2017 IEEE 44th Photovoltaic Specialist Conference (PVSC 2017)

Washington, DC, USA
25-30 June 2017

Pages 1-705

IEEE Catalog Number: CFP17PSC-POD
ISBN: 978-1-5090-5606-4

**Copyright © 2017 by the Institute of Electrical and Electronics Engineers, Inc.
All Rights Reserved**

Copyright and Reprint Permissions: Abstracting is permitted with credit to the source. Libraries are permitted to photocopy beyond the limit of U.S. copyright law for private use of patrons those articles in this volume that carry a code at the bottom of the first page, provided the per-copy fee indicated in the code is paid through Copyright Clearance Center, 222 Rosewood Drive, Danvers, MA 01923.

For other copying, reprint or republication permission, write to IEEE Copyrights Manager, IEEE Service Center, 445 Hoes Lane, Piscataway, NJ 08854. All rights reserved.

*** This is a print representation of what appears in the IEEE Digital Library. Some format issues inherent in the e-media version may also appear in this print version.

IEEE Catalog Number:	CFP17PSC-POD
ISBN (Print-On-Demand):	978-1-5090-5606-4
ISBN (Online):	978-1-5090-5605-7
ISSN:	0160-8371

Additional Copies of This Publication Are Available From:

Curran Associates, Inc
57 Morehouse Lane
Red Hook, NY 12571 USA
Phone: (845) 758-0400
Fax: (845) 758-2633
E-mail: curran@proceedings.com
Web: www.proceedings.com

2017 IEEE 44th Photovoltaic Specialist Conference (PVSC 2017)

Washington, DC, USA
25-30 June 2017

Pages 1-705

IEEE Catalog Number: CFP17PSC-POD
ISBN: 978-1-5090-5606-4

TABLE OF CONTENTS

OPEN CIRCUIT VOLTAGE CALCULATION USING TEMPERATURE AND IRRADIANCE 1
Andrew Melvin

EFFECT OF CL-DOPING IN ZNTEO ON PHOTOLUMINESCENCE AND PHOTOVOLTAIC
PROPERTIES OF ZNTEO-BASED INTERMEDIATE BAND SOLAR CELLS 3
T. Tanaka ; S. Tsutsumi ; Y. Okano ; K. Matsuo ; K. Saito ; Q. Guo ; M. Nishio ; T. Tayagaki ; K. M. Yu ; W. Walukiewicz

TOWARD LEAD HALIDE PEROVSKITE-BASED INTERMEDIATE BAND ABSORBERS 6
Matthew D. Sampson ; Ji-Sang Park ; Richard D. Schaller ; Maria K. Y. Chan ; Alex B. F. Martinson

TYPE-II QUANTUM DOTS FOR APPLICATION TO PHOTON RATCHET INTERMEDIATE
BAND SOLAR CELLS 10
Ryo Tamaki ; Yasushi Shoji ; Yoshitaka Okada

AN INVESTIGATION OF THE ROLE OF RECOMBINATION PROCESSES IN THE
OPERATION OF INAS/GAASL-XSBX QUANTUM DOT SOLAR CELLS 14
Y. Cheng ; A. J. Meleco ; A. J. Roeth ; V. R. Whiteside ; M. C. Debnath ; M. B. Santos ; T. D. Mishima ; S. Hatch ; H.Y. Liu ; I. R. Sellers

TEMPERATURE AND VOLTAGE-BIAS DEPENDENT TWO-STEP PHOTON ABSORPTION IN
INAS/GAASL AL0.3GAAS QUANTUM DOT IN A WELL SOLAR CELLS 18
Yushuai Dai ; Brittany L. Smith ; Michael A. Slocum ; Zachary S. Bittner ; Hyun Kum ; Julia D'Rozario ; Seth M. Hubbard

INCREASING CURRENT GENERATION BY PHOTON UP-CONVERSION IN A SINGLE-
JUNCTION SOLAR CELL WITH A HETERO-INTERFACE 23
Shigeo Asahi ; Kazuki Kusaki ; Toshiyuki Kaizu ; Takashi Kita

RTP-ASSISTED EX-SITU ANALYSIS OF (AG,CU)(IN,GA)SE2FORMATION USING
SELENIZATION 26
Sina Soltanmohammad ; William N. Shafarman

ROLE OF EV+0.98 EV TRAP IN LIGHT SOAKING-INDUCED SHORT CIRCUIT CURRENT
INSTABILITY IN CIGS SOLAR CELLS 30
P. K. Paul ; T. Jarmar ; L. Stolt ; A. Rockett ; A. R. Arehart

STUDY OF DEFECT PROPERTIES IN CUGASE2THIN-FILM SOLAR-CELLS USING
ADMITTANCE SPECTROSCOPY 33
Muhammad Monirul Islam ; Shogo Ishizuka ; Hajime Shibata ; Shigeru Niki ; Katsuhiro Akimoto ; Takeaki Sakurai

TRANSMISSIVE SPECTRUM-SPLITTING CONCENTRATOR PHOTOVOLTAIC CELLS AND
MODULES 37
Yaping Ji ; Qi Xu ; Brian Riggs ; John Robertson ; Kazi Islam ; Vince Romanin ; Dimitri D. Krut ; Jim H. Ermer ; Matthew D. Escarra

ALGAINP/GAAS TANDEM SOLAR CELLS FOR POWER CONVERSION AT 400 C AND 1000X
CONCENTRATION 42
Myles A. Steiner ; Emmett E. Perl ; John Simon ; Daniel J. Friedman ; Nikhil Jain ; Paul Sharps ; Claiborne Mcpheeters ; Minjoo L. Lee

GALNASP SOLAR CELLS GROWN BY HYDRIDE VAPOR PHASE EPITAXY FOR ONE-SUN &
LOW-CONCENTRATION III-V/SI PHOTOVOLTAICS 46
Nikhil Jain ; John Simon ; Kevin L. Schulte ; David R. Diercks ; Corinne E. Packard ; David Young ; Aaron J. Ptak

PHOTO-ELECTROCHEMICAL HYDROGEN GENERATION FROM INVERTED
METAMORPHIC MULTIJUNCTION III-VS 47
Todd G. Deutsch ; James L. Young ; Myles A. Steiner ; Henning Döscher ; Ryan M. France ; John A. Turner

ADVANCED SILICON THIN FILMS FOR HIGH-EFFICIENCY SILICON HETEROJUNCTION-
BASED SOLAR CELLS 50
A. Descoeudres ; C. Allebe ; N. Badel ; L. Barraud ; J. Champliaud ; G. Christmann ; L. Curvat ; F. Debrot ; A. Faes ; J. Geissbiihler ; J. Horzel ; A. Lachowicz ; J. Levrat ; S. Martin De Nicolas ; S. Nicolay ; B. Paviet-Salomon ; L.-L. Senaud ; A. Tomasi ; C. Ballif ; M. Despeisse

MOOXAND WOXBASED HOLE-SELECTIVE CONTACTS FOR WAFER-BASED SI SOLAR
CELLS 55
Stephanie Essig ; Julie Dréon ; Jérémie Werner ; Philipp Löper ; Stefaan De Wolf ; Mathieu Boccard ; Christophe Ballif

METAL NANOPARTICLE HOLE CONTACTS FOR SILICON SOLAR CELLS 59
James Bullock ; Zhaoran Xu ; Mark Hettick ; Yimao Wan ; Ali Javey

NEAR-FIELD TRANSPORT IMAGING APPLICATION OF PHOTOVOLTAIC MATERIALS................................62
Chuanxiao Xiao ; Chun-Sheng Jiang ; John Moseley ; John Simon ; Kevin Schulte ; Aaron J. Ptak ; Steve Johnston ; Brian Gorman ; Mowafak Al-Jassim ; Nancy M. Haegel ; Helio Moutinho

APPLICATIONS OF DMD-BASED INHOMOGENEOUS ILLUMINATION PHOTOLUMINESCENCE IMAGING FOR SILICON WAFERS AND SOLAR CELLS................................66
Yan Zhu ; Mattias Klaus Juhl ; Ziv Hameiri ; Thorsten Trupke

NUMERICAL MODEL TO EXTRACT MATERIALS PROPERTIES MAP FROM SPECTRALLY RESOLVED LUMINESCENCE IMAGES................................70
Nicolas Paul ; Vincent Le Guen ; Daniel Ory ; Laurent Lombez

NON-DESTRUCTIVE CONTACT RESISTIVITY MEASUREMENTS ON SOLAR CELLS USING THE CIRCULAR TRANSMISSION LINE METHOD................................74
Geoffrey Gregory ; Andrew M. Gabor ; Andrew Anselmo ; Rob Janoch ; Zhihao Yang ; Kristopher O. Davis

RADIATION RESISTANCE OF LOW COST HIGH EFFICIENCY TRIPLE JUNCTION SOLAR CELLS................................76
Roberta Campesato ; Erminio Greco ; Giuseppe Gabetta ; Mariacristina Casale ; Gabriele Gori ; M. Sankaran ; Suresh E. Puthanveettil ; B. R. Uma ; M. Ravindra ; Sheeja Krishnan

AMORPHOUS SILICON CARBIDE REAR-SIDE PASSIVATION AND REFLECTOR LAYER STACKS FOR MULTI-JUNCTION SPACE SOLAR CELLS BASED ON GERMANIUM SUBSTRATES................................83
Stefan Janz ; Charlotte Weiss ; Christian Mohr ; Rufi Kurstjens ; Bruno Boizot ; Bianca Fuhrmann ; Victor Khorenko

HOT CARRIER TRANSPORTATION DYNAMICS IN INAS/GAAS QUANTUM DOT SOLAR CELL................................85
Tomah Sogabe ; Kohdai Nii ; Katsuyoshi Sakamoto ; Koichi Yarnaquchi ; Yoshitaka Okada

INTEGRATION OF CRACK-TOLERANT COMPOSITE GRIDLINES ON TRIPLE JUNCTION PHOTOVOLTAIC CELLS................................88
Omar K. Abudayyeh ; Geoffrey K. Bradshaw ; Steven Whipple ; David M. Wilt ; Sang M. Han

SUBCELL LIGHT CURRENT- VOLTAGE CHARACTERIZATION OF IRRADIATED MULTIJUNCTION SOLAR CELL................................93
Don Walker ; John Nocerino ; Yao Yue ; Colin J. Mann ; Simon H. Liu

ANALYTICAL METHOD FOR PREDICTING SPACECRAFT POWER GENERATION ON PARTIALLY SHADED SOLAR PANELS................................96
Gordon Wu ; Bao Hoang

EVALUATING THE EMISSIVITY OF PSEUDOMORPHIC GLASS (PMG)................................102
Ryan D. Beauchemin ; David M. Wilt ; Paul E. Hausgen

CHARACTERIZING THE IMPACT OF SOLAR SPECTRAL IRRADIANCE ON PV MODULE OUTPUT................................107
M. Schweiger ; W. Herrmann

USE OF MEASURED AEROSOL OPTICAL DEPTH AND PRECIPITABLE WATER TO MODEL CLEAR SKY IRRADIANCE................................110
Mark M. Mikofski ; Clifford W. Hansen ; William F. Holmgren ; Gregory M. Kimbal

RECENT ADVANCEMENTS IN THE NUMERICAL SIMULATION OF SURFACE IRRADIANCE FOR SOLAR ENERGY APPLICATIONS................................116
Yu Xie ; Manajit Sengupta ; Chris Deline

OPTIMAL IRRADIANCE SENSOR PLACEMENT FOR PHOTOVOLTAIC SYSTEMS USING MUTUAL INFORMATION BASED GREEDY ALGORITHM IN GAUSSIAN PROCESS................................120
Lian Lian Jiang ; R. Srivatsan ; Douglas L. Maskell

EVALUATING DIFFERENT UPSCALING APPROACHES TO DERIVE THE ACTUAL POWER OF DISTRIBUTED PV SYSTEMS................................126
Sven Killinger ; Björn Müller ; Bernhard Wille-Haussmann ; Russell Mckenna

ADVANCES IN LONG-TERM SOLAR ENERGY PREDICTION AND PROJECT RISK ASSESSMENT METHODOLOGY................................132
Alemu Tadesse ; Adam Kankiewicz ; Alex Kubiniec ; Richard Perez ; John Dise ; Thomas Hoff

DECOUPLING THIN FILM CDTE GROWTH FROM PACKAGING: TOWARD RECORD SPECIFIC POWER IN LOW COST POLYCRYSTALLINE PV................................138
D. Clayton-Warwick ; M.D. Kempe ; M. S. Dabney ; T. M. Barnes ; C. A. Wolden ; M. O. Reese

JUNCTION ACTIVATION OF CDTE/CDS SOLAR CELL USING MGCL2................................142
G. Angeles-Ordóñez ; E. Regalado-Pérez ; M.G. Reyes-Banda ; N. R. Mathews ; X. Mathew

VARIATION OF CU CONTENT OF SPRAYED CU(IN, GA)(S,SE)2 SOLAR CELLS BASED ON A THIOL-AMINE SOLVENT MIXTURE................................146
Panagiota Arnou ; Sona Ulicná ; Alexander Eeles ; Mustafa Togay ; Lewis D. Wright ; Andrei V. Malkov ; John M. Walls ; Jake W. Bowers

CUINSE2 ABSORBER LAYER GROWN UNDER COPPER EXCESS WITH A COPPER POOR SURFACE FORMED BY A KF POST DEPOSITION TREATMENT 151

Finn Babbe ; Hossam Elanzeery ; Michele Melchiorre ; Susanne Siebentritt

CU2ZNSNSE4SOLAR CELLS ONTO POLYIMIDE SUBSTRATES FABRICATED AT LOW TEMPERATURE 155

Ignacio Becerril-Romero ; Simón Lopez-Marino ; Moisés Espíndola-Rodríguez ; Laura Acebo ; Markus Neuschitzer ; Yudania Sánchez ; Edgardo Saucedo ; Paul Pistor

AN OPTIMIZED PHOTOLITHOGRAPHY RECIPE FOR CU(IN1-X,GAX)(SY,SE1-Y)2(CIGSSE) SOLAR CELLS 160

Xia Hao ; Shenghao Wang ; Katsuhiro Akimoto ; Takuya Kato ; Hiroki Sugimoto ; Takeaki Sakurai

EFFECTS OF CDCL2PASSIVATION ON THIN CDTE ABSORBERS FABRICATED BY CLOSE-SPACE SUBLIMATION 164

Anna Wojtowicz ; Alexandra M. Huss ; Jennifer A. Drayton ; James R. Sites

CDS1-XSEXWINDOW LAYER FOR CDTE PREPARED BY THE EXCHANGE OF S WITH SE IN CDS FILMS 170

Geethika K. Liyanage ; Adam B. Phillips ; Zhaoning Song ; Suneth C. Watthage ; Ramez H. Ahanzhamejhad ; Michael J. Heben

EFFECT OF ILLUMINATION ON THERMAL CDCL2TREATMENT OF CDTE 175

Sudhajit Misra ; Carina E. Hahn ; Vasilios Palekis ; Christos Ferekides ; Michael A. Scarpulla

CHALLENGES IN THE INDUSTRIAL PRODUCTION OF CZTS MONOGRAIN SOLAR CELLS 178

Gerhard Peharz ; Valentin Satzinger ; Sandra Pötz ; Gemot Oreski ; Theodoros Dimopoulos ; Stefan Edinger ; Wolfeanz Hackl ; Hannes Starkl ; Parichehr Esfandiari ; Peter Krabb ; Stefan Gahr ; Lukas Plessing ; Dieter Meissner

UNDERSTANDING INSTABILITIES AND DEGRADATION DUE TO MOISTURE INGRESS IN CU(IN, GA)SE2SOLAR CELLS 182

Grace Rajan ; Shankar Karki ; Isaac Butt ; Krishna Aryal ; Tyler J. Grassman ; Angus Rockett ; Sylvain Marsillac

CONTROL OF MOSE2 FORMATION IN HYDRAZINE-FREE SOLUTION-PROCESSED CIS/CIGS THIN FILM SOLAR CELLS 186

Sona Ulicná ; Panagiota Arnou ; Alexander Eeles ; Mustafa Togay ; Lewis D. Wright ; Ali Abbas ; Andrei V. Malkov ; John M. Walls ; Jake W. Bowers

GROWTH AND PROPERTIES OF EPITAXIAL CU(IN, GA)SE2THIN FILMS DEPOSITED BY THE THREE-STAGE PROCESS FOR SOLAR CELLS 192

Takeru Yamagami ; Yuta Ando ; Ishwor Khatri ; Mutsumi Sugiyama ; Tokio Nakada

IMPROVEMENT OF CIS SOLAR CELLS WITH KF POSTDEPOSITION FOLLOWING A SIMPLE TWO-STEP SELENIZATION PROCESS 195

Yang Zhang ; Robert E. Bartolo ; Sang Jik Kwon ; Mario Dagenais

THE TWINS STRUCTURE, ELECTRICAL PROPERTIES AND CELL PERFORMANCE OF MAGNETRON SPUTTERING DEPOSITED CHLORINE DOPED CDTE 198

Ziyao Zhu ; Fu-Kuo Chiang ; Zhongming Du ; Yufeng Zhang ; Xiangxin Liu

INVESTIGATION AND MITIGATION OF SHUNTS FOR HIGHER EFFICIENCY EPITAXIAL GASB/GASB AND GASB/GAAS SOLAR CELLS 202

George T. Nelson ; Bor-Chau Juang ; Steve Johnston ; Michael A. Slocum ; Zachary S. Bittner ; Ramesh B. Lagumavarapu ; Diana Huffaker ; Seth M. Hubbard

DEVELOPMENT OF GASB SOLAR CELLS ON GAAS BY MOVPE VIA INTERFACE MISFIT TECHNIQUE 206

Michael A. Slocum ; Alessandro Giussani ; Emily Kessler ; Phil Ahrenkiel ; George T. Nelson ; Seth M. Hubbard

FABRICATION OF INGAASP SOLAR CELLS FOR CONCENTRATOR APPLICATIONS 210

Mitchell F. Bennett ; Matthew P. Lumb ; Kenneth J. Schmieder ; Brent Fisher ; Eric A. Armour ; Robert J. Walters

DETAILED CHARACTERIZATION FOR TCAD SIMULATIONS OF GAAS0.76P0.24/SI1-YGEY/SI SINGLE JUNCTION SOLAR CELLS 213

Sabina Abdul Hadi ; Timothy Milakovich ; Eugene A. Fitzgerald ; Ammar Nayfeh

COMPARATIVE STUDY OF >2 EV LATTICE-MATCHED AND METAMORPHIC (AL)GAINP MATERIALS AND SOLAR CELLS GROWN BY MOCVD 215

Daniel J. Chmielewski ; Christine Jackson ; Jacob Boyer ; Daniel Lepkowski ; John A. Carlin ; Aaron R. Arehart ; Tyler J. Grassman ; Steven A. Ringel

PERFORMANCE OF GASB PHOTOVOLTAICS WITH GRAPHENE COATING 219

Benjamin P. Conlon ; Daniel J. Herrera ; Shaimaa A. Abdallah ; Jonathan O. Okafor ; Luke F. Lester

HIGH EFFICIENCY SINGLE-JUNCTION INGAP PHOTOVOLTAIC DEVICES UNDER LOW INTENSITY LIGHT ILLUMINATION 222

Yushuai Dai ; Hyun Kum ; Michael A. Slocum ; George T. Nelson ; Seth M. Hubbard

RADIATION RESISTANT OF UPRIGHT METAMORPHIC GAINP/GAINAS/GE TRIPLE JUNCTION SOLAR CELLS FOR SPACE USE 226

Liang Fang ; Abuduwayiti Aierken ; Zhen Pan ; Qiming Zhang ; Zhanhang Li ; Heini Maliya ; Wei Gao ; Hui Gao ; Ronghua Wan ; Bao Zhang ; He Wang ; Qi Guo

HIGH EFFICIENCY GLASS WAVEGUIDING SOLAR CONCENTRATOR 229

Chehao Hu ; Yusuf Dogan ; Matthew Morrison ; A. Nanda ; D. Ma ; R. Atkins ; C. K. Madsen

GAINASP/GAINAS TANDEM SOLAR CELL WITH 32.6% ONE-SUN EFFICIENCY 232

Nikhil Jain ; Kevin L. Schulte ; John F. Geisz ; Ryan M. France ; Myles A. Steiner

EVALUATION OF TANDEM EFFICIENCIES: DILUTE NITRIDE P-I-N (BULK OR MQWS) IN CONJUNCTION WITH PRACTICAL SI SOLAR CELLS 236

Khim Kharel ; Alexandre Freundlich

GALLIUM PHOSPHIDE NANOSTRUCTURE ON SILICON BY SILICA NANOSPHERES LITHOGRAPHY AND METAL ASSISTED CHEMICAL ETCHING 240

Sangpyeong Kim ; Chaomin Zhang ; Som Dahal ; Stuart Bowden ; Christiana B. Honsberg

EFFICIENCY ENHANCEMENT OF INGAP/INGAAS/GE SOLAR CELLS WITH GRADUALLY DOPED P-N JUNCTION ACTIVE LAYERS 244

Youngjo Kim ; Sang Hyun Jung ; Chang Zoo Kim ; Kangho Kim ; Hyun-Beom Shin ; Kyung Ho Park ; Won-Kyu Park ; Jaejin Lee ; Ho Kwan Kang

ANALYSIS OF INGAP OXIDE GROWTH RATE AT HIGH TEMPERATURES AND AMBIENT CONDITIONS FOR TERRESTRIAL PHOTOVOLTAIC APPLICATIONS 247

Nicole A. Kotulak ; Matthew P. Lumb ; Raymond Hoheisel ; Erin Cleveland ; Mitchell Bennett ; Phillip P. Jenkins ; Robert J. Walters

GRAIN BOUNDARIES IN THIN-FILM POLYCRYSTALLINE GAAS SOLAR CELLS: A SIMULATION STUDY 251

Khushboo Kumari ; Sushobhan Avasthi

TIME-RESOLVED PL MEASUREMENTS IN THE GROWTH OF HIGH VOLTAGE (AL)GAINP/GAAS SOLAR CELLS 255

Xinyi Li ; Wei Zhang ; Hongbo Lu

LOW-RESISTANCE AND HIGHLY-TRANSPARENT GASB-BASED TUNNEL JUNCTIONS 259

Matthew P. Lumb ; Shawn Mack ; Maria Gonzalez ; Kenneth J. Schmieder ; Mitchell F. Bennett ; Chaffra A. Affouda ; James E. Moore ; Robert J. Walters

MODULATED PHOTOCURRENT MEASUREMENTS IN DOUBLE JUNCTION SOLAR CELLS 263

Nicolás Márquez Peraca ; Behrang H. Hamadani

EFFECT OF ATMOSPHERIC ABSORPTION BANDS ON THE OPTIMAL DESIGN OF MULTIJUNCTION SOLAR CELLS 268

William E. Mcmahon ; Daniel J. Friedman ; John F. Geisz

EFFECTS OF CONTACT CONFIGURATION AND PERIMETER RECOMBINATION ON OPTIMAL CELL SIZE FOR HIGH CONCENTRATION PHOTOVOLTAICS 272

James E. Moore ; Matthew P. Lumb ; Kenneth J. Schmieder ; Robert J. Walters ; Brent Fisher ; Matt Meitl ; Scott Burroughs

NUMERICAL SIMULATION OF DEFECTS IN III-V PV CELLS: THE EFFECT OF VOLTAGE BIAS AND DOPING CONCENTRATION 276

Vasiliki Paraskeva ; Constantinos Lazarou ; Andreas Livera ; Venizelos Venizelou ; Maria Hadjipanayi ; George E. Georghiou

IMPROVEMENT OF OPEN-CIRCUIT VOLTAGE IN METAMORPHIC GASB CELLS GROWN ON GAAS SUBSTRATES BY USING AN INTERFACIAL MISFIT ARRAY AND AN ALSB BLOCKING LAYER 281

E. J. Renteria ; S. J. Addamane ; D. M. Shima ; A. Mansoori ; A. L. Soudachanh ; G. Balakrishnan

ENERGY YIELD EVALUATION FOR FIELD OPERATION OF SOLAR CELLS IN SINGAPORE: GAAS/GAAS TANDEM VS. GAAS SINGLE-JUNCTION SOLAR CELLS 284

Maung Thway ; Zekun Ren ; Kevin Nay Yaung ; Haohui Liu ; Zhe Liu ; Samuel Raj ; Soo Jin Chua ; Armin G. Aberle ; Tonio Buonassisi ; Ian Marius Peters ; Fen Lin

SIMULATION OF THE PERFORMANCES OF MULTIJUNCTION SOLAR CELLS WITH IMPROVED VOLTAGE BY TRANSFER AND SCATTERING MATRIX METHODS 290

Gianluca Timò ; Lucio Andreani

OPTIMIZED DESIGN OF BACK-CONTACT THIN-FILM GAAS SOLAR CELLS 294

Jia-Ling Tsai ; Chung-Yu Hong ; Tien-Chien Zhan ; Yuh-Renn Wu ; Albert Lin ; Peichen Yu

DESIGN CONSIDERATIONS ON GAINNAS SOLAR CELLS WITH BACK SURFACE REFLECTORS 297

Antti Tukiainen ; Arto Aho ; Timo Aho ; Ville Polojärvi ; Mircea Guina

QUANTITATIVE ELECTROLUMINESCENCE ANALYSIS OF TRIPLE JUNCTION SOLAR CELLS TO DETERMINE SUBCELL VOLTAGE-TEMPERATURE COEFFICIENTS 301

Kevin Tyler ; Geoffrey K. Bradshaw ; Sam Wilt ; David M. Wilt ; Richard R. King

PROGRESS TOWARDS DOUBLE-JUNCTION INGAN SOLAR CELL 305

Ehsan Vadiee ; Evan A. Clinton ; Heather Mcfavilen ; Alec M. Fischer ; Yi Fang ; Joshua J. Williams ; Christiana B. Honsberg ; William A. Doolittle ; Stephen M. Goodnick

A PHYSICS-BASED SIMULATION TOOL FOR LEAKAGE CURRENTS IN C-SI PV MODULES 309

John M. Waddle ; Saroj Dahal ; Marco Nardone

BROADBAND TA2O5 MOTH-EYE ANTIREFLECTION COATINGS FOR TANDEM SOLAR CELLS ON SI 315

Bo Yuan ; Brian Thibeault ; David Payne ; James Mutitu ; Ivan Perez-Wurfl ; Kevin Dobson ; Brianna Conrad ; Allen Barnett ; Robert L. Opila

CARRIER TRANSPORT IN POLYCRYSTALLINE SILICON AT HIGH OPTICAL INJECTION: TRANSIENT PHOTOCONDUCTANCE VS. NUMERICAL MODELING 319

Uchechi Anyanwu ; Christian Harris ; Andrey Semichaevsky

IMPROVING SILICON SURFACE PASSIVATION WITH A SILICON OXIDE LAYER GROWN VIA OZONATED DEIONIZED WATER 322

Sara Bakhshi ; Ngwe Zin ; Kristopher O. Davis ; Marshall Wilson ; Ismail Kashkoush ; Winston V. Schoenfeld

DEPOSITION OF SIOC BY PLASMA-FREE ULTRA-LOW-TEMPERATURE ALD (ULT-ALD) AND ITS PASSIVATION ON P-TYPE SILICON 326

Meixi Chen ; Naoto Noda ; Raphael Rochat ; Abhishek Iyer ; James H. Hack ; Changhee Ko ; Christian Dussarrat ; Robert L. Opila

A METHOD FOR QUANTITATIVELY INVESTIGATING THE REAR-SIDE PASSIVATION PERFORMANCE OF PERC CELLS 329

Tsung-Cheng Chen ; Yung-Sheng Lin ; Chen-Hao Ku ; Ting-Wei Kuo ; Cheng-Shun Hu ; Ching-Chang Wen

FIELD-EFFECT PASSIVATION BY NEGATIVE CHARGE ON BORON EMITTER AND BORON-DOPED SURFACES BY A NOVEL LOW-COST PLASMA CHARGE INJECTION 333

Eunhwan Cho ; Young-Woo Ok ; James Hwang ; Aditi Jain ; Vijay D. Upadhyaya ; John Keith Tate ; Ajeet Rohatgi

INDUSTRY RELEVANT RIE TEXTURING FOR MC-SI DIAMOND WIRE OR DIRECT WAFER® PRODUCT: OPTIMIZED REFLECTIVITY, UNIFORMITY, AND THROUGHPUT 337

Jose Luis Cruz-Campa ; Ray Fraser ; Rob Steeman ; John Linton

SHORT-CIRCUIT CURRENT-DENSITY ENHANCEMENT OF SILICON SOLAR CELLS USING PLASMONICS ANTIREFLECTIVE COATING AND LUMINESCENT DOWNSHIFTING 343

Sheng-Kai Feng ; Wen-Jeng Ho ; Guan-Yi Li ; Jheng-Jie Liu ; Hao-Yu Yang ; Ta-Wei Chuang

EXTREMELY LOW REFLECTIVITY NANOPOROUS BLACK SILICON SURFACE BY COPPER CATALYZED ETCHING FOR EFFICIENT SOLAR CELLS 346

K A S M Ehteshamul Haque ; Wenqi Duan ; Fatima Toor

IMPACT OF FRONT SIDE PYRAMID SIZE ON THE LIGHT TRAPPING PERFORMANCE OF WAFER BASED SILICON SOLAR CELLS AND MODULES 352

Oliver Höhn ; Nico Tucher ; Benedikt Bläsi

A STUDY OF BLISTER CONTROL OF AL2O3 THIN FILM DEPOSITED BY PLASMA-ASSISTED ATOMIC LAYER DEPOSITION AFTER FIRING PROCESS 356

Min Gu Kang ; Jeong In Lee ; Hee-Eun Song ; Myeong Sangjeong ; Kyung Taekjeong ; Hyo Sikchang

PYPVCELL: AN OPEN-SOURCE SOLAR CELL MODELING LIBRARY IN PYTHON 359

Kan-Hua Lee ; Kenji Araki ; Omar Elleuch ; Nobuaki Kojima ; Masafumi Yamaguchi

IMPROVEMENT IN SURFACE PASSIVATION OF C-SI USING GRADIENT-LAYERED A-SI:H FILM FOR HIGH EFFICIENCY SILICON HETEROJUNCTION SOLAR CELLS 363

Soonil Lee ; Leo Mathew ; Rajesh Rao ; Jae Hyun Kim ; Sanjay K. Banerjee ; Edward T. Yu

PHOTOVOLTAIC PERFORMANCE ENHANCEMENT OF TEXTURED SILICON SOLAR CELLS USING LUMINESCENT DOWN-SHIFTING METHYLAMMONIUM LEAD TRIBROMIDE PEROVSKITE NANOPHOSPHORS 367

Guan-Yi Li ; Wen-Jeng Ho ; Sheng-Kai Feng ; Hao-Yu Yang ; Ta-Wei Chuang ; Bang-Jin You ; Zong-Xian Lin ; Zong-Liang Tseng ; Lung-Chien Chen

SINX THIN FILMS WITH APPROPRIATE ANTIREFLECTION AND SHIFT-CONVERSION PROPERTIES FOR SILICON SOLAR CELLS 370

E. Men-Pérez ; J. Salazar ; A. Dutt ; J. Santoyo-Salazar ; G. Santana

NUMERICAL SIMULATION OF CRYSTALLINE SILICON SOLAR CELLS WITH FULL AREA METAL OXIDE REAR CONTACTS 373

James E. Moore ; Woojun Yoon ; Phillip P. Jenkins ; Robert J. Walters

INTERDIGITATED BACK CONTACT SILICON SOLAR CELL WITH PEROVSKITE LAYER FOR FRONT SURFACE PASSIVATION AND ULTRAVIOLET RADIATION STABILITY 377

Rahul Pandey ; Shivam Gupta ; Trijul Khatri ; Rishu Chaujar

POTENTIAL OF A-SI:H/C-SI HETEROJUNCTION SOLAR CELLS WITH VERY THIN WAFERS 381
Hitoshi Sai ; Hiroshi Umishio ; Takuya Matsui ; Shota Nunomura ; Tomoyuki Kawatsu ; Hidetaka Takato ; Koji Matsubara

MANIPULATING FIXED CHARGES IN ZRO2 BY DOPING FOR PASSIVATION AND ANTIREFLECTION ON WAFER-SI SOLAR CELLS 385
Woo Jung Shin ; Laidong Wang ; Wen-Hsi Huang ; Meng Tao

LOW TEMPERATURE ANTIREFLECTION COATING FOR SILICON SOLAR CELLS 389
O. S. Shinde ; Ej Schneller ; N. Dhere ; S. V. Ghaisas

RELATIONSHIP BETWEEN POWER LOSS AND VOLTAGE APPLIED TO SOLAR CELLS IN PID-AFFECTED SOLAR MODULES 392
Fumei Wang ; Baosong Duan ; Wenshuang He ; He Wang ; Hong Yang ; Chengfeng Su ; Bojie Su ; Xue Zhang ; Yunxue Cao ; Hui Zhao

A NEW LOW-COST AND LOW-TEMPERATURE CHEMICAL PASSIVATION PROCESS FOR LARGE AREA INDUSTRIAL SINGLE CRYSTALLINE SILICON WAFERS 396
Tarun S. Yadav ; K. Sandeep ; Ashok K. Sharma ; B. Spandana ; K.L. Narasimhan ; B.M. Arora ; Anil Kottantharayil ; Prabir K. Basu

EVALUATION OF ALD PASSIVATION LAYERS FOR INDUSTRIAL PERC PROCESS 399
Chang Youn Yoo ; Keunkee Hong ; Jisun Kim ; Eunjoo Lee ; Dong Seop Kim

QUANTITATIVE ANALYSIS OF ELECTROLUMINESCENCE AND INFRARED THERMAL IMAGES FOR AGED MONOCRYSTALLINE SILICON PHOTOVOLTAIC MODULES 402
Irene Berardone ; Juan Lopez Garcia ; Marco Paggi

GAP PASSIVATION STRUCTURE FOR SCALABLE N-TYPE INTERDIGITATED ALL BACK CONTACT SILICON HETERO-JUNCTION SOLAR CELL 408
Lei Zhang ; Ujjwal Das ; Steven Hegedus

PROPOSAL OF THE BANDGAP DESIGN USING THE SUN HEIGHT OF THE CULMINATION ON THE WINTER SOLSTICE 412
Kenji Araki ; Kan-Hua Lee ; Masafumi Yamaguchi

PHOTOEXCITED CARRIERS, PHONONS, AND THEIR SCATTERING MEASURED IN SEMICONDUCTOR JUNCTIONS BY TRANSIENT EXTREME ULTRAVIOLET SPECTROSCOPY 417
Scott K. Cushing ; Brett M. Marsh ; Mihai E. Vaida ; Lucas M. Carneiro ; Ilana J. Porter ; Angela Lee ; Stephen R. Leone

ON THE USE OF VOLTAGE MEASUREMENTS FOR DETERMINING CARRIER LIFETIME AT HIGH ILLUMINATION INTENSITY 420
Robert Dumbrell ; Mattias K. Juhl ; Thorsten Trupke ; Ziv Hameiri

HIGH RESOLUTION 3D CHEMICAL CHARACTERISATION OF A CADMIUM TELLURIDE SOLAR CELL BY DYNAMIC SIMS 424
Thomas Fiducia ; Kexue Li ; Chris Grovenor ; Kurt Barth ; Walajabad Sampath ; Michael Walls

HARSH OUTDOOR EVALUATION SETUP AND FIRST POWER PRODUCTION RESULTS FOR SI MINI-MODULES COVERED BY EU3+-BASED DOWN CONVERTERS 429
Benjamín González-Díaz ; Carlos Montes ; Joaquín Sanchiz ; Luis Ocaña ; Carlos Quinto ; Cecilio Hernández-Rodríguez ; Mari Paz Friend ; Manuel Cendagorta-Galarza ; David Cañadillas ; Ricardo Guerrero-Lemus

STUDY OF MICRO-STRUCTURAL PROPERTIES OF ZNO AND TIO2THIN FILM GROWN BY SPRAY PYROLYSIS 433
G. Gordillo ; J.M. Correa ; A.A. Ramirez ; E. A. Ramírez

NONLINEAR RESPONSE OF SILICON SOLAR CELLS 437
Behrang H. Hamadani ; Andrew Shore ; Howard W. Yoon ; Mark Campanelli

EXTENDED LINEAR INTERPOLATION/EXTRAPOLATION PROCEDURE FOR ACCURATE AND VERSATILE TRANSLATION OF THE I-V CURVES OF PV CELLS AND MODULES 441
Y. Hishikawa ; H. Ohshima ; M. Higa ; K. Yamagoe ; T. Takenouchi ; T. Doi

SEVERITY TEST WITH UNEVEN LOAD DUE TO WIND ACTION ON PHOTOVOLTAIC MODULE 445
Shu-Tsung Hsu

STANDARDIZED DURABILITY TEST FOR ORGANIC PHOTOVOLTAIC AND DYE SENSITIZED SOLAR CELL 448
Shu-Tsung Hsu ; Yean-San Long ; Teng-Chun Wu

SPATIAL THICKNESS UNIFORMITY AND STRUCTURAL EVALUATION OF RF SPUTTERED ZNO THIN FILMS FOR SOLAR CELL 451
Babar Hussain ; Taj M. Khan

LOCAL MEASUREMENTS OF SURFACE CAPACITANCE BY ELECTROSTATIC FORCE MICROSCOPY ON CU(IN, GA)SE2MATERIALS 455
Tomoaki Ishii ; Takashi Minemoto ; Takuji Takahashi

A COMPARISON OF SI-BASED CAMERAS FOR IMAGING LUMINESCENCE FROM PHOTOVOLTAIC MATERIALS AND DEVICES 459

Steve Johnston

BLISTERING OF AL2O3/A-SINX:H STACKS: ANALYSIS OF THE SUBMERGED PART OF THE ICEBERG BY COLORED PICOSECOND ACOUSTIC MICROSCOPY 464

Fabien Lebreton ; Arnaud Devos ; Etienne Drahi ; Patricia De Coux ; François Silva ; Sergej Filonovich ; Pere Roca I Cabarrocas

SELF-REFERENCE PROCEDURE TO REDUCE UNCERTAINTY IN MODULE CALIBRATION 467

D.H. Levi ; C.R. Osterwald ; S. Rummel ; L. Ottoson ; A. Anderberg

UNCERTAINTY EVALUATION OF PRIMARY REFERENCE PHOTOVOLTAIC CELL CALIBRATION UNDER OUTDOOR CONDITION IN TIBET 472

Haitao Liu ; Shiyu Sang ; Guomin Zhou ; Yonghui Zhai

REQUIREMENT OF ARTIFICIAL LIGHTING SIMULATOR FOR EVALUATION EMERGING PV PERFORMANCE RATING UNDER INDOOR ENVIRONMENT 476

Yean-San Long ; Shu-Tsung Hsu ; Teng-Chun Wu

NON-CONTACT VOLTAGE MEASUREMENT OF SOLAR CELL WITH ELECTROSTATIC VOLTMETER 480

Sakutaro Miyajima ; Kensuke Nishioka ; Yoshihiro Hishikawa

NREL'S CELL AND MODULE PERFORMANCE GROUP'S ASYMPTOTIC PMAX PROTOCOL FOR PEROVSKITE DEVICES 483

Tom Moriarty ; Dean Levi

OUTDOOR OPERATING TEMPERATURE MODELING OF PHOTOVOLTAIC MODULES INCLUDING TRANSIENT EFFECT 487

Soo-Young Oh ; Min-Soo Kim ; Won-Shup So ; Woo Kyoung Kim ; Jae Hak Jung ; Chinho Park ; Benazzouz Aboubakr ; Ikken Badr ; Naimi Zakaria ; Benlarabi Ahmed

PRIMARY REFERENCE CELL CALIBRATIONS WITH REDUCED MEASUREMENT UNCERTAINTY 490

C.R. Osterwald ; L. Ottoson ; R. Williams ; C. Mack ; T. Moriarty ; K.A. Emery ; D.H. Levi

IMPLEMENTATION OF NOVEL PIN CONNECTION AND TEST ROUTINE FOR IMPROVED ACCURACY IN I-V MEASUREMENTS 496

Samuel Raj ; Johnson Kai Chi Wong ; Mohan Krishan Bhan ; Evan Palmer ; Jian Wei Ho ; Sumukh Ramprasad ; Wang Junci ; Thomas Mueller ; Armin G. Aberle

A NEW METHOD TO QUANTIFY CONTACT RESISTANCE USING LOCALIZED-ILLUMINATION PHOTOLUMINESCENCE TECHNIQUE IN A SOLAR CELL 499

Amit Singh Rajput ; Samuel Raj ; Johnson Wong ; Armin G. Aberle

IMPROVEMENT OF THE PROPERTIES OF CZTS THIN FILMS PREPARED BY SPRAY PYROLYSIS USING DMSO IN ACETONE AS SOLVENT 503

E. A. Ramírez ; A. Ramírez ; G. Gordillo

ASSESSMENT OF CARRIER LIFETIMES AND SURFACE RECOMBINATION VELOCITY THROUGH SPECTRAL MEASUREMENTS 508

John Roller ; Behrang H. Hamadani

IMPACT OF SPACE RADIATION ENVIRONMENT ON CONCENTRATOR PHOTOVOLTAIC SYSTEMS 512

Pilar Espinet-Gonzalez ; Tatiana Vinogradova ; Michael D. Kelzenberg ; Alexander Messer ; Emily C. Warmann ; Chris Peterson ; Nina Vaidya ; Ali Naqavi ; Jing-Shun Huang ; Samuel P. Loke ; Don Walker ; Colin J. Mann ; Sergio Pellegrino ; Harry A. Atwater

EXTRACTING THE FIXED CHARGE DENSITY IN HFOX FILMS GROWN ON HIGHLY-DOPED P-SI SAMPLES 517

Alexander To ; Jie Cur ; Bram Hoex

NEAR-UNITY ULTRA-WIDEBAND THERMAL INFRARED EMISSION FOR SPACE SOLAR POWER RADIATIVE COOLING 521

Ali Naqavi ; Samuel P. Loke ; Michael D. Kelzenberg ; Emily C. Warmann ; Pilar Espinet-González ; Nina Vaidya ; Jing-Shun Huang ; Tatiana A. Roy ; Alexander J. Messer ; Tatiana G. Vinogradova ; Ali Hajimiri ; Sergio Pellegrino ; Harry A. Atwater

LINE-FOCUS AND POINT-FOCUS SPACE PHOTOVOLTAIC CONCENTRATORS USING ROBUST FRESNEL LENSES, 4-JUNCTION CELLS, & GRAPHENE RADIATORS 525

Mark O'Neill ; A.J. Mcdanal ; Michael Piszczor ; Matt Myers ; Paul Sharps ; Claiborne Mcpheeters ; Jeff Steinfedt

SIMULATION OF LIGHT TRAPPING STRUCTURES FOR ENHANCING RADIATION HARDNESS IN SPACE SOLAR CELLS 531

Nizami Z. Vagidov ; Kyle H. Montgomery ; Geoffrey K. Bradshaw ; David M. Wilt

AN ALTERNATIVE METHOD FOR SOLAR CELL INTEGRATION 537

Jessica Buckner ; Tracy Davis ; Eric Muskovin ; Bernard Carpenter

NIEL DOSE ANALYSIS ON TRIPLE JUNCTION CELLS 30% EFFICIENT AND RELATED SINGLE JUNCTIONS 541

Roberta Campesato ; Erminio Greco ; Mariacristina Casale ; Massimo Gervasi ; P.G. Rancoita ; Davide Rozza ; Mauro Tacconi ; Enos Gombia ; Aldo Kingma ; Carsten Baur

THIN AND FLEXIBLE TRIPLE JUNCTION CELLS 30% EFFICIENT: QUALIFICATION RESULTS AND FUTURE SPACE APPLICATIONS 545

Roberta Campesato ; Mariacristina Casale ; Giuseppe Gabetta ; Emilio Fernandez Lisbona ; Laurent D'Abrigeon

PRINTED ASSEMBLIES OF MICROSCALE TRIPLE-JUNCTION (3J) INVERTED METAMORPHIC (IMM) GAINP/GAAS/INGAAS SOLAR CELLS 549

Boju Gai ; John Geisz ; Daniel Friedman ; Jongseung Yoon

COMPARATIVE STUDY ON NONRADIATIVE RECOMBINATION CENTERS IN PROTON IRRADIATED INAS/GAAS QUANTUM DOT STRUCTURE BY TWO WAVELENGTH EXCITED PHOTOLUMINESCENCE 552

M. D. Haque ; N. Kamata ; S-I. Sato ; S. M. Hubbard

DESIGN AND PROTOTYPING EFFORTS FOR THE SPACE SOLAR POWER INITIATIVE 558

Michael D. Kelzenberg ; Pilar Espinct-Gonzalez ; Nina Vaidya ; Tatiana A. Roy ; Emily C. Warmann ; Ali Naqavi ; Samuel P. Loke ; Jing-Shun Huang ; Tatiana G. Vinogradova ; Alexander J. Messer ; Christophe Leclerc ; Eleftherios E. Gdoutos ; Fabien Royer ; Ali Hajimiri ; Sergio Pellegrino ; Harry A. Atwater

DEFECT CHARACTERIZATION OF III-V QUANTUM STRUCTURE SOLAR CELLS USING PHOTO-INDUCED CURRENT TRANSIENT SPECTROSCOPY 562

Shin-Ichiro Sato ; Takeyoshi Sugaya ; Tetsuya Nakamura ; Takeshi Ohshima

EFFECT OF LUMINESCENCE COUPLING BETWEEN INGAP AND GAAS SUBCELLS TO EXTERNAL QUANTUM EFFICIENCY IN TRIPLE-JUNCTION SOLAR CELLS 567

Mitsunobu Suga ; Mitsuru Imaizumi ; Tetsuya Nakamur ; Takeshi Ohshima

LIGHTWEIGHT CARBON FIBER MIRRORS FOR SOLAR CONCENTRATOR APPLICATIONS 572

Nina Vaidya ; Michael D. Kelzenberg ; Pilar Espinet-Gonzalez ; Tatiana G. Vinogradova ; Jing-Shun Huang ; Christophe Leclerc ; Ali Naqavi ; Emily C. Warmann ; Sergio Pellegrino ; Harry A. Atwater

GAAS SOLAR CELLS ON V-GROOVED SILICON VIA SELECTIVE AREA GROWTH 578

Michelle Vaisman ; Nikhil Jain ; Qiang Li ; Kei May Lau ; Adele C. Tamboli ; Emily L. Warren

HIGH TEMPERATURE ANNEALING OF INI-XGAXN MQW SOLAR CELLS 582

Joshua J. Williams ; Heather Mcfavilen ; Steven Young ; Christiana B. Honsberg ; Stephen M. Goodnick

SOLAR PROBE PLUS ARRAY RELIABILITY 585

Anton Yanchilin ; Edward Gaddy

PHOTOVOLTAIC TEMPERATURE ESTIMATION MODEL FOR RAPID IRRADIANCE CHANGE CONDITIONS IN TROPICAL REGIONS USING HEURISTIC ALGORITHMS 589

R. Srivatsan ; Lian L. Jiang ; Douglas L. Maskell

ACCURACY OF CDTE PV ENERGY PREDICTIONS USING SPECTRAL CORRECTIONS 595

Mitchell Lee ; Kendra Passow ; Paul Wolffersdorff

PLANTPREDICT: SOLAR PERFORMANCE MODELING MADE SIMPLE 600

Kendra Passow ; Lauren Ngan ; Geoffrey Rich ; Mitch Lee ; Stephen Kaplan

INTEGRABILITY COMPARISON BETWEEN BIPV AND BAPV IN TROPICAL CONDITIONS: A BANGALORE CASE-STUDY 604

Gayathri Aaditya ; Roshan R Rao ; Monto Mani

A NEW PHOTOVOLTAIC SYSTEM TOPOLOGY THROUGH LOAD MANAGEMENT 608

Joseph A. Azzolini ; Meng Tao

FIRST STEP FOR POWER GENERATION AMOUNT ESTIMATION OF SOLAR MATCHING SYSTEM 613

Kazuya Hosokawa ; Toshiaki Yachi ; Yoichi Hirata ; Yasuyuki Watanabe

IRRADIANCE AND TEMPERATURE DISTRIBUTIONS AT HIGH LATITUDES: DESIGN IMPLICATIONS FOR PHOTOVOLTAIC SYSTEMS 619

Anne Gerdimenes ; Josefine Sclj

STEP-BY-STEP EVALUATION OF PHOTOVOLTAIC MODULE PERFORMANCE RELATED TO OUTDOOR PARAMETERS: EVALUATION OF THE UNCERTAINTY 626

Anne Migan Dubois ; Jordi Badosa ; Fausto Calderón-Obaldía ; Olivier Atlan ; Vincent Bourdin ; Marko Pavlov ; Dae Young Kim ; Yvan Bonnassieux

PERFORMANCE COMPARISONS OF A PV SYSTEM BY MONITORING SOLAR IRRADIANCE WITH DIFFERENT PYRANOMETERS 632

Yasuhiro Matsumoto ; J. Antonio Urbano ; Ramón Peña ; María De La Luz Olvera ; Nun Pitalúa ; Miguel A. Luna ; René Asomoza

FINANCIAL ANALYSIS OF A GRID-CONNECTED PHOTOVOLTAIC SYSTEM IN SOUTH FLORIDA 638

Hadis Moradi ; Amir Abtahi ; Ali Zilouchian

STUDY OF PHOTOVOLTAIC SYSTEMS MONITORING METHODS............643
E. Ortega ; G. Aranguren ; M.J. Sáenz ; R. Gutiérrez ; J.C. Jimeno

GLOBAL DESIGN ASPECTS OF PERSISTENT AND AUTONOMOUS PV POWERED SYSTEMS 648
I. M. Peters ; S. Watson ; N. Sahraei ; T. Buonassisi

HOW TO CHOOSE THE BEST EMPIRICAL MODEL FOR OPTIMUM ENERGY YIELD PREDICTIONS 652
Steve Ransome ; Juergen Sutterlueti

MODELING AND ANALYSIS OF PHOTOVOLTAIC ELECTROCHEMICAL SYSTEM USING MODULE-LEVEL POWER ELECTRONICS............658
Gowri M. Sriramagiri ; Nuha Ahmed ; Kevin D. Dobson ; Steven S. Hegedus

BETAVOLTAIC GENERATION FUNCTION IN SILICON............663
A.V. Sachenko ; I.O. Sokolovskyi ; M. Evstigneev

MULTI-OBJECTIVE OPTIMIZATION FOR COLOR-TUNABILITY AND TRANSPARENCY IN COLLOIDAL QUANTUM DOT SOLAR CELLS............667
Ebuka S. Arinze ; Botong Qiu ; Nathan Palmquist ; Yan Cheng ; Yida Lin ; Gabrielle Nyirjesy ; Gary Qian ; Susanna M. Thon

CUBIC PHASE INXGA1-XN/GAN QUANTUM WELLS FOR THEIR APPLICATION TO TANDEM SOLAR CELLS............670
C. A. Hernández-Gutiérrez ; Y. L. Casallas-Moreno ; Dagoberto Cardona ; Yu. Kudriavtsev ; A. Morales-Acevedo ; G. Santana-Rodríguez ; M. López-López

MODELING OF P-I-N GAASPN/GAP MQWS SOLAR CELL: TOWARDS LATTICE MATCHED III-V/SI TANDEM............673
Khim Kharel ; Alexandre Freundlich

INP QUANTUM DOT INTERMEDIATE BAND SOLAR CELL GROWN VIA MOCVD 677
Hyun Kum ; Yushuai Dai ; Michael Slocum ; Zachary Bittner ; Seth Hubbard

MODIFIED LIMITING EFFICIENCY FOR MULTIPLE EXCITON GENERATION SOLAR CELLS............681
Jongwon Lee ; Christiana B. Honsberg

A SIMPLE MONTE CARLO MODEL OF A HOT CARRIER CELL............685
Tor Oskar Saetre

OPTIMIZATION OF SEMICONDUCTOR QUANTUM DOTS FOR LUMINESCENT SOLAR CONCENTRATORS: MINIMIZING REABSORPTION LOSSES............690
Anatoli I. Shkrebtii ; Anatoliy V. Sachenko ; Igor O. Sokolovskyi ; Vitaliy P. Kostylyov ; Mykola R. Kulish ; Denis V. Khomcnko ; Mykhaylo A. Evstigneev

DEVELOPMENT OF ABSORBER AND ENERGY SELECTIVE CONTACTS FOR HOT CARRIER SOLAR CELLS............696
Santosh Shrestha ; Simon Chung ; Yuanxun Liao ; Wenkai Cao ; Neeti Gupta ; Yi Zhang ; Xiaoming Wen ; Gavin Conibeer

GAASBI DEVICES FOR THERMAL ENERGY CONVERSION............701
Margaret Stevens ; Abigail Licht ; Nicole Pfiester ; Emily Carlson ; Kevin Grossklaus ; Thomas E. Vandervelde

ANALYTIC JV-CHARACTERISTICS OF IDEAL IMPURITY PV-CELLS 706
Rune Strandberg

PHOTOLUMINESCENCE PROPERTIES OF IN-PLANE ULTRAHIGH-DENSITY INAS QUANTUM DOTS ON GAASSB/GAAS(001) FOR SOLAR CELL APPLICATIONS............712
Ryo Sugiyama ; Naoki Akimoto ; Tomah Sogabe ; Koichi Yamaguchi

CARRIER SELECTIVE BACK CONTACT (CSBC) SOLAR CELL USING TRANSITION METAL OXIDES............716
Astha Tyagi ; Kunal Ghosh ; Anil Kottantharayil ; Saurabh Lodha

ANALYSIS OF OPEN-CIRCUIT VOLTAGE AND CONVERSION EFFICIENCY IN QUANTUM-DOT SOLAR CELLS VIA DETAILED-BALANCE-LIMIT THEORY............721
Lin Zhu ; Hidefumi Akiyama ; Yoshihiko Kanemitsu

ZINC SELENIDE SURFACE PASSIVATION LAYER FOR SINGLE-CRYSTALLINE CZTSE SOLAR CELLS............726
Michael A. Lloyd ; Douglas Bishop ; Brian E. Mccandless ; Robert Birkmirc

USE OF SINGLE WALL CARBON NANOTUBE FILMS DOPED WITH TRIETHYLOXONIUM HEXACHLORANTIMONATE AS A TRANSPARENT BACK CONTACT FOR CDTE SOLAR CELLS............730
Fadhil K. Alfadhili ; Jacob M. Gibbs ; Geethika K. Liyanage ; Patrick W. Krantz ; Suneth C. Watthage ; Zhaoning Song ; Adam B. Phillips ; Michael J. Heben

GRAIN AND GRAIN BOUNDARY GEOMETRICAL SHAPE CONSIDERATIONS ON SODIUM AND POTASSIUM DIFFUSION THROUGH MOLYBDENUM FILMS............735
Orlando Ayala ; Chinedum Akwari ; Tasnuva Ashrafee ; Shankar Karki ; Grace Rajan ; Sylvain Marsillac

SOLUTION-PROCESSED NICKEL-ALLOYED IRON PYRITE THIN FILM AS HOLE TRANSPORT LAYER IN CADMIUM TELLURIDE SOLAR CELLS 738

Ebin Bastola ; Khagendra P. Bhandari ; Randy J. Ellingson

USE OF CDS:O AND CDSE AS WINDOW LAYERS FOR CDTE PHOTOVOLTAICS 742

Tom Baines ; Guillaume. Zoppi ; Ken Durose ; Jonathan D. Major

APPLICATIONS OF HYBRID ORGANIC-INORGANIC METAL HALIDE PEROVSKITE THIN FILM AS A HOLE TRANSPORT LAYER IN CDTE THIN FILM SOLAR CELLS 748

Khagendra P. Bhandari ; Suneth C. Watthage ; Zhaoning Song ; Adam Phillips ; Michael J. Heben ; Randy J. Ellingson

MAGNESIUM-DOPED ZINC OXIDE AS A HIGH RESISTANCE TRANSPARENT LAYER FOR THIN FILM CDS/CDTE SOLAR CELLS 752

Francesco Bittau ; Elisa Artegiani ; Ali Abbas ; Daniele Menossi ; Alessandro Romeo ; Jake W. Bowers ; John M. Walls

INVESTIGATION OF ZNL-XMGXO:A1 FILM BY RATIO FREQUENCY MAGNETRON CO-SPUTTERING AS TRANSPARENT CONDUCTIVE OXIDE LAYER 757

Jakapan Chantana ; Yuya Ishino ; Takashi Minemoto

A NEW TCO/WINDOW-BUFFER FRONT STACK FOR CDTE SOLAR CELLS AND ITS IMPLEMENTATION 761

Alan E. Delahoy ; Xuehai Tan ; Akash Saraf ; Payal Patra ; Surya Manda ; Yunfei Chen ; Krishnakumar Velappan ; Bastian Siepchen ; Shou Peng ; Ken K. Chin

SYNTHESIS OF HIGH-QUALITY AZO POLYCRYSTALLINE FILMS VIA TARGET BIAS RADIO FREQUENCY MAGNETRON SPUTTERING 767

Zhongming Du ; Yufeng Zhang ; Xiangxin Liu

CLOSE-SPACE SUBLIMATED CDTE SOLAR CELLS WITH CO-SPUTTERED CDSXSE1-XALLOY WINDOW LAYERS 771

Corey R. Grice ; Maxwell Junda ; Alexander Archer ; Jian Li ; Yanfa Yan

EFFECTS OF GRAPHENE OXIDE BARRIER ON CU2ZNSNSXSE4-XTHIN FILM SOLAR CELLS 777

Woo-Lim Jeong ; Jung-Hong Min ; In-Young Kim ; Hae-Sun Kim ; Jin-Hyeok Kim ; Dong-Seon Lee

13% CDS/CDTE SOLAR CELL USING A NANOCOMPOSITE (CUS)X(ZNS)1-X THIN FILM HOLE TRANSPORT LAYER 781

Kamala Khanal Subedi ; Khagendra P. Bhandari ; Ebin Bastola ; Randy J. Ellingson

MOLYBDENUM OXIDE AND MOLYBDENUM NITRIDE BACK CONTACTS FOR THIN-FILM CDTE SOLAR CELLS 785

Anna Kindvall ; Jason Kephart ; Walajabad Sampath

INVESTIGATION AND OPTIMIZATION OF CD-FREE BUFFER LAYERS IN2S3 AND ZN(O, S) FOR CU2ZNSN(S, SE)4-BASED SOLAR CELLS 791

Willi Kogler ; Thomas Schnabel ; Andreas Bauer ; Stefanie Spiering ; Erik Ahlswede ; Michael Powalla

REAR CONTACT PASSIVATION FOR HIGH BANDGAP CU(IN, GA)SE2 SOLAR CELLS WITH VARYING ABSORBER THICKNESS AND FLAT GA PROFILE 796

Dorothea Ledinek ; Pedro Salome ; Carl Hägglund ; Marika Edoff

LASER ANNEALED BACK CONTACTS FOR CDTE SOLAR CELLS 802

Vasilios Palekis ; Shamara Collins ; Imran Khan ; Vamsi Evani ; Sudhajit Misra ; Michael A. Scarpulla ; Mark Lonergan ; Don Morel ; Chris Ferekides

ENHANCED ANTI-REFLECTIVE COATING FOR THIN FILM SOLAR CELLS 807

Grace Rajan ; Shankar Karki ; Robert W. Collins ; Sylvain Marsillac

INFLUENCE OF AGS LAYER INSERTION AT ABSORBER/ITO INTERFACE ON STRUCTURAL AND PHOTOVOLTAIC PROPERTIES OF ULTRATHIN CU(IN,GA)SE2 SOLAR CELLS 810

Muhammad Saifullah ; Jihye Gwak ; Kihwan Kim ; Joo Hyung Park ; Junsik Cho ; Jae Ho Yun

NOVEL, FACILE BACK SURFACE TREATMENT FOR CDTE SOLAR CELLS 815

Suneth C. Watthage ; Geethika K. Liyanage ; Zhaoning Song ; Fadhil K. Alfadhili ; Rabee B. Alkhayat ; Khagendra P. Bhandari ; Randy J. Ellingson ; Adam B. Phillips ; Michael J. Heben

OPTIMIZING CDS BUFFER LAYER FOR CIGS BASED THIN FILM SOLAR CELL 820

Weijie Zhang ; Korhan Demirkan ; Geordie Zapalac ; David Spaulding ; Jochen Titus ; Neil Mackie

INVESTIGATION OF INP DEFECT CHARACTERISTICS GROWN USING NOVEL TF-VLS TECHNIQUE 823

Abhinav Chikhalkar ; Alec Fischer ; Mark Hettick ; Ali Javey ; Richard R. King

INVESTIGATION OF FAST GROWTH GAAS-BASED SOLAR CELL ON REUSABLE SUBSTRATE BY METALORGANIC CHEMICAL VAPOR DEPOSITION 827

Chaomin Zhang ; Abhinav Chikhalkar ; Ehsan Vadiee ; Richard King ; Christiana Honsberg ; Eric Armour ; Yeongho Kim

DEVELOPMENT OF ALUMINUM EPILAYERS AS BUFFERS FOR GAINAS 831
Phil Ahrenkiel ; Nathan Smaglik ; Nikhil Pokharel ; Alessandro Giussani ; Michael A. Slocum ; Seth M. Hubbard

LASER CRYSTALLIZATION OF AMORPHOUS GERMANIUM ON TITANIUM NITRIDE-COATED STEEL FOR LOW-COST GAAS SOLAR-CELLS 837
Saloni Chaurasia ; Srinivasan Raghavan ; Sushobhan Avasthi

HIGH QUALITY EPITAXIAL GERMANIUM ON SI (100) FOR LOW -COST III–V SOLAR-CELLS 841
Saloni Chaurasia ; Srinivasan Raghavan ; Sushobhan Avasthi

CRYSTALLINITY CONTROL IN LOW-TEMPERATURE GROWTH OF POLY-CRYSTALLINE GE BY ION BEAM DEPOSITION 845
S. I. Maximenko ; N. A. Mahadik ; P. P. Jenkins ; R. J. Walters ; A. Giussani ; E. L. Mcclure ; S. M. Hubbard ; C. Bailey

HIGH EFFICIENCY GAINP/GAAS DOUBLE JUNCTION SOLAR CELL ON SI SUBSTRATE ASSISTED BY THE ELECTRON BEAM TREATMENT 849
Hyo Jin Kim ; Yong Whan Kim

ANALYSIS OF DEPOSITED RESIDUES AND ITS CLEANING PROCESS ON GAAS SUBSTRATE AFTER EPITAXIAL LIFT-OFF 854
Tatsuya Nakata ; Kentaroh Watanabe ; Hassanet Sodabanlu ; Daiki Kimura ; Naoya Miyashita ; Yoshitaka Okada ; Yoshiaki Nakano ; Masakazu Sugiyama

ULTRATHIN SILICON-AN-INSULATOR (SOI) WAFER FOR COMPLIANT SUBSTRATE 858
Shinyoung Noh ; Anita Ho-Baillie ; Stephen Bremner ; Martin A. Green ; Xiaojing Hao

CHARACTERIZATION OF GAAS SOLAR CELLS GROWN BY HYDRIDE VAPOR PHASE EPITAXY IN HORIZONTAL REACTOR 861
Ryuji Oshima ; Kikuo Makita ; Takeyoshi Sugaya ; Akinori Ubukata

FLEXIBLE GAAS SINGLE-JUNCTION SOLAR CELLS BASED ON SINGLE-CRYSTAL-LIKE THIN-FILM MATERIALS DIRECTLY GROWN ON METAL TAPES 866
Sara Pouladi ; Monika Rathi ; Mojtaba Asadirad ; Pavel Dutta ; Seung Kyu Oh ; Devendra Khatiwada ; Shahab Shervin ; Yao Yao ; Venkat Selvamanickam ; Jae-Hyun Ryou

REDUCED DEFECT DENSITY IN SINGLE-CRYSTALLINE-LIKE GAAS THIN FILM ON FLEXIBLE METAL SUBSTRATES BY USING SUPERLATTICE STRUCTURES 869
M. Rathi ; P. Dutta ; D. Khatiwada ; Y. Yao ; Y. Gao ; Y. Li ; S. Sun ; S. Pouladi ; S. Reed ; A. Khadimallah ; J. Ryou ; V. Selvamanickam ; N. Zheng ; P. Ahrenkiel

ECONOMIC ANALYSIS OF TRANSFER PRINTED III–V VIRTUAL SUBSTRATES 873
Kenneth J. Schmieder ; Matthew P. Lumb ; Michael K. Yakes ; Shawn Mack ; Mitchell F. Bennett ; Sergey I. Maximenko ; Laura B. Ruppalt ; Michael A. Meeker ; Chase T. Ellis ; Matthew Meitl ; Joseph G. Tischler ; Robert J. Walters

THIN FILMS OF ZINC-DOPED GAAS BY RF MAGNETRON SPUTTERING FOR USE IN PHOTOVOLTAIC CELLS 876
Kirby Simon ; Kyle Cepeda ; Nishit Shetty ; Elijah Thimsen

SELF ALIGNED ALUMINUM SELECTIVE EMITTER FOR N-TYPE SI CELLS 881
San Theigi ; Robert C. Reedy ; Vincenzo Lasalvia ; Paul Stradins ; Benjamin G. Lee

HOW TO REALIZE SOLAR CELLS WITH LASER STRUCTURED PLATED NI-CU-CONTACTS WITH EXCELLENT ADHESION AND HIGH FILL-FACTORS WITHOUT PARASITIC PLATING 884
A. Büchler ; S. Kluska ; J. Bartsch ; B. Grübel ; A.A. Brand ; S. Gutscher ; M. Glatthaar

EXPLOITING THE POTENTIALS OF THE FRONT SURFACE FIELD (FSF) INDUSTRIAL SILICON SOLAR CELL 888
Ahrar Ahmed Chowdhury ; Yu -Chen Hsu ; Veysel Unsur ; Abasifreke Ebong

PHOTOVOLTAIC PERFORMANCE OF SILICON SOLAR CELLS ENHANCED BY PLASMONIC SILVER NANOPARTICLES OF VARIOUS DIMENSIONS DEPOSITING THROUGH ANODIC ALUMINUM OXIDE TEMPLATE 893
Ta-Wei Chuang ; Wen-Jeng Ho ; Sheng-Kai Feng ; Jheng-Jie Liu ; Guan-Yi Li ; Hao-Yu Yang ; Yun-Chie Yang ; Cho-Chun Chiang ; Yao-Hui Chen

MITIGATION OF POTENTIAL-INDUCED DEGRADATION 896
Orry Faur ; Maria Faur

ELECTRODEPOSITION OF SI-LAYER THROUGH REDUCTION OF DIATOMACEOUS EARTH FOR THE APPLICATION OF SOLAR-CELLS 900
Muhammad Monirul Islam ; Imane Abdellaoui ; Takeaki Sakurai ; Saad Hamzaoui ; Katsuhiro Akimoto

EFFECT OF SI CONTENT IN Al PASTE ON LOCAL Al REAR CONTACTS IN PERC CELL 904
Supawan Joonwichien ; Katsuhiko Shirasawa ; Satoshi Utsunomiya ; Hidetaka Takato

NEW SILVER PASTE METALLIZATION APPROACH ON P+ DIFFUSION ZONES OF SILICON SOLAR CELLS 907
Yunjun Li ; Mohshi Yang ; Igor Pavlovsky ; Guoping Zeng

INFLUENCES OF ANNEALING AND DEFECT LIMITATION ON P-TYPE SILICON SOLAR CELL 911
Yu-Hsuan Lin ; Sung-Yu Chen ; Kuen-Yi Wu ; Chien-Hsun Chen ; Chen-Hsun Du ; Chun-Ming Yeh

REDUCED TEMPERATURE SILVER PASTE WITH LOW CONTACT RESISTANCE FOR ADVANCED SOLAR CELL APPLICATIONS 914
Ryan Mayberry ; Daniel Holzmann ; Gerd Schulz ; Lindsey Karpowich ; Mark Naylor ; Matthias Hoerteis

BSF ISLANDS FOR REDUCED RECOMBINATION IN IBC CELLS 917
Agnes A. Mewe ; Nicolas Guillevin ; Ilkay Cesar ; Antonius R. Burgers

THERMAL STABILITY OF HYDROGENATED BORON EMITTERS 921
Khaja H. Mohammed ; Larry C. Cousar ; Philip A. Mcmeans ; Garrett Z. Evans ; Douglas A. Hutchings ; Hameed A. Naseem ; Sergiu C. Pop

LIGHT INDUCED PLATING OF SILICON SOLAR CELLS USING BORIC ACID-FREE NICKEL CHEMISTRY 925
Krystal Munoz ; Lynne Michaelson ; Joseph Karas ; Tom Tyson ; James Rand ; Stuart Bowden

BAKING TEMPERATURE DEPENDENCE OF CU PASTE ON A1-BSF CELL PROPERTIES 931
Tomohiro Saito ; Tetsuya Fukuda ; Hoang Tri Hai ; Yuji Kurimoto ; Daisuke Ando ; Yuji Sutou ; Katsuhiko Shirasawa ; Junichi Koike

THE SILVER CONTACT AND FORMATION MECHANISM OF THE BORON EMITTER AND THE CURRENT FLOW MECHANISM OF THE SOLAR CELL ELECTRODE 935
Seunghyun Shin ; Soohyun Bae ; Sungeun Park ; Yoonmook Kang ; Hae-Seok Lee ; Donghwan Kim

LASER ANNEALING TO ENHANCE PERFORMANCE OF ALL-LASER-BASED SILICON BACK CONTACT SOLAR CELLS 937
Zeming Sun ; Mool C. Gupta

LARGE AREA N-TYPE SELECTIVE EMITTER CELLS USING LASER DOPING THROUGH BORON DOPED SCREEN PRINTED PASTE 940
Ajay D Upadhyaya ; Vijaykumar D Upadhyaya ; Brian Rounsaville ; Keeya Madani ; Ajeet Rohatgi ; Toru Hanada

METALLIZED BORON-DOPED BLACK SILICON EMITTERS FOR FRONT CONTACT SOLAR CELLS 944
Guillaume Von Gastrow ; Hele Savin ; Eric Calle ; Pablo Ortega ; Ramón Alcubilla ; Andreana Daniil ; Elias Z. Stutz ; Anna Fontcuberta I Morral ; Sebastian Husein ; Tara Nietzold ; Mariana Bertoni

CONTACT RESISTANCE MEASUREMENT FOR THERMALLY DIFFUSED POINT CONTACT BY LOCALIZED DIELECTRIC BREAKDOWN SOLAR CELLS 948
Qilin Ye ; Ned J. Western ; Anqi Liao ; Stephen P. Bremner

LOW TEMPERATURE REAR SURFACE METALLIZATION OF MULTI-CRYSTALLINE SILICON SOLAR CELLS FOR IMPROVED BULK LIFETIME 953
N. J. Western ; S. P. Bremner

INVESTIGATION OF HIGH PERFORMANCE PEROVSKITE-BASED SOLAR CELLS GROWN BY HYBRID CHEMICAL VAPOR DEPOSITION TECHNIQUE 958
Huseyin Cem Gokkaya ; Shen Qian ; Zhiwei Ren ; Annie Ng ; Charles Surya

ENHANCED PEROVSKITE SOLAR CELL PERFORMANCE USING FULL SPACE DEVICE OPTIMIZATION 963
Ahmer A.B. Baloch ; Shahzada P. Aly ; Mohammad I. Hossain ; Raka Jovanovic ; Nouar Tabet ; Fahhad H. Alharbi

MEASURING OPTICAL ABSORPTION IN ORGANIC PHOTOVOLTAICS USING MONOCHROMATED ELECTRON ENERGY-LOSS SPECTROSCOPY 966
Jessica A. Alexander ; Frank J. Scheltens ; David W. Mccomb ; Lawrence F. Drummy ; Michael F. Durstock ; James B. Gilchrist ; Sandrine Hentz

ADVANCED DEPOSITION OF PHOTO-CATALYTIC TIO2 FILM BY ATMOSPHERIC SPPS FOR DYE SENSITIZED SOLAR CELLS 970
Ifeanacho Anyadiegwu ; Dickson Kindole ; Geoffrey Kibiegon Ronoh ; Yoshimasa Noda ; Yasutaka Ando

CH3NH3PBI3-XBRXPEROVSKITE SOLAR CELLS VIA SPRAY ASSISTED TWO-STEP DEPOSITION: INFLUENCE OF BROMIDE ON THE DEVICE PERFORMANCE 976
Gaoda Chai ; Shiqiang Luo ; Shizhen Wang ; Hang Zhou

MODULATED STRUCTURE TO MAXIMIZE THE OPEN-CIRCUIT VOLTAGE WITH MODERATE BAND-GAP OF SMALL MOLECULE ORGANIC SOLAR CELLS-DFT APPROACH 980
Saravanan Chinnusamy ; Amita Munshi ; Sukanya Santhosh Kumar ; W. S. Sampath ; Milind S. Dangate

PEROVSKITE GRAIN SIZE MODULATION BY ANNEALING IN METHYL-AMINE ENVIRONMENT 986
Arun Singh Chouhan ; Naga Prathibha Jasti ; Srinivasan Raghavan ; Sushobhan Avasthi ; Shreyash Hadke

FE2O3AS AN ELECTRON TRANSPORT MATERIAL FOR ORGANO-METAL HALIDE PEROVSKITE SOLAR CELLS 989
Dallas Fisher ; Pravakar P. Rajbhandari ; Tara P. Dhakal

OPTICAL EVALUATION OF PEROVSKITE FILMS IN AND FOR SOLAR CELL DEVICE STRUCTURES 993

Kiran Ghimire ; Dewei Zhao ; Changlei Wang ; Yanfa Yan ; Nikolas J. Printraza

HYBRID ORGANIC-INORGANIC SOLAR CELLS WITH A BENZOQUINONE PASSIVATING LAYER 999

James Hack ; Abhishek Iyer ; Meixi Chen ; Nicole Kotulak ; Akirt Sridharan ; Robert Opila

PRECISE 1-V CURVE MEASUREMENT PROCEDURE FOR PEROVSKITE SOLAR CELLS: APPLICATION TO VARIOUS TYPES OF DEVICES 1003

Y. Hishikawa ; M. Yoshita ; H. Shimura ; A. Sasaki ; T. Ueda

ENHANCING THE CRYSTALLINE OF PLANAR-STRUCTURE CH3NH3PBI3PEROVSKITE SOLAR CELLS VIA SANDWICH EVAPORATION TECHNIQUE 1006

Po-Tsun Kuo ; Shang-Pang Lin ; Cheng-Shian Lin ; Ching-Fuh Lin

TOWARD HIGH PERFORMANCE ORGANIC-SILICON HYBRID SOLAR CELLS 1009

Yi Lai ; Hong-Jhang Syu ; Ching-Fuh Lin

NICKEL OXIDE THIN FILMS BY RADIO FREQUENCY SPUTTER FOR INVERTED PEROVSKITE SOLAR CELLS 1012

Hyeonseok Lee ; Yu-Ting Huang ; Shien-Ping Feng

ANOMALOUS EFFICIENCY SCALING WITH DARK CURRENT IN PEROVSKITE SOLAR CELLS 1015

Vikas Nandal ; Pradeep R. Nair

NUMERICAL SIMULATION AND PERFORMANCE OPTIMIZATION OF PEROVSKITE SOLAR CELL 1018

Sai Naga Raghuram Nanduri ; Mahbube K. Siddiki ; Ghulam M. Chaudhry ; Yahya Z. Alharthi

PERFORMANCE PREDICTION FOR LARGE AREA PEROVSKITE SOLAR CELLS 1022

Yojak Raote ; Hitarth Choubisa ; Pradeep R. Nair

PHOTOCONVERSION EFFICIENCY MODELING IN PEROVSKITE SOLAR CELLS 1025

A.V. Sachenko ; V.P. Kostylyov ; A.V. Bobyl ; V.M. Vlasiuk ; I.O. Sokolovskyi ; E.I. Terukov ; M. Evstigneev

INFLUENCE OF MONO- AND DI-VALENT METAL ADDITIVES ON MORPHOLOGY AND CHARGE CARRIER DYNAMICS OF CH3NH3PBI3PEROVSKITE 1030

Niraj Shrestha ; Suneth C. Watthage ; Zhaoning Song ; Paul J. Roland ; Adam B. Phillips ; Michael J. Heben ; Randall J. Ellingson

EFFECT OF DUAL CATHODE BUFFER LAYER ON TERNARY ORGANIC SOLAR CELL 1034

Ashish Singh ; T. Bhim Raju ; Anamika Dey ; Ritesh Kant Gupta ; Parameswar K. Iyer

COPPER PLATED TOP ELECTRODE FOR AN INVERTED ORGANIC PHOTOVOLTAIC 1037

Malia Steward ; Zhan Shi ; Kyoung- Tae Kim ; Seungkeun Choi

INTERFACE BAND GAP AND CHARGE TRAPPING IN BULK HETEROJUNCTION SOLAR CELLS 1040

Marian Tzolov ; Maxwell Mcintyre

FABRICATION OF EFFICIENT CH3NH3PBI3 SOLAR CELLS IN AMBIENT AIR 1044

Feng Wang ; Ye Zhongbiao ; Hojjatollah Sarvari ; Somin Park ; Kenneth Graham ; Yuetao Zhao ; Zhi David Chen

HIGH EFFICIENCY PEROVSKITE SOLAR CELLS BY A MODIFIED LOW-TEMPERATURE SOLUTION PROCESS INTER-DIFFUSION METHOD 1048

Yangyi Yao ; Wei-Lun Hsu ; Mario Dagenais

INTERFACIAL MODIFICATION OF SOL-GEL ZNO/AZO BILAYER AS HIGHLY EFFICIENT ELECTRON TRANSPORT LAYER FOR PEROVSKITE SOLAR CELLS 1051

Shang-Hsuan Wu ; Ming-Yi Lin ; Sheng-Hao Chang ; Wei-Chen Tu ; Chi-Wei Chu ; Via-Chung Chang

THE POTENTIAL OF BIFACIAL PHOTOVOLTAICS: A GLOBAL PERSPECTIVE 1055

Xingshu Sun ; Mohammad R. Khan ; Amir Hanna ; Muhammad M. Hussain ; Muhammad A. Alam

PERFORMANCE ASSESSMENT OF STAND ALONE BIFACIAL SOLAR PANEL UNDER REAL TIME CONDITIONS 1058

Ahmer A.B. Baloch ; Maher Armoush ; Basel Hindi ; Abdelkader Bousselham ; Nouar Tabet

OPERATION AND PERFORMANCE ASSESSMENT OF GRID-CONNECTED PV SYSTEMS IN OPERATION IN MAUI, HAWAII 1061

Severine Busquet ; Jonathan Kobayashi ; Richard E. Rocheleau

A NOVEL MULTILEVEL SOLAR PANEL SYSTEM: IMPLEMENTATION AND VERIFICATION 1067

Tanmoy Debnath ; Syed N. Imtiaz ; Syed F. Nawaz ; Abdullah Al Mahmud ; Mosaddequr Rahman

PREDICTING POWER LOSS DUE TO MODULE MISMATCH IN UTILITY-SCALE PHOTOVOLTAIC SYSTEMS 1071

Stephen Kaplan ; Kendra Passow

APPLICATION OF SHAPED REFLECTORS TO INCREASE THE ENERGY HARVEST OF BIFACIAL PV SYSTEMS - ANALYZED WITH A MINIATURIZED TEST ARRAY 1077

Hartmut Nussbaumer ; Markus Klenk ; Nico Keller ; Dominic Heller ; Remo Kaslin ; Thomas Baumann ; Franz Baumgartner

TOWARDS NEW MODULE AND SYSTEM CONCEPTS FOR LINEAR SHADING RESPONSE 1081

Kostas Sinapis ; Tom T.H. Rooijakkers ; Lenneke H. Slooff ; Lars A.G. Okel ; Mark J. Jansen ; Anna J. Carr

PARTIAL SHADING ABATEMENT THROUGH CASCADED H-BRIDGE TOPOLOGY 1086

Steven Tidwell ; Joseph Latham ; Michael Mcintyre

DATA ANALYSIS FOR EFFECTIVE MONITORING OF PARTIALLY SHADED PHOTOVOLTAIC SYSTEMS 1090

Odysseas Tsafarakis ; Kostas Sinapis ; Wilfried G.J.H.M. Van Sark

BIFACIAL PHOTOVOLTAIC MODULE ENERGY YIELD CALCULATION AND ANALYSIS 1094

Christopher E. Valdivia ; Chu Tu Li ; Annie Russell ; Joan E. Haysom ; Rui Li ; David Lekx ; Mohsen M. Sepeher ; Dan Henes ; Karin Hinzer ; Henry P. Schriemer

DESIGN AND DEVELOPMENT OF A SOLAR PHOTOVOLTAIC MODULE DETECTION CONTROL SYSTEM BASED ON PLC 1100

Yiwang Wang ; Jili Zhang ; Kanglin Liu ; Houjun Tang ; Hui Pan ; Yan Lin ; Peter Yang ; Rui Wang

DETECTING CALIBRATION DRIFT AT GROUND TRUTH STATIONS A DEMONSTRATION OF SATELLITE IRRADIANCE MODELS' ACCURACY 1104

Richard Perez ; James Schlemmer ; Adam Kankiewicz ; John Dise ; Alemu Tadese ; Thomas Hoff

PERFORMANCE OF SOLAR RESOURCE MONITORING STATIONS IN HOT CLIMATE REGIONS 1110

Yahya Z. Alharthi ; Mahbube K. Siddiki ; Ghulam M. Chaudhry ; Saad Muaddi ; Ahmed Alahmed

FIRST RESULTS OF A LOW COST ALL-SKY IMAGER FOR CLOUD TRACKING AND INTRA-HOUR IRRADIANCE FORECASTING SERVING A PV-BASED SMART GRID IN LA GRACIOSA ISLAND 1116

David Cañadillas ; Walter Richardson ; Benjamín Gonzalez-Díaz ; Les E. Shephard ; Ricardo Guerrero Lemus

STATISTICAL ANALYSIS OF PV INSOLATION DATA 1122

Abdulmunim Guwaeder ; Rama Ramakumar

A COMPARISON OF PV POWER FORECASTS USING PVLIB-PYTHON 1127

William F. Holmgren ; Antonio T. Lorenzo ; Clifford Hansen

COMPARING THE TYPICAL GHI YEAR VS TYPICAL POWER YEAR 1132

Alex Kubiniec ; Adam Kankiewicz ; Alemu Tadesse

THE HOLY GRAIL OF RESOURCE ASSESSMENT: LOW COST GROUND-BASED MEASUREMENTS WITH GOOD ACCURACY 1134

Bill Marion ; Benjamin Smith

GLOBAL COMPARISON OF THE IMPACT OF TEMPERATURE AND PRECIPITABLE WATER ON CDTE AND SILICON SOLAR CELLS 1140

I. M. Peters ; L. Haohui ; T. Reindl ; T. Buonassisi

ESTIMATION OF MEAN MONTHLY GLOBAL SOLAR RADIATION USING MODEL BASED ON SUNSHINE HOURS FOR COLOMBIA 1143

Diego J. Rodríguez ; Johan Hernández ; Adolfo Jaramillo

IMPLEMENTATION OF SOLAR DIFFUSE CIE MODEL IN RAY TRACING PROGRAM FOR IRRADIANCE CALCULATIONS 1147

Liliana Ruiz Diaz ; Pierre-Alexandre Blanche ; Robert A. Norwood

INVESTIGATION OF CITY-LEVEL SITE-PAIR CORRELATIONS OF SOLAR VARIABILITY USING EMPIRICAL SATELLITE DATA 1151

Rhythm Singh ; Rangan Banerje

ULTRA-SHORT-TERM PHOTOVOLTAIC GENERATION FORECASTING MODEL BASED ON WEATHER CLUSTERING AND MARKOV CHAIN 1158

Jin Tan ; Changhong Deng

DAILY SOLAR IRRADIANCE PROFILE CHARACTERIZATION AND RAMP RATE ANALYSIS AT DIFFERENT TIME RESOLUTIONS 1163

Spyros Theocharides ; Venizelos Venizelou ; George Makrides ; George E. Georghiou

COMPARISON AND ANALYSIS OF INSTRUMENTS MEASURING PLANE OF ARRAY IRRADIANCE FOR ONE-AXIS TRACKING PV SYSTEMS 1169

Frank Vignola ; Chun-Yu Chiu ; Josh Peterson ; Michael Dooraghi ; Manajit Sengupta

A SKY IMAGE ANALYSIS SYSTEM FOR SUB-MINUTE PV PREDICTION 1175

Rodrigo Verschae ; Li Li ; Shohei Nobuhara ; Takekazu Kato

LARGE AREA NANOSTRUCTURE INTEGRATION FOR BROAD-SPECTRUM, OMNIDIRECTIONAL ANTIREFLECTION IMPROVEMENTS ON POLYMER PACKAGED, MECHANICALLY FLEXIBLE, EPITAXIAL LIFT-OFF III-V SOLAR CELLS 1181

Gabriel Cossio ; Jihwan Lee ; Gautham Ragunathan ; Andre Wibowo ; Sudersena Rao Tatavarti ; Kimberly Sablon ; Edward T. Yu

DEVELOPMENT OF BACK SURFACE TEXTURE FOR LIGHT MANAGEMENT IN EPITAXIAL LIFT OFF (ELO) QUANTUM DOT SOLAR CELLS 1184

Brittany L. Smith ; George T. Nelson ; Yushuai Dai ; Michael A. Slocum ; Andre Wibowo ; Rao Tatavarti ; Seth M. Hubbard

ENABLING HIGH-EFFICIENCY INAS/GAAS QUANTUM DOT SOLAR CELLS BY EPITAXIAL LIFT-OFF AND LIGHT MANAGEMENT 1189

F. Cappelluti ; A. P. Cédola ; A. Khalili ; Farid Elsehrawy ; G. Bauhuis ; P. Mulder ; J. Schermer ; G. Bissels ; T. Aho ; T. Niemi ; M. Guina ; D. Kim ; J. Wu ; H. Liu

CHARACTERIZATION OF ARSENIC DOPED CDTE LAYERS AND SOLAR CELLS 1193

Sachit Grover ; Xiaoping Li ; Wei Zhang ; Ming Yu ; Gang Xiong ; Markus Gloeckler ; Roger Malik

ENHANCING P-TYPE DOPING IN POLYCRYSTALLINE CDTE FILMS 1196

Brian Mccandless ; Wayne Buchanan ; Gowri Sriramagiri ; Christopher Thompson ; Joel Duenow ; David Albin ; Soren Jensen ; John Moseley ; M. Al-Jassim ; Wyatt K. Metzger

SPECTRAL AND CONCENTRATION SENSITIVITY OF MULTIJUNCTION SOLAR CELLS AT HIGH TEMPERATURE 1201

Daniel J. Friedman ; Myles A. Steiner ; Emmett E. Perl ; John Simon

ON THE USE OF TRANSPARENT CONDUCTIVE OXIDES IN HIGH CONCENTRATOR III-V MULTIJUNCTION SOLAR CELLS 1204

Ignacio Rey-Stolle ; Yeonbae Lee ; Iván Garcia ; Luis Cifuentes ; Kin Man Yu ; Carlos Algora ; Wladek Walukiewicz

COMPONENT INTEGRATION EFFECTS IN 4-JUNCTION SOLAR CELLS WITH DILUTE NITRIDE 1EV SUBCELL 1210

I. García ; M. Ochoa ; I. Lombardero ; L. Cifuentes ; P. Caño ; M. Hinojosa ; I. Rey-Stolle ; C. Algora ; A. D. Johnson ; J. I. Davies ; K.H. Tan ; W.K. Loke ; S. Wicaksono ; S. F. Yoon

BISMUTH SURFACTANT-MEDIATED GROWTH OF GANASSB(BI) SOLAR CELLS 1215

Aymeric Maros ; Chaomin Zhang ; Jongwon Lee ; Hongfeng Wang ; Stephen Bremner ; Nikolai Faleev ; Christiana B. Honsberg ; Richard. R. King

AMORPHOUS SILICON CARBIDE FOR SILICON SURFACE PASSIVATION IN CARRIER-SELECTIVE CONTACT DEVICES 1220

Mathieu Boccard ; Christophe Ballif ; Zachary C. Holman

SURFACE PASSIVATION OF BORON DIFFUSED JUNCTIONS BY BOROSILICATE GLASS AND IN SITU GROWN SILICON DIOXIDE INTERFACE LAYER 1222

Valentin D. Mihailetchi ; Haifeng Chu ; Jan Lossen ; Radovan Kopecek

IMPROVED LIGHT INCOUPLING IN PLANAR SOLAR CELLS VIA IMPROVED TEXTURE MORPHOLOGY OF PDMS SCATTERING LAYER 1228

Salman Manzoor ; Zhengshan J. Yu ; Asad Ali ; Waqar Ali ; Zachary C. Holman

DAMAGE-FREE LASER ABLATION FOR EMITTER PATTERNING OF SILICON HETEROJUNCTION INTERDIGITATED BACK-CONTACT SOLAR CELLS 1233

Menglei Xu ; Twan Bearda ; Miha Filipic ; Hariharsudan Sivaramakrishnan Radhakrishnan ; Maarten Debucquoy ; Ivan Gordon ; Jozef Szlufcik ; Jef Poortmans

BENEFITS OF A THERMAL DRIFT DURING ATOMIC LAYER DEPOSITION OF AL2O3 FOR C-SI PASSIVATION 1237

Fabien Lebreton ; Andy Zauner ; Pavel Bulkin ; Francois Silva ; Sergej Filonovich ; Pere Roca I Cabarrocas

GROWTH DIFFERENCE OF AMORPHOUS SILICON BETWEEN PLASMA ENHANCED AND CATALYTIC CVD BASED ON SILICON HETEROJUNCTION SOLAR CELLS 1241

Liping Zhang ; Renfang Chen ; Zhuopeng Wu ; Chenguang Sun ; Fanying Meng ; Zhengxin Liu

DEVELOPING AN UNDERSTANDING-BASED SELECTION OF HYBRID-PEROVSKITE COMPOUNDS AND THE CU-IN HYBRID-PEROVSKITE (CIHP) FAMILY 1245

Alex Zunger ; G. Dalpian ; Qihang Liu ; L.B Abdalla ; L.L. Kazmerski

EFFECTS OF ELECTRON AND PROTON RADIATION ON PEROVSKITE SOLAR CELLS FOR SPACE SOLAR POWER APPLICATION 1248

Jing-Shun Huang ; Michael D. Kelzenberg ; Pilar Espinet-González ; Colin Mann ; Don Walker ; Ali Naqavi ; Nina Vaidya ; Emily Warmann ; Harry A. Atwater

TOWARDS PEROVSKITE SILICON TANDEM SOLAR CELLS WITH OPTIMIZED OPTICAL PROPERTIES 1253

Jan Christoph Goldschmidt ; Alexander J. Bett ; Patricia S.C. Schulze ; Nico Tucher ; Martin Bivour ; Markus Kohlstädt ; Seunghun Lee ; Simone Mastroianni ; Laura Mundt ; Markus Mundus ; Paul Ndione ; Karl Wienands ; Kristina Winkler ; Uli Würfel ; Martin Hermle ; Stefan W. Glunz

FIRST-PRINCIPLES DENSITY FUNCTIONAL THEORY CALCULATION OF METAL-SUBSTITUTED LEAD HALIDE PEROVSKITE 1256

Ji-Sang Park ; Matthew D. Sampson ; Alex B.F. Martinson ; Maria K.Y. Chan

ESTIMATING THE EFFECTS OF MODULE AREA ON THIN-FILM PHOTOVOLTAIC SYSTEM COSTS 1259

Kelsey A. W. Horowitz ; Ran Fu ; Xingshu Sun ; Tim Silverman ; Michael Woodhouse ; Muhammad A. Alam

COST ANALYSIS OF TANDEM MODULES 1264

Sarah E. Sofia ; Jonathan Mailoal ; Dirk Weiss ; Tonio Buonassisi ; Ian Marius Peters

CAUSE OF CURRENT-COLLECTION FAILURE OBSERVED INISC-REDUCTION PHASE OF PV CELLS AND MODULES EXPOSED TO ACETIC ACID 1268

Tadanori Tanahashi ; Norihiko Sakamoto ; Hajime Shibata ; Atsushi Masuda

COMPARISON OF PV MODULE PERFORMANCE BEFORE AND AFTER 11, 20, AND 25.5 YEARS OF FIELD EXPOSURE 1271

Jacob Rada ; Charles Chamberlin ; Peter Lehman ; Arne Jacobson

MARRYING QUALITY ASSURANCE WITH DESIGN ENGINEERING – A WINNING PARTNERSHIP! BUT, A CULTURAL DIVIDE? 1275

Sarah Kurtz ; Govind Ramu ; Robert Cornell ; Sumanth Lokanath ; Edward Hsi ; Tony Sample ; Masaaki Yamamichi ; George Kelly ; Ted Spooner ; Jonathan Previtali ; John Wohlgemuth

UPDATED EVALUATION OF SHOCK HAZARDS TO FIREFIGHTERS WORKING IN PROXIMITY OF PV SYSTEMS 1280

Olga Lavrova ; Jimmy E. Quiroz ; Jack Flicker ; Renee Gooding

GROWTH AND OPTIMIZATION OF GAINP/INP NANOWIRE TUNNEL DIODE 1286

Xulu Zeng ; Gaute Otnes ; Magnus Heurlin ; Magnus T Borgström

CATHODOLUMINESCENCE MAPPING FOR THE DETERMINATION OF N-TYPE DOPING IN SINGLE GAAS NANOWIRES 1289

Hung-Ling Chen ; Chalermchai Himwas ; Andrea Scaccabarozzi ; Pierre Rale ; Fabrice Oehler ; Aristide Lemaître ; Laurent Lombez ; Jean-François Guillemoles ; Maria Tchemycheva ; Jean-Christophe Harmand ; Andrea Cattoni ; Stéphane Collin

OPTICAL OPTIMIZATION OF PASSIVATED GAAS NANOWIRE SOLAR CELLS 1294

Kyle W. Robertson ; Ray R. Lapierre ; Jacob J. Krich

HIGH EFFICIENCY GAN NANOWIRE/SI PHOTOCATHODE FOR PHOTOELECTROCHEMICAL WATER SPLITTING 1299

Srinivas Vanka ; Sheng Chu ; Yichen Wang ; Ishiang Shih ; Hong Guo ; Zetian Mi

ANALYTIC DESCRIPTION OF THE IMPACT OF GRAIN BOUNDARIES ON VOC 1303

Paul Haney ; Benoit Gaury

ROLE OF TELLURIUM BUFFER LAYER ON CDTE SOLAR CELLS' ABSORBER/BACK-CONTACT INTERFACE 1308

Tao Song ; James R. Sites

SIMULTANEOUS EXAMINATION OF GRAIN-BOUNDARY POTENTIAL, RECOMBINATION, AND PHOTOCURRENT IN CDTE SOLAR CELLS USING DIVERSE NANOMETER-SCALE IMAGING 1312

C.S. Jiang ; H.R. Moutinho ; J. Moseley ; A. Kanevce ; J.N. Duenow ; E. Colegrove ; C. Xiao ; W.K. Metzger ; M.M. Al-Jassim

NANOPARTICLE/METAL REAR REFLECTORS FOR LOW- AND HIGH-TEMPERATURE SILICON SOLAR CELLS 1317

Syeda Qudsia ; Farah Qazi ; Mehwish Azher Javed ; Mathieu Boccard ; Zhengshan J. Yu ; Peter Firth ; Jonathan Bryan ; Zachary C. Holman

ABSORPTION IN EACH LAYER OF A SILICON HETEROJUNCTION SOLAR CELL 1322

Keith R. Mcintosh ; Malcolm D. Abbott ; Benjamin A. Sudbury ; Salman Manzoor ; Zhengshan J. Yu ; Mehdi Leilaeioun ; Jiatiwei Shi ; Zachary C. Holman

INVESTIGATIONS ON PLASMONIC COLOR TUNING COATING ON C-SI SOLAR CELLS 1329

Gerhard Peharz ; Wolfgang Waldhauser ; Christine Prietl ; Bettina Großschädl ; Martin C. Schubert ; Bernhard Michl

INVESTIGATION OF INTERFACE AND BULK LOCALIZED STATES IN A-SI:H SOLAR CELLS 1333

Adrien Bidiville ; Takuya Matsui ; Hitoshi Sai ; Koji Matsubara

EXPERIMENTAL AND THEORETICAL STUDY OF THE INFRARED EMISSIVITY OF CRYSTALLINE SILICON SOLAR CELLS 1339

Alberto Riverola ; Alexander Mellor ; Diego Alonso Alvarez ; Lourdes Ferre Llin ; Ilaria Guarracino ; Christos N. Markides ; Douglas Paul ; Daniel Chemisana ; Ned Ekins-Daukes

HIGH PERFORMANCE MOLECULAR DONORS FOR ORGANIC SOLAR CELLS, MATERIALS DESIGN AND DEVICE OPTIMIZATION 1342

Paul Geraghty ; Haotian Wang ; Calvin Lee ; Jegadesan Subbiah ; David Jones

ADVANCED OPTICAL MODELLING OF MICRO-TEXTURED SOLUTION-PROCESSED SOLAR CELLS WITH CONSIDERATION OF SMALL-AREA EFFECTS..1346

Benjamin Lipovšek ; Marko Jošt ; Andrej Campa ; Fei Gu ; Christoph J. Brabec ; Karen Forberich ; Janez Krc ; Marko Tonic

IDENTIFICATION OF DEGRADATION PATHWAYS OF ORGANIC SOLAR CELLS USING INFRARED SPECTROSCOPY 1350

S. Shah ; R Biswas ; T. Koschny ; V L Dalal

A DEVICE-INDEPENDENT SCREENING TECHNIQUE FOR RAPIDLY IDENTIFYING NEXT GENERATION OPV MATERIALS..1354

Bryon W. Larson ; Andrew J. Ferguson ; Bertrand J. Tremolet De Villers ; Ross E. Larsen

NOVEL ANTHANTHRONE AND ANTHANTHRENE CO-POLYMERS AS P-TYPE CONJUGATED SEMICONDUCTORS FOR ORGANIC PHOTOVOLTAICS..1360

Suru Vivian John ; Patrick Denk ; Christoph Ulbricht ; Herwig Heilbrunner ; Jean-Benoit Giguère ; Antoine Lafleur-Lambert ; Jean-Francois Morin ; Emmanuel Iwuoha ; Daniel Ayuk Mbi Egbe

REDUCING UV INDUCED DEGRADATION LOSSES OF SOLAR MODULES WITH C-SI SOLAR CELLS FEATURING DIELECTRIC PASSIVATION LAYERS..1366

Robert Witteck ; Henning Schulte-Huxel ; Boris Veith-Wolf ; Malte Ruben Vogt ; Fabian Kiefer ; Marc Kontges ; Robby Peibst ; Rolf Brendel

LARGE-AREA JUNCTION DAMAGE IN POTENTIAL-INDUCED DEGRADATION OF C-SI SOLAR MODULES 1371

Chuanxiao Xiao ; Chun-Sheng Jiang ; Steve Johnston ; Steve P. Harvey ; Peter Hacke ; Brian Gorman ; Mowafak Al-Jassim

SEARCH FOR MICROSTRUCTURAL DEFECTS AS NUCLEI FOR PID-SHUNTS IN SILICON SOLAR CELLS 1376

Volker Naumann ; Otwin Breitenstein ; Jan Bauer ; Christian Hagendorf

INVESTIGATING PID SHUNTING IN POLYCRYSTALLINE SILICON MODULES VIA MULTI-SCALE, MULTI-TECHNIQUE CHARACTERIZATION 1381

Steven P. Harvey ; John Moseley ; Adam Stokes ; Andrew Norman ; Brian Gorman ; Peter Hacke ; Steve Johnston ; Mowafak Al-Jassim

POTENTIAL-INDUCED DEGRADATION OF A SI NITRIDE/CRYSTALLINE SI INTERFACE OBSERVED THROUGH MINORITY CARRIER LIFETIME MEASUREMENT..1385

Naoyuki Nishikawa ; Seira Yamaguchi ; Keisuke Ohdaira

FIELD INSPECTION OF PV MODULES: QUANTIFICATION OF EVA BROWNING LEVEL USING AN IMAGE PROCESSING TOOL..1389

Sushanth Gudla ; Govindasamy Tamizhmani

PREVENTING POTENTIAL-INDUCED DEGRADATION IN CRYSTALLINE SILICON PV MODULES: RELATIONSHIP BETWEEN DEGRADATION AND BILL OF MATERIAL 1395

Alessandro Virtuani ; Eleonora Annigoni ; Christophe Ballif

IDENTIFYING REVERSE-BIAS BREAKDOWN SITES IN CUINXGA(1-X)SE2 1400

Steve Johnston ; Elizabeth Palmiotti ; Andreas Gerber ; Harvey Guthrey ; Lorelle Mansfield ; Timothy J. Silverman ; Mowafak Al-Jassim ; Angus Rockett

HIMAWARI-8 ENABLED REAL-TIME DISTRIBUTED PV SIMULATIONS FOR DISTRIBUTION NETWORKS 1405

Nicholas A. Engerer ; Jamie M. Bright ; Sven Killinger

REDUCED MEASUREMENT UNCERTAINTY IN PV MODULE BATCH TESTING 1411

Blagovest Mihaylov ; Bengt Jaeckel ; Juergen Arp ; Ralph Gottschalg

CLOUD MOTION IDENTIFICATION ALGORITHMS BASED ON ALL-SKY IMAGES TO SUPPORT SOLAR IRRADIANCE FORECAST..1415

Lydie Magnone ; Fabrizio Sossan ; Enrica Scolari ; Mario Paolone

AUTOMATIC DETECTION OF INACTIVE SOLAR CELL CRACKS IN ELECTROLUMINESCENCE IMAGES 1421

Sergiu Spataru ; Peter Hacke ; Dezso Sera

APPLYING SPATIAL DOWNSCALING AND SMART PERSISTENCE TO PROVIDE AN IMPROVED SOLAR FORECAST TO REDUCE COMMERCIAL DEMAND CHARGES..1427

Alex Kubiniec ; Ted Belanger ; Adam Kankiewicz ; Skip Dise ; Nate Glasgow ; Alemu Tadesse

THERMAL CHARACTERISTICS OF PID-AFFECTED MONOCRYSTALLINE SILICON SOLAR MODULES UNDER ILLUMINATED AND DARK CONDITIONS 1430

Pan Zhao ; Shuwen Guo ; He Wang ; Hong Yang ; Dengyuan Song ; Shiyu Sang ; Bojie Su ; Xue Zhang ; Yunxue Cao ; Hui Zhao

TARGETED EVALUATION OF UTILITY-SCALE AND DISTRIBUTED SOLAR FORECASTING 1435

Matthew Lave ; Robert J. Broderick ; Laurie Burnham

RECORD EFFICIENCIES FOR SELENIUM PHOTOVOLTAICS AND APPLICATION TO INDOOR SOLAR CELLS 1441

Douglas M. Bishop ; Teodor Todorov ; Yun Seog Lee ; Oki Gunawan ; Richard Haight

CLOSE-SPACED SUBLIMATION FOR SB2SE3SOLAR CELLS 1445

Laurie J. Phillips ; Peter Yates ; Oliver S. Hutter ; Tom Baines ; Leon Bowen ; Ken Durose ; Jonathan D. Major

FABRICATION OF COPPER ARSENIC SULFIDE THIN FILMS FROM NANOPARTICLES FOR APPLICATION IN SOLAR CELLS 1449

Scott A. Mcclary ; Joseph Andler ; Carol A. Handwerker ; Rakesh Agrawal

ORIENTATION CONTROLLED GE THIN FILMS ON GLASS BY AL-INDUCED CRYSTALLIZATION 1452

Kaveh Shervin ; Khim Kharel ; Alexandre Freundlich

IN-LINE POTASSIUM FLUORIDE TREATMENT OF CIGS ABSORBERS DEPOSITED ON FLEXIBLE SUBSTRATES IN A PRODUCTION-SCALE PROCESS TOOL 1455

Ryan Kaczynski ; Jinwoo Lee ; Jane Van Alsburg ; Baosheng Sang ; Urs Schoop ; Jeffrey Britt

LIGHT-SOAK AND DARK-HEAT INDUCED CHANGES IN CU(IN, GA)SE2 SOLAR CELLS: A MACROSCOPIC TO MICROSCOPIC STUDY 1459

Rouin Farshchi ; Benjamin Hickey ; Dmitry Poplavskyy

A NEW MODEL TO DETERMINE INSTALLED SYSTEM COST AND LCOE FOR ARPA-E'S MOSAIC MICRO-CONCENTRATOR PV PROGRAM 1463

Ran Fu ; Kelsey A.W. Horowitz ; Daniel W. Cunningham ; James Zahler

FIXED-TILT 660 × CONCENTRATING PHOTOVOLTAIC SYSTEM WITH 30% EFFICIENCY 1469

Alex J. Grede ; Jared S. Price ; Baomin Wang ; Michael V. Lipski ; Brent Fisher ; Kyu-Tae Lee ; Junwen He ; Gregory S. Brulo ; Xiaokun Ma ; Scott Burroughs ; Christopher D. Rahn ; Ralph G. Nuzzo ; John A. Rogers ; Noel C. Giebink

WAFER INTEGRATED MICRO-SCALE CONCENTRATING PHOTOVOLTAICS 1473

Tian Gu ; Duanhui Li ; Lan Li ; Bradley Jared ; Gordon Keeler ; Bill Miller ; William Sweatt ; Scott Paap ; Michael Saavedra ; Ujjwal Das ; Steve Hegedus ; Anna Tanke-Pedretti ; Juejun Hu

TOWARD STATIONARY CONCENTRATOR PHOTOVOLTAIC PANELS 1476

Peter Kozodoy ; Christopher Gladden ; Michael Pavilonis ; Tobias Wheeler ; Christopher Rhodes ; Chadwick Casper ; Kevin Schneider

CPV TECHNOLOGIES NOT RELYING ON PERFECTION OF TRACKERS 1479

Kenji Araki ; Yasuyuki Ota ; Kan-Hua Lee ; Kensuke Nishioka ; Masafumi Yamaguchi

THE GETTERING EFFECT OF DIELECTRIC FILMS FOR SILICON SOLAR CELLS 1485

A. Y. Liu ; C. Sun ; V. P. Markevich ; A. R. Peaker ; J. D. Murphy ; D. Macdonald

TABULA RASA: OXYGEN PRECIPITATE DISSOLUTION THOUGH RAPID HIGH TEMPERATURE PROCESSING IN SILICON 1491

Erin E. Looney ; Hannu S. Laine ; Mallory A. Jensen ; Amanda Youssef ; Vincenzo Lasalvia ; Paul Stradins ; Tonio Buonassisi

TOWARD EFFECTIVE GETTERING IN BORON-IMPLANTED SILICON SOLAR CELLS 1494

Hannu S. Laine ; Ville Vähänissi ; Zhengjun Liu ; Ernesto Magaña ; Ashley E. Morishige ; Jan Krügener ; Kristian Salo ; Hele Savin ; Barry Lai ; David P. Fenning

IMPACT OF THE INITIAL GROWTH INTERFACE ON THE GRAIN STRUCTURE IN HPMC-SI INGOT 1498

Giri Wahyu Alam ; Etienne Pihan ; Benoit Marie ; Nathalie Mangelinck-Noël

EFFECT OF CARBON CONCENTRATION AND GROWTH CONDITIONS ON OXYGEN PRECIPITATION BEHAVIOR IN N-TYPE CZ-SI 1504

Takuto Kojima ; Ryota Suzuki ; Kosuke Kinoshita ; Kyotaro Nakamura ; Atsushi Ogura ; Yoshio Ohshita ; Isao Masada ; Shoji Tachibana

NANO-IMAGING OF PERFORMANCE IN PHOTOVOLTAICS 1508

Elizabeth M. Tennyson ; Marina S. Leite

IMPLICATIONS OF CONDUCTIVE GRAIN BOUNDARIES IN CHLORINE-TREATED CDTE SOLAR CELLS 1511

Mohit Tuteja ; Vasilios Palekis ; Allen Hall ; Scott Maclaren ; Chris S. Ferekides ; Angus A. Rockett

IMAGING THE MULTI-TEMPORAL PHOTO-CARRIER DYNAMICS AT THE NANOMETER SCALE IN ORGANIC AND INORGANIC SOLAR CELLS 1516

Pablo A. Fernández Garrillo ; Lukasz Borowik ; Florent Caffy ; Renaud Demadrille ; Benjamin Grévin

NANOSCALE TOMOGRAPHIC CHARGE TRANSPORT IN POLYCRYSTALLINE CHALCOGENIDE ABSORBERS: CDTE VERSUS CIGS 1522

Justin L. Luria ; Andrew Moore ; Sun Yu ; Mark Aindow ; Bryan D. Huey

IMPROVING THE PV MODULE SINGLE-DIODE MODEL ACCURACY WITH TEMPERATURE DEPENDENCE OF THE SERIES RESISTANCE 1526

Kyumin Lee

CELL-TO-MODULE (CTM) ANALYSIS FOR PHOTOVOLTAIC MODULES WITH SHINGLED SOLAR CELLS 1531
Max Mittag ; Tobias Zech ; Martin Wiese ; David Blasi ; Matthieu Ebert ; Harry Wirth

A PRACTICAL IRRADIANCE MODEL FOR BIFACIAL PV MODULES 1537
Bill Marion ; Sara Macalpine ; Chris Deline ; Amir Asgharzadeh ; Fatima Toor ; Daniel Riley ; Joshua Stein ; Clifford Hansen

A DETAILED MODEL OF REAR-SIDE IRRADIANCE FOR BIFACIAL PV MODULES 1543
Clifford W. Hansen ; Renee Gooding ; Nathan Guay ; Daniel M. Riley ; Johnson Kallickal ; Donald Ellibee ; Amir Asgharzadeh ; Bill Marion ; Fatima Toor ; Joshua S. Stein

VIEW FACTOR MODEL AND VALIDATION FOR BIFACIAL PV AND DIFFUSE SHADE ON SINGLE-AXIS TRACKERS 1549
Marc Abou Anoma ; David Jacob ; Ben C. Bourne ; Jonathan A. Scholl ; Daniel M. Riley ; Clifford W. Hansen

A FAST QUASI-STATIC TIME SERIES (QSTS) SIMULATION METHOD FOR PV IMPACT STUDIES USING VOLTAGE SENSITIVITIES OF CONTROLLABLE ELEMENTS 1555
Xiaochen Zhangl ; Santiago Grijalva ; Matthew J. Reno ; Jeremiah Deboever ; Robert J. Broderick

FAST DETERMINATION OF DISTRIBUTION-CONNECTED PV IMPACTS USING A VARIABLE-TIME-STEP QUASI-STATIC TIME-SERIES APPROACH 1561
Barry Mather

SCALABILITY OF THE VECTOR QUANTIZATION APPROACH FOR FAST QSTS SIMULATION 1567
Jeremiah Deboever ; Santiago Grijalva ; Matthew J. Reno ; Xiaochen Zhang ; Robert J. Broderick

MACHINE LEARNING FOR RAPID QSTS SIMULATIONS USING NEURAL NETWORKS 1573
Matthew J. Reno ; Robert J. Broderick ; Logan Blakely

ALGORITHMIC ASPECTS OF A COMMERCIAL-GRADE DISTRIBUTION SYSTEM LOAD FLOW ENGINE 1579
Francis Therrien ; Marc Belletête ; Jean-Sébastien Lacroix ; Matthew J. Reno

RESONANT AND NON-RESONANT DIELECTRIC COATINGS FOR HIGH EFFICIENCY SOLAR CELLS 1585
Dongheon Ha ; Chen Gong ; Marina S. Leite ; Jeremy N. Munday

ENHANCED LIGHT TRAPPING IN THIN SILICON SOLAR CELLS USING EFFECTIVELY TRANSPARENT CONTACTS (ETCS) 1589
Rebecca Saive ; André Augusto ; Stuart G. Bowden ; Harry A. Atwater

ENHANCED POWER CONVERSION EFFICIENCY IN SINGLE NANOWIRE DEVICES THROUGH SYMMETRY BREAKING DESIGN 1594
Jian Zhou ; Yonggang Wu ; Zihuan Xia ; Xuefei Qin ; Zongyi Zhang

CDSE(TE)/CDS/CDSE RODS VS. CDTE/CDS/CDSE SPHERES: MORPHOLOGY-DEPENDENT CARRIER DYNAMICS FOR PHOTON UPCONVERSION 1598
Eric Y. Chen ; Zhuohui Li ; Christopher C. Milleville ; Kyle R. Lennon ; Matthew F. Doty

DRIFT-DIFFUSION INGAN/GAN SOLAR CELL SIMULATOR WITH OPTICAL MANAGEMENT 1603
Y. Fang ; D. Guo ; A. Fischer ; E. Vadiee ; C. Zhang ; J. Williams ; S. M. Goodnick ; D. Vasileska

PERFORMANCE ENHANCEMENT OF A GAAS SOLAR CELL WITH COLLOIDAL QUANTUM DOTS EMBEDDED IN TRENCHES 1606
Chia-Jhe Shu ; Yu-Ming Huang ; Shun-Chieh Hsu ; Jinn-Kong Shu ; Jia-Lin Tsai ; Pei-Chen Yu ; Yung-Jr Hung ; Chien-Chung Lin

ENHANCED PHOTORESPONSE OF INN DEVICES USING INDIUM-TIN OXIDE NANORODS 1610
Lung-Hsing Hsu ; Yuh-Jen Cheng ; Peichen Yu ; Hao-Chung Kuo ; Chien-Chung Lin

PLASMONIC SILVER STRUCTURES FOR IMPROVED PEROVSKITE PHOTOVOLTAIC PERFORMANCE 1614
Arul Varman Kesavan ; Arun D Rao ; Praveen C Ramamurthy

QUANTUM CUTTING LUMINESCENT PMMA FILMS CONTAINING CE3+ - YB3+ CODOPED YAG PHOSPHOR FOR SI CONCENTRATOR SOLAR CELLS 1619
Lu Li ; Chaogang Lou ; Huihui Cao

NUMERICAL EVALUATION ON THE NANO-ROD ARRAY ON A N-SIDE-UP THIN-FILM GAAS SOLAR CELLS 1623
Po-Ching Wu ; Yan-Zhang Lin ; Shun-Chieh Hsu ; Chia-Jhe Hsu ; Chien-Chung Lin

DOWN SHIFTED CONVERSION FOR ENHANCED HIT SOLAR CELL EFFICIENCY 1627
Albert S. Lin ; Parag Parashar ; Wei-Ming Huang ; Yi-Wen Huang ; Ding-Rung Jian ; Ming-Hsuan Kao ; Shi-Wei Chen ; Chang-Hong Shen ; Jia-Min Shieh ; Tzu-Yu Chen ; Chien-Chung Lin ; Hao-Chung Kuo

THE PLANAR THERMOPHOTOVOLTAIC SELECTIVE NEARLY-PERFECT ABSORBERS/EMITTERS 1631
Parag Parashar ; Ding-Rung Jian ; Weiming Huang ; Vi-Wen Huang ; Albert Lin

HYBRID PEDOT:PSS SILICON SOLAR CELLS WITH PENCIL ROD STRUCTURES 1635
Ruei-Ying Wu ; Liang-Chian You ; Hsin-Fei Meng ; Chun-Chi Chen ; Peichen Yu

PL STUDY OF PHOSPHORUS-DOPED CDTE EVT FILMS 1638
Shamara Collins ; Imran Khan ; Vamsi Evani ; Chih An Hsu ; Vasilios Palekis ; Don Morel ; Chris Ferekides

CHARACTERIZATION OF SINGLE-SOURCE DEPOSITED CLOSE-SPACE SUBLIMATION CDTEXSE1-XTHIN FILM SOLAR CELLS 1643
Corey R. Grice ; Jian Li ; Yanfa Yan

THE INFLUENCE OF THE CU-RICH/CU-POOR SEQUENCE ON THE PROPERTIES OF CU(IN, GA)SE2 FILMS DEPOSITED BY IN-LINE CO-EVAPORATION PROCESS 1648
He Wang ; Fang Fang Liu ; Yi Tong Yang ; Li You Yao ; Peng Gao ; Zhi Bin Xiao ; Qiang Sun

DETERMINATION AND MODELING OF INJECTION DEPENDENT SERIES RESISTANCE IN CIGS SOLAR CELLS 1651
Vito Huhn ; Bart E. Pieters ; Andreas Gerber ; Yael Augarten ; Uwe Rau

LARGE GRAIN GROWTH IN CU2ZNSNS4 THIN FILMS IN THE ABSENCE OF NA USING RAPID THERMAL ANNEALING 1656
J. L. Johnson ; A. Bhatia ; J. G. Bolke ; M. A. Scarpulla

CU2ZNSNS4THIN FILMS SYNTHESIZED BY COSPUTTERING AND RAPID THERMAL ANNEALING: EFFECTS OF COMPOSITION AND TEMPERATURE 1661
J.L. Johnson ; W.M. Hlaing Oo ; M. Karmarkar ; M.A. Scarpulla

EARTH-ABUNDANT CZTSSE THIN FILM SOLAR CELLS ON FLEXIBLE STAINLESS STEEL FOIL SUBSTRATES 1665
Hae-Sun Kim ; Woo-Lim Jeong ; Dong-Seon Lee

COMPARISON OF MGCL2AND CDCL2ACTIVATION TREATMENT FOR CDTE SOLAR CELLS: RECRYSTALLIZATION AND DEFECTS 1669
Daniele Menossi ; Elisa Artegiani ; Ivan Rimmaudo ; Alessia Le Donne ; Simona Binetti ; Juan Luis Pena ; Fabio Piccinelli ; Alessandro Romeo

CHARACTERIZATION OF CDTE PHOTOVOLTAIC DEVICES PASSIVATED USING HYDROGEN PLASMA 1674
Amit Munshi ; Piotr Kaminski ; Ali Abbas ; Shiva Tarun Chenna ; Sreeram Chandralal ; John Walls ; Walajabad Sampath

GROUP-V DOPING IMPACT ON CD-RICH CDTE SINGLE CRYSTALS GROWN BY TRAVELING-HEATER METHOD 1679
Akira Nagaoka ; Kenji Yoshino ; Yoshitaro Nose ; Darius Kuciauskas ; Michael A. Scarpulla

BAND-GAP ENGINEERING IN CU2ZNSN(S,SE)4SOLAR CELLS BY POST-SULPHURIZATION OF SELENIZED ABSORBER LAYERS 1682
Markus Neuwirth ; Elisabeth Seydel ; Heinz Kalt ; Michael Hetterich

IMPACT OF GA/III PROFILE ON VOLTAGE-DEPENDENT COLLECTION LOSSES IN CIGS SOLAR CELLS 1686
Dmitry Poplavskyy ; Jeff Bailey ; Rouin Farshchi ; David Spaulding

CL DIFFUSION IN CDTE SOLAR CELLS ACTIVATED BY GASEOUS CHCLF2ATMOSPHERE 1691
I. Rimmaudo ; R. Mis Fernandez ; V. Rejon ; A. Abbas ; F. Lisco ; J.M. Walls ; J.L. Peña

STABILITY OF CD1-XZNXTE ALLOYS UNDER CDTE PROCESSING CONDITIONS 1697
Yegor Samoilenko ; Colin A. Wolden

CIGSE ABSORBER PREPARATION: AN ALTERNATIVE TO H2SE 1701
O.S. Shinde ; E.J. Schenller ; S.R. Jadkar ; S.V Ghaisas ; N. Dhere

CHARGE CONTROLLED SEQUENTIAL ELECTRODEPOSITION FOR SYNTHESIS OF CU2ZNSNS4ON MO-COATED GLASS SUBSTRATE 1704
Ashish K. Singh ; Rajiv Dubey ; Manoj Neergat ; Kavaipatti R. Balasubramaniam

EFFECT OF DEPOSITED PRESSURE ON THE CDTE THIN FILMS BY CLOSED SPACE SUBLIMATION METHOD 1707
Yufeng Zhang ; Zhongming Du ; Xiangxin Liu

ANALYZING THE COST REDUCTION POTENTIAL OF III-V/SI HYBRID CONCENTRATOR PHOTOVOLTAIC SYSTEMS 1711
Kan-Hua Lee ; Kenji Araki ; Masafumi Yamaguchi

GENERALIZED NUMERICAL DESIGN OF AXIALLY-ASYMMETRICAL AND GRID-ARRANGED STATIC CPV ARRAY FOR MAXIMIZING ANNUAL ENERGY GENERATION 1714
Kenji Araki ; Kan-Hua Lee ; Masafumi Yamaguchi

SPECTRAL TRANSMITTANCE ANALYSIS OF LIQUIDS FOR HIGH CONCENTRATION III-V PHOTOVOLTAIC IMMERSION COOLING APPLICATIONS 1719
Xinyue Han ; Yongjie Guo

OPTICAL DESIGN FOR 2-TERMINAL III-V/SI SMAC MODULE 1724
Masaaki Baba ; Kikuo Makita ; Hidenori Mizuno ; Hidetaka Takato ; Takeyoshi Sugaya ; Noboru Yamada

DESIGN OF OPTICAL ELEMENTS FOR LOW PROFILE CPV PANEL WITH SUN TRACKING FOR ROOFTOP INSTALLATION 1728

Xinbing Liu ; Zhou Lu ; Riccardo Leto ; Carlton Brule ; Nanu Brates

MICRO CHIPLET PRINTER DEVELOPMENT FOR MOSAIC PROGRAM 1733

P.Y. Maeda ; Y. D. Wang ; S. Raychaudhuri ; J. Kalb ; D. K. Biegelsen ; R. Lujan ; Q. Wang ; Y. Wang ; J. Bert ; B. Rupp ; I. Matei ; L. Crawford ; A. Plochowietz ; E.M. Chow ; J.P. Lu ; V. Gupta

MICRO-OPTICAL TANDEM LUMINESCENT SOLAR CONCENTRATOR 1737

David R. Needell ; Zach Nett ; Ognjen Ilic ; Colton R. Bukowsky ; Junwen He ; Lu Xu ; Ralph G. Nuzzo ; Benjamin G. Lee ; John F. Geisz ; A. Paul Alivisatos ; Harry A. Atwater

INCREASE IN MAXIMUM POWER OF A-SI, C-SI AND GAAS.76P.24 SOLAR CELLS UNDER LOW CONCENTRATION 1741

Hiba Riaz ; Sabina Abdul Hadi ; Ammar Nayfeh

DESIGN AND EVALUATION OF PARTIAL CONCENTRATION III-V/SI MODULE WITH ENHANCED DIFFUSE SUNLIGHT TRANSMISSION 1743

Daisuke Sato ; Noboru Yamada ; Kan-Hua Lee ; Kenji Araki ; Masafumi Yamaguchi

CONTAMINATION CONTROL CHALLENGES ON SHJ SOLAR CELL PROCESSING 1747

G. Condorelli ; P. Rotoli ; A. Canino ; A. Battaglia ; W. Favre ; A. -S. Ozanne ; A. Moustafa ; A. Danel ; D. Muñoz ; P. -J. Ribeyron ; C. Gerardi

>23% SILICON HETEROJUNCTION SOLAR CELLS IN MEYER BURGER'S DEMO LINE: RESULTS OF PILOT PRODUCTION ON MASS PRODUCTION TOOLS 1752

J. Zhao ; M. König ; A. Wissen ; V. Breus ; D. Deckerl ; M. Fritzsche ; M. Schorch ; H. J. Nonnenmacher ; M. Leonhardt ; T. Große ; J. Hausmann ; A. Waltmger ; D. Landgraf ; S. Burkhardt ; H. Mehlich ; E. Vetter ; F. Schitthelm ; Y. Yao ; T. Söderström ; A. Richter ; D. Habermann ; S. Leu

EXPERIMENTAL AND SIMULATION STUDIES ON TIO2/SILICON HETEROJUNCTION DIODES 1755

Swasti Bhatia ; Neha Raorane ; Nimisha Sreekumar ; Pradeep R. Nair ; Aldrin Antony

A STUDY ON BLISTER FORMATION AND ELECTRICAL PROPERTIES UNDER VARIOUS ANNEALING CONDITION FOR TUNNELING OXIDE PASSIVATION LAYER 1758

Sungjin Choi ; Ka-Hyun Kim ; Min Gu Kang ; Jeong In Lee ; Donghwan Kim ; Hee-Eun Song

PROCESSING APPROACHES AND CHALLENGES OF INTERDIGITATED BACK CONTACT SI SOLAR CELLS 1761

Ujjwal Das ; Lei Zhang ; Steven Hegedus

FABRICATION OF CUI/A-SI:H/C-SI STRUCTURE FOR APPLICATION TO HOLE-SELECTIVE CONTACTS OF HETEROJUNCTION SI SOLAR CELLS 1765

Kazuhiro Gotoh ; Min Cui ; Nguyen Cong Thanh ; Koichi Koyama ; Isao Takahashi ; Yasuyoshi Kurokawa ; Hideki Matsumura ; Noritaka Usami

CHARACTERISTICS OF THIN CRYSTALLINE SILICON SOLAR CELLS WITH RIB STRUCTURE 1769

Yukimi Ichikawa ; Shuhei Yoshiba ; Masakazu Hirai ; Makoto Konagai

MEASUREMENT OF TIO2/P-SI SELECTIVE CONTACT PERFORMANCE USING A HETEROJUNCTION BIPOLAR TRANSISTOR WITH A SELECTIVE CONTACT EMITTER 1773

Janam Jhaveri ; Alexander Berg ; Sigurd Wagner ; James C. Sturm

EFFECT OF GROWTH AND POST-OXIDATION ANNEALING TEMPERATURE OF THERMALLY GROWN TUNNELING SIOX, ON THE IIMPLIED VOCOF PASSIVATED CONTACTS FOR C-SI BASED SOLAR CELLS 1777

Abhijit S. Kale ; William Nemeth ; Matthew Page ; Sumit Agarwal ; Paul Stradins

PARTIALLY CONTACTED SURFACES WITH CONTACT SIZE IN THE 1 μM RANGE FOR C-SI PERC SOLAR CELLS 1781

R. Khoury ; I. Martín ; G. López ; C. Jin ; J.M. López-González ; L. Zeyu ; P. Bulkin ; E.V. Johnson ; R. Alcubilla

ENTRANCE OF LOW COST FABRICATION OF BACK-CONTACT HETEROJUNCTION SOLAR CELLS BY USING PLASMA ION IMPLANTATION 1787

Koichi Koyama ; Keisuke Ohdaira ; Hideki Matsumura

TLM MEASUREMENTS VARYING THE INTRINSIC A-SI:H LAYER THICKNESS IN SILICON HETEROJUNCTION SOLAR CELLS 1790

Mehdi Leilaeioun ; William Weigand ; Pradyumna Muralidharan ; Mathieu Boccard ; Dragica Vasileska ; Stephen Goodnick ; Zachary Holman

SOLAR CELLS APPLICATION OF P-TYPE POLY-SI THIN FILM BY ALUMINUM INDUCED CRYSTALLIZATION 1794

Shota Masuda ; Kazuhiro Gotoh ; Isao Takahashi ; Kyotaro Nakamura ; Yoshio Ohshita ; Noritaka Usami

A SELF - CONSISTENTLY COUPLED DRIFT DIFFUSION AND MONTE CARLO SIMULATOR TO MODEL SILICON HETEROJUNCTION SOLAR CELLS 1797

Pradyumna Muralidharan ; Stuart Bowden ; Stephen M. Goodnick ; Dragica Vasileska

DOPANT PATTERNING BY PECVD AND MECHANICAL MASKING FOR PASSIVATED TUNNELING CONTACT IBC CELL ARCHITECTURES 1801

William Nemeth ; Vincenzo Lasalvia ; Benjamin G. Lee ; Abhijit Kale ; Paul Stradins

ALD ALUMINUM OXIDE AS A HOLE SELECTIVE TUNNELING CONTACT FOR CRYSTALLINE SILICON SOLAR CELLS 1804

Kortan Ögütman ; Kristopher O. Davis ; Winston V. Schoenfeld ; Michael Haslinger ; Sofie Robert ; Emanuele Cornagliotti ; Joachim John

SCREEN PRINTED, LARGE AREA BIFACIAL N-PERT CELLS WITH TUNNEL OXIDE PASSIVATED BACK CONTACT 1807

Young-Woo Ok ; Ajay D Upadhyaya ; Brian Rounsaville ; Ying-Yuan Huang ; Vijaykumar D Upadhyaya ; Ajeet Rohatgi

CORRELATION BETWEEN ELECTROLUMINESCENCE AND PHOTOCONVERSION EFFICIENCY IN A-SI:H/C-SI HETEROJUNCTION SOLAR CELLS 1811

A.V. Sachenko ; A.V. Bobyl ; V.N. Verbitskiy ; V.M. Vlasyuk ; D.M. Zhigunov ; V.P. Kostylyov ; I.O. Sokolovskyi ; E.I. Terukov ; P.A. Forsh ; M. Evstigneev

AN ISOTOPE STUDY OF HYDROGEN PASSIVATION OF POLY-SI/SIOXPASSIVATED CONTACTS FOR SI SOLAR CELLS 1817

Manuel Schnabel ; William Nemeth ; Bas W.H. Van De Loo ; Bart Macco ; Wilhelmus M.M. Kessels ; Paul Stradins ; David L. Young

ALLEVIATING HYDROGEN PLASMA DAMAGE TO AMORPHOUS/CRYSTALLINE SILICON INTERFACE PASSIVATION 1820

Jianwei Shi ; Zachary C. Holman

LARGE-AREA N-TYPE TOPCON CELLS WITH SCREEN-PRINTED CONTACT ON SELECTIVE BORON EMITTER FORMED BY WET CHEMICAL ETCH-BACK 1824

Yuguo Tao ; Felix Book ; Barbara Terheiden ; Viiaykumar Upadhvaya ; Keeya Madani ; Brian Rounsaville ; Eunhwan Cho ; Ajeet Rohatgi

HYDROGEN PLASMA POST-DEPOSITION TREATMENT FOR PASSIVATION OF A-SI/C-SI INTERFACE FOR HETEROJUNCTION SOLAR CELL BY CORRELATING OPTICAL EMISSION SPECTROSCOPY AND MINORITY CARRIER LIFETIME 1828

Anishkumar Soman ; Ugochukwu Nsofor ; Lei Zhang ; Ujjwal Das ; Tingyi Gu ; Steve Hegedus

MEASURING DIODE RESISTIVITY OF PASSIVATED CONTACTS 1832

San Theingi ; William Nemeth ; David L. Young ; Paul Stradins ; Benjamin G. Lee

ULTRA-THIN CRYSTALLINE SILICON SOLAR CELLS WITH NICKEL OXIDE INTERLAYER AS HOLE-SELECTIVE CONTACT 1835

Muyu Xue ; Raisul Islam ; Junyan Chen ; Zheng Lyu ; Yusi Chen ; Daniel Dewitt ; Albert Pleus ; Christian Tae ; Ching-Ying Lu ; Kai Zang ; Jieyang Jia ; Yijie Huo ; Ted Kamins ; Krishna Saraswat ; James Harris

CRYSTALLINE SI SOLAR CELLS WITH PASSIVATING, CARRIER-SELECTIVE NICKEL OXIDE CONTACTS 1838

Woojun Yoon ; James Moore ; David Scheiman ; Eunhwan Cho ; Young-Woo Ok ; Nicole Kotulak ; Phillip P. Jenkins ; Ajeet Rohatgi ; Robert J. Walters

GAP/SI HETEROJUNCTION SOLAR CELLS GROWN BY MOLECULAR BEAM EPITAXY 1841

Chaomin Zhang ; Ehsan Vadiee ; Richard R. King ; Christiana B. Honsberg

SPIN COATED NICKEL OXIDE AND VANADIUM OXIDE LAYERS ON SILICON FOR A CARRIER SELECTIVE CONTACT SOLAR CELL 1845

Jing Zhao ; Fa-Jun Ma, Jae-Yun ; Anita Ho-Baillie ; Stephen Bremner

QUANTIFICATION OF PV MODULE DISCOLORATION USING VISUAL IMAGE ANALYSIS 1850

Shashwata Chattopadhyay ; Chetan Singh Solanki ; Anil Kottantharayil ; K.L. Narasimhan ; Juzer Vasi ; Sai Tatapudi ; Govindasamy Tamizhmani

TEMPERATURE AND POWER STUDY OF ADHERED AND RACKED DOUBLE GLASS PHOTOVOLTAIC MODULES 1855

Volker Beutner ; Rubina Singh ; Cameron Stark

FIELD INSPECTION OF PV MODULES: QUANTITATIVE DETERMINATION OF PERFORMANCE LOSS DUE TO CELL CRACKS USING EL IMAGES 1858

Carlos A. Rodríguez Castañeda ; Shashwata Chattopadhyay ; Jaewon Oh ; Sai Tatapudi ; Govindasamy Tamizhmani ; Hailin Hu

SCALE UP DESIGNS FOR HAND-HELD LIGHT-WEIGHT TPV DC POWER SUPPLY 1863

L. M. Fraas ; J. E. Avery ; L. Minkin ; Hui She ; L. Ferguson

HIGH EFFICIENCY ANTI-REFLECTIVE COATING FOR PV MODULE GLASS 1869

Brennen M. Freiburger ; Corey S. Thompson ; Robert A. Fleming ; Douglas Hutchings ; Sergiu C. Pop

INVESTIGATION OF EFFICIENCY FOR PID-AFFECTED SOLAR MODULE AT NONSTANDARD TEST CONDITIONS 1873

Shuwen Guo ; Pan Zhao ; Weijing Huang ; Jipeng Chang ; He Wang ; Hong Yang ; Chengfeng Su ; Bojie Su ; Xue Zhang ; Yunxue Cao ; Hui Zhao

THERMAL UNIFORMITY MAPPING OF PV MODULES AND PLANTS..........1877
Ashwini Pavgi ; Jaewon Oh ; Joseph Kuitche ; Sai Tatapudi ; Govindasamy Tamizhmani

CLIMATE-SPECIFIC THERMAL MODEL COEFFICIENTS FOR C-SI AND THIN-FILM PV MODULES..........1883
Ashwini Pavgi ; Joseph Kuitche ; Jaewon Oh ; Govindasamy Tamizhmani

EFFECT OF THE THERMOPHYSICAL PROPERTIES OF A PHASE CHANGE MATERIAL ON THE ELECTRICAL OUTPUT OF A CONCENTRATED PHOTOVOLTAIC SYSTEM..........1888
Jawad Sarwar ; Ahmed E. Abbas ; Konstantinos E. Kakosimos

PASSIVE COOLING OF PHOTOVOLTAICS WITH DESICCANTS..........1893
Lin J. Simpson ; Jason Woods ; Nicolas Valderrama ; Alex Hill ; Nina Vincent ; Timothy Silverman

MODIFIED MAXIMUM POWER EXTRACTION TECHNIQUE FOR RAPIDLY CHANGING NUI AND DYNAMIC LOADS..........1898
U Aswani ; S.P. Duttagupta ; T.I. Eldho ; B.V. Rao

REAL-TIME MONITORING OF PHOTO VOLTAIC RELIABILITY ONLY USING MAXIMUM POWER POINT - THE SUNS-VMP METHOD..........1904
Xingshu Sun ; Haejun Chung ; Raghu Vamsi Krishna Chavali ; Peter Bermel ; Muhammad Ashraful Alam

PHOTOVOLTAIC MODULE DURABILITY AND RELIABILITY: ANALYSIS OF A 23-YEAR-OLD ARRAY OPERATING IN QUEBEC, CANADA..........1908
Christopher Baldus-Jeursen ; Alexandre Côté ; Naveen Goswamy ; Tanya Deer ; Yves Poissant

ARE E-W TRACKERS A BETTER OPTION FOR FUTURE INVESTMENTS IN PV SECTOR-A DETAILED TECHNO-COMMERCIAL STUDY..........1912
Rakesh Bohra ; Ramesh Rame Gowda ; Mani R. Krishnan

EXPERIMENTAL EVALUATION OF THE PERFORMANCE OF CRYSTALLINE SI PV MODULE DEGRADATION AFTER 15-YEARS OF FIELD EXPOSURE..........1917
Denio A. Cassini ; Antonia Sônia A. C. Diniz ; Marcelo Machado Viana ; Michele C. C. De Oliveira ; F. C. Lins Vanessa De ; Roberto Zilles ; Lawrence L. Kazmerski

FIELD INVESTIGATIONS OF POTENTIAL-INDUCED DEGRADATION (PID) FOR CRYSTALLINE SILICON PV PANELS IN DIFFERENT CLIMATES..........1922
Yifeng Chen ; Peter Hacke ; Yong Sheng Khoo ; Kaitlyn Vansant ; Zigang Wang ; Wei Luo ; Jing Chai ; Chris Deline ; Yan Wang ; Armin G. Aberle ; Pietro P. Altermatt ; Zhiqiang Feng ; Sarah Kurtz ; Pierre J. Verlinden

DETERMINING THE POWER RATE OF CHANGE OF 353 PLANT INVERTERS TIME-SERIES DATA ACROSS MULTIPLE CLIMATE ZONES, USING A MONTH-BY-MONTH DATA SCIENCE ANALYSIS..........1927
Alan J. Curran ; Yang Hu ; Rojiar Haddadian ; Jennifer L. Braid ; David Meakin ; Timothy J. Peshek ; Roger H. French

PHOTOVOLTAIC ARRAY DIFFERENTIAL BACKSIDE EXPOSURE CONDITIONS: BACKSHEET DEGRADATION AND SITE DESIGN..........1933
Andrew Fairbrother ; Julien Avenet ; Yadong Lyu ; Matthew Boyd ; Scott Julien ; Kai-Tak Wan ; Liang Ji ; Kenneth Boyce ; Sebastien Merzlic ; Amy Lefebvre ; Greg O'Brien ; Yu Wang ; Laura Bruckman ; Roger French ; Michael Kempe ; Brian Dougherty ; Xiaohong Gu

STUDY ON RANDOM FAILURE OF CRYSTALLINE SILICON SOLAR MODULES IN THE FIELD..........1937
Xuefang Jiang ; Fumei Wang ; Ao Wang ; Hong Yang ; He Wang ; Jie Ding ; Junjun Zhang ; Jingsheng Huang

POTENTIAL INDUCED DEGRADATION (PID) POWER LOSS CORRELATION TO LEAKAGE AND REVERSE BIAS CURRENTS..........1941
Michalis Florides ; Georgios Konstantinou ; Venizelos Venizelou ; George Makrides ; George E. Georghiou

PERFORMANCE STUDY OF VARIOUS PV MODULE TECHNOLOGIES IN DESERT CONDITIONS..........1946
Jim J John ; Ammar Elnosh ; Anwar Almheiri ; Wadhah Alzahmi ; Marco Stefancich ; Pedro Banda

HIGH-SPEED MEASUREMENTS OF GENERATED POWER AND ITS RELATIONSHIP TO WEATHER OBSERVATIONS AT YOSHINOGARI MEGA SOLAR POWER PLANT..........1950
Makoto Kasu ; Shigeomi Hara ; Takumi Uematsu

IMPACT OF MISSING DATA ON THE ESTIMATION OF PHOTOVOLTAIC SYSTEM DEGRADATION RATE..........1954
Andreas Livera ; Alexander Phinikarides ; George Makrides ; George E. Georghiou

FIELD DEGRADATION AND FAILURES OF AGED CRYSTALLINE SILICON PV MODULES IN MEXICO..........1959
D. Martínez Escobar ; P. A. Sánchez-Pérez ; Rocío De La Luz Santos Magdaleno ; José Ortega Cruz ; Sai Tatapudi ; Aarón Sánchez Juárez ; Govindasamy Tamizhmani

RAPID SHUTDOWN WITH PANEL LEVEL ELECTRONICS-A SUITABLE SAFETY MEASURE?..........1965
Adam Cordova ; Christopher Merz ; Gerd Bettenwort ; Markus Hopf ; Hannes Knopf ; Joachim Laschinski

INVESTIGATING A NEW OPERATING POINT FOR PV PANELS SEEKING MAXIMUM LIFE SPAN..1968

Bechara Nehme ; Nacer K. M'sirdi ; Tilda Akiki

POWER GENERATION EVALUATION OF LARGE-SCALE PHOTOVOLTAIC SYSTEMS LOCATED ON INCLINED PLANE 1973

Naotaka Oka ; Yasuhito Takahashi ; Koji Fujiwara ; Kazuyuki Hidaka ; Hiroshi Morita

INVESTIGATING THE IMPACT OF SOLAR CELLS PARTIAL SHADING ON PHOTOVOLTAIC MODULES BY THERMOGRAPHY...1979

David Pera ; José A. Silva ; Sara Costa ; João M. Serra

ANNUAL DEGRADATION RATE AND ITS LINEARITY ANALYSIS USING METERED KWH DATA...1984

Christopher Raupp ; Govindasamy Tamizhmani

ELECTRICAL PERFORMANCE ANALYSIS OF A 27 KW GRID-CONNECTED PV SYSTEM WITH SOILING AND SHADING IN MORELOS MEXICO..1990

P. A. Sánchez-Pérez ; D. Martínez Escobar ; E. O. Ángel Ruiz ; R. Santos Magdaleno ; José Ortega Cruz ; A. Sánchez Juárez

MODIFIED STC CORRECTION PROCEDURE FOR ASSESSING PV MODULE DEGRADATION IN FIELD SURVEYS 1995

Hemant K. Singh ; R. Dubey ; S. Zachariah ; K. L. Narasimhan ; B. M. Arora ; A. Kottantharayil ; J. Vasi

DEGRADATION MODELS OF PHOTOVOLTAIC MODULE BACKSHEETS EXPOSED TO DIVERSE REAL WORLD CONDITION..2000

Yu Wang ; Sebastien Merzlic ; Andrew Fairbrother ; Scott Julien ; Lucas Fridman ; Camille Loyer ; Amy L. Lefebvre ; Gregory O'Brien ; Xiaohong Gu ; Liang Ji ; Ken Boyce ; Michael Kempe ; Kai-Tak Wan ; Roger H. French ; Laura S. Bruckman

ADDRESSING HOTSPOTS IN THE PRODUCT ENVIRONMENTAL FOOTPRINT OF CDTE PHOTOVOLTAICS..2005

Parikhit Sinha ; Andreas Wade

PHOTOVOLTAIC SMART HOME SYSTEM - DUBAI CASE STUDY 2011

Ammar Natsheh ; Marwa Aljaziri ; Maitha Moosa ; Gharibah Essa ; Hassa Moosa

DIRECT DRIVE PHOTOVOLTAIC MILK CHILLING EXPERIENCE IN KENYA............2014

Robert Foster ; Brian Jensen ; Brian Dugdill ; Wendy Hadley ; Bruce Knight ; Abudul Faraj ; Johnson Kyalo Mwove

COST OPTIMIZATION OF DECOMMISSIONING AND RECYCLING CDTE PV POWER PLANTS 2019

V. Fthenakis ; Z. Zhang ; J. -K Choi

CHALLENGES FOR DECISION MAKERS WHEN FEED-IN TARIFFS OR NET METERING SCHEMES CHANGE TO INCENTIVES DEPENDENT ON A HIGH SHARE OF SELF-CONSUMED ELECTRICITY 2025

Mattias Gustafsson

PROCEDURES TO MAKE PROJECTS ABOUT RENEWABLE ENERGY GENERATION CONNECTED TO THE GRID IN COLOMBIA..2031

J. A. Hernandez ; C. A. Arredondo ; D. J. Rodriguez

A CRITICAL ANALYSIS ON THE THIN CRYSTALLINE SILICON PV MODULE OF THE LIGHTWEIGHT PV SYSTEM...2035

Meixi Chen ; Abhishek Iyer ; Cheng-Hao Shih ; Lado Kurdgelashvili ; Robert Opila

PHOTOVOLTAIC MODULE MANUFACTURING COSTS, AVERAGE PRICES AND INDUSTRY BALANCE 2006–2016..2039

Paula Mints ; Zhengshan J Yu

SOLAR CELL AND WIND ENERGY REPLACEMENT OF POWER PLANTS GLOBALLY 2042

Larry Partain ; Shirley Hansen ; Dirk Bennett ; Richard Hansen ; Allan Newlands ; Lewis Fraas

ANALYSIS OF LIGHT ENVIRONMENT UNDER SOLAR PANELS AND CROP LAYOUT 2048

Deng Wang ; Yaojie Sun ; Yandan Lin ; Yuan Gao

INTERFACE EFFECTS OF ALKALI TREATMENT ON CU-RICH THIN FILM SOLAR CELLS 2054

Hossam Elanzeery ; Finn Babbe ; Anastasiya Zelenina ; Michele Melchiorre ; Susanne Siebentritt

INCREASEDVOCAND FF IN ZN01-XSX-BUFFERED CUIN1-XGAXSE2SOLAR CELLS BY CADMIUM PARTIAL ELECTROLYTE TREATMENT 2058

Andreas Bauer ; Dimitrios Hariskos ; Wiltraud Wischmann

PASSIVATING AND CARRIER-SELECTIVE CONTACTS - BASIC REQUIREMENTS AND IMPLEMENTATION 2064

S.W. Glunz ; M. Bivour ; C. Messmer ; F. Feldmann ; R. Müller ; C. Reichel ; A. Richter ; F. Schindler ; J. Benick ; M. Hermle

FIRST-PRINCIPLES MODELING OF ALKALI METAL POST DEPOSITION TREATMENT EFFECTS IN CIGS SOLAR CELLS..2070

Maria Fedina ; Hannu-Pekka Komsa ; Ville Havu ; Martti J. Puska

EXPLORING SILICON CARBIDE- AND SILICON OXIDE-BASED LAYER STACKS FOR PASSIVATING CONTACTS TO SILICON SOLAR CELLS..2073

P. Löper ; G. Nogay ; P. Wyss ; M. Hyvl ; P. Procel ; J. Stuckelberger ; A. Ingenito ; I. Mack ; Q. Jeangros ; M. Ledinsky ; A. Fejfar ; C. Allebé ; J. Horzel ; M. Despeisse ; F. Crupi ; F.-J. Haug ; C. Ballif

EFFICIENT ELECTRON CONTACTS FORN-TYPE SILICON SOLAR CELLS USING MAGNESIUM METAL, OXIDE, AND FLUORIDE..2076

Yimao Wan ; Chris Samundsett ; James Bullock ; Di Yan ; Thomas Allen ; Jun Peng ; Jie Cui ; Mark Hettick ; Ali Javey ; Andres Cuevas

GRADED (ALZGA1-Z)XIN1-XP WINDOW-EMITTER STRUCTURES FOR IMPROVED SHORT-WAVELENGTH RESPONSE..2079

Jacob T. Boyer ; Daniel L. Lepkowski ; Daniel J. Chmielewski ; Steven A. Ringel ; Tyler J. Grassman

INTEGRATION OF QUANTUM DOTS AND QUANTUM WELLS INTO INGAAS METAMORPHIC SUBCELL FOR RADIATION HARD 3-J ELO IMM PHOTOVOLTAICS..................2084

Zachary S. Bittner ; Hyun Kum ; Michael A. Slocum ; George T. Nelson ; Rao Tatavarti ; Andre Wibowo ; Seth M. Hubbard

PROTON IRRADIATION OF 3J SOLAR CELLS AT LOW TEMPERATURE....................................2087

Seonyong Park ; Jacques C. Bourgoin ; Olivier Cavani ; Sandrine Picard ; Jérôme Bourcois ; Victor Khorenko ; Carsten Baur ; Bruno Boizot

ULTRA-THIN GAAS SOLAR CELLS: RADIATION TOLERANCE AND SPACE APPLICATIONS............2091

Louise C. Hirstl ; Michael K. Yakes ; Jeffery. H. Warner ; Mitchell F. Bennett ; Kenneth J. Schmieder ; Stephanie Tomasulo ; Erin Cleveland ; Sergey Maximenko ; James Moore ; Robert J. Walters ; Phillip P. Jenkins

LARGE AREA MULTIJUNCTION III-V SPACE SOLAR CELLS OVER 31% EFFICIENCY........................2094

X.Q. Liu ; C. Fetzer ; P. Chiu ; M. Haddad ; X. Zhang ; R. Cravens ; D. Law ; J. Ermer ; J. Krogen ; S. Sharma ; J. Hanley

ADVANCED-ARCHITECTURE HIGH-EFFICIENCY SOLAR CELLS FOR LOW IRRADIANCE LOW TEMPERATURE (LILT) APPLICATIONS..2099

Andreea Boca ; Jonathan Grandidier ; Claiborne Mcpheeters ; Paul Sharps ; Philip Chiu ; Xing-Quan Liu ; James Ermer

ULTRA-LIGHTWEIGHT PV MODULE DESIGN FOR BUILDING INTEGRATED PHOTOVOLTAICS..2104

Ana C. Martins ; Valentin Chapuis ; Alessandro Virtuani ; Christophe Ballif

DESIGN IT WITH LSCS; AN EXPLORATION OF APPLICATIONS FOR LUMINESCENT SOLAR CONCENTRATOR PV TECHNOLOGIES..2109

Wouter Eggink ; Angèle Reinders

INVESTIGATING PV-BATTERY 3-TERMINAL INTEGRATION CONCEPT AS A SELF-SUSTAINING POWER SOLUTION..2114

Solomon N. Agbo ; Oleksandr Astakhov ; Uwe Rau ; Tsvetelina Merdzhanova

PERFORMANCE ASSESSMENT OF A BIPV ROOFING TILE IN OUTDOOR TESTING 2118

Cristina S. Polo Lopez ; Pierluigi Bonomo ; Francesco Frontini ; Vasco Medici ; Lorenzo Nespoli

LIFE CYCLE ASSESSMENT OF TRANSPARENT ORGANIC PHOTOVOLTAIC FOR WINDOW APPLICATIONS..2124

Annick Anctil ; Eunsang Lee ; Jack Stephan ; Anjali Munasinghe ; Christopher Traverse ; Richard R. Lunt

A REDUCED ORDER MODEL FOR A TOV STUDY IN A SOLAR PV PROJECT.................................2128

Ahmad Abdullah ; Billy Yancey

CYBER SECURITY ASSESSMENT OF DISTRIBUTED ENERGY RESOURCES.................................2135

Cedric Carter ; Ifeoma Onunkwo ; Patricia Cordeiro ; Jay Johnson

EVALUATION OF FAST-FREQUENCY SUPPORT FUNCTIONS IN HIGH PENETRATION ISOLATED POWER SYSTEMS..2141

Mohamed Elkhatib ; Jason Neely ; Jay Johnson

LOSS OF UTILITY DETECTION CAPABILITIES FOR TODAY'S UTILITY INTERCONNECTED PHOTOVOLTAIC INVERTERS..2147

Sigifredo Gonzalez ; Gregory Kern ; Michael Ropp

PARAMETRIC PV GRID-SUPPORT FUNCTION CHARACTERIZATION FOR SIMULATION ENVIRONMENTS..2153

Javier Hernandez-Alvidrez ; Jay Johnson

COST ANALYSIS AND COST REDUCTION OPPORTUNITIES OF RESIDENTIAL PV SYSTEM IN THE JAPAN..2159

Izumi Kaizuka ; Haruki Yamaya ; Takashi Ohigashi ; Risa Kurihara ; Osamu Ikki

SUPPLY AND DEMAND CONSTRAINTS ON FUTURE PV POWER IN THE USA..........................2163

Paul A. Basore ; Wesley J. Cole

RESIDENTIAL PHOTOVOLTAIC ELECTRICITY GENERATION IN THE EUROPEAN UNION 2017-OPPORTUNITIES AND CHALLENGES 2167
Arnulf Jäger-Waldau ; Thomas Huld ; Sandor Szabo

INVESTIGATING NANOSCALE DETERMINANTS OF CHARGE COLLECTION IN QUASI-2D PEROVSKITE SOLAR CELLS 2170
Yanqi Luo ; Xueying Li ; Bat-El Cohen ; Barry Lai ; Lioz Etgar ; David P Penning

RECENT DEVELOPMENTS OF SOLAR PHOTOVOLTAIC SYSTEMS IN INDIA 2172
Saravanan Vasudevan ; Arumugam Murugesan

OPERANDO X-RAY DIFFRACTION FOR CHARACTERIZATION OF PHOTOVOLTAIC MATERIALS 2176
Laura T Schelhasl ; Jeffrey A. Christians ; Joseph J. Berry ; Michael F. T Oney ; Christopher J. Tassone ; Joseph M. Luther ; Kevin H. Stone

X-RAY BEAM INDUCED VOLTAGE: A NOVEL TECHNIQUE FOR ELECTRICAL NANOCHARACTERIZATION OF SOLAR CELLS 2179
Michael E. Stuckelberger ; Tara Nietzold ; Bradley M. West ; Barry Lai ; Jörg M. Maser ; Volker Rose ; Mariana I. Bertoni

ELECTRO-LUMINESCENT REFRIGERATION ENABLED BY HIGHLY EFFICIENT PHOTOVOLTAICS 2185
T. Patrick Xiao ; Kaifeng Chen ; Parthiban Santhanam ; Shanhui Fan ; Eli Yablonovitch

MULTIPLE QUANTUM WELLS AS SLOWED HOT CARRIER COOLING ABSORBERS IN HOT CARRIER CELLS 2186
Gavin Conibeer ; Yi Zhang ; Simon Chung ; Yuaxun Liao ; Stephen Bremner ; Santosh Shrestha

QUANTITATIVE OPTOELECTRONIC MEASUREMENTS OF CARRIER THERMODYNAMICS PROPERTIES IN QUANTUM WELL HOT CARRIER SOLAR CELL 2192
Dac-Trung Nguyen ; Laurent Lombez ; François Gibelli ; Soline Boyer-Richard ; Alain Le Corre ; Olivier Durand ; Jean-François Guillemoles

ABSORPTION ENHANCEMENT IN INGAASP/INGAP QUANTUM WELL SOLAR CELLS 2195
Islam E.H. Sayed ; Nikhil Jain ; Myles A. Steiner ; John F. Geisz ; Salah M. Bedair

CARRIER COLLECTION MODEL AND DESIGN RULE FOR QUANTUM WELL SOLAR CELLS 2201
Kasidit Toprasertpong ; Boram Kim ; Yoshiaki Nakano ; Masakazu Sugiyama

INFLUENCE OF CONDUCTION BAND OFFSETS AT WINDOW/BUFFER AND BUFFER/ABSORBER INTERFACES ON THE ROLL-OVER OF J-V CURVES OF CIGS SOLAR CELLS 2205
Giovanna Sozzi ; Simone Di Napoli ; Roberto Menozzi ; Florian Werner ; Susanne Siebentritt ; Philip Jackson ; Wolfram Witte

OVERVIEW OF SURFACE PASSIVATION SCHEMES FOR THIN FILM SOLAR CELLS 2209
Ratan Kotipalli ; Bart Vermang

TOWARDS 10% STATE-OF-THE-ART PURE SULFIDE CU2ZNSNS4 SOLAR CELL BY MODIFYING THE INTERFACE CHEMISTRY 2213
Kaiwen Sun ; Jialiang Huang ; Steve Johnston ; Chang Yan ; Fangyang Liu ; Xiaojing Hao ; Martin Green

BAND GAP CHANGES OF THE CDS BUFFER INDUCED BY POST-ANNEALING OF CU2ZNSN(S,SE)4SOLAR CELLS 2216
Mario Lang ; Nicolas Schäfer ; Christian Huber ; Thomas Schnabe ; Heinz Kalt ; Michael Hetterich

22.61 % EFFICIENT FULLY SCREEN PRINTED PERC SOLAR CELL 2220
Weiwei Deng ; Feng Ye ; Ruimin Liu ; Yunpeng Li ; Haiyan Chen ; Zhen Xiong ; Yang Yang ; Yifeng Chen ; Yongqian Wang ; Pietro P. Altermatt ; Zhiqiang Feng ; Pierre J. Verlinden

HOW TO ACHIEVE 23% EFFICIENT LARGE-AREA CU PLATED N-PERT CELLS? 2227
Monica Aleman ; Angel Uruena ; Emanuele Cornagliotti ; Patrick Choulat ; Joachim John ; Richard Russell ; Sukvhinder Singh ; Loic Tous ; Wen-Cheng Sun ; Filip Duerinckx ; Jozef Szlufcik

MICROSTRUCTURE AND RECOMBINATION ACTIVITY OF GRAIN BOUNDARIES FROM FRONT AND REAR SIDE DURING A LID-CYCLE OF MC-PERC SOLAR CELLS 2232
Tabea Luka ; Marko Turek ; Stephan Großer ; Christian Hagendorf

THERMODYNAMIC EFFICIENCY LIMIT OF BIFACIAL SOLAR CELLS FOR VARIOUS SPECTRAL ALBEDOS 2236
Thomas C.R. Russell ; Rebecca Saive ; Harry A. Atwater

PROCESS-INDUCED DEGRADATION RESISTANT N-CZ WAFERS THROUGH TABULA RASA DEFECT ENGINEERING 2242
Vincenzo Lasalvia ; William Nemeth ; Matthew Page ; Wooseok Nam ; Youngsik Han ; Sungsun Baik ; Amanda Youssef ; Tonio Buonassisi ; Paul Stradins

DETECTION OF A SHIFTING BROMINE CONCENTRATION IN HYBRID PEROVSKITES BY X-RAY FLUORESCENCE MICROSCOPY 2245
Yanqi Luo ; Parisa Khoram ; Sarah Brittman ; Barry Lai ; Erik C. Garnett ; David P. Fenning

INFLUENCE OF GRAIN SIZE AND INTERFACES ON PHOTO-STABILITY OF PEROVSKITE SOLAR CELLS 2247

Istiaque Hossain ; Liang Zhang ; Ranjith Kottokkaran ; Mohamed El-Henawey ; Pranav Joshi ; Max Noack ; Vikram Dalal

COLD THOUGHTS ON PEROVSKITE FEVER 2251

Tao Xu ; Jue Gong

LBIC ANALYSIS OF PEROVSKITE BASED SOLAR CELLS STABILITY 2255

Carmen M. Ruiz ; Javier Ramos ; Richard Garuz ; Damien Barakel ; Jean Reusser ; Judikaël Le Rouzo

ASSESSING JOB GROWTH AND SUSTAINABILITY IN THE US PV INDUSTRY 2258

Brion Bob

ENSURING THE RELIABILITY OF PHOTOVOLTAIC POWER SYSTEMS USING INTERNATIONAL STANDARDS AND THE IECRE CONFORMITY ASSESSMENT SYSTEM 2263

George Kelly ; Adrian Häring ; Ted Spooner ; Greg Ball ; Sarah Kurtz ; Matthias Heinze ; Masaaki Yamamichi ; Govind Ramu

A FRAMEWORK TO CALCULATE UNCERTAINTIES FOR LIFETIME ENERGY YIELD PREDICTIONS OF PV SYSTEMS 2267

Bjorn Muller ; Peter Bostock ; Boris Farnung ; Christian Reise

INTEGRATED PV-RECYCLING-MORE EFFICIENT, MORE EFFECTIVE 2272

Wolfram Palitzsch ; Ulrich Loser

ANALYSIS OF GAINP SOLAR CELLS GROWN BY HYDRIDE VAPOR PHASE EPITAXY 2275

Kevin L. Schulte ; John Simon ; David L. Young ; Aaron J. Ptak

INVESTIGATION OF ADHESION FORCES BETWEEN DUST PARTICLES AND SOLAR GLASS 2280

H.R. Moutinho ; C.-S. Jiang ; B. To ; C. Perkins ; M. Muller ; M.M. Al-Jassim ; L. Simpson

ANTI-REFLECTIVE AND ANTI-SOILING PROPERTIES OF A KLEANBOOST™, A SUPERHYDROPHOBIC NANO-TEXTURED COATING FOR SOLAR GLASS 2285

Illya Nayshevsky ; Qianfeng Xu ; Gil Barahman ; Alan Lyons

MULTILAYER-GROWN ULTRATHIN NANOSTRUCTURED GAAS SOLAR CELLS 2291

Boju Gai ; Yukun Sun ; Minjoo Lee ; Jongseung Yoon

LABORATORY STUDIES OF PARTICLE CEMENTATION AND PV MODULE SOILING 2294

Craig L. Perkins ; Matthew Muller ; Lin Simpson

VIRTUAL SUBSTRATES FOR LOW-COST HIGH EFFICIENCY III-V PHOTOVOLTAICS 2298

Sean J. Babcock ; Marlene L. Lichty ; Shankar Karki ; Grace Rajan ; Sylvain Marsillac ; Elisabeth L. Mcclure ; Seth M. Hubbard ; Christopher G. Bailey

SEASONAL TRENDS OF SOILING ON PHOTOVOLTAIC SYSTEMS 2301

Leonardo Micheli ; Daniel Ruth ; Matthew Muller

INTERRELATIONSHIPS AMONG NON-UNIFORM SOILING DISTRIBUTIONS AND PV MODULE PERFORMANCE PARAMETERS, CLIMATE CONDITIONS, AND SOILING PARTICLE AND MODULE SURFACE PROPERTIES 2307

Lawrence L. Kazmerski ; Antonia Sonia A.C. Diniz ; Daniel Sena Braga ; Cristiana Brasil Maia ; Marcelo Machado Viana ; Suellen C. Costa ; Pedro P. Brito ; Cláudio Dias Campos ; Sergio De Morais Hanriot ; Leila R. De Oliveira Cruz

PV MODULE DURABILITY -CONNECTING FIELD RESULTS, ACCELERATED TESTING, AND MATERIALS 2312

T. John Trout ; W. Gambogi ; T. Felder ; K. R. Choudhury ; L. Garreau-Iles ; Y. Heta ; K. Stika

FEMTOSECOND VS NANOSECOND: AN ANALYSIS ON THE LASER ABLATION PROPERTIES OF DIELECTRIC LAYERS FOR SOLAR CELLS 2318

Jaffar Moideen Yacob Ali ; Vinodh Shanmugam ; Carlos D. Rodríguez-Gallegos ; Bianca Lim ; Armin Aberle ; Thomas Mueller

GROWTH OF MOS2 THIN FILMS WITH MICRODOME TEXTURE AS OMNIDIRECTIONAL LIGHT TRAP FOR SOLAR CELL APPLICATIONS 2324

Hussain M. Abouelkhair ; Nina A. Orlovskaya ; Robert E. Peale

STUDY OF SPATIAL DISTRIBUTION OF ELECTRICAL, OPTICAL AND STRUCTURAL PROPERTIES OF MAGNETRON SPUTTERED AZO THIN FILMS 2330

Mohit Agarwal ; Rajiv O Dusane

MULTIBAND FORMATION IN CR DOPED CUGAS2 THIN FILMS SYNTHESIZED BY CHEMICAL SPRAY PYROLYSIS 2334

Nazmul Ahsan ; Sivaperuman Kalainatharr ; Naoya Miyashita ; Takuya Hoshii ; Yoshitaka Okada

EFFECTS OF ANNEALING AND SUBSTRATE TEMPERATURE FOR SN-S THIN FILMS 2338

Yoji Akaki ; Kazuya Iwasaki ; Shigeyuki Nakamura ; Hideaki Araki

MOLYBDENUM OXIDE THIN FILMS FOR HETEROJUNCTION SOLAR CELLS 2342

A. Dominguez ; Ateet Dutt ; O. De Melo ; G. Santana

DUAL ION BEAM SPUTTERED TCO THIN FILMS: SPUTTER-INSTIGATED PLASMONIC FEATURES FOR ULTRATHIN PHOTOVOLTAICS 2345

Vivek Garg ; Brajendra S. Sengar ; Vishnu Awasthi ; Shailendra Kumar ; Shaibal Mukherjee

COMBINATORIAL STUDY OF SN-TI-W-O TRANSPARENT CONDUCTING OXIDE THIN FILMS FOR PHOTOVOLTAIC APPLICATIONS 2349

Michael N. Gona ; Patrick J. M. Isherwood ; Jake W. Bowers ; John M. Walls

BANDGAP AND ELECTRON AFFINITY OPTIMIZATION OF ZINC OXIDE FOR N-ZNO/P-SI SINGLE HETEROJUNCTION SOLAR CELL 2355

Babar Hussain ; Aasma Aslam

MODELING AND OPTIMIZING THE EFFICIENCY OF A ZNO/ZNTE SOLAR CELL USING SCAPS SOFTWARE 2358

Amal Kabalan ; Sam Roy ; Benjamin Chen

TERNARY PHOSPHIDE SEMICONDUCTOR INMG/ZN3P2SOLAR CELLS 2361

Ryoji Katsube ; Kenji Kazumi ; Yoshitaro Nose

NUMERICAL MODELING OF WSE2SOLAR CELLS 2364

H. Kyureghian ; M. Hilfiker ; E. Ediger ; V. Medic ; N.J. Ianno

BIAXIAL-TEXTURED TITANIUM NITRIDE THIN FILMS ON LOW-COST, FLEXIBLE METAL SUBSTRATE AS A CONDUCTIVE BUFFER LAYER FOR THIN FILM SOLAR CELLS 2368

Yongkuan Li ; Yao Yao ; Ying Gao ; Sicong Sun ; Pavel Dutta ; Monika Rathi ; Jae-Hyun Ryou ; Venkat Selvamanickam

SNS BY IONIZED JET DEPOSITION FOR PHOTOVOLTAIC APPLICATIONS 2372

Daniele Menossi ; Simone Di Mare ; Ivan Rimmaudo ; Elisa Artegiani ; Giampiero Tedeschi ; Juan Luis Pena ; Fabio Piccinelli ; Andrei Salavei ; Alessandro Romeo

EFFECT OF VALENCE BAND SPLITTING ON THE ABSORPTION SPECTRA OF MONOLAYER MOS2 IN PRESENCE OF SULPHUR VACANCIES 2376

Himani Mishra ; Sitangshu Bhattacharya

THE STUDY OF SOME MATERIALS AS BUFFER LAYER IN COPPER ANTIMONY SULPHIDE (CUSBS2) SOLAR CELL USING SCAPS 1-D 2381

Muteeu Olopade ; Adeyinka Adewoyin ; Michael Chendo ; Adewumi Bolaji

INFLUENCE OF HETERO-INTERFACES ON PHOTOVOLTAIC PERFORMANCE IN SOLAR CELLS BASED ON ZNSNP2BULK CRYSTAL 2385

Shigeru Nakatsuka ; Shunsuke Akari ; Jakapan Chantana ; Takashi Minemoto ; Yoshitaro Nose

JUNCTION BY DIFFUSION OF ELEMENTAL SODIUM ALONE INTO BRIDGMAN CU(IN, GA) SE2 2388

S. Park ; C. H. Champness ; S. Vanka ; Z. Mi ; I. Shih

OXYGEN SUBSTITUTION AND SULFUR VACANCIES IN NABIS2: A PB-FREE CANDIDATE FOR SOLUTION PROCESSABLE SOLAR CELLS 2392

Robert J Patterson ; Hongze Xia ; Long Hu ; Zhilong Zhang ; Lin Yuan ; Jianfeng Yang ; Weijian Chen ; Zihan Chen ; Yijun Gao ; Yicong Hu ; Binesh Puthen Veettil ; John A. Stride ; Gavin Conibeer ; Shujuan Huang

EFFECT OF ANNEALING ON PERFORMANCE OF SOLAR CELLS WITH NEW OXIDE ABSORBER MN2V2O7 2395

Pramod Ravindra ; Eashwer Athresh ; Rajeev Ranjan ; Srinivasan Raghavan ; Sushobhan Avasthi

ELECTRO-OPTICAL PROPERTIES OF ZN2MO3O8THIN-FILMS: A NOVEL LOW-BANDGAP SOLAR ABSORBER 2399

Pramod Ravindra ; Eashwer Athresh ; Rajeev Ranjan ; Srinivasan Raghavan ; Sushobhan Avasthi

LOW TEMPERATURE SOLUTION PROCESS FOR RANDOM HIGH ASPECT RATIO SILVER NANOWIRE AS PROMISING TRANSPARENT CONDUCTIVE LAYER 2403

Arastoo Teymouri ; Supriya Pillai ; Zi Ouyang ; Xiaojing Hao ; Martin Green

OXYGEN INCORPORATION INTO SI NANOCRYSTAL/SIC MULTILAYERS 2407

Charlotte Weiss ; Andreas Reichert ; Johannes Hofmann ; Stefan Janz

DESIGN OF CASCADED HETEROSTRUCTURED P-I-I-N CDS/CDSE LOW COST SOLAR CELL 2411

M. Zinaddinov ; S. Mil'shtein

FAST C-V METHOD TO MITIGATE EFFECTS OF DEEP LEVELS IN CIGS DOPING PROFILES 2414

P. K. Paull ; J. Bailey ; G. Zapalac ; A. R. Arehart

CRYSTAL GROWTH PHENOMENA IN POLYCRYSTALLINE (CU)ZNTE/CDTE/CDS VIA MOLECULAR DYNAMICS 2419

Rodolfo Aguirre ; Jose J. Chavez ; Xiao W. Zhou ; David Zubia

USING HIGH-RESOLUTION ANOMALOUS-SCATTERING X-RAY DIFFRACTION TO OBSERVE OFF-STOICHIOMETRIC CU2ZNSNS4CRYSTAL STRUCTURES 2423

Christopher J. Bosson ; Max T. Birch ; Douglas P. Halliday ; Chiu C. Tang ; Peter D. Hatton

SIMULATION OF ZNMGO AS THE WINDOW LAYER FORCDTESOLAR CELLS 2427

Yunfei Chen ; Shou Peng ; Xin Cao ; Alan E. Delahoy ; Ken K. Chin

MODELING EFFECT OF DEFECTS ON EFFICIENCY OF NANOWIRE CDS-CDTE SOLAR
CELLS 2432

Hongmei Dang ; Esther Ososanya ; Nian Zhang ; Xiaohui Wang ; Hojjatollah Sarvari ; Vijay P. Singlr

ANALYTICAL DESCRIPTION OF CHARGED GRAIN BOUNDARY RECOMBINATION IN
POLYCRYSTALLINE THIN FILM SOLAR CELLS 2438

Benoit Gaury ; Paul M. Haney

IMAGING THE EFFECT OF CDSE WINDOW LAYERS IN CDTE PHOTOVOLTAICS 2443

John M. Howard ; Elizabeth M. Tennyson ; William B. Gunnarsson ; Naba R. Paudel ; Yanfa Yan ; Marina S.
Leite

INVESTIGATION OF TRAPS DENSITY AND POSITION IN ALKALI TREATED CU(IN,GA)SE2
THIN FILMS AND SOLAR CELLS 2446

Shankar Karki ; Pran K. Paul ; Grace Rajan ; Chinedum Akwari ; Angus Rockett ; Steven A Ringel ; Aaron R.
Arehart ; Sylvain Marsillac

THE EFFECT OF DEPOSITION STOICHIOMETRY AND POST-DEPOSITION TREATMENTS
ON DEEP DEFECTS IN CDTE 2449

Imran S. Khan ; Vamsi Evani ; Shamara Collins ; Chih An Hsu ; Vasilis Palekis ; Chris Ferekides

TESTING THE LIMITS OF MECHANICALLY-SCRIBED CIGS MICROCELLS 2453

Ombline Lafont ; Nicolas Vandamme ; Leia Ruffini ; Jia Yu ; Philip Jackson ; Jose Alvarez ; Daniel Lincot

PHOTOLUMINESCENCE IMAGING ANALYSIS OF DOPING IN THIN FILM CDS AND
CDS/CDTE DEVICES 2457

C. Potamialis ; F. Lisco ; B. Maniscalco ; M. Togay ; A. Abbas ; M. Biiss ; J.W. Bowers ; J.M. Waiis ; I.
Rimmaudo ; R. Mis Fernandez ; V. Rejon ; J.L. Peña

APPLICATION OF MAPPING SPECTROSCOPIC ELLIPSOMETRY FOR CDSE/CDTE SOLAR
CELLS: OPTIMIZATION OF LOW-TEMPERATURE PROCESSED DEVICES WITH ALL-
SPUTTERED SEMICONDUCTORS 2462

Mohammed A. Razooqi ; Adam B. Phillips ; Geethika K. Liyanage ; Fadhil K. Al-Fadhili ; Maxwell M. Junda ;
Nikolas J. Podraza ; Michael J. Heben ; Robert W. Collins ; Prakash Koirala

ASSESSING THE VALIDITY AND ACCURACY OF EFFECTIVE ELECTRONIC MATERIALS:
CAN 1D SIMULATIONS PREDICT POLYCRYSTALLINE DEVICE PERFORMANCE? 2467

Yubo Sun ; Allison Perna ; Sudhajit Misra ; Vasilios Palekis ; Chris Ferekides ; Jeffrey Aguiar ; Peter Bermel ;
Michael A. Scarpulla

CHARACTERIZING RECOMBINATION IN CDTE-BASED SOLAR CELLS BY THE
TEMPERATURE AND EXCITATION DEPENDENCE OF OPEN-CIRCUIT VOLTAGE AND
PHOTOLUMINESCENCE 2473

Craig H. Swartz ; Sanjoy Paul ; Corey R. Grice ; Yanfa Yan ; Lorelle Mansfield ; Sachit Grover ; Gang Xiong ;
Jian V. Li

EXPERIMENTAL EVIDENCE FOR CDS-RELATED TRANSPORT BARRIER IN THIN FILM
SOLAR CELLS AND ITS IMPACT ON ADMITTANCE SPECTROSCOPY 2478

Florian Werner ; Anastasiya Zelenina ; Susanne Siebentritt

TRANSPARENT CONDUCTIVE ADHESIVES FOR TANDEM SOLAR CELLS 2482

Talysa R. Klein ; Benjamin G. Lee ; Manuel Schnabel ; Emily L. Warren ; Pauls Stradins ; Adele C. Tamboli ;
Maikel F.A.M. Van Hest

MODELING THREE-TERMINAL III- V LSI TANDEM SOLAR CELLS 2488

Emily L. Warren ; Michael G. Deceglie ; Paul Stradins ; Adele C. Tamboli

WAFER BONDING APPROACHES FOR III-V ON SI MULTI-JUNCTION SOLAR CELLS 2492

Laura Vauche ; Elias Veinberg-Vidal ; Clément Weick ; Christophe Morales ; Vincent Larrey ; Christophe
Lecouvey ; Mickaël Martin ; Jérémy Da Fonseca ; Christophe Jany ; Thibaut Desrues ; Céline Brughera ;
Philippe Voarino ; Thierry Salvetat ; Frank Fournel ; Mathieu Baudrit ; Cécilia Dupré

DESIGN ARITHMETIC OF THE LATERAL III-V / SI HYBRID MODULE 2498

Kenji Araki ; Kyotaro Nakamura ; Kan-Hua Lee ; Takefumi Kamioka ; Yu-Cian Wang ; Nobuaki Kojima ; Yoshio
Ohshita ; Masafumi Yamaguchi

GAASP NANOWIRE SOLAR CELL DEVELOPMENT TOWARDS NANOWIRE/SI TANDEM
APPLICATIONS 2502

Enrique Barrigon ; Yang Chen ; Gaute Otnes ; Vilgaile Dagyte ; Nicklas Anttu ; Lars Samuelson ; Magnus
Borgström

DEMONSTRATION OF GAINP2/SI VOLTAGE MATCHED TANDEM SOLAR CELLS 2506

David C. Bobela ; Kenneth J. Schmieder ; Matthew P. Lumb ; James E. Moore ; Robert J Walters ; Eric A. Armour
; Leo Matthew ; Rajesh Rao ; Angelo Mascarenhas ; Kirstin Alberi

WAFER BONDED III–V ON SILICON MULTI -JUNCTION CELL WITH EFFICIENCY
BEYOND 31% 2511

Romain Cariou ; Jan Benick ; Paul Beutel ; Nico Tucher ; Martin Graf ; David Lackner ; Martin Hermle ; Stefan
W. Glunz ; Andreas W. Bett ; Frank Dimroth

INTEGRATION OF THIN AL FILMS ON IN0.18GA0.82AS METAMORPHIC GRADE STRUCTURES FOR LOW-COST III- V PHOTOVOLTAICS 2514

Alessandro Giussani ; Michael A. Slocum ; Seth M. Hubbard ; Nathan Smaglik ; Nikhil Pokharel ; S. Phillip Ahrenkiel

TEMPERATURE DEPENDENT CHARACTERISTICS OF GAINP/GAAS/GAINNASSB SOLAR CELL UNDER SIMULATED AM0 SPECTRA 2520

Riku Isoaho ; Arto Aho ; Antti Tukiainen ; Mircea Guina

EFFICIENCY OF GAAS P/SI TWO-JUNCTION SOLAR CELLS WITH MULTI-QUANTUM WELLS: A REALISTIC MODELING WITH CARRIER COLLECTION EFFICIENCY 2524

Boram Kim ; Kasidit Toprasertpong ; Oliver Supplie ; Agnieszka Paszuk ; Thomas Hannappel ; Yoshiaki Nakano ; Masakazu Sugiyama

INVERSE METAMORPHIC III-V/EPI-SIGE TANDEM SOLAR CELL PERFORMANCE ASSESSED BY OPTICAL AND ELECTRICAL MODELING 2528

Raphaël Lachaurne ; Martin Foldyna ; Gwénaëlle Hamon ; Nicolas Vaissiére ; Jean Decobert ; Romain Cariou ; Pere Roca I Cabarrocas ; José Alvarez ; Jean-Paul Kleider

TOWARDS MONOLITHICALLY INTEGRATED GAAS ON SI TANDEM SOLAR CELL 2532

Zhen Liu ; Zekun Ren ; Haohui Liu ; Tonio Buonassisi ; Ian Marius Peters

ZNSIP2 THIN FILM GROWTH FOR SI-BASED TANDEM PHOTOVOLTAICS 2536

Aaron D. Martinez ; Elisa M. Miller ; Andrew G. Norman ; Paul Stradins ; Eric S. Toberer ; Adele C. Tamboli

IN SITU CONTROL OVER THE SUBLATTICE ORIENTATION OF GAP/SI(100): AS VIRTUAL SUBSTRATES FOR TANDEM ABSORBERS 2538

Aznieszka Paszuk ; Oliver Supplie ; Sebastian Brückner ; Matthias M. May ; Anja Dobrich ; Andreas Nägelein ; Boram Kim ; Yoshiaki Nakano ; Masakazu Sugiyama ; Peter Kleinschmidt ; Thomas Hannappel ; Thomas Hannappel

III-V/SI TANDEM CELL TO MODULE INTERCONNECTION - COMPARISON BETWEEN DIFFERENT OPERATION MODES 2543

Henning Schulte-Huxel ; Emily L. Warren ; Manuel Schnabel ; Paul Stradins ; Daniel Friedman ; Adele C. Tamboli

INGAP/GAAS/ITO/SI HYBRID TRIPLE-JUNCTION CELLS WITH GAAS/ITO BONDING INTERFACES 2548

Naoteru Shigekawa ; Tomoya Hara ; Tomoki Ogawa ; Jianbo Liang ; Takefumi Kamioka ; Kenji Araki ; Masafumi Yamaguchi

MEASUREMENTS OF POTENTIALS AT TAP CONTACTS AND ESTIMATION OF RESISTANCE ACROSS BONDING INTERFACES IN INGAP/GAAS/SI HYBRID TRIPLE-JUNCTION CELLS 2551

Naoteru Shigekawa ; Jianbo Liang

OPTIMIZATION OF A GAASP TOP CELL FOR IMPLEMENTATION IN A III-V/SI TANDEM STRUCTURE 2554

Amber C. Silvaggio ; Daniel L. Lepkowski ; Daniel J. Chmielewski ; Jacob T. Boyer ; Steven A. Ringel ; Tyler J. Grassman

THEORETICAL DESIGN OF PEROVSKITE/CDTE FOUR-TERMINAL TANDEM SOLAR CELLS 2558

Tao Tang ; Huan Zhang ; Xingzhi Du ; Yiming Lnr ; Hang Zhou

WAFER-BONDED ALGAAS///SI DUAL-JUNCTION SOLAR CELLS 2562

Elias Veinberg-Vidal ; Laura Vauche ; Clément Weick ; Jérémy Da Fonseca ; Christophe Jany ; Christophe Morales ; Christophe Lecouvey ; Thibaut Desrues ; Philippe Voarino ; Frank Fournel ; Anne Kaminski-Cachopo ; Alejandro Datas ; Pablo Garcia-Linares ; Mathieu Baudrit ; Pierre Mur ; Cécilia Dupré

ENHANCEMENT OF SI PHOTOVOLTAIC MODULE BY INTRODUCING III-V/SI HYBRID CONFIGURATIONS AND COST EVALUATIONS UNDER VARIOUS COST RATIOS OF III-V/SI PHOTOVOLTAICS 2566

Yu-Cian Wang ; Kenii Araki ; Kyotaro Nakamura ; Kan-Hua Lee ; Takefumi Kamioka ; Nobuaki Kojima ; Yoshio Ohshita ; Masafumi Yamaguchi

NUMERICAL SIMULATION OF P-TYPE FRONT JUNCTION PERL SILICON CELL FOR III-V LSI TANDEM DEVICES 2569

Chuqi Yi ; Fa-Jun Ma ; Anita Ho-Baillie ; Stephen Bremner

EPITAXIAL GAP LAYERS GROWN ON SI SUBSTRATES USING MIGRATION ENHANCED AND MOLECULAR BEAM EPITAXY 2573

Chaomin Zhang ; Allison Boley ; Nikolai Faleev ; David J. Smith ; Christiana B. Honsberg

INVESTIGATION OF CARRIER-INDUCED DEFECT BEHAVIOR IN P-TYPE MULTICRYSTALLINE SILICON 2576

Catherine E. Chan ; Tsun H. Fung ; David N.R. Payne ; Daniel Chen ; Malcolm D. Abbott ; Alison M. Ciesla ; Ran Chen ; Brett J. Hallam ; Stuart R. Wenham

MAGNETRON SPUTTERED HYDROGENATED SILICON THIN FILMS: ASSESSMENT FOR APPLICATION IN PHOTOVOLTAICS 2582

Dipendra Adhikari ; Maxwell M. Junda ; Sylvain X. Marsillac ; Robert W. Collins ; Nikolas J. Podraza

HIGH QUALITY AND THIN SILICON WAFER FOR NEXT GENERATION SOLAR CELLS 2588

Yoshio Ohshita ; Takuto Kojima ; Ryota Suzuki ; Kosuke Kinoshita ; Tomoyuki Kawatsu ; Kyotaro Nakamura ; Atsushi Ogura

FIRST DEMONSTRATION OF RADIAL JUNCTION SILICON NANOWIRE SOLAR MINI-MODULES PREPARED BY PECVD AND LASER SCRIBING 2593

Mutaz Al-Ghzaiwat ; Martin Foldyna ; Takashi Fuyuki ; Wanghua Chen ; Erik V. Johnson ; Jacques Meot ; Pere Roca I Cabarrocas

IMPACT OF INDUCED DEFECTS ON DEVICE PERFORMANCE IN SILICON HETEROJUNCTION SOLAR CELLS 2596

Pradeep Balaji ; André Augusto ; Stuart G. Bowden

LASER HYDROGENATION ON HEAVILY DISLOCATED CAST-MONO SILICON CELLS 2600

Alison M. Ciesla ; Catherine E. Chan ; Sisi Wang ; Malcolm D. Abbott ; Cheemun Chong ; Stuart R. Wenham

PERFORMANCE OPTIMIZATION OF SEMI-TRANSPARENT THIN-FILM AMORPHOUS SILICON CELLS 2605

Yuan Gao ; Fai Tong Si ; Olindo Isabella ; Rudi Santbergen ; Guangtao Yang ; Jianfei Dong ; Guoqi Zhang ; Miro Zeman

LOW TEMPERATURE SPALLING OF SILICON: A CRACK PROPAGATION STUDY 2610

Pablo Guimera Coll ; Tine Uberg Nærland ; Nathan Stoddard ; Michael Stuckelberger ; Mariana Bertoni

NEW FINDINGS OF THERMAL EFFECT ON PM-SI:H SOLAR CELLS OPTOELECTRONIC PROPERTIES 2614

L. Hamui ; L. A. Górnez-González ; G. Santana

STUDY OF PV MODULE DEGRADATION RATE PREDICTION THROUGH CORRELATION OF FIELD-AGED AND ACCELERATED-AGED MODULE DEGRADATION DATA 2618

Babak T. Hamzavy ; William J. Grieco ; Brian J. Fields ; Cara S. Libby ; William B. Hobbs ; Olga Lavrova ; C. Birk Jones

ADVANCED ANALYSIS OF MULTI WIRE WAFERING PROCESSES 2622

Ringo Koepgel ; Samuel Brinnig ; Felix Kaule ; Hartmut Schwabe ; Stephan Schoenfelder

CONSIDERATION ON OPEN-CIRCUIT VOLTAGE OF SI HETEROJUNCTION SOLAR CELLS UNDER LOW CONCENTRATION CONDITION 2627

Makoto Konagai

CHARACTERIZATION OF MICROCRYSTALLINE SILICON THIN FILM SOLAR CELLS PREPARED BY HIGH WORKING PRESSURE PLASMA-ENHANCED CHEMICAL VAPOR DEPOSITION 2631

Jung-Dae Kwon ; Dong-Ho Kim ; Ji-Hoon Lee ; Myungkwan Song ; Myunghun Shin

ATOMIC-LAYER-DEPOSITEDV2O5-XFILMS AS A HIGHLY-EFFICIENT P-TYPE LAYER FOR THIN FILM A-SI SOLAR CELLS 2634

Ji-Hoon Lee ; Myungkwan Song ; Dong-Ho Kim ; Jung-Dae Kwon

A NOVEL DEFECT PASSIVATION METHOD FOR MULTICRYSTALLINE SI WAFER BY H2S REACTION 2637

Hsiang-Yu Liu ; Ujjwal K. Das ; Robert W. Birkmire

CARRIER TRANSPORTATION AT NOVEL SILVER PASTE CONTACT 2642

Takefumi Kamioka ; Satoshi Kamevama ; Kazuo Muramatsu ; Aki Tanaka ; Naotaka Iwata ; Kyotaro Nakamura ; Atsushi Ogura ; Yoshio Ohshita

INFLUENCE OF DEPOSITION PARAMETERS ON SILICON THIN FILMS DEPOSITED BY MAGNETRON SPUTTERING 2646

Grace Rajan ; Tejaswini Miryala ; Shankar Karki ; Robert W. Collins ; Nikolas Podraza ; Sylvain Marsillac

MINORITY CARRIER LIFETIME VARIATIONS IN MULTICRYSTALLINE SILICON WAFERS WITH TEMPERATURE AND INGOT POSITION 2651

Sissel Tind Søndergaard ; Jan Ove Odden ; Rune Strandberg

CUO NANOWIRES-BASED RADIAL HETERO-JUNCTION THIN FILM SILICON SOLAR CELLS WITH A HIGH OPEN-CIRCUIT VOLTAGE 2656

Xiaolin Sun ; Jiawen Lu ; Fan Yang ; Linwei Yu ; Jun Xu ; Ling Xu ; Kunji Chen

THE EFFECT OF CHEMICAL COMPOSITION ON POROUS ETCHING FOR EPI AND LIFT-OFF WAFER PROCESS 2660

Teng-Yu Wang ; Peng-Wei Chen ; Han-Wen Liu

ELECTRICAL AND OPTICAL PERFORMANCE OF SILICON SOLAR CELLS USING PLASMONICS INDIUM NANOPARTICLES LAYER EMBEDDED IN SIO2ANTIREFLECTIVE COATING 2664

Hao-Yu Yang ; Wen-Jeng Ho ; Sheng-Kai Feng ; Jheng-Jie Liu ; Ta-Wei Chuang ; Guan-Yi Li ; Yun-Chie Yang ; Cho-Chun Chiang ; Yao- Hui Chen

ELECTROLUMINESCENCE ANALYSIS FOR SEPARATION OF SERIES RESISTANCE FROM RECOMBINATION EFFECTS IN SILICON SOLAR CELLS WITH INTERDIGITATED BACK CONTACT DESIGN 2667

Nuha Ahmed ; Lei Zhang ; Ujjwal Das ; Steven Hegedus

INDOOR MEASUREMENT OF ANGLE RESOLVED LIGHT ABSORPTION BY BLACK SILICON 2672

Mekbib W. Amdemeskel ; Beniamino Iandolo ; Rasmus S. Davidsen ; Ole Hansen ; Gisele A. Dos Reis Benatto ; Nicholas Riedel ; Peter B. Poulsen ; Sune Thorsteinsson ; Anders Thorseth ; Carsten Dam-Hansen

IMPACT OF NON- FLAT PHOTOGENERATION AND CARRIER PROFILES ON THE LUMINESCENT EMISSION AND DETECTION OF SILICON SOLAR CELLS 2677

Nekane Azkona ; Federico Recart ; Pedro Rodríguez ; Vanesa Fano ; Aloña Otaegi ; Juan Carlos Jimeno

DEVELOPMENT OF OUTDOOR LUMINESCENCE IMAGING FOR DRONE-BASED PV ARRAY INSPECTION 2682

Gisele A. Dos Reis Benatto ; Nicholas Riedel ; Sune Thorsteinsson ; Peter B. Poulsen ; Anders Thorseth ; Carsten Dam-Hansen ; Claire Mantel ; Soren Forchhammer ; Kenn H. B. Frederiksen ; Jan Vedde ; Michael Petersen ; Henrik Voss ; Michael Messerschmidt ; Harsh Parikh ; Sergiu Spataru ; Dezso Sera

CLIMBING DRUM PEEL (CDP) TEST METHOD FOR CHARACTERIZING ADHESION IN FLEXIBLE PV MODULES 2688

Venkata Bheemreddy ; Kedar Hardikar

ACCURACY OF SOLAR SIMULATOR SPECTRAL DETERMINATION USING BAND-PASS FILTERING METHOD 2692

Weston Dobson ; Harrison Wilterdink ; Cassidy Sainsbury ; Adrienne Blum ; Justin Dinger ; Ronald A. Sinton ; Karsten Bothe ; David Hinken ; Martin Wolf

CORRELATION OF I-V CURVE PARAMETERS WITH MODULE-LEVEL ELECTROLUMINESCENT IMAGE DATA OVER 3000 HOURS DAMP-HEAT EXPOSURE 2697

Justin S. Fada ; Andrew J. Loach ; Alan J. Curran ; Jennifer L. Braid ; Shuying Yang ; Timothy J. Peshek ; Roger H. French

A NOVEL METHOD TO INVESTIGATE STOICHIOMETRY AND PERFORMANCE OF BURIED PASSIVATED CONTACTS UTILIZING TIME-OF-FLIGHT SIMS 2702

Steven P. Harvey ; William Nemeth ; Jeff Aguiar ; Craig Perkins ; Pauls Stradins

A COMPARISON BETWEEN QUASI-STEADY STATE AND TRANSIENT PHOTOCONDUCTANCE LIFETIMES IN SILICON INGOTS: SIMULATIONS AND MEASUREMENTS 2707

Mohsen Goodarzi ; Ronald Sinton ; Daniel Chung ; Bernhard Mitchell ; Thorsten Trupke ; Daniel Macdonald

NEW DEVELOPMENT IN GLOW DISCHARGE OPTICAL EMISSION SPECTROMETRY FOR THE CHARACTERIZATION AND THE THICKNESS MEASUREMENT OF LAYERS FOR PHOTOVOLTAIC APPLICATIONS 2711

Philippe Hunault ; Matthieu Chausseau ; Patrick Chaporr ; Sofia Gaiaschi ; Anais Loubar ; Muriel Bouttcmy ; Arnaud Etcheberry

DEEP LEVEL TRANSIENT SPECTROSCOPY MEASUREMENTS OF SILICON HETEROJUNCTION CELLS 2716

Sanchit Khatavkar ; C. V. Kannan ; Vijay Kumar ; P. R. Nair ; B. M. Arora

CHARACTERIZATION OF MODULES AND ARRAYS WITH SUNS VOC 2719

Alex Killam ; Stuart Bowden

A STUDY OF PERFORMANCE CHARACTERIZATION WITH REAR LIGHT SOURCE IN CONVENTIONAL BIFACIAL SOLAR CELLS 2723

Soo Min Kim ; Sang Hoon Jung ; Rae-Won Choi ; Yong Bae Kim ; Min Gu Kang ; Hee-Eun Sonp ; Gyu-Seok Choi

ELECTRICAL CHARACTERIZATION OF THE CARRIER TRANSPORT PROPERTIES IN ACU(IN,GA)SE2SOLAR CELL 2728

Roberto Lopez ; Sanjoy Paull ; Ingrid Repins ; Jian V. Li

SYSTEMATIC THERMALPHOTOVOLTAIC SOLAR CELL OPTIMIZATION 2732

Zheng Lyu ; Muyu Xue ; Junyan Chen ; Jieyang Jia ; Shanhui Fan ; James Harris

CHARACTERIZATION OF TELLURIUM AS A BACK CONTACT FOR CDTE SOLAR CELLS 2736

C.E. Moffett ; W.S. Sampath

ON THE DIFFERENT EXPLANATIONS OF THE RECOMBINATION CURRENTS WITH HIGH IDEALITY FACTOR IN SILICON SOLAR CELLS 2740

A. Otaegi ; V. Fano ; N. Azkona ; J. R. Gutiérrez ; J. C. Jimeno

IDENTIFICATION OF SHUNTS IN A MONOLITHIC MULTIJUNCTION GAAS/GAAS DEVICE BY SPECTROMETRIC CHARACTERIZATION 2744

Felipe Oviedo ; Liu Zhe ; Zekun Ren ; Kevin Nay Yaung ; Maung Thway ; Liu Haohui ; Tonio Buonassisi ; Ian Marius Peters

A SIMULATION STUDY ON RADIATIVE RECOMBINATION ANALYSIS IN CIGS SOLAR CELL 2749

Sanjoy Paul ; Roberto Lopez ; Md Dalim Mia ; Craig H. Swartz ; Jian V. Li

SIMULATION AND SPECTROSCOPY OF CARRIER RELAXATION IN GASB AND GAAS 2755

A.C. Scofield ; A.I. Hudson ; B.L. Liang ; B.C. Juang ; D.L. Huffaker ; S.M. Hubbard ; W.T. Lotshaw

COMPUTATIONAL DESIGN OF DOPANTS IN CDTE GRAIN BOUNDARIES FOR EFFICIENT PHOTOVOLTAICS 2759

Fatih G. Sen ; Tadas Paulauskas ; Ce Sun ; Moon Kim ; Robert F. Klie ; Maria K.Y. Chan

ANALYSES OF PHOTOVOLTAIC POWER PLANT PERFORMANCE ESTIMATES BASED ON DETAILED LABORATORY MODULE CHARACTERIZATIONS AND TYPICAL REAL-WORLD INPUT DATA SOURCES 2762

Rajeev Singh ; John L.R. Watts ; Kellen Gillispie

CRITICAL EVALUATION OF THE FOUNDATIONS OF SOLAR SIMULATOR STANDARDS 2765

Ronald A. Sinton ; Harrison Wilterdink ; Justin Dinger ; Adrienne L. Blum ; Weston Dobson ; Cassidy Sainsbury

IMPACT OF INFRARED OPTICAL PROPERTIES ON CRYSTALLINE SI AND THIN FILM CDTE SOLAR CELLS 2771

Indra Subedi ; Timothy J Silverman ; Michael Deceglie ; Nikolas J. Podraza

THE IMPACT OF IMPURITIES ON THE RELATIVE EFFICIENCIES OF SOLAR CELLS FROM DIFFERENT SILICON FEEDSTOCKS 2776

Muhammad Tayyib ; Aleksandr Dobroliubov ; Zekija Ramic ; Muhammad Nadeem Akarm ; Jan Ove Odden

ACCURACY EVALUATION OF ABSOLUTE ELECTROLUMINESCENCE-EFFICIENCY MEASUREMENTS OF SOLAR CELLS USING A SENSITIVITY-CALIBRATED-PHOTODETECTOR CONTACT METHOD 2781

Masahiro Yoshita ; Yoshihiro Hishikawa ; Yoshihiko Kanemitsu ; Hidefumi Akiyama

NANOMETER-SCALE CARRIER IMAGING OF POTENTIAL-INDUCED DEGRADATION IN C-SI SOLAR CELLS 2785

C.-S. Jiang ; C. Xiao ; H.R. Moutinho ; S. Johnston ; M.M. Al-Jassim ; X. Yang ; Y. Chen ; J. Ye

NREL EFFORTS TO ADDRESS SOILING ON PV MODULES 2789

Lin J. Simpson ; Matthew Muller ; Michael Deceglie ; Helio Moutinho ; Craig Perkins ; C. S. Jiang ; David C. Miller ; Leonardo Micheli ; Govindasamy Tamizhmani ; Sai Ravi Vasista Tatapudi ; Mowafak Al-Jassim

MODELING POTENTIAL-INDUCED DEGRADATION (PID) OF FIELD-EXPOSED CRYSTALLINE SILICON SOLAR PV MODULES: FOCUS ON A REGENERATION TERM 2794

Eleonora Annigoni ; Alessandro Virtuani ; Fanny Sculati-Meillaud ; Christophe Ballif

SOILING LOSS ON PV MODULES AT TWO LOCATIONS IN INDIA STUDIED USING A WATER BASED ARTIFICIAL SOILING METHOD 2799

Sonali Bhaduri ; Sachin Zachariah ; Lawrence L. Kazmcrski ; Balasubramaniam Kavaipatti ; Anil Kottantharayil

QUANTIFYING YEAR-TO-YEAR VARIATIONS IN SOLAR PANEL SOILING FROM PV ENERGY-PRODUCTION DATA 2804

Michael G. Deceglie ; Leonardo Micheli ; Matthew Muller

ACCURATELY MEASURING PV SOILING LOSSES WITH SOILING STATION EMPLOYING PV MODULE POWER MEASUREMENTS 2808

Michael Gostein ; Bill Stueve ; Mandy Chan

PERFORMANCE OF MONOCRYSTALLINE SILICON SOLAR CELL- INFLUENCE OF DUST ON ULTRA-VIOLET AND VISIBLE REGION DURING EARLY STAGE OF DEPOSITION 2811

Hemaprabha Elangovan ; Upasna Ranjan ; A K Jagdish ; Praveen C. Ramamurthy ; Kamanio Chattopadhyay

A COMPREHENSIVE STUDY OF LIGHT SOAKING EFFECT IN CDTE SOLAR CELLS 2816

D. Guo ; A. Moore ; D. Krasikov ; I. Sankin ; D. Vasileska

CORRECTION FOR METASTABILITY IN THE QUANTIFICATION OF PID IN THIN-FILM MODULE TESTING 2819

Peter Hacke ; Sergiu Spataru ; Steve Johnston

A FINE MODEL OF POWER DEGRADATION FOR CRYSTALLINE SILICON SOLAR MODULES 2823

Wenshuang Hea ; Baosong Duan ; Fumei Wang ; Ao Wang ; Jipeng Chang ; He Wang ; Hong Yang ; Jie Ding ; Junjun Zhang ; Jingsheng Huang

TEST METHODS FOR HYDROPHOBIC COATINGS ON SOLAR COVER GLASS 2827

Kenan Isbilir ; Biancamaria Maniscalco ; Ralph Gottschalg ; John Michael Walls

IMPACT OF DEGRADATION RATES ON SOLAR PV FINANCING FOR PROJECTS LOCATED IN THE UNITED STATES 2833

Rounak A. Kharait ; Phil Stiles ; Jarrett Carriere ; Larry Mcclung

ANALYSIS OF WIND DIRECTION AND SPEED MEASUREMENTS IN ARID REGION - A SITE EVALUATION USING DATA WITH LOW TEMPORAL RESOLUTION 2836

Elisabeth Klimm ; Felix Guischard ; Karl-Anders Weiss

FORECASTING ENVIRONMENTAL DEGRADATION POWER LOSS IN SOLAR PANELS WITH A PREDICTIVE CRACK OPENING TEST 2839

Jason L. Lincoln ; Andrew M. Gabor ; Eric J. Schneller ; Hubert Seigneur ; Joseph Walters ; Rob Janoch ; Andrew Anselmo ; Victor Huayamave ; Winston Schoenfeld

FLUORESCENCE IMAGING ON THE CROSS-SECTION OF PHOTOVOLTAIC LAMINATES AGED UNDER DIFFERENT UV INTENSITIES 2844

Yadong Lyu ; Jae Hyun Kim ; Xiaohong Gu

STATISTICAL ANALYSIS OF DEGRADATION DATA FOR C-SI MODULES OBSERVED IN INDIA IN 2016 2849

Chiranjibi Mahapatra ; Rajiv Dubey ; Shashwata Chattopadhyay ; Sachin Zachariah ; Sanjeev Sabnis

PROCESS INDUCED DEFLECTION AND STRESS ON ENCAPSULATED SOLAR CELLS 2854

Xiaodong Meng ; Michael Stuckelberger ; Peter Hacke ; Mariana Bertoni

A UNIFIED GLOBAL INVESTIGATION ON THE SPECTRAL EFFECTS OF SOILING LOSSES OF PV GLASS SUBSTRATES: PRELIMINARY RESULTS 2858

Leonardo Micheli ; Eduardo F. Fernández ; Greg P. Smestad ; Hameed Alrashidi ; Nabin Sarmah ; Nazmi Sellami ; Ibrahim A. I. Hassan ; Amal Kasry ; Gustavo Nofuentes ; Neeru Sood ; Bala Pesala ; S. Senthilarasu ; Florencia Almonacid ; K.S. Reddy ; Matthew Muller ; Tapas K. Mallick

REFERENCE: PROCEEDINGS OF THE IEEE PVSC CONF., 2017 THE DEVELOPMENT OF A DC BREAKDOWN VOLTAGE TEST FOR PHOTOVOLTAIC INSULATING MATERIALS 2864

David C. Miller ; Bernt Ake-Sultan ; Axel Borne ; Rene Eugen ; Bradley L. Givot ; Jiirgen Jung ; Steven W. Macmaster ; Byron K. Mcdanold ; Ulf H. Nilsson ; Nancy H. Phillips ; Ian A. Tappan ; Nick S. Bosco

FIELD-EVALUATION OF ELECTRODYNAMIC SCREENS FOR MAINTAINING HIGH OPTICAL EFFICIENCY OPERATION OF SOLAR COLLECTORS 2870

Cristian Morales ; Annie Bernard ; Ryan Eriksen ; Julius Yellowhair ; Sean Garner ; Ricci La Centra ; Alecia Griffin ; Alexis Lloyd ; Yujie Gao ; Ramakrishnan Lakshmanan ; Mark Horenstein ; Malay Mazumder

EFFECT OF REVERSE BIAS VOLTAGES ON SMALL SCALE GRIDDED CIGS SOLAR CELLS 2875

Soheyl Mortazavi ; Klaas Bakker ; Jome Carolus ; Michael Daenen ; Gabriela De Amorim Soares ; Henk Steijvers ; Arthur Weeber ; Mirjam Theelen

A METHOD TO EXTRACT SOILING LOSS DATA FROM SOILING STATIONS WITH IMPERFECT CLEANING SCHEDULES 2881

Matthew Muller ; Leonardo Micheli ; Alfredo A. Martinez-Morales

ANALYTICAL (S)TEM STUDIES OF DEFECTS ASSOCIATED WITH PID IN STRESSED SI PV MODULES 2887

Andrew Norman ; Adam Stokes ; John Moseley ; Steven Harvey ; Steve Johnston ; Harvey Guthrey ; Mowafak Al-Jassim

DESIGN, DEVELOPMENT, AND EVALUATION OF ELECTRODYNAMIC SCREENS FOR SELF-CLEANING SOLAR PANELS AND CONCENTRATING MIRRORS 2891

Annie Bernard ; Cristian Morales ; Ryan S. Eriksen ; Alecia C. Griffin ; Yujie Gao ; Ramakrishnan Lakshmanan ; Ricci La Centra ; Arash Sayyah ; Julius E. Yellowhair ; Sean M. Garner ; N Mark Horenstein ; Malay K. Mazumder

EVALUATING SOLAR CELL FRACTURE AS A FUNCTION OF MODULE MECHANICAL LOADING CONDITIONS 2897

Eric J. Schneller ; Andrew M. Gabor ; Jason Lincoln ; Rob Janoch ; Andrew Anselmo ; Joseph Walters ; Hubert Seigneur

COMPUTATIONAL STUDY OF THE EFFECT OF PHOTOVOLTAIC (PV) MODULE PARAMETERS ON STRESS DEVELOPMENT IN SILICON UNDER STATIC LOADING 2902

Saurabh Sethia ; Karan Shishir Yadav ; Sudharm Rathore ; Abhishek Shubhrant ; Aparna Singh

A SIMPLE METHOD FOR MEASURING SOLAR RADIATION INTENSITY BY IMAGE ANALYSES 2906

Akiko Takahashi ; Akinori Moriki ; Nobuyuki Yamada ; Jun Imai ; Shigeyuki Funabiki

DEGRADATION OF SOLDER BONDS IN FIELD AGED PV MODULES: CORRELATION WITH SERIES RESISTANCE INCREASE 2912

Abhishiktha Tummala ; Jaewon Oh ; Sai Tatapudi ; Govindasamy Tamizhmani

PERFORMANCE OF LIGHT AND DARK CURRENT-VOLTAGE CHARACTERISTICS FOR PID-AFFECTED MONOCRYSTALLINE SILICON SOLAR MODULES 2918

He Wang ; Pan Zhao ; Shuwen Guo ; Hong Yang ; Weijing Huang ; Shiyu Sang ; Bojie Su ; Xue Zhang ; Yunxue Cao ; Hui Zhao

SOILING RATES OF PV MODULES VS. THERMOPILE PYRANOMETERS 2923

Martin Waters ; Tejas Tirumalai ; Michael Gostein ; Bill Stueve

GRID INTEGRATION OF BUILDING SYSTEMS AND 1 MW PHOTOVOLTAIC ARRAY USING VOLTTRON 2926

David Raker ; Andrew Sellers ; Roshan Kini ; Michael Green ; Thomas Stuart ; Randall Ellingson ; Raghav Khanna ; Michael Heben

INTERCONNECTION STUDY OF DISTRIBUTED PV SYSTEMS BY INTERFACING MATLAB WITH OPENDSS AND GIS 2931

Joseph A. Ahamioje ; Hariharan Krishnaswami

NOVEL MPPT ALGORITHM FOR ACTIVE POWER CONTROL OF MULTI-LEVEL DUAL-ACTIVE BRIDGE PV CONVERTER IMPLEMENTED IN NI MYRIO 2936

Shilpa Marti ; Hariharan Krishnaswami

MODELING A GRID-CONNECTED PV/BATTERY MICROGRID SYSTEM WITH MPPT CONTROLLER 2941

Genesis Alvarez ; Hadis Moradi ; Mathew Smith ; Ali Zilouchian

>94.5%REDUCTION IN GRID-BUY ELECTRICITY AND ELIMINATION OF AM & PM ENERGY PEAKS/SPIKES BY OPTIMIZING ENERGY USAGE AND INTEGRATION OF CUSTOMER SELF-SUPPLY ROOFTOP SOLAR PV WITH ELECTRICAL & THERMAL (HOT & COLD) STORAGE BATTERIES: A CASE STUDY FOR RESIDENTIAL HAWAII 2947

John Borland ; Jay Moore ; Corpuz Poncho ; Takahiro Tanaka ; Harumi Mcclure

A SINGLE-STAGEC"UK-BASED TRANSFORMERLESS INVERTER FOR 1-Φ GRID-CONNECTED PV SYSTEMS 2952

Phani Kumar Chamarthi ; Amit Kumar Gupta ; Madhuwanti S. Joshi ; Vivek Agarwal

A STATE SPACE AVERAGE MODEL FOR DYNAMIC MICROGRID BASED SPACE STATION SIMULATIONS 2957

Rachid Darbali-Zamora ; Eduardo I. Ortiz-Rivera

BUCK CONVERTER AND SEPIC BASED ELECTRONIC POWER SUPPLY DESIGN WITH MPPT AND VOLTAGE REGULATION FOR SMALL SATELLITE APPLICATIONS 2963

Rachid Darbali-Zamora ; Nicolás Cobo-Yepes ; John E. Salazar-Duque ; Eduardo I. Ortiz-Rivera ; Amilcar A. Rincon-Charris

VIRTUAL POWER PLANT FEEDBACK CONTROL DESIGN FOR FAST AND RELIABLE ENERGY MARKET AND CONTINGENCY RESERVE DISPATCH 2969

Mohamed Elkhatib ; Jay Johnson ; David Schoenwald

INTELLIGENT SAMPLING OF PERIODS FOR REDUCED COMPUTATIONAL TIME OF TIME SERIES ANALYSIS OF PV IMPACTS ON THE DISTRIBUTION SYSTEM 2975

Jason Galtieri ; Matthew J. Reno

A PWM SCHEME TO REALISE TWO TIMES EFFECTIVE SWITCHING FREQUENCY WITH CONSTANT COMMON MODE VOLTAGE AND REACTIVE POWER CAPABILITY IN 1- Φ GRID-TIED TRANSFORMERLESS H6 PV INVERTER 2981

Amit Kumar Gupta ; Madhuwanti S. Joshi ; Vivek Agarwal

A SOLAR PV RETROFIT SOLUTION FOR RESIDENTIAL BATTERY INVERTERS 2986

Amit Kumar Gupta ; Vaibhav Pawar ; Madhuwanti S. Joshi ; Vivek Agarwal ; Deepak Chandran

COST BENEFIT AND ALTERNATIVES ANALYSIS OF DISTRIBUTION SYSTEMS WITH ENERGY STORAGE SYSTEMS 2991

Tom Harris ; Adarsh Nagarajan ; Murali Baggu ; Tom Bialek

EVALUATION OF PV HOSTING CAPACITIES OF DISTRIBUTION GRIDS WITH UTILIZATION OF SOLAR-ROOF-POTENTIAL-ANALYSES 2996

Gerd Heilscher ; Falko Ebe ; Basem Idlbi ; Jeromie Morris ; Florian Meier

EXPERIMENTAL DISTRIBUTION CIRCUIT VOLTAGE REGULATION USING DER POWER FACTOR, VOLT-VAR, AND EXTREMUM SEEKING CONTROL METHODS 3002

Jay Johnson ; Sigifredo Gonzalez ; Daniel B. Arnold

DYNAMIC SETPOINT CONTROL OF ELECTRIC HOT WATER HEATER TANKS FOR INCREASED INTEGRATION OF SOLAR PHOTOVOLTAIC SYSTEMS 3008

C. Birk Jones ; Monte Lunacek ; Matthew Lave ; Jay Johnson ; Robert Broderick

SPATIAL ANALYSIS OF RESIDENTIAL COMBINED PHOTOVOLTAIC AND BATTERY POTENTIAL: CASE STUDY UTRECHT, THE NETHERLANDS 3014

Geert Litjens ; Bala Bhavya Kausika ; Ernst Worrell ; Wilfried Van Sark

POWER BALANCE REQUIREMENTS FOR SUSTAINED ISLANDING OF INVERTER BASED DISTRIBUTED GENERATION 3020

Gregory A. Kern ; Michael Ropp ; Sigifredo Gonzalez

FULL-SCALE DEMONSTRATION OF DISTRIBUTION SYSTEM PARAMETER ESTIMATION TO IMPROVE LOW-VOLTAGE CIRCUIT MODELS 3025

Matthew Lave ; Matthew J. Reno ; Robert J. Broderick ; Jouni Peppanen

CREATION AND VALUE OF SYNTHETIC HIGH-FREQUENCY SOLAR INPUTS FOR DISTRIBUTION SYSTEM QSTS SIMULATIONS 3031

Matthew Lave ; Matthew J. Reno ; Robert J. Broderick

A DIRECT MAXIMUM POWER POINT SEARCH USING CURRENT-VOLTAGE BASED POWER-LAW RELATION FOR PHOTOVOLTAIC SYSTEM UNDER UNIFORM IRRADIANCE 3038
Hitesh K. Mehta ; Ashish K. Panchal

PASSIVITY BASED CONTROLLER FOR PHOTOVOLTAIC MODULES USING ZETA CONVERTER 3044
Daniel A. Merced Cirino ; Rachid Darbali Zamora ; Eduardo I. Ortiz Rivera

SIC SWITCH BASED SINGLE-STAGE BUCK-BOOST TRANSFORMERLESS MINI INVERTER WITH LOW LEAKAGE CURRENT AND NEGLIGIBLE DC INJECTION 3050
Soumya Ranjan Mohapatra ; Amit Kumar Gupta ; Madhuwanti S. Joshi ; Vivek Agarwal

OPEN SOURCE TOOLS FOR HIGH PERFORMANCE QUASI-STATIC-TIME-SERIES SIMULATION USING PARALLEL PROCESSING 3055
Davis Montenegro ; Roger C. Dugan ; Matthew J. Reno

MAXIMUM POWER POINT TRACKING OF PV MODULE BASED ON NEW EXPLICIT I-V RELATION 3061
Tejeswar Nukala ; A. K. Panchal

AN AUTOCORRELATION-BASED COPULA MODEL FOR PRODUCING REALISTIC CLEAR-SKY INDEX AND PHOTOVOLTAIC POWER GENERATION TIME-SERIES 3067
Joakim Munkhammar ; Joakim Widén

DYNAMIC RESPONSE OF MAXIMUM POWER POINT TRACKING USING PERTURB AND OBSERVE ALGORITHM WITH MOMENTUM TERM 3073
Gautam A. Raiker

A FRAMEWORK FOR COMPARING THE ECONOMIC PERFORMANCE AND ASSOCIATED EMISSIONS OF GRID-CONNECTED BATTERY STORAGE SYSTEMS IN EXISTING BUILDING STOCK: A NYISO CASE STUDY 3077
Julian Do Nascimento Ricardo ; Vasilis Fthenakis

IMPROVING ANY ARBITRARY MPPT HILL CLIMBER WITH ANN ESTIMATIONS 3083
Jesse Roberts ; Indranil Bhattacharya

INCREASING SOLAR PHOTOVOLTAIC PENETRATION USING THERMAL ENERGY STORAGE 3088
Alexander F. Routhier ; Christiana Honsberg

MODEL PREDICTIVE CONTROL OF GRID CONNECTED MODULAR MULTILEVEL CONVERTER FOR INTEGRATION OF PHOTOVOLTAIC POWER SYSTEMS 3092
Amir Shahirinia ; Amin Hajizadeh

MAXIMIZATION OF SELF-SUFFICIENCY WITH GRID CONSTRAINTS: PV GENERATORS, WIND TURBINES AND STORAGE TO FEED TERTIARY SECTOR USERS 3096
Filippo Spertino ; Jawad Ahmad ; Alessandro Ciocia ; Paolo Di Leo ; Francesco Giordano

SWITCHES CONTROLLING TO IMPLEMENT ADAPTIVE MULTILEVEL INVERTER ON PV SYSTEM 3102
Hadi Suhana ; Ngapuli I Sinisuka ; Muhammad Nurdin ; Yvon Besanger ; Vincent Debusschere

DEMAND RESPONSE FOR THE PROMOTION OF PHOTOVOLTAIC PENETRATION 3107
Venizelos Venizelou ; Spyros Theocharides ; George Makrides ; Venizelos Efthymiou ; George E. Georghiou

GRIDDLER AI: NEW PARADIGM IN LUMINESCENCE IMAGE ANALYSIS USING AUTOMATED FINITE ELEMENT METHODS 3113
Johnson Wong ; Percis Teena ; Daniel Inns

INTERACTION OF O2IDIMERS WITH GA IN SI AND IMPLICATIONS FOR A COMPREHENSIVE MODEL OF LIGHT- INDUCED DEGRADATION 3119
Yu Jin ; Scott T. Dunham

NUMERICAL SIMULATION OF EBIC FOR ANALYSIS OF EXTENDED DEFECTS 3123
Marco Nardone ; John Moseley ; Saroj Dahal ; Anuja V. Parikh ; John M. Waddle

COLLOIDAL QUANTUM DOT SOLAR CELL ELECTRICAL PARAMETER IMAGING USING CAMERA-BASED HIGH-FREQUENCY HETERODYNE LOCK-IN CARRIEROGRAPHY 3129
Lilei Hu ; Mengxia Liu ; Andreas Mandelis ; Qiming Sun ; Alexander Melnikov ; Edward H. Sargent

A NEW PERSPECTIVE ON POTENTIAL-INDUCED DEGRADATION OF THE SHUNTING TYPE BY MICRO RAMAN-SPECTROSCOPY AND MICRO LIGHT-BEAM-INDUCED CURRENT 3135
A. Büchler ; H. Nagel ; M. Breitwieser ; S. Kluska ; F. D. Heinz ; M. C. Schubert ; M. Glatthaar ; S. Glunz

NANOSCALE DETECTION OF DEEP LEVELS IN CIGS USING ELECTRON ENERGY LOSS SPECTROSCOPY 3139
Julia I. Deitz ; Pran K. Paul ; Shankar Karki ; Sylvain Marsillac ; Aaron R. Arehart ; Tyler J. Grassman ; David W. Mccomb

MEASUREMENT OF CARRIER DYNAMICS IN PHOTOVOLTAIC CZTSE BY TIME-RESOLVED TERAHERTZ SPECTROSCOPY 3143

Siming Li ; Michael A. Lloyd ; Andrew A. Golembeski ; Brian E. Mccndless ; Jason B. Baxter

DECOUPLING GRAIN-BOUNDARY, GRAIN-INTERIOR, AND SURFACE RECOMBINATION WITH CATHODOLUMINESCENCE 3147

John Moseley ; Pierre Rale ; Stéphane Collin ; Ana Kanevce ; Eric Colegrove ; Joel Duenow ; Soren Jensen ; Wyatt K. Metzger ; Mowafak M. Al-Jassim

HIGH RESOLUTION THZ SCANNING FOR OPTIMIZATION OF DIELECTRIC LAYER OPENING PROCESS ON DOPED SI SURFACES 3150

P. Spinelli ; F.J.K. Danzl ; D. Deligiannls ; N. Guillevin ; A.R. Burgers ; S. Sawallich ; M. Nage ; I. Cesar

DEGRADATION ASSESSMENT OF FIELDED CIGS PHOTOVOLTAIC ARRAYS 3155

Bruce H. King ; Joshua S. Stein ; Daniel Riley ; C. Birk Jones ; Charles D. Robinson

APPLICATION OF IEC 61724 STANDARDS TO ANALYZE PV SYSTEM PERFORMANCE IN DIFFERENT CLIMATES 3161

Katherine A. Klise ; Joshua S. Stein ; Joseph Cunningham

EFFECTS OF URBAN ENVIRONMENT ON SOLAR PV PERFORMANCE 3167

Panagiotis Moraitis ; Bala Bhavya Kausika ; Wilfried G.J.H.M. Van Sark

IRRADIANCE MEASUREMENT CONSIDERATIONS FOR SYSTEM PERFORMANCE ASSESSMENT WHEN MANAGING FLEETS OF PHOTOVOLTAIC ASSETS ACROSS ASIA 3172

André M. Nobre ; Shravan Karthik ; Chenxi Liu ; Rohit Jaswal ; Rupesh Baker ; Raghav Malhotra ; Alan Khor

MACHINE LEARNING IN PV FAULT DETECTION, DIAGNOSTICS AND PROGNOSTICS: A REVIEW 3178

Sandy Rodrigues ; Helena Geirinhas Ramos ; F. Morgado-Dias

OUTDOOR FIELD PERFORMANCE FROM BIFACIAL PHOTOVOLTAIC MODULES AND SYSTEMS 3184

Joshua S. Stein ; Daniel Riley ; Matthew Lave ; Clifford Hansen ; Chris Deline ; Fatima Toor

DEFINING THRESHOLD VALUES OF ENCAPSULANT AND BACKSHEET ADHESION FOR PV MODULE RELIABILITY 3190

Nick Bosco ; Joshua Eafanti ; Sarah Kurtz ; Jared Tracy ; Reinhold Dauskardt

CHARACTERIZATIONS OF AGED GLASS/ETHYLENE VINYL ACETATE/GLASS USING FLUORESCENCE SPECTROSCOPY AND INSTRUMENTED INDENTATION 3195

Jae Hyun Kim ; Yadong Lyu ; David C. Miller ; Xiaohong Gu

ENCAPSULANT ADHESION TO SURFACE METALLIZATION ON PHOTOVOLTAIC CELLS 3200

Jared Tracy ; Nick Bosco ; Reinhold Dauskardt

IMPACT OF UV LIGHT INTENSITY ON PHOTODEGRADATION OF PV BACKSHEETS 3204

Xiaohong Gu ; Li-Chieh Yu ; Yadong Lyu ; Jae Hyun Kim ; Andrew Fairbrother ; Tinh Nguyen

SURVEY OF MECHANICAL DURABILITY OF PV BACKSHEETS 3208

Michael D. Kempe ; David C. Miller ; Allen Zielnik ; Daniel Montiel-Chicharro ; Jiang Zhu ; Ralph Gottschalg

SOLAR VARIABILITY REDUCTION USING OFF-MAXIMUM POWER POINT TRACKING AND BATTERY STORAGE 3214

Jason Galtieri ; Philip T. Krein

INTEGRATION OF ELECTROCHEMICAL CAPACITORS ON SILICON PHOTOVOLTAIC MODULES FOR RAPID-RESPONSE POWER BUFFERING 3220

Yu Jiang ; Xuanyi Shi ; Derwin Lau ; Da-Wei Wang ; Zi Ouyang ; Alison Lennon

DESIGN & EVALUATION OF A HYBRID SWITCHED CAPACITOR CIRCUIT WITH WIDE-BANDGAP DEVICES FOR COMPACT MVDC PV POWER CONVERSION 3224

J. Stewart ; J. Delhotal ; J. Richards ; J. Neely ; L. Rashkin ; J. D. Flicker ; R. Kaplar ; S. Gonzalez ; J. Lehr

SOLAR ENERGY FOR CLEAN AND AFFORDABLE WATER DESALINATION 3230

V. M. Fthenakis ; Adam A. Atia

GLOBAL RESIDENTIAL AIR-CONDITIONING SECTOR AS A DRIVER FOR PHOTOVOLTAIC INDUSTRY GROWTH DURING THE 21ST CENTURY 3236

Hannu S. Laine ; Jyri Salpakari ; Marius Peters ; Erin E. Looney ; Ashley E. Morishige ; Hele Savin ; Gregory Wilson ; Tonio Buonassisi

MEASURES TO REMOVE ECONOMIC NON-MARKET FAILURE AND INSTITUTIONAL BARRIERS THAT RESTRICT PHOTOVOLTAICS SELF-CONSUMPTION AND NET-METERING IN SPAIN 3240

Enrique Rosalcs-Ascnsio ; Juan A. Méndez ; Benjamín Gonzálcz-Díaz ; Ricardo Guerrero Lemus

COST COMPETITIVE CONCENTRATOR PHOTOVOLTAICS FOR SOLAR THERMAL APPLICATIONS 3245

Brian C. Riggs ; Richard E. Biedenham ; Chris Dougher ; Yaping Vera Ji ; Qi Xu ; Vince Romanin ; Daniel S. Codd ; James M. Zahler ; Matthew D. Escarra

PREDICTING THE EFFICIENCY OF THE SILICON BOTTOM CELL IN A TWO-TERMINAL TANDEM SOLAR CELL 3250

Zhengshan J. Yu ; Zachary C. Holman

MECHANICALLY STACKED 4-TERMINAL III-V/SI TANDEM SOLAR CELLS 3254

Stephanie Essig ; Christophe Allebe ; John F. Geisz ; Myles A. Steiner ; Loris Barraud ; Antoine Descoeudres ; J. Scott Ward ; Manuel Schnabel ; David L. Young ; Matthieu Despeisse ; Christophe Ballif ; Adele Tamboli

PEROVSKITE/SILICON TANDEM SOLAR CELLS: CHALLENGES TOWARDS HIGH-EFFICIENCY IN 4-TERMINAL AND MONOLITHIC DEVICES 3256

Jérémie Werner ; Florent Sahli ; Brett Kamino ; Davide Sacchetto ; Matthias Bräuninger ; Arnaud Walter ; Christophe Ballif ; Matthieu Despeisse ; Sylvain Nicolay ; Bjoern Niesen ; Raphaël Monnard ; Stefaan De Wolf ; Soo-Jin Moon ; Loris Barraud ; Bertrand Paviet-Salomon ; Jonas Geissbuehler ; Christophe Allebé

THE OUTCOME OF REPLACING SN COMPLETELY BY GE IN KESTERITE CU2ZNSNSE4SOLAR CELLS 3260

S. Sahayaraj ; G. Brammertz ; B. Vermang ; T. Schnabel ; E. Ahlswede ; Z. Huang ; S. Ranjbar ; M. Meuris ; J. Vleugels ; J. Poortmans

TRANSITION METAL OXIDES NANO-LAYERS AS EFFICIENT BACK ELECTRON REFLECTORS FOR CU2ZNSNSE4SOLAR CELLS 3265

Sergio Giraldo ; Moisés Espíndola-Rodríguez ; Florian Oliva ; Víctor Izquierdo-Roca ; Alejandro Pérez-Rodríguez ; Edgardo Saucedo

MIXED SULFUR AND SELENIUM ANNEALING STUDY OF COMPOUND-SPUTTERED BILAYER CU2ZNSNS4/ CU2ZNSNSE4PRECURSORS 3269

N. Ross ; S. Grini ; L. Vines ; C. Platzer-Björkman

REVEALING THE ROLE OF MN INCORPORATION IN CU2ZNSN(S, SE)4PHOTOVOLTAIC ABSORBER LAYER 3275

Stener Lie ; Joel M. R. Tan ; Wenjie Li ; Shin Woei Leow ; Oki Gunawan ; Doug Bishop ; Lydia H. Wong

NON-VACUUM SINGLE STEP SYNTHESIS OF LARGE-GRAIN SIZE CZTS PHOTO ABSORBER FOR THIN FILM SOLAR CELLS BY FLUX ASSISTED CHEMICAL SPRAY 3279

Ratheesh R. Thankalekshmi ; Navjot Kaur Sidhu ; A.C. Rastogi

RAMAN SCATTERING ASSESSMENT OF POINT DEFECTS IN KESTERITE SEMICONDUCTORS: UV RESONANT RAMAN CHARACTERIZATION FOR ADVANCED PHOTOVOLTAICS 3285

Florian Oliva ; Laia Arqués Farré ; Sergio Giraldo ; Mirjana Dimitrievska ; Paul Pistor ; Alejandro Martínez-Pérez ; Lorenzo Calvo-Barrio ; Edgardo Saucedo ; Alejandro Pérez-Rodríguez ; Victor Izquierdo-Roca

ASSESSING THE DEFECT RESPONSIBLE FOR LETID: TEMPERATURE- AND INJECTION-DEPENDENT LIFETIME SPECTROSCOPY 3290

Mallory A. Jensen ; Yan Zhu ; Erin E. Looney ; Ashley E. Morishige ; Carlos Vargas ; Ziv Hameiri ; Tonio Buonassisi

MICROSCOPIC DISTRIBUTION OF LUMINESCENCE FROM DISLOCATION CLUSTERS IN MULTICRYSTALLINE SILICON WAFERS 3295

H. T. Nguyen ; M. A. Jensen ; L. Li ; C. Samundsett ; H. C. Sio ; B. Lai ; T. Buonassisi ; D. Macdonald

DO GRAIN BOUNDARIES MATTER? ELECTRICAL AND ELEMENTAL IDENTIFICATION AT GRAIN BOUNDARIES IN LETID-AFFECTED P-TYPE MULTICRYSTALLINE SILICON 3300

Mallory A. Jensen ; Ashley E. Morishige ; Sagnik Chakraborty ; Romika Sharma ; Hang Cheong Sio ; Chang Sun ; Barry Lai ; Volker Rose ; Amanda Youssef ; Erin E. Looney ; Sarah Wieghold ; Jeremy Poindexter ; Juan-Pablo Correa-Baena ; Daniel Macdonald ; Joel B. Li ; Tonio Buonassisi

PERC SOLAR CELL PERFORMANCE PREDICTIONS FROM MULTICRYSTALLINE SILICON INGOT METROLOGY DATA 3304

Bernhard Mitchell ; Daniel Chung ; Qiuxiang He ; Hua Zhang ; Zhen Xiong ; Pietro P. Altermatt ; Peter Geelan-Small ; Thorsten Trupke

PHOTOLUMINESCENCE-IMAGING-BASED EVALUATION OF NON-UNIFORM CDTE DEGRADATION 3305

Steve Johnston ; David Albin ; Peter Hacke ; Steven P. Harvey ; Helio Moutinho ; Mowafak Al-Jassim ; Wyatt K. Metzger

MACHINE LEARNING AND CORRELATIVE MICROSCOPY: HOW 'BIG DATA' TECHNIQUES CAN BENEFIT THIN FILM SOLAR CELL CHARACTERIZATION 3309

Bradley M. West ; Michael Stuckelberger ; Tara Nietzold ; Barry Lai ; Jörg Maser ; Mariana I. Bertoni

METAL INDUCED CONTACT RECOMBINATION MEASURED BY QUASI-STEADY-STATE PHOTOLUMINESCENCE 3315

Robert Dumbrell ; Mattias K. Juhl ; Mengjie Li ; Thorsten Trupke ; Ziv Hameiri

USING TIME-OF-FLIGHT SIMS TO INVESTIGATE GROUP V DOPANT DISTRIBUTION IN CDTE 3319

Steven P. Harvey ; Eric Colegrove ; Brian Mccandless ; David Albin ; Mowafak Al-Jassim ; Wyatt K. Metzger

QUANTITATIVE ANALYSIS OF ACTIVE DOPANT DISTRIBUTION AND ESTIMATION OF EFFECTIVE DIFFUSIVITY IN PHOSPHORUS- IMPLANTED EMITTER OF SI SOLAR CELL USING SCANNING NONLINEAR DIELECTRIC MICROSCOPY 3323

Kotaro Hirose ; Katsuto Tanahashi ; Hidetaka Takato ; Yasuo Cho

SIMULATION OF DRIVE-LEVEL CAPACITANCE PROFILING TO INTERPRET MEASUREMENTS ON CU(IN, GA)SE2SCHOTTKY DEVICES 3327

Geordie Zapalac ; Jeff Bailey

ANALYSIS OF THE IMPACT OF INSTALLATION PARAMETERS AND SYSTEM SIZE ON BIFACIAL GAIN AND ENERGY YIELD OF PV SYSTEMS 3333

Amir Asgharzadeh ; Tomas Lubenow ; Joseph Sink ; Bill Marion ; Chris Deline ; Clifford Hansen ; Joshua Stein ; Fatima Toor

DEPENDENCE OF STRING POWER ON ITS HEIGHT IN THE ARRAY IN YOSHINOGARI MEGA SOLAR POWER PLANT 3339

Shigeomi Hara ; Makoto Kasu ; Yasuki Masutomi

A BOTTOM-UP ENERGY SIMULATION FRAMEWORK TO ACCURATELY COMPARE PV MODULE TOPOLOGIES UNDER NON-UNIFORM AND DYNAMIC OPERATING CONDITIONS 3343

Patrizio Manganiello ; Maro Baka ; Hans Goverde ; Tom Borgers ; Jonathan Govaerts ; Arvid Van Der Heide ; Eszter Voroshazi ; Francky Catthoor

A PERFORMANCE MODEL FOR BIFACIAL PV MODULES 3348

Daniel Riley ; Clifford Hansen ; Joshua Stein ; Matthew Lave ; Johnson Kallickal ; Bill Marion ; Fatima Toor

ACCURATE MODELING OF PARTIALLY SHADED PV ARRAYS 3354

Bennet Meyers ; Mark Mikofski

EVALUATION OF UNCERTAINTY IN PV PROJECT DESIGN: DEFINITION OF SCENARIOS AND IMPACT ON ENERGY YIELD PREDICTIONS 3360

Giorgio Belluardo ; Magnus Herz ; Ulrike Jahn ; Mauricio Richter ; David Moser

MONOCRYSTALLINE 1.7 EV MGCDTE DOUBLE-HETEROSTRUCTURE SOLAR CELL WITH 11.2% EFFICIENCY 3366

Calli M. Campbell ; Xin-Hao Zhao ; Yuan Zhao ; Mathieu Boccard ; Cheng- Ying Tsai ; Jacob J. Becker ; Zachary Holman ; Yong-Hang Zhang

MBE GROWTH OF 1.7EV AL0.2GA0.8AS AND 1.42EV GAAS SOLAR CELLS ON SI USING DISLOCATIONS FILTERS: AN ALTERNATIVE PATHWAY TOWARD III-V/ SI SOLAR CELLS ARCHITECTURES 3370

Arthur Onno ; Mingchu Tang ; Mu Wang ; Yurii Maidaniuk ; Mourad Benamara ; Yuriy I. Mazur ; Gregory J. Salamo ; Lars Oberbeck ; Jiang Wu ; Huiyun Liu

III- V/SI TANDEM CELLS UTILIZING INTERDIGITATED BACK CONTACT SI CELLS AND VARYING TERMINAL CONFIGURATIONS 3371

Manuel Schnabel ; Michael Rienacker ; Agnes Merkle ; Talysa R. Klein ; Nikhil Jain ; Stephanie Essig ; Henning Schulte-Huxel ; Emily Warren ; Maikel F.A.M. Van Hest ; John Geisz ; Jan Schmidt ; Rolf Brendel ; Robby Peibst ; Paul Stradins ; Adele Tamboli

TOWARDS HIGH-EFFICIENCY GAASP/SI TANDEM CELLS 3376

S. Fan ; M. Vaisman ; K. Nay Yaung ; E. Perl ; D. Martín-Martín ; M. Leilaeioun ; Z. C. Holman ; M. L. Lee

CHARACTERIZATION OF HETEROEPITAXIAL GAAS FILMS GROWN ON SI USING SELECTIVE AREA NUCLEATION 3381

Emily L. Warren ; Emily A. Makoutz ; Michelle Vaisman ; Benjamin F. Bachman ; William E. Mcmahon ; Jeramy D. Zimmerman ; Adele C. Tamboli

EFFICIENT PHOTON UPCONVERSION IN SEMICONDUCTOR NANOSTRUCTURES: CONSTRAINTS AND OPPORTUNITIES 3384

Matthew F. Doty ; Eric Y. Chen ; Jing Zhang ; Diane G. Sellers ; Zhuohui Li ; Christopher C. Milleville ; Kyle Lennon ; Joshua M. O. Zide

ENHANCED ULTRA-THIN A-GE:H SOLAR CELLS BY PLASMONIC NANOPARTICLES EMBEDDED IN THE OPTICAL RESONANT CAVITY 3388

Brendan Brady ; Volker Steenhoff ; Benedikt Nickel ; Martin Vehse ; Alexander G. Brolo

NATIVE-METAL-OXIDE-COATED PLASMONIC ELECTRODE METASURFACES FOR NANOPHOTONIC LIGHT TRAPPING AND EFFICIENT CHARGE COLLECTION 3393

Deirdre M. O'Carroll ; Christopher E. Petoukhoff ; Zhongkai Cheng ; Zeqing Shen ; Catrice M. Carter

IN-GA PRECURSOR ISLANDS FOR CU(IN, GA)SE2MICRO-CONCENTRATOR SOLAR CELLS 3396

Katharina Eylers ; Franziska Ringleb ; Berit Heidmann ; Sergiu Levcenco ; Thomas Unold ; Hagen W. Klemm ; Gina Peschel ; Alexander Fuhrich ; Thomas Teubner ; Thomas Schmidt ; Martina Schmid ; Torta Boeck

ADVANCES IN SILICON SURFACE TEXTURIZATION BY METAL ASSISTED CHEMICAL ETCHING FOR PHOTOVOLTAIC APPLICATIONS 3402

Sylvain Le Gall ; Raphaël Lachaume ; Encarnacion Torralba ; Mathieu Halbwax ; Vincent Magnin ; Taha El Assimi ; Marin Fouchier ; Joseph Harari ; Jean-Pierre Vilcot ; Christine Cachet-Vivier ; Stéphane Bastide

SINGLE CRYSTALLINE SUBSTRATES FOR III- V GROWTH VIA EXFOLIATION OF BULK SINGLE CRYSTALS 3406

Celeste L. Melamed ; Brenden R. Ortiz ; Aaron D. Martinez ; William E. Mcmahon ; Adele C. Tamboli ; Andrew G. Norman ; Eric S. Toberer

CUZNS HOLE CONTACTS ON MONOCRYSTALLINE CDTE SOLAR CELLS 3410

Jacob J. Becker ; Xiaojie Xu ; Rachel Woods-Robinson ; Calli M. Campbell ; Maxwell Lassise ; Joel Ager ; Yong-Hang Zhang

THE EFFECT OF THE CDCL2 HEAT TREATMENT ON CDSEXTE1-X SOLAR CELLS 3413

Chih An Hsu ; Vasilios Palekis ; Imran Khan ; Shamara Collins ; Don Morel ; Chris Ferekides

EFFECTS OF CDCL2TREATMENT ON THE LOCAL ELECTRONIC PROPERTIES OF POLYCRYSTALLINE CDTE MEASURED WITH PHOTOEMISSION ELECTRON MICROSCOPY 3417

Morgann Berg ; Jason M. Kephart ; Walajabad S. Sampath ; Taisuke Ohta ; Calvin Chan

POINT DEFECTS IN CDTE BULK SINGLE CRYSTALS GROWN IN CD-RICH CONDITIONS 3422

Tursun Ablekim ; Santosh K. Swain ; Teresa M. Barnes ; Kelvin G. Lynn

OPTICAL PROPERTIES OFCDSE1-XSXANDCDSE1-YTEYALLOYS AND THEIR APPLICATION FOR CDTE PHOTOVOLTAICS 3426

Maxwell M. Junda ; Corey R. Grice ; Prakash Koirala ; Robert W. Collins ; Yanfa Yan ; Nikolas J. Podraza

BLISTERING OF MAGNETRON SPUTTERED THIN FILM CDTE DEVICES 3430

P.M. Kaminski ; S. Yilmaz ; A. Abbas ; F. Bittau ; J.W. Bowers ; R.C. Greenhalgh ; J.M. Walls

ENERGY YIELD IN HOT & SUNNY CLIMATES: IMPACT OF SILICON SOLAR CELL ARCHITECTURE AND CELL INTERCONNECTION 3435

Jan Haschke ; Johannes P. Seif ; Yannick Riesen ; Andrea Tomasi ; Jean Cattin ; Loïc Tous ; Patrick Choulat ; Monica Aleman ; Emanuele Comagliotti ; Angel Uruena ; Richard Russell ; Filip Duerinckx ; Jonathan Champliaud ; Jacques Levrat ; Amir A. Abdallah ; Brahim Aïssa ; Nouar Tabet ; Nicolas Wyrsch ; Matthieu Despeisse ; Jozef Szlufcik ; Stefaan De Wolf ; Christophe Ballif

NOVEL REAR SIDE METALLIZATION ROUTE FOR SI SOLAR CELLS USING A TRANSPARENT CONDUCTING ADHESIVE 3439

Manuel Schnabel ; Talysa R. Klein ; Benjamin G. Lee ; William Nemeth ; Vincenzo Lasalvia ; Maikel F.A.M. Van Hest ; Paul Stradins

MULTILAYER FOIL METALLIZATION FOR ALL BACK CONTACT CELLS 3442

David H. Levy ; David E. Carlson

ELECTROLUMINESCENCE EXCITATION SPECTROSCOPY: A NOVEL APPROACH TO NON-CONTACT QUANTUM EFFICIENCY MEASUREMENTS 3448

Kristopher O. Davis ; Greg S. Horner ; Joshua B. Gallon ; Leonid A. Vasilyev ; Kyle B. Lu ; Antonius B. Dirriwachter ; Terry B. Rigdon ; Eric J. Schneller ; Kortan Ogutman ; Richard K. Ahrenkiel

ILLUMINATED OUTDOOR LUMINESCENCE IMAGING OF PHOTOVOLTAIC MODULES 3452

Timothy J Silverman ; Michael G. Deceglie ; Kaitlyn Vansant ; Steve Johnston ; Ingrid Repins

ELECTROLUMINESCENT IMAGE PROCESSING AND CELL DEGRADATION TYPE CLASSIFICATION VIA COMPUTER VISION AND STATISTICAL LEARNING METHODOLOGIES 3456

Justin S. Fada ; Mohammad A. Hossain ; Jennifer L. Braid ; Shuying Yang ; Timothy J Peshek ; Roger H. French

TOWARDS DEVELOPING A STANDARD FOR TESTING BIFACIAL PV MODULES: SINGLE-SIDE VERSUS DOUBLE-SIDE ILLUMINATION METHOD I-V MEASUREMENTS UNDER DIFFERENT IRRADIANCE AND TEMPERATURE 3462

Stefan Roest ; Witek Nawara ; Bas B. Van Aken ; Elias Garcia Goma

ELECTRICAL TRANSPORT PROPERTIES FROM LONG WAVELENGTH ELLIPSOMETRY 3468

Prakash Uprety ; Maxwell M. Junda ; Indra Subedi ; Michael A. Slocum ; David V. Forbes ; Seth M. Rubbard ; Nikolas J. Podraza

IN SITU RAMAN MONITORING OF KESTERITE CU2ZNSNS4 PHASE FORMATION FROM SULFURIZATION OF SOL-GEL OXIDE PRECURSORS 3473

Osama Awadallah ; Joseph Hernandez ; Andriy Durygin ; Zhe Cheng

PERFORMANCE OF FIELD-AGED PV MODULES IN INDIA: RESULTS FROM 2016 ALL INDIA SURVEY OF PV MODULE RELIABILITY 3478

Rajiv Dubey ; Sachin Zachariah ; Shashwata Chattopadhyay ; Vivek Kuthanazhi ; Sugguna Rambabu ; Sonali Bhaduri ; Hemant K. Singh ; Archana Sinha ; Birinchi Bora ; Rajesh Kumar ; O. S. Sastry ; Chetan S. Solanki ; Anil Kottantharayil ; Brij M. Arora ; K. L. Narasimhan ; Juzer Vasi

INFERRING THE PERFORMANCE RATIO OF PV SYSTEMS DISTRIBUTED IN AN REGION: A REAL-CASE STUDY IN SOUTH TYROL 3482

Marco Pierro ; Giorgio Belluardo ; Philip Ingenhoven ; Cristina Cornaro ; David Moser

QUANTIFY PHOTOVOLTAIC MODULE DEGRADATION USING THE LOSS FACTOR MODEL PARAMETERS 3488

C. Birk Jones ; Bruce H. King ; Joshua S. Stein ; Justin S. Fada ; Alan J. Curran ; Roger H. French ; Erdmut Schnabel ; Michael Koehl ; Olga Lavrova

SIMULATING PV SYSTEM PERFORMANCE WITH COMPONENT RELIABILITY DISTRIBUTIONS 3494

Geoffrey T. Klisel ; Janine M. Freeman ; Olga Lavrova

LIFETIME AND DEGRADATION OF PRE-DAMAGED PV-MODULES – FIELD STUDY AND LAB TESTING 3500

Claudia Buerhop ; Sven Wirsching ; Simon Gehre ; Tobias Pickel ; Thilo Winkler ; Andreas Bemrrr ; Julia Merghcim ; Christian Camus ; Jens Hauch ; Christoph J. Brabec

IMM TRIPLE-JUNCTION SOLAR CELLS AND MODULES OPTIMIZED FOR SPACE AND TERRESTRIAL CONDITIONS 3506

Tatsuya Takamoto ; Hiroyuki Juso ; Kohsuke Ueda ; Hidetoshi Washio ; Hiroshi Yamaguchi ; Mitsuru Imaizumi ; Taishi Sumita ; Tetsuya Nakamura

VERY HIGH SPECIFIC POWER ELO SOLAR CELLS (>3 KW/KG) FOR UAV, SPACE, AND PORTABLE POWER APPLICATIONS 3511

D. Cardwell ; A. Kirk ; C. Stender ; A. Wibowo ; F. Tuminello ; M. Drees ; R. Chan ; M. Osowski ; N. Pan

ENHANCED ENDURANCE OF A UNMANNED AERIAL VEHICLES USING HIGH EFFICIENCY SI AND III-V SOLAR CELLS 3514

David Scheiman ; Raymond Hoheisel ; Daniel J Edwards ; Andrew Paulsen ; Justin Lorentzen ; Steve Carruthers ; Sam Carter ; Matthew Kelly ; Phillip Jenkins ; Robert Walters

HIGH PERFORMANCE, LIGHTWEIGHT GAAS SOLAR CELLS FOR AEROSPACE AND MOBILE APPLICATIONS 3520

Aarohi Vijh ; Lori Washington ; Robert C. Parenti

THROUGH-EPITAXIAL-VIA BACK-CONTACT MULTI-JUNCTION SOLAR CELLS FABRICATED USING EPITAXIAL LIFT-OFF 3524

Rekha Reddy ; Marilyn L. Nowakowski ; David Rowell ; Christopher L. Stender ; Christopher Youtsey

DESIGN OF INGAP/GAAS/LNGAAS MULTI-JUNCTION CELLS WITH REDUCED LAYER THICKNESSES USING LIGHT-TRAPPING REAR TEXTURE 3528

Lin Zhu ; Anurag Reddy ; Kentaroh Watanabe ; Masakazu Sugiyama ; Yoshiaki Nakano ; Hidefumi Akiyama

Author Index

Open Circuit Voltage Calculation Using Temperature and Irradiance

Andrew Melvin

Ulteig Engineering, Inc. 5575 DTC Parkway Suite 200. Greenwood Village, CO 80111

Abstract — A module's voltage changes based on the environment the module is in. The common method for calculating system voltage relies on the modules voltage temperature factor. This factor is applied for the coldest temperature to the Open Circuit Voltage, as calculated at STC conditions of 1000 W/m2. However, many of these coldest conditions occur at very low light levels. While voltage is largely temperature dependent, it is not solely temperature dependent and irradiance has an impact particularly in low light conditions. This finding is substantiated by both manufacturers and field measurements from operational systems. By using publicly available, multiyear satellite derived data sets, designers are able to show realistic Voc's far lower than the traditional method allowing in some cases several extra modules per string without the risk of exceeding IEC/NEC voltage limitations. Longer strings equate to considerable savings in a project be it through reduced home run quantities, more modules per tracker/racking, reduced DC line losses and keeping voltage higher and in the inverter's power point window for later in the projects life.

Index Terms — solar modeling, open circuit voltage

I. INTRODUCTION

The traditional method for calculating maximum system voltage, as outlined in the National Electric Code, involves adjusting the standard test condition (STC) calculated open circuit voltage (Voc) to the ASHRAE minimum temperature. Basically, account for the maximum voltage of a system at the expected minimum temperature sunlight is present. Though sufficiently ensuring that a system does not exceed voltage allowances, using the STC voltage as calculated at 1000W/m2 leads to a very conservative calculated voltage. This is because a module's voltage, while being mostly dependent on temperature, is also dependent on irradiance levels. Otherwise, a module would be generating voltage and thus the potential to do work without sunlight present. As a majority of the colder conditions occur during low light levels this can lead to misrepresented maximum voltage and unnecessarily limit system options. Instead, a method utilizing the module's IV (Voltage/current) curve at various irradiances can be employed along with long term, site specific data sets to create a statistically significant maximum voltage.

II. OVERVIEW OF STUDY

12 module strings of 275W and 285W Modules mounted on a Fixed Tilt racking system were set to Open Circuit over a course of One (1) Year. One (1) Minute data was collected for voltage, module temperature and POA Irradiance. Current/Voltage (IV) curves for a wide range of irradiances were created from 3rd party lab test results. The open circuit voltage (Voc) is then extracted from this dataset for performing open circuit calculations as shown in figure 1.

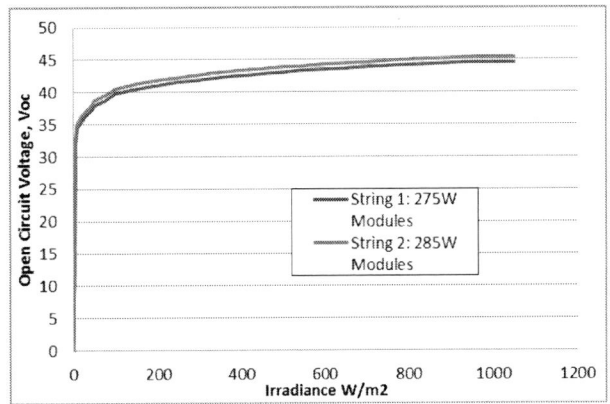

Figure 1: Voltage vs Irradiance Curve at 25°C

Voltage is dependent on both temperature and irradiance. By isolating the voltage values for a single temperature, the non-linear relationship between voltage and irradiance can be derived. This Voc value is then modified using the modules temperature coefficient for the measured temperature using equation 1, the standard equation for temperature correction. These calculated values can then be compared to the measured values to determine how accurate the proposed calculation method is.

$$Voc(T) = Voc(ref) * (1-\alpha(T-Tref)) \qquad [1]$$

Where
Tref = 25°C
T = Recorded Cell Temperature
Voc(Ref) = Irradiance Adjusted Open Circuit Voltage
α = Module's Temperature Coefficient (%/°C)

III. FINDINGS

Using the collected data, it was shown that the calculated method utilizing the Voc binning respective of irradiance was consistently approximating the measured data with +/- 5% whereas the calculation method using only the STC Voc was consistently over approximating the low irradiance Voc values by a considerable margin. Typical calculated values using just the Voc at STC method were over approximating by 30%,

978-1-5090-5606-4/17 $31.00 © 2017 IEEE

sometimes as much as 55%. For the same data points, the alternate method presented here was no more than 5% over approximating. During periods of higher insolation, all methods were very similar and trended very close to the measured value. However, as this represented the lowest voltages due to higher module temperatures, is less significant than the morning values. Figure 2 represents the three voltages in a single day.

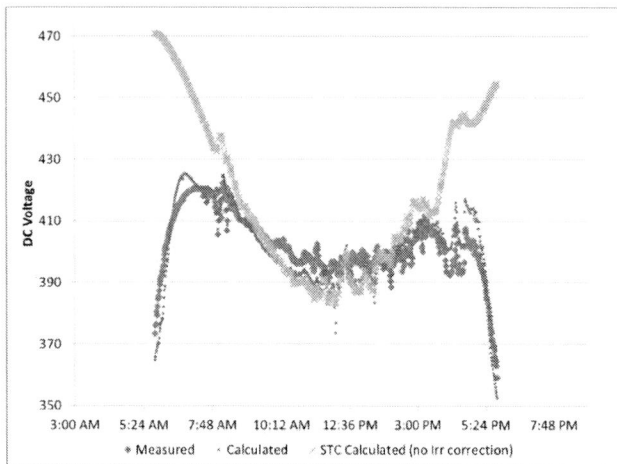

Figure 2: Measured and Calculated Voc for Single Day

To determine the potential impact of the revised voltage calculation, a long term weather data set is used and voltage calculated for each point in the yearly set. For this example, 12 years' worth of site-specific data was pulled off of SolarAnywhere[1], a subscription based service. However, Solar Prospector[2] can be used to generate the same length dataset without the need for a subscription. These yearly datasets are run through a program such as PVSyst or NREL SAM to generate 8760 hourly output tables for each year including module temperature and POA irradiance. Similar to the real world data, irradiance binning is used along with the temperature coefficient to calculate maximum system voltage for all irradiance values.

The results are shown in figure 3. Maximum VOC, as calculated using the ASHRAE value of -8°C and the Voc at STC for 20 module strings, comes out to 1054Vdc. The maximum voltage using irradiance adjusted Voc values yields a maximum of 997Vdc with an average of just 989Vdc.

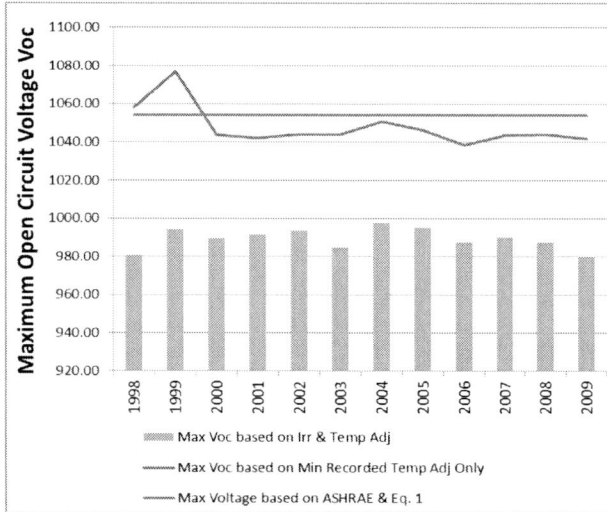

Figure 3: Long Term Voc Calculation

IV. CONCLUSION

While the traditional method of maximum system voltage in a PV system is adequate for most designs, it can be limiting and overestimate the maximum voltage a system will experience. This can lead to unnecessary cost adders in racking, wiring and larger voltage drop resulting in poorer financial returns on a system. It has been shown that an alternative method for calculating the maximum system voltage can be achieved by using open circuit voltages values for various irradiance levels, not just STC, and applying temperature corrections over a long term dataset. The result may be longer strings and improved economics without sacrificing safety, Authority Having Jurisdiction or manufacturer requirements.

REFERENCES

[1] Solar Anywhere, https://www.solaranywhere.com/Public/SelectData.aspx 1/27/2016
[2] Solar Prospector, http://maps.nrel.gov/prospector 1/27/2016

Effect of Cl-doping in ZnTeO on Photoluminescence and Photovoltaic Properties of ZnTeO-based Intermediate Band Solar Cells

T. Tanaka[1], S. Tsutsumi[1], Y. Okano[1], K. Matsuo[1], K. Saito[1], Q. Guo[1], M. Nishio[1], T. Tayagaki[2], K. M. Yu[3], and W. Walukiewicz[4,5]

[1]Department of Electrical and Electronic Engineering, Saga University, Saga 840-8502, Japan.
[2]National Institute of Advanced Industrial Science and Technology, Tsukuba 305-8568, Japan.
[3]Department of Physics and Materials Science, City University of Hong Kong, Kowloon, Hong Kong.
[4]Materials Sciences Division, Lawrence Berkeley National Laboratory, Berkeley, California 94720, U.S.A.
[5]Department of Materials Science and Engineering, University of California at Berkeley, Berkeley, CA 94720, U.S.A.

Abstract — **We report the effect of Cl-doping in ZnTeO alloys on the photoluminescence (PL) and photovoltaic (PV) properties of ZnTeO-based intermediate band solar cells (IBSCs) with *n*-ZnS layer. The Cl-doping was done by supplying ZnCl$_2$ flux during the growth of ZnTeO on ZnTe substrate by molecular beam epitaxy to introduce electrons into the intermediate band (IB) of ZnTeO that is required to be half-filled with electrons for the efficient operation of an intermediate band solar cell. The ZnCl$_2$ cell temperature was varied between 70 and 250 °C. Low temperature PL spectra indicated that the doped Cl atoms acts as donors in ZnTeO. The PV properties of IBSCs using ZnTeO layer with and without Cl-doping were investigated. The external quantum efficiencies (EQE) of the IBSC with Cl-doped ZnTeO are higher than those without Cl-doping in whole energy range between 1.5 and 3.8 eV. The photocurrent generated by two-step photon absorption through IB was observed at room temperature in the IBSC using Cl-doped ZnTeO whereas it was not detected in the IBSC without Cl-doping. The results indicate that the Cl-doping is effective to improve the PV properties of ZnTeO-based IBSC through the introduction of electron into IB.**

I. Introduction

The concept of multiband or intermediate band solar cell (IBSC) has recently attracted a renewed attention as a viable approach to achieving high solar power conversion efficiencies [1]. Several approaches have been employed to demonstrate the concept of IBSC including quantum dot superlattices [2,3] and highly mismatched alloys (HMAs) [4-7]. HMAs is a class of materials whose electronic band structures are dramatically modified through the substitution of a relatively small fraction of host atoms with an element of very much different electronegativity/size, for example N in GaAs [8] and O in ZnTe [4].

As has been previously reported highly mismatched ZnTe$_{1-x}$O$_x$ (ZnTeO) is one of the promising absorber materials for IBSCs because ZnTeO has a narrow O-derived intermediate band (E_-) located well below the conduction band (E_+) edge of the ZnTe [4]. The energy position of E_- and E_+ bands in ZnTeO can be tuned by adjusting the O content as described

by the band anticrossing (BAC) model [8]. The three absorption edges of ZnTeO cover the entire solar spectrum providing a material envisioned for the multi-band, single junction, high efficiency photovoltaic (PV) devices.

So far, several research groups have grown ZnTeO layers by molecular beam epitaxy (MBE) [9-11] and pulsed laser deposition [10]. Using radio frequency plasma-assisted MBE (RF-MBE), we have previously grown ZnTeO layers on ZnTe substrates and reported the dependence of the energy position of E_+ and E_- bands on the O composition [11]. Also, we have demonstrated the generation of photocurrent by two-step photon absorption through the intermediate E_- band in ZnTeO-based IBSC with *n*-ZnO window layer [12].

In order to enhance the two-step photon absorption, the intermediate band is required to be half-filled with electrons according to the theoretical simulation [13]. As a donor impurity, Cl is one of the most promising elements because the successful *n*-type doping in ZnTe was reported by Tao *et al.* [14] and we have also reported the preliminary results on the growth and photoluminescence (PL) properties of Cl-doped ZnTeO layers, showing the possible formation of Cl-donors in ZnTeO [15]. Here, we report the detailed PL analyses on Cl-doped ZnTeO layers and the effect of Cl-doping in ZnTeO on the PV properties of ZnTeO-based IBSCs.

II. Experimental

Cl-doped ZnTeO layers were grown on ZnTe(001) substrates by a conventional MBE system with a radio frequency radical cell for O. 7N Zn and 6N Te were used as source materials while 6N ZnCl$_2$ was used as a dopant source. ZnTe(001) substrates were ultrasonically cleaned in organic solvents and were wet-etched using Br-methanol solution. The substrate temperature was set to 400 °C during growth. The ZnCl$_2$ cell temperature was varied between 70 and 250 °C. The O radical supply conditions were kept constant as the RF power and the O$_2$ flow rate of 100 W and 0.2 sccm,

978-1-5090-5606-4/17 $31.00 © 2017 IEEE

respectively, throughout the growth experiments. The growth time was 60 min.

O composition x was estimated from a relaxed lattice constant a_{ZnTeO} of ZnTeO assuming Vegard's law. a_{ZnTeO} was obtained from the in-plane and out-of-plane lattice constants determined by high-resolution X-ray diffraction (HR-XRD). The O composition x in the Cl-doped ZnTeO was in the range between 0.51 and 0.99%.

The IBSCs with an n-ZnS/ZnTe/Cl-doped ZnTeO/ZnTe/p-ZnTe (referred as "Cl-doped IBSC") and an n-ZnS/ZnTe/undoped ZnTeO/ZnTe/p-ZnTe (referred as "undoped IBSC") structures were fabricated to clarify the effect of Cl-doping on the solar cell performances.

III. RESULTS AND DISCUSSION

A. Photoluminescence properties of Cl-doped ZnTeO films

Fig. 1 shows the PL spectra at 6 K for Cl-doped ZnTeO films grown at various $ZnCl_2$ cell temperatures. A PL spectrum for an undoped ZnTeO with O concentration of 0.99% is also shown in Fig. 1 for comparison. In all samples, PL peaks are observed at the photon energy range between 1.5 and 2.0 eV and the spectra are fitted with Gaussian function as shown by dashed lines in Fig. 1. In the undoped ZnTeO, two PL peaks were observed at around 1.80 (P_3) and 1.62 eV (P_4) that are attributed to a band-to-band luminescence between E_- and valence bands and a luminescence peak related to O-related localized defect [16], respectively.

In Cl-doped ZnTeO grown at $ZnCl_2$ cell temperature of 100 °C, a new luminescence peak at 1.7 eV (P_2) becomes dominant and the peak shifts to the higher energy side with increasing $ZnCl_2$ cell temperature. The band-to-band luminescence (P_1) and an O-related luminescence peaks (P_4) are also detected in the Cl-doped ZnTeO samples.

In order to clarify the origin of P_2 peak, the PL spectra of the sample grown at $ZnCl_2$ cell temperature of 250 °C were measured at various excitation powers. It was found that the peaks shift to higher energy side as the excitation power increases. This indicates that the origin of this peak is due to the donor-acceptor pair (DAP) emission. This suggests that Cl introduces donor state in ZnTeO.

B. Photovoltaic properties of IBSCs

The PV performances of the Cl-doped IBSC were V_{OC}=0.84V, J_{SC}=1.21mA/cm^2, FF=0.41 and η=0.42% while those for undoped IBSC were V_{OC}=0.78V, J_{SC}=0.89mA/cm^2, FF=0.60 and η=0.42%. The Cl-doped IBSC showed higher V_{OC} and J_{SC} than the undoped IBSC but the overall efficiency is same because of low FF, which is probably due to a high resistivity of the Cl-doped ZnTeO layer resulting in high series resistance.

Fig. 2 shows the external quantum efficiency (EQE) curves for the Cl-doped and undoped IBSCs at room temperature. The Cl-doped IBSC showed higher EQE than the undoped IBSC in the whole photon energy range between 1.6 and 3.7 eV. If the IB is filled partly by electrons, the recombination of electrons in the CB through the IB will be reduced, leading to the increase of EQE. The obtained result for the Cl-doped IBSC might imply the formation of the partially filled IB by Cl-doping.

In order to further clarify the effect of Cl-doping, we measured photocurrents induced by two-step photon absorption. The infrared (IR) light with photon energy less than 1.2 eV was irradiated simultaneously during the EQE measurement and the difference of EQE (ΔEQE) with and without IR irradiation was recorded, as shown in Fig. 3. In the undoped IBSC, almost no difference was observed, indicating that no photocurrent was induced by two-step photon

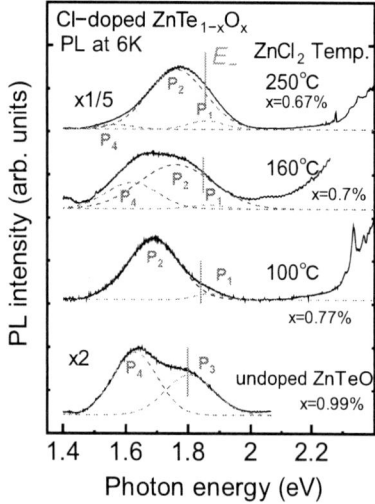

Fig. 1 PL spectra at 6 K for Cl-doped ZnTeO films grown at various $ZnCl_2$ cell temperatures.

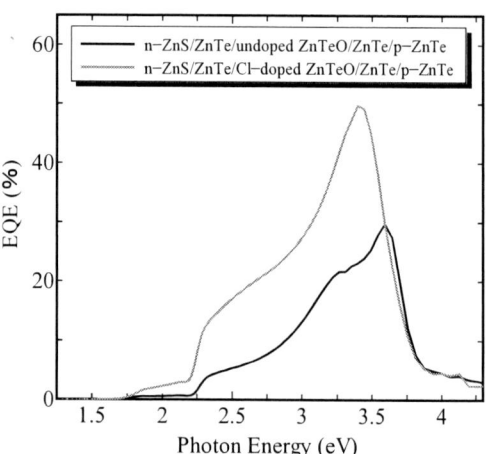

Fig. 2 External quantum efficiency curves for the Cl-doped and undoped IBSCs at room temperature.

absorption and that the excited electrons in IB were lost or escaped from IB. Because the photocurrent induced by two-step photon absorption was detected previously in the ZnTeO IBSCs using n-ZnO layer [12], the increase of the internal electric field in the depletion layer including ZnTeO layer by using n-ZnS layer might be the reason for the disappearance of the two-step excitation current.

On the other hand, the increase of ΔEQE was observed in the Cl-doped IBSC at the energy region between 1.6 and 2.25 eV where the electron transition from VB to IB takes place. This result also supports the introduction of electrons in the IB by Cl-doping that can enhance two-step photon absorption. The further increase of ΔEQE at the photon energy region above 2.25 eV where the electron transition from VB to CB takes place is due to the reexcitation of electrons trapped into IB from CB. These results indicate that the Cl-doping is effective to improve the PV properties of ZnTeO-based IBSC through the introduction of electron into IB.

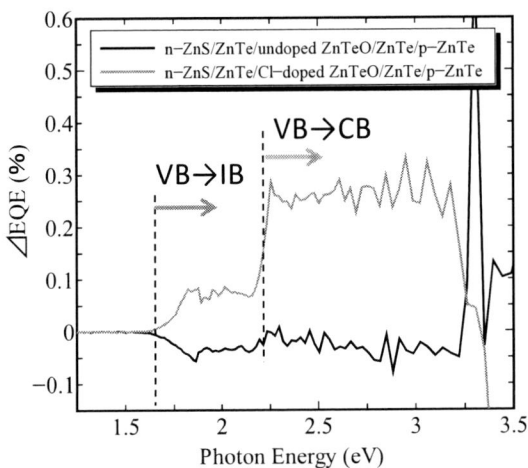

Fig. 3 Difference of EQE (ΔEQE) with and without an additional IR irradiation.

VI. CONCLUSIONS

We have grown Cl-doped ZnTeO films by MBE for the application of IBSCs. Low temperature PL spectrum for undoped ZnTeO shows the band-to-band luminescence together with a luminescence peak related to the O-related localized defect. The DAP emission becomes dominant after Cl-doping, indicating the formation of Cl-related donors in ZnTeO. In the IBSC using Cl-doped ZnTeO layer, the EQE was improved in the wide photon energy range and a photocurrent induced by two-step photon absorption through IB was detected. These results indicate that the Cl-doping is effective to improve the PV properties of ZnTeO-based IBSC through the introduction of electron into IB.

REFERENCES

[1] A. Luque and A. Martí, Phy. Rev. Lett. **78** (1997) 5014.
[2] A. Marti, E. Antolin, C. R. Stanley, C. D. Farmer, N. Lopez, P. Diaz, E. Canovas, P. G. Linares, and A. Luque, Phys. Rev. Lett. **97** (2006) 247701.
[3] R. Oshima, A. Takata, and Y. Okada, Appl. Phys. Lett. **93** (2008) 083111.
[4] K. M. Yu, W. Walukiewicz, J. Wu, W. Shan, J. W. Beeman, M. A. Scarpulla, O. D. Dubon, and P. Becla, Phys. Rev. Lett. **91** (2003) 246403.
[5] N. Lopez, L. A. Reichertz, K. M. Yu, K. Campman, and W. Walukiewicz, Phys. Rev. Lett. **106** (2011) 028701.
[6] W. Wang, A. S. Lin, and J. D. Phillips, Appl. Phys. Lett. **95** (2009) 011103.
[7] N. Ahsan, N. Miyashita, M. M. Islam, K. M. Yu, W. Walukiewicz, and Y. Okada, Appl. Phys. Lett. **100** (2012) 172111.
[8] W. Shan, W. Walukiewicz, J. W. Ager III, E. E. Haller, J. F. Geisz, D. J. Friedman, J. M. Olson, and S. R. Kurtz, Phys. Rev. Lett. **82** (1999) 1221.
[9] Y. Nabetani, T. Okuno, K. Aoki, T. Kato, T. Matsumoto, and T. Hirai, *Phys. Stat. Sol.* A **203** (2006) 2653.
[10] W. Wang, W. Bowen, S. Spanninga, S. Lin, and J. Phillips, J. Electron. Mater. **38** (2009) 119.
[11] T. Tanaka, S. Kusaba, T. Mochinaga, K. Saito, Q. Guo, M.Nishio, K. M. Yu, and W. Walukiewicz, Appl. Phys. Lett. **100** (2012) 011905.
[12] T. Tanaka, M. Miyabara, Y. Nagao, K. Saito, Q. Guo, M. Nishio, K. M. Yu, and W. Walukiewicz, Appl. Phys. Lett. **102** (2013) 052111.
[13] L. Cuadra, A. Marti, and A. Luque, Thin Solid Films, **451-452** (2004) 593.
[14] I. W. Tao, M. Jurkovic, and W. I. Wang, Appl. Phys. Lett. **64** (1994) 1848.
[15] T. Tanaka, S. Tsutsumi, Y. Okano, K. Saito, Q. Guo, M. Nishio, K. M. Yu, and W. Walukiewicz, Proc. of 43rd IEEE Photovoltaic Specialist Conference, 2016, p. 2830.
[16] Y. C. Lin, M. J. Tasi, W. C. Chou, W. H. Chang, W. K. Chen, T. Tanaka, Q. Guo, M. Nishio, Appl. Phys. Lett. **103** (2013) 261905.

978-1-5090-5606-4/17 $31.00 © 2017 IEEE

Toward Lead Halide Perovskite-Based Intermediate Band Absorbers

Matthew D. Sampson,[†] Ji-Sang Park,[§] Richard D. Schaller,[§] Maria K. Y. Chan,[*,§] and Alex B. F. Martinson[*,†]

[†]Materials Science Division, Argonne National Laboratory, Argonne, Illinois 60439, United States
[§]Center for Nanoscale Materials, Argonne National Laboratory, Argonne, Illinois 60439, United States

Abstract — **Lead halide perovskites allow easy access to the long minority carrier lifetimes and diffusion lengths desirable for traditional photovoltaics. We explore the extent to which these semiconductors may be translated to a substitutionally-doped intermediate band absorber. We computationally and experimentally explore the substitution of transition metals on the Pb site of MAPbX$_3$ (MA = methylammonium, X = Br or Cl) to achieve a tunable density of states within the parent gap. First-principles density functional theory (DFT) calculations support the existence of intermediate bands upon Co incorporation as do an experimentally observed sub-gap absorption confirmed by UV-visible-NIR absorption spectroscopy.**

I. INTRODUCTION

Intermediate band (IB) PVs are a class of multi-junction devices that have the theoretical potential to exceed the Shockley-Queisser limit for single p-n junction cells.[1] IB PVs require an IB material sandwiched between two conventional (n- and p-type) semiconductors, which serve as selective contacts to the conduction band (CB) and valence band (VB), respectively. These PVs are designed to retain the high output voltages characteristic of large band gap semiconductors, while harvesting more energy from the available solar spectrum by absorbing pairs of lower energy sub-band gap photons to produce additional high-energy carriers. Only a handful of examples of materials that exhibit the properties necessary for IB PV operation have been reported to date.[2,3] Thus far, GaAs is the most successful parent material owing to its excellent photophysical properties, to which perovskite halides have been compared, as well as the ability to form quantum wells with smaller gap. While the GaAs-quantum wells operate as IB PVs, the contribution to the photocurrent enabled by the intermediate band is modest, as it is limited by the achievable spatial density of the structurally strained quantum wells. The band gap of GaAs is also not ideal, as the optimal total band gap of an IB material has been calculated to be approximately 2.0 eV, which is split by a mid-gap level into two sub-band gaps of approximately 0.7 eV and 1.3 eV (Figure 1).[2] In this arrangement, the maximum photovoltage is the difference between the quasi-Fermi levels (i.e. the electrochemical potentials of the electrons) in the CB and VB of the n- and p-side electrodes (2.0 eV).

We identify lead halide perovskites as prime candidates for IB PVs due to their outstanding photophysical properties, likely structural tolerance for metal substitution, and tunable band gap. Compared to conventional inorganic binary semiconductors like GaAs (band gap = 1.4 eV), a large compositional space has been identified for APbX$_3$ perovskites (A = cations such as Cs$^+$, methylammonium [MA], or formamidinium [FA] and X = Cl, Br, and/or I). We have identified MAPbBr$_3$ as a promising parent material with nearly ideal parent band gap (~2.2 eV) and hypothesize that partial substitution of Pb for earth-abundant transition metals may provide stable mid-gap states with tunable energy level, while retaining many of the desirable optoelectronic attributes of the parent material.

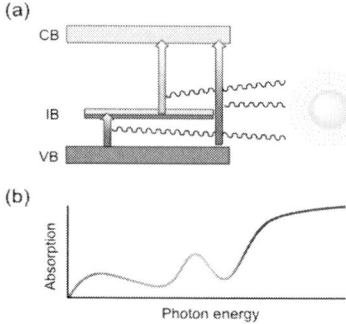

Figure 1. (a) Idealized schematic energy level diagram illustrating the working principle of an IB PV. (b) Corresponding idealized absorption spectrum of an IB photovoltaic.

In order to screen the widest range of substitutional possibilities for Pb in APbX$_3$, we have utilized computational high throughput screening with density functional theory (DFT) calculations to select promising substituents that might be incorporated into the hybrid perovskite halide framework. We report details of the synthesis and characterization of Co-substituted MAPbBr$_{3-y}$Cl$_y$ thin films, prepared via one-step solution processing. Thin films are characterized by X-ray diffraction (XRD), UV-Vis absorption spectroscopy, and photoluminescence (PL) spectroscopy to understand both the structural and photophysical dependences on the transition metal substitution level (x).

II. EXPERIMENTAL METHODS

Thin film fabrication. Thin films of Co-substituted MAPb$_{1-x}$Br$_{3-y}$Cl$_y$ were prepared by simple one-step solution spin coating, following a method adapted from previous

978-1-5090-5606-4/17 $31.00 © 2017 IEEE

reports.[4] In each case, the appropriate ratio of 0.25 M MABr, MACl, $PbBr_2$, $PbCl_2$, and/or $CoBr_2$ was heated in dimethylformamide (DMF) at 100 °C for 4 hrs. A 3% excess of MABr/MACl was added to each solution to ensure all $PbBr_2$/$PbCl_2$ was converted to perovskite.

Characterization. XRD measurements were performed with a Bruker D2 Phaser diffractometer with Cu Kα radiation. Optical absorption spectra were measured using a Varian Cary 5000 UV-Vis-NIR spectrophotometer equipped with an integrating sphere diffuse reflectance accessory. PL emission spectra and time-resolved PL were obtained using a 405 nm, 35 ps pulse width diode laser operated at 1 MHz.

First principles modeling. To investigate the optoelectronic property of $MAPb_{1-x}Co_xBr_3$, calculations were performed within the density functional theory (DFT) framework by using the generalized gradient approximation (GGA) parameterized by Perdew, Burke, and Ernzerhof (PBE) for the exchange correlation potential.[5] The projector-augmented wave (PAW) potentials were used, as implemented in the VASP code.[6] Calculations were performed both with and without spin-orbit coupling. The implementation of Dudarev for the Hubbard U correction was used.[7]

III. RESULTS AND DISCUSSION

Thin Film Fabrication. Thin films of Co-substituted $MAPbBr_{3-y}Cl_y$ and $MAPb_{1-x}Co_xBr_{3-y}Cl_y$, were prepared on either fused quartz (FQ) or fluorine-doped tin oxide (FTO) by simple one-step spin coating, following a method adapted from previous reports.[4] Attempts to prepare Co-substituted perovskite films via two-step deposition or adding solvents or additives to single-step deposition did not result in any observable improvements in film morphology or grain size as compared to one-step deposition. Attempts were made to soak $MAPbBr_{1-y}Cl_y$ films in $CoBr_2$ solutions in isopropanol in order to explore the ability for transition metal substitution to occur in pre-fabricated films; however, in most cases, this resulted in destruction of the films.

Initially, Co-substituted films were only prepared by substituting the Br-only parent material, $MAPbBr_3$. Cl was later introduced in order to better support Co^{2+} ions in the perovskite framework, as Cl is expected to give a more desirable perovskite 'tolerance factor' and 'octahedral factor'.[8] Together, the tolerance factor and octahedral factor have been established as widely accepted metrics for determining perovskite stability. Perovskite halides possess tolerance and octahedral factors in the ranges of 0.813 – 1.107 and 0.442 – 0.895, respectively.[8] Qualitatively, as Co^{2+} (r_{Co} = 90 pm) has a much smaller ionic radii than Pb^{2+} (r_{Pb} = 129 pm), one would expect smaller Cl^- (r_{Cl} = 181 pm) to better support Co^{2+} in the perovskite structure as compared to Br^- (r_{Br} = 196 pm). For Co^{2+}, t = 0.930 and 0.942 and μ = 0.459 and 0.497 for Br^- and Cl^-, respectively. For Co^{2+}, the octahedral factor for Br^- yields a μ on the border of predicted stability for perovskite halides.

This prediction is borne out in films of the mixed Br/Cl perovskite that appear more uniform by optical microscopy and SEM than Br-only perovskites; however, both Br-only and mixed halide films possess a homogeneous perovskite structure, as evidenced by X-ray diffraction (XRD) studies (*vide infra*).

Structural Prediction and Characterization. DFT calculations predict that 1/8 Co and Fe substitution for Pb in $MAPbBr_3$ and $MAPb_{1-x}Co_xBr_{3-y}Cl_y$ will not change the lattice constant appreciably. Using PBE, the optimized lattice constant of $MAPbBr_3$ is 6.065 Å with an insignificant change in lattice constant (Δ_{lat} = -0.05% for Co) to 6.062 Å upon Co substitution ($MAPb_{0.87}Co_{0.13}Br_3$. Comparing calculations performed with and without spin-orbit coupling, the difference in lattice parameters changes upon doping is of order 0.1%. For Co, the lattice constant is also largely insensitive to the U value selected. In all calculations, the change of the lattice constant upon Co substitution is less than 0.3%, which will be shown to be consistent with experimental findings below.

Experimentally, all levels of Co substitution (x = 0.13 to 0.5 for Co) result in films that exhibit the expected perovskite structure, as evidence by XRD patterns, Figure 2. From the observed reflections, it is evident that each film formed the same *Pm3m* cubic crystal structure characteristic of $MAPbBr_3$.

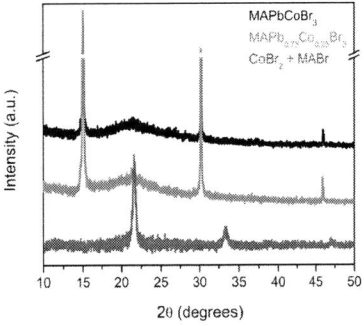

Figure 2. XRD pattern of a control film consisting of a 1:1 ratio of $CoBr_2$ to MABr (blue), with XRD patterns for $MAPbBr_3$ (black) and $MAPb_{0.75}Co_{0.25}Br_3$ (red) thin films.

We find no evidence for free $PbBr_2$/$PbCl_2$ in thin films for substitution levels from x = 0.13 to 0.5. Crystallinity of the films decreases considerably at substitution levels above x = 0.25, and thus, for simplicity, we have only included levels of Co substitution of x = 0.13 and 0.25 in these studies. As a control, we prepared a film consisting of a 1:1 ratio of $CoBr_2$ and MABr (i.e. substitution level of x = 1, no Pb), and this film no longer possesses the perovskite structure of $MAPbBr_3$.

Electronic Structure and Optical Properties. Reflectance-corrected, steady state absorption spectra were obtained for the Co-substituted $MAPb_{1-x}Co_xBr_{3-y}Cl_y$ films. For each Co-substituted film, the absorption spectrum shows a sharp band edge and excitonic peak that is characteristic of

MAPbBr$_3$ thin films. Co-substituted samples display an additional sub-band gap absorption feature between 1.65 – 2.0 eV, Figure 3. This feature is made up of at least four closely spaced peaks and becomes broader with a shift to slightly higher energy in the Cl-containing films. The multiplet centered at ~1.8 eV is characteristic of CoBr$_2$ in a Br-rich environment, commonly represented as either [CoBr$_4$]$^{2-}$ or [CoBr$_6$]$^{4-}$.[9,10] For comparison, this feature is also present in a control sample of CoBr$_2$ and MABr (no Pb, x = 1; see Figure 3). However, the observation of a shift of this Co-related absorption in the Cl-containing films, as well as shift in band gap of the MAPbBr$_{1.5}$Cl$_{1.5}$ film, in addition to the shift in the

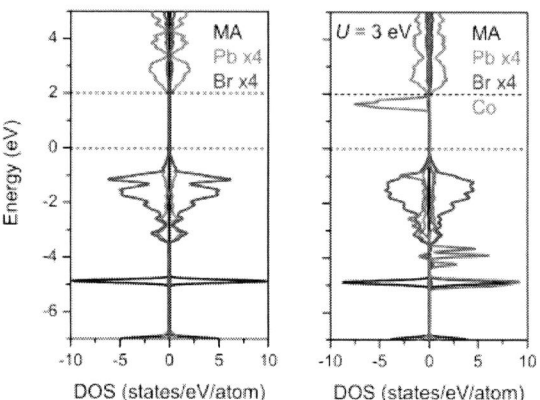

Figure 4. DFT-calculated electronic structure of MAPbBr$_3$ (left) and MAPb$_{0.87}$Co$_{0.13}$Br$_3$ (right).

Acknowledgement. This material is based upon work supported by Laboratory Directed Research and Development (LDRD) funding from Argonne National Laboratory, provided by the Director, Office of Science, of the U.S. Department of Energy under Contract No. DE-AC02-06CH11357.

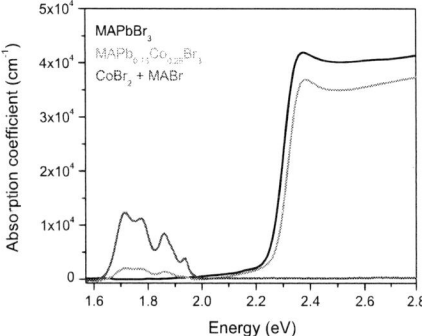

Figure 3. Absorption spectrum of a control film consisting of a 1:1 ratio of CoBr$_2$ to MABr (blue), with the absorption spectra for MAPbBr$_3$ (black) and MAPb$_{0.75}$Co$_{0.25}$Br$_3$ (red).

lattice parameters and lack of CoBr$_2$ in XRD upon adding CoBr$_2$ provide strong evidence that Co is chemically incorporated into the perovskite structure.

The computationally predicted electronic densities of states and absorption coefficient for MAPbBr$_3$, MAPb$_{0.87}$Co$_{0.13}$Br$_3$, MAPbBr$_{1.5}$Cl$_{1.5}$, and MAPb$_{0.87}$Co$_{0.13}$Br$_{1.5}$Cl$_{1.5}$ are in reasonable agreement with those derived from experiment, Figure 4. Additional details are published elsewhere.[11]

To determine the effect that Co substitution has on photoexcited carrier lifetimes, we obtained time-resolved PL data (data not shown). For each film studied, the PL intensity displays a decay that was sufficiently modeled by a two-component exponential. For the Br-only containing perovskites, the first component possesses a lifetime of $\tau_1 =$ ~10 ns, and the second component possesses a lifetime of $\tau_2 =$ ~26 ns. Substitution with Co did not have a significant effect on either of these lifetimes for the Br-only films. For applications in IB PVs, it's important to retain the long-lived photoexcited state lifetimes inherent of perovskite films, and therefore, this result demonstrates promise for creating quality IB materials from the Co-substituted films. The first transient absorption measurements have also been made which reveal parent band edge bleaching upon excitation of the 1.7 eV transition, which suggests good communication between the parent and intermediate band DOS.

REFERENCES

(1) Okada, Y.; Ekins-Daukes, N. J.; Kita, T.; Tamaki, R.; Yoshida, M.; Pusch, A.; Hess, O.; Phillips, C. C.; Farrell, D. J.; Yoshida, K.; Ahsan, N.; Shoji, Y.; Sogabe, T.; Guillemoles, J.-F.: Intermediate band solar cells: Recent progress and future directions. *Appl. Phys. Rev.* 2015, *2*, 021302.

(2) Luque, A.; Marti, A.; Stanley, C.: Understanding intermediate-band solar cells. *Nature Photon.* 2012, *6*, 146-152.

(3) Luque, A.; Martí, A.; Stanley, C.; López, N.; Cuadra, L.; Zhou, D.; Pearson, J. L.; McKee, A.: General equivalent circuit for intermediate band devices: Potentials, currents and electroluminescence. *J. Appl. Phys.* 2004, *96*, 903-909.

(4) Jeng, J.-Y.; Chiang, Y.-F.; Lee, M.-H.; Peng, S.-R.; Guo, T.-F.; Chen, P.; Wen, T.-C.: CH$_3$NH$_3$PbI$_3$ Perovskite/Fullerene Planar-Heterojunction Hybrid Solar Cells. *Adv. Mater.* 2013, *25*, 3727-3732.

(5) Perdew, J. P.; Burke, K.; Ernzerhof, M.: Generalized Gradient Approximation Made Simple. *Phys. Rev. Lett.* 1996, *77*, 3865-3868.

(6) Kresse, G.; Furthmüller, J.: Efficient iterative schemes for *ab initio* total-energy calculations using a plane-wave basis set. *Phys. Rev. B* 1996, *54*, 11169-11186.

(7) Dudarev, S. L.; Botton, G. A.; Savrasov, S. Y.; Humphreys, C. J.; Sutton, A. P.: Electron-energy-loss spectra and the structural stability of nickel oxide: An LSDA+*U* study. *Phys. Rev. B* 1998, *57*, 1505-1509.

(8) Li, C.; Lu, X.; Ding, W.; Feng, L.; Gao, Y.; Guo, Z.: Formability of ABX$_3$ (X = F, Cl, Br, I) halide perovskites. *Acta Crystallogr. Sect. B* 2008, *64*, 702-707.

(9) Fine, D. A.: Halide Complexes of Cobalt(II) in Acetone Solution. *J. Am. Chem. Soc.* 1962, *84*, 1139-1144.

(10) Stanescu, G.; Trutia, A.: Cobalt Complexes Formation in CoBr$_2$/PEG Systems. *Rom. Rep. Phys.* 2005, *57*, 223022.

(11) Sampson, M. D.; Park, J. S.; Schaller, R. D.; Chan, M. K. Y; Martinson, A. B. F. Transition metal-substituted lead halide perovksite absorbers. *J. Mater. Chem. A* 2017, *5*, 3578-3588.

Type-II Quantum Dots for Application to Photon Ratchet Intermediate Band Solar Cells

Ryo Tamaki, Yasushi Shoji, and Yoshitaka Okada

Research Center for Advanced Science and Technology (RCAST), The University of Tokyo,
4-6-1 Komaba, Meguro-ku, Tokyo 153-8904, Japan

Abstract — In intermediate band solar cells (IBSCs), long photo-carrier lifetime in IB is indispensable to obtain appropriate carrier filling ratio in IB and achieve efficient two-step photon absorption. In this paper, type-II quantum dot (QD) heterostructures, electron-confinement InGaAs/GaAsSb QDSC and hole-confinement GaSb/GaAs QDSC, have been investigated as candidates for photon ratchet IBSCs. Fourier transform infrared (FTIR) photocurrent spectroscopy with absolute calibration revealed that the quantum efficiency was ~10^{-5} for IR photon absorption in QDSCs at 9 K. Furthermore, higher operation temperature was determined in type-II QDSCs, which is a more desirable feature as photon ratchet IBSCs.

Index Terms — III-V semiconductor materials, indium gallium arsenide, gallium arsenide, antimony, quantum dots, photovoltaic cells, Fourier transform infrared spectroscopy.

I. INTRODUCTION

Novel physical concepts have been proposed to overcome the limiting efficiency of single-junction solar cells, which is 41% under maximum concentration [1]. In intermediate band solar cells (IBSCs), an additional sub-bandgap level, that is the IB, in the host single-junction solar cell can reduce transmission losses while preserving the output voltage. Detailed balance calculations predict the limiting efficiency of 63% on IBSCs with a single IB under sun-light concentration [2]-[4]. In addition to valence band (VB) to conduction band (CB) transitions, two-step photon absorption via IB can convert below-bandgap infrared (IR) photons to photocurrent [5]-[13]. However, to realize efficient two-step photon absorption in IBSCs, an appropriate carrier filling ratio in IB is indispensable [14]-[16], therefore, both the long photo-carrier lifetime and the large absorption coefficient should be

realized simultaneously [4]. In recent years, photon ratchet IBSCs have been proposed to overcome the above requirement with introducing a ratchet band (RB) as depicted in Fig. 1 [17]. In ideal conditions, radiative recombination from RB to VB is forbidden, so that the long photo-carrier lifetime in RB can improve two-step photon absorption even at low filling conditions. To this end, we have explored various semiconductor nanostructures based on III-V quantum dots (QDs) to implement IB. In this study, type-II QD heterostructures were investigated for application to photon ratchet IBSCs.

II. EXPERIMENTAL

Multi-stacked (10-layers) InAs/GaAs [9], $In_{0.4}Ga_{0.6}As$ (InGaAs) / $GaAs_{0.86}Sb_{0.14}$ (GaAsSb) [10], and GaSb/GaAs [13] QD layers were embedded in the host GaAs *p-i-n* single-junction solar cells fabricated on *n*-GaAs substrates by solid-source molecular beam epitaxy (MBE). As shown in Fig. 2(a), InAs/GaAs QDs form a type-I band alignment. On the other hand, as shown in Figs. 2(b) and (c), type-II heterostructures can be obtained in InGaAs/GaAsSb and GaSb/GaAs QDs. In type-I InAs/GaAs heterostructure, both electrons and holes are confined in the QDs, while in the type-II InGaAs/GaAsSb or GaSb/GaAs heterostructures, only electrons or holes are confined in QDs. As schematically shown in Figs. 2(b) and (c), electrons and holes in QD layers are spatially separated by the type-II band alignment, therefore, long photo-carrier lifetime is expected in these QD systems [18], [19].

In addition to standard external quantum efficiency (EQE) spectroscopy at visible to near-IR spectral region, we have

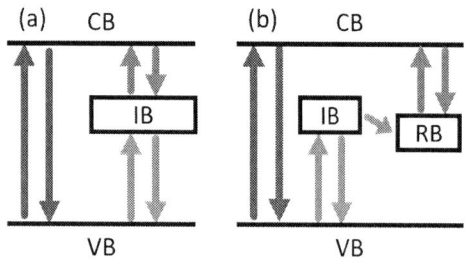

Fig. 1. Schematic band diagrams of (a) conventional IBSC and (b) photon ratchet IBSC with a ratchet band (RB).

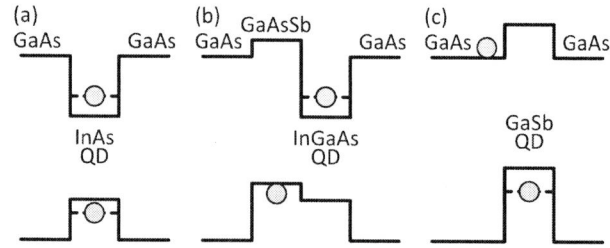

Fig. 2. Schematic illustration of band diagrams of (a) InAs/GaAs, (b) InGaAs/GaAsSb, and (c) GaSb/GaAs QDs.

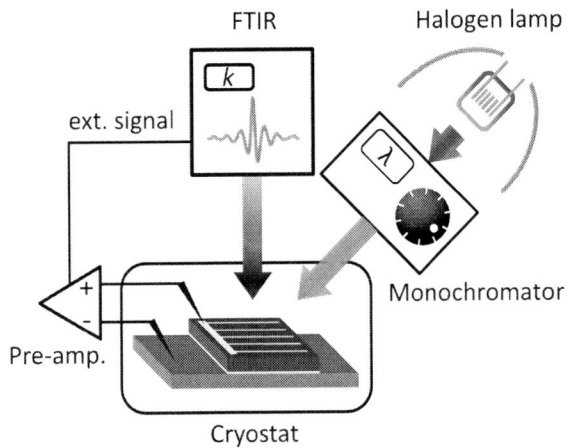

Fig. 3. A schematic illustration of low temperature Fourier transform infrared (FTIR) photocurrent spectroscopy setup.

investigated the second step IR photon absorption by applying Fourier transform infrared (FTIR) photocurrent spectroscopy [20], [21]. A schematic illustration of the optical setup is shown in Fig. 3. A monochromatic light was used as the bias light source to generate photo-carriers in QDs by the first step interband transitions. An IR lamp equipped with a FTIR spectrometer was utilized as the IR light source to excite photo-carriers out of QDs and corrected as photocurrent. Photocurrent signal was converted to voltage signal by a low noise transimpedance pre-amplifier, and it was connected to the external signal input of the FTIR spectrometer. The interferogram synchronized with the interferometer in FTIR spectrometer was Fourier transformed to get signal intensity spectra at IR region extended to 0.1 eV (12.4 μm). On the other hand, the power intensity spectrum of the IR light irradiating on the sample was measured by a wavelength insensitive DLaTGS pyroelectric detector, and the signal intensity spectra were normalized by the irradiated IR photon flux spectrum. Finally, the absolute intensity was calibrated by EQE at 1.2 eV (1.0 μm) for the respective samples at room temperature (RT). As the result, the EQE spectra of the second step IR photon absorption of two-step photon absorption can be evaluated. The benefit of FTIR technique is quick acquisition of the whole IR range with fine wavelength resolution, which is suitable to systematic investigation on series of samples. All measurements were conducted at short-circuit conditions for both standard EQE and FTIR photocurrent spectroscopy.

III. RESULTS AND DISCUSSION

EQE spectra at RT of InAs/GaAs, InGaAs/GaAsSb, and GaSb/GaAs QDSCs are summarized in Fig. 4. In three types of samples, QD layers were embedded in the i-layer of GaAs single-junction solar cells, and the host GaAs absorption

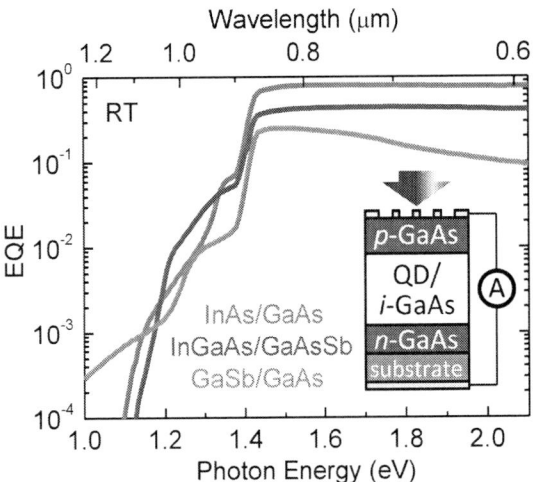

Fig. 4. EQE spectra of InAs/GaAs (red), InGaAs/GaAsSb (blue), and GaSb/GaAs (green) QDSCs at room temperature. Inset: A schematic of the device structure of the QDSCs.

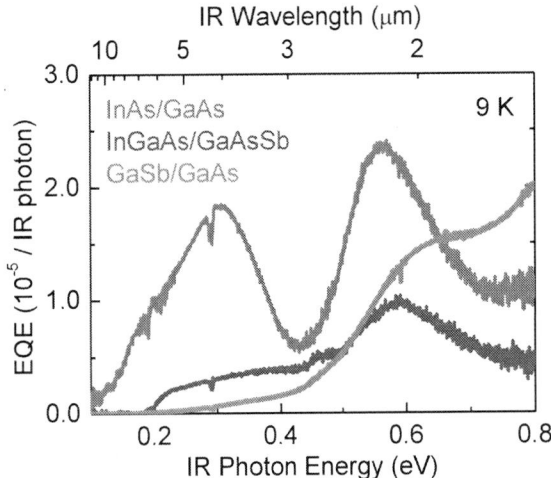

Fig. 5. Mid-IR EQE spectra of InAs/GaAs (red), InGaAs/GaAsSb (blue), and GaSb/GaAs (green) QDSCs at 9 K. The measurements were performed at short-circuit conditions under 800 nm bias light.

appeared at photon energy above 1.42 eV (873 nm). The longer wavelength region below the bandgap of GaAs was attributed to the contribution from QD heterostructures. It is noted that the accessible wavelength range in conventional EQE apparatus is limited due to the light source and optical components. Therefore, VB to IB and VB to CB transitions can be evaluated, but not for IR photon absorption by the IB to CB transitions. To investigate spectral response of photocurrent production via longer wavelength transitions in two-step photon absorption processes, we performed FTIR photocurrent spectroscopy with absolute intensity calibration.

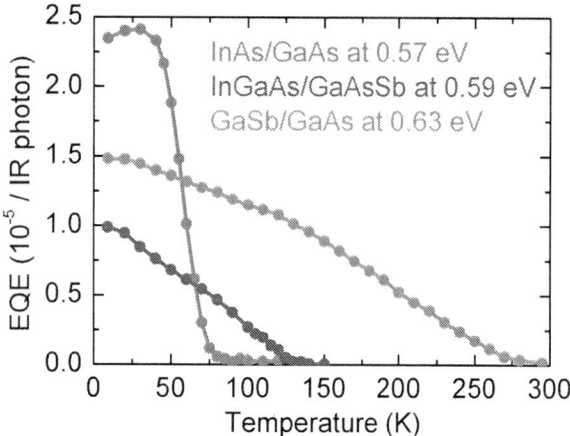

Fig. 6. Temperature dependence of EQE at respective IR peak energies of InAs/GaAs (red), InGaAs/GaAsSb (blue), and GaSb/GaAs (green) QDSCs.

As shown in Fig. 5, we clarified low temperature EQE spectra of IR photon absorption in two-step photon absorption processes in QDSCs. The dip appeared at 0.3 eV was due to atmospheric absorption of CO_2. Monochromatic bias light with a wavelength at 800 nm was applied to generate photo-carriers via VB to CB interband transitions in GaAs. Part of the photo-carriers was captured by QDs and re-excited by IR photo-irradiation, contributing photocurrent production via two-step photon absorption. In this experiment, the samples were mounted on a closed-cycle helium cryostat and the measurements were performed at 9 K to eliminate thermionic emission of photo-carriers out of QDs and enhance carrier filling ratio in the QDs. The IR EQE on the other of $\sim10^{-5}$ per IR photons was obtained in QDSCs under study. Two peak features, at around 0.3 and 0.6 eV, were determined in three types of QDSCs. Mid-IR transition at 0.3 eV was obtained in both InAs/GaAs and InGaAs/GaAsSb QDSCs, but not in GaSb/GaAs QDSC. These results suggest that the low energy transition is originated from "electron" excitation from QDs. On the other hand, the assignment of transitions at 0.6 eV is unclear at this stage. In GaSb/GaAs QDSC, "hole" excitation out of QDs is a possible interpretation [11], [12]. However, in InAs/GaAs and InGaAs/GaAsSb QDSCs, the hole confinement is smaller than that of electrons, hence such higher energy transitions can be attributed to "electron" bound-to-continuum transitions [7], [8]. Further investigation is necessary to conclude the origins of IR transition features.

It is noted that EQE in type-II QDSCs at low temperature was the same order of magnitude as the type-I QDSC. In two-step photon absorption processes, long photo-carrier lifetime can enhance the carrier filing ratio in IB, therefore, large EQE was expected to obtain in type-II heterostructures. On the other hand, the oscillator strength of IR transitions in type-I and type-II QD heterostructures would be different. Nevertheless, in our future study, carrier filling ratio and

photo-carrier lifetime in QDs will be determined directly by applying time-resolved spectroscopy technique in these devices, and discuss in detail about the photocurrent enhancement by two-step photon absorption in type-II QDSCs.

To investigate further insights of the influence of band alignment at type-I and type-II QD heterostructures on two-step photon absorption processes, temperature dependence of EQE at IR peak energies were evaluated as shown in Fig. 6. In the type-I InAs/GaAs QDSC, IR EQE rapidly decreased by increasing temperature, and almost zero at 80 K. Meanwhile in type-II InGaAs/GaAsSb and GaSb/GaAs QDSCs, EQE at IR peak gradually decreased with increasing temperature, and finite signal could be observed upto 130 and 280 K, respectively. The operation temperature of the GaSb/GaAs QDSC was comparable or even higher than the other wide bandgap host materials, such as AlGaAs or InGaP [21]. These results clearly demonstrated the higher operation temperature in type-II QD heterostructures for photocurrent production via two-step photon absorption and desirable feature as photon ratchet IBSCs by enhancing carrier filling ratio against thermionic emission from QDs.

IV. CONCLUSION

In this study, type-II QD heterostructures were investigated for application to photon ratchet IBSCs. EQE spectra of IR photon absorption in two-step photon absorption processes were evaluated by using FTIR photocurrent spectroscopy with absolute intensity calibration in type-I and type-II QD heterostructures embedded in GaAs p-i-n single-junction solar cells. We have investigated a prototypical type-I InAs/GaAs QDSC, type-II InGaAs/GaAsSb and GaSb/GaAs QDSCs with electron- and hole-confinement, respectively. In all of type-I, electron-confinement type-II, and hole-confinement type-II QDSCs, EQE of $\sim10^{-5}$ per IR photon was obtained at 9 K. IR spectral response reflected transition features in respective QD heterostructures. Furthermore, IR EQE gradually decreased with increasing temperature in the type-II QDSCs, therefore, type-II QD heterostructures would be beneficial to reduce thermionic emission of photo-carriers out of QDs. Long photo-carrier lifetime is expected in type-II heterostructures while the absolute intensity of IR EQE was comparable as the type-I QDs. These results suggested the feasibility of type-II heterostructures as photon ratchet IBSCs.

ACKNOWLEDGEMENT

This work is performed under Research and Development of ultra-high efficiency and low-cost III-V compound semiconductor solar cell modules supported by National Research and Development Agency, New Energy and Industrial Technology Development Organization (NEDO), and Ministry of Economy, Trade and Industry (METI), Japan.

References

[1] W. Shockley and H. J. Queisser, "Detailed balance limit of efficiency of *p-n* junction solar cells," *Journal of Applied Physics*, vol. 32, pp. 510-519, 1961.

[2] A. Luque and A. Martí, "Increasing the efficiency of ideal solar cells by photon induced transitions at intermediate levels," *Physical Review Letters*, vol. 78, pp. 5014-5017, 1997.

[3] A. Luque, A. Martí, and C. Stanley, "Understanding intermediate-band solar cells," *Nature Photonics* vol. 6, pp. 146-152, 2012.

[4] Y. Okada, N. J. Ekins-Daukes, T. Kita, R. Tamaki, M. Yoshida, A. Pusch, O. Hess, C. C. Phillips, D. J. Farrell, K. Yoshida, N. Ahsan, Y. Shoji, T. Sogabe, and J.-F. Guillemoles, "Intermediate band solar cells: Recent progress and future directions," *Applied Physics Reviews*, vol. 2, pp. 021302, 2015.

[5] A. Martí, E. Antolín, C. R. Stanley, C. D. Farmer, N. López, P. Díaz, E. Cánovas, P. G. Linares, and A. Luque, "Production of photocurrent due to intermediate-to-conduction-band transitions: A demonstration of a key operating principle of the intermediate-band solar cell," *Physical Review Letters*, vol. 97, pp. 247701, 2006.

[6] Y. Okada, T. Morioka, K. Yoshida, R. Oshima, Y. Shoji, T. Inoue, and T. Kita, "Increase in photocurrent by optical transitions via intermediate quantum states in direct-doped InAs/GaNAs strain-compensated quantum dot solar cell," *Journal of Applied Physics*, vol. 109, pp. 024301, 2011.

[7] R. Tamaki, Y. Shoji, Y. Okada, and K. Miyano, "Spectrally resolved intraband transitions on two-step photon absorption in InGaAs/GaAs quantum dot solar cell," *Applied Physics Letters* vol. 105, pp. 073118, 2014.

[8] R. Tamaki, Y. Shoji, Y. Okada, and K. Miyano, "Spectrally resolved interband and intraband transitions by two-step photon absorption in InGaAs/GaAs quantum dot solar cells," *IEEE Journal of Photovoltaics* vol. 5, pp. 229-233, 2014.

[9] T. Sogabe, Y. Shoji, M. Ohba, K. Yoshida, R. Tamaki, H.-F. Hong, C.-H. Wu, C.-T. Kuo, S. Tomić, and Y. Okada, "Intermediate-band dynamics of quantum dots solar cell in concentrator photovoltaic modules," *Scientific Reports* vol. 4, pp. 4792, 2014.

[10] Y. Shoji, K. Akimoto, and Y. Okada, "InGaAs/GaAsSb type-II quantum dots for intermediate band solar cell," *the 38th IEEE Photovoltaic Specialists Conference* vol. 2, pp. 1-4, 2012.

[11] J. Hwang, K. Lee, A. Teran, S. Forrest, J. D. Phillips, A. J. Martin, and J. Millunchick, "Multiphoton sub-band-gap photoconductivity and critical transition temperature in type-II GaSb quantum-dot intermediate-band solar cells," *Physical Review Applied* vol. 1, pp. 051003, 2014.

[12] I. Ramiro, E. Antolín, J. Hwang, A. Teran, A. J. Martin, P. G. Linares, J. Millunchick, J. Phillips, A. Martí, and A. Luque, "Three-bandgap absolute quantum efficiency in GaSb/GaAs quantum dot intermediate band solar cells," *IEEE Journal of Photovoltaics* vol. 7, pp. 508-512, 2016.

[13] Y. Shoji, R. Tamaki, and Y. Okada, "Multi-stacked GaSb/GaAs type-II quantum nanostructures for application to intermediate band solar cells," accepted for *AIP Advances*.

[14] R. Strandberg and T. W. Reenaas, "Photofilling of intermediate bands," *Journal of Applied Physics* vol. 105, pp. 124512, 2009.

[15] K. Yoshida, Y. Okada, and N. Sano, "Self-consistent simulation of intermediate band solar cells: Effect of occupation rates on device characteristics," *Applied Physics Letters* vol. 97, pp. 133503, 2010.

[16] K. Yoshida, Y. Okada, and N. Sano, "Device simulation of intermediate band solar cells: Effects of doping and concentration," *Journal of Applied Physics* vol. 112, pp. 084510, 2012.

[17] M. Yoshida, N. J. Ekins-Daukes, D. J. Farrell, and C. C. Phillips, "Photon ratchet intermediate band solar cells," *Applied Physics Letters*, vol. 100, pp. 263902, 2012.

[18] K. Nishikawa, Y. Takeda, T. Motohiro, D. Sato, J. Ota, N. Miyashita, and Y. Okada, "Extremely long carrier lifetime over 200 ns in GaAs wall-inserted type II InAs quantum dots," *Applied Physics Letters*, vol. 100, pp. 113105, 2012.

[19] F. Hatami, M. Grundmann, N. N. Ledentsov, F. Heinrichsdorff, R. Heitz, J. Böhrer, D. Bimberg, S. S. Ruvimov, P. Werner, V. M. Ustinov, P. S. Kop'ev, and Zh. I. Alferov, "Carrier dynamics in type-II GaSb/GaAs quantum dots," *Physical Review B*, vol. 57, pp. 4635-4641, 1998.

[20] M. Vanecek and A. Poruba, "Fourier-transform photocurrent spectroscopy of microcrystalline silicon for solar cells," *Applied Physics Letters* vol. 80, pp. 719-721, 2002.

[21] R. Tamaki, Y. Shoji, T. Sugaya, and Y. Okada, "Universal linear relationship on two-step photon absorption processes in In(Ga)As quantum dot solar cells," in *the 43rd IEEE Photovoltaic Specialist Conference*, 2016.

An investigation of the role of recombination processes in the operation of InAs/GaAs$_{1-x}$Sb$_x$ quantum dot solar cells

Y. Cheng[1], A. J. Meleco[1], A. J. Roeth[1], V. R. Whiteside[1], M. C. Debnath[1], M. B. Santos[1], T. D. Mishima[1], S. Hatch[2], H-Y. Liu[2], and I. R. Sellers[1]

[1]Homer L. Dodge Department of Physics & Astronomy, University of Oklahoma, Norman, OK

[2]Department of Electrical & Electronic Engineering, University College London, Torrington Place. London.   WC1E 7JE. U. K

Abstract — **The electroluminescence and photoluminescence from an InAs/GaAs$_{1-x}$Sb$_x$ quantum dot solar cell are investigated as a function of temperature and correlated to the PV characteristics of the cell over the same temperature range. Analysis of the dominant recombination mechanism is shown to change from radiative to non-radiative above ~ 150 K, which is consistent with a *reduction* in the J_{sc} (and V_{oc}) at elevated temperatures in these devices.**

I. Introduction

Single band gap solar cells are limited to efficiencies on the order of 30% as described by the Shockley-Queisser limit [1]. Intermediate band solar cells (IBSC) [2] have been proposed as a candidate system to overcome this limit through the absorption of sub-gap photons with a potential efficiency of greater than 60 % under ideal conditions [2][3] . The creation of an intermediate band (IB) requires the formation of an isolated band within the band gap of the absorber material. The 3-dimensional confinement of semiconductor quantum dots (QDs) makes them a candidate system to form such IBs. Several QD systems have been studied in the past decade [4]-[9]; InAs/GaAs QDs being the most well investigated material system for IBSC applications [4]-[6][9][11]. However, by replacing the GaAs matrix material with GaAs$_{1-x}$Sb$_x$, a higher density of QDs has been demonstrated [12][13] with a better spectral match to the solar resource [14].

II. Experiment Details

The InAs/GaAs$_{1-x}$Sb$_x$ QD solar cell studied in this work was grown by molecular beam epitaxy (MBE). The InAs/GaAs$_{1-x}$Sb$_x$ QD structure, which serves as the intrinsic region, is repeated 5 times and sandwiched within p- and n-type GaAs layers to form a p-i-n diode architecture. The composition of the the GaAs$_{1-x}$Sb$_x$ matrix was $x = 0.14$ to achieve a *quasi-flat* valence band (VB) as determined for this system previously [12][13].

III. Results and Discussions

Figure 1(a) compares the photoluminescence (*PL*) and electroluminescence (*EL*) for the QD region at various temperatures. The *EL* displays a single peak at 1050 nm (77 K), 1130 nm (210 K), and 1153 nm (298 K), respectively. The *PL*

at 77 K and 210 K display much broader emission consistent with non-idealities and bimodality in the system; which are presumably quenched or saturated at the higher injection levels present in *EL*.

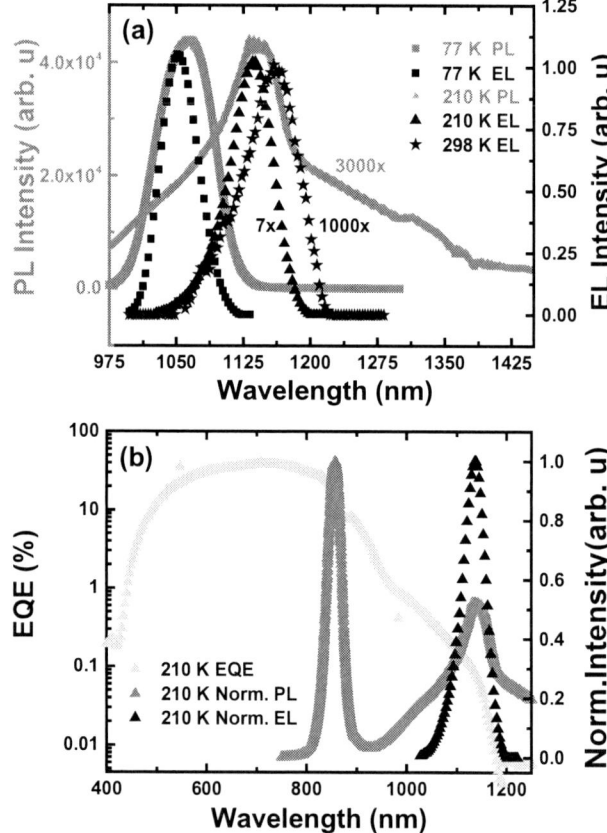

Figure 1 (a) *PL* **(red symbols) and** *EL* **(black symbols) spectra at 77 K (squares), 210 K (triangles), and** *EL* **at room temperature (stars); (b) Normalized** *PL* **(red triangles),** *EL* **(black triangles), and** *EQE* **(green triangles) at 210 K.**

Figure 1(b) shows a comparison of the external quantum efficiency (*EQE*) of the device at 210 K to illustrate that the single peak evident in the *EL* measurements is indeed related to the QD transition; i.e., the VB to IB transition in the active region of the device. Figure 1(b) also shows the normalized *PL* measured at 210 K, which displays two additional features not evident in the *EL*: at 870 nm and 970 nm - correlating to absorption in the GaAs and GaAsSb matrix, respectively. The absence of these features in the *EL* measurements reflects the direct injection of the carriers into the QDs and the separation of the quasi-fermi level that is set by the difference in energy of the QD transitions in *EL*. In the PL measurements the GaAs emitter and QDs are probed simultaneously [15] which, with the combination of the longer radiative lifetime of photogenerated carriers in the type-II QDs, contributes to significant *PL* from the continuum regions.

By *T* = 210 K, the *PL* shows a three order of magnitude reduction in intensity and is similar in strength to a broad (900 – 1500 nm) defect band; which dominates the *PL* as the temperature is increased above 210 K (not shown here). This defect band reflects the contribution of dislocations in the matrix region that result during growth due to the significant lattice mismatch (1.4 %) between GaAs and GaAsSb regions [12] [16]. At low temperatures, photogenerated carriers (electrons) in the QDs are isolated from these defect centers due to the strong confinement of the QDs. However, at elevated temperatures the increased thermal energy not only facilitates the photogenerated carriers to transfer between QDs but also, the defect states now serve as efficient single particle non-radiative recombination centers for these mobile carriers - dominating the *PL* spectra. The rapid loss of the photogenerated carriers indicates the absence of a well isolated intermediate band in the system. Indeed, evidence of carrier extraction at 77 K (not shown, see [12]) indicates the absence of such an IB at all temperatures measured (above 77 K) in the current series of samples.

The PL recombination processes at low and high temperature have been presented previously and appear to indicate two regimes: (1) a low temperature regime, in which photogenerated carriers are isolated and recombine directly in the QDs.; (2) a higher temperature (> 160 K) region within which carriers have enough thermal energy to escape from the QDs where they experience considerable non-radiative loss due to exposure to defects and traps. These non-radiative recombination process serve to simultaneously increase the reverse saturation (dark) current in current-voltage (*J-V*) measurements and reduce the open circuit voltage.

The non-radiative recombination of the QDs is also observed to be correlated to a reduction in the *EL* intensity *except* this reduction occurs at slightly higher temperatures than that of the *PL* (above 210 K). This may be explained by a partial saturation of defects in the matrix at increasing injection levels during *EL*. Previous measurements of temperature dependent *J-V* and *EQE* [12] also indicated a transition in which carrier recombination significantly affects the performance of the device with large reductions in open circuit voltage (*V_oc*) with increasing temperature. At T < 160 K, the radiative recombination of the carriers served as the limiting effect in the extraction of the photogenerated carriers; while non-radiative recombination, facilitated by thermally activated non-radiative centers, was deemed responsible for poor carrier extraction and large increase (decrease) in the dark current (*V_oc*) at higher temperature (*T* > 210 K).

These two regimes displayed very unconventional (non-monotonic) photocurrent behavior whereby, a slightly increasing short circuit current density (*J_sc*) was observed between 77 K and 160 K reaching a minimum point at 170 K; above 170 K, the *J_sc* fluctuated (see Figure 2 (b)). This complex behavior is attributed to a transition from the dominance of radiative to non-radiative processes at elevated temperatures, along with the competing processes of carrier extraction (increasing *J_sc*) and recombination (decreasing *J_sc*).

To further investigate these hypotheses, here, we present a comparison of the spontaneous emission in *EL* measurements to determine the nature and dominance of the recombination processes under different conditions and correlate these findings to the PV analysis. The current injected into a device can be approximated by [17]:

$$I = eV(An + Bn^2 + Cn^3) + I_{leak} \ . \quad (1)$$

Where *A*, *B*, and *C* represent: single carrier (defect-mediated) recombination, radiative recombination, and Auger-mediated process coefficients, respectively [17]. *V and e* are the active region volume and electronic charge, respectively.

Single carrier recombination is associated with the recombination through traps and defects (Shockley-Reed-Hall) [17]; the radiative recombination is related to the *spontaneous emission (EL)*. The total integrated spontaneous emission rate *L* is proportional to the radiative recombination n^2. Equation (1) can be further simplified to $I \propto n^z \propto L^{1/2^z}$ such as to allow a correlation between current injection (*I*), spontaneous emission (*L*), and the dominant recombination mechanism (*n*) [17] to be determined. When radiative recombination dominates the current, the *z*-factor will be close to 2 ($\propto n^2$). However, if *z* is closer to 1 ($\propto n$), this indicates the current is dominated by non-radiative recombination centers - defects and traps.

By plotting $\ln(I) - \ln(L^{1/2})$, *z* can be extracted from the slope to determine the nature of the recombination processes. Figure 2 (a) shows data along with a linear fit for 4 different temperatures from 77 K to 298 K. The slope becomes shallower as the temperature increases, which indicates a reduction of *z*. The temperature dependent behavior of *z* is shown in Figure 2 (b). Between *T* = 77 K and *T* = 150 K, the *z*-factor remains above 2. This indicates that the current is dominated (predominately) by radiative recombination, with some contribution from Auger-processes, at higher injection. Losses due to defect-mediated non-radiative processes, however, appear limited below *T* = 150 K.

As the temperature is increased above 150 K, *z* decreases towards ~1 at room temperature, indicating the increasing dominance of non-radiative processes at elevated temperatures. The decrease of *z* corresponds directly to the temperature dependent (TD) reduction in *EL* intensity (or radiative efficiency). This is qualitatively consistent to that evident in the TD *PL* and steady decrease in *V_oc* evident in the TD *J-V* under 1-sun illumination [12] [16]. These results serve to further reflect the thermally activated behavior of the non-radiative changes as a function of temperatures.

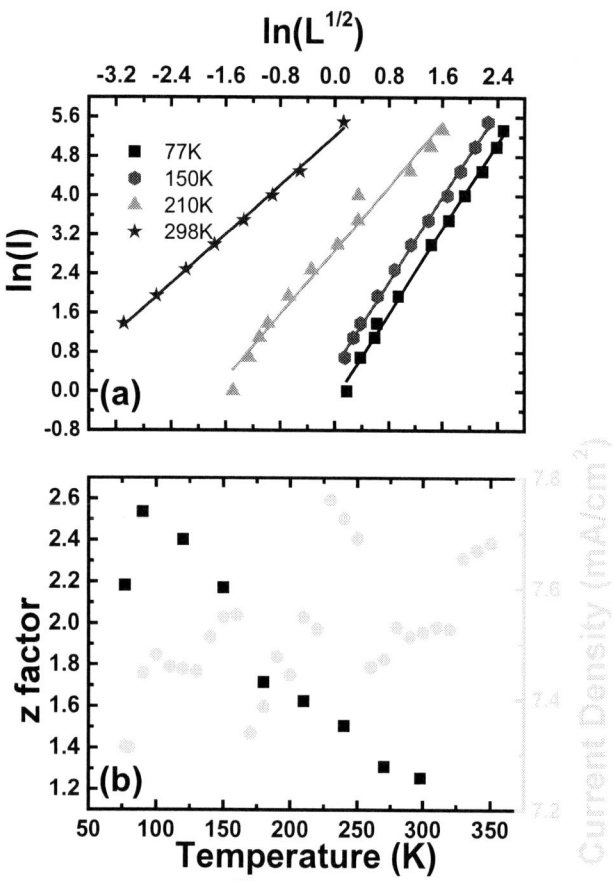

Figure 2 (a) ln (I)- ln(L$^{1/2}$) at 77 K, 150 K, 210 K and 298 K; (b) Temperature dependent behavior of Z-factor (black squares)，Jsc extracted from the J-V measurements (green circles).

IV. CONCLUSION

An InAs/GaAs$_{0.86}$Sb$_{0.14}$ QDSC was investigated using complementary *PL* and *EL* measurements to support previous measurements on a series of similar structures that show unusual escape and transport mechanisms. A rapid quenching of the *PL* and *EL* intensity, along with a transition (above 150 K) in the dominant recombination process (*z*-factor) in high temperature *EL* measurements further indicate the prevalence of non-radiative processes at elevated temperatures in these systems. This correlates qualitatively with TD *EQE* and *J-V* measurements - supporting the conclusion that non-radiative processes introduced by the lattice mismatch between the GaAsSb matrix and GaAs perturbs the transport behavior of the QDs; suggesting further improvements in material growth and quality are required before these systems can be considered practical for applications in IBSCs.

V. ACKNOWLEDGMENT

The authors acknowledge financial support through the NASA EPSCoR program, Grant(s) #NNX15AM75A, and #NNX13AN01A. Support is also acknowledged from the state of Oklahoma OCAST program, Grant No. #AR14-041.

REFERENCES

[1] W. Shockley and H. J. Queisser, "Detailed balance limit of efficiency of p‑n junction solar cells." *Journal of applied physics* 32.3 (1961): 510-519.

[2] A. Luque, and A. Martí. "Increasing the efficiency of ideal solar cells by photon induced transitions at intermediate levels." *Physical Review Letters* 78.26 (1997): 5014.

[3] A. Luque and A. Martí, "The intermediate band solar cell: progress toward the realization of an attractive concept." *Advanced Materials* 22, no. 2 (2010): 160-174.

[4] S. M. Hubbard, C. D. Cress, C. G. Bailey, R. P. Raffaelle, S. G. Bailey, and D. M. Wilt, "Effect of strain compensation on quantum dot enhanced GaAs solar cells." *Applied Physics Letters* 92, no. 12 (2008): 123512.

[5] G. Jolley, H. F. Lu, L. Fu, H. H. Tan, and C. Jagadish, "Electron-hole recombination properties of In$_{0.5}$Ga$_{0.5}$ As/GaAs quantum dot solar cells and the influence on the open circuit voltage." *Applied Physics Letters* 97, no. 12 (2010): 123505.

[6] K. A. Sablon, J. W. Little, K. A. Olver, Z. M. Wang, V. G. Dorogan, Y. I. Mazur, G. J. Salamo, and F. J. Towner, "Effects of AlGaAs energy barriers on InAs/GaAs quantum dot solar cells." *Journal of Applied Physics* 108, no. 7 (2010): 074305.

[7] P. J. Simmonds, R. B. Laghumavarapu, M. Sun, A. Lin, C. J. Reyner, B. Liang, and D. L. Huffaker, "Structural and optical properties of InAs/AlAsSb quantum dots with GaAs (Sb) cladding layers." *Applied Physics Letters* 100, no. 24 (2012): 243108.

[8] R. B. Laghumavarapu, A. Moscho, A. Khoshakhlagh, M. El-Emawy, L. F. Lester, and D. L. Huffaker, "GaSb⁄ GaAs type II quantum dot solar cells for enhanced infrared spectral response." *Applied Physics Letters* 90, no. 17 (2007): 173125.

[9] S. Hatch, J. Wu, K. Sablon, P. Lam, M. Tang, Q. Jiang, and H. Liu, "InAs/GaAsSb quantum dot solar cells." *Optics express* 22, no. 103 (2014): A679-A685.

[10] F. K. Tutu, I. R. Sellers, M. G. Peinado, C. E. Pastore, S. M. Willis, A. R. Watt, T. Wang, and H. Y. Liu, "Improved performance of multilayer InAs/GaAs quantum-dot solar cells using a high-growth-temperature GaAs spacer layer." *Journal of Applied Physics* 111, no. 4 (2012): 046101.

[11] A Martí, E. Antolín, C. R. Stanley, C. D. Farmer, N. López, P. Díaz, E. Cánovas, P. G. Linares, and A. Luque. "Production of photocurrent due to intermediate-to-conduction-band transitions: a demonstration of a key operating principle of the intermediate-band solar cell." *Physical Review Letters* 97, no. 24 (2006): 247701.

[12] Y. Cheng, M. Fukuda, V. Whiteside, M. Debnath, P. Vallely, T. Mishima, M. Santos, K. Hossain, S. Hatch, H. Liu, et al., M. Fukuda, V. R. Whiteside, M. C. Debnath, P. J. Vallely, T. D. Mishima, M. B. Santos et al. "Investigation of InAs/GaAs$_{1-x}$Sb$_x$ quantum dots for applications in intermediate band solar cells." *Solar Energy Materials and Solar Cells* 147 (2016): 94-100.

[13] Debnath, M. C., T. D. Mishima, M. B. Santos, Y. Cheng, V. R. Whiteside, I. R. Sellers, K. Hossain, R. B. Laghumavarapu, B. L. Liang, and D. L. Huffaker. "High-density InAs/GaAs1− x Sb x quantum-dot structures grown by molecular beam epitaxy for use in intermediate band solar cells." *Journal of Applied Physics* 119, no. 11 (2016): 114301.

[14] M. Y. Levy and C. Honsberg, "Nanostructured absorbers for multiple transition solar cells." *IEEE Transactions on Electron Devices* 55, no. 3 (2008): 706-711.

[15] I. Ramiro, E. Antolín, P. G. Linares, E. Hernández, A. Martí, A. Luque, C. Farmer, and C. Stanley. "Application of photoluminescence and electroluminescence techniques to the characterization of intermediate band solar cells." *Energy Procedia* 10 (2011): 117-121.

[16] Y. Cheng, M. Fukuda, V. R. Whiteside, M. C. Debnath, P. J. Vallely, A. J. Meleco, A. J. Roeth et al. "Investigation of InAs/GaAs$_{1-x}$Sb$_x$ quantum dots for applications in intermediate band solar cells." *Photovoltaic Specialists Conference (PVSC)*, 2016 IEEE 43rd, pp. 0005-0008, 2016

[17] A. F. Phillips, S. J. Sweeney, A. R. Adams, and P. J. Thijs. "The temperature dependence of 1.3-and 1.5-um compressively strained InGaAs (P) MQW semiconductor lasers." *IEEE Journal of selected topics in quantum electronics* 5, no. 3 (1999): 401-412.

Temperature and voltage-bias dependent two-step photon absorption in InAs/GaAs/Al$_{0.3}$GaAs quantum dot in a well solar cells

Yushuai Dai, Brittany L. Smith, Michael A. Slocum, Zachary S. Bittner, Hyun Kum, Julia D'Rozario and Seth M. Hubbard

NanoPower Research Laboratory, Rochester Institute of Technology, 111 Lomb Memorial Drive, Rochester, NY 14623

Abstract — **The realization of the concept of the intermediate band solar cell (IBSC) requires that two-step photon absorption dominates at room temperature. To increase two step photon absorption (TSPA), an InAs/GaAs/Al$_{0.3}$GaAs quantum dot in a well (Dwell) structure is used to reduce thermal escape of carriers. The Dwell structure shows decreased TSPA with increasing temperature because of faster thermal escape. With an increased electron ground state barrier height relative to Al$_{0.3}$GaAs, the observable TSPA occurs up to 80K. The extracted thermal activation energy from temperature dependent TSPA is between 80-95 meV, which is associated with holes escape processes. Charge separation along the growth direction reduces the recombination rate in the IB. The stable TSPA observed at -2V reverse bias may be a balance between the reduced recombination rate and increased tunneling rate.**

Index Terms — **quantum dot, intermediate band solar cell, AlGaAs, charge separation.**

I. INTRODUCTION

The concept of the intermediate band solar cell (IBSC) [1] has been proposed as a method to surpass the Shockley-Queisser limit in a single junction solar cell. The ideal IBSC reduces transmission loss without degradation in output voltage due to the two-photon absorption process, which enables optical transition from the valence band (VB) to the IB, and the IB to the conduction band (CB). Due to their discrete density of states, quantum dots (QDs) are considered a potential candidate to form an IB. So far, InAs/GaAs quantum dot solar cells (QDSCs) have been widely studied as a prototype of IBSC because they have a matured growth technique. However, the small difference between the IB and the CB in the InAs/GaAs QDSCs introduces fast thermal escape from the IB and reduces the probability of two step photon absorption (TSPA) [2].

To reduce thermal escape from QDs and achieve the highest IBSC conversion efficiency of 63% under maximum solar concentration, the optimized band gap values are 1.96 eV [3] for the host material, and 1.24 eV and 0.72 eV between the VB-IB and IB-CB, respectively. It has been proposed to use wide-band-gap materials including In$_{0.5}$GaP (1.9 eV at 300K)[4], [5] or Al$_{0.3}$GaAs (1.84 eV at 300K) [6], [7] as the host material. Because InAs/In$_{0.5}$GaP QDs have deeper confinement than the InAs/Al$_{0.3}$GaAs QD system [8], a large loss in carrier collection was observed due to hole capture and recombination in the QDs [5], [8].

The ideal use of Al$_{0.3}$GaAs barriers would be without the integration of any GaAs wells within the superlattice (SL), however that is challenging since Al$_{0.3}$GaAs often getters oxygen, which creates trap states at the QD interface if QDs are grown directly on Al$_{0.3}$GaAs [9]. Furthermore, the Al$_{0.3}$GaAs growth temperature necessary for direct growth of InAs QDs is lower than the optimal temperature for Al$_{0.3}$GaAs, so the optical and electrical properties are degraded [10]. To achieve high quality crystalline films, InAs/GaAs/Al$_{0.3}$GaAs Dwell SL was used, which shows suppressed thermal escape in previous studies [8], [10]. This paper demonstrates temperature and voltage-bias dependent TSPA in the Dwell structure by directly measuring TSPA-introduced photocurrent. To further improve device design, the paper also discusses the roles of absorption, recombination and charge separation on TSPA.

II. EXPERIMENTAL

Figure 1 shows the investigated cell structure of 10-layers of InAs QDs embedded in an *n-i-p* Al$_{0.3}$GaAs solar cell. This cell was grown in a 3×2" Aixtron close-couple showerhead metal organic vapor phase epitaxy (CCS-MOVPE) reactor on a Zinc-doped GaAs substrate with a 2° offcut towards the (110) orientation. The metalorganic precursors used were trimethylindium, trimethygallium, trimethyaluminum, and arsine, with disilane as the n-type dopant for the 50 nm (1×10^{18} cm^{-3}) Al$_{0.3}$GaAs emitter and diethylzinc as the p-type dopant for the 1500 nm (5×10^{16} cm^{-3}) Al$_{0.3}$GaAs base. The average built-in electric field across the 10-layer QD superlattice in the 600 nm intrinsic region is 14kV/cm based on Sentaurus Device™ simulations. In the QD superlattice, prior to the formation of InAs QDs, 3 nm of GaAs was grown after 3 nm of Al$_{0.3}$GaAs to prevent the intermixing of aluminum near the InAs QDs [11]. As in our previous work, InAs QDs are formed by the strain-driven Stranski-Krastanow growth mode with InAs coverage between 1.8 ML-2.0 ML. After QD formation, the QD layer is capped with a low temperature 3 nm GaAs layer to maintain the InAs QD height during the following high temperature Al$_{0.3}$GaAs layer. The 4.6 nm Al$_{0.3}$GaAs barrier layer is grown at 620°C to reduce aluminum and oxygen bonding, and is followed by a GaP strain-balancing layer. The solar cell was fabricated using standard III–V processing techniques. Individual cells were

978-1-5090-5606-4/17 $31.00 © 2017 IEEE

isolated using wet chemical etching techniques. No anti-reflective coating was applied. Measurements were performed on 1×1 cm^2 mesa-isolated device without grid fingers.

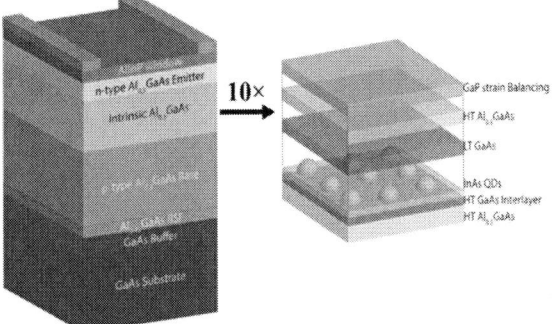

Fig.1. Structural layout for InAs/GaAs/Al$_{0.3}$GaAs Dwell solar cell.

Figure 2 shows a schematic of the TSPA-introduced photocurrent measurement setup. A Cryoindustries 10K M-22 cryo-system, with a CaF$_2$ window to reduce the loss of infrared (IR) input light, was used to cool the sample to 30K. A continuous monochromatic light generated by a Tungsten lamp was used to pump carriers from the VB to IB, while a second IR photon source (Omega 800^{0}C blackbody radiation sources with 1500 nm filter) is coincident upon the sample to enable the second photon excitation from the IB to CB. A chopper is positioned in front of the IR light source in order to generate an alternating current (AC) signal of the carrier collection from the IB-CB optical transition.

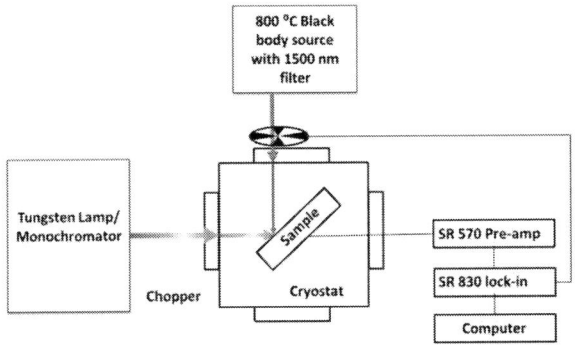

Fig. 2. Schematic of TSPA introduced photocurrent measurement setup.

The low temperature external quantum efficiency (EQE) data was collected via an OL750 spectroradiometric measurement system. Additional temperature-dependent photoluminescence (TDPL) experiments were completed by pumping the sample with a 100 mW 532 nm laser, and the PL signal was detected with an InGaAs detector and a Princeton Instrument monochromator. Atomic force microscopy (AFM) measurements were completed with a Bruker Dimension 3100 scanning probe micrometer. The statistical analysis of the surface QDs was achieved with the image recognition software SPIP^{TM} by Image Metrology.

III. RESULTS AND DISCUSSION

Figure 3(a) shows the measured temperature-dependent TSPA-introduced photocurrent. With increasing temperature, the TSPA signal decreases due to increasing carrier thermal escape from the confined level. The increased electron barrier height [8] relative to Al$_{0.3}$GaAs allows TSPA-introduced photocurrent to be observed even at 80K. Figure 3(b) shows the 40K EQE and PL measurements, indicating the band-edges of Al$_{0.3}$GaAs and GaAs are around 640 nm (1.93 eV) and 840 nm (1.48 eV) respectively, while the QD ground state PL emission is around 920 nm (1.35 eV). TSPA via filtered 800^{0}C black body excitation is mainly from the GaAs/Al$_{0.3}$GaAs quantum well (650-850 nm). There is almost zero TSPA photocurrent from QD optical transitions (>850 nm) in Figure 3(a). This was also observed by Asahi *et al.* [10] in their Dwell structure. The elimination of TSPA at the ground state may be due to limited optical generation rates from the VB to IB, where the optical generation rate depends on the product of the incident photon flux and the optical cross-section of the investigated QDs [12], [13]. The incident photon flux is limited by the filtered black body light sources: integrated band radiance (1500 nm-10 μm) of 2.17×10^4 W/m^2/Sr is two orders of magnitude lower than the integrated band radiance of 2.56×10^6 W/m^2/Sr from the Sun (a 6000K black body). Meanwhile the optical cross-section is affected by the QD absorption coefficient that strongly depends on the QD surface density. A 1×1 μm^2 AFM image of InAs QDs is shown in the inset of Figure 3(a). A QD density of 8.4×10^9 cm^{-2} explains the small QD absorption.

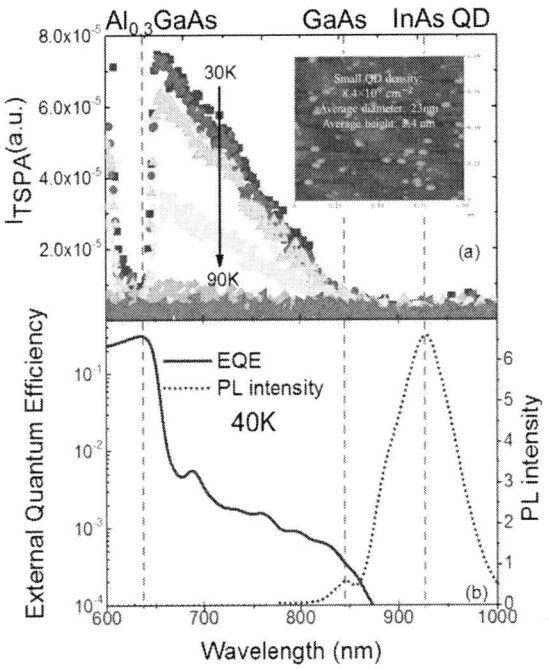

Fig. 3. (a) Temperature dependent TSPA, with inset of $1 \times 1 \mu$m^2 AFM. (b) 40K EQE and PL measurements.

978-1-5090-5606-4/17 $31.00 © 2017 IEEE

The experimental TSPA performance in terms of working temperature and optical generation rate can be improved by increasing incident light concentration, photon recycling, improved QD surface morphology, and number of layers of QDs. On the other hand, Figure 3(a) also shows that TSPA decreases at longer wavelengths. This interesting phenomenon was also observed in other types of QD systems [14], [15], which contradicts the argument that fast escape via thermal means or tunneling eliminates the TSPA from shallow confined levels (650-850 nm). To quantitatively analyze temperature-dependent TSPA, Figure 4 shows the normalized TSPA intensity at 700 nm and 800 nm with temperature varied from 30K to 90K. Normalized TSPA is used to exclude the factor of carrier density in the confined levels that depends on the absorption from the VB to a certain confined level. The normalized TSPA intensity is a ratio between the optical generation rate (R_{TSPA}) over the other competing processes including thermal escape, recombination, and tunneling, which can be summarized in the rate equation (1)

$$I_{TSPA_{normalized}} = \frac{R_{TSPA}}{R_{TSPA} + R_{r\&t} + Aexp(\frac{-E_a}{kT})} \quad (1)$$

where $Aexp(-E_a/kT)$ is the term for thermal escape rate that exponentially increases with temperature, A is the constant of thermal escape depending on the mass of the escaped carriers and average height of the QDs, k is the Boltzmann constant, T is the temperature and E_a is activation energy. $R_{r\&t}$ is the total rate of recombination and tunneling, which is considered stable with temperature. The extracted activation energies are 80±3 meV and 94 ±3 meV for 700 nm (1.77 eV, 160 meV band-offset from $Al_{0.3}GaAs$) and 800 nm (1.55eV, 380 meV band-offset from $Al_{0.3}GaAs$), respectively. These small activation energies may be due to fast hole escape [16]. Fast hole escape extends the electron lifetimes in confinement levels [17].

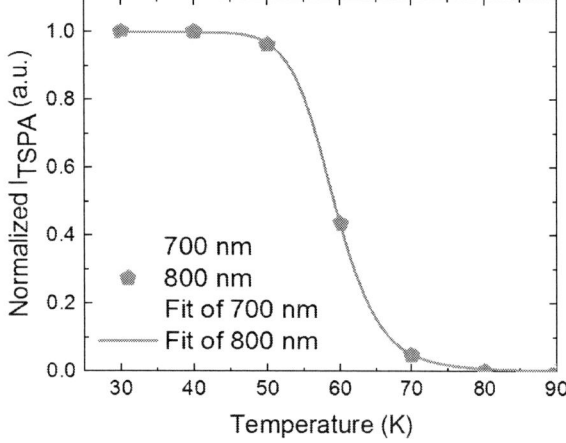

Fig. 4. Normalized TSPA intensity from InAs/GaAs/Al0.3GaAs Dwell IBSC with varied temperature from 30K to 90K at 700 nm and 800 nm.

Also the extract ratio of R_{r+t}/R_{TSPA} is below 0.02, which indicates slow recombination and tunneling at the TSPA-active QD layer. To explain the dynamic carrier processes occurring in the temperature-dependent TSPA measurements towards an understanding that leads to design optimization, Figure 5 shows a simplified band structure of the *n-i-p* Dwell IBSC that depicts absorption, recombination, and charge separation. Absorption of photons with energy lower than the $Al_{0.3}GaAs$ bandgap generates electron-hole pairs in the confined levels formed by $Al_{0.3}GaAs/GaAs$ quantum wells and the InAs wetting layer on GaAs. The photo-excited carriers in the shallow confined levels can then escape, either by thermal escape or tunneling. After a carrier escapes, it is possible that it could be recaptured to the confined states and repeat the above process. The photo-excited carriers left in the confined states can either be excited by TSPA, relax to a lower confined energy level, or recombine.

Due to a built-in electric field across the intrinsic region, the number of electrons may be not equal to the number of holes in a given layer [18]. This charge separation along the growth direction reduces the carrier recombination rate inside the QDs [16] and increases the electron lifetime [17]. This is significant at low temperature, where the TSPA rate mainly competes with the radiative recombination rate because of the reduced thermal escape resulting from deeper QD confinement [19]. Despite the lower TSPA rate at the shallow confined states, TSPA can still pump carriers out after relaxation into the deep confined level and be detected. Furthermore, because of the lower density of states in deeper confinement, the number of photo-excited carriers decreases with longer wavelength below the bandgap. The total number of carriers that relax into deeper confined levels may also decrease, which decreases state filling and the TSPA absorption coefficient. Therefore, TSPA signal decreases towards longer wavelengths (deeper confinement) as shown in Figure 3(a).

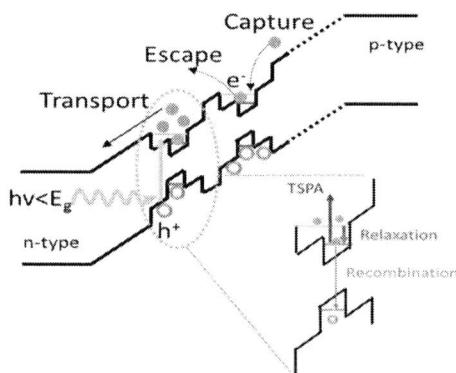

Fig. 5. Simplified band diagram of the *n-i-p* InAs/GaAs/Al$_{0.3}$GaAs Dwell solar cell at 30K.

To verify the process discussed above, Figure 6 shows the voltage-biased TSPA-introduced photocurrent measurements at 40K, which is consistent with the voltage bias results of the InGaAs/GaAs QDSC [14]. It was observed that the TSPA signal is stable with reverse bias from 0V to -2 V. The increasing electric field with reverse bias improves the charge separation that increases the recombination lifetime but also increases the tunneling rate. Stable TSPA is a balance between the rates of recombination and tunneling. Increased tunneling enables faster carrier escape from shallow confined states, which reduces the number of carriers relaxing into the QDs and ultimately decreases TSPA photocurrent, so TSPA is quenched with further reverse bias at -3 V. On the other hand, forward bias reduces both the electric field around the QD region as well as charge separation, so fast recombination in the shallow confined states limits TSPA photocurrent.

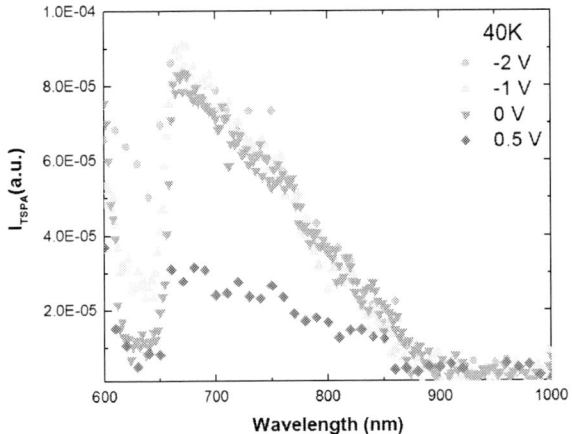

Fig. 6. Voltage bias TSPA photocurrent at 40K.

IV. CONCLUSION

The temperature-dependent and voltage bias-dependent two-step photon absorption processes were investigated in the MOCVD-grown InAs/GaAs/Al$_{0.3}$GaAs Dwell-IBSC. It was found that the working temperature of TSPA is up to 80K under IR (1500 nm) filtered 800°C black body illumination. Due to the limited QD ground state absorption resulting from low QD surface density, QD ground state TSPA was not observed, though could be improved by increasing incident photon density (sun concentration). Instead, TSPA was observed from shallow confined levels formed by the Al$_{0.3}$GaAs/GaAs wells, which is mainly due to the combination of enhanced absorption and charge separation. The enhanced absorption of the wells introduces higher carrier concentrations in the shallow confined levels, and the charge separation results in longer radiative lifetimes.

Based on the observed experimental results, three areas for improvement can be addressed. First, the first step of two-step photon absorption (between the deep confined levels) should be enhanced to improve direct TSPA, which could be achieved by improving QD surface density, the number of QD layers, and photon recycling [20]. Second, the QD capture process affects carrier collection and distribution in a given QD layer. Novel QD systems like InP/In$_{0.5}$GaP [21] or GaSb/Al$_x$GaAs could be used to reduce the hole or electron capture/recombination in the QD, respectively, by careful QD growth optimization and device design. Third, efficient TSPA

requires a reduced recombination rate and escape rate. The electric field decreases the recombination rate via charge separation, while it also increases the escape rate. To balance the effect of electric field and finally achieve a high efficiency room temperature IBSC, it is important to investigate the magnitude of the electric field required at P_{max} condition for a specific design in order to achieve a photon ratchet for a given IBSC design [22], [23].

ACKNOWLEDGEMENT

This work was supported by the Air Force Research Laboratory through STTR FA9453-15-C-0404 and the US Government.

REFERENCES

[1] A. Luque and A. Martí, "Increasing the Efficiency of Ideal Solar Cells by Photon Induced Transitions at Intermediate Levels," *Phys. Rev. Lett.*, vol. 78, no. 26, pp. 5014–5017, Jun. 1997.

[2] Y. Dai *et al.*, "Effect of electric field on carrier escape mechanisms in quantum dot intermediate band solar cells," *J. Appl. Phys.*, vol. 121, no. 1, p. 013101, Jan. 2017.

[3] A. Datas *et al.*, "Intermediate Band Solar Cell with Extreme Broadband Spectrum Quantum Efficiency," *Phys. Rev. Lett.*, vol. 114, no. 15, p. 157701, Apr. 2015.

[4] I. Ramiro *et al.*, "Wide-Bandgap InAs/InGaP Quantum-Dot Intermediate Band Solar Cells," *IEEE J. Photovolt.*, vol. 5, no. 3, pp. 840–845, May 2015.

[5] Y. Dai, S. Polly, S. Hellstroem, D. V. Forbes, and S. M. Hubbard, "Carrier collection in quantum dots solar cells with barrier modification," in *2015 IEEE 42nd Photovoltaic Specialist Conference (PVSC)*, 2015, pp. 1–5.

[6] I. Ramiro *et al.*, "InAs/AlGaAs quantum dot intermediate band solar cells with enlarged sub-bandgaps," in *2012 38th IEEE Photovoltaic Specialists Conference (PVSC)*, 2012, pp. 000652–000656.

[7] H. Xie *et al.*, "Improved optical properties of InAs quantum dots for intermediate band solar cells by suppression of misfit strain relaxation," *J. Appl. Phys.*, vol. 120, no. 3, p. 034301, Jul. 2016.

[8] Y. Dai, M. A. Slocum, Z. Bittner, S. Hellstroem, D. V. Forbes, and S. M. Hubbard, "Optimization in wide-band-gap quantum dot solar cells," in *Photovoltaic Specialists Conference (PVSC), 2016 IEEE 43rd*, 2016, pp. 0151–0154.

[9] R. Jakomin *et al.*, "InAs quantum dot growth on Al$_x$Ga1−xAs by metalorganic vapor phase epitaxy for intermediate band solar cells," *J. Appl. Phys.*, vol. 116, no. 9, p. 093511, Sep. 2014.

[10] S. Asahi, H. Teranishi, N. Kasamatsu, T. Kada, T. Kaizu, and T. Kita, "Suppression of thermal carrier escape and efficient photo-carrier generation by two-step photon absorption in InAs quantum dot intermediate-band solar cells using a dot-in-well structure," *J. Appl. Phys.*, vol. 116, no. 6, p. 063510, Aug. 2014.

[11] S.-K. Park, J. Tatebayashi, and Y. Arakawa, "Formation of ultrahigh-density InAs/AlAs quantum dots by metalorganic chemical vapor deposition," *Appl. Phys. Lett.*, vol. 84, no. 11, pp. 1877–1879, Mar. 2004.

[12] J. Hwang, A. J. Martin, J. M. Millunchick, and J. D. Phillips, "Thermal emission in type-II GaSb/GaAs quantum dots and prospects for intermediate band solar energy conversion," *J. Appl. Phys.*, vol. 111, no. 7, p. 074514, Apr. 2012.

[13] V. Aroutiounian, S. Petrosyan, and A. Khachatryan, "Studies of the photocurrent in quantum dot solar cells by the application of a new theoretical model," *Sol. Energy Mater. Sol. Cells*, vol. 89, no. 2–3, pp. 165–173, Nov. 2005.

[14] Y. Shoji, K. Akimoto, and Y. Okada, "Self-organized InGaAs/GaAs quantum dot arrays for use in high-efficiency intermediate-band solar cells," *J. Phys. Appl. Phys.*, vol. 46, no. 2, p. 024002, 2013.

[15] J. Hwang *et al.*, "Multiphoton Sub-Band-Gap Photoconductivity and Critical Transition Temperature in Type-II GaSb Quantum-Dot Intermediate-Band Solar Cells," *Phys. Rev. Appl.*, vol. 1, no. 5, p. 051003, Jun. 2014.

[16] A. Creti *et al.*, "Role of charge separation on two-step two photon absorption in InAs/GaAs quantum dot intermediate band solar cells," *Appl. Phys. Lett.*, vol. 108, no. 6, p. 063901, Feb. 2016.

[17] A. Cedola, F. Cappelluti, and M. Gioannini, "Dependence of quantum dot photocurrent on the carrier escape nature in InAs/GaAs quantum dot solar cells," *Semicond. Sci. Technol.*, vol. 31, no. 2, p. 025018, 2016.

[18] S. Asahi, H. Teranishi, N. Kasamatsu, T. Kada, T. Kaizu, and T. Kita, "Saturable Two-Step Photocurrent Generation in Intermediate-Band Solar Cells Including InAs Quantum Dots Embedded in Al Ga As/GaAs Quantum Wells," *IEEE J. Photovolt.*, vol. 6, no. 2, pp. 465–472, Mar. 2016.

[19] G. Jolley, L. Fu, H. F. Lu, H. H. Tan, and C. Jagadish, "The role of intersubband optical transitions on the electrical properties of InGaAs/GaAs quantum dot solar cells," *Prog. Photovolt. Res. Appl.*, vol. 21, no. 4, pp. 736–746, Jun. 2013.

[20] B. L. Smith *et al.*, "Inverted growth evaluation for epitaxial lift off (ELO) quantum dot solar cell and enhanced absorption by back surface texturing," in *2016 IEEE 43rd Photovoltaic Specialists Conference (PVSC)*, 2016, pp. 1276–1281.

[21] T. Tayagaki, Y. Nagato, Y. Okano, and T. Sugaya, "A proposal for wide-bandgap intermediate-band solar cells using type-II InP/InGaP quantum dots," in *2016 IEEE 43rd Photovoltaic Specialists Conference (PVSC)*, 2016, pp. 0160–0162.

[22] O. J. Curtin *et al.*, "Quantum Cascade Photon Ratchets for Intermediate-Band Solar Cells," *IEEE J. Photovolt.*, vol. 6, no. 3, pp. 673–678, May 2016.

[23] A. Pusch *et al.*, "Limiting efficiencies for intermediate band solar cells with partial absorptivity: the case for a quantum ratchet," *Prog. Photovolt. Res. Appl.*, vol. 24, no. 5, pp. 656–662, May 2016.

Increasing Current Generation by Photon Up-Conversion in a Single-Junction Solar Cell with a Hetero-Interface

Shigeo Asahi, Kazuki Kusaki, Toshiyuki Kaizu, and Takashi Kita

Kobe University, 1-1 Rokkodai, Nada, Kobe 657-8501, Japan

Abstract — The up-conversion of below-gap photons is very promising for breaking the Shockley–Queisser limit restricting the conversion efficiency of single-junction solar cells (SCs). In this work, we demonstrate drastic increase in the photocurrent generated by a two-step photon up-conversion process in a single-junction SC with a hetero-interface comprising different band-gaps of $Al_{0.3}Ga_{0.7}As$ and GaAs. Below-gap photons for $Al_{0.3}Ga_{0.7}As$ excite GaAs and create electrons and holes. Electrons are densely accumulated at the hetero-interface and are pumped upwards into the $Al_{0.3}Ga_{0.7}As$ barrier by other below-gap photons. We observe a dramatic increase in the additional photocurrent.

Index Terms — intermediate-band solar cells, photovoltaic cells, GaAs, hetero-interface.

I. INTRODUCTION

It is well known that the intermediate-band solar cell (IBSC) is one of the promising candidates of next-generation, ultrahigh-efficiency photovoltaic cells [1]. IBSC contains intermediate states in the band gap and covers a broadband solar spectrum by absorbing multi-color photons corresponding to three optical transitions; the valence band (VB) to the conduction band (CB), the VB to the intermediate band (IB), and the IB to the CB. Sequential absorption of two below-gap photons via the IB up-converts the energy of excited carriers. This two-step photon up-conversion (TPU) produces extra photocurrent without decreasing the output photovoltage. According to an ideal theoretical prediction, the conversion efficiency of IBSC is expected to be greater than 60% under the maximum concentration and 48% under one-sun illumination. A large amount of effort has been dedicated to realize IBSCs, especially for improving the efficiency of the TPU. Generally, the absorption strength of the intraband transition is insufficient, and quick energy relaxation of excited electrons into the IB reduces the photocurrent. Therefore, improving the second-excitation efficiency in the TPU process has been a key issue [2], [3]. In order to overcome this issue, we have proposed a new type single-junction SC structure, called TPU-SC with a hetero-interface of semiconductors [4]. In this work, we fabricated two types of single-junction SCs incorporating $Al_{0.3}Ga_{0.7}As$/GaAs hetero-interfaces of which the interface positions are different. The detailed optical responses of these SCs were characterized. TPU-SCs demonstrate increasing current generation by TPU.

II. CONCEPT OF TWO-STEP PHOTON UP-CONVERSION SOLAR CELLS

Figure 1 illustrate the schematic device structure of the TPU-SCs. The TPU-SC comprises a diode structure of n-$Al_{0.3}Ga_{0.7}As$/$Al_{0.3}Ga_{0.7}As$/GaAs/p-GaAs on a p^+-GaAs(001) substrate. High-energy photons are absorbed in $Al_{0.3}Ga_{0.7}As$, and excited electrons and holes drift in the opposite directions toward n-$Al_{0.3}Ga_{0.7}As$ and p-GaAs, respectively. Here, excited electrons can reach the n-$Al_{0.3}Ga_{0.7}As$ layer without being captured. Below-gap photons passing through $Al_{0.3}Ga_{0.7}As$ directly excite the GaAs layer. Excited carriers in GaAs drift in the opposite directions in the internal electric field. Excited holes reach the p-GaAs contact layer, while electrons are accumulated at the $Al_{0.3}Ga_{0.7}As$/GaAs interface. As electrons accumulated at the hetero-interface are separated from holes, these electron's life time are extended. Such long-lived electrons improve the intraband absorption strength for below-gap photons for GaAs and are efficiently pumped upwards into the $Al_{0.3}Ga_{0.7}As$ barrier. Thus, the TPU process occurring at the hetero-interface increases the photocurrent.

We fabricated two types of TPU-SCs on a p^+-GaAs (001) substrate by using solid-source molecular beam epitaxy, as shown in Fig. 1. A 150-nm-thick p-GaAs (Be: 2×10^{18} cm^{-3}) layer was grown over a 400-nm-thick p^+-GaAs (Be: 1×10^{19} cm^{-3}) buffer layer at a substrate temperature of 550 °C. Subsequently, a 1400-nm-thick i-layer was deposited. Finally, n^+-GaAs (Si: 2.5×10^{18} cm^{-3}), n^+-$Al_{0.3}Ga_{0.7}As$ (Si: 2.5×10^{17} cm^{-3}), and n-$Al_{0.3}Ga_{0.7}As$ (Si: 1×10^{17} cm^{-3}) layers were fabricated on the SC structure. Then, metal Au/Au-Ge and Au/Au-Zn contacts were formed on the top and the bottom surfaces, respectively. The dimensions of the SC were 4×4 mm^2. The difference of the two types of TPU-SCs is the position of the $Al_{0.3}Ga_{0.7}As$/GaAs interface in the i-layer. The thickness of the i-$Al_{0.3}Ga_{0.7}As$ layer from the n-side is 1380 nm or 250 nm. In this paper, we labeled these SCs as p- and n-side TPU-SCs, respectively. By comparing the optical responses of the two SCs, we elucidate the influence of electron accumulation at the hetero-interface on the TPU efficiency. We conducted external quantum efficiency (EQE) and ΔEQE measurements at room temperature. The excitation light was produced by a tungsten halogen lamp, passed through a 140-mm single monochromator, and chopped by an optical chopper with a frequency of 800 Hz. The excitation power density depends on the wavelength and the integrated power density was approximately 2 mW/cm², which is much lower than that of the one-sun solar irradiance of 100 mW/cm². The beam diameter of the monochromatic light was 1.2 mm on the SC surface. The photocurrent was detected by a lock-in amplifier synchronized with the optical chopper. The photocurrent was measured under the short-circuit condition without an external bias voltage. Here, the EQE was defined as the efficiency of the photocurrent generation under monochromatic excitation. The TPU was demonstrated by measuring the change in the EQE signal amplitude under two-color excitation using different light sources. The first interband-excitation light source was the monochromated tungsten halogen lamp described above. The second intraband light source was a continuous-wave laser diode (LD) with the 1300-nm emission, which was used for pumping electrons accumulated at the $Al_{0.3}Ga_{0.7}As$/GaAs interface into the $Al_{0.3}Ga_{0.7}As$ barrier. The 1300-nm LD wavelength was sufficiently long to prevent interband transitions. The excitation power density of the 1300-nm LD was 360 mW/cm². ΔEQE was defined as the difference between the EQE obtained with and without the 1300-nm LD illumination.

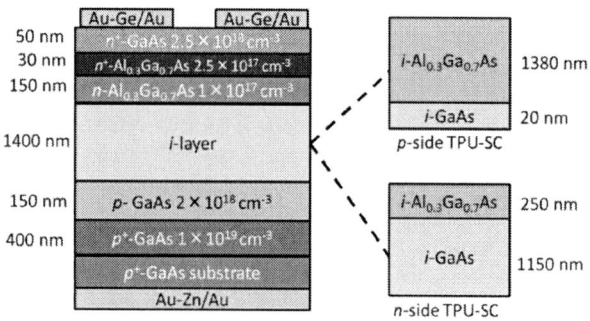

Fig. 1.   Schematic device structure of the *p*-side and *n*-side TPU-SCs.

III. RESULTS AND DISCUSSION

Figure 2(a) shows the EQE spectra for the *p*-side TPU-SC at 290 K. Without the 1300-nm LD illumination, two clear absorption edges appear at 685 nm and 875 nm in the EQE spectrum (black color in Fig. 2(a)); these edges correspond to the band gaps of $Al_{0.3}Ga_{0.7}As$ and GaAs, respectively. When excited above the band gap of $Al_{0.3}Ga_{0.7}As$, the excited electrons and holes are collected at the corresponding electrodes. However, the behavior of carriers generated by below-gap photons for $Al_{0.3}Ga_{0.7}As$ is different. The below-gap photons are absorbed in *i*-GaAs and generate carriers at that place. The excited holes drift towards the *p*-layer of GaAs. On the other hand, drifting electrons are partially accumulated at the $Al_{0.3}Ga_{0.7}As$/GaAs interface, resulting in a significant drop of the EQE signal below the band gap of $Al_{0.3}Ga_{0.7}As$. In this wavelength region, the EQE signal also shows a gradual decrease with increasing wavelength because the optical absorption coefficient becomes small with increasing wavelength. In the near-infrared wavelength region below the band gap of GaAs, the EQE signal decreases drastically and becomes extremely low below the detection limit.

Next, we discuss the EQE spectrum measured under the 1300-nm LD illumination. In this case, the spectrum was substantially changed. It must be noted that the EQE signal (drawn by red color in Fig. 2(a)) increases in the wavelength region between the band gaps of $Al_{0.3}Ga_{0.7}As$ and GaAs. We defined ΔEQE (blue color in Fig. 2(b)) as the difference between the EQE obtained with and without the 1300-nm LD illumination. The ΔEQE signal was approximately 1% in that wavelength region. The generated electrons in *i*-GaAs drift toward n-$Al_{0.3}Ga_{0.7}As$ and accumulated at the $Al_{0.3}Ga_{0.7}As$/GaAs interface. These electrons are spatially separated from holes and are expected to exhibit extended lifetime. Thus, the EQE signal drastically increases with the additional 1300-nm LD illumination at room temperature.

Figures 2(c) and 2(d) show the EQE and ΔEQE spectra for the *n*-side TPU-SC. The EQE signal intensity improves as compared to the *p*-side TPU-SC in the wavelength region below the band gap of $Al_{0.3}Ga_{0.7}As$ while the EQE decreases above the $Al_{0.3}Ga_{0.7}As$ band gap. As the GaAs layer of the *n*-side TPU-SC is much thicker than that of the *p*-side TPU-SC, sufficient electrons are generated in GaAs. Moreover, the EQE signal drop is very steep at the absorption edge of GaAs, suggesting that electrons in the *i*-GaAs layer are effectively separated from holes, and the electron lifetime is extended. These separated electrons are densely accumulated at the $Al_{0.3}Ga_{0.7}As$/GaAs interface. With the 1300-nm LD illumination, the ΔEQE dramatically increases and reaches 10 % which is greater than that of the *n*-side TPU-SC by one order of magnitude. The densely accumulated long-lived electrons are easily pumped into the $Al_{0.3}Ga_{0.7}As$ barrier by the 1300-nm LD light, which attains efficient TPU at the hetero-interface. The optical selection rule of the intersubband transition of electrons in an ideal two-dimensional structure is forbidden for light irradiating the two-dimensional plane perpendicularly. The finite thickness of the accumulation layer relaxes the selection rule. Thus, efficient TPU occurs at the $Al_{0.3}Ga_{0.7}As$/GaAs interface in single-junction SCs.

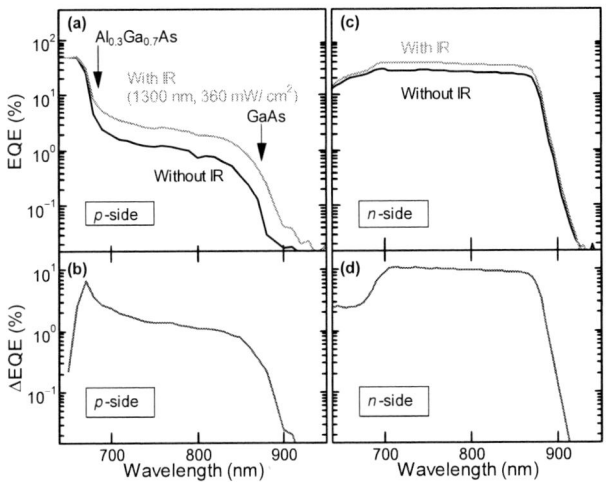

Fig. 2 EQE and ΔEQE spectra obtained at 290 K. (a) and (c) show EQE spectra for the *p*-side and *n*-side TPU-SC, respectively. The black and red lines represent the EQE spectra measured with and without the additional 1300-nm LD illumination, respectively. (b) and (d) show ΔEQE spectra for the *p*-side and *n*-side TPU-SC, respectively. ΔEQE is defined as the difference between the EQE signals measured with and without the additional 1300-nm LD illumination.

IV. CONCLUSION

We fabricated TPU-SCs with a hetero-interface comprising the different band-gaps of $Al_{0.3}Ga_{0.7}As$ and GaAs and investigated the optical responses using the two-color photo-excitation EQE measurement. We observed a dramatic increase in the additional photocurrent generated by the TPU at the $Al_{0.3}Ga_{0.7}As$/GaAs interface. The electron density accumulated at the hetero-interface critically influences the TPU efficiency. The obtained results suggest that the TPU-SC has a high potential for implementation in the next-generation high-efficiency solar cells.

ACKNOWLEDGEMENT

This work was partially supported by the Incorporated Administrative Agency New Energy and Industrial Technology Development Organization (NEDO) and the Japan Society for the Promotion of Science (JSPS) KAKENHI.

REFERENCES

[1] A. Luque and A. Martí, "Increasing the efficiency of ideal solar cells by photon induced transitions at intermediate levels," *Phys. Rev. Lett*, vol. 78, pp. 5014–5017, 1997.
[2] Y. Okada *et al.*, "Intermediate band solar cells: Recent progress and future directions," *Appl. Phys. Rev.*, vol. 2, pp. 021302-1–021302-48, 2015.
[3] T. Kada *et al.*, "Two-step photon absorption in InAs/GaAs quantum-dot superlattice solar cells," *Phys. Rev. B*, vol. 91, no. 20, pp. 201303-1–201303-6, 2015.
[4] S. Asahi, H. Teranishi, K. Kusaki, T. Kaizu and T. Kita, "Two-step photon up-conversion solar cells," *Nat. Commun.*, vol. 8, p. 14962, 2017.

RTP-assisted Ex-situ Analysis of $(Ag,Cu)(In,Ga)Se_2$ Formation using Selenization

Sina Soltanmohammad[1,2] and William N. Shafarman[1,2]

[1]Institute of Energy Conversion, University of Delaware, Newark, Delaware 19716, USA
[2]Department of Materials Science & Engineering, University of Delaware, Newark, Delaware 19716, USA

Abstract — The addition of Ag to Cu-Ga-In precursors for reaction to form $(AgCu)(InGa)Se_2$ (ACIGS) has shown benefits including improved adhesion, greater process tolerance and potential for improved device performance. In this study, metal precursors were sputtered with $Ag/(Ag+Cu) \approx 0.25, 0.75, 1.00$ and $Ga/(Ga+In) \approx 0.25$. Chemical pathways of formation of the ACIGS were studied by rapid thermal processing assisted *ex-situ* time-progressive experiments at $450°C$ under 5% Ar/H_2Se atmosphere. The reaction time was varied from 2–45 min. Composition and structure were characterized by energy dispersive x-ray spectroscopy (EDS), x-ray fluorescence (XRF), x-ray diffraction (XRD) and Raman spectroscopy. Reaction pathways analysis for low $Ag/(Ag+Cu) \approx 25\%$ indicate that multiple stable intermetallic phases were formed during the reaction including κ-$(Cu,Ga)_2(Ag,In)$, ζ-$(Ag,Cu)_3(In,Ga)$ and $Cu_9(In,Ga)_4$; and the reaction to form ACIGS was complete after 10 min. However XRD analysis showed that Ag tends to stay in the stable ζ-$(Ag,Cu)_3(In,Ga)$ phase during the reaction at higher $Ag/(Ag+Cu)$ $\approx 75\%$ and 100% and this causes the longer reaction time and the non-uniformity. With Ag ratio $=100\%$, the reaction was completed with formation of $Ag(In,Ga)Se_2$ and $AgIn_5Se_8$ within 20 min.

I. INTRODUCTION

Silver alloying of $Cu(InGa)Se_2$ to form $(AgCu)(InGa)Se_2$ (ACIGS) has been investigated for thin film photovoltaics. Introducing silver into the CIGS lattice during co-evaporation resulted in improved V_{OC} and diminished defect concentration, attributed to the lower melting point of the alloy [1]. Also, Ag alloying gave significant improvement in the hydride gas reaction of metal precursors by altering the morphology of the metal precursors and enhancing adhesion to enable a wider process window including higher reaction temperature and improved device performance [2].

Knowledge of the reaction process during the absorber formation is essential to develop an absorber compatible with high performance solar cells. Reaction chemistry of the formation of ternary and quaternary chalcopyrites from metal precursors has been studied by different groups using time-progressive reactions [3]–[5] or *in-situ* analysis [6]–[9]. Verma et al.[3] and Orbey et al.[4] studied the chemical formation of $CuInSe_2$ using time-progressive reactions of Cu-In precursors with H_2Se at $250°C$, $325°C$ and $400°C$ in a tubular reactor. Brummer et al. [6] and Hergert et al. [7] performed *in-situ* x-ray diffraction analysis during annealing of Cu/In/Se precursors with Se excess in a temperature range from 25 to $550°C$. Similar studies also used *in-situ* energy dispersive XRD (EDXRD) [8], [9].

Hergert suggested a two-step process for the formation of the chalcopyrite phase [7]. In the first step, $CuInSe_2$ forms by reaction of InSe and CuSe at the Se melting point ($221°C$). This is followed by decomposition of CuSe into Cu_2Se and then a fast reaction of InSe and $Cu_{2-x}Se$ to form $CuInSe_2$. While different reactions have been reported for the former steps, the final reaction has been verified by other studies. This could originate from different processing conditions, uncertainties in the identification of Cu-In phases particularly with rapid *in-situ* XRD scans and possible low temperature phase transformations in the quenched samples at different stages of the *ex-situ* reaction process [6].

Previously, we studied the effect of Ag-alloying on the precursor structure using co-sputtered or stacked layer deposition [10]. Here, we report a comprehensive study on chemical pathways during the formation of ACIGS using *ex-situ* time-progressive experiments. This will lead to development of an advanced precursor reaction process with a controlled composition profile.

II. EXPERIMENTAL PROCEDURES

Ag-Cu-Ga-In metal precursors were deposited onto Mo/soda-lime glass substrates by dc magnetron sputtering at room temperature using $Cu_{0.77}Ga_{0.23}$, $Ag_{0.77}Ga_{0.23}$, and In targets. Samples were deposited with Mo/Cu-Ga/In/Ag-Ga stacked layers, as described previously [11]. Sputtering parameters were determined to give $Ag/(Ag+Cu) \approx 0.25, 0.75, 1$ and $Ga/(Ga+In) \approx 0.25$ with an average thickness ≈ 500 nm. Selenization was done using rapid thermal processing (RTP) at atmospheric pressure in a tubular quartz reactor charged with 5% H_2Se in Ar. The reaction temperature was $450°C$ and the reaction time was varied from 2 – 45 min. The temperature ramp up was set to ~ 1 second and the cool-down was ~ 30 sec to $250°C$.

Composition of the samples was measured by x-ray fluorescence (XRF) and energy dispersive x-ray spectroscopy (EDS). The XRF measurements sample the entire film. EDS with excitation voltage 20 keV, however, gives a composition value weighted toward the top ~ 0.3 μm of the film. Cross-section EDS mapping was performed with 10 keV in order to decrease matrix effects. The crystal structures of the films were evaluated using symmetric x-ray diffraction (XRD) and glancing incidence XRD (GIXRD) with Cu Kα radiation.

III. RESULTS AND DISCUSSION

Fig. 1 compares the Ga/(Ga+In) and Se/M ≡ Se/(Ag+Cu+Ga+In) ratios measured by EDS and XRF versus reaction time for all films. For films with Ag/(Ag+Cu) ≈ 0.25 and 0.75, Ga grading and Se uptake occur in the first 10 minutes

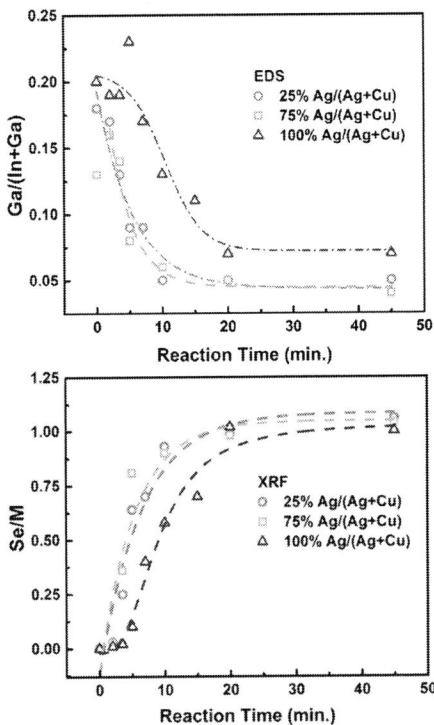

Fig. 1. Ga/(Ga+In) and Se/Metals ratios measured by EDS and XRF versus reaction time.

of reaction; however, those happen after 20 minutes for the Ag/(Ag+Cu) ~ 1 sample, indicating slower reaction rate of Ag(In,Ga)Se$_2$ compared to the other two compounds.

Symmetric XRD scans of reacted films with identified phases are shown in Fig. 2. XRD analysis of the Ag/(Ag+Cu) ≈ 25% samples (Fig. 2(a)) indicated that the ACIGS phase mostly formed by 10 min through reaction of multiple intermetallic phases of In, (Ag,Cu)In$_2$, Ag$_3$In, Cu$_9$(In,Ga)$_4$ and a recently reported κ-(Cu,Ga)$_2$(Ag,In) phase [12]. This phase which has a cubic structure in the space group of $Fd\bar{3}m$ forms with disappearance of the Cu$_3$Ga phase and is stable up to 7 min reaction. Cross section EDS map analysis of the 7 min reacted film (not shown here) shows localized intermetallic compounds that are fully separated from the top surface of the film. No significant changes are seen in XRD patterns of samples selenized from 10-45 min, indicating that the chalcogenide phase is fully formed after 10 min reaction. Lattice constants from Rietveld analysis ($a = 5.83\text{Å}, c = 11.73\text{Å}$) are consistent with those reported for the (Ag$_{0.25}$Cu$_{0.75}$)InSe$_2$ phase [13]. SEM/EDS and Raman studies on the back side of the films (not shown here) showed incorporation of Ga into CuGaSe$_2$ and GaSe phases.

XRD analysis of the samples with Ag/(Ag+Cu) ≈ 75% (Fig. 2(b)) indicated that the ACIGS phase formed between ~ 3.5-10 min similar to the 25% samples. Cross-section EDS maps (not shown here) showed that a Cu-In-Se layer formed on the surface after 3.5 min reaction. GIXRD analysis at 0.5° and Raman analysis (not shown here), show that the peak belongs to a CuInSe$_2$ phase on the top surface of the film. By increasing the reaction time to 10 min, all of the intermetallic and InSe peaks disappeared and only chalcogenide peaks are seen. Apart from improving the crystallinity of the chalcogenide phase, XRD patterns of 20 and 45 min reacted films do not show any

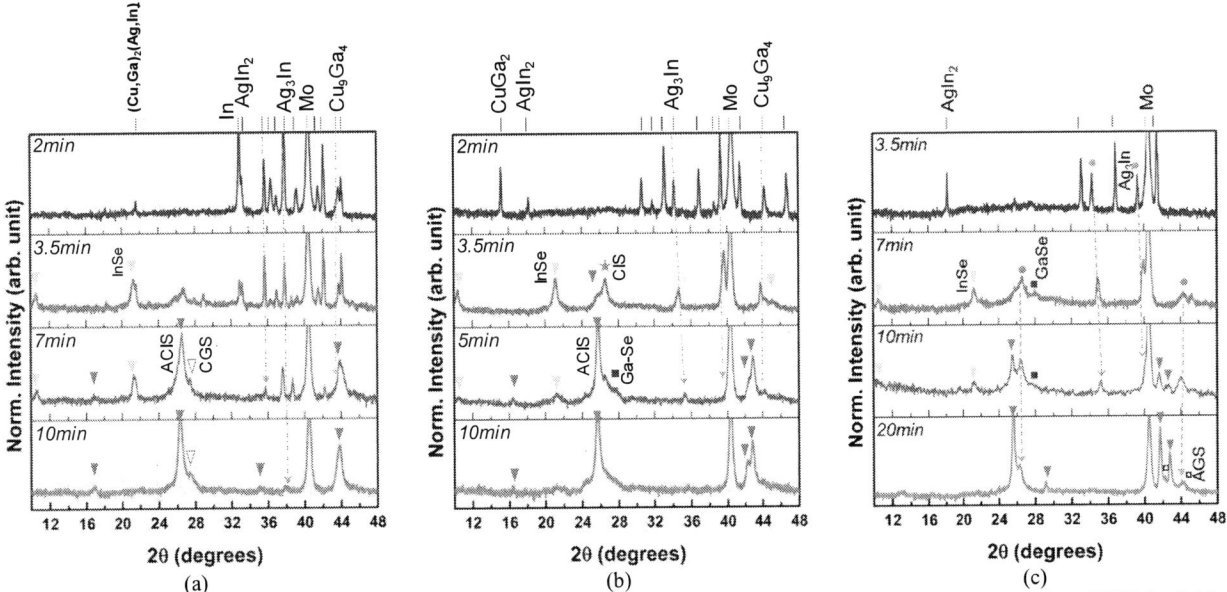

Fig. 2. X-ray diffraction patterns (square-root scale) of the reacted films with Ag/(Ag+Cu) ≈ (a) 0.25, (b) 0.75 and (c) 1.00 at 450°C for 2-45 min reaction time. Phases are indicated in the figures.

978-1-5090-5606-4/17 $31.00 © 2017 IEEE

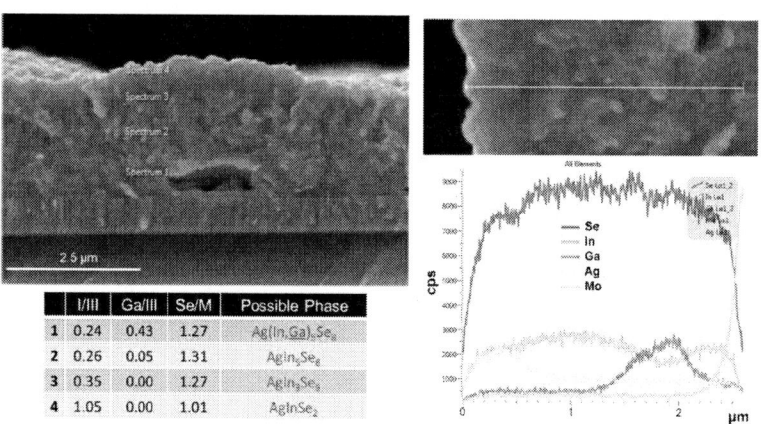

Fig. 3. Cross-sectional SEM images and point-EDS analysis (left) and line-EDS analysis (right) of 45 min reacted sample with Ag/(Ag+Cu) ≈ 1.0.

TABLE II
SUMMARIZED CHEMICAL PATHWAY FORMATION OF THE (AG,CU)(IN,GA)SE₂ ABSORBER LAYER WITH AG/(AG+CU) ≈ 0.25, 0.75 AND 1.00.

Time	Detected phases
Ag/(Ag+Cu) ≈ 0.25	
0	In, $AgIn_2$, Cu_3Ga, $Cu_9(In,\underline{Ga})_4$
2	In, $AgIn_2$, Cu_3Ga, $Cu_9(In,\underline{Ga})_4$, Ag_3In, $(Cu,Ga)_2(Ag,In)$, $InSe$, In_4Se_3
3.5	In, $AgIn_2$, $Cu_9(In,\underline{Ga})_4$, Ag_3In, $(Cu,Ga)_2(Ag,In)$, $Ag_9(\underline{In},Ga)_4$, $InSe$, In_4Se_3, $(Ag,\underline{Cu})InSe_2$
5	$Cu_9(In,\underline{Ga})_4$, Ag_3In, $(Cu,Ga)_2(Ag,In)$, $Ag_9(\underline{In},Ga)_4$, $InSe$, $(Ag,\underline{Cu})InSe_2$, $GaSe/CuGaSe_2$
10	$Cu_9(In,\underline{Ga})_4$, $(Ag,\underline{Cu})InSe_2$, $GaSe/CuGaSe_2$
20 - 45	$(Ag,\underline{Cu})InSe_2$, $GaSe/CuGaSe_2$

Time	Detected phases
Ag/(Ag+Cu) ≈ 0.75	
0	$AgIn_2$, $Ag_9(\underline{In},Ga)_4$, $CuGa_2$, $Cu_9(In,\underline{Ga})_4$
2	$AgIn_2$, $Ag_9(\underline{In},Ga)_4$, $CuGa_2$, $Cu_9(In,\underline{Ga})_4$, Ag_3In, $InSe$, In_2Se_3
3.5	$Cu_9(In,\underline{Ga})_4$, Ag_3In, $InSe$, In_2Se_3, $CuInSe_2$, $(\underline{Ag},Cu)InSe_2$
5	Ag_3In, $InSe$, In_2Se_3, $CuInSe_2$, $(\underline{Ag},Cu)InSe_2$, $GaSe/(\underline{Ag},Cu)Ga_5Se_8$
10	Ag_3In, $InSe$, $(\underline{Ag},Cu)InSe_2$, $GaSe/(\underline{Ag},Cu)Ga_5Se_8$
20 - 45	$(\underline{Ag},Cu)InSe_2$, $GaSe/(\underline{Ag},Cu)Ga_5Se_8$

Time	Detected phases
Ag/(Ag+Cu) ≈ 1.00	
0	$AgIn_2$, $Ag_9(\underline{In},Ga)_4$
3.5	$AgIn_2$, $Ag_3(\underline{In},Ga)$, $(\underline{In},Ga)_2Se_3$
7	$Ag_3(\underline{In},Ga)$, $(In,\underline{Ga})Se$, $InSe$, $Ag(\underline{In},Ga)_5Se_8$
10	$Ag_3(\underline{In},Ga)$, $(In,\underline{Ga})Se$, $InSe$, $Ag(\underline{In},Ga)_5Se_8$, $AgInSe_2$
20	$(In,\underline{Ga})Se$, $InSe$, $Ag(\underline{In},Ga)_5Se_8$, $AgInSe_2$, $AgGaSe_2$
45	$Ag(\underline{In},Ga)_5Se_8$, $AgInSe_2$, $AgGaSe_2$

significant changes. However, point-EDS analysis from the remnants on the Mo-Side of the 20 min reacted film after delaminaiton (not shown here) indicated existence of unreacted ζ-$(Ag,Cu)_3(\underline{In},Ga)$ or α-(Ag) phases. Also, remnants with composition close to the $Ag(In,\underline{Ga})_5Se_8$ (ODC) phase were detected.

By increasing the Ag to 100%, the initial formation of crystalline In-Se and the chalcopyrite phases was delayed. As seen in the XRD analysis in Fig. 2(c), for up to 3.5 min reaction only intermetallic phases including $AgIn_2$ and ζ-$(Ag,Cu)_3(\underline{In},Ga)$ phases were detected while InSe phase was first seen after 5 min. An In-rich ordered defect compound (ODC) $AgIn_5Se_8$ appeared in the XRD pattern of the sample reacted for 7 min. The reaction completed with formation of $Ag(\underline{In},Ga)Se_2$ and $AgIn_5Se_8$ after 20 min. Based on the XRD analysis, besides the $AgIn_2$ phase that disappears between 5-7 min, ζ-$(Ag,Cu)_3(\underline{In},Ga)$ phase was the main intermetallic phase seen during the reaction. EDS mapping analysis of the 45 min reacted film (Fig. 3) shows a Ag-rich layer ($AgInSe_2$) on the top surface and the ODC phase in the rest of the film while Ga remains at the back interface. SEM/EDS analysis on the Mo remnants after delamination (not shown here) also indicated existence of $AgGaSe_2$ and Ga-Se features at the back interface. Table 1 summarizes the reaction pathway analysis of $(Ag_xCu_{1-x})(In,Ga)Se_2$ absorber materials at 450°C.

IV. CONCLUSIONS

In this study, the chemical pathway for formation of $(Ag,Cu)(In,Ga)Se_2$ absorber materials was studied using RTP reaction. *Ex-situ* time progressive experiments were performed at 450°C in a tubular reactor charged with Ar/5% H_2Se. Ag-Cu-In-Ga metal precursors were sputtered with sequential layer structures of Cu-Ga/In/Ag-Ga with Ag/(Ag+Cu) ≈ 25%, 75%, and 100% and Ga/(Ga+In) ≈ 0.25 ratios. Composition and structure were characterized by energy dispersive x-ray spectroscopy (EDS), x-ray fluorescence (XRF), x-ray diffraction (XRD) and Raman spectroscopy. Reaction pathways analysis for films with Ag/(Ag+Cu) ≈ 25% indicated that multiple stable intermetallic phases were formed during the reaction including κ-$(Cu,Ga)_2(Ag,In)$, ζ-$(\underline{Ag},Cu)_3(\underline{In},Ga)$ and $Cu_9(\underline{In},Ga)_4$. The reaction was complete after 10 min. However, by increasing the Ag ratio to 100%, the initial formation of the chalcopyrite phase was delayed and the reaction was completed with formation of $Ag(\underline{In},Ga)Se_2$ and $Ag(\underline{In},Ga)_5Se_8$ within 20 min. XRD analysis showed that Ag tends to stay in the stable ζ-$(Ag,Cu)_3(In,Ga)$ phase during the reaction at higher $Ag/(Ag+Cu) \approx 75\%$ and 100% and this causes the longer reaction time and the non-uniformity. We also show that Ga accumulated in $AgGaSe_2$ and Ga-Se phases at the back-side of all films regardless of the amount of Ag-alloying.

ACKNOWLEDGEMENTS

The authors acknowledge the technical support of John Elliott, Kevin Hart and Wayne Buchanan. This work is based in part upon work supported by the U.S. Department of Energy SunShot Initiative, under Award Numbers DE-EE0005407 and DE-EE0007542.

Disclaimer: "This report was prepared as an account of work sponsored by an agency of the United States Government. Neither the United States Government nor any agency thereof, nor any of their employees, makes any warranty, express or implied, or assumes any legal liability or responsibility for the accuracy, completeness, or usefulness of any information, apparatus, product, or process disclosed, or represents that its use would not infringe privately owned rights. Reference herein to any specific commercial product, process, or service by trade name, trademark, manufacturer, or otherwise does not necessarily constitute or imply its endorsement, recommendation, or favoring by the United States Government or any agency thereof. The views and opinions of authors expressed herein do not necessarily state or reflect those of the United States Government or any agency thereof."

REFERENCES

[1] P. Erslev, G. M. Hanket, W. N. Shafarman, and D. J. Cohen, "Characterizing the effects of silver alloying in chalcopyrite CIGS with junction capacitance methods," in *MRS Proceedings*, 2011, vol. 1165, pp. 1165-M01-7.

[2] Y. Tauchi, K. Kim, H. Park, and W. Shafarman, "Characterization of $(AgCu)(InGa)Se_2$ absorber layer fabricated by a selenization process from metal precursor," *IEEE J. Photovoltaics*, vol. 3, no. 1, pp. 467–471, 2013.

[3] S. Verma and N. Orbey, "Chemical reaction analysis of copper indium selenization," *Prog. Photovoltaics Res. Appl.*, vol. 4, pp. 341–353, 1996.

[4] N. Orbey, H. Hichri, R. Birkmire, and T. Russell, "Effect of temperature on copper indium selenization," *Prog. Photovoltaics Res. Appl.*, vol. 5, pp. 237–247, 1997.

[5] G. M. Hanket, W. N. Shafarman, B. E. McCandless, and R. W. Birkmire, "Incongruent reaction of Cu–(InGa) intermetallic precursors in H_2Se and H_2S," *J. Appl. Phys.*, vol. 102, no. 7, p. 74922, 2007.

[6] A.Brummer, V. Honkimäki, P. Berwian, V. Probst, J. Palm, and R. Hock, "Formation of $CuInSe_2$ by the annealing of stacked elemental layers—analysis by in situ high-energy powder diffraction," *Thin Solid Films*, vol. 437, no. 1–2, pp. 297–307, 2003.

[7] F. Hergert, R. Hock, A. Weber, M. Purwins, J. Palm, and V. Probst, "In situ investigation of the formation of $Cu(In,Ga)Se_2$ from selenised metallic precursors by X-ray diffraction—The impact of Gallium, Sodium and Selenium excess," *J. Phys. Chem. Solids*, vol. 66, no. 11, pp. 1903–1907, 2005.

[8] E. Rudigier, J. Djordjevic, C. Von Klopmann, B. Barcones, A. Pérez-Rodríguez, and R. Scheer, "Real-time study of phase transformations in Cu–In chalcogenide thin films using in situ Raman spectroscopy and XRD," *J. Phys. Chem. Solids*, vol. 66, no. 11, pp. 1954–1960, 2005.

[9] J. Djordjevic, E. Rudigier, and R. Scheer, "Real-time studies of phase transformations in Cu–In–Se–S thin films—3: Selenization of Cu–In precursors," *J. Cryst. Growth*, vol. 294, no. 2, pp. 218–230, 2006.

[10] S. Soltanmohammad, L. Chen, B. McCandless, and W. N. Shafarman, "Ag–Cu–In–Ga Metal Precursor Thin Films for $(Ag,Cu)(In,Ga)Se_2$ Solar Cells," *IEEE J. Photovoltaics*, vol. 7, no. 1, pp. 273–280, 2017.

[11] S. Soltanmohammad, D. M. Berg, L. Chen, K. Kim, H. Simchi, and W. N. Shafarman, "Effect of sputtering sequence on the properties of Ag-Cu-In-Ga metal precursors and reacted $(Ag,Cu)(In,Ga)Se_2$ films," in *2014 IEEE 40th Photovoltaic Specialist Conference (PVSC)*, 2014, pp. 1707–1711.

[12] S. Soltanmohammad, B. Mccandless, and W. N. Shafarman, "A quaternary Laves-type phase in Ag-Cu-In-Ga thin films," *J. Alloys Compd.*, vol. 710, no. 2017, pp. 819–824, 2017.

[13] J. H. Boyle, B. E. McCandless, W. N. Shafarman, and R. W. Birkmire, "Structural and optical properties of $(Ag,Cu)(In,Ga)Se_2$ polycrystalline thin film alloys," *J. Appl. Phys.*, vol. 115, no. 22, 2014.

Role of $E_V+0.98$ eV trap in light soaking-induced short circuit current instability in CIGS solar cells

P. K. Paul[1], T. Jarmar[2], L. Stolt[2], A. Rockett[3], and A. R. Arehart[1]

[1]Electrical and Computer Engineering, The Ohio State University, Columbus, OH USA
[2]Solibro Research AB, Uppsala, Sweden
[3]Department of Metallurgical and Materials Engineering, Colorado School of Mines, Golden, CO USA

Abstract — **Light-induced instabilities/degradation in Cu(In,Ga)Se₂ (CIGS) solar cells are a prevalent and urgent issue to resolve to improve performance, uniformity, and reliability. Here, mechanisms contributing to light–induced instabilities are identified focusing on an observed short circuit current (J_{SC}) reduction. External quantum efficiency measurements before and after light soaking identified a reduction in long wavelength photon carrier collection efficiency in the CIGS absorber layer. Using deep level optical spectroscopy (DLOS), the concentration of CIGS $E_V+0.98$ eV deep level is correlated with the amount of J_{SC} degradation, Finally, capacitance voltage (C-V) measurements reveal light induces a large reduction in the depletion depth and reduction of carrier collection and are all correlated with the J_{SC} reduction. Finally, the $E_V+0.53$ eV trap concentrations are shown to correlate with V_{OC} instability but not the J_{SC} reduction confirming that multiple trap-induced mechanism are responsible for the light-induced instabilities.**

I. INTRODUCTION

CIGS is an established absorber material for thin-film solar cells due to its high optical absorption, tunable bandgap, and low manufacturing cost [1]. The record efficiency of CIGS thin film solar cell is 22.6%, which is the highest among all thin-film solar [2]. In spite of high initial efficiency, CIGS solar cell instabilities due to the effects of light, temperature and moisture is an active research area [3]. Previously, CIGS solar cells have shown both beneficial and detrimental effects after extended light soaking (LS) [4]. Typically, CIGS solar cells subjected to LS exhibit increased effective p-type doping due to changes in trap occupancy [5]. Additionally, deep levels and interface traps are believed to be responsible for loss of solar cell performance [6], but the actual mechanisms and the knowledge about the individual traps responsible for LS-induced degradation are still not well understood. In CIGS, light-soaking can impact open circuit voltage (V_{OC}), Jsc fill factor (FF), and efficiency (η), but since it influences so many cell parameters it suggests more than one mechanism is responsible for all these changes [7]. To understand the mechanisms of LS- induced instability, we investigated solar cells with different LS responses.

II. APPROACH

In this study, three CIGS solar cells with varying degrees of instability grown by Solibro Research AB were used. First, Mo metal back contacts were deposited on a soda lime glass substrate and then the CIGS absorber layer was deposited by a 2-stage vacuum co-evaporation process. Then the CdS buffer layer was deposited by chemical bath deposition. The front contact was formed by sputtering intrinsic ZnO (IZO) and Al-doped ZnO (AZO) layers. The cells were characterized then ~1 mm² sub-cells were physically circumscribed to provide devices with suitable capacitance for the capacitance-based measurements.

To understand and identify the role of deep traps responsible for the light-induced J_{SC} degradation, capacitance–voltage (C-V) profiling and deep level transient and optical spectroscopies (DLTS/DLOS) were all performed with external quantum efficiency (EQE) measurements. The DLTS transients were analyzed using the double boxcar method with rate windows from 0.8 to 2000 s⁻¹, and the equipment and additional details are described in Ref. 8. For the DLTS and DLOS measurements, the traps were filled with 10 ms and 10 s with +0.4 V pulses, respectively, and the trap emission was recorded at -1.0 V. The DLOS capacitance transients were recorded for 350 s at room temperature with monochromatic light incident with photon energies from 0.5 to 1.4 eV in 0.02 eV steps. C-V measurements were performed using a Boonton 7200 capacitance meter and the net doping profiles (N) were calculated using [9]

$$N = \frac{-C^3}{q\varepsilon_s A^2 \left(\frac{dC}{dV}\right)}$$

where A is the device area, ε_s is the permittivity, and q is the elementary charge.

III. RESULTS AND DISCUSSION

Table I shows the impact of light soaking at 1000 W/m² at approximately 25⁰ C for 24 h on the three samples. After light soaking, Sample 1 shows the smallest J_{SC} reduction and Samples 3 show ~6X larger J_{SC} reduction. Additionally, Samples 1 and 2 show reduced open circuit voltage (V_{OC}), which is a different pattern of instability than the J_{SC} reduction. To understand the light-induced J_{SC} reduction, EQE

TABLE I
Light-induced changes for the three CIGS samples

	Sample 1	Sample 2	Sample 3
η change (%)	-0.6	-1.16	-0.6
V_{OC} change (V)	-0.01	-0.02	+0.005
Jsc change (mA/cm²)	**-0.20**	**-0.85**	**-1.25**
FF change (%)	-1.4	-1.8	0.0
[$E_V + 0.98$ eV] (cm⁻³)	1.5×10^{15}	5.0×10^{15}	6.0×10^{15}
C-V N_A change (cm⁻³)	1.0×10^{15}	3.5×10^{15}	5.1×10^{15}
[$E_V + 0.53$ eV] (cm⁻³)	1.3×10^{13}	2.3×10^{13}	BD

BD = below detection

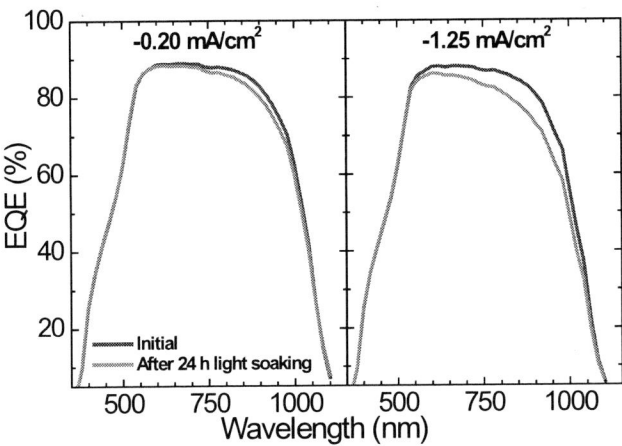

Figure 1: EQE spectra before and after light soaking (left) Sample 1 (-0.20 mA/cm²) with the lowest instability and (right) Sample 3 (-1.25 mA/ cm² J_{SC} degradation) with the highest instability. Larger reduction in EQE is observed in the long wavelengths of Sample 3, consistent with the J_{SC} degradation and indicating the CIGS layer is responsible for the J_{SC} reduction.

measurements were performed before and after LS. From Fig. 1, the EQE measurements show a clear decrease in carrier collection efficiency at longer wavelengths after LS. The longer wavelength photons are absorbed deeper in the sample (i.e. in the CIGS absorber layer), this indicates that the CIGS layer is responsible for the loss of carrier collection. The carrier collection efficiency loss is clearly correlated with the measured LS-induced J_{SC} reduction. The collection efficiency is dictated by the minority carrier diffusion length L_n and the depletion depth W, which is given by [9]

$$W = \sqrt{\frac{2\varepsilon_s(V_{bi}-V)}{qN_A}}$$

where V is the applied voltage, V_{bi} is the built-in potential, and N_A is the carrier concentration. In the case of CIGS, the trap

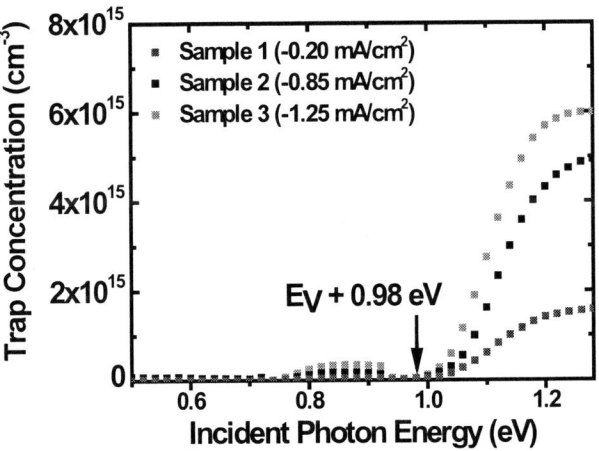

Figure 3: DLOS steady state photocapacitance spectra showing a correlation between the E_V+0.98 eV concentration and J_{SC} reduction. The trap concentrations are summarized in Table I.

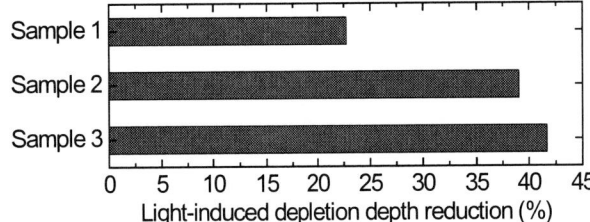

Figure 2: Depletion depth reduction for Sample 1 (-0.2 mA/cm²), Sample 2 (-0.85 mA/cm²), and Sample 3 (-1.25 mA/cm²) after–LS soaking. The samples with higher J_{SC} instability have the highest relative reduction of the depletion region leading to higher J_{SC} reductions.

densities can be larger than the doping concentration, which results in large changes in the depletion depth, and hence is a likely source of the reduced carrier collection efficiency.

To understand the J_{SC} reduction likely due to the trap-induced depletion depth changes, C-V measurements were performed: in the dark to maintain the equilibrium trap occupancy (e.g. traps filled with holes), and with 1.3 eV illumination light to empty all traps in the bandgap. Because the monochromatic light intensity (~10^{14}-10^{15} photons/cm²/s) is much less than the typical above bandgap solar spectra, any changes observed under this illumination will also be observed under normal solar cell operation. From the C-V measurements, the carrier concentration vs. depth was extracted [9]. From Fig. 2, LS caused a significant reduction in the depletion width for all three samples. Importantly, after LS Samples 2 and 3, which exhibit the highest J_{SC} reduction, have the highest relative change of the depletion width where after LS Samples 2 and 3 have a depletion depth (0.35-0.39 μm) that is less than half that of Sample 1 (0.85 μm). Thus, the light-induced depletion depth reduction is the likely source of the carrier collection efficiency reduction, reduction in the long wavelength EQE, and also the reduction in J_{SC}. The precise source of this depletion depth reduction can be easily identified in the defect spectroscopy measurements.

The DLOS measurements in Fig. 3 show a trap at E_V + 0.98eV in all samples, which was previously suggested to be the $V_{Cu}+V_{Se}$ defect [10]. Table I shows a clear trend between the increase in effective doping measured by C-V (Fig. 2) due to 1.3 eV illumination and the concentration of the E_V+0.98 eV level indicating the E_V+0.98 eV trap is likely responsible for the light-induced Jsc reduction in these samples and that reduction of the $V_{Cu}+V_{Se}$ divacancy concentration would be necessary to alleviate the loss of J_{SC}.

To complete the defect study throughout the CIGS bandgap, DLTS measurements were performed. Fig. 4 shows DLTS measurement spectra where a trap at E_V+0.53 eV is present in all samples. The trap concentrations are listed in Table I, but it is evident that there is no clear correlation between the trap concentration and light-induced J_{SC} reduction. However, there is a clear correlation between the V_{OC} change and the E_V+0.53 eV concentration. Because the E_V+0.53 eV trap is near mid-gap, it might be an efficient recombination center and efficiently reduce V_{OC}, but this mechanism is still under investigation.

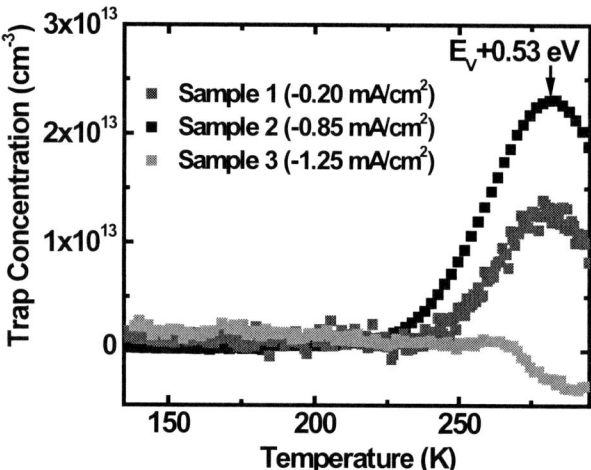

Figure 4: DLTS spectra showing trap with activation energy E_V+ 0.53 eV. Concentration of E_V+0.53 eV shows no correlation with J_{SC} reduction indicates E_V+0.53 eV might not be the cause of J_{SC} reduction.

IV. CONCLUSIONS

CIGS samples subjected to light soaking results in changes in solar cell performance that are correlated with the E_V+0.98 eV level in the case of the observed J_{SC} reduction and E_V+0.53 eV level for the V_{OC} changes. The J_{SC} reduction results from loss of carrier collection in the CIGS layer where the E_V+0.98 eV level causes a large reduction in the depletion depth when light above 1.0 eV is incident on the sample (i.e. normal solar cell conditions), which is the likely source of the J_{SC} reduction. On the other hand, the V_{OC} change is correlated with the concentration of the near mid-gap E_V+0.53 eV level, which may be an efficient recombination center. This suggests there is indeed more than one mechanism for the light-soaking-induced changes in cell performance and reducing the concentration of these two deep levels would help mitigate the light-soaking induced J_{SC} and V_{OC} instabilities.

Acknowledgements

The authors would like to thank the Department of Energy (Contract #DE-DD0007141) for partial financial support.

V. REFERENCES

1. K. L. Chopra, P. Paulson and V. Dutta, "Thin film solar cells and overview," *Progress in Photovoltaics*, vol. 12, pp. 69-92, 2004.
2. Pv-tech.org "ZSW achieves world record CIGS lab cell efficiency of 22.6%,"[Online] Available: http://www.pv-tech.org/news/zsw-achieves-world-record-cigs-lab-cell-efficiency-of-22.6G. [Accessed: 15- Jun- 2016].
3. P. Reinhard, et al., "Review of progress toward 20% efficiency flexible CIGS solar cells and manufacturing issues of solar modules," *IEEE J. Photovoltaics*, vol. 3, pp. 572–580, 2013.
4. S. Chen, T. Jarmar, Sven Södergren, U. Malm, E. Wallin, O. Lundberg, S. Jander, R. Hunger, and L. Stolt, "Light soaking induced doping increase and sodium redistribution in Cu(In,Ga)Se2-based thin film solar cells", *Thin Solid Films*, vol. 582, pp. 35-38, 2015.
5. R. Farshchi, et al., "Light soaking effects on photovoltaic modules: overview and literature review" 2016 IEEE 43rd Photovoltaic Specialists Conference (PVSC), Portland, OR, 2016, pp. 2157-2160.
6. J. Bailey, G. Zapalac and D. Poplavskyy, "Metastable defect measurement from capacitance-voltage and admittance measurements in Cu(In, Ga)Se2 Solar Cells," 2016 IEEE 43rd Photovoltaic Specialists Conference (PVSC), Portland, OR, 2016, pp. 2135-2140.
7. T. Ishii, K. Otani, T. Takashima and K. Ikeda, "Change in I-V characteristics of thin-film photovoltaic (PV) modules induced by light soaking and thermal annealing effects", *Progress in Photovoltaics: Research and Applications*, vol. 22, no. 9, pp. 949-957, 2013.
8. A. R. Arehart, A. A. Allerman, and S. A. Ringel, "Electrical characterization of n-type Al$_{0.30}$Ga$_{0.70}$N Schottky diodes," *Journal of Applied Physics*, vol. 109, pp. 114506, 2011.
9. D. K. Schroeder, *Semiconductor Material and Device Characterization*. Hoboken, New Jersey: John Wiley & Sons, Inc., 2006.
10. S. Lany and A. Zunger, "Light- and bias-induced metastabilities in Cu(In,Ga)Se2 based solar cells caused by the (V_{Se}-V_{Cu}) vacancy complex," *Journal of Applied Physics*, vol. 100, p. 113725, 2006.

Study of Defect Properties in CuGaSe$_2$ Thin-film Solar-cells Using Admittance Spectroscopy

Muhammad Monirul Islam[1,2], Shogo Ishizuka[3], Hajime Shibata[3], Shigeru Niki[3], Katsuhiro Akimoto[1], and Takeaki Sakurai[1]

[1]Division of Applied Physics, Faculty of Pure and Applied Sciences, University of Tsukuba, Tsukuba, Ibaraki 305-8573, Japan.

[2]Alliance for Research on North Africa (ARENA), Faculty of Pure and Applied Sciences, University of Tsukuba, Tsukuba, Ibaraki 305-8572, Japan.

[3]Research Center for Photovoltaics, National Institute of Advanced Industrial Science and Technology (AIST), Tsukuba, Ibaraki 305-8568, Japan.

Abstract — **Defect study of the CuGaSe$_2$ thin-films deposited with three-stage evaporation process under different Se-flux (P_{Se}) conditions were performed using admittance spectroscopy (AS) technique. A dominant defect, A2 has been identified around 230~350 meV above the valance band (E_V) of the CuGaSe$_2$. In general, density of defects was found to decrease with an increase in the P_{Se} condition during deposition of the CuGaSe$_2$ layer. Effect of the Se-flux on the performance of the fabricated solar-cells were discussed in relation to the defect-properties of the corresponding CuGaSe$_2$ absorber-layer identified by admittance spectroscopy.**

Index Terms — **CuGaSe$_2$ materials, thin-films, solar-cells, admittance spectroscopy, defect properties.**

I. INTRODUCTION

Chalcopyrite Cu(In$_{1-x}$Ga$_x$)Se$_2$ (CIGS) alloy has got much attention due to its potentials to realize high-efficiency thin-film solar-cells. Moreover, high optical absorption of the CIGS makes it possible to get required efficiency withing thickness limit of thin-films, thus, reduce consumption of the material. Quaternary CIGS is a pseudo-binary alloy of the ternary CuInSe$_2$(x = 0) and the CuGaSe$_2$ (x = 1.0). Band-gap of the CIGS can be controlled by varying the Ga-content, x = [Ga]/([Ga]+[In]) = Ga/III in the material, while it becomes 1.68 eV for CuGaSe$_2$. Band-gap of the CuGaSe$_2$ is close to the ideal band-gap of the absorber-layer to achieve highest possible efficiency with single junction solar cell under AM 1.5 sunlight [1]. Moreover, larger band-gap makes CuGaSe$_2$ suitable for the top-cell in the tandem structure together with CuInSe$_2$ as the bottom-cell. Nevertheless, so far, CuGaSe$_2$ solar cells with a CdS buffer have achieved efficiency of around 11% for thin-film [2]. Therefore, to achieve the efficiency beyond current limit, an extensive study of the CuGaSe$_2$-material including defect-study with various compositions is indispensable. In this study, we have used admittance spectroscopy (AS) to investigate the defect properties of CuGaSe$_2$ thin-film solar-cell structure, where

CuGaSe$_2$ absorber layers were grown with various Se-flux (P_{Se}) conditions. AS has been extensively used to determine and characterize deep-level defects in the semiconductors including activation energy, pre-exponential factor, density, and capture cross section of defects *etc.* [3]. For the AS measurement, we considered the depletion region as a parallel combination of capacitance, C_p and conductance, G_p, thus admittance becomes, $Y = G_p + i\omega C_p$. Since, imaginary part C_p and real part G_p in the admittance term are related explicitly by Kramers- Kronig relations, similar information can be obtained from both C_p and G_p. Here, we used both parameters to investigate the defect properties in CuGaSe$_2$ based solar-cells.

II. EXPERIMENTAL

Polycrystalline CuGaSe$_2$ thin films with the typical thickness of 2 μm were grown over Mo-coated soda lime glass (SLG) substrates through a three-stage co-evaporation process using molecular beam epitaxy system. Evaporation of CuGaSe$_2$ was done from three Knudsen–cells (K-cells) that were the respective Cu, Ga, and Se sources. Beam flux of each material was measured as a beam equivalent pressure (BEP), which is the difference of material flux before and after the opening of the shutter. Beam flux of the Cu and Ga was kept constant for all the samples by keeping the similar K-cell temperature. However, to study the effect of Se-flux over the defect formation in the CuGaSe$_2$ films, several samples were grown under different beam flux of Se, i.e., P_{Se} ranging from 2.36 × 10^{-3} to 3.6 × 10^{-3} Pa. To fabricate solar cell structure, a 50-nm-thick CdS buffer layer was deposited through a chemical bath process followed by the deposition of *i*-ZnO and Al:ZnO transparent conductive oxide (TCO) by sputtering. For the measurement of admittance spectroscopy, sample was mounted on a cold finger inside Janis closed cycle refrigerator (CCR) system. Admittance spectroscopy was carried out in the dark within the temperature range, 10 ~ 350 K using Agilent

978-1-5090-5606-4/17 $31.00 © 2017 IEEE

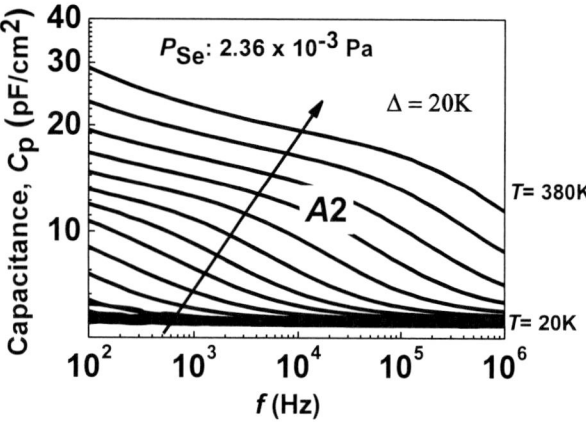

Fig. 1. Temperature dependent junction-capacitance of a CuGaSe₂-based solar-cell measured as a function of the modulation frequency.

4284A LCR meter. Amplitude of the AC modulation voltage was kept as small as 25 mV (rms) to maintain linear response, while modulation frequency was varied from 1 kHz to 1 MHz.

III. RESULTS AND DISCUSSIONS

AS measures thermal emission rates of trapped carriers from a localized defect-states from the variation of the junction-capacitance or conductance in response to a applied modulation voltage as a function of temperature and frequency. The characteristic frequency, ω_0, that is, frequency-normalized conductance or differential-capacitance peak is related to the emission rate (τ_0^{-1}) of trapped carriers according to the equation [4],

$$\omega_0 = 2\tau_0^{-1} = 2N_{C,V}v_{th}\sigma_t \exp(-E_A/kT) \qquad (1)$$

Here, $N_{C,V} \propto T^{3/2}$ is the effective density of the band involved, $v_{th} \propto T^{1/2}$ is the average thermal velocity, σ_t is the capture cross section of trap centers, and E_A denotes activation energy of the defect. Considering the temperature dependences of $N_{C,V}$ and v_{th}, Eq. (1) can be written as

$$\omega_O = 2\tau_0^{-1} = 2\xi_0 T^2 \exp(-E_A/kT), \qquad (2)$$

where, $\xi_0 = (\sigma_t v_{th} N_{C,V})/T^2 \qquad (3)$

is the temperature independent pre-exponential factor. Thus, thermal capture cross section, σ_t calculated from Eq. (3) becomes a temperature-independent single quantity.

Shown in Fig. 1 is the temperature dependent junction capacitance of a CuGaSe₂ solar-cell structure where CuGaSe₂ absorber -layer was deposited with low Se-flux of $P_{Se} \sim 2.36 \times 10^{-3}$ Pa. Capacitance was measured as a function of the

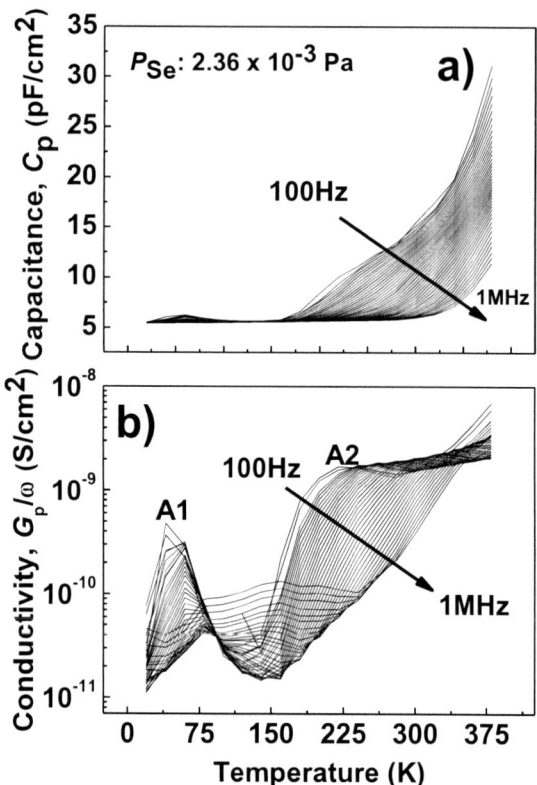

Fig. 2. (a) Temperature dependent steady state junction-capacitance of a CuGaSe₂-based solar cell showing steps low and high temperature region; (b) Frequency-normalized junction conductance showing peaks corresponding to each capacitance step.

modulation frequency obtained at 0 bias voltage. At a temperature, when applied modulation frequency, ω becomes greater than the characteristic frequency, $\omega0$ ($\sim 2e_T$) of the emission rate of the trapped carriers, corresponding junction capacitance, C decreases as carriers from the trap level cannot follow AC voltage, thus making a step at capacitance value in the frequency scale. As temperature is increased further, the steps (*i.e.,* characteristic frequency, $\omega0$) in the capacitance move to the higher frequency, as because emission rate becomes higher at higher temperature. These, steps in the *C-f* spectra along the temperature indicate the electrical response of the thermally activated defects in the CuGaSe₂ material. Admittance spectroscopy shows a prominent hole trap center, *A2* (steps in the *C-f* curve) in all the sample.

To get more insight of the defect response, junction capacitance has been measured in the temperature scale at various frequencies ranging from 100 Hz to 1 MHz. Figure 2 shows AS of the similar CuGaSe₂ solar-cell structure ($P_{Se} \sim 2.36 \times 10^{-3}$ Pa), measured with 0 bias voltage. Steady-state capacitance, C_p at various frequencies exhibited well defined steps (Fig. 2(a)). While, conductance, G_p/ω of the structure showed peaks *A*1, and *A*2 (Fig. 2(b)) corresponding to the each capacitance-step suggesting the electrical response from the trap-levels in the CuGaSe$_2$. Activation energy of the defects were calculated form the Arrhenius plot of the corresponding defect obtained from conductance peaks appeared as a function of temperature and frequency. Activation energy of $A1$ and $A2$ was calculated around 50 and 350 meV above the valance band (E_V) edge of the CuGaS$_2$, respectively. We assigned $A1$ as a shallow acceptor level, while $A2$ is considered as majority carrier trap centers.

To study the effect of P_{Se}-condition on the defect properties of the CuGaSe$_2$ samples, AS of several solar-cell structures were compared where CuGaSe$_2$ absorber layer was deposited under different P_{Se}-condition. As seen from the comparative AS spectra in the Fig. 3, in general, intensity of the defect-peak, $A2$ increases with a decrease in the P_{Se}-condition during growth of the CuGaSe$_2$ films.

Shown in Fig. 4 is the defect density spectrum of a representative sample with $P_{Se} = 2.36 \times 10^{-3}$ Pa. Defect density of each defect has been calculated from the Gaussian fitting of the spectrum which was calculated using the following equation proposed by Walter et al [5].

$$N_t(E_A) = - \frac{2V_{bi}^{3/2}}{wkT\sqrt{q}\sqrt{qV_{bi} - (E_g - E_A)}} \frac{dC}{d\ln\omega} \quad (5)$$

Here, $N_t(E_A)$ is the defect density of the trap with activation energy E_A, while V_{bi}, W, E_g, K and q being the built-in-potential, depletion width, band-gap, Boltzmann constant and elementary charge, respectively.

Capture cross-sections (σ_t) extrapolated to T = ∞; were calculated from Eq. (1), once ξ_0 is known. Then, carrier life-time, τ (*i.e.*, life time before captured by traps) could be obtained as,

$$\tau = 1/(\sigma_t v_{th} N_t), \quad (4)$$

N_t is the defect density. Capture cross section for $A2$ defect in all the samples were found in the range of $10^{-16} \sim 10^{-18}$ cm^2.

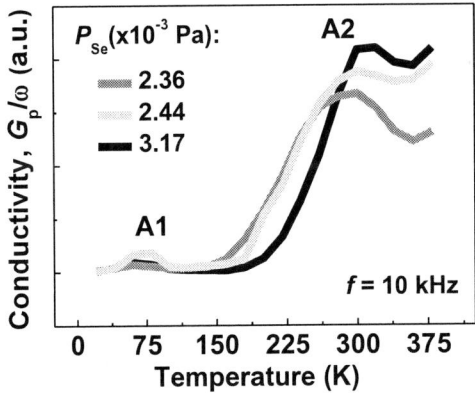

Fig. 3. Admittance spectra of CuGaSe$_2$-based solar cell structures with various P_{Se} condition during CuGaSe$_2$ growth.

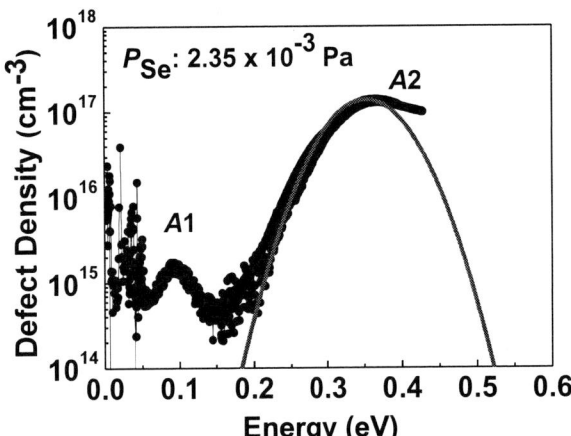

Fig. 4. Defect spectrum of a CuGaSe$_2$ solar-cell obtained using Eq. (5).

Table-1 shows summary of the defect parameters of peak-$A2$. Performance of the solar-cells fabricated with corresponding CuGaSe$_2$ absorber was also shown in the table. From the table it is clear that in general cell-performance becomes better mainly due to increase in the short-circuit current-density (J_{SC}) with an increase in the P_{Se} during growth of the CuGaSe$_2$ films. It indicates material properties get improved with increasing Se-flux, presumably due to decreased defect density with an increase in Se-flux; in consistent to our previous study of defects for CIGS (x~ 0.5) solar cells [6,7]. Moreover, it has been clear from the Fig. 5 that in fact, J_{SC} of the solar-cells get improved with a decrease in the $A2$-defect, although we could not find any clear correlation with the defect density and open circuit voltage (V_{OC}) in the solar-cells.

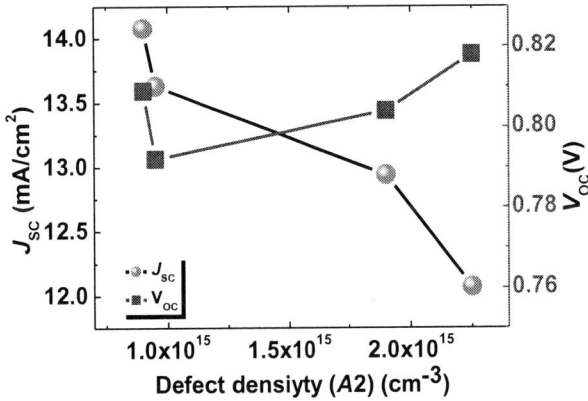

Fig. 5 Performance of CuGaSe$_2$ solar-cells as a function of A2-defect determined by AS.

TABLE I

SUMMARY OF DEFECT PARAMETERS FOR $A2$-DEFECT AS CALCULATED FROM AS DATA

Se temp (^{O}C)	Se-flux ($P_{Se} \times 10^{-3}$ Pa)	E_A (meV)	$N_{t(A2)}$ (cm^{-3})	J_{sc} (mA/cm^2)	V_{oc} (V)	FF	*Eff*
178	*2.36*	*353*	*2.3 $\times 10^{15}$*	*12.1*	*0.82*	*0.53*	*5.2*
182	*2.44*	*231*	*1.9 $\times 10^{15}$*	*12.9*	*0.80*	*0.52*	*5.4*
186	*3.17*	*322*	*9.4 $\times 10^{14}$*	*13.6*	*0.79*	*0.59*	*6.4*
190	*3.58*	*360*	*9.0 $\times 10^{14}$*	*14.1*	*0.81*	*0.59*	*6.7*

Nevertheless, regarding defect-$A2$, smaller capture cross section in the range of $10^{-16} \sim 10^{-18}$ cm^2 accompanied by relatively larger capture life time in the μs order suggests that thermally active defect-$A2$ as probed by AS technique in this study might not so effective to act as trapping or recombination center.

It is worthy to be mentioned that an increasing tendency of conductivity at higher temperature (Fig. 2(b), Fig. 3) above 350°C hints presence of a more deep-level defect in the CuGaSe$_2$ which might not be possible to be probed by the AS techniques. There is an upper limit of the depth (E_A) of the thermally active defect to be detected by AS technique as imposed by the Eq. 2, *i.e.*, this upper limit depends on the measurement temperature and frequency. Thus, optical characterization, *viz.*, transient photocapacitance (TPC) might be helpful to probe deep-level defect beyond the limit by AS. Previously, we reported using two wavelength photoluminescence (PL) technique that 0.8 eV-defect in CIGS as probed by TPC works as a recombination center [8]. More recently, Conrad Spindler et al. [9] also reported about 600 meV defect in epitaxial-grown CuGaSe$_2$ using PL which might be powerful candidate as a recombination center.

IV. CONCLUSION

Defect properties in the CuGaSe$_2$ solar cells where studied by AS technique, where CuGaSe$_2$ absorber layers were evaporated with various P_{Se}-conditions. A bulk type defect, $A2$ has been identified in all the samples within $230 \sim 350$ meV above E_v of the CuGaSe$_2$ absorber layer. In general, CuGaSe$_2$ samples grown with higher P_{Se}-condition show lower defect density (peak- $A2$), and improved solar cell performances. Relatively smaller capture cross section and larger capture life time of the defects suggest that they are not so effective as trap center or recombination center in the CuGaSe$_2$ absorber layers.

ACKNOWLEDGEMENT

This work is supported by the New Energy and Industrial Technology Development organization (NEDO) and the Ministry of Economy, Trade and Industry (METI), Japan.

REFERENCES

[1] W. Shockley, H. Queisser, "Detailed Balance Limit of Efficiency of p-n Junction Solar Cells," *Journal of Applied Physics*, Vol. 32, PP. 510-519,1961.

[2] S. Ishizuka, A. Yamada, P. Fons, H. Shibata, S. Niki, "Impact of a binary Ga$_2$Se$_3$ precursor on ternary CuGaSe$_2$ thin-film and solar cell device properties," *Applied Physics Letter*, Vol. 103, PP.143903 (1-5), 2013.

[3] M. M. Islam et al. "Identification of defect types in moderately Si-doped GaInNAsSb layer in p-GaAs/n- GaInNAsSb/n-GaAs solar cell structure using admittance spectroscopy", *Journal of Applied Physics*, Vol. 112, pp 114910 (1-9), 2012.

[4] D. L. Losee, "Admittance spectroscopy of impurity levels in Schottky barriers", *Journal of Applied Physics*, Vol. 46, pp 2204-2214, 1975.

[5] T. Walter al. "Determination of defect distributions from admittance measurements and application to Cu(In,Ga)Se$_2$ based heterojunctions", *Journal of Applied Physics*, Vol. 80, pp 4411-4420, 1996.

[6] T. Sakurai, M. M. Islam, S. Ishizuka, A. Yamada, H. Shibata, K. Sakurai, K. Matsubara, S. Niki, K. Akimoto, Dependence of Se beam pressure on defect states in CIGS solar cells, *Solar Energy Materials & Solar Cells*, Vol. 95, pp. 227-230, 2011.

[7] M. M. Islam et al. "Impact of Se flux on the defect formation in polycrystalline Cu(In,Ga)Se$_2$ thin films grown by three stage evaporation process", *Journal of Applied Physics*, Vol. 113, pp 064907(1-7), 2013.

[8] A.Gupta, N.Hiraoka, T.Sakurai, A.Yamada, S.Ishizuka, S.Niki, and K.Akimoto,"Characterization of Cu(In,Ga)Se$_2$ grown by MBE by two-wavelength excited photoluminescence spectroscopy", *Journal of Crystal Growth*, Vol. 378, pp. 162-164, 2013.

[9] C. Spindler, D. Regesch, and S. Siebentritt, "Revisiting radiative deep-level transitions in CuGaSe$_2$ by photoluminescence", *Applied Physics Letters*, Vol. pp. 032105 (1-4), 2016.

Transmissive Spectrum-Splitting Concentrator Photovoltaic Cells and Modules

Yaping Ji[1], Qi Xu[1,2], Brian Riggs[1], John Robertson[1], Kazi Islam[1], Vince Romanin[3], Dimitri D. Krut[4], Jim H. Ermer[4], and Matthew D. Escarra[1]

[1]Tulane University, New Orleans, LA 70118, USA; [2]Stion Corp., San Jose, CA 95119, USA; [3]Otherlab, 3101 20th Street, San Francisco, CA 94110, USA; [4]Boeing-Spectrolab, Inc., Sylmar, CA 91342, USA

Abstract — Novel transmissive concentrator multijunction (TCMJ) solar cells, and their corresponding module, are here introduced and demonstrated as a promising component of a hybrid concentrator photovoltaic/concentrated solar thermal power (CPV/CSP) system. Using this system, incoming concentrated light can be effectively split into two parts; "in-band", shorter-wavelength light is absorbed into the TCMJ solar cells to generate electricity, and "out-of-band" light transmits through the cells and module to a thermal receiver where photons are collected as heat that may be used for a variety of purpose. When on-substrate cells are employed, the in-band power conversion efficiencies measured under 1 sun at room temperature are 34.7% and 35.8% and the energy-weighted out-of-band light transmission is 68.6% and 60.8% for the TCMJ cell and module, respectively. We also demonstrate the stability of TCCMJ cells and modules via thermal cycling and show that the temperature of the TCMJ module can be controlled under 105C via transparent active cooling microfluidic channels. Thin-film cells fabricated by means of the epitaxial lift-off (ELO) technique are being developed to enhance both the spectrum splitting efficiency and cost effectiveness of this TCMJ module.

Index Terms — III-V materials, concentrated solar cells, transmissive CPV module

I. INTRODUCTION

A hybrid concentrator photovoltaic (CPV) /concentrated solar power (CSP) system, taking advantage of both CPV and CSP system components, can generate electrical and thermal energy simultaneously. The CPV cells directly convert sunlight to electrical energy, while the heat energy is captured and stored in a thermal receiver for future use. Generally, CPV is more efficient at converting higher energy parts of the solar spectrum, while CSP systems use the entire solar spectrum with relatively uniform efficiency, providing an opportunity for a hybrid CPV/CSP systems to split the solar spectrum into two different parts, one of which is directed to solar cells where high power conversion efficiency (PCE) is achieved, while the other spectral segment is diverted to a thermal receiver. Most conventional spectrum splitting techniques, such as heat reflectors [1], holographic filters [2], and liquid absorption filters [3], require extra apparatus and complicated optical design. Alternatively, transmissive concentrator multijunction (TCMJ) solar cells (See Figure 1(a)) and modules provides a novel way to act as beam splitter, effectively dividing incident solar radiation into two parts. Photons with energy above the lowest bandgap of the TCMJ cells are absorbed and converted to electricity, while lower energy photons pass through to the

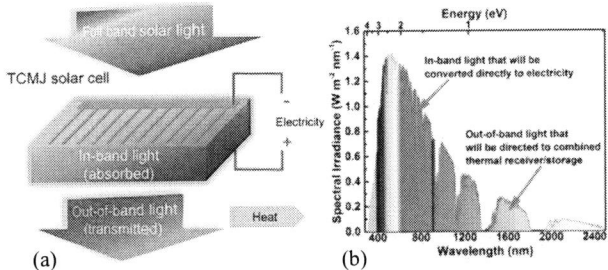

(a)   (b)

Fig. 1. (a) TCMJ solar cells acts as a beam splitter –high energy photons are absorbed in the cell, while low energy photos are passing through [4]. (b) The ASTGM173-03 reference spectra derived from SMARTS v.2.9.2, with the division of the spectrum between in-band electrical conversion and out-of-band thermal energy capture noted.

thermal receive and are is captured and stored as thermal energy. Here we refer to those two light spectra as "in-band" and "out-of-band" light (see Figure 1(b)). In this paper, PCE is used to evaluate the quality of in-band light conversion, and out-of-band light transmission is similarly used to evaluate the effectiveness of the spectrum-splitting technology. The electrical and optical performance of the TCMJ cells and modules are studied independently. Since the CPV/CSP system presented here needs to maintain reasonably low cell temperature, the thermal performance is also investigated.

II. TCMJ SOLAR CELLS

The TCMJ solar cells are epitaxially grown by metal-organic vapor phase epitaxy (MOVPE) and fabricated using standard III-V solar cell fabrication techniques [5]. They are designed with three bandgaps at 1.410eV or higher: GaAs at 1.410eV for the bottom cell (C1), $Al_{0.18}Ga_{0.82}As$ at 1.675eV for the middle cell (C2), and $Al_{0.23}Ga_{0.26}In_{0.51}P$ at 2.098eV for the top cell (C3) (see Fig. 2(a)). The polarity reversing (PR) tunnel junction below the bottom cell enables the use of an n-type GaAs growth substrate with doping concentration around $1E17cm^{-3}$, which is chosen for epitaxial growth due to its lower free carrier absorption [6] as well as low resistivity semiconductor-metal contacts obtained [7], which doesn't harm the electrical power output much and increases the out-of-band light transmission of TCMJ solar cells. A unique characteristic of these TCMJ solar cells is the rear contact design. It is composed of a busbar and sparse contact grid, which enables the out-of-band light to transmit through the cells. Two anti-reflection coatings are fabricated on the front and back side of the cells to decrease the light reflection at both interfaces.

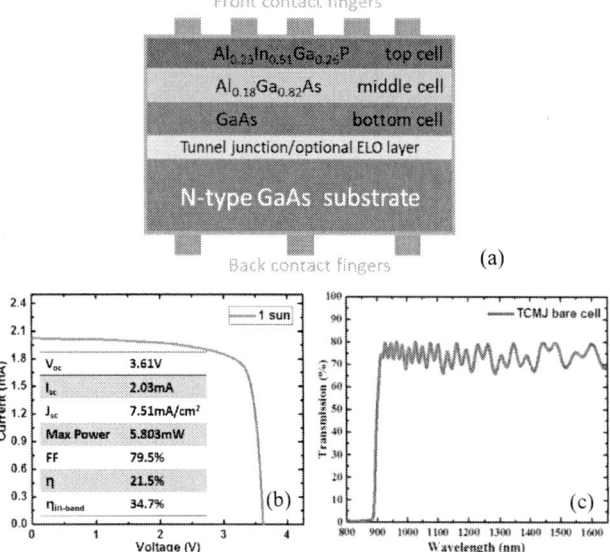

Fig. 2 (a) Structure of the TCMJ solar cells. (b) Measured illuminated I-V characteristics for the TCMJ solar cells under 1 sun AM1.5D spectrum at room temperature [3]. (c) Measured transmission through a TCMJ solar cell using a broadband (185nm to 1700nm) spectrometer (Ocean Optics).

The electrical performance is measured under 1 sun at room temperature (~25°C) using a solar simulator (Unisim, TS-Space Systems), with a total intensity of 0.09W/cm². A broadband UV/Vis/NIR spectrometer is also used to characterize the out-of-band light transmission of the TCMJ solar cells. The illuminated I-V curves derived from the TCMJ solar cell and the out-of-band light transmission are shown in Fig. 2 (b) and (c). The PCE is 21.5%, which corresponds to 34.7% in-band efficiency with a bandgap cutoff of 1.41eV. The transmission is almost zero at wavelengths below 880nm, and then dramatically increases to around 75%. The interference fringe pattern is due to slight differences in index for III-V cell epi-layers. The energy-weighted transmission efficiency of out-of-band light (from 874nm to 1650nm) is calculated to be 68.6%. This value is promising for TCMJ solar cells to act as beam splitters.

III. TCMJ SOLAR MODULE

A module was designed to protect and cool the TCMJ cells while still permitting high out-of-band transmission, as shown in Fig. 3(a). The module consists of eight layers in total: two pieces of glass as top and bottom cover material, two layers of encapsulant to provide adhesion between cover layers and cells, a piece of sapphire as a thermal conduction layer, water to remove the waste heat, and a PDMS layer containing a cooling water microchannel pattern. A previous study [8] showed the

incident power distribution from a parabolic dish on the TCMJ solar module is a Gaussian distribution. Considering the non-uniform light intensity distribution, all cells are connected in parallel instead of in series, taking advantage of low variability in voltage in order to reduce electrical mismatch losses. The TCMJ solar module contains 49 cells, composed of 7 columns of 7 cells (See Fig. 3(b) and (c)). All cells in a column are wired to be electrically in parallel.

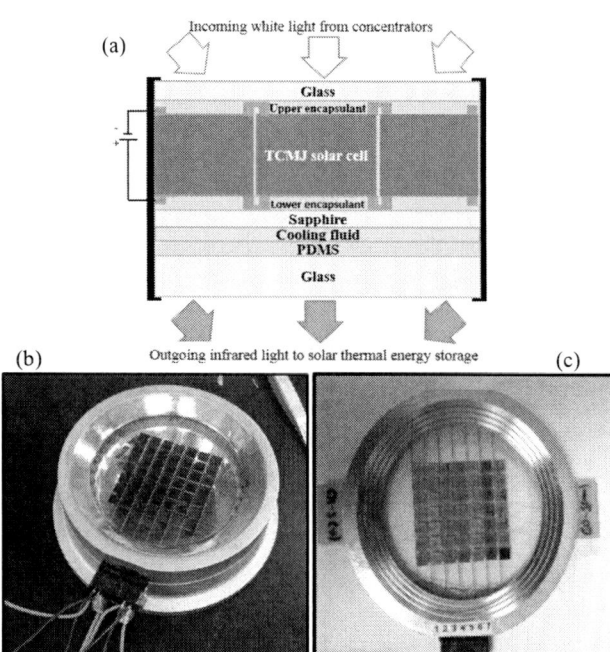

Fig. 3. (a) The TCMJ solar module consists of eight layers: glass, upper encapsulant, a TCMJ solar cell, lower encapsulant, sapphire, cooling fluid, and glass. (Dimensions not to scale) (b) Full TCMJ solar module assembly with water cooling. (c) Forward bias test on the solar module. Cell in bottom right is dimmer than the rest, but still works.

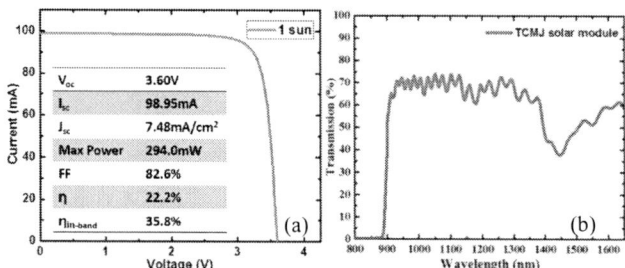

Fig.4. (a) Measured illuminated I-V characteristics for the TCMJ solar modules under 1 sun AM1.5D spectrum at room temperature (b) Measured transmission through a TCMJ solar module using a broadband (185nm to 1700nm) spectrometer.

The TCMJ module was tested under 1 sun to measure its I-V curve as shown in Fig. 4(a). The open circuit voltage of the 7×7 module is nearly the same as an individual cell. While the short circuit current of the module is 48.7 times larger than that of a single cell, a valid number considering the module contains 49

978-1-5090-5606-4/17 $31.00 © 2017 IEEE   38

cells. The total efficiency of the module is 22.2%, showing a better performance than individual cells; that can be explained by a higher fill factor in the module, and less reflection in the interface between upper encapsulant and the solar cell. The front side anti-reflection coating of the solar cell is specially designed assuming light comes from encapsulant, not air. Fig. 4(b) shows the out-of-band light transmission spectrum through the TCMJ solar module. There are dips near 1200nm and 1450nm. The first one is attribute to C-H absorption in the encapsulant. The second dip is due to both encapsulant and water absorption. The energy-weighted out-of-band light transmission is calculated to be 60.8% for the module.

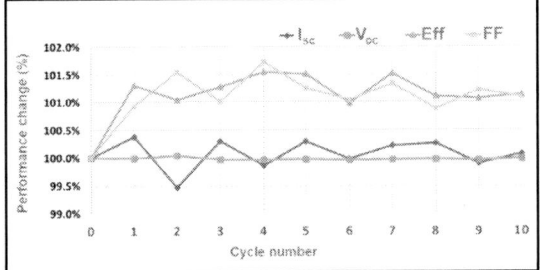

Fig. 5 Electrical performance under repeated thermal cycling tests, ramping the temperature of 3-cell mini module up to 150°C and back down to room temperature on each cycle.

The TCMJ solar cells are required to cycle from low night-time temperatures to temperatures as high as 105°C under concentrated light during the day. A mini-module was built with a similar structure as the full module except consisting of only three TCMJ cells. The thermal cycling stability test process is as follows, the mini-module is heated up to 150°C on a hotplate, held at 150°C for 10 minutes, and then cooled naturally until it reaches room temperature. The hot plate was heated to 150°C so that the cell temperature reaches at least 105°C. After each cycle of heating and cooling, the electrical performance of the TCMJ mini-module was measured until 10 cycles were finished. As can be seen (Fig. 5), after the first thermal cycle, the change of open circuit voltage and short circuit current are negligible while PCE and fill factor show a slight increase of about 1 to 1.5% compared to the initial state of the TCMJ cells. Cell performance is stable with successive cycles.

Our TCMJ module uses active cooling to maintain a reasonable temperature range for cells in order to maximize performance and minimize degradation of materials. A coolant is pumped through microfluidic channels on the backside of the module, taking the waste heat away from the bottom of the CPV cells (see Figure 6(a)). Water is chosen as the coolant due to its low viscosity, high optical transmission, and environmental friendliness. To investigate the cooling performance, the temperature distribution along the cooling flow direction is calculated using a 1-D model with water velocity of 2.61m/s. In this model, only four layers are considered: the TCMJ cells, encapsulant, sapphire, and water with thicknesses of 450μm, 20

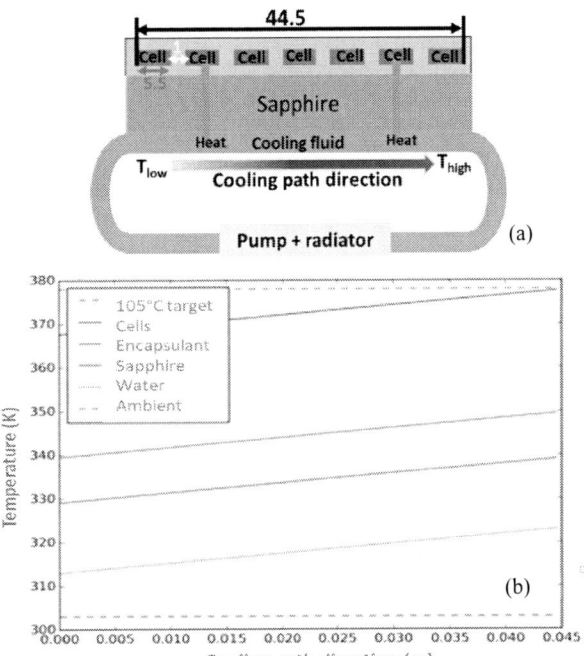

Fig. 6. (a) The active cooling diagram (Unit: mm). The water flows under the solar cell, encapsulant, and sapphire layer to take away the waste heat from the TCMJ cells. The widths of cell and the gap are 5.5mm and 1mm, respectively. The total distance between the cells at two edges is 44.5mm. (b) the temperature distribution in each layer along the cooling path direction.

μm, 1mm, and 100μm, respectively. The size of individual TCMJ cells is 5.5mm × 5.5mm and the 1mm-wide gap between two cells is filled with encapsulant. As seen in Fig. 6(b), the temperature of each layer is rising along the cooling path direction, or the direction of water flow, as the temperature gradually rises. This 1D model is corroborated by a finite element method 3D model [8], a thermal testbed [9], and preliminary full module testing up to 50 suns, indicating that the cells' temperature can be well controlled under our target temperature (105°C). The encapsulant makes up >50% of the total temperature drop between the cell and the cooling water, which could be reduced by decreasing the encapsulant layer's thickness or embedding sapphire microparticles in the encapsulant layer.

IV. TCMJ ELO CELLS

The epitaxial lift-off (ELO) technique, in which thin-film III-V solar cells are separated from a GaAs substrate using HF acid etchant and an AlGaAs release layer, was first described by Konagai et al. in 1978 [10]. The technique results in a potentially significant cost reduction of III-V devices due to the ability to re-use GaAs substrates, allowing for the repeated growth of thin film solar cells on the same wafer after the lift-off process. By removing the GaAs substrate, the ELO process

978-1-5090-5606-4/17 $31.00 © 2017 IEEE

helps enhance the transmission of our TCMJ solar cells by removing parasitic absorption near the bandgap and the free carrier absorption at longer wavelengths in GaAs. According to our transmission model, the out-of-band light transmission of a TCMJ ELO cell will be around 6% (absolute) more than that of an on-substrate cell. We are working to first demonstrate the ELO process in the fabrication of single junction transmissive thin-film GaAs cells. Additional junctions are not expected to significantly affect this process.

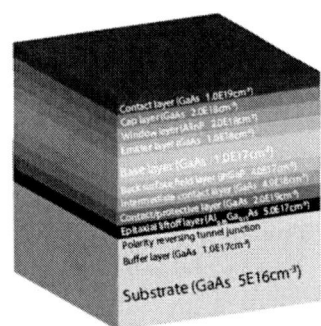

Fig. 7. Single junction GaAs solar cell structure for ELO process.

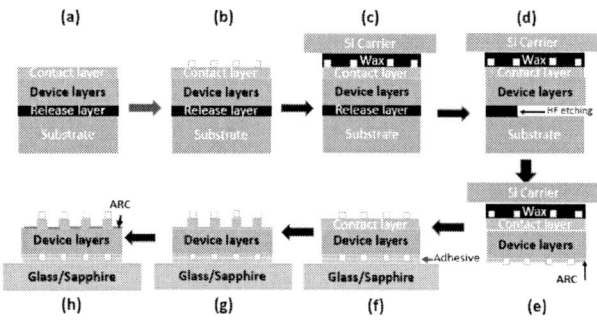

Fig. 8. The transplantation process of the ELO thin film solar cell from GaAs substrate to a glass/sapphire substrate. (a) As grown solar cell device. (b) N-type contact fabrication. (c) Mounting to a Si carrier. (d) Liftoff etching process. (e) P-type contact fabrication and anti-reflection deposition. (f) Si carrier removal. (g) Contact layer etching. (h) Antireflection coating deposition.

The overall ELO device structure is designed as an n-on-p cell, as shown in Fig.7. A buffer layer was grown before the solar cell structure to preserve the original GaAs wafer surface in its epi-ready condition, thereby providing a high-quality regrowth interface without polishing. An $Al_{0.85}Ga_{0.15}As$ layer with a thickness of 10nm was employed as a release layer. A contact/protective layer with high doping concentration is employed for two purposes: helping the formation of ohmic p-type backside contacts and preventing the etching into the following layers. The next two layer are in the intermediate contact layer and the back surface field layer, continuing from bottom to top. The n-type emitter layer is heavily doped and 200nm thick, whereas the lightly doped base layer is over an order of magnitude thicker at 3.5μm. A higher bandgap AlInP layer acts as a window layer. The cap layer and n-type contact layer both use GaAs to minimize resistivity in the top contact.

The fabrication and transplantation process for the single junction thin-film GaAs from the GaAs substrate to the glass/sapphire module substrate is schematically shown in Fig.8. After growth, the n-side metal contact of Pd/Ge/Au (10/20/200nm) is deposited, patterned, and annealed at 450°C for 30 seconds. Then a Si carrier is attached on top of the sample to help supply tension to lift the thin film structure. After putting the cell in a mixture of HF and DI-water (1:1) for 24 hours, the epi-cell is separated from its substrate. After the liftoff process (Fig. 7d) the back side of the cell structure is accessible for the deposition and patterning of the p-side contact, which requires annealing at a relatively low temperature (300~350°C). The rear contact has a grid pattern similar to the pattern of the front-side contact. Following metallization, an antireflection coating composed of two layers (TiO_2 and Al_2O_3, 140nm and 220nm respectively) is deposited. The thin film cell is then mounted on a glass or sapphire substrate using a commercially available silicone-based adhesive (Dow Corning Sylgard 184 silicone), followed by the removal of the Si carrier by immersion in trichloroethylene (TCE). The thin film cell is soaked in the mixed solution of $NH_4OH:H_2O_2:H_2O$ to etch away the contact layer. Finally, a second anti-reflection coatings consisting of TiO_2/Al_2O_3 (58nm/100nm) is deposited by e-beam evaporation. We are currently working to experimentally demonstrate these transmissive spectrum-splitting thin-film GaAs cells and an associated transmissive module with improved performance.

V. SUMMARY

We have presented a novel beam splitting technology for hybrid solar electricity and thermal energy conversion utilizing, transmissive concentrator multijunction solar cells, and a corresponding module. TCMJ cells and modules effectively split the incoming concentrated light into in-band and out-of-band light, where the former part is converted directly to electricity and the latter part is captured as thermal energy for a variety of application. The in-band power conversion efficiencies measured under 1 sun at room temperature are 34.7% and 35.8% and the energy-weighted out-of-band light transmission is 68.6% and 60.8% for the TCMJ cell and module, respectively. Thermal cycling teste of the TCMJ module shows a stable electrical output, and the temperature of the TCMJ module can be effectively controlled below 105°C by active cooling in microfluidic channels. ELO technology, in which the thin film cell structure is "peeled off" from the substrate, will increase the TCMJ solar cells and modules' out-of-band light transmission by nearly 6% (absolute) and reduce system cost.

This information, data, or work presented herein was funded in part by the Advanced Research Projects Agency-Energy (ARPA-E), U.S. Department of Energy, under Award Number DE-AR0000473.

REFERENCES

[1] D. E. Soule, S. E. Woods, *Proceedings of the SPIE,* Vol. 653, 1986, pp. 172-180

[2] M. Escarra, S. Darbe, E.C. Warmann, H. A. Atwater, *Proceedings of the 40th IEEE Photovoltaic Specialists Conference* (2013)

[3] M. Sabry, et al, *Proceedings of the 19th IEEE PVSC*, New Orleans, LA, 2002, pp. 1588-1591

[4] Q. Xu, Y. Ji, D. Krut, J. Ermer, and M. Escarra, *Appl. Phys. Lett.* 109, 193905 (2016)

[5] J. F. Geisz, D. J. Friedman, J. S. Ward, A. Duda, W. J. Olavarria, T. E. Moriarty, J. T. Kiehl, M. J. Romero, A. G. Norman, and K. M. Jones, *Appl. Phys. Lett.* 93, 123505 (2008).

[6] H. Casey Jr, D. Sell, and K. Wecht, *J. Appl. Phys.* 46, 250-257 (1975).

[7] J. Kwak, H. Baik, J. Lee, C. Park, H. Kim, and K. Suh, *Thin Solid Films 290*, 497-502 (1996).

[8] Q. Xu, Y. Ji, B. Riggs, A. Ollanik, N. Farrar-Foley, J. Ermer, V. Romanin, P. Lynn, D. Codd, and M. Escarra, *Solar Energy* 137 (2016) 585-593

[9] B. Riggs, N. Farrar-Foley, S. Deckoff-Jones, Q. Xu, V. Romanin, D. Codd, and M. Escarra, *Proceedings of the 43rd IEEE Photovoltaic Specialists Conference* (2016)

[10] M. Konagai, M. Sugimoto, K. Takahashi, *Journal of Crystal Growth*, Volume 45,277-280(1978)

AlGaInP/GaAs Tandem Solar Cells For Power Conversion At 400°C And 1000X Concentration

Myles A. Steiner[1], Emmett E. Perl[1], John Simon[1], Daniel J. Friedman[1], Nikhil Jain[1], Paul Sharps[2], Claiborne McPheeters[2], Minjoo L. Lee[3]

[1]National Renewable Energy Laboratory, Golden, Colorado, United States
[2]SolAero Technologies Corp., Albuquerque, New Mexico, United States
[3]University of Illinois at Urbana-Champaign, Urbana, Illinois, United States

Abstract — We demonstrate dual junction (Al)GaInP/GaAs solar cells that are designed to operate at 400°C and 1000X concentration. The cells have stable front metallization and anti-reflection coatings at 400°C. The choice of bandgaps was first estimated using ideal solar cell characteristics and then refined based on empirical data. We show power conversion efficiency of ~15±1% for AR-coated cells at 400°C and high concentration, with pathways to improved performance.

Index Terms — multijunction; III-V solar cell; high temperature.

I. INTRODUCTION

The efficiency of a solar cell decreases as the temperature rises, due to the loss in voltage, and most solar cell systems are therefore designed to operate as close to room temperature as possible. Some applications, however, may *require* high temperature operation. Consider a hybrid system combining concentrator photovoltaics (CPV) and concentrated solar thermal power (CSP), which would in principle allow for direct electricity generation along with storage [1]. In such a system, the CPV portion would convert the short wavelength light directly to electricity, while the CSP portion would capture longer wavelength light (and any heat dissipated by the CPV cells) in a thermal fluid, for power generation after sunset. Since the CSP portion of the system operates most efficiently at high temperature, one implementation of a combined system would have the photovoltaics also operating at high temperature [1]. If properly designed and integrated, a hybrid system could lead to higher power production and lower levelized cost of electricity (LCOE) compared to either individual component.

We have designed and developed dual junction solar cells for part of such a hybrid system. These cells could also find application in, for example, satellite missions toward the sun and inner planets where the operating temperature is at times well above 25°C [2]. Here we report on (Al)GaInP/GaAs tandem cells that are designed to operate at 400°C and ~1000X, an operating condition that is representative of a potential hybrid system.

II. MODEL

The initial approach to modeling the optimum bandgap combination was to use ideal QEs where every absorbed photon is converted to photocurrent, and room-temperature values of 400 and 370 mV for W_{oc} ($=E_g/q - V_{oc}$) for the top and bottom junctions respectively. The efficiency contours were then calculated for the tandem, at 400°C and 1000X, as a function of the two room-temperature bandgaps. Those contours are shown as the black dashes in Fig. 1. The optimum combination of ~1.9/1.3 eV (point A in Fig. 1) can be realized with $Ga_{0.5}In_{0.5}P/Ga_{0.9}In_{0.1}As$, where the GaInAs lower junction is lattice-mismatched with respect to a GaAs substrate. With only slightly lower projected efficiency, however, a lattice-matched $Al_xGa_{0.5-x}In_{0.5}P/GaAs$ combination of 2.0/1.42 eV (point B) was deemed simpler and thus a more suitable first design target. AlGaInP solar cells were developed and demonstrated with a range of Al compositions *x* up to 24% [3] as can be seen in the inset of Fig. 2, with a focus on the 12% Al composition that corresponded to a bandgap of ~2.0 eV. Those cells were used to determine an empirical relationship for W_{oc} as a function of bandgap, shown in Fig. 2, and together with the measured external quantum efficiencies (EQEs) were then used to re-compute the efficiency contours, as shown in blue in Fig. 1. The realistically achievable efficiencies are lower than for the idealized cells. The optimum bandgap shifts to a lower Al top cell (point C) and the desired lattice-matched solution has a top cell with lower Al content of ~6% (point D). The GaInP/GaAs tandem is shown by point X, which represents

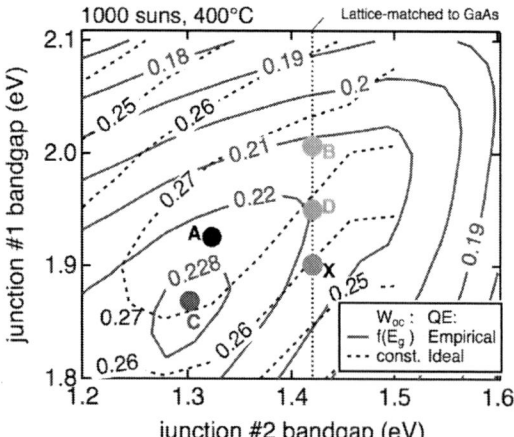

Figure 1 Efficiency contours, at 1000X and 400°C, of a two junction solar cell. The modeling assumptions are shown in the legend. Points A, B, C, D and X are described in the text.

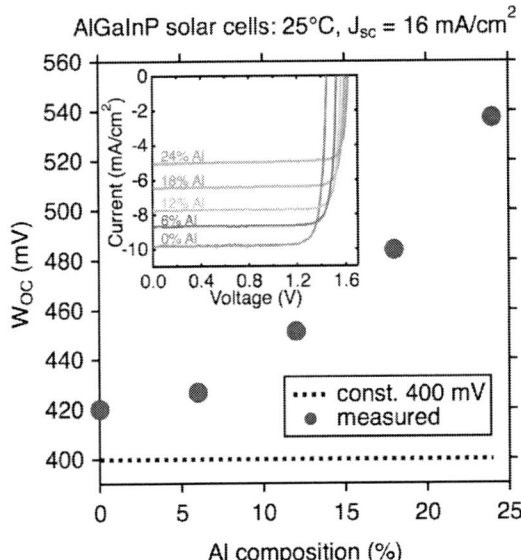

Figure 2 Bandgap-voltage offset W_{oc} ($=E_g/q-V_{oc}$) for AlGaInP solar cells, evaluated at 16 mA/cm^2 and 25°C. The inset shows IV curves at 1-sun, 25°C, with different Al compositions [3].

just a slight theoretical drop from point D, but eliminates the use of Al in the top cell absorber.

For operation in the intended high-concentration application, minimizing series resistance is also critical. We find that the lower Al content cells provide much lower series resistance in lateral current transport in the emitter layer, attributed to the boost in mobility. Because of both the lower series resistance and the lower W_{oc}s, the lower Al-content cells provide our highest demonstrated efficiencies at present.

III. EXPERIMENT

Cells were fabricated by atmospheric pressure metalorganic vapor phase epitaxy (MOVPE) and processed into devices using standard cleanroom photolithography and wet chemical etching techniques. Details can be found elsewhere [3,4].

Several front metallization schemes were tested for stability at 400°C. As shown in Fig. 3, electroplated gold clearly degrades at high temperatures, as evidenced by the increased room-temperature dark IV curves after 30 minutes of annealing at progressively higher temperatures. Alloyed stacks of Ti/WTi/Ag/Ti and Ti/Pt/Al/Pt, both fabricated by electron beam deposition, are more stable high-temperature metallizations with very little evidence of degradation in similar experiments. Electroplated gold was used as the back contact to the substrate for all cells, since the thick substrate separates the metal from the active semiconductor layers by >300 μm.

We also tested three different anti-reflection coatings for stability at 400°C. Figure 4 shows the room-temperature EQE of AR-coated (Al)GaInP top cells before and after heating to 400°C. Bilayer coatings of ZnS/MgF$_2$ and Ta$_2$O$_5$/MgF$_2$ both

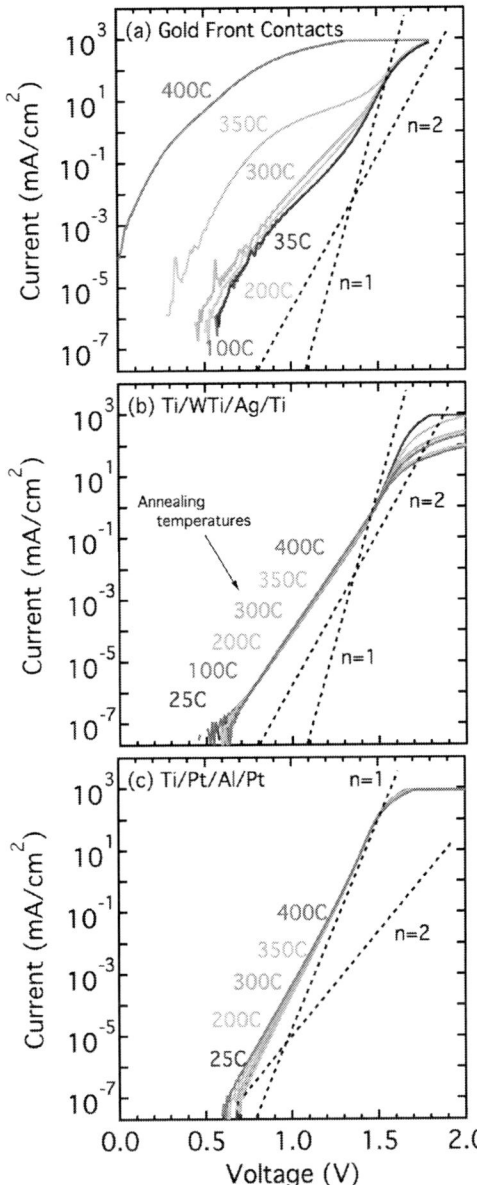

Figure 3 Room-temperature dark IV curves for AlGaInP cells with different front metalizations, after annealing at successively higher temperatures: (a) gold, (b) Ti/WTi/Ag/Ti and (c) Ti/Pt/Al/Pt. All cells have gold back contacts on the back side of the substrate. The cells were heated to the indicated temperature (100°C, 200°C...etc.) for 30 minutes and then cooled back to room temperature and measured, then heated again to the next temperature. The dashed lines are guides to ideality factors of 1 and 2.

show significant degradation, whereas TiO$_2$/Al$_2$O$_3$ shows good stability.

Figure 5a shows the measured EQE for a GaInP/GaAs tandem cell at temperatures up to 400°C. Because of the high saturation dark current, the GaAs subcell EQE could not be measured above 300°C, but previous measurements of

978-1-5090-5606-4/17 $31.00 © 2017 IEEE

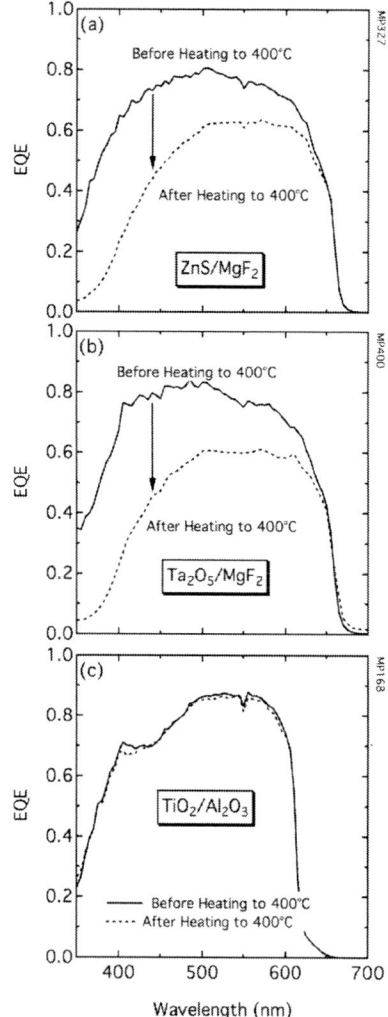

Figure 4 EQE of an (Al)GaInP top cell with different anti-reflection coatings: (a) ZnS/MgF$_2$; (b) Ta$_2$O$_5$/MgF$_2$; (c) TiO$_2$/Al$_2$O$_3$. The EQEs were measured at 25°C before and after heating to 400°C.

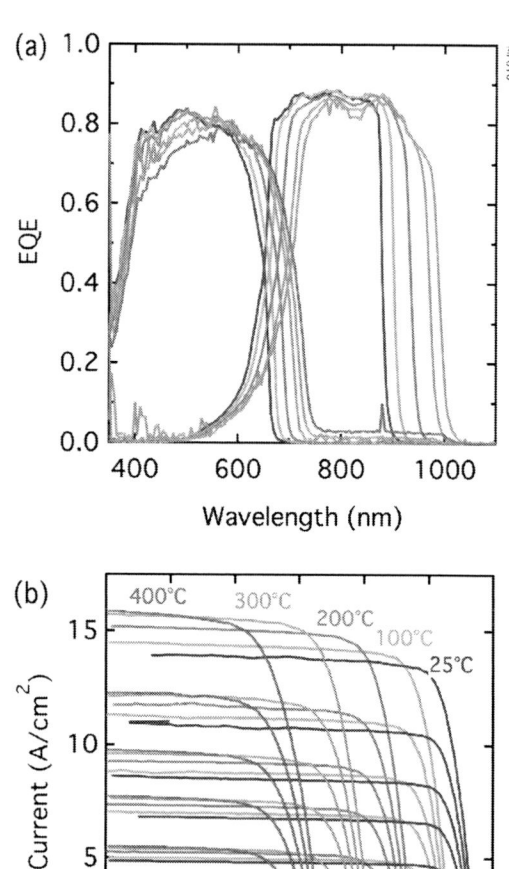

Figure 5 Characteristics for an AR-coated GaInP/GaAs solar cell. (a) EQE measured at temperatures from 25-400°C. The EQE of the bottom cell could not be measured at 400°C due to the high saturation dark current. (b) IV curves taken on the High Intensity Pulsed Solar Simulator (HIPSS) at the indicated temperatures. The spectrum was not re-adjusted at each temperature, so some error may result.

filtered, single-junction cells indicate that there will be no measurable decrease in the J$_{SC}$ of the GaAs subcells as the temperature is increased to 400°C [4]. The cell was measured using a temperature-controlled Linkam Instruments stage capable of heating up to 600°C [4]. Though the J$_{sc}$ in the tandem is significantly top-limited at room temperature, the shifts in bandgap with temperature lead to a more current-matched device at 400°C. Current-voltage curves taken on a High Intensity Pulsed Solar Simulator (HIPSS) under concentration are shown in Fig. 5b. The cells are stable and the data are reproducible at temperatures up to 400°C, and the tunnel junction is clearly able to handle the current at 400°C. The drop in V$_{oc}$ with temperature, which dominates the efficiency loss, is also apparent.

Based on these IV curves, the efficiencies as a function of concentration and temperature are shown in Fig. 6. At 400°C, the efficiency reaches 15±1% over the range of 300-1000 suns. At a lower temperature of 300°C, the efficiency reaches 20±1% at 730 suns. The one-sun characteristics were measured separately, on a class A adjustable solar simulator, and used to determine the concentration ratios under the assumption of linearity. Since the HIPSS spectrum was not carefully adjusted at each temperature, some systematic error is expected in these un-certified data but is accounted for by the relatively large error bars.

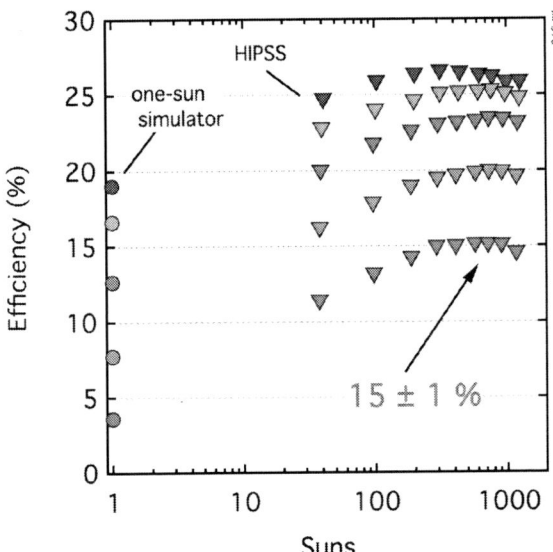

Figure 6 Efficiency of an AR-coated GaInP/GaAs tandem solar cell at high concentration and temperature. Data were taken on the HIPSS and are based on Figure 3b. The one-sun IV characteristics were measured separately on an adjustable solar simulator and used to determine the concentration, based on the assumption of linearity.

IV. DISCUSSION

Comparing the data in Fig. 6 to the model in Fig. 1, there is clearly headroom to improve the performance of these cells. Though the fill factor (FF) has an upper limit of only ~76% at 1000X and 400°C, based on the diode equation in the absence of series resistance [5], the cell in Figs. 5 and 6 is limited by series resistance that further reduces the fill factor to ~65%. Some of this FF loss can be mitigated by a denser grid with tall, narrow fingers. Narrow grids will also reduce shadow losses, which are largely responsible for low EQE plateaus in Fig. 5a, especially in the GaAs cell. The cumulative effects of repeated thermal cycling over the course of the cell processing and suite of measurements have yet to be fully assessed, and if necessary, mitigated. Along with improvements to the top cell that will enable better current-matching and a higher overall J_{sc}, and the development of a more transparent tunnel junction, we expect to demonstrate an efficiency >17% at 400°C and 1000X concentration.

V. SUMMARY

(Al)GaInP/GaAs tandem cells have been designed and demonstrated at temperatures up to 400°C and concentrations up to 1000X. We have developed anti-reflection coatings and metallizations for the front grids that are stable at annealing temperatures as high as 400°C. The best cell to date demonstrates a peak efficiency of 15±1% over the range of 300-1000 suns, with several pathways to improved performance.

ACKNOWLEDGEMENTS

The authors are pleased to thank W. Olavarria, M. Young and C. Beall at NREL for processing work, and John Geisz for useful conversations. This work was supported by the the U.S. Department of Energy through the ARPA-E FOCUS program under Award DE-AR0000508, and through contract DE-AC36-08GO28308 with the National Renewable Energy Laboratory. The U.S. Government retains, and the publisher by accepting the article for publication acknowledges that the U.S. Government retains, a nonexclusive, paid up, irrevocable, worldwide license to publish or reproduce the published form of this work, or allow others to do so, for U.S. Government purposes.

REFERENCES

[1] H. M. Branz, W. Regan, K. J. Gerst, J. B. Borak, and E. A. Santori, "Hybrid solar converters for maximum exergy and inexpensive dispatchable electricity," *Energy Environ. Sci.,* vol. 8, p. 3083, 2015.

[2] G. A. Landis, D. Merritt, R. Raffaelle, and D. Scheiman, "High-temperature solar cell development," in *Proc. 18th Space Photovoltaic Res. Technol. Conf.,* 2005, pp. 241-247.

[3] E. E. Perl, J. Simon, J. F. Geisz, W. Olavarria, M. Young, A. Duda, *et al.,* "Development of high-bandgap AlGaInP Solar Cells Grown by organometallic vapor-phase epitaxy," *J. Photovoltaics,* vol. 6, p. 770, 2016.

[4] E. E. Perl, J. Simon, J. F. Geisz, M. L. Lee, D. J. Friedman, and M. A. Steiner, "Measurements and modeling of III-V solar cells at high temperatures up to 400°C," Journal of Photovoltaics " *J. Photovoltaics,* vol. 6, p. 1345, 2016.

[5] A. L. Fahrenbruch and R. H. Bube, *Fundamentals of Solar Cells Photovoltaic Solar Energy Conversion.* New York: Academic Press, 1983.

GaInAsP Solar Cells Grown by Hydride Vapor Phase Epitaxy for One-Sun & Low-Concentration III-V/Si Photovoltaics

Nikhil Jain[1], John Simon[1], Kevin L. Schulte[1], David R. Diercks[2], Corinne E. Packard[1,2], David Young[1] and Aaron J. Ptak[1]

[1]National Renewable Energy Laboratory, Golden, Colorado, 80401, USA

[2]Colorado School of Mines, Golden, Colorado, 80401, USA

Abstract — **Dynamic hydride vapor phase epitaxy (D-HVPE) has recently reemerged as a low-cost alternative to metalorganic chemical vapor deposition (MOCVD) for the growth of high-efficiency III-V solar cells. Quaternary GaInAsP solar cells in the bandgap range of ~1.6-1.8 eV are promising top-cell candidates for III-V/Si tandem solar cells. In this work, we report on the development of lattice-matched GaInAsP (Eg~1.66 eV) solar cells grown via low-cost HVPE at very high growth rates of ~ 0.7 μm/min (~ 42 μm/h). We demonstrate for the first time HVPE grown passivated GaInAsP homojunction solar cells that show substantial improvement in the short-wavelength photoresponse attributed to the incorporation of a GaInP window layer. The heterointerfaces in these multilayer devices were characterized by transmission electron microscopy. The best device achieved a certified one-sun efficiency of 18.7% under AM1.5G, demonstrating the viability of HVPE to grow multilayered structures comprising ternary and quaternary alloys. This work represents a promising step towards low-cost III-V/Si tandem photovoltaics with one-sun efficiencies exceeding 30%.**

I. INTRODUCTION

The efficiency of market incumbent silicon (Si) photovoltaic (PV) technology is approaching its theoretical limit (29.4%) [1]. Any substantial performance improvement beyond the recent record efficiency of 26.3% [2] will be increasingly challenging. As single-junction Si solar cells approach their practical limits, Si-based tandem solar cells, in both current-matched and electrically independent configurations, have attracted immense interest to influence the balance-of-system economics [3-4]. Recently, mechanically stacked dual- and triple-junction III-V/Si tandem solar cells with one-sun efficiency of 30.4% (four-terminal) [3] and 30.2% (two-terminal) [4], respectively have been demonstrated. However, the bandgap of the top-cell was not optimal for III-V/Si tandems. Furthermore, the III-V solar cells in these tandem designs were grown by conventional MOCVD, which is costly owing to the expensive metalorganic precursors, low throughput and typical growth rates of 1 to 6 μm/h. These cumulative costs make it imperative to develop an alternative low-cost III-V deposition technique that will allow III-V/Si tandems to become economically viable. HVPE can attain significantly higher growth rates than typical MOCVD (up to 300 μm/h) and uses lower Group-V over pressures, while still achieving high-quality heterointerfaces and devices. Quaternary GaInAsP

solar cells with bandgap tunability in the range of ~1.6-1.8 eV are optimal for III-V/Si tandem cells. Fig. 1(a) shows the projected one-sun and low-concentration efficiencies for III-V/Si tandem solar cells, while the proposed integration scheme for realizing a GaInAsP/Si tandem stack is illustrated in Fig. 1(b).

In this work, we report on the development of GaInAsP homojunction cells grown lattice-matched to GaAs, wherein the GaInAsP base layer is grown at ~ 0.7 μm/min, about 10x faster than typical MOCVD reports [5]. We present GaInAsP solar cells with a passivating GaInP window layer that show substantial improvement in the short-wavelength photoresponse and achieve an efficiency of 18.7% under AM1.5G. Key factors limiting the performance are identified and directions for future improvements are discussed.

II. EXPERIMENTAL METHODS

The GaInAsP solar cells were grown in a custom-built dual-chamber HVPE reactor [6]. Metal chlorides (GaCl, InCl), which are formed upstream by the reaction of HCl gas with elemental Ga or In at 800°C, were used as Group-III sources. Arsine and phosphine gases were used as Group-V precursors. All epitaxial layers were grown on (001) GaAs:Zn substrates with a 4° miscut toward (111)B at a growth temperature of 625°C. The device schematics of unpassivated GaInAsP heterojunction and passivated GaInAsP homojunction cells are shown in Fig. 2(a) and (b), respectively. All the cells were processed with electroplated Au as the front and back Ohmic contact layer. A bi-layer MgF₂/ZnS (92nm/51nm) anti-

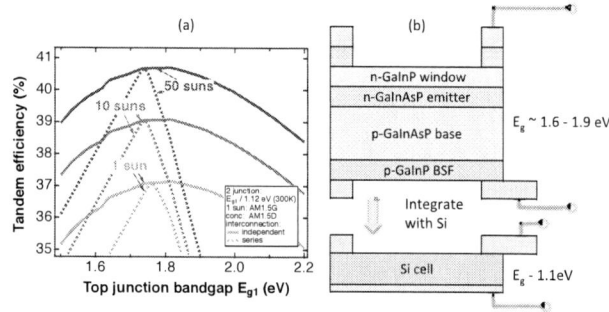

Fig. 1 (a) Projected efficiencies for III-V/Si tandems, (b) Integration scheme for realizing mechanically stacked GaInAsP solar cell on bottom Si subcell.

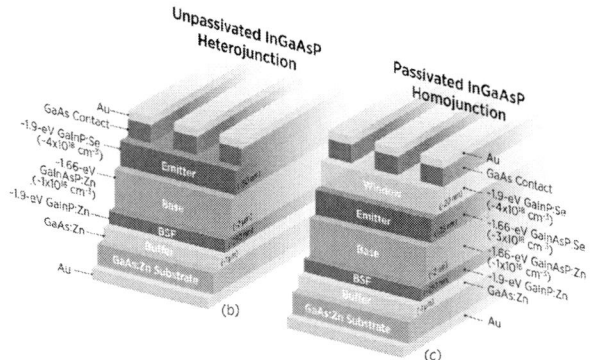

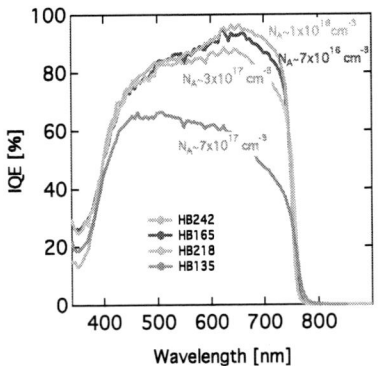

Fig. 2 Device schematic of (a) unpassivated GaInAsP heterojunction and (b) passivated GaInAsP homojunction solar cell with a GaInP window layer.

Fig. 4 Internal quantum efficiency (IQE) of passivated GaInAsP homojunction cells with varying base doping. (No ARC).

reflection coating (ARC) was deposited using thermal evaporation. The quantum efficiency (QE) and specular reflectance were measured on a custom-built QE set-up, and current density-voltage (JV) measurements were performed on an XT10 simulator, tuned to simulate the AM1.5G spectrum.

III. RESULTS AND DISCUSSION

Preliminary solar cell designs were focused on front heterojunction GaInAsP devices with a p-type GaInAsP:Zn base (E_g~1.66 eV) and an n-type GaInP:Se emitter (E_g~1.90 eV) (see Fig. 2(a)). The base thickness was fixed to ~2 μm because nearly 95% of the incident photons are expected to be absorbed in a 2 μm thick GaInAsP layer, based on the absorption coefficient determined from ellipsometry measurements [5]. Representative EQE and JV characteristics are shown in Fig. 3(a) and (b), respectively. These heterojunction GaInAsP solar cells suffered from poor short-wavelength photoresponse given the lack of front-passivation. This limited the short-circuit current density (J_{SC}) to 16.4 mA/cm^2 and the one-sun efficiency to 15.6%.

Homojunction GaInAsP solar cells with GaInP window layers were developed in order to improve the short-wavelength carrier collection. We grew a set of samples with varying base doping (~1x10^{16} to ~7x10^{17} cm^{-3}) to investigate the carrier collection in the GaInAsP base layer. Reducing the base doping led to an improvement in the long-wavelength photoresponse, as evident from Fig. 4. The device with the

lowest base doping exhibited the best performance and was the only device that showed a transition to an ideality factor of n=1 at the maximum power point. Fig. 3 also shows the performance of the homojunction GaInAsP cell compared to heterojunction GaInAsP cells with similar base doping (N_A~1x10^{16} cm^{-3}). It is worth noting that we saw ~20 meV drop in open-circuit voltage (V_{OC}) for the homojunction cell. We attribute this drop to the lower bandgap emitter layer in homojunction cells, translating to higher diffusion-region limited dark current (J_{01}). More importantly, the passivated homojunction cell showed a substantial improvement in the short-wavelength photoresponse, translating to an ~18% boost in the J_{SC} (16.4 mA/cm^2 to 19.2 mA/cm^2), indicative of the effectiveness of front surface passivation. The heterointerfaces in the GaInAsP homojunction solar cells were characterized using TEM. Fig. 5 (a) shows a high-resolution {011} bright field image showing nearly flat interface between the GaInP BSF and GaInAsP base layer. Individual elemental maps were also acquired using nano-energy dispersive x-ray spectroscopy (EDS). The elemental maps (see Fig. 5(b)) suggest minimal chemical intermixing between the respective layers, highlighting the viability of the HVPE technique to grow these multilayer structures.

The best device achieved a certified efficiency of 18.7% with a fill-factor of ~85.4% (see Fig. 6(d)). It is worth mentioning that a $W_{OC} = E_g/q - V_{OC}$ ~0.52 V suggests significant

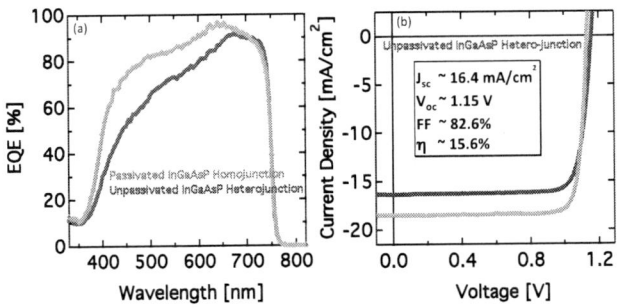

Fig. 3 (a) EQE and (b) JV characteristics of passivated and unpassivated GaInAsP cells measured under AM1.5G. The certified JV for passivated GaInAsP is included in Fig. 6.

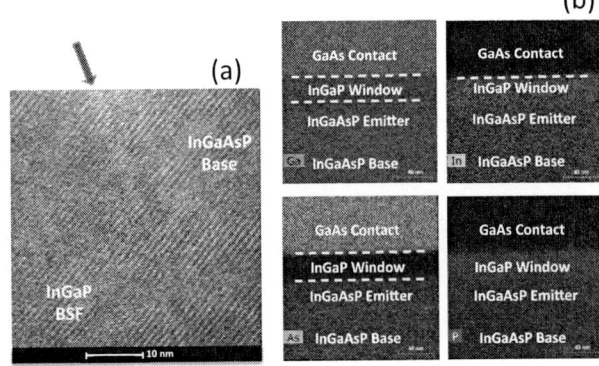

Fig. 5(a) High-resolution {011} bright field TEM image of GaInP BSF & GaInAsP base heterointerface, (b) EDS elemental map showing minimal chemical intermixing between the device layers.

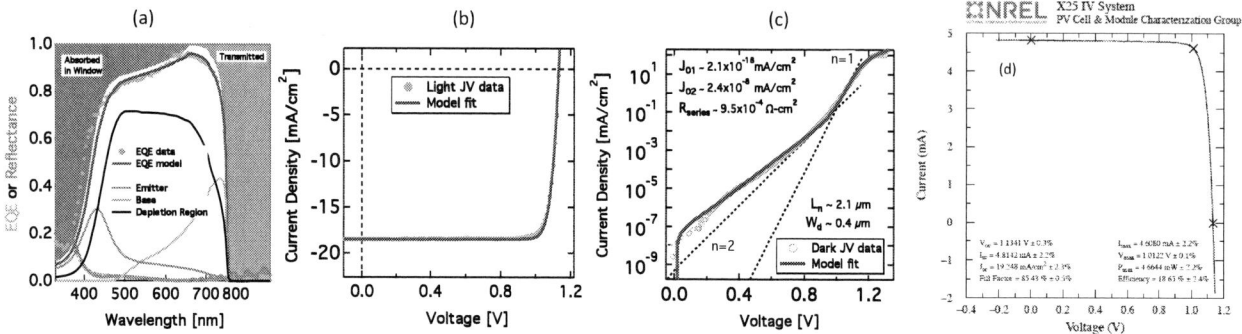

Fig. 6 Modeled (blue line) and experimental data (green dots) for 18.65% passivated GaInAsP homojunction solar cell representing (a) EQE, (b) light J-V and (c) dark J-V plots. Certified LIV measurement shown in (d).

room for further improvement. We modeled the EQE and JV plots using Hovel's drift-diffusion equations (see Fig. 6) in order to gain insight into carrier collection. A minority carrier diffusion length (L_n) of ~2.1 μm in the GaInAsP:Zn base layer was estimated at N_A~1x10^{16} cm^{-3}. Although the peak EQE exceeds 95%, the back surface field (BSF) layer could be influencing the bandedge collection. It is also worth noting that the dark JV plots (Fig. 6(c)) indicate the diode transition from an ideality factor of n=2 to n=1 behavior at the maximum power point.

Realistic efficiencies exceeding 35% can be realized by operating III-V/Si tandem under concentrations (~50 suns). We measured the performance of selected cells under concentration on a High Intensity Pulsed Solar Simulator. No attempts were made to optimize the grid-design and/or sheet resistance for concentrator operation. Consequently, the efficiency of these devices peaked below 200 suns owing to high series resistance. Nonetheless, both devices show a logarithmic dependence of V_{OC} with increasing concentration as expected (see Fig. 7). We anticipate optimization of design trade-offs between shadow losses and series resistance should improve concentrator cell performance.

Future efforts to improve performance should predominantly focus on improving the V_{OC}. Exploring growth

conditions (temperature, V/III ratio) that could influence the bulk GaInAsP material quality could be a promising avenue. Additionally, development of a higher bandgap GaInAsP emitter could reduce the dark current and potentially improve the V_{OC} and FF. Incorporation of an even wider bandgap window layer (such as AlInP, E_g~2.2eV) is expected to improve the short-wavelength carrier collection.

IV. CONCLUSION

We show the first demonstration of HVPE grown GaInAsP (Eg~1.66 eV) homojunction solar cells that incorporate a GaInP window layer to enhance the short-wavelength collection. The GaInAsP (Eg~1.66 eV) absorber layer was grown at a high growth rate of ~0.7 μm/min, about 10x faster than the growth of similar alloys reported by the MOCVD technique. The successful demonstration of HVPE grown GaInAsP homojunction solar cells with ~18.7% not only demonstrates the effectiveness of the front surface passivation, but also highlights the viability of HVPE to grow complex multilayer devices. This work shows a promising path towards realizing high-efficiency top cells grown via low-cost HVPE for III-V/Si tandems with one-sun efficiencies exceeding 30%.

REFERENCES

[1] A. Richter, et al, "Reassessment of the Limiting Efficiency for Silicon Cells," *IEEE J. Photovoltaics*, vol. 3, pp. 1184, 2013.

[2] M. A. Green, et al., "Solar cell efficiency tables (version 49)," *Prog. Photovolt: Res. Appl.*, vol. 25, pp. 3-13, 2017.

[3] S. Essig, et al., Dual-Junction III-V/Si solar cells with over 30% one-sun efficiency" under review.

[4] R. Cariou, et al., "Monolithic Two-Terminal III-V//Si Triple-Junction Solar Cells With 30.2% Efficiency Under 1-Sun AM1.5g," *IEEE J. Photovoltaics*, vol. 7, pp. 367-373, 2017.

[5] N. Jain, et al., "Enhanced Current Collection in 1.7 eV GaInAsP Solar Cells Grown on GaAs by Metalorganic Vapor Phase Epitaxy", *IEEE Journal of Photovoltaics*, in press.

[6] D. L. Young, et al., "High throughput semiconductor deposition system," U.S. Patent US 20 130 309 848, 2013.

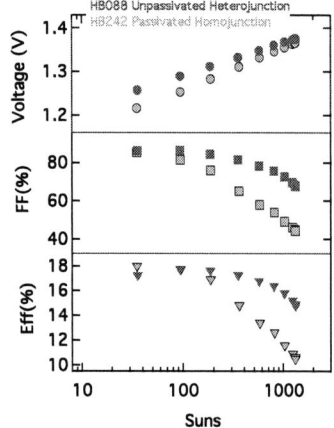

Fig. 7 Performance of GaInAsP solar cells under concentration.

Photo-Electrochemical Hydrogen Generation from Inverted Metamorphic Multijunction III-Vs

Todd G. Deutsch[1], James L. Young[2], Myles A. Steiner[2], Henning Döscher[2,3], Ryan M. France[2], and John A. Turner[2]

[1] National Renewable Energy Laboratory, Todd.Deutsch@nrel.gov, Golden, Colorado, USA

[2] National Renewable Energy Laboratory, Golden, Colorado, USA

[3] Philipps-Universität Marburg, Marburg, Germany

Abstract — **Our goal is to improve solar-to-hydrogen (STH) efficiency from just over 10% to over 20% via novel tandem semiconductor materials and configurations. Our primary focus is to develop inverted metamorphic multijunction (IMM) III-V semiconductors that have bandgaps optimized for water splitting. We will also discuss measurement challenges in appraising STH efficiency, some of which are specific to tandem absorbers. Using more stringent measurement standards, we have confirmed 16.2% STH on our most advanced IMM devices.**

I. Introduction

In order to economically generate renewable hydrogen fuel directly from solar energy using semiconductor-based photoelectrochemical (PEC) devices, the U.S. Department of Energy Fuel Cells Technology Office has established technical targets of over 20% solar-to-hydrogen (STH) efficiency with several thousand hours of stability under operating conditions [1]. Techno-economic sensitivity analysis shows that increasing STH efficiency has the largest effect on the levelized cost of hydrogen [2]. Here we report on our progress towards the 20% goal by improving STH efficiency from just over 10% to over 16% via novel tandem semiconductor materials and configurations. Hydrogen via solar driven electrolysis can proceed via PEC routes or through PV coupled to an electrolyzer; we will discuss cells that apply to both approaches.

The requirements for a PEC device are more stringent than a PV-coupled electrolyzer because of the fixed energetics of the band edges at the semiconductor/electrolyte interface. The electrochemical concepts of flat band potentials and reference electrodes as well as kinetic overpotentials for the water reduction and oxidation half-reactions must be considered and are not intuitive. Flat band potentials are a measurement of the Fermi level when the bands are flat (i.e. no electric field is present) and are used to determine the potentials at which the valence and conduction band edges are pinned at the interface with the electrolyte. These potentials are measured vs. a reference electrode potential (e.g. Ag/AgCl) which allows them to be placed on the electrochemical scale for determining what electrochemical reactions are possible. Electrons can only drive reduction half-reactions that have reversible potentials below the conduction band edge potential (more

positive on an electrochemical scale). Conversely, holes can only affect oxidation half-reactions with potentials above the valence band edge potential (more negative on an electrochemical scale). Most III-V semiconductors have a conduction band edge that can drive the water reduction half-reaction, but are incapable of oxidizing water to oxygen without applying a bias. Even the III-Vs that can generate a V_{oc} well above the thermodynamic minimum required for water electrolysis ($\Delta G° = 1.23$ V) cannot split water because of their insufficiently positive valence band edge position. Tandem cells, which are optically and electrically in series, can eliminate the need for an external bias. By monolithically integrating a lower bandgap p/n junction below a PEC junction, the oxidative potential of the hole from the wider bandgap PEC junction can be augmented via generation of a second electron hole pair and recombination at the tunnel junction in a process depicted in Figure 1. The graphic in Figure 1 is based on lattice-matched GaInP₂/GaAs cell reported by Khaselev and Turner and was the first demonstration of unbiased PEC water splitting at a reasonable efficiency [3]. While the reported 12.4% STH efficiency was impressive, the device they used was not current matched and significantly current-limited by the bottom GaAs cell.

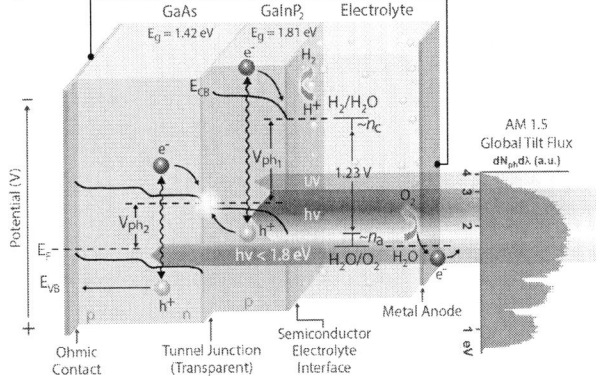

Fig. 1. Schematic depicting some of the concepts of a PEC water splitting cell. The potential of the hole generated in the p-GaInP₂ PEC is not positive enough to perform the water oxidation half-reaction. A second electron hole pair generated in the p/n-GaAs junction and accompanying recombination at the tunnel junction provides this second hole with a sufficiently positive potential for water oxidation and allows unbiased photoelectrolysis.

Although iso-efficiency contour modeling of tandem bandgap combinations for photoelectrolysis has been known for several years [4], systems that are lattice matched (LM) to GaAs are unable to access the optimal combinations. Our primary focus is to develop inverted metamorphic multijunction (IMM) III-V semiconductors with bandgaps that maximize attainable water splitting efficiency.

II. EXPERIMENTAL DETAILS

The IMM devices were grown at NREL by metalorganic vapor phase epitaxy using a GaAs substrate and use a transparent step-graded buffer layer to allow lattice-mismatched growth. Details of the growth and processing [5] as well as photoelectrode characterizations [6] have been previously reported.

II. RESULTS

We used IMM growth to achieve a high-quality, lattice-mismatched lower bandgap (1.2 eV) InGaAs bottom cell that captures a larger fraction of the solar spectrum. Representations of the various device structures appear in Figure 2. This 1.8/1.2 eV bandgap combination of the GaInP$_2$/InGaAs configuration is able to generate higher photocurrents, and thus efficiencies, than the 1.8/1.4 eV bandgap combination of the lattice-matched GaInP$_2$/GaAs configuration. Our first attempts to use this IMM configuration with a p-GaInP$_2$/electrolyte junction was unable to split water because the device produced insufficient photovoltage to overcome losses associated with the PEC junction. We incorporated an n-type GaInP$_2$ layer to move the space charge region away from the electrolyte interface and through this "buried junction" were able to improve the photocurrent onset (V_{oc}) by 700 mV. This translation of the J-V curve by 700 mV allowed us to drive water splitting at the light-limited photocurrent of our semiconductor device without any additional bias and achieve 14% STH under 1-sun AM1.5 D illumination. The STH efficiency was improved to over 16% STH by passivating surface recombination with an AlInP window layer and an n-GaInP$_2$ encapsulating layer (Figure 3). The largest loss channel in this system is photon reflection at the semiconductor/electrolyte interface. We are currently investigating photon management strategies to minimize this loss and achieve over 20% STH from these IMM device configurations. Device stability also needs to be improved to thousands of hours to be commercially viable.

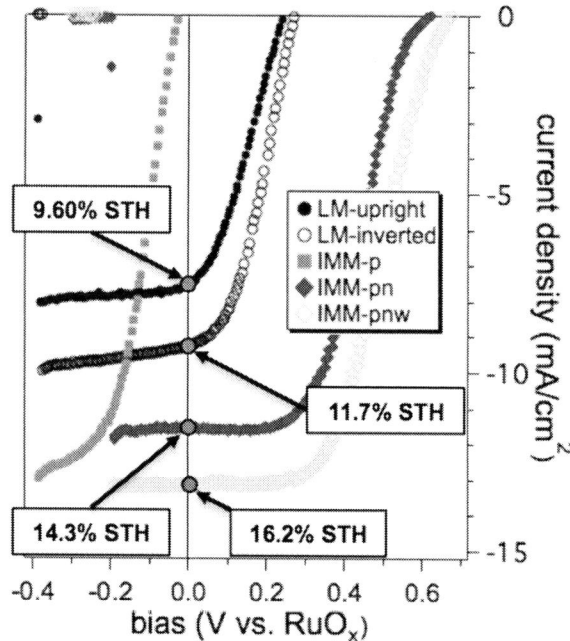

Fig. 3. J-V curves of various tandem configurations measured under natural sunlight with a spectral correction factor applied, showing the progression of water-splitting efficiency.

This work is significant because it represents the highest reported STH efficiency for an immersed photoelectrode. The reported 12.4% by Khaselev and Turner was considered the record for PEC water splitting from 1998 until it was recently bested by an upright metamorphic tandem cell [7]. This work surpasses the upright metamorphic tandem cell by at least two absolute efficiency points in STH as well as demonstrating the advantage of placing the metamorphic junction between the two absorbers, which permits independent tuning of bandgaps without requiring them to be lattice matched.

Even if 20% STH can be sustained for several thousand hours, semiconductor absorber costs need to be significantly reduced, down to ~$150/m^2, to achieve DOE hydrogen production cost targets. Optical concentration is another way to defray prohibitively expensive absorber costs. PEC systems are limited to low concentrations due to bubble evolution off the light-absorbing surface as well as unsustainable potential drop (iR losses) across the electrolyte at high currents due to long ionic transport distances typical to PEC systems. Tembhurne and Haussener have proposed an integrated system [8] that directly couples high (1000x) concentration photovoltaics with a polymer electrolyte membrane electrolysis device that is designed to be operated at high current densities. Their initial cost modeling suggests this might be a viable route to economically produce hydrogen from renewable resources.

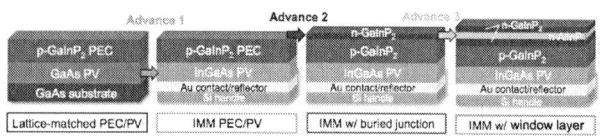

Fig. 2. Cartoon representations of the device structural advances that led to improved STH efficiency.

REFERENCES

[1] U.S. Department of Energy June 2015. http://energy.gov/sites/prod/files/2014/10/f19/fcto_myrdd_production.pdf

[2] B. A. Pinaud, J. D. Benck, L. C. Seitz, A. J. Forman, Z. Chen, T. G. Deutsch, B. D. James, K. N. Baum, G. N. Baum, S. Ardo, H. Wang, E. Miller, and T. F. Jaramillo, "Technical and economic feasibility of centralized facilities for solar hydrogen production via photocatalysis and photoelectrochemistry," *Energy Environ. Sci.*, vol. 6, pp. 1983-2002, 2013.

[3] O. Khaselev and J. A. Turner, "A monolithic photovoltaic-photoelectrochemical device for hydrogen production via water splitting," *Science*, vol. 280, pp. 425-427, 1998.

[4] R. E. Rocheleau and E. L. Miller, "Photoelectrochemical production of hydrogen: Engineering loss analysis," *Int. J. Hydrogen Energy*, vol. 22, pp. 771-782, (1997).

[5] J. L. Young, M. A. Steiner, H. Döscher, R. M. France, J. A. Turner, and T. G. Deutsch, "Direct solar-to-hydrogen conversion via inverted metamorphic multi-junction semiconductor architectures," *Nature Energy,* vol. 2, 17028, 2017.

[6] H. Döscher[†], J. L. Young[†], J. F. Geisz, J. A. Turner, and T. G. Deutsch, "Solar to hydrogen efficiency: Shining light on photoelectrochemical device performance," *Energy Environ. Sci.*, vol. 9, pp. 74-80, 2016.

[7] M. M. May, H.-J. Lewerenz, D. Lackner, F. Dimroth, and T. Hannappel, "Efficient direct solar-to-hydrogen conversion by *in situ* interface transformation of a tandem structure," *Nature Communications*, vol. 6, 8286, 2015.

[8] S. Tembhurne and S. Haussener, "Integrated photo-electrochemical solar fuel generators under concentrated irradiation," *J. Electrochem. Soc.*, vol. 163, H999-H1007, 2016.

Advanced Silicon Thin Films for High-Efficiency Silicon Heterojunction-Based Solar Cells

A. Descoeudres[1], C. Allebé[1], N. Badel[1], L. Barraud[1], J. Champliaud[1], G. Christmann[1], L. Curvat[1], F. Debrot[1], A. Faes[1], J. Geissbühler[1], J. Horzel[1], A. Lachowicz[1], J. Levrat[1], S. Martin de Nicolas[1], S. Nicolay[1], B. Paviet-Salomon[1], L.-L. Senaud[1], A. Tomasi[2], C. Ballif[1,2], M. Despeisse[1]

[1]CSEM, PV-Center, CH-2002 Neuchâtel, Switzerland

[2]EPFL, IMT, PV-Lab, CH-2002 Neuchâtel, Switzerland

Abstract — The drop in passivation usually observed after the deposition of the *p*-doped amorphous silicon (a-Si:H) layer on top of the passivating intrinsic buffer a-Si:H layer during the fabrication of silicon heterojunction (SHJ) solar cells is shown to be mostly related to the properties of the *i*-layer itself. After optimization of the *i*-layer to reduce this loss, minority carrier lifetimes above 50 ms were achieved on very lowly doped wafers, and close to 18 ms on actual SHJ cell precursors with *i*-layers as thin as 4 nm. These films were integrated into SHJ solar cells fabricated with industry-compatible processes, yielding efficiencies up to 23.1% on large-area devices and up to 23.9% on 4 cm² devices. In addition, the developed a-Si:H layers were also used as key building blocks in more advanced high-efficiency solar cell architectures, such as interdigitated back-contacted SHJ solar cells (>23%), III-V//SHJ tandems (>30%), and perovskite//SHJ tandems (>25%), for example.

Index Terms — amorphous silicon, crystalline silicon, heterojunction, photovoltaic cells, surface passivation.

I. INTRODUCTION

In order to reach high conversion efficiencies with crystalline silicon (c-Si) solar cells, one key requirement is a high passivation quality of the wafer surfaces. Hydrogenated amorphous silicon (a-Si:H) films are known for decades to yield excellent surface passivation on c-Si substrates [1]. The integration of such films in silicon heterojunction (SHJ) solar cells is the root cause of their record open-circuit voltages (V_{oc}'s) [2]. Combined with the interdigitated back-contacted (IBC) architecture, IBC SHJ solar cells are nowadays the most efficient single-junction c-Si devices, with 1-sun efficiencies exceeding 26% [3].

Although a-Si:H passivation layers have been studied and developed for years, continuous improvement is still made, showing their remarkable potential. In this work, it is shown that the optimization of the intrinsic a-Si:H passivation layers in SHJ solar cells should take into account the impact of the overlaying *p*- and *n*-doped a-Si:H layers, which are forming the hole and electron selective contacts, respectively (Fig. 1). After such an optimization, very high minority carrier lifetimes were demonstrated, with remarkably thin passivation layers (< 5 nm).

Fig. 1. Schematic structure of a standard SHJ solar cell.

The integration of these ultra-thin films into standard SHJ solar cells resulted in a significant improvement yielding device fill factors (FF) exceeding 81.5% and efficiencies exceeding 23.9% using industrial fabrication processes. Finally, it is shown that these films are also useful building blocks for advanced high-efficiency SHJ-based devices, such as IBC SHJ, III-V//SHJ and perovskite//SHJ tandem solar cells.

II. RESULTS AND DISCUSSION

A. High-Quality a-Si:H Passivation Layers

During the fabrication of SHJ solar cells, one usually observes a drop in passivation with the deposition of the *p*-doped a-Si:H layer on top of the passivating intrinsic buffer a-Si:H layer. Depending on the layer properties, the difference in minority carrier lifetime of a wafer passivated with only *in/i* a-Si:H stacks compared to the same wafer with the subsequent addition of the *p*-layer (i.e. passivated with *in/ip* a-Si:H stacks) can be really significant: a drop from almost 20 ms to 1.3 ms has been observed at CSEM (Fig. 2). One often attributes this loss in passivation mainly to the highly defective nature of *p*-doped a-Si:H layers, and also to defect equilibration mechanisms [2, 4]. Defects are indeed induced in the *i*-layer by the doped overlayer, via Fermi-level dependent Si-H bond breakage at the c-Si/a-Si:H interface [4].

978-1-5090-5606-4/17 $31.00 © 2017 IEEE

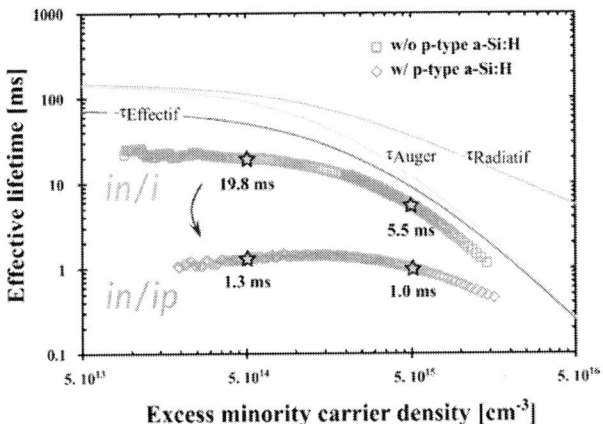

Fig. 2. Example of drop in minority carrier lifetime observed after the deposition of the *p* a-Si:H layer in a SHJ solar cell precursor (FZ textured c-Si wafer, 3 Ω·cm, 240 μm, passivated with ~20 nm *in/ip* a-Si:H stacks).

Ratios of lifetimes after / before the deposition of the *p*-layer were measured for different intrinsic and *p*-doped a-Si:H layers, as shown in Fig. 3. It appears that the main driver of the passivation drop are the properties of the intrinsic buffer layer rather than those of the *p*-layer itself. Indeed, the lifetime ratios depend only weakly on the *p*-layer doping (different activation energies in Fig. 3) or on the *i*-layer thickness, for example. In addition, potential passivation degradation eventually induced by the *p*-layer plasma processes can be excluded as well (see data points denominated "intrinsic overlayer" in Fig. 3, obtained with layers deposited with the exact same conditions as one of the *p*-layers but without dopants in the gas mixture). On the other hand, different *i*-layers (obtained from different deposition conditions) produce significantly different results, with lifetime ratios ranging from 1.2 down to 0.15, although lifetimes measured before the *p*-layer depositions are all in the same range (2–6 ms).

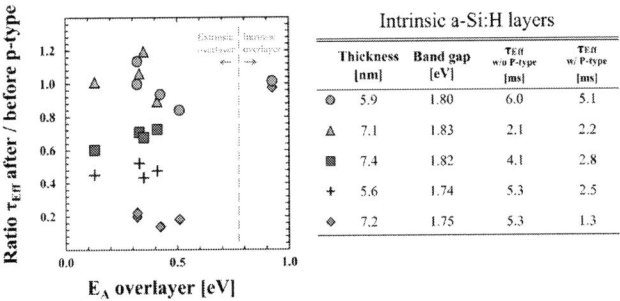

Intrinsic a-Si:H layers			
Thickness [nm]	Band gap [eV]	τ_{Eff} w/o P-type [ms]	τ_{Eff} w/ P-type [ms]
⊙ 5.9	1.80	6.0	5.1
△ 7.1	1.83	2.1	2.2
▦ 7.4	1.82	4.1	2.8
+ 5.6	1.74	5.3	2.5
◇ 7.2	1.75	5.3	1.3

Fig. 3. Ratio of minority carrier lifetimes measured on SHJ cell precursors after/before the *p*-layer deposition, as a function of the *p* layer activation energy E_A (related to doping level). Different symbols represent different *i*-layers (see table).

Surprisingly, as shown in Fig. 4, the analysis of the *i*-layer properties with Fourier transform infrared (FTIR)

spectroscopy [5] shows that the drop in lifetime after the addition of the *p*-layer is *reduced* when using *i*-layers characterized by an *increased* microstructure factor R* (calculated from intensities of high and low Si-H stretching modes [6]). Consequently, the V_{oc} of the finished solar cells increases with R*. In addition to V_{oc}, the ideality factor *n* of the final devices are also found to improve (i.e. lowered towards 1) with *i*-layers having increased R*, leading to improved FF as well (Fig. 5).

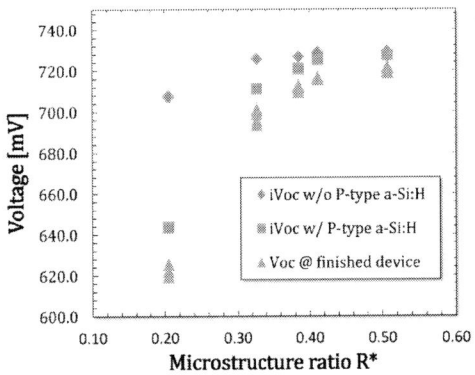

Fig. 4. Implied V_{oc} and V_{oc} for different *i*-layers, characterized by different microstructure factors R* (obtained from FTIR measurements [5]).

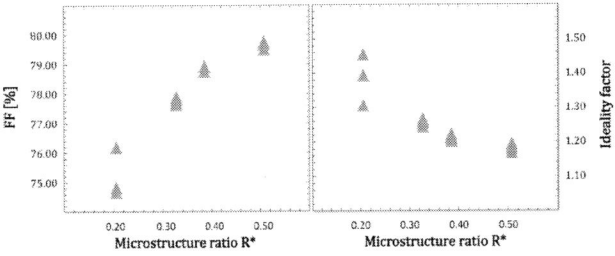

Fig. 5. Fill factor and ideality factor for different *i*-layers, characterized by different microstructure factors R* (obtained from FTIR measurements [5]).

High R* values reflect an increase in the a-Si:H bulk disorder, in the form of microvoids for instance. Such disordered a-Si:H materials are known to show significant light-induced degradation when incorporated in thin film solar cells, for example [6, 7]. Similarly, it has been reported that the use of *i*-layers with high Urbach energy (i.e. high disorder) in SHJ solar cells leads also to a decrease in passivation quality, with or without the presence of a doped overlayer [8]. The results of the present work (disordered *i*-layers give better results) seem thus contradictory to what is generally believed, and further work is still needed to have a complete understanding of the observed behavior. In any case, these results emphasize that defect equilibration between the intrinsic and the doped layers plays most likely a major role in

SHJ solar cells. Additionally, from a practical point of view, it shows also that full *ip*-stacks should be deposited for the *i*-layer development in the hole selective hetero-contact (and not solely the *i*-layer), so as to efficiently probe the *i*-layer characteristics that will drive the final device performance.

The deposition conditions for the *i*-layer were optimized by taking into account this passivation drop after *p*-layer deposition and by trying to minimize it. As a result (see Table I), SHJ cell precursor lifetimes close to 18 ms and 10 ms were obtained on FZ and CZ c-Si wafers, respectively, using *i*-layers of only ~4 nm and doped layers of each type of ~10 nm in thickness. In more favorable conditions (very lowly doped and double-side polished wafers, thicker a-Si:H stacks), minority carrier lifetimes above 50 ms were even reached using these advanced passivation layers. This emphasizes once more the outstanding potential of a-Si:H thin films for very high-quality surface passivation.

TABLE I
BEST MINORITY CARRIER LIFETIMES OBTAINED IN THIS WORK

Wafer type	a-Si:H layers	Lifetime	Implied V_{oc}
FZ *n*-type, 20 kΩ·cm, double-side polished, (111), 500 µm	*in/in* (10+50 nm)	51.1 ms*	701 mV
FZ *n*-type, 3 Ω·cm, textured, 230 µm	*in/ip* (4+10 nm)	17.9 ms**	734 mV
CZ *n*-type, 3 Ω·cm, textured, 150 µm	*in/ip* (4+10 nm)	9.6 ms**	738 mV

* measured at 10^{14} cm^{-3} ** measured at $5 \cdot 10^{14}$ cm^{-3}

B. Layer Integration into Industrial Silicon Heterojunction Solar Cells

Large-area SHJ solar cells were fabricated using CSEM's R&D platform applying industry-compatible processes, with the following steps: wafer texturing and cleaning by wet-chemistry, a-Si:H depositions on both sides in a large-area PECVD cluster tool, transparent conductive oxide (TCO) depositions on both sides and full-area back side Ag metallization by sputtering, and screen-printing of the front metallization grid using a low-temperature Ag paste. The integration of the very thin and highly passivating a-Si:H layers described above into such SHJ devices helped to increase their conversion efficiencies, due to a significant increase in FF (up to 81.7% in some cases) while maintaining V_{oc}'s at a high level. Fig. 6 shows that FF above 80% are regularly obtained.

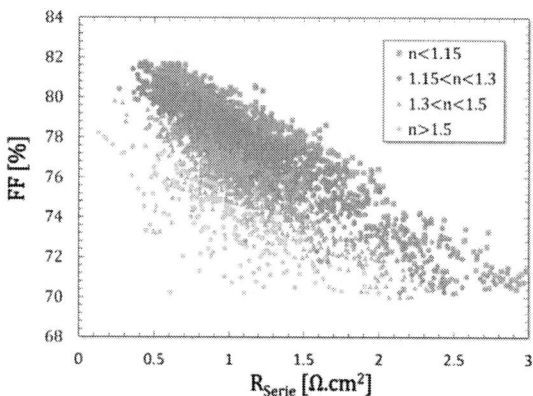

Fig. 6. Fill factors of SHJ solar cells produced at CSEM as a function of device series resistance and ideality factor *n* (extracted with the method described in [9]).

As shown in Fig. 7, the best efficiency measured on a large-area SHJ solar cell is 23.1%, using a rather simple and standard cell architecture (both-sides contacted, CZ *n*-type 150 µm thick 6" wafer, full a-Si:H layers, front emitter configuration, ITO front side, 3-busbars grid design). Further efficiency increase can thus be expected by introducing several improvement in material properties and cell architecture. On smaller devices (4 cm^2) and using the exact same layers and cell structure, best efficiencies exceed 23.9% (FZ *n*-type 230 µm thick 4" wafer, see Fig. 7).

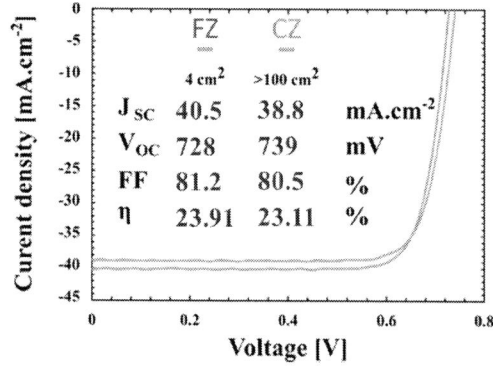

Fig. 7. Illuminated IV curves and cell parameters of the best both-sides-contacted SHJ solar cells (in-house measurements).

C. Layer Integration into Advanced Solar Cell Architectures

In addition to standard SHJ solar cells, the developed a-Si:H layers turn to be useful building blocks also for other and more advanced high-efficiency cell structures. The first example is IBC SHJ solar cells. Although this type of c-Si devices is currently the most efficient, the technology sophistication needed to realize carrier selective contacts for both polarities on the rear side may limit their appeal and spreading in industrial mass production. The approach followed jointly by

EPFL, CSEM and Meyer Burger Research is therefore focused on the simplification of the fabrication process, with a photolithography-free patterning process of the backside, based on hard-masking for a-Si:H layers and inkjet printing combined with wet etching for selective formation of TCO/metal electrodes [10]. A further significant simplification was achieved with the introduction of an interband silicon tunnel junction at the backside (Fig. 8), requiring only one hard-masking step for the silicon thin films patterning instead of two [11]. In addition to the passivating intrinsic a-Si:H layers, doped nanocrystalline layers (nc-Si:H) were integrated as well, allowing for decreased contact resistances and FF improvement. Using the improved hetero-contacts developed, a photolithography-free medium-sized 25 cm^2 IBC SHJ device with a conversion efficiency of 23.2% could be fabricated [12].

Fig. 8.   Schematic structure of the tunnel IBC SHJ solar cell [11].

The second example of advanced solar cells benefitting from the developed a-Si:H layers is tandem devices. Thanks to their good response in the red part of the spectrum and under reduced (< 1 sun) illumination while keeping high V_{oc}'s, the SHJ solar cells developed in this work are indeed excellent candidates for bottoms cells, in combination with higher band-gap top cells. As an example, after a further specific optimization, SHJ bottom cells were successfully integrated in 4-terminal tandem devices with GaInP or GaAs top cells developed at NREL, making 1-sun certified efficiencies above 30% possible [13-15]. In a more cost competitive approach, efficiencies above 25% have also been demonstrated using perovskite top cells, in collaboration with EPFL (4-terminal tandem devices) [16, 17].

III. CONCLUSIONS

Very high passivation levels have been achieved with the developed a-Si:H layers, demonstrated by minority carrier lifetimes above 50 ms on lowly-doped wafers and close to 18 ms on actual SHJ solar cell precursors. Such highly passivating ultra-thin layers are not only interesting for standard SHJ solar cells (>23% demonstrated on large-area industrial devices), but are also key components once integrated into more advanced high-efficiency devices, such as IBC SHJ and tandem cells. Further improvements in the performance of such devices will be pursued in the future,

while continuing to reduce the fabrication cost at the same time.

REFERENCES

[1] J. I. Pankove and M. L. Tarng, "Amorphous silicon as a passivant for crystalline silicon", *Appl. Phys. Lett.*, vol. 34, pp. 156-157, 1979.

[2] S. De Wolf, A. Descoeudres, Z. C. Holman, C. Ballif, "High-efficiency Silicon Heterojunction Solar Cells: A Review", *Green*, vol. 2, pp. 7-24, 2012.

[3] K. Yoshikawa, H. Kawasaki, W. Yoshida, T. Irie, K. Konishi, K. Nakano, T. Uto, D. Adachi, M. Kanematsu, H. Uzu, K. Yamamoto, "Silicon heterojunction solar cell with interdigitated back contacts for a photoconversion efficiency over 26%", *Nature Energy*, vol. 2, pp. 17032, 2017.

[4] S. De Wolf and M. Kondo, "Nature of doped a-Si:H/c-Si interface recombination", *J. Appl. Phys.*, vol. 105, pp. 103707, 2009.

[5] L. Barraud, C. Allebé, G. Christmann, A. Descoeudres, A. Faes, S. Nicolay, B. Paviet-Salomon, C. Ballif, M. Despeisse, "Defect creation in silicon heterojunction solar cells", *7th International Conference on Crystalline Silicon Photovoltaics (SiliconPV)*, April 2017, Freiburg, Germany.

[6] E. Bhattacharya and A. H. Mahan, "Microstructure and the light-induced metastability in hydrogenated amorphous silicon", *Appl. Phys. Lett.*, vol. 52, pp. 1587, 1988.

[7] A. Shah (ed.), Thin-film silicon solar cells. Lausanne, EPFL Press, 2010.

[8] T. F. Schulze, C. Leendertz, N. Mingirulli, L. Korte, and B. Rech, "Impact of Fermi-level dependent defect equilibration on V_{oc} of amorphous/crystalline silicon heterojunction solar cells", *Energy Procedia*, vol. 8, pp. 282-287, 2011.

[9] S. Bowden and A. Rohatgi, "Rapid and accurate determination of series resistance and fill factor losses in industrial silicon solar cells", *7th European Photovoltaic Solar Energy Conference and Exhibition*, October 2001, Munich, Germany.

[10] A. Tomasi, B. Paviet-Salomon, D. Lachenal, S. Martin de Nicolas, A. Descoeudres, J. Geissbühler, S. De Wolf and C. Ballif., "Back-Contacted Silicon Heterojunction Solar Cells With Efficiency >21%", *IEEE J. Photovoltaics*, vol. 4, pp. 1046-1054, 2014.

[11] A. Tomasi, B. Paviet-Salomon, Q. Jeangros, J. Haschke, G. Christmann, L. Barraud, A. Descoeudres, J. Seif, S.Nicolay, M. Despeisse, S. De Wolf, C. Ballif, "Simple processing of back-contacted silicon heterojunction solar cells using selective-area crystalline growth", *Nature Energy*, vol. 2, pp. 17062, 2017.

[12] A. Tomasi, B. Paviet-Salomon, L. Barraud, A. Descoeudres, J. Seif, Q. Jeangros, J. Geissbühler, G. Christmann, N. Badel, A. Faes, S. Nicolay, D. Lachenal, B. Strahm, M. Despeisse, S. De Wolf, and C. Ballif, "Simple Interdigitated Back-Contacted Silicon Heterojunction Solar Cells with an Interband Silicon Tunnel Junction", *7th International Conference on Crystalline Silicon Photovoltaics (SiliconPV)*, April 2017, Freiburg, Germany.

[13] S. Essig, M. A. Steiner, C. Allebé, J. F. Geisz, B. Paviet-Salomon, S. Ward, A. Descoeudres, V. LaSalvia, L. Barraud, N. Badel, A. Faes, J. Levrat, M. Despeisse, C. Ballif, P. Stradins, and D. L. Young, "Realization of GaInP/Si dual-junction solar cells with 29.8% 1-sun efficiency", *IEEE J. Photovoltaics*, vol. 6, pp. 1012-1019, 2016.

[14] S. Essig, C. Allebé, J. F. Geisz, M. A. Steiner, L. Barraud, J. S. Ward, M. Schnabel, A. Descoeudres, D. L. Young, M.

Despeisse, C. Ballif, A. Tamboli, "Mechanically stacked 4-terminal III-V/Si tandem solar cells ", this conference.

[15] S. Essig C.Allebé, T. Remo, J. F. Geisz, M. A. Steiner, L. Barraud, J. S. Ward, M. Schnabel, K. Horowitz, A. Descoeudres, D. L. Young, M. Woodhouse, M. Despeisse, C. Ballif, A. Tamboli, *Nature Energy*, accepted for publication.

[16] J.Werner, L. Barraud, A. Walter, M. Bräuninger, F. Sahli, D. Sacchetto, N. Tétreault, B. Paviet-Salomon, S.-J. Moon, C. Allebé, M. Despeisse, S. Nicolay, S. De Wolf, B. Niesen, C. Ballif, "Efficient Near-Infrared-Transparent Perovskite Solar Cells Enabling Direct Comparison of 4‑Terminal and Monolithic Perovskite/Silicon Tandem Cells", *ACS Energy Lett.*, vol. 1, pp. 474-480, 2016.

[17] J. Werner, B. Kamino, F. Sahli, D. Sacchetto, M. Brauninger, A. Walter, S.-J. Moon, L. Barraud, B. Paviet-Salomon, J. Geissbühler, C. Allebé, S. De Wolf, M. Despeisse, S. Nicolay, B. Niesen, C. Ballif, "Perovskite/Silicon Tandem Solar Cells: Challenges Towards High-Efficiency in 4-Terminal and Monolithic Devices", this conference.

MoO$_X$ and WO$_X$ based hole-selective contacts for wafer-based Si solar cells

Stephanie Essig[1], Julie Dréon[1], Jérémie Werner[1], Philipp Löper[1],
Stefaan De Wolf[2], Mathieu Boccard[1], Christophe Ballif[1]

[1] École Polytechnique Fédérale de Lausanne (EPFL), Institute of Microengineering (IMT),
Photovoltaics and Thin-Film Electronics Laboratory, Neuchâtel, Switzerland
[2] King Abdullah University of Science and Technology (KAUST), Thuwal, Saudi Arabia

Abstract — **Highly-transparent carrier-selective front contacts open a pathway towards entirely dopant free Si solar cells. Hole-selective a-Si:H/MoO$_x$/ITO front contact stacks were already successfully applied in such novel devices. However, for optimum device performance, further improvements are required: We evaluate the use of the high-work-function material WO$_X$ as a replacement for MoO$_x$ in an attempt to reduce optical absorption losses. In addition, we investigate the use of thin hydrogenated SiO$_X$ instead of a-Si:H, and the impact of the residual pressure for MoO$_x$ evaporation.**

Index Terms — **Silicon heterojunction solar cell, carrier-selective contacts, transition metal oxides**

I. INTRODUCTION

The development of dopant-free Si solar cells in which carrier extraction is provided by electron and hole transport layers (ETL and HTL), integrated into passivating contacts, is a promising route to reach efficiencies close to 27% [1]. A typical carrier-selective contact employs a broadband optically transparent material which provides both a chemical passivation of Si surface states and a high (low) work function, inducing in dark and at equilibrium an electrical potential at the Si wafer surface, yielding hole (electron) collection.

A fully doping-free cell with efficiency of 19.4% was realized in ref. [2] by employing a transparent MoO$_X$-based hole-selective contact on the front side and a LiF-based electron-selective stack on the rear side of an n-type Si wafer. As MoO$_X$ and LiF themselves do not passivate electronic defects at the c-Si surface, thin ~5 nm thick intrinsic hydrogen-rich amorphous silicon (i)a-Si:H interlayers have to be inserted to achieve a high V$_{OC}$. Our present research aims to further optimize the hole-selective front contact, as already used in ref. [2] and [3]. The motivation for this is that the proposed (i)a-Si:H/MoO$_X$/ITO front contact stack comes with an important drawback: parasitic light absorption [3] in both the MoO$_X$, mainly due to transparent conductive oxide (TCO) sputter-induced damage [4], and the ~5 nm thick (i)a-Si:H layer. Furthermore, MoO$_X$ based contact stacks suffer from a fill factor degradation upon annealing at 190°C [5], required for curing of screen printed contacts. This effect could recently be attributed to the release of hydrogen from adjacent

layers [6]. In our experiments, we evaluated the possible replacement of MoO$_X$ by WO$_X$ layer. WO$_X$ provides a high work function similar to MoO$_X$ [7, 8] but has a higher transparency both before and after TCO sputter-deposition [4]. Furthermore we tested the impact of the residual base pressure in the evaporation chamber on the performance of solar cells with MoO$_X$-based front contacts and investigated the use of thermal silicon oxide as an alternative passivation layer with higher transparency and improved thermal resilience.

II. EXPERIMENTAL

Float-zone (100) Si wafers with a thickness of 240 μm and resistance of 3 Ωcm (n-type) were textured and cleaned. For solar cell (Fig. 1a) fabrication, 5-nm-thick (i)a-Si:H layers were applied on the front and rear sides by plasma enhanced chemical vapor deposition (PECVD). On the rear side an about 10 nm thick phosphorous doped a-Si:H layer was deposited in the same tool. After loading in an evaporation chamber, an either ~7-nm-thick MoO$_X$ (x<3) or 5-nm-thick WO$_X$ (x< 3) layer was deposited on the wafer front side by thermal evaporation using stoichiometric powder (MoO$_3$ or WO$_3$). Then, 2-cm^2-sized pads of ITO or IZO/ITO were sputter-deposited on the front side and a full-area ITO/Ag rear contact stack was applied. Front metal grids were deposited by Ag screen-printing and cured 20 minutes at 190 °C.

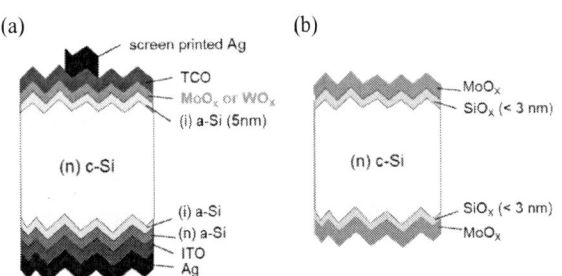

Fig. 1. Schematic sketches of (a) our solar cells with MoO$_x$ or WO$_x$ based hole selective front contact, (b) symmetrical structure to test silicon-oxide buffer layers.

Symmetrical test structures (Fig. 1b) were fabricated to evaluate the passivation of thin thermal silicon oxide layers (1.5-2.5 nm) as buffer layers underlying the MoO$_X$. For this, textured 200 μm thick (100) n-type Si wafers (~2 Ωcm) and double side polished (DSP) (100) n-type Si wafers (280 μm, ~3 Ωcm) were cleaned in standard wet-chemical solutions. Thin thermal oxide layers were grown in a Rapid Thermal Processing (RTP) furnace using different peak temperatures T_{peak} and oxidation times. Part of the samples were hydrogenated *in-situ* by a Forming Gas Annealing (FGA) process (30 min, 500 °C). Afterwards, all wafers were loaded in a vacuum chamber to evaporate MoO$_X$ layers which had on glass thicknesses of about 15 nm. Finally, the minority carrier lifetimes were determined by QSSPC (Quasi Steady State Photo Conductance) measurements.

III. CELLS WITH WO$_X$ BASED FRONT CONTACTS

Figure 3 shows the JV-curve of our so far best cell with WO$_X$-based hole-selective contact and its comparison to a cell with an unoptimized MoO$_X$-based contact fabricated in the same evaporation chamber at low base pressures (10^{-6} mbar). In both cases, sputtered IZO [9] was used as the front TCO layer, deposited in the same chamber immediately after the MoO$_X$ or WO$_X$ thermal evaporation without breaking the vacuum. The cells were characterized after annealing the screen-printed Ag front contact at 190°C. Even though the WO$_X$ based cell achieves a higher short-circuit current density J$_{SC}$, its V$_{OC}$ equals only 638 mV (Table 1) compared to 705 mV for the MoO$_X$ based cell. Both cells suffer from an S-shaped JV-curve close to the V$_{OC}$, leading to fill factors of only 70% (WO$_X$) and 73% (MoO$_X$). Increasing the WO$_X$ thickness leads to an even stronger S-shape and further reduction in FF (data not shown here). Most likely, the observed low V$_{OC}$ and efficiencies of cells with WO$_X$ based front contact are due to an insufficient band bending provided by our thermally evaporated WO$_X$ material, as discussed in ref. [10].

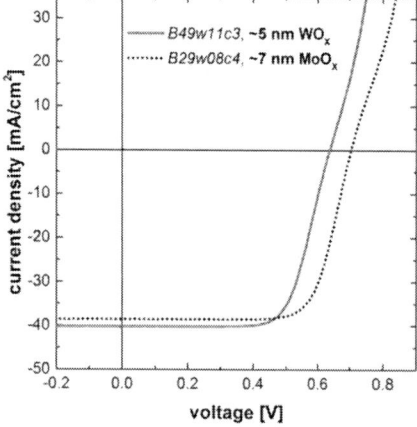

Fig. 2. Light JV-curves of solar cells with MoO$_X$ and WO$_X$ based hole-selective front contact measured after annealing at 190 °C

Table 1. JV parameters of cells of the cells with MoO$_X$ and WO$_X$ based hole-selective contacts from Figure 2.

cell-ID	V$_{oc}$ [mV]	J$_{sc}$ [mA/cm^2]	FF [%]	eff. [%]
B29w08c4, ~7nm MoO$_X$	705	38.6	72.8	19.8
B49w11c3, ~5nm WO$_X$	638	40.2	70.1	18.0

IV. IMPACT OF BASE PRESSURE ON MoO$_X$ BASED CONTACTS

Figure 2 compares JV curves of cells with MoO$_X$-based hole-selective front contacts in which the MoO$_X$ layers were evaporated in vacuum chambers pumped down to different base pressures. We specifically investigated the influence of the water partial pressure by comparing results using a tool equipped with a glovebox and a transfer chamber as a water-free deposition tool (labelled N$_2$), and a tool opened to air before pumping down (thus with most residual pressure being water vapor, labelled as H$_2$O). Pumping times were adjusted in both tools to reach a base pressure around 10^{-5} mbar (labelled high p) or 10^{-6} mbar (labelled low p). ITO was then sputter-deposited on all samples, a silver grid was screen printed, and samples were cured at a low temperature of 130 °C. All JV curves show very similar V$_{OC}$ and J$_{SC}$. A slight S-shape is observed in all cases, yet much more pronounced for the sample prepared with a high base pressure in the tool vented to atmosphere (high p H$_2$O). This suggests that residual water during evaporation can impact negatively the performance of MoO$_X$-based devices.

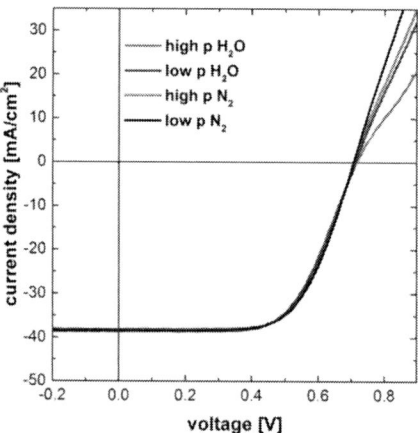

Fig. 3. JV curves of SHJ cells with (i)a-Si:H/MoO$_X$/ITO hole-selective front contact. The MoO$_X$ layers were deposited after evacuating the chamber to a high (~10^{-5} mbar) or low (10^{-6} mbar) base pressure. MoO$_X$ layers evaporated in the chamber installed in a N$_2$ glovebox lead to superior cell performance compared to those deposited in the chamber which was vented in air before each deposition (labelled "H$_2$O").

978-1-5090-5606-4/17 $31.00 © 2017 IEEE

IV. ALTERNATIVE BUFFER LAYERS FOR MoO$_X$ BASED HOLE-SELECTIVE CONTACTS

Figure 4 compares the minority carrier lifetimes of symmetrical test structures with different buffer layers underneath the thermally evaporated MoO$_x$. The investigated silicon oxide layers were generated by two different recipes leading to oxide thicknesses of 2.0 nm and 2.4 nm, as summarized in Table 2.

The thin thermal oxide layers result in higher lifetimes than achieved with a reference sample that was HF-dipped prior to the MoO$_x$ deposition (no buffer). However their lifetimes of 100 µs to 170 µs at an injection level of 1.10^{15} cm^{-3} are one order of magnitude lower than achieved with an (i)a-Si:H buffer layer (~6 ms). Our data indicate that the surface passivation increases with increasing oxide thickness and hydrogenation. Hydrogenation of the oxide layer 2 (DSP wafer) prior to MoO$_x$ deposition resulted in a higher lifetime (Fig. 4) which translates into an increase in implied V$_{OC}$ (iV$_{OC}$) from 603 mV to 621 mV. The same effect was observed on textured (TXT) wafers which achieved an iV$_{OC}$ up to 650 mV (hydrogenated oxide 2, Fig. 4).

Further analysis is necessary to investigate the carrier transport in SiO$_X$/MoO$_X$/TCO contact stacks and to achieve higher passivation levels comparable to (i)a-Si:H. We expect that there will be a tradeoff between passivation and series resistance for the oxide layer thickness when applied in a real solar cell device.

Table 2. Parameters used for the different rapid thermal oxidations. The oxide thicknesses were determined by spectral ellipsometry on reference DSP Si (100) wafers.

Oxide recipe	T$_{peak}$ [°C]	oxidation time [s]	Oxide thickness measured on DSP [nm]
Oxide 1	700	90	2.0
Oxide 2	750	90	2.4

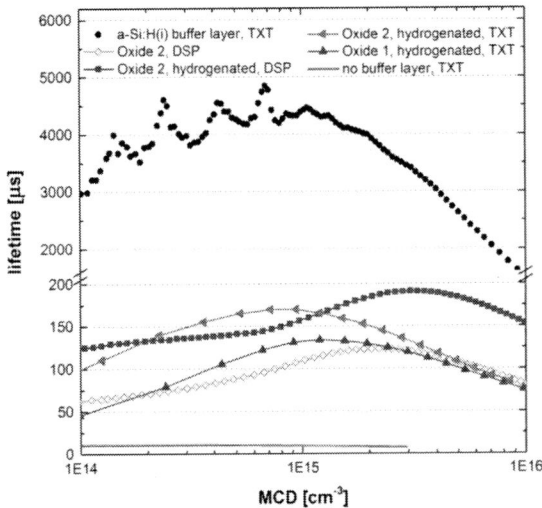

Fig. 4. Carrier lifetime versus minority carrier density (MCD) of symmetrical test structures (Fig 1b) with different buffer layers as described in the text.

V. CONCLUSION AND OUTLOOK

Our experimental results show that both MoO$_X$ and WO$_X$ can be used in hole-selective contacts. However, so far WO$_X$-based hole-selective contacts resulted in lower V$_{OC}$. The performance of a-Si:H/MoO$_X$/TCO hole-selective contact stacks depends critically on the base pressure and residual water vapor in the deposition chamber. Symmetrical test structures with stacks of 2.4 nm thick hydrogenated SiO$_x$ layers and MoO$_x$ resulted in iV$_{OC}$ of 650 mV. Further investigation is required to evaluate its applicability in a real device and its optimal properties for best performance.

ACKNOWLEDGEMENTS

The authors would like to thank Raphaël Monnard and Guillaume Charitat from EPFL and Nicolas Badel from CSEM for work performed in the context of this publication. S. Essig holds a Marie Skłodowska-Curie Individual Fellowship from the European Research Council (ERC) under the European Union's Horizon 2020 research and innovation programme (grant agreement No: 706744, action acronym: COLIBRI). Part of this work was supported by the European Union's Horizon 2020 Programme for research, technological development and demonstration Grant Agreements no. 727529 (project DISC).

REFERENCES

1. Battaglia, C., A. Cuevas, and S. De Wolf, *High-efficiency crystalline silicon solar cells: status and perspectives.* Energy & Environmental Science, 2016. **9**(5): p. 1552-1576.

2. Bullock, J., et al., *Efficient silicon solar cells with dopant-free asymmetric heterocontacts.* Nature Energy, 2016. **1**: p. 15031.

3. Geissbühler, J., et al., *22.5% efficient silicon heterojunction solar cell with molybdenum oxide hole collector.* Applied Physics Letters, 2015. **107**(8): p. 081601.

4. Werner, J., et al., *Parasitic Absorption Reduction in Metal Oxide-Based Transparent Electrodes: Application in Perovskite Solar Cells.* ACS Applied Materials & Interfaces, 2016. **8**(27): p. 17260-17267.

5. Geissbühler, J., et al., *Silicon Heterojunction Solar Cells With Copper-Plated Grid Electrodes: Status and Comparison With Silver Thick-Film Techniques.* IEEE Journal of Photovoltaics, 2014. **4**(4): p. 1055-1062.

6. Essig, S., et al., *MoO$_X$-based hole-selective contacts for Si photovoltaics: hydrogen-rich layers as fill-factor killers.* submitted.

7. Gerling, L.G., et al., *Transition metal oxides as hole-selective contacts in silicon heterojunctions solar cells.* Solar Energy Materials and Solar Cells, 2016. **145, Part 2**: p. 109-115.

8. Bivour, M., et al., *Molybdenum and tungsten oxide: High work function wide band gap contact materials for hole selective contacts of silicon solar cells.* Solar Energy Materials and Solar Cells, 2015. **142**: p. 34-41.

9. Morales-Masis, M., et al., *Low-Temperature High-Mobility Amorphous IZO for Silicon Heterojunction Solar Cells.* IEEE Journal of Photovoltaics, 2015. **5**(5): p. 1340-1347.

10. Mews, M. et al., *Oxygen vacancies in tungsten oxide and their influence on tungsten oxide/silicon heterojunction solar cells.* Solar Energy Materials & Solar Cells 158, pp. 77–83 (2016).

Metal Nanoparticle Hole Contacts for Silicon Solar Cells

James Bullock, Zhaoran Xu, Mark Hettick, Yimao Wan, Ali Javey

Department of Electrical Engineering and Computer Sciences, University of California, Berkeley, California 94720, USA.

Materials Sciences Division, Lawrence Berkeley National Laboratory, Berkeley, California 94720, USA

Abstract — **Recent years have seen the development of a broad range of novel contacting strategies for crystalline silicon (c-Si) solar cells. This study is focused on the use of a sparse layer of platinum nanoparticles to enhance the hole contact characteristics of a c-Si / transparent conductive oxide hole contact. It is shown that the addition of the Pt nanoparticle layer results in a 2 order of magnitude reduction in contact resistivity on both lightly and heavily doped p-type surfaces with values of ~3 and ~0.2 mΩcm², respectively. A high transparency can be maintained for such contacts from strict control of the nanoparticle deposition process with a predicted J_{sc} loss of just 0.15 mA/cm² as a result of the Pt nanoparticle layer. Finally, the thermal and damp heat stability of the Pt nanoparticle contact is investigated, revealing promising initial results.**

Index Terms — **Silicon, metal nanoparticles, photovoltaic cells, contact resistivity.**

I. INTRODUCTION

In recent times, there has been a renewed interest in the development of novel contacting structures for high efficiency crystalline silicon (c-Si) solar cells. This push has been motivated by performance limiting issues with conventional screen-printed doped contacts. Included in the growing list of alternative contacting materials are doped silicon layers [1], [2], metal oxides [3]–[5], alkali and alkaline earth metal salts [6], [7], organic materials [8], and high/low work function metals [9]. These rely on favorable band offsets or band bending induced by strong work functions to selectively extract electrons or holes.

In this study, we explore the use of high work function metal nanoparticles as an effectively transparent interface modification layer to improve hole contact characteristics of c-Si solar cells. The study is split into two consecutive sections focused first on the demonstration of the proof-of-concept, followed by an investigation of the thermal and damp heat stability.

II. PROOF-OF-CONCEPT PT NANOPARTICLE CONTACT

The basic concept of this study, as highlighted in Figure 1a, is the use of a sparse, transparent layer of platinum nanoparticles to enhance the hole contact characteristics of the indium tin oxide (ITO) / c-Si interface. In these initial experiments, a focus is made on the ability of such a scheme to improve the contact resistivity. Future studies could focus on looking at factors such as contact recombination with, for example, the integration of a thin passivating interlayer [10], [11] or any potential additional light management benefits of metal nanoparticles.

To first examine the influence of the Pt deposition on ρ_c, simple transfer length method (TLM) contact structures are employed. These structures are fabricated on ~1 Ωcm, p-type, float-zone, (100), wafers with a thickness of ~250 μm. Pt nanoparticles are deposited by electron beam evaporation

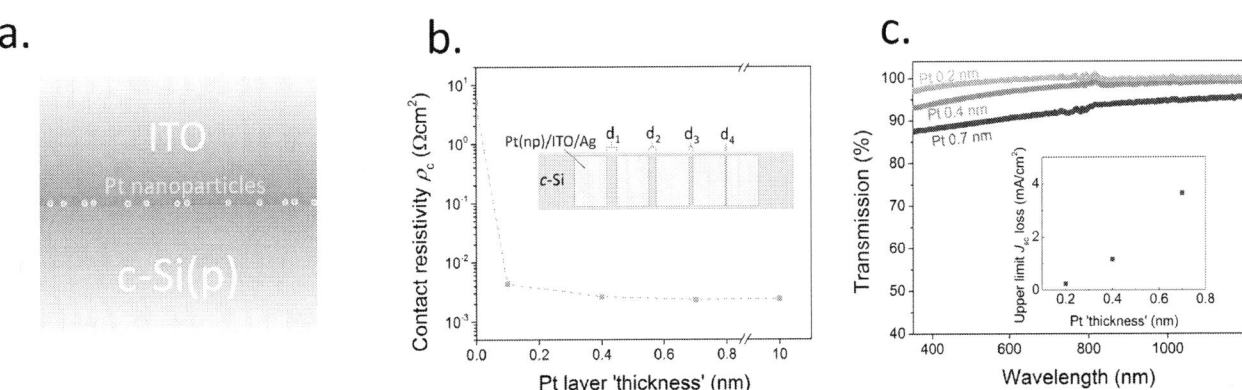

Fig. 1. a.) concept diagram of Pt nanoparticle enhanced hole contact; b.) dependence of contact resistivity on Pt layer 'thickness' for c-Si(p) / Pt / ITO contacts. c.) transmission spectrum of different Pt layer 'thicknesses' across the UV, visible and near IR portions of the spectrum. The inset shows the estimated upper limit current loss for different Pt layer thicknesses.

978-1-5090-5606-4/17 $31.00 © 2017 IEEE

through a TLM shadow mask to a layer 'thickness' in the 0.1-0.7 nm range (as judged using a crystal monitor). Such deposition conditions have been previously shown to produce a discontinuous layer of nanoparticles. On top of the Pt nanoparticles, an ITO (~120 Ω/sq., ~55 nm) / Ag (~200 nm) stack is deposited through the same TLM shadow mask via RF sputtering and thermal evaporation, respectively. Figure 1b shows the extracted ρ_c, dependence on Pt layer thickness, where values for samples without Pt (c-Si / ITO) and continuous layers of Pt (c-Si / 10nm Pt / ITO) have been included for reference. It should be noted that the ρ_c value provided for the c-Si / ITO sample should be taken as an estimate only due to difficulties with accurately extracting ρ_c values in this range. As can be seen, even with a 'thickness' of 0.1 nm, a reduction in the hole contact ρ_c of more than two orders of magnitude can be obtained. The addition of a thicker and even fully continuous Pt layer does not significantly decrease the ρ_c further, which instead plateaus at a value of ~2 mΩcm².

As discussed earlier, a second important factor to the proof-of-concept for this contact is its transparency. To investigate this, platinum nanoparticles are deposited on thin glass slides (~200 μm) and the transmission (corrected using a blank glass slide background) is measured across the near ultra-violet (UV), visible and near infrared (IR) portions of the spectrum. A summary of these measurements is provided in Figure 1c. The expected decrease in transmission with 'thicker' Pt deposition is clearly seen, however, a 0.2 nm layer results in almost no transmission loss across the full UV-visible-IR range studied. To further quantify this loss, the inset of Figure 1c provides an upper-limit estimation of J_{sc} loss, assuming such a contact was used on the front sunward side of a solar cell. This plot is prepared using the AM 1.5G spectrum to spectrally weigh the transmission loss. From this analysis, it can be seen that for nanoparticle 'thicknesses' of 0.2 nm or less the J_{sc} loss is limited to 0.15 mA/cm² – which is acceptable even in high efficiency solar cell designs.

The above extracted ρ_c of ~2 mΩcm² and the predicted J_{sc} loss of only 0.15 mA/cm² suggest that this contact strategy may be suitably applied as the rear contact of a bifacial p-type partial rear contact (PRC) cell, which could present significant efficiency advantages over conventional mono-facial p-type PRC cells.

To extend the above analysis to heavily p-type doped surfaces, ITO (~55 nm) / Ag (~200 nm) TLM contact structures are prepared both with and without a 0.2 nm Pt interfacial layer. These are fabricated on high resistivity silicon wafers with boron diffused surfaces (~200 Ω/sq, N_{surf} ~ 1.6×10^{19} cm⁻³). The ρ_c comparison of these two structures again reflects an improvement in contact performance by more than 2 orders of magnitude from 26 mΩcm² to 0.25 mΩcm² for the structures without and with the Pt nanoparticle interlayer, respectively. This result suggests that the use of Pt nanoparticles may also be useful as an interlayer in monolithic 2-terminal tandem solar cell devices employing a silicon n/p^+ homojunction bottom cell.

III. CONTACT STABILITY

An additional important demonstration in the proof-of-concept for this contact structure is to investigate its stability under thermal and damp heat stresses. To achieve this a series of c-Si(p) / Pt / ITO / Ag TLM contact structures, identical to those described above, are fabricated and subjected to worst case scenario, unencapsulated environmental stressing.

Figure 2a shows the measured thermal stability in ρ_c for contacts with Pt nanoparticle layers of different thickness, as assessed by annealing samples on a hotplate in sequential 10 minute anneals at increasing temperatures between 100°C and 300°C. From this plot, it can be seen that a greater thermal stability is afforded by thicker layers, with 0.7 nm layers resulting in negligible change over the measurement range. Even for lower thicknesses reasonable ρ_c values can be maintained if temperatures are kept to less than 200°C.

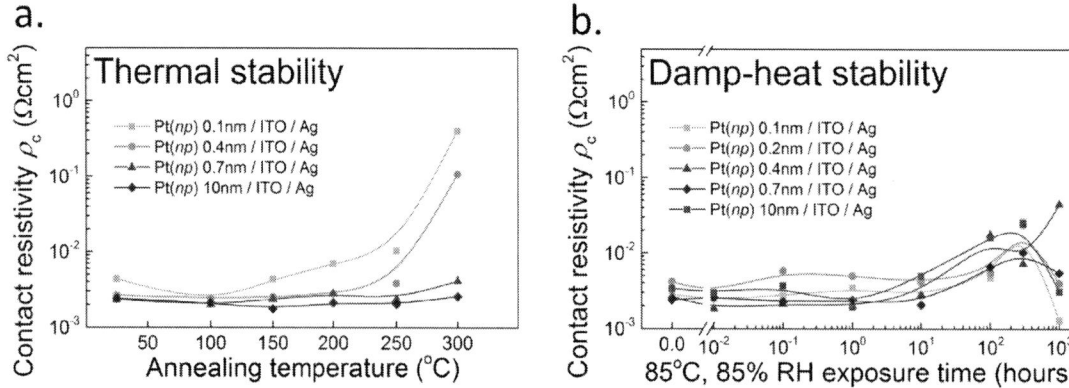

Fig. 2.   a.) Thermal and; b.) damp-heat stability of c-Si(p) / Pt nanoparticle / ITO contact resistivity samples with different Pt nanoparticle deposition conditions.

Figure 2b shows an analogous set of results for contact structures subjected to standard damp-heat conditions (85°C, 85% relative humidity) for up to 1000 hours. While no thickness dependence can be seen, after 1000 hours most of the contacts still have ρ_c values on the mΩcm^2 scale suggesting that damp-heat stability is not a fundamental issue for this contact scheme.

IV. CONCLUSION

This study has been focused on the development of a transparent Pt nanoparticle / ITO hole contact to c-Si. It is found that with the addition of a Pt nanoparticle interfacial layer, a reduction in contact resistivity of more than two orders of magnitude can be achieved without significantly affecting the transparency of the contact. Further, the stability of this contact is explored under both thermal and damp-heat stressors.

While the specific demonstration in this study has been limited to a contact resistivity analysis, a number of potential additions to the concept are highlighted. In particular, the addition of thin passivating interlayers to improve surface recombination, or the possibility of integrating particles with favorable nanophotonic properties such as engineered light scattering could be studied in the future.

ACKNOWLEDGEMENT

Device design, fabrication and characterization were funded by the Bay Area Photovoltaics Consortium (BAPVC).

REFERENCES

[1] K. Yoshikawa *et al.*, "Silicon heterojunction solar cell with interdigitated back contacts for a photoconversion efficiency over 26%," *Nat. Energy*, vol. 2, no. 5, p. nenergy201732, Mar. 2017.

[2] S. W. Glunz *et al.*, "The Irresistible Charm of a Simple Current Flow Pattern – 25% with a Solar Cell Featuring a Full-Area Back Contact," *31st Eur. Photovolt. Sol. Energy Conf. Exhib.*, pp. 259–263, Nov. 2015.

[3] C. Battaglia *et al.*, "Hole Selective MoOx Contact for Silicon Solar Cells," *Nano Lett.*, vol. 14, no. 2, pp. 967–971, 2014.

[4] J. Geissbühler *et al.*, "22.5% efficient silicon heterojunction solar cell with molybdenum oxide hole collector," *Appl. Phys. Lett.*, vol. 107, no. 8, p. -, 2015.

[5] S. Avasthi, W. E. McClain, G. Man, A. Kahn, J. Schwartz, and J. C. Sturm, "Hole-blocking titanium-oxide/silicon heterojunction and its application to photovoltaics," *Appl. Phys. Lett.*, vol. 102, no. 20, 2013.

[6] Y. Wan *et al.*, "Magnesium Fluoride Electron-Selective Contacts for Crystalline Silicon Solar Cells," *ACS Appl. Mater. Interfaces*, vol. 8, no. 23, pp. 14671–14677, Jun. 2016.

[7] J. Bullock *et al.*, "Efficient silicon solar cells with dopant-free asymmetric heterocontacts," *Nat. Energy*, no. 1, p. 15031, 2016.

[8] D. Zielke, A. Pazidis, F. Werner, and J. Schmidt, "Organic-silicon heterojunction solar cells on n-type silicon wafers: The BackPEDOT concept," *Sol. Energy Mater. Sol. Cells*, vol. 131, pp. 110 – 116, 2014.

[9] T. G. Allen *et al.*, "A Low Resistance Calcium/Reduced Titania Passivated Contact for High Efficiency Crystalline Silicon Solar Cells," *Adv. Energy Mater.*, p. n/a–n/a, Feb. 2017.

[10] J. Bullock, D. Yan, and A. Cuevas, "Passivation of aluminium n+ silicon contacts for solar cells by ultrathin Al2O3 and SiO2 dielectric layers," *Phys. Status Solidi RRL Rapid Res. Lett.*, vol. 7, no. 11, pp. 946–949, 2013.

[11] D. Zielke, J. H. Petermann, F. Werner, B. Veith, R. Brendel, and J. Schmidt, "Contact passivation in silicon solar cells using atomic-layer-deposited aluminum oxide layers," *Phys. Status Solidi RRL Rapid Res. Lett.*, vol. 5, no. 8, pp. 298–300, 2011.

Near-Field Transport Imaging Application of Photovoltaic Materials

Chuanxiao Xiao,[1,2] Chun-Sheng Jiang,[1] John Moseley,[1] John Simon,[1] Kevin Schulte,[1] Aaron J. Ptak,[1] Steve Johnston,[1] Brian Gorman,[2] Mowafak Al-Jassim,[1] Nancy M. Haegel,[1] and Helio Moutinho[1]

[1] National Renewable Energy Laboratory, Golden, CO 80401 USA
[2] Colorado School of Mines, Golden, CO 80401 USA

Abstract — **We applied a novel analytical technique—near-field transport imaging (TI)—to photovoltaic materials for charge-carrier transport mapping in nanometer-scale. We measured the diffusion length of a well-controlled gallium arsenide (GaAs) thin-film samples and it agrees well with the results calculated by time-resolved photoluminescence. We report for the first time on TI experiments on thin-film cadmium telluride, including the effective carrier diffusion length, as well as the first near-field imaging of the effect of a single small defect on carrier transport and recombination in a GaAs sample. Furthermore, by changing the scanning setup, we were able to do near-field cathodoluminescence (CL), and correlated the results with standard CL results. The TI technique shows great potential for high spatial resolution mapping transport properties in solar cell materials.**

Index Terms — Transport imaging, nanometer-scale, effective diffusion length, near-field, single defect.

I. INTRODUCTION

Measurement of the minority-carrier diffusion, especially localized transportation, is vital to improving solar cell device performance. Transport imaging (TI) is an innovative approach that integrates near-field scanning optical microscopy (NSOM) with the imaging and highly localized excitation capabilities provided by incident electrons in a scanning electron microscope (SEM), to enable "seeing" carrier transport by imaging light associated with local recombination [1, 2].

In this work, we applied near-field TI technique on gallium arsenide (GaAs) and cadmium telluride (CdTe) solar cell materials. We use a Matlab data-fitting procedure to extract diffusion length (Ld) values from the TI measurements and demonstrate good agreement between the Ld values determined from time-resolve photoluminescence (TRPL) and TI data on GaAs samples. Most importantly, we report for the first time measurements on thin-film CdTe using the TI technique, including the determination of effective carrier diffusion length, as well as the first near-field cathodoluminescence (CL) imaging of the effect of a single localized defect on carrier transport and recombination in a GaAs heterostructure.

II. EXPERIMENT DETAILS

The TI technique combines two microscopes—a Hitachi S4000-N SEM and a Nanonics Multiview 2000 NSOM with an atomic force microscope (AFM) apparatus [3]. In this technique, the focused electron beam of SEM is used to generate excess carriers within a micro- to nano-sized volume of subsurface material while the beam is fixed at a point or in a linescan mode. Carrier diffusion is tracked by an NSOM probe with a nano-size aperture scanning above the surface and detecting the local light emission due to carrier radiative recombination.

Near-field CL can be performed with both the electron beam and NSOM probe fixed at a certain distance and letting the sample move. In the measurement, the light is collected at a given distance from its generation point, after part of the generated carrier population has already recombined.

III. RESULTS AND DISCUSSIONS

A. Measuring Diffusion Lengths of GaAs and CdTe

We used TI to investigate the transport properties of heterostructure GaAs and large-grain CdTe samples (see Fig. 1). The electron beam is fixed at the center of the lower edge of the image—with the SEM set in spot mode and in a voltage of 20 kV—while the optical fiber is scanned to collect the spatial distribution of the luminescence. Fig. 1a shows the result from the GaAs sample, and Fig. 1b shows the intensity variation marked by the green line in Fig. 1a; clear decay was observed in the image as the diffusion and recombination of the excess minority carriers occurred. On the lower edge of the image, the dark spot is because of the NSOM probe shading the electron beam. Because the experimental geometry is a point source on a thin film (point source in quasi 2-D film) [1], we developed a data-fitting procedure for extracting reliable diffusion length values by a Bessel function,

$$y = k \cdot \text{Besselk}(0, x/L_d) + b,$$

where Besselk is a modified Bessel function of the second kind, and k, L_d, and b are fitting parameters: k is the scaling factor and b is the offset of background. The estimated diffusion length is $5.24 \pm 0.03\ \mu m$. The other two white lines were also fitted in the same way and the results are similar.

This TI-measured value of L_d was compared with the one obtained indirectly from the well-established TRPL measurement. The TRPL-measured value of τ is 15.7 ns, and the value of the diffusion coefficient, D, is estimated from the doping (3×10^{16} cm^{-3}) and the majority-carrier electron mobility in GaAs (5,000 cm^2/Vs). According to the relation between L_d and τ, $L_d = \sqrt{D \cdot \tau}$, the calculated value for L_d is 4.9 μm. This value is very close to the TI-measured L_d (5.2 μm). However, the calculation from TRPL also includes

978-1-5090-5606-4/17 $31.00 © 2017 IEEE

uncertainty associated with majority versus minority carrier mobility estimates. It is important to note that the TI provides a direct image of where the carriers go—including all factors such as bulk behavior, surface recombination, and the extent to which ambipolar diffusion plays a role. Thus, the TI image shows the actual behavior of interest in a given structure.

Similarly, TI was applied on a large-grain CdTe sample. The TI image is showed in Fig 1c, and Fig. 1d shows the intensity variation of the green line in Fig. 1c. One can see that

the decay is much faster than the one of high-quality GaAs. We used the same fitting procedure and found $L_d = 0.70 \pm 0.07\,\mu m$, which is within the range generally expected for similar CdTe samples. Different directions of diffusion (marked by white lines) on the same image were also examined, and L_d values are similar. To our knowledge, this is the first time that near-field TI measurements, including calculation of L_d values, have been reported for a polycrystalline thin film.

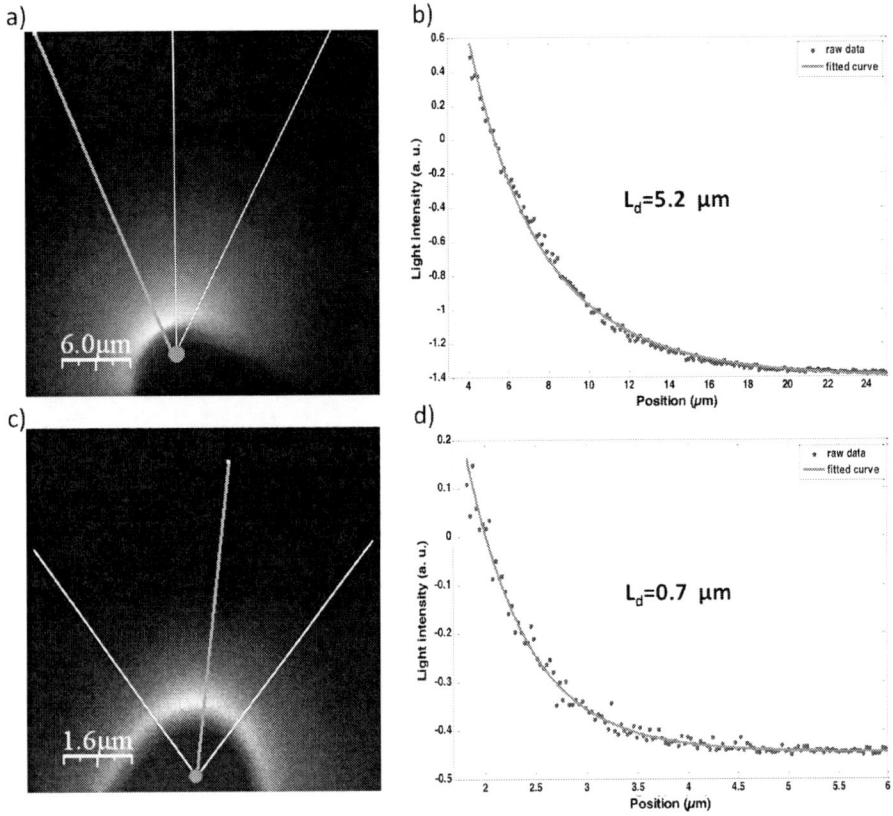

Fig. 1. a) TI image of GaAs sample; b) light intensity decay and fitting of the green line in a); c) TI image of CdTe sample; d) light intensity decay and fitting of the green line in c).

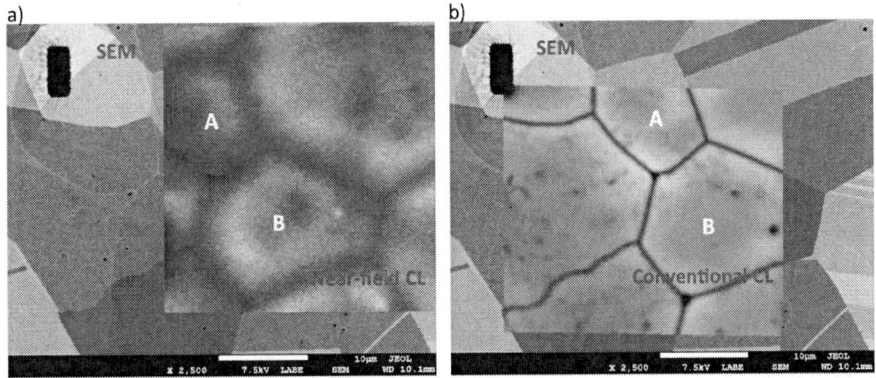

Fig. 2. Comparison between near-field CL (a) and conventional CL (b) analyses on the previous TI-analyzed CdTe thin film. Same grains on both images are marked A and B.

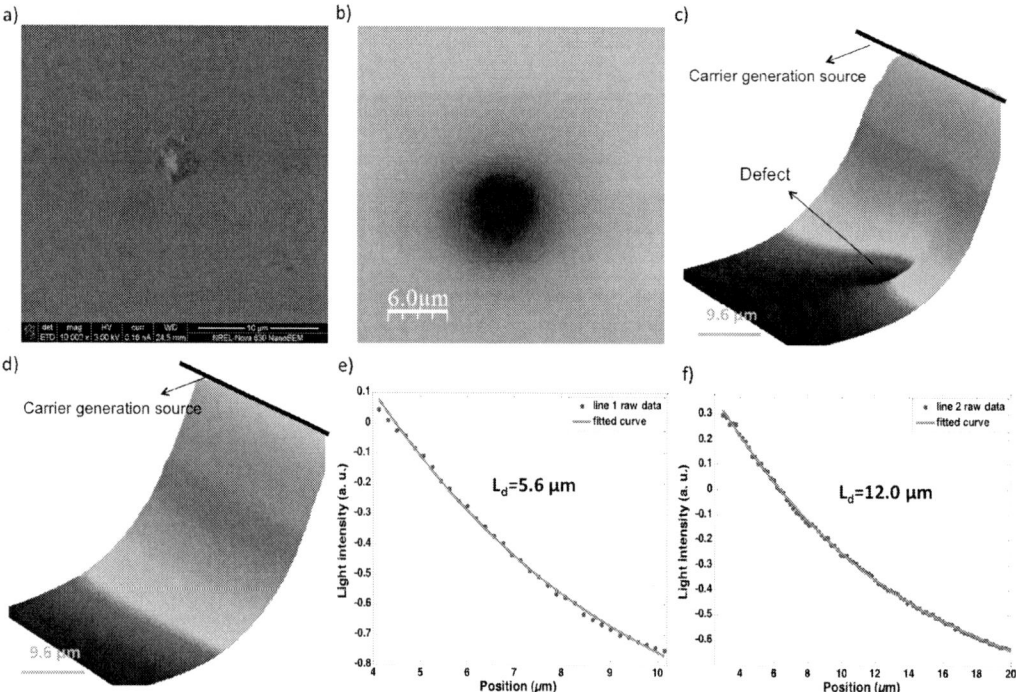

Fig. 3. a) An SEM image of the defective area; b) Near-field CL of the defective area; c) 3-D transport image with the line generation source in the vicinity of the defect; d) 3-D transport image away from the defect; e) Light intensity decay and fitting in c) around the defect; f) Light intensity decay and fitting in d).

B. Near-Field CL of CdTe

In this TI setup, a near-field CL measurement can be performed with the NSOM probe and the electron beam stationary, about 1 μm away from each other, and with the sample moving. Fig. 2a shows the near-field CL map, and Fig. 2b shows the conventional CL map (technique detail see Ref. [4]); the two measurements were done on the same location with the SEM backscattered image in the background. All the same grains and grain boundaries can be identified in the two images, where "A" and "B" in the images indicate the same grains.

The images are very similar because with the proximity between the electron beam and NSOM probe in TI, we probe the recombination signal close to the generation point (whereas conventional CL measures the total far-field luminescence mapped to the point of excess carrier generation). Note that this is an additional mode of analysis that makes the TI technique more versatile.

C. Transport and Near-Field CL Imaging of a Defective GaAs

Near-field CL (Fig. 3b) was first performed in a region with a clear defect as indicated by the SEM image (Fig. 3a), and we found that the defective area has weaker luminescence intensity. The dark region is larger than the area indicated by SEM. This is not surprising because the material around the observed defective area is normally not as high quality as material away from the defect.

The TI imaging was acquired with the electron beam operating in linescan mode. It is easier to observe the defect on local transport by using a line source for carrier generation, which results in a net 2-D carrier diffusion perpendicular to the line source, rather than random diffusion in every direction with the electron beam in a fixed point. Two images were collected and are 3-dimensional: in Fig. 3c, the excitation source is in the vicinity of the defect, whereas the one in Fig. 3d is away from the defect as a reference. The black lines indicate electron-beam source generation. The defect significantly affects the local carrier transportation, as nonuniform decay was observed in Fig. 3c. As carriers diffuse, when the carriers "see" the defect, the luminescence is much weaker (orange color indicates highest luminescence intensity). Similar to the previous fitting procedure, a line near the defect of Fig. 3c was fitted by Matlab, and the diffusion length value is $5.6 \; \mu m$ (see Fig. 3e). In Fig. 3d, the carrier diffusion length is $12 \; \mu m$ (see Fig. 3f). The TI mapping results image the local carrier diffusion process, rather than giving a single averaged value of diffusion length.

IV. CONCLUSION

We applied the near-field TI technique to two important photovoltaic materials, GaAs and CdTe. The diffusion lengths were measured directly, and the value from the well-controlled GaAs sample agrees with TRPL results. Near-field CL was also conducted on the large-grain CdTe sample and clearly identifies grains and grain boundaries. Most importantly, a defective GaAs sample was investigated by TI and the results showed significant local carrier transportation variations. The developed TI technique will be a powerful tool to investigate localized transport properties of photovoltaic materials in high spatial resolution.

ACKNOWLEDGEMENTS

The authors thank Dr. David Albin (National Renewable Energy Laboratory) for providing the CdTe sample. C. X. thanks Dr. Xianguang Yang (Jinan University, P. R. China) for fruitful discussion on NSOM scanning, and Dr. Mingjian Cui and Yuanzhi Liu (The University of Texas at Dallas, USA) for helpful discussion of Matlab fitting. This work was supported by the U.S. Department of Energy under Contract No. DE-AC36-08GO28308 with the National Renewable Energy Laboratory. The U.S. Government retains and the publisher, by accepting the article for publication, acknowledges that the U.S. Government retains a nonexclusive, paid up, irrevocable, worldwide license to publish or reproduce the published form of this work, or allow others to do so, for U.S. Government purposes.

REFERENCES

[1] N. M. Haegel, Nanophotonics 3(1–2), 75–89, 2014.

[2] D. R. Luber, F. M. Bradley, N. M. Haegel, M. C. Talmadge, M. P. Coleman, and T. D. Boone, Appl. Phys. Lett. 88, 163509, 2006.

[3] C. Xiao, C.-S. Jiang, J. Moseley, J. Simom, K. Schulte, A. J. Ptak, S. Johnston, B. Gorman, M. Al-Jassim, N. M. Haegel, H. Moutinho, Solar Energy 153, 134-141 (2017).

[4] J. Moseley, W. K. Metzger, H. R. Moutinho, N. Paudel, H. L. Guthrey, Y. Yan, R. K. Ahrenkiel, and M. M. Al-Jassim, J. Appl. Phys. 118, 025702, 2015.

Applications of DMD-based Inhomogeneous Illumination Photoluminescence Imaging for Silicon Wafers and Solar Cells

Yan Zhu, Mattias Klaus Juhl, Thorsten Trupke, and Ziv Hameiri

The University of New South Wales, Sydney, NSW, 2052, Australia

Abstract — Under inhomogeneous illumination, a silicon wafer or solar cell can generate lateral current flow due to gradients in the quasi Fermi energy levels. This lateral current flow is impacted by a few of the key electrical parameters of the sample, such as emitter sheet resistance, series resistance and diffusion length. This lateral current flow can be directly correlated to the spatially resolved luminescence intensity from the photoluminescence image. Therefore, photoluminescence imaging with inhomogeneous illumination can provide additional information, which is usually difficult to extract from a conventional photoluminescence image at open circuit voltage.

Index Terms — photoluminescence image, DMD, solar cells, silicon wafers, series resistance, diffusion length.

I. INTRODUCTION

Photoluminescence (PL) imaging is a fast and powerful characterization technique for silicon wafers and solar cells [1]. In conventional PL imaging, uniform illumination is used. The regions with lower quality (low lifetime or shunts) can be directly identified from the lower counts in the PL image. In order to obtain resistance information from luminescence imaging based techniques, current flows must be stimulated, which is commonly done by electroluminescence (EL) [2] or PL with current extraction (PLCE) [3]. Both techniques require electrical contact. The requirement of contact increases the possibility of damaging the sample and impedes the application to "on-the-fly" inline inspection. One way to induce lateral current flow without making electrical contact is using inhomogeneous generation across the sample. By applying inhomogeneous generation, Kasemann *et al.* [4] have proposed a contactless technique to extract qualitative series resistance (R_s) image. The idea of inhomogeneous generation PL imaging was also applied by Juhl *et al.* [5] to extract emitter sheet resistance (ρ_{she}) of diffused silicon wafers.

In the two cases mentioned above, the inhomogeneous generation was achieved by a combination of conventional homogeneous illumination with an opaque mask or optical filter placed on top of the sample. These methods can be limited by the physical geometry of samples and still require contact of the mask with the sample. These disadvantages can be easily solved by using a digital micromirror device (DMD) to achieve an inhomogeneous illumination light source. A DMD consists of a rectangular array of micrometer sized mirrors which can be controlled individually between on and off states. Each micromirror is imaged onto a different location in the sample plane. When in the on state a single mirror reflects light onto the sample, and when in the off state the light is directed away from the sample. In this study, a DMD-based PL setup is presented and the two aforementioned applications for resistance measurements will be demonstrated. Additionally, the capability of inhomogeneous illumination PL for measuring diffusion length will be presented.

II. EXPERIMENTS

A. Experimental setup

A sketch of the DMD-based PL imaging setup is shown in Fig. 1. An ultra-high-performance (UHP) lamp is used as the light source and the light is firstly guided to a DMD chip. The light is then either reflected into a light sink or projected to the sample. A 950 nm short pass filter is mounted after the projection lens to filter out light with wavelength above 950 nm. The intensity of the illumination light was measured to be between 0.1 and 0.4 suns, depending on the working distance. The emitted PL is captured by a silicon charge-coupled device (CCD) camera with a long pass filter in the front of the lens.

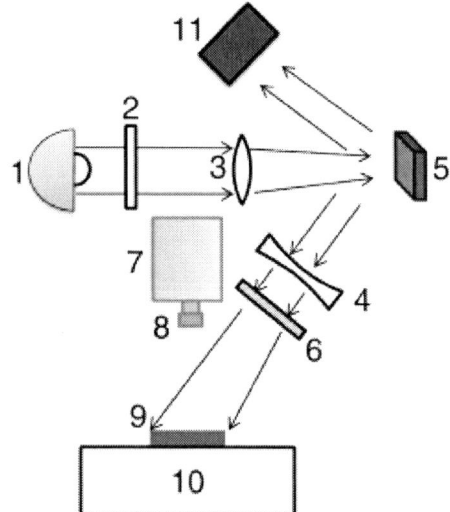

Fig. 1. Sketch of the DMD-based PL imaging setup with: (1) UHP lamp and concave mirror; (2) beam homogenizer; (3) and (4) lenses; (5) DMD; (6) 950 nm short pass filter; (7) silicon CCD

camera; (8) long pass filter; (9) sample; (10) stage; and (11) light sink.

B. Contactless series resistance imaging

We first apply this DMD-based PL imaging setup to contactless R_s imaging following the method of Kasemann et al. [4] A mono-crystalline silicon aluminum back-surface-field solar cell with three busbars is used in this measurement.

To implement the method of Kasemann et al., three PL images are needed: one with full area homogeneous illumination and two with inhomogeneous illumination [4]. The three resulting images are shown in Figs. 2(a)-(c). Only half of the cell is displayed. In Fig. 2(b), only the top part of the cell is illuminated and carriers generated in this part flows to the bottom non-illuminated region, while in Fig. 2(c) the opposite happens. According to Kasemann et al. [4], the illuminated areas of these two partially illuminated PL images [top part of Fig. 2(b) and bottom part of Fig. 2(c)] are combined together and then divided by the full area illuminated image. The resulting image is shown in Fig. 2(d). This ratio image is proportional to the voltage difference of the two operation points, which is correlated to the local series resistance.

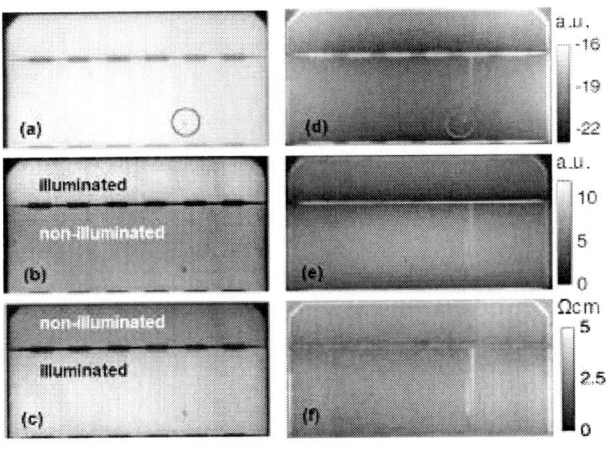

Fig. 2. (a) PL image of a solar cell with homogeneous illumination; (b) and (c) inhomogeneously illuminated PL images of the same cell; (d) contactless R_s image by the DMD-based PL; (e) contactless R_s image by the DMD-based PL with a modified method; and (f) R_s image obtained using the method of Kampwerth et al. [6].

Two finger breaks can be observed in Fig. 1(d). However, due to the relatively low intensity of the light source used in this study, the contrast of the resulting image is not high enough to suppress the lifetime artifact marked by the red circle. This can be solved by another advantage of the DMD-based method: the PL from the non-illuminated area can be also captured by the camera, whereas the PL from non-illuminated area will be blocked by the opaque mask in the original method of Kasemann et al. For the illuminated area, the captured image is actually PLCE, and high R_s regions

appear to be brighter. For the non-illuminated area, the image is actually EL, and high R_s image regions appear to be darker. Therefore, direct division of the two partially illuminated images with opposite illumination patterns can achieve a higher contrast R_s image. The result of this modified method is shown in Fig 2(e), where the top part is obtained by dividing Fig. 2(b) by Fig. 2(c), and the bottom part is the reverse. For comparison, the R_s image measured by the contact method of Kampwerth et al. [6] is presented in Fig. 2(f). As can be seen, the R_s image obtained by the modified method not only clearly indicates the high R_s area caused by broken fingers, but also significantly suppresses the lifetime artifact.

C. Contactless emitter sheet resistance measurement

We then use the setup to demonstrate the contactless method proposed by Juhl et al. [5] for ρ_{she} measurement of diffused silicon wafers. The set of samples with different ρ_{she} used in this section went through a similar process as Juhl et al. [5].

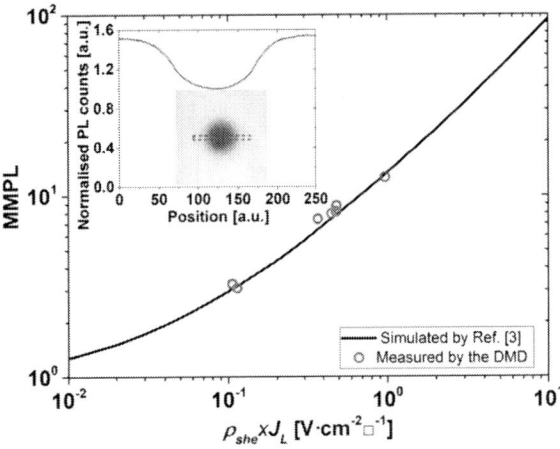

Fig. 3. Maximum to minimum PL counts ratio (MMPL) obtained by the DMD-based method as a function of the product of emitter sheet resistance and generation current, compared to the theoretical relationship from Ref. [5].

With the DMD-based PL setup, the inhomogeneous illumination was achieved by setting all the mirormorrors to the on state except for a small circular region. The resulting image and cross section PL counts of a representative sample are shown in Fig. 3. The diameter of the circle was chosen to be same as the diameter of the circular filter used by Juhl et al. [5]. The ratio between the maximum and minimum PL counts (MMPL) measured by the DMD-based method is plotted against the product of the photo-generation current (J_L) and ρ_{she} for each sample. The ρ_{she} of each sample was measured by four-point probe and the generation current was calculated by the integration of the sample absorption and the spectrum of the illumination light. A relationship between MMPL and $\rho_{she} \times J_L$ product developed by Juhl et al. [5] from simulation for this particular geometry is also shown in a solid back line. As can be seen in Fig. 2, the measured data agree well (in the range of 12%) with the predicated results. This is a clear demonstration of the ability of the proposed method to extract the ρ_{she} from inhomogeneous illumination PL imaging. By using the DMD-based method, the size of the circular pattern is no longer limited by the physical dimensions of the circular filter. A much smaller diameter can be used and a new correlation between MMPL and $\rho_{she} \times J_L$ can be easily obtained by repeating the simulation of Juhl *et al.* [5] Therefore, ρ_{she} scanning or even simultaneous measurement of multiple spots across a wafer can be easily implemented.

D. Diffusion length measurement

The lateral carrier diffusion induced by inhomogeneous illumination can also be used for the determination of the diffusion length of a high lifetime silicon wafer. For a silicon wafer with length and width much larger than its thickness, the excess carriers distribution from an illuminated area to a non-illuminated area can be calculated as [7]:

$$\Delta n(x) = A e^{\frac{-x}{L_{\text{eff}}}}, \qquad (1)$$

where Δn is the excess carrier concentration, x is the distance along the direction perpendicular to the edge of illumination in the sample surface, L_{eff} is the effective diffusion length. Under low injection, the PL intensity can be calculated as: $PL = C \cdot N_{dop} \cdot \Delta n$, where N_{dop} is the bulk doping concentration and C is a constant. The correlation between the PL intensity and the effective diffusion length can be obtained:

$$\left| \frac{\partial}{\partial x} \ln PL \right| = \left| \frac{\partial}{\partial x} \ln \Delta n \right| = \frac{1}{L_{\text{eff}}} \cdot \qquad (2)$$

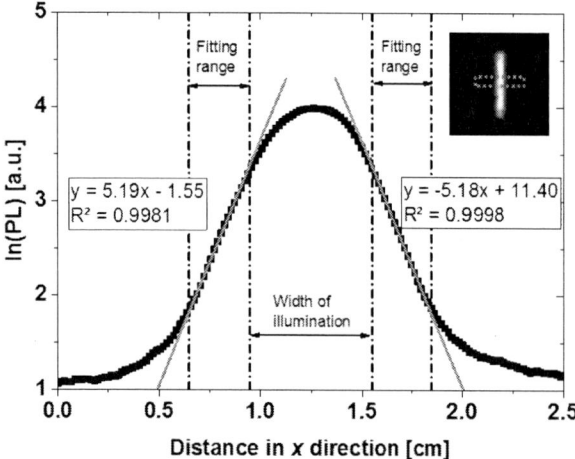

Fig. 4. Linear fit of the logarithm of the PL counts as a function of wafer distance. The diffusion length of the sample can be extracted from the inverse of the fitted slope. The inset presents the PL imaging of a sample by the DMD based setup.

Therefore, by inhomogeneously illuminating and measuring the gradient of the PL intensity, the diffusion length of the wafer can be extracted.

With the DMD-based method, the inhomogeneous illumination can be easily achieved by projecting a white strip on a dark background. In this experiment, a set of float-zone wafers with high quality amorphous silicon or aluminum oxide surface passivation was used. The diffusion length of the wafers was calculated to be between 0.18 to 0.29 cm based on effective lifetime measurements using a photoconductance-based system [8]. Fig. 4 presents the inhomogeneous illumination PL image of a representative wafer (inset) and a line scan of the logarithm PL counts. The dotted red box indicates the selected region used for the line scan. The black dashed lines indicate the width of the illumination strip and the range used for the linear fitting. The two red solid lines are fitting of the cross section PL counts just outside the illuminated region. According to Eq. (2), if the logarithm of the PL counts is plotted against the distance in the x direction, the effective diffusion length can be estimated as the inverse of slope. Using this method, a diffusion length of 0.19 cm was calculated for the sample presented in Fig. 4. The reduction of the slope in the region further away from the illuminated area is a result of the PL counts reaching the noise level of the image.

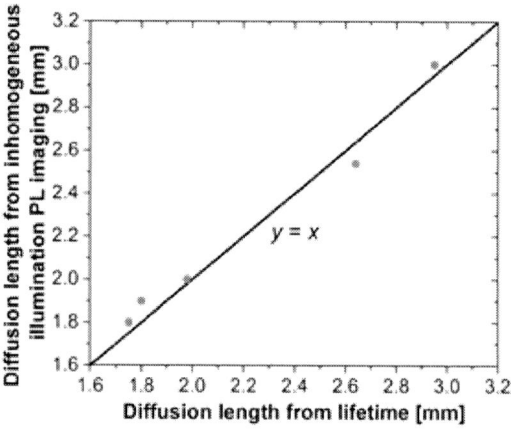

Fig. 5. Result of the diffusion length measurement by the DMD-based method.

Similarly, the diffusion lengths of other wafers were measured and the results are plotted in Fig. 5 against the diffusion length calculated from effective lifetime measurements. As can be seen, for all the wafers, the diffusion length extracted by the DMD based method is in good agreement (within 3% of relative difference) with the diffusion length calculated based on the effective lifetime.

III. CONCLUSIONS

This contribution presented a novel DMD-based PL imaging setup which is able to achieve inhomogeneous illumination. The inhomogeneous illumination PL imaging method extends the applications of conventional PL imaging to series resistance imaging, emitter sheet resistance, and diffusion length measurement, while maintaining the advantages of being contactless. The DMD-based PL setup can be further used for other applications involving inhomogeneous illumination.

ACKNOWLEDGEMENT

This work was supported by the Australian Government through the Australian Renewable Energy Agency under grant 2014/RND097 Z. Hameiri was supported by the Australian Research Council (ARC) through the Discovery Early Career Researcher Award under grant DE150100268).

The authors would like to thank Dr. Catherine Chan, Dr. Hongzhao Li, Chang-Yeh Lee and Kyung Hun Kim for kindly providing the samples. The authors also wish to thank Dr. David Payne for the help and discussion on image deconvolution.

REFERENCES

[1] T. Trupke, R. A. Bardos, and J. Nyhus, "Photoluminescence characterization of silicon wafers and silicon solar cells," in *18th workshop on Crystalline Silicon Solar cells &*

Modules, 2008.

[2] D. Hinken, K. Ramspeck, K. Bothe, B. Fischer, and R. Brendel, "Series resistance imaging of solar cells by voltage dependent electroluminescence," *Appl. Phys. Lett.*, vol. 91, no. 18, p. 182104, Oct. 2007.

[3] T. Trupke, E. Pink, R. A. Bardos, and M. D. Abbott, "Spatially resolved series resistance of silicon solar cells obtained from luminescence imaging," *Appl. Phys. Lett.*, vol. 90, no. 9, p. 93506, 2007.

[4] M. Kasemann, L. M. Reindl, B. Michl, W. Warta, A. Schütt, and J. Carstensen, "Contactless qualitative series resistance imaging on solar cells," *IEEE J. Photovoltaics*, vol. 2, no. 2, pp. 181–183, 2012.

[5] M. Juhl, T. Trupke, and Y. Augarten, "Emitter sheet resistance from photoluminescence images," in *39th IEEE Photovolt. Spec. Conf.*, 2013, pp. 198–202.

[6] H. Kampwerth, T. Trupke, J. W. Weber, and Y. Augarten, "Advanced luminescence based effective series resistance imaging of silicon solar cells," *Appl. Phys. Lett.*, vol. 93, no. 20, p. 202102, 2008.

[7] O. V. Sorokin, "Measurement of Surface Recombination Rates in Thin Semiconductor Specimens with Qualitatively Different Boundaries," *Sov. physics. Tech. phisics*, vol. 1, no. 11, pp. 2384–2389, 1956.

[8] R. A. Sinton, A. Cuevas, and M. Stuckings, "Quasi-steady-state photoconductance, a new method for solar cell material and device characterization," in *25th IEEE Photovoltaic Specialists Conference*, 1996, pp. 457–460.

Numerical model to extract materials properties map from spectrally resolved luminescence images

Nicolas Paul[1], Vincent Le Guen[1], Daniel Ory[1,2], Laurent Lombez[2,3]

[1]EDF R&D, 6 quai Watier, 78400 CHATOU Cedex, FRANCE

[2]Institute of Research and Development on Photovoltaic Energy (IRDEP), UMR 7174 CNRS-EDF- Chimie ParisTech, EDF R&D Chatou, France

[3]Institut Photovoltaïque d'Ile-de-France (IPVF), 8 rue de la Renaissance - 92160 Antony, France

Abstract — **We present a numerical model to investigate spectrally resolved photoluminescence images. For each image location, the complete luminescence spectrum is modeled to extract absorption properties as well as thermodynamic properties such as the temperature and the quasi-Fermi level splitting. Not only the whole spectral information is used but also the spatial information is taken into account to reduce the uncertainty in the determination. The validity of the method is discussed to get rid of any ambiguity that could be inherent to fitting procedures with multi-parameters. We apply the method to CIGS polycrystalline thin films solar cells. Results will be discussed (such as correlation between parameters) and compared to previous results and other methods.**

I. Introduction

The characterization of solar cells based on luminescence allows the understanding of the physical and chemical phenomena that are involved in the operation of photovoltaic devices. In fact, the emission properties are a function of the two thermodynamic parameters of a reservoir, namely the temperature and the (electro-) chemical potential. The latter, sometimes called the chemical potential of radiation, represents the difference between the electrochemical potentials of electrons and holes and reflects the maximum open circuit voltage one can expect from the investigated material.

The Planck's law, which describes the emission of a black body at a certain temperature, has been generalized by Kirchhoff for every body by taking into account the absorptivity in equilibrium with its environment. For a non-equilibrium situation as it is the case on photoluminescence experiment, the description of the emission process was developed by Lasher and Stern and then by De Vos and Würfel. This law is written [1,2]:

$$\varphi(E, \theta) = \frac{CE^2 a(E, \theta)}{\exp\left(\dfrac{E - \Delta\mu}{kT}\right) - 1}$$

where a(E) is the absorption probability of the incident flux at the energy E and $C = \dfrac{\cos\theta}{4\pi^3 c_0^2}$ The separation of the quasi-Fermi levels $\Delta\mu$ is assumed to be constant in the thickness of the material, in which case it is equal to the open circuit voltage.

As a first approximation, it is possible to determine the thermodynamic parameters T and $\Delta\mu$ by observing the slope of the high energy luminescence spectrum which is written:

$$ln\left(\frac{\varphi(E)}{a(E)E^2}\right) = \frac{-E}{kT} + \frac{\Delta\mu}{kT}$$

On a logarithmic scale, the slope reflects the temperature, from which the value $\Delta\mu$ is determined. This method has been applied in the literature but might suffer from severe consideration such as the knowing of the absorptivity a(E).

The scientific community recently questioned the validity of this method [1]. The sensitivity on the value of a(E) is not critical because a 50 \% error only results in about 25 meV error on the $\Delta\mu$ value. However, the spectral variation of a(E) and/or the precise value of the temperature can be critical. Part of the paper presented here aims at discussing that point.

Assuming a known temperature value, Katahara and Hillhouse have demonstrated the critical knowledge of a(E) in thin-film PV materials and considered sub-bandgap gap variation of a(E) [1]. Our work is based on their funding. Reader is referred to Ref [1] for details. Briefly, the generalized Planck's law has been rewritten including sub-band gap properties. A general form of a(E) is proposed. We recall the expression:

$$a(E) = (1 - \text{Exp}(-\alpha_{0R}(E)d))(1 - \frac{2}{\text{Exp}(\frac{E - \Delta\mu}{2kT}) + 1})$$

$$\alpha(E, 0, \gamma) = \alpha_0\sqrt{\gamma}G(\frac{E - E_g}{\gamma}, 0)$$

Where α_0: absorption strength (as defined in [2] supplementary information…), d the sum of absorption length and diffusion length, E_g the bandgap, γ the sub-bandgap tail state width and

θ the sub-bandgap tail state exponent.

We present inversion of the PL spectrum in order to determine the thermodynamics properties T, $\Delta\mu$ and the absorption properties θ, σ and Eg. A model is developed that takes the spectral variation as well as the spatial variation to properly map all the properties. The validity of the model is discussed. In particular we show that the Quasi-Fermi level splitting and the temperature can be estimated without ambiguity for a noise to N/S ratio inferior to 2%. More generally it allows us to properly evaluate any spatial correlation between parameters. Application is possible for any polycrystalline materials. Example is shown on CIGS microcell.

II. Validity of the model

We propose to test the robustness of the proposed method as a function of the noise to signal ratio one can add when dealing with experimental data. Therefore we simulate one PL spectrum and add a Gaussian noise (N/S ratio = NSR). As we said, our work is based on the model proposed in [1] to fit the simulated luminescence spectra and estimate the complete cell characteristics.

We note $\varphi(E)$ the measured luminescence spectrum and $\varphi_{model}(E, p)$ the luminescence spectrum model described as it is done in [1]. $p = (T, \Delta, \alpha_0 d, E_g, \gamma, \theta)$ is a vector gathering the different searched parameters. $\varphi_{model}(E, p)$ can be expressed as :

$$\varphi_{model}(E, p) = \frac{Ce^2 a(E, p)}{\exp\left(\dfrac{e - \Delta\mu}{kT}\right) - 1}$$

where $C = 2\pi / h^3 c^2$ and $a(Ep)$ is the absorptivity function modeled by :

$$a(E, p) = 1 - \exp\left(-\alpha_0 dG(E, p)\left(1 - \frac{2}{\exp\left(\dfrac{E - \Delta\mu}{2kT}\right) + 1}\right)\right)$$

and $G(E, p)$ is the absorption coefficient modeled by the convolution of two densities of states :

$$G(E, p) = g_{sub-bandgap}(E, p) \otimes g_{above-bandgap}(E, p)$$

The sub-bandgap density of states is described by:

$$g_{sub-bandgap}(E, p) = \exp\left(-\left(\frac{e - E_G}{\gamma}\right)^{\theta}\right)$$

The above bandgap density of states is the ideal square root absorption coefficient:

$$g_{above-bandgap}(E, p) = \sqrt{E - E_G} \; for \; E > E_G$$
$$g_{above-bandgap}(E, p) = 0 \; for \; E < E_G$$

Our notation is general and valid for any energy, which enables to express the absorption coefficient as a convolution product and keep infinite limits for the integration:

$$G(E, p) = g_{sub-bandgap}(E, p) \otimes g_{above-bandgap}(E, p)$$

The least-square estimation of the cell characteristic consists in finding the vector **p** which minimizes the quadratic distance between the measured spectrum and the model:

$$r(p) = \sum_e \left(\varphi(E) - \varphi_{model}(E, p)\right)^2$$

At first, the model is too complex to get an analytical expression of the solution. Therefore some optimization algorithm has to be used to minimize $r(p)$. In [1] a Levenberg-Markhard algorithm is used; however it is an unconstrained optimization algorithm. We propose to use some *constrained optimization algorithms (interior point algorithm)* in order to limit the parameter search to some physical coherent area. It accelerates the optimization algorithm and prevents from the convergence to non-physical solution.

An important question is the estimation error, one can wonder if the optimization problem has a unique solution or if some different set of parameters would produce the same fit quality. This potential fit ambiguity can be evaluated with synthetic data. We start by generating synthetic spectrum $\varphi_{model}(E, p_{true})$ with parameter p_{true} and synthetic Poisson noise at different noise to signal ratio (NSR).

For each NSR value, 1000 different simulations are generated with different noise realizations. The inversion algorithm is used for each simulation. For NSR=1%. The estimation bias (average error) is only 0.8 K for the temperature and 1.2 meV for $\Delta\mu$. Relative RMSE error for all parameters as a function of the NSR is shown in Fig 1.

The spatial information is also taken into account when dealing with spectrally resolved images (not discussed here). The method finally includes a spectral *inversion* and a spatial *deconvolution.* The complete method will be described in the final paper. We present here results on CIGS microcell.

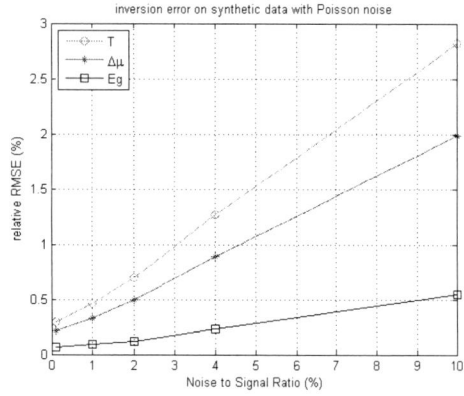

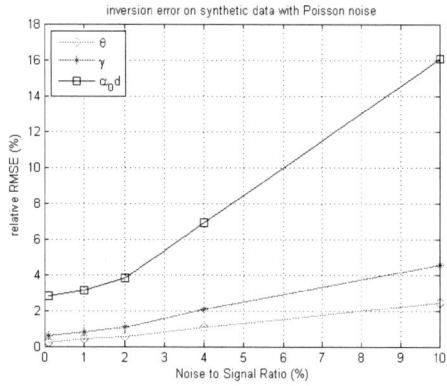

Figure 1 : (up) estimation performances on T, $\Delta\mu$ and Eg (down) estimation performances on θ, γ and $\alpha 0d$

III. CIGS cell investigation

We analyze a CIGS microcell with a diameter of 35µm. To obtain spectrally resolved photoluminescence images we use a hyperspectral imaging system. The description of the system as well at the classical treatment of the cell have been published elsewhere [3]. It allows one to obtain absolute intensity measurement of the PL spectrum at each pixel. Following the linearized generalized Planck's law, quasi-Fermi levels splitting and temperature maps have been obtained and analyzed. In these previous results we used to consider that a(E) is spatially constant as well as the temperature. The method exposes in this paper does not suffer from such hypothesis but allows verifying/infirming it.

In this paper we only present two maps representing the spatial variation of T and $\Delta\mu$ (See Fig2). One can see a clear anti-correlation that is also displayed in Fig 3 when plotting

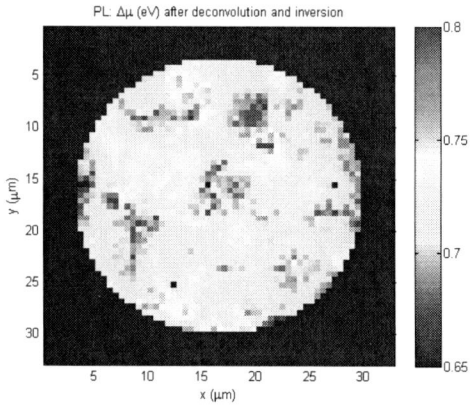

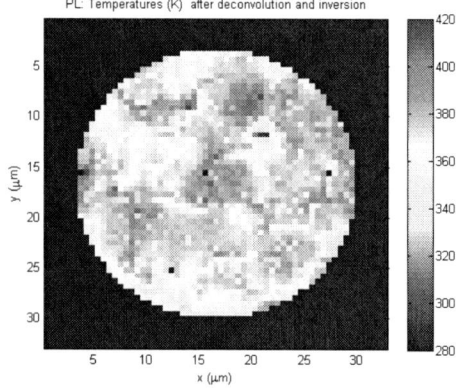

Figure 2 : Spatial fluctuations on CIGS microcell (up) quasi Fermi level splitting $\Delta\mu$ and (down) temperature

$\Delta\mu(T)$; each point is a given spatial location. In a first approximation this plot could be seen as a voltage-temperature plot from one can extract the activation energy (i.e. the bandgap) from the ordinate intercept. Indeed we found the value of E_g. Nerveless it is important to discuss the real origin of this anti-correlation as it can also be due to the mathematical model we use. In fact we observe the same anti-correlation by using results obtained by fitting 1000 spectra with a N/S ratio of 2%. Therefore those two variables are not independent.

Besides those observations we also looked at other possible correlation. Among many correlations we found it

978-1-5090-5606-4/17 $31.00 © 2017 IEEE

seems that experimentally the maximum PL intensity is weakly correlated to the QFLS which is not the case with synthetic data. Therefore it appears that the spatial variations of PL spectra observed in our CIGS cell are due to spatial fluctuations of the QFLs but with a much less contribution as compare to previous studies. Fluctuations of several absorptions properties also influence PL spectra spatial variations.

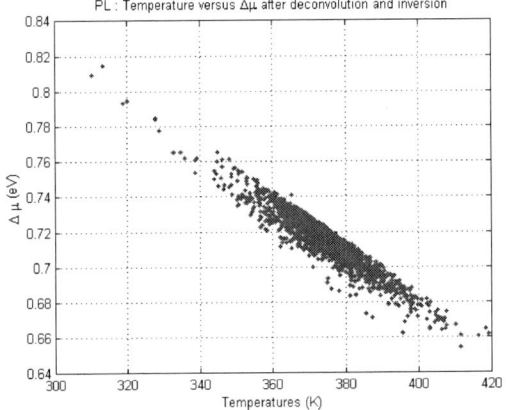

Figure 3 : Quasi Fermi levels Δμ as a function of temperature T. Each point correspond to a specific location of maps in Fig 2

IV. Conclusion

We developed a numerical model to extract several physical parameters from spectrally resolved luminescence images. The complete luminescence spectrum and the spatial information are taken into account to make the model robust and beyond the state of the art. The validity of the method is discussed to get rid of any ambiguity and to quantify the uncertainty as a function of the experimental noise to signal ratio. On CIGS cell, (anti)correlations between extracted parameters are observed mainly due to independent variables.

References

[1] J. K. Katahara and H. W. Hillhouse J. Appl. Phys. 116, 173504 (2014).
[2] P. Würfel, Sol. Energy Mats. and Sol. Cells., 46, (1997).
[3] A. Delamarre et al. Prog. Photovolt: Res. Appl., 23: 1305–1312 *(2015)*

This project has been supported by the French Government in the frame of the program of investment for the future (Programme d'Investissement d'Avenir – ANR-IEED-002-01)

Non-Destructive Contact Resistivity Measurements on Solar Cells Using the Circular Transmission Line Method

Geoffrey Gregory[1], Andrew M. Gabor[2], Andrew Anselmo[2], Rob Janoch[2], Zhihao Yang[3], Kristopher O. Davis[1]

[1] Florida Solar Energy Center, Orlando, FL, 32826, USA

[2] BrightSpot Automation, Westford, MA, 01886, USA

[3] School of Materials Science and Energy Engineering, Foshan University, Guangdong, China, 528000

Abstract — **Several methods have been determined for measuring the contact resistivity of solar cell devices, but these methods are all either destructive in nature or require the fabrication of special metal contacts. In this paper, we present a non-destructive method for measuring the contact resistivity of commercial grade solar cells using the circular transmission line method. We first determine an optimal method for probing the total resistance of these structures on a solar cell and investigate the importance of measuring the exact dimensions of each contact. Then, by comparing the results of the measurement to traditional TLM results, we select a proper geometry for the circular patterns so that they can be hidden within the busbars of finished cells and not affect cell efficiency or aesthetic. Good correlation is demonstrated between automated circular TLM measurements performed on a new tool, the *ContactSpot-PRO*, and traditional TLM measurements performed manually. The implementation of this high-speed tool can allow contact resistance to be measured on every R&D or production cell with only a minor change in front silver paste screen artwork.**

I. INTRODUCTION

One critical component of series resistance in solar cell devices is the contact resistance R_C (ohms) or contact resistivity ρ_c (ohms-cm^2) between the metallization and the semiconductor. In order to optimize the performance of these devices it is important to accurately measure ρ_c after metallization. While customized test structures can be made for all of these measurements, other tests can also be performed on actual commercial-grade solar cells in a variety of ways, some being destructive in manner while others allow the cell to remain intact. A popular method of contact resistance measurement uses the Corescan tool from Sunlab [1]. This tool is most appropriate for spot checks as it has low throughput and is destructive since it uses a probe to scratch across the surface. A method for measuring ρ_c during production on a large set of cells is desirable as it would allow for the optimization and the monitoring of front metallization processes.

The transmission line method (TLM) [2, 3], is another common way of measuring ρ_c as well as the sheet resistance R_{sh} (ohms/sq) of the underlying doped Si layer and the transfer length L_T on commercial grade solar cells [4, 5]. When applying this technique to cells, the devices are cut into strips parallel to their busbars so that current flow can be isolated between incrementally spaced contact pairs and the resultant resistance can be measured. From there ρ_c is extrapolated from a plot of total resistance R_T versus contact spacing d. This method is often chosen over others because of its accuracy, but its destructive nature prohibits the use of the solar cell in its intended commercial application.

A variation of the TLM, known as the circular transmission line method (cTLM) [6], has also been used to measure ρ_c and R_{sh} non-destructively on semiconductor devices. Fig. 1 shows the general form of a cTLM structure with conducting inner regions of varying radius L, non-conducting regions with fixed radius r, and a conducting outer region with a variable width w. The difference between dimension r and L is designated as the gap size d.

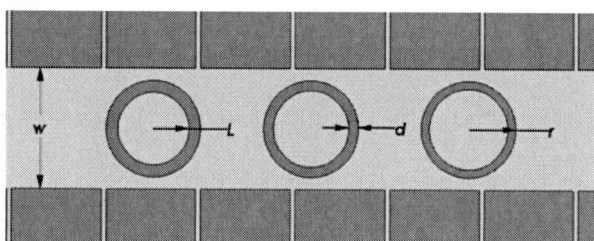

Fig. 1.   cTLM structure with conducting inner region of radius L, non-conducting region of radius r and gap size d.

This method works in a similar manner to the TLM; R_T is measured between the outer conducting region and each of the inner dots while the incremental contact spacings are created using the variable dimension L. While this method is commonly used in many semiconductor applications, we have seen no published reference of its application to solar cell devices, although researchers from Suniva and the Rochester Institute of Technology have independently been exploring these approaches simultaneously with our work [7]. Due to the geometry of the cTLM structures there is no need for destructive edge isolation as in the TLM. Therefore, the method presents an opportunity for manufacturers and researchers to assess ρ_c on finished cells in a non-destructive manner.

In this paper, we detail a set of experiments in which cTLM structures were screen printed within the busbars of standard crystalline silicon (c-Si) solar cells and analyzed in an attempt to exploit the non-destructive nature of the method.

We detail the investigation of different front-side metallization designs driven by cTLM theory [8], as well as appropriate methods for probing R_T of each dot structure. The *ContactSpot-PRO*, an automated multiplexing tool with image recognition capabilities and a translation stage was built in conjunction with these experiments by BrightSpot Automation for in-line ρ_c characterization purposes with a takt time as short as 3 seconds. The importance of the accurate measurement of circular radii for parameter extraction is highlighted along with its incorporation into the tool. The results of the cTLM measurements are compared to the results of linear TLM measurements performed on the same set of cells to validate this novel solar cell characterization method.

II. EXPERIMENT

A. Metallization Design

The cTLM technique calculates the values of ρ_c and R_{sh} using different contact geometries, each with a slight variation in the dimensions L and r. The basis for the method, proposed by Schroder [8], defines R_T as a function of $\ln(1+d/L)$:

$$ R_T = \frac{R_{SH}}{2\pi}\left[\frac{L_T}{L}\frac{I_0(L/L_T)}{I_1(L/L_T)} + \frac{L_T}{L+d}\frac{K_0(L/L_T)}{K_1(L/L_T)} + \ln\left(1+\frac{d}{L}\right)\right], (1) $$

where I and K are the modified Bessel functions of the first kind and their subscripts denote their order. For situations where $L \gg 4L_T$, I_0/I_1 and K_0/K_1 approach unity and (1) can be simplified as:

$$ R_T = \frac{R_{SH}}{2\pi}\left[\frac{L_T}{L} + \frac{L_T}{L+d} + \ln\left(1+\frac{d}{L}\right)\right]. (2) $$

The cTLM structures designed for these experiments employed 6-circle arrays, where each circle within each array had unique values of r, L, and d. An R_T value was measured from each dot of the cTLM arrays and a Nelder-Mead solver was used to calculate the values of L_T and R_{sh} that corresponded to each data point measured.

To simplify the calculation of ρ_c and R_{sh}, the metallization screen designed for these experiments employed large circular contact radii that would meet the requirement $L \gg 4L_T$ so that (2) could be used as a solution method. This led to circular contact designs that were larger than many standard busbar widths (~1200 µm). Therefore, the busbars had to take on a unique shape near the cTLM contacts in order to compensate for the large radii. Fig. 2 shows one of these printed cTLM structures as well as a TLM structure used for validation of measured ρ_c.

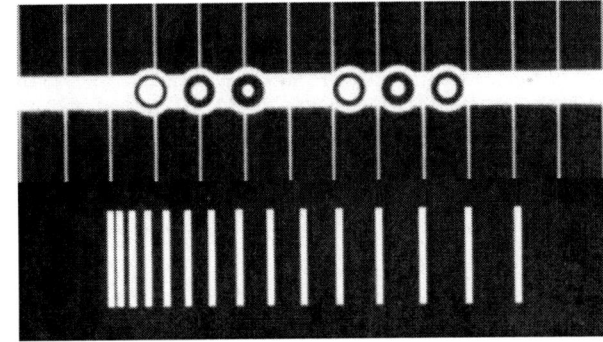

Fig. 2. cTLM contact on 1200 µm busbar after firing.

B. Metal Resistance

Different methods of probing the cTLM structures were analyzed to determine if additional line resistance was introduced when measuring R_T. The front side metallization was deposited and fired by Gonda Electronic Technology, Co. Ltd. using experimental Ag pastes at different peak firing temperatures. A Keithley 2400 Sourcemeter was used in combination with a micro-probing station to measure R_T; a current probe and voltage probe were then placed onto the inner dot of each circular contact while one voltage probe and one, two, or three current probes were placed on the outer region of the contact as shown in Fig. 3. This was done to assess the uniformity of current distribution along the busbar regions of the cTLM structures.

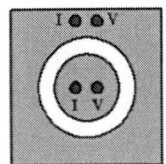 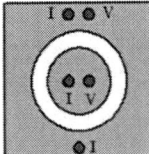 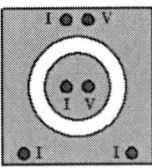

Fig. 3. Position of current and voltage probes on cTLM structures using four, five, and six probes.

The effect of metal resistance on the measurement technique was also assessed by varying the width w of the busbars. R_T was measured on several circular structures deposited within busbars of varying widths. This was done to assess whether using a targeted busbar width of 1200 µm would introduce unnecessary line resistance to the cTLM measurement.

C. Contact Dimension Measurement

During the printing and firing process of the front side contacts, the dimensions of the metallization are often different from the screen artwork design due to factors such as paste slumping, temperature variations, paste viscosity variations, and screen wear. These dimensions can vary spatially within a cell and from cell to cell. The cTLM contacts used in this study were designed with only small (~ 50 μm) incremental changes from one contact radius to the next, so slight changes in the cTLM dimensions had the potential to significantly impact the measurement results. Because of this, special attention was given to the actual contact dimensions after firing.

To study this, a confocal scanning microscope was used to record the actual dimensions of several cTLM contacts. An optical inspection camera with image processing software was integrated into the *ContactSpot-PRO* and then calibrated with the data taken from the confocal microscope. The focus, brightness, and ambient light intensity were all optimized on the optical unit until accurate dimension measurement was possible.

III. RESULTS AND DISCUSSION

A. Metal Resistance

A plot of R_T versus $\ln(1+d/L)$ for the three studied probing schemes shows that the metal resistance has a significant effect on the total measured resistance when only four probes were used. Fig. 4 shows such a plot. By adding just one extra current probe to the outer ring of the cTLM structure, the measured values of R_T decreased by 35.8%. The remainder of experiments that were performed used five probes to measure the total resistance of each circular structure. Additionally, the *ContactSpot-PRO* probe head was designed with multiple current probes in order to minimize this added resistance.

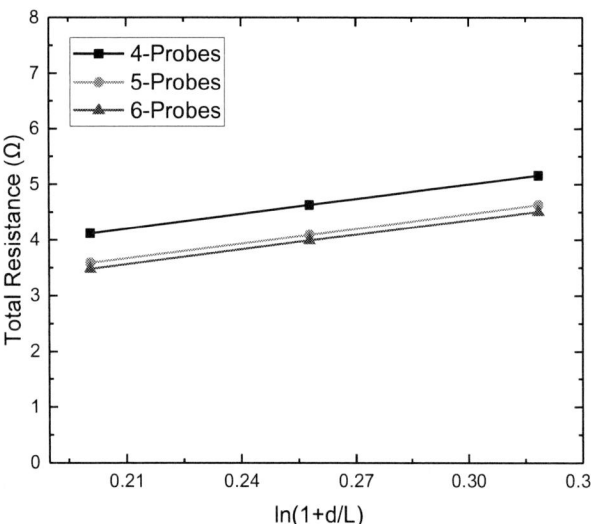

Fig. 4.   Plot of R_T versus $\ln(1+d/L)$ for a sample fired at 855 ℃.

Fig. 5 shows a cTLM plot of two circular arrays, each with different busbar widths that were printed on the same cell. The first busbar had a width of 1200 μm and the second had a width of 4000 μm. Despite this, the resistance values

gathered from each array were in good agreement with each other. A line of best fit was generated for the data taken from this cell with a coefficient of determination of 0.999. This test was repeated on several cells and the results were similar. Since the busbar width w showed little effect on the measured resistance values, 1200 μm busbars were used for the remainder of these experiments. In order to minimize shading losses and the front side contact fraction, the value of w should be small (≤ 1200 μm) in future applications.

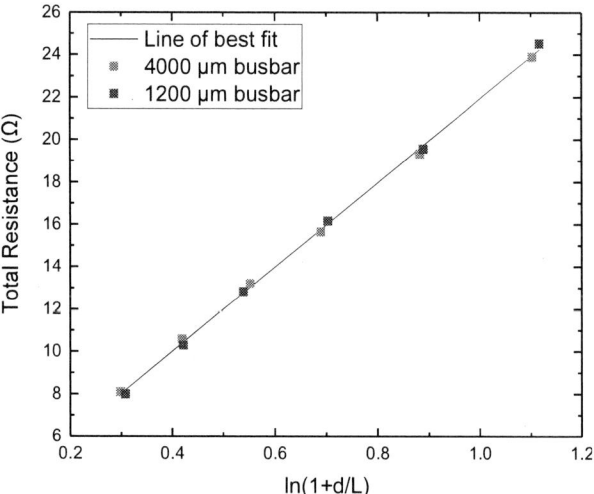

Fig. 5.   cTLM plot of two circular arrays with different busbar widths from the same cell.

B. Results of Dimension Measurement Study

An error analysis was performed using the Nelder-Mead solver to investigate the magnitude of error that would result in the final value of ρ_c if the cTLM dimensions were systematically under-measured or over-measured. Typical values of ρ_c and R_{sh} (5 mΩ-cm^2 and 125 Ω/□) were implemented into the solver along with the designed cTLM dimensions, and theoretical R_T values were generated. The analysis showed that if the circular dimensions were systematically over-measured by just 1% then the error in ρ_c would be on the order of 11.4%. Similarly, the resultant error in ρ_c would be on the order of 10.8% if the circular dimensions were under-measured by the same magnitude.

The image recognition software and optical unit on the *ContactSpot-PRO* were calibrated with the measurements taken on the confocal microscope so that they would accurately measure the dimensions r and L on all of the cTLM structures used in this study. A statistical analysis was performed on the dimensions measured on 20 individual cTLM arrays by the tool and their deviation from their expected artwork values. The results of the analysis are shown in Table 1. Since the range of deviation was well above 1%, all the measured dimensions were used for the calculation of ρ_c and R_{sh} in order to minimize the error in the results.

TABLE 1
ANALYSIS OF cTLM DIMENSIONS

	Expected Dimension (μm)	Dimension Deviation (%)	Range of Deviation (%)
L_1	200	2.49%	4.99%
L_2	250	1.49%	3.29%
L_3	300	1.18%	3.55%
L_4	350	1.17%	2.19%
L_5	400	1.16%	3.08%
L_6	450	0.88%	1.96%
r	600	4.41%	4.11%

C. Results of cTLM Measurements

With the parasitic resistances of the measurement technique minimized and the image recognition software calibrated, ρ_c and R_{sh} were measured using both the linear TLM and the cTLM. This set of measurements was performed on p-type multi-crystalline silicon solar cells from Gonda with a range of sheet resistances and with different experimental front-side Ag pastes fired at 4 different peak temperatures over a range of 60C. The linear TLM was performed using the Keithley 2400 Sourcemeter after laser scribing the patterns from the cells. The cTLM was performed on the *ContactSpot-PRO*, shown in Fig. 6, using the calibrated optical inspection camera and the multiplexing probe head.

Fig. 6. The *ContactSpot-PRO* tool.

Fig. 7 shows that the values of R_{sh} that were calculated using (2) did not correlate well with the values of R_{sh} measured from the linear TLM structures. Since R_{sh} was over measured in this method, the values of ρ_c were consequently

under measured. It is possible that the values of L may have not actually met the requirement for simplification, that $L>>4L_T$.

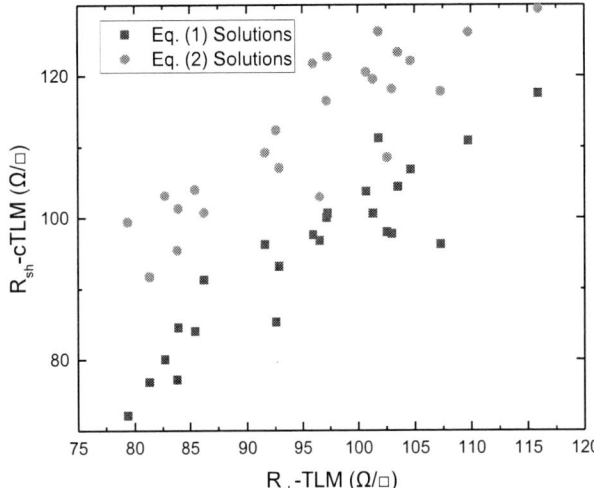

Fig. 7. Correlation plot of R_{sh} calculated using equations (1) and (2) from *ContactSpot-PRO* data versus R_{sh} calculated using the linear TLM.

The values of R_{sh} calculated using (1) did correlate well with the linear TLM data and with sufficient computational power, this calculation method will not affect takt times significantly. Fig. 8 shows promising correlation between ρ_c measured on the *ContactSpot-PRO* using the cTLM and the TLM data. The cTLM tended to undermeasure the values of ρ_c. This difference may be due to remaining inaccuracies in the measurement of circular dimensions. Implementation of a correction factor could also help improve the correlation.

Without the need to maximize the values of L to meet the requirements of (2), inner dot radii can be made to fit well within the bounds of narrow busbars. Thus, the contact fraction and shading losses of these types of cells can be even further minimized.

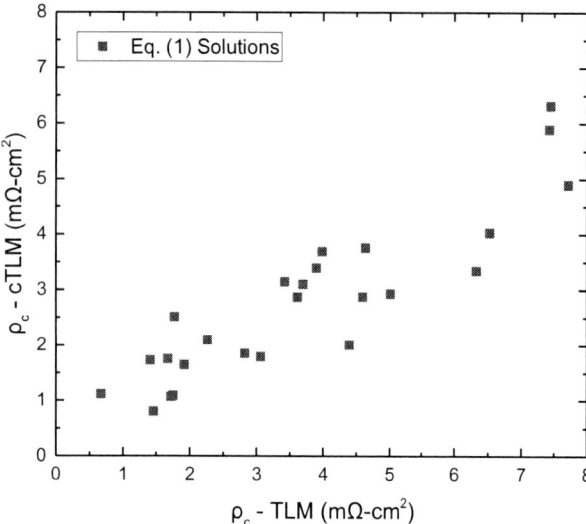

Fig. 8. Correlation plot of ρ_c calculated using equation (1) from *ContactSpot-PRO* data versus ρ_c calculated using the linear TLM.

IV. CONCLUSIONS

In these studies, we demonstrated a successful application of the circular TLM method on commercial grade solar cells. This nondestructive method allows measurement of cells for both research & development as well as production lines with only a minor change in screen artwork. The present semiautomatic tool allows the easy and fast testing of every R&D cell for improved experimentation. An in-line version of the *ContactSpot-PRO* tool located near the standard IV testing equipment could enable the testing of contact resistance and sheet resistance on every cell in a production line for improved factory quality control. Since the circular structures were hidden within the busbars, these structures will have essentially no impact on module aesthetics or power after soldering of the interconnect wires. It was determined that in order to account for the line resistance of these structures without compromising the functionality of the measurement, three current probes and two voltage probes should be used to measure the R_T of each dot. An optimal method for calculating the parameters of interest was identified by comparing its results to traditional TLM values. A specific cTLM geometry was then chosen to fit within the bounds of a 1200 μm busbar, with the potential to fit well within even narrower busbars in future photovoltaic applications. In cases where a floating busbar paste is used, the portion of the busbar that is printed with the cTLM patterns should be printed along with the fingers.

Future work will explore cTLM application to alternate metallization schemes such as 1) plating, 2) to rear side Al paste contacts to silicon through dielectric openings (PERC), 3) to metal on TCO application such as heterojunction cells,

and 4) to thin-film applications for example by masking the metal deposition onto CdTe layers.

ACKNOWLEDGEMENTS

The authors would like to thank Gonda Electronic Technology, Co. Ltd. for providing the test cells and metallization for these experiments. This material is based upon work supported in part by the U. S Department of Energy's Office of Energy Efficiency and Renewable Energy, in the Solar Energy Technologies Program, under Award Number DE-EE0004947.

REFERENCES

1. Heide, A.S.H.v.d.S., A.; Wyers, G.P.; Sinke, W.C., *Mapping of contact resistance and locating shunts on solar cells using resistance anaysis by mapping of potential (RAMP) techniques.* 16th European Photovoltaic Solar Energy Conference and Exhibition 2000: p. 4.

2. Murrmann, H. and D. Widmann, *Current crowding on metal contacts to planar devices.* IEEE Transactions on Electron Devices, 1969. **16**(12): p. 1022.

3. Berger, H.H., *Models for contacts to planar devices.* Solid-State Electronics, 1972. **15**: p. 145-158.

4. Andrew M. Gabor, G.G., Adam M. Payne, Rob Janoch, Andrew Anselmo, Vijay Yelundur, Kristopher O. Davis, *Dependence of Solar Cell Contact Resistivity Measurements on Sample Preparation Methods.* IEEE Photovoltaic Specialists Conference,, 2016. **43**.

5. Guo, S., et al., *Detailed investigation of TLM contact resistance measurements on crystalline silicon solar cells.* Solar Energy, 2017. **151**: p. 163-172.

6. Reeves, G.K., *Specific contact resistance using a circular transmission line model.* Solid-State Electronics, 1980. **23**: p. 487-490.

7. *Conversations with Adam Payne and Satosh Kurinec, 2016.*

8. Schroder, D.K., *Semiconductor material and device characterization.* 1990, New York: Wiley.

Radiation Resistance of Low Cost High Efficiency Triple Junction solar cells

Roberta Campesato [1], Erminio Greco[1], Giuseppe Gabetta[1], Mariacristina Casale[1], Gabriele Gori[1], M. Sankaran[2], Suresh E. Puthanveettil[2] , B.R.Uma[2] , DR.M.Ravindra[2], Sheeja Krishnan[3]

[1]CESI, 20134 Milan, via Rubattino 54 , Italy, [2] ISRO Satellite Centre, Bengaluru, India, [3] Shree Devi Institute of Technology,Mangalore,India

Abstract — **Low Cost solar cells based on the high efficiency structure InGaP/GaAs/Ge have been manufactured and verified for space application. These solar cells are characterized by a AM0 BOL efficiency of 28% and very high radiation resistance to proton and electron irradiation thanks to their simplified structure. These features make them really appealing for the new market of mega constellation of minisatellites. In this paper the results of the electron and proton irradiation campaign in the energies range from 0.5 MeV to 10 MeV are reported and compared with standard CTJ30 solar cells.**

Index Terms— **Photovoltaic cells, III-V semiconductor materials**

I. INTRODUCTION

Triple junction solar cells based on III-V compound semiconductors have been the core of space solar panels for more than 10 years. These solar cells are characterized by high conversion efficiency, radiation hardness and reliability in the space environment.

The space market, for many applications (e.g. "cubesats", commercial constellations of mini satellites), is currently facing the need of combining the high efficiency and the reliability with much lower prices than the standard triple junction devices.

Silicon solar cells are rather inexpensive, but their efficiency is too low and their radiation hardness is unsatisfactory for main space market demands.

CESI has combined high efficiency, radiation hardness and reliability with the low cost in a peculiar triple junction solar cell type InGaP/GaAs/Ge, characterized by the following elements:

- Efficiency 28% in AM0, 25 °C (just 1 point lower than standard TJ solar cells)
- High radiation resistance (typical of III-V solar cells)
- Cost about 30-50% less than the standard triple junction solar cells
- High production capacity (obtained reducing the production time).

These features were obtained at CESI through the optimization of the whole production chain of the III-V solar cells, namely:

- MOCVD growth at very high rates combined with a reduction of precursor consumption

- Post growth process sequence optimization and simplification of the process steps
- High rate metal deposition and reduction of the metal quantities.

At the end of the development phase, 100 low cost cell prototypes were manufactured and submitted to an intensive irradiation campaign at ISRO using both protons and electrons to fully characterize the EOL behavior of such components.

II. LOW COST TRIPLE JUNCTION SOLAR CELLS

To date, the commercial space market demand is oriented to reduce the cost of the overall satellite and also in particular of the solar generators.

The use of the electrical propulsion and the satellite lifetime extension in LEO orbit above 5 years, does not allow the use of the cheaper silicon solar cells derived from the terrestrial application.

CESI has developed a new concept for a III-V based triple junction solar cell, characterized by:

- a cost lower than the state of the art triple junction solar cells such as CESI CTJ30 solar cell,
- a slightly reduced efficiency BOL (28% instead of 29.5%)
- a good radiation resistance to the charged particles
- temperature coefficients in line with the CTJ30 standard solar cells.

CESI low cost triple junction solar cell, named CTJ-LC, has the structure reported in Figure 1. It is composed by 3 junctions and 2 tunnel diodes and, with respect to the standard CESI CTJ30 structure:

- the top junction is InGaP with a reduced thickness;
- the MID junction is InGaAs with a reduced thickness and the InGaAs layers are grown at very high growth rate;
- the germanium junction is unchanged but requirements on germanium wafers are relaxed;
- the front contact has a thinner gold layer to protect silver from oxidation;

978-1-5090-5606-4/17 $31.00 © 2017 IEEE

- the post growth processes were simplified to significantly reduce the manufacturing time.

Figure 2 reports a picture of the CESI low cost solar cell with an area of 26.5 cm^2. With respect to a standard CTJ30, the noticeable feature is the slightly whiter grid and pads appearance.

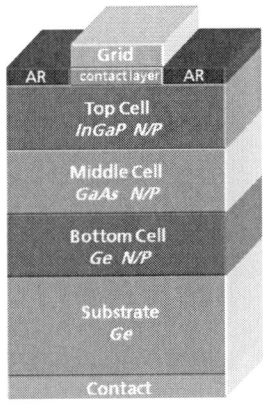

Fig. 1: Scheme of a low cost TJ

Fig. 2: Picture of a CTJ-LC, area 26.5 cm^2

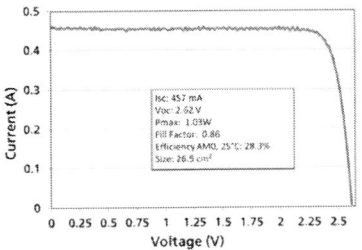

Fig. 3: I-V measurement under AM0, 25 °C

The electrical performances on the 100 samples manufactured at the end of the development phase are reported in table 1 and compared to the results achieved on the CESI standard triple junction solar cells CTJ30.

An average efficiency of 28% was obtained and the cost of these devices is more than 30% lower with respect to the CTJ30 solar cells.

TABLE I

ELECTRICAL PERFORMANCE OF CTJ-LC SOLAR CELLS UNDER AM0, 25 °C BOL CONDITIONS..

@ AM0, 25°C	Jsc (mA/cm^2)	Voc (V)	FF	Eff (%)
CTJ 30 BOL	18.0	2.60	0.86	29.5
CTJ LC BOL	17.3	2.62	0.85	28.0

III. EXPERIMENTAL IRRADIATION RESULTS

CESI Low cost solar cells were submitted by ISRO to a irradiation campaign using protons and electrons with different energies and fluencies.

In particular electrons with energies 1 MeV and 10 MeV were used at fluencies from 1e14 e$^-$/cm^2 up to 3e15 e$^-$/cm^2. Proton energies of 2 MeV and 10 MeV were used.

After irradiation, the solar cells were measured at ISRO and CESI "as is" and after annealing. The annealing consisted of 8 hours under AM0 light with a temperature of 60 °C (typical of in orbit condition).

The solar cells generally recovered a portion of the irradiation damage after annealing, especially when the solar cells got irradiated by electrons.

For the low cost solar cells after annealing, figure 4 shows the remaining power factors, i.e. the ratio between the maximum power after irradiation and before irradiation.

Generally the remaining power factor of low cost solar cell resulted higher with respect to standard triple junction solar cells to all energies and fluencies [1].

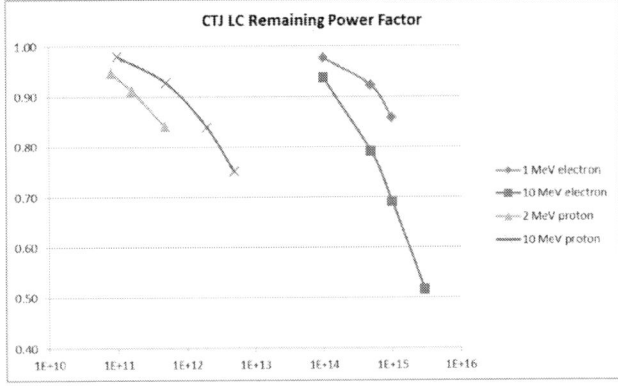

Fig. 4.   Remaining power factors of CTJ-LC solar cells vs fluence.

In figure 5 the remaining power factor (RPF) of CTJ-LC solar cell is compared to the RPF of standard CTJ-30 solar cells after 10 MeV proton irradiation and 10 MeV electron irradiation. Very high RPF are obtained on low cost devices

978-1-5090-5606-4/17 $31.00 © 2017 IEEE

with respect to standard triple junction solar cells with BOL efficiency higher than 29% (CTJ30).

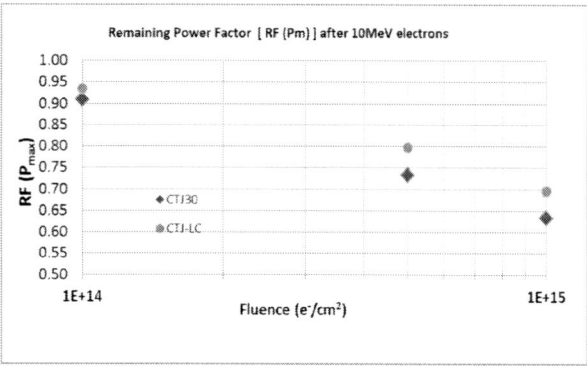

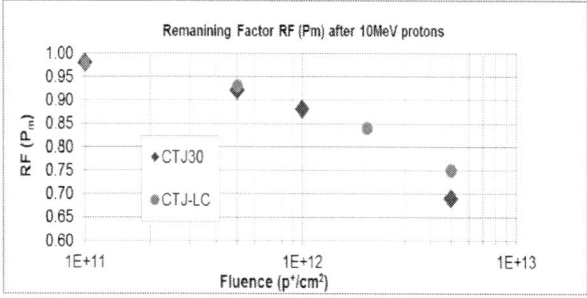

Fig. 5. Remaining power factors of CTJ-LC solar cells vs fluency compared with standard CTJ-30 devices after irradiation with 10 MeV electrons and 10 MeV protons

The remaining factors for Isc, Voc and Pm are reported in table 2 for 1MeV electrons and table 3 for protons with different energies. The Remaining Factors at 10 MeV electrons are reported in table 4 of the next paragraph.

TABLE II
REMAINING FACTORS OF CTJ LC (4 CM^2 AREA) BEFORE AND AFTER 1 MEV ELECTRONS IRRADIATION

Electron Energy	Fluence (e/cm2)	Isc	Voc	Pm
1MeV	1E14	0.99	0.97	0.98
1MeV	5E14	0.98	0.95	0.92
1MeV	1E15	0.92	0.94	0.86

TABLE III
REMAINING FACTORS OF CTJ LC (4 CM^2 AREA) BEFORE AND AFTER PROTONS IRRADIATION

Proton Energy	Fluence (p/cm2)	Isc	Voc	Pm
700KeV	2E+11	0.93	0.90	0.80
700KeV	3E+11	0.88	0.87	0.69
700KeV	5E+11	0.82	0.86	0.67

1MeV	5E+10	0.99	0.95	0.90
1MeV	2E+11	0.90	0.90	0.78
2MeV	8E+10	1.00	0.96	0.95
2MeV	2E+11	1.00	0.94	0.91
2MeV	5E+11	0.98	0.91	0.84
10 MeV	1E+11	1.00	0.98	0.98
10 MeV	5E+11	1.00	0.95	0.93
10 MeV	2E+12	0.98	0.90	0.84
10 MeV	5E+12	0.94	0.86	0.75

IV DATA ANALYSIS

In order to understand the behavior of the solar cells, the I-V curves obtained before and after irradiation have been simulated using a mathematical code developed at CESI and based on the Hovel theoretical model for solar cells.

In such model the triple junction solar cell is represented by three subcells connected with a series resistance representing the tunnel diode. The Top and Middle subcells are represented with the three diode model considering the diffusion dark current, the recombination current and the tunneling effect (generally unnoticeable at room temperature) [2].

Using this model, once the solar cell structure is known (composition, thickness and doping of each layer in the full TJ stack), it is possible to evaluate the minority carrier lifetimes.

The CTJ LC irradiated with 10 MeV electrons at different fluencies were considered: their electrical performances before and after irradiation without and with annealing are reported in table 4.

TABLE IV
ELECTRICAL PERFORMANCES OF CTJ LC (4 CM2 AREA) BEFORE AND AFTER 10 MEV ELECTRONS IRRADIATION.

	Isc [A]	Voc [V]	Pmax [W]	F.F.	Eff. [%]
Cell no.	BOL MEASUREMENT				
LC02-057	0.069	2.606	0.151	0.84	27.6
LC01-191	0.068	2.601	0.151	0.85	27.7
LC02-088	0.068	2.636	0.149	0.83	27.2
LC03-056	0.067	2.638	0.149	0.85	27.3
Fluences	10 MeV electrons "as is"				
1E+14	0.068	2.459	0.137	0.82	25.1
5E+14	0.062	2.327	0.116	0.81	21.1
1E+15	0.056	2.235	0.097	0.78	17.8
3E+15	0.046	2.107	0.073	0.74	13.3
	10MeV electrons after annealing				
1E+14	0.069	2.478	0.141	0.82	25.8
5E+14	0.062	2.359	0.120	0.82	22.0
1E+15	0.059	2.278	0.105	0.79	19.3
3E+15	0.049	2.141	0.078	0.75	14.3

From these data, the minority carrier lifetimes for electrons in the bases of the top junction (InGaP), mid junction (InGaAs) and bottom junction (Ge) were evaluated as reported in table 5. The Lifetime dependance to irradiation was found in line with the predictions of the Aspnaugh model [3]:

$$\frac{1}{\tau} \equiv \frac{1}{\tau 0} + K \cdot Fluence$$

The K value for InGaP and GaAs were also determined.

TABLE V

MINORITY CARRIER LIFETIMES EVALUATION

	No Annealing after irradiation		
CTJ_LC	Minority carrier lifetimes (bases)		
	τ_top	τ_mid	τ_bot
BOL	2.5E-08	6.0E-08	1.0E-07
1E+14	1.5E-09	6.0E-09	1.0E-08
5E+14	3.0E-10	8.0E-10	1.0E-09
1E+15	1.0E-10	3.0E-10	5.0E-10
3E+15	5.0E-11	9.0E-11	2.0E-10
	Annealing after irradiation		
BOL	Minority carrier lifetimes (bases)		
1E+14	τ_top	τ_mid	τ_bot
5E+14	2.5E-08	6.0E-08	1.0E-07
1E+15	3.0E-09	6.5E-09	2.0E-08
3E+15	5.0E-10	8.5E-10	2.0E-09
	2.5E-10	3.5E-10	6.0E-10
	7.0E-11	1.2E-10	3.0E-10

V. CONCLUSIONS AND FUTURE WORK

Low cost solar cells have been developed and the electron and proton irradiation results show how these devices are resistant to space environment.

In particular results after 10 MeV proton and 10 MeV electrons at high doses show that the efficiency of low cost solar cells is higher with respect to CTJ30 solar cells as reported in table 6.

The CTJ-LC solar cells qualification campaign at bare and SCA level is currently running with the objective to provide the availability of such components for the future market of low cost solar arrays whenever the requirements on reliability

and on end of life performances hinder the use of silicon devices.

TABLE VI

ELECTRICAL PERFORMANCES OF CTJ-LC AND CTJ30 BOL AND EOL AFTER IRRADIATION

Cell ID		Isc [mA]	Voc[V]	F.F.	Eff. [%]
	10 MeV electron				
CTJ30	BOL	72	2.616	0.84	29.13
	1E+14	70	2.475	0.83	26.50
	5E+14	61	2.338	0.81	21.33
	1E+15	56	2.259	0.78	18.45
CTJ-LC	BOL	69	2.620	0.84	28.00
	1E+14	67	2.495	0.84	26.18
	5E+14	62	2.359	0.82	22.32
	1E+15	59	2.278	0.79	19.50
	10 MeV proton				
CTJ30	BOL	72	2.616	0.84	29.13
	1E+11	72	2.554	0.83	28.38
	5E+11	71	2.493	0.83	27.06
	5E+12	62	2.279	0.77	20.22
CTJ-LC	BOL	69	2.620	0.84	28.00
	1E+11	0.069	2.565	0.83	27.48
	5E+11	0.069	2.486	0.82	26.26
	5E+12	0.063	2.298	0.79	21.26

A further simplification of the post growth process with the use of large area solar cells using 6" germanium wafers is envisaged to further decreasing the solar cell cost reaching 50% of standard triple junction solar cells for space while maintaining the electrical performances EOL.

REFERENCES

[1] R. Campesato, G.Gori, M. Casale, G. Gabetta, M. Sankaran, E.P. Suresh , B.R.Uma, "Radiation Effects on Advanced Multi Junction Solar Cells for Space Mission" 30th EUPVSEC 2016
[2] Hovel Semiconductors and semimetals vol. 11
[3] Anspaugh "GaAs Solar Cell Radiation Handbook" NASA 1996

Amorphous Silicon Carbide Rear-Side Passivation and Reflector Layer Stacks for Multi-Junction Space Solar Cells based on Germanium Substrates

[1]Stefan Janz, [1]Charlotte Weiss, [1]Christian Mohr, [2]Rufi Kurstjens, [3]Bruno Boizot, [4]Bianca Fuhrmann and [4]Victor Khorenko

[1]Fraunhofer Institute for Solar Energy Systems, Freiburg, 79110, Germany; [2]Umicore Electro-optic Materials, Watertorenstraat 33, 2250 Olen, Belgium; [3]Laboratoire des Solides Irradiés, CNRS-UMR 7642, CEA- DRF-IRAMIS, Ecole Polytechnique, Université Paris-Saclay, Palaiseau Cedex, 91120, France; [4]AZUR SPACE Solar Power GmbH, Theresienstr. 2, 74072 Heilbronn, Germany

Abstract — **New developments for space solar cells mainly address efficiency improvements and weight reduction. In this paper we developed amorphous SiC based layer stacks for passivation and enhanced reflection on thin and lowly doped ($2x10^{16}$ - $1x10^{17}$ at/cm^3) Ge wafers. Passivated Ge samples with minority carrier lifetimes of more than 300 µs and surface recombination velocities of just 17 cm/s are presented. Thermal annealing at 400 °C and additional "mirror" layer deposition do not harm the minority carrier lifetimes or lead to an even slight increase. Electron irradiation with fluences of $1x10^{15}$ e/cm^2 and more lead to strong material degradation and lifetimes of just 5 µs.**

I. INTRODUCTION

Recent developments in space solar cell research are focusing on higher efficiencies and thinner cells to increase the power-to-weight (W/g) ratio [1]. In addition, the End-Of-Life performance is important especially for satellites using electric orbit raising technology. The latter technology results in a higher radiation load during the transfer time of the satellite [2]. Today's state-of-the-art technology for space is a triple-junction GaInP/GaInAs/Ge cell in which the Ge bottom cell generates a significant excess current. A promising candidate for the next generation space solar cells is a four-junction metamorphic device with a Ge bottom sub-cell [3] in which all junctions are current matched. It is therefore important to understand how the photocurrent in the Ge bottom cell can be increased and how it reacts to the radiation environment in space. In order to increase the current generated by the longer wavelength photons, the bulk minority carrier lifetime needs to be increased. Furthermore applying a mirror layer to the back surface of the Germanium is needed in order to significantly enhance the light path and absorption in this wavelength range. This is especially important as Germanium is an indirect semiconductor with low absorption between 1600-1850 nm. Besides that such a mirror reflects photons with energy below the bandgap back into space which allows reaching lower cell operating temperatures and therefore higher efficiency.

In this paper we are investigating layer stacks of amorphous silicon carbide (a-Si$_x$C$_{1-x}$:H / a-SiC:H) which should provide both, excellent surface passivation and enhanced reflection beyond 1800 nm (see Fig. 1). In order to investigate the right combination of diffusion length of carriers, passivation performance and current generation in the Ge we fabricate wafers in a wide doping range from $2x10^{16}$ at/cm^3 to $1x10^{17}$ at/cm^3. The investigations of plasma cleaning and passivation are discussed and compared to literature values [2, 3]. Furthermore post processing like thermal annealing and accelerated aging with electrons is investigated.

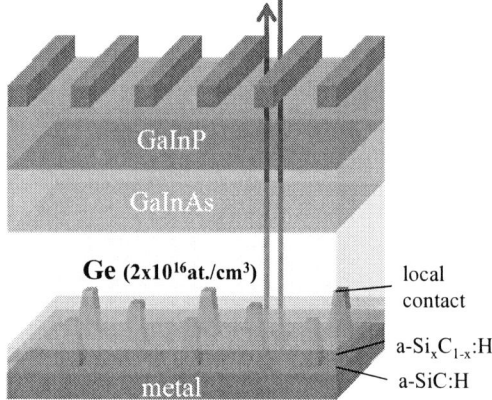

Fig. 1. Lattice-matched triple-junction solar cell with a-Si$_x$C$_{1-x}$:H / a-SiC:H back side structure for excellent surface passivation and enhanced reflection including local contact points.

II. EXPERIMENTAL

Dislocation free Czochralski pulling of Ge substrates for solar cells is a well-established technology. However, most activities have been focusing on the doping range of interest for lattice matched triple-junction cells ($1x10^{17}$ – $1x10^{18}$ at/cm^3) with a process optimization towards stable and uniform doping throughout the crystal. In order to investigate which doping range is optimal, a crystal with a doping gradient was pulled in the frame of this work. Taking wafers from such an ingot allows an investigation of different doping concentrations where all wafers have otherwise similar crystal quality. Ge ingots were pulled with a target gradient in doping from $1 \times 10^{16} - 1 \times 10^{17}$ at/cm³ enabling good material quality over a wide doping range. Out of these ingots 4" wafers with thicknesses from 150 to 650 μm were produced.

After the pre-conditioning and shipping of the wafers they were unpacked and stored for one day under clean-room conditions to allow a reproducible oxidation of the surface. This step is necessary to avoid undefined surface conditions of the samples depending on the interval between unpacking and subsequent processing. In this way, a reasonable comparison and analysis of the passivation quality of different Ge samples should be assured. Depending on further processing some 4" wafers were broken or cut into smaller pieces simply to allow for efficient use of the precious material.

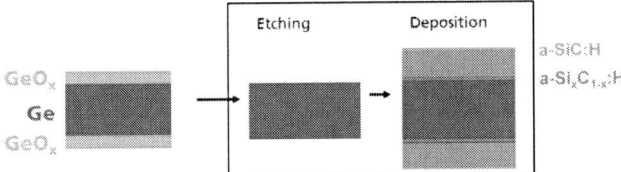

Fig. 2. Process flow for the symmetrically processed "lifetime" samples starting with the oxidized Ge wafer, the plasma etching, the passivation (a-Si$_x$C$_{1-x}$:H) and mirror layer (a-SiC:H) deposition.

The removal of the oxide formed on the Ge surface was done by a dry etching step under vacuum conditions inside a plasma enhanced chemical vapor deposition (PECVD) reactor using a H$_2$/Ar gas mixture. The exposure time of this etching process has to be carefully optimized to achieve (i) a complete removal of the native oxide on the Ge surface without (ii) damaging the Ge surface. The plasma cleaning process duration for oxygen removal was varied between 0 and 50 s. In an in-situ process an amorphous silicon carbide a-Si$_x$C$_{1-x}$:H layer (passivation layer) with (x ≈ 0.95) was deposited using methane (CH$_4$), silane (SiH$_4$) and hydrogen (H$_2$) as precursor gases at a temperature of 270 °C. For some samples the second stoichiometric a-SiC:H layer ("mirror" layer) was deposited in-situ using the same precursors. As for further solar cell processing p-doping of the film is essential, H$_2$ is substituted by diborane (B$_2$H$_6$) in order to dope the layers (see Fig. 2). Associated to plasma related processes we focused on the investigation of the plasma pre-conditioning of the Ge surface. In addition to that the temperature stability of the passivation layers was tested by exposing it to an annealing step on a hotplate at air ambient between 400 °C and 500 °C for 5 to 30 min. These thermal budgets are representative for further processes like e.g. metal/Ge contact annealing (400 °C). The prepared Ge samples for lifetime evaluation were processed exactly the same way on both sides (see Fig. 2).

The electrical quality of the surface passivation layer was characterized using microwave-detected photoconductance decay (μPCD) [4] technique which is a purely transient method measuring the exponential decay of excess carriers immediately after a short laser illumination. In this system, a short laser pulse (904 nm) on top of a steady state bias light generates excess carriers within the investigated wafer. This leads to an increase of the wafer conductance. After termination of each pulse, the excess carriers diffuse and recombine within the bulk and at both surfaces of the wafer and the photo-conductance decreases exponentially to its initial value. For the detection of the conductance, microwaves generated in a phase-locked microwave oscillator are directed through a waveguide and the reflected microwaves are redirected towards the detector. Depending on the sample conductance, the signal is reflected with varying intensity. The time dependence of this quantity is recorded by a digital storage oscilloscope and then analyzed.

The measured minority carrier lifetime τ_{eff} is an effective differential lifetime combining the carrier lifetime within the bulk τ_{bulk} and a surface carrier lifetime, characterized by the parameter surface combination velocity (S_{eff}):

$$\frac{1}{\tau_{eff}} = \frac{1}{\tau_{bulk}} + \frac{1}{\tau_{surface}} = \frac{1}{\tau_{bulk}} + \frac{2}{W} \times S_{eff}. \quad (1)$$

For this experiment we varied the Ge wafer thickness (150 μm, 300 μm and 500 μm) and plotted the inverse effective differential lifetime versus two divided by the thickness W of the wafer to easily determine the carrier bulk lifetime as y-axis intersection and the surface recombination velocity S_{eff} as slope of the applied fit.

Finally, electron accelerated ageing at different fluences was done. For these experiments the "lifetime" samples were coated with three different mirror layers in order to investigate if the radiation damage can be influenced by them. Two batches with intrinsic mirror layers 100 nm or 200 nm thick and a third batch with a doped mirror layer (200 nm) were prepared. After annealing all wafers (4") were cut to pieces of 20x20 mm². For this irradiation campaign, the electron 1 MeV beam surface on the sample was a 26 mm diameter disk measured on a microscope glass lamella using a circular 23 mm diaphragm before the window of the irradiation chamber. During the irradiation, the sample and diaphragm currents were recorded and the electron charge at the back of the sample using an ORTEC integrator was integrated. The sample temperature was around 17 °C. In total more than 100 samples were irradiated between 3×10^{13} and 10^{16} e/cm². All irradiated samples were measured before and after irradiation in order to analyze the changes in the effective minority carrier lifetimes (τ_{eff}) as a function of the total electron fluence.

III. RESULTS AND DISCUSSION

The current crystal pulling hardware and process is optimized towards reducing the doping gradient within one crystallization run. Some of the process steps could be modified to increase the gradient but still a second crystal had to be pulled to be able to provide enough material. The first crystal that was pulled (cz102/803) had a steep gradient but a doping range that was above target (see Fig. 3). Some parts of this crystal (from head to middle) were used to supply wafers with doping concentration in the range of 1×10^{17} at/cm³. The second crystal that was pulled (cz102/804) had the same steep gradient and followed the target more closely. Material from this ingot was used to supply wafers of 2×10^{16} at/cm³ and 5×10^{16} at/cm³. For further experiments we merged wafers into three doping categories ("high" 9.7×10^{16}-1.1×10^{17}, "middle" 5.2-6.2×10^{16}, "low" $\approx 2.3 \times 10^{16}$ at/cm³).

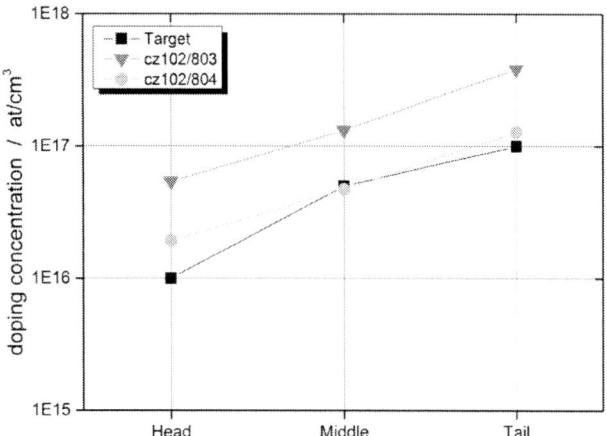

Fig. 3. Overview of the doping target and results for the two Ge crystals pulled (grey) including the target values (black).

The normalized minority carrier lifetime values for Ge samples of all three doping categories cleaned with the Ar/H₂ plasma are presented in Fig. 4. τ_{eff} improves with reduced bulk doping concentration which is not surprising considering stronger Auger recombination contribution with increasing dopant concentration. Furthermore the values clearly improve with longer plasma exposure. This parallel behavior for all doping levels can be explained with an advancing removal of the oxide layer up to process durations of 30 s. The saturation or even a slight decrease in lifetimes can be due to starting bulk damage for even longer exposure times. However, this damage behavior can be concluded from the data for the lowly doped samples only. In contrast to this results Fernandez *et al.* [5] presented a much stronger degradation of passivation performance with longer plasma exposure. So far it is not clear why, with comparable plasma parameters and using the same reactor the samples presented here seem to show almost no plasma related bulk damage effects. Either a plasma generator adaptation in between the experiments leads to "softer" plasma with lower energy densities in the current setup or the oxidized Ge surfaces differ significantly in composition and/or thickness.

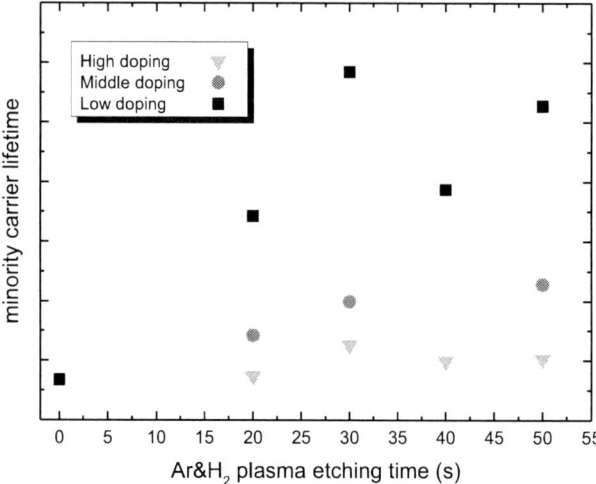

Fig. 4. Relative changes in minority carrier lifetimes of Ge samples with different doping concentrations after plasma cleaning and passivation with 30 nm of a-Si$_x$C$_{1-x}$:H on both sides of the wafer.

Based on the cleaning experiments a plasma exposure of >30 s was set for all following deposition experiments. The effective lifetime values for all three doping levels of Ge can be found in Fig. 5. From [5] we know that maximum effective lifetimes of 200 µs for highly and 400 µs for lowly doped Ge wafers can be expected. Measured lifetimes of $\tau_{eff} > 300$ µs for the lowly doped samples are the best lifetime values reported so far on such Ge wafers and proof that cleaning and passivation procedure are performing very well.

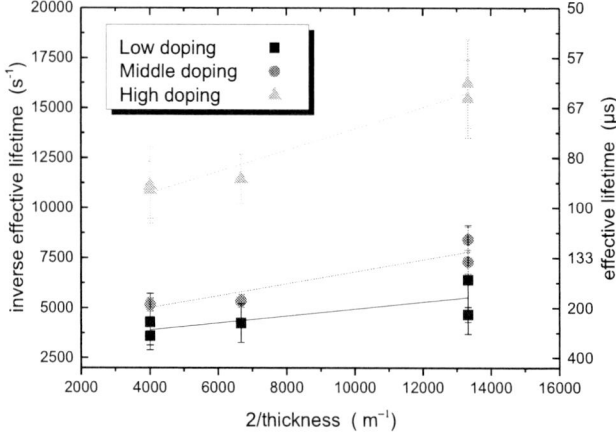

Fig. 5. Inverse effective lifetime versus 2/thickness for three different Ge doping levels and three wafer thicknesses of 150 µm – 500 µm (right to left).

Plotting the inverse effective differential lifetime versus two divided by the thickness W of the wafer (see Fig. 5) leads to carrier bulk lifetimes of 116 µs, 261 µs and 311 µs, respectively. Surface recombination velocities S_{eff} of 17 to

53 cm/s (see Table 1) confirm the excellent passivation performance of the a-Si$_x$C$_{1-x}$:H layer. So far it is not clear why the S_{eff} values are increasing with doping level, but higher defect concentrations and/or other band structures at the interface are possible explanations.

Table 1   Bulk lifetime and surface recombination velocity for the three doping levels.

Doping level (at/cm^3)	τ_{bulk}	S_{eff} (cm/s)
2x10^{16}	311	17
5x10^{16}	261	30
1x10^{17}	116	53

After an annealing of samples at the maximum temperature of 500°C for more than 5 min the minority carrier lifetimes are reduced by around 100 µs. As blistering is observed, which is typical for hydrogen effusion from amorphous hydrogenated layers, we suppose that this out-diffusion of hydrogen is the reason for the degradation. We can therefore conclude that using already passivated Ge substrates for the epitaxial growth of the III-V top absorber layers seems to be impossible due to the degradation of the passivation performance with temperatures of ≥500 °C. However, in the further solar cell process the maximum temperature applied will be much below 500 °C. Therefore we annealed parallel samples at just 400 °C, which is e.g. a standard temperature for contact annealing. The minority carrier lifetimes for all three doping levels are (after 15 min), within the accuracy of the measurement, unchanged compared to not annealed samples.

In Fig. 6 a µPCD mapping of two lowly doped 4" wafers with passivation and additional stoichiometric a-SiC:H layers (mirror layers) are shown. The mirror layers are 200 nm thick and during deposition diborane was added for the sample on the right. After annealing at 400 °C some inhomogeneities of τ_{eff} values over the wafers can be found. As plasma process inhomogeneity below 10 %, at least for the deposition, could be proven by ellipsometry measurements of the layers' thicknesses we think that the surrounding ring can be explained by edge effects (measurement artefacts). Other regions with lower lifetime values show tweezer prints and other handling traces. Further features like the ring-shaped structures on the wafer could have their origin in deviations of the bulk quality or dopant inhomogeneities in the bulk. For solar cell integration it is important to note that the additional mirror layer deposition leads to no degradation of the lifetime values (τ_{eff} > 300 µs). In addition to that we find even increased lifetime values for the sample with the boron doped

mirror layers (see Fig. 6, right). One possible explanation could be the higher amount of charges in the doped layers leading to field effect passivation. However, so far it is not clear if saturation of dangling bonds or any charge related effect are leading to the excellent passivation performance of the a-Si$_x$C$_{1-x}$:H layers on Ge. Charging experiments and the measurement of surface voltage are needed to further investigate this point.

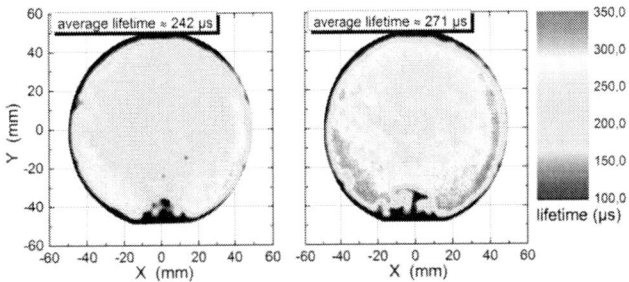

Fig. 6.   µPCD mappings of two different 4" Ge wafers (doping concentration ≈2x10^{16} cm^{-3}) after plasma cleaning, passivation with a-Si$_x$C$_{1-x}$:H, coating with stoichiometric intrinsic SiC:H (left) and p-doped SiC:H (right) and annealing.

As expected the electron irradiations show an increasingly detrimental effect on the samples' lifetimes with increasing fluences for all three doping levels of Ge (Fig. 7). The lowest total fluence of 3x10^{13} e/cm^2 reduces the lifetime from 206 µs, 118 µs and 54 µs to 181 µs, 100 µs and 41 µs for the three doping levels. Until a total fluence of 3x10^{14} e.cm^{-2} the three doping levels can still be distinguished from each other (34 µs, 19 µs and 12 µs) but for 1x10^{15} e/cm^2 and 1x10^{16} e/cm^2 all three doping levels show the same lifetime values of 5 µs and 2 µs, respectively. We could find no influence of passivation layer thickness or doping level on the lifetime values after irradiation which leads to the conclusion that doping levels in Ge alone determine the radiation hardness of the structure. So far it is not clear which fluences have to be considered in order to represent the launching conditions through Van-Allen's belt and if the Ge bulk can regenerate under irradiation (photon annealing). However, as mobilities in Ge are very high even lifetimes of just 5 µs in lowly doped material correspond to diffusion lengths for minority carriers of around 200 µm. With a final Ge wafer thickness in the range of 70 µm this can still lead to a significant efficiency improvement in the final solar cell.

978-1-5090-5606-4/17 $31.00 © 2017 IEEE

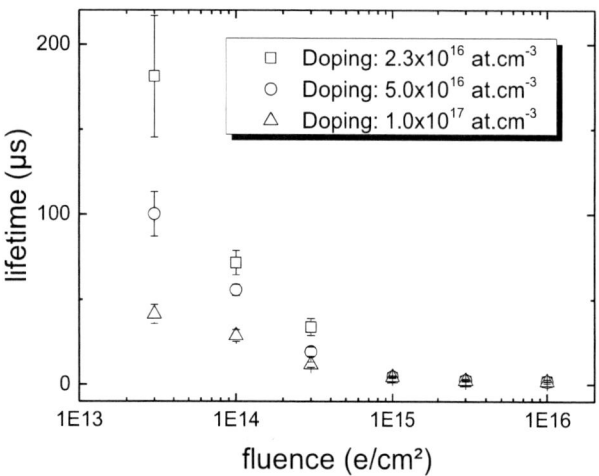

Fig. 7. Influence of electron irradiation on the lifetime degradation for the three different Ge doping levels.

experiments showed that with increased total electron fluences the lifetime values strongly decrease ending up at just 5 μs for fluences of 1×10^{15} e/cm^2 independent of the substrate doping level. Layer composition, thickness or annealing seem to have no influence on the degradation behavior. However, as mobilities in Ge are very high, even this reduced lifetime values of just 5 μs in lowly doped material correspond to diffusion lengths for minority carriers of around 200 μm and can therefore lead to an significant efficiency improvement in the final solar cell.

ACKNOWLEDGEMENT

The authors would like to express their gratitude to F. Dimroth and E. Oliva at ISE for their support and input in many valuable discussions. This work has received funding from the European Union's Horizon 2020 research and innovation programme within the project SiLaSpaCe under grant agreement No 687336.

IV. CONCLUSIONS

We investigated bulk and surface properties of Germanium for the use in next generation space solar cell devices. In two crystallization runs, Ge substrates with a wide doping concentration range from 2×10^{16} to 1×10^{17} at/cm^3 were produced. An in-situ plasma cleaning and passivation procedure for Ge wafers was developed applying H$_2$/Ar plasma for 30 s and depositing an a-Si$_x$C$_{1-x}$:H passivation layer. For lowly doped samples champion lifetimes of $\tau_{eff} > 300$ μs could be achieved which is, considering the theoretical maximum value of 400 μs and reported values in literature of well below 300 μs, an excellent result. The achieved lifetimes correspond to τ_{bulk} between 116 and 311 μs depending on Ge bulk dopant concentration and surface recombination velocities (S_{eff}) between 17 and 53 cm/s. After annealing at 400 °C the τ_{eff} or the deposition of additional optical layers on top of the passivation layers lifetime values do not decrease or even slightly increase. Electron irradiation

REFERENCES

[1] Strobl, G. F. X. *et al.,* "European roadmap of multijunction solar cells and qualification status," in *Proceedings of the 21st European Photovoltaic Solar Energy Conference,* Hofmann, 2006, pp. 1793–1796.

[2] N. E. Posthuma, G. Flamand, W. Geens, and J. Poortmans, "Surface passivation for germanium photovoltaic cells," *Sol. Energy Mater. Sol. Cells,* vol. 88, no. 1, pp. 37–45, 2005.

[3] J. Fernandez, "Development of Germanium for TPV and High-Efficiency Multi-Junction Solar Cells," Dissertation, Fakultät für Physik, Konstanz, Universität, Konstanz, 2010.

[4] J. Schmidt and A. G. Aberle, "Accurate method for the determination of bulk minority-carrier lifetimes of mono- and multicrystalline silicon wafers," *J. Appl. Phys.,* vol. 81, no. 9, pp. 6186–6199, 1997.

[5] J. Fernández, F. Dimroth, E. Oliva, and A. W. Bett, "Development of germanium tpv cell technology," in 2007, pp. 516–519.

Hot Carrier Transportation Dynamics in InAs/GaAs Quantum Dot Solar Cell

Tomah Sogabe[1,2,3], Kohdai Nii,[2] Katsuyoshi Sakamoto[2], Koichi Yamaguchi[2], Yoshitaka Okada[3]

[1]i-Powered Energy Research Center (i-PERC), The University of Electro-Communications, Tokyo, Japan
[2]Department of Engineering Science, The University of Electro-Communications, Tokyo, Japan
[3]Research Center for Advanced Science and Technology (RCAST), The University of Tokyo, Tokyo, JAPAN

ABSTRACT

The hot carrier dynamics and its effect on the device performance of GaAs solar cell and InAs/GaAs quantum dot solar cell (QDSC) was investigated. At first, the fundamental operation feature of conventional hot carrier solar cell was simulated based on the detailed balance thermodynamic model. Then we investigated the hot carrier dynamics in the normal pn junction based solar cell using hydrodynamic/energy Boltzmann transportation model (HETM) where the two temperature (carrier temperature and lattice temperature are treated separately. For the first time, we report an inherent quasi-equivalence between the detailed balance model and HETM model. The inter-link revealed here addresses the energy conservation law used in the detailed balance model from different angle and it paves a way toward an alternative approach to curtail the selective contact constraints used in the conventional hot carrier solar cell. In simulation, a specially designed InAs/GaAs quantum dot solar cell was used in the simulation. By varying the hot carrier energy relaxation time τ_w, an increase in the open circuit voltage V_{oc} was clearly found with the increase of τ_w. Detailed analysis was presented regarding the spatial distribution of hot carrier temperature and its interplay with electric field and three hot carrier recombination processes (Auger, SRH and radiative)

INTRODUCTION

Hot carrier solar cell has drawn a lot of attentions due to its high theoretical efficiency limit of more than 80% [1,2]. Conventionally, implementation of hot carrier solar cell requires (1) a photoactive material where cooling is slower than the transport to contacts; (2) a contact material which allows to selective extraction of electron or holes through a narrow energy band [2]. Since this model is based on thermodynamics, it lacks of the interlink to the semiconductor device. Especially the implementation of selective contact in terms of semiconductor material is severely hindered due to the insufficient interpretation of the energy conservation thermodynamic constraints from the semiconductor device point of view. This is one of the main targets of the current work i.e. addressing the hot carrier operation principle from closed form of Boltzmann device transportation model.

Meanwhile, in a typical bulk semiconductor device, cooling occurs in less than 10ps, while carrier extraction may take nanoseconds or longer. A lot of efforts have been devoted to slow the cooling process by well controlling the phonon scattering such as employing special bulk materials in which the energy coupling between phonon and electron could hardly occur. Recently, quantum well and quantum dot have also showed potentials to slow the hot carrier cooling rate [3]. An alternative way to slow down the cooling rate is to enhance the carrier separation rate. As having been reported in dye sensitized solar cell, an interfacial charge

transfer can be fast and in less than a picosecond be exploited [4]. In this work, we focus on the investigation of the hot carrier separation/transportation dynamics in InAs/GaAs QDSC.

The paper is arranged as follows: 1) At first, the hot carrier solar cell operation principle was simulated based on the thermodynamic detail balance model containing a 'cooling' and 'hot' competing process. 2) A hydrodynamic model including the carrier energy ω ($\omega = 3/2 k T_H$ and T_H denotes the hot carrier temperature) was used to simulate a specially designed GaAs$_{n\text{-emitter}}$/AlGaAs$_{barrier}$/ GaAs$_{intrinsic}$ / GaAs$_{p\text{-base}}$ solar cell. The results were compared with the simulation results using the drift-diffusion model (DDM) by varying internal electric filed through adjusting GaAs intrinsic layer thickness, hot carrier relaxation time τ_w as well as the AlGaAs barrier thickness. 3) Detailed analysis regarding the spatial distribution of hot carrier temperature and the interplay with electric field and recombination was presented.

'COOL' AND 'HOT' CARRIER COMPETING MODEL

Figure 1(a) shows the thermodynamic principle involving 'cool' and 'hot' carriers. The hot carriers are assumed to reach a self-equilibrium between temperature T_H and chemical potential μ_{cool}. For a hot carrier solar cell, T_H and μ_H are competing and regulated based on the following equations [1]:

$$\mu_{out} = \mu_H \left(\frac{T_a}{T_H}\right) + E_{hot}\left(1 - \frac{T_a}{T_H}\right) \tag{1}$$

$$J = q\{Xf_S N(E_g,\infty, T_{sun},0) - N(E_g,\infty, T_H, \mu_H)\} \tag{2}$$

$$J * E_{hot} = q\{Xf_S L(E_g,\infty, T_{sun},0) - L(E_g,\infty, T_H, \mu_H)\} \tag{3}$$

where T_a is the ambient temperature and E_{hot} is the energy separation between the extracted electron and hole and in physics it equals the kinetic energy ω; $N(\ )$ is the photon flux density and $L(\ )$ is the energy flux density, X and f_S are the parameters related to light concentration. If the hot carrier reaches the thermal equilibrium with

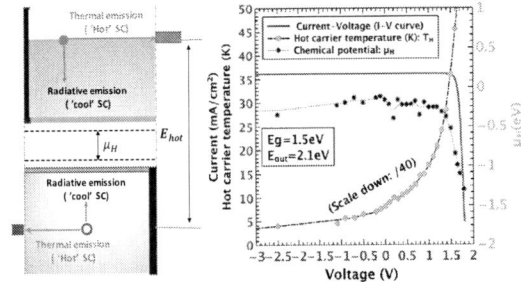

Figure 1 (a) the sketch of the cooling and 'hot' competing process model. (b) Calculated results based on detailed balance principle together with energy conservation constraints. Here, $E_g = 1.5eV$ and $E_{hot} = 2.1eV$, V=μ_{out}/q. note: T_H was scaled down by 40 to fit on the same figure.

ambient temperature T_a, then the hot carrier solar cell

978-1-5090-5606-4/17 $31.00 © 2017 IEEE

becomes the conventional solar cell and the output potential equals to $\mu_{out} = \mu_H$ (the quasi-Fermi level splitting). The J-V relation is derived by considering both the particle conservation (number of absorbed photons = number of emitted photons) described in equation (2) and the energy conservation in equation (3). Figure 1(b) shows the results of a hot carrier solar cell with bandgap $E_g = 1.5eV$ and the extraction energy separation E_{hot} was set at $2.1eV$ (note: this value can be varied to sort the maximum conversion efficiency). It is clearly seen here that by varying the output voltage V= μ_{out}/q the temperature of carriers T_H and chemical potential μ_H showed opposite trend indicating the competing behavior. As V increases, the output current decreases due to the increased photon emission which follows the Planck's radiation law $n(E) \propto E^2 \exp\{-\left(\frac{E-\mu_H}{k\,T_H}\right)\}$. Here we refer the photon emission controlled by μ_H as a 'cool' process and ones induced by temperature T_H as 'hot' process. The word 'cool' was so chosen as to reflect its similarity to the normal solar cell where the photo emission is solely controlled by chemical potential μ because T_H is usually assumed to be constant as T_a. The word 'cool' was also chosen to reflect its value is much less than its counterpart in the normal solar cell and becomes zero or negative at the vicinity of V_{oc}. The existence of the 'cool' and 'hot' competition lies in the fact that for a hot carrier solar cell, it allows both T_H and μ_H to vary the number of emitted photons $n(E)$ so as to reach the detailed balance with the totally absorbed photons. When V reaches the V_{oc} regime, the carrier temperature T_H increases dramatically while μ_H goes even to negative due to the stringent constraints of particle conservation and energy conservation. In other words, conventional solar cell at V_{oc} can be viewed approximately as an 'emitting diode' at ambient temperature T_a while the hot carrier solar cell is exemplified as 'hot (high temperature T_H) thermal engine with μ_H being zero or even negative. as shown in the Figure 1(b).

HOT CARRIER DYNAMICS SIMULATION USING HYDRODYNAMIC / ENERGY TRANSPORTATION MODEL

After having investigated the competing dynamic behaviour of the 'cool' and 'hot' carrier using the thermodynamic model, we further studied the hot carrier relaxation and extraction effect on solar cell device performance. This was implemented by converting the conventional DDM into the hydrodynamic model where the hot carrier energy $\omega = 3/2kT_H$ was included. We have tackled this issue from two approaches: (i) a hydrodynamic model including hot carrier energy transportation using the commercial software, the Crosslight APSYS; (ii) extension of the self-developed DDM code to include the effects such as the electric field dependent mobility and diffusion coefficient as well as cooling and 'hot' carrier competing dynamics illustrated in equation (1) [5,6,7]. The main formula used in approach (i) are given as follows [8]:

$$\nabla \cdot (\epsilon_s \nabla \psi) = -q(p - n + N_D^+ - N_A^-) \tag{4}$$

$$\frac{1}{q} \nabla \cdot J_n = -G + R \tag{5}$$

$$\frac{1}{q} \nabla \cdot J_p = G - R \tag{6}$$

$$\frac{J_n}{q} = \mu_n n \nabla \epsilon_c + D_n \nabla n + S_n \nabla T_{n,H} \tag{7}$$

$$\frac{J_p}{q} = \mu_p p \nabla \epsilon_c - D_p \nabla p - S_p \nabla T_{p,H} \tag{8}$$

$$\frac{\partial}{\partial t}(nw) + \nabla \cdot S - \frac{J_n}{q} \cdot \nabla \epsilon_c = \frac{\partial}{\partial t}(nw)\Big|_c - w(R - G) \tag{9}$$

where the Here equation (4), (5), (6), (7), (8) are the familiar Poisson's equation, current continuity equation

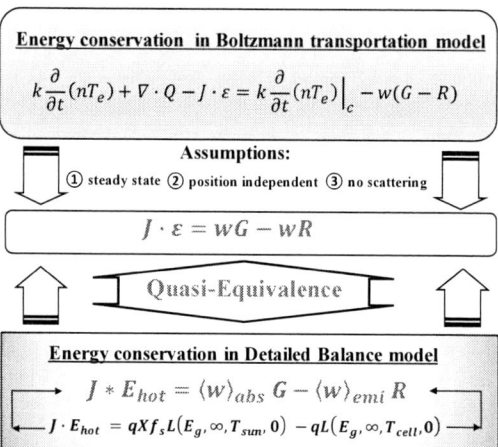

Figure 2 Sketch of the derivation of quasi-equivalence between the energy conservation in detailed balance model and energy conservation in Boltzmann transportation model.

and current density. Note that in (7) and (8), there is additional contribution to the current density due to the hot carrier temperature gradient when compared to the normal drift-diffusion simulation (DDM). These formulae are physically corresponding to the particle conservation equation (2) in thermodynamic detailed balance model.

The key feature in hydrodynamic/energy transportation model is the energy balance equation (9). This equation, based some assumptions, can be simplified to a similar form of the energy conservation equation (3), as shown in Figure 2. $\nabla \epsilon_c$ here represents the electric field ε. If we assume that in a conceptual thermodynamic device, it is reasonable to assume that there is no position dependence, thus $\nabla \cdot S = 0$ and there is no scattering $(\tau_w \rightarrow \infty)$, thus $\frac{\partial}{\partial t}(nw)\Big|_c = 0$. Equation (9) can then be further simplified as:

$$J \cdot \varepsilon = wG - wR. \tag{10}$$

Meanwhile, the formula (3) in the thermodynamic detailed balance model is rewritten so that the photon currents can be related to the energy currents via the average energies of the absorbed $\langle w_{abs} \rangle$ and emitted $\langle w_{emi} \rangle$ photons. These average photon energies are given by the absorbed or emitted energy currents divided by the appropriate photon currents G and R. The formula (3) was finally changed to:

$$J \cdot E_{hot} = \langle w_{abs} \rangle G - \langle w_{emi} \rangle R. \tag{11}$$

As shown in Figure 2, it is apparent that a quasi-equivalence is built between the formula of (10) and formula (11). It is interesting to note that the most intriguing and challenging parameter in the thermodynamic model of hot carrier solar cell is the E_{hot}, which can be found corresponding exactly to the electric field ε, as shown in Figure 2.

Figure 3 shows the simulated results where a specially designed GaAs$_{n-emitter}$/AlGaAs$_{barrier}$/ GaAs$_{intrinsic}$/GaAs$_{p-base}$ solar cell was used as reference sample. A narrow n-type emitter was chosen and the intrinsic layer (i-GaAs) was set at 100nm to enhance the electric field effect. AlGaAs barrier was inserted between the i-GaAs and n-GaAs emitter to tune the hot carrier dynamics. The band diagram at thermal equilibrium and short circuit current J_{sc} as well as the open circuit voltage V_{oc} were illustrated

978-1-5090-5606-4/17 $31.00 © 2017 IEEE

in Figure 3(c) and 3(d). In order to further reveal the hot carrier transportation dynamics, energy dependent relaxation time τ_w was varied from $1\times10^{-14}s$ to $1\times10^{-11}s$ and the results were compared with those calculated by conventional DDM in which the hot carrier effect is ignored.

As it can be seen from 3(c), the J_{sc} tends to decrease with the increase of the relaxation time τ_w. We attributed the reduction of J_{sc} to the hot electrons induced current leak to the back electrode. An electron block layer acting as back surface field is expected to suppress the current reduction. Interestingly, we found that V_{oc} simulated under the HETM model showed much higher value than those simulated by DDM. Detailed analysis regarding the V_{oc} variation with τ_w will be given in the next section together with the interpretation of the hot carrier temperature distribution. In addition, the fill-factor (FF) was found improved in the hot carrier dynamics simulation model from 83.5% at $1\times10^{-14}s$ to 84.8 at $1\times10^{-11}s$. The FF increases with elongated carrier relaxation time indicating the enhanced carrier extraction under 'hot' states, which will be discussed in more detail in the next section.

DISCUSSIONS

a): Simplified J-V relation:
One naïve approximation for the J-V relation two-temperature HETM can be made by taking the analogue of conventional one-temperature J-V model while inserting the temperature dependence for the related parameters. The formula can be approximated as:

$$V_{oc,HETM} \approx m(T_H)kT_H \ln\left\{\left(\frac{J_{sc}}{J_0(T_H)}\right) + 1\right\} \qquad (12)$$

where the $m(T_H)$ is the temperature dependent ideality factor; $J_0(T_H)$ is the temperature dependent recombination current, which corresponds to the SRH, Auger and radiative recombination mentioned previously. We have found that when at the V_{oc}, the J_0 calculated by HETM was higher than the ones by DDM. Meanwhile, the device temperature T_H is almost equal to $T_L = 300K$, so the only parameters which contribute to the increase of V_{oc} is the ideality factor $m(T_H)$. Based on a simple relation between FF and ideality factor derived by Green[9], increase of FF calculated by using the HETM should be originated from a decrease of ideality factor. From the analysis above, we found the formula (12) failed to interpret the increase of V_{oc} calculated by using the HETM. In other words, a simplified J-V diode model is not applicable to analyse the HETM results.

b): Energy conservation analysis:
In principle, the difference between the DDM and HETM is the applied energy conservation law in the HETM model. The increase of V_{oc} should be fully understood through the closed form of formula (9). At V_{oc}, the current $J_n = 0$, formula (9) then reduced to:

$$-\frac{\partial}{\partial t}(nw)\Big|_c = (G - R) \approx \frac{1}{\tau_w}\left(1 - \frac{T_l}{T_e}\right) \qquad [13]$$

This is very significant results for understanding HETM model. We found $(R - G)$ iis depending on the carrier temperature T_e and relaxation time τ_w. Although further investigation is needed to gain deeper insight on this issue, $(R - G)$ remains nonzero at V_{oc} is considered to be the most decisive factor which altered the V_{oc} value in the HETM simulation. We will show more detailed analysis regarding the T_e distribution and (G-R) in the conference.

Our further investigation will be focused on the incorporation of carrier separation term E_{hot} and InAs QD into the current simulation model. Meanwhile, an InAs/GaAs QDSC will be fabricated based on the the simulated device structure to verify the hot carrier transportation effects by varying internal electric filed as well as the AlGaAs barrier height.

CONCLUSIONS

Hot carrier dynamics in GaAs based solar cell was studied by using HETM by varying the hot carrier relaxation time and the results were compared to those simulated by using DDM. It was found that the I_{sc} tends to decrease with the increase of hot carrier relaxation time and showed lower value in the energy transport model than the DDM model. The V_{oc}, on the contrary, showed much higher value and which was analyzed based on both thermodynamic model and energy conservation in HETM simulation. Simulations are undergoing by incorporating carrier separation term E_{hot} and InAs QD into the current simulation model. Meanwhile experimental characterization of InAs/GaAs QDSC will be used to further examine the simulation results. Note the reduction of V_{oc} at very short relaxation time is mainly due to the reduction of I_{sc}. The fill-factor (FF) was found improved in the hot carrier dynamics simulation model. The FF increases with elongated carrier relaxation time indicating the enhanced carrier extraction under 'hot' states. The final conversion efficiency decreases mainly due to the reduction in I_{sc}.

REFERENCES

[1] J. Nelson, The Physics of Solar Cells (Imperial College Press, London, (2003).

[2] P. Würfel, "Solar energy conversion with hot electrons from impact ionisation", Sol. Energy Mater. Sol. Cells **46**, 43 (1997).

[3] A. Nozik, "Spectroscopy and hot electron relaxation dynamics in semiconductor quantum wells and quantum dots," Ann. Rev.Phys.Chem. **52**, 193 (2001)

[4] A.Hagfeldt and M.Graetzel, "molecular photovoltaics", Acc, Chem,Res,**33**,269 (2000).

[5] T. Sogabe et al., "Intermediate-band dynamics of quantum dots solar cell in concentrator photovoltaic modules," Sci. Rep., **4**, 4792 (2014).

[6] S. Tomić, T. Sogabe and Y. Okada, "In-plane coupling effect on absorption coefficients of InAs/GaAs quantum dots arrays for intermediate band solar cell," Prog. Photovolt: Res. Appl., **23**,546 (2014).

[7] T. Sogabe, T. Kaizu, Y. Okada and S. Tomić, "Theoretical analysis of GaAs/AlGaAs quantum dots in quantum wire array for intermediate band solar cell", J.Renew. Sustain. Ener., **6**, 011206 (2014)

[8] E.M. Azoff. "Energy transport numerical simulation of graded AlGaAs/GaAs heterojunction bipolar transistors," IEEE Trans. ED, **36**, 609 (1989).

[9] M. A. Green, "Solar cell fill factors: General graph and empirical expressions", Solid-State Electronics, vol. 24, pp. 788 – 789(1981).

Integration of Crack-Tolerant Composite Gridlines on Triple Junction Photovoltaic Cells

Omar K. Abudayyeh[1], Geoffrey K. Bradshaw[2], *Member, IEEE*, Steven Whipple[3], David M. Wilt[2], and Sang M. Han[1]

[1]University of New Mexico, Albuquerque, NM, 87131, USA
[2]Air Force Research Laboratory, KAFB, NM, 87123
[3]SolAero Technologies, Albuquerque, NM, 87123

Abstract—**Metal matrix composites consisting of multi-walled carbon nanotubes embedded in a sliver matrix are successfully integrated onto commercial triple junction photovoltaic cells. The performance of triple junction cells with composite gridlines is analyzed; the fill factor and efficiency closely match those of cells with standard evaporation-based metallization. The cells are then intentionally cracked, using external mechanical stress. We observe substantially enhanced crack tolerance in composite-enhanced cells in comparison to standard triple junction cells. Upon introducing cracks, the control cell result in a loss of 54% in the short circuit current, whereas no significant loss is observed for the composite-enhanced cells. The composite metal gridlines show strong potential to improve the lifetime of space photovoltaic cells against stress-induced fracture.**

Index Terms—**Carbon nanotubes, composite materials, cracking, metallization, photovoltaic cells, reliability, solar cells.**

I. INTRODUCTION

Multi-junction (MJ) solar cells have been used almost exclusively for space vehicles due to their high efficiency [1] and high radiation hardness [2]. The efficiency of state-of-practice (SOP) space triple-junction (TJ) cells today are approximately 30% under 1-sun Air Mass 0 (AM0) spectrum [3]. While MJ cells provide a high efficiency, microcracks can develop in crystalline photovoltaic (PV) cells due to a variety of reasons: e.g., growth defects, film stress by lattice mismatch, and mechanical stress introduced during shipping, installation, and operation. Microcracks can electrically isolate fractured portions of the cell, which can lead to a substantial power loss.

To mitigate the performance degradation due to microcracks, we have investigated the use of multi-walled carbon nanotubes (MWCNTs) as mechanical reinforcement to metal gridlines, where CNTs are embedded in the metal to form metal matrix composites (MMC). Our previous work [4] has shown that MMC gridlines (1-mm wide, 6-μm-thick) can electrically bridge > 40-μm-wide fractures and repeatedly "self-heal" when the lines are strained to failure and brought back to close proximity. Fig. 1 shows SEM micrographs of fractured MMC gridlines (~9-μm-wide gap) with MWCNTs bridging the gap while providing redundant electrical conduction pathways.

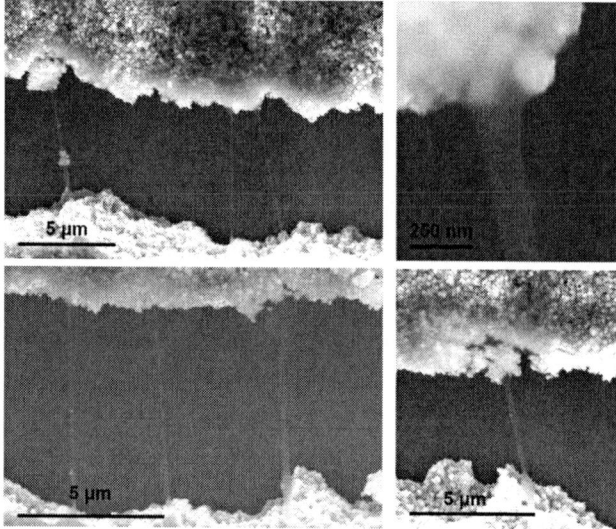

Fig. 1. SEM micrographs of CNTs anchored in Ag layers, bridging 9-μm-wide gaps [4].

The CNTs are deposited on a layer of electrodeposited silver (Ag), and a second layer of Ag is electrodeposited on top of the CNT layer, forming a layer-by-layer microstructure of the MMCs. The MMC gridlines are then integrated onto commercial TJ cells and later fractured to assess the ability of MMC to maintain electrical conductivity. After cracks are introduced, we observe that the cells with MMC gridlines exhibit higher performance than the cells with standard metallization.

II. EXPERIMENTAL METHODS

A. Materials

Two sets of 2 cm x 2 cm TJ cells with a modified metallization pattern were used to produce control and MMC-enhanced devices. The gridline width and the front contact pad size are made larger than production cells in order to facilitate MMC integration and testing. Cyanide-free alkaline silver plating solution (E-Brite 50/50 RTP, *Electrochemical Products Inc.*) is used for Ag plating. Low-cost, dry MWCNTs purchased from *SWeNT* are functionalized and suspended in water. For a

978-1-5090-5606-4/17 $31.00 © 2017 IEEE

more detailed description on CNT solution preparation and Ag plating optimization, refer to Ref. [4]. The CNTs are spray-coated on top of the plated Ag, and another layer of Ag is electroplated on top of the CNT layer. A full description of the CNT deposition method can be found in Ref.[5].

Following the gridline deposition, the cells are characterized through light current-voltage (*LIV*) sweeps and electroluminescence (EL) measurements. An X-25 solar simulator (Mark II, *Spectrolab Inc.*) is used to perform the *LIV* sweeps. The beam is first calibrated using a standard TJ cell, and values are reported under AM0 spectrum. EL measurements are performed by forward biasing the cells at 30 mA (~ 7.5 mA/cm^2). The top subcell (GaInP) strongly emits in the visible region ($\lambda = 683$ nm) [6] when forward-biased and can be easily examined with the naked eye to visually locate cracks in the substrate.

B. Silver Plating

Silver can be deposited through a variety of techniques, including screen printing, electro and electroless deposition, and vacuum deposition techniques (e.g., chemical/physical vapor deposition and ion sputtering). Electroplating was chosen as the method of deposition because it is a simple and inexpensive technique capable of producing homogenous, highly reflective, thin-film Ag deposits. Metal electroplating is a process that coats conductive or semi-conductive objects with a thin metal film. The process uses an electrical current to reduce cations of a desired metal from a solution. The process is stable and suitable for growth of thin films and/or nanostructures with potentially enhanced thermoelectrical properties [7, 8].

In our previous work [4, 9], we were able to determine a window of operation for depositing smooth, coherent, and compact silver films in a range of current densities using the aforementioned Ag plating solution. We further optimize the previously developed recipe to plate Ag into the 50-μm-wide, 7-μm-deep recess patterns defined for gridlines. Through careful and systematic manipulation of deposition rates (operating current density), we are able to reproduce high-quality silver gridlines. The first Ag layer is deposited using a two-step process, starting with a slow deposition rate (low current density) for a short period followed by a high deposition rate (ramping up the current density). Ag is plated at -1.7 mA/cm^2 for 400 s followed by -6.7 mA/cm^2 for 800 s, resulting in a 2 to 3-μm-thick Ag film (Fig. 2A). The initial deposition at a low current density creates Ag nucleation sites on the seed layer. The proper formation of these nucleation sites allows for the second-stage growth of uniform Ag layers at a high deposition rate. Without the proper formation of initial nucleation sites, dendrites start to form (Fig. 2B) with coarse surface morphology. The high growth rate corresponds to fast reduction of Ag ions at the surface owing to high charge density. A similar behavior is reported in the literature [10-12]. Yasnikov *et al.* [11] reported the formation of various microstructures (e.g., pentagonal, regular shaped, planar, and dendritic) under different potentiostatic conditions as the overpotential increased from 160 to 200 mV. The formation of such dendrites occurs as a result of severe depletion of silver ions near the cathode.

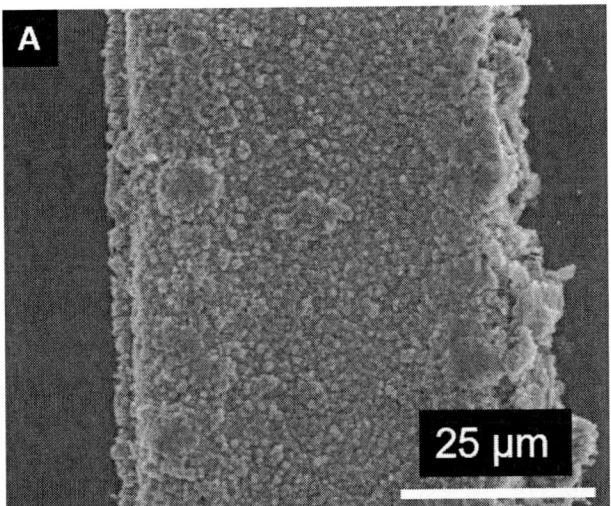

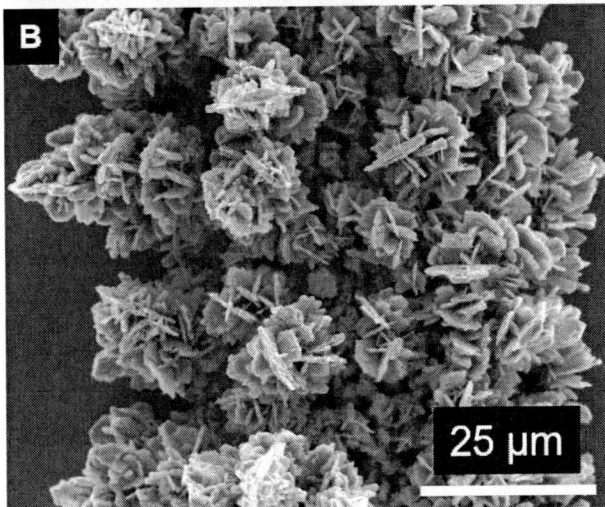

Fig. 2. SEM micrograph (top view) of a gridline after Ag deposition at (A) -1.7 mA/cm^2 for 400 s followed by -6.7 mA/cm^2 for 800 s, resulting in a smooth, dendrite-free gridline; and (B) -6.7 mA/cm^2 for 1200 s, where dendrites form on top of the seed layer creating "desert rose" structures.

C. MMC Integration on TJ Cells

Following the first electroplated Ag layer on TJ cells, the entire cell is spray-coated with an aqueous CNT solution (1.3g/L). The cells are placed on a moving stage (3.5 mm/s) and sprayed 15 times in a repeat cycle, resulting in complete surface coverage of CNTs on Ag. By controlling the moving speed of the sample stage under the spray nozzle and the substrate temperature, we are able to deposit thin, uniform layers of CNTs across the entire surface of the cell. The solvent evaporates quickly from the heated substrate, leaving behind functionalized MWCNTs. The cells are then transferred back into the electrolytic solution for the final deposition of Ag. The second Ag layer is electroplated at -6.7 mA/cm^2 for 500 s, resulting in a ~ 1-μm-thick layer. A short plating time for the second layer is chosen in order to selectively deposit Ag mostly on the gridline pattern and not on the photoresist covered with conductive CNTs. Samples are then rinsed with water to

978-1-5090-5606-4/17 $31.00 © 2017 IEEE

remove residual plating solution and soaked in acetone to remove the photoresist and CNTs on its top by liftoff.

To verify integration of CNTs within the metal matrix and their homogenous dispersion without agglomeration, the MMC gridlines are examined by scanning electron microscopy (SEM) using a *Hitachi S-4300* at 20 KeV. Figure 3 shows a cross sectional-view of two MMC gridlines. These images show that we can deposit thick (~15-μm) as well as thin (~ 6-μm) gridlines by adjusting the deposition time. CNTs can be seen in the enlarged SEM view (Fig. 3, red arrows) as veiny threads. The CNTs are localized within the stack at a specified depth and appear to adhere well to the surrounding metal matrix. We do not observe agglomeration or bundling of CNTs within the stack; CNTs are well dispersed due to surface functionalization.

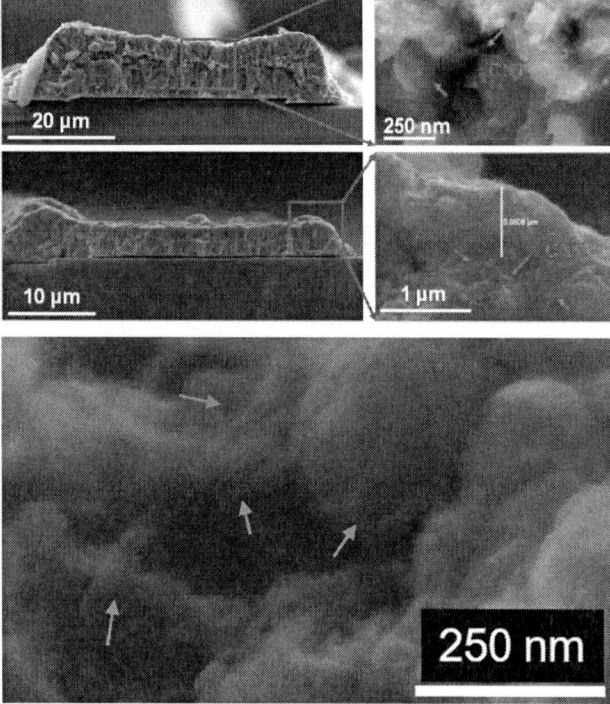

Fig. 3. Cross-sectional SEM micrographs of MMC gridlines on TJ cells. CNTs indicated by red arrows.

III. RESULTS AND DISCUSSION

A. MMC Integration

A TJ control sample with standard metallization is used as a baseline for comparing *LIV*, *DIV* and EL characteristics to cells with MMC gridlines. Fig. 4A shows the summary of *LIV* data and EL images of the control sample and five test cells (MMC 1 − 5). The initial attempts of integration resulted in poor cell performance (Fig. 4A, MMC 1 − 3). However, the integration process is later optimized by modifying the electroplating step which resulted in significant improvement in cell performance (Fig. 4A, MMC 4 and 5). The most significant improvement is seen in the *FF* and η reaching 86% and 26%, respectively. Early electroplating attempts resulted in the etching of the cell edges, which can be seen in the EL images of MMC 1 − 3 (Fig.

4A inset), where a non-uniform EL response is observed around the cell edges. This is a result of the plating solution used, which is alkaline with a trace amount of KOH and is a known etchant for most III-V materials [13, 14]. The electroplating step is optimized by properly sealing the cell edges with photoresist prior to electroplating. The dark spots observed are due to KOH etching of the exposed semiconductor material. In comparison, a MMC-integrated TJ cell whose edges are properly sealed during Ag plating (MMC 5 in Fig. 4A) shows a uniform EL response without any dark spots. The result from MMC 5 closely matches the control sample.

To confirm successful integration of MMC gridlines, we examine the diode properties through *DIV* measurements. The inset of Fig. 4B shows the dark diode characteristics of the MMC samples, plotted on semi-log scale. The *DIV* characteristics improve going from MMC 1 to MMC 5 as is particularly evident by the reduced ideality factor at higher voltages. Above 1.5 V, MMC 4 and 5 samples show the diode ideality factor similar to the control sample. For the applied voltage below 1.5 V, the performance of MMC 4 and 5 cells appears to deviate from the control cell. This may be attributed

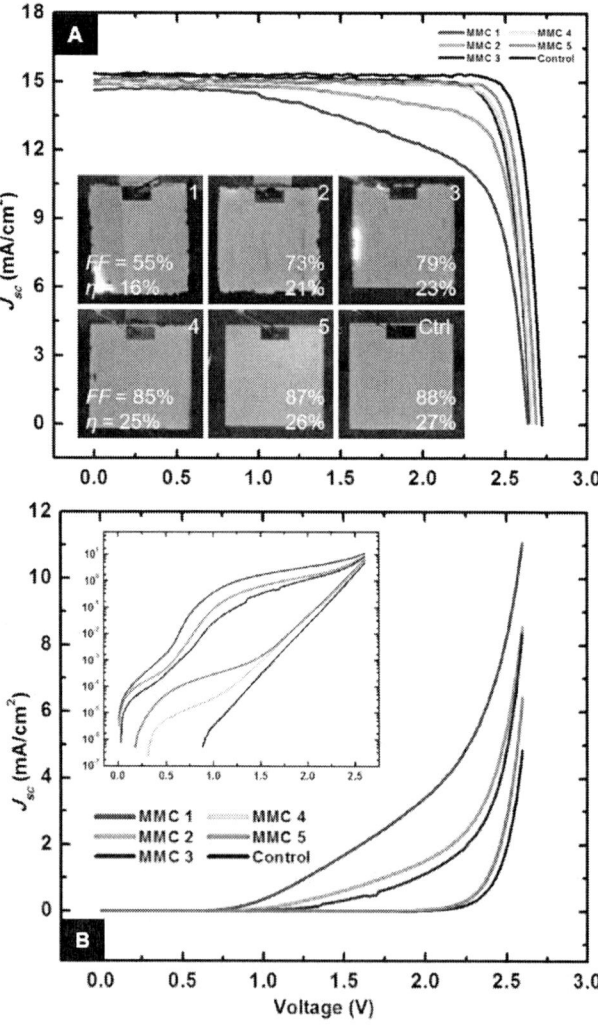

Fig. 4. (A) *LIV* characteristics with EL images of control and 5 MMC test cells, (B) *DIV* scans on linear scale and semi-log scale (inset).

to a small decrease in the shunt resistance of the cells, most likely the result of processing and handling, although the overall cell performance of MMC 4 and 5 is very close to the control sample (Fig. 4A) without any noticeable slope in the current response. In contrast, MMC 1 – 3 samples exhibit poor diode characteristics with a pronounced deviation from the control sample. The non-ideal diode characteristics of MMC 1 – 3 cells are consistent with the poor *LIV* performance, which is largely due to the non-radiative recombination on the exposed areas of the cells that are etched away during Ag deposition.

B. Cell Performance Degradation due to Microcracks

We evaluate the effects of microcracks on the performance of typical TJ cells with standard metallization vs. TJ cell with MMC gridlines. The gridlines on the control cell consist of 100% evaporated Ag, where no CNTs are embedded. *LIV* scan is performed before and after cracking. Cracks are generated by resting cells against a curved surface ($r = 6$ cm) while applying an external mechanical stress on the cell's top surface. Fig. 5 summarizes the *LIV* and EL results of both control and test samples.

The EL response of the control sample shows a large dark region after introducing cracks. This dark region is the result of electrical isolation from the busbar, where the cracks propagated through both semiconductor and metal gridlines. The illuminated regions correspond to the remaining active cell area that contributes to current generation in the *LIV* scan (Fig. 5). The control sample suffered a significant loss in all cell parameters. A degradation in both V_{oc} (5.1%) and J_{sc} (53%) is clearly observed, demonstrating a possible outcome of cell cracking with Ag gridlines. The *FF* and η also decreased by 29% and 68%, respectively due to cracking. In addition, we note a decrease in shunt resistance and an increase in series resistance of the control cell as a result of cracking.

In contrast, the test sample with MMC gridlines is capable of maintaining electrical continuity even in the presence of cracks.

The test cell with MMC gridlines generates nearly the same J_{sc} (0.78% loss) after being fractured. Additionally, the test sample is able to maintain a good EL response after fracturing without any visible dark spots. While both gridlines and substrate are completely fractured, the CNTs appear to provide redundant electrical conduction pathways.

While J_{sc} remains virtually unchanged, we observe brighter EL intensity at and around the fracture locations than the regions that remain intact. The increased brightness suggests an increased number of recombination sites for minority carriers at and around the fracture location. These recombination sites would reduce V_{oc}, as clearly shown in Fig. 5 for both control (5.1% loss) and MMC test (6.1% loss) samples. The V_{oc} loss observed is attributed to the fracturing process and not due to the MMC gridline integration. In addition, the series resistance of the cell with MMC gridlines appears to increase after fracturing. As the gridlines get severed due to microcracks developing in the underlying substrate, the electrical connection in that region is sustained only through the embedded CNTs. Thus, an overall increase in series resistance is observed. This increase in the series resistance ultimately reduces the *FF* and η by 26% and 35%, respectively.

Although the MMC integration on cells may not mitigate against all the loss mechanisms due to fracturing, the test sample with MMC gridlines maintains the same current generation after fracturing. Current preservation is critically important as cells are connected in series on a string, and current loss in one cell can decrease the overall module output.

IV. CONCLUSION

In this study, we demonstrate the successful integration of MMC gridlines onto commercial TJ cells. These composite lines show strong potential to replace conventional metal ones on space photovoltaic cells. Upon fracturing, the test cells integrated with our MMC gridlines show clearly increased fracture tolerance than the control cell with evaporated Ag gridlines. That is, embedding CNTs into metal mitigates the

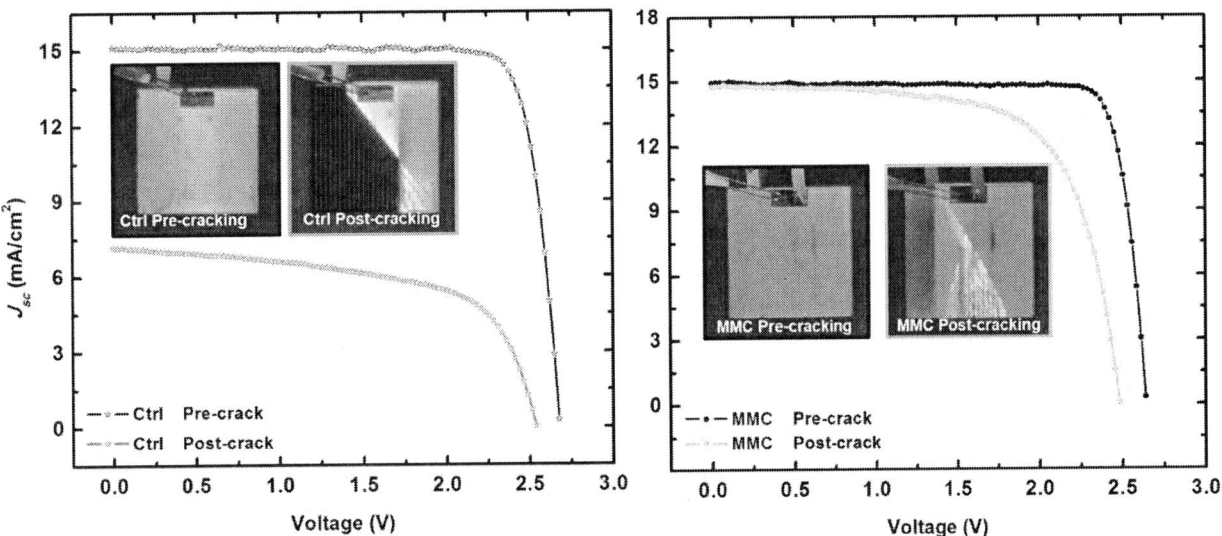

Fig. 5. *LIV* and EL measurement of control (left) composed of bare Ag gridlines, and test sample (right) composed of MMC gridlines.

electrical disconnect due to microcracks. The continuous areal EL response and the preservation of J_{sc} after substrate fracture are strong evidences that MMC lines are more resilient to microcracks developed in the semiconductor substrate than 100% metal gridlines. However, we acknowledge that the cracks may introduce other loss mechanisms that MMCs cannot fully counter, leading to an unavoidable loss in V_{oc}, FF, and η. This demonstration supports that our MMC gridlines are suitable to replace traditional gridlines and to help mitigate the loss in cell performance as microcracks develop in cells.

ACKNOWLEDGMENT

We acknowledge a generous financial support from the Air Force Research Laboratory (FA9453-14-1-0242) to conduct this engineering research.

REFERENCES

[1] M. Stan, D. Aiken, B. Cho, A. Cornfeld, J. Diaz, V. Ley, *et al.*, "Very high efficiency triple junction solar cells grown by MOVPE," *Journal of Crystal Growth,* vol. 310, pp. 5204-5208, Nov 2008.

[2] N. Dharmarasu, M. Yamaguchi, A. Khan, T. Yamada, T. Tanabe, S. Takagishi, *et al.*, "High-radiation-resistant InGaP, InGaAsP, and InGaAs solar cells for multijuction solar cells," *Applied Physics Letters,* vol. 79, pp. 2399-2401, Oct 2001.

[3] D. C. Law, J. C. Boisvert, E. M. Rehder, P. T. Chiu, S. Mesropian, R. L. Woo, *et al.*, "Recent Progress of Spectrolab High-Efficiency Space Solar Cells," in *Nanophotonics and Macrophotonics for Space Environments Vii.* vol. 8876, E. W. Taylor and D. A. Cardimona, Eds., ed Bellingham: Spie-Int Soc Optical Engineering, 2013.

[4] O. K. Abudayyeh, N. D. Gapp, C. Nelson, D. M. Wilt, and S. M. Han, "Silver-Carbon-Nanotube Metal Matrix Composites for Metal Contacts on Space Photovoltaic Cells," *IEEE Journal of Photovoltaics,* vol. 6, pp. 337-342, 2016.

[5] O. K. Abudayyeh, N. D. Gapp, G. K. Bradshaw, D. M. Wilt, and S. M. Han, "Spray-coated carbon-nanotubes for crack-tolerant metal matrix composites as photovoltaic gridlines," in *2016 IEEE 43rd Photovoltaic Specialists Conference (PVSC),* 2016, pp. 0835-0839.

[6] L. J. Kong, Z. M. Wu, S. S. Chen, Y. Y. Cao, Y. Zhang, H. Li, *et al.*, "Performance evaluation of multi-junction solar cells by spatially resolved electroluminescence microscopy," *Nanoscale Research Letters,* vol. 10, pp. 1-7, Feb 2015.

[7] Q. Y. Feng, T. J. Li, Z. T. Zhang, J. Zhang, M. Liu, and J. Z. Jin, "Preparation of nanostructured Ni/Al2O3 composite coatings in high magnetic field," *Surface & Coatings Technology,* vol. 201, pp. 6247-6252, Apr 2007.

[8] N. V. Dziomkina, M. A. Hempenius, and G. J. Vancso, "Towards true 3-dimensional BCC colloidal crystals with controlled lattice orientation," *Polymer,* vol. 50, pp. 5713-5719, 2009.

[9] O. K. Abudayyeh, C. Nelson, S. M. Han, N. Gapp, and D. M. Wilt, "Silver-carbon-nanotube metal matrix composites for metal contacts on space photovoltaic cells," in *2015 IEEE 42nd Photovoltaic Specialist Conference (PVSC),* 2015, pp. 1-3.

[10] J. X. Fang, H. Hahn, R. Krupke, F. Schramm, T. Scherer, B. J. Ding, *et al.*, "Silver nanowires growth via branch

fragmentation of electrochemically grown silver dendrites," *Chemical Communications,* pp. 1130-1132, 2009.

[11] I. S. Yasnikov, Y. D. Gamburg, and P. E. Prokhorov, "Peculiarities of morphology of silver microcrystals electroplated under potentiostatic conditions from ammonium solutions," *Russian Journal of Electrochemistry,* vol. 46, pp. 524-529, May 2010.

[12] Y. H. Cheng and S. Y. Cheng, "Nanostructures formed by Ag nanowires," *Nanotechnology,* vol. 15, pp. 171-175, Jan 2004.

[13] J. R. Mileham, S. J. Pearton, C. R. Abernathy, J. D. MacKenzie, R. J. Shul, and S. P. Kilcoyne, "Patterning of AlN, InN, and GaN in KOH-based solutions," *Journal of Vacuum Science & Technology a-Vacuum Surfaces and Films,* vol. 14, pp. 836-839, May-Jun 1996.

[14] C. Youtsey, I. Adesida, and G. Bulman, "Highly anisotropic photoenhanced wet etching of n-type GaN," *Applied Physics Letters,* vol. 71, pp. 2151-2153, Oct 1997.

Subcell Light Current-Voltage Characterization of Irradiated Multijunction Solar Cell

Don Walker, John Nocerino, Yao Yue, Colin J. Mann, and Simon H. Liu

The Aerospace Corporation, El Segundo, CA 90245, U.S.A

Abstract — The degradation of individual subcell J-V parameters, such as short circuit current, open circuit voltage, fill factor, and power of a GaInP/GaInAs/Ge triple junction solar cell by 1 MeV electrons were derived utilizing the spectral reciprocity relation between electroluminescence and external quantum efficiency. After exposure to a fluence of 1 x 10^15 1 MeV electrons, it was observed that up to 67% of the voltage loss is from the middle, GaInAs subcell. Also, the dark saturation current of the Ge and GaInAs subcells increased but a simultaneous decrease in ideality factor caused a reduction of the open circuit voltage. The reduced ideality factor further indicates a change in the primary recombination mechanism.

I. INTRODUCTION

Understanding the effects that influence the current-voltage (I-V) characteristics of space photovoltaics is critical to predicting their on-orbit performance[1-3]. By investigating the effects of space radiation on the underlying device parameters we can better understand and predict the performance of on-orbit solar arrays. Current state-of-the-art solar cells are made from direct bandgap III-V materials because of their high efficiency and radiation hardness. Multijunction solar cells are more radiation hard than previous generation single junction gallium arsenide (GaAs) solar cells due to the radiation hardness of the current limiting indium gallium phosphide based top cell[4]. While current degradation can be tuned for and characterized by quantum efficiency measurements, it has only been recently discovered that the open circuit voltage of each subcell of a monolithic GaInP/GaInAs/Ge can be derived along with the partial dark and light current-voltage parameters. By utilizing the reciprocity theorem between electroluminescence and external quantum efficiency [5], the subcell dark and light current-voltage curves can be derive before and after irradiation [6-8]. From the subcell dark and light current-vottage curves we can determine the radiation induced degradation of each subcell's short circuit current, open circuit voltage, fill factor, power, ideality, and dark saturation current by 1 MeV electrons at a fluence of 1 x 10^15 electrons/cm^2. The ability to measure all the degradation current-voltage parameters of each subcell can lead to better degradation modelling. Understanding how each subcell degrades in voltage adds a new dimension of potentially tuning for less voltage degradation after particle irradiation

II. EXPERIMENT

Triple-junction ATJ solar cells of GaInP/GaInAs/Ge were purchased from SolAero Technologies®. Light current-voltage, quantum efficiency, and electroluminescence

measurements were performed before and after electron radiation at The Aerospace Corporation's Photovoltaic Evaluation And Research Laboratory (PEARL). The short circuit current density (J_{sc}), open circuit voltage (V_{oc}), fill factor, and efficiency were determined under a calibrated AM0 solar simulator on a temperature-controlled stage at 301.15 K (28°C). The simulator was calibrated using ATJ primary, JPL balloon flown standards. Quantum efficiency measurements were conducted using a Newport monochromator that was calibrated using a silicon photodiode and InGaAs diode that are National Institute of Standards and Technology traceable. Electroluminescence measurements were also performed at 301.15 K using a Horiba monochromator with a 2d, thermoelectrically cooled silicon CCD, and a temperature controlled, liquid nitrogen cooled extended InGaAs linear array. Electron radiation was performance at the National Institute of Standards and Technology radiation facility. The triple junction cells were exposed to electrons at an energy of 1 MeV and a fluence of 1 x 10^15 electrons/cm^2.

III. RESULTS AND DISCUSSION

The spectral reciprocity relation as defined by Eq. 1 relates the external luminescence (ϕ_{em}) and quantum efficiency (Q_e) to the voltage (V) of a solar cell, where ϕ_{bb} is the black body photon flux, q is the electron charge, k is the Boltzmann constant, and T is temperature. By taking electroluminescence spectra of each junction in a multijunction solar at different injection currents you can calculate the voltage of each subcell at each injection current. The open circuit voltage of each subcell is derived from the voltage calculated using Eq. 1 at an injection current equal to the short circuit current of the individual subcell.

$$\phi_{em} = Q_e \phi_{bb} \left[e^{\frac{qV}{kT}} - 1 \right] \tag{1}$$

The electroluminescence of each junction was taken before and after irradiating the triple junction cell with a fluence of 1 x 10^15 1 MeV electrons/cm^2 as can be seen in Fig. 1. The electroluminescence spectra of each spectra was taken at injection current densities ranging from 25 µA/cm^2 to 35 mA/cm^2. After exposure the electron radiation it is observed in Fig. 1a that the electroluminescence of GaInP and GaInAs is reduced by over two orders of magnitude. Also, the relative peak intensity between GaInP and GaInAs changes such that GaInP is brighter than GaInAs. The electroluminescence of the Ge subcell also decreases, losing about half of its electroluminescence. Also, the luminescence below 1400nm is quenched after electron radiation. The luminescence below 1400 is believed to be from direct transfer states in the Ge.

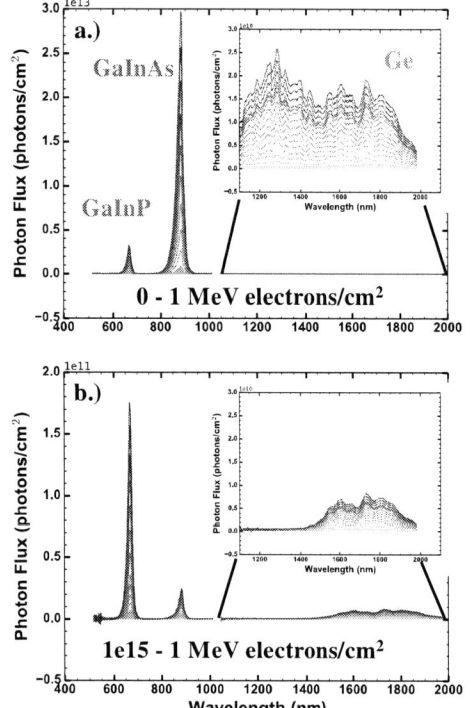

Fig. 1. Electroluminescence of GaInP, GaInAs, and Ge subcells in a triple junction solar cell before irradiation with 1 MeV electrons (a.) and after a fluence of 1 x 10^{15} 1 MeV electrons/cm^2 (b.).

By using the reciprocity relation we can calculate the voltage of each subcell for each injection current and derive subcell dark current-voltage characteristics for each subcell before and after electron radiation as seen in Fig 2. The most obvious observed change is an increase in the dark current for each subcell at any given voltage. Less obvious, is the change in ideality of each subcell as well as the change in the saturation current density. By fitting the subcell dark current-voltage curves, we can see that ideality becomes more ideal, closer to 1 from the Ge and GaInAs subcell (Tab. 1).

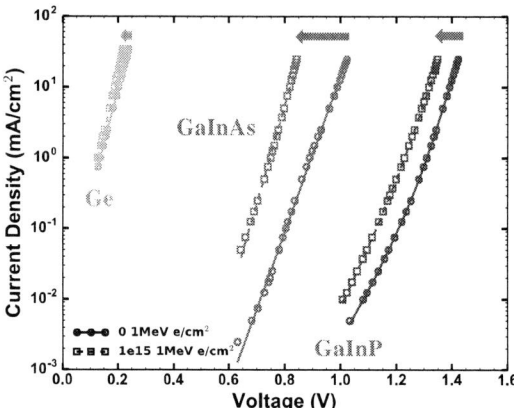

Fig. 2. Dark current-voltage properties before and after irradiation with 1 x 10^{15} 1 MeV electrons/cm^2. The dark current-voltage curves were derived by using the spectral reciprocity relationship. The GaInAs subcell showed the greatest increase in dark current. Also, the GaInAs and Ge subcells had a decrease in their respective ideality factor.

The decrease in ideality indicates that the subcells are becoming more dominated by radiative recombination after irradiation, albeit there is much less radiative recombination as indicated by the electroluminescence spectra.

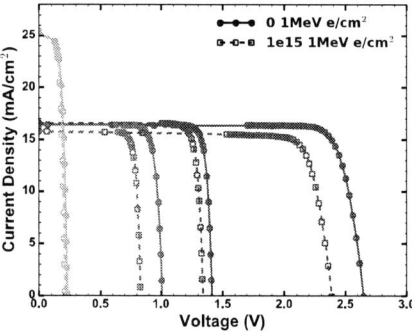

Fig. 3. The subcell light current-voltage properties are derived by using superposition. The photocurrent derived from the respective quantum efficiency spectrum and the AM0 spectra were used to derive the subcell light current-voltage properties pre and post irradiation with 1 x 10^{15} 1 MeV electrons/cm^2.

Finally, by offsetting the dark current-voltage curves using the derived photocurrent from integrating the AM0 spectrum with the quantum efficiency spectrum

Table 1

1 MeV e$^-$	Cell	J_{sc} (mA/cm^2)	ff	P_{max} (W)	V_{oc} (volts)	V_{loss} (%)	J_o (mA/cm^2)	n
0	Full	16.42	0.84	0.147	2.65			
	J_1	17.30	0.88	0.086	1.41		3.75E-19	1.20
	J_2	17.26	0.83	0.058	1.01		1.50E-10	1.52
	J_3	26.44	0.64*	0.016*	0.23		8.88E-03	1.11
1 x 10^{15}	Full	15.78	0.83	0.124	2.39			
	J_1	17.16	0.87	0.080	1.34	26.45	2.11E-17	1.26
	J_2	16.34	0.85	0.046	0.83	67.06	4.44E-11	1.21
	J_3	26.35	0.65*	0.015*	0.21	6.49	6.85E-03	1.00

Table I. The measured and derived light current-voltage properties of a multijunction solar cell (Full) its subcells(J_1, J_2, and J_3) before and after irritation with 1 x 10^{15} 1 MeV electrons/cm^2. The dark current-voltage parameters for each subcell are also presented. (*These values were calculated and presented for completion, but have error associated with not being able to capture the full 'knee' of the subcell current-voltage curve.

we can produce light current-voltage curves for each subcell. As can be seen in Fig. 3 and Tab 1, the subcell fill factors, Voc, Jsc, and power. Previously this has not been accomplished, as the lower injection currents needed to obtain the knee of the current-voltage curve had not been obtained[7, 8]. Utilizing the derived subcell voltages we can now determine where the greatest loss in voltage comes from in the multijunction solar cell by evaluating the $V_{i_{loss}}$ factor as described in Eq. 2.

$$V_{i_{loss}} = \frac{\Delta V_i}{\Delta V_{Total\ Loss}} = \frac{V_{i_0} - V_{i_{1e15}}}{V_0 - V_{1e15}} \qquad (2)$$

In Eq. 2, $V_{i_{loss}}$ is the normalized voltage loss for each junction (i). $V_{i_{loss}}$ is simply a way to determine what subcell contributes to the greatest loss in voltage of the multijunction solar cell. In this case the GaInAs contributes 67.06% of the voltage loss, GaInP contributes 26.45%, and Ge contributes 6.49% of the loss.

IV. CONCLUSION

By deriving the subcell current-voltage properties and characterizing their radiation induced degradation, we can now identify where most of the voltage is coming from, radiation effects on fill factor, as well as the root cause for voltage degradation in multijunction solar cells. The ability to measure the degradation of subcell current-voltage properties has implications in better predicting radiation degradation of space photovoltaics, as well as developing more radiation hard solar cells.

V. REFERENCES

[1] H. Y. Tada, J. R. Carter, B. E. Anspaugh, and R. G. Downing, "Solar cell radiation handbook," Jet Propulsion Laboratory, California Institute of Technology 82-69, 1982.

[2] B. E. Anspaugh, "Gaas solar cell radiation handbook," Jet Propulsion Laboratory, California Institute of Technology 96-9, 1996.

[3] K. C. Reinhardt, C. S. Mayberry, B. P. Lewis, and T. L. Kreifels, "Multijunction solar cell iso-junction dark current study," in *Photovoltaic Specialists Conference, 2000. Conference Record of the Twenty-Eighth IEEE*, 2000, pp. 1118-1121.

[4] K. C. Reinhardt, Y. K. Yeo, P. H. Ostdiek, and R. L. Hengehold, "Junction characteristics of electron-irradiated ga0.5in0.5pn+-p diodes and solar cells," *Journal of Applied Physics*, vol. 81, pp. 3700-3706, 1997.

[5] T. Kirchartz and U. Rau, "Detailed balance and reciprocity in solar cells," *Physica Status Solidi a-Applications and Materials Science*, vol. 205, pp. 2737-2751, Dec 2008.

[6] T. Kirchartz, U. Rau, M. Hermle, A. W. Bett, A. Helbig, and J. H. Werner, "Internal voltages in gainp/gainas/ge multijunction solar cells determined by electroluminescence measurements," *Applied Physics Letters*, vol. 92, Mar 2008.

[7] S. Roensch, R. Hoheisel, F. Dimroth, and A. W. Bett, "Subcell i-v characteristic analysis of gainp/gainas/ge solar cells using electroluminescence measurements," *Applied Physics Letters*, vol. 98, Jun 2011.

[8] R. Hoheisel, F. Dimroth, A. W. Bett, S. R. Messenger, P. P. Jenkins, and R. J. Walters, "Electroluminescence analysis of irradiated gainp/gainas/ge space solar cells," *Solar Energy Materials and Solar Cells,* vol. 108, pp. 235-240, Jan 2013.

Analytical Method for Predicting Spacecraft Power Generation on Partially Shaded Solar Panels

Gordon Wu, Bao Hoang

Space Systems Loral (SSL), Palo Alto, CA 94303, USA

Abstract — **Telecommunication satellites in Earth geostationary orbit (GEO) often have large communication antennas and other external appendages. Throughout the 24-hour day in GEO orbit, and depending on the orbit seasons, the external appendages may cast various changing shadows onto the satellite's solar arrays. The geometry and composition of the external appendages result in a complex shadowing profile involving soft and hard shadows. These conditions have a nontrivial impact on the solar array power generation and the spacecraft payload power management. Space Systems Loral (SSL) has developed an analytical tool to accurately predict the changes to array power using a combination of CAD modeling, raytracing and simulation of solar cell performance under various levels of illumination. The predictions from the analytical tool have been shown to correlate well with on-orbit solar array telemetry; in particular, shadow correlation data from a 12-m unfurlable mesh antenna will be presented.**

I. INTRODUCTION

In this paper, we will demonstrate our analytical methodology to predict solar array power in an event of partial shading of the solar cells within an array string. The geometry of the spacecraft, its external appendages, and the resulting shadow on the solar array panels including individual solar cells are modeled in a CAD software package as part of the tool. Within this tool, the solar flux throughout the 24-hour GEO orbit is simulated using a SSL-designed raytracing analysis software package with ability to simulate different types of orbits, and orbital maneuvers such as yaw steering. Optical properties for the external appendages are included to capture the soft shadows from semi-transparent reflector meshes. The subtended angle of the sun is included to capture the umbra and penumbra of completely opaque objects.

The tool includes computational logic to calculate voltage, current and power contribution from individual solar cells within an array string at any given time in the orbit based on the illumination data from the raytracing model. The tool also includes the effects from electrical current flowing through the individual solar cell's bypass diode in the event of shadowing on part of the array string.

II. ANALYTICAL METHODOLOGY

An overview of the power prediction tool work flow diagram can be seen in Fig. 1.

Each solar cell in the array string is modeled in the geometry model and given a unique node identification to capture illumination data in the form of absorbed solar flux. All of the other spacecraft geometry is modeled to capture potential obstructions to each solar cell's view to the sun.

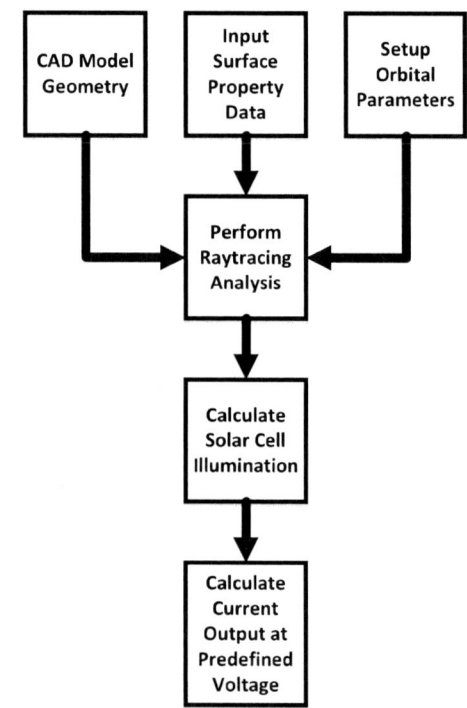

Fig. 1. Diagram of work flow in analytical tool

All the objects or nodes in the computer-aided design (CAD) model are assigned thermal optical properties associated with the outer surface finish. For semi-transparent objects such as antenna mesh or polyimide tape, transmissivity terms at various angle of incidence based on ground test data are used to capture the non-opaque shadowing effects.

Orbital parameters such as orbit inclination, orbital period, body steering maneuvers, beta-angle, subtended angle of the sun and solar flux are included in the raytracing software to simulate the location of the sun relative to the spacecraft at any given time.

The raytracing software gathers the orbital parameters, the CAD model geometry and thermal optical properties to calculate each solar cell's view to the sun at each position in the orbit. Factoring in opaque obstructions, semi-transparent materials and the subtended angle of the sun, the analysis records the solar flux that would reach each solar cell. The resulting solar flux data is used to calculate the illumination.

The illumination data is used as an input to calculate I-V curves for each array circuit and therefore the current at the predefined voltage of interest. For a multifunction solar cell used on the SSL array, the current and voltage characteristic of the cell is typically represented as a calculated current as a function of voltage, shown in equation (1) [1].

$$I = I_{sc} - I_s \times (e^{\frac{qV}{k_BT}} - 1) \qquad (1)$$

In a string of solar cells connected in series, the voltage contribution of each cell is additive at a given current. Therefore it is of interest to determine the voltage as a function of current, shown in equation (2).

$$V = \frac{k_BT \times \ln(\frac{I_{sc}-I}{I_s} + 1)}{q} \qquad (2)$$

Not shown in the equations above are the solar cell series and shunt resistance and a set of constants are used for beginning of life (BOL) performance based on predicted temperature and radiation exposure. For end of life (EOL), another separate set of constants are used to account for radiation damage over the on-orbit mission life. For this particular array design, performance data of an array string of multiple coverglass-interconnector-cells (CICs) in series is used to calculate the contribution of an individual CIC. The solar cells are state-of-practice multi-junction GaAs/Ge solar cells.

Each CIC has an associated silicon bypass diode electrically connected in parallel. As the CIC is shaded and cannot generate current, current generated by other in-series CICs in the string flows through the bypass diode of the shadowed CIC and to the next in-series CIC in the string. There is a voltage drop across the bypass diode that has an impact on overall power generated on the string. The effects of current routed through the bypass diode while some CICs are shadowed is modeled in the power prediction tool. An illustration of the difference in the power prediction tool's I-V curve calculation with and without modeling the bypass diode can be seen in Fig. 2.

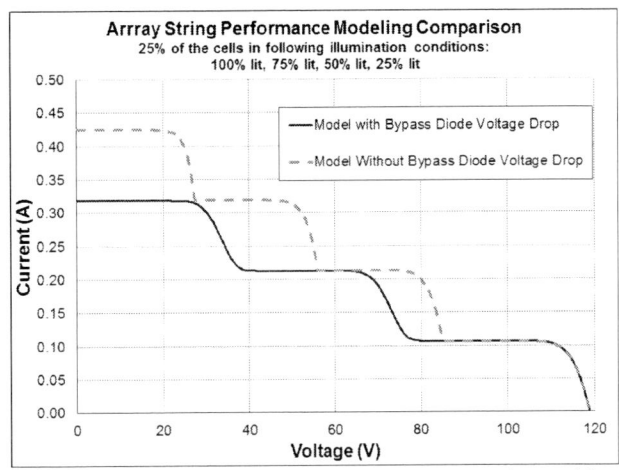

Fig. 2. Power prediction tool's I-V curve comparison with and without bypass diode modeling

Figure 3 shows a comparison of I-V curve predictions for multiple array strings under three different illumination conditions. There is a string that has all of its CICs completely illuminated by the sun. There is a second string that has approximately half the CICs completely shaded. There is a third string that has all the CICs shaded by an object with 50% transparency. Referencing the current level of a fully illuminated string, the string with half the cells completely shaded can generate full current up to 20V and drops to zero current generation at 55V. The string with uniform reduction in illumination generates half the current up to 100V before current begins to drop with no current generated at 118.5V.

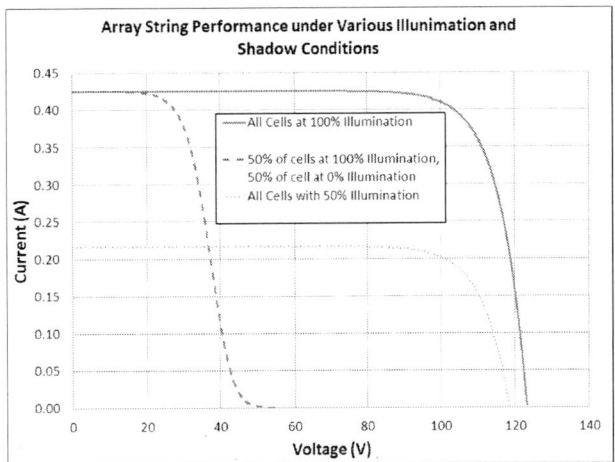

Fig. 3. I-V curve comparison of strings with three levels of illumination, calculated in the model

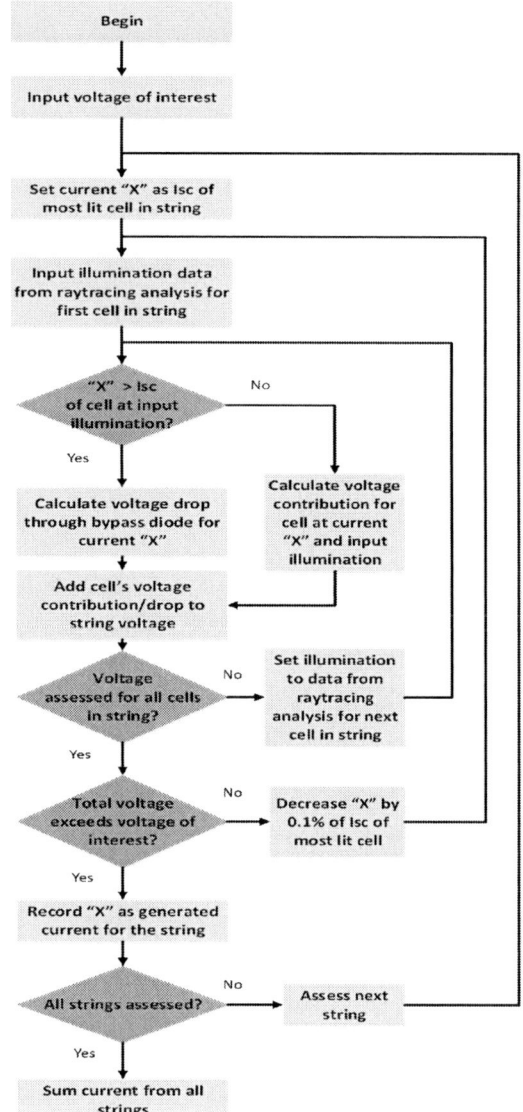

Fig. 4. Logic flowchart for current calculation at predefined voltage of interest

The response shown in Fig. 3 are two of numerous shading conditions that can occur simultaneously on the solar array during any given moment in a shadowing event. The model has capability to calculate of current generation at the bus voltage of interest. Tediously manual calculation of current and voltage performance of multiple CICs within a string at a particular shadowing condition is not difficult given the illumination state of each CIC. The challenge arise from the need to calculate CIC performance for hundreds of strings at one-minute intervals over a five hour shadowing event. Over 3 million calculations are required which is nearly impossible to compute by hand. Therefore a software was written to take in all the illumination inputs and calculate the current generation at the voltage of interest.

The overall logic for the software computation is shown in Fig. 4. The voltage of interest is specified at the beginning of the calculations. Since the current generated by the string cannot exceed the current generated by the CIC with highest illumination, the short circuit current for the CIC with higher illumination in the string is used as the initial current. This current value is compared with the short circuit current of the first CIC in the string, given the illumination.

If the computed current value exceeds the short circuit current of the CIC, then the CIC is fully reversed bias and it will not have any voltage contribution. A voltage drop is calculated for the current that is going through the parallel bypass diode of the affected CIC. If the computed current value is below the short circuit current of the CIC, then the voltage contribution is calculated based on equation (2) using a set of constants that factor in radiation effects and predicted temperature. This voltage calculation is performed for all the CICs in the string. The summation of the voltages is compared with the voltage of interest. If the calculated string voltage exceeds the voltage of interest, the current value is recorded for that string. If the calculated string voltage is below the voltage of interest, the iterated current value is decreased by 0.1%. The voltage calculations are iterated until the voltage of interest is reached. The calculations are repeated for all the strings in the solar array. The summation of all the current values provides the total current generated by the solar array at the voltage of interest.

The geometry model of the unfurlable mesh antenna was simplified to only represent the bulk structure: structural ribs, boom, hub and antenna mesh. Miscellaneous items such as clips, pass-thrus and fittings shown on Fig. 5 were not included in the model and therefore the shadowing effects of small components were not completely captured in the power prediction.

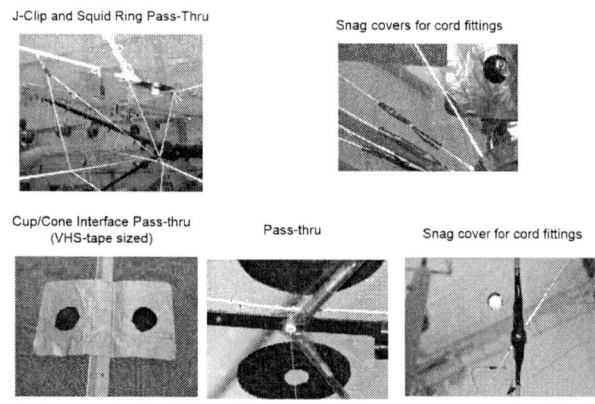

Fig. 5. Miscellaneous unfurlable mesh antenna geometry not captured in analytical model.

A solar cell's performance is dependent on its temperature. Equations (1) and (2) are used to calculate the I-V curve characteristics at a given temperature. Solar cell temperature can change by more than 150°C in a shadowing event. Currently, the transient temperature effects are not captured in the tool which accounts for some discrepancy between the analytical prediction and flight data. An example of this is shown in Fig. 6 where the analytical prediction exiting the Earth seasonal eclipse shows an immediate recovery in current but the flight data shows that the panel takes about 10 minutes for the current generation to recover as the sun warms the solar cells.

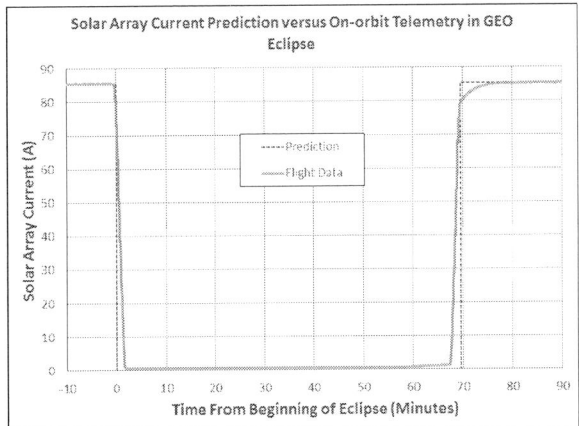

Fig. 6.   Solar cell current response after Earth seasonal eclipse

III. FLIGHT CORRELATION

The analytical methodology was validated by applying it to an existing SSL-built spacecraft and comparing the prediction with solar array flight data. Figure 7 show the GEO communication spacecraft and the large 12-m unfurlable mesh antenna. Figs. 8 and 9 show the geometry models of the SSL built spacecraft and its solar arrays that was used for this analytical study. The 12-m unfurlable mesh antenna casts a complex shadow that varies in illumination on the solar array throughout the orbit. The antenna boom and hub structure cast hard shadows that are wide enough to temporarily drop power generation from an entire string whereas the soft shadows cast by the antenna ribs and the antenna mesh reduce the power generation of each string depending on the angle of the mesh relative to the sun during the orbit.

A comparison of the predicted and actual on-orbit solar array available current throughout a 24-hour period in GEO orbit is shown in Fig. 10. The changes to the array electrical load current were due to the shadow of the 12-m unfurlable mesh antenna.

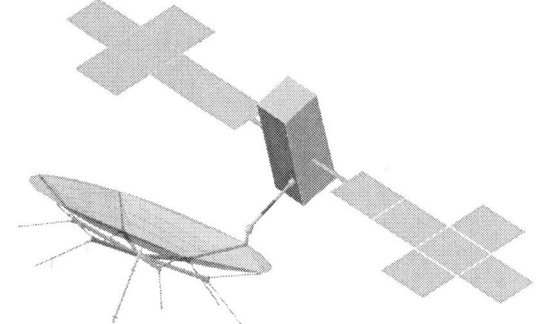

Fig. 7.   View of analytical model of the spacecraft with a 12-m unfurlable mesh antenna and solar array

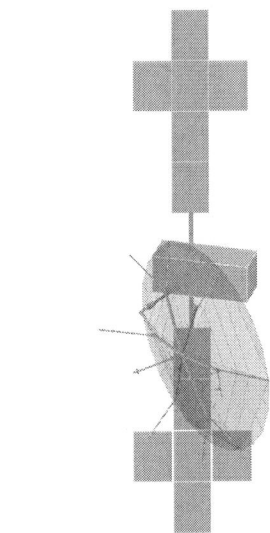

Fig. 8.   Analytical model viewing from the sun at the time of day with the deepest shadow

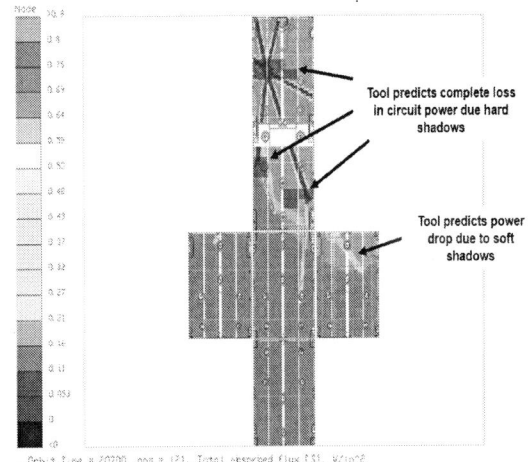

Fig. 9.   Analytical results from power prediction tool in the geometry model at the time of day with the deepest shadow

The analytical prediction matches the flight data within 2% during the shadowing event as shown in Fig. 10. The maximum discrepancy of 2% is observed at the peak of the array shadow, at approximately 05:30 local spacecraft time. The small discrepancy, observed throughout the shadow period, can be attributed to the geometry modeling of the antenna in the analytical model as previously discussed in Paragraph II and Fig. 5. The peak discrepancy, which occurs at 05:30 local time, may be attributed to the transient temperatures of the array.

Taking a look at the total power generation during the approximately 5-hour shadowing event, the difference between the analytical tool and the flight data is 3Ah with a total of 488Ah generated based on flight data. This is equivalent to over-predicting the impact of the shadowing by 300W-hr in total energy.

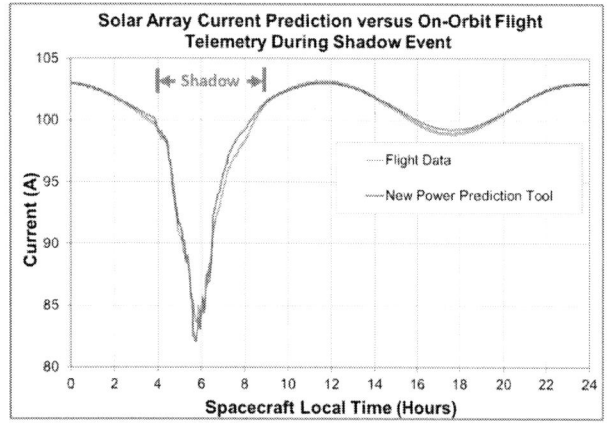

Fig. 10. Solar Array analytical prediction compared with solar array on-orbit flight data for the 24-hour orbit during seasonal shadow event

IV. DISCUSSION

Prior to creation of the power prediction tool, analytical power prediction during the complex shadow event was simply predicted. A "snap-shot" was taken at an assumed worst case shadow during the event. An example of a "snap-shot" can be seen on Fig. 11. Power drop was assessed based on graphical data of the shadowing for each string for that specific orbital position in the shadowing event, as shown in Fig. 12. The resulting power impact was applied across the entire duration of the shadowing event. The former power prediction method had a 67Ah difference compared to the flight data, over-predicting the shadowing impact by 6700W-hr. As previously discussed, the new method over-predicted only by 300W-hr.

The former power prediction method led to multiple concerns. First of all, without a mathematical model incorporating the performance of each solar cell, it was nearly

impossible to ascertain the orbital position of the worst case power drop during the shadowing event. Secondly, an analytical calculation of the power drop was not possible based on a graphical map of the drop in illumination. Therefore analytical uncertainty needed to be applied to the estimate for conservatism. Lastly, applying the worst case power drop to the entire shadowing event resulted in a very conservative estimate of the power impact from the shadowing event. A comparison of the power prediction between the former and new methods can be seen in Fig. 13.

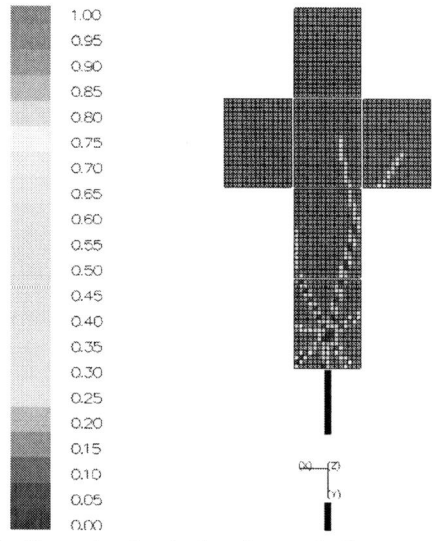

Fig. 11. Example of a shadow "snap-shot" assessed with the former power prediction method

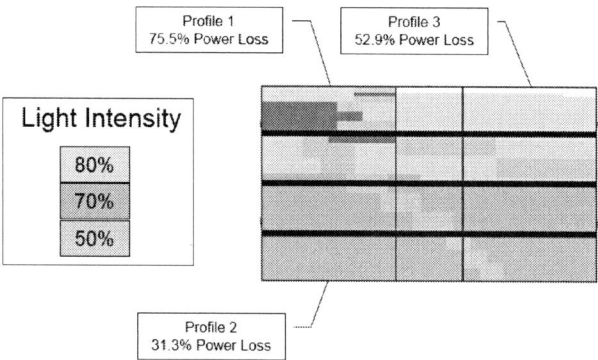

Fig. 12. Assessment of power drop for individual strings in the former power prediction method

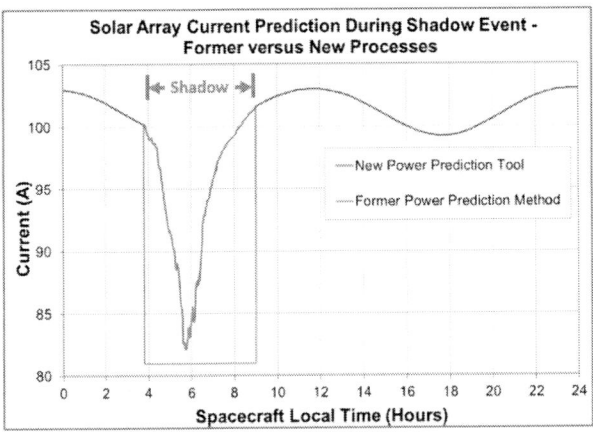

Fig. 13. Power prediction comparison of the present power prediction tool with the former power prediction method.

In the case of a 12-meter unfurlable mesh antenna shadowing event, the difference in total predicted power generation during the event is approximately 6700W-hr in total energy between the two different methods. Assuming the power deficit during the shadowing event is compensated by battery power, an additional 370W – 400W would need to be booked to recharge the batteries during non-shadowing periods. This deficit is equivalent to two or three more typical radio frequency (RF) transponders that could be added to the communications payload using predictions from the new model. Using a metric of $1M revenue per RF transponder per year, the estimated impact is $30M - $45M in additional revenue for the spacecraft operator during the 15 year operating life using results from SSL's new prediction tool.

V. Summary

An analytical tool to predict photovoltaic power drop during transient shadowing of solar panels was created using CAD geometry modeling, raytracing analysis and SSL-designed software based power analysis. The analytical predictions have been shown to correlate well with on-orbit data.

The power prediction tool's ability to use different solar cell performance models allows studies to be performed to optimize the length of the array strings based on the shadow profile expected from external appendages, the mission life, and the electrical power requirements.

The high fidelity of the analytical power prediction tool removes unnecessary conservatism when sizing the power subsystem during the design phase of the spacecraft. The resulting benefit is substantial and quantifiable. This can enable additional transponders and increased revenue for the spacecraft operators.

VI. Acknowledgements

The authors wish to acknowledge Mr. Robert Neff and Mr. Tod Redick, both formerly at SSL, for their contribution in this paper.

VII. References

[1] H. Rauschenbach, *Solar Cell Array Design Handbook*, Van Nostrand Reinhold Company, 1980

Evaluating the Emissivity of Pseudomorphic Glass (PMG)

Ryan D. Beauchemin[1], David M. Wilt[2], Paul E. Hausgen[2]

[1]Applied Technology Associates (ATA)

[2]Space Vehicles Directorate, Air Force Research Laboratory, Albuquerque, NM 87117

Abstract — A flexible replacement for solar cell coverglass called Pseudomorphic Glass (PMG) has been previously characterized for relevant properties including stability under UV and particle radiation and electrical conductivity. PMG is composed of small beads of coverglass material suspended in a space grade silicone adhesive. The material has shown good environmental stability and offers several benefits for space applications. Recently, the spectral emittance of PMG was characterized and demonstrated to have superior behavior compared to conventional coverglass. This data suggests that the PMG may enable the underlying solar cell to operate at lower temperatures, and therefore increased cell efficiency. PMG configurations with different types of bead materials and silicone adhesives were selected for examination. Using measured spectral emittance data, an on-orbit model was developed to predict the temperature of the various PMG materials compared with coverglass under AM0 solar flux in a vacuum environment. The calculated temperature differences between the PMG materials and coverglass were then used to predict efficiency changes, using reported values of cell efficiency temperature coefficients. The calculations suggest that PMG coated solar cells would run ~5°C cooler than coverglass coated cells, leading to a ~0.3% absolute increase in cell efficiency. To validate this claim, an experiment was setup in which the temperature of two types of cells, one coated in coverglass and the other with PMG, were measured in response to receiving AM0 radiation at open circuit. The PMG coated cells operated at slightly lower temperatures than the coverglass coated cells.

Index Terms — Photovoltaic cells, spectral emissivity, multi-junction cells, coverglass.

I. INTRODUCTION

Multi-junction photovoltaic cells play a critical role in meeting spacecraft power needs. Harnessing the sun's power as a widely available and accessible source of energy, it is desirable to come up with designs that maximize the amount of energy being collected in a given area while remaining lightweight and cost effective at the same time [1]. Multi-junction cells achieve optimal energy collection by using three or more solar cells layered on top of each other such that each absorbs and efficiently converts a portion of the sun's emissions [2]. Although multi-junction cells achieve higher levels of efficiency they are prohibitively expensive for terrestrial use and are primarily implemented in space power applications.

In order to protect the cells from damaging radiation in space, a rigid, ceria-doped borosilicate coverglass is adhered to the top of the cell using space grade silicone adhesives. Conventional solar cell coverglasses work quite well, but have challenges in device fabrication and providing full panel encapsulation [3].

Pseudomorphic Glass (PMG) is a flexible solar cell coverglass replacement option. PMG is a composite of small beads of coverglass materials (e.g. fused silica or ceria-doped borosilicate glass) suspended in silicone adhesive combining the optical properties of coverglass materials with the flexibility of the silicone [1]. With its flexibility, PMG is less prone to cracking, can easily provide complete electrical encapsulation and can be adapted to a wide array of satellite configurations. In addition, PMG also has the advantage of having a higher emissivity in the infrared spectrum. A higher emissivity is desirable for a coverglass material since it radiates heat more effectively, resulting in a lower solar cell operating temperature and an improved cell efficiency. With this in mind, the emissive properties of PMG are investigated by characterizing a variety of PMG samples along with traditional coverglass for comparison. By measuring the spectral emittance for each material, the operating temperature can be calculated and used to estimate the increase in solar cell efficiency.

II. MEASUREMENT OF RADIATIVE AND OPTICAL SURFACE PROPERTIES

When characterizing the optical properties of the materials, two variations of PMG (DC93-500 and SCV1-2590 adhesives with fused silica beads) along with coverglass were selected for testing. A Surface Optics Corporation 400T reflectometer was used to measure the reflectance of the samples from 2 microns to 25 microns with two different backings; a cavity absorber and a gold reflector [1]. For the spectral range where the two measurements agreed, it was concluded that the films were fully absorptive and thus the spectral absorptance (α_λ) could be calculated as:

$$\alpha_\lambda = 1 - \rho_\lambda \qquad (1)$$

Applying Kirchhoff's law ($\varepsilon_\lambda = \alpha_\lambda$) yields the spectral emittance:

$$\varepsilon_\lambda = 1 - \rho_\lambda \qquad (2)$$

These spectral emittance values were graphed over the 5 to 25-micron range as seen in Fig. 1. It can be noted that the emissivity of cover glass drops far below that of the PMG's in the 9-micron range, which is important as this is near the peak in the 300K blackbody spectrum (Fig. 1).

To calculate the predicted temperature of each cell, the total frontal emittances (ϵ_f) for the cover materials were required. To do this, the black body curve at a temperature of 300 K between the wavelengths (λ) 4 and 25 microns was obtained by

978-1-5090-5606-4/17 $31.00 © 2017 IEEE

calculating the irradiance at each wavelength. This was done using Plank's formula [3]:

$$S_\lambda = \frac{8\pi hc}{\lambda^5} \frac{1}{e^{\frac{hc}{\lambda kT}}-1} \qquad (3)$$

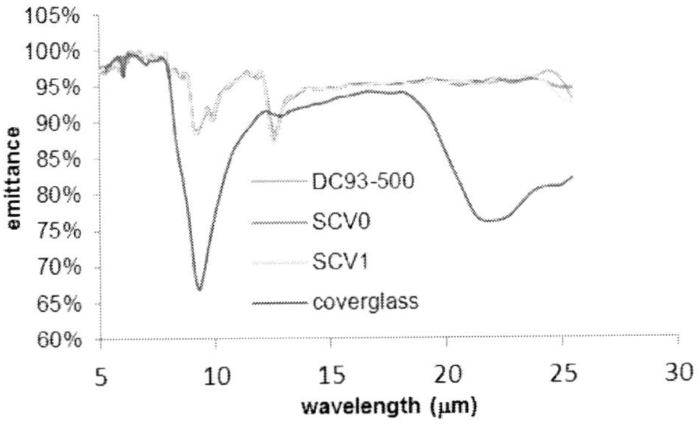

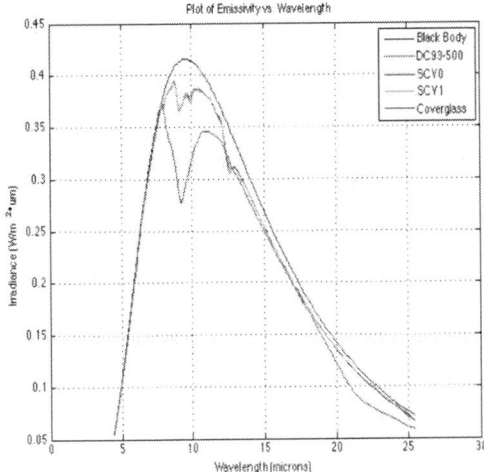

Fig. 1. The graph compares the spectral emittance of different materials including the several PMG samples and coverglass. The emissivity for each material varies over wavelength, and compared to the PMG, coverglass shows a significant drop in emissivity in the 9-micron range[1].

The blackbody irradiances obtained in (3) were multiplied by the measured spectral emittance values for each cell sample and are plotted for comparison in Fig. 2. It can be noted that the curves for all the samples fall within the black body curve and that the curve for coverglass has a significant drop within the 8 to 11 micron range.

Fig. 2. Plot showing calculated irradiance curves of coverglass and PMG samples and a black body at 300°K.

Next, the spectral irradiance curves for the black body and the samples, seen in Fig. 3, were numerically integrated to find the total radiated flux. For each sample, its integrated area

under the curve was divided by the total emissive power of the black body (E_b) (found to be 494.75 W/m²) [5] giving the total emittance (ϵ_f , PMG – 0.95, Coverglass – 0.88). This calculation is shown in the equation below:

$$\epsilon_f = \frac{\int_{\lambda_1}^{\lambda_2} \epsilon_{\lambda f}(\lambda,T)S_\lambda(\lambda,T)d\lambda}{\int_{\lambda_1}^{\lambda_2} S_\lambda(\lambda,T)d\lambda} \qquad (4)$$

Using a similar process, the spectral reflectance data over the relevant solar wavelengths for the cell cover materials was multiplied by the AM0 radiation spectrum, integrated to find the total energy absorbed, and then divided by the total solar flux (1366 W/m²) to obtain their total solar reflectance values (PMG - 0.169, Coverglass - 0.154).

III. CALCULATION OF PREDICTED ON-ORBIT TEMPERATURES

Predicted on-orbit solar array temperatures were calculated to assess the effect on solar cell conversion efficiency resulting from using the PMG cover material compared to standard coverglass. The radiative and optical properties obtained in section II were used in this calculation. To calculate the predicted on-orbit temperature, an energy balance was applied to the solar array in the space environment, which yields the following equation (assuming steady state conditions, no spatial temperature variations across or through solar array panel, thermal interactions with the spacecraft are neglected).

$$q_s + q_{alb} + q_{ir} - q_{emm\,f} - q_{emm\,b} - q_{ref} - q_{elec} = 0 \qquad (5)$$

Where q_s is the incident solar energy, q_{alb} is the Earth albedo absorbed by the solar array, q_{ir} is the thermal radiation emitted from the earth and absorbed by the solar array, q_{emmf} is the thermal radiation emitted from the front side of the array, q_{emmb} is the thermal radiation emitted from the backside of the array, q_{ref} is the reflected incident solar energy, and q_{elec} is the electrical power produced by the solar array. The energy balance is show graphically in Fig. 3.

Using the Stefan-Boltzmann Law for thermal emission and expressing the reflected and generated electrical power in terms of the solar reflectivity (ρ_s) and electrical conversion efficiency (η), the terms in the energy balance can be written in the following forms:

$$q_{emm\,f} = A\sigma\epsilon_f T^4 \qquad (6)$$
$$q_{emm\,b} = A\sigma\epsilon_b T^4 \qquad (7)$$
$$q_{ref} = A\rho_s q_s'' \qquad (8)$$
$$q_{elec} = Aq_s''\eta PF \qquad (9)$$
$$q_s = Aq_s'' \qquad (10)$$

Where the incident solar flux at AM0 conditions ($q_{s''}$) is assumed to be 1366 W/m^2 , Stefan-Boltzmann's constant (σ) to be 5.6703E-8 W/m^2·K^4, the solar cell packing factor (PF) is assumed to be 0.88, and the baseline solar cell efficiency (η) to be 29.5% at 25 °C [6].

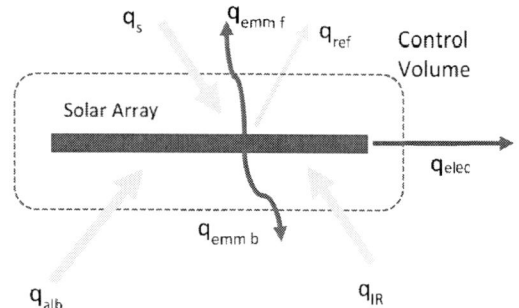

Fig. 3. Energy balance (W) of a solar array in the space environment

The energy balance in (5) can be expanded by inserting equations (6), (7), (8), (9), and (10). Because the areas are assumed to be the same, they can be factored out leaving the equation in terms of fluxes (W/m^2). For both types of panels (PMG and standard coverglass), the emittance of the backside (ϵ_b) is assumed to be the same leaving the emittance and solar reflectance of the front side as the variables of interest. The assumed values for the remaining terms in the energy balance are shown in Table I. The values in Table I were obtained using data and procedures outlined in [7], [8], and [9].

TABLE I
ASSUMED SOLAR ARRAY PROPERTIES

Backside Emittance	Backside Solar Absorptance	Absorbed Earth Albedo (W/m^2)	Absorbed Earth IR (W/m^2)
0.85	0.47	103	207

The temperature dependence of the solar cell conversion efficiency is incorporated using the measured maximum power temperature coefficient (-85.7 µW/cm^2-°C, [6]). The result is two simultaneous equations (one nonlinear, with temperature and efficiency as unknowns) that were solved using Mathematica. The predicted on-orbit solar array temperature and corresponding solar cell conversion efficiency are shown in Table II.

TABLE II
CALCULATED ON-ORBIT TEMPERATURE AND EFFICIENCY

Sample	Frontside Emittance	Solar Reflectance	Array Temperature (°C)	Efficiency (%)
93-500	0.95	0.169	50	27.9
SCV1	0.95	0.169	50	27.9
Coverglass	0.88	0.154	55	27.6

IV. EXPERIMENTAL SETUP

With the on-orbit model suggesting that PMG coated cells would run 5°C cooler than coverglass coated cells, the next step was to build an experiment to test this claim. Thermal performance testing for cells is typically done in a cryo-cooled thermal balance chamber, but because one was not available, an alternative approach was used. For the test, the cells were mounted on an insulating pad made from layers of Pyrogel insulating fiber to prevent conductive heat transfer to the underlying stage, simplifying the experiment while eliminating

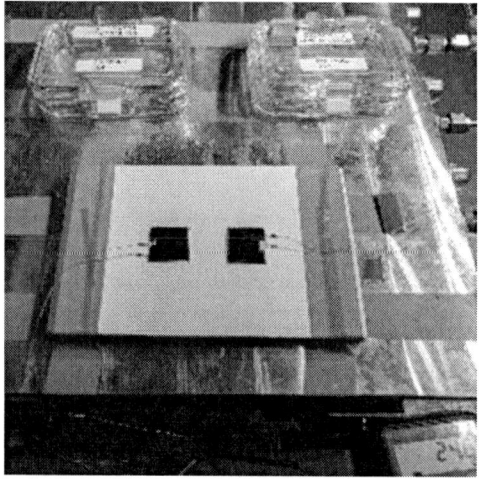

Fig. 4. Experimental setup consisting of solar cells (CMG-left and PMG-right) on a thermal insulating pad.

unwanted variables in the calculations. The insulating pad was placed on a water cooled chuck which was set to 24.8°C to keep the surface at ambient air temperature. A picture of the experimental setup is shown in Fig. 4.

During the experiment, the cells were exposed to AM0 solar radiation using an X-25 solar simulator and their open circuit voltages were recorded over time and used to determine the solar cell temperature, given the known 6.3 mV/°C temperature coefficient [6]. Assuming the cells were at room temperature (25°C) when they are first illuminated, the cell temperature could be calculated from the drop in open circuit voltage as the cell heats. The energy balance for the experimental setup is pictured in Fig. 5.

Each term for heat transfer can be added to obtain the following equation:

$$q_{in} = q_{cond} + q_{ref} + q_{conv} + q_{emm} \qquad (11)$$

By writing each heat flux in terms of its constants and assuming the heat conducted through the insulator to be negligible (9.375x10^{-4} W/cm^2), the equation can be rewritten in the form below:

$$q_{in} = \rho q_{in} + h(T_1 - T_s) + \epsilon\sigma(T_1^4 - T_s^4) \qquad (12)$$

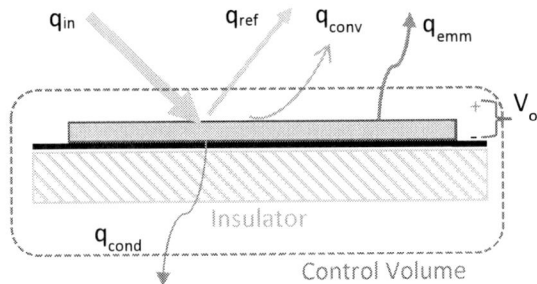

Fig. 5. Energy balance for the experimental setup where q_{in} is radiation from the simulator, q_{ref} is radiation reflected by the cell, q_{conv} is heat lost due to convection, q_{cond} is heat conducted through the insulator, and q_{emm} is radiation emitted by the cell cover.

For the constants written in [12], h is a convective coefficient for air, T_1 is the cell's operating temperature, T_s is the surrounding air and enclosure temperature, and ρ is the solar reflectance.

Since the cells are close enough to be in the same ambient air environment, their convective coefficients (h = 8.86 W/m²K) and surrounding temperatures (T_s) are assumed to be the same. This leaves the operating temperature as being a function of the reflectance and emittance of the cover materials. Having a higher reflectance, higher emittance, or both would result in a lower operating temperature, however, having a high reflectance would also result in less solar radiation received by the cell so being able to radiate heat without being reflective to the incident solar wavelength range in which the solar cell responds would be ideal.

The tests involved the same set of cells mounted on white Gylon reaching a steady state operating temperature of around 62°C. When calibrating the radiation flux of the X-25, it was discovered that the light intensity received by each cell varied by ~3% with the location they were placed. To account for this variation, the cells were tested separately in each test position.

A duplicate set of cells also covered in coverglass and PMG were measured to ensure consistent, reproducible behavior and to adjust for slight variations found in each cell. Overall, two samples of PMG and CMG (coverglass) coated cells were measured in two different test positions.

V. Results and Discussion

The experimental results and calculations were tabulated and graphed to be analyzed and compared with predictions made by the on-orbit model. The voltages recorded in each test were plotted over time intervals of roughly 20 minutes to allow for a steady state to be reached. The graph shown in Fig. 6 is a sample of one of these plots.

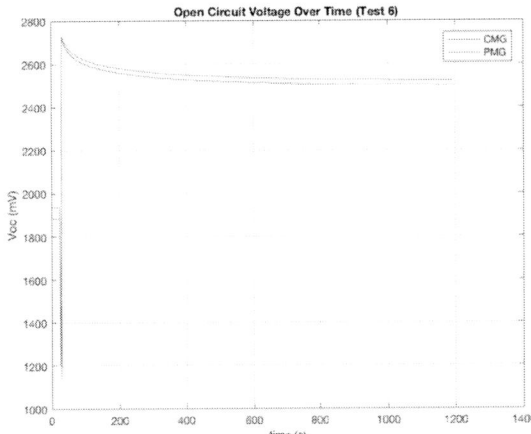

Fig. 6. Plot of open circuit voltages for PMG and Coverglass (CMG) changing over time.

When receiving AM0 radiation from the simulator, the cells initially developed a maximum voltage as the simulator shutter was opened, but as heating occurred due to absorption of the light within the cells, the voltages began to drop asymptotically until reaching a steady state. Exploiting the linear relationship between voltage and temperature change, the drop in voltage was used to calculate the operating temperatures of the cells as seen in Fig. 7.

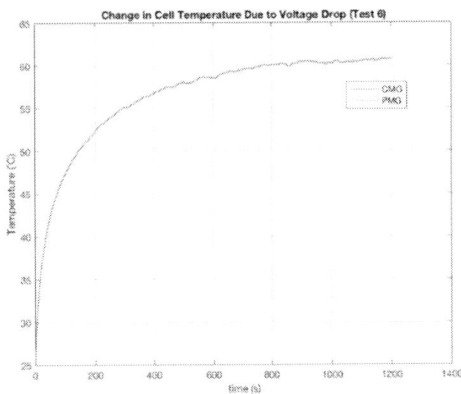

Fig. 7. Plot of temperatures for the PMG and Coverglass (CMG) coated cells corresponding to their voltage drops.

For calculating the temperatures, the maximum voltages for each cell was set to equal the ambient room temperature of 25°C. The voltages shown in Fig. 6 were subtracted by the maximum values and divided by a temperature coefficient, rated at -6.3 mV/°C for the type of cells used, to find changes in temperature which were then added to the baseline of 25°C. The steady state temperatures for the cells mounted on white Gylon were consistently found to be close to 60°C. By looking at where the temperature curves, like the one in Fig. 7, flattened out, the steady state temperatures for each test were estimated and are listed in Table III.

TABLE III
OPERATING TEMPERATURES OF CELLS AT DIFFERENT TEST CONDITIONS

Cell Condition	Sample 1 (°C)	Sample 2 (°C)	Average (°C)	CMG-PMG Difference
PMG Spot 1	61.41	60.61	61.01	0.25
CMG Spot 1	60.81	61.71	61.26	
PMG Spot 2	60.22	61.30	60.76	0.71
CMG Spot 2	62.05	60.89	61.47	

Because the differences in radiative intensity due to cell placement proved to cause the most variation in results, the temperature measurements were grouped by this test condition so that relevant comparisons could be made between the two cover materials. To test for consistency and account for variations in performance found in each individual cell, a different set of cell samples were tested under the same conditions and the measurements for each sample were averaged. As seen in the last column of Table III, PMG had an operating temperature 0.25°C lower than Coverglass (CMG) for cells placed in spot 1 and 0.75°C lower for cells placed in spot 2. The thermal performance benefit of PMG shown by the experiment was lower than predicted by the on-orbit model. This is believed to be due to the presence of different heat transfer modes, which are not found in space.

VII. SUMMARY

Pseudomorphic glass (PMG) is a new coverglass approach that combines the optical properties of traditional coverglass with the flexibility of silicone adhesives. PMG has the additional advantage of having a higher emittance, which should reduce the on-orbit operating temperature of the underlying solar cell. The emissive properties of PMG were characterized in order to determine these temperatures and a model developed to predict how they correspond to an increase in solar cell efficiency. In addition, an experiment was setup in which two cells, one coated in PMG and another in coverglass, were mounted on an insulated pad and were exposed to simulated AM0 radiation. The open circuit voltages for each cell were simultaneously measured and recorded over time while the change in temperature from the cells heating up was calculated using the voltage drop. To account for unwanted variables, tests were repeated with changing the placement of the cells, and using a different set of samples to test for reproducibility. In the end, it was found that PMG operated at a temperature of between 0.25 and 0.71°C lower than coverglass in a terrestrial, air environment. This is compared to a predicted 5°C temperature reduction using PMG in an on-orbit, space environment.

REFERENCES

[1] M. Armbruster, D.M. Wilt, R.J. Cooper, and R.C. Hoffman, "Optical and Electrical Properties of Pseudomorphic Glass," Proceedings of the IEEE 43rd Photovoltaic Specialist Conference, 2016.

[2] A.W. Bett, S.P. Phillips, S. Essig, S. Heckelmann, R. Kellenbenz, V. Klinger, M. Niemeyer, D. Lackner, and F. Dimroth, "Overview About Technology Perspectives For High Efficiency Solar Cells For Space And Terrestrial Applications," Proceedings of the 28th European Photovoltaic Solar Energy Conference and Exhibition, Paris, France, 2013.

[3] D.M. Wilt, "Advanced Photovoltaic Development at Air Force Research Laboratory," SPIE Photonics West, San Francisco, 2012.

[4] "Blackbody Radiation" Hyperphysics, 18 Jan. 2016, http://hyperphysics.phy-astr.gsu.edu/hbase/mod6.html

[5] B.V., Karlekar, and R.M. Desmond. *Heat Transfer.* 2nd ed., West Publishing Company, 1977

[6] "ZTJ Space Solar Cell" Solaero Technologies, 4 May. 2016 http://solaerotech.com/wp-content/uploads/2016/10/ZTJ-Datasheet-Updated-2016

[7] D. H. Nguyen, L. M. Skladany, and B. D. Prats. "Thermal Performance of the Hubble Space Telescope Solar Array-3 During the Disturbance Verification Test (DVT)," 4th International Symposium on Environmental Testing for Space Programmes; 12 Jun. 2001; Liege; Belgium. Available at https://ntrs.nasa.gov/search.jsp?R=20010070991

[8] B. J. Anderson and R. E. Smith, "Natural Orbital Environment Guidelines for Use in Aerospace Vehicle Development," NASA Technical Memorandum 4527, June 1994, available at https://ntrs.nasa.gov/archive/nasa/casi.ntrs.nasa.gov/199400316 68.pdf

[9] B. J. Anderson, C. G. Justus, and G. W. Batts, "Guidelines for Selection of Near-Earth Thermal Environment Parameters for Spacecraft Design," NASA Technical Memorandum 2001-211221, Oct 2001. Available at https://ntrs.nasa.gov/archive/nasa/casi.ntrs.nasa.gov/200200043 60.pdf

Characterizing the Impact of Solar Spectral Irradiance on PV Module Output

M. Schweiger, W. Herrmann

TÜV Rheinland Energy GmbH, 51105 Cologne, Germany, www.tuv.com/solarpower

Abstract — **The energy delivery of PV devices for a specific location is determined by the interaction of the electrical performance of the PV device and the site specific environmental and meteorological conditions. The spectral response curve of a PV device is a determining factor for its energy yield performance with regard to spectral effects. For a given spectral irradiance distribution the generated photocurrent results out of the product of both curves. Deviations to AM 1.5 of measured real sun spectra can be either subject to a blue shift when the composition of the spectrum shows a higher intensity in the low wavelength range or a red shift when a higher intensity for high wavelength range is observed. Accordingly, PV devices with a narrow band of spectral response will benefit from a blue shift of spectral irradiance. Global solar irradiance is commonly measured with spectrally neutral pyranometers. Therefore, the effective irradiance for photocurrent generation of a PV device can be higher or lower depending on the spectral composition of solar irradiance. This relation is described by the spectral mismatch correction factor (MMF), which is a direct measure for spectral gains (MMF > 1) or losses (MMF <1). We analyzed spectral shifts in the solar spectral irradiance using the average photon energy (APE) and quantify the impact on the performance of eight PV technologies in Arizona and Italy using the spectral mismatch factor.**

Index Terms — **spectral effects, spectral mismatch factor, average photon energy, energy yield.**

I. INTRODUCTION

The energy delivery of PV devices for a specific location is determined by the interaction of the electrical performance of the PV device and the site specific environmental and meteorological conditions. Besides the nominal output power, the temperature coefficients, the efficiency at low irradiance and the angular response for large angles of incidence, the spectral response curve of a PV device is a determining factor for the energy yield performance of a PV device [1]. For a given spectral irradiance distribution curve (i.e. sunlight) the generated photocurrent is the product of both curves.

The nominal output power of PV devices is commonly related to standard test conditions which also define the AirMass 1.5 (AM1.5) reference spectral irradiance distribution [2]. Deviations of measured real sun spectra can be either subject to a blue shift when the composition of the spectrum shows a higher intensity in the wavelength range below 720 nm and a red shift when a higher intensity above 720 nm is observed. Accordingly, PV devices with a narrow band of spectral response will benefit from a blue shift of spectral irradiance.

Fig. 1. Setup for spectral irradiance measurements: Input optic in the plane of the array, Wavelength range 300 – 1600 nm

Global solar irradiance is commonly measured with spectrally neutral pyranometers. Therefore, the effective irradiance for photocurrent generation of a PV device can be higher or lower depending on the spectral composition of solar irradiance. This relation is described by the spectral mismatch factor (MMF), which is a direct measure for spectral gains (MMF > 1) or losses (MMF <1).

In this study we analysed the impact of spectral shifts in the solar spectrum on the spectral mismatch factor for eight PV technologies and two test locations in Tempe (Arizona) and Ancona (Italy). We measure the spectral irradiance in the wavelength range 300 – 1600 nm. In combination with spectral response measurements it is possible to evaluate the influence of spectral effects on the energy yield performance of different cell technologies (a-Si, CIS, CdTe, CIGS, c-Si and HJT).

II. AVERAGE PHOTON ENERGY

The average photon energy (APE) according to [3] is a factor to rate the spectral shift of a measured solar spectrum compared to AM1.5 reference spectral distribution, which has an APE of 1.65eV for the given wavelength range of 300 – 1600nm. The APE factor is calculated out of the spectral irradiance and the flux density according to

$$APE = \int_{\lambda_{Min}}^{\lambda_{Max}} E_i(\lambda)d\lambda / (q_e \int_{\lambda_{Min}}^{\lambda_{Max}} \Phi_i(\lambda)d\lambda) \cdot \qquad (1)$$

An APE factor greater than 1.65 eV means a blue shift compared to AM1.5 and an APE factor smaller 1.65eV means a red shift compared to AM1.5. The measurements reveal a

978-1-5090-5606-4/17 $31.00 © 2017 IEEE

blue shift in summer and a red shift in winter for the northern hemisphere as illustrated in figure 2 for monthly averages.

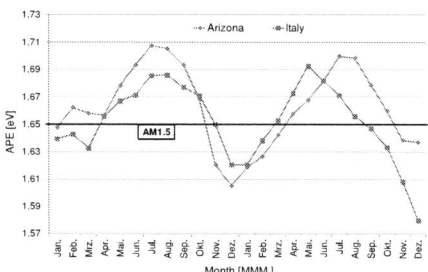

Fig. 2. Average Photon Energy (APE) on a monthly basis in Arizona and Italy

III. SPECTRAL RESPONSE OF PV DEVICES

We measure the spectral response of single- and multi-junction PV modules without destroying the samples with a setup developed by AIST [4]. The results for eight representative PV module types are plotted in figure 3.

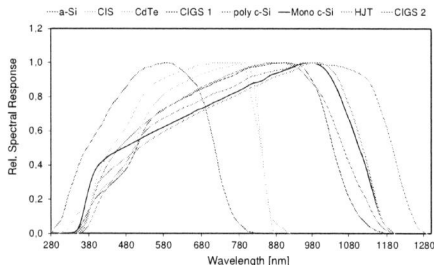

Fig. 3. Relative spectral response of six different PV module technologies normalized to peak value

IV. SPECTRAL MISMATCH FACTOR

We calculated the spectral mismatch factor [5] using measured spectral irradiance curves (E_{Sun}) according to

$$MMF = \frac{\int E_{Sun}(\lambda) \cdot SR_{Mod}(\lambda) d\lambda}{\int E_{STC}(\lambda) \cdot SR_{Mod}(\lambda) d\lambda} \times \frac{\int E_{STC}(\lambda) \cdot SR_{Det}(\lambda) d\lambda}{\int E_{Sun}(\lambda) \cdot SR_{Det}(\lambda) d\lambda}. \quad (2)$$

The input parameters include the measured spectral response of the PV device under test SR_{Mod} (as shown in figure 3), the spectral response of the reference irradiance detector (SR_{Det}) and the AM1.5 reference spectral irradiance (E_{STC}). The reference irradiance detector is a pyranometer with a spectral response curve assumed to be independent from wavelength and equal to unity.

The annual average solar spectral irradiance distribution can be used to calculate spectral gains or losses for a specific PV technology in the laboratory [6] or in the field. Figure 4 shows

the results for PV modules of different technology and test locations in Arizona and Italy. Annual gains or losses are more pronounced in the hot and dry climate of Arizona. Further results indicate greater impact of spectral effects for CdTe and a-Si in tropic climates like in south-east India. The monthly averages of spectral mismatch factors are shown in Fig. 5. The diagrams reveal seasonal variations for all PV technologies.

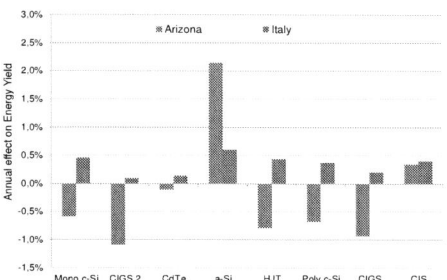

Fig. 4. Annual spectral gains or losses for different PV modules and test locations in Italy and Arizona

V. CONCLUSIONS

Our analysis is focusing on the impact of spectral effects on the performance of different PV module technologies. The analysis of measured spectral irradiance data revealed a blue shift in summer and a red shift in winter for the northern hemisphere compared to AM1.5. Having measured the spectral response curves of eight representative cell technologies the spectral mismatch factor is calculated for different technologies and climates. The analysis revealed gains in winter for cell technologies with a broad band of spectral responsivity up to 1280 nm like c-Si, HJT and CIGS. For cell technologies with a narrow band of spectral responsivity ending below 900 nm, gains can be achieved in summer times when there is a blue shift of solar irradiance.

The average annual energetic impact of spectral effects is low in Arizona and Italy. Further data from India, Germany and Saudi-Arabia is still under investigation.

REFERENCES

[1] M. Schweiger, W. Herrmann: Comparison of Energy Yield Data of Fifteen PV Module Technologies Operating in Four Different Climates, 2015, 42nd IEEE Photovoltaic Specialists Conference, New Orleans, Louisiana.

[2] IEC 60904-3, Photovoltaic devices – Part 3: Measurement principles for terrestrial photovoltaic (PV) solar devices with reference spectral irradiance data, 2015.

[3] T. Betts, Investigation of Photovoltaic Device Operation under Varying Spectral Conditions, Loughborough University, 2004.

[4] Y. Tsuno, Y. Hishikawa and K. Kurokawa, "A Method for Spectral Response Measurements of Various PV Modules," in 23rd European Photovoltaic Solar Energy Conference and Exhibition, Valencia, 2008.

[5] IEC 60904-7, Photovoltaic devices - Part 7: Computation of the spectral mismatch correction for measurements of photovoltaic devices, 2009.

[6] C. Monokroussos et al., IEC 60904-9 spectral classification and impact on industrial rating of c-Si devices, WCPEC-6, 2014

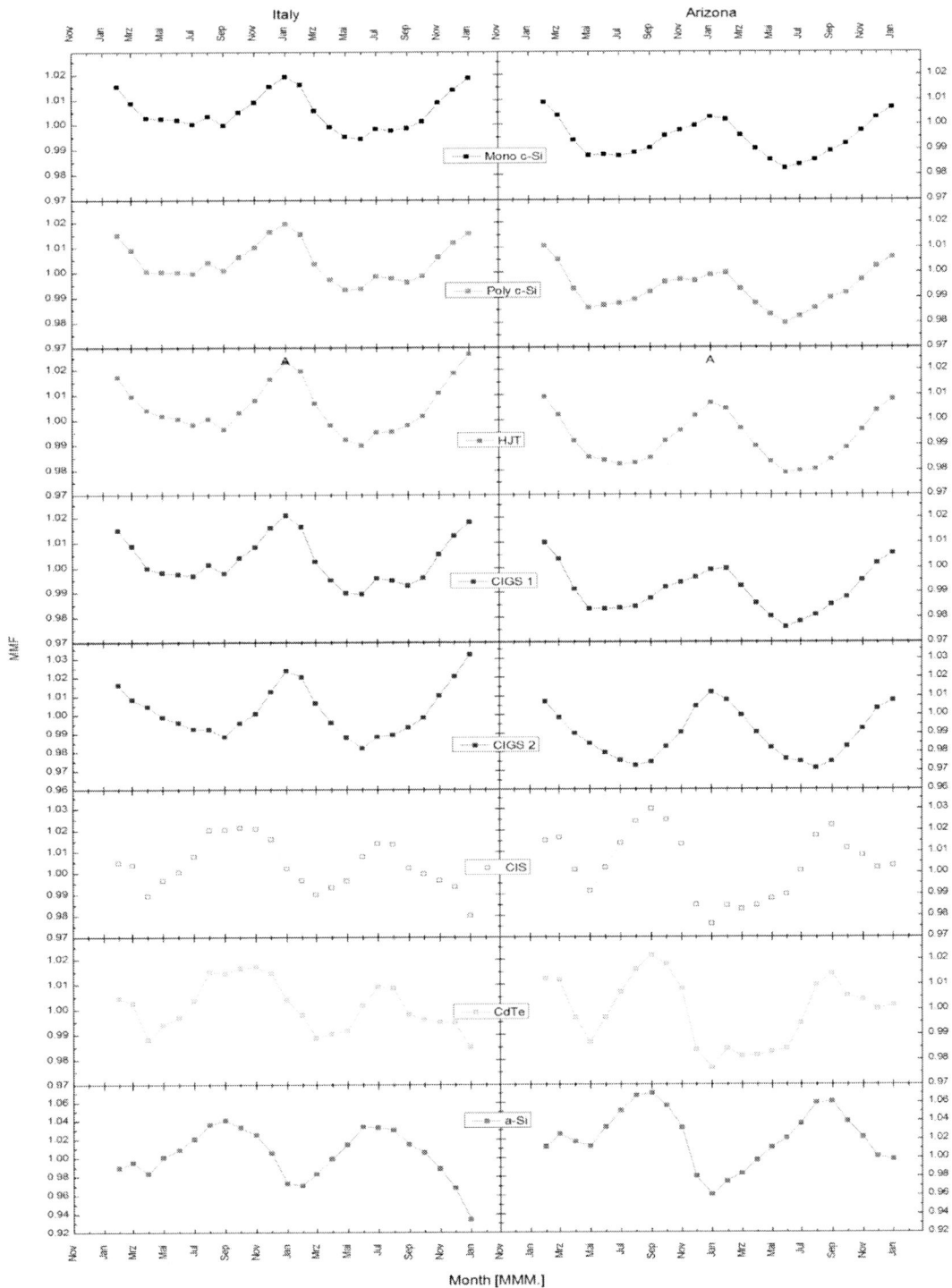

Fig. 5. Time series of monthly spectral mismatch factor, calculated for eight different PV technologies and test locations in Arizona and Italy

Use of Measured Aerosol Optical Depth and Precipitable Water to Model Clear Sky Irradiance

Mark M. Mikofski[1], Clifford W. Hansen[2], William F. Holmgren[3] and Gregory M. Kimball[1]

[1]SunPower Corporation, Richmond, CA, 94804, USA
[2]Sandia National Laboratories, Albuquerque, NM, 87185, USA
[3]University of Arizona, Tucson, AZ, 85721, USA

Abstract — Predicted clear sky irradiance depends on atmospheric composition as well as solar position and extra-terrestrial irradiance. The effects on clear sky irradiance of year to year variations in atmospheric composition were studied using measurements of aerosol optical depth (AOD) and precipitable water (P_{wat}) at seven locations in the United States. Three clear sky models were evaluated, including one that uses Linke turbidity (T_L). This model was evaluated using historical, static T_L as well as updated values derived from real-time AOD and P_{wat} measurements. The average annual error in predicted clear sky irradiance using static T_L did not differ significantly from year to year. Annual average error in predicted GHI was less than 5% for all models with no significant difference between models. The model with static T_L had the lowest DNI errors, and the Bird model had the smallest GHI error but the largest DNI error. On average DNI and GHI were under-predicted.

Index Terms — clear sky, irradiance, aerosol optical depth, precipitable water.

I. INTRODUCTION

Predicting clear sky irradiance is important for estimating energy generation by solar power systems. Clear sky models predict the direct normal (DNI), diffuse horizontal (DHI) and global horizontal (GHI) components of irradiance on a cloudless day. Since the concentration of aerosol and water vapor in the atmosphere can affect all three of these irradiance components, they can also influence power production. We analyzed the effects of aerosol optical depth (AOD) and precipitable water (P_{wat}) on irradiance predictions from three clear sky models by comparing them with irradiance measurements at seven US locations. This paper is a report on our analysis.

II. METHODS

This section describes the differences between the clear sky models and the sources of atmospheric composition and irradiance measurements used in our analysis.

A. Clear Sky Models

Several numerical models are available for prediction of clear sky irradiance. Ineichen recently published a study of seven clear sky models [1], evaluating them using atmospheric data from the Monitoring Atmospheric Composition and Climate (MACC) project of the Copernicus Atmospheric Monitoring Service (CAMS). This data is provided by the European Center for Medium-Range Weather Forecasts (ECMWF). Ineichen concluded that the Simplified Solis model [2] demonstrated the smallest long term variance from measurements at twenty two irradiance stations mostly in Europe over an 8 year period. The National Renewable Energy Laboratory (NREL) performed a similar study [3] with irradiance and atmospheric data from the National Oceanic and Atmospheric Administration (NOAA) Earth System Research Laboratory (ESRL) Surface Radiation Network (SURFRAD) and found the Bird model [4]–[7] to be a better fit. We analyzed these models as well as the Ineichen-Perez model, popular due to its long-established implementation in PVsyst and in the PVLIB MATLAB and Python modeling libraries [8]–[10].

We compared the accuracies of Bird, Simplified Solis, and Ineichen-Perez models using PVLIB-Python. The Bird and Simplified Solis models take inputs of P_{wat} and broadband AOD measurements directly, but the Ineichen-Perez model [11], [12] uses Linke turbidity (T_L) [13] as a parameter to represent both components of the atmosphere. PVLIB-Python provides a gridded static set of monthly T_L values from 2003, obtained from the SoDa Pro website. We re-calculated the T_L values from AOD and P_{wat} measurements using the method described in the next section and compared irradiance predictions from both the static and re-calculated T_L values to demonstrate year to year variability.

B. Measurements of Atmospheric Composition

We used measurements of AOD and P_{wat} from the CAMS MACC project provided by ECMWF. This data is derived from an atmospheric model that assimilates satellite data from MODIS and is calibrated with independent ground measurements from AERONET [14]. Aerosol data at several wavelengths and total column water vapor are available over the entire globe at 0.75° increments every 3 hours from 2003 to 2012.

In our analysis, we calculated T_L from broadband AOD and P_{wat} using Eq. (1) in which AM is airmass, calculated using the NREL solar position algorithm (SPA) form PVLIB, and δ_{total}.is the total atmospheric attenuation, derived in Eq. (2).

$$T_L = -\frac{-(9.4 + 0.9AM)\log(\exp(-AM\delta_{total}))}{AM} \quad (1)$$

The method developed by Kasten [17], [18] is explained in detail by Ineichen and Perez [12], [19]. The contributions from pure Rayleigh scattering, $\delta_{Rayleigh}$, through a hypothetical "clean dry atmosphere" are combined with water absorption, δ_{water}, and the broadband AOD to get the total atmospheric attenuation, δ_{total}, in Eq. (2).

$$\delta_{total} = \delta_{Rayleigh} + \delta_{water} + \tau_{aerosol} \quad (2)$$

There are several options for determining the broadband AOD, $\tau_{aerosol}$. Molineaux [20] proposed using a single AOD measurement at 700 nm which is used in the Simplified Solis model. For the Bird model, Bird and Hulstrom [21] suggested two AOD measurements at 380 nm and 500 nm correlated by the expression in Eq. (3) where τ is AOD and λ is wavelength.

$$\tau_{aerosol} = 0.27583\tau_{\lambda=380nm} + 0.35\tau_{\lambda=500nm} \quad (3)$$

To calculate AOD at 380 nm, 500 nm and 700 nm, we obtained AOD at 550 nm and 1240 nm from the ECMWF MACC data. Then, assuming AOD is related to wavelength by the Angstrom turbidity model [15], [16], we calculated the Angstrom exponent, α, from AOD at the two wavelengths, and used α to obtain AOD at the desired wavelengths. This is demonstrated in Eq. (4).

$$\tau = \tau_0 \left(\frac{\lambda}{\lambda_0}\right)^{-\alpha} \quad (4)$$

C. Measurements of Clear Sky Irradiance

To evaluate the clear sky irradiance models and the measurement sources of AOD and P_{wat}, predictions of DNI, DHI and GHI were compared to SURFRAD measurements of irradiance, ambient temperature, relative humidity and pressure at either 1-minute or 3-minute intervals. The SURFRAD stations listed in Table I were used for the years from 2003 to 2012. Down-sampled measurements at 3-minute intervals were filtered for clear sky conditions using PVLIB-Python with a 30-minute window and clear sky calculated using Simplified Solis. Measurements below a GHI threshold of 200 W/m² were also removed.

Mean bias error (MBE) was calculated between the filtered measured data and the predictions using the formula in Eq. (5) in which N is the number of measurements. Relative error was obtained by dividing the calculated MBE by the average of the measurements. The analysis was done in a Python notebook that can be accessed from an online repository at https://github.com/mikofski/pvsc44-clearsky-aod.

$$MBE = \frac{\sum_{n=1}^{N}(predicted_n - measured_n)}{N} \quad (5)$$

TABLE I
SURFRAD SURFACE RADIATION STATIONS

Station Name	Station ID	Latitude	Longitude	Elevation (m)
Bondville, IL	bon	40.05	-88.37	213
Table Mountain, CO	tbl	40.13	-105.24	1689
Desert Rock, NV	dra	36.62	-116.02	1007
Fort Peck, MT	fpk	48.31	-105.10	634
Goodwin Creek, MS	gwn	34.25	-89.87	98
Penn State, PA	psu	40.72	-77.93	376
Sioux Falls, SD	sxf	43.73	-96.62	473

III. RESULTS

Fig. 1 to 4 compare monthly static T_L at the Bondville, Fort Peck, Table Mountain and Desert Rock stations, monthly average T_L values calculated from AOD and P_{wat} using Eq. (1) and (2) for all years in the study and the monthly average over all years of the calculated T_L. Atmospheric data was filtered for clear sky and low light before calculating T_L. The magnitude and shape of the historical and calculated T_L were similar for Bondville, Sioux Falls, Goodwin Creek and Penn State, but deviated for Fort Peck, Table Mountain and Dessert Rock. For all stations, T_L was greater in summer than winter.

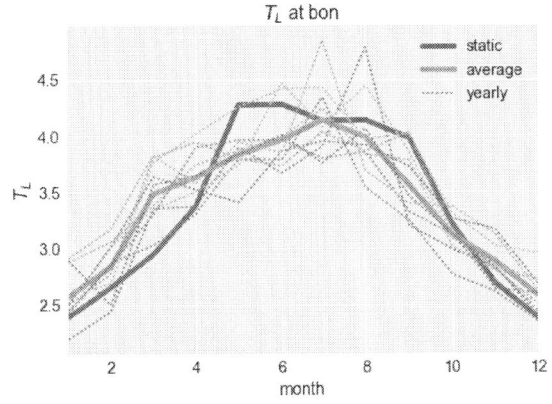

Fig. 1. Linke turbidity at Bondville, IL, from 2003 to 2012 calculated using filtered AOD and P_{wat} as dotted lines, the average for all years as solid green line and the 2003 historical values as solid blue line.

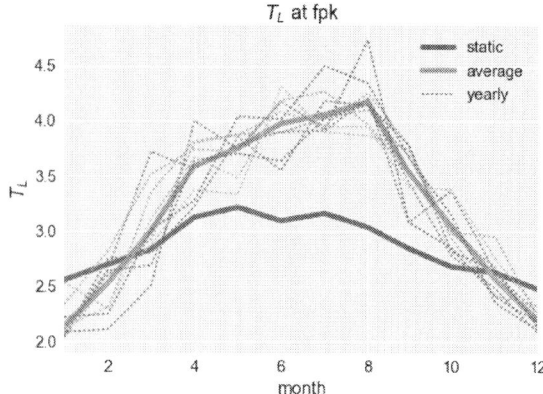

Fig. 2. Linke turbidity at Fort Peck, MT, from 2003 to 2012 calculated using filtered AOD and P_{wat} as dotted lines, the average for all years as solid green line and the 2003 historical values as solid blue line.

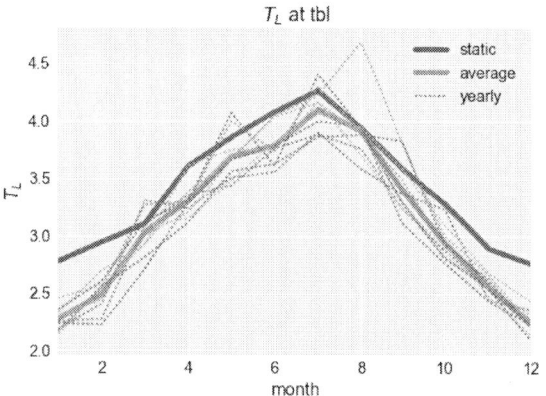

Fig. 3. Linke turbidity at Table Mountain, CO, from 2003 to 2012 calculated using filtered AOD and P_{wat} as dotted lines, the average for all years as solid green line and the 2003 historical values as solid blue line.

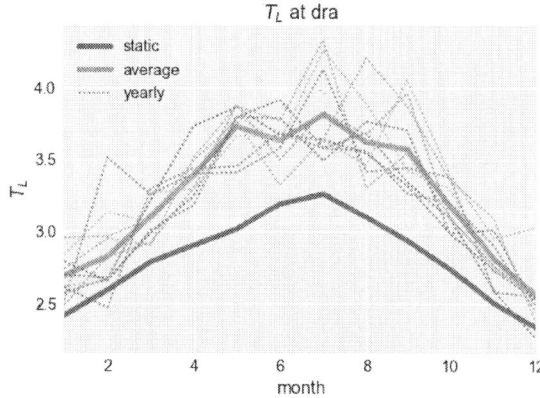

Fig. 4. Linke turbidity at Desert Rock, NV, from 2003 to 2012 calculated using filtered AOD and P_{wat} as dotted lines, the average for all years as solid green line and the 2003 historical values as solid blue line.

In Fig. 5 and 6, different clear sky models are compared to measured data at Bondville, IL on July 16th, 2006.

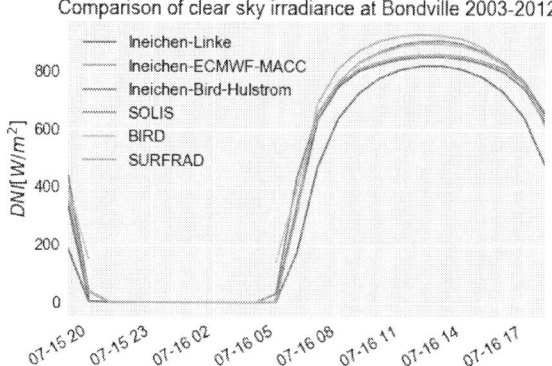

Fig. 5. Comparison of DNI at Bondville, IL on 7/16/2006 shows good agreement with Bird and Simplified Solis and poorer agreement with Ineichen-Perez using either historic or calculated T_L.

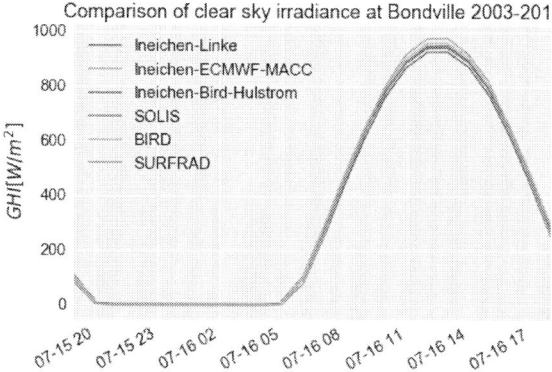

Fig. 6. Comparison of GHI at Bondville, IL on 7/16/2006. Bird and Simplified Solis with MACC data are slightly better than the historic or calculated T_L.

Fig. 7 to 14 show box plots of the distributions of average monthly relative errors, calculated using Eq. (5), for each clear sky model. The box bounds the the 2nd and 3rd quartiles, the whiskers show the 5% and 95% confidence bounds, the dashed red line is the mean, the solid black line is the median, and the flyers are values that fall outside of the confidence bounds. Fig. 7 and 8 show comparisons between clear sky models by year in different colors. Fig. 7 shows that there are no significant long term trends in DNI errors and no significant differences between models. The Ineichen-Perez model with static T_L shows no trend from year to year while the models using ECMWF MACC data show an increasing negative error with time. The Bird model had the largest mean yearly error. Fig. 8 shows that there are no significant long term trends in GHI errors and no significant differences between models. The Ineichen-Perez model with static T_L shows no trend from year to year while the models using ECMWF MACC data show an increasing negative error with time. The Simplified Solis model had the largest mean yearly error.

From the year to year comparison, there does not appear to be significant difference between the use of static and real-time atmospheric data in clear sky predictions. The increasing yearly mean bias observed in DNI and GHI year to year box plots for models using real-time AOD and P_{wat} may be an artifact of the measured atmospheric data. For GHI the increase in relative mean bias error is less than 5%.

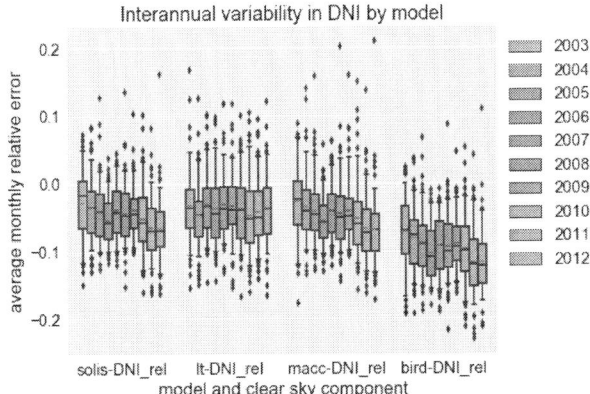

Fig. 7. Comparison of DNI errors at all stations by year (colors) and by model shows no statistical year to year variation and no statistical variation between models. The Ineichen-Perez model with static T_L has no trend from year to year while the models using ECMWF MACC show increasing mean error over time. The Bird model has the largest mean yearly error.

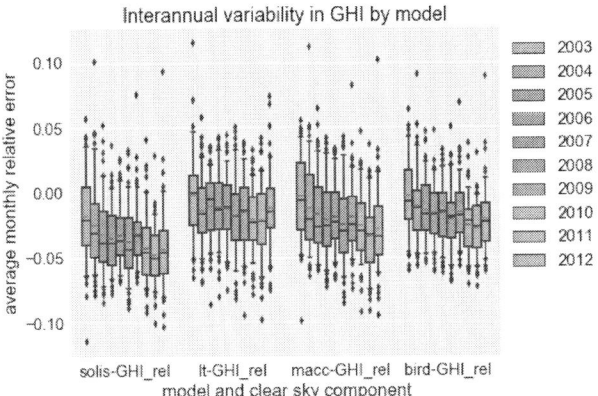

Fig. 8. Comparison of GHI errors at all stations by years (colors) and by model show no statistical year to year variation and no statistical variation between models. The Ineichen-Perez model with static T_L has no trend from year to year while the models using ECMWF MACC show increasing mean error over time. The Simplified Solis has the largest mean yearly error.

Fig. 9 and 10 show seasonal variations in error between clear sky models by month in different colors. The average monthly relative errors are grouped by month across all years. There are no significant differences in DNI errors by month or by model. The Bird model has the largest mean monthly

errors. There is a seasonal bias in GHI errors for all models, including the Ineichen-Perez model with static T_L, so the seasonal bias cannot be an artifact of the AOD and P_{wat} measurements unless it arises from a common instrument error. The seasonal bias under-predicts GHI in summer, with a delta between the summer and winter mean error of less than 5%.

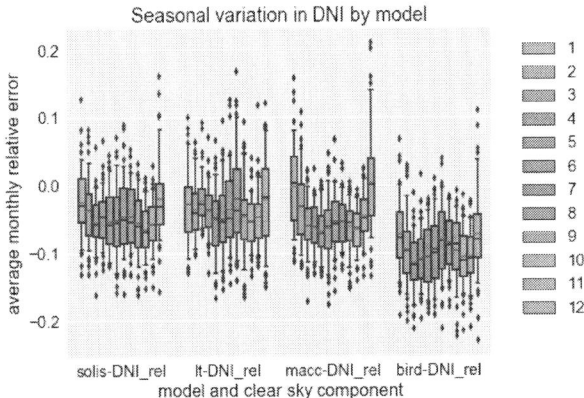

Fig. 9. Comparison of DNI errors at all stations by months (colors) shows no statistical differences by month or model.

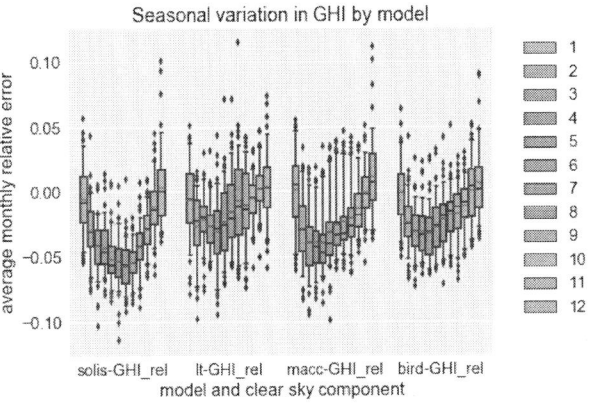

Fig. 10. Comparison of GHI errors at all stations by months (colors) show a seasonal bias, with a delta between summer and winter mean error of less than 5%.

Fig. 11 and 12 show regional variations in error between clear sky models by stations in different colors. The average monthly errors are grouped by station for all months and years. Fig. 8 shows the errors in DNI by station. The stations that had small differences between calculated T_L and static values have roughly consistent errors for all models. Two of the stations that had calculated T_L that deviated from the static values, Fort Peck and Desert Rock, show lower errors in both DNI and GHI with the Ineichen-Perez model using static T_L. The other station with calculated T_L that differed from the static values was the Table Mountain station, and it shows lower errors in both DNI and GHI with the Ineichen-Perez model using ECMWF MACC data. These three stations, Fort Peck, Desert

Rock and Table Mountain, were also the stations with the highest elevation and highest average DNI.

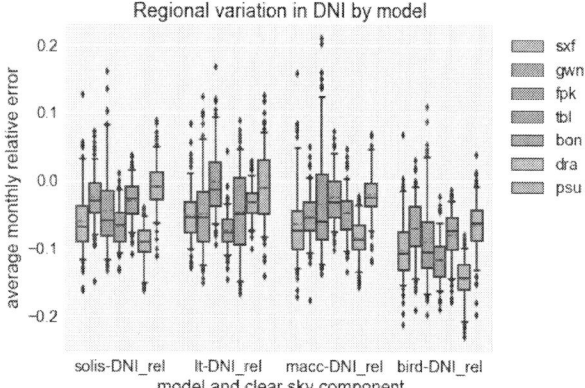

Fig. 11. Comparison of monthly DNI errors grouped by station (colors) show roughly the same mean error except for Fort Peck, Desert Rock, and Table Mountain, which were also the stations that had calculated T_L that differed from static values.

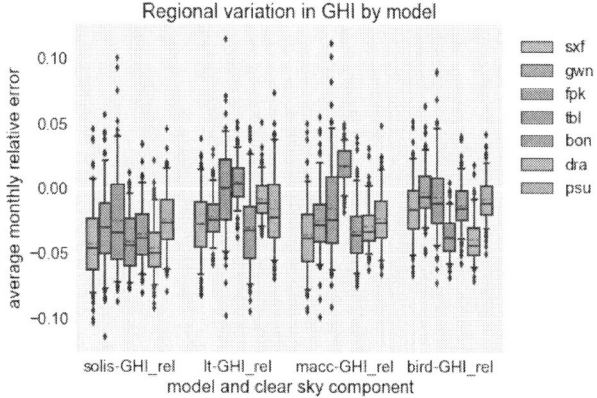

Fig. 12. Comparison of monthly GHI errors grouped by station (colors) show the mean error is less than 5% for all stations and models.

Fig. 13 and 14 show average relative error in DNI and GHI for all stations sorted by model. There were significant differences between GHI errors, although all models had average errors less than 5%. The Simplified Solis had the largest GHI error, and the Bird model had the lowest median error in this study but was not significantly different from the Ineichen-Perez model with either historical or real-time T_L. The Ineichen-Perez model with static T_L had the smallest errors in DNI, but was not statistically different from the Ineichen-Perez model with ECMWF MACC data or the Simplified Solis model. The Bird model had the largest DNI errors.

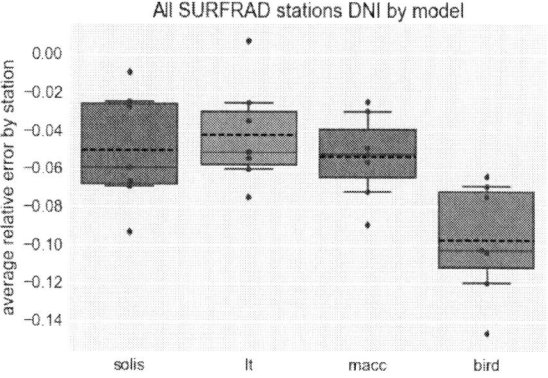

Fig. 13. Comparison of DNI errors for all stations by model show significant difference between Bird and other models.

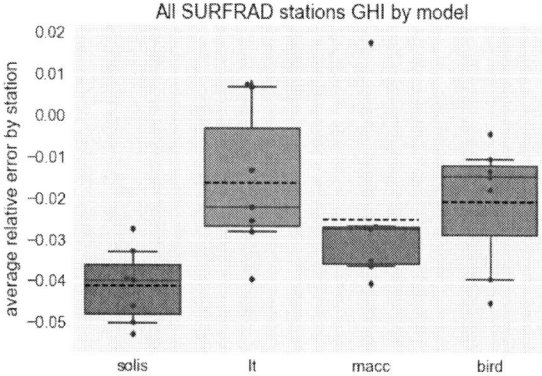

Fig. 14. Comparison of GHI errors for all stations by model show significant differences between the models in this study but all errors are less than 5%.

IV. CONCLUSIONS

A study of variations in clear sky irradiance due to AOD and P_{wat} has shown that for the seven stations and the ten-year period examined in this study there is no significant improvement in model accuracy when using real-time AOD and P_{wat} measurements. There is a seasonal bias in the GHI error that does not appear to be caused by the real-time AOD and P_{wat} measurements because it also appears in the errors from model using static T_L. The average monthly errors in GHI were not significantly different between models and were all less than 5%. The Ineichen-Perez model with static T_L had the lowest errors for DNI, but were not significantly different than the Simplified Solis model. The Bird model had significantly larger errors for DNI, but had the lowest GHI median error.

978-1-5090-5606-4/17 $31.00 © 2017 IEEE

ACKNOWLEDGMENTS

Sandia National Laboratories is a multi-mission laboratory managed and operated by National Technology and Engineering Solutions of Sandia, LLC., a wholly owned subsidiary of Honeywell International, Inc., for the U.S. Department of Energy's National Nuclear Security Administration under contract DE-NA0003525.

REFERENCES

[1] P. Ineichen, "Validation of models that estimate the clear sky global and beam solar irradiance," *Sol. Energy*, vol. 132, pp. 332–344, 2016.

[2] P. Ineichen, "A broadband simplified version of the Solis clear sky model," *Sol. Energy*, vol. 82, no. 8, pp. 758–762, 2008.

[3] M. Sengupta and P. Gotseff, "Evaluation of Clear Sky Models for Satellite-Based Irradiance Estimates," 2013.

[4] R. E. Bird and R. L. Hulstrom, "Simplified Clear Sky Model for Direct and Diffuse Insolation on Horizontal Surfaces," 1981.

[5] R. E. Bird and R. L. Hulstrom, "Review, Evaluation, and Improvement of Direct Irradiance Models," *J. Sol. Energy Eng.*, vol. 103, no. 3, p. 182, 1981.

[6] D. R. Myers, "Solar radiation modeling and measurements for renewable energy applications: data and model quality," *Energy*, vol. 30, no. 9, pp. 1517–1531, Jul. 2005.

[7] R. L. Hulstrom, *Solar Resources*. MIT Press, 1989.

[8] W. F. Holmgren, R. W. Andrews, A. T. Lorenzo, and J. S. Stein, "PVLIB Python 2015," in *Photovoltaic Specialists Conference (PVSC), 2015 IEEE 42nd*, 2015, pp. 1–5.

[9] R. W. Andrews, J. S. Stein, C. Hansen, and D. Riley, "Introduction to the open source PV LIB for python Photovoltaic system modelling package," *2014 IEEE 40th Photovolt. Spec. Conf. PVSC 2014*, pp. 170–174, 2014.

[10] J. S. Stein, W. F. Holmgren, J. Forbess, and C. W. Hansen, "PVLIB: Open Source Photovoltaic Performance Modeling Functions for Matlab and Python," *IEEE 43rd Photovolt. Spec. Conf.*, pp. 3–8, 2016.

[11] R. Perez, P. Ineichen, K. Moore, M. Kmiecik, C. Chain, R. George, and F. Vignola, "A New Operational Satellite-to-Irradiance Model - Description and Validation," *Sol. Energy*, vol. 73, no. 5, pp. 307–317, 2002.

[12] P. Ineichen and R. Perez, "A new airmass independent formulation for the Linke turbidity coefficient," *Sol. Energy*, vol. 73, no. 3, pp. 151–157, Sep. 2002.

[13] F. Linke, "Transmissions-Koeffizient und Trubungsfaktor," *Beitrage zur Phys. der Atmosphare*, vol. 10, pp. 91–103, 1922.

[14] V. Cesnulyte, A. V. Lindfors, M. R. A. Pitkänen, K. E. J. Lehtinen, J. J. Morcrette, and A. Arola, "Comparing ECMWF AOD with AERONET observations at visible and UV wavelengths," *Atmos. Chem. Phys.*, vol. 14, no. 2, pp. 593–608, 2014.

[15] A. Angstrom, "On the Atmospheric Transmission of Sun Radiation and On Dust in the Air," *Geogr. Ann.*, vol. 11, pp. 156–166, 1929.

[16] A. ÅNGSTRÖM, "Techniques of Determinig the Turbidity of the Atmosphere," *Tellus A*, vol. 13, no. 2, pp. 214–223, 1961.

[17] F. Kasten, "A simple parameterization of the pyrheliometric formula for determining the Linke turbidity factor," *Meteorol. Rundschau*, vol. 33, pp. 124–127, 1980.

[18] F. Kasten, "The linke turbidity factor based on improved values of the integral Rayleigh optical thickness," *Sol. Energy*, vol. 56, no. 3, pp. 239–244, Mar. 1996.

[19] P. Ineichen, "Conversion function between the Linke turbidity and the atmospheric water vapor and aerosol content," *Sol. Energy*, vol. 82, no. 11, pp. 1095–1097, Nov. 2008.

[20] B. Molineaux, P. Ineichen, and N. O'Neill, "Equivalence of pyrheliometric and monochromatic aerosol optical depths at a single key wavelength.," *Appl. Opt.*, vol. 37, no. 30, pp. 7008–18, Oct. 1998.

[21] R. E. Bird and R. L. Hulstrom, "Direct Insolation Models," 1980.

Recent Advancements in the Numerical Simulation of Surface Irradiance for Solar Energy Applications

Yu Xie, Manajit Sengupta, and Chris Deline

National Renewable Energy Laboratory, Golden, CO, 80401, USA

Abstract—This paper briefly reviews the National Renewable Energy Laboratory's recent efforts to develop all-sky solar irradiance models for solar energy applications. The Fast All-sky Radiation Model for Solar applications (FARMS) uses the simulation of clear-sky transmittance and reflectance and a parameterization of cloud transmittance and reflectance to rapidly compute broadband irradiances on horizontal surfaces. The accuracy of FARMS is comparable to that of two-stream approximation, but it is approximately 1,000 times faster. A FARMS for Narrowband Irradiance over Tilted surfaces (FARMS-NIT) has been developed to compute spectral irradiances on photovoltaic (PV) panels in 2,002 wavelength bands. FARMS-NIT has been extended to bifacial PV panels by accounting for solar radiation reaching the backside of a PV panel.

I. INTRODUCTION

Multiple radiative transfer models that simulate atmospheric radiation under all-sky conditions have been developed and used in a broad range of applications, such as climate and weather studies [1]. Compared to those applications, radiative transfer models for solar energy applications have unique requirements; thus, specific prerequisites are inherent in the model design. For instance, radiative transfer models for climate and weather studies provide broadband irradiances only in direct, upwelling, and downwelling directions or radiances in narrow-wavelength bands [2, 3]. In contrast, solar energy applications require solar irradiances over inclined surfaces because solar systems track the sun on multiple axes. Because of the spectral response of photovoltaic (PV) panels, irradiances in numerous narrow-wavelength bands are particularly desired for solar cell research. In addition, solar resource assessment and forecasting studies require extremely fast and efficient computations because irradiance changes rapidly during time and space, and fast computations at high resolutions are required for large geographic areas.

During recent decades, a significant number of solar irradiance models have been developed for and actively used by solar energy applications [4-6]; however, many of them lack the ability to simulate solar irradiance under clouds, which cover approximately 70% of the Earth's surface at any given time [7]. Solar irradiance models that have taken clouds into account in broadband or spectral irradiance simulations [8] have represented the complex processes of absorbing and scattering by clouds by using empirically determined cloud-modification factors that are less accurate compared to the physical solutions of radiative transfer models for climate and weather studies. Thus, there is a critical need for advanced models that can bridge the special demands of solar energy applications and the advantage of radiative transfer models for climate and weather studies.

This study briefly introduces the National Renewable Energy Laboratory's (NREL's) recent advancements in rapid broadband and spectral radiative transfer models to meet the requirements of solar energy applications. The rest of this paper is organized as follows. Section 2 summarizes the rapid simulation of solar irradiance over horizontal surfaces. Section 3 discusses the efforts to simulate narrowband irradiances over inclined PV panels and those consisting of bifacial PV modules. The last section summarizes the results and future studies.

II. RAPID SIMULATION OF SHORTWAVE IRRADIANCES OVER HORIZONTAL SURFACES

A Fast All-sky Radiation Model for Solar applications (FARMS) [9] was developed to compute broadband diffuse horizontal irradiance and direct normal irradiance (DNI). The cloud transmittance and reflectance of solar irradiance are precomputed by the Rapid Radiation Transfer Model [1] for all possible cloud conditions and solar and viewing geometries. They are parameterized using exponential functions, and are combined with clear-sky transmittances and reflectances of irradiance to rapidly compute broadband irradiances for all-sky conditions. Details of FARMS are not restated here because they are extensively given in [9].

The clear-sky tranmisttance and reflectance are computed using the clear-sky irradiance model REST2 [4]. The accuracy of FARMS is comparable to or better than the two-stream approach, but it is approximately 1,000 times more efficient. We recently implemented the Bird Clear-Sky Model [6] (hereafter referred to as BIRD) in FARMS. Our investigation indicated that using FARMS with BIRD is consistently accurate as validated by surface measurements, but the computational efficiency increases by more than 100% compared to using FARMS with REST2 as the clear-sky model.

978-1-5090-5606-4/17 $31.00 © 2017 IEEE

III. NARROWBAND IRRADIANCES OVER INCLINED PV PANELS

To extend the capability of FARMS, we recently developed the FARMS for Narrowband Irradiances over Tilted surfaces (FARMS-NIT) to efficiently compute spectral plane-of-array (POA) irradiances received by PV panels. A comprehensive lookup table of cloud bidirectional transmittance distribution functions (BTDFs) was developed using LibRadtran [10]. To account for multiple reflections between cloud and land surface, we also computed the cloud transmittance and reflectance of spectral irradiances. They were then combined with a clear-sky model, SMARTS [5], to compute all-sky radiances.

A. Narrowband Irradiances over Monofacial PV Panels

From the simulation of radiances, solar irradiance in the POA can be given by:

$$POAI = POAI_d + POAI_{u,sky} + POAI_{u,ground} \quad (1)$$

where $POAI_d$, $POAI_{u,sky}$, and $POAI_{u,ground}$ represent the POA irradiances from direct solar radiation, diffuse sky radiation, and solar radiation reflected by the land surface that reaches the PV panel, respectively.

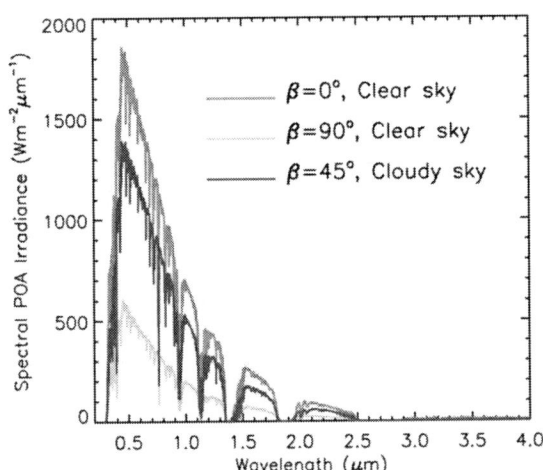

Fig. 1. Spectral POA irradiance for a solar zenith angle of 15°.

$POAI_d$ is given by the direct solar radiation in the normal direction of the PV panel, as follows:

$$POAI_d = DNI \cos \theta' \quad (2)$$

where θ' is the angle between the direct solar radiation and the normal direction of the PV panel. $POAI_{u,sky}$ can be given by the integration of radiances in the perpendicular direction to the tilted PV panel:

$$POAI_{u,sky} = \int_0^{2\pi} \int_0^{\Theta} I \cos \theta' \sin \theta d\theta d\varphi \quad (3)$$

where I is the radiance, and Θ denotes the upper limit of θ. The contribution from the reflected solar radiation by land surface can be given by:

$$POAI_{u,ground} = \int_0^{2\pi} \int_0^{\frac{\pi}{2}-\Theta} I_r \cos \theta' \sin \theta d\theta d\varphi \quad (4)$$

where I_r is reflected radiance by land surface.

In FARMS-NIT, we follow the wavelengths from SMARTS because they are used to compute the clear-sky transmittance and reflectance; thus, 2,002 narrow-wavelength bands are considered. An example output from FARMS-NIT is shown in Fig. 1, where β represents the tilt angle of the PV panel. For the cloudy sky, we assume a water cloud with an optical thickness of 3 and an effective particle diameter of 20 μm.

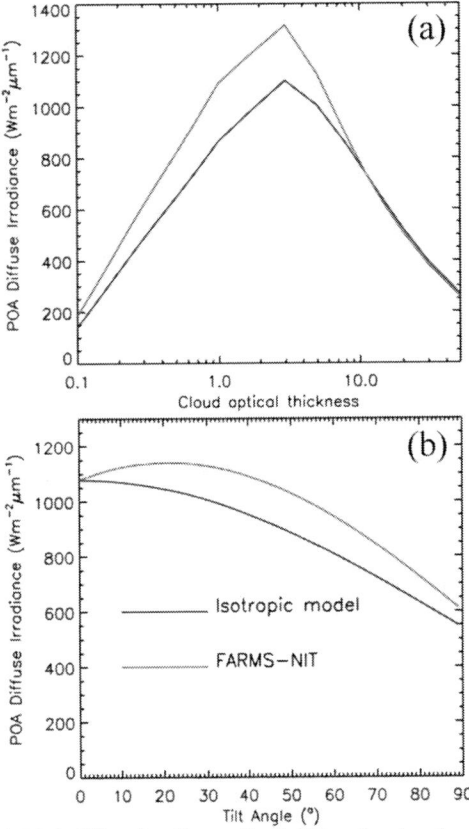

Fig. 2. (a) POA diffuse irradiances for β=30° and water clouds with various cloud optical thickness and (b) those for a cloud optical thickness of 5 and various β. The wavelength λ=0.6 μm. We assume a solar zenith angle of 30° and water clouds with an effective particle diameter of 20 μm.

Compared to FARMS-NIT, models assuming isotropic distribution of diffuse radiation might dramatically underestimate POA irradiance due to the neglect of the stronger forward scattering by clouds (see Fig.2a). The underestimation increases with cloud optical thickness, but it rapidly decreases when the clouds are thick. For λ=0.6 μm and a solar zenith angle of 30°, the underestimation can reach more than 20%. Figure 2b shows that the underestimation by the isotropic model reaches the maximum when the tilt angle is around the solar zenith angle because the maximum of direct POA irradiance received by the PV panel.

978-1-5090-5606-4/17 $31.00 © 2017 IEEE

B. Narrowband Irradiances over Bifacial PV Panels

Because radiances in the atmosphere are computed by FARMS-NIT, this model can be extended for the simulation of POA irradiances on bifacial PV panels. Similar to (1), the POA irradiance on the backside of a bifacial PV is:

$$POAIB = POAIB_d + POAIB_{u,sky} + POAIB_{u,ground} \qquad (5)$$

$POAIB_d$, $POAIB_{u,sky}$, and $POAIB_{u,ground}$ are the backside irradiances from the direct solar radiation, diffuse sky radiation, and land-surface reflection, respectively. With the radiation in the perpendicular directions on the backside of the PV panel, $POAIB_d$, $POAIB_{u,sky}$, and $POAIB_{u,ground}$ can be derived as follows:

$$POAIB_d = \begin{cases} -DNI\cos\theta' & for \ \cos\theta' < 0 \\ 0 & for \ \cos\theta' \ge 0 \end{cases} \qquad (6)$$

$$POAIB_{u,sky} = -\int_0^{2\pi}\int_\Theta^{\frac{\pi}{2}} I \cos\theta' \sin\theta \, d\theta \, d\varphi \qquad (7)$$

$$POAIB_{u,ground} = \int_0^{2\pi}\int_{\frac{\pi}{2}-\Theta}^{\frac{\pi}{2}} I_r \cos\theta' \sin\theta \, d\theta \, d\varphi \qquad (8)$$

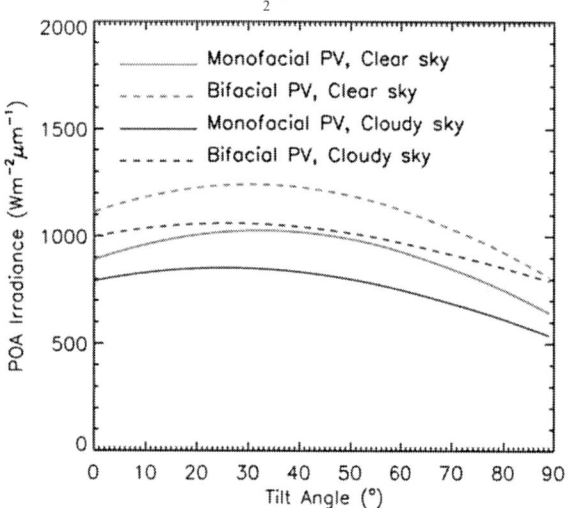

Fig. 3. POA irradiances over monofacial and bifacial PV panels for a solar zenith angle of 30° and a land-surface albedo of 0.25. For the clear-sky condition, AOD is 0.5. For the cloudy-sky condition, a water cloud with cloud optical thickness of 3 and effective particle diameter of 10 μm is assumed.

We consider a single monofacial or bifacial PV panel where solar shadow from the PV panel itself and nearby PV panels is neglected. The surface-reflected radiance can then be approximated as

$$I_r = \frac{GHI\sigma}{\pi} \qquad (9)$$

where σ denotes land-surface albedo.

Figure 3 shows the POA irradiances simulated for clear- and cloudy-sky conditions when the solar zenith angle is 30° and the land-surface albedo is 0.25. For the clear-sky condition, the aerosol optical depth (AOD) is 0.5. When clouds are present, a water cloud with the optical thickness of 3 and effective particle diameter of 10 μm is assumed. The POA irradiances are computed for the 2,002 spectral bands and integrated to give the broadband irradiances from 0.28-4.0 μm as shown in Fig. 3. Also shown is that bifacial PV panels receive significantly more solar radiation than monofacial PV panels because of the radiation reaching the backside of the PV panels after being scattered in the atmosphere or reflected by the land surface. For the clear-sky condition, the bifacial PV panel receives 21.88% more solar radiation compared to the monofacial PV panel. For the cloudy-sky condition, the bifacial PV panel receives 28.42% more solar radiation because of greater diffuse irradiance scattered by clouds.

Note that greater land-surface albedo will further increase the POA irradiances, especially for bifacial PV panels; however, solar shadows because of PV panels decrease the global horizontal irradiance (GHI) reaching the land surface and thus more significantly affect bifacial PV panels receiving much more reflected solar radiation than monofacial PV panels. This effect depends on the size, geometry, and tilt angles of the PV panels as well as their density and displacement over the area, which should be investigated by a parallel research effort.

IV. CONCLUSIONS

Conventional radiative transfer models for climate and weather studies differ from solar irradiance models for solar energy. Although the former can precisely simulate solar irradiance for all-sky conditions by solving the radiative transfer equation, they do not meet the requirements for solar energy applications. In this paper, we briefly reviewed NREL's recent efforts to improve the radiative transfer models for efficient solar energy applications. We developed FARMS using a clear-sky model and a parameterization of cloud transmittance and reflectance. We discovered that the clear-sky models REST2 and BIRD provide comparable accuracy when used by FARMS. The computational efficiency of FARMS increased by more than 100% when REST2 was replaced with BIRD. Because FARMS provides rapid and accurate solutions of GHI and DNI, it has been used in a number of applications since its development (e.g., NREL's National Solar Radiation Database and the Weather Research and Forecasting model for solar energy forecasts). We extended the capability of FARMS by developing FARMS-NIT to compute spectral irradiances over both horizontal and tilted surfaces. The irradiances on tilted surfaces are provided by a lookup table of cloud BTDFs and a spectral irradiance model for clear-sky conditions. We also demonstrated the capability of FARMS-NIT to compute POA irradiances over bifacial PV panels.

FARMS and FARMS-NIT as a whole have bridged the special demands of solar energy and the advantages of radiative transfer models for climate and weather studies. Further improvements to and the applications of these models will provide opportunities to improve the availability and accuracy of solar resource assessments and forecasts.

ACKNOWLEDGEMENT

This work was supported by the U.S. Department of Energy under Contract No. DE-AC36-08GO28308 with the National Renewable Energy Laboratory. Funding provided by U.S. Department of Energy Office of Energy Efficiency and Renewable Energy Solar Energy Technologies Office.

The U.S. Government retains and the publisher, by accepting the article for publication, acknowledges that the U.S. Government retains a nonexclusive, paid-up, irrevocable, worldwide license to publish or reproduce the published form of this work, or allow others to do so, for U.S. Government purposes.

REFERENCES

[1] E. J. Mlawer, S. J. Taubman, P. D. Brown, M. J. Iacono, and S. A. Clough, "RRTM, a validated correlated-k model for the longwave," *J. Geophys. Res.,* vol. 102, pp. 16663-16682, 1997.

[2] P. Jimenez, P. Hacker, J. Dudhia, S. Haupt, J. Ruiz-Arias, C. Gueymard, G. Thompson, T. Eidhammer, and A. Deng, "WRF-Solar: Description and clear-sky assessment of an augmented NWP model for solar power prediction," *Bull. Amer. Meteor. Soc.,* vol. 97, pp. 1249-1264, 2016.

[3] Y. Xie, P. Yang, G. W. Kattawar, P. Minnis, Y. X. Hu, and D. Wu, "Determination of ice cloud models using MODIS and MISR data," *Int. Remote Sens.,* vol. 33, pp. 4219-4253, 2012.

[4] C. Gueymard, "REST2: High-performance solar radiation model for cloudless-sky irradiance, illuminance, and photosynthetically active radiation - Validation with a benchmark dataset," *Sol. Energy,* vol. 82, pp. 272-285, 2008.

[5] C. Gueymard, "SMARTS2: a simple model of the atmospheric radiative transfer of sunshine: algorithms and performance assessment," *Florida Solar Energy Center,* vol. Cocoa, FL, 1995.

[6] R. Bird and R. Hulstrom, "A simplified clear sky model for direct and diffuse insolation on horizontal surfaces," Solar Energy Research Institute, Golden, CO1981.

[7] C. Stubenrauch, W. Rossow, S. Kinne, S. Ackerman, G. Cesana, H. Chepfer, L. Di Girolamo, B. Getzewich, A. Guignard, A. Heidinger, and B. Maddux, "Assessment of global cloud datasets from satellites: Project and database initiated by the GEWEX radiation panel," *Bull. Amer. Meteor. Soc.,* vol. 94, pp. 1031-1049, 2013.

[8] R. Perez, P. Ineichen, K. Moore, M. Kmiecik, C. Chain, R. George, and F. Vignola, "A new operational model for satellite-derived irradiances: Description and validation," *Sol. Energy,* vol. 73, pp. 307-317, 2002.

[9] Y. Xie, M. Sengupta, and J. Dudhia, "A Fast All-sky Radiation Model for Solar applications (FARMS): Algorithm and performance evaluation," *Sol. Energy,* vol. 135, pp. 435-445, 2016.

[10] B. Mayer and A. Kylling, "Technical note: The libRadtran software package for radiative transfer calculations: description and examples of use," *Atmos. Chem. Phys.,* vol. 5, pp. 1855-1877, Jul 26 2005.

Optimal Irradiance Sensor Placement for Photovoltaic Systems Using Mutual Information Based Greedy Algorithm in Gaussian Process

Lian Lian Jiang [1], R. Srivatsan [1], and Douglas L. Maskell [2]

[1]Energy Research Institute @NTU, Singapore, 637553

[2]School of Computer Science and Engineering, Nanyang Technological University Singapore, 639798

Abstract—This paper proposes the application of the mutual information criteria based greedy algorithm to place the irradiance sensor for photovoltaic (PV) systems. There is little information about irradiance sensor placement for PV systems in literature because of the complexity caused by the variability of the irradiance distribution. Existing methods in the literature are not able to provide good accuracy due to either their experience based characteristic or the low resolution in the satellite data used to determine the irradiance distribution. In this work, to get the near optimal sensor placement, that is, the best *b* locations out of *n* possible sensor locations, a mutual information based greedy algorithm is used to maximize the mutual information increase (MII) in the unsensed locations. The kriging interpolation technique is then used to predict the irradiance values at these unsensed locations. The effectiveness of the greedy algorithm based on the mutual information criteria is verified using experimental datasets measured at two different locations, namely at Nanyang Technological University, Singapore and NREL Oahu, Hawaii. The results show that the proposed method can provide a near optimal sensor placement which provides close performance to the minimum average root mean square error (RMSE) of the irradiance values at the unsensed locations achieved using an exhaustive search.

Index Terms—Optimal irradiance sensor placement, mutual information, kriging interpolation, greedy algorithm, Gaussian process.

I. INTRODUCTION

Photovoltaics (PVs) represent one of the more promising renewable energy technologies, with exponential growth worldwide over the past two decades. The improved efficiency in PV materials, advancements in power electronic devices, feed-in tariffs from governments, large-scale production, etc., have reduced the installed price of U.S. solar PV. The irradiance monitoring is critical for applications in PV systems such as system modeling, power management, fault control, power quality control, and battery integration to the grid, etc. In energy prediction or estimation, the classical approach is to determine the irradiance value from satellite data collected from meteorological geostationary satellites. However, despite the advantage of a broad geographical coverage, the irradiance values estimated from satellite datasets suffer from very low resolution [1–3]. An alternative method to solve this problem is to measure the irradiance values using ground based sensors [4–6]. The irradiance values of locations where there is no sensors deployed can then be interpolated with irradiance values taken from nearest sensors. This method usually gives better resolution than that using satellite maps. However, the position of these sensors greatly impacts the interpolation

results due to variable environmental conditions. One of the major factors affecting the interpolation results is the partial shading caused by passing clouds, which results in significant irradiance variations and thus causes inaccurate power predictions [7–11].

Deploying more irradiance sensors can improve the prediction accuracy, but it increases system costs due to the high cost of the irradiance sensors. Alternatively, the irradiance sensors placed at the extremities of the PV array and interpolation is used to get the irradiance profile wherever necessary. However, this sensor placement may not be the optimal choice when large irradiance variations exist. Thus, in order to measure the irradiance profile accurately with a minimized system cost, it is necessary to investigate the optimal PV sensor placement. While there is considerable research on optimal sensor placement in applications such as temperature monitoring [12–14], wind distribution measurement for the water quality detection [15], parametric identification of structural systems [16], etc., there is little on optimal irradiance sensor placement. One method is to consider the irradiance time series at each spatial location as another dimension to the attributes of interest [17], then group the neighboring spatial locations with similar temporal characteristics together, and finally place a monitoring sensor at the center of each clustered region. For the clustering problem, various techniques can be used such as *k*-means [18] or the variance Quadtree algorithm [19]. However, these methods require long-enough datasets to train the classifiers, which may not be universally available. In addition, the clustering results depend on the initial centroid selections and the outliers [19], and thus, it is difficult for users to decide the algorithm initial conditions to achieve a given error. In [20], the optimal sensor placement is found by minimizing the distances between each point of the plant and the closest sensor for the PV panels using the genetic algorithm. However, this method is only based on the geographic arrangement but does not account for the irradiance distribution over the area of interest.

Another popular method is to greedily add the placements at each iteration by maximizing the joint entropy of the Gaussian process [21]. However, this entropy criterion suffers from the problem of placing the resulting sensors along the boundary of the space and thus part of the information collected by the sensors at the borders is wasted [22]. A mutual information criterion was applied to the Gaussian process in [13] to reduce the temperature uncertainty at the unsensed locations. The

978-1-5090-5606-4/17 $31.00 © 2017 IEEE

resulting temperature sensor placement for measuring indoor temperature showed superior prediction accuracy by reducing the probability of selecting a suboptimal sensor location compared to the conventional chain-rule of entropies and random placement strategies [13].

In this paper, a mutual information based greedy algo-rithm is proposed to find a near-optimal placement for irradiance sensors. The mutual information, which defines the reduction in the entropy for all the unsensed locations and is defined in Section II.B, is used to quantify how informative the selected set of sensor placements is. By greedily adding a sensor in sequence to the existing sensor set, the optimal sensor placement set is ultimately found according to the mutual information increase (MII). The result shows that the application of greedy algorithm by maximizing MII shows a very low average root mean square error (RMSE) which is very close to the minimum average RMSE, that is, the ideal case obtained using an exhaustive search of all possible sensor combinations. The detailed method is described in the following sections.

II. Problem Statement

In this section, the fundamental optimization problem for the irradiance sensor placement and definition of the sensing quality are introduced.

A. Optimization Problem in Sensor Placement

The sensor placement problem is to find a finite subset of locations A from a set of possible locations V ($V \subset R^2$, where R is the symbol of a set of real numbers.) so that the sensing quality is the highest. More generally, the sensor placement can be standardized as solving the optimization problem of the form

$$\max_{A \subseteq V} F(A)$$
$$\text{Subject to } c(A) \leqslant B_g \quad (1)$$

where $F(A)$ is the sensing quality of the sensor placement A, $c(A)$ is the corresponding cost, and B_g is the budget limit which is larger than zero. This maximization problem aims to find the most informative sensor placement providing the best sensing quality subject to the cost budget limit.

B. Definition of Sensing Quality

To quantify the sensing quality, the irradiance value measured at each location is assumed as a variable with a multivariate Gaussian joint distribution. Considering a sensor network with n possible positions, a set of random variables chosen from the complete set V is defined as $X_S \subseteq X_V$, with location $S \subseteq V$, $S = \{s_j | j = (1, 2, \cdots, n)\}$. For a subset $A \subseteq V$, $X_A \subseteq X_V$ and $X_{V-A} \subseteq X_V$ denote a set of random variables at corresponding sensor locations (A) and unsensed (V-A) positions, respectively. It is assumed that the measured irradiance values at all the locations have a Gaussian distribution. For a set of n random variables X, we have the distribution in the form of

$$P(X = x) = \frac{1}{\sqrt{(2\pi)^n |\Omega|}} e^{-\frac{1}{2}(x-\mu)^T \Omega^{-1}(x-\mu)} \quad (2)$$

where μ and Ω are the mean vector and the positive definite covariance matrix of all the possible sensors, respectively. Thus, for a given set of known sensor locations A, the conditional distribution $P(X_s = x_s | X_A = x_A)$ is also a Gaussian distribution, where the mean $\mu_{s|A}$ and variance $\sigma^2_{s|A}$ are calculated by

$$\mu_{s|A} = \mu_s + \Omega_{sA} \Omega_{AA}^{-1}(x_A - \mu_A) \quad (3)$$

$$\sigma^2_{s|A} = \sigma^2_s - \Omega_{sA} \Omega_{AA}^{-1} \Omega_{sA}^T \quad (4)$$

where Ω_{AA}^{-1} is the inverse of the covariance matrix for the sensed locations, Ω_{sA}^T is the row vector of the covariance of X_s with all the sensed locations X_A, σ^2_s is the variance of X_s, μ_s and μ_A are the means of X_s and X_A, respectively. Thus, after getting the measurements from the set of sensed sensors X_A, the irradiance distribution at any unsensed position can be predicted using Eqs.(3) and (4).

The sensing quality is usually defined as the uncertainty of unsensed locations. In our work, the uncertainty of the sensor placement is given by the mutual information of the Gaussian random variable X_s conditioned on a known set A. It is a monotonic function of its variance and represented as

$$H(V - A|A) = 0.5 log_2((2\pi e)^n |\Omega_{V-A|A}|) \quad (5)$$

where $|\Omega_{V-A|A}|$ denotes the determinant of the conditional covariance matrix $\Omega_{V-A|A}$ which can be calculated from Eq. (4). Thus, when $n = 1$ and $|\Omega_{V-A|A}| = \sigma^2_{S|A}$, the conditional mutual information value for each newly added sensor at location s can be calculated by rewritting Eq.(5) to

$$H(S|A) = 0.5 log_2(2\pi e \sigma^2_{S|A}). \quad (6)$$

Optimizing the mutual information criteria using a lazy greedy algorithm has been proved to be a poly-time algorithm with a constant factor approximation guarantee [14, 15]. The mutual information as the target function is defined as

$$F(A) = I(A; V - A) = H(V - A) - H(V - A|A). \quad (7)$$

This mutual information represents the reduction in the entropy of all unsensed locations (V-A), conditioned by the measurements of the sensed locations A. Thus, the goal is to find a set b out of a finite possible n locations so that the errors in the predicted irradiance for the unsensed locations are minimized. This goal is achieved by maximizing the mutual information in Eq. (7). The maximization problem is given as

$$\varphi(A) = \underset{A \subseteq V : |A| = k}{\arg\max} \{H(V - A) - H(V - A|A)\}. \quad (8)$$

However, as indicated in [23] this maximization problem is NP-complete and it can be reduced to a poly-time algorithm with a constant factor approximation guarantee using the method given in [13]. Thus, Eq.(8) is rewritten as

$$
\begin{aligned}
\varphi(A) &= \underset{A \subseteq V:|A|=k}{\operatorname{argmax}} \{F(A \cup S) - F(A)\} \\
&= \underset{A \subseteq V:|A|=k}{\operatorname{argmax}} \ H(A,S) - H((A,S)|\overline{A \cup S}) \\
&\quad - [H(A) - H(A|\overline{A \cup S} \cup S)] \\
&= \underset{A \subseteq V:|A|=k}{\operatorname{argmax}} \{H(S|A) - H(S|\overline{A \cup S})\}
\end{aligned}
\tag{9}
$$

The covariance matrix is required to calculate the mutual information as in Eq. (7). The method in [26] is used to calculate the covariance matrix. Suppose we have a set of measurements during time $1 \leq t \leq T_0$ from n irradiance sensors, $G_t = (G_t(s_1), G_t(s_2), \cdots, G_t(s_n))^T$, $(t = 1, 2, \cdots, T_0)$, the estimated spatial mean vector for each sensor position can be calculated by

$$
\overline{G}(s_i) = \frac{1}{T_0} \sum_{n=1}^{T_0} G_t(s_i), i = (1, 2, \cdots, n) \tag{10}
$$

Let $C(G(s_1), G(s_2), \cdots, G(s_n))$ be the covariance matrix of all the sensor locations and then it can be estimated by

$$
\begin{aligned}
\Omega &= C(G(s_1), G(s_2), \cdots, G(s_n)) \\
&= \frac{1}{T_0} \sum_{i=1}^{T_0} (G_i - \overline{G})(G_i - \overline{G})^T
\end{aligned}
\tag{11}
$$

where $\overline{G} = (\overline{G}(s_1), \overline{G}(s_2), \cdots, \overline{G}(s_n))^T$. The The mutual information is obtained by substituting Eqs. (4) and (8) into Eq. (9). Subsequently, the optimization problem of Eq. (11) can be solved by the greedy algorithm which will be introduced in Section III.

In fact, there are three ways to calculate the covariance matrix. The first is to use the stationary or nonstationary Gaussian kernel models as indicated in [12] to estimate the covariance matrix. However, the isotropy and stationarity assumptions are usually not suitable for the practical case because it does not take into account the surrounding shading conditions. This inhomogeneity process assumption generally can match with the real data, however it is more suitable for a large area with a large number of known sensors. At least 10 sensors are recommended in [24] for fitting the stationary model locally. For a small area (as in our case), the nonstationary covariance matrix we obtained from the result is negtive-definite, and thus cannot be used due to the requirement of a positive-definite matrix. The second way is to calculate the covariance matrix directly with the measured data from all the sensors using Eq. (10)-(11). This requires that all the sensors are available. However, it is usually not practical to have all the sensors at the initial stage of design. The third way is to calculate the covariance matrix based on the interpolated data. Assume that there are k sensors available in the initial deployment which are randomly placed within the test area. The kriging interpolation technique can then be used to predict the irradiance values at unsensed locations. Thus, the covariance matrix can be easily calculated from Eq. (10)-(11) using the predicted and the sensed datasets. In this case, only

a small number of irradiance sensors are required ($k < n$). In this work, we use the third method.

III. OVERVIEW OF THE PROPOSED METHOD

For a particular number of sensors, a general procedure to find the optimal sensor placement is proposed as follows.
1. *Collect data and calculate the covariance matrix*: We collect the irradiance data from k (k=10) sensors initially deployed in the region of the experimental setup. We then predict the irradiance values at the unsensed locations in the initial deployment using the kriging interpolation technique.
2. *Predict the sensing quality*: After obtaining the covariance matrix, we compute the sensing quality indicated by the mutual information criteria for any set of available sensors.
3. *Find the near-optimal sensor placement*: By greedily adding the next sensor which gives the maximum increase in the mutual information, the near-optimal set of sensed locations, for a particular number of sensors, will be eventually found.
4. *Evaluate the sensor placement*: To evaluate the resulted sensor placement by the mutual information based greedy algorithm, the kriging interpolation technique is then used to predict the irradiance at the unsensed locations.

A. Kriging interpolation

The geostatistical knowledge based kriging technique is a tool for interpolation purposes [25]. In this work, we use the ordinary kriging method, which assumes a constant unknown mean value, to conduct the interpolation for different sensor placement. The basic idea of ordinary kriging is to predict the value of a function at a given point s_0 by computing a weighted combination of the observed values in the neighborhood. In our case, the estimated irradiance at location s_0, assuming there are b available sensors, is written as

$$
\widetilde{G}(s_0) = \sum_{i=1}^{b} \lambda_i G(s_i) \tag{12}
$$

where λ_i is the weight of location s_i, $G(s_i)$ is the observed irradiance value at location s_i, and b is the number of available sensors. Thus, the key point of kriging interpolation is to calculate the weights λ_i. To guarantee an unbiased estimation, firstly the expected error should be zero. That is $E[\widetilde{G}(s_0) - G(s_0)] = 0$, which means that the sum of weights should be equal to one, that is $\sum_{i=1}^{b} \lambda_i = 1$. Secondly, the variance between the estimated irradiance value and the original value should be minimized. This variance is represented by

$$
\begin{aligned}
\sigma^2(s_0) &= E\{[G(s_0) - \widetilde{G}(s_0)]^2\} \\
&= 2 \sum_{i=1}^{b} \lambda_i \gamma(s_i, s_0) - \sum_{i=1}^{b} \sum_{j=1}^{b} \lambda_i \lambda_j \gamma(s_i, s_j)
\end{aligned}
\tag{13}
$$

where $\gamma(s_i, s_j)$ is the semi-variance function between s_i and s_j. The Lagrange multiplier method is then used to find the optimal weights. We define a function $\Psi = \sigma^2(s_0) - 2\rho(\sum_{i=1}^{b} \lambda_i - 1)$, where ρ denotes the Lagrange multiplier. To minimize the covariance, the partial derivative of this function with respect to the weights λ_i and Lagrange multiplier ρ need to be zero. By rearranging the equation, we get

$$
\sum_{j=1}^{b} \lambda_j \gamma(s_i, s_j) + \rho = \gamma(s_i, s_0), \sum_{i=1}^{b} \lambda_i = 1 \tag{14}
$$

978-1-5090-5606-4/17 $31.00 © 2017 IEEE

By rewriting Eq. (14) in matrix form, we have

$$A\lambda = B \qquad (15)$$

$$A = \begin{bmatrix} \gamma(s_1,s_1) & \gamma(s_1,s_2) & \cdots & \gamma(s_1,s_b) & 1 \\ \gamma(s_2,s_1) & \gamma(s_2,s_2) & \cdots & \gamma(s_2,s_b) & 1 \\ \vdots & \vdots & \vdots & \cdots & \vdots \\ \gamma(s_b,s_1) & \gamma(s_b,s_2) & \cdots & \gamma(s_b,s_b) & 1 \\ 1 & 1 & \cdots & 1 & 0 \end{bmatrix},$$

$$\lambda = \begin{bmatrix} \lambda_1, \lambda_2, \ldots, \lambda_b, \rho \end{bmatrix}^T,$$

$$B = \begin{bmatrix} \gamma(s_1,s_0), \gamma(s_2,s_0), \ldots, \gamma(s_b,s_0), 1 \end{bmatrix}^T.$$

Thus, when the inverse matrix of A is calculated, the weights and Lagrange multiplier values can be obtained using $\lambda = A^{-1}B$. Therefore, by substituting the weights in Eq. (14), the irradiance value at an unsensed location can be estimated with a variance of

$$\sigma^2 = \sum_{i=1}^{b} \lambda_i \gamma(s_i,s_0) - \gamma(s_0,s_0) + \rho \qquad (16)$$

In order to evaluate the result of interpolation, for a period of T_1 seconds, the normalized RMSE between the predicted irradiance value $\widetilde{G}(s_i), s_i = (b+1, b+2, \cdots, n)$ and the measured value $G(s_i)$ over the target locations is defined as

$$RMSE(s_i) = \sqrt{\frac{1}{T_1} \sum_{j=1}^{T_1} [\frac{G_j(s_i) - \widetilde{G}_j(s_i)}{G_j(s_i)}]^2} \qquad (17)$$

The average RMSE of all the unsensed locations, which represents quality of the interpolation for the unsensed locations, is calculated by:

$$\overline{RMSE} = \frac{1}{n-b} \sum_{i=b+1}^{n} RMSE(s_i) \qquad (18)$$

B. The mutual information based greedy algorithm

In this section we describe how the greedy algorithm is applied to get the optimal sensor placement. The basic idea of the greedy algorithm is as follows:

1. *Initialization*: Prepare the covariance matrix of the whole space as the input for the greedy algorithm and start the optimization with an empty set for the intial number of sensors. Initialize the MII as a positive infinite number for a sensor set.

2. *Adding a new sensor*: To find the best location to put the new sensor, firstly sort the MII in descending order. This means that it is not necessary to check all sensor locations due to the monotonic characteristics of the mutual information function, thus accelerating the search process. Next, calculate the MII for each newly sorted location and add a new sensor to the position which has the largest MII, as indicated in Eq.(9).

3. *Termination*: Continue to add additional sensors in sequence to the sensor group until the MII values, for all positions, are negative or until the required number of b sensors is reached. Otherwise, go back to step 2. A negative value of MII is considered a termination condition because after this threshold, increasing sensors will not reduce the uncertainty in the irradiance value at the unsensed positions. The detailed

---

Algorithm 1: The greedy algorithm for maximizing the MII.

Input: Irradiance covariance matrix Ω_V.
Output: Optimal sensor positions, $\{p_1, p_2, \cdots, p_b\}$.
Begin A = Empty set;
Initialize the current MI ($MI_{current} = 0$) and MII ($MII = Inf$);
for *i=1 to b* **do**
    Initialize the best MII for the ith step: $B\_MII(i,0) = 0$;
    Sort MII vector;
    for *j=1 to n* **do**
        if $MII(i-1,j) >= B\_MII(i,j-1)$, **then**
            Calculate MII for jth position:
                $MII(i,j) = H(s|A) - H(s|\overline{A \cup S})$;
            Calculate MII: $\Delta MII(i,j) =$
                $MII(i,j) + MI_{current} - B\_\Delta MII(i-1)$;
            Find the best MII: $B\_MII(i,j) =$
                $max(B\_MII(i,j-1), \Delta MII(i,j))$;
        else
            break;
        end
    end
    Find the best MII for jth location:
    $B\_MII(i) = max\{MII(i,j)|j = 1, 2, \cdots, n\}$;
    if $B\_MII(i) > 0$ **then**
        $p_i \leftarrow \underset{A \subseteq V : |A| = k}{\operatorname{argmax}} \{\Delta MII(i,j)|j = 1, 2, \cdots, n\}$;
        Update values for set A and
        $B\_\Delta MII(i) = B\_\Delta MII(i-1) + B\_MII(i)$;
        Update $MI_{current}$ using Eqs.(5) and (7).
    else
        break;
    end
end

---

greedy algorithm for maximizing mutual information is given in Algorithm 1.

Note that in the greedy algorithm, $\Delta MII(i,j)$ is the MII for jth sensor at the ith step, $MII(i,j)$ is individual MII calculated by substituting Eq. (6) into equation $MII(i,j) = H(s|A) - H(s|\overline{A \cup S})$, whereas the MII $\Delta MII(i,j)$ is calculated by $\Delta MII(i,j) = MII(i,j) + MI_{current} - B\_\Delta MII(i-1)$, where $MI_{current}$, calculated by substituting Eq.(5) into Eq. (7), is the mutual information between the unsensed and sensed locations after obtaining the best locations in the last step, and $B\_\Delta MII(i-1)$ is the summation of the best MII at the *(i-1)*th step.

IV. EXPERIMENTAL RESULTS AND DISCUSSION

In order to verify the feasibility of optimal sensor placement using the mutual information criteria based greedy algorithm, we use two experimental datasets measured from different locations and spatial scales. One is an irradiance dataset collected over a small area located at Nanyang Technological University (NTU) in Singapore (103.6829^oE, 1.34653^oN). The irradiance sensor array consists of 16 sensors, with a 4 meter spacing, in a predesignated square matrix, as shown in Fig. 1. The global horizontal irradiance (GHI) values from 8:00am to 6:00pm, with a 1 second sampling interval, are used for this study. The other dataset (also with 1s resolution) is downloaded from the NREL Oahu network [26]. This sensor network, consisting of 17 sensor locations, covers a relatively larger region (about 1km×1.2km), as shown in Fig. 2.

For finding the optimal sensor placement, we randomly choose the NTU data for 3rd Dec 2015 and NREL data for

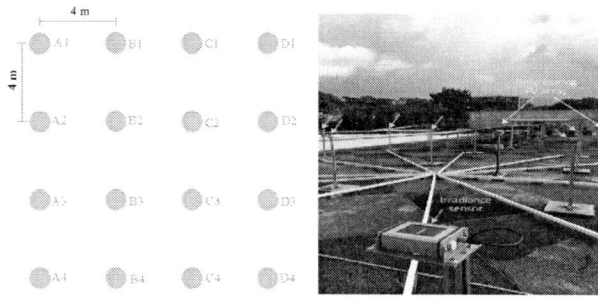

Fig. 1. The irradiance sensor array (Left). The layout of irradiance sensors on rooftop of the NTU (Right).

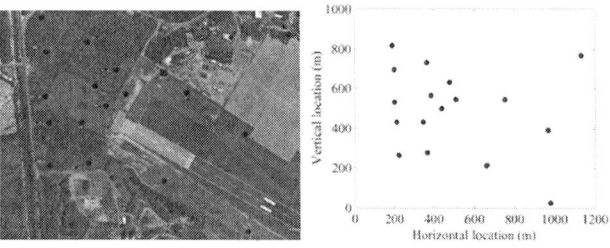

Fig. 2. The 17 irradiance stations at the NREL Oahu Network [26] (Left). The station locations on a converted plane coordinate (Right).

22nd March 2011 to calculate the covariance matrix. After the kriging interpolation process based on the irradiance values measured by the available sensors, the covariance matrix is obtained and input to the greedy algorithm to optimize the sensor locations based on the mutual information criterion in Eq.(9). This is then compared to the optimal placement, for the selected number of sensors, determined from the actual global horizontal insolation (GHI) data for all sensors (16 for NTU and 17 for NREL) using an exhaustive search. At this point, it should be stressed that the GHI data for all sensors would not normally be known *a priori*, and hence cannot be used to assist with determining optimal or near-optimal sensor location. In these calculations, the latitude and longitude coordinates of the sensor locations are converted to the universal transverse Mercator plane coordinate system.

By applying the MII based greedy algorithm to both cases, the best positions for different numbers of sensors are found. Because the mutual information based objective function for the greedy algorithm is only monotonic for sets of size up to $2b$ [12], we only show results up to a maximum 7 available sensors. As an example, the best locations for 5 available sensors, for the two different locations, are shown in Fig. 3. After finding the optimal positions for different numbers of sensors, the ordinary kriging algorithm is used to predict the irradiance values at the unsensed locations. The average RMSE of the unsensed locations is then calculated.

This exhaustive search is conducted by placing all combinations of the different numbers of available sensors at the possible sensor locations and calculating the average RMSE for each of these combinations. For example, using four available sensors in a 16 sensor array, there will be 1820 different

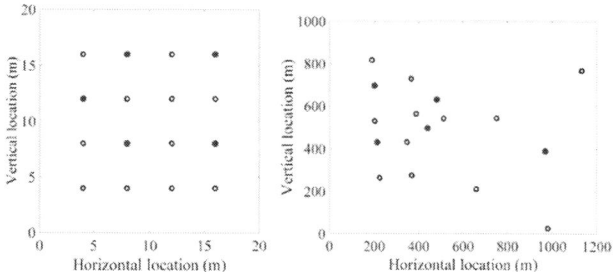

Fig. 3. The best locations of available sensors found using the greedy algorithm based on the mutual information criteria for the test regions at NTU (left) and NREL Oahu (right).

placement cases for 4 sensors ($C_{16}^4 = \frac{16!}{4!(16-4)!} = 1820$). Due to the large amount of data, the irradiance interpolation is subsampled to every 10 mins for the exhaustive search. After calculating the average RMSE values for all sensor permutations, we find both of the best and worst locations which provide minimum and maximum average RMSE of the unsensed locations, respectively. Finally, the minimum and maximum average RMSE are recalculated using GHI data subsampled to a 1 minute sampling interval to give the minimum and maximum limit representing the best and worst possible sensor locations.

Fig.4 and Fig.5 show the relationship between the number of sensors and the average RMSE of the unsensed locations for the NTU dataset and the Oahu network dataset, respectively. In Fig.4 and Fig.5, the upper line represents the sensor choice which gives the worst average RMSE of the unsensed locations and the lower line represents the sensor choice which gives the best average RMSE of the unsensed locations, both determined using an exhaustive search technique. The middle (red) line in both figures is the average RMSE obtained by the proposed method. From Fig.4, we can see that the average RMSE of the unsensed locations using the proposed method is reasonably close to the minimum average RMSE obtained from an exhaustive search, and approaches the optimal sensor location, with a low average RMSE, for 6 or 7 sensors. Similarly, Fig.5 shows the average RMSE of the proposed method and that by the exhaustive search for the dataset collected in NREL Oahu. Here the proposed method is very close to the optimal sensor location for all sensor sets. The difference in the average RMSE between these two locations is expected and is due to the different scale of the measured regions.

V. CONCLUSION

In this work, we propose to use a mutual information criteria to optimize the placement of irradiance sensors for PV systems. By assuming a Gaussian process for the problem, the optimal sensor locations are achieved by greedily adding the next sensor which maximizes the MII until the specified number of sensors are reached or until there is no MII for all the locations. The effectiveness of the proposed method is verified using two different datasets, measured at two different regions with different areas. The experimental results in both cases show that the mutual information based greedy algorithm results in a near optimal sensor placement which

978-1-5090-5606-4/17 $31.00 © 2017 IEEE

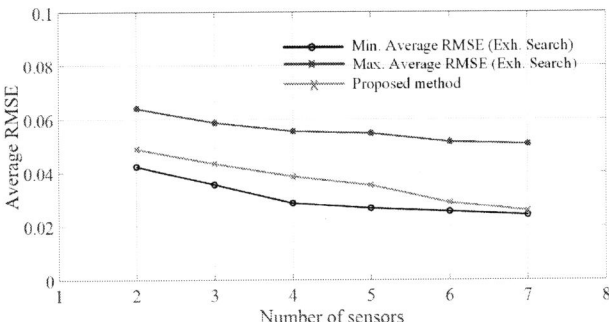

Fig. 4. The average RMSE of the estimated irradiance at unsensed locations using the proposed method and an exhaustive search at NTU.

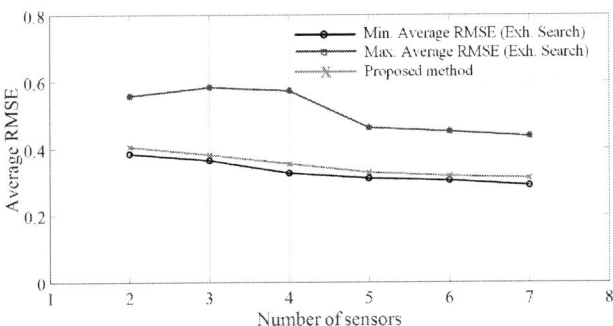

Fig. 5. The average RMSE of the estimated irradiance at unsensed locations using the proposed method and an exhaustive search at NREL Oahu.

provides a low average RMSE close to the minimum average RMSE determined using an exhaustive search. As there is little information in the research literature on how to optimally place irradiance sensors in PV systems, this work provides general guidance for users to wisely locate their available minimum number of sensors so that the interpolation error can be minimized. Our future work include improving the performance of the estimation by examining other methods for calculating the covariance matrix, the influence of the covariance matrix on the final sensor placement and a method to determine the minimum number of sensors required to achieve a predefined error.

ACKNOWLEDGMENT

This research is supported by the Singapore National Research Foundation under NRF2012EWT-EIRP001.

REFERENCES

[1] D. W. McKenney, S. Pelland, Y. Poissant, R. Morris, M. Hutchinson, P. Papadopol, K. Lawrence, and K. Campbell, "Spatial insolation models for photovoltaic energy in canada," *Solar Energy*, vol. 82, no. 11, pp. 1049–1061, 2008.

[2] H. T. Nguyen and J. M. Pearce, "Estimating potential photovoltaic yield with r.sun and the open source geographical resources analysis support system," *Solar Energy*, vol. 84, no. 5, pp. 831–843, 2010.

[3] M. Alasdair and L. Ben, "Utility scale solar power plants: A guide for developers and investors," South asia Department of International Finance Corporation World Bank Group, Report, 2012.

[4] G. Graditi, S. Ferlito, and G. Adinolfi, "Comparison of photovoltaic plant power production prediction methods using a large measured dataset," *Renewable Energy*, vol. 90, pp. 513–519, 2016.

[5] G. Graditi, S. Ferlito, G. Adinolfi, G. M. Tina, and C. Ventura, "Energy yield estimation of thin-film photovoltaic plants by using physical

approach and artificial neural networks," *Solar Energy*, vol. 130, pp. 232–243, 2016.

[6] L. L. Jiang, D. L. Maskell, and J. C. Patra, "Chebyshev functional link neural network-based modeling and experimental verification for photovoltaic arrays," in *IEEE International Joint Conference on Neural Network (IJCNN)*, 2012, Conference Proceedings, pp. 1–8.

[7] L. L. Jiang, D. L. Maskell, S. Rama, and Q. Xu, "Power variability of small scale pv systems caused by shading from passing clouds in tropical region," in *IEEE Photovoltaic Specialists Conference*, 2016.

[8] W. Jewell and R. Ramakumar, "The effects of moving clouds on electric utilities with dispersed photovoltaic generation," *IEEE Transactions on Energy Conversion*, vol. EC-2, no. 4, pp. 570–576, 1987.

[9] E. C. Kern and M. C. Russell, "Spatial and temporal irradiance variations over large array fields," in *The 20th IEEE Photovoltaic Specialists Conference*, vol. 2, 1988, Conference Proceedings, pp. 1043–1050.

[10] L. L. Jiang, D. L. Maskell, and J. C. Patra, "A novel ant colony optimization-based maximum power point tracking for photovoltaic systems under partially shaded conditions," *Energy and Buildings*, vol. 58, pp. 227–236, 2013.

[11] L. L. Jiang, D. R. Nayanasiri, D. L. Maskell, and D. M. Vilathgamuwa, "A hybrid maximum power point tracking for partially shaded photovoltaic systems in the tropics," *Renewable Energy*, vol. 76, pp. 53–65, 2015.

[12] A. Krause, A. Singh, and C. Guestrin, "Near-optimal sensor placements in gaussian processes: theory, efficient algorithms and empirical studies," *The Journal of Machine Learning Research*, vol. 9, pp. 235–284, 2008.

[13] C. Guestrin, A. Krause, and A. Singh, "Near-optimal sensor placements in gaussian processes," in *The 22nd International Conference on Machine Learning (ICML)*, 2005, pp. 265–272.

[14] R. Jedermann and W. Lang, *The minimum number of sensors interpolation of spatial temperature profiles in chilled transports*. Springer Berlin Heidelberg, 2009, vol. 5432, pp. 232–246.

[15] W. Du, Z. Xing, M. Li, B. He, L. H. C. Chua, and H. Miao, "Optimal sensor placement and measurement of wind for water quality studies in urban reservoirs," in *Proceedings of the 13th international symposium on Information processing in sensor networks*, 2014, pp. 167–178.

[16] C. Papadimitriou, "Optimal sensor placement methodology for parametric identification of structural systems," *Journal of Sound and Vibration*, vol. 278, no. 45, pp. 923–947, 2004.

[17] D. Yang and T. Reindl, "Solar irradiance monitoring network design using the variance quadtree algorithm," *Renewables: Wind, Water, and Solar*, vol. 2, no. 1, pp. 1–8, 2015.

[18] A. Zagouras, R. H. Inman, and C. F. M. Coimbra, "On the determination of coherent solar microclimates for utility planning and operations," *Solar Energy*, vol. 102, pp. 173–188, 2014.

[19] B. Minasny, A. B. McBratney, and D. J. J. Walvoort, "The variance quadtree algorithm: Use for spatial sampling design," *Computers & Geosciences*, vol. 33, no. 3, pp. 383–392, 2007.

[20] M. Pau, N. Locci, and C. Muscas, "A tool to define the position and the number of irradiance sensors in large pv plants," in *IEEE International Energy Conference*, 2014, pp. 374–379.

[21] M. D. McKay, R. J. Beckman, and W. J. Conover, "A comparison of three methods for selecting values of input variables in the analysis of output from a computer code," *Technometrics*, vol. 21, no. 2, pp. 239–245, 1979.

[22] N. Ramakrishnan, C. Bailey-Kellogg, S. Tadepalli, and V. N. Pandey, *Gaussian processes for active data mining of spatial aggregates*, 2005, pp. 427–438.

[23] C. W. Ko, J. Lee, and M. Queyranne, "An exact algorithm for maximum entropy sampling," *Operations Research*, vol. 43, no. 4, pp. 684–691, 1995.

[24] D. J. Nott and W. T. M. Dunsmuir, "Estimation of nonstationary spatial covariance structure," *Biometrika*, vol. 89, no. 4, pp. 819–829, 2002.

[25] S. H. Monger, E. R. Morgan, A. R. Dyreson, and T. L. Acker, "Applying the kriging method to predicting irradiance variability at a potential pv power plant," *Renewable Energy*, vol. 86, pp. 602–610, 2016.

[26] M. Sengupta and A. Andreas, "Oahu solar measurement grid (1year archive)," National Renewable Energy Laboratory Technical Report, Report, 2010.

Evaluating different upscaling approaches to derive the actual power of distributed PV systems

Sven Killinger[*†], Björn Müller[*], Bernhard Wille-Haussmann[*], Russell McKenna[†]

[*]Fraunhofer Institute for Solar Energy Systems ISE, 79110 Freiburg, Germany

[†]Chair for Energy Economics, Karlsruhe Institute of Technology (KIT), 76187 Karlsruhe, Germany

Abstract—The power generated by distributed PV systems within a given region is an important information e.g. for grid operators. But PV power measurements are often only available for a small fraction of systems. Hence, various upscaling approaches estimate the power of a large number of unmeasured photovoltaic systems based on these measured reference units. This paper evaluates the application of different methods for upscaling PV power. Results show overall improvements up to $\approx 13\%$ in terms of $RMSE_{rel}$, mostly by the consideration of the module orientation and a spatial interpolation. Additionally a significant decline in the variance of MBE_{rel} is encountered due to a calibration of all PV systems. The evaluation is extended by an economic assessment, which assumes a quadratic relation between the simulation error and the costs. When compared to a standard approach, the presented methods reach cost savings of $15\text{-}25\%$ on average.

I. Introduction

A precise knowledge of the power generated by PV systems is important for grid operators to reduce the amount of balancing power and ensure grid stability. Various approaches exist which either estimate the current state (nowcasting) or predict a future value (forecasting) by applying physical or statistical techniques or a combination of both. A good overview about that topic is provided in [1], [2], [3], [4]. Upscaling is one technique within this context and typically employs information from a subset of systems to define the total quantity of regional PV power. The analyzed methods are based on (multiple) data sources such as measured reference units [5], [6], [7], [8], [9], [10], [11], [12], [13], [14], irradiance from numerical weather models [10], [11], [12] and satellite pictures [10], [12], [13], [15], [16]. Some of these approaches take the module orientation into account by deriving it either from a database [7], [8], [10], [11], applying statistical assumptions [17] or parameterizing azimuth and inclination angles with LiDAR data [9], [16].

Even though the literature review reveals a huge variety of approaches, a comprehensive assessment of different methodologies for upscaling PV power is missing. Apart from typical error metrics, it is often unclear what the economic benefits of such approaches are. Hence, the objective of this paper is to present and evaluate a holistic approach in order to quantify the statistical and economic value of several (optional) advancements.

The paper is organized as follows: the methodology and scenarios are introduced in section II together with an approach to quantify the economic importance of upscaling. In section III the different scenarios are evaluated within a statistical analysis and economic benefits assessed. In addition to that, the whole evaluation is critically discussed and future advancements presented. The paper concludes with a short summary and outlook in section IV.

II. Methodology

A. Definition of Scenarios

The upscaling approach (UA) presented in this paper is mainly based on previous work done by the authors [9], [18], [19], [20] and combines several independent methodologies. In order to evaluate their interaction, different scenarios are defined within this section and are displayed in Table I. All power values within this contribution are normalized by their installed capacity.

A standard approach "STD" applies the methodology currently used by many grid operators and serves as a benchmark. It takes the power of reference units P_{Ref}, in order to estimate the power of the target units P_{Tar}. No spatial interpolation is conducted in this scenario, but in case multiple reference systems are used, their average value is taken.

In contrast to "STD", the upscaling approach of the authors applies the power projection algorithm presented in [18],

TABLE I: Upscaling methodologies differing in the spatial interpolation, consideration of the module orientation and calibration. Methods are as follows: IDW is the inverse distance weighting method with an exponent of 2.0 and LF a loss factor.

Scenario	Methodology	Spatial interpolation	Module orientation	Calibration reference system	Calibration target system
STD	$P_{Ref} \rightarrow P_{Tar}$	Average	-	-	-
UA1	$P_{Ref} \rightarrow G_h \rightarrow P_{Tar}$	Average	From ref. system	-	-
UA2	$P_{Ref} \rightarrow G_h \rightarrow P_{Tar}$	IDW 2.0	From ref. system	-	-
UA3	$P_{Ref} \rightarrow G_h \rightarrow P_{Tar}$	Average	From tar. system	-	-
UA4	$P_{Ref} \rightarrow G_h \rightarrow P_{Tar}$	IDW 2.0	From tar. system	-	-
UA5	$P_{Ref} \rightarrow G_h \rightarrow P_{Tar}$	IDW 2.0	From tar. system	with LF	-
UA6	$P_{Ref} \rightarrow G_h \rightarrow P_{Tar}$	IDW 2.0	From tar. system	with LF	with LF

which derives the global horizontal irradiance G_h as an intermediate value. Although the power projection is designed to consider the module orientation, only the information provided from the reference systems is used in the upscaling approaches "UA1" and "UA2", in order to independently evaluate the influence of the spatial interpolation. "UA3" and "UA4" then consider the correct module orientation of each target system individually and again compare the different spatial interpolation scenarios. Similar to the definition in "STD", "average" takes the mean value of G_h derived from all reference units, whereas "IDW 2.0" is the inverse distance weighting method presented in [21] with an exponent of 2. In addition to that, a calibration by a loss factor LF [19], [20] is applied only for the reference systems in "UA5" and for all systems in "UA6".

B. Economic Assessment of Upscaling

Weather forecasts have a strong impact on the economy and [22], [23] assume that the benefits of such forecasts are 20 times higher than their costs. Power forecasts for both wind and solar power strongly depend on such weather forecasts and increase the overall economic value [24], [25] of both technologies by reducing the uncertainty in the predicted generation. Various publications discuss the economic value in the context of power forecasting. Since the upscaling methods presented in this paper can be used for both nowcasting and forecasting, assumptions from the literature can be used to frame the assessment in this section. While [1], [26], [27], [28], [29], [30], [31] focus on solar energy, [32], [33], [34], [35], [36], [37], [38], [39] describe the benefits of forecasting in the context of wind energy and [40], [41], [42], [43], [44], [45] for both technologies. The general approaches of these publications are similar, and either based on empirical analyses or unit-commitment models.

However, the analysis of this work is restricted to a small portfolio of 45 PV systems in southern Germany (see section III). Using a unit-commitment model for such a regional evaluation adds additional uncertainty and seems to be only of limited use. In fact, general evaluation criteria are needed and an evaluation of empirical analyses with a focus on the German market is suitable to do define them.

In Germany and many other European countries, forecast errors on the day-ahead market can be compensated through the intraday market or with balancing power. Hence, economic benefits can be quantified by differences between the prices on the day-ahead and intraday market as well as the day-ahead market and balancing power. [44] shows that the difference between the prices of the intraday (p_{ID}) and day-ahead market (p_{DA}) as well as the forecast error x_{ID} have a correlation of 0.58[1]. As a consequence, the compensating trade on the intraday market leads to additional costs. Based on a linear regression and a factor describing the gradient of the prices g_{ID}, p_{ID} can be expressed by:

$$p_{ID} = p_{DA} + g_{ID} \cdot x_{ID}. \tag{1}$$

[1]The correlation might be limited due to further uncertainties such as forecasting the load [26], [27].

The overall costs C_{ID} due to this trade depend on the price difference as well as the forecast error and are defined by:

$$C_{ID} = x_{ID} \cdot (p_{ID} - p_{DA}). \tag{2}$$

Combining (1) and (2) shows a quadratic relation between the costs and the forecast error, also reported by [34], [35]:

$$C_{ID} = x_{ID}^2 \cdot g_{ID}. \tag{3}$$

Analyzing the correlation between the difference in the prices of balancing power and the day-ahead market as well as the forecast error does not show such a strong link. However, [44], [34], [35], [42] conclude that costs for balancing power are much higher than for trading on the intraday market and [42] approximates them by a quadratic model.

Even though such correlations with market prices are not found for the regional portfolio in this paper, the presented methods can easily be applied to a larger portfolio. Hence, using a quadratic relation between the simulation error and costs[2] seems to be a realistic and maybe even conservative assumption to assess the potential economic benefits.

III. CROSS-VALIDATION AND DISCUSSION

The cross-validation in this section is conducted for 5 years of data between 2010 and 2014 from 45 PV systems in the region of Freiburg, Germany, with a temporal resolution of 5 minutes. The module orientation, defined by the tilt and azimuth angle, is known for these systems. Two different evaluation schemes (ES) are defined:

- **ES1:** Similar to the analysis of [17], a fixed number of 30 systems is drawn randomly from the 45 systems and treated as target systems. Subsequently, a number between 1 and 15 system(s) is drawn from the remaining ones and used as reference system(s). For each of these scenarios 150 variations for the selection of reference and target systems are randomly sampled. Then, the average normalized power values of the target systems P_{Tar} from the different scenarios in Table I are compared to the measurements.
- **ES2:** In a second evaluation, the leave-one-out cross-validation is applied. Within this validation 44 reference systems are used to estimate the power of one target system. Hence, 45 variations are possible.

A. Statistical Analysis

Fig. 1 shows the relative root-mean-square error $RMSE_{rel}$ (top) and relative mean bias error MBE_{rel} (bottom) of P_{Tar} for a varying number of reference systems. Every boxplot visualizes the range of 150 randomly chosen variations.

In general, aggregation effects due to an increasing number of reference systems significantly improve the results of all scenarios. All the scenarios considering the module orientation outperform the standard approach. As expected, the spatial interpolation is more important with a higher number of

[2][42], [28] discuss whether there is an exponential relation.

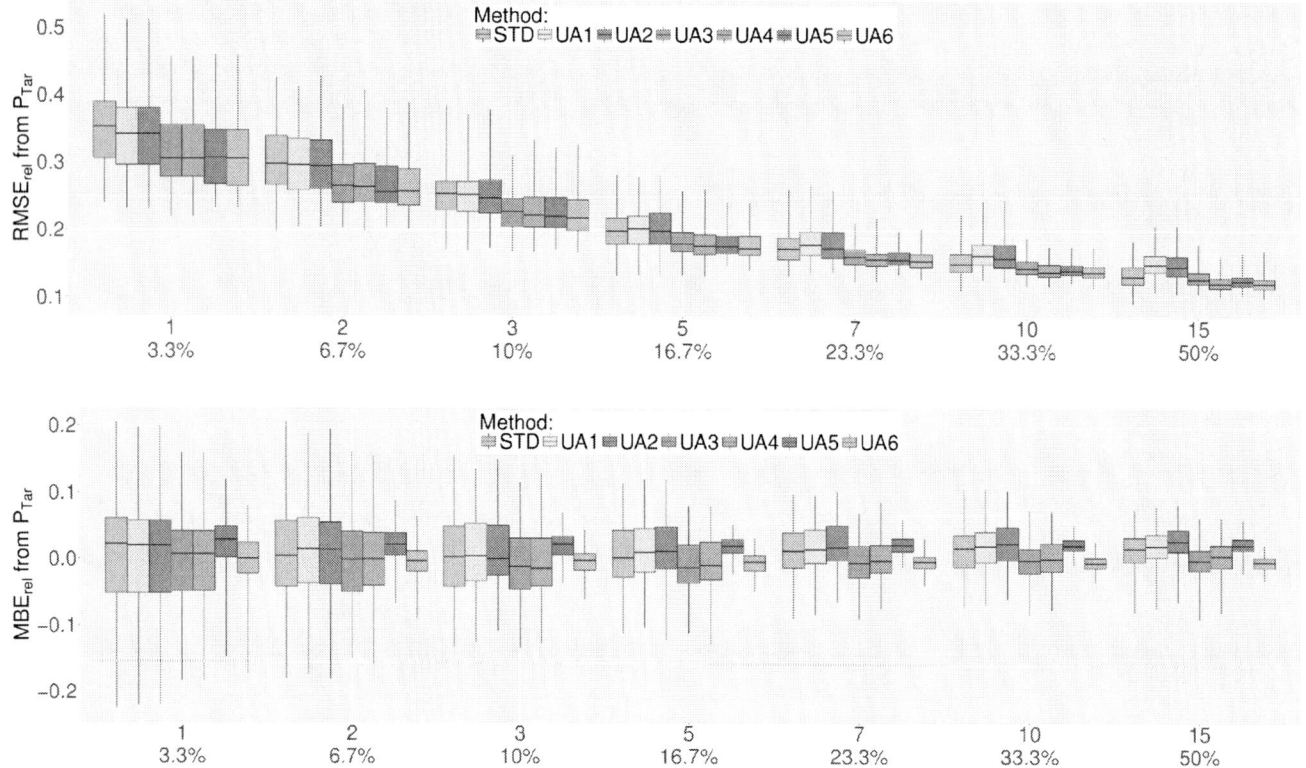

Fig. 1: Relative root-mean-square error $RMSE_{rel}$ (top) and relative mean bias error MBE_{rel} (bottom) of P_{Tar} for different methods of Table I as well as a varying number of reference systems and their relative share compared to the number of target systems as defined in "ES1". The boxplots for the different scenarios depict the 25 % and 75 % quantiles at the bottom and top edges of the box and the median as the band inside the box. The ends of the whiskers represent the minimum and maximum of the results.

reference systems and the "IDW 2.0" shows improvements if 5 or more reference systems are used, as can be seen by comparing "UA1" and "UA2" or "UA3" and "UA4".

Using G_h as an intermediate without considering the module orientation in "UA1" and "UA2" slightly worsens the $RMSE_{rel}$ values due to the the complex simulation chain involved in this procedure [18]. However, the difference is not large and in order to improve it the simulated G_h could be combined with G_h derived from numeric weather models, satellite pictures or ground measurements.

Calibrating the PV systems shows a small improvement for the $RMSE_{rel}$ and a significant reduction in the variance of MBE_{rel}. Whereas calibrating only the reference systems leads to a positive MBE_{rel}, calibrating reference and target systems leads to a median of MBE_{rel} values around 0. In general, a reduction in the variance of MBE_{rel} can be observed with each of the three presented advancements (considering the module orientation, applying a spatial interpolation and calibrating PV systems) in the upscaling approach between "UA1" and "UA6".

Fig. 2 compares the performance of the different scenarios, when 44 reference systems are taken to estimate the power of one target system. In general, the trends observed from

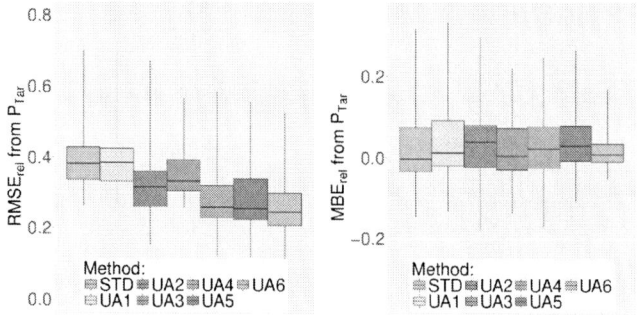

Fig. 2: Relative root-mean-square error $RMSE_{rel}$ (left) and mean bias error MBE_{rel} (right) of P_{Tar} for 44 reference systems predicting one target system as defined in "ES2". The different methods are defined in Table I.

the previous evaluation continue but differences are more strongly pronounced. As expected, the $RMSE_{rel}$ values are significantly higher for a single system, but MBE_{rel} stays in a similar range compared to Fig. 1.

Again, "STD" and "UA1" both lead to similar results. How-

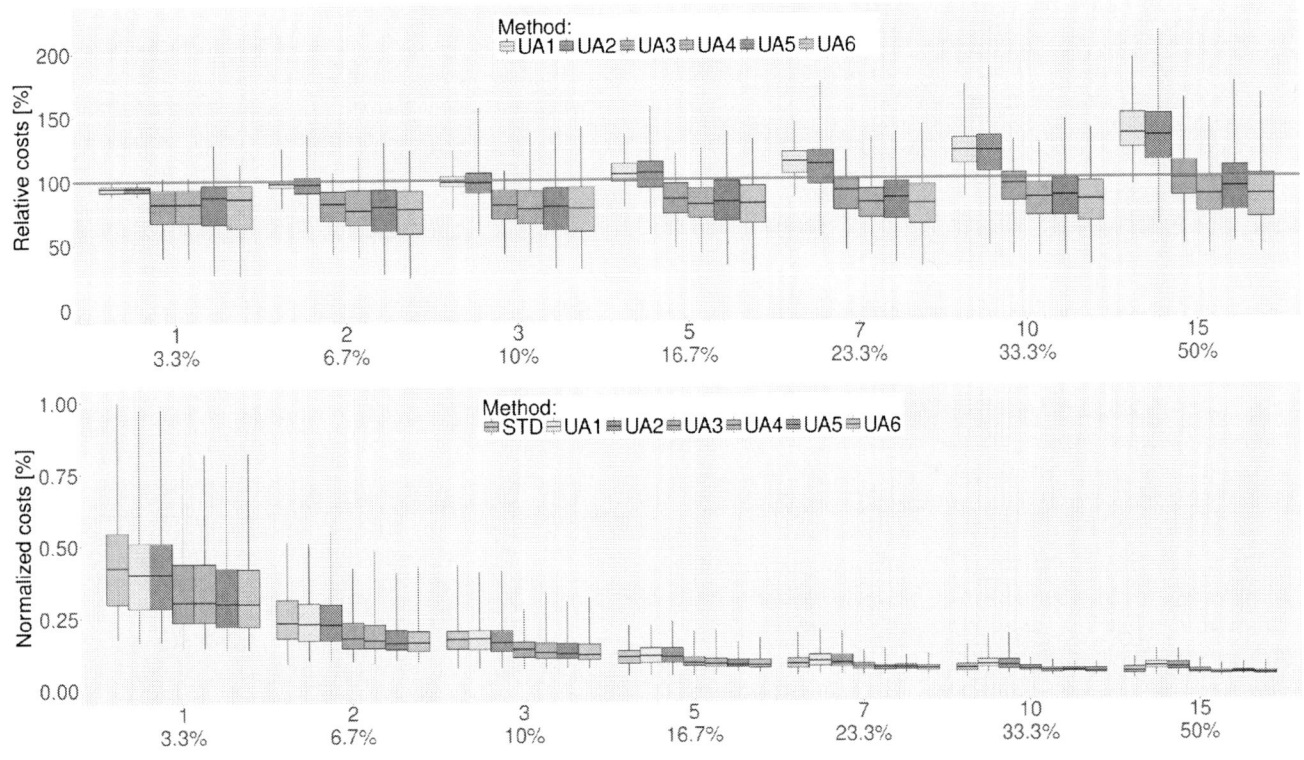

Fig. 3: Relative (top) and normalized (bottom) costs determined by the quadratic error for different scenarios given in Table I as well as a varying number of reference systems as defined in the "ES1".

ever, the additional value due to a spatial interpolation with "IDW 2.0" is tremendously improved due to the high number of reference systems, as can be seen in comparison of "UA1" and "UA2" or "UA3" and "UA4". Both the consideration of the module orientation and the calibration by LF further improve the error metrics.

The significant reduction in the variance of MBE_{rel} due to a calibration of reference and target systems in "UA6" strongly reduces the uncertainty in the upscaling approach and emphasizes its importance. In real applications where no measurements of target systems are known to determine LF, other information such as the annual yield of the systems can be employed for an individual calibration.

In summary, results are promising and show improvements due to the scenarios up to $\approx 13\%$ in Fig. 2 and $\approx 6\%$ in Fig. 1 in terms of the medians of the $RMSE_{rel}$ values.

B. Assessment of the Economic Value

Based on the assumptions in section II-B, costs are determined by the quadratic error and both terms are used as synonyms within this paper. Costs are calculated by the "ES1" and are visualized for all scenarios in Fig. 3. The graphic on top shows the costs in relation to the standard approach, which is displayed by the red 100%-line. The graphic on the bottom shows the absolute value of the costs, normalized with respect to the highest occurring value. Hence, both figures are independent of the gradient g_{ID} in (3).

Especially the decrease in the normalized costs due to an increasing number of reference systems is significant in Fig. 3 (bottom). For example the costs in "UA6" are 8 times higher, if only one instead of 15 reference systems is available. Similar to the $RMSE_{rel}$ in Fig. 1 and observations in [46], [47], the marginal use of each additional reference systems reduces, but absolute differences are more strongly pronounced due to the missing root in the quadratic error. In summary, the number of reference systems might be an inexpensive and very effective approach to improve upscaling approaches, especially if their overall number is low.

In contrast, the relative costs stay on a similar level for an increasing number of reference systems within each scenario. The scenarios "UA1" and "UA2", which do not take the module orientation into consideration, cannot be recommended and are more expensive than "STD", especially for a higher number of reference systems. However, in many variations "UA3-6" stay significantly under the costs of "STD". The consideration of the module orientation in "UA3-6", the spatial interpolation in "UA4-6", as well as the additional calibration of the reference and target systems in "UA6", are apparently valuable features. In summary, the scenarios "UA4-6" show a relative decrease in the costs of 15-25 % when compared to "STD" and the advantages of the presented (optional) methods are almost independent from the number of reference systems.

Assuming average costs of 3 €/MWh [44] and an annual energy production of 37.53 TWh from all PV systems in

Germany [48], the overall economic benefits account for $\approx$16.9-28.1 million €/a. Even though this is only a strong simplification, it helps to estimate the range of potential economic savings due to the presented methods.

C. Critical Reflections on the Presented Approach

Within this section, a critical reflection discusses the consideration of the module orientation, the calibration, potential improvements due to a quality control as well as the transferability of the results to other regions.

The module orientation of the target systems in "UA3-6" might not be known in real applications. Hence, [49], [9], [50] present GIS-based approaches which allow a parametrization of the inclination and azimuth angles. Since a slight error can be expected by these approaches, the improvements being achieved due to the consideration of the module orientation in "UA3-6" can be seen as an upper limitation.

Similarly, the calibration of reference and target systems is conducted with the approach of [19] requiring the measured PV power as an input. However, in real applications this information might not be available for most target systems. Hence, other approaches are needed for a calibration, e.g. by using the annual energy generation given in registers such as [51]. Since individual differences in real PV systems are expected to be much higher than in the systems used in this paper, potential improvements might be even higher even though less information can be used for a calibration.

In [19] a routine to quality control power measurements from PV systems (QCPV routine) is presented. Since upscaling approaches are based on a minority of reference systems, these systems have a strong influence on the estimation of the regional PV power. Hence, an automated quality control which decides if a system is appropriate to be used as a reference unit is expected to reach significant improvements. Before the QCPV routine is applied to an upscaling approach, it must be clarified how to differentiate automatically between measurement errors and atypical behavior which is not representative for a regional portfolio. Furthermore it must be discussed how to prevent a bias such as in "UA5", if only reference systems are quality controlled. These topics as well as the compensation of systematic influences by a tuning approach will be the focus of future work.

Even though many scenarios and variations are evaluated within this paper, it remains unclear if results will be similar in other regions. It can be stated that:

- Results are very sensitive to the choice, number and spatial distribution of reference and target systems. This might be similar in other regions too.
- It is expected that potential improvements due to a spatial interpolation, calibration and quality control might be higher in a real application and larger portfolio.
- If reference systems cover different module orientations and are similar to the target systems, improvements due to their consideration might be small. However, if systematic differences occur such as in [9], the consideration of the module orientation is recommended.

- Potential improvements must be discussed for each individual case. A simulation which allows an analysis for different temporal and spatial aggregations can support such an evaluation and will be the topic of future work.

IV. SUMMARY AND OUTLOOK

This paper evaluates the additional value of considering the module orientation, applying a spatial interpolation and calibrating PV systems in the context of upscaling approaches. The impact due to these advancements strongly depends on the choice, number and spatial distribution of reference and target systems. In general, considering the module orientation and applying a spatial interpolation achieves the highest improvements in terms of $RMSE_{rel}$ of up to $\approx 13\%$, whereas a calibration significantly reduces the MBE_{rel} to one third of the range between the 25% and 75% quantiles of a standard approach.

Additionally, the economic value due to these improvements is assessed by assuming a quadratic relation between the simulation error and the costs. When compared to a standard approach, the presented methods promise average savings of 15-25%. Another cost effective approach might be to use a higher number of reference systems.

Future work is needed to assess further methodologies such as a quality control in the context of upscaling approaches, derive recommendations for grid operators and evaluate the role of the temporal and spatial resolution.

V. ACKNOWLEDGMENT

The first author would like to thank the Nagelschneider Foundation for partially supporting this work. Furthermore the authors thank their colleagues of the Department Quality Assurance PV Modules and Power Plants at Fraunhofer ISE for their support and provision of monitoring data.

REFERENCES

[1] J. Antonanzas, N. Osorio, R. Escobar, R. Urraca, F. J. Martinez-de Pison, and F. Antonanzas-Torres, "Review of photovoltaic power forecasting," *Solar Energy*, vol. 136, pp. 78–111, 2016.

[2] R. H. Inman, H. T. Pedro, and C. F. Coimbra, "Solar forecasting methods for renewable energy integration," *Progress in Energy and Combustion Science*, vol. 39, no. 6, pp. 535–576, 2013.

[3] M. Paulescu, E. Paulescu, P. Gravila, and V. Badescu, *Weather modeling and forecasting of PV systems operation*, ser. Green Energy and Technology. Dordrecht: Springer, 2012.

[4] M. Q. Raza, M. Nadarajah, and C. Ekanayake, "On recent advances in PV output power forecast," *Solar Energy*, vol. 136, pp. 125–144, 2016.

[5] M. T. Beck, de Meer, Hermann de, S. Schuster, and M. Kreuzer, Eds., *Energy-efficient data centers: Estimating photovoltaic power supply without smart metering infrastructure.* Berlin, Heidelberg: Springer-Verlag, 2014.

[6] R. J. Bessa, "Solar power forecasting for smart grids considering ICT constraints," in *4th Solar Integration Workshop: International Workshop on Integration of Solar Power into Power*, Berlin, Germany, 2014.

[7] N. A. Engerer, "City-wide simulations of distributed photovoltaic array power output," PhD Thesis, The Australian National University, Canberra, Australia, 2015.

[8] A. Golnas, J. Bryan, R. Wimbrow, C. Hansen, and S. Voss, "Performance assessment without pyranometers: Predicting energy output based on historical correlation," in *37th IEEE Photovoltaic Specialists Conference (PVSC)*, Seattle, USA, 2011, pp. 2006–2010.

978-1-5090-5606-4/17 $31.00 © 2017 IEEE

[9] S. Killinger, P. Guthke, A. Semmig, B. Müller, B. Wille-Haussmann, and W. Fichtner, "Upscaling PV power considering module orientations," *IEEE Journal of Photovoltaics*, 2017.

[10] J. Kühnert, "Development of a photovoltaic power prediction system for forecast horizons of several hours," PhD Thesis, Universität Oldenburg, Oldenburg, Germany, 2016.

[11] E. Lorenz, T. Scheidsteger, J. Hurka, D. Heinemann, and C. Kurz, "Regional PV power prediction for improved grid integration," *Progress in Photovoltaics: Research and Applications*, vol. 19, no. 7, pp. 757–771, 2011.

[12] V. P. Lonij, A. E. Brooks, A. D. Cronin, M. Leuthold, and K. Koch, "Intra-hour forecasts of solar power production using measurements from a network of irradiance sensors," *Solar Energy*, vol. 97, pp. 58–66, 2013.

[13] Y.-M. Saint-Drenan, S. Bofinger, B. Ernst, T. Landgraf, and K. Rohrig, "Regional nowcasting of the solar power production with PV-plant measurements and satellite images," in *30th ISES Solar World Congress*, Kassel, Germany, 2011.

[14] S. Schierenbeck, D. R. Graeber, A. Semmig, and A. Weber, "Ein distanzbasiertes Hochrechnungsverfahren für die Einspeisung aus Photovoltaik," *Energiewirtschaftliche Tagesfragen*, vol. 60, no. 12, pp. 60–64, 2010.

[15] L. Grossi, G. Wirth, E. Lorenz, A. Spring, and G. Becker, "Simulation of the feed-in power of distributed PV systems," in *29th European PV Solar Energy Conference and Exhibition*, Amsterdam, Netherlands, 2014.

[16] H. Ruf, "Computation of the load flow at the transformer in distribution grids with a significant number of photovoltaic systems using satellite-derived solar irradiance data," Ph.D. dissertation, University of Agder, Kristiansand, Norway, 2016.

[17] Y. M. Saint-Drenan, G. H. Good, M. Braun, and T. Freisinger, "Analysis of the uncertainty in the estimates of regional PV power generation evaluated with the upscaling method," *Solar Energy*, vol. 135, pp. 536–550, 2016.

[18] S. Killinger, F. Braam, B. Müller, B. Wille-Haussmann, and R. McKenna, "Projection of power generation between differently-oriented PV systems," *Solar Energy*, vol. 136, pp. 153–165, 2016.

[19] S. Killinger, N. Engerer, and B. Müller, "QCPV: A quality control algorithm for distributed photovoltaic array power output," *Solar Energy*, vol. 143, pp. 120–131, 2017.

[20] S. Killinger, N. A. Engerer, and B. Müller, "Identification of typical quality control issues in distributed PV power measurements," in *Asia Pacific Solar Research Conference 2016*, Canberra, Australia, 2016.

[21] D. Shepard, "A two-dimensional interpolation function for irregularly-spaced data," in *23rd ACM National Conference*, Las Vegas, USA, 1968, pp. 517–524.

[22] B. J. Mason, "The role of meteorology in the national economy," *Weather*, vol. 21, no. 11, pp. 382–393, 1966.

[23] T. N. Palmer, "The economic value of ensemble forecasts as a tool for risk assessment: From days to decades," *Quarterly Journal of the Royal Meteorological Society*, vol. 128, no. 581, pp. 747–774, 2002.

[24] E. Baker, M. Fowlie, D. Lemoine, and S. S. Reynolds, "The economics of solar electricity," *Annual Review of Resource Economics*, vol. 5, no. 1, pp. 387–426, 2013.

[25] T. Jónsson, P. Pinson, and H. Madsen, "On the market impact of wind energy forecasts," *Energy Economics*, vol. 32, no. 2, pp. 313–320, 2010.

[26] C. Brancucci Martínez-Anido, A. Florita, and B.-M. Hodge, "The impact of improved solar forecasts on bulk power system operations in ISO-NE," in *4th Solar Integration Workshop: International Workshop on Integration of Solar Power into Power*, Berlin, Germany, 2014.

[27] C. Brancucci Martinez-Anido, B. Botor, A. R. Florita, C. Draxl, S. Lu, H. F. Hamann, and B.-M. Hodge, "The value of day-ahead solar power forecasting improvement," *Solar Energy*, vol. 129, pp. 192–203, 2016.

[28] S. A. Fatemi and A. Kuh, "Solar radiation forecasting under asymmetric cost functions," in *International Joint Conference on Neural Networks (IJCNN)*, Beijing, China, 2014.

[29] A. Kaur, L. Nonnenmacher, H. T. Pedro, and C. F. Coimbra, "Benefits of solar forecasting for energy imbalance markets," *Renewable Energy*, vol. 86, pp. 819–830, 2016.

[30] A. Mills, A. Botterud, J. Wu, Z. Zhou, B.-M. Hodge, and M. Heaney, "Integrating solar PV in utility system operations," Argonne, USA, 2013. [Online]. Available: https://emp.lbl.gov/sites/all/files/lbnl-6525e.pdf

[31] R. Perez, T. Hoff, J. Dise, D. Chalmers, and S. Kivalov, "The cost of mitigating short-term PV output variability," in *ISES Solar World Congress*, vol. 57, Cancún, Mexico, 2013, pp. 755–762.

[32] R. J. Barthelmie, F. Murray, and S. C. Pryor, "The economic benefit of short-term forecasting for wind energy in the UK electricity market," *Energy Policy*, vol. 36, no. 5, pp. 1687–1696, 2008.

[33] A. Fabbri, T. GomezSanRoman, J. RivierAbbad, and V. H. MendezQuezada, "Assessment of the cost associated with wind generation prediction errors in a liberalized electricity market," *IEEE Transactions on Power Systems*, vol. 20, no. 3, pp. 1440–1446, 2005.

[34] D. Graeber, *Handel mit Strom aus erneuerbaren Energien: Kombination von Prognosen.* Springer Gabler, 2013.

[35] D. Graeber and A. Kleine, "The combination of forecasts in the trading of electricity from renewable energy sources," *Journal of Business Economics*, vol. 83, no. 5, pp. 409–435, 2013.

[36] B. M. Hodge, A. Florita, J. Sharp, M. Margulis, and D. Mcreavy, "Value of improved short-term wind power forecasting: NREL/TP-5D00-63175," Golden, USA, 2015. [Online]. Available: http://www.nrel.gov/docs/fy15osti/63175.pdf

[37] D. Lew, M. Miligan, G. Jordan, and R. Piwko, "The value of wind power forecasting," in *91st American Meteorological Society Annual Meeting, the Second Conference on Weather, Climate, and the New Energy Economy*, Washington D.C., USA, 2011.

[38] M. R. Milligan, A. H. Miller, and F. Chapman, "Estimating the economic value of wind forecasting to utilities," in *Windpower '95*, Washington D.C., USA, 1995.

[39] G. P. Swinand and A. O'Mahoney, "Estimating the impact of wind generation and wind forecast errors on energy prices and costs in Ireland," *Renewable Energy*, vol. 75, pp. 468–473, 2015.

[40] L. Hirth and I. Ziegenhagen, "Balancing power and variable renewables: Three links," *Renewable and Sustainable Energy Reviews*, vol. 50, pp. 1035–1051, 2015.

[41] L. Hirth, F. Ueckerdt, and O. Edenhofer, "Integration costs revisited – An economic framework for wind and solar variability," *Renewable Energy*, vol. 74, pp. 925–939, 2015.

[42] J. Mueller, M. Hildmann, A. Ulbig, and G. Andersson, "Grid integration costs of fluctuating renewable energy sources," in *IEEE Conference on Technologies for Sustainability (SusTech)*, Portland, USA, 2014.

[43] S. von Roon, "Empirische Analyse über die Kosten des Ausgleichs von Prognosefehlern der Wind- und PV-Stromerzeugung," in *7. Internationalen Energiewirtschaftstagung (IEWT)*, Vienna, Austria, 2011.

[44] ——, "Auswirkungen von Prognosefehlern auf die Vermarktung von Windstrom," Ph.D. dissertation, Technische Universität München, München, 2012.

[45] F. Ueckerdt, L. Hirth, G. Luderer, and O. Edenhofer, "System LCOE: What are the costs of variable renewables?" *Energy*, vol. 63, pp. 61–75, 2013.

[46] E. Lorenz, J. Hurka, G. Karampela, D. Heinemann, H. G. Beyer, and M. Schneider, "Qualified forecast of ensemble power production by spatially dispersed grid-connected PV systems," in *23rd European Photovoltaic Solar Energy Conference and Exhibition*, Valencia, Spain, 2008, pp. 3285–3291.

[47] E. Lorenz, J. Hurka, D. Heinemann, and H. G. Beyer, "Irradiance forecasting for the power prediction of grid-connected photovoltaic systems," *IEEE Journal of Selected Topics in Applied Earth Observations and Remote Sensing*, vol. 2, no. 1, pp. 2–10, 2009.

[48] Fraunhofer ISE, "Energy Charts: Net installed electricity generation capacity in Germany," 2017. [Online]. Available: https://www.energy-charts.de/power_inst.htm

[49] S. Killinger, L. Burckhardt, R. McKenna, and W. Fichtner, "GIS-basierte Parametrierung der Modulorientierung von Photovoltaik-Anlagen," in *VDI Wissensforum - Optimierung in der Energiewirtschaft*, vol. 2266, Düsseldorf, Germany, 2015, pp. 131–136.

[50] K. Mainzer, D. Schlund, S. Killinger, R. McKenna, and W. Fichtner, "Rooftop PV potential estimations: Automated orthographic satellite image recognition based on publicly available data," in *32nd European PV Solar Energy Conference and Exhibition (EU PVSEC)*, Munich, Germany, 2016.

[51] German society for solar energy (DGS), "Map of renewable energies," 2015. [Online]. Available: http://www.energymap.info/

Advances in long-term solar energy prediction and project risk assessment methodology

Alemu Tadesse[1], Adam Kankiewicz[1], Alex Kubiniec[1], Richard Perez[2], John Dise[1] & Thomas Hoff[1]

[1]Clean Power Research, Napa, California, 94558, USA

[2]Atmospheric Sciences Research Center, SUNY, Albany, New York, 12203, USA

Abstract — In this article, we present a state-of-the-art long-tern energy prediction methodology to minimize risk in the project investment. We applied the algorithm to the SolarAnywhere data over 19 years period. he algorithm systematically creates synthetic years from the period of dataset to draw distribution that best describes the data. We conducted the experiment for projects under different climatic conditions. We compared the result with several theoretical distribution models. The results show that the method is so robust that it can predict the real production data at a site with high level of accuracy and provide confidence in the investment.

Index Terms— **solar resource, satellite, modeling, irradiance, benchmarking, risk, investment, PVsyst , energy**

I. INTRODUCTION

Before investing in a solar energy project, developers, investors, and owners perform long-term energy prediction for the project and conduct the associated risk analysis to decide whether to invest or not. Traditionally, commercial software like PVsyst, SAM, or similar models are used for long term energy prediction. A file called Typical Meteorological Year (TMY) or Typical GHI Year (TGY) are used in Photo Voltaic (PV) production simulation software to calculate long-term energy prediction. These TMY3 and TGY files contain a one year typical (median) Global Horizontal, Direct Normal and Diffuse Irradiance (components of solar irradiance reaching the sensors such as Ground sensors and Satellite sensors), ambient temperature and wind speed along with other weather variables. The use of the typical weather values in the PV simulation software gives a median (commonly known as P50) value of long term energy production values. However, in order to make risk assessment on the production estimate over a long-term, a time series data is required along with an appropriate algorithm that helps to calculate energy prediction values at different risk level (commonly represented as Pxx). Pxx, is the probability of exceedance, where P stands for Probability and xx stands for the position of a value in the resource distribution, and xx% of the data values are above the value at position xx. For example, P90 stands for probability where 90% of the resources are above the resource value represented by P90. In general, the

Pxx values are estimated from certain distribution that the resource data values are drawn from. For example, in PVsyst the Pxx values are calculated by assuming that the solar resource (solar irradiance values in TMY or TGY files) are drawn from a normal distribution (shown in Fig.1). PVsyst states that the probabilistic values are "based on several hypotheses which requires some decisions of the user". Therefore, the user needs to understand the nature of the data that is used in the simulation process.

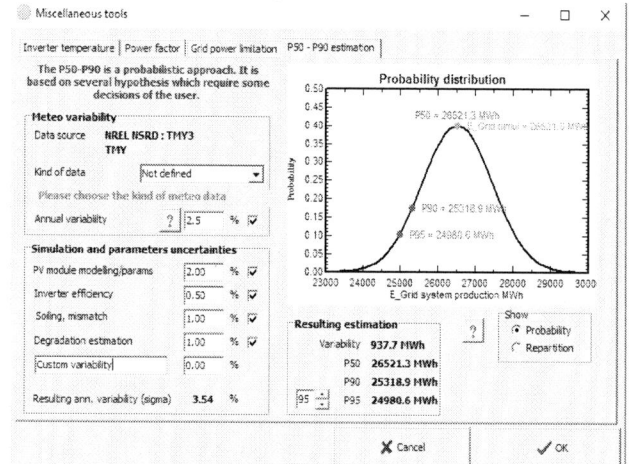

Fig.1.   PVsyst probability distribution

But, research shows that the solar resource distributions are not always normal and it deviates away from normality as the temporal resolution increases. In the following sections, we show distributions of E Grid (the AC power from the inverter to the grid). The following analysis is based on a 13000kWp (Kilowattpeak) solar project over anonymous sites located in different climatic regimes. kilowattpeak is the power of PV installation or the power that solar panels generate under standard conditions. In the following sections, we present Data and Methodology, , Results and Discussions and Conclusions.

II. DATA AND METHODOLOGY

We selected several locations with distinct climates and varying surface roughness condition. We based our choices of locations on the Köppen climate classification (Fig.2. We chose four climates: Dry such as Hot semi-arid climate (Bsh) and Cold desert climate (Bwk), and Temperate: such as Humid subtropical (Cfa) and Hot summer and Mediterranean climate(Csa). However, the ground measurement data is not available for every location we selected. Fig.3, shows the spatial distribution of National Oceanographic and Atmospheric Administration (NOAA) ground stations (Left) and Clean Power Research's SolarAnywhere data (Right). From the figure, we can see that there are very limited ground measurement stations (maximum of 14 stations with high quality ground data and a long-term period of record). It is very difficult to get consistent ground data over a long period of time due to sensor drift or calibration error. A well calibrated and validated satellite data with 20 years of consistent hourly solar irradiance data is preferred for reliable resource assessment. Satellite data is becoming popular due to its quality and spatial coverage (available over any place under the satellite's Field of View) and provides historical and consistent dataset over several years. Clean Power Research produces satellite data, in TGY format and timeseries data for over 19 years, for use in PV simulation, plant performance monitoring and related applications.

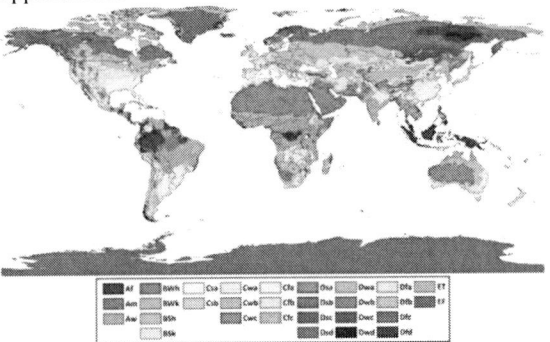

Fig.2   Köppen climate classification (source -
https://en.wikipedia.org/wiki/K%C3%B6ppen_climate_classification)

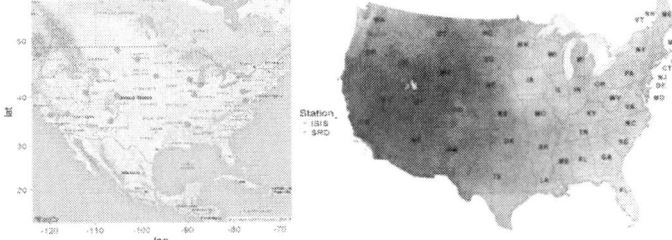

Fig.3   NOAA's stations (on the left) and CPR's satellite data on the right

The SolarAnywhere data is used by several solar energy companies for variety of applications including resource assessment, plant performance monitoring and project risk evaluation. Recently, the data has been used to also help detect ground measurement data sensors calibration issues [5].

In this paper, we investigated TGY based hourly energy distribution, and SolarAnywhere historical data based annual energy distribution to investigate if the energy values follow certain theoretical distribution and can be modeled.

a.   TGY HOURLY DISTRIBUTION

We ran PVsyst for several locations under different climatic conditions. We have also investigated the resource variability under the same climatic condition but with varying surface roughness. We used SolarAnywhere TGY data to investigate if there is certain theoretical distribution that may fit the hourly energy that is injected to the grid (E Grid). Understanding the hourly energy distribution will help to characterize the risk associated with the energy resource for certain obligations of the power purchase agreement process to meet power dispatch at certain times of the day. Figures 4 to 7 show the density distribution of hourly E Grid for two climatic conditions during winter and summer months (January for winter and July for summer). The density distributions of the hourly E Grid did not follow any certain theoretical distribution for any time of the day, season and climatic conditions. From Fig. 4, we see that certain hours of the day have a bi-modal distribution of different shapes (Skewed to the left or skewed to the right or with an oval shape). There is no one model that fits these distributions. Fig. 5 shows an hourly energy profile for July, for the same location and climate as in Fig. 4. The energy distributions across the hours of the day are not similar. Moreover, distributions for corresponding hours of the day between Fig. 4 and Fig. 5 are also very different. This observation leads to a need to design an approach by which hourly distributions can be modeled (without the use of any assumption for an underlying distribution). Fig. 6 and Fig. 7 show hourly energy distributions for January and July months for a location with a climate type of Bwk. Most Bwk climatic regions zones are usually located far from oceanic influences into the deep interiors of continental areas. Therefore, they usually remain dry and moisture-bearing air masses rarely reach them. The Bwk regions are also influenced by orography and several of such climatic regions are located on the leeward side of major orographic belts, and therefore in the rain shadow of the mountains.

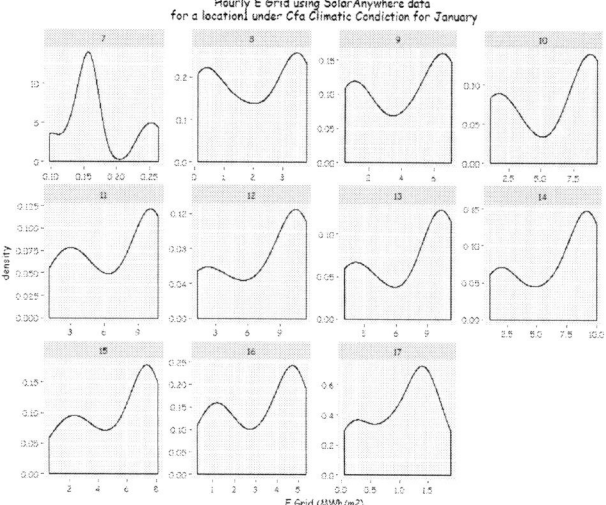

Fig.4.   Hourly E Grid distribution for January for climate type Cfa

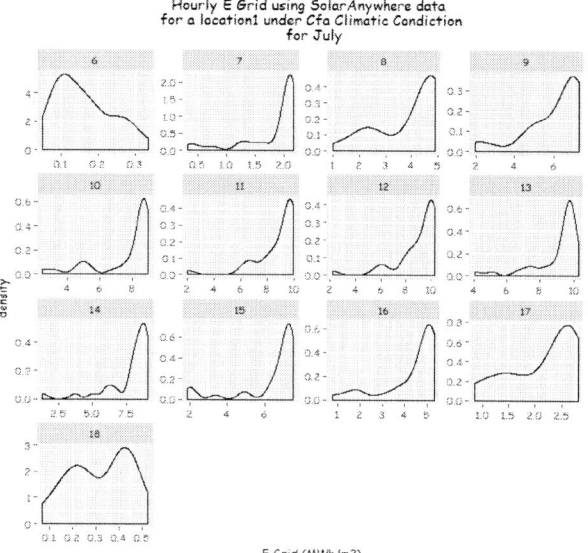

Fig.5.   Hourly E Grid distribution for July for a climate type Cfa

It is very interesting that the January and July months of most of corresponding hours have similar hourly energy distribution as opposed to what has been seen in the Cfa case in the previous section. Even though the Bwk climate type (for this location) shows some kind of similarity in the distribution, there is conclusive evidence that will lead us to believe that we can model hourly energy distributions of certain location with a theoretical model with certainty. In part b of this section, we present resource distribution characterization on an annual basis. Annual energy resource characterization provides information on inter-annual variability and also to predict the

probability of lower resource value for investment risk evaluation.

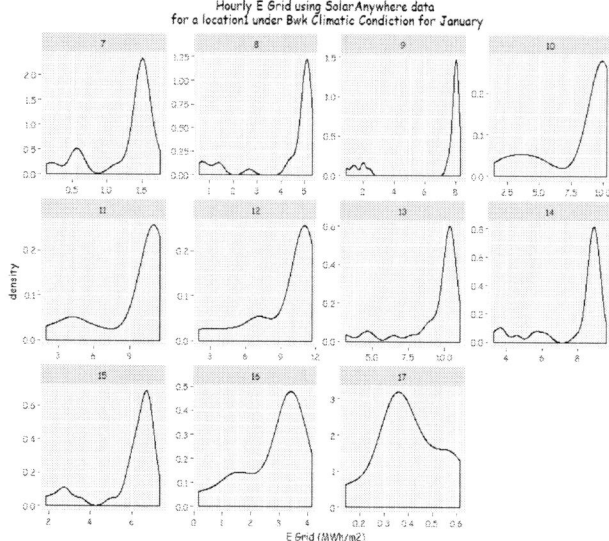

Fig.6.   Hourly E Grid distribution for January for a climate type Bwk

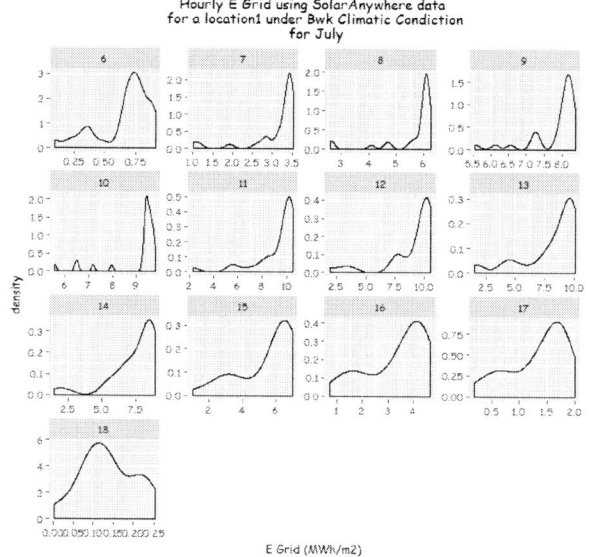

Fig.7.   Hourly E Grid distribution for July for a climate type Bwk

b.   E GRID HISTORICAL DATA

In this analysis, we used SolarAnywhere time series data and we run PVsyst for several locations over 19 years period to investigate annual energy distribution. In this section we present annual energy distribution for four locations in the United States. Fig. 8, shows the annual energy distribution over these locations. The top left panel shows a distribution that is skewed to the left and the top right panel shows a bimodal

978-1-5090-5606-4/17 $31.00 © 2017 IEEE   134

distribution. The lower left panel and lower right panel also show a very different distribution from the top panel locations.

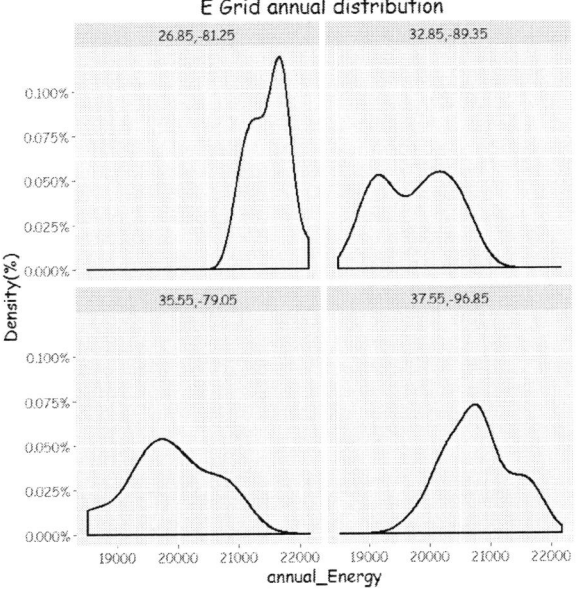

Fig.8. Hourly E Grid distribution for January for a climate type Bwk

Therefore, there is no evidence of any theoretical distribution that can best explain these distributions.

III. Results & Discussions

We fitted several models to the data shown in Fig. 8. We conducted the Akaike Information Criterion (AIC) as a way of selecting a few models that we presented below. We chose, Weibull, Lognormal and Gamma distributions, and presented the results in Fig. 9 to Fig 12 (corresponding to each location in Fig.8). The histograms in the figures show the distribution of the actual data (19 years of E Grid simulated using SolarAnywhere data as input in PVsyst). The Figures in topright panels represent empirical cumulative distributions created from an actual data. The bottom line represents quantiles and probability plots. Q-Q plots take the sample data (the 19 years E Grid values), sort it in ascending order, and then plot them versus quantiles calculated from a theoretical distribution. The number of theoretical quantiles are quantiles selected to match the size of the number of years data we have. If the quantiles from the sample match the quantiles from the chosen theoretical distribution, then the points fall along the straight line. A P-P plot compares the empirical cumulative distribution function of the E Grid dataset with a specified theoretical cumulative distribution function. If the theoretical probability values match that of the sample data, then the points

fall along the straight line. In all of these cases, we see that the density functions derived from the models missing the histogram and also Q-Q and P-P plots not being aligned with each other. Therefore, we can conclude that neither of these models were able to simulate the distribution of the E Grid data on an annual basis.

Due to the capital-intensiveness of renewable energy projects, the cost of capital is a crucial element in every renewable energy investment decision and can significantly influence the business case of a project. Therefore, there is a risk associated with any underlying distribution assumptions for a project investment risk assessment. There are a number of methods that are available for best model selection that gives the best balance between model fit and complexity [2 & 3].

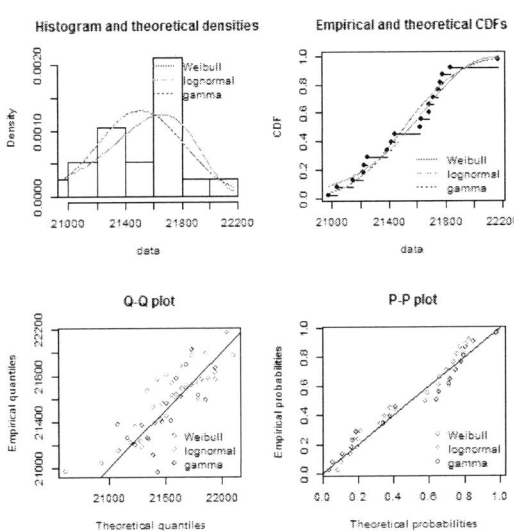

Fig.9 Location Lat=26.85 & Lon=-81.25

However, such approach requires to have several models for each location, climate, month and hour and may lead to a more complex scenario and computationally expensive situation.
In this paper, we used SolarAnywhere data from 1998 to 2016 and ran PVsyst for each year. We assumed the same system size, module and inverter technology for all simulations. We employed the following methodology in order generate synthetic years from which we calculated the Pxx values. If x is the number of experimental data years available and 1/n is the considered yearly fraction, the number of possible composite years is equal to x^n. In the present case, we started from 19 experimental data-years and assembled 6859 possible composite years by producing all possible combinations of individual 3-month periods (n=4), effectively increasing the data sample by an order of magnitude. This method is based on

the method that Perez R. has developed [1] for resource estimation.

Based on this method, we created 6859 years of data from which we constructed a probability distribution (Fig. 13). This distribution is closer to a normal distribution from which we can draw information on the historical annual energy distribution for risk assessment applications.

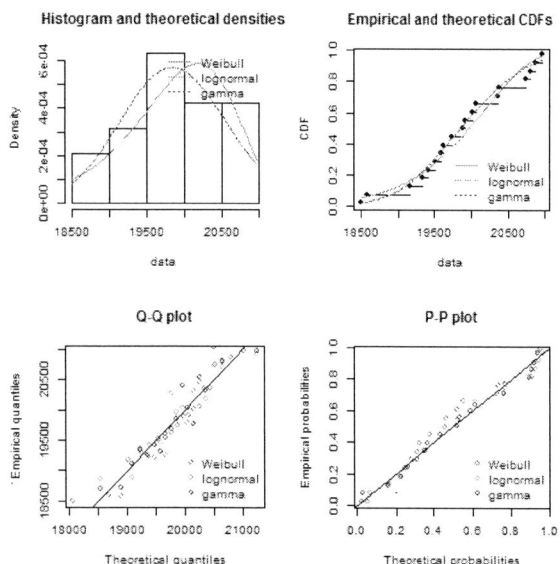

Fig.10.   Lat=32.85 & Lon=,-89.35

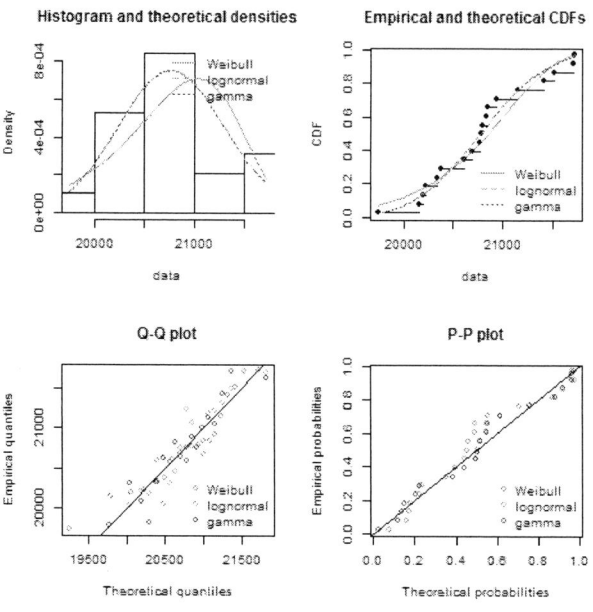

Fig.12.   Lat=37.55 & Lon=-96.85

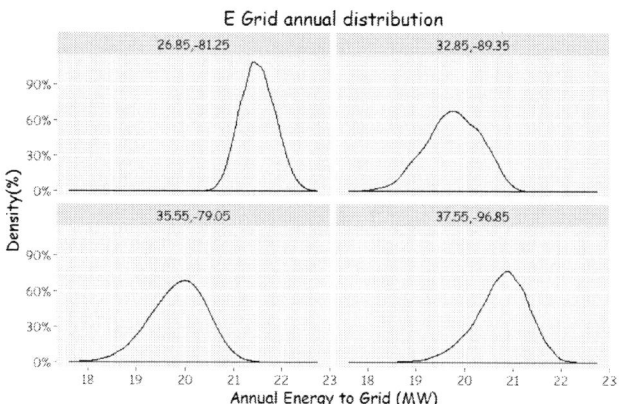

Fig.13.   E Grid distribution created using the methodology described in this section

The method presented in this paper indicates the importance of using long-term satellite data in the PV system simulation software to perform an accurate long-term energy prediction at a project site and also to conduct risk assessment of the investment by providing accurate Pxx values. In addition to the method we described here, in this paper, to estimate the probability of exceedance values we have also calculated 6859 data points from a normal distribution assumption (using mean and standard deviation of annual E Grid data from the 19 years PVsyst simulation) to investigate the deviation in the Pxx

Fig.11.   Lat=35.55 & Lon=-79.05

values from our new method. We calculated Pxx values from the 19 years of data (with normal distribution assumption) and compared the results to the Pxx values calculated from the synthetic years generation approach we described above. We denote the Pxx calculated from the normal distribution assumption as Pxx_{19} and the Pxx calculated from the synthetic years as $Pxx_{synthetic}$. We compared the results from the two approaches as shown in equation 1 below. The results are shown in Fig. 14 and Fig. 15.

$$Pxx\ diff = \frac{(Pxx_{19} - Pxx_{synthetic})}{Pxx_{synthetic}} * 100 \quad (1)$$

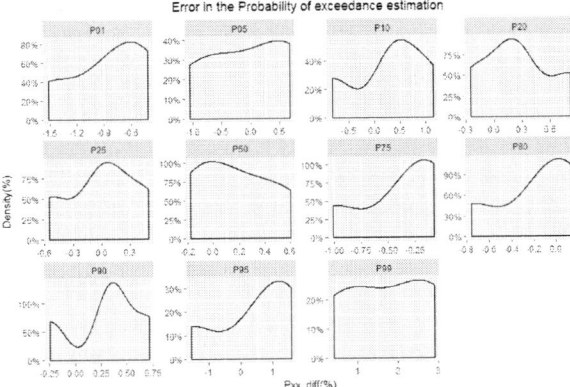

Fig. 14.   Differences in Pxx annual E Grid values

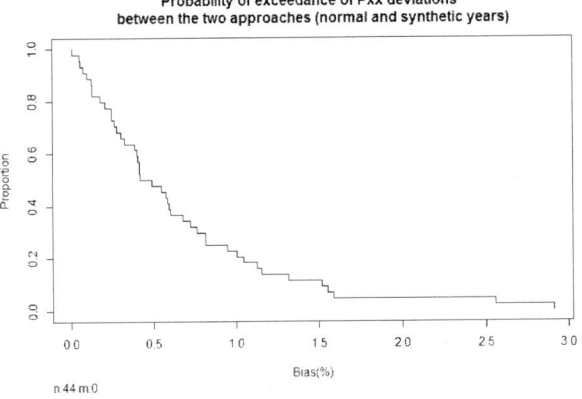

Fig.15.   Probability of exceedance of absolute values of Pxx_diff

Fig. 14, shows that the synthetic year approach and the normal distribution assumption approach have lower error for the P50 values. However, the differences can be as high as 1 to 3% for P99 and P95 values. The P99 and P95 values are the most important values for the loan lenders and project developers. In Fig. 15, we can see that for over 10% of the time the differences in the Pxx values can exceed 1%. A 1% in

resource estimation uncertainty can cost the project, over its life time, millions of dollars.

IV. CONCLUSIONS

An approach to a resource assessment can make or break to a project feasibility. Not giving enough scrutiny to resource assessment process, early-on in the project feasibility studies, may lead to the chance of failure or financial losses. Risk assessment—by properly and carefully calculating the Pxx values —helps to avoid an avoidable cost associated with bad assumptions. In the approach discussed in this paper, we have demonstrated the use of consistent historical data for resource assessment and investment risk quantification (in the form of Pxx values). In this proposed approach, the actual energy data at the project site (calculated using consistent satellite data - SolarAnywhere data) is not changed, which gives confidence in the results obtained.

We will conduct similar studies for all of the climatic conditions to see if there is a pattern (as depicted by the location chosen from Bwk climate) for other climate classifications zones to generalize the result of this study.

REFERENCES

1. Perez. R., has developed this probability determination technique for several of his customers in the US including Solar Millennium and RW Beck-SAIC and results have been received satisfactorily by financing parties. The technique shall be describing in a planned 2012 Solar PACES conference article
2. Burnham K. P. & Anderson D. R. (2002): " Model selection and multimodel inference: A practical information-theoretic approach." Springer
3. Hastie T., Tibshirani R. & Friedman J. (2009): "The 0.01 elements of statistical learning: Data mining, inference, andprediction". Springer.
   3. Shao J. (1993):"Linear model selection by cross-validation". Journal of American Statistical Association, 88, 486-494
4. Perez R, 2017: "SolarAnywhere detects calibration error in nation's most trusted ground site"-
   blog- https://www.cleanpower.com/2017/solaranywhere-detects-calibration-error/

Decoupling Thin Film CdTe Growth from Packaging: Toward Record Specific Power in Low Cost Polycrystalline PV

D. Clayton-Warwick[1,2], M. D. Kempe[1], M. S. Dabney[1], T. M. Barnes[1], C. A. Wolden[2], M. O. Reese[1]

[1]National Renewable Energy Laboratory, Golden, CO 80401, USA
[2]Colorado School of Mines, Golden, CO 80401, USA

Abstract — **There are critical material and scientific barriers to producing high efficiency, flexible, lightweight solar cells that maximize specific power in a cost-effective manner. III-V solar cells have produced the highest specific power of any photovoltaic technology due to their extremely high efficiency, however, they are also among the most expensive to produce. This work explores a novel lift-off approach with thin-film solar cells, particularly CdTe, to achieve high specific power at low costs. Thin-film devices can be delaminated from their heavy, glass growth substrates post-growth by exploiting a mismatch in coefficient of thermal expansion between the two. This allows thin-film PV to be decoupled from high temperature growth requirements and repackaged using lightweight materials. In this work we evaluate different materials and approaches for achieving large area delamination of CdTe films from their glass superstrates. It is found that epoxy bonded to a rigid handle provides robust and reproducible delamination at the TCO/CdS interface upon submersion in liquid nitrogen. Following delamination, a transparent front contact is re-deposited and selectively etched to expose the back contact to produce a functioning device. Effects of delamination and contributions to device functionality from $CdCl_2$ and CdS layers are discussed. Reproducible delamination of high efficiency, thin-film solar cells over large areas will enable work to further improve device efficiency through passivation and reconstruction of the previously buried interface between the transparent contact and CdS.**

I. INTRODUCTION

The specific power, or power-to-weight ratio, of a solar cell is determined by the efficiency of the photovoltaic (PV) device and the weight of the final package (substrate, encapsulation, etc.). Currently, GaAs solar cells produce the highest specific power of any PV technology due to their high efficiency and ability to reduce weight via substrate removal. Additionally, removing and reusing expensive single crystal substrates lowers production costs for III-V based PV. However, they are still extremely expensive compared with other PV technologies [1]. In particular, the production cost per area of thin films, such as CdTe and $CuIn_xGa_{1-x}Se_2$ (CIGS), are two orders of magnitude less than GaAs and other III-V's [2, 3]. At current thin-film PV production module efficiencies of 14-17% [4, 5, 6], specific powers approaching

500 W/kg hypothetically could be achieved by reducing thin-film package weight from 16.7 kg/m^2 to 350 g/m^2. Package weights as low as this have been reported for commercial III-V products [7]. The bulk of a conventional module's weight is due to the double glass module construction. Thus, post-growth delamination and repackaging using lightweight materials can theoretically make high specific powers achievable with the low production costs from thin-film PV growth. Other attempts have been made to decrease the weight of thin-film devices by growing directly on lightweight polymeric films or ultra-thin metal foils. However, because of high-temperatures (500-600°C) used in conventional processing steps, this approach can be complicated and requires extensive development to decrease growth temperatures, to manage the dimensional stability of the substrate, and limit impurity diffusion.

In this paper, we explore the thermo-mechanical delamination/lift-off of completed CdTe devices by exploiting a mismatch in coefficient of thermal expansion (CTE) between the thin film and growth substrate. This produces separation at or near the front contact/CdS interface (Figure 1a-b), which is particularly beneficial because of its proximity to the p-n heterojunction. Recombination near this junction is thought to be a major source of voltage loss in CdTe devices, which can be detrimental to device performance [8, 9]. By gaining access to this interface, post-growth reconstruction and passivation can be performed to reduce recombination and improve device efficiency. However, before this can be addressed, delamination of high-efficiency thin films must be made reproducible over large areas. This requires a careful understanding and control of adhesion at the interface at which lift-off occurs. While delamination of epitaxial materials may be well understood [10, 11], the polycrystalline nature of thin films introduces new challenges. This paper aims to initiate the study of delamination in thin films, which will enable reductions in package weight while maintaining efficiency. Additionally, by altering the newly exposed interface, we can probe the material system in new ways. This type of knowledge will lay the groundwork in improving device efficiency through interface modifications.

978-1-5090-5606-4/17 $31.00 © 2017 IEEE

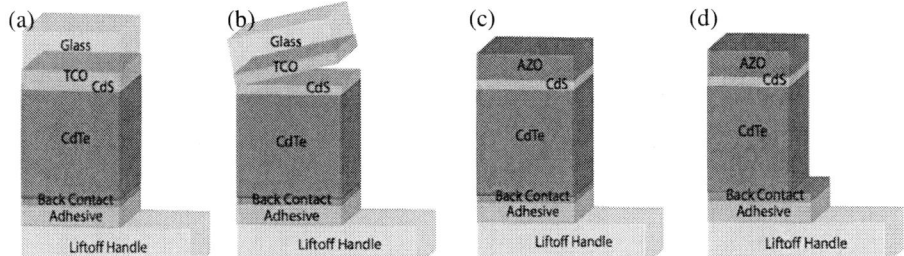

Figure 1: (a) CdTe device structure prior to lift-off; (b) Delamination occurs at the front contact (TCO)/CdS interface; (c) New front contact material (AZO) is regrown; (d) Selective etch performed to partially expose back contact.

II. MATERIALS AND METHODS

All CdTe devices were grown in the superstrate configuration using Corning® 7059 glass coated with a fluorinated tin oxide (FTO) using chemical vapor deposition. 100 nm of oxygenated cadmium sulfide (CdS:O) was deposited via rf magnetron sputtering in an oxygen/argon ambient. CdTe (~4 μm) was then deposited using close-spaced sublimation (CSS) with the substrate held at 600 °C in an oxygen/helium ambient. Vapor-phase $CdCl_2$ annealing was also performed in a CSS configuration with the substrate temperature fixed at 400 °C in an oxygen/helium ambient. Finally, a Cu/Au back contact was evaporated and annealed post-deposition to promote Cu diffusion [12].

Figure 1 displays the current process of delaminating and re-contacting devices. First, a glass handle was adhered to the Cu/Au back contact using several types of adhesives, including silver paint, pressure sensitive adhesives, epoxy (Hysol® 1C™), thermoplastic polyurethane (TPU), and ethylene-vinyl acetate (EVA). The structure was then submerged in liquid nitrogen to induce delamination. To first order, it is believed that the primary cause for delamination is a mismatch in CTE, which causes each layer to contract by varying amounts. This strains the interfaces and leads to fracture along either the weakest interface or the region with the highest strain.

After lift-off, the front contact was regrown (Figure 1c) using aluminum-doped zinc oxide (AZO), a transparent conducting oxide (TCO) that can be deposited at low temperatures (~60°C) via sputtering. Because the Cu/Au back contact was made inaccessible by the adhesive and handle, it must be subsequently exposed. A selective etch was performed to remove all device layers down to the back contact (Figure 1d) using Kapton® stickers (4 mm diameter) as an etch mask and ferric chloride as the etchant (Figure 2).

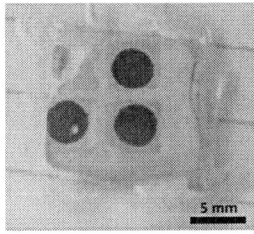

Figure 2: CdTe is in the substrate configuration after lift-off. Selective etch performed to access Cu/Au back contact (gold squares) for performance measurements.

III. RESULTS AND DISCUSSION

Key factors affecting uniformity of lift-off were determined by testing various types of adhesive. Silver paint, which has a grainy surface texture made up of silver particles after solvent evaporation, resulted in a rough, blistered surface of CdTe. This transfer of texture from the adhesive to film was also seen with various tapes (carbon, 3M™ XYZ-axis electrically conductive, Kapton®). In addition, tapes provided inconsistent and partial lift-off that was dependent on pressure and method of application.

Of the adhesives tested, epoxy provided the most consistent lift-off, with clean delamination of more than 1 cm² areas. Using epoxy, a functioning CdTe device has been delaminated and re-contacted to make a working device. Before lift-off, this device had an efficiency of 12.3% and its J-V curve (black) can be seen in Figure 3. After delamination and regrowth of the front contact, the efficiency was ~7.5% (red curve in Figure 3). Solar cell efficiency is defined as $\eta = V_{OC}J_{SC}FF/P_{in}$ where V_{OC} is the open-circuit voltage (x-intercept in Figure 3), J_{SC} is the short-circuit current density (y-intercept in Figure 3), FF is the fill factor, and P_{in} is the incident power.

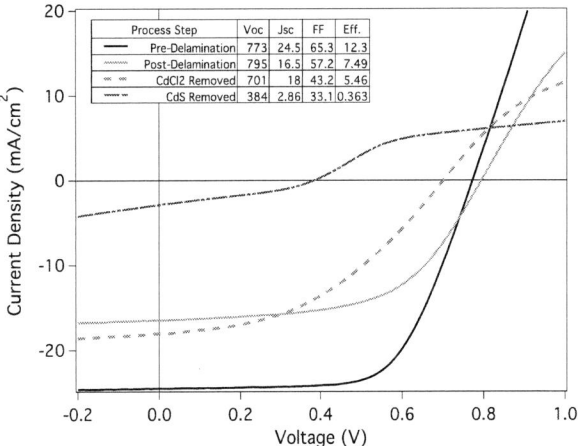

Figure 3: J-V curves for a CdTe device before lift-off (black curve), after lift-off (red), after rinsing with DI water to remove $CdCl_2$ (dashed green), and after rinsing with HCl acid to remove CdS (dashed blue).

By comparing the black and red J-V curves, it can be seen that the decrease in efficiency following delamination is largely due to a considerable decrease in J_{SC} as well as an increase in series resistance. The increased series resistance is largely due to a poorly optimized front contact (~80 Ω/sq). The decrease in current density is likely caused by large cracks that formed during AZO deposition (Figure 4). Because epoxy has a much higher CTE than CdTe, it expands to a greater degree when heated during sputtering. This, in turn, applies enough tensile stress on the CdTe film to cause cracking. A minor problem that results is the creation of shunt pathways through AZO deposition within the cracks. As seen in Figure 3, this results in a slight decrease in shunt resistance.

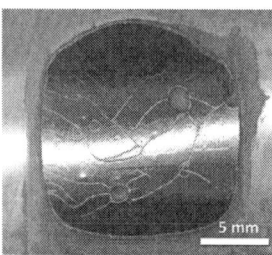

Figure 4: Large cracks seen in films after AZO deposition and before selective etching.

More importantly, these cracks can drastically reduce the contacted cell's effective area, making the area used for J_{SC} calculations (that of the 4 mm Kapton® sticker) too large. The efficiency measurements reflect these limitations and current work is focused on addressing the source of such cracking. The measurement of interest then, is V_{OC} because it does not depend on area. Of particular interest is the increase in V_{OC} after lift-off. Immediately after the growth of the superstrate device, higher efficiency/voltage was seen. Over the course of months it dropped from ~850 mV to the pre-delamination voltage of 773 mV. This is likely due to Cu migration, which can be a source of metastability that can be reflected in a voltage loss [13]. It is commonly observed that this effect can be modestly reversed with elevated temperatures, such as that seen during regrowth of the front contact (~60°C).

To roughly determine the importance of the layers near the TCO/CdS interface, some further processing was done. Because CdCl$_2$ is a water-soluble salt, this layer was removed by rinsing with DI water after delamination and before AZO deposition (the accumulation of a monolayer of CdCl$_2$ at the CdS/TCO interface is supported in detail by a comprehensive photoemission study reported in a separate paper [14]). The removal of this layer resulted in an apparent decrease in V_{OC} of roughly 100 mV (dashed green curve in Figure 3). This seems to suggest that CdCl$_2$ plays an important role in maintaining voltage in CdTe devices. On a separate sample, the CdS layer was removed by swabbing the newly exposed interface with 10% HCl acid. By removing this layer, the heterocouple between CdTe (p-type) and CdS (n-type) breaks down and the V_{OC} decreases by about a factor of two (dashed blue curve in Figure 3).

V_{OC} measurements provide an assessment of the quality of the heterojunction, and it is clear that altering the interface also alters the performance of the solar cell. Thus, by improving the previously buried TCO/CdS interface, such as through surface reconstruction and passivation, it is possible to improve the efficiency of CdTe devices using this lift-off technique. Additionally, it is possible that parameters of a chosen adhesive may be tuned to access other buried interfaces, allowing for a much more robust understanding of CdTe devices.

While epoxy may serve as a good adhesive to demonstrate the reliability of thermo-mechanical lift-off, it is not practical for long-term studies. This is due not only to the aforementioned cracking, but also its rigidity. To move toward flexible and lightweight devices, a different adhesive must be used. TPU has been shown to result in clean delamination (Figure 5a), but this adhesive has its own complications. Namely, its adhesion to glass is weak such that when the CdTe/TPU/glass handle structure is submerged in liquid nitrogen, the TPU detaches from the glass handle while simultaneously lifting off the CdTe. Because the CTE of TPU is much higher than CdTe, the isolated CdTe/TPU multilayer curls up, leaving the CdTe film susceptible to cracking and tearing. Possible solutions to this issue are being explored in current research. Finally, EVA has shown some promising results (Figure 5b), yet delamination is inconsistent and not reproducible. Future work will focus on understanding why this is the case and how to correct it.

(a)   (b)

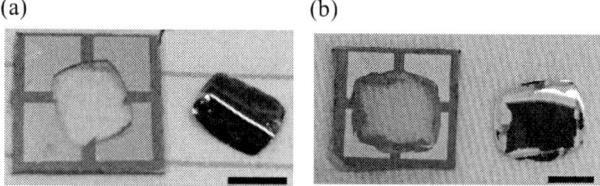

Figure 5: Delaminated CdTe film using (a) TPU (which has detached from its glass handle and curled up) and (b) EVA with a small plastic handle (5 mm scale bars).

IV. SUMMARY AND CONCLUSIONS

In this paper, we examined the effects of various types of adhesives (silver paint, pressure sensitive adhesives, epoxy, TPU, EVA) on delamination of CdTe thin-film solar cells. Qualitative results are summarized in Table 1. We found that epoxy results in the most consistently clean lift-off over large areas (larger than 1 cm^2). Using epoxy as the adhesive, we delaminated CdTe films, regrew the front contact (by sputtering AZO), and selectively etched to make complete devices. Because of cracks in the film, efficiency was reduced relative to the starting value. The partial recovery of V_{OC} showed us that delamination might be a tool to modify the interface and junction in delaminated cells. A thin layer of CdCl$_2$ at the interface appears to help reduce voltage loss. While epoxy provides consistent and clean lift-off that has

TABLE I
SUMMARY OF ADHESIVES USED FOR DELAMINATION

Adhesive	Level of Delamination	Appearance
Silver paint	Complete	Blistered surface
Pressure sensitive adhesives	Partial	Textured surface
Epoxy	Complete	Rigid, cracks after AZO deposition
Thermoplastic polyurethane (TPU)	Varied	Curled, easily torn/cracked
Ethylene-vinyl acetate (EVA)	Varied	Smooth, some partial lift-off

been useful in initial studies, its rigidity precludes it from use in flexible delaminated devices. Therefore, future research will be done to identify, understand, and optimize other promising adhesives and handles. Once reliable delamination is achieved, improvements to the interface can be investigated. Examples include regrowth of a CdS layer after its removal, passivation of the exposed interface prior to depositing the front contact, and methods to eliminate cracking in delaminated films.

ACKNOWLEDGEMENTS

This work was supported by the U.S. Department of Energy under Contract No. DE-AC36-08GO28308 with Alliance for Sustainable Energy, LLC, the Manager and Operator of the National Renewable Energy Laboratory. Funding provided by Office of Naval Research. The U.S. Government retains and the publisher, by accepting the article for publication, acknowledges that the U.S. Government retains a nonexclusive, paid-up, irrevocable, worldwide license to publish or reproduce the published form of this work, or allow others to do so, for U.S. Government purposes.

REFERENCES

[1] K.A.W. Horowitz, M. Woodhouse, H. Lee, G.P. Smestad "A bottom-up cost analysis of a high concentration PV module" *AIP Conf. Proc.* 1679, 100001 (2015).

[2] K.A.W. Horowitz, R. Fu, M. Woodhouse "An analysis of glass-glass CIGS manufacturing costs" *Sol. Energy. Mat. Sol. Cells* 154, 1 (2016).

[3] R. Jones-Albertus, D. Feldman, R. Fu, K. Horowitz, M. Woodhouse "Technology advances needed for photovoltaics to achieve widespread grid price parity" *Prog. Photovolt.: Res. Appl.* 24, 1272 (2016).

[4] M.A. Green, K. Emery, Y. Hishikawa , W. Warta, E.D. Dunlop "Solar cell efficiency tables (version 47)" *Prog. Photovolt.: Res. Appl.* 24, 3 (2016).

[5] http://www.firstsolar.com/Home/Technologies-and-Capabilities/PV-Modules/First-Solar-Series-3-Black-Module, accessed January 25, 2017.

[6] http://www.solar-frontier.com/eng/solutions/products/index.html, accessed January 25, 2017.

[7] http://mldevices.com/index.php/product-services/photovoltaics, accessed January 26, 2017.

[8] T.A. Gessert, S.H. Wei, J. Ma, D.S. Albin, R.G. Dhere, J.N. Duenow, D. Kuciauskas, A. Kanevce, T.M. Barnes, J.M. Burst, W.L. Rance, M.O. Reese, H.R. Moutinho, "Research strategies toward improving thin-film CdTe photovoltaic devices beyond 20% conversion efficiency" *Sol. Energy Mater. Sol. Cells* 119, 149 (2013).

[9] A. Kanevce, M.O. Reese, T.M. Barnes, S.A. Jensen, W.K. Metzger "The roles of carrier concentration and interface, bulk, and grain-boundary recombination for 25% efficient CdTe solar cells" *J. Appl. Phys.* (Online, DOI: 10.1063/1.4984320).

[10] G.J. Bauhuis, P. Mulder, E.J. Halverkamp, J.J. Shermer, E. Bongers, G. Oomens, W. Kolster, G. Strobl, "Wafer reuse for repeated growth of III–V solar cells" *Prog. Photovolt.: Res. Appl.* 18, 155 (2010).

[11] J.J. Schermer, G.J. Bauhuis, P. Mulder, E.J. Haverkamp, J. van Deelen, A.T.J. van Niftrik, P.K. Larsen "26.1% thin-film GaAs solar cell using epitaxial lift-off" *Thin Sol. Films* 511, 645-653 (2006).

[12] J.N. Duenow, R.G. Dhere, J.V. Li, M.R. Young, T.A. Gessert, "Effects of back-contacting method and temperature on CdTe/CdS solar cells" *Proc. 35th PVSC*, 1001 (2010).

[13] D.S. Albin "Accelerated stress testing and diagnostic analysis of degradation in CdTe solar cells" *SPIE Optics+Photonics Meeting* (2008).

[14] C.L. Perkins, C. Beall, J.M. Burst, A. Kanevce, M.O. Reese, T.M. Barnes, "Two dimensional cadmium chloride nanosheets in cadmium telluride solar cells" *ACS Appl. Mater. Interfaces* (Online, DOI: 10.1021/acsami.7b03671).

Junction Activation of CdTe/CdS Solar Cell Using MgCl$_2$

G. Angeles-Ordóñez, E. Regalado-Pérez, M.G. Reyes-Banda, N.R. Mathews, X. Mathew*

Instituto de Energías Renovables, Universidad Nacional Autónoma de México, Temixco,
Morelos, 62580 México

* Corresponding author (X. Mathew); e-mail: xm@ier.unam.mx

Abstract:- **A comparative study of CdTe/CdS junction activation using MgCl$_2$ and CdCl$_2$ is discussed. V$_{oc}$, J$_{sc}$, FF and efficiency of the devices were comparable in both cases indicating that CdCl$_2$ can be avoided in the processing of CdTe solar cells. From a survey of the device parameters of small area devices at different locations of the large substrate it was observed that MgCl$_2$ activation results in higher spatial uniformity. Extent of S diffusion was significantly higher for CdCl$_2$ activated devices at the same conditions, indicating that either the recrystallization process or the grain boundary passivation is distinct for both cases.**

Keywords: CdTe/CdS; MgCl$_2$ junction activation

1. INTRODUCTION

Activation of CdTe/CdS junction in presence of a chloride compound is a key step in obtaining high efficiencies, and CdCl$_2$ is the common and widely accepted chloride source for this purpose. Also it is known that this annealing process promotes CdTe recrystallization and grain growth resulting in reduced defect densities, diffusion of S and Te at the interface, passivation of grain boundaries, suppression or segregation of electrically active species, Cl doping of CdTe, etc. [1]-[6]. The notable benefits are gain in open-circuit voltage, short-circuit current and increase in fill factor. There exists a good amount of literature regarding the attempts to experiment different chloride sources in activation process including NaCl, KCl, MnCl$_2$, HCl, NH$_4$Cl, ZnCl$_2$, CHF$_2$Cl, etc. [7]-[12]. There is also report about a two-stage process in which CdCl$_2$ and MgCl$_2$ were employed in the device processing, however, the purpose of this process was to look into the formation of electron reflector at the back [13]. Among the mentioned materials only CHF$_2$Cl treatment was promising [12], however its use is restricted due to environmental considerations. Recently Major et al. reported promising results using MgCl$_2$ as junction activation source, obtaining efficiencies comparable to that obtained with CdCl$_2$ [10]. The promise of this result was the substitution of toxic CdCl$_2$ with a risk-free chloride source in the cell fabrication.

A comparative study of MgCl$_2$ activated devices with the control samples activated using CdCl$_2$ in our laboratory showed that the cell voltage, current, and fill factor are comparable or slightly higher for devices activated with MgCl$_2$ [14]. In addition, the spatial uniformity in efficiency of small area devices fabricated on large area substrate was also better in devices activated with MgCl$_2$. In this presentation, we are discussing the activation of CdTe/CdS junction using MgCl$_2$ as the chloride source. The control samples were junction-activated with CdCl$_2$ under exactly the same conditions. In order to minimize the errors and make the comparison as accurate as possible, we used dip-coating technique for chloride loading and same molar concentrations of MgCl$_2$ and CdCl$_2$ were used.

2. THIN FILM DEPOSITION AND JUNCTION ACTIVATION PROCEDURE

The thin film layers CdS and CdTe were deposited by close spaced sublimation (CSS) over Tec 15 glass substrates, the corresponding substrate temperatures were 530 and 550 °C respectively. The thickness of CdS film was about 100 nm and that of CdTe was in the range 3 to 4 μm. In order to perform the junction activation the CdTe/CdS heterostructure was immersed in 0.07 M solution of MgCl$_2$ in methanol for 60 s and dried in air. The annealing was at 410 °C in an ambient of Ar and O$_2$. The reference sample was junction activated using 0.07 M solution of CdCl$_2$ in identical conditions. The molarity of 0.07 was chosen since it corresponds to the saturation of CdCl$_2$ in methanol. After annealing, the samples were etched with nitric-phosphoric (N-P) acid. The electrical contact was made by vacuum depositing Cu/Au (3/30 nm) and later annealed at 150 °C in N$_2$ atmosphere to allow Cu diffusion. The area of the devices was 0.255 cm^2.

The structural properties and the morphological features of CdTe films exposed to annealing were studied. SIMS (Cameca IMS 4f) technique was utilized to evaluate the extent of elemental diffusion across the device layers. I-V and EQE data were analyzed to compare the device performance of both batches of devices. I-V data was collected with 100 mW/cm^2 incident power under AM1.5 illumination from an Oriel Class AAA solar simulator (Oriel 94043A, Newport Corp.). EQE spectra were recorded using an Oriel QEPVSI-B equipped with a 300-W xenon arc lamp. The incident intensity at each wavelength was calibrated with a UV-enhanced Si photodetector (Newport 71889).

3. DISCUSSION OF THE DATA

The XRD patterns of the as-deposited and annealed films showed the characteristic reflections of CdTe with the intense reflection corresponding to (111) plane. In the XRD reflections of MgCl$_2$ treated sample very weak signals of Mg$_3$(OH)$_5$Cl·4H$_2$O and Mg(ClO4)$_2$·3H$_2$O were also detected. No significant changes in crystallite size (25.4 – 26.5 nm) was observed due to annealing in both chloride salts. The SEM images of three representative films; as-deposited (a), annealed in CdCl$_2$ (b), and in MgCl$_2$ (c) are shown in Fig. 1. In order to make the comparison realistic the annealed films were cut from the same source sample as that of the as-deposited film. The annealing conditions are given in section 2 above. It is clear that the surface morphology of the film exposed to CdCl₂ is more or less identical to that of the virgin film, however, the morphology is significantly changed in the case of film annealed with MgCl₂. The EDXS data showed that the film composition is similar in all cases, slightly rich in Te.

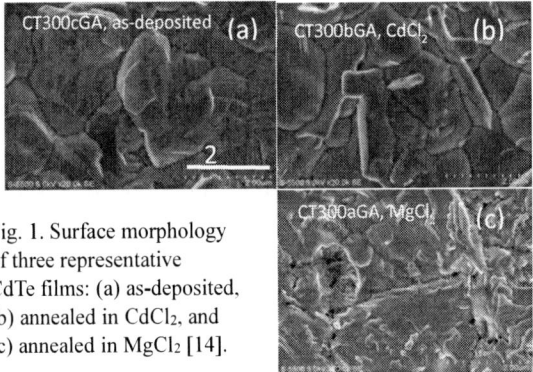

Fig. 1. Surface morphology of three representative CdTe films: (a) as-deposited, (b) annealed in CdCl₂, and (c) annealed in MgCl₂ [14].

Prototype devices were developed by activating the junction in the chloride vapor environment at different time durations. It was observed that the optimum time duration was different for CdCl₂ and MgCl₂ activation. Our preliminary studies showed that the junction activation in MgCl₂ vapor requires only half of the time that required with CdCl₂ at the same temperature. For a comparison the device parameters (V_{oc}, J_{sc}, FF and efficiency) of CdTe/CdS cells activated with both methods are shown in Fig. 2. The data corresponds to the average of values measured on six cells. In the case of MgCl₂ treatment best results were obtained for 10 min activation while for CdCl₂ activation 20 min. was the adequate time duration. The devices activated with CdCl₂ for 10 min. duration was poor in performance.

SIMS analysis was performed on different sets of devices to get an insight into the elemental profile across the cross-section. The Cu profile was observed to be identical in both cases indicating that Cu diffusion is independent of the activation agent used, as expected the concentration is much higher at the back contact region. However, Mg concentration at the back was lower than that in the bulk and identical in both cases till a depth of 0.75 μm from surface, which was un-expected. The higher Mg concentration towards heterojunction region was interpreted as due to diffusion from glass substrate [10].

Special attention was given to S diffusion, a qualitative information about the diffusion of S through the CdTe was obtained by analyzing the SIMS data of the completed devices (CdTe/CdS/Cu-Au) activated using both CdCl₂ and MgCl₂. In SIMS measurement the scan was started from the back contact and stopped at the glass substrate. Fig. 3 shows the amount

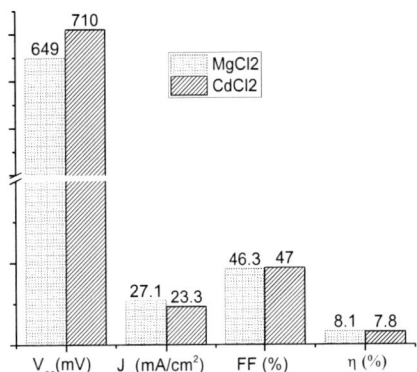

Fig. 2. A comparison of the device parameters of CdTe/CdS solar cells activated using CdCl₂ and MgCl₂. The data corresponds to the average of measurements on six cells. Junction activation was at 410 °C, and time duration was 10 min for MgCl₂ and 20 min for CdCl₂. Activation time of 10 min. was not enough to produce comparable efficiency for CdCl₂ case.

(atoms/cc) of S atoms in CdTe at distances 0.5, 1, and 2 μm from the CdTe surface, which gives a qualitative information about the extent of diffusion of S from CdS into the bulk of CdTe. It is clear that the S diffusion is significant in all cases across the bulk of CdTe, and S diffusion is significantly higher for devices activated with CdCl₂. This observation is validated by the EQE data discussed below. It should be noted that the above observation is valid for different activation durations.

The EQE spectra of two representative devices, one activated with MgCl₂ and the other with CdCl₂ at identical conditions (410 °C, 10 min.) are shown in Fig. 4. Two important features are (i) sharp CdS absorption edge at about 510 nm of the MgCl₂ treated device, and (ii) high blue response of the CdCl₂ treated device. The sharp absorption edge in MgCl₂ treated device indicates that the CdS film is almost intact (no S loss) which is in agreement with the SIMS data that S diffusion is less (Fig. 3). On the other hand the enhanced blue response in CdCl₂ device shows the evidence of thinning of CdS as a result of S diffusion which is again in agreement with SIMS data (Fig.3, S diffusion is high for CdCl₂ activation). The shallow absorption edge (500 to 650 nm) of CdS in this case indicates the intermixing and the formation of a ternary compound CdTe₁₋ₓSₓ at the interface with band gap slightly lower than that of CdTe. This explains the slight displacement of EQE curve at the long wavelength end for CdCl₂ case. The above observations indicate that further studies are needed to identify the optimum MgCl₂ activation conditions to produce adequate intermixing which reduces the lattice mismatch and the interface recombination center density.

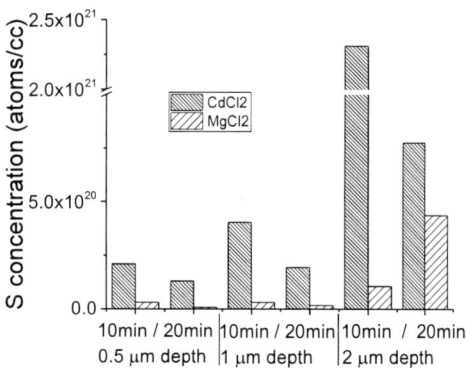

Fig. 3. Quantitative information about the concentration of S atoms at different locations (0.5, 1, and 2 μm from CdTe surface) in the CdTe film. The data corresponds to two pairs of devices, one activated with $CdCl_2$ and the other with $MgCl_2$. The first pair was annealed for 10 min and the second pair for 20 min. The annealing temperature was at 410 °C in all cases.

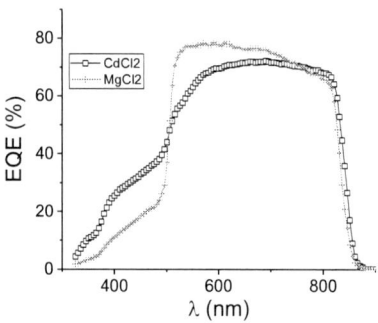

Fig. 4. EQE data of two CdTe/CdS devices activated using $MgCl_2$ (CT-277GA) and $CdCl_2$ (CT-276GA). The temperature was 410°C and duration 10 min.

CONCLUSION

The study reveal that $MgCl_2$ can be a substitute to $CdCl_2$ in junction activation giving equally good results in CdTe/CdS device performance. The experimental data on S diffusion and junction alloying suggest that the recrystallization process depends on the type of the chloride salt used. The observed slow diffusion of S during $MgCl_2$ activation indicates that the temperature and time window is not as narrow as in the case of $CdCl_2$ activation, hence $MgCl_2$ can give better control of junction activation. Further studies are needed to understand the elemental diffusion, alloying at the CdTe/CdS interface, and to optimize the junction activation parameters using $MgCl_2$.

ACKNOWLEDGEMENTS

The authors thank María Luisa Ramón García for the XRD measurements, Rogelio Morán Elvira for the SEM images, and Gildardo C. Segura for technical support in the CSS laboratory. This work is part of the project CeMIE-Sol 207450/P25. The data analysis and write-up of this work was performed during the sabbatical period of X. Mathew and served as the foundation for the alternative thermal processing procedure for the nano-wire CdTe project. X. Mathew acknowledge the ECE department at UKy for hosting his sabbatical year and the support of CONACyT through the project 0265494.

REFERENCES

[1] S. P. Harvey, G. Teeter, H. Moutinho, M.M. Al-Jassim, "Direct evidence of enhanced chlorine segregation at grain boundaries in polycrystalline CdTe thin films via three-dimensional TOF-SIMS imaging", *Prog. Photovolt.: Res. Appl.*, vol. 23, pp. 838–846, 2015.

[2] N. Spalatu, J. Hiie, V. Mikli, M. Krunks, V. Valdna, N. Maticiuc, T. Raadik, M. Caraman, "Effect of $CdCl_2$ annealing treatment on structural and optoelectronic properties of close spaced sublimation CdTe/CdS thin film solar cells vs deposition conditions", *Thin Solid Films*, vol. 582,pp. 128–133, 2015.

[3] S.A. Galloway, A.W. Brinkman, K. Durose, P.R. Wilshaw, A.J. Holland, "A study of the effects of post-deposition treatments on CdS/CdTe thin film solar cells using high resolution optical beam induced current", *Appl. Phys. Lett.*, vol. 68, pp. 3725–3727, 1996.

[4] W.K. Metzger, D. Albin, M.J. Romero, P. Dippo, M. Young, "$CdCl_2$ treatment, S diffusion, and recombination in polycrystalline CdTe", *J. Appl. Phys.*, vol. 99, pp. 103703, 2006.

[5] M. Emziane, K. Durose, N. Romeo, A. Bosio, D.P. Halliday, D.P., "A combined SIMS and ICPMS investigation of the origin and distribution of potentially electrically active impurities in CdTe/CdS solar cell structures", *Semicond. Sci. Technol.*, vol. 20, pp. 334–442, 2005.

[6] K. Durose, M.A. Cousins, D.S. Boyle, J. Beier, D. Bonnet, "Grain boundaries and impurities in CdTe/CdS solar cells", *Thin Solid Films*, vol. 403,pp. 396–404, 2002.

[7] T.X. Zhou, N. Reiter, R.C. Powell, R. Sasala, P.V. Meyers, "Vapor chloride treatment of polycrystalline CdTe/CdS films", *in: Conference Record of the 1994 IEEE First World Conference on Photovoltaic Energy Conversion*, vol. 1, pp. 103-106, 1994.

[8] Y, Qu, P.V. Meyers, B.E. McCandless, "HCl vapor post-deposition heat treatment of CdTe/CdS films", *in: Conference Record of the 1996 IEEE Twenty Fifth Conference on Photovoltaic Energy Conversion*, pp. 1013-1016, 1996.

[9] J.D. Major, L. Bowen, R.E. Treharne, L.J. Phillips, K. Durose, "NH_4Cl alternative to the $CdCl_2$ treatment step for CdTe thin-film solar cells", *IEEE J. Photovolt.*, vol. 5, pp. 386–389, 2015.

[10] J.D. Major, R.E. Treharne, L.J. Phillips, K. Durose, K., "A low-cost non-toxic post-growth activation step for CdTe solar cells", *Nature*, vol. 511, pp. 334–337, 2014.

[11] C. Drost, B. Siepchen, V. Krishnakumar, B. Späth, C. Kraft, T. Modes, O. Zywitzki, "Activation of CdTe-based thin films with zinc chloride and tetrachlorozincates", *Thin Solid Films*, vol. 582, pp. 100–104, 2015.

[12] S. Mazzamuto, L. Vaillant, N. Romeo, A. Bosio, N. Armani, G. Salviati, "A study of the CdTe treatment with a Freon gas such as CHF_2Cl", *Thin Solid Films*, vol. 516, pp. 7079–7083, 2008.

[13] J. Drayton, R. Geisthardt, J. Raguse, J.R. Sites, "Metal chloride passivation treatments for CdTe solar cells", *in: Proceedings of the 2013 MRS Spring Meeting*, Cambridge Univ. Press, vol. 1538, pp. 269–274, 2013.

[14] G. Angeles-Ordóñez, E. Regalado-Pérez, M.G. Reyes-Banda, N.R. Mathews, X. Mathew, "CdTe/CdS solar cell junction activation: Study using $MgCl_2$ as an environment friendly substitute to traditional $CdCl_2$", *Solar Energy Mater. & Solar Cells*, vol. 160, pp. 454–462, 2017

Variation of Cu Content of Sprayed Cu(In,Ga)(S,Se)$_2$ Solar Cells Based on a Thiol-Amine Solvent Mixture

Panagiota Arnou [1*], Soňa Uličná [1], Alexander Eeles [1], Mustafa Togay [1], Lewis D. Wright [1], Andrei V. Malkov [2], John M. Walls [1] and Jake W. Bowers [1]

[1] CREST, Wolfson School of Mechanical, Electrical and Manufacturing Engineering, [2] Department of Chemistry, Loughborough University, Leicestershire, LE11 3TU, UK

Abstract— Cu(In,Ga)(S,Se)$_2$ (CIGS) thin films were formed by a low cost solution-based approach using metal sulfide precursors. The stoichiometry of the absorber layer is tailored in order to improve film morphology and electrical properties. Cu$_y$In$_{0.7}$Ga$_{0.3}$Se$_2$ films were prepared with a varied Cu content (0.8>y>1.1) and were completed in solar cell devices. The compositional, structural and electrical properties of the devices were investigated. Increased Cu content improves lateral crystallization, but results in the formation of Cu-rich secondary phases in-between CIGS grain boundaries. Characterization of the completed devices shows that Cu content has an important effect on the device electrical properties and the dominant recombination mechanisms.

Index Terms —CIGS, low cost, solar cells, solution processing, stoichiometry

I. INTRODUCTION

The chalcopyrite semiconductor CuInSe$_2$, along with its related alloys (CIGS), is a promising light absorbing material commonly used in thin film solar cells. Due to desirable material properties, such as high optical absorption, tunable bandgap and high stability, CIGS solar cells have the highest performance among thin film technologies [1]. CIGS solar modules are conventionally fabricated using well-established vacuum-based techniques, such as multi-stage co-evaporation or sputtering [1, 2]. Recently, there has also been increasing attention in atmospheric processes which are highly attractive for low cost production of photovoltaics.

The development of a hydrazine-based method was a breakthrough in the solution processing of CIGS [3]. High quality absorbers can be fabricated using this method, owing to the solvent properties of hydrazine and the excellent solubility of metal chalcogenides [3]. Nonetheless, the high toxicity and explosive nature of hydrazine raise safety concerns which hinder the potential for commercialization. A safer alternative solvent combination of a diamine and a dithiol has recently been found to effectively dissolve metal chalcogenides [4]. Following this work, molecular-based approaches have been developed for CIGS solar cells based on the amine-thiol system [5-8].

In previous work, we presented a straight-forward deposition technique for CIGS thin films, starting from metal sulfides dissolved in a mixture of 1,2-ethylenediamine (EDA) and 1,2-ethanedithiol (EDT) [5]. Addition of Ga metal in the starting solution allowed a fine bandgap adjustment and improved photovoltaic performance [6]. These devices were limited by the incomplete crystallization of the absorber and the formation of two separate layers after selenization, which is a common problem found in solution-based CIGS [5, 9].

The ratio of Cu to Ga+In (CGI) is known to have a strong effect on grain size in vacuum deposited films. Films with CGI>1 show much larger grains than films with CGI<1. This is thought to be due to the effect of Cu$_x$Se forming a quasi-liquid surface layer which acts as a fluxing agent [10]. Devices with CGI>1 however are usually dominated by interface recombination and have much lower efficiencies than devices with CGI<1 [11]. For this reason, most high performance devices undergo a Cu rich (CGI>1) stage during the film formation in order to promote grain growth, but are then eventually finished as Cu-poor with CGI in the range of 0.88 to 0.95 [2]. Alternatively, CIGS can be made as Cu-rich, followed by a chemical wet etching step to selectively remove Cu$_x$Se phases [11].

In this work we varied CGI in an effort to improve the crystal quality and fully recrystallize the absorber layer. The impact of the absorber composition on the film microstructure and solar cell properties is investigated.

II. EXPERIMENTAL

CIGS thin films were spray-deposited in ambient atmospheric conditions, using a similar approach to what has been reported previously [6]. Here, a constant GGI and a varied CGI were attempted. Different stock solutions were prepared for In$_2$S$_3$ (717mg In$_2$S$_3$, 10ml EDA, 1ml EDT), Cu$_2$S (350mg Cu$_2$S, 10ml EDA, 1ml EDT) and Ga precursor (107mg Ga, 243mg Se, 7ml EDA, 0.7ml EDT). In$_2$S$_3$ and Cu$_2$S precursor solutions were dissolved at room temperature, whilst the Ga precursor required mild heating at ~50°C. After dissolution, the three component solutions were mixed in predetermined ratios to form the CIGS precursor solution with the desired composition. The GGI ratio was fixed to 0.3 and the CGI ratio was varied from 0.8 to 1.1. The mixed precursor solution was left stirring for 3-4 hours. Before deposition, the precursor solution was diluted with ethyl acetate (2:1 v/v) and then filtered (0.45 μm PTFE).

The films were sprayed onto molybdenum-coated soda lime glass substrates placed on a hot plate, controlled at 310°C. A deposition/drying cycle was repeated 6 times. Unless otherwise stated, the same precursor solution was used for all the spray cycles. In one case, however, the first 3 sprayed layers were performed using the solution with CGI=1.0, followed by 3 layers with CGI=0.8. Layers of different stoichiometry were combined in order to promote elemental interdiffusion and improve recrystallization during the subsequent selenization step. A graded CGI profile can also result in a combination of the favourable interface properties of Cu-poor material and the improved crystal growth of Cu-rich material [11].

After the last deposition/drying cycle, a selenization step was performed inside a tube furnace, where two 2.5x2.5cm samples were placed inside a graphite box with Se pellets. The tube was first purged with nitrogen, after which the pressure was set to 450 Torr. The tube remained sealed during selenization in order to allow a higher Se partial pressure. A total heating time of 50min (~35°C/min) and a final temperature of 540°C resulted in a final pressure of ~770 Torr and evaporation of the entire Se amount (~300 mg).

Two sister samples of each composition were deposited and selenized in the same run. One sample of each pair was chemically etched for 30sec in a 10% KCN aqueous solution immediately before the CdS buffer layer deposition. Although the formation of Cu_xSe phases is unlikely for CGI<1, the KCN etch can have additional beneficial effects, such as recovery of the minority carrier lifetime for air-exposed samples [11]. The CdS layer (~60nm) was deposited by chemical bath deposition. The intrinsic ZnO and Al doped ZnO layers (~80nm and 500nm, respectively) were both deposited using RF sputtering. No contact grid or anti-reflective coatings were used in this configuration. Mechanical scribing was performed to define individual cells of ~0.25cm^2 area. Sodium is only unintentionally supplied from diffusion from the glass substrate.

Device J-V characterization was performed using an in-house solar simulator under 1000W/m^2 illumination. The film morphology was investigated using a Carl Zeiss 1530 VP field emission gun scanning electron microscope (FEGSEM) with 30 μm aperture size and 5 kV operating voltage. The grain size was measured offline from the SEM images using AxioVision software (release 4.9.1, Zeiss). Energy Dispersive X-ray spectroscopy (EDS) was used for compositional analysis, with an aperture size of 60 μm and 20 kV operating voltage. A Bruker D2 phaser X-ray diffractometer was used for X-ray diffraction (XRD) analysis, using a Cu-Kα X-ray source and a Lynxeye detector. Capacitance-voltage profiling (CV) was carried out using a Keysight E4990 Impedance Analyzer and four-point probes. The temperature was adjusted with a LakeShore 335 Temperature Controller through a Janis CCS-150 closed cycle helium cryostat. The current density-voltage (JVT) characteristics were measured at different temperatures using a Keysight B2902A unit under 500W/m^2 illumination.

III. RESULTS

Five CIGS thin films were synthesized targeting a varied Cu content, as described in the experimental section. The targeted CGI varied from 0.8 to 1.1, with 0.1 increments. Additionally, one sample was prepared by combining solutions with CGI=1.0 and 0.8 (0.8/1.0).

The microstructure of the films was investigated. Fig. 1 shows the top view SEM images (left column), as well as the cross sectional images (right column) for each sample. For the Cu-rich sample (CGI=1.1), the SEM images after the KCN etch are also shown (1.1 E). The images of the etched samples for the rest of the compositions are omitted, as no influence is visible on the film microstructure.

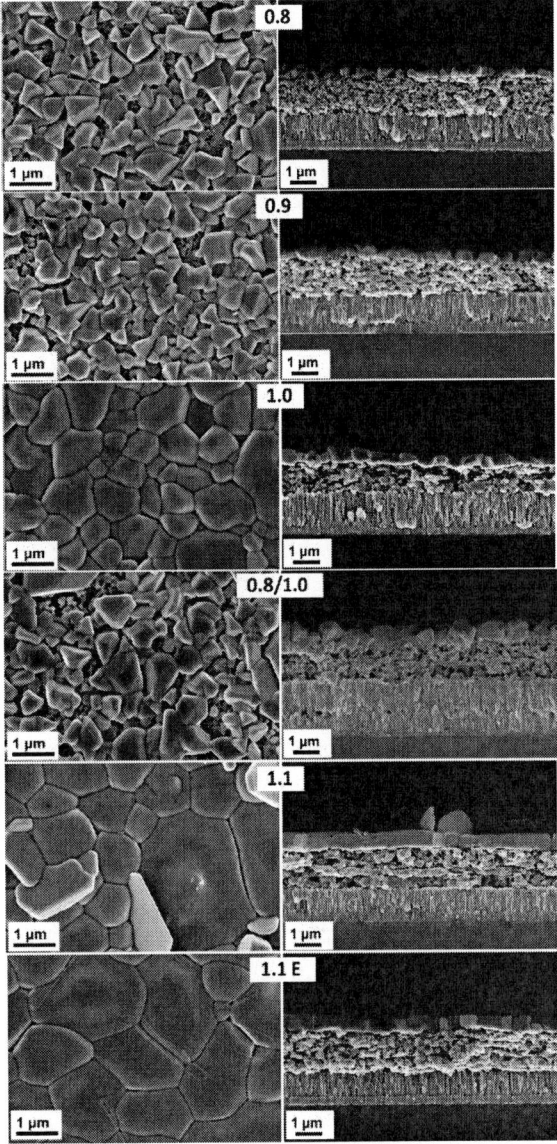

Fig. 1   Top-view (left) and cross section (right) SEM images of selenized CIGS layers with different CGI ratio.

The KCN etch dissolves the Cu_xSe secondary phases initially present on the Cu-rich sample. Surprisingly, this leaves behind a "chalk outline" on the film surface. It is still unclear whether this feature degrades device performance. Nonetheless, the KCN etch does not seem to form significant voids, which would be deleterious to the device performance.

In terms of the crystal quality, as anticipated, grain growth is significantly improved with CGI. The lateral grain size was increased from ~1.1 μm (CGI=0.8) to >3μm (CGI=1.1). The cross sectional images, however, show that the crystallized depth of the absorber remains fairly constant (500-700nm), despite the increased lateral grain size. Fig. 2 (top) shows the influence of the CGI ratio on the lateral and vertical grain size. The constant crystalline depth could suggest that the recrystallization is more likely limited due to non-optimum selenization or due to the presence of oxides/residual carbon in the film. Total carbon, hydrogen and nitrogen contents were determined for an as-deposited and a selenized sample with CGI=0.9, using a CE-440 CHN Elemental Analyzer (Exeter Analytical Inc., Europe). The as deposited sample contained a carbon, hydrogen and nitrogen content of 4.1, 0.7 and 2.1 at.% respectively, whilst the corresponding contents in the selenized sample were 1.1, 0.0 and 0.8 at.%. The low C content suggests that this is unlikely to cause the incomplete crystallization, as opposed to other atmospheric techniques with a C content of up to 60 at.% in the uncrystallised bottom layer [7]. Fig. 2 (bottom) shows the CGI and GGI ratios for each sample, as determined by EDS analysis. It is confirmed that the GGI ratio is fixed to ~0.3 and that the CGI ratio is increased.

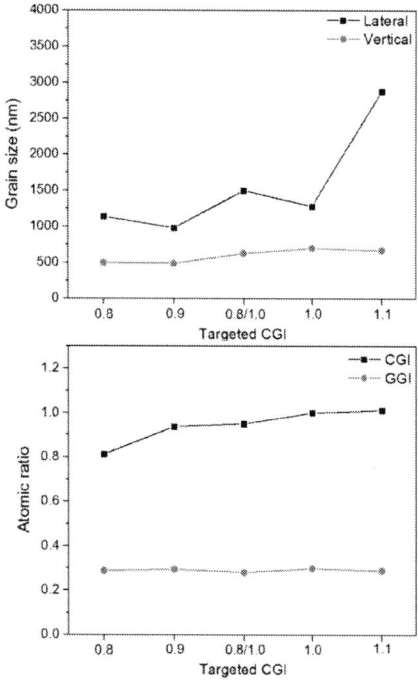

Fig. 2 Top: Lateral and vertical grain size for each sample. Bottom: CGI and GGI ratios for each sample, as determined by EDS.

The effect of the composition on the structural properties of CIGS films was characterized by XRD. Fig. 3 shows the XRD pattern of each sample. Each pattern consists of the same distinct peaks associated with the chalcopyrite structure of $CuIn_{0.5}Ga_{0.5}Se_2$ (JCPDS 40-1488), as well as peaks which correspond to Mo and $MoSe_2$. An additional peak is only evident at ~31.3° and it is removed after the KCN etch. This peak is likely associated with the Cu_xSe secondary phases. The inset table summarizes the position and the full width at half maximum (FWHM) of the (1 1 2) peak for each sample. The peak position remained unchanged, which suggest that the variation in the Cu content did not change the lattice parameters of the chalcopyrite structure. Finally, there is small decrease of the FWHM of the peak with CGI, which is consistent with the increase of the grain size, as shown by SEM imaging.

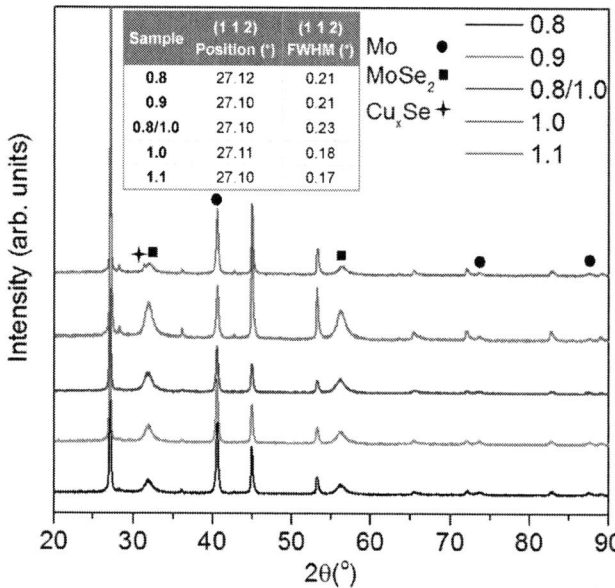

Sample	(1 1 2) Position (°)	(1 1 2) FWHM (°)
0.8	27.12	0.21
0.9	27.10	0.21
0.8/1.0	27.10	0.23
1.0	27.11	0.18
1.1	27.10	0.17

Fig. 3 XRD patterns of each selenized sample. Inset table: Summary of the (1 1 2) peak position and FWHM.

Fig. 4 shows the J-V curves of the highest performing cell of each sample. The J-V curves are considerably varied with Cu content. The Cu-poor samples have similar V_{oc} and J_{sc}, whilst these values are lower in the stoichiometric sample (CGI=1.0). This could be due to higher porosity in the bulk of the absorber, as seen in the cross sectional image of the stoichiometric sample, which could cause shunting losses. Additionally, the lower V_{oc} could be related to inferior interface properties. As expected, the Cu-rich sample exhibits a lower efficiency. This could be caused by shunting induced by the presence of a large amount of Cu_xSe secondary phases. However, the shunt resistance measured on the dark IV curve is not significantly high. This effect, as well as the fact that the KCN etch (designed to remove Cu_xSe) only marginally improves the performance, suggest that the performance loss is mostly caused by the inferior electronic properties of the Cu-rich material. The best result was obtained for the graded sample (0.8/1.0 CGI) with an efficiency (η), fill factor (FF), open circuit voltage (V_{OC}) and short circuit current density (J_{SC}) of 9.0%, 50.2%, 547mV and 32.6mA/cm^2, respectively. This demonstrates the possibility of this technique to combine the larger grains from the stoichiometric layer with a Cu-poor overall composition. As previously discussed however there is another factor which limits the vertical grain growth. It should also be noted that FF is relatively low for all the samples due to series resistance losses caused by the thick MoSe$_2$ layer, the incompletely crystallized absorber, and the lack of a metallic collection grid. A barrier layer at the back contact is currently under development to control the MoSe$_2$ layer formation. This is expected to improve the FF of the devices and make the effect of the CGI clearer.

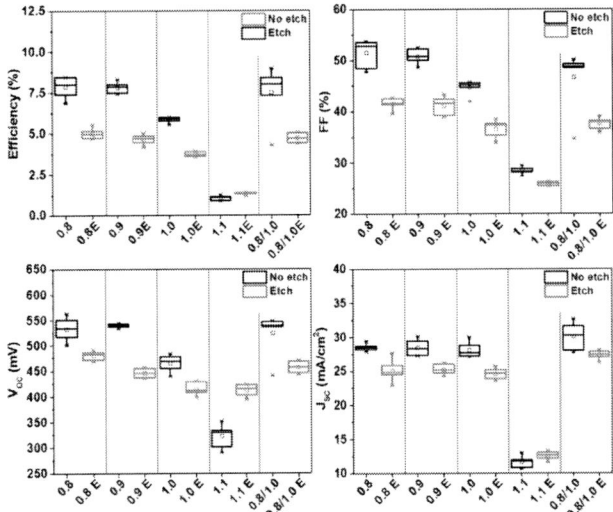

Fig. 5   Box plots summarizing the results from J-V measurements for each sample, before and after KCN etch.

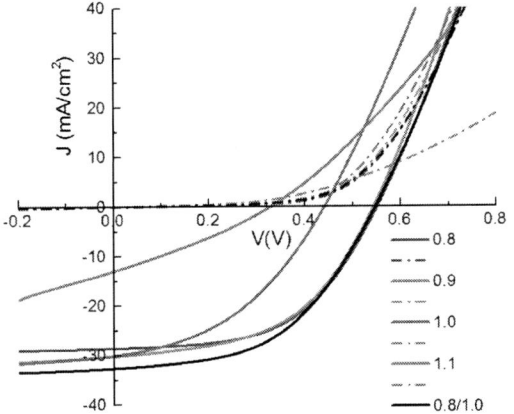

Fig. 4   J-V curves for the highest performing CIGS cells for each sample.

Statistical analysis was performed by measuring 6 adjacent cells on each sample. The J-V characteristics (η, FF, J_{SC}, V_{OC}) are summarized in the box plots of Fig. 5. It is evident that both η and FF values are decreased with CGI, as also seen in Fig. 4. J_{SC}, on the other hand, remains fairly constant, with the exception of the Cu-rich sample. The performance parameters of each sample after the 30sec etch are also included. Interestingly, the performance decreases for all the samples, apart from the Cu-rich, for which the improvement is marginal. This suggests that longer etching times could be required, which could also be associated with the presence of the outline features on the surface of the etched sample.

Fig. 6 shows the light and dark J-V curves at different temperatures for the sample with CGI=0.8. The light measurements were performed at a constant light intensity of 500W/m^2, using a halogen light source. The J-V curves for the rest of the samples are omitted as they have a similar behaviour. No roll-over of the J-V is observed at low temperatures indicating that there are no significant diode current barriers.

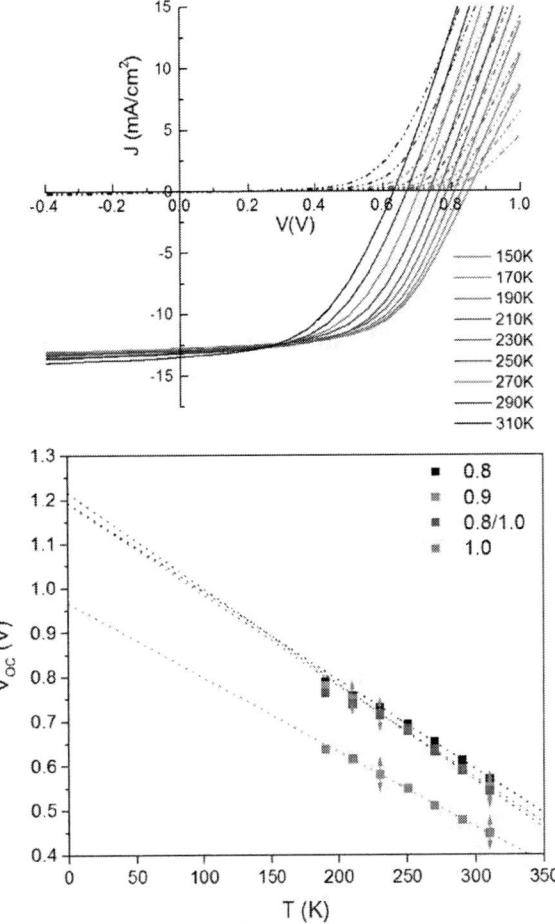

Fig. 6   Top: The light and dark J–V curves at varying temperatures from 150 to 310K for the sample with CGI=0.8. Bottom: Temperature dependence V_{OC} data for each sample, determined by the light J–V curves.

Fig. 6 shows the open circuit voltage as a function of temperature for each device. Extrapolation of Voc to 0K gives an activation energy for recombination $E_a=qV$ of ~1..19 eV for CGI=0.8 and 0.9 and ~1..21 eV for CGI=0.8/1.0 [12]. These values are very similar to the expected bandgap of the material based on the empirical formula for $CuIn_{1-x}Ga_xSe_2$: $E_g=1.65\times+1.01(1-x)-0.151(1-x)x$, with x=0.3 [13]. This indicates that the main recombination mechanism in these devices occurs in the bulk of the absorber, which is common for CIGS solar cells [3, 12]. In contrast, the extrapolation for the sample with CGI=1.0 gives a lower E_a of about ~0.97 eV. The fact that the E_a value is smaller than the bandgap of the absorber confirms that the device is limited by interface recombination rather than recombination in the bulk, which is the common result for stoichiometric devices [11]. The doping profile of each device was extracted from each CV curve at 300 K. These profiles (Fig. 7) indicate that the net doping density is higher in the stoichiometric film than in the Cu-poor, which is consistent with vacuum processed devices [11].

Based on these results, it is shown that the Cu content has a significant effect on the structural properties of the absorber film and the electrical properties of the solar cell device.

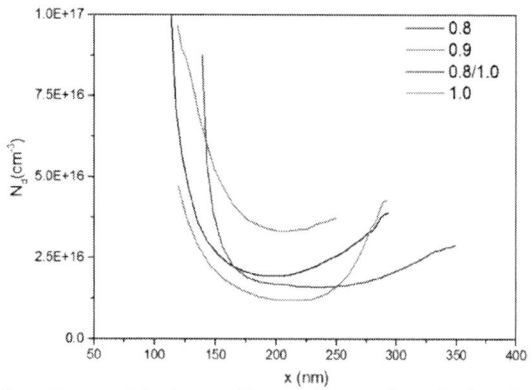

Fig. 7 Extracted doping profiles vs. distance from the junction for each CIGS sample, measured at 300 K.

III. CONCLUSIONS

The effects of CGI on the material and device properties of solution-processed CIGS were investigated. It was shown that good compositional control is possible using this deposition approach. The effect of CGI on the electrical properties of the devices was found to be similar to that seen in vacuum processes, with stoichiometric and Cu-rich devices dominated by interface recombination. Interestingly, the improved grain growth anticipated for higher CGI samples was observed only in the lateral direction, with the crystalline depth remaining fairly constant. Depositing a Cu-poor layer on top of a Cu-rich layer produced the highest efficiency by combining the larger crystals of the Cu-rich material with the favorable interface properties of a Cu-poor film. This device however was still limited by incomplete crystallization through the depth of the absorber.

ACKNOWLEDGEMENT

The authors are grateful for funding from the EPSRC grant (EP/N026438/1) to support this work.

REFERENCES

[1] R. Kamada, T. Yagioka, S. Adachi, A. Handa, K. F. Tai, T. Kato, and H. Sugimoto, "New world record Cu(In, Ga)(Se, S)2 thin film solar cell efficiency beyond 22%," *2016 IEEE 43rd Photovoltaic Specialists Conference*, pp. 1287–1291, 2016.

[2] P. Jackson, D. Hariskos, R. Wuerz, O. Kiowski, A. Bauer, T. M. Friedlmeier, and M. Powalla, "Properties of Cu(In,Ga)Se2 solar cells with new record efficiencies up to 21.7%," *Phys. Status Solidi RRL*, vol. 9, no. 1, pp. 28–31, Dec. 2014.

[3] T. K. Todorov, O. Gunawan, T. Gokmen, and D. B. Mitzi, "Solution-processed Cu(In,Ga)(S,Se)2 absorber yielding a 15.2% efficient solar cell," *Prog. Photovolt Res. Appl.*, vol. 21, no. 1, pp. 82–87, 2012.

[4] D. H. Webber and R. L. Brutchey, "Alkahest for V2VI3 chalcogenides: dissolution of nine bulk semiconductors in a diamine-dithiol solvent mixture.," *J. Am. Chem. Soc.*, vol. 135, no. 42, pp. 15722–15725, Oct. 2013.

[5] P. Arnou, M. F. A. M. Van Hest, C. S. Cooper, A. V. Malkov, J. M. Walls, and J. W. Bowers, "Hydrazine-Free Solution-Deposited CuIn(S,Se)2 Solar Cells by Spray Deposition of Metal Chalcogenides," *ACS Appl. Mater. Interfaces*, vol. 8, no. 19, 2016.

[6] P. Arnou, C. S. Cooper, S. Uličná, A. Abbas, A. Eeles, L. D. Wright, A. V. Malkov, J. M. Walls, and J. W. Bowers, "Solution processing of CuIn(S,Se)2 and Cu(In,Ga)(S,Se)2 thin film solar cells using metal chalcogenide precursors," *Thin Solid Films*, 2016.

[7] X. Zhao, M. Lu, M. Koeper, and R. Agrawal, "Solution-Processed Sulfur Depleted Cu(In,Ga)Se2 Solar Cells Synthesized from a Monoamine-Dithiol Solvent Mixture," *J. Mater. Chem. A*, vol. 4, no. 19, pp. 7390–7397, 2016.

[8] D. Zhao, Q. Tian, Z. Zhou, G. Wang, Y. Meng, D. Kou, W. Zhou, D. Pan, and S. Wu, "Solution-deposited pure selenide CIGSe solar cells from elemental Cu, In, Ga, and Se," *J. Mater. Chem. A Mater. energy Sustain.*, vol. 3, no. 38, pp. 19263–19267, 2015.

[9] I. Klugius, R. Miller, A. Quintilla, T. M. Friedlmeier, D. Blázquez-Sánchez, E. Ahlswede, and M. Powalla, "Growth mechanism of thermally processed Cu(In,Ga)S2 precursors for printed Cu(In,Ga)(S,Se)2 solar cells," Phys. status solidi – Rapid Res. Lett., vol. 6, no. 7, pp. 297–299, 2012.

[10] R. Klenk, T. Walter, H.-W. Schock, and D. Cahen, "A model for the successful growth of polycrystalline films of CuInSe2 by multisource physical vacuum evaporation," *Adv. Mater.*, vol. 5, no. 2, pp. 114–119, 1993.

[11] S. Siebentritt, L. Gütay, D. Regesch, Y. Aida, and V. Deprédurand, "Why do we make Cu(In,Ga)Se2 solar cells non-stoichiometric?," *Sol. Energy Mater. Sol. Cells*, vol. 119, pp. 18–25, 2013.

[12] S. S. Hegedus and W. N. Shafarman, "Thin-film solar cells: device measurements and analysis," *Prog. Photovoltaics Res. Appl.*, vol. 12, no. 2–3, pp. 155–176, 2004.

[13] B. J. Stanbery, "Copper Indium Selenides and Related Materials for Photovoltaic Devices," *Crit. Rev. Solid State Mater. Sci.*, vol. 27, no. 2, pp. 73–117, Apr. 2002.

978-1-5090-5606-4/17 $31.00 © 2017 IEEE

CuInSe$_2$ absorber layer grown under copper excess with a copper poor surface formed by a KF post deposition treatment

Finn Babbe, Hossam Elanzeery, Michele Melchiorre and Susanne Siebentritt

Laboratory for Photovoltaics, Physics and Material Science Research Unit, University of Luxembourg, 41, Rue du Brill, L-4422 Belvaux, Luxembourg

Abstract — **The post deposition treatment (PDT) of chalcopyrite absorber layers with heavy alkalis leads to an increase in power conversion efficiencies in CI(G)S solar cells. Here, we present an in-situ treatment with etching step on CuInSe$_2$ absorber layers grown under copper excess, which improves the open circuit voltage (V$_{OC}$), the fill factor (FF) and power conversion efficiency. Low temperature photoluminescence measurements show that besides the stoichiometric absorber a second copper deficient phase is present near the surface after the PDT. Calibrated photoluminescence measurements reveal that there is an improvement in quasi Fermi level splitting after treatment contributing to the improved V$_{OC}$.**

I. INTRODUCTION

The post deposition treatment (PDT) of copper indium gallium di-selenide Cu(In,Ga)Se$_2$ (CIGS) chalcopyrite absorber layers with heavy alkalis [1], led to an increase in power conversion efficiencies reaching up to 22.6 % on the laboratory scale [2]. The effects of the PDT on the absorber layer are manifold and a sophisticated understanding is still missing. The ternary copper indium di-selenide (CIS) shows lower power conversion efficiencies than CIGS but with its less complex structure it is a good material to help unveil the effects of KF PDT. The copper content in CIS can vary over a broad range. All record cells are grown non stoichiometric with copper over indium ratio well below 1 (Cu-poor). From the phase diagram [3] it can be seen that absorbers grown under copper excess (Cu-rich) form stoichiometric CIS as well as a secondary copper selenide phase. For solar cell application this secondary phase has to be removed by potassium cyanide (KCN) etching, since it has highly conductive properties. After etching, all Cu-rich absorbers have a stoichiometric composition.

There are two different methods reported to perform a KF PDT. The in-situ treatment leading to the highest power conversion efficiency [1][2] is done directly after the absorber growth. The cooldown from growth temperature is stopped at an elevated temperature and KF is deposited within the same vacuum chamber. The second method is the ex-situ treatment. For this method the sample is taken out of the vacuum chamber to deposit the KF in another machine. Afterwards the sample is placed back into vacuum to be annealed at elevated

temperature under selenium atmosphere [4][5][6]. The first method cannot be used as described for CIS absorber grown under copper excess, since the secondary copper selenide has to be removed before the deposition of KF. Using the second method, improvements in the open circuit voltage (V$_{OC}$) have been reported [4][5], but at decreased fill factor (FF) and lower power conversion efficiency for both Cu-poor and Cu-rich material. The reduced parameters have been attributed to the hygroscopic nature of KF, which enables it to adsorb water from ambient air during the transfer between KF deposition and subsequent annealing [4]. To overcome those problems a third method for the PDT was developed and will be presented in this contribution. We label this method "in-situ PDT with etching step": it consists of the following steps. Similar to the ex-situ treatment, the samples are taken out of the vacuum chamber after growth and a KCN etch is used to remove secondary phases. Immediately after etching, the samples are transferred back into the vacuum chamber. There the sample is heated up and, similar to the in-situ PDT, the KF is deposited while the sample is annealed. To avoid the loss of selenium from the absorber at elevated temperatures, a constant background pressure of selenium is used. The rest of the baseline process is the same as for untreated solar cells and can be found in the experimental section. Using this method, parameters like annealing temperature, annealing duration and KF flux have been varied and studied on a total of 28 Cu-rich and Cu-poor absorbers. This proceedings paper concentrates on one treatment condition used on a Cu-rich CIS sample and the comparison to a reference sample without a PDT. Further results can be found more detailed in [7]. The samples are compared with respect to their current voltage characteristics as well as their spectral response measured by external quantum efficiency. Photoluminescence (PL) spectra at low temperatures are used to probe the electronic structure. CIS samples grown under copper excess show sharp transition lines corresponding to the defect states involved in the transition [8] whereas Cu-poor samples show broad peaks due to their high degree of compensation. In a second proceedings paper as well as in the more detailed paper [7][9] the advanced electrical characterizations are presented using the same in-situ PDT with etching step.

II. EXPERIMENTAL DETAILS

The CIS absorbers have been prepared in a single stage co-evaporation process on molybdenum coated soda lime glass substrates in a molecular beam epitaxy (MBE) system [10]. The samples are grown under copper excess and have a final copper over indium ratio of 1.3 after growth, determined by energy dispersive x-ray spectroscopy (EDX). Before the PDT, the Cu₂Se secondary phase is removed by a 5 minute etching step in 10% aqueous KCN solution. A reference sample from the same process is directly processed into a solar cells. After the etching step, the samples are immediately transferred back into the MBE system. During the PDT, the samples are heated up under selenium atmosphere to a temperature of 350 °C measured by a pyrometer and annealed under KF flux corresponding to roughly 1 nm/minute for 6 minutes. During the cooldown, the samples are still under selenium flux to avoid losses of it from the absorber layer. A small piece of the treated absorber is mechanically split off for analysis whereas the rest is used for solar cell fabrication. After rinsing the absorber layers with deionized water, the solar cells are finished with a roughly 50 nanometer thick CdS buffer layer deposited by chemical bath deposition, a sputtered zinc oxide window layer grown under bias voltage [11] and nickel/aluminum grids.

Current voltage characteristic (IV) are measured under the illumination of an AAA solar simulator. The external quantum efficiency (EQE) is measured by means of spectral response in a home built setup. Photoluminescence measurements are carried out under 660 nanometer wavelength continuous excitation from a diode laser. The emitted luminescence is collected by two off axis mirrors, focused into a fiber, spectrally resolved by a monochromator and detected by a InGaAs-CCD array. The calibration steps needed for photon counting, necessary for the extraction of the quasi Fermi level splitting (qFLs), are described elsewhere [12][13][supplement in 14]. For measurements at 10 K, the samples are introduced into a liquid Helium flow cryostat and are measured by the same optical set up.

III. EXPERIMENTAL RESULTS

The current voltage characteristic of the untreated reference solar cell (black) and treated solar cell (red) are depicted on the left side of Fig. 1. The PDT has several effects on the IV curve. The most prominent one is the increase in V_{OC} by about 50 mV from 343 mV to 393 mV. The V_{OC} gain is more distinct than in the former ex-situ PDT (30 mV - 40 mV) [4]. Furthermore the ex-situ PDT resulted in a smaller FF, higher diode factor and larger series resistance when compared to the reference sample. For the in-situ PDT with etching step the opposite is observed: A lower diode factor (decreased from 1.9 to 1.7) and an increased shunt resistance lead to a better fill factor (increase from 52% to 59%). The current density decreases slightly after the treatment, which can be mostly attributed to the reduced EQE response shown in Fig 2. The

loss in the short wavelength region is attributed to a thicker CdS layer [15]. The efficiency increases from 7.1% to 9.0% with the PDT. The breakdown behaviour seen in the untreated cell at reverse voltage above 0.2 V is not observed in the treated cell, like already seen in the ex-situ treatment [4]. The V_{OC} of absorbers grown under Se excess is known to be limited by recombination near the interface. Temperature dependent V_{OC} measurements show that this is still the case for the untreated cell but that in the treated cell the interface recombination is strongly reduced [9].

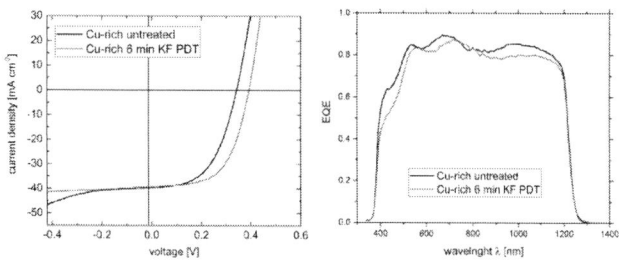

Fig. 1. Left-hand side: Current density plotted over the applied voltage under AM1.5 illumination condition measured for a standard Cu-rich cell (black) and a Cu-rich cell with KF PDT (red). Right-hand side: EQE measurement of the same solar cells.

To investigate the defect structure of the samples, low temperature PL measurements at 10 K are carried out on the untreated reference and on the treated absorber layer. The normalized PL spectra for both samples under equivalent illumination conditions are shown in Fig. 2 (a) and (b). A PL spectrum under lower excitation for the treated absorber is also included Fig. 2 (c). The energetic position and the excitation dependency of the reference Cu-rich absorber (a) agree well with the known results for CIS absorber grown under high Cu-excess [8]. The peak at 0.97 eV can be assigned to the donor-acceptor (DA) pair transition commonly denoted as DA2. The peaks around 0.94 eV and 0.91 eV are phonon replica of the DA transition. The distance between the replica peaks is 27 meV corresponding to the LO phonon involved which agrees well with literature data. At 1.04 eV, excitonic luminescence is observed with a low PL yield. The transition around 0.99 eV, commonly described as DA1, is only visible as a small shoulder. The excitonic luminescence as well as the fact that two phonon replica are visible indicate a good electronic structure of the absorber layer. The PL spectrum at low temperature of the absorber with PDT is more involved but can be understood when measured at various excitation powers. At low excitation (c) a broad asymmetric peak around 0.91 eV and a second peak around 0.96 eV are visible. With increasing excitation the broad peak at lower energy (dashed red line) shifts to higher energies with 17 meV per decade of excitation intensity. When the integrated peak intensity is plotted double logarithmically over the excitation power, an exponent of 1 can be extracted [16]. The shift with excitation intensity is quite large and indicates a high degree of compensation. This behaviour as well as the asymmetric shape are expected from an absorber layer with Cu-poor composition [8]. The second peak at higher energy (solid red line) shows a lower shift with excitation (4 meV per decade) and has a higher exponent (1.1). The characteristics match the DA2 transition plus the phonon replica known from the reference CIS sample grown under Cu-excess (part a of the figure). The only difference is a small shift of the peaks towards lower energies by 10 meV. Furthermore, the spectrum at high excitation (b) shows excitonic luminescence at 1.04 eV. From this, it can be deduced that within the probed surface region there is a stoichiometric ("Cu-rich") as well as Cu-poor phase that are both PL active. Given the fact that the samples with PDT show strongly reduced interface recombination (shown by IV(T) [9]) we propose that those phases are at least partly in a layered form, creating a Cu-poor surface.

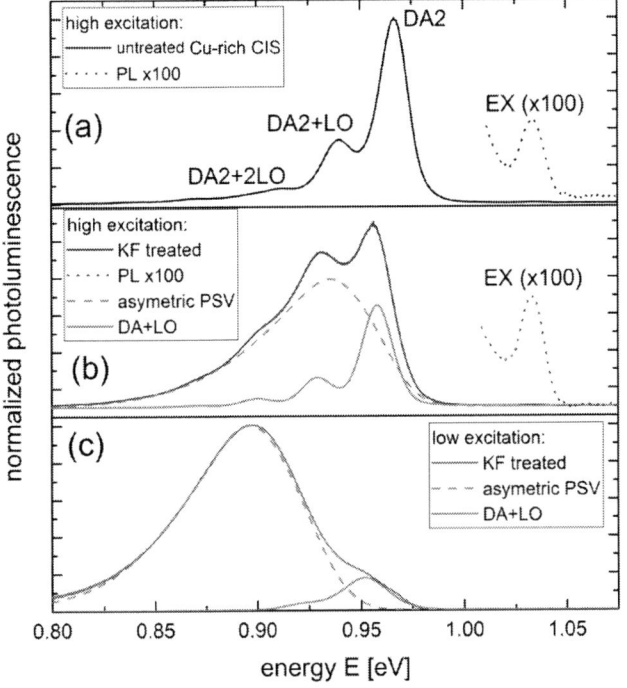

Fig. 2. Photoluminescence spectra measured at 10 K of the reference Cu-rich absorber (black) as well as a KF treated absorber at high (blue) and low excitation (green). The red curves represent the fitting functions used. The peak around 1.04 eV is magnified by a factor of 100 for better visibility.

From photoluminescence measurements at room temperature the quasi Fermi level splitting can be derived. To avoid possible errors from degraded surfaces due to oxidation, the qFLs is measured on absorber layers covered with CdS [14][17]. On the standard absorber a qFLs of roughly 465 meV is measured under the equivalent illumination of one sun whereas the absorber with PDT shows a qFLs of 485 meV. This indicates that the absorber and its interface with CdS are

improved by the PDT. The V_{OC} improvement is higher than the qFLS improvement, indicating that part of the V_{OC} loss in the untreated absorber occurs only after the complete formation of the pn-junction by the ZnO deposition. This is in agreement with previous observations and simulations [18] that indicated that the high electric field in the absorber is responsible for the V_{OC} loss.

IV. CONCLUSION

The presented in-situ PDT with etching step greatly increased the power conversion efficiency from 7.1 % to 9.0 %. This increase is driven by an increase in V_{OC} and FF, which is higher due to a higher shunt resistance as well as a lower diode factor. EQE measurements show a reduced spectral response in the short wavelength region attributed to a thicker CdS layer. Low temperature PL measurements of the reference sample confirm the favorable electronic properties of a CIS absorber layer grown under copper excess. The known DA2 transition with two phonon replica as well as excitonic luminescence is observed. The sample after treatment shows the spectrum of an absorber grown under copper excess with its sharp transition peaks superimposed by a broad asymmetric peak known from copper deficient CIS. From this it can be deduced that the probed surface region has two different phases which are assumed to be at least partly in a layered structure. The copper poor phase is proposed to make a good interface with the CdS, shifting the main recombination path from the interface (untreated) into the bulk after treatment. Calibrated PL measurements at room temperature show an increase in the qFLs by 20 meV with the treatment. This accounts only for half of the increase in V_{OC}, indicating that part of the V_{OC} loss in (untreated) Cu-rich solar cells is due to the high electric field in the absorber.

The authors gratefully acknowledge the Luxembourgish Fonds National de la Recherche (FNR) for funding

REFERENCES

[1] F. Pianezzi, P. Reinhard, A. Chirilă, B. Bissig, S. Nishiwaki, S. Buecheler, and A. N. Tiwari, "Unveiling the effects of post-deposition treatment with different alkaline elements on the electronic properties of CIGS thin film solar cells," *Phys. Chem. Chem. Phys.*, vol. 16, no. 19, p. 8843, 2014.

[2] P. Jackson, R. Wuerz, D. Hariskos, E. Lotter, W. Witte, and M. Powalla, "Effects of heavy alkali elements in Cu(In,Ga)Se2 solar cells with efficiencies up to 22.6%," *Phys. status solidi - Rapid Res. Lett.*, vol. 4, pp. 1–4, 2016.

[3] T. Gödecke, T. Haalboom and F. Ernst, "Phase equilibria of Cu-In-Se I. The In2Se3-Se-Cu2Se subsystem," *Zeitschrift für Met.*, vol. 91, no. 8, pp. 622–634, 1948.

[4] H. Elanzeery, F. Babbe, M. Melchiorre, A. Zelenina, and S. Siebentritt, "Potassium Fluoride Ex Situ Treatment on Both Cu-Rich and Cu-Poor CuInSe2 Thin Film Solar Cells," *IEEE*

J. Photovoltaics, vol. 7, no. 2, pp. 684–689, Mar. 2017.

[5] P. Pistor, D. Greiner, C. a. Kaufmann, S. Brunken, M. Gorgoi, A. Steigert, W. Calvet, I. Lauermann, R. Klenk, T. Unold, and M.-C. Lux-Steiner, "Experimental indication for band gap widening of chalcopyrite solar cell absorbers after potassium fluoride treatment," *Appl. Phys. Lett.*, vol. 105, no. 6, p. 63901, Aug. 2014.

[6] R. Kamada, T. Yagioka, S. Adachi, A. Handa, K. F. Tai, T. Kato, and H. Sugimoto, "New world record Cu(In,Ga)(Se,S)$_2$ thin film solar cell efficiency beyond 22%," in *2016 IEEE 43rd Photovoltaic Specialists Conference (PVSC)*, 2016, pp. 1287–1291.

[7] F. Babbe, H. Elanzeery, M. Melchiorre, A. Zelenina, and S. Siebentritt, "In-situ potassium fluoride treatment of on both Cu rich and Cu poor CuInSe$_2$ thin film solar cells." - submitted

[8] S. Siebentritt, N. Rega, A. Zajogin, and M. C. Lux-Steiner, "Do we really need another PL study of CuInSe$_2$?," *Phys. status solidi*, vol. 1, no. 9, pp. 2304–2310, Aug. 2004.

[9] H. Elanzeery, F. Babbe, A. Zelenina, M. Melchiorre, and S. Siebentritt, "Interface effects of alkali treatment on Cu-rich thin film solar cells," *IEEE 44th Photovolt. Spec. Conf.*, 2017.

[10] V. Deprédurand, T. Bertram, D. Regesch, B. Henx, and S. Siebentritt, "The influence of Se pressure on the electronic properties of CuInSe$_2$ grown under Cu-excess," *Appl. Phys. Lett.*, vol. 105, no. 2014, p. 172104, 2014.

[11] M. Hála, H. Kato, M. Algasinger, Y. Inoue, G. Rey, F. Werner, C. Schubert, T. Dalibor, and S. Siebentritt, "Improved environmental stability of highly conductive nominally undoped ZnO layers suitable for n-type windows in thin film solar cells," *Sol. Energy Mater. Sol. Cells*, vol. 161, pp. 232–239, Mar. 2017.

[12] P. Wurfel, "The chemical potential of radiation," *J. Phys. C Solid State Phys.*, vol. 15, no. 18, pp. 3967–3985, 1982.

[13] T. Unold and L. Gütay, "Photoluminescence Analysis of Thin-Film Solar Cells," in *Advanced Characterization Techniques for Thin Film Solar Cells*, Weinheim, Germany: Wiley-VCH Verlag GmbH & Co. KGaA, 2011, pp. 151–175.

[14] F. Babbe, L. Choubrac, and S. Siebentritt, "Quasi Fermi level splitting of Cu-rich and Cu-poor Cu(In,Ga)Se$_2$ absorber layers," *Appl. Phys. Lett.*, vol. 109, no. 8, p. 82105, Aug. 2016.

[15] A. Chirilă, P. Reinhard, F. Pianezzi, P. Bloesch, A. R. Uhl, C. Fella, L. Kranz, D. Keller, C. Gretener, H. Hagendorfer, D. Jaeger, R. Erni, S. Nishiwaki, S. Buecheler, and A. N. Tiwari, "Potassium-induced surface modification of Cu(In,Ga)Se$_2$ thin films for high-efficiency solar cells.," *Nat. Mater.*, vol. 12, pp. 1107–11, 2013.

[16] S. Siebentritt and U. Rau., *Wide-Gap Chalcopyrites*, vol. 86, no. 6058. Berlin/Heidelberg: Springer-Verlag, 2006.

[17] D. Regesch, L. Gütay, J. K. Larsen, V. Deprédurand, D. Tanaka, Y. Aida, and S. Siebentritt, "Degradation and passivation of CuInSe$_2$," *Appl. Phys. Lett.*, vol. 101, no. 11, 2012.

[18] Y. Aida, V. Depredurand, J. K. Larsen, H. Arai, D. Tanaka, M. Kurihara, and S. Siebentritt, "Cu-rich CuInSe$_2$ solar cells with a Cu-poor surface," *Prog. Photovoltaics Res. Appl.*, vol. 23, no. 6, pp. 754–764, Jun. 2015.

Cu₂ZnSnSe₄ Solar Cells onto Polyimide Substrates Fabricated at Low Temperature

Ignacio Becerril-Romero,[a] Simón López-Marino,[a] Moisés Espíndola-Rodríguez,[a] Laura Acebo,[a] Markus Neuschitzer,[a] Yudania Sánchez,[a] Edgardo Saucedo,[a] and Paul Pistor[a]

[a] IREC, Jardin de les Dones de Negre 1, 08930, Sant Adrià del Besòs, Spain

Abstract — **Polyimide is an interesting substrate for thin-film photovoltaics as corroborated by recent results obtained with CIGS. However, little work has been done to use this polymer for CZTSSe. This work proves the feasibility of fabricating flexible and efficient CZTSe devices on polyimide. The low thermal robustness and the lack of alkali of this material are tackled in this work. First, the effect of temperature on CZTSe devices fabricated on SLG at low temperature (< 500°C) is studied. Then, solar cells are fabricated on polyimide using Na and K doping strategies. A 4.9% efficiency device on polyimide is reported.**

Index Terms — **alkali doping, CZTSe, kesterite solar cells, low temperature, polyimide substrate, SnSe₂ secondary phase.**

I. INTRODUCTION

A very attractive way of exploiting the capabilities of thin-film photovoltaics (PV) is through the use of flexible and light-weight substrates. Besides widening the range of applications of PV (e.g. portable electronics, BIPV, space applications, automotive industry, etc.), flexible substrates are expected to reduce fabrication costs through roll-to-roll high-throughput industrial processes. Cu₂ZnSn(S₁₋ₓSeₓ)₄ (CZTSSe) based solar cells are very well-suited for mass production since they are formed mainly by non-toxic and earth-abundant elements. Yet, soda-lime glass (SLG) is still the quintessential substrate for kesterites due to its favorable mechanical properties, availability, and its beneficial composition.

Polyimide (PI) has proved to be a very promising substrate yielding very high efficiency devices for more mature thin-film technologies, especially CIGS [1]. PI features several advantageous properties like high mechanochemical stability (for a polymer) or the absence of harmful impurities (it is mainly formed by C, H, N and O) that combined with its insulating nature avoid the use of additional barrier layers and allow monolithic interconnection. These characteristics make PI a very interesting substrate for flexible PV. Still, the kesterite community appears to be turning their back on this material. Only a very scarce number of publications report kesterite solar cells fabricated on polyimide substrates [2]-[4].Thus, in this work, we investigate the potential of this polymer as a low-cost, low-weight and flexible alternative to soda-lime glass for earth-abundant Cu₂ZnSnSe₄-based (CZTSe) solar cells. Two main concerns arise when working with polyimide: 1) its low thermal robustness limits process temperatures below 500°C and 2) its lack of alkalis in contrast

to conventional SLG necessary for high-efficiency CZTSe devices [5]-[7].

This work aims to tackle both issues. First, an experiment is carried out in which CZTSe devices are fabricated on SLG by sequential precursor sputtering and selenization at different temperatures (450-490°C). The effect of temperature and the viability of obtaining efficient working devices at such low temperatures are analyzed. Then, CZTSe devices are fabricated on polyimide by an identical procedure and different alkali doping strategies are investigated for the incorporation of Na and K into the kesterite absorber: pre-absorber synthesis (PAS) and post-deposition treatment (PDT). Finally, through further experimentation and the addition of a Ge nanolayer we report a 4.9% efficiency device on PI that sets a record for kesterite on a polymer substrate.

II. EXPERIMENTAL

Two different substrates are employed in this work: 3 mm soda-lime glass and 50 μm polyimide foil (Upilex). A 670 nm trilayer Mo back contact was deposited on the clean substrates by DC-magnetron sputtering (*Alliance Concept CT100*). CZTSe absorbers were synthesized by a sequential process in which a metallic precursor stack was deposited by DC-magnetron sputtering (*Alliance Concept Ac450*) onto the Mo back contact and calibrated to obtain Cu-poor and Zn-rich absorbers, and then selenized through a thermal reactive annealing.

A three-zone tubular furnace was used to synthesize the CZTSe absorber. Samples were placed inside a graphite box together with 100 mg of Se and 5 mg of Sn. A two-step thermal annealing was carried out: first at 400°C and 1.5 mbar for 30 min and then at 450°C-490°C (depending on the experiment) and 1 bar for 15 min.

The absorbers were subjected to three chemical etchings to remove (Zn,Sn,Cu)Se secondary phases: first in KMnO₄/H₂SO₄, then in (NH₄)₂S and finally in diluted KCN. A CdS buffer layer (~40 nm) was deposited by chemical bath deposition. The cells were then completed with i-ZnO (50 nm) and ITO (200 nm) layers deposited by DC-magnetron sputtering (*Alliance Concept CT100*). Individual 3×3 mm² solar cells were isolated using a manual microdiamond scriber (*MR200 OEG*).

In the case of polyimide samples, an additional step was performed in which a 0-15 nm layer of NaF or KF was thermally evaporated (*Oerlikon Univex 250*) to act as a doping layer. Two different approaches were employed in this regard: pre-absorber synthesis (PAS) and post-deposition treatment (PDT). In PAS, the NaF/KF layer was deposited on the Mo back contact prior to the deposition of the metallic precursor stack. The diffusion of the alkali into the absorber takes place during the synthesis of the CZTSe absorber. In the PDT approach, the NaF/KF layer was deposited on top of the as-synthesized CZTSe absorber. The PDT samples were then submitted to an additional annealing at 325°C and 1.5 mbar for 12.5 min in order to induce the diffusion the alkali elements into the absorber.

Scanning electron microscopy (SEM) (*ZEISS Series Auriga*) was used to analyze the morphology of the surface and the bulk of the CZTSe absorbers. Energy-dispersive X-ray spectroscopy (EDX) (*Oxford Instruments, XMax*) was used to investigate the composition of the chemically etched surface of the absorbers using an acceleration voltage of 15 kV. Capacitance-Voltage (C-V) measurements were carried out at 126 kHz to evaluate the carrier concentration of the full devices. The JV-curves of the finished solar cells were measured (AM1.5, 1 Sun, 25°C) using a pre-calibrated Sun 3000 Class AAA solar simulator (*Abet Technologies*).

III. RESULTS

A. Low Temperature CZTSe on glass

Solar cells were fabricated on SLG at different annealing temperatures (450-490°C) in order to shed light on the impact of low process temperatures on CZTSe.

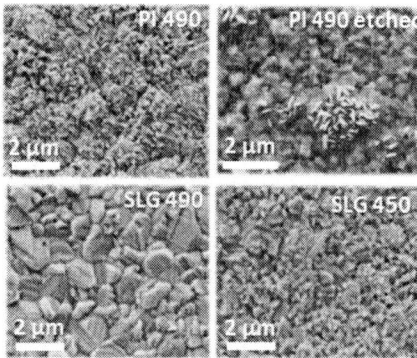

Fig. 1.   SEM top view images of CZTSe absorbers: on PI at 490°C (top-left), on PI at 490°C after chemical etchings (top-right), on SLG at 490°C (bottom-left) and on SLG at 450°C (bottom-right).

Fig. 1 (bottom) shows SEM top view images of the as-annealed CZTSe absorbers. At 450°C the surface of the absorber is covered by a layer of crystallites in which some elongated grains can also be observed. The small grains are identified as ZnSe and elongated ones as a $SnSe_x$ phase [8], [9]. At 470°C, micron-size grains mixed with smaller crystals

can be clearly spotted throughout the whole absorber. As the temperature increases up to 490°C the presence of small crystallites vanishes and a well-packed surface remains.

Fig. 2 shows the J-V parameters of the finished devices. The performance of the devices is highly enhanced with the increasing annealing temperature. Voc is the dominant parameter and exhibits an almost linear behavior. This behavior correlates with the SEM observations. After an air reannealing of the full 490°C device at 200°C for 30 min on a hot plate, a 6.4% efficiency is achieved. However, the reannealing was detrimental for the rest of the samples. This is probably due to remaining $SnSe_2$ after chemical etchings as will be discussed later on. Despite this phase, these results prove the feasibility of obtaining efficient cells at temperatures <500°C and, thus, the transferability of the process to PI substrates.

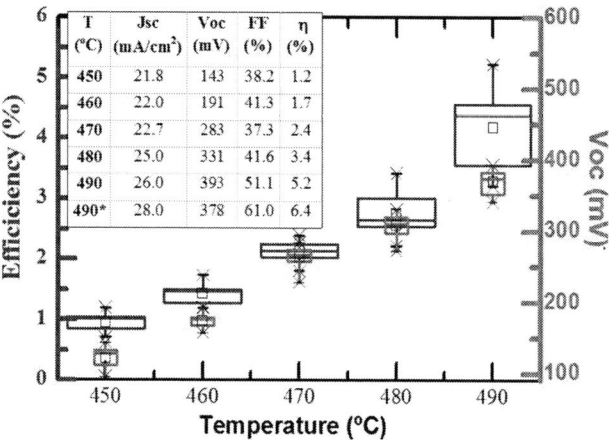

T (°C)	Jsc (mA/cm²)	Voc (mV)	FF (%)	η (%)
450	21.8	143	38.2	1.2
460	22.0	191	41.3	1.7
470	22.7	283	37.3	2.4
480	25.0	331	41.6	3.4
490	26.0	393	51.1	5.2
490*	28.0	378	61.0	6.4

Fig. 2.   Efficiency and Voc of the devices prepared on SLG at different temperatures. Inset: table with main optoelectronic parameters of the record cells of each sample.*after hot plate

B. CZTSe on polyimide: doping strategies

Solar cells were produced on polyimide at a moderate temperature of 470°C to ensure substrate stability. Na and K PAS and PDT doping strategies were studied. A SLG reference was processed in the same batch. The full devices were reannealed in a hot plate at 180°C in air during 30 min. Table 1 shows the main optoelectronic parameters of the record cells obtained in each of the samples. The PDT samples exhibit a poorer performance than the alkali-free reference (1.3%). While Jsc and FF seem unaffected by the PDT doping, the Voc is severely reduced. This effect intensifies when using KF. On the other hand, PAS enhances all the cell parameters when employing either NaF or KF. In the case of NaF, Voc is found to increase when using thicker doping layers (up to +85 mV with respect to the undoped reference). As for KF, 10 nm increase the Voc similarly to NaF but FF further improves than with the latter alkali salt (+18% absolute with respect to the undoped references). Going beyond this KF thickness affects the cell negatively. The best cell is obtained using 10 nm of KF with a 3.6% efficiency. Nevertheless, all the PI samples perform below the SLG reference (5.9%) which exhibits higher Voc, Jsc, and FF.

Sample			Jsc (mA/cm^2)	Voc (mV)	FF (%)	η (%)
SLG ref			29.9	344	57.6	5.9
Undoped			18.6	231	31.3	1.3
PAS	NaF	10 nm	24.8	257	43.5	2.8
		15 nm	24.0	303	44.3	3.2
	KF	10 nm	24.7	282	51.2	3.6
		15 nm	22.8	255	42.0	2.4
PDT	NaF	10 nm	18.9	147	32.2	0.9
		15 nm	19.5	180	31.6	1.1
	KF	10 nm	18.7	116	31.0	0.7
		15 nm	10.6	76.2	27.2	0.2

Table 1. Main solar cell parameters of the record cells obtained on PI at 470°C with different doping strategies.

The samples were inspected by SEM both in top view (not shown) and cross-section configuration. The PDT samples do not show any differences with respect to the undoped reference. This result was expected since the incorporation on the alkalis takes place after the crystallization of the CZTSe absorber.

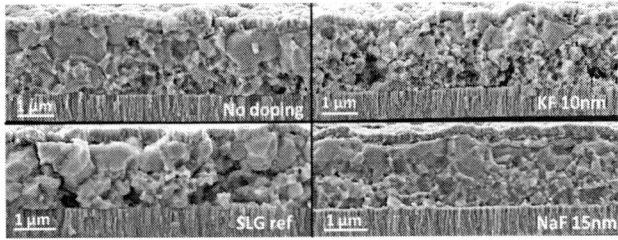

Fig. 3. SEM cross-sectional images of the full CZTSe devices on PI at 470°C and different doping strategies.

Fig. 3 shows cross-section images of the best performing samples: the soda-lime glass reference, the undoped PI reference and the PI samples doped with 15 nm of NaF and 10 nm of KF by PAS. No morphological differences could be spotted between samples with different thicknesses of the same dopant, therefore these can be taken as representative for the whole batch of samples. In top view configuration, while the SLG reference in this experiment shows a clear surface with well-defined grains, PI samples display a surface covered by a layer of small crystallites analogous to the one observed before on the 450°C SLG sample (Fig. 1 bottom) regardless the doping strategy. Cross-section images reveal a bilayer structure in every sample (Fig. 3). The largest grains are observed in the undoped and the SLG samples. Na doping does not modify the grain morphology significantly, contrarily

to most of the reports [5], [10]-[12]. Doping with K, on the other hand, hinders crystal growth and displays a predominant presence of small crystals.

C-V measurements were performed on these devices in order to estimate their charge carrier concentration. The measurements provide evidence of an increased carrier concentration by the PAS doping approach. The cell without intentional doping shows the lowest carrier concentration with $\sim 4 \cdot 10^{15}$ cm^{-3}. The addition of 15 nm of NaF and 10 nm of KF (the best cases) increase the carrier concentration up to $7 \cdot 10^{15}$ cm^{-3} and $1 \cdot 10^{16}$ cm^{-3} respectively indicating an effective incorporation of the dopants into the CZTSe absorber. This augmented carrier concentration might explain, at least partially, the increased Voc of the doped samples with respect to the undoped reference.

C. Further Experiments, Record Cell and SnSe$_2$

Further experiments were carried out in order to optimize the thermal annealing for PI at temperatures up to 490°C. In addition, the effect of introducing a Ge nanolayer was also studied. We managed to improve previous results significantly but observed repeatability issues, too. Specifically, the behaviour of the samples during air reannealing was complex and did not follow clear trends. For the sake of simplicity, we only report here the best results as summarized inn Fig. 4. A remarkable 4.2% efficiency was achieved at 490°C without any doping after reannealing at 300°C. A 4.4% efficiency was achieved at 480°C employing PAS NaF doping without any reannealing. The main effect of NaF is observed in the FF. Nevertheless, the record cell on PI was achieved by combining 10 nm NaF and 10 nm Ge doping at 480°C with a reannealing at 300°C. A 4.9% efficiency cell was obtained due to increased Voc and FF. This result sets a record for a kesterite device developed on a polymer substrate.

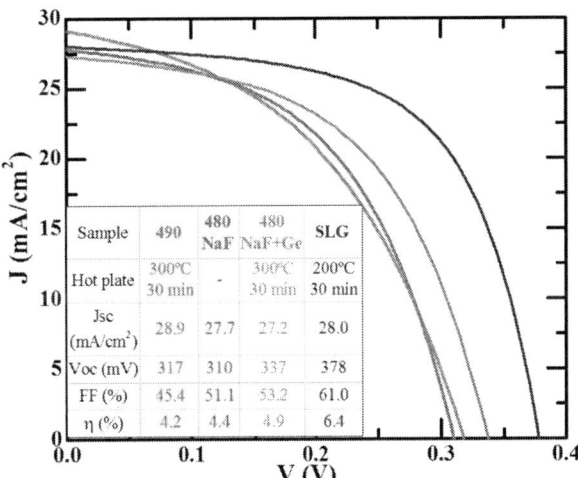

Fig. 4. J-V curves of the record cells achieved on PI together with the record obtained in SLG at 490°C. Inset: main cell parameters

However, contrarily to SLG samples, working at higher temperatures on PI did not avoid the superficial layer of secondary phases (ZnSe and SnSe$_x$) on top of the absorber, even on the best samples. Taking a look at Fig. 1, it is clear that the PI sample processed at 490°C (top-left) does not resemble its SLG counterpart processed at the same temperature (bottom-left) but rather looks like the SLG sample processed at 450°C (bottom-right). This suggests that PI behaves thermally differently to SLG during annealing hindering heat transfer from the graphite box into the absorber and resulting in a lower effective annealing temperature under the same processing conditions compared to SLG. In principle, the top layer of secondary phases should not be a concern since the chemical etchings applied before the deposition of the CdS remove almost completely any ZnSe and SnSe surface crystals as shown in Fig. 1 (top-right). However, after the etchings, clusters of crystals of another compound are revealed underneath the top layer (Fig.1 bottom-right). This phase seems deeply attached to the CZTSe structure and is found randomly distributed throughout the whole surface of the absorbers. An EDX analysis (Fig. 5) shows that this phase corresponds to SnSe$_2$. This phase is not observed on high-temperature SLG samples so its formation seems to be inherent to low-temperature CZTSe synthesis. This phase was not observed either on the record cells. We suspect, thus, that SnSe$_2$ is the main responsible for the low performance and the low repeatability observed in the different experiments and that it is related to the odd behaviour of the devices during air reannealing processes.

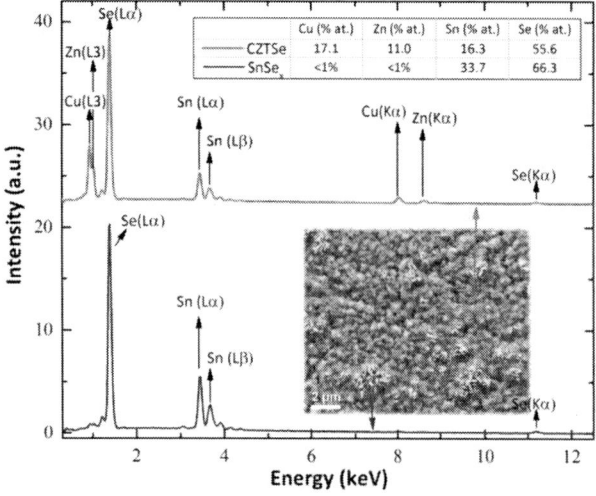

Fig. 5. EDX analysis of secondary phase found on the CZTSe absorbers synthesized on PI after chemical etchings.

IV. SUMMARY

The feasibility of producing efficient CZTSe devices by sequential precursor sputtering and selenization at annealing temperatures below 500°C has been demonstrated both on SLG (6.4% efficiency) and on PI (4.9% efficiency) substrates setting a record for CZTSe devices fabricated on a polymer substrate. Different doping strategies were investigated to introduce Na and K on CZTSe devices fabricated on PI. The results lay bare the difficulty of the PDT doping approach. On the contrary, PAS is an effective doping strategy that causes important improvements in solar cell performance, carrier concentration as well as modifying the grain morphology when doping either with NaF or KF. Further experimentation led to a 4.9% record sample on PI by combining NaF and Ge doping. However, we came across the formation of a detrimental SnSe$_2$ secondary phase apparently induced by low process temperatures. This effect is aggravated on PI devices due to a hindered heat absorption of this substrate during annealing compared to SLG. However, we strongly believe that SnSe$_2$ can be almost completely avoided by a deep optimization of the annealing process which will allow to further improve the results reported in this work.

ACKNOWLEDGEMENT

This research was supported by MINECO (Ministerio de Economía y Competitividad de España) under the ECOART project (RTC-2014-2426-7) and the NASCENT project (ENE2014-56237-C4-1-R), and by European Regional Development Funds (ERDF, FEDER Programa Competitivitat de Catalunya 2007–2013), and CERCA Programme / Generalitat de Catalunya. Authors from IREC and the University of Barcelona belong to the M-2E (Electronic Materials for Energy) Consolidated Research Group and the XaRMAE Network of Excellence on Materials for Energy of the "Generalitat de Catalunya". P.P. thanks the European Union for the JUMPKEST Marie Curie Individual Fellow (FP7-PEOPLE-2013-IEF- 625840).

REFERENCES

[1] A. Chirilă, P. Reinhard, F.Pianezzi, P. Bloesch, A. R. Uhl, C. Fella, L. Kranz, D. Keller, C. Gretener, H. Hagendorfer, D. Jaeguer, R. Erni, N. Nishiwaki, S. Buecheler and A. N. Tiwari, "Potassium-induced surface modification of Cu(In,Ga)Se$_2$ thin films for high-efficiency solar cells," *Nat. Mater.*, vol. 12, no. 12, pp. 1107–1111, Nov. 2013.

[2] M. Boshta, S. Binetti, A. Le Donne, M. Gomaa, M. Acciarri, A chemical deposition process for low-cost CZTS solar cell on flexible substrates, *Materials Technology.* (2016) 1–5.

[3] Z. Zhou, Y. Wang, D. Xu, Y. Zhang, Fabrication of Cu$_2$ZnSnS$_4$ screen printed layers for solar cells, *Solar Energy Materials and Solar Cells.* 94 (2010) 2042–2045.

[4] J. Xu, Z. Cao, Y. Yang, Z. Xie, Fabrication of Cu$_2$ZnSnS$_4$ thin films on flexible polyimide substrates by sputtering and post-sulfurization, *Journal of Renewable and Sustainable Energy.* 6 (2014) 053110.

[5] S. López-Marino, Y. Sánchez, M. Espíndola-Rodríguez, X. Alcobé, H. Xie, M. Neuschitzer, I. Becerril, S. Giraldo, M. Dimitrievska, M. Placidi, L. Fourdrinier, V. Izquierdo-Roca, A. Pérez-Rodríguez, E. Saucedo, Alkali doping strategies for flexible and light-weight $Cu_2ZnSnSe_4$ solar cells, *J. Mater. Chem. A.* 4 (2016) 1895–1907.

[6] J.V. Li, D. Kuciauskas, M.R. Young, I.L. Repins, Effects of sodium incorporation in Co-evaporated $Cu_2ZnSnSe_4$ thin-film solar cells, *Applied Physics Letters.* 102 (2013) 163905.

[7] Z. Tong, C. Yan, Z. Su, F. Zeng, J. Yang, Y. Li, L. Jiang, Y. Lai, F. Liu, Effects of potassium doping on solution processed kesterite Cu2ZnSnS4 thin film solar cells, *Applied Physics Letters.* 105 (2014) 223903.

[8] S. López-Marino, Y. Sánchez, M. Placidi, A. Fairbrother, M. Espíndola-Rodríguez, X. Fontané, V. Izquierdo-Roca, J. López-Martinez, L. Calvo-Barrio, A. Pérez-Rodríguez and E. Saucedo, "ZnSe Etching of Zn-Rich $Cu_2ZnSnSe_4$: An Oxidation Route for Improved Solar-Cell Efficiency," *Chem. - Eur. J.*, vol. 19, no. 44, pp. 14814–14822, Oct. 2013.

[9] H. Xie, Y. Sánchez, S. López-Marino, M. Espíndola-Rodríguez, M. Neuschitzer, D. Sylla, A. Fairbrother, V. Izquierdo-Roca, A. Pérez-Rodríguez and E. Saucedo, "Impact of Sn(S,Se) Secondary Phases in $Cu_2ZnSn(S,Se)_4$ Solar Cells: a Chemical Route for Their Selective Removal and Absorber Surface Passivation," *ACS Appl. Mater. Interfaces*, vol. 6, no. 15, pp. 12744–12751,Aug.2014

[10] T. Gershon, B. Shin, N. Bojarczuk, M. Hopstaken, D.B. Mitzi, S. Guha, The Role of Sodium as a Surfactant and Suppressor of Non-Radiative Recombination at Internal Surfaces in Cu_2ZnSnS_4, *Advanced Energy Materials.* 5 (2015).

[11] K. Sun, F. Liu, C. Yan, F. Zhou, J. Huang, Y. Shen, R. Liu, X. Hao, Influence of sodium incorporation on kesterite Cu_2ZnSnS_4 solar cells fabricated on stainless steel substrates, *Solar Energy Materials and Solar Cells.* 157 (2016) 565–571.

[12] M. Johnson, S.V. Baryshev, E. Thimsen, M. Manno, X. Zhang, I.V. Veryovkin, C. Leighton, E.S. Aydil, Alkali-metal-enhanced grain growth in Cu_2ZnSnS_4 thin films, *Energy Environ. Sci.* 7 (2014) 1931–1938.

An optimized photolithography recipe for $Cu(In_{1-x},Ga_x)(S_y,Se_{1-y})_2$ (CIGSSe) solar cells

Xia Hao[1], Shenghao Wang[1*], Katsuhiro Akimoto[1], Takuya Kato[2], Hiroki Sugimoto[2], Takeaki Sakurai[1]

[1]Institute of Applied Physics, University of Tsukuba, Tsukuba, Ibaraki 305-8573, Japan

[2] Atsugi Research Center, Solar Frontier K. K., Atsugi, Kanagawa, 243-0206, Japan

Abstract — **In order to characterize solar cell modules, patterning a sub-module into small-sized devices by photolithography are required. For CIGSSe sub-modules fabricated by sulfurization-after-selenization (SAS), the small-size device performance was considerably deteriorated (from 19.3 % to 11.3 % for power conversion efficiency (PCE)) as compared to the sub-module after traditional photolithography process. We found that, undesirable $Cu_{2-x}Se$ phase was introduced into the film by the traditional photolithography process. By utilizing an optimized photolithograph recipe, the PCE as high as 19.5 % (nearly the same as the sub-module) for the patterned small-size unit cells were achieved. This allows us to perform specific characterizations to the small-size device in order to evaluate the properties of the sub-modules.**

Key words — **Solar cells, CIGSSe thin films, photolithography.**

I. INTRODUCTION

$Cu(In_{1-x},Ga_x)(S_ySe_{1-y})_2$ (CIGSSe) is an excellent material for fabricating thin film solar cells and the power conversion efficiency (PCE) of the best researched CIGSSe solar cell has been achieved to 22.6 %, [1] which is the highest one among thin film photovoltaic (PV) devices. It is almost close to the conventional Si solar cell. [2-3] For commercial application of CIGS PV products, the PCE should be further improved and the cost should be reduced. Therefore, it is essential to characterize the device or module properties (e.g., defect levels) to further understand the device physics and find the efficient way to further improve the performance.

Large-areared PV module is the basic unit of PV array. Generally, CIGSSe PV module performance can be easily characterized by solar simulator to obtain the information of short circuit current density (J_{SC}), open circuit voltage (V_{OC}), fill factor (FF) and PCE. However, for some other characteristics, such as admittance spectroscopy, capacitance-voltage measurements and so on, they can only be performed on a small-sized unit cell. Therefore, patterning a sub-module into small-size devices by photolithography are required. Photolithography can achieve small-sized unit cells with specific device area (e.g., circular unit cells with a diameter of 0.1 cm, resulting in 0.007854 cm^2 of device area).

In the present work, we studied the effect of photolithography process on the properties of CIGSSe solar cells. By applying a traditional photolithography recipe, the PCE of the patterned cells are severely deteriorated as compared to the original sub-module and the main reason of the performance degradation is ascribed to the appearance of $Cu_{2-x}Cu$ secondary phase. The recipe was then modified by utilizing a manual scribing process prior to photolithography and controlling the acid concentration. Then small-sized unit cells with nearly the same performance as the CIGSSe sub-module was achieved. The results suggested that the manual scribing prior to photolithography and controlling the acid etching condition are very important for maintaining the device performance.

II. EXPERIMENTS

CIGSSe thin films were grown on Mo-coated soda-lime glass substrates by using a two-step process, where a sputter-deposited Cu-In-Ga precursor was prepared and followed by selenizanition and sulfurization (SAS) process. Films with a Ga contents (x) of 0.14 was prepared. To form device structure, $Zn(O,S,OH)_x$ buffer layers were prepared by chemical bath deposition (CBD), and ZnO/ZnO:B (BZO) was prepared by chemical vapor deposition (CVD). The formed device structure is glass/Mo/CIGSSe/$Zn(O,S,OH)_x$/ZnO/ZnO:B.

The current density versus voltage (J-V) characteristics of the devices were measured with a J-V source meter (Advantest R6245) in the dark and under AM 1.5 G 100 mW/cm^2 simulated solar light. Raman measurement was carried out by using a 532 nm laser.

OFPR-800 LB (Ethyl lactate (90 %~54 %) butyl acetate (10 %~5 %)) and NMD-W 2.38 (Tetramethyl ammonium Hydroxide) were used as ultraviolet light (UV) resist and developer, respectively. The power of the UV light is 30 mW/cm^2 and the exposure time is 60 second. Patterned unit cells were achieved by using hydrochloric acid (HCl) for etching. In this work, two recipes were used. In Recipe 1, the photolithography was carried out directly on the sub-module. The device was etched by using HCl with a weight concentration of 0.07 % for 60 second. And in Recipe 2, a manual scribing was applied before photolithography. Then the HCL concentration and etching time were optimized.

III. RESULTS AND DISCUSSION

The process of Recipe 1 is schematically shown in Fig. 1(a) and the J-V curves of the devices before and after patterning by using Recipe 1 are shown in Fig. 1(b). The photovoltaic parameters are shown in Table 1. After photolithography, the J_{SC}, V_{OC} and the FF were all severely decreased. As a result, device performance was considerably deteriorated. The degradation of the device performance implies that the photolithography process should be optimized.

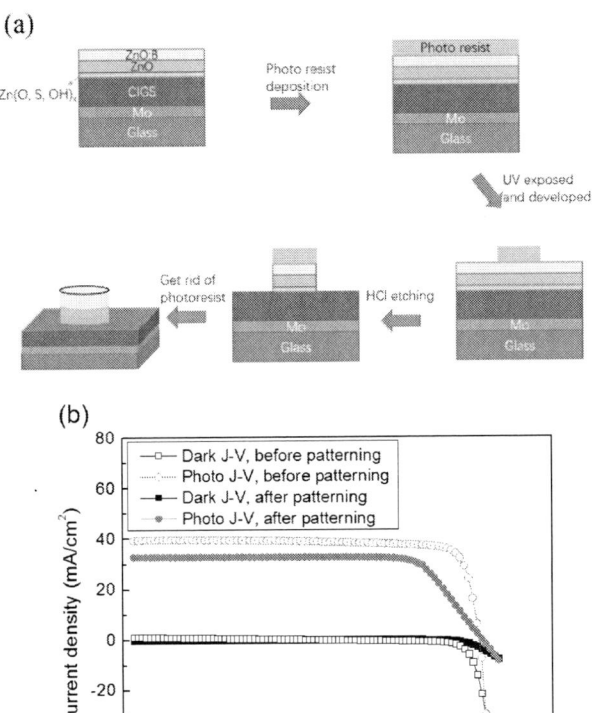

(a)

(b)

Fig. 1. (a) Schemes of the photolithography processes of Recipe 1 and (b) J-V characteristics of the devices before and after photolithography by using Recipe 1.

Table 1. Photovoltaic parameters for the devices before and after photolithography by using Recipe 1.

Devices	J_{SC} (mA/cm^2)	V_{OC} (mV)	FF (%)	PCE (%)
Before patterning	38.8	666	74.6	19.3
Patterned by Recipe 1	32.7	660	52.3	11.3

To optimize the patterning process, we modified Recipe 1 by changing the concentration of HCl. Note that the etching time was primarily kept as 60 s. Firstly, we decreased the HCl concentration from 0.07 wt.% to 0.04 wt.%, and we found that the buffer layers still largely remained on top of CIGSSe layer, resulting in failure of device performance measurement (data not shown here). Then we increased the concentration of HCl solution from 0.07 wt.% to wt.0.14 %. The corresponding J-V curve of unit cell patterned by modified HCl concentration was shown in Fig. 2, marked by filled squares. From the J-V parameters (shown in Table 2), one can tell the device performance was also severely deteriorated, even worse than

that of with 0.07 wt.% HCl concentration. Then we kept the HCl concentration as 0.14 wt.% and further shortened the etching time to 40 s and 20 s. The J-V curves of the unit cells with these processes are also shown in Fig. 2, marked as filled circles and filled triangles, respectively. Even though the PCE in case of 20 s etching time (13.3 % of PCE) seems to be higher than the case of 40 s etching time (16.3 % of PCE), the J_{SC} of the unit cell in case of 20 s was actually over estimated due to the exaggeration of the total device area. This was evidenced by the very low resistance (caused by BZO part), checked by multi-meter simply at the range around the circular cell. This suggests that the BZO was not completely removed out of the shadow mask range. And in the case of 40 s etching time, the PCE of the patterned unit cell is 13.3%, and it is still lower than that of the original sub-module. The degraded performance maybe imply that damages were introduced.

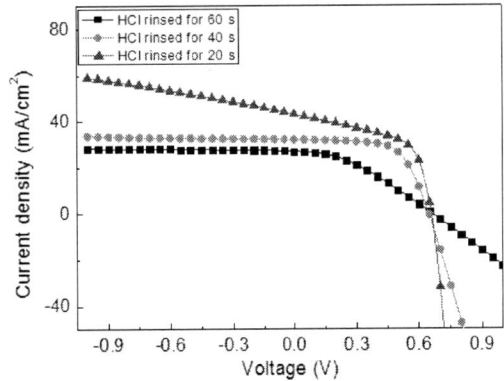

Fig. 2. J-V characteristics of the unit cells after photolithography by using modified Recipe 1, with HCl solution concentration of 0.14 wt.% and etching time for 60, 40 and 20 s.

Table 2. Photovoltaic parameters for the unit cells after photolithography by using modified Recipe 1 (0.14 wt.% HCl concentration) with different rinse times.

Devices	J_{SC} (mA/cm^2)	V_{OC} (mV)	FF (%)	PCE (%)
Etched for 60 s	26.9	663	35.5	6.5
Etched for 40 s	31.5	653	64.6	13.3
Etched for 20 s	43.1	659	57.6	16.3

The recipe of photolithography was then further optimized by introducing a step of manual scribing prior to photolithography and setting the HCl rinse time as 40 s (so-called Recipe 2). The schematic view of the process of Recipe 2 is shown in Fig. 3(a) and the J-V curves of the unit cell before and after patterning with Recipe 2 are shown in Fig. 3(b). The shape of the J-V curve and the photovoltaic parameters were almost the same for the unit cell before and after photolithography by using Recipe 2 (see Table 3). This suggests

978-1-5090-5606-4/17 $31.00 © 2017 IEEE

that no damage was introduced by patterning sub-module into small-sized unit cells by using Recipe 2.

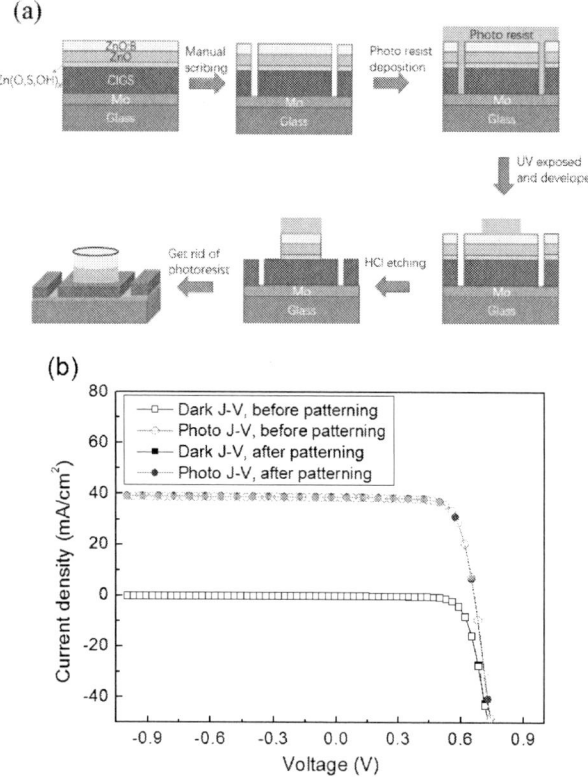

Fig. 3. (a) Schemes of the photolithography processes of Recipe 2 and (b) J-V characteristics of the devices before and after photolithography by using Recipe 2. In (b), the *J-V* curves before and after photolithography are overlapped.

Table 3. Photovoltaic parameters for the devices before and after photolithography.

Devices	J_{SC} (mA/cm²)	V_{OC} (mV)	FF (%)	PCE (%)
Before patterning	38.8	666	74.6	19.3
Patterned by Recipe 2	38.6	665	76.3	19.5

To reveal the influence induced by photolithography, we characterized the devices by using Raman spectroscopy. The results (see Fig. 4) suggest that before photolithography, there is a distinct peak around 170 cm⁻¹ indicating the CIGSSe films. [4,5] However the peaks ranging from 250 to 270 cm⁻¹ for the device patterned by Recipe 1 (HCl concentration of 0.07 % and rinse time of 60 s) suggests the existence of low resistance $Cu_{2-x}Se$ compounds. [6,7] It has been reported that in both $CuInSe_2$ and $CuGaSe_2$ thin films, the introduction of oxygen will generate oxides (*i.e.*, In_2O_3, Ga_2O_3, SeO_2 and Cu_xO) and $Cu_{2-x}Se$ compound or $Cu_{2-x}Se$ inter layer between the CIGS and oxide layers. [8,9] Therefore, it is reasonable that $Cu_{2-x}Se$

compound appears on the top of bare CIGSSe because CIGSSe itself is ease to degrade in the ambient air. This could be the main cause of the degradation of device performance. The $Cu_{2-x}Se$ compound observed in the device region could be caused by over etching of the top transparent conductive oxides (BZO), as well as the buffer layer and intrinsic ZnO layer. In that case, some damage was introduced to the unit cell and as a result, the locally "buried" CIGSSe was exposed to the air. Thus, the chemical reactions between CIGSSe and oxygen would occur and as a result the $Cu_{2-x}Se$ Raman peak was also observed at the unit cell region.

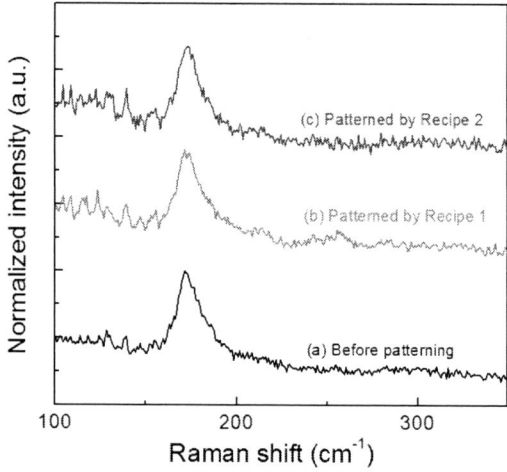

Fig. 4. Raman spectra of the device (a) before and (b, c) after photolithography. The Raman measurement results were obtained on BZO part of the unit cells. For (b), the unit cell was photolithographied by Recipe 1 (0.07 wt.% HCl and rinse time of 60 s). For (c), the unit cell was patterned by Recipe 2 with manual scribe process and optimized etching condition.

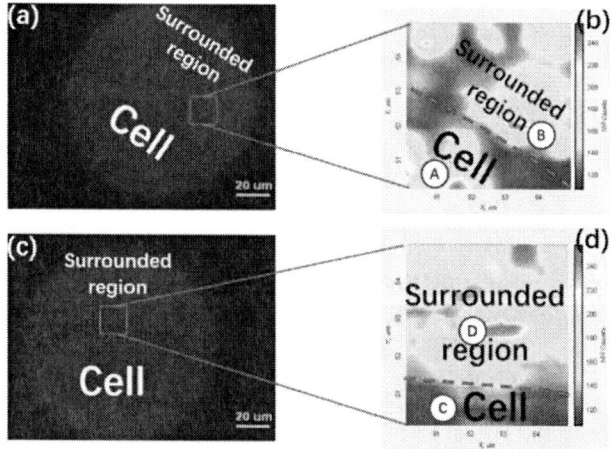

Fig. 5. Microscopic images of the boundary of the circular cells patterned by (a) Recipe 1 and (c) Recipe 2. The corresponding Raman mapping results for $Cu_{2-x}Se$ peak are shown in (b) and (d), respectively. The A, B, C, D are used for marking different

positions at the BZO parts and the surrounded regions of the circular devices.

On the contrary, in the case of Recipe 2 (with manual scribe process and optimized etching condition), there is no peak indicating $Cu_{2-x}Se$ phase on the top of BZO. This implies a well protection for the device.

The Raman mapping was then carried out to compare the $Cu_{2-x}Se$ distribution of the patterned devices by using different recipes. The results are shown in Fig. 5. As shown in Fig. 5(b), for the device pattered by Recipe 1, the $Cu_{2-x}Se$ phase distributes in both the unit cell part and the surrounding part of the unit cell; while the $Cu_{2-x}Se$ phase could be only detected at the regions surrounding the unit cell for the device patterned by Recipe 2 (Fig. 5d). This implies that Recipe 2 is an optimized process of photolithography. As a result, the damage to the unit cell can be avoided and the device performance of the unit cell nearly keeps the same as the sub-module.

IV. CONCLUSIONS

In this work, we clarified the reason of the performance degradation for photo-etched CIGSSe solar cells. The traditional recipe introduced damages on the surface of the CIGSSe solar cell, and the formation of low resistance $Cu_{2-x}Se$ was responsible for the degradation of device performance. After applying a manual scribing process and optimizing acid etching duration time, the device was refrained from over etching and the performance can be maintained. The optimized recipe of the photo-etching process for CIGSSe solar cell from sub-module to unit cell allows us to extensively understand the performance and properties of modules by means of charactering the patterned small unit cells.

ACKNOWLEDGEMENT

This work was partially supported by the New Energy and Industrial Technology Development Organization (NEDO).

The authors would like to thank Dr. Shogo Ishizuka from National Institute of Advanced Industrial Science and Technology (AIST), Japan for his support with sample scribing.

REFERENCES

[1] P. Jackson, R. Wuerz, D. Hariskos, E. Lotter, W. Witte and M. Powalla, "Effects of heavy alkali elements in Cu(In,Ga)Se2 solar cells with efficiencies up to 22.6%", *Physica Status Solidi Rapid Research Letters*, vol. 10, pp. 583–586, 2016.

[2] K. Kushiya, "CIS-based thin-film PV technology in solar frontier K.K.," *Solar Energy Materials & Solar Cells*, vol. 122, pp. 309–313, 2014.

[3] K. Masuko, M. Shigematsu, T. Hashiguchi, D. Fujishima, M. Kai, N.Yoshimura, T. Yamaguchi, Y. Ichihashi, T. Mishima, N. Matsubara, T. Yamanishi, T. Takahama, M. Taguchi, E. Maruyama, and S. Okamoto "Achievement of more than 25% conversion efficiency with crystalline silicon heterojunction solar cell," *IEEE Journal of Photovoltaic.*, vol. 4, pp. 1433–1435, 2014.

[4] J. Parravicini1, M. Acciarri1, A. Lomuscio, M. Murabito1, A. Donne1, A. Gasparotto and S. Binetti1, "Gallium In-depth profile in Bromine etched Copper–Indium–Galium–(Di)selenide (CIGS) thin films inspected using Raman spectroscopy", *Applied Spectroscopy*, vol. 71, pp.1334-1339, 2017.

[5] J. Xiang, X. Huang, G. Lin, J. Tang and C. Ju and X. Miao, "CIGS thin films for Cd-free solar cells by one-step sputtering process", *Journal of Electronic Materials*, vol. 43, pp. 2658-2666, 2014.

[6] B. Jheng, M. Wang, M. Wu, and D. Jan, "Enhanced conversion efficiencies of Cu(In,Ga)Se2-based thin film solar cells using new precursors by sputtering", *Science of Advanced Materials*, vol. 8, pp. 1464-1469, 2016.

[7] T. Hsieh, C. Chuang, C. Wu, J. Chang, J. Guo and W. Chen, "Effects of residual copper selenide on CuInGaSe2 solar cells", *Solid-State Electronics*, vol. 56, pp. 175–178, 2011.

[8] L. L. Kazmerski, O. Jamjoum, J. F. Wager, and P. J. IrelandK. J. Bachmann, "Summary Abstract: Oxidation of CuInSe2", Journal of Vacuum Science & Technology A", vol. 1, pp.668-669, 1983.

[9] R. Würza, A. Meedera, D. Fuertes Marróna, Th. Schedel-Niedriga, A. Knop-Gerickeb and K. Lipsc, "Native oxidation of CuGaSe2 crystals and thin films studied by electron paramagnetic resonance and photoelectron spectroscopy", Physical Review B, vol. 70, pp. 205321-1-10, 2004.

Effects of CdCl₂ Passivation on Thin CdTe Absorbers Fabricated by Close-Space Sublimation

Anna Wojtowicz, Alexandra M. Huss, Jennifer A. Drayton, and James R. Sites

Colorado State University Department of Physics, Fort Collins, CO, 80523, USA

Abstract — **Thin CdTe solar cells with absorber thicknesses between 0.6 and 1.2 μm were passivated with varying amounts of CdCl₂. Current densities and voltages were relatively flat with CdCl₂ dose for the larger doses, but decreased markedly for smaller amounts. The voltages of the cells with sufficient CdCl₂ were near 750 mV; the current densities were near 23 mA/cm², and decreased only slightly for the thinner absorbers. Spatial characterization indicates that devices with optimal CdCl₂ dosing are uniform.**

Index Terms — **thin CdTe, CdCl₂ treatment, close-space sublimation, processing, optimization.**

I. INTRODUCTION

Cadmium telluride (CdTe) is currently the second most common photovoltaic (PV) technology worldwide, and the only technology cost competitive with crystalline silicon in commercial manufacturing [1, 2]. This is due to its favorable optical and electrical properties which make CdTe nearly ideal for terrestrial applications.

One of the largest hurdles thin-film PV technology faces is trap-assisted carrier recombination, which becomes considerably enhanced by the short carrier lifetimes of p-type CdTe (<10 ns) [3, 4]. By thinning the absorber, devices become more fully-depleted, reducing recombination in the space-charge region by quickly separating carriers. Due to the direct bandgap nature of CdTe and the material's favorable properties, modeling predicts that a thin CdTe absorber layer less than 1 μm is viable [5]. It has also been demonstrated that sputtered CdTe absorbers as thin as 0.25 μm yielded a reasonable performance level [6].

To assess the benefits of thin CdTe absorber devices, it is necessary to establish an optimal fabrication process in which reliable thin CdTe devices can be made. Cadmium chloride (CdCl₂) is a passivation treatment commonly used in CdTe device fabrication. When sublimated onto polycrystalline CdTe and annealed, the chloride species will nucleate in grain boundaries and voids encouraging recrystallization, grain coalescence, passivation of dangling bonds, and improved bulk electronic properties [7]. In this paper, we provide a systematic approach to the optimization of CdCl₂ passivation by close space sublimation (CSS) deposition for CdTe thicknesses ranging from 0.6-1.2 μm. Our results show obvious negative effects of under-passivation and the more subtle effects of over-passivation.

II. EXPERIMENTAL DETAILS

Devices were fabricated on commercial 3.2-mm soda-lime glass coated with a 400-nm SnO₂:F thin-conducting oxide (TEC10 made by Pilkington). (Mg,Zn)O (MZO) was selected as a window layer for its tunable bandgap and superior transparency at lower wavelengths compared to the traditional choice of cadmium sulfide (CdS) [8]. 100-nm n-type MZO was deposited by RF magnetron sputtering at room temperature directly onto the SnO₂:F layer. The CdTe and CdCl₂ were deposited using the CSS system built at Colorado State University [9]. The p-type CdTe layer was deposited onto the MZO in a range of thicknesses by controlling the temperature of the sublimating material. CdTe thicknesses of 0.6 ± 0.1, 0.8 ± 0.05, 1.0 ± 0.15 and 1.2 ± 0.05 μm were fabricated. CdTe thickness was measured using an Alpha Step surface profilometer.

The CdCl₂ dose was defined as the time the CdTe film was exposed to the sublimating CdCl₂ material [9]. CdCl₂ dose times were 38, 58, 78, 98, 118, 138, 158 and 180 seconds and an annealing process at 400°C for 180 seconds followed. The CdCl₂ dose times were chosen to encompass poor and well-passivated devices based on the system-optimized CdCl₂ dose for standard 2.2-2.5 μm CdTe devices. After cooling and removal from the vacuum system, the CdTe surface was rinsed with deionized water to remove any residual CdCl₂ material. A 40-nm Te buffer layer was then evaporated onto the CdTe film using a Cooke MK VII Evaporator. The back contact was finished by applying approximately 140-μm colloidal Ni onto the Te layer. After the Ni layer cured, 25 small-area devices were delineated for each superstrate using a syphon-type bead blaster and stainless steel mask [9]. The finished cells were approximately 0.6-0.7 cm² in area. The completed superstrate device structure is shown in Fig. 1.

There was no intentional Cu used in the fabrication of these devices, however, the CSS system has been used for Cu doping in the past.

978-1-5090-5606-4/17 $31.00 © 2017 IEEE

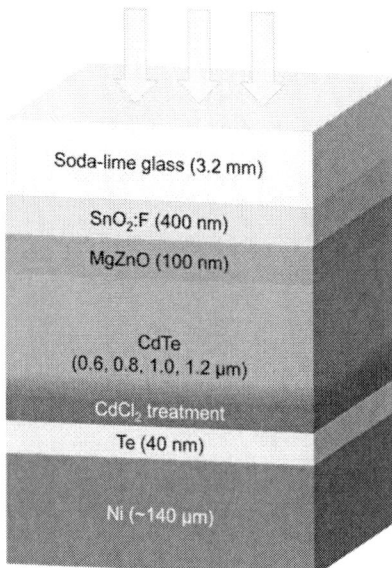

Fig. 1. Device superstrate structure (not to scale) with the direction of incident light indicated. MZO, Te and Ni layers have consistent thickness on all devices, while CdTe thickness and CdCl₂ dose time are varied.

Completed devices were characterized under standard test conditions and included measurement of current density-voltage (J-V) with a Solar Light Co. XPS 400 xenon lamp AM 1.5 solar simulator, external quantum efficiency (EQE), and optical reflection using a Perkin Elmer Lambda 2 spectrometer. Additional characterization included room-temperature photoluminescence (PL), electroluminescence (EL), and light beam induced current (LBIC) measurements.

The room-temperature PL measurement utilized a 520-nm excitation source laser which illuminates the device through the glass. EL measurements were conducted in a light-tight enclosure using a Finger Lakes MicroLine ML8300 Si CCD camera operated at -25°C to minimize thermal noise [10]. The exposure time for the EL images was 100 seconds, and an injection current density of 40 mA/cm² was used on all devices. Images were background-corrected and normalized by exposure time and injection current density, scaled logarithmically and given false coloring. LBIC was also measured in a light-tight enclosure with the system setup described by [11]. The measurement used a 638-nm diode laser light source focused to a 100 μm spot size with no voltage bias applied. The LBIC image is normalized to the measured EQE at 638 nm to produce a spatial mapping of the device's QE.

III. DEVICE PERFORMANCE VS. CdCl₂ PASSIVATION

A. Current Density-Voltage (J-V)

The light J-V curves of the best performing devices for each absorber thickness at 118 seconds CdCl₂ dose are shown in Fig. 2. The devices were well behaved and demonstrated good junction quality with average short-circuit current density $J_{sc} \approx$ 23 mA/cm² and average open-circuit voltage $V_{oc} \approx 0.75$ V.

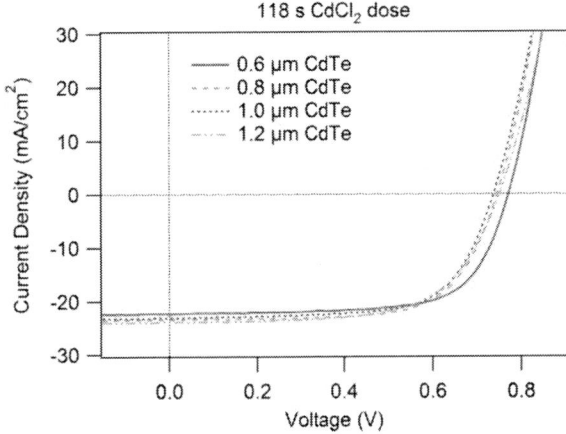

Fig. 2. Light J-V curves for the best performing devices of each absorber thickness at 118 seconds CdCl₂ dose time.

A comparison of J-V metrics (V_{oc}, J_{sc}, fill factor, efficiency) of best cells for each absorber thickness across all CdCl₂ dose times are shown in Fig. 3. Worth noting in all four plots is that overall device performance significantly improves up to 60-80 seconds CdCl₂ dose. This indicates that any benefits of CdCl₂ passivation occur only after a minimum passivation point is achieved regardless of CdTe absorber thickness.

Fig. 3(a) shows that after achieving minimum passivation, V_{oc} generally increases with CdCl₂ dose. Fig. 3(b) shows that J_{sc} rapidly increases before leveling out over a wide range of CdCl₂ doses, indicating J_{sc} experiences only marginal losses at higher doses. Fig. 3(c) shows that after a rapid increase, fill factor decreases at longer CdCl₂ dose times. Fig. 3(d) demonstrates that overall device efficiency is reduced at longer CdCl₂ doses and shows the optimal dose time is approximately 118 seconds regardless of the CdTe absorber thickness.

The trends observed in Fig. 3 have been reproduced on separate occasions for different, nominally identical superstrate devices with good agreement.

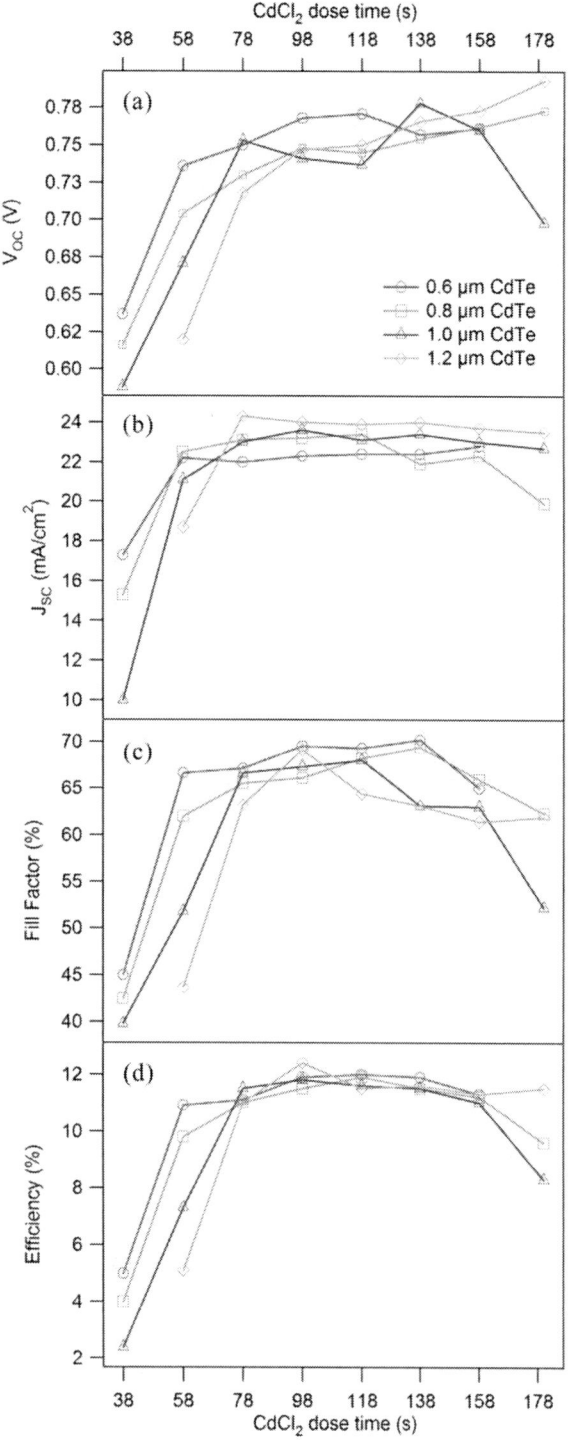

Fig. 3. J-V metrics as a function of $CdCl_2$ dose time for best cells of each absorber thickness: (a) V_{oc}, (b) J_{sc}, (c) Fill Factor and (d) Efficiency.

B. Quantum Efficiency (QE) and Reflection

Fig. 4 shows the external QE and reflection data for the highest efficiency devices of each absorber thickness at 118 second $CdCl_2$ dose. As shown in Table 1, the integrated J_{sc} values agree well with the J_{sc} values measured from J-V. There is increased current collection near the CdTe absorption edge (~830 nm) as the CdTe thickness increases, but the difference in overall current densities is not significant as demonstrated by the values given below.

TABLE I
J_{sc} COMPARISON OF BEST DEVICES AT 118S $CdCl_2$ DOSE

CdTe Thickness (µm)	Jsc from J-V (mA/cm²)	Jsc from QE (mA/cm²)
0.6	22.4	22.5
0.8	23.5	23.3
1.0	23.3	23.9
1.2	24.0	24.3

Optical measurements show that there is approximately a 10% loss due to reflection on the fully-completed devices. This loss is fairly consistent across all absorber thicknesses; since the interfaces between glass and films are alike on all devices, the thin-film structures share the same indices of refraction resulting in similar front-reflection profiles [11]. The signal loss between QE and reflection corresponds to photons absorbed by the window layers preceding CdTe, recombination losses, and photons transmitted through the absorber layer to the back contact. As CdTe becomes thinner, the effect of unabsorbed photons increases near the CdTe absorption edge due to insufficient material for complete absorption.

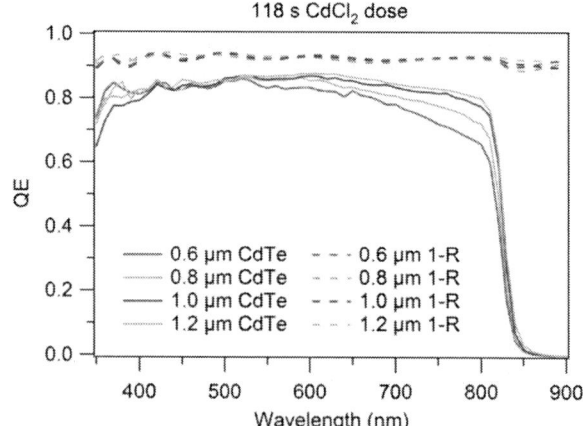

Fig. 4. QE and reflection of best performing cells at 118 seconds $CdCl_2$ dose for each absorber thickness. No light or voltage bias was applied.

The small dip in the reflection just past the CdTe absorption edge in Fig. 4 indicates that the Te buffer layer could be acting as an optical reflector in the thinner devices.

C. Additional Characterization

Fig. 5 shows room temperature PL for the highest efficiency devices of each absorber thickness at an 118-second $CdCl_2$ dose. The PL peak is near the CdTe bandgap for all of the absorber thicknesses, and this bandgap agrees with the QE absorption edge of approximately 1.50 eV (~830 nm). PL intensity is approximately proportional to thickness.

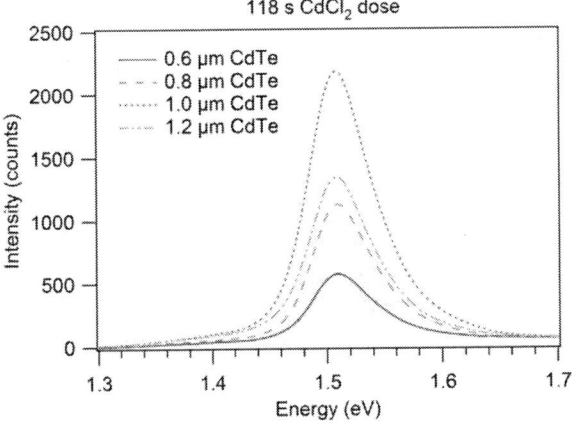

Fig. 5. Room temperature PL of best performing devices at 118 seconds $CdCl_2$ dose for each absorber thickness.

Electroluminescence images of various $CdCl_2$ dose times are shown in Fig. 6 for 0.8-µm CdTe devices. The features highlighted in these images were also observed for the other absorber thicknesses.

Fig. 6(a) shows a device with a $CdCl_2$ dose time of 78 seconds. As implied by [10], darker regions indicate a weaker EL signal and thus lower performance in that region. The gradient across the device in Fig. 6(a) spans approximately a decade of EL intensity, illustrating the effects of under-passivation. In comparison, Fig. 6(b) illustrates an optimized $CdCl_2$ dose of 118 seconds with a high degree of uniformity across the device. Fig. 6(c) demonstrates the effects of over-passivation with a $CdCl_2$ dose time of 180 seconds. Like Fig. 6(a), the overpassivated device has a gradient of non-uniform intensity across the device, in addition to spot-like defects not seen in the other images.

In addition to the effects of passivation, the EL images in Fig. 6 also reveal defects that appear as dark spots. Defects such as these have the potential to limit performance, for example, by creating electrical shunts in the diode structure. However, these defects can often be present and benign in devices with good performance.

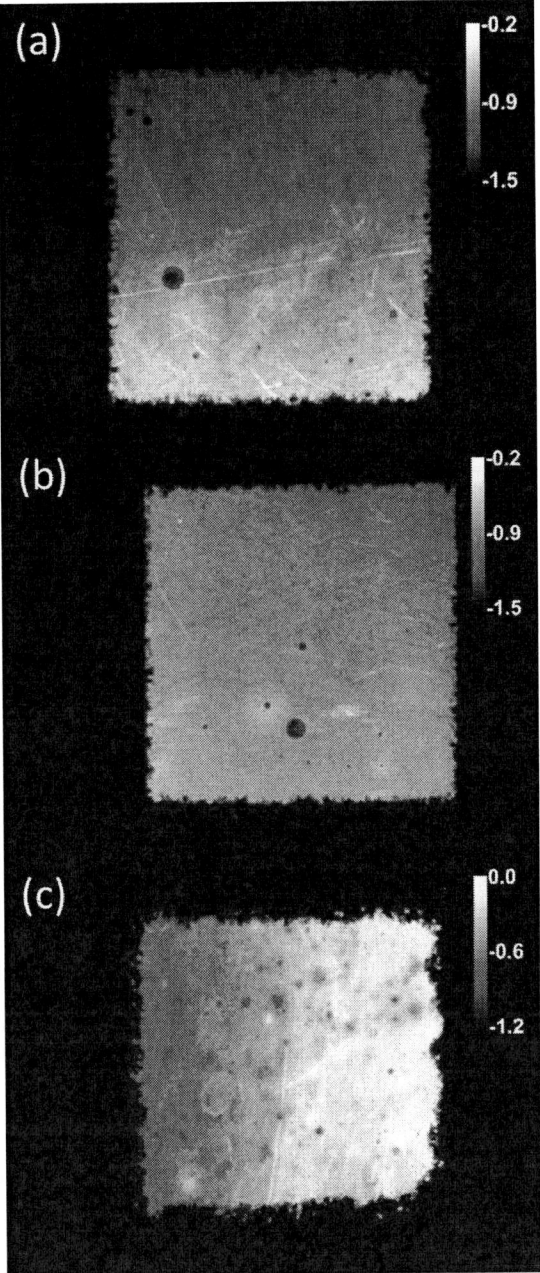

Fig. 6. EL images of 0.8 µm CdTe devices of varying $CdCl_2$ doses illustrating: (a) under-passivation (78 s dose), (b) optimal passivation (118 s dose), and (c) over-passivation (180 s dose). The scales shown are $\log_{10}$ intensity with arbitrary zero.

Fig. 7 shows an LBIC image of the same optimally-passivated device as depicted in Fig. 6(b). When compared to EL, the circular defect labeled 'A' appears less prominent in LBIC imaging. Conversely, defect 'B' appears more prominent in LBIC imaging with an 18% relative drop in QE from the device average.

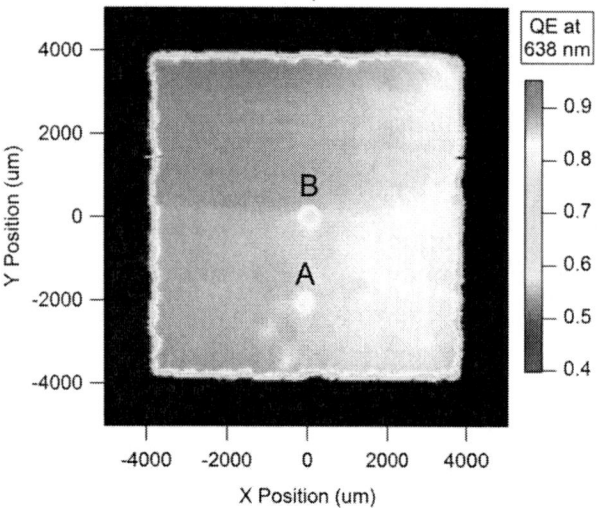

Fig. 7. LBIC of 0.8 μm CdTe device at 118 second CdCl$_2$ dose time. Defects labeled A and B have lower QE than the average device QE.

Fig. 8(a) and (b) show linescans performed across the width of the device at the defect positions for EL and LBIC respectively. Comparing the relative size of the defects in both plots, defect B widens more significantly under illumination than defect A. Due to the nature of EL and LBIC measurements, this suggests that defect B is modestly detrimental to J$_{sc}$, while defect A has more effect on V$_{oc}$ [11]. However, both defects are relatively small in both measurements; since the device is uniform and has good performance, these defects appear to be benign.

IV. DISCUSSION AND CONCLUSIONS

The data indicates that there is a range of CdCl$_2$ doses within which thin CdTe devices fabricated by CSS will have consistent performance. This is illustrated in Table 2 which provides data for the three best performing cells of each absorber thickness. The efficiencies differ only slightly (<1% relative) for each absorber thickness suggesting that good, comparable devices can be fabricated over a moderate range of CdCl$_2$ doses.

Our results demonstrate the effects of under-passivation and that there exists a minimum passivation point in the range of 60-80 seconds of CdCl$_2$ dosing after which cell performance drastically improves. This is supported by Fig. 6(a) depicting EL of 78-seconds CdCl$_2$ dose; the performance was acceptable but the device response was not uniform.

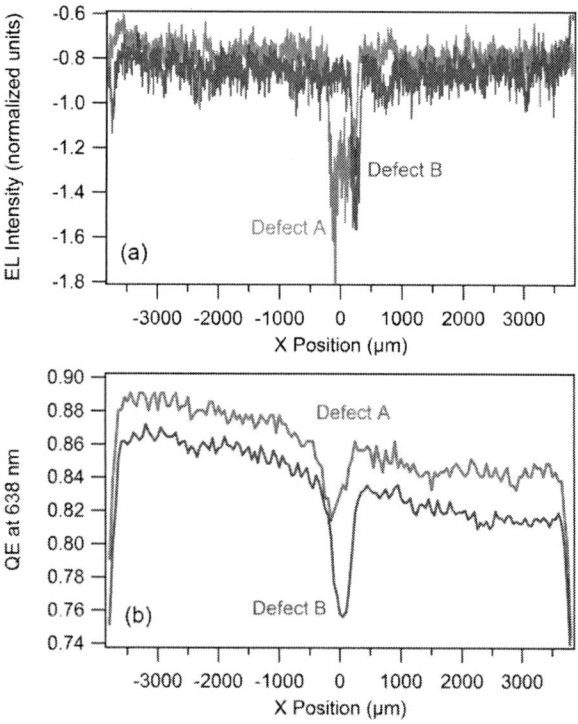

Fig. 8. Linescans performed across defects labeled A and B for the (a) EL image, and (b) LBIC image of the optimally-passivated device.

Our devices also show the subtle effects of over-passivation. This is shown in the non-uniformity of EL depicted in Fig. 6(c), and a general decline in J-V performance for higher CdCl$_2$ doses shown in Fig. 3. The fill factor appeared most susceptible with an average relative decrease of 13% for all absorber thicknesses at the higher CdCl$_2$ doses. The simultaneous improvement of V$_{oc}$ and degradation of fill factor at higher CdCl$_2$ dose times indicates that carrier lifetimes and densities are likely changing at each CdCl$_2$ dose.

TABLE II
SUMMARY OF CDCL$_2$ DOSE RANGE OPTIMIZATION

CdTe Thickness (μm)	Efficiency (%)	CdCl$_2$ Dose Time (s)
0.6	11.9	98
	12.0	**118**
	11.9	138
0.8	11.0	78
	11.5	98
	11.9	**118**
1.0	11.5	78
	11.8	**98**
	11.6	118
1.2	**12.4**	**98**
	11.5	118
	11.6	138

In addition to having the least variation in cell efficiency, the 0.6 and 0.8 μm devices have very similar efficiencies to the 1.0 and 1.2 μm devices despite having a significantly thinner absorber. We believe that the Te buffer layer may be acting as a partial optical reflector at the back of the thinner devices.

Optimal CdCl₂ passivation time varies slightly with absorber thickness but produces like-performing devices over a moderate span of CdCl₂ doses. Fig. 3(c) and (d) demonstrate the detrimental effects of longer CdCl₂ dose times on fill factor and—in turn—efficiency, with devices in this series suffering a relative efficiency loss up to 30%. Our data strongly implies that a thin CdTe absorber makes devices especially susceptible to over-passivation resulting in noticeably decreased performance.

ACKNOWLEDGEMENTS

The authors would like to thank Professor W.S. Sampath for use of the deposition systems, Kevan Cameron for system support, and Drew Swanson and Andrew Moore for helpful discussions. This work has been funded by the US DOE Photovoltaic Research and Development (PVRD) SunShot Initiative program DE-EE0007543.

REFERENCES

[1] M. Woodhouse , R. Jones-Albertus , D. Feldman , R. Fu , K. Horowitz, D. Chung , D. Jordan, and S. Kurtz, "On the Path to SunShot: The Role of Advancements in Solar Photovoltaic Efficiency, Reliability, and Costs," Technical report. Golden, CO: National Renewable Energy Laboratory, 2016.

[2] C. Helbig, A. M. Bradshaw, C. Kolotzek, A. Thorenz, and A. Tuma, "Supply risks associated with CdTe and CIGS thin-film photovoltaics," *Applied Energy*, vol. 178, pp. 422-433, 2016.

[3] C. Sah, R. Noyce, and W Shockley, "Carrier Generation and Recombination in P-N Junctions and P-N Junction Characteristics," *Proceedings of the IRE*, vol. 45, issue 2, pp. 1228-1243, 1957.

[4] T.A. Gessert, "Cadmium Telluride Photovoltaic Thin Film: CdTe," in *Earth Systems and Environmental Sciences*, Elsevier, 2012, pp. 423-438.

[5] L. Kuhn, U. Reggiani, L. Sandrolini, and N.E. Gorji, "Physical device modeling of CdTe ultrathin film solar cells," *Solar Energy*, vol. 132, pp.165-172, 2016.

[6] N.R. Paudel, K.A. Wieland, and A.D Compaan, "Ultrathin CdS/CdTe solar cells by sputtering," *Solar Energy Materials & Solar Cells*, vol. 105, pp.109-112, 2012.

[7] S.S. Hegedus and B.E. McCandless, "Processing options for CdTe thin film solar cells," *Solar Energy*, vol. 77, pp. 839-856, 2004.

[8] J. Kephart, "Optimization of the front contact to minimize short-circuit current losses in CdTe thin-film solar cells," Ph.D. Dissertation, Spring 2015, Colorado State University.

[9] D. Swanson, J. Kephart, P. Kobyakov, K. Walters, K. Cameron, K. Barth, W.S. Sampath, J. Drayton, and J. Sites, "Single vacuum chamber with multiple close space sublimation sources to fabricate CdTe solar cells," *Journal of Vacuum Science & Technology A*, vol. 34, issue 2, 021202, 2016.

[10] J. Raguse, J.T. McGoffin, and J. Sites, "Electroluminescence system for analysis of defects in CdTe cells and modules," in *Proc. 38th IEEE Photovoltaic Spec. Conf.*, 2012, pp. 448-451.

[11] R. Geisthardt "Device Characterization of Cadmium Telluride Photovoltaics," Ph.D. Dissertation, Summer 2014, Colorado State University.

CdS$_{1-x}$Se$_x$ Window Layer for CdTe Prepared by the Exchange of S with Se in CdS Films

Geethika K. Liyanage, Adam B. Phillips, Zhaoning Song, Suneth C. Watthage, Ramez H. Ahangharnejhad, and Michael J. Heben

Wright Center for Photovoltaics Innovation and Commercialization, Department of Physics and Astronomy, University of Toledo, Toledo, OH, 43606, USA

Abstract — Recent studies show that using CdSe as a window material for CdTe solar cells can enhance the current density in both short and long wavelength regions. However, to preserve the high open circuit voltage in this structure, a thin CdS layer at the transparent conducting oxide layer interface is still required. As an alternative to the CdS/CdSe bilayer, we attempted to use a CdS$_{1-x}$Se$_x$ window layer to fabricate CdTe devices. Preparation of the CdS$_{1-x}$Se$_x$ film was done by heat treating pure CdS films with Se vapor, converting a fraction of the CdS to CdSe by an exchange reaction. The degree of conversion increased as the selenization time was increased. The resultant films showed a mixture of CdSe and CdS phases at short times, and the formation of CdS$_{1-x}$Se$_x$ phases at longer times. To study the effect of the selenized window on device performance, sputtered CdTe films were prepared and devices were finished. Current-voltage characteristics and external quantum efficiency measurements showed that the selenized CdS did not perform as well as either sputtered CdS or sputtered CdSe windows. Better control of the intermixing with CdTe and the defect physics could lead to higher performance devices in the future.

Index Terms — CdSe, CdTe, CdS$_{1-x}$Se$_x$, CdTe Quantum efficiency, selenization

I. INTRODUCTION

CdTe is a promising absorber material for thin film photovoltaics (PV) due to its direct band gap and high optical absorption coefficient [1]. The record efficiency of CdTe has been dramatically increased over the past five years, reaching a power conversion efficiency (PCE) of 22.1% for research scale devices [2]. Recent increases in the record PCE are mainly due to the increases in the short circuit current density (J_{SC}) which has been achieved by increasing the current collection in both the short- and long-wavelength regions.

CdS has been extensively used as the heterojunction partner for CdTe devices. However, photons absorbed in the CdS generally do not contribute to the photo-generated current and, due to its fairly small band gap (2.4 eV), the lost current density can be substantial. To overcome this, various methods have been investigated, including reducing the thickness of the CdS layer [3],[4], and using wide band gap window layers such as CdS:O [1],[5],[6], and ZnS [7].

Recently, Paudel and Yan [8] showed that using CdSe (bandgap ~ 1.7 eV) as the window material can lead to an enhancement of J_{SC} in both short and long wavelengths. The high solubility of Se in CdTe lead to the formation of a graded CdTe$_x$Se$_{1-x}$ layer during high temperature absorber preparation

or post-deposition treatments, reducing the effective bandgap of a portion of the absorber layer [8]-[10]. Preparation of a properly intermixed CdTe$_x$Se$_{1-x}$ layer can enhance the current collection in both the long and short wavelength regions. Here, one goal is to prepare the CdTe$_x$Se$_{1-x}$ material in the photoactive zincblende phase [10]. The loss of open circuit voltage (V_{OC}) and fill factor (FF) observed in these CdSe/CdTe devices was attributed to the lower carrier lifetime in the intermixed CdTe$_x$Se$_{1-x}$ layer [10]. By incorporating a thin CdS layer at the SnO$_2$ interface prior to depositing CdSe, the values of these parameters were recovered while retaining the high J$_{SC}$ [8].

Preparation of a CdS/CdSe window has been explored by depositing layers of CdS and CdSe by sputtering [8],[9] and pulsed laser deposition [11],[12]. In the present study, we prepared CdS$_{1-x}$Se$_x$ thin films by an exchange reaction which allowed Se to replace S in the CdS thin films. CdS samples were heated in a closed box in the presence of Se vapor for various periods of time. Se-exchanged CdS$_{1-x}$Se$_x$ films were optically and structurally characterized by spectrophotometry, X-ray diffraction (XRD), and scanning electron microscopy (SEM). Finally, these films were used as the window layer in CdTe PV devices, and the current density-voltage (J-V) and external quantum efficiency (EQE) characteristics were examined.

II. EXPERIMENTAL DETAILS

100 and 10 nm thick CdS thin films were prepared on commercially available TEC 15M (SnO$_2$:F/SnO$_2$; Pilkington NA) substrates by sputtering at a substrate temperature of 250 °C (0.41 W/cm^2, 15 mTorr and 23 sccm of Ar). An aluminum box with inner dimensions of 50 mm × 50 mm × 5 mm was used to perform the Se exchange. The CdS samples and excess Se powder (Sigma Aldrich, 99.99%) were loaded in a glove box, and the Al box was sealed with a graphite gasket. Films were heated to 350 °C for 30 min, 60 min, 90 min, 120 min, or 300 min. After cooling to the room temperature, the films were rinsed with methanol to remove any Se residue. During the exchange reaction, the vapor pressure of Se was ~1 Torr as determined from the phase diagram.

The optical transmittance of the films was measured using a Perkin-Elmer Lambda 1050 spectrophotometer. Surface images of these films were acquired using a Hitachi S-4800

978-1-5090-5606-4/17 $31.00 © 2017 IEEE

UHR-Scanning Electron Microscope, and the composition analysis was done using energy dispersive x-ray spectroscopy (EDS). To examine the structure of these films, XRD patterns were obtained using a Rigaku Ultima III X-ray diffractometer in parallel beam mode with a fixed angle of 1°.

To fabricate CdTe solar cells, 2.1 µm thick CdTe films were deposited by sputtering on the prepared $CdS_{1-x}Se_x$ films at a substrate temperature of 250 °C (0.41W/cm^2, 10 mTorr and 23 sccm of Ar). As-grown CdTe devices were activated by heat treating with $CdCl_2$ at 387 °C for 30 min in dry air. $CdCl_2$ activations for longer durations (up to 45 min) were also performed on similar films to explore the intermixing of the $CdS_{1-x}Se_x$ with CdTe. After removing the excess $CdCl_2$ with methanol, a standard back contact processing was performed by evaporating 3 nm of Cu and 40 nm of Au followed by a heat treatment at 150 °C in air for 30 min. The prepared devices were laser scribed to define a cell area of 0.08 cm^2. J-V characteristics were measured under simulated AM1.5 illumination. The photocurrent generation was characterized by EQE.

III. RESULTS AND DISCUSSION

The evolution of the optical properties was examined by measuring the transmittance spectra. As shown in Fig. 1a, as-deposited CdS films show a clear absorption edge at the wavelength of ~ 512 nm corresponding to a 2.42 eV band gap. As the selenization time increases, the absorption feature at 512 nm decreases. At the same time, there is an increase in absorption over the range of 520 to 710 nm, likely due to the formation of $CdS_{1-x}Se_x$ and/or CdSe in the films. With increased selenization time, further Se exchange resulted the reduction in the fraction of pure CdS. However, evidence for

Fig. 2. (a) Compositional variation of $CdS_{1-x}Se_x$ film with time, SEM images of (b) a clean TEC 15M glass after selenization at 350 °C for 120 min, (c) prepared $CdS_{1-x}Se_x$ film by selenization at 350 °C for 120 min, and (d) an as-deposited CdS film.

the CdS absorption edge was still found for selenization times up to 120 min. After 300 min, a distinct CdS edge was absent. Fig. 1b shows the evolution of the physical appearance of the selenized samples. The color of these films changed from light yellow (CdS) to reddish-brown (expected for CdSe) with increasing selenization time.

EDS was performed to investigate the composition of the films as a function of the selenization time. Fig. 2a shows that the S content drops and the Se content increases with selenization time. Note that the electron beam was scanned over the image area (~ 4 x 4 µm) and the excitation depth was large compared to the film thickness, so that EDS data indicates the film's average composition. After 120 minutes of selenization, the Se:S ratio was close to 1:1.

SEM images revealed surface morphology changes due to selenization. Fig. 2b shows selenized TEC 15M sample after rinsing with methanol and evidence of Se was found as flakes condensed on the bare SnO_2 surface. These flakes were not present in the selenized CdS sample (Fig. 2c) even after 120 min of exposure to Se vapor. This result supports the conclusion that the change in the optical properties of the selenized CdS films is due to Se exchange, and not Se deposition. It is interesting to note that the selenized CdS film shows smoother edges in comparison to the as-deposited CdS (Fig. 2d). One explanation is that the rough surfaces of the as-deposited CdS grains were more readily reactive toward Se exchange and that selenization resulted in surfaces with lower energy.

XRD patterns were obtained for the $CdS_{1-x}Se_x$ films in an effort to fully understand the details of the selenization phase (Fig. 3). As-deposited CdS shows the characteristic peaks for the hexagonal wurtzite structure (h-CdS). After 30 min of selenization diffraction intensities associated with the h-CdS phase decreased, while peaks belonging to the wurtzite CdSe

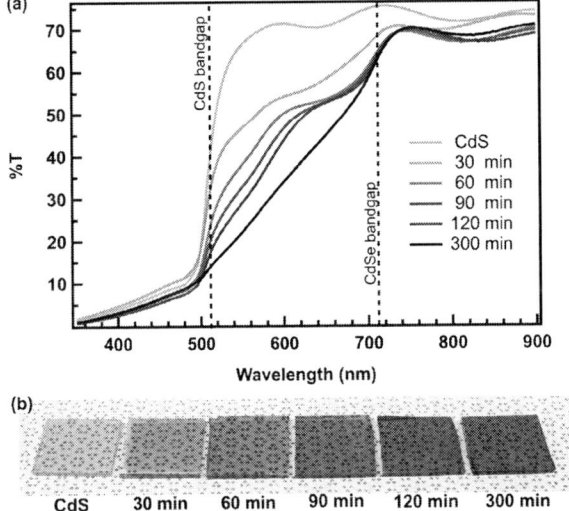

Fig. 1. (a) Optical transmittance spectra and (b) optical images of the CdS films after interacting with Se vapor for different times.

978-1-5090-5606-4/17 $31.00 © 2017 IEEE

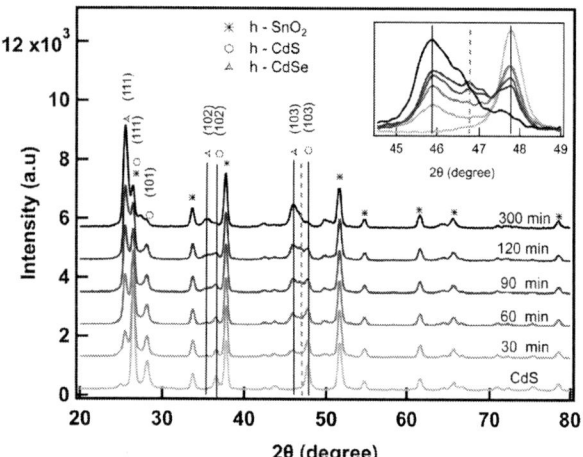

Fig. 3. XRD pattern data for CdS films after Se exchange for different exposure times to Se vapor.

(h-CdSe) started to appear. This change is most clearly seen in the range of 2θ from 45° to 49° where there is no overlapping with the SnO_2 signals (inset of Fig. 3). After 30 min of selenization time, two distinct peaks at $2\theta = 45.9°$ and $2\theta = 47.8°$ associated with diffractions from (103) planes belonging to h-CdSe and h-CdS, respectively, were visible. This indicated that interaction with Se vapor for this relatively short time period resulted in a structurally distinct CdSe phase rather than a $CdS_{1-x}Se_x$ alloy. This is consistent with Se vapor readily interacting with the high-energy surfaces of CdS grains (Fig. 2d) and the formation of a CdSe shell on the exterior of the grain. This process could lead to the smoother grain surfaces seen in Fig. 2c. The (103) diffraction intensity for h-CdS further decreased with increasing selenization time while the intensity of the companion peak belonging to h-CdSe increased with no shift in the peak position. At the same time, diffraction intensity between $2\theta = 45.9°$ and $2\theta = 47.8°$ increased, suggesting the formation of intermediate $CdS_{1-x}Se_x$ phases. For longer selenization times (300 min), the (103) diffraction for h-CdSe was prominent while the companion peak from CdS was absent. However, there was still a significant shoulder on the high 2θ side, indicating the presence of $CdS_{1-x}Se_x$ phases. Note that the diffraction intensities from the SnO_2 layer were unchanged during the selenization, indicating that the structure of the SnO_2 film was not significantly changed.

To test the photovoltaic performance of thin film solar cells prepared with the selenized CdS window layer, we deposited CdTe layers by sputtering and finished the devices with back contacts, as described above. Two control devices, one with 100 nm CdS and one with 100 nm CdSe, were also fabricated for comparison. Fig. 4 shows the J-V characteristics and EQE of several devices.

The PV performance (Fig. 4a) of the devices prepared with the $CdS_{1-x}Se_x$ window layers was worse than that of either of the control devices. The losses in V_{OC}, J_{SC}, and FF increased as the selenization time was increased. The EQE (Fig. 4b) showed an increase response in the long wavelength region (>850 nm) when Se was present. This is consistent with a reduction in the CdTe band gap with Se incorporation [13]. For longer selenization times, increase in the wavelength at which a photoresponse occur is consistent with further Se diffusion into the grains. While the device made with sputtered CdSe shows an enhanced photocurrent generation in both short and long wavelength regions, as well a flat generation profile across the intermediate wavelengths, devices made with a selenized layer show improved long wavelength response but poor response in the 350 to 700 nm regions.

The devices fabricated with the control CdSe and $CdS_{1-x}Se_x$ window layers showed a lower V_{OC} and FF as compared to the control CdS device, with the values for the $CdS_{1-x}Se_x$ layers being below those for the control CdSe device and being worse with increasing selenization time. Poplawsky et al. [10] attributed the loss in V_{OC} with Se incorporation to lower carrier lifetimes that arise from the intermixing of Se with CdTe during the high temperature deposition/post-deposition

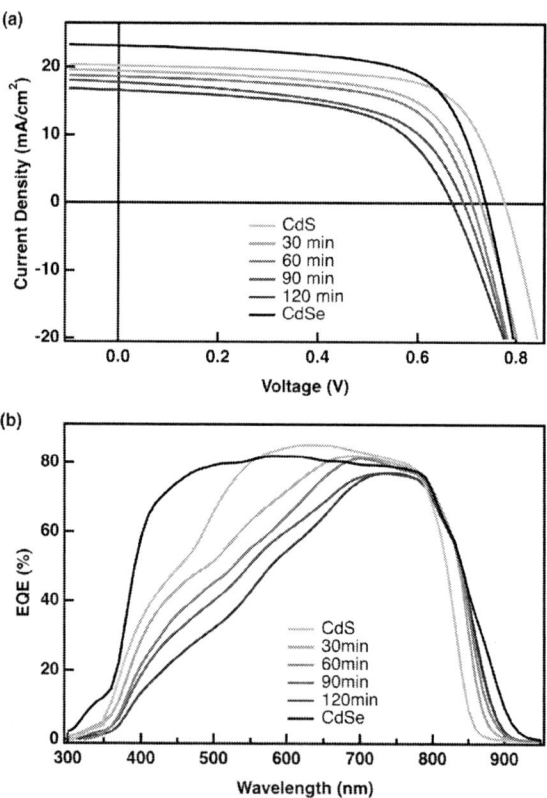

Fig. 4. (a) J-V characteristics and (b) EQE measurements of CdTe devices made on window layer comprised with 30 min, 60 min, 90 min, and 120 min Se exchange compared to CdTe devices made on 100 nm CdS (control CdS), and 100 nm CdSe (control CdSe).

978-1-5090-5606-4/17 $31.00 © 2017 IEEE

processes used in that work. The evolution of the EQE with selenization time and the inflection point around 500 nm suggests the presence of CdS for both 30 and 60 min of selenization time, consistent with the XRD data. At longer selenization times the shape of the EQE curve becomes more triangular, with very little response on the short wavelength side. At the longest selenization time the EQE data resembles the data found for the 400 nm thick sputtered CdSe layer in the work of Poplawsky et al. [10]. In that work, the relatively thick 400 nm CdSe layer produced a wurtzite CdTe-CdSe alloy that was photo-inactive, while thinner CdSe layers led to Se-poor zincblende CdTe-CdSe alloys that were photoactive. We note that our highest processing temperature (387 °C) occurred during the $CdCl_2$ treatment while Poplawsky et al. grew the CdTe layer by closed space sublimation at 610 °C. Therefore, the degree of intermixing of Se and Te in the CdTe is likely to be different despite the similarity between the EQE data.

To fully investigate the role of Se exchange with Te in CdTe, we varied the $CdCl_2$ treatment time. Experiments were performed on CdTe devices prepared with 100 nm thick CdS samples that had been selenized for 60 min. Fig. 5a shows that the V_{OC} and J_{SC} were adversely impacted, suggesting that the devices were in fact over-treated, perhaps even at after 30 min.

sufficient to nearly fully selenize the thin CdS layer to produce a CdSe layer with a small S content. After selenization, 90 nm of CdSe was subsequently sputtered on top. This stack and two other stacks, comprised of 100 nm CdS and 100 nm CdSe layers on TEC15M, were finished into devices by sputtering 2.1 μm of CdTe, performing a $CdCl_2$ treatment (30 min at 387 °C), followed by standard back contact processing. Fig. 6 shows the J-V and EQE performance of these devices. Once again, we see characteristic beahavior for the CdS and CdSe devices. The J-V data shows a higher V_{OC} and lower J_{SC} for the CdS device, while the reverse is true for the CdSe device. Turning to the EQE, the CdS device shows a notch associated with absorption at wavelengths < 500 nm that does not contribute to the photocurrent, while the CdSe device has a broader, flatter response that extends to longer wavelengths. The device with the $CdS_{1-x}Se_x$/CdSe bilayer has V_{OC} and J_{SC} values that are smaller than those of the control devices. The J-V curves also shows roll-over in forward bias, which suggests the formation of an additional junction in the device, perhaps at the $CdS_{1-x}Se_x$/CdSe interface.

In this device configuration the Se-Te interdiffusion is expected to be governed by the properties of the CdSe/CdTe interface. Consequently, it is somewhat surprising that the inclusion of a thin, nearly fully selenized $CdS_{1-x}Se_x$ layer could

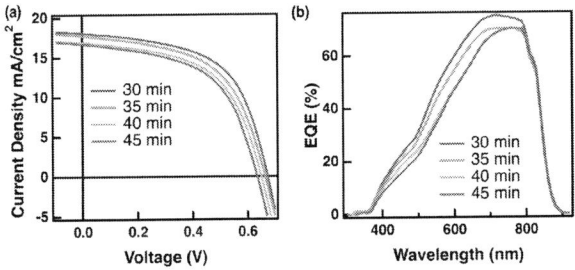

Fig. 5.   (a) J-V characteristics and (b) EQE measurements of CdTe devices prepared on $CdS_{1-x}Se_x$ (60 min selenization time at 350 °C) finished with different $CdCl_2$ activation times.

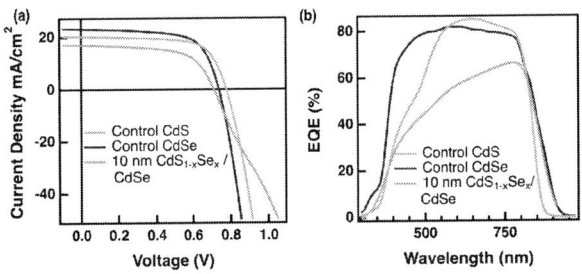

Fig. 6.   (a) J-V characteristics and (b) EQE measurements of CdTe devices made on 100 nm CdS (control-CdS), 100 nm CdSe (control-CdSe), and 10 nm selenized CdS (60 min selenization time) with additional 90 nm CdSe.

Consistently, the EQE data (Fig. 5b) shows loss in current generation across the range 400 – 700 nm with increased treatment time, indicating increasing recombination losses for light absorbed near the junction. It is interesting to note that the long wavelength response was not effected.

We orginally anticipated that a fully selenized CdS layer could perform as well as a sputtered CdSe layer and, if some CdS was retained, it might be possible to obtain the short circuit current enhancements while maintaing a high V_{OC}. However, the selenized CdS samples, in all cases, lead to poorer performance than sputtered CdS alone or pure sputtered CdSe. In another approach, we sputtered a 10 nm CdS film on TEC15M and selenized it for 60 min at 350 °C to prepare a thin $CdS_{1-x}Se_x$ film. From the transmission and EDS data presented earlier we concluded that 60 min would be

introduce such significant losses in the EQE. Evidently, the defect distribution within the $CdS_{1-x}Se_x$ layer and its structure is significantly different from that of the sputtered CdSe film. We note that the EQE data for the $CdS_{1-x}Se_x$/CdSe bilayer and the CdSe devices overlaps in the long wavelength region indicating that the photocurrent generation and collection mechanisms are the same at these wavelengths. In contrast, the EQE at shorter wavelengths is severely impacted. The data suggests that the recombination rates are much higher for carriers that are generated near, or can diffuse in to, the $CdS_{1-x}Se_x$ layer.

The poor performance of devices with $CdS_{1-x}Se_x$ layers is most likely due to the defect structure in the layers. Additionally, the XRD data indicated that the selenization

978-1-5090-5606-4/17 $31.00 © 2017 IEEE   173

does not proceed in a homogeneous fashion, particularly when the as-deposited CdS layer consists of small grain material. Even in the limit of long-time selenization there is still evidence of S content and the grain structure suggests that the S is not uniformly distributed. This lack of homogeneity is very likely translated to the Se-Te exchange that has been shown to be beneficial at the CdSe/CdTe interface. For example, one can imagine a situation where CdSe on grain surfaces in the initial window layer may intermix with the CdTe during the $CdCl_2$ activation, while S and Se content in elsewhere in the heterogeneous layer may not participate in the intermixing. These issues have been highlighted by Grice *et al.* [9] where it was determined that CdSe/CdTe interface mixing of low temperature sputtered devices occurs only during the $CdCl_2$ activation step while, in contrast, significant intermixing occurs during a high temperature deposition processes such as close space sublimation. A similar approach may promote intermixing in these $CdS_{1-x}Se_x$/CdTe devices to obtain a better junction quality.

IV. CONCLUSION

$CdS_{1-x}Se_x$ thin films were prepared by selenization of CdS films in a closed aluminum box. The degree of selenization was controlled by adjusting the exposure time to Se vapor at 350 °C. X-ray diffraction data indicates that films produced with 30 min of selenization consist of grains with a CdSe shell CdS interior. Increasing selenization time resulted in the further diffusion of Se in to the grains and a more complete conversion to CdSe. The $CdS_{1-x}Se_x$ films were used as a window layer in sputtered CdTe devices. Prepared devices showed poor performance compared to standard CdTe devices fabricated on either CdS or CdSe. EQE results show a significant photocurrent loss in the 400-700 nm range in these devices due to poor intermixing of the $CdS_{1-x}Se_x$ with CdTe. A high temperature absorber preparation method could be used to promote this intermixing to obtain better junction quality.

ACKNOWLEDGEMENTS

Authors would like to thank NSG (Pilkington NA) Toledo, Ohio for providing SnO_2:F/SnO_2 coated soda lime glass substrates. The work was funded by the Air Force Research Laboratory, Space Vehicles Directorate (contract # FA9453-11-C-0253) and the Wright Center Endowment for Photovoltaics Innovation and Commercialization.

REFERENCES

[1] X. Wu, "High-efficiency polycrystalline CdTe thin-film solar cells," *Solar Energy,* vol. 77, no. 6, pp. 803-814, 2004.

[2] NREL. Solar Efficiency Chart [Online]. Available: http://www.nrel.gov/pv/assets/images/efficiency-chart.png (Accessed : May, 2017).

[3] B. E. McCandless and K. D. Dobson, "Processing options for CdTe thin film solar cells," *Solar Energy,* vol. 77, no. 6, pp. 839-856, 2004.

[4] A. Bosio, N. Romeo, S. Mazzamuto, and V. Canevari, "Polycrystalline CdTe thin films for photovoltaic applications," *Progress in Crystal Growth and Characterization of Materials,* vol. 52, no. 4, pp. 247-279, 2006.

[5] X. Wu, Y. Yan, R. G. Dhere, Y. Zhang, J. Zhou, C. Perkins, and B. To, "Nanostructured CdS:O film: preparation, properties, and application," *Physica Status Solidi (c),* vol. 1, no. 4, pp. 1062-1066, 2004.

[6] N. R. Paudel and Y. Yan, "Fabrication and characterization of high-efficiency CdTe-based thin-film solar cells on commercial SnO_2:F-coated soda-lime glass substrates," *Thin Solid Films,* vol. 549, pp. 30-35, 2013.

[7] C. Xiao, N. R. Paudel, C. R. Grice, Y. Yu, and Y. Yan, "CdTe solar cells using combined ZnS/CdS window layers," in *40th Photovoltaic Specialist Conference,* 2014, pp. 2428-2430.

[8] N. R. Paudel and Y. Yan, "Enhancing the photo-currents of CdTe thin-film solar cells in both short and long wavelength regions," *Applied Physics Letters,* vol. 105, no. 18, p. 183510, 2014.

[9] C. R. Grice, A. Archer, S. Basnet, N. R. Paudel, and Y. Yan, "Characterization of CdS/CdSe window layers in CdTe thin film solar cells," in *42th Photovoltaic Specialist Conference,* 2016, pp. 1459-1463.

[10] J. D. Poplawsky , W. Guo, N. Paudel, A. Ng, K. More, D. Leonard, and Y. Yan, "Structural and compositional dependence of the $CdTe_xSe_{1-x}$ alloy layer photoactivity in CdTe-based solar cells," *Nature Communications,* Article vol. 7, p. 12537, 2016.

[11] X. Yang , Z. Bao, R. Luo, P. Tang, B. Li, B. Liu, J. Zhang, W. Li, L. Wu, and L. Feng, "Preparation and characterization of pulsed laser deposited CdS/CdSe bi-layer films for CdTe solar cell application," *Materials Science in Semiconductor Processing,* vol. 48, pp. 27-32, 2016.

[12] X. Yang , B. Liu, J. Zhang, W. Li, and L. Feng, "Preparation and characterization of pulsed laser deposited a novel CdS/CdSe composite window layer for CdTe thin film solar cell," *Applied Surface Science,* vol. 367, pp. 480-484, 2016.

[13] S. H. Wei, S. B. Zhang, and A. Zunger, "First-principles calculation of band offsets, optical bowings, and defects in CdS, CdSe, CdTe, and their alloys," *Journal of Applied Physics,* vol. 87, p. 1304, 2000.

Effect of Illumination on Thermal CdCl₂ Treatment of CdTe

Sudhajit Misra[1], Carina E. Hahn[1], Vasilios Palekis[2], Christos Ferekides[2] and Michael A. Scarpulla[1]

[1]University of Utah, Salt Lake City, Utah, 84112, USA
[2]University of South Florida, Tampa, Florida, 33620, USA

Abstract — **Photon absorption is a critical process for photovoltaic operation, yet the effect of illumination on semiconductor growth and processing is rarely investigated. In this work we investigate the effect of illumination during the cadmium chloride (CdCl₂) annealing treatment of cadmium telluride (CdTe) thin films for photovoltaic applications while keeping the sample temperature constant. Illumination at 808 nm with varying intensity (0, 0.5, 1 and 1.5 W) when incident on CdTe during thermal CdCl₂ treatment of CdTe lead to enhancement of grain size and optoelectronic properties of CdTe. However photo-assisted etching of the CdTe surface is also observed at illuminations sufficiently low that the sample temperature was barely affected during annealing. Additionally, annealing under high illumination sufficient to drive CdCl₂ annealing in the absence of a furnace has no such effect on the CdTe surface. We attempt to reconcile these observations and provide rationale for further experiments.**

I. INTRODUCTION

In recent years, cadmium telluride (CdTe) has emerged as a leading thin film photovoltaic (PV) technology. CdTe is a direct bandgap semiconductor with a bandgap of (1.45-1.5 eV at 300 K), has high absorption coefficient ($>10^4/cm^2$), which makes it an ideal material for PV applications. A very thin layer (1-3 μm) is sufficient for maximum absorption of incident solar radiation. Rapid improvements in CdTe efficiency has led to successful commercialization in the PV industry and current research efficiency have surpassed 22% [1].

Chlorine thermal treatment of CdTe layers after deposition is a crucial step manufacturing high efficiency solar cells; it is used to improve structural and optoelectronic properties of CdTe [2-5]. Generally, a thin layer of cadmium chloride is deposited on top of the CdTe film either by evaporation or by drip-coating a solution of CdCl₂ in methanol followed by a drying step and then annealing is done or close space sublimation (CSS) is used for the whole process. The CdCl2 treatment is carried out typically at a temperature of 400-430 °C and times of 10's of minutes, although it is believed that the takt time in manufacturing is much shorter. Rapid thermal annealing (RTP) may be used to reduce the time of thermal treatment process. This process involves the use of a high-intensity lamps to rapidly heat the sample. The CdTe device stack may or may not be directly exposed to the incident illumination in RTP – susceptors and proximity caps may be used. Additionally, we have demonstrated infrared laser annealing as a means of rapidly accomplishing CdCl₂ treatment although to date the device results have not proven quite as high as for traditional CSS based CdCl₂ treatment. These illuminated annealing processes, experiments proving

differences in diffusion and surface point defect concentrations, and recent theory that established the role of excess carriers in determining point defect thermodynamic stability suggest that experiments decoupling the roles of temperature, time, heating and cooling rates, and illumination are of interest for fully understanding and perhaps further optimizing CdCl₂ processes.

In this preliminary study, we investigated the effect of isothermally annealing CdTe thin films with varying light bias intensity. An 808 nm laser with varying fluence was directed onto the samples during cadmium chloride (CdCl₂) thermal treatment in a tube furnace as shown in Fig.1a. The sample temperature was measured by a thermocouple in contact with the back of the glass substrate. The sample temperature was kept constant by reducing the power supplied to the furnace to compensate for the power added by the laser.

We found an unanticipated photo-etching effect on the surface of the CdTe films for the small photon fluxes used. The etching appears to target the grain boundary regions and in some areas lead to creation of etch pits. Other observations include an increase in grain size, as is typical in CdCl₂ thermal treatment of CdTe.

II. EXPERIMENT

Transparent fluorinated tin oxide coated glass substrates (2 mm thick) were cut to fit inside the 1" tube furnace standing up such that the sample normal was along the tube axis. A thin

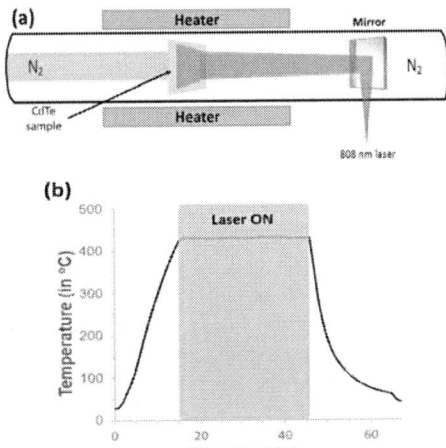

Figure 1(a). Setup of tube furnace (b) temperature profile of CdCl₂ treatment measured with thermocouple in contact with back of sample.

layer (50-80 nm) of cadmium sulphide (CdS) was deposited by evaporation. The 3-5 μm thick CdTe layer was deposited on top of the CdS layer by CSS [6]. For CSS the temperature of source and substrate were maintained at 600 °C and 475 °C respectively, in an atmosphere of helium at 5 Torr. The samples were then drip coated with a saturated solution of $CdCl_2$ in methanol at room temperature followed by ambient evaporation to leave a layer of $CdCl_2$ on top of the CdTe.

For chlorine activation, the $CdCl_2$ coated samples were annealed in tube furnace at 430 °C for 30 mins in dry nitrogen

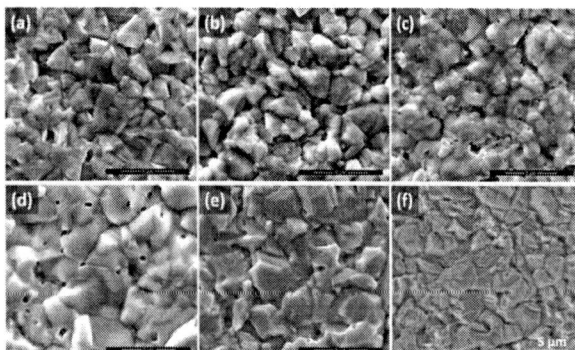

Fig2 - SEM images of (a) reference sample, (b) dark annealed control sample, and samples annealed in furnace at surface illumination of (c) 0.5 W (d) 1.5 W (e) 2.5 W (f) sample annealed by 808 nm laser. Samples b, c, d, e were annealed in a tube furnace at 430°C.

atmosphere. A continuous-wave 808 nm fiber-coupled diode laser (Coherent FAP) was used to illuminate the samples at 3 different powers of 0.5 W, 1.2 W and 2.5W. The experiment setup of the tube furnace is shown in Fig. 1(a). A thermocouple attached to the back of the sample was used to keep the temperature constant during the experiments. We know from pervious laser annealing experiments that the time constant for establishment of steady state temperature profile for such samples is approximately one minute. Additionally, in steady state, because the overwhelming majority of the thermal resistance comes from the thick glass substrate, the temperature in the device stack is within 1 °C even when all heating power comes from a laser and the back of the substrate is heat-sunk to room temperature. Therefore any explanation of the results based on local overheating of the CdTe surface should be discounted. The power to the furnace was reduced when the laser illumination was turned on in order to ensure that heating of the CdTe film does not occur solely due to the incident laser beam. The temperature profile of the annealing process is shown in Fig. 1(b). Two more samples, one thermally annealed in dark at same conditions and another laser annealed in argon atmosphere were used to compare the effects of illumination on surface morphology of the samples. The laser power for the laser annealed sample was chosen so as to maintain a similar temperature at the CdTe surface as the thermally treated samples.

Low temperature wavelength resolved photoluminescence spectroscopy (PL) was performed on the samples to observe the opto-electronic properties of the samples. A helium cryostat was used to cool the samples to 10 K. A 532 nm solid state laser operating at 30mW was used excitation source. The PL emission form the CdTe film was wavelength resolved using a 0.5m SPEX monochromator and detected using a silicon CCD line camera.

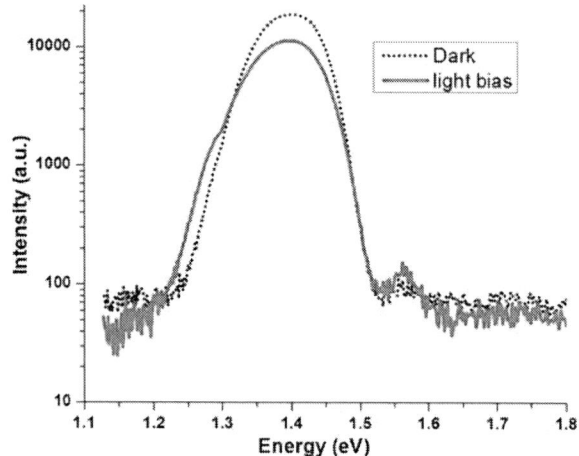

Figure 3 - Comparison of low temperature PL (10 K) of samples annealed in dark and under illumination. An exciton peak (1.57 eV) is observed only in sample annealed under illumination in a furnace.

III. RESULTS

The surface morphology of all the annealed samples were compared with respect to a reference sample (Fig. 2.a) in a SEM. Minor growth in grain size was observed in the sample annealed in dark (Fig. 2.b) and sample annealed with 500mW light bias(Fig. 2.c). Significant grain growth was observed in samples that received an illumination of 1.5 W (Fig. 2.d) and 2.5 W (Fig. 2.e). Largest increase in grain size however was observed in the sample that did not receive a furnace annealing but was annealed solely with an 808 nm laser.

An interesting feature was observed in the samples that were annealed in the furnace with additional illumination (Fig. 2c, d, e). Features that look like surface etching mainly concentrated around some grain boundaries were observed in these samples. These observations led us to believe that the presence of chlorine and photons during the annealing process leads to outgassing of $CdCl_2$ from the surface leaving behind the etch-pit structures. However, these etch pits were not observed under a high photon flux annealing (laser annealing sample, Fig. 2.f). Further investigation is needed to better understand this effect.

Annealing in presence of light has a different effect than annealing in dark on opto-electronic property of the CdTe film. Low temperature spectrally resolved photoluminescence spectra (figure on left) of the samples annealed at 430 °C shows that annealing in light bias has a lower defect peak at 1.4 eV and a excitonic peak at higher energy than the sample annealed in dark (Fig. 3) [7, 8].

IV. DISCUSSION

The primary motivation behind the work was to understand the effect of photo-carriers on grain growth and point defects in CdTe during the annealing process. Our preliminary observations suggest, under low illumination annealing in a furnace, the opto-electronic property of the film in enhanced as is evidenced by the appearance of the excitonic peak in the sample annealed under illumination. This is beneficial to device fabrication.

However, a photo-etching effect is also observed on the CdTe surface but only under low illumination annealing. This effect is not observed when the sample is annealed under high illumination (laser annealed sample, Fig. 2f). Since the mechanics of annealing by laser is similar to rapid thermal annealing process (RTP), investigations of RTP treatment of CdTe do not show this effect. Further investigation is needed to better understand the origin and mechanism of this photo-etching effect.

V. SUMMARY

In this work we demonstrated the effect of illumination on thermal $CdCl_2$ treatment of CdTe thin films. Even though at low illumination conditions the opto-electronic property and grain size of CdTe film seems to be enhanced, a photo-etching effect is observed at the surface of CdTe. This etching is deleterious to device performance. However, high illumination annealing (laser annealing and RTP) do not show this effect. Further investigation is needed to understand the phot-etching effect.

ACKNOWLEDGEMENT

This work was supported in full by the Department of Energy through the Bay Area Photovoltaic Consortium under award DE-EE0004946.

REFERENCES

[1] M. A. Green, K. Emery, Y. Hishikawa, W. Warta, E. D. Dunlop, D. H. Levi, and A. W. Ho-Baillie, "Solar cell efficiency tables (version 49): Solar cell efficiency tables (version 49)," Progress in Photovoltaics, vol. 25, no. NREL/JA--5J00-67687, 2016.

[2] B. E. McCandless, L.V. Moulton and R. W. Birkmire, "Recrystallization and sulfur diffusion in $CdCl_2$-treated CdTe/CdS thin films," Progress in Photovoltaics: Research and Applications, vol. 5, issue 4, pp. 249-260, 1997.

[3] M. Terheggen, H. Heinrich, G. Kostorz, A. Romeo, D. Baetzner, A. N. Tiwari, A. Bosio and N. Romeo, "Structural and chemical interface characterization of CdTe solar cells by transmission electron microscopy," Thin Solid Films, vol. 431, pp. 262-266, 2003.

[4] W. K. Metzger, D. Albin, M. J. Romero, P. Dippo and M. Young, "CdCl2 treatment, S diffusion, and recombination in polycrystalline CdTe," Journal of Applied Physics, vol. 99, issue 10, 103703, 2006.

[5] B. E. McCandless, M. G. Engelmann, and R. W. Birkmire, "Interdiffusion of CdS/CdTe thin films: Modeling x-ray diffraction line profiles," Journal of Applied Physics, vol. 89, pp. 988-994, 2001.

[6] C. S. Ferekides, D. Marinskiy, V. Viswanathan, B. Tetali, V. Palekis, P. Selvaraj, and D. Morel, "High efficiency CSS CdTe solar cells," Thin Solid Films, vol. 361, pp. 520-526, 2000.

[7] C. Kraft, H. Metzner, M. Hädrich, U. Reislöhner, P. Schley, G. Gobsch, and R. Goldhahn A. L. Fahrenbruch and R. H. Bube, "Comprehensive photoluminescence study of chlorine activated polycrystalline cadmium telluride layers," Journal of Applied Physics, vol. 108, pp. 124503, 2010.

[8] P. L. Roland, N. R. Paudel, C. Xiao, Y. Yan and R. J. Ellingson, "Photoluminescence spectroscopy of cadmium telluride deep defects," in 40th Photovoltaic Specialist Conference (PVSC), 2014, p. 3266-3271.

Challenges in the industrial production of CZTS monograin solar cells

[1]Gerhard Peharz, [1]Valentin Satzinger, [2]Sandra Pötz, [2]Gernot Oreski, [3]Theodoros Dimopoulos, [3]Stefan Edinger, [4]Wolfgang Hackl, [4]Hannes Starkl, [5]Parichehr Esfandiari, [5]Peter Krabb, [5]Stefan Gahr, [5]Lukas Plessing and [5,6]Dieter Meissner

[1]JOANNEUM RESEARCH, Weiz, 8160, Austria; [2]PCCL, Leoben, 8700, Austria
[3]AIT, Vienna, 1210, Austria, [4]Forster, Waidhofen, 3340, Austria; [5]crystalsol, Vienna, 1110, Austria;
[6]Tallinn University of Technology, Tallinn, 19086, Estonia

Abstract — The material class comprising of $Cu_2ZnSn(S,Se)_4$ (CZTS) is non-toxic and comprises of abundant elements, which makes it to be very interesting for the application in solar cells. Recent progress on the understanding of the materials and devices resulted in increasing efficiencies and it is expected that these will increase further. For a commercial success of CZTS solar cells an industrial low-cost production is required. The CZTS monograin solar cell technology allows developing a solar cell production process independently from material and device research. Challenges and latest results on the roll-to-roll production of monograin CZTS solar cells are discussed and presented.

Keywords — CZTS, module, production, roll-to-roll

I. INTRODUCTION

In the recent years a substantial increase in research and developments in the field of $Cu_2ZnSn(S,Se)_4$ (CZTS) solar cells has been recognized. The main reasons are that CTZS is a non-toxic material and comprises of abundant elements. Those are considered to be long-term advantages in particular when comparing to CIGS and/or CdTe thin film solar cells. However, compared to CIGS and CdTe the efficiency of CZTS solar cells is still substantially lower. In particular, the highest certified solar cell efficiency until now is 12.6% [1].

One of the main reasons for the relatively poor efficiency of CZTS devices (e.g. compared to CIGS) is a lack in open circuit voltage (V_{OC}) [1],[2]. That decrease in V_{OC} is considered to be related to material defects, and in the recent years very good progress has been made in understanding those material defects in CZTS [1],[3] – [7]. In particular by carefully controlling the annealing and the Cu-Zn ordering, recently a V_{OC} 784 mV was achieved with a Cu_2ZnSnS_4 monograin layer [6].

CZTS monograin solar cells comprise of individual CZTS crystals coated with a buffer layer. These are arranged in a monolayer [8],[9]. In contrast to (vacuum and/or solution based) thin-film processes, used typically producing CZTS solar cells, the monograin approach allows to separate the CZTS material production (including buffer) from the processes for making solar cells and modules. Consequently a side-by-side development is enabled, where the CZTS material is independently optimized from developing a module manufacturing process. Improved CTZS material quality and

better (hetero-)junction properties can be quickly transferred to the solar cell and/or module level by replacing the CZTS powder used for making the monograin devices (drop-in solution).

Since monograin production technology is promising but new approach for solar cells and/or modules, one of the challenges is to develop a low-cost and high-throughput roll-to-roll production process. The scope of this paper is to present latest results on processes for the roll-to-roll production of CZTS monograin solar cells and discuss their feasibility in terms of industrial production.

II. MONOGRAIN PRODUCTION

The aim of the CZTS monograin production is to obtain a monolayer of small (around 50 μm) crystal grains (including buffer layer) assembled by a non-conductive polymer adhesive. Contacts are formed with a transparent front side contact (TFC) on top and a back-side electrode on the rear-side. The process scheme of the most important process steps considered in the work presented is shown in Fig. 1.

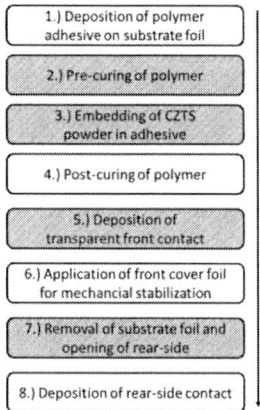

Fig. 1: The process scheme for making CZTS monograin solar cells is sketched.

Some of the processes shown in Fig. 1 are already state of the art in industrial roll-to-roll production. For instance printing of polymer layers (1), post-curing (4), laminating

978-1-5090-5606-4/17 $31.00 © 2017 IEEE

front-covers (6) and printing of electrodes (8) are industrial standard processes. However, about half of the remaining processes (indicated by grey boxes in Fig. 1) are challenging in terms of high throughput roll-to-roll production. In the following sections the key requirements of those critical processes are discussed in terms of an industrial production of CZTS monograin solar modules and achievable web-speed in a roll-to-roll production. Please note that in the considerations on the web-speed shown below a web-width of 0.5 m was taken into account.

For the qualification of the processes test-modules ranging from a size of 10x10 to 20x20 cm² have been produced. The CZTS material used for making those modules is of decent but not very high quality. Solar cells and modules made of the CZTS powder used for the test trials typically show efficiencies well above 5%. That efficiency is considered to be the benchmark for the alternative processes discussed below.

III. PRE-CURING OF POLYMER

As already mentioned above roll-to-roll printing of polymer adhesives is industrial state-of-the-art and high throughput rates of >100 m/min can be achieved. For that high-speed printing the printed adhesive usually requires to have a rather low viscosity. When trying to embed (process 3) a monolayer of crystalline (CZTS) grains (diameter about 50 μm) in a liquid resin layer with such low viscosity, one faces the problem of capillary effects (between the grains) and unwanted wetting of the grains at the top-surface. In particular no decent front contact formation is possible when the grains are wetted by the adhesive. If the polymer is wetting only parts of the grains top-surface, that would result in a reduction of the active area. Moreover a high wetting of the grains can result in reduced thickness of the polymer in-between the grains and the risk of shunt formation would be increased. The application of an adhesive which intrinsically shows a low wetting of the CZTS grains is no feasible option, as the mechanical integrity of the monograin film would be weakened and the risk of shunt formation would be increased.

In a pre-curing process the viscosity (hardness) of the liquid resin layer is increased. An optimized pre-curing process enables a subsequent embedding of a monolayer of crystal grains where only half of the grain volume is sticking in the partially cured adhesive (no contamination of top-surface).

Epoxy resins are widely used in electronics industry and these materials typical have very good insulating properties, long-term reliability and are not critical in terms of (environmental) toxicity. For printing the liquid epoxy resin is mixed with a hardener and the viscosity is tuned by fillers. The cross-linking typically is triggered thermally and also the speed of the reaction can be increased by elevating the temperature. The viscosity of the epoxy resin increases with the degree of cross-linking. At the optimum degree of cross-linking the CZTS powder can be well embedded into the partially cured resin layer. If the resin is too hard, embedding of the grains into the membrane will be prevented. In alternative, if the viscosity of the epoxy resin is too low the complete surface of the grains will be covered and no feasible front contact can be created.

For an industrial implementation the main challenge is to achieve a high level of control of the cross-linking degree of the polymer. The most important process parameter is temperature, which needs to be controlled very well during the pre-curing process. For this purpose the substrate foil with the uncured epoxy film is passed through an oven. The industrial feasibility of using epoxy resins and thermal pre-curing were evaluated in pre-tests and a continuous production of monograin foils for about 2 hours has been successfully demonstrated. A minimum curing time of about 1.5 minutes was achieved when pre-curing was carried out at 95°C. Consequently the maximum web-speed during production is limited by the oven length and a web-speed of 4 m/min can be achieved when using a standard oven with a length of 6 m.

IV. EMBEDDING OF CZTS POWDER INTO THE POLYMER

The main requirements for the embedding process are a high packing density of the grains in the monolayer. In particular at least 80% of the cell area should be covered with crystal grains. Moreover, the grains must not be damaged or contaminated during the embedding process. Two processes have been investigated in detail until now:

A. Mechanical embedding by gravitational forces

A straight-forward method for embedding the CZTS powder is to use a funnel and create a curtain of trickling grains falling onto the moving web. That method was tested to be feasible and CZTS monograin modules with efficiencies >5% were produced in a roll-to-roll process. Moreover, investigations on increasing the speed of the embedding process have been conducted. Currently a web-speed of 10 m/min is validated to result in high packing densities of >80% and a further increase of web-speed is being tested.

B. Embedding by electrostatic forces

Since it is anticipated that the limit of the mechanical embedding process described in the sub-section above will be limited to be in the range of 40 – 60 m/min, an alternative process was tested. An electrostatic field is applied to accelerate the CZTS powder when the web with the pre-cured polymer is passing by in order to embed the grains in the adhesive. Test on a roll-to-roll set-up show decently high packing densities of about 80% and can be achieved also at high web-speeds of 40 m/min; potentially even 120 m/min can be achieved. In addition it was found that the grains are not damaged in the electrostatic embedding process and CZTS monograin membranes can be produced.

V. DEPOSITION OF TFC

Once the CZTS powder is decently embedded and impressed the adhesive is post-cured. In contrast to the pre-curing the post-curing is less time-critical, since it can be performed with a coiled web and therefore does not require to be done in the roll-to-roll production. More critical is the subsequent process of depositing a transparent front contact (TFC). A common TFC for CZTS solar cells is based on transparent conductive oxides (TCO). Such TCOs are typically deposited in a sputtering process and the current state-of-the art for making CZTS monograin solar modules is based on sputtered TFCs.

Inline sputtering of TCOs can be done in a roll-to-roll process [10]; however, high investment costs are required for the sputtering equipment when aiming at roll-to-roll production with high web-speeds. Consequently, a solution based process for depositing the TFC was developed which allows the fabrication of CZTS monograin modules. In particular a three step process was applied for creating the solution based front contacts:

- In a first step, a seeding layer of ZnO nanoparticles is deposited on top of the CZTS grains embedded in the polymer.
- In a second step, a thin layer (< 100 nm) of doped ZnO is grown on the seeding layer by means of chemical bath deposition (see Fig. 2).
- Finally, the conductivity of the front contact is enhanced by depositing silver nanowires.

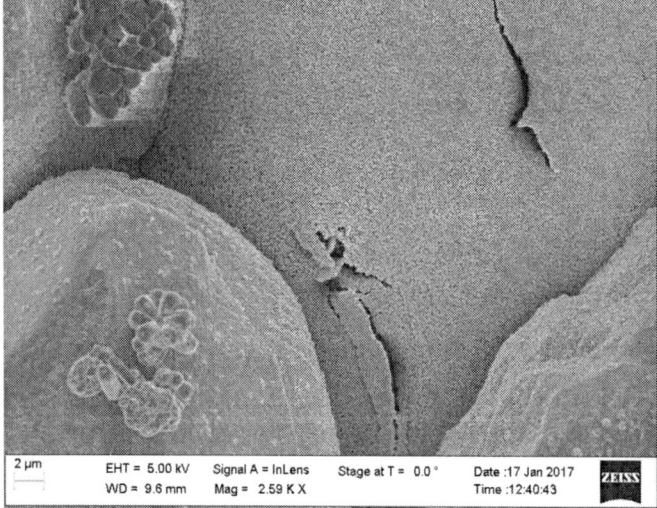

Fig. 2: A SEM image of the monograin after the CBD coating is shown.

The initial performance of test-modules with the "all-solution-based" front-contact is found to be equal to that made by sputtering and is well above 5% for both types of TFCs. Moreover, degradation studies show that solar cells with these TFCs are at least as stable as the reference devices with sputtered TFCs.

The main advantage of the solution based TFC process is that it can easily be integrated in a roll-to-roll production. The main limiting factor for the web-speed is the chemical bath deposition. It was found that less than 1 min is sufficient for the chemical bath deposition process. Thus for a web-speed of 4 m/min a chemical bath deposition unit with an effective process length of 4 m would be required, which is feasible.

VI. REAR-SIDE CONTACT PREPARATION

After a transparent front-cover foil is laminated, the substrate foil is removed. Subsequently about 5-15 μm of the rear side of the monograin solar cell needs to be removed in order to expose the core of the CZTS grains. That can be done by polishing the rear-side of the monograin and decent solar cell efficiencies can be achieved by applying that process. However, in a roll-to-roll process well-defined polishing of a monograin is not trivial. Therefore an alternative process has been developed which is relying on laser-ablation.

The laser-ablation process developed for CZTS monograin solar cells allow a very high control of the ablation depth of less than 1 μm. That very high process control is achieved also on module level as can be seen in the very homogeneous electro-luminescence image shown in Fig. 3. The efficiencies of test-modules made by applying the laser-ablation process are found to be well above 5% and are equal to those made by polishing the rear-side.

Fig. 3: Electro-luminescence image of a 20x20 cm² CZTS monograin module made by applying laser-ablation for the rear-side contacting.

Laser-ablation processes are already state-of-the art also in roll-to-roll production. However, until now mainly the scribing of lines for the monolithic series connection is done in the production of photovoltaic devices. For the monograin production a full area ablation is required which limits the achievable process speed. The laser equipment used for processing the module shown in Fig. 3 requires less than 1 minute to ablate an area of 20x20 cm². That would correspond to a web-speed of about 0.1 m/min for a 50 cm wide web. However, industrial laser sources with at least 10 times more power are already available and based on conservative extrapolations of already achieved results a web speed of a few m/min should be achievable

VII. DISCUSSION AND CONCLUSIONS

The results shown above indicate that an industrial roll-to-roll production of CZTS monograin modules is feasible and can be realized without vacuum processes. In terms of process speed the most limiting process is that of the rear-side ablation which is required for contact formation. The equipment used for the laser ablation is proven to be very precise and homogeneous; however, it only allows achieving a web-speed in the range of 0.1 m/min. Considering a 24 hour production such a web-speed would correspond to 720 m² of CZTS monograin module are produced per day. An increase of web-speed would require to use laser sources with higher average or to use several laser sources in parallel. That would allow increasing the web-speed to be in the range of a few m/min.

The other processes investigated can be conducted at a web-speed of at least 4 m/min. In particular the embedding of the crystal powder was identified to be non-critical in terms of process speed. Also the first process step – the printing of the polymer – was shown to be feasible for a roll-to-roll web-speed of 4 m/min.

Very promising is the solution process allowing to produce decent transparent front contacts. That process is limited by the time required for the chemical bath deposition, which was found to be about 1 minute. For a web-speed of 4 m/min a roll-to-roll chemical bath deposition reactor with an effective length of 4 m would be required. Such a reactor is realistic and consequently also a corresponding production of CZTS monograin modules at a roll-to-roll web-speed of a few m/min can be realized. That is very encouraging for the further development of the CZTS material (powder). An improved understanding of the material and the solar cell architecture will result in performance improvements. Those efficiency improvements on CZTS material and on the device junction can be quickly transferred to a roll-to-roll produced monograin module. In particular the main advantage of the monograin roll-to-roll production is that by "drop-in" of a new generation of crystal grain powder the efficiency modules produced a low costs can be improved and reduces the time to marked of future CZTS solar cells with improved material and junction quality.

ACKNOWLEDGEMENTS

This work was partly funded by the Austrian Climate and Energy Fund within the project print.PV in the frame of the e!MISSION.at program (project number: 845017). One of us (DM) gratefully acknowledges support by the European Regional Development Fund, Project TK141.

REFERENCES

[1] W. Wang et al., "Device Characteristics of CZTSSe Thin-Film Solar Cells with 12.6% Efficiency," Adv. Energy Mater., vol. 4, no. 7, p. n/a-n/a, May 2014.

[2] D. B. Mitzi, O. Gunawan, T. K. Todorov, K. Wang, and S. Guha, "The path towards a high-performance solution-processed kesterite solar cell," Sol. Energy Mater. Sol. Cells, vol. 95, no. 6, pp. 1421–1436, Jun. 2011.

[3] S. Siebentritt and S. Schorr, "Kesterites—a challenging material for solar cells," Prog. Photovolt. Res. Appl., vol. 20, no. 5, pp. 512–519, Aug. 2012.

[4] X. Liu et al., "The current status and future prospects of kesterite solar cells: a brief review," Prog. Photovolt. Res. Appl., vol. 24, no. 6, pp. 879–898, Jun. 2016.

[5] S. Bourdais et al., "Is the Cu/Zn Disorder the Main Culprit for the Voltage Deficit in Kesterite Solar Cells?," Adv. Energy Mater., vol. 6, no. 12, p. n/a-n/a, Jun. 2016.

[6] K. Timmo et al., "Influence of order-disorder in Cu2ZnSnS4 powders on the performance of monograin layer solar cells," Thin Solid Films, doi: 10.1016/j.tsf.2016.10.017, 2016

[7] M. Kauk-Kuusik, K. Timmo, M. Danilson, M. Altosaar, M. Grossberg, and K. Ernits, "p–n junction improvements of Cu2ZnSnS4/CdS monograin layer solar cells," Appl. Surf. Sci., vol. 357, Part A, pp. 795–798, Dec. 2015.

[8] D. Meissner: " Photovoltaics Based on Semiconductor Powders" in: A. Méndez-Vilas (Ed.): Materials and processes for energy: communicating current research and technological developments, Energy Book Series #1, Formatex Research Center, Badajoz, Spain, 2013 (ISBN: 978-84-939843-7-3), pp. 126 - 141

[9] Enn Mellikov, et al.: " Growth of CZTS-Based Monograins and Their Application to Membrane Solar Cells", chapter 13 in: Kentaro Ito (ed.): "Copper Zinc Tin Sulfide-Based Thin-Film Solar Cells", John Wiley & Sons Ltd, Chichester, West Sussex, UK, 2015, doi: 10.1002/9781118437865.ch13, pp. 289 - 309

[10] Y.-S. Park, H.-K. Kim, S.-W. Jeong, and W.-J. Cho, "Highly flexible indium zinc oxide electrode grown on PET substrate by cost efficient roll-to-roll sputtering process," Thin Solid Films, vol. 518, no. 11, pp. 3071–3074, Mar. 2010.

Understanding Instabilities and Degradation due to Moisture Ingress in Cu(In,Ga)Se₂ Solar Cells

Grace Rajan[1], Shankar Karki[1], Isaac Butt[1], Krishna Aryal[1], Tyler J. Grassman[2,3], Angus Rockett[4], Sylvain Marsillac[1]

[1] Virginia Institute of Photovoltaics, Old Dominion University, Norfolk, VA, USA
[2] Dept. of Materials Science & Engineering, The Ohio State University, Columbus, OH, 43210, USA.
[3] Dept. of Electrical & Computer Engineering, The Ohio State University, Columbus, OH, 43210, USA.
[4] Metallurgical and Material Engineering, Colorado School of Mines, Golden, CO, USA

Abstract — **In order to understand the role of moisture ingress in the degradation of Cu(In,Ga)Se₂ solar cells, different samples were soaked in water at different stages during the fabrication of the device. The samples along with reference devices were characterized by current density–voltage (J-V) and external quantum efficiency (QE) measurements to obtain an outline of the deteriorating effect. The device performance was observed to be the worst for the water soaked CIGS thin film. Raman and Secondary ion mass spectroscopy analysis were done on the same sample. More chemical analysis will be done in the near future.**

I. INTRODUCTION

Thin film solar cells based on Cu(In,Ga)Se₂ (CIGS) have advanced into a new realm of high efficiencies in the past few years. With major revisions in the deposition process, different research groups around the world have reached efficiencies over 22% [1]. As the technology matures to production on an industrial scale, the long-term stability and degradation process of these solar cells are of great interest to attain market success [2]. The susceptibility of the CIGS solar cells to moisture ingress and its effects on the chemical, structural, electronic properties and device efficiency is investigated and the preliminary results are presented in this study.

II. EXPERIMENTAL DETAILS

The devices were grown on molybdenum coated soda lime glass substrates. CIGS absorber layers were deposited by 3-stage process in a high vacuum co-evaporation chamber. After the CIGS process, the samples were dipped into a chemical bath to form a thin layer of CdS buffer layer. High resistive ZnO layers along with ZnO:Al layers were deposited by RF sputtering to obtain a transparent window layer. Finally, Ni/Al/Ni grids were evaporated by e-beam evaporation for the electrical contacts.

We generated four different types of samples by soaking them in water (WS) during different stages of device fabrication: after Mo (labeled A), after CIGS (labeled B), after CdS (labeled C) and after device completion (labeled D). The samples were soaked in water at 50°C for 24 hours in all cases. A reference device was prepared alongside the sample that underwent soaking in water (labeled with subscript "Ref" (e.g.: A_{Ref})).

The electrical characterization including current density-voltage measurement (J-V) and external quantum efficiency measurement (QE) were obtained for all the different WS devices and reference devices. Raman and Secondary ion mass spectroscopy analysis were performed on specific samples.

III. RESULTS

III.1. Electrical Characterizations

The open circuit voltage (Voc), short-circuit current density (Jsc), fill factor (FF) and the power conversion efficiency (η) of the various samples are shown in Figure 1. The diode analysis of the light I-V curves (using a single diode equation model) for the best samples in each category is shown in Figure 2.

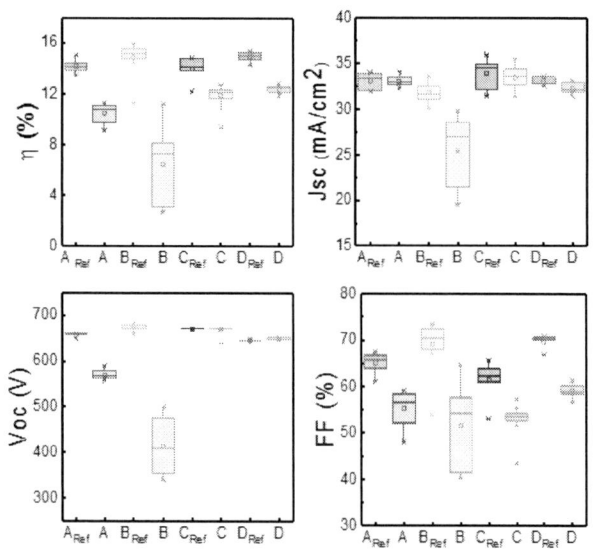

Fig 1. Box plots showing (a) Efficiency (b) short circuit current density, Jsc (c) fill factor, FF and (d) open circuit voltage, Voc of the devices soaked in water for 24 hours at 50°C at various stages compared with the reference devices.

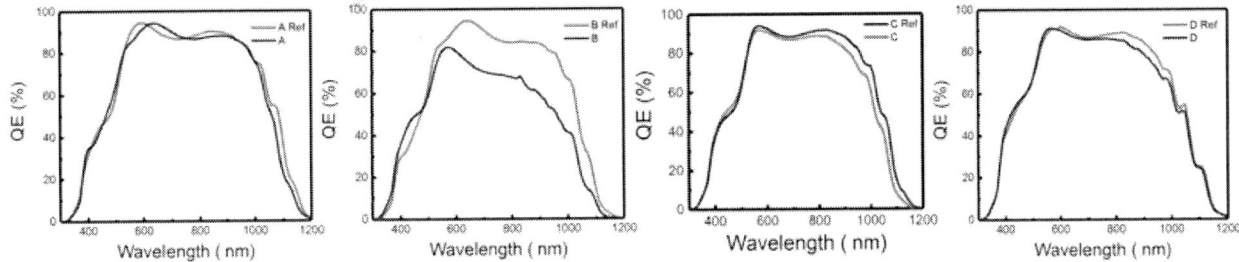

Fig. 3 QE plots for the best devices of various WS samples

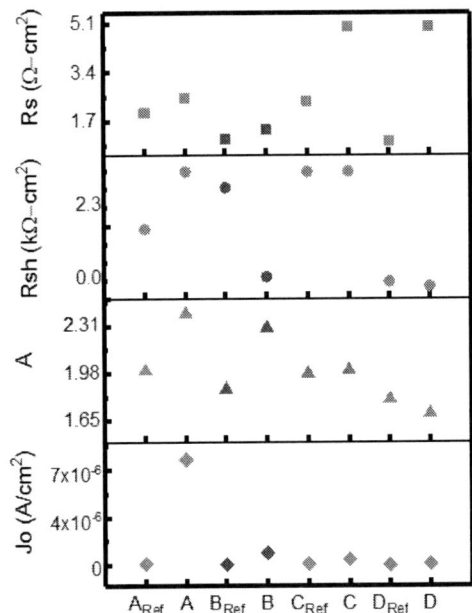

Fig. 2 Diode analysis (light I-V curves) of the best devices using single diode model

A. Samples of type A (Mo treated)

The reverse saturation current (J_0) in device A is one order of magnitude higher in the WS samples when compared to the reference device. There is also an increase in the diode quality factor (A), shunt resistance (R_{SH}) and series resistance (R_S) after WS treatment. The increase in R_S and A causes the deterioration in FF, while the change in J_0 leads to the loss of Voc. The change in Jsc is negligible, which was validated by the QE measurements (Figure 3).

B. Samples of type B (CIGS treated)

All the device parameters in this set of samples worsened after the WS treatment, most notably the reverse saturation current, which increases, and the shunt resistance which decreases from 3.5 to 0.7. However, no real shunt is observed in the devices, and it actually is a voltage dependent current collection

problem. The QE spectrum also shows that there is a loss of current after WS treatment, due to a lack of collection at all wavelengths.

C. Samples of type C (CdS treated)

Although there is not much noticeable change in the Voc, the FF of the device is significantly affected due to the increase in series resistance. There is almost a difference of 10% in the median value. There is also a slight loss in the QE spectra over the wavelength range of 500 nm to 1100 nm.

D. Samples of type D (Device treated)

The deterioration in Sample D is similar to Sample C. There is a clear indication of diminution of FF thus deteriorating the device efficiency. This is due mostly to the increase in series resistance of the sample after soaking in water. For sample D, there is little change in Voc and Jsc, and the values are much more consistent from one sample to the other.

III.2. Other Characterizations

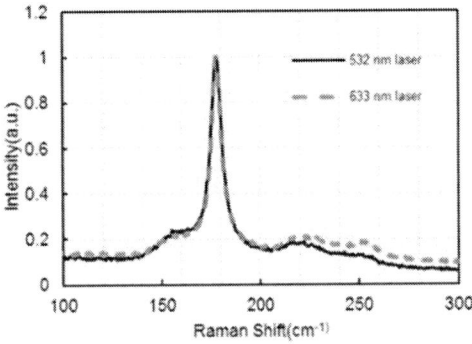

Fig. 4 Raman spectra of the water soaked CIGS (sample B).

At this stage, we focused our attention on the samples of type B (CIGS treated), as these are the ones showing the most drastic changes. Raman spectroscopy was performed on the water soaked CIGS (Figure 4). The Raman spectra were measured with the 532 nm and 633 nm excitation wavelength. The Raman spectrum exhibits a peak at 177 cm^{-1}, which is attributed to the A1 mode of the CIGS compound. The broad shoulder at around 150 cm^{-1} is generally attributed to the ordered vacancy compound like $Cu(In,Ga)_3Se_5$ or $Cu_2(In,Ga)_4Se_7$ [3]. The

978-1-5090-5606-4/17 $31.00 © 2017 IEEE   183

weaker peak at 260 cm^{-1} is assigned to the $Cu_{2-x}Se$ compounds [3]. The observed peaks in the Raman spectra were identified in earlier measurements [3, 4]. Thus, the effect of water soaking on CIGS probed by the Raman spectroscopy did not show any specific change. We also performed SIMS measurements on the surface of the CIGS water soaker sample. As one can see, there is a clear hydrogen peak at the surface that decreases quickly in the film. Another interesting feature seems to be the constant Ga concentration throughout the film, despite this being a 3-stage deposited film. We are in the process of repeating this analysis and will perform further analysis by XPS and TEM to assess the chemical bounding and structural effect of water soaking on each layer in the structure.

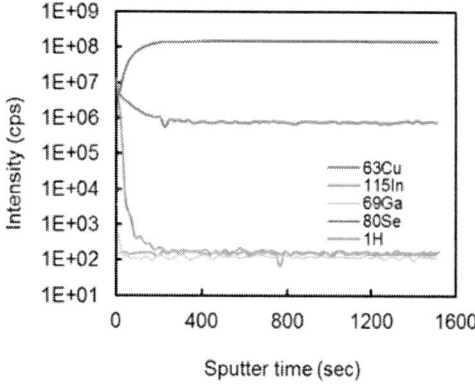

Fig. 5 SIMS profile of the water soaked CIGS (sample B).

IV. CONCLUSIONS

The effects of moisture ingress on each layer of the $Cu(In,Ga)Se_2$ solar cell was investigated and the preliminary results of the detrimental performance was presented in this paper. More samples will be generated to ensure that these changes observed are statistically significant. Once this is demonstrated, we will pursue more detailed analysis of the samples to see where and how the reaction occurs within the samples. The further development on this study is directed to examine deterioration in chemical, structural and electronic properties in detail.

ACKNOWLEDGEMENT

This research was supported by the Department of Energy Contract No. DE-EE0007141.

DISCLAIMER

This report was prepared as an account of work sponsored by an agency of the United States Government. Neither the United States Government nor any agency thereof, nor any of their employees, makes any warranty, express or implied, or assumes any legal liability or responsibility for the accuracy, completeness, or usefulness of any information, apparatus, product, or process disclosed, or represents that its use would not infringe privately owned rights. Reference herein to any specific commercial product, process, or service by trade name, trademark, manufacturer, or otherwise does not necessarily constitute or imply its endorsement, recommendation, or favoring by the United States Government or any agency thereof. The views and opinions of authors expressed herein do not necessarily state or reflect those of the United States Government or any agency thereof.

REFERENCES

[1] Jackson P, Wuerz R, Hariskos D, Lotter E, Witte W, Powalla M,. "Effects of heavy alkali elements in $Cu(In,Ga)Se_2$ solar cells with efficiencies up to 22.6%", *Phys. Status Solidi RRL.* 2016; 1-4.

[2] Pern F.J, Egaas B, To B, Jiang C.S, Li J.V, Glynn S, DeHart C,. "A study on the Humidy Susceptibility of thin Film CIGS Absorber", 34th IEEE PVSC, Pennsylvania, USA,2009.

[3] W. Witte, R. Kniese, and M. Powalla, "Raman investigations of Cu(In,Ga)Se2 thin films with various copper contents," *Thin Solid Films,* vol. 517, pp. 867-869, 11/28/ 2008.

[4] J. Han, L. Ouyang, D. Zhuang, C. Liao, J. Liu, M. Zhao, *et al.*, "Raman and XPS studies of CIGS/Mo interfaces under various annealing temperatures," *Materials Letters,* vol. 136, pp. 278-281, 12/1/ 2014.

This Page Intentionally Left Blank.

Control of MoSe$_2$ formation in hydrazine-free solution-processed CIS/CIGS thin film solar cells

Soňa Uličná[1*], Panagiota Arnou[1], Alexander Eeles[1], Mustafa Togay[1], Lewis D. Wright[1], Ali Abbas[1], Andrei V. Malkov[2], John M. Walls[1] and Jake W. Bowers[1]

[1]CREST, Wolfson School of Mechanical, Electrical and Manufacturing Engineering, [2]Department of Chemistry, Loughborough University, Loughborough, Leicestershire, LE11 3TU, UK

Abstract — This study investigated an approach to control the MoSe$_2$ layer formation at the Mo/CIGS interface of hydrazine-free solution-processed CIGS solar cells. The MoSe$_2$ layer thickness reduction was achieved by deposition of a MoN$_x$ back contact barrier layer, which effectively acts as a diffusion barrier against selenium (Se). The resulting Mo/MoN$_x$/Mo multilayer was applied in a CIGS device as the back contact. The electrical performance of this device was compared to our baseline approach with bare Mo as the back contact. The MoSe$_2$ layer formed after selenization was dramatically reduced when the barrier layer was present and the corresponding device exhibited a power conversion efficiency (PCE) of 8.2%. More importantly, the application of the barrier layer as an intermediate layer within the Mo back contact allows for longer, or even multiple selenization steps. A longer or a multiple selenization was shown to improve the absorber grain growth and consequently result in higher PCEs.

Index Terms — CIGS, diffusion barrier, MoN$_x$, MoSe$_2$, selenization, solution-process.

I. INTRODUCTION

Cu(In,Ga)Se$_2$ is one of the best performing thin-film photovoltaic technologies [1]. High efficiencies are however achieved using expensive and sophisticated vacuum-based equipment. To reduce the production costs, non-vacuum solution-based deposition approaches for the absorber layer are of increased popularity. These techniques promise many potential advantages. As well as the lower capital cost, solution approaches offer process simplicity, straightforward compositional control, large area uniformity and the possibility for flexible substrate application. So far, the best performing true solution-based CIGS solar cell with a PCE of 15.2% was developed using hydrazine as the solvent [2]. This fabrication method has overcome some of the limitations of non-vacuum techniques, such as phase impurity and incomplete grain growth. However the large scale implementation of this method is difficult due to hydrazine being an extremely hazardous solvent.

A hydrazine-free solvent combination consisting of 1,2-ethanedithiol/1,2-ethylenediamine (eth/en) in a 1:10 volumetric ratio was found to effectively dissolve metal chalcogenides [3]. In our previous work, this diamine/dithiol solvent mixture was used as a safer alternative to hydrazine to prepare CIGS precursor solutions by dissolving copper and indium sulfides, as well as elemental gallium and selenium. The solution was spray-coated in air onto molybdenum (Mo) coated substrates followed by post-deposition selenization. This method resulted in PCEs up to 8% for CIS and 9.8% for CIGS solar cells [4]. A similar molecular precursor route using amine-thiol mixture was developed to fabricate CIGSe from elemental Cu, In, Ga and Se with a reported PCE of 9.5% [5] and from a combination of metal salts and chalcogenides reporting PCE of 12.2% [6]. Although these methods are very promising for a scalable industrial application, there is still a large room for further improvement in terms of device performance.

Currently, one of the limiting factors of these devices is the excessive MoSe$_2$ formation during the high temperature selenization step. A thin MoSe$_2$ layer is beneficial as it forms an ohmic contact at the Mo/CIGS interface. However, excessive formation of MoSe$_2$ can have detrimental effects on the device performance, by decreasing the fill factor (FF) and causing adhesion problems [7]. Some of the factors that can affect the MoSe$_2$ formation are the sputtering conditions for Mo, residual stress in the film, selenization conditions or presence of sodium (Na) [8]. Selenium diffusion barriers have been previously reported to hinder the excessive transformation of Mo into MoSe$_2$. These include TiN, molybdenum oxide (MoO$_x$) and molybdenum nitride (MoN$_x$) [9]-[11].

The purpose of this work is to investigate the impact of the MoN$_x$ diffusion barrier and the selenization configuration on the MoSe$_2$ layer formation. Subsequent CIGS devices were fabricated using these substrates and were compared with a baseline sample without the barrier layer.

II. EXPERIMENTAL DETAILS

A. Molybdenum deposition

A MoN$_x$ thin film of ~30 nm thickness was deposited onto a 600 nm thick Mo coated soda-lime glass (SLG) substrate. The Mo layer had a bilayer structure, as this is optimized for high quality CIGS solar cells [12]. The MoN$_x$ barrier layer was deposited using DC magnetron sputtering at a base pressure lower than 3 x 10^{-6} Torr. A mixture of Ar/N$_2$ sputtering gases (10/5 sccm) was introduced into the sputtering chamber, resulting in a working pressure of 2.4 mTorr. The deposition was carried out using a power density of 4 W/cm². Finally a ~50 nm thick Mo layer was deposited on top of the MoN_x layer, using 2 sccm of Ar and a sputter power and pressure of 4 W/cm² and 1.2 mTorr respectively. The sheet resistance of the final $Mo/MoN_x/Mo$ multilayer remained unchanged compared to the Mo single layer due to the MoN_x film being relatively thin.

B. Deposition of CIGS absorber films and fabrication of solar cells

$Cu(In,Ga)(S,Se)_2$ films were prepared in two steps. First, the precursor solution was deposited onto the modified Mo coated substrates. Secondly, the as-deposited film was thermally annealed in Se atmosphere to recrystallize the absorber layer.

Metal chalcogenides (copper and indium sulfides, elemental gallium in presence of excess selenium) were dissolved in the eth/en solvent mixture as described in our previous work [4]. The $Cu_{0.9}In_{0.7}Ga_{0.3}Se_2$ precursor solution was diluted with ethyl acetate (2:1 v/v), filtered (0.45 μm PTFE) and subsequently sprayed in layers onto the Mo coated substrates placed on a preheated hot plate. Each sprayed layer was immediately dried to evaporate the excess solvent. The final film consisted of 6 sprayed layers in total to obtain a film thickness of 2-3 μm. The precursor film was then selenized in a tube furnace. The sample is placed together with Se pellets in a partially closed graphite box heated at 540°C for 50-90 minutes, including the ramping (~40°C/min), at a starting pressure of 450 Torr. The single selenized samples on bare Mo are denoted as 'S-50' and 'S-90', with 50 and 90 indicating the annealing time. The single selenized samples that contain MoN_x barrier layer are denoted as 'SB-50' and 'SB-70'. In a separate approach, a thinner absorber (3 sprayed layers) was selenized using the same conditions. The spraying and selenization procedure was then repeated in the same way. The final device consists of 6 sprayed layers and was selenized for 100 minutes in total. The double selenized sample is denoted as 'D-50-50'.

CIGS devices were completed by chemical bath deposition of CdS buffer layer and sputtering of intrinsic ZnO and Al doped ZnO (AZO). A top contact silver grid was evaporated. Mechanical scribing was performed to delimit cells of a total area of ~0.25 cm².

C. Characterization

Transmission Electron Microscopy (TEM) was carried out using FEI Tecnai F20 (S)TEM equipped with an Oxford Instruments X-Max 80 silicon drift detector (SDD) Energy Dispersive X-ray detector (EDX). The TEM samples were prepared by Focused Ion Beam (FIB) milling using a dual beam FEI Nova 600 Nanolab. The absorber microstructure was observed using a JEOL JSM-7800F Field Emission Scanning Electron Microscope (FE-SEM) equipped with EDX. X-ray diffraction (XRD) was performed using a Bruker D2 Phaser X-ray diffractometer equipped with a Lynxeye™

detector and Cu-Kα X-ray source. The current density/voltage (JV) characteristics of the individual cells were measured using AM1.5G simulated sunlight from a dual source solar simulator (Wacom, Japan) under 100mWcm⁻², using a calibrated Si reference cell. Prior to the JV measurements, the cell area was measured using a digital microscope. The external quantum efficiency (EQE) spectra were acquired with chopped light using a Bentham PVE300 system. The measurements were performed at 0 V bias with a spectral resolution of 5 nm. Temperature-dependent current density/voltage (JVT) measurements were performed using a Lakeshore 335 temperature controller by heating or cooling through a Janis CCS150 closed cycle helium cryostat. Capacitance-Voltage (CV) measurements were performed using a Keysight E4990A impedance analyzer and four-point probe at room temperature.

III. RESULTS

The biggest obstacles against achieving higher PCEs for the solution-processed CIGS solar cells are the non-optimized back contact and a poorly recrystallized absorber. It is expected that a reduction in the $MoSe_2$ layer thickness can lead to lower series resistance (R_S) and therefore increased FF.

The TEM cross-section of the $Mo/MoN_x/Mo$ multilayer after 50 minutes-long selenization is shown in Fig. 1. EDX elemental maps show that Se diffusion is effectively blocked from migrating towards the substrate and converting the entire Mo layer into $MoSe_2$. The top 50 nm thick Mo was all converted into a thick $MoSe_2$ with a thickness of >200 nm. This confirms the role of the MoN_x as an effective diffusion barrier against Se migration.

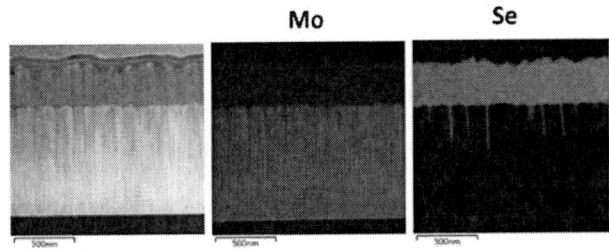

Fig. 1. TEM cross-section and EDX elemental maps of the $Mo/MoN_x/Mo$ multilayer after selenization.

A more aggressive selenization is often required in order to fully recrystallize the absorber. However, this may cause delamination due to formation of a thick $MoSe_2$ layer. Therefore, a compromise needed to be made in the choice of the selenization conditions, resulting in incompletely crystallized absorbers. A bilayer is typically formed after selenization of solution-processed CIGS, consisted of an uncrystallized part at the bottom and larger grains on the top. This is suspected to be a limiting factor towards achieving higher efficiencies [13]. Longer dwell times and higher selenium partial pressures during selenization are expected to improve the crystal quality of the absorber. However, the process window for the selenization step is limited by the excessively thick $MoSe_2$ layer that may be formed, causing delamination issues.

The $MoSe_2$ layer was effectively controlled by introducing the MoN_x barrier layer. The XRD patterns in Fig. 2 show distinct peaks corresponding to the chalcopyrite structure of CIGS (JCPDS 40-1488 of $CuIn_{0.5}Ga_{0.5}Se_2$).

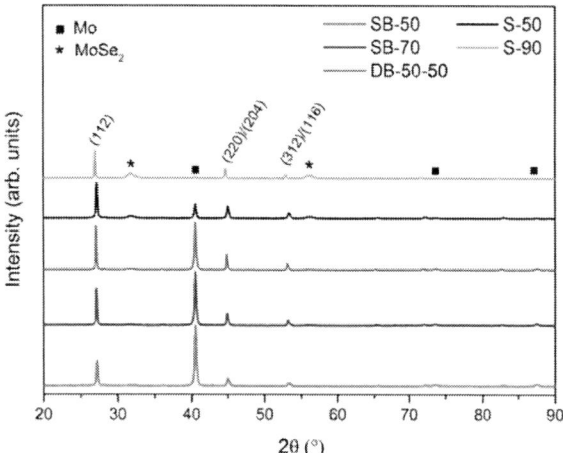

Fig. 2.   XRD patterns of the CIGS devices showing the increase of $Mo/MoSe_2$ ratio with introducing the MoN_x barrier layer.

The intensity of the $MoSe_2$ (100) peak at 2θ ~32° is substantially decreased when the barrier layer is present. This peak is more pronounced for the barrier-free samples. Table I summarizes the Mo (110)/$MoSe_2$ (100) peak intensity ratio extracted from the XRD data. The ratio is substantially higher when the barrier layer is present, 41.92 ('SB-50') compared to 4.35 for 'S-50'. However it decreases with longer selenization durations, from 41.92 to 24.91 for the 'D-50-50' sample. It was shown in Fig. 1 that the barrier layer allows locally some selenium to go through the barrier layer and form $MoSe_2$. It is likely that Se diffuses through the barrier to a bigger extend for the longer selenized samples.

TABLE I
FWHM AND INTENSITY RATIOS OF THE XRD PEAKS

| | FWHM (°) | | | Intensity |
	CIGS (112)	Mo (110)	$MoSe_2$ (100)	Mo (110) / $MoSe_2$ (100)
SB-50	0.254	0.317	NA	41.92
SB-70	0.193	0.316	NA	30.07
DB-50-50	0.161	0.317	NA	24.91
S-50	0.249	0.333	NA	4.35
S-90	0.161	NA	1.09	0.42

The extracted full width at half maximum (FWHM) of the dominant (112) CIGS peak gives an indication of the crystal growth during selenization. The crystal growth does not seem to be significantly affected by the presence of the barrier layer with a FWHM value of ~0.25° for both 'SB-50' and 'S-50'

samples. On the other hand the FWHM decreases to 0.193° with longer selenization times and even to 0.161° for the double selenized absorber.

The SEM images of the absorber surface and cross-sections displayed in Fig. 3 are in agreement with the XRD observations. First, the reduced $MoSe_2$ thickness can be clearly seen on the cross-section of the sample with the barrier layer. The sample can withstand even longer or multiple selenizations, without formation of cracks or delamination at the back contact. On the contrary, delamination is evident for the barrier-free sample selenized for longer times 'S-90'. Secondly, the bilayer structure of large/small crystals is seen in all single (S) selenized absorbers. The grain size in the top crystallized layer increases with longer selenization time ('SB-70'). Larger grains are present in the bulk rather than on the surface after the double selenization ('DB-50-50').

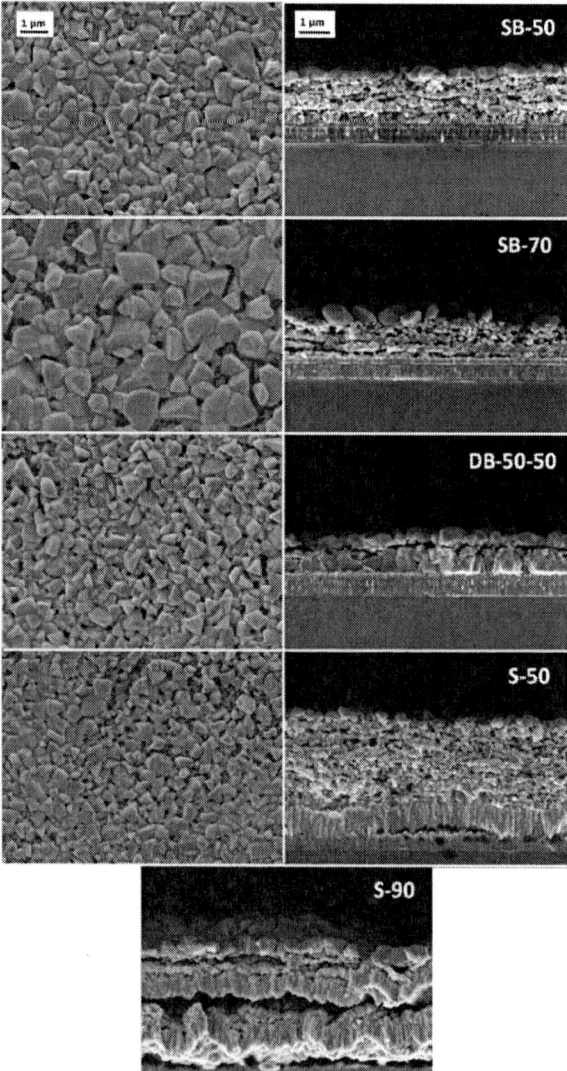

Fig. 3.   SEM surface and cross-section images of the samples.

The light and dark JV characteristics of the champion cell for each sample are displayed in Fig. 4. The key performance indicators corresponding to each of the JV curves are summarized in Table II. The best performing cell is the 'SB-70' with a PCE of 9.0 %. This confirms that the presence of the MoN_x barrier layer does not detriment the device properties. This device has the highest open circuit voltage (V_{OC}) (622 mV) and the highest short circuit current (J_{SC}) (24.3 mA/cm^2) among the compared samples. Surprisingly, the 'DB-50-50' device has a lower performance considering the improved crystal growth seen by SEM (Fig. 3). However, this device has the highest FF, exceeding 64 %. The increase in FF is most likely associated with the improved crystallization in the bulk of the absorber causing lower R_S losses. The low performance of the 'S-90' cell can be attributed mainly to the poor quality of the back contact. A 1.5% increase in the PCE was obtained for the 'SB-50' sample in comparison to the 'S-50', showing the beneficial effect of the barrier layer. The unintentional thickness variation between the two samples can also affect the device performance.

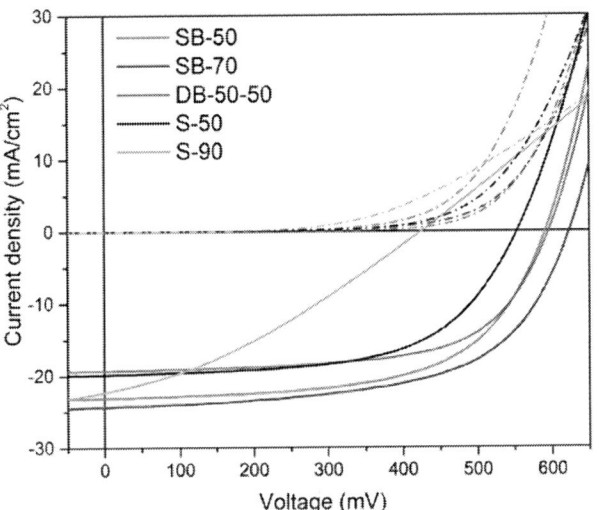

Fig. 4.   Light and dark JV curves of representative devices.

TABLE II
KEY PERFORMANCE INDICATORS OF EACH OF THE DISPLAYED
JV CURVES IN FIG. 4

	Efficiency (%)	FF (%)	V_{OC} (mV)	J_{SC} (mA/cm^2)
SB-50	8.2	60	590	23.0
SB-70	9.0	59	622	24.3
DB-50-50	7.3	64	595	19.2
S-50	6.5	59	553	19.8
S-90	3.1	32	425	22.2

To further investigate the effect of the barrier layer on the device properties, EQE, EDX, CV and JVT characterization were performed. The presence of the barrier layer can also affect the doping density of the device, by limiting Na

diffusion from the SLG substrate. There is also a possibility that the barrier can affect the absorber composition, as it may prevent Cu migration into the $MoSe_2$. The Cu out-diffusion into the back contact may be possible in barrier-free CIGS solar cells [14]. Fig. 5 shows the extracted doping profiles from the CV measurements at 300 K.

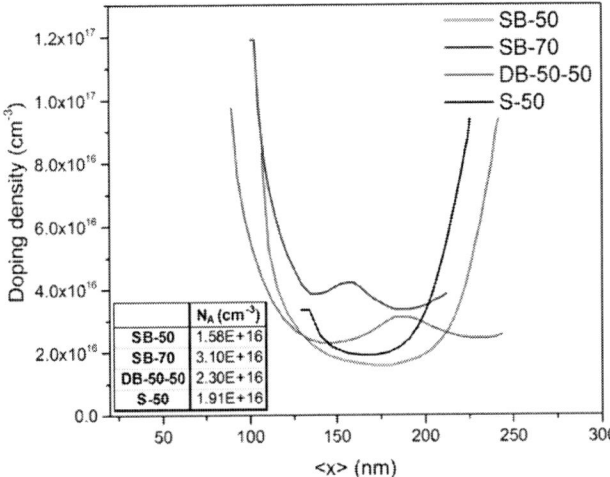

Fig. 5.   Room temperature doping profiles for the devices with and without the MoN_x barrier layer.

Both samples selenized for 50 min display U-shape profiles. The doping densities were estimated from the minima of these curves. The net doping density is the lowest for the 'SB-50' device. This could be caused by reduced sodium diffusion from SLG due to the presence of barrier, but these changes are too small to be conclusive on the Na-blocking effect of the barrier layer. Moreover, the double and longer selenized devices on the barrier layer show higher doping densities than the device on bare Mo. This indicates that a sufficient amount of Na could be diffused through discontinuities in the MoN_x, as seen in Fig. 1. Alternatively, higher Cu amounts could be present in the absorber, either introduced unintentionally or due to MoN_x barrier layer that would block the Cu migration into the $MoSe_2$ [14]. A further analysis is needed in order to quantify the Na and Cu contents in the samples.

The doping profiles of these two samples have an unusual shape presenting a local maximum and two minimums. It seems reasonable to assume that the unusual doping density profile is connected to the double layer structure of the devices. It could represent a genuine doping profile or an artifact caused roughness and incomplete coverage of the large crystal layer. From the SEM cross section (Fig. 3), the devices which display this double dip characteristic have a much larger top crystal region (~500nm) compared to the devices which do not (~200nm). We interpret this as showing the large crystal region is fully depleted in the devices with short selenizations, whereas for the long and double selenizations the depletion width crosses the interface between the large and small layers

978-1-5090-5606-4/17 $31.00 © 2017 IEEE   189

during the voltage sweep. This is not consistent with the measured profile depth <x> however this measure is strongly affected by deep defects and interface states, which could have artificially lowered its value [15].

The EDX data summarized in Table III show that the Ga/In+Ga (GGI) ratio agrees well with the targeted values. No significant Ga loss is observed during longer selenizations. The deviation of Cu and Se contents might be related to unintentional deposition variations rather than the effect of the barrier or the selenization duration.

TABLE III
ELEMENTAL COMPOSITION OF EACH FILM COMPARED TO THE TARGETED CIGS COMPOSITION

	Targeted	SB-50	SB-70	DB-50-50	S-50
GGI	0.3	0.30	0.30	0.29	0.30
CGI	0.9	0.87	0.84	0.84	0.90
Se/GI	2	1.87	2.18	1.90	1.88

The EQE spectra of the three single selenized devices are shown in Fig. 6. The devices on the barrier layers have a higher collection compared to the barrier-free counterpart. The 'SB-70' device has the best collection, just above 80% between 540 and 570 nm. This device also exhibited the highest J_{SC} value of 24.3 mA/cm^2. A gradual decay of the QE is observed in longer wavelengths for all the devices. This is likely attributed to recombination losses in the fine-grained part of absorber layer. The small decay below 530 nm is due to the absorption in the CdS layer. The inset of Fig. 6 shows the extracted band gaps (E_g) from the EQE curves. The band gap is slightly lower for the longer selenization, which is likely due to variation of S/Se, given that the GGI is constant. The sulphur content was difficult to be quantified with EDX due to a peak overlap with Mo.

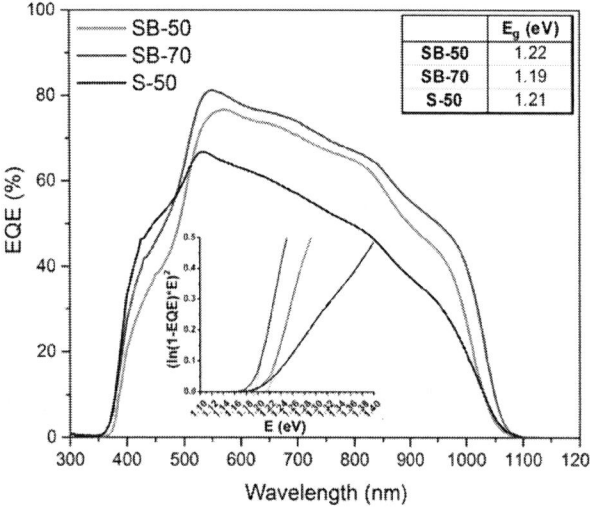

Fig. 6.   EQE spectra of the CIGS solar cells. The inset shows the absorber band gap, as extracted from the EQE.

A method typically used to determine the dominating recombination path is JVT. The activation energy for recombination (E_A) can be estimated from the JVT measurement. In the plot of the V_{OC} vs. Temperature (Fig. 7), the linear extrapolation to T = 0 K gives the E_A for each sample. Activation energy equal or close to the band gap indicates that the SRH recombination in the bulk is dominant. Values lower than the band gap indicate that the major recombination occurs at the heterojunction interface. For both devices (i.e. with and without MoN$_x$ barrier) the extracted E_A is smaller than the band gap, suggesting that the main recombination path is interface recombination. As seen from the SEM images, larger grains cover the porous fine-grained bottom layer. However the crystallized absorber layer does not fully cover the surface and so the porous absorber may come in contact with the CdS buffer layer. This could be responsible for the junction recombination losses.

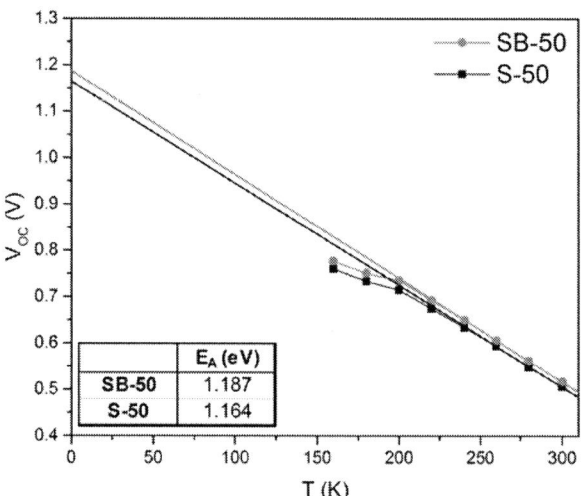

Fig. 7. V_{OC} vs. Temperature obtained from the JVT measurement.

IV. CONCLUSIONS

The MoN$_x$ barrier layer was effectively introduced at the back contact for hydrazine-free solution-processed CIGS solar cells. The barrier layer was shown to effectively block Se diffusion during selenization, hence controlling the MoSe$_2$ formation. Excessive MoSe$_2$ formation was shown to cause delamination problems. CIGS solar cells with the barrier layer reached comparable or even higher efficiencies to the baseline device on bare Mo. Moreover, the barrier layer allows for longer or multiple selenizations, resulting in better crystallization and consequently higher J_{SC} and V_{OC} values. FF was improved, especially for the double selenized device, where the fine-grain layer thickness was substantially reduced. The QE loss in the red portion of the spectrum for all the single selenized samples indicates that there is still room for improvement in terms of the absorber quality. Nonetheless, this work shows that the application of the barrier layer allows a broader process window for the selenization step. Further work is required to test whether the barrier hinders Na diffusion from the substrate or causes compositional variations.

ACKNOWLEDGEMENTS

The authors would like to thank Edgardo Saucedo and Sergio Giraldo, IREC, Spain for EQE measurements. The authors are also grateful for funding from the EPSRC (EP/N026438/1) to support this work.

REFERENCES

[1] P. Jackson, R. Wuerz, D. Hariskos, E. Lotter, W. Witte, and M. Powalla, "Effects of heavy alkali elements in Cu(In,Ga)Se 2 solar cells with efficiencies up to 22.6%," *Phys. status solidi - Rapid Res. Lett.*, vol. 4, 2016.

[2] T. K. Todorov, O. Gunawan, T. Gokmen, and D. B. Mitzi, "Solution-processed Cu(In,Ga)(S,Se)$_2$ absorber yielding a 15.2% efficient solar cell," *Prog. Photovoltaics Res. Appl.*, vol. 21, 2013.

[3] D. H. Webber and R. L. Brutchey, "Alkahest for V2VI3 chalcogenides: dissolution of nine bulk semiconductors in a diamine-dithiol solvent mixture.," *J. Am. Chem. Soc.*, vol. 135, 2013.

[4] P. Arnou, C. S. Cooper, S. Uličná, A. Abbas, A. Eeles, L. D. Wright, A. V. Malkov, J. M. Walls, and J. W. Bowers, "Solution processing of CuIn(S,Se)$_2$ and Cu(In,Ga)(S,Se)$_2$ thin film solar cells using metal chalcogenide precursors," *Thin Solid Films*, 2016.

[5] D. Zhao, Q. Tian, Z. Zhou, G. Wang, Y. Meng, D. Kou, W. Zhou, D. Pan and S. Wu, "Solution-deposited pure selenide CIGSe solar cells from elemental Cu, In, Ga, and Se," *J. Mater. Chem. A*, 2015.

[6] X. Zhao, M. Lu, M. Koeper, and R. Agrawal, "Solution-Processed Sulfur Depleted Cu(In,Ga)Se$_2$ Solar Cells

Synthesized from a Monoamine-Dithiol Solvent Mixture," *J. Mater. Chem. A*, 2016.

[7] D. Abou-Ras, G. Kostorz, D. Bremaud, M. Kalin, F. V Kurdesau, A. N. Tiwari, and M. Dobeli, "Formation and characterisation of MoSe$_2$ for Cu(In,Ga)Se$_2$ based solar cells," *Thin Solid Films*, vol. 480, 2005.

[8] X. L. Zhu, Z. Zhou, Y. M. Wang, L. Zhang, a M. Li, and F. Q. Huang, "Determining factor of MoSe$_2$ formation in Cu(In,Ga)Se$_2$ solar Cells," *Sol. Energy Mater. Sol. Cells*, vol. 101, 2012.

[9] T. Schnabel and E. Ahlswede, "On the interface between kesterite absorber and Mo back contact and its impact on solution-processed thin-film solar cells," *Sol. Energy Mater. Sol. Cells*, vol. 159, 2017.

[10] A. Duchatelet, G. Savidand, R. N. Vannier, and D. Lincot, "Optimization of MoSe$_2$ formation for Cu(In,Ga)Se$_2$-based solar cells by using thin superficial molybdenum oxide barrier layers," *Thin Solid Films*, vol. 545, 2013.

[11] C. W. Jeon, T. Cheon, H. Kim, M. S. Kwon, and S. H. Kim, "Controlled formation of MoSe$_2$ by MoN$_x$ thin film as a diffusion barrier against Se during selenization annealing for CIGS solar cell," *J. Alloys Compd.*, vol. 644, 2015.

[12] J. H. Scofield, a. Duda, D. Albin, B. L. Ballard, and P. K. Predecki, "Sputtered molybdenum bilayer back contact for copper indium diselenide-based polycrystalline thin-film solar cells," *Thin Solid Films*, vol. 26, 1995.

[13] D. Zhao, Q. Fan, Q. Tian, Z. Zhou, Y. Meng, D. Kou, W. Zhou, S. Wu, J. Mater, D. Zhao, A. Qingmiao Fan, A. Qingwen Tian, A. Zhengji Zhou, A. Yuena Meng, A. Dongxing Kou, A. Wenhui Zhou, and S. Wu, "Eliminating Fine-Grained Layer in Cu(In,Ga)(S,Se)$_2$ Thin Films for Solution-Processed High Efficient Solar Cells," *J. Mater. Chem. A*, 2013.

[14] J. F. Guillemoles, L. Kronik, D. Cahen, U. Rau, A. Jasenek, and H.-W. Schock, "Stability Issues of Cu(In,Ga)Se$_2$-Based Solar Cells," *J. Phys. Chem. B*, vol. 104, 2000.

[15] J. T. Heath, J. D. Cohen, and W. N. Shafarman, "Bulk and metastable defects in CuIn$_{1-x}$Ga$_x$Se$_2$ thin films using drive-level capacitance profiling," *J. Appl. Phys.*, vol. 95, 2004.

Growth and properties of epitaxial Cu(In, Ga)Se₂ thin films deposited by the three-stage process for solar cells

Takeru Yamagami[1], Yuta Ando[1], Ishwor Khatri[2], Mutsumi Sugiyama[1,2], and Tokio Nakada[2]

[1]Faculty of Science and Technology, [2]Research Institute for Science and Technology,
Tokyo University of Science, 2641 Yamazaki Noda, Chiba, 278-8510, Japan

Abstract — **Epitaxial CIGS thin films were deposited by the modified three-stage process on epitaxially-grown Mo {110}/sapphire {0001} substrates. The crystallographic properties were investigated by means of scanning electron microscopy (SEM), transmission electron microscopy (TEM), and X-ray diffraction (XRD). The 2 µm-thick epitaxial CIGS thin films showed high flatness and a few grain boundaries with voids at the CIGS/MoSe₂ interface. The CIGS {112} plane grew parallel to hexagonal MoSe₂ {0001}/Mo {110}/sapphire {0001}. The misfit dislocations were not found in the CIGS or MoSe₂ films. Time resolved photo-luminescence (TR-PL) measurements revealed that the PL lifetime of the epitaxial CIGS thin films did not show a significant difference before or after the CBD-CdS process. This result contrasts clearly to polycrystalline CIGS thin films grown on soda-lime glass substrates.**

Index Terms — **Cu(In, Ga)Se₂, Epitaxial growth, Three-stage process, TEM analysis, PL.**

INTRODUCTION

Epitaxial growth of CIGS-based chalcopyrite thin films has been reported on single-crystal semiconductors such as Si [1], GaP [2], and GaAs [3] by various deposition techniques including MBE and MOCVD. In these works, the crystallographic and optical properties as well as the growth kinetics of epitaxially-grown CIGS films have been investigated. Recently, epitaxial CIGS films were grown on n-type (100)-Ge substrates using a pulsed-electron deposition (PED) technique and the photovoltaic properties of CIGS/Ge hetero-junction diodes has been reported [4].

In our previous reports, we have tried to grow the epitaxial CIGS thin films on Mo/sapphire substrates for solar cells by the three-stage co-evaporation process. The epitaxial CIGS thin films have been grown on Mo {110}/sapphire {0001} at relatively high substrate temperatures above 600°C [5, 6].

Usually, high-temperature growth is required to achieve high-quality thin films. However, serious problems occur due to re-evaporation of high vapor-pressure elements such as In and Se. We thus tried to modify the three-stage process, where Cu and Se are deposited before the subsequent three-stage process to prevent the re-evaporation of the (In, Ga)₂Se₃ layer during the first stage.

In the present contribution, we report the crystallographic and optical properties of epitaxial CIGS thin films grown by the modified three-stage process.

EXPERIMENTAL

Approximately 2 µm-thick epitaxial CIGS thin film was deposited by the modified three-stage co-evaporation process using a water-cooling molecular beam epitaxy system (EW-500, EIKO Engineering Co., Ltd.) at a maximum substrate temperature of 625°C. For this purpose, we first realized a Mo epitaxial layer on single crystal sapphire (0001) substrates with a thickness of 400 nm by DC-magnetron sputtering. Before the subsequent three-stage co-evaporation process, Se was first deposited to enhance the adhesion between the CIGS and Mo layer. In the second step, Cu and Se were deposited onto a thin MoSe₂ layer at the CIGS/Mo interface. Following that, the three-stage co-evaporation process was performed. As-deposited epitaxial CIGS thin films were rinsed in ammonia (1.0 M) for 10 min before the CBD-CdS process. The CBD-CdS process was performed on the epitaxial CIGS thin films using $CdSO_4$ (1.4×10^{-3} M)–ammonia (1.0 M)–Thiourea (0.14 M) aqueous solutions at bath temperatures up to 80°C.

Inductively coupled plasma spectroscopy (SPS 7700, Seiko, Japan) was used to measure the average film composition. Epitaxial relations were investigated by selected area electron diffraction (SAED) and X-ray diffraction (XRD) pole figure (SmartLab, Rigaku) for CdS {220}, CIGS {220}, MoSe₂ {11$\bar{2}$4}, Mo {110}, and sapphire {11$\bar{2}$3}. The microstructure of the film was investigated by scanning electron microscopy (S-4800 FE-SEM, Hitachi) and high resolution transmission electron microscopy (JEM-2100F, JEOL). A HR-TEM sample was prepared by a focused ion beam (FIB) with a thickness of approximately 100 nm. The photoluminescence (PL) of the CIGS thin film and CdS/CIGS structure was measured for both the epitaxial and polycrystalline CIGS thin films at room temperature using a time resolved photo-luminescence (TR-PL) system (C12132, Hamamatsu Photonics K. K., Japan) with a YAG laser (λ = 532 nm, 1.1 mW for excitation) for comparative studies.

RESULTS AND DISCUSSION

Fig. 1 shows (a) surface and (b) cross-sectional (tilted 13°) SEM photographs of the epitaxial CIGS thin film. The CIGS thin film was flat and shiny. The cross-sectional photograph shows the presence of a few grain boundaries. The deposition condition has been described in the experimental section. Pre-deposition of Cu and Se was performed before the subsequent three-stage co-evaporation process. Absence of pre-deposition conditions reduces the thickness of the epitaxial CIGS thin films. Therefore, it is assumed that the pre-deposition of Cu and Se before the subsequent three-stage co-evaporation process suppresses the re-evaporation of $(In, Ga)_2Se_3$ layer at the first stage of the co-evaporation process.

Voids were observed at the CIGS/Mo interface (Fig. 1 (b)), which might have been formed because of the re-evaporation of pre-deposited Cu-Se binary compounds. The adhesion was sufficient not to be peeled off the CIGS thin films after the CBD-CdS process. Additional experiments are needed to conclude the influence of voids on epitaxial growth of the CIGS thin films.

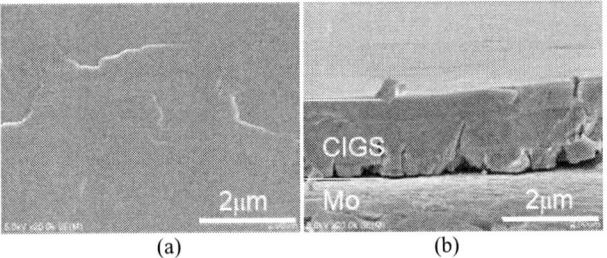

(a)   (b)

Fig. 1. SEM photographs of epitaxial CIGS thin film (a) surface and (b) cross-section (tilted 13°)

As can be seen in Fig. 2 (a) and (b), dislocation was not observed in either the epitaxial CIGS thin film or epitaxial $MoSe_2$ layer. The spacing of d_{CIGS} was 0.32 nm, which is close to the equivalent {112} plane of $Cu(In_{0.7}, Ga_{0.3})Se_2$ that is 0.331 nm [7]. On the other hand, the spacing of $d_{MoSe_2} = 0.64$ nm is approximately half of the {0001} plane of $MoSe_2$ [8]. Therefore, it was found that the c-axis of $MoSe_2$ was perpendicular to the epitaxial Mo {110} plane. In contrast, for the polycrystalline CIGS solar cell, the c-axis of $MoSe_2$ is parallel to the Mo {110} plane [9].

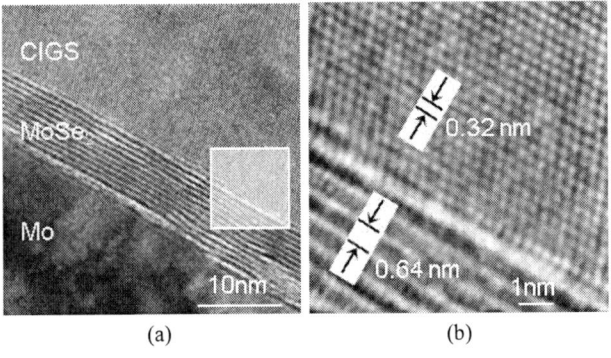

(a)   (b)

Fig. 2. Lattice image of epitaxial CIGS/epitaxial $MoSe_2$/epitaxial Mo interfaces (a) low magnification and (b) high magnification part indicated by white square in Fig. 2 (a)

Fig. 3 shows a cross sectional high resolution TEM image of an epitaxial CdS/epitaxial CIGS interface and its corresponding

diffraction pattern along $[01\bar{1}]$ for CdS and $[02\bar{1}]$ for CIGS. A twin was observed in the CdS thin film.

In our previous work, we grew CdS epitaxially without the twins in a polycrystalline CIGS solar cell [10]. This apparent inconsistency is due to the difference in the CBD-CdS recipe. Thus, it is possible to suppress twins in the CdS buffer layer by controlling the CBD-CdS condition.

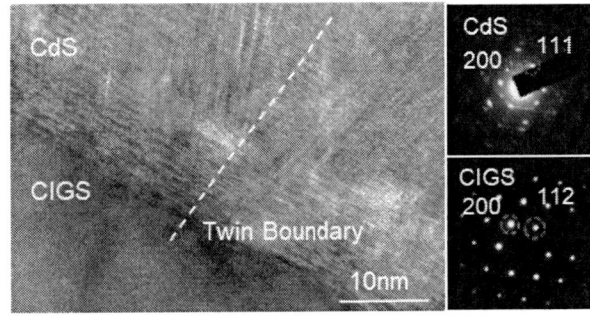

Fig. 3. Lattice image of epitaxial CdS/epitaxial CIGS interface and its corresponding diffraction patterns along the $[01\bar{1}]$ and $[02\bar{1}]$ zone axes of CdS and CIGS, respectively.

TABLE 1
The summary of epitaxial relations of each layer

	Epitaxial relation	Misfit (%)
CdS/CIGS	CdS{111}<110>//CIGS{112}<110>	1.5
CIGS/MoSe$_2$	CIGS{112}<110>//MoSe$_2${0001}<11$\bar{2}$0>	-17.8
MoSe$_2$/Mo	MoSe$_2${0001}<11$\bar{2}$0>//Mo{110}<110>	4.0
Mo/Sapphire	Mo{110}<110>//Sapphire{0001}<11$\bar{2}$0>	-6.4

The epitaxial relations and the misfit at 300 K of the CdS/CIGS/MoSe$_2$/Mo/sapphire stacked layers observed in this experiment are summarized in Table 1.

In-plane epitaxial relations and domains were confirmed by an XRD pole figure. CdS had two domains, which had 60° rotations about {111} poles. CIGS also had two domains, which had 60° rotations about {112} poles. The $MoSe_2$ layer had a single domain. Mo had three domains, which had 60° rotations about {110} poles on the sapphire {0001} substrate. Despite a large misfit between each layer, we have obtained epitaxially-grown CdS/CIGS/MoSe$_2$/Mo/sapphire stacked layers. Additional experiments are required in order to suppress the multi domain.

Fig. 4 shows the PL decay curves of the (a) epitaxial and (b) polycrystalline CIGS thin films [11] before (black lines) and after (red lines) the CBD-CdS process. No significant difference in the PL decay curve was observed in the epitaxial CIGS thin film before or after the CBD-CdS process (Fig. 5 (a)). In contrast, the PL lifetime of the polycrystalline CIGS thin film significantly increased after the CBD-CdS process. This result is consistent with those obtained in the previous work [12].

The difference may be due to the surface degradation owing to alkali-metals. Polycrystalline CIGS thin film contains alkali-metals such as Na or K at the surface, which diffuse from SLG during CIGS deposition. Alkali-metals absorb or trap O and OH at the CIGS surface. Therefore, the CIGS surface is degraded after air-exposure.

It has been reported that during the CBD-CdS process, alkali-metals as well as O and OH are removed from the surface region due to the ammonia etching [13], thereby enhancing the PL intensity and lifetime. In contrast, the PL decay curve was not significantly changed for epitaxial CIGS thin film before or after the CBD-CdS process, which is expected due to the lack of alkali-metals in the epitaxial CIGS thin film.

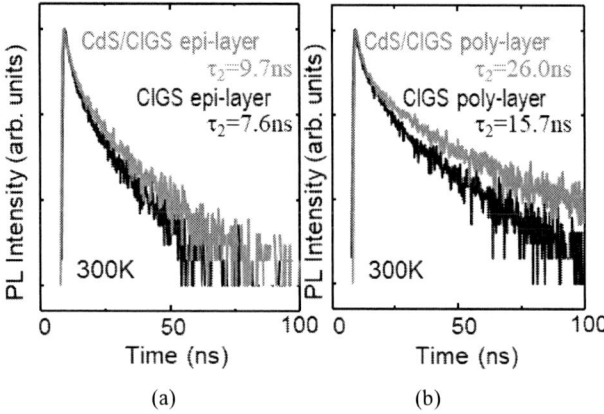

Fig. 4. PL decay curves of (a) epitaxial and (b) polycrystalline CIGS thin films before (black lines) and after (red lines) the CBD-CdS process.

CONCLUSION

Crystallographic and optical properties of the epitaxial CIGS thin films grown by the modified three-stage co-evaporation process were investigated using SEM, TEM, and TR-PL. We found that the Cu and Se deposition prior to the three-stage process could suppress the re-evaporation of the $(In, Ga)_2Se_3$ layer during the first stage. The epitaxial CIGS thin films showed a high flat surface and a few grain boundaries with relatively large voids at the $CIGS/MoSe_2$ interface.

TEM analysis revealed that the misfit dislocations were not observed in either the epitaxial CIGS thin film or epitaxial $MoSe_2$ layer grown on the Mo/sapphire substrates.

The epitaxial relations were confirmed by the lattice images and the selective area diffraction (SAED) patterns as CdS $\{111\}<110>//CIGS$ $\{112\}<110>//MoSe_2$ $\{0001\}<11\bar{2}0>//Mo$ $\{110\}<110>//sapphire$ $\{0001\}<11\bar{2}0>$.

TR-PL measurements confirmed that the carrier lifetime of the epitaxial CIGS thin films did not show any significant

difference before or after the CBD-CdS process, which is expected owing to the absence of alkali-metals in the epitaxial CIGS thin film.

ACKNOWLEDGEMENTS

This work was supported by New Energy and Industrial Technology Development Organization (NEDO). The authors would like to express sincere appreciation to Dr. T. Ichihashi and Prof. Y. Idemoto for their help with TEM analysis.

REFERENCES

[1] A. N. Tiwari, S. Blunier, K. Kessler, V. Zelezny, and H. Zogg, "Direct growth of heteroepitaxial $CuInSe_2$ layers on Si substrates," *Applied physics Letters*, vol. 65, pp. 2299, 1994.

[2] O. Igarashi, "Epitaxial growth of $CuGaSe_2$ and $CuInSe_2$ single crystals by halogen transport method using $Se(CH_3)_2$," *Journal of Crystal Growth*, vol. 143, pp. 213-220, 1994.

[3] N. Rega, S. Siebentritt, I. Beckers, J. Beckmann, J. Albert, and M. Lux-Steiner, "MOVPE of epitaxial $CuInSe_2$ on GaAs," *Journal of Crystal Growth*, vol. 248, pp. 169-174, 2003.

[4] S. Rampino, M. Bronzoni, L. Colace, P. Frigeri, E. Gombia, C. Maragliano, F. Mezzadri, L. Nasi, L. Seravalli, F. Pattini, G. Trevisi, M. Motapothula, T. Venkatesan, and E. Gilioli, "Low-temperature growth of single-crystal $Cu(In, Ga)Se_2$ films by pulsed electron deposition technique," *Solar Energy Materials & Solar Cells*, vol. 133, pp. 82-86, 2015.

[5] H. Masuko and T. Nakada, "Epitaxial growth of $Cu(In, Ga)Se_2$ thin films for solar cells on Mo thin layers epitaxially-grown on sapphire substrates," in *12th International Photovoltaic Science and Engineering Conference Technical Digest*, 2001, pp-126.

[6] H. Matsumori and T. Nakada, "Epitaxial growth of CIGS thin films on Mo-coated sapphire substrates," in *19th International Conference on Ternary and Multinary Compounds*, 2014, P4-131.

[7] JCPDS Powder Diffraction File Card No. 35-1102.

[8] JCPDS Powder Diffraction File Card No. 29-0914.

[9] D. Abou-Ras, G. Kostorz, D. Bremaud, M. Kälin, F. V. Kurdesau, A. N. Tiwari, M. Döbeli, "Formation and characterisation of $MoSe_2$ for $Cu(In, Ga)Se_2$ based solar cells," *Thin Solid Films*, vol. 480–481, pp. 433-438, 2005.

[10] T. Nakada, "Nano-structural investigations on Cd-doping into $Cu(In, Ga)Se_2$ thin films by chemical bath deposition process," *Thin Solid films*, vol. 361-362, pp. 346-352, 2000.

[11] I. Khatri, H. Fukai, H. Yamaguchi, M. Sugiyama, and T. Nakada, "Effect of potassium fluoride post-deposition treatment on $Cu(In, Ga)Se_2$ thin films and solar cells fabricated onto sodalime glass substrates," *Solar Energy Materials & Solar Cells*, vol. 155, pp. 280-287, 2016.

[12] S. Shirakata, H. Ohta, K. Ishihara, T. Takagi, A. Atarashi, and S. Yudate, "Photoluminescence characterization of surface degradation mechanism in $Cu(In, Ga)Se_2$ thin films grown on Mo/soda lime glass substrate," *Japanese Journal of Applied Physics*, vol. 53, pp. 05FW11, 2014.

[13] R. Hunger, T. Schulmeyer, A. Lcbedev, A. Klein, W. Jaegermann, R. Kniese, M. Powalla, K. Sakurai, and S. Niki, "Removal of the surface inversion of $CuInSe_2$ absorbers by $NH_{3, aq.}$ etching," in *3rd World Conference on Photovoltaic Energy Conversion*, 2003, 2LN-C-04.

Improvement of CIS Solar Cells with KF Postdeposition Following a Simple Two-Step Selenization Process

Yang Zhang, Robert E. Bartolo, Sang Jik Kwon, and Mario Dagenais

Department of Electrical Engineering, University of Maryland, College Park, MD 20742 USA

Abstract — A CuInSe2 (CIS) thin film solar cell (without Ga) with 14.7% efficiency is demonstrated based on the use of a potassium fluoride (KF) post-deposition treatment (PDT). By comparison, a CIS solar cell with 10.3% efficiency is obtained without KF PDT. No anti-reflection (AR) coating was used. KF PDT helps to form a hole-blocking layer between CdS/CIS interface and reduce the symmetry of the non-ohmic shunt leakage at low forward bias and reverse bias. The hole-blocking layer reduces the tunneling recombination and significantly increases the open circuit voltage. A space charge limited (SCL) current model is used to explain the symmetry feature of the current vs voltage, which is indicative of metal-semiconductor-metal (MSM) shunts. To understand possible defects and shunt path, some discussion is presented.

I. INTRODUCTION

Several research laboratories in the world have reported CuInGaSe$_2$-based (CIGS) solar cells with efficiencies above 22% in recent years. This includes the Centre for Solar Energy and Hydrogen Research Baden-Württemberg (ZSW) [1], and Solar Frontier [2]. Literature shows that many aspects of CIGS are still under intense investigation, like the grain boundaries, defects, the phase diagram, the role of sodium and phosphorus content, all impacting the efficiencies of CIGS solar cells. This indicates that there is still room for further progress in improving the efficiency of CIGS solar cells.

To pursue a low-cost and large-area approach with high yield and reproducibility for depositing a CuInSe2 (CIS) absorber layer, the two-step selenization process is preferred as compared to the three-stage co-evaporation process. We present a simple two-step process by selenization of stacked elemental layers (SEL) under selenium vapor in atmospheric ambient within a simple graphite box. The highest efficiency CIS solar cells have exhibited an efficiency of about 15.4% at a short circuit current density of about 41.2 mA/cm^2 [3]. CIS has a very high optical absorption coefficient, which makes it able to absorb more than 90% of the incident photons with energies higher than 1.0 eV within 1-2 μm thickness. Because of the high absorption coefficient and low bandgap, high quality CIS solar cells can have a very high short circuit current compared with other thin film materials or other type of solar cells. With our simple fabrication approach, our conversion efficiency is near record, at a value of 14.7%, and was obtained without anti-reflection (AR) coating. An AR coating would typically raise the solar cell efficiency by about 1.2%. We are able to steadily achieve above 12% conversion efficiency in CIS cells. Great control of the Cu/In ratio in the

precursor and Se amount provided during selenization of CIS thin film ensures that single phase chalcopyrite CIS films with grain sizes larger than 2 μm, which is the key for us to make high efficiency solar cells [4] [5]. In this paper, a variety of characterization techniques are used to discuss the properties of high-efficiency CIS solar cells both with and without the potassium fluoride (KF) post-deposition treatment (PDT), with an emphasis on the current-voltage (IV) characteristics. A hole-blocking layer between CdS/CIS interface is formed with KF PDT in a Se vapor environment, which reduces the tunneling recombination and significantly increases the open circuit voltage. In measurements of dark I-V characteristics, non-ohmic shunt leakage at low forward bias and reverse bias is observed, which can be explained by a space charge limited (SCL) current model. KF PDT clearly reduces the symmetry of shunt current around zero bias, which can be explained by the improvement of surface morphology. Discussion is presented to understand possible defects introduced during our processing steps and the physical origin of the shunt paths.

II. EXPERIMENTAL DETAILS AND RESULTS

Our CIS thin film solar cells were fabricated by using a two-step sequential process. Cu and In were sequentially deposited by sputtering to form the SEL, and then the SEL was selenized in a partially closed graphite box under Se vapor and converted to a CIS film in an optimized thermal process. Heterojunction devices using the structure ZnO/CdS/CuInSe$_2$ were fabricated. Details of the process are presented in this paper [6]. Something worth to mention is to indicate that the KF PDT process was done before the deposition of CdS layer, so it is not possible to study pre- and post-deposition of CIS devices using the same CIS device.

Fig. 1. shows the I-V characteristics of the best cells with and without KF PDT. The cell without KF PDT has V_{oc} = 0.423 V, J_{sc} = 40.5 mA/cm^2, a fill factor of 60.0%, and a power conversion of 10.3%. The cell with KF PDT has V_{oc} = 0.516 V, J_{sc} = 40.2 mA/cm^2, a fill factor of 70.7%, and a power conversion of 14.7%. Both cells have quite similar short-circuit current density. The increase in open-circuit voltage and Fill Factor is the main reason that leads to the 4.4% absolute increase in efficiency. To determine whether the bandgap of CIS film is increased after applying

978-1-5090-5606-4/17 $31.00 © 2017 IEEE

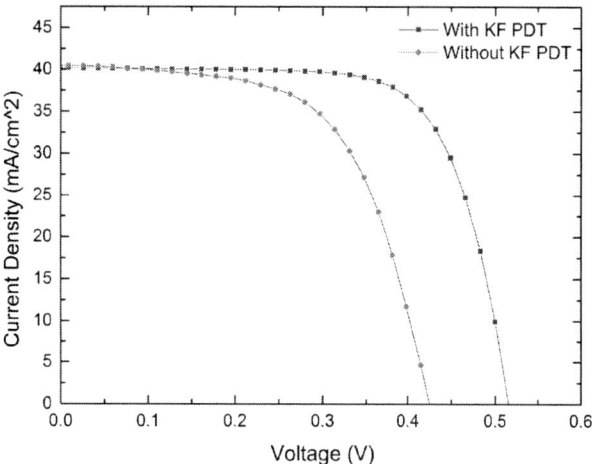

Fig. 1. Current density versus voltage for best CIS solar cells with and without KF PDT.

KF PDT, an external quantum efficiency (EQE) measurement is done on both cells, see Fig. 2. According to the measurement, both film have a similar bandgap of about 1 eV. This means that other factors result in the increase of V_{oc} and FF. In fact, the CdS/CIGS interface was improved during KF PDT, and reduced recombination at the interface was found using temperature-dependent J-V measurements [7]. Cu can diffuse into bulk during PDT in a Se environment since the Cu-Se bond is weaker than the In-Se bond [8]. So a hole-blocking layer at the interface is formed and the chance of tunneling of holes across n-ZnO into p-CIS becomes smaller, which reduces the surface tunneling recombination and results in the increase of V_{oc}.

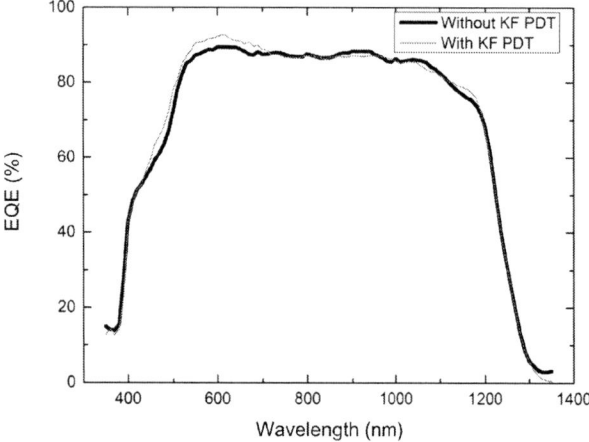

Fig. 2. ECE results of the best CIS solar cells without applied bias.

III. DARK IV CHARACTERIZATIONS

From the dark IV curves of CIS thin film solar showed in Fig. 3., two distinct regions can by identified. One is the high forward bias region ($V >\sim 0.05V$), where both cells show an exponential forward current behavior

$$I \propto e^{qV/nk_BT}.$$

A good fit for both cells with a power exponent n = 1.5 - 2.5 is obtained within the voltage range of -6 V to 0 V by fitting the reverse current with a power law. The traditional equivalent circuit model cannot explain the non-linear dependence of their reverse saturation current on the applied voltage. Another region is around V = 0 in the low bias region ($|V| <\sim 0.05V$), the symmetry of current in a I vs |V| plot is clearly observed in Fig. 3(b), while the curve with KF PDT (Fig. 3(c)) shows almost no symmetry. A space charge limited (SCL) current model was assumed [9] [10] to explain both the power law voltage dependence and the symmetry of the current around V = 0.

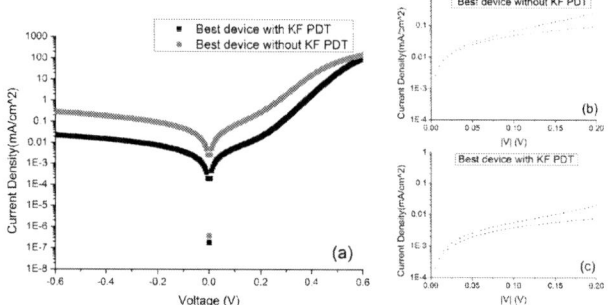

Fig. 3. (a) Dark IV characteristics of CIS solar cells. (b), (c) dark current plotted in absolute value terms, i.e. |I| versus |V|.

In our CIS solar cells, metal-semiconductor-metal (MSM) structure may exist in certain localized regions, to explain the symmetric and power law voltage dependent SCL current. Any pinholes introduced during the deposition of the thin window and buffer layers can cause Ni/Al layer contact directly with the CIS thin film layer. Unlike co-evaporation method, two-step selenization could cause the nonuniform distribution of Cu, In and Se in the CIS film. Thus, the electronic properties vary through the large area, which means certain regions may behave as intrinsic material. The reduction of the symmetry nature of current with KF PDT indicates that KF treatment helps reduce the shunt current in the SCL model. To prove this, in Fig. 4., we compare the top view SEM images of CIS films before and after KF PDT. Fewer pinholes were observed on surface after KF PDT and the surface smoothness was also improved. This enhances the following deposition quality of thin window and buffer layer, and then reduce the possibility of forming local MSM structure.

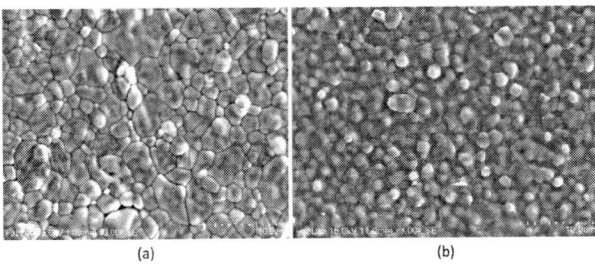

(a)   (b)

Fig. 4. Top view SEM images of CIS thin films. (a) Before KF PDT. (b) After KF PDT.

IV. SUMMARY

A very simple two-step process by selenization of SEL under selenium vapor in atmosphere with in a partially closed graphite box is used to obtain high efficiency CIS solar cells. By applying KF PDT, a CIS solar cell with a conversion efficiency of 14.7% can be obtained, compared with 10.3% efficiency solar cell without KF PDT. The dark IV characteristics of the CIS solar cells show that the shunt current leads to a symmetric I-V characteristics at low voltage and to a power law voltage dependence, which can be explained by the SCL current model and likely formation of MSM shunts. CIS solar cell with KF PDT clearly indicates a sizable reduction of the SCL current, presumably due to a reduction of the density of pinholes.

REFERENCES

[1] S. GmbH, "ZSW sets European CIGS efficiency record of 22%," PV Mag. [On-line]. Available: http://www.pv-magazine.com/news/details/beitrag/zsw-sets-european-cigs-efficiency-record-of-22_100023961/.

[2] S. GmbH, "Solar frontier hits 22.3% on CIGS cell," PV Mag. [On-line]. Available: http://www.pv-magazine.com/news/details/beitrag/solar-frontier-hits-223-on-cigs-cell_100022342/.

[3] J. Hedstrom et al., "ZnO/CdS/Cu(In,Ga)Se2 thin film solar cells with im- 419 proved performance," in Conf. Rec. 23rd IEEE Photovoltaic Spec. Conf., 420 1993, pp. 364–371.

[4] F. O. Adurodija, M. J. Carter, and R. Hill, "Synthesis and characterization of CuInSe2 thin films from Cu, In and Se stacked layers using a closed graphite box," Sol. Energy Mater. Sol. Cells, vol. 40, no. 4, pp. 359–369, Aug. 1996.

[5] J. López-García and C. Guillén, "Adjustment of the selenium amount provided during formation of CuInSe 2 thin films from the metallic precursors," Phys. Status Solidi Appl. Mater. Sci., vol. 206, no. 1, pp. 84–90, 2009.

[6] Y. Zhang, R. E. Bartolo, S. J. Kwon, M. Dagenais, "High Short-Circuit Current Density in CIS Solar Cells by a Simple Two-Step Selenization Process With a KF Postdeposition Treatment," in IEEE Journal of Photovoltaics , vol.PP, no.99, pp.1-8

[7] F. Pianezzi, P. Reinhard, A. Chirilă, B. Bissig, S. Nishiwaki, S. Buecheler, and A. N. Tiwari, "Unveiling the effects of post-deposition treatment with different alkaline elements on the electronic properties of CIGS thin film solar cells," Phys. Chem. Chem. Phys., vol. 16, no. 19, p. 8843, 2014.

[8] Y. M. Shin, C. S. Lee, D. H. Shin, H. S. Kwon, B. G. Park, and B. T. Ahn, "Surface modification of CIGS film by annealing and its effect on the band structure and photovoltaic properties of CIGS solar cells," Curr. Appl. Phys., vol. 15, no. 1, pp. 18–24, Jan. 2015.

[9] S. Dongaonkar, J. D. Servaites, G. M. Ford, S. Loser, J. Moore, R. M. Gelfand, H. Mohseni, H. W. Hillhouse, R. Agrawal, M. A. Ratner, T. J. Marks, M. S. Lundstrom, and M. A. Alam, "Universality of non-Ohmic shunt leakage in thin-film solar cells," J. Appl. Phys., vol. 108, no. 12, pp. 124509-124509–10, Dec. 2010.

[10] S. Dongaonkar, S. Loser, E. J. Sheets, K. Zaunbrecher, R. Agrawal, T. J. Marks, and M. A. Alam, "Universal statistics of parasitic shunt formation in solar cells, and its implications for cell to module efficiency gap," Energy Environ. Sci., vol. 6, no. 3, pp. 782–787, Mar. 2013.

The Twins Structure, Electrical Properties and Cell Performance of Magnetron Sputtering Deposited Chlorine Doped CdTe

Ziyao Zhu[1,3], Fu-Kuo Chiang [2], Zhongming Du[1,3], Yufeng Zhang[1,3], Xiangxin Liu[1,3] *

[1] The Key Laboratory of Solar Thermal Energy and Photovoltaic System, Institute of Electrical Engineering, Chinese Academy of Sciences, Beijing 100190, China.

[2] National Institute of Clean and Low Carbon Energy, Beijing 102211, China
[3] University of Chinese Academy of Sciences, Beijing 100049, China

Abstract — We studied the twins structure, optical properties, electrical properties and cell performance of chlorine doped CdTe (CdTe:Cl) polycrystalline film deposited by magnetron sputtering. Serried twins structure consisting of only several atomic layers can be found in both as-deposited CdTe:Cl and CdTe:Cl with annealing. Relatively heavy chlorine doping by magnetron sputtering may also cause rapid donor-acceptor pair (DAP) recombination. These cause the series resistance of CdTe:Cl film to be too high compared to undoped CdTe film with CdCl₂ treatment.

Index Terms — chlorine doped CdTe, CdCl₂ treated CdTe, photovoltaic cells, high-revolution transmission electron microscope, conductive atomic force microscope.

I. INTRODUCTION

CdTe is a promising material for fabricating high-efficient and low-cost thin film solar cell [1]. To achieve high energy conversion efficiency, polycrystalline CdTe films must go through an annealing process in an atmosphere containing $CdCl_2$ [2]. This is a thermal equilibrium process, in which the final distribution of Cl is determined by the solubility of Cl in CdTe. Numerous researches have been done to study the mechanisms of CdCl₂ treatment. For example, Chen et al reported that in CdCl₂ treated CdTe film, Cl accumulated at grain boundaries of CdTe and form localized p-n-p junction [3]. Meanwhile, first-principle calculation claimed that Cl atom formed V_{Cd}-Cl_{Te} complex, and provide extra shallow p-doping energy level [4]. Both Cl_{Te} donor at $E_c - 0.014$ eV and A-center acceptor at $E_v + 0.12$ eV has been observed by photoluminescence of Cl ion implanted CdTe crystal [5], a non-equilibrium doping process. It seems both Cl atoms segregation and Cl atoms doping can enhance cell performance of CdTe.

For better understanding of the mechanism of CdCl₂ treatment, we have fabricated Cl doped CdTe film by magnetron sputtering to study the effects of chlorine doping in CdTe lattice. Lattices structure, electrical properties and device performance of CdTe:Cl films and those of undoped CdTe film with CdCl₂ treatment have been investigated. Results showed chlorine doping by magnetron sputtering can

cause series resistance increasing drastically, and an explanation is proposed to explain why heavy chlorine doping may be hazardous to CdTe solar cell performance.

II. EXPERIMENT RESULTS

To fabricate homemade sputtering target, CdCl₂ powder was dissolved in ethanol and mixed with micron CdTe powder. We use ultra-dry CdCl₂ (99.998%) fabricated by Alfa Aesa and CdTe (99.999%) provided by Sichuan Xinlong. Then the mixed powder is ball-milled and dried by infrared heating to remove ethanol. The mixed powder is then cold-pressed in graphite die to fabricate the green body of chlorine doped CdTe ceramic target. Finally, the green body together with the graphite die is hot-press sintered to form ceramic target. The chamber of hot pressing furnace is vacuumized to 10^{-2}Pa then filled with flowing Ar to maintain micro-positive pressure. The green body is heated with the heating rate of 4°C/min and held at 600°C for 180min under a pressure of 5MPa. It has been reported that the overall concentration of Cl in vapor deposited CdTe film post-CdCl₂ treatment is 3-5 $\times 10^{18}$ at/cm³ (about 100 ppm), but Cl atoms accumulation at grain boundaries up to 10^{19} at/cm³ level can be detected [6]. Therefore, the amount of Cl atoms in the CdTe:Cl target is controlled to 100 ± 5 ppm. As a reference, undoped CdTe ceramic target was milled and hot-press sintered from the same CdTe powder as the CdTe:Cl target.

SnO₂:F (FTO) coated soda lime glass - TEC15 (Pilkington), as well as, Corning 7059, are used as substrate. Glass substrates were cleaned by IPC Micor-90 solution and dried by flow N₂ before deposition. All CdTe and CdTe:Cl films are sputtered at 267°C with RF power density of 1.32 W/cm² in flowing Ar at 2Pa . During the deposition, film thickness was measured in situ by a home-made optical transmission film thickness acquisition apparatus.

We fabricated two types of CdTe:Cl thin film solar cell both with a structure of TEC15 glass/FTO/80nm CdS:O/2.3μm CdTe:Cl/2nm Cu/20nm Au. The 1ˢᵗ type is that after CdTe:Cl is deposited, it goes through an annealing process without CdCl₂. The other kind is that as-deposited CdTe:Cl is directly used as absorb layer, without annealing. We also fabricated

978-1-5090-5606-4/17 $31.00 © 2017 IEEE

normal CdTe thin film solar cell of the same structure with CdCl₂ treatment. The undoped CdTe film is treated at 400°C for 30 min in the atmosphere of CdCl₂ and flowing gas of 4% O₂ and 96% N₂ mixture, in an E-Star RTP600 rapid thermal process equipment. In this process, the CdTe film is capped by a piece of CdCl₂ coated 7059 glass in the distance around 1 mm, with CdTe film facing to CdCl₂ source. The CdTe:Cl film is annealed at the same configuration and environment, without the CdCl₂ source. To avoid unintentional contamination from CdCl₂, before annealing the CdTe:Cl films, the RTP600 chamber was heated to 600°C for 30min with a dry air flow of 1.5L/min to purge remaining CdCl₂, since the melting point of CdCl₂ is 569°C, and CdCl₂ is also prone to sublimation.

Figure 1 gives the cross-sectional SEM image of conventional CdTe cell with CdCl₂ treatment. Not every CdTe grain contains twins and quenching of twin boundaries in some grains can be observed. The width of twins is usually wider than 2nm. Serried twins can only be discovered at CdTe/CdS interface. However, in CdTe:Cl films, serried twins exist in both as-deposited and annealed films, as shown in figure 2. Some twins only contain several atomic layers. And we did not observe Cl segregation in CdTe:Cl films whether annealing or not

Fig. 1. (a) the cross-sectional TEM image of CdTe thin film solar cell with CdCl₂ treatment; (b) the TEM image of twins at CdTe/CdS interface in CdTe thin film solar cell with CdCl₂ treatment.

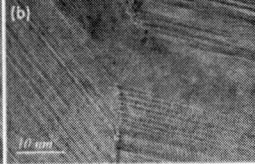

Fig. 2. (a) the cross-sectional TEM image of CdTe:Cl thin film solar cell without annealing process; (b) the cross-sectional TEM image of CdTe:Cl thin film solar cell with annealing process.

The serried ultra-thin twins may be generated due to the non-equilibrium doping process of magnetron sputtering. For undoped CdTe film, Cl atoms segregate at grain boundaries after CdCl₂ treatment and the diffusion coefficient of Cl in CdTe lattice is small [7]-[8]. Cl concentration in CdTe lattice of magnetron sputtering films could be higher than thermodynamic equilibrium solubility, because of the non-equilibrium doping in sputtering, which will be discussed in detail later We speculate that these excess Cl atoms may distract normal CdTe lattice structure and induce twins boundaries.

Shift of optical band gap E_g in CdTe:Cl post-annealing is observed, as shown in figure 5, which does not occur in CdTe after CdCl₂ treatment. The E_g of as-deposited CdTe:Cl increased to 1.53 eV, comparing to 1.5 eV of undoped CdTe. It decreases to 1.515 eV after annealing, but still higher than CdTe. The change of optical band gap may be the result of degeneration of chlorine related states. 100 ppm of Cl in CdTe equals to concentration of $N_D = 3.22 \times 10^{18}/cm^3$, assuming all chlorine forms Cl_{Te}. The concentration of ionization Cl_{Te}^+ is $2.44 \times 10^{17}/cm^3$, based on optical band gap of as-deposited CdTe:Cl and (1) [9]:

$$n_D^+ = \frac{N_D}{1 + 2\exp(\frac{E_F - E_D}{k_0 T})} \qquad (1)$$

The minimum concentration of weak n-type degeneration of Cl_{Te} states is $1.03 \times 10^{17}/cm^3$, based on (2) [9]:

$$N_D = \frac{2N_C}{\sqrt{\pi}}\left[1 + 2\exp\left(\frac{E_F - E_C}{k_0 T}\right)\exp\left(\frac{\Delta E_D}{k_0 T}\right)\right]F_{1/2}\left(\frac{E_F - E_C}{k_0 T}\right) \qquad (2)$$

The concentration of Cl atoms is sufficient to cause weak n-type degeneration. Magnetron sputtering is a non-equilibrium doping process, the concentration of Cl in film is determined by the concentration of target rather than thermodynamic equilibrium solubility. We speculate that once Cl atom entered CdTe lattice, it's hard to move, based on following four facts. First, Cl atom has different diffusion mechanisms in CdTe lattice from other halogen elements, since the diffusion energy is between 0.63eV and 1.32eV, much higher than other halogen atoms of Br (0.14-0.26eV) and I (0.21-0.28eV) [10]. Second, the diffusion coefficient of Cl atom in CdTe lattice is $6 \times 10^{-25}cm^{-2}s^{-1}$, which is much lower than other atoms. For example, the diffusion coefficient of I atom in CdTe lattice is ~$10^{-14}cm^{-2}s^{-1}$ [11]. Third, the two-phase diagram of CdTe and CdCl₂ shows their solid phases are completely immiscible [12]. And the last, we did not observe Cl atoms segregation in the CdTe:Cl films either annealing or not. After annealing, partial Cl_{Te} dopants attract V_{Cd} intrinsic defects and form A-centers ($Cl_{Te} - V_{Cd}$). V_{Cd} diffuses relatively easier in CdTe lattice than Cl_{Te}. Formation of A-centers during annealing should reduce the concentration of Cl_{Te} and E_g as well. While in CdCl₂ treated CdTe, Cl atoms may segregate along grain boundaries rather than form dopants in lattices so the E_g of CdTe doesn't change after CdCl₂ treatment.

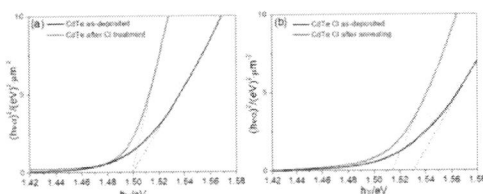

Fig. 3. (a) is the optical band gap of as-deposited CdTe film and CdTe film after CdCl₂ treatment; (b) is the optical band gap of as-deposited CdTe:Cl film and CdTe:Cl film after annealing.

As shown in table 1, Hall measurement indicates dramatic improvement of mobility but significant drop of carrier density in CdTe:Cl post-annealing, which lead to even more resistive film. One speculated reason causing such high resistivity in as-deposited CdTe:Cl film is the high density of ultra-thin twins. These dense twins can scatter carriers and led to low carrier mobility. After annealing, the donor-and-acceptor pair (DAP) as effective traps for free carriers which significantly reduce the carrier density may be the main reason for higher resistivity. Cl_{Te} can bind with V_{Cd} forming A-center when Cl concentration increases [5]. Cl atoms are uniformly distributed and pined in CdTe lattice by non-equilibrium doping, but V_{Cd} is an intrinsic defect and can move at elevated temperature. With the concentration of Cl in CdTe lattice is high, the distance between Cl_{Te} donors and A-center acceptors is closer, and the carrier recombination rate is higher. So CdTe:Cl exhibit intrinsic feature after annealing.

Tab. 1.   Resistivity, carrier mobility and carrier concentration of CdTe film before and after CdCl₂ treatment and those of CdTe:Cl film before and after annealing.

Sample	ρ (Ω·cm)	Mob (cm²/V·s)	N (/cm³)
CdTe as-deposited	2.54E+6	26.8	-3.99E+10
CdTe with CdCl₂ treatment	4.87E+5	1.18	-5.41E+12
CdTe:Cl as-deposited	7.99E+5	0.282	-1.39E+13
CdTe:Cl with annealing	2.66E+6	22.6	5.20E+10

C-AFM and SKPM scan are shown in figure 4 and 5. In CdTe:Cl, there are no obvious difference between grain and grain boundary. But in CdTe with CdCl₂ treatment grain is the conduction channel of holes and grain boundary is the conduction channel of electrons. The serried and dense twins in CdTe:Cl film can cause severe carrier scattering for carriers not transport along the twin-boundaries. This could explain the low carrier mobility of 0.28cm²/V·s but reasonable carrier density of 10^{13}/cm³ in as-deposited CdTe:Cl, as listed in table 1. In previous work [13]-[14], we have observed electrons transport along grain boundaries and holes transport within intra-grain in conventional CdTe film post chloride annealing. Assuming this is the same case in CdTe:Cl films with ultra-thin twins, the distance between channels of electrons and holes could reduce to couple nanometers, much smaller than in conventional CdTe films, which could lead to high recombination possibility.

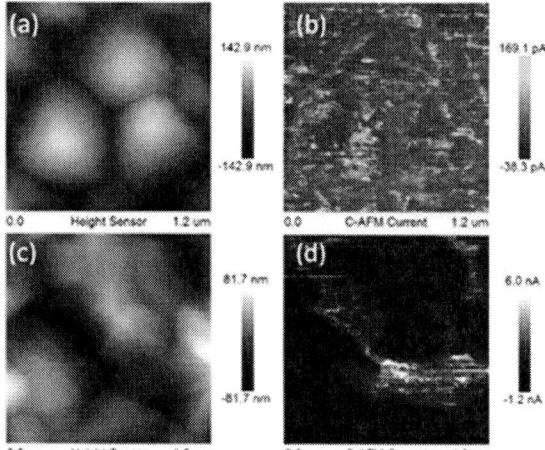

Fig. 4.   (a) is the AFM image of CdTe:Cl; (b) is the C-AFM image of the same area in (a); (c) is the AFM image of CdTe after CdCl₂ treatment; (d) is the C-AFM image of the same area in (c).

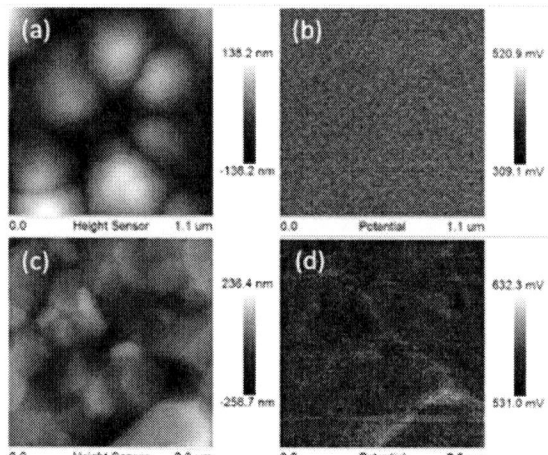

Fig. 5.   (a) is the AFM image of CdTe:Cl; (b) is the SKPM image of the same area in (a); (c) is the AFM image of CdTe after CdCl₂ treatment; (d) is the SKPM image of the same area in (c).

Previous research on chlorine doped single crystal CdTe found that if the doping concentration is higher than 10^{17}/cm³, severe donor-acceptor-pair (DAP) recombination will happen [15]. In this work, 100ppm chlorine doped CdTe:Cl corresponds to concentration of 10^{18}/cm³ scale. Therefore, the distance between Cl_{Te} and A-center is short, and DAP recombination rate in CdTe:Cl can be quick when doping concentration is high, which makes DAP effective traps for both free electron and holes. This could explain the significantly reduction of the carrier concentration in CdTe:Cl to intrinsic level after annealing.

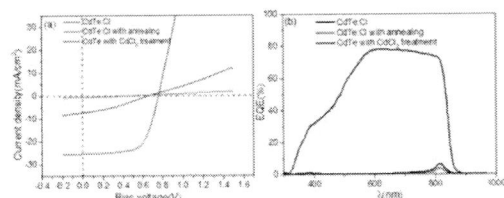

Fig. 6.   (a) the J-V curves of CdTe:Cl cell, CdTe:Cl cell with annealing and CdTe cell with CdCl$_2$ treatment; (b) is the EQE curves of CdTe:Cl cell, CdTe:Cl cell with annealing and CdTe cell with CdCl$_2$ treatment.

Tab. 2.   Cell performance of CdTe:Cl cell, CdTe:Cl cell with annealing and CdTe cell with CdCl$_2$ treatment

Sample	Voc(V)	Jsc(mA/cm^2)	Rs(Ω•cm^2)	Rsh(kΩ•cm^2)	FF(%)	Eff(%)
CdTe:Cl without annealing	0.65	7.36	75.56	0.18	32.27	1.55
CdTe:Cl with annealing	0.40	0.51	696.50	0.93	26.81	0.06
CdTe with CdCl$_2$ treatment	0.76	25.31	5.45	0.78	64.78	12.49

The J-V curve, external quantum efficiency (EQE) curve and cell performance of CdTe:Cl cell and CdTe cell with CdCl$_2$ treatment are shown in figure 6 and table 2. The series resistance of CdTe:Cl cell without annealing is about one magnitude higher than CdTe cell with CdCl$_2$ treatment. And if CdTe:Cl went through annealing before deposit back contact deposition, the series resistance can increase one more magnitude. The EQE peaks of CdTe:Cl at 390nm and 810nm also indicate poor carrier collection. Only collection of electrons near p-n junction and holes near back contact is visible in EQE, but much lower than that in conventional cells. This is also because of the high resistance of CdTe:Cl film.

III. CONCLUSION

CdTe:Cl film deposited by magnetron sputtering contains dense ultra-thin twins and the structure remains stable after annealing. Resistance of devices composed of CdTe:Cl film increases drastically compared to that composed of CdCl$_2$ treated CdTe film solar cell with the same cell structure. And series resistance of CdTe:Cl film solar cell can further increase if CdTe:Cl film is annealed. Such serried twins are speculated to cause severe carrier scattering and recombination. We also speculate donor-acceptor-pair recombination in annealed CdTe:Cl could effectively trap both carriers and reduce carrier density. Heavy Cl doping by non-equilibrium process may carrier traps causing high resistivity by reducing the carrier density and reduce cell performance.

Investigation on lower Cl concentration is still necessary to draw a conclusion.

REFERENCES

[1] X. Wu, "High-efficiency polycrystalline CdTe thin-film solar cells," *Solar energy*, vol. 77, pp. 803-814, 2004.

[2] McCandless, E. Brian, and W. A. Buchanan, "High throughput processing of CdTe/CdS solar cells with thin absorber layers," in *33rd IEEE Photovoltaic Specialists Conference*, 2008, p. 1.

[3] C. Li, Y. Wu, J. Poplawsky, T. J. Pennycook, N. Paudel, W. Yin, and S. J. Pennycook, "Grain-boundary-enhanced carrier collection in CdTe solar cells," *Physical review letters*, vol. 112, pp. 156103, 2014.

[4] H. Zhu, M. Gu, L. Huang, J. Wang, and X. Wu, "Structural and electronic properties of CdTe:Cl from first-principles," *Materials Chemistry and Physics*, vol. 143, pp. 637-641, 2014.

[5] X. Liu, and A. D. Compaan, "Photoluminescence from Ion Implanted CdTe Crystals," *MRS Proceedings*, vol. 865, pp. F5-25, 2005.

[6] D. Mao, C. E. Wickersham, and M. Gloeckler, "Measurement of chlorine concentrations at CdTe grain boundaries," *IEEE Journal of Photovoltaics*, vol. 4, pp. 1655-1658, 2014.

[7] A. Abbas, G. D. West, J. W. Bowers, P. Isherwood, P. M. Kaminski, B. Maniscalco, and K. L. Barth, "The effect of cadmium chloride treatment on close-spaced sublimated cadmium telluride thin-film solar cells." *IEEE Journal of Photovoltaics*, vol. 3, pp.1361-1366, 2013

[8] D. Shaw, and E. Watson, "The diffusion of chlorine in CdTe," *Journal of Physics C: Solid State Physics*, vol. 17, pp.4945, 1984.

[9] E. Liu, B. Zhu, and J. Luo, *Semiconductor Physics*, Beijing, Beijing: Electronics Industry Press, 2003.

[10] J. Malzbender, E. D. Jones, N. Shaw, and J. B. Mullin, "Studies on the diffusion of the halogens into CdTe," *Semiconductor science and technology*, vol. 11, pp. 741, 1996.

[11] E. D. Jones, J. Malzbender, J. B. Mullin, and N. Shaw, "The diffusion of Cl into CdTe," *Journal of Physics: Condensed Matter*, vol. 6, pp. 7499, 1994.

[12] H. Tai, and S. Hori, "Equilibrium Phase Diagrams of the CdTe-CdCl$_2$ and CdTe-CdBr$_2$ Systems," *Journal of the Japan Institute of Metals*, vol. 40, pp. 722-725, 1976.

[13] H. Li, X. Liu, Y. S. Lin, B. Yang, and Z. M. Du, "Enhanced electrical properties at boundaries including twin boundaries of polycrystalline CdTe thin-film solar cells," *Physical Chemistry Chemical Physics*, vol. 17, pp. 11150-11155, 2015.

[14] H. Li, X. Liu, Z. M. Du, and B. Yang, "Twin Boundaries Enhanced Current Transport in 14.4%-Efficient CdTe solar Cells by RF Sputtering," in *42nd IEEE Photovoltaic Specialists Conference*, 2015, p. 1.

[15] X. Liu, "Photoluminescence and extended x-ray absorption fine structure studies on CdTe material," *Ph. D. Dissertation*, 2006.

Investigation and Mitigation of Shunts for Higher Efficiency Epitaxial GaSb/GaSb and GaSb/GaAs Solar Cells

George T. Nelson*, Bor-Chau Juang†, Steve Johnston‡, Michael A. Slocum*, Zachary S. Bittner*, Ramesh B. Lagumavarapu†, Diana Huffaker†, and Seth M. Hubbard*

*Rochester Institute of Technology, Rochester, NY, USA
†University of California, Los Angeles, CA, USA
‡National Renewable Energy Labs, Golden, CO, USA

Abstract—Multi-junction cells using GaAs/GaSb interface misfit arrays for metamorphic growth can achieve comparable efficiency to conventional inverted metamorphic multijunction cells without the need for costly graded buffer layers. In this work, GaSb single junction cells were grown via molecular beam epitaxy on both GaSb and GaAs substrates to compare homoepitaxial and heteroepitaxial GaSb material quality. The homoepitaxial cell achieved 4.1% efficiency under 1 sun AM1.5g while the IMF cell achieved 0.70%. The IMF cell had higher non-radiative dark current likely caused by defects related to the mismatched growth. Lock-in thermography was used to explore shunt causing defects and sidewall shunting was passivated by dielectric deposition on device sidewalls. Homoepitaxial cells with sidewalls passivated by Al_2O_3 had ~5x higher average shunt resistance compared to unpassivated devices. Reverse bias thermograms indicated that MBE-related defects filled with front grid metal were the cause of device-crippling shunts which lowered device yield and limited growth of large-area devices.

Index Terms—III-V semiconductors, photovoltaic cells, epitaxy, multijunction solar cells

I. INTRODUCTION

The use of interfacial misfit arrays (IMF) enables hetero-epitaxial growth of III-Sb materials directly on Si [1] or GaAs [2] substrates with potentially low threading dislocation density and without the need for a step-graded buffer. These properties make IMF growth an attractive low-cost alternative to the current state of the art multi-junction solar cell, the inverted metamorphic (IMM) cell, which relies on thick and costly metamorphic layers. In addition to the cost benefit, very high efficiency is possible with III-Sb IMF growth as it can employ an assortment of well-developed and understood lower bandgap materials, such as AlGaSb or InAsSb, that spectrally compliment the GaAs family. Our physics-based simulations predict that a GaSb/AlGaSb/GaAs/AlGaAs/AlInGaP 5-J cell of sufficient material quality can achieve 43% efficiency under AM1.5g 1-sun illumination.

During IMF growth, high amounts of strain allow for relaxation to occur via 90° misfit dislocations which are confined to the thin interface layer [3]. IMF technology has been used in the past to grow InGaAsSb photodetectors on GaAs with comparable results to detectors grown natively on GaSb [4]. In this work, we have continued to make progress on GaSb photovoltaic cells grown on either GaSb and GaAs

substrates, previously reported on in [5]. We have also studied some of the complications that arise when growing GaSb epitaxially via molecular beam epitaxy (MBE).

There has been successful work on diffused junction GaSb photovoltaic cells [6], [7] and epitaxial cells grown by metal-organic chemical vapor deposition (MOCVD) have reached as high as 10% efficient under AM1.5g [8]. However, only preliminary progress has been made on MBE-grown GaSb cells [5], [7], [9], [10], [11]. One significant impediment are the reported problems with growing large area GaSb devices due to defects originating from the substrate or from undesired growth complications [7], [12]. We have taken two approaches to this issue: 1. Fabricate small area cells that will achieve higher yield by avoiding defects altogether, and 2. Search for and mitigate the defects that shunt large area devices. For the large area cells, we have characterized a similar defect found in [12] which we suspect originates during MBE growth. We have used dark lock-in thermography (DLIT) to show that the defect only becomes a shunt path when it is covered with grid metal. Small area cells unaffected by defects were found to be subject to perimeter effects from the GaSb native oxide formed on the unpassivated device sidewalls. The Al_2O_3 passivation scheme for GaSb/InAs detectors found in [13] was implemented and increased shunt resistance of these cells by a factor of ~5 compared to the unpassivated baseline. These cells compare favorably to previous large-area results [11].

II. METHODS

Samples were grown via MBE in a Veeco Gen 930 solid-source reactor. The epitaxial GaSb layers were grown at a growth temperature of 500° C with a V/III beam equivalent pressure of 6. The n- and p-type dopants were Te and Be, respectively. Control cells were grown on a p-type GaSb substrate after the growth of a GaSb buffer layer. IMF cells were grown on a p-type GaAs substrate and the details for the IMF array are reported in previous work [2]. The epitaxial structure can be seen in the Figure 1, which depicts a fabricated cell.

IMF and control cells were processed in the same batch with identical procedure. Two wet mesa-etch chemistries were

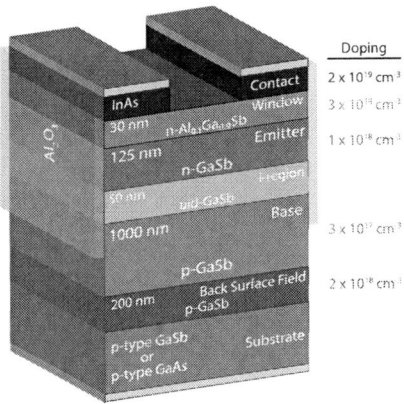

Fig. 1: Layer schematic for the sidewall passivated GaSb cells.

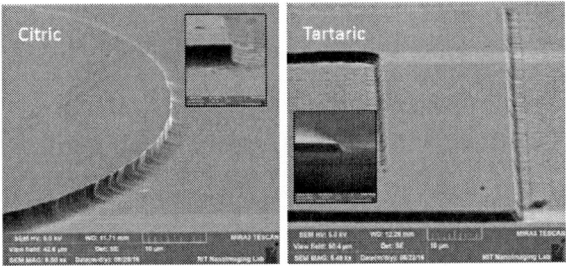

Fig. 2: SEM images sidewalls etched with (a) the citric-based solution and (b) the tartaric-based solution. The insets show the sidewall profile.

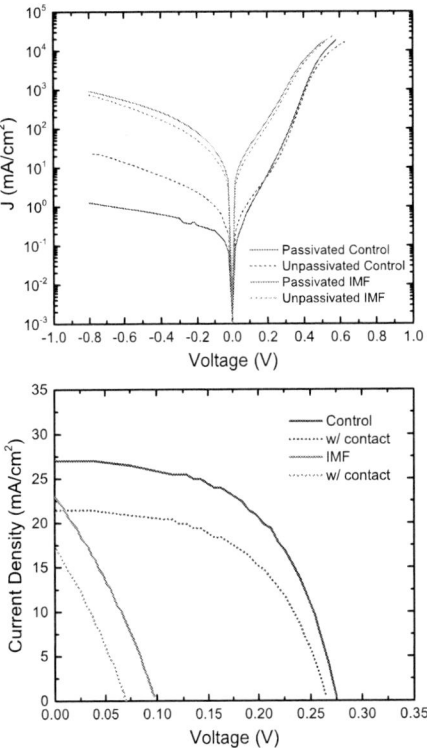

Fig. 3: (a) Dark I-V for unpassivated (dotted) and passivated (solid) homoepitaxial (control) and IMF cells. (b) AM1.5G illuminated I-V results at 1 sun for cells with contact layer (dotted) and contact layer removed (solid).

developed for device isolation, $C_6H_8O_7$/HF/H_2O_2/H_2O ('citric solution', 4 g:4 µl:1 mL:4 mL, etch rate of ~120 nm/min) and KNa-$C_4H_4O_6$-4H_2O/HCl/H_2O_2/H_2O ('tartaric solution', 12 g:33 mL:9 mL:500 mL, etch rate of ~130 nm/min). Testing indicated that that for these two solutions the etched surfaces were smooth and the difference in perimeter effects was negigible (see Figure 2). The citric solution was chosen for the vertical sidewalls it produced. After the mesa etch, a passivating layer of 100 nm thick Al_2O_3 was deposited by atomic layer deposition (ALD) using a 2nd generation Cambridge Nanotech Savannah. Before passivation, the native oxide was removed using 1:1 HCl:H_2O [14] and transferred to the ALD reactor covered in methanol to prevent re-oxidation. Roughly 100 nm of Al_2O_3 was deposited using TMA and water precursors at 150° C with 5 sec wait time between pulses. The passivation layer was patterned and then etched using a dilute HF solution. Standard deposition techniques were used to deposit the back contact metal stack of Pt/Ti/Au and front grid metal stack of Ti/Au. The 50 nm InAs contact layer was etched using a citric acid and peroxide mixture (1:1) for 50 sec. The tested cell sizes were circular with 400 µm and 500 µm diameter.

Dark current-voltage (I-V) measurements were performed with a Kiethley Source Meter 2440-C. A TSS Space Systems two zone solar simulator with AM1.5G filters was used for illuminated I-V. The two lamps were calibrated to InGaP and Ge reference cells. Concentration measurements up to 45 suns took place in the simulator with the use of concentrating optics. Dark lock-in thermography (DLIT) was used to determine the cause of the shunt seen in the defective devices that led to low yield, the experimental setup for which is detailed elsewhere [15].

III. RESULTS

Dark I-V results (Figure 3a) were used to compare passivated and unpassivated homoepitaxial and IMF devices with 250 µm radius. Homoepitaxial cells with the Al_2O_3 coating had an average factor of ~5 improvement in shunt resistance over unpassivated devices. Current under reverse bias was also significantly reduced; it may be that the dielectric prevented a surface tunneling current mechanism. The surface treatment had no beneficial effect on the IMF devices as these devices were bulk limited.

I-V results under 1-sun AM1.5g illumination for 250 µm-radius control and IMF cells are shown in Figure 3b. The GaSb control cell was 4.1% efficient at 1 sun, with a fill factor (FF) of 56%, an open-circuit voltage (V_{OC}) of 275 mV, and a short-circuit current (J_{SC}) of 27.0 mA/cm^2. These results compare favorably to reported MBE-grown GaSb photovoltaic cells [5], [9], [10], [11]. The IMF cell efficiency was

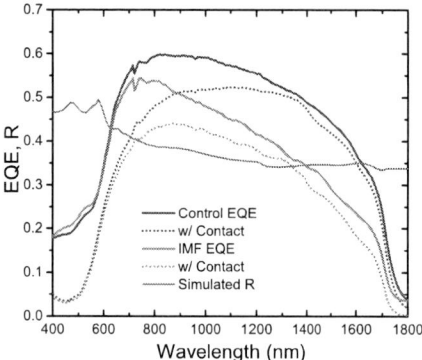

Fig. 4: EQE of homoepitaxial (control) and IMF cells with (solid) and without (dotted) the contact layer and with simulated reflectance (R).

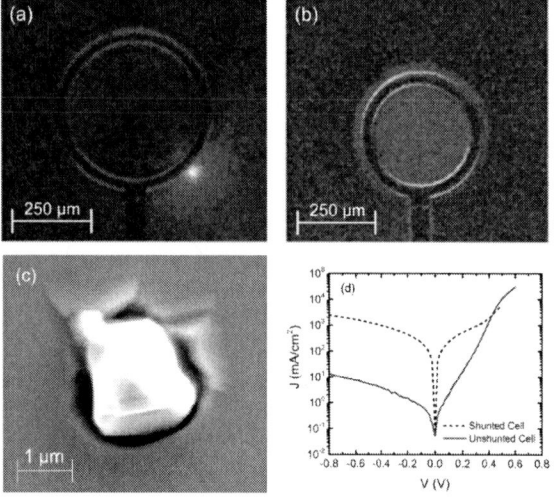

Fig. 5: Reverse bias thermograms of two cells, a shunted cell (a) and an unshunted cell (b). In (c) is an SEM image of a Ga-rich GaSb surface defect believed to be the cause of the shunt. In the dark I-V results for these cells (d), the dotted line is the cell in (a) and the blue line is the cell in (b).

0.70%, the FF was 32%, the V_{OC} was 99 mV, and the J_{SC} was 22.9 mA/cm^2 1-sun J_{SC}. Compared to the previously reported results [5]. The higher current here was due to optimization of the cell design for the expected IMF diffusion lengths, *i.e.* by thinning the emitter. Fitting the I-V in Figure 3 revealed that the IMF cell was shunted and had high dark saturation current, likely due to defects caused by the mismatched growth. Work on the IMF material quality is ongoing.

The measured EQE results are shown in Figure 4. Using a drift-diffusion model, minority carrier diffusion lengths in the cell base were fit for the control and IMF devices. The extracted minority carrier diffusion length (MCDL) for the control cell base was at least 6 µm, close to the 8 µm value calculated from literature source [16]. The IMF cell had a

degraded base MCDL of 0.8 µm, consistent with increased recombination seen in the I-V results.

Dark lock-in thermograms (Figure 5) correlated well to dark I-V results for shunted and unshunted homoepitaxial cells. The cell in (a) of the figure was known to be shunted as demonstrated by the black dotted dark I-V curve in (d). The current under reverse bias flowed nearly exclusively through a small area in the bottom right part of the image (bright spot in thermogram). The unshunted cell in (b) of the figure had uniform illumination indicative of normal diode leakage current under reverse bias. The trend from these cells and thermograms of other shunted cells indicated that the shunt was related to the front grid metal. In part (c) we have imaged, via scanning electron microscope (SEM), similar surface defects to those found by Romero *et al.* [12]. The defects in the reference were caused by Ga spitting from the MBE solid source. If the defects imaged here are also from spitting, then the DLIT results are further evidence that a shunt is formed when a surface defect void is filled with metal. It is possible that the use of a dual-filament Ga cell would reduce or eliminate the incidence of these defects and enable the growth of large-area cells [12].

IV. CONCLUSION

Homoepitaxial and IMF heteroepitaxial GaSb solar cells were grown via MBE. The ALD Al$_2$O$_3$ passivated homoepitaxial and IMF cells achieved 4.1% and 0.70% efficiency, respectively, under 1-sun AM1.5g illumination, with 275 mV and 99 mV V_{OC} 56% and 32% FF, and 27.0 and 22.9 mA/cm^2 J_{SC}, respectively. Shunting and high non-radiative dark current were main cause of FF and efficiency loss in the IMF devices. Severe shunts in the homoepitaxial cells were investigated by DLIT and it was discovered that the shunts were caused by front grid metal deposited into a void, likely caused by MBE Ga-rich surface defects found by SEM.

ACKNOWLEDGMENT

This work was supported by grants from the National Science Foundation (ECCS-1509468) and the Bay Area Photovoltaic Consortium (DE-EE0004946 / 60964954-51077).

REFERENCES

[1] A. Jallipalli, G. Balakrishnan, S. H. Huang, A. Khoshakhlagh, L. R. Dawson, and D. L. Huffaker, "Atomistic modeling of strain distribution in self-assembled interfacial misfit dislocation (IMF) arrays in highly mismatched IIIV semiconductor materials," *Journal of Crystal Growth*, vol. 303, no. 2, pp. 449–455, May 2007.

[2] S. Huang, G. Balakrishnan, and D. L. Huffaker, "Interfacial misfit array formation for GaSb growth on GaAs," *Journal of Applied Physics*, vol. 105, no. 10, p. 103104, May 2009.

[3] C. J. Reyner, J. Wang, K. Nunna, A. Lin, B. Liang, M. S. Goorsky, and D. L. Huffaker, "Characterization of GaSb/GaAs interfacial misfit arrays using x-ray diffraction," *Applied Physics Letters*, vol. 99, no. 23, p. 231906, Dec. 2011.

[4] K. C. Nunna, S. L. Tan, C. J. Reyner, A. R. J. Marshall, B. Liang, A. Jallipalli, J. P. R. David, and D. L. Huffaker, "Short-Wave Infrared GaInAsSb Photodiodes Grown on GaAs Substrate by Interfacial Misfit Array Technique," *IEEE Photonics Technology Letters*, vol. 24, no. 3, pp. 218–220, Feb. 2012.

[5] B.-C. Juang, R. B. Laghumavarapu, B. J. Foggo, P. J. Simmonds, A. Lin, B. Liang, and D. L. Huffaker, "GaSb thermophotovoltaic cells grown on GaAs by molecular beam epitaxy using interfacial misfit arrays," *Applied Physics Letters*, vol. 106, no. 11, p. 111101, Mar. 2015.

[6] L. Fraas, J. Avery, V. Sundaram, V. Dinh, T. Davenport, J. Yerkes, J. Gee, and K. Emery, "Over 35% efficient GaAs/GaSb stacked concentrator cell assemblies for terrestrial applications," in *21st IEEE Photovoltaic Specialists Conference*, May 1990, pp. 190–195 vol.1.

[7] A. W. Bett and O. V. Sulima, "GaSb photovoltaic cells for applications in TPV generators," *Semiconductor Science and Technology*, vol. 18, no. 5, p. S184, 2003.

[8] T. Schlegl, "GaSb-Photovoltaikzellen für die Thermophotovoltaik," PhD Dissertation, Universität Regensburg, Jan. 2006. [Online]. Available: https://epub.uni-regensburg.de/10412/

[9] D. DeMeo, C. Shemelya, C. Downs, A. Licht, E. S. Magden, T. Rotter, C. Dhital, S. Wilson, G. Balakrishnan, and T. E. Vandervelde, "GaSb Thermophotovoltaic Cells Grown on GaAs Substrate Using the Interfacial Misfit Array Method," *Journal of Electronic Materials*, vol. 43, no. 4, pp. 902–908, Feb. 2014.

[10] E. J. Renteria, A. Mansoori, S. J. Addamane, D. M. Shima, C. P. Hains, and G. Balakrishnan, "Development of thin film metamorphic GaSb cells by epitaxial lift-off from GaAs substrates," in *43rd IEEE Photovoltaic Specialists Conference*, Jun. 2016, pp. 2310–2312.

[11] G. T. Nelson, B. C. Juang, M. Slocum, Z. Bittner, R. B. Lagumavarapu, D. Huffaker, and S. Hubbard, "GaSb on GaAs interfacial misfit solar cells," in *43rd IEEE Photovoltaic Specialists Conference*, Jun. 2016, pp. 2349–2353.

[12] O. S. Romero, A. A. Aragon, N. Rahimi, D. Shima, S. Addamane, T. J. Rotter, S. D. Mukherjee, L. R. Dawson, L. F. Lester, and G. Balakrishnan, "Transmission Electron Microscopy-Based Analysis of Electrically Conductive Surface Defects in Large Area GaSb Homoepitaxial Diodes Grown Using Molecular Beam Epitaxy," *Journal of Electronic Materials*, vol. 43, no. 4, pp. 926–930, Mar. 2014.

[13] O. Salihoglu, A. Muti, K. Kutluer, T. Tansel, R. Turan, C. Kocabas, and A. Aydinli, "Atomic layer deposited Al2o3 passivation of type II InAs/GaSb superlattice photodetectors," *Journal of Applied Physics*, vol. 111, no. 7, p. 074509, Apr. 2012.

[14] A. Nainani, T. Irisawa, Z. Yuan, B. R. Bennett, J. B. Boos, Y. Nishi, and K. C. Saraswat, "Optimization of the Interface and a High-Mobility GaSb pMOSFET," *IEEE Transactions on Electron Devices*, vol. 58, no. 10, pp. 3407–3415, Oct. 2011.

[15] S. Johnston, I. Repins, N. Call, R. Sundaramoorthy, K. M. Jones, and B. To, "Applications of imaging techniques to Si, Cu(In,Ga)Se2, and CdTe and correlation to solar cell parameters," in *35th IEEE Photovoltaic Specialists Conference*, Jun. 2010, pp. 001 727–001 732.

[16] O. V. Sulima and A. W. Bett, "Fabrication and simulation of GaSb thermophotovoltaic cells," *Solar Energy Materials and Solar Cells*, vol. 66, no. 14, pp. 533–540, Feb. 2001.

Development of GaSb solar cells on GaAs by MOVPE via interface misfit technique

Michael A. Slocum*, Alessandro Giussani*, Emily Kessler*, Phil Ahrenkiel[†], George T. Nelson* and Seth M. Hubbard*

*Rochester Institute of Technology, Rochester, NY, 14623
[†]South Dakota School of Mines & Technology, Rapid City SD, 57701

Abstract—Current research in photovoltaics has pushed lattice matched triple junction solar cells on Ge to average efficiencies of 31% under one sun AM0. Further improvements are possible through an inverted metamorphic (IMM) design, departing from the lattice matching constraint to achieve improved current matching with a 1eV bottom cell. Even with these improvements the record one sun efficiencies fall short of the detail balance limit for three junctions of 48% largely due to the predominant use of two materials, (In)GaAs and InGaP, while neglecting the remainder of possible III-V semiconductor materials. It is proposed that including Sb-based materials will significantly expand the opportunities for high efficiencies. This is done through the development of an interfacial misfit (IMF) growth of highly mismatched materials with low defect densities by MOVPE. The development of IMF growth has led to the direct growth of GaSb on GaAs with a rocking curve FWHM of 307 arcsec, with the potential to reduce that further with alternate precursors. Single junction GaSb IMF and homoepitaxial solar cells have been grown and will be presented to demonstrate the potential of this technique. This enables triple and five junction solar cells with ideal bandgap combination that would result in efficiencies over 40% under 1 sun AM0.

I. INTRODUCTION

Over the past 30 years, the development of multi-junction (MJ) solar cells was based on lattice-matched (LM) materials to provide efficient operation and low defect densities. The LM triple junction solar cell (TJSC) is comprised of InGaP$_2$ (1.85 eV), (In$_{0.01}$)Ga$_{0.99}$As (1.40 eV) and Ge (0.66 eV), which are all lattice matched to the commonly available Ge substrate. Optimization of the device design has resulted in manufactured, 1 sun, lot efficiencies near 28-30% AM0 and wide-spread application to both satellite power systems and terrestrial concentrating photovoltaics (CPV). However, these LM materials are not the best combination for current matching and thus maximal efficiency. The optimal set of bandgaps for a 1 sun monolithic device under an AM0 spectrum is 1.90 eV, 1.35 eV, and 0.92 eV, which gives a detailed balance efficiency of over 47% at one sun (AM0) and logarithmically higher under solar concentration.

A major technology shift in MJ solar cells became possible in 2006 when Geisz, et al. [1] demonstrated the technology to current match a TJSC by replacing the Ge bottom subcell with a lattice mismatched (LMM) 1.0 eV InGaAs subcell. This is completed by growing a step graded buffer from GaAs (5.65 Å) and 1.0 eV InGaAs (5.77 Å). The higher voltage InGaAs subcell provided improved current-matching and resulted in cell efficiency over 30% under one-sun AM0 illumination. In this device, known as the inverted metamorphic (IMM) solar cell, the subcells are grown in reverse order on a GaAs substrate. The GaAs substrate is subsequently removed and the device is bonded to a handle and inverted for operation. While the IMM approach has achieved impressive results, the thick metamorphic (MM) buffer needed between GaAs and lattice mismatched InGaAs requires significant effort and time to grow and still retains a fairly high defect density with threading dislocation densities (TDD) approaching 5×10^6 cm^{-2}. While this level of TDD has allowed for high efficiency operation under concentration, the overall process is complex, time consuming and results in limited minority carrier diffusion length and enhanced non-radiative recombination

in the bottom junction. Additionally, it has been shown that the bottom junction in the IMM is quite susceptible to degradation due to radiation seen in the space environment [2].

An interface misfit (IMF) dislocation layer developed initially by MBE [3] enables the growth of high quality GaSb films grown on GaAs without the need for a buffer layer. The high degree of strain (7.8%) between GaAs and GaSb allows for strain relaxation to occur solely by laterally propagating 90 misfit dislocations, which are confined to the GaAs-GaSb interface. The growth of dislocation free GaSb on GaAs enables a much broader parameter space given the direct bandgap materials lattice matched to GaSb range from InAsSb at 0.28 eV to AlGaAsSb at 1.3 eV, and indirect bandgap materials up to AlAsSb at 1.64 eV. Given this material parameter space, perfectly current matched triple junctions can be designed. This technology can set the stage for future 5 and 6-junction devices comprised of lattice matched (Al)InGaP2, (Al)GaAs and GaAs as the top cells, a single IMF transition, and lattice matched AlGa(As)Sb with varying composition for the bottom cells as depicted in Figure 1a. These 5 or 6 junction devices would be capable of significant advances in efficiency, with the potential for nearing or breaking the 45% mark at one sun AM0. The I-V curve shown in Figure 1b demonstrates a potential efficiency of 44.1%, which was simulated with a drift-diffusion model in Synopsys Sentaurus, a physics-based device simulation program.

In this work, we will demonstrate the growth of GaSb directly on GaAs by metallic-organic vapor phase epitaxy (MOVPE) without the use of a graded buffer layer. Although others have demonstrated GaSb IMF by MOVPE [4], [5], the majority of the work has been done by MBE and significant development is still required by MOVPE. Additionally, we will demonstrate the growth and performance of GaSb solar cells grown by IMF, as a first step towards the development of 3J or 5J solar cells with IMF GaSb based subcells.

II. EXPERIMENTAL

The GaSb films was grown on (001) oriented GaAs substrates with a 2° offcut towards <110> in an Aixtron Close Coupled Showerhead reactor. Source gases were Trimethylgallium (TMGa) and Arsine (AsH$_3$) for the GaAs buffer, and Triethylgallium (TEGa) and Trimethylantimony (TMSb) were primarily used for the GaSb layers, although tests were also completed with alternate sources TMGa and Triethylantimony (TESb). For doped layers Diethylzinc (DEZn) and Diethyltellerium (DETe) were chosen for the p-type and n-type dopants respectivly. For these studies a baseline condition was chosen, which consisted of a GaAs buffer growth at 620°C, a temperature ramp to 520°C followed by a 30s AsH$_3$ pause and a 60s TMSb preflow at 40 μmol/min. The TMSb flow and chamber temperature are kept constant for the growth of GaSb using TMSb and TEGa at a V/III of 1.2 with a final thickness of 500 nm.

Multiple test structures have been grown to evaluate the x-ray diffraction peak broadening and photoluminescence from IMF grown GaSb. Four series of growths were completed to study multiple variables in forming an ideal IMF transition. First the AsH$_3$ pause

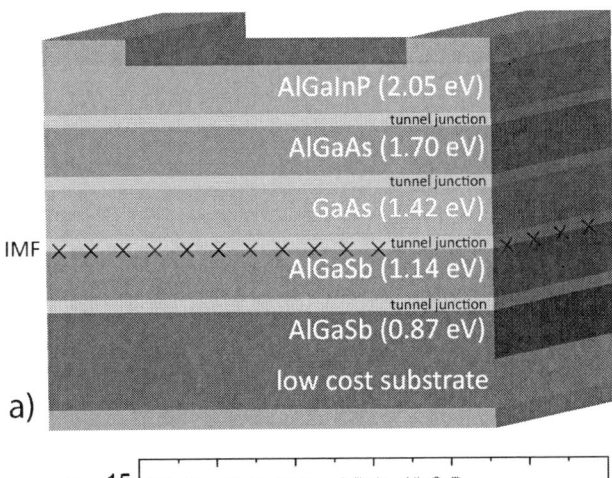

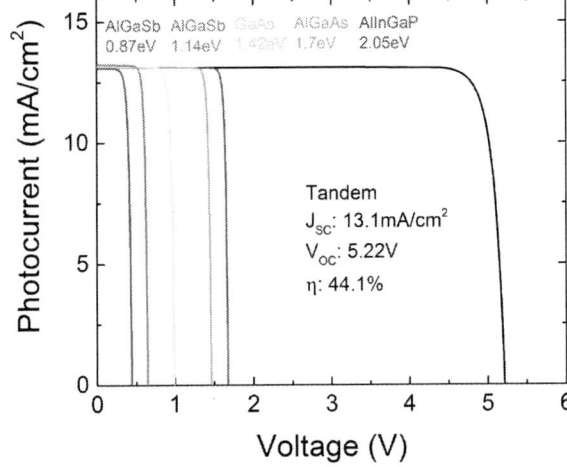

Fig. 1. (a)Current matched 5 junction solar cell design with subcells lattice matched to both GaAs and GaSb, and (b) AM0 one sun IV for each 5J subcell and the tandem device with a simulated efficiency of 44.1%.

and Sb pre-flow time was evaluated. For the AsH_3 pause study no Sb preflow was used, while the AsH_3 pause was varied from 0 to 50 s. With the AsH_3 pause fixed at 30 s the TMSb pre-flow was varied from 0 to 60s with a flow of 40 μmol/min. To study the effect of precursor selection the standard baseline growth conditions were used and each combination of TESb/TMSb and TEGa/TMGa was used for both the Sb pre-flow and GaSb growth. Finally, three varied conditions were also chosen which include: (a) a low temperature buffer with an initial 3 monolayer GaSb film grown at 490°C and a V/III of 10, prior to ramping to 520°C and a V/III of 1.2 over 36 s to grow the 500 nm GaSb buffer, (b) including a 200 nm GaSb initial IMF layer, then a 200 nm $In_{0.01}$GaSb buffer layer, followed by the 500 nm GaSb buffer, and finally (c) GaSb growth at a V/III of 2.5 with the TMSb preflow molar flow increased to 80 μmol/min.

X-ray diffraction (XRD) measurements were taken on each of the test structures, with full width half max (FWHM) extracted from $2\Theta - \omega$ and rocking curve scans taken with a 4-bounce source monochromator and 3-bounce analyzer crystal on the detector.

III. RESULTS & DISCUSSION

The XRD FWHM of the $2\Theta - \omega$ and rocking curve scans tended together, so Figure 2 only displays the FWHM of the rocking curve. In evaluating the AsH_3 pause, the inclusion of a pause did improve the FWHM, which is expected given the findings from MBE based

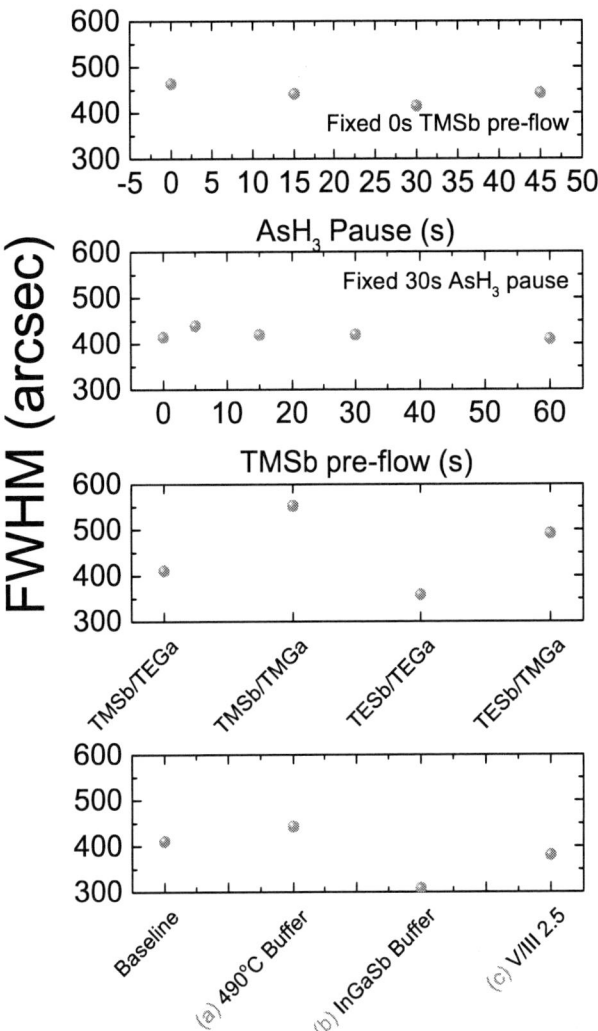

Fig. 2. Full width half max of XRD rocking curves for GaSb IMF growths with varying growth conditions

IMF growth where the As flow is stopped to cause a (4x2) Ga rich reconstruction prior to Sb flow. It is unclear whether the reconstruction in MOVPE grown IMF is identical to MBE due to the lack of surface reconstruction characterization such as RHEED, however the improvement with the introduction of an AsH_3 pause would indicate some similarities. In MBE IMF growth the introduction of Sb flux transforms the RHEED pattern to a (2x8) reconstruction to cause the IMF array, however there is no strong impact from the TMSb pre-flow time, only indicating a slight decrease in FWHM as pre-flow time is increased with a minimum of 410.1 arcsec with a 60 s pre-flow time.

It was unknown whether precursor selection would have a major impact on the IMF growth, however others have indicated that TESb and TEGa would be preferred due to complete pyrolization at lower growth temperatures below 550°C [6]. The XRD FWHM analysis agreed with the earlier conclusions, where the FWHM was decreased from 492 to 358 arcsec when switching from TMGa to TEGa with TESb as the group V source. Similarly the FWHM dropped from 411 to 358 arcsec when switching from TMSb to TESb with TEGa as the group III source. It is however unclear at this point whether

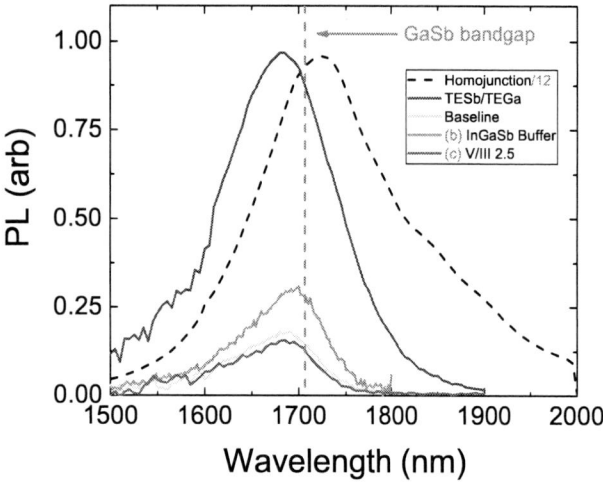

Fig. 3. Photoluminescence of GaSb IMF growths, as well as a homoepitaxial GaSb growth. Homojunction GaSb PL was reduced by a factor of 12 for comparison.

the reduction in FWHM comes from an improvement in the IMF transition and lower TDD density, or a reduction in mosaicity in the GaSb film grown following the IMF transition.

The additional conditions described in Section II as (a), (b) and (c) demonstrated further improvements in FWHM. Condition (a) degraded the FWHM relative to the baseline slightly, potentially due to the increased V/III resulting in residual Sb droplets or the lower pyrolization efficiency at the reduced temperature leading to non-uniformity due to small temperature gradients in the kinetically limited regime. The introduction of an InGaSb buffer layer in (b) was intended to allow complete strain relaxation for a partially strained film by increasing the in-plane lattice constant to that of GaSb, similar to the methodology used for the overshoot layer of a metamorphic grade [1]. The result was a decrease in FWHM to 307 arcsec, which was the lowest broadening found in this study. Relaxation for the baseline IMF was calculated to be 99.7% from reciprocal space maps, which was increased to 99.9% with the InGaSb interlayer. A similar improvement was seen when AlSb/GaSb compressive superlattices were grown after an IMF transition for MBE grown GaSb where it is expected that the SL traps dislocations and prohibits threading [7], it is possible the improvement from adding InGaSb is due to a similar effect. The total thickness of the (In)GaSb stack was 900 nm and only 500 nm with the GaSb IMF only, however MBE grown IMF GaSb has shown a quadrupling of the GaSb thickness is required to increase relaxation from 99.7% to 99.9% [8]. Also, increased layer thickness can reduce XRD FWHM, however this effect is more prevalent in $2\Theta - \omega$ scans and not rocking curves. Finally an increase in V/III ratio in (c) does also improve the FWHM from the baseline by 8.7%, however as with the alternate precursor growth it is unclear whether the improvement comes from an improved bulk crystal or IMF transition.

Photoluminescence measurements shown in Figure 3 demonstrate that we can achieve PL from the IMF GaSb at room temperature, which others have shown is challenging to realize above cryogenic temperatures [9]. For comparison PL from homoepitaxial GaSb was included, which is a factor of 12 higher intensity than the most luminescent IMF growth which was grown with TESb/TEGa precursors. Adding the InGaSb buffer in (b) did improve PL intensity relative to the baseline by 70%, however the improvement in PL from switching

Sb precursor to TESb relative to the baseline is 4,349%. It is expected that including a InGaSb buffer could further improve PL and material quality for GaSb grown with TESb/TEGa. Also, a shift in PL peak position was measured between the IMF and homoepitaxial growths of 19 meV. This has been seen elsewhere, and it was thought to be due to an in-plane tensile stress from the differential in thermal expansion between epi-layer and substrate [9]. To confirm this, and evaluate other material difference photoreflectance (PR) and deep level transient spectroscopy (DLTS) will be used to measure the band energy and trap states respectively.

IV. CONCLUSION

The growth of GaSb directly on GaAs has been demonstrated by MOVPE for the intended application of the monolithic growth of a three to five junction solar cell with subcells lattice matched to both GaAs and GaSb. Analysis of AsH_3 pause, TMSb pre-flow, precursor selection and initial buffer layers was conducted in order to determine the impact of these variables on the XRD peak broadening, as well as PL intensity. Rocking curve FWHM was reduced to 307 arcsec with a measured relaxation of 99.9% when a $In_{0.01}$GaSb layer was added as a buffer layer prior to the GaSb growth. This demonstrates the potential for use of GaSb growth directly on GaAs, while minimizing the total buffer layer thickness.

V. ACKNOWLEDGMENTS

This research was supported by an appointment to the Intelligence Community Postdoctoral Research Fellowship Program at the Rochester Institute of Technology, administered by Oak Ridge Institute for Science and Education through an interagency agreement between the U.S. Department of Energy and the Office of the Director of National Intelligence.

REFERENCES

[1] J. F. Geisz, S. Kurtz, M. W. Wanlass, J. S. Ward, A. Duda, D. J. Friedman, J. M. Olson, W. E. McMahon, T. E. Moriarty, and J. T. Kiehl, "High-efficiency GaInPGaAsInGaAs triple-junction solar cells grown inverted with a metamorphic bottom junction," *Applied Physics Letters*, vol. 91, no. 2, p. 023502, 2007.

[2] P. Patel, D. Aiken, A. Boca, B. Cho, D. Chumney, M. B. Clevenger, A. Cornfeld, N. Fatemi, Y. Lin, J. McCarty, F. Newman, P. Sharps, J. Spann, M. Stan, J. Steinfeldt, C. Strautin, and T. Varghese, "Experimental Results From Performance Improvement and Radiation Hardening of Inverted Metamorphic Multijunction Solar Cells," *IEEE Journal of Photovoltaics*, vol. 2, no. 3, pp. 377 –381, 2012.

[3] S. Huang, G. Balakrishnan, and D. Huffaker, "Characterization of Interfacial Misfit Array Formation for GaSb Growth on GaAs by Transmission Electron Microscopy," *Microscopy and Microanalysis*, vol. 15, no. Supplement S2, pp. 1062–1063, 2009.

[4] J. Wu, S. Liu, W. Zhou, W. Tang, and K. M. Lau, "Effect of antimony ambience on the interfacial misfit dislocations array in a GaSb epilayer grown on a GaAs substrate by MOCVD," in *2012 International Conference on Optoelectronics and Microelectronics (ICOM)*, 2012, pp. 5–9.

[5] W. Zhou, X. Li, S. Xia, J. Yang, W. Tang, and K. M. Lau, "High Hole Mobility of GaSb Relaxed Epilayer Grown on GaAs Substrate by MOCVD through Interfacial Misfit Dislocations Array," *Journal of Materials Science & Technology*, vol. 28, no. 2, pp. 132–136, 2012.

[6] C. A. Wang, "Progress and continuing challenges in GaSb-based IIIV alloys and heterostructures grown by organometallic vapor-phase epitaxy," *Journal of Crystal Growth*, vol. 272, no. 14, pp. 664–681, 2004.

[7] R. Hao, S. Deng, L. Shen, P. Yang, J. Tu, H. Liao, Y. Xu, and Z. Niu, "Molecular beam epitaxy of GaSb on GaAs substrates with AlSb/GaSb compound buffer layers," *Thin Solid Films*, vol. 519, no. 1, pp. 228–230, 2010.

[8] C. J. Reyner, J. Wang, K. Nunna, A. Lin, B. Liang, M. S. Goorsky, and D. L. Huffaker, "Characterization of GaSb/GaAs interfacial misfit arrays using x-ray diffraction," *Applied Physics Letters*, vol. 99, no. 23, p. 231906, 2011.

[9] E. T. R. Chidley, S. K. Haywood, A. B. Henriques, N. J. Mason, R. J. Nicholas, and P. J. Walker, "Photoluminescence of GaSb grown by metal-organic vapour phase epitaxy," *Semiconductor Science and Technology*, vol. 6, no. 1, p. 45, 1991.

Fabrication of InGaAsP Solar Cells for Concentrator Applications

Mitchell F. Bennett[*], Matthew P. Lumb[†], Kenneth J. Schmieder[‡],
Brent Fisher[§], Eric A. Armour[¶], Robert J. Walters[‡]

[*]Sotera Defense Solutions, Annapolis, MD 20701
[†]George Washington University, Washington, DC 20052
[‡]Naval Research Laboratory, Washington, DC 20375
[§]Semprius Inc., Durham, NC 27713
[¶]Veeco MOCVD, Somerset, NJ 08873

Abstract—**Transfer printing is an emerging technology for the mechanical stacking of III-V multijunction solar cells, yielding ultra high efficiency devices under concentration while avoiding existing epitaxial growth limitations. This commercial technique allows for energy harvesting of the entire solar spectrum through combinations of devices grown on GaAs, InP and GaSb substrates. In this paper, individual single junction materials outlined as transfer printing candidates in an InP-based tandem were investigated to optimize a fabrication process and to study electrical device performance. InGaAsP single junction solar cells grown by MOCVD at compositions of $In_{0.9}Ga_{0.1}As_{0.21}P_{0.79}$ (corresponding to a 1.2 eV band gap) and $In_{0.76}Ga_{0.24}As_{0.52}P_{0.48}$ (1.0 eV band gap) were demonstrated. Process optimization, including particular mesa isolation challenges, and device characterization are discussed.**

Index terms— **fabrication, photovoltaic cells, III-V semiconductor materials**

I. INTRODUCTION

Full harvesting of the solar spectrum has been a recent target for obtaining high solar cell conversion efficiencies. Innovative splitting of available photons to III-V multijunction solar cells (MJSCs) can provide the means to device performances greater than 50% efficiency under concentrations over 500X AM1.5D illumination. MJSCs for concentrator photovoltaics (CPV) have been demonstrated through advantageous techniques such as metamorphic growth and wafer bonding to obtain high cell efficiencies under concentration, although these methods can introduce critical growth and processing challenges [1]–[4].

Transfer printing is an alternative approach that allows for heterogeneous integration of devices grown on different substrates, greatly increasing the available combinations of materials, and enabling high-throughput CPV arrays while still taking advantage of substrate recycling [5]–[7]. Furthermore, printing with elastomeric stamps and utilizing thin, transparent adhesives allows for a high yield transfer and provides additional applications for other established technologies [8]. Transfer printing can ultimately realize full energy harvesting of up to 5 and 6 junction devices. A long-term roadmap for transfer printed mechanically stacked architectures on GaAs, InP and GaSb has been presented in detail in reference [9]. Figure 1(a) shows a simple representation of a GaAs solar cell being printed onto an InP solar cell. The bottom device has contacts embedded in a thick lateral conduction layer, which can be patterned using wet or dry etch chemistries to ensure

a planar front surface [10]. A candidate for the bottom cell of a high efficiency transfer printed MJSC based on InP is an InGaAsP/InGaAsP/InGaAs 3J with band gaps of 1.2/1.0/0.74 eV. In this paper, we characterize progress in the experimental development for the fabrication of InGaAsP solar cells lattice matched to InP, with structures related to Figure 1(b). A particular challenge for processing this quaternary is its complex relationship with standard wet etchants used for mesa isolation, especially when used in conjunction with InP cladding layers as shown here.

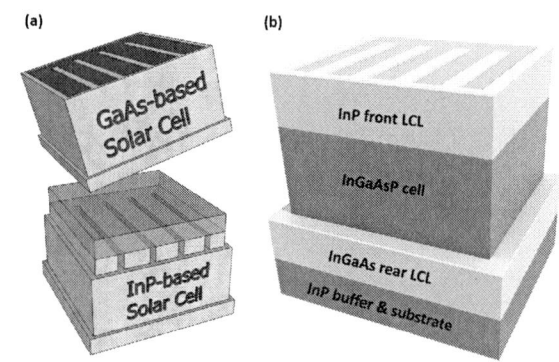

Figure 1. (a) Transfer printing schematic depicting mechanical stacking of a GaAs-on-InP MJSC; (b) Representative device structure for InGaAsP solar cells.

II. EXPERIMENTAL METHODOLOGY

InGaAsP test structures were grown by MOCVD at compositions lattice-matched to InP with band gaps of 1.2 eV ($In_{0.9}Ga_{0.1}As_{0.21}P_{0.79}$) and 1.0 eV ($In_{0.76}Ga_{0.24}As_{0.52}P_{0.48}$). In order to determine suitably balanced mixtures for etching and selectivity, test structures of InP and InGaAsP were used to evaluate a variety of etch chemistries. $HCl:H_3PO_4$ mixed at a ratio of 1:3 (referred to as etchant A), a standard etchant for InP that minimizes lateral undercut, was used to determine selectivity between InP and the quaternary materials. Two sulfuric acid based etchants, mixed at ratios of 1:1:5 and 3:2:2 $H_2SO_4:H_2O_2:H_2O$ (etchants B and C, respectively), were used both to establish etch rates for the quaternary materials and to determine relative selectivity to InP. Also, attention was paid to selectivity of these etchants to photoresist, as

Table I

ETCH RATES FOR InP AND InGaAsP COMPOUNDS IN VARIOUS WET ETCHANTS, GIVEN IN NM/MIN. RELEVANT SELECTIVITY BETWEEN EPI LAYERS IS ALSO INFERRED FROM THE TABLE.

Recipe	Mixture	InP	InGaAsP - 1.0 eV	InGaAsP - 1.2 eV
A	$HCl:H_3PO_4$ [1:3]	1000	Non-appreciable	Appreciable
B	$H_2SO_4:H_2O_2:H_2O$ [1:1:5]	Non-appreciable	108	18
C	$H_2SO_4:H_2O_2:H_2O$ [3:3:2]	Appreciable	N/A	69

higher concentrations of H_2SO_4 can increase the etch rate to ideal values but lead to aggravation of an organic etch mask. Alternatively, hard masks like SiN_x can be used, which don't etch in these chemistries, but require additional process steps and time to etch and remove. The established etchants were then used in the fabrication of 1.0 eV and 1.2 eV InGaAsP solar cells with structures represented in Figure 1(b). Annealed Ti/Pt/Au contacts were deposited via e-beam evaporation in a dual front-side contact pattern and no anti-reflection coatings were applied.

III. RESULTS AND DISCUSSION

A. Etching Study

Measured etch rates for the study described above are summarized in Table I. An appreciable etch indicates an etch that, while small but not insignificant, was not fast enough to determine a reliable rate. A non-appreciable etch had no measureable effect on a particular material after a minimum of 10 minutes. Overall, the tests identified a good selectivity between InP and 1.0 eV InGaAsP, but conversely implied a low selectivity between InP and 1.2 eV InGaAsP. It was also noted that the solutions for recipes B and C are volatile, leading to a decreased etch rate over a few hours and creating problems obtaining a consistent etch rate. This can cause complications in situations where several samples are etched successively over a long period of time. This issue was eliminated during these tests by (1) waiting 30 minutes after adding the H_2SO_4 to H_2O for the highly exothermic mixture to stabalize in temperature, and (2) adding H_2O_2 and then immediately etching all test pieces simultaneously in the same bath. Alternatively, the etchant could be held at a constant elevated temperature to investigate whether these etch chemistries are reaction limited or otherwise affected by chemical decomposition.

The rates shown in Table I provide a good basis for wet etching of InGaAsP solar cells. An etch rate for 1.0 eV InGaAsP of 108 nm/min (etchant B) is suitable for typical solar cell structures of a few μm thickness. However, this etchant performed much slower in 1.2 eV InGaAsP at a rate of only 18 nm/min. Etchant C gave a higher etch rate of 69 nm/min, which could lower the etching time considerably. However, due to low selectivity between InP and 1.2 eV InGaAsP care must be taken when timing the etches of this material. A combination of both etchants B and C is suggested to competently etch thicker 1.2 eV InGaAsP structures.

B. InGaAsP single junction solar cells

Figure 2 compares experimental quantum efficiency of the 1.0 eV and 1.2 eV InGaAsP solar cells. These spectra were taken on a gridless QE test structure on the same sample as the one used for I-V measurements. An important note is that these cells are not isotype structures that fully account for the correct optical filtering that these junctions would observe in a multijunction stack. Thus, they receive a larger amount of high energy photons, leading to augmented integrated short circuit current values for the single junction InGaAsP solar cells. Corrected optical filtering would require a thick InP lateral conduction layer on top of the 1.2 eV InGaAsP solar cell to simulate the low sheet resistance contact layer necessary for transfer printing, and an additional 1.2 eV junction on top of the 1.0 eV solar cell to simulate the InP-based tandem. The external quantum efficiency is provided with no antireflection coatings. It should also be noted that the short-wavelength response observed in Figure 2 is not a failure of high surface recombination velocity, but is instead due to cell design preventing the removal of the contact layer. Furthermore, in a mechanically stacked configuration, not all available photons will be transmitted to the lower cell. For instance, a separate tandem grown on GaAs, when stacked on an InP LCL, will absorb the higher energy photons.

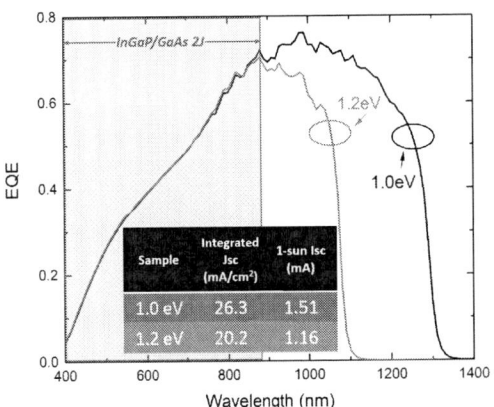

Figure 2. Measured quantum efficiency for single junction InGaAsP solar cells. The embedded table summarizes integrated J_{sc} values obtained from EQE spectra as well as derived one-sun I_{sc} values used to approximate concentration factors.

Convolution of the EQE data with the AM1.5D spectra

gives a quantifiable value for the short circuit current density. These values are tabulated in the inset of Figure 2. The 1.2 eV InGaAsP cell has an integrated J_{sc} of 20.2 mA/cm^2, while the 1.0 eV InGaAsP cell has an integrated J_{sc} of 26.3 mA/cm^2. Although these values will be substantially larger than those from comparable isotype devices because they collect more high-energy photons, the number was used to calculate a one-sun short circuit current value used for calculating concentration factors in I-V measurements. The one-sun I_{sc} values were calculated using appropriate area and shading figures, resulting in 1.16 mA for the 1.2 eV InGaAsP cell and 1.51 mA for the 1.0 eV InGaAsP cell.

Figure 3 shows the illuminated current-voltage characteristics of the InGaAsP concentrator cell measured by changing the height of a singlet lens to defocus light from a single source Xenon lamp calibrated with a Si reference cell. Increasing concentration, as noted in the figure, increases both the open circuit voltage and short circuit current of both the 1.0 eV InGaAsP (Figure 3a) and 1.2 eV InGaAsP (Figure 3b) solar cells. V_{oc} is plotted as a function of concentration factor in Figure 3(c). This factor, denoted concentration ratio (CR), is measured in suns and calculated using the integrated J_{sc} values determined from EQE measurements. A useful metric in comparing the quality of solar cells is band gap voltage offset, W_{oc}, which is the difference between the band gap and the V_{oc}. At the highest measured CR values of 61 suns, the V_{oc} was 704 mV for the 1.0 eV cell and 868 mV for the 1.2 eV cell. This corresponds to W_{oc} values of 296 mV and 332 mV, respectively, indicative of high material quality.

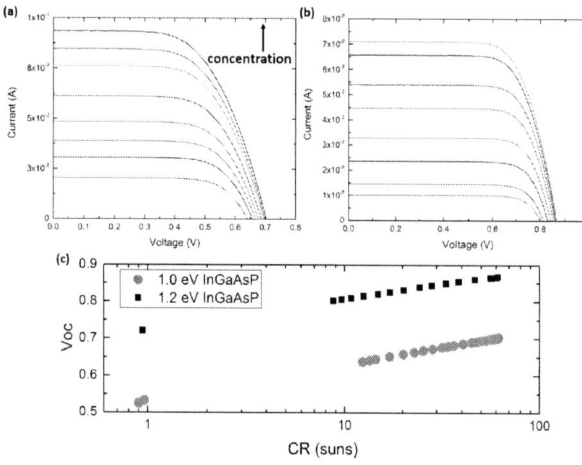

Figure 3. Light I-V characteristics under increasing concentration for single junction InGaAsP solar cells at (a) 1.0 eV and (b) 1.2 eV band gap; (c) Open-circuit voltage as a function of increasing intensity and concentration ratio.

IV. CONCLUSION AND FUTURE WORK

InGaAsP single junction solar cells at compositions of $In_{0.9}Ga_{0.1}As_{0.21}P_{0.79}$ (corresponding to a 1.2 eV band gap)

and $In_{0.76}Ga_{0.24}As_{0.52}P_{0.48}$ (1.0 eV band gap) were grown by MOCVD. Mesa isolation strategies were investigated using HCl:H$_3$PO$_4$ [1:3] and H$_2$SO$_4$:H$_2$O$_2$:H$_2$O [1:1:5] and [3:2:2]. Etch rates identified from the wet etch tests were used to fabricate solar cells from these materials. Quantum efficiency and light I-V measurements were used to determine V_{oc} of the solar cells as a function of increasing concentration, highlighting W_{oc} values of 296 mV and 332 mV for the 1.0 eV and 1.2 eV cells, respectively. These metrics are important to further classify these materials as favorable subjunctions for a fully transfer-printed MJSC for CPV.

REFERENCES

[1] S. Kurtz and J. Geisz, "Multijunction solar cells for conversion of concentrated sunlight to electricity," *Optics Express*, vol. 18, no. S1, pp. A73–A78, 2010.

[2] M. Stan, D. Aiken, B. Cho, A. Cornfeld, V. Ley, P. Patel, P. Sharps, and T. Varghese, "High-efficiency quadruple junction solar cells using OMVPE with inverted metamorphic device structures," *Journal of Crystal Growth*, vol. 312, no. 8, pp. 1370–1374, 2010.

[3] D. C. Law, R. R. King, H. Yoon, M. J. Archer, A. Boca, C. M. Fetzer, S. Mesropian, M. Isshiki, M. Haddad, K. M. Edmondson, D. Bhusari, J. Yen, R. A. Sherif, H. A. Atwater, and N. H. Karam, "Future technology pathways of terrestrial IIIV multijunction solar cells for concentrator photovoltaic systems," *Solar Energy Materials and Solar Cells*, vol. 94, no. 8, pp. 1314–1318, 2010.

[4] F. Dimroth, M. Grave, P. Beutel, U. Fiedeler, C. Karcher, T. N. D. Tibbits, E. Oliva, G. Siefer, M. Schachtner, A. Wekkeli, A. W. Bett, R. Krause, M. Piccin, N. Blanc, C. Drazek, E. Guiot, B. Ghyselen, T. Salvetat, A. Tauzin, T. Signamarcheix, A. Dobrich, T. Hannappel, and K. Schwarzburg, "Wafer bonded four-junction GaInP/GaAs//GaInAsP/GaInAs concentrator solar cells with 44.7% efficiency," *Progress in Photovoltaics: Research and Applications*, vol. 22, no. 3, pp. 277–282, 2014.

[5] M. P. Lumb, M. Meitl, B. Fisher, S. Burroughs, K. J. Schmieder, M. Gonzalez, M. K. Yakes, S. Mack, R. Hoheisel, M. F. Bennett, C. W. Ebert, D. V. Forbes, C. G. Bailey, and R. J. Walters, "Transfer-printing for the next generation of multi-junction solar cells," in *Photovoltaic Specialist Conference (PVSC), 2015 IEEE 42nd*, Jun. 2015, pp. 1–6.

[6] S. Burroughs, R. Conner, B. Furman, E. Menard, A. Gray, M. Meitl, S. Bonafede, D. Kneeburg, K. Ghosal, R. Bukovnik, W. Wagner, S. Seel, and M. Sullivan, "A New Approach For A Low Cost CPV Module Design Utilizing MicroTransfer Printing Technology," vol. 1277, 2010, pp. 163–166.

[7] X. Sheng, C. A. Bower, S. Bonafede, J. W. Wilson, B. Fisher, M. Meitl, H. Yuen, S. Wang, L. Shen, A. R. Banks, C. J. Corcoran, R. G. Nuzzo, S. Burroughs, and J. A. Rogers, "Printing-based assembly of quadruple-junction four-terminal microscale solar cells and their use in high-efficiency modules," *Nature Materials*, vol. 13, no. 6, pp. 593–598, 2014.

[8] M. A. Meitl, Z.-T. Zhu, V. Kumar, K. J. Lee, X. Feng, Y. Y. Huang, I. Adesida, R. G. Nuzzo, and J. A. Rogers, "Transfer printing by kinetic control of adhesion to an elastomeric stamp," *Nature Materials*, vol. 5, no. 1, pp. 33–38, Jan. 2006.

[9] M. P. Lumb, K. J. Schmieder, M. Gonzlez, S. Mack, M. K. Yakes, M. Meitl, S. Burroughs, C. Ebert, M. F. Bennett, D. V. Forbes, X. Sheng, J. A. Rogers, and R. J. Walters, "Realizing the next generation of CPV cells using transfer printing," vol. 1679, 2015, p. 040007.

[10] M. F. Bennett, M. P. Lumb, M. Gonzlez, K. J. Schmieder, S. Mack, J. A. Nolde, and R. J. Walters, "Development of recessed contacts for mechanical stacking of GaSb solar cells," 2016, pp. 2323–2326.

Detailed Characterization for TCAD Simulations of GaAs$_{0.76}$P$_{0.24}$/Si$_{1-y}$Ge$_y$/Si Single Junction Solar Cells

Sabina Abdul Hadi[a], Timothy Milakovich[b], Eugene A. Fitzgerald[b] and Ammar Nayfeh[a]

[a] Department of Electrical Engineering and Computer Science (EECS), Khalifa University of Science and Technology, Masdar Institute, Masdar City PO Box. 54224, Abu Dhabi, United Arab Emirates

[b] Department of Materials Science and Engineering, MIT, Cambridge, MA 02139, USA

Abstract — In this work a TCAD simulation model for GaAs$_{0.76}$P$_{0.24}$ single junction solar cells grown on Si substrate via SiGe buffer layers is presented, along with experimental characterization of the cells. Simulations and experimental results are used to identify recombination factors present in the cells. GaAsP lifetime is estimated to be between 1-10 ns, depending on the window layer lifetime and interface recombination. Results show very good fit for current-voltage characteristics, while spectral response of simulated cells shows that more specific recombination definitions and accurate optical data are required to further improve the accuracy of simulation estimated recombination factors.

Keywords: III-V, solar cell, simulations, TCAD, recombination

I. INTRODUCTION AND BACKGROUND

III-V solar cells hold the record for the highest reported efficiencies for single and multi-junction solar cells [1], due to the direct bandgap (E_g) and high absorption coefficients. However, the growth of III-V materials is costly so use of these cells is limited mostly to space and niche terrestrial applications. In order to bring these cells to a wider pool of consumers, researchers are manufacturing III-V cells on inexpensive substrates, such as Si. Due to lattice mismatch between III-V materials and Si, graded buffer layers (such as Si$_{1-x}$Ge$_x$ or GaAs$_{1-x}$P$_x$) are used to minimize the threading dislocation defects that can severely decrease the performance of the solar cell [2,3]. Understanding the recombination parameters in these cells for the purpose of accurate device modeling can decrease the costs of III-V solar cells and aid innovation process.

In this paper we analyze by experiment and simulations the single junction (SJ) GaAs$_{76}$P$_{24}$ solar cell, grown on Si substrate via graded Si$_{1-x}$Ge$_x$ buffer. The cross-section of the analyzed cell is shown in Fig. 1. Fabrication details and performance parameters for the best of these cells are previously presented in [4], where state-of-the art bandgap-voltage offset, W$_{oc}$ ~0.48 eV and corresponding threading dislocation density, TDD~ $2 \pm 0.5 \times 10^6$ cm^{-2} were reported. Such level of TDD is expected to result in minority carrier lifetime of ~7 ns [3]. However, some of the cells fabricated using the same stack and growth method as in [4] can exhibit open-circuit voltage (V$_{oc}$) ranging between ~1 V to 1.23 V, suggesting that recombination mechanisms can vary from cell to cell. Analysis of current-voltage data along with quantum efficiency (QE) and reflectance (R), can help identify non-reflection related losses. The loss factors could be an insufficient absorber thickness, short diffusion length (carrier lifetime), interface and surface recombination, non-ohmic contacts or other parasitic optical losses (i.e. in a window layer).

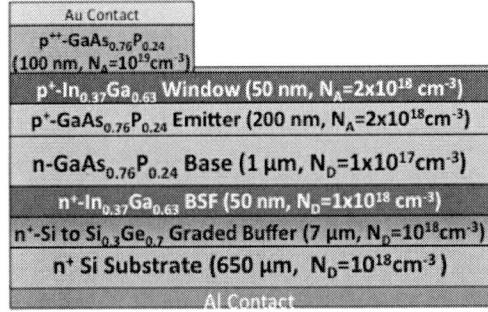

Fig. 1. Cross-section of simulated SJ GaAs$_{76}$P$_{24}$ cell on SiGe/Si substrate, as in [4], with 8.7% contact shading. Figure not to scale.

In this work, the simulation model using Sentaurus TCAD by Synopsys ® is developed in order to observe the effects of different parameters, such as the window or absorber carrier lifetime and surface recombination at different interfaces. Experimentally extracted cell parameters are used to fine-tune simulation model and to quantify carrier lifetime or surface recombination velocity in the device, as well as to identify the reasons for simulation shortcomings.

II. SIMULATIONS

A. Model

The cross-section of the simulated SJ GaAs$_{76}$P$_{24}$ solar cell is shown in Fig. 1, along with the thickness and carrier concentrations, based on the cell structure reported in [4]. Physics based TCAD model is used for electrical and optical simulation of the device. For optical simulations, transfer matrix method (TMM) is used with AM 1.5G standard spectra [5]. Activated physics models include *Shockley Read Hall* (SRH) recombination (represented by effective lifetime, τ) and *Auger* and *Radiative* recombination models using default parameters. Furthermore, thermionic emission, heterointerface carrier transport models and Fermi-Dirac statistics are used. Minority carrier lifetime, τ, of Si substrate and SiGe layers are fixed at 10^{-5} and 10^{-7} s, respectively [6,7], while τ of GaAsP

978-1-5090-5606-4/17 $31.00 © 2017 IEEE

base, emitter and InGaP material are treated as variables. Additionally, surface recombination velocity, S (cm/s), is considered at all interfaces and is treated as a variable.

Bandgap of $GaAs_{0.76}P_{0.24}$ material is selected to be 1.71 eV [4], a value that is extracted from measured internal quantum efficiency (IQE) using the method based on an Urbach's rule [8]. However, measured optical properties of $GaAs_{0.76}P_{0.24}$ alloy revealed that E_g is lower than 1.71 eV observed by IQE measurements. The reason for this may be the difference in the growth between the stack used for optical parameters extraction (GaAsP layers without impurity doping and without InGaP) and the stack used for the cell fabrication. Hence, for simulation purposes, extinction coefficient, k, was adjusted such that E_g equals 1.71 eV, while using Urbach's tail extracted from measured data. InGaP optical properties are utilized from [9] and E_{g_InGaP}=2 eV. $Si_{1-y}Ge_y$ graded buffer layer is modeled as a stack of 1 μm layers with Ge fraction increasing from 0 to 70% (10% / μm). Optical properties and E_g for SiGe alloys are used from [10] and [11].

Band diagram for simulated GaAsP cell in equilibrium (structure in Fig.1) is shown in Fig.2. Space charge region (SCR) is around 120 nm at 0V and dark, while built-in potential is ~ 1.61 eV. These values are comparable to experimentally extracted SCR width and built-in potential, presented later in this text.

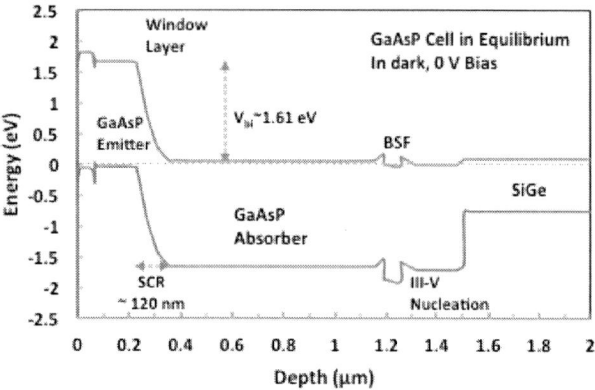

Fig. 2. Band-diagram of simulated SJ $GaAs_{76}P_{24}$ cell grown on SiGe/Si substrate, in equilibrium conditions. SCR is calculated to be ~ 120 nm, while built-in potential is ~ 1.61 eV.

Performance parameters of the best and the worst performing cells (stack shown in Fig. 1) that are used for simulation data fitting are listed in Table 1.

TABLE I PERFORMANCE PARAMETERS OF GAAS $_{76}$P$_{24}$ SJ CELLS

Cell	J_{sc} (mA/cm²)	V_{oc} (V)	Fill Factor (FF)	Efficiency (%)	Area (cm²)	Shading (%)
Upper Limit [4]	11.00	1.22	0.829	11.20	0.01	8.7
Lower limit	11.97	1.04	0.754	9.94	0.16	7.4

B. Simulation Results

In order to estimate the recombination mechanisms in analyzed GaAsP cell, these recombination factors are treated

as variables. Fig.3 shows simulated J_{sc} (left axis) and V_{oc} (right axis) as a function of effective $GaAs_{76}P_{24}$ minority carrier lifetime, τ_{GaAsP}. For these results, base and emitter lifetime are set to be equal, while surface recombination, S, at all interfaces is set to 0 cm/s. Simulation results are also shown for three different InGaP minority carrier lifetime values, τ_{InGaP}, while dashed lines indicate the range of experimentally observed J_{sc} and V_{oc}. Since V_{oc} is sensitive to recombination, we can estimate that effective GaAsP lifetime ranges from ~0.1 ns to ~50 ns, which also depends on the τ_{InGaP} value. On the other hand, simulations show that J_{sc} is more sensitive to InGaP lifetime, which narrows $GaAs_{76}P_{24}$ effective lifetime between 0.5-50 ns.

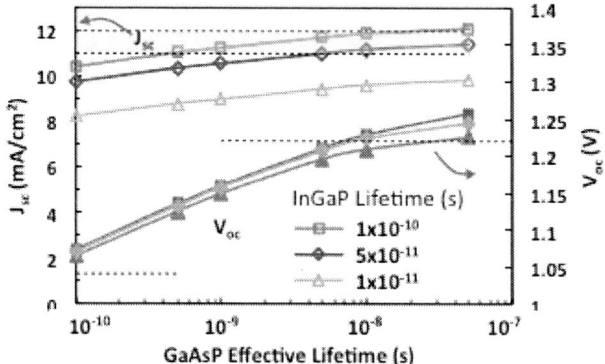

Fig. 3. Simulated J_{sc} (left) and V_{oc} (right) vs. effective GaAsP minority carrier lifetime, shown for different InGaP lifetime values. Also shown are the ranges for experimental J_{sc} and V_{oc}.

The results shown in Fig. 3 also imply that the window layer can contribute to optically generated current or it can cause parasitic optical losses if τ_{InGaP} is too low. This can also be seen from external quantum efficiency (EQE) characteristics for different InGaP lifetime values, as shown in Fig. 4, where GaAsP absorber and emitter lifetimes are set to 1 ns and 0.5 ns, respectively.

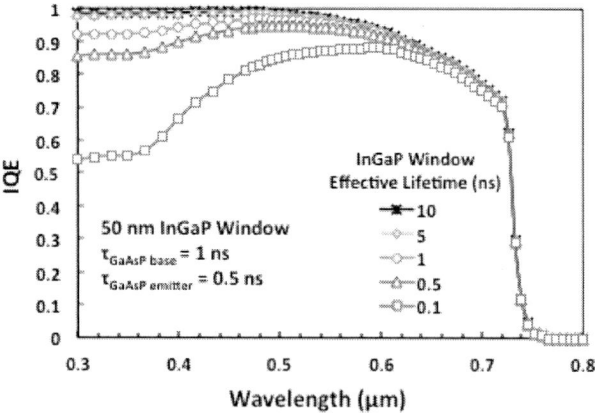

Fig. 4. Simulated EQE for variable τ_{InGaP}= 0.1-10 ns, and GaAsP absorber lifetime =1ns, emitter lifetime = 0.5ns, and all S=0 cm/s.

Simulation results shown in Fig. 4 show that InGaP window layer can contribute in photogenerated current as high as 40% in wavelength range between 300-400 nm, explaining the strong effect InGaP lifetime on Jsc shown in Fig. 3. These results show importance of having optimum thickness and high quality material for window layers.

In order to correctly estimate recombination factors, the analysis of QE data is needed when varying all the parameters. Fig. 5 shows simulated EQE, IQE and R data for a structure with fixed τ_{InGaP}=0.1 ns and τ_{GaAsP} varied from 0.1 to 10 ns. Also shown in Fig. 5 is experimentally measured data (curves with symbols [4]). The plot shows that τ_{GaAsP} affects the collection at the long wavelengths, as expected, while short wavelength response is limited by the collection in InGaP window layer, for given τ_{InGaP}. For the wavelength range between 400-600 nm there is a good agreement between experimental and simulated values for EQE data. Furthermore, there is a slight offset between the simulated and measured reflectance, R, which results in a larger mismatch between the simulated and measured IQE. The mismatch between the experiment and simulation can be due to the difference between the actual and simulated optical parameters and the layer thicknesses.

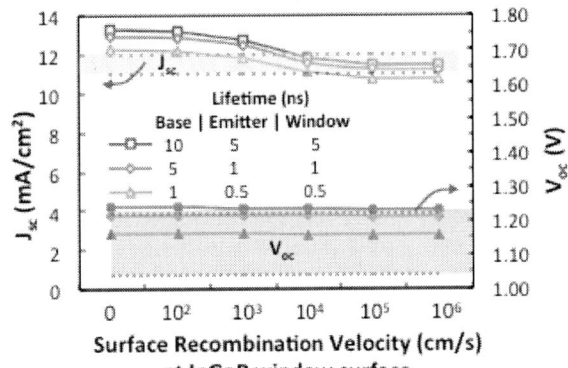

Fig. 6. J_{sc} (left axis) and V_{oc} (right axis) vs. surface recombination velocity S (cm/s) at the window layer interface, for base τ_{GaAsP} =1-10 ns. Also shown is the range of experimental J_{sc} and V_{oc}.

III. EXPERIMENTAL ANALYSIS

A fabricated GaAs$_{76}$P$_{24}$ solar cell on Si [4] (4 mm × 4 mm cell) is characterized using the Sol3A Class AAA solar simulator under Standard Test Conditions (25°C, 1 sun at AM1.5G). Furthermore, the probe stage temperature is varied from 10-40°C, for both dark and light measurements. Capacitance-voltage (CV) measurements are taken to estimate space charge region width (W$_{SCR}$), base region doping and built-in potential in the juntion.

A. Experimental Results

CV measurements at 100 kHz are used to estimate the width of space charge region (W$_{SCR}$) using relation $W_{SCR}(V) = (\varepsilon_0 \, \varepsilon_r \, A)/C(V)$, where ε_0 is vacuum permittivity, A is the area of the cell and GaAs$_{76}$P$_{24}$ electric permittivity is assumed to be equal to that of GaAs (ε_r=12.9). Further, using the abrupt junction (p$^+$-n) approximation, the carrier concentration in the base region, N$_D$, is estimated from the slope of the linear region of the $1/C^2(V)$ curve. Fig.7 shows W$_{SCR}$ (a) and $1/C^2$ as a function of voltage bias for GaAs$_{76}$P$_{24}$ cell.

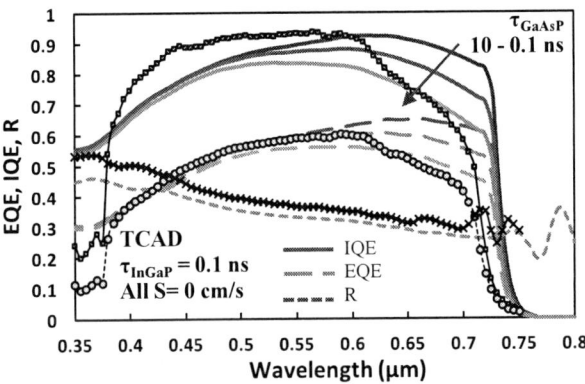

Fig. 5. Simulated EQE, IQE and R for τ_{GaAsP} = 0.1 to 10 ns, τ_{InGaP}=0.1 ns and all S=0 cm/s. Also shown are experimentally measured data reproduced from [4] (curves with symbols).

The results shown in Fig.5 do not account for lifetime doping dependence or for the interface surface recombination. Separating these factors from effective lifetime helps better quantify bulk τ_{GaAsP}. Fig.6 shows J_{sc} and V_{oc} vs. S at the InGaP window interface, for structures with base τ_{GaAsP}=1-10 ns and window and emitter lifetime set to 0.5-5 ns. Results show that J_{sc} is more susceptible to surface recombination velocity at window interface, while V_{oc} is affected more by the base τ_{GaAsP}. Comparing simulated results to the range of experimental values, τ_{GaAsP} is estimated to be 5 ns, while S at the top interface can be between 10^4-10^6 cm/s.

To better differentiate between diffusion and interface based recombination, or other loss factors, and to better understand simulation uncertainty, a feedback loop between simulation model and experimental observations is required.

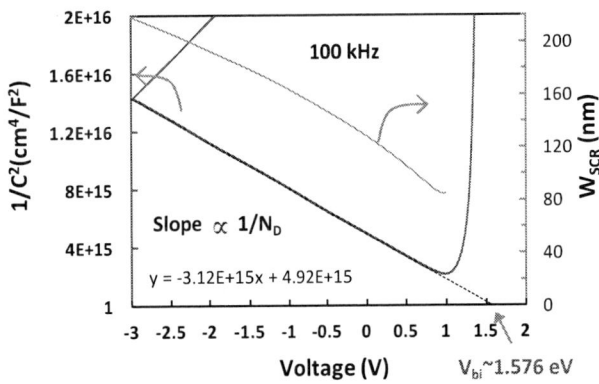

Fig. 7. Measured $1/C^2$ (left axis) and extracted space charge region W$_{SCR}$ (right axis) vs. voltage bias for GaAs$_{76}$P$_{24}$ cell.

The built-in voltage is estimated to be ~1.576 eV, an upper limit for V_{oc} of this particular cell design. Built-in potential of simulated cell is ~1.61 eV, which is close to experimentally extracted value. The base layer doping N_D is extracted at ~1.37×10^{17} cm^{-3}, close to the expected value (specified in Fig.1). Extracted W_{SCR} increases with reverse bias, ranging from 128 nm at 0V to ~220 nm at -3V reverse bias. Experimentally extracted and simulated SCR at 0V bias are at a close agreement, verifying the simulated design.

The analysis of V_{oc} temperature dependence (shown in Fig. 8 below) for one of the low performing GaAs$_{76}$P$_{24}$ cells showed that linearly interpolated V_{oc} at T=0 K is ~1.321 eV, also called an activation energy [12]. For cells with dominant band-to-band recombination, activation energy is equal to bandgap, while for the cells with dominant interface recombination, this energy would indicate barrier potential [12]. Comparing extrapolated $V_{oc}(0K)$ value to built-in potential, we can estimate that for a measured cell, an interface recombination factor is present. Hence, recombination at the interfaces, including junction interface should be taken into account in the simulation model.

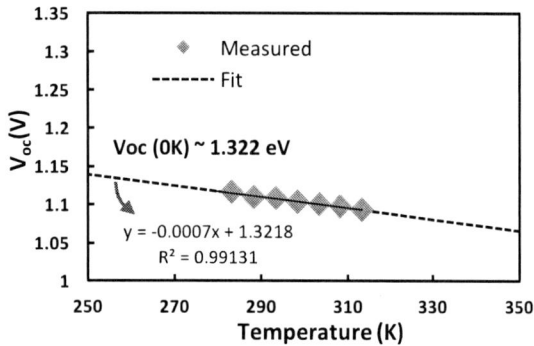

Fig. 8. Measured V_{oc}(V) under 1 sun as a function of ambient temperature for GaAs$_{76}$P$_{24}$ cell.

B. Refining Simulation Model

Simulation parameters are updated based on the experimentally extracted values for the base doping and layer thicknesses estimated from device TEM image [4]. The lifetime and surface recombination S values are varied within previously anticipated range, considering all the interfaces with III-V materials. Fig. 9 compares experimental [4] and updated simulation results for JV (a) and QE (b) data, along with the set of updated parameters. The absorber GaAsP layer minority carrier lifetime, τ_{GaAsP}, is estimated to be 10 ns, which is in agreement with lifetime expected from reported TDDs value [3] [4] [13] [14]. The interface recombination at the emitter/base interface equal to $S_{base}=10^5$ cm/s, based on the experimental extraction of Voc at 0K. The GaAsP emitter and InGaP lifetime are both set to 5 ns, while surface recombination at the interface between emitter and window is modeled as 100 cm/s. Note that simulated EQE is larger than experimental EQE across most wavelengths, yet simulated and experimental J_{sc} match quite well. Closer inspection showed

8.6% difference between J_{sc} extracted from simulated EQE and JV curves. This is due to the reflectance mismatch and contact shading (8.74 %) which is not accounted for in simulated EQE. This also shows that further model improvements could be implemented using more accurate optical data and specific definition of recombination mechanisms at the interface.

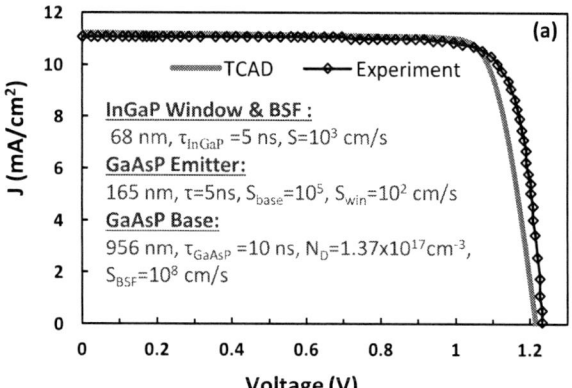

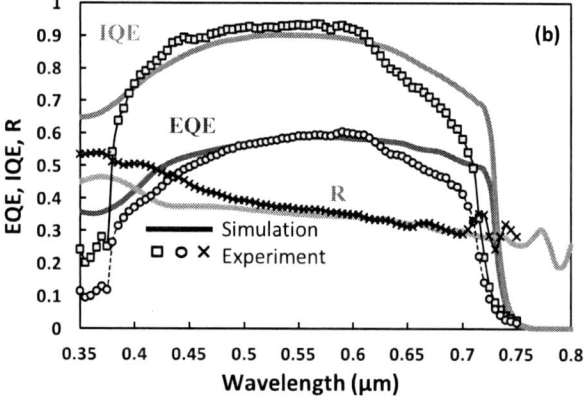

Fig. 9. Comparison of updated simulations and experimental results (reproduced from [4]) for (a) JV and (b) EQE, IQE and R data. Also shown are updated simulation parameters.

IV. CONCLUSION

In this paper a TCAD simulation model for GaAs$_{76}$P$_{24}$ single junction solar cells grown on Si substrate is presented. Simulations results compared to detailed experimental characterization help identify and estimate recombination factors in the cell. Using this model and results, the lifetime of GaAsP base layer is estimated to be equal to ~10 ns, which is in agreement with reported TDDs for analyzed cells. InGaP lifetime is estimated to be 5 ns and simulations imply that window layer can contribute to photogenerated current, given sufficiently high carrier lifetime. Refined TCAD model presented here provides a good current-voltage characteristics fit, but modeling more accurate spectral response requires further and more detailed analysis of the recombination physic while also using accurate material optical properties.

REFERENCES

[1] M. A. Green, et. al, "Solar cell efficiency tables (version 47)," *Prog. Photovolt: Res. Appl.* vol. 24, pp. 3-11, 2016.

[2] E. Fitzgerald et.al, "Totally relaxed Ge_xSi_{1-x} layers with low treading dislocation densities grown on Si substrates," *Applied Physics Letters*, vol. 59, no. 7, p. 811, 1991.

[3] K. Hayashi, T. Soga, H. Nishikawa and T. J. a. M. Umeno, "MOCVD growth of GaAsP on Si for tandem solar cell applications," 1st World Conf. on PV Energy Conversion, 1994.

[4] T. Milakovich et.al, "Growth and characterization of GaAsP top cells for high efficiency III-V/Si tandem PV", 42nd IEEE PVSC, p.1-4, 2015.

[5] National Renewable Energy Laboratory (NREL), "Renewable Resource Data Center," [Online]. Available: http://rredc.nrel.gov/solar/spectra/am1.5/. [Accessed 26.02.2012].

[6] T. Ghani, J. L. Hoyt, D. B. Noble, J. F. Gibbons, J. E. Turner, and T. I. Kamins, "Effect of Oxygen on Minority Carrier Lifetime and Recombination Currents in $Si_{1-x}Ge_x$ Heterostructure Devices," *Applied Physics Letters* 58(12), 1317–1319 (1991).

[7] S. Abdul Hadi, P. Hashemi, N. DiLello, E. Polyzoeva, A. Nayfeh and J. L. Hoyt, "Effect of germanium fraction on the effective minority carrier lifetime in thin film amorphous-Si/crystalline-Si1-xGex/crystalline-Si heterojunction solar cells," *AIP Advances*, vol. 3, no. 5, pp. 052119-052119-6, 2013.

[8] H. Helmers, C. Karcher and A. W. Bett, "Bandgap determination based on electrical quantum efficiency," *Applied Physics Letters*, vol. 103, 2013.

[9] B. Conrad et.al, "Optical characterisation of III-V alloys grown on Si by spectroscopic ellipsometry", *Solar Energy Materials and Solar Cells*, vol. 162, 7–12, 2017.

[10] E. Kasper, Properties of Strained and Relaxed SiGe, London: Institution of Electrical Engineers, 1995.

[11] R. Braunstein A. R. Moore, and F. Herman, "Intrinsic Optical Absorption of Germanium-Silicon Alloys," *Physical Review*,109, 3, pp. 695-710, 1958.

[12] S. Grover, J. V. Li, D. L. Young, P. Stradins, and H. M. Branz, "Reformulation of solar cell physics to facilitate experimental separation of recombination pathways", *Applied Physics Letters*, 103, 093502, 2013.

[13] M. Yamaguchi and C. Amano, "Efficiency calculations of thin-film GaAs solar cells on Si substrates," *Journal of Applied Physics* , vol. 58, no. 9, pp. 3601-3606, 1985.

[14] C. L. Andre, J. J. Boeckl, D. M. Wilt, A. J. Pitera, M. L. Lee, E. A. Fitzgerald, B. M. Keyes and S. A. Ringel, "Impact of dislocations on minority carrier electron and hole lifetimes in GaAs grown on metamorphic SiGe substrates," *Applied Physics Letters*, vol. 84, no. 18, 2004.

978-1-5090-5606-4/17 $31.00 © 2017 IEEE

Comparative Study of >2 eV Lattice-Matched and Metamorphic (Al)GaInP Materials and Solar Cells Grown by MOCVD

Daniel J. Chmielewski, Christine Jackson, Jacob Boyer, Daniel Lepkowski, John A. Carlin, Aaron R. Arehart, Tyler J. Grassman, and Steven A. Ringel

The Ohio State University, Columbus, Ohio, 43210, United States

Abstract — **This work investigates >2 eV (Al)GaInP alloys grown via MOCVD for application as the top junction of IMM solar cells. We explore balancing Al content and lattice constant as two complementary variables to achieve target bandgaps in one material system. Prototype solar cell performance and defect spectroscopy (DLTS/DLOS) are used to evaluate the various alloys. Results thus far suggest that metamorphic GaInP is the most promising route for a high-performance top cell up until metamorphic GaInP becomes indirect.**

I. Introduction

Inverted metamorphic multijunction (IMM) solar cells have reached AM0 efficiencies above 34% [1], and by increasing the number of junctions, even higher efficiencies are predicted due to further idealized partitioning of the solar spectrum. These IMMs may include 6+ junctions, and such designs require top junctions with bandgaps of 2.05-2.4 eV. Even so, the 4J IMM can improve substantially by improving material properties of the top 2.05 eV junction [2]-[3].

We are engaged in a program to explore and compare two strategies that enable these bandgap targets. Figure 1 outlines the general approach, which involves optimizing Al content and misfit within the (Al)GaInP system such that maximum material quality over a range of bandgaps between 2.05-2.4 eV can be identified. Two broad studies are currently ongoing; the first at the 2.05 eV bandgap target and the second at the 2.2 eV bandgap target. The 2.05 eV study compares lattice-matched (LM) $(Al_{0.25}Ga_{0.75})_{0.51}In_{0.49}P$ to metamorphic (MM) $Ga_{0.63}In_{0.37}P$ to identify the impacts of Al content vs. misfit, respectively, where lattice-matching is defined with respect to GaAs. The 2.2 eV study explores these two approaches with adjusted compositions for the necessary bandgap—LM $(Al_{0.40}Ga_{0.60})_{0.52}In_{0.48}P$ and MM $Ga_{0.76}In_{0.24}P$—as well as MM $(Al_{0.12}Ga_{0.88})_{0.64}In_{0.36}P$ to explore the impact of a combination of Al content and misfit. Table I lists all target compositions within these two studies, as well as the nomenclature used to distinguish these compositions in the remainder of this paper.

Both studies build upon previous work at 2.05 eV [4], which enabled the identification of an optimized device structure and doping profile to utilize for this subsequent work for a fair comparison between devices. In addition, previous results led to a primary focus on materials and devices grown by metalorganic chemical vapor deposition (MOCVD) due to the apparent superior material quality compared with molecular beam epitaxy (MBE)-grown material and device structures.

Here, solar cells were fabricated for all target compositions for the 2.05 eV and 2.2 eV studies and characterized for direct comparison. The analysis of cell performance metrics was aided through the use of both deep level transient spectroscopy (DLTS) and deep level optical spectroscopy (DLOS). This information is used to build correlations between specific defect states and cell characteristics as a function of alloy composition.

II. Approach

(Al)GaInP n^+p diodes and prototype solar cell structures were fabricated for all test conditions. Metamorphic solar cells were grown on tensile, step-graded $GaAs_yP_{1-y}$ buffers on GaAs. Electron beam induced current (EBIC) revealed that the GaAsP virtual substrates possessed a threading dislocation density (TDD) of about 1×10^6 cm^{-3}. A growth temperature of 625°C was used for all devices. All diodes and cells were designed with a 1.5 µm p-base and 50 nm n$^+$-emitter doped with Zn and

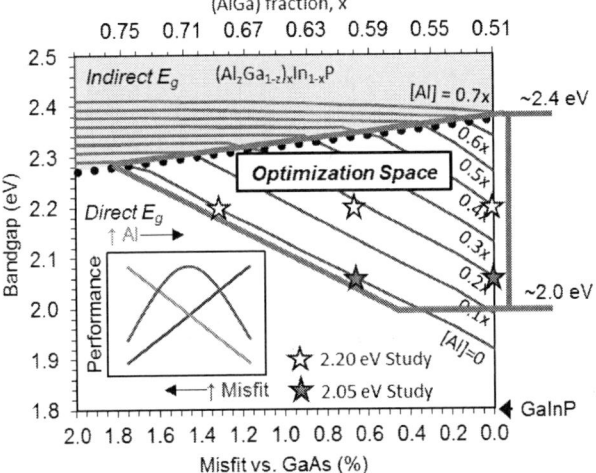

Fig. 1. Bandgap vs. misfit plot of the AlGaInP optimization space. The inset depicts the general approach of optimizing misfit vs. Al content.

TABLE I
TARGET COMPOSITIONS OF 2.05 eV AND 2.20 eV STUDIES

Target E_g (eV)	Composition	Nomenclature
2.05	$Ga_{0.63}In_{0.37}P$	2.05eV MM GaInP
2.05	$(Al_{0.25}Ga_{0.75})_{0.51}In_{0.49}P$	2.05eV LM AlGaInP
2.20	$Ga_{0.76}In_{0.24}P$	2.2eV MM GaInP
2.20	$(Al_{0.40}Ga_{0.60})_{0.52}In_{0.48}P$	2.2eV LM AlGaInP
2.20	$(Al_{0.12}Ga_{0.88})_{0.64}In_{0.36}P$	2.2eV MM AlGaInP

Si, respectively. A 20 nm n⁺-$Al_xIn_{1-x}P$ window layer was grown internally lattice-matched to the respective (Al)GaInP alloys, along with a highly doped GaAs(P) cap for ohmic contact.

Lighted current-voltage (LIV) measurements were performed with a single-zone, Xe lamp solar simulator filtered for AM0. Capacitance-voltage (CV) measurements were used to characterize base doping. External quantum efficiency (EQE) measurements were performed on a custom-built small spot system. Deep levels within the base layer of each type of device were characterized by DLTS and DLOS, yielding quantitative information about each type of detected trap state—trap density (N_T), energy level (E_T), and capture cross section (σ). DLTS, which utilizes thermally-stimulated emission of carriers from bandgap states, can identify states existing within ~1 eV of the majority carrier band edge. DLOS, which is based on optically stimulated emission, allows probing of the remainder of the bandgap. Full details of DLTS and DLOS, and their use to explore the entire range of states in the bandgap, have been reported previously [5].

III. RESULTS AND DISCUSSION

A. 2.05 eV Study

In the pursuit of this effort, detailed materials and device characterization revealed that the original (Gen1) 2.05 eV bandgap MM GaInP and LM AlGaInP cells that were previously reported [4] were not directly comparable to one another due to differences in their doping profiles. Namely, the Gen1 MM GaInP window and emitter were doped well above the design targets of 2×10^{18} cm⁻³ and 5×10^{18} cm⁻³, respectively, while the Gen1 LM AlGaInP base was doped 2.5× greater than the 1×10^{17} cm⁻³ target. While the general trends in the DLTS/DLOS study reported previously are expected to still hold, the Gen1 cells cannot be directly compared due to these differences.

As such, improved doping calibrations were incorporated into a new round of growths. These Gen2 devices were fabricated with matching structures and doping profiles, ensuring the target doping profile was achieved. Figure 2

presents LIV and EQE on the Gen1 and Gen2 solar cells for comparison. A significant improvement in the short wavelength (λ_{short}) EQE response is seen in the Gen2 2.05 eV MM GaInP cell compared to its Gen1 counterpart. Similarly, the reduction in base doping in the Gen2 2.05 eV LM AlGaInP cell resulted in a significantly improved long wavelength (λ_{long}) EQE response compared to its Gen1 counterpart.

The EQE of the Gen2 devices reveals an important optical property related to the $Al_xIn_{1-x}P$ windows of these cells. A key observation is that the peak in the Gen2 2.05 eV LM AlGaInP EQE occurs at 490 nm, while in the Gen2 2.05 eV MM GaInP the peak occurs at 455 nm. These wavelengths correspond to the direct bandgap of the $Al_xIn_{1-x}P$ windows of each of these cells, as confirmed by optical transmission measurements on isolated $Al_xIn_{1-x}P$ epilayers. The ability for the 2.05 eV MM GaInP cell to utilize a wider bandgap $Al_xIn_{1-x}P$ window compared to the 2.05 eV LM AlGaInP cell enables an enhancement in the λ_{short} response, as observed in the comparison of the Gen2 EQE. A comparison of the Gen2 λ_{long} response suggests improved base material quality using 2.05 eV LM AlGaInP, evident by both the improved EQE at wavelengths longer than 490 nm (ignoring the discrepancy in cutoff wavelengths due to slight deviations in alloy compositions) and the reduced slope in the EQE in this region, indicative of a longer base diffusion length.

The overall improvement in the EQE of the Gen2 devices compared to their Gen1 counterparts is also evident in the comparison of the short-circuit current densities (J_{SC}) in the LIV of each device. The higher J_{SC} of the Gen2 2.05 eV MM GaInP compared to the Gen2 2.05 eV LM AlGaInP comes from the improved λ_{short} response due to the wider bandgap window (even after accounting for the additional current due to the slight discrepancy in bandgaps).

Another key observation from the LIV is dependence of the slope in the LIV data between 0 V and the max power point (V_{MPP}). A non-zero slope in this region (assuming no shunt issues) is due to voltage-dependent carrier collection effects that are characteristic of non-ideal base diffusion lengths. Figure 2(a) shows that the MM and LM approaches for the

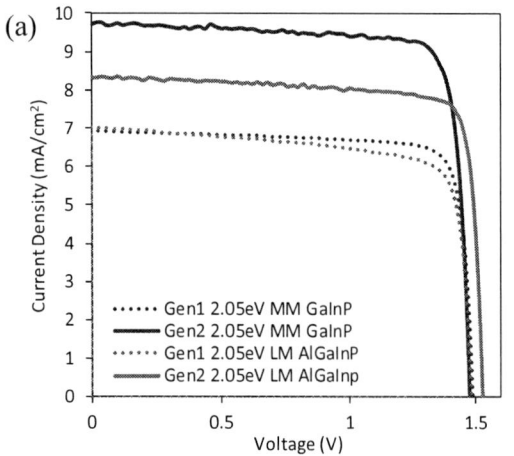

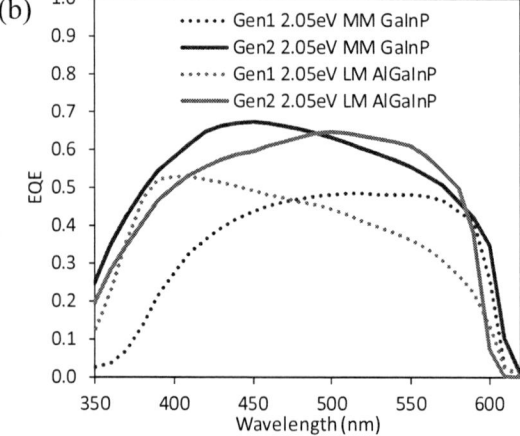

Fig. 2. (a) AM0 LIV and (b) EQE of MOCVD-grown 2.05 eV MM GaInP and LM AlGaInP cells without ARCs.

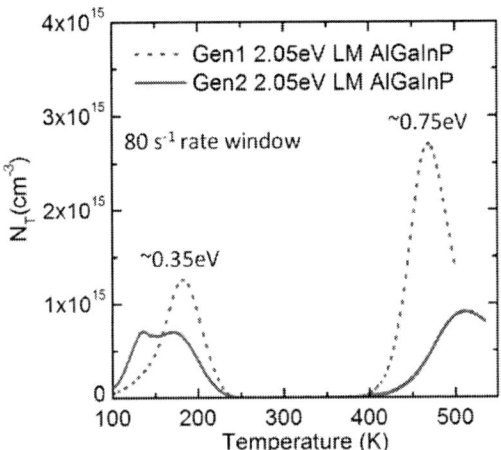

Fig. 3. DLTS on the Gen1 and Gen2 2.05 eV LM AlGaInP devices, revealing trap levels and concentrations present within the Zn-doped p-type base layers of the cells.

Gen2 series have nearly identical slopes, implying similar base diffusion lengths for both approaches to achieve the same bandgap. This is in contrast with the Gen1 series where an improved (reduced) slope for the 2.05 eV MM GaInP device compared with 2.05 eV LM AlGaInP is evident, and this is consistent with the above-mentioned diffusion length reduction presumably due to the higher base doping for the latter. These trends are indeed supported by and consistent with a comparison of λ_{long} EQE responses in Figure 2(b).

Since the cell results imply a significant difference in the Gen1 and Gen2 LM AlGaInP 2.05 eV p-type base material quality, DLTS and DLOS measurements were used to explore potential linkages with high densities of bandgap states. Figure 3 shows the DLTS results obtained for the Gen1 and Gen2 2.05 eV LM AlGaInP cells. An immediate observation is the reduction in concentrations of both the Ev + 0.35 eV and Ev + 0.75 eV traps between Gen1 and Gen2, consistent with the lower base quality for Gen1 2.05 eV LM AlGaInP, and implies a possible connection between these defect states and diffusion length, likely through increased Shockley-Read-Hall (SRH)

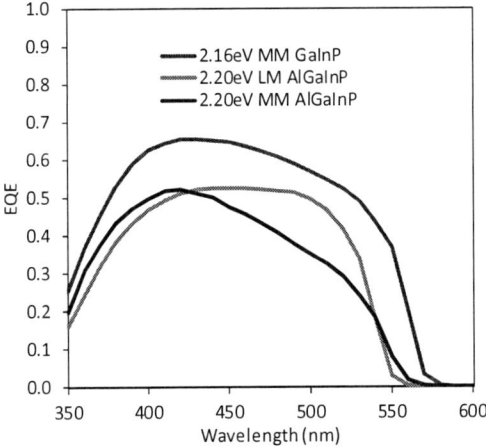

Fig. 4. EQE of various 2.2 eV (Al)GaInP cells without ARCs.

recombination lifetimes. The question as to if and how this may be connected with the Zn base doping under the current growth conditions (e.g. 625°C growth temperature) would require a deeper study to confirm.

B. 2.2 eV Study

The 2.2 eV bandgap study is ongoing in parallel with the 2.05 eV study. Three 2.2 eV bandgap target devices have been fabricated and characterized via LIV and EQE. Figure 4 presents EQE on the preliminary 2.2 eV devices. Thus far, the MM GaInP cell has demonstrated the best performance. However, it has been observed that the bandgap of MM GaInP begins to deviate from the direct GaInP tie-line as the composition approaches that required for 2.2 eV (which is why the MM GaInP is in fact only 2.16 eV, as seen in the EQE). This may be related to either approaching the direct-indirect transition or to sub-lattice ordering effects. In the case of 2.2 eV LM AlGaInP, an overall absolute reduction in EQE is observed, likely due to high Al content. The preliminary 2.2 eV MM AlGaInP cell demonstrates a similar overall reduction as the 2.2 eV LM AlGaInP, in addition to a similar slope in the λ_{long} EQE response compared to the MM GaInP cell. This suggests the base diffusion length is limited by the TDD of the tensile-graded metamorphic buffer. The TDD can be reduced by optimization of the buffer, which should minimize/eliminate this slope in the response. DLTS and DLOS are ongoing for the 2.2 eV devices.

IV. Conclusions

Prototype lattice-matched and metamorphic 2.05 eV and 2.2 eV (Al)GaInP devices have been produced and characterized to assess and compare the impact of misfit and Al content on material quality for the top junction of current and future IMM cells. Improvements to the doping profile have led to significantly better performance in the Gen2 2.05 eV devices compared to their Gen1 counterparts. The Gen2 devices revealed that a major benefit of MM GaInP device is that the $Al_xIn_{1-x}P$ window has a significantly wider bandgap than on the LM AlGaInP device, leading to improvements in the λ_{short} EQE response. DLTS/DLOS has also demonstrated that the base material quality of 2.05 eV LM AlGaInP is highly sensitive to base doping. Parallel studies on 2.2 eV material have revealed challenges due to perhaps the GaInP direct-indirect transition on the tie-line and/or ordering, as well as difficulties with growth of high Al-content AlGaInP. Currently, MM GaInP appears to be a promising choice for future IMM solar cells.

Acknowledgment

This work was funded by the Air Force Research Laboratory, Space Vehicles Directorate (contract# FA9453-14-C-0373).

References

[1] A. B. Cornfeld, D. Aiken, B. Cho, A. V. Ley, P. Sharps, M. Stan, and T. Varghese, "Development of a four sub-cell inverted metamorphic multijunction (IMM) highly efficient AM0 solar cell," in *35th IEEE Photovoltaic*

Specialists Conference, 2010, p. 105.

[2] D. J. Aiken, A. B. Cornfeld, M. A. Stan, and P. A. Sharps, "Consideration of high bandgap subcells for advanced multijunction solar cells," in *4th World Conference on Photovoltaic Energy Conversion*, 2006, p. 838.

[3] P. Patel, D. Aiken, D. Chumney, A. Cornfeld, Y. Lin, C. Mackos, J. McCarty, N. Miller, and P. Sharps, "Initial results of the monolithically grown six-junction inverted metamorphic multi-junction solar cell," in *38th IEEE Photovoltaic Specialist Conference*, 2012, p. 2.

[4] D. J. Chmielewski, K. Galiano, P. Paul, D. Cardwell, S. Carnevale, J. A. Carlin, A. R. Arehart, T. J. Grassman, and S. A. Ringel, "Comparative study of 2.05 eV AlGaInP and metamorphic GaInP materials and solar cells grown by MBE and MOCVD," *Proc. 43rd IEEE Photovoltaic Spec. Conf.*, 2016.

[5] M. González, A. M. Carlin, C. L. Dohrman, E. A. Fitzgerald, and S. A. Ringel, "Determination of bandgap states in p-type $In_{0.49}Ga_{0.51}P$ grown on SiGe/Si and GaAs by deep level optical spectroscopy and deep level transient spectroscopy," *J. Appl. Phys.*, vol. 109, p. 063709, 2011. (and references within)

Performance of GaSb Photovoltaics with Graphene Coating

Benjamin P. Conlon[1], Daniel J. Herrera[1], Shaimaa A. Abdallah[1], Jonathan O. Okafor[2] & Luke F. Lester[1]

[1]Virginia Tech Bradley Department of Electrical and Computer Engineering, Blacksburg, VA, 24060, USA

[2]Norfolk State University, Norfolk, VA, 23504, USA

Abstract — **GaSb cells grown by MBE have high quality material growth, but struggle to overcome high sheet resistance in p-type emitter designs. To overcome this issue, GaSb photovoltaic devices with a p-type emitter were fabricated and coated with a single layer of graphene. Graphene's intrinsic transport properties and low optical transparency make it a suitable candidate for use as a transparent conducting electrode. By adding single layer graphene to the surface of the fabricated GaSb photovoltaics, the series resistance and shunt resistance are improved, with an associated increase in fill factor and efficiency.**

Index Terms – **gallium antimonide, graphene, molecular beam epitaxy, photovoltaic, transparent conductive electrode.**

I. INTRODUCTION

As the need for lightweight micro-autonomous systems increases, so does the requirement for on-board power sources to fuel their operation. Thermophotovoltaics (TPV) are an attractive technology which can supply a higher energy density than rechargeable batteries that would typically be used for this application. TPVs are photovoltaic cells composed of low bandgap materials, which allow them to absorb long wavelength radiation from a blackbody emitter. TPVs are typically composed of III-V semiconductors with bandgap energies between 0.5-0.75 eV such as InGaAs, InGaAsSb, and GaSb.

GaSb offers the flexibility of epitaxial methods, such as metal-organic chemical vapor deposition (MOCVD) or molecular beam epitaxy (MBE), and non-epitaxial methods, such as Zn diffusion and Be ion implantation, all of which are available on commercial n-type GaSb substrates. Due to the ultra-high-vacuum purity conditions of MBE, GaSb TPVs grown by this method have the potential to exhibit longer minority carrier lifetimes than similar cells realized via other methods such as MOCVD and ion implantation. Recent progress has resulted in the ability to fabricate GaSb devices as large as 10x10 mm by reducing the total amount of epitaxy used [1]. This was previously prevented by the formation of growth defects that adversely affect the device's performance by increasing the shunt resistance [2].

Although the n-type substrate of GaSb is technologically preferred for the purposes of crystal quality and substrate consistency, the p-on-n device design has an inherent challenge because of the low hole mobility in the extrinsic p-type emitter and the minority carrier holes in the n-type base region. The former increases series resistance, while the latter decreases collection efficiency. Even though a low sheet resistance p-type emitter could be designed, it would require an extremely high doping level that would degrade the minority electron lifetime. To deal with this obstacle, a separate method for improving the emitter sheet resistance is desirable. Two-dimensional materials are a possible solution.

Graphene, when first discovered by Geim and Novoselov in 2004 as large-scale flakes, sparked an explosion of interest into thin carbon films and materials [3]. The transport characteristics of graphene are some of the highest ever recorded, and given its unique band structure, graphene also has the ability to function as a material with ambipolar transport characteristics [4]-[5]. Coupled with graphene's small and wavelength-independent optical transparency, these properties imply that graphene has the ability to be incorporated into any device that needs to transport photo-generated carriers [6].

One of the major hurdles involving graphene following its discovery was that large area single layer graphene could not be easily and consistently grown. This problem was solved when it was found that graphene could be grown via chemical vapor deposition onto a metal foil [7]. Transferring graphene from these metal foils proved to be a relatively simple process, but refinements to it are still being attempted to overcome the defects that are inherent to an atomically thin 2D material [5].

Adding a conductive layer which could improve the poor lateral mobility of p-GaSb would inherently improve the emitter characteristics. In this work, single layer graphene (SLG) was transferred onto previously processed GaSb photovoltaics with an emitter metallization 400 nm thick. The goal of this application was to reduce the series resistance of the MBE-grown cells by improving the lateral current path of photo-generated carriers.

II. FABRICATION

The p-on-n epitaxial layer structure shown in Fig. 1 was grown via MBE. Though a more optimal emitter thickness has been shown to be 200 nm [8], an initial thickness of 670 nm was chosen to target an emitter sheet resistance of 100 Ω/sq. After growth, a back-metal contact was deposited over the entire back surface via e-beam evaporation and annealed at 290 °C for 45 seconds. The front metal contact was deposited via standard contact photolithography and an additional e-beam deposition over a total device area of 5x5 mm. The area of the cell covered by the front metal contact was calculated to be 9.8%, containing few fingers with a large spacing between them. A wet mesa etch was then done using an $HCl:H_2O_2:H_2O$ solution (50:1:150) at 15°C for 15 minutes before dicing with a diamond blade in an automated saw.

978-1-5090-5606-4/17 $31.00 © 2017 IEEE

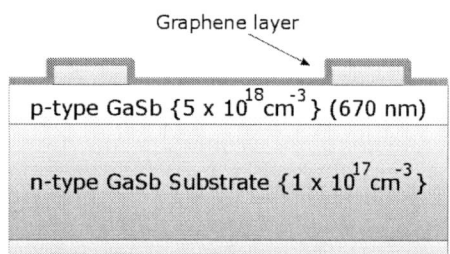

Fig. 1. Device structure for graphene-assisted GaSb solar cell grown by MBE. The emitter is an epitaxially-grown p-type layer of GaSb, while the n-type substrate is the base region of the device. A monolayer of CVD-grown graphene is placed on the front surface of the device, acting as a transparent electrode to improve lateral conduction between the metal gridlines via a wet PMMA transfer process.

Graphene was then transferred onto the cell through a modified procedure based on the standard PMMA-in-anisole transfer process [7]. Graphene-on-copper foil was cut into pieces which were roughly the same size as the cell to be covered (5x5 mm). The graphene was then attached to a glass slide by a water droplet method, wherein a small drop of water is placed on a glass slide, and the graphene is pressed down onto this water droplet to create a seal, enabling the foil to have PMMA spun on it without vacuum damage. The PMMA was spun at 500 rpm for 70 seconds to ensure a thick coating that made the graphene easier to maneuver throughout the transfer process, and was followed by an 80 °C bake for 5 minutes. The PMMA-coated graphene was then placed in an $FeCl_2$ bath to remove the copper and subjected to the transfer process detailed in [7].

The graphene structure was verified using a Witec Raman spectroscopy system with a 512 nm wavelength laser. As shown in Fig. 2, the GaSb provides a clean Raman spectroscopy image of graphene, and by observing the G band it is possible to infer that the transfer was of a single layer of graphene. The ratio of 2D to G band signal heights being 1.3 indicates an acceptable transfer with some defects present [9]. Given repeated Raman scans of each fabricated and coated device, it was estimated that 80% of each cell was covered by graphene.

The GaSb cell was tested using an Oriel IV Test Station and Keithley 2400 SourceMeter under AM 1.5G conditions. Although the cell would perform better under longer wavelength conditions, these test conditions were chosen as an initial test, because using AM 1.5 results are easier to compare to previously reported works.

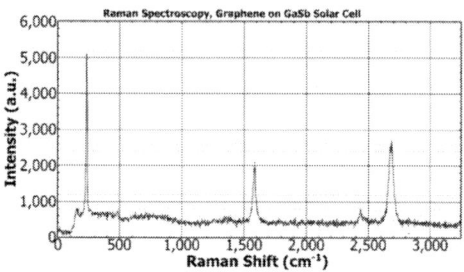

Fig. 2. Raman spectroscopy results for graphene transferred onto a GaSb cell via a wet PMMA transfer process. The presence of the graphene G band at 1650 cm^{-1} and 2D band at 2600 cm^{-1} is consistent with single layer graphene spectra. The GaSb substrate gives good results as compared to SiO_2 for detection of graphene, as the characteristic Raman peaks for GaSb (200 cm^{-1}) are far from the characteristic peaks of graphene.

III. RESULTS

Table I shows solar characteristics for the GaSb cell before and after transferring graphene onto its surface. These results must be interpreted through the lens of the single diode equation. The single-diode solar cell equation is described as follows:

$$ J = J_{ph} - J_0 \exp\left(\frac{V + AJR_s}{nV_t}\right) - \frac{V + AJR_s}{R_p} \qquad (1) $$

where R_s and R_p are the series and shunt resistances, respectively and n is the ideality factor, estimated at 1.1 [10]. The most prominent characteristic of these cells are their high R_s values. Also of note are low R_p values, attributed to lingering growth defects in the epitaxial layer. Normally in photovoltaics, factors like R_s and R_p (as well as their ratio) are inconsequential. In GaSb cells with p-type emitters, these values are substantial in their effect on the diode characteristics of the cell. The effect of R_s is exacerbated by the non-optimal value of R_p.

As evidenced by the cell characteristics in Table I, the graphene significantly reduced R_s of the cell, equaling a 50% decrease. This showcases graphene's ability to act as a highly conductive layer to mitigate the inherently poor lateral sheet resistance. Furthermore, the use of graphene does not exhibit the tradeoff of higher shading losses, typically associated with reducing the contribution to R_s due to finger resistance. The R_p increase exhibited by the device suggests that adding graphene improved the lateral resistance path, which mitigated the effect of growth defects on R_p.

	V_{oc} (V)	J_{sc} $\left(\frac{mA}{cm^2}\right)$	R_s (Ω)	R_p (Ω)	Fill Factor (%)
GaSb cell (uncoated)	0.200	32.3	14.3	31.0	30.8
GaSb cell (coated)	0.203	39.7	7.58	44.0	34.6

Table I. AM1.5 behavior of GaSb cells pre and post graphene coating. The graphene improved all characteristics of cell performance.

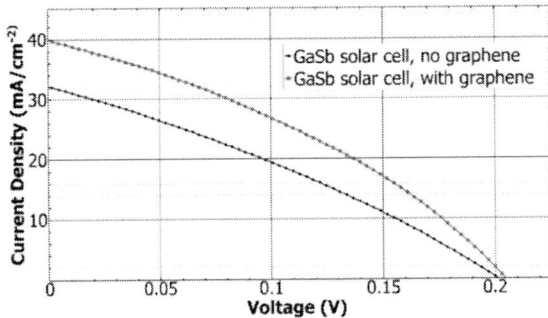

Fig. 3. J-V plot for GaSb solar cell under an AM1.5G solar spectrum, before and after adding graphene to the front surface of the device. Shading losses due to the front metal gridlines were removed from this plot. Adding graphene positively impacted the series resistance of the cell, also resulting in an increase of the short circuit current density.

This decrease in R_s and increase in R_p combined to raise the short circuit current by 7 mA/cm^2, while in some separate calculations, photocurrent density remains unchanged. Fig. 3 shows J-V characteristics for the cell shown in Table I under an AM1.5G solar spectrum, before and after the addition of graphene, with front metal shading losses compensated for. As shown, the addition of graphene increased the short circuit current density to 39.7 mA/cm^{-2}. The V_{oc} should have improved considerably by shunt resistance increases, but pinhole growth defects in the epitaxy could have been filled with evaporated metal to form a Schottky diode with the underlying n-type base layer [2]. This would mask any meaningful V_{oc} change that would be brought on the addition of the graphene. The data for each cell shows that with the application of graphene, the current generation characteristics increase by way of improving the series and shunt resistance values. Other cells that were coated did display an increase in photocurrent density by as much as 2%, which is an indicator that graphene also has the capability of passivating the GaSb surface.

IV. CONCLUSIONS & FUTURE WORK

As evidenced by the substantial reduction in series resistances as well as the increase in shunt resistance and J_{sc}, graphene acts as an effective transparent conductive electrode for p-type emitter GaSb cells. Though AM 1.5 is a good indicator of potential TPV performance, testing the device under longer wavelength conditions would give a great deal of understanding as to how the impressive phonon transport characteristics of graphene could be put to good use in standard TPV conditions. Removing the front metal contact pattern and replacing it entirely with graphene as the front electrode could potentially result in a cell with near zero shading losses and low series resistances.

ACKNOWLEDGEMENTS

The authors would like to thank Dr. Xiaoting Jia for providing training and guidance on graphene transfer, Dr. Wei Zhou and Dr. Rober Bodnar for the use of their Raman systems, Micron Inc. for providing facilities to produce our devices, as well as Dr. Mantu Hudait and ADSEL for providing facilities to test our devices.

REFERENCES

[1]   N. Rahimi et al., "GaSb thermophotovoltaics: current challenges and solutions," 2015, vol. 9358, pp. 935816-935816–10.

[2]   O. S. Romero et al., "Transmission Electron Microscopy-Based Analysis of Electrically Conductive Surface Defects in Large Area GaSb Homoepitaxial Diodes Grown Using Molecular Beam Epitaxy," J. Electron. Mater., vol. 43, no. 4, pp. 926–930, Apr. 2014.

[3]   K. S. Novoselov et al., "Electric Field Effect in Atomically Thin Carbon Films," Science, vol. 306, no. 5696, pp. 666–669, 2004.

[4]   L. Banszerus et al., "Ballistic Transport Exceeding 28 μm in CVD Grown Graphene," Nano Lett., vol. 16, no. 2, pp. 1387–1391, Feb. 2016.

[5]   D. R. Cooper et al., "Experimental Review of Graphene," ISRN Condens. Matter Phys., pp. 1–56, Jan. 2012.

[6]   R. R. Nair et al., "Fine Structure Constant Defines Visual Transparency of Graphene," Science, vol. 320, no. 5881, pp. 1308–1308, Jun. 2008.

[7]   A. Reina et al., "Large area, few-layer graphene films on arbitrary substrates by chemical vapor deposition," Nano Lett., vol. 9, no. 1, pp. 30–35, 2008.

[8]   S. A. Abdallah, D. J. Herrera, B. P. Conlon, N. Rahimi, and L. F. Lester, "Emitter thickness optimization for GaSb thermophotovoltaic cells grown by molecular beam epitaxy," 2015, vol. 9562, p. 95620L–95620L–7.

[9]   M. G. Babenco, L. Tao, and D. Akinwande, "Graphene Raman imaging and spectroscopy processing: characterization of graphene growth," 2012, vol. 8466, p. 84660O–84660O–7.

[10]   N. Rahimi et al., "Epitaxial and non-epitaxial large area GaSb-based thermophotovoltaic (TPV) cells," in Photovoltaic Specialist Conference (PVSC), 2015 IEEE 42nd, 2015, pp. 1–3.

High efficiency single-junction InGaP photovoltaic devices under low intensity light illumination

Yushuai Dai, Hyun Kum, Michael A. Slocum, George T. Nelson and Seth M. Hubbard

NanoPower Research Laboratory, Rochester Institute of Technology, 111 Lomb Memorial Drive, Rochester, NY 14623

Abstract — The fabricated single junction InGaP photovoltaic devices show an overall 30% conversion efficiency under 1.27 uW/cm^2 illumination. The development not only enables long lifetime radio-isotope based batteries but also, more important for the daily life, has the potential to promote the concept of the internet of things by efficiently powering indoor wireless sensors. To reduce the dark current and increase the absorption at longer wavelengths (>550 nm), several parameters including doping and thickness are optimized for the device design. Additional current-voltage characteristics under dark conditions and external quantum efficiency were also performed in order to evaluate the performance of the InGaP photovoltaic cells.

Index Terms —high efficiency, single junction, InGaP, internet of things.

I. INTRODUCTION

The development of photovoltaic (PV) device under low intensity illuminations is not only useful for radioisotope batteries [1] but also for new markets emerging from internet of things (IoT). The IoT refers to a network of nodes and sensors (embedded in home appliances, buildings, vehicles, etc.), which are connected wirelessly and communicate functionally to complete tasks. Most sensors are off grid devices, which require long-term energy sources to function. PV devices are as one of the methods to harvest ambient energy have been considered as replacement for less sustainable battery technology with limited charge cycles [2]–[4]. Some of the sensors are for indoor applications, so they require the photovoltaics devices output high power under low light illuminations.

Traditional terrestrial solar cells are optimized under standard test conditions (STC) to simulate outdoor illumination during the day. The power intensity (100 mW/cm^2 [5]) of STC with AM 1.5 sun spectrum is much higher than the power intensity (1uW/cm^2-1mW/cm^2 [6]) of an indoor fluorescent lamp, or light emitting diodes, or phosphor radioluminescence with a narrow spectrum within the visible region (400-700 nm). The low intensity narrower wavelength light affects PV devices conversion efficiency due to an increasing importance of the dark current [6] and shunt resistance [7]. To optimize PV devices performance under indoor illumination, the PV devices should be designed and evaluated under these specific light sources.

To date, several photovoltaic material have been investigated for the indoor light applications, including silicon[7], [8], GaAs [9], AlGaAs [4], pervoskite [2], organic solar cells [10] and so on. Based on the detailed balance model, Freunek *et al.* theoretically predicated an optimal band gap of 1.9 eV -2.0 eV with a power conversion efficiency up to 60% under narrow band artificial indoor light sources [11], so wide-band-gap III-V materials have drawn attention for higher conversion efficiency for indoor applications. However, the characterization and design of single junction InGaP, as one of the III-V material with a band gap of 1.9 eV at 300 K, has not been fully investigated under low light illuminations. In this paper, the design and performance of single junction InGaP photovoltaic devices is reported. The effect of parameters, including doping and thickness in the emitter and base on the efficiency of *n-i-p* InGaP PV devices has been evaluated. Additional current-voltage characteristics under dark and external quantum efficiency were also performed.

II. SAMPLE AND EXPERIMENT

Doping dependent recombination and diffusion length of minority carriers affect PV devices' dark current, while the total thickness of the devices determine the total number of photons absorption, so four parameters including the n-emitter doping/thickness, p-base doping/thickness are evaluated based on efficiency simulated from Sentaurus TCAD. Figure 1 shows the layer structure for the two optimized InGaP *n-i-p* photovoltaic devices grown by 3×2" Aixtron close-couple showerhead metal organic vapor phase epitaxy (CCS-MOVPE). Thin (50 nm) InGaP back surface field layer was employed on a p-type GaAs substrate to reduce interface recombination, followed by a p-type base of 1500 nm with a doping of 5×10^{16} cm^{-3}. Then a 10 nm intrinsic region was grown to improve the charge separation at the junction, followed by a 100 nm Si-doped (1.6×10^{18} cm^{-3}) InGaP emitter for design A. The 20 nm InAlP front window layer was used to reflect minority holes in the emitter and passivate the device leading to a low dark current [12]. Finally a heavily doped GaAs contact layer was incorporated for ohmic contact formation. Additionally, a double layer ZnS (56 nm)/MgF$_2$ (110 nm) were designed via TFcalc software to minimize the reflection between 400-700 nm. Compared to design A, design B has a longer base of 2000 nm and a shorter emitter of 50

978-1-5090-5606-4/17 $31.00 © 2017 IEEE

nm, respectively. The increased total length of design B is used to reduce the transmission loss of wavelength above 550 nm according to Beer-Lambert law. The samples were then fabricated using standard III–V processing and microlithography techniques. Individual cells (1×1 cm^2) with less than 4% grid fingers were isolated using wet chemical etching techniques.

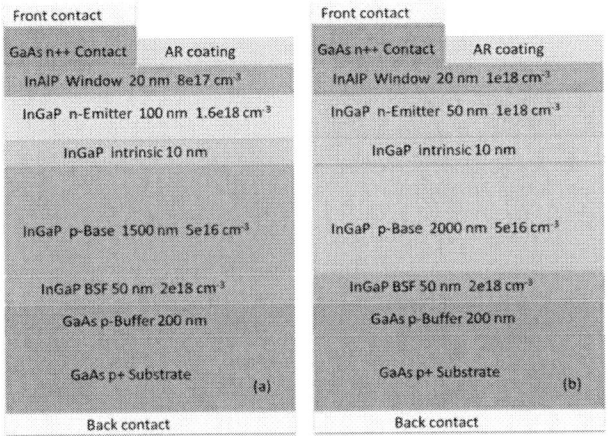

Fig.1 Layer structure layout for InGaP n-i-p solar cell devices.

Reflection measurements were completed using Filmetrics F20. Room temperature EQE measurements were taken with a Newport IQE-200 Spectroradiometric Measurement system. The low power illuminated J-V characteristics were completed with OL750 spectroradiometric measurement system with wavelength fixed at 600 nm. The wavelength of 600 nm correlates the peak spectrum from the cool and warm white LED [5]. Furthermore, a 600 nm peak also matches with phosphor emission from Tritium vials. The power of the light sources (1.27 μW/cm^2) matches scintillation light from phosphors for tritium powered sensor network [6]. The dark J-V measurements were completed with a Keithley source meter.

III. RESULTS AND DISCUSSION

Figure 2 shows the light IV curves of the Design A and Design B under 1.27 uW/cm^2 illuminations. Such low power light allows observing the effect of the shunt resistance on the efficiency. Both cells show a high efficiency over 30%. Both cells have a fill factor around 80% that indicates the shunt resistance is not the dominant effect on the devices performance. The short circuit current density of the design A (491 nA/cm^2) is lower than that (511 nA/cm^2) in the design B. The open circuit voltage (V_{oc}) in the design B is 0.04 V less than the Design A. The reduction in V_{oc} indicates larger dark current from recombination process.

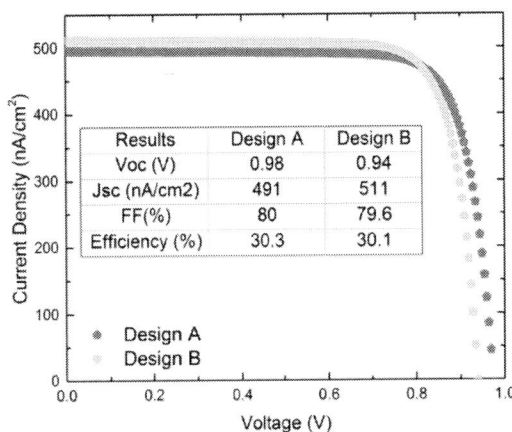

Fig.2. Measured current density vs. voltage curves 1.27 uW/cm^2 illuminations.

To detect the wavelength dependent carrier collection of the InGaP devices with different designs, Figure 3 shows the room temperature EQE measurements. Both designs show similar EQE between 400 and 550 nm. Because the reflectance (inset) is lower at wavelength between 400-450 nm in the design B, the EQE is higher in Design B than that in design A. The difference of EQE above 600 nm between design A and B may be the combination effect of reflection difference and longer base enhanced absorption.

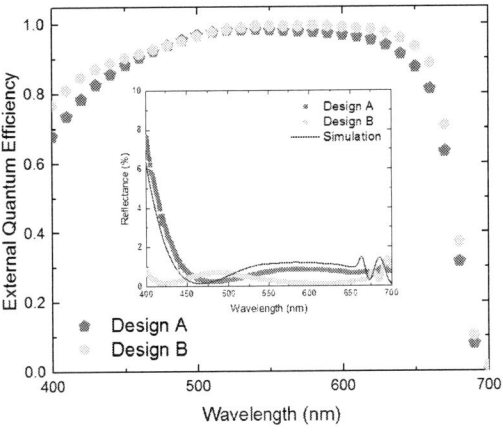

Fig.3. Room temperature EQE at zero voltage bias of the investigated design A and B InGaP photovoltaic devices. The inset is reflectance from both simulation and measurements.

To investigate the recombination process in the design A and B, Figure 4 shows the dark IV measurements. Design A shows a generally lower dark current than design B at each forward bias voltage, which is correlated to higher V_{oc} in the light IV measurements. To separate the recombination components, the dark IV curve was fit by a double diode Equation as shown in Equation (1).

$$J = J_{o1}\left\{exp\left[\frac{q(V-JR_s)}{n_1 kT}\right] - 1\right\} + J_{o2}\left\{exp\left[\frac{q(V-JR_s)}{n_2 kT}\right] - 1\right\} + \frac{V-JR_s}{R_{shunt}} \qquad (1)$$

Here J is the measured current densities and V is the bias voltage. J_{o1} and J_{o2} are the reverse saturation current densities of diode 1 (recombination in the quasi-neutral region [9]) and diode 2 (junction recombination at low voltage), respectively. The ideality factor of a diode associates different recombination process [13]. n_1 (usually equals to 1, SRH recombination in the bulk or surface recombination) and n_2 (usually equals to 2, two-carrier recombination) refer the ideality factor of this diode. R_s is the series resistance and R_{shunt} represents the shunt resistance. The extracted parameters are shown in the inset table. At lower bias voltage, the junction recombination dominates the dark current in the photovoltaic devices. Under the order of $1uW/cm^2$ illumination, the V_{oc} (around 1V) mainly dominates the J_{o1}. Because design A has a shorter base of 1500 nm than design B, design A shows a lower J_{o1} than design B due to reduced bulk recombination.

To increase the diffusion length of electrons in the n-type base, higher quality InGaP crystal via MOVPE growth is required for a longer carrier lifetime. The length of the intrinsic region should also be considered to further separate charge along the device growth direction [12]. Additionally, the shunt resistance is lower in the design A than that in design B, which is associated with a slightly reduced FF in device B as shown Figure 1, so path of shunting introduced during growth (such as point defects) and fabrication (such as sidewall recombination [14]) should be eliminated. The difference in series resistance may be due to anti-reflection coating not fully removed on the top of the contact metal finger during the dark IV measurements, which only affects photovoltaic device performance under high intensity light illumination.

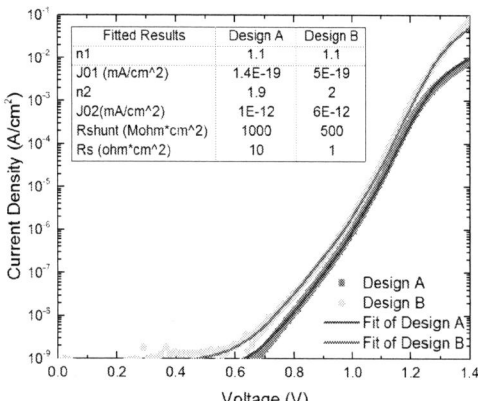

Fig.4. Measured and simulated dark current –voltage characteristic for the Design A and B

IV. CONCLUSION

This paper demonstrates the performance on two designs of InGaP PV devices with varied thickness in base and emitter for low intensity light applications. It shows reduced front surface reflection (below 1% between 450-700 nm) using MgF_2/ZnS as anti-reflection coating. Both devices show over 30% high efficiencies with an open circuit voltage toward 1 V. Meanwhile, due to the recombination in the surface, bulk and sidewall, the devices with a total length over 2 μm shows slightly V_{oc} (0.04V) drop, compared to the device with a shorter length. Side wall passivation during the fabrication and epi-growth related crystal quality can be further optimized to improve device performance in different designs with InGaP based low intensity light energy harvester.

ACKNOWLEDGEMENT

This work was supported by the Army Research Laboratory and the US Government.

REFERENCES

[1] C. D. Cress, B. J. Landi, R. P. Raffaelle, and D. M. Wilt, "InGaP alpha voltaic batteries: Synthesis, modeling, and radiation tolerance," *J. Appl. Phys.*, vol. 100, no. 11, p. 114519, Dec. 2006.

[2] C.-Y. Chen, J.-H. Chang, K.-M. Chiang, H.-L. Lin, S.-Y. Hsiao, and H.-W. Lin, "Perovskite Photovoltaics for Dim-Light Applications," *Adv. Funct. Mater.*, vol. 25, no. 45, pp. 7064–7070, Dec. 2015.

[3] I. Mathews, P. J. King, F. Stafford, and R. Frizzell, "Performance of III #x2013;V Solar Cells as Indoor Light Energy Harvesters," *IEEE J. Photovolt.*, vol. 6, no. 1, pp. 230–235, Jan. 2016.

[4] A. S. Teran *et al.*, "AlGaAs Photovoltaics for Indoor Energy Harvesting in mm-Scale Wireless Sensor Nodes," *IEEE Trans. Electron Devices*, vol. 62, no. 7, pp. 2170–2175, Jul. 2015.

[5] M. Kasemann, K. Rühle, K. M. Gad, and S. W. Glunz, "Photovoltaic energy harvesting for smart sensor systems," 2013, vol. 8763, p. 87631T–87631T–6.

[6] M. S. Litz, J. A. Russo, and D. Katsis, "Tritium-powered radiation sensor network," 2016, vol. 9824, pp. 982412-982412–12.

[7] G. E. Bunea, K. E. Wilson, Y. Meydbray, M. P. Campbell, and D. M. D. Ceuster, "Low Light Performance of Mono-Crystalline Silicon Solar Cells," in *2006 IEEE 4th World Conference on Photovoltaic Energy Conference*, 2006, vol. 2, pp. 1312–1314.

[8] E. Moon, D. Blaauw, and J. D. Phillips, "Small-Area Si Photovoltaics for Low-Flux Infrared Energy Harvesting," *IEEE Trans. Electron Devices*, vol. 64, no. 1, pp. 15–20, Jan. 2017.

[9] A. S. Teran *et al.*, "Energy Harvesting for GaAs Photovoltaics Under Low-Flux Indoor Lighting Conditions," *IEEE Trans. Electron Devices*, vol. PP, no. 99, pp. 1–6, 2016.

[10] H. K. H. Lee, Z. Li, J. R. Durrant, and W. C. Tsoi, "Is organic photovoltaics promising for indoor applications?," *Appl. Phys. Lett.*, vol. 108, no. 25, p. 253301, Jun. 2016.

[11] M. Freunek, M. Freunek, and L. M. Reindl, "Maximum efficiencies of indoor photovoltaic devices," *IEEE J. Photovolt.*, vol. 3, no. 1, pp. 59–64, Jan. 2013.

[12] P. Cabauy, L. C. Olsen, and N. Pan, "Tritium direct conversion semiconductor device," US8487507 B1, 16-Jul-2013.

[13] K. Ishaque, Z. Salam, and H. Taheri, "Simple, fast and accurate two-diode model for photovoltaic modules," *Sol.*

Energy Mater. Sol. Cells, vol. 95, no. 2, pp. 586–594, Feb. 2011.

[14] T. Gu, M. A. El-Emawy, K. Yang, A. Stintz, and L. F. Lester, "Resistance to edge recombination in GaAs-based dots-in-a-well solar cells," *Appl. Phys. Lett.*, vol. 95, no. 26, p. 261106, Dec. 2009.

Radiation resistant of upright metamorphic GaInP/GaInAs/Ge triple junction solar cells for space use

Liang Fang[1], Abuduwayiti Aierken[2], Zhen Pan[1], Qiming Zhang[1], Zhanhang Li[2], Heini Maliya[2], Wei Gao[1], Hui Gao[1], Ronghua Wan[1], Bao Zhang[1], He Wang[1], Qi Guo[2]

1Tianjin Institute of Power Sources, Tianjin 300384, China

2Xinjiang Technical Institute of Phys.& Chem., Chinese Academy of Sciences, Urumqi 830011, China

ABSTRACT — The electrical parameters and EQE spectral response of 1 MeV electron beam irradiated MOVPE grown lattice-matched (LM) and upright metamorphic (UMM) GaInP/GaInAs/Ge triple-junction solar cells have been investigated with different flux density and fluence range from 5×10^{14} to $1.5 \times 10^{15} cm^{-2}$. The overall degradation of electrical parameters of UMM cell was higher than LM cell due to the existence of dislocation within the epitaxy active layers. It was observed that the EQE spectra degrade mainly in top and middle subcells, and top GaInP subcell in UMM cell showed better radiation resistance due to the higher InP composition. An artifact phenomenon was observed in bottom Ge subcell due to the low shunt resistance.

Index Terms — radiation resistant, upright metamorphic, triple junction solar cells, III-Ⅴsemiconductor.

Ⅰ. INTRODUCTION

Presently, the state-of-art triple-junction solar cells consists of two III-Ⅴsemiconductor junctions epitaxially grown lattice matched (LM) to Ge bottom junction, with the band gap combination of 1.86, 1.42, and 0.67 eV, have been widely adopted for space use due to high-efficiency, radiation hardness, and mature fabrication process[1]. However, the constrained band gap combination of LM configuration is not optimal for maximum conversion efficiency of the solar spectrum. In order to enhance conversion efficiency an upright metamorphic (UMM) triple junctions solar cells has been proposed which consists of two III-Ⅴsemiconductor junctions connected to Ge bottom junction by a compositionally graded buffer layer[2], with the band gap combination of 1.80, 1.30, and 0.67 eV, respectively. However, radiation effects and radiation mechanism of space solar cells with UMM solar cells have not been studied as extensively as that of LM solar cells.

In this paper the radiation effects of UMM solar cells are investigated by irradiated with 1 MeV electron beam with different flux density, and LM solar cells with similar conversion efficiency are used as reference cells. The current-voltage and external quantum efficiency are measured before and after the irradiation. The effects of device configuration and irradiation flux density on device performance have been discussed.

Ⅱ. EXPERIMENTAL

The samples used in this work were fabricated by using atmosphere-pressure metal-organic vapor phase exitaxy (MOVPE) system on a single-crystal (001) Ge substrates miscut by 6° toward (111). Lattice matched $Ga_{0.51}In_{0.49}P/Ga_{0.99}In_{0.01}As/Ge$ and upright metamorphic $Ga_{0.43}In_{0.57}P/Ga_{0.92}In_{0.08}As/Ge$ solar cells with initial conversion efficiency 29.0-30.0% have been studied. The external quantum efficiency (EQE) was measured by an alternative current type (Enlitech QE-R3018), and the current-voltage (IV) characteristic of solar cells were taken by a solar simulator (Spectrosun X25A) under AM0 solar spectrum (135.3 mW/cm^2) at 25℃±2℃. Electron beam irradiation was carried out by an ELV-8 vertical electron accelerator at room temperature. Two kind of electron flux densities, 5×10^{10} e/(cm^{2}s) and 1×10^{11} e/(cm^{2}s), were chosen. Solar cells were measured prior to irradiation and at total irradiated electron fluence of 5×10^{14} e/cm^2, 1×10^{15} e/cm^2, and 1.5×10^{15} e/cm^2, respectively.

Ⅲ. RESULTS AND DISCUSSION

Fig. 1 shows the degradation of short circuit current (I_{sc}), open current voltage (V_{oc}), and maximum output power (P_{max}) of two different types of solar cells against the electron irradiation fluence with two different flux densities. All I_{sc}, V_{oc} and P_{max} values decreased with the increase of irradiated electron fluence amount in both cases, and degradation of V_{oc} is bigger than that of I_{sc}. Comparing the effect of different flux density, the degradation is slightly larger in 5×10^{10} e/(cm^{2}s) flux density than that of 1×10^{11} e/(cm^{2}s), but the difference of degradation is not significant. However, when the electron fluence increased to 1.5×10^{15} e/cm^2, the P_{max} value of UMM cell dropped notably in larger flux density irradiation, whereas the LM cell showed better performance.

Comparing LM and UMM cells, LM cell showed better radiation hardness in all parameters. At the irradiated electron fluence of 1×10^{15} e/cm^2 with flux density of 5×10^{10} e/(cm^{2}s), the value of P_{max} decreased to 83% of its initial value in LM cell, whereas it dropped to 81% in UMM cell. This difference is caused by the cell structure, where the UMM cell has threading dislocations (TDs) due to lattice constant mismatch

978-1-5090-5606-4/17 $31.00 © 2017 IEEE

(a)

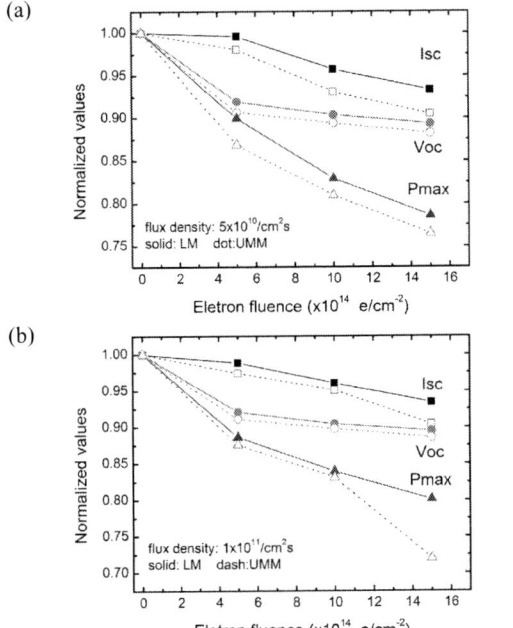

(b)

Fig. 1. Normalized I_{sc}, V_{oc}, and P_{max} values of LM and UMM solar cells irradiated by 1 MeV electron beamwith flux density of (a) 5×10^{10} e/(cm²s) and (b) 1×10^{11} e/(cm²s), respectively.

between the Ge substrate and two metamorphic junctions. The TDs cannot been fully eliminated by compositionally graded buffer layer, and still exists within active layers of subcells.

(a)

(b)

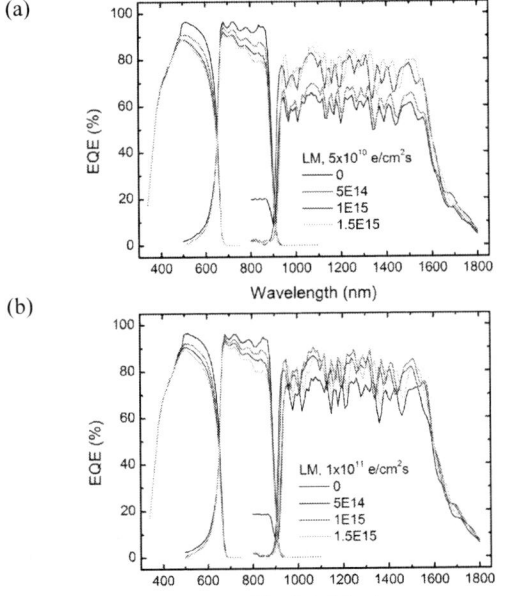

Fig. 2. EQE spectra of three subcells of LM cell against the 1 MeV electron beam irradiation with flux density of (a) 5×10^{10} e/(cm²s) and (b) 1×10^{11} e/(cm²s), respectively.

Fig. 2 (a) shows the degradation of EQE spectral response of each subcell of LM cell against the irradiated electron beam fluence with two different flux densities. EQE values decreased with the increase of electron fluence in top GaInP and middle GaInAs cells, but no clear differences were observed comparing the degradation against two different flux densities. However, abnormal EQE response was observed from Ge bottom cell. This phenomenon was observed by some other groups as artifact EQE response, and the main reasons are low shunt resistance of the Ge subcell and Luminescence coupling effects [3-5]. Similar EQE response against irradiated electron fluence was observed in UMM cell as well (Fig. 3).

(a)

(b)

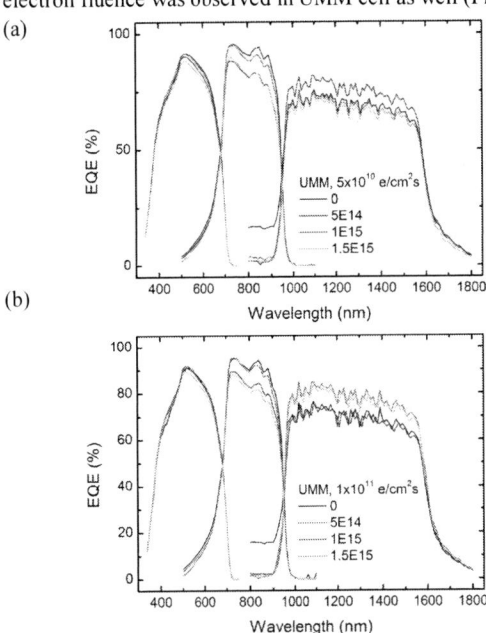

Fig. 3. EQE spectra of three subcells of UMM cell against the 1 MeV electron beam irradiation with flux density of (a) 5×10^{10} e/(cm²s) and (b) 1×10^{11} e/(cm²s), respectively.

As shown in Fig 2 and Fig 3, the degradation of EQE spectra happened in top and middle junctions in both LM and UMM solar cells, furthermore, some differences exist due to different cell structures and material composition. For LM cells, the degradation of EQE spectra happened at longer wavelength regions from 500 to 600nm in top GaInP subcell, and from 700 to 900nm in middle GaInAs subcell, which indicates the radiation induced defects are mainly exiting within the base region of subcells. For UMM cells, the degradation of EQE spectra in top GaInP subcell was rather smaller even with electron irradiation fluence of 1e15cm⁻², and the degradation of EQE spectra in middle GaInAs subcell shows similar phenomena at 700-900nm wavelength regions with relative larger decrease compared to LM solar cells. The better radiation resistance of GaInP subcell for UMM solar

978-1-5090-5606-4/17 $31.00 © 2017 IEEE

cells is related to the increase of InP composition[6], and the relatively weak radiation resistance of GaInAs subcell is related to TDs within active layers due to lattice constant mismatch between Ge substrate and epitaxy layers of subcells. The compositionally graded buffer layers will be optimized to reduce threading dislocation density further and increase radiation resistance of GaInAs subcell of UMM cells.

(a)

(b)

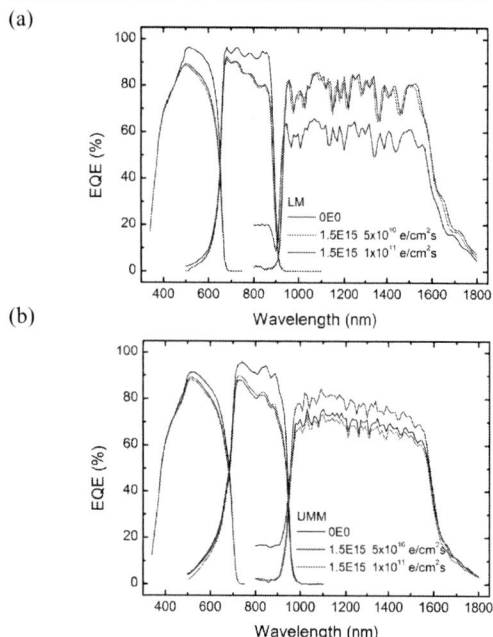

Fig. 4. EQE spectra of three subcells of (a) LM and (b) UMM cells irradiated by 1 MeV electron beam with flux density of 5×10^{10} e/(cm^{2}s) and 1×10^{11} e/(cm^{2}s), respectively.

IV. CONCLUSION

Fig.4 shows the comparison of EQE spectra of LM and UMM cells before and after electron irradiation with fluence of 1.5×10^{15} e/cm^2 with different flux density, respectively. No clear difference was observed in both shape and intensity of EQE spectra in top GaInP and middle GaInAs subcells. However, it is difficult to compare the EQE spectra of bottom Ge subcells due to the artifact EQE spectra.

The electrical and EQE spectral response of 1 MeV electron beam irradiated MOVPE grown LM and UMM GaInP/GaInAs/Ge triple junction solar cells have been investigated. The degradation of electrical parameters of UMM cell is slightly larger than that of LM cell due to lattice constant mismatch between Ge substrate and epitaxy layers of subcells. Comparing the effect of different flux density, the degradation is slightly larger in 5×10^{10} e/(cm^{2}s) flux density than that of 1×10^{11} e/(cm^{2}s). The degradation of EQE spectra were observe in top GaInP subcell and middle GaInAs subcell in both LM and UMM configurations, however, top GaInP

subcell in UMM structure showed better radiation resistance due to the higher InP composition. Artifact EQE spectra were observed in bottom Ge subcell due to the low shunt resistance. No significant difference was observed on EQE spectra degradation with different irradiation flux density.

Acknowledgements

Abuduwayiti Aierken would like to thank National Natural Science Foundation of China (No. 61534008) for the financial support.

REFERENCES

[1] R. R. King, D. C. Law, K. M. Edmondson, C. M. Fetzer, G. S. Kinsey, et al., "Advances in High- Efficiency III-V Multijunction Solar Cells", Advances in OptoElectronics, Vol. 2007, pp. 29523, 2007

[2] R. R. King, D. C. Law, K. M. Edmondson, et al., "40% efficient metamorphic GaInP/GaInAs/Ge multijunction solar cells," Applied Physics Letters, vol. 90, no. 18, pp.183516, 2007

[3] M. Sugai, J. Harada, M. Imaizumi, S. Sato, and T. Ohshima, "A study of the artifact external quantum efficiency on Ge-bottom sub cell in triple-junction solar cell", 39th PVSC, June 16-21,2013, Tampa, Florida, USA.

[4] S. H. Lim, J. J. Li, E. H. Steenbergen, Y. H. Zhang, "Luminescence coupling effects on multijunction solar cell external quantum efficiency measurement", Progress in Photovoltaics, Volume 21, Issue 3, May 2013

[5] M. Meusel, C. Baur1, G. Letay, A.W. Bett, W. Warta and E. Fernandez "Spectral Response Measurements of Monolithic GaInP/Ga(In)As/Ge Triple-Junction Solar Cells: Measurement Artifacts and their Explanation", Progress in Photovoltaics: Research and Applications 11(8): pp. 499-514, 2003.

[6] M. Yamaguchi, "Radiation-resistant solar cells for space use",Solar energy materials & solar cells 68 (2001) 31-53

High Efficiency Glass Waveguiding Solar Concentrator

Chehao Hu, Yusuf Dogan, Matthew Morrison, A. Nanda, D. Ma, R. Atkins and C. K. Madsen

Texas A&M University, College Station, Texas, 77843, USA

Abstract — **A new waveguiding concentrator design is presented that does not suffer from decoupling losses common to other waveguiding architectures. The geometric optical efficiency (η_{geo}), before considering absorption and scattering, is higher than 99.9%. For a tracking (incidence) angle tolerance of ±0.8°, the efficiency only drops to 95.6%. The concentration optics are transparent, allowing collection of the diffuse sunlight with a low-cost single-junction solar cell. Glass waveguides are used so that absorption losses are minimized and the design can be scaled to large areas. We present simulation results using Zemax and TracePro to achieve high optical path efficiencies for both concentrated (direct) and diffuse light and energy harvesting efficiencies over 30% in both high and low DNI conditions using currently achievable PV cell efficiencies. We also discuss our first measurement results on fabricated glass waveguides and for a first prototype measurement with an efficiency of 90% for a single lens-to-waveguide optical path.**

Index Terms — **CPV, diffuse light, high efficiency, solar concentrator, waveguide.**

I. INTRODUCTION

Concentrating photovoltaic (CPV) systems employ optical architectures to focus the direct sunlight onto a small area multi-junction PV (MJPV) cell with the main design goals being to achieve high efficiencies at minimal cost. Our previous work [1-2] focused on a waveguiding architecture for concentrations up to 1000x, but was limited in scaling due to decoupling losses. MJPV efficiencies have been reported up to 43.2% under 1000X concentration [3]. Since the MJPV is a costly component, high concentration designs allow a reduction in the required area and thus a substantial cost benefit. To reap the benefits, the optical path efficiency must be very high. We present simulation results using Zemax and TracePro to achieve optical path efficiencies over 90% for the direct normal irradiance (DNI).

Various CPV architectures have been explored, including Flat Reflector, V-trough, Linear Fresnel Reflector [4] and Planar Waveguide Concentrators [1-2,5]. Each type has its own tradeoffs. The Planar Waveguide Concentrator was first proposed by Karp et al. [5] in 2010. To improve upon these designs, we report on our novel design that avoids decoupling losses, which are inherent in other published architectures.

II. DESIGN OVERVIEW

A small, 3x5 (lens element) prototype architecture is shown in Fig. 1 to illustrate the basic components. It is composed of a lens array, waveguide array, MJPV cells and an underlying Si PV cell. The red and yellow lines represent direct and diffuse sunlight, respectively. The direct sunlight is focused by each len element onto a waveguide turning-mirror facet, whereby the direct light propagates through the waveguide to an MJPV cell. The diffuse light is not concentrated, but passes through the waveguide array and is collected by Si PV cells. Other low-cost, single-junction PV cells could be used for diffuse light collection as well.

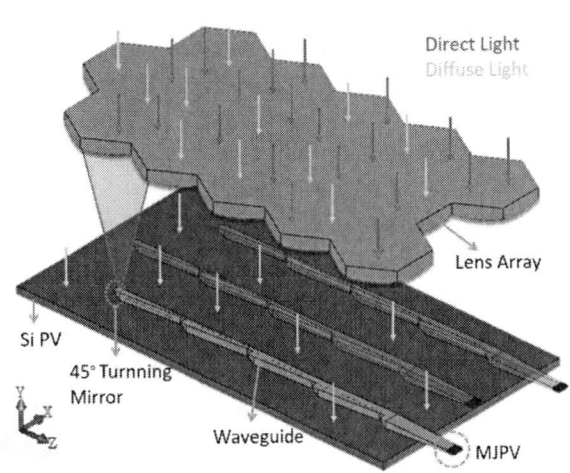

Fig. 1. Schematic of a small, 3x5 lens-to-waveguide array prototype system, so that the components are easily visualized.

A. Lens Design Tradeoffs

An ideal material for the lens array is low-loss glass. However, the cost and complexity of fabrication for a molded glass lens are relatively higher than PMMA. Therefore, the PMMA is still an attractive option. The PMMA has the disadvantage of higher absorption loss, which reduces the optical throughput. We maximized transmittance through the PMMA lens by reducing the necessary thickness of the lens. For a given lens element area of 100 mm², the central thickness is reduced from 2.33mm to 1.84mm for a square and hexagonal lens shape (see Fig. 2), respectively. This 21% reduction in thickness allows a 2.3% increase in lens transmittance for the optical path.

978-1-5090-5606-4/17 $31.00 © 2017 IEEE

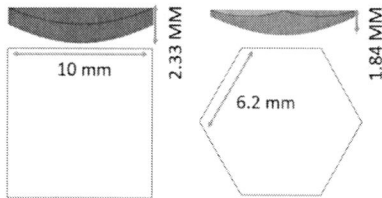

Fig. 2. Original design (left) used a square lens, while our improved design uses a hexagonal lens (right).

B. Waveguide design

The waveguide includes three components: a first coupler, a slab channel and a second coupler (see Fig. 3). The first coupler is tapered in the X direction, while the second coupler is tapered in the Y direction to achieve maximum concentration without violating a TIR condition. A gap is located between the first coupler and slab channel that prevents decoupling losses. In our previous design [2], the waveguide array required a slightly larger area than the lens array. Our new design scales to larger array sizes as shown in Fig. 6 with the full lens array aperture captured in the waveguide layer that is confined within the lens array area.

The maximum number of coupling elements combined in a row will depend on the achievable waveguide loss. Using fused silica, absorption losses are negligible and we expect scattering loss to dominate. Fig. 4 shows the simulation result using a scattering model, with parameters from our surface roughness measurements, in Zemax to model efficiency versus waveguide length. There is no geometric loss, because there is no de-coupling mechanism in our design. These simulations predict that long lengths (up to 300mm) could be viable with high efficiencies (>95%) for the waveguide path.

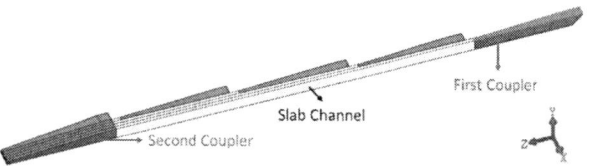

Fig. 3. The red, white and blue color represents the first coupler, slab channel and second coupler.

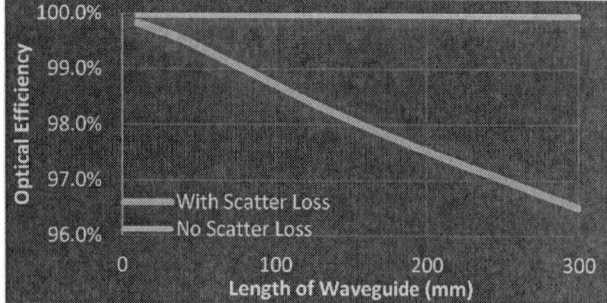

Fig. 4. Comparison of the waveguide path optical efficiency without scatter loss and with scatter loss.

The distance between the lens and waveguide arrays is about 20mm, so the design is quite compact (<1inch total thickness). Under these constraints, we optimized the lens design and show an improved incidence angle (tracking) tolerance compared to our previous design [2] in Fig. 5. The blue line and orange line represent the new and old design, respectively. The incident angle tolerance is almost ±1°at 90% optical efficiency.

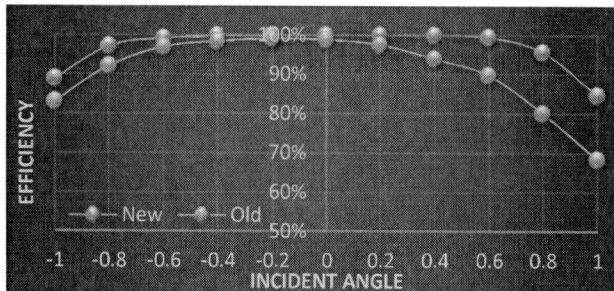

Fig. 5. Incident angle tolerance for the new and previous [2] designs.

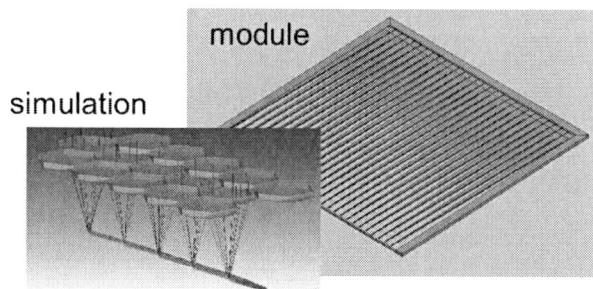

Fig. 6. Simulation of one row of 5 lens elements coupling into a common waveguide section. A module view showing the waveguides (in pink) coupled to a 28x28 lens element array.

C. Diffuse Light Collection

Diffuse light constitutes a significant portion of total sunlight. Hence, higher total solar harvesting efficiency requires collection of diffuse light. We the Zemax and TracePro to simulate the performance of diffuse light capture in our optical system. We followed two steps to simulate diffuse sunlight: 1) we used a swept-angle source approach in Zemax, and 2) an atmospheric model in TracePro to get an angular distribution.

1) Swept-angle source: Based on Fig. 7, we collect the rays which hit detector 2, and increased the tilt angle of the source from 0° to 90°. The average loss is around 14% due to the reflection of the lens.

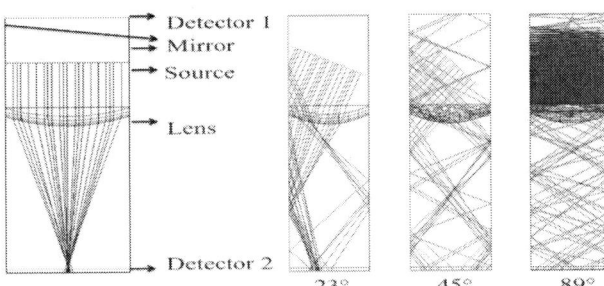

Fig. 7. Swept-angle diffuse light simulation.

2) To simulate the angular distribution of a given atmospheric model, we used TracePro's "Solar Emulator" tool, which can simulate the sun in any location and weather. We tested for Singapore under partly cloudy conditions with the Igawa All sky model. The simulation structure is shown in Fig. 8(a). Applying the sky model inside of the red circle, we placed a mask under the source to block the direct light, and built a wall around the system to make sure the only entrance of diffuse light is the lens. Fig. 8(b) shows the light that reaches the Si PV.

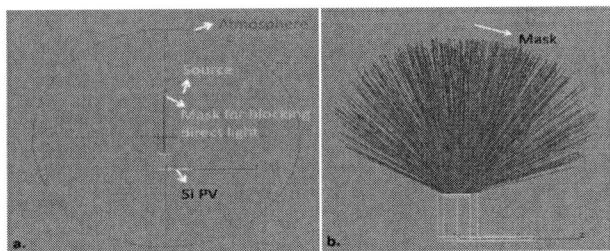

Fig. 8. (a) Simulation structure setting. (b) Visualization of the angular distribution.

Collecting the diffuse light on the Si PV provides the angular distribution, shown in Fig. 9. Then, we weighted the efficiency versus tilt angle from step 1 by this angular distribution. The average diffuse optical path loss is estimated at 5.7% using this approach.

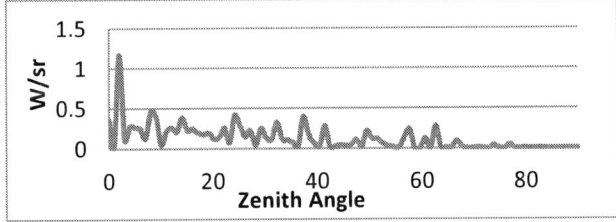

Fig. 9. The angular distribution of diffuse sunlight.

III. MEASUREMENT & FABRICATION

For a prototype demonstration, we had a custom-designed acrylic lens array fabricated. We have recently fabricated fused silica waveguides with a 1mm×1mm cross-section in lengths from 5mm to 15mm, for a first demonstration. Our measured values using an AFM for surface roughness are as low as Rq=1.5nm (RMS), which represents a precision quality surface finish. Measuring optical throughput at 638nm, we have achieved excess losses down to 0.11 dB after accounting for Fresnel reflection losses on a 5mm length sample. We have also fabricated 45degree turning mirror surfaces on fused silica waveguides, which are needed for the first coupler sections. Our first measurements using a lens element coupled to a TIR mirror and single waveguide yielded an optical path loss of 90%, after accounting for Fresnel reflection losses. Thus, we are on the path to demonstrating very high overall optical path efficiencies for integrated waveguide structures with further development.

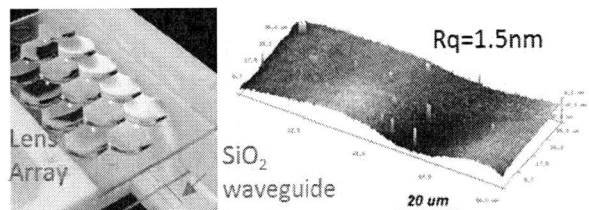

Fig. 10. Prototype assembly and AFM measurement of glass waveguide surface.

IV. CONCLUSION

Our new design provides high concentration (1000x) and almost perfect geometric optical efficiency, 99.9%. In addition, this architecture can collect substantial diffuse light (94.3% simulated for one location so far). If operating this system with a 43% efficiency MJPV and 20% efficiency Si PV, the maximum solar harvesting efficiency can achieve 33% under 30% diffuse conditions (70% DNI). We have demonstrated the basic building blocks to achieving an optical path with 90% efficiency or more.

Acknowledgement: We thank the DOE ARPA-E MOSAIC program for funding on this work.

REFERENCES

[1] Y. Liu, R. Huang, C. K. Madsen. "Design of a lens-to-channel waveguide system as a solar concentrator structure," *Optics Express*, vol. 22, pp. A198, 2014.

[2] Y. Liu, R. Huang, C. K. Madsen. "Design optimizations for a lens-to-channel waveguide solar concentrator," *Photovoltaic Specialist Conference*, vol. 10, pp. 1109, 2015.

[3] *AZUR SPACE Solar Power GmbH*, Cell Type: 3C44 – 3x3mm², http://www.azurspace.com/images/products/DB_435 7-00-00_3C44_AzurDesign_3x3_2015-04-02.pdf.

[4] K. Shanks, S. Senthilarasu, T. K. Mallick, "Optics for concentrating photovoltaics: Trends, limits and opportunities for materials and design," *Renewable and Sustainable Energy Reviews*, vol. 60, pp. 394-407, 2016.

[5] J. H. Karp, E. J. Tremblay, and J. E. Ford, "Planar micro-optic solar concentrator," *Optics Express*, vol. 18, pp. 1122-1133, 2010.

GaInAsP/GaInAs Tandem Solar Cell with 32.6% One-Sun Efficiency

Nikhil Jain, Kevin L. Schulte, John F. Geisz, Ryan M. France, Myles A. Steiner

National Renewable Energy Laboratory, Golden, CO 80401

Abstract — We demonstrate two-junction, inverted metamorphic GaInAsP/GaInAs solar cells grown by MOVPE on GaAs substrates. The bandgaps of the two junctions were ~1.7 and ~1.1 eV, respectively; the second junction was grown monolithically on a transparent AlGaInAs compositionally graded buffer layer. This tandem architecture achieved a one-sun efficiency of $(32.6 \pm 1.4)\%$ under the AM1.5 G173 global spectrum at 1000 W/m^2, highlighting the promise of newly developed high quality alloys in combination with highly transparent metamorphic buffers with low dislocation density.

I. INTRODUCTION

Multijunction solar cells enable increased power conversion efficiencies compared to single junction cells, primarily due to the reduction in thermalization losses. While the best single junction cell to date is a GaAs cell with an efficiency of 28.8%, the best dual junction cell is a GaInP/GaAs cell with an efficiency of 31.6% [1]. Though both cells can absorb the same fraction of the solar spectrum, the tandem converts the shorter wavelength photons at the higher junction voltage of the GaInP cell; the current is halved but the voltage is more than doubled.

The ideal bandgap combination for a two junction solar cell, operating under the global solar spectrum, would be ~1.6/0.9 eV. These band gaps are not attainable in the III-V materials palette in a lattice-matched, monolithic configuration, however. With bandgaps of ~1.8 eV and 1.4 eV, respectively, the GaInP/GaAs tandem is well away from the optimum bandgap configuration for a two-junction cell. And yet, first on its own and then in conjunction with a germanium third junction, this tandem has enjoyed tremendous commercial success in the photovoltaics industry because of the high material quality that results from the lattice-matched epitaxial growth.

Advances in lattice-mismatched epitaxy and development of transparent buffers have enabled design flexibility to achieve bandgap combinations spanning the entire solar spectrum. Recent efforts have led to mismatched GaInAs alloys with material quality nearly as high as GaAs and GaInP [2]. Using compositionally step-graded AlGaInP and AlGaInAs buffer layers (CGBs), the in-plane lattice constant is stepwise increased, with the misfit dislocations that mediate the strain relaxation largely confined to the interfaces between steps. These CGBs can be engineered to minimize threading dislocation density by maximizing the threading dislocation glide length, with measured threading dislocation densities ~10^6 cm^{-2} in structures with over 2% lattice mismatch [2]. Lattice-mismatched epitaxy has enabled three-junction and

four-junction inverted metamorphic multijunction (IMM) concentrator solar cells to reach efficiencies in excess of 45%[2].

GaInAsP solar cells with a bandgap of ~1.7 eV offer an attractive aluminum-free alternative for integration as the second junction in next generation of five and six-junction IMM solar cells, that have a realistic potential to exceed 50% efficiency at high concentration.

Two junction IMM cells have more recently been demonstrated for solar-to-hydrogen water splitting applications [3]. Electrolysis of water requires a minimum potential of 1.23 V. A high quality, practical device requires an additional ~0.5 V to overcome surface kinetic overpotentials, for a total required voltage of ~1.8 V, and any additional voltage is effectively wasted. GaInP/GaAs tandems have been used for water splitting with demonstrated solar-to-

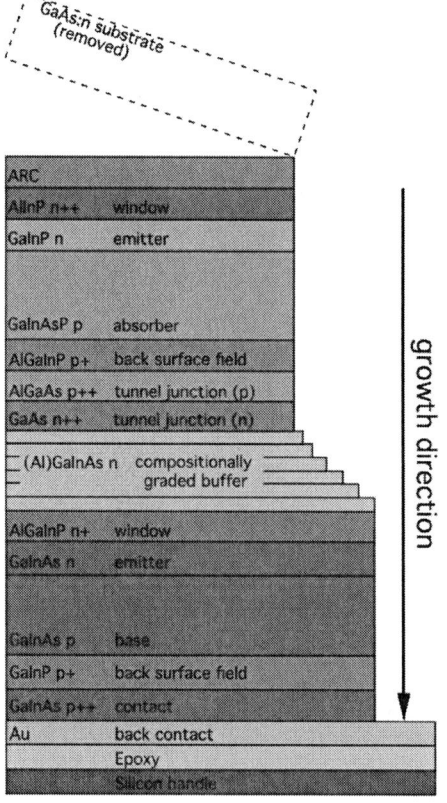

Figure 1 Schematic of the layer structure of the 1.7/1.1 eV GaInAsP/GaInAs tandem solar cell, not drawn to scale. The lateral direction represents the lattice constant.

978-1-5090-5606-4/17 $31.00 © 2017 IEEE

hydrogen efficiencies of ~10%, but as shown in Ref. [3], a more optimized device would evolve the bandgaps toward a ~1.7 eV top junction and a 1.1-1.2 eV bottom junction, for a total voltage of ~1.9-2.0 V and a solar-to-hydrogen conversion efficiency >16%. This bandgap combination is also relevant to the silicon community since a tandem cell based on a silicon bottom junction (1.1eV) would optimally be designed with a ~1.7 eV top junction. Several technical challenges must be overcome to realize such a device, but the results discussed here for high quality III-V tandems indicate that efficiencies >30% may be realistically achievable for III-V/Si two-terminal tandem cells.

Here we report the photovoltaic characteristics of a GaInAsP/GaInAs monolithic, tandem cells with bandgaps of approximately 1.7 eV and 1.1 eV, respectively. The independently confirmed efficiency of 32.6% reported here exceeds the previous dual-junction tandem record of 31.6% efficiency of the GaInP/GaAs, demonstrating the power and utility of newly developed high quality alloys in combination with high transparent metamorphic buffers. Key factors limiting the performance are identified and directions for future improvements are discussed.

II. EXPERIMENTAL METHODS

GaInAsP and AlGaAs (~23% Al) alloys were both developed for potential use as a 1.7 eV junction. Performance of both junctions has been promising, but at one-sun current densities the GaInAsP has slightly better recombination characteristics and for that reason was included in the present tandem design for one-sun. A schematic of the tandem device is shown in Figure 1. The development of the $Ga_{.68}In_{.32}As_{.34}P_{.66}$ cell was discussed in detail in Ref. [4]. The AlGaInAs CGB was designed to be transparent to wavelengths beyond 700 nm, by varying the aluminum fraction of the layers from 31% to 48% as the indium fraction was stepped forward, such that a constant band gap was maintained throughout the graded structure [5]. A strain grading rate of $0.8\%/\mu m$ was employed, and a composition step back was used for the final layer in order to achieve strain free growth at the end of the grade.

Cells were grown at NREL by atmospheric pressure metalorganic vapor phase epitaxy (MOVPE), on a custom built vertical reactor. Trimethylgallium, triethylgallium, trimethylaluminum and triethylindium precursors were used as source material for the Group-III atoms; arsine and phosphine sources were used for the Group-V atoms; and disilane, hydrogen selenide and diethylzinc sources were used for the dopants. Samples were grown on (001) GaAs substrates, miscut 6 degrees toward (111)A. The A miscut substrate suppresses phase separation during growth of the GaInAsP, as was discussed in [4]. The reactor was equipped with a multibeam optical stress sensor (MOS) for *in-situ* evaluation of the film stress at growth temperature.

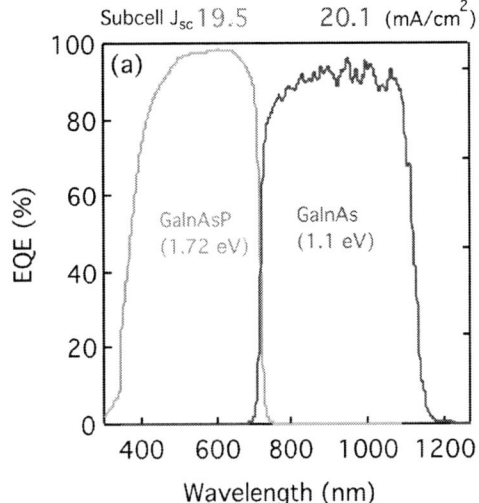

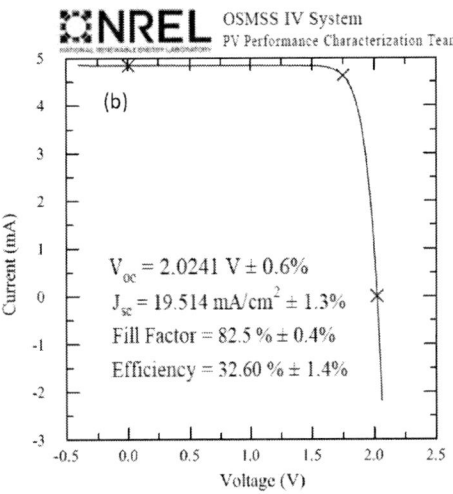

Figure 2 (a) External quantum efficiency and (b) light JV curves for a 1.7/1.1 eV GaInAsP/GaInAs tandem solar cell, measured under AM1.5 G173 (one-sun) illumination at 1000 W/m².

The epitaxial structure was processed into individual devices using standard photolithography and wet-chemical etching techniques. Details can be found elsewhere [2, 4]. Briefly, the cells were grown inverted so that the back contact layer was the last layer grown. After growth, a gold back contact was electroplated to the back contact layer and the structure was bonded with epoxy to a silicon handle. The substrate was then etched away in an alkaline solution of $NH_4OH:H_2O_2$ (1:2 by volume). Gold contacts were electroplated to the front surface, the cells were isolated using

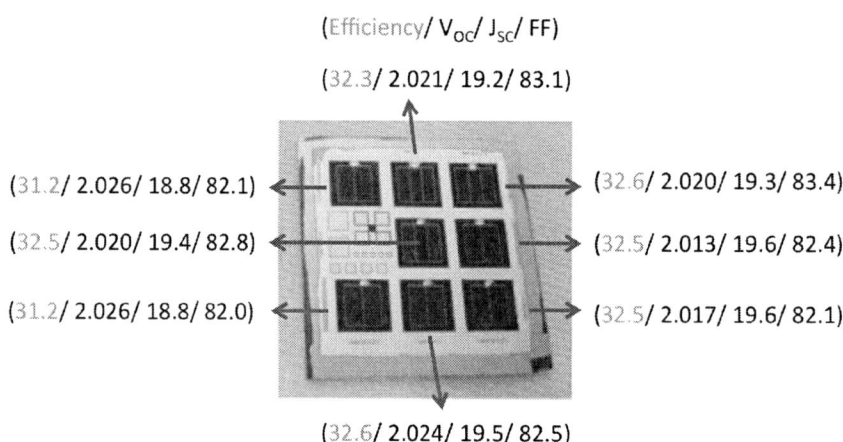

(Efficiency/ V_{OC}/ J_{SC}/ FF)

(32.3/ 2.021/ 19.2/ 83.1)

(31.2/ 2.026/ 18.8/ 82.1)　　　　(32.6/ 2.020/ 19.3/ 83.4)

(32.5/ 2.020/ 19.4/ 82.8)　　　　(32.5/ 2.013/ 19.6/ 82.4)

(31.2/ 2.026/ 18.8/ 82.0)　　　　(32.5/ 2.017/ 19.6/ 82.1)

(32.6/ 2.024/ 19.5/ 82.5)

Figure 3 Performance parameters for all eight fabricated devices of 1.7/1.1 eV GaInAsP/GaInAs tandem solar cells. Each cell is ~0.25 cm^2 in area and the entire sample size is about ~4 cm^2. The primary source of uncertainty in the reported efficiency is from the photocurrent measurement. All the measurements reported here are independently certified by the NREL PV Cell & Module Characterization Group.

a combination of selective and non-selective etchants, and finally a bilayer MgF$_2$/ZnS (98nm/46nm) anti-reflection coating was deposited on the front surface. The cell area was measured to be 0.248 cm^2 with ~2% shadowing by the contact pad and grid fingers.

The external quantum efficiency (EQE) was measured on custom-built instrumentation using a tungsten-halogen lamp and a 1-nm resolution monochromator, and high brightness LEDs to overdrive the non-limiting cell. Current-voltage (JV) characteristics were measured on NREL's one sun multisource simulator (OSMSS), using a primary calibrated reference cell to set the correct intensity.

III. RESULTS AND DISCUSSION

Figure 2a shows the external quantum efficiency of the tandem. The bandgaps of the two subcells are 1.720 and 1.095 eV, as determined by the detailed-balance equivalence method [6]. There appears to be some absorption on the short-wavelength side of the bottom cell, indicating that either the CGB or the tunnel junction is not as transparent as desired, but the loss is small, and the integrated current densities (shown in Figure 2a) indicate that the tandem is marginally current-limited by the top junction.

The certified one-sun JV curve for the best device is shown Figure 2b. Based on the measured bandgaps, the total bandgap-voltage offset W_{oc} = 0.790 V, for an average of 0.395 V. The tandem cell achieved a conversion efficiency of (32.6 ± 1.4)% under the AM1.5 G173 global spectrum at 1000 W/m^2, which is highest efficiency to date for a monolithic two-junction tandem solar cell under one-sun. Both the individual junctions transition from ideality factors of n=2 to n=1 near the one-sun current density (~20 mA/cm^2) and the

junction voltages estimated from electroluminescence and dark JV measurements (not shown) were ~1.33 V for the GaInAsP and 0.71 V for the GaInAs, which again indicates a W_{OC}~0.39 V for each junction. These are very promising W_{OC}s given the propensity for phase separation in 1.7 eV GaInAsP alloys and 1.1 eV GaInAs subcell being a lattice-mismatched junction.

It is also worth noting that controlling the composition and achieving good uniformity in mixed arsenide-phosphide alloys can be challenging. We measured eight devices, each ~0.25 cm^2 in area, to assess the uniformity across the entire sample area (~4cm^2). The solar cell performance parameters for these eight devices are summarized in Figure 3. Two (left side of image in Figure 3) of the eight devices exhibited slightly lower efficiency attributed to unexpected drop in the J_{SC}. Nonetheless, overall the consistency of the parameters achieved for these tandems is encouraging. Future efforts to improve the device performance should predominantly focus on improving the fill factor, which is limited by the low bandgap, metamorphic bottom GaInAs subcell. Evolving both alloys toward the global optimum of 1.6/0.9 eV may yet lead to even higher efficiency.

IV. CONCLUSION

In summary, we have demonstrated a series-connected, 1.7/1.1 eV, GaInAsP/GaInAs IMM solar cell with an independently measured one-sun efficiency of 32.6 ±1.4 %. The material quality, as determined by the individual junction voltages, is excellent in both alloys, despite the propensity for phase separation in the GaInAsP, and the lattice-mismatch in the GaInAs alloys.

ACKNOWLEDGMENTS

The authors would like to acknowledge T. Deutsch, J. Young, H. Doescher, and J. Turner for useful conversations, and T. Moriarty for certified measurements. Michelle Young and Waldo Olavarria assisted in cell processing and epitaxial growth. Andrew Norman and Steve Harvey performed TEM and CL characterization. Alfred Hicks at NREL assisted with graphical device schematics. This work was supported by the U.S. Department of Energy under Contract No. DE-AC36-08GO28308 with the National Renewable Energy Laboratory. Funding provided by U.S. DOE Office of Energy Efficiency and Renewable Energy Solar Energy Technologies Program.

The U.S. Government retains and the publisher, by accepting the article for publication, acknowledges that the U.S. Government retains a nonexclusive, paid up, irrevocable, worldwide license to publish or reproduce the published form of this work, or allow others to do so, for U.S. Government purposes.

REFERENCES

[1]	M. A. Green, K. Emery, Y. Hishikawa, W. Warta, E. D. Dunlop, D. H. Levi, *et al.*, "Solar cell efficiency tables (Version 49)," *Prog. Photovolt.*, vol. 25, pp. 3-13, 2017.

[2]	R. M. France, J. F. Geisz, I. Garcia, M. A. Steiner, W. E. McMahon, D. J. Friedman, *et al.*, "Design flexibility of ultrahigh efficiency four-junction inverted metamorphic solar cells," *IEEE J. Photovoltaics*, vol. 6, pp. 578-583, 2016.

[3]	J. L. Young, M. A. Steiner, H. Doescher, R. M. France, J. A. Turner, and T. G. Deutsch, "Direct solar-to-hydrogen conversion via inverted metamorphic multi-junction semiconductor architectures," *Nature Energy*, vol. 2, p. 17028, 2017.

[4]	N. Jain, J. F. Geisz, R. M. France, A. G. Norman, and M. A. Steiner, "Enhanced current collection in 1.7 eV GaInAsP solar cells grown on GaAs by metalorganic vapor phase epitaxy," *IEEE J. Photovoltaics*, to appear, 2017.

[5]	K. L. Schulte, R. M. France, and J. F. Geisz, "Highly transparent compositionally graded buffers for new metamorphic multijunction solar cell designs," *IEEE J. Photovoltaics*, vol. 7, pp. 347-353, 2017.

[6]	J. F. Geisz, M. A. Steiner, I. Garcia, R. France, W. E. McMahon, C. R. Osterwald, *et al.*, "Generalized optoelectronic model of series-connected multijunction solar cells," *IEEE J. Photovoltaics*, vol. 5, p. 1827, 2015.

Evaluation of Tandem efficiencies: Dilute nitride p-i-n (bulk or MQWs) in conjunction with practical Si Solar cells.

Khim Kharel and Alexandre Freundlich

Center for Advanced Materials, University of Houston

Houston, Texas, 77096, USA

Abstract --- Tandem silicon solar cells are purposed to lower the levelized cost of solar electricity by increasing cell's efficiencies more than 30%. In reality, dual junction thin film III-V/Si cell's efficiencies are limited to 20% because of high dislocation densities ($>10^6/cm^2$) due to lattice mismatch. GaAsyP1-x-yNx has a lattice constant matched with silicon at y=4.7*x-0.1 and has an optimal bandgap for tandem III-V/Si solar cells. In term of top subcell of silicon tandem, we have designed p-i-n (GaAsPN/GaP MQWs) or GaAsPN bulk solar cells lattice matching with silicon and simulate their solar cell properties using realistic drift-diffusion approach. Using experimental spectral responses including with other solar cells properties of various silicon solar cells such that HIT, PERL, PERC, PERT, IBC etc., we have evaluated tandem efficiencies of designed p-i-n bulk or MQWs in conjunction with different silicon solar cells. Tandem efficiencies of optimized p-i-n MQWs and bulk under AM 1.5G spectrum maximum at 34% and 32% in conjunction with 25.6% efficient HIT silicon solar cell.

Index Terms — III-V-N semiconductor materials, Resonant thermo-tunneling design, Tandem solar cell, Drift-diffusion model.

I. INTRODUCTION

To get high efficiency solar cells and reduce the cost at the same time are always important in photovoltaic technologies. Moreover, single junction silicon solar cell's efficiencies are almost saturated. Thus, to reduce levelized cost of solar electricity, we need to integrate high efficient III-V with cheap silicon solar cell for the increment of efficiency more than 30%. Maximum radiative efficiency under AM 1.5G spectrum of silicon based tandem (1.74eV/1.12eV) is 41.9% [1]. Similarly, optimized top cell thickness of series connected III-V/Si (1.7eV/1.12eV) tandem has a maximum efficiency at 37% under AM1.5G spectrum [2]. There are two major approaches to integrate III-V on silicon: a) hetero-epitaxial growth or monolithic b) mechanical stacking or wafer bonding technique. Monolithic integration technique is designed to integrate high efficient III–V solar cells on Si using a single substrate and single epitaxial process. However realistic efficiencies of such dual junction solar cells are limited to 20% because of high dislocation densities ($>10^6 cm^{-2}$), due to lattice mismatch. The performance of such solar cells can be improved by growing

metamorphic graded buffer layers to bridge the lattice constant between III-V and Si. This technique has created a lot of attention in the recent year [3, 4], but the major drawback of this method is that it increases the overall cost and time of the epitaxial growth. Another approach which could tolerate a large lattice mismatch is to develop III-V/Si by mechanical stacking using a transparent adhesive between two subcells. The measured efficiencies, under AM 1.5G spectrum, of a four terminal mechanically stacked GaInP/Si are up to 30% [5, 6]. Despite promising results in the literature, challenges such as matching the bonding temperature with III-V and Si, bonding layer should be thin, optically transparent and electrically conductive. Promising but less explored approach is to design dilute nitride (due to their abnormal properties such that bandgaps shrinkage and increase in effective mass via absorption with nitrogen) solar cells lattice matching with silicon. In this paper, we study the possible integration of p-i-n $GaAs_{0.09}P_{0.87}N_{0.04}$/GaP MQWs or bulk $GaAs_{0.09}P_{0.87}N_{0.04}$ solar cells with various practical silicon solar cells. We have also applied resonant thermotunneling designed purposed by A.Freundlich and A. Alemu (US pattern 20130186458A1) in MQWs solar cells to extract electrons from deep quantum well in a picosecond time before recombine with holes.

II. MODELING APPROACH

At the beginning, we worked on the estimation of the performance of designed p-i-n bulk or MQWs solar cells using realistic drift-diffusion approach. Dark currents in both p-i-n bulk and MQWs solar cells are derived from the parameters extracted from references [7] as well as other past experiments in the literature. Similarly, minority charge carrier's diffusion lengths are derived as a function of doping concentrations. Bandgap evolution of purposed design is calculated using 8 band k.p Hamiltonian with Band Anti-crossing model to the conduction band to take account for lesser amounts of nitrogen impurities [8]. Optical absorption in a coupled quantum wells and bulk including the effect of excitons are evaluated using confinement energies and bandgaps. The calculated value of absorption corresponds with the maximum experimental

978-1-5090-5606-4/17 $31.00 © 2017 IEEE

absorption coefficient of bulk GaAsPN below bandgap of GaP i.e about 40000/cm [9, 10]. We have also evaluated internal quantum efficiencies of the resonantly coupled quantum wells in MQWs solar cells using thermionic escape time for carriers excited to higher confined states and tunneling escape time for resonantly coupled states and recombination time which we have taken 100ps by experience for MQWs [11]. For practical silicon solar cells, short circuit current is derived from EQE measurement is given by

$$ J_{sc} \equiv q \int \phi(\lambda) EQE(\lambda) d\lambda \qquad (1) $$

$\phi(\lambda)$ Is the solar spectrum.

Similarly, EQE values from SR measurement is given by

$$ EQE \equiv \frac{hc * SR}{q\lambda} \qquad (2) $$

Dark currents and ideality factors for silicon solar cells are extracted from experimental values of open circuit voltages and fill factors. Finally, the performance of dual junction series connected solar cell is determined under the assumption that both junctions operate at same current.

III. RESULT AND DISCUSSION:

Based on the k.p calculation of band structures of GaAsPN lattice matching on silicon, we have evaluated the absorption coefficient as a function of energies which is as shown in the inlet of the figure [1].

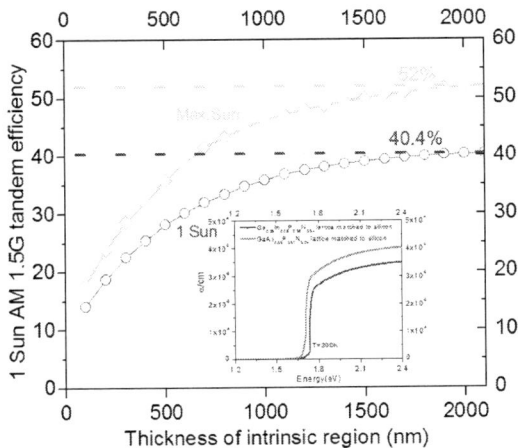

Fig 1: Detail balanced calculation for tandem efficiencies of p-i-n bulk GaAsPN/Si solar cell using calculated values of absorption coefficient of GaAsPN as shown in the inlet of the figure.

Using calculated value of absorption, we have evaluated the detail balanced efficiency under AM 1.5G spectrum of tandem junction p-i-n $GaAs_{0.09}P_{0.87}N_{0.04}$ with silicon for 1 sun and maximum concentration. Ideally, we can get efficiency of about 40% for 1 sun and 52% under maximum concentration using 2000 nm bulk GaAsPN in the intrinsic region of p-i-n GaP based solar cells as shown in figure [1].

In a purposed resonantly coupled quantum well design which consists of three asymmetric coupled quantum wells of width

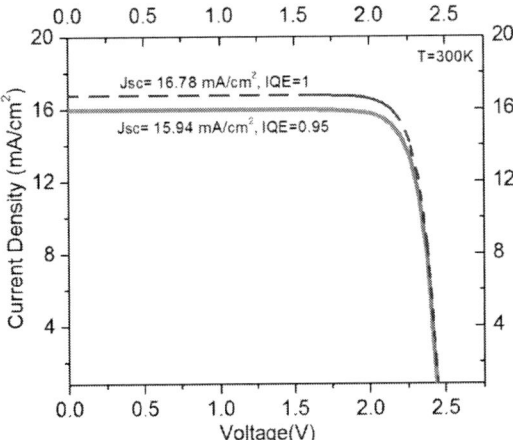

Fig 2: JV characteristics of dual junction tandem p-i-n $GaAs_{0.09}P_{0.87}N_{0.04}/GaP$ coupled MQWs in conjunction with HIT silicon solar cell with interdigitated back contacts [12].

of 12 nm 5.9 nm and 2.5 nm of $GaAs_{0.09}P_{0.87}N_{0.04}$ separated by 4 nm width of GaP barrier, we have calculated IQE=0.9983 for a single coupled well and 0.95 for thirty periods resonantly coupled wells. Bandgap of such design calculated from confinement energies is 1.72eV contributed from 12 nm quantum well. Simulation of current voltage characteristics of dual junction optimized p-i-n MQWs in conjunction with 25.6% efficient HIT solar cells with interdigitated back contacts [12] is as shown in the Fig [2]. Resonant thermo-tunneling design in the top sub cell improve both Jsc and Voc. This is because, faster collection of charge carriers prevents from recombination losses. Similarly, Figure [3] shows the calculation of the dual junction, series connected tandem efficiencies of p-i-n MQWs under AM 1.5G spectrum in conjunction with HIT silicon solar cell having interdigitated back contacts with respect to the number of coupled quantum wells of MQWs solar cells. Optimized efficiency of such tandem junction solar cell is maximum at 33.80% under AM 1.5G spectrum.

Similarly, we have also done similar type of calculation for other several types of silicon solar cells such that HIT only, IBC, PERL, PERT and PERC solar cells using their spectral responses, Voc and FF. We have reported the performance of tandem junction in conjunction with HIT silicon solar cell having interdigitated back contacts as shown in Fig [2] and Fig [3] because it has highest spectral response towards band edge

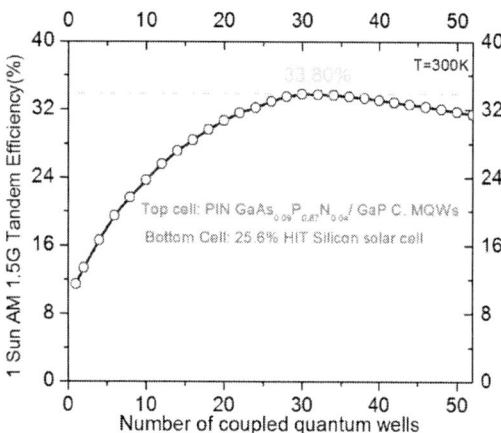

Fig 3: Tandem efficiencies of p-i-n $GaAs_{0.09}P_{0.87}N_{0.04}$ /GaP coupled MQWs solar cell in conjunction with HIT silicon solar cell having interdigitated back contacts [12].

of silicon. Despite highest Voc in HIT silicon solar cell reported in reference [14], it's not possible for monolithic integration of III-V-N with this silicon solar cell due to the presence of a-Si layer on top of crystalline silicon. Among rest of them, PERC

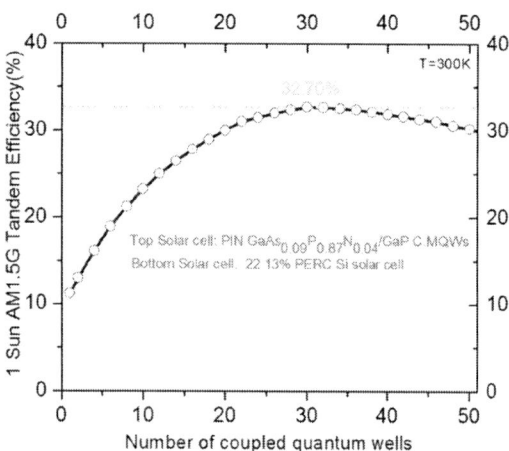

Fig 4: Tandem efficiencies of p-i-n $GaAs_{0.09}P_{0.87}N_{0.04}$ /GaP C. MQWs solar cell in conjunction with PERC silicon solar cell [13].

solar cell is one of the best candidate for the integration of III-V-N with silicon because of its suitable design. Simulation of tandem efficiencies of p-i-n MQWs in conjunction with 22.13% PERC solar cells [13], under AM 1.5G spectrum with respect

to the number of such coupled quantum wells is as shown in the Fig [4]. Maximum efficiency of the optimized tandem solar cell is 32.70% under AM 1.5G spectrum.

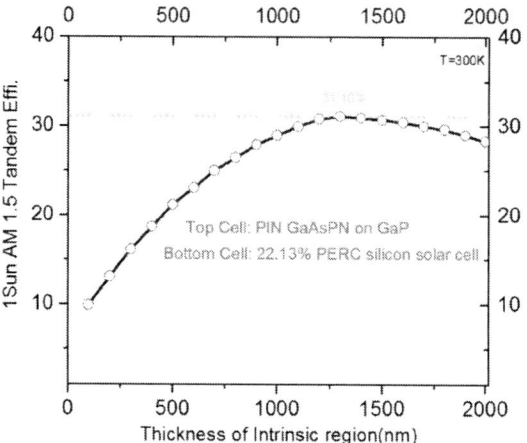

Fig 5: Tandem efficiencies of p-i-n bulk $GaAs_{0.09}P_{0.87}N_{0.04}$ solar cell in conjunction with PERC silicon solar cell [13].

On the other hand, we have also calculated the performance of p-i-n bulk $GaAs_{0.09}P_{0.87}N_{0.04}$ (having bandgap 1.71eV) solar cell taking $10^{14}/cm^3$ (nominal) background doping density in the intrinsic region. Optimized bulk p-i-n solar cell can contribute short circuit current of about $18mA/cm^2$ and efficiency about 20% under AM 1.5G spectrum. Simulation of tandem junction efficiencies of p-i-n bulk in conjunction with 22.13% PERC silicon solar cells [13], under AM 1.5G spectrum with respect to the thickness of intrinsic region i.e. thickness of GaAsPN absorbing layers is as shown in the Fig [5]. Maximum efficiency of the optimized tandem solar cell is about 31% under AM 1.5G spectrum using 1300 nm thickness of the intrinsic region.

IV. CONCLUSION

In this paper, we have evaluated the tandem solar cell's efficiencies of p-i-n bulk or coupled MQWs of GaAsPN in conjunction with various high efficient silicon solar cells using experimental measured values of spectral response, Voc and FF. Obtaining such promising performance is a very important motivation for monolithic integration of lattice matching III-V-N on silicon solar cells and achieving practical value of silicon based tandem cell's efficiency above 30%. Designing of tunnel junction between III-V-N and silicon tandem solar cells is explained in reference [15].

REFERENCES

[1] James P. Connolly, Denis Mencaraglia, Charles Renard and

978-1-5090-5606-4/17 $31.00 © 2017 IEEE

Daniel Bouchier, Prog. Photovolt: Res. Appl. 2014; 22:810–820.

[2] S.R. Kurtz, P. Faine, J.M. Olson, J.Appl.Phys.68, 4(1990).

[3] Tyler J. Grassman, Mark R. Brenner, Maria Gonzalez, Andrew M. Carlin, Raymond R. Unocic, Ryan R. Dehoff, Michael J. Mills, and Steven A. Ringel, IEEE Transactions on Electron Devices, 57, 10 (2010).

[4] M. R. Lueck, C. L. Andre, A. J. Pitera, M. L. Lee, E. A. Fitzgerald, and S. A. Ringel, IEEE Electron Device Letters 27, 3 (2006).

[5] Stephanie Essig, Myles A. Steiner, Christophe Allebe, John F. Geisz, Bertrand Paviet-Salomon, Scott Ward, Antoine Descoeudres, Vincenzo LaSalvia, Loris Barraud, Nicolas Badel, Antonin Faes, Jacques Levrat, Matthieu Despeisse, Christophe Ballif, Paul Stradins, and David L. Young, IEEE Journal of Photovoltaic, 6, 4 (2016).

[6] Stephanie Essig, Scott Ward, Myles A. Steiner, Daniel J. Friedman,John F. Geisz, Paul Stradins, David L. Young., Energy Procedia 77 (2015) 464 – 469.

[7] Xuesong Lu, Susan Huang, Martin B. Diaz, Nicole Kotulak, Ruiying Hao, Robert Opila, and Allen Barnett, IEEE Journal of Photovoltaic, 2, 2 (2012).

[8] L. Bhusal and A. Freundlich, Physical Review B 75, 075321(2007).

[9] J.F. Geisz, D.J. Friedman, Semicond. Sci. Technol., 17 (2002), pp. 769–777.

[10] S. Ilahi, Samy Almosni, Fares Chouchane, Mathieu Perrin, K. Zelazna, Solar Energy Materials and Solar Cells, Elsevier, 2015, 141, pp.291- 298.

[11] A. Freundlich, A. Alemu, IEEE Journal of Photovoltaics 2, 3 (2012)

[12] Keiichiro Masuko, Masato Shigematsu, Taiki Hashiguchi, Daisuke Fujishima, Motohide Kai, Naoki Yoshimura, Tsutomu Yamaguchi, Yoshinari Ichihashi, Takahiro Mishima, Naoteru Matsubara, Tsutomu Yamanishi, Tsuyoshi Takahama, Mikio Taguchi, Eiji Maruyama, and Shingo Okamoto, IEEE JOURNAL OF PHOTOVOLTAICS, 4, 6 (2014).

[13] Feng Ye1, Weiwei Deng, Wangwu Guo, Ruiming Liu, Daming Chen1, Yifeng Chen, Yang Yang,Ningyi Yuan, Jianning Ding, Zhiqiang Feng, Pietro P. Altermatt, Pierre J. Verlinden, Conference Record of the IEEE PVSC (2016).

[14] Mikio Taguchi, Ayumu Yano, Satoshi Tohoda, Kenta Matsuyama, Yuya Nakamura, Takeshi Nishiwaki,Kazunori Fujita, and Eiji Maruyama, IEEE JOURNAL OF PHOTOVOLTAICS, 4, 1 (2014).

[15] Alain Rolland, Laurent Pedesseau, Jacky Even,Samy Almosni, Cedric Robert , Charles Cornet, Jean Marc Jancu, Jamal Benhlal, Olivier Durand,Alain Le Corre,Pierre Rale,Laurent Lombez, Jean-Francois Guillemoles , Eric Tea ,Sana Laribi, Opt Quant Electron (2014) 46:1397–1403.

Gallium Phosphide nanostructure on Silicon by Silica nanospheres lithography and Metal Assisted Chemical Etching

Sangpyeong Kim, Chaomin Zhang, Som Dahal, Stuart Bowden, and Christiana B. Honsberg

School of Electrical, Computer and Energy Engineering, Arizona State University, Tempe, Arizona, 85284, U.S.A.

Abstract - **This work focuses on exploring the ways to integrate III-V nanostructures on silicon to enhance the current silicon cell's efficiency by effective absorption and possibly using quantum mechanical phenomena such as MEG (multiple exciton generation) and effective absorption of light above the band gap in the nanostructures' confined states. The gallium phosphide (GaP) is integrated with silicon wafers in the form of nanostructures. GaP is epitaxially grown on silicon, and the nanostructures are fabricated by silica nanospheres lithography (SNL) with metal assisted chemical etching (MACE). With optimized MACE condition, we are able to fabricate these nanopillars on GaP with controlled size and aspect ratios. Silica nanospheres are coated on the GaP is on Si by spin-coating, and their size is modified by a reactive ion etching (RIE) process. Palladium (Pd) is deposited on the sample by electron beam deposition for MACE process. In this MACE process, we control the MACE solution by changing the ration of HF and H_2O_2 concentrations. The GaP nanostructures created from MACE solutions with high HF and H_2O_2 concentrations is taller and possesses a more well-defined shape. Also, GaP nanostructures created from different metal of MACE such as Pd, Au and Ti/Au. MACE with Pd shows the better shape nanostructure than others.**

I. Introduction

As single junction silicon solar cells are approaching the Shockley-Queisser (SQ) limit, exploring routes and approaches to go beyond this limit is essential. In this work, we explore the ways to integrate III-V nanostructures on silicon which not only enhances the path length of photons but also has possibility of effectively using light with wavelengths above the band gap for power conversion. Especially for thin silicon with implied Voc of up to 750 mV [1], the efficiency is limited by the short circuit current, and the integration of nanostructures with these solar cells is one of the options to approach or exceed the SQ limit. III-V semiconductor nanostructures can improve optical and electrical performance of photovoltaic devices. Nanostructures potentially improve light trapping effects for solar cells. The literature describes many different ways to fabricate III-V nanostructures [2]-[9]. Mostly, III-V nanostructures are fabricated by a vapor-liquid-solid (VLS) mechanism [10] or reactive ion etching (RIE) [11]. However, these mechanisms by their very nature generate defects such as metal impurities, surface damage and crystallographic defects [12]. These defects have detrimental effect on the electrical property of nanostructures. Conversely, the metal assisted chemical etching (MACE) process offers a defect-free method for creation of III-V nanostructures. Defect-free GaAs p-i-n nanopillar arrays have been fabricated

by MACE [4]. In the MACE process, the nanostructure is fabricated by patterning with a noble metal film [13] or metal particle such as Ag, Au or Pt [14]-[16] in a mixed solution with HF, H_2O_2 and H_2O. In this experiment, we use a Pd layer for MACE [12] and silica nanospheres lithography (SNL) for patterning [17]. For the SNL process, silica nanospheres are coated on the target wafer by spin-coating. Before spin-coating, the nanospheres are prepared with DMF (N,N-dimethyl-formamide) solvent because this enables a more stable coating condition [17]. This method provides a uniform and fast silica nanosphere coating on the target wafer. For III-V nanostructures on Si, we chose gallium phosphide (GaP) because of several advantages. First, GaP has very low lattice mismatch (<0.4%) with Si and can be grown epitaxially on Si [18]-[20]. Second, the free carrier mobility of GaP is higher comparable to Si than mobility with amorphous Si. Third, the bandgap of GaP (2.26eV) is much higher than that of Si (1.12eV). This means that GaP can be expected to be a good minority carrier blocker in a heterostructure with a silicon solar cell [21]. In this work, we fabricated GaP nanostructures using SNL and MACE. To optimize the results, we changed the mixed ratio of the MACE solution. Different MACE solution recipes have different HF and H_2O_2 concentrations, and these results will be compared. Also, we compare GaP nanostructure from different metals (Palladium (Pd), Gold (Au) and Titanium (Ti)/Au) of MACE.

II. Experiments and Results

A. Fabrication of gallium phosphide (GaP) nanostructures on Si

We used a solid-source Veeco GEN-III MBE system, equipped with a phosphorus valved-cracker cell and a gallium source to grow GaP. GaP of 1 μm and 2 μm thicknesses was epitaxially grown on the Si (001) substrates with a growth rate of 0.52 um/hr at 580 °C.

The GaP nanostructure was fabricated by silica nanospheres lithography (SNL) and metal assisted chemical etching (MACE). The nanospheres were prepared using a DMF solution to implement SNL. After they are mixed with the DMF solution an ultrasonic treatment is used for a few hours. After the ultrasonic treatment, spin-coating on the GaP layer is performed at 2000 RPM for 200sec. Fig. 1 schematically depicts the whole process flow for nanostructure fabrication.

Spin-coated nanospheres are etched by a florine-based reactive ion etching (RIE) process for size modification. After the RIE etching, the mean size of nanospheres is a 300 nm in diameter. For MACE, we deposited palladium (Pd) to a thickness of 30 nm using electron beam deposition. The MACE solution is comprised of 2 parts of HF, 1 part of H_2O_2 and 1 part of H_2O. The samples were subsequently etched in the solution for 10min. Fig. 2 shows the spin-coated nanospheres on a wafer. The mean size of the nanospheres was determined to be 520 nm in diameter. We success the conformal coating of nanospheres on Si by spin-coating.

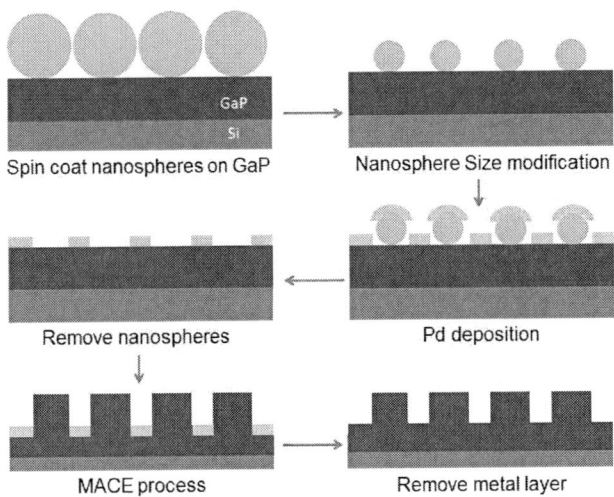

Fig. 1. Schematic of process flow – GaP layer is grown on Si and nanospheres are spin-coated and then modified in size by RIE. Metal layers are deposited by electron beam evaporation. After that, sample is given the MACE process with several solution recipes.

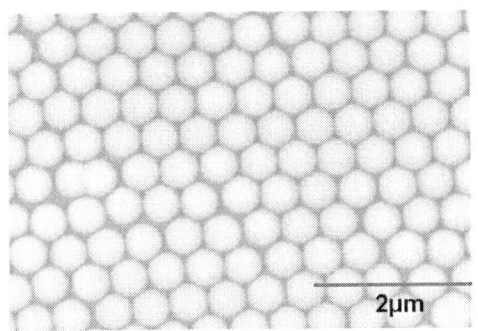

Fig. 2. SEM image of nanospheres on GaP dispersed by spin coating

Fig. 3 shows the GaP nanostructure after MACE. The nanopillars as seen in Fig. 3, have average height and diameter of 165 nm and 300 nm respectively.

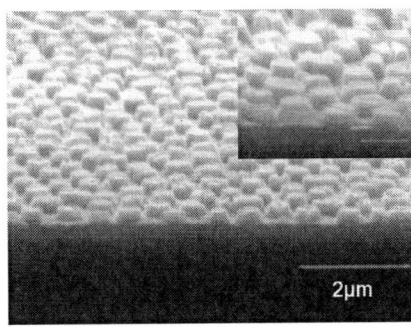

Fig. 3. SEM image of GaP nanopillar structure after the SNL and MACE processes. After MACE process, metal layers are removed by Aqua Regia. Inset scale bar is 500 nm.

B. *GaP nanostructure using different MACE solution recipes*

GaP nanostructures were fabricated using several MACE solutions which have different mixing ratios of HF, H_2O_2 and H_2O. One mixing ratio explored was 2 parts of HF, 1 part of H_2O_2 and 1 part of H_2O [12]. This mixing ratio has the fastest MACE speed, but the GaP layer is underetched a lot. Another solution recipe tested was 6 parts of HF, 1 part of H_2O_2 and 10 parts of H_2O. This recipe exhibited more gentle MACE conditions and avoided critical side etching of the GaP layer. A third recipe consisted of 5 parts of HF, 1 part of H_2O_2 and 44 parts of H_2O [17]. We tried to use this recipe for more controlled MACE conditions. During the MACE process, bubbles were observed on the samples. Due to these bubbles, the Pd layer peeled off partially. Fig. 4 compares the GaP nanostructure from different solution recipes. All samples were etched for 10 min at room temperature.

(a)

(b)

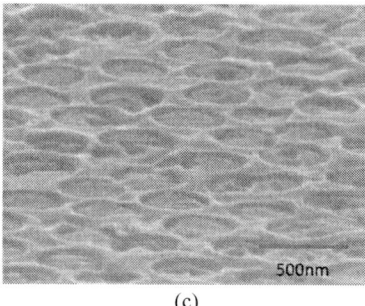
(c)

Fig. 4. SEM image of GaP nanostructure fabricated with different MACE solution recipes: (a) HF:H₂O₂:H₂O (2:1:1) solution (b) HF:H₂O₂:H₂O (6:1:10) solution (c) HF:H₂O₂:H₂O (5:1:44) solution.

We can see a trend that higher HF concentrations yield taller nanostructures for the same processing time and temperature. Figure 4(c) indicates that the GaP is not etched enough for nanostructure creation. However, the high HF concentration recipe shows side etching of the GaP layer.

C. GaP nanostructures using different metals for MACE

GaP nanopillar structures were also fabricated by the MACE process with different metal layers such as Pd, Au and Ti/Au. We compared these results for improving the GaP MACE process. Fig. 5 shows the MACE results for these different metal layers. Every MACE process used the 2:1:1 ratio solution for 10 min. Fig 5(a) is a SEM image showing the best results from all process splits. During the MACE process, the reaction was very aggressive which produced a lot of bubbles. After the MACE process, metal layers are lifted off by chemical reaction. Fig. 5(c) indicates that a nanostructure is not formed because the Ti/Au layers were peeled off quickly during the MACE process.

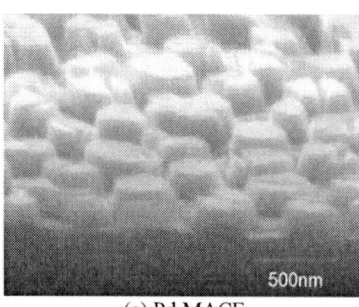
(a) Pd MACE

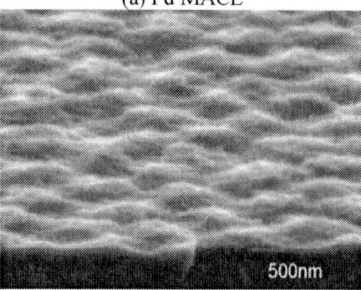
(b) Au MACE

(c) Ti/Au MACE

Fig. 5. SEM image of MACE results using several different metal layers. (a) The best results, MACE with Pd. (b) Poorly-defined nanostructures using MACE with Au. (c) MACE with Ti/Au exhibits no nanostructure. Every MACE process was performed for 10 min with 2:1:1 mixed ratio solution.

III. Conclusion

We demonstrated successful fabrication of GaP nanostructure using SNL and MACE. GaP nanostructures were fabricated by different MACE solution recipes with Pd layer, and high HF concentration has better etching speed. Also, structures were fabricated using different metal layer for MACE with 2:1:1 ratio MACE solution. MACE with a Pd layer yielded better results than observed for the other metal layers (Au and Ti/Au). Further optimization is necessary to produce nanostructures in GaP using SNL and MACE. Going forward, the MACE recipe will be optimized and the resulting nanostructure will be fully characterized and different schemes of surface passivation will be implemented to study the effectiveness of these structures in resulting solar cells.

References

[1] A. Augusto, K. Tyler, S. Y. Herasimenka and S. G. Bowden, "Flexible modules using <70µm thick silicon solar cells", *Energy Procedia*, vol. 92, pp. 493-499, 2016

[2] Y. Yasukawa, H. Asoh and S. Ono, "Site-Selective Metal Patterning/Metal-Assisted hemical Etching on GaAs Substrate Through Colloidal Crystal Templating", *Journal of the Electrochemical Society*, vol. 156, pp. H777-H781, 2009

[3] M. DeJarld *et al.*,"Formation of High Aspect Ratio GaAs Nanostructure with Metal-Assisted Chemical Etching", *Nano Letter*, vol. 11, pp. 5259-5263, 2011

[4] P. K. Mohseni *et al.*, "GaAs pillar array-based light emitting diodes fabricated by metal-assisted chemical etching", *Journal of Applied Physic*, vol. 114, pp. 064909, 2013

[5] L. Xiuling, *et al.*, "In-plane bandgap control in porous GaN through electroless wet chemical etching", *Applied Physics Letters*, vol. 80, pp. 980, 2002

[6] X. Y. Guo *et al.*, "Enhanced ultraviolet photoconductivity in porous GaN prepared by metal-assisted electroless etching", *Solid State Communications*, vol. 140, pp. 159-162, 2006

[7] F. K. Yam and Z. Hassan, "Structural and potical characteristics of porous GaN generated by electroless checmical etching", *Materials Letters*, vol. 63, pp. 724-727, 2009

[8] B. K. Duan and P. W. Bohn, "High sensitivity hydrogen sensing with Pt-decorated porous gallium nitride prepared by metal-assisted electroless etching", *Analyst*, vol. 135, pp. 902-907, 2010

[9] T. Yokoyama, H. Asoh and S. Ono, "Site-selective anodic etching of InP substrate using self-organized spheres as mask", *Physica status solidi. A*, vol. 207, pp. 943-946, 2010

[10] R. E. Algra *et al.*, "Paired Twins and {112} Morphology in GaP Nanowires", *Nano Letters*, vol. 10, pp. 2349-2356, 2010

[11] C. Battaglia *et al.*, "Enhanced Near-Bandgap Response in InP Nanopillar Solar Cells", *Advanced Energy Materials*, pp. 1400061, 2014

[12] J. Kim and J. Oh, "Formation of GaP nanocones and micro-mesas by metal-assisted chemical etching", *Physical Chemistry Chemical Physics*, vol. 18, pp. 3402-3408, 2016

[13] K. Peng et al.," Ordered silicon nanowire arrays via nanoshpere lithography and metal-indued etching", *Applied Physics Letters*, vol. 90, pp. 163123, 2007

[14] X. Li and P. W. Bohn, "Metal-assisted chemical etching in HF/H_2O_2 produces porous silicon", *Applied physics Letters*, vol. 77, pp. 2572-2574, 2000

[15] K. Tsujino and M. Matsumura, "Morphology of nanoholes formed in silicon by wet etching in solutions containing HF and H2O2 at different concentrations using silver nanoparticles as catalysts", *Electrochimica Acta*, vol. 53, pp. 28-34, 2007

[16] X. Li, et al., "Fast electroless fabrication of uniform mesoporous silicon layers", *Electrochimica Acta*, vol. 94, pp. 57-61, 2013

[17] J.-Y. Choi, T. L. Alford and C. B. Honsberg, "Fabrication of Periodic Silicon Nanopillars in a Two-Dimensional Hexagonal Array with Enhanced Control on Structural Dimension and Period," *Langmuir*, vol.31, pp. 4018-4023, 2016

[18] J. P. Andre, J. Hallais and C. Schiller, "Heteroepitaxial growth of GaP on Silicon", *Journal of Crysta Growth*, vol. 31, pp. 147-157, 1975

[19] S. L. Wright, H. Kroemer and M. Inada, "Molecular beam epitaxial growth of GaP on Si", *Journal of Applied Physics*, vol. 55, 1983

[20] J. T. Kelliher et al., "Low temperature chemical beam epitaxy of gallium phosphide/silicon heterostructures", *Materials Science and Engineering: B*, vol. 22, pp. 97-102, 1993

[21] H. Wagner et al., "A numerical simulation study of gallium-phosphide/silicon heterojunction passivated emitter and rear solar cell", *Journal of Applied Physics*, vol. 115, 2014

Efficiency Enhancement of InGaP/InGaAs/Ge Solar Cells with Gradually Doped P-N Junction Active Layers

Youngjo Kim[1], Sang Hyun Jung[1], Chang Zoo Kim[1], Kangho Kim[1,2], Hyun-Beom Shin[1], Kyung Ho Park[1],
Won-Kyu Park[1], Jaejin Lee[2], and Ho Kwan Kang[1,*]

[1]Korea Advanced Nano Fab. Center, Suwon, 16229, Korea

[2]Department of Electrical and Computer Engineering, Ajou University, Suwon, 16499, Korea

Abstract — **Graded doping techniques have been studied to improve the power conversion efficiency of the InGaP/InGaAs/Ge solar cells. We have investigated the efficiency enhancement of the triple-junction solar cells with gradually doped p-n junction active layers. The triple-junction epi-structures are grown by metalorganic chemical vapor deposition on Ge (100) substrates. The photovoltaic devices are fabricated and characterized under AM1.5 global illuminations. It was found that the graded doping profiles in InGaP emitter and InGaAs base layers can contribute to the efficiency improvement of the InGaP/InGaAs/Ge solar cells. The open circuit voltage and the short circuit current were improved up to 3.7 and 1.3 %, respectively, by employing the graded doping technique.**

I. INTRODUCTION

III-V multi-junction solar cells promise very high power conversion efficiencies beyond 40% under concentrated illuminations [1]. Recently, the InGaP/GaAs/Ge triple-junction solar cells have been used for various space and terrestrial applications due to their high efficiency and reliability [2–4]. However, III-V solar cell based photovoltaic systems are still suffering from the higher cell cost compared to Si based systems. Therefore, many research groups have been trying to reduce the fabrication cost and improve the power conversion efficiency.

Optimizing the doping profile in the p-n junction active layers is one possible method to achieve higher efficiencies for the III-V multi-junction solar cells. It is known that the open circuit voltage and the short circuit current density can be improved by employing graded doping profiles. However, there are few studies on the device characterization for the InGaP/GaAs/Ge solar cells, which have gradually doped p-n junction structures.

In this study, we demonstrate the efficiency enhancement of the InGaP/InGaAs/Ge solar cells with various graded doping profiles in p-n junction active layers. The triple-junction solar cells are grown by metalorganic chemical vapor deposition (MOCVD) on Ge (100) substrates and the fabricated device characteristics are investigated with a solar simulator and a quantum efficiency measurement system under AM1.5G illuminations.

II. EXPERIMENTAL

Lattice matched InGaP/InGaAs/Ge triple-junction solar cells are grown by MOCVD (AIXTRON, AIX2600 G3) on Ga doped p-type Ge (100) substrates with a misorientation of 6° toward [111]. Fig. 1 shows the schematic diagrams of the solar cells with three different p-n junction structures: sample A has uniformly doped p-n junction layers, sample B has gradually doped InGaP emitter and InGaAs base layers, and sample C has four gradually doped p-n junction layers.

During the epitaxial growth, the doping concentrations are kept at 1×10^{18} and 1×10^{17} cm^{-3} in the uniformly doped emitter and base layers, respectively. The doping level in the gradually doped emitter layer increases from 1×10^{18} to 8×10^{18} cm^{-3}, whereas that in the gradually doped based layer decreases from 8×10^{18} to 1×10^{17} cm^{-3}.

Phosphine (PH$_3$), arsine (AsH$_3$), Trimethylindium (TMIn), and trimethlygallium (TMGa) are used as the reactant sources for P, As, In, and Ga, respectively. Silane (SiH$_4$) and diethyltelluride (DETe) are used as n-type dopant sources, whereas diethylzinc (DEZn) and carbontetrabromide (CBr$_4$) are used as p-type dopant sources. The planetary reactor pressure is fixed at 50 mbar and palladium-diffused ultra-high-purity hydrogen (H$_2$) is used as a carrier gas with a total flow of 50,000 sccm. The reactor temperature is varied from 550 to 680 °C for the epitaxial growth.

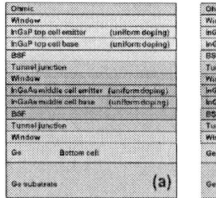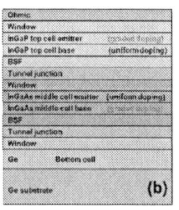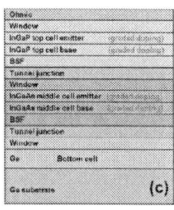

Fig. 1. Schematic diagrams of the solar cells with three different p-n junction structures: (a) sample A has uniformly doped p-n junction layers, (b) sample B has gradually doped InGaP emitter and InGaAs base layers, and (c) sample C has four gradually doped p-n junction layers.

The InGaP/InGaAs/Ge solar cells are fabricated by a conventional photolithography, metal evaporation, annealing, wet etching, anti-reflection coating, and back-end processes. The n- and p-type contact metals consist of AuGe/Ni/Au and Ti/Pt/Au, respectively. The ohmic contact layer is selectively etched in a $NH_4OH:H_2O_2:H_2O$ (2:1:10) solution. The solar cells are isolated by a conventional mesa isolation process using wet chemical etching and MgF_2/ZnS double-layers are deposited as the anti-reflection coating (ARC) on the top surface by an e-beam evaporator. Each solar cell device is divided by a dicing saw system and mounted on a printed circuit board (PCB) using silver paste and gold wires. The cell area of the solar cell is 0.3025 cm².

The photovoltaic J-V characteristics of the fabricated InGaP/InGaAs/Ge solar cells are measured using a class A solar simulator (WACOM, WXS-220S-L2) under AM1.5G illuminations. The external quantum efficiencies of the GaAs solar cells are measured with a solar cell quantum efficiency measurement system (PV Measurements, QEX7).

III. RESULTS AND DISCUSSION

The photovoltaic J-V characteristics of the fabricated InGaP/InGaAs/Ge solar cells with the three different p-n junction doping profiles are shown in Fig. 2. The inset shows the detailed J-V curves around 2.5 V to identify the difference of the open circuit voltage (V_{oc}). The specific device parameters of the fabricated solar cells under AM1.5G illuminations are summarized in Table I.

TABLE I
SUMMARY OF PHOTOVOLTAIC DEVICE PARAMETERS

Sample	Efficiency (%)	V_{oc} (V)	J_{sc} (mA/cm²)	Fill factor
A	26.87	2.44	12.97	0.849
B	28.26	2.51	13.14	0.856
C	28.00	2.53	12.89	0.859

The highest efficiency of 28.26% was achieved by the sample B, which has gradually doped InGaP emitter and InGaAs base layers, with a V_{oc} of 2.51 V, a short circuit current density (J_{sc}) of 13.14 mA/cm², and a fill factor (FF) of 0.856. The V_{oc}, J_{sc}, and FF were improved up to 2.9, 1.3, and 0.8 % compared to the sample A, which has the uniformly doped p-n junction layers. This means that the efficiency enhancement of the sample B is mainly due to the improved V_{oc}.

The additional electric field in the p-n junction layers, which is induced by the graded doping profile, could contribute to the improved carrier collection efficiency [5, 6]. In this reason, the sample C, which has four gradually doped p-n junction layers, shows the highest V_{oc} of 2.53 V. However, the J_{sc} of the sample C was degraded compared to other samples. This indicates that employing graded doping profiles in InGaP base and InGaAs emitter layers can be rather detrimental on current generation.

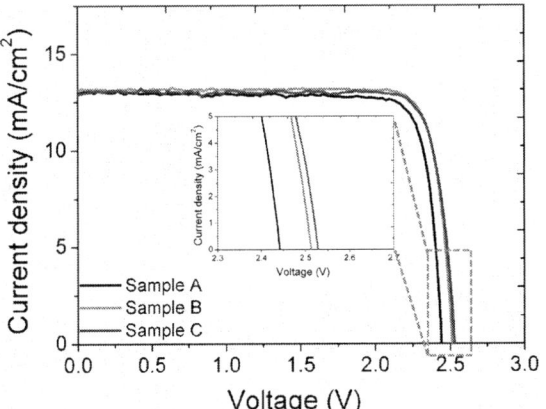

Fig. 2. Photovoltaic J-V characteristics of the fabricated InGaP/InGaAs/Ge triple-junction solar cells with three different p-n junction structures: sample A has uniformly doped p-n junction layers, sample B has gradually doped InGaP emitter and InGaAs base layers, and sample C has four gradually doped p-n junction layers.

Fig. 3. External quantum efficiencies of InGaP and InGaAs subcells in the fabricated InGaP/InGaAs/Ge triple-junction solar cells with three different p-n junction structures: sample A has uniformly doped p-n junction layers, sample B has gradually doped InGaP emitter and InGaAs base layers, and sample C has four gradually doped p-n junction layers.

Spectral responses of the InGaP/InGaAs/Ge solar cells were investigated with external quantum efficiency (EQE) measurements under AM1.5G illuminations to find out the reason that the J_{sc} changes with the graded doping profiles. The measured EQEs of the InGaP/InGaAs/Ge solar cells are shown in Fig. 3. The sample B, which has gradually doped InGaP emitter and InGaAs base layers, shows the higher EQEs in a broad range of measured wavelengths. This result is in good agreement with the highest J_{sc} of the sample B shown in Table I.

The InGaP top cell (TC) EQE of the sample B is higher than that of the sample A, which has uniformly doped p-n junction layers, especially, in the short wavelengths below 560 nm. This is evidence that the graded doped InGaP emitter layer can improve the carrier collection efficiency in the short wavelength range. In contrast, the TC EQE of the sample C was degraded in the wavelength range from 500 to 700 nm. This means that the minority carrier life time in the base layer becomes shorter by employing gradually doped InGaP base layer [7]. An increase of the doping level in the p-type InGaP base layer might affect to the change of electron diffusion length.

On the other hand, InGaP/InGaAs/Ge solar cells show relatively small changes in the InGaAs middle cell (MC) EQE. This could be attributed to the different absorption condition because the InGaAs MC absorbs filtered light by the 1 μm thick InGaP TC. Although the difference of the MC EQE affects to the InGaAs MC current, it cannot result in changes of the J_{sc} values in Table I because the total current is limited by the InGaP TC current.

IV. CONCLUSION

Lattice matched InGaP/InGaAs/Ge triple-junction solar cells were grown by MOCVD on Ge (100) substrates with different doping profiles. Gradually doped p-n junction structures were investigated to improve the power conversion efficiency of the triple-junction solar cell. The efficiency was improved up to 5.2% by employing gradually doped InGaP emitter and InGaAs base layers. It was demonstrated that the efficiency enhancement is mainly due to an increase of the V_{oc} attributed to an additional electric field in the gradually doped active layers. It was also found that graded doping profiles in InGaP base and InGaAs emitter layers can be rather detrimental on current generation.

ACKNOWLEDGEMENT

This work was supported by the Korea Institute of Energy Technology Evaluation and Planning (KETEP) and the Ministry of Trade, Industry & Energy (MOTIE) of the Republic of Korea (No. 20163030013980).

REFERENCES

[1] M. A. Green, K. Emery, Y. Hishikawa, W. Warta, and E. D. Dunlop, "Solar cell efficiency tables (version 48)", *Prog. Photovolt: Res. Appl.*, vol. 24, pp. 905–913, 2016.

[2] R. R. King, D. C. Law, K. M. Edmondson, C. M. Fetzer, G. S. Kinsey, H. Yoon, R. A. Sherif, and N. H. Karam, "40% efficient metamorphic GaInP/GaInAs/Ge multijunction solar cells", *Appl. Phys. Lett.*, vol. 90, p. 183516, 2007.

[3] W. Guter, J. Schone, S. P. Philipps, M. Steiner, G. Siefer, A. Wekkeli, E. Welser, E. Oliva, A. W. Bett, and F. Dimroth, "Current-matched triple-junction solar cell reaching 41.1% conversion efficiency under concentrated sunlight", *Appl. Phys. Lett.*, vol. 94, p. 223504, 2009.

[4] H. Helmers, M. Schachtner, and A. W. Bett, "Influence of temperature and irradiance on triple-junction solar subcells", *Sol. Energy Mater. Sol. Cells*, vol. 116, pp. 144–152, 2013.

[5] H. J. Hovel, R. K. Willardson, and A. C. Beer, *Semiconductors and semimetals.* New York: Academic Press, vol. 11, pp. 17–20, 1975.

[6] A. Fahrenbruch and R. Bube, *Fundamentals of solar cells: photovoltaic solar energy conversion.* New York: Academic Press, pp. 83–85, 1983.

[7] H. Cotal, C. Fetzer, J. Boisvert, G. Kinsey, R. King, P. Hebert, H. Yoon, and N. Karama, "III–V multijunction solar cells for concentrating photovoltaics," *Energy Environ. Sci.*, vol. 2 pp. 174-192, 2009.

Analysis of InGaP Oxide Growth Rate at High Temperatures and Ambient Conditions for Terrestrial Photovoltaic Applications

Nicole A. Kotulak*, Matthew P. Lumb[†], Raymond Hoheisel[†], Erin Cleveland[‡], Mitchell Bennett[§], Phillip P. Jenkins[‡], Robert J. Walters[‡]

*NRC Postdoctoral Associate residing at the U.S. Naval Research Laboratory, Washington, DC 20375 USA
[†]George Washington University, Washington, DC 20052 USA
[‡]U.S. Naval Research Laboratory, Washington, DC 20375 USA
[§]Sotera Defense Solutions, Inc., Herndon, VA 20171

Abstract—To evaluate the effects of high temperature terrestrial ambient conditions on InGaP films for use in PV devices, InGaP with a bandgap of 1.87 eV was grown on GaAs by MOVPE. The films were annealed in ambient conditions at 400°C and 450°C for a cumulative duration of 16 hours, respectively. At intervals of 1, 6, and 16 hours of annealing, VASE measurements were taken to determine the composition and thickness of the developing oxide layer. The oxide was found to be like GaP-oxide in nature, with parallel growth rates over 16 hours of annealing time. This trend was independent of temperature. Final oxide thickness was dependent on annealing temperature, with InGaP annealed at 450°C having a thicker oxide. Surface roughness also increased after annealing, as shown by AFM.

Index Terms—III-V semiconductor, InGaP, oxide, high temperature, ellipsometry

I. INTRODUCTION

Semiconductor devices have long been used in extreme environments, where the ability of the materials to withstand environmental stressors and maintain high levels of performance is mandatory.

The performance of photovoltaic (PV) devices at high temperatures is of great interest for terrestrial applications, and a robust theoretical body of work from the space sector presents ideal materials and device design considerations to ensure successful implementation of PV in extreme environments [1]–[5]. Drawing from theory, as well as state-of-the-art space PV technology, the characteristics of many materials have been evaluated at high temperature conditions, including element diffusivity, contact stability, and the behavior of defects at high temperatures [6]–[13]. Device-level analysis has led to suggested cell design alterations, process changes to prevent degradation from adhesives in the high temperature and high intensity space environment, and a deeper understanding of the temperature coefficients as a function of concentration [2], [3], [10], [12]–[15]. Consistently, materials with higher bandgaps are shown to better tolerate high temperatures, leading to both research and commercial space photovoltaics employing InGaP as one of the junctions [2]–[5], [12], [13], [15], [16].

For this work, terrestrial applications where the devices will continuously function at temperatures greater than, or equal to, 400°C are considered. These devices are intended as topping cells for solar thermal systems, which operate by collecting heat in the range of 200°C to 600°C [17]. This hybrid system design, combining photovoltaics with solar thermal technology, addresses the growing importance of energy storage, which is a necessary component of a grid with increasing renewable energy sources of supply. Previous studies have evaluated materials at temperatures much higher than 400°C, where materials were annealed for short bursts of time, on the order of 5 minutes under non-ambient gas flow (such as N_2) or vacuum [7]. This is not representative of the operating conditions of the proposed terrestrial hybrid system, therefore, longer annealing times under ambient conditions are required for this study. In addition, other methods used for space applications to externally divert direct solar insolation on the structures to reduce temperature-related damage is not feasible for this application [5], [13]. As InGaP has already been shown to perform well at high temperatures, evaluating the effects of long annealing times in ambient conditions to simulate extreme terrestrial conditions will identify potential sources of material failure.

II. EXPERIMENTAL METHODS

The InGaP film was grown by MicroLink Devices on an Aixtron 2600 MOVPE reactor. On a 625 μm semi-insulating GaAs substrate with a 6° offcut toward the (111) plane, a 500 Å undoped GaAs buffer layer was grown, followed by 5000 Å of undoped $In_xGa_{1-x}P$, where $x = 0.49$, with a resulting bandgap of 1.87 eV, which implies ordered material [18], [19].

Two InGaP films were annealed, one at 400°C and one at 450°C, using a Thermo Scientific Thermolyne benchtop muffle furnace, in an ambient atmosphere. Both samples were annealed in additive time intervals of 1 hr, 5 hrs, and 10 hrs, which led to measurements of oxide growth after 1, 6, and 16 total hours of annealing. During annealing, the samples were housed in a covered Pyrex petri dish, epi-side up, to minimize any particulate contamination of the surface during the anneal.

Characterization was performed with a Digital Instruments Dimension 3100 Series microscope for AFM analysis, and a J. A. Woollam Co., Inc. VASE system for ellipsometry. VASE measurements were taken at angles of 65°, 70°, and 75° over a photon energy range of 1 to 6 eV in steps of 0.03 eV.

978-1-5090-5606-4/17 $31.00 © 2017 IEEE

Fitting of ellipsometric data requires simultaneous analysis of all layers in the structure. The oxide optical parameters used a GaP oxide material file provided in the WVASE software. The GaP oxide was compared with two other oxides – InP oxide and a weighted Bruggeman EMA of InP and GaP oxides – and found to provide the best fit for the data. GaP oxide was used for all ellipsometric data presented in this work.

Data fitting of the InGaP films utilized the Herzinger–Johs parameterized semiconductor oscillator (Psemi) function [20]–[22]. This model was developed to analyze the dielectric functions of compound semiconductor films. In order to maintain Kramers–Kronig consistency, the model simulates the semiconductor critical point structures through the use of highly flexible oscillator lineshapes. The features of the oscillators were then used to provide the critical points (CPs) of the material, as well as amplitude and broadening characteristics.

III. RESULTS AND DISCUSSION

A. InGaP Optical Constants

The fitted real and imaginary dielectric functions of the InGaP film and their second derivatives are shown in Fig. 1(a) and Fig. 1(b), respectively. The critical points were determined by the energies of each Psemi oscillator function. The Psemi oscillator critical point values, which include the energy, amplitude, and broadening, are shown in Table I. The final column of the table shows the corresponding literature value for each critical point location for $x = 0.49$ InGaP.

TABLE I
TABLE OF PARAMETERS FOR CRITICAL POINT OSCILLATORS OF 1.87 EV
INGAP.

InGaP Critical Points

Oscillator	Energy (eV)	Amp.	Broadening	Adachi (eV) [23]
E_0	1.87	0.18	20	1.91
$E_0+\Delta_0$	2.02	0.44	100	2.00
E_1	3.28	28.50	171	3.23
E_0'	4.63	13.76	235.24	4.72
E_2	4.88	45.90	231.27	5.19

As can be seen in Table I, the critical points from the fit are in close agreement with the literature values from Adachi [23], with E_0 differing due to ordering [18], [19]. Proper fitting of the optical constants of the InGaP film provides the foundation for fitting the oxide layers and determining the oxide growth trend as a function of annealing time and temperature.

B. Oxide Growth Trend

Oxide thicknesses at each stage of the annealing study as described in Section II are shown in Fig. 2(a). The best fit of the data occurred when the oxide layer was modeled as GaP oxide, and an example of fit quality is shown in Fig. 2(b).

The thickness of the oxide from the as-grown InGaP material is nominally 18 Å, and is observed to increase in thickness after annealing, and with each subsequent annealing step (Fig. 2(a)). The anneal temperature impacts initial oxide thickness after 1 hr, with a temperature increase of 50°C resulting in an oxide that is approximately double the thickness of

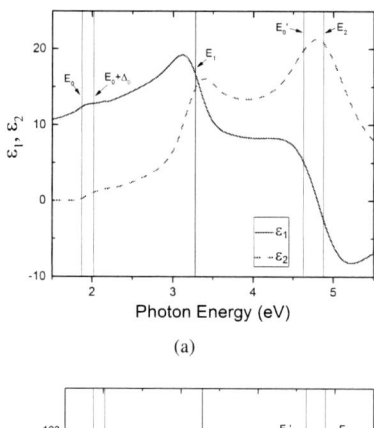

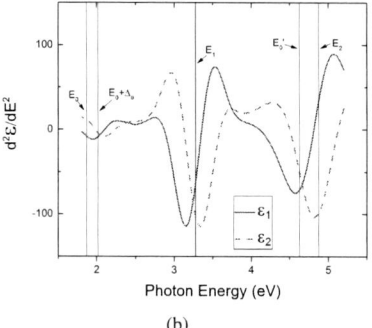

Fig. 1. Fitted *a)* Real and imaginary dielectric functions and *b)* their second derivatives for 1.87 eV InGaP material with critical points overlayed.

that which was grown at 400°C. This temperature dependence only appears to affect the growth rate within the first hour. During subsequent additive anneal times, shown at 6 and 16 hrs, the thickness of the oxide increased at similar parallel rates that were independent of annealing temperature. The overall thickness of the oxide at the end of the annealing test, therefore, maintained a ~40Å difference between the samples annealed at 400°C and at 450°C.

Although InGaP oxide growth rates appear to decrease over time, a device functioning under high temperature for a longer duration will likely be subject to extensive compounding oxide growth – the magnitude of which being determined by the operating temperature. Thus, high temperature applications for InGaP will require design considerations to either counteract oxide growth or eliminate its effect on device performance.

C. Surface Morphology

During the annealing process, changes in surface morphology are expected to occur as the oxide layer develops. To quantify this change in surface morphology, AFM scans of the as-grown and annealed samples are shown in Fig. 3.

No development of large or clustered surface features is aparent in the AFM scans. The surface remains uniform as the samples are annealed. Increased roughening of the surface does occur, however, as a function of temperature, as indicated by the increasing RMS surface roughness (R_q) values. The

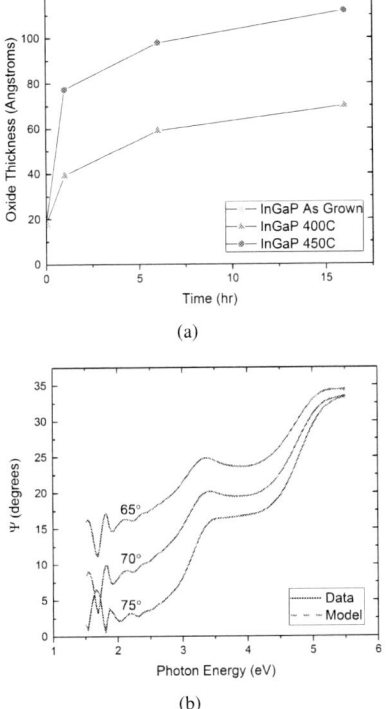

(a)

(b)

Fig. 2. *a)* Oxide thickness before annealing, and after each subsequent annealing step. *b)* An example fit of the data using a Psemi model for the InGaP film and GaP oxide for the oxide layer.

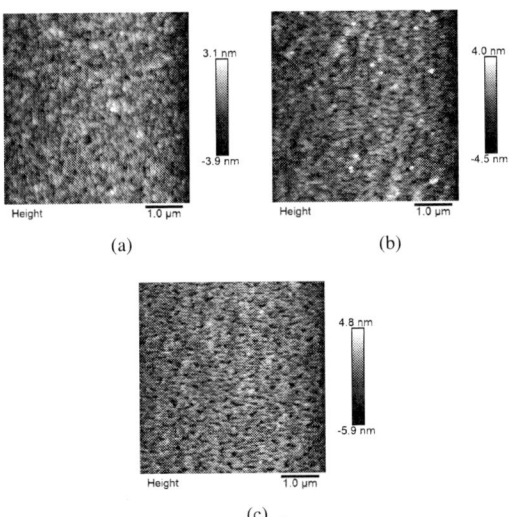

(a)　　　　　　　　(b)

(c)

Fig. 3. AFM on an InGaP surface when *a)* pre-annealed ($R_q = 0.944\ nm$), *b)* annealed at 400°C for 16 hours ($R_q = 1.17\ nm$), and *b)* annealed at 450°C for 16 hours ($R_q = 1.52\ nm$).

as-grown sample (Fig. 3(a)) has an R_q of $0.944\ nm$, with

R_q increasing to $1.17\ nm$ after 16hrs at 400°C (Fig. 3(b)) and $1.52\ nm$ after 16hrs at 450°C (Fig. 3(c)). This surface roughness increase is not an unexpected effect of oxide growth and possible desorption of P, which is known to occur at the annealing temperatures used for this work [24].

IV. CONCLUSIONS

In this work, the growth of InGaP native oxide in an ambient high-temperature environment for 1, 6, and 16 hours, during which oxide growth continuously occurs, was characterized. Annealing temperature strongly impacts the initial oxide growth rate, with a 450°C anneal resulting in an oxide ~40Å thicker than at 400°C after 1 hour. After the initial 1 hour, oxide growth rates decrease and maintain the ~40Å thickness difference between the two annealing temperatures, resulting in parallel growth curves. Surface morphology shows increased surface roughness after 16 hrs at both temperatures. Based on these results, device design considerations will be required to mitigate the effects of oxide growth due to the environmental conditions of a photovoltaic device used as a topping cell on a solar thermal system.

ACKNOWLEDGEMENT

The authors thank MicroLink Devices, Inc. for supplying the MOVPE-grown InGaP material. This work was performed while N. A. Kotulak held a National Research Council Research Associateship Award at the U.S. Naval Research Laboratory.

REFERENCES

[1] M. F. Lamorte and D. H. Abbott *IEEE Trans. Electron Devices*, vol. 27, no. 4, pp. 831–840, 1980.

[2] G. Siefer and A. W. Bett *Prog. Photovolt: Res. Appl.*, vol. 22, no. 5, pp. 515–524, 2014.

[3] G. A. Landis, D. J. Belgiovane, and D. A. Scheiman in *Proc. of 37th IEEE PVSC*, (Seattle), June 2011.

[4] D. A. Scheiman, G. A. Landis, and V. G. Weizer in *AIP Conf. Proc.*, (Albuquerque), Jan.–Feb. 1999.

[5] G. Landis, P. Jenkins, D. Scheiman, and R. Rafaelle in *2nd IECEC*, (Providence), Aug 2004.

[6] S. Gasner, G. Pack, M. Gates, and R. Given in *Proc. of 21st IEEE PVSC*, (Kissimmee), May 1990.

[7] M. B. Spitzer, J. Dingle, R. P. Gale, and R. H. Morrison *IEEE Trans. Electron Devices*, vol. 38, no. 8, pp. 1787–1791, 1991.

[8] P. Sharps, N. Fatemi, and D. Thang *IEEE 4th World Conf. Photovolt. Energy Conv.*, May 2006.

[9] S. P. Tobin, M. B. Spitzer, C. Bajgar, L. Geoffroy, and C. J. Keavney in *Proc. of 19th IEEE PVSC*, (New Orleans), May 1987.

[10] C. G. Zimmermann, C. Nömayer, M. Kolb, and A. Caon in *Proc. of 37th IEEE PVSC*, June 2011.

[11] R. P. Raffaelle, S. G. Bailey, P. Neudeck, R. Okojie, C. M. Schnabel, M. Tabib-Azar, D. Scheiman, P. Jenkins, and S. Hubbard in *Proc. of 28th IEEE PVSC*, (Anchorage), September 2000.

[12] C. G. Zimmermann *Appl. Phys. Lett.*, vol. 102, p. 233506, 2013.

[13] C. G. Zimmermann, C. Nömayer, M. Kolb, and A. Rucki *Prog. Photovolt: Res. Appl.*, vol. 21, pp. 420–435, 2013.

[14] O. V. Sulima, P. E. Sims, J. A. Cox, M. G. Mauk, R. L. Mueller, R. C. R. Jr., A. M. Khammadov, P. D. Paulson, and G. A. Landis *IEEE 3rd World Conf. Photovolt. Energy Conv.*, May 2003.

[15] H. Helmers, M. Schachtner, and A. W. Bett *Sol. Energy Mater. Sol. Cells*, vol. 116, pp. 144–152, 2013.

[16] T. Shimada, H. Toyota, A. Kukita, M. Imaizumi, K. Hirose, M. Tajima, H. Ogawa, H. Hayakawa, A. Okamoto, Y. Nozaki, H. Watabe, and T. Hisamatsu in *Proc. of 35th IEEE PVSC*, (Honolulu), June 2010.

[17] H. M. Branz, W. Regan, K. J. Gerst, J. B. Borak, and E. A. Santori *Energy Environ. Sci.*, vol. 8, pp. 3083–3091, 2015.

[18] J. K. Shurtleff, R. T. Lee, C. M. Fetzer, and G. Stringfellow *Appl. Phys. Lett.*, vol. 75, no. 13, 1999.

[19] T. Suzuki, A. Gomyo, S. Iijima, K. Kobayashi, S. Kawata, I. Hino, and T. Yuasa *Jpn. J. Appl. Phys.*, vol. 27, no. 11, 1988.

[20] C. M. Herzinger, B. Johs, W. A. McGahan, J. A. Woollam, and W. Paulson *J. Appl. Phys.*, vol. 83, p. 3323, 1998.

[21] C. M. Herzinger and B. D. Johs *U.S. Patent*, no. 5,796,983, 1998.

[22] B. Johs, C. Herzinger, J. Dinan, A. Cornfeld, and J. Benson *Thin Solid Films*, vol. 313–314, pp. 137–142, 1998.

[23] S. Adachi, *Properties of Semiconductor Alloys: Group-IV, III-V and II-VI Semiconductors*, chapter 6, p. 156. Materials for Electronic & Optoelectronic Applications, West Sussex, UK: Wiley, 2009.

[24] M. Heinrich, C. Domke, P. Eber, and K. Urban *Phys. Rev. B*, vol. 53, pp. 894–897, 1996.

Grain boundaries in Thin-Film Polycrystalline GaAs Solar Cells: A Simulation Study

Khushboo Kumari and Sushobhan Avasthi

Centre for Nano Science and Engineering, Indian Institute of Science, Bengaluru, India.

Abstract — There is a demand of highly efficient, thin, light and flexible solar cells. Polycrystalline GaAs could satisfy this need. However, effect of grain size on performance has to be evaluated. Here, the effect of grain boundaries under dark and illuminated conditions is quantified. A 2-D simulation of single junction GaAs solar cell was done using SILVACO software. Columnar grain structures were assumed for simplicity and also because these kinds of grain structures give maximum efficiency. Device performance was simulated for various grain sizes, assuming Gaussian distribution and varying density of defect states at the grain boundary. Carrier lifetime of at least 10 ns was required to have proposed efficiency of greater than 20% for a grain size lesser than 100 µm.

Index Terms — Gallium Arsenide, Grain boundary, Polycrystalline, Solar cell, Simulation.

I. INTRODUCTION

GaAs solar cells have unrivalled efficiency of 28.8 % [1]. Recently, a 22.08 % efficient GaAs thin-film solar cell on flexible substrate was also demonstrated [2]. However, all the demonstrations use monocrystalline GaAs which requires expensive substrates, such as Ge and GaAs. One way to reduce cost is to use polycrystalline GaAs for thin-film solar cells [3]. Polycrystalline GaAs will suffer grain boundary defects but hopefully the performance penalty can be reduced by intelligent device design. In this work, we report a 2D-simulation study of thin-film polycrystalline GaAs solar cell. Using a SILVACO TCAD software and published data on distribution and density of trap and defect states at GaAs grain boundaries, maximum potential efficiency of GaAs solar cells is calculated. We find that efficiency > 20% are possible in polycrystalline solar cells, as long as GaAs has columnar grains with ~100 um grain size, base thickness, and emitter doping are optimized. Effect of recombination is also studied, which provides a rough measure of maximum bulk contamination that can be tolerated in GaAs layers.

II. THEORY AND DISCUSSION

The grain boundaries act as recombination centers and hence decrease the efficiency of the solar cells. The effect is decided by the distribution (uniform, delta or Gaussian), types (acceptors or donors) and density of these defects at the grain boundaries. We assumed a Gaussian distribution of defects for our analysis and simulation.

$$D_{it}(E) = N_{it} e^{\frac{(E-E_t)^2}{\frac{2S^2}{\sqrt{2\pi}S}}} \tag{1}$$

where D_{it} (E) is the energy distribution of defect states at the grain boundary (per unit area per eV), N_{it} is the total density of defect states per unit area, S is the Gaussian distribution parameter and E_t is the energy position of the mean value of interface states from the valence band edge [4].

These defects trap the majority carriers and hence a depletion width or a potential barrier is formed around the grain boundary like the Schottky junction. We assumed Shockley-Read-Hall recombination (eqn.2) at the grain boundary defects which causes reduction in the barrier height under optical illumination.

$$U_{SRH} = \frac{pn - n_i^2}{\tau_{po}\left[n + n_i e^{\frac{E_i-E_t}{kT}}\right] + \tau_{no}\left[p + n_i e^{\frac{-(E_i-E_t)}{kT}}\right]} \tag{2}$$

where U_{SRH} is the net recombination rate and E_i is the intrinsic fermi level.

III. DEVICE MODELLING AND SIMULATION

The analysis for evaluating the effect of grain boundaries on the performance of polycrystalline GaAs solar cells was done using SILVACO software. GaAs thin-films with columnar grains do not present grain-boundaries in the current path of a solar cell. Hence, in terms of efficiency, columnar films have the highest potential for thin-film solar cells. In the present work, GaAs thin-films with columnar grains were simulated.

Two models have been done for simulating the grain boundary. The values of various parameters used in the simulation are listed in Table 1. The cell is illuminated by AM 1.5 G spectrum from the top.

Table 1: Various Parameters used during the simulation

Parameter	Optimized Value
Eg (GaAs)	1.42 eV
N_C	4.35×10^{17}
N_V	1.29×10^{19} cm^{-3}
ni (GaAs)	2.67×10^6 cm^{-3}
Emitter Doping (p$^+$)	10^{19} cm^{-3}
Base Doping (n)	10^{16} cm^{-3}

978-1-5090-5606-4/17 $31.00 © 2017 IEEE

BSF Doping (n^+)	10^{19} cm^{-3}
Interface Recombination Velocity (S_{GB})	10^6 cms^{-1}
Electron Lifetime	10ns
Hole lifetime	200 ns
Defect Distribution	Gaussian
Total density of defect states, N_{it}	10^{13} cm^{-2}
Mobility, μ_n	8000 cm^{2}V^{-1}S^{-1}
Mobility, μ_p	400 cm^{2}V^{-1}S^{-1}

Simplistic Model: In this model, grain boundary is simulated as an interface with an effective surface recombination velocity (SRV), S_{GB}. S_{GB} has been reported to be between 10^5 and 10^7 cms^{-1} [5]. In this work, we took the average S_{GB} to be 10^6 cm/s. The electron (τ_n) and hole (τ_p) lifetimes are assumed to be equal to their values in bulk. Fig. 1(a) shows the schematic of the simulated structure.

In the simulations, the grain size, base thickness, doping of emitter, BSF (Back Scattered Field) and base layer, carrier lifetime and surface recombination velocity at the grain boundary interface were varied. The effects of these variations are explained in the following sections.

Single crystalline GaAs solar cell was also simulated for comparison by keeping the same device parameters and the following values are obtained:

Table 2: Various solar cell parameter values for simulated single crystalline solar cell

Parameter	Values
J_{SC} (mA/cm^2)	24.076
V_{OC} (V)	0.952
FF	0.863
Efficiency (%)	19.833

A. Effect of Grain Size

With an increase in grain size, all the four parameters — J_{SC}, V_{OC}, FF and efficiency – increase. However, the marginal gains reduce as the grain-size increases beyond 20 um. Efficiency saturates at ~20 % (J_{SC}=26.347 mA/cm^2, V_{OC}=0.936 and FF=0.835) for n-type base solar cells with grain size> 64 μm (Fig. 1(b)). The diffusion length of the carriers in these calculations is around 4.56 μm. Once the grain size becomes 3-4x bigger than the diffusion length, the effect of grain boundary recombination becomes insignificant.

B. Effect of Base thickness

Increase in base thickness from 1μm to 3μm leads to increase in the short circuit current, due to enhanced absorption. Further increasing base thickness to 5 μm does not change J_{SC} and power conversion efficiency too much (Fig.1(b)), because most of the solar spectrum gets absorbed in 3 um thick GaAs. However, this increase in J_{SC} is substantial only for grain size > 5 μm. For grain size smaller than 5 μm, base thickness of 1 μm

provides better J_{SC} than base thickness of 3 and 5 μm. This anomalous behavior is a direct result of grain-boundary recombination. Increasing the base does not only increase absorption but also provides more opportunity for carrier to recombine. So there is an optimum thickness of the base, beyond which the loss in recombination is way more than the gain in absorption.

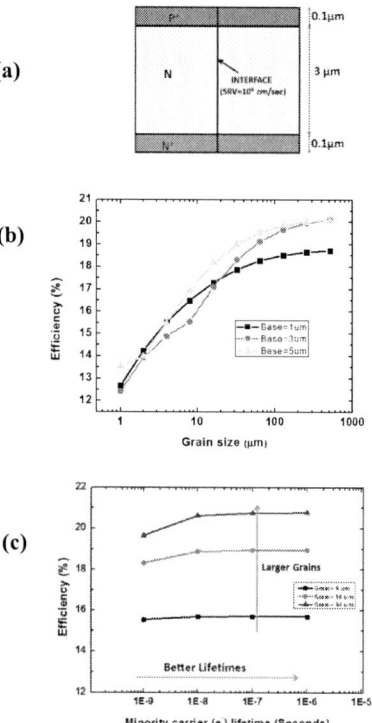

Fig.1: (a) Schematic (b) Efficiency of GaAs solar cells for different base thicknesses (c) for different lifetimes as a function of grain size

C. Effect of Carrier Lifetime

It has been reported that carrier lifetime for polycrystalline GaAs varies from few μs to few ns depending on the fabrication techniques and contamination. Device efficiency was calculated for a range for electron (hole) carrier lifetimes: τ_n=1ns (τ_p=20ns), 10ns (200ns), 100ns (2000ns) and 1μs (20μs). As expected, all the performance parameters increased with increasing carrier lifetimes. We find that for an efficiency >20 % and grain size ~ 100 μm, the carrier lifetime has to be > τ_n =10 ns (τ_p = 200 ns) (Fig.1(c)).

Thus, the simplistic model showed us the effect of recombination. Recombination can happen either in the bulk or at the interface. The effect of interface can be seen in Fig. 1(b). For different base thicknesses, the efficiency of polycrystalline solar cell varied from 12 % to 20 % depending on the grain size. Based on this, a base thickness of 5 μm was picked. The recombination can also occur in the bulk of the material as thin film devices tend to have contamination. That contamination decides the highest efficiency of the solar cell. For example, Fig. 1(c) shows how the carrier lifetimes affect the efficiency of polycrystalline solar cell as a function of grain size.

The simplistic model showed us the worst case scenario. It assumed the grain boundary to be a perfect sink. However, the reality is not that bad. The effective S_{GB} comes out to be $< 10^6$ cms^{-1}.

Hence, a more detailed model is needed to capture the complete picture of grain boundary defect states rather than simply assuming an effective SRV.

Detailed Model: The simplistic model did not consider any band bending, types of defects, distribution and density of these defect states. In reality, grain boundaries have donors and acceptor types of defect states. These defect states, as discussed earlier give rise to band bending at these grain boundaries. (Fig.2 (a))

We assumed a Gaussian distribution of defect states at the grain boundary (as shown in Fig.2 (b)). The band bending reduces under light illumination as compared to the dark condition. (as shown in Fig.3. and 4) Typically, the defect densities are in the range of 10^{10}-10^{15} cm^{-2} [6]. The efficiency decreases with increase in defect density and saturates at 10^{16} cm^{-2} (Fig.5.)

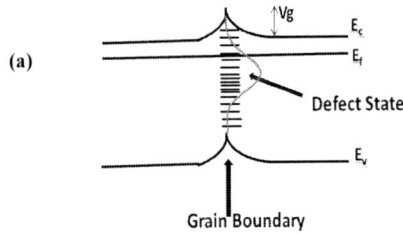

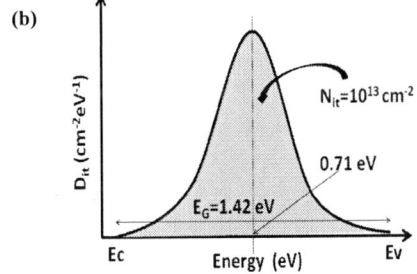

Fig.2. (a) Band diagram and (b) Distribution of defect states at the grain boundary.

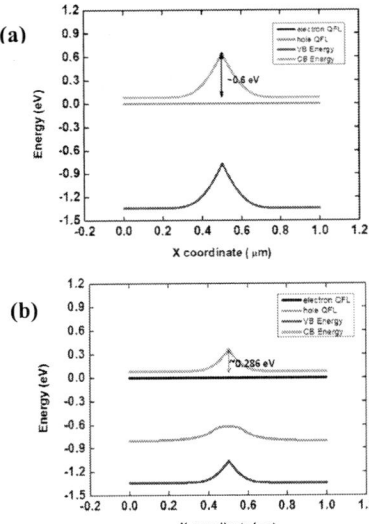

Fig.3. Band bending at the grain boundary under (a) Dark (b) Illumination (Grain Size =1 μm)

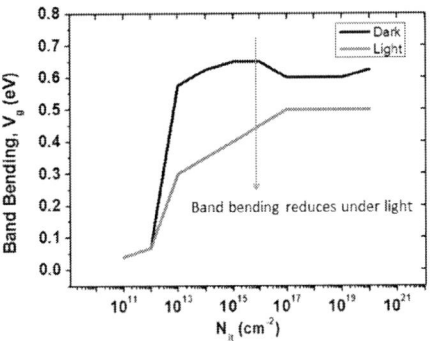

Fig.4. Decrease in band bending under light at various defect densities.

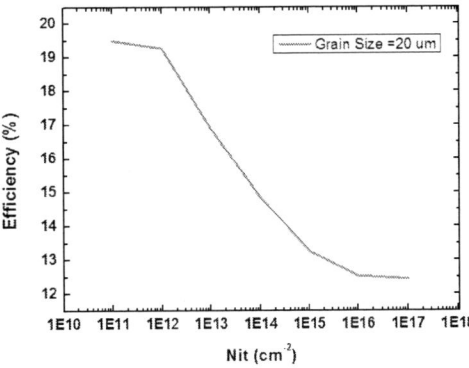

Fig.5. Decrease in efficiency of polycrystalline GaAs solar cell as the defect density increases

III. CONCLUSION

The simulation of thin film GaAs solar cells with columnar grains without anti reflection coatings were carried out. To attain an efficiency> 20 %, the thickness of GaAs thin-films should be 3-5 um. Further, lateral grain-size larger than 60 um is essential to achieve such efficiencies. The GaAs layers must also be free from contamination, with an electron recombination lifetime of at least 10 ns. This lifetime approximately corresponds to a defect density in the range of 10^{12}-10^{13} cm^{-2}. These numbers can be used as guidelines by materials growth specialists to fabricate better thin-film GaAs solar cells. This efficiency can be improved by ~2% by using p type base and ~3% by the addition of AlGaAs window layer. This will be the subject of our future work.

REFERENCES

[1] Green, M. A., et al. "Solar cell efficiency tables (version 46)." Prog. Photovoltaic: Res Appl 23 (2015): 805-812.

[2] Moon, Sunghyun, et al. "Highly efficient single-junction GaAs thin-film solar cell on flexible substrate." Scientific Reports 6 (2016).

[3] Venkatasubramanian, R., et al. "18.2%(AM1. 5) efficient GaAs solar cell on optical-grade polycrystalline Ge substrate."Photovoltaic Specialists Conference, 1996., Conference Record of the Twenty Fifth IEEE. IEEE, 1996.

[4] Joshi, Dinesh Prasad, and Devesh Prasad Bhatt. "Theory of grain boundary recombination and carrier transport in polycrystalline silicon under optical illumination." IEEE Transactions on Electron Devices 37.1 (1990): 237-249

[5] Lanza, C., and H. J. Hovel. "Efficiency calculations for thin-film polycrystalline semiconductor pn junction solar cells." IEEE Transactions on Electron Devices 27.11 (1980): 2085-2088.

[6] Kazmerski, Lawrence L. "The effects of grain boundary and interface recombination on the performance of thin-film solar cells." Solid-State Electronics 21.11-12 (1978): 1545-1550

Time-resolved PL measurements in the growth of high voltage (Al)GaInP/GaAs solar cells

Xinyi Li*, Wei Zhang, Hongbo Lu

State Key Laboratory of Space Power, Shanghai Institute of Space Power-sources, Shanghai, PRC, 200245

Abstract — **Aluminum doped GaInP, (Al)GaInP hereafter, is used as absorber in double-junction (Al)GaInP/GaAs solar cells, to enhance the open-circuit voltage and further the efficiency. By investigating the time-resolved PL of (Al)GaInP/AlGaInP heterojunction, whose interface is equivalent to that of base/back-surface-field (bsf), the bulk lifetime and interface recombination velocity are obtained. With further improvements, a recombination velocity of 139 cm/s is achieved at the interface between the base and bsf. A preliminary double-junction solar cell is grown and fabricated, showing an open-circuit voltage (Voc) of 2506 mV and a short-circuit current (Jsc) of 15.6 mA/cm^2 only. With further current matching, including thickening the base, employing transparent tunnel diode and improving the anti-reflection coating, this solar cell could be used as top-cell in Ge-based solar cell, IMM solar cells and SBT solar cells, to approach higher conversion efficiency.**

Index Terms — **time-resolved PL, III-V semiconductors.**

I. INTRODUCTION

GaInP sub-cells are usually used as top sub-cells in high efficiency III-V group solar cells. The bandgap of GaInP is about 1.88-1.90 eV, related to the degree of structure disorder. Since the efficiency of traditional Ge-based 3J solar cell has almost reached its limit, new mutlijunction solar cells, such as IMM solar cells and SBT solar cells, have been widely studied to boost the conversion efficiency [1]-[3]. To match the solar spectra, sub-cells of higher bandgap than 1.90 eV are required to effectively utilize the high energy photons. 3-5% aluminum doped GaInP, (Al)GaInP, whose bandgap is around 1.95 eV, is quite a convenience substitute material as top sub-cell. Such (Al)GaInP sub-cells have even been employed in some Ge-based 3J solar cells to enhance the Voc.

It is always an issue to improve the quality of Al-contained semiconductors because of notable Al-O related defects. In this case, besides the(Al)GaInP as base layer, the quality of the interface between (Al)GaInP and high Al-composition (Al$_{0.5}$Ga$_{0.5}$)InP (AlGaInP hereafter) bsf, namely interface recombination velocity, is crucial for carrier collection and surface passivation. Analyzing the photoluminescence (PL) lifetime of the bulk and/or the interface is the most efficient approach to improve the growth condition. Without time-consuming and costly fabrication process, information such as lifetime, surface recombination, carrier drift etc. could be extracted from simple samples [4]-[7]. One can use these knowledges to guide the both structure design and epitaxial growth.

In this work, the lifetime decay of (Al)GaInP/AlGaInP DHs are measured by time-resolved PL (TRPL) technique, and then bulk lifetime and interface recombination velocity are obtained. By adjusting the V/III ratio during the growth, the in-

terface is quite improved. A preliminary (Al)GaInP/GaAs double-junction solar cell is grown and fabricated. IV shows Voc = 2506 mV, Jsc = 15.6 mA/cm^2 and *ff* = 0.845. With further current matching the Jsc could reach 16.5 mA/cm^2, making it a promising top sub-cell for high efficiency multijunction solar cell.

II. EXPERIMENTAL CONDITIONS

(Al)GaInP/AlGaInP isotype DHs are grown at 700 °C under 50 mBar in an AIX G3 MOVPE reactor. The carrier gas used during the growth is Pd-purified hydrogen. The TMGa, TMAl and TMIn sources are held at 5.0 ± 0.1 °C, 17.0 ± 0.1 °C and 17.0 ± 0.1 °C, respectively. The TMGa and TMAl flows are controlled by mass flow controllers, and the TMIn flow is controlled by real-time concentration measuring apparatus. The phosphorus source used is pure phosphine (PH$_3$). Hydrogen diluted DEZn is employed as *p*-type dopant. The doping concentration in the confinement layer is about 4E17 cm^{-3} while the active layer is only slightly doped, both of which are similar to the situation in the solar cell.

TRPL measurement is performed to evaluate the recombination of interface. The decay of excess minority-carrier density as a function of time, is recorded through time correlated single-photon counting (TCSPC) technique. A picosecond Fianium supercontinuum whitelaser combined with a scanning monochromator (Ex-mono) to choose appropriate wavelength is employed to excite the excess minority-carriers in the sample, and the photons emitted by the sample are focused on the input slit of a scanning monochromator (Em-mono), which is tuned to the appropriate wavelength corresponding to the band-to-band recombination. The system uses a Hamamatsu R3809U-50 ultra-fast microchannel plate detector which has a transit time spread of 25 *ps* at full-width-at-half-maximum (FWHM). The detector is followed by a high-speed amplifier which is connected to the photon-counting motherboard. The measurements are performed at room temperature. Before measuring the decay, steady PL is taken to obtain the wavelength of the band-to-band recombination luminescence.

(Al)GaInP/GaAs solar cells employing AlGaAs/AlGaInP tunnel diode is grown and fabricated. The IV curve of the cell is measured under the illumination of Spectrolab X-25 AM0 simulator. Quantum efficiency (QE) in the wavelength range from 300 – 970 nm is taken with an EnLi QE apparatus.

978-1-5090-5606-4/17 $31.00 © 2017 IEEE

III. Results and Discussions

Fig.1 shows the measurements of the DHs with various thickness. The steady PL shows that the wavelength of band-to-band recombination luminescence in the DHs is located at 632 nm. Therefore, for the TRPL, the Em-mono is tuned to 632 nm while the Ex-mono is tuned to 610 nm to avoid the unintended excitation via spontaneous emission from the confinement layers. The repetition rate of the laser is set to 20 MHz. In the time range, no additional features have been observed for times later than shown. Two single-exponential decays are observed in time-resolved PL, Table 1. Fast decay with time constant τ_1 smaller than 0.2 ns are noticed. With regard to the mechanisms governing the PL decays, it is possible that this fast decay is related to the swept-out of excess minority-carriers by internal electric field, which quickly diminishes PL. The origin of this decay need be further investigated. The slow decay τ_2 is considered to be proportional to the band-to-band recombination in the active layer. Neglecting photon recycling, the PL lifetime of a DH is linear according to the expression

$$\frac{1}{\tau_{PL}} = \frac{1}{\tau_B} + \frac{2S}{d} \qquad (1)$$

where τ_{PL} is the time constant of lifetime decay, τ_B the bulk lifetime, S the interface recombination velocity, and d the thickness of active layer. By plotting $1/\tau_{PL}$ vs. $1/d$, S is deduced from the slope and the τ_B from the intercept. Therefore, the interface recombination velocity between (Al)GaInP/AlGaInP is about 257 cm/s, which is also the the recombination velocity between base and bsf in solar cell. The bulk lifetime, τ_B, of active layer is about 82.2 ns, but it should be pointed out that without taking any lifetime-enhancing effects such as photon recycling in consideration, τ_B would be much larger than its intrinsic bulk lifetime.

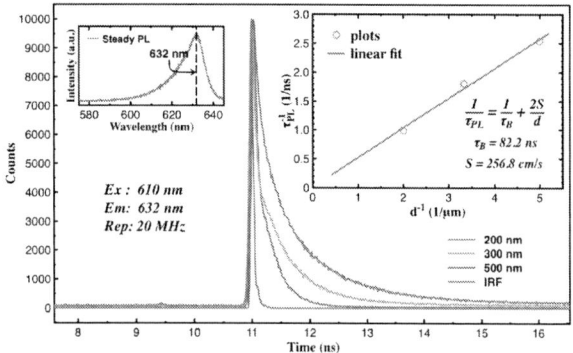

Fig. 1. Decays of the DHs with various thickness d of active layer (200 nm, 300 nm, and 500 nm, respectively). DHs are excited with 20 MHz laser repetition rate at 610 nm. The lifetimes τ_{PL} are extracted by multi single-exponential fitting, and the bulk lifetime τ_B and interface recombination velocity S are then obtained from linear fitting of plots $1/\tau_{PL}$ vs. $1/d$.

TABLE I
SUMMARY OF EXTRACTED TIME CONSTANT

Active Layer (nm)	τ_1 (ps)	τ_2 (ps)
200	10.50 *invalid	393.41
300	34.92	553.71
500	141.46	1008.27

Fig.2 shows the measurements of the DHs (d = 400 nm) with various V/III ratio. No additional PL features have been observed for times later than shown. Also, one fast decay τ_1 and one slow decay τ_2 are observed in TRPL. Since the bulk lifetime τ_B is larger than PL lifetime τ_{PL}, it could be simplified that $1/\tau_{PL} \approx 2S/d$. Therefore, time constant τ_2 indicates the quality of the interface, or the effectiveness of surface-field in the solar cells. The inset of Fig.2 shows the dependence between τ_2 and V/III ratio during the growth, suggesting that a V/III ratio of 127 produces a much lower interface recombination velocity around 139 cm/s. It is also noticed that the τ_1 follows the same trend as the τ_2.

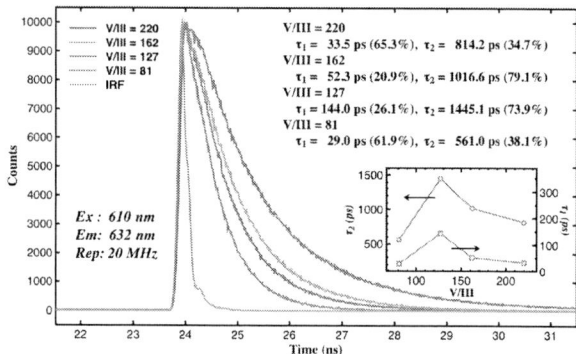

Fig. 2. Decays of the DHs grown with various V/III ratio (220, 162, 127, and 81, respectively). DHs are excited with 20 MHz laser repetition rate at 610 nm. The lifetimes τ_{PL} are extracted by multi single-exponential fitting. Two decays are observed in time-resolved PL.

A preliminary (Al)GaInP/GaAs double-junction solar cell, Fig.3, employing optimized τ_B and surface recombination velocity was grown and fabricated into size of 3 × 4 cm². The J-V characteristics of the double-junction solar cell is illustrated under one sun AM0 condition, Fig.4. The Voc and Jsc are 2506 mV and 15.6 mA/cm², respectively. Considering the Voc of GaAs sub-cell is normally about 1040-1045 mV in our early work, a voltage of ~1466 mV is estimated for (Al)GaInP sub-cell, resulting the [Eg/q-Voc] of 495 mV.

QE response of the solar cell, Fig.5, shows the current mismatch between sub-cells. The integrated current to the AM0 spectra of the top and bottom sub-cells are 15.70 and 17.52 mA/cm², respectively. It is obvious that the (Al)GaInP sub-cell holds the total Jsc through the solar cell. Such mismatch mainly is caused by the thin base in the (Al)GaInP sub-cell. It could be noticed that, even before the bandgap edge of the top sub-cell, the QE drops drastically while the QE of bottom subcell is booming. By thickening the base of (Al)GaInP, the mismatch would be improved. Since an AlInP window layer of 50 nm in thickness is used in the solar cell, obvious absorption in short wavelength range is unavoidable. By thinning the window layer, which might introduce wet-etch difficult in fabrication, the QE in short wavelength range could approach 90% and further improve the Jsc. Also QE drops slightly near 650 nm. This is because the tunnel diode employed in the solar cell is tellurium doped GaInP junction. Although the heavily doped GaInP layer is quite thin (15-20 nm), more than 10% of photons are lost near the bandgap edge of GaInP. By changing the GaInP with AlGaInP, this part of photon loss could be avoided. Meanwhile, a new anti-reflection coating is needed for the cell to minimize the reflection loss of incident photons. In general, by further improving and current-matching, the Jsc could reach 16.5 mA/cm^2 without any cost of Voc, producing an efficiency beyond 25.8%.

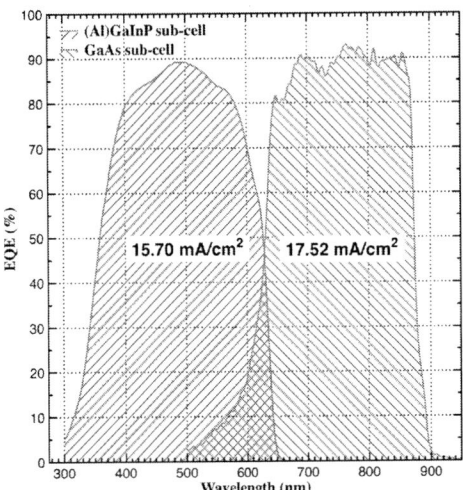

Fig. 5.   EQE of sub-cells in fabricated (Al)GaInP/GaAs solar cell.

IV. CONCLUSION

With the help of the time-resolved PL, both bulk and hetero-interface quality are optimized to approach their best photoelectrical characteristics. A preliminary (Al)GaInP/GaAs double-junction solar cell is fabricated, with Voc of 2506 mV, Jsc of 15.6 mA/cm^2 and fill factor of 0.845. With further improving and current-matching, the Jsc could reach at least 16.5 mA/cm^2, producing a conversion efficiency near 26%. Such double-junction sub-cell could be employed as top-cell in Ge-based solar cells, IMM solar cells and SBT solar cells, to achieve higher efficiency of sunlight harvesting.

	Gold Front Contact	
	n$^+$-GaAs Contact Layer	
Window	n-AlInP:Si	30 nm
Emitter	n-(Al)GaInP:Si	90 nm
Base	p-(Al)GaInP:Zn	550 nm
BSF	p-AlGaInP:Zn	50 nm
p^{++}	p-AlGaAs:C	20 nm
n^{++}	n-AlGaInP:Te	20 nm
Window	n-AlInP:Si	50 nm
Emitter	n-GaAs:Si	100 nm
Base	p-GaAs:Zn	3000 nm
BSF	p-AlGaAs:Zn	50 nm
	GaAs Tunnel Diode	
	15°A n-type GaAs Substrate	
	Gold Back Contact	

Fig. 3.   Schematic showing the structure of the epitaxially growth layers and metallization of the (Al)GaInP/GaAs solar cell. The diagram is not to scale.

REFERENCES

[1] P. T. Chiu, D. C. Law, S. B. Singer, D. Bhusari, A. Zakaria, X. Q. Liu, S. Mesropian, and N. H. Karam, "High performance 5J and 6J direct bonded (SBT) space solar cells", in *IEEE 42nd Photovoltaic Specialists Conference*, pp.1-3, 2015.

[2] P. T. Chiu, D. C. Law, R.L. Woo, S. B. Singer, D. Bhusari, W. D. Hong, A. Zakaria, J. Boisvert, S. Mesropian, R. R. King, and N. H. Karam, "35.8% space and 38.8% terrestrial 5J direct bonded cells", in *IEEE 40th Photovoltaic Specialists Conference*, pp.11-13, 2014.

[3] A. B. Cornfeld, P. Patel, J. Spann, D. Aiken, and J. McCarty, "Evolution of a 2.05eV AlGaInP Top sub-cell for 5 and 6J-IMM Applications", in *IEEE 38th Photovoltaic Specialists Conference*, pp.2788-2791, 2012.

[3] A. B. Cornfeld, P. Patel, J. Spann, D. Aiken, and J. McCarty, "Evolution of a 2.05eV AlGaInP Top sub-cell for 5 and 6J-IMM Applications", in *IEEE 38th Photovoltaic Specialists Conference*, pp.2788-2791, 2012.

[4] D. M. Tex, T. Ihara, H. Akiyama, M. Imaizumi, and Y. Kanemitsu, "Time-resolved photoluminescence measurements for determing voltage-dependent charge-separation efficiencies of subcells in triple-junction solar cells", *Applied Physics Letters*, vol. 106, 013905, 2015.

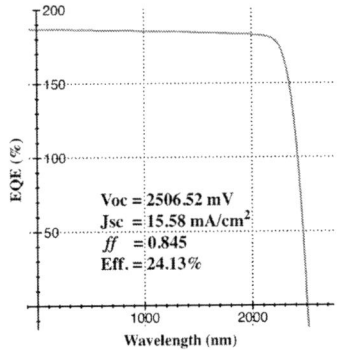

Fig. 4.   J-V characteristics for fabricated (Al)GaInP/GaAs solar cell (AM0, one sun).

[5] J. Afalla, M. H. Balgos, A. Garcia, J. J. Ibanes, A. Salvador, and A. Somintac, "Observation of picosecond carrier lifetimes in GaAs/AlGaAs single quantum wells grown at 630°C", Journal of Luminescence, vol. 143, pp.538-541, 2013.

[6] R. K. Ahrenkiel, and S. W. Johnston, "An optical technique for measuring surface recombination velocity", *Solar Energy Materials & Solar Cells*, vol. 93, pp.645-649, 2009.

[7] M. A. Steiner, J. F. Geisz, I. Garcia, D. J. Friedman, A, Duda, W. J. Olavarria, M. Young, D. Kuciauskas, and S. R. Kurtz, "Effects of Internal Luminescence and Internal Optics on Voc and Jsc of III-V Solar Cells", IEEE Journal of Photovoltaics, vol. 3, pp.1437-1442, 2013.

Low-resistance and highly-transparent GaSb-based tunnel junctions

Matthew P. Lumb[1], Shawn Mack[2], Maria Gonzalez[3], Kenneth J. Schmieder[2], Mitchell F. Bennett[3], Chaffra A. Affouda[2], James E. Moore[1] and Robert J. Walters[2]

[1]The George Washington University, Washington, DC, 20037, USA
[2]US Naval Research Laboratory, Washington, DC, 20375, USA
[3]Sotera Defense Solution, Annapolis Junction, MD, 20701, USA

Abstract — **Tunnel junctions with low differential resistance and high transparency have been grown by molecular beam epitaxy on GaSb substrates. The resulting devices have been characterized and analyzed using a combination of electrical measurements and modeling. These devices have importance to multi-junction solar cells grown on GaSb substrates. The structures contain an n-type InAs quantum well embedded in a GaSb p/n junction, exploiting the high tunnel probability at the broken gap interface between p-type GaSb and n-type InAs, whilst having a minimal impact on the transparency of the device.**

I. INTRODUCTION

Materials grown on GaSb substrates are important for a broad range of optoelectronic device applications at near to mid-infrared wavelengths [1]. An emerging application for GaSb-based materials is in multi-junction solar cells (MJSCs) for concentrator photovoltaics (CPV) applications, as these enable access to a wavelength range not achievable using lattice matched (LM) alloys on either InP or GaAs substrates. Therefore, GaSb-based absorbers unlock the potential to harvest the entire solar spectrum using lattice-matched, direct-bandgap materials as part of mechanically stacked architectures.

Previous studies [2] have shown that, using the alloy InGaAsSb, a lattice-matched material with a bandgap of 0.5 eV can be grown on GaSb. This is significant as the optimum bottom cell bandgap for terrestrial CPV cells with four or more junctions is ~0.5 eV (~2500 nm), defined by a significant spectral absorption band beginning at this wavelength. As a result, single, two and three-junction solar cells on GaSb have the potential to be extremely efficient components of mechanically stacked solar cells when combined with wider bandgap materials grown on InP or GaAs substrates using heterogeneous integration [2-5]. For MJSCs on GaSb, interband tunnel junctions (TJs) with low electrical resistance and high optical transparency are very important for interconnecting the subcells.

A particular candidate material combination for high performance GaSb-based MJSCs is a GaSb/InGaAsSb 2J device. A solar cell with this bandgap combination can be combined with GaAs or InP-based solar cells in a mechanically stacked architecture to give extremely high conversion efficiency. In this case, the TJ interconnecting GaSb and InGaAsSb subcells would ideally be transparent for all photons of energy less than the bandgap of GaSb to avoid filtering light reaching the InGaAsSb cell.

The conventional approach to creating tunnel junctions between III-V subcells in an MJSC on GaAs or InP is to use a highly-doped homojunction Esaki diode [6, 7]. However, molecular beam epitaxy (MBE) grown GaSb materials pose a particular challenge for the development of an Esaki diode, due to the electrically active donor concentration in Te-doped GaSb being limited to approximately $1\text{-}3\times10^{18}$ cm^{-3} [8-10]. This limitation prevents highly degenerately-doped n-type GaSb from being produced, which creates a severe limitation in the performance of the device as a TJ.

An alternative approach to the conventional Esaki diode is a TJ device formed from a p-type GaSb and n-type InAs heterostructure, which has been demonstrated in the literature to have very low resistance [11]. The key aspect of the bandstructure of the device leading to low resistance is that InAs and GaSb have a broken-gap band alignment, which facilitates efficient interband tunneling across the forbidden gap. However, the inclusion of thick InAs or InAsSb layers is detrimental to the transmission of the TJ below the GaSb bandgap. In this paper, we describe an alternative approach: the broken-gap quantum-well tunnel junction (BG-QWTJ). In this structure, a single InAs QW is employed at the interface of a GaSb p/n homojunction, creating a low resistance path for carriers to tunnel across the forbidden gap yet maintaining a high degree of transmission required for efficiency MJSC operation. Quantum wells with a type-I band alignment have been shown previously to significantly improve the tunnel probability across the forbidden region in a device with only a small impact on transparency [12-14], and this work extends this concept to a new set of materials and investigates QWs with a different band alignment type.

II. EXPERIMENTAL

The devices in this study were grown by solid-source MBE, with Te-doping achieved using a GaTe solid source compound and either Si or Be for p-doping. Si was also used as an n-type dopant in the InAs QW. The devices were grown on quarters of 2-inch GaSb wafers in the (001) direction and processed into fully-metallized, circular mesas 1 mm in diameter using standard photolithography techniques. The bulk TJ was grown on an n-GaSb substrate with the front contact formed using an unannealed TiAu stack and the back contact using a CrSnPtAu stack, annealed at 300 °C for 90 seconds under a nitrogen atmosphere. The BG-QWTJ sample was grown on a p-GaSb substrate, and the front and back metal contacts were formed using unannealed TiAu. The current-voltage characteristics of the devices were then characterized using a four-point-probe technique. The devices were simulated using the optoelectronic device modeling package NRL MultiBands® [6], which includes a drift-diffusion model incorporating non-local tunneling and an exhaustive library of III-V band parameters and optical constants.

III. RESULTS AND DISCUSSION

Figure 1 shows the equilibrium band diagram at the junction region for structure A, a GaSb n/p^{++} homo-junction TJ. The band diagram demonstrates that the limited electrically-active n-type doping results in a wide depletion region with little degeneracy in the n-type layer, creating a very low probability for interband tunneling in forward bias.

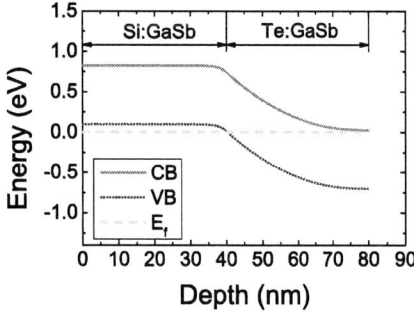

Figure 1. Equilibrium band diagram of structure A, the GaSb homojunction tunnel junction.

Figure 2 shows structure B, where a thin (80 Å) Te-doped InAs quantum well (QW) is situated at the interface of a GaSb homo-junction. The broken-gap alignment promotes extremely efficient tunneling across the p-GaSb/n-InAs interface, and the weak absorption associated with the broken-gap QW maintains the high transmission required for photons below the GaSb bandgap.

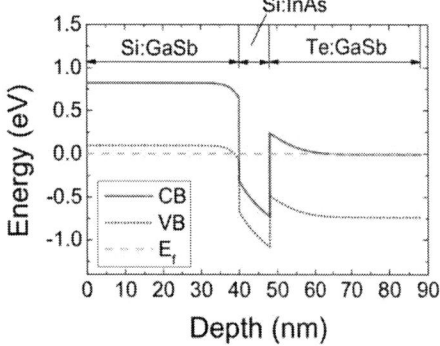

Figure 2. Equilibrium band diagram of structure B, the broken-gap quantum well tunnel junction.

The measured current density-voltage (JV) curve of a typical device homo-junction is shown in Figure 3 (dashed line) and demonstrates rectifying behavior, with little evidence of tunneling behavior in forward bias, as expected. The JV curve of the broken-gap quantum-well tunnel-junction (BG-QWTJ), also shown Figure 3 (solid line), was found to be ohmic with a resistance of 1.7×10^{-3} Ω·cm^2. The low resistance of the TJ leads to a significant improvement in performance over the homo-junction device, incurring a much smaller voltage penalty under high photocurrent conditions.

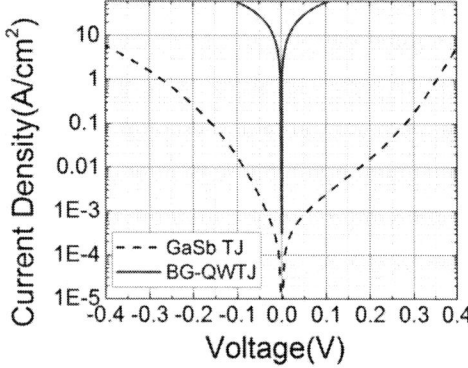

Figure 3. Absolute values of measured JV curves for structures A (homo-junction) and B (broken gap quantum well tunnel junction).

The impact of the QW on the transparency of the device was estimated using the transfer-matrix technique, assuming the bulk optical constants of the layers in structures A and B. This approximation does not take into account potential band-filling effects in the highly doped layers, which can result in a significant reduction in absorption near the band-edge of the layers, nor any modification to the density of states due to quantum confinement effects. However, the technique does give a reasonable estimate of the upper limit of absorption to be expected from the structure, which is shown in Figure 4. The average transmission loss in the region below the onset of absorption in the GaSb at approximately 1800 nm is less than 1%, which reveals that the high electrical performance can be achieved without sacrificing optical transparency.

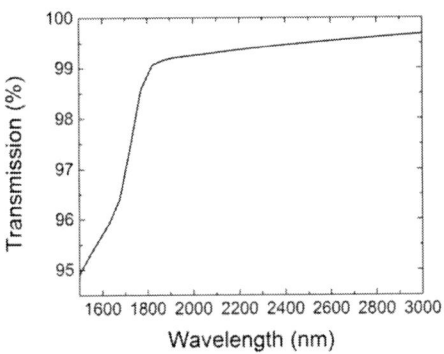

Figure 4. Calculated transmission through a BG-QWTJ with an 8 nm InAs QW.

Figure 5(a) shows the impact of QW thickness on the electrical performance of the BG-QWTJ. Three different designs with InAs QW thickness varied between 20-80 Å were grown and the differential resistance at $V = 0$ evaluated. Typically, 10 to 12 devices were tested on each wafer and the mean differential resistance and standard deviation calculated.

(a)

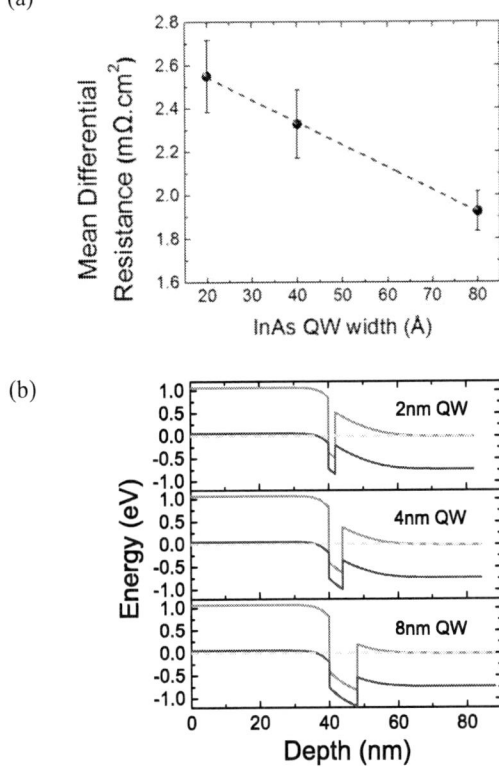

(b)

Figure 5. (a) Mean differential resistance of BG-QWTJs with different InAs QW thicknesses. The average values were calculated from measurements of multiple devices on each separate design. The error bars show the standard deviation. (b) Equilibrium band diagrams for BG-QWTJs with InAs QW thicknesses ranging from 2 nm to 8 nm.

The mean differential resistance was found to increase with decreasing QW thickness in a linear fashion. The proposed explanation for this effect is the increasing conduction band barrier formed by the n-GaSb layer as the InAs well thickness decreases, shown in the equilibrium band diagrams in Figure 5(b), introducing a greater series resistance for majority carriers. However, the exact nature of the electrical transport through the QW structure and surrounding barriers is not fully understood at this time.

IV. CONCLUSION

In this paper, we have presented electrical characterization of BG-QWTJ devices designed to operate as low resistance electrical interconnections between subcells of GaSb-based multi-junction solar cells. Including an InAs quantum well at the interface of a GaSb-based TJ creates a high probability of interband tunneling and overcomes the difficulties associated achieving high TJ performance with the limited electrical activity in Te-doped GaSb.

V. ACKNOWLEDGMENTS

This work was supported by the Office of Naval Research.

REFERENCES

[1] A. G. Milnes and A. Y. Polyakov, "Gallium antimonide device related properties," *Solid-State Electronics,* vol. 36, pp. 803-818, 1993.

[2] M. P. Lumb, M. Meitl, K. J. Schmieder, M. González, S. Mack, M. Yakes, M. F. Bennett, J. Frantz, M. A. Steiner, J. F. Geisz, D. J. Friedman, M. A. Slocum, S. M. Hubbard, B. Fisher, S. Burroughs, and R. J. Walters, "Towards the Ultimate Multi-Junction Solar Cell Using Transfer Printing," in *43rd IEEE Photovoltaics Specialists Conference*, Portland, OR, USA, 2016.

[3] M. P. Lumb, M. Meitl, J. W. Wilson, S. Bonafede, S. Burroughs, D. V. Forbes, C. G. Bailey, N. M. Hoven, M. Gonzalez, M. K. Yakes, S. J. Polly, S. M. Hubbard, and R. J. Walters, "Development of InGaAs Solar Cells for >44% Efficient Transfer-Printed Multi-junctions," in *40th IEEE Photovoltaics Specialists Conference*, Denver, CO, USA, 2014.

[4] M. P. Lumb, M. Meitl, B. Fisher, S. Burroughs, K. J. Schmieder, M. Gonzalez, M. K. Yakes, S. Mack, R. Hoheisel, M. F. Bennett, C. W. Ebert, D. V. Forbes, C. G. Bailey, and R. J. Walters, "Transfer-printing for the next generation of multi-junction solar cells," in *42nd IEEE Photovoltaic Specialist Conference*, New Orleans, LA, USA, 2015, pp. 1-6.

[5] M. P. Lumb, K. J. Schmieder, M. González, S. Mack, M. K. Yakes, M. Meitl, S. Burroughs, C. Ebert, M. F. Bennett, D. V. Forbes, X. Sheng, J. A. Rogers, and R. J. Walters, "Realizing the next generation of CPV cells using transfer printing," in *CPV 11*, Aix Les Bains, France, 2015, p. 040007.

[6] M. P. Lumb, M. González, M. K. Yakes, C. A. Affouda, C. G. Bailey, and R. J. Walters, "High temperature current–voltage characteristics of InP-based tunnel junctions," *Progress in Photovoltaics: Research and Applications,* vol. 23, pp. 773-782, 2014.

[7] J. F. Wheeldon, C. E. Valdivia, A. W. Walker, G. Kolhatkar, A. Jaouad, A. Turala, B. Riel, D. Masson, N. Puetz, S. Fafard, R. Arès, V. Aimez, T. J. Hall, and K. Hinzer, "Performance comparison of AlGaAs, GaAs and InGaP tunnel junctions for concentrated multijunction solar cells," *Progress in Photovoltaics: Research and Applications,* vol. 19, pp. 442-452, 2011.

[8] A. Chandola, R. Pino, and P. S. Dutta, "Below bandgap optical absorption in tellurium-doped GaSb," *Semiconductor Science and Technology,* vol. 20, pp. 886-893, 2005.

[9] T. H. Chiu, J. A. Ditzenberger, H. S. Luftman, W. T. Tsang, and N. T. Ha, "Te doping study in molecular beam epitaxial growth of GaSb using Sb2Te3," *Applied Physics Letters,* vol. 56, pp. 1688-1690, 1990.

[10] G. W. Turner, S. J. Eglash, and A. J. Strauss, "Molecular beam epitaxial growth of high-mobility n-GaSb," *Journal of Vacuum Science & Technology B,* vol. 11, pp. 864-867, 1993.

[11] K. Vizbaras, M. Törpe, S. Arafin, and M.-C. Amann, "Ultra-low resistive GaSb/InAs tunnel junctions," *Semiconductor Science and Technology,* vol. 26, p. 075021, 2011.

[12] M. P. Lumb, M. K. Yakes, M. Gonzalez, I. Vurgaftman, C. G. Bailey, R. Hoheisel, and R. J. Walters, "Double quantum-well tunnel junctions with high peak tunnel currents and low absorption for InP multi-junction solar cells," *Applied Physics Letters,* vol. 100, p. 213907, 2012.

[13] J. P. Samberg, C. Zachary Carlin, G. K. Bradshaw, P. C. Colter, J. L. Harmon, J. B. Allen, J. R. Hauser, and S. M. Bedair, "Effect of GaAs interfacial layer on the performance of high bandgap tunnel junctions for multijunction solar cells," *Applied Physics Letters,* vol. 103, p. 103503, 2013.

[14] M. P. Lumb, M. K. Yakes, M. González, M. F. Bennett, K. J. Schmieder, C. A. Affouda, M. Herrera, F. J. Delgado, S. I. Molina, and R. J. Walters, "Wide bandgap, strain-balanced quantum well tunnel junctions on InP substrates," *Journal of Applied Physics,* vol. 119, p. 194503, 2016.

Modulated Photocurrent Measurements in Double Junction Solar Cells

Nicolás Márquez Peraca and Behrang H. Hamadani

National Institute of Standards and Technology, Gaithersburg, Maryland, 20899, United States

Abstract — Frequency dependent external quantum efficiency (EQE) measurements were performed on double junction solar cells by a custom-designed system consisting of an array of various monochromatic LEDs. LEDs were operated both at constant intensity and pulsed at various frequencies to explore the frequency response of each junction under various conditions. An equivalent circuit model, incorporating the effects of shunt resistances, junction capacitances, optical light coupling and the series resistance was then used to explain the various features and findings obtained from these measurements.

I. INTRODUCTION

With significant improvements in design, fabrication, and performance of multijunction solar cells [1,2], it becomes necessary to establish more advanced opto-electronic characterization techniques to explore the characteristics of these solar cells. In recent years, extensive light bias and voltage bias dependent external quantum efficiency (EQE) measurements have been performed to elucidate artifacts and phenomena such as low shunt resistance effects [3,5-8,10,11], reverse breakdown voltage [3,4], light coupling between junctions [9,10,12], etc. in these devices. Most EQE measurements are performed using a differential spectral response system where a monochromatic light source incident upon the cell is chopped at a certain frequency creating an AC signal in the measurement junction of interest, while a DC light bias is applied to the other junctions. Although there has been much work discussing the effects observed under these circumstances, very little work has been dedicated to the frequency response of the AC photocurrent extracted from the current limited junction. One can think of this type of measurement as a frequency-dependent EQE, since the internal junction capacitances and resistances of each junction affect the extracted photocurrent magnitude and phase in the frequency domain.

In this work, we describe the result of our modulated photocurrent measurements in a simple double junction solar cell and show that an equivalent circuit model can be used to describe the unique features observed in both the amplitude and the phase response of the normalized photocurrent or the EQE of these solar cells. In particular, it is demonstrated that EQE shows a significant frequency dependence based on each junction's bias current and capacitive effects.

II. EXPERIMENTAL DETAILS

A diagram of the experimental setup used for performing the

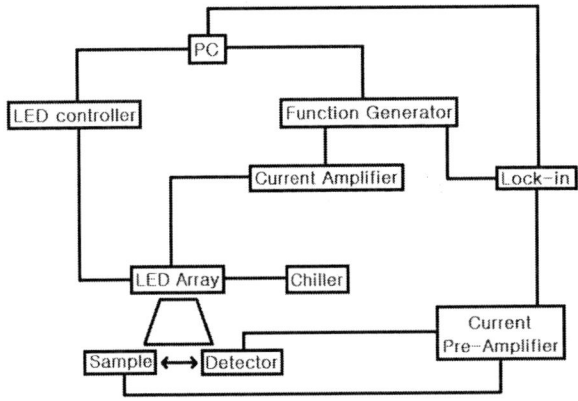

Fig. 1 Experimental setup used for the measurements. Both an AC and DC source are taken as an input of the LED array, which illuminates the sample through a quartz light pipe. A high-speed lock-in amplifier measures the amplitude and relative phase of the signal, and those values are then recorded on a computer.

modulated photocurrent measurements is shown in Fig. 1. A function generator is used in conjunction with a custom current amplifier to provide a pulsed AC signal to the LED array, while an LED controller provides the DC input. A solid glass light pipe in the form of a frustum is mounted in front of the LED array, which allows for a uniform illumination spot at the sample location (the exit port of the light pipe) for each LED type used. The cell's output is connected to a high-speed current to voltage pre-amplifier, which in turn is connected to a lock-in amplifier, providing amplitude and relative phase of the signal. This lock-in is synchronized with the function generator, and the whole system is controlled and automated by a computer program. Amplitude and phase dependence of the photocurrent on the frequency of the modulated light can then be found by changing the pulsed LED frequency. To provide stable operation of the LEDs, a water chiller is used to cool down the LED array plate to approximately 15 °C, as shown in Fig. 1.

Fig. 2. shows an actual photo of the optical segment of the setup. The LED array plate can be seen on the left side of the image. It has 12 LEDs of different wavelengths ranging from 460 nm to 928 nm. In this case, both the 460 nm and 623 nm LEDs are turned on, producing a purple color on the sample mounting plate due to the homogenizing of the blue and the red colors passing through the light pipe. The light pipe is in the center, encased in a 3D printed holder.

978-1-5090-5606-4/17 $31.00 © 2017 IEEE

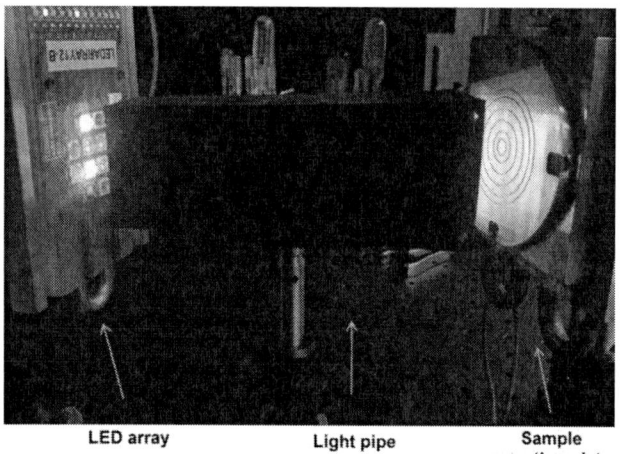

LED array   **Light pipe**   **Sample mounting plate**

Fig. 2. Photo of the experimental setup. Both 460 nm (blue) and 623 nm (red) LEDs are turned on, producing a purple color at the sample mounting plate.

The solar cell used for this study was a *GaInP/GaAs* cell, with an illuminated active area of $0.2533 \, \text{cm}^2$. The top *GaInP* junction is $0.9 \, \mu\text{m}$ thick with a bandgap around 1.84 eV, and the bottom junction is $3.5 \, \mu\text{m}$ thick. The active region of the top and bottom junctions were characterized by performing spectral response measurements on the cell by use of a monochromator-based system (see Fig. 3). Then, 460 nm and 850 nm LEDs were selected for the setup in Fig. 1, the former used as the pulsed light and the latter as bias light. Throughout the measurements the intensity of the pulsed 460 nm LED was kept fixed at $0.5 \, W / m^2$.

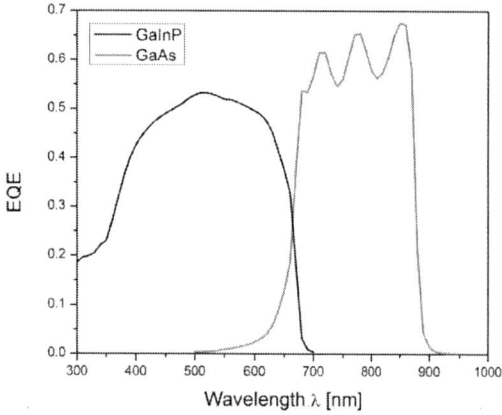

Fig. 3. External quantum efficiency as a function of wavelength for the *GaInP/GaAs* cell, obtained by performing spectral response measurements on the cell by use of a monochromator-based system.

III. THEORETICAL MODEL

An equivalent circuit for the AC light excitation measurements on the double junction solar cell can be seen in

Fig. 4. $\tilde{I}_T$ and $\tilde{I}_B$ represent the AC currents generated in each junction, R_T and R_B their dynamic resistances (which might depend on the DC light bias current), and C_T and C_B the depletion region capacitances. The dependent current source $\eta_1 \tilde{I}_r$ models any possible light coupling from the top junction to the bottom. The general solution for the circuit-extracted current, $\tilde{I}_{SC}$ in this model can be found in [9]. In the case where the pulsed light is applied to the top junction while the bottom junction is DC light biased, i.e., similar to EQE measurement conditions for the top junction, this result simplifies to:

$$\frac{\tilde{I}_{SC}}{\tilde{I}_T} = \frac{Z_T}{Z_T + Z_B + R_S} = X(\omega) + jY(\omega), \qquad (1)$$

when there is no light coupling from the top junction to the bottom, and:

$$Z_i = \frac{R_i}{1 + j\omega C_i R_i}, \qquad (2)$$

$$R_B = \frac{nk_B T}{q} \frac{1}{I_B} = \frac{nV_T}{I_B}, \qquad (3)$$

where n is the diode ideality factor, $k_B T / q$ is the thermal voltage V_T (≈ 25 mV at room temperature), Z_i (for i=T, B) is the complex impedance for the top and bottom junction, I_B is the DC current generated in the bottom junction, and $X(\omega), Y(\omega)$ are the real and imaginary parts of $\tilde{I}_{SC}/\tilde{I}_T$, respectively. It is noted that the ratio $\tilde{I}_{SC}/\tilde{I}_T$ actually represents the internal quantum efficiency (IQE) of this cell because $\tilde{I}_T$, the AC photocurrent generated in the top junction, is proportional to the modulated light intensity, $\tilde{E}_T$. Therefore, the ratio $\tilde{I}_{SC}/\tilde{I}_T \propto \tilde{I}_{SC}/\tilde{E}_T = \tilde{R}_T \propto IQE$, $\tilde{R}_T$ being the internal spectral responsivity of the junction. We have multiplied this value by a fixed constant before comparing the model to the experimental data to include reflectance effects, so that it can represent the EQE at the excitation wavelength probed.

It can be easily seen from (2) that in the low frequency limit where $\omega \rightarrow 0$, $Z_i \rightarrow R_i$ and so:

$$\frac{\tilde{I}_{SC}}{\tilde{I}_T} \rightarrow \frac{R_T}{R_T + R_B + R_S} \approx 1 \qquad (4)$$

where the last step follows from the approximation $R_T \gg R_B, R_S$ when the top cell is in reverse bias.

On the other hand, in the high frequency limit $\omega \rightarrow +\infty$, $Z_i \rightarrow 1 / j\omega C_i$ and if we consider a negligible series resistance, then we find:

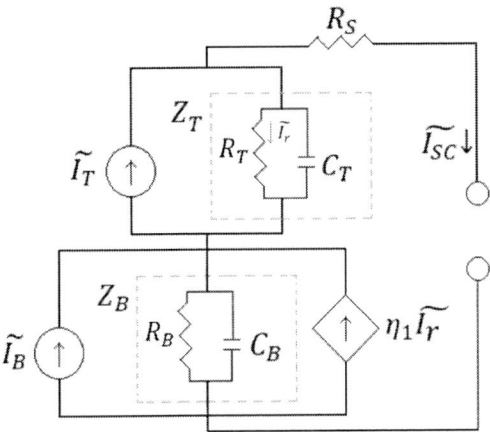

Fig. 4. Equivalent circuit for the AC measurements performed on the double junction cell. The two AC sources represent the currents generated on the junctions, whilst the dependent current source accounts for any possible light coupling.

$$\frac{\tilde{I}_{SC}}{\tilde{I}_T} = \frac{1}{1+\dfrac{Z_B}{Z_T}} \rightarrow \frac{1}{1+\dfrac{C_T}{C_B}} \ . \tag{5}$$

The ratio C_T / C_B can then be obtained from the EQE(ω) plot.

The phase $\theta(\omega) = \text{Arctan}(Y(\omega)/X(\omega))$ presents a resonant behavior and, as it can be seen by taking the quotient of the imaginary and real parts of (1), when there is no series resistance present it goes to zero both in the high and low frequency limits. The value for which this resonant peak happens is found to be:

$$\omega_{\min} = \frac{\sqrt{1+R_B/R_T}}{R_B C_B \sqrt{1+\dfrac{C_T}{C_B}}} \approx \frac{1}{R_B C_B} \propto I_B \ , \tag{6}$$

a result that can be verified by setting the first derivative of $\theta(\omega)$ to zero, and where the last step follows from (3).

Furthermore, the real part of $\tilde{I}_{SC}/\tilde{I}_T$ evaluated at this frequency is equal to:

$$\left.\frac{\tilde{I}_{SC}}{\tilde{I}_T}\right|_{\omega=\omega_{\min}} = 2\left[\left(\frac{\tilde{I}_{SC}}{\tilde{I}_T}(\omega\to 0)\right)^{-1} + \left(\frac{\tilde{I}_{SC}}{\tilde{I}_T}(\omega\to +\infty)\right)^{-1}\right]^{-1} \tag{7}$$

On the other hand, a non-zero series resistance causes a drop in the amplitude from the value in (5) to zero, and makes the phase rotate from 0 to -90° in the high frequency limit. Even so, in the region where the condition $\omega_{\min}(I_B) < 1/R_S C_T$ is met the previous analysis continues to be approximately valid.

These features can be seen in Fig. 5, where the left Y-axis represents the magnitude $IQE(\omega) = \sqrt{X(\omega)^2 + Y(\omega)^2}$ and the right Y-axis the phase $\theta(\omega)$, expressed in degrees. Here R_S was kept fixed at $90\,\Omega$ (see next section) while I_B was increased from $0.1\,\mu A$ to $100\,mA$ for exemplification purposes. The curve for $I_B = 100\,mA$ shows the case where $1/R_S C_T < \omega_{\min}$, and so no resonance is present in the phase plot and the IQE shows a sudden drop to zero at $\omega \sim 1/R_S C_T$.

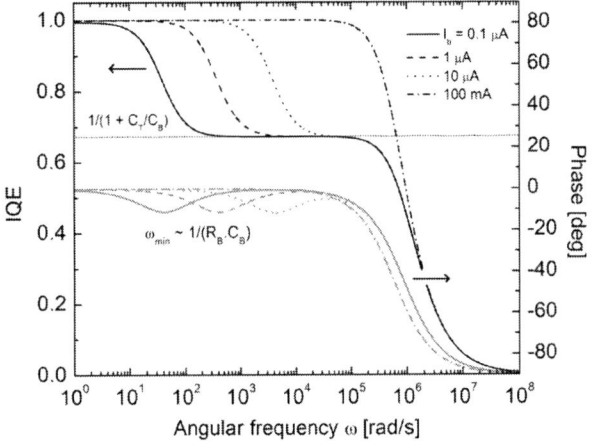

Fig. 5. Predicted internal quantum efficiency as a function of frequency. Here R_S was kept fixed at 90Ω, while I_B was increased from $0.1\mu A$ to $100mA$. At the frequency $\omega_{\min}$, both a resonant behavior for the phase and a sudden drop in the IQE to the value in (5) occur. The high and low frequency limits of (4) and (5) can also be seen in this figure.

IV. DISCUSSION AND RESULTS

Starting from the amplitude and phase data obtained from the lock-in measurements in the experimental setup, both for the solar cell and the reference detector, the external quantum efficiency (EQE) and net phase can be calculated. Fig. 6 shows the results obtained from these measurements (scatter points), as well as the model predictions of (1) (solid lines). The left Y-axis represents the external quantum efficiency and the right Y-axis is the phase, expressed in degrees. Both the model predictions and the measurements were scaled to the EQE value obtained from the monochromator setup at 460 nm, whilst a fixed $\approx 1°$ was subtracted from the phase data to account for the unphysical non-zero phase at low frequencies related to a small phase lag in the instrumentation. The series resistance used in the model was $90\,\Omega$: $50\,\Omega$ corresponding to the pre-amplifier's input impedance (from the specification data) and $40\,\Omega$ obtained through I-V measurements from the cell itself. Setting 1 to Setting 4 correspond to different DC light bias conditions which, expressed in terms of the LED controller current values, are 1, 2, 5, and 10 mA, respectively.

978-1-5090-5606-4/17 $31.00 © 2017 IEEE

As Fig. 6 shows, the EQE drop occurs around the same frequency where the phase minimum happens and, as expected from (6), it shifts towards higher frequencies when the DC current generated on the bottom junction increases. This effect suggests that when performing monochromator-based differential spectral response measurements for determining the steady-state EQE of the cell, the chopper's frequency should satisfy the condition $\omega_{meas} < I_B / nV_T C_B$. Otherwise, one risks underestimating the correct magnitude of the EQE for a given junction. In general, it is recommended to perform EQE measurements under the lowest frequencies possible, particularly when light bias conditions are low.

The little discrepancy between the model and the measurements in the high frequency region in Fig. 6 is explained by the fact that our current to voltage pre-amplifier has a strong bandwidth dependence with the source capacitance. For the measured cells having capacitances around 30 nF, the bandwidth drops to approximately 100 kHz to 200 kHz from its maximum value of ≈ 100 MHz.

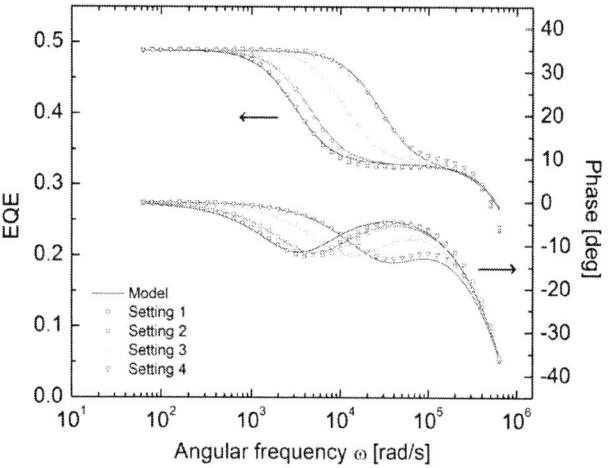

Fig. 6.   Results obtained from the measurements superposed to the theoretical model predictions, both being scaled to the EQE reported by the monochromator system. As shown, the optimal measurement frequency depends on the light bias intensity and on the bottom junction capacitance (see text). The values used for the fits were initially estimated from published works, and then adjusted through the model to obtain $C_T = 18.7$ nF , $C_B = 38.4$ nF , $R_S = 90\ \Omega$, $R_T \approx 10^8\ \Omega$, and $I_B = 8.1\ \mu A$, 11.93 μA, 30.51 μA, 74.78 μA for Settings 1-4, respectively.

V. CONCLUSIONS

Measurements of the amplitude and phase dependence of the external quantum efficiency on frequency for a double junction cell were performed and compared against the predictions of an equivalent circuit model. It was shown that in general the optimal measurement frequency will depend both on the light bias intensity levels and the capacitance of the junction that is in forward bias. Our recommendation is to use a measurement frequency as low as possible, while increasing the light bias. The frequency-dependent photocurrent measurements also allow for the determination of the internal capacitances and resistances of each junction by fitting the described model to a large set of data.

ACKNOWLEDGEMENT

The authors would like to thank Dr. Daniel Friedman of NREL for graciously providing the solar cells used in this study. N. M. P. would like to thank the Solar Energy Laboratory (LES, Uruguay) and the Technological Laboratory of Uruguay (LATU) for their support of the research projects on which he was selected to participate, and to NIST for their hospitality throughout this stay. The authors also gratefully acknowledge the support of the NIST International and Academic Affairs Office.

REFERENCES

[1]   H. Yoon *et al.*, "Recent advances in high-efficiency III-V multi-junction solar cells for space applications: Ultra triple junction qualification," in *Progress in Photovoltaics: Research and Applications*, 2005, vol. 13, no. 2, pp. 133–139.

[2]   M. A. Green, K. Emery, Y. Hishikawa, W. Warta, and E. D. Dunlop, "Solar cell efficiency tables (version 48)," *Prog. Photovoltaics Res. Appl.*, vol. 24, no. 7, pp. 905–913, 2016.

[3]   J. P. Babaro, K. G. West, and B. H. Hamadani, "Spectral response measurements of multijunction solar cells with low shunt resistance and breakdown voltages," *Energy Sci. Eng.*, vol. 4, no. 6, pp. 372–382, 2016.

[4]   M. Meusel, C. Baur, G. Létay, A. W. Bett, W. Warta, and E. Fernandez, "Spectral Response Measurements of Monolithic GaInP/Ga(In)As/Ge Triple-Junction Solar Cells: Measurement Artifacts and their Explanation," *Prog. Photovoltaics Res. Appl.*, vol. 11, no. 8, pp. 499–514, 2003.

[5]   J.-J. Li, S. H. Lim, and Y.-H. Zhang, "A novel method to eliminate the measurement artifacts of external quantum efficiency of multi-junction solar cells caused by the shunt effect," *Proc.SPIE*, vol. 8256, pp. 825616–825623, 2012.

[6]   J.-J. Li and Y.-H. Zhang, "Elimination of Artifacts in External Quantum Efficiency Measurements for Multijunction Solar Cells Using a Pulsed Light Bias," *IEEE J. Photovoltaics*, vol. 3, no. 1, pp. 364–369, 2013.

[7]   V. Paraskeva, M. Hadjipanayi, M. Norton, M. Pravettoni, and G. E. Georghiou, "Voltage and light bias dependent quantum efficiency measurements of GaInP/GaInAs/Ge triple junction devices," *Sol. Energy Mater. Sol. Cells*, vol. 116, pp. 55–60, 2013.

[8]   G. Siefer, C. Baur, and A. W. Bett, "External quantum efficiency measurements of Germanium bottom subcells: Measurement artifacts and correction procedures," in *Conference Record of the IEEE Photovoltaic Specialists Conference*, 2010, pp. 704–707.

[9]   M. A. Steiner, S. R. Kurtz, J. F. Geisz, W. E. McMahon, and J. M. Olson, "Using phase effects to understand measurements of the quantum efficiency and related

luminescent coupling in a multijunction solar cell," *IEEE J. Photovoltaics*, vol. 2, no. 4, pp. 424–433, 2012.

[10] J.-J. Li, S. H. Lim, C. R. Allen, D. Ding, and Y.-H. Zhang, "Combined Effects of Shunt and Luminescence Coupling on External Quantum Efficiency Measurements of Multijunction Solar Cells," *IEEE J. Photovoltaics*, vol. 1, no. 2, pp. 225–230, 2011.

[11] M. Pravettoni, R. Galleano, A. Virtuani, H. Müllejans, and E. D. Dunlop, "Spectral response measurement of double-junction thin-film photovoltaic devices: the impact of shunt resistance and bias voltage," *Meas. Sci. Technol.*, vol. 22, no. 4, p. 45902, 2011.

[12] M. A. Steiner *et al.*, "Measuring IV curves and subcell photocurrents in the presence of luminescent coupling," *IEEE J. Photovoltaics*, vol. 3, no. 2, pp. 879–887, 2013.

Effect of Atmospheric Absorption Bands on the Optimal Design of Multijunction Solar Cells

William E. McMahon, Daniel J. Friedman, John F. Geisz

National Renewable Energy Laboratory, Golden, Colorado, 80401, USA

Abstract — The optimization of subcell bandgaps for series-connected multijunction solar cells is complicated by atmospheric absorption bands in terrestrial spectra, and this complexity increases with the number of subcells. We systematically characterize the resulting design space, and introduce a taxonomy for MJ cells based upon the position of subcell bandgaps with respect to atmospheric absorption bands. Once this is done, the full set of local efficiency optima can be quickly identified and described for arbitrarily large numbers of junctions, and deviations from current-matched designs can be more easily characterized and understood.

I. Introduction

The optimization of subcell bandgaps for terrestrial multijunction (MJ) solar cells becomes increasingly complex with increasing number of junctions, primarily because atmospheric absorption bands create a family of optimal and near-optimal designs. Although there has been considerable work published for specific cases [1-4], a holistic understanding and classification of the resulting design space is lacking.

To provide such a framework, we have conducted a comprehensive study of terrestrial MJ design space. This produced a new taxonomy for MJ cells which enables the full set of local efficiency optima to be quickly identified and understood, for arbitrarily large numbers of junctions. Deviations from the "current-matching" rule are discussed, along with implications for numerical convergence.

II. Background

All terrestrial spectra (global or direct) contain the atmospheric absorption bands exhibited by the G173d spectrum shown in Figure 1(left). Because an efficient design will never place a bandgap in the middle of an absorption band, these absorption bands divide design space into multiple local efficiency maxima. This can be seen in Figure 1(right), where two designs with similar efficiencies have been plotted.

If only these pairs of designs existed, the labels "higher-bandgap design" and "lower-bandgap design" might be sufficient, but design space actually contains many more local efficiency maxima (not shown in Figure 1), so these descriptors are inadequate. Because many of these designs have similar efficiencies and design space becomes enormous as the number of junctions increases, the goal of this study is to provide better methods to find, describe, visualize, and understand the best designs.

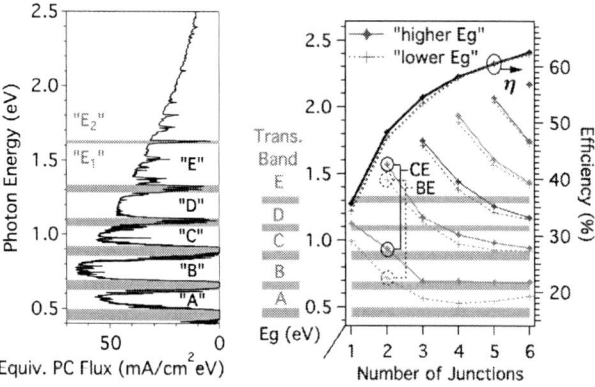

Fig. 1. (Left) The G173d spectrum, converted into equivalent photocurrent flux using a quantum efficiency (QE) equal to unity. The labels "A"-"E" divide the spectrum into "transmission bands". (Right) As the number of junctions increases, the efficiency difference between competing designs decreases. "CE" and "BE" are "MJ types" within a taxonomy which connects each subcell bandgap to a transmission band. This method provides a unique name for each of the designs in this plot.

III. Modeling Methods

For this work, we needed to comprehensively understand and visualize large regions of multi-dimensional parameter space in a reasonable amount of computer time, while still capturing all of the essential physics. This demands a very fast, computationally simple cell model, and effective search algorithms.

All multijunction solar cells in this study were modeled at 300K and 1000 suns concentration under the direct terrestrial AM1.5 standard spectrum (G173d), using the methods described in Reference [5]. The external radiative efficiency (ERE) was set to 0.01 for all subcells, and the subcell-to-subcell optical coupling set to zero. By assuming a single diode model with ideality factor n=1, J_i^{01} effectively becomes J_i^{db}/ERE_i. This creates a set of subcells of similar, but realistic, quality while quenching luminescent coupling to simplify the understanding of current-matching.

Simple "boxcar" EQEs approximate ideal, optically thick subcells, and the effects of shunt resistance and reverse-bias breakdown were neglected. This set of approximations was enabling for this work, because it reduced the computation time by about a factor of 100, while still retaining all of the salient points.

978-1-5090-5606-4/17 $31.00 © 2017 IEEE

Our computations were done in three steps. First, a simulated annealing algorithm [6] was performed multiple times to find all significant local optima, including the global optimum. Then, the location of each resulting optima was further refined using a gradient-based method.

It should be emphasized that our intent is to perform this comprehensive survey once, then use the results to develop more deterministic (and thereby faster) methods for subsequent cell optimization, enabling the use of more detailed, computationally intensive models for subsequent work.

IV. TAXONOMY AND NOMENCLATURE

To facilitate the description and comparison of the myriad possible designs, we will introduce a taxonomical structure based upon "transmission bands" and "MJ types".

"Transmission bands" are portions of the spectrum separated by absorption bands. As seen in Figure 1(left), we have chosen to label them in order of increasing photon energy from "A" to "E". When helpful, further subdivision by minor absorption bands is done with subscripts (i.e. "E_1" and "E_2").

A "MJ type" describes a given MJ design by indicating which transmission band contains each subcell, starting with the bottom subcell. For example, "CE" in Figure 1(right) indicates a lower subcell bandgap in transmission band "C" and an upper subcell bandgap in transmission band "E". In "BE", the bottom bandgap has moved down to transmission band "B". By extension, this naming scheme gives each (locally-optimized) design in Figure 1 (and subsequent designs) a unique name.

Parentheses will be used to indicate current-matched subcells, i.e. "(CE)". If a MJ design is *not* current matched, a "+" superscript indicates a subcell receiving "extra" photocurrent. For example, for an "$A^+(DE)$" 3J design, "D" is current-limiting, "E" is current-matched to "D", and "A" is over-driven.

V. RESULTS

Because current-matching is such a strong driver for the efficiency of series-connected MJ devices, all viable designs are *nearly* current-matched. However, the word "*nearly*" is very important, because none of the optimal designs in this study is *exactly* current-matched. Furthermore, there are *two distinct types* of non-current-matched designs. One of these will be called a "current-graded" (CG) design, and it deviates only slightly from a "current-matched" (CM) design. The other will be called an "absorption-band constrained" (ABC) design, and it differs qualitatively from both the CM and CG designs.

Because the existence of these three design types affects every aspect of MJ cell design, a complete understanding of them is quite valuable.

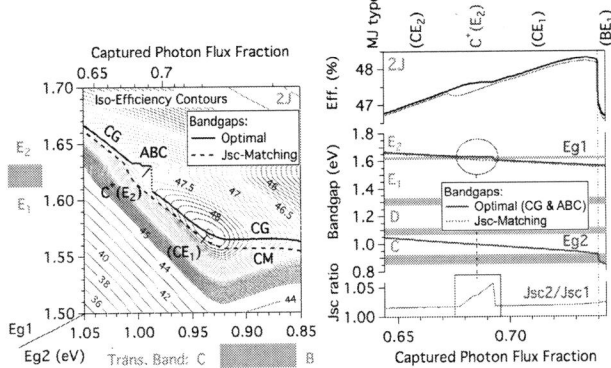

Fig. 2. (Left) Iso-efficiency contours for a 2-junction series-connected solar cell, with CM, CG and ABC designs labeled. The "absorption-band constrained" (ABC) local maximum "$C^+(E_2)$" places Eg1 just above the small absorption band dividing transmission band E into E_1 and E_2 (gray bar on left axis). (Right) A parametric plot can be extended from the 2J case shown here to an arbitrary number of junctions. The anomalous Jsc ratio (bracketed) serves as a "fingerprint" for the "$C^+(E_2)$" ABC design.

A. Description of Design Types

i. Current-Matched (CM) Designs

In a current-matched design ("CM" in Figure 2), each of the subcell Jsc values is equal. Because all viable designs are "near" a CM design, the set of CM designs serves as a useful and relatively easy-to-compute landmark in parameter space.

ii. Current-Graded (CG) Designs

In a CG design ("CG" in Figure 2), the bandgaps of the upper subcell(s) are slightly raised, increasing both the open-circuit voltage (Voc) and fill factor (FF) of the MJ device (because a CG design is not current-matched). Although the resulting increase in efficiency can be very small, it is important to understand CG designs so as to distinguish them from other effects. In particular, it is important to emphasize that *CG designs result from an operating point shift* [7]. Disruption of current-matching by terrestrial atmospheric absorption bands (next) is a completely independent effect.

iii. Absorption-Band Constrained (ABC) Designs

ABC designs occur where *an atmospheric absorption band disrupts current matching* ("ABC" in Figure 2). This happens when the CM design places an upper-subcell bandgap just below an absorption band, where the thermalization losses are large. An ABC design reduces these thermalization losses (and breaks the current-matching rule) by increasing the upper-subcell bandgap to place it just above the absorption band.

B. Visualization of Designs

Traditionally, parameter space for MJ cell optimization is presented as 2D plots of efficiency versus pairs of bandgaps (Figure 2 left). However, the number of 2D plots and associated efficiency data become large as the number of junctions is increased.

Fortunately, parametric plots of the CM, CG and ABC cell efficiencies and associated subcell bandgaps offer an effective alternative (Figure 2 right). In our parametric plots, we have found it to be useful to plot everything against a "captured photon flux fraction" (CF), where CF is simply the fraction of the incident photons with energies above the bottom-cell bandgap. When plotted in this way, the horizontal axis roughly corresponds to current.

To better understand and differentiate between CM, CG and ABC designs, Figure 2b includes a "Jsc ratio". This is the Jsc for each subcell divided by the current-limiting Jsc for the MJ device (Jsc1). The Jsc ratios indicates that CG designs are slightly top-cell limited, becoming more so as the bandgaps decrease. The Jsc ratio for an ABC design is noticeably larger, and can be used as a "fingerprint" for ABC designs. Although the differences between CM, CG and ABC designs are very small for a 2J device, a 2J device serves as an excellent example of all fundamental concepts.

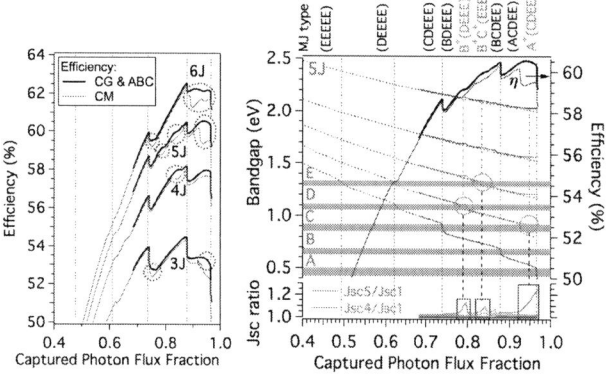

Fig. 3. (Left) MJ cell efficiency increases as the number of junctions increases, but with additional structure due to the spectral absorption bands. The dashed vertical lines indicate where the bottom subcell bandgap crosses an absorption band. ABC designs are marked with circles. (Right) Modeled results for a 5J device plotted parametrically. ABC designs occur wherever an upper subcell bandgap for the CM design crosses an absorption band (circled). For an ABC design, the bandgap remains above the absorption band and becomes current-limiting (bracketed Jsc ratios).

C. Extension to 6 Junctions

Figure 3 extends these basic concepts to 6 junctions. In Figure 3(left), ABC designs (circled) become more prevalent and deviate more significantly from the CM design as the number of junctions increases. This is because more upper-subcells encounter absorption bands, as can be seen in the 5J example shown in Figure 3(right). Three ABC designs are found, and each occurs where a (circled) subcell bandgap encounters an absorption band.

Once this is understood, all of the complexity associated with ABC designs can be explained and described using "MJ type" nomenclature. Furthermore, the fact that each MJ type supports a single efficiency maximum has important implications for numerical optimization of MJ cells.

This is because all of the feasible CG and ABC MJ types can be deduced from a parametric plot of CM designs, which are relatively fast and easy to calculate. Once this is done, it should be possible to compute the efficiency maximum for each MJ type using a simple and fast gradient solver.

VI. CONCLUSIONS

A classification system was developed to explain the complexity of MJ design space under terrestrial spectra. The results reveal two types of designs which can improve upon a "current-matched" (CM) design (for which all subcells are Jsc-matched): 1) The first is a "current-graded" (CG) design which deviates only slightly from the CM design. 2) The second is an "absorption-band constrained" (ABC) design which forms wherever an upper-subcell bandgap encounters an absorption band in the terrestrial spectrum. An understanding of these design types should facilitate the development of more computationally-efficient MJ cell optimization algorithms.

ACKNOWLEDGMENT

Funding for this work was provided by DOE EERE through contract SETP DE-AC36-08GO28308. The U.S. Government retains and the publisher, by accepting the article for publication, acknowledges that the U.S. Government retains a nonexclusive, paid up, irrevocable, worldwide license to publish or reproduce the published form of this work, or allow others to do so, for U.S. Government purposes. The authors would like to thank Myles Steiner and Iván García for contributing to the ERE-based cell-efficiency model used in these calculations.

REFERENCES

[1] P. Patel *et al.*, "Initial results of the monolithically grown six-junction inverted metamorphic multi-junction solar cell," in *38th IEEE PVSC*, vol. **2** (2012).

[2] P. T. Chiu *et al.*, "Direct Semiconductor Bonded 5J Cell for Space and Terrestrial Applications," *IEEE Journal of Photovoltaics* **4**, 493 (2013).

[3] S. P. Bremner, M. Y. Levy, and C. B. Honsberg, "Analysis of tandem solar cell efficiencies under AM1.5G spectrum using a rapid flux calculation method," *Progress in Photovoltaics: Research and Applications* **16**, 225 (2008).

[4] M. D. Yandt, K. Hinzer, and H. Schriemer, "Efficient Multijunction Solar Cell Design for Maximum Annual Energy Yield by Representative Spectrum Selection," *IEEE Journal of Photovoltaics*, available online (2017).

[5] J. F. Geisz *et al.*, "Generalized Optoelectronic Model of Series-Connected Multijunction Solar Cells," *IEEE Journal of Photovoltaics* **5**, 1827 (2015).

[6] W. H. Press, S. A. Teukolsky, W. T. Vetterling, and B. P. Flannery, *Numerical Recipes in C*, 2nd ed., Cambridge University Press (1992).

[7] M. W. Wanlass and D. S. Albin, "A Rigorous Analysis of Series-Connected, Multi-Bandgap, Tandem Thermophotovoltaic (TPV) Energy Converters," AIP Conference Proceedings **738**, 462 (2004).

Effects of Contact Configuration and Perimeter Recombination on Optimal Cell Size for High Concentration Photovoltaics

James E. Moore[1,2], Matthew P. Lumb[1,2], Kenneth J. Schmieder[2], Robert J. Walters[2], Brent Fisher[3], Matt Meitl[3], Scott Burroughs[3]

[1]The George Washington University, Washington, DC 20037, USA
[2]U.S. Naval Research Laboratory, Washington, DC 20375, USA
[3]Semprius Inc., Durham, NC 27713

Abstract — **Simulations were performed with LT-SPICE™ to evaluate the limiting effects of cell size, perimeter recombination, and sheet resistance on the performance of a solar cell under 500 sun light concentration. A circuit model was generated using parameters extracted from a 3 junction InGaP/GaAs/InGaAsN device and simulated with various contact configurations. The results of the simulation were then used to determine the optimal cell size for a gridless cell and the conditions under which a gridless cell is more efficient than a cell with grid fingers.**

I. INTRODUCTION

High efficiency multijunction solar cells are commonly designed to operate under high light concentration in order to both reduce the required surface area of the expensive solar cell material itself and to enhance the open circuit voltage, which increases logarithmically with light concentration [1]. Reducing the size of the cell also has the added benefit of reducing the series resistance from surface current. Solar cells operating under high concentration typically experience very high lateral surface currents as electrons spread out to reach the contacts, which can reduce the fill factor (FF) if the sheet resistance is significant [2].

The solar cells on which the simulation is based are designed with a metal busbar at the perimeter of the cell and a centered aperture where the light is concentrated. The aperture area of concentrator cells typically contains a series of narrow metal grid-fingers, which help spread the photocurrent laterally for efficient collection at the contacts and improved FF, at the expense of increased shadowing losses. As the surface area of the cell is reduced, the need for grid fingers to maintain high FF is also reduced. In fact, if the cells are small enough, lateral current spreading in the semiconductor layers alone is sufficient to achieve high FF and the grid fingers can be eliminated entirely and improve overall efficiency.

However, we must also take into account the effects of recombination at the edges of the cell, which become more prominent as the size of the cell is reduced. For a device with a non-zero perimeter recombination rate there must therefore be an optimum size at which the highest efficiency can be achieved. Previous studies have simulated the lateral current flow effects [2-4], but to the authors' knowledge none have included the effects of perimeter recombination. Limitations

on the useable fraction of semiconductor active area due to constraints on the minimum sizes of metallization features also impose a penalty on device efficiency for micro-scale cells. In this study, we will evaluate the optimum cell size for a high concentration solar cell designed to operate at 500 suns, taking into account these different losses.

How the addition of grid fingers will impact the efficiency of the cell depends on both the surface sheet resistance and the size of the cell. We will therefore look at cell performance as a function of cell size and sheet resistance for several different contact configurations to find which configuration works best under each set of conditions.

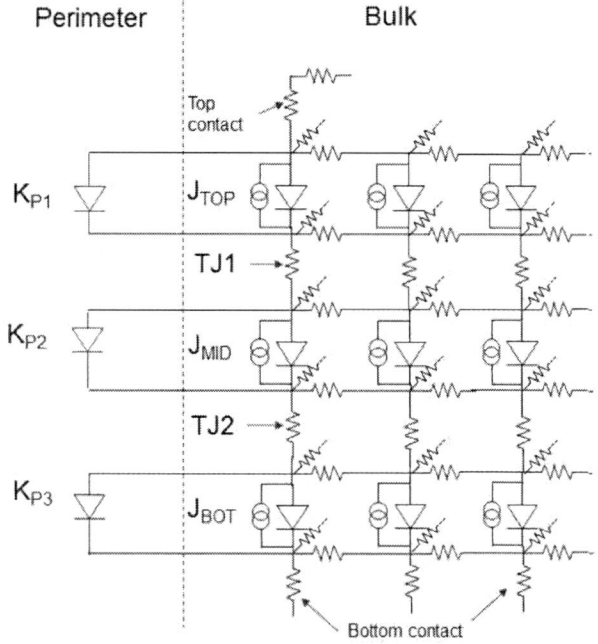

Fig. 1. Example circuit diagram of netlist used in LT-SPICE™ simulation. The nodes in the bulk, consisting of 3 diodes in series with 2 tunnel junctions, are arranged in a 2D array connected laterally by sheet resistance, and a set of 3 diodes are attached at each edge node along the perimeter.

978-1-5090-5606-4/17 $31.00 © 2017 IEEE

TABLE 1
LIST OF SIMULATION PARAMETERS

J_{TOP}	J_{MID}	J_{BOT}	K_{P1}	K_{P2}	K_{P3}	J_{sc}	R_{sheet}	R_{tun}
10^{-26} A/cm^2	10^{-20} A/cm^2	10^{-14} A/cm^2	10^{-17} A/cm	10^{-15} A/cm	10^{-15} A/cm	14.5 mA/cm^2	600 Ω/□	10^{-5} Ωcm^2

II. SIMULATION METHOD

We have written a C# program that automatically generates a netlist from an input of cell parameters, which can then be run using LT-SPICE™. The metal grid geometry and cell size are specified using a bitmap input file with pixels corresponding to individual circuit elements in the netlist. As shown in Fig. 1, the netlist uses a model consisting of multiple diode and current source elements arranged in a 3 dimensional array to simulate a 3 junction device. The bulk of the model consists of an array of nodes with 3 diodes in series representing the top, middle, and bottom junctions, with a current source representing the light generated current. Several of these nodes are distributed in parallel on a 2D grid separated by a sheet resistance. The full list of input parameters used in the simulation can be found in Table 1.

The perimeter recombination is modeled using a diode connected to each of the three junctions along each edge node following the same conventions used in [3]. The value of the perimeter recombination current was estimated to be 10^{-17} A/cm for the top cell (K_{p1}) and 10^{-15} A/cm for the bottom two cells (K_{p2}, K_{p3}), and these numbers were used as a baseline for the simulation. These values are taken from light and dark IV measurements of differently sized GaAs and InGaP cells. Similar estimates of the recombination current for the perimeter of GaAs and InGaP can be found in literature [5-6].

III. RESULTS

Several simulations were performed using the model described in the previous section, focusing on the effects of three different variables: perimeter recombination, cell size, and sheet resistance. Additionally, 5 different contact configurations were investigated to determine the effect these had on the optimum cell size.

A. Perimeter Recombination

We simulated the IV characteristics of a gridless device for four cases: without perimeter recombination, with the baseline perimeter recombination current (K_p), and with 10, 100 and 1000 times the baseline current value. As expected, when there is no perimeter recombination current the efficiency increases as the cell size decreases, although shading becomes more of a factor for a fixed busbar width in the smallest cells. Using the estimated perimeter recombination of the measured device, we get a maximum efficiency for an edge length of 50μm as shown

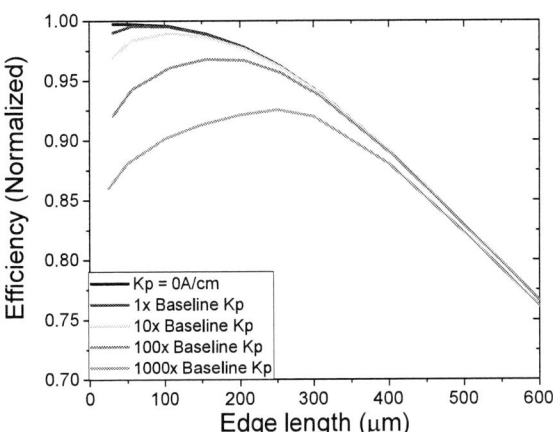

Fig. 2. Plot showing normalized efficiency vs. edge length for 4 values of edge recombination rate.

in Fig. 2. If the perimeter recombination rate is increased, the optimum cell size increases as well.

B. Cell Size and Sheet Resistance

In the baseline simulations we assumed a sheet resistance of 600 Ω/□. We also find that for higher sheet resistances (R_{sheet}) the cell size that gives us the highest efficiency decreases. In Fig. 3 we have plotted contour maps of the simulated efficiency vs. edge length and sheet resistance for the baseline K_p and a perimeter recombination rate 100 times higher. The effects of sheet resistance on the optimum cell size can most clearly be seen in Fig. 3b, where the edge length with the highest efficiency decreases from about 300μm when R_{sheet} is low to 150μm when R_{sheet} is high.

C. Contact Configuration

The efficiency was evaluated for different cell sizes with edge lengths ranging from 25μm to 1000μm using five different contact configurations. The netlist program generates a metal contact grid using a bitmap input by the user such as those shown in Fig. 4, which were used for this simulation. Each of the bitmap images is 13x13 pixels, where each pixel corresponds to one node in the circuit model. The black pixels tell the program where to put the metal contacts in the netlist.

In each simulation, all gridlines were assumed to be 3μm in width. For 600Ω/□ sheet resistance and using the baseline perimeter recombination, we find that the best overall efficiency can be achieved with a gridless cell that has an edge length of 50μm (Fig. 5). For cells with edge lengths between

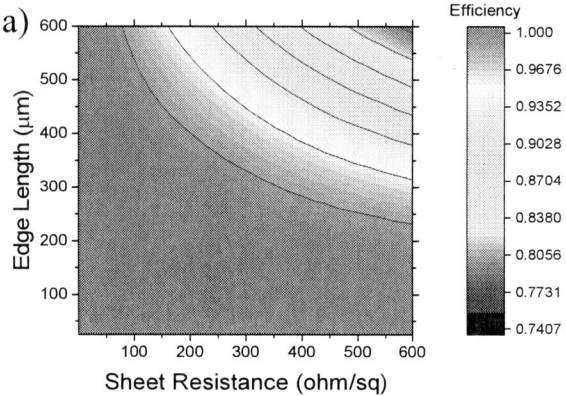

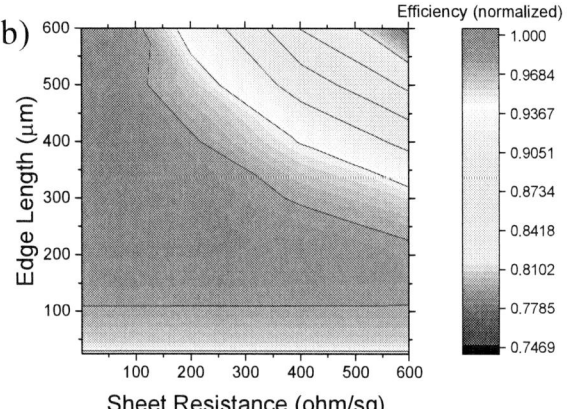

Fig. 3. Contour plot of normalized efficiency for a) cell with baseline K_p and b) a cell with 100x the baseline K_p

200 and 300μm 1 grid finger gives us the highest efficiency, while above 300μm 2 grid fingers show the best performance. However, these larger cell sizes still show lower efficiency overall than a small gridless cell

As shown in the previous sections, raising K_p increases the optimum cell size, so one might ask at what point does it become preferable to use a contact grid? We find that even by increasing K_p up to 1000 times the baseline value, although the optimum edge length increases to 250μm for a gridless cell, it is still slightly preferable to the best cell with grid fingers. We therefore conclude that for the majority of cells, further investigation of contact structures is probably unnecessary, as a gridless design is preferable for any reasonable value of K_p.

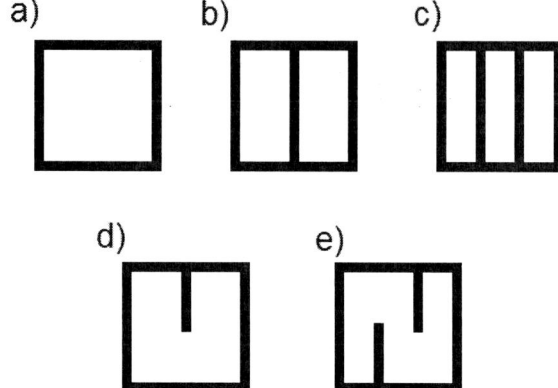

Fig. 4. Grid contact patterns used in simulation.

IV. CONCLUSIONS

In this manuscript we have demonstrated a simulation tool which can predict the IV characteristics for cells under high light concentration as a function of sheet resistance, cell size, perimeter recombination, and contact pattern. We found that the highest efficiencies are generally achieved using the smallest cell sizes without a contact grid. Although in this study we assumed uniform light concentration, this program also has the capability to simulate non-uniform intensity profiles. Simulations with non-uniform light intensity may be included in future studies.

We have also shown that the perimeter recombination rate can play an important role in determining the optimum cell size for a solar cell under high light concentration. Although the estimated perimeter recombination current for our cell is not large enough to greatly limit our minimum optimal cell size, for higher recombination currents we find that this effect does significantly increase the cell size required to get the highest efficiency. If perimeter recombination can be accurately measured, this tool therefore promises to be invaluable in assisting in determining the size limitations of high concentration solar cells.

978-1-5090-5606-4/17 $31.00 © 2017 IEEE

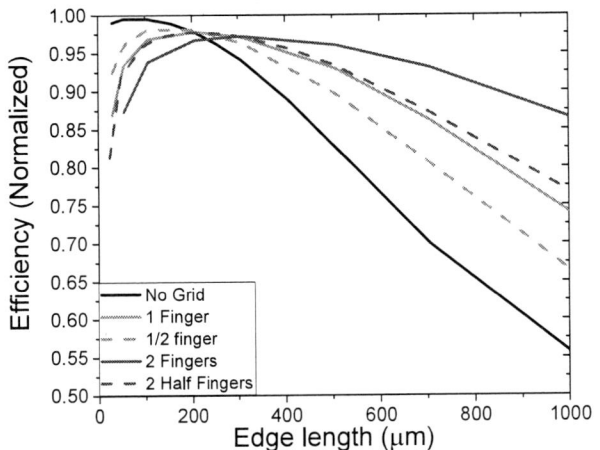

Fig. 5. Results for different grid patterns for a cell with 600 $\Omega/\square$ sheet resistance and using baseline perimeter recombination.

REFERENCES

[1] P. Perez-Higueras, E. Munoz, G. Almonacid, P.G. Vidal "High Concentrator PhotoVoltaics efficiencies: Present status and forecast" *Renewable and Sustainable Energy Reviews,* vol. 15, pp. 1810-1815, 2011

[2] L. Nielsen, "Distributed series resistance effects in solar cells," *Electron Devices, IEEE Transactions on,* vol. 29, pp. 821–827, 1982.

[3] A. Antonini, M. Stefancich, D. Vincenzi, C. Malagu, F. Bizzi, A. Ronzoni, G. Martinelli "Contact grid optimization methodology for front contact concentration solar cells" *Solar Energy Materials and Solar Cells,* vol. 80, pp. 155-166, 2003

[4] A.W. Haas, J.R Wilcox, J.L. Gray, "Numerical Modeling of loss mechanisms resulting from the distribute emitter effect in concentrator solar cells", *Proc. 34th IEEE Photovoltaic Specialists Conference* 2009

[5] P. Espinet-Gonzales, I Rey-Stolle, M. Ochoa, C. Algora, I. Gracia, and E. Barrigon "Analysis of perimeter recombination in the subcells of GaInP/GaAs/Ge triple junction solar cells," *Prog. in Photovoltaics,* vol. 23, pp. 874-88

[6] IG Vara "Development of GaInP/GaAs dual-junction solar cells for high light concentrations" PhD Thesis, University of Madrid 2010

Numerical simulation of defects in III-V PV cells: the effect of voltage bias and doping concentration

Vasiliki Paraskeva, Constantinos Lazarou, Andreas Livera, Venizelos Venizelou, Maria Hadjipanayi, and George E.Georghiou

FOSS Research Centre for Sustainable Energy, PV Technology, University of Cyprus, 75 Kallipoleos St., Nicosia, 1678, Cyprus

Abstract — The presence of defects in a solar cell device affects the electrical parameters causing possible performance deterioration. Point-like defects are expected to have significant impact on their local surroundings. Their effect weakens away from them. For this reason a physical model of a GaAs device taking into consideration the impact of local ohmic shunts was developed in order to examine the effect of the shunts on the electrical and physical parameters. Defect states have been treated as degenerate semiconductors and the impact of voltage bias and degeneracy on the main electrical parameters of the cell have been investigated.

I. INTRODUCTION

An ideal solar cell is assumed to show a homogeneous current flow across the whole area, both under illumination and in the dark [1]. However it has been observed that in real cells, there are local defects (also known as shunts) which affect the operation of the device. All types of defects act as internal short-circuits degrading the electrical parameters of the cell and often even preventing their applicability [2]. Real cells behave very differently from ideal ones because recombination in the depletion region due to shunts reduces significantly the total external current of the cell. Knowledge of the influence of shunts on their local surroundings is therefore required for a deeper understanding of shunt phenomena. Numerical modelling provides a very useful way of studying the processes behind shunts and can thus help in understanding their effect on cell operation.

In a previous work of the authors, shunts in the depletion region were introduced and a physical model of a defect in a GaAs solar cell device was developed in an attempt to investigate the impact of local ohmic shunts in the cell [3]. In this paper the local changes of the electrical and physical parameters in the shunt region in the presence of varying voltage bias and degeneracy are investigated. The current-continuity equations for the description of electrons and holes coupled with Poisson's equation for the electric field have been considered and the resulting simulations have provided detailed results about the chemical and electrical potential, in the region inside and outside defects. The simulation results demonstrated that the voltage bias has a greater influence in the region of a defect when smaller forward voltage bias is applied on the cell. Furthermore, the degeneracy of the defect

state was found to determine the electrical and physical parameters around shunts.

II. DESCRIPTION OF THE MODEL

In an attempt to develop a realistic physical model for GaAs PV cells in the presence of shunts, the detailed doping concentration of the emitter and base layers as well as their heights are firstly required. The schematic diagram of a real GaAs junction with the doping concentrations that exist in the structure is given in Fig. 1.

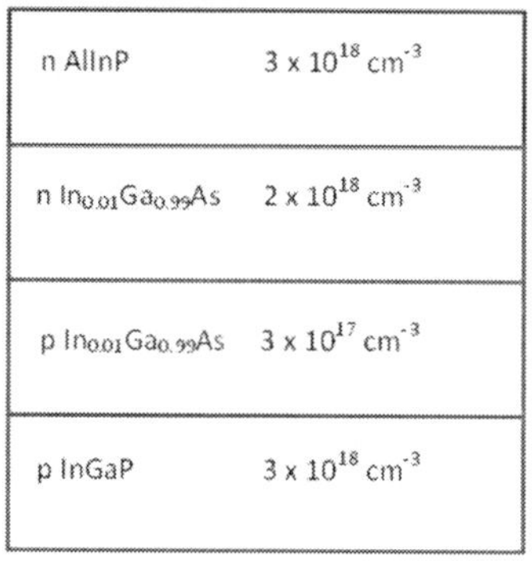

n AlInP	3×10^{18} cm^{-3}
n In$_{0.01}$Ga$_{0.99}$As	2×10^{18} cm^{-3}
p In$_{0.01}$Ga$_{0.99}$As	3×10^{17} cm^{-3}
p InGaP	3×10^{18} cm^{-3}

Fig. 1. Schematic illustration of the investigated GaAs junction.

The cell, primarily, is made up of the emitter, the base and the barrier layer. The emitter is 200 nm thick while the base has a much larger height (1600 nm). The back surface layer located at the bottom of the base is very thin (200 nm) and it is of higher concentration compared to the base. In order to take into account the defect, a tunnel junction was introduced across the p-n junction to act effectively as an ohmic shunt in contact with the neutral regions on both sides. Fig. 2 presents a two-dimensional axisymmetric schematic of the simulated GaAs junction in the presence of a single shunt in the depletion region. The simulated junction is identical with the real device. The modelled junction is p-n junction with an overall height of 2 μm as in the real device presented in Fig. 1. The radius of the simulated junction was chosen to be 15 μm for computational efficiency and is much smaller compared to

978-1-5090-5606-4/17 $31.00 © 2017 IEEE

the dimensions of the real device. Spatially-resolved Electroluminescence measurements have been undertaken for the investigation of defects in the solar cell. The measurements revealed that a reasonable length of the shunt could be 7.5 μm and that value was chosen for the radius of the shunt.

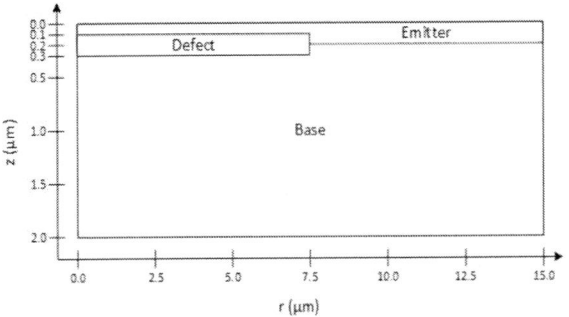

Fig. 2. Modelling structure: z=0 μm corresponds to the GaAs surface and z=0.2 μm corresponds to the p-n junction interface (depletion region). The defect has a radius of 7.5 μm and height of 0.2 μm.

The defect state in the material was chosen to be located across the depletion region and as observed in Fig. 2 lies between z=0.1 μm and z=0.3 μm from the surface. As it is well known, inside the defects the conductivity of the material increases locally and so does the forward current. Therefore, one reasonable assumption is the treatment of the defects as degenerate semiconductors with metallic behaviour (tunnel junctions). For that reason strongly p-doped acceptor states (p^{++}GaAs) were located on the p-side of the junction and n-doped donor states (n^{++}GaAs) on the n-side. The higher doping densities inside the defects were simulated with two Gaussian distributions in each region (n and p) which overlap in the depletion region. The Gaussian peaks in the n and p regions of the junction were selected to be at z=0.15 μm and z=0.25 μm from the surface respectively. Both Gaussian distributions were selected to have the same doping concentration to resemble a uniform defect on both sides of the junction. The standard deviation of the Gaussian peaks is 0.02 μm and the height of the defect was set to 0.2 μm (from z=0.1 μm to z=0.3 μm from the surface). In this way, defects cross the depletion region and are in contact with neutral regions on both sides as shown in Fig. 2. Direct recombination was assumed in the bulk (region outside defects) while inside the defects, Auger recombination was considered as the latter is observed for higher doping densities.

An axisymmetric two-dimensional model (r-z) was used to study the effect of the defect. The electrical and physical characteristics such as chemical and electric potential have been investigated at a radial distance of r=3 μm inside the defect and r=9 μm outside the defect. The upper and lower layers of the shunt in the z-axis are affected significantly by

the defect state and the impact of the shunt in those regions merits investigation. The results inside and outside the defect have been considered. No light was applied on the simulated p-n junction and thus the generation of carriers due to light is zero. The generation of carriers originates from the biasing of the junction with voltage. For the simulation of the p-n junction the current-continuity equations for the description of electrons and holes coupled with Poisson's equation for the solution of the electric field have been solved. Drift-diffusion equations define the electron-hole current densities

$$J_n = -q\mu_n n\nabla V + qD_n\nabla n \qquad (1)$$

$$J_p = -q\mu_p p\nabla V - qD_p\nabla p \qquad (2)$$

where n and p are the electron and hole concentrations respectively, D_n, μ_n and D_p, μ_p are the diffusion coefficients and mobilities of electrons and holes respectively and V is the electric potential. The current continuity equations are given by:

$$\frac{\partial n}{\partial t} = -\frac{1}{q}(\nabla.J_n) - U_n \qquad (3)$$

$$\frac{\partial p}{\partial t} = -\frac{1}{q}(\nabla.J_p) - U_p \qquad (4)$$

where U_n, U_p represent the net electron and hole recombination rate. The electric field is given from Poisson's equation:

$$\nabla.(\varepsilon\nabla V) = -q(p - n + N_D^+ - N_A^-) \qquad (5)$$

where N_A and N_D are the doping concentrations of the ionized acceptors and donors in the p- and n-type side respectively and ε is the material permittivity. The Newmann boundary conditions are:

$$\hat{n}.\overrightarrow{J_n} = 0 \qquad (6)$$

$$\hat{n}.\overrightarrow{J_p} = 0 \qquad (7)$$

Continuous hetero-junction exists at the interface between defect and actual semiconductor material. The same holds for the interface between the n- and p-side of the junction.

$$D_1 = D_2 \qquad (8)$$

$$E_1 = E_2 \qquad (9)$$

TABLE I

BASIC MODEL PARAMETERS USED IN SIMULATION.

Model parameters	Value
Base doping (N_A)	$3 \times 10^{17}\,cm^{-3}$
Emitter doping (N_D)	$2 \times 10^{18}\,cm^{-3}$
Defect thickness (d)	$0.2\,\mu m$
Electron mobility (μ_n)	$8500\,cm^2/V.s$
Hole mobility (μ_p)	$400\,cm^2/V.s$

where D and E represent the electric displacement and the electric field respectively. A constant mobility for the electrons and holes has been taken into consideration. The diffusion coefficients are estimated from Einstein's relation. The model parameters are provided in Table I.

III. RESULTS AND DISCUSSION

Initially, the impact of voltage bias on the potential around the defect sites was examined. Bias voltage dependence of the local shunt was found in previous works using the Light Beam Induced Current [4] and Electroluminescence-Photoluminescence methods [5]. The simulated results are expected to provide valuable information about the local changes of the potential around the defects as a function of the applied voltage bias. An investigation of the chemical potential of the solar cell over a broad voltage range has also been explored. The built-in potential of the junction is around 1.4 V in the region outside the defect and 1.9 V in the region inside the defect since the higher dopant concentration in the defect region increases the built-in potential. Voltage bias in the range between 0.1 - 0.9 V was applied to the junction and the chemical potential was examined in the local surroundings of the defects and outside them. The temperature of the device was set to 298 K and the doping concentration of the defect to $10^{22}\,cm^{-3}$. Fig. 3(a) and 3(b) present the chemical potential across the z-axis and at radial distances of r=9 μm (outside) and r=3 μm (inside) at different voltage biases.

Just inside the depletion region (z=0.2 μm) the quasi-Fermi levels are flat and the chemical potential equals the applied voltage in both radial distances under investigation. According to Fig. 3(a) the chemical potential in the absence of defect states remains almost constant along the z-axis and has values roughly equal to the applied voltage. However, Fig. 3(b) demonstrates that at distance of r=3 μm and at 0.1<z<0.3 μm where the defect exists, the chemical potential reduces significantly due to the presence of the defect. As a result in the upper and lower regions of the defect across the z-axis the chemical potential is low. The reduction is higher on the emitter side indicating that the potential is affected more on the surface compared to the rear regions of the solar cell.

A reduction of roughly 0.2 eV is apparent between the chemical potential value at z=0.2 μm and z=0.5 μm almost at all voltage biases under examination. The point z=0.2 μm lies inside the depletion region and the point z=0.5 μm lies outside the depletion region. Even if similar chemical potential reduction appears for all voltage biases, the amount of reduction compared to the initial chemical potential is larger at lower voltage biases indicating that in those conditions the impact of defects in their local surroundings is more pronounced. This is in agreement with the literature which states that the effect of the shunt on the local potential is more pronounced for lower applied voltages [6]. Specifically, the chemical potential reduction in the presence of 0.9 V, 0.5 V and 0.1 V is around 19%, 34.5% and 97%. The reduction of the chemical potential in the region of the defect was found to follow a polynomial function of the form of $-144x^3+367x^2-335x+128$ with voltage bias (see Fig. 4). For comparison purposes the chemical potential reduction against voltage bias at radial distance of r=9 μm outside the defect is plotted on the same graph. From the simulated results it can be obtained that the chemical potential reduction in the region outside the defect is much lower compared to the chemical potential reduction inside at all voltage biases under investigation.

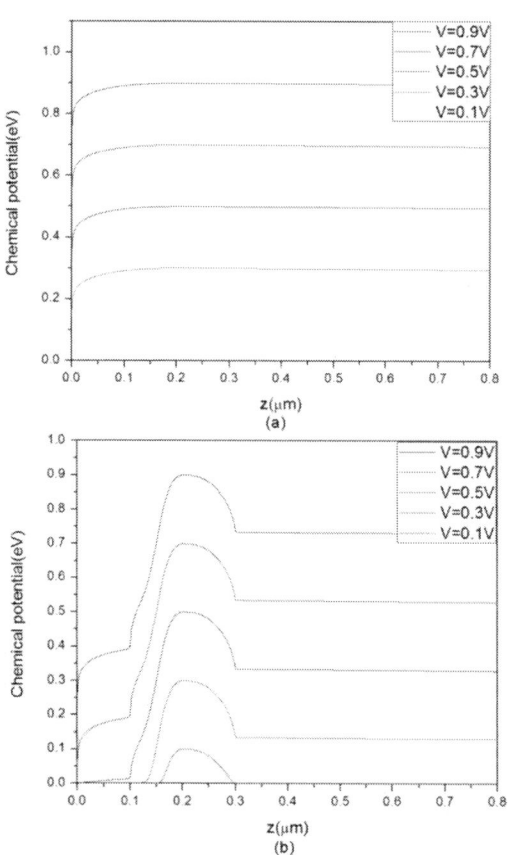

Fig. 3. Chemical potential along the z-axis (a) outside the defect (r=9 μm) and (b) in the region of the defect (r=3 μm) with the application of different forward voltage bias.

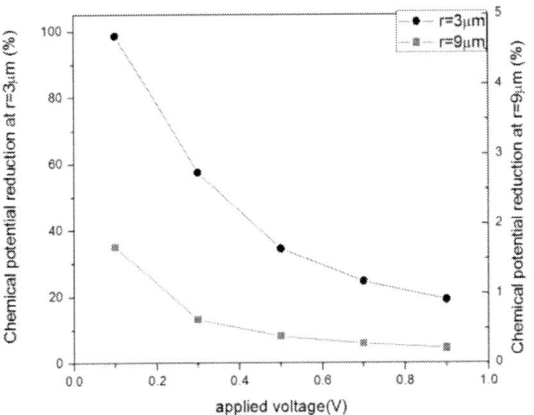

Fig. 4. Chemical potential reduction along the z-axis at different forward voltage bias in the region inside and outside the defect. It is clearly observed that the impact of the defect on the chemical potential becomes more pronounced at lower voltage bias.

Different doping densities of defects were used next in order to examine the effect of doping on the potential around the defects. Since the model used for the simulation of defects treats the defect states as degenerate semiconductors then the doping density determines the degeneracy of the defect state and the impact of the defect on its surroundings. A reasonable assumption is that weak shunts correspond to degenerate semiconductors with lower doping densities (10^{19} cm^{-3}) while stronger ones correspond to semiconductors with higher doping concentration (10^{23} cm^{-3}). Defects with five different doping concentrations were investigated (10^{19}, 10^{20}, 10^{21}, 10^{22} and 10^{23} cm^{-3}). Apart from these doping concentrations, the concentration of 10^{18} was also examined. This concentration approaches the case where no defect is introduced in the device. The results for the chemical and electric potential in the region of the defect are depicted in Fig. 5. The temperature of the device was set to 298 K and the applied voltage bias on the device was 0.9 V in order to excite the junction close to its built-in potential.

The chemical potential reduces outside the depletion across the z-axis at the radial distances where the shunt is present (r<7.5 μm). The chemical potential reduction as shown in 5(a) is more pronounced at a higher doping density of defects indicating that as the doping density increases the influence of the defect on its surroundings becomes more pronounced. Particularly, reduction of the chemical potential along the z-axis is greatly enhanced at a doping density of 10^{23} cm^{-3}. Chemical potential reduction in that case reaches 46% and thus the impact of shunts in the device becomes more important. The chemical potential reduction is correlated with

the quasi-Fermi levels E_{Fn} and E_{Fp}. According to the literature in degenerate semiconductors, the difference of the quasi-Fermi levels becomes smaller in agreement with the simulation results [7]. Also, the electric potential in the depletion region increases due to the presence of defects and this is more severe in the presence of shunts with high degeneracy. Higher built-in potential is exhibited in the depletion region in the presence of defects with higher doping profile resulting in higher electric potential. An increase of the electric potential of 0.6 V is obtained in the presence of a defect with strong metal properties. Although these changes in the fields are local they affect the total potential of the device.

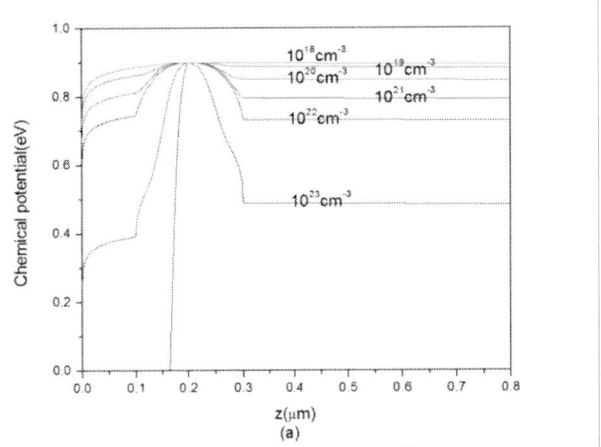

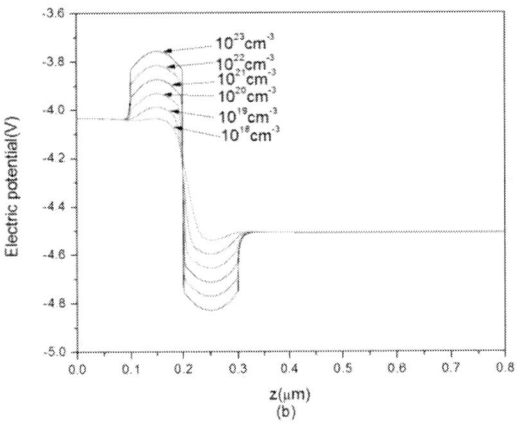

Fig. 5. Modelling results for the (a) chemical potential and (b) electric potential along the z-axis in the region of defects (r=3 μm) for different doping concentrations.

In order to investigate in detail the chemical potential reduction with doping density, the chemical potential is plotted against degeneracy (see Fig. 6). The chemical potential reduction was estimated in relation to the values at z=0.2 μm and z=0.5 μm as discussed below. The chemical potential reduction increases parabolically with degeneracy indicating the strong effect of the metal properties of the defect on the local potential of the device.

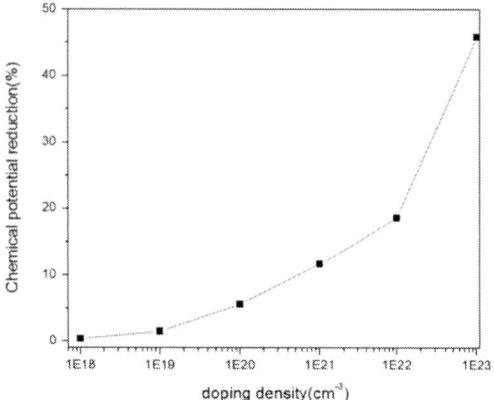

Fig. 6. Chemical potential reduction along the z-axis at different doping densities of the defect at radial distance of r=3 μm.

IV. CONCLUSIONS

A physical model for a GaAs solar cell, taking into consideration the impact of local ohmic shunts, has been developed. Defects have been treated as degenerate semiconductors with higher doping density. The effect of voltage bias and doping concentration has been investigated in order to demonstrate the impact of those parameters on the electrical parameters around the defect. The level of voltage bias applied on the device determines the chemical potential reduction occurring in the vicinity of the defect. Furthermore, the doping density of the defects determines both the electrical and chemical potential of the solar cell device.

ACKNOWLEDGMENTS

This work has been co-financed by the European Regional Development Fund and the Republic of Cyprus in the framework of the project 'H-Volt PV' with grant number 'KOINA/SOLAR-ERA.NET/1214/09'.

REFERENCES

[1]   A. Schenk and U. Krumbein, "Coupled defect-level recombination: Theory and application to anomalous diode characteristics," *J. Appl. Phys.*, vol. 78, no. 5, p. 3185, 1995.

[2]   O. Breitenstein, J. Bauer, and J. P. Rakotoniaina, "Material-induced shunts in multicrystalline silicon solar cells," *Semiconductors*, vol. 41, no. 4, pp. 440–443, 2007.

[3]   V. Paraskeva, C. Lazarou, M. Hadjipanayi, M. Norton, M. Pravettoni, G. E. Georghiou, M. Heilmann, and S. Christiansen, "Photoluminescence analysis of coupling effects: The impact of shunt resistance and temperature," *Sol. Energy Mater. Sol. Cells*, vol. 130, pp. 170–181, 2014.

[4]   M. Nguyen, A. Scutt, J. Carstensen, and H. Foll, "Quantitative Defect Analysis on Solar Cells by Laser Beam Induced Current (LBIC) Measurements and 3D Network Simulations," *Proc. MRS Fall Meet.*, vol. 1493, 2012.

[5]   O. Breitenstein, J. Bauer, T. Trupke, and R. A. Bardos, "On The Detection of Shunts in Silicon Solar Cells by Photo- and Electroluminescence Imaging," *Prog. Photovoltaics Res. Appl.*, vol. 16, pp. 325–330, 2008.

[6]   D. Abou-Ras, T. Kirchartz, and U. Rau, *Advanced Characterization Techniques for Thin Film Solar Cells*. 2011.

[7]   S. Sze, *Physics of Semiconductor Devices*. 2006.

Improvement of open-circuit voltage in metamorphic GaSb cells grown on GaAs substrates by using an interfacial misfit array and an AlSb blocking layer

E. J. Renteria, S. J. Addamane, D. M. Shima, A. Mansoori, A. L. Soudachanh, and G. Balakrishnan

Center for High Technology Materials, University of New Mexico, Albuquerque, NM, 87106, USA

Abstract — The performance of metamorphic GaSb solar cells grown on GaAs substrates has been improved by adding an AlSb layer that blocks the propagation of threading dislocations into the cell. The threading dislocation density has been reduced from 2 x 10^{18} cm^{-2} to 3.6 x 10^{17} cm^{-2}. This increases the open circuit voltage from 0.13 V to 0.16 V and decreases the leakage current by 77%. Transmission Electron Microscope data and J-V characteristics under dark and illumination are presented.

Index Terms — current-voltage characteristics, degradation, electron microscopy, materials reliability, photovoltaic cells, III-V semiconductor materials.

I. INTRODUCTION

In 1989, a mechanically stack GaAs/GaSb solar cell with over 35% efficiency was demonstrated by Fraas and et al [1]. Since then, a great effort has been devoted to the study of GaSb/GaAs heterojunctions for the applications in thermophotovoltaics (TPVs) and multi-junction photovoltaics [2-6]. In addition to the advantage of monolithically integrating narrow band gap cells with wide band gap cells, the growth of III-Sb based layers on GaAs substrates is an attractive idea due to lower cost and larger wafer sizes for GaAs substrates compared to GaSb substrates [6]. Also, the availability of semi-insulating (SI) GaAs substrates is valuable for TPVs as it facilitates the architecture of monolithically interconnected modules where TPVs are connected in series to build up voltage [7-8].

However, the 7.78 % lattice mismatch between GaSb and GaAs can result in a significant amount of threading dislocations in the crystalline structure. Such dislocation density can be reduced by inducing arrays of 90° interfacial misfit dislocations (IMF) at the GaSb/GaAs interface [9]. This technique reduces the threading dislocation density (TDD) in the GaSb epitaxial layer to the low 10^8 defects/cm^2, which has been sufficient to demonstrate a wide range of devices [10-12]. However, for photon-absorption applications, a higher crystal quality is required [13]. The feasibility of making GaSb cells on GaAs substrates has already been explored. GaSb photodiodes of 400 μm and 500 μm in diameter were used for optical measurements and it was concluded that GaSb cells on GaAs substrates is a promising idea if the threading dislocation density in the GaSb epitaxial layers is further reduced [13-14].

In this work, we present the progress in attempting to reduce the number of threading dislocation density in metamorphic GaSb solar cells.

II. EXPERIMENT

For this study, three samples were grown by means of molecular beam epitaxy. The first sample is a lattice matched GaSb solar cell grown on a GaSb substrate to use as a control sample (LM sample). The other two samples are metamorphic GaSb solar cells grown on GaAs substrates by inducing arrays of 90° interfacial misfit dislocations at the GaSb/GaAs interface. One sample has the AlSb blocking layer and other does not.

A. Growth

The p-on-n GaSb solar cell structure consists of a 50 nm thick p-type (5 x 10^{18} cm^{-3}) GaSb contact layer, a 500 nm p-type (5 x 10^{17} cm^{-3}) GaSb emitter layer, a 2000 nm thick n-type (4 x 10^{17} cm^{-3}) base layer, and a 200 nm thick n-type (5 x 10^{18} cm^{-3}) GaSb contact layer. The IMF sample with the AlSb blocking layer (IMF BL sample) has a 500 nm GaSb layer after the GaAs buffer layer and then it is followed by a 250 nm AlSb layer and then the GaSb solar cell layers. For the regular IMF sample, the cell's layers are grown after the GaAs buffer layer. The growth conditions for the IMF samples are established as described by Huang, et al [9].

B. Processing and characterization of devices

Device fabrication was performed using standard photolithography techniques. Metals were deposited via e-beam evaporation. For the top grid, p-type Ti/Pt/Au ohmic contacts were deposited. For the bottom contacts, a metal sequence of Ni/Ge/Au/Pt/Au was deposited if the material was n-GaSb and Ge/Au/Ni/Au if it was n-GaAs. To make the n-contacts ohmic, the n-GaSb contacts were annealed at 290°C for 45 seconds and the n-GaAs at 380°C for 60 seconds. The devices were isolated by etching the mesas below the p-n junction using an inductively coupled plasma reactor with BCl$_3$ gas.

III. RESULTS AND DISCUSSION

Fig. 1 shows a cross-sectional transmission electron microscopy (XTEM) image of a GaSb membrane grown on a GaAs substrate with a 250 nm thick AlSb layer in between. The AlSb layer helps to block the propagation of threading dislocations to the GaSb epitaxial layer as it can be seen in Fig. 1. To calculate the TDD in the 1 μm thick GaSb epitaxial layer plan view TEM was done. The calculated TDD was 3.6×10^7 defects/cm^2 which is almost an order of magnitude less than the IMF samples without an AlSb blocking layer.

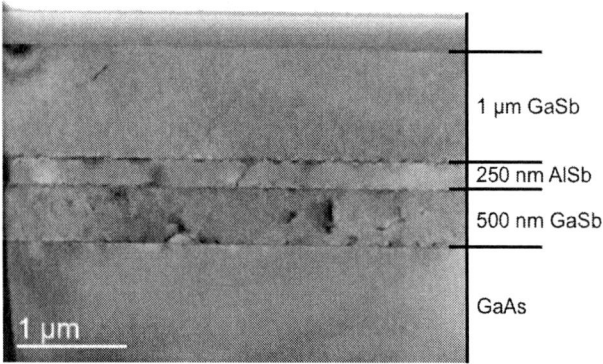

Fig. 1.   XTEM images of GaSb epitaxial layer grown on a GaAs substrate with an AlSb layer that blocks the propagation of threading dislocations

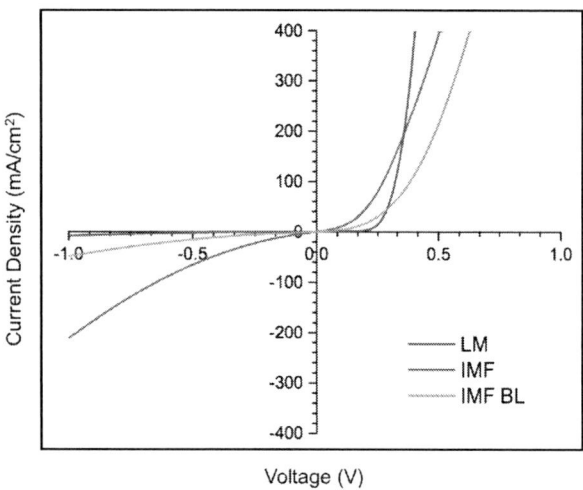

Fig. 2.   Dark J-V characteristics of solar cells

Fig. 2 shows the dark current density – voltage (J-V) characteristics of the three samples. It can clearly be seen that the reverse-bias leakage current of the IMF BL device has been reduced significantly in comparison to IMF device without BL. At -1V, the dark current is -8 mA/cm^2 for the LM sample, -48 mA/cm^2 for the IMF BL sample, and -210 mA/cm^2 for the IMF sample. The one order of magnitude reduction in TDD has reduced the leakage current by about 77 %.

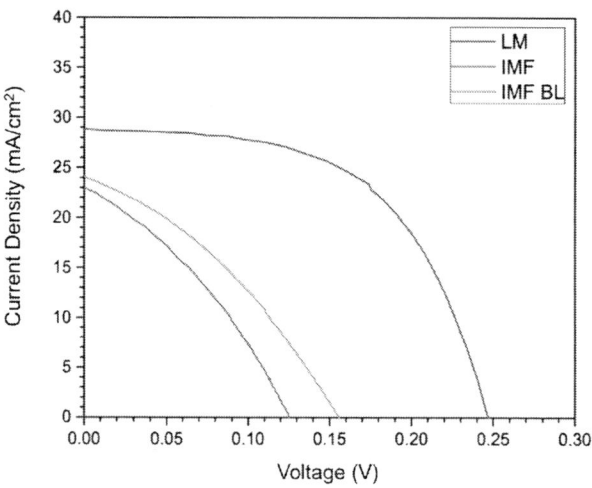

Fig. 3.   J-V characteristics of GaSb cells under AM0 illumination

TABLE I
SUMMARY OF SOLAR CELL'S PARAMETERS

	LM	IMF	IMF BL
J_{sc} (mA/cm^2)	28.72	22.93	23.91
V_{oc} (V)	0.25	0.13	0.16
FF (%)	57.25	33.58	34.53
η (%)	2.98	0.71	0.95
R_s (Ωcm^2)	6	12	22
R_{sh} (Ωcm^2)	1000	44	47

Fig. 3 shows the J-V characteristics of the 3 samples measured under air mass zero (AM0) illumination. The 0.5 cm x 0.5 cm control cell exhibits an open-circuit voltage (V_{OC}) of 0.25 V, a short-circuit current density (J_{SC}) of 28.72 mA/cm^2, a fill factor (FF) of 57.25%, and an efficiency (η) of 2.98%. The same size IMF cell shows a V_{OC} of 0.13 V, a J_{SC} of 22.93 mA/cm^2, a FF of 33.58%, and a η of 0.71%. The IMF BL cell shows a V_{OC} of 0.16 V, a J_{SC} of 23.91 mA/cm^2, a FF of 34.53%, and a η of 0.95%. Table I has a summary of these parameters. Comparing the IMF-based cells, it is observed that the IMF BL sample has a higher V_{oc} than the IMF sample. This means that the crystal quality of the IMF BL sample has increased and that indeed the threading dislocation density in the sample has been reduce. Also, the J_{SC} has increased slightly. This increase in J_{SC} can be observed in the external quantum efficiency (EQE) measurements shown on Fig. 4. Throughout the wavelength range, the EQE curve for the IMF BL sample is higher than the IMF sample. Hence, the IMF BL sample was expected to have a higher J_{SC} than the IMF sample. The short circuit currents calculated from EQE measurements are 29.25 mA/cm^2 for the LM sample, 20.94 mA/cm^2 for the IMF BL sample, and 18.76 mA/cm^2 for the IMF sample.

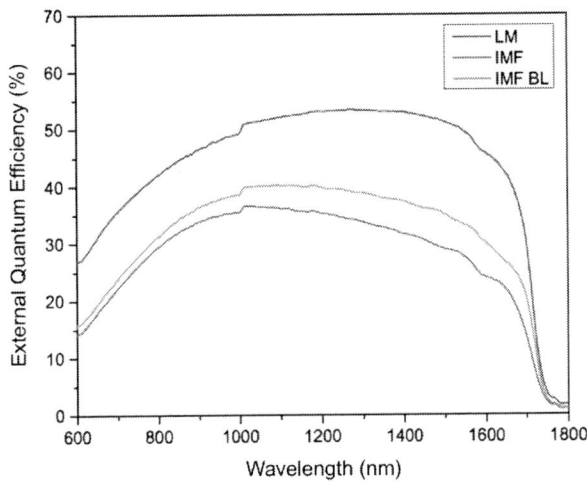

Fig. 4.   External quantum efficiency of solar cells

IV. SUMMARY

In conclusion, we have shown that the performance of metamorphic GaSb solar cells grown on GaAs substrates by IMF technique can be improved by using AlSb layers that block the propagation of treading dislocations to the p-n junction. The leakage current has been reduced by 77 % and the V_{oc} has been increased by 23 %. These two parameters are dependent on the crystal quality of the material. Hence, the improvement of these parameters verifies that the amount of threading dislocations in the sample with the AlSb blocking layer has been reduced.

ACKNOWLEDGEMENT

This material is based upon work primarily supported by the National Science Foundation (NSF) and the Department of Energy (DOE) under NSF CA No. EEC-1041895. Any opinions, findings and conclusions or recommendations expressed in this material are those of the author(s) and do not necessarily reflect those of NSF or DOE. Also, the authors thank the advanced space power group at the Kirtland Air Force Research Laboratory for allowing us to use their I-V and EQE setup.

REFERENCES

[1]   L. M. Fraas, J. E. Avery, J. Martin, V. S. Sundaram, G. Girard, V. T. Dinh, T. M. Davenport, J. W. Yerkes, and M. J. O'neil. "Over 35-percent efficient GaAs/GaSb tandem solar cells," *IEEE Transactions on Electron Devices*, vol. 37, pp. 443-449, 1990.

[2]   Haywood, S. K., C. G. Scott, G. M. Sweileh, M. Lakrimi, N. J. Mason, P. J. Walker, and L. Zheng. "Effect of GaSb growth temperature on p-GaSb/n-GaAs diodes grown by MOVPE." *IEE Proceedings-Optoelectronics* 145, no. 5 (1998): 287-291.

[3]   Toušková, J., D. Kindl, E. Samochin, J. Toušek, E. Hulicius, J. Pangrác, T. Šimeček, and Z. Výborný. "Charge transport study

and spectral response of GaSb/GaAs heterojunctions prepared by MOVPE." *Solar energy materials and solar cells* 76, no. 2 (2003): 135-145.

[4]   Bumby, C. W., Q. Fan, P. A. Shields, R. J. Nicholas, S. K. Haywood, and L. May. "InAs passivated GaSb thermo-photovoltaic cells on a GaAs substrate grown by MOVPE." *International journal of ambient energy* 25, no. 2 (2004): 73-78.

[5]   Aroutiounian, V. M., G. Sh Shmavonyan, O. A. Zadoyan, K. M. Gambaryan, and A. M. Zadoyan. "Investigation of photoelectrical properties of epitaxially grown heterojunction thermophotovoltaic cells." *Journal of Contemporary Physics (Armenian Academy of Sciences)* 49, no. 6 (2014): 258-263.

[6]   Zheng, L., S. K. Haywood, N. J. Mason, and G. Verschoor. "p-GaSb/n-GaAs heterojunction diodes for TPV and solar cell applications." *IEE Proceedings-Optoelectronics* 147, no. 3 (2000): 205-208.

[7]   Wang, C. A., R. K. Huang, D. A. Shiau, M. K. Connors, P. G. Murphy, P. W. O'Brien, A. C. Anderson, D. M. DePoy, G. Nichols, and M. N. Palmisiano. "Monolithically series-interconnected GaInAsSb/AlGaAsSb/GaSb thermophotovoltaic devices with an internal backsurface reflector formed by wafer bonding." *Applied physics letters* 83, no. 7 (2003): 1286-1288.

[8]   Kim, Jung Min, Partha S. Dutta, Eric Brown, Jose M. Borrego, and Paul Greiff. "Wafer-scale processing technology for monolithically integrated GaSb thermophotovoltaic device array on semi-insulating GaAs substrate." *Semiconductor Science and Technology* 28, no. 6 (2013): 065002.

[9]   S. H Huang, G. Balakrishnan, A. Khoshakhlagh, A. Jallipalli, L. R. Dawson, and D. L. Huffaker. "Strain relief by periodic misfit arrays for low defect density GaSb on GaAs." *Applied physics letters* vol. 88, pp. 131911, 2006.

[10]   Mehta, M., G. Balakrishnan, S. Huang, A. Khoshakhlagh, M. N. Kutty, L. R. Dawson, and D. L. Huffaker. "Demonstration of GaSb QW-based" Buffer-Free" LED on GaAs Substrate." In *Device Research Conference, 2006 64th*, pp. 135-136. IEEE, 2006.

[11]   Mehta, Manish, Ganesh Balakrishnan, Maya N. Kutty, Pravin Patel, Larry R. Dawson, and Diana L. Huffaker. "1.55 μm GaSb/AlGaSb MQW diode lasers grown on GaAs substrates using interfacial misfit (IMF) arrays." In *Conference on Lasers and Electro-Optics*, p. CThK6. Optical Society of America, 2007.

[12]   Plis, Elena, Jean-Baptiste Rodriguez, G. Balakrishnan, Y. D. Sharma, H. S. Kim, T. Rotter, and Sanjay Krishna. "Mid-infrared InAs/GaSb strained layer superlattice detectors with nBn design grown on a GaAs substrate." *Semiconductor Science and Technology* 25, no. 8 (2010): 085010.

[13]   D. DeMeo, C. Shemelya, C. Downs, A. Licht, E. S. Magden, T. Rotter, C. Dhital, S. Wilson, G. Balakrishnan, and T. E. Vandervelde, "GaSb Thermophotovoltaic Cells Grown on GaAs Substrate Using the Interfacial Misfit Array Method," *Journal of Electronic Materials,* vol. 43, pp. 902-908, 2014.

[14]   B. C. Juang, R. B. Laghumavarapu, B. J. Foggo, P. J. Simmonds, A. Lin, B. Liang, and D. L. Huffaker, "GaSb thermophotovoltaic cells grown on GaAs by molecular beam epitaxy using interfacial misfit arrays," *Applied Physics Letters,* vol. 106, pp 111101, 2015.

Energy yield evaluation for field operation of solar cells in Singapore: GaAs/GaAs tandem vs. GaAs single-junction solar cells

Maung Thway[1,2], Zekun Ren[3], Kevin Nay Yaung[3], Haohui Liu[1], Zhe Liu[1,4], Samuel Raj[1], Soo Jin Chua[2], Armin G. Aberle[1,2], Tonio Buonassisi[3,4], Ian Marius Peters[4], Fen Lin[1]

[1] Solar Energy Research Institute of Singapore (SERIS), National University of Singapore, 7 Engineering Drive 1, 117574, Singapore

[2] Department of Electrical and Computer Engineering, National University of Singapore, 4 Engineering Drive 3, 117583, Singapore

[3] Singapore-MIT Alliance for Research and Technology (SMART), 1 CREATE Way, 138602, Singapore

[4] Massachusetts Institute of Technology (MIT), 77 Massachusetts Avenue, Cambridge, MA 02139, United States

Abstract — **Tandem solar cells are one of the most promising ways to exceeding the one-sun efficiency limit of single-junction (SJ) solar cells. However, compared to SJ solar cells, tandem solar cells often experience larger energy yield losses in the field, due to variations in operating conditions (i.e. spectrum, irradiance and operating temperature). In this work, we determine the energy yield losses of dual-junction and SJ GaAs solar cells, which are potential candidates for thin-film on silicon tandem solar cells. Our measurements show that a GaAs SJ cell has an annual energy yield of 3.0% more than with standard testing conditions (STC), whereas the investigated GaAs/GaAs dual-junction cell has an annual energy yield loss of 11.2% compared with STC.**

Index Terms — **Energy yield, GaAs, GaAs/GaAs, irradiance effect, multijunction, spectra effect, tandem solar cells, temperature effect.**

I. INTRODUCTION

Electricity production from solar cells will play an important role in reducing climate change by not contributing to greenhouse gases (e.g. CO_2). The cost of PV electricity needs to be further reduced to ensure wider adoption. Improving solar cell efficiencies is one key lever to reduction of levelised cost of electricity (LCOE) [1]. III-V/Si tandem (or multijunction) solar cells are one of the most promising concepts for high-efficiency solar cells [2]. Among the different materials for tandem solar cells, the GaAs with Si combination is attractive because of the established fabrication processes and proven long-term stability of GaAs single-junction (SJ) solar cells [3]-[5]. The GaAs/Si tandem solar cell can potentially achieve more than 30% flat-plate non-concentrating PV module efficiency [6]. Since GaAs does not have an ideal top cell bandgap on a Si bottom cell, current matching between the top and bottom cells in an integrated two-terminal structure becomes challenging. An alternative way is to mechanically stack GaAs on Si and operate the tandem in a 4-terminal configuration [7]. However, the four-terminal configuration is challenging to integrate at module level and possibly incur additional cost compared to a 2-terminal configuration. In order to achieve current matching in the 2-terminal configuration, we have proposed replacing the GaAs top cell with a GaAs/GaAs tandem [8]. This configuration can better balance the current generated by each subcell, and thus achieve a current-matched 2-terminal tandem. The radiative 1-sun efficiency limit of a GaAs/GaAs/Si tandem solar cell is 41.7%, and theoretical calculations based on realistic material parameters of record GaAs and Si cells show a lower efficiency potential of 33.0% [8]. There were no reports of a tandem (dual-junction; DJ) GaAs cell until the concept was introduced by Ren et al. [8].

To achieve current matching, the thicknesses of the absorbers in a tandem are usually designed based on the standard AM1.5G spectrum. However, in actual field operations, the solar illumination received by the tandem experiences spectrum and irradiance variation [9]. Low irradiance conditions frequently occur during field operations and have a noticeable impact on tandem energy yield. In addition, it has been shown that tandem cells generally are more sensitive to spectrum variations than SJ solar cells [10]-[11]. However, Liu et al. showed that the losses due to spectrum variation can be minimized by optimizing the thickness of the top cell [12]. Therefore, in addition to using the standard testing condition (STC) efficiency alone as performance indicator, the energy yield of a tandem solar cell needs to be evaluated using realistic outdoor spectra [12]-[13]. Spectrum effect studies were conducted extensively for concentrator photovoltaic (CPV) solar cells [14]-[16]. These studies used simulated spectra based on atmospheric parameters (such as, air mass, optical depth and ambient moisture), and indicated that the spectrum effect could be significant in tandem structures. In this work, the spectrum, irradiance and operating temperature effects for non-concentrating tandem structures were studied experimentally.

II. EXPERIMENTAL DETAILS

We compare the energy yield of a GaAs SJ solar cell and a GaAs/GaAs tandem. The GaAs solar cell was exposed to the same illumination conditions in either the GaAs SJ, or GaAs/Si tandem configurations. Likewise, the light input for the GaAs/GaAs solar cell will be the same in both GaAs/GaAs DJ

978-1-5090-5606-4/17 $31.00 © 2017 IEEE

and GaAs/GaAs/Si triple-junction solar cells. Hence, the losses and the yield calculated in this work can be directly correlated to GaAs/Si and GaAs/GaAs/Si tandem studies. Figure 1 shows the schematic structures of the GaAs/GaAs tandem solar cell and the GaAs SJ solar cell used in this experiment. The top GaAs/GaAs tandem [8] and the GaAs SJ cells were fabricated on epi-ready *n*-type GaAs on-axis wafers (400 μm, <100> oriented) using an AIXTRON Cirus Metal Organic Chemical Vapour Deposition (MOCVD) reactor. The measured STC efficiencies were 11.7% for the GaAs/GaAs tandem cell without antireflective coating (ARC) and 17.8% for the GaAs SJ cell with ARC. A LED based *I-V* tester (WaveLabs, Sinus-220) was used in this study [17]. The 21 LEDs in this *I-V* tester can be controlled independently to mimic the shape of a target spectrum. The actual field spectra was extracted from the spectroradiometer placed at an outdoor module testing facility of SERIS [18]. The spectra were recorded in 1-minute intervals. The one-year data collection period was from February 2013 to January 2014 in Singapore. The spectrum data discussed in this work were obtained by choosing 5-minute intervals within the recorded period. This time resolution is necessary for Singapore as its solar spectrum and illumination intensity vary rapidly due to the movement of clouds.

$$APE = \frac{\int_{\lambda_1}^{\lambda_2} I(\lambda)d\lambda}{\int_{\lambda_1}^{\lambda_2} \phi(\lambda)d\lambda} = \frac{\int_{\lambda_1}^{\lambda_2} I(\lambda)d\lambda}{\int_{\lambda_1}^{\lambda_2} \frac{I(\lambda)}{hc/\lambda}d\lambda} \quad (1)$$

where $I(\lambda)$ is the intensity distribution for each wavelength of a spectrum ($W/m^2 \cdot nm$), $\phi(\lambda)$ photon flux density for each wavelength of the same spectrum ($number/m^2.nm$), and hc is the product of Planck's constant and the speed of light [12]. APE is used as a metric to define the spectral composition of solar illumination. Spectra with the same APE values were averaged to obtain a representative spectrum for each APE value. Figure 2 represents the averaged representative solar spectrum with each APE value for Singapore over the wavelengths from 300 nm to 1100 nm. The input target spectra of the *I-V* tester have APE values ranging from 1.70 to 2.10 eV, with 0.05 eV intervals [12].

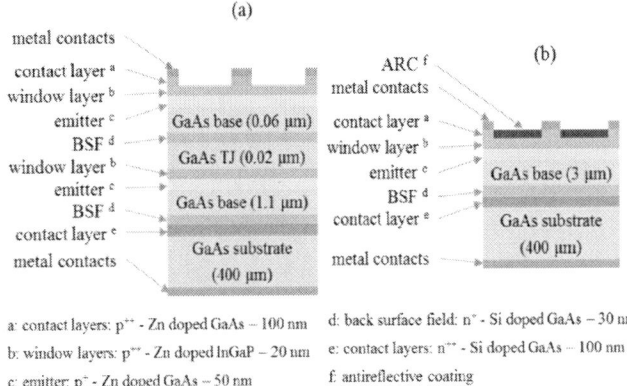

(a)

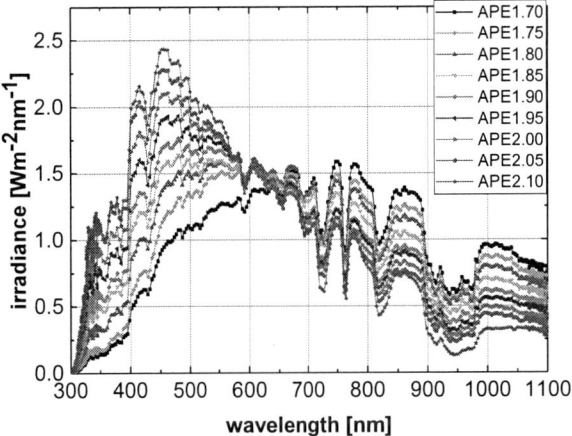

Fig. 2. Averaged representative spectrum for each APE value (from 1.70 eV to 2.10 eV with 0.05 eV interval) for Singapore over the wavelength range of 300 nm to 1100 nm.

Fig. 1. Schematic structures of (a) GaAs/GaAs dual-junction tandem solar cell. (b) GaAs SJ solar cell.

Spectrum and irradiance data were collected using a spectroradiometer, while the operating temperature data were collected using a temperature sensor. The *I-V* characteristics of the solar cells under various operating conditions are experimentally determined using an LED-based *I-V* tester. The *I-V* tester has been programmed to simulate different solar spectra with suitable light intensities and operating temperatures in Singapore. The STC efficiency of the test structures will be compared with a realistic measured power conversion efficiency due to three dominant effects (i.e., spectrum, irradiance, and operating temperature).Average photon energy (APE) is the average of the energy carried by all the photons in a certain wavelength range. APE of a spectrum within a defined wavelength range can be calculated by dividing the integrated irradiance by the total number of photons:

Figure 3 shows the target spectrum and the *I-V* tester simulated spectrum with an APE of 1.70. The simulated spectrum (with LED tester) was carefully adjusted to power match to the target spectrum at each 100 nm wavelength interval. The mismatch percentage for spectral power is maintained below 5% (Class A) for each wavelength interval to meet the classification requirement by IEC 60904-9 [19]. The overall mismatch APE value of every simulated spectra was also kept below 0.5%. Table I summarizes the power matching per 100 nm wavelength interval and the obtained APE value for an APE 1.70 target. It is important that all APE spectra have the same total power input so that the spectral and irradiance effects are independent. Therefore, all APE spectra were rescaled to match the power of AM1.5G (802.3 Wm⁻²) in the wavelength range 300 to 1100 nm. From the data collected, the contribution of each APE spectrum with specific irradiance and temperature was tabulated. This will be referred to as the measurement matrix throughout the paper. The detailed energy yield measurement methodology will be reported elsewhere.

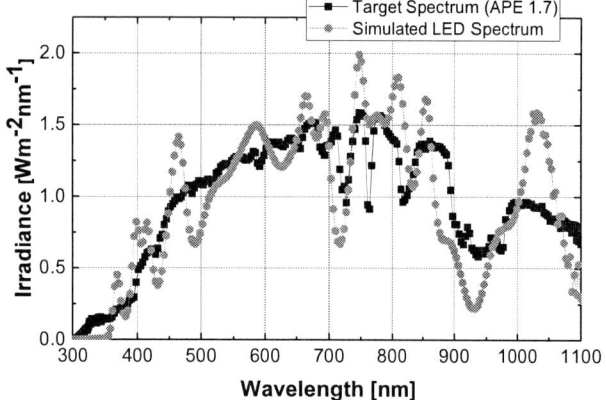

Fig. 3. APE 1.70 target spectrum (black) and LED *I-V* tester's simulated spectrum (red).

TABLE I
POWER AND APE MISMATCH OF SIMULATED SPECTRUM

Wavelength Range [nm]	Target Power [Wm⁻²]	Simulated Power [Wm⁻²]	Relative Mismatch [%]
300 – 400	16.18	16.14	0.2%
400 – 500	84.34	84.34	0.0%
500 – 600	127.11	127.28	0.1%
600 – 700	145.81	145.41	0.3%
700 – 800	136.76	137.04	0.2%
800 – 900	128.06	124.30	2.9%
900 – 1100	164.09	161.42	1.6%
Total	802.34	795.92	0.8%
APE	1.7000	1.7070	0.4%

III. PRELIMINARY RESULTS

First, we measured the *I-V* characteristics of the GaAs/GaAs and GaAs SJ cells under standard testing condition (AM1.5G spectrum with 1 sun illumination at 25°C operating temperature), as shown in Fig. 4. Ideally, the GaAs/GaAs tandem cell has half of the short-circuit current density (J_{SC}) of the GaAs SJ cell. However, in this case the thickness of the top GaAs cell in the GaAs/GaAs tandem cell is 110 nm. It is challenging to produce a very thin top cell (ideally 85 nm) that is not shunted in the GaAs/GaAs tandem cell [8]. In addition, the GaAs/GaAs tandem cell does not have an ARC. The fill factor (FF) increased in GaAs/GaAs tandem cell due to lower J_{SC} compared to that of GaAs SJ cell. However, both the GaAs and the GaAs/GaAs tandem cell have high series resistance. Both devices are still undergoing an efficiency optimization process. The solar cells used in this work are the best cells among similar cells produced at the time of measurement.

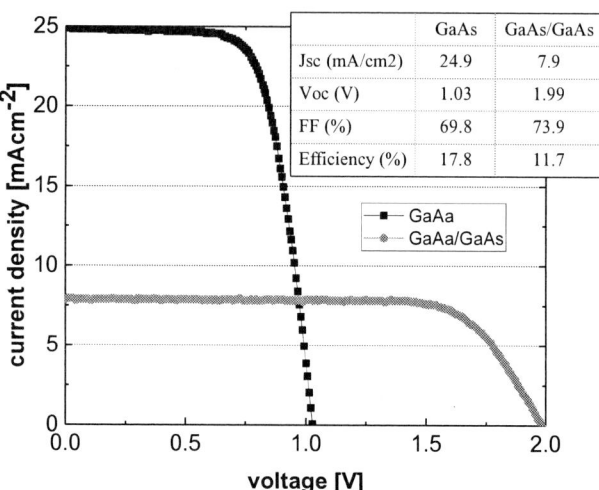

Fig. 4. *I-V* characteristics of GaAs/GaAs and GaAs structure under AM1.5G at 1 sun intensity and 25°C operating temperature.

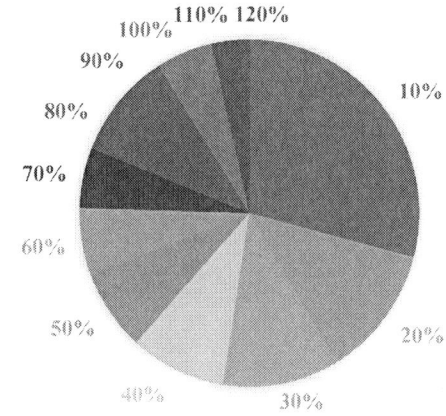

(a) Irradiance frequency distribution

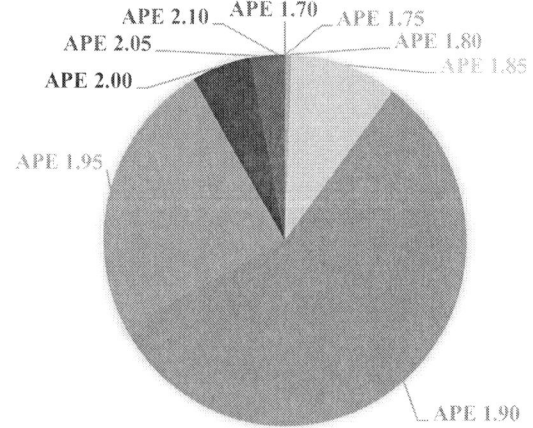

(b) APE frequency distribution

(c) Temperature frequency distribution

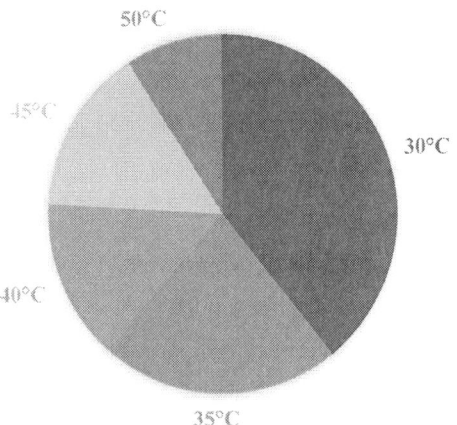

Fig. 5. Frequency distribution of three components in the collected data in Singapore. (a) irradiance frequency distribution ranging from 10% to 120% irradiance level; (b) APE (spectrum) frequency distribution ranging from APE 1.70 to APE 2.10; (c) operating temperature frequency distributions ranging from 30°C to 50°C.

The distribution of irradiance, APE and temperature was collected over one year in Singapore. The annual frequency distribution of each component is shown in Fig. 5. Figure 5(a) shows the frequency distribution of different irradiance levels with an interval of 10%. It can be observed that an irradiance level of more than 1 sun) was observed 4% of the time during the recorded year. Low irradiance level (10% irradiance) is the most frequent and comprise of more than one quarter of the total frequency distribution (see Fig. 5(a)). The APE frequency distribution is summarized in Fig. 5(b). The STC spectrum (AM1.5G) has an APE equivalent of 1.85 eV. However, it can be observed from Fig. 5(b) that the most frequent spectrum in Singapore throughout the data collection period has APE of 1.90. More than half of the spectra recorded corresponds to the APE 1.90 spectrum. On the other hand, APE 1.70 and APE 1.75 spectra are less likely to occur in Singapore as they contribute less than 0.2% in the frequency distribution pie chart (see Fig. 5(b)). In Fig. 5(c), the frequency distribution of the operating temperature recorded from 30°C to 50°C is shown. As expected, low temperature condition (30°C) occurs more frequently than the others.

The irradiance, spectrum and temperature effects were studied individually. The flow chart of calculating the performance ratio (PR) of each component (irradiance, spectrum, and temperature) for efficiency under APE 1.80 spectrum at 10% irradiance at 30°C is shown in Fig. 6. The irradiance loss/gain was calculated by measuring the efficiency of the cells at different irradiance levels of STC condition. Performance ratio is defined as the relative percentage increase/decrease of efficiencies under different conditions, irradiance for example, compared to that obtained under STC condition. The frequency distribution of irradiance (see Fig. 5(a)) is taken into account in this calculation. The GaAs SJ cell

shows a relative PR increase of +3.34%, while the PR of GaAs/GaAs tandem cell is -2.45%. The GaAs SJ cell performs relatively better under most irradiance levels compared to STC conditions. It is mainly due to the increment in FF at lower irradiance levels, despite the decrease in open-circuit voltage (V_{OC}) and short-circuit current (J_{SC}). On the other hand, the GaAs/GaAs tandem suffers from reduced overall performance under lower irradiance levels.

The spectra effect on both cells was calculated by varying the spectrum (APE) but maintaining the irradiance level of the AM1.5G spectrum and the operating temperature. As this comparison is limited to the same irradiance level and same temperature, the difference calculated between these efficiencies are due to the spectrum effects. These efficiency differences are multiplied with the APE frequency distribution (shown in Fig. 5(b)) to obtain the spectral loss/gain. It is observed that both GaAs SJ and GaAs/GaAs tandem cells decreases -1.13% and -4.87%, respectively. At lower APE values, there are more photons in the longer wavelength of the spectrum which is beyond the wavelength that GaAs can absorb. Therefore, at lower APE spectra, the efficiencies of both the GaAs and GaAs/GaAs cells are reduced. Additionally, different spectra cause different current generation in the subcells of the GaAs/GaAs tandem cell. Hence, the PR drop of the GaAs/GaAs tandem cell is higher than that of the GaAs SJ cell.

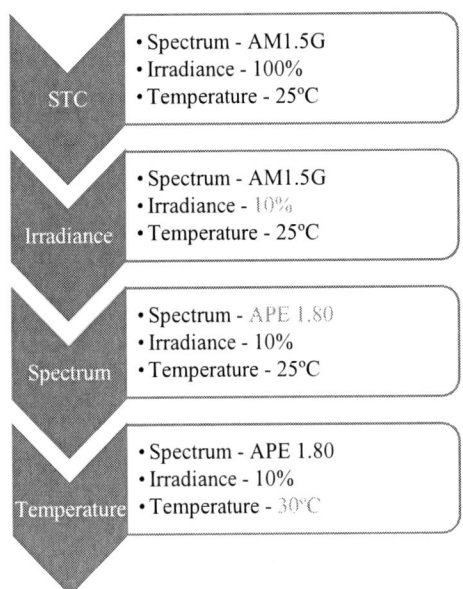

Fig. 6. Flow chart describing an example of calculating the PR breakdown for each component (irradiance, spectrum and operating temperature) for APE 1.80 spectrum at 10% irradiance, and at 30°C condition. At each step only one component is changed (written in red), while keeping the others constant.

Similarly, the effect of operating temperature is calculated. We compared the efficiencies of the cells under varying temperatures but with the same spectrum and irradiance conditions. The efficiency differences are then multiplied with the frequency distribution of temperature (shown in Fig. 5(c))

978-1-5090-5606-4/17 $31.00 © 2017 IEEE

to obtain the temperature loss/gain. The GaAs SJ cell shows an increment in PR (+0.83), while the GaAs/GaAs tandem cell results in PR drop of -3.91%. Even though the V_{OC} of the GaAs SJ cell reduces with increasing temperature, the efficiency of GaAs SJ cell increases due to J_{SC} and FF. On the other hand, the efficiency of GaAs/GaAs tandem cell decreases with increasing temperature. This might be due to the difference in temperature coefficient of these cells. The authors are aware that the sequence of the breakdown (i.e. the order of calculation for all components: irradiance, spectrum, and operating temperature) has an impact on the PR distribution of loss/gain among the individual components. However, the consistent use of the same sequence among different cells will help in comparing and optimizing the cells. A comparison between different categories will be reported in a future publication.

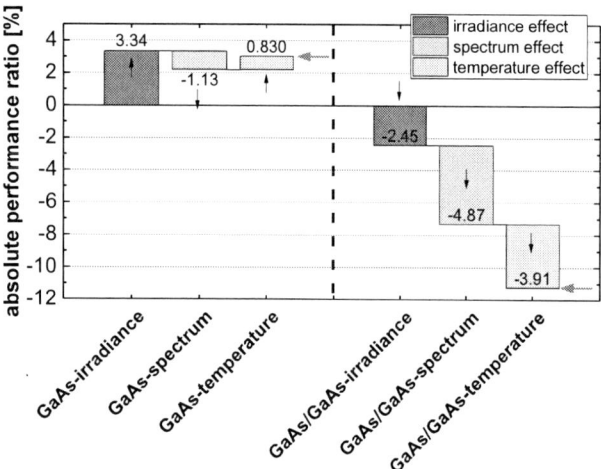

Fig. 7. Measured performance ratio after taking into account the spectrum effect, the irradiance effect, and the operating temperature effect for a SJ GaAs and dual-junction GaAs/GaAs operating in Singapore.

In order to perform a complete energy yield analysis, there is a need to create an efficiency measurement matrix of different spectra, irradiances and temperatures. By repeating the above mentioned methods for different spectra, irradiances and temperature, an efficiency measurement matrix was obtained. In the efficiency measurement matrix, not all possible combinations of APEs, irradiance levels, and operating temperature values are included as some combinations were not observed throughout the year of data collection, for example, APE 1.70 spectrum at 100% irradiance. By multiplying the efficiency measurement matrix and the frequency distribution matrix, we calculated the energy yield of the GaAs/GaAs tandem cell and GaAs SJ cell for Singapore. The annual performance ratio (PR) of the GaAs/GaAs tandem cell is 88.8%, while that of GaAs SJ cell is 103.0%. This means the annualized harvesting efficiency has a drop of 11.2% for the GaAs/GaAs tandem cell, while there is a gain of 3.0% for the GaAs SJ cell, relative to their respective STC performance. Figure 7 shows the PR calculated for both cells with a breakdown of the effects of irradiance, spectrum, and operating

temperature. The rightmost bar in Fig. 7 for each solar cell (in cyan colour) shows the PR change due to the temperature effect and it also represents the annual overall PR of each cell (indicated by the red arrow). It should be noted that in Fig. 7, individual effects such as spectrum effect, irradiance effect and temperature effect shown are in the sequence of breakdowns mentioned earlier. The irradiance effect was calculated first, followed by spectrum, and operating temperature.

IV. CONCLUSION

We evaluated experimentally the energy yield of solar cells under specific weather conditions in Singapore. A STC *I-V* tester with an LED-based solar simulator was configured to generate different operating conditions of irradiance, spectrum and operating temperature, simulating the specific weather conditions in Singapore. With these indoor measurements, the actual annual energy yields of a GaAs/GaAs tandem cell and a GaAs SJ cell in Singapore were predicted. Our study showed that the GaAs SJ cell had a PR increase of 3.0% and is more robust to various outdoor operation conditions in Singapore than the GaAs/GaAs tandem cell, which showed a PR decrease of 11.2% even though a GaAs/GaAs cell stack can be current matched to the Si bottom cell in a GaAs/GaAs/Si configuration [8]. Our study suggests that the GaAs/GaAs DJ should be redesigned to be more resistant to the weather conditions in Singapore. This study shows that addition to the STC efficiency, energy yield should be applied for evaluation of tandem solar cells in the field

ACKNOWLEDGEMENT

SERIS is sponsored by the National University of Singapore (NUS) and Singapore's National Research Foundation (NRF) through the Singapore Economic Development Board (EDB). This research is supported by the National Research Foundation, Prime Minister's Office, Singapore under its Campus for Research Excellence and Technological Enterprise (CREATE) program and under its Energy Innovation Research Program (EIRP-13, Award No. NRF2015EWT-EIRP003-004).

REFERENCES

[1] D. M. Powell, R. Fu, K. Horowitz, P. A. Basore, M. Woodhouse, and T. Buonassisi, "The capital intensity of photovoltaics manufacturing: barrier to scale and opportunity for innovation," *Energy and Environmental Science*, vol. 8, pp. 3395-3408, 2015.

[2] M. Yamaguchi, T. Takamoto, K. Araki, and N. Ekins-Daukes, "Multi-junction III-V solar cells: current status and future potential," *Solar Energy*, vol. 79, pp. 78-85, 2005.

[3] T. P. White, N. N. Lal, K. R. Catchpole, "Tandem solar cells based on high-efficiency c-Si bottom cells: top cell requirements for >30% efficiency," *IEEE Journal of Photovoltaics*, vol. 4, pp. 208-214, 2014.

[4] A. Pozza, and T. Sample, "Crystalline silicon PV module degradation after 20 years of field exposure studied by electrical tests, electroluminescence, and LBIC," *Progress in Photo-voltaics*, vol. 24, pp. 368-378, 2016.

[5] F. Fertig, S. Nold, N. Wohrle, J. Greulich, I. Hadrich, K. Kraub, M. Mittag, D. Biro, S. Rein, R. Preu, "Economic feasibility of bifacial silicon solar cells," *Progress in Photovoltaics*, vol.24, pp.800-817, 2016.

[6] Z. Ren, J. P. Mailoa, Z. Liu, H. Liu, S. C. Siah, T. Buonassisi, and I. M. Peters, "Numerical analysis of radiative recombination and reabsorption in GaAs/Si tandem," *IEEE Journal of Photovoltaics*, vol. 5, pp. 1079-1086, 2015.

[7] M. Thway, Z. Ren, Z. Liu, S. J. Chua, A. G. Aberle, T. Buonassisi, and I. M. Peters, "Sensitivity analysis for III-V/Si tandem solar cells: a theoretical study," accepted in *Japanese Journal of Applied Physics*, vol. 56, no. 8S2, 2017.

[8] Z. Ren, H. Liu, Z. Liu, C. S. Tan, A. G. Aberle, T. Buonassisi, and I. M. Peters, "The GaAs/GaAs/Si solar cell – Towards current matching in an integrated two terminal tandem," *Solar Energy Materials & Solar Cells*, vol. 160, pp. 94-100, 2017.

[9] H. Liu, Z. Ren, Z. Liu, A. G. Aberle, T. Buonassisi, and I. M. Peters, "The realistic energy yield potential of GaAs-on-Si tandem solar cells: a theoretical case study," *Optic Express*, vol. 23, pp. A382-A390, 2015.

[10] S. R. Kurtz, J. M. Olson, and P. Faine, "The difference between standard and average efficiencies of multijunction compared with single-junction concentrator solar cells," *Solar Energy Materials and Solar Cells*, vol. 31, pp. 501-513, 1991.

[11] P. Faine, S. R. Kurtz, C. Riordan, and J. M. Olson, "The influence of spectral solar irradiance variations on the performance of selected single-junction and multijunction solar cells," *Solar Cells*, vol. 31, pp. 259-278, 1991.

[12] H. Liu, Z. Ren, Z. Liu, A. G. Aberle, T. Buonassisi, and I. M. Peters, "Predicting the outdoor performance of flat-plate III–V/Si tandem solar cells," *Solar Energy*, vol. 149, pp. 77–84, 2017.

[13] H. Liu, A. G. Aberle, T. Buonassisi, and I. M. Peters, "On the methodology of energy yield assessment for one-Sun tandem solar cells," *Solar Energy*, vol. 135, pp. 598–604, 2016.

[14] N. L.A. Chan, T. B. Young, H. E. Brindley, N. J. Ekins-Daukes, K. Araki, Y. Kemmoku, and M. Yamaguchi, "Validation of energy prediction method for a concentrator photovoltaic module in Toyohashi Japan," *Progress in Photovoltaics*, vol. 21, pp. 1598-1610, 2013.

[15] X. Wang and A. Barnett, "The effect of spectrum variation on the energy production of triple-junction solar cells," IEEE *Journal of Photovoltaics*, vol. 2, pp. 417-423, 2012.

[16] E. F. Fernandez, F. Almonacid, J. A. Ruiz-Arias, and A. Soria-Moya, "Analysis of the spectral variations on the performance of high concentrator photovoltaic modules operating under different real climate conditions," *Solar Energy Materials and Solar Cells*, vol. 127, pp. 179-187, 2014.

[17] Wavelabs, Sinus-220, [Available at http://www.wavelabs.de/en]

[18] Solar Energy Research Institute of Singapore (SERIS), Outdoor Module Testing (OMT) Facility, [Available at http://www.seris.nus.edu.sg/activities/outdoor-module-performance-testing.html/]

[19] J. H. Wohlgemuth, "Standards for PV modules and components – recent developments and challenges," in *27th European Photovoltaic Solar Energy Conference and Exhibition*, 2012.

Simulation of the performances of multijunction solar cells with improved voltage by transfer and scattering matrix methods

Gianluca Timò[1,2] and Lucio Andreani[2]

[1]RSE, Strada Torre della Razza, le Mose, 29100 Piacenza, Italy

[2]Department of Physics, University of Pavia, Via Bassi 6, I-27100 Pavia, Italy

Abstract — So far the transfer matrix method (TMM) has been extensively used for the calculation of the propagation of an electromagnetic wave through planar stratified media and in particular for the calculation of the reflection coefficient and generation function in III-V solar cells. TMM however can present numerical instability when applied to MJ solar cells whose structures allow improving the cell photovoltage. It is shown that for such structures, in order to overcome the numerical instability of TMM, a simplified scattering matrix method can be successfully applied.

I. INTRODUCTION

Multijunction (MJ) solar cells allow reaching very high conversion efficiency owing to the ability to better exploit the solar spectrum. Further improvement in the solar cell performances are possible by applying proper strategies for increasing the cell voltage [1]. For the simulation of the MJ cell performances adopting advanced structures, proper calculation methods are essential. The transfer matrix method (TMM) has been extensively used for analyzing the propagation of an electromagnetic wave through planar stratified media and in particular for the calculation of the reflection coefficient and the generation function in III-V solar cells [2]. So far, as the best of the authors knowledge, no mention has been reported about the possible failure of the TMM method when applied to calculate the reflectivity of MJ junction solar cells. TMM method can present numerical instability in particular when applied to triple junction solar cells whose structure allows improving the cell photovoltage. For such structures, in order to overcome the numerical instability of TMM, a simplified scattering matrix (S-matrix) method reported in [3] has been implemented and successfully applied.

II. THEORETICAL BACKGROUND

A. Transfer matrix method

To study the reflection and the transmission of electromagnetic radiation through a MJ solar cell with the TMM, we consider a structure as reported in Fig.1. The MJ cell structure present several layers with different refraction indexes and thicknesses comparable with light wavelength, therefore, multiple reflections and interference phenomena take place inside the device.

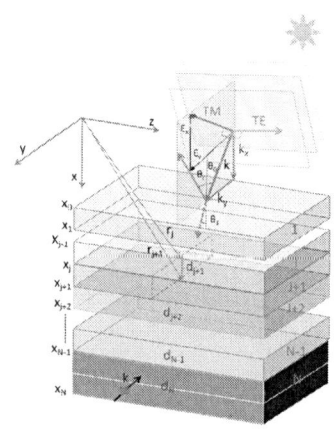

Fig. 1. Schematic of a MJ structure. Solar Radiation propagates like plane waves coming from the #0 layer (air). The solar intensity is not polarized and it can be thought as equally distributed in the TM and TE polarization modes. When the substrate is very thick, the backward propagating wave coming from the substrate-air interface can be neglected.

In order to calculate the intensity of the forward and backward propagating waves in each layer of the MJ solar cell structure, the Maxwell equations are solved with proper boundary conditions (which assure the continuity of the electric and magnetic fields at the interfaces) and by considering the following assumption: i) each layer of the MJ structure is thought as homogeneous and isotropic, ii) the layers and interfaces are source-free (J=0 and ρ=0), iii) the Y and Z dimensions of the MJ solar cell are much greater than X dimension (1D problem). The complex amplitudes of the forward χ_j^+ and backward χ_j^- waves at the end of a generic layer "j" can be shown related to the complex amplitudes of the forward χ_{j+1}^+ and backward χ_{j+1}^- waves at the end of the adjacent layer "j+1" through a two column "transfer matrix" M_{j+1} (see Eq. 1 and 2).

$$\begin{pmatrix} \chi_j^+ \\ \chi_j^- \end{pmatrix} = M_{j+1} \begin{pmatrix} \chi_{j+1}^+ \\ \chi_{j+1}^- \end{pmatrix}; \quad M_{j+1} =$$

$$\frac{1}{2w_j} \begin{bmatrix} \psi_{j+1}^-(w_j + w_{j+1}) & \psi_{j+1}^+(w_j - w_{j+1}) \\ \psi_{j+1}^-(w_j - w_{j+1}) & \psi_{j+1}^-(w_j + w_{j+1}) \end{bmatrix} \quad (1)$$

$$\psi_{j+1}{}^{\pm} = e^{\pm i k_{x,j+1} d_{j+1}}; \; w_j = \frac{\varepsilon_j}{k_{x,j}}; \; k_{x,j} = \frac{2\pi}{\lambda_0} n_j \cos\theta_j \qquad (2)$$

Where: d_{j+1} = thickness of the j+1 layer, $k_{x,j+1}$= component of the wave vector along the "x" direction of the j+1 layer, ε_j= complex dielectric constant of the layer j, n_j= complex refractive index of the layer j, ϑ_j= propagation angle in the layer j and λ_0= wavelength in the vacuum.

Once all the interfaces of the MJ solar cell structure are considered, it is possible to connect the complex amplitude of the forward and backward waves at the two ends of the overall medium by a single matrix, that is the matrix product, as schematically reported in Fig.2 and showed in Eq.3.

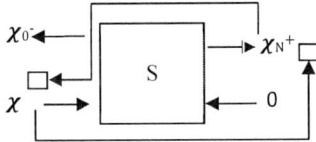

$$M = M_1 \, M_2 \, .. \, M_{j+1} \, ... M_N$$

Fig. 2. Tracking the complex amplitudes of the forward and backwards waves from the layer "0" to the layer "N" with the transfer matrix M. The backward propagating wave coming from the last layer-air interface is considered null.

$$\begin{pmatrix} \chi_0{}^+ \\ \chi_0{}^- \end{pmatrix} = M \begin{pmatrix} \chi_N{}^+ \\ 0 \end{pmatrix} \qquad (3)$$

While the TMM is very powerful, it presents an intrinsic numerical instability, since the left half of the transfer matrix M grows exponentially as the number of layer increases (see M_{j+1} in Eq.1). We can then expect to find a numerical instability when the MJ device reaches a certain total thickness.

B. S-matrix method

The simple S-matrix formulation suggested in [3] is an alternative method that we have applied to MJ solar cells in order to avoid the numerical instabilities found applying the TMM. In the S-matrix method the outgoing waves are connected to the incoming waves as reported in Fig.3 and Eq.4.

Fig. 3. Tracking the complex amplitudes of the outgoing waves and incoming waves of the layer "0" and layer "N" connected by the S-matrix. The backward propagating wave coming from the last layer-air interface is considered null.

$$\begin{pmatrix} \chi_0{}^- \\ \chi_N{}^+ \end{pmatrix} = S \begin{pmatrix} \chi_0{}^+ \\ 0 \end{pmatrix} \qquad (4)$$

To derive the S-matrix, the two linear equation systems reported in Eq.3 and Eq.4 are forced to be compatible. Furthermore, the terms that grows exponentially are eliminated, obtaining solutions numerical stable.

III. MJ SOLAR CELL FOR HIGH PHOTOVOLTAGE

As an example of MJ solar cell, we have considered the InGaP/InGaAs/Ge triple junction (TJ) structure. One strategy to increase the photovoltage in such a device it is to make thinner the Ge substrate (see Fig.4). In this way the light can be recycled and the recombination losses can be reduced. The optimized substrate thickness, as part of the active layer of the solar cell, can be determined by simulating the performance of the MJ solar cell by applying the TMM or the S-matrix method.

However, while the semi-infinite substrate structure allows neglecting the coherent reflections in the substrate, which in turn, it is equivalent to consider a substrate with zero thickness, once the substrate is thinned, coherent reflections inside this layer have to be taken into account. This means that a further layer has to be included in the MJ cell structure for calculating the intensity of the forward and backward propagating waves. We pointed out that in this situation the TMM can become numerically instable.

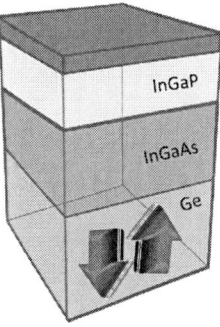

Fig. 4. Schematic of the MJ cell structure for getting high photovoltage. By thinning the Ge substrate and considering a perfect mirror on the back side, it is possible to recycle part of the light and reduce the recombination losses owing to the lower cell volume.

III. SIMULATION RESULTS

The reflection coefficient of a InGaP/InGaAs/Ge solar cell structure with a semi-infinite substrate calculated by TMM and S-scattering methods is reported in Fig. 5.

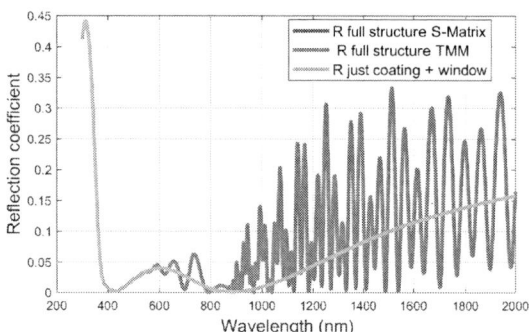

Fig.5. Reflection coefficient calculated by TMM and S-matrix method for the full InGaP/InGaAs/Ge cell structure and for a simplified structure which considers only the coherent reflection in the coating and solar cell window.

Both methods allow an accurate determination of the reflection coefficient considering all the interference phenomena taking place in the structure. The determination of the reflection coefficient only considering the coherent reflections in the coating and in the solar cell window would bring to under evaluate this parameter in wide range of wavelength.

The calculation of the reflection coefficient has then been carried out by considering a TJ solar cell structure in which the Ge substrate has been thinned to 6 μm. The results are showed in Fig.6.

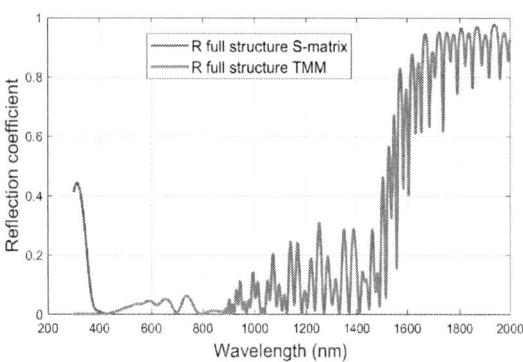

Fig.6. Reflection coefficient calculated by TMM and S-matrix method for the full InGaP/InGaAs/Ge solar cell structure considering a 6 μm thick Ge substrate. Al mirror was considered as last layer.

The TMM fails in the calculation of the reflection coefficient for wavelength lower than 400 nm. It can be shown that the numerical instability of TMM gets worse, that is, the calculus starts to be interrupted at longer wavelength, as the thickness of the substrate is increased. On the contrary, the S-scattering method is stable in the whole wavelength range.

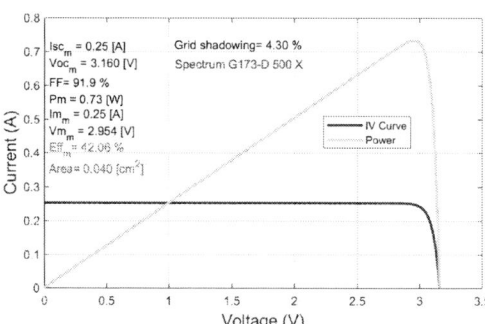

Fig.7. Ideal I-V curve of the InGaP/InGaAs/Ge solar cells with a semi-infinite Ge substrate.

The improvement in the solar cells voltage owing to the thinning of the substrate (only considering the reduction of the recombination losses) can be assessed by comparing the IV curve reported in Fig.7 and Fig.8.

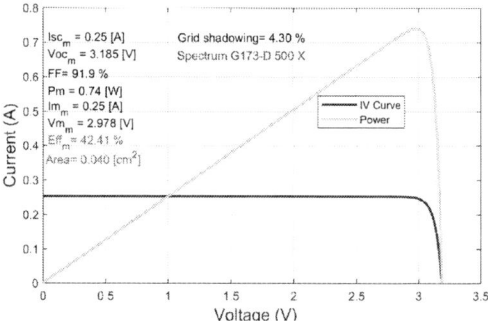

Fig. 8. Ideal I-V curve of the InGaP/InGaAs/Ge solar cells with a 6 μm thick Ge substrate.

The increase in the open circuit voltage is even higher if the different optical losses are considered.

IV. CONCLUSION

With this contribution it is shown that the TMM is not adequate to calculate the reflection coefficient of InGaP/InGaAs/Ge solar cell structures when the substrate is thinned in order to improve the solar cell photovoltage. As alternative to the TMM, a simple S-matrix method can be applied in order to avoid numerical instabilities in the whole wavelength range of interest.

REFERENCES

[1] Owen D. Miller, Eli Yablonovitch, and Sarah R. Kurtz, "Strong Internal and External Luminescence as Solar Cells Approach the

Shockley–Queisser Limit" *IEEE Journal of Photovoltaics*, vol.2.No3.July 2012, 303-310.

[2] G. Letay, M. Breselge, A.W. Bett, "Calculating the generation function of III_V solar cells" in *3rd World Conference on Photovoltaic energy conversion*, May-11-18, 741-744, 2003 Osaka, Japan 25.

[3] Alex J. Yuffa and John A. Scales, "Object-oriented electrodynamic S-matrix code with modern applications", *Journal of Computational Physics* 231 (2012) 4823–483

Optimized Design of Back-Contact Thin-Film GaAs Solar Cells

Jia-Ling Tsai[1], Chung-Yu Hong[1, 3], Tien-Chien Zhan[1], Yuh-Renn Wu[2], Albert Lin[4], and Peichen Yu[1]

[1]Department of Photonic & Institute of Electro-Optical Engineering, National Chiao Tung University, 1001 Ta Hsueh Road, Hsinchu 300, Taiwan

[2] Graduate Institute of Photonics and Optoelectronics, National Taiwan University, Taipei, Taiwan

[3] Arima Photovoltaic & Optical Corporation, 119 Guangfu North Road, Hukou Township, Hsinchu 303, Taiwan

[4]Department of Electronics Engineering, National Chiao Tung University, 1001 Ta Hsueh Road, Hsinchu 300, Taiwan

(*E-mail: yup@faculty.nctu.edu.tw)

Abstract —In this work, we attempt to find the optimal design parameters for the doping concentration, thickness of the base region, the length and pitch of back electrodes by employing a validated Sentaurus TCAD model. Through current-voltage characteristics, one could determine best power conversion efficiency and the correlation between design parameters. While optical shadowing is eliminated in the back-contact design, it is found that electrical shading affects the cell performance and is sensitive to the electrode length and pitch. Through minimizing the length of anode, the efficiency can be elevated significantly. Moreover, a relatively wide cathode results in efficient carrier separation and superior transportation properties. Based on these simulation outcomes, we are able to optimize the design of back-contact back-junction GaAs solar cells

Index Terms — GaAs solar cells, back-contacted, thin film.

I. INTRODUCTION

Back-contact back-junction (BC-BJ) solar cells, also named interdigitated back contact (IBC), were first proposed in 1975 by Schwartz and Lammert [1]. As an innovation for concentration applications, BC-BJ has a variety of advantages like completely eliminating the shadowing effect of the contact fingers at the front surface and lowering the series resistance. Besides, this architecture simplifies interconnection between cells at the module level. These benefits makes BC-BJ also valuable for nowadays pursuing higher conversion efficiency at one-sum illumination. In 2014, K.Masuko et al improved the conversion efficiency of a single junction silicon-based solar cells to 25.6% by adopting the IBC concept [2]. In late 2016, this record was further updated as 26.3% by Kaneka Corporation [3]. Compared to their previous result, which didn't utilize back-contact structure, the short-circuit current is elevated from 40.8 to 42.25 mA/cm². These results reveal that BC-BJ is an indispensable part of high-efficiency solar cells.

Not like most researchers studying about silicon back-contact solar cells, we select GaAs as absorbing material. Since Shockley-Queisser limit was published, GaAs has always been considered to have better conversion efficiency due to its desirable bandgap [4]. And recently, Alta devices improved the single-junction GaAs world record from about 25 to 28.8% [5]. This improvement is attributed to a strengthened photon-recycling effect caused by the removal of substrate and the high reflective back contact. As a result, the incident photons can be absorbed effectively and the loss by radiative recombination can be minimized. In addition, owing to its large absorption coefficient, it only needs merely a few micrometers to absorb most of the incident photons. Thus, the cell can be made both ultra-thin and high-efficiency, which allows it to have wider range of applications. These benefits not only cause thin-film the growing trend for GaAs solar cells but also make GaAs a leading material for high efficiency.

We believe that higher efficiency can be achieved if the concept of back-contact is applied to thin-film GaAs solar cells. But only few experiments have been conducted because of the complicated structure and fabricated challenges. In 2010, some researchers reported such a structure in a small size [6]. Additionally, we fabricated a slightly larger back-contact GaAs solar cells with 6.5% conversion efficiency in 2016 [7]. Neither of the results show a breakthrough performance, therefore, we study how it is affected by the structural and fabricated parameters with the help of SENTAURUS TCAD. In this work, we aim to search for the optimal design parameters.

II. SIMULATION METHODOLOGY

A schematic cross-section of the back-contact solar cells in our simulation is shown in Fig 1. The design is based on what we used previously and slightly modified. To have a better performance, a double layer anti-reflection coating composed by silicon oxide (SiO_2) and silicon nitride (SiN_x) is added to the surface. Moreover, we add a back-surface field above the anode to avoid electrons flowing through.

Each simulation considers the standard AM1.5G solar spectrum. As for the recombination loss, band to band recombination and trap-assisted recombination are included in our simulation. To predict realistic results, many doping dependent models are adopted, including the SRH lifetime model, since it is strongly related to the doping density.

Similarly, the model for deterioration in mobility and bandgap narrowing effect while doping concentration increases are considered.

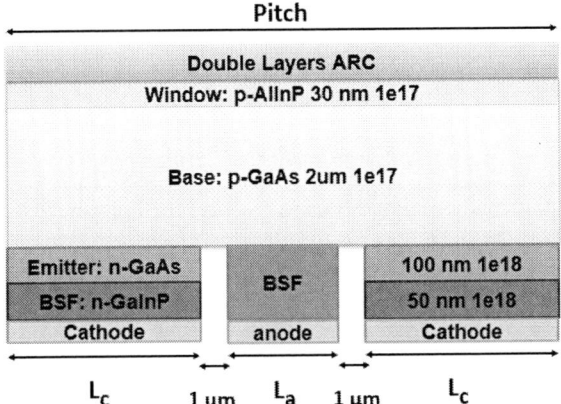

Fig. 1. Schematic cross-section of the back-contact GaAs solar cell in the simulation.

Because base region is where major absorption, generation and recombination happens, we take thickness and doping concentration of base region as the most important parameters. Therefore, we first investigate the influence of base thickness on the current-voltage characteristic, and then repeat it under different base doping density. Through the sequence of simulations, we can acquire appropriate fabrication parameters.

From the similar research about silicon BC-BJ [8], there is an effect named electrical shading which locates in non-collecting region. In our simulation, it is especially intense above the anode. This effect replaces the optical shading loss caused by front contact and becomes a major issue for back-contact solar cells. Consequently, the next step is to determine the layout of contact. In order to express the influence of anode width, we introduce a parameter called anode ratio (ρ), which is defined by the formula

$$\rho = \frac{L_a}{L_a + 2L_c} \qquad (1)$$

By varying the composition of electrodes under different pitches, we expect to observe how the electrical shading degrades the cell. After series of simulations, we are able to find which design is better and its correlation between contacts length and pitch.

III. RESULT AND DISCUSSION

A. base region

At the beginning, we study the current-voltage characteristic variation with thickness of base and show it in Fig 2. As the base gets thicker, the photon current density (J_{ph}), which stands for the absorption in the entire cell, increases without a doubt.

On the contrary, the short-circuit current density (J_{sc}) has the same trend with J_{ph} when base is too thin to have compelling recombination, and starts to reduce when the recombination cannot be neglected.

There is a strong correlation between the open-circuit voltage (V_{oc}) and the recombination. That the extent of recombination keeps growing while the thickness increases leads to the tendency.

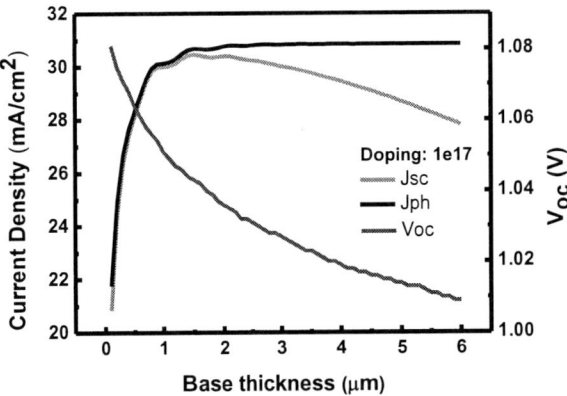

Fig. 2. Photon current density (Jph), short-circuit current density (Jsc), and open-circuit voltage (Voc) with base thickness of 0.1 to 6 μm.

The fill factor (FF) will not turn to an influential level if base thickness changes, so the value of power conversion efficiency (PCE) will be determined by the product of Jsc and V_{OC} only. Therefore, we can find out a balance between the two factors when the value of base thickness is 1.5 μm and causes the PCE to reach maximum. The relevant content is illustrated in Fig 3. Also, the results of different base doping concentrations are plotted.

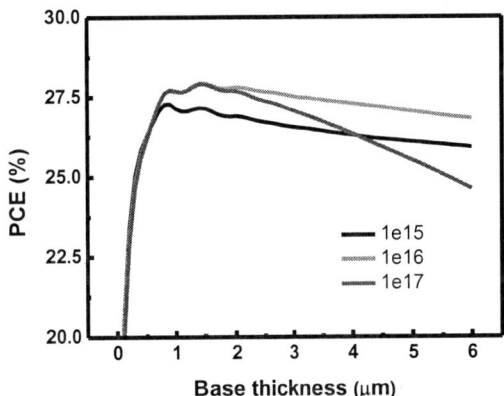

Fig. 3. Power conversion efficiency (PCE) under different doping density with base thickness of 0.1 to 6 μm.

As shown in Fig 3, the interaction between thickness and recombination becomes stronger when the doping concentration increases. After conducting detail simulations, we set the base doping for 5×10^{16} cm^{-3} and keep on the following optimization about contact design.

B. contact design

Fig 4 shows PCE as a function of pitch and ρ for fixed height (1.5μm) and doping concentration (5×10^{16} cm^{-3}). It can be observed that higher efficiency values are corresponding to narrower anodes.

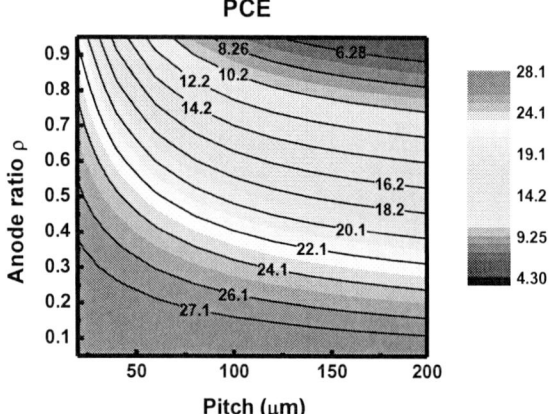

Fig. 4. Power conversion efficiency (PCE) as a function of pitch and anode ratio ρ

The p-n junction, which can separate electrons and holes, only distributes above the cathode. Namely, a wider anode will give less chance of collecting carriers. Fig 5 illustrate the spatial distribution of SRH recombination in the back-contact solar cell. As just mentioned, there is a serious recombination called electrical shading right above the anode. Hence, we ought to design the anode to be comparatively narrow.

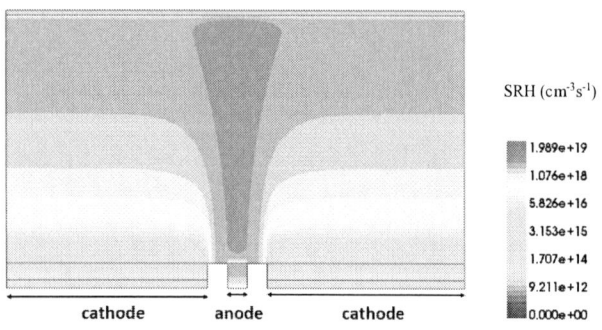

Fig. 5. Spatial distribution of SRH recombination

As for the pitch, we can speculate wider pitch is favorable to carriers' separation for the same reason above. But it won't last permanently during the width keeps increasing. When the pitch broadens to a specific extent, the benefit of carriers' separation will be less than the harm that transportation brings. In other words, the reduction in FF will be dominant if pitch is too wide. For a cell with 1μm anode, the best length of pitch will be about 100μm for our structure.

IV. CONCLUSION

In this simulation work, we consider 1.5μm and 5×10^{16} cm^{-3} base region to yield a better efficiency. Through minimizing the length of anode, the efficiency can be significantly improved due to the weakened electrical shading effect. In addition, a relatively wide cathode results in efficient carrier separation and transportation properties. Based on these outcomes, we are able to optimize the design of back-contact back-junction GaAs solar cells.

REFERENCES

[1] R.J. Schwartz and M.D. Lammert, Silicon solar cells for high concentration applications, in Technical Digest of the International Electron Devices Meeting, Washington, DC, 350-2, 1975

[2] K. Masuko; M. Shigematsu; T. Hashiguchi; D. Fujishima; M. Kai; N. Yoshimura; T. Yamaguchi; Y. Ichihashi; T. Mishima; N. Matsubara; T. Yamanishi; T. Takahama; M. Taguchi; E. Maruyama; S. Okamoto, " Achievement of More Than 25% Conversion Efficiency With Crystalline Silicon Heterojunction Solar Cell," *IEEE Journal of Photovoltaics*, vol 4, issue 6, pages 1433-1435, 2014

[3] M. A. Green, K. Emery, Y. Hishikawa, W. Warta, E. D. Dunlop, D. H. Levi, and Anita W. Y. Ho-Baillie, "Solar Cell Efficiency Tables (version 49)," Progress in Photovoltaics, vol 25, issue 1, pages 3-13, 2017

[4] W.Shockley and H.J. Queisser, "Detailed Balance Limit of Efficiency of p-n Junction Solar Cells," *Journal of Applied Physics*, 32, 510-519, 1961

[5] L.S. Mattos, S.R. Scully, M. Syfu, E. Olson, L. Yang, C. Ling, B.M. Kayes, and G. He, "New Module Efficiency Record:23.5% under 1-Sun Illumination Using Thin-film Single-junction GaAs Solar Cells, " in *38th IEEE Photovoltaic Specialists Conference*, 2012

[6] J. L. Cruz-Campa, G. N. Nielson, M. Okandan, M. W. Wanlass, C. A. Sanchez, P. J. Resnick, P.J. Clews, T. Pluym, and V. P. Gupta, "Back-Contacted and Small Form Factor GaAs Solar Cell," *Proceedings of the 35nd Photovoltaics Specialists Conference,* " Honolulu, 1248-1252, 2010.

[7] C.Y. Hong, Y.C. Lin, K.Y. Ho, J.L. Tsai, Z.T. Chien, Y.R. Wu, A. Lin, W.Y. Uen, G.C. Chi, and P.Yu, "Back-Contacted Thin-Film GaAs Solar Cells, " in 43*th IEEE Photovoltaic Specialists Conference*, 2016

[8] M. Hermle, F. Granek, O. Schutlz-Wittmann, and S. W. Glunz " Shading effects in back-junction back-contacted silicon solar cells, " in 33*th IEEE Photovoltaic Specialists Conference*, 2008

Design considerations on GaInNAs solar cells with back surface reflectors

Antti Tukiainen, Arto Aho, Timo Aho, Ville Polojärvi, Mircea Guina

Optoelectronic Research Centre, Faculty of Science and Engineering, Tampere University of Technology,
P.O. Box 692, FIN-33100 Tampere, Finland

Abstract — We report on design considerations for developing dilute nitride solar cells with back surface reflectors. Two different scenarios including specular and diffuse reflectors were modeled and their effects on solar cell characteristics were compared with cells without reflectors. We show that for optimal performance of the solar cell with back surface reflectors, the layer structure of the cell has to be optimized taking into account the reflector properties. Using high quality reflectors the usable background doping range for GaInNAs sub-junctions can be extended to above 1×10^{16} cm^{-3}.

I. INTRODUCTION

Dilute nitrides (GaInNAs) based materials are offering unique properties for developing high-efficiency III-V multijunction solar cell [1]. Efficiency levels exceeding 44% have already been demonstrated using dilute nitrides fabricated by molecular beam epitaxy (MBE) as bottom junction in triple junction solar cells [2].

While they offer widely tuneable bandgaps and lattice matching to GaAs, dilute nitrides often exhibit short carrier lifetimes and short minority carrier diffusion lengths, as well as high background doping concentrations, prompting for extensive optimization of the fabrication conditions. It has been shown that the carrier lifetime in low quality GaInNAs is only a few tens of picoseconds and while it can approach ns level when the fabrication conditions are optimized [3]. Short carrier lifetimes with rather low electron and hole mobilities lead to the fact that the carrier collection probability in GaInNAs n-i-p solar cells is largely dependent on the width of the depletion region and thus on the background doping of the unintentionally doped intrinsic GaInNAs layer. To reach depletion widths larger than 1 μm, thus allowing high photocurrent generation and at the same time high carrier collection probability, the GaInNAs layer usually should have background doping level below 10^{16} cm^{-3} [5]. Such low background doping levels can be controllably achieved by using MBE but not so easily by using MOCVD [6]. Therefore, to alleviate at least partially the requirement for low background doping level, structural modifications are needed. For example, the optical path of photons inside the GaInNAs layer can be extended using more complex optical designs [7-9]. Metallic back surface reflectors and distributed Bragg reflectors (DBR) inserted below the GaInNAs junctions have already been used for increasing the photogeneration in

GaInNAs solar cells. [10-12]. To this end, it is however, important to understand how the choice of the reflector affects the optimal layer design. Here, we present results of a study aiming to understand how the use of back reflector affects the design constrains of GaInNAs solar cell.

II. EXPERIMENTAL

The solar cells comprised an unintentionally doped GaInNAs layer sandwiched between n-GaAs and p-GaAs layers. The cell contained a p-GaInP back surface field (BSF) layer and a thin AlInP window layer. .

We studied two reflector concepts, shown in Fig. 1. The first reflector is a highly reflecting metal mirror producing a specular reflection at the rear side of the GaInNAs junction. The second reflector scheme provides a highly efficient diffuse reflection at the rear of the junction.

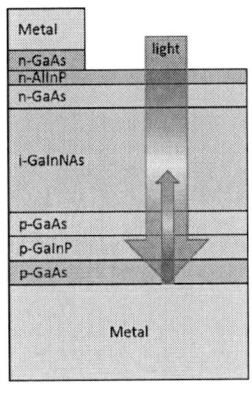 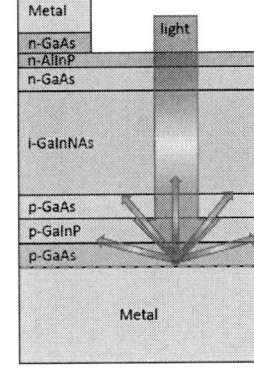

Specular reflector   Diffuse reflector

Fig 1. Generic test structures for the GaInNAs back surface reflector solar cell with different reflector types.

The operation of the cells was first simulated using PC1D for which we have developed a model adapted to GaInNAs [13, 14].

First, the electrical parameters of a standard GaInNAs junction were modeled and the results were checked against experimental results from similar structures to validate the model. The front side of the cell was assumed to be antireflection (AR) coated. For the specular and diffuse reflectors, we assumed a lump reflectance for all the wavelengths to allow for simpler comparison between the two approaches.

The short circuit current density (J_{sc}), open circuit voltage (V_{oc}), and efficiency (η) were modeled for junctions with specular and diffuse reflectors. GaInNAs layer thickness, doping level and backside reflectance were set as variables in the simulations.

III. RESULTS AND DISCUSSION

The modeled J_{sc} values of solar cells without reflector and with specular or diffuse reflectors as a function of GaInNAs thickness are shown in Fig. 2. The cell structure was considered similar to what was studied in [14]. Both reflectors were assumed to provide a very high reflectance, which is feasible for high quality metal reflectors. We considered an AM1.5G spectrum and a 900 nm long-pass filter that was inserted between the light source and the solar cell to mimic the conditions below the GaAs sub-junction of a GaInP/GaAs/GaInNAs triple junction or GaInP/GaAs/ GaInNAs/Ge four junction solar cell.

The cell without a reflector produces a J_{sc} of only 12.4 mA/cm² , which is well in line with real measurements of similar cells. Therefore, the GaInNAs sub-junction would limit the current in triple junction cells. However, when a specular reflector is used, the maximum J_{sc} increases to 13.9 mA/cm². The corresponding value for a diffuse reflector is 14.4 mA/cm². Such parameters would already be closely current matched in triple junction cells [15, 16]. For the four junction cells, J_{sc} without reflector would be enough for current matching.

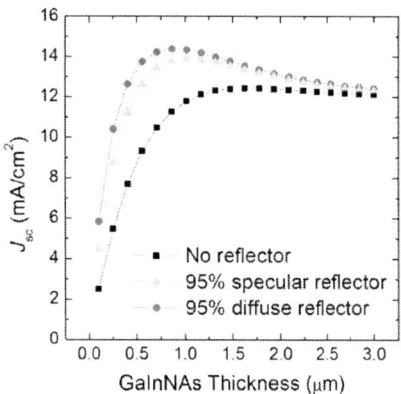

Fig. 2. Comparison of calculated J_{sc} vs. GaInNAs thickness for solar cells with and without reflectors. The simulation parameters for the cell are taken from [14]. The background doping level of the GaInNAs layer is 2.2×10^{16} cm⁻³.

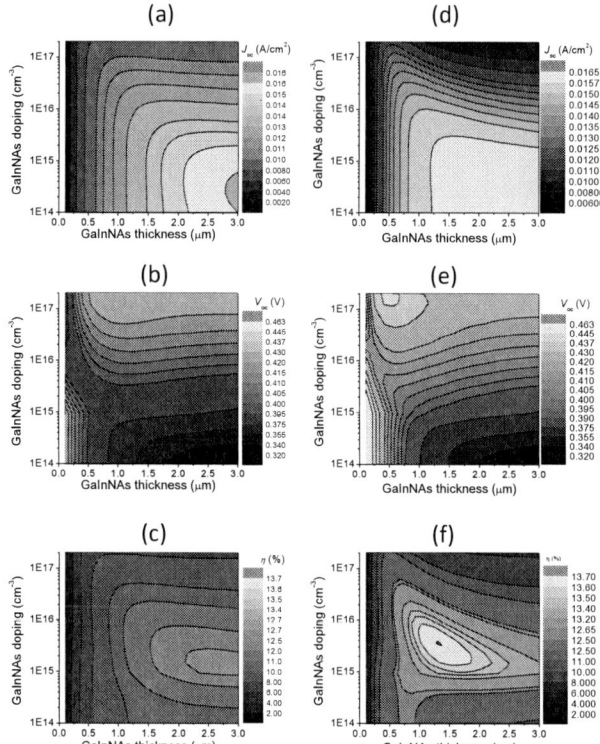

Fig. 3. (a)-(c) Modeled GaInNAs n-i-p solar cell characteristics without backside reflector. (d)-(e) Modeled GaInNAs n-i-p solar cell characteristics with a 95% diffuse backside reflector.

The J_{sc} results show that there is clearly an optimal thickness for the GaInNAs layer for each type of reflector. Another important detail is that the optimal GaInNAs thickness is considerably reduced for the cells having a backside reflector. For this particular cell type, the optimum GaInNAs thickness with a 95% specular reflector is at about 1 µm, and for a 95% diffuse reflector the optimal thickness is reduced even further down to 850 nm. The ability to reduce the GaInNAs layer thickness without losing its functionality has favorable effects on the fabrication costs of the multijunction solar cells due to increased throughput and reduced material costs.

Next, different combinations of thickness and the doping level of GaInNAs were modeled for cells with and without diffuse reflector and the results are shown in Fig. 3. It is evident, that the highest J_{sc} for both samples is obtained when low doping level and rather thick GaInNAs layers are used. However, the V_{oc} behaves the differently, i.e., it has a minimum when the doping level is low and the GaInNAs thickness is large. As consequence, the efficiency of the GaInNAs solar cell is maximized at a unique combination of GaInNAs thickness and doping level, which for the cell without reflector is at 2.75 µm and 1.5×10^{15} cm⁻³, respectively. For the 95% diffuse reflector the maximum efficiency is obtained when GaInNAs thickness is 1.3 µm and

978-1-5090-5606-4/17 $31.00 © 2017 IEEE

the doping level is 4×10^{15} cm^{-3}. Using the latter values, the cell made of GaInNAs would have potential for J_{sc} of 15.5 mA/cm^2 and V_{oc} of 0.402 V when a highly reflective diffuse reflector is inserted at the back of the junction. This indicates, that for optimal behavior, the background doping level of GaInNAs should be reduced from 2.2×10^{16} cm^{-3} down to 4×10^{15} cm^{-3}.

With the GaInNAs solar cell material discussed in [14] as an initial point, the thickness of the GaInNAs layer can also be optimized for various reflectance values of the back surface reflector. This is depicted in Fig. 4, in which the efficiency of a GaInNAs solar cell is shown as functions of backside reflectance and GaInNAs layer thickness. The results indicate that for each backside reflectance value, a unique GaInNAs layer thickness is required to maintain the highest possible performance of the solar cell. For this particular cell, the highest efficiency without reflector is obtained with GaInNAs thickness of about 1.7 μm.

In addition, as the backside reflectance is varied, the J_{sc} and V_{oc} are not maximized at the same GaInNAs thickness. When designing the use of GaInNAs sub-junctions with backside reflectors, it is important to pay attention to the fact that regardless of the backside reflectance, the maximum J_{sc} is always obtained with thicker GaInNAs layer than what is needed to obtain the maximum V_{oc}. For example, in Fig. 4 the maximum J_{sc} of a junction with a 100% reflector is obtained for GaInNAs thickness of 0.87 μm and the maximum V_{oc} is obtained at 0.6 μm.

The model used here does not take into account multiple reflections at the back and front surfaces. This would most likely have some effect on the modeled optimal layer thicknesses. However, because the AR-coating has already reflectance of only a few percentage points the amount of light for the second roundtrip within the cell is very small and thus we have neglected this possibility.

IV. Conclusion

We have used a PC1D based model for simulation of the effect of back surface reflectors on GaInNAs single junction n-i-p solar cells. The results show that using diffuse reflectors it is possible to extend the usable background doping range for GaInNAs to levels above 1×10^{16} cm^{-3} and yet maintain high current production needed for the bottom sub-junction of high-efficiency triple junction solar cell.

Acknowledgment

The authors want to thank European Research Council for financial support under the ERC Advanced Grant AMETIST #695116.

References

[1] D. J. Friedman, J. F. Geisz, S. R. Kurtz, and J. M. Olson "1-eV GaInNAs Solar Cells for Ultrahigh-Efficiency Multijunction Devices, " in *2nd World Conference and Exhibition on Photovoltaic Solar Energy Conversion*; 1998.

[2] Green MA, Emery K, Hishikawa Y, Warta W, DunlopED. Solar cell efficiency tables (version 41). Progress in Photovoltaics: Research and Applications 2013; 21:1. DOI:10.1002/pip.2352.

[3] A. Gubanov, V. Polojärvi, A. Aho, A. Tukiainen, N. V. Tkachenko and M. Guina, "Dynamics of time-resolved photoluminescence in GaInNAs and GaNAsSb solar cells", Nanoscale Research Letters, 9(1):80, 2014.

[4] A. Aho, V. Polojärvi, V.-M. Korpijärvi, J. Salmi, A. Tukiainen, P. Laukkanen and M. Guina, "Composition dependent growth dynamics in molecular beam epitaxy of GaInNAs solar cells," *Solar Energy Materials & Solar Cells*, vol. 124, p. 150-158, 2014.

[5] A. J. Ptak, D. J. Friedman, S. Kurtz, and R. C. Reedy, "Low-acceptor-concentration GaInNAs grown by molecular-beam epitaxy for high-current p-i-n solar cell applications ", *Journal of Applied Physics*, vol. 98, issue 9, p. 094501, 2005.

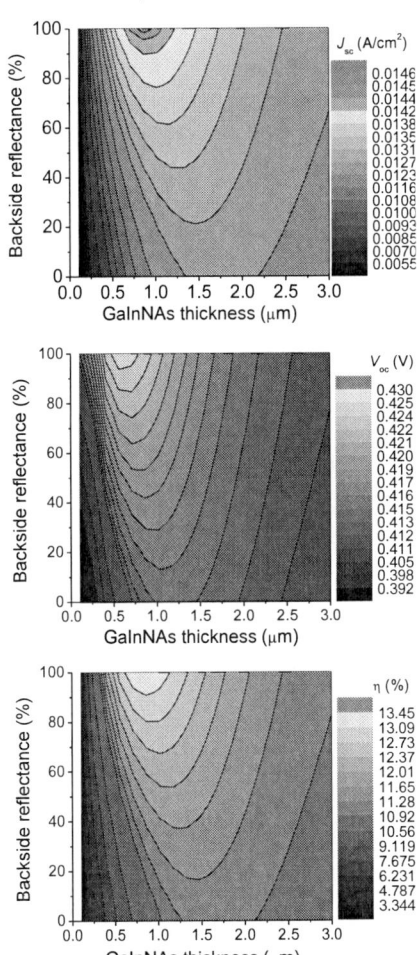

Fig. 4 J_{sc}, V_{oc} and η of a n-i-p solar cell as functions of reflectance of the diffuse reflector and GaInNAs thickness.

[6] S. Kurtz, J. F. Geisz, D. J. Friedman, W. K. Metzger, R. R. King, and N. H. Karam, "Annealing-induced-type conversion of GaInNAs", Journal of Applied Physics 95 (5), pp. 2505-2508, 2004.

[7] J. J. Schermer, G. J. Bauhuis, P. Mulder, E. J. Haverkamp, J. van Deelen, A. T. J. van Niftrik, P. K. Larsen, "Photon confinement in high-efficiency, thin-film III–V solar cells obtained by epitaxial lift-off", Thin Solid Films, 511-512, p. 645-653, 2006.

[8] N. Vandamme, H.-L. Chen, A. Gaucher, B. Behaghel, A. Lemaitre, A. Cattoni, C. Dupuis, N. Bardou, J.-F. Guillemoles and S. Collin, "Ultrathin GaAs Solar Cells With a Silver Back Mirror," IEEE Journal of photovoltaics, vol. PP, issue 99 IEEE Early Access Articles, p. 1-6, 2014.

[9] G. J. Bauhuis, P. Mulder, E.J. Haverkamp, J. C. C. M. Huijben and J. J. Schermer, "26,1% thin film GaAs solar cell using epitaxial lift-off", Solar energy and materials & Solar Cells, vol. 93, p. 1488-1491, 2009.

[10] T. Aho, A. Aho, A. Tukiainen, V. Polojärvi, J-P. Penttinen, M. Raappana, and M. Guina, "GaInNAs Solar Cell with Back Surface Reflector", 42nd IEEE Photovoltaic Specialists Conference (PVSC), 2015. IEEE, 4 p.

[11] T. Aho, A. Aho, A. Tukiainen, V. Polojärvi, T. Salminen, M. Raappana, and M. Guina, "Enhancement of Photocurrent in GaInNAs Solar Cells using Ag/Cu Double-Layer Back Reflector", Applied Physics Letters. 109, 251104, 2016

[12] A. Tukiainen, A. Aho, V. Polojärvi, and M. Guina, "Improving the current output of GaInNAs solar cells using distributed Bragg reflectors" IEEE 43rd Photovoltaic Specialists Conference (PVSC). IEEE, p. 0368-0371 4 p.

[13] A. Tukiainen, A. Aho, V. Polojärvi, and M. Guina, "Modeling of MBE-Grown GaInNAs Solar Cells", 10th European Space Power Conference ESPC 2014, European Space Agency, p. 1-4 4 p. (European Space Agency - Special Publication (ESA - SP); vol. 719), 2014.

[14] A. Tukiainen, A. Aho, V. Polojärvi, R. Ahorinta, and M. Guina, "High efficiency dilute nitride solar cells: Simulations meet experiments", Journal of Green Engineering 5, 3-4, p. 113-132 20 p. 8, 2016.

[15] A. Aho, A. Tukiainen, V. Polojärvi, M. Guina, "Performance assessment of multijunction solar cells incorporating GaInNAsSb", Nanoscale Research Letters, 9(1):61, 2014.

[16] D. B. Jackrel, S. R. Bank, H. B. Yuen, M. A. Wistey, J. S. Harris Jr., A. J. Ptak, S. W. Johnston, D. J. Friedman, and S. R. Kurtz, "Dilute nitride GaInNAs and GaInNAsSb solar cells by molecular beam epitaxy", Journal of Applied Physics, vol. 101, issue 11, p. 114916, 2007.

Quantitative Electroluminescence Analysis of Triple Junction Solar Cells to Determine Subcell Voltage-Temperature Coefficients

Kevin Tyler[1], Geoffrey K. Bradshaw[2], Sam Wilt[3], David M. Wilt[2], and Richard R. King[1]

[1]Arizona State University, Tempe, AZ, 85281, USA
[2]United States Air Force Research Laboratory, Albuquerque, NM, 87117, USA
[3]University of New Mexico, Albuquerque, NM, 87131, USA

Abstract — **Solar cells in space experience varying temperatures throughout their lifetime on spacecraft and satellites. At higher temperatures, photovoltaic performance decreases. In multijunction cells, a reduction in open-circuit voltage occurs across each subcell. This voltage-temperature coefficient is crucial for design purposes. The voltage of subcells within multijunction cells has traditionally been difficult to characterize, requiring the use of isotype cells for each subcell. However, the reciprocity relation allows the voltage characteristics of a cell to be correlated with its electroluminescent emission and quantum efficiency, even within a multijunction cell, providing an alternative to the use of isotypes. This research presents the steps and accuracy of this method.**

I. Introduction

Multijunction (MJ) solar cells remain the dominant technology for power generation onboard spacecraft and satellites. In order to fully optimize the performance of these photovoltaic (PV) devices, it is necessary not only to characterize the full MJ cell, but also to understand the characteristics of each subcell that together form the MJ cell. While measuring subcell parameters has traditionally been much harder than characterizing single-junction cells or MJ cells as a whole, new measurement methods are now being developed that are faster and more accurate.

The reciprocity relation [1] is the backbone of this approach, providing the fundamental physical principle that allows efficient subcell characterization using the full MJ cell. Previously, subcells had to be analyzed using "isotype" cells, one for each subcell. These isotype cells contain active layers of the base, emitter, and other layers of the subcell being characterized, and inactive layers for the other subcells which serve as an optical filter above the active subcell. By using the reciprocity relation, the quantitative electroluminescence, the quantum efficiency, and the light current-voltage (LIV) curve of a full MJ cell, the voltages of each individual subcell within the device can be calculated [1-6]. This is a far more time efficient and cost-effective proposition than acquiring and measuring isotypes for each subcell.

In this paper, we demonstrate how the voltage-temperature coefficients of each subcell may be extracted from a full MJ cell. This is useful for the Air Force mission, as solar cells experience frequent and large variations in temperature during operation in space. The reciprocity relation is used in this powerful method, demonstrating accurately and efficiently the

ability to extract information from a full MJ cell in lieu of acquiring isotype cells.

II. Experimental Results and Discussion

A. Electroluminescence Results

When an electrical current is applied across a solar cell, radiative and non-radiative recombination occurs. The radiative recombination can be seen as the emission of light and therefore can be measured and quantified. Photons are emitted in this process when electrons and holes recombine, and the photon energy is equal to the energy difference between the electrons and holes. It is this electroluminescence (EL) that is used in the reciprocity relation from Rau [1], shown below in Equation 1, to determine the individual subcell voltages:

$$\emptyset_{EL}(E) = EQE_{PV}(E) \; \emptyset_{bb}(E) \left[exp\left(\frac{qV}{kT} \right) - 1 \right] \quad (1)$$

where $\emptyset_{EL}(E)$ is the electroluminescence emission photon flux density of a subcell per unit photon energy, normal to the surface of the subcell, at photon energy E; $EQE_{PV}(E)$ is the external photovoltaic quantum efficiency of a subcell measured at normal incidence; $\emptyset_{bb}(E)$ is the photon flux density of a blackbody at the cell temperature T per unit photon energy; V is the cell voltage; q is the electron charge; and k is the Boltzmann constant. For a given subcell j in a multijunction cell, the voltage V_j can be found from the measured electroluminescence and external quantum efficiency of the subcell through [6]:

$$V_j = \frac{kT}{q} ln \left[\frac{\phi_{EL}(E)}{EQE_{PV,j}(E) \; \phi_{bb}(E)} \right] \quad (2)$$

The electroluminescent photon flux is measured using an ASD Inc. FieldSpec4 spectroradiometer while a Keithley 2400 SMU injects current into the cell. A white body reference is used to calibrate the data to produce absolute electroluminescence intensities. The fiber-optic cable within the FieldSpec4 is coupled to three detectors covering the ranges of 350-1000 nm, 1001-1800 nm, and 1801-2500 nm. These detectors have a hard time picking up the signal from the germanium bottom cell, which is already small and occurs around 1800 nm. Because this is also where the detectors

978-1-5090-5606-4/17 $31.00 © 2017 IEEE

switch, the germanium peak may contain a small offset within it.

The EL demonstrates a very good response for the top two subcells (GaInP/GaAs) while showing a weak response for the bottom cell, germanium. This occurs because germanium is an indirect bandgap material, while GaInP and GaAs are direct bandgap materials. For this experiment, EL was taken for a range of temperatures. Fig. 1(a) shows the lowest and highest temperature EL scans to provide the sharpest contrast between the various EL curves. All other EL measurements lie in between the two EL curves shown. The top and middle subcell peaks both show a slight shift to longer wavelengths and a slightly lower intensity at 80°C compared to 10°C. Fig. 1(b) shows the change in wavelength of the GaAs peak to be approximately 20 nm across the temperature range. This shift occurs because the bandgap decreases with increasing temperature. Fig. 1(c) shows the general decreasing trend in peak intensity between 10°C and 80°C which, through the reciprocity relation, is representative of the decreasing voltage as temperature increases.

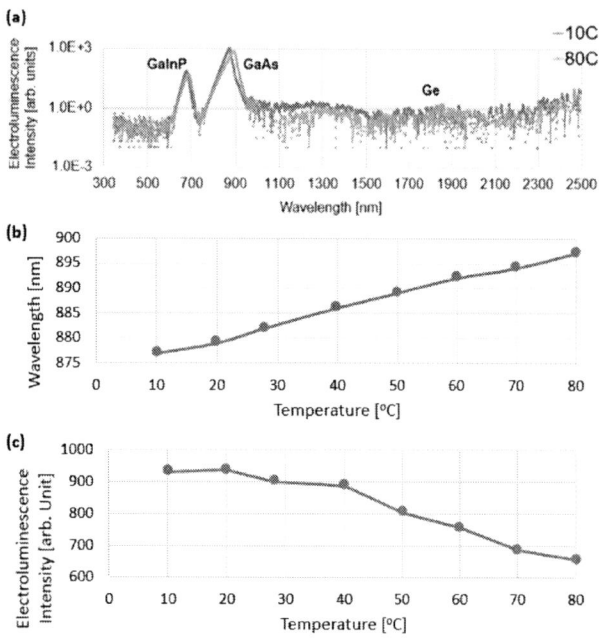

Fig. 1. (a) Electroluminescence intensity of the cell at 10°C and 80°C. (b) Shift in wavelength of the GaAs peak with temperature. (c) Decrease in electroluminescence intensity as temperature increases.

B. External Quantum Efficiency Results

The external quantum efficiency (EQE) can be measured by illuminating the solar cell with a monochromatic light source and then measuring the current produced by the cell. The ratio of electrons collected at a given wavelength to the amount of incident photons is then used to determine the EQE. This is done across a range of wavelengths. EQE measurements were performed using a Newport Cornerstone monochromator with a Xenon light source and standard lock-in techniques.

In addition, a light bias must be introduced to activate subcells other than the device under test. This light bias increases the current of the cells that are not being measured in order to ensure that the subcell being measured is the current-limiter at the time of the measurement. Finally, a voltage bias must be introduced when measuring the bottom germanium subcell in order to better reproduce short circuit current conditions. This bias can be found by performing a voltage (V_{bias}) sweep while the light bias is active and finding the point in which the EQE signal is maximized.

The external quantum efficiency measurements in Fig. 2 show a shift of approximately 20 nm to longer wavelengths in both the top and middle subcells as the temperature is increased from 28°C and 80°C. While the germanium may also have this trend, the data, especially at higher temperatures, is much noisier and thus the wavelength shift is more difficult to determine. This noise comes from the relatively weak response of the indirect gap germanium cell, and because the wavelength of the Ge band edge is near the limit of the detector used to calibrate the incident light. The largest change in EQE signal can be seen in the reduction of the bottom germanium subcell between 28°C and 80°C, showing that this subcell leaks progressively more current as the temperature rises.

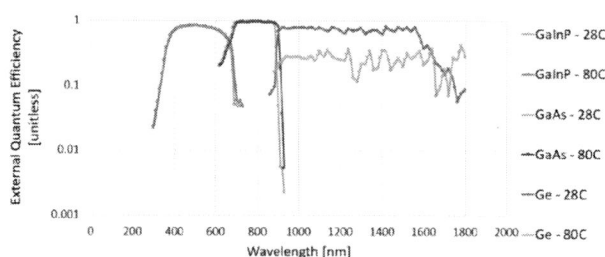

Fig. 2. External quantum efficiency of each subcell at 28°C and 80°C.

C. LIV Results

The light current-voltage (LIV) characteristics are determined for the full multijunction cell using a Spectrolab Inc. X-25 solar simulator, calibrated with a triple-junction flight standard, to simulate the AM0 solar spectrum. The LIV results provide a clear picture of how the full cell behaves under varying temperatures. Fig. 3(a) shows the measured and calculated decrease in open-circuit voltage (V_{OC}) as the temperature increases. There is a linear relationship across the full range of temperatures tested, with a calculated voltage-temperature constant of -6.107 mV/°C. According to the manufacturer of the cells being tested, the voltage-temperature constant should be approximately -5.9 mV/°C. Fig. 3(b) demonstrates that the short circuit current (J_{SC}) increases by only 4.4% over the temperature range, while the V_{OC} decreases by 15.7%. Fig. 3(c) shows the linear decrease of AM0 efficiency of the full multijunction cell, from approximately 28% to 24% from 10°C to 80°C.

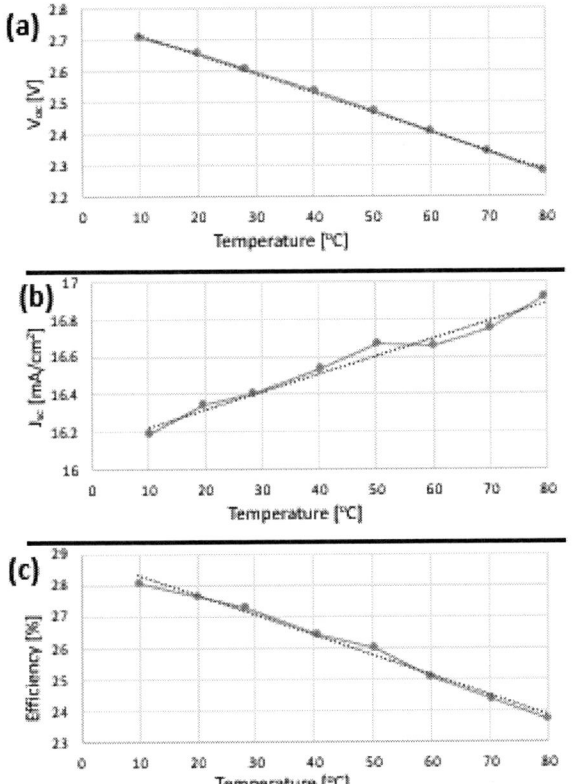

Fig. 3. Measured and calculated light I-V parameters for a 3-junction GaInP/GaAs/Ge solar cell: (a) the drop in open-circuit voltage with increasing temperature, (b) increase in short-circuit current with increasing temperature, (c) decrease in overall efficiency with increasing temperature. Solid circles represent measured data, while the dashed line shows the calculated dependence.

Ideally, the subcell voltages extracted from electroluminescence measurements and the reciprocity relation will sum to be the total voltage acquired from the LIV measurements on the full cell. However, due to the angular dependence of the electroluminescence emission and the external quantum efficiency measurements, the V_{OC} from the full cell must be used to determine a voltage offset to correct for the angular differences. This value can then be used to correct the values obtained in Eq. 1. This offset is determined as follows:

$$\delta V = \frac{(V_{OC} - \Sigma V_{Subcell\,i})}{Total\ number\ of\ subcells} \qquad (3)$$

D. Reciprocity Relation Results

The reciprocity relation, when used on the data gathered above, provides the voltage-temperature coefficients for each subcell, shown in Fig. 4. Each subcell provides approximately the expected voltage at room temperature [2, 4-6]. In addition, the sum of subcell voltage-temperature coefficients is -5.846

mV/°C, with the theoretical value being -5.9 mV/°C and the experimentally determined full cell value being -6.107 mV/°C. The measured value of -6.107 mV/°C for the full cell is most likely high due to running the measurements on separate days with slightly different calibration values. This value could be brought closer to the theoretical value and the sum of the individual subcells value by performing the measurements on the same day. The individual subcell temperature voltage coefficients are approximately the same as those seen in literature for single cells [7]. The differences can be attributed to many factors. For one, the subcells are connected in series, causing current limiting effects that can affect voltage in a way that will not be seen in stand-alone materials. In addition, other layers in the single cells as well as different experimental set ups can cause uncertainty when measuring the voltage-temperature coefficient of a single subcell material. Overall though, the voltage-temperature coefficients obtained are comparable to those in literature.

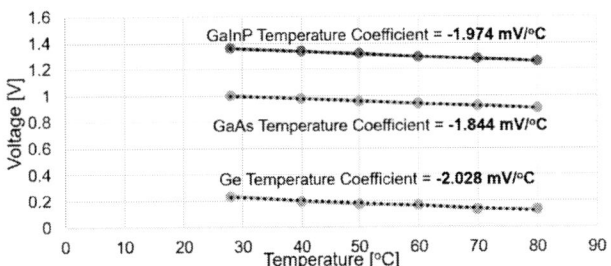

Fig. 4. Voltages of individual subcells within a multijunction cell as a function of temperature, experimentally determined from EL measurements using the reciprocity relation.

The measured V_{OC} value minus the V_{OC} predicted from the fit, or residual, for each subcell and temperature was plotted in Fig. 5, revealing some information about experimental error in the study. The residual plot for each subcell, when analyzed individually, shows both randomness in the error as well as low absolute values of the residuals. However, the trends in the GaInP and GaAs residual plots are nearly identical, while the Ge residuals generally follow the inverse of the trends for GaInP and GaAs.

A likely explanation for this anomaly is the inexactness in determining the temperature of the cell during measurement. A thermocouple is used to measure the temperature during LIV, EL, and EQE measurements, and it is placed on the chuck that holds the cell, rather than directly on the cell. Therefore, the actual cell temperature may have some random variation about the set values assumed in the study, such that the voltage of the cell may have been obtained at 39.8 °C rather than 40 °C, for example.

Logically, this would lead the residuals of the subcells to match each other. If, taking the example above, the desired temperature is 40°C while the actual cell temperature is slightly lower, this would lead to a larger measured voltage than the true voltage that would be represented at 40°C. Because the data for all subcells is collected at the same time, this would cause the residuals of each subcell to match each other, as is seen for the GaInP and GaAs cells.

This also explains the inversed residuals in the Ge plot. Because the Ge cell EL and EQE data are too noisy, the program used to implement the reciprocity relation determines the Ge subcell voltage by subtracting the determined GaInP and GaAs subcell voltages from the full MJ cell voltage determined by the LIV measurements. Thus, there is some internal correction done by the program when the first two subcell voltages are high by determining the voltage of the Ge subcell to be lower than its true value, so that the total sum of subcell voltages is closer to the full multijunction cell voltage at that temperature.

One positive conclusion about this error analysis is that if the source of error is in fact the inexactness of the temperature measurements, then a fixture better designed to measure the precise temperature of the cell will yield a fit with even less error when determining the voltage-temperature coefficients of each subcell.

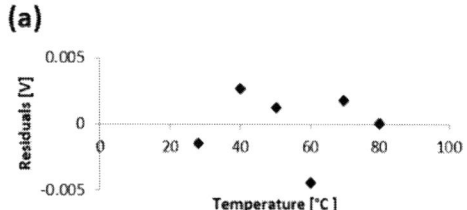

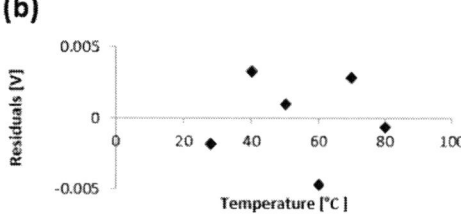

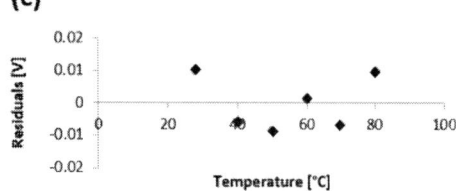

Fig. 5.   Residual subcell voltage plots for (a) GaInP (b) GaAs (c) Ge.

III. SUMMARY AND SIGNIFICANCE

The method of using the reciprocity relation on a full multijunction solar cell to determine the voltage-temperature coefficients has been shown to be accurate and efficient. The individual subcell coefficients, when added together, are very

close to the given full cell voltage-temperature coefficient. The slight discrepancy could be a result of doing the measurements on separate days with slightly different calibrations or the cells themselves slowly degrading from the time they were produced. In addition, error can be largely reduced by precisely controlling the temperature during measurement. The final results are shown below in Table 1.

Method	Voltage-Temperature Coefficient (mV/°C)
Full MJ Theoretical	-5.9
Full MJ Measured LIV	-6.107
Sum of subcell values	-5.846
Reciprocity - GaInP	-1.974
Reciprocity - GaAs	-1.844
Reciprocity - Ge	-2.028

Table 1. Comparison of different techniques to achieve full cell and subcell voltage-temperature coefficients.

The method used in this paper can be extended to include multijunction cells of any number of junctions and across larger temperature ranges if needed. Isotypes are no longer necessary to determine the voltage-temperature coefficients of each subcell, ultimately saving time and resources.

ACKNOWLEDGEMENTS

This work was supported by the Air Force Research Laboratory, Space Vehicles Directorate (Phillips Summer Scholars program). Special thanks to Paul Roland for providing the software used in this study.

REFERENCES

[1] U. Rau, "Reciprocity relation between photovoltaic quantum efficiency and electroluminescent emission of solar cells," *Physics Review B,* 76 (8), 2008.

[2] S. Chen, L. Zhu, M. Yoshita, T. Mochizuki, C. Kim, H. Akiyama, M. Imaizumi and Y. Kanemitsu, "Thorough subcells diagnosis in a multi-junction solar cell via absolute electroluminescence-efficiency measurements," *Sci. Rep.*, vol. 5, p. 7836, 2015.

[3] M. Zazoui and J. Bourgoin, "Space degradation of multijunction solar cells: An electroluminescence study," *Appl. Phys. Lett.*, vol. 80, no. 23, p. 4455, 2002.

[4] S. Roensch, R. Hoheisel, F. Dimroth and A. Bett, "Subcell I-V characteristic analysis of GaInP/GaInAs/Ge solar cells using electroluminescence measurements," *Appl. Phys. Lett.*, vol. 98, no. 25, p. 251113, 2011.

[5] R. Hoheisel, F. Dimroth, A. Bett, S. Messenger, P. Jenkins and R. Walters, "Electroluminescence analysis of irradiated GaInP/GaInAs/Ge space solar cells," *Solar Energy Materials and Solar Cells,* vol. 108, pp. 235-240, 2013.

[6] T. Kirchartz, U. Rau, M. Hermle, A. Bett, A. Helbig and J. Werner, "Internal voltages in GaInP/GaInAs/Ge multijunction solar cells determined by electroluminescence measurements," *Appl. Phys. Lett.*, vol. 92, no. 12, p. 123502, 2008.

[7] N. Kensuke, T. Tatsuya, A. Takaaki, *et. al.*, "Evaluation of temperature characteristics of high-efficiency InGaP/InGaAs/Ge triple-junction solar cells under concentration," *Solar Energy Materials and Solar Cells*, 85(3): 429-436, 2005.

Progress Towards Double-Junction InGaN Solar Cell

Ehsan Vadiee[1], Evan A. Clinton[2], Heather McFavilen[3], Alec M. Fischer[1], Yi Fang[1], Joshua J. Williams[1], Christiana B. Honsberg[1], William A. Doolittle[2], and Stephen M. Goodnick[1]

[1] Arizona State University, Tempe, AZ 85280, USA

[2] Georgia Institute of Technology, Atlanta, GA 30332, USA

[3] Photonitride Devises Inc., Tempe, Arizona, 85284, USA

Abstract—**This paper investigates the performance of a tandem indium gallium nitride/gallium nitride (InGaN/GaN) solar cell, comprised of two single-junction solar cells which are monolithically connected through a heavily doped MBE grown p-GaN/n-GaN tunnel junction (TJ). The higher band gap cell has a multi-quantum well (MQW) InGaN/GaN structure grown by metalorganic chemical vapor deposition (MOCVD). The lower band gap cell has a dual heterojunction (DHJ) InGaN/GaN p-i-n structure grown by a MBE. The MQW structure was chosen to improve open-circuit voltage (Voc) through delaying the onset of the InGaN relaxation, resulting in a high crystal quality. The DHJ cell was grown by the MBE in order to improve the material quality at higher indium contents, achieved by a low growth temperature under nitrogen rich condition. The tandem cell structure was developed for the possible operation at high temperature regimes of photovoltaic-thermal hybrid converters. The tandem cell shows a Voc ~2.22 V and a short-circuit current (Jsc) of ~2.2 mA/cm². The dark J-V results are also presented in order to have an insight into the origin of the performance degradation.**

Index Terms— **Epitaxial layers, indium gallium nitride, multi-junction solar cell**

I. INTRODUCTION

Nitride-based devices have a mature technology in high power electronics and LED industry. It is possible to use these existing platforms to develop high efficiency InGaN solar cells. Current single-junction InGaN solar cells exhibit low efficiencies, due to multiple material quality issues, such as indium phase separation, induced polarization band bending, and low p-type doping concentration. However, theoretical calculations show that InGaN multijunction solar cells can reach efficiencies up to ~46% without considering any performance degradation caused by polarization-induced electric fields and potential barriers, resulting from InGaN/GaN heterointerface [1].

One of the major applications of InGaN solar cells is in photovoltaic-thermal (PV-T) hybrid collectors that can operate in high temperature environments [2]. PV-T hybrid systems can utilize both solar radiation and heat by combining a photovoltaic device as a topping cell, and a thermal absorber as a bottoming cell. PV-T hybrid systems can transfer heat to the thermal absorber and provide dispatchability at costs comparable to traditional sources, whether the sun is shining or not [3]. InGaN material system is favorable for PV-T applications due to its tunable band gap (~0.7 - ~3.4 eV),

covering from UV to IR wavelength regime, high radiation resistance, high thermal resistance, and high absorption coefficient [4], [5]. Motivated by that, a single-junction MQW InGaN solar cell has been successfully shown to operate at high temperatures (T ≈ 450 °C) with good robustness and durability [6]. However, it has been a long-standing goal to increase the absorption band of InGaN solar cell by using multijunction approach. This requires an implementation of a tunnel junction (TJ) between the InGaN subcells.

In this work, we have designed, grown, and fabricated a double-junction cell for the purpose of operation at high temperatures. The tandem cell is comprised of a MOCVD-grown InGaN/GaN MQW and a MBE-grown dual-heterojunction (DHJ) InGaN/GaN subcells with different indium compositions. These two subcells are connected with a degenerately doped p-n GaN TJ grown by the plasma assisted MBE. We present the material and electrical properties of this tandem device here.

II. EXPERIMENTAL DETAILS

The InGaN/GaN MQW p-i-n subcell was grown on a (001) sapphire substrate. The structure consists of a ~2.5 μm GaN: Si buffer layer, followed by a MQW region, consisting of 40 quantum wells (InGaN (3 nm)/GaN: Si (2 nm)). A thicker InGaN barrier was chosen to increase the absorption while maintaining the crystal quality. The MQW layers were followed by a 150 nm thick p-GaN and a thin 10 nm p+GaN, as a capping layer. Si and Mg were used as n-type and p-type doping, respectively. The XRD measurements revealed that the indium content is ~15% which corresponds to a band gap of ~2.8 eV at room temperature. The high indium content MQW structure was successfully implemented for this subcell because the thin InGaN layers can be grown through strain compensation, remaining fully strained to the underlying GaN layers (verified by XRD) [7]. The lower bandgap DHJ subcell with a structure of i-InGaN (100 nm)/p-GaN (50 nm) was grown on a heavily doped p+GaN (25 nm)/n+GaN (150 nm) tunnel junction with the doping concentrations of 3×10^{19} and 1×10^{20} cm⁻³, respectively, by a Riber 32 MBE system. The tunnel junction acts as an intermediary layer to transfer solar generated carriers between the conduction band of one junction and the valence band of the next junction. The p-n tunnel junction is expected to provide a low electrical resistance between the two InGaN subcells without absorbing most of the solar spectrum due to its high band gap of ~3.4 eV. The XRD

978-1-5090-5606-4/17 $31.00 © 2017 IEEE

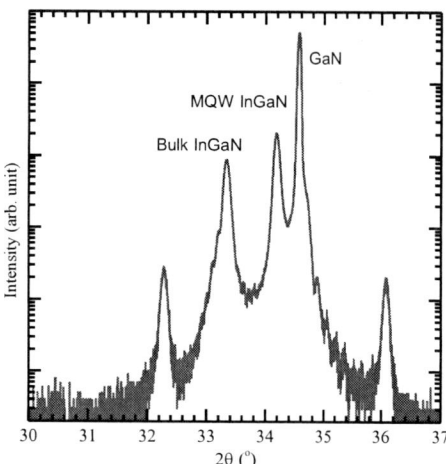

Fig. 1. (002) 2θ–ω diffraction scans of the tandem cell showing the absence of phase separation and the presence of superlattice peaks. The MBE-grown InGaN (002) and (105) ω rocking curves show values of 317 arcsec and 771 arcsec, respectively.

measurements revealed that the indium content is ~25% which corresponds to a band gap of ~2.4 eV at room temperature. These InGaN band gaps are chosen based on a detailed balance calculation with consideration of the InGaN material quality limitations [8].

The tandem structure was fabricated into 1×1 mm² solar cells using photolithography and inductively coupled plasma etching processes. Ti/Al/Ti/Au and Ni/Au layers were deposited by e-beam evaporation as contacts to n-GaN and p-GaN. Ni (5 nm)/Au (5 nm) were also deposited on top of the p-GaN layer as a current spreading layer to increase the lateral conductivity of the p-GaN and decrease the p-GaN contact resistance.

The light current density-voltage (J-V) of the tandem cell was measured under one-sun AM1.5G condition with irradiation intensity of ~1000 W.m⁻² by a solar simulator equipped with a Xenon ARC lamp manufactured by PV Measurements Inc.

III. DEVICE PERFORMANCE AND DISCUSSION

$2\theta - \omega$ scan and reciprocal space maps (RSM), measured by x-ray diffraction (XRD), were used to evaluate the crystal quality and composition of the alloys. Figure 1 shows the double-crystal $2\theta - \omega$ scan in the vicinity of GaN and InGaN (002) reflections. The results show the absence of phase separation and the presence of single phase InGaN layers. In addition, the superlattice peak positions were used to calculate the period of the MQWs, showing a period of ~5 nm which agrees well with the targeted InGaN/GaN QW thickness. Fig. 2 exhibits the RSM in the vicinity of the (105) reflection. The low relaxation degree of InGaN film in the DHJ structure is ascribed to the nitrogen rich and low temperature MBE growth of InGaN, which can improve the crystal quality and indium incorporation into the film [9]. Other methods have been shown in the literature to further improve the crystal quality, such as using nano-patterned sapphire substrates and silica microsphere templates [17], [18]. However, the slight relaxation of the $In_{0.25}Ga_{0.75}N$ peak can degrade the crystal homogeneity and consequently reduce the V_{OC} value. Since even a slight relaxation of InGaN film can cause V_{OC} degradation through formation Shockley-Read-Hall (SRH) recombination centers, the growth of high quality InGaN films with no strain is crucial for future InGaN tandem cells. This can be achieved by using a thick homo-junction InGaN cell grown on an InGaN template. However, up to now the quality of InGaN templates are not good compared to the current low defect density substrates, required for high efficiency III-V solar cells.

The tandem solar cell has a photovoltaic response with a Voc value of ~2.22 V, J_{SC} value of 2.2 mA/cm², fill factor (FF) of ~83%, as shown in Fig. 3. These results show that the carriers are collected from the both cells through the tunnel junction. It is noteworthy that the tandem cell was designed to reach a theoretical efficiency of above 17% under 200x, as demonstrated by Fang *et al.* [10]. Single-junction MQW InGaN and MBE grown DHJ solar cells with similar structures but slightly different indium contents were reported in our earlier works [11].

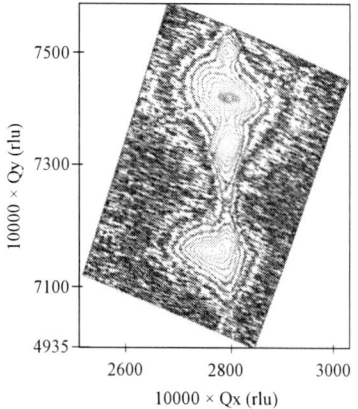

Fig. 2. RSM along the (105) reflection of GaN showing partially relaxed $In_{0.25}Ga_{0.75}N$ peak and fully strained $In_{0.15}Ga_{0.85}N$.

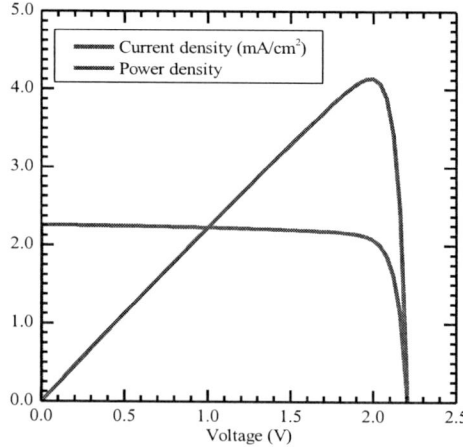

Fig. 3. The light J-V and power density characteristics of the InGaN tandem cell.

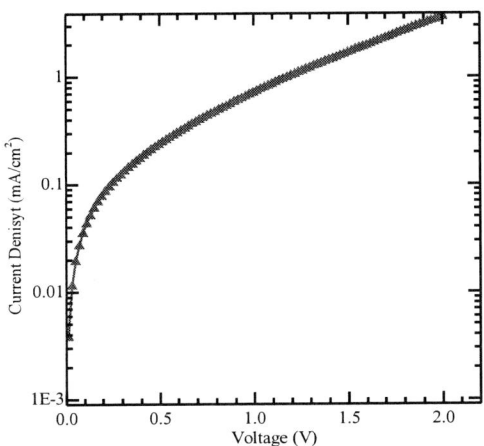

Fig. 4. Experimental and fitted dark J–V curves of tandem MQW InGaN solar cell. The curve was fitted with a single-diode model with R_{SH}, R_S, and one recombination term (no lifetime-minority correction factor was considered).

Recently, nitride-based tunnel junctions were successfully demonstrated by introducing an InGaN layer in a p-n GaN junction to reduce depletion width and increase tunneling probability [12], [13]. Using InGaN in a GaN-based tunnel junction causes undesirable photon absorption due to its low band gap, which can reduce the total photogenerated current density of the tandem cell. However, in this work, a highly doped p-n GaN junction was used in order to reduce the tunneling distance and also to avoid any absorption at the junction. The high p-type doping of GaN (3×10^{19} cm^{-3}) with a low activation energy was achieved by metal modulated epitaxy (MME) growth method [14].

The dark J–V data of the best performing tandem cell was fitted with a diode model, as shown in Fig. 4. The dark J-V data shows a single-diode model. The dark current density (J_0) and ideality factor have values of ~4.6×10^{-18} mA/cm^2 and ~1.2, respectively. The small slope at J_{SC} was fitted to a shunt resistance (R_{SH}) of ~2.9 K$\Omega\cdot$cm^2, indicative of no significant shunting problem in the device. Furthermore, the slope around the V_{OC} was fitted to series resistance (R_S) of ~137 $\Omega\cdot$cm^2. Thus, a carrier transport (carrier collection) issue and not a shunting problem is possibly responsible for the performance degradation. The high R_{SH} and low J_0 can be an indicative of a low density of non-radiative recombination centers in the bulk of the cell [15]. The threading dislocation density (TDD) in the subcells and their effects on the total R_{SH} is difficult to parse. Thus, to understand more thoroughly about the effect of the TDs on the performance of the tandem cell, more optical and electrical measurements for the individual subcells are necessary. The high R_S can be related to a high p-GaN resistivity, high contact resistivity of p-GaN to Ni/Au current spreading layer, and most importantly the polarization induced band bending at the InGaN/GaN interfaces and high TJ resistivity.

The resulting tandem cell was expected to have high voltage and relatively low current. However, the W_{OC} (bandgap-voltage offset) is dramatically higher than the widely used value of W_{OC} ~0.4 [19]. This can be an indication of low tunneling probability and presence of defects in the structure. Given this device was known to be current mismatched, the expectation is to see a light J-V with the small response from one of the cells. Additionally, as these are preliminary results and have only been tested under a UV deficient lamp at present, we express caution in the results.

Finally, the tandem cell performance presented in this work can be used as the groundwork for future development of tandem InGaN solar cells. These are preliminary results providing our initial assumptions that the TJ is successfully implemented in the tandem device. Thus, more measurements are necessary to provide an insight into the TJ behavior.

IV. SUMMARY

In an attempt to investigate the possibility of using III-N material systems for PV-T hybrid solar converter systems, we have designed, grown, and fabricated a double-junction InGaN/GaN MQW solar cell with different indium contents. The device shows a photovoltaic response with a V_{OC} of ~2.22 V. However, due to the presence defects and low tunneling probability (high resistivity in TJ), the V_{OC} value is not significantly higher than the MOCVD-grown MQW single-junction solar cell. We believe by optimizing the crystal quality and increasing the tunnel junction conductivity, we can improve the cell performance.

ACKNOWLEDGMENT

The information, data, or work presented herein was funded in part by the Advanced Research Projects Agency Energy (ARPA-E), U.S. Department of Energy, under Award Number DE-AR0000470.

REFERENCE

[1] J.Y. Chang, S.-H. Yen, Y.-A. Chang, B.-T. Liou, and Y.-K. Kuo, "Numerical Investigation of High-Efficiency InGaN-Based Multijunction Solar Cell," *IEEE Trans. Electron Devices*, vol. 60, no. 12, pp. 4140–4145, Dec. 2013.

[2] H. M. Branz, W. Regan, K. J. Gerst, J. B. Borak Ac, and E. A. Santori, "Hybrid solar converters for maximum exergy and inexpensive dispatchable electricity," *Energy Environ. Sci. Energy Environ. Sci*, vol. 8, no. 8, pp. 3083–3091, 2015.

[3] Y. Vorobiev, J. González-Hernández, P. Vorobiev, and L. Bulat, "Thermal-photovoltaic solar hybrid system for efficient solar energy conversion," 2005.

[4] J. Wu, "When group-III nitrides go infrared: New properties and perspectives," *J. Appl. Phys.*, vol. 106, no. 1, p. 11101, 2009.

[5] E. Trybus, G. Namkoong, W. Henderson, S. Burnham, W. A. Doolittle, M. Cheung, and A. Cartwright, "InN: A material with photovoltaic promise and challenges," *J. Cryst. Growth*, vol. 288, no. 2, pp. 218–224, 2006.

[6] J. J. Williams, H. McFavilen, A. M. Fischer, D. Ding, S. R.

Young, E. Vadiee, F. A. Ponce, C. Arena, C. B. Honsberg, and S. M. Goodnick, "Development of a high-band gap high temperature III-nitride solar cell for integration with concentrated solar power technology," in *2016 IEEE 43rd Photovoltaic Specialists Conference (PVSC)*, 2016, pp. 0193–0195.

[7] M. Leyer, "The critical thickness of InGaN on (0 0 0 1)GaN," *J. Cryst. Growth*, vol. 310, pp. 4913–4915, 2008.

[8] S. P. Bremner, M. Y. Levy, and C. B. Honsberg, "Analysis of tandem solar cell efficiencies under AM1.5G spectrum using a rapid flux calculation method," *Prog. Photovoltaics Res. Appl.*, vol. 16, no. 3, pp. 225–233, May 2008.

[9] C. A. M. Fabien, B. P. Gunning, W. Alan Doolittle, A. M. Fischer, Y. O. Wei, H. Xie, and F. A. Ponce, "Low-temperature growth of InGaN films over the entire composition range by MBE," *J. Cryst. Growth*, vol. 425, pp. 115–118, 2015.

[10] Y. Fang, D. Vasileska, C. Honsberg, and S. M. Goodnick, "High temperature InGaN solar cell modeling," in *2015 IEEE 42nd Photovoltaic Specialist Conference (PVSC)*, 2015, pp. 1–5.

[11] J. J. Williams, H. McFavilen, A. M. Fischer, D. Ding, S. R. Young, E. Vadiee, F. A. Ponce, C. Arena, C. B. Honsberg, and S. M. Goodnick, "Development of a high-band gap high temperature III-nitride solar cell for integration with concentrated solar power technology," in *2016 IEEE 43rd Photovoltaic Specialists Conference (PVSC)*, 2016, pp. 0193–0195.

[12] F. Akyol, S. Krishnamoorthy, Y. Zhang, J. Johnson, J. Hwang, and S. Rajan, "Low-resistance GaN tunnel homojunctions with 150 kA/cm2 current and repeatable negative differential resistance," *Cit. Appl. Phys. Lett.*, vol. 108, 2016.

[13] S. Krishnamoorthy, D. N. Nath, F. Akyol, P. S. Park, M. Esposto, and S. Rajan, "Polarization-engineered GaN/InGaN/GaN tunnel diodes Tunneling-based carrier regeneration in cascaded GaN light emitting diodes to

overcome efficiency droop Polarization-engineered GaN/InGaN/GaN tunnel diodes," *Cit. Appl. Phys. Lett. J. Vac. Sci. Technol. BGaN heterojunctions Appl. Phys. Lett. Phys. Lett. Appl. Phys. Lett. Appl. Phys. Lett*, vol. 97, no. 99, pp. 203502–62203, 2010.

[14] G. Namkoong, E. Trybus, K. K. Lee, M. Moseley, W. A. Doolittle, and D. C. Look, "Metal modulation epitaxy growth for extremely high hole concentrations above 1019cm−3 in GaN," *Appl. Phys. Lett.*, vol. 93, no. 17, p. 172112, Oct. 2008.

[15] D. Cherns, S. J. Henley, F A Ponce, S. J. Henley, and F. A. Ponce, "Edge and screw dislocations as nonradiative centers in InGaN/GaN quantum well," *Cit. Appl. Phys. Lett. Appl. Phys. Lett. Appl. Phys. Lett. Appl. Phys. Lett. GaN Appl. Phys. Lett. J. Appl. Phys.*, vol. 78, no. 100, pp. 2691–2701, 2001.

[16] C. A. M. Fabien, A. Maros, S. Member, C. B. Honsberg, S. Member, W. A. Doolittle, and S. Member, "III-Nitride Double-Heterojunction Solar Cells With High In-Content InGaN Absorbing Layers : Comparison of Large-Area and Small-Area Devices," *IEEE J. Photovoltaics*, vol. 6, no. 2, pp. 460–464, Mar. 2016.

[17] Y-K. Ee, J. M. Biser, W. Cao, H. M. Chan, R. P. Vinci, and N. Tansu, "Metalorganic Vapor Phase Epitaxy of III-Nitride Light-Emitting Diodes on Nanopatterned AGOG Sapphire Substrate by Abbreviated Growth Mode," *IEEE J. Sel. Top. Quantum Electron.*, vol. 15, no. 4, pp. 1066–1072, Jul. 2009.

[18] Q. Li, J. J. Figiel, and G. T. Wang, "Dislocation density reduction in GaN by dislocation filtering through a self-assembled monolayer of silica microspheres Dislocation density reduction in GaN by dislocation filtering through a self-assembled monolayer of silica microspheres," *Cit. Appl. Phys. Lett*, vol. 94, 2009.

[19] R. R. King, D. Bhusari, A. Boca, D. Larrabee, X. Q. Liu, W. Hong, C. M. Fetzer, D. C. Law, and N. H. Karam, "Band gap-voltage offset and energy production in next-generation multijunction solar cells," *Prog. Photovoltaics Res. Appl.*, vol. 19, no. 7, pp. 797–812, Nov. 2011.

A Physics-Based Simulation Tool for Leakage Currents in c-Si PV Modules

John M. Waddle, Saroj Dahal, Marco Nardone

Dept. of Physics and Astronomy, Bowling Green State University, Bowling Green, Ohio, 43403

Abstract—**This project encompasses the development of a simulation tool for leakage currents in c-Si photovoltaic modules using the finite-element method. It accounts for variable material and surface properties due to changes in temperature, relative humidity, and surface wetting. Moisture diffusion from the module perimeter into the encapsulant and its resultant impact on charge transport is calculated. The electric field in the SiN$_x$ antireflection coating, which may affect shunting due to potential induced degradation, is also quantified; it can be greater than 10 kV/cm depending on temperature, relative humidity, and the SiN$_x$ properties. By simulating time-varying weather conditions, leakage currents in fielded modules can be predicted.**

I. INTRODUCTION

In order to reduce losses from high current in photovoltaic (PV) arrays, as many cells as possible will be arranged in series, leading to a high voltage difference between cells and the grounded frame. The resulting leakage current and associated detrimental effects, such as Potential Induced Degradation (PID) will only become more prevalent as the size of module strings continues to increase. Sustained efficiency and power output are critical to the economic viability of PV systems; therefore, understanding and mitigating these effects is of considerable importance.

To this end, the leakage current has been extensively studied [1]. However, related modeling tools have been based on empirical fits [2], equivalent circuit methods [3], or finite element analysis [4] without consideration of detailed charge/ion/moisture transport physics. In this work we develop a leakage current simulation tool that incorporates the best available knowledge on the properties of the packaging materials. Our finite element models—utilizing COMSOL Multiphysics®—calculate leakage current along surfaces and through the bulk materials, water diffusion from the module perimeter into the encapsulant and its resultant impact on charge transport. The electric field distribution in the SiN$_x$ anti-reflective coating is also calculated, which has important implications for PID of the shunting type. This work will focus on typical Si-based PV, but our approach is applicable to any type of module.

An overview of the simulation tool and lifetime prediction method is shown in Fig. 1. The primary inputs include the geometry of the module, packaging material properties, and operating conditions. Geometry includes various thicknesses and lateral spacings, but it may also include 2D geometries when certain symmetries can be exploited. There are several material parameters that pertain to electronic, ionic, and moisture transport processes in the various packaging components,

as described below. Operating conditions can include laboratory stress test settings or local weather information. The transport models are comprised of the governing equations for charge and moisture movement; details are provided in Sec. II. For the given input parameters and physical models, time-dependent simulation using the finite element method is used to calculate: (1) the leakage current (from which total charge transfer can be obtained by integration over time), (2) moisture concentration; and (3) the electric field (in particular, across the SiN$_x$ layer) as functions of space and time. If test data are used to determine the maximum allowable charge transfer or moisture content for a particular module design, then lifetime can be predicted as the time at which those maximums are attained. In what follows, we provide some details on the model components and example results.

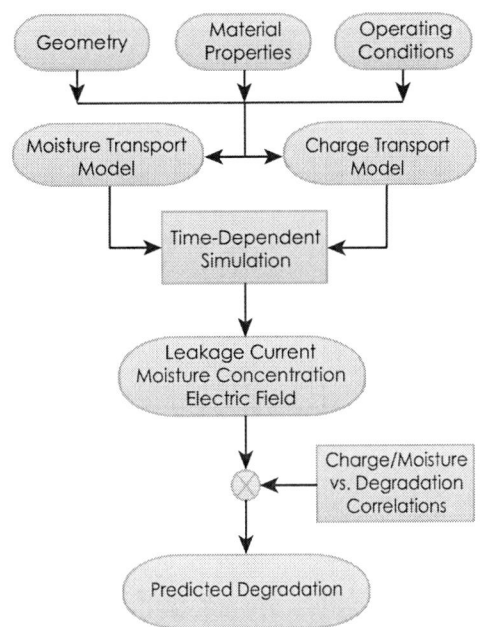

Fig. 1: Overview of the simulation approach and module lifetime prediction method.

In this study, we consider the typical module packaging arrangement: glass, polymer laminate, Si cells, Tedlar-Polyethylene-Tedlar (TPT) backsheet, and edge sealant. Our calculations are based on a 2D cross-section geometry of a module, as shown in Fig. 2; extension to 3D is also possible. As preliminary results with and without metallic fingers

978-1-5090-5606-4/17 $31.00 © 2017 IEEE

suggest that at they have little effect on the leakage current and potential distribution—since the Si layer is practically an equipotential under high voltage stress—the metallic fingers were omitted in our calculations. Material parameter values are provided in Table I. The operating conditions are either fixed temperatures and RH, as would be expected in an environmental chamber, or the specific Typical Meteorological Year-3 (TMY-3) data for Denver, CO [5].

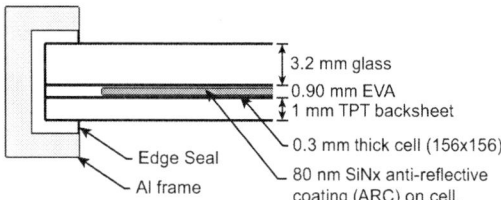

Fig. 2: Schematic of the module cross-section used in this study. Not to scale. All units in mm. The thickness of the edge seal is variable and the Al frame is treated as a $V = 0$ (grounded) boundary condition. The cell is held at a negative bias for leakage current calculations.

II. TRANSPORT MODELS

Leakage current, I_L, between the cell contacts and the grounded module components depends on various fluxes, $\mathbf{J}$, which are governed by conservation equation $\nabla \cdot \mathbf{J} = Q$ where Q is a source term. The flux may be comprised of electronic charge, ions, water vapor, or thermal energy, all of which influence I_L. Furthermore, the properties of the encapsulant can be influenced by the degree of moisture diffusion through the edge seals. Charge and moisture transport models are described below. Dependencies on temperature and relative humidity (RH) are critical environmental parameters.

A. Charge Transport

We consider first the electric current in each of the module packaging materials (ionic or electronic current depending on the material). Charge flux is given by Ohm's Law, $\mathbf{J} = -\sigma \nabla \varphi$, and conservation with $Q = 0$ yields,

$$-\nabla \cdot (\sigma \nabla \varphi) = 0, \qquad (1)$$

which is solved for the electric potential φ. Depending on the material, the conductivity σ can depend strongly on temperature, water content, and/or electric field, as described below. The boundary conditions include specific values for φ at contacts or $\hat{\mathbf{n}} \cdot \mathbf{J} = 0$ (zero normal current density) at insulation boundaries.

1) Bulk Electrical Conductivities: Glass conductivity is ionic in nature with Na ions being the predominant charge carriers in soda-lime glass [6]. Ohmic conduction is typically valid for electric field $\mathcal{E} < 10^4 \, \mathrm{V/cm}$ [7] and follows Arrhenius behavior with respect to temperature,

$$\sigma_g = \sigma_{0g} \exp\left(-\frac{E_g}{kT}\right). \qquad (2)$$

For typical PV glass, we use $\sigma_{0g} = 25.2$ S/cm and $E_g = 0.814$ eV [3]. There are many subtleties with glass conduction, such as current fluctuations and decay with time due to space charge build-up. For our present purposes, we assume that Eq. (2) is valid for bulk glass, is constant in time, and does not depend on RH.

The conductivity of ethylene vinyl acetate (EVA, most common laminate) depends on vinyl acetate content, absorbed water, other chemical properties, and aging. Mon et. al. [8] provided a relation for the bulk EVA conductivity as a function of the relative permittivities of the polymer and water, ϵ_{EVA} and ϵ_w, along with water content, c (mass water per unit mass of polymer),

$$\sigma_{EVA} = \sigma_{0EVA} \exp\left[\left(-\frac{E_{EVA}}{kT}\right)\left(\frac{\epsilon_{EVA}}{\epsilon_w}\right)^c\right]. \qquad (3)$$

The water content, c, can be based on absorption isotherms for the relevant material or it can be obtained from the solution of the moisture transport model described below. In the latter case the charge and moisture transport models are coupled.

For the TPT back sheet, a very low electrical conductivity of $\sigma_{TPT} = 3 \times 10^{-16}$ S/cm was used [9]. Measured conductivities for various edge seal materials, σ_{edge} (such as silicone and PIB) are not readily available. The conductivity of a few sealants in the range of $\sigma_{edge} \approx 10^{-11} - 10^{-10}$ S/cm at $T = 295$ K and $50\% RH$, were provided in Ref. [10], but temperature and moisture dependencies were not provided. Therefore, we used σ_{edge} as an independent variable in this study and determined that the edge material can play an important role in controlling leakage current.

2) Thin SiN$_x$ Layer: As meshing high aspect ratio geometry presents difficulties for the finite-element-method, we assume a linear voltage as a function of space (constant electric field) across the thin SiN$_x$ film of conductivity σ_f and thickness δ_f, modeling it as an internal 2D boundary with the following conditions on the flux components normal to the surface,

$$\hat{\mathbf{n}} \cdot \mathbf{J_1} = \frac{\sigma_f}{\delta_f}(V_1 - V_2) \text{ and } \hat{\mathbf{n}} \cdot \mathbf{J_2} = \frac{\sigma_f}{\delta_f}(V_2 - V_1). \qquad (4)$$

The potentials on the upper and lower faces of the boundary are given by V_1 and V_2, respectively, resulting in an electric field, $E = (V_2 - V_1)/\delta_f$; $\delta_f \approx 80$ nm for SiN$_x$.

We consider two Si$_3$N$_4$ charge transport mechanisms identified by Sze [11] as relevant for typical PV conditions: a moderate temperature, low-field ohmic regime and a moderate temperature, high-field Poole-Frenkel (PF) regime. Ohmic conduction by hopping between isolated states is given by [12],

$$\sigma = \sigma_0 \exp\left(-\frac{E}{kT}\right), \qquad (5)$$

where E is the activation energy, k is Boltzmann's constant, and T is temperature. PF conductivity is described by,

$$\sigma_{PF} = \sigma_{0PF} \exp\left(-\frac{E_{PF}}{kT} + \sqrt{\frac{\mathcal{E}}{\mathcal{E}_0}}\right), \qquad (6)$$

with $\mathcal{E}_0 = \dfrac{\pi \epsilon_r \epsilon_0 (kT)^2}{q^3}$,

where E_{PF} is the activation energy, $\mathcal{E}$ is the local electric field, and $\mathcal{E}_0$ is a characteristic field with relative permittivity ϵ_r, vacuum permittivity ϵ_0, and elementary charge q. The total conductivity is the sum of (5) and (6):

$$\sigma_{SiN} = \sigma + \sigma_{PF}, \qquad (7)$$

which is used in Eq. (4) in place of σ_f.

SiN$_x$ in PV cells can exhibit both types of conductivity behavior shown in Eq. (7) depending on the Si:N ratio, deposition method, and resulting refractive index. Naumann et. al. [13] found that the conductivity was greater in SiN$_2$ (with a refractive index of $n = 2.32$) compared to SiN$_1$ ($n = 1.93$). Our fit of their current-voltage data showed that the SiN$_2$ exhibited both ohmic and PF conduction. The characteristic field obtained from that fit was $\mathcal{E}_0 = 1.2 \times 10^4$ V/cm which, from Eq. (6), yields $\epsilon_r \approx 10$, in reasonable agreement with typical measured values of $\epsilon_r \sim 4 - 7$. Only ohmic behavior was exhibited by the SiN$_1$ material. Our simulations allow for calculation of the field across the SiN$_x$ layer in different scenarios.

3) Surfaces and Interfaces: In the cases of surfaces and material interfaces, we solve for the potential along the directions tangential to the surface with,

$$-\nabla_t \cdot (\delta_s \sigma_S \nabla_t \varphi) = \delta_s Q, \qquad (8)$$

where ∇_t is the tangential derivative and δ_s is the layer thickness. The validity of this 2D approach was tested by comparing it to calculations with a 3D thin layer included in the model as a domain representing the interface between the glass and EVA. Indistinguishable potential distributions and leakage currents were obtained for both scenarios over a wide range of thickness and conductivity parameters.

Surface conductivity (units of S or $1/\Omega$) of the glass depends on RH and the fraction of the surface covered by water, f_w according to [8],

$$\sigma_{S,g} = \sigma_{0S,g} \exp \left\{ \left[-\frac{E_{S,g}}{kT} \right] \left[1 - f_w RH \left(1 - \frac{\epsilon_g}{\epsilon_w} \right) \right] \right\}. \qquad (9)$$

The same expression is used for EVA surfaces with the subscript 'g' replaced by 'EVA'. Conduction data was fit with $f_w \approx 0.1$ for glass and $f_w \approx 1$ for EVA in Ref. [8]. At very high values of $RH > 95\%$, we assume that the outer glass surface is fully covered with water and $f_w = 1$. Coincidentally, the magnitude of the conductivity given by Eq. (9) when using the parameters in Table I and $f_w RH = 1$, is close to that of rainwater (1–100 µS/cm).

At the interface of glass and EVA, the conductivity is given by,

$$\sigma_I = \sigma_{S,EVA} + \sigma_{S,g}, \qquad (10)$$

where it is assumed that the two materials are well-bonded.

B. Moisture Transport

The water concentration, c, as a function of time is determined by the diffusion equation,

$$\frac{\partial c}{\partial t} = \nabla \cdot (D \nabla c), \qquad (11)$$

where D is the diffusivity of water. With this approach, diffusion can be Fickian or non-Fickian ($D = D(c)$), but here we assume the former. At the external boundaries, the concentration is due to the solubility S of water in the specific edge material, which is proportional to relative humidity RH [8], [14]. Both D and S are thermally activated (see Table I for parameter values):

$$D = D_0 \exp \left(-\frac{E_{a_D}}{k_B T} \right), \; S = S_0 RH \exp \left(-\frac{E_{a_S}}{k_B T} \right). \qquad (12)$$

Eqs. (12) apply both to the edge seal and encapsulant with material specific values (see Table I).

TABLE I:
PARAMETER LIST

Symbol	Unit	Value	Reference
σ_0	S/cm	1.4×10^{-12}	[11], [13]
E	eV	0.1	[11], [13]
σ_{0PF}	S/cm	9.9×10^4	[11], [13]
E_{PF}	eV	1.3	[11], [13]
ϵ_r	–	7	[11], [13]
σ_{0g}	S/cm	25.2	[3]
E_g	eV	0.814	[3]
ϵ_{eva}	–	3.8	[8]
ϵ_w	–	82	[8]
σ_{0eva}	S/cm	2.24×10^{-4}	[8]
E_{eva}	eV	0.59	[8]
$\sigma_{0S,eva}$	S/cm	1.7×10^{12}	[8]
$E_{S,eva}$	eV	0.28	[8]
$\sigma_{0,g}$	S/cm	0.02775	[8]
$E_{S,g}$	eV	0.68	[8]
σ_{TPT}	S/cm	3×10^{-16}	[9]
σ_{PIB}	S/cm	var.	NA
$E_{a_D,eva}$	kJ/mol	38.1	[14]
$D_{0,eva}$	cm^2/s	2.310	[14]
$E_{a_S,eva}$	kJ/mol	16.7	[14]
$S_{0,eva}$	g/cm^3	1.81	[14]
$E_{a_D,pib}$	kJ/mol	58.4	[14]
$D_{0,pib}$	cm^2/s	17	[14]
$E_{a_S,pib}$	kJ/mol	5	[14]
$S_{0,pib}$	g/cm^3	0.036	[14]

III. Results and Discussion

The primary outputs considered here are the leakage current, I_L, water concentration, c (in the EVA and edge seal), and the electric field across the SiN$_x$ layer, E_{SiN}. We use the following criteria to specify on-set of significant degradation: (1) total charge transfer, $Q_{tot} = 0.02$ Coulomb per cm of module perimeter [15] (this is the integral of I_L over time); (2) maximum moisture content in the EVA, $c_{max} = 0.003$ g water/g EVA; and (3) a maximum field of $E_{SiN,max} = 5 \times 10^4$ V/cm across the SiN$_x$ layer. These criteria are simple examples that can be determined for specific module designs through controlled laboratory stress tests, field tests, and material specific measurements. All calculations assume that delamination does not occur.

Considering the geometry and layer thicknesses shown in Fig. 2 with an edge seal thickness of 0.4 mm, leakage current was calculated as a function of $50\,°C < T < 90\,°C$, $20\% < RH < 90\%$, and 10^{-14} S/cm$< \sigma_{edge} < 10^{-10}$ S/cm, assuming the constant conditions of an environmental chamber and bias of -1000 V at the cell. In this case, it was further assumed that the water content of the EVA was fixed at $c = 0.001$ g water/g EVA (i.e. the edge seal did not allow moisture ingress). Therefore, the humidity only affected the glass surface conductivity [Eq. (9)]. The results shown in Fig. 3 indicate that the leakage current increases with T and RH, with a stronger dependence on T.

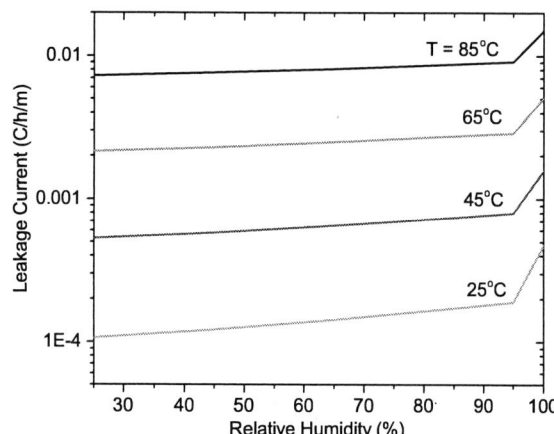

Fig. 4: Leakage current as a function of RH with $\sigma_{edge} = 10^{-11}$ S/cm at -1000 V bias and the temperatures show. Model geometry from Fig. 2 and parameter values from Table I.

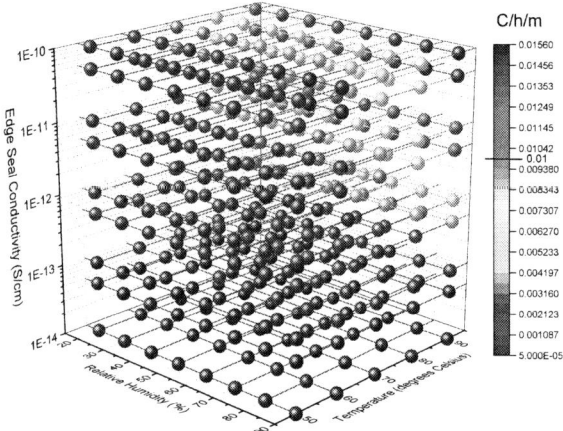

Fig. 3: Leakage current as a function of T, RH and edge seal conductivity under constant stress conditions at -1000 V. Model geometry from Fig. 2 and parameter values from Table I. Criteria of 0.01 C/h/m shown on the legend.

If we assume that 200 hours before on-set of power loss under these constant stress conditions is a reasonable time for module reliability, our criteria of $Q_{tot} = 0.02$ C/cm converts to $I_L < Q_{tot}/200 = 0.01$ C/h/m. Fig. 3 suggests that this criteria is satisfied for $T < 90\,°C$ and $\sigma_{edge} < 10^{-11}$ S/cm. Note that for $\sigma_{edge} < 10^{-11}$ S/cm, the criteria is satisfied for all T and RH.

When water covers most of the glass surface and contacts the grounded frame, the leakage current can increase dramatically because the effective electrode area increases. Here we assume that the factor $f_w = 1$ for glass in Eq. (9) when $RH > 95\%$. The significant increase in leakage current can be clearly seen in Fig. 4. It should noted that when water covers the entire front surface, the leakage current will scale with the area of the module rather than the perimeter. Therefore, the increase in leakage current close to $RH = 100\%$ will be more dramatic in large area modules (the model used here had an equivalent area of a single cell coupon, approx. 20 x 20 cm^2.).

The maximum electric field across the SiN$_x$ layer as a function of σ_{edge} is shown in Fig. 5 for various T/RH values at a bias of -1000 V (the field is always strongest closer

to the grounded frame, except for when water covers the glass surface). In all cases, $E_{SiN,max}$ increases with σ_{edge} since more of the voltage drop is taken up by the SiN$_x$ layer as the resistance of other elements in the system decreases. Interesting behavior is observed with respect to temperature: $E_{SiN,max}$ increases with T up to 65 °C, then decreases as T approaches 85 °C. This is likely due to the different conductivity temperature dependencies of the materials involved and the electrical field dependence of the SiN$_x$ conductivity via the Poole-Frenkel effect. In any case, if the criteria of $E_{SiN,max} \leq 5 \times 10^4$ V/cm is set, it is always satisfied for 25 °C/25% and otherwise $\sigma_{edge} < 5 \times 10^{-12}$ S/cm is required. Such high fields could act as driving forces for ions across the SiN$_x$, which may lead to PID shunting. The time it takes for sodium to diffuse through the SiN$_x$ at a given field depends on the mobility.

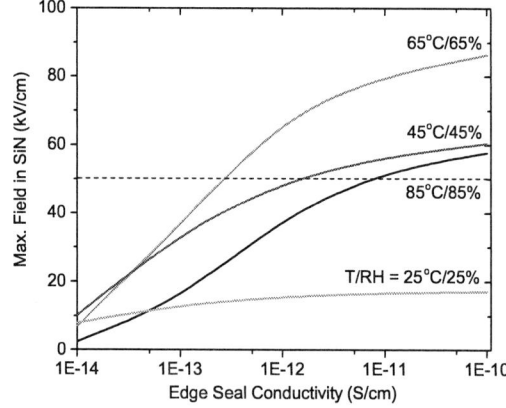

Fig. 5: Maximum electric field in the SiN$_x$ layer at the T/RH combinations shown and bias of -1000 V. The arbitrary criteria of $E_{SiN,max} \leq 5 \times 10^4$ V/cm is shown by the dashed line.

Next we study moisture ingress for the same model geometry as above (but now with variable edge seal thickness) and the parameter values from Table I, assuming PIB edge seal. The initial water vapor concentration in the EVA was set to zero. Fig. 6 shows the maximum water concentration in the EVA layer as a function of time for damp heat conditions of 85 °C/85%RH and edge seal thicknesses of 1, 5, and 10 mm. 85% RH provides a boundary concentration at the edge seal of 0.0051 g/cm^3 due to Eq. (12). The criteria of $c_{max} = 0.003$ g water/g EVA is met at approximately 100, 300, and 1000 h for edge seal thicknesses of 1, 5, and 10 mm, respectively.

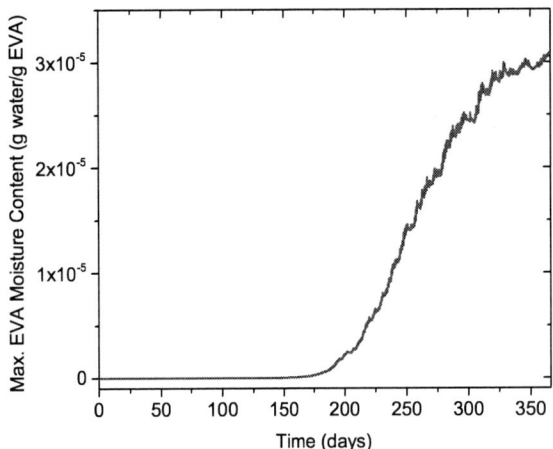

Fig. 7: Maximum moisture content in the EVA layer over a one-year period using the TMY-3 data for Denver, CO. Edge seal thicknesses is 10 mm.

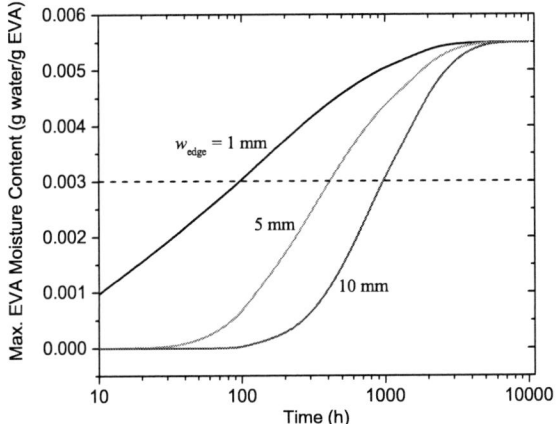

Fig. 6: Maximum moisture content in the EVA layer over time at 85 °C/85%RH for the three edge seal thicknesses shown. An arbitrary criteria of 0.003 g water/g EVA is indicated.

As an example of extending the model to field conditions, we used hourly TMY-3 weather data (including estimated module temperature) for Denver, CO [5]. The resultant c_{max} in the EVA layer over a one-year period time is shown in Fig. 7 for an edge seal width of 10 mm. c_{max} is well below the criteria of 0.003 g water/g EVA after one year. This approach can be extended to multiple years.

IV. SUMMARY

A module packaging simulation tool is being developed that can predict leakage current and moisture ingress while accounting for changes in material parameters due to either static or time-dependent environmental factors. Critical factors such as the total leaked charge, moisture content in the encapsulant, and electric field across the SiN$_x$ layer can be predicted for specific module designs. Our results suggest that the electrical resistance of the edge seal can play an important role in limiting the leakage current and related detrimental effects. Reducing the electrical connectivity between the grounded frame and surface water at high RH could significantly reduce leakge current in fielded modules. This modeling approach is generally applicable to any type of module.

ACKNOWLEDGMENTS

This conference paper was developed based upon funding from the Alliance for Sustainable Energy, LLC, Managing

and Operating Contractor for the National Renewable Energy Laboratory (NREL) for the U.S. Department of Energy (DOE).

REFERENCES

[1] W. Luo, Y. S. Khoo, P. Hacke, V. Naumann, D. Lausch, S. P. Harvey, J. P. Singh, J. Chai, Y. Wang, A. G. Aberle, and others, "Potential-induced degradation in photovoltaic modules: a critical review," *Energy & Environmental Science*, 2017. [Online]. Available: http://pubs.rsc.org/-/content/articlehtml/2017/ee/c6ee02271e

[2] P. Hacke, S. Spataru, K. Terwilliger, G. Perrin, S. Glick, S. Kurtz, and J. Wohlgemuth, "Accelerated Testing and Modeling of Potential-Induced Degradation as a Function of Temperature and Relative Humidity," *IEEE Journal of Photovoltaics*, vol. 5, no. 6, pp. 1549–1553, Nov. 2015.

[3] H. Nagel, M. Glatthaar, and S. Glunz, "Quantitative assessment of the local leakage current in PV modules for degradation prediction," in *Proceedings of the 31st European Photovoltaic Solar Energy Conference and Exhibition*, Sep. 2015, pp. 1825 – 1829.

[4] N. Shiradkar, E. Schneller, and N. G. Dhere, "Finite element analysis based model to study the electric field distribution and leakage current in PV modules under high voltage bias," in *SPIE Solar Energy+ Technology*. International Society for Optics and Photonics, 2013, pp. 88 250G–88 250G.

[5] "TMY-3 Data." [Online]. Available: http://rredc.nrel.gov/solar/olddata/nsrdb/1991-2005/tmy3/

[6] F. V. Natrup, H. Bracht, S. Murugavel, and B. Roling, "Cation diffusion and ionic conductivity in soda-lime silicate glasses," *Physical chemistry chemical physics*, vol. 7, no. 11, pp. 2279–2286, May 2005. [Online]. Available: http://pubs.rsc.org/en/content/articlelanding/2005/cp/b502501j

[7] R. J. Maurer, "Deviations from Ohm's Law in Soda Lime Glass," *The Journal of Chemical Physics*, vol. 9, no. 8, pp. 579–584, Aug. 1941. [Online]. Available: http://aip.scitation.org/doi/abs/10.1063/1.1750958

[8] G. Mon, L. Wen, and R. Ross Jr, "Encapsulant free-surfaces and interfaces: critical parameters in controlling cell corrosion," in *19th IEEE Photovoltaic Specialists Conference*, vol. 1, 1987, pp. 1215–1221.

[9] F.-J. Pern, *Module encapsulation materials, processing and testing*. National Renewable Energy Laboratory, 2008.

[10] G. R. Mon, L. C. Wen, R. S. Sugimura, and R. G. Ross, "Reliability studies of photovoltaic module insulation systems," in *Electrical Electronics Insulation Conference, 1989. Chicago '89 EEIC/ICWA Exposition., Proceedings of the 19th*. IEEE, 1989, pp. 324–329.

[11] S. M. Sze, "Current Transport and Maximum Dielectric Strength of Silicon Nitride Films," *Journal of Applied Physics*, vol. 38, no. 7, pp. 2951–2956, Jun. 1967. [Online]. Available: http://scitation.aip.org/content/aip/journal/jap/38/7/10.1063/1.1710030

[12] N. F. Mott and W. D. Twose, "The theory of impurity conduction," *Advances in Physics*, vol. 10, no. 38, pp. 107–163, Apr. 1961. [Online]. Available: http://dx.doi.org/10.1080/00018736100101271

[13] V. Naumann, K. Ilse, and C. Hagendorf, "On the discrepancy between leakage currents and potential induced degradation of crystalline silicon modules," in *Proceedings of the 28th European Photovoltaic Solar Energy Conference*, 2013.

[14] M. D. Kempe, A. A. Dameron, and M. O. Reese, "Evaluation of moisture ingress from the perimeter of photovoltaic modules," *Progress in Photovoltaics: Research and Applications*, vol. 22, no. 11, pp. 1159–1171, Nov. 2014.

[15] P. Hacke, M. Kempe, K. Terwilliger, S. Glick, N. Call, S. Johnston, S. Kurtz, I. Bennett, and M. Kloos, "Characterization of multicrystalline silicon modules with system bias voltage applied in damp heat," *Proceedings of 25th EUPVSEC, Valencia, Spain*, pp. 3760–3765, 2010.

Broadband Ta_2O_5 Moth-eye Antireflection Coatings for Tandem Solar Cells on Si

Bo Yuan[1], Brian Thibeault[2], David Payne[3], James Mutitu[4], Ivan Perez-Wurfl[3], Kevin Dobson[4], Brianna Conrad[3], Allen Barnett[3] and Robert L. Opila[1,5]

1. Department of Chemistry and Biochemistry, University of Delaware, Newark, DE 19716 USA
2. Department of Electrical and Computer Engineering, University of California Santa Barbara, Santa Barbara, CA 93106 USA
3. School of Photovoltaic and Renewable Energy Engineering, University of New South Wales, Sydney NSW 2052, Australia
4. Institute of Energy Conversion, University of Delaware, Newark, DE 19716 USA
5. Department of Materials Science and Engineering, University of Delaware, Newark, DE 19716 USA

Abstract — **Wafer-scale Ta_2O_5 moth-eye antireflection coatings are directly patterned on dual junction GaAsP/SiGe solar cells on Si substrates using deep UV photolithography and plasma etching. These sub-wavelength structures have an aspect ratio of 1.2, which results in excellent antireflection properties with an average reflectance of 7% over the entire 400-1100nm. Further optimizations of moth-eye and traditional double layer antireflection coatings (DLARCs) on the device using finite difference time domain (FDTD) method are performed. Optimized moth-eye structures outperform optimized traditional DLARCs with average reflection as low as 2.2% from 400-1100nm.**

I. INTRODUCTION

Due to the ability to exploit multiple absorption bands, III-V multi-junction solar cells (MJSCs) are the most efficient solar cells ever developed [1][2]. Nevertheless, due to the high cost of lattice-matched substrates and deposition processes, III-V devices are extremely expensive and mainly used for space and highly concentrated terrestrial solar power plants. Si solar cells hold dominant position in PV industry [3], which is largely due to its natural abundance and cost-effective growth and fabrication processes. It is in this context that research on fabricating multi-junction devices directly on Si platforms has gained increasing interest [4][5]. By combining the high performance of III-V devices and well-developed Si-based fabrication technologies, a potential 1-sun efficiency of 40% is expected to be achieved [5].

As on single junction solar cells where antireflection coatings (ARCs) are used to couple the maximum amount of light into the device, ARCs are also indispensable for tandem solar cells to achieve broadband antireflection. However, optimization of ARCs for tandem cells is more challenging than single junction solar cells [6]. Overall performance of the ARCs is evaluated by the current-limiting subcell. Step-down interference-based coatings are one of the most common ARCs used in tandem solar cells [7]. In this case, however,

any deviation of coatings from optimum might increase current mismatch of the device [6]. Recently, antireflective nanostructures have been proposed as a solution to achieve low reflection across a broad spectral band, which is more forgiving to current mismatch subcells [8]-[11]. The sub-wavelength structures provide a gradient change rather than abrupt steps in the effective index of refraction from air to the top semiconductor layer. Reflection thus can be efficiently decreased over a broad wavelength range and at all angles of incidence [12]. In this paper, we first present fabrication of broadband Ta_2O_5 moth-eye ARCs for tandem solar cells on Si. Ta_2O_5 has been chosen as the material of interest because of its relatively high refractive index in addition to low extinction coefficient across the spectrum. Finite-difference time-domain (FDTD) was then used as simulation method to further optimize the design geometry.

II. EXPERIMENTAL DETAILS

A. Fabrication of moth-eye ARCs

Fabrication details for the dual junction GaAsP/SiGe solar cell have been presented previously [13]. 25nm GaInP was deposited on GaAsP top cell as a window layer. Therefore, directly patterning moth-eye structures by using this window layer on our device is not feasible. The cell was first treated by NH_4OH: H_2O_2: H_2O (7:35:160) to remove the uppermost GaAsP contact layer. Cone-shaped moth-eye structures were then fabricated in the following fashion. First, Ta_2O_5 was deposited onto the etched solar cell structure using a Veeco Nexus Dual-Beam ion beam deposition system. The deposition beam (aimed at the Ta target) parameters were set to 1200 V, 420 mA, with Xe at 5.2 sccm. The assist beam (aimed at the substrate) parameters were set to 50 V, 310 mA, with O_2 at 20 sccm. Deposition was carried out for 85 minutes on the sample at 40 degrees relative to normal. The resulting films were approximately 580 nm in thickness as measured by a Woollam M2000DI Variable Angle Spectroscopic Ellipsometer system over the wavelength range of 400-1100 nm. After deposition, the samples were patterned using 248 nm deep UV (DUV) lithography on an ASML 5500/300 system. The full wafer pattern was made by stitching together 4mm x 4mm areas of dense HCP arrays consisting of 150 nm dots spaced with a 350 nm pitch. The lithography process proceeded as follows. A DUV antireflective coating, AR2-600, was first spun at 3.5 krpm for 30 s and baked for 60 s at 210°C on a hot plate. Shipley UVN2300-0.5 was then coated at 2.5krpm for 30s and baked for 60 s at 110°C on a hot plate. The pattern was exposed with a dose of 109mJ using a system NA of 0.63 with annular illumination conditions (S_o=0.8, S_i=0.5). The sample was developed for 20 s using AZ300MIF developer. Following lithography, the sample was etched in a Panasonic E640 ICP etching system at 15°C substrate temperature. The sample was mounted to a 6" silicon wafer carrier using Santovac 5 oil to provide thermal contact between the sample and wafer carrier. Etching was executed in 3 steps. Gases, powers, and material etched are indicated in table I. During the etching of the Ta_2O_5 in step 2, the PR mask was also etched into a cone shape due to physical sputtering. CHF_3 polymer sidewall passivation and the PR ablation together led to the final structures in the Ta_2O_5 shown in Figure 1. The cone height of these nanostructures is 425 nm and their base diameter is 350 nm, which gives an aspect ratio (cone height/base diameter) of 425/350≈1.2. Closely packed nanocones indicate there is no obvious discontinuity of refractive index between air and the top of the semiconductor layer.

B. Optical simulations

The device structure was optically modelled using finite difference time domain (FDTD) modelling technique, in this case the commercial software package, 'FDTD Solutions' by *Lumerical* (Lumerical Solutions, Inc. Canada) was used. When modelling a 3D structure using FDTD, the geometry and optical properties (n & k) of each different material involved, along with any optical input sources and monitors, must be clearly specified and the simulation region and boundary types must also be defined. The geometry is then discretised into a mesh of cuboids known as Yee cells [14] and Maxwell's time-dependent curl equations are solved at nodes across these cells. In this work, simulations were first validated by comparing with experimental results and parameter sweeps were then used to further optimize the device geometries and explore the limits of the antireflection approaches. Two antireflection schemes were simulated, one of the device structure with conventional double layer antireflection coatings (DLARCs) and one of device structure with Ta_2O_5 moth-eye antireflection hemispheres on top.

The initial material geometries were chosen to match those of the physical device and the optical properties of each material were determined either using *Lumerical*'s present material database or from ellipsometry measurements and model fitting [15]. A plane wave source was placed at normal incidence to the front surface of the modelled structure and a 2D frequency-domain field and power (FDFP) monitor was placed behind the source in order to calculate reflectance. The simulation boundaries were specified as perfectly matched layers (PML) in the +/- z direction, resulting in a semi-infinite SiGe substrate. In the +/- x and y directions the simulation boundaries were specified as periodic, creating an effectively infinitely wide periodic device. The wavelength range of the results was set from 400nm to 1100nm with 300 frequency points recorded by the monitor.

TABLE I
ICP etching parameters for preparation of Ta_2O_5 moth-eye structures

Etch Step	1	2	3
Material	AR2 between dots	Ta_2O_5	Residual PR Mask
Gas	O_2	CHF_3	O_2
Flow (sccm)	40	40	40
ICP Power (W)	75	500	200
Sample Power (W)	75	200	100
Time (m:s)	0:40	4:40	3:20

Fig.1 SEM image of Ta_2O_5 moth-eye structures with 50° relative to normal. Dimensions are already corrected for tilt angle.

III. RESULTS AND DISCUSSIONS

A. Optical characterization of Ta_2O_5 moth-eye structures

The total optical reflectivity of the solar cell with the patterned moth-eye structures was measured from 400-1100 nm by using a UV-visible-IR spectrometer fitted with an integrating sphere (Perkin-Elmer Lambda 750) at an angle of incidence of 8 degrees from normal. Figure 2 compares the reflectance from a bare dual junction cell without ARCs and the same cell with patterned Ta_2O_5 moth-eye structures. The bare solar cell shows an average reflectance of 35% over the measured spectral range. However, the cell with Ta_2O_5 moth-eye structures shows an average absolute reflectance decrease of 28% over 400-1100 nm.

For wavelengths above 700 nm, since the absorption coefficient of the GaAsP layer drops off rapidly [15], interference patterns occur because the light reaches various interfaces throughout the III-V stack and most prominently between the III-V and SiGe. A proof of concept Ta_2O_5 moth-eye structures fabricated directly on III-V/SiGe solar cell on Si for ARCs has thus been shown here.

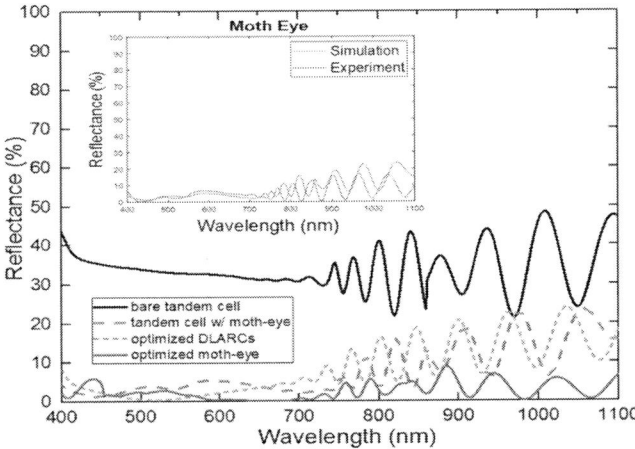

Fig.2 Measured total reflection for bare tandem cell (solid black), tandem cell with fabricated moth-eye (dotted red) and simulated total reflection of tandem cell with optimized DLARCs (dotted orange) and optimized moth-eye structures (solid green). Top left plot shows comparison of simulation and experimental results for moth-eye structures with base diameter of 350 nm and height of 425 nm.

B. Optical simulations

The FDTD model and simulation technique was first validated by comparison of Ta_2O_5 moth-eye ARCs simulation results with experimental results at fabricated dimensions: base diameter of 350 nm and height of 425 nm. (Figure 2 top left). FDTD simulation gave reasonably good agreement with experimental results. Some of the reasons for their differences are that simpler hemisphere structures are used in the model; there could also be error in the ellipsometric fitting of the layers, etc. The DLARCs were optimized using a dual parameter sweep of the SiO_2 and SiN_x layer thicknesses. For each layer the thickness was varied between 0 and 160 nm, in intervals of 10 nm, resulting in 289 full simulations in total. Broadband reflectance result was acquired for each case and the total weighted absorption was then calculated to conduct comparison by using Equation 1. The Ta_2O_5 moth-eye antireflection hemispheres were optimized by sweeping the height and diameter of the hemispheres. Hemisphere height varied from 100 to 600 nm in interval steps of 25 nm while hemisphere diameter changed from 120 to 620 nm in interval steps of 100 nm. The modeled hemispheres were arranged in a hexagonal pattern, matching that of the experimental device. Throughout optimization, the spacing of the x and y boundaries was adjusted such that the pitch of the simulated hemisphere pattern was equal to the hemisphere diameter, creating a close packed array. For each parameter tested,

broadband reflectance result was acquired and total weighted absorption was then calculated. Contour plots showing the resulting weighted absorption for both moth-eye and DLARCs parameter sweeps are shown in Figure 3 (a) and (b) respectively.

$$A = 1 - R = 1 - \frac{\sum_{400}^{1100}(R(\lambda) * AM1.5(\lambda))}{\sum_{400}^{1100} AM1.5(\lambda)} \quad (1)$$

Where A is weighted absorption; R is weighted reflectance; $R(\lambda)$ is reflectance at wavelength λ; AM1.5 spectrum is taken from tabulated values of the ASTM G173-03.

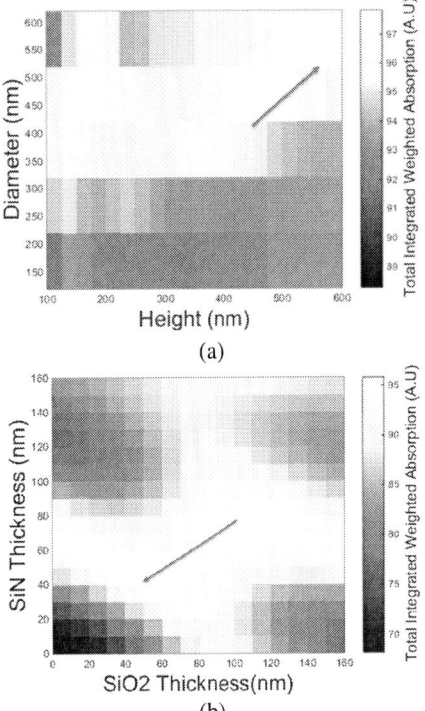

Fig.3 Contour plot for total weighted absorption as a function of diameter and height of moth-eye structures and DLARCs thickness. Blue arrows denote parameters that produced highest weighted absorption. For moth-eye structures (a), base diameter of 520 nm with cone height of 550 nm gives the best result (absorption of 97.8%). For DLARCs (b), 40 nm SiN_x with 50 nm SiO_2 gives the highest absorption (95.8%).

Figure 2 also shows broadband reflectance results for the optimized DLARCs and moth-eye structures. The optimized DLARCs design of 50 nm SiO_2 with 40 nm SiN_x gives the best weighted absorption result of 95.8% and base diameter of 520 nm with height of 550nm gives the best moth-eye structures with weighted absorption of 97.8% from 400-1100 nm. An absolute reflectance decrease of 9.7% for the optimized moth-eye structure is shown compared with optimized DLARC from 770-1100 nm, which is within the absorption range of the current-limiting SiGe bottom cell. More light will then be absorbed in the device, which will lead to an increase of cell current. Fabrication of this optimized moth-eye geometry can be achieved by tuning parameters of the template used for plasma etching.

IV. CONCLUSIONS

In summary, a simple, cost-effective approach to fabricate sub-wavelength structured Ta_2O_5 moth-eye ARCs on dual junction solar cells on Si was developed. This type of ARCs shows excellent broadband antireflective performance: an average reflectance of 7% was achieved from 400-1100 nm. FDTD simulations were used to optically optimize moth-eye structures and traditional DLARCs. With simulated parameters sweeps, further optimization of the moth-eye pattern geometry can achieve reflection as low as 2.2% from 400-1100 nm. It shows that optimized moth-eye structures outperform traditional DLARCs in longer wavelength from 550-1100 nm. Future work will focus on testing antireflective performances of moth-eye structures at different incident angles and improving tandem cells performances by patterning Ta_2O_5 ARCs.

ACKNOWLEDGEMENT

This work has been supported by Australian Renewable Energy Agency (ARENA). Part of this work was carried out at Nanofabrication facility in University of California, Santa Barbara, supported by the National Science Foundation and the National Nanofabrication Infrastructure Network (NNIN).

REFERENCES

[1] R. R. King, D. C. Law, K. M. Edmondson, C. M. Fetzer, G. S. Kinsey, H. Yoon, R. A. Sherif and N. H. Karam, "40% efficient metamorphic GaInP/GaInAs/Ge multijunction solar cells," *Appl. Phys. Lett.*, vol. 90, pp. 183516, 2007.

[2] W. Guter, J. Schone, S. Philipps, M. Steiner, G. Siefer, A. Wekkeli, E. Welser, E. Oliva, A. Bett and F. Dimroth "Current-matched triple-junction solar cell reaching 41.1% conversion efficiency under concentrated sunlight," *Appl. Phys. Lett.*, vol. 94, pp. 223504, 2009.

[3] Fraunhofer ISE, "Photovoltaics Report," 2015.

[4] K. J. Schmieder, A. Gerger, M. Diza, Z. Pulwin, C. Ebert, A. Lochtefeld, R. L. Opila and A. Barnett "Analysis of tandem III-V/SiGe devices grown on Si," in *38th IEEE Photovoltaic Specialists Conference*, 2012, pp. 968–973.

[5] M. Diaz, L. Wang, D. Li, X. Zhao, B. Conrad, A. Soeriyadi, A. Gerger, A. Lochtefeld, C. Ebert, R. L. Opila, I. Perez-Wurfl and A. Barnett "Tandem GaAsP/SiGe on Si solar cells," *Sol. Energy Mater. Sol. Cells*, vol. 143, pp. 113–119, 2015.

[6] D. J. Aiken, "Antireflection coating design for series interconnected multi-junction solar cells," *Prog. Photovoltaics Res. Appl.*, vol. 8, no. 6, pp. 563–570, 2000.

[7] S. Saylan, T. Milakovich, S. A. Hadi, A. Nayfeh, E. A. Fitzgerald and M. S. Dahlem, "Multilayer antireflection coating design for $GaAs_{0.69}P_{0.31}$/Si dual-junction solar cells," *Solar Energy*, vol. 122, pp. 76-86, 2015.

[8] S. L. Diedenhofen, G. Grzela, E. Haverkamp, G. Bauhuis, J. Schermer and J. G. Rivas, "Broadband and omnidirectional anti-reflection layer for III/V multi-junction solar cells," *Solar Energy Mater. Solar Cells*, vol. 101, pp. 308, 2012.

[9] E. E. Perl, W. E. McMahon, J. E. Bowers and D. J. Friedman, "Design of antireflective nanostructures and optical coatings for next-generation multijunction photovoltaic devices," *Opt. Express*, vol. 22, no. S5, pp. A1243–A1256, 2014.

[10] Y. M. Song, S. J. Jang, J. S. Yu and Y. T. Lee, "Bioinspired Parabola Subwavelength Structures for Improved Broadband Antireflection," *Small* vol. 6, pp. 984, 2010.

[11] S. A. Boden and D. M. Bagnall, "Tunable reflection minima of nanostructured antireflective surfaces" *Appl. Phys. Lett.*, vol. 93, pp. 133108, 2008.

[12] Q. Chen, G. Hubbard, P. A. Shields, C. Liu, D. W. E. Allsopp, W. N. Wang and S. Abbott, "Broadband moth-eye antireflection coatings fabricated by low-cost nanoimprinting" *Appl. Phys. Lett.* vol. 94, pp. 263118, 2009.

[13] K. J. Schmieder, "III-V/SiGe Tandem Solar Cells on Si Substrates," University of Delaware, 2013.

[14] K. Yee, "Numerical solution of initial boundary value problems involving maxwell's equations in isotropic media" *IEEE Transactions on Antennas and Propagation*, vol.14, pp.302-307, 1966.

[15] B. Conrad, T. Zhang, A. Lochtefeld, A. Gerger, C. Ebert, M. Diaz, L. Wang, I. Perez-Wurf and A. Barnett "Double layer antireflection coating and window optimization for GaAsP/SiGe tandem on Si," in *40th IEEE Photovoltaic Specialists Conference*, 2014, pp.1143–1147.

Carrier transport in polycrystalline silicon at high optical injection: transient photoconductance vs. numerical modeling

Uchechi Anyanwu, Christian Harris, Andrey Semichaevsky

Department of Chemistry, Physics, and Engineering, Lincoln University, Pennsylvania, 19352, USA

Abstract — This paper discusses our experimental results obtained for several p-type polycrystalline Si wafers using transient photoconductance decay method in combination with a numerical model that accounts for the carrier density dependence of recombination lifetime. Using peak illumination intensities of up to 65 Suns, we are able to estimate the recombination rates for the various carrier recombination mechanisms separately. The model with such realistic recombination parameters produces a realistic diffusion length of minority carriers in Si at high optical injection. Our approach can be applied to the characterization of semiconductor materials for concentrator photovoltaics as well as to design of optoelectronic devices.

Keywords — Silicon, non-radiative recombination, high optical injection, transient photoconductance decay, diffusion-reaction equation.

I. INTRODUCTION

Photovoltaic technologies that reduce the demand for device fabrication and material processing, such as concentrator photovoltaics, have become popular in recent years [1], [2]. Silicon is widely used in solar cell manufacturing, including solar cells that work under concentrated sunlight.

Carrier recombination in Si at room temperature occurs to a large extent due to non-radiative (NR) mechanisms, since the energy bandgap for this material is much larger than the thermal energy, kT at operating temperatures.

Defects in silicon, such as interstitial metal atoms, for instance, Fe and Ti, are known to facilitate charge carrier trapping that can result in the increased non-radiative recombination [1], [2]. When polycrystalline silicon is used to fabricate solar cells, this trap-assisted recombination can reduce the cell fill factors at high optical injection [2], [3].

Previously, carrier density- and irradiance-dependent recombination processes in polycrystalline silicon were studied experimentally [1], [3], [4], [5] and also assessed using computer models of carrier transport [6]. As our previous results [3] may suggest, the shunt resistances found experimentally for some PV cells decrease rapidly as the optical injection increases, which may suggest that some of the recombination processes may involve more than two charge carriers, as in the case of Auger recombination [5].

In this study, we have combined the transient photoconductance (TPC) experiments with computational modeling using experimental carrier lifetimes, in order to explain the recombination processes in polycrystalline silicon at high injection levels.

II. THEORETICAL BACKGROUND

For a semiconductor wafer illuminated by pulsed light, as in TPC, one can describe the carrier transport by a set of diffusion-generation-recombination equations (1a) and (1b) as follows:

$$\frac{\partial n}{\partial t} = G(t) - R(t) + \nabla \cdot [-\mu_n n \nabla \Phi + \frac{\mu_n kT}{e} \nabla n] \quad (1a)$$

$$\frac{\partial p}{\partial t} = G(t) - R(t) - \nabla \cdot [-\mu_p p \nabla \Phi - \frac{\mu_p kT}{e} \nabla p], \quad (1b)$$

where n and p are electron and hole concentrations, μ_n and μ_p are the electron and hole mobilities, e is the electron charge, Φ is the electric potential, k is the Boltzmann's constant, T is temperature, $G(t)$ and $R(t)$ are the time-dependent electron-hole pair generation and recombination rates, respectively. Recombination rate due to all mechanisms is given by (1c).

$$R = R_{SRH} + R_{rad} + R_{Auger} \quad (1c)$$

The carrier recombination rate due to the SRH mechanism at high injection can be approximated as:

$$R_{SRH} = \frac{np}{\tau_n p + \tau_p n}, \quad (1d)$$

where τ_n and τ_p are electron and hole lifetimes associated with this carrier recombination mechanism, and the Auger recombination rate for high injection can be assumed to be proportional the third power of injected carrier concentration:

$$R_{Auger} = (C_n n + C_p p)np, \quad (1e)$$

where C_n and C_p are Auger coefficients for electrons and holes in silicon, respectively.

For our numerical model, the parameters associated with the carrier generation and with SRH (1d) and Auger (1e) recombination can be found experimentally.

For practical purposes, for p-type Si samples, in the absence of applied electric field, equations (1a) and (1b) can be simplified to equation (2) that describes the transport of minority carriers (electrons):

$$\frac{\partial n}{\partial t} = G(t) - \frac{n}{\tau(n)} + \frac{\mu_n kT}{e} \frac{\partial^2 n}{\partial r^2}, \quad (2)$$

where r is a spatial coordinate.

978-1-5090-5606-4/17 $31.00 © 2017 IEEE

III. EXPERIMENTAL RESULTS

Two sets of polycrystalline p-type Si wafers were characterized using the WCT-120 carrier lifetime tester from Sinton, Inc. [7]. This instrument is capable of producing optical pulses of up to 65 Suns in intensity and of conducting time-dependent carrier transport measurements. Using the raw data to describe time-dependent irradiance, we have estimated density-dependent minority carrier (electron) lifetimes. Optical pulse FWHM duration was set at 5.5 ms, and a flash lamp with a broad light spectrum that covers the absorption wavelength range of silicon was used. One set of wafers consists of samples that come from the same batch, and the other set has multiple wafers that were made by different companies under different conditions.

Figure 1 shows the experimental carrier lifetime dependences on injected carrier concentrations for a set of three p-type polycrystalline Si wafers that come from the same batch, provided by Solar World, at T=290 K.

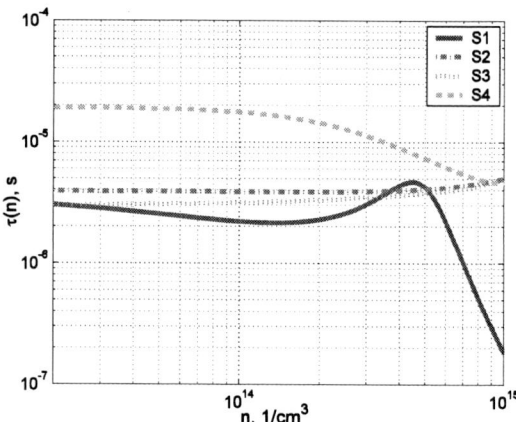

Fig. 2. Electron recombination lifetimes as functions of optically-injected density n, for a set of polycrystalline Si wafers from multiple batches.

As one can see from Figure 2, the electron lifetimes increase slightly with the carrier density and irradiance in samples S2 and S3, while in sample S4, the lifetime decreases (recombination increases). Sample S1, on the other hand exhibits a lifetime dependence with a maximum at n≈4.5×10^{14} 1/cm^3.

IV. NUMERICAL ANALYSIS

Commonly, the carrier lifetime due to all mechanisms of recombination is approximated [1], [2] as:

$$\frac{1}{\tau} = \frac{1}{\tau_{rad}} + \frac{1}{\tau_{SRH}} + \frac{1}{\tau_{Auger}} \approx \frac{1}{\tau_{SRH}} + \frac{1}{\tau_{Auger}}, \quad (3)$$

where τ_{rad}, τ_{SRH}, and τ_{Auger} are the carrier lifetimes associated with radiative, and non-radiative SRH and Auger recombination mechanisms, respectively. The approximation in (3) holds well due to relatively wide bandgap of Si.

The polynomial fitting for the inverse injection-dependent lifetimes presented in Figures 1, 2 is carried out using equation (4):

$$\frac{1}{\tau(n)} = An^3 + Bn^2 + Cn + D \quad (4)$$

The R^2 coefficient for the fitting of the inverse lifetime by (4) varied among the samples in the range of 0.96-0.99.

Our numerical calculations are carried out using a simplified version of equation (2), assuming that the generation rate is given by (5):

$$G(t) = \frac{G_{max}}{\sqrt{2\pi}\tau} e^{\frac{(t-t_0)^2}{2\tau^2}}, \quad (5)$$

where is G$_{max}$ is the peak generation rate in a pulse, τ is the optical pulse width.

Equation (2) is discretized in accordance with the finite-difference approach, applying Dirichlet boundary conditions

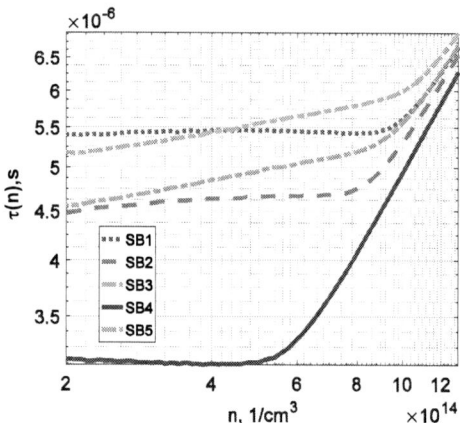

Fig. 1. Electron recombination lifetimes as functions of optically-injected density n, for a set of polycrystalline Si wafers from same batch.

As one can see from Figure 1, the electron lifetime dependences on illumination are very similar among all samples, especially when the electron concentration is around 10^{15} 1/cm^3 or higher. Some differences in low-injection lifetime values, e.g., for sample SB4, may be due to specific microstructure or defect density.

Figure 2 presents electron lifetime dependences on the optically-injected carrier density for the set of polycrystalline silicon wafers provided by Sinton, Inc. and taken from multiple batches.

for the carrier concentration unmodified by the optical injection at the edges of the 1-D domain. A source condition for the photogeneration rate, in accordance with (5) is applied at r=0. The difference equation (2) is solved in space and in time.

Figure 3 shows how the predicted one-dimensional concentration profile changes with position and time when carriers are optically generated over a small area near r=0. Figure 3a presents the calculation results for an assumed carrier lifetime that approaches infinity, and Figures 3 b,c,d present the carrier density for samples S4, S2, and S3 from Figure 2, respectively.

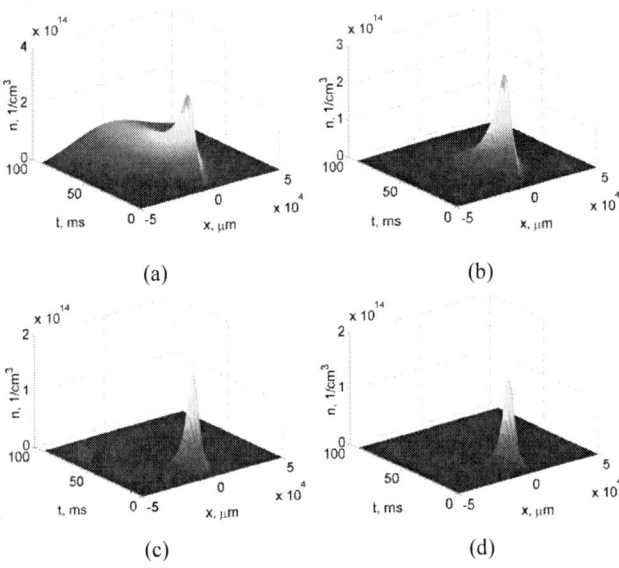

(a)　　　　　　　　　(b)

(c)　　　　　　　　　(d)

Fig. 3. Electron density in p-type Si samples predicted numerically from (2), a) for $\tau \to \infty$, b) for $\tau(n)$ of S4, c) for $\tau(n)$ of S2, d) for $\tau(n)$ of S3 from Figure 2.

As one can see from Figures 3a-d, the NR recombination present in the simulations for samples S2 and S3 shown in Figures 3c and 3d greatly reduces the distance that photoexcited electrons travel by diffusion in poly-Si before they recombine. The electrons in sample S4 have longer lifetimes, so the NR recombination has less effect on carrier transport than for samples S2 and S3 with shorter lifetimes. Using the same approach, one can also evaluate the diffusion length dependence on the optical irradiance.

VI. SUMMARY OF THE WORK

In this paper we present and discuss our transient photoconductance measurements at relatively high peak optical intensities for polycrystalline p-type silicon wafers. This is particularly useful for the development and parameterization of a semiclassical carrier transport model that includes realistic lifetime dependences on optical injection. Therefore, one can consider various NR recombination mechanisms, such as Shockley-Reed-Hall and Auger as well as spatial distribution of recombination lifetimes in simulations. The polycrystalline silicon samples that we included in this study show significant variation in minority lifetime dependences on carrier density. However, this variability is much less remarkable when Si samples come from the same fabrication and processing batch. This can suggest that microstructural properties of polycrystalline material strongly affect recombination at high optical injection levels. Results of diffusion length modeling with experimental electron lifetimes included also reveal big differences in spatiotemporal carrier concentrations among various poly-Si samples.

VII. ACKNOWLEDGEMENT

This work was supported through the NSF-RIA award HRD-1505377.

REFERENCES

[1] D. McDonald, Sinton, R.A., Cuevas A., "Recombination in highly injected silicon," *Journal of Applied Physics*, vol. 89, pp. 2772-2777, 2001.

[2] D. Macdonald and Cuevas A., "Reduced fill factors in multicrystalline silicon solar cells due to injection-level dependent bulk recombination lifetimes," *Progress in Photovoltaics: Research and Applications*, vol. 8, pp. 363-375, 2000.

[3] E. Connell, Semichaevsky, A.,"Degradation of polycrystalline Si solar cell efficiency with increased incident optical power – experiments and theory," *43rd IEEE PVSC*, Portland, OR, 2016.

[4] J. Schmidt, "Temperature- and injection-dependent lifetime spectroscopy for the characterization of defect centers in semiconductors," *Appl. Phys. Lett.*, vol. 82, pp. 2178, 2003.

[5] Pang S.K., A. W. Smith, A. Rohatgi, "Effect of trap location and trap-assisted Auger recombination on silicon solar cell performance," *IEEE Trans. on Electron Devices*, vol. 42, pp. 662-668, 2002.

[6] S. K. Saha, A.M. Farhan, S. I. Reba, et al., "An analytical model of dark saturation current of silicon solar cell considering both SRH and Auger recombination," *IEEE Regional Symposium on Micro- and Nanoelectronics (RSM)*, Malaysia, 2011.

[7] Sinton, Inc., "WCT-120 lifetime tester manual", 2016.

Improving Silicon Surface Passivation with a Silicon Oxide Layer Grown via Ozonated Deionized Water

Sara Bakhshi[1,2], Ngwe Zin[1,2], Kristopher O. Davis[1,2], Marshall Wilson[3], Ismail Kashkoush[4], and Winston V. Schoenfeld[1,2]

University of Central Florida, Orlando, Florida, 32816, United States

Florida Solar Energy Center, Orlando, Florida, United States

SemilabSDI, Tampa, Florida, 33617, United States

Akrion Systems, Allentown, PA

Abstract — Passivation quality of silicon nitride (SiNx), aluminum oxide (AlOx) and a stack of AlOx/SiNx has been investigated in the presence and absence of a thin silicon oxide (SiOx) layer formed using ozonated deionized water. Lifetime measurements show ≈3ms effective carrier lifetime (τeff) for the stack of AlOx/SiNx in the presence of the oxide. Low saturation current density (Jo) and interfacial trap density (D_{it}) confirm and explain the high τ_{eff} for this sample. The stack of AlOx/SiNx can offer excellent surface passivation because of both field-effect passivation and chemical passivation. Note that the presence of oxide also shows a crucial impact in achieving a good passivation.

I. INTRODUCTION

Excellent surface passivation is essential to improve the efficiency of crystalline silicon (c-Si) solar cells. A dielectric passivation layer at the front and rear of the cell brings higher performance due to lower surface recombination velocities [1-5].Good chemical passivation is achieved by the saturation of dangling bonds resulting in a decrease in interfacial trap densities (D_{it}). While field effect passivation decreases the surface concentration of minority carriers by repelling electrons (in p-type surfaces) or holes (in n-type surfaces) with a built in electric field coming from electrostatic charges (Q_f) near the silicon surface.

Silicon nitride (SiN$_x$) is the industry standard for antireflection coatings (ARC) in c-Si solar cells, due to its appropriate refractive index and limited parasitic optical absorption. Additionally, SiN$_x$ is a very effective passivation layer [6-8]for n-type surfaces. However, it is not very effective for p-type silicon due to the large positive Q_f near the Si/SiN$_x$ interface. This positive Q_f introduces an inversion layer underneath of the SiN$_x$, resulting in parasitic shunting[9]. Aluminum oxide (AlO$_x$) is able to solve this problem by inducing large negative Q_f near the Si/AlO$_x$ interface. Aluminum oxide is a great passivation material for p-type surfaces in high-efficiency solar cells [4, 10],coming from the combination of field effect passivation (i.e., large, negative Q_f) and good chemical surface passivation (i.e., low D_{it}). Several methods of depositing AlO$_x$ have been explored, such as plasma-enhanced chemical vapor deposition(PECVD)[11],

atmospheric pressure chemical vapor deposition (APCVD)[12-13] , reactive sputtering[14],and atomic layer deposition(ALD)[15].It has been demonstrated that ALD AlO$_x$ offers a low D_{it} in the range of 10^{11} eV^{-1}.cm^{-2} at the Si surface and negative Q_f in the 10^{12} -10^{13} cm^{-2} range. The AlO$_x$/SiN$_x$ stack also exhibits a negative fixed charge density [16]. This stack offers several advantages in comparison to the single layer of AlO$_x$ or SiN$_x$, such as higher chemical and thermal stability as well as enhanced internal reflectivity[17].

Here we try to investigate the passivation qualities of AlO$_x$, SiN$_x$ and the stack of AlO$_x$ /SiN$_x$, as well as the effect of thermal annealing on effective carrier life time improvement. We also have studied how the growth of a thin SiO$_x$ layer, formed using ozonated deionized water, at the Si surface prior to AlO$_x$ deposition influences the effective carrier lifetime (τ_{eff}) and thermal stability.

II. EXPERIMENTAL APPROACH

Symmetrical lifetime samples were fabricated using 5Ω·cm n-type Cz Si wafers with a 200 μm thickness. All wafers received an RCA clean, followed by a dip in hydrofluoric (HF) acid, and then the growth of a ≈1-2 nm SiO$_x$ layer using ozonated deionized water. Passivation layers were then deposited directly onto a group of the SiO$_x$ coated wafers. Other wafers went through an additional HF dip and ozonated deionized water rinse just before the deposition of the passivation layer(s).

For the passivation layers, three different experimental groups were used: (1) SiN$_x$ only; (2) AlO$_x$; and (3) AlO$_x$/SiN$_x$ stack. For the SiN$_x$ group, an 80 nm layer of SiN$_x$ was deposited using PECVD at a temperature of 300°C with a SiH$_4$/NH$_3$/N$_2$ratio of 250/5/25. For the AlO$_x$ group, ALD (Cambridge Savanah 100) was used to deposit15 nm Al$_2$O$_3$ at a temperature of 200°C. This process is split up into two self-limiting reactions consisting of trimethylaluminium (TMA) exposure until it is saturated. The deposition rate is only0.09 nm/cycle and a very thin layer of AlO$_x$ can be achieved using

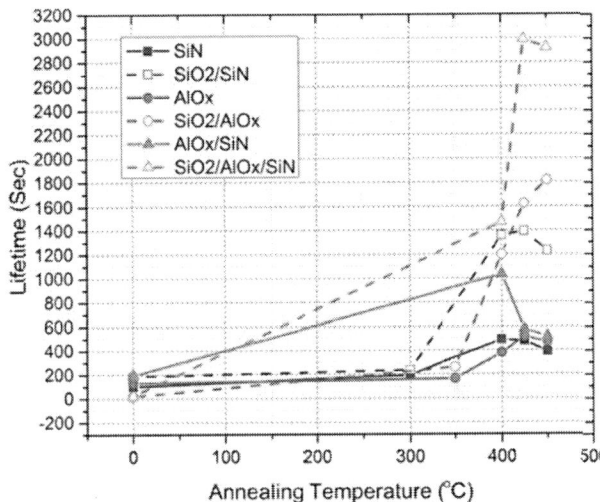

Fig.1. Carrier effective lifetime versus annealing temperature

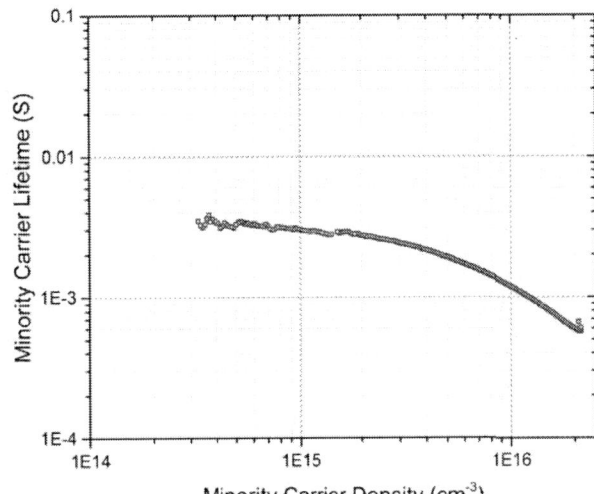

Fig.2.Carrier effective lifetime versus injection level

this technique. The AlO_x/SiN_x stack was fabricated using the same processes described above sequentially.

The samples were annealed at various temperatures (300-450°C) in a tube furnace with N_2 ambient for 30 minutes. The injection-level dependent τ_{eff} was measured at each step using a Sinton Instruments WCT-120. The highest τ_{eff} and lowest saturation current density (J_o) were achieved with either the 425°C or 450° C anneal for different samples.

The electrical properties and surface passivation of the samples were measured by means of corona-voltage measurements (SemilabSDI PV-2000). Corona charging is a contactless method which induces an electric charge on the silicon surface to change the electric field in the dielectric film and the semiconductor itself. The surface band bending on silicon wafer will be varied and thus the flat-band voltage (V_{fb}) can be attained when there is no band bending observed [18-20].

III. RESULT AND DISCUSSION

It can be observed in figure 1 that passivation quality is enhanced upon annealing, leading to a higher τ_{eff}. Thermal annealing under certain condition is critical for full activation of field-effect passivation. This can be attributed to local reconstruction of the AlO_x which increases the negative built-

Table I
Effective carrier life time and current density for 6 various samples

	$\tau_{eff}(\mu Sec)$	Jo ($10^{15}A/cm^2$)
SiN_x	490	77
SiO_x/SiN_x	1391	31
AlO_x	516	44
$SiOx/AlO_x$	1818	20
$AlOx/SiN_x$	1033	19
$SiO_x/AlOx/SiN_x$	2991	16

in potential and hydrogen diffusion from SiN_x or AlO_x bulk to the interface and provides chemical passivation.

The result shows that all samples that have the ozonated SiO_x provide a τ_{eff} that is more than two times higher than the same wafers where the oxide has been removed. Hence, the

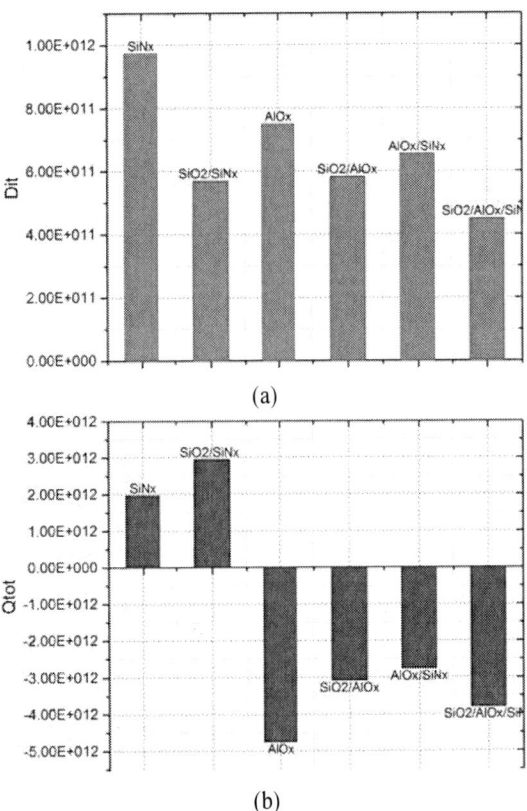

Fig. 3. (a) Interfacial state density (b) total charge for all 6 samples

4.49×10^{11} cm^{-2} eV^{-1} and confirms the effective carrier lifetime results.

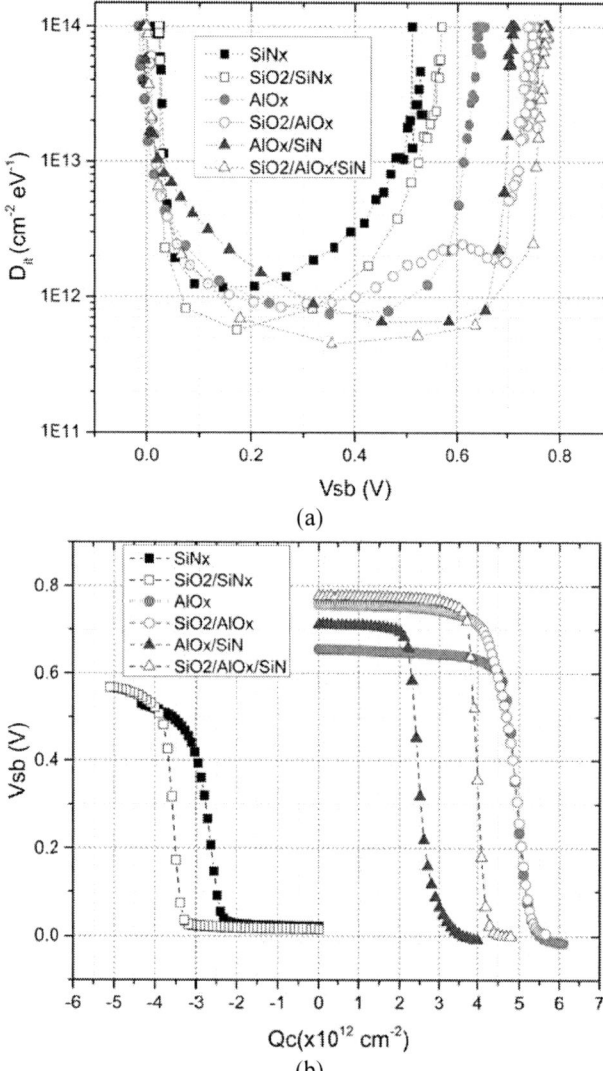

Fig. 4. Interfacial state density versus substrate voltage (a), and charge versus substrate voltage (b) for 6 different samples.

stack of SiO$_x$/ AlO$_x$ /SiN$_x$ with ~3 ms of τ_{eff} and 16.3×10^{15} A/cm^2 of J_0 depicts the best results in this contribution. The effective life time versus the injection level for this sample can be seen in figure 2. It depicted the effective life time is high and almost independent of the injection level.

Figure 3 is a comparison of Q_c and D_{it} for between all samples while figure illustrates the interface state defect density (D_{it}) versus voltage, and charge (Q_c) versus voltage for each sample. As we expected, AlO$_x$ provided the highest amount of negative charge density equal to 4.7×10^{12} cm^{-2}, while the stack of SiO$_x$/ AlO$_x$ /SiN$_x$ has a negative charge density of 3.8×10^{12} cm^{-2}. It is due to the compensation of the negative charge of the AlO$_x$ layer with the positive charge that the SiN$_x$ layer provides. Obviously, this stack brings the lowest interface defect density among these samples which is

IV. CONCLUSION

In this paper we have studied three different passivation techniques, SiNx, AlO$_X$, and the stack of AlO$_X$/SiN$_x$. SiNx deposition was achieved by using PECVD, for the deposition of the AlO$_X$, an ALD system has been utilized. For all three groups, the effect of the presence of the ozonated oxide has been examined. Effective carrier Lifetime and corona-voltage measurement have been performed to investigate the passivation quality and electrical properties of these layers.

The highest τ_{eff} of ~3mSec belongs to the stack of AlO$_X$ /SiN$_x$ in presence of ozonated oxide underneath of the stack. The lowest D_{it} among these 6 samples is 4.49×10^{11} cm^{-2} eV^{-1}, which belongs to the same sample and is compatible with the lifetime results.

REFERENCES

[1] Blakers, A.W., et al., 22.8% efficient silicon solar cell. Applied Physics Letters, 1989. 55(13): p. 1363-1365.

[2] Hoex, B., et al., On the c-Si surface passivation mechanism by the negative-charge-dielectric Al2O3. Journal of Applied Physics, 2008. 104(11): p. 113703.

[3] Schmidt, J., et al., Surface passivation of high-efficiency silicon solar cells by atomic-layer-deposited Al2O3. Progress in Photovoltaics: Research and Applications, 2008. 16(6): p. 461-466.

[4] Schmidt, J., et al., Advances in the Surface Passivation of Silicon Solar Cells. Energy Procedia, 2012. 15: p. 30-39.

[5] Werner, F., et al., Electronic and chemical properties of the c-Si/ Al2O3 interface. Journal of Applied Physics, 2011. 109(11): p. 113701.

[6] Hezel, R. and R. Schörner, Plasma Si nitride—A promising dielectric to achieve high-quality silicon MIS/IL solar cells. Journal of Applied Physics, 1981. 52(4): p. 3076-3079.

[7] Lauinger, T., et al., Record low surface recombination velocities on 1 Ω cm p-silicon using remote plasma silicon nitride passivation. Applied Physics Letters, 1996. 68(9): p. 1232-1234.

[8] Soppe, W., H. Rieffe, and A. Weeber, Bulk and surface passivation of silicon solar cells accomplished by silicon nitride deposited on industrial scale by microwave PECVD. Progress in Photovoltaics: Research and Applications, 2005. 13(7): p. 551-569.

[9] Dauwe, S., et al., Experimental evidence of parasitic shunting in silicon nitride rear surface passivated solar cells. Progress in Photovoltaics: Research and Applications, 2002. 10(4): p. 271-278.

[10] Dingemans, G. and W.M.M. Kessels, Status and prospects of Al2O3-based surface passivation schemes for silicon solar cells. Journal of Vacuum Science & Technology A: Vacuum, Surfaces, and Films, 2012. 30(4): p. 040802.

[11] Töfflinger, J.A., et al., PECVD-AlOx/SiNx Passivation Stacks on Silicon: Effective Charge Dynamics and Interface Defect State Spectroscopy. Energy Procedia, 2014. 55: p. 845-854.

[12] Black LE, McIntosh KR. Surface passivation of c-Si by atmospheric pressure chemical vapor deposition of Al2O3.

Applied Physics Letters 2012; 100: 202107-1-202107-5. DOI: 10.1063/1.4718596.

[13] Davis KO, Jiang K, Wilson M, Demberger C, Zunft H, Haverkamp H, Habermann D, Schoenfeld WV. Influence of precursor gas ratio and firing on silicon surface passivation by APCVD aluminium oxide. physica status solidi (RRL) – Rapid Research Letters 2013; 7: 942-945. DOI: 10.1002/pssr.201308092.

[14] Li, T.T. and A. Cuevas, Effective surface passivation of crystalline silicon by rf sputtered aluminum oxide. physica status solidi (RRL) - Rapid Research Letters, 2009. 3(5): p. 160-162.

[15] Hoex, B., et al., Ultralow surface recombination of c-Si substrates passivated by plasma-assisted atomic layer deposited Al2O3. Applied Physics Letters, 2006. 89(4): p. 042112.

[16] Mack, S., et al., Silicon Surface Passivation by Thin Thermal Oxide/PECVD Layer Stack Systems. IEEE Journal of Photovoltaics, 2011. 1(2): p. 135-145.

[17] Saint-Cast, P., et al., High-Efficiency c-Si Solar Cells Passivated With ALD and PECVD Aluminum Oxide. IEEE Electron Device Letters, 2010. 31(7): p. 695-697.

[18] Wilson M, D'Amico J, Savtchouk A, Edelman P, Findlay A, Jastrzebski L, Lagowski J, Kis-Szabo K, Korsos F, Toth A, Pap A, Kopecek R, Peter K. Multifunction metrology platform for photovoltaics. 37th IEEE Photovoltaic Specialists Conference, Seattle, WA, 2011; 1748-1753. DOI: 10.1109/PVSC.2011.6186292.

[19] Wilson M, Hameiri Z, Nandakumar N, Duttagupta S. Application of non-contact corona-Kelvin metrology for characterization of PV dielectrics on textured surfaces. Photovoltaic Specialist Conference (PVSC), 2014 IEEE 40th, 2014; 0680-0685. DOI: 10.1109/PVSC.2014.6925012.

[20] Wilson M, Marinskiy D, Byelyayev A, D'Amico J, Findlay A, Jastrzebski L, Lagowski J. The Present Status and Recent Advancements in Corona-Kelvin Non-Contact Electrical Metrology of Dielectrics for IC-Manufacturing. ECS Transactions 2006; 3: 3-24. DOI: 10.1149/1.2355694.

[21]

Deposition of SiOC by plasma-free ultra-low-temperature ALD (ULT-ALD) and its passivation on p-type silicon

Meixi Chen[a], Naoto Noda[b], Raphael Rochat[b], Abhishek Iyer[a], James H. Hack[a],
Changhee Ko[b], Christian Dussarrat[b], Robert L. Opila[a]

[a] Department of Materials Science and Engineering, University of Delaware, Newark, DE, 19716, United States
[b] K.K. Air Liquide Laboratories, 28 Wadai, Tsukuba, Ibaraki, 300-4247, Japan

Abstract—SiOC thin films have been successfully deposited using carbosilane compounds with ultra-low-temperature ALD (ULT-ALD) around 50°C to 70°C. This plasma-free ULT-ALD eliminates the damage to the silicon surfaces due to plasma and high temperature. Silicon substrates with native oxide, as well as hydroxyl terminated silicon, without a pre-cleaning, can be passivated by this method , which greatly shortens the processing time in fabrication.

Lifetime tests show that the SiOC film deposited by precursors R#2 (Air Liquide proprietary) and a BTCSM precursor can passivate the p-Si substrates giving a lifetime of up to 349μs. XPS and FTIR analysis shows that the film is approximately 10%C, 40%O and 50%Si with trace amounts of N and Cl. Most of the films' lifetime performances follow an exponential decay with time when exposed to air. Some of the relatively thinner samples do have good stability in that they can maintain 80% passivation performance for two days. Moreover, an increase in lifetime is observed when samples are heated to 100°C, which might be related to hydrogen migration in the film.

Index Terms—Plasma-free, ultra-low-temperature ALD, Silicon passivation

Fig. 1. PREC structure

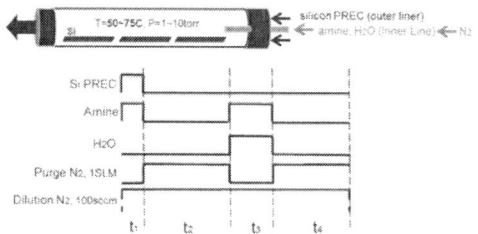

Fig. 2. The home-built ALD reaction system and gas delivery sequences

I. INTRODUCTION

Effective surface passivation has become one of the biggest challenges in making thinner and more efficient solar cells nowadays. The conventional passivation layer is silicon oxide or silicon nitride. Silicon oxide is normally thermally grown, and it requires a high thermal budget using temperatures greater than 1000°C. PECVD is widely used for silicon nitride deposition, but one disadvantage is plasma induced surface damage, which can cause a strong decrease of lifetime due to the defects created on the surface [1]-[5]. Hence, passivation materials are needed that can be deposited without plasma and at relatively low temperatures.

In this work, we deposited SiOC films using carbosilane compound precursors with plasma-free ultra-low-temperature ALD. The elimination of plasma and high temperature ensures the high quality of the substrate surface and also greatly decreases the thermal budget. Composition and bonding of the SiOC films are studied using XPS and FTIR. The surface passivation by the SiOC film, as well as its in-air and thermal stability are also studied.

II. EXPERIMENT

Substrates used in this work are p-type Silicon (100) Cz wafers, with a resistivity of 1-100 Ω-cm, and thickness of 700-750 μm. Two types of surface treatments were performed

prior to ALD: Hydrogen termination is obtained by 1 wt% HF, and OH termination by SPFM solution (details described in reference [6]). Samples were then loaded into the reactor immediately for the SiOC deposition. Three precursors (PRECs) were deposited in ULT-ALD in this work: BTCSM and BDCSM (Gelest 95% and Aldrich 97%, respectively) and R#2 shown in Fig. 1, among which R#2 is a novel, proprietary compound synthesized in Air Liquide Laboratories.

The home-built ALD system and the gas supply sequences are shown in Fig. 2. The silicon precursors and water are introduced into the reactor in sequence with a precursor to water ratio around 1:15. Amines (pyridine or trimethylamine), introduced with the precursor serve as the catalyst for both precursor and water steps. The reactants are supplied through bubblers where the flow rates of each gas supply line are controlled separately by the flow rate and temperature of the nitrogen bubbling gas. Deposition rate of R#2, BTCSM and BDCSM were found to be around 0.7Å/min, 1.5Å/min and 3Å/min respectively, under these conditions.

After deposition, samples were loaded into an X-ray photoelectron spectrometer (XPS, Thermo scientific K-alpha) or FT-IR spectrometer (Thermo Fisher Nicolet) immediately. Lifetime data was acquired with a Sinton WCT-120 Lifetime tester at MCD=1x10^{14} cm^{-3} using quasi-steady-state conductance decay (QSSCD). The thickness of the SiOC film is measured

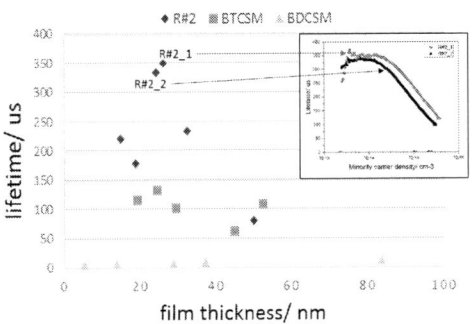

Fig. 3. ALD mechanism

by an ellipsometer (Semilab, SE-2000).

III. RESULTS AND DISCUSSION

A. Influence of surface treatment prior to ALD and its effect on deposition mechanism

It is essential to study different surface features since both ALD and passivation are surface sensitive processes [7]. Three surfaces are studied in this work, hydrogen terminated Si (H-Si), hydroxyl terminated Si (OH-Si) and silicon with native oxide surfaces.

BTCSM was deposited on these three types of wafers using the same ALD conditions. It was found that the deposition rate for H-Si substrates is 15% lower than for OH-Si and Si-native oxide substrates. This can be explained by the proposed mechanism in Fig. 3. The oxygen on the substrate surfaces is the active site that bonds with silicon precursors to deposit the first monolayer, with amine as the catalyst; as water exposure follows, more chlorine is substituted by oxygen creating reaction sites for the next layer deposition. BDCSM and R#2 presumably follow a similar mechanism based on the precursor structures. For the rest of this paper, the OH-Si surface is used, unless otherwise specified. The activity of Si-native oxide surfaces in this ULT-ALD method indicates that, in production, wafers can be loaded directly into the fabrication line without a surface precleaning, which will greatly shorten the processing time.

Lifetime tests show that the films deposited on both OH-Si and Si native oxide substrates can passivate the surface with an increase in lifetime from less than $20\mu s$ to around $120\mu s$, while BTCSM films deposited on H-Si do not passivate the surface. It needs to be noted that the SiOC depositions here have not been optimized for passivation, and it will be shown in the following sections that the longest carrier lifetime observed was $349\mu s$.

B. Overall lifetime performances

The deposition conditions for R#2 and BTCSM are optimized, while BDCSM is more of a CVD-like process, and is still under study. It can be seen from Fig. 4 that, comparing with bare silicon, which has a lifetime under $20\mu s$, the SiOC film deposited by both R#2 and BTCSM can passivate the p-Si substrates; the film deposited by R#2 with a thickness of 26.2nm, noted as sample R#2_1, gives the highest lifetime of $349\mu s$. The lifetime has a weak dependence on thickness for all samples.

BDCSM might be expected to give a higher lifetime than BTCSM because of a higher hydrogen incorporation in the

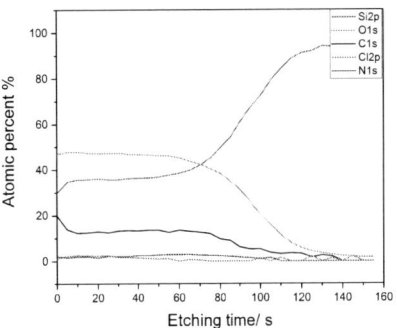

Fig. 4. Lifetime performance of the films deposited by three PRECs as a function of film thickness (nm).

Fig. 5. XPS depth profiles of R#2_1

PREC molecules. However, the opposite results are shown here, and it may be because the deposition conditions for BDCSM are not yet optimized. In this process the PREC vapors are mixed in the gas phase and then deposited without further reaction on the silicon substrates; as a result, the reactive surface defects sites are less likely to be saturated by PREC molecules. FTIR of the deposited film shows the presence of unreacted precursor. (data not shown here)

The study of the composition as a function of depth in Fig. 5 shows that the best passivation sample R#2_1, has around 10%C, 40%O and 50%Si. Si-O, C-O and Si-C bonding are observed in all films from XPS and FTIR.

C. Stability – in-air degradation

The in-air stability of SiOC was examined. R#2 samples were kept in ambient room condition for up to three weeks after deposition. The change in lifetime is recorded in Fig. 6.

Fig. 6 indicates that most films' lifetime follow an exponential decay when the films are exposed to air. Some samples degrade much slower than others; R#2_4($220\mu s$ initially), can maintain 80% performance for 2 days. The degradation rate slows in samples with relatively thin SiOC layers. One hypothesis is that among the thinner samples, oxygen diffuses to the interface more easily, which introduces negative charges and decreases the number of interface state defects [8]-[10].

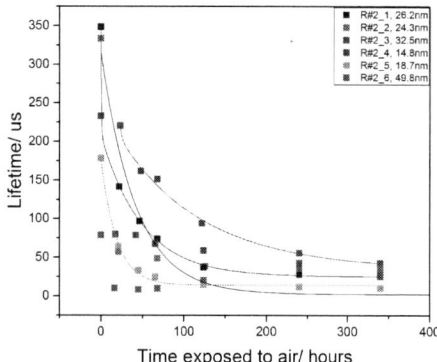

Fig. 6. Lifetime as a function of time in air of R#2 samples

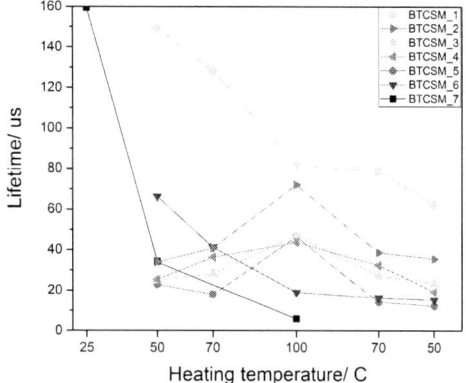

Fig. 7. Lifetime performances in high temperature

D. Stability – heating effect

To characterize the temperature dependence of the degradation, BTSCM samples were heated up to $100°C$ on a Sinton tester (WC-120 TS) equipped with a thermal probe in ambient room condition, and then cooled down. The heating takes place in a period of a few minutes, and the cooling requires several hours. The corresponding lifetimes are shown in Fig. 7.

It can be seen here that the lifetime of four samples increased when heating the films to $100°C$, and decreased when cooled down. One hypothesis is that heating facilitates hydrogen diffusion from dielectric films to the interface, and terminates Si dangling bonds, which decreases the interface state density and thus gives a higher lifetime [8]-[10]. Fig. 7 also indicates that the impact of hydrogen or oxygen is more significant on relatively low lifetime samples; for samples with a lifetime longer than $80\mu s$, the improvement is not strong enough to compensate the oxidation of the passivation layer, and the overall lifetime degrades dramatically at an elevated temperature, like for samples BTSCM_1,6,7.

IV. CONCLUSIONS

SiOC thin films have been successfully deposited by carbosil precursors using ULT-ALD at $50°C$ to $70°C$. The

plasma-free ultra-low temperature process eliminates the damage of silicon surfaces due to plasma and high temperature processing. This ALD can be performed on both OH terminated Si and Si-native oxide surfaces. This demonstrates another advantage of the SiOC film: wafers can be loaded directly into the deposition chamber without a surface cleaning, which will greatly shorten the processing time.

Lifetime tests show that the SiOC film deposited by R#2 and BTCSM can passivate the p-Si substrates; the film deposited by R#2 with a thickness of 26nm gives the highest lifetime of $349\mu s$. Surface analysis shows that the film has around 10%C, 40%O and 50%Si with trace amounts of N and Cl.

Most of the films lifetime performances follow an exponential decay when exposed to air. Some of the relatively thinner samples degrade much slower than other thicker films. This can be explained by the diffusion of oxygen to the interface more easily in thinner samples, which introduces negative charges and lowers the interface state defect density.

Moreover, an increase in lifetime is observed with an elevated storage temperature up to $100°C$, which may be related to hydrogen migration.

Overall, SiOC has been found to be a good alternative in passivating silicon solar cells. This SiOC plasma-free ULT-ALD process can also be used in thin film fabrication where plasma and high temperature are critical factors.

ACKNOWLEDGMENT

This material is based upon work primarily supported by K.K. Air Liquide Laboratories, Japan and the National Science Foundation (NSF) and the Department of Energy (DOE) under NSF CA No. EEC-1041895. Any opinions, findings and conclusions or recommendations expressed in this material are those of the author(s) and do not necessarily reflect those of K.K. Air Liquide Laboratories, NSF or DOE.

REFERENCES

[1] J. R. Elmiger, "Investigations of Silicon Nitride Films for Silicon Solar Cells, *MRS Proceedings, Cambridge University Press*, Vol. 426, , 1996.

[2] J. Schmidt and A. G. Aberle, "Carrier recombination at siliconsilicon nitride interfaces fabricated by plasma-enhanced chemical vapor deposition, *J. Appl. Phys.*, 85, 3626, 1999.

[3] J. Schmidt, F. M. Schuurmans, W. C. Sinke, S. W. Glunz, and A. G. Aberle, "Observation of multiple defect states at siliconsilicon nitride interfaces fabricated by low-frequency plasma-enhanced chemical vapor deposition, *Appl. Phys. Lett.*, 71, 252, 1997.

[4] S. Steingrube, P. P. Altermatt, D. S. Steingrub, J. Schmidt, and R. Brendel, "Interpretation of recombination at c-Si/SiNx interfaces by surface damage, *J. Appl. Phys.*, 108, 014506, 2010.

[5] Tachibana T, Takai D, Kojima T, et al, "Minority Carrier Recombination Properties of Crystalline Defect on Silicon Surface Induced by Plasma Enhanced Chemical Vapor Deposition, *ECS Journal of Solid State Science and Technology*, 5(9), 253-256, 2016.

[6] K. Ljungberg, U. Jansson, S. Bengtsson, A. Sderbrg, "Modification of silicon surfaces with H2SO4:H2O2:HF and HNO3: HF for wafer bonding applications, *J. Electrochem. Soc.*, Vol. 143, No.5, May 1996.

[7] H. Moshe, Y. Mastai, "Atomic Layer Deposition on Self Assembled Monolayers, *Materials Science Advanced Topics*, InTech, 6384, 2013.

[8] C Zhou, J Zhu, et al. "SiOyNx/SiNx stack: a promising surface passivation layer for high-efficiency and potential-induced degradation resistant mc-silicon solar cells, *Prog. Photovolt: Res. Appl.*, 10.1002, 2803, 2016.

[9] S. Keipert-Colberg, N. Barkmann, et al., "Investigation of a PECVD Silicon Oxide/Silicon Nitride Passivation System Concerning Process Influences, *26th EU PVSCE*, 1770, 2011.

[10] J. Schmidt, F. Werner F, B. Veith , et al., "Advances in the surface passivation of silicon solar cells, *Energy Procedia*, 15: 30-39, 2012.

A Method for Quantitatively Investigating the Rear-Side Passivation Performance of PERC cells

Tsung-Cheng Chen[*], Yung-Sheng Lin, Chen-Hao Ku, Ting-Wei Kuo,

Cheng-Shun Hu, and Ching-Chang Wen

E-ton Solar Tech Co., LTD., No.498, Sec. 2, Bentian Rd., Annan Dist., Tainan City 709, Taiwan, R.O.C.

Abstract — **A method that can have insight into the rear-side passivation performance of passivated emitter and real cells (PERC) is crucial for further improving the cell efficiency. In this paper, we propose a method based on high-resolution laser beam induced current (HR-LBIC) for investigating the rear-side passivation performance of PERC cells. Using this method, the performance of local back surface field (LBSF) regions and rear dielectric passivation regions can be distinguished quantitatively, and it helps easily comparing the cells with different rear-side passivation performance. Most importantly, the potential V_{oc} of the PERC cells under current used aluminum paste can be estimated as well.**

I. INTRODUCTION

PERC cells are expected to gain significant market share over traditional full-area aluminum back surface field (Al-BSF) cells for the next few years; meanwhile, the efficiency of PERC cells needs to be continuously improved to sustain its competitiveness for the demand of ever-higher module output power. Several researchers have showed that the top two recombination losses of today's state-of-the-art PERC cells or of maximal simulated efficiency PERC cells are emitter (included front passivation) loss and rear-Al contact loss, respectively [1 – 4], implying there is still room to improve in these two losses.

LBSF quality, which is related to rear-Al contact loss, has been generally analyzed by one of or the combination of the following methodologies: scanning acoustic microscopy (SAM), laser beam induced current (LBIC), scanning electron microscope (SEM), electroluminescence (EL), and photoluminescence (PL) [5 – 7]. These studies mainly focused on investigating the impact of different kind of voids and LBSF thickness on local internal quantum efficiency (IQE) performance. Their results showed that voids do not necessarily exhibit increased recombination compared to filled contacts and the risk for enhanced recombination increases only for BSF thickness below around 1 μm.

Among abovementioned methodologies, LBIC analysis, especially HR-LBIC analysis, is more informative since it can acquire local IQE information which is the final result of local electrical performance affected by local carrier recombination property and local carrier collection probability. Therefore, the performance between rear-Al contact regions (i.e. laser ablation regions or LBSF regions) and passivation regions are easily distinguished by using a long-wavelength HR-LBIC analysis. An example of such analysis is illustrated in Fig. 1.

There are three samples with the same passivation quality but with different LBSF quality which could be caused from different Al pastes or different laser ablations or different co-firing conditions or all. The upper part of Fig.1 shows the IQE mapping of these three samples. The perpendicular lines are the locations of laser ablation lines; the horizontal lines are the locations of front silver fingers. A comparison of IQE line scan (along dash line) of these three samples is also shown at the bottom part of Fig.1. It can be seen that the IQE of passivation regions (plateau regions in line scan) are comparable among samples, but the IQE of LBSF regions (valley regions in line scan) have notable difference. Furthermore, the full width at half maximum (FWHM) of these valleys is a measure of what length the rear-Al contact loss can influence and is related to the diffusion length of minority carrier.

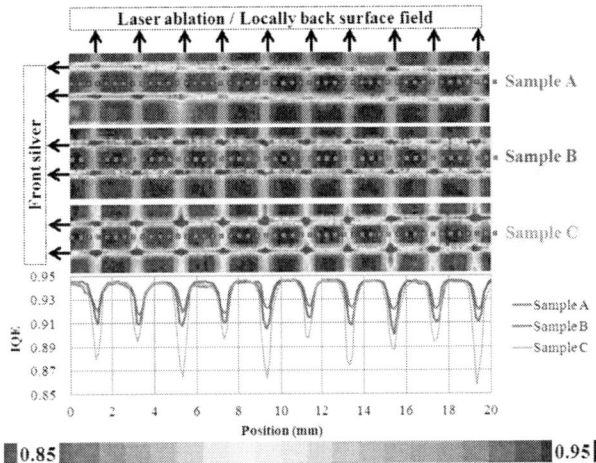

Fig. 1. An example of HR-LBIC analysis. The upper part: the IQE mapping of three samples with the same passivation quality but with different LBSF quality. The bottom part: the corresponding IQE line scan of these three samples.

The potential V_{oc}, the implied V_{oc}, of cell is usually evaluated by quasi-steady-state photoconductance (QSSPC) method [8]. However, the method has some restrictions and is only for a well-passivated and non-metalized sample. The aluminum paste used in PERC cells needs to well anchor to the rear dielectric layer. Inevitably, some damages to the rear dielectric layer happen. Such effect of damage caused by aluminum paste on cell V_{oc} cannot be evaluated directly by

QSSPC method. In this study, a method extended from abovementioned HR-LBIC local IQE analysis was proposed for further quantitatively investigating the correlation between LBSF quality and cell V_{oc} and for estimating the potential V_{oc} of the PERC cells under current used aluminum paste. Please note that the LBIC/QE performance is usually more relative to short-circuit current density (J_{sc}). However, the uncertainty during acquiring I_{sc} (short-circuit current) in a realistic cell IV equipment is around 2 %, which is much higher than that of V_{oc} around 0.5 %. Therefore, we mainly analyzed cell V_{oc} in this study.

II. EXPERIMENTAL

The PERC cells used in this study were produced in our ExcelTon^{TM}-Cell-III mass production line [9]. To demonstrate the impact of LBSF quality on local IQE performance and the corresponding cell V_{oc}, we chose laser ablation and wet deposit of aluminum paste as experimental variables. The detail conditions of the experimental variables are shown in Fig. 2. The experimental cells, i.e. the semi-finished cells before laser ablation and final metallization, were well shuffled into each experimental lot to diminish lot-to-lot and cell-to-cell variance.

Items	Parameters
Laser ablation -dotted line space	25 μm, 50 μm, 64 μm, 120 μm
Wet deposit of advanced rear aluminum paste	1 g/pcs, 2 g/pcs

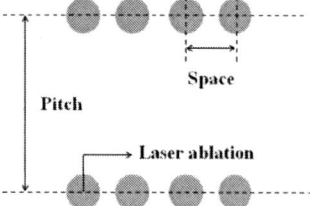

Fig. 2. The detail conditions of the experimental variables and the schematic diagram of laser ablation parameters. Note that the pitch of laser ablation was fixed as 2000 μm in this study.

After finishing the final metallization process, an IV measurement was used to extract the cell electrical parameters. The HR-LBIC measurement of each cell center with 6 mm x 20 mm area was also carried out to acquire IQE mapping data by using Semilab WT-2000. To get the information of rear-side response, the measuring wavelength was set at 1013 nm, and the highest resolution of 62.5 μm was implemented.

III. RESULT AND DISCUSSION

A. Cell V_{oc} versus LBSF quality

The correlation of cell V_{oc} among different experimental variables is shown in Fig. 3. Firstly, the wet deposit of

aluminum paste is a key factor whether optimizing V_{oc} or enlarging the process window of laser ablation is achieved. Because the experimental variables chosen were all from LBSF-related process step, the trend in Fig. 3 can be attributed to the difference of LBSF quality. The phenomenon of lower aluminum wet deposit getting higher V_{oc} (better LBSF quality) is ascribed to that the lower volume of aluminum layer can limit the lower total amount of silicon diffused into aluminum layer during co-fire process, as a result of reducing Kirkendall effect and enhancing LBSF quality [10]. As for laser ablation space, larger space stands for less laser damage and less metal recombination center, theoretically resulting in increasing cell V_{oc}. In fact, larger space makes Kirkendall void ratio higher as well. However, this issue has been gradually improving by the revolution of aluminum paste over the past few years. Currently, the mainstream of laser ablation design is dotted line with laser-spot space smaller than 60-80 μm suggested by our paste vendor, which is consistent with the result of this experiment.

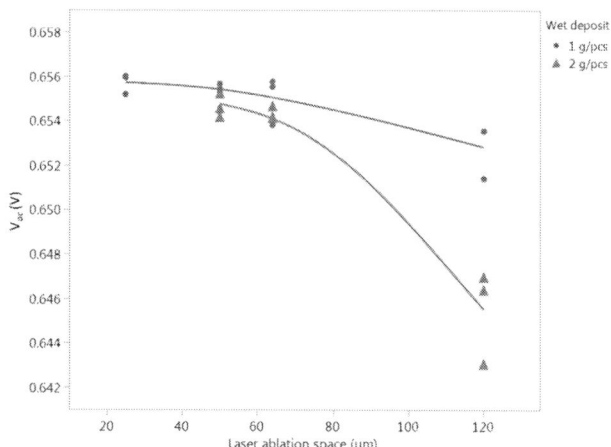

Fig. 3. The cell V_{oc} with different experimental variables.

B. HR-LBIC analysis and results

Refer to IQE line scan in Fig. 1, each valley value is related to LBSF quality and each plateau value represents the passivation quality. To obtain statistically meaningful IQE values of LBSF regions and passivation regions, the HR-LBIC mapping data of each cell was processed using a procedure as illustrated in Fig.4. At step B, we excluded the data of silver finger regions to avoid misleading the analysis. Only 64 rows were remained for all cells. Note that the row width was equal to 62.5μm (resolution of mapping) and 4 rows adjacent to each finger (total in 8 rows per finger) were excluded. At step C, each raw were further divided equally into ten columns denoted as C1 to C10, resulting in 640 bricks per cell. Then, the maximum (represented potential passivation quality) and the minimum (related to LBSF quality) in each brick were calculated, and finally the statistical values (the average and the standard deviation) of those maxima and minima were calculated for representing the IQE performance at rear side, i.e. the potential passivation quality and the LBSF quality of each cell, respectively.

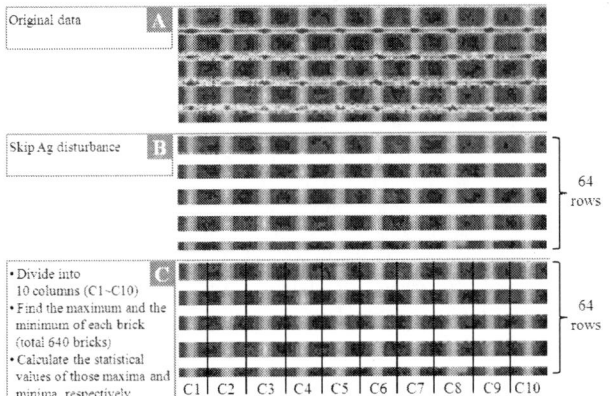

Fig. 4. The process flow from HR-LBIC mapping raw data to the IQE performance of the cell. Note: the maxima and the minima are related to passivation regions and LBSF regions, respectively

The correlation of the IQE performance at rear side of each cell and its corresponding cell V_{oc} are shown in Fig. 5. The average IQE related to potential passivation quality is denoted as $IQE_{Pass.}$ and blue filled circles; the average IQE related to LBSF quality is denoted as IQE_{LBSF} and red filled triangles; the standard deviation is denoted as error bars. As expected, the cell V_{oc} is mainly dominated by IQE_{LBSF}. V_{oc} is a measure of the amount of recombination in the device and can be determined from the carrier concentration [8]:

$$V_{oc} = \frac{kT}{q} \ln \left[\frac{(N_A + \Delta n)\Delta n}{n_i^2} \right] \quad (1)$$

where kT/q is the thermal voltage, N_A is the doping concentration, Δn is the excess carrier concentration and n_i is the intrinsic carrier concentration. Generally, N_A is much greater than Δn, resulting in V_{oc} is proportional to $\ln(\Delta n)$, where Δn is related to IQE performance. Taking the LBSF area ratio into consideration, the normalized IQE of the cell, denoted as $IQE_{Norm.}$, can be written as

$$IQE_{Norm.} = \frac{IQE_{Pass.} \times (1 - f_{LBSF}) + IQE_{LBSF} \times (f_{LBSF})}{IQE_{Pass.}}$$

$$= 1 - f_{LBSF} + \frac{IQE_{LBSF}}{IQE_{Pass.}} \times f_{LBSF} \quad (2)$$

where f_{LBSF} is estimated by the FWHM of LBSF valley in line scan and the number of LBSF lines. Therefore, V_{oc} is a function of $\ln(IQE_{Norm.})$.

Fig. 6 shows that cell V_{oc} is linearly proportional to $\ln(IQE_{Norm.})$ and an empirically linear equation can be obtain. The intercept ($IQE_{Norm.}$ is equal to 1) of the linear equation

represents the potential V_{oc} of the PERC cells under current used aluminum paste. Note that the potential V_{oc} may change as processes change (for example, emitter quality change, bulk quality change, front-side or rear-side passivation quality change, and aluminum paste change), which would result in a change in the potential V_{oc}. In this study, the potential cell V_{oc} was 0.6593 V. As for the slope of the linear equation, represented $\Delta V_{oc}/\Delta IQE_{Norm}$, can be widely used to estimate how much V_{oc} can be improved compared to the potential V_{oc} because of its normalized feature.

We also found that all cells featured plateaus with comparable values in line scan and the FWHM of LBSF valleys was independent of the LBSF quality (around 485 μm), implying the diffusion length of minority carrier of our PERC cells after metallization was limited to 485 μm in this study. The width of LBSF influenced region, i.e. double the FWHM (970 μm), is quiet consistent with the optimized pitch of laser ablation in mainstream PERC design when taking the tradeoff between V_{oc} and FF into consideration.

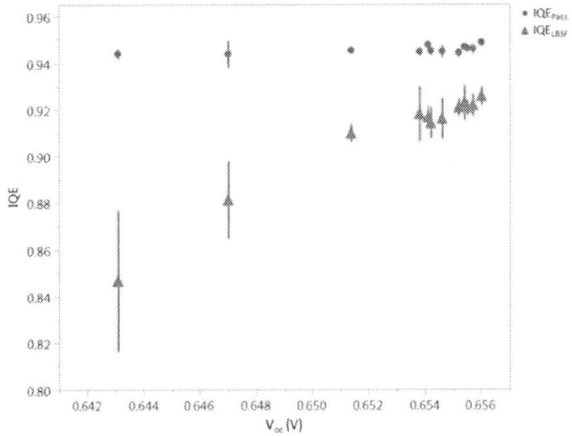

Fig. 5. The correlation of the IQE performance at rear side of each cell and its corresponding cell V_{oc}.

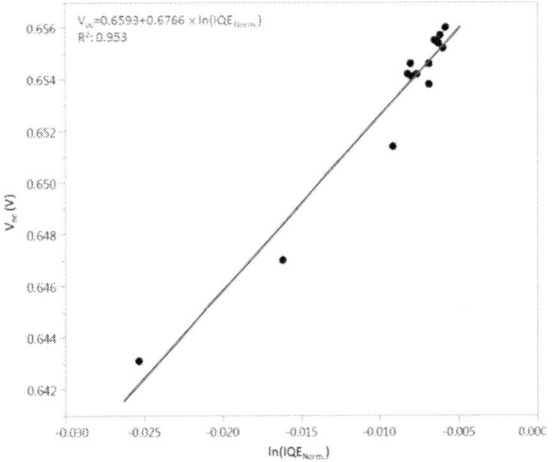

Fig. 6. The correlation of cell V_{oc} and $\ln(IQE_{Norm.})$.

IV. CONCLUSION

In this study, we proposed a method to evaluate the effect of LBSF quality on cell V_{oc} by using HR-LBIC analysis. We demonstrated that an empirically linear equation between cell V_{oc} and $\ln(IQE_{Norm})$ can be obtained and the potential V_{oc} of the studied cells can be estimated by the intercept of this linear equation. This intercept may change as processes change. However, the slope of this linear equation can be widely used because of its normalized feature. In summary, this method has the following advantages.

1. Provide a quantitatively method for comparing the rear-side performance of PERC cells.
2. Provide a method for estimating the potential V_{oc} of the PERC cells under current used aluminum paste and how much V_{oc} can be improved compared to the potential V_{oc}.

REFERENCES

[1] C. Kranz, J. H. Petermann, T. Dullweber, and R. Brendel, "Simulation-Based Efficiency Gain Analysis of 21.2 %-Efficient Screen-Printed PERC Solar Cells," *Energy Procedia*, vol. 92, pp. 109-115, 2016.

[2] D. C. Walter, B. Lim, and J. Schmidt, "Realistic Efficiency Potential of Next-Generation Industrial Czochralski-Grown Silicon Solar Cells after Deactivation of the Boron–Oxygen-Related Defect Center," *Progress in Photovoltaics: Research and Applications*, vol. 24, pp. 920-928, 2016.

[3] B. Min, H. Wagner, M. Muller, H. Neuhaus, R. Brendel, and P. P. Altermatt, "Incremental Efficiency Improvements of Mass-Produced PERC Cells UP to 24 %, Predicted Solely with Continuous Development of Existing Technologies and Wafer Materials," in *31st*

European Photovoltaic Solar Energy Conference and Exhibition, pp. 473-476, 2015.

[4] D. Chen et al., "21.40 % Efficient Large Area Screen Printed Industrial PERC Solar Cell," in *31st European Photovoltaic Solar Energy Conference and Exhibition*, pp. 334-340, 2015.

[5] R. Horbelt, G. Hahn, R. Job, and B. Terheiden, "Void Formation on PERC Solar Cells and Their Impact on the Electrical Cell Parameters Verified by Luminescence and Scanning Acoustic Microscope Measurements," *Energy Procedia*, vol. 84, pp. 47-55, 2015.

[6] S. Großer, S. Swatek, J. Pantzer, M. Turek, and C. Hagendorf, "Quantification of Void Defects on PERC Solar Cell Rear Contacts," *Energy Procedia*, vol. 92, pp. 37-41, 2016.

[7] R. Horbelt, G. Micard, P. Keller, R. Job, G. Hahn, and B. Terheiden, "Surface Recombination Velocity of Local Al-Contacts of PERC Solar Cells Determined from LBIC Measurements and 2D Simulation," *Energy Procedia*, vol. 92, pp. 82-87, 2016.

[8] R. A. Sinton and A. Cuevas, "Contactless Determination of Current-Voltage Characteristics and Minority-Carrier Lifetimes in Semiconductors from Quasi-Steady-State Photoconductance Data," *Applied Physics Letters*, vol. 69, no. 17, pp. 2510-2512, 1996.

[9] T.-cheng Chen et al., "E-Ton's Printed-AlO$_x$ PERC Cells: Efficiencies beyond 21 % with a Next-Generation AlO$_x$ Paste," in *32nd European Photovoltaic Solar Energy Conference and Exhibition*, pp. 680-682, 2016.

[10] E. Urrejola, K. Peter, H. Plagwitz, and G. Schubert, "Understanding and Avoiding the Formation of Voids for Rear Passivated Silicon Solar Cells," in *3rd Workshop on Metallization for Crystalline Silicon Solar Cells*, pp. 1-22, 2011.

Field-Effect Passivation by Negative Charge on Boron Emitter and Boron-doped Surfaces by a Novel Low-cost Plasma Charge Injection

Eunhwan Cho[1], Young-Woo Ok[1], James Hwang[2], Aditi Jain[1] Vijay D. Upadhyaya[1], John Keith Tate[1] and Ajeet Rohatgi[1]

[1]Georgia Institute of Technology, Atlanta, Georgia, 30332, USA
[2]Amtech Systems Inc., Tempe, Arizona, 85281, USA

Abstract — Field effect passivation by negative charge is important for back surface of p-type PERC and front surface of n-type PERT with boron emitter. Currently, Al_2O_3 passivation is widely used for both surfaces due to its negative charge. However, Al_2O_3 tool has high operation cost and safety related issues due to the use of TMA precursor. In this paper, a novel plasma charging tool is used to inject negative charge into the SiN_x/SiO_2 passivation stack to passivate boron emitter and boron-doped silicon surface. Large area bifacial n-type PERT and p-type PERC cells were fabricated with controlled negative charge injection to demonstrate the effectiveness of this technique. Field effect passivation by injecting $1E13cm^{-2}$ negative charge by this plasma charging method gave ~1.35% and ~0.36% increases in absolute cell efficiencies for n-type PERT and p-type PERC cells, respectively. Detailed characterization and analysis show that these improvements are due to enhanced surface passivation resulting from the formation of accumulation layer on p-type surfaces which reduces the SRH recombination and shunting associated with depletion and inversion layers.

Index Terms — Negative charge, charge injection, field-effect passivation, PECVD SiNx, PERT, PERC, silicon solar cell.

I. INTRODUCTION

Dielectric passivation quality is becoming more important to reduce loss mechanisms in the solar cells. For example, currently full Al-BSF solar cells are being replaced by local Al-BSF contact cell, known as passivated emitter rear contact cell (PERC) which reduces back metal contact fraction from 100% to ~6% and increases efficiency by ~1% over full Al-BSF cell due to better rear passivation and reflection. Similarly, n-type passivated emitter rear totally diffused (PERT) solar cells have B-doped emitter and local contacts with dielectric passivation for higher efficiency. Negatively charged Al_2O_3 is widely used in PV industry not only to enhance passivation quality of B-doped surfaces but also to remove shunt mechanisms due to induced electron inversion layer by positive charge present in SiO_2/SiN_x or SiN_x dielectrics. However, Al_2O_3 has high operation cost and safety concern due to the use of TMA precursor. In this paper, a novel plasma charging method has been used to inject negative charge in the SiO_2/SiN_x stack to improve passivation quality by forming an accumulation layer.

II. EXPERIMENT

In this study, large area (239 cm^2) bifacial n-type PERT and p-type PERC cells were fabricated with SiO_2/SiN_x stacks on both surfaces for dielectric passivation as shown in Fig. 1. Then, a novel low-cost plasma tool, invented by Amtech (Patent, US 8,338,211 B2), was applied to inject negative charge in the SiO_2/SiN_x passivated back surface of the p-type PERC and front surface of the n-type PERT cells to enhance passivation quality and increase cell efficiency. Details of the charging tool and C-V measurements are described in the reference [1]. In the p-type PERC cells, negative charge injection transforms inversion layer due to initial positive charge in the dielectric stack to accumulation layer. In the case of n-type PERT, depletion layer in the p+ emitter is transferred to accumulation layer. Thus, in both cases, SRH surface recombination is reduced in addition to the improvement in minority carrier shunting because of significant reduction in minority carrier concentration at the surface.

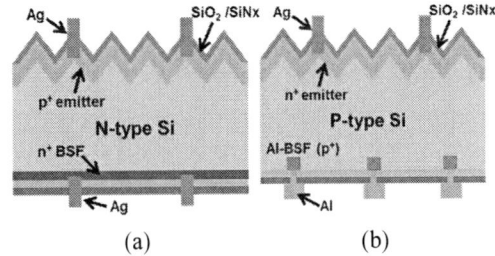

Fig.1 Schematic picture of (a) bifacial n-type PERT and (b) bifacial p-type PERC solar cells.

III. RESULT AND DISCUSSION

A. Field Effect Passivation by Negative Charging on N-type PERT Solar Cells

Table I shows LIV data of n-type PERT cells before and after negative charge injection.

978-1-5090-5606-4/17 $31.00 © 2017 IEEE

TABLE I
MEASURED N-TYPE PERT CELL LIV DATA BEFORE AND AFTER NEGATIVE
CHARGING ON FRONT $SiO_2/SiNx$ STACK($-1E13CM^{-2}$)

Cell Type	AVG of 4	Voc (mV)	Jsc (mA/cm²)	FF (%)	η (%)
n-type PERT Cell	Pre-charging	636	37.9	79.0	19.00 ±0.15
	Post-charging	648	39.5	79.5	20.35 ±0.13

In case of n-type PERT cells, cell efficiency increased by 1.35% absolute after -1E13 cm^{-2} negative charge injection in $SiO_2/SiNx$ stack on the B-doped emitter. The efficiency boost comes from all cell parameters including Voc (+12mV), Jsc (+1.6mA/cm²) and FF (+0.5%) due to enhanced front surface passivation. This is because B-doped emitter surface state is changed from depletion to accumulation. Prior to charge injection, B-doped emitter surface is depleted due to high boron concentration ($\sim$1e19cm^{-3}) and positive charge in the SiO_2/SiN stack. Negative charge injection attracts more holes to the surface and drive it into accumulation. Internal quantum efficiency (IQE) measurements in Fig. 2 show that short wavelength responses (400nm~800nm) increased appreciably after negative charge injection in $SiO_2/SiNx$ stack, supporting that front surface passivation improved by field effect passivation.

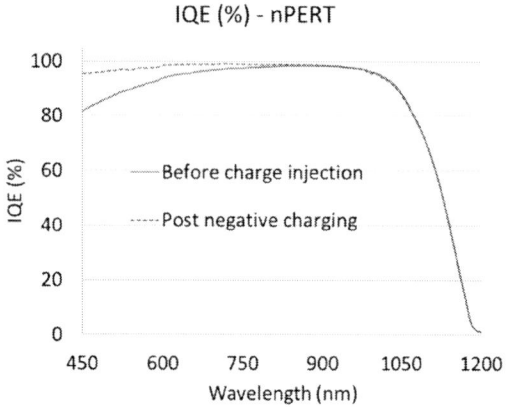

Fig. 2 Measured Internal Quantum Efficiency (IQE) on n-type PERT solar cell before and after charge injection.

This was also further supported by Suns-Voc measurements in Fig. 3, which showed that effective lifetimes increased after negative charge injection at all the injection levels.

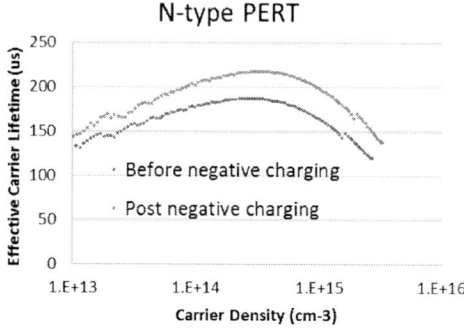

Fig. 3 Measured effective lifetimes before and after negative charge injection (net charge: -1E13 cm^{-2}) as a function of injection level on n-type PERT cell by Sinton's Suns-Voc tool.

B. Field Effect Passivation by Negative Charge in $SiO_2/SiNx$ Passivated P-type Rear Surface of PERC Solar Cells

Table II shows light I-V data of p-type PERC cells before and after $\sim$1e13cm^{-2} negative charge injection in the $SiO_2/SiNx$ stack on the back surface.

In case of p-type PERC cells, cell efficiency increased by

TABLE II
MEASURED P-TYPE PERC CELL LIV DATA BEFORE AND AFTER NEGATIVE
CHARGING ON FRONT $SiO_2/SiNx$ STACK($-1E13CM^{-2}$)

Cell Type	AVG of 3	Voc (mV)	Jsc (mA/cm²)	FF (%)	η (%)
p-type PERC Cell	Pre-charging	656	38.4	78.1	19.68 ±0.07
	Post-charging	659	38.6	78.8	20.04 ±0.04

0.36% absolute due to higher Voc (+3mV), FF (+0.7%) and Jsc (+0.2mA/cm²). Notice that there was a smaller increase in efficiency of p-type PERC compared to the n-type PERT cells. This is because p-PERC back surface is inverted due to low boron doping (1E16 cm^{-3}) and high positive charge ($\sim$5E12cm^{-2}) in the $SiO_2/SiNx$ stack. This inversion layer significantly reduces surface recombination velocity (SRV). Therefore, formation of accumulation layer after negative charge injection has little impact on the SRV and Voc. This can be explained quantitatively on the basis of SRV which is defined by Equation (1) [2, 3]:

$$S_{eff}\left(\frac{cm}{s}\right) = \frac{U_s}{\Delta n}$$

$$U_s\left(\frac{1}{cm^2 \cdot s}\right) = \frac{n_s p_s - n_i^2}{\frac{n_s + n_1}{S_{po}} + \frac{p_s + p_1}{S_{no}}} \quad (1)$$

where S_{eff} is effective SRV, Δn is electron injection level, n_s and p_s are electron and hole concentrations at the surface, n_1 and p_1 are equilibrium electron and hole concentrations, and

S_{n0} and S_{p0} are fundamental surface recombination velocities for electrons and holes, respectively. According to Equation (1) SRV decreases when one type of carrier concentration far exceeds the other type of carrier concentration at the surface and it is maximum when $n_s \approx p_s$ (if $S_{n0} \approx S_{p0}$). Table III shows the calculated SRV for the Si-SiO$_2$ passivated 1.8 ohm-cm Si surface from Equation (1) as a function of charge density (first n_s and p_s values were calculated as a function of surface charge and band bending, and SRV was calculated using Si-SiO$_2$ interface parameters (D_{it}, σ_n, σ_p, n_1, p_1 and E_t) from Glunz et al's paper [4]).

TABLE III
CALCULATED SRV AS A FUNCTION OF NET CHARGE AMOUNT ON SI-SIO$_2$ SURFACE.

Injected Charge Amount (cm^{-2})	-1E13	-1E12	-2E11	0	2E11	1E12	1E13
SRV (cm/s) @ 1E15 cm^{-3} injection level	0.5	45	761	954	11	0.2	1E-3

The SRV calculations in Table III clearly show that injection of ~1E13cm^{-2} positive or negative charge lowers the SRV of 1.8 ohm-cm Si surface below 1 cm/s. Next, we performed device simulation using Sentaurus 2D model, which allows us to place desired amount of charge on the back of the p-PERC cell (Fig. 4). Slight asymmetry of efficiency curve in Fig. 4 is due to different S_{n0} and S_{p0} values from different electron and hole capture cross section values (σ_n, σ_p) on Si-SiO$_2$ surface [5].

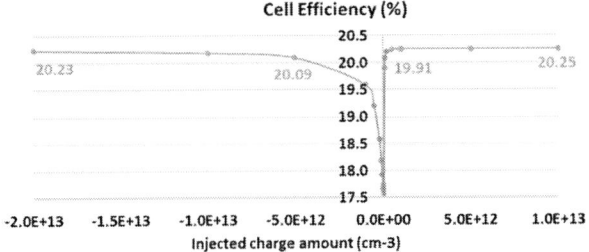

Fig. 4 Sentaurus 2D device simulation on p-type bifacial PERC as a function of net charge amount on back SiO$_2$/SiNx stack.

Table III and device modeling showed that with no surface charge or field effect passivation SRV is ~1000 cm/s and efficiency drops to 17.5% but with $\geqq$5E12 cm^{-2} positive or negative charge efficiency jumps to ~20.2%. Since initial charge in the SiO$_2$/SiNx stack is ~5E12cm^{-2} and post negative charging provided ~1E13 cm^{-2} negative charge, there should be virtually no change in efficiency according to Fig. 4.

However, p-type PERC did show an increase in efficiency by 0.36% after charge injection. This is because negative charge injection removed the minority carrier shunting or leakage at the back surface by creating accumulation layer.

This gave an increase in efficiency. This is supported by IQE measurements with and without the light bias in Fig. 5.

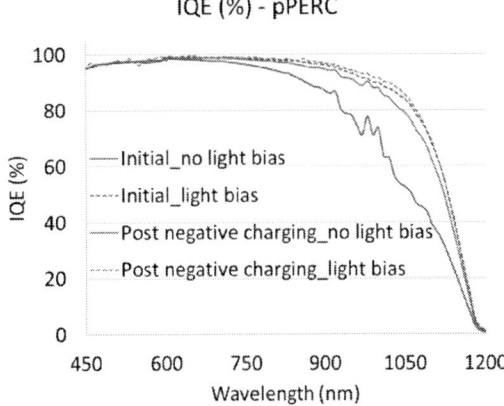

Fig. 5 Measured Internal Quantum Efficiency (IQE) on p-type PERC solar cell before and after charge injection.

Without light-bias IQE response in long wavelength range (650nm~1200nm) increased significantly after negative charging. This increase is because negative charging of PERC back surface switches the inversion layer to accumulation layer, removing parasitic shunt mechanisms between minority carrier inversion layer and metal contact such as pinholes in the positive charged SiO$_2$/SiNx passivation or insufficiently formed back surface field in Al-BSF [6].

IV. CONCLUSION

This paper demonstrated field effect passivation by negative charging on boron emitter (n-type PERT front surface) and boron-doped surface (p-type PERC back surface) compared to as-grown positive fixed charge in SiO$_2$/SiNx passivation. Negative charge injection was performed by a novel low cost plasma charging method developed by Amtech company. Large area (239 cm^2) n-type PERT and p-type PERC cells efficiencies increased by 1.35% and 0.36% absolute, respectively, after 1E13 cm^{-2} negative charge injection in SiO$_2$/SiNx stack. N-type PERT cell efficiency increased due to all LIV parameters including Voc (+12mV), Jsc (+1.6mA/cm^2) and FF (+0.5%). This is because heavily doped boron (~1e19cm^{-3}) emitter surface on n-type PERT cell changed from depletion mode due to positive charge in initial SiO$_2$/SiNx to accumulation mode after negative charging. In case of p-type PERC cell, efficiency increased only by 0.36% absolute due to higher Voc (+3mV), FF (+0.7%) and Jsc (+0.2mA/cm^2). This is because negative charging does not alter passivation quality much but removes parasitic shunt mechanisms in the inversion layer on back surface of p-PERC.

978-1-5090-5606-4/17 $31.00 © 2017 IEEE

ACKNOWLEDGEMENT

The authors would like to thank to all other UCEP group members for their technical support. This program has been supported by the Department of Energy (Award number: DE-EE0007189).

REFERENCES

[1] E. Cho, Y.-W. Ok, J. Hwang, A. D. Upadhyaya, J. K. Tate, F. Zimbardi, *et al.*, "Field-effect passivation by charge injection into SiNx using a novel low-cost plasma charging method," in *Photovoltaic Specialists Conference (PVSC), 2016 IEEE 43rd*, 2016, pp. 2874-2877.

[2] A. G. MARTING, "Operating Principles Technology and System Applications," *Solar Cells,* vol. 93, 1998.

[3] W. Shockley and W. Read Jr, "Statistics of the recombinations of holes and electrons," *Physical review,* vol. 87, p. 835, 1952.

[4] S. Glunz, D. Biro, S. Rein, and W. Warta, "Field-effect passivation of the SiO 2 Si interface," *Journal of Applied Physics,* vol. 86, pp. 683-691, 1999.

[5] R. B. Girisch, R. P. Mertens, and R. De Keersmaecker, "Determination of Si-SiO/sub 2/interface recombination parameters using a gate-controlled point-junction diode under illumination," *IEEE Transactions on Electron Devices,* vol. 35, pp. 203-222, 1988.

[6] A. Lorenz, J. John, B. Vermang, E. Cornagliotti, and J. Poortmans, "Comparison of illumination level dependency and rear internal reflectance of PERC-type cells with different dielectric passivation stacks," in *Proc. 26th Eur. Photovoltaic Sol. Energy Conf,* 2011, pp. 1486-1488.

Industry relevant RIE texturing for mc-Si diamond wire or Direct Wafer® product: optimized reflectivity, uniformity, and throughput

Jose Luis Cruz-Campa, Ray Fraser, Rob Steeman, John Linton

1366 Technologies, Bedford, MA, 01730, USA

Abstract — **Low-cost wafering processes such as diamond wire sawn and Direct Wafer® technology are excellent candidates for emerging texturing techniques. Reactive Ion Etch enables enhanced light-trapping with respect to traditional acid texturing techniques. Simulations predict that narrow pyramids have the lowest reflectance. The experiments in an industrial RIE machine demonstrate that the lowest reflectance etch is not necessarily the most uniform or the fastest etch. Fast-forming texture with low reflectivity and high uniformity were obtained using a ratio of Cl_2 to SF_6 between 0.35 and 0.65, a ratio of O_2 to SF_6 between 1.2 and 1.7, power of 18-22 kW, and pressure of 19-25 Pa.**

I. Introduction

Multicrystalline-silicon based solar panels are the dominant technology worldwide for photovoltaic power. Wafers contribute as much as 40% of the cost to the panel [1]. The need to further reduce the cost of solar modules has given rise to novel wafering techniques for silicon such as Diamond Wire Sawn (DWS) and Direct Wafer® (DW) technology. Direct Wafer products from 1366 Technologies have the advantage of 50% lower production cost and 66% reduction in energy use. These wafers are grown directly from molten silicon rather than diced from silicon ingots. The method enables features such as 3D thickness control and internal doping gradient control [2 , 3] which are impossible through conventional wafer manufacturing. Furthermore DW have certified cell efficiencies (Fraunhofer ISE CalLab) up to 19.9% fabricated with commercial partners [4].

The surfaces of wafers produced with new technologies such as DWS and DW are smoother and less defective than for slurry sawn wafers. Because of this, texturing using conventional methods such as acidic chemical etching (ISO-texture) is not possible. There are several emerging texturing techniques including Metal Assisted Etching (MAE) [5], Atmospheric Dry Etching (ADE) [6], and laser texturing [7]. Reactive Ion Etch (RIE) is a commercially available technology to texture silicon wafers. RIE can boost solar cell performance compared to standard ISO texture. This improved light trapping in the cell can enhance photon collection and electrical current and thus efficiency.

II. Methods: Reactive Ion Etch

RIE relies on a chemically reactive plasma (containing molecules and atoms partially or totally ionized) to both create a mask and to remove material from the wafer surface. Fig. 1a

illustrates how the non-volatile components formed during etching create a polymer mask ($Si_xO_yF_z$ or $Si_xO_yCl_z$) that impedes the etching of certain parts of the wafer. The unmasked areas of the wafer are etched more than the ones with polymer masking. If the plasma composition is right, this process creates pyramid shapes.

The etching of the wafer is affected by multiple variables: gas kinetics, the transport of charged particles, electrodynamics, surface kinetics, ratio of ions in plasma and temperature. In general, increasing O_2 content favors polymer creation while increasing SF_6 increases etching in open areas and decreases polymer formation. The reflectivity is mostly controlled by the pyramid aspect ratio, which in turn is a function of the plasma energy of the ions [8, 9].

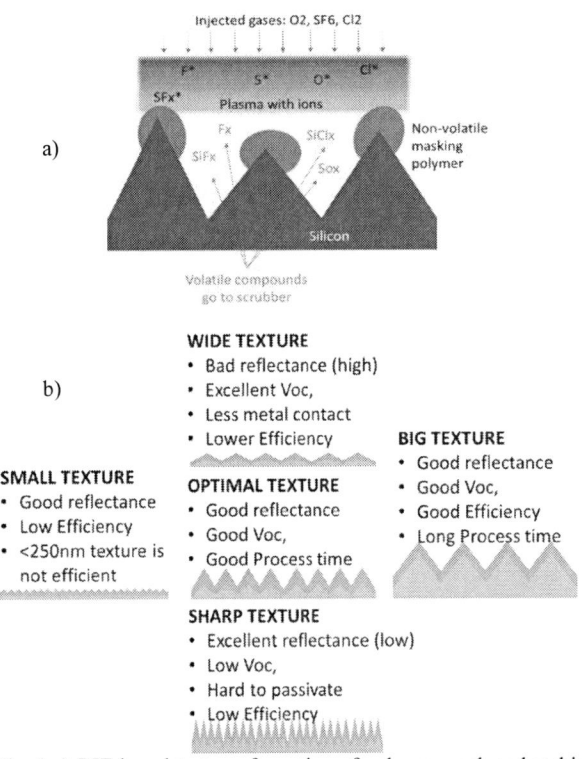

Fig. 1 a). RIE based texture: formation of polymer mask and etching of Silicon by ions in plasma b) Diagram explaining the different implications of size/shape of texture in the performance of solar cells.

The texture in a solar cell is important in several aspects: 1) changes reflection, 2) can change IQE in long wavelength, 3)

978-1-5090-5606-4/17 $31.00 © 2017 IEEE

changes contacting of emitter, 4) creates more surface that enables more nitride to be deposited and thus a higher hydrogen source, 5) depending on size, it modifies the shape of the emitter, 6) it changes the firing profile of the metal/nitride stack and 7) it changes cell to module (CTM) gains. Different textures are available depending on the technique and variables used. Figure 1b shows in a generalized diagram illustrating the benefits and detriments of different shapes/sizes of texture.

III METHODS: CHAMBER AND MEASUREMENT TECHNIQUES

An industrial scale, single chamber, reactive ion etch, silicon texturing machine from Wonik IPS capable of 50 MW/year of production was used for the experiments shown here. Fig. 2 shows a picture of the RIE machine used. The design of this machine is based on Kyocera's design for RIE texturing [10]. One of the key features of this machine is the use of a perforated parallel plate between the top electrode and the product that confines the plasma reaction in order to increase the texturing speed.

Fig. 2. Industrial RIE chamber/machine used for the experiments

DW product wafers with standard size of 156.75 mm × 156.75 mm and an average thickness of 180 μm were used. To calculate the amount of Si etched, the wafers were weighed before and after texturing.

The reflectivity of the wafers was measured using integrating spheres (ISP-REF) and spectrophotometers (USB4000) from Ocean Optics. Each wafer was measured at nine equidistant points through the wafer to obtain an average of the reflectance and the coefficient of variance throughout the wafer in wavelengths from 400-1000 nm. The reflectance reported here is the reflectance at 650 nm.

For scanning electron microscope (SEM) pictures, a JEOL 6510 SEM with a tilt of 30 degrees and a magnification of 20,000 × was used.

IV. SIMULATIONS

Two short simulations using Opal 2 [11] (available on PV Lighthouse) were used to quantify the influence of texture on light reflected on a silicon substrate and in photocurrent in a cell. The simulations shown assume geometric optics and do not consider effects from nano-scale texture.

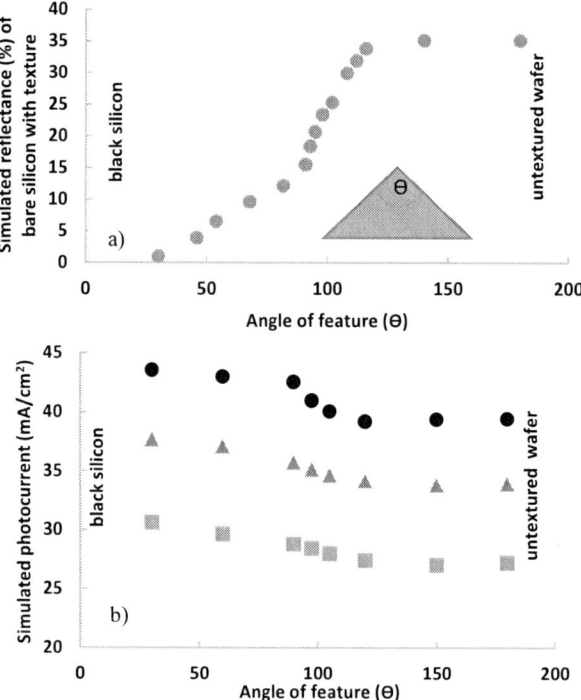

Fig. 3. a) Simulated reflectance of bare silicon b) Simulated photocurrent of a cell. Sharper features are better at achieving lower reflectance and higher photocurrents.

We simulated a 180 μm thick silicon with an incident AM 1.5G spectrum normal to the surface, with a planar fraction of 5% and a max:min height ratio of 5. All other variables are set to default. The angle of the texture was varied from a flat surface to a sharp feature of 30° as defined by the internal angle of the triangle peak. Fig. 3 a) shows the effect of different feature angles on bare silicon substrates on reflectance. Fig. 3 a) also shows that as the internal angle of the texture increases, the reflectance increases saturating past 110° where the light angle inside the silicon is not sharp enough to achieve total internal reflection. Fig. 3 b) shows the simulated photocurrent of a solar cell (with 70 nm of nitride anti-reflection coating) with respect to the angle of the feature. Fig 3. b) also shows that most of the photocurrent is gained from taking the angle from 110° to less than 80°. Past 80° the gains are smaller.

V. EXPERIMENTAL RESULTS, ANALYSIS AND DISCUSSION

Five experiments were conducted: the variables explored were O_2 to SF_6 ratio, power, Cl_2 to SF_6 ratio, pressure, and etch time. The measured outputs were etch rate, reflectivity and uniformity. SEM and an optical pictures are shown for each experiment.

A. Oxygen content dependence

The first experiment explored the behavior of different ratios of O_2 in SF_6. The fixed variables were: etch time (150 seconds), pressure (19 Pa), power (19 kW) and a fixed ratio of $Cl_2:SF_6$ of 3:5 In Fig. 4, the left SEM picture shows the case for a low ratio of O_2 in SF_6 (90%), it can be seen that the texture is very small, rounded and wide. The right picture shows the case for high O_2 content (225% O_2 in SF_6), this picture shows the pyramid shape with sharper angles than at 90% O_2.

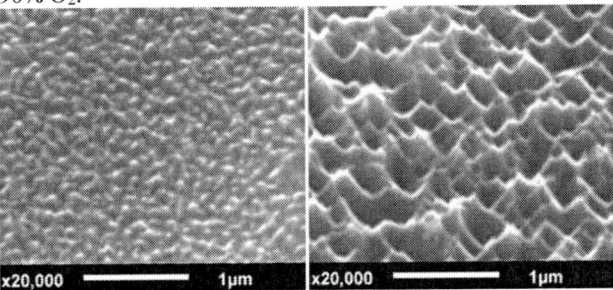

Fig. 4. 30° tilted SEM pictures at 20,000× magnification. Left 90% O_2 in SF_6 right 225% O_2 in SF_6.

Fig. 5 a) shows the relationship between % of O_2 and amount of silicon etched: the etch rate is increased slightly up to a feed gas composition of 25% O_2, then the etch rate diminishes for greater than 25% O_2. Maximum etch rates of 0.42 μm/min were obtained at 25% O_2 in SF_6. Minimum etch rates of to 0.24 μm/min were achieved at 270% O_2 in SF_6.

Other research [12] with different reactors (without the plasma concentration plate) has shown a stronger etch rate dependence with O_2. An important observation is that the internal composition of gasses between the concentration plate and the wafers appears to change the dependence on O_2

Fig. 5 b) shows the relationship between reflectivity for the wafers as the percentage of O_2 increases. Comparing graphs in Fig. 5 a) and 5 b), it can be seen that while the material etches with zero oxygen in the chamber, the texture does not form. At least 100% O_2 content in SF_6 is necessary to start the formation of masking material that will then create the texture.

Fig. 5 c) shows the relationship between non-uniformity within the wafer as the percentage of O_2 increases. Also, Fig. 5 c) shows the shades of the wafers along this process. It can be seen that between 0% and 60% O_2 content the wafers are very uniform. However, Fig. 5b), it can be observed that texture does not form with those concentrations. Concentrations of O_2 between 60% and 120% yield highly non-uniform textures. Concentrations >120% of O_2 are recommended for better uniformity. A fast texture with low reflectivity and high uniformity can be obtained using percentages of O_2 to SF_6 between 120%-170%.

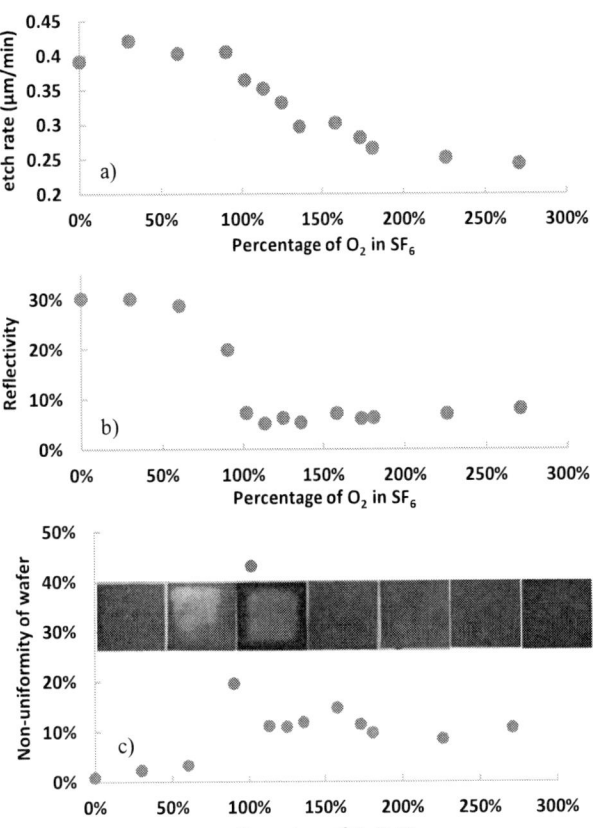

Fig. 5. a) Etch rate b) reflectivity and c) non-uniformity of wafers for different percentage of content of O_2 in SF_6. For Fig. 5 c), each square in the picture is associated to each one of the red points in the graph in the order presented according to percent O_2. A fast texture with low reflectivity and high uniformity can be obtained using percentages of O_2 in SF_6 between 120%-170%.

B. RF Power dependence

The second experiment explored the behavior of different RF power. The fixed variables were: etch time (150 s), pressure (19 Pa), and a ratio of $O_2:Cl_2:SF_6$ of 3:6:5. In Fig. 6, the left SEM picture shows the case for a low RF power (14kW) where it can be seen that the texture is short and wide. The right picture shows the case for high RF power (24 kW), this picture shows the deeper features with sharper angles than at 14kW. Also the presence of two sizes of texture is appreciable at higher power, a small texture feature riding along of a bigger texture.

Fig. 7 a) shows the relationship between RF power and amount of silicon etched: the etch rate remains relatively constant until 22kW, after then, the etch rate decreases. Maximum etch rates of 0.325 μm/min were obtained at 18kW.

Fig. 7 b) shows the relationship between reflectivity for the wafers as the percentage of O_2 increases. Comparing graphs in Fig. 7 a) and 7 b), it can be seen that while the material etches and forms texture at lower power, the reflectivity is higher.

Fig. 7 c) shows the relationship between non-uniformity within the wafer as the RF power increases. Also, Fig. 7c) shows the shades of the wafers along this process. Even though the lowest reflectivity is obtained at the highest power, the best uniformity is achieved at the lowest power. RF power between 17 and 21 kW observe good etching rates with low reflectance and uniform texture across the wafer.

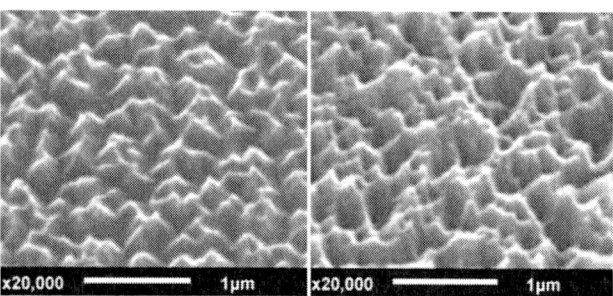

Fig. 6. 30° tilted SEM pictures at 20,000× magnification. Left, power at 14kW, right power at 24kW.

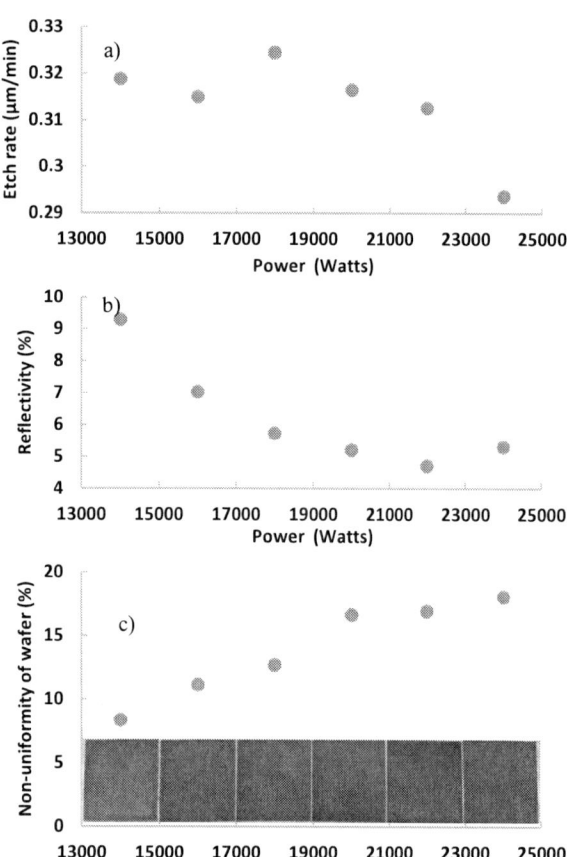

Fig 7 a) Etch rate b) reflectivity and c) non-uniformity of wafers for different RF power. For Fig. 5 c), each square (wafer picture) in the graph is associated to each one of the points in the graph in the order presented according to power. A fast texture with low reflectivity and high uniformity can be obtained using RF power between 17 and 21kW.

C. Chlorine content dependence

The third experiment explored the behavior of different ratios of Cl_2 in SF_6. The fixed variables were: etch time (150 seconds), pressure (19 Pa), power (19 kW) and a fixed ratio of $O_2:SF_6$ of 6:5.

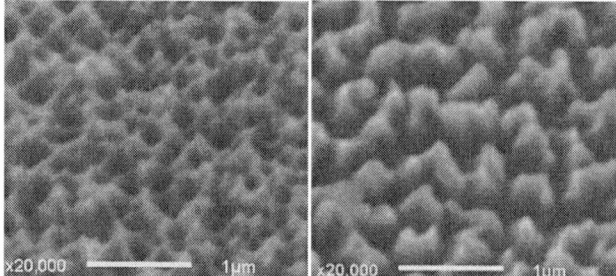

Fig. 8. 30° tilted SEM pictures at 20,000× magnification. Left, Cl_2 content of 21% in SF_6, right Cl_2 content of 83% in SF_6

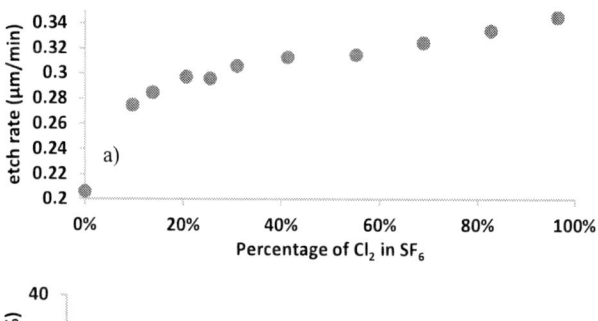

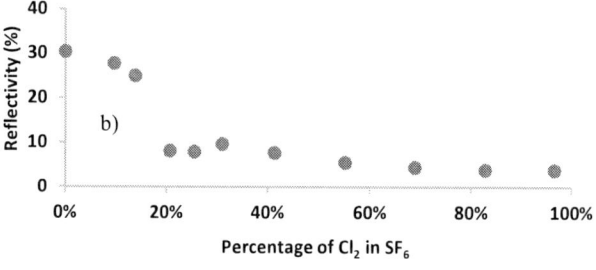

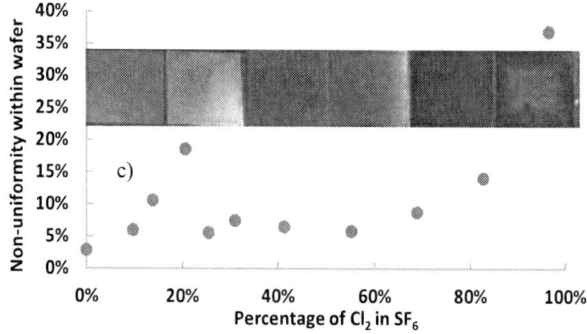

Fig. 9. a) Etch rate b) reflectivity and c) non-uniformity of wafers for different percentage of content of Cl_2 in SF_6. For Fig. 9 c), each square in the picture is associated to each one of the red points in the graph in the order presented according to percent Cl_2. A fast texture with low reflectivity and high uniformity can be obtained using percentages of Cl_2 in SF_6 between 30%-65%.

In Fig. 8, the left SEM picture shows the case for a low ratio of Cl_2 in SF_6 (21%), it can be seen that the texture is small and sharp. The right picture shows the case for high Cl_2 content (83% Cl_2 in SF_6), this picture shows tall/wide pyramid shapes.

Fig. 9 a) shows the relationship between % of Cl_2 and amount of silicon etched: the etch rate is increased rapidly at the beginning as composition goes from 0% Cl_2 to >10%, after that, the etch rates continues to increase almost linearly in relationship with the amount of Cl_2 in the chamber. Maximum etch rates of 0.34 µm/min were obtained at 100% Cl_2 in SF_6. Minimum etch rates of to 0.2 µm/min were achieved at 0% Cl_2 in SF_6.

Fig. 9 b) shows the relationship between reflectivity vs Cl_2 in the chamber. Comparing graphs in Fig. 9 a) and 9 b), it can be seen that while the material etches with zero chlorine in the chamber, the texture does not form. At least 20% Cl_2 content in SF_6 is necessary to start the formation of masking material that will then create the texture.

Fig. 9c) shows the relationship between non-uniformity within the wafer as the percentage of Cl_2 increases and the shades of the wafers along this process. It can be seen that between 0% and 20% and above 80% Cl_2 content in SF_6 the wafers increase non-uniformity as Cl_2 content increases. It is only between 20-80% Cl_2 content that the wafers are uniform. A fast texture with low reflectivity and high uniformity can be obtained using percentages of Cl_2 to SF_6 between 30%-65%.

D. Pressure dependence

The fourth experiment explored the behavior of different pressure in the chamber. The fixed variables were: etch time (150 seconds), power (19 kW), and a fixed ratio of $O_2:Cl_2:SF_6$ of 3:6:5. In Fig. 10, the left SEM picture shows the case for a low pressure (17 Pa) where it can be seen that the texture is sharp, pointy and tightly packed. The right picture shows the case for high pressure (29 Pa), this picture shows rounder, less dense, and shorter texture with wider angles than at lower pressure.

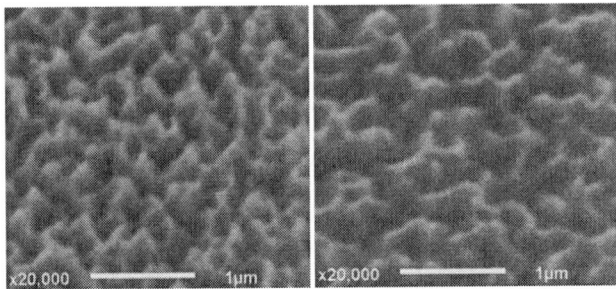

Fig. 10. 30° tilted SEM pictures at 20,000× magnification. Left image, 17 Pa, right image, 29 Pa.

Fig 11a shows the relationship between reflectivity for the wafers as the pressure increases. Lower pressure reduces the reflectivity, presumably by increasing the acceleration and mean free path of ions. Fig. 11b shows that reflectivity increases as the pressure increases.

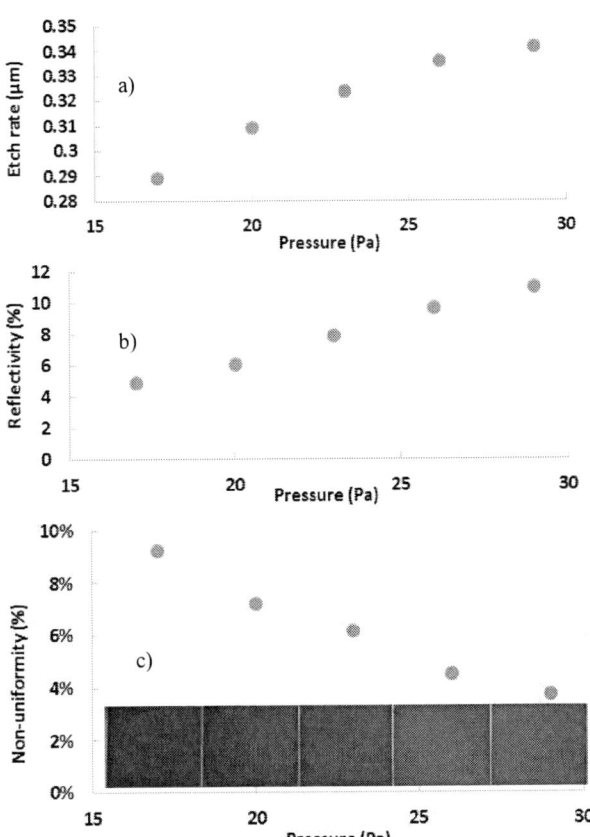

Fig. 11. a) Etch rate b) reflectivity and c) non-uniformity of wafers for different chamber pressure scenarios. For Fig. 11 c), each square in the picture is associated to each one of the blue points in the graph in the order presented according to pressure. A fast texture with low reflectivity and high uniformity can be obtained using pressure between 20 and 25 Pa.

Fig 11c shows the relationship between non-uniformity within the wafer for the wafers as the power increases. Also, Fig. 11c shows the shades of the wafers along this process. To the eye, the variation in uniformity is not seen easily but the measurements indicate that the lower the pressure, the less uniform the texture is over the wafer. Overlaying the results from figures 11a, b and c a fast texture with low reflectivity and high uniformity can be obtained using RF powers between 20 and 25 Pa.

E. Time dependence

The fifth experiment explored the behavior of different process times. The fixed variables were: power (19 kW), pressure (19 Pa), and a fixed ratio of $O_2:Cl_2:SF_6$ of 3:6:5. In Fig. 12, the left SEM picture shows the case for wafers processed for 75s where it can be seen that the texture is barely forming. The right picture shows the case for 150s process time, where it shows a sharp and tall texture. Fig. 13a shows the linear relationship between time and amount of silicon etched, the etch rate is about 4.2 nm /s or 0.25 μm/min. Fig 13b shows the relationship between reflectivity for the wafers as time advances. A hyperbolic decay behavior is observed. Also, Fig. 13b shows the colors of the wafers along this process. Most of the change in reflectivity/color occurs in the first 80-100 seconds of etching. Non-uniformity is not a function of time and had a value around 10 % for all times.

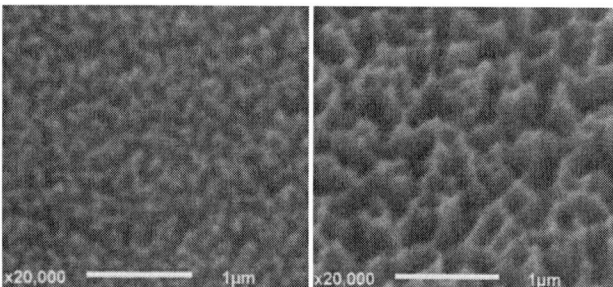

Fig. 12. 30° tilted SEM pictures at 20,000× magnification. Left image, wafers processed for 75s, right image, with 150s process time.

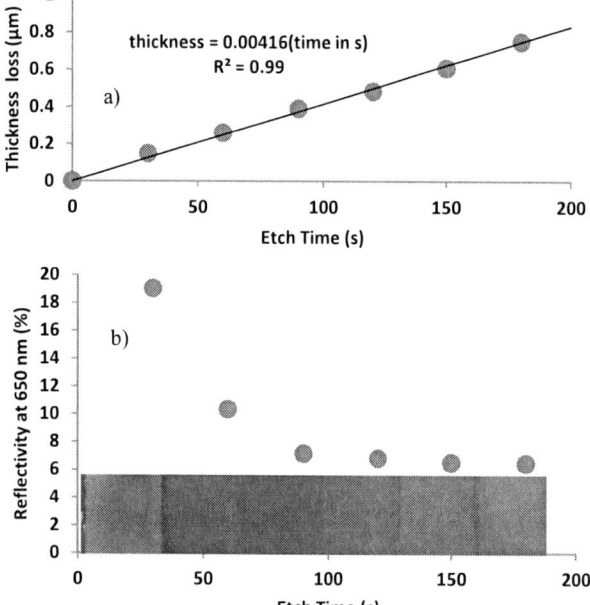

Fig. 13. a) Etch rate and b) reflectivity of wafers for different process times. For Fig. 13 c), each square in the picture is associated to each one of the blue points in the graph in the order presented according to pressure. The etch rate is linear with time and the reflectivity reaches and stays constant at about 7% after 100s.

V SUMMARY

This paper presented simulations and experimental work about the optical behavior of different RIE-based textures on smooth wafers including DW technology. Simulations predict the influence of angle of texture in reflectance demonstrating that narrow pyramids are best at absorbing light. Five

experiments showed the influence of varying O_2 to SF_6 ratio, power, Cl_2 to SF_6 ratio, pressure, and etch time. The measured outputs were etch rate, reflectivity and uniformity. SEM and optical pictures were shown for each experiment. The results, graphs, optical and SEM pictures show that a fast forming texture with low reflectivity and high uniformity can only be obtained in a narrow range of parameters: between gas compositions of Cl_2 to SF_6 in the range of 35%-65%, O_2 to SF_6 ratios between 120%-170%, power ranging between 18-22 kW, and pressures between 19 - 25 Pa. Overall, we showed the optimal industrial-scale RIE parameters needed to achieve a great texture that can be attained fast, with low reflectivity, and high uniformity on smooth wafers such as Direct Wafer product or diamond sawn.

REFERENCES

[1] http://pvinsights.com/
[2] A. Lorenz, J. Hofstetter, H. Malkasian, L. Sanderson, F. van Mierlo "3 Dimensional Direct Wafer Product with Locally-Controlled Thickness", in *33rd European Photovoltaic Solar Energy Conference and Exhibition* , 2BO.2.5 3, 2016.
[3] R. Jonczyk, A. Lorenz, A. Ersen, J. Hofstetter, K. Hubener, K. Dunker, J. Scharf, L. Neibergall, K. Petter, J Muller, D. Jeong, "Low-cost kerfless wafers with gradient dopant concentration exceeding 19% cell efficiency in PERC production line", *Energy Procedia*, vol. 92, pp. 822-827, 2016.
[4] http://1366tech.com/2017/03/08/1366-technologies-achieves-19-9-efficiency-using-direct-wafer-hanwha-q-cells-q-antum-cell-technologies/
[5] F, Toor, J. B. Miller, L. M. Davidson, L. Nichols, W. Duan, M. P. Jura, J. Yim, J. Forziati and M. R Black "Nanostructured silicon via metal assisted catalyzed etch (MACE): chemistry fundamentals and pattern engineering", *Nanotechnology*, vol. 27, pp. 15448-15466, 2016.
[6] B. Kafle, T. Freund, A. Mannan, L. Clochard, E. Duffy, S. Werner, P. Saint-Cast, M. Hofmann, J. Rentsch, R. Preu "Plasma-free Dry-chemical Texturing Process for High-efficiency Multicrystalline Silicon Solar Cells", *Energy Procedia*, vol. 92, pp. 359–368, 2016.
[7] http://sionyx.com/pdf/solarcellperformancewhitepaper.pdf
[8] M. Sugawra, *Plasma etching fundaments and applications*. New York, New York: Oxford Science Publications, 1998.
[9] R. J. Shul and S.J. Pearton, *Handbook of advanced plasma processing techniques.*, New York, New York, 2002
[10] O. Yosuke Inomata, Y. Youko Fukawa, "Method for producing a solar cell", *US Patent* 7,556, 741 B2, 2009
[11] K. R. McIntosh and S. C. Baker-Finch, "OPAL 2: Rapid optical simulation of silicon solar cells," in *38th IEEE Photovoltaic Specialists Conference*, pp. 265-271, 2012.
[12] R. Legtenberg, H. Jansen, M. de Boer, and M. Elwenspoek "Anisotrapic Reactive Ion Etching of Silicon Using SF6/O2/CHF3 Gas Mixtures", *J. Electrochem. Soc.*, vol. 142, pp. 2020-2028, 2012.

Short-Circuit Current-Density Enhancement of Silicon Solar Cells Using Plasmonics Antireflective Coating and Luminescent Downshifting

Sheng-Kai Feng, Wen-Jeng Ho*, Guan-Yi Li, Jheng-Jie Liu, Hao-Yu Yang, and Ta-Wei Chuang

Department of Electro-Optical Engineering, National Taipei University of Technology, Taipei 10608, Taiwan, R.O.C. *: wjho@ntut.edu.tw

Abstract — The combination of plasmonic scattering of silver nanoparticles (Ag-NPs) embedded in SiO₂ layer and luminescent downshifting (LDS) of Eu-doped phosphor layer applied on silicon solar cells to enhance photovoltaic performances was demonstrated. By annealing 3, 5, and 7 nm thick silver films at 200 °C for 30 min under ambient H₂ to form Ag NPs of various dimensions were obtained, which corresponding to the average dimensions of Ag-NPs are 20.13, 25.03, and 32.14 nm determining by SEM images. The optical and electrical properties of the cells with Ag-NPs embedded in a SiO₂ antireflection coating (ARC) were characterized firstly by optical reflectance, absorbance, and external quantum efficiency (EQE) measurements. Larger Ag-NPs dimension of 32.14 nm exhibited a larger short-circuit current-density (J_{sc}) enhancement of 31.24 %, which is higher than that of 30.90% for Ag-NPs dimension of 25.03 nm and 28.64% for Ag-NPs dimension of 20.13 nm. The combined effects of plasmonic scattering and LDS are studied by applying Ag-NPs and Eu-doped phosphor particles within SiO₂ ARC, which the J_{sc} enhancement can be further enhanced from 28.64% to 29.37%.

I. INTRODUCTION

Silver nanoparticle (Ag-NP) has been widely used as a promising material to capture and couple incident light into solar cells. By forming Ag nanoparticles on top of dielectric layer, a strong scattering have been observed [1, 2]. In general, the surface plasmon resonance (SPR) of metallic NPs is the reason for causing this phenomenon and is very sensitive to the size, shape, pacing, and dielectric environment of metallic NPs [3]. Comparing to the most commonly used structure of plasmonic solar cells, forming metallic NPs on top of solar cells, the metallic NPs embedded in antireflection coating (ARC) would be more suitable for enhancing conversion efficiency of solar cell applications. Locating metallic nanoparticle in dielectric layer can not only enhance the SPR effect but also form a protective layer to keep metallic NPs from being exposed to the air [2]. In addition, a problem of lower conversion efficiency of silicon solar cell in short wavelength region is always an obstacle to be overcome. Because absorption coefficient of silicon is quite high in UV-blue region, meaning that UV-blue photons will be absorbed near the surface which cause higher recombination loss and lower conversion efficiency. Recently, innovative ways to improve this issue by using spectral conversion of various phosphor materials have been demonstrated [4, 5].

Luminescent downshifting (LDS) phosphors can transfer high-energy photons into low-energy photons, which avoid the recombination loss on the surface and enhance the conversion efficiency in short wavelength region.

In this study, we have proposed and demonstrated that photovoltaic performances of silicon solar cell can be enhanced by using the combination of plasmonic scattering of silver nanoparticles (Ag-NPs) and luminescent downshifting (LDS) of Eu-doped phosphor particle. The experiment is designed into three groups, solar cell with a pure SiO₂ antireflection layer (ARC), cells with various dimensions of Ag-NPs embed in SiO₂ ARC (Ag-NPs ARC), and cells with Ag-NPs embedded in the mixed phosphor-SiO₂ ARC (P-ARC). The optical and electrical properties including optical reflectance, absorbance, external quantum efficiency (EQE), and photovoltaic current-voltage (J-V) are measured and compared to examine plasmonics scattering and LDS effects.

II. EXPERIMENT

A 20 nm thick SiO₂ film was deposited on the bare solar cells using e-beam evaporation. Next, 3, 5, and 7 nm silver films were deposited on the bare solar cells with a 20 nm SiO₂ space layer by e-beam evaporation and subsequently annealed at 200 °C for 30 min under ambient H₂ to form Ag-NPs of various dimensions. Then, a 75 nm thick SiO₂ film was deposited on the surface of Ag-NPs layer using spin-on film process, which Ag-NPs was embedded in SiO₂ layer, and the fabricated cell was called the cell with Ag-NPs ARC. Finally, we prepared silicate phosphors $(Sr_{1-x}Ba_x)_2SiO_4:Eu^{2+}F$ of 3 wt.% mixing with silica solution instead of pure SiO₂ solution spinning on the cell with Ag-NPs upon a 20 nm SiO₂ space layer, and the fabricated cell was called P-ARC.

The dimension and profile of deposited Ag-NPs are determined by scanning electron microscope (SEM). The reflectance (R) and transmittance (T) of proposed samples were measured by integrating sphere, and the absorption (A) was calculated by the formula A=1-R-T. The EQE response was characterized over a range of wavelengths from 350 nm to 1100 nm. The photovoltaic current density-voltage (J-V) characteristics were measured under 1-sun AM 1.5G solar simulation.

978-1-5090-5606-4/17 $31.00 © 2017 IEEE

III. RESULTS AND DISCUSSION

Fig. 1(a), (b) and (c) show the SEM images of Ag NPs after annealing the deposited Ag films with the thickness of 3, 5, and 7 nm, respectively. The coverage and average diameter were calculated by ImageJ software. The average diameter was 20.13 nm, 25.03 nm, and 32.14 nm corresponding to the deposited thickness of 3, 5, and 7 nm respectively. The coverage of Ag NPs would also increase along with the thickness.

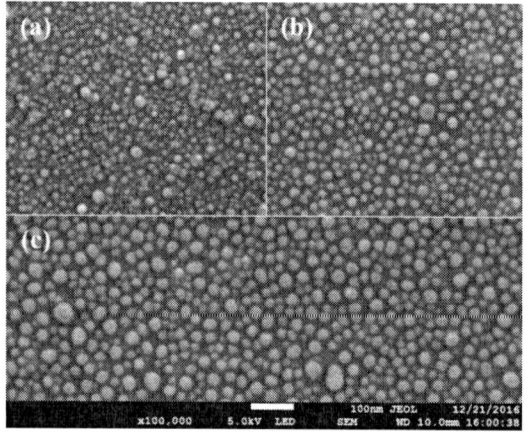

Fig. 1. SEM images of Ag NPs annealed from the deposition Ag film thickness of (a) 3 nm, (b) 5 nm, and (c) 7 nm.

Optical reflectance of the reference solar cell, the cell with Ag-NPs ARC and the cell with P-ARC are presented in Fig. 2. The reflectance of the cells with Ag-NPs embedded in SiO_2 ARC was lower than the cell with an only SiO_2 ARC one. Notice that there is a strong reduction of reflectance in the wavelength from 400 to 650 nm, which is the SPR region of Ag-NPs causing by parasitic absorption [6].

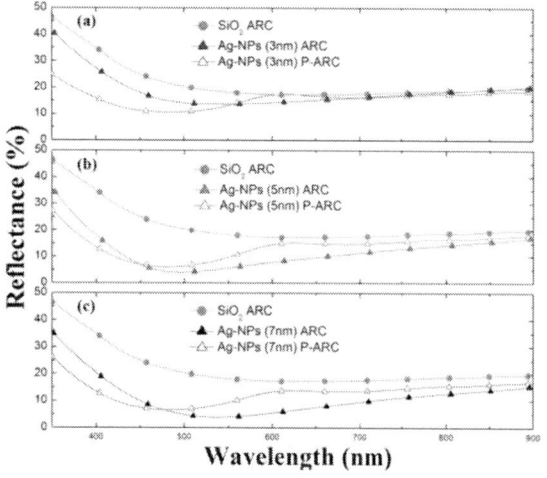

Fig. 2. The reflectance of all the cells tested in this study.

In addition, an lower reflectance in 350-450 nm and an higher reflectance at around 610 nm can be observed for the cell with P-ARC compared to the cell with Ag-NPs ARC, which be attributed to the LDS effects of Eu-doped phosphor on the silicon solar cells. In this study, the reflectance of cells with P-ARC exhibited a lower reflectance in short wavelength region, particularly for the cell with smaller Ag-NPs particles. The optical absorption spectrum of Ag-NPs with various particles dimension is shown in Fig. 3. The absorption range of 350-550 nm and the peak of absorption at 450-460 nm can be observed. Although the resonant absorption of Ag NPs seems a negative effect, it provided a strong forward scattering in the off-resonant region.

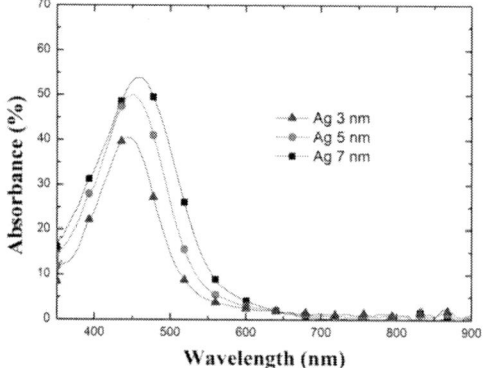

Fig. 3. The absorption spectrum of Ag NPs of various sizes embedded in SiO_2 ARC.

Fig. 4 shows the EQE spectrum of the cell with SiO_2 ARC, the cell with Ag-NPs ARC and the cell with P-ARC. The EQE values decreased from 400 to 550 nm because the resonant absorption of Ag-NPs. However, the EQE values were increased beyond 550 nm wavelength due to forward scattering by Ag-NPs. The reduction of reflectance and increase of the effective diffusion length of minority carriers of solar cells has been reported by Thouti , et al. [7].

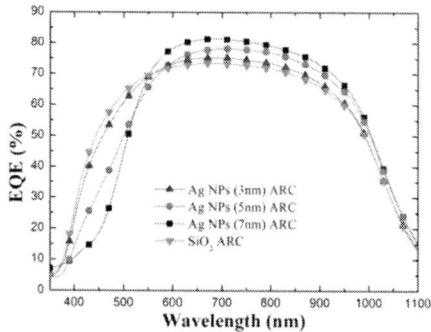

Fig. 4. The EQE response of the cell with SiO_2 ARC and the cell with Ag-NPs ARC. The deposited Ag film thickness of 3, 5, and 7 nm.

Fig. 5 shows the EQE response and EQE enhancement of the cell with a pure SiO_2 ARC, the cell with Ag-NPs (the

978-1-5090-5606-4/17 $31.00 © 2017 IEEE

deposited film thickness of 3 nm) embedded in pure SiO_2 ARC, and the Ag-NPs (the deposited film thickness of 3 nm) embedded in phosphor-SiO_2 ARC (P-ARC). The plasmonic Ag-NPs cell with phosphor has pronounced EQE enhancement in 350-400 nm wavelength region due to LDS effects. Besides, the EQE enhancement of the cell with Ag NPs embedded in phosphor-SiO_2 ARC was increased at the wavelength beyond 550 nm, due to the light scattering induced by Ag-NPs and Eu-doped phosphor particles.

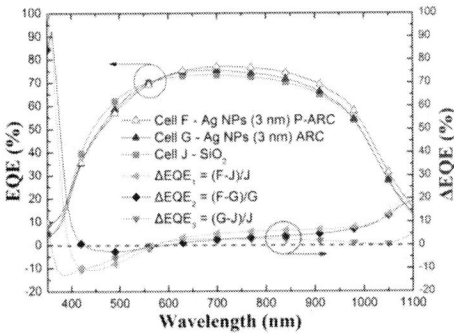

Fig. 5. The EQE response and ΔEQE of the cell with Ag-NPs ARC and cell with P-ARC. The deposited Ag film thickness of 3nm.

Fig. 6 shows the photovoltaic J-V characteristics of the bare cell, the cell with SiO_2 ARC, and the cell with Ag-NPs (the thickness of deposited Ag-films of 3, 5, and 7 nm) embedded in phosphor-SiO_2 ARC (P-ARC). The short circuit current density (J_{sc}) of the cells with P-ARC was higher than the cell with SiO_2 ARC. In addition, the short circuit current density enhancement (ΔJ_{sc}) of the cell with Ag-NPs SiO_2 ARC and cell with P-ARC as a function of the thickness of the deposited Ag film are also shown in the inset of Fig. 6. The large dimension Ag-NPs (the film thickness of 7 nm) exhibited a larger ΔJ_{sc} of 31.24 %, which is higher than that of 30.90% for medium dimension Ag-NPs (the film thickness of 5 nm) and 28.64% for small dimension Ag-NPs (the film thickness of 3 nm). The combine effects of plasmonic scattering and LDS are studied by applying LDS Eu-doped phosphor particles and Ag-NPs within SiO_2 ARC, which the J_{sc} can be further enhanced from 28.64% to 29.37%.

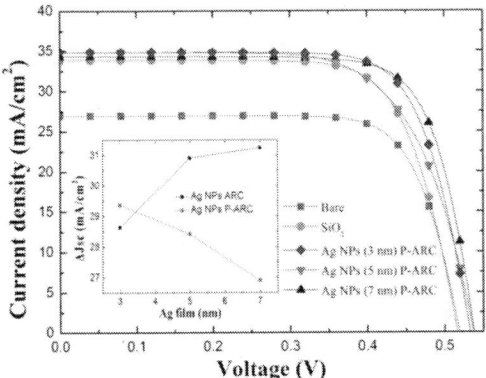

Fig. 6. J_{sc} of the bare cell, the cell with SiO_2 ARC, and the cell with P-ARC. The deposited Ag film thickness of 3, 5, and 7 nm.

IV. CONCLUSION

The plasmonic effects of Ag-NPs of various dimensions embedded in SiO_2 ARC were demonstrated. The reflection, absorption, and EQE spectrum were used to reveal plasmonic scattering of Ag-NPs. Larger Ag-NPs exhibited a high plasmonic light scattering as well as a larger J_{sc} enhancement. Moreover, the combination of plasmonic scattering and LDS effects was demonstrated by applying Ag-NPs and Eu-doped phosphor particles within a SiO_2 ARC, and the J_{sc} was further enhanced. The optimal combined condition of plasmonic scattering and LDS is undergoing study in our laboratory.

ACKNOLOWLEDGEMENTS

The authors would like to thank the Ministry of Science and Technology (MOST) of the Republic of China for financial support under Grant MOST 103-2221-E-027-049-MY.

REFERENCES

[1] Ta-Wei Chuang, Wen-Jeng Ho, Chia-Hua Hu, Wei-Lien Wang, Jian-Cheng Lin, Sheng-Kai Feng, Su-Han Weng, Han-Chung Huang, Guan-Yi Li and Hao-Yu Yang, "Plasmonic Scattering Comparing of Silicon Solar Cells Coated with Silver, Aluminum, and Indium Nanoparticles Based on Similar Diameter and Coverage," in *43th IEEE PVSC Conference*, 2016.

[2] Jae-Phil Shim, Sang-Bae Choi, Duk-Jo Kong, Dong-Ju Seo, Hyung-Jun Kim, and Dong-Seon Lee, "Ag nanoparticles-embedded surface plasmonic InGaN-based solar cells via scattering and localized field enhancement," *Optics Express*, vol. 24, pp. A1176-A1178, 2016.

[3] Shi-e Yang, Ping Liu, Yu-jie Zhang, Qiao-Neng Guo, and Yong-sheng Chen. "Effects of silver nanoparticles size and shape on light scattering," *Optik Optics*, vol. 127, pp. 5722-5728, 2016.

[4] Wen-Bin Hung, and Teng-Ming Chen. "Efficiency enhancement of silicon solar cells through a downshifting and antireflective oxysulfide phosphor layer," *Solar Energy Materials & Solar Cells*, vol. 133, pp. 379-47, 2015.

[5] Guo-Chang Yang, Wen-Jeng Ho, Chien-Wu Yeh, Yu-Tang Shen, Ruei-Siang Sue, Chia-Hua Hu, and Yu-Jie Deng, "Electrical and Optical Characterization of Crystalline-Si Solar Cell with Down Shifting Eu-Doped Silicate Phosphors Film Depending on the Coverage of Phosphors Particles Using Spin-on Film Process," in *42th IEEE PVSC*, 2015.

[6] Eshwar Thouti, and Vamsi K. Komarala, "Investigation of parasitic absorption and charge carrier recombination losses in plasmonic silicon solar cells using quantum efficiency and impedance spectroscopy," *Solar Energy*, vol. 132, pp. 143-149, 2016.

[7] Eshwar Thouti, Ashok K Sharma, Sanjay K Sardana, and Vamsi K Komarala1, "Internal quantum efficiency analysis of plasmonic textured silicon solar cells: surface plasmon resonance and off-resonance effects," *Journal of Physics D: Applied Physics*, vol. 47, pp. 422001-425501, 2014.

Extremely Low Reflectivity Nanoporous Black Silicon Surface by Copper Catalyzed Etching for Efficient Solar Cells

K A S M Ehteshamul Haque, Wenqi Duan and Fatima Toor

Electrical and Computer Engineering Department, University of Iowa, Iowa City, IA 52242, USA

Abstract — In this work, we report on a process based on copper (Cu) catalyzed etching of silicon (Si) to obtain extremely low reflectivity nanoporous 'black silicon' (bSi) surface. We explore both one-step and two-step etching process, and find that one-step etching results in a uniformly etched surface reproducibly, compared with the two-step process. Adding ascorbic acid ($C_6H_8O_6$) in the one-step process recipe results in inverted pyramid shaped pores that enhance light trapping and lowers the bSi surface reflectivity. We observe that the optimum concentration of hydrogen peroxide (H_2O_2) in the etching solution depends on the sample size. The lowest spectrum-weighted-average reflectivity (R_{avg}) obtained from the optimized one-step etching process is 3.36%.

Index terms — black silicon; metal assisted catalyzed etching; antireflective porous silicon surface; light trapping

I. INTRODUCTION

Metal assisted catalyzed etching (MACE) of Si [1] results in extremely low reflectivity over a broad spectrum of wavelengths (300-1000 nm). This eliminates the need for creating an expensive vacuum-deposited antireflection (AR) layer (such as silicon nitride (SiN_x)) in Si solar cells and reduces the overall manufacturing cost. Contrary to the conventional vacuum deposited AR coating, MACE creates a graded-index nanostructured surface, which gives ultra-low spectrum-weighted-average reflectivity (R_{avg}) [2]. The antireflective nanoporous silicon surface formed using the MACE process is typically termed as 'black Si' (bSi) [3]. Several research groups are developing nanostructured bSi based solar cells due to their low cost potential at commercial scale [3-6]

Forming bSi by MACE can be a one- or two-step process [7]. In the one-step process, the metal precursor, the etchant and the oxidant are present in the same solution. In this case, the metal film formation and the etching processes occur simultaneously. In the two-step process, the Si wafer is first dipped in a plating solution which contains the metal ions. This creates a coating of metal nanoparticles on the Si surface. Then, this coated wafer is dipped in an etching solution that contains an etchant (usually hydrofluoric acid (HF)) in the presence of an oxidant (usually hydrogen peroxide H_2O_2) [4]. The metal nanoparticles act as local catalyst for the oxidation of Si in the presence of H_2O_2. Then the oxidized Si is etched by HF, resulting in a nanoporous Si surface.

Metals commonly used in the MACE process are gold (Au), silver (Ag), platinum (Pt) and palladium (Pd) [1]. Using Cu instead of Au [3] or Ag [8] for the MACE process gives several advantages. Firstly, Cu is orders of magnitude less expensive than Au or Ag. Secondly, Si nanowires grown using Cu catalyst demonstrated longer minority carrier diffusion length (~10 μm) than those using Au catalyst (~4 μm) [9]; longer minority carrier diffusion length results in improved solar cell performance.

In this work, various parameters of the Cu MACE process were optimized to obtain a functional and reproducible process. Reproducibility is an issue in Cu MACE when the process parameters are not optimized, because excess Cu (relative to the oxidant) may create a continuous Cu film on Si, which results in minimal etching of Si [10]; while with less Cu (relative to the oxidant), the MACE may not occur successfully over the entire Si surface, which may result in craters or shallow pits instead of nanopores [11]. Our initial plan was to optimize the parameters of the two-step Cu MACE process. However, from further observation and analysis, we moved to the one-step process, and optimized its parameters. The optimized one-step process produced reproducible results. The minimum spectrum-weighted-average-reflectivity (R_{avg}) obtained from one-step Cu MACE was 3.36%.

II. EXPERIMENTAL METHODS

Single-side polished two-inch mono-crystalline Si (100) Czochralski (CZ) wafers (p-type, 1-10 Ω-cm, University Wafer) were cleaved into 1-cm by 1-cm square samples. These samples were first cleaned in 10% HF solution (in deionized (DI) water) for 1-min to remove the native oxide on the surface. Next, we performed the one- or two-step MACE process.

For the one-step process, the MACE solution contained copper (II) sulfate pentahydrate ($CuSO_4.5H_2O$), ascorbic acid ($C_6H_8O_6$), HF and H_2O_2 in DI water. Concentrations of all these reactants were varied independently and optimized. The solution was kept in a water bath at 55 °C.

For the two-step MACE process, the plating solution contained 50 vol% ammonium fluoride (NH_4F), 30 vol% methanol and 20 vol% DI water, $CuSO_4.5H_2O$, $C_6H_8O_6$ and 0.05 M sodium potassium tartrate ($KNaC_4H_4O_6.4H_2O$). $C_6H_8O_6$ increases the adhesion of the Cu film onto the Si substrate [12]. In fact, the adhesion of Cu film to the substrate is strongly dependent upon the presence of $C_6H_8O_6$ in the

978-1-5090-5606-4/17 $31.00 © 2017 IEEE

plating solution. Cu film formed without $C_6H_8O_6$ fails the standard scotch tape test [12]. Methanol and $KNaC_4H_4O_6 \cdot 4H_2O$ are added to the plating solution in order to reduce the internal stress of the Cu film [12].

Concentrations of $CuSO_4.5H_2O$, $C_6H_8O_6$ and the plating time were varied independently and optimized, as these parameters directly control the distribution of Cu nanoparticles on the Si surface. For the etching solution, 7.2 mL of 48% HF was mixed with 5 mL of 30% H_2O_2. Then, DI water was added to make a solution of 50 mL. This solution has a molar ratio ρ of 80%, where ρ is the molar percentage of the etchant in the solution (ρ = [HF] / ([HF] + [H_2O_2]); here [HF] and [H_2O_2] represent the molar concentrations of HF and H_2O_2, respectively). Toor $et\ al.$ reported that ρ ~80% results in the lowest R_{avg} utilizing the two-step Cu MACE process [13]. The etching solution was kept in a water bath at 55 °C during the etching process.

After the one- or two-step MACE processes, the residual Cu nanoparticles in the nanoporous bSi samples were removed by submerging the sample in 70% nitric acid (HNO_3) for 20 minutes with ultrasonication. Next, the bSi samples were dipped in 5% HF solution (in DI water) for 90 sec, rinsed in DI water and blow dried using a nitrogen (N_2) gun.

II. RESULTS AND DISCUSSIONS

A. Two-step MACE Process

We systematically analyzed the electroless plating time, and the concentrations of Cu^{2+}, $C_6H_8O_6$, HF and H_2O_2 of the two-step MACE process to determine the optimum process parameters that result in the formation of a low reflectivity nanoporous bSi surface. Initially, the concentrations of Cu^{2+}, $C_6H_8O_6$, HF and H_2O_2 were set at 0.01 M, 0.01 M, 4 M and 1 M, respectively. When one parameter was varied, the other parameters were kept fixed at the values specified above.

(i) Electroless Plating Time

The electroless plating time for Cu nanoparticles was varied from 10 to 60 seconds (s). From the scanning electron microscopy (SEM) images of the top surface (Fig 1), we observed that plating time greater than 10 s created a continuous Cu film on parts of the Si surface. For effective etching, there should be 20-40% exposed Si surface among the Cu nanoparticles [3]. So, the plating time was kept at 10 s for future experiments.

(ii) Cu^{2+} ($CuSO_4.5H_2O$) Concentration

Cu^{2+} molar concentration ([Cu^{2+}]) was varied from 0.005 M to 0.015 M. From the observation of top surface SEM images (Fig. 2), [Cu^{2+}] of 0.01 M was chosen for future experiments, as it gave the optimum covering of the Si surface by Cu nanoparticles as mentioned earlier [3].

(iii) $C_6H_8O_6$ Concentration

$C_6H_8O_6$ molar concentration ([$C_6H_8O_6$]) was varied from 0.01 to 0.03 M. Optimum distribution of Cu nanoparticles on

the Si surface was observed for [$C_6H_8O_6$] = 0.01 M (Fig. 3). So, [$C_6H_8O_6$] was chosen as 0.01 M for future experiments.

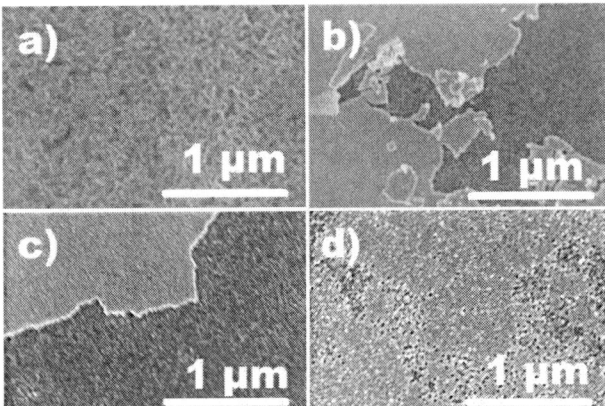

Fig. 1. Top surface SEM images of Si after Cu plating of a) 10 s b) 20 s c) 30 s and d) 1 min

Fig. 2. Top surface SEM images of Si after plating with a) 0.005 M [Cu^{2+}] b) 0.01 M [Cu^{2+}] and c) 0.015 M [Cu^{2+}]

(iv) HF and H_2O_2 Concentrations

Earlier work suggested that obtaining nanoporous bSi surface depends only on the relative concentration (ρ) of HF and H_2O_2 in the etching solution [14]. However, Cao $et\ al.$ [10] reported recently that varying HF and H_2O_2 concentrations in the etching solution (while keeping ρ = 92%) produces significant changes in the surface morphology of Si. To investigate this, we carried out experiments varying [HF] from 5.8 M to 11.6 M, while the ρ was kept fixed at 92%. It should be noted that as the ρ was kept constant, [H_2O_2] was also varied proportionally (from 0.5 M to 1 M).

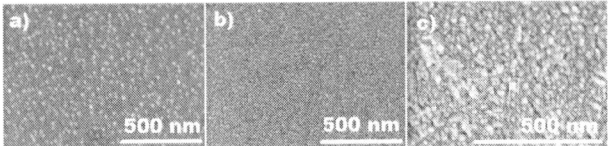

Fig. 3. Top surface SEM images of Si after Cu plating with a) [$C_6H_8O_6$] = 0.01 M b) [$C_6H_8O_6$] = 0.02 M and c) [$C_6H_8O_6$] = 0.03 M in the plating solution.

It was found that with increasing [HF] and [H_2O_2], reflection from the Si surface decreases, and the etching time also decreases (Fig. 4), even though the ρ was unchanged. The lowest R_{avg} (8.46%) obtained in this process was with [HF] = 11.6 M, with an etching time of 20 minutes.

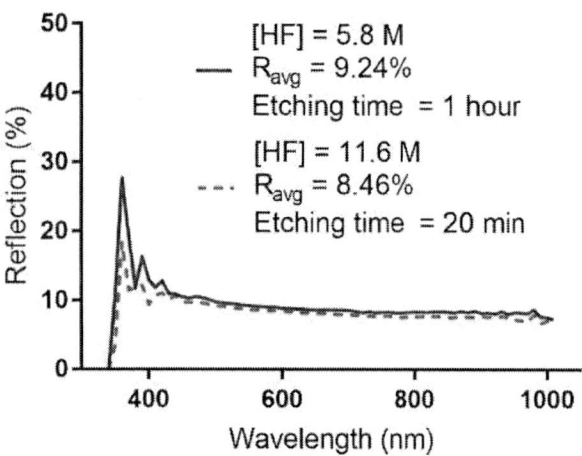

Fig. 4. Reflection as a function of wavelength of nanoporous bSi surface after two-step Cu MACE with different HF concentrations.

B. One-step MACE Process

Even though the R_{avg} became lower (<10%) with higher [HF] and [H$_2$O$_2$] in the two-step MACE process, top surface SEM images revealed that there is non-uniform etching over the Si surface, which is due to the non-uniform distribution of Cu nanoparticles over the surface (Fig. 5a). This explains the irreproducibility in the two-step Cu MACE process. In two-step etching, the Cu plating occurs separately, and there is a chance of large variation in Cu nanoparticle distribution over the sample surface. This may result in non-uniform etching across the surface. On the other hand, we found that the one-step etching is more uniform (Fig. 5b), because the Cu nanoparticles interact with H$_2$O$_2$ continuously, and there is always Cu^{2+} ions present in the solution to replenish any deficit. Due to these process considerations, we utilized the one-step Cu MACE for the next set of experiments.

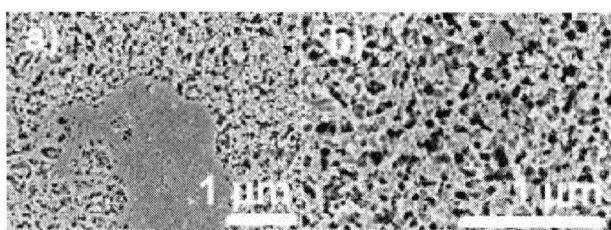

Fig. 5. Top surface SEM images of a) non-uniform etching of Si utilizing two-step Cu MACE process and b) uniform etching of Si utilizing one-step Cu MACE process.

(i) Optimization of Process Parameters in the One-step MACE Process

Top surface SEM images of bSi samples etched with [HF] = 5.8 M and [HF] = 11.6 M show that the ones etched with 11.6 M [HF] resulted in a more uniform porous structure (Fig 6a), and hence, lower R_{avg} than the ones etched with 5.8 M [HF]

(Fig. 6b). Therefore, we kept the [HF] at 11.6 M for future one-step etching experiments.

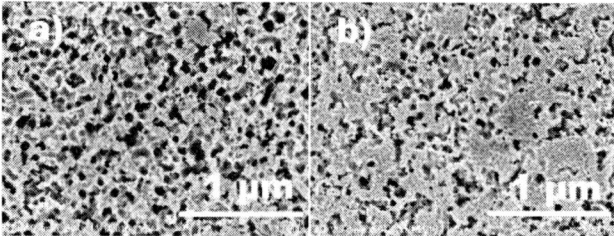

Fig. 6. Top surface SEM images of nanoporous Si etched in the one-step Cu MACE process utilizing a) [HF] = 11.6 M b) [HF] = 5.8 M.

Next, keeping [HF] = 11.6 M and ρ = 92%, the concentration of CuSO$_4$.5H$_2$O in the one-step solution (and hence [Cu^{2+}]) was varied from 0.01 M to 0.02 M, in steps of 0.005 M. We found that the lowest R_{avg} of 4.65% was obtained when [Cu^{2+}] was 0.015 M (Fig. 7), while the R_{avg} increases for higher or lower [Cu^{2+}]. So, the optimum [Cu^{2+}] was taken as 0.015 M for future one-step experiments. Below the optimum Cu^{2+} concentration, the catalytic rate of Cu goes down. Above the optimum concentration, there is a chance of continuous Cu film creation on Si surface, which also inhibits the etching. This explains the surface reflection trend in Fig. 7. It should be noted that the optimum [Cu^{2+}] in the one-step MACE process is higher than the one in the two-step process.

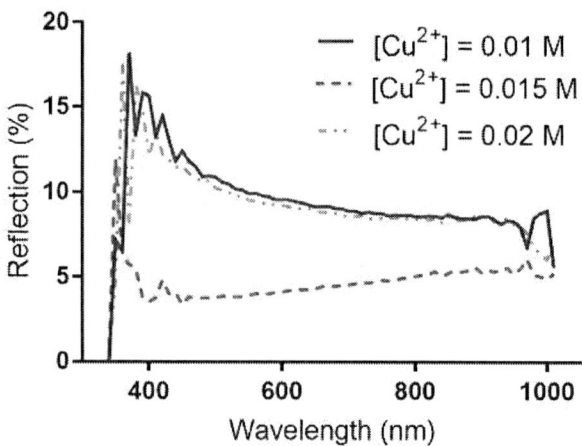

Fig. 7. Reflection as a function of wavelength for bSi for different Cu^{2+} concentrations utilized in the one-step MACE process.

In addition, [H$_2$O$_2$] in the one-step MACE solution was varied from 0.8 M to 1.4 M, while [HF] and [Cu^{2+}] were kept at 11.6 M and 0.015 M, respectively (Fig. 8). Etching time for each case was 20 minutes. We see from Fig. 8 that the lowest R_{avg} (3.36%) was obtained for [H$_2$O$_2$] = 1 M (data point circled in the inset of Fig. 8), while the R_{avg} increases for higher or lower H$_2$O$_2$ concentrations. This happens because there is a competition between H$_2$O$_2$ and Cu in terms of reacting with Si and catalyzing the etching reaction [10]. Higher than optimum H_2O_2 slows down the etching reaction rate by diminishing the Cu catalysis, while lower than optimum H_2O_2 reduces the etching rate by minimizing the oxidation of Si. So, the optimum $[H_2O_2]$ in the one-step MACE solution was chosen as 1 M for etching 1 cm²-area samples.

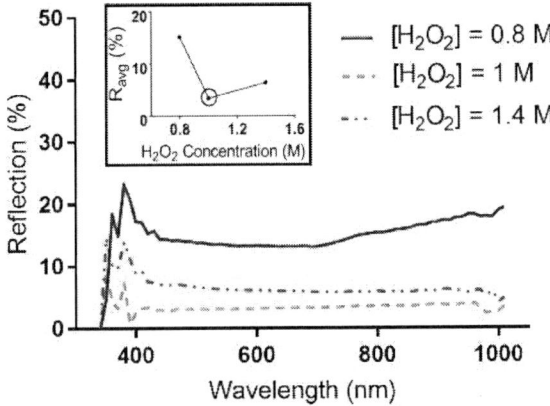

Fig. 8.   Reflection as a function of wavelength for 1 cm²-area bSi samples at different H_2O_2 concentrations utilized in the one-step MACE process. The inset shows the R_{avg} values as a function of H_2O_2 concentration.

(ii) Role of $C_6H_8O_6$ in One-Step MACE

In previous works, $C_6H_8O_6$ was used in the two-step Cu MACE process [12, 13], but it has not yet been used in the one-step process [10]. In order to determine whether the increased adhesion of Cu film due to the presence of $C_6H_8O_6$ would help in the one-step etching process, we prepared bSi samples with (0.01 M) and without $C_6H_8O_6$ in the one-step solution. Sample made with $C_6H_8O_6$ resulted in slightly lower R_{avg} (and lower reflection for wavelengths > 750 nm) than the one made without $C_6H_8O_6$ (Fig. 9). Top surface SEM images show that without $C_6H_8O_6$, circular pores are formed on the surface, while with $C_6H_8O_6$, much larger pores are formed, which have the shape of inverted pyramids (inset of Fig. 9). Such inverted pyramid shaped pores have been reported to result in enhanced light trapping [15], and hence, lower surface reflection.

(iii) Dependence of Optimum $[H_2O_2]$ on Sample Size

The samples used in all of the above experiments were 1 cm² in area. For making solar cells, we would use larger samples (area = 5.06 cm²). For these samples, we observed that etching failed to occur with 1 M $[H_2O_2]$ as the H_2O_2 got depleted before significant etching could take place. So, in order to find the optimum $[H_2O_2]$ for our large-area solar cell samples, we varied $[H_2O_2]$ in the one-step etching solution within 1 - 1.5 M (Fig. 10). Effective etching took place for $[H_2O_2] \geq 1.4$ M. The lowest R_{avg} of 3.65% was obtained at $[H_2O_2] = 1.5$ M (inset of Fig. 10). Etching time was kept at 20 minutes. It should be

mentioned that the HF concentration was kept constant (11.6 M) during this set of experiments, so the ρ got changed. The optimum $[H_2O_2] = 1.5$ M gives a ρ of 89%.

From our analysis, it can be concluded that for effective etching, $[H_2O_2]$ needs to be increased as the sample area increases. Our next step is to utilize the optimized one-step Cu MACE process to fabricate bSi solar cells.

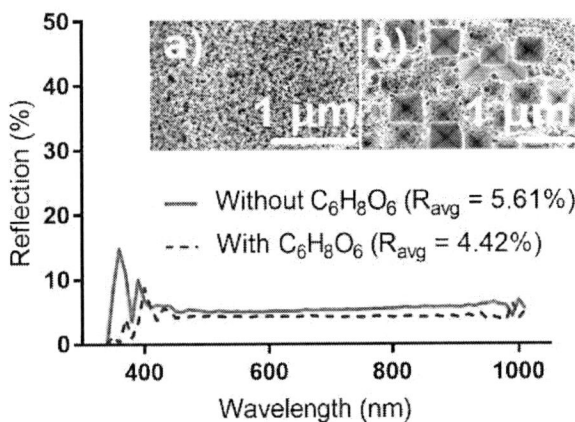

Fig. 9.   Reflection as a function of wavelength for bSi after one-step Cu MACE (inset shows top surface SEM image a) without $C_6H_8O_6$ and b) with $C_6H_8O_6$).

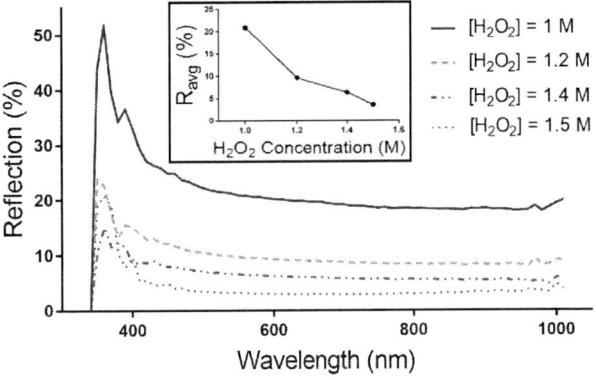

Fig. 10.   Reflection as a function of wavelength for 5.06 cm²-area bSi samples with varying H_2O_2 concentrations in the one-step MACE process. The inset shows the R_{avg} values as a function of H_2O_2 concentration.

(iv) Fabricating Cu MACE bSi solar cells

Two-inch single-side polished mono-crystalline Si (100) Czochralski (CZ) wafers (p-type, 1-10 Ω-cm, University Wafer) were cut into 5.06 cm²-area samples. Each sample was cleaned utilizing the standard RCA1 and RCA2 cleaning processes, followed by acetone and isopropanol (IPA) cleaning, and a 1 min dip in 10% HF (in DI water) solution. Then the samples were converted to black Si (bSi) utilizing the optimized one-step Cu MACE process described above. In order to create a p-n junction, n-type doping was performed by ammonium dihydrogen phosphate (ADP) solution (15% in water). Before doping, the samples were dipped into 30% H_2O_2 solution at room temperature for 5 minutes to increase hydrophilicity [16]. Then, the ADP solution was spin coated on p-type bSi at 1000 rpm. Subsequently, the bSi samples were put on a hotplate (at 100 °C for 8 minutes) and then inside an open tube furnace (950 °C) for 90 minutes to create the n-type emitter. The rear contact was created initially by e-beam deposition of 2 μm aluminum (Al) on the back of each sample. The samples were then placed in the tube furnace (closed with 300 sccm N_2 flow) for 10 minutes for the annealing of the back contact and the creation of back surface field (BSF). Then, another 2 μm of Al was e-beam deposited on the back of each sample to create low resistance back contact. Front contact grid was patterned by photolithography, and the front contact was made by subsequent e-beam deposition of 50 nm titanium (Ti) and 1 μm silver (Ag). Edge isolation was performed in each cell by reactive ion etching of Si (etching depth = 3.22 μm). For front surface passivation, a SiN_x layer (81 nm) was deposited on the front surface. In this batch, the highest efficiency obtained from a Cu MACE bSi cell was 9.1%. The output characteristics of this bSi cell and a planar non-bSi cell from the same batch (both measured under AM1.5G) are shown in table 1.

TABLE I

OUTPUT CHARACTERISTICS OF FABRICATED SOLAR CELLS

Cell type	Short circuit current density, J_{sc} (mA/cm^2)	Open circuit voltage, V_{oc} (V)	Fill factor, FF (%)	Efficiency, η (%)
bSi cell	29.74	0.55	55.85	9.1
Planar non-bSi cell	26.17	0.55	57.41	8.3

The light J-V curves of the cells (under AM1.5G) are shown in Fig. 11. The external quantum efficiency (EQE) curves of the cells, plotted along with the normalized photon flux in AM1.5G is shown in Fig. 12. We can see that the short wavelength response (blue response) of the planar non-bSi cell is better than that of the bSi cell. This is due to the fact that bSi nanostructure acts as a recombination-active zone because of its high surface area (surface recombination) and heavy emitter doping (Auger recombination). The latter is caused by the heavy in-diffusion of n-type dopant from the high surface area of the nanostructure. These factors damage the blue response of the bSi cell [17].

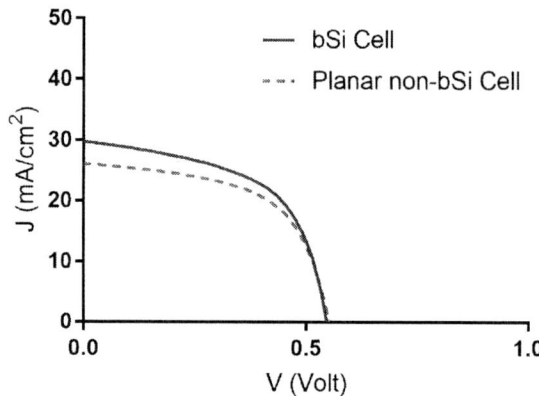

Fig. 11. Light J-V curves of the fabricated solar cells.

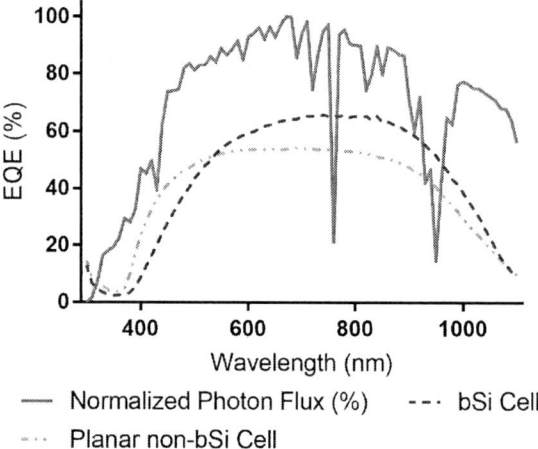

Fig. 12. EQE curves of the fabricated solar cells, plotted along with the normalized photon flux.

We are currently working on improving the fill factor, since it is low, by improving the shunt and series resistances of our solar cells. We calculated the shunt resistance values and those were 124.6 and 164 Ω-cm^2 for the bSi cell and the planar non-bSi cell mentioned above, respectively. We can see that the shunt resistance values are low, and we plan to increase those by improving edge isolation of the solar cells. To minimize the series resistance of the top contacts, we plan to utilize the rapid thermal processing (RTP) tool to induce alloying between the top contact metals (Ti/Ag) and Si that can reduce the series resistance.

III. CONCLUSIONS

This work analyzes the role of different process parameters in one- and two-step Cu nanoparticle-based MACE of Si and provides the details of an optimized one-step Cu MACE process. We also present results on the advantages of using $C_6H_8O_6$ in one-step Cu MACE. We have found that the optimum H_2O_2 concentration in the one-step solution depends on sample size, and it increases for larger samples. Finally, the output characteristics of a Cu MACE bSi cell from our initial batch have been reported. At present, we are working on improving the solar cell output characteristics. The findings of this work will help establish a cost-effective Cu MACE process to replace the currently utilized technologies for obtaining AR in Si solar cells.

ACKNOWLEDGEMENTS

We gratefully acknowledge the Iowa Energy Center Grant Number: OG-16-019, the Internal Funding Initiative (IFI) grant sponsored by the University of Iowa Vice President of Research Office, and the 2015 Old Gold Summer Fellowship awarded by the University of Iowa to Fatima Toor.

REFERENCES

[1] F. Toor, J. B. Miller, L. M. Davidson, L. Nichols, W. Duan, M. P. Jura, et al., "Nanostructured silicon via metal assisted catalyzed etch (MACE): chemistry fundamentals and pattern engineering," *Nanotechnology*, vol. 27, p. 412003, Oct 14 2016.

[2] R. B. Stephens and G. D. Cody, "Optical reflectance and transmission of a textured surface," *Thin Solid Films*, vol. 45, pp. 19-29, 1977/08/15/ 1977.

[3] S. Koynov, M. S. Brandt, and M. Stutzmann, "Black nonreflecting silicon surfaces for solar cells," *Applied Physics Letters*, vol. 88, pp. 203107 - 203107-3, 2006.

[4] X. Li and P. W. Bohn, "Metal-assisted chemical etching in HF/H2O2 produces porous silicon," *Applied Physics Letters*, vol. 77, pp. 2572-2574, 2000/10/16 2000.

[5] K. Tsujino, M. Matsumura, and Y. Nishimoto, "Texturization of multicrystalline silicon wafers for solar cells by chemical treatment using metallic catalyst," *Solar Energy Materials and Solar Cells*, vol. 90, pp. 100-110, 1/6/ 2006.

[6] M. Kulakci, F. Es, B. Ozdemir, H. E. Unalan, and R. Turan, "Application of Si Nanowires Fabricated by Metal-Assisted Etching to Crystalline Si Solar Cells," *IEEE Journal of Photovoltaics*, vol. 3, pp. 548-553, 2013.

[7] Z. Huang, N. Geyer, P. Werner, J. de Boor, and U. Gösele, "Metal-Assisted Chemical Etching of Silicon: A Review," *Advanced Materials*, vol. 23, pp. 285-308, 2011.

[8] J. Oh, H.-C. Yuan, and H. M. Branz, "An 18.2%-efficient black-silicon solar cell achieved through control of carrier recombination in nanostructures," *Nat Nano*, vol. 7, pp. 743-748, 11//print 2012.

[9] M. C. Putnam, D. B. Turner-Evans, M. D. Kelzenberg, S. W. Boettcher, N. S. Lewis, and H. A. Atwater, "10µm minority-carrier diffusion lengths in Si wires synthesized by Cu-catalyzed vapor-liquid-solid growth," *Applied Physics Letters*, vol. 95, pp. 163116 - 163116-3, 2009.

[10] Y. Cao, Y. Zhou, F. Liu, Y. Zhou, Y. Zhang, Y. Liu, et al., "Progress and Mechanism of Cu Assisted Chemical Etching of Silicon in a Low Cu2+ Concentration Region," *ECS Journal of Solid State Science and Technology*, vol. 4, pp. P331-P336, 2015.

[11] K. Q. Peng, J. J. Hu, Y. J. Yan, Y. Wu, H. Fang, Y. Xu, et al., "Fabrication of Single-Crystalline Silicon Nanowires by Scratching a Silicon Surface with Catalytic Metal Particles," *Advanced Functional Materials*, vol. 16, pp. 387-394, 2006.

[12] L. Magagnin, R. Maboudian, and C. Carraro, "Selective deposition of thin copper films onto silicon with improved adhesion," *Electrochemical and Solid-State Letters*, vol. 4, pp. C5-C7, 2001.

[13] F. Toor, J. Oh, and H. M. Branz, "Efficient nanostructured 'black' silicon solar cell by copper-catalyzed metal-assisted etching," *Progress in Photovoltaics: Research and Applications*, vol. 23, pp. 1375-1380, 2015.

[14] C. Chartier, S. Bastide, and C. Lévy-Clément, "Metal-assisted chemical etching of silicon in HF–H2O2," *Electrochimica Acta*, vol. 53, pp. 5509-5516, 7/1/ 2008.

[15] S. Eyderman, S. John, M. Hafez, S. S. Al-Ameer, T. S. Al-Harby, Y. Al-Hadeethi, et al., "Light-trapping optimization in wet-etched silicon photonic crystal solar cells," *Journal of Applied Physics*, vol. 118, p. 023103, 2015.

[16] Y. Tang, C. Zhou, W. Wang, Y. Zhao, S. Zhou, J. Fei, et al., "N+ emitters realized using Ammonium Dihydrogen Phosphate for silicon solar cells," *Solar Energy*, vol. 95, pp. 265-270, 2013.

[17] F. Toor, H. M. Branz, M. R. Page, K. M. Jones, and H.-C. Yuan, "Multi-scale surface texture to improve blue response of nanoporous black silicon solar cells," *Applied Physics Letters*, vol. 99, p. 103501, 2011/09/05 2011.

Impact of front side pyramid size on the light trapping performance of wafer based silicon solar cells and modules

Oliver Höhn, Nico Tucher, Benedikt Bläsi

Fraunhofer Institute for Solar Energy Systems ISE, Heidenhofstraße 2, 79110 Freiburg, Germany

Abstract — **Smaller pyramid sizes for Silicon solar cell front side textures attract more and more interest. At the same time a very good optical performance of the front side texture is crucial to achieve high PV module efficiencies. In this paper an optical study of the impact of periodically arranged front side pyramids on the useful absorption in a Silicon solar cell is presented. Also a way to account for random pyramids is described. Results for solar cells facing semi-infinite EVA/glass are compared to the modeling results of a full solar module stack with the help of the OPTOS formalism. We found that the impact of different pyramids differs depending on the quality of the planar rear side reflector. It is shown that the impact of the module case is crucial and that it has to be accounted for when predicting the optimal pyramid sizes that are relevant for cell and module manufacturers. We also demonstrate that the use of simple parameters such as a single pass light path enhancement factor for the structured surface can lead to wrong conclusions, and a full modeling is required to predict the real module performance. The results indicate that the overall optical performance of a solar module does almost not vary for pyramid sizes above 600 nm.**

I. INTRODUCTION

As silicon solar cell efficiency improves further, the optical performance – the quality of the light-trapping – becomes more and more important. Random pyramids at the front side of silicon solar cells are the industrial standard for wafer based monocrystalline silicon solar cells. These pyramids address two different needs for the optical performance: They decrease the reflection loss on the one hand and allow for a path length enhancement and therefore a better absorption inside the Silicon solar cells on the other hand. In state of the art industrial textures, typical pyramid sizes are in the range of several micrometers [1]. In recent research and development pyramids with smaller sizes in the range of several 100nm gained more attention due to different reasons. From a production point of view the realization of smaller pyramids can improve the process chain, as there is. a cost reduction potential via a process time reduction and also a reduction in material consumption. But even more importantly, the evaluation of pyramid sizes gains more interest in order to maximize the light-trapping performance of the structure and with this to maximize module efficiency. While recent literature [2–4] evaluates smaller pyramids mainly on cell level, in this work also the optical influence of the module stack is accounted for. With the help of the OPTOS formalism [5, 6] the influence of periodically arranged pyramids with different periods – and therefore different pyramid sizes – to the light trapping performance is

systematically evaluated. We choose a periodic arrangement as it is straightforward to model wave optically e.g. with the help of rigorous coupled wave analysis (RCWA). Even more important is that we can obtain basic physical insight into the different effects that influence the light-trapping performance. In the end, a way to account for randomness with the help of scattering assumptions is briefly described and preliminary conclusions of first estimations are shortly summarized.

Within this work we show that predicting the system performance just with the help of a few parameters or even only one single parameter can lead to wrong conclusions. Here, we especially focus on the light path enhancement factor (LPEF) and the ratio of modes in the superstrate to modes within Silicon (R_{trap}). We depict that it is important to also account for the module stack and not only focus on the cell level.

II. OPTOS FORMALISM

The OPTOS formalism [5] was developed by Tucher et al. and allows for the optical modelling of solar module stacks [6] while also light-trapping within the module glass is accounted for. OPTOS has the advantage that different interfaces acting in different optical regimes can be combined in an efficient way. Changes in parameters such as the cell thickness or the specular reflectance of the assumed rear side mirror can be realized very efficiently so that such parameter variations can be performed with low computational effort. A detailed description of the OPTOS formalism can be found in [5, 6].

III. COMPARISON OF DIFFERENT PERIODIC FRONT SIDE PYRAMIDS

We evaluate three simple parameters to explain the behavior of the structure:

1. The **light path enhancement factor (LPEF)** of a single pass is typically defined as

   $$\text{LPEF} = \frac{\Sigma_{pq} \beta(\theta_{pq}) \cdot \eta_{pq}}{\Sigma_{pq} \eta_{pq}},$$

   where η_{pq} is the diffraction efficiency in to the diffraction order (p,q) and $\beta(\theta_{pq}) = 1/\cos\theta_{pq}$ is the light path enhancement factor of the specific diffraction order of a given surface texture. It describes how strongly the weighted light path is increased in comparison to a perpendicular pass by a given surface texture.

2. The **reflectance** that is weighted with the AM1.5g solar spectrum from 300nm to 1200nm, which is the relevant wavelength range for a Silicon solar cell.
3. The mode ratio R_{mode} as number of propagating diffraction orders M in silicon compared to the number of propagating diffraction orders in the superstrate (EVA/Air):

$$R_{mode} = M_{Silicon}/M_{EVA/Air}$$

A. Solar cells facing semi-infinite glass/EVA

Silicon solar cells are encapsulated in almost all relevant applications. Therefore, the behavior with air as surrounding medium is of minor importance and we start the evaluation with a solar cell facing a semi-infinite glass respectively EVA encapsulation. Between the EVA/glass and the Silicon solar cell an antireflective coating (ARC) made from SiNx [7] with a thickness of 70nm is assumed. At the rear side of the 180μm thick silicon solar cell a perfect mirror is assumed. It is important to account for the ARC as it does not only change

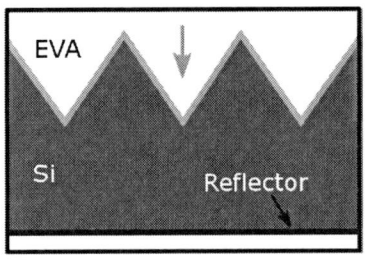

Fig. 1   Sketch of the modeled structure. A 70nm thick ARC made from Si3N4 is assumed at the Si/EVA interface. At the rear side an ideal mirror is assumed. The Silicon solar cell thickness was set to 180μm.

reflectance, but also has a significant influence to the light distribution within the Silicon solar cell that is caused by diffraction at the pyramids.

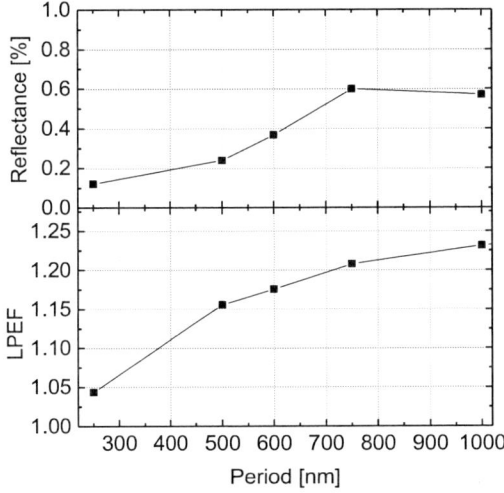

Fig. 2   Weighted reflectance and LPEF of the structure shown in Fig. 1

At first the spectrally weighted reflectance and the spectrally weighted LPEF are considered. Having a closer look at the

reflectance, depicted in Fig. 2, one can see that it is extremely low for all assumed pyramid sizes. The LPEF increases with increasing structure period, which means that the texture redistributes the light into larger angles. Therefore, one could argue that due to these two parameters the bigger pyramids behave slightly better for wavelengths close to the band gap as the longer light path within the structure could lead to higher absorption.

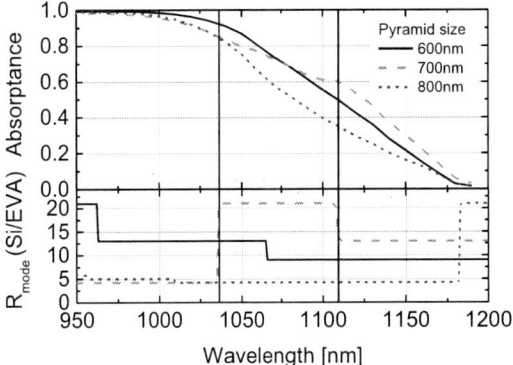

Fig. 3   Absorptance and R_{mode} of the system shown in Fig. 1

In order to verify this preliminary conclusion, the OPTOS formalism was used to model the described structure. The respective results are shown in Fig. 3. As one can see, this first conclusion could not be confirmed: The system with the 800nm pyramids absorbs less light in the long wavelength region than with the 600nm pyramids.

The LPEF only assumes the path length enhancement within one light path. Usually photons at wavelengths close to the band gap need multiple light paths before being absorbed.

So, the transmission loss at the front side at each interaction will also play an important role. Due to this reason the above mentioned mode ratio R_{mode} was calculated and is also shown in Fig. 3. Comparing this ratio for the 600nm and 800nm pyramids one can see that the mode ratio is higher in the 800nm case in the relevant wavelength regime. For a good absorption it therefore seems to be more important to trap light over many interactions. The relative escape is highly relevant for each interaction with the front side, so a higher mode ratio can lead to a better trapping of light over many interactions. Having a look at the 700nm periods within the marked regime, it becomes obvious that also this simple parameter of the mode ratio alone does not allow for a reliable prediction of the optimal structure behavior. Probably not all diffraction orders play the same role integrated over all relevant interactions of light at the front side of the Silicon solar cell.

Hence, it must be concluded that the only reliable way of predicting the optimal pyramid size at the moment is to model the full system, e.g. using RCWA in combination with OPTOS.

978-1-5090-5606-4/17 $31.00 © 2017 IEEE   353

B. Solar cells in a solar module stack

Having this first conclusion in mind, we extended the investigation to the full system – meaning adding a physical

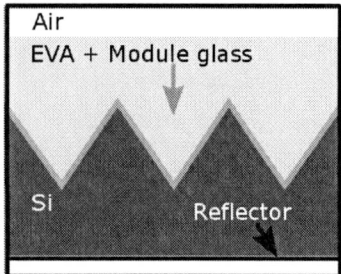

Fig. 4 Sketch of the modeled full module stack. A 70nm thick ARC made from Si3N4 is assumed at the Si/EVA interface. At the rear side an ideal mirror is assumed. The Silicon solar cell thickness was set to 180μm.

interface between the module glass and air. This interface can, as is well known from literature [6, 8–10], have a significant influence to the overall light trapping properties of the system. The result is shown in Fig. 5 again together with the ratio R_{mode}.

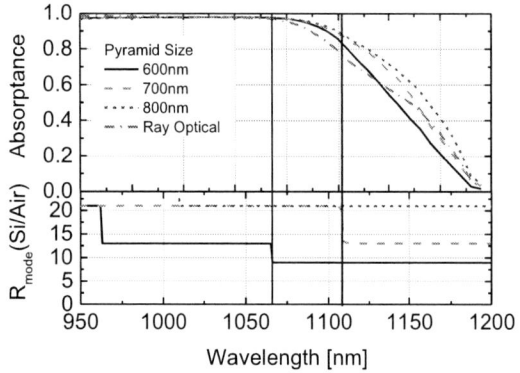

Fig. 5 Absorptance and Rmode of the system shown in Fig. 4

Comparing these curves to the results with the semi-infinite EVA/glass above, one can see that accounting for a real glass/air interface changes the performance. Again the above defined mode ratio gives a good indication that one important factor is to have lots of trapped modes. But still, this is not the only important parameter.

Here, additionally the results for regular pyramids that are sufficiently large for being treated ray optically are shown. It becomes obvious that small pyramids show a different behavior, so a ray optical treatment of the small pyramids is not sufficient.

C. Influence of the mirror reflectivity

Up to this point, the rear side of the Silicon solar cell was assumed to be perfectly reflective. In a last evaluation step the reflectivity of the rear side was changed continuously from 0 to 1 in order to estimate the importance of the rear side mirror quality. Again OPTOS modeling was performed to calculate the absorbed photon current in the wavelength range of 800-1200nm, where light trapping plays an important role.

The integrated current for the three different pyramid sizes as well as the ray optical limit is shown in Fig. 6. For very weak rear side reflectance there is hardly any difference between the shown pyramid sizes.

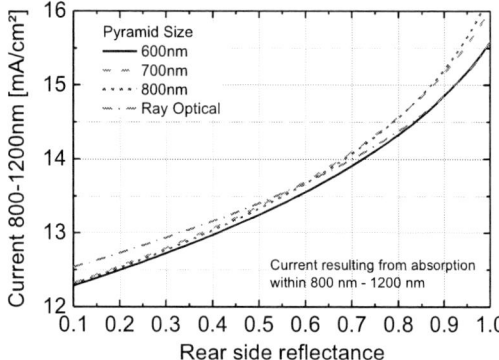

Fig. 6 Current resulting from absorption within 800-1200nm of the system shown in Fig. 4 for different pyramid sizes and rear side mirror reflectance

The only influences in this case are the slightly different reflectance and the slightly different LPEF. For high rear side reflectivity the ratio R_{trap} gains more importance. As this influence is non-linear, the slopes of the three curves are slightly different. One can see that for very good rear side reflectance a difference between the three pyramid sizes and the ray optical limit occurs. Still this difference is small. The maximal difference here is calculated to be 0.67 mA/cm² for perfect rear side reflectivity, 0.35 mA/cm² for 90% rear side reflectivity und 0.23 mA/cm² for 80%.

D. Influence of scattering

Typically, in industrial solar cells no regular but random pyramids are used. This irregularity could in principle be accounted for by implementing scattering into the front surface. First results indicate that this scattering even decreases the small difference between the different pyramid sizes in the case of perfect rear side reflection, meaning that in a real solar module hardly any difference will occur for different pyramid sizes assumed in this study.

IV. CONCLUSION

From the described broad analysis of complete solar module stacks with a variation of the texture dimensions, two general conclusions can be drawn:

1. The investigated simple parameters, e.g. LPEF, allow for an explanation and comparison of the different effects in a solar module stack. However, they do not allow for a reliable conclusion about the overall performance of a given front side texture. To find the optically optimal texture a full modeling of the module stack, e.g. with the help of OPTOS, is necessary.

2. In the investigated regime of pyramid sizes hardly any difference in absorption will be measurable, as the small

occurring effects will most likely vanish due to imperfect rear side reflectors and the influence of scattering especially in the case of random pyramids. It is likely that in a solar module any pyramid size above 600nm will lead to very similar measured absorption and therefore current.

V. ACKNOWLEDGEMENTS

This work was supported by the German Federal Ministry for Economics and Energy in frame of the GROSCHEN project (Contract no. 0324012B)

VI. REFERENCES

[1] P. Campbell and M. A. Green, "Light trapping properties of pyramidally textured surfaces," *J. Appl. Phys.*, vol. 62, no. 1, p. 243, 1987.

[2] C. Trompoukis *et al.*, "Photonic nanostructures for advanced light trapping in thin crystalline silicon solar cells," *phys. status solidi a*, vol. 212, no. 1, pp. 140–155, 2015.

[3] S. Sivasubramaniam and M. M. Alkaisi, "Inverted nanopyramid texturing for silicon solar cells using interference lithography," *Microelectronic Engineering*, vol. 119, pp. 146–150, 2014.

[4] P. Wang *et al.*, "Periodic Upright Nanopyramids for Light Management Applications in Ultrathin Crystalline Silicon Solar Cells," *IEEE J. Photovoltaics*, pp. 1–9, 2017.

[5] N. Tucher *et al.*, "3D optical simulation formalism OPTOS for textured silicon solar cells," *Opt. Express*, vol. 23, no. 24, pp. A1720, 2015.

[6] N. Tucher *et al.*, "Optical simulation of photovoltaic modules with multiple textured interfaces using the matrix-based formalism OPTOS," *Opt. Express*, vol. 24, no. 14, pp. A1083, 2016.

[7] S. Duttagupta, F. Ma, B. Hoex, T. Mueller, and A. G. Aberle, "Optimised Antireflection Coatings using Silicon Nitride on Textured Silicon Surfaces based on Measurements and Multidimensional Modelling," *Energy Procedia*, vol. 15, pp. 78–83, 2012.

[8] S. C. Baker-Finch, K. R. McIntosh, and M. L. Terry, "Isotextured Silicon Solar Cell Analysis and Modeling 1: Optics," *IEEE J. Photovolt.*, vol. 2, no. 4, pp. 457–464, 2012.

[9] J. Gjessing and E. S. Marstein, "An Optical Model for Predicting the Quantum Efficiency of Solar Modules," *IEEE J. Photovoltaics*, vol. 4, no. 1, pp. 304–310, 2014.

[10] I. Geisemeyer *et al.*, "Angle Dependence of Solar Cells and Modules: The Role of Cell Texturization," *IEEE J. Photovoltaics*, vol. 7, no. 1, pp. 19–24, 2017.

A Study of Blister Control of Al_2O_3 Thin Film Deposited by Plasma-assisted Atomic Layer Deposition after Firing Process

Min Gu Kang[1], Jeong In Lee[1], Hee-eun Song[1], Myeong Sang Jeong[1], Kyung Taek Jeong[1], Hyo Sik Chang[2]

[1]Korea Institute of Energy Research, Daejeon, 34129, South Korea

[2]Chungnam National University, Daejeon, 34134, South Korea

Abstract — **Al_2O_3 was deposited by RF plasma-assisted atomic layer deposition method. Al_2O_3 thin films having various OH group density were fabricated. These Al_2O_3 thin films were characterized by quasi-steady state photoconductance decay method, X-ray photoelectron spectroscopy, optical microscopy and capacitance-voltage characterization. The blisters were observed at the samples having low passivation properties after firing process. These samples had high OH group density and reducing OH group could suppress degrading passivation quality.**

I. INTRODUCTION

Crystalline silicon solar cell has dominated the photovoltaic market. Passivated emitter rear contact(PERC) cell has been studied by many groups. To achieve high efficiency of silicon solar cell, increasing light absorption and decreasing carrier recombination should be achieved. At the point of PERC cell, the cell has a passivation layer of rear surface. Recombination velocity of rear surface is reduced. To achieve the passivation properties, there are two kind of passivation methods which are chemical passivation method and field effective passivation method. At the Si surface, unpaired Si atoms are existed. These unpaired bonds are dangling bonds. These dangling bonds increase the recombination velocity. Chemical passivation is neutralizing dangling bonds at the silicon surface. One of chemical passivation methods is thermal oxidation process[1]. Thermally grown SiO_2 is known as the best surface passivation layer. The other passivation method is field effective passivation. Field effective passivation is repelling the minority carriers by the fixed charge in the passivation layer.

For the p type silicon surface which has electrons as the minority carriers, passivation layer with negative fixed charge is need. Annealed Al_2O_3 thin film which contains negative fixed charges is one of the p type silicon passivation layers.

For fabricating crystalline solar cells, high temperature firing process should be carried out to form electrode using metal paste. However, blistering could be formed during the firing process because Al_2O_3 thin film contains hydrogen atoms and hydrogen gas forms at the high temperature process.

Generally, Al_2O_3 thin film is deposited by thermal atomic layer deposition method. Water vapor as O source is used to deposit Al_2O_3 thin film. Water vapor is decomposed of OH group and H atom by heat, and OH group remains in Al_2O_3

thin film. This OH group in Al_2O_3 thin film forms H atoms and O interstitial atoms during the annealing process. H atoms passivate the silicon surface by effusion. O interstitial atoms make negative fixed charges. In case of plasma assisted atomic layer deposition (PAALD), O_2 gas as O source is used to deposit Al_2O_3 thin film. O_2 plasma is effective oxidants which can react with CH_3 ligand of trimethylaluminum (TMA) by O radical driven combustion-like reaction. Also, O radical and CH_3 form OH groups which are contained in Al_2O_3 thin film by PAALD [2].

In this study, Al_2O_3 film was deposited by PAALD. Electrical properties of Al_2O_3 thin film were characterized by quasi-steady-state photoconductance(QSSPC), capacitance voltage(C-V), conductance-voltage(G-V), and X-ray photoelectron spectroscopy(XPS).

II. EXPERIMENTAL PROCEDURE

In this experiment, two types of substrates were used. To measure effective minority carrier lifetime, solar grade Si wafer with thickness of 200μm and resistivity of 1-3 Ω cm was used. To characterize interface trap density (D_{it}), fixed charge (Q_{tot}) and chemical states, electronic grade one side polished Si wafer with 660μm and resistivity of 5-10 Ω cm was used.

Samples for characterizing effective minority carrier lifetime have a sandwich-like structure of SiNx/Al_2O_3/p type Si/ Al_2O_3/SiNx. P type Si substrate was texturized by KOH/IPA mixture. And Al_2O_3 thin film of 10nm thickness was deposited by PAALD. The RF powers were varied from 200W to 800W. TMA as Al source and O_2 gas as O source were used. The deposition temperature, process pressure, O_2 gas flow and process time of O_2 plasma condition were 250 °C, 1.0 Torr, 1000 sccm and 1.5 seconds per 1 cycle, respectively. TMA source was vaporized during 1 second per 1 cycle. N_2 gas flow rate and time at purge cycle were 1000 sccm and 6 seconds. Silicon nitride of 80 nm for capping layer was deposited by plasma enhanced chemical vapor deposition (PECVD). Deposition temperature, process pressure, SiH_4 gas flow rate, NH_3 flow rate, N_2 gas flow rate, RF power and process time were 400 °C, 0.8 Torr, 80 sccm, 130 sccm, 1700 sccm, 400 W and 84 seconds, respectively.

978-1-5090-5606-4/17 $31.00 © 2017 IEEE

Samples for X-ray photoelectron spectroscopy (XPS), capacitance-voltage measurement (C-V) and conductance-voltage measurement (G-V) had a structure of Al$_2$O$_3$/p type Si. All samples were annealed at 425 °C for 15 minutes in a N$_2$ atmosphere and fired at 870 °C.

The minority carrier lifetime of the sample was characterized by quasi-steady-state photoconductance decay method (QSSPC). Qtot and Dit were calculated from C-V measurement and G-V measurement, respectively. Mercury probe was used for the C-V and G-V measurements to contact the samples. To characterize the concentration of OH groups of as-deposited, annealed and fired samples, angle-resolved XPS system with AlKα as excitation source was used.

III. RESULT AND DISCUSSION

Fig 1 shows the QSSPC result when subjected to the respective steps of the Al$_2$O$_3$/SiNx:H stack structure. As the RF power was increased, lifetime was increased in the annealing process. OH is generated by the combustion effect, it is included in the Al$_2$O$_3$ film by chemical absorption at the deposition in layer by layer. Lifetime of radical increases as plasma power is increased [3]. Accordingly, as plasma power is increased, it is expected that the number of radicals and OH concentration are increased. OH in Al$_2$O$_3$ film is decomposed into O and H during the annealed. The hydrogen is moved in the Al$_2$O$_3$/Si interface by the capping layer. Hydrogen which has moved to the interface is attached to a dangling bond. Depending on increase of OH concentration, chemical passivation was increased by annealing process. On the other hand, remaining O radicals have negative fixed charge in Al$_2$O$_3$ bulk within interstitial. Depending on the increase of OH, field effect passivation also increases. Lifetime and iVoc were increased depending on the OH concentration when annealing temperature was lower than the hydrogen evolution temperature. However, it is larger than the Si-H bond energy, hydrogen is released at the interface. In this case, the hydrogen is released in the form of molecule. If the number of released hydrogen was increased, it is to give damage to the Al$_2$O$_3$ film. This phenomenon is called as blister, which is formed at the Al$_2$O$_3$/SiNx:H stacks after firing process. Lifetime was decreased above 800 W during firing process, this result was expected due to blister. From this result, RF power of 700 W has the most appropriate OH concentration. After firing process, lifetime was increased in 100-700 W. It was expected by an influence of the fixed charge.

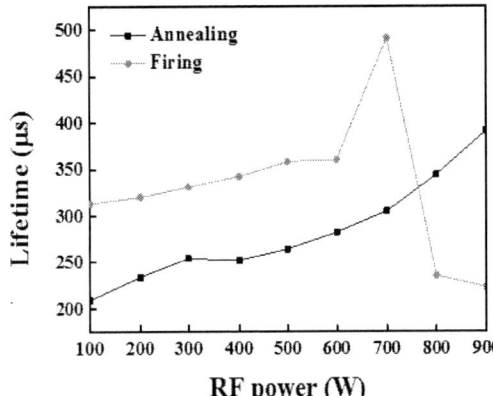

Fig. 1. Characterization of passivation property with different RF powers

Table 1 shows the fixed charge density and interface properties of Al$_2$O$_3$ thin film. Al$_2$O$_3$ film was annealed for 15 minutes in N$_2$ ambient at 425 °C. Electrode in the MOS structure was used for mercury (99.999%). This mercury electrode was used to avoid the annealing process for the metal contact.

SiNx:H is known to have a positive fixed charge, SiNx:H makes recombination of holes and electrons at the surface of n-type semiconductor. On the other hand, Al$_2$O$_3$ has a negative fixed charge. Al$_2$O$_3$ film makes recombination of holes and electrons at the surface of p-type substrate because this negative fixed charge repels electron at the surface. The flat band voltage(V_{FB}) was increased as the RF power was increased. Flat band voltage was shifted that it changes the fixed charge. Also, the fixed charge density(Q_{tot}) was increased as the RF power was increased. Flat band voltage was increased when RF power increased, which showed that negative fixed charge was increased. Therefore, fixed charge increased as RF power increased.

TABLE I
FIXED CHARGE DENSITY AND INTERFACE PROPERTIES OF AL$_2$O$_3$ THIN FILM

RF power (W)	After annealing process		
	$V_{FB}(V)$	$Q_{tot}(cm^{-2})$	$D_{it}(eV^{-1}cm^{-2})$
200	1.31	-9.34×10^{12}	5.84×10^{11}
500	1.34	-9.52×10^{12}	2.29×10^{11}
800	1.89	-1.22×10^{13}	2.23×10^{11}

The surface was observed by OM. Fig 2 shows that Al$_2$O$_3$/SiNx stack with thickness of 10/80 nm was observed by OM after firing. Blistering was observed after the firing process at RF power over 800 W. $Al_2O_3/SiNx$ film was delaminated by blistering, as a result, silicon surface was exposed by the blister. Silicon surface has a lot of dangling bonds and dangling bonds make the trap sites in the bandgap, therefore, these trap sites causes the recombination center of electrons-hole pair.

The number of blister at the fired sample deposited with RF power under the 700 W was very small or non-existent. There was no damage to the capping layer by hydrogen release at the Al_2O_3 film containing the concentration of OH below 4.01 %. Blister was generated when RF power was over 800 W. Because Al_2O_3 layer deposited with RF power over 800 W contains a lot of hydrogen, the number of the blister was increased as RF power was increased. From this result, the optimized RF power was 700 W.

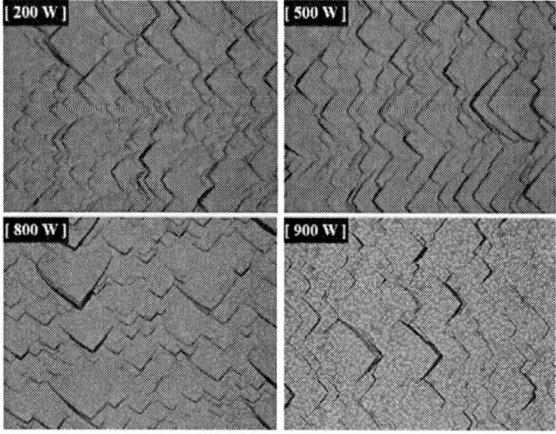

Fig. 2. OM images of $Al_2O_3/SiNx$ stack layer on c-Si after firing process

IV. CONCLUSION

The objective of this work is reduction of back surface recombination velocity and improvement of back surface passivation quality of crystalline silicon solar cells. In order to improve the back surface passivation quality, PAALD deposited Al_2O_3 film was introduced on back surface of solar cells. Al_2O_3 is promising material as an alternative of aluminium back surface field (Al-BSF), providing better Voc compared to that of conventional screen-printed Al-BSF. Al_2O_3 film forms negative fixed charge on surface of crystalline silicon, resulting in much reduced surface recombination.

As-deposited Al_2O_3 film does not form negative fixed charge on crystalline silicon surface. The negative change can be formed by thermal annealing process. As-deposited Al_2O_3 film consists of OH groups, and the OH groups are disassociated and produce atomic hydrogen by thermal annealing. Thermal annealing generates atomic hydrogen and leads passivation of dangling bond on crystalline silicon surface. As a result, Al_2O_3 film chemically passivates crystalline silicon surface. Furthermore, thermal annealing introduces Al_2O_3 transform from octahedral to tetrahedral structure and from O interstitial. Since O interstitial has negative charge, negative fixed charges are formed in entire Al_2O_3 film. Therefore, Al_2O_3 film also acts as field effect passivation.

Comprehensive study of C-V and G-V measurement, demonstrated that improvement of back surface passivation of Al_2O_3 film was consequences of both chemical passivation and field effect passivation. XPS was used for determining OH content of Al_2O_3 film. OM reveals that Al_2O_3 films deposited at RF power higher than 800 W accompany blisters. Therefore, we suggest Al_2O_3 film should have decent hydrogen content for making chemical passivation, but any excess hydrogen content leads to deterioration of passivation caused by hydrogen effusion and blister formation after firing. Finally, we obtained optimized RF power for Al_2O_3 deposition of the crystalline silicon back surface passivation.

REFERENCES

[1] A. G. Aberle, et al., "High - efficiency silicon solar cells: Si/SiO_2, interface parameters and their impact on device performance," *Progress in Photovoltaics: Research and Applications*, vol. 2, p. 265-273, 1994.

[2] S. B. S. Heil, et al., "Reaction mechanisms during plasma-assisted atomic layer deposition of metal oxides: A case study for Al_2O_3," *Journal of Applied Physics e*, vol. 103, p. 103302, 2008.

[3] T. O. KÄÄRIÄINEN, D. C. Cameron, "Plasma-Assisted Layer Deposition of Al_2O_3 at Room Temperature," *Plasma Processes and Polymers*, vol. 6, p.S237-S241, 2009.

Pypvcell: An Open-Source Solar Cell Modeling Library in Python

Kan-Hua Lee, Kenji Araki, Omar Elleuch, Nobuaki Kojima and Masafumi Yamaguchi

Abstract—We announced a open source solar cell modeling and analysis toolkit written in Python. The standard off-the-shelf solar cell simulation software is often difficult to modify or reuse some of its functionality into new solar cell models. In this software package, we wrap the solar cell simulations into individual modules and application programming interfaces to make them very user-friendly. This library contains a wide range of functions to do the heavy-lifting, error-prone jobs of modeling solar cells, such as unit conversions and arithmetic operations of spectrum data, absorption-emission reciprocity or solving the I-Vs of multi-junction cells. This allows the users to rapidly adapt and build their own models with these modularized and verified components. Source codes, detailed documentation and examples can be found in https://kanhua.github.io/pypvcell.

Index Terms—solar cells, simulation

I. INTRODUCTION

Modeling solar cells is an important part in designing and optimizing solar cell structure. The most common option is using off-the-shelf solutions to model the solar cells, such as PC1D and its descendants [1] [2] [3]. Their graphical user interface lowers the slope of learning curve, however, it is difficult to automate large amount of iterative simulation processes or tailor the model to suit their research purposes. Researchers therefore choose to develop their own codes from scratch. However, this slows down the development of solar cell modeling because significant amount efforts were spent on reinventing the wheels that may have been built by others. Therefore, in our opinion, an ideal solar cell modeling software package for advanced users should meet the following requirements:

1) It is formed by a rich set of carefully designed application programming interfaces (APIs), so the users can easily plug any parts of the functionalities of the software to their own model.
2) The code is well-documented and open-sourced, allowing everyone to validate and adapt the codes for their own research
3) Each functions are validated and tested comprehensively.

Pypvcell was therefore designed and implemented to meet these requirements. Our goal is to define a standard of APIs so that everyone can build their own solar cell model rapidly and share their models under a common framework.

Pypvcell is written in Python [4], a general-purpose programming language that recently become very popular in scientific computing. Implementing the solar cell modeling library in Python also makes it easy to streamline the energy yield prediction with PVLIB-Python [5], a renowned Python softwarepackage for predicting the energy yields.

This paper will present an overview of the design concept, followed by the progress and the validation of Pypvcell. We would like to emphasize that although these design concepts are implemented in python in this work, these can also be implemented in other modern programming languages that supports object-oriented programming.

II. DESIGN OF THIS PACKAGE

A. Architecture of Pypvcell

The design architecture of Pypvcell is illustrated in Fig. 1. Pypvcell focuses on the simulation of the solar cell. Most of the low-level numerical computation such as matrix operations and interpolation are rely on Numpy [6] and Scipy [7], which are the definitive packages for numerical calculation in Python. A rich set of functions that performs the operations related to solar cells, including optics, photogeneration and carrier recombination, were built upon Numpy and Scipy. These function are further encapsulated into different solar cell models. In Pypvcell, every solar cell model is defined as a class. For example, the following codes initializes an instance of solar cell object that implements the radiative-limit model:

```
gaas=SQCell(eg=1.42,cell_T=292)
```

The above code initialize a radiative-limit cell with a band gap of 1.42 eV. The solar cell models that are currently implemented in Pypvcell are listed in TABLE I.

These solar cell classes are all inherited from the same base class and share the same interfaces of retrieving the electrical and optical output. These standardized interface let the switching between different the solar cell model very easy.

Pypvcell currently simulate two-terminal series connected multi-junction solar cell by solving the I-V of series connected individual subcells. One can establish this multi-junction cell by putting different solar cell instances in a pipeline. This approach can accommodate subcells with different models. For instance, a multi-junction solar cell can be set up like this:

```
gaas=SQCell(eg=1.42,cell_T=292)
si=DBCell(qe,rad_eta)
mj=MJCell([gaas,si])
```

In this example, the multi-junction cell has a GaAs top cell at Shockley-Queisser limit, and a silicon bottom cell with user-defined quantum efficiency and radiative efficiencies. Users can easily write the model for their own solar cell and incorporate it into the multi-junction cell as long as they use the same API interface to name the functions.

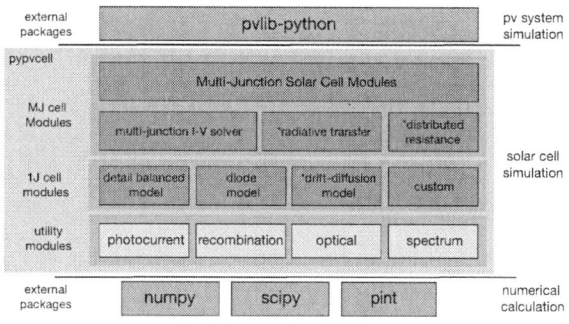

Fig. 1. The architecture of Pypvcell.

B. Data Structure

In general, Pypvcell uses `ndarray` defined in Numpy as the main data structure in the calculation. `ndarray` is the standard data structure for storing the values of vector and matrix. Internally, the physical quantities are converted to SI units for calculation. In the current version, the user has to explicitly convert the physical quantities to SI units when inputting the values to the APIs. The capability of handling physical quantities with dimensions will be supported in future versions.

Modeling solar cells involves a lot of arithmetic operations, interpolation and unit conversions of spectrum data, such as solar spectrum, quantum efficiency, absorption coefficients and so on. These operations can be very troublesome and error prone. We therefore implemented a class called `Spectrum` to do these operations. An instance of `Spectrum` class stores two arrays, the abscissa and the ordinate of a spectrum, which are denoted as `x` and `y` in later discussion, respectively. Despite that it may seem inconvenient to bundle the x and y data in a class instance, it can save tremendous efforts of debugging especially when doing unit conversions, since the values of y are often coupled with x. Also, we have created many functions in `Spectrum` to reduce these inconveniences and make the syntax as intuitive as possible.

An instance of the class can be initialized by the following code:

```
sp=Spectrum(x_data=wavelength,
            y_data=spec,x_unit='nm')
```

Arithmetic operations in `Spectrum` is very intuitive, for example,

```
# multiply every element of y by 0.8
new_sp=sp*0.8

# sum up the spectrum of sp1 and sp2
sp_sum=sp1+sp2
```

In the above code snippet, if x in `sp1` and `sp2` do no match, `Spectrum` class will first interpolates `sp2` and perform then perform addition on `sp1`.

The value of `spectrum` is retrieved by `get_spectrum` method for plotting or exporting. When retrieving the values of x and y, the user has to specify the units of x. The user

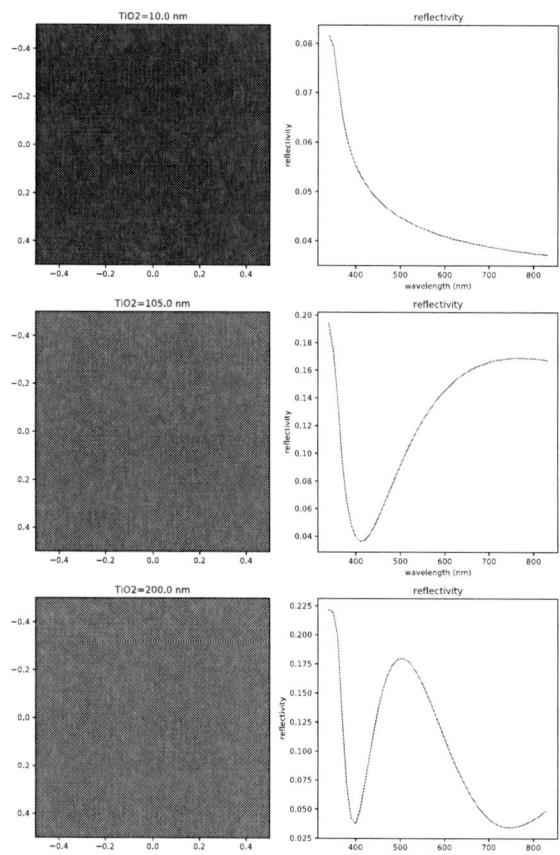

Fig. 2. The modelled color appearence of the surface of glass substrates with different thicknesses of TiO2 coated on the top. The left column shows the colors and the right column is the reflectivity.

can also elect to convert the spectrum from energy to photon flux by specifying the parameter `to_photon_flux`:

```
x1,y1=sp.get_spectrum(to_x_unit='eV',
            to_photon_flux=True)
```

`Spectrum` class can also deal with y that comes with the unit of per wavelength or per photon energy, such as W/m^2. In other words, the integral of the spectrum $\int y(x)dx$ would give a legitimate physical quantity, such as total power.

III. OPTICAL MODEL

Pypvcell implements a transfer matrix model to calculate the transmission, reflection and absorption described in [8] and [9]. As an example, Fig. 2 models the reflectivity of a glass substrate with different thicknesses of TiO2 along with it color appearances. In this example, pypvcell calculated the reflected spectrum of the TiO2-coated glass substrate and then feed the reflected spectrum into python-colormath to calculate the color. This example demonstrates the flexibility and extendability of using a script-based solar cell model package instead of a graphic-user-interface-based one.

TABLE I
SOLAR CELL MODELS IMPLEMENTED IN PYPVCELL

Model	Description
SQCell	Solar cell at Shockley-Queisser limit.
DBCell	Generalized detailed balance model with arbitrary absorptivity and external radiative efficiency.
MJCell	Multi-junction solar cell by solving I-V of individual cell.
TwoDiodeCell	Two-diode model with specified J_{01}, J_{02}, n_1, n_2, R_s and R_{sh}
AnaPNCell	Solar cell model with analytical solver of drift-diffusion equations (in progress).
NumPNCell	Solar cell model with numerical drift-diffusion equation solver (in progress).

IV. SOLAR CELL MODELS AND VALIDATIONS

In this section we briefly describe underlying algorithms of several the models that are implemented in Pypvcell, which is then followed by some validations by comparing Pypvcell's results to the value reported in the literature.

A. Generalized detailed balance model

Pypvcell has a detailed balance model for modeling single junction solar cells [10] [11]. This model is widely used for predicting the efficiency limits of solar cells. Implementations of this model can be found on any code repository, such as EtaOpt [12]. Most of these packages can only deal with bulk material with stepped absorption. Our implementation generalized this model for arbitrary absorptivity or quantum efficiency, making it easy to project the efficiency limit of solar cell with indirect band gaps or nanostructure. Mathematical details of the model can be found elsewhere, such as [12].

B. Series-connected-diode multi-junction model

A series-connected-diode model is currently implemented in Pypvcell. In this model, the I-V characteristics of each subcell is first calculated individually. After that, an I-V solver is launched to find the I-V characteristics of the multi-junction cell from the I-V characteristics of the individual subcells. The details of these two steps are described as below.

The illumination of the i-th junction is assumed to be the sum of the transmitted photons from its upward junction and other contribution, including the reflection from its downward junction and the radiative emission from its upward junction, which can be formulated by:

$$\phi_i(E) = \phi_{i-1}(E)(1 - \text{EQE}_{i-1}(E)) + \phi_i'(E) \quad (1)$$

where $\phi_{i-1}(E)$ and $\text{EQE}_{i-1}(E)$ are the incident photons and the EQE of the junction stacked above the i-th junction. $\phi_i'(E)$ accounts for those photons contributions from the reflection of $i+1$-th junction or the radiative emission from $i-1$-th junction. Note that this assumption only applies to thick subcells. For thin subcells, more accurate model such transfer matrix model should be used to calculate the illumination of each subcell properly. Once the illumination is determined, the I-V characteristics of i-th subcell $V_i(J)$ can then be calculated

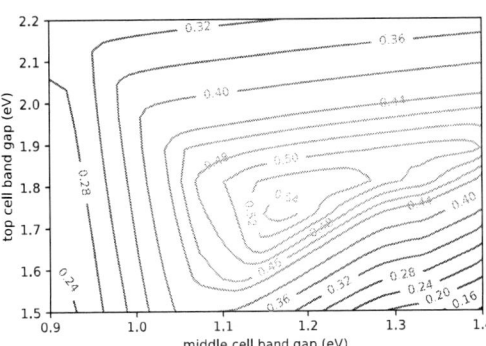

Fig. 3. Radiative-limit efficiency contour of 3J $Eg_x/Eg_y/0.67$eV against Eg_x and Eg_y. This figure matches Figure 1 of [13].

by any single-junction solar cell model provided in Pypvcell or a custom model.

Pypvcell provides two different modes to calculate the efficiency of the multi-junction cells. One is assuming a mechanical stack cell, in which case the efficiency is simply calculated by adding up the maximum power of the individual subcell. Another one is the two-terminal, series-connected multi-junction cell. To solve this numerically, a suitable range of J is discretized to get a series values of current, that is,

$$\{j_m\}, m = 0...N - 1 \quad (2)$$

The I-V of the multi-junction device is thus solved by interpolating the voltages for discretized current density j_m for every subcell i and adding up the subcell voltages to obtain the I-V characteristics of the multi-junction cell $V_{tot}(j_m)$, namely,

$$V_{tot}(j_m) = \sum_{i=1}^{N} V_i(j_m) \quad (3)$$

where $V_i(J)$ is the voltage of the i-th subcell at the current density J.

C. Validation of the models

Our results of radiative-limit solar cells show reasonable agreements with others' work. We caluclated the efficiency contour of three-junction cell X/Y/Ge(0.67eV) cell against the band gap of subcell X and Y, as shown in Fig. 3. This calculation assumes all the cells at Shockley-Queisser limits. This figure also match nicely with the same calculation presented in [13]. The results shows that the detailed balance model and the I-V solver implemented in Pypvcell is robust and reliable. Other validations and test of the functions and components of Pypvcell can also be found in the package.

V. CONCLUSION AND FUTURE DEVELOPMENT

We presented the design concepts, data structure and implementation details of a solar cell modeling library in Python. The design of Pypvcell allows great extensibility for incorporating new or user-defined solar cell models. A number of solar cell models were implemented in Pypvcell, featuring a generalized detailed balance model and a robust multi-junction I-V solver. The development of more sophisticated model such as drift-diffusion model is currently underway.

ACKNOWLEDGMENT

The authors would like to thank Japan New Energy and Industrial Technology Development Organization (NEDO) for supporting this research (NEDO 15100731-0).

REFERENCES

[1] D. A. Clugston and P. A. Basore, "PC1D version 5: 32-bit solar cell modeling on personal computers," in *Photovoltaic Specialists Conference, 1997., Conference Record of the Twenty-Sixth IEEE.* IEEE, 1997, pp. 207–210.

[2] H. Haug, J. Greulich, A. Kimmerle, and E. S. Marstein, "PC1Dmod 6.1 – state-of-the-art models in a well-known interface for improved simulation of Si solar cells," *Solar Energy Materials and Solar Cells,* vol. 142, no. C, pp. 47–53, Nov. 2015.

[3] H. Haug, A. Kimmerle, J. Greulich, A. Wolf, and E. S. Marstein, "Implementation of Fermi–Dirac statistics and advanced models in PC1D for precise simulations of silicon solar cells," *Solar Energy Materials and Solar Cells,* vol. 131, no. C, pp. 30–36, Dec. 2014.

[4] "Official Python Website."

[5] W. F. Holmgren, R. W. Andrews, A. T. Lorenzo, and J. S. Stein, "PVLIB Python 2015," in *2015 IEEE 42nd Photovoltaic Specialists Conference (PVSC).* IEEE, 2015, pp. 1–5.

[6] [Online]. Available: http://www.numpy.org

[7] [Online]. Available: http://www.scipy.org

[8] L. A. A. Pettersson, L. S. Roman, and O. Inganäs, "Modeling photocurrent action spectra of photovoltaic devices based on organic thin films," *Journal of Applied Physics,* vol. 86, no. 1, pp. 487–496, Jul. 1999.

[9] P. Peumans, A. Yakimov, and S. R. Forrest, "Small molecular weight organic thin-film photodetectors and solar cells," *Journal of Applied Physics,* vol. 93, no. 7, pp. 3693–3723, Apr. 2003.

[10] G. L. Araújo and A. Martí, "Absolute limiting efficiencies for photovoltaic energy conversion," *Solar Energy Materials and Solar Cells,* vol. 33, no. 2, pp. 213–240, 1994.

[11] W. Shockley and H. J. Queisser, "Detailed Balance Limit of Efficiency of p-n Junction Solar Cells," *Journal of Applied Physics,* vol. 32, no. 3, pp. 510–519, 1961.

[12] G. Létay and A. Bett, "EtaOpt–a program for calculating limiting efficiency and optimum bandgap structure for multi-bandgap solar cells and TPV cells," in *The 17th EC-PVSEC,* 2001.

[13] R. R. King, D. C. Law, K. M. Edmondson, C. M. Fetzer, G. S. Kinsey, H. Yoon, R. A. Sherif, and N. H. Karam, "40GaInP/GaInAs/ Ge multijunction solar cells," *Applied Physics Letters,* vol. 90, no. 18, pp. 183 516–3, 2007.

Improvement in Surface Passivation of c-Si Using Gradient-Layered a-Si:H Film for High Efficiency Silicon Heterojunction Solar Cells

Soonil Lee,[a] Leo Mathew,[b] Rajesh Rao,[b] Jae Hyun Kim,[a] Sanjay K Banerjee,[a] and Edward T. Yu[a]

[a]Microelectronics Research Center, University of Texas at Austin, 10100 Burnet Rd, Building 160, Austin, Texas 78758, United States
[b]Applied Novel Devices Inc., 15844 Garrison Circle, Austin, Texas 78758, United States

Abstract — Gradient-layered *i* a-Si:H passivation is shown to enhance minority carrier lifetimes in silicon heterojunction solar cells compared to single or dual-layer passivation structures. The gradient layered passivation contains optimized interfacial and capping layers for passivating the *i* a-Si:H/c-Si interface and a gradient layer for suppressing charge traps in bulk *i* a-Si:H. Gradient layers were obtained by changing the hydrogen dilution ratio continuously during deposition of *i* a-Si:H. The minority carrier lifetime of the c-Si substrate using gradient-layered passivation was as high as 2451 µs, compared to 1831 µs for dual-layered passivation at 2×10^{15} cm^{-3} injection level. JV characteristics of Silicon heterojunction (SHJ) solar cell employing these gradient passivation schemes will also be discussed. When the gradient-layered passivation was implied to interdigitated back contact (IBC) SHJ solar cells, there was an improvement in efficiency, compared to other passivation schemes such as single and dual passivation.

I. INTRODUCTION

SHJ solar cells have attracted much recent interest due to their high efficiency and tunable band offsets. In 2014, Panasonic achieved 25.6% energy conversion efficiency using heterojunctions in combination with an heterostructure with intrinsic thin layer (HIT) solar cell structure [1]. In SHJ solar cells, surface passivation is a key step for improving solar cell efficiency. In particular, passivation quality affects minority carrier lifetime and open circuit voltage (V_{OC}). [2] Intrinsic a-Si:H (*i* a-Si:H) passivation layers have been considered to be highly effective on c-Si surfaces since they can efficiently remove dangling bonds at the c-Si surface and thereby improve the minority carrier lifetime. [3] Usually, *i* a-Si:H films have been deposited using chemical vapor deposition (CVD). [4] In the CVD process for depositing *i* a-Si:H film, the passivation quality can be enhanced by optimizing parameters such as processing temperature, [5] pressure, [6] H_2 dilution ratio (R), [7] and film thickness. [8] Among these parameters, the crystallinity and defect density of *i* a-Si:H film are mainly controlled by R, the ratio of H_2 to SiH_4 gas flow rates (R = $[H_2]/[SiH_4]$). [7] Lee *et al.* [9] reported that there was an improvement of surface passivation on c-Si surface by depositing an additional *i* a-Si:H layer with high R on the interfacial *i* a-Si:H layer deposited with optimized R. This dual-layered *i* a-Si:H passivation showed 33.4% enhancement

of the effective lifetime comparing to single-layered *i* a-Si:H passivation.

Herein, we demonstrate gradient-layered passivation with *i* a-Si:H film for improving passivation quality at the *i* a-Si:H/c-Si interface. The gradient-layered passivation, which consists of 3 layers (interfacial, gradient, and capping layers), was obtained by depositing *i* a-Si:H film with varying R between interfacial and capping *i* a-Si:H layers. Compared to dual-layer passivation, due to the gradual change in R during *i* a-Si:H deposition, gradient-layered passivation yielded lower defect density at the interface between interfacial and capping layers and better passivation quality. Therefore, the gradient-layered passivation leads to suppress not only surface recombination on *i* a-Si:H/c-Si interface but also bulk recombination in *i* a-Si:H film. The passivation quality of gradient layered *i* a-Si:H film was evaluated by measuring minority carrier lifetime and compared to single and dual layered passivation. Additionally, the IBC structured SHJ solar cells were fabricated using gradient passivation and the current density-voltage (*JV*) characteristics were observed to figure out the enhancement in electrical properties of solar cell.

II. EXPERIMENTAL DETAIL

A. Saw Damage Removing (SDR) Process

As-cut n-type (100) c-Si wafers (thickness t ~ 180 µm and resistivity ρ ~ 2.0 Ωcm) of commercial solar grade for mass production were prepared with standard RCA cleaning. To reduce the surface defects, as-cut c-Si wafers were etched in 20% KOH solution at 70 °C for 40 min. The etched c-Si wafers were dipped in HCl:H_2O_2:H_2O 5:1:1 solution to remove K^+ ions from the surface.

B. i a-Si:H Deposition

The *i* a-Si:H layers were deposited on SDR processed c-Si substrate by remote plasma-enhanced CVD (RPCVD). During the deposition process, the processing conditions were maintained at 250 °C processing temperature, 30 W plasma power, and 100 sccm Ar gas flow. Detailed deposition conditions for each *i* a-Si:H layers are presented in Table I.

C. i a-Si:H Film Characterization

To measure the film thickness, *i* a-Si:H layers were deposited on an Si/SiO₂ wafer (SiO₂ thickness t ~ 200 nm), followed by film thickness measurement using spectroscopic ellipsometry. Fourier transform infrared (FTIR) measurements were performed on *i* a-Si:H/c-Si substrates which were deposited with different R values.

The passivation quality of the *i* a-Si:H films was evaluated by measuring minority carrier lifetime of the substrate. Each *i* a-Si:H passivation layer was deposited on both sides of c-Si wafer and the effective minority carrier lifetimes for various passivation methods were measured by photoconductance decay method using Sinton lifetime tester (WCT 120).

To estimate the effects of passivation layers in IBC cells, commercial IBC cells were prepared. 10 nm of a-Si:H films were deposited on the top surfaces of prepared IBC structured solar cells with different passivation schemes using RPCVD processes. *J-V* characteristics of these IBC solar cells were measured under AM 1.5G illumination.

III. RESULTS AND DISCUSSION

A. Optimizing Interfacial and Capping Layers

In the *i* a-Si:H passivation scheme, the interfacial layer serves to passivate dangling bonds at the *i* a-Si:H/c-Si interface. Low hydrogen dilution increases the intensity of Si-H_2 bonds at the *i* a-Si:H/c-Si interface which leads to a high density of micro-voids and defects. Conversely, high hydrogen dilution causes epitaxial growth of the a-Si:H film so that the passivation quality is degraded. Therefore, optimizing the R value in the a-Si:H deposition process is essential to improving passivation quality.

Fig. 1 (a) shows the FTIR absorption spectra in the 1900-2200 cm⁻¹ region of single-layered *i* a-Si:H films which were deposited with different R value in depositing interfacial layer (R_I). As R_I values increase from from 0 to 12.5, an obvious decrease in the Si-H_2 stretching mode was observed at 2090 cm⁻¹. As shown in Fig. 1 (b), a sufficiently reduced intensity for the Si-H_2 stretching mode at R_I=5 leads to enhanced passivation quality of *i* a-Si:H film, so that its minority carrier lifetime was dramatically increased, to 1388 µs. However, after the transition point at R_I=5, there was a sudden decline, which is believed to be due to epitaxial Si growth in the *i* a-Si:H film on c-Si substrate. As a result, the interfacial layer showed optimized passivation quality with highest minority

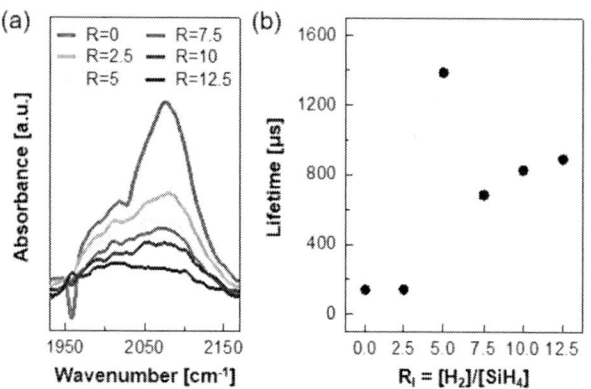

Fig. 1. (a) FTIR Spectra for 10 nm single-layered *i* a-Si:H films on glass substrates with different R values in RPCVD processes. (b) Minority carrier lifetime measurements for c-Si substrates passivated by 10 nm single-layered *i* a-Si:H films as a function of R values in depositing interfacial layer (R_i).

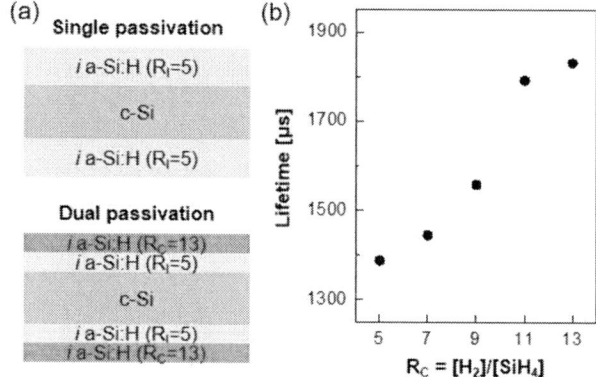

Fig. 2. (a) Schematics of single-layered and dual-layered *i* a-Si:H films on c-Si substrates. (b) Minority carrier lifetime measurements for c-Si substrates passivated by 10 nm dual-layered *i* a-Si:H films (interfacial layers with R_I=5) as a function of R values in depositing capping layer (R_C).

TABLE I
CONDITIONS OF RPCVD PROCESS

Section	Chamber Condition			Interfacial Layer			Gradient Layer			Capping Layer			Minority Carrier Lifetime
	RF Power	Temp. (°C)	Ar flow (sccm)	Thic. (nm)	Gas flow		Thic. (nm)	Gas flow		Thic. (nm)	Gas flow		
					SiH₄ (sccm)	H₂ (sccm)		SiH₄ (sccm/nm)	H₂ (sccm)		SiH₄ (sccm)	H₂ (sccm)	
Best single passivation	30 W	250	100	10	5	25	-	-	-	-	-	-	1388 µs
Best dual passivation	30 W	250	100	5	5	25	-	-	-	5	5	25	1831 µs
Gradient passivation 1	30 W	250	100	3	5	25	4	+0.10	25	3	5	25	2003 µs
Gradient passivation 2	30 W	250	100	2	5	25	6	+0.15	25	2	5	25	2127 µs
Gradient passivation 3	30 W	250	100	1	5	25	8	+0.20	25	1	5	25	2451 µs

carrier lifetime at $R_I=5$.

To optimize the value of R value in depositing interfacial layer (R_C) for the capping layer in the gradient passivation scheme, dual–layered i a-Si:H films were deposited on c-Si substrates with $R_I=5$ and different R_C. (Fig. 2 (a)) In the case of capping an i a-Si:H layer, the phase transition of a-Si:H occurs at higher hydrogen dilution since the capping layer is deposited not on a c-Si surface but on an a-Si:H surface. [10] Therefore, i a-Si:H deposition with high R for capping layer shows highly effective passivation and low hydrogen concentration without crystallization. As shown in Fig. 2. (b), with increasing R_C value, the minority carrier lifetime gradually increases and the best performance is obtained at $R_C=13$.

B. Passivation Properties of Gradient-Layered i a-Si:H Films

In the RPCVD process for dual-layered i a-Si:H films, there is an abrupt change of R value between deposition of the interfacial and capping layers. To overcome the bulk recombination in dual-layered i a-Si:H films caused by this abrupt change, a gradient layer was added between the interfacial and capping layers as illustrated in Fig. 3. (a). As shown in Table I, the total thicknesses of a series of i a-Si:H films were fixed to 10 nm and three kinds of gradient layers were applied with different rates of change of R (Gradient passivation 1, 2, and 3). According to the minority carrier lifetime results (Fig. 3. (b)), the best gradient passivation scheme showed 33.9 % improvement in passivation quality compared to the best dual-layer passivation. The gradual change of R lowers the charge trapping probability in the bulk i a-Si:H film, so that the changing ratio of R affects the film quality. The minority carrier lifetime of gradient-layered i a-Si:H film increased as the rate of change of R across the passivation layer decreased. As a result, it was investigated that the moderate change of R in i a-Si:H film enhanced the passivation quality of gradient-layered i a-Si:H films

C. Effect of Gradient Passivation in IBC Solar Cell Device

To investigate the effect of the gradient-layered i a-Si:H passivation film on improvement of electrical performance of SHJ solar cells, each passivation scheme was implied to prepared IBC structured SHJ solar cells. The prepared IBC solar cells were passivated by 10 nm i a-Si:H films with single-, dual-, and gradient-layered passivation schemes. Fig. 4 represents measured JV characteristics for each IBC solar cells with different passivation methods and without any passivation. Before passivation IBC cell showed lowest performance with J_{sc} of 15.2 mA cm^{-2}, V_{oc} of 355.3 mV and FF of 67%, yielding a PCE of 3.8%. However, they were improved after every passivation schemes. Especially, the IBC cell with gradient-layered i a-Si:H passivation film showed the most improved performance with J_{sc} of 30.8 mA cm^{-2}, V_{oc} of 405.2 mV and FF of 72%, yielding a PCE of 8.7%. In the case of IBC cell, at the top surface of c-Si substrate, most carriers are generated under light illumination. Therefore, the electrical performance of IBC cell is highly affected by the surface passivation quality on c-Si substrate. Therefore, these results explains that the gradient-layered passivation leaded to enhance the surface passivation quality and, therefore, the electrical performance of passivated IBC cell.

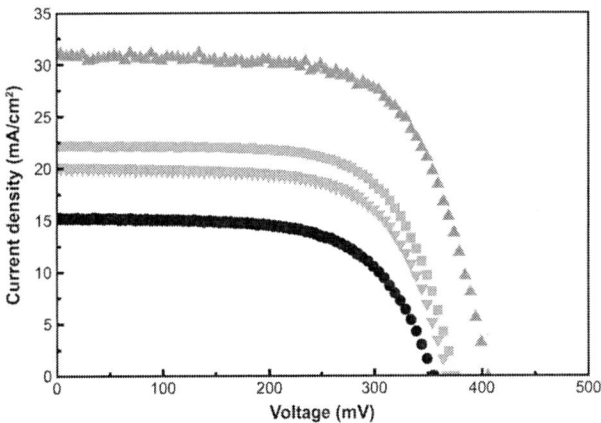

Fig. 4. Black circle, green reversed triangle, blue square, and red triangle indicate JV characteristics of IBC structured SHJ solar cells with no-, single-, dual-, and gradient-layered i a-Si:H films on the top surface.

IV. CONCLUSION

In conclusion, we demonstrate an improvement in passivating c-Si for high efficiency SHJ solar cells using gradient-layered i a-Si:H films. The gradient a-Si:H passivation layer enables grading of the interface regions within the passivation layer by adding the gradient layer between interfacial and capping a-Si:H layers. The gradual change in R value in the gradient layer can reduce the defect density in the a-Si:H film compared to that present with an abrupt change. Therefore, the gradient-layered a-Si film

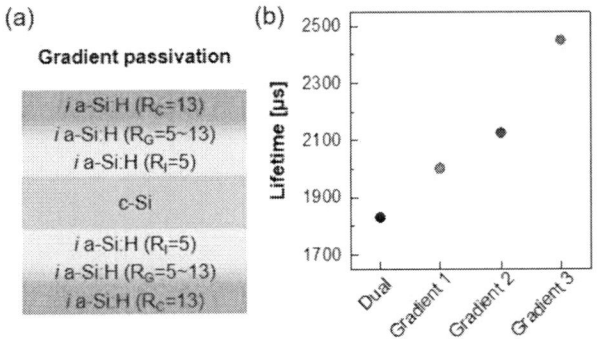

Fig. 3. (a) Schematics of gradient-layered i a-Si:H film on c-Si substrate. (b) Minority carrier lifetime measurements for c-Si substrates passivated by 10 nm dual-layered (Best dual passivation) and gradient-layered i a-Si:H films with different gradient structures (Gradient passivation 1, 2, and 3).

enables suppression not only of surface recombination at the c-Si/a-Si:H interface by interfacial and capping layers but also of bulk recombination in the a-Si:H film. The improved passivation quality of gradient-layered a-Si:H films is expected to significantly improve minority carrier lifetime and *JV* characteristics of gradient passivated SHJ solar cell devices.

ACKNOWLEDGMENTS

This material is based upon work supported by the Army Contracting Command (ACC) under Contract No. W911QX-15-C-0035.

REFERENCES

[1] K. Masuko, M. Shigematsu, T. Hashiguchi, D. Fujishima, M. Kai, N. Yoshimura, T. Yamaguchi, Y. Ichihashi, T. Mishima, N. Matsubara, T. Yamanishi, T. Takahama, M. Taguchi, E. Maruyama, and S. Okamoto, "Achievement of more than 25% conversion efficiency with crystalline silicon heterojunction solar cell," *IEEE Journal of Photovoltaics*, vol. 4, pp. 1433-1435, 2014.

[2] S. Y. Herasimenka, W. J. Dauksher, and S. G. Bowden, " >750 mV open circuit voltage measured on 50 μm thick silicon heterojunction solar cell," *Applied Physics Letters*, vol. 103, p. 053511, 2013.

[3] T. Mishima, M. Taguchi, H. Sakata, and E. Maruyama, "Development status of high-efficiency HIT solar cells," *Solar Energy Materials and Solar Cells*, vol. 95, pp. 18-21, 2011.

[4] A. J. Flewitt, and W. I. Milne, "Low-temperature deposition of hydrogenated amorphous silicon in an electron cyclotron resonance reactor for flexible displays", *IEEE*, vol. 93, pp. 1364-1373, 2005.

[5] T. Schutz-Kuchly, and A. Slaoui, "Double layer a-Si:H/SiN:H deposited at low temperature for the passivation of N-type silicon", *Applied Physics A*, vol. 112, pp. 863-867, 2013.

[6] T. Hirao, K. Setsune, M. Kitagawa, T. Kamada, K. Wasa, K. Tsukamoto, and T. Izumi, "Hydrogen concentration and bond configurations in silicon nitride films prepared by ECR plasma CVD method", *Japanese Journal of Applied Physics*, vol. 27, pp. 30, 1988.

[7] T. Mishima, M. Taguchi, H. Sakata, and E. Maruyama, "Development status of high-efficiency HIT solar cells," *Solar Energy Materials and Solar Cells*, vol. 95, pp. 18-21, 2011.

[8] D. Deligiannis, V. Marioleas, R. Vasudevan, C. C. G. Visser, R. A. C. M. M. Swaaij, and M. Zeman, and E. Maruyama, "Understanding the thickness-dependent effective lifetime of crystalline silicon passivated with a thin layer of intrinsic hydrogenated amorphous silicon using a nanometer–accurate wet-etching method," *Journal of Applied Physics*, vol. 119, p. 235307, 2016.

[9] K. Lee, C. Yeon, S. Yun, K. Jung, and J. Lim, "Improved surface passivation using dual-layered a-Si:H for silicon heterojunction solar cells," *ECS Solid State Letters*, vol. 3, pp. 33-36, 2014.

[10] J. Koh, Y. Lee, H. Fujiwara, C. R. Wronski, and R. W. Collins, "Optimization of hydrogenated amorphous silicon *p-i-n* solar cells with two-step *i* layers guided by real-time spectroscopic ellipsometry," *Applied Physics Letters*, vol. 73, pp. 1526-1528, 1998.

Photovoltaic Performance Enhancement of Textured Silicon Solar Cells Using Luminescent Down-Shifting Methylammonium Lead Tribromide Perovskite Nanophosphors

Guan-Yi Li, Wen-Jeng Ho*, Sheng-Kai Feng, Hao-Yu Yang, Ta-Wei Chuang, Bang-Jin You, Zong-Xian Lin, Zong-Liang Tseng, and Lung-Chien Chen

Department of Electro-Optical Engineering, National Taipei University of Technology, Taipei 10608, Taiwan, R.O.C.

*: wjho@ntut.edu.tw

Abstract — The methylammonium lead tribromide ($MAPbBr_3$) perovskite nanophosphors deposited by spin-on film technique on the textured silicon solar cells with a SiN_x anti-reflection coating to enhance photovoltaic performances are demonstrated. The coverage and dimension of $MAPbBr_3$ particles were 1.6% and 20-50 nm, respectively, determined by SEM images. The luminescent downshifting (LDS) characteristics of $MAPbBr_3$ perovskite nanophosphors were examined by the measurements of photoluminescence (PL), optical reflectance and external quantum efficiency (EQE). The peak of PL signal and photons emitted range are 530 nm and 500-575 nm. LDS effects can be clear displayed on the reflective spectrum and EQE response curve. An increasing in absolute conversion efficiency of 0.54% (from 15.16% to 15.70%) was obtained when the cell with $MAPbBr_3$ nanophosphors layer, compared to the reference cell without $MAPbBr_3$ nanophosphors layer.

Index Terms —nanophosphors, luminescent downshifting (LDS), textured silicon solar cells.

I. INTRODUCTION

The conversion efficiency of crystalline-silicon (C-Si) solar cells is relatively low, due to high reflectance and low spectral response at ultraviolet (UV) and blue wavelengths (300–450 nm). A higher recombination loss would be exhibited in the surface of photovoltaic devices because the incident photons of higher energy were absorbed within a short distance from the surface. The luminescent downshifting (LDS) phosphors materials can absorb high-energy photons and re-emitted lower-energy photons for the applications of solar cells to improve low spectral response at short wavelength band [1-3]. Recently, the nanometer-sized, high quantum yield and organolead halide methylammonium lead tribromide perovskite ($MAPbBr_3$) had been reported [4]. These $MAPbBr_3$ nanoparticles can be maintained stable in the solid state as well as in concentrated solutions for more than three months, without requiring a mesoporous material. This makes it possible to prepare homogeneous thin films of these nanoparticles on the solar cells by spin-on technique.

In this study, the uniform $MAPbBr_3$ nanophosphor layer deposited by spin-on film on the textured C-Si solar cells is demonstrated. The surface morphology and the coverage of nanophosphors are examined by scanning electron microscopy (SEM) images and J-image software. Photoluminescence (PL), optical reflectance and external quantum efficiency (EQE) measurements are used to examine the effectiveness of LDS of $MAPbBr_3$ nanophosphors. Furthermore, the photovoltaic performance enhancements of C-Si solar cells with and without $MAPbBr_3$ nanophosphors layer are characterized by photovoltaic current-voltage (I-V) measurement.

II. EXPERIMENT

In this work, the $MAPbBr_3$ nanophosphors of 10 mg/ml mixing with toluene solutions are prepared for spin-on film processes. Then, the $MAPbBr_3$ nanophosphors solutions were coated on the fabricated textured C-Si solar cells with a SiN_x anti-reflective coating (ARC) by spin-on technique using two-step spinning rate. Finally, surface morphology of the sample with nanophosphors was examined by SEM image and the fluorescence emission spectrum was characterized by PL measurement at room temperature.

To characterize electrical and optical properties of textured C-Si solar cells coated with LDS $MAPbBr_3$ nanophosphors material, the textured C-Si solar cells are fabricated as following process. First, the C-Si wafer was cut into chips in size of 1×1 cm^2. The saw damaged surface C-Si sample was removed by dipping the sample in a H_2O/KOH solution. The surface of the sample was then etched by dipping in a solution of H_2O/KOH/IPA. After standard cleaning, an n^+-Si emitter layer with a sheet resistance of approximately 80 Ω/sq was applied to the textured wafer using a $POCl_3$ diffusion process in a tube diffusion chamber at 850 °C for 3 min. After phosphorus diffusion, the phosphorus glass on the surface of the wafer was removed using hydrogen fluoride (HF) solution, and a SiN_x layer of one-quarter wavelength thick was deposited by PECVD on the emitter surface. Next, Al paste was printed on the rear side and dried at 200 °C .The grid pattern of Ag was screen-printed on the front side, then, the Ag and Al contacts were co-fired at a peak temperature of 750 °C to form a good ohmic contact with p-Si and n^+-Si semiconductor. After annealing, the bare textured C-Si solar cells with SiN$_x$ ARC (which called the reference cells) were obtained. Finally, the 10 mg/ml MAPbBr$_3$ nanophosphors solutions were coated on the bare solar cells. The schematic diagram of the fabricated textured silicon solar cell with a MAPbBr$_3$ nanophosphors layer is shown in Fig. 1. The contribution of down shifting MAPbBr$_3$ nanophosphors is confirmed and characterized using EQE response and photovoltaic I-V measurement under one-sun AM 1.5G illuminations.

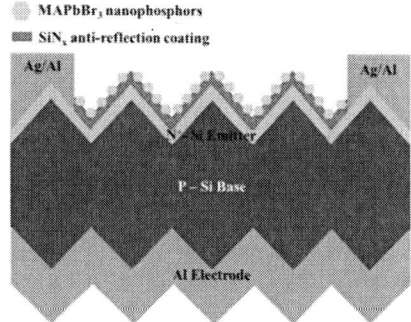

Fig. 1. Schematic diagram of a textured C-Si with SiN$_x$ ARC solar cells coated with a MAPbBr$_3$ nanophosphors layer.

III. RESULTS AND DISCUSSION

Figures 2(a) and (b) show the top-view SEM images of textured C-Si substrates with and without a MAPbBr$_3$ nanophosphor layer. MAPbBr$_3$ nanophosphors were uniform distributed over textured surface using our proposed two-step spinning rate spin-on technique. The coverage of nanophosphor particles was 1.6% and the dimension of nanophosphors particles was approximately 20-50 nm which are examined by SEM images and J-image software.

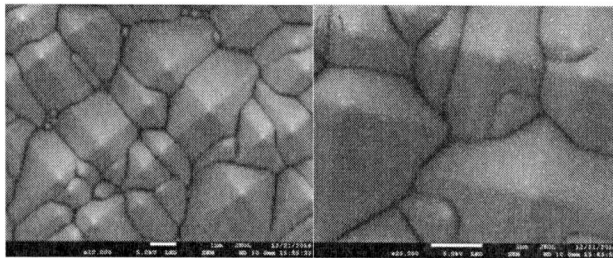

Fig. 2. Top-view SEM images of the cell (a) without, (b) with MAPbBr$_3$ nanophosphors layer.

Figure 3 shows the PL fluorescence spectrum of perovskite MAPbBr$_3$ nanophosphors at room temperature. The excited source was a semiconductor laser with 442 nm wavelength. The peak of PL signal and photons emitted range are 530 nm and 500-575 nm for MAPbBr$_3$ nanophosphors layer. It exhibited superior LDS characteristics of MAPbBr$_3$ nanophosphors that it absorbed approximately the blue-band

photons and converting them into the visible photons with wavelengths of 500-575 nm.

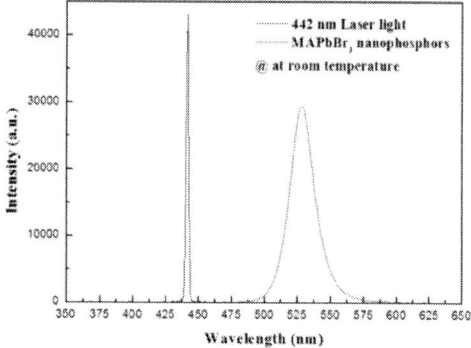

Fig. 3. The PL spectrum of the MAPbBr$_3$ nanophosphors.

The optical reflectance of textured C-Si solar cells with and without MAPbBr$_3$ nanophosphors layer are show in Fig. 4. Two important optical properties can found in the reflection spectrum. First, the reflectance of the cell with MAPbBr$_3$ nanophosphors layer was lower than that of the cell without MAPbBr$_3$ nanophosphors layer at UV-blue wavelength range, due to the absorption of MAPbBr$_3$ nanophosphors particles. The other is that the reflectance decreased in the full wavelength range which can be ascribed to the light scattering of the deposited MAPbBr$_3$ nanophosphors particles. In addition, the weighted average reflectivity of the cell with and without MAPbBr$_3$ nanophosphors layer calculated from 350 to 1000 nm wavelengths are 3.92% and 4.41%, respectively. The reduction in reflection of 0.49% was obtained, which it will provide much incident photons into the cells.

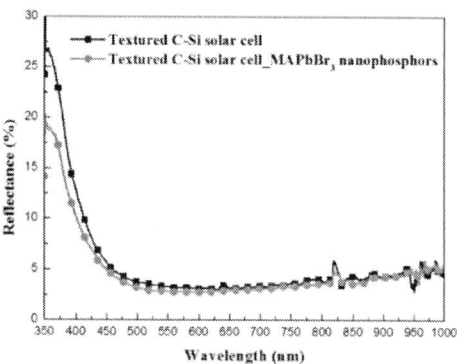

Fig. 4. Optical reflectance of textured C-Si solar cell with and without MAPbBr$_3$ nanophosphors layer.

Figure 5 shows the EQE response of the textured C-Si solar cells with and without MAPbBr$_3$ nanophosphors layer. Comparing with EQE response curves, high EQE level is presented in the cell with MAPbBr$_3$ nanophosphors layer. Particularly, higher EQE values from the wavelength range of 325 to 475 nm can be observed, which suggest the LDS effects of MAPbBr$_3$ nanophosphors particles. The weighted average EQE of the cell with and without MAPbBr$_3$ nanophosphors layer calculated from 325 to 1100 nm wavelengths are 88.01% and 86.62%, respectively. The improved in EQE of the cell with a MAPbBr$_3$ nanophosphors layer was 1.49%. Thus, the obtained EQE result is in agreement with the obtained optical reflectance result.

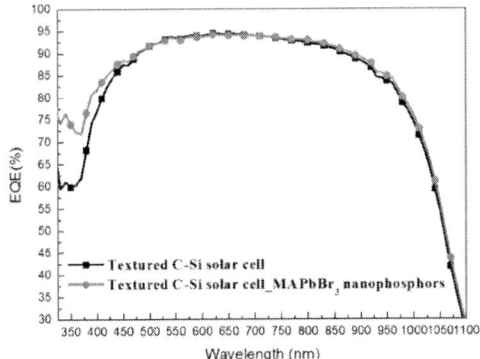

Fig. 5. EQE responses of textured C-Si solar cell with and without MAPbBr$_3$ nanophosphors layer.

Figure 6 displays the photovoltaic J-V characteristics of the textured C-Si solar cells with and without MAPbBr$_3$ nanophosphors layer and the photovoltaic performances are summarized in Table I. The short-circuit current density (J$_{sc}$) and open-circuit voltage (V$_{oc}$) are increased when the cell coated with a MAPbBr$_3$ nanophosphors layer. The J$_{sc}$ enhancement of 2.14% was obtained, compared to the cell without a MAPbBr$_3$ nanophosphors layer. The increasing in J$_{sc}$ is well agreed to that of EQE one. Therefore, an increasing in absolute conversion efficiency of 0.54% (from 15.16% to 15.70%) was obtained when the cell with MAPbBr$_3$ nanophosphors layer, compared to the reference cell without MAPbBr$_3$ nanophosphors layer.

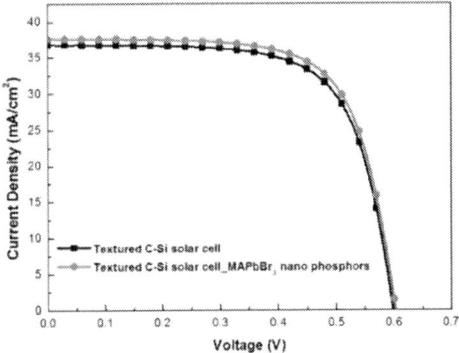

Fig. 6. The photovoltaic J-V of textured C-Si solar cell with and without MAPbBr$_3$ nanophosphors layer.

TABLE I
THE PHOTOVOLTAIC PERFORMANCES OF THE CELL WITH AND WITHOUT MAPbBr$_3$ NANOPHOSPHORS LAYER.

	Reference	MAPbBr$_3$ nanophosphors layer
V$_{oc}$ (mV)	598.2	602.5
J$_{sc}$ (mA/cm^2)	36.83	37.62
F.F (%)	68.82	69.30
η (%)	15.16	15.70
ΔJ$_{sc}$ (%)	2.14 %	
Δη (%)	3.56 %	

IV. CONCLUSION

The electrical and optical properties of the textured C-Si solar cells with a MAPbBr$_3$ perovskite nanophosphors layer using spin-on film technique were demonstrated. The LDS effects of MAPbBr$_3$ perovskite nanophosphors were examined using optical reflectance and EQE measurements. An increasing in absolute conversion efficiency of 0.54% (from 15.16% to 15.70%) was obtained when the cell with MAPbBr$_3$ perovskite nanophosphors layer. To obtain a higher absolute conversion efficiency for commercial textured C-Si solar cell applications, the coverage and long-time stable operation of MAPbBr$_3$ perovskite nanophosphor particles are undergoing study in our laboratory.

ACKNOLOWLEDGEMENTS

The authors would like to thank the Ministry of Science and Technology (MOST) of the Republic of China for financial support under Grant MOST 103-2221-E-027-049-MY.

REFERENCES

[1] W.B. Hung, et. al., "Enhanced conversion efficiency of crystalline Si solar cells via luminescent down-shifting using Ba$_2$SiO$_4$:Eu^{2+} phosphor," *J. of Ceramic Processing Research.* vol. 15(3), pp. 157-161, 2014.

[2] Guojian Shao, et. al., "Enhancing the efficiency of solar cells by down shifting YAG:Ce^{3+} phosphors," *J. Lumin.* vol. 157, pp. 344-348, 2015.

[3] S.D. Hodgson, et. al., "Increased conversion efficiency in cadmium telluride photovoltaics by luminescent downshifting with quantum dot/poly (methyl methacrylate) films," *Prog. Photovolt. Res. Appl.* vol. 23, pp. 150-159, 2015.

[4] Luciana C. Schmidt, et. al., "Nontemplate Synthesis of CH$_3$NH$_3$PbBr$_3$ Perovskite Nanoparticles," *J. Am. Chem. Soc.* vol. 136, pp. 850-853, 2014.

SiN$_x$ thin films with appropriate antireflection and shift-conversion properties for silicon solar cells

E. Mon-Pérez[1], J. Salazar[1], A. Dutt[1*], J. Santoyo-Salazar[2] and G. Santana[1,*]

[1] Instituto de Investigaciones en Materiales, Universidad Nacional Autónoma de México. A.P. 70-360, Coyoacán, C.P. 04510, México, D.F.

[2] Departamento de Física, CINVESTAV-IPN, A. P. 14-740, C. P. 07000, México, D. F., Mexico

* Corresponding authors: Guillermo Santana, gsantana@iim.unam.mx, Ateet Dutt, adutt@cinvestav.mx

Abstract —

In the present work, silicon nitride thin films has been deposited by using dichlorosilane in plasma enhanced chemical vapor deposition (PECVD). Adequate deposition condition has been found to obtain the minimal reflectance corresponding to one particular wavelength at the suitable value of refractive index and thickness of the thin film. Furthermore, using transmission electron microscopy (TEM) presence of silicon quantum dots (QD's) in the size regime of 3-4 nm has been found. Strong visible photoluminescnce observed from the as deposited film is due to the well-known quantum confinement effect. Intense visible pholuminescence observed in the present work could be used for the shift conversion (downshift) which could subsequently improve the efficiency in silicon solar cells. Optical property studies carried out in the present work highlight the prospects of SiN$_x$ thin films for down shifting, antireflection and as an passivation coating in silicon solar cells.

I. INTRODUCTION

Antireflection coating is one of the techniques to reduce major reflection loss in solar cells [1-2]. An efficient anti-reflection coating diminishes the reflection at the front surface of the solar cells, thereby, increasing the photocurrent and thus the photovoltaic efficiency.

SiN$_x$ deposited by plasma enhanced chemical vapor deposition (PECVD) is one of the most used candidates for anti-reflection coating because of tunable refractive index, as well it provides an excellent level of passivation, and high transparency in the visible region due to high band gap [3-4]. PECVD is well suited for large area solar cell fabrication due to lower deposition temperature. In case of PECVD, film properties could be varied by changing different deposition parameters like flow rate of gases, pressure inside the chamber, temperature and RF power.

Apart from the vast study made on the antireflection layer to improve the efficiency of solar cells several other investigations has also been carried out. In current years, lot of study has been done to improve the conversion efficiency of silicon solar cells by employing different kind of structures like microstructure and nanostructures [5]. Recently, using nanostructures a new kind of generation of solar cells has been investigated where higher energy photons could be absorbed and will be converted to lower energy photons when absorbed by silicon solar cells [6]. Along with this other kind of mechanism to achieve maximum efficiency is also been investigated and is called as multiple exciton generation. Using new generation of solar cells, increment in open circuit voltage (VOC) and short circuit current has been already reported [7].

In our previous reports, using chlorinated silanes, i.e., SiH$_2$Cl$_2$ lot of research has been made on the development of polymorphous silicon thin films [8, 9]. Advantages of this precursor material are less presence of weak Si-H bonding peaks, and it has the possibility to produce nanoparticles in the as-deposited thin films. Similarly, in the previous published reports control over the phase transition has also been established as in response to various deposition conditions.

In the present work, we have investigated the formation of quantum dots related to silicon in non-stoichiometric SiN$_x$ thin films. By TEM, we have confirmed the formation of small nanoparticles in the range of 3-4 nm and afterwards we have carried out optical study on the samples. Finally, depending on the formation of nanoparticles in both cases, photoluminescence and reflectance characteristics obtained in this work could be used for a good anti-reflection coating as well as for down shifting in silicon-based solar cells. The present research could result in the economic bulk fabrication of SiN$_x$ thin films, which could be used for multiple applications. Multiple applications with additional feature of cost sustainability could promote the development of thin films proposed in the present work.

II. EXPERIMENTAL

The SiN$_x$ thin films were deposited on high-resistivity n-type [1 0 0] monocrystalline silicon substrates by using *PECVD* reactor operating at *RF* of 13.56 MHz with parallel plates of 128 cm^2 in area and 1.5 cm apart. Hyrogen-diluted SiH_2Cl_2 gas was used as the silicon precursor. Turbomolecular pump was used to obtain a high vaccum of 10^{-6} Torr before starting the deposition. For different characterizations, depositions were made on different substrates like fused silica (*quartz*), high resistivity (100) single crystalline silicon, glass and *NaCl*. The SiH_2Cl_2 and Ar flow rates were fixed at 10 sccm and 40 sccm, respectively, while the NH_3 flow rate was kept constant at 10 sccm. H_2 flow rate was kept at 20 scmm. In all cases, the substrate temperature, deposition time, the RF plasma power and the deposition pressure were fixed at 150 °C, 30 min, 50 W and 250 mTorr, respectively.

The chemical bonding structure of the films was analysed by using a FTIR spectrometer (Nicolet- 560) in the range of 400 to 4000 cm^{-1}. The size and structure of Si-QD's were confirmed by TEM images. Samples were observed in a TEM, JEOL (JEM-2010) with LaB6 filament with an acceleration voltage at 200 kV and a wavelength of 0.0027 nm. Photoluminescence (PL) was measured using a He-Cd laser (Kimmon Koha Co., Ltd., Centennial, CO, USA) with an excitation wavelength of 325 nm. Filmetrics brand model F10-RT-UV was used for the antireflectance measurement. It has a double light source: a deuterium lamp for the UV range and a halogen for the visible region. This technique allows to obtain spectra of transmittance and specular reflectance in the range of 190-1100 nm.

III. Results and discussion

Figure 1 shows the representative FTIR spectrum of the as deposited SiN_x sample. The spectrum consisted of one strong absorption band that corresponds to the Si-N stretching vibration around at 860 cm^{-1}. As can be seen from Figure 1, hydrogen can be found mainly bonded with nitrogen and silicon species at N-H bending (stretching) modes at 1180 cm^{-1} (3360 cm^{-1}) and Si-H stretching mode near 2220 cm^{-1}.

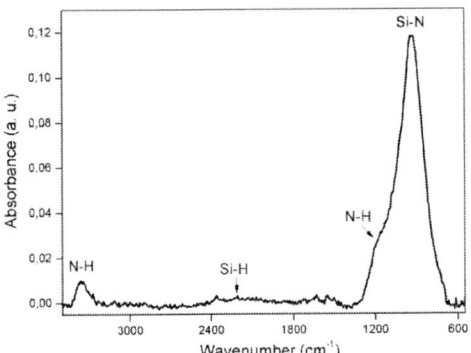

Figure1. Representative FTIR spectra of as-deposited SiN_x thin film.

Additionally, the lack of the Si-O stretching mode around 1070 cm^{-1} confirms the stability and resistance to oxidation of the deposited thin film. Even more, precursor dichlorosilane avoids

the formation of SiH_n hydrogen species, which could result in degradation of the deposited film during course of the time.

TEM image displayed in Figure 2 evidence the presence of Si-QD's embedded in the amorphous matrix of SiNx. In addition, the size distribution and density of Si-QD's in the films was obtained using the Digital Micrograph 3.7.0, software by Gatan, Inc. USA.

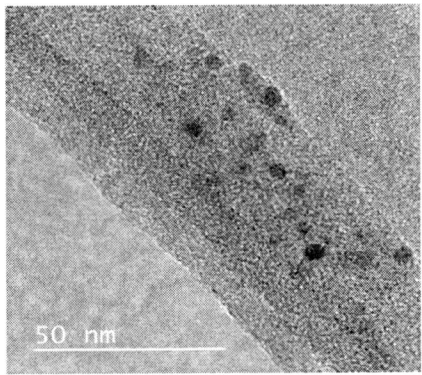

Figure 2. TEM micrograph of the Si-QD's embedded in the amorphous SiNx matrix corresponds to sample.

For the sample, using TEM analysis the average size and density are in the order of 4.31 ± 1.26 nm and 2.67 $X10^{11}$ Si-QDs/cm^2, respectively.

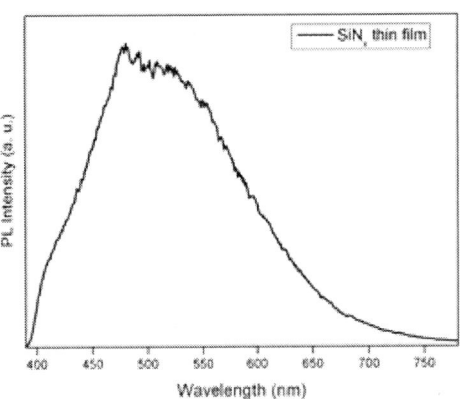

Figure 3. Visible PL spectrum of the as-deposited sample when excited with He-Cd laser (325 nm) at room temperature.

Figure 3 displays the PL spectra of the as deposited sample. As can be seen, the PL main peak can be found at around 480 nm with a broad shoulder from around 400 to 700 nm. The presence of wide distribution of QD's (Figure 2) could provoke this strong emission by quantum confinement. Hence, it can be illustrated that in the present case we have been able to deposit thin film with bright and stable emission in the visible region, which further can be used in the shift conversion property in the silicon solar cells. Furthermore, the passivation provided in the present case due to embedding of quantum dots in the amorphous matrix ensures the stability of the material in the practical application.

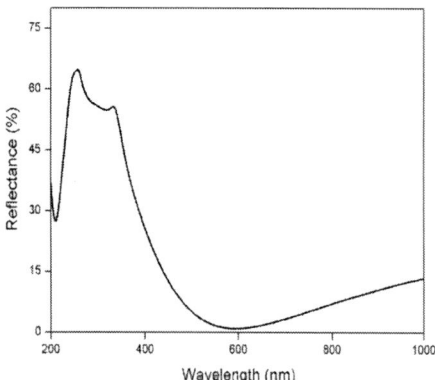

Figure 4. Reflectance spectrum of the as-deposited sample.

Figure 4 shows the reflectance spectrum of the sample. As can be observed, the sample shown minimum total reflection by the adjustment of the refraction index and thickness corresponding to one particular wavelength. These suitable properties felicitate the use of deposited film as antireflection coating in solar cells.

IV. Conclusions

In this work, we have obtained the presence of silicon quantum dots in nonstoichiometric SiN_x ambience grown using dichlorosilane as a precursor gas by PECVD. Deposited thin films shown the intense photoluminescence in the visible region, which remarks the shift conversion property from this material. In addition, the reflectance observed for the thin films shown adequate value as for their usage as an antireflection coating in silicon solar cells.

V. Acknowledgements

The authors acknowledge financial support for this project from DGAPA-UNAM PAPIIT Projects IN108215 and IN107017. E. Mon-Pérez is grateful for CONACYT scholarship 517361. We want to thank Carlos Ramos Vilchis and Cain Gonzalez for their technical assistance.

VI. References

[1] D. Chen, "Anti-reflection (AR) coatings made by sol–gel processes: A review" Sol. Energy Mater. Sol. Cells vol. 68, pp. 313-336, 2001.

[2] S. Chattopadhyay, Y. F. Huang, Y. J. Jen, A. Ganguly, K. H. Chen, and L. C. Chen, "Anti-reflecting and photonic nanostructures" Mater. Sci. Eng. R Vol.69, pp. 1-35, 2010.

[3] H. Nagel, AG. Aberle, R. Hezel "Optimised antireflection coatings for planar silicon solar cells using remote PECVD silicon nitride and porous silicon dioxide" Prog. Photovoltaics, vol. 7, pp. 245-60, 1999.

[4] P. Doshi, GE Jellison, A. Rohatgi. "Characterization and optimization of absorbing plasma-enhanced chemical vapor deposited antireflection coatings for silicon photovoltaics". Appl. Opt., vol.36, pp. 7826-37, 1999.

[5] G. Conibeer, M. Green, R. Corkish, Y. Cho, E. Cho, Chu-Wei Jiang, T. Trupke, B. Richards, A. Shalav, and K. Lin, "Silicon nanostructures for third generation photovoltaic solar cells" Thin Solid Films, vol. 511, pp. 654-662, 2011.

[6] T. Trupke, M.A. Green, and P. Wurfel, " Improving solar cell efficiencies by down-conversion of high-energy photons" J. Appl. Phys., Vol. 92, pp.1668-1674, 2002.

[7] H. Shpaisman, O Niitsoo, I. Lubomirsky and D. Cahen," Up-and Down-Conversion,and Multi-Exciton Generation for Improving Solar Cells:A Reality Check" vol. 1101, 1101-KK05-14, 2008.

[8] B. Monroy, G. Santana, J. Aguilarhernandez, a Benami, J. Fandino, a Ponce, G. Contreraspuente, a Ortiz, and J. Alonso, "Photoluminescence properties of SiNx/Si amorphous multilayer structures grown by plasma-enhanced chemical vapor deposition," J. Lumin., vol. 121, no. 2, pp. 349–352, Dec. 2006.

[9] E Mon-Pérez, J Salazar, E Ramos, J Santoyo Salazar, A López Suárez, A Dutt, G Santana and B Marel Monroy, "Experimental and theoretical rationalization of the growth mechanism of silicon quantum dots in non-stoichiometric SiN x : role of chlorine in plasma enhanced chemical vapour deposition", Nanotechnology, vol 27, pp. 455703, 2016.

Numerical Simulation of Crystalline Silicon Solar Cells with Full Area Metal Oxide Rear Contacts

James E. Moore[1,2], Woojun Yoon[2], Phillip P. Jenkins[2], and Robert J. Walters[2]

[1]The George Washington University, Washington, DC 20037, USA
[2]U.S. Naval Research Laboratory, Washington, DC 20375, USA

Abstract — In this study, we investigate the effectiveness of MoO_x and TiO_2 layers as minority carrier selective contacts in crystalline silicon solar cells. We use numerical simulation to show that it is possible to achieve an improvement in open circuit voltage by up to 70 mV for TiO2 and 100 mV for MoO_x with one-sun efficiencies exceeding 20%. We further determine the material parameters of primary importance to each of the materials in determining their effectiveness at surface passivation, finding that the density of surface defects is most important for TiO_2, while the electron affinity predominates for MoO_x.

I. INTRODUCTION

High efficiency crystalline silicon (c-Si) solar cells have achieved recent improvements in material quality and surface passivation to reach efficiencies exceeding 20% [1]. As a result, one current limiting factor in c-Si solar cell performance is now carrier recombination at the back surface, which has renewed interest in finding better materials for creating back surface fields (BSF). Some technologies such as passivated emitter rear contacts (PERC) and passivated emitter rear totally diffused (PERT) contacts have shown that they can achieve very high efficiencies, but usually require localized contacts created by cost intensive and complex processes [2]

In contrast, carrier selective contacts based on metal oxides can be deposited as full-area rear contacts, thereby bypassing the need for such complex processes. Recent interest in metal oxide passivation layers used as BSFs have shown remarkable promise, with materials such as TiO_2 on n-type substrates above 22% efficiency [3] and MoO_x on p-type substrates above 20% efficiency [4]. Numerical simulation, however, indicates that still higher efficiencies can be achieved. The purpose of this study is to examine the material properties of these metal oxides to which the passivation effects are most sensitive to help researchers better understand how to further improve the use of these materials in passivation layers.

II. SIMULATION METHOD

Simulations were performed using 2D numerical simulation with the Advanced Physical Models of Semiconductor Devices (APSYS) modeling tool developed by Crosslight [5]. This software was chosen for its ability to model both interband quantum tunneling as well as intraband tunneling through a

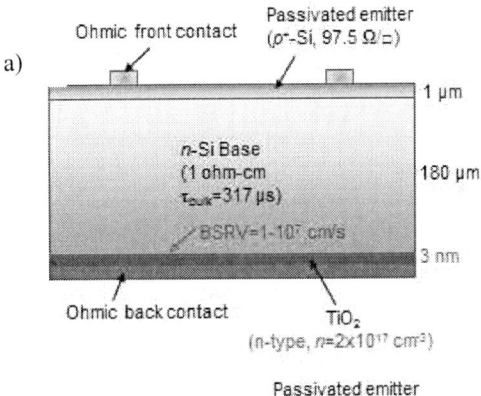

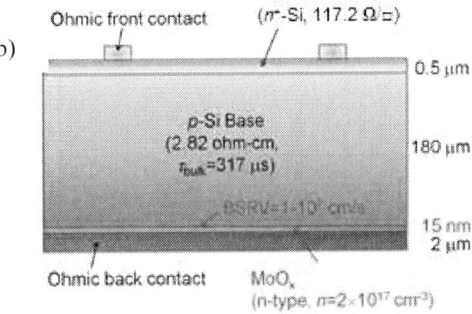

Fig. 1. Simulated device models and parameters for a) n-type Si base with TiO_2 BSF and b) p-type Si base with MoO_x BSF. Parameters which were varied in simulation are shown in red [6].

potential barrier, which is required to accurately model the carrier transport through very thin layers of metal oxides.

The simulated structures and associated material parameters are shown in Fig. 1. The emitter and base parameters were kept constant in all simulations to better focus on the effects of changing the back contact parameters. The base and emitter use standard material parameters for c-Si. Both devices have a 180 µm base, with doping and material characteristics for the emitter and base taken from literature [3-4].

For the metal oxides, we simulated a variation in several different material parameters, but only a few were found to have a significant effect on the quality of the surface passivation and device performance. The thickness of the metal oxide layers were varied in simulation, but both have large enough potential barriers that only extremely thin layers (<1 nm) were seen to have tunneling currents large enough to noticeably

978-1-5090-5606-4/17 $31.00 © 2017 IEEE

effect the performance of the cell. Likewise, adjusting the doping in the metal oxide layers also had little noticeable effect on both the MoO$_x$ and TiO$_2$ model devices.

III. RESULTS

The most important parameters that affect performance were found to be the quality of the n-Si/TiO$_2$ and p-Si/MoO$_x$ interface in terms of back surface recombination velocity (BSRV) and the electron affinity of the metal oxide. The electron affinity plays a particularly important role for the MoO$_x$ and will be discussed in detail in the following section.

A. n-Si/TiO$_2$

Of the various material parameters examined for the device with the TiO$_2$ BSF, the BSRV had the largest effect on performance. As shown in Fig. 2, the open circuit voltage (V_{oc}) drops sharply for BSRV above 100 cm/s. Although TiO$_2$ can enhance the V_{oc} by as much as 70 mV, this means that the device can only benefit from high level of surface passivation, i.e. BSRV<10 cm/s. This is in agreement with the experimental results reported in [7] which reports an implied V_{oc} above 680 mV after TiO$_2$ deposition with a very low recombination current density (J_0) of 20 fA/cm^2 at n-Si/TiO$_2$, but the V_{oc} falls to 639 mV after high temperature annealing due to passivation degradation caused by phase transformation in the TiO$_2$.

We also did not observe a reduction in V_{oc} for TiO$_2$ thicknesses below 4.5nm as was reported in [7]. We had initially assumed that this loss in V_{oc} was due to quantum tunneling in the ultrathin TiO$_2$ layer, but based on our simulation the tunneling current does not become large enough to adversely affect the device performance until the TiO$_2$ thickness becomes less than 1 nm. We therefore conclude that the experimentally measured V_{oc} loss for the thinnest TiO$_2$ layers must arise from another mechanism that was not included in this simulation.

B. p-Si/MoO$_x$

In addition to the BSRV, the electron affinity (χ) was also found to play an important role in the performance of the MoO$_x$ device. The electron affinity of MoO$_x$ is known to vary from 4.7 eV to 6.9 eV depending on several conditions during deposition and post-deposition processes [8-9]. We found that MoO$_x$ devices with χ<5.5eV were severely limited in V_{oc}, having a large impact on efficiency (Fig. 3). The BSRV was also found to have an important impact on V_{oc}, but is of secondary importance to the electron affinity and only has a significant effect for χ>5.5eV where the electron affinity is high enough not to limit the performance of the cell.

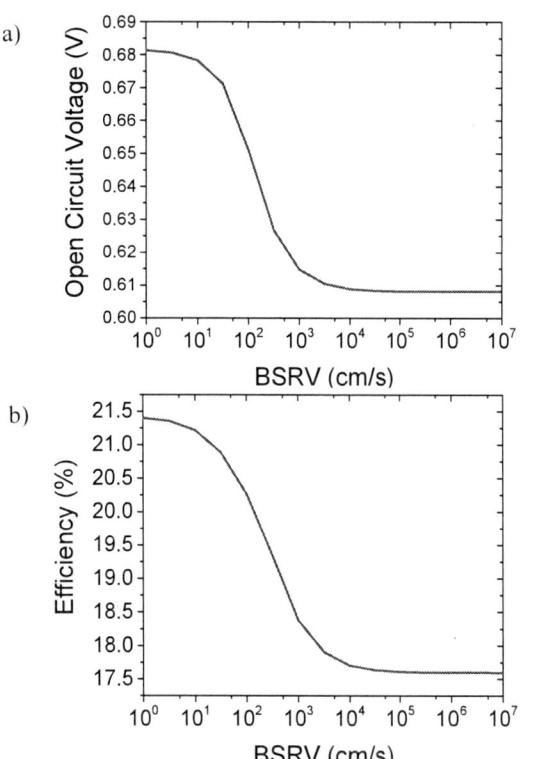

Fig. 2. Plots of a) the open circuit voltage and b) the efficiency for TiO$_2$ BSF with different n-Si/ TiO$_2$ interface BSRV [6]

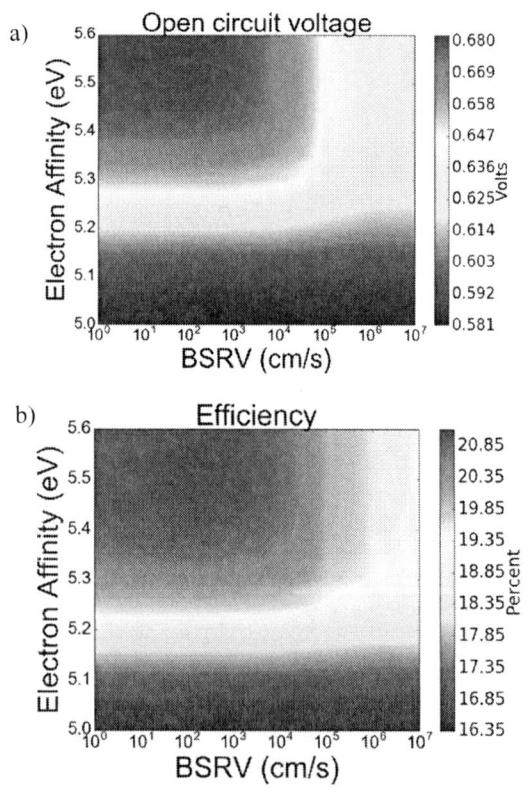

Fig. 3. Contour plot of a) open circuit voltage and b) efficiency as a function of MoO$_x$ electron affinity and BSRV at p-Si/MoO$_x$ interface [6].

a)

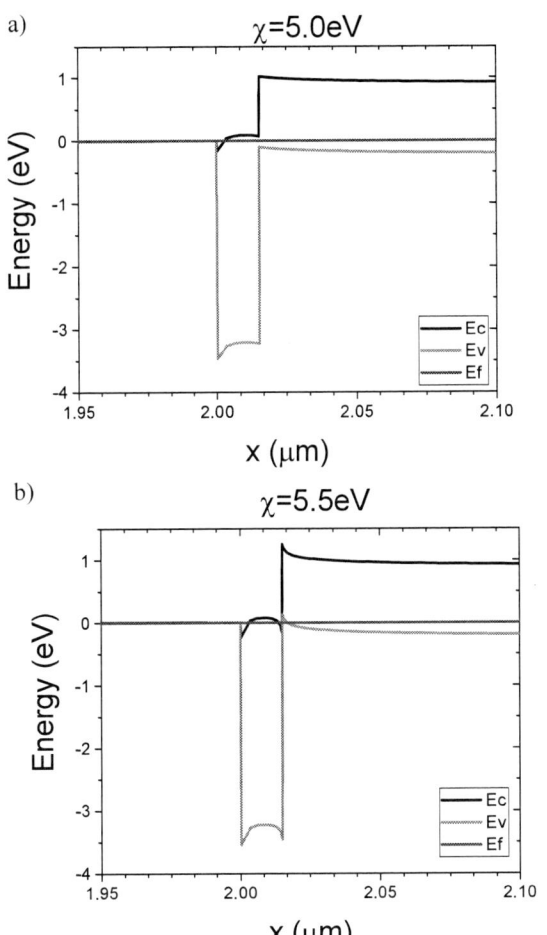

conditions when creating the device. TiO₂ was shown to be quite sensitive to recombination at the n-Si/TiO₂ interface, indicating that the V_{oc} enhancement could be significantly reduced if defects are present at this interface when it is used in an actual solar cell. MoO$_x$ was not quite as sensitive to the effects of BSRV, but was shown to be very sensitive to the electron affinity and band offset with silicon if it was too small. If these potential pitfalls can be avoided during device fabrication, however, both of these materials offer promising alternatives to conventionally used localized contact methods in silicon solar cells.

Fig. 4 Energy band plots at MoO$_x$/p-Si interface for a) χ= 5.0 eV and b) χ=5.5 eV [6].

The reason for the large role that the value of χ plays in the MoO$_x$ contacted cell's performance can be seen from the band diagrams shown in Fig. 4. When χ is too small as in Fig. 4a, the conduction band spike that is meant to block the minority carrier electrons is too small to be effective. Furthermore, the majority carrier holes cannot easily reach the back contact due to the large barrier in the valance band. In contrast, when χ is large enough as in Fig. 4b, the conduction band spike blocks the electrons, and p-Si and MoO$_x$ form a type III heterojunction, allowing the holes to pass through the junction and reach the back contact unimpeded.

IV. CONCLUSIONS

Through numerical simulation we have demonstrated the potential passivation effects of two metal oxides, MoO$_x$ and TiO₂, as BSF layers on c-Si solar cells with the passivated emitters. [10, 11] Our model shows that both materials are good candidates for forming carrier selective contacts, but could also be limited by the quality of the material and processing

REFERENCES

[1] W. Yoon, E. Cho, J. D. Myers, Y.-W. Ok, M. P. Lumb, J. A. Frantz, N. A. Kotulak, D. Scheiman, P. P. Jenkins, and A. Rohatgi, "Transparent conducting oxide-based, passivated contacts for high efficiency crystalline Si solar cells," in *Photovoltaic Specialist Conference (PVSC), 2015 IEEE 42nd*, 2015, pp. 1-4.

[2] J. Berrick, B. Steinhauser, R. Muller, J. Bartsch, M. Kamp, A. Mondon, A. Richter, M. Hermie, S. Glunz, "High efficiency n-type PERT and PERL solar cells," *Proceeding of the 40th IEEE PVSC*, pp. 3637-3640, 2014.

[3] X. Yang, Q. Bi, H. Ali, K. Davis, W.V. Shoenfeld, and K. Weber "High-Performance TiO2-Based Electron Selective Contacts for Crystalline Solar Cells," *Advanced Materials,* vol. 28, pp. 5891-5897, 2016

[4] J. Bullock, C. Samundsett, A. Cuevas, D. Yan, Y. Wan, and T. Allan "Proof of concept p-type silicon solar cells with molybdenum oxide partial rear contacts," *Proceeding of the 42nd IEEE PVSC*, 2015

[5] Crosslight Device Simulation Software User Manual ©2010 Crosslight Software Inc.

[6] W. Yoon, J. E. Moore, E. Cho, D. Scheiman, N. A. Kotulak, E. Cleveland, Y.-W. Ok, P. P. Jenkins, A. Rohatgi, and R. J. Walters, "Hole-selective Molybdenum Oxide as a Full-area Rear Contact to Crystalline p-type Si Solar Cells," Japanese Journal of Applied Physics, vol. 56, 2017.

[7] X. Yang and K. Weber "N-type silicon solar cells featuring an electron selective TiO2 contact" *Proceeding of the 42nd IEEE PVSC*, 2015

[8] J. Meyer, S. Hamwi, M. Kroger, W. Kowalsky, T. Riedl, A. Khan, "Transition Metal Oxides for Organic Electronics: Energetics, Device Physics and Applications" Advanced Materials vol. 24 pp. 5408-5427, 2012

[9] J.J. Jasieniak, J. Seifter, J. Jo, T. Mates, A.J. Heeger "A Solution-Processed MoOx Anode Interlayer for Use within Organic Photovoltaic Devices," Advanced Functional Materials, vol. 22, pp. 2594-2605, 2012

[10] M. P. Lumb, W. Yoon, C. G. Bailey, D. Scheiman, J. G. Tischler, and R. J. Walters, "Modeling and analysis of high-performance, multicolored anti-reflection coatings for solar cells," Optics express, vol. 21, pp. A585-A594, 2013.

[11] W.Yoon, J. E. Moore, A. Lochtefeld, N. A. Kotulak, D. Scheiman, A. Barnett, P. P. Jenkins, R. J. Walters, "Surface Passivation of Epitaxial Boron Emitters for High-efficiency Ultrathin Crystalline Si Solar Cells," Japanese Journal of Applied Physics, vol. 56, 2017.

978-1-5090-5606-4/17 $31.00 © 2017 IEEE

978-1-5090-5606-4/17 $31.00 © 2017 IEEE

Interdigitated Back Contact Silicon Solar Cell with Perovskite layer for Front Surface Passivation and Ultraviolet Radiation Stability

Rahul Pandey, Shivam Gupta, Trijul Khatri and Rishu Chaujar*

Department of Engineering Physics, Delhi Technological University,
Main Bawana Road, New Delhi, Delhi 110042
E-mail: rishu.phy@dce.edu*

Abstract — **In this work, advanced surface passivation technique has been discussed for the n-type wafer based interdigitated back contact silicon (IBC-Si) solar cells. Perovskite ($CH_3NH_3PbI_3$) has been used as a front surface passivation layer. Results show that the presence of perovskite as a passivation layer improves the photovoltaic behavior of the device by preventing the surface recombination of minority carriers (holes), due to the creation of positive electric field at the perovskite/silicon interface. UV stability at higher surface recombination velocity (SRV) has also been observed in the surface passivation technique. The impact of SRV on optical as well as electrical behavior has also been obtained, and results are compared with conventional antireflective (AR) coating based cell. Power conversion efficiency of the perovskite passivated solar cell is 9.5% higher compared to AR coating based solar cell. All the simulations have been done using SILVACO ATLAS device simulator.**

Index term- Perovskite, solar cell, recombination, interdigitated back contact.

I. INTRODUCTION

Advancement in crystalline Silicon solar cells by passivation of both cell surfaces led power conversion efficiencies above 20% for the first time. With modern industries tending towards thinner wafers and higher power conversion efficiency, the passivation of the front and rear surfaces has now become vitally important for commercially sold silicon cells [1]. Thus, to improve the UV stability and performance of the devices, the passivation of the front and rear surfaces are required to reduce the concentration of charge carriers, electrons or holes. Any excess energy beyond the semiconductor band gap is lost and released as heat, which can deteriorate the performance due to over-heating [2]. Overheating causes a rise in temperature of the module which leads to a decrease in power conversion efficiency(PCE) as high as 0.08% $°C^{-1}$ [3]. An optimized AR coating, (to reduce reflection losses) exhibits a low reflectivity only at relatively narrow wavelength range ratio compared to the broad solar spectrum [4]. The region of low reflectivity should be chosen to coincide with the wavelength range across which the semiconductor can harvest the largest energy from sunlight [4]. The AR coating is usually optimized for red light (for Si_3N_4 it is 80nm, at a wavelength of 700nm) and becomes highly reflective in blue part of the spectrum. In this work, we have deposited organometal trihalide perovskite layer on Silicon wafer to enhance the performance of back contact Silicon cell.

II. STRUCTURE AND SIMULATION

SILVACO ATLAS device simulator has been used to simulate the devices. Fig. 1 shows the simulated device structure. $CH_3NH_3PbI_3$ layer of 20nm is deposited on top surface of the silicon wafer. The 20μm-wide $p+$ regions and the 20μm-wide $n+$ regions are used. The substrate is n-type, with a doping density of 3×10^{15} cm^{-3}. $p+$ region is doped with boron (4×10^{18} cm^{-3}, depth of 1 μm). The $n+$ region is doped with phosphorous (4×10^{18} cm^{-3}, depth of 1 μm). All contacts are of aluminum. The above cell is compared with a cell replacing $CH_3NH_3PbI_3$ with Si_3N_4(80nm thick) as the top layer.

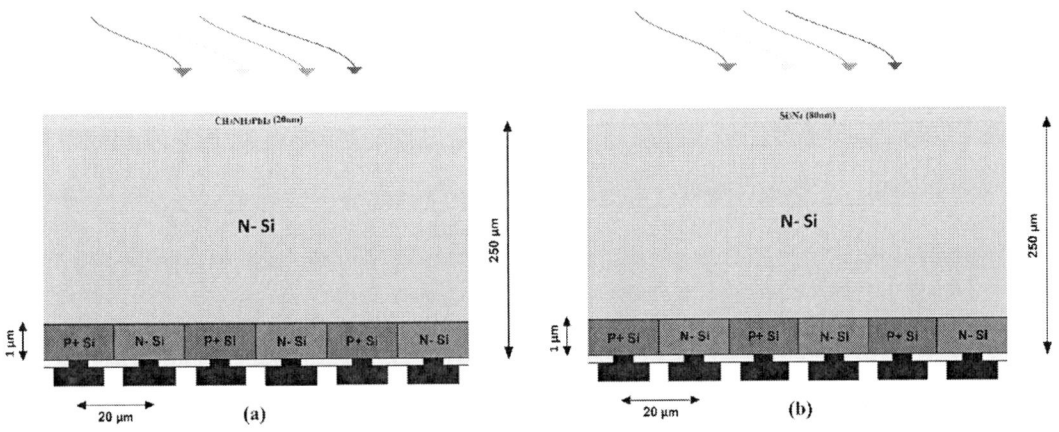

Fig. 1   Structure used: Interdigitated back contact Silicon cell with (a) $CH_3NH_3PbI_3$ passivation layer and, (b) Si_3N_4 AR-layer.

978-1-5090-5606-4/17 $31.00 © 2017 IEEE

Bandgap narrowing, concentration-dependent Shockley Read Hall (SRH), concentration-dependent auger, concentration dependent mobility, and field dependent mobility models are used during simulation analysis [5]. Poisson's equation together with the continuity equation for electrons and holes are solved simultaneously to obtain the current density– voltage (J–V) characteristics under illumination of the AM1.5G solar spectrum.

III. RESULTS

To obtain high PCE, passivation of the front surface is very important. The cell performance highly depends on surface recombination velocity of carriers in the conventional IBC-Si solar cell.

The external quantum efficiencies (EQEs) of conventional AR based IBC-Si solar cell and, $CH_3NH_3PbI_3$ layer based IBC-Si solar cell is shown in Fig. 2a, and Fig. 2b respectively. Results show that EQE of the conventional AR based IBC-Si solar cell decreases drastically from 88% to 5% at a wavelength of 600nm, with changing SRV from 1 to 10^4 cm/s, while, $CH_3NH_3PbI_3$ layer based IBC-Si solar cell shows negligible effect, shown in Fig. 2b, with EQE only changing from 96% to 90% at the wavelength of 600nm with a change in SRV. This is due to the front surface field (FSF) created by $CH_3NH_3PbI_3$ layer, that creates the potential barrier for minority carrier holes and thus decreases the surface

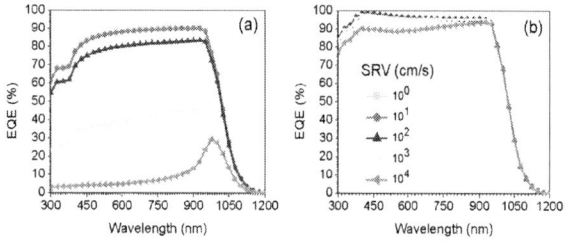

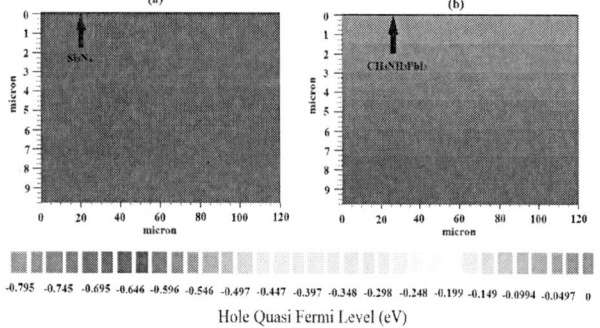

Fig. 2   EQEs of (a) Si_3N_4 AR-based IBC-Si solar cell and (b) $CH_3NH_3PbI_3$ passivation layer based IBC-Si solar cell for different surface recombination velocities.

Fig. 3   Hole quasi fermi level near the interfaces, (a) Si_3N_4 /n-Si, and (b) $CH_3NH_3PbI_3$ /n -Si.

concentration, which consequently reduces surface recombination. The barrier created, can be seen in Fig.3, shows hole quasi-fermi level (HQFL) in both cells. The Fig. 3b shows decreases in HQFL sharply at the surface in

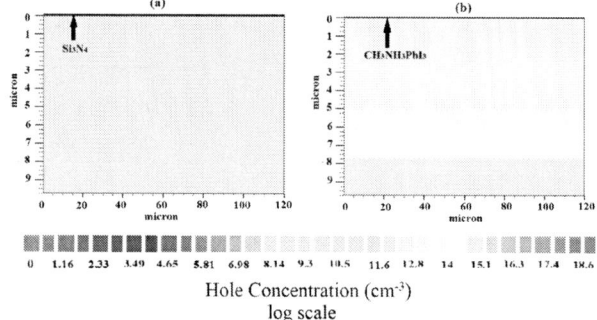

Fig. 4   Hole concentration near the interfaces, (a) Si_3N_4 /n-Si, and (b) $CH_3NH_3PbI_3$ /n -Si.

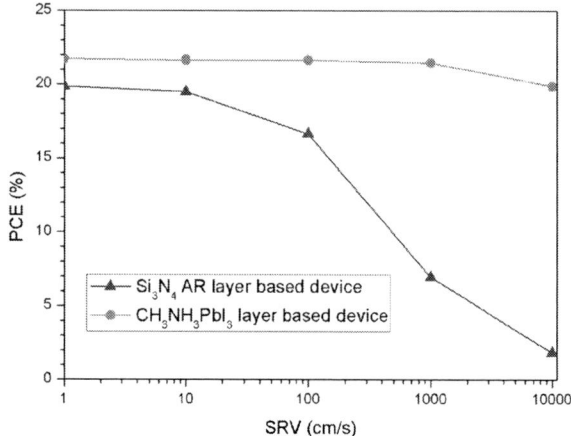

Fig. 5   Performance of both the cells for different SRVs.

$CH_3NH_3PbI_3$ layer based IBC-Si solar cell, due to the presence of $CH_3NH_3PbI_3$ layer, which creates the barrier talked about earlier. Fig. 3a shows no such barrier in conventional AR based IBC-Si solar cell. In support to this decrease in hole concentration at the $CH_3NH_3PbI_3$ /n -Si interface of the $CH_3NH_3PbI_3$ layer-based IBC-Si solar cell, can be seen in Fig. 4b, while no reduction in hole concentration, in Fig. 4a is seen at Si_3N_4 /n-Si interface of conventional AR based IBC-Si solar cell. Results have shown PCE increasing from 19.86% to 21.75%, i.e. an increase of nearly 9.5% with the introduction of the $CH_3NH_3PbI_3$ layer for surface passivation. This is due to increase in short-circuit current density (J_{SC}), in $CH_3NH_3PbI_3$ layer based IBC-Si solar cell, while there is a negligible change in the open circuit voltage (V_{OC}) with the introduction of the $CH_3NH_3PbI_3$ layer for front surface passivation. Fig.5 shows that perovskite cell performs better in comparison to the conventional Si_3N_4 AR based IBC-Si solar cell at higher SRVs. This is due to the FSF produced at the interface of $CH_3NH_3PbI_3$ layer-based IBC-Si solar cell. Maximum carriers

978-1-5090-5606-4/17 $31.00 © 2017 IEEE   378

are generated at the surface, and their recombination is limited by the FSF formed. Further results suggest that conventional Si_3N_4 AR coating is not suitable for SRV greater than 100 cm/s while $CH_3NH_3PbI_3$ passivated IBC-Si solar cell performs better at all SRVs.

Table 1 Results obtained for conventional cell and with perovskite passivation layer for different surface recombination velocity.

Top layer	SRV (cm/s)	J_{SC} (mA)	V_{OC} (mV)	FF (%)	PCE (%)
Si_3N_4	10^0	34.75	679.73	84.12	19.86
	10^1	34.25	677.83	84.04	19.50
	10^2	29.92	664.80	83.67	16.64
	10^3	13.52	620.30	82.95	6.95
	10^4	3.84	579.28	82.18	1.83
$CH_3NH_3PbI_3$	10^0	37.89	681.54	84.27	21.75
	10^1	37.75	681.47	84.26	21.67
	10^2	37.72	681.45	84.26	21.65
	10^3	37.43	681.28	84.24	21.47
	10^4	34.80	679.54	84.11	19.88

Further, Table 1 shows the photovoltaic parameters i.e. J_{SC}, V_{OC}, fill factor (FF), and PCE obtained for different values of SRV for both the cells. With the increase in SRV, a slight reduction in PCE is observed in $CH_3NH_3PbI_3$ passivated IBC-Si solar cell than in Si_3N_4 AR based IBC-Si solar cell, suggesting superior passivation compared to the Si_3N_4 AR based IBC-Si solar cell. The J-V characteristics of both Si_3N_4 AR based IBC-Si solar cell and $CH_3NH_3PbI_3$ passivated IBC-Si solar cell with a variation of SRV can be seen in Fig. 6a,

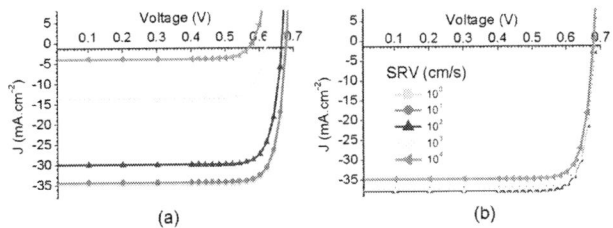

Fig. 6 J-V characteristics under illumination for (a) Si_3N_4 AR based IBC-Si solar cell, and (b) $CH_3NH_3PbI_3$ passivated IBC-Si solar cell for different surface recombination velocities.

and Fig. 6b respectively. The reason for a change in J_{SC} in $CH_3NH_3PbI_3$ passivated IBC-Si solar cell can be explained, with the help of Fig. 7. More recombination in a $CH_3NH_3PbI_3$ passivated IBC-Si solar cell can be seen in Fig. 7d with SRV of 10^4 cm/s than in Fig. 7c with SRV of 1 cm/s. The radicle change in current density of conventional Si_3N_4 AR based IBC-Si solar cell can be explained by an increase in surface recombination with an increase in SRV from 1cm/s to 10^4 cm/s as shown in Fig. 7a and Fig. 7b respectively. Results also show higher optical coupling and improved surface passivation quality in $CH_3NH_3PbI_3$ passivated IBC-Si solar cell. $CH_3NH_3PbI_3$ layer based device shows good UV stability

under high SRV condition. EQE of $CH_3NH_3PbI_3$ passivated IBC-Si solar cell at 300nm is 85%, which is quite high compared to the conventional AR-based device, i.e. 62% at 300nm. At high SRV it can be seen that EQE of the conventional Si_3N_4 AR-based device decreases to 3% at 300nm, while EQE of $CH_3NH_3PbI_3$ layer based device only decreases to 77%, making it a UV stable device.

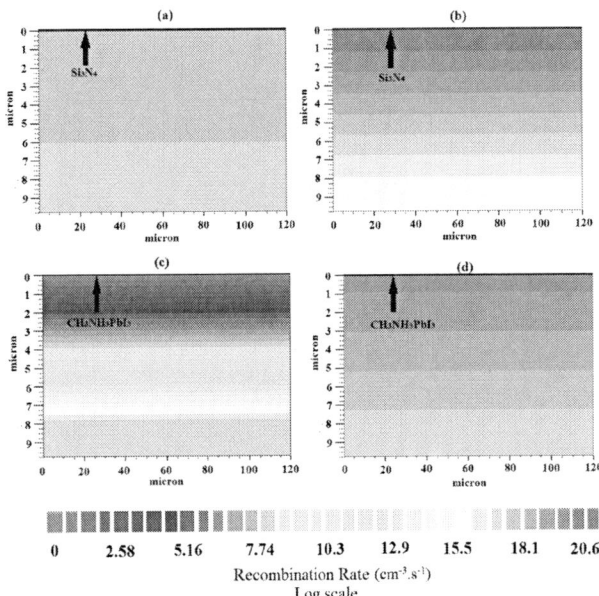

Fig. 7 Recombination rate at surface for (a) Si_3N_4 AR based IBC-Si solar cell at SRV 1cm/s, (b) Si_3N_4 AR based IBC-Si solar cell at SRV 10^4 cm/s, (c) $CH_3NH_3PbI_3$ passivated IBC-Si solar cell at SRV 1cm/s and, (d) $CH_3NH_3PbI_3$ passivated IBC-Si solar cell at SRV 10^4 cm/s.

IV. Conclusion

The introduction of $CH_3NH_3PbI_3$ as a front surface passivation layer on the n-type wafer based crystalline IBC-Si solar cell, enhances the photovoltaic behavior, by creating a FSF at perovskite/silicon interface. This further, decreases the surface recombination of minority carrier (holes) compared to conventional AR based solar cell, thus resulting in 21.75% PCE, which is 9.5% higher compared to the AR-based device. UV stability has also been observed in perovskite passivated device i.e. EQE of the device is less prone to higher SRV, throughout the spectrum range. Thus, high performance, $CH_3NH_3PbI_3$ layer based interdigitated back contact silicon solar cell, with UV stability is proposed with PCE as high as 21.75%.

Acknowledgements

The authors would like to thank Microelectronics Research Lab, Department of Engineering Physics, Delhi Technological University to carry out this work.

REFERENCES

[1] A. G. Aberle, "Surface passivation of crystalline silicon solar cells: a review," Prog. Photovoltaics Res. Appl., vol. 8, no. 5, pp. 473–487, 2000.

[2] W. Shockley, H.J, Queisser, J.Appl. Phys. 196, 32, 510.

[3] E. Radziemska, Renewable Energy 2003, 28, 1.

[4] The Light-Induced Field-Effect Solar Cell Concept – Perovskite Nanoparticle Coating Introduces Polarization Enhancing Silicon Cell Efficiency." DOI: 10.1002/adma.201606370.

[5] Software, "ATLAS User ' s Manual," no. 408, pp. 567–1000, 2013.

Potential of a-Si:H/c-Si heterojunction solar cells with very thin wafers

Hitoshi Sai[1,2], Hiroshi Umishio[1,3], Takuya Matsui[1,2], Shota Nunomura[1,2], Tomoyuki Kawatsu[4], Hidetaka Takato[2], and Koji Matsubara[1,2]

[1]Research Center for Photovoltaics, National Institute of Advanced Industrial Science and Technology (AIST), Tsukuba, Japan. [2]Renewable Energy Research Center, AIST, Koriyama, Japan. [3]Tsukuba University, Tsukuba, Japan. [4]Komatsu NTC Ltd. Nanto, Japan.

Abstract — **Potential of a-Si:H/c-Si heterojunction cells with very thin wafers was examined from the optical and electrical points of view. Optical characterization of dummy c-Si cell structures shows that a realistic light trapping scheme, i.e., pyramidally textured Si wafers with a dielectric anti-reflection coating and a back reflector, realizes an efficient quasi-Lambertian light absorption enhancement, even for very thin wafers. The potential photocurrent densities evaluated using absorption spectra suggests that J_{SC} reduction by decreasing the wafer thickness can be mitigated by using the realistic light trapping scheme, suppose that the parasitic absorption loss is minimized. The potential of higher V_{OC} in thin c-Si cells was investigated using thin c-Si wafers passivated with intrinsic a-Si:H thin layers. A steady increase of implied V_{OC} with decreasing the wafer thickness was experimentally confirmed in a wide range of wafer thickness down to 30 μm. The V_{OC} enhancement was experimentally confirmed in a SHJ cell with a thickness below 60 μm.**

I. INTRODUCTION

In the field of crystalline silicon (c-Si) solar cells, it is expected that the thickness of c-Si wafers is steadily decreased for reducing the material cost. To meet this requirement, wafer slicing equipment manufacturers are developing slicing technologies which are capable to produce thinner wafers with less kerf losses. It is also expected that the main-stream solar cell structure in the industry changes from the common Al-BSF (Back surface filed) structure to other advanced structures such as PERC (Passivated emitter rear cell) in the near future, for improving the conversion efficiency. In particular, a-Si:H/c-Si heterojunction (SHJ) solar cell structure, which utilizes an excellent surface passivation of c-Si with intrinsic (i) hydrogenated amorphous silicon (a-Si:H) thin layers, has been actively researched because of its potential for high-efficiency cells exceeding 25 % [1-3]. In the high-efficiency cell structures including SHJ cells, the surface recombination loss can be substantially reduced. In such a situation, using thinner wafers contributes to the reduction of the bulk recombination loss, and therefore the conversion efficiency of c-Si cells can be improved, suppose that the reduction of light absorption is sufficiently suppressed by applying light management techniques. Taguchi et al. reported a notably high open circuit voltage (V_{OC}) of 0.750 V as well as an excellent efficiency of 24.7% in a SHJ cell with a 100-μm-thick wafer [4]. For much thinner wafers, a very high V_{OC}

of 0.753 V was reported by Herasimenka et al. in a 50-μm-thick SHJ cell structure, although other parameters were not reported [5]. Another notable thin c-Si solar cell was developed by Solexel using epitaxial lift-off technology and diffusion-based interdigitated back contact (IBC) design [6]. They reported a 21.2% efficient c-Si cell having a V_{OC} of 0.687 V, with a wafer thickness of only 35 μm [7]. However, to date, there are few systematic experimental reports on the potential of very thin c-Si solar cells. In this study, the impact of wafer thickness on the optical and electrical properties in c-Si solar cells are characterized in a wide range of wafer thickness from 300 μm down to 40 μm, with a focus on SHJ solar cells.

II. EXPERIMENTAL

P-doped n-type CZ-grown monocrystalline Si wafers (~2 Ωcm, <100> oriented) with a wide variety of thicknesses from 50 to 300 μm were prepared from the same ingot. The wafer thickness was controlled by mechanical thinning and polishing. These wafers were cut into 5 cm × 5 cm in size due to the limitation in our experimental facility. Then they were subjected to wet-chemical etching with a KOH-based solution to form random pyramidal textures on both surfaces. In this process, the wafers were etched by 5 – 10 μm in thickness on each surface, and the average thickness after texturing was determined from their weight, with an accuracy of ± 0.5 μm. The average texture size was approximately 5 μm. One group of the wafers was subjected to parallel plate plasma enhanced chemical vapor deposition (PECVD) to form SiN$_X$ films (100

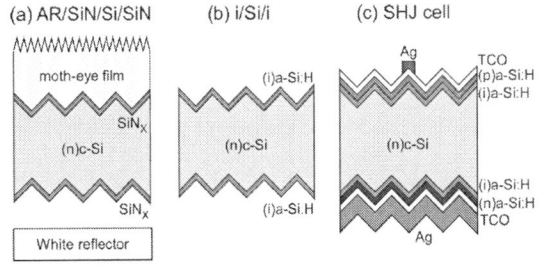

Fig. 1. Sample structures prepared in this study. (a) dummy cell structure for optical characterization (AR/SiN/Si/SiN), (b) passivated wafer (i/Si/i), and (c) completed SHJ solar cell.

nm) on both sides, followed by attachment of an anti-reflection film with moth-eye feature, as shown in Fig. 1(a). These wafers (AR/SiN/c-Si/SiN), which can be regarded as dummy cell structures with minimal parasitic absorption loss, were used for optical measurement. Total light absorption of the AR/SiN/c-Si/SiN structures was evaluated by the spectrometer with an integral sphere (Perkin Elmer, Lambda-950) and a detached white back reflector (Labsphere, Spectralon). Another group of the wafers was wet-chemically cleaned with our standard cleaning process followed by the PECVD deposition (60 MHz) of 10-nm-thick (i)a-Si:H layers on both sides with a SiH_4/H_2 gas mixture and pressures of 0.1 – 0.2 torr for surface passivation. These samples are referred as i/Si/i structures, as schematically drawn in Fig. 1(a). Two sets of i/Si/i were prepared with respect to the growth temperature (T_g) and the post annealing temperature (T_{ann}): (a) processed at T_g = 150°C followed by a post annealing at T_{ann} = 240°C for 2h in a vacuum oven, and (b) processed at T_g = 200 °C without post annealing. The minority carrier lifetime and implied V_{OC} (iV_{OC}) of the i/Si/i structures were characterized by quasi-steady state photo-conductance (QSSPC) technique (Sinton Instruments, WCT-120). For some wafers, (p) and (n)a-Si:H were successively deposited with PECVD followed by In_2O_3:Sn (ITO) film and metal electrode formation with sputtering to complete SHJ solar cells, as illustrated in Fig. 1(c). The current density-voltage (J-V) characteristics of these cells were evaluated using a solar simulator under AM 1.5 100 mW/cm^2 illumination. The cell area was 1 cm^2 designated by a shadow mask.

III. RESULTS AND DISCUSSION

A. Potential of J_{SC}

The potential of thin c-Si cells in terms of short circuit current density (J_{SC}) was investigated experimentally via optical characterization of AR/SiN/Si/SiN dummy cell structures. Figure 2 shows the optical absorptance of AR/SiN/Si/SiN structures, measured with a white diffuse reflector on the backside. AR/SiN/Si/SiN structures exhibit almost ideal absorption within the visible region, while the optical absorption in the near infrared (NIR) region decreases steadily with decreasing the wafer thickness, as seen in the solid lines in Fig. 2. The dashed line in Fig. 2 shows the absorptance of a 280-μm-thick c-Si slab with ideal Lambertian interfaces, calculated using the formula reported by Green [8]. It can be seen that, even with the relatively simple structure, the absorptance of the AR/SiN/Si/SiN is rather close to the ideal case, confirming that the random pyramid structure obtained with anisotropic etching is very efficient in light trapping.

In order to quantify the maximum achievable J_{SC}, an implied J_{SC} (iJ_{SC}) were calculated using the absorptance spectra shown in Fig. 2, by simply assuming that internal

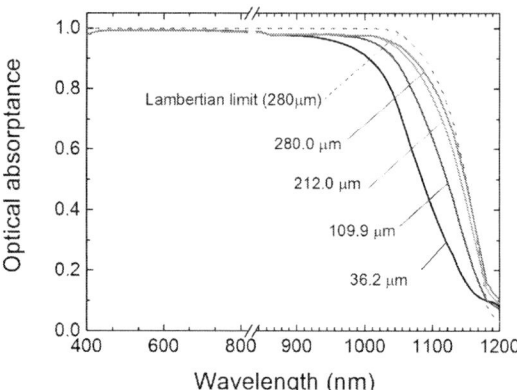

Fig. 2. Optical absorptance spectra of AR/SiN/Si/SiN structures with a variety of wafer thicknesses from 36 to 280 μm (solid lines). Optical absorption of a Si slab with ideal Lambertian surfaces and a backside mirror is also plotted for comparison (dashed line).

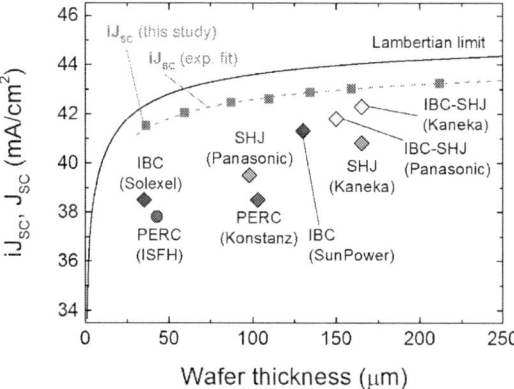

Fig. 3. Implied J_{SC} calculated from the optical absorptance spectra shown in Fig. 2. Several J_{SC} in the high-efficiency c-Si cells [1-4,7,9,10] are plotted for comparison. The solid curve shows iJ_{SC} expected from Lambertian limit, and the dashed curve shows the exponential fit of the experimentally determined iJ_{SC} data.

quantum efficiency is unity in the wavelength range up to 1200 nm. The result is summarized in Fig. 3, with the J_{SC} data of several high-efficiency c-Si cells found in the literature [1-4,7,9,10]. From Fig. 3, we can confirm the following issues: First, the wafer-thickness dependence of the experimentally obtained iJ_{SC} is very similar to that of the Lambertian limit. This implies that the optical structure in the standard c-Si cells (pyramidal textures with AR coatings) behaves like quasi-Lambertian, even for very thin wafers below 50 μm. Second, in spite of its quasi-Lambertian behavior, iJ_{SC} of AR/SiN/Si/SiN is still smaller than that of the ideal Lambertian case by approximately 1 mA/cm^2, regardless of wafer thickness at least up to ~30 μm. This is mainly originated from insufficient absorption in the NIR region, as seen in Fig. 2, indicating that there is room for improvement from the optical point of view. Third, iJ_{SC} obtained in this study is higher than the experimentally confirmed J_{SC} in high-efficiency cells. The gap between them tends to increase with decreasing the wafer thickness. This result suggests that, with decreasing the wafer thickness, the parasitic absorption loss becomes more pronounced, and/or the fabrication of efficient light trapping structures becomes more difficult. Nevertheless, the high iJ_{SC} confirmed in this study indicates that a high current density exceeding 40 mA/cm^2 is experimentally feasible for thin (~50 μm) c-Si cells with a realistic optical design.

B. Potential of V_{OC}

The potential of thin c-Si cells in terms of V_{OC} was investigated experimentally via characterizing iV_{OC} of i/Si/i structures. Figure 4 shows the measured iV_{OC} of the i/Si/i structures as a function of wafer thickness. It is found that iV_{OC} shows a relatively large scattering, especially for those processed at T_g = 200°C, likely due to the insufficient reproducibility in our process. However, it is clearly seen that the maximum iV_{OC} in each thickness increases steeply and monotonically with decreasing the wafer thickness down to 30 μm, in agreement with the previous reports [5,11]. It is also found that two different series of i/Si/i samples show almost equivalent maximum iV_{OC}, although those with T_g = 200°C shows a larger scattering than the other. The highest iV_{OC} of 0.763 V was experimentally confirmed with a 32-μm-thick textured wafer. The thickness dependence in iV_{OC} observed here can be well reproduced by model simulation with a fixed bulk lifetime of 3 msec and a surface saturation current density of 1.8 fA/cm^2, as shown as the solid line in Fig. 4. This result indicates that the bulk lifetime and surface passivation quality of the i/Si/i structures were maintained in our relatively "good" samples, in a wide range of wafer thicknesses from 30 to 300 μm.

It should be noted that, iV_{OC} obtained in i/Si/i structures can be deteriorated during the following process including doped layer and TCO film formations to complete SHJ cells. We found that the iV_{OC} deterioration seems to be more pronounced in post-annealed i/Si/ structures, despite that the post annealing is effective to improve iV_{OC}. This issue is limiting the performance of SHJ cells fabricated in this study, as shown in later.

D. SHJ solar cells

Figure 5 shows the J-V curve of thin (59 μm) and standard (184 μm) SHJ solar cells fabricated in this study. As seen in Fig. 5, the SHJ cell properties obtained are not comparable to the previously reported best-efficiency cells, because our process has not yet sufficiently matured for achieving a high V_{OC} in SHJ cells close to iV_{OC} in i/Si/i structures, as well as a

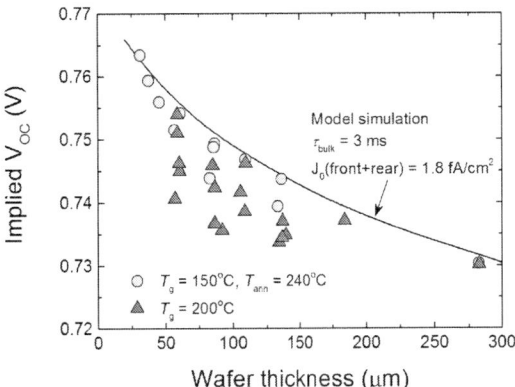

Fig. 4. Measured implied V_{OC} of i/Si/i structures as functions of wafer thickness (symbol). Solid line shows ideal V_{OC} variations against wafer thickness, assuming a bulk lifetime of 3 ms and a surface saturation current density of 1.8 fA/cm^2. Simulation was performed using PC1D.

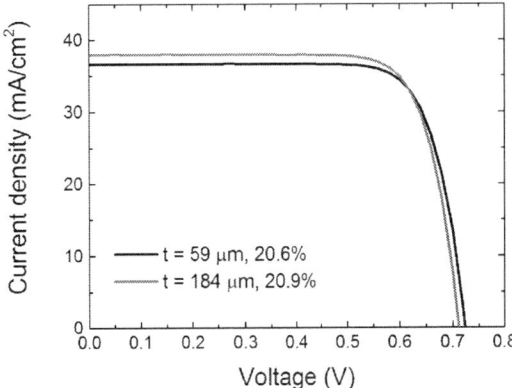

Fig. 5. J-V curves of SHJ solar cells with wafer thicknesses of 59 μm (J_{SC} = 36.6 mA/cm^2, V_{OC} = 0.724 V, FF = 0.779, Eff. = 20.6%) and 184 μm (37.9 mA/cm^2, 0.712 V, 0.776, 20.9%).

high J_{SC} close to iJ_{SC}. In addition, there is also significant room for improvement in FF. Nevertheless, it is clearly demonstrated in Fig. 5 that SHJ cells with almost equivalent efficiencies can be realized in spite of the significant reduction in the wafer thickness by a factor of 1/3, since the drop in J_{SC} is partly compensated by the gain in V_{OC}, as expected in Fig. 4. The thin SHJ cell yields a higher V_{OC} by 12 mV (~ 1.7%) than the other, which is in good agreement with the relative iV_{OC} gain shown in Fig. 4. However, this gain is over-compensated by its smaller J_{SC} by 1.3 mA/cm^2 (~ 3.3%), resulting in a slightly lower conversion efficiency. The relatively large J_{SC} decrease can be ascribed to parasitic absorption loss by the supporting layers (intrinsic and doped a-Si:H layers, TCO layers, and metal electrode), which will become more severe with decreasing the thickness.

IV. SUMMARY

Potential of a-Si:H/c-Si heterojunction cells with very thin wafers was examined from the optical and electrical points of view. It was experimentally confirmed that a realistic light trapping scheme, i.e., pyramidally textured Si wafers with a dielectric anti-reflection coating and a back reflector, shows an efficient light absorption enhancement which is very close to that predicted by the Lambertian limit, even for very thin wafers down to 30 µm. This indicates that, in principle, J_{SC} reduction by decreasing the wafer thickness can be mitigated with such a structure, suppose that the parasitic absorption loss by the supporting layers is minimized. It was also experimentally demonstrated that excellent surface passivation by (i)a-Si:H thin layers contributes to iV_{OC} enhancement with decreasing the wafer thickness. A very thin SHJ solar cell with a thickness of 59 µm was also developed, and a conversion efficiency of 20.6% was obtained. It was found that the efficiency decrease by reducing the thickness from 184 µm to 59 µm was very small, owing to the trade-off between J_{SC} and V_{OC}, indicating the high potential of very thin c-Si cells. More systematic investigation on the cell performances will be the future work.

ACKNOWLEDGEMENT

The authors acknowledge Mr. Oku, Mr. Sato, and Ms. Tanabe for the technical supports in this work A part of this work was conducted with the financial support from New Energy and Industrial Technology Development Organization (NEDO) under the Ministry of Economy, Trade and Industry (METI), Japan.

REFERENCES

[1] K. Yoshikawa, H. Kawasaki, W. Yoshida, T. Irie, K. Konishi, K. Nakano, T. Uto, D. Adachi, M. Kanematsu, H. Uzu and K. Yamamoto, "Silicon heterojunction solar cell with interdigitated back contacts for a photoconversion efficiency over 26%," *Nature Energy* vol. 2, art.no. 17032, 2017.

[2] K. Masuko, M. Shigematsu, T. Hashiguchi, D. Fujishima, M. Kai, N. Yoshimura, T. Yamaguchi, Y. Ichihashi, T. Mishima, N. Matsubara, T. Yamanishi, T. Takahama, M. Taguchi, E.

Maruyama, and S. Okamoto, "Achievement of More Than 25% Conversion Efficiency With Crystalline Silicon Heterojunction Solar Cell," *IEEE J. Photovolt.* vol. 4, no. 6, pp.1433-1435, 2015.

[3] D. Adachi, J.L. Hernandez, and K. Yamamoto, "Impact of carrier recombination on fill factor for large area heterojunction crystalline silicon solar cell with 25.1% efficiency," *Appl. Phys. Lett.* vol. 107, art.no. 233506, 2015.

[4] M. Taguchi, A. Yano, S. Tohoda, K. Matsuyama, Y. Nakamura, T. Nishiwaki, K. Fujita, and E. Maruyama, "24.7% Record Efficiency HIT Solar Cell on Thin Silicon Wafer," *IEEE J. Photovolt.* vol. 4, no. 1, pp.96-99, 2014.

[5] S.Y. Herasimenka, W.J. Dauksher, and S.G. Bowden, ">750 mV open circuit voltage measured on 50 µm thick silicon heterojunction solar cell," *Appl. Phys. Lett.* vol. 103, art.no. 053511, 2013.

[6] MM. Moslehi, P. Kapur, J. Kramer, V. Rana, S. Seutter, A. Deshpande, T. Stalcup, S. Kommera, J. Ashjaee J, A. Calcaterra, D. Grupp, D. Dutton, and R. Brown "World-record 20.6% efficiency 156mm× 156mm full-square solar cells using low-cost kerfless ultrathin epitaxial silicon & porous silicon lift-off technology for industry-leading high-performance smart PV modules," *PV Asia Pacific Conference (APVIA/PVAP)*, 24 October 2012.

[7] M.A. Green, K. Emery, Y. Hishikawa, W. Warta, E.D. Dunlop, D. H. Levi and A. W. Y. Ho-Baillie, "Solar cell efficiency tables (version 49)," *Prog. Photovolt: Res. Appl.* vol. 25, pp. 3-13 2017.

[8] M.A. Green, "Lambertian Light Trapping in Textured Solar Cells and Light-Emitting Diodes: Analytical Solutions," *Prog. Photovolt: Res. Appl.* vol. 10, pp. 235-241, 2002.

[9] B. Terheiden, T. Ballmann, R. Horbelt, Y. Schiele, S. Seren, J. Ebser, G. Hahn, M. Vertens, M. B. Koentopp, M. Scherff, J. W. Müller, Z. C. Holman, A. Descoeudres, S. De Wolf, S. Martin de Nicolas, J. Geissbuehler, C. Ballif, B. Weber, P. Saint-Cast, M. Rauer, C. Schmiga, S. W. Glunz, D. J. Morrison, S. Devenport, D. Antonelli, C. Busto, F. Grasso, F. Ferrazza, E. Tonelli, and W. Oswald, "Manufacturing 100-µm-thick silicon solar cells with efficiencies greater than 20% in a pilot production line," *Phys. Status Solidi A* vol. 22, no. 1, pp. 13-24, 2015.

[10] J. H. Petermann, D. Zielke, J. Schmidt, F. Haase, E. Garralaga Rojas and R. Brendel, "19%-efficient and 43µm-thick crystalline Si solar cell from layer transfer using porous silicon," *Prog. Photovolt: Res. Appl.* vol. 20, pp. 1-5, 2012.

[11] S. Tohoda, D. Fujishima, A. Yano, A. Ogane, K. Matsuyama, Y. Nakamura, N. Tokuoka, H. Kanno, T. Kinoshita, H. Sakata, M. Taguchi, and E. Maruyama, "Future directions for higher-efficiency HIT solar cells using a Thin Silicon Wafer," *J. Non-Crystalline Solids* vol. 358, pp. 2219-2222, 2012.

[12] A. Richter, M. Hermle, and S.W. Glunz, "Reassessment of the Limiting Efficiency for Crystalline Silicon Solar Cells," *IEEE J. Photovolt.* vol. 3, no. 4, pp. 1184-1191, 2013.

Manipulating Fixed Charges in ZrO₂ by Doping for Passivation and Antireflection on Wafer-Si Solar Cells

Woo Jung Shin,[a] Laidong Wang,[b] Wen-Hsi Huang,[a] and Meng Tao[b]

[a] School for Engineering of Matter, Transport and Energy and [b] School of Electrical, Computer and Energy Engineering, Arizona State University, Tempe, Arizona, 85287, USA

Abstract — **In this study, we attempted to manipulate fixed charges within ZrO₂ films by doping for effective field-effect passivation. ZrO₂ films doped with Cr, Mn or Zn (1–10 at. %) were grown on p-type Si (100) wafers by spray pyrolysis, followed by rapid thermal annealing at 700–850°C in air. Flat-band voltage shifts from capacitance-voltage measurements indicate that incorporation of dopants into Zr⁺ sites leads to changes in sign and density of fixed charges in ZrO₂. The optical properties of doped ZrO₂ are comparable to, and in some cases slightly better than, PECVD SiN_x. These results suggest that spray-deposited doped ZrO₂ could be a suitable material for antireflection and passivation on p-type Si in wafer-Si solar cells.**

I. INTRODUCTION

Previously, our studies on un-doped spray-deposited ZrO₂ have shown that the optical properties of ZrO₂ films are comparable to those of SiN_x films by PECVD [1], which makes it a potential candidate for antireflection on wafer-Si solar cells. Zr is an earth-abundant element allowing terawatt-scale production of wafer-Si solar cells with a ZnO₂ coating [2]. For low-cost processing, spray pyrolysis was chosen to deposit ZrO₂ films due to its versatility, simplicity, low cost and high throughput.

Un-doped ZrO₂ films have a negative fixed charge density of ~9.5×10^{11} cm^{-2} for a 75-nm film, which is still low for effective field-effect passivation on p-type Si [3]. Therefore, manipulation of fixed charges within the ZrO₂ film is required in order to increase the negative charge density to ~1×10^{13} cm^{-2} for sufficient field-effect passivation. In this paper, we report doping in spray-deposited ZrO₂ to manipulate the density and sign of the fixed charges in ZrO₂. Our objective is to identify a dopant that generates a high negative charge density so ZrO₂ can serve for both antireflection and passivation on p-type Si in wafer-Si solar cells. This would simplify the current Al₂O₃/SiN_x stack on p-type Si.

II. EXPERIMENTAL

Cr, Mn or Zn doped ZrO₂ films were synthesized by low-cost open-air spray pyrolysis. Zr acetylacetonate, Zr(C₅H₇O₂)₄, dissolved in dimethylformamide (C₃H₇NO) was used as the spray solution. CrCl₆, MnCl₂ and Zn acetate (Zn(O₂CCH₃)₂) were used as the dopant precursors. These elements were chosen because they have similar atomic size to Zr, but have different numbers of valence electrons. The concentration of Zr(C₅H₇O₂)₄ was fixed to 0.05 M in the spray solution and the

dopant/Zr ratio in the solution was varied from 1–10 at. %. Pulsed spray was employed for film deposition, with 10 s on cycle and 50 s off cycle. The films were deposited at 500°C, followed by rapid thermal annealing at 700–850°C for 1 min in air. The process parameters for doped ZrO₂ thin films are listed in Table I.

TABLE I
OPTIMIZED PARAMETERS FOR SPRAY DEPOSITION AND POST
ANNEALING OF DOPED ZIRCONIUM OXIDE

Nozzle-substrate distance	15 cm
Solution flow rate	1 ml/cycle
Carrier gas	air
Atomization/piston pressure	60/40 pa
Deposition temperature	500°C
Post annealing Temperature	700–850°C
Post annealing time	1 min

The wafers used in this work were 10 Ω-cm p-type Si (100) wafers and soda lime glass slides. Prior to deposition, Si wafers were cleaned with diluted HF to remove native oxide from the surface. The glass slides were ultrasonically cleaned with acetone, methanol and water respectively. Doped ZrO₂ films deposited on Si wafers were used to characterize electrical and structural properties; and those on glass slides were used to characterize optical properties of the films.

The capacitance-voltage (C-V) measurement was carried out using a MDC C-V system equipped with a mercury probe. The thickness and refractive index of the doped ZrO₂ films was measured by a VASE ellipsometer. Transmittance and reflectance of the films was characterized by a JASCO V-670 spectrophotometer. Surface morphology and roughness of the film was characterized by a Park System atomic force microscope (AFM) using contact mode.

III. RESULTS AND DISCUSSION

A. C-V Characteristics of Cr, Mn, Zn-doped ZrO₂

Fig. 1 presents the retraced C-V characteristics of un-doped ZrO₂ and Cr-doped ZrO₂ films as a function of post annealing temperature and doping concentration. All of the films used for C-V measurement have a thickness of 75 nm. The flat-band voltage of un-doped ZrO₂ film is located at about +0.6V corresponding to a negative charge density of ~9.5×10^{11} cm^{-2}. A slight hysteresis of 0.3V is observed in un-doped ZrO₂ indicating that the charges in un-doped ZrO_2 is not perfectly stable. However, after doping ZrO_2 with Cr the C-V hysteresis is decreased significantly. A positive flat-band voltage shift is observed with Cr, and the magnitude of the flat-band shift is strongly dependent on doping concentration and post annealing temperature.

(a)

(b)

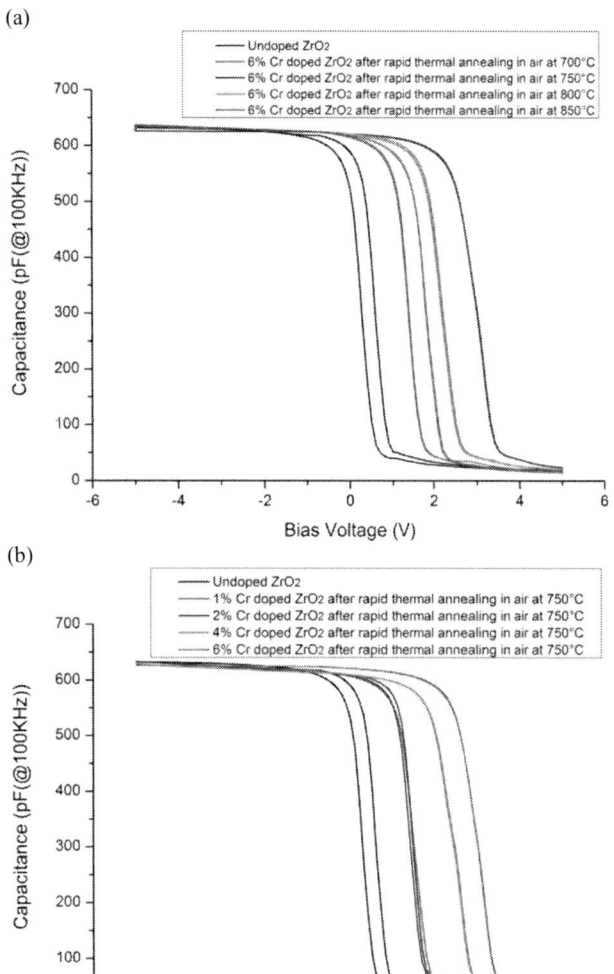

Fig. 1. (a) C-V of un-doped and 6% Cr-doped ZrO_2 films after different post annealing temperatures from $700^\circ C$ to $850^\circ C$ and (b) C-V of un-doped and Cr-doped ZrO_2 films with different Cr concentrations. The post annealing conditions are fixed at $750^\circ C$ for 1 min.

Fig. 1(a) shows the effect of annealing temperature on the magnitude of the positive flat-band shift in Cr-doped ZrO_2 films. The amount of shift increases with post annealing temperature between $700^\circ C$ and $750^\circ C$, and then decreases above $800^\circ C$. The maximum positive flat-band shift is observed at annealing temperature of $750^\circ C$. Fig. 1(b) shows

the effect of doping concentration on its flat-band shift. The amount of positive flat-band shift increases linearly as Cr concentration increases from 1% to 6%. Further increasing Cr concentration above 6% does not result in a larger shift. The maximum flat-band shift of +2.5V is observed at 6% Cr doping with rapid thermal annealing at $750^\circ C$ for 1 min. Under these conditions, the flat-band voltage and net negative charge density are calculated to be 3.14 V and 4.07×10^{12} cm^{-2}, respectively.

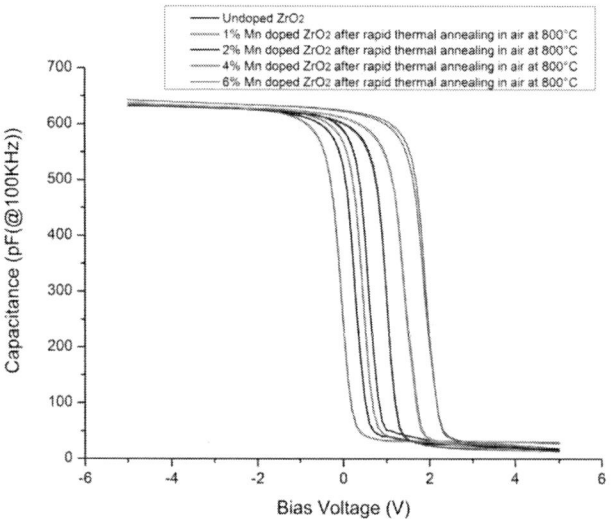

Fig. 2. C-V of un-doped and Mn-doped ZrO_2 films with different Mn concentrations. The post annealing conditions are fixed at $800^\circ C$ for 1 min.

For Mn-doped ZrO_2 films, the maximum positive flat-band shift is observed at annealing temperature of $800^\circ C$ for all doping concentrations. Fig. 2 compares the magnitude of the positive flat-band shift for Mn-doped ZrO_2 films as a function of doping concentrations. There is a negligible shift when ZrO_2 is doped with a low concentration of Mn, 1%, and the retraced C-V still shows a hysteresis. However, the magnitude of the positive flat-band shift increases linearly from 2% to 6%. The maximum flat-band shift of +1.35V is observed at 6% Mn doping after rapid thermal annealing at $800^\circ C$. No significant C-V hysteresis is observed above doping concentration of 4%. Under these conditions, the flat-band voltage and net negative charge density are calculated to be 2.0 V and 2.83×10^{12} cm^{-2}.

In case of Zn-doped ZrO_2 films, a negative shift in flat-band voltage is observed instead of a positive shift. The maximum negative flat-band shift is observed at annealing temperature of $850^\circ C$ for all doping concentrations. Fig. 3 shows the negative flat-band in Zn-doped ZrO_2 films as a function of doping concentration after post annealing at $850^\circ C$ for 1 min. A maximum flat-band shift of -1.6V is observed with 6% Zn doping concentration. Under these conditions, the flat-band voltage and net positive charge density are calculated to be

−1.0 V and 8.7×10^{11} cm^{-2}. These results prove that doping can change not only the density of the fixed charge but also the sign of fixed charge in ZrO$_2$ films.

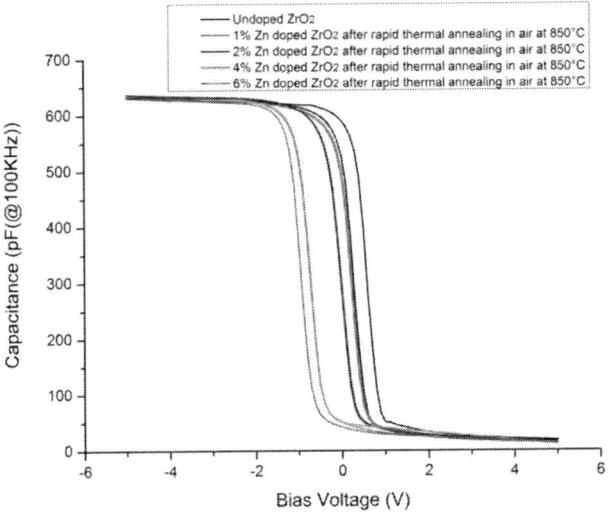

Fig. 3. C-V of un-doped and Zn-doped ZrO$_2$ films with different Zn concentrations. The post annealing conditions are fixed at 850°C for 1 min.

B. Optical and Structural Properties of Doped ZrO$_2$

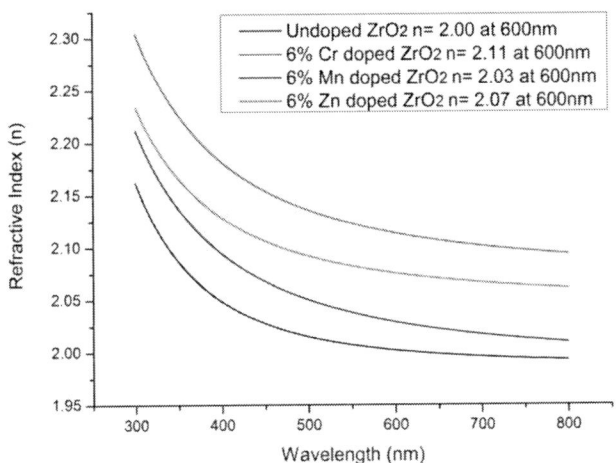

Fig. 4. Comparison of refractive index between un-doped and Cr, Mn or Zn-doped ZrO$_2$.

Fig. 4 presents the refractive index of un-doped ZrO$_2$ and 6% Cr, Mn or Zn-doped ZrO$_2$ films. The refractive index of doped ZrO$_2$ films are found to be higher than un-doped ZrO$_2$. Among the doped ZrO$_2$ films, Cr-doped ZrO$_2$ has the highest value of n = 2.11 at 600 nm. This is higher than the PECVD SiN$_x$ which has favorable value of n = 2.0 although it can be varied in a range of 1.8 and 2.3 by means of the silane/ammonia ratio [4]. This result show that Cr-doped ZrO$_2$

can be a potential candidate for antireflection and passivation on p-type Si in wafer-Si solar cells.

Fig. 5 compares the total transmittance of PECVD SiN$_x$ with un-doped and doped ZrO$_2$ on glass substrate. The total transmittance is measured with an integrating sphere. The thickness of all the films is 75 nm. At long wavelengths, all the films have similar transmittance above 83%, but doped ZrO$_2$ films have slightly higher transmittance below 400 nm. Moreover, there is no change in absorption edge among un-doped and doped ZrO$_2$ films indicating that the band gap of doped ZrO$_2$ does not change due to doping.

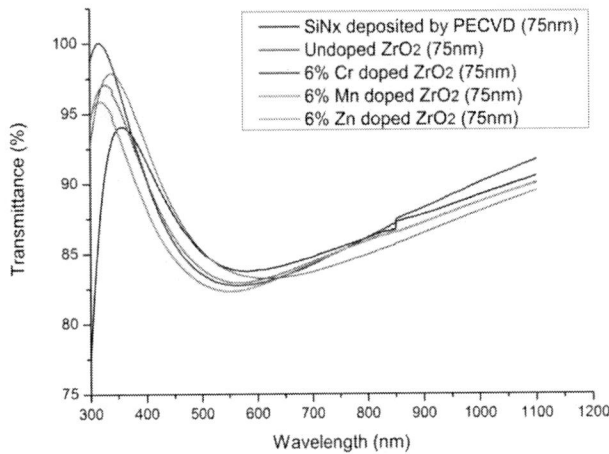

Fig. 5. Comparison of total transmittance spectra for SiN$_x$, un-doped ZrO$_2$ and Cr, Mn or Zn-doped ZrO$_2$.

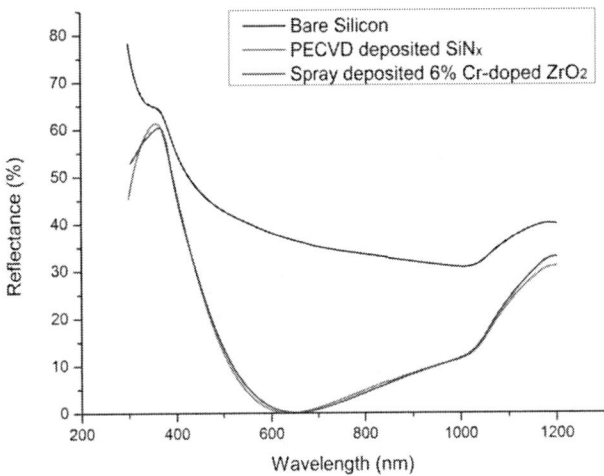

Fig. 6. Comparison of reflectance spectra for PECVD SiN$_x$ and spray-deposited 6% Cr-doped ZrO$_2$ films on Si wafers.

Reflectance spectra of a bare Si wafer, a Si wafer coated with PECVD SiN$_x$ and a Si wafer with spray deposited 6% Cr-doped ZrO$_2$ are shown in Fig. 6. These films were deposited on polished p-type Si (100) wafers and the

978-1-5090-5606-4/17 $31.00 © 2017 IEEE

thickness of SiN_x and Cr-doped ZrO_2 films were 75nm. They have similar reflectance trends with minimum reflectance located at ~640 nm, confirming that they have similar refractive indices.

(a)

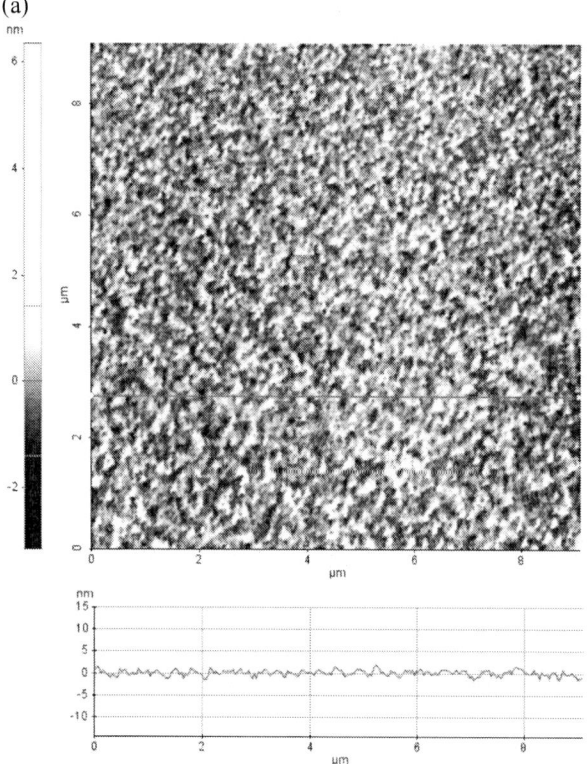

(b)

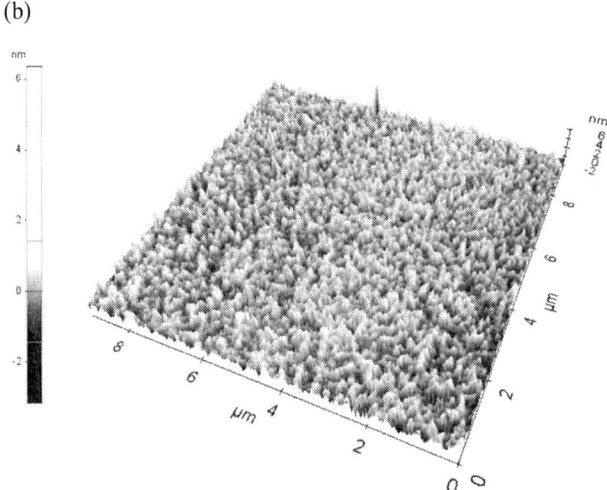

Fig. 7. (a) Top view AFM image and (b) 3-D AFM image of a 6% Cr-doped ZrO_2 film on a Si wafer.

Contact mode AFM was used to study surface morphology and roughness of spray-deposited Cr-doped ZrO_2 films, as shown in Fig. 7. The scan area on the sample is 9×9 μm^2.

The surface roughness keeps decreasing as the deposition temperature increases. At deposition temperature of 500˚C, the root mean square value of surface roughness is found to be 0.7 nm in the ZrO_2 film doped with 6% Cr.

IV. CONCLUSIONS

Optical, electrical and structural properties of Cr, Mn or Zn doped ZrO_2 films are studied for antireflection and passivation on p-type Si in wafer-Si solar cells. C-V results indicate that manipulation of fixed charges in ZrO_2 films, including both density and sign, is possible by doping. Among the three dopants, Cr doping shows the largest flat-band voltage shift of +2.5V in the positive direction, i.e. a significant increase in negative charge density up to 4.07×10^{12} cm^{-2}. Optical properties such as total transmittance, reflectance and refractive index of Cr-doped ZrO_2 are found to be comparable to or slightly better than PECVD SiN_x. Moreover, the surface roughness of Cr-doped ZrO_2 is very small having a root mean square value of only 0.7 nm. These results indicate the possible use of low-cost spray deposited Cr-doped ZrO_2 films as an antireflection and passivation layer to simplify the current Al_2O_3/SiN_x stack on p-type Si.

ACKNOWLEDGEMENT

Authors thank Wayne Paulson (Arizona State University NanoFab) for discussions and technical assistance on C-V measurements. Financial support for this project is provided in part by the U.S. National Science Foundation under grant no. DMR-1306542.

REFERENCES

[1] W. J. Shin, L. Wang, and M. Tao. "Low-cost spray deposited ZrO_2 for passivation and antireflection on p-type Si." In *43rd IEEE Photovoltaic Specialist Conference,* pp. 2971-2974. 2016.

[2] M. Tao, *Terawatt Solar Photovoltaics – Roadblocks and Opportunities.* London: Springer, 2014.

[3] B. Hoex, S. B. S. Heil, E. Langereis, M. C. M. van de Sanden, and W. M. M. Kessels. "Ultralow surface recombination of c-Si substrates passivated by plasma-assisted atomic layer deposited Al_2O_3." *Applied Physics Letters* 89, no. 4 (2006): 042112.

[4] Duerinckx, F., and J. Szlufcik. "Defect passivation of industrial multicrystalline solar cells based on PECVD silicon nitride." *Solar energy materials and solar cells* 72.1 (2002): 231-246.

Low temperature antireflection coating for silicon solar cells

O. S. Shinde[1,2], EJ Schneller[1], N. Dhere[1], S. V. Ghaisas[3]

[1]Florida Solar Energy Center, University of Central Florida, FL 32922-5703, USA
[2]Deparment of Electronics Science, SP Pune University, Pune-411007, India
[3]IIT Bombay, Mumbai-400076, India

Abstract — **for potential reduction in the economical budget of silicon solar cells requires the low temperature manufacturing process. Normally, at industrial level dielectric materials such as SiO₂, Al₂O₃ and a-SiNx:H used for passivation as well as antireflection coating. These materials requires higher process temperatures in the range of 200-900⁰c. Here, in this paper, we have discussed a simple method for the room temperature epoxy antireflection coating for the AL-BSF silicon solar cells. It is found that both open circuit voltage (Voc) and short circuit current (Isc) is increased after applying the coating on the solar cell. Further, comparison is made between the standard antireflection a:SiNx coatings and epoxy coating.**

Index Terms — **Anti reflection, epoxy, Silicon nitride, current voltage, efficiency increment, low temperature.**

I. INTRODUCTION

Cost dynamics of a solar panel fabrication depends on the processes involve in manufacturing i.e wafer to module. The requirement of technological advancement and simplicity in fabrication processes is must. However, process should be cost effective [1, 2]. Here in this paper, we are focusing on the process related to the antireflection coating. Generally, on the manufacturing bed passivation/antireflection is done by depositing the dielectric materials such as a-SiNx:H, SiO₂ ,Al₂O₃ and TiO₂ on the top of emitter [3-7]. In this process, reduction in carrier recombination achieve by removing the surface states by applying the different passivation/ARC layers. Silicon photovoltaic industry adopted the deposition techniques such as PECVD, thermal oxidation atomic layer deposition and APCVD to deposit these layers [8-10]. These processes are well developed and reliable for high efficiency solar cells. However, process requires the higher temperatures in the range of 200 to 900 ⁰c and costly instruments[11].

To resolve the problems related to higher passivation/antireflection temperature, attempts are being made to develop a low temperature passivation/antireflection processes. Usually, epoxy is the organic coating and for low temperature passivation, organic coatings of Oley amine and 9,10-phenanthrenequinone or iodine are used in many laboratories across the world [12-16]. Further, for antireflection coatings no attempts were made by using organic coatings. On the same approach, in present work, we have applied the epoxy coatings on the AL-BSF solar cell and discuss its antireflection and current voltage characteristics.

II. EXPERIMENTAL

All experiments are done on commercial available multicrystalline 156 mm X 156 mm AL-BSF silicon solar cells provided by University of Central Florida, Orlando, FL, USA. First, Solar cell with 1.75 cm² samples created by TYKMA laser system model Minilase OH, USA . Then all samples were subjected to the standard current voltage and reflectance characterization.

Then passivation/ARC layer removed by chemical reaction with HF 49%v/v. After that epoxy coating was applied on the top of above samples (emitter side) at room temperature. Same set of measurements and characterization repeated on these samples.

All current voltage characterizations were done with the help well equipped Oriel Class 2A solar simulator (Newport Corporation).

III. RESULT AND DISCUSSION

A. *Reflectance measurements:*

To investigate antireflection property of epoxy coating, the reflection measurements were performed over the wide range of wavelength 400-1200nm as shown in figure 1. Measurements were carried out on the same sample at three distinct stages i) solar cell with ARC, a-SiNx: H, ii) solar cell without ARC and iii) solar cell without ARC but coated with epoxy. From figure 1 it is depicted that solar cell with SiNx acts as efficient anti reflection layer in the range of 550-1000nm well below the 10% of reflection (black). After removal of a-SiNx:H the antireflection coating, the reflection increases up to 25-30% in the full range of wavelength spectrum (wine) and this is standard reflection spectra of silicon surface[17].

After epoxy coating on the top of emitter of a solar cell, reflection profile mainly changes in two regions of the spectrum. In the first region, in the range 400-550nm, observation made as reflection improved (cyan) and shown 10 % constant profile as compared to the a:SiNx:H ARC (black) coating which was the linear decrement in the reflectance spectra in the same region. The second region 1000-1200nm, attributed to the escape reflectance which was also improved with the coating of the epoxy layer. Improvement in reflection in above two regions contributes in the increment of power conversion efficiency.

The red curve in the graph shows the only epoxy reflectance spectra with/almost 8% reflection in the spectrum range 450-1000nm.

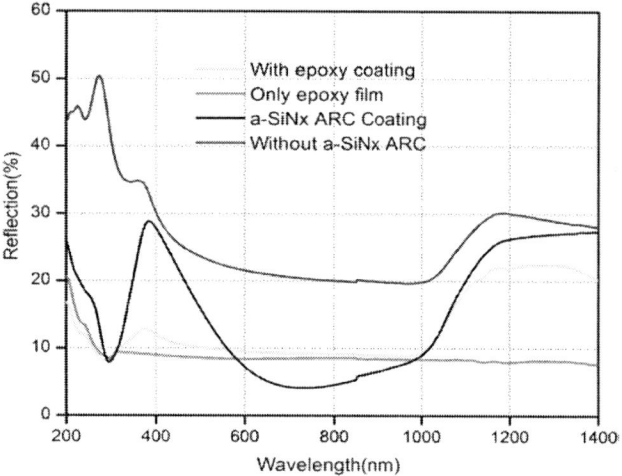

Fig1: Reflectance spectra of a-SiNx:H ARC coated solar cell(Black), Without ARC layer solar cell(Wine), With epoxy coated solar cell(Black) and only epoxy (red).

B. Current voltage characteristics:

Current voltage characteristics are measured in three different stages with ARC, without ARC and samples with the epoxy coatings shown in figure.

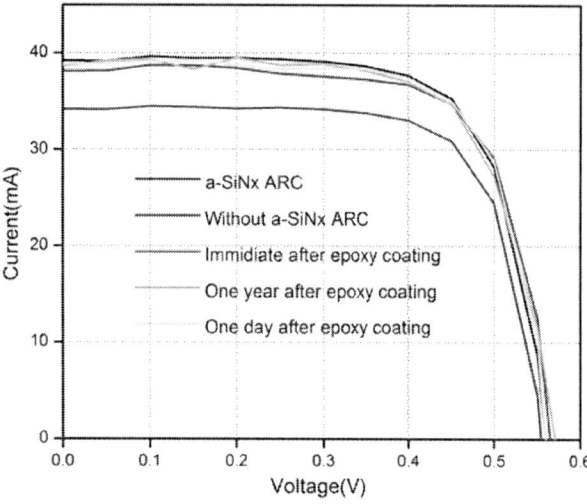

Fig 2: Current voltage characteristics of a-SiNx:H ARC coated solar cell(black), Without ARC layer solar cell (wine), With epoxy coated solar cell(cyan), after one year of epoxy coating(magenta)

From current voltage characteristics it is observed that a sample with a-SiNx: H has efficiency 15.77% (black), without a-SiNx:H the efficiency becomes 13.9%(wine) and after coating of the epoxy efficiency incremented to 15.72%(cyan). This increment in efficiency is due to increment in Isc. LT magenta curve shows the IV

characteristics of the same sample after one year with efficiency 15.62%.

C. Jsc (mA/cm²) analysis:

We have measure and the calculated the current density (Jsc) values form quantum efficiency curves and quantify them in three batches. First is epoxy coated Solar cell(cyan) ,second, Bare solar cell (wine) and third one is a-SiNx coated Solar cell (Black) as shown in figure 3.

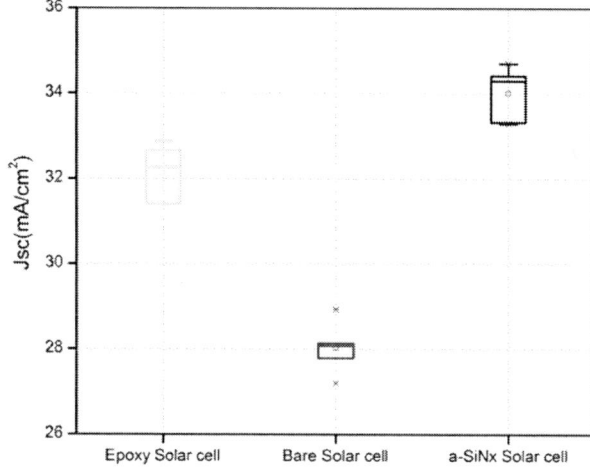

Fig 3: Reflectance spectra of a-SiNx:H ARC coated solar cell(Black), Without ARC layer solar cell(Wine), With epoxy coated solar cell(Black) and only epoxy (red).

It is observed that a-SiNx coated solar cell samples shows current density values higher than the epoxy coated samples. This increment attributed to well maintained a-SiNx coatings thickness in nanometer range required for ARC coatings and in case of the epoxy coating thickness management is very critical. In the case of epoxy coating the thickness control study is not being conducted yet and this will be the future prospective of the research.

IV. CONCLUSION

Two advantages of epoxy coatings are observed, first. It acts as antireflection coating on the n-type emitter of multicrystalline AL-BSF Silicon solar cells. Second, it doesn't require the high temperature to get coated on the Silicon sample. We can easily deposit epoxy layer at room temperature. These two advantages will help to reduce the heat and economical budget of solar cells/modules.

ACKNOWLEDGMENT

This research is based upon work supported in part by the Solar Energy Research Institute for India and the U.S. (SERIIUS) funded jointly by the U.S. Department of Energy subcontract DE AC36-08G028308 (Office of Science, Office of Basic Energy Sciences, and Energy Efficiency and Renewable Energy, Solar Energy Technology Program, with support from the Office of International Affairs) and the Government of India subcontract IUSSTF/JCERDC-SERIIUS/2012 dated 22nd Nov. 2012.

REFERENCES

[1] O. Inganäs and V. Sundström, "Solar energy for electricity and fuels," *Ambio,* journal article vol. 45, no. 1, pp. 15-23, 2016.

[2] T. Saga, "Advances in crystalline silicon solar cell technology for industrial mass production," *NPG Asia Mater,* vol. 2, pp. 96-102, 2010.

[3] A. G. Aberle, "Surface passivation of crystalline silicon solar cells: a review," *Progress in Photovoltaics: Research and Applications,* vol. 8, no. 5, pp. 473-487, 2000.

[4] B. S. Richards, "Comparison of TiO2 and other dielectric coatings for buried-contact solar cells: a review," *Progress in Photovoltaics: Research and Applications,* vol. 12, no. 4, pp. 253-281, 2004.

[5] N. B. S. Q. H. C. Park, "Surface Passivation Schemes for High-Efficiency c-SiSolar Cells - A Review."

[6] A. Rohatgi, P. Doshi, J. Moschner, T. Lauinger, A. G. Aberle, and D. S. Ruby, "Comprehensive study of rapid, low-cost silicon surface passivation technologies," *IEEE Transactions on Electron Devices,* Article vol. 47, no. 5, pp. 987-993, 2000.

[7] K. O. Davis *et al.,* "Manufacturing metrology for c-Si module reliability and durability Part II: Cell manufacturing," *Renewable and Sustainable Energy Reviews,* vol. 59, pp. 225-252, 2016.

[8] H. C. Kim and V. M. Fthenakis, "Comparative life-cycle energy payback analysis of multi-junction a-SiGe and nanocrystalline/a-Si modules," *Progress in Photovoltaics: Research and Applications,* vol. 19, no. 2, pp. 228-239, 2011.

[9] P. Saint-Cast *et al.,* "High-Efficiency c-Si Solar Cells Passivated With ALD and PECVD Aluminum Oxide," *IEEE Electron Device Letters,* vol. 31, no. 7, pp. 695-697, 2010.

[10] K. O. Davis *et al.,* "Influence of precursor gas ratio and firing on silicon surface passivation by APCVD aluminium oxide," *physica status solidi (RRL) – Rapid Research Letters,* vol. 7, no. 11, pp. 942-945, 2013.

[11] S. W. Glunz, "High-efficiency c-Si solar cells passivated with ALD and PECVD aluminum oxide," 2012.

[12] O. S. Shinde *et al.,* "Construing the interaction between solar cell surface and fatty amine for the room temperature passivation," *Solar Energy,* vol. 135, pp. 359-365, 2016.

[13] O. S. Shinde, A. M. Funde, S. R. Jadkar, R. O. Dusane, N. G. Dhere, and S. V. Ghaisas, "Reliability and efficacy of organic passivation for polycrystalline silicon solar cells at room temperature," 2016, vol. 9938, pp. 99380G-99380G-5.

[14] S. Avasthi, Y. Qi, G. K. Vertelov, J. Schwartz, A. Kahn, and J. C. Sturm, "Electronic structure and band alignment of 9,10-phenanthrenequinone passivated silicon surfaces," *Surface Science,* vol. 605, no. 13–14, pp. 1308-1312, 2011.

[15] N. Batra *et al.,* "A comparative study of silicon surface passivation using ethanolic iodine and bromine solutions," *Solar Energy Materials and Solar Cells,* vol. 100, pp. 43-47, 2012.

[16] P. R. B. Sopori, J. Appel, and and D. Guhabiswas, "Light-Induced Passivation of Si by Iodine Ethanol Solution," 2008.

[17] "http://www.pveducation.org/pvcdrom/materials/optical-properties-of-silicon."

Relationship between Power Loss and Voltage Applied to Solar Cells in PID-affected Solar Modules

Fumei Wang[a], Baosong Duan[b], Wenshuang He[a], He Wang[a]*, Hong Yang[a], Chengfeng Su[c], Bojie Su[d], Xue Zhang[d], Yunxue Cao[e], Hui Zhao[e]

[a]MOE Key laboratory for Nonequilibrim Synthese and Modulation of Condensed Matter, School of Science,

Xi'an Jiaotong University, Xi'an 710049, People's Republic of China

[b]Xi'an Communications Institute, Xi'an 710106, People's Republic of China

[c]Taizhou Chisolar Co., Ltd., Taizhou 318020, People's Republic of China

[d]China Quality Certification Center, Beijing 100070, People's Republic of China

[e]SPIC Power Plant Operation Technology Co., Ltd, Beijing 100190, People's Republic of China

*Corresponding author: He Wang, hw69cn@126.com

Abstract —In this paper, the electric field of modules in a photovoltaic (PV) system was simulated by ANSYS finite element software. The relationship between power loss and voltage applied to solar cells in potential induced degradation (PID)-affected solar modules was revealed for the first time. The electric field of solar module is related to module sequence installed in module string. And the electric field applied to solar cell is the key factor for the power loss. Meanwhile the solar modules exposed to lower voltage are also affected by PID.

Index terms —power loss, p-type crystalline silicon solar module, potential induced degradation, ANSYS finite element software

I. INTRODUCTION

The effect of potential induced degradation (PID) which reduced the output power has been discovered in the field [1-5]. And the crystalline silicon solar modules which subjected to PID were investigated by different methods. In large scale photovoltaic (PV) systems, the high voltage occurs in PV sub-array because of a large number of solar modules interconnected serially. According to various investigations in different levels about the reason that cause PID, the high system voltage seems to be the primary factor. TEM/EDX measurements reveal that stacking faults crossing the p-n junction are decorated with Na causing PID [6.7]. The calculation by first principles calculations based on density functional theory show that the presence of sodium (Na) leads to a substantial degradation [8-10]. Even an acceleration model as a function of time and of temperature and humidity was applied to power performance of crystalline silicon (Si) solar modules [11-14]. Various accelerated testing by applying a high voltage were operated to study the mechanism of potential induced degradation. Full recovery of power performance through applying to reversed voltage could not

be achieved. According to a number of studies about the effect of potential induced degradation of p-type crystalline silicon solar cells in cooperated power plant, the power loss presents certain regularity [2]. But how the high system voltage affects the power loss is uncertain.

This paper focuses on the simulation of electric field of p-type single crystalline silicon solar module installed on photovoltaic system. The simulated results vividly explained the relationship between power loss and voltage applied to solar cells in PID affected solar modules. The power loss is related to the direction and intensity of electric field between glass and solar cell in solar module. This investigation lays the foundation for the precaution and recovery of modules with PID.

II. EXPERIMENT

TABLE I
THE PARAMETER OF PV MODULE AT STANDARD TEST CONDITIONS (STC)

Rated Maximum Power (Pmax)	195 W
Voltage at Pmax (Vmp)	36.5 V
Current at Pmax (Imp)	5.34 A
Open-Circuit Voltage (Voc)	44.5 V
Short-Circuit Current (Isc)	5.73A
Dimensions (mm)	1580×808×40

STC: 1 kW/m^2 irradiance, 25°C module temperature and AM1.5 global spectrum.

In this work, the electric field of p-type crystalline silicon solar modules in module string which consists of 17 modules

TABLE II
THE PARAMETERS OF SOLAR MODULE AND CORRESPONDING REDUCTION RATE AFTER ONE YEAR

Module 1	Voc (V)	Isc (A)	Vm (V)	Im (A)	Rsh (Ω)	Pmax (W)
Before Installation	44.52	5.73	36.51	5.34	396.25	195.36
After One Year's Grid	44.47	5.62	36.41	5.28	383.37	192.08
Degradation Rate (%)	0.11	1.92	0.27	1.12	3.25	1.68
Module 8	Voc (V)	Isc (A)	Vm (V)	Im (A)	Rsh (Ω)	Pmax (W)
Before Installation	44.46	5.69	36.48	5.41	435.87	197.32
After One Year's Grid	44.02	5.58	36.24	5.37	422.27	194.67
Degradation Rate (%)	0.09	1.93	0.67	0.71	3.12	1.34
Module 10	Voc (V)	Isc (A)	Vm (V)	Im (A)	Rsh (Ω)	Pmax (W)
Before Installation	44.54	5.71	36.53	5.39	429.16	196.81
After One Year's Grid	43.89	5.54	35.75	5.36	393.93	191.73
Degradation Rate (%)	1.46	2.98	2.14	0.49	8.21	2.58
Module 17	Voc (V)	Isc (A)	Vm (V)	Im (A)	Rsh (Ω)	Pmax (W)
Before Installation	44.49	5.73	36.49	5.33	389.24	194.37
After One Year's Grid	30.56	5.61	19.21	5.20	166.36	99.89
Degradation Rate (%)	31.31	2.09	47.36	2.44	57.26	48.61

were simulated using ANSYS finite element software by changing the inputted system voltage. The parameters of solar module before installed on PV power plant at STC were shown in table I. Each module contains 72 pieces p-type crystalline silicon solar cells which have a dimension of 125 mm ×125 mm.

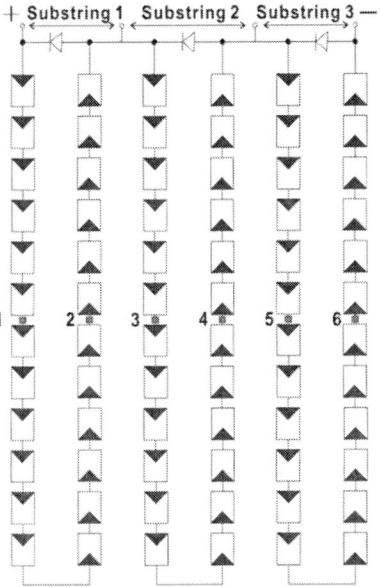

Fig. 1.   The circuit diagram of the simulated solar module

In one module string, the solar module exposed to different system voltage with the difference of module sequence. And the voltage of each solar cell in one solar module is difference. Given the small size at door of 17 modules in one module string, the slightly difference of environmental factor of one module string can be ignored. For the modules in one module string installed on PV power plant, the only difference is the negative system voltage. Fig. 1 gives out the six different positions simulated by ANSYS finite element software.

From the previous investigation [2], the power loss shows certain regularity with the module sequence variation. The performance variation of partial module was shown in table II. Some parameters have no obviously variation. Compared with the module before installed on power plant, the power of module 1 and module 8 hardly shows degradation. Then module 10 and module 17 show obvious degradation. The degradation rate about the power loss of module 17 is up to 48.61%. This has a severe influence on the long term stability and reliability. Meanwhile the unusual phenomenon that solar modules exposed to lower negative voltage are affected by PID occurs. In order to explain the law, the electric field of module with different module sequence was simulated. The potential of each point in different modules was shown in table III.

TABLE III
THE POTENTIAL OF DIFFERENT SOLAR MODULE AT SIX POINTS

Potential	Module 1	Module 8	Module 10	Module 17
Point 1	345.08 V	58.08 V	-23.88 V	-310.92 V
Point 2	338.24 V	51.24 V	-30.72 V	-317.76 V
Point 3	331.4 V	44.4 V	-37.56 V	-324.6 V
Point 4	324.56 V	37.56 V	-44.4 V	-331.44 V
Point 5	317.72 V	30.72 V	-51.24 V	-338.28 V
Point 6	310.88 V	23.88 V	-58.08 V	-345.12 V

III. RESULTS AND DISCUSSION

The electric field of solar module 1 in one module string was simulated using ANSYS finite element software. The lateral view after superposition, the partial figure after 10 times magnification and the intensity of electric field are present in Fig. 2. From the description of the previous chapter, the potential difference between aluminum (Al) frame and solar cells in module 1 is positive and higher. The intensity and the direction of the electric field are shown vividly by Fig. 2. The electric field is mainly concentrated on the edge of the module after superposition. The intensity of the electric field near the Al frame is higher. From the partial figure after 10 times magnification, intensity of electric field from solar cells to becomes stronger. The direction of electric field perpendicular to solar cell points to glass from solar cell. The maximum intensity of electric field ups to 4.05×10^4 V/m.

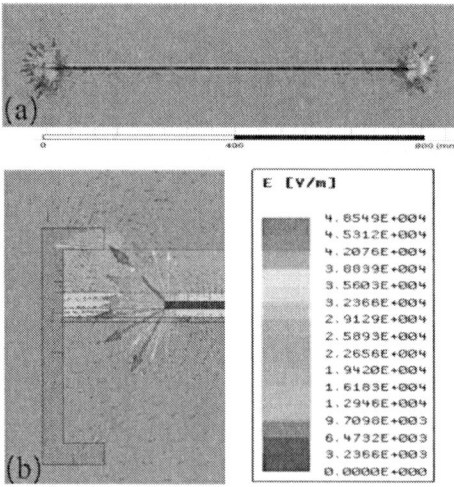

Fig. 2. The simulation of electric field of solar module 1 in one module string

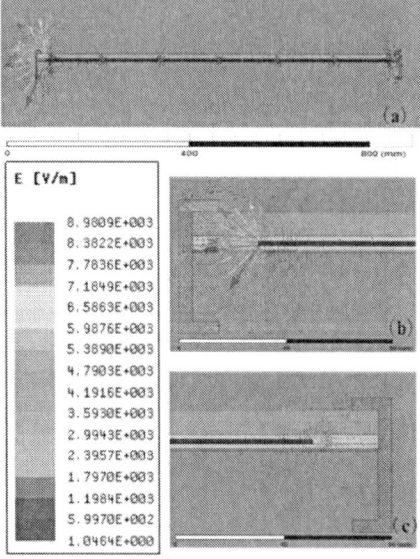

Fig. 3. The simulation of electric field of solar module 8 in one module string

From Fig. 3, the maximum electric field intensity of module 8 is about an order of magnitude less than module 1. With the decrease of voltage applied to the solar cells, the intensity of electric field after superposition becomes weaker. From the partial figure after 10 times magnification, the intensity of electric field near the positive pole is stronger than the electric field near the negative pole. Although the intensity is weaker, the direction perpendicular to solar cell still points to glass.

According to the installation of the power plant, the potential between Al frame and solar cell of module 10 to module 17 is negative. So the direction of electric field is shown in Fig. 4. Obviously the Al frame seems to be the source of electric field. The intensity of electric field is relative weak. The distribution of electric field is corresponding to the power loss. The electric field points to solar cells made the power loss. Meanwhile the electric field of module 17 (Lower) is also simulated using ANSYS finite element software in Fig. 4. The electric field intensity of module 17 (Right) is equal to approximatively the module 8 (Left). But the electric field direction perpendicular to solar cell is points to solar cells. The adverse electric field direction makes the power loss. Compared with the module 10 (Upper) with same electric field direction, the electric field intensity is stronger. The stronger negative potential difference between Al frame and solar cells make the power loss seriously.

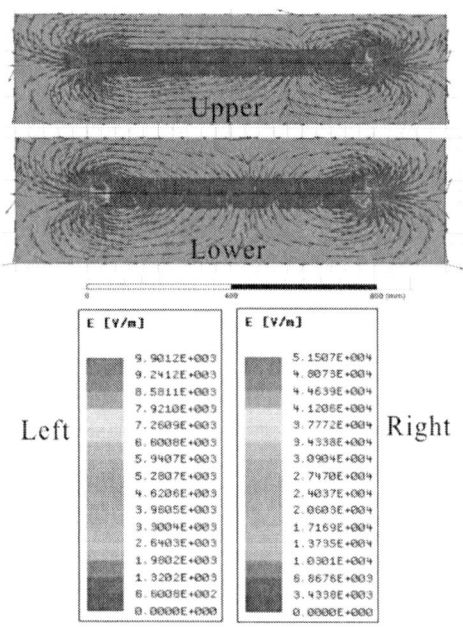

Fig. 4. The simulation of electric field of solar module 17 affected by PID

According to table III, the electric field intensity of each point from module 10 to module 17 was simulated. The distribution of electric field was obtained from ANSYS finite element software. The electric field intensity of each point from module 10 to module 17 was shown by Fig. 5. The results show that with the increase of potential, the distribution of electric field intensity of each module appears to be the structure of the basin.

Obviously, the electric field intensity of the sixth point in each module is the strongest. This is in accordance with the distribution of the potential at the sixth point. However, compared with the electric field intensity at the second point, the electric field intensity at the first point is stronger. This is attributed to the distance between Al frame and solar cell. On the other hand, the distribution of electric field does verify the power loss of three substrings in one module in Ref. [4]. Meanwhile, the electric field between Al frame and solar cell results in the migration of metal ion.

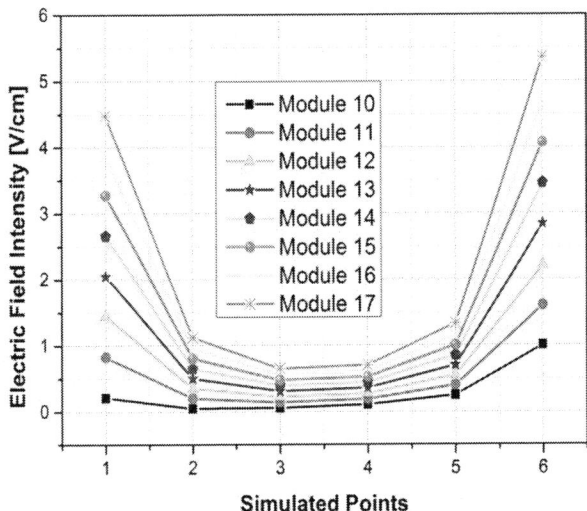

Fig. 5. The distribution of electric field of each point from module 10 to module 17

IV. Conclusion

In this paper, the performance of solar module installed on the power plant for one year was measured. The power loss presents certain regularity. For this phenomenon, the electric performance of solar module was further studied. The electric field intensity of solar module was simulated using ANSYS finite element software. The electric field intensity of solar module with different module sequence was simulated by changing the inputted system voltage. The results reveal the relationship between power loss and voltage applied to solar cells in PID affected module. Whether the power loss happens depends on the direction of electric field perpendicular to solar cell. And the degree of power loss is in accordance with the electric field intensity.

Acknowledgement

The authors would like to thank the support of Natural Science Foundation of China (Grant No. 61376067 and 61274050). This study was also supported by the National High Technology Research and Development Program of China (Grant No.2015AA050301). This study was also

supported by the Bureau of Science and Technology of Taizhou City (2016023).

References

[1] R. Swanson, M. Cudzinovic, D. DeCeuster, "The surface polarization effect in high-efficiency silicon solar cells," in *Proceedings 15th International Photovoltaic Science and Engineering Conference (PVSEC-15)*, Shanghai, China, 2005.

[2] Fumei Wang, He Wang,Hong Yang, Jipeng Chang, Pan Zhao, Ao Wang, Dengyuan Song, Effect of Potential Induced Degradation on Crystalline Silicon Solar Modules in Photovoltaic Power Plant, in： Proceedings 43rd IEEE PVSC,2016, pp. 1752-1756

[3] A. Ndiaye, et al., Degradations of Silicon Photovoltaic Modules: A Literature Review, Sol. Energy 96 (2013) 140–151.

[4] Hong Yang, Fumei Wang, He Wang, Jipeng Chang, Dengyuan Song, Chengfeng Su, Performance deterioration of p-type single crystalline silicon solar modules affected by potential induced degradation in photovoltaic power plant, Microelectronics Reliability 72 (2017) 18–23.

[5] K. Hara, S. Jonai and A. Masuda, Potential-induced degradation in photovoltaic modules based on n-type single crystalline Si solar cells. Sol. Energy Mater. Sol. Cells 140 (2015) 383–389.

[6] Volker Naumann, Dominik Lausch, and Martina Werner, "Explanation of potential-induced degradation of the shunting type by Na decoration of stacking faults in Si solar cells," *Solar Energy Materials and Solar Cells*," 120 (2014) :383-389.

[7] V. Naumann, et al., The role of stacking faults for the formation of shunts during potential-induced degradation of crystalline Si solar cells. Physica Status Solidi RRL 7 (2013) 315–318.

[8] Pan Zhao, He Wang, Hong Yang, Fumei Wang, Ao Wang and Dengyuan Song, "Mechanism Analysis of Potential-induced Degradation of P-type Crystalline Si Solar Cells," in: Proceedings 43rd IEEE Photovoltaic Specialists Conference, Portland USA, June 2016, pp. 2756-2760.

[9] J. Bauer, et al. On the mechanism of potential-induced degradation in crystalline silicon solar cells. Physica Status Solidi RRL 6 (8) (2012) 331–333.

[10] Benedikt Ziebarth, Mayous Mrovec, Christian Elasser, and Peter Gumbsch, "Potential –induced Degradation in solar cells: Electronic structure and diffusion mechanism of sodium in stacking faults of silicon," *Journal of Applied Physics* 116. 093510 (2014).

[11] Peter Hacke, Sergiu Spataru, Kent Terwilliger, Greg Perrin, "Accelerated testing modeling of Potential-induced degradation as a function temperature and relative humidity," *IEEE Journal of Photovoltaics*, 2015, pp.1549-1553.

[12] S. Pingel, et al., Potential induced degradation of solar cells and panels, in: Proceedings of the 35th IEEE PVSC, Honolulu, HI, USA, 2010, pp. 2817–2822.

[13] S. Hoffmann and M. Koehl, Effect of humidity and temperature on the potential-induced degradation, Prog. Photovolt: Res. Appl 22 (2014) 173–179.

[14] S. Spataru, et al., Temperature-dependency analysis and correction methods of in situ power-loss estimation for crystalline silicon modules undergoing potential-induced degradation stress testing, Prog. Photovolt: Res. Appl 23 (2015) 1536–1549.

978-1-5090-5606-4/17 $31.00 © 2017 IEEE

A New Low-Cost and Low-Temperature Chemical Passivation Process for Large Area Industrial Single Crystalline Silicon Wafers

Tarun S. Yadav[1,2], Sandeep K.[1], Ashok K. Sharma[1], Spandana B.[1], K.L. Narasimhan[1,2], B.M. Arora[1,2], Anil Kottantharayil[1,2], Prabir K. Basu[1]

[1]National Centre for Photovoltaics for Research and Education (NCPRE) and [2]Department of Electrical Engineering, Indian Institute of Technology Bombay, Mumbai, Maharashtra, 400076, India

Abstract — In this paper, a unique low-cost, low-temperature chemical passivation process (named as NCPRE-oxide) is investigated. This process uses only a single chemical to grow an ultrathin oxide layer of thickness ~1.7 nm on the silicon surface. Being a single component chemical process, the solution is homogeneous which helps in uniform growth of silicon oxide layer. Our study establishes that the introduction of this chemical oxide layer in between silicon and silicon nitride improves passivation on both n-type and p-type surfaces. An enhancement in the effective minority carrier lifetime (τ_{eff}) is also observed on the pyramidal textured surface with chemical oxide/silicon nitride stack for both the p-base and phosphorous diffused silicon surface. The versatility of the NCPRE-oxide is verified by X-ray photoelectron spectroscopy (XPS), carrier lifetime measurements and photoluminescence (PL) imaging.

Index Terms — chemical oxide, low-temperature, passivation, silicon, ultrathin oxide.

I. INTRODUCTION

Silicon solar cells suffer from different types of losses like electrical, optical and recombination losses [1]. Surface recombination is one of the major recombination losses in thin Si wafers [2]. To reduce the surface recombination loss, various passivation layers are used on the solar cell surfaces. Hydrogenated silicon nitride (SiN_x:H) layer is most commonly used for the passivation of the n^+ emitter layer in conventional Al-BSF technology. Additionally, it acts as an anti-reflection layer to reduce optical losses. SiN_x:H has a high density of positive fixed charge that prevents it from being used as a surface passivation layer on p^+ layer [3]. This problem can be addressed by introducing a thin silicon oxide layer between the SiN_x:H and crystalline silicon [3, 4].

Thermally grown silicon dioxide (SiO_2) layer is an excellent surface passivation layer for silicon. It is already shown in the literature that oxide layer stacked with silicon nitride layer provide good passivation in high-efficiency solar cells [2, 4, and 5]. However, thermally grown silicon dioxide is not viable for industrial production due to low throughput and high cost. Nitric acid [6] and ozone [7] are also used for growing thin oxide on silicon wafers.

In this work, we have investigated a new low-temperature cost effective chemical process to grow ultrathin silicon oxide ($SiO_x \sim 1.7$ nm) on the large area single crystalline silicon

wafers. The passivation quality for both p-type and n-type surface is discussed in detail.

II. EXPERIMENTAL DETAILS

In this work, pseudo-square (125 mm × 125 mm), Cz, p-type, monocrystalline Si wafers with a bulk resistivity of 1 - 3 Ω-cm, the thickness of ~180 μm, and (100) surface orientation were used for lifetime sample preparation. Single side polished (SSP) 4" diameter, Cz, p-type Si wafers with a bulk resistivity of 1 - 3 Ω-cm, the thickness of 285 μm, and (100) surface orientation were used for ellipsometry and X-ray photoelectron spectroscopy (XPS) of the chemically grown SiO_x layer.

The pseudo square wafers were saw damage etched in a KOH/NaOCl/DI water solution at 80°C for 15 minutes as described by Basu et al. [8] to planarize the surface by removing approximately 3 μm silicon. Subsequently, alkaline texturing in KOH/IPA/K$_2$SiO$_3$ water solution was used to form random pyramids [9]. After texturization, some of the wafers were treated in sodium hypochlorite (NaOCl) solution at 40°C for few minutes to grow ultrathin oxide. This was followed by a treatment with hydrochloric acid to neutralise the sodium ions [8]. These wafers are referred to as non-diffused wafers and this ultrathin chemical oxide growth process is referred to as NCPRE-oxide process. The thickness of the film was measured by a spectroscopic ellipsometer (Sentech) at several points and the measured refractive index and thickness were 1.46 and ~1.7 nm respectively. Phosphorous was diffused into some of the wafers after texturization to form n^+pn^+ structures using phosphorus oxychloride (POCl$_3$) in a tube furnace (Protemp, Sirius Pro 200). The sheet resistance of these samples was approximately 50 Ω /sq. measured by using a four probe system. These wafers are referred to as diffused wafers. Subsequently, phosphosilicate glass (PSG) was removed and ultrathin oxide was grown on some of these diffused wafers to find out the passivation quality of NCPRE-oxide process on the n-type surface.

To compare the quality of ultrathin oxide with low-temperature thermal oxide, some of the diffused wafers were oxidised in nitrogen (N$_2$) and oxygen (O$_2$) ambient at 600°C in a tube furnace. The refractive index and thickness were measured as 1.46 and 3 nm.

978-1-5090-5606-4/17 $31.00 © 2017 IEEE

Layers of SiN$_x$:H using PECVD (Oxford Instruments Plasmalab System100) at 380°C were deposited on both sides of test wafers referred to as (i) double side diffused only (D-1); (ii) double side diffused with ultrathin oxide (D-2); (iii) non-diffused only (ND-1) and (iv) non-diffused with ultrathin oxide (ND-2). The refractive index and thickness of the SiN$_x$:H were measured as 2 and 95 nm.

Silicon nitride SiN$_x$:H was used as a capping layer to protect ultrathin oxide and also used as an anti-reflection coating.

All samples were subjected to post-deposition anneal in the rapid thermal processing (RTP) unit (Allwin21 Corp., AW 610) for few seconds at ~800°C.

III. CHARACTERIZATION

Carrier lifetime measurements were done using contactless flash-based photoconductance decay (PCD) tester (WCT-120, Sinton Instruments).

Photoluminescence (PL) Imaging was done with a customised PL setup at the wavelength of 630 nm to evaluate the quality of passivation over 125 mm x 125 mm large area.

X-ray photoelectron spectroscopy (XPS) was used for the detailed analysis of the oxide stoichiometry (Si$^+$, Si^{2+} and Si^{3+} suboxides species) [7]. The analysis was carried out with a PHI 5000 Versa Probe-II instrument at an emission angle of 45°, a pass energy of 11.75 eV and a step size of 0.05 eV. Al Kα source with an excitation energy of 1486.6 eV was used for the analysis. XPS survey spectrum of NCPRE-oxide sample is shown in Fig. 1. The highest peak is at ~99 eV that represents the silicon substrate (Si^{0+}). The smallest peak at ~103 eV in XPS spectrum represents the SiO$_2$ (Si^{4+}) [7]. These large and small peaks are having several smaller peaks of sub-stoichiometric oxide species like Si$_2$O (Si$^+$), SiO(Si^{2+}) and Si$_2$O$_3$(Si^{3+}).

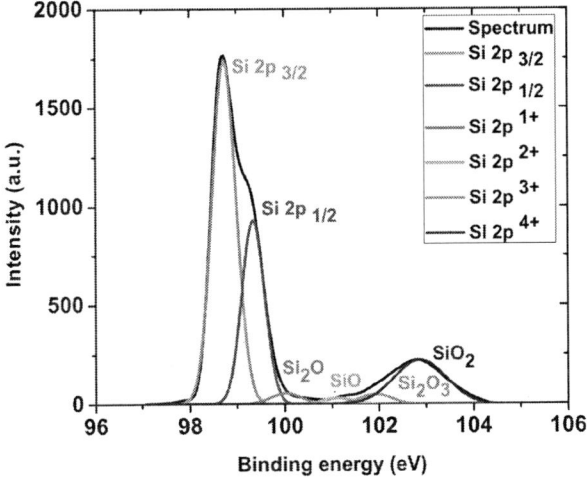

Fig. 1. XPS survey spectrum of NCPRE oxide sample with silicon and SiO$_2$ peaks.

IV. RESULTS AND DISCUSSION

Fig. 2 shows the Auger corrected inverse effective minority carrier lifetime (τ_{eff}) with the function of injected minority carrier density. From Fig. 2, it can be concluded that NCPRE-oxide process improves the passivation for both phosphorous

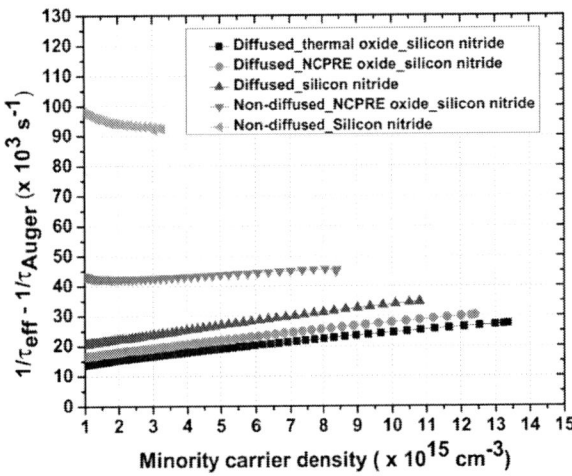

Fig. 2. Auger corrected inverse minority carrier lifetime (τ_{eff}) as a function of minority carrier density for phosphorous diffused and non-diffused wafers with and without ultrathin NCPRE-oxide.

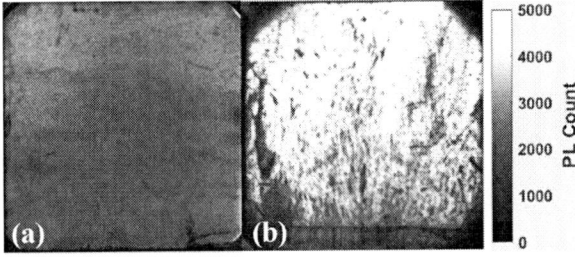

Fig. 3. Spatially resolved photoluminescence (PL) images of non-diffused textured wafers (a) Silicon nitride (SiN$_x$:H) only (ND-1) and (b) Silicon nitride on ultrathin oxide (ND-2).

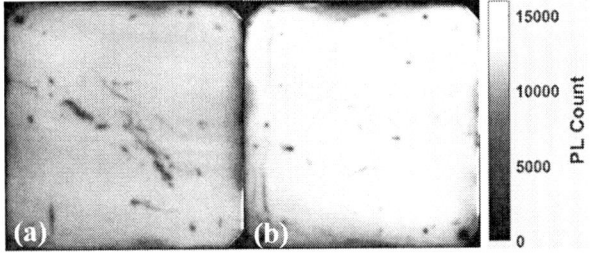

Fig. 4. Spatially resolved photoluminescence (PL) images of diffused textured wafers (a) Silicon nitride (SiN$_x$:H) only (D-1) and (b) Silicon nitride with ultrathin oxide (D-2).

diffused and non-diffused wafers. The amount of improvement in carrier lifetime is almost same for diffused and non-diffused samples. However, thermal oxide process diffused samples have the highest minority carrier lifetime compared to other samples due to better surface passivation. Minority carrier lifetime is low due to the quality of Cz wafers.

NaOCl ionises in hypochlorite ions (OCl^-) and hypochlorous (HOCl) acid; these hypochlorite ions (OCl^-) further form silicon dioxide (SiO_2) and chlorine ions upon reaction with silicon [8, 10]. These chlorine ions (Cl^-) react with potentially unwanted metallic impurities and dissolve them into the solution as soluble chlorides. The chemical reactions are as follows [8]:

$$NaOCl \leftrightarrow Na^+ + OCl^- \qquad (1)$$

$$2OCl^- + H_2O \leftrightarrow HOCl + OH^- + OCl^- \qquad (2)$$

$$Si + 2OCl^- \leftrightarrow 2Cl^- + SiO_2 \qquad (3)$$

$$Na^+ + OH^- \leftrightarrow NaOH \qquad (4)$$

The resulting NaOH is a silicon etchant, but due to the low temperature used for the process as well as the formation of oxide described in equation (3), it is not able to etch the silicon.

Photoluminescence (PL) images shown in Fig. 3(a) and 3(b) shows the spatially resolved PL image of the non-diffused wafers with silicon nitride coating only and silicon nitride capping layer on ultrathin NCPRE-oxide, respectively. It is clearly visible from Fig. 3(b) that the PL count is higher on most of the sample ND-2 in comparison to ND-1. However, there are several dark marks visible on the wafer as shown in Fig. 3(b). A similar observation can be made for diffused samples shown in Fig. 4 where sample D-2 had higher PL count than D-1. In addition to this, there is a difference of approximately 10,000 of PL count between non-diffused and diffused samples. This could be partly due to gettering of impurities during diffusion and also because of nitride capping layer, which may lead to surface inversion on p-type silicon. In Fig. 4, PL count is large and uniform in diffused wafer D-2(with ultrathin oxide and silicon nitride) in comparison to only silicon nitride deposited diffused wafer D-1. This indicates that the chemical oxide provides better passivation than SiN_x:H alone.

V. CONCLUSION

Significant carrier lifetime improvement is observed on the n-type and p-type silicon surface by using ultrathin NCPRE-oxide/silicon nitride stack instead of the only silicon nitride passivation layer. PL images show good uniformity and higher PL count on the n-type surface with NCPRE-oxide. It eventually confirms its utilisation for n^+ Si surface passivation. Due to the low cost of NaOCl and low-temperature process, it

can be implemented at industrial scale. This process can also become a candidate to grow tunnel oxide in TOPCON (Tunnel oxide passivated contact) solar cells.

ACKNOWLEDGEMENT

We would like to acknowledge the efforts of R. Manoj Kumar, Pradeep P. for their assistance in characterization and processing. This work was carried out at the National Centre for Photovoltaic Research and Education (funded by the Ministry of New and Renewable Energy) and IIT Bombay Nanofabrication Facility at IIT Bombay.

XPS measurement was done at the Centre of Excellence in Nanoelectronics (funded by Ministry of Electronics & Information Technology) at IIT Bombay.

REFERENCES

[1] A. G. Aberle, W. Zhang, and B. Hoex, "Advanced loss analysis method for silicon wafer solar cells," *Energy Procedia*, vol. 8, pp. 244–249, 2011.

[2] S. W. Glunz, "High-Efficiency Crystalline Silicon Solar Cells," *Advances in OptoElectronics*, vol. 2007, pp. 1–15, 2007.

[3] S. Duttagupta, F.-J. Ma, B. Hoex, and A. G. Aberle, "Excellent surface passivation of heavily doped p silicon by low-temperature plasma-deposited SiOx/SiNy dielectric stacks with optimised antireflective performance for solar cell application," *Solar Energy Materials and Solar Cells*, vol. 120, pp. 204–208, 2014.

[4] G. Dingemans, M. M. Mandoc, S. Bordihn, M. C. M. Van De Sanden, and W. M. M. Kessels, "Effective passivation of Si surfaces by plasma deposited SiOx/a-SiNx:H stacks," *Applied Physics Letters*, vol. 98, no. 22, p. 222102, 2011.

[5] M. Hofmann, S. Janz, C. Schmidt, S. Kambor, D. Suwito, N. Kohn, J. Rentsch, R. Preu, and S. W. Glunz, "Recent developments in rear-surface passivation at Fraunhofer ISE," *Solar Energy Materials and Solar Cells*, vol. 93, no. 6-7, pp. 1074–1078, 2009.

[6] H. Kobayashi, K. Imamura, W.-B. Kim, S.-S. Im, and A., "Nitric acid oxidation of Si (NAOS) method for low temperature fabrication of SiO2/Si and SiO2/SiC structures," *Applied Surface Science*, vol. 256, no. 19, pp. 5744–5756, 2010.

[7] A. Moldovan, F. Feldmann, G. Krugel, M. Zimmer, J. Rentsch, M. Hermle, A. Roth-Fölsch, K. Kaufmann, and C. Hagendorf, "Simple Cleaning and Conditioning of Silicon Surfaces with UV/Ozone Sources," *Energy Procedia*, vol. 55, pp. 834–844, 2014.

[8] P. K. Basu, D. Sarangi, and M. B. Boreland, "Single-Component Damage-Etch Process for Improved Texturization of Monocrystalline Silicon Wafer Solar Cells," *IEEE Journal of Photovoltaics*, vol. 3, no. 4, pp. 1222–1228, 2013..

[9] P. K. Basu, D. Sarangi, K. D. Shetty, and M. B. Boreland, "Liquid silicate additive for alkaline texturing of mono-Si wafers to improve process bath lifetime and reduce IPA consumption," *Solar Energy Materials and Solar Cells*, vol. 113, pp. 37–43, 2013.

[10] U. Gangopadhyay, S. K. Dhungel, K. Kim, U. Manna, P. K. Basu, H. J. Kim, B. Karunagaran, K. S. Lee, J. S. Yoo, and J. Yi, "Novel low cost chemical texturing for very large area industrial multi-crystalline silicon solar cells," *Semiconductor Science and Technology*, vol. 20, no. 9, pp. 938–946, Jan. 2005.

Evaluation of ALD Passivation Layers for Industrial PERC Process

Chang Youn Yoo, Keunkee Hong, Jisun Kim, Eunjoo Lee and Dong Seop Kim

Technical Research Center, Shinsung E&G Co., Ltd.,
#8 Daewangpangyo-ro 395 Bundang-gu Seongnam-si Gyeonggi-do, 463-420, Korea

Abstract — The passivation quality of batch type thermal atomic layer deposition (ALD) is investigated in the industrial production line. The surface morphology, reflectance, and carrier lifetime of samples are measured and analyzed based on the deposited Al_2O_3 film and SiNx capping layer thickness. Front layer control with SiNx/SiON layers further boosts the level of passivation. Our results show thin (~5nm) layer of Al_2O_3 on both side of wafer could potentially improve the Passivated Emitter and Rear Cell (PERC) performance.

Index Terms — PERC, Mono-crystalline Silicon, ALD, Passivation, Carrier lifetime

I. INTRODUCTION

Passivated Emitter and Rear Cell (PERC) technology has recently demonstrated its capability for high efficiency mass production. With decades of research the majority of PV industry has adopted Al_2O_3 as rear passivation material, for its dielectric layer contains sufficient amount of negative fixed charge and provides outstanding field-effect passivation on p-type substrate. The chemical passivation quality of Al_2O_3 to reduce dangling bonds of the surface is also comparable with conventional oxidation [1]. The method to deposit Al_2O_3 is Plasma Enhanced Chemical Vapor Deposition (PECVD) for most manufactures today and others also use Atomic Layer Deposition (ALD) for high film quality and good step coverage [2]. The optimum field-effect passivation of Al_2O_3 can be realized at minimum thickness between 5-6 nm with minimum vacancy between the bulk and oxide layer. High quality passivation requires the integration of rear surface polishing and film deposition with conventional process of the Aluminum Back Surface Field (Al-BSF) cell production lines.

In order to compete with PECVD system, high throughput is required for ALD process. The Spatial ALD has been introduced for in-line process which is more suitable for PV industry [3,4]. The drawback has been a process pressure being at atmosphere unlike batch type, and the relative preventive maintenance period is short for this reason. With recent development, the batch type ALD has gained improved throughput in production greater than that of PECVD. In this paper, we highlight the feature of ALD process with its precise control of uniform film thickness resulting higher carrier lifetime. The carrier lifetime has been a criterion for PERC structure next to the final cell efficiency. By evaluating its surface morphology, surface reflection, thickness of Al_2O_3 layer and front/rear SiNx layers, the process and carrier lifetime have been optimized.

II. SAMPLE FABRICATION

Industrial 6-inch monocrystaline boron doped p-type silicon wafers were used for each process evaluation. All samples were alkaline textured and doped by $POCl_3$ diffusion furnace, followed by PSG removal via wet isolation etching. For PERC structure with improved rear surface condition, the chemical process was adjusted to enable inline polishing step simultaneously with wet isolation. A thin layer of thermal oxide was grown subsequently, and SiNx layer or SiNx/SiON layer was deposited on front surface with PECVD process.

Al_2O_3 rear surface passivation was applied using batch type ALD system employing deposition rate of 1.2Å/cycle at temperature of 250°C. To obtain and characterize optimal Al_2O_3 layer, samples were split into groups of 5 nm, 7 nm and 10nm deposition thickness. To compensate the change of total film thickness and refractive index, both front and rear passivation layers were varied. The SiNx on rear side is used to cap Al_2O_3 layer with PECVD process for thickness between 80nm and 140nm.

Another sample batch was deposited with Al_2O_3 using PECVD process for thickness range from 10nm to 30nm. The SiNx capping layer was maintained near 120nm thick to compare the relative compatibility of ALD and PECVD process. All layer thickness was confirmed by ellipsometer. The effective minority carrier lifetime was measured and evaluated using QSSPC technique after fast firing process having peak temperature near 850°C used in production line. The overall fabrication steps are summarized in Fig. 1.

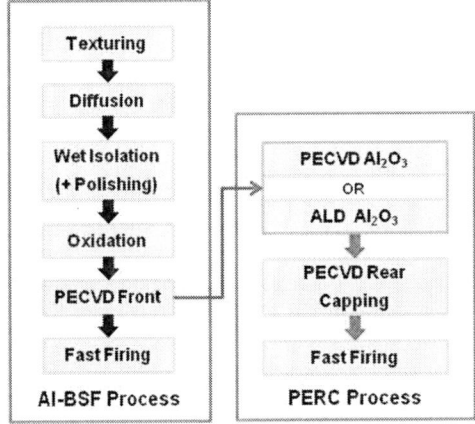

Fig. 1.   Flow chart for test process.

III. RESULT AND DISCUSSION

A. Front Surface Reflectance

The batch type ALD process used in the experiment performs double side deposition which adds 5-10nm of Al_2O_3 layer on top of SiNx or SiNx/SiON layer at front surface. This may not have huge impact on passivation quality, but it moderately changes the front surface reflectance which determines the amount of light entered into the substrate and the internal reflection of the rear surface. As shown in Fig. 2, the reflectance of both SiNx layer and SiNx/SiON layer for conventional process and ALD process are compared. Additional layer of Al_2O_3 improves short wavelength reflectance for SiNx but not much for SiNx/SiON layer. The SiNx/SiON layer passivation has been applied previously on Al-BSF structure to improve short wavelength (<450nm) reflectance in exchange of slight loss in the long wavelength (>1100nm) by its thicker film and lower refractive index property. However, as for PERC structure, lower reflectance in the long wavelength is more favorable to maximize the rear internal reflection gain. For short wavelength, front surface engineering such as nano-texturing can be considered instead of having SiNx/SiON layer.

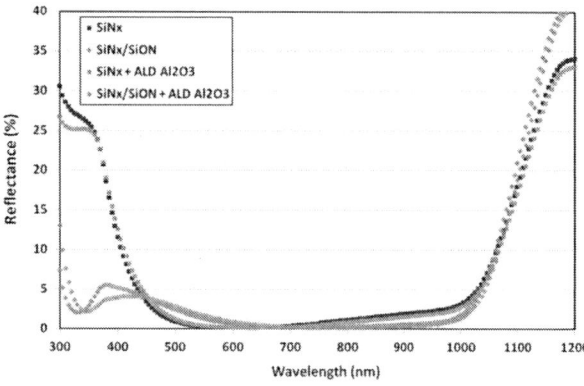

Fig. 2. Reflectance of front surface passivated with and without ALD process for SiNx layer and SiNx/SiON layer.

B. Rear Surface Morphology and Lifetime

Fig.3 shows the rear surface of silicon wafer after wet isolation process with and without polishing. The polishing removal thickness is estimated to be 3-5μm by measuring its mass change. This is well matched with the optimum etching discussed by others [5,6]. Our conventional process creates rough textured surface while tips are still visible, and polishing process appropriate for PERC process rounds the surface. The average surface roughness (Rs) of rear side is measured to be 0.666μm and 0.268μm for conventional and polishing process respectively. The planar surface is responsible for higher current density (Jsc) by increased rear internal reflectance and

removed damage layer improves open circuit voltage (Voc). The step coverage of Al_2O_3 layer deposition by ALD expects to be more uniform and less sensitivity to the change in roughness compared to the PECVD process.

The effective carrier lifetime of samples after firing process is illustrated in Fig. 4. The highest carrier lifetime of 215μs, a corresponding implied Voc of 678mV, was achieved when SiNx/SiON layer was combined with ALD and polishing process. The carrier density vs. carrier lifetime curve at one-sun near the peak is shifted upward by upgrading passivation layers from conventional to ALD to SiNx/SiON. Finally the combination of ALD and SiNx/SiON layers with improved polishing surface has shifted the curve outward at greater extend. The polished samples have reduced rear surface area which lowers the surface recombination velocity by far the most. The surface geometry different from the first three cases clearly separates its carrier lifetime at higher carrier density.

C. Al_2O_3 Passivation Layer

Batches of samples were run varying Al_2O_3 layer thickness and SiNx capping layer thickness is controlled at constant to demonstrate the strength of ALD passivation. The implied Voc

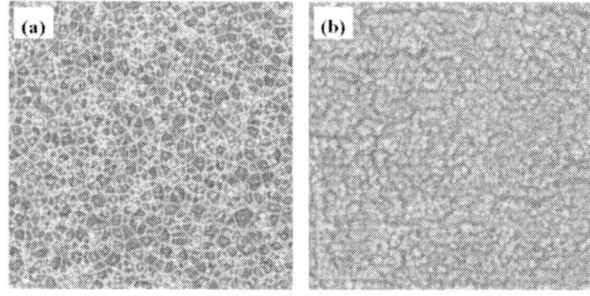

Fig. 3. Rear surface morphology of wet isolated etching with (a) conventional process and (b) polishing process.

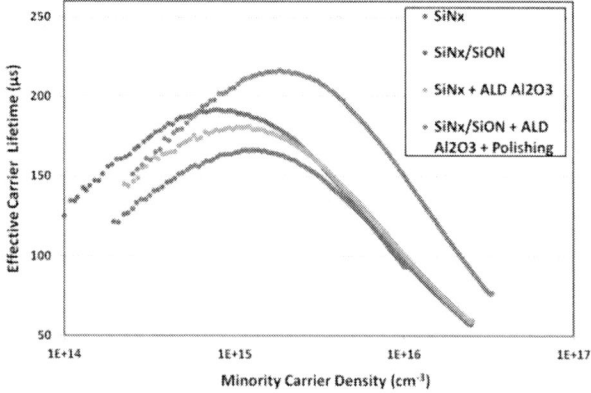

Fig. 4. Effective minority carrier lifetime as function of carrier density of passivated samples after fast firing.

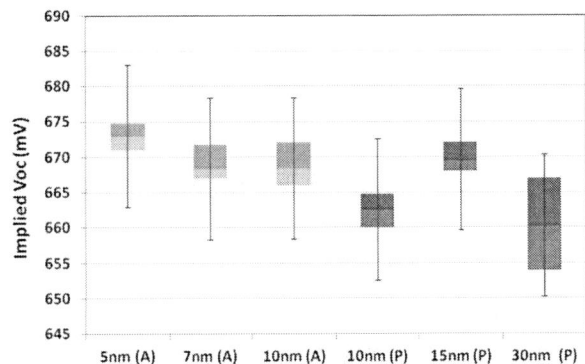

Fig. 5. Implied Voc after fast firing for varying Al_2O_3 layer thickness by ALD (A) and PECVD (P) process, with the SiNx capping layer thickness maintained between 100nm and 120nm.

(iVoc) of PECVD process is compared as shown in Fig. 5. Typical Al_2O_3 deposition range for PECVD is 10-30nm in industry. Moving from 30nm to 15nm in thickness, the passivation quality and the uniformity have been improved and average iVoc is increased by 10mV after series of optimization. Further reduction of thickness by PECVD process is limited at 10nm which makes it difficult to produce same quality as before. In case of ALD process, the baseline of iVoc is already similar to optimized PECVD process at 15nm thickness. Al_2O_3 as thin as 5nm is deposited by ALD to result the highest iVoc of 673mV on average and 683mV for the best value. The reliable control of film thickness explains the potential of ALD process for quality improvement.

Analyzing the economic value of batch type ALD process relative to PECVD process also provides several advantages from production standpoint. The reduced consumption of trimethylamine (TMA), the precursor for Al_2O_3, can save the cost and depreciation of equipment for a long run. The process time from 10nm to 5nm deposition is reduced by 28% which makes more competitive in manufacturing production.

D. Rear Capping Layer

The role of rear capping layer is to prevent aluminum paste to penetrate and damage Al_2O_3 layer for possible junction shunt after firing step. To find necessary capping condition, the SiNx thickness is varied from 80nm to140nm by PECVD while Al_2O_3 is fixed at 7nm by ALD as shown in Fig. 6. From industrial market perspective, thick SiNx capping layer is not economic because of higher cost of precursors supplied and longer process time in mass production. Although the carrier lifetime is similar for all thickness except 140nm and above, below 80nm is not desired as it is susceptible to aluminum paste penetration at 850°C peak firing step. This peak firing temperature is fixed for proper contact formation of the silver paste in the front surface. Considering metallization and final cell efficiency, 100nm-120nm thickness of capping layer is recommended and also considered by PV industry report [2].

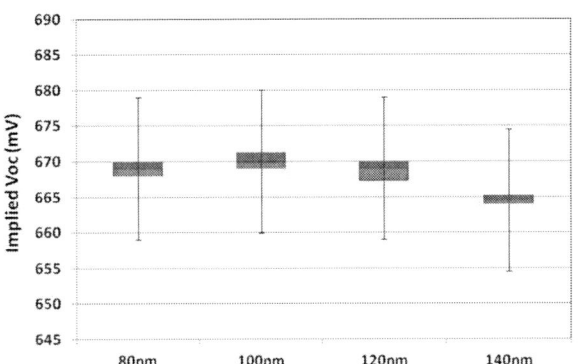

Fig. 6. Implied Voc after fast firing for varying SiNx capping layer thickness on top of Al_2O_3 ALD passivation.

The carrier lifetime is heavily depends on the thickness and the quality of Al_2O_3 layer rather than SiNx capping layer thickness.

IV. CONCLUSION

The batch type ALD passivation is evaluated and discussed through the combined effect of rear surface morphology, front surface reflectance, and carrier lifetime/iVoc measurement. Although PECVD process currently dominate the commercial market, batch type ALD process with aids of ultra-thin layer deposition technology and in-line integration has benefit of lower manufacturing cost. The SiON/SiNx layer demonstrates its strength as strong absorber in the short wavelength, and the overall passivation quality is improved if rear passivation is well balanced with the front side.

ACKNOWLEDGEMENT

This work was supported and funded by the Small and Medium Business Administration (SMBA, Korea) under the contract No. S2419224.

REFERENCES

[1] Schmidt, J., et al. "Surface passivation of silicon solar cells using plasma-enhanced chemical-vapour-deposited SiN films and thin thermal SiO2/plasma SiN stacks." Semiconductor Science and Technology 16.3 (2001): 164.

[2] Shravan., C., et al. "PERC Solar Cell Technology." TaiyangNews (2013).Semiconductor Science and Technology 27.7 (2012): 074002.

[3] Van Delft, J., et al. "Atomic layer deposition for photovoltaics: applications and prospects for solar cell manufacturing." Semiconductor Science and Technology 27.7 (2012): 074002.

[4] Granneman, E., et al. "Spatial ALD, Deposition of Al2O3 Films at Throughputs Exceeding 3000 Wafers per Hour." ECS transactions 61.3 (2014): 3-16.

[5] Kranz, C., et al. "Impact of the Rear Surface Roughness on Industrial-Type PERC Solar Cells" 27th European Photovoltaic Solar Energy Conference and Exhibition, 2012.

[6] Kranz, C., et al. "Wet chemical polishing for industrial type PERC solar cells." Energy Procedia 38 (2013): 243-249.

Quantitative analysis of electroluminescence and infrared thermal images for aged monocrystalline silicon photovoltaic modules

Irene Berardone[1], Juan Lopez Garcia[2], and Marco Paggi[1]

[1]IMT School for Advanced Studies Lucca, Piazza San Francesco 19, 55100 Lucca, Italy

[2] European Commission, DG JRC, Directorate C - Energy, Transport and Climate, Energy efficiency and Renewables Unit, Via Fermi 2749, 21027 Ispra (VA), Italy

Abstract — A quantitative statistical analysis of infrared images (IR), under forward bias and under illuminated condition, and electroluminescence (EL) images for damaged monocrystalline photovoltaic (PV) modules after 20 years, is herein proposed. The proposed methodology relies in the analysis frequency histograms of red, blue and green channel of IR images and the intensity of EL images. Dark IR images provide local information, useful to distinguish disconnected cell interconnect (DCI), humidity corrosion (HC) and finger interruption (FC) in highly damaged solar cells. This extends and generalizes the correlation between dark IR and EL established in the literature for thin-film solar cells also to monocrystalline silicon photovoltaic modules.

Index Terms — electroluminescence, thermal infrared imaging, crystalline silicon photovoltaics, image analysis.

I. INTRODUCTION

Quality control and performance monitoring are strictly connected to photovoltaic costs reduction. For this reason, Scientific Community collects and analyzes data on PV technical performance and failures to study and understand challenges involved in damage phenomena.

IR thermography and EL are the most employed non-destructive techniques for quality control of PV module. Both optical techniques provide real-time two-dimensional images and detect type of defects, malfunctioning, or even failures.

EL imaging is a high-resolution technique that provides a map of intensities over the domain of the solar cells. From the EL intensities, arbitrarily set in the range from 0 to 255, it is possible to assess the voltage drop from a point to another [1-3]. This drop is usually caused by defects or cracks, which induce a localized additional resistance. Therefore, intrinsic (grain boundaries, dislocation, shunts or process failures) and extrinsic defects (cracks, interrupted contacts and TCO corrosion for thin films) can be visually identified. In the EL test, modules operate under forward bias like a light emitting diode (LED), and therefore have to be power supplied. Excitation current can be lower or equal to I_{SC}. The detected emission is due to radiative interband recombination of charge carriers (near-IR in the range from 300 up to 1000 nm, depending on the semiconductor type). An EL drawback is its operating condition, that requires dark environment, thus confining mostly to the lab. However, progresses have recently been made to perform EL images in the field, with the use of shrouds to cover the module, or using cameras able to work also in nearly dark environments at sunsets, or integrating the camera in drones for aerial inspection.

Although some practical limitations, mainly related to power supply and module orientation, in-field EL inspection techniques are gradually growing. To date, EL data up to 1 MWp are acquired only in one night [4].

On the other hand, IR thermography is a cost-effective and time saving method able to detect mostly general problems related to electric circuit (electrical disconnections, bypassed strings, faulty soldering and short-circuited cells) [6]. IR thermography can be performed in dark or illuminated conditions [7]. In the former instance, this procedure shares some similarities with EL technique, being the module subjected to a forward voltage. In the latter, the illuminated IR thermography is similar to luminescence, where the current flowing through cells increases PV module temperature.

Although a correlation between increase of temperature and power reduction of a module has been demonstrated [5], a quantitative interpretation of IR images is under debate. Moreover, even if a large number of works analyzes EL and IR techniques [6-16], only in few cases quantitative conclusions are drawn. The majority of works on EL is focused on the determination of physical proprieties of solar cells, such as local effective diffusion length l_{eff} [11], carrier collection length l_c [12], local junction voltage [10], series resistance [14, 15]. Zamini et al. [5] proposed preliminary qualitative comparisons between EL and IR images for crystalline silicon and thin film technologies. They found a spatial correlation between dark IR measurement and EL images for the specific case of thin film solar cells.

To quantitatively compare IR and EL methods and understand challenges involved in damage phenomena, we propose a statistical analysis of the data from EL, dark-IR and illuminated IR imaging of crystalline silicon PV modules. The importance of the results relies in the possibility to implement the identified correlations in machine learning software for the automatic inspection of PV modules in real time, very important for a fast and accurate detection and classification of the various forms of damage affecting PV modules durability.

I. METHODS AND MATERIALS

To assess the ability of EL and IR imaging techniques to detect a wide range of defects and forms of damage in solar cells caused by a significant time exposure in the field, four crystalline silicon PV modules manufactured in the mid-1980s have been considered. The modules were installed on a solar car (Mazzieri model, equipped with an electric motor of 7 cV and a maximum speed of 70 km/h) and exposed to outdoor condition for 20 years. They were named IY41, IY42, IY43 and IY44. Different colors have been attributed to them in the present study, namely orange, blue, green and gray, respectively (see Tab.1). Three modules were placed on the roof (module IY42, IY43, IY44) and the one on the bonnet (IY41) of the car. They were connected to a battery. The disposition of module on the solar car was the following: module IY41 had a tilt angle of 12°, IY42 and IY44 of 8°, while IY43 was installed horizontally on the roof. Details on module manufacture and characteristic are also summarized in Tab. 1.

The vehicle was a service car at JRC in Ispra and it was driven in that region for a period of about 20 years, without module cleaning. Ispra is in the North of Italy, at 220 m above sea level and belongs to moderate subtropical climate zone (temperature excursion during the year from -10°C up to +35°C, relative humidity <90%) [17].

TABLE I
LIST OF PV MODULE ANALYZED IN THE PRESENT STUDY

Module	Producer	Model	Front Glass	N. Cells	Year outdoor
IY41	BP	1242	Flat	36	20
IY42	Arcosolar	M75	Flat	33	20
IY43	Arcosolar	M75	Flat	33	20
IY44	Arcosolar	M75	Flat	33	20

After removal of the PV modules from the car, they were cleaned manually for 10 minutes with a soft sponge, a standard commercial glass cleaning detergent and finally dried with a cloth. Afterwards, the modules were characterized in the JRC facilities through visual characterization, I-V measurement, electroluminescence and infrared thermography.

Visual characterization showed delamination, cracked cells degradation and discoloration of cells interconnect ribbons and yellowing of the center or whole cells. In this study, each cell is identified with a code composed by module name, a letter and a number. The letter identifies the row, instead the number identifies the column of each defective cell, see Fig.1 (EL images).

Electroluminescence images were performed with a Sensovation digital camera SVSB14-M under forward bias with a current equal to I_{SC} and exposure time of 300 s (see Fig. 1).

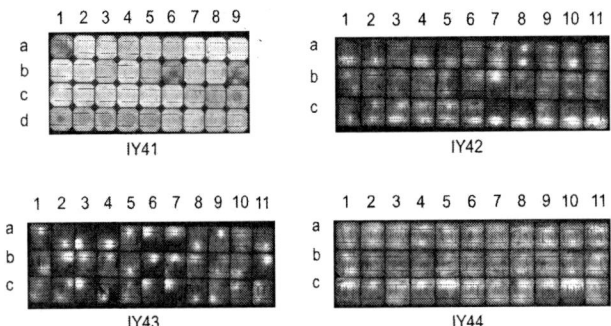

Fig. 1.   Electroluminescence images for modules IY41, IY42, IY43 and IY44.

Original dark IR images of the PV modules were taken at JRC using a Fluke Ti55 camera under forward bias, with a current equal to I_{SC} (see Fig.2). Illuminated IR images were taken under continuous illumination (≈ 1000 W/m^2), provided by a steady state solar simulator and short circuited with a 15Ω resistance (see Fig.3).

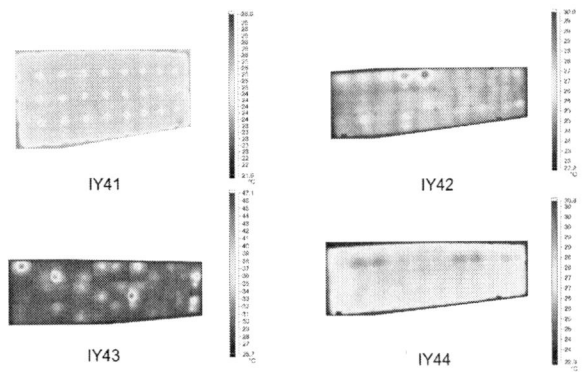

Fig. 2.   Dark IR images of the modules IY41, IY42, IY43 and IY44.

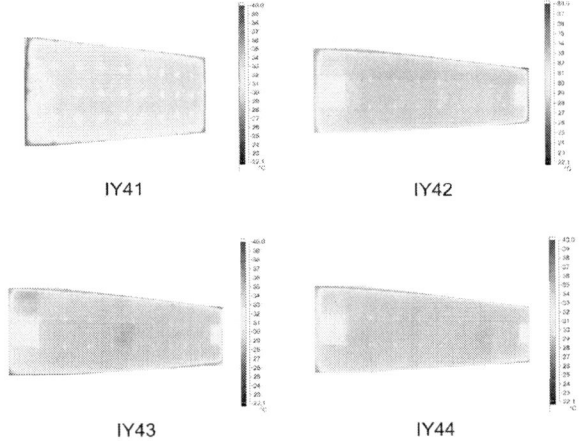

Fig. 3.   Illuminated IR images of the modules IY41, IY42, IY43 and IY44.

To perform a statistical analysis of EL and IR signal, we modified the initial images reducing distortions in IR images. According to the International Energy Agency classification reported in [18], we identified and classified defective cells of the analyzed modules from the EL images. After this EL-based classification, we analyzed IR and EL signals through a user-defined subroutine written in MATLAB.

Preliminary and automatic operations of the developed subroutine involve the extraction of all the individual solar cells and the conversion of the signals into multi-dimensional matrices of real numbers. For IR images, the multi-dimensional matrix collects the x and y position coordinates of each pixel and the corresponding RGB channels. Each channel is normalized in the range from 0 to 1. To synthetically characterize the information contained in the RGB channels, these data are also converted to real values ranging from 255 (code number for blue) to 16711680 (code number for red) and then transformed in a temperature scale, of easy visualization to draw thermal maps for each module under forward bias or illumination. Similarly, for EL images, a three-dimensional matrix is introduced by collecting the x and y coordinates of each pixel and the corresponding EL signal ranging from 0 to 255. Such a range has been normalized from 0 to 1 for easy of visualization. The first operation for the statistical analysis of the data was the computation of frequency histograms for R, G, B and EL signal (using 50 bins for each image). Quantity of interest is the position of the histogram interval, with occurrences higher than a prescribed minimum tolerance. To minimize the noise this threshold is set equal to the 80% of the average of the occurrences. This procedure is applied to each cell and the interval position depends both on defect type and on module condition.

III. RESULTS AND DISCUSSION

From the analysis of the EL signal, damaged cells can be classified depending on the type of defect (see Tab. 2), according to the classification in [18]. The first column of Tab. 2 is reporting code of the defect, followed by the nomenclature, the EL image of a corresponding solar cell and finally in the last column the module belonging to each class of defect. The severity of defects increases from SR to DCL based on the brightness of the EL image of the cells that becomes dimmer and dimmer from SR to DCL. The range of intensity of a defect class is calculated by averaging the intervals corresponding to all solar cells of the corresponding defect class. Such a range depends both on the defect class and on module condition. The range of intensity for EL signal is shown in Fig. 4.

The analysis of the EL signal shows an increasing trend in the range of intensities by moving from HC 2 up to SR (Fig. 4). Due to frequent overlapping between the intervals corresponding to different defect classes, visual inspection seems to be required for their careful discrimination. Due to overlapping between to different defect classes intervals, visual inspection seems required for their careful discrimination.

TABLE II

CLASSES OF DEFECTS IN TESTED SAMPLES ACCORDING TO [18]

Module	Name		Module / cells
SR	Striation Ring		IY41
SH	Sucker handling		IY41
CFb	Contact forming		IY41-IY44
HC1	Humidity corrosion in a low damaged solar cell		IY41
HC2	Humidity corrosion in a high damaged solar cell		IY43
CC1	Cross Crack in a low damaged solar cell		IY43
CC2	Cross Crack in a high damaged solar cell		IY41
Fc	Finger interruption C		IY43, IY44
DClxs1	Disconnected cell interconnect (spot) in a high damaged solar cell		IY43
DClxs2	Disconnected cell interconnect (spot) in a medium damaged solar cell		IY42
DClxs3	Disconnected cell interconnect (spot) in a low damaged solar cell		IY44
DClxxs1	Disconnected cell interconnect (spot) in a low damaged solar cell		IY42, IY43
DClxxs2	Disconnected cell interconnect (spot) in a high damaged solar cell		IY44
DClxxl1	Disconnected cell interconnect (line) in a high damaged solar cell		IY43, IY42 (1A)
DClxxl2	Disconnected cell interconnect (line) in a low damaged solar cell		IY42
DClxxl3	Disconnected cell interconnect (line, extended to a full busbar)		IY44
DClxxxl1	Disconnected cell interconnect (line) in a high damaged solar cell		IY42 (10C, 9C,3C)
DClxxxl2	Disconnected cell interconnect (line) in a low damaged solar cell		IY43, IY42 (8C,11C,7C)
MC	Micro-crack C (isolating part)		IY43

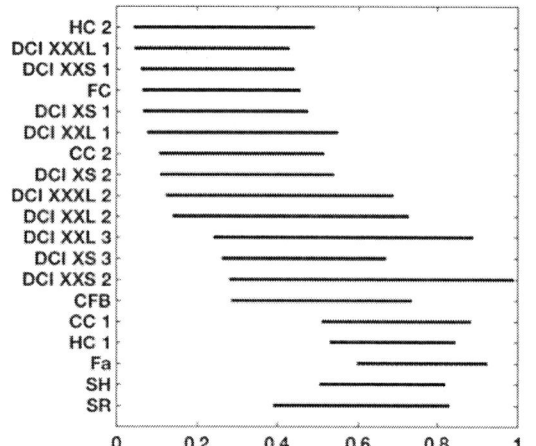

Fig. 4. X-range of intensity histogram of EL signal for the different classes of defects in Tab.2.

We applied the same approach to IR histograms under biased and illuminated condition. Average values of the x-range of intensity for red, green and blue are shown in Fig. 5 for biased condition and in Fig. 6 for illuminated condition. The order sequence adopted is the same chosen in Fig. 4.

For IR images taken in forward bias condition, the red and blue channels have an abrupt and opposite change in the ranges from the class DCI XXL 2 onwards. For defect classes from HC 2 up to DCI XXL 2, the range in in the interval between zero and 0.23. From DCI XXL 3 to the SR classes, on the other hand, the interval shifts in the range from 0.78 and unity.

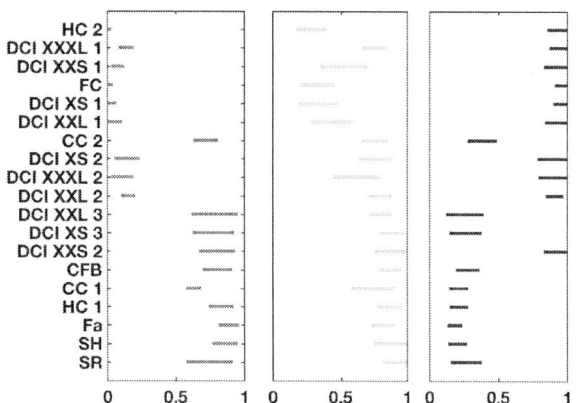

Fig. 5. X-range of intensities histogram for red, green and blue channels of the IR signal under forward bias condition, for the different classes of defects in Tab. 2.

The opposite trend is observed for the blue channel, while the green channel has a trend similar to the red one, but without a sharp shift. Only the CC class (CC 1 for the red channel and CC 2 for the red and the blue channels) does not fulfill the observed trend, which can be motivated by the fact that the detected cross cracks are sources of minor defects for the examined cells. From the physical point of view, these results imply that it is possible to distinguish disconnected cell interconnect (DCI),

humidity corrosion (HC) and finger interruption (FC) in highly damaged solar cells. The same analysis repeated for the histograms of the IR images taken under illumination (Fig. 12) do not show any shift in the interval ranges, thus making nearly impossible to distinguish between the different classes of defects.

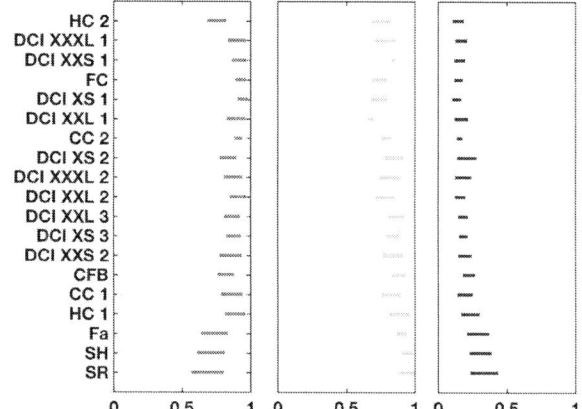

Fig. 6. X-range of intensities for red, green and blue channels of the IR signal under illuminated condition, for the different classes of defects in Tab. 2.

IV. CONCLUSION

We proposed a comparative statistical analysis of IR images under forward bias or illuminated conditions and of EL images for monocrystalline PV modules, after 20 years of exposure. This study represents an attempt to quantitatively establish correlations between the information that can be gained from EL and IR images, to draw guidelines for their combined use. The following main results indicated that:

(1) From visual inspection of EL and IR images, most of the defects detected in dark IR images under forward bias condition can also be identified in the EL images.

(2) Frequency histograms of the EL and R, G, B channel intensities, normalized in the range from zero to unity, provide useful information and should be considered as a tool for carrying our quantitative analyses.

(3) The range of intensities in the frequency histograms can be correlated to the classes of defects for IR images taken in forward bias conditions. In such a case, an abrupt change of the range for red and blue channels towards higher intensities is observed from class DCI XXL2 to HC2 as compared to class from SR to DCI XX3. Correspondingly, an increase of the EL intensities is observed.

(4) The proposed analysis confirms the qualitative relationship between EL and IR images under forward bias condition established for thin-film solar cells by Zamini et al. [6] and Erber at al. [7-8], and it is herein extended to monocrystalline solar cells.

(5) Machine learning algorithm could take advantage from the observed trends that can be implemented as criteria for real time identification of some classes of defects.

ACKNOWLEDGEMENT

The authors would like to acknowledge the European Research Council for supporting the ERC Grant Agreement No. 306622 (ERC Starting Grant "Multi-field and multi-scale Computational Approach to Design and Durability of PhotoVoltaic Modules" – CA2PVM) within the European Union's Seventh Framework Programme (FP/2007–2013).

REFERENCES

[1] M. Köntges, et al. "The risk of power loss in crystalline silicon based photovoltaic modules due to micro-cracks," *Solar Energy Materials and Solar Cells*, vol. 95, pp.1131-1137, 2011.

[2] M. Paggi, I. Berardone, A. Infuso, and M. Corrado, "Fatigue degradation and electric recovery in Silicon solar cells embedded in photovoltaic modules," *Scientific Reports*, vol. 4, pp.1-7, 2014.

[3] M. Paggi, I. Berardone, and M. Martire, "An Electric Model of Cracked Solar Cells Accounting for Distributed Damage Caused by Crack Interaction," *Energy Procedia*, vol. 92, pp. 576-584, 2016.

[4] S. Koch, T. Weber, C. Sobottka, A. Fladung, P. Clemens, and J. Berghold, "Outdoor electroluminescence imaging of crystalline photovoltaic modules: comparative study between manual ground-level inspections and drone-based aerial surveys," in *32nd European Photovoltaic Solar Energy Conference and Exhibition*, pp. 1736-1740, 2016.

[5] C. Buerhop, D. Schlegel, C. Vodermayer, and M. Nieβ, "Quality control of PV-modules in the field using infrared-thermography," in *26th European Photovoltaic Solar Energy Conference and Exhibition*, pp. 3894-3897, 2011.

[6] S. Zamini, "Non-destructive-techniques for quality control of photovoltaic modules: electroluminescence imaging and infrared thermography," *Photovoltaics International Journal, PV-Tech PRO*, vol. 15, pp. 127-203, 2012.

[7] R. Ebner, B. Kubicek, G. Újvári, S. Novalin, M. Rennhofer, and M. Halwachs, "Optical Characterization of Different Thin Film Module Technology," *International Journal of Photoenergy*, pp.1-12, 2015.

[8] R. Ebner, "Defect analysis in different photovoltaic modules using electroluminescence (EL) and infrared (IR)-thermography," in *25th European Photovoltaic Solar Energy Conference and Exhibition*, 2010.

[9] O. Breitenstein, J. Bauer, K. Bothe, D. Hinken, and J.Müller, "Luminescence imaging versus thermography on solar cells and wafers," in *26th European Photovoltaic Solar Energy Conference and Exhibition*, 2011.

[10] V. Gade, N. Shiradkar, M. Paggi, and J. Opalewski, "Predicting the long term power loss from cell cracks in PV modules," *IEEE 42nd Photovoltaic Specialist Conference*, pp. 1-6, 2015.

[11] P. Würfel, et al. "Diffusion lengths of silicon solar cells from luminescence images," *Journal of Applied Physics* 101, pp. 123110, 2007.

[12] K. Bothe, et al. "Combined quantitative analysis of electroluminescence images and LBIC mappings," in *22nd European Photovoltaic Solar Energy Conference*, pp. 1673, 2007.

[13] K. Bothe, et al. "Electroluminescence imaging as an in-line characterization tool for solar cell production," in *21st European Photovoltaik Solae Energy Conference*, pp. 597-600 2006.

[14] D. Hinken, et al. "Series resistance imaging of solar cells by voltage dependent electroluminescence," *Applied Physics Letters*, 91, pp. 182104, 2007.

[15] J. Haunschild, et al. "Fast series resistance imaging for silicon solar cells using electroluminescence," *Physica status solidi* (RRL)-Rapid Research Letters vol. 3, pp. 227-229, 2009.

[16] J. Lopez, A. Pozza, and T. Sample. "Analysis of crystalline silicon PV modules after 30 years of outdoor exposure," in *31st European Photovoltaic Solar Energy Conference and Exhibition*, pp. 1839-1845, 2015.

[17] A. N. Strahler, A. H. Strahler, M. Barrutia, and P. Sunyer. *Geografía física*, Barcelona: Omega, 1981

[18] Review of Failures of Photovoltaic Modules. IEA PVPS Task 13. External final report IEA-PVPS. March 2014. ISBN 978-3-906042-16.

Gap passivation structure for scalable n-type interdigitated all back contact silicon hetero-junction solar cell

Lei Zhang[1,2], Ujjwal Das[2], Steven Hegedus[1,2]

[1]Department of Electrical & Computer Engineering, University of Delaware, Newark, DE 19716, USA
[2]Institute of Energy Conversion, University of Delaware, Newark, DE 19716, USA

Abstract — **Minimizing the lateral transport loss over the rear surface gap region between emitter and base contacts is crucial for a scalable interdigitated all back contact silicon heterojunction solar cell (IBC-SHJ) process with gap width ≥ 100 μm. In this work, we investigated four variations of gap passivation structure having a wide range of interface defect density (D_{it}) and interface charge (Q_{pass}) and demonstrated surface recombination velocity (SRV) < 5cm/s. 2-D simulations indicate > 21.5% IBC-SHJ efficiency is achievable by the simplified cell process with proper light management. However, results of IBC-SHJ cell featuring two different gap structures fabricated via simplified processes suggest that a gap structure which induces inversion at n-type c-Si surface should be avoided.**

Index Terms — **gap passivation, interface defects density, interface charge density, IBC-SHJ, inversion effects, simplified fabrication process, scalable process**

I. INTRODUCTION

N-type IBC-SHJ solar cell has been leading the silicon based single-junction terrestrial solar cell performance with excellent surface passivation, no metal grids shading loss, well-designed light trapping and contacting schemes [1]. Recently a record efficiency of 26.6% has been reported in a practical size wafer (180 cm^2) [1]. An inherent feature of IBC-SHJ cells is the two-dimensional (2D) carrier transport. Some work has been done to understand the lateral transport loss mechanisms with respect to the rear surface patterning and pitch design [2]. In this work, we have investigated the effect of the mulit-layer stacked structure in the 100 μm wide gap on the lateral transport and device performance with a relatively large pitch dimension (1850 μm). Large pitch dimension with a wide gap between emitter and base contacts reduces limitation on patterning and loosens tolerance for alignment in a manufacturing environment. First, we studied four possible gap structure candidates for surface passivation and adopted extended Shockley-Read-Hall (SRH) formalism [2] to extract two key parameters: 1) passivation material/silicon interface defect density (D_{it}) and 2) passivation material surface area charge density (Q_{pass}). Next we evaluated the impact of the different D_{it} & Q_{pass} parameters on cell efficiency in a 2D Sentaurus TCAD model previously developed and validated for IBC-SHJ [2] [3] with new features which account for Q_{pass} and gap width. Then, we tried to correlate with the simulation results via simplified IBC-SHJ fabrication processes to implement two different gap passivation structures. Cell results will be discussed in section III C. The goal of this work is to identify how D_{it} and Q_{pass} of different gap passivation

structures determine the gap recombination and utilize this information to down-select usable gap passivation candidates to develop an industrially-scalable process of IBC-SHJ with gap width ⩾ 100 μm.

II. EXPERIMENTAL

Four groups of stack layers, as listed below, were symmetrically deposited by DC plasma enhanced chemical vapor deposition (PECVD) under 300°C on both surfaces of 150 μm 8.5 ohm.cm n-type CZ c-Si wafers. The wafer cleaning process before deposition is described elsewhere [2]. The stacks are listed in order of the deposition on Si.

1. Intrinisic i.a-Si:H (8nm)/ a-SiNx:H (80nm)
2. i.a-Si:H (8nm)/ phosphorus doped n.a-Si:H (30nm)
3. i.a-Si:H (8nm) only
4. i.a-Si:H (8nm)/ boron doped p.a-Si:H

Effective lifetimes, τ_{eff} were measured as a function of excess minority carrier density (Δn) using Sinton Quasi-steady-state-photoconductance (QSSPC) lifetime tester and translated into effective surface recombination velocity, SRV(Δn), using equation (1):

$$\frac{1}{\tau_{eff}(\Delta n)} \equiv \frac{1}{\tau_{bulk}(\Delta n)} + \frac{2 \times SRV(\Delta n)}{w} \tag{1}$$

Table I lists the relevant parameters and models used to calculate $\tau_{bulk}(\Delta n)$. D_{it} and Q_{pass} of optimized condition from each group (four sets in total) were extracted by fitting the experimental SRV (Δn) curves using extended SRH model.

TABLE I
PARAMETERS & MODELS USED FOR CALCULATION OF $\tau_{bulk}(\Delta n)$

Parameters/Models	Value
Radiative recombination	[6]
Auger recombination	[7]
c-Si bulk SRH recombination	Single trap level SRH Et=Ei=Eg/2=0eV
c-Si bulk majority carrier lifetime τ_{n0}	5ms
c-Si bulk minority carrier lifetime τ_{p0}	5ms

We then performed 2D numerical simulations of IBC-SHJ unit cell design depicted in Fig. 1. The emitter half-width and base half-width were fixed at 700 μm and 125 μm respectively. D_{it} and Q_{pass} were varied in a range that covered

978-1-5090-5606-4/17 $31.00 © 2017 IEEE

all the four sets of extracted data for simulations with a fixed gap width of 100 μm. In addition, simulations of constant gap SRV=3cm/s and 30cm/s were also performed as a function of gap width to show the dependence of device performance on gap passivation as the cell design became more industrially scalable gap width of $\geqslant$ 100 μm. Planar devices are simulated for simplicity. A simplified photolithography (PLith) free process was developed for fabrication of the device (1"×1") with the gap structures in study. Current-voltage (JV) characteristics were measured with a AM1.5G spectra class AAA simulator from OAI.

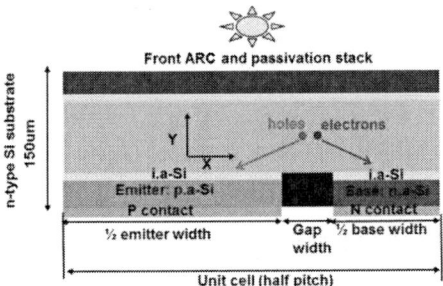

Fig. 1. Unit cell layout as used in simulations. The gap between emitter and base regions has various multi-layer stack structures.

III. RESULTS AND DISCUSSIONS

A. SRV (Δn) model & Extraction of D_{it} and Q_{pass}

In the extended SRH model that we used, the calculation of injection dependent SRV (Δn) can be expressed as:

$$SRV(\Delta n) \equiv \frac{v_{th}(p_s n_s - n_i^2)}{\Delta n} \int_{Ev}^{E_c} \frac{D_{it}(E)}{\frac{(p_s + p_1)}{\sigma_n(E)} + \frac{(n_s + n_1)}{\sigma_p(E)}} dE \quad (2)$$

$$Q_{pass} + Q_{si}(\Delta n, \psi_s) = 0$$
$$(Q_{pass} = Q_{aSi} + Q_{it}) \quad (3)$$

n_s, p_s are the carrier concentrations of electrons and holes at the surface of bulk c-Si, which are governed by the band bending ψ_s at the surface resulting from the electric charge distributions in the passivation structure (Q_{pass}) as shown in Fig. 4, and can be solved numerically by (3). Q_{pass} represents the effective area charge for the passivation layer. p1, n1 are the thermally emitted carrier densities from recombination centers. v_{th} is the thermal velocity and σ_n, σ_p are the energy dependent capture cross sections of electrons and holes and we assigned their relations as commonly used: $\sigma_p = 10\sigma_n$ where $\sigma_{charged} = 10\sigma_{neutral}$ for each of the structure. In our model, we also assigned the energy dependent $D_{it}(E)$ to have the same distribution functions as the a-Si:H bulk defect model described in [3], which composed of two Gaussians

near mid-gap and two exponential band-tails representing both acceptor and donor defects.

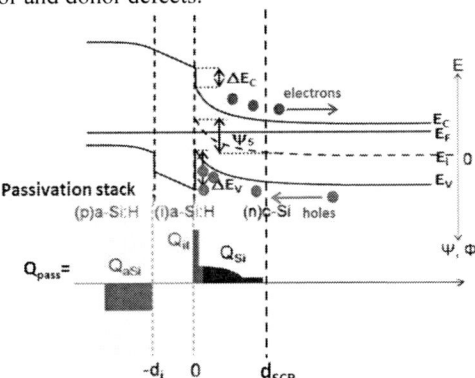

Fig. 2. Schematic of band-alignment and charge distributions at SHJ surface for i./p.a-Si passivation stack where Q_{pass} is negative.

Fig. 3 demonstrates good agreement between the experimental data and corresponding fitting curves for four types of gap passivation structure under study. The inset table shows the integrated D_{it} and Q_{pass} for the four gap structures estimated from the fitting. SRV (Δn=1×1015 cm-3) $\leqslant$ 5cm/s were achieved on all of the structures. Extremely low SRV (< 2 cm/s) are found for i./a-SiNx and i./n.a-Si:H structures due to simultaneous reduction of Dit and high positive Qpass, which repel holes, the minority carriers, away from the surface.

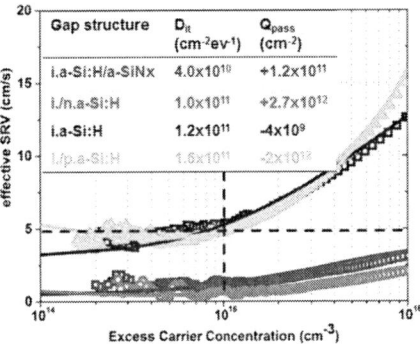

Fig. 3. SRV (Δn) data: experimental (open symbols), model fitting (lines). The table lists D_{it} and Q_{pass} for 4 gap structures.

B. 2D IBC-SHJ simulations with different D_{it} and Q_{pass}

Fig. 4 shows simulated efficiencies of planar IBC-SHJ solar cells (Fig. 1) as a function of (a) gap width, (b) varying Dit with low Qpass and (c) varying Q_{pass} with high D_{it}. The IBC-SHJ cell efficiency decreases strongly with increase in gap SRV and/or width due to lateral carrier transport losses. Even with an extremely low SRV of 3 cm/s, efficiency decreases by an absolute 1% when gap width increases from 10 μm to 100 μm (Fig. 4(a)). Two parameters that determine the SRV, namely, D_{it} and Q_{pass} were then varied with a gap width of 100 µm. Fig. 4(b) shows that with low Q_{pass}, as estimated for our i.a-Si:H layer, D_{it} needs to be $<1\times10^{11}$ cm^{-2}ev^{-1} to maintain high efficiency with 100 µm gap. Fig. 4(c) shows that either high positive or negative Q_{pass} ($\sim 10^{12}$ cm^{-2}) can maintain high efficiency even with high Dit=6×10^{12} cm^{-2}ev^{-1} and gap width of 100 µm because they repel one type of carrier (electron or hole for negative or positive Q_{pass} respectively) hence yielding difference in n_s vs. p_s. This result is important and could guide us to design a scalable patterning process with less stringent alignment requirement by relying on a high negative or positive charge in the gap stack.

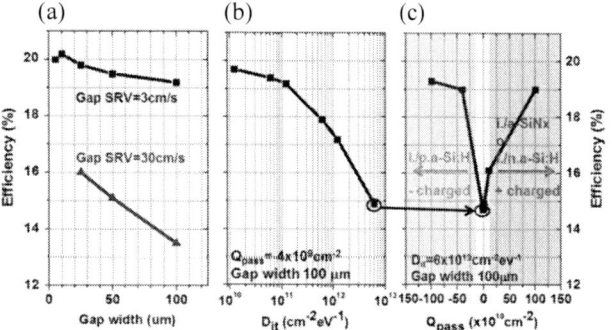

Fig. 4.   Simulated planar IBC-SHJ efficiencies as a (a): function of gap width with two constant SRVs of 3cm/s and 30cm/s. (b): function of D_{it} with low Q_{pass} = -4$\times10^9$cm^{-2}; (c): function of Q_{pass} with high D_{it} =6$\times10^{12}$cm^{-2}ev^{-1}. Gap width = 100 µm in (b) and (c).

C. Simplified IBC-SHJ fabrication with 100 µm gap

Following the simulation results, IBC-SHJ solar cells were fabricated using two simplified PLith free processes with 1) i./p.a-Si:H as gap passivation stack (Cross section is shown in the inset of Fig. 5(b)); 2) i. /a-SiNx:H as gap passivation stack (Cross section is shown in the inset of Fig. 5(d)). Gap width of the two device structures was designed to be 100 µm.

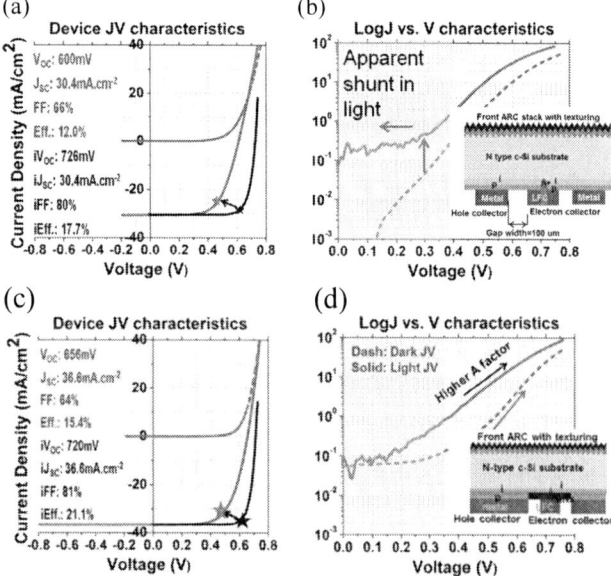

Fig. 5 (a) Measured dark (red dashed), light (red solid) J-V curves and re-constructed pseudo-JV (black) for i./p.a-Si:H gap passivation. (b) Measured dark (red dashed) and light (red solid) J-V plotted in log scale for same cell shown in (a), where the light curve was shifted upward from 4th quadrant to 1st quadrant by the measured J_{SC} value. (c) Measured dark (red dashed), light (red solid) J-V curves and re-constructed pseudo-JV (black) for i./a-SiNx:H gap passivation device. (d) Measured dark (red dashed) and light (red solid) J-V plotted in log scale for same cell shown in (c), where the light curve was shifted upward from 4th quadrant to 1st quadrant by the measured J_{SC} value.   Maximum power points are marked as stars on the graphs (a) and (c). Inset tables in (a) and (c) list all the relevant JV parameters; Insets of (b) and (d) show the schematic of IBC-SHJ cell structures fabricated with the simplified processes.

The patterning process for device structure 1) required only two steps of metallization using two sets of shadow mask after all PECVD thin film depositions. The shadow masks were designed in a way that misalignment was minimized to < 10 µm in the sample holder. Al was used for emitter contact and a metal stack (Ti/Sb/Al) was optimized for laser fired contact (LFC) formation with minimized passivation loss (< 15 mV) as electron collector as reported in [4]. The gap structure was i./p.a-Si stack with an emitter contact width of 1400 µm and base contact width of 250 µm. It is noted that the experimental cell structure differs from simulated structure shown in Fig. 1 in the base region structure, where no p.a-Si:H layer is present under the base metal contact in simulated structure. The patterning process for device structure 2) required one step of masked PECVD deposition and following two steps of masked metallization where the metal contacts are same as those used in device structure 1). The gap structure was i. /a-SiNx:H stack with an emitter contact width of 1300 µm and base contact width of 400µm, and the device structure also required LFC for base contact formation.

Fig. 5(a) & (c) compare the measured JV with the implied-JV constructed from QSSPC data assuming the measured short-circuit current density J_{SC} value for both device structures. Implied-JVs for both device structures had iV_{OC} of ~ 720 mV and iFF=80%. However, for the device with i./p.a-Si:H as gap passivation structure the measured values were much lower with V_{OC} = 600 mV, FF = 66% and J_{SC} = 30.4mA.cm^{-2}, indicating a significant recombination loss that was absent in implied-JV measurement. The device with i. /a-SiNx:H as gap passivation, on the other hand, shows much better V_{OC} = 656mV and J_{SC} = 36.6mA.cm^{-2}. The measured dark and light JV curves plotted on a log scale in Fig. 5(b) for i./p.a-Si:H as gap passivation structure show a significant upward shifting from dark to light JV with 'apparent' shunt-like behavior in low bias (< 0.3 V) in light JV, while shunting is absent in the dark curve. Fig. 5 (d), device with a-Si:H/a-SiNx:H as gap passivation, does not show obvious upward shifting in low bias region. We attribute this phenomenon to the presence of inversion layer in i./p.a-Si:H gap region ($p_s \gg n_s$) as shown in the band-alignment in Fig. 2, which creates a conductive channel for light-generated holes to be favorably transported to the local defects created by the LFC regions.

978-1-5090-5606-4/17 $31.00 © 2017 IEEE   410

This shunting path is more pronounced due to base contact metal being in direct contact with emitter p.a-Si:H where SRV > 10^7 cm/s. This induced inversion effect has been investigated by several other groups [5] [6] with capacitance studies and Deceglie et al. [7] utilized the scanning laser-beam-induced current to show how the induced inversion layer at front emitter junction can degrade the SHJ device's carrier collection if any localized defects exist, especially at forward bias, consistent with our results and interpretation. The low FF of IBC-SHJ device with i. /a-SiNx:H stack as gap passivation can be related to degradation of surface passivation due to non-ideal LFC process, and this can also translate into a V_{OC} loss from iV_{OC} to device V_{OC}. Also, the higher ideality factor of light-JV curve compared to that of the dark-JV curve shown in Fig. 8(d) (inverse to the slope) at bias > 0.4V indicates that the majority carriers, namely the electrons, collecting base region has high recombination at maximum power point that could reduce the V_{OC} and therefore FF. More in-depth investigation of the above mentioned phenomena have been discussed in another of our papers to be published at this conference [8].

IV. CONCLUSIONS

Simulations and measurements on test structures suggests how to optimize the gap passivation to achieve >20% with a scalable IBC-SHJ fabrication processes for manufacturable gap width ≥ 100 μm on n-type wafer. We achieved SRV < 5 cm/s on four types of low temperature gap passivation meeting the range of D_{it} and Q_{pass} required to achieve >21.5% cell efficiency on textured substrate. We investigated simplified processes to fabricate IBC-SHJ with different gap passivation structures, namely i. /a-SiNx:H stack with fixed positive charge and i./p.a-Si:H stack with induced negative charge. Device results suggest gap structure needs to have fixed positive charge to avoid forming an inversion layer.

ACKNOWLEDGEMENT

This work is funded by US Department of Energy PVRD program under contract number DE-EE-0007534.

Disclaimer: "This report was prepared as an account of work sponsored by an agency of the United States Government. Neither the United States Government nor any agency thereof, nor any of their employees, makes any warranty, express or implied, or assumes any legal liability or responsibility for the accuracy, completeness, or usefulness of any information, apparatus, product, or process disclosed, or represents that its use would not infringe privately owned rights. Reference herein to any specific commercial product, process, or service by trade name, trademark, manufacturer, or otherwise does not necessarily constitute or imply its endorsement, recommendation, or favoring by the United

States Government or any agency thereof. The view and opinions of authors expressed herein do not necessarily state or reflect those of the United States Government or any agency thereof."

REFERENCES

[1] Kaneka Corporation, "World's highest conversion efficiency of 26.33% achieved in a crystalline silicon solar cell" News Release, 2016.

[2] Z. Shu, U. Das, J. Allen, R. Birkmire and S. Hegedus, "Experimental and simulated analysis of front vs. all-back-contact silicon heterojunction solar cells: effect of interface and doped a-Si:H layer defects", Prog. Photovoltaics: Res. Appl., 23:0178-93, 2015.

[3] L. Zhang, U. K. Das, Z. Shu, H. Liu, R. Birkmire and S. Hegedus, "Experimental and simulated analysis of p a-Si:H defects on silicon heterojunction solar cells: trade-offs between VOC and FF," 2015 IEEE 42nd Photovoltaic Specialist Conference (PVSC), DOI: 10.1109/PVSC.2015.7356043, 2015.

[4] J. He, S. Hegedus, U. Das, Z. Shu, M. Bennett, L. Zhang and R. Birkmire, "Laser Fired Contact for n-Type Crystalline Si Solar Cells", Prog. Photovoltaics: Res. Appl., 23:1091-1099, 2015.

[5] O.Maslova, A. Brézard-Oudot, M. E. Gueunier-Farret, J. Alvarez, W. Favre, D. Muñoz, and J. P. Kleider, "Understanding inversion layers and band discontinuities in hydrogenated amorphous silicon/crystalline silicon heterojunctions from the temperature dependence of the capacitance", Appl. Phys. Lett., 103, 183907, 2013.

[6] Jian V. Li, Richard S. Crandall, David L. Young, Matthew R. Page, Eugene Iwaniczko, and Qi Wang, "Capacitance study of inversion at the amorphous-crystalline interface of n-type silicon heterojunction solar cells", Journal of App. Phsy., 110, 114502, 2011.

[7] Michael G. Deceglie, Hal S. Emmer, Zachary C. Holman, Antoine Descoeudres, Stefaan De Wolf, Christophe Ballif and Harry A. Atwater, "Scanning laser-beam-induced current of lateral transport near-junction defects in silicon heterojunction solar cells", IEEE Journal of Photovoltaics, vol.4, no.1, pp.154-159, 2014.

[8] N. Ahmed, L. Zhang, U. Das, S. Hegedus, "Electroluminescence analysis for separation of series resistance from recombination effects in silicon solar cells with interdigitated back contact design", to be published in this conference, 2017

Proposal of the Bandgap Design Using the Sun Height of the Culmination on the Winter Solstice

Kenji Araki, Kan-Hua Lee and Masafumi Yamaguchi

Toyota Technological Institute, Nagoya, 468-8611 Japan

Abstract — **Calculation of the optimized bandgap combination at the specific site required many annually-observed atmospheric parameters. For evaluating the impact of the atmospheric conditions, both the best and the worst distributions were taken from the measurement data. It was found that the optimization of the bandgaps combination considering ever-changing spectrum, may be determined by the matching condition to the sun height at the culmination on the winter solstice. Considering influence on the annual energy yield of the CPV multi-junction cell, the best number of junction was 3 and more than 4 junctions was not better for the annual energy yield.**

Index Terms —**Spectrum, Multi-junction cell, Optimization, Bandgap engineering, Outdoor measurement.**

I. INTRODUCTION

Multi-junction solar cells are sensitive to the spectrum change and its impact has been studied by many scientists [1-3]. Bandgap optimization of the multi-junction solar cells has been calculated and realized by many scientists. It was applied to various types of solar cells, including III-V multi-junction cell [4-8], thin film cells [9-12], and advanced cells even considering luminescence coupling effects [13-14].

It is true that optimized bandgap combination varies by sites. But the above-mentioned calculations of the bandgap optimization required many annually-observed atmospheric parameters and would not be realistic to most of the installation site in the world, unless a database is established by sophisticated observation of atmospheric parameters in worldwide [13-14]. It is also recognized that the atmospheric conditions, especially aerosol depth has a great influence on the performance of the multi-junction cells [13-14]. The impact of the extreme atmospheric conditions to the optimum bandgap combinations was not discussed before.

The question is how the bandgap is optimized in the area without annual spectrum information. One possibility is observation of the power generation trend at least one-year [15-16]. The prediction of annual energy yield of spectrum sensitive multi-junction cells especially concentrator photovoltaic (CPV) has been progressed [17-18]. However, the prediction still depends on precise observation of annual spectrum measurement. Alternative method based on a single day or at least several days of measurement is expected.

Supposing that the sky is clear or at least the minimum DNI divided by averaged DNI is more than 30 %, the spectrum-mismatching loss calculated from the classical Bird's model [19] was small. In this condition, it was shown that the

calculated 3-J and 4-J cell with optimized bandgap combination has wide range of the high-performance plateau in the range of the sun height at around the culmination on the winter solstice [20]. This fact implies a possibility of the simple and robust design rule of the bandgap optimization at the specific site. Namely, the bandgaps of multi-junction solar cells may be determined at the sun height of the culmination on the winter solstice (Fig. 1).

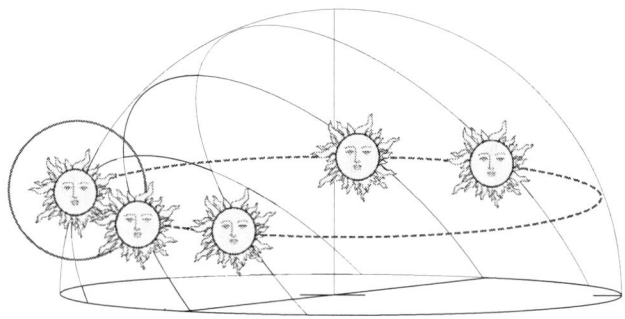

Fig. 1. Proposed representative sun height for optimization of annual energy yield to multi-junction solar cells.

This paper proposes a simple method to determine the optimized bandgap combination by a single point of the spectrum data available to all year, using numerical model considering modulation and variation (given by a random number) of atmospheric parameters, sunshine durations and other irradiance conditions.

II. METHOD

It was known that aerosol depth and water precipitation were the two main parameters that impacted to the performance of multi-junction cells [13-14].

For the evaluation of the impact of the aerosol depth fluctuation, it was assumed that its distribution is the logarithmic normal distribution. The aerosol transmittance is given by Bird's model [19].

Fig. 2 (left chart) shows the distribution of the aerosol depth in several sites in the world from AERONET database [13-14]. Note that some sites, like Kanpur in India and Solar Village in the Middle East, have often more aerosol depth than the value in the AM1.5D reference spectrum. Figure 2 (bottom-left) corresponds to the generated aerosol depth distribution. The top chart of Figure 2 is normalized frequency distribution for aerosol optical depth at 500 nm for various locations, extracted from AERONET database. In this chart, the grey vertical line indicates the value (0.084) used to generate the AM1.5D reference spectrum. The same procedure was applied to the water precipitation. The water vapor transmittance is given by Bird's model [19].

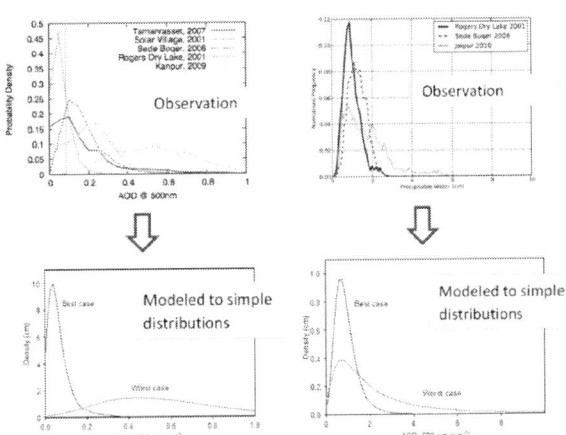

Fig. 2. Distribution of aerosol optical depth (left) and water precipitation (right) used in this calculation.

For the evaluation of the impact by the sunshine duration (ex., sunshine in summer or sunshine in winter), the collections of daily spectrum of various levels of the sun height were weighted by sinusoidal function with the modulation.

After the latitude, seasonal modulation of the sunshine duration and the distributions of atmospheric parameters were given, the set of the both direct normal irradiation and global irradiation on the sloped surface spectrum consists of 365 days and time divisions from sunrise and sunset of each day were calculated by the Bird's model [19] and prepared for bandgap optimization of the multi-junction cells for maximizing annual energy yield. Shading loss related with the air mass variation was not considered [20].

For the behavior of the multi-junction solar cells, we used a simple and ideal model. The output current of the multi-junction cell impacted by the spectrum change was assumed that it would be constraint by the minimum current output from all the series-connected junctions, considering that each sub-cell is connected in series. The luminescent coupling effect [21] was not considered in this model. The external quantum efficiency was assumed as the rectangular windows with the cutoff wavelength corresponded to the band edge. The peak efficiency was assumed as unity.

The output open-circuit voltage at one sun was simply assumed as minus 0.3 V from the bandgap [22-23]. Bandgap and dark current dependence from the junction temperature was ignored. Diode ideality factor of each sub-cell was assumed to unity.

Fill factor (FF) was modeled by a simple linear approximation from the degree of current mismatching among

sub-cells. At first, both bandgap combination and spectrum conditions (sun-height, aerosol depth and water precipitation) were simultaneously given by random numbers. The I-V curve of the cell of each combination was generated and FF was calculated by solving maximum value of the product of I and V in each time. It is shown the relationship between the current mismatching among three junctions (min(Isc) / mean(Isc)) and FF. Both bandgap combination and spectrum parameters were simultaneously given by random numbers. Internal resistance components in the cell were ignored.

The annual energy yield was calculated by integration of dates and times. The sampled sun height was done by the equal seven division starting at the 5° of the sun height (morning) and ending at 5° of the sun height (evening) regardless of the season.

The combination of the bandgap of sub-cells was numerically calculated for maximizing the annual energy yield per cell aperture area using different combinations of atmospheric parameters and other factors. This optimization calculation was repeated more than 100 times to each condition. The distribution of the resolved optimum bandgap was compared. At the same time, the set of the optimized bandgap calculated only by the culmination of the winter solstice was compared to the one that maximized the annual energy generation to see if the hypothesis was practically valid.

It is true these assumptions are too idealistic. However, it is useful for proving the hypothesis that optimization of the bandgap combination at the specific site may be determined by the matching condition to the sun height at the culmination on the winter solstice. The realistic model such as non-ideal junction behavior, incomplete absorption around the band-edge, and recombination loss in the shorter wavelength region tends to reduce the sensitivity to the spectrum change.

III. RESULTS

Among the atmospheric parameters, the main impact is given by the aerosol depth and the water precipitation [13-14]. At first, the direct spectrum was examined. The aerosol depth has an influence on the shorter wavelength region by scattering. The water precipitation affects the longer wavelength region by absorption. Both factors directly impact the performance of concentrator photovoltaic that only uses the direct sunlight. At first, the impact through the direct sunlight was examined.

Figure 3 indicates the calculated results with both the best and the worst distributions of aerosol depth and water precipitation. The 3-junction cells were operated in 1,000 x concentration using only the direct sunlight. The solid lines were optimized bandgaps by maximizing annual energy generation. The dashed lines were optimized bandgaps only by the sun height at the culmination on the winter solstice. The latitude is N40° and the seasonal modulation of the sunshine duration factor m = 0.

978-1-5090-5606-4/17 $31.00 © 2017 IEEE

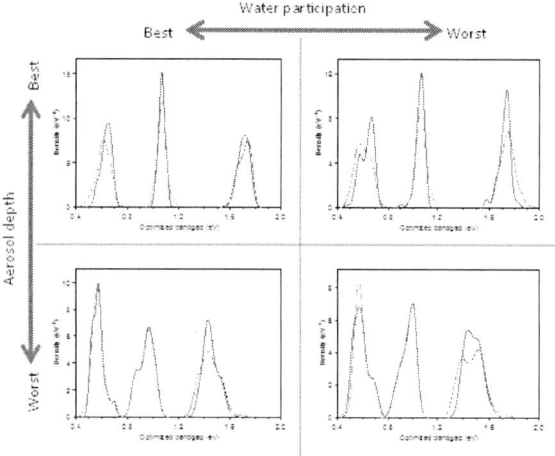

Fig. 3. The optimum bandgap distribution of the 3-junction cell impacted by the aerosol density and water precipitation.

Apparently, the peak position of the top cell in the worst aerosol depth distribution moved significantly to the lower bandgap direction because the top junction needed to expand the absorption region to reach to the current matching condition impacted by the insufficient blue component of the light by scattering. This phenomenon was also discussed by Chen for seeking deployment of CPV to the north India [13-14]

The influence by the water precipitation was opposite but not as obvious as the aerosol depth. The influence was seen in the bottom cell. The peak position of the bottom cell in the worst water precipitation distribution moved slightly to the lower bandgap because the bottom junction needed to expand the absorption region to reach to the current matching condition impacted by the insufficient infrared (IR) component by absorption of water in the sky.

In both cases, the distribution shape of the optimum bandgaps became wider either in the worst case of the aerosol depth and the water precipitation, indicating that the fluctuation of the atmospheric conditions significantly affecting the bandgap optimization results.

The difference of the optimized bandgaps between the ones by annual optimization and the ones only by the sun height at the culmination of the winter solstice was insignificant compared with the inherent variance of the bandgap by the fluctuation of the atmospheric conditions. Practically, the combination of the optimized bandgap can be designed by the spectrum of the sun height at the culmination of the winter solstice within the error of atmospheric parameter fluctuation.

The same kinds of the Monte Carlo simulations were repeated to the influence of (1) Direct spectrum vs. global spectrum on the sloped surface, (2) Impact of latitude of the installation sites, (3) Impact of the seasonal modulation of the sunshine duration, and (4) Number of junctions up to 6. Like the examination of the atmospheric parameters, the difference of the optimized bandgaps between the ones by annual optimization and the ones only by the sun height at the

culmination of the winter solstice was insignificant compared with the inherent variance.

IV. OPTIMIZED NUMBER OF JUNCTIONS

It has been believed for a long time that the increase of number of junctions will simply improve the performance. However, this dream was denied by the successive conference presentations in 2003 and 2004, on the precise calculation of the annual energy field considering spectrum variations by the sun-height [24-25]. In these studies, the annual energy yield increases by the number of junctions up to 3. However, its increase leveled off at 4 or 5 junctions. The number of junctions of the practical CPV system is still 3. More recently, it was implied that these calculations were too pessimistic by ignorance of the luminescence coupling [11,13]. However, its effects have not been clarified. At the same time, thanks to the development of the research on the global warming, fluctuation of atmospheric parameters has been quantitatively recognized [11-12]. It is high time to re-evaluate the problem of the best number of junctions with consideration of the fluctuation of the atmospheric parameters.

The above calculation was expanded to other levels of the number of junctions. For saving calculation time and increasing the size of random numbers for accurate searching for optimized combinations of bandgap energy combination, the result of the previous section was used, namely, the combination of the optimum bandgap energy can be obtained by the sun-height at the culmination of the winter solstice. For practical calculations, the series resistance was assumed 5 $m\Omega cm^2$ for 1,000 x concentration and 5 Ωcm^2 for flat-plate non-concentration operation using global irradiance on the fixed and sloped surface. The relation of FF to ($\min(Isc)$ / $mean(Isc)$) of sub-cells for each junction level was recalculated considering series resistance loss. For the model of the global irradiance spectrum was Bird's [19]. The latitude was represented as N20°. For the evaluation of the impact by the sunshine duration (ex., sunshine in summer or sunshine in winter), the collections of daily spectrum of various levels of the sun height were weighted by sinusoidal function with the modulation factor. It was for modeling seasonal fluctuation of the irradiance, if the irradiance varies sinusoidally with date. In this case of the calculation, we assumed the pattern of "sunshine in summer and zero irradiance at the winter solstice" for weighting yearly-averaged efficiency.

The calculated result is shown in Fig. 4. Surprisingly, the annual energy yield drops at more than 5 junctions. Namely, 4 is maximum number and the extra number of junctions to CPV application will be useless, considering terrestrial direct spectrum change influenced by the fluctuation of atmospheric parameters. This trend may be released by the increase of luminescent coupling and redundancy of the absorption band like the bottom cell in the lattice-matched 3-junction cell.

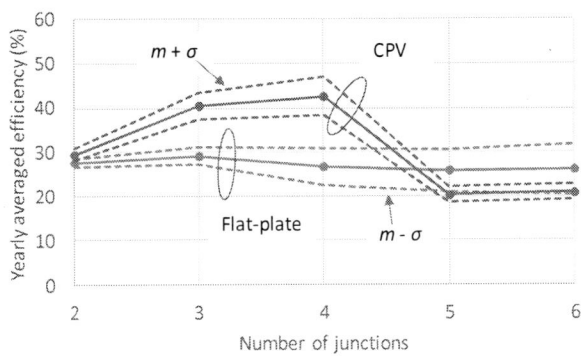

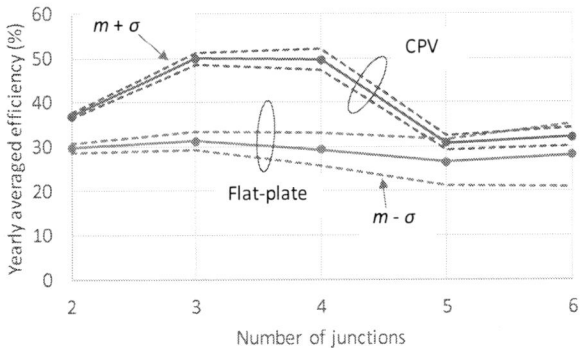

Fig. 4. Yearly averaged efficiency of multi-junction cells influenced by fluctuation of atmospheric parameters. The top chart is the area of the worst aerosol density and the best water precipitation like northern India. The bottom is the area of the best aerosol density and the best water precipitation like middle-west area of USA.

V. CONCLUSION

The hypothesis that optimization of the bandgaps combination of multi-junction solar cells in view of the maximization of the annual energy yield may be determined by the matching condition to the sun height at the culmination on the winter solstice was examined and it was practically applied within the best and worst cases of the atmospheric conditions, latitude (0° to 60° in the northern and southern hemisphere), meteorological conditions (modulation factor of the sunshine duration m = ±1), number of junctions (2 to 6) and the types of the irradiance (the global irradiance in sloped surface and the direct irradiance).

Considering this enormous influence on the annual energy yield of the CPV multi-junction cell, the best number of junction was 3 and more than 4 junctions was not better for the annual energy yield.

ACKNOWLEDGEMENT

This work has been partially supported by NEDO in Japan.

REFERENCES

[1] K. Araki and M. Yamaguchi, "Influences of spectrum change to 3-junction concentrator cells", Sol. Energy Mater. Sol. Cells 75, 2003, 707-71.

[2] C. Domínguez, I. Antón, G. Sala, and S. Askins, "Current-matching estimation for multijunction cells within a. CPV module by means of component cells", Prog. Photovolt. Res. 21, 2013. 1478-1488.

[3] R. Núñez, C. Domínguez, S. Askins, M. Victoria, R. Herrero, I Antón, and G. Sala, "Concentration photovoltaic optical system irradiance distribution measurements and its effect on multi - junction solar cells", Prog. Photovolt. Res. Appl. 24, 2016, 663.

[4] I. Antón, M. Martínez, F. Rubio, R. Núñez, R. Herrero, C. Domínguez, M. Victoria, S. Askins, and G. Sala, "Power rating of CPV systems based on spectrally corrected DNI", AIP Conf. Proc. 2012, 331–335.

[5] S. Fafard, "Low-Cost High-Performance Concentrated PhotoVoltaic (CPV) Solar Cell Production and Optimization using III-V Quantum Dot Material Band-Gap Engineering,", in Optics for Solar Energy 2012.

[6] D. C. Law, R. R. King, H. Yoon, M. J. Archer, A. Boca, C. M. Fetzer, S. Mesropian, T. Isshiki, M. Haddad, K.M. Edmondson, D. Bhusari, J. Yen, R.A. Sherif, H.A. Atwater, N.H. Karam, "Future technology pathways of terrestrial III–V multijunction solar cells for concentrator photovoltaic systems." Sol. Energy Mater. Sol. Cells, 94(8), 1314-1318. 2010.

[7] J. H. Ermer, R. K. Jones, P. Hebert, P. Pien, R. R. King, D. Bhusari, R. Brandt, O. Al-Taher, C. Fetzer, G. S. Kinsey, and N. Karam, "Status of C3MJ+ and C4MJ production concentrator solar cells at spectrolab." IEEE J. Photovolt., 2(2), 209-213.

[8] M. McDonald and C. Barnes, "Spectral optimization of CPV for integrated energy output", Proc. SPIE 7046, 2008, 704604.

[9] T.N.D. Tibbits, M. P. Lumb and A. Dobbin, "Quantum wells in multiple junction photovoltaics", Proc. SPIE 7933, 2011, 793303.

[10] T. Minemoto, M.Toda, S. Nagae, M. Gotoh, A. Nakajima, K. Yamamoto, H. Takakura and Y. Hamakawa, "Effect of spectral irradiance distribution on the outdoor performance of amorphous Si//thin-film crystalline Si stacked photovoltaic modules.", Sol. Energy Mater. Sol. Cells, 91, 2-3, 2007, 120-122.

[11] S. Reynolds and V. Smirnov, "Modelling performance of two- and four-terminal thin-film silicon tandem solar cells under varying spectral conditions" Energy Procedia 84, 2015, 251 – 260.

[12] R. Gottschalg, T. R. Betts, D. G. Infield, and M. J. Kearney, "The effect of spectral variations on the performance parameters of single and double junction amorphous silicon solar cells." Sol. Energy Mater. Sol. Cells, 85, 2005, 415–428.

[13] N. L. A. Chan, T. Thomas, M. Fuhrer, and N. J. Ekins - Daukes, "Practical limits of multijunction solar cell performance enhancement from radiative coupling considering realistic spectral conditions", J. Photovolt. 4.5 (2014), 1306-1313.

[14] N. L. A. Chan, H. E. Brindley, and N. J. Ekins - Daukes, "Impact of individual atmospheric parameters on CPV system power, energy yield and cost of energy.", Prog. Photovolt. Res. 22.10, 2014, 1080-1095.

[15] K. Araki Y. Kemmoku, and M. Yamaguchi, "A simple rating method for CPV module and systems", Conf. Proc. CPV-5, 2008.

[16] P. J. Verlinden, and J. B. Lasich, "Analysis and Simulation of Performance of CPV Systems with Multi - Junction Solar Cells." in AIP Conf. Proc. Vol. 1277, No. 1, pp. 277-280. 2010.

978-1-5090-5606-4/17 $31.00 © 2017 IEEE

[17] S. Kurtz, M. Muller, B. Marion, K. Emery, R. McConnell, S. Surendran, and A. Kimber, "Considerations for How to Rate CPV", AIP Conf. Proc. 1407, 2011, 25-29.

[18] P. Besson C. Domínguez and M. Baudrit, "Contributions to reproducible CPV outdoor power ratings", AIP Conf. Proc. 2014, 1616,167-173.

[19] R. E. Bird and C. Riordan, "Simple solar spectral model for direct and diffuse irradiance on horizontal and tilted planes at the earth's surface for cloudless atmospheres", Journal of Climate and 1Applied Meteorology, 25(1), 87-97. 1986.

[20] K. Araki, H. Nagai, K. Ikeda, K. H. Lee, and M. Yamaguchi, "Optimization of Land Use for a Multitracker System Using a Given Geometrical Site Condition", IEEE Journal of Photovoltaics, 6(4), 960-966. 2016.

[21] M. A. Steiner, and J. F. Geisz, "Non-linear luminescent coupling in series-connected multijunction solar cells." Applied Physics Letters, 100(25), 251106.2012.

[22] R. R, King, D. Bhusari, D. Larrabee, X.-Q. Liu, E. Rehder, K. Edmondson, H. Cotal, R. K. Jones, J. H. Ermer, C. M. Fetzer, D. C. Law and N. H. Karam. "Solar cell generations over 40% efficiency", Prog. Photovolt: Res. Appl. (2012), Published on-line. 2011.

[23] E.S. Toberer, A.C. Tamboli, M. Steiner, and S. Kurtz, "Analysis of solar cell quality using voltage metrics." Photovoltaic Specialists Conference (PVSC), 2012 38th IEEE, 1327-1331.01. 2012.

[24] K. Araki, N. J. Ekins-Daukes, M. Yamaguchi, "Which is the best number of junctions under ever-changing solar spectrum?", Proc. 3rd WCPEC Conf., 307, 2003.

[25] G. Létay, C. Baur, A. W. Bett, "Theoretical investigations of III-V multi-junction concentrator cells under realistic spectrum conditions", Proc. 19th European PVSEC. 2004.

Photoexcited Carriers, Phonons, and their Scattering Measured in Semiconductor Junctions by Transient Extreme Ultraviolet Spectroscopy

Scott K. Cushing[1,2], Brett M. Marsh[1], Mihai E. Vaida[1], Lucas M. Carneiro[1], Ilana J. Porter[1], Angela Lee[1], Stephen R. Leone[1,2,3]

[1]*Department of Chemistry, University of California, Berkeley, CA 94720, USA.*
[2]*Chemical Sciences Division, Lawrence Berkeley National Laboratory, Berkeley, CA 94720, USA.*
[3]*Department of Physics, University of California, Berkeley, CA 94720, USA.*

Abstract — **We use transient XUV spectroscopy to measure the carrier and phonon dynamics for each atomic-species in a semiconductor or semiconductor heterojunction, giving the technique potential for table-top photovoltaic device characterization. First, the electron populations, phonon populations, and their scattering pathways are measured in the Δ, L, and Γ valleys of Si(100) from femtoseconds to hundreds of picosecond. Second, the Si(100) is coated with TiO_2 and Ni to form a metal-insulator-semiconductor (MIS) junction. The Si L_{23}, Ni M_{23}, and Ti M_{23} edges are probed using XUV radiation from 20 eV to 150 eV, measuring the photoexcited charge-transfer, equilibration, and lattice dynamics for each component of the MIS junction.**

I. INTRODUCTION

The measurement of the valley-specific electron-phonon and phonon-phonon scattering pathways that follow photoexcitation is critical to predicting the performance of photovoltaic devices. Currently, the possible scattering processes are determined by aggregating measurements from multiple instruments with incommensurate time scales [1]. This has often limited modeling to the lowest-lying valleys of crystalline semiconductors, even though the photoexcited carrier distributions in a solar cell can span several eV.

Transient extreme ultraviolet (XUV) spectroscopy is a powerful new spectroscopic technique that allows probing of elemental x-ray absorption edges in a table-top package [2]-[4]. Time-resolved electron and hole distributions can be measured by the change in the occupancy on the core-level transition. Lattice dynamics can be measured by the local structural information imparted on the XUV absorption by the core-hole. The complexity of the core-hole screening effects in the XUV absorption previously limited the technique to measuring the coupled carrier and lattice dynamics in atomic and molecular systems.

Here, we discuss the recent breakthroughs in measuring the coupled carrier and lattice dynamics in semiconductors and semiconductor heterojunctions. First, the differential XUV absorption is measured in Si(100) from femtoseconds to hundreds of picoseconds after photoexcitation of the Δ, L, and Γ valleys. The coupled lattice and vibrational dynamics are separated using a single plasmon pole model and the Bethe-Salpeter equation (BSE) with density functional theory (DFT),

allowing the valley-dependent electron populations, phonon populations, and their scattering to be determined. Second, using the element-specificity of the XUV probe, the photoexcited charge transfer is measured between each component of a Si-Ni-TiO_2 metal-insulator-semiconductor junction. The measured absorption changes of the Si L_{23}, Ni M_{23}, and Ti M_{23} edges reveal that photo-excitation of the Si initially tunnels carriers into the Ni layer, followed by charge-equilibration through the TiO_2 insulating layer.

II. RESULTS AND ANALYSIS

The differential XUV absorption of Si(100) following photoexcitation with 800 nm, 500 nm, and 266 nm light is shown in Figure 1a-c, respectively. The measured ground state XUV absorption is theoretically predicted by a single plasmon pole and BSE-DFT calculation. The excited state absorption is then predicted by changing the valence and conduction carrier densities to match the photoexcited carrier density. The change in screening renormalizes and broadens the XUV absorption. The change in occupancy of the valence and conduction bands allows or blocks the possible XUV transition pathways. Finally, electron-phonon and phonon-phonon scattering creates an expansion of the lattice, which depends on the anharmonicity of the excited phonon modes.

Immediately following excitation, inter-valley electron-phonon scattering leads to the redistribution of photoexcited carriers between valleys. The phonon modes involved in the scattering have a momentum determined by the difference in the location of the initial and final valley in the Brillouin zone [1]. The symmetry and momentum restrictions on the possible excited phonon modes means that an anisotropic strain of the lattice occurs, energetically shifting and breaking the degeneracy of the critical points in the XUV absorption [5]. On longer time scales, intra-valley electron-phonon scattering (>100 fs in Figure 1a-c) and phonon-phonon scattering (>2-4 ps in Figure 1a-c) excite a more random distribution of phonon momenta than inter-valley scattering, leading to an isotropic expansion of the lattice. On long time scales the isotropic expansion converges to a simple heating of the lattice.

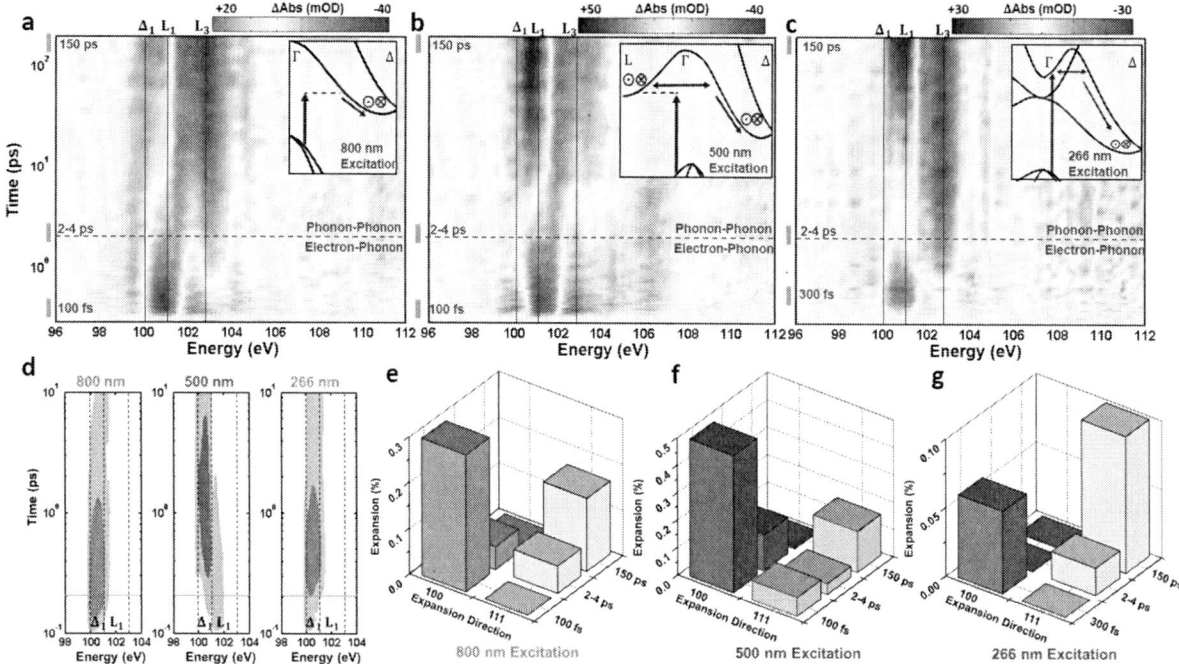

Fig. 1 Differential XUV absorption following photoexcitation into the **(a)** Δ, **(b)** L, and **(c)** Γ valley with 800 nm, 500 nm, and 266 nm light, respectively. The inset shows the possible scattering pathways. In and out of plane arrows indicate inter-valley scattering between degenerate valleys. The approximate cross-over timescale between electron-phonon and phonon-phonon scattering is marked as a dashed horizontal line. Key time points are denoted by grey squares. The Δ_1, L_1, and L_3 critical points are marked by solid vertical lines. **(d)** The electron distribution extracted from **(a)-(c)**. The grey horizontal line indicates the pulse-width. **(e)-(f)** The anisotropic and isotropic lattice expansions extracted from **(a)-(c)** are shown for key time scales following **(e)** 800 nm excitation, **(f)** 500 nm excitation, and **(g)** 266 nm excitation. The measured lattice expansions identify the scattering process and phonon modes at each time scale.

The state-filling induced changes in the XUV spectrum mainly occur below 102 eV. The lattice-induced changes in the XUV spectrum mainly occur above 102 eV and are distinct for the involved phonon modes. Using these spectral regions and the single plasmon pole and BSE-DFT model allows the time-resolved carrier (Figure 1d) and phonon (Figure 1e-g) populations to be determined from the measured differential absorptions.

Following 800 nm excitation into the Δ valley (Figure 1a), state-blocking is measured at the Δ_1 critical point (Figure 1d), and an initial anisotropic lattice expansion is measured in <60 fs due to the [100] optical phonon mode involved in the g- and f-type inter-valley scattering (Figure 1e). By ~2-4 ps after excitation, intra-valley scattering has thermalized the excited carrier distribution, and the more-random acoustic phonon distribution that is produced by the electron-phonon scattering begins to expand the lattice isotropically (Figure 1e). Phonon-phonon thermalization continues to expand the lattice isotropically until the ~150 ps timescale, converting the anisotropic strain of the optical phonons excited in inter-valley scattering into low-momentum acoustic phonons (Figure 1e).

Following 500 nm excitation into the L valley (Figure 1b), state-blocking is measured at the L_1 critical point (Figure 1d) in agreement with the excited carrier density, and a mixture of

[100] and [111] type phonons are measured from the resulting inter-valley L-L' and L-X scattering (Figure 1f). The L-L' inter-valley scattering occurs by a [100] phonon on a 20-100 fs timescale, while the L-X inter-valley scattering occurs by a [110] to [111] type phonon on a 100-200 fs timescale. Between the initial excitation and ~2-4 ps, the change in state-blocking represents the thermalization of the excited carriers from the L_1 to the Δ_1 point (Figure 1d). The increased isotropic expansion relative to the anisotropic expansion on this timescale also results from the intra-valley thermalization process (Figure 1f). On long time-scales an isotropic expansion is measured similar to 800 nm excitation, in agreement with the relative valley-independence of the phonon-phonon scattering process.

For 266 nm excitation into the Γ valley, state-blocking is not measured immediately following excitation because the Si $2p$ core level transition is dipole forbidden to the Γ_2' point. Instead, a delayed state-filling is measured in the Δ_1 critical point (Figure 1d). The rise-time corresponds to the 200-300 fs Γ-X inter-valley scattering as measured by the [100] lattice strain. The subsequent intra-valley scattering is again measured by the increase of the isotropic lattice expansion (Figure 1g). A larger number of intra-valley scattering events are necessary to thermalize the excess carrier energy of the 266 nm pump. This

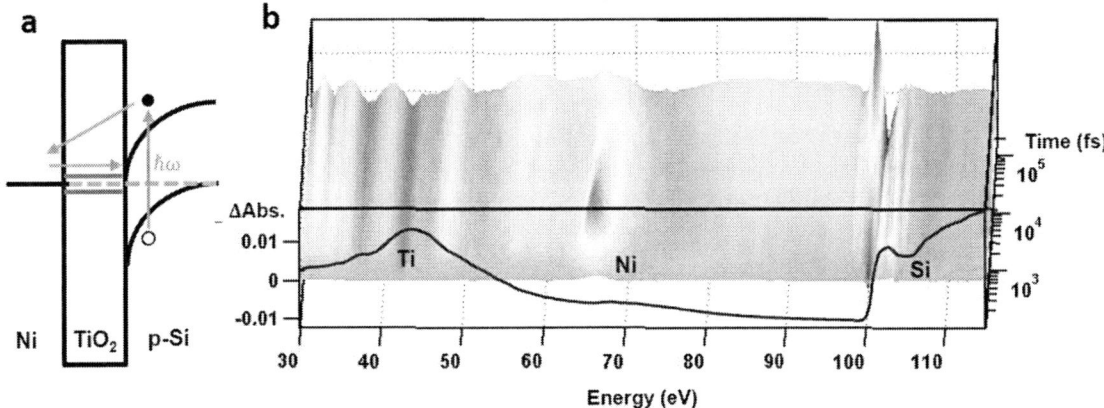

Fig. 2 **(a)** Diagram of the Si-Ni-TiO$_2$ metal-insulator-semiconductor junction. **(b)** Differential XUV absorption of the Ti M$_{23}$, Ni M$_{23}$, and Si L$_{23}$ edges following photoexcitation of the Si with 800 nm light. The overlay shows the XUV absorption.

is reflected by the higher final isotropic strain amount compared to the initial anisotropic strain amount.

The Si(100) is next coated with a ~10 nm TiO$_2$ thin film and a ~10 nm Ni layer (Figure 2a). High harmonic generation in Ar and He allow the the Si, Ni, and Ti edges to be measured (absorption overlay in Figure 2b). Photoexcitation with 800 nm light excites carriers into the Δ valley of Si, and a differential absorption is measured in Figure 2b similar to Figure 1a. Within 100 fs, a subsequent absorption rise is measured at the Ni edge. Within 1 ps, the Ni edge absorption begins to decrease while the differential absorption is measured to increase in magnitude across the Ti edge. No photoexcited signal is measured at this excitation energy for reference samples of TiO$_2$ and Ni without the Si base layer.

The changes in absorption for the MIS junction suggest an initial photoexcited tunneling between the Si and Ni layers, followed by a subsequent equilibration or back-transfer in the TiO$_2$ insulating layer. Applying a forward or reverse bias to the junction alters the magnitude of the charge-transfer measured in the Si, Ni, and Ti. By examining the Si edge, the effect of the junction on the electron, hole, and scattering processes, including both electron-phonon and phonon-phonon scattering, can be determined similar to Figure 1.

III. SUMMARY

We use transient XUV spectroscopy to measure the carrier and lattice dynamics of each atomic-species in a semiconductor or semiconductor heterostructure. Specifically, the carrier and phonon populations, as well as their scattering pathways, are measured for photoexcitation into the Δ, L, and Γ valleys of Si. Next, photoexcited charge-transfer is measured for the metal, insulator, and semiconductor layers in a Si-TiO$_2$-Ni MIS

junction. The versatility of transient XUV spectroscopy for measuring coupled electron and lattice dynamics may allow rapid characterization of excited state dynamics and scattering processes in photovoltaic devices.

ACKNOWLEDGEMENTS

The Si experiments were supported by the U.S. Department of Energy, Office of Science, Office of Basic Energy Sciences, Materials Sciences and Engineering Division, under Contract No. DE-AC02-05-CH11231 within the Physical Chemistry of Inorganic Nanostructures Program (KC3103). The Si-TiO$_2$-Ni experiments were supported by the U.S. Air Force Office of Scientific Research (Grant FA9550-14-1-0154). SKC acknowledges a postdoctoral fellowship through the Office of Energy Efficiency and Renewable Energy.

REFERENCES

[1] J. Shah, *Ultrafast Spectroscopy of Semiconductors and Semiconductor Nanostructures.* Berlin, Springer Berlin Heidelberg, 1999.

[2] T. Brabec and F. Krausz, "Intense few-cycle laser fields: frontiers of nonlinear optics," *Rev. Mod. Phys.,* vol. 72, pp. 545, 2000.

[3] E. Goulielmakis, Z.-H. Loh, A. Wirth, R. Santra, N. Rohringer, V. S. Yakovlev, S. Zherebtsov, T. Pfeifer, A. M. Azzeer, M. F. Kling, S. R. Leone, and F. Krausz, "Real-time observation of valence electron motion," *Nature,* vol. 466, pp. 739, 2010.

[4] E. R. Hosler and S. R. Leone, "Characterization of vibrational wave packets by core-level high-harmonic transient absorption spectroscopy," *Phys. Rev. A,* vol. 88, pp. 023420. 2013.

[5] C. Euaruksakul, Z. W. Li, F. Zheng, F. J. Himpsel, C. S. Ritz, B. Tanto, D. E. Savage, X. S. Liu, and M. G. Lagally, "Influence of strain on the conduction band structure of strained silicon nanomembranes," *Phys. Rev. Lett.,* vol. 101, pp. 147403, 2008.

On The Use of Voltage Measurements For Determining Carrier Lifetime at High Illumination Intensity

Robert Dumbrell, Mattias K. Juhl, Thorsten Trupke, Ziv Hameiri

University of New South Wales, Sydney, Australia

Abstract — **Illumination intensity dependent open circuit voltage measurements (commonly known as Suns-V_{oc}) are often used to measure the current-voltage characteristic of a solar cell without the impact of series resistance. Deviations have previously been reported between Suns-V_{oc} measurements and contactless measurements, such as injection-dependent photoluminescence (Suns-PL) at high illumination levels. These deviations are analyzed in detail in this paper and shown to cause significant errors when converting Suns-V_{oc} data to injection-dependent minority carrier lifetimes. Experimental data are used to demonstrate the magnitude of this effect for a range of different solar cell types.**

I. INTRODUCTION

Two common measurements performed on photovoltaic devices are illumination intensity dependent open circuit voltage (V_{oc}) measurements, also known as Suns-V_{oc} [1]; and injection-dependent lifetime measurements. Lifetime measurements can be used to predict the final cell V_{oc} (known in the literature as either implied open circuit voltage [iV_{oc}] [2] or as Suns-photoluminescence [3]); while V_{oc} measurements of cells can be used to determine the effective lifetime of the underlying material structure [1]. This conversion procedure is robust if: (a) the sample has a small proportion of voltage independent carriers [4]; and (b) the carrier density, and thus the local diode voltage, is laterally uniform. This study focuses on the impact of the latter issue, which can result in large errors if not considered.

Deviations between Suns-V_{oc} and iV_{oc} measurements of the same device have previously been reported in the literature at high illumination intensities (1 to 10 Suns). These include well known effects such as the impact of a Schottky diode at the rear of a cell [5] and other subtler effects that have been observed but not investigated in detail [3], [6]–[10]. Conversion between the two measurands has recently been suggested as a means to estimate the surface recombination saturation current density J_{0s} of finished cells [11]. As it will be shown, this process is likely to lead to large errors.

This study presents measurements of a range of solar cell types that exhibit the above deviation at high illumination intensities. The conversion of Suns-V_{oc} data into injection-dependent lifetimes and subsequent analysis of the high injection data according to the method of Kane and Swanson [12] is shown to lead to significant errors.

II. THEORY

Injection-dependent lifetime measurements are commonly used on unmetallized test structures to characterize the bulk or surface recombination of a particular sample [13]. Typical measurement techniques such as quasi-steady-state photoconductance (PC) [6] or photoluminescence (PL) [7] measure the average excess carrier density (Δn_{av}) in the volume.

The Δn_{av} can be interpreted either as an effective lifetime τ_{eff} using generalized analysis [14] via

$$\tau_{eff} \cong \frac{\Delta n_{av}}{G_{av} - \frac{d\Delta n_{av}}{dt}} \quad (1)$$

or as iV_{oc} via

$$iV_{oc} \cong \frac{kT}{q}\ln\left(\frac{\Delta n_{av}[N_{dop} + \Delta n_{av}]}{n_i^2}\right) \quad (2)$$

where G_{av} is the average generation rate, t is time, N_{dop} is the ionized dopant concentration, n_i is the intrinsic carrier concentration, and kT/q is the thermal voltage.

When Δn_{av} is interpreted as an iV_{oc} the implicit assumptions are that: (a) Δn_{av} is approximately equal to the excess carrier density at the junction [$\Delta n(0)$] in the illuminated region; and (b) $\Delta n(0)$ in the illuminated region is the same as $\Delta n(0)$ underneath the metal contacts.

Suns-V_{oc} measurements are commonly used to determine device and material properties of metallized solar cells. In these measurements, the voltage represents the separation of quasi Fermi levels at the junction underneath the probed metal contact, which can be equated with the junction excess carrier density underneath the contact [$\Delta n(0)_{probe}$] as

$$V_{oc} = \frac{kT}{q}\ln\left(\frac{\Delta n(0)_{probe}[N_{dop} + \Delta n(0)_{probe}]}{n_i^2}\right) \quad (3)$$

Under the same assumptions above, a Suns-V_{oc} measurement can also be converted into an implied effective lifetime measurement

$$\tau_{eff,implied} \cong \frac{\Delta n(0)_{probe}}{G_{av} - \frac{d\Delta n(0)_{probe}}{dt}} \quad (4)$$

As will be shown, the second assumption becomes invalid for standard front junction devices at high illumination intensities. The front contacts of the device create laterally non-uniform generation and increased local recombination. This results in carriers moving laterally at open circuit, which are opposed by the lateral series resistance. High illumination intensities result in larger lateral currents which eventually results in a measurable voltage drop from the non-metallized region to the metalized region.

In this case, interpreting the V_{oc} as τ_{eff} will result in an underestimation of τ_{eff}. Since this deviation between V_{oc} and iV_{oc} is strongly dependent on the illumination intensity, the deviations between the actual minority carrier lifetime and the lifetime inferred from measured V_{oc} increases with illumination intensity, which in turn causes errors in the common Kane and Swanson analysis [12] up to a factor of five.

III. METHOD

The theoretically predicted deviations between V_{oc} and iV_{oc} are now demonstrated by measurements on different types of solar cells. The investigated cells are: (1) a cell with non-optimized laser fired contacts (LFC), causing a Schottky diode at the rear contact (obtained from the cells presented in Ref. [5]); (2) a 30×30 mm^2 token cut from a standard screen printed back surface field (BSF) cell; and (3) a 30×30 mm^2 token cut from a commercially-produced passivated emitter and rear cell (PERC).

The measurement setup simultaneously measures absolute V_{oc}, relative PL intensity (I_{PL}) and relative illumination intensity under pulsed illumination by a xenon flash lamp, filtered such that light in the silicon band-to-band emission spectrum is blocked.

The iV_{oc} was calculated from I_{PL} using

$$iV_{oc} = \frac{kT}{q}\ln(C \cdot I_{PL}) \tag{5}$$

and adjusting the proportionality constant C such that the iV_{oc} agrees with the V_{oc} measurements at low light intensity. The relative illumination intensity was converted to absolute units using a separate measurement of the Suns-V_{oc} using a laser system for which the incident photon flux is known accurately.

IV. RESULTS AND DISCUSSION

The Suns-V_{oc} and Suns-iV_{oc} (also known as Suns-PL) plots of the three solar cells are displayed in Fig. 1 (a)-(c). There is a clear deviation between the V_{oc} and iV_{oc} for all three cells at high light intensities. Even at one sun illumination intensity there is a 6-mV difference for the LFC and the PERC solar cells and a 3-mV difference for the BSF sample. This demonstrates that this effect can be significant for the most common cell structures.

For all three measured cells, the Suns-iV_{oc} curves can be modelled with a single diode with ideality factors n in the range 1.2-1.4. Only at very high voltages does the Suns-iV_{oc} curve deviate upwards away from the single diode behavior, possibly indicating onset of the dominance of Auger

recombination at high carrier injection levels. The Suns-V_{oc} curves, however, are not readily modelled as single diodes. This deviation from the one diode model is an indication that carriers are no longer able to reach the metal contacts of the device without a significant voltage loss.

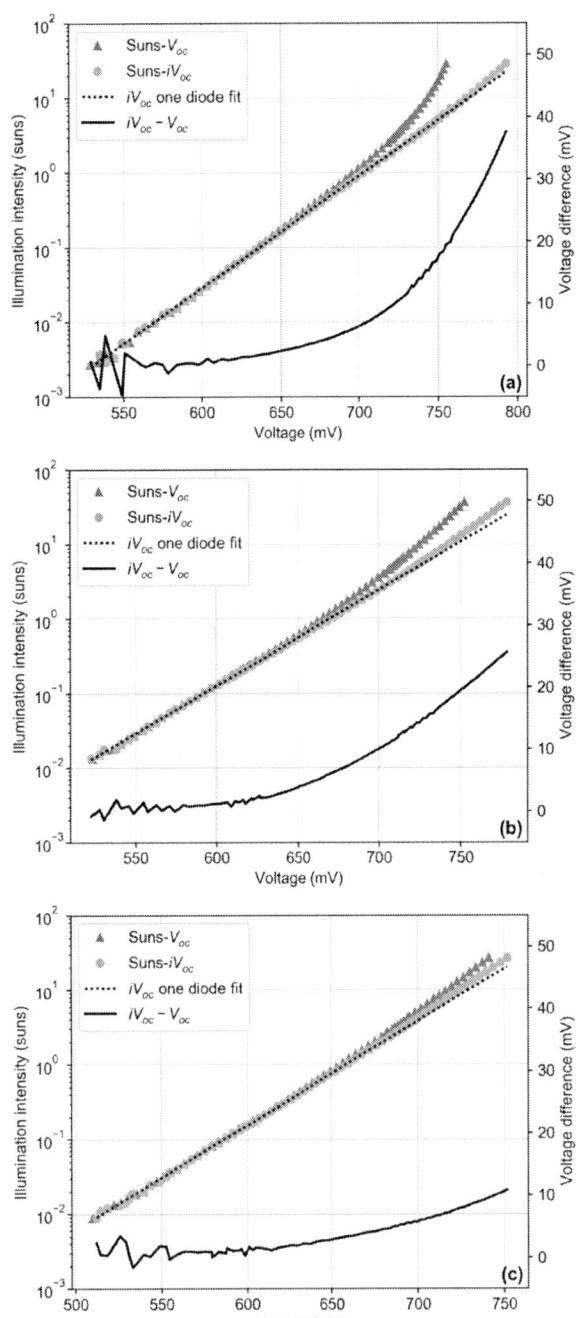

Fig. 1. Suns-V_{oc} and Suns-iV_{oc} data measured on: (a) the LFC sample, (b) the PERC cell, (c) the BSF cell. The difference between iV_{oc} and V_{oc} (solid black line) is plotted on the right vertical axis against iV_{oc}.

In the original study from which the LFC sample was taken, some cells exhibited even more extreme behavior than the deviation measured here, with a reduction in V_{oc} observed with increasing illumination intensity. However, the Suns-iV_{oc} data indicate that the recombination activity and the implied current-voltage (I-V) curve of the cell have normal behavior. In this specific cell the deviations between V_{oc} and iV_{oc} are known to be strongly dominated by the Schottky contact on the rear surface, but it is noteworthy that an impact of the effects of lateral current flow as described above could not be ruled out based on this measurement alone.

The BSF sample and PERC sample are both cut from commercially manufactured cells but because they are cut from a region including a segment of busbar, the metal shading area fraction is higher than for a typical full-size industrial cell, and thus can be expected to exaggerate the shading effects discussed in the previous section. However, given that the shading fraction of the two samples is similar, a comparison between the two samples is still instructive of the difference between the two cell types. For the same illumination intensity, the PERC cell has a higher voltage than the BSF cell, which indicates there is less recombination occurring in the PERC cell.

The difference between the iV_{oc} and V_{oc} is higher for the PERC cell. This difference is caused by the different rear contact design on the two cells. The BSF cell has a uniform rear contact which forms a low resistance path across the entire rear of the device. In contrast, the local rear contacts in the PERC cell create a laterally non-uniform resistance and rear surface recombination rate and therefore cause a further voltage loss, which further contributes to the increased deviation between iV_{oc} and V_{oc} that was measured.

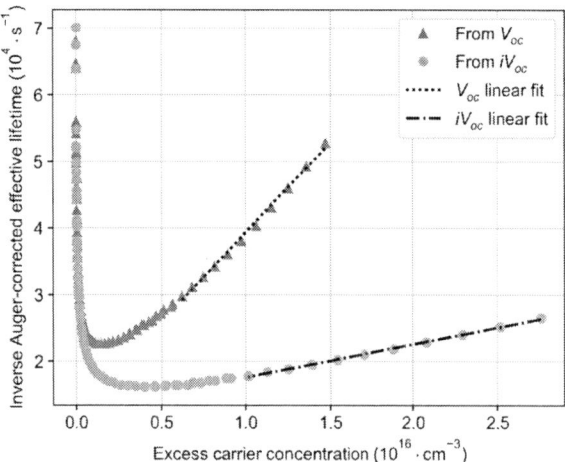

Fig. 2. Inverse effective lifetimes obtained from V_{oc} and iV_{oc} measurements of the PERC solar cell. Linear fits result in significantly different slopes.

V. J_{0s} ERROR CALCULATION

The deviations demonstrated in the measurements are consequential when calculating J_{0s} from the V_{oc} measurements. In this method, the implied injection-dependent τ_{eff} is calculated using (4) and J_{0s} is determined by fitting the inverse Auger corrected implied τ_{eff} against Δn_{av} at high injection [11] [12]. The procedure was performed on the PERC cell and the data is shown in Fig. 2. The slopes of the linear fits in Fig. 2 clearly differ for the iV_{oc} and V_{oc} data, demonstrating that the V_{oc} data will overestimate J_{0s}.

The J_{0s} values were determined from the V_{oc} and iV_{oc} measurements for both the BSF and PERC cells. For the PERC cell, the J_{0s} was 139 fA/cm² from the iV_{oc} measurement and 757 fA/cm² from the V_{oc} measurement—more than five times larger. The BSF cell J_{0s} was 282 fA/cm² from the iV_{oc} measurement and 642 fA/cm² from the V_{oc} measurement—more than two times larger.

It should be noted that strictly, the J_{0s} extraction method requires high injection level conditions, which could not be achieved due to limitations of the experimental setup. But the values obtained here are still indicative of the resulting relative error from applying the method to V_{oc} based as opposed to PL based iV_{oc} measurements.

VI. CONCLUSION

Deviation between the measured terminal voltage and implied voltage from PL data at high illumination intensities was observed on two typical cell structures and a third cell, with a known Schottky contact. Interpretation of terminal voltage data in terms of injection dependent minority carrier lifetime data and subsequent calculation of J_{0s} was shown to result in relative errors up to a factor of five when compared to the PL measurements.

ACKNOWLEDGEMENT

The authors acknowledge support from the Australian Government through the Australian Renewable Energy Agency (ARENA, Project 2014/RND097). The views expressed herein are not necessarily the views of the Australian Government, and the Australian Government does not accept responsibility for any information or advice contained herein. Z. Hameiri acknowledges the support of the Australian Research Council (ARC) through the Discovery Early Career Researcher Award (DECRA, Project DE150100268).

REFERENCES

[1] R. A. Sinton and A. Cuevas, "A quasi-steady-state open-circuit voltage method for solar cell characterization," in *16th European Photovoltaic Solar Energy Conference*, 2000, pp. 1152–1155.

[2] R. A. Sinton and A. Cuevas, "Contactless determination of

current – voltage characteristics and minority carrier lifetimes in semiconductors from quasi-steady-state photoconductance data," *Applied Physics Letters*, vol. 69, no. 17, pp. 2510–2512, 1996.

[3] T. Trupke, R. A. Bardos, M. D. Abbott, and J. E. Cotter, "Suns-photoluminescence: Contactless determination of current-voltage characteristics of silicon wafers," *Applied Physics Letters*, vol. 87, p. 93503, 2005.

[4] M. K. Juhl and T. Trupke, "The impact of voltage independent carriers on implied voltage measurements on silicon devices," *Journal of Applied Physics*, vol. 120, no. 16, 2016.

[5] S. W. Glunz, J. Nekarda, H. Mäckel, and A. Cuevas, "Analyzing back contacts of silicon solar cells by Suns-Voc measurements at high illumination densities," in *22nd European Photovoltaic Solar Energy Conference and Exhibition*, 2007, pp. 849–853.

[6] A. Cuevas and S. López-Romero, "The combined effect of non-uniform illumination and series resistance on the open-circuit voltage of solar cells," *Solar Cells*, vol. 11, no. 2, pp. 163–173, 1984.

[7] N. P. Harder, A. B. Sproul, T. Brammer, and A. G. Aberle, "Effects of sheet resistance and contact shading on the characterization of solar cells by open-circuit voltage measurements," *Journal of Applied Physics*, vol. 94, no. 4, pp. 2473–2479, 2003.

[8] J. Greulich, M. Glatthaar, A. Krieg, G. Emanuel, and S. Rein, "JV characteristics of industrial silicon solar cells: Influence of distributed series resistance and Shockley Read Hall recombination," in *24th European Solar Energy Conference and Exhibition*, 2009, pp. 2065–2069.

[9] J. A. Giesecke, M. C. Schubert, and W. Warta, "Carrier lifetime from dynamic electroluminescence," *IEEE Journal of Photovoltaics*, vol. 3, no. 3, pp. 1012–1015, 2013.

[10] J. Wong, S. Raj, J. W. Ho, J. Wang, and J. Lin, "Voltage loss analysis for bifacial silicon solar cells: Case for two-dimensional large-area modeling," *IEEE Journal of Photovoltaics*, vol. 6, no. 6, pp. 1421–1426, 2016.

[11] A. L. Blum, R. A. Sinton, W. Dobson, H. Wilterdink, and J. H. Dinger, "Lifetime and substrate doping measurements of solar cells and application to in-line process control," in *43rd IEEE Photovoltaic Specialists Conference*, 2016, pp. 3534–3537.

[12] D. E. Kane and R. M. Swanson, "Measurement of the emitter saturation current by a contactless photoconductivity decay method," in *18th IEEE Photovoltaic Specialists Conference*, 1985, pp. 578–583.

[13] A. Cuevas and D. H. Macdonald, "Measuring and interpreting the lifetime of silicon wafers," *Solar Energy*, vol. 76, no. 1–3, pp. 255–262, 2004.

[14] H. Nagel, C. Berge, and A. G. Aberle, "Generalized analysis of quasi-steady-state and quasi-transient measurements of carrier lifetimes in semiconductors," *Journal of Applied Physics*, vol. 86, no. 11, p. 6218, 1999.

High Resolution 3D Chemical Characterisation of a Cadmium Telluride Solar Cell by Dynamic SIMS

Thomas Fiducia[1], Kexue Li[2], Chris Grovenor[2], Kurt Barth[3], Walajabad Sampath[3], and Michael Walls[1]

[1]Loughborough University, Loughborough, Leicestershire, LE11 3TU, United Kingdom

[2]Materials Department, Oxford University, Oxford, OX1 3PH, United Kingdom

[3]Colorado State University, Fort Collins, Colorado, 80523, USA

Abstract — **Impurity elements such as chlorine and sulphur can have significant effects on the electrical performance of cadmium telluride (CdTe) solar cells. Here, the 3D distribution of such elements in a cadmium chloride treated CdTe device has been determined by high resolution dynamic SIMS, a novel technique that has not been applied to thin-film PV cells. It is found that as well as segregating to grain boundaries following treatment, chlorine also segregates to the CdS/CdTe interface. Conversely, sulphur shows a U-shaped diffusion profile. These results have potential implications for the processing thin-film CdTe devices.**

I. Introduction

It is well-known that an annealing treatment in the presence of cadmium chloride ($CdCl_2$) is necessary for the production of high efficiency CdTe solar cells. However, the treatment causes complex microstructural and chemical changes to the cell, including: CdTe recrystalisation and grain growth [1], the removal of stacking faults from CdTe grains [2], [3], CdTe grain boundary passivation with chlorine, and interdiffusion of sulphur and tellurium at the CdS/CdTe interface [1] (each of which has at one time or other been given as a reason for the positive effects of the $CdCl_2$ treatment). Keeping track of the chemical effects of the treatment is therefore important in determining the optimal cell processing conditions for maximum efficiency. Here, the 3-dimensional distribution of elements in a $CdCl_2$ treated device is determined by high resolution dynamic SIMS, a technique that has not been used before to characterise thin-film CdTe solar cells.

Two dimensional elemental characterisation of CdTe has been performed by Mao et al [4], who measured the planar distribution of chlorine and sulphur in a $CdCl_2$ treated device by Time-of-Flight (ToF) SIMS. The 3D distribution of chlorine and sulphur in a *substrate* CdTe device has been measured by Kranz et al [5], also by ToF SIMS. Harvey et al performed 3D ToF SIMS in the absorber layer of a superstrate CdTe device [6]. The current technique combines the high resolution of ToF SIMS with the superior sensitivity of dynamic SIMS.

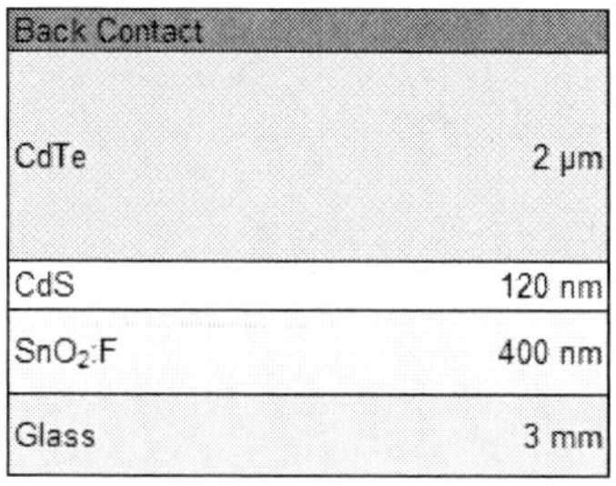

Fig. 1   Schematic of the superstrate CdTe device structure

II. Experimental

A plate of superstrate cadmium telluride (CdTe) devices was fabricated in an all-in-one vacuum process at Colorado State University [7]. In the process, ~120 nm of cadmium sulphide (CdS) was sublimated onto an NSG Plikington, TEC12D glass substrate. This was followed by ~2 μm of CdTe, resulting in a (glass/F:SnO_2/SnO_2/CdS/CdTe) device structure, as seen in figure 1. The stack was then exposed to process of reference $CdCl_2$ and CuCl treatments. The $CdCl_2$ treatment lasted a total of 6 minutes, and was carried out in an atmosphere of nitrogen (with 2% oxygen). The stack was contacted with nickel-carbon paint, then sand-blasted to create 9 separate devices. These were JV tested, and a 12.01% efficient device selected for further characterisation.

Following the removal of the back contact with acetone, the device was characterised with a Cameca NanoSIMS 50/50L Secondary Ion Mass Spectrometer (SIMS). Since the depth resolution of SIMS depth profiles are affected by initial surface topography, a region of the CdTe back surface was first polished in a dual beam SEM with a low current ion beam (polishing parallel to the substrate, with a beam current of 0.5nA). In addition, a layer of platinum was deposited to help smooth the polishing of the surface, as is done in preparation of samples for Transmission Electron Microscopy. Mass spectrometry measurements were taken over a 5 x 5 micron beam raster area on the polished surface. This area was then ion eroded with a second ion beam, leaving the layer beneath,

978-1-5090-5606-4/17 $31.00 © 2017 IEEE

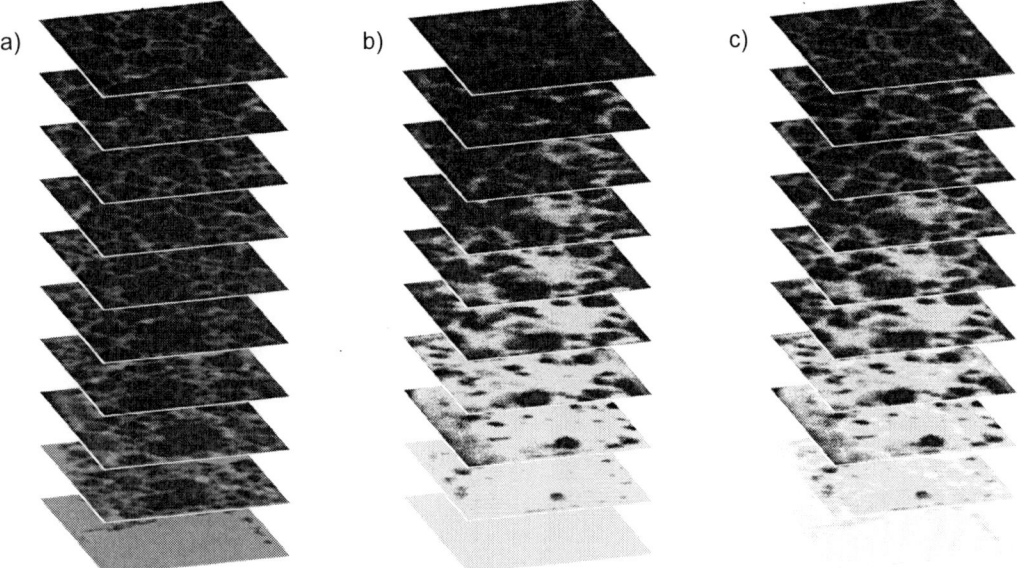

Fig. 2   Image stacks showing the distribution of a) chlorine and b) sulphur through the CdTe absorber, up to the CdS interface. Images in stacks a) and b) are combined to produce those in c) (with yellow indicating regions containing both chlorine and sulphur counts). The 10 images in each stack are planes 1, 7, 13, 19, 25, 31, 37, 43, 49 and 55. The field of view of each image is 5 x 5 microns.

around 10 nm below, to be measured. This process was then repeated through the depth of the cell. The spectrometer was tuned for Cl, S, Te, Cu, and F. Results for copper and fluorine are not included in the analysis.

III. RESULTS

The device chosen for elemental characterisation had a measured efficiency of 12.01%, with a short-circuit current of 22.7 mA/cm^2, an open circuit voltage of 789 mV, and a fill factor of 67.1%. These are in the range of typical values for CSU's basic, process of reference CdS cells. This indicates that there was nothing unusual with the deposition or processing of the measured cell.

NanoSIMS characterisation produced multiple stacks of 2D images, with each stack showing the distribution of a different element through the depth of the measured volume (one stack for each of the 5 elements). Chlorine, sulphur, and 'combined' stacks are shown in figure 2. Plotting the average number of counts per pixel in each image versus the plane number of that image produces a depth profile of the 1D distribution of the element through the cell. Each data point in a profile therefore corresponds to a 2D image for that element. This has been performed for Te, S and Cl in figure 3. As such, the 2D images shown in figures each correspond to a point on a profile in figure 3.

Chlorine – Figures 2, 3, 4 and 5 display the important features of the distribution of chlorine through the CdTe and CdS layers. Figure 2 a), showing the Cl distribution towards the back of the CdTe layer (plane number 3), clearly shows

that chlorine is located mainly at the grain boundaries (GBs). Some bright spots corresponding to the highest chlorine counts can be seen at the confluence points of multiple GBs, and/or where the grain size – as seen from this plane – is small. There is no signal from the grain interiors (GIs).

These general patterns continue through the depth of the CdTe layer. Figure 3 shows a steady increase in average chlorine counts from planes 35 to 50. This coincides with a decrease in apparent grain size closer to the CdS interface (and hence an increase in total GB length). This can be seen clearly in stack a) of figure 2.

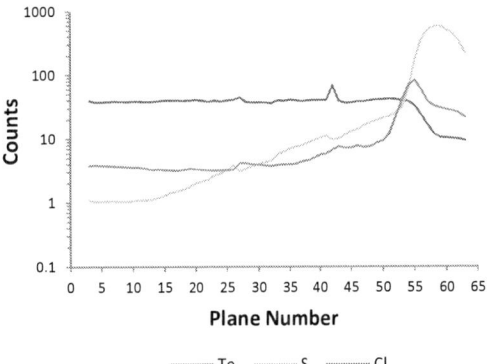

Fig. 3   1D profile of average chlorine, sulphur and tellurium counts through the CdTe and CdS layers. Plane number 1 is close to the back of the CdTe layer and plane number 65 is close to the transparent conductor

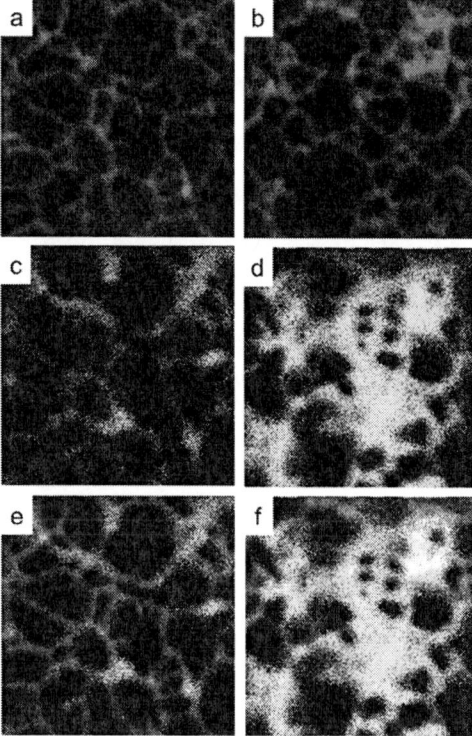

Fig. 4   NanoSIMS images of the chlorine (a & b) and sulphur (c & d) distributions in planes 3 and 26 of the image stack (LHS and RHS respectively). Images e) and f) are combinations of the S and Cl images above. The grain structures in planes 3 and 26 do not align because of slight sample drift during the data acquisition. The field of view in each image is 5 x 5 microns.

Figure 1 also shows a sharp rise in average chlorine counts at the CdS/CdTe interface (the position of the interface is indicated by the spike in the S profile and drop in the Te profile at around plane 55). Although in 1D the chlorine concentrations appear very high in this region, it is not clear how much this is caused by geometry or change in matrix, rather than genuinely higher levels of Cl segregation at this

interface than at grain boundaries (i.e. the chlorine-containing interface is now parallel to or in the plane of the beam raster, meaning chlorine counts are collected over all areas of the image). This point is well illustrated in figure 3. The images clearly show that chlorine is segregated along the CdS/CdTe interface. While the tellurium counts are reducing going from planes 50 to 57, indicating the end of the CdTe and beginning of the CdS, chlorine counts are increasing. Planes 56 and above in figures 3 and 5 both show high levels of chlorine counts in the CdS layer.

Sulphur – Figures 2, 3 and 4 display the important features of the distribution of sulphur through the CdTe and CdS layers. Figure 4 c), showing the S distribution towards the back of the CdTe layer, indicates some moderate S segregation at the grain boundaries. The GB segregation is qualitatively weaker than for chlorine in the same plane, as indicated by the combined image in part e) of the figure. Unlike with chlorine, average sulphur concentrations rise steadily from this point up until the CdS/CdTe interface. Figure 4 d) shows that this is down to increased counts from the grain interiors, with sulphur signal now emanating from most areas but the middle of the larger grains. It should be noted that this is at the stage before apparent grain sizes have started to become significantly smaller, shown by the Cl in image 4 b).

IV. DISCUSSION

The acquisition of high resolution elemental data in 3D allows predictions to be made in two key areas: 1) the kinetics of diffusing impurity species; and 2) what affect the species may be having on device performance. These can then be used to inform changes in device processing to improve performance.

It is clear from the data that chlorine strongly segregates to grain boundaries in CdTe. This is in agreement with other experiments [8]. The data also suggests that during treatment chlorine travels down the GBs to the CdS/CdTe interface, and continues into the CdS. While the segregation of chlorine at grain boundaries (and its pacifying effects) have been relatively well studied, chlorine segregation at the CdS interface has not. Chlorine has been seen at the interface in

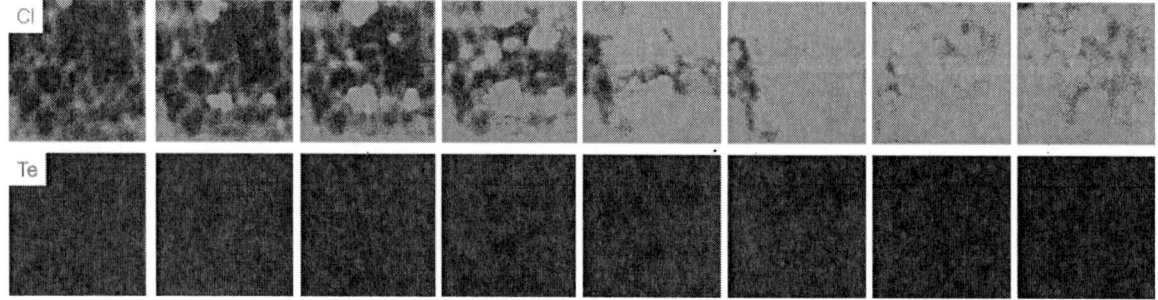

Fig. 5   Series of NanoSIMS images crossing the CdS/CdTe interface (planes 50 to 57, left to right). The top row shows chlorine and the bottom tellurium. The field of view in each image is 3.4 x 3.4 microns.

978-1-5090-5606-4/17 $31.00 © 2017 IEEE

cross-sectional TEM studies [8], [9] and an atom probe tomography study [10], but not in a plan-view series moving across the junction. The electrical effects of chlorine at the CdS/CdTe junction are not clear, but are likely to be significant given the key location at the p-n junction and the known electronic and band structure effects of chlorine in CdTe. It should also be noted that chlorine is likely to be present at the back surface of the CdTe, which is removed here in the process of polishing.

The series view of sulphur distribution through the CdTe – showing progressively more S in the CdTe grain interiors as the CdS layer is approached – is consistent with a type B diffusion regime in the Harrison classification system [11]. This is where fringes of sulphur develop alongside grain boundaries due to out-diffusion from the boundaries, but do not overlap with one another [12]. The different diffusion regimes are shown in figure 6 in the appendix. The presence of sulphur in the CdTe layer – forming a $CdTe_{1-x}S_x$ alloy – has potential microstructural and electrical effects. In terms of the microstructural effects, one theory is that the alloy reduces the C.10% lattice mismatch between the two crystals, resulting in fewer misfit dislocations at the interface. Sulphur also decreases the band gap of CdTe due to band gap bowing.

V. CONCLUSIONS

High resolution 3D dynamic SIMS measurements have been carried out on a $CdCl_2$ treated CdTe solar cell. Chlorine is observed to segregate strongly at grain boundaries and at the CdS/CdTe interface, and is present in the CdS layer. Sulphur shows weaker segregation to grain boundaries, and plan-view slices through the depth of the CdTe reveal a U-shaped diffusion profile within grains, suggesting a mixed grain boundary and lattice diffusion regime. These measurements give useful insights into the complex chemical effects of the $CdCl_2$ annealing treatment, and are especially useful as through-thickness, plan-view images of this kind have rarely been employed in the chemical characterisation of CdTe. In future, the technique could be used to compare device sets with a range of different processing conditions and/or layer structures, and could include detailed studies of other elements such as copper, fluorine, sodium and oxygen.

ACKNOWLEDGEMENTS

One of the authors (TF) is grateful to the EPSRC SuperSolar Hub for funding an International Engagement secondment at Colorado State University.

REFERENCES

[1]   B. E. Mccandless, Ã. L. V Moulton, and R. W. Birkmire, "Recrystallization and Sulfur Di   usion in CdCl 2 -Treated CdTe / CdS Thin Films," vol. 5, no. March, pp. 249–260, 1997.

[2]   A. Abbas, G. D. West, J. W. Bowers, P. Isherwood, P. M. Kaminski, B. Maniscalco, P. Rowley, J. M. Walls, K. Barricklow, W. S. Sampath, and K. L. Barth, "The effect of cadmium chloride treatment on close-spaced sublimated cadmium telluride thin-film solar cells," *IEEE J. Photovoltaics*, vol. 3, no. 4, pp. 1361–1366, 2013.

[3]   T. Fiducia, A. Abbas, K. Barth, W. Sampath, and M. Walls, "Intragranular Defects in As-Deposited and Cadmium Chloride-Treated Polycrystalline Cadmium Telluride Solar Cells," in *2016 IEEE 43rd Photovoltaic Specialist Conference (PVSC)*, 2016.

[4]   D. Mao, G. Blatz, C. E. Wickersham, and M. Gloeckler, "Correlative impurity distribution analysis in cadmium telluride (CdTe) thin-film solar cells by ToF-SIMS 2D imaging," *Sol. Energy Mater. Sol. Cells*, vol. 157, pp. 65–73, 2016.

[5]   Lukas Kranz, Christina Gretener, Julian Perrenoud, Dominik Jaeger, Stephan S. A. Gerstl,Rafael Schmitt, Stephan Buecheler, and Ayodhya N. Tiwari "Tailoring Impurity Distribution in Polycrystalline CdTe Solar Cells for Enhanced Minority Carrier Lifetime," *Adv. Energy Mater.*, 2014.

[6]   S. P. Harvey, G. Teeter, H. Moutinho, and M. M. Al-Jassim, "Direct evidence of enhanced chlorine segregation at grain boundaries in polycrystalline CdTe thin films via three-dimensional TOF-SIMS imaging," *Prog. Photovoltaics Res. Appl.*, vol. 23, no. 7, pp. 838–846, Jul. 2015.

[7]   D. E. Swanson, J. M. Kephart, P. S. Kobyakov, K. Walters, K. C. Cameron, K. L. Barth, W. S. Sampath, J. Drayton, and J. R. Sites, "Single vacuum chamber with multiple close space sublimation sources to fabricate CdTe solar cells," *J. Vac. Sci. Technol. A Vacuum, Surfaces, Film.*, vol. 34, no. 2, p. 021202, 2016.

[8]   A. Abbas, D. Swanson, A. Munshi, K. L. Barth, W. S. Sampath, J. W. Bowers, P. M. Kaminski, and J. M. Walls, "The Effect of a Post-Activation Annealing Treatment on Thin Film CdTe Device Performance," 2015.

[9]   A. Abbas, G. . West, J. W. Bowers, P. M. Kaminski, B. Maniscalco, J. M. Walls, K. L. Barth, and W. S. Sampath, "Cadmium chloride assisted re-crystallization of CdTe: The effect of annealing over-treatment," in *2014 IEEE 40th Photovoltaic Specialist Conference (PVSC)*, 2014, pp. 0701–0706.

[10]   J. D. Poplawsky, C. Li, N. R. Paudel, W. Guo, Y. Yan, and S. J. Pennycook, "Nanoscale doping profiles within CdTe grain boundaries and at the CdS/CdTe interface revealed by atom probe tomography and STEM EBIC," *Sol. Energy Mater. Sol. Cells*, vol. 150, no. March, pp. 95–101, 2016.

[11]   L. G. Harrison, "Influence of dislocations on diffusion kinetics in solids with particular reference to the alkali halides," *Trans. Faraday Soc.*, vol. 57, p. 1191, 1961.

[12]   H. Mehrer, *Diffusion in Solids*. 2007.

APPENDIX

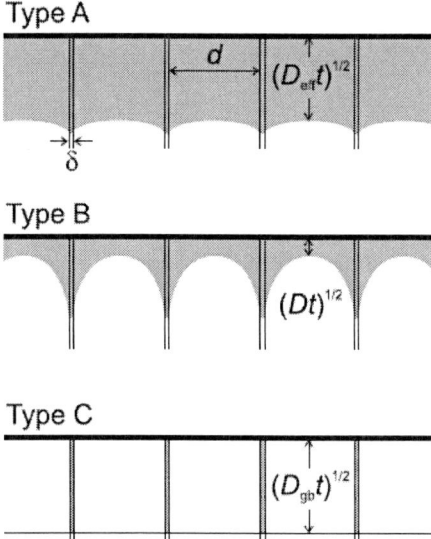

Fig. 6   Schematic of the type A, B, and C diffusion regimes [12]

Harsh Outdoor Evaluation Setup and First Power Production Results for Si Mini-Modules Covered by Eu³⁺-Based Down Converters

Benjamín González-Díaz[1], Carlos Montes[2], Joaquín Sanchiz[3], Luis Ocaña[2], Carlos Quinto[2], Cecilio Hernández-Rodríguez[3], Mari Paz Friend[2], Manuel Cendagorta-Galarza[2], David Cañadillas[4] and Ricardo Guerrero-Lemus[4].

[1]Departamento de Ingeniería Industrial. Universidad de La Laguna. Avenida Astrofísico Francisco Sánchez S/N 38206 S/C de Tenerife. Spain.

[2]Instituto Tecnológico y de Energías Renovables, S. A. (ITER), Pol. Industrial de Granadilla, s/n. 38600 Granadilla de Abona, Spain.

[3]Departamento de Química. Universidad de La Laguna. Avenida Astrofísico Francisco Sánchez S/N 38206 S/C de Tenerife. Spain.

[4]Departamento de Física. Universidad de La Laguna. Avenida Astrofísico Francisco Sánchez S/N 38206 S/C de Tenerife. Spain.

Abstract — **A new outdoor evaluation setup has been installed in a harsh location close to the Sahara Desert, in South Tenerife (28°04'16.48"N 16°30'48.15"W). The environmental conditions are unique in terms of very high sun irradiation, salinity, thermal cycles and dust. The setup is composed by 8 mini-module positions. Some mini-modules have been covered with different down-converters synthesized in our lab, in order to evaluate improvements in efficiency compared to mini-modules without any treatment. Results show the down-conversion layers increase efficiency in a first step but also suffers early degradation that requires further analysis.**

I. INTRODUCTION

One of the most interesting strategies to increase efficiency in solar cells is to cost-effectively increase its external quantum efficiency (EQE) in the ultraviolet (UV) spectral range. Mainly due to absorption and reflection losses in the front glass, encapsulation material, antireflective coatings and heavily doped emitters, the UV photons are not completely used [1]. One of the technologies to produce the EQE enhancement in the UV range is based on the down-conversion and down-shifting of UV photons to longer wavelengths in the visible range, where the EQE is closer to 100%.

In several studies, down-converters and down-shifters have been applied to solar cells and the enhancement of the EQE in the 280 - 400 nm spectral range has been reported [2]. However, none of them has tested the converters placed on PV modules and tested under ambient conditions. Our group has synthesized down-converters from commercial chemicals and applied to photovoltaic (PV) cells and modules, resulting in a cost-effective procedure to increase the energy efficiency of these devices.

First efficiency results for PV minimodules covered by Eu³⁺-based down-converters and tested under harsh ambient conditions are presented in this work. Firstly, a description of the chemical compounds used for synthesizing the converter is included (Fig. 1), and the different procedures to place the down-converter on the PV module. Secondly, a description of the testbed and the ambient conditions is exposed in order to show the harsh environment prevalent in the area. Finally, first results in terms of power with temperature correction for the different PV minimodules are exposed and a slight increase is observed for the minimodules covered with the down-converters.

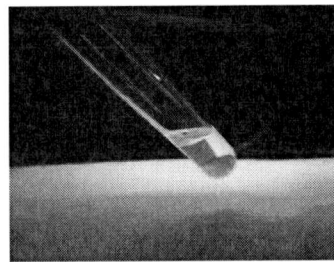

Fig. 1.   Solution of the active specie of the down-converter under UV light.

II. TESTBED AND AMBIENT CONDITIONS

The testbed has been placed outdoor in the Instituto Tecnológico de Energías Renovables SA (ITER), in the south of Tenerife (28°04'16.48"N 16°30'48.15"W). Average daily global horizontal irradiation reaches 4.822 kWh/kW (1,761 kWh/kW·yr), and average temperature reaches 19.8° C.(Fig. 2). Average wind speed in the area is 7.58 m/s (80 m in height) [3].

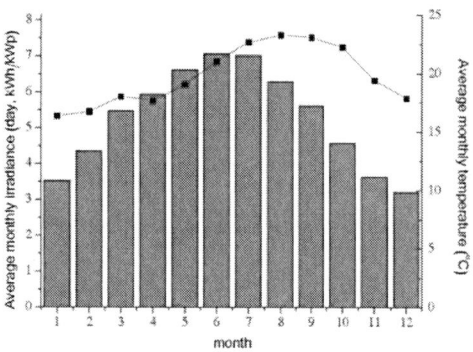

Fig. 1. Average monthly irradiance (bars) and temperature (symbols) in the testbed location.

The Sahara Desert is the world's largest source of Aeolian soil dust [4], and affects air temperatures through the absorption and scattering of solar radiation. Saharan haze dust can have variable duration in terms of a few days or extended to weeks. The aerosol optical thickness (a quantitative measure of the extinction of solar radiation by aerosol scattering and absorption between the point of observation and the top of the atmosphere), haze dust density and total dust load are the parameters used for characterizing the hazes dust events. Aerosol optical thicknesses at 550 nm up to 0.8, haze dust densities up to 320 g/m^3 and total dust load up to 500 mg/m^2 has been measured in the area [5]. Moreover, contrary to expected, during the Saharan hazes dust invasions the relative humidity increases (up to more than 90%, while in non-hazes days the average humidity is around 65%). This increment is interpreted as an adhesion of water molecules to the Saharan haze dust during its trip over the Atlantic Ocean, acting as seed for cloud formation [4].

Fig. 2. Image of the minimodules manufacture (left) and testbed in ITER (right).

The testbed is composed by 8 minimodule positions (Fig. 2). Each minimodule (260 x 260 mm) has been produced in ITER installations and is composed by a mc-Si solar cell encapsulated in EVA and framed in aluminum. Every minimodule is connected to a data acquisition card, placed inside of the junction-box, where the electrical performance, I-V curve and module temperature are been recorded. The tilt and azimuthal angles of the minimodules can be manually adjusted.

The testbed is connected to a Supervisory Control And Data Acquisition system (SCADA), where solar parameters and I-V curves are recorded every minute. All components and SCADA system have been developed in ITER.

Thus, this testbed has been designed to be an excellent system to foretell the behavior of any PV component or installation that may be placed in harsh conditions in any desert.

III. DOWN CONVERTER SYNTHESIS AND DEPOSITION PROCEDURE ON THE MINIMODULES

Three minimodules have been covered in the laboratory with a down shifter layer containing 90% of PMMA and 10% of active species which consisted of an Eu(III).

The down converter layers have been prepared from Europium(III) nitrate pentahydrate (99.99%), 4,7-biphenyl-1,10-phenanthroline (bphen, 97%), thenoyltrifluoroacetone (Htta, 99%), triethylamine (99%), poly(methylmethacrylate), (PMMA, average Mw 996,000, ref. 182265). All reagents for the synthesis of the active species have been used as received to preserve their properties.

Film casting or spin coating methods were used for film deposition on commercial PV glass substrates (Pilkington). The first module (Module A) has been covered with a layer with 10% [Eu(phen)(tta)3] and a thickness between 50 and 100 μm. The second module (Module B) has been covered with 10% [Eu(bath)(tta)3] and a thickness between 100 and 150 μm. Finally, the last module (Module C) have been covered with 10% [Eu(bath)(tta)3] with a thickness between 50 and 100 μm.

The other minimodules are maintained without any treatment to be used as reference for the characterization of the minimodules covered with the down-converter layers.

In order to compare the PV production, the solar panels performance have been corrected with the temperature and referred to 25°C.

The solar minimodules have been integrated in a house called "La Geria" (Fig. 3), one of the 25 bioclimatic buildings located also at ITER. Currently, a 3.06 kW_p PV plant on the roof of this house, oriented to the SE and 10° tilt is providing 5,202 kWh/yr. The aim of this project is to evaluate the benefits to cover the installed PV power plant with down converter layers to maximize the energy production, without increase the weight of the installation on the roof.

Fig. 3. Bioclimatic house "La Geria", where the minimodules with down-converters and showing the best performance have been integrated.

III. RESULTS AND DISCUSSION

Average power produced by the PV minimodules slightly increases when are covered with the down-converters layers deposited.

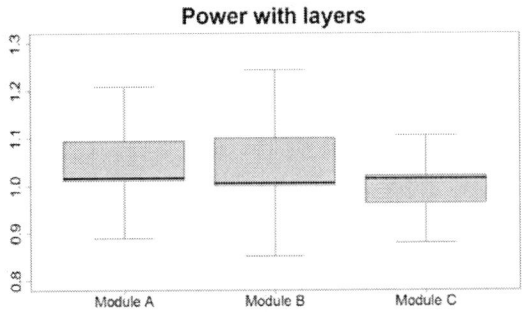

Fig. 4. Boxplot of the normalized and temperature corrected output power from the solar minimodules with down-converter layers.

In figure 4, the normalized out power of three solar minimodules (A, B and C) during the first operation day with the down converter layers are shown.

The statistical analysis reveals that, in comparison with the average reference panel, the Module A presents an increment of 1.6%, Module B of 0.497% and Module C of 1.47%.

The statistical deviations during the day are attributed to the variations detected with diffuse radiation because the down converter layers are not flat when they are deposited on the surface of the minimodules.

These increments are directly related with the improvement of the photocurrent generated in the solar cells due to the EQE increment in the UV range [1]. To corroborate this effect, the IV curve of the minimodules has been obtained using the AMSE [6] procedure. All the curves have been obtained simultaneously, in order to assure the same radiation level and therefore, to obtain an accurate comparative of the electrical performance of the minimodules. In the figure 5, the IV curves from the three modules under study and the reference cell are shown.

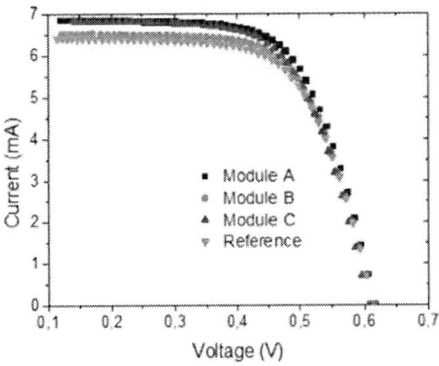

Fig. 5. IV curves from the modules with the down converter layers compared with the reference cell.

The modules with down converter layers present major short current intensity in comparison with the reference module. The increments obtained from the statistical analysis of the power match with the I_{sc} increment detected in the IV curves from the minimodules. However, after aging for some days the conversion layers show early degradation that requires further analysis.

IV. CONCLUSIONS

In this work, the behavior of down-converter layers tested on harsh outdoor conditions is reported for first time. Two different compounds and thicknesses have tested in order to increase the EQE values and therefore the outpower of the solar minimodules. Power increment have detected in all the samples when thin down converter layers have been placed. After the exposure to harsh outdoor conditions for some days, the minimodules with down-converter present early degradation of the conversion layer that requires further analysis.

978-1-5090-5606-4/17 $31.00 © 2017 IEEE

V. ACKNOWLEDGEMENTS

This work has been supported by the Ministerio de Ciencia e Innovación, Spain (Project ENE2013-41925R), co-supported by the European Social Fund and by the Fundación Cajacanarias (Project ENER10).

REFERENCES

[1] T. Monzón-Hierro, J. Sanchiz, S. González-Pérez, B. González-Díaz, S. Holinski, D. Borchert, Cecilio Hernández-Rodríguez and Ricardo Guerrero Lemus, "A new cost-effective polymeric film containing an Eu(III) complex acting as UV protector and down-converter for Si-based solar cells and modules," *Solar Energy Materials & Solar Cells* vol. 136, pp. 187 - 192, 2015.

[2] S. González-Pérez, J. Sanchiz, B. González-Díaz, S. Holinski, D. Borchert and R. Guerrero-Lemus, "Luminescent polymeric film containing Eu(III) complex acting as UV protector and down-converter for Si-based solar cells and modules," Surface & Coatings Technology vol. 271, pp. 106-111, 2015.

[3] Sistema de Información Territorial de Canarias. Available on 16 January 2017 at http://visor.grafcan.es/visorweb/default.php?svc=svcStatISTAC &lat=28.3&lng=-15.799999999999954&zoom=8&lang=es

[4] A.S. Goudie, N.J. Middleton, Earth-Science Reviews vol. 56, pp. 179-204, 2001.

[5] C. Montes, B. González-Díaz, A. Linares, E. Llarena, O. González, D. Molina, A. Pío, M. Friend, M. Cendagorta,, J.P. Díaz and F.J. Expósito, "Effects of the Saharan dust in the performance of multi-MW PV grid connected facilities in the Canary Islands (Spain), 25th European Photovoltaic Solar Energy Conference and Exhibition, pp. 5046-5049, 6-10 September 2010, Valencia, Spain.

[6] B. Paviet-Salomon, J. Levrat, V. Fakhfouri, Y. Pelet, N. Rebeaud, M. Despeisse, and C. Ballif, "New guidelines for a more accurate extraction of solar cells and modules key data from their current-voltage curves," Prog. Photovoltaics Res. Appl., 2017.

Study of micro-structural properties of ZnO and TiO₂ thin films grown by spray pyrolysis

G. Gordillo, J.M. Correa , A.A. Ramirez and E. A. Ramírez

National University of Colombia, Bogotá, ggordillog@unal.edu.co, Colombia

Abstract — **Micro structural and optical properties of ZnO and TiO2 films prepared by spray pyrolysis were studied by XRD (X-ray diffraction) analysis and measurements of transmittance and Urbach Energy. The influence of preparation conditions on the micro structural properties of ZnO and TiO2 films was investigated throug the evaluation of the X-ray peak broadening. The Williamson Hall (W-H) analysis and size strain plot method were used to study the individual contributions of crystallite sizes and lattice strain on the peak broadening. The physical parameters such as strain, stress and energy density values were calculated for all the reflection peaks of XRD corresponding to both type of samples from the modified form of the W-H plot assuming a uniform deformation model (UDM), uniform stress deformation model (USDM) and uniform deformation energy density model (UDEDM). It was found that the particle mean size of the ZnO and TiO2 films estimated from the W-H method agree quite good with AFM results. Further, information regarding the influence of preparation conditions on the formation of structural defects was achieved through Urbach energy measurements.**

Index Terms — ZnO, TiO2, thin films, spray pyrolysis, Urbach energy , micro-structural properties.

I. Introduction

Thin films of ZnO and TiO₂ have been used for many applications including solar cells [1,2]. Recently these compounds also began to be used as electron selective (electron transport/hole blocking layer) layer in inverted organic solar cells [3] and as electron transport layer in perovskite based hybrid solar cells [4,5]. These layers are usually prepared by spin-coating or spray-pyrolysis using a titanium or zinc precursor solutions [6-8]. The hole-blocking layers, which contribute to transport electron and block hole to electrode, are commonly applied to solar devices to avoid recombination at the interface between the electrode and light absorber. Optimization of the hole-blocking layer has drawn increasing attention as it can significantly improve the performance of perovskite solar cell.

In the present work, we have investigated the effect of deposition conditions such as the nature of precursors and substrate temperature on the formation of structural defects as well as on the structural and optical properties of sprayed ZnO and TiO₂ films.

II. Experimental

ZnO and TiO₂ films were deposited inside a glove box by spray pyrolysis using nitrogen as gas of transport. The experimental setup designed and implemented for this study, includes the following units:

i) Nozzle with facilities to perform a fine control of the solution flow.

i) Electronic control of amplitude and oscillatory frequency of the nozzle

ii) Nitrogen Generator with facilities to control the pressure of the transport gas that is injected to the nozzle

iii) PID substrate temperature controller.

The best conditions for the synthesis of ZnO and TiO2 films were found through a parameter study carried out following a methodology which includes initially a study to optimize the molar concentration of the precursor salts ((Zn(Ac)₂ and ZnCl₂ for ZnO and TiCl4 for TiO2), using as reference data of synthesis parameters reported in the literature. Subsequent to the optimization of the chemical composition of the work solution, in a second phase, there was realized a study that allowed to optimize the substrate – nozzle separation, the solution flow, the pressure of transport gas and the frequency of the oscillatory displacement of the nozzle. The optimization of synthesis parameters was achieved correlating these parameters varied in a wide range with measurement of x-rays diffraction and spectral transmittance.

Finally there was optimized the substrate temperature (T_s), that is the parameters that more critically affect the properties of the ZnO and TiO2 films. This parameter study was done varying T_S between 350 and 450°C.

The best conditions to grow ZnO and TiO2 films with good properties, which resulted from the parameters study are the following:

i) The ZnO films were grown using a solution prepared from both zinc acetate (Zn(Ac)₂) and ZnCl₂ as precursor salts. The solution includes 0.1 M zinc acetate (or 0.1 M ZnCl₂) in a mixture of deionised water and ethanol with the ratio of (2:3) in 100 ml. The TiO2 films were deposited from a solution prepared dissolving 1.5 g of TiCl4 in a solution containing 72.94 g of deionized water, 5.5 g of peroxide (H_2O_2) and 1.65 g of $H_2C_2O_4$ to obtain a solution 0.1 m of $Ti4^+$.

ii) Substrate temperature of 400° C, nozzle height of 20 cm, solution flow of 0,75 ml/s, transport gas pressure of 4 bar and a frequency of the oscillatory displacement of the nozzle of 0.6 cps, interruption of solution supply during 60 seconds after every cycle.

Transmittance and reflectance measurements were performed using a Varian – Cary 5000 spectrophotometer and the resistivity was done using the four probe method with contacts placed on the surface of the sample and the film thickness was determined using a Veeco Dektak 150 surface profiler. The XRD (x-ray diffraction) measurements were performed in a Shimadzu-6000 diffractometer.

978-1-5090-5606-4/17 $31.00 © 2017 IEEE

III. RESULTS AND DISCUSSION

The influence of substrate temperature T_s, on the optical, electrical, structural and morphological properties of both ZnO and TiO2 films was evaluated through measurements of transmittance, resistivity, XRD and AFM. Fig. 1 show the influence of Ts and film thickness on the transmittance spectra of ZnO and TiO$_2$ films deposited by spray pyrolysis, using solutions containing ZnCl$_2$ and Zn(Ac)$_2$ as precursor salts for the ZnO films and TiCl$_4$ for the TiO$_2$ films. The following facts can be highlighted from the results shown in figure 1:

i) All the transmittance spectra of the ZnO films deposited by spray pyrolysys from solutions based on both ZnCl$_2$ and Zn(Ac)$_2$ present the same cutoff wave length, indicating that the energy gap Eg of the ZnO films is not affected by the substrate temperature, as well as nor by the thickness and the type of precursor salt used. It was found that the ZnO films exhibit an Eg value (obtained from the intercept with the axis hv of the curve of $(\alpha h\nu)^2$ vs hv) of about 3,3 eV.

ii) The ZnO films deposited from a solution based on (Zn(Ac)$_2$) varying both the substrate temperatures and film thickness present high transmittance (around 90%) and high slope indicating that in this type of samples the absorption of photons occur mainly via fundamental transitions and that the ZnO films grow with a low density of intrinsic and structural defects. In contrast, the slope of the transmittance curves of ZnO films deposited from ZnCl$_2$ significantly decreases by increasing of film thickness; however the slope is slightly affected by the substrate temperature. The effect of thickness on the slope of transmittance curves of ZnO films grown from ZnCl$_2$, seem to be caused by absorption of photons in the states of tails of bands associated with structural defects.

iii) The TiO$_2$ films exhibit an energy gap Eg of 3.5 eV; this value is not affected by the substrate temperature, nor by their thickness.

iv) In general, the TiO$_2$ films present high transmittance (around 90%) and high slope independently of the deposition temperature (in the studied range) indicating that in TiO$_2$ films the absorption of photons occur via fundamental transitions as it happens in sprayed ZnO films. However, the slope of the transmittance curves of TiO$_2$ films significantly decreases by increasing of film thickness, indicating that the increase of the thickness of the TIO$_2$ films induces an increase in the density of structural defects, generating in this way absorption via states of tail of bands.

Absorption via transitions between states of tails of bands was evaluated through the calculation of the absorption coefficient near the band edge (α_U) that has an exponential dependence on photon energy and Urbach energy [9] ($\alpha_U = \alpha_0 exp[h\nu - E_i/E_u]$). Where E_U is the Urbach energy, E_1 and α_0 are constant. Thus, a plot of lnα vs. hv should be linear and Urbach energy can be obtained from the slope. In Table 1 are given results that show the effect of substrate temperature and film thickness on the Urbach energy of both ZnO and TiO2

films. Taking into account that Urbach energy is related to density of localized states in the band gap induced by structural disorder (stress and dislocation) [9], can be concluded that the ZnO films prepared from (Zn(Ac)$_2$ present an improved crystallinity, as compared with those of ZnO films deposited from ZnCl2 and TiO2 films. It is also observed that the substrate temperature affects very little the Urbach energy of both the ZnO and TiO2 layers, while the value of E_U increases significantly when its thickness increases. These results allow to conclude that the crystallographic properties of the ZnO and TiO2 films are slightly affected by the temperature of synthesis, whereas an increase in thickness induces significant deterioration of these properties.

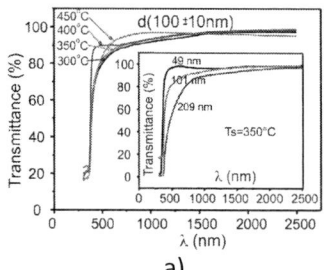
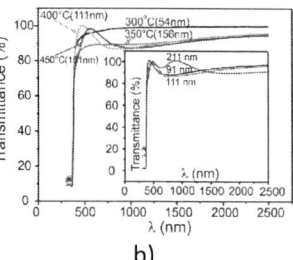

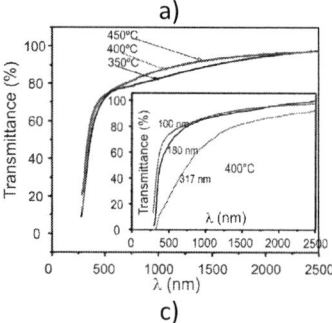

Fig. 1: Influence of substrate temperature on the transmittance of ZnO films a) deposited from ZnCl$_2$, b) deposited from Zn(Ac)$_2$ and c) of TiO$_2$ films. In inset are displayed Transmittance spectra of ZnO and TiO2 films showing the effect of film thickness.

The structural properties of the ZnO and TiO$_2$ films were also evaluated through XRD measurements. In Fig. 2 are displayed XRD spectra of ZnO and TiO$_2$ films prepared by spry pyrolysis varying the substrate temperature maintaining constant the rest of parameters whose optimized values are given above. It is observed that the ZnO films prepared using both type of precursor salts present Bragg peaks of (002),(100), (101) ,(102), (103) and (112) which belong to the ZnO phase with hexagonal structure (PDF data file # 01-0790208), whereas the TiO2 films present reflections along the,(101), (101) ,(002), (105) and (211) planes which belong to the TiO2 anatase phase with tetragonal structure (pdf card #01-0841286). It is observed that the ZnO films prepared from ZnCl$_2$ based solution grow with preferred orientation along c-axis, i.e. (002) plane and that the substrate temperature does not induce changes in the orientation of the planes of reflection; On the contrary, the orientation of the planes of reflection in samples of ZnO prepared from (Zn(Ac)$_2$) is affected by the deposition temperature. On the other hand however, at above 400°C the ZnO films prepared

TABLE 1: Influence of substrate temperature and film thickness on the Urbach Energy of ZnO and TiO2 films prepared by spray pyrolysis.

	ZnO from ZnCl$_2$						ZnO From Zn(Ac)$_2$	TiO$_2$					
	T$_S$ (°C) Film thickness = 50 nm			Thickness (nm) T$_S$=350°C				T$_S$ (°C) Film thickness = 100 nm			Thickness (nm) T$_S$=400°C		
	350	400	450	50	100	200		350	400	450	100	180	317
E$_U$ (meV)	151.3	105.6	124.5	165.1	266.2	602.3	54.7	704.4	687.4	688.4	687.4	893.9	1125.1

from (Zn(Ac)$_2$) tend to grow with preferential orientation along the (002) plane.

It was found that the full width at half maximum (FWHM) value of the studied ZnO and TiO2 films is affected by the deposition temperature, indicating that this parameter affect their crystallinity. XRD can be utilized to evaluate peak broadening with crystallite size D and lattice strain ε due to dislocation [10]. The particle size of the ZnO and TiO2 can be determined by the X-ray line broadening method using the Scherrer equation [11]: $D = \frac{K\lambda}{\beta_{hkl}cos\theta_{hkl}}$, where D is the particle size in nanometers, λ is the wavelength of the radiation for CuKα radiation), k is a constant equal to 0.94, β_{hkl} is the peak width at half-maximum intensity and Θ_{hkl} is the peak position. The XRD peak broadening is due to both instrument and sample dependent effects. The contributions of each of these effects are convoluted causing an overall broadening of the diffraction peaks. To decouple these contributions, it is necessary to correct the instrumental effect from the line broadening of a standard material to determine the instrumental broadening. The instrumental corrected broadening β_{hkl} was estimated using the equation [12]:

$$\beta_{hkl} = \left[(\beta_{hkl})^2_{measured} - (\beta)^2_{instrumental} \right]^{1/2} \quad (1)$$

In addition to the instrumental X-ray peak broadening, lattice strain ε and crystallite size are the other two independent factors that contribute to the total peak broadening. The strain induced line broadening β_s is given by the relation $\beta_s = 4\varepsilon tan\theta_{hkl}$. Now the total peak broadening is represented by the sum of the contributions of crystallite size and strain present in the material. Assuming that the strain present in the material is uniform in all crystallographic directions, the **Williamson-Hall equations** [13] for the total peak broadening is given by:

$$\beta_{hkl}cos\theta_{hkl} = \frac{K\lambda}{D} + 4\varepsilon sin\theta_{hkl} \quad (2)$$

Therefore, Eq. (2) represents the UDM (Uniform Deformation Modell), where the strain was assumed to be uniform in all crystallographic directions. The term $\beta cos\theta$ was plotted with respect to $4sin\theta$ for the preferred orientation peaks of ZnO and TiO$_2$ films. Accordingly, the slope and y-intersect of the fitted line represent strain and particle size, respectively.

In the Uniform Stress Deformation Model (USDM), a generalized Hooke's law refers to the strain, keeping only the linear proportionality between the stress and strain as given by $\sigma = Y\varepsilon$, where σ is the stress of the crystal and Y is the modulus

of elasticity or Young's modulus. This equation is an approximation that is valid for a significantly small strain. Applying the Hooke's law approximation to Eq. (2) yields:

$$\beta_{hkl}cos\theta_{hkl} = \frac{K\lambda}{D} + \frac{4\sigma sin\theta_{hkl}}{Y_{hkl}} \quad (3)$$

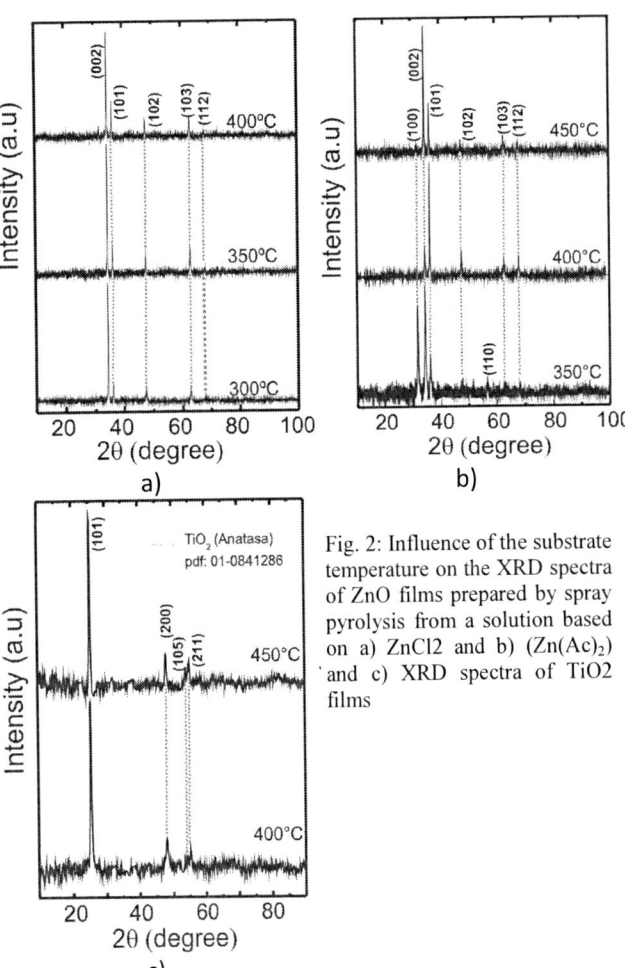

Fig. 2: Influence of the substrate temperature on the XRD spectra of ZnO films prepared by spray pyrolysis from a solution based on a) ZnCl2 and b) (Zn(Ac)$_2$) and c) XRD spectra of TiO2 films

There is another model that can be used to determine the energy density of a crystal called the Uniform Deformation Energy Density Model (UDEDM). When the strain energy density u is considered the constants of proportionality associated with the stress-strain relation are no longer independent. For an elastic system that follows Hooke's law,

TABLE 2: Influence of substrate temperature on the grain size, stress, strain and energy density of ZnO and TiO$_2$ films prepared by spray pyrolysis.

| Sample | Temp. substrate | Scherrer method | Williamson - Hall method | | | | | | | | | | AFM |
| | | | UDM | | USDM | | | UDEDM | | | | |
		D (nm)	D (nm)	ϵx10^{-3}	D (nm)	ϵx10^{-3}	σ (MPa)	D (nm)	ϵx10^{-3}	σ (MPa)	u (KJm^{-3})	D (nm)
ZnO (ZnCl$_2$)	350	170.36	163.58	1.4	145.05	1.2	156.0	144.12	1.2	156.0	93.60	188 ± 66
	400	192.94	183.24	0.9	171.37	0.9	117.0	169.80	1.0	130.0	75.00	202 ± 56
	450	219.17	210.50	0.2	196.83	0.2	26.0	192.32	0.2	26.0	2.60	208 ± 70
ZnO (Zn(Ac)$_2$)	350	122.99	112.98	2.1	108.38	2.0	260.0	108.01	1.9	247.0	235.60	145 ± 39
	400	152.78	138.21	1.6	131.92	1.5	195.0	129.53	1.5	195.0	146.20	169 ± 42
	450	172.58	155.88	1.2	160.08	1.1	143.0	155.14	1.1	143.0	78.65	186 ± 42
TiO$_2$	400	148.56	146.85	2.3	145.97	2.2	481.8	143.52	2.2	481.8	529.98	160 ± 60
	450	152.34	144.58	2.0	142.51	2.1	459.9	141.76	2.0	438.0	438.00	176 ± 45

the energy density u can be calculated from u = ($\epsilon^2 Y_{hkl}$)/2. Then Eq. (3) can be rewritten according the energy and strain relation.

$$\beta_{hkl} cos\theta_{hkl} = \frac{K\lambda}{D} + \left(4 sin\theta \left(\frac{2u}{Y_{hkl}} \right)^{1/2} \right) \qquad (4)$$

Plots of $\beta_{hkl}cos\theta$ versus 4 $sin\Theta/(Y_{hkl})^{1/2}$ were constructed and the data fitted to lines. The anisotropic energy density u was estimated from the slope of these lines, and the crystallite size D from the y-intercept.

The results obtained from the Scherrer method, UDM and USDM, UDEDM, models and AFM are summarized in Table 2. For that, we used values of Y= 130 GPa and Y= 219 GPa reported in the literature for ZnO hexagonal [14] and for TiO2 films grown with anatasa structure respectivelly [15].

IV. CONCLUSIONS

The influence of deposition temperature on the optical and structural properties of ZnO and TiO2 thin films prepared sprary pyrolysis were evaluated through transmittance and X-ray diffraction measurements using Williamson–Hall-isotropic strain model and Urbach energy estimated from absorption coefficient calculated near the band edge. It was found that ZnO films prepared from (Zn(Ac)$_2$ present an improved crystallinity, as compared with those of ZnO films deposited from ZnCl$_2$ and TiO$_2$ films. It is also observed that the substrate temperature affects very little the Urbach energy of both the ZnO and TiO$_2$ layers, while the value of E$_U$ increases significantly when its thickness increases. These results allow to conclude that the crystallographic properties of the ZnO and TiO$_2$ films are slightly affected by the temperature of synthesis, whereas an increase in thickness induces significant deterioration of these properties. It was also found that the strain and stress values decreased but the particle size increased as subatrate temperature was increased.

ACKNOWLEDGEMENTS: This work was supported by Colciencias (Contract #184/2016) and Universidad Nacional de Colombia, Sede Bogotá, Facultad de Ciencias, Departamento de Física, Grupo GMS&ES, K30 #45-03, Bogotá DC, Colombia.

REFERENCES

[1] M. A. Green , K. Emery , Y. Hishikawa , W. Warta and E. D. Dunlop, *Prog. Photovolt.Res. Appl.*Vol 23, no. 7, pp 805-812, 2015, 2015.

[2] L. Mingzhen, M. B. Johnston and H. J. Snaith, Efficient planar heterojunction perovskite solar cells by vapour deposition , Nature , **501**, 395-398 (2013).

[3] Chien-Hung Chiang and Chun-Guey Wu, Bulk heterojunction perovskite–PCBM solar cells with high fill factor, *Nature Photonics*, **10**,196–200(2016).

[4] G. Hodes, Perovskite-based solar cells, Science 2013; 342:317–318.

[5] O. Malinkiewicz, A. Yella, Y. Hui Lee, M. Graetzel, M. K. Nazeeruddin & H. J. Bolink, Perovskite solar cells employing organic charge-transport layers, *Nature Photonics* 8, 128–132 (2014)

[6] F. Zahedi, R.S. Dariani, S.M. Rozati, Effect of substrate temperature on the properties of ZnO thin films prepared by spray pyr olysi, Materials Science in Semiconductor Processing 16 (2013) 245–249

[7] J. Duan, J. Wu, J. Zhang , Y. Xu , H. Wang, D. Gao and P. D. Lund, TiO2/ZnO/TiO2 sandwich multi-layer films as a holeblocking layer for efficient perovskite solar cells, Int. J. Energy Res. 2016; 40:806–813.

[8] X. Yin, Zh. Xu, Y. Guo, P. Xu, and M. He, Ternary Oxides in the TiO2−ZnO System as Efficient ElectronTransport Layers for Perovskite Solar Cells with Efficiency over 15% , ACS Appl. Mater. Interfaces 2016, 8, 29580−29587.

[9] M.V. Kurik, Review of Urbach´s tail. Phys. *Status Solidi A*, vol. 8, No. 9, 1971.

[10] R. Yogamalar, R. Srinivasan, A. Vinu, K. Ariga, A.C. Bose, Solid State Commun. 149 (2009) 1919

[11] P. Scherrer, Nachrichten von der Gesellschaft der Wissenschaften zu Göttingen, Mathematisch-Physikalische Klass, PHIS Z. Volume: 1918, Page 98-100

[12] R. Srinivasan, R. Yogamalar, R.J. Josephus, A.C. Bose, Funct. Mater. Lett. 2 (2009) 1.

[13] G.K. Williamson,W.H. Hall, X-Ray line Broadening from filed Aluminium and Wolframium, Acta Metall. 1 (1953) 2231

[14] A. Khorsand , W.H. Abd. Majida, M.E. Abrishami , R. Yousefi, X-ray analysis of ZnO nanoparticles by Williamson Hall and size estrain plot methods, Solid State Sciences 13 (2011) 251-256

[15] A. K. Prasad , R. Jha , R. Ramaseshan , S. Dash , I. Manna and A. K. Tyagi, Comparison of microstructure and electronic properties of TiO2 thin films grown by different techniques, Surface Engeneering Vol.27, **5**, (2011), 350-354

Nonlinear Response of Silicon Solar Cells

Behrang H. Hamadani[1], Andrew Shore[1], Howard W. Yoon[1], and Mark Campanelli[2]

[1]National Institute of Standards and Technology, 100 Bureau Drive, Gaithersburg, MD 20899
[2]Intelligent Measurement Systems LLC, Bozeman, MT 59715

Abstract — We used an LED-array-based combinatorial flux addition method to explore the wavelength and the intensity-dependence of the spectral responsivity in silicon solar cells. Many types of silicon cells, whether mono- or multi-crystalline type, exhibit notable nonlinear behavior of current with light intensity at illumination intensities below 0.01-sun equivalent levels. This effect is particularly pronounced when exposed to near-infrared light close to the peak of the spectral responsivity. However, some cases show the perseverance of nonlinearity all the way up to 1-sun intensity levels. In such cases, use of these cells as reference cells has implications for the accuracy of electrical performance measurements.

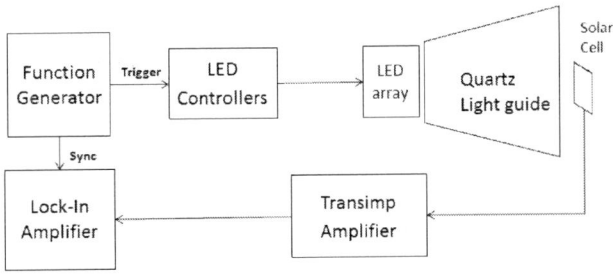

Fig. 1. Schematic of the nonlinearity measurement apparatus

I. INTRODUCTION

In an *ideal* photovoltaic (PV) solar cell, a linear relationship exists between the incident irradiance flux on the solar cell and the resultant photogenerated current output. Therefore, increasing the total irradiance by a factor of x should also result in a factor of x increase in the short circuit current (I_{sc}) of the PV device. However, most *real-world* PV devices do not follow this simple rule; rather, they show a nonlinear behavior that is both dependent on the spectrum of the incident radiation as well as the intensity of the illumination source itself. For most silicon solar cells that were evaluated in our laboratory, the nonlinear behavior described here appears at longer wavelengths (i.e., 600 nm to 1000 nm) and even then, it is generally revealed at very low intensities. However, there are unusual cases where a significant nonlinear behavior is observed even at incident intensities close to the standard reporting conditions (SRC), i.e., 1-sun equivalent intensities.

Traditionally, nonlinear behavior in solar cells and optical detectors have been evaluated by well-established methods such as the differential spectral responsivity (DSR) or AC-DC method [1], and the superposition method such as the two-lamp flux addition technique [2]. The DSR method can indeed be used to determine the actual mathematical relationship between the signal s (generally the I_{sc}) and the flux φ. However, the technique is time-consuming and the AC/DC current separation, especially at large ratios, presents a significant electrical engineering challenge. Also, the DSR method is usually implemented by use of a monochromator and it is difficult to change or modify the intensity of the monochromatic beam. Some superposition methods, such as the two-lamp method, based on comparing the ratio of the individually-added photocurrents obtained from two light sources to their combined two-source output, can reveal device

nonlinearity, but do not provide any insight on the actual nonlinear relationship.

In this work, we demonstrate the feasibility of using a combinatorial flux addition technique [3] based on the use of two sets of light emitting diode (LED) arrays to accurately measure the nonlinear behavior of a variety of silicon based solar cells, over a large range of signals (by controlling the intensities) and wavelength. Our results clearly indicate that linearity should not be automatically assumed when evaluating the performance of a solar cell under a given light intensity, or when using it as a reference cell for intensity measurements at conditions other than what it was originally calibrated or tested.

II. EXPERIMENTAL DETAILS

A schematic of the LED-based combinatorial approach is shown in Fig. 1. The measurement requires illumination of the cell by two *identical* LEDs (or sets of LEDs) in singular and a combinatorial fashion. To achieve this, commercial LED controllers with multiple independent current source channels were used to run each LED according to a prescribed schedule [4]. For each LED setting, the I_{sc} from the device under test (DUT) is recorded by way of a transimpedance preamplifier connected to a lock-in amplifier. All LEDs were operated

Fig. 2. The LED array as seen through the light guide

978-1-5090-5606-4/17 $31.00 © 2017 IEEE

in pulsed mode through a trigger signal applied to the LED controllers by a function generator. Pulsing ensures better measurement stability and less drift in the LED signal during the course of the measurement. The light from the LEDs is coupled into a solid-glass, tapered light guide to ensure homogeneity and uniform illumination at the cell location. The nonuniformity for each LED at the cell location across the 50 mm exit port is up to 10 %. Fig. 2 shows an image of one such LED array as seen looking into the light guide from the front of the exit port. Both the circuit-mounted LED array and the light guide are custom designed and fabricated for our unique measurement needs.

For the data presented here, 15 unique current levels are sourced to LED 1 in a sequential way while the I_{sc} is recorded for each. This sequence is followed by 15 other current levels to LED 2 in a similar manner along with recording of the I_{sc}. Then combinations of these currents are applied to both LEDs simultaneously, usually in a manner whereby each current from source 1 is paired with 3 currents from source 2. Therefore, 45=15×3 combination fluxes are supplied and 45 signals are

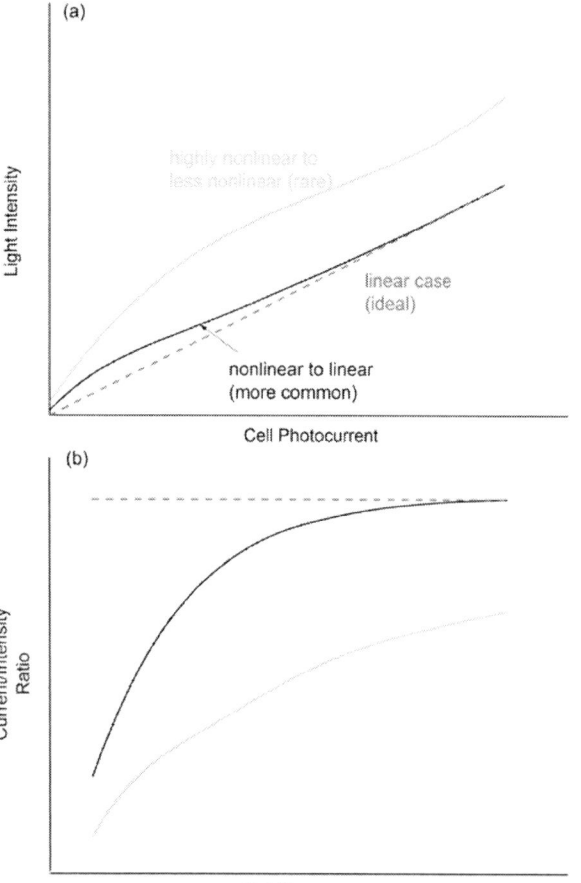

Fig. 3 (a) Two common forms of nonlinear behavior observed with Si cells (b) Representing nonlinearity with ratio plots.

obtained from the cell. Thus, 75=15+15+45 total data points are measured in one run.

III. ANALYSIS AND RESULTS

A. Mathematical Framework

The objective is to use the combinatorial signal data to construct a linear system of equations relating signals and fluxes and solve for the unknown flux values and coefficients. This system of equations is overdetermined because the total number of unknowns is far smaller than the total number of equations (or signals). Assuming an Nth-degree polynomial model for flux ϕ (say, in W/m^2) as a function of short-circuit current signal s (say in A DC), i.e.,

$$\phi = r_0 + r_1 s + r_2 s^2 + ... + r_N s^N \quad , (1)$$

we can then write K equations for the K distinct fluxes from LED source 1, J equations for the J distinct fluxes from LED source 2, and M equations for the M combinations of the fluxes from both sources 1 and 2. As described in detail previously [4], these equations can be made *related* and solved by finding a linear least squares solution (or can be solved using a matrix-based approach) and the fluxes and the ratio of signal to flux values can be calculated for each signal value. Depending on the severity of the nonlinear behavior, it may become necessary to solve for the r_N coefficients up to $N = 5$ or more. The residuals of the fit determine the order, N. This solution remains unscaled. However, it is possible to obtain a scaled relationship if one or more calibrated flux values were known. Flux calibration can be achieved by using a reference photodetector with a known irradiance spectral responsivity. For most practical cases, it is sufficient to plot the ratio of signal to flux as a function of signal and observe whether it is constant or not. Constant cases correspond to a linear response whereas a changing ratio corresponds to a nonlinear behavior. It is noted that a signal to flux ratio is essentially an unscaled spectral responsivity.

B. Nonlinear vs Linear Behavior

Fig. 3 demonstrates the concept of nonlinearity and a convenient way to plot such data. In Fig. 3a, we show an exaggerated schematic of 3 typical situations observed with silicon solar cells in our measurements. In the ideal case, we observe a linear relationship between the light intensity and the cell photocurrent. If plotted as a function of the ratio of current to light intensity vs. current, a constant plot is observed as shown by the dotted red line in Fig 3b. A majority of Si solar cells that we have measured, however, behave as shown by the black curve. For these cells, the light intensity rises super-linearly with respect to the current at low intensity but then trends linear at higher intensities. The point of transition from nonlinear to linear behavior varies among different cells and is likely related to the material quality and the influence of charge carrier recombination mechanisms. The light intensity for this transition can be as low as 0.0001 sun-equivalent intensity or as high as 0.1 (or more) sun-equivalent intensity. The ratio plot in this case (black curve in Fig 3b) shows a ratio that increases with current and reaches steady state above a certain current or intensity value. Finally, there are rare cases represented by the green curve where one could see a ratio plot that initially rises rapidly with current, followed by a more moderate rise at higher intensities. Some of these cases never show a leveling-off behavior even at the highest intensities at which we have irradiated the cells. Also, these types of cells generally have lower I_{sc} output and inferior current-voltage performance, although clearly these characteristics improve with intensity.

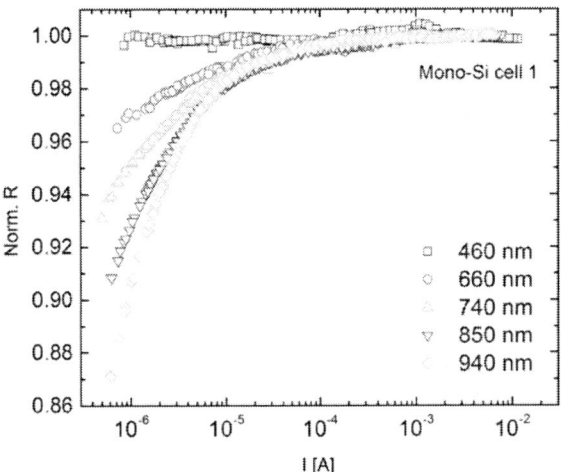

Fig. 5. The ratio of signal to flux plotted as a function of signal for the mono-Si cell 1 at multiple wavelengths.

10^{-6} A to 10^{-4} A, there is approximately a 10 % nonlinearity, meaning the output of the device is 10 % lower at the lower signal level. Furthermore, the data show that at higher light intensities, the spectral responsivity (which is proportional to the external quantum efficiency of these devices) is larger than that at lower intensities. This finding has major implications for spectral response measurements. Most research groups perform such measurements using the differential spectral response method, where a monochromator and mechanical chopper are used to sweep the wavelength across a spectral range and measure the I_{sc} of the cell in response to this modulated excitation. The monochromatic light intensity in most setups is very low, typically on the order of 1 µW to 20 µW of incident power. These conditions will yield a device spectral response/EQE that is not representative of the spectral response under the standard reporting conditions, i.e., air mass 1.5, 1 kW/m^2 intensity. In such cases, a DC-operated light bias needs to accompany the modulated light source to ensure that the cell is operating within the linear regime.

For the combinatorial measurements presented here, the main uncertainty component is associated with the repeatability of the current sourcing to the LEDs and the resulting signal from the cell [4]. This repeatability uncertainty is roughly 1 %, therefore allowing this method to be used to examine nonlinear relations on the order of a 1 % change. All the other sources of uncertainty, including the LED stability and statistical variations are minimal.

Fig. 5 shows the nonlinear behavior of the mono-crystalline cell 1 for different excitation wavelengths. These data indicate that nonlinearity is dependent upon the wavelength of light. At lower wavelengths, e.g., 460 nm, the nonlinearity is nonexistent over the entire range probed here, but the nonlinearity becomes more severe at low signal levels under higher wavelength LED light. Although the nonlinearity for this cell occurs under very low light conditions and improves at higher intensities (cell

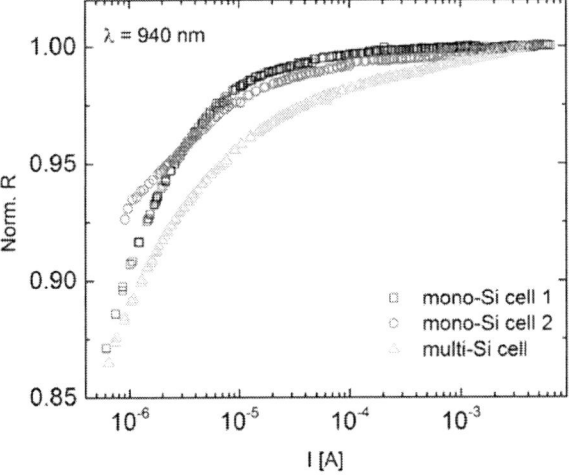

Fig. 4. The ratio of signal to flux plotted as a function of signal for 3 different types of silicon solar cells, with an excitation wavelength of 940 nm. Each cell is 2 cm × 2 cm in dimensions.

C. Measurement Results

We examined the relationship between current and intensity of a large variety of solar cells, including some reference solar cells available for purchase from private calibration laboratories. The results indicate large variations in nonlinear behavior among nominally similar silicon solar cells. Fig. 4, for example, shows the normalized ratio of signal-to-flux plotted as a function of current over a large signal range on a linear-logarithmic scale for measurements with two 940 nm LEDs. Here, the data were collected in multiple overlapping segments of signal so that a large signal range can be probed and plotted. The two mono-crystalline cells behave slightly differently, particularly under the low signal/ low intensity regime, but they both reach a saturating value, i.e., linear behavior at current values $> \approx 10^{-4}$ A. The multi-crystalline Si cell measurements, however, reveal significantly more nonlinear response at lower signals and do not appear to reach a steady state behavior even at signal values $\approx$ 10 mA (roughly 0.1 sun-equivalent for these cells). This particular cell shows that, between the signal levels

978-1-5090-5606-4/17 $31.00 © 2017 IEEE   439

becomes linear at all wavelengths), the implications of these results in practical applications are notable. For example, if such a cell were to be employed in indoor PV-powering applications [5], one indeed should be concerned with nonlinear behavior under low light conditions, particularly with light sources that contain more near-IR components in their emission spectra. As for outdoor installed PV, such types of cells would obviously have lower outputs/power conversion efficiency under low light conditions, i.e., cloudy skies, and early or late in the day, than they would closer to the SRC illumination conditions.

IV. CONCLUSIONS

Nonlinear behavior of current with light intensity in various types of silicon solar cells were measured by an LED-based combinatorial flux addition method, taking advantage of spectral and light intensity level control afforded by use of LED sources. The combinatorial measurements allow for calculation of the incident flux, and hence the ratio of signal to flux, by setting and solving an overdetermined linear system of equations. Our results indicate that linearity is strongly dependent on both the intensity of the light source and the wavelength of the illumination.

REFERENCES

[1] J. Metzdorf, "Calibration of solar cells. 1: The differential spectral responsivity method.," *Appl. Opt.*, vol. 26, no. 9, pp. 1701–1708, 1987.

[2] K. Emery, S. Winter, S. Pinegar, and D. Nalley, "Linearity testing of photovoltaic cells," *Conf. Rec. 2006 IEEE 4th World Conf. Photovolt. Energy Conversion, WCPEC-4*, vol. 2, pp. 2177–2180, 2007.

[3] D. R. White, M. T. Clarkson, P. Saunders, and H. W. Yoon, "A general technique for calibrating indicating instruments," *Metrologia*, vol. 45, no. 2, pp. 199–210, 2008.

[4] B. H. Hamadani, A. Shore, J. Roller, H. W. Yoon, and M. Campanelli, "Non-linearity measurements of solar cells with an LED-based combinatorial flux addition method," *Metrologia*, vol. 53, no. 1, pp. 76–85, 2016.

[5] R. Haight, W. Haensch, and D. Friedman, "Solar-powering the Internet of Things," *Science (80-.).*, vol. 353, no. 6295, pp. 124–125, 2016.

Extended linear interpolation/extrapolation procedure for accurate and versatile translation of the I-V curves of PV cells and modules

Y. Hishikawa[1], H. Ohshima[1], M. Higa[1], K. Yamagoe[1], T. Takenouchi[1], T. Doi[1]

[1]National Institute of Advanced Industrial Science and Technology (AIST), Japan

Abstract — Modification of the procedure of linear interpolation method is investigated, in order to improve its precision and applicability to practical I-V curve translation of PV devices for irradiance and temperature. The present study comprises of modified and extended choice of data on the I-V curves for the linear interpolation. It is demonstrated that the present procedure is especially useful for the translation of I-V curves whose voltage ranges are not wide enough, the experimental noise is relatively large, or extrapolative translation is required, which are important for practical translation.

I. INTRODUCTION

Photovoltaic (PV) devices are designed to operate under various irradiance and temperature conditions, whereas their specification is usually based on STC (Standard Test Conditions; irradiance of 1 kW/m², spectrum of air mass 1.5G, and device temperature of 25 °C) [1]. Therefore, it is practically important to translate the current-voltage (I-V) characteristics under STC to different irradiance and temperature conditions or vice versa. Various formulas to translate the I-V curves for irradiance and/or temperature were investigated so far [2]-[6]. Three kinds of procedures are described in a current international standard [7]. Among them, the linear interpolation method [6], which is described as "correction procedure 3" in ref. [7], has a feature that precise translation is possible with no predetermined parameters such as the temperature coefficient or series resistance. It is also applicable to various kinds of PV technologies [8]. However, its translation results were sometimes sensitive to experimental noise and steps in the I-V curve, which possibly resulted in increased translation error. This study proposes and demonstrates that slight modification of the procedure substantially improves its versatility and precision of the translation of those I-V curves.

II. TRANSLATION PROCEDURE

1. Basic procedure of the linear interpolation method

Translation of the I-V curves by the linear interpolation method is illustrated in Fig. 1. Its detail is described in previous publications such as [6] and [7]. Two experimental reference I-V curves are used for the translation, which are denoted as reference curves 1 (blue line) and 2 (black line) in the figure. Data points (V_1, I_1) and (V_2, I_2) are chosen on the reference curves 1 and 2, respectively, so that the difference in

their current is equal to the difference in the short circuit current, as shown in the figure,

$$I_2 - I_1 = I_{SC2} - I_{SC1} = \Delta I . \tag{1}$$

Then the data points on the translated curve, shown by a red line in the figure, is calculated [7];

$$\begin{aligned} V_3 &= V_1 + a \cdot (V_2 - V_1) \\ I_3 &= I_1 + a \cdot (I_2 - I_1) \end{aligned} \tag{2}$$

The translation is carried out by scanning the data points throughout the reference curves. Here, a is a constant which determines the ratio of interpolation. For example, when the (irradiance G, temperature T) of the reference curves 1 and 2 are (0 kW/m², 20 °C) and (1 kW/m², 30 °C) and a is 0.6, the (G, T) of the translated curve is (0.6 kW/m2, 26 °C). Since Eq. (2) is very simple and straightforward, substantive procedure to affect the precision of the translation is to choose the reference points (V_1, I_1) and (V_2, I_2), which satisfy the relation of Eq. (1).

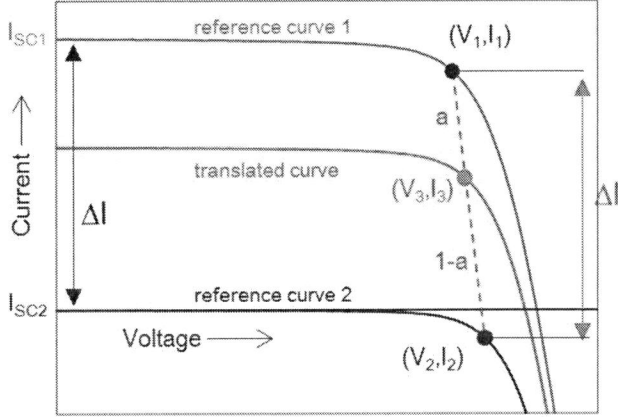

Fig. 1 Schematic basic procedure of the translation of I-V curves by the linear interpolation method.

The basic precision of the linear interpolation is very good, which usually agrees with experiments well within 1% in wide irradiance and temperature ranges, as long as precise values of irradiance, device temperature, and reference I-V curves are available [8]. However, the reference I-V curves are often affected by measurement errors such as the electrical and optical noise, nonuniform irradiance, and misestimation in the irradiance and temperature. These errors possibly result in irregular choice of the reference points, which is sensitive to the noise and steps in the reference I-V curves, as shown in Fig. 2. No significant distortion of the translated I-V curve, shown by a red line, is visible from the figure, thanks to the flexibility of the linear interpolation method. However, the irregular distribution of the reference points results in irregular voltage interval and sometimes reversed voltage interval of the translated curve, which are problematic when extrapolation is necessary, or when it is used as a reference curve for the subsequent interpolation. Also, the voltage range of the translated curve was usually smaller than the reference curves as shown in Fig. 2.

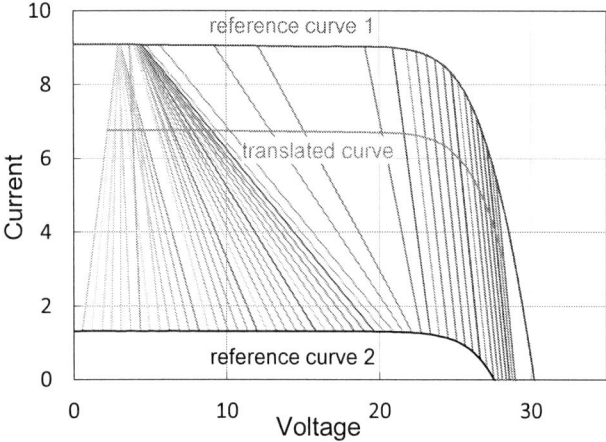

Fig. 2 Irregular interval of reference data points which occur when the noise and steps in the reference I-V curves are significant. Corresponding reference data points are connected by lines. About 1/3 of the lines are shown for visual clarity. It is noted that the appearance of the lines are dependent on the detailed algorithm of translation calculation.

2. Extended linear interpolation with modified choice of reference data points

Slight modification of the choice of reference points is proposed in the present study, as shown in Fig. 3, in order to solve the problems discussed above. Corresponding reference data are connected by lines, in order to illustrate the way of choosing the reference data points of the present study.
(1) The conventional procedure is used as is, when the interval of reference data points are uniform (region A in Fig. 3).
(2) The slope between the reference points is kept constant, when the interval of reference data as determined by the conventional procedure is irregular (region B in Fig. 3)

(3) The translated curve is calculated by only one reference curve while the slope is fixed to the same value, if one of the reference curves is missing. Reference curves 2 and 1 are used in the regions C and D of Fig. 3, respectively.

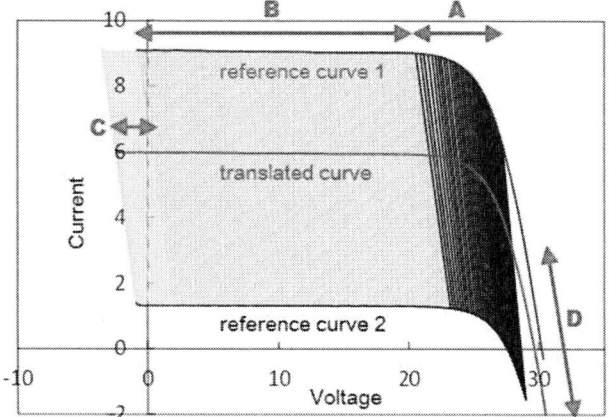

Fig. 3 Modified choice of reference data of the present study for the linear interpolation / extrapolation.

The slope between the reference points as illustrated in Figs, 1 and 2 is usually interpreted as the series resistance R_s of the device, and not actually constant. There have been many studies on the variation of R_s at different current or voltage levels in an I-V curve [10]-[12]. Therefore, the constant slope in the region B of Fig. 3 is not a theoretical requirement or assumption, but an approximation to mitigate the problem caused by the irregular interval of reference data.

3. Examples of translation calculations

Conventional procedure of linear interpolation is useful enough, when precise reference I-V curves are available over the whole range of irradiance and temperatures of interest. On the other hand, the translation procedure of the present study is practically useful when experimental noise or steps are not negligible in the reference I-V curves, or when extrapolation is necessary. Figures 4(a) and 4(b) shows examples of the preset procedure to calculate the STC I-V curve.

Translated curve 2 in Fig. 4(b) is compared with experimental STC I-V curve in Fig. 5. Results of translations based on other sets of reference I-V curves are also shown, all of which include extrapolative calculations. Relative differences of the translated curves with experiment were within about 1% for both current and voltage around the maximum operation conditions, as shown by the inset. The differences in P_{max}, I_{sc} and V_{oc} were also within about 1%, 1.5% and 1%, respectively. These results demonstrate the precision of the present procedure even when substantial extrapolation is included. Since the translation results are possibly affected by the detailed choice of reference I-V curves and data points as well as the type of the device, further

978-1-5090-5606-4/17 $31.00 © 2017 IEEE   442

confirmation is necessary to clarify the detailed uncertainty of the procedure.

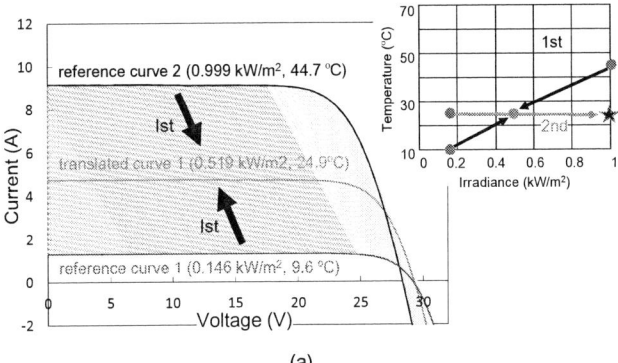

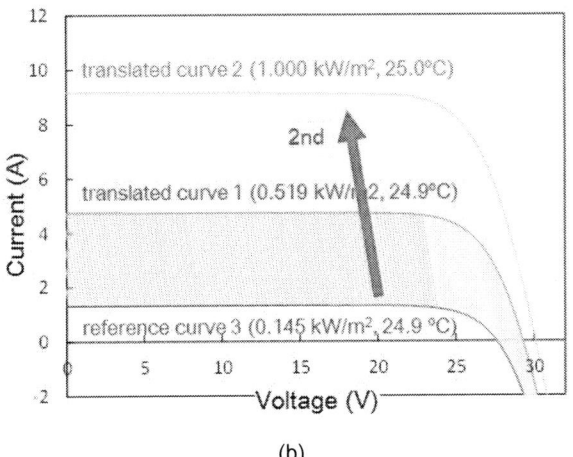

(b)

Fig. 4 Examples of the translation of the present study to calculated the STC I-V curve from three reference curves where extrapolation is necessary. (a) Interpolation to calculate translated curve 1 from reference curves 1 and 2. Irradiances and temperatures of the reference curves and translated curves are schematically shown in the inset. (b) Extrapolation to calculate the STC curve from the translated curve 1 and reference curve 3. Regions A, B, and C are shown by green, gray, and red lines.

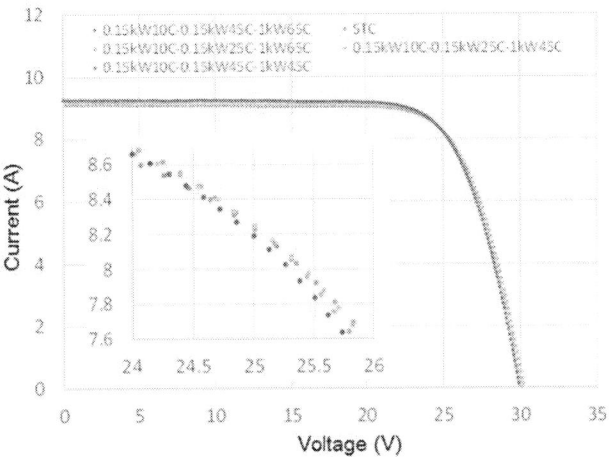

Fig. 5 Examples of the translation of the I-V curves to STC which include extrapolative calculation. The result of Fig. 4(b) (brown symbols) as well as results based on different sets of reference curves are shown by different colors. Experimental STC I-V curve is shown by orange symbols.

III. SUMMARY

Translation of the I-V curves by linear interpolation method has an advantage that precise translation is possible with no predetermined parameters such as the temperature coefficient or series resistance. The present study has investigated a modified procedure, in order to improve its precision even when the experimental noise and steps in the reference I-V curves are not negligible, or extrapolative translation is required, which are important for practical translation. Relative differences of the P_{max}, I_{sc} and V_{oc} of the translated curves were within about 1% - 1.5%, which demonstrate the precision of the present procedure even when substantial extrapolation is included in the translation procedure.

Acknowledgement: This study is supported in part by NEDO under METI. The authors are thankful to Dr. Blagovest Mihaylov for discussion on the basic nature of the I-V curves and series resistance of PV devices.

REFERENCES

[1] IEC 60904-3:2016, "Photovoltaic devices - Part 3 Measurement principles for terrestrial photovoltaic (PV) solar devices with reference spectral irradiance data "

[2] G. Blaesser, " PV system measurements and monitoring the European experience ", Sol. Energy Mat. Sol. Cells 47, 167 (1997)

[3] A. J. Anderson, " PV translation equations a new approach" AIP Conference Proceedings 353-1, 604, The 13th NREL photovoltaics program review meeting (1996) , Lakewood http://doi.org/10.1063/1.49391

[4] W. Herrmann and W. Wiesner, " Current-Voltage Translation Procedure for PV Generators in the German 1,000 Roofs-

Programme" Eurosun Proceedings; 2 (1996) 701-705; Internationales Sonnenforum; EuroSun '96, Freiburg

[5] B. Marion, S. Rummel and A, Anderberg, "Current-Voltage Curve Translation by Bilinear Interpolation" Prog. Photovolt: Res. Appl. 12 (2004) 593-607

[6] Y. Tsuno, Y. Hishikawa and K. Kurokawa, " Modeling ofthe I–V curves of the PV modules using linear interpolation/extrapolation", Solar Energy Materials & Solar Cells 93 (2009) 1070–1073

[7] IEC 60891:2009, "Photovoltaic devices - Procedures for Temperature and Irradiance Corrections to Measured I-V Characteristics of Photovoltaic Devices"

[8] Y.Tsuno, Y.Hishikawa and K.Kurokawa, "Temperature and irradiance dependence of the I–V curves of various kinds of solarcells", Technical Digest of the PVSEC, vol.15, Shanghai, 2005, pp.422–423

[9] Y. Hishikawa, Y. Tsuno and K. Kurokawa, "Translation of the I-V curves of various solar cells by improved linear

interpolation", Proceedings of the 21st EU PVSEC (2006) Dresden; 2093-2096

[10] G. L. Araujo, A. Cuevas and J. M. Ruiz, "The effect of distributed series resistance on the dark and illuminated current-Voltage characteristics of solar cells " IEEE Trans. Electron Devices 33-3 (1986) 391-401

[11] P. P. Altermatt, G. Heiser, A. G. Aberle, A. Wang, J. Zhao, S. J. Robinson, S. Bowden and M. A. Green, "Spatially resolved analysis and minimization of resistive losses in High-efficiency Si solar cells", Prog. Photovol. 21-4 (2011) 490-499

[12] K. C. Fong, K. McIntosh and A. W. Blakers, " Accurate series resistance measurement of solar cells ", Prog. Photovol. 4-5 (1996) 399-414

[13] O. Breitenstein, "Understanding the current-voltage characteristics of industrial crystalline silicon solar cells by considering inhomogeneous current distributions", Opto Electronics Review 21-3 (2013), 259–282

Severity Test with Uneven Load due to Wind Action on Photovoltaic Module

Shu-Tsung Hsu

Center for Measurement Standards, Industrial Technology Research Institute, Hsinchu, TAIWAN
Rm.401, Bldg.53, No. 195, Sec. 4, Chung Hsing Rd., Chutung, Hsinchu, Taiwan 31040,
andersonhsu@itri.org.tw

Abstract — The issue of typhoon has received considerable critical attention since the associated strong winds generally damage Photovoltaic (PV) modules severely. Previous IEC standards examined the effect of static or dynamic uniform-loads for PV module, but overlooked the wind actions with moment effect or non-uniform loads on PV module. In this paper, we developed an analytical model named extended wind-pressure test method considering wind actions. Therefore, the non-uniform wind-loads on PV module can be simulated directly by mechanical loads in labs. Results of this study also provided a severity test and revealed that such simulated non-uniform mechanical loads relied on the environmental condition like wind velocity and inclined angle of PV platform.

I. INTRODUCTION

Typhoon is one of the world's most frequently occurring natural disaster that causes severe damage. The losses caused by typhoons are not only related to the strength and structure of a particular typhoon, but also to the population density, home range, and type of economy in the affected area. Taiwan, situated between 22-25° N and 120-121° E, is centrally situated in the main path arc of typhoons generated in the Northwest Pacific Ocean. On average, 3~4 typhoons approach or make landfall in Taiwan yearly. The severity of a typhoon is associated with its wind strength and moisture content. Thus, the installation of a photovoltaic (PV) module is of necessity concerned with the wind uplift produced by typhoons. Breakage of the glass covers of PV modules and frames is mainly due to the strong wind uplift and the associated large deflection. Fig. 1 showed Meranti typhoon had numerous PV modules or trackers damaged at Southern Taiwan in September 2016. In addition, the trend of module technologies includes plastic frame or frameless module, lightweight of metal frame (0.5-0.3 kg/m), thin glass for packaging (3.3-2.0 mm) and high- power module (60-72 cells), etc. These changes of components will increase the difficulties of PV modules to withstand wind action. Therefore, a good simulation of the wind resistance by mechanical load is essential for preventing such damage.

In this work, a new concept of test method was introduced for uneven surface-loads on single PV module due to wind pressure, in order to simulate the non-uniform loads induced by wind flow through along the surfaces of PV module,

especially at some worst cases with torsion moment effect under specific wind direction and inclined angle of PV module.

Fig.1 One example of typhoon damaged solar trackers. Resource: http://news.ftv.com.tw/NewsContent.aspx?ntype=class&sno=201691 7U13M1

II. CURRENT STANDARDS OF MECHANICAL LOADS FOR PV MODULE

Currently, the IEC Standards still lack of technical documents to define wind test for PV module, especially without considering the non-uniform pressure and moment effect. Below are three standards defined mechanical loads on PV module.

A. IEC 61215:2016 [1] – Static uniform mechanical load

IEC 61215 is designed for the type approval of PV modules, and contain the static mechanical load test in C11.16. This test defined three cycles of 2400 Pa uniform load, applied for one hour to front and back surfaces in turn. Optional snow load of 5400 Pa during the last front cycle. To pass the tests, the modules must withstand a uniform test stress of 2400 Pa, and even 5400 Pa in cases of higher snow load classes.

B. IEC TS 62782:2016 [2] – Dynamic uniform mechanic load

IEC TS 62782 provides a test method (± 1000 Pa, 1000 cycles, 1 to 3 cycles per minute) for performing a (cyclic) dynamic mechanical loading test in which the module is supported at the design support points and a uniform load normal at the module surface is cycled in alternating negative and positive directions. This test may be utilized to evaluate if components within the module including solar cells, interconnect ribbons, and electrical bonds within the module are susceptible to breakage, or if edge seals are likely to fail due to the mechanical stresses encountered during installation and operation.

C. IEC 62938 Ed 1[3] - Static non-uniform snow load

IEC 62938 recommends a consistent pulling force of 2400 Pa or 5400 Pa in three one-hour cycles. The load to be applied to the PV module and its distribution by separate weight elements are determined as a function of the characteristic snow load Sk. High snow load frequently causes damage to photovoltaic systems. Systems on slanted roofs are most affected, due to the uneven distribution of the overall snow load. When snow slides down the PV module and accumulates on the lower frame, a particularly critical snow load occurs, which could lead to an overload of the entire mounting system. Until now, standards for the type approval of photovoltaic systems only included load tests carried out horizontally.

III. NBE-AE 88 [4] WITH INCLINED ANGLE

NBE-AE 88 presents a model of wind action on inclined open surfaces very similar with the module for PV platform with inclined angle. This standard can be used to determine the wind load on PV platforms, even if it is not especially dedicated. NBE-AE 88 defines the whole range of tilt angles (0-90 degrees), with values close to those from Eurocode [5]-[6], for canopies with tilt angles up to 30 degrees. Values of the pressure coefficients (C_p) are given for different wind direction angles β (Fig. 2) and wind angles α (Table I). Since the wind direction angle assumes parallel with ground, so the wind angle equal to the inclined angle of PV module.

Wind action is one of the most important loads of a mechanical system such as wind, rain, snow, earthquake and own weight. An important issue is to establish the moment acting on PV module due to wind action. Standards [5]-[6] present calculus procedures related to the wind loads' determinations. In the general case of trapezoidal distribution of wind pressure on PV module (Fig. 2), and a torsion moment effect appears [7]. The edge of a module faced the upstream wind velocity generally experienced the bigger value of the pressure coefficient.

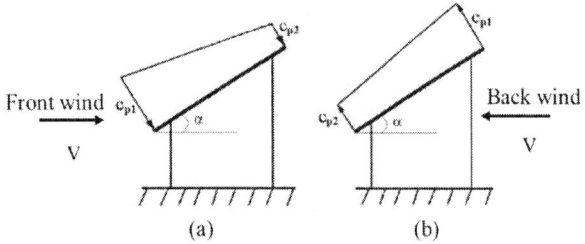

Fig.2 Trapezoidal distribution of wind-pressure due to (a) front wind ($\beta = 180°$) (b) back wind ($\beta = 0°$).

TABLE I
PRESSURE COEFFICIENT C_p FOR OPEN PLATFORMS [8] AND EXTENDED COEFFICIENT C_α [9]

α (°)	C_{p1}	C_{p2}	C_{p1}/C_{p2}	F_I/F_{II}	C_α
0	0	0			
10	0.8	0			
20	1.2	0.4	3	1.0	2.3
30	1.6	0.8	2	0.5	1.7
40	1.6	0.8	2	0.5	1.7
50	1.4	1.0	1.4	0.2	1.3
60-90	1.2	1.2	1	NA	1

IV. EXTENDED WIND-PRESSURE TEST METHOD [9]-[10]

In this paper, we developed a dichotomy method for surface-pressure on PV module due to wind loads with the moment effect. The surface-pressure estimated along the longitudinal direction on PV module, no matter the wind direction either from upstream (front wind) or downstream (back wind). For the general case of trapezoidal distribution of pressure (Fig. 2), a torsion moment appears. It can be evaluated by the arm e of the resultant force. Therefore, the calculate moment M equal to the resultant force F from triangle region I multiplied by arm distance e (Fig. 3). F_{II} means the total force of rectangle region II, F_I is of region I, F_{III} is of rectangle region III to instead of triangle region I with the same moment effect, and L is the dimension of longitudinal length of a test module. Equations (1)-(3) define each load in detail.

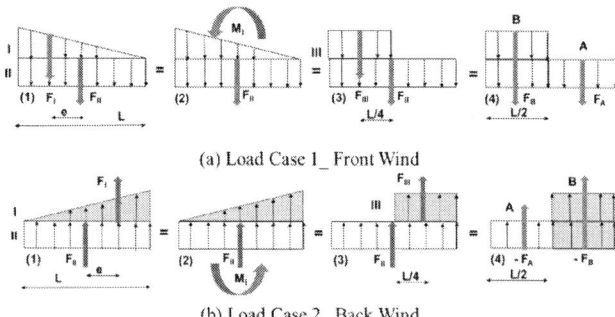

Fig.3 Wind load diagram for different load case.

$$F_{III} = 2/3 \times F_I \qquad (1)$$

$$F_B = F_{III} + 1/2 \times F_{II} \qquad (2)$$

$$F_A = 1/2 \times F_{II} \qquad (3)$$

Consequently, if we consider using P_A and P_B to simulate uneven pressure on both half-side of module (Fig. 3(a)(4), 3(b)(4), equations (4)-(5)) coherently, the pressure P_A is the design wind pressure covering safety factor 3 and C_{p2}, and then P_B is the extended wind pressure covering the moment

TABLE II

THE DESIGN WIND-PRESURE P_A AND EXTENDED WIND-PRESSURE P_B WITH SEVERITY OF INCLINED ANGLE α AND WIND VELOCITY V (OR WIND PRESSURE W_p) FOR BEAUFORT SCALE (BS) 15-17 [9]-[10]

BS	V(m/s)	W_p(Pa)	α = 20°		α = 30°		α = 40°		α = 50°		α = 60° - 90°	
			P_A(Pa)	P_B(Pa)	P_A(Pa)	P_B(Pa)	P_A(Pa)	P_B(Pa)	P_A(Pa)	P_B(Pa)	P_A(Pa)	P_B(Pa)
15	50.9	1674	2009	4621	4018	6831	4018	6831	5022	6529	6026	6026
16	56.0	2026	2431	5591	4862	8265	4862	8265	6078	7901	7294	7294
17	61.2	2420	2904	6679	5808	9874	5808	9874	7260	9438	8712	8712

effect and C_{p1}. The choice extended coefficient C_α means the ratio of P_B/P_A (or F_B/F_A) with severity of inclined angle α. On the other hand, in the general case of trapezoidal distribution of pressure, e.g., the limit pressure C_{p1} near two times of C_{p2} with relation at the inclined angle 30 degree (Table I). The more value of C_α means the more severity chosen to withstand moment effect from design view. Base on the Bernoulli's principle, wind pressure W_p (Pa) measured by equation (6) with wind velocity V (m/s) and air density ρ (e.g., 1.2923 kg/m^3 at 25 °C).

$$P_A = C_{p2} \times 3 \times W_p = F_A / (0.5 \times \text{module area}) \quad (4)$$

$$P_B = C_\alpha \times P_A = F_B / (0.5 \times \text{module area}) \quad (5)$$

$$W_p = V^2/1.6 \quad (6)$$

V. SEVERITY TEST WITH UN-EVEN WIND LOADS FOR PV MODULE [9]-[10]

Inclined angle, wind directional angle and wind velocity are studied together (Table I and Table II) to get the optimum design for installation of the PV modules. Furthermore, the effect of wind-pressure on PV module can be simulated by both P_A and P_B, which are determined in the choices of different wind velocity and inclined angle. In addition, the designed wind-pressure P_A is determined in C_{p2}, inclined angle α, wind pressure W_p and safety factor 3. Base on the moment effect, the extended wind-pressure P_B is determined in an extended coefficient C_α, inclined angle α and radio of C_{p1}/C_{p2}. In Table II, the value of C_α decreases while inclined angle (α) increases (e.g., α = 20°, P_B = 2.3P_A ; α = 40°, P_B = 1.7P_A).

VI. CONCLUSION

Taiwan (22-25°N, 120-121°E) situate in the main path of typhoons generated in the western North Pacific Ocean (19-28°N, 117-125°E). On average, 3~4 typhoons approach or landfall in Taiwan yearly. The issue of typhoon has received considerable critical attention since the associated strong winds generally damage PV modules severely. Previous IEC standards have examined the effect of static (IEC 61215) or dynamic (IEC 62782) uniform-loads for PV module, and have

overlooked the wind loads with moment effect or non-uniform loads on PV module indeed.

The purpose of this study is to understand the relationship between wind-velocity and wind-pressure on PV module under different orientations. In this paper, we developed an analytical model about extended wind-pressure test method, which can successfully simulate the non-uniform wind-loads with moment effect on PV module. All the surface-pressure pattern need to correlate with wind-tunnel experiment or CFD simulation. Results of this study revealed that such simulated mechanical loads relied on the choices of environmental condition like wind velocity and inclined angle of PV module. Furthermore, this paper contributes to ongoing discussions about providing one severity test considering wind-load and different orientations for PV module.

REFERENCES

[1] IEC 61215-2:2016, Crystalline Silicon Terrestrial Photovoltaic (PV) Modules - Design Qualification and Type Approval.

[2] IEC TS 62782:2016, Photovoltaic (PV) Modules – Cyclic (Dynamic) Mechanical Load Testing.

[3] IEC 62938 Ed. 1, Non-uniform Snow Load Testing for PV Modules.

[4] NBE-AE/88, Actions on Structures, 1988 (in Spanish).

[5] EN 1991-1-4:2005, Eurocode 1: Actions on Structures - Part 1-4: General Actions -Wind Actions.

[6] ASCE/SEI 7-05, Minimum Design Loads For Buildings and Other Structures.

[7] I. Scaletchi, I. Vişa, R. Velicu and M. Moldovan, "Torsion Moment from Wind Action on PV Platforms", 26th European Photovoltaic Solar Energy Conference and Exhibition, pp 3556-3559, 2010.

[8] Puneeth kumar H P, S B Prakash, "CFD Analysis of Wind Pressure over Solar Panels at Different Orientations of Placement", International Journal of Advanced Technology in Engineering and Science, 02(07), 2014.

[9] S. T. Hsu, "Testing Requirement for Non-Uniform Dynamic (Cyclic) Mechanical Load on PV Module", 33rd National Conference on Mechanical Engineering of CSME, 2016.

[10] S. T. Hsu, T. C. Wu, S. Y. Ting, "Simulated Wind Action on Photovoltaic Module by Non-uniform Dynamic Mechanical Load and Mean Extended Wind Load", SNEC 11th (2017) International Photovoltaic Power Generation Conference.

978-1-5090-5606-4/17 $31.00 © 2017 IEEE

Standardized Durability Test for Organic Photovoltaic and Dye Sensitized Solar Cell

Shu-Tsung Hsu[1]*, Yean-San Long[1], Teng-Chun Wu[1]

[1]Center for Measurement Standards, Industrial Technology Research Institute, Hsinchu, TAIWAN
Rm.401, Bldg.53, No. 195, Sec. 4, Chung Hsing Rd., Chutung, Hsinchu, Taiwan 31040,
*andersonhsu@itri.org.tw

Abstract — This paper aims to propose a new durability test that towards a more complete simulation or test requirements for OPV/DSSC devices, which having the portfolio for energy harvesting in low-light application. Base on harmonizing both SEMI PV57 (I-V) and SEMI PV69 (SR) in solving the hysteresis problem due to capacity effect and getting the reliable I-V characteristics, this paper focused on the detailed guidelines of durability test with severity I & II for OPV/DSSC under different scenarios. The main findings from this work were applied for SEMI Doc. 5598, and successfully released as SEMI PV76 (Durability) in Jan. 2017.

I. INTRODUCTION

OPV/DSSC is a high-potential product used for energy harvesting, especially in the context of indoor illumination applications. It is important to enhance the quality and reliability of such products, and to overcome the measurement error caused by capacity effect like the hysteresis problem. Industry needs guidelines of test specifications in order to get the reliable measurement. Therefore, it is necessary to establish I-V (current – voltage) and SR (spectrum response) standardized methods of measurement for OPV/DSSC, and based on this basis, to establish the durability test method is applicable. The detailed documents and industry standards for OPV/DSSC were developed by SEMI OPV and DSSC TF [1]-[2] in Taiwan. The key TF members include ITRI, EVERLIGHT, King Design, TDP, INER, NDHU, NCKU, MUST and NCU, etc. The work focused on the performance evaluation method of I-V [3]-[5] and SR [6] for OPV/DSSC during 2014-2015, to fulfill the PV industry or any other party interested, can thus have the common testing standards to refer to when desired.

The current technology challenge for OPV/DSSC products still focused on a serious packaging problem that certainly appears in subtropical environments. In general, environmental degradation factors (humidity, temperature and light) would seriously make OPV/DSSC with poor performance and reliability, and shorten its lifetime as well. Therefore, this paper aims to propose a new durability test, which towards a more complete simulation or test requirements, for OPV/DSSC devices as follows.

II. SEMI PV76 [7]

A. Aging Tests with severity I & II

This work of PV76 proposed three aging tests (TC, LS and DH) and one combined test (TC50 + DH), according to the specific test condition (see Table I) designed for durability test with two severities I & II, and compared with IEC 61646 [8].

TABLE I
DURABILITY TEST METHOD FOR PV CELL AND MODULE

Item	SEMI PV76	IEC 61646
Test sample	OPV/DSSC cell and module	thin-film module
Pre-LS	1 h	NA
Severity	I & II	NA
LS (by lighting exposure model)	1000 Wm^{-2} (*300 Wm^{-2}) ; 60 °C ± 1 °C ; 1000 h(I)/ 3000 h(II)	1000 Wm^{-2} ; 50 °C ± 10 °C ; at least 43 kWhm^{-2}
DH (by inverse power model)	65 °C ± 2 °C ; 65 %RH ± 10 %RH ; 600 h(I)/1800 h(II)	85 °C ; 85 %RH ; 1000 h
TC (by Arrhenius model)	-10 °C ~ 60 °C ; TC50 (I)/ TC150 (II)	−40 °C ~ 85 °C ; TC50/ TC200
Combined test	TC50 + DH	NA
Life-time estimated	3 Yr (I)/ 10 Yr (II)	20-25 Yr
Criteria	△ P$_{max}$ < 5 %	△ P$_{max}$ < 5 %
I-V requirement	SEMI PV57	IEC 60904-1[9]

Note: *only for low-lighting, h (hour), Yr (year)

1) Light Soaking (LS): LS test can stabilize the electrical characteristics of OPV/DSSC by natural sunlight or simulated solar irradiation or artificial lighting source. In order to simulate the life time of OPV/DSSC under T5 lighting with irradiance 1 Wm^{-2} and room temperature 60 °C, stress condition (1000 Wm^{-2}, 60 °C) of LS refer to lighting exposure model [10]. e.g., for T5 device, 60 °C, 1 Wm^{-2}, intensity ratio = 1/1000 x100 % = 0.1 %, lamp power 80 % (assumption), exposure ratio = 100 − 74.8 = 25.2. The life-time of product is estimated as 3/10 years while setting stress time is 1043/3476 hours. The designed minimum stress time is 1050/3000 hours.

978-1-5090-5606-4/17 $31.00 © 2017 IEEE

Therefore, if OPV/DSSC merely tests with irradiance less than 1000 Wm⁻² due to some specific materials or device only used for low-light application, LS test needs to adjust the stress specification in order to fulfil the same life-time estimation.

2) Thermal Cycle (TC): TC test can determine the ability of OPV/DSSC to withstand thermal mismatch, fatigue and other stresses caused by repeated changes of temperature. Arrhenius model [18] with the degradation factor of temperature, can estimate the life-time of OPV/DSSC is 3/10 years, according to different stress time (TC50/TC150) and temperature (-10 °C ~ 60 °C), which differ with the stress condition of IEC 61646 C10.11 (-40 °C ~ 90 °C and TC200).

3) Damp Heat (DH): DH test can determine the ability of OPV/DSSC to withstand the effects of long-term penetration of humidity. The life-time, estimated by inverse power model [18] with multi-degradation factors (65 °C, 65 %RH), is 3/10 years while setting minimum stress time is 600/1800 hours, which differ with stress condition (85 °C, 85 %RH, 1000 hour) of IEC 61646 C10.13.

4) Combined Test (TC50+DH): This test is to determine the final packaging ability of OPV/DSSC to withstand thermal mismatch, fatigue and other stresses caused by repeated changes of temperature, and the effects of long-term penetration of humidity. Before installing the product at room temperature in the chamber of damp heat, shall be introduced into the chamber with thermal cycle preconditioning repeat change temperature by 50 cycles (TC50). The following test conditions are applied.

- test temperature: 85 °C ± 2 °C
- relative humidity: 85 %RH ± 5 %RH
- test sequence refers to Fig. 1.

Final measurements shall operate both visual inspection and I-V measurement before a minimum recovery time of one hour.

B. Test process

Nine test samples shall be divided into five-group and subjected to the qualification test sequences in Fig. 1, carried out in the order laid down. Each group refers to the corresponding sub-clause in this test method. Test procedures and severities, including initial and final measurements where necessary. However, with regard to the tests of initial and final measurements, it should be noted that the procedures laid down in SEMI PV57, these tests shall be carried out within ± 5 % of the specified irradiance and within ± 2 °C of the specified temperature. Any single test, executed independently of a test sequence, shall be preceded by the initial tests of visual inspection and I-V measurement.

C. Standardization by global voting

In order to develop test methods for OPV/DSSC, SEMI OPV and DSSC TF in Taiwan was held since 2013. ITRI coordinated this TF to set up a series of experiments [11]-[12], and made discussions which focused on the characteristics' methods of I-V, SR and durability. I-V evaluation was applied

to SEMI PV57 [3]. SR evaluation was applied to SEMI PV69 [6]. Based on these standards, the work continually develops a durability test method for OPV/DSSC, especially in low-light environment. Results were applied for SEMI Doc. 5598 [15]-[17] and released as SEMI PV 76 after successfully voting globally in Sep. 2016.

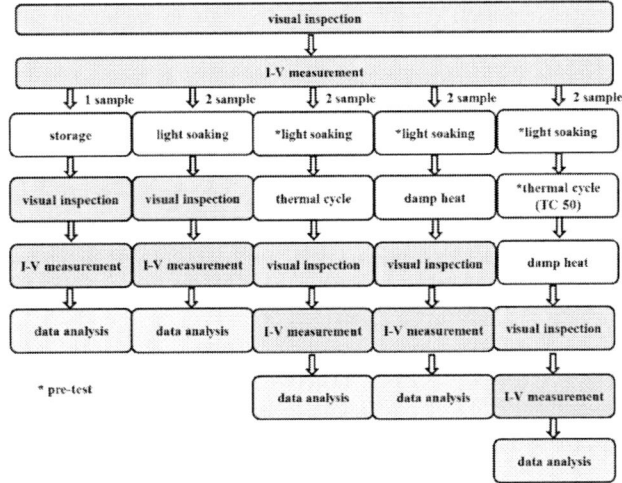

Fig. 1.   Test flow in parallel [7].

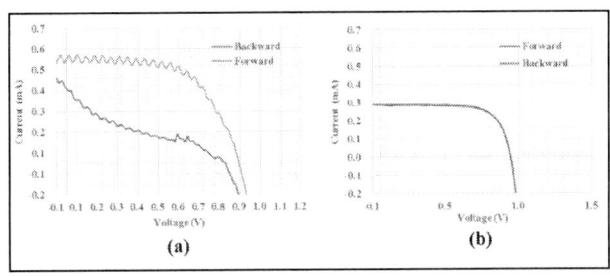

Fig. 2.   I-V measurement (at STC) using the (a) sweeping rate method and (b) asymptotic I-V method [4].

III. DISCUSSION

1) The asymptotic I-V method can provide stable characteristics for emerging PV, unlike the sweeping rate method. Reliable measurement of I-V [11] and SR [14] has become important tools to evaluate the durability of OPV/DSSC properly. All I-V tests need to follow the test procedure defined in SEMI PV57, and remove the capacity effect during measurement (see Fig. 2).

2) This study of SEMI PV76 explores different aspects of OPV/DSSC by means of standardizing a new durability test method, which extends previous work in I-V (SEMI PV57) and SR (SEMI PV69). In addition, the main benchmark documents for these three SEMI standards in order are IEC 61646, IEC 60904-1 and IEC 60904-8 [18]. SEMI PV76 aims to propose a durability test method to evaluate OPV/DSSC.

For other PV technologies, the proposed qualification would be helpful to refer to in similar indoor application like OPV/DSSC. Furthermore, SEMI PV76 especially addresses the test requirement for application scenario indoors or under low light. Therefore, such performance measurement has been estimated to keep expanded measurement uncertainty less than 2.7 %. Then the final criteria should be able to fulfill 5 %.

3) Upon for the low-light energy harvesting application and differentiating with standard test condition used for c-Si devices, there are two new indexes were defined in SEMI PV57. One is very low irradiance condition (VLIC: AM1.5G, 25 °C, 60 Wm^{-2}), and the other is ultra-low irradiance condition (ULIC: AM1.5G, 25 °C, 1 Wm^{-2}). Base on the low-light application indoors, the efficiency of OPV/DSSC module shall be defined as (P_{max}/module area)/input power. The value of input power depends on the choice of irradiation type or artificial lighting source, and keeps 1000 Wm^{-2} in using natural sunlight or simulated solar irradiation. Furthermore, there are two ongoing drafts are developing by OPV and DSSC TF in Taiwan since Jan. 2017, to define the specification of indoor lighting simulator (SEMI Doc. 5979 [19]) and current-voltage (I-V) measurement in indoor lighting (SEMI Doc. 5980 [20]).

4) For LS test (see Table I), in order to anneal and stabilize the electrical characteristics of test sample by simulated solar irradiation approach from 10 minutes to 30 minutes, the irradiance (a) for outdoor application, between 600 Wm^{-2} and 1000 Wm^{-2}, or (b) for low lighting application less than 300 Wm^{-2}.

5) Due to the demand of different consumer products, SEMI PV76 defines durability test according with two severities I & II (see Table I), which refer to the products' life-year, and fulfill the application scenario even happened indoor environment with low lighting.

IV. CONCLUSION

This paper aims to improve the existing durability tests, such as IEC 61646 for thin-film PV product, towards a more complete simulation or test requirements for OPV/DSSC devices, which having the portfolio for energy harvesting in low-light application. Base on harmonizing both SEMI PV57 (I-V) and SEMI PV69 (SR) in solving the hysteresis problem due to capacity effect first then getting the reliable I-V characteristics, this paper focused on the detailed guidelines of durability test including severity I & II for OPV/DSSC. The main findings from this work were applied for SEMI Doc. 5598, and successfully released as SEMI PV76 in Jan. 2017. Through this standardization of SEMI PV76, the work is helpful to alleviate the early performances of new product for OPV/DSSC, and influence the relevant reliability aspect in a long-term perspective.

REFERENCES

[1] S. T. Hsu, Y. S. Long, T. C. Wu, H. H. Hsieh, "A Case Study of Developing SEMI OPV/DSSC Standards in Taiwan", 32nd EU PVSEC, 2016.

[2] S. T. Hsu, Y. S. Long, T. C. Wu, "A Case Study of Developing SEMI PV Standards in Taiwan", *Applied Mechanics and Materials* (accept).

[3] SEMI PV57, Test Method for Current-Voltage (I-V) Performance Measurement of Organic Photovoltaic (OPV) and Dye-Sensitized Solar Cell (DSSC).

[4] Y. S. Long, Keith Emery, S. T. Hsu, T. C. Wu, "I-V Characteristics of Emerging PV under Indoor and Outdoor Lighting Conditions", *Journal of Photovoltaics* (accept).

[5] Y. S. Long, S. T. Hsu, T.C. Wu, "Induction of Internal Capacitance Effect in Performance Measurement of Organic Photovoltaic Device (OPV) by Real-Time One-Sweep Method (RTOSM)", *Journal of Energy and Power Engineering*, 8 (2014) 1059-1066.

[6] SEMI PV69, Test Method for Spectrum Response (SR) Measurement of Organic Photovoltaic (OPV) and Dye-Sensitized Solar Cell (DSSC).

[7] SEMI PV76, Test Method for Durability of Low Light Intensity Organic Photovoltaic (OPV) and Dye Sensitized Solar Cell (DSSC).

[8] IEC 61646, Thin-Film Terrestrial Photovoltaic (PV) Modules – Design Qualification and Type Approval.

[9] IEC 60904-1, Photovoltaic Devices – Part 1: Measurements of Photovoltaic Current-Voltage Characteristics.

[10] IEC 60068-1, Environmental Testing - Part 1: General and Guidance.

[11] T. C. Wu, S. T. Shu, Y. S. Long, "New Set-up Procedures and Integrated Measurement System for Organic Photovoltaic (OPV) Module", International Photovoltaic Science and Engineering Conference (PVSEC23), 2013.

[12] Y. S. Long, S. T. Hsu, T. C Wu, "Induction of Internal Capacitance Effect in Performance Measurement of OPV (Organic Photovoltaic) Device by RTOSM (Real-Time One-Sweep Method)", *Journal of Energy and Power Engineering,* 8:1059-1066, 2014.

[13] S. T. Hsu, Y. S. Long, T. C. Wu, "Standardization of Current - Voltage Test Method for DSSC Products", 12th ISMTII, 2015.

[14] S. T. Hsu, Y. S. Long, T. C. Wu, Jay Lin, "Standardization of Spectrum Response Measurement for OPV and DSSC", SNEC 10th (2016) International Photovoltaic Power Generation Conference.

[15] S. T. Hsu, Y. S. Long, T. C. Wu, J. C. Chou, "Durability Test for Organic Photovoltaic (OPV) and Dye Sensitized Solar Cell (DSSC)", PVSEC26, 2016.

[16] S. T. Hsu, Y. S. Long, T. C. Wu, J. C. Chou, "Standardized Durability Test for Emerging Photovoltaic in Low Light Intensity Environment", 4th IMETI, 2016.

[17] S. T. Hsu, Y. S. Long, T. C. Wu, H. H. Hsieh, "Requirement of Durability Test for DSSC and OPV", 32nd EU PVSEC, 2016.

[18] IEC 60904-8, Photovoltaic devices – Part 8: Measurement of spectral response of a photovoltaic (PV) device.

[19] SEMI Doc. 5979, Specification of Indoor Lighting Simulator Requirements for Emerging PV.

[20] SEMI Doc. 5980, Test Method of Current-Voltage (I-V) Measurement in Indoor Lighting for DSC and OPV.

Spatial Thickness Uniformity and Structural Evaluation of RF Sputtered ZnO Thin Films for Solar Cell

Babar Hussain[1,3,4]* and Taj M. Khan[2,3]

[1]Energy Production and Infrastructure Center, Department of Electrical and Computer Engineering,
University of North Carolina at Charlotte, Charlotte, NC, 28223, USA
[2]School of Physics, Trinity College Dublin, Dublin, Ireland
[3]National Institute of Lasers and Optronics, Nilore 45650, Islamabad, Pakistan
[4]Intel Corporation, Rio Rancho, NM, USA

*Corresponding Author: babar.hussain@intel.com

Abstract—**Spatial uniformity in thickness and other characteristics over the surface of thin films plays a fundamental role in characterization and device fabrication particularly in the solar cells. We report preparation of zinc oxide thin films on Sapphire substrates by RF magnetron sputtering and characterization by interferometric spectral reflectance and atomic force microscopy. The results of the sample prepared by RF magnetron sputtering are analyzed over the surface of samples at different spots to compare with the results previously reported of zinc oxide thin films grown by metal organic chemical vapor deposition. The results show that ZnO thin films prepared by sputtering are superior in thickness uniformity. Furthermore, the importance of thickness uniformity specifically with reference to solar cell performance has been discussed which has been so far overlooked by research community.**

Index Terms—**Zinc oxide, thin film, sputtering, spectral reflectance, AFM, solar cells.**

I. INTRODUCTION

Zinc oxide (ZnO) is a II-VI compound semiconductor with wide bandgap tunable from 3 to 5 eV. Along with several other potential applications [1–3], ZnO has unique advantages such as natural abundance, lower toxicity, and lower processing cost. The common methods to grow or deposit doped and undoped ZnO films include; pulsed laser deposition (PLD) [4], molecular beam epitaxy (MBE) [5], ion plating [6], metal organic chemical vapor deposition (MOCVD) [7], and RF magnetron sputtering [8]. ZnO can be grown by MOCVD at moderate temperatures with high growth rate. This technique is considered as a best crystal growth method with potential at large scale; however, high-cost maintenance and the use of toxic metal-organic materials are the essential drawbacks of MOCVD. RF sputtering has been used extensively. The main advantage of sputtering is that high melting points materials such as ZnO can be easily sputtered with better thickness uniformity where evaporation of such materials is difficult.

We have recently proposed an n-ZnO/p-Si solar cell structure [7]. The degradation in device performance based on thin film due to thickness non-uniformity has so far remained overlooked in semiconductor industry. Spatial thickness uniformity is an important parameter because it directly affects physical properties and performance of thin film device as well as plays important role in characterization of the material/device. From device point of view, there is a direct relation between thickness uniformity and performance of the thin film solar cell device because spatial thickness distribution influences the electrical and optical properties of thin films.

In this paper, we present an experimental study of spatial thickness uniformity and optical analysis over the surface of ZnO thin films deposited by RF magnetron sputtering and a comparison is made with the MOCVD grown ZnO films. There are several studies reporting sample-to-sample analysis of ZnO films prepared by sputtering for thickness uniformity and reproducibility but to the best of our knowledge, in literature no one has reported a study describing thickness uniformity over the surface of ZnO film prepared by sputtering and to have a comparison with the samples prepared by MOCVD which is important for device performance. In general, material science/engineering community has a consensus that MOCVD produces thin films with overall better quality as compared to sputtering. But it depends on material, precursors used, and precision in calibration of the systems. We have prepared ZnO thin films using sputtering and performed measurements at several points over the film surface using spectral reflectance and atomic force microscopy (AFM). Our results establish that the ZnO films prepared by sputtering are more uniform in thickness as compared to those prepared by MOCVD.

II. ZINC OXIDE DEPOSITION BY SPUTTERING

ZnO thin films were grown on sapphire substrates of diameter of 2 inch by using the AJA International Inc. sputtering system. The substrates were initially cleaned by acetone for 10 min in an ultrasonic bath and subsequently cleaned by methanol, isopropanol, and deionized water. The sputtering pressure was set to 3×10^{-3} torr in argon working gas environment. Additionally, the suceptor rotational speed was set to 20 rpm. The RF power and frequency used for sputtering was set to 180 W and 13.56 MHz respectively.

It was determined that thin film of room temperature RF sputtered ZnO does not crystallize. We prepared three samples at different temperatures and examined their crystallinity. We observed an improvement in film crystallinity with increase in deposition temperature as shown in Fig. 1 [9]. It is important to note that there is a possibility to grow ZnO films with even better crystallinity at higher deposition temperatures; however, the facility available in our cleanroom at UNC- Charlotte has an upper temperature limit of 300 °C.

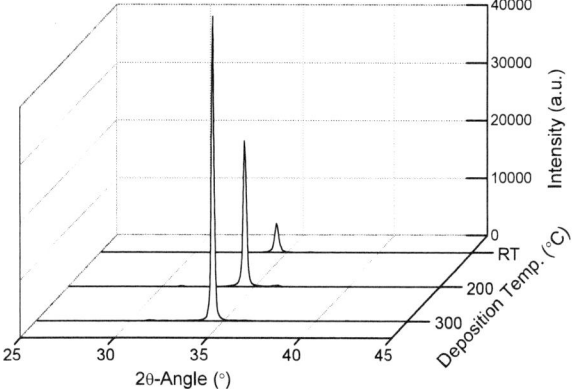

Fig. 1. XRD rocking curves showing 002 reflections of the ZnO samples deposited at room temperature (RT), 200 °C, and 300 °C [9].

III. CHARACTERIZATION RESULTS AND ANALYSIS

A. Thickness Measurements by Spectral Reflectance

The thickness measurements were carried out by utilizing the Filmetrics F20-UV system that exploits spectral reflectance data to calculate thickness, optical constants, and roughness of the sample. The measurements performed at nine distinct points are plotted on a 3-D histogram as illustrated in Fig. 2.

TABLE 1. Comparison of thickness uniformity over the surface of ZnO samples grown by sputtering and MOCVD.

Growth Method	Growth Temperature (°C)	Average Thickness (nm)	Standard Deviation (nm)
Sputtering	300	350.78	2.08
MOCVD	550	483.40	36.62

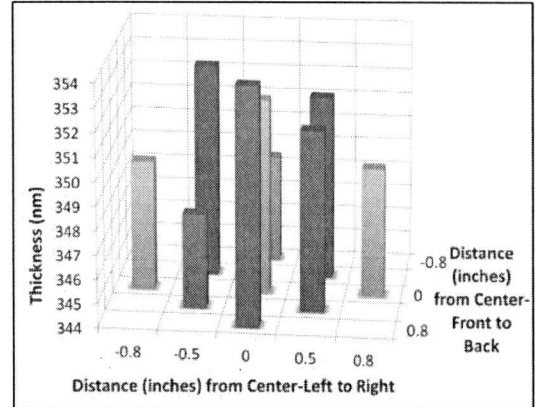

Fig. 2. Thickness measurements using spectral reflectance data at nine distinct points over the surface of ZnO thin film sputtered on sapphire substrate of diameter of 2 inches. The inset of Fig. 3 depicts the positions front, back, left, and right.

Overall, the thickness for the ZnO sample grown with sputtering technique is highly uniform. The mean value of thickness is 350.78 nm with a standard deviation of 2.08 nm and thicknesses over the surface are in the range of 347.91 to 353.89 nm. Previously, the reported samples grown by MOCVD had thickness measurements varied over the range of ~70 nm [10]. Table 1 shows the average thicknesses and standard deviation for the ZnO samples grown by sputtering and MOCVD. It is evident that thickness uniformity over the surface of ZnO thin films is significantly higher for the sample prepared by RF sputtering as compared to those prepared by MOCVD.

B. Atomic Force Microscopy (AFM)

The AFM measurement was carried out by NT-MDT AFM system by Spectrum Instruments and image data was processed and analyzed using the image analysis software provided by the same company. The Fig. 5 depicts AFM images at four different spots of ZnO thin film sputtered on sapphire substrate having 2" diameter at a temperature of 300 °C and RF power of 180 W. The table appended with each AFM image contains summary of the grain analysis results. The average grain size is 50–100 nm which is coherent with the value calculated by Debye-Scherrer equation using XRD data. The grain size is much larger in the vicinity of top-right region showing better crystal quality. It is noteworthy that crystal quality varies significantly at large scale distance over the sample surface that can play an important role in performance of the final device. This difference in crystal quality cannot be observed in small scale mapping of the samples. This consideration is very important in large area devices such as solar cells.

IV. DISCUSSION

Thickness uniformity of the film is an important parameter because thickness directly affects physical properties of the single- or multi-layer structure and thus performance of the device [11, 12]. Also, it plays important role in sample evaluation especially in electrical characterizations like Hall and 4-probe measurements because thickness is a fundamental factor which is used to calculate carrier concentration, lifetime, resistivity, and mobility of the sample.

The thickness uniformity of the film prepared by magnetron sputtering is primarily dictated by the geometry of the target relative to substrate, applied electric power, gas temperature, and erosion zone of the target ends. The thickness uniformity and deposition rate both increase with the decrease in target-substrate distance as well as with increase in power [13]. Jiang et al. have recently proposed a theoretical model to improve uniformity in thickness of the films grown on a large substrate by employing a step-moving target [14]. There were two critical parameters, the target stay time and the target moving step, that affect film thickness distribution in the model.

We have experimentally proved that uniformity in thickness distribution over the surface of ZnO films prepared by sputtering at lower temperature is better than the films produced by MOCVD at higher temperature. The high temperature processing of samples can degrade the quality of substrate and thus performance of the device. This is specifically applicable to ZnO/Si heterojunction solar cell because high temperature causes solar cell performance degradation mainly correlated with oxygen precipitation [15]. Also, there is a direct relation between spatial thickness uniformity and solar cell performance. The behavior of absorption of light has been attributed to normalized standard deviation of thickness [16]. Karpov et al. in 2002, modeled non-uniformity effects in terms of the equivalent circuit of a system of many interacting random diodes [17]. The several weak diodes were associated with a non-uniform thickness which consumed most of the photogenerated current. The non-uniform thickness was believed to influence the optical and electrical properties of thin films.

There are some additional advantages associated with sputtering as compared to MOCVD. It is possible to implement shadow-mask lithography in sputtering if ZnO (or any other material) needs to be deposited on selected areas. Also, source vapors can penetrate below the substrate in case of MOCVD causing unwanted growth of material at back side of substrate where this is not possible in sputtering. We believe that this work will help to improve performance of thin film devices especially solar cells [7]. Further details are provided somewhere else [18]. The development of ZnO/Si solar cell is reported elsewhere [19, 20].

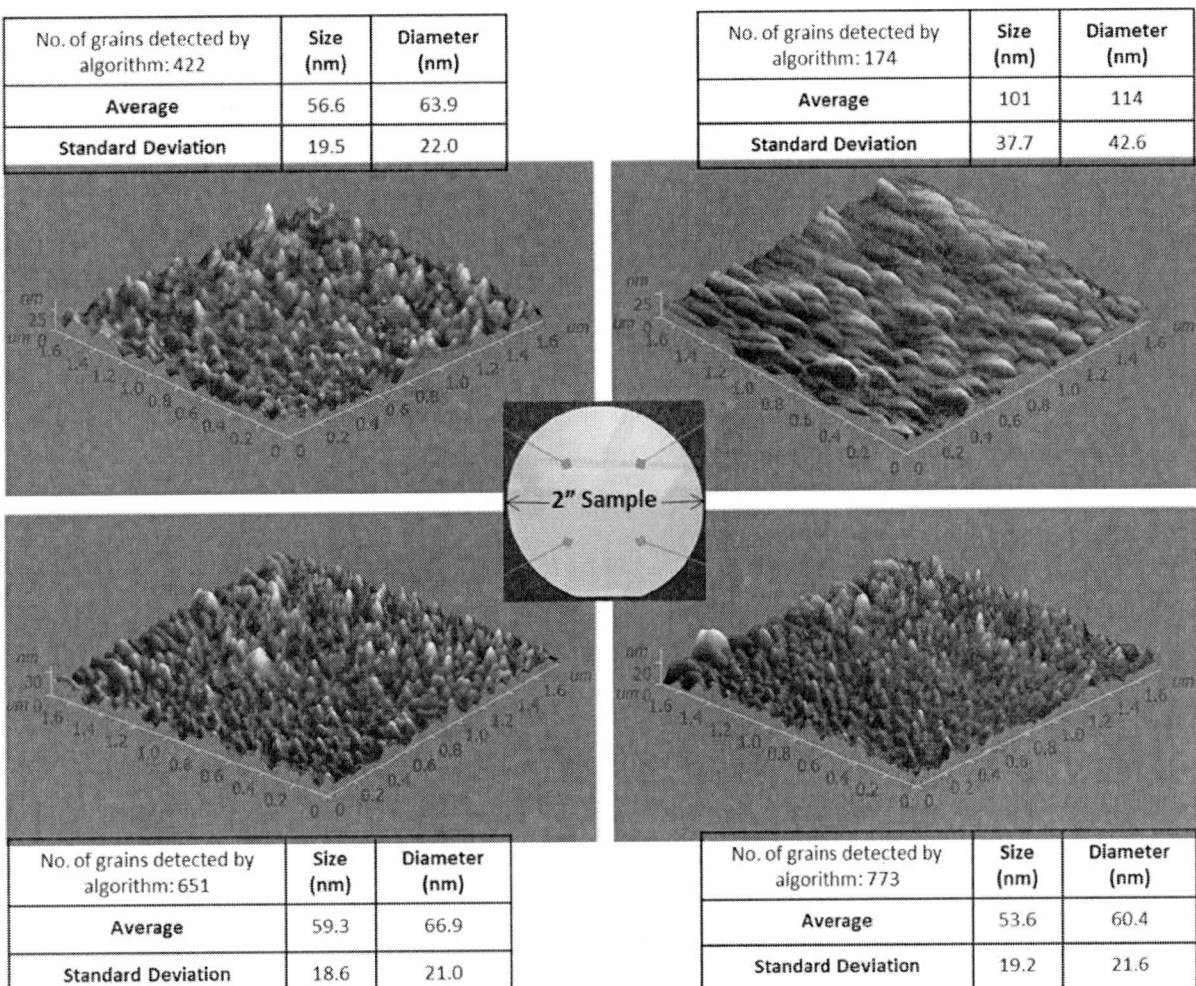

Fig. 5. AFM images at four different spots of n-ZnO grown on sapphire substrate using magnetron sputtering at 300 °C with RF power of 180 W. The table appended with each AFM image contains summary of the grain analysis results.

V. CONCLUSION

We have experimentally evaluated spatial thickness uniformity and optical quality of ZnO thin films sputtered on sapphire substrates. The results confirmed that the ZnO thin films produced by magnetron sputtering are more uniform in spatial thickness distribution as compared to ZnO thin films grown the same substrates by MOCVD. Based on our analysis, overall sputtered ZnO films exhibited better and an improved optical quality over the MOCVD films. These findings can help in making thin film based solar cells and other optical devices with improved performance. Our research work is in progress with extension to make n-ZnO/p-Si solar cells by sputtering ZnO on Si substrates.

ACKNOWLEDGEMENTS

We are grateful to Dr. Edward Stokes for letting us use his equipment to measure thickness.

REFERENCES

[1] E. Fortunato, P. Barquinha, and R. Martins, "Oxide semiconductor thin-film transistors: a review of recent advances," Adv. Mater., vol. 24, pp. 2945–2986, 2012.

[2] H. Tampo et al., "Polarization-induced two-dimensional electron gases in ZnMgO/ZnO heterostructures," Appl. Phys. Lett., vol. 93, pp. 202104-1–3, 2008.

[3] K. Remashan, Y. S. Choi, S. J. Park, and J. H. Jang, "High performance MOCVD-grown ZnO thin-film transistor with a thin MgZnO layer at channel/gate insulator interface," J. Electrochem. Soc., vol. 157, pp. H1121–H1126, 2010.

[4] S. J. Henley, M. N. R. Ashfold, and D. Cherns, "The growth of transparent conducting ZnO films by pulsed laser ablation," Surf. Coat. Tech., vols. 177-178, pp. 271-276, 2004.

[5] A. Tsukazaki et al., "High electron mobility exceeding 10^4 cm^{2}V^{-1}s^{-1} in Mg$_x$Zn$_{1-x}$O/ZnO single heterostructures grown by molecular beam epitaxy," Appl. Phys. Express, vol. 1, pp. 055004-1-3, 2008.

[6] T. Yamada, H. Makino, N. Yamamoto, and T. Yamamoto, "Ingrain and grain boundary scattering effects on electron mobility of transparent conducting polycrystalline Ga-doped ZnO films," J. Appl. Phys., vol. 107, pp. 123534-1-8, 2010.

[7] B. Hussain, A. Ebong, and I. Ferguson, "Zinc oxide as an active n-layer and antireflection coating for silicon based heterojunction solar cell," Solar Energy Materials & Solar Cells, vol. 139, pp. 95–100, 2015.

[8] X. Yu et al., "Preparation and properties of ZnO:Ga films prepared by r.f. magnetron sputtering at low temperature," Appl. Surf. Sci., vol. 239. pp. 222–226, 2005.

[9] B. Hussain and A. Ebong, "Improvement in open circuit voltage of n-ZnO/p-Si solar cell by using amorphous-ZnO at the interface," under review.

[10] P. Mishra, R. Patel, B. Hussain, J. Stansell, B. Kucukgok, M. Y. Raja, N. Lu, and I. Ferguson, "Spatial Analysis of ZnO Thin Films Prepared by Vertically Aligned MOCVD," Proc. IEEE 10th International HONET Conference, DOI: 10.1109/HONET.2014.7029363, 2014.

[11] S. H. Mohamed and M. Raaif, "Effects of thickness and rf plasma oxidizing on structural and optical properties of SiO$_x$N$_y$ thin films," Surface & Coating Technology, vol. 205, pp. 525–532, 2010.

[12] N. Choudhary, D. K. Kharat, and D. Kaur, "Structural, electrical and mechanical properties of magnetron sputtered NiTi/PZT/TiO$_x$ thin film heterostructures," Surface & Coating Technology, vol. 205, pp. 3387–3396, 2011.

[13] Z. Yichen, S. Qingzhu, and S. Zhulai, "Research on thin film thickness uniformity for deposition of rectangular planar sputtering target," Physics Procedia, vol. 32, pp. 903–913, 2012.

[14] C. Z. Jiang, J. Q. Zhu, J. C. Han, P. Lei, and X. B. Yin, "Uniform film in large areas deposited by magnetron sputtering with a small target," Surface & Coating Technology, vol. 229, pp. 222–225, 2013.

[15] G. Gaspar et al., "Identification of defects causing performance degradation of high temperature n-type Czochralski silicon bifacial solar cells," Solar Energy Materials & Solar Cells, vol. 153, pp. 31–43, 2016.

[16] Y. O. Choi, N. H. Kim, J. S. Park, and W. S. Lee, "Influences of thickness-uniformity and surface morphology on the electrical and optical properties of sputtered CdTe thin films for large-area II-VI semiconductor heterostructured solar cells," Material Science and Engineering B, vol. 171, pp. 73–78, 2010.

[17] V. G. Karpov, A. D. Compaan, and D. Shvydka, "Effects of nonuniformity in thin-film photovoltaics," Applied Physics Letters, vol. 80, pp. 4265–4258, 2002.

[18] T. Ortiz, C. Conde, T. M. Khan, and B. Hussain, "Thickness uniformity and optical/structural evaluation of RF sputtered ZnO thin films for solar cell and other device applications," Applied Physics A, 123:280, 2017; DOI: 10.1007/s00339-017-0909-2.

[19] B. Hussain, "Improvement in open circuit voltage of n-ZnO/p-Si solar cell by using amorphous-ZnO at the interface," Prog Photovolt Res Appl. 0:1–9, 2017.

[20] B. Hussain, "Development of n-ZnO/p-Si single heterojunction solar cell with and without interfacial layer," The University of North Carolina at Charlotte, 2017, 154; 10258481.

Local Measurements of Surface Capacitance
by Electrostatic Force Microscopy on Cu(In,Ga)Se₂ Materials

Tomoaki Ishii [1], Takashi Minemoto [3] and Takuji Takahashi [1,2]

[1] Institute of Industrial Science, The University of Tokyo, Tokyo 153-8505, Japan
[2] Institute for Nano Quantum Information Electronics, The University of Tokyo, Tokyo 153-8505, Japan
[3] Graduate School of Science and Engineering, Ritsumeikan University, Shiga 525-8577, Japan

Abstract — We have performed electrostatic force microscopy (EFM) on Cu(In,Ga)Se₂ thin films with different Ga contents to examine their surface depletion capacitance locally. Especially from the analyses of the various frequency components in electrostatic force, influence of Cd diffusion into CIGS as well as its uniformity on a granular surface of CIGS were investigated. In addition, relationship between such properties and conversion efficiency in solar cell has been discussed.

Index Terms — electrostatic force microscopy, surface capacitance, depletion capacitance, Cd diffusion, CIGS.

I. INTRODUCTION

A Cu(In,Ga)Se₂ [CIGS] material is very promising for a thin film solar cell owing to its unique properties like bandgap tunability and very high absorption coefficient, and the conversion efficiency over 22 % has already been achieved [1][2]. However, some inconsistency between the theoretically and experimentally optimum bandgaps remains. If this inconsistency can be solved, we have more room for improving the efficiency. In addition, most CIGS materials used in solar cells have a microcrystalline structure that includes many small grains and their boundaries. Therefore microscopic investigations of material properties in CIGS by scanning probe microscopies are very informative, and, up to now, several works have been reported [3]-[10]. In this study, we have performed local electrostatic force measurements on bare CIGS and CdS/CIGS layers by electrostatic force microscopy (EFM) to examine their surface depletion capacitance, and have discussed an influence of Cd diffusion into CIGS.

II. EXPERIMENTAL

Figure 1 shows our experimental setup for EFM based on a commercial atomic force microscopy (AFM) system (NanoNavi/ E-sweep, Hitachi High-Tech Science Corp., Japan) operated in high vacuum condition (~ 10⁻⁵ Pa) at room temperature with a Pt-coated AFM cantilever whose spring constant and fundamental resonant frequency were about 2 N/m and 70 kHz, respectively. Between this AFM tip and a sample, an a.c. bias voltage ($V_{AC} \cos 2\pi ft$) and d.c. offset bias voltage (V_{DC}) were applied to generate an electrostatic force. The induced electrostatic force F_{ES} is given by the following equation in the assumption that a capacitance C between the tip and sample is independent of the external bias:

$$F_{ES} = \frac{\partial C}{\partial z}[(V_{DC} - \Delta\varphi) + V_{AC}\cos 2\pi ft]^2$$
$$= F_{ES,\,DC} + F_{ES,\,f} + F_{ES,\,2f} \tag{1}$$

where z and $\Delta\varphi$ are distance and intrinsic contact potential difference between the tip and sample, respectively, and $F_{ES,DC}$, $F_{ES,f}$ and $F_{ES,2f}$ are d.c., f- and $2f$-components in the electrostatic force. Now, $F_{ES,f}$ and $F_{ES,2f}$ are given by

$$F_{ES,\,f} = \frac{\partial C}{\partial z}[2V_{AC}(V_{DC} - \Delta\varphi)\cos 2\pi ft] \tag{2}$$

$$F_{ES,\,2f} = \frac{\partial C}{\partial z}[\frac{V_{AC}^2}{2}\cos 2\pi(2f)t], \tag{3}$$

and those components can be separately extracted by a lock-in amplifier as shown in Fig. 1.

In the normal operation mode in EFM, $F_{ES,f}$ or $F_{ES,2f}$ at a certain V_{DC} are used to construct electrostatic force images on a scanned area. Since an amplitude of $F_{ES,2f}$ simply depends on the capacitance as shown in (1) under a constant V_{AC}, we can investigate distribution of capacitance from $2f$-component image, which is an indication of the uniformity of depletion in CIGS. In addition, EFM enables us to record the dependence of $F_{ES,f}$ on V_{DC} given by (2), which is referred to as an electrostatic force spectrum in this study, at a certain point on the sample surface. When C is independent of V_{DC} as assumed above, linear

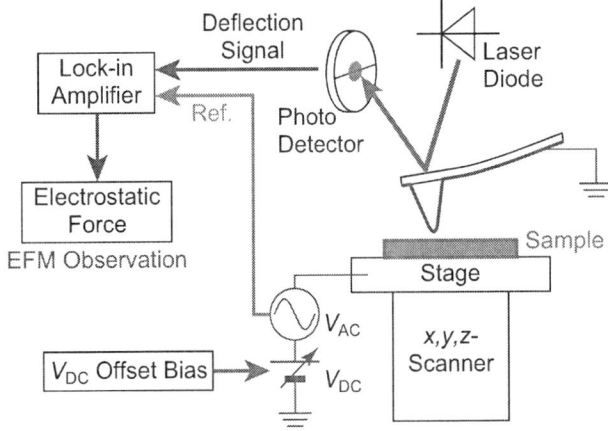

Fig. 1   Experimental setup for EFM operated in high vacuum condition (~ 10⁻⁵ Pa) at room temperature.

Fig. 2 (a) Topographic image and (b) and (c) images of 2f- and 3f-components in the electrostatic force, respectively, taken by EFM on a bare CIGS thin film with Ga content of 31 %.

spectrum should be obtained. If, to the contrary, C depends on V_{DC} like in the case of the depletion capacitance, non-linearity should appear in the spectrum, and triple frequency 3f-component should appear in the electrostatic force [11].

As for the samples in this study, we first prepared CIGS thin films with different Ga contents by the three-stage co-evaporation method on Mo-coated soda lime glass (SLG). Then, a CdS buffer layer was formed on each CIGS thin film by the chemical bath deposition, and subsequently the CdS layer was etched out by HCl solution. By adopting this etching process, we expect that a series capacitance originating from the CdS layer will be excluded and consequently that an influence of Cd diffusion into CIGS will be apparently observed through the depletion capacitance measurements by EFM.

III. RESULTS AND DISCUSSION

Figure 2(a) shows an AFM topographic image on the bare CIGS thin film with Ga content [= Ga/(Ga+In)] of 31 %, and Figs. 2(b) and (c) show images of 2f- and 3f-components in the electrostatic force, respectively, on the identical area, observed by EFM. Here, V_{DC} (sample bias) and V_{AC} were set at -0.5 and 0.5 V, respectively, for 2f-component measurements and at -0.4 and 2.0 V for 3f-component measurements. A frequency f for the a.c. bias was tuned so that a value of 2f or 3f was almost equal to the second mode resonant frequency of the cantilever (typically, 360 kHz) in order to enhance the response to the 2f- or 3f-component of the electrostatic force, respectively. The same series of images as those in Fig. 2 were taken on the CdS/CIGS

sample with Ga content of 48 % after CdS removal, as shown in Figs. 3(a)-(c). By comparing Fig. 2(c) and Fig. 3(c), we found that the 3f-component signal on the bare CIGS was very weak, while apparent signals with a certain intensity were clearly observed on the CdS/CIGS sample. The latter result means that the surface capacitance can be varied by the external bias voltage, indicating that the CIGS surface is depleted in some degree. The depletion at the CIGS surface is attributable to the Cd diffusion into CIGS and resulting formation of Cd_{Cu} as an anti-site donor which compensates Cu vacancy V_{Cu} as a main acceptor in CIGS [12].

In addition, Fig. 3 shows that some grain interiors (GIs) exhibited strong 3f-component signals as indicated by red ellipses, while some others do not. These results suggest that the depletion capacitance was fluctuated among various grains, which is attributable to non-uniformity in the acceptor density and/or the Cd diffusion level. At grain boundaries (GBs), on the other hand, small 3f-component signals were acquired as indicated by blue ellipses. This result indicates that the depletion capacitance hardly varies by the external bias, implying strong pinning of surface Fermi level. As an origin of the strong Fermi level pinning at GB, we consider that the Cd diffusion may be enhanced near GB and consequently the anti-site defect Cd_{Cu} acting as a donor may densely exist [12].

We also acquired the electrostatic force spectra, as shown in Fig. 4, at Points A and B indicated in Fig. 3. Figure 4(a) indicates that the decrease of capacitance which corresponds to the broadening of the depletion layer appeared at negative sample bias at Point A. This tendency is a typical characteristics of the

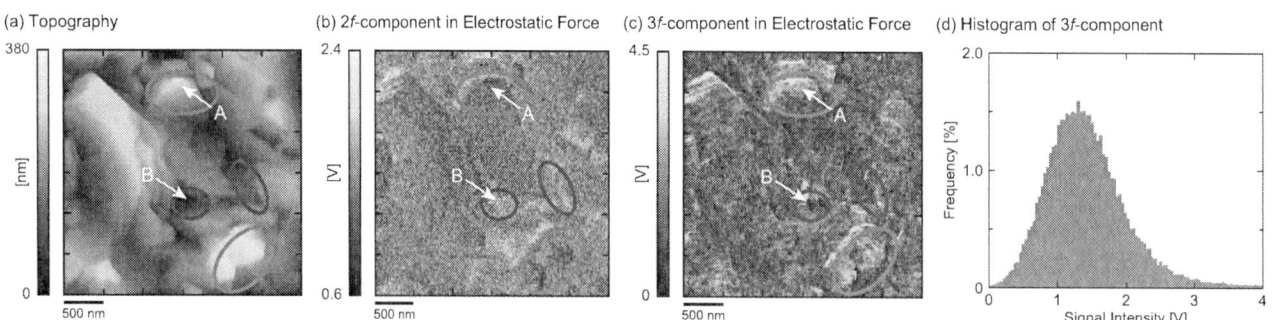

Fig. 3 (a) Topographic image, (b) and (c) images of 2f- and 3f-components in the electrostatic force, respectively, taken by EFM on a CdS/CIGS sample with Ga content of 48 % after CdS removal, and (d) histogram of 3f-components. Points A and B indicate the positions where the electrostatic force spectra shown in Fig. 4 were taken.

978-1-5090-5606-4/17 $31.00 © 2017 IEEE

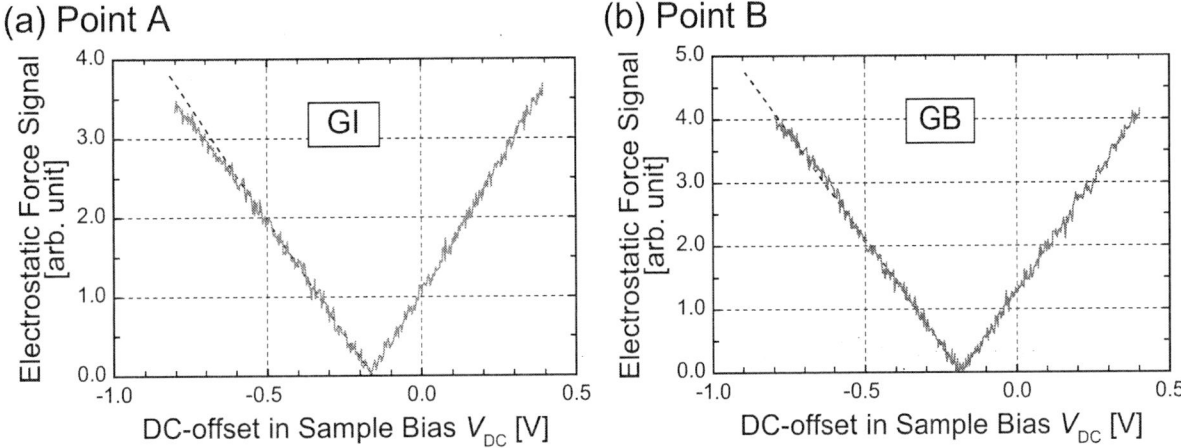

Fig. 4　(a) and (b) Electrostatic force spectra taken at Points A and B, respectively, indicated in Fig. 3. Broken straight lines are just eye-guides.

depletion capacitance in *p*-type semiconductor like CIGS. To the contrary, almost linear response was observed at Point B as shown in Fig. 4(b), suggesting the strong pinning of surface Fermi level, which is very consistent with the above deduction from the EFM images in Fig. 3.

On the other hand, Figs. 5(a)-(c) indicate the same series of images as those in Figs. 2 and 3 taken on the CdS/CIGS sample with Ga content of 31 % after CdS removal. As shown in this figure, both *2f*- and *3f*-component signals were relatively small and uniform, indicating the uniformity of the acceptor density. Now, Figs. 3(d) and 5(d) indicate histograms of the *3f*-components in the electrostatic force shown in Figs. 3(c) and 5(c), respectively, and we can surely confirm that the *3f*-component signal was small and uniform on the sample with low Ga content. As a matter of fact, a conversion efficiency of solar cell fabricated on CIGS with Ga content of 31 % was better than that fabricated on CIGS with Ga content of 48 %. Therefore, we consider that the non-uniformity of net acceptor density after the compensation by Cd_{Cu} anti-site donor which is deduced from the results shown in Fig. 3 is one cause to degrade the conversion efficiency in CIGS solar cell.

IV. CONCLUSIONS

We performed EFM on the CIGS thin films with different Ga contents and the CdS/CIGS layers after CdS removal. As a result, we found that the surface depletion of CIGS was changed after the deposition process of the CdS layer, being attributed to the Cd diffusion into CIGS layer and consequent formation of Cd_{Cu} aniti-site donor. At some grain boundaries, on the other hand, the depletion capacitance was almost independent of the external bias, and this result suggests the strong pinning of surface Fermi level due to the enhanced Cd diffusion. In addition, the CIGS layer with high Ga content exhibited non-uniform distribution of the *3f*-components in the electrostatic force, implying that the surface depletion capacitance are not uniform and consequently that the conversion efficiency is degraded.

ACKNOWLEDGEMENTS

This work was partly supported by a Grand-in-Aid for Scientific Research from Japan Society for the Promotion of Science (JSPS), and by Project for Developing Innovation Systems of the Ministry of Education, Culture, Sports, Science and Technology (MEXT), Japan.

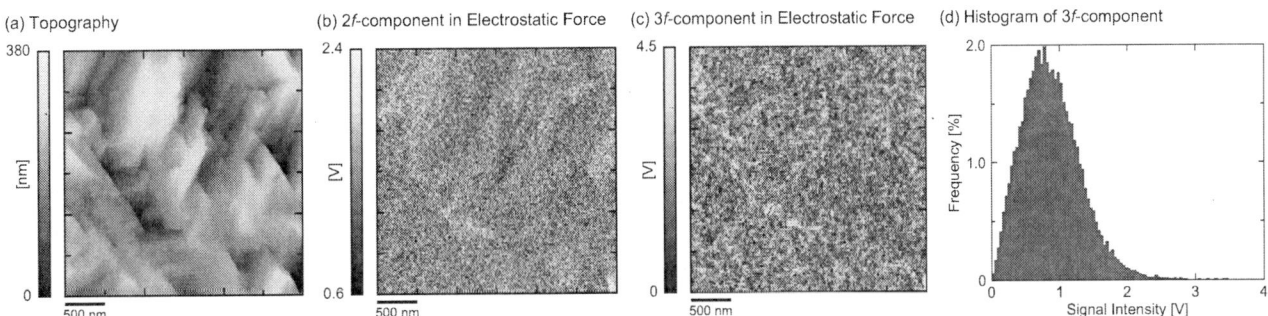

Fig. 5　(a) Topographic image, (b) and (c) images of *2f*- and *3f*-components in the electrostatic force, respectively, taken by EFM on a CdS/ CIGS sample with Ga content of 31 % after CdS removal, and (d) histogram of *3f*-components.

REFERENCES

[1] M.A. Green, K. Emery, Y. Hishikawa, W. Warta, E.D. Dunlop, D.H. Levi, and W.Y. Ho-Baillie, "Solar cell efficiency tables (version 49)," *Prog. Photovolt: Res. Appl.*, **25**, pp. 3-13, 2017.

[2] https://www.zsw-bw.de/fileadmin/user_upload/PDFs/Pressemitteilungen/2016/pr09-2016-ZSW-WorldRecord CIGS.pdf (accessed 25 October 2016).

[3] U. Rau, K. Taretto, and S. Siebentritt, "Grain boundaries in Cu (In,Ga)(Se,S)$_2$ thin-film solar cells," *Appl. Phys. A*, **96**, pp. 221–234, 2009.

[4] C.S. Jiang, R. Noufi, J.A. AbuShama, K. Ramanathan, H.R. Moutinho, J. Pankow, and M. Al-Jassim, "Local built-in potential on grain boundary of Cu(In,Ga)Se$_2$ thin films," *Appl. Phys. Lett.*, **84**, pp. 3477–3479, 2004.

[5] C.S. Jiang, R. Noufi, K. Ramanathan, J.A. AbuShama, H.R. Moutinho, and M. Al-Jassim, "Does the local built-in potential on grain boundaries of Cu(In,Ga)Se$_2$ thin films benefit photovoltaic performance of the device?" *Appl. Phys. Lett.*, **85**, pp. 2625–2627, 2004.

[6] G. Hanna, T. Glatzel, S. Sadewasser, N. Ott, H.P. Strunk, U. Rau, and J.H. Werner, "Texture and electronic activity of grain boundaries in Cu(In,Ga)Se$_2$ thin films," *Appl. Phys. A*, **82**, pp. 1–7, 2006.

[7] D.F. Marrón, S. Sadewasser, A. Meeder, T. Glatzel, and M.C. Lux-Steiner, "Electrical activity at grain boundaries of Cu(In,Ga) Se$_2$ thin films," *Phys. Rev. B*, **71**, p. 033306, 2005.

[8] M. Takihara, T. Minemoto, Y. Wakisaka, and T. Takahashi, "An investigation of band profile around the grain boundary of Cu(InGa)Se$_2$ solar cell material by scanning probe microscopy," *Prog. Photovolt: Res. Appl.*, **21**, pp. 595-599, 2013.

[9] H. Yong, Y. Nakajima, T. Minemoto, and T. Takahashi, "Photovoltage decay processes in Cu(In,Ga)Se$_2$ solar cells studied by photo-assisted Kelvin probe force microscopy", *Proceedings of the 39th IEEE Photovoltaic Specialists Conference*, 2013.

[10] Y. Hamamoto, K. Hara, T. Minemoto, and T. Takahashi, "Photothermal spectroscopy by atomic force microscopy on Cu(In,Ga)Se$_2$ solar cell materials", *Solar Energy Materials & Solar Cells*, **141**, pp. 32–38, 2015.

[11] K. Kobayashi, H. Yamada, and K. Matsushige, "Dopant profiling on semiconducting sample by scanning capacitance force microscopy", *Appl. Phys. Lett.*, **81**, pp. 2629-2631, 2002.

[12] O. Cojocaru-Miredin, P. Choi, R. Wuerz, and D. Raabe, "Exploring the p-n junction region in Cu(In,Ga)Se$_2$ thin-film solar cells at the nanometer-scale", *Appl. Phys. Lett.*, **101**, p. 181603, 2012.

A Comparison of Si-based Cameras for Imaging Luminescence from Photovoltaic Materials and Devices

Steve Johnston

National Renewable Energy Laboratory, Golden, CO, 80401, U.S.A.

Abstract — **Si cameras for luminescence imaging of photovoltaic devices are compared. The Princeton Instruments PIXIS 1024BR camera is a scientific, cooled, Si-based camera that has back-illumination and deep-depletion for improved near-infrared sensitivity. This camera is compared to a consumer-grade Nikon D5100 camera with the internal short-pass filter removed. For a fraction of the cost, the Nikon camera shows reasonable comparison in image quality, but variations of luminescence intensity and exposure time give a less linear response than that of the PIXIS camera. For cost and performance between these examples, other camera choices, such as the Thor Labs 1501M, can provide good image quality and linearity.**

Index Terms — **imaging, cameras, photovoltaic cells, photoluminescence, electroluminescence, silicon.**

I. INTRODUCTION

Electroluminescence (EL) imaging [1] and photoluminescence (PL) imaging [2]-[3] are valuable characterization tools for the photovoltaics community. Silicon cells and modules compose a vast majority of the market. Thin films such as $CuIn_xGa_{(1-x)}Se_2$ (CIGS) and CdTe are currently the other significant contributors. Si, CIGS, and CdTe can all be imaged at various process steps using Si-based cameras. While InGaAs cameras have more sensitivity to collect luminescence from Si and CIGS, the larger cost and lower resolution currently limit the implementation of these cameras for such luminescence applications.

PL imaging has been shown to be useful and applicable along the manufacturing line from bricks to wafers to finished cells for silicon cell production and also at various process steps for CIGS cell production [4]-[6]. While some process steps demand highly sensitive cameras due to weak PL signals, such as for as-cut or textured Si wafers, other steps have passivated surfaces that lead to stronger luminescence. At these steps, such as after diffusion, after anti-reflective coating, at the finished cell, or module-level assembly and testing, a moderately priced camera may sufficiently collect useful images for process monitoring, defect control, and quality assurance. Comparisons of a high-end, cooled, scientific-grade camera to an uncooled scientific camera and a consumer-grade camera with its internal short-pass filter removed are presented.

II. EXPERIMENT

Besides using an InGaAs camera, the most sensitive cameras for imaging in the near-infrared for Si and CIGS luminescence are Si cameras with detectors that are back-illuminated and use deep-depletion. Cooling of the detector reduces dark counts and allows for longer exposure times to improve the image signal-to-noise ratio. Many options and various camera models are available, but a comparison is made here between a Princeton Instruments scientific-grade, back-illuminated, deep-depletion, cooled Si camera; an uncooled Thor Labs scientific Si camera; and a Nikon consumer-grade Si camera.

The consumer-grade camera is a Nikon D5100 with up to 4,928 x 3,264 pixels in the large image size format. This camera was disassembled to remove the built-in short-pass filter that would normally block near-infrared light, and then reassembled. A similar example is given in [7]. The camera is operated using Nikon Control Pro software. The settings include monochromatic, fixed ISO 6400 for gain, but other automatic processing features are turned off, if possible. The exposure time can be set from fractions of a second up to 30 s. The Nikon camera uses a 50 mm F/1.2 manual focus lens. A photograph of this camera and its lens is shown in Fig. 1(a).

The uncooled scientific camera is a 1.4-megapixel Thor

Fig. 1. Lab photographs show the cameras: (a) Nikon D5100 camera, (b) Thor Labs 1501M camera, and (c) Princeton Instruments PIXIS 1024BR camera.

Labs model 1501M having 1,392 x 1,040 14-bit pixels. The pixels are 6.45 μm square. This camera, as shown in Fig. 1(b), has a user-removable infrared filter, which was removed for maximum sensitivity. This camera model also uses a near-infrared enhanced (boost) mode that is selected via the software. The lens used with this camera is a c-mount Schneider 50 mm F/2.8 lens.

The cooled scientific camera is a Princeton Instruments PIXIS 1024BR having a 1,024 x 1,024 imaging array with 13-μm x 13-μm 16-bit pixels. The camera is cooled to -60°C and operated using WinView software. The PIXIS camera uses the same c-mount Schneider 50 mm F/2.8 lens. A photograph of this camera and its lens is shown in Fig. 1(c). For all cameras, images were collected in a dark cabinet.

III. RESULTS

A comparison of EL images of a multi-crystalline Si cell is shown in Fig. 2. The left column shows images from the Nikon camera, while the right column shows images from the

PIXIS camera. The top row shows EL images with 1 A of current flowing in the cell in forward bias, and the exposure time for each camera is 8 s. The bottom row shows images for 4 A of current and 1 s exposure times. The images appear comparable when looking at general features such as grain boundaries and clusters of grain-related defects, back-contact soldering structures, cracks, and degradation in the center area. A red box in Fig. 2(d) identifies a relatively uniform region to be used for more quantitative comparison of signal sensitivity and linearity.

EL images were collected for both cameras using a range of currents through the cell (1 to 5 A) and a range of camera exposure times (0.5 to 256 s). The image area within the red box of Fig 2(d) was averaged. This average value of intensity (average number of pixel counts) is summarized for the camera comparison plots in Fig. 3. The plots in Fig. 3 show the increasing counts as the exposure time is lengthened. The EL signal from the cell increases when driving more current through the device. For weak signals, the readout noise of the camera can be comparable to the signal. For every exposure

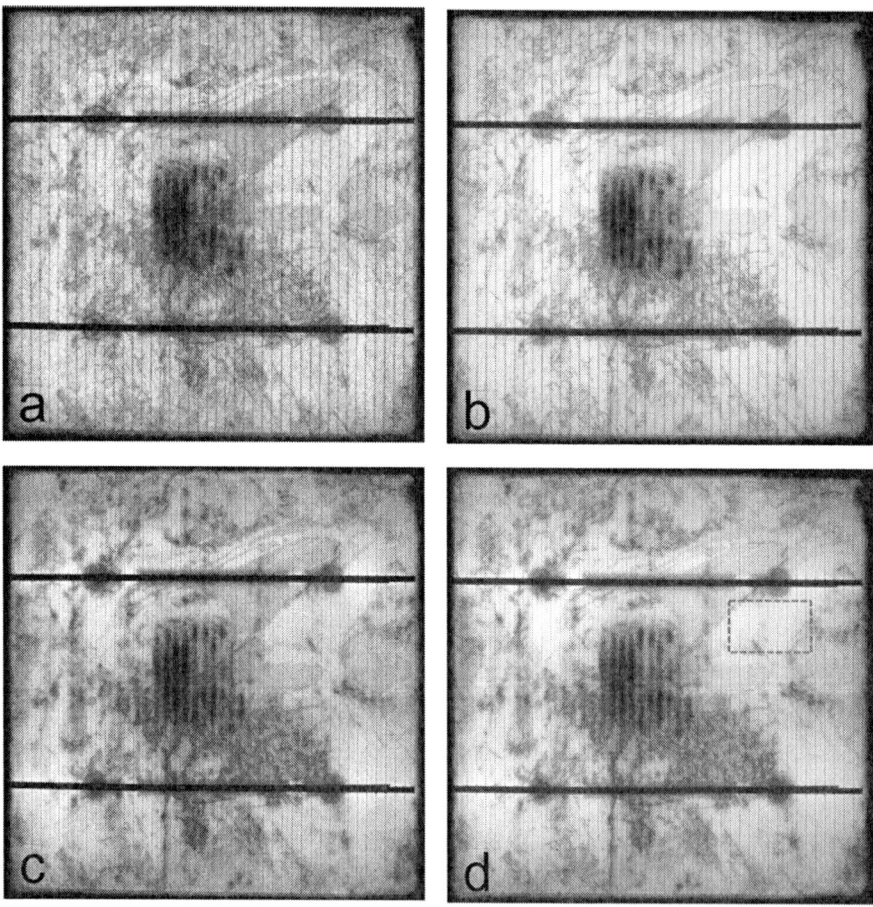

Fig. 2. EL images collected using 1 A of current and an 8-second exposure time for a Nikon D5100 camera (a) and a PIXIS camera (b). EL images for 4 A of current and a 1-second exposure time for the Nikon camera (c) and the PIXIS camera (d). The red box highlights the region used to average intensity for comparison.

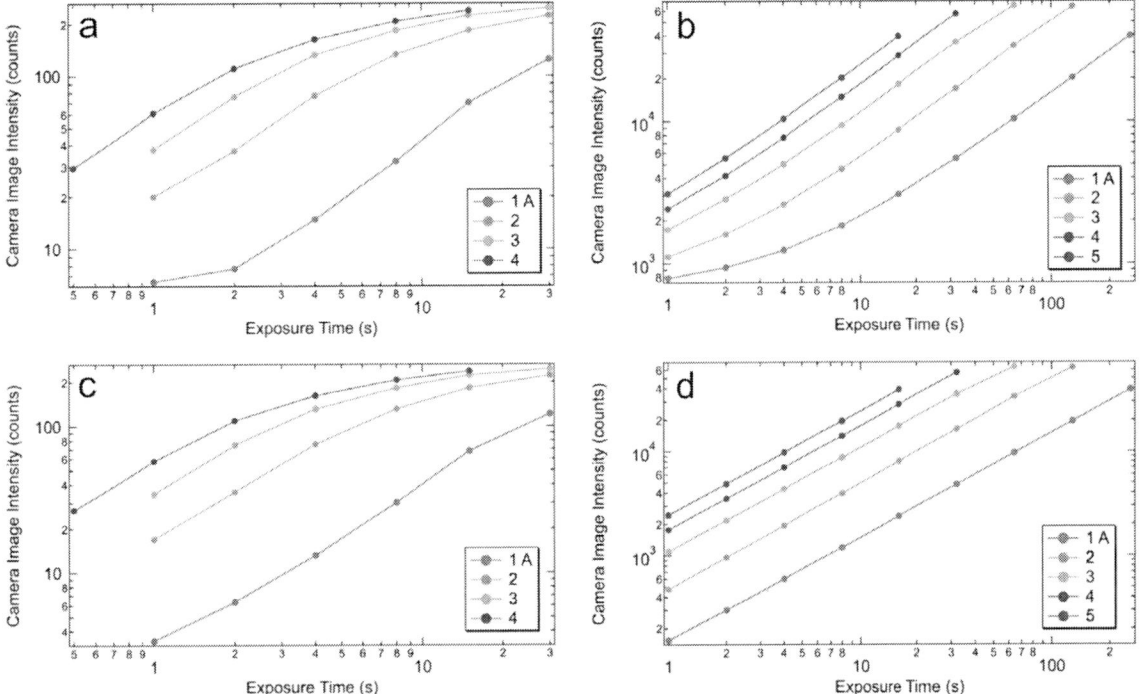

Fig. 3. Average camera pixel counts from the red box area of Fig. 2 are plotted as a function of camera exposure times for varying EL currents flowing through the cell. The top graphs, (a) and (b), represent data without background subtraction, while the lower graphs, (c) and (d), include background subtraction. Data from the Nikon D5100 camera are plotted in (a) and (c). Data from the PIXIS camera are plotted in (b) and (d).

time, a dark background image is collected with no current running through the device. In Fig. 3(c) and 3(d), the respective dark background image is subtracted from the EL image, giving average pixel counts that are more linear at low signal strength.

The same plotting procedures are used for both the Nikon camera (Fig. 3(a) and 3(c)) and the PIXIS camera (Fig. 3(b) and 3(d)). Average pixel counts are plotted as a function of exposure time and cell current. Fig. 3(d) shows the linear response of the PIXIS camera over the entire range when dark background images are subtracted to account for readout noise. The PIXIS camera collects images with 16-bit depth, so pixel intensity ranges from 0 to 65,535. This range gives the PIXIS camera a larger dynamic range of counts and nearly three orders of magnitude of linear response as shown in Fig. 3(d). The images from the Nikon camera are output in an 8-bit format, so the pixel intensity ranges from 0 to 255. Features of the Nikon camera's hardware and software appear to prevent the camera from providing a linear response near the upper range of the bit depth. The camera appears to have roughly an order of magnitude linear response region as shown in Fig. 3(c). If long exposure times are used and counts are in the upper portion of the 8-bit range, then the image contrast or ratio between good, high-performance regions of the cell and poor, defect-limited regions will be reduced due to the non-linear response.

The images in Fig. 4 compare the PIXIS camera to the 1501M camera from Thor Labs, also using a multicrystalline Si cell as an example. Fig. 4(a) shows an EL image collected from the PIXIS camera using an exposure time of 16 s with 4.4 A of current flowing through the cell. The exposure time of 16 s allowed the 16-bit pixels to nearly approach saturation (65,535 counts) in the brightest areas. A dark background image with 0 A of current flowing through the cell and the same exposure time of 16 s has been subtracted from the EL image, even though it has negligible effect for this cooled detector when the counts per pixel are large.

For comparison, an EL image, also using 4.4 A of current flowing through the cell, is collected by the Thor Labs 1501M camera and is shown in Fig. 4(b). For the 1501M camera, the near-infrared boost mode was active, and the gain was set to its highest value of 1,023. Because the pixels are 14-bit, they approach saturation when the counts reach 16,383. A shorter exposure time of 4 s was used to prevent saturation. The camera's software also allowed for up to 32 averages of the images to improve the signal-to-noise ratio and help compensate for the lack of significant detector cooling. A dark background image was also collected for subtraction from the EL image. So, a total acquisition time of about 4 minutes was used for the 1501M image, and, as shown in Fig. 4, the EL images from each camera appear very similar.

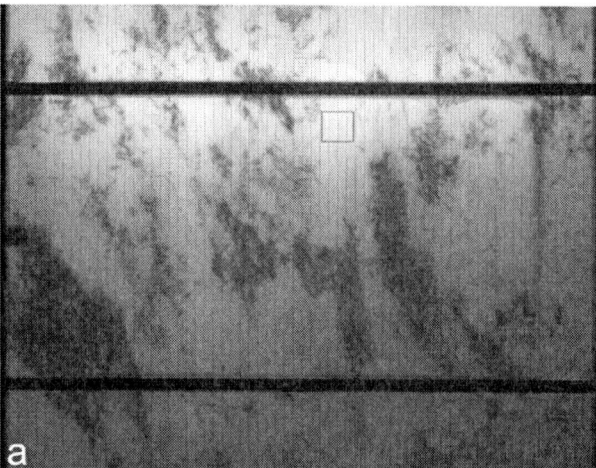

Fig. 4. EL images collected using 4.4 A of current for the PIXIS camera (a) and the Thor Labs 1501M camera (b). The PIXIS camera image was acquired with a 16 s exposure, and the 1501M camera used an average of 32 images with 4-second exposure times. The red box in (a) corresponds to the area used for plotting average intensity as a function of camera exposure time.

While Fig. 4 illustrates the maximum exposure time (and averaged images for the 1501M camera) for each camera without saturation, other images with shorter exposure times were also collected to provide data for plotting linearity. For both the PIXIS and 1501M cameras, EL images were collected using exposure times of 125 ms, 250 ms, 500 ms, 1 s, 2 s, and 4 s. For the 1501M camera, a 32-image average was used for each exposure time setting, and both cameras used a dark, zero-current image of the same exposure time (and averages for the 1501M camera) for background subtraction. After the dark background image was subtracted, the area within the red box shown in Fig. 4(a) was averaged to give an average intensity value for each camera exposure time. The results are plotted in Fig. 5. For the PIXIS camera, a linear relationship is observed for average EL intensity as a

function of camera exposure time. These values are represented by the solid dark-blue circles. The highest count values extend toward the maximum values of the 16-bit pixels, which is 65,535. For the 1501M camera, a linear relationship is also observed. The average EL intensity values are represented by the red, open-circle symbols. These values extend to about the maximum for the 14-bit pixels, which is 16,383.

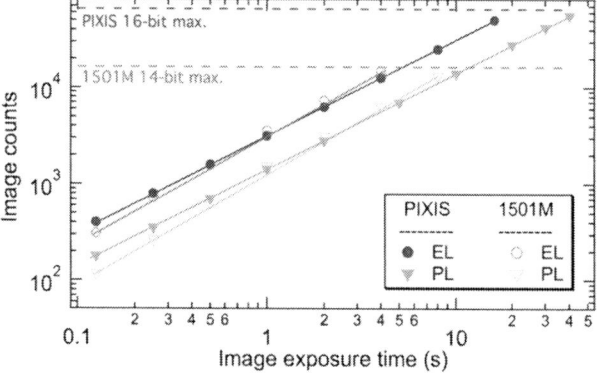

Fig. 5. Average camera pixel counts from the red box area of Fig. 4 are plotted as a function of camera exposure times for a current of 4.4 A flowing through the cell for the EL curves. PL curves use image data as shown in Fig. 6. Data from the PIXIS camera are plotted using solid symbols, while data from the 1501M camera use open symbols.

PL imaging is also used to compare the PIXIS and 1501M cameras using a multicrystalline wafer that has only been partially processed into a cell. Fig. 6 shows PL images and corresponding histograms of the PL intensities. Similar to the EL images, the PL images are collected over a range of exposure times with dark background images of the same exposure times subtracted. The PIXIS camera uses one exposure, while the 1501M averages 32 images of that exposure time. The area of the red box shown in Fig. 6(a) is averaged and plotted in Fig. 5 to show linearity of response from short exposure times with small filling of the pixels to long exposure times with near-saturated pixels. Again, as seen with the EL images, the PL images show a linear relationship of average pixel counts as a function of the camera image exposure time for each camera's range of pixel bin depth.

IV. SUMMARY

For luminescence imaging of photovoltaic materials and devices, a large range of camera models and capabilities exists. A comparison of a high-performance cooled scientific camera to both a low-cost consumer-grade camera and an uncooled scientific camera has shown that comparable images can be acquired. While high sensitivity in the near-infrared and image-sensor cooling are likely necessary for imaging

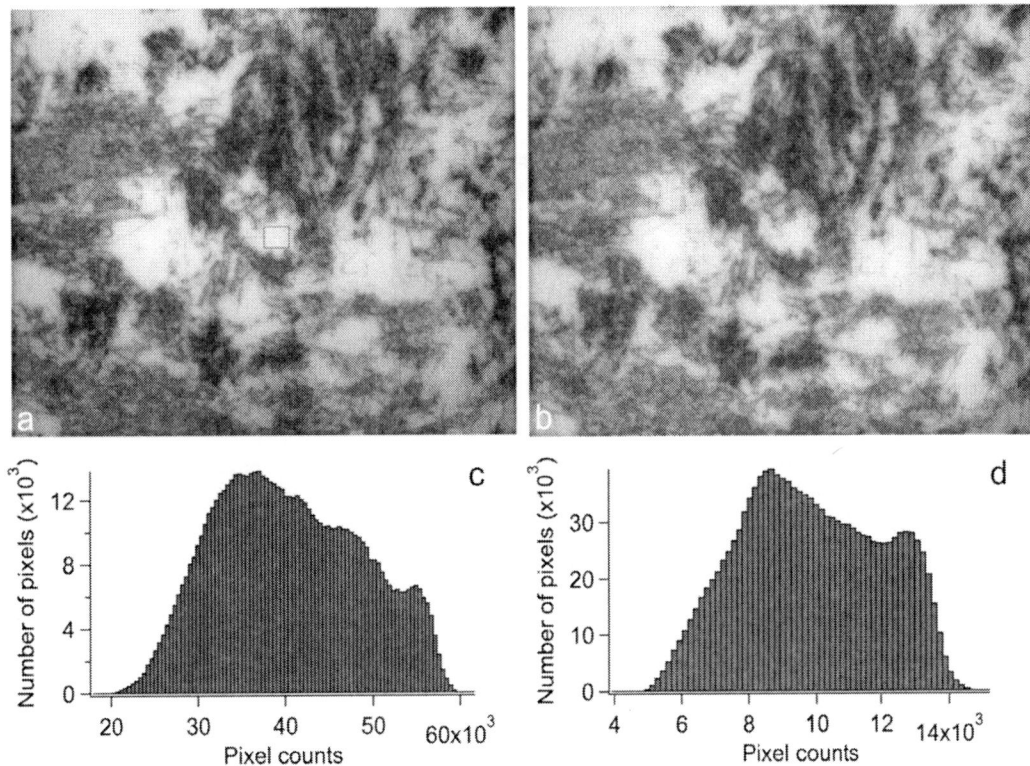

Fig. 6. PL images of a partially-processed multicrystalline Si cell collected with the PIXIS camera (a) and the Thor Labs 1501M camera (b). The PIXIS camera image was acquired with a 40 s exposure, and the 1501M camera used an average of 32 images with 8-second exposure times. Histograms (c) and (d) correspond to the full images (a) and (b), respectively. The red box in (a) corresponds to the area used for plotting average PL intensity as a function of camera exposure time. Data from the red box area is plotted in Fig. 5.

unpassivated or poor-quality materials, a low-cost camera may provide reasonable images on materials and cells when possible limitations of sensitivity or linearity are considered.

ACKNOWLEDGEMENT

The U.S. Government retains and the publisher, by accepting the article for publication, acknowledges that the U.S. Government retains a nonexclusive, paid-up, irrevocable, worldwide license to publish or reproduce the published form of this work, or allow others to do so, for U.S. Government purposes. This work was supported by the U.S. Department of Energy under Contract No. DE-AC36-08GO28308 with the National Renewable Energy Laboratory. Funding provided by the U.S. Department of Energy Office of Energy Efficiency and Renewable Energy Solar Energy Technologies Office.

REFERENCES

[1] T. Fuyuki, H. Kondo, T. Yamazaki, Y. Takahashi, and Y. Uraoka, "Photographic surveying of minority carrier diffusion length in polycrystalline silicon solar cells by electroluminescence," *Appl. Phys. Lett.*, vol. 86, pp. 262108, 2005.

[2] T. Trupke, R. A. Bardos, M. C. Schubert, and W. Warta, "Photoluminescence imaging of silicon wafers," *Appl. Phys. Lett.*, vol. 89, pp. 044107, 2006.

[3] T. Trupke, R. A. Bardos, M. D. Abbott, F. W. Chen, J. E. Cotter, and A. Lorenz, "Fast photoluminescence imaging of silicon wafers," in *32nd IEEE PVSC and WCPEC-4*, pp. 928-931, 2006.

[4] S. Johnston, F. Yan, K. Zaunbrecher, M. Al-Jassim, O. Sidelkheir, and K. Ounadjela, "Quality characterization of silicon bricks using photoluminescence imaging and photoconductive decay," in *38th IEEE Photovoltaics Specialist Conference*, pp. 406-410, 2012.

[5] S. Johnston, F. Yan, D. Dorn, K. Zaunbrecher, M. Al-Jassim, O. Sidelkheir, and K. Ounadjela, "Comparison of photoluminescence imaging on starting multi-crystalline silicon wafers to finished cell performance," in *38th IEEE Photovoltaic Specialist Conference*, pp. 2161-2166, 2012.

[6] S. Johnston, T. Unold, I. Repins, A. Kanavce, K. Zaunbrecher, F. Yan, J. V. Li, P. Dippo, R. Sundaramoorthy, K. M. Jones, and B. To, "Correlations of Cu(In, Ga)Se$_2$ imaging with device performance, defects, and microstructural properties," *J. Vac. Sci. Technol. A*, vol. 30, pp. 04D111-1-04D111-6, 2012.

[7] J. S. Fada, N. R. Wheeler, D. Zablyaka, N. Goel, T. J. Peshek, and R. H. French, "Democratizing an electroluminescence imaging apparatus and analytics project for widespread data acquisition in photovoltaic materials," *Rev. Sci. Instrum.*, vol. 87, pp. 085109-1-085109-6, 2016.

Blistering of Al₂O₃/a-SiN$_X$:H stacks: analysis of the submerged part of the iceberg by colored picosecond acoustic microscopy

Fabien Lebreton[1,2,3], Arnaud Devos[4], Etienne Drahi[1,3], Patricia de Coux[1], François Silva[2,3], Sergej Filonovich[1,3], Pere Roca i Cabarrocas[2,3]

[1]Total S.A. Renewables, 2 Place Jean Millier, 92078 Paris La Défense cedex, France

[2]LPICM, CNRS, Ecole Polytechnique, Université Paris-Saclay, 91128 Palaiseau, France

[3]Institut Photovoltaïque d'Ile-de-France (IPVF), 8 rue de la renaissance, 92160 Antony, France

[4]IEMN UMR 8520, Dpt ISEN, 41 Bd Vauban, 59046 Lillle Cedex, France

Abstract — **Blistering of Al₂O₃/SiN$_X$:H stacks has been regularly reported over the last decade. Despite several studies, it has not been possible to link blistering density and lifetime degradation. In this work we demonstrate the use APiC technique to probe the c-Si/Al₂O₃ interface. It allows us to show that it might not be the delaminated surface fraction which rules the lifetime degradation but most likely the interface quality of non-delaminated areas. Blistering is the top of the iceberg; characterizing the interface with the high sensitivity provided by APiC is a breakthrough.**

I. INTRODUCTION

Aluminum oxide (Al₂O₃) capped with amorphous silicon nitride (a-SiN$_X$:H) is nowadays a common passivation stack for p-type surfaces [1]. The combination of atomic layer deposition (ALD) for Al₂O₃ with plasma-enhanced chemical vapor deposition (PECVD) for the capping made its success. It is due to very high passivation properties provided by the first one and the good thermal stability ensured by the second one [2]. Performances of such stacks seem to be ascribed to the large amount of hydrogen released by the stack during post-deposition thermal treatments [3]. This hydrogen can easily diffuse to the c-Si/Al₂O₃ interface passivating remaining dangling bonds. The main drawback of this gas release is the well-known blistering of the passivation stacks, which can deteriorate their optical and electrical properties [4, 5]. Up to now, it has not been possible to establish a direct link between blistering density and passivation quality [5, 6]. One way to limit blistering is to grow a thin silicon oxide layer prior to ALD deposition [7]. This layer is a good nucleation surface for the ALD process allowing a direct layer by layer growth mode rather than the island growth during the first ten cycles [8].

These observations reported in the literature and our own observations of non-homogeneously delaminated blisters (Fig. 1) let us to suggest an important contribution of the adhesion parameter. The proportion of delaminated surface due to blistering is generally lower than 10% [5], but it seems to be enough to deteriorate the whole minority carrier lifetime. We think that there is no direct relation between blistering density and lifetime. Passivation quality might be ruled by surface located between blisters, *i.e.* within supposed "healthy" areas. In this work we propose to assess these interface properties by colored picosecond acoustic microscopy (APiC) [9].

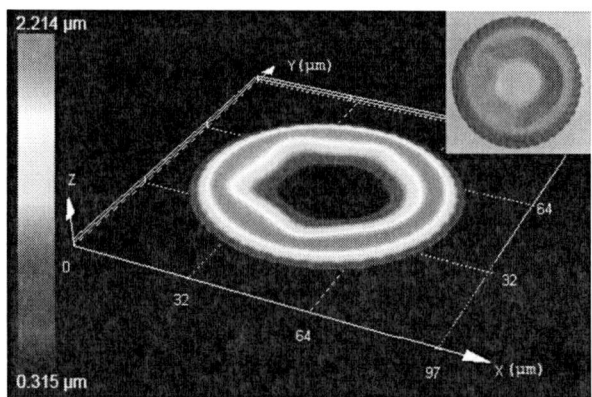

Fig. 1. Topography of a non-uniformly delaminated blister obtained by confocal microscopy. The colored scale indicates the height in the Z direction. Inset: Top view of the blister.

TABLE I

SUMMARY OF SAMPLES PARAMETERS AND MEASURED PROPERTIES. LIFETIME VALUES ARE DETERMINED BY QSS-PC. BLISTERING DATA ARE ACQUIRED BY OPTICAL MICROSCOPY.

Sample	Al₂O₃ thickness (nm)	Post-Deposition Annealing		τ_{eff} (µs)	Blistering		
		Duration (min)	Temperature (°C)		Density (blister/cm²)	Average diameter (µm)	Delaminated area (%)
#1	30	-	-	1633	2760 ± 420	60 ± 7	7.8 ± 3.4
#2	30	10	380	1612	1820 ± 60	21± 4	0.6 ± 0.3
#3	6	10	380	3860	0	0	0

II. EXPERIMENTAL APPROACH

Al_2O_3/a-SiN_X:H symmetrically processed c-Si samples were produced. Their process parameters and main numerical results are summarized in Table I. Al_2O_3 is deposited on both sides of the c-Si substrate in the same run thanks to a specific substrate holder which allows diffusion of precursors to both surfaces. An Al_2O_3 thickness of 30 nm was chosen to promote blistering on samples #1 and #2. A reference sample using a thickness of 6 nm to avoid blistering was chosen according to previous studies [10]. The c-Si wafers were double-side polished, float-zone, p-type, <100>, with a resistivity of 1–5 Ω.cm and a thickness of 280 µm. After HF dip, Al_2O_3 deposition was carried out at 250 °C. Some samples received post-deposition annealing (PDA) in a forming gas atmosphere to activate hydrogen desorption. After this optional PDA, all samples received a 75 nm PECVD a-SiN_X:H capping at a substrate temperature of 300 °C. A post-capping annealing for 20 min at 380 °C in forming gas atmosphere was then carried out to mimic the thermal budget of Ni/Cu plating metalization developed at IPVF [11].

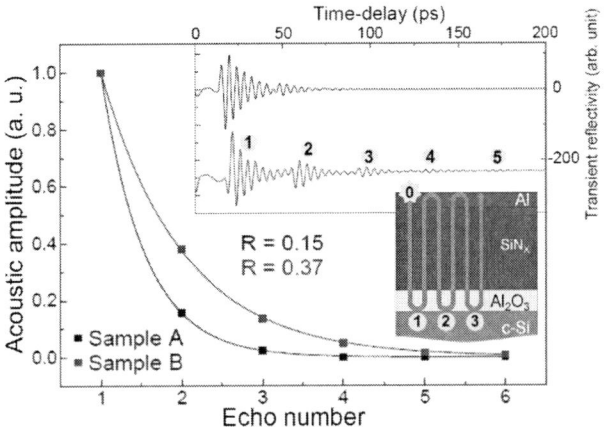

Fig. 2. Principle of APiC measurement. Inset graph and sample sketch: a thin aluminum layer is evaporated on top of the SiN_X:H/Al_2O_3 passivation stack. An acoustic wave is generated by a laser pulse on this Al layer (point 0). The wave propagates through the dielectric stack and is reflected back at the Al_2O_3/c-Si interface (point 1). Depending of the adhesion between c-Si and the passivation stack, several reflections can occur (points 2, 3 and more). Main graph: acoustic amplitude as a function of the echo number. The reflection coefficient R is determined by fitting the mean echo number (n) with a power law R^n. The lower R, the better the adhesion. In this example c-Si/Al_2O_3 adhesion of sample A is better than that of sample B.

Lifetime measurements have been done by photoconductance decay in quasi-steady-state mode (QSS-PC). Lifetimes are reported in this paper for carrier densities of 10^{15} cm^{-3}. High resolution lifetime maps have been obtained by QSS-PC calibrated high magnification photoluminescence. Blistering density has been determined by optical microscopy. The delaminated area has been estimated

assuming circular blisters [5]. In order to probe the c-Si/Al_2O_3 interface between blisters, high resolution APiC microscopy mapping was used. This technique, rather new in the field of photovoltaics, is described in Fig. 2 [9].

III. RESULTS AND DISCUSSION

The high lifetime of sample #3 (3.8 ms) coated by 6 nm of Al_2O_3 could be partially ascribed to its blister-free surface. The presence of blisters on samples #1 and #2 directly decreases their lifetime by more than a factor of 2 compared to sample #3 (see table I). However samples #1 and #2 have different blisters sizes, densities and so delaminated surfaces by more than one order of magnitude.

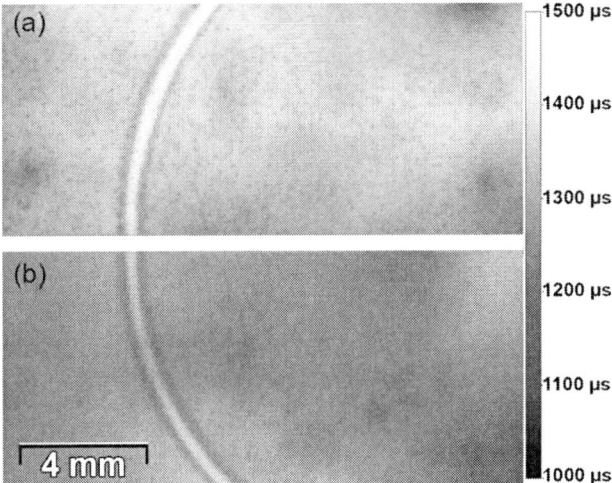

Fig. 3. High resolution QSS-PC calibrated photoluminescence of (a) sample #1 and (b) sample #2. Visible arc on both images is the Sinton stage sensor.

High magnification photoluminescence (Fig. 3) highlights a quite different response of these two samples. The sample having the lowest blistering density (Fig. 3-b) shows a rather homogenous photoluminescence response over its entire surface while on the other sample a spotted surface is visible (Fig. 3-a). This interesting pattern might not be visible on sample #2 because of the lower blistering density but also due to the average blister size, slightly smaller than a pixel of the detector. However, the average lifetime between blisters seems higher on samples #1 than on sample #2.

High resolution APiC map shows a very low and uniform reflection coefficient (R) for the reference sample (sample #3, Fig. 4-c). Sample #2 also presents a rather homogeneous R between blisters (Fig. 4-b) but its value is higher than for sample #3, highlighting a weaker interface. Sample #1 shows a mixed pattern of strong and weaker adhesion between its blisters. This contrast between delaminated and non-delaminated areas might be responsible for the higher photoluminescence signal observed on Fig. 3-a. In terms of

978-1-5090-5606-4/17 $31.00 © 2017 IEEE

delamination, 4.5% and 0.45% of the APiC probed area are considered as defective for samples #1 and #2 respectively. These values are in good agreement with these determined by optical microscopy. Statistically, it can be seen on Fig. 4-d that the median value of the R is lower for sample #1 than for sample #2, underlining a better interfacial property between its blisters. However, the average value of these samples is similar, reflecting maybe more the QSS-PC lifetime which is an average value over the probed surface.

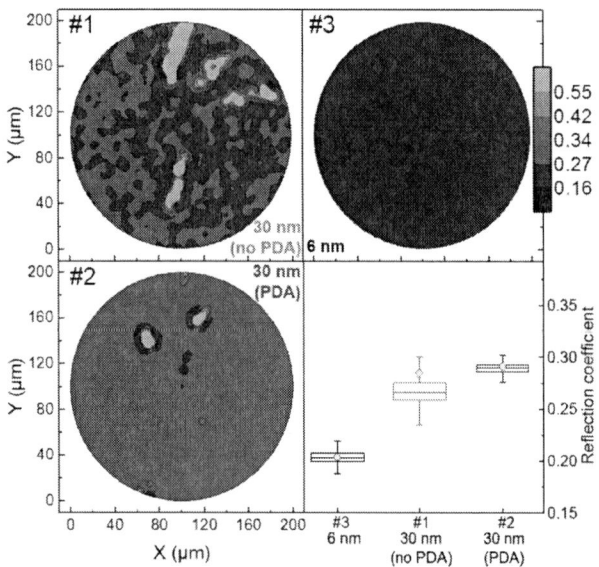

Fig. 4. High resolution APiC mapping of the reflection coefficient of samples #1, #2 and #3. Orange areas are blisters. Bottom right: Statistic distribution of the adhesion coefficient for the 1330 measured points of each map. Lines in the middle of boxes represent the median values. Green dots represent the average value.

IV. CONCLUSION

Blistering is a visible macroscopic phenomenon highlighting delamination of the films. Like the visible part of an iceberg, blisters do not represent what is happening under the sea. Blisters are just a local manifestation of a broader interface degradation. Our results show that the acoustic reflection coefficient is the relevant parameter to evaluate passivation quality of a blistered surface. Its median value is representative of the percentage of delaminated area, while its average value seems to be more representative of the effective lifetime. APiC microscopy seems to be the key characterization to dive deeper into the understanding of blistering. As APiC requires metalization of the passivation stack, it can be the perfect technique to assess the passivation quality of PERC/PERT solar cells backside with full metalization.

ACKNOWLEDGMENT

This work was carried out in the framework of a project A from IPVF (Institut Photovoltaïque d'Ile-de-France). This project has been supported by the French Government in the frame of Programme d'Investissement d'Avenir – ANR-IEED-002-01. Fabien Lebreton thanks ANRT for CIFRE scholarship.

REFERENCES

[1] V. I. Kuznetsov, M. A. Ernst, and E. H. A. Granneman, "Al$_2$O$_3$ surface passivation of silicon solar cells by low cost ald technology," in *40th IEEE Photovoltaic Specialists Conference*, 2014, pp. 0608-0611.

[2] B. Veith, F. Werner, D. Zielke, R. Brendel, and J. Schmidt, "Comparison of the thermal stability of single Al$_2$O$_3$ layers and Al$_2$O$_3$/SiN$_X$ stacks for the surface passiviation of silicon," *Energy Procedia,* vol. 8, pp. 307-312, 2011.

[3] A. Richter, J. Benick, M. Hermle, and S. W. Glunz, "Excellent silicon surface passivation with 5 Å thin ALD Al$_2$O$_3$ layers: Influence of different thermal post-deposition treatments," *physica status solidi (RRL) – Rapid Research Letters,* vol. 5, no. 5-6, pp. 202-204, 2011.

[4] B. Vermang, H. Goverde, V. Simons, I. De Wolf, J. Meersschaut, S. Tanaka, J. John, J. Poortmans, and R. Mertens, "A study of blister formation in ALD Al$_2$O$_3$ grown on silicon," in *38th IEEE Photovoltaic Specialists Conference*, 2012, pp. 001135-001138.

[5] L. Hennen, E. Granneman, and W. Kessels, "Analysis of blister formation in spatial ALD Al$_2$O$_3$ for silicon surface passivation," in *38th IEEE Photovoltaic Specialists Conference*, 2012, pp. 001049-001054.

[6] P. Saint-Cast, D. Kania, R. Heller, S. Kuehnhold, M. Hofmann, J. Rentsch, and R. Preu, "High-temperature stability of c-Si surface passivation by thick PECVD Al$_2$O$_3$ with and without hydrogenated capping layers," *Applied Surface Science,* vol. 258, no. 21, pp. 8371-8376, 2012.

[7] Y. Bao, S. Li, G. von Gastrow, P. Repo, H. Savin, and M. Putkonen, "Effect of substrate pretreatments on the atomic layer deposited Al$_2$O$_3$ passivation quality," *Journal of Vacuum Science & Technology A: Vacuum, Surfaces, and Films,* vol. 33, no. 1, pp. 01A123, 2015.

[8] V. Naumann, M. Otto, R. B. Wehrspohn, and C. Hagendorf, "Chemical and structural study of electrically passivating Al$_2$O$_3$/Si interfaces prepared by atomic layer deposition," *Journal of Vacuum Science & Technology A,* vol. 30, no. 4, pp. 04D106, 2012.

[9] A. Devos, "Colored ultrafast acoustics: From fundamentals to applications," *Ultrasonics,* vol. 56, pp. 90-97, 2015.

[10] A. Zauner, F. Lebreton, P. Saint-Cast, M. Hofmann, J.-Y. Letellier, E. Urrejola, Y. Marot, F. Gouhinec, C. Charpentier, J. Hong, F. Coeuret, and S. Pouliquen, "PERC solar cells comparison of Al precursors for rear-side surface passivation," in *29th European Photovoltaic Solar Energy Conference*, 2014, pp. 1413-1416.

[11] J. Couderc, H. e. Belghiti, D. Aureau, J. Dupuis, P.-P. Grand, É. Delbos, A. Etcheberry, and D. Lincot, "Study of one-step annealing for plated nickel-copper contacts on n-pert bifacial monocrystalline silicon solar cells," in *32nd European Photovoltaic Solar Energy Conference*, 2016, pp. 697-702.

978-1-5090-5606-4/17 $31.00 © 2017 IEEE

Self-Reference Procedure to Reduce Uncertainty in Module Calibration

D.H. Levi, C.R. Osterwald, S. Rummel, L. Ottoson, A. Anderberg

National Renewable Energy Laboratory, Golden, CO, USA 80401, etc.

Abstract — **This paper presents a composite measurement approach that capitalizes on complimentary strengths of 3 different test beds to significantly reduce uncertainty in I-V parameters for secondary module calibration. This approach addresses PV manufacturers' need for reduced uncertainty in the calibration modules that provide the basis for their module power ratings. This new method enables NREL to reduce uncertainty for secondary module calibration in commercial c-Si modules from ±3.2% to ±0.7% for I_{SC} and from ±3.3% to ±1.1% for P_{MAX}. This new module self-reference, or MSR procedure is based on the sensitivity of module voltage to temperature plus the uniformity of outdoor sunlight. By calibrating module V_{OC} in thermal equilibrium on a flash simulator we provide an accurate gauge of junction temperature for subsequent measurements. By calibrating I_{SC} outdoors in natural sunlight we enable the module to serve as its own reference device in setting the intensity of a continuous simulator, effectively eliminating the impact of spatial non-uniformity and spectral mismatch on the I-V measurement. These procedures significantly reduce measurement errors due to temperature uncertainty, spatial non-uniformity, and spectral mismatch.**

Index Terms — **photovoltaic module, calibration, measurement uncertainty.**

I. INTRODUCTION

PV module manufacturers rely on calibration traceability to accredited test labs for reliable power measurements of their products. The financial impact of each additional 1% uncertainty in module power is up to $3.5M USD for each 1 GW of production at $0.35/W module sales price. In today's PV industry with hyper-thin operating margins, manufacturers can ill afford to relinquish any profit to less than optimal calibrations. Each of the tens of millions of modules produced per year must be flash tested at the end of the production line to determine its power output. The accuracy of this measurement depends directly on the calibration uncertainty of the calibration module used to set the irradiance level of the simulator. PV manufacturers depend on calibration laboratories to provide secondary calibration modules with minimal uncertainty. In response to this need the Cell and Module Performance laboratory at the National Renewable Energy Laboratory in Golden, CO has recently instituted a new procedure to minimize I_{SC} and P_{MAX} uncertainties for secondary calibration modules by combining measurements from three different test beds to achieve expanded uncertainty (k=2) in P_{MAX} as low as ±1.1% on commercial c-Si modules.

A. Module calibration considerations

Leading calibration laboratories such as NREL, Fraunhofer-ISE, AIST, and JRC-ESTI calibrate PV modules under standard test conditions (STC: 25°C, 1000 W/m², AM1.5G spectrum) following IEC 60904 and/or ASTM E44 standards under ISO17025 quality standards [1]. IEC 60904-1 specifies methods and equipment for measurement of PV cell and module I-V data. The standard allows measurement under solar simulators or natural sunlight. It requires that irradiance be measured using a pyranomter or PV reference device, which can be either a packaged reference cell or a module. If the reference device is not spectrally matched to the device under test (DUT) the standard requires that a spectral mismatch correction be applied. Spectral mismatch corrections require accurate measurement of the spectral response of the reference device, the DUT, and the spectrum of the simulator or sunlight during the test. Such measurements require costly equipment and specialized expertise and are rarely performed outside of advanced calibration laboratories. Module manufacturers generally avoid the need for spectral corrections by utilizing calibration modules that are closely matched to the module under test in both size and spectral response.

Because calibration laboratories test many dozens of different types of modules it is not practical to use matched modules for measurement of irradiance. Calibration labs typically use specially made and calibrated primary reference cells to measure irradiance [2]. Primary reference cells have the advantage of advanced calibration procedures that minimize uncertainty in I_{SC}, typically below ±1%. One significant disadvantage of primary reference cells is that they are much smaller in size than commercial PV modules, hence if the light source has non-uniform irradiance there is an uncertainty in the module current, and subsequently module power, that is proportional to the non-uniformity of the light source. [3] This is frequently a dominant source of uncertainty in module calibration, as even a Class A simulator can have nonuniformity of up to ±3.0%.

Another significant potential source of uncertainty in module calibrations is temperature. Typical P_{MAX} temperature coefficients for c-Si modules are on the order of 0.5%/°C. Hence an error of 2°C in module temperature could result in an error of 1.0% in the measured P_{MAX}. While temperature uncertainty is minimal under pulsed simulators if the pulse duration is short enough to minimize module heating, there can be significant heating under continuous simulators or natural sunlight. When temperature sensors are attached to the back of the module the measured temperature is influenced by the thermal conductivity of the encapsulant and backsheet

978-1-5090-5606-4/17 $31.00 © 2017 IEEE

materials. This issue is aggravated by non-uniform temperature distributions among cells in the module, making average cell temperature a difficult parameter to accurately measure. Fortunately V_{OC} has a linear dependence on temperature and logarithmic dependence on irradiance, making it an accurate surrogate for temperature measurements, as described in IEC 60904-5 for determining equivalent cell temperature using the open-circuit voltage method.

II. SELF-REFERENCE METHOD

A. NREL Module Calibration Lab

The NREL module calibration laboratory utilizes 3 different module test beds. These include a Spectrolab model X200 class AAA continuous solar simulator, a Spire model 5600 class AAA pulsed solar simulator, and an outdoor module test bed using natural sunlight. Each of these test beds has strengths and weaknesses. The Spectrolab X200 has the advantage of continuous illumination, which eliminates concerns of inaccurate IV curve measurement due to hysteresis effects in high capacitance modules. The spatial non-uniformity of this test bed is ±3.0%, making it less accurate for I_{SC} and P_{MAX} measurements. The Spire 5600 has a flash duration of ~ 100 ms. This is short enough to minimize heating during the flash, making this test bed very accurate for V_{OC} measurements as long as the module is in thermal equilibrium prior to the measurement. The short pulse causes concerns with hysteresis effects in high capacitance modules due to the high scan rate required to complete the IV scan in 100 ms. The outdoor test bed has the advantage of highly uniform light due to the uniformity of natural sunlight. The challenge of using natural sunlight is that most IV measurements must be scaled to 1000 W/m^2 because of the continually changing intensity of sunlight. This scaling can introduce undesirable uncertainties.

B. Strategic Combination of Test Bed Measurements

We have developed a procedure that utilizes the strengths of each test bed to minimize the impact of weaknesses of the others, resulting in the minimum P_{MAX} uncertainty reported by any internationally recognized test lab. First, the module is brought to thermal equilibrium at 25°C indoors. V_{OC} vs. irradiance is measured at 800, 900, and 1000 W/m^2 using the Spire 5600. The light level in the simulator is set using a c-Si reference cell in a module package with I_{SC} uncertainty of ±1.4%. This procedure provides a calibration of V_{OC} vs. irradiance within 0.3°C of 25°C.

This V_{OC} calibration is then used as an accurate measure of junction temperature in the module during measurements in the outdoor test bed and Spectrolab X200, significantly reducing errors due to temperature variations. Prior to outdoor measurement the module is cooled to ~15°C, then mounted on the outdoor test bed and allowed to heat under natural sunlight while V_{OC} is monitored. When V_{OC} indicates the module temperature is 24.5°C the IV curve is measured. Irradiance is

measured using a c-Si primary reference cell calibrated at NREL with a U95 expanded uncertainty in I_{SC} of ±0.42%. The spectrum is measured from 350 – 2400 nm using an ASD FieldSpec3 spectroradiometer. Because of the uniformity of natural sunlight, this technique minimizes uncertainty in Isc related to spatial non-uniformity of the irradiance. The value of I_{SC} from the outdoor test bed is adjusted for spectral mismatch and scaled to irradiance of 1000 W/m^2 [4].

After the outdoor measurement the module is again cooled to ~15°C, then mounted on the Spectrolab X200 continuous simulator. Instead of a reference cell the module itself is used to set the light level based on the value of I_{SC} from the outdoor test bed, hence the term *module self-reference*. This procedure minimizes uncertainty in P_{MAX} by eliminating spectral error between the module and reference cell, and also minimizes error in I_{SC} due to spatial non-uniformity of the light source. After mounting, the module is illuminated and V_{OC} is monitored until it indicates the module is at 24.5°C. The IV sweep is then performed, providing a measurement within 0.5°C of 25°C with minimal error due to uncertainties in temperature, spectrum, and non-uniformity of irradiance. Lastly, the measured current data points are scaled so that the I_{SC} is equal to the outdoor calibrated I_{SC}.

C. Uncertainty Calculation

The uncertainty calculation presented here is based on the methods detailed in the ISO document "Guide to the Expression of Uncertainty in Measurement" or GUM [5]. In the GUM, the standard deviation, σ, of a quantity x is termed the standard uncertainty, $u(x)$, and guidance is provided for obtaining $u(x)$ values with statistical analysis. The expanded uncertainty equals the standard uncertainty multiplied by a coverage factor that depends on the statistical distribution of the quantity in question. In this paper we are assuming the k=2 coverage means 95% coverage without having specific knowledge of the distributions. All uncertainties in this discussion are expressed in relative percent. Note that we utilize a 'sample specific' uncertainty analysis where the uncertainty is calculated for each measurement. Hence the numerical values discussed herein are based on a specific measurement of a 1.58 m^2 mono-Si commercial module.

For V_{OC}, the calibration consists of fitting voltage vs. irradiance at 800, 900, and 1000 W/m^2 measured on the Spire 5600 at 25 ± 0.3°C. Assuming ideal single diode behavior and the light-generated current directly proportional to the irradiance E, a logarithmic fit provides a way to calculate the expected V_{OC} at 25°C and a given E. The uncertainty in V_{OC} is estimated with three components: 1) the long-term V_{OC} stability of the Spire 5600, ($u(V_{LT})$, measured to be 0.162% over three years), 2) the manufacturer's uncertainty specification of the V channel, $(U(V_{5600}) = ±0.25\%$ with a 95% coverage factor $k = 2$), and 3) the uncertainty of the logarithmic fit, $u(\ln(E))$.

At 1000 W/m^2, $u(\ln(E))$ is taken as half the difference between the upper and lower prediction bounds of the logarithmic least-squares regression. From the fit for a commercial 1.58 m² mono-Si module, $u(\ln(E)) = 0.183\%$. Using the root-sum-square (RSS) method from the GUM [5] we determine $U_{95}(V_{OC}) = \pm 0.27\%$, as shown in (1).

$$
\begin{aligned}
U_{95}\left(V_{OC}\right) &= k\,u\left(V_{OC}\right) \\
&= k\left[\left(\frac{u\left(V_{LT}\right)}{1}\right)^2 + \left(\frac{U\left(V_{5600}\right)}{2}\right)^2 + \left(\frac{u\left(\ln\left(E\right)\right)}{1}\right)^2\right]^{1/2} \quad (1) \\
&= (2)\left[\left(0.162\right)^2 + \left(\frac{0.25}{2}\right)^2 + \left(0.183\right)^2\right]^{1/2} = \pm 0.27\%
\end{aligned}
$$

The outdoor I_{SC} calibration is an implementation of Eq. 3 from [4], which is a translation of PV effective irradiance ratio (measured with a primary reference cell) from one spectral and temperature condition to another. Solving that equation for the calibrated module short circuit current we derive the equation for calibration of module I_{SC} in the outdoor test bed:

$$
I_{0,TD} = I_{TD}\,\frac{I_{0,RC}}{I_{RC}}\,\frac{1}{M\left(T_{TD}, T_{RC}\right)} \quad (2)
$$

where I_{TD} = measured test device (module) short circuit current, I_{RC} = measured reference cell short circuit current, $I_{0,RC}$ = reference cell calibration value, $M(T_{TD}, T_{RC})$ = temperature dependent spectral correction, and $I_{0,TD}$ = spectral- and temperature-corrected module short circuit current.

Based on the terms in (2) and our measurement procedures there are eleven individual error sources that contribute to the uncertainty in the outdoor calibration of module short circuit current. These 11 terms are described in Table I. Capital $U_{95}(x)$ refers to the 95% confidence expanded uncertainty of

TABLE I
UNCERTAINTY TERMS IN CALCULATION OF I_{SC}

Term	Description
$U_{95}(M)$	T-dependent spectral mismatch correction
$U_{95}(I_{0,RD})$	Primary reference device calibration
$u(S)$	Spatial non-uniformity of irradiance
$u(PM_{TD,RD})$	Planar misalignment of test and reference devices
$u(I_{T,TD})$	Current error due to temperature change before/after I-V sweep
$u(R_{RD})$	Current sense resistor for reference device
$u(R_{TD})$	Current sense resistor for test device
$\sigma(I_{SC0,RD})$	Fit of I-V data for reference device I_{SC}
$\sigma(I_{SC0,TD})$	Fit of I-V data for test device I_{SC}
$u(VM_{RD})$	Voltmeter for reference device current measurement across sense resistor
$u(VM_{TD})$	Voltmeter for test device current measurement across sense resistor

variable x, while small u(x) represents the unexpanded uncertainty of variable x, and $\sigma(x)$ denotes the standard deviation of variable x.

The largest single error source is the uncertainty in the primary reference cell calibration value, with an expanded uncertainty $U_{95}(I_{0,RD}) = \pm 0.42\%$. The second largest factor is the temperature dependent spectral mismatch correction $U_{95}(M) = \pm 0.3\%$. Although sunlight is expected to have minimal spatial non-uniformity, previous work has estimated the standard (unexpanded) uncertainty due to this factor $u(S) = 0.25\%$ (rectangular distribution) [6].

An easy-to-overlook error source in outdoor solar measurements is planar misalignment between the reference and test devices. For the measurement of the 1.58 m² mono-Si module with a 1° misalignment between module and reference cell and an 8° solar incidence angle, the error $u(PM_{TD,RD}) = 0.242\%$, also with rectangular distribution. Smaller contributions include the test and reference devices current sense resistors, the linear fits used to calculate test and reference device I_{SC} from the I-V data points, and the voltmeters used to measure voltage across the current sense resistors for test and reference devices. Applying root sum square analysis with appropriate coverage factors the full uncertainty calculation for outdoor calibration of module short circuit current is shown in (3), with an expanded U95 uncertainty of $U(I_{0,TD}) = \pm 0.684\%$.

$$
\begin{aligned}
u\left(I_{0,TD}\right) &= \left[\begin{array}{l}
\left(U_{95}\left(M\right)/2\right)^2 + \left(U_{95}\left(I_{0,RD}\right)/2\right)^2 + \\
u^2\left(S\right)/3 + u^2\left(PM_{TD,RD}\right)/3 + u^2\left(I_{T,TD}\right)/3 + \\
u^2\left(R_{RD}\right) + u^2\left(R_{TD}\right) + \sigma^2\left(I_{SC0,RD}\right) + \\
\sigma^2\left(I_{SC0,TD}\right) + u^2\left(VM_{RD}\right) + u^2\left(VM_{TD}\right)
\end{array}\right]^{0.5} \\[6pt]
&= \left[\begin{array}{l}
\left(0.3/2\right)^2 + \left(0.422/2\right)^2 + \\
\left(0.25\right)^2/3 + \left(0.242\right)^2/3 + \left(0.0185\right)^2/3 + \\
\left(0.0074\right)^2 + \left(0.0240\right)^2 + \left(0.0678\right)^2 + \\
\left(0.0623\right)^2 + \left(0.0141\right)^2 + \left(0.0139\right)^2
\end{array}\right]^{0.5} \quad (3) \\[6pt]
&= 0.342\%
\end{aligned}
$$

$$
U_{95}\left(I_{0,TD}\right) = k \times u\left(I_{0,TD}\right) = (2)(0.342) = \pm 0.684\%
$$

The final step is measurement of the full I-V curve using the module as the reference device to set the irradiance level in the Spectrolab X200 continuous simulator. As noted in section B, using this method eliminates several significant error sources in determination of P_{MAX}. Errors due to simulator spatial non-uniformity, previously $\pm 3\%$, are almost completely eliminated. The one remaining effect of non-uniformity is the potential I-V curve distortion [7]. This effect produces a reduction in the difference between I_{SC} and I_{MAX}, flattening the I-V curve. By comparing multiple indoor and outdoor I-V curves on a group of control modules, this distortion has been

978-1-5090-5606-4/17 $31.00 © 2017 IEEE

estimated to be a maximum of 0.5% (rectangular distribution). By using the V_{OC} versus E curve to monitor module temperature, excursions away from 25°C are minimized, which reduces voltage uncertainty. Remaining sources of error include V_{OC} uncertainty from the Spire calibration, I_{SC} uncertainty from the outdoor calibration, voltage error due to temperature change before/after the I-V sweep, and multiple small errors due to measurement hardware and data fitting.

As expressed previously in Table I for the terms contributing to uncertainty in the I_{SC} calibration, the terms contributing to uncertainty in calibration of P_{MAX} are summarized in Table II. In this case dominant error sources include uncertainty in I_{SC} calibration from the outdoor measurement, equal to 0.342%, I-V curve distortion due to spatial non-uniformity, estimated as 0.5% (rectangular distribution), and uncertainty in the V_{OC} calibration from the Spire, equal to 0.276%.

TABLE II
UNCERTAINTY TERMS IN CALCULATION OF P_{MAX}

Term	Description
$u(V_{OC,0})$	Calibrated V_{OC} at 25°C, 1000 W/m² (Spire)
$u(V_{V,TD})$	Voltage error due to temperature change before/after I-V sweep
$u(VM_{V,TD})$	Voltmeter error for voltage measurement
$u(I_{0,TD})$	Calibrated I_{SC} from outdoor measurement
$u(I_{T,TD})$	Current error due to temperature change before/after I-V sweep
$u(VM_{I,TD})$	Voltmeter error for current measurement across current sense resistor
$u(R_{TD})$	Current sense resistor for test device
$\sigma(I_{SC,TD})$	Fit of I-V curve for test device I_{SC}
$u(I\text{-}V)$	Maximum bounds of I-V curve distortion from spatial non-uniformity
$\sigma(P_{MAX,TD})$	Uncertainty of P_{MAX} curve fit using prediction intervals

The root sum square analysis of uncertainty in P_{MAX} is shown in (4). Numerical values shown in the second expression are again from the 1.58 m² mono-Si module. In this case we determine an expanded U95 (k=2) uncertainty in P_{MAX} of ±1.14%. To our knowledge this is the lowest reported uncertainty for a module of this size for any accredited calibration laboratory. Note that previous measurements of module P_{MAX} in the X200 with standard procedures using a reference cell to set the irradiance in which the spatial non-uniformity error dominates [2] results in $U_{95}(P_{MAX}) = \pm3.3\%$. Hence this new procedure provides nearly a factor of 3 reduction in uncertainty for module power calibrations.

$$u(P_{MAX}) = \begin{bmatrix} u^2(V_{OC0}) + u^2(V_{T,TD})/3 + u^2(VM_{V,TD}) + \\ u^2(I_{0,TD}) + u^2(I_{T,TD})/3 + u^2(VM_{I,TD}) + \\ u^2(I\text{-}V)/3 + \sigma^2(I_{SC,TD}) + \sigma^2(P_{MAX,TD}) + \\ u^2(R_{TD}) \end{bmatrix}^{0.5}$$

$$= \begin{bmatrix} (0.276)^2 + (0.0018)^2/3 + (0.0088)^2 + \\ (0.342)^2 + (0.0382)^2/3 + (0.0173)^2 + \\ (0.5)^2/3 + (0.0759)^2 + (0.2075)^2 + \\ (0.0340)^2 \end{bmatrix}^{0.5} \quad (4)$$

$$= 0.572\%$$

$$U_{95}(P_{MAX}) = k \times u(P_{MAX}) = (2)(0.572) = \pm1.14\%$$

III. DISCUSSION AND CONCLUSIONS

An uncertainty analysis is a useful exercise for gauging the accuracy of a measurement procedure, but will remain somewhat academic until statistics have been obtained through multiple measurements on identical samples, and through comparisons with results by other laboratories. We have recently completed an intercomparison with certified module calibration laboratories at Fraunhofer-ISE, JRC-ESTI, and AIST. In measurements on commercial mono-Si and multi-Si modules all laboratories' I_{SC} values are within ±0.5% of the mean. This is a clear indication that not only NREL, but all 4 of these laboratories have significantly reduced errors relative to previous intercomparisons. A manuscript detailing the results of this international intercomparison has been accepted for publication in the journal Solar Energy and should be appearing in the journal later this year. [8]

In order to develop a statistical evaluation of the accuracy of this method we have begun repeated measurements on a set of 16 modules of different sizes using the MSR procedure. The results of these measurements to date are presented in Fig.1. The plot shows the deviation of each measurement from the mean value of I_{SC} for each module in percent. Different colored symbols represent different modules. While the statistics for a single module are weak with between 2 and 7 measurements per module, the overall statistics begin to present a valid evaluation of the accuracy of this method. The black horizontal lines at ±0.68% represent the two standard deviation, 95% confidence interval calculated for I_{SC} in the text. There are 51 total measurements. Of these, 48, or 94% fall within this confidence level. Statistical analysis calculates a standard deviation for these measurements of 0.358%, which is only slightly larger than the 0.342% calculated for the 1.58 m² mono-Si module evaluated in the discussion on uncertainty analysis. The precise uncertainty is specific to each module and each measurement; hence this evaluation is only approximate.

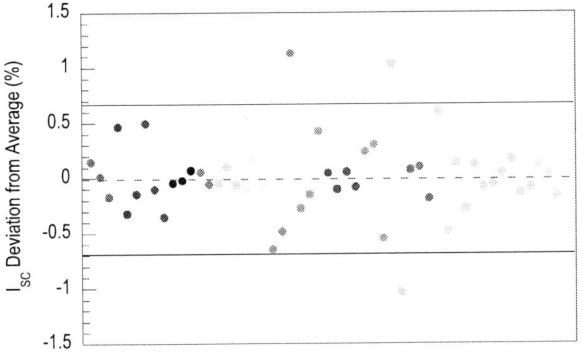

Fig. 1 I_{SC} data collected on 16 modules using the module self reference procedure. Each color represents a different module. Data is plotted as the percent difference between each measured I_{SC} value and the average I_{SC} for that module. Horizontal lines at ±0.68% show the U95 (k=2), two standard deviation 95% confidence range.

Yet, this initial evaluation demonstrates that the uncertainty analysis corresponds well to the experimental results up to this point.

IV. SIGNIFICANCE AND SUMMARY

Reducing uncertainty in module calibrations is one of the key challenges facing certified calibration laboratories because of the financial impact of uncertainty on PV manufacturers. This new self-reference procedure enables NREL's Cell and Module Performance lab to overcome relatively high spatial non-uniformity in our X200 continuous simulator to cut P_{MAX} uncertainty by a factor of 3. This paper has presented the details of the procedure and the uncertainty analysis behind it such that manufacturers and other calibration laboratories can understand and repeat the procedure themselves. At this time the procedure currently has only been applied to mono-Si and multi-Si modules. Thin film modules such as CdTe, CIGS, and a-Si present additional technical difficulties that must be solved to extend the application of this method to those technologies.

ACKNOWLEDGEMENT

This work was supported by the U.S. Department of Energy under Contract No. DE-AC36-08-GO28308 with the National Renewable Energy Laboratory. The U.S. Government retains and the publisher, by accepting the article for publication, acknowledges that the U.S. Government retains a nonexclusive, paid-up, irrevocable, worldwide license to publish or reproduce the published form of this work, or allow others to do so, for U.S. Government purposes.

REFERENCES

[1] Standard ISO/IEC 17025, General requirements for the competence of testing and calibration laboratories (2005 edition 2). Geneva, Switzerland.
[2] C. R. Osterwald, S. Anevsky, A. K. Barua, P. Chaudhuri, J. Dubard, K.Emery, B. Hansen, D. King, J. Metzdorf, F. Nagamine, R. Shimokawa, Y.X.Wang, T.Wittchen,W. Zaaiman, A. Zastrow, and J. Zhang, "The world-photovoltaic scale: An international reference cell calibration program," in Proc. IEEE 26th Photovoltaic Spec. Conf., Anaheim, CA, USA, 1997,pp. 1209–1212.
[3] Emery, K., "Uncertainty Analysis of Certified Photovoltaic Measurements at the National Renewable Energy Laboratory," August 2009, NREL Technical Report NREL/TP-520-45299
[4] M.B. Campanelli and C.R. Osterwald, Effective Irradiance Ratios to Improve I–V Curve Measurements and Diode Modeling Over a Range of Temperature and Spectral and Total Irradiance, *IEEE Journal of Photovoltaics* **6**(1), pp. 48-55, 2016. DOI 10.1109/JPHOTOV.2015.2489866.
[5] Procedure ISO GUM, Guide to the Expression of Uncertainty in Measurement. ISO: Geneva, 1995, ISBN 92-67-10188.
[6] K. Emery, internal NREL communication, 2015.
[7] H.W. Wilterdink, A.L. Blum, C.L. Sainsbury, and R.A. Sinton, "Spatial non-uniformity of irradiance and uncertainty in power rating for c-Si modules," in the NREL PV Reliability Workshop, Golden CO, 2016.
[8] E. Salis, D. Pavanello, M. Field, U. Kraeling, F. Neuberger, K. Kiefer, C. Osterwald, S. Rummel, D. Levi, Y. Hishikawa, K. Yamagoe, H. Ohshima, M. Yoshita, and H. Müllejans, "Improvements in world-wide intercomparison of PV module calibration," accepted for publication, Solar Energy.

978-1-5090-5606-4/17 $31.00 © 2017 IEEE

Uncertainty Evaluation of Primary Reference Photovoltaic Cell Calibration under Outdoor Condition in Tibet

Haitao Liu[1], Shiyu Sang[1], Guomin Zhou[2], Yonghui Zhai[1]

[1]Key Laboratory of Solar Thermal Energy and Photovoltaic System, Institute of Electrical Engineering, Chinese Academy of Sciences, Beijing, China, 100190

[2]Tibet New Energy Research and Demonstration Centre, Lhasa, China, 850000

Abstract — **A primary reference cell calibration testbed using direct normal irradiance calibration method is developed to establish calibration traceability chain of PV scale by Institute of Electrical Engineering (IEECAS) and New Energy Research and Demonstration Centre (NERDC) of Tibet. The testbed is designed to carry out PV reference cell primary calibration under direct irradiance condition, which meets the requirement of IEC 60904-4. The calibration value and uncertainty analysis of direct normal irradiance, spectral mismatch correction, and short-circuit current measurement are discussed and evaluated for photovoltaic metrology calibration reference. The calibration method is based on absolute cavity radiometer which is calibrated against world radiation scale. The calibration result shows expanded uncertainty of 0.91% and desired results reproducibility for establishing on-site photovoltaic metrology traceability chain in China.**

Index Terms — **uncertainty, reference cell, primary calibration, spectral mismatch factor.**

I. INTRODUCTION

Photovoltaic reference cell is used for rating irradiance of natural sunlight or artificial light source. The reference cells are also reference solar devices to measure current-voltage characteristics of photovoltaic cells, modules, and arrays. The primary calibration value of PV cell is the ratio of short circuit current and incident irradiance. As there is spectral distribution difference between natural sunlight and reference solar spectrum, it is important to define a rating method to obtain incident irradiance value. As primary calibration is to obtain short circuit current of PV cell under irradiance of $1000 W/m^2$ with 25 ℃ cell temperature, the measurement methods of incident irradiance, spectral distribution and cell temperature are key factors to influence the results and uncertainty. There are two components of uncertainty source during the primary calibration of PV cells: (1) the short circuit current measurement, and (2) incident irradiance measurement. In general, the uncertainty of short circuit current (Isc) measurement is influenced by accuracy of electronic digital multi-meters, which has better uncertainty than irradiance measurement component. To obtain natural sunlight irradiance, pyranometer and pyrheliometer are often used to measure global irradiance and direct normal irradiance, but both pyranometer and pyrheliometer have large uncertainty of

measurement (2%~5%). The equipments above are not suitable for primary calibration. According to the requirement of IEC 60904-4[1], the absolute cavity radiometer is the recommended equipment to measure incident irradiance, which has uncertainty of less than 1%, and can be traced world PV scale.

II. MATHEMATICAL MODELS OF PRIMARY CALIBRATION

According to the PV cell short circuit current calibration theory, the mathematical model of primary calibration is defined as follows:

(i) Primary Calibration under Reference Spectral Irradiance

$$C_R = \frac{i_R}{E_{e,R}} = \frac{\int_{300nm}^{4000n} E_R(\lambda)R(\lambda)d\lambda}{\int_{300nm}^{4000n} E_R(\lambda)d\lambda} \qquad (1)$$

where:

C_R is the calibration value of reference cell

$R(\lambda)$ is the absolute spectral responsivity of reference cell

$E_R(\lambda)$ is the reference solar spectral distribution (IEC 60904-3[2])

i_R is the short circuit current of reference cell

$E_{e,R}$ is the reference solar irradiance (IEC 60904-3)

The equation (1) is the function expression of PV cell primary calibration under reference (theory) condition. As there is spectral irradiance distribution deviation between natural sunlight and reference sunlight, the spectral mismatch factor should be included in the following definitions.

(ii) Primary Calibration under Sunlight Spectral Irradiance

We assume that the cell temperature is a constant. The direct normal irradiance is measured by absolute cavity radiometer. The natural sunlight spectral irradiance distribution is expressed by $E_M(\lambda)$, the measured short circuit current of reference cell is expressed by i_M, the measured incident irradiance is expressed by $E_{e,M}$.

$$C_M = \frac{i_M}{E_{e,M}} \qquad (2)$$

$$i_M = \int E_M(\lambda) R(\lambda) d\lambda \qquad (3)$$

As the reference solar irradiance is expressed by $E_{e,R}$, the spectral mismatch factor k_c can be expressed by:

$$k_c = \frac{C_R}{C_M} = C_R \cdot \frac{\int E_M(\lambda) d\lambda}{\int E_M(\lambda) R(\lambda) d\lambda} \qquad (4)$$

$$k_c = \frac{\int E_M(\lambda) d\lambda}{\int E_M(\lambda) R(\lambda) d\lambda} \cdot \frac{\int E_R(\lambda) R(\lambda) d\lambda}{\int E_R(\lambda) d\lambda} \qquad (5)$$

The equation (4) can be expressed by following equation:

$$C_R = \frac{i_M}{E_{e,M}} \cdot \frac{\int E_M(\lambda) d\lambda}{\int E_M(\lambda) R(\lambda) d\lambda} \cdot \frac{\int E_R(\lambda) R(\lambda) d\lambda}{\int E_R(\lambda) d\lambda} \qquad (6)$$

The calibration value (C_R) is the product of the spectral mismatch factor k_c and the ratio of measured Isc and irradiance. The mismatch factor k_c equals 1 when the two spectral irradiance distributions coincide. The integral limit in equation (3) ~ (6) is the same as that in equation (1).

III. DIRECT SUNLIGHT PRIMARY CALIBRATION METHOD

To obtain satisfied meteorological condition, the outdoor calibration test bed is installed in Lhasa, Tibet. The test bed consists of outdoor solar tracker, meteorological parameter measurement equipment and indoor measurement devices. The detailed equipments of test bed are as follows:

(a) two-axis solar tracker loaded 4 PV cells and collimator tubes
(b) solar tracker (kipp&zonen SOLYS-2) loaded HF absolute cavity radiometer (Eppley Lab)
(c) Agilent 34410 multi-meters and Agilent 34970 data acquisition equipment
(d) Cell temperature control and measurement device
(e) Pyranometer, pyrheliometer, wind velocity and other meteorological devices
(f) Spectro-radiometer with tracker and collimator tube

Fig.1 shows the two reference cells with collimator tube installed on the two-axis tracker. The short circuit current of reference cell, incident direct irradiance and spectral irradiance distribution are measured simultaneously. The HF absolute cavity radiometer traced to world radiometric reference (calibrated against National Center for Meteorological Metrology of China) is used to measure direct normal irradiance. The reference cell is located below the collimator tube, which limits the field of view to 5.0° during experiment implementation [3]. An optical research MSR-7000 spectro-radiometer with 5.0° field of view collimating tube is used for measuring spectral irradiance between 300nm to 2200nm. The cell temperature is controlled to 25±1℃ by water cycling facility. Each instrument and PV cell to be calibrated is mounted to a separate dual axis solar tracker.

During the calibration time period, The cell temperature can be controlled with 25±0.2℃, the temperature coefficient is not used for temperature correction computation. Spectral irradiance correction is applied using relative spectral response which is measured by CEP-25/CH spectral response measuring equipment.

Fig. 1. Direct sunlight primary calibration test bed in Tibet

The reference cell size to be calibrated is 20mm×20mm with 4 wires connection design. The Pt100 temperature sensor is attached to the backside of reference cell. The stability, linearity and package design meet the requirement of IEC 60904-2 [4,5]

Fig.2. Crystalline silicon reference PV cell to be calibrated

The spectral irradiance distribution, direct normal irradiance $E_{e,M}$, short circuit current $i_R(t)$ and cell temperature t are measured simultaneously over direct normal irradiance range of 800~1100W/m². The measured $C_R(t)$ can be corrected to $C_R(25)$ based on equation (7).

$$i_R(t) = i_R(25) \cdot [1 + A_{coeff}(t-25)] \qquad (7)$$

$$C_R(25) = \frac{C_R(t)}{1 + A_{coeff}(t-25)} \qquad (8)$$

where, the measured cell temperature is expressed by t, the temperature coefficient of Isc is expressed by A_{coeff}

IV. UNCERTAINY EVALUATION OF CALIBRATION VALUE

The uncertainty analysis of is focused on calibration value parameter $C_R(t)$, temperature coefficient, and spectral mismatch factor k_c. The uncertainty evaluation is based on ISO/IEC Guide 98-3 [6].

(i) Uncertainty Evaluation of $C_R(t)$

The equation (1) above can be expressed following equation under desired temperature:

$$C_R(t) = \frac{i_R(t)}{E_{e,R}} \qquad (9)$$

The standard uncertainty of calibration value, short circuit current and normal irradiance is expressed by $u(C_R)$, $u(i_R)$, $u(E_e)$. The parameters $u(i_R)$ and $u(E_e)$ are uncorrelated variable. So following equation can be obtained:

$$u^2(C_R) = [\frac{1}{E_{e,R}} \cdot u(i_R)]^2 + [\frac{-i_R}{E_{e,R}^2} \cdot u(E_e)]^2 \qquad (10)$$

The equation above also can be expressed by relative standard uncertainty $u_r(C_R)$, $u_r(i_R)$, $u_r(E_e)$:

$$u_r^2(C_R) = u_r^2(i_R) + u_r^2(E_e) \qquad (11)$$

The combined standard uncertainty $u_r(C_R)$ of calibration value $C_R(t)$ can be expressed by:

$$u_r(C_R) = \sqrt{u_r^2(i_R) + u_r^2(E_e)} \qquad (12)$$

The relative expanded uncertainty $U_r(C_R)$ of calibration value $C_R(t)$ can be expressed by:

$$U_r(C_R) = 2 \cdot u_r(C_R) \qquad (13)$$

Short circuit current measurement is implemented by Agilent 34410 multi-meter and 10Ω standard resistance. The calculated expanded uncertainty caused by multi-meter is

0.01%, the expanded uncertainty caused by standard resistance is 0.01%. The HF absolute cavity radiometer is traced to National Center for Meteorological Metrology of China, the calibration report shows the uncertainty of 0.45% with rectangle distribution. After collecting each component of uncertainty of calibration value, type A, type B and combined standard uncertainty are calculated and shown in table I.

TABLE I
STANDARD UNCERTAINTY SUMMARY

Source of uncertainty	Type A	Type B	combined
$u_{r,1}(i_R)$	0.11%		
$u_{r,2}(i_R)$ (digital multi-meter)		0.01%	
$u_{r,3}(i_R)$ (10Ω)		0.01%	
Combined $u_r(i_R)$			0.11%
$u_{r,1}(E_e)$	0.06%		
$u_{r,2}(E_e)$ (absolute cavity radiometer)		0.21%	
$u_r(E_e)$ combined			0.22%
$u_r(C_R)$ combined			0.22%
$U_r(C_R)$ expanded			0.44%

(ii) Uncertainty Evaluation of Temperature Coefficient A_{coeff}

The cell temperature is controlled by water cycling method. The uncertainty caused by temperature is based on temperature coefficient of cell and only related to temperature range. The cell temperature coefficient is measured over the range of 15~45℃ under outdoor condition. The calculated relative A_{coeff} is within the range of 0.0427%/℃ ~ 0.0753%/℃. As the cell temperature range is (25±2)℃, the absolute temperature fluctuation range is 4℃. The relative temperature range can be calculated according to following equations:

$$0.0427\% / \,^\circ C \times 4\,^\circ C = 0.17\% \qquad (14)$$
$$0.0753\% / \,^\circ C \times 4\,^\circ C = 0.30\% \qquad (15)$$

The measured short circuit current value influenced by temperature varies over the range of 0.17% ~ 0.30%. The distribution of temperatue coefficient is taken to be rectangular in the absence of further information. The expanded uncertainty of A_{coeff} is calculated as follows:

$$U_{A_{coeff}} = 030\% - 0.17\% = 0.13\% \qquad (16)$$

(iii) Uncertainty Evaluation of Spectral Mismatch Factor k_c

The reference spectral irradiance distribution $E_R(\lambda)$ is constant according to equation (5). The uncertainty of kc consists of natural sunlight spectral distribution $E_M(\lambda)$ and relative spectral responsivity $R(\lambda)$. The kc can be expressed by following equation:

$$k_c = \frac{\int E_M(\lambda)d\lambda}{\int E_M(\lambda)R(\lambda)d\lambda} \cdot \frac{\int E_R(\lambda)R(\lambda)d\lambda}{\int E_R(\lambda)d\lambda} \qquad (17)$$

The relative spectral responsivity is measured by grating monochromator measurement system. As measurement results are relative value, the uncertainty component source can be eliminated. The spectral irradiance distribution is measured by Optical Research MSR-7000 spectro-radiometer, which gives calibration report to evaluate uncertainty of type B. The uncertainty source of kc is shown in Table II.

TABLE II
UNCERTAINTY OF SPECTRAL MISMATCH FACTOR

Source of uncertainty	Type A	Type B	Distribution
Monochromatic light fluctuation	0.5%	0.3%	Rectangular
Silicon photodiode		0.3%	Gauss
Non-uniformity of monochromatic light		0.3%	Rectangular
Spectro-radiometer (MSR-7000)		0.3%	Rectangular

$$U_{kc} = 2 \left[\begin{array}{l} \left(\dfrac{0.5}{\sqrt{10}}\right)^2 + \left(\dfrac{0.3}{\sqrt{3}}\right)^2 + \left(\dfrac{0.3}{2}\right)^2 \\ + \left(\dfrac{0.3}{\sqrt{3}}\right)^2 + \left(\dfrac{0.3}{\sqrt{3}}\right)^2 \end{array} \right]^{0.5} = 0.74\% \quad (18)$$

(iv) Uncertainty Evaluation of Final Calibration Value

After summarizing each uncertainty component $C_R(t)$, A_{coeff}, and k_c, the overall uncertainty evaluation is shown as follows

Table III
UNCERTAINTY RESULTS OF PRIMARY CALIBRATION

Source of uncertainty	Uncertainty		Distribution
$U_r(C_R)$	0.44%		Rectangular
U_{Acoeff}	0.13%		Rectangular
U_{kc}	0.74%		Gauss
$U_{<CV>}$	0.91%		

$$U_{<CV>} = 2 \left[\left(\frac{U_r}{\sqrt{3}}\right)^2 + \left(\frac{U_{A_{coeff}}}{\sqrt{3}}\right)^2 + \left(\frac{U_{kc}}{2}\right)^2 \right]^{0.5} = 0.91\% \quad (19)$$

V. CONCLUSIONS

The reference cell primary calibration test bed is based on absolute cavity radiometer traced to world radiometric reference [7]. The uncertainty of absolute cavity radiometer is the most important factor to influence the overall uncertainty result. As the absolute cavity radiometer has time constant, it is important that measured data from cell current, spectral irradiance, temperature and irradiance should be acquired simultaneously.

The overall uncertainty of calibration value is 0.91%, which is obtained under cell temperature condition of 25±2℃. But the effective data is hard to be acquired over 3 days according to the requirement of IEC 60904-4 even though the direct normal irradiance meets the experiment requirement. As the natural sunlight spectrum varies in different region and time. A vehicle-mounted calibration facilities is equipped with this primary cell calibration test bed to compare and improve calibration uncertainty of primary reference cell in different region of China.

ACKNOWLEDGMENTS

This work was supported by key project in the National Science & Technology Pillar Program of Ministry of Science and Technology of China under contract No. 2015BAA09B02.

REFERENCES

[1] IEC 60904-4 Ed.1.0, Photovoltaic devices- Part 4: Reference solar devices-Procedures for establishing calibration traceability [S]. 2009.
[2] IEC 60904-3 Ed.2.0, Photovoltaic devices - Part 3: Measurement principles for terrestrial photovoltaic (PV) solar devices with reference spectral irradiance data [S]. 2008.
[3] ASTM E1125-10, Standard Test Method for Calibration of Primary Non-Concentrator Terrestrial Photovoltaic Reference Cells Using a Tabular Spectrum, American Society for Testing Materials [S]. 2010.
[4] IEC 60904-2 Ed.2.0, Photovoltaic devices- Part 2: Requirements for reference solar devices [S]. 2007.
[5] Osterwald C R, Anevsky S, Bücher K, et al. The world photovoltaic scale: an international reference cell calibration program [J]. *Progress in Photovoltaic Research and Applications*, 1999, 7(4): 287-297.
[6] ISO/IEC Guide 98-3, Uncertainty of measurement – Part 3: Guide to the expression of uncertainty in measurement.
[7] Reda I. Calibration of a Solar Absolute Cavity Radiometer with Traceability to the World Radiometric Reference[R].NREL/TP-463-20619, 1996.

Requirement of Artificial Lighting Simulator for Evaluation Emerging PV Performance Rating under Indoor Environment

Yean-San Long[1], Shu-Tsung Hsu[1] and Teng-Chun Wu[1]

[1]Center for Measurement Standards, Industrial Technology Research Institute, Hsinchu, Taiwan

Abstract—The measured I-V hysteresis is complicated by the vast array of different materials, lighting sources, and device architectures. In our study, we used RTOS method for I-V then the results showed better accuracy by eliminating in real time the acceptance effect. We also used this method to compare emerging PV cells with the latter cells exhibited promising power conversion efficiency and no hysteresis behavior under indoor lighting simulator. In this study, we defined a performance ratio to compare between indoor lighting and solar and the results will be show different performance trend for the different type PV in indoor/outdoor, which depends on device. That's will be shown what's type PV have the application of energy harvesting for indoor.

Index Terms—RTOS Method, Indoor Lighting, Solar simulator, Performance ratio.

I. INTRODUCTION

In general indoor lighting, a lot of different light sources exist. The International Commission on Illumination (CIE) is responsible for publishing binding standards for different types of light sources, called illuminants. A standard illuminant represents a mathematical table of relative energy versus wavelength, used for colorimetric calculations. It is a (theoretical) source of visible light with a set spectrum, determined by convention and therefore provides a worldwide basis for comparing images under different lighting. Which illuminants are useful standards for indoor illumination ? We consider some typical, widely used, light sources for indoor environments (residential and commercial) and relate them to an appropriate illuminant (standard). First, we will propose a standard for the common incandescent light bulb. The CIE has agreed upon a standard for the incandescent light bulb, called "illuminant A". This illuminant is intended to represent typical, domestic, tungsten-filament lighting. Its relative spectral power distribution is that of a Planckian radiator at a temperature of approximately 2856 K. In other words, a typical light bulb can be characterized by a black body at temperature T = 2856 K. Technically, illuminant A is only defined over the spectral region from 300 nm to 830 nm. However, solar cells often absorb longer wavelengths. Therefore, we will not use illuminant A itself to represent the spectrum of the incandescent light bulb, but we will use Planck's law for a black body at a temperature of 2856 K. In the 300–830 nm range, the difference between Planck's law at this temperature and illuminant A never exceeds 1%.

However, In indoor lighting, the International Commission on Illumination is an organization devoted to international cooperation and exchange of information between its member countries on all matters relating to the science and art of lighting. In accordance with an agreement between ISO and CIE, standards are published as double logo standards by ISO. Standards produced by the CIE are a concise documentation of data defining aspects of light and lighting, for which international harmony requires a unique definition. Therefore, the standard follow that the CIE has defined some lighting spectrums for different indoor colorful applications and situations. However, the requirements for color testing do not apply to related indoor energy harvesting under different lighting uniformity and stability, for performance measurement of emerging PV and device. Furthermore, countless types of indoor lighting patterns are required to do performance characteristics measurement for different application contexts. The most common performance indicator is the PV efficiency under standard test conditions (STC). It is the maximum electrical power divided by the total irradiance. For indoor lighting illuminance values are measured used a lux meter or spectradiometer. Therefore, we find (a) Lux is a typical unit for measuring indoor lighting. (b) For outdoors, irradiance is measured in W/m^2 and there is an approximate conversion of 0.0079 $W/m^2/$ Lux for the sun. Irradiance includes the power from all wavelengths weighted equally, whereas illuminance weights the power from each wavelength in proportion to the sensitivity of the human (Fig. 1). Although these conditions are seldom achieved in practice (except in the lab), this characterization provides a reasonable guideline for comparing different solar cell types under outdoor conditions. However, the STC are not relevant for indoor applications. Accurate characterization of emerging PV requires level lighting consideration on each very slow temporal response in the I-V curves of the emerging PV are clearly dependent on the voltage sweep direction, even when the sweep time is the order of seconds [1]. Furthermore, the temporal response is dependent on different level lighting consideration. In our study, RTOS (real-time one-sweep) method can real-time remove the capacity effect under solar simulator for I-V successfully and proposes the forward/backward schematic reaction mechanism for DSC [2]. In this study, RTOS method [3, 4] practices for measuring perovskite solar cell performance showed having better accuracy results by eliminating acceptance effect instantly. Additionally, RTOS method will be useful to measuring cell performance more accurately and rapidly when evaluating emerging PV performance in indoor lighting condition. Therefore, we also defined a performance ratio to compare between indoor lighting and solar simulator.

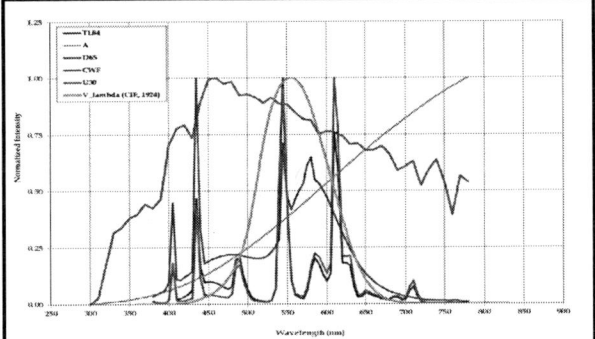

Fig. 1. Illumination spectra of CIE Standard Lighting [11, 12]

II. EXPERIMENTS

The samples Sample size ranges from 0.1 cm² to 1 cm × 5 cm. These I-V curves were done on a Keithley Source Meter (Model 2460), AM1.5G solar simulator, and Indoor Multi-Light Source System (see Fig. 4). Its meet follow as:

1) Spectral match, available methods are the use of spectroradiometer. Standard lighting spectrum is defined by the deviation from CIE[12]. For three wavelength intervals of interest, the percentage of total irradiance is specified in Table I. Calculate the spectral match for each wavelength interval, which is the ratio of calculated percentage for the CIE spectrum and the lighting simulator spectrum (Table I). The data comparison with the solar spectrum shall indicate the spectral match classification as per the following as Table II.

2) Non-uniformity of irradiance in the test plane, the lux meter is recommended to be used as uniformity detector for determining the non-uniformity of irradiance in the test area of the lighting simulator. The uniformity detector shall have a spectral response appropriate for the simulator. The linearity and time response of the uniformity detector shall conform to the characteristics of the simulator being measured. And the designated test area divided by 25. The area covered by the detector measurements should be 100% of the designated test area. The measurement positions should be distributed uniformly over the designated test area. Therefore, the maximum and minimum irradiance are those measured with the detector(s) over the designated test area. The calculation of non-uniformity is defined as Eq. (1).

3) Temporal instability of irradiance, the lux meter is recommended to be used as temporal instability detector for determining the temporal instability of irradiance in the

test area of the lighting simulator. For lighting simulator the temporal instability is related to the irradiance change of measured data sets during the time of data acquisition (Fig. 2).

The Indoor Multi-Light Source included D65 (6500 K, Average North USA sky daylight) / TL84 (4100 K, European shop fluorescent) / CWF (4150 K, cool white fluorescent, shop lighting) / U30 (3000 K, shop lighting) / A (2856 K, typical home lighting), meeting CIE Standard illuminants. The adjustable light intensity (ranging from 0 to 2500 lux) is supplied with an ultraviolet light source for aging testing, with non-uniformity of less than 2% (at 20 cm × 20 cm) and temporal instability of less than 2%. During the I-V measurement in the dark and under Indoor Multi-Light Source System, we followed the testing flowchart in Fig. 5; the scan direction was forward and backward, the sample temperature should be stable at 25 °C with a fluctuation of less than 1 °C, and the irradiance and luminance were determined using a NIST-traceable spectroradiometer (StellarNet) and an NML (National Measurement Laboratory, Taiwan)-traceable lux meter (HIOKI, model FT3424).

TABLE I THE RATIO OF CALCULATED PERCENTAGE FOR THE CIE SPECTRUM

Wavelength	Bandwidth	Percentage of total irradiance (300 nm – 900 nm)				
Band		A	D65	U30	CWF	TL84
1	300 - 450	3.78	20.31	11.40	17.99	15.18
2	450 - 650	40.15	45.35	81.72	76.73	79.13
3	650 - 900	56.06	34.34	6.88	5.29	5.69

TABLE II SPECTRAL DEFINITION OF LIGHTING SIMULATOR CLASSIFICATIONS

Classification	Percentage
A	1 %
B	3 %
C	5 %

$$Non - uniformity(\%) \equiv \frac{(max - min)}{(max + min)} \times 100\% \qquad (1)$$

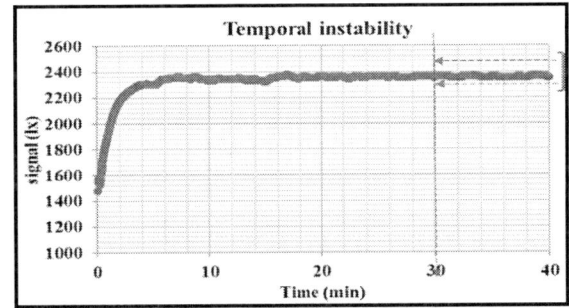

Fig. 2. Evaluation of temporal instability

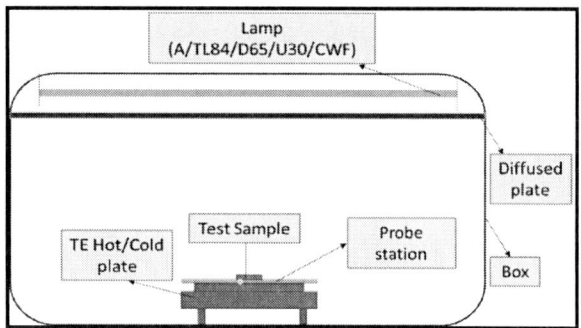

Fig. 3. Indoor multi-light source system custom-made by CMS/ITRI for solar cell measurement.

III. TEST AND ANALYSIS

In this study, we follow the testing flow [3,4] in the I-V measurement. We used real-time removal of capacity effect; the RTOS method shows the best repeatability. A proposed list of steps [2,5] for the measurement of a sample is given in Fig. 8. For most traditional semiconductor PV devices, the current will respond to small changes in bias voltage in less than a millisecond. Sweeping the voltage from open-circuit voltage (V_{oc}) to short-circuit current (I_{sc}) in a few seconds at a constant illumination for such a device will yield accurate currents. Many emerging PV cells respond far more slowly, and consequently, the sweep time must be increased.

Therefore, we used the RTOS Method [2] following the testing flowchart (see Fig. 4) in the I-V measurement. applied bias voltage is changed from I_{sc} to V_{oc} (forward scan) or in the reverse direction (backward scan, V_{oc} to I_{sc}). When a capacitance effect occurs, this measurement process produces the phenomenon. In the forward scan, an overshot current appears immediately after an abrupt increase of applied voltage and then the current gradually decreases to an equilibrium state. Sweeping to determine maximum power (P_{max}) may not be valid if the device is metastable because of the long times that it sits near I_{sc} or V_{oc} before getting to P_{max} confirm the "near-true" P_{max} of emerging PV, especially for samples that depend strongly on sweep direction or rate. This also shows that the asymptotic I-V method may be the only reliable way to obtain efficiency. In the light intensity under artificial lighting conditions found in offices and factories is less than 5 W/m^2 as compared to 100–1000 W/m^2 under outdoor conditions, depending on the type of light source and its distance. Moreover, the spectrum can be totally different from the outdoor solar reference spectrum (e.g. AM1.5G). The spectrum depends not only on the type of light source, but also on the presence of reflected and diffused light. Therefore, we shown a definition of performance (η) between indoor lighting and solar simulator below as

$$\eta \equiv (P_{max}/Area)/P_{in}. \qquad (2)$$

Where (a) P_{max} means maximum power of device, (b) Area means area of device, (c) P_{in} means Illuminant power of lighting source. Therefore, we also defined a normalized power density (performance ratio, η_n) to compare between indoor lighting and solar simulator below as

$$\eta_n \equiv \eta_{Indoor}/\eta_{AM1.5G} \qquad (3)$$

Where (a) η_{Indoor} means performance of device at each indoor lighting (A/TL84/D65/U30/CWF), (b) $\eta_{AM1.5G}$ means performance of device at solar simulator.

IV. DISCUSSION

The characterization after testing showed that one must measure performance under all typical artificial light sources. Therefore, we propose that each solar cell should be measured varied indoor and outdoor lighting condition for its characterization by RTOS method. In Fig. 5a, we will be compare the P_{max} of the different solar cells in light levels in indoor environments to show linearity under near range lighting level, which depends on device. In Fig. 5b shown nearly a constant from 400 lux to 1000 lux, it will be get an average of normalized power density and follow the defined of equation (3). Therefore, the results will be show the different solar cells in light levels in indoor environments to show linearity under near range lighting level, which depends on device. From Fig. 10b, it is also shown the different solar cells in indoor lighting and solar environments to show different performance, ex. DSSC have more 0.26 times for TL84 than D65. We are also known STC under outdoor application, therefore, we also define a STC for indoor application, that is " LS/ 25 °C/ 1000 lx " , LS mean: lighting source, ex. A/D65/TL84/U30/CWF.

V. CONCLUSION

In this work, we defined performance and normalized power density, is called performance ratio, shown a comparison for different emerging PV in indoor lighting and solar environments. Therefore, the results also find DSSC have more 0.26 times for TL84 than D65, but OPV/a-Si have less 0.5 times for TL84 than D65. That's shown DSSC have more indoor application than others in this case study. When we will be follow through CIE of the defined some lighting spectrums for different indoor colorful applications and situations. However, the requirements for color testing do not apply to related PV indoor performance measurement under different lighting uniformity and stability, for performance measurement of emerging PV and device, and Its characteristics will be having a classification. Furthermore, countless types of indoor lighting patterns are required to do performance characteristics measurement for different application contexts. We should be define a different STC than STC in AM1.5G, which we will be define a STC in indoor lighting as below as "I-V measurement for Indoor STC condition: 1000 lx @ 25 °C and Illuminant A". Furthermore, our goal is to determine which light source is relevant for a new international standard test method. Upon the indoor characterization of solar cells, it is the spectrum range of the light source that decides the lighting performance, not the light intensity, which is present and affects the cell equally through different light sources.

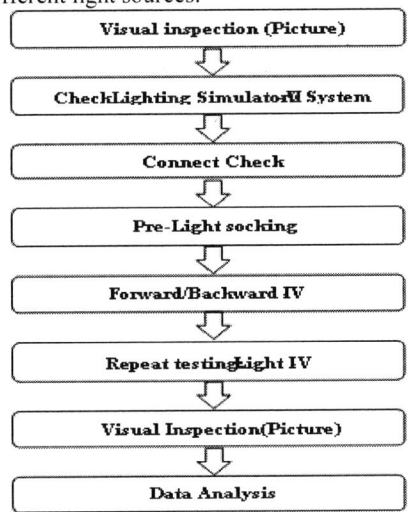

Fig. 4. Testing flowchart [2,5].

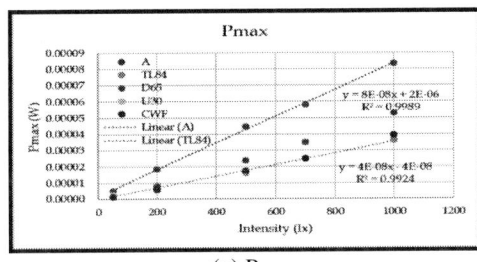

(a) P_{max}

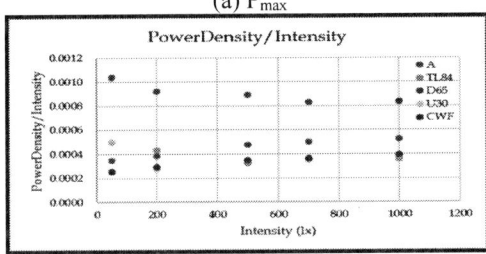

(b) Normalized Power density

Fig. 5. Light linearity of DSSC measured by indoor multi-light source system under varied lighting condition (I-V using RTOS method).

REFERENCES

[1] T. C. Wu, S. T. Hsu, and Y. S. Long, "New Set-up Procedures and Integrated Measurement System for Organic Photovoltaic (OPV) Module," *International Photovoltaic Science and Engineering Conference (PVSEC23)*, 2013.

[2] Y. S. Long, S. T. Hsu, and T. C. Wu, "Induction of Internal Capacitance Effect in Performance Measurement of OPV (Organic Photovoltaic) Device by RTOSM (Real-Time One-Sweep Method)," *Journal of Energy and Power Engineering*, Vol 8, pp1059-1066, 2014.

[3] Y. S. Long, S. T. Hsu, and T. C. Wu, "Capacitance Effect from High to Low Level Lighting by Real-Time One-Sweep Method for OPV Device," *IEEE PVSC42*, 2015.

[4] Y. S. Long, S. T. Hsu, and T. C. Wu, "A Study of Capacitance Effect in DSC from High to Low Level Lighting by Real-Time One-Sweep Method," *OPTIC2014*.

[5] SEMI PV57-1214, "Test Method for Current-Voltage (I-V) Performance Measurement of Organic Photovoltaic (OPV) and Dye-Sensitized Solar Cell (DSSC)."

[6] US PTO. 8224598, Method for Forming Optimal Characteristic Curves of Solar Cell and System Thereof.

[7] ASTM E490, "Standard Solar Constant and Zero Air Mass Solar Spectral Irradiance Tables."

[8] Handbook of Photovoltaic Science and Engineering, Chapter 16

[9] IEC 60904-1, "Photovoltaic Devices – Part 1: Measurement of Photovoltaic Current-Voltage Characteristics," 2006.

[10] IEC 60904-3, "Photovoltaic Devices – Part 3: Measurement Principles for Terrestrial Photovoltaic (PV) Solar Devices with Reference Spectral Irradiance Data," 2008.

[11] ISO 10526, "CIE Standard Illuminants for Colorimetry," 2007.

[12] CIE S 014-2, "CIE Standard Illuminants for Colorimetry," 2006.

[13] K. Emery, S. Winter, S. Pinegar, and D. Nalley, "Linearity Testing of Photovoltaic Cells," *4th World PV Conf.*, 2006.

[14] CIE S 014-2, "CIE Colorimetry - Part 2: Standard Illuminants for Colorimetry "

[15] ISO 11664-2:2007, "Colorimetry -- Part 2: CIE standard illuminants"

[16] ASTM D 1729, "Standard Practice for Visual Appraisal of Colors and Color Differences of Diffusely-Illuminated Opaque Materials";

[17] SEMI Doc. 5979, "Specification of indoor lighting simulator requirements for emerging PV "

[18] Handbook of Photovoltaic Science and Engineering, Chapter 16

Non-contact Voltage Measurement of Solar Cell with Electrostatic Voltmeter

Sakutaro Miyajima[1], Kensuke Nishioka[1] and Yoshihiro Hishikawa[2]

[1]Faculty of Engineering, University of Miyazaki, Miyazaki, 1-1, Gakuen Kibanadai Nishi, 889-2192, Japan

[2] Research Center for Photovoltaics, National Institute of Advanced Industrial Science and Technology (AIST), Tsukuba, Central 2, 1-1-1, Umezono, Tsukuba 305-8568, Japan

Abstract — **An accurate and quick performance measurement of PV modules and cells are very important. Non-contact measurements of PV characteristics without interruption of the operation of the PV system are desired. A high accuracy electrostatic voltmeter was used for the non-contact measurement of voltage of solar cell in a module structure. The values of the voltage measured with the voltage tester and electric potential measured with the electrostatic voltmeter agreed well. The non-contact electrostatic voltmeter can measure the voltage of solar cell in the module structure with a margin of error within 0.01 V.**

Index Terms — **electrostatic voltmeter, non-contact, photovoltaic cell, photovoltaic module, temperature, voltage.**

I. INTRODUCTION

An accurate and quick performance measurement of PV modules and cells are very important to their quality control, characterization of PV performance, and detection of degradation and failure [1]. Normally, PV performance is evaluated with current-voltage (I-V) measurement. However, I-V measurements require the insertion of an I-V tester and eventually interrupt the operation of the PV system. Therefore, non-contact measurements of PV characteristics are strongly desired.

The electrostatic potential or voltage of PV modules and cells is an important parameter that directly reflects their performance and operating conditions. If a method to measure the electrostatic potential or voltage without contact is established, it is very useful for the evaluation of PV modules and cells under their operations. The principle of the non-contact voltage measurement is based on the detection of the static electric field by periodically modulating the electric field at its electrode [2-6]. However, the characterization of PV modules and cells covered with a multilayer structure has not been well investigated and the accuracy of the measured value with the non-contact voltage measurement is not enough. In this study, a high accuracy electrostatic voltmeter was used for the non-contact measurement of voltage of solar cell covered with a multilayer structure.

The conversion efficiency of solar cells decreases with increasing temperature due to the decrease of voltage [7-9]. In order to develop the simple PV performance characterization technologies, the development of simple measurement technologies of PV temperature is very important. If we can measure the voltage of PV modules and cells without contact, we can easily estimate the solar cell temperature in PV module.

II. EXPERIMENT

A single-cell module was fabricated. Figure 1 displays the photograph and schematic diagram of the module.

(a)

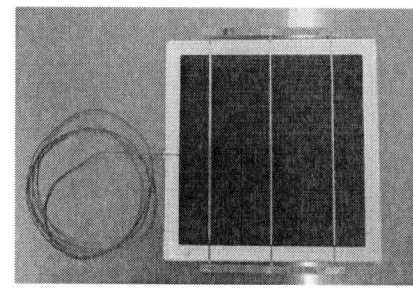

(b)

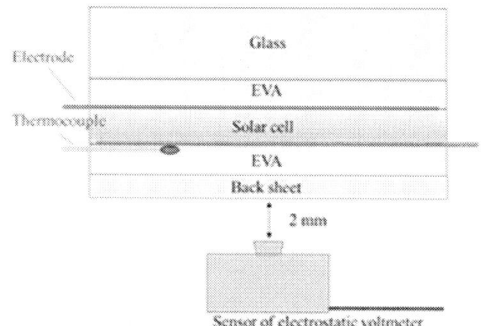

Fig. 1. (a) Photograph and (b) schematic diagram of the single-cell module. A single crystalline silicon solar cell was covered with ethylene-vinyl acetate (EVA), glass and back sheet. The sensor of the electrostatic voltmeter was set under the back sheet. The distance from back sheet to sensor was 2 mm.

In the module, a single crystalline silicon solar cell was covered with ethylene-vinyl acetate (EVA). The module front and back surface consisted of glass and back sheet, respectively. In order to measure the accurate cell temperature, a thermocouple was set just below the cell. A high accuracy electrostatic voltmeter (TREC JAPAN, TREC-320C) was used for measuring the potential of the module. The electrostatic voltmeter measures the potential of the module by measuring the field generated between the module and the sensor. The sensor of the electrostatic voltmeter was set under the back sheet. Figure 2 displays the photograph of the sensor. The distance from back sheet to sensor is an important factor for the measurement of potential. Therefore it was strictly set at 2 mm. A voltage tester (voltmeter) was connected to the electrode and the actual voltage of solar cell was also measured. Light from a solar simulator was irradiated to the module.

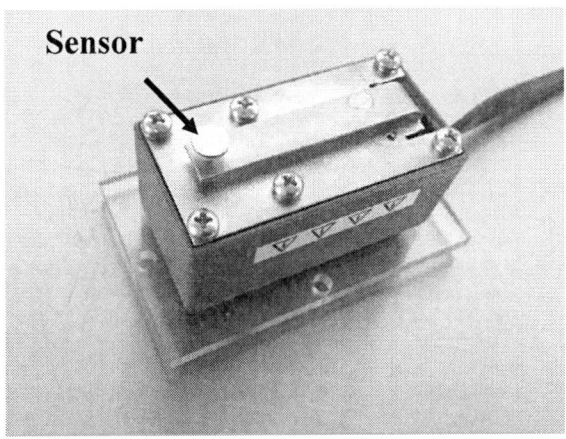

Fig. 2. Photograph of the sensor.

III. RESULTS

Figure 3 shows the relationship between the voltage measured with the voltage tester and electric potential measured with the electrostatic voltmeter. The intensity of light from solar simulator was varied to change the voltage of the solar cell. Temperature of the solar cell during the measurement was 25°C. The values of the voltage measured with the voltage tester and electric potential measured with the electrostatic voltmeter agreed well. The non-contact electrostatic voltmeter can measure the voltage of solar cell in module structure with a margin of error within 0.01 V. At the time of measurement, it is necessary to connect the electrostatic voltmeter to a ground as a reference of potential. When the earth of the module and the electrostatic voltmeter were at the same potential, the electrostatic potential could be

measured stably. When the module was not grounded, the electrostatic potential fluctuated.

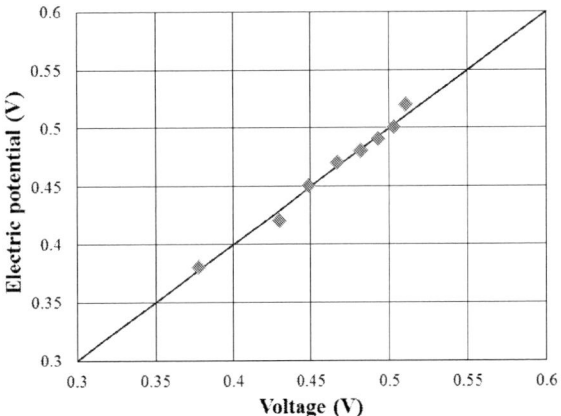

Fig. 3. Relationship between the voltage measured with the voltage tester and electric potential measured with the electrostatic voltmeter. The intensity of light from solar simulator was varied to change the voltage of solar cell.

Figure 4 shows the temperature dependence on the electric potential measured with the electrostatic voltmeter. The solar cell temperature was varied from 25 to 60 °C. The intensity of irradiated light was 1 kW/m^2. Measurement was repeated 5 times and the maximum, minimum, average values were shown in this figure. The electric potential of solar cell decreased with increasing temperature. In a general way, the voltage of solar cell decreases with increasing temperature. The decrease of voltage of solar cell with increasing temperature was successfully detected by using the non-contact electrostatic voltmeter.

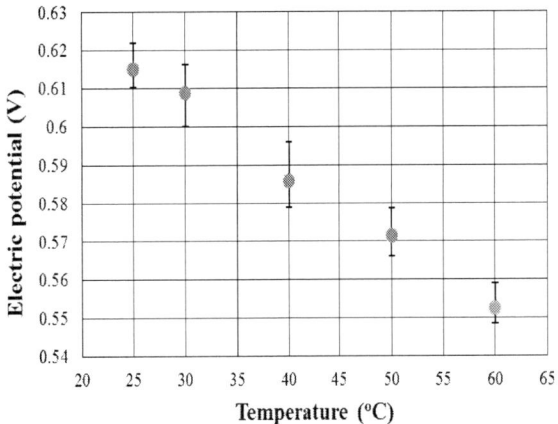

Fig. 4. Temperature dependence on the electric potential measured with the electrostatic voltmeter. The solar cell temperature was varied from 25 to 60 °C. The intensity of irradiated light was 1 kW/m^2.

Figure 5 shows the temperature dependence on the voltage measured with the voltage tester and the averaged electric potential measured with the electrostatic voltmeter. The linear approximations by least-square method were also shown in this figure. The values and linear approximations for the voltage and electric potential agreed well. By using this technique, we may estimate the solar cell temperature in PV module with simple method.

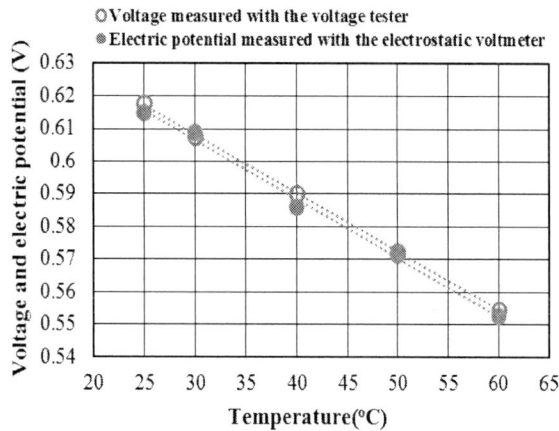

Fig. 5. Temperature dependence on the voltage measured with the voltage tester and the averaged electric potential measured with the electrostatic voltmeter. The linear approximations by least-square method were also shown.

IV. CONCLUSION

The electrostatic potential or voltage of PV modules and cells is an important parameter that directly reflects their performance and operating conditions. If a method to measure the electrostatic potential or voltage without contact is established, it is very useful for the evaluation of PV modules and cells. In this study, a high accuracy electrostatic voltmeter was used for the non-contact measurement of voltage of solar cell in PV module structure. The values of the voltage measured with the voltage tester and electric potential measured with the electrostatic voltmeter agreed well. The non-contact electrostatic voltmeter can measure the voltage of solar cell in module structure with a margin of error within

0.01 V. Moreover, temperature dependence on the voltage measured with the voltage tester and the electric potential measured with the electrostatic voltmeter was assessed. The values and linear approximations for the voltage and electric potential agreed well.

ACKNOWLEDGEMENT

A part of this work was supported by the incorporated administrative agency New Energy and Industrial Technology Development Organization (NEDO) under the Ministry of Economy, Trade and Industry (METI), Japan.

REFERENCES

[1] H. Kawamura, K. Naka, N. Yonekura, S Yamanaka, H. Kawamura, H. Ohno, and K. Naito, "Simulations of I-V characteristics of a PV module with shaded PV cells," *Sol. Energy Mat. Sol. cells*, vol.75, pp.613-621, 2003

[2] R. E. Vosteen, "DC Electrostatic voltmeters and fieldmerter," *in 9th Annu. Meet. IEEE Industry Application Society*, 1974

[3] M. N. Horenstein, "Measuring Isolated Surface Charge with a Noncontacting Voltmeter," *J. Electrost.*, vol. 35, pp. 203-213, 1995.

[4] Y. Hishikawa, K. Yamagoe and T. Onuma, "Non-Contact Measurement of the Electric Potential of PV Modules," *Proc. JSES/JWEA Joint Conference*, vol 40, pp282, 2014

[5] Y. Hishikawa, K. Yamagoe and T. Onuma, "Non-contact measurement of electric potential of PV cells in a module and novel characterization technologies," *Tech. Dig. WCPEC-6*, 2014, p. 1121.

[6] Y. Hishikawa, K. Yamagoe, and T. Onuma, "Non-contact measurement of electric potential of photovoltaic cells in a module and novel characterization technologies," *Japanese Journal of Applied Physics*, vol. 54, 08KG05, 2015.

[7] K. Nishioka, T. Hatayama, Y. Uraoka, T. Fuyuki, R. Hagihara, and M. Watanabe, " Field-test Analysis of PV System Output Characteristics Focusing on Module Temperature," *Sol. Energy Mat. Sol. cells*, vol. 75, pp. 665-671, 2003.

[8] K. Nishioka, T. Takamoto, T. Agui, M. Kaneiwa, Y. Uraoka and T. Fuyuki, " Evaluation of temperature characteristics of high-efficiency InGaP/InGaAs/Ge triple-junction solar cells under Concentration," *Sol. Energy Mat. Sol. cells*, vol. 85, pp. 429-436, 2005.

[9] K. Nishioka, T. Sueto, M. Uchida, and Y. Ota, " Detailed Analysis of Temperature Characteristics of an InGaP/InGaAs/Ge Triple-Junction Solar Cell," *Journal of Electronic Materials*, vol. 39, pp. 704-708, 2010.

NREL's Cell and Module Performance group's asymptotic Pmax protocol for perovskite devices.

Tom Moriarty, Dean Levi

National Renewable Energy Laboratory, Golden, CO 80401 USA

Abstract — **Perovskite photovoltaic devices have demonstrated a very rapid rise in reported efficiencies. Potentially extreme artifacts which tend to manifest themselves as large hysteresis effects, depending on scan rates and directions and light bias exposure history are often seen. This can make characterization difficult and may have resulted in some exaggerated efficiency reports. NREL's Cell and Module Performance (CMP) group's approach to standardized measurements of perovskites is discussed.**

I. INTRODUCTION

Perovskite solar cells have shown a rapid advance in reported efficiencies in the last several years. Certified record efficiencies have gone form 14% to 22% in only four years. Compare this to single crystal Si non-concentrators, which took 16 years to advance through the same range of efficiencies (1977-1993) and over 20 years to rise from 22% to 25%. [1]

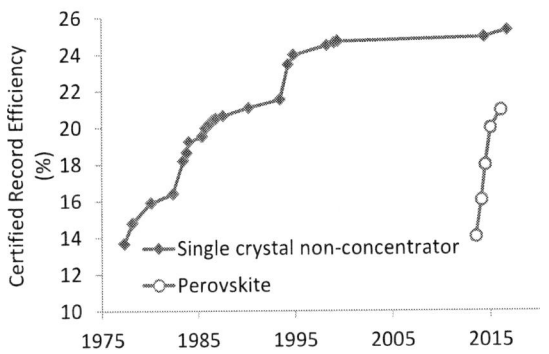

Figure 1. Comparison of record efficiency advances for single crystal Si non-concentrators to perovskites. [1]

This rapid advance has been accompanied by some unique and extreme characterization artifacts. Some measurement artifacts have resulted in reported efficiencies that have engendered doubts in the photovoltaic community [2][3]. A 2015 online ChemistyWorld.com article quoting several prominent players in the characterization of perovskite devices noted "the very nature of efficiency testing, as well as the questionable stability of perovskites themselves, is only serving to exaggerate device performance [4]." Chief among the artifacts is hysteresis effects that depend on scan rates, scan directions and light bias history [5][6]. It was under this complicated testing environment that NREL's CMP group ventured into the characterization of perovskite devices. The

goal was to create a characterization procedure that could be applied to any perovskite cell and provide an unbiased "level playing field" efficiency measurement.

The National Renewable Energy Laboratory's CMP group is part of NREL's National Center for Photovoltaics. We measure the quantum efficiency and current-voltage performance parameters of devices with wide ranges of areas, material technologies, packaging, irradiances, etc. Our clients include domestic and international commercial sources, research labs, universities, etc. Our goal is to provide accurate characterization of any PV technology.

We are accredited to ISO 17025 standards by the American Association of Laboratory Accreditation for secondary module measurements and primary and secondary reference cell calibrations. This accreditation technically does not cover perovskite measurements or even the majority of our overall measurements, which fall outside of the three categories mentioned above. However, we endeavor to apply the same group standards for hardware development, technique development, software development and data handling applied to our ISO accredited measurements to all of our measurements. The same hardware used for the secondary reference cell calibrations is used for the perovskite IV measurements.

However, the measurement protocol for perovskites is significantly different. Usually, the parameter of ultimate concern is the efficiency or Pmax, but the hysteresis artifacts may be particularly pronounce in the region of maximum power (Pmax). While there might be much useful information to be gained by sweeping IV curves at various bias rates in each direction with various dwell times and light soaking strategies, it is often a poor strategy for determining Pmax and efficiency for a perovskite device. A better approach is to determine steady state currents for constant voltages in the region of the voltage (Vmax) that yields the maximum power. The CMP group's solution is find the time asymptotic Pmax.

II. AREA AND QE

In order to overcome the measurement artifacts frequently seen when characterizing perovskites using the standard measurement procedures which work well for most other conventional cells, NREL's CMP group has developed a new protocol for measuring perovskites.

The basics of CMP group's standard procedure for more conventional cells are well known [7] and used with variation

978-1-5090-5606-4/17 $31.00 © 2017 IEEE

by many labs: measure cell area, measure quantum efficiency, derive the spectral mismatch correction factor and apply it to the adjustment of a simulator's irradiance, and measure the IV curve of the test device.

A frequently underappreciated major source of error in Jsc and efficiency is area error. The CMP group's cell measurement lab uses a Nikon microscope with optics and hardware that allow the measurement of near micron sized features. It is difficult to see a sharp edge of the active area of some of many solar cells, including perovskites, and light piping artifacts (typically from sandwiching glass edges, etc.) are encountered [8]. The best practice is to define the cell area with a mask. We encourage clients submitting cells for measurement to apply their own mask, otherwise the CMP group will apply a mask to the client's cell or characterize the cell without a mask and not report J_{sc} or efficiency. If the client provides their own attached mask, then they have control of defining the cell area and there tends to be less controversy over J_{sc} or efficiency results.

Our primary use of QE data is the calculation of the spectral mismatch correction factor,[9] which is used in conjunction with a primary calibrated reference cell to adjust the irradiance of the IV solar simulator such that the test device will yield the same Jsc under the simulator's spectrum that it would have yielded under the reference spectrum. The quantum efficiency of small, single junction cells of conventional materials are usually measured without complications beyond the usual handling, probing, temperature control, and light biasing, issues.

The quantum efficiency measurements of perovskites are often problematic due to the same response time and bias history issues that lead to high hysteresis in IV sweeps. Bias light level and chopping frequency can have strong effects on the measured quantum efficiency of perovskites. Depending on the particular cell, the effects can be seen in the magnitude and/or shape of the cell QE. It is important to look for these effects when setting up a QE measurement. The chopping frequency and bias light level must be adjusted for each device to insure as square a chopped response as possible. It is desirable to use a high chopping frequency if a square wave can be achieved, because this allows a faster scan rate which reduces bias light exposure time. The CMP group's rapid scan grating monochrometer based QE system typically chops at about 275 Hz. However, we are equipped to chop at sub-Hz rates for devices that have extremely long response times, but lower chopping frequencies require longer bias light exposure and risk damaging the test device.

Ideally, light bias level during QE should be 37% of one sun. In practice, for most devices, it is adequate to apply a white bias light such that the test device yields about 37% of the approximate expected one sun Jsc. [10] It is useful to measure the QE of a perovskite device at multiple light bias levels, including zero and the approximate Isc, if possible.

This will bracket the variation of the shape of the QE and the resulting spectral mismatch correction factor.

Over the last two years, 85% of the about 75 perovskite QEs that have been successfully measured have resulted in spectral mismatches of less than 4% under our solar simulator using a colored-glass filtered Si reference cell. In some extreme cases the measured quantum efficiency is so dubious that it is better to assume a unity spectral mismatch and increase the uncertainty calculations for the measured IV parameters rather than to apply a spectral mismatch correction. In other cases where a very low chopping frequency is required, and consequently long exposure to bias light, it is often desirable to measure the IV response the cell before measuring the quantum efficiency, then apply a spectral mismatch correction to the IV data in post processing.

III. CURRENT VS. VOLTAGE

Current vs. voltage measurement of many perovskite devices show pronounced scan rate and hysteresis effects. Figure 2 shows an example where the conventional protocol for IV measurement was applied to a 0.1 cm^2 perovskite device with varying sweep times. If taken at face value, the 2.2 second sweeps indicate an 8.1% efficient cell and the 412 second sweep indicates a 3.9% efficient cell. The sweep time required to get an accurate Pmax and efficiency may be impractically long, especially for research devices that that decay with light and air exposure.

Our practical solution is to do time asymptotic current measurements with a set of constant voltages in the region of

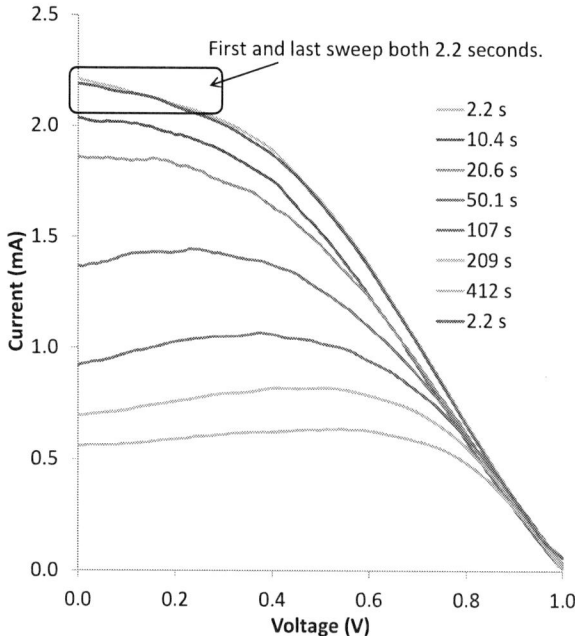

Figure 2. Current vs. voltage sweeps of 0.1 cm2 perovskite device. Voltage bias is from Voc to Isc. Cell manufacturer has given permission to use this data, but does not want to be identified.

978-1-5090-5606-4/17 $31.00 © 2017 IEEE

Vmax. Preliminary Voc to Isc and Isc to Voc fast IV sweeps give an indication of the magnitude of the hysteresis effect and help determine the voltage window in the region of Vmax that will be examined. The perovskite cell is voltage biased to a small set of voltages in that window and each selected voltage is held constant until the current settles to an asymptotic level. Additionally, the cell is held at 0 volts until the asymptotic Isc is achieved and at 0 amps until an asymptotic Voc is achieved. All of these measurements are done with continuous temperature monitoring via thermocouples or RTDs attached to or near the cell and temperature control through stage temperature, fans and venturi cooled air.

The approach to an asymptotic current typically appears to be exponential. Figure 3 shows the current vs time at voltages near the maximum power point of a very slow responding cell. Figure 4 shows an enlargement of the 0.76 volt data from figure 3, where the near exponential nature of the asymptotic approach and about 1% noise can be seen.

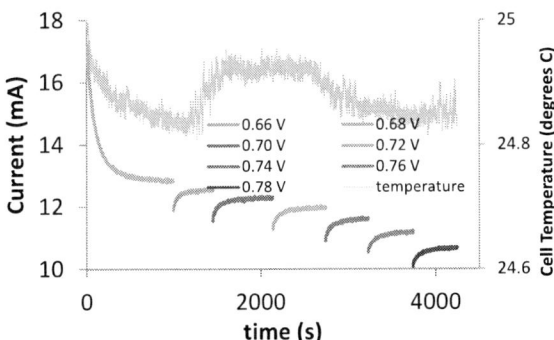

Figure 3. Current as a function of time for a very slow responding perovskite device. Voltage is held constant until an asymptotic current derived.

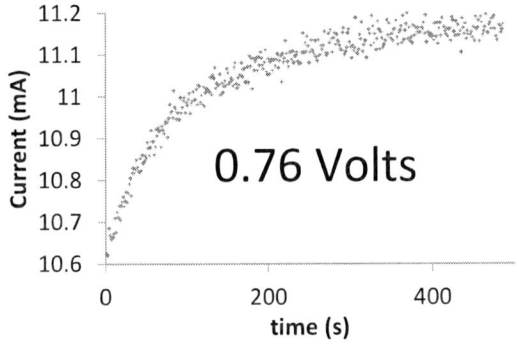

Figure 4. Enlargement of 0.76 volt data from figure 3. 1% noise can be seen in the data.

As the data is collected, a sliding window encompassing about the last 25% of the collected data is fit to a line. When the slope of that line reaches some acceptance threshold, typically about 0.1% of the cells estimated Isc per minute, the cell is biased to the next voltage and the process begins again.

As the threshold is satisfied at each voltage the current vs. time for that voltage is fit to an exponential of the form

$$I(t) = Ae^{\beta(t-t_0)} + C$$

where C is the asymptotic limit.

Figure 5 shows the slopes associated with several iterations of the sliding window during the current tracking of the 0.76 volt case shown in figure 4. In this example, once the slope threshold of 0.1% of the approximate Isc was reached, fitting the above equation for the time and current data shown in figure 5 yielded an asymptotic limit of 11.18 mA for 0.76 volts.

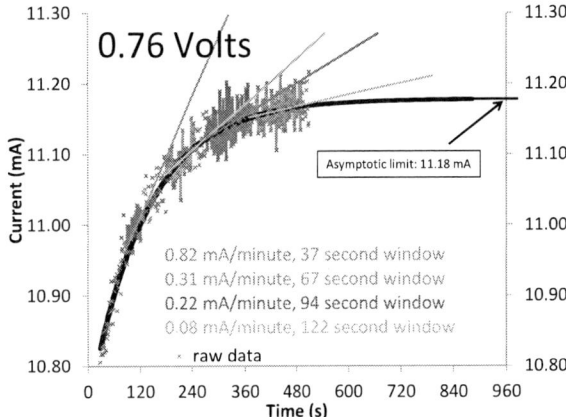

Figure 5. Once the slope of the data in the sliding window achieves the acceptance threshold the current vs. time data is fit

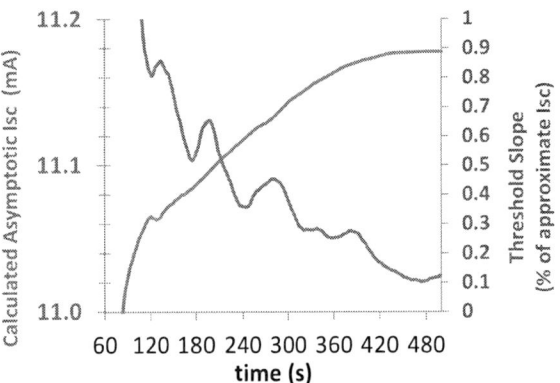

Figure 6. Threshold slopes and their associated calculated asymptotic limits the 0.76 volt case shown in figure 3.

There is a tradeoff between the acceptance threshold and the measurement time. A looser acceptance threshold means a shorter measurement time, but also a greater uncertainty in the asymptotic currents, and consequently in Pmax. Figure 6 shows the asymptotic limits for the Isc that would have been calculated for thresholds ranging from slopes of 0.1% to 1.0%. Had the threshold slope been set to 0.5% of the approximate Isc, the threshold would have been achieved in about 180 seconds and the calculated asymptotic current would have been about 11.10 mA (0.7% lower than derived with a 0.1% threshold).

IV. PMAX

Our validated code for extracting Pmax from current vs. voltage data is common to both our module and conventional cell measurements. It takes the entire set of current vs. voltage data from an IV curve as input, isolates points near the maximum power point, applies a polynomial fitting routine to those points, and finds a root of the derivative of that polynomial. This effectively interpolates between discrete voltage and current pairs. The same validated code is used to find the maximum power point from the voltage and asymptotic current pairs found near the maximum power point of perovskite cells.

In some instances, a client of the CMP may be interested in verifying the belief that their device has achieved a record efficiency or beaten some efficiency threshold. In this case, it is legitimate for that client to specify a single voltage (say, V_{proxy}) as a proxy for Vmax. Instead of examining a set of voltages in the expected region of Vmax and interpolating the resulting voltage/asymptotic current pairs with our standard algorithm, we may simply measure the asymptotic current at V_{proxy}. If

$$\frac{V_{proxy} I_{asymtotic}}{Area} > \text{record efficiency or efficiency threshold}$$

then a valid claim to a new record has been established or a desired efficiency threshold satisfied. The short light exposure and simplicity of this approach are obvious.

V. CONCLUSION

The example used in this paper was an extreme case, requiring 4500 seconds to measure seven time asymptotic currents near the maximum power point. It took an additional 1500 seconds to measure an asymptotic Isc

Changing sweep direction, varying sweep rates, dwell times and light soaking strategies often causes perovskite device measurements to yield different current vs. voltage parameters. The asymptotic Pmax protocol is an attempt to get Pmax and efficiency of a perovskite under continuous illumination and standard testing conditions and to put the characterization of such devices on a level playing field.

ACKNOWLEDGEMENTS

This work was supported by the U.S. Department of Energy under Contract No. DE-AC36-08-GO 28308 with the National Renewable Energy Laboratory

REFERENCES

[1] NREL Best Research-Cell Efficiency chart. http://www.nrel.gov/pv/assets/images/efficiency-chart.png

[2] M. Gratzel, 2014, "The light and shade of perovskite solar cells," *Nature Materials*, Sept 2014, **13**, pp838-842

[3] N. J. Jeon, J. H.Noh, Y. C. Kim,W. S. Yang, S. Ryu and S. I. Seok, "Solvent engineering for high-performance inorganic organic hybrid perovskite solar cells," *Nature Materials*, Sept 2014, **13**, pp897–903

[4] M Gunther, "Meteoric rise of perovskite solar cells under scrutiny over efficiencies," ChemistryWorld.com, 2 March 2015

[5] E. L. Unger, E. T. Hoke, C. D. Bailie, W. H. Nguyen, A. R. Bowring, T. Heumüller, M. G. Christoforod and M. D. McGehee, "Hysteresis and transient behavior in current–voltage measurements of hybrid-perovskite absorber solar cells," *Energy Environ. Sci.*, 2014, 7, pp3690-3698

[6] W. Tress, N. Marinova, T. Moehl, S. M. Zakeeruddin, M. K. Nazeeruddin and M. Grätzel, "Understanding the rate-dependent J–V hysteresis, slow time component, and aging in CH3NH3PbI3 perovskite solar cells: the role of a compensated electric field," *Energy Environ. Sci.*, 2015, **8**, pp995-1004

[7] Keith Emery, Allan Anderberg, Mark Campanelli, Paul Ciszek, Charles Mack, Tom Moriarty, Carl Osterwald, Larry Ottoson, Steve Rummel, and Rafell Williams, "Rating Photovoltaics," *Proceedings 39th IEEE Photovoltaic Specialists Conference*, Tampa Florida, June 16-21, 2013

[8] S. Ito, M. K. Nazeeruddin, P. Liska, P. Comte, R. Charvet, P. Pechy, M. Jirousek, A. Kay, S. M. Zakeeruddin and M. Gratzel, "Photovoltaic Characterization of Dye-sensitized Solar Cells: Effect of Device Masking on Conversion Efficiency," *Prog. Photovolt: Res. Appl.* 2006, **14**, pp589–601

[9] C.R. Osterwald, "Translation of device performance measurements to reference conditions," *Solar Cells*, 1986, **18**, pp 269-279

[10] Keith Emery, "Research Level I-V and QE Measurements," CRSP Annual Meeting, Golden Colorado, August 2013

Outdoor Operating Temperature Modeling of Photovoltaic Modules including Transient Effect

Soo-Young Oh, Min-Soo Kim, Won-Shup So, Woo Kyoung Kim, Jae Hak Jung, Chinho Park
Yeungnam University, Gyeongsan, Gyeongbuk, Korea,
Benazzouz Aboubakr, Ikken Badr, Naimi Zakaria, Benlarabi Ahmed
Institut de Recherche en Energie Solaire et Energies Nouvelles (IRESEN), Morocco

ABSTRACT — The conventional temperature model of PV modules is developed for the steady state condition. In order to improve the accuracy, transient effect due to the heat capacity of the PV module should be included, thus a new model is developed based on the heat transfer equation including the transient effect. Test sites are set up, and the new model has been benchmarked with these field data. The RMSE(Root-Mean-Square-Error) of the new transient model is reduced to almost half of those of the conventional steady-state model.

I. INTRODUCTION

It is important to predict the power production accurately in design, installation, operation and management of the PV power plant to reduce LCOE (Levelized Cost Of Energy). The name-plate value of the PV module power is at STC condition (T_{mod} = 25 °C, Solar irradiation = 1000 W/m^2). However, actual power production in the field is less than the name-plate value. One of the reasons is the solar irradiance is below the STC condition during most of the day. Another reason is the PV module temperature. Since not all of the solar irradiance is converted to the electrical power, the residual energy increases the PV module temperature, usually above the STC condition. The increase of the module temperature causes the decrease in the PV module power.

King [1] and Faiman [2] recently reported a series of work targeted to predict the temperature of PV modules accurately. However, both neglected the heat capacity of the PV modules in their models assuming steady-state condition. On the other hand, solar irradiance and ambient temperature are both continuously varying in the real situation, especially the solar irradiance. Recently, Cuiffi [3] reported that the difference between the field measurement of PV module temperature and King's model prediction increases, when the measurement time interval is reduced. In order to improve the accuracy, instantaneous temperature should be calculated including the transient effect due to the heat capacity of the PV module. In this paper, therefore, the temperature of the PV module is modeled using the heat transfer equation which includes the transient effect.

II. TEMPERATURE MODELING OF PV MODULES

The temperature of PV modules is determined by the incoming solar irradiance, the electrical, optical, and thermal properties of the PV module, and its heat exchange with the surroundings. Up to now, the PV module temperature has been modeled at the steady state, neglecting the heat capacity of PV modules. In the real situation, however, instantaneous temperature is significantly different from its steady-state value. In order to improve the accuracy of the temperature modeling of the PV module, and thus to improve the prediction of the PV module power, its instantaneous temperature should be modeled as a transient phenomena including the heat capacity of the PV module. Cuiffi [3] tried to model the instantaneous temperature including the heat capacity of the PV module using a simple empirical equation without any physical basis.

In this paper, the instantaneous temperature of the PV module is physically modeled (YU model) as a transient phenomenon using the heat transfer equation as below:

$$I_r(t_n) - P_m(t_n) - h \left[T_m(t_n) - T_A(t_n) \right] = m\,C\,\left[T_m(t_{n+1}) - T_m(t_n) \right] / (t_{n+1} - t_n) \quad (1)$$

$$T_m(t_{n+1}) - T_m(t_n) = \Delta t / (m\,C) \left\{ I_r(t_n) - P_m(t_n) - h \left[T_m(t_n) - T_A(t_n) \right] \right\} \quad (2)$$

where t_n : nth time

    $I_r(t_n)$: Net solar irradiance at t = t_n (W/m^2)

    $P_m(t_n)$: Module power at t = t_n (W/m^2)

    $T_m(t_n)$: Module temperature at t = t_n

    $T_A(t_n)$: Ambient temperature at t = t_n

    h   : heat transfer coefficient (W/m^2 °C)

    Δt = (t_{n+1} - t_n)

    n   = $\Delta t / (m\,C)$

    m   : Mass of the module

    C   : Heat capacity of the module

Eq. (1) is based on the heat transfer equation. The left hand side of Eq. (1) is the residual power of the PV module. It is the net solar irradiation (Ir) minus the sum of the electrical power generated by the PV module (Pm) and the heat transferred to the ambient. The heat transfer coefficient, h includes the conduction, convection, and radiation heat exchange with the surroundings. The module temperature will be increased by the residual power as shown in the right hand side of Eq. (1). Eq. (1) is then rearranged to Eq. (2) to explicitly express the increment of the temperature change of the PV module. Eq. (2) has two empirical coefficients, h and n. Both can be extracted from the field measurement data of the PV module, which will be illustrated in Section IV.

III. EXPERIMENTAL SETUP

Two experimental test beds have been set up and operating to benchmark and prove the accuracy of our temperature model (YU model), Eq. (2). One test bed is located at the

Yeungnam University, Gyeongsan, Korea (Latitude = 35.82 °N, Longitude = 128.75 °E, Elevation = 65 m). The test set up is shown in Figure 1. The stand contains 8 multi-crystalline silicon PV modules with the conventional 60-cell and 250 Wp name-plate power. All modules have glass cover and Tedlar^{TM} back sheet. The tilt angle of the PV modules is 30 degree. Each module is connected to an active load which tracks its maximum power point and dissipates the power. The stand also contains a thermopile pyranometer (Kipp&zonen CHP1) for monitoring plane of array (POA) irradiance. To the rear of each module, one thermocouple (TC) is attached by double sided bonding tape. The temperature distribution within the modules is expected to be less than 1~2 °C. A data logger samples and stores all 8 TCs and the pyranometer every 30 seconds. The ambient temperature and wind speed are monitored at the meteorological station close to the test bed.

Figure 1. Test bed set up at Yeungnam University, Gyeongsan, Korea.

The other test bed is located at Green Energy Park in Benguerir, Morocco (Latitude = 32.12 °N, Longitude = -7.94 °E. Elevation = 450 M). The test set up is shown in Figure 2. The stand contains 8 multi-crystalline silicon PV modules with the conventional 72-cell and 300 Wp name-plate power. The tilt angle of the PV modules is 32 degree. The stand also contains a thermopile pyranometer (Hukseflux SR20) for monitoring plane of array (POA) irradiance. To the rear of each module, one thermistor (SOL Connect Sensor T) is attached by a double sided bonding tape. The temperature distribution within the modules is expected to be less than 1~2 °C. The modules temperatures and the POA irradiances are measured continuously and synchronously every 10 seconds. The ambient temperature and wind speed are as well monitored every 60 seconds at the meteorological station next to the test bed.

Figure 2. Test bed set up at Green Energy Park in Benguerir, Morocco.

IV. RESULTS AND DISCUSSION

The PV module temperature model, Eq. (2) has two empirical coefficients (h and n). First, the field data are collected at the day with the clear and low wind speed (< 2 m/sec) day as shown in Figure 3. The period is selected when the temperature is steady like part B just after the noon. $\Delta T_m = 0$ in this steady-state period, and Eq. (1) is reduced to Eq. (3).

$$I_r(t_n) - P_m(t_n) - h\,[T_m(t_n) - T_A(t_n)] = 0 \qquad (3)$$

h can be expressed as below to include the effect of the wind speed (WS).

$$h = h_0 + h_1 * WS \qquad (4)$$

The data in Figure 3 is collected in the low wind speed day. The second term can be neglected. h_0 is determined using Eq. (3) with the data in part B. After h_0 is determined, h_1 can be calculated with the different wind speed case. After h is determined, n can be determined using Eq. (2) and the data in part A.

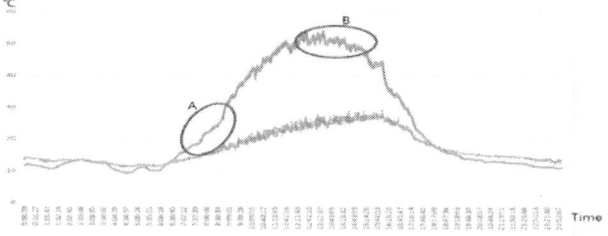

Figure 3. Ambient & module temperature variations

Using these empirical temperature coefficients, h_0, h_1, and n, the PV module temperatures are calculated using YU model and compared with the measurements. The simulations with the Faiman model [2] with their parameters are also compared in Figure 4 for YU site.

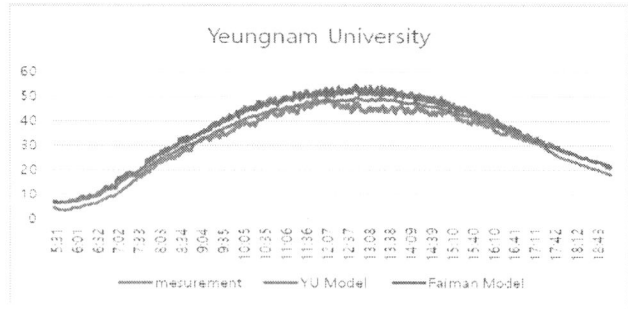

Figure 4. Comparison of the measurements and simulations of the PV module temperature at YU site. (10-05-2015)

The average and maximum RMSE (Root-Mean-Square-Error) between the measurements and simulations at YU site are shown in Table 1. As shown in the Figure 4 and Table 1, our model (Average RMSE=2.40 °C) has much better accuracy when compared with Faiman model (average RMSE = 4.52 °C). The reason is because Faiman model does not include the transient effect due to the heat capacity of the PV module. It is more evident in the maximum RMSE differences in Table 1.

	YU model	Faiman Model
Average of RMSE	2.40	4.52
Maximum of RMSE	5.52	9.70

Table 1. Average and maximum RMSE between the measurements and simulations at YU site

The average RMSE of the two models are compared for the different time intervals and shown in Table 2. For a larger time interval like 1 hour time interval, the average RMSE of YU model is comparable to the RMSE of Faiman model. However, when the time interval is reduced, the average RMSE of Faiman model increases. On the other hand, the average RMSE of YU model decreases and becomes almost half of that of Faiman model. This clearly shows that the steady-state model has bigger error than the transient model (e.g. YU model), when the time interval becomes shorter.

Time Interval	Average RMSE of YU model	Maximum RMSE of YU model	Average RMSE of Faiman model	Maximum RMSE of Faiman model
1 hour	3.69	5.63	3.75	6.81
15 min	2.91	5.39	3.91	7.67
30 sec	2.40	5.81	4.52	9.71

Table 2. The average RMSE between the measurements and simulations at YU site for different time intervals.

Using these empirical temperature coefficients, h_o, h_1, and n, the PV module temperatures are simulated using our model (YU model) and compared with the measurements at Green Energy Park in Benguerir. The simulations with Faiman model [2] with their parameters are also compared in Figure 5.

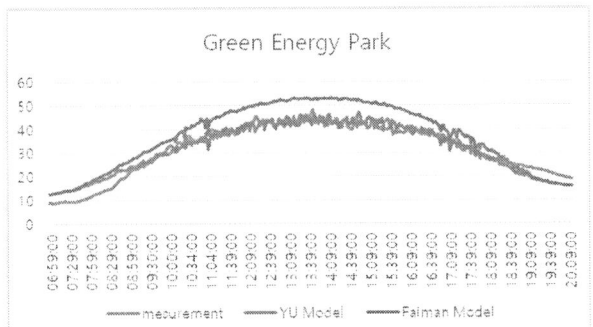

Figure 5. Comparison of the measurements and simulations of the PV module temperature at Morocco. (4-12-2016)

The average and maximum RMSE between the measurements and simulations at Green Energy Park are shown in Table 3. As shown in the Figure 4 and Table 3, YU model (Average RMSE=2.84 ° C) has much better accuracy compared with Faiman model (Average RMSE=6.72 ° C).

	YU model	Faiman Model
Average of RMSE	2.84	6.72
Maximum of RMSE	7.04	12.24

Table 3. The average and maximum RMSE between the measurements and simulations at Green Energy Park.

V. SUMMARY

Due to the massive installation of PV power plants around the world, it becomes important to predict the power production and the PV module temperature accurately. The conventional temperature model of PV modules is based on the steady state condition. In order to improve the accuracy, transient effect due to the heat capacity of the PV module should be included. YU model is based on the heat transfer equation including the transient effect due to the heat capacity of the PV module. Two test sites are set up at Yeungam University in Korea and at Green Energy Park in Morocco. YU model has been benchmarked with these field data. It is also compared with Faiman model. YU model has reduced the RMSE almost half compared with Faiman model at both sites. It is also demonstrated that the transient model like YU model shows the comparable RMSE with the steady-state model like Faiman model at the longer time interval. However, the RMSE of the transient model is reduced to almost half of the RMSE of the steady-state model with the shorter time interval.

REFERENCES

[1] D.L. King, W.E. Boyson, J.A. Kratochvil, Photovoltaic Array Performance Model, Sandia National Laboratories, Albuquerque, (2004).
[2] David Faiman, "Assessing the Outdoor Operating Temperature of Photovoltaic Modules," Prog. Photovolt: Res. Appl. 2008; 16:307–315.
[3] Joseph Cuiffi et al, "Smart Solar Field Instrumentation for Development of Site-Specific Irradiance to Power Models," EPRI-Sandia PV Systems Symposium, May 9th, 2016.

Primary Reference Cell Calibrations with Reduced Measurement Uncertainty

C.R. Osterwald, L. Ottoson, R. Williams, C. Mack, T. Moriarty, K.A. Emery, and D.H. Levi

National Renewable Energy Laboratory, Golden, Colorado, 80401, USA

Abstract — This paper presents an overview of the improvements made since 2014 to the primary reference cell calibrations performed at the National Renewable Energy Laboratory (NREL). Reducing the measurement uncertainty of the calibrations was the primary goal of the improvements, which were used during the recent calibrations conducted in the autumn of 2016. Changes and improvements have been made to the spectral irradiance and short-circuit current measurements, spectral irradiance modeling, and reference cell temperature corrections. Almost 6000 points were collected on 38 reference cells. Analysis of the results demonstrated reductions of expanded uncertainty from ±1.0-1.2% in previous years down to ±0.4-0.5% for crystalline-Si reference cells.

Index Terms — photovoltaic reference cells, primary, calibration, measurement uncertainty.

I. INTRODUCTION

The Cell and Module Performance (CMP) group at NREL has conducted primary reference cell calibrations since the mid-1980s; these calibrations are performed in direct sunlight in accordance with ASTM E1125 [1]. The CMP group holds a third-party laboratory accreditation for these calibrations from the American Association for Laboratory Accreditation (A2LA). Briefly, the short-circuit current (I_{SC}) of a cell under calibration is compared with the incident total irradiance (E_T) as measured with an absolute cavity radiometer (ACR), and a spectral correction to the hemispherical (i.e. global) reference spectral irradiance [2] is performed. The calibration equation is (1), in which F is the spectral correction factor (analogous but not identical to a spectral mismatch correction, M), and CV is the calibration value in mA/W/m^2:

$$CV = \frac{I_{SC}}{E_T}\left(\frac{1}{F}\right).$$

(1)

An ACR measures direct solar irradiance because it incorporates a collimator that limits its field-of-view to 5° [3]. Thus, the instrument must be mounted on a two-axis tracker that points toward the sun during measurements. So that all three measurements experience the same illumination, the cell under calibration and the spectral irradiance measurement (spectroradiometer, necessary for the spectral correction) are also on solar trackers with identical fields-of-view.

Details of the calibrations have been previously published [4]-[5], including the atmospheric transmittance model used to extend the spectral irradiance measurement beyond the upper and lower limits of spectroradiometer wavelength range [6].

Typically NREL's primary reference cells are calibrated annually in autumn when clear, stable weather conditions prevail. Multiple CV results obtained on multiple days are averaged to obtain a final reference cell calibration result.

Beginning in 2014 an effort was initiated to improve the group's reference cell calibration program through changes to both hardware and software, with the goal of reducing the uncertainty. The secondary goal was to enable calculation of device-specific calibration uncertainties.

A. Measurement Uncertainty Overview

Laboratory accreditations require documentation of the measurement uncertainty according to the international Guide to the Expression of Uncertainty in Measurement, commonly referred to as the GUM [7]. In the GUM, the standard deviation, σ, of a quantity x is termed the standard uncertainty, $u(x)$, and guidance is provided for obtaining $u(x)$ values with statistical analysis. There are two methods of standard uncertainty evaluations, Type A and Type B. Type A are obtained "*...by the statistical analysis of series of observations*" (GUM 2.3.2), and Type B are "*...by means other than the statistical analysis of series of observations*" (GUM 2.3.3). Because standard uncertainty is an expression of standard deviation, a Type A value is obtained from multiple measurements.

By multiplying $u(x)$ times a coverage factor, k, the expanded uncertainty is obtained (GUM 6.2.1):

$$U(x) = \pm k \times u(x).$$

(2)

For a Normal probability distribution of x values, using $k = 2$ is termed to provide 95% coverage. Lacking detailed knowledge of the distributions of CV, in this work a $k = 2$ will be assumed without any statement of the exact coverage (see GUM 6.3.3) and denoted as $U_{k=2}$.

When a quantity is the result of multiple quantities, the combined standard uncertainty is calculated from the individual $u(x)$ values. For a measurement function in which all quantities are independent products raised to the +1 or −1 power, as in (1), and the $u(x)$ values expressed in relative percentages, a linearization using a Taylor series expansion allows the combined u to be computed as the square root of the sum of the squares (see GUM 5.1.5 and [8]). Thus, for (1), with all standard uncertainties as relative quantities in percent:

$$u(CV) = \sqrt{u(I_{SC})^2 + u(E_T)^2 + u(F)^2}.$$

(3)

978-1-5090-5606-4/17 $31.00 © 2017 IEEE

For the CMP group's primary cell calibrations this was previously documented in [5], where Si solar cells were estimated to have a $U = \pm 0.91\%$ (with $k = 2$). The analysis determined that the uncertainty is dominated by two sources: the cavity radiometer and the spectral correction factor. It should be noted that this $U_{k=2}$ was determined for one particular Si reference cell, and is not reflective of cells with different bandgaps such as GaAs, InP, and GaInP, or silicon cells with colored-glass windows to simulate amorphous-Si quantum efficiency (QE).

II. CALIBRATION PROCESS IMPROVEMENTS

A. Spectral Irradiance Measurements

Until 2014, the calibration used a LI-COR Inc. model LI-1800 spectroradiometer, introduced in 1982 and was the first compact, portable instrument intended for outdoor operation. The LI-1800 had a mechanically scanned diffraction grating with a Si photodiode detector, and required about 30 s to scan the 350-1100 nm range. Because of signal-to-noise levels, the usable wavelength range was 400-1000 nm. LI-COR discontinued the product in 2002 and ended technical support in 2012.

The obsolete LI-1800 was replaced with an ASD Inc. FieldSpec 3 spectroradiometer, a triple grating instrument that spans 350-2400 nm. The full-width-half maximum (FWHM) bandwidth of the visible grating (350-1050 nm) is specified as 3 nm at 700 nm, while the infrared gratings (1000-2400 nm) are 10 nm at 1400 and 2100 nm; the FWHM bandwidths increase to about 12 nm at the endpoints of the gratings' ranges. Spectroradiometer calibrations are performed by the NREL Metrology group against National Institute of Standards and Technology (NIST) standard irradiance lamps.

For each reference cell calibration point, the ASD is programmed to collect 120 spectra during a 15 s scan time; these are averaged to produce a single spectral irradiance. This is unlike the LI-1800 that could not average small irradiance changes during a scan. I_{SC} and E_T are simultaneously sampled 40 times during the scan time.

B. Short-Circuit Current Measurement

The reference cell tracker is equipped with four collimators and can measure up to four cells simultaneously. Two potential sources of error were identified in this system. First, the design of the tubular collimators could not block all stray light around the bottoms from reaching the reference cells, and second, the 40 mm-diameter illumination area was unable to completely fill the 50×50 mm square windows of many reference cell packages.

Stray light had initially been reduced with black cloth wrapped around the collimator bottom ends, with questionable effectiveness. About 3% of the total window area at the corners was not illuminated; attempts were made to measure the magnitude of the effect were inconclusive and seemed to be reference cell-dependent. Thus, the collimators were replaced with a new design that eliminated both problems.

First, the receiving area diameter was increased to 80 mm. This completely illuminates all reference cells and permits cell misalignments up to several mm inside the area. Inner aperture locations were redesigned to minimize off-angle illumination from multiple reflections. For stray light minimization, the distance from the reference cell top surface to the last aperture was increased, and a gasket around the collimator bottom edges makes a conformal seal to the temperature control plate and reference cell cables.

C. Spectral Irradiance Extension and Modeling

The wavelength of the infrared response limit of an ACR is important because the spectral correction factor relies on the convolution integral of the wavelength response with the incident spectral irradiance, E_λ, being proportional to the measured total irradiance, or:

$$E_T \propto \int_{\lambda 1}^{\lambda 2} Z_\lambda E_\lambda d\lambda ,\qquad (4)$$

where $\lambda 1$ and $\lambda 2$ are the short and long wavelength limits of the ACR response, Z_λ, which is assumed to be unity from $\lambda 1$ to $\lambda 2$, and zero elsewhere. The LI-1800 model used $\lambda 2 = 4\ \mu m$ because this is where the ASTM G173 tables end [2] ($\lambda 1$ was 0.3 μm). ASTM E1125 estimates there are 2.9 W/m^2 beyond 4 μm, and provides an extension to the G173 tables to 10 μm [1]. Accordingly, the new model expands the ASD data from 0.28 μm out to 10 μm.

The variable FWHM bandwidths of the new ASD spectroradiometer differ from those of the single, fixed FWHM bandwidth of the LI-1800; this precluded reuse of the same constants used to calculate the atmospheric transmittance parameters necessary for extending the spectral irradiance range to match the response of the ACR [6]. Using the additional infrared data, it became possible to improve the accuracy of the fitting.

After recalculating the atmospheric optical depth constants, all the fitting algorithms were rewritten and improved. New absorption bands now fitted include H_2O vapor bands at 1120, 1370, and 1840 nm, and the 1270 nm O_2 band. The 2048 nm CO_2 band is used to calculate a carbon dioxide concentration. The aerosol scattering calculation was also improved.

A qualitative way to compare the performance of the spectral model against the ACR is to rearrange (4) as the ratio of total irradiance ratios, R_E, or:

$$R_E = \int_{\lambda 1}^{\lambda 2} E_\lambda\, d\lambda \Big/ E_T .\qquad (5)$$

During the 2013 and 2014 annual calibrations, the ASD was operated in parallel with the LI-1800, which produced two independent spectra for each data point and allowed a direct comparison of the two instruments. Fig. 1 shows the histograms of R_E for both spectra sets, and it is evident that the mean of the ASD distribution is considerably closer to one— 1.0009 versus 0.9827. Note, however, that the offset between the two instruments does not represent an error in *CV* because multiplicative constants on the spectral quantities cancel from the spectral correction factor in (6) below.

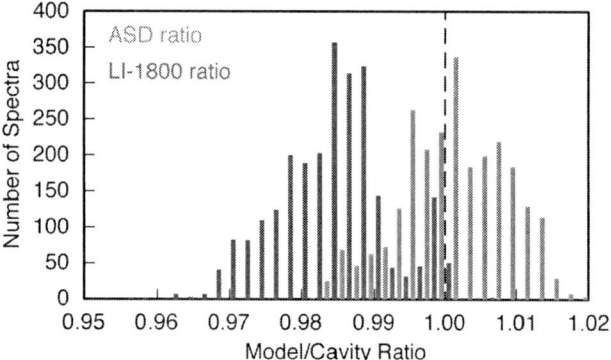

Fig. 1.   Dual histogram of total irradiance ratios R_E from (5), for spectra measured simultaneously with the LI-1800 and the ASD spectroradiometers (2013 and 2014 data, 2500 spectra).

D. Reference Cell Temperature

Solar cell temperature variations have traditionally been corrected with I_{SC} temperature coefficients (α) [3]. However, these coefficients are only an approximation of the variation of I_{SC} with temperature, principally because they are functions of spectral irradiance and are difficult measurements [9]. Therefore, they have been replaced with temperature-dependent QE translations that combine spectral and temperature corrections into a single function [1,9].

In simplified form, the spectral correction factor is (see [1] for the wavelength limits of these definite integrals):

$$F = \frac{\int \lambda Q_\lambda (T_c) E_{S\lambda}\, d\lambda}{\int E_{S\lambda}\, d\lambda} \times \frac{\int E_{0\lambda}\, d\lambda}{\int \lambda Q_\lambda (T_0) E_{0\lambda}\, d\lambda}, \quad (6)$$

where λ is wavelength, T_0 is the reference temperature of the *CV* (i.e. 25°C), T_C is the reference cell temperature during calibration, $E_{0\lambda}$ the reference spectral irradiance, $E_{S\lambda}$ the extended ASD spectral irradiance, and $Q_\lambda(T)$ the reference cell quantum efficiency, at both temperatures. From (6), F can be seen as the ratio of two *CV*s in integral form, one at the test condition and the other at the reference condition.

To avoid the need for QE measurements at every potential cell measurement temperature, the top-left integral in (6) can be well-approximated using a linearization as (with $\Delta T = T_C - T_0$) [9]:

$$F = \frac{\int \lambda Q_\lambda (T_0) E_{S\lambda}\, d\lambda + \Delta T \int \lambda \frac{\partial Q_\lambda}{\partial T} E_{S\lambda}\, d\lambda}{\int E_{S\lambda}\, d\lambda} \\ \times \frac{\int E_{0\lambda}\, d\lambda}{\int \lambda Q_\lambda (T_0) E_{0\lambda}\, d\lambda} . \quad (7)$$

This expanded form needs only two spectral quantities for the reference cell: $Q_\lambda(T_0)$, and the partial derivative of QE with temperature, $\partial Q_\lambda/\partial T$. Reference cell QEs are measured at 15, 25, and 40°C, and the QE temperature dependence calculated as $\partial Q_\lambda/\partial T$ using line fits of QE versus T at each wavelength. Fig. 2 shows an example of a temperature-dependent QE characterization.

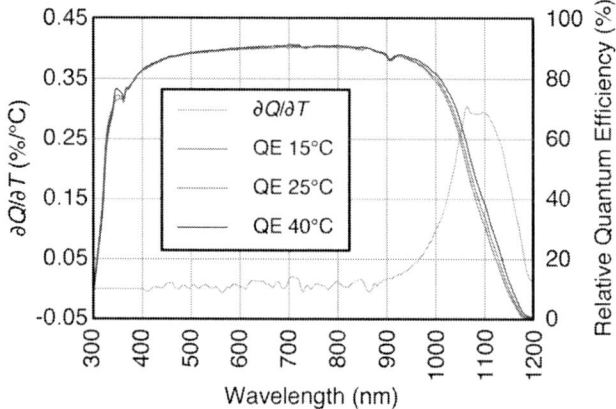

Fig. 2.   Plot of relative QE versus wavelength for a crystalline-Si reference cell, at three temperatures, and the resulting $\partial Q_\lambda/\partial T$.

E. Calibration Data Processing

One complete calibration data set of 20-40 reference cells can exceed 2000 data points which are post-processed to reduce the final spectral- and temperature-corrected calibration values. Previously this processing could only be done as a single batch operation, and detecting bad data points was time-consuming and error-prone. New software was developed to automate the processing and provide visualization for any individual reference cell using control charts of calibration value versus data point number. Individual data points are now filtered by the ranges of I_{SC} and E_T, the wind speed (important for ACR data integrity), and the total irradiance ratio of R_E.

III. UNCERTAINTY ANALYSIS

Returning to the three terms in (3), the u values are now analyzed in further detail. First, the reference cell I_{SC} during calibration is measured with an Agilent 34401 6-½ digital voltmeter (DVM) connected across a 10Ω, 4-terminal precision current sense resistor with a temperature coefficient of resistance equal to ±5 ppm/°C. An analog feedback network automatically keeps the voltage across the reference cell under test to within a few millivolts of zero (the sense resistor and feedback network combination functions as a transimpedance amplifier). Sense resistor and DVM calibrations are performed at one-year intervals by the NREL Metrology group. With V_R the voltage across the resistor and R_S the resistor calibration, the I_{SC} is:

$$I_{SC} = V_R / R_S . \qquad (8)$$

For the sense resistor, considering the calibration value uncertainty, year-to-year calibration stability history, and a maximum temperature range of 15°C, $u(R_S)$ was calculated to be 0.0086%. For the DVM uncertainty, the manufacturer's specifications were consulted and a lookup table constructed from which the voltage measurement is calculated for a voltage range, reading, and instrument temperature (this is a Type B standard uncertainty evaluation, see [5] for an example calculation). Using an average $I_{SC} = 120$ mA, V_R is 1.2 V and $u(V_R) = 0.011\%$.

Second, the ACR is calibrated at least annually by intercomparison against a group of four NREL ACRs, which provides a numeric transfer factor (TF) to the World Radiometric Reference (WRR), and the uncertainty of total irradiance measurements for the instrument [10]. Our ACR has participated in 33 pyrheliometer intercomparisons over 21 years, with a mean $TF = 0.99824$ and $\sigma(TF) = 0.043\%$. From the 2016 intercomparison, $U(E_T)$ was given as ±0.37% with $k = 1.96\%$. Thus, $u(E_T) = 0.37 \div 1.96 = 0.189\%$.

Finally, obtaining uncertainty values for the spectral correction factor is not an easy task, even though multiplicative calibration values for the spectral quantities cancel in (6) and (7). In [5], it was assumed that $u(F) \approx 0.2 \times F = 0.4\%$, based on experience and Monte Carlo studies of spectral mismatch studies in solar simulators, as a Type B evaluation. With the improvements to the spectral correction factor, however, this assumption was likely no longer valid. As an alternative, $u(F)$ has been changed to a Type A evaluation by studying the variability of calibration data over time.

Several NREL reference cells have calibration histories going back as far as 1990, using the same procedures, and these data can be viewed in a control chart. Fig. 3 shows over 1000 calibration points for a crystalline-Si reference cell, arranged in chronological order and normalized by the mean value from the most recent calibration in 2016. The standard deviation for all points from 1990 through 2014 was 0.58% for 881 points, while for the new 2016 calibration σ has been reduced to just 0.083% for 139 points.

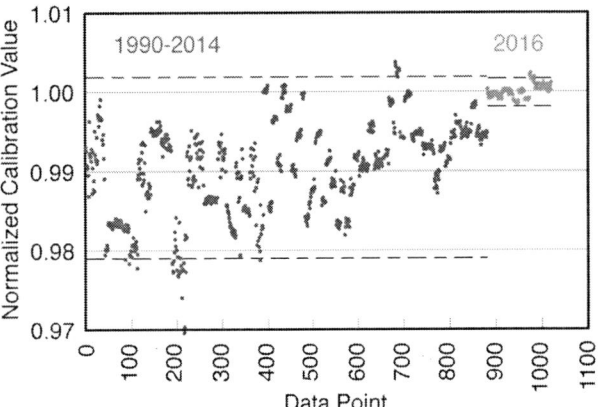

Fig. 3. Control chart showing every data point for NREL mono-Si reference cell S01 starting in 1990. Each point corresponds to one spectral irradiance measurement. Red points were measured in 2016 with the hardware and software improvements; calibration values are normalized to the mean 2016 value. The dashed lines are ±2 standard deviations.

The uncertainty of the spectral correction factor is estimated by the standard deviation of the repeated measurements of the calibration value. Normally, for Type A evaluations, the standard deviation is divided by the square root of the number of points used to calculate the mean, $\sqrt{n}$. For several reasons, we have chosen to assume that $u(F) = \sigma(CV)$ instead of $\sigma(CV) / \sqrt{n}$. First, the GUM states this is done when multiple observations have been obtained under the same conditions of measurement [3]. Because the spectral irradiance differs for every point, the measurement conditions are constantly changing. Second, using standard deviation as a direct estimate of uncertainty is consistent with ASTM E2554 [11]. Third, it is a more conservative estimate when the number of points in a sample is high (i.e. 100 or more). Fourth, the statistical distributions of CV tend to vary from cell-to-cell, and are not always normal; some are rectangular and some even bimodal.

Combining the individual contributions for the cell in Fig. 3, the standard uncertainty is:

$$u(CV) = \left[u^2(E_T) + u^2(F) + u^2(R_S) + u^2(V_R) \right]^{0.5}$$
$$= \left[0.189^2 + 0.083^2 + 0.0086^2 + 0.011^2 \right]^{0.5} \quad (9)$$
$$= 0.207\%.$$

The expanded uncertainty is then:

$$U_{k=2} = k \times u(CV) = \pm 0.41\%. \quad (10)$$

Using $\sigma = 0.58\%$ from the 1990-2014 data, $U_{k=2} = \pm 1.22\%$.

IV. RESULTS

Table 1 lists the $U_{k=2}$ results for the 32 NREL reference cells recalibrated during 2016 as ranges sorted by cell type. Si cells have the lowest uncertainties overall, which is consistent with the low sensitivity to spectral variations. Overall, $U_{k=2}$ increases with decreasing wavelength response range. Here, 'III-V isotype' is a broad category that includes cells such as GaInAs underneath GaAs windows to simulate the QEs of component subcells in series-connected multijunction devices; some of these have very narrow response ranges.

TABLE I
UNCERTAINTY RESULTS FOR NREL REFERENCE CELLS

Type	$U_{k=2}$ (±%)	Number
Mono-Si	0.40-0.45	11
GaAs	0.42-0.51	5
GaInP	0.55-0.61	3
InP	0.57	1
KG5-filtered Si	0.49	3
III-V isotype	0.41-0.68	9

Reanalysis of historic data for the III-V isotypes, as was done for S01 in Fig. 3, revealed that $U_{k=2}$ exceeded ±2% in several cases.

The offset between the red points and the more recent blue points in Fig. 3 illustrate that the *CV* for cell S01 has increased by about 0.5%. Overall, the calibration values for Si and GaAs increased by 0.5 to 0.9%, which will translate to higher I_{SC} values for test devices measured against these reference cells.

Uncertainty analyses are worthwhile efforts for establishing a laboratory's competency, but until they are verified through testing with other laboratories, they remain a somewhat intellectual exercise alone. Fig. 4 shows an intercomparison with five crystalline-Si cells that were calibrated with the updated NREL procedures described here and by Physikalisch-Techniche Bundesanstalt (PTB) using the differential spectral responsivity (DSR) method [12]. This indicates that the *CV*s of all five cells are within the error bars quoted by both laboratories and lends credence to the uncertainty calculations.

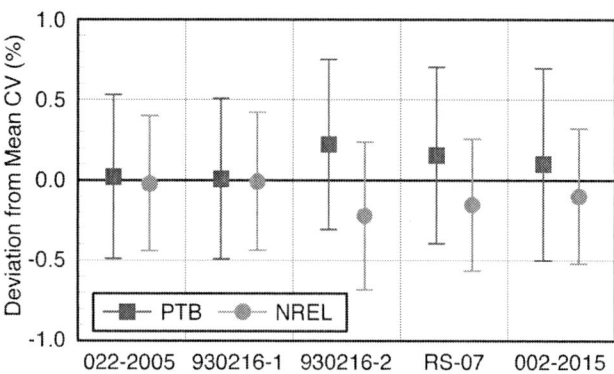

Fig. 4. Comparison of primary calibration results on five crystalline-Si reference cells calibrated by both NREL and Physikalisch-Techniche Bundesanstalt (PTB), using calibration values normalized by the mean *CV* reported by both laboratories. The error bars are expanded uncertainties.

Looking at Fig. 3 and the historical data for cell S01 again, a group of points exceeding two standard deviations can be seen around data point number 200. This is an example where control charting might have highlighted these points and caused them to be investigated as possible outliers.

V. SIGNIFICANCE AND SUMMARY

Beginning in 2014, the primary reference cell calibrations conducted by the CMP group at NREL have been improved in several key areas, including: 1) spectral irradiance, 2) reference cell I_{SC}, 3) temperature-dependent spectral correction factor, and 4) data processing. Results of the calibrations indicate greatly reduced standard deviations, and $U_{k=2}$ uncertainty values of ±0.40-0.45% for crystalline-Si reference cells. The improved uncertainties will propagate to cell and module measurements performed against these primary cells and manifest as lower I_{SC} uncertainties.

ACKNOWLEDGEMENTS

We gratefully acknowledge the contributions by Stefan Winter and the staff of PTB in providing intercomparison reference cell calibrations over many years. This work was supported by the U.S. Department of Energy under Contract No. DE-AC36-08-GO28308 with the National Renewable Energy Laboratory. The U.S. Government retains and the publisher, by accepting the article for publication, acknowledges that the U.S. Government retains a nonexclusive, paid-up, irrevocable, worldwide license to publish or reproduce the published form of this work, or allow others to do so, for U.S. Government purposes.

978-1-5090-5606-4/17 $31.00 © 2017 IEEE

REFERENCES

[1] ASTM E1125–16, "Standard Test Method for Calibration of Primary Non-Concentrator Terrestrial Photovoltaic Reference Cells Using a Tabular Spectrum," in ASTM International Annual Book of ASTM Standards, vol. 12.02, West Conshohocken, PA, 2016, DOI: 10.1520/E1125-16.

[2] ASTM G173–12, "Standard Tables for Reference Solar Spectral Irradiances: Direct Normal and Hemispherical on 37° Tilted Surface," in ASTM International Annual Book of ASTM Standards, vol. 14.04, West Conshohocken, PA, 2013.

[3] World Meteorological Organization, WMO-No. 8, "Guide to Meteorological Instruments and Methods of Observation," 7th ed., 2008, Geneva, available from http://www.wmo.int.

[4] C.R. Osterwald, K.A. Emery, D.R. Myers, and R.E. Hart, "Primary Reference Cell Calibrations at SERI: History and Methods," in 21st IEEE Photovoltaic Specialist Conference, 1990, pp. 1062-1067.

[5] K. Emery, "Uncertainty Analysis of Certified Photovoltaic Measurements at the National Renewable Energy Laboratory," Technical Report NREL/TP-520-45299, National Renewable Energy Laboratory, Golden CO, USA, 2009.

[6] C.R. Osterwald and K.A. Emery, "Spectroradiometric Sun Photometry," Journal of Atmospheric and Oceanic Technology, vol. 17, pp.1171–1188, 2000.

[7] Bureau International des Poids et Mesures, Joint Committee for Guides in Metrology, "Evaluation of measurement data — Guide to the expression of uncertainty in measurement," JCGM 100:2008, available from http://www.bipm.org.

[8] M. Campanelli, "Special Case Formula for Root-Sum-of-Squares," unpublished manuscript, June 8, 2017.

[9] C.R. Osterwald, M. Campanelli, T. Moriarty, K.A. Emery, and R. Williams, "Temperature-Dependent Spectral Mismatch Corrections," IEEE Journal of Photovoltaics, vol. 5, no. 6, 2015, DOI 10.1109/JPHOTOV.2015.2459914.

[10] I. Reda, M. Dooraghi, A. Andreas, and A. Habte, "NREL Pyrheliometer Comparisons: September 26 – October 7, 2016 (NPC-2016)," Technical Report NREL/TP-3B10-67311, National Renewable Energy Laboratory, Golden CO, USA, 2016.

[11] ASTM E2554–13, "Standard Practice for Estimating and Monitoring the Uncertainty of Test Results of a Test Method Using Control Chart Techniques", in ASTM International Annual Book of ASTM Standards, vol. 14.05, West Conshohocken, PA, 2016, DOI: 10.1520/E2554-13.

[12] S. Winter, T. Wittchen, J. Metzdorf, "Primary Reference Cell Calibration at the PTB Based on an Improved DSR Facility," in 16th European Photovoltaic Solar Energy Conference and Exhibition, 2000 pp. 2198-3001.

978-1-5090-5606-4/17 $31.00 © 2017 IEEE

Implementation of novel pin connection and test routine for improved accuracy in I-V measurements

Samuel Raj[1], Johnson Kai Chi Wong[1], Mohan Krishan Bhan[3], Evan Palmer[3], Jian Wei HO[1],

Sumukh Ramprasad[1], Wang Junci[2], Thomas Mueller[1], Armin G. Aberle [1,2]

[1]Solar Energy Research Institute of Singapore (SERIS), National University of Singapore, Singapore

[2]Department of Electrical and Computer Engineering, National University of Singapore, Singapore

[3]OAI-Optical Associates, INC. 685 River Oaks Parkway, San Jose, CA 95134 USA

Abstract — **Customary four-wire I-V testing of solar cells involve multiple voltage sense points on the cell front plane that are electrically connected. This work explores an alternative configuration in which the sense pins are isolated by high value resistors. This configuration allows the distribution of sense voltages to be read out during current extraction, as a way to check that the pin contacts are even. Simulations also show that voltage sensing after the isolation resistors leads to reduced influence of uneven pin contacts on the maximum power point, and therefore improved measurement accuracy.**

I. INTRODUCTION

I-V testing of solar cells is as standard as currency counting: it is necessary to develop the equipment that has high precision and accuracy. Accuracy is important not only so that different cells can be compared on a tight basis, but also so that the results can be reliably used in solar cell loss analysis [1, 2]. In the solar cell measurement laboratory in the Solar Energy Research Institute of Singapore (SERIS), we have identified several routine issues that influence the accuracy in I-V measurement, such as broken pins from probe bars, uneven pressure distribution along the probe bars, and misalignment of probe bars with respect to the cell busbars. All of these issues lead to a spread in voltage under the current extraction points along the cell plane, which distorts the cell current flow and makes the measured I-V curve not repeatable.

Our techniques focus on improving the measurement accuracy of the I-V curve and therefore the maximum power point and fill factor (FF). The next section outlines the proposed electrical connection configuration.

II. EXPERIMENTAL DETAILS

A. Experimental Setup

As shown in figure 1, I-V testing is a four-wire measurement, with the source and sense wires to the top side of the solar cell electrically connected to the cell plane at multiple points along the cell busbars. The perfect connection is one in which each contact point can be repeatably set to the control voltage. Unfortunately, such a level of detailed control is impractical, and any inequality in the resistance along the wires going to the busbars, or at the pin contacts, will lead to the voltages at the point contacts having a finite spread. In traditional probe bars,

which form the electrical connections to the cell busbars, the voltage sense pins are shorted together to the same metal trace. In contrast, in the proposed probe bar configuration, the sense pins are isolated by high value resistors, thus offering the possibility to monitor the sense pin voltages individually as a way to check the degree of spread in the voltages at the current extraction points.

Fig. 1. Chuck and probe bars connection setup of this technique

The new probe bar consists of 15 pairs of pins (sense and source). The voltage sense pins are connected to separate copper traces on the probe bar. Each trace passes through a 1kΩ isolation resistor, after which the traces merge into a single conductor out to the sense wire (B). Before the isolation resistors, each trace voltage can be individually monitored by a multichannel DAQ card (NI-6225) as shown in figure 2 (A).

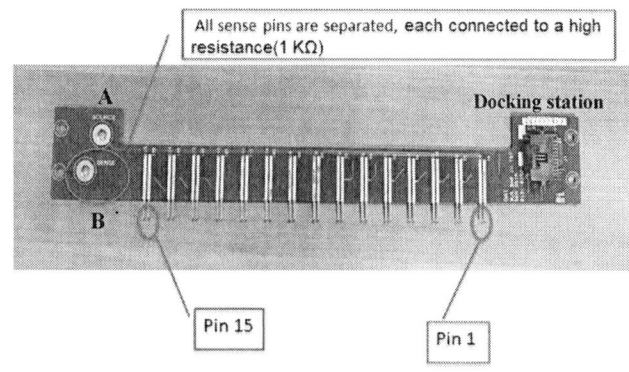

Fig. 2. Probe Bar with individual sense pin output

B. Analysis

There are a total of 45 sense pins input which acquire voltages from the three probe bars. The individual sense pin voltage monitoring data acquisition takes place when about 9A of current is extracted out of the solar cell under 1-Sun illumination, and involves calculating the mean of 100 samples in a time series at each sense pin. This is implemented in a Labview program. Figure 3 shows the voltage-spread graph in the program for one of the test that we had performed using this technique. The voltage at sense pins 1 to 15 for each probe bar can be monitored. The monitoring allows the voltage spread to be examined and is carried out before the I-V measurement starts, as a way to check that the pin contacts are even. If there is an abnormality, an instance like a small "spike" or sharp "cone" on the graph of voltage distribution can be seen. Adjustments will be required on the sense pins by replacing it or tuning the probe bar position until a constant voltage distribution in the graph is obtained as shown in figure 4. If one of the sense pin is faulty or broken, that pin will have a floating voltage and at the same time, the particular faulty pin's data will be highlight on the program. If there is poor source pin contact, or a broken source pin, that could also lead to a local swelling in voltage which will be detected. Based on this data, it is easy to locate the faulty probe pin in the probe bar.

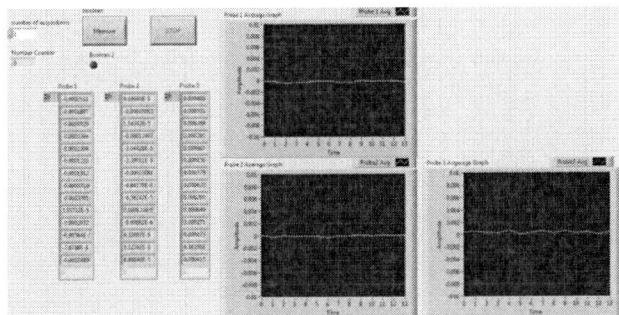

Fig. 3. Software: LabVIEW Interface design.

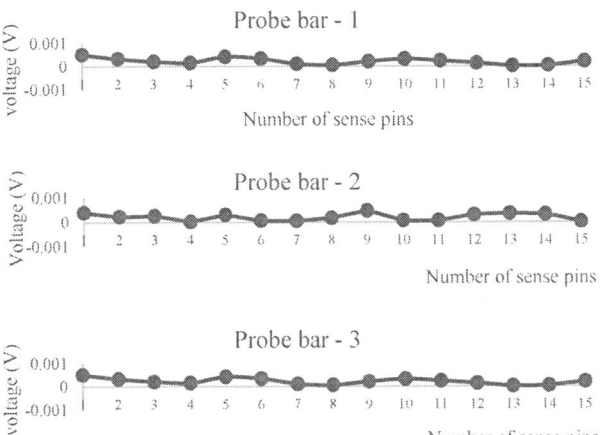

Fig. 4. Sense voltage distribution along each probe bar.

We implement this technique to test cells with different numbers of busbar and to compare with traditional way of measurement. In addition, we configured the source pins to extract current from both ends of the probe bar to avoid the current sag inside the probe bar circuit.

The results of the two experiments are analyzed ANOVA (Analysis of Variance). The statistical value F as defined in Equation 1 was applied in this comparison. It represents the ratio of two variances. The larger the values, the greater the dispersion in the data from the mean value.

F = variation between group sample means / variation within the group samples (1)

Since the testing environment for both old and new probe bar configuration was maintained the same, the contribution of Voc and Isc to the results can be treated as constant. Therefore, the significant data is Vmp measured by sense pin.

Multiple measurements were performed with both old and new probe bar configuration. Experimental results shows the F-value for the new probe bar is much smaller than the old probe bars.

This indicates that the variance between the testing group sets under the new probe bar condition is smaller. The improvement in the performance can be attributed to the monitoring system, which provided a straightforward visual indication on whether there is an abnormity at any of the sense pins before IV testing. This indication was effective, as during the experiment, from time to time, even without external interference, the probe bars might have missed aligned with the solar cell and lose contact. This will be reflected on the graph as a pin error such that the user can adjust the connection and alignment accordingly to ensure three smooth graphs for each probe bar are generated before proceeding to IV testing.

Although, F-value for the new probe bar is smaller, it is rather high and should be lowered as much as possible in the future experiment. One of the possible ways to improve can be further improve the structure of the probe bars body. We observed that the probe bar is made with PCB board and it could be bended without much force required. If possible, some effort to be do to make the bar to be more resilient to force and hopefully, when moving up and down, it would not be tilted away easily.

With the implementation of this new probe bars technique, unstable factors such as misaligned probes with solar cells or faulty pins etc are eliminated, since LabVIEW program could detect those problems and allow the user to correct the faults. Therefore, 98% of the errors came from the probe bars contacts will eliminated, remaining errors will occur from the cable connection that connects all the source pins together, this can eliminated by using a low resistive cable path for the source current to extract.

An added advantage to the new probe bar configuration, that simulations show it produces I-V curves that are less likely influence by uneven sense pin contacts. Comparison between the experimental and simulations are well explain in this paper.

III. SIMULATION

Over 90 pins spread over three-busbar increase the measuring sensitivity of the probe bar, but this may create measurement inaccuracy in the measurement. There are multiple reasons such as uneven pressure distribution on the probe bars, misalignment of the probe bars, unawareness of the broken pins, inability to monitor individual sense pins. The aim is to find whether there is a correlation between the sense pins contact resistance to FF and how it affects the I-V measurement. We randomly assign different pins contact resistance values to all 90 pins in the software over the three number of busbars and repeatedly run in Griddler 2D finite element simulation software as shown in figure 5.

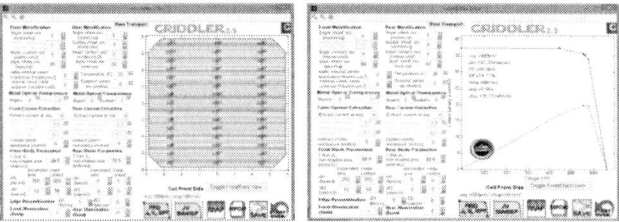

Fig.5. Griddler 2D finite element simulation software with probe pins numbering layout [5] [9].

After a few run of data analysis with repeated measurements trial, achieved the best FF spread as shown in figure 6.

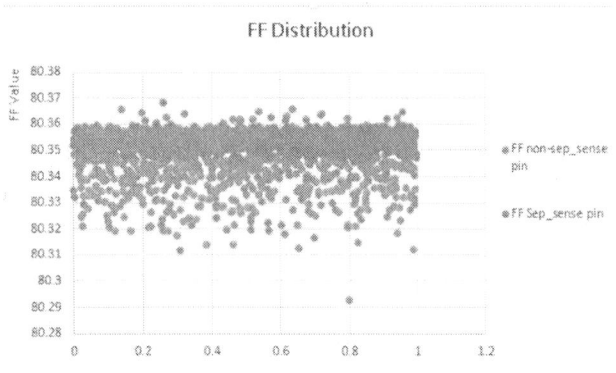

Fig.6. Shows the best FF distribution vs perfect separated sense pin graph

FF distribution graph for separate sense pins provides a better approach and best fit FF of 80.36. All points are majorly indicated the best-fitted FF. A good correlation between sense pin voltage and FF is achieved. This is because separated sense pins allows the measuring probe bar to read the voltage individually and allow the users to check the measurement errors before start the standard I-V measurement.

IV. SUMMARY

Significantly improved measurement of fill factor has been achieved in this advanced technique compared to the conventional method. Provision of a straightforward graphical representation of sense pin contacts can easily identify whether there is an abnormalities on the sense pins. Faulty sense pins can be conveniently located and adjusted before the start of I-V testing. Our next step involves further I-V testing using different types of cells and with different bus bar patterns to analyze the impact of FF losses due to probe pins contact, as well as checking the repeatability and accuracy of measurement.

ACKNOWLEDGEMENT

This research was supported by the Solar Energy Research Institute of Singapore (SERIS). SERIS is sponsored by the National University of Singapore (NUS) and Singapore's National Research Foundation (NRF) through the Singapore Economic Development Board (EDB).

REFERENCES

[1] "A Fill Factor Analysis Method for Silicon Wafer Solar Cells"Ankit Khanna, Thomas Muller, Rolf A. Stagl, Bram Hoex, Prabir K. Basu and Armin G. Aberle, IEEE Journal of Photovoltaics, Vol.3, No.4, October 2013.

[2] "Fill Factor analysis of solar cells' current-voltage curves", Johanness Greulich, Markus Glatthaar and Stefan Rein, Fraunhofer Institute for Solar Energy Systems (ISE), Heidenhofstrabe 2, D-79110 Freiburg, Germany

[3] "Solar Simulators and I-V Measurement Methods" K.A.EMERY, Solar Energy Research Institute, 1617 Cole Boulevard, Golden, CO 80401 (U.S.A)

[4] "Effect of MS contact on the electrical behavior of solar cells" Journal of solid-state Electronics, Vol.31, Jan 1988

[5] J. Wong, "Griddler: Intelligent computer aided design of complex solar cell metallization patterns," *39th IEEE Photovolt. Spec. Conf.*, pp. 933–938, 2013.

[6] J. Wong, R. Sridharan, and V. Shanmugam, "Quantifying Edge and Peripheral Recombination Losses in Industrial Silicon Solar Cells," *IEEE Trans. Electron Devices*, vol. 62, no. 11, pp. 3750–3755, 2015.

[7] IEC, "IEC 60904-7 Edition 3.0 Part 7: Computation of the spectral mismatch correction for measurements of photovoltaic devices," 2008.

[8] IEC, "IEC 60904-3 Edition 2.0 Part 3: Measurement principles for terrestrial photovoltaic (PV) solar devices with reference spectral irradiance data," 2008.

[9] J. Wong and R. Sridharan, "Griddler 2: Two Dimensional Solar Cell Simulator with Facile Definition of Spatial Distribution in Cell Parameters and Bifacial Calculation Mode," *42th IEEE Photovolt. Spec. Conf.*, 2015.

[10] A. Cuevas and S. López-Romero, "The combined effect of non-uniform illumination and series resistance on the open-circuit voltage of solar cells," *Sol. Cells*, vol. 11, no. 2, pp. 163–173, Mar. 1984.

A new method to quantify Contact Resistance using Localized-illumination Photoluminescence technique in a Solar Cell

Amit Singh Rajput [1,2], Samuel Raj [2], Johnson Wong [2], Armin G. Aberle [1,2]

[1]Department of Electrical and Computer Engineering, National University of Singapore, Singapore

[2]Solar Energy Research Institute of Singapore (SERIS), Singapore

Abstract — **An approach to derive the contact resistance of solar cell using localized-illumination photoluminescence (PL) imaging technique is presented. In this work, we used advanced finite-element simulations in order to understand dependency of PL intensity with variation in contact resistance. The technique involves the comparative analysis of PL images taken at different contact resistance using finite-element simulation software. These data will be then compare with the experimental values in order to quantify this parameter. For this localized–illumination, confined-concentrated light beam using array of high power LEDs to generate electron-hole pairs at one part of the solar cell, thus injecting photocurrent, which travels to the dark area of the cell. The interpretation of this comparative work results in evaluation of contact resistance for solar cell.**

I. INTRODUCTION

In recent years, the use of non-contact characterization techniques like Electroluminescence (EL) and Photo-luminescence (PL) imaging techniques have opened up new possibilities for inline cell metrology or even module inspection in the field. The advantages like easy to carry and fast data acquisition time make these techniques more popular amongst the industries. However, these non-contact techniques received substantial attention in this decade due to the photovoltaic (PV) module inspection under field condition. These techniques give a picture of the voltage profile, saturation current density profile, series resistance across the cell plane, aside from the general ability to highlight defects like cracks in the solar cell [1]–[5].

While there has been plenty of published methods to extract the distribution of series resistance across the solar cell plane from luminescence data, virtually all of them make use of an independent diode equivalent circuit model [6]–[8], which produces an effective term for series resistance that cannot be traced to physical origins. As of today, if one were interested to determine a specific source of series resistance, such as the contact resistance, one may have to use destructive techniques like the transfer length method (TLM).

As the non-contact, non-destructive characterization techniques has been gaining more attentions from researchers based on its advantages. The motivation behind this work is to combine photoluminescence with a detailed finite-element model of the solar cell plane, with the power to predict spatial distributions in the luminescence, to derive a tool which can accurately determine the cell contact resistance without having to destroy or even physically probe the sample.

The purpose of this paper to investigate and summarize the findings obtained from investigation of contact resistance using PL technique.

II. EXPERIMENTAL DETAILS

In the following, we outline the basic theory of proposed method based on luminescence technique, which we use for our simulations.

A. Details on measurements

The solar cell used for this investigation was n-type multicrystalline (15.6 × 15.6 cm^2), equipped with five bus bar and industrially fabricated. For simulation purpose, we used H-patterned three bus bar monocrystalline solar cell. This design allows us to analyze total PL counts coming from the cell when the light source is illuminated. In order to take PL images, we adopted BTI-Imaging tool.

B. Luminescence image analysis

In luminescence imaging, the local intensity at each pixels is exponential function of voltage at corresponding pixel. Mathematically,

$$I_i = C_i \cdot exp\left(\frac{qV_i}{k_B T}\right) \qquad (1)$$

Where C_i is the calibration constant, V_i is the voltage at that pixel positioned at 'i', q is the elementary charge, T is solar cell temperature and k_B is the Boltzmann constant.

By knowing the quantity C_i, it is possible to get voltage drop across the cell. Several methods have been published to get voltage from luminescence images [3], [4], [9]. In most of the methods, calibration constant can be calculated using low voltage image. Then the voltage drop have been investigated using higher voltage image [3]. However, the calibration constant cannot be obtained in the same fashion using PL imaging technique.

In this work, we assume that C_i is independent of the pixel, and then the luminescence image can be converted into a voltage image with an offset. This allow us to determine the voltage drops across the cell by knowledge of the luminescence distribution, which is the basis behind the contact resistance parameter extraction by comparison of experimental images to simulated luminescence data.

978-1-5090-5606-4/17 $31.00 © 2017 IEEE

C. Experimental Setup

In order to induce current to move laterally in the cell plane and to generate a luminescence pattern that is sensitive to the contact resistance, we have used power LEDs in order to create a confined light beam that illuminates only part of the cell plane. This system is based on a 620-630 nm wavelength light from LED's as an excitation source and the illumination intensity can reach up to 12-13 Suns (locally).

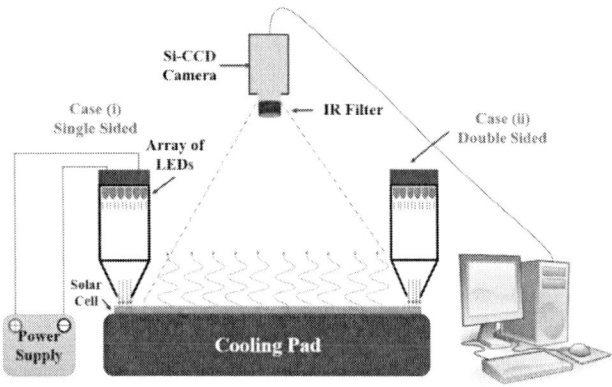

Fig. 1. Schematic diagram of localized illuminated PL setup.

TABLE I
DEATAILS OF LIGHT SOURCE USED

Total number of LED's used	Power of LED's	Wavelength Range	Total Light Intensity on solar cell
10	100 Watt each	620-630 nm	12-13 Suns

The 100 watts power LEDs calibrated to maintain continuous light intensity. For the detailed analysis of the device design parameters, refer to table 1. However, the light intensity is varying from 12-13 suns because of the heat dissipation from semiconductor device. We have been used high power exhaust fans to minimize the temperature effect.

The "Localized-illumination Photoluminescence technique" process is shown in Fig. 1. There are two sets of LED's array, externally powered by power supply. In order to maintain the constant temperature of the cell, we used cooling chuck.

As shown, an external power supply was used to generate a constant luminescence beam of light. The light beam from array of power LED's, are guided and inserted into the solar cell from one side (case (i)) and both sides (case (ii)) as well. This light beam is responsible for the electron-hole pair generation in solar cell and thus current flows from one side to the other. The recombination of these carriers are captured using silicon charge-coupled device (Si-CCD) camera in wavelength region of 800-1100 nm.

D. Procedure to take PL Images

Two different arrangements were used for the measurements: First, when one LED source is used and second when both LED sources are on. This approach requires measurement on finished solar cells as shown in Fig. 2. In top view and side view, LED source is mounted on five bus bar solar cell. A sample cell is used and illuminated light source from one side of the solar cell.

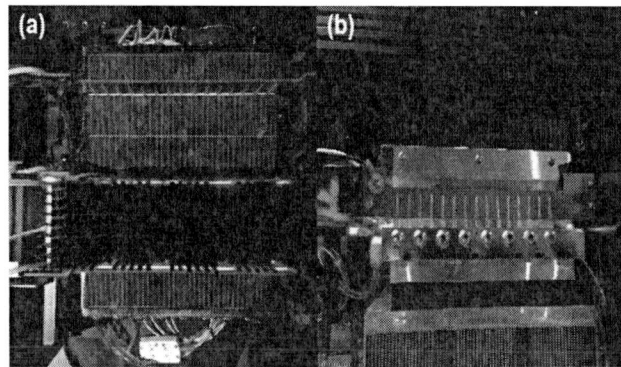

Fig. 2. (a) Top view and (b) side view of our measurement setup.

For this measurement, we have used 156 mm × 156 mm n-type standard solar cell. The spatially resolved PL image was recorded. The PL image of p-type multi-crystalline Si solar cell is shown in Fig. 3. The position of LED source is clearly shown in the image. It shows the how far the current can travel inside the solar cell. The electron-hole pair generation rate near by the LED Source are higher as compared to the other side. The decreasing trend in PL counts have been observed as we go along the other side. Based on this experiment we analyzes the curvature, slope and height of the PL curve relative to the horizontal distance. All these factors leads us to study the behavior of luminescence intensity with variable contact resistance in the sample.

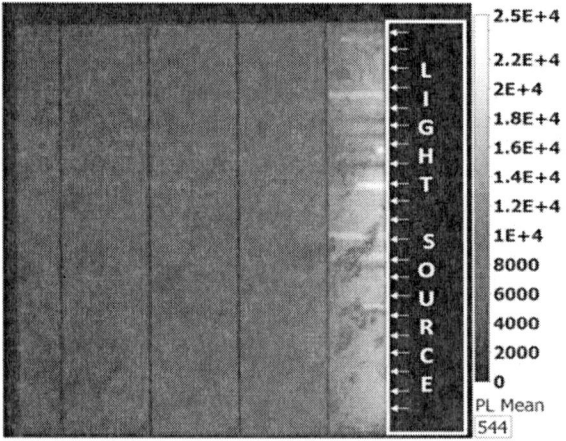

Fig. 3. PL image when light source was placed at one side on a p-type multi-crystalline Si solar cell.

III. SIMULATION RESULTS AND DISCUSSION

In order to quantitatively estimate the contact resistance using PL imaging technique, it is necessary to look out the behavior of radiation patterns coming from solar cell when the light is illuminated from one side as well as from both sides of the solar cell. We have performed simulation using 'Griddler 2.5 Pro' finite-element simulation software [10], [11]. Some of the simulated results are shown here in Fig. 4 & 5.

A. Case (i) Single-Sided Illumination

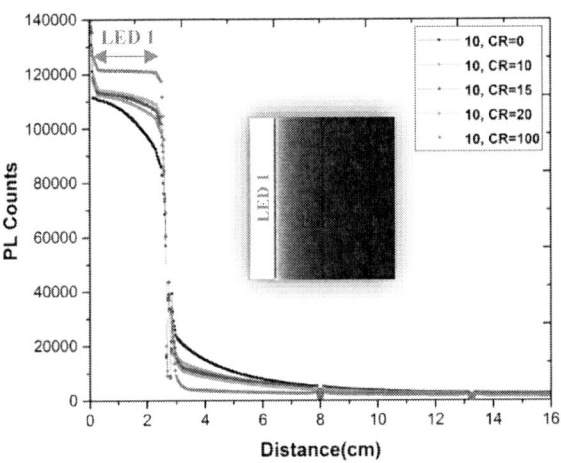

Fig. 4. Variation in illumination intensity as a function of distance and contact resistance when only one side is illuminated.

As our device is able to produce more than 10 suns light intensity, so we carried out our simulation with minimum intensity set as 10. All the simulations are carried out with 10 Suns illuminations in order to verify our process to investigate contact resistance.

Inset image of fig.4 shows the simulated PL image when one light source is on. We took the line scan of this PL image and plotted in a graph as function of contact resistance (in mΩ/cm^2) and horizontal distance. The PL counts near at the device position is very high as shown in graph. As you move further, The PL counts started exponentially decreasing. Furthermore, we repeated this process for different values of contact resistance and we noticed the same decreasing patterns with different slopes as a function of distance as well as contact resistance. This is the main observation from this simulation, which lead us to perform actual experiments.

Same trend have been observed for the other simulation case when both light sources are on. The positions of the LED's sources are shown in fig. 5. The same trend have been observed with PL counts as a function of contact resistance and horizontal distance. From fig.5, the slope and curvature of the PL counts gives more information about the variation in contact resistance. As contact resistance goes up, the PL counts are getting saturation value. In other words, nearby light source, the

slope of the curvature is very high when contact resistance is very high and vice-versa.

B. Case (ii) Double-Sided Illumination

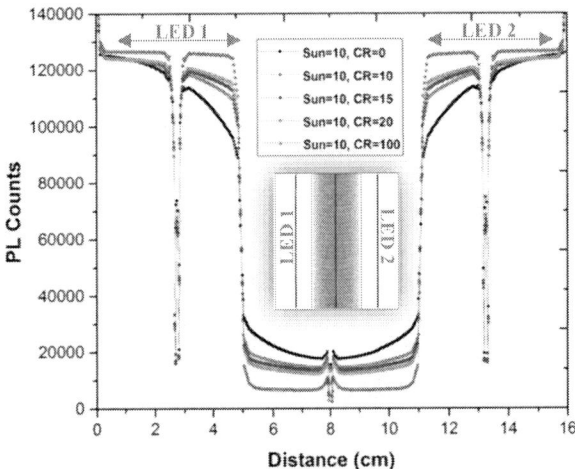

Fig. 5. Variation in illumination intensity as a function of distance and contact resistance when two sides are illuminated.

In order to demonstrate the effectiveness of this method, we performed simulation as shown in Fig. 4 and 5 for different values of contact resistance (in mΩ/cm^2). Until now we know that the contact resistance depends upon series resistance and sheet resistivity [12]. In order to investigate another way to quantify this limiting parameter, we are introducing this new PL based method.

Furthermore, we carried out this simulation with variations in other limiting parameters; one of them is area of illumination and the positions of light source in a solar cell.

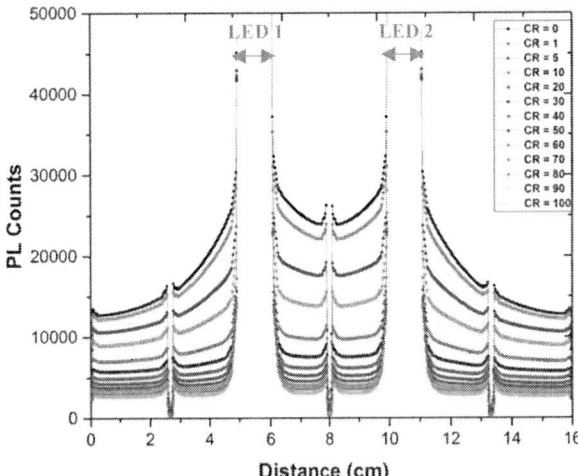

Fig. 6. Variation in illumination intensity as a function of distance and contact resistance when two LED sources, placed very near to each other, are illuminated.

978-1-5090-5606-4/17 $31.00 © 2017 IEEE

As area of illumination is fixed at 15.6 cm², variation in PL counts with respect to distance and varying contact resistance can be seen clearly in fig. 6. Our area of interest lies between the two light source where the height, slope of the curve and the curvature can clearly been differentiated for different values of contact resistance.

The luminescence patterns changes as the cell metallization contact resistance is varied. These PL intensity Vs distance graph shows the strong dependency of contact resistance on illuminated intensity.

III. SUMMARY

This study shows the possibility of investigating contact resistance using non-contact method photoluminescence technique. Currently we are investigating and trying to match this simulation results with the actual experimental one. Although there is strong dependency of contact resistance on PL intensity. The height, slope and curvature of PL counts depends upon the contact resistance and area of illumination. So our next step is to reliable means of deducing contact resistance based on luminescence images. We are planning to formulate this hypothesis mathematically and will present as soon as possible.

ACKNOWLEDGEMENT

This research was supported by the Solar Energy Research Institute of Singapore (SERIS). SERIS is sponsored by the National University of Singapore (NUS) and Singapore's National Research Foundation (NRF) through the Singapore Economic Development Board (EDB).

REFERENCES

[1] O. Breitenstein, A. Khanna, Y. Augarten, J. Bauer, J. M. Wagner, and K. Iwig, "Quantitative evaluation of electroluminescence images of solar cells," *Phys. Status Solidi - Rapid Res. Lett.*, vol. 4, no. 1–2, pp. 7–9, 2010.

[2] T. Fuyuki and A. Kitiyanan, "Photographic diagnosis of crystalline silicon solar cells utilizing electroluminescence," *Appl. Phys. A Mater. Sci. Process.*, vol. 96, no. 1, pp. 189–196, 2009.

[3] J. Haunschild, M. Glatthaar, M. Kasemann, S. Rein, and E. R. Weber, "Fast series resistance imaging for silicon solar cells using electroluminescence," *Phys. Status Solidi - Rapid Res. Lett.*, vol. 3, no. 7–8, pp. 227–229, 2009.

[4] M. Glatthaar, J. Haunschild, R. Zeidler, M. Demant, J. Greulich, B. Michl, W. Warta, S. Rein, and R. Preu, "Evaluating luminescence based voltage images of silicon solar cells," *J. Appl. Phys.*, vol. 108, no. 1, pp. 1–5, 2010.

[5] W. Luo, P. Hacke, J. P. Singh, J. Chai, Y. Wang, S. Ramakrishna, and A. G. Aberle, "In-Situ Characterization of Potential-Induced Degradation in Crystalline Silicon Photovoltaic Modules Through Dark I – V Measurements," vol. 7, no. 1, pp. 104–109, 2017.

[6] M. D. Trupke, T. and Pink, E. and Bardos, R. A. and Abbott, "Spatially resolved series resistance of silicon solar cells obtained from luminescence imaging," *Appl. Phys. Lett.*, vol.

90, no. 93506, 2007.

[7] M. Glatthaar, J. Haunschild, M. Kasemann, J. Giesecke, W. Warta, and S. Rein, "Spatially resolved determination of dark saturation current and series resistance of silicon solar cells," *Phys. Status Solidi - Rapid Res. Lett.*, vol. 4, no. 1–2, pp. 13–15, 2010.

[8] K. Bothe and D. Hinken, *Quantitative Luminescence Characterization of Crystalline Silicon Solar Cells*, 1st ed., vol. 89. Elsevier Inc., 2013.

[9] T. Trupke, B. Mitchell, J. W. Weber, W. McMillan, R. A. Bardos, and R. Kroeze, "Photoluminescence Imaging for Photovoltaic Applications," *Energy Procedia*, vol. 15, no. 2011, pp. 135–146, 2012.

[10] J. Wong, "Griddler: Intelligent computer aided design of complex solar cell metallization patterns," in *39th IEEE Photovolt. Spec. Conf.*, pp. 933–938, 2013.

[11] J. Wong, R. Sridharan, "Griddler 2: Two Dimensional Solar Cell Simulator with FacileDefinition of Spatial Distribution in Cell Parameters and Bifacial Calculation Mode," in *42th IEEE Photovolt. Spec. Conf.*, 2015.

[12] D. Schroder, "Solar cell contact resistance — A review," *IEEE Trans. Electron Devices*, vol. ED-31, no. June 1984, 1984.

Improvement Of the Properties Of CZTS Thin Films Prepared by Spray Pyrolysis Using DMSO in Acetone As Solvent

E. A. Ramírez, A. Ramírez, G. Gordillo.

Facultad de Ingeniería – UNAL, Dep. de Física – UNAL, , Universidad Nacional de Colombia, Bogotá,
edaramirezpe@unal.edu.co, Colombia

Abstract — **This work reports results of a study carried out to improve the optical, electrical and structural properties of CZTS films grown by spray pyrolysis in a one-step process using a precursor solution prepared dissolving thiourea and salts of Cu, Sn and Zn in a solvent constituted by a mixture of dimethyl sulfoxide (DMSO) and acetone. The improvement of the properties of the CZTS films was achieved through a parameters study performed by using a factorial experimental design 2^3 face centered with reply in the central point. Special emphasis was done on studying the influence of substrate temperature (T_s), carrier gas pressure (P_g) and spray pulse time (t_{sp}) on the optical, electrical and structural properties of the CZTS films. The study revealed that the t_{sn} and T_s as well as their interaction are the parameters that most critically affect the above mentioned properties.**

Index Terms — **CZTS, thin films, Spray pyrolysis, DMSO route, parameter study.**

I. INTRODUCTION

Efforts are currently being made to fabricate solar cells with high conversion efficiency, low cost and based on abundant and pollution free materials. Materials like Cu_2ZnSnS_4 (CZTS), $Cu_2ZnSnSe_4$ (CZTSSe) are considered promising materials for low-cost thin film solar cell. A variety of routes has been undertaken for thin-film deposition; these include vacuum and solution based deposition approaches [1,2]. Among these, SP deposition process is also a suitable technique for CZTS thin films deposition due to the facility to adjust the composition just tuning the precursor solution and to the conceptual simplicity of the process [3]. In the year 2013 were reported efficiencies of 7,5% for solar cells based on CZTS deposited by SP using DMSO as solvent which allow propose this route as a substitute of the hydrazine toxic route [4]. To prevent incorporation of oxygen during deposition, the CZTS films were deposited inside a hermetic glove box under nitrogen atmosphere and using nitrogen as carrier gas. In addition, instead of water we used DMSO as a solvent.

In this work we used a factorial design 2^3 face centered with reply in the central point to evaluate the effect of the main SP parameters in the electrical, optical and structural properties of CZTS thin films.

II. EXPERIMENTAL

CZTS films were prepared inside a glove box connected to a nitrogen generator that provides a permanent flow of nitrogen; these were grown by spray pyrolysis technique using nitrogen as carrier gas and a precursor DMSO/acetone solution containing cupric chloride (0.01 M), zinc acetate (0.005 M), stannic chloride (0.005 M) and thiourea (0.03 M).

Considering that substrate temperature (T_s), spray pulse time (t_{sp}) and carrier gas input pressure (P_g) are the parameters that more critically effect the CZTS films properties, a parameter study was done varying T_s between 350 and 450°C, t_{sp} between 0,4 and 1.2s and P_g between 2 and 4 bar. The direct effect and the interaction effect of these parameters onto the film properties were evaluated by means of a factorial design 2^3 face centered with reply in the central point. After deposition, the samples were annealed at 500°C during 30 min maintaining the chamber at nitrogen partial pressure of 20 mbar.

Transmittance and reflectance measurements were performed using a Varian – Cary 5000 spectrophotometer and the resistivity was done using the four probe method with contacts placed on the surface of the sample. The film thickness was determined using a Veeco Dektak 150 surface profiler. The XRD measurements were performed in an Shimadzu-6000 diffractometer and Raman spectroscopy in an Alpha 300 Raman spectrometer of WITec with a high resolution confocal optical microscope using excitation source of 785 nm.

III. RESULTS AND DISCUSSION

A. Influence of Ts, tsp and Pg on the film resistivity

The influence of T_s, t_{sp} and P_g on the film resistivity was studied using a precursor solution prepared as described above and keeping the rest of synthesis parameters constant (Substrate-nozzle distance =20 cm, flow rate of precursor solution =3 ml/min, nozzle diameter =0.6 mm and waiting time between pulses =90 s). From this study, the following facts are highlighted:

i) Substrate temperature significantly affect the resistivity under the conditions of $t_{sp} < 0.8$ s and $P_g > 3$ bar. In this region the resistivity decreases in the range of 95.3-11.3 Ω.cm. T_s values around 400°C gives raise to high homogeneity and low resistivity. The lowest resistivity (11.3 Ω.cm) was achieved using T_s =400 °C, t_{sp} =0.4 s and P_g =4 bar. In contrast, at T_s around 450 °C was observed an increased resistivity. Therefore 400 °C was chosen as the most adequate substrate temperature.

ii) In the range of $t_{sp} \geq 0.8$ s and $P_g \leq 3$ bar, t_{sp} is the parameter that most strongly affects resistivity and the homogeneity. Furthermore, it was observed that the small droplets (when Pg is increased) contribute to improve the crystallinity of the films, which was confirmed by X-ray diffraction analysis.

B. Influence of Ts, tsp and Pg on the optical properties

The influence of T_s, t_{sp} and P_g on the optical properties was also evaluated through transmittance and reflectance measurements. Fig. 1 shows typical transmittance spectra of CZTS films deposited varying T_s, t_{sp} and P_g. It is clear from Fig. 1 that the best set of SP parameters is: T_s =400 °C, t_{sp}= 0.4 s and P_g =4 bar, which were established considering that the samples prepared under these conditions present both high transmittance (around 70%) and high slope indicating that this type of samples grow with a low density of intrinsic and structural defects. In contrast, the CZTS film deposited under the parameter set: T_s =350 °C, t_{sp}= 1.2 s and P_g =2 bar are characterized by having both low transmittance (around 5%) and low slope, indicating that these type of samples grow with a high density of intrinsic and structural defects.

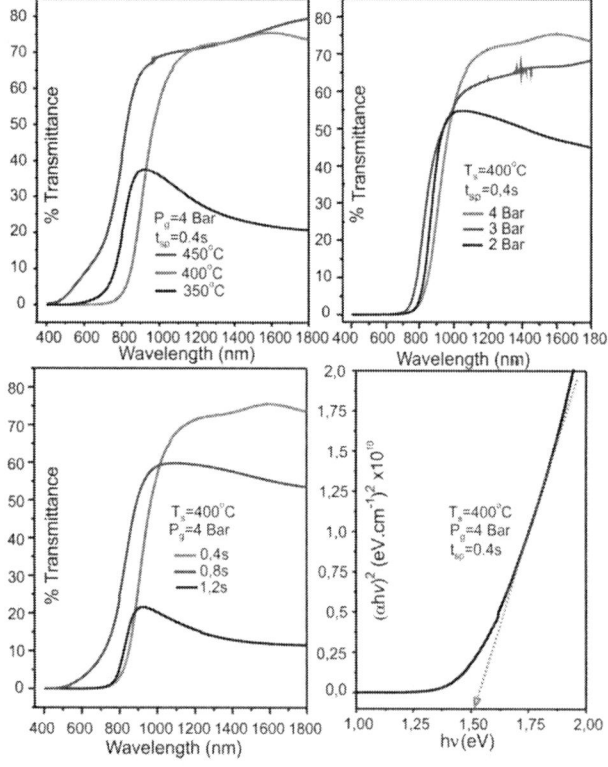

Fig. 1. Transmittance spectra of CZTS films deposited varying the parameters T_s, P_g and t_{sp}. The curve of $(\alpha h\nu)^2$ vs hν correspond to a typical CZTS film prepared under the best parameters.

Intrinsic defects generate absorption centers within the band gap which contribute to a reduction of the transmittance in the visible and infrared regions; on the other hand,

structural defects induce distortion of the bands, generating tails of states that extend the bands within the gap causing a decrease of the slope of the transmittance curves. Absorption via transitions between states of tails of bands was evaluated through the calculation of the absorption coefficient near the band edge (α_U) that has an exponential dependence on photon energy and Urbach energy [5] ($\alpha_U = \alpha_0 exp\left[\frac{h\nu - E_i}{Eu}\right]$).Where E_1 and α_0 are constant and E_U is the Urbach energy which is weakly dependent on temperature and is often interpreted as the width of the tail of localized states in the band gap. Thus, a plot of lnα vs. hν should be linear and Urbach energy can be obtained from the slope. In Table 1 are listed values of Urbach energy calculated from the slope of the curves of lnα_u vs hν.

Taking into account that Urbach energy is related to density of localized states in the band gap induced by structural disorder (stress and dislocation) [6], can be concluded that CZTS films prepared around 400°C under low t_{sp} and high P_g values present an improved crystallinity. The excess of precursor solution on the substrate by the increase of the t_{sp} contributes to the formation of localized states, associated to structural defects and presence of secondary phases. It is observed in Table 1 that the samples prepared at t_{sp}> 0.4 s exhibit increased Eu values and present a mixture of the CZTS, Cu_2SnS_3 and CuS phases. On the other hand, it was found that the variation of P_g does not affect the Urbach energy nor gives place to the formation of secondary phases.

TABLE 1
E_u VALUES OF CZTS FILMS PREPARED VARYING T_S, t_{sp} And P_g.

	t_{sp}=0.4 s, P_g=4 bar			T_S=400 °C, P_g=4 bar		T_S=400 °C, t_{sp}=0.4 s	
	T_S (°C)			t_{sp}(s)		P_g (bar)	
	350	400	450	0.8	1.2	2	3
E_u (meV)	192.1	135.9	220.0	230.5	196.4	133.1	137.6

C. Influence of Ts, tsp and Pg on the structural properties

The structural properties of the CZTS films were also evaluated through XRD measurements. Fig. 2 shows XRD spectra corresponding to CZTS films prepared varying T_s, t_{sp} y P_g. The analysis of these diffractograms reveal that it is possible to have thin films of CZTS with the presence of a single phase of Cu_2ZnSnS_4 (JCPDS file # 00-021-0883), but also samples with presence of secondary phases of CuS (JCPDS file #. 00-001-1281) and Cu2SnS3 (JCPDS file # 00-019-0412) depending on deposition parameters used (see Table2).

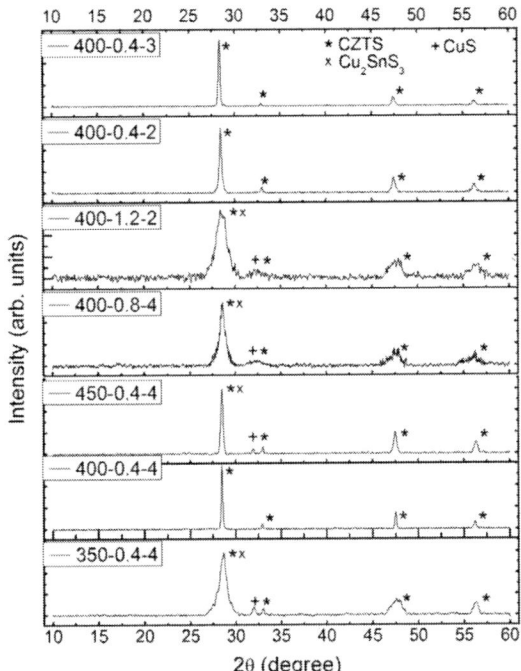

Fig. 2: X-ray diffraction spectra of CZTS films prepared varying T_s, t_{sp}, and P_g; the phase associated to each peak is indicated.

In general, thin films of CZTS with presence of single phase of Cu_2ZnSnS_4 can be obtained using values of $T_s=400\,°C$, $T_{sp}=0.4$ s and P_g between 2 and 4 bar. The increase of t_{sp} promotes the formation of secondary phases of CuS and Cu_2SnS_3 due to presence of residual solvent and excessive reduction of the substrate temperature caused by excess solvent sprayed to the substrate surface. The use of substrate temperatures around 350 °C or 450 °C (keeping $t_{sp}=0.4$ s and $P_g=4$ bar) also gives rise to the formation of secondary phases of CuS and Cu_2SnS_3, apparently as a result of incomplete reaction (when deposition is done at T =350 °C) and excess of evaporation of Zn and S (when deposition is done at 450 °C). Based on the XRD study, can be concluded that the best conditions to grow thin films with single phase of Cu_2ZnSnS_4 was using values of $T_s=400\,°C$, $t_{SP}=0.4$ s and P_g between 2 and 4 bar and additionally leaving an interval of time of 90 s between pulse and pulse.

TABLE 2
PHASES IDENTIFIED IN CZTS FILMS PREPARED VARYING T_s, t_{sp} And P_g.

	$t_{sp}=0.4$ s, $P_g=4$ bar			$T_s=400\,°C$, $P_g=4$ bar		$T_s=400\,°C$, $t_{sp}=0.4$ s	
	T_s (°C)			t_{sp} (s)		P_g (bar)	
	350	400	450	0.8	1.2	2	3
	Phases						
CZTS	x	x	x	x	x	x	x
Cu_2SnS_3	x		x	x	x		
CuS	x		x	x	x		

It was also found that the full width at half maximum (FWHM) value of the Cu_2ZnSnS_4 is affected by the P_g value, indicating that this parameter affect their crystallinity. Considering that the peak broadening is affected by crystallite size D and lattice strain $\mathcal{E}$ due to dislocation [7], these two parameters can be determined by the X-ray line broadening method using the Williamson-Hall equations [8] given by:

$$\beta_{hkl}cos\theta_{hkl} = \frac{K\lambda}{D} + 4\varepsilon sin\theta_{hkl} \qquad (1)$$

Where D is the particle size, λ is the wavelength of the radiation for CuKα radiation), k is a constant equal to 0.94, β_{hkl} is the peak width at half-maximum intensity and Θ_{hkl} is the peak position. The XRD peak broadening is due to both instrument and sample dependent effects. The contributions of each of these effects are convoluted causing an overall broadening of the diffraction peaks. To decouple these contributions, it is necessary to correct the instrumental effect from the line broadening of a standard material to determine the instrumental broadening. The instrumental corrected broadening β_{hkl} was estimated using the equation [9]:

$$\beta_{hkl} = \left[\left(\beta_{hkl}\right)^2_{measured} - \left(\beta\right)^2_{instrumental}\right]^{1/2} \qquad (2)$$

If the strain is assumed to be uniform in all crystallographic directions, Eq. (1) represents the UDM model (Uniform Deformation Modell). The term $\beta cos\theta$ was plotted with respect to $4sin\theta$ for the preferred orientation peaks of CZTS films. Accordingly, the slope and y-intersect of the fitted line represent strain and particle size, respectively.

In the Uniform Stress Deformation Model (USDM), a generalized Hooke's law refers to the strain, keeping only the linear proportionality between the stress and strain as given by $\sigma=Y\mathcal{E}$, where σ is the stress of the crystal and Y is the modulus of elasticity or Young's modulus. This equation is an approximation that is valid for a significantly small strain. Applying the Hooke's law approximation to Eq. (1) yields:

$$\beta_{hkl}cos\theta_{hkl} = \frac{K\lambda}{D} + \frac{4\sigma sin\theta_{hkl}}{Y_{hkl}} \qquad (3)$$

There is another model that can be used to determine the energy density of a crystal called the Uniform Deformation Energy Density Model (UDEDM). When the strain energy density u is considered the constants of proportionality associated with the stress-strain relation are no longer independent. For an elastic system that follows Hooke's law, the energy density u can be calculated from u = $(\mathcal{E}^2Y_{hkl})/2$. Then Eq. (3) can be rewritten according the energy and strain relation.

$$\beta_{hkl}cos\theta_{hkl} = \frac{K\lambda}{D} + \left(4sin\theta\left(\frac{2u}{Y_{hkl}}\right)^{\frac{1}{2}}\right) \qquad (4)$$

TABLE 3.
INFLUENCE OF SUBSTRATE TEMPERATURE ON THE GRAIN SIZE, STRESS, STRAIN AND ENERGY DENSITY OF CZTS FILMS PREPARED BY SPRAY PYROLYSIS.

| Sample | P_g (Bar) | Williamson - Hall method | | | | | | | | | AFM |
| | | UDM | | USDM | | | UDEDM | | | | D |
		D (nm)	ε x 10^{-3}	D (nm)	ε x 10^{-3}	σ GPa	D (nm)	ε * 10^{-3}	σ GPa	u (KJm^{-3})	(nm)
CZTS	2	34.8	1.95	34.8	1.95	152.1	34.8	1.95	152.1	148.8	29 ±3
T_s= 400 °C	3	66.6	2.69	66.6	2.69	209.6	66.6	2.69	209.6	284.4	59±4
t_{sp}= 0.4 s	4	84.0	1.48	84.0	1.48	115.2	84.0	1.48	115.2	85.29	86±7

Plots of $\beta_{hkl}\cos\theta$ versus $4\sin\Theta/(Y_{hkl})^{1/2}$ were constructed and the data fitted to lines. The anisotropic energy density u was estimated from the slope of these lines, and the crystallite size D from the y-intercept.

The results obtained from the UDM, USDM, UDEDM, models and AFM are summarized in Table 3; in this case we used a value of Y =77.8 GPa reported in the literature for kesterite-type Cu_2ZnSnS_4 films, which was calculated using density-functional theory (DFT) [10].

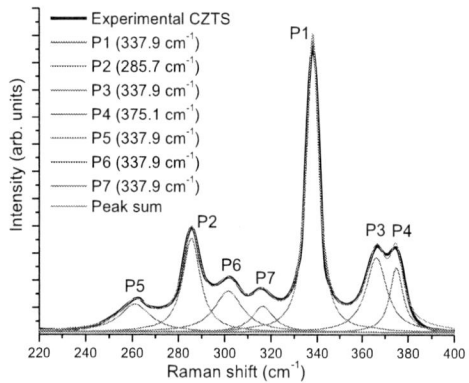

Fig. 3: Raman spectrum of a typical CZTS sample prepared at T_s=400 °C, t_{sp}=0.4 s and P_g=4 bar, together with the fitting of the peaks with lorentzian curves.

X-ray diffraction analysis which is mostly used for the phase identification cannot clearly distinguish the formation of secondary phases such as Cu_2SnS_3 and CuS, since these compounds have similar diffraction pattern as Cu_2ZnSnS_4; therefore Raman scattering analysis was used to distinguish these phases with a higher degree of reliability. In Fig. 3 is displayed a Raman spectrum of a typical CZTS sample prepared at T_s= 400 °C, t_{sp}= 0.4 s and P_g= 4 bar, together with the fitting of the peaks with lorentzian curves. The spectrum shows a dominant peak P1 located at 337–338 cm^{-1} which have been identified as the main vibrational A symmetry mode from kesterite CZTS [11-15]. Besides, the peaks located in the Raman bands P2, P3, P4, P5, P6 and P7, found by fitting the spectra with Lorentzian curves have also been associated to single phase kesterite type CZTS films [14-16]. In CZTS

samples prepared under these conditions were not identified phases corresponding to the compounds CuS and Cu_2SnS_3.

IV. CONCLUSIONS

Through a study of synthesis parameters carried out with the help of a factorial experimental design 2^3 face centered with reply in the central point, conditions were found to grow by spray pyrolysis, single phase Cu_2ZnSnS_4 thin films with improved structural, optical and electrical properties, from a precursor solution prepared dissolving thiourea and salts of Cu, Sn and Zn in a solvent constituted by a mixture of DMSO and acetone. The study revealed that these properties are affected by most of the synthesis parameters being T_s and t_{sp}, the parameters that most critically affect them.

CZTS films prepared under optimized conditions grow with a low density of intrinsic and structural defects and present an energy band gap close to 1.5 eV, resistivities ranging from 10 to 30 Ωcm and transmittances around 70%, which can be considered adequate to be used as absorber layer in solar cells.

ACKNOWLEDGEMENTS: This work was supported by Colciencias (Contract #184-2016) and Universidad Nacional Facultad de Ciencias, Grupo GMS&ES, Bogotá DC, Colombia (Proy. 28135supported by DIB).

REFERENCES

[1] K. Wang, O. Gunawan, T. Todorov, B. Shin, S.J. Chey, N.A. Bojarczuk, D. Mitzi, S. Guha, "Thermally evaporated Cu2ZnSnS4 solar cells", Appl. Phys. Lett., vol. 97, pp. 143508-14311, 2008.

[2] J. Kim, H. Hiroi, T. K. Todorov, O. Gunawan, M. Kuwahara, T. Gokmen, D. Nair, M. Hopstaken, B. Shin, Y. S. Lee, W. Wang, H. Sugimoto, and D. B. Mitzi, High Efficiency Cu2ZnSn(S,Se)4 Solar Cells by Applying a Double In2S3/CdS. Emitter. Adv. Mater. vol. 26, pp. 7427–7431, 2014.

[3] Y. B. Kishore Kumar, G. Suresh Babu, P. UdayBhaskar, and V. Sundara Raja, Effect of starting-solution pH on the growth of Cu2ZnSnS4 thin films deposited by spray pyrolysis. Phys. Status Solidi A.vol206, p.p.1525–1530, 2009

[4] T. Schnabel, M. Löw, E. Ahlswede, Vacuum-free preparation of 7.5% efficient Cu2ZnSn(S,Se)4 solar cells based on metal salt precursors. Sol Energ Mat Sol C. vol. 117, p.p. 324–328, 2013

[5] J. Tauc. Amorphous and Liquid Semiconductors, Plenum Press, London, New York, Chapter 4, ISBN 10: 0306307774. 1974

[6] M.V. Kurik, Review of Urbach's tail. Phys. Status Solidi A., vol. 8, p. 9, 1971.

[7] R. Yogamalar, R. Srinivasan, A. Vinu, K. Ariga, A.C. Bose, Solid State Commun, vol. 149, p. 1919, 2009

[8] G.K. Williamson,W.H. Hall, Acta Metall, vol. 1, p. 2231, 1953

[9] R. Srinivasan, R. Yogamalar, R.J. Josephus, A.C. Bose, Funct. Mater. Lett, vol. 2, p. 1, 2009

[10] X. He and H. Shen, "First-principles study of elastic and thermo-physical properties of kesterite-type Cu_2ZnSnS_4, Physica B: Condensed Matter, vol. 406, pp. 4604–4607, 2011.

[11] Xavier Fontané, " Caracterización por espectroscopía Raman de semiconductores, Cu_2ZnSnS_4 para nuevas tecnologías fotovoltaicas", PhD Tesis, Universidad de Barcelona, 2013

[12] K. Wang, B. Shin, K. B. Reuter, T. Todorov, D. B. Mitzi, and S. Guha, Structural and elemental characterization of high efficient Cu_2ZnSnS_4 solar cells, Appl. Phys. Lett, vol. 98, p. 051912, 2011

[13] F. Biccari, Fabrication of Cu2ZnSnS4 solar cells by sulfurization of evaporated precursors, Energy Procedia, vol.10, p.p. 187-191, 2011

[14] P. Fernandes, P.M.P. Saloméa, A.F.D. Cunha, Study of polycrystalline Cu2ZnSnS4 films by Raman scattering, J. Alloys Compound, vol. 509, p.7600, 2011

[15] X. Fontané, L. Calvo-Barrio, V. Izquierdo-Roca, E. Saucedo, A. Pérez-Rodriguez, J.R. Morante, D.M. Berg, P.J. Dale, S. Siebentritt, In-depth resolved Raman scattering analysis for the identification of secondary phases: characterization of Cu2ZnSnS4 layers for solar cell applications, Appl. Phys. Lett, vol. 98, p. 181905, 2011

[16] P. A. Fernandes, P. M. P. Salomé, and A. F. Cunha, Study of polycrystalline Cu2ZnSnS4 films by Raman scattering. J. Alloys Compd., vol. 509, no. 28, pp. 7600-606, 2011.

Assessment of Carrier Lifetimes and Surface Recombination Velocity through Spectral Measurements

John Roller and Behrang H. Hamadani

National Institute of Standards and Technology, Gaithersburg, MD, 20899, USA

Abstract — Charge carrier lifetimes and surface recombination velocities were measured on photovoltaic-grade silicon wafers using a spectral-dependent method. Narrow bandwidth light emitting diodes (LEDs) were used to generate excess charge carriers inside the material. The effective lifetimes were accurately measured and then analyzed in conjunction with an analytical model relating the bulk lifetime and surface recombination velocity with effective lifetime and excitation wavelength. The agreement between the model and the measured data validates the ability to ascertain information about the bulk lifetime and surface recombination velocity through these measurements. The limitations of the model are also discussed in detail.

I. INTRODUCTION

Charge carrier lifetime is a useful physical parameter that helps characterize the quality of a semiconductor material. Longer carrier lifetimes in bulk material have been shown to correlate to higher efficiency in completed solar cells, thus there is a need for a reliable measurement of bulk lifetime [1]. Surface effects can influence the measurement of the effective lifetime, and can have a dominant effect in thin samples where the charge carrier diffusion length is on the order of the material thickness. With strong surface effects, materials will show carrier lifetimes orders of magnitude lower than the bulk lifetime without proper passivation.

PV module manufacturers want to be able to extract information on the bulk lifetime and surface passivation of their materials; therefore, this study focuses on silicon, a material which is widely used in the industry. Existing approaches to determine this information include contactless photoconductance-based Quasi-Steady State (QSS) measurements of several wafers with either varying thickness [2] or varying levels of processing [3], transient decay experiments which utilize the decay profile of a probe signal [4], or microwave photoconductivity decay which measures the charge carrier concentration through a proportional relation to the power of reflected microwaves [5]. These techniques require expensive specialized equipment or specialized samples, and are therefore not widely used by manufacturers.

This work assesses the bulk lifetime and surface recombination velocity of several wafers using a spectral technique that utilized the wavelength dependence of the absorption coefficient of silicon. An array of monochromatic LEDs was used to generate carriers with differing profiles as a function of depth into the material. The resulting photoconductive response was then obtained through a radio frequency (RF) photoconductance (PC) measurement apparatus, and could then be used to extract information about the material. The well-established QSSPC method [6] was used to evaluate the effective lifetimes, which were then used in conjunction with a recently developed analytical model[2] to evaluate the bulk lifetime and surface recombination velocity of a wafer from wavelength dependent measurements. Previous reports on the spectral dependence of lifetimes utilize either filtered broad-spectrum flash lamps [7], which are not as easily controllable or scalable as LEDs, or tunable laser pulses, which are used mainly for transient studies.

Various samples were measured with this technique, and the spectral dependence was more evident on poorly passivated wafers where the surface effects had a major effect. The exact values for bulk lifetime and surface recombination velocity could not always be extracted from the model due to the sensitivity of the measurement. When the surface recombination effects were significantly more dominant than bulk effects, only a lower bound for the bulk lifetime could be found.

II. EXPERIMENTAL DETAILS

The measurements in this work implemented a modified version of the Sinton* WCT-120 instrument, which used a custom-built LED array as an input light source instead of the provided broadband flashlamp. Nine types of narrow band LEDs were used ranging from 456 nm to 1032 nm, and were controlled via a waveform generator in conjunction with custom-built current amplifiers. The waveform generator could be controlled in such a way that QSS conditions could be met for the lifetimes of the various wafers, which approached 1 ms. In addition to this setup helping with the measurement of long lifetime samples, it could also be utilized in a way to more effectively measure short lifetime samples. By increasing the frequency of measurement, the QSS measurement could be repeated several times in quick succession, and averaging the measurements resulted in a useful signal to noise ratio for these samples which usually have a low output signal.

Most of the measurements in this study were done using a waveform frequency of 0.97 Hz, though some of the low lifetime measurements were sped up to 9.7 Hz or even

978-1-5090-5606-4/17 $31.00 © 2017 IEEE

97 Hz. The input waveform used for these measurements was a sinusoidal signal, which allowed for easy determination of transient effects in the output signals. These transients would appear in the form of a phase lag in the output signal, as compared against the incident optical signal, which indicated a need for a new measurement at a slower frequency. Figure 1 shows the setup of the modified instrument, which includes a quartz light pipe that provides for a uniform input intensity in the range of 10 W/m² to 1000 W/m². This range is comparable to the typical exposure of installed solar modules.

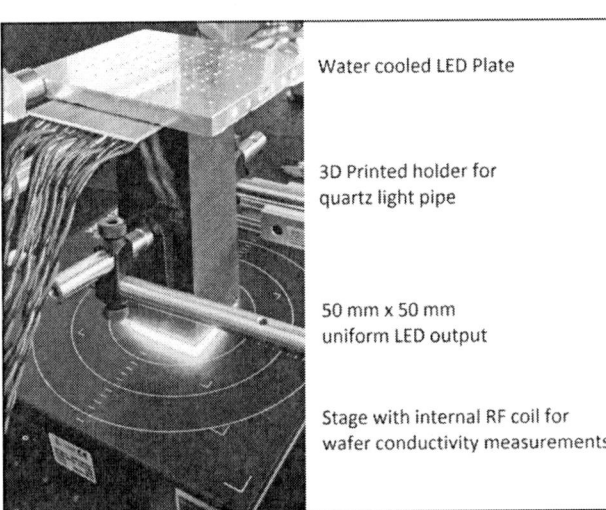

Fig. 1. Measurement setup of the LED array coupled with the Sinton WCT-120 stage. The quartz light pipe focuses the output of the LEDs to a uniform 50 mm x 50 mm area. In this image, a 456 nm LED source was used for the illumination.

The density of excess carriers in the wafer was measured by the photoconductive coil, having been previously calibrated against known resistivity wafers, and the steady state generation rate was determined from the measurement of incident light by a calibrated photodetector at the exit plane. Using these two measurements, the lifetimes were calculated by using the continuity equation:

$$\tau_{eff} = \frac{\Delta n}{g - \frac{d\Delta n}{dt}}, \qquad (1)$$

The effective lifetime, τ_{eff}, is the quotient of the excess charge carrier density, Δn, and the generation rate, g. The QSS nature of the measurement makes the transient change in carrier density, $d\Delta n/dt$, negligible. The generation rate, g, was calculated using the equation:

$$g = \int_0^w \int_{\lambda_{min}}^{\lambda_{max}} \alpha(\lambda)\, \phi(\lambda)\, e^{-\alpha(\lambda)z} \left(1 - r(\lambda)\right) d\lambda\, dz \quad (2)$$

The total generation rate in the material is a function of the absorption coefficient of the material, α, the incident flux, ϕ, the extinction through the material and the reflection coefficient, r, integrated over the wavelength spectrum of the

LED, λ_{min} to λ_{max}, and the thickness of the wafer, w. A calibrated photodetector was used to calculate the incident flux, and a separate monochromator-based measurement was performed on each wafer to independently obtain the reflection coefficient. This equation assumes that reflection losses are solely at the front surface, which is valid for most wavelengths considered here; however, the LEDs with wavelengths above 900 nm required a pathlength enhancement factor, a scalar modification to the thickness value, to compensate for the effects of backside reflection and surface texturing.

Trap states in the measured wafers caused issues for the appropriate measurement of the excess charge carrier density. It has been shown that trap states can have sizable photoconductive responses with a small amount of incident light [8]. To compensate for this effect, which would cause an artificially high photoconductive response, a bias term is used to subtract out the trap conductivity.

There are several sources of uncertainty for these calculations including nonuniformity of the incident LED light of around 7 % to 11 %, instrument resolution of the RF coil, and nonuniformity of the measured wafer. To estimate this uncertainty, a series of measurements was performed with a set of LEDs at a representative wavelength to see the variations in calculated effective lifetime. The standard deviations from these measurements range from 5 % to 33 % depending on the signal level of the samples, with higher deviations on the shorter lifetimes.

III. EXPERIMENTAL RESULTS AND ANALYSIS

In a recent study, Turek et al. found an analytical solution to the diffusion equation in steady state [2]. (3) is the diffusion equation in one dimension where D is the diffusion coefficient, τ_b is the bulk lifetime, z is the depth into the material and $n(z)$ and $g(z)$ are the carrier concentration and generation at z respectively.

$$D \frac{d^2 n(z)}{dz^2} - \frac{n(z)}{\tau_b} = -g(z) \qquad (3)$$

The analytical solution takes the form of (4), where l is the diffusion length defined as $\sqrt{D\tau_b}$, S is the surface recombination velocity, and w is the total thickness of the material.

$$\tau_{eff,\lambda} = \frac{\tau_b}{1 - \alpha^2 l^2} \left[1 - \alpha l \, \frac{\alpha l + \frac{Sl}{D} \coth \frac{\alpha w}{2}}{1 + \frac{Sl}{D} \coth \frac{w}{2l}} \right] \qquad (4)$$

While Turek et al. focused on the surface passivation and wafer thickness to determine the relationship between effective and bulk lifetimes, the measurements in this study use the dependence of the absorption coefficient, α, on the wavelength, λ, to determine the bulk lifetime from several measured effective lifetimes. This dependence is shown in Figure 2.

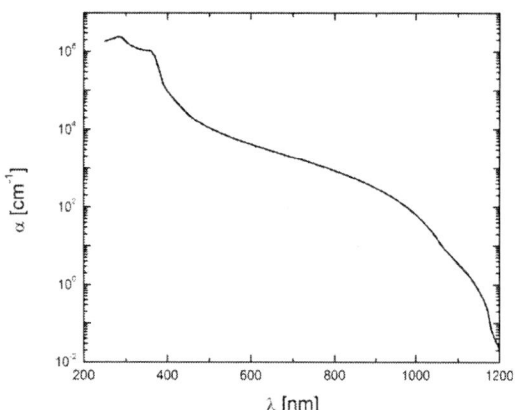

Fig. 2. Absorption coefficient for crystalline silicon at 300 K as a function of wavelength. This data was taken from PC1D, a commonly used solar cell modeling program.

For this model to give conclusive results, the surface recombination cannot mask other recombination mechanisms. Figure 3 shows the measured effective lifetimes of an unpassivated 1 mm polished n-type monocrystalline silicon wafer overlaid on the analytical model evaluated at several values for bulk lifetime. The data are in a regime where the model has effectively saturated, meaning further increases in the τ_b values will not shift the curves higher along the τ_{eff} axis; therefore, only a minimum bound for the bulk lifetime can be assessed. In this case, although the model fits well to the data and there is good confidence over the extracted $S = 10\,600$ cm/s, we observe that we can only state a lower bound value of $\tau_b \geq 2$ ms for the bulk lifetime because the data are beyond the sensitivity range of the model.

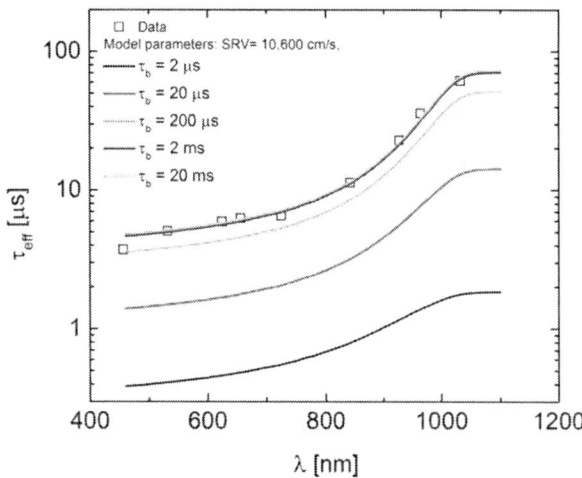

Fig. 3 Measured data and modeled behavior of effective lifetime for a 1mm thick, high resistivity n-type wafer. For the determined surface recombination velocity, S, of 10 600 cm/s and a diffusion constant, D, of 12 cm²/s, the model was calculated at five separate values for bulk lifetime.

Figure 4 shows the measured effective lifetimes of an unpassivated 550 µm n-type polished monocrystalline silicon wafer overlaid on the analytical model evaluated at several values for bulk lifetime. Notice how for this sample, the data fall below this saturation regime and a value for the bulk lifetime can be calculated. From the best fit of the model to these data, $\tau_b = 63$ µs and $S = 8800$ cm/s were extracted at an excess carrier density of $\approx 10^{13}$ cm⁻³.

The data taken in Figures 3 and 4 show that information about the properties of the wafer can be assessed when the surface recombination velocity is influential on the effective lifetime. When bulk recombination is dominant, other factors need to be considered to glean information about a wafer.

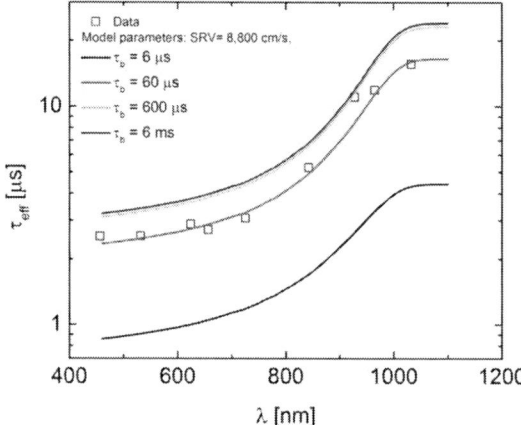

Fig 4. Measured data and modeled behavior of effective lifetime for a 550 µm thick, low resistivity n-type wafer. The model was calculated at four separate values for bulk lifetime using a S of 8800 cm/s and a D of 12 cm²/s.

This measurement technique can be used to verify the quality of wafers throughout the cell production process. Wafers can be measured at various stages of the production line to calculate changes in the surface recombination velocity [9]. Figure 5 shows two sets of data collected on 160 µm n-type wafers at two stages of the production process for Heterojunction with Intrinsic Thin Layer (HIT) cells. For the wafer that has undergone a passivation process, the bulk recombination is the dominant mechanism. This means neither the bulk lifetime or the surface recombination velocity can be extracted reliably because there is no wavelength dependence of the data. However, using the assumption that the bulk lifetime of the wafer does not change through the production process, values from the unpassivated wafer can be used to assess values for the passivated wafer.

The unpassivated data is unfortunately in the lifetime saturation regime, meaning that the calculated values are $S = 33\,300$ cm/s and $\tau_b \geq 1$ ms. Although there is no definite lifetime calculated for the wafer, the bounds $\tau_b = 1$ ms and

$\tau_b \to \infty$ can be used to calculate bounds on the surface recombination of the unpassivated wafer.

Figure 5 also shows the modeled behavior of effective lifetime for several values of surface recombination velocity while holding the bulk lifetime constant at the minimum bound of $\tau_b = 1$ ms. From these data, the minimum bound of the surface recombination velocity of the passivated sample is found to be $S \geq 5.7$ cm/s. A similar method can be used with the bulk lifetime set to $\tau_b \to \infty$, which gives an upper bound of $S \leq 13.5$ cm/s. Through using information garnered from the unpassivated sample, the properties of the passivated sample were able to be reasonably assessed.

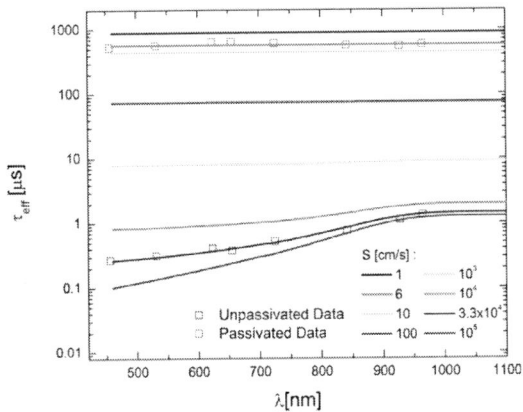

Fig 5. Measured data and modeled behavior of effective lifetime for two 160 µm thick, n-type wafers at two stages in the HIT cell production process. The model was calculated at eight values of surface recombination velocity using a D of 18 cm²/s and a τ_b of 1 ms, the lower bound of bulk lifetime for the unpassivated wafer.

IV. CONCLUSIONS

An apparatus was developed that allows for measurement of effective carrier lifetimes by way of optical excitation from narrow band illumination sources. The measurements taken agree with a recently developed analytical model that shows the wavelength dependence of lifetime data. From these measurements, information can be gathered on the bulk and surface properties of the material, and, at a minimum, bounds on the bulk lifetime and surface recombination velocity can be estimated.

The strength of the measurement lies in the similarities to PV cell use conditions, namely being quasi steady state with light intensities on the order of or below one sun. The measurements in this study were done primarily on silicon as a proof of concept but can be done on any material with a wavelength dependent absorption coefficient.

REFERENCES

[1] D. Klein, F. Wuensch, and M. Kunst, "The determination of charge-carrier lifetime in silicon," *Phys. Status Solidi Basic Res.*, vol. 245, no. 9, pp. 1865–1876, 2008.

[2] M. Turek, "Interplay of bulk and surface properties for steady-state measurements of minority carrier lifetimes," *J. Appl. Phys.*, vol. 111, no. 12, p. 123703, 2012.

[3] K. Bothe, R. Krain, R. Falster, and R. Sinton, "Determination of the bulk lifetime of bare multicrystalline silicon wafers," *Prog. Photovoltaics Res. Appl.*, vol. 18, no. 3, pp. 204–208, 2010.

[4] R. K. Ahrenkiel and S. W. Johnston, "An optical technique for measuring surface recombination velocity," *Sol. Energy Mater. Sol. Cells*, vol. 93, no. 5, pp. 645–649, 2009.

[5] O. Palais and A. Arcari, "Contactless measurement of bulk lifetime and surface recombination velocity in silicon wafers," *J. Appl. Phys.*, vol. 93, no. 8, pp. 4686–4690, 2003.

[6] R. a Sinton, T. Mankad, S. Bowden, and N. Enjalbert, "Evaluating silicon block and ingots with quasi-steady-state lifetime measurements," *Proc. 19th Eur. Photovolt. Sol. Energy Conf. , Paris, Fr.*, pp. 520–523, 2004.

[7] H. Mäckel and A. Cuevas, "Determination of the surface recombination velocity of unpassivated silicon from spectral photoconductance measurements," *Proc. 3rd World Conf. Photovolt. Energy Convers.*, no. 4, p. , 2003.

[8] D. Macdonald, R. A. Sinton, and A. Cuevas, "On the use of a bias-light correction for trapping effects in photoconductance-based lifetime measurements of silicon," *J. Appl. Phys.*, vol. 89, no. 5, pp. 2772–2778, 2001.

[9] J. F. Roller, Y. Li, M. Dagenais, and B. H. Hamadani, "Spectral dependence of carrier lifetimes in silicon for photovoltaic applications," *J. Appl. Phys.*, vol. 120, no. 23, p. 233108, 2016.

Impact of Space Radiation Environment on Concentrator Photovoltaic Systems

Pilar Espinet-Gonzalez[1,*], Tatiana Vinogradova[2], Michael D. Kelzenberg[1], Alexander Messer[2], Emily C. Warmann[1], Chris Peterson[2], Nina Vaidya[1], Ali Naqavi[1], Jing-Shun Huang[1], Samuel P. Loke[1], Don Walker[3], Colin J. Mann[3], Sergio Pellegrino[1] and Harry A. Atwater[1]

[1] California Institute of Technology, Pasadena, CA 91125

[2] Northrop Grumman Aerospace Systems, Azusa, CA 91702

[3] The Aerospace Corporation, El Segundo, CA 90245

* Corresponding author e-mail address pespinet@caltech.edu

Abstract — Concentrator photovoltaic systems can provide supplementary shielding against high energy particles. In this paper we compare the radiation environment that the same solar cell would experience in a flat-plate module versus in a parabolic mirror concentrator system. We have observed that the shielding provided by the concentrator system is remarkable.

In order to obtain an accurate prediction of the overall shield needed in our concentrator system triple-junction space solar cells have been irradiated on the edge with 350-keV protons at a fluence of 10^{12} p$^+$cm^{-2}. A mild degradation of the open circuit voltage was measured (~70 mV).

Index Terms — Concentrator photovoltaic systems, concentrator solar cells, space radiation modeling, radiation tests, space solar power, shielding.

I. INTRODUCTION

Concentrator photovoltaic systems for space applications have the potential of significantly reducing the specific power (W/g) of the solar panels. This is achieved by replacing most of the area of the solar cells with lighter-weight concentrator optics. This is beneficial even when the cells themselves are lightweight, because the cells must be covered by heavy cover glass to protect them against damage from high energy particles. In fact, low concentration in space is not a new concept. Several reflective [1-2] and refractive [3] designs have been developed, and some of them have been successfully tested in space [4-5]. However, concentrator systems make more challenging the thermal management of the system, the pointing sensitivity, and the overall system reliability; thus flat panels remain as the state of the art for space solar power. In order to fulfill all the requirements for operating in space while maintaining a significant advantage in specific power over the latest flexible flat plate modules [6], it is essential in designing the concentrator system to carry out a global optimization, taking into account the different parameters that have an impact on the thermal management, optical performance and radiation hardness of the system.

In this paper we have studied how the radiation of high energy particles impacts the optimization of a reflective concentrator system that we are designing. We have compared the radiation environment experienced by the solar cells in a flat plate configuration versus that in our concentrator. Additionally, we have studied the impact of radiation on small (~mm size) solar cells with a high perimeter-to-area ratio, which are a characteristic feature of our concentrator design. In order to carry out this analysis, we have manufactured solar cells of 1 mm x 10 mm area, and irradiated them from their front, back, and side, with 350-keV protons. They have been characterized before and after the radiation exposure in order to determine the impact that the irradiation has on their photovoltaic performance.

II. ANALYSIS OF THE SHIELDING PROVIDED BY THE CONCENTRATOR OPTICS

Concentrator systems can contribute to protecting the solar cells against omnidirectional radiation from protons and low energy electrons, thus reducing the amount of cover glass required for a given mission.

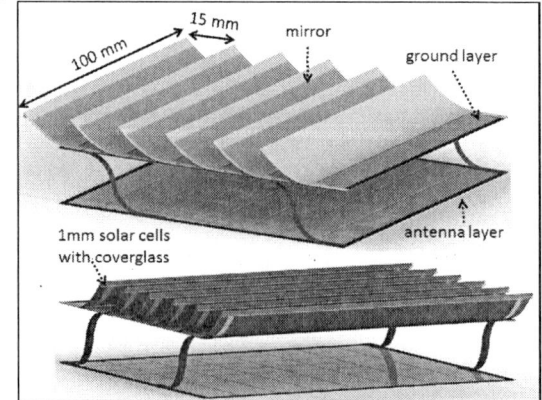

Figure 1. Sketch of the concentrator system simulated with NOVICE.

To compare the radiation environment that the solar cells experience during 10 years in geostationary orbit (GEO) in a flat plate module versus in the concentrator system that we are designing, we have simulated both (Figure 2 and 3) using NOVICE, a 3D adjoint (reverse) Monte Carlo transport simulation program [7]. The incident trapped electron and protons spectrums are from the AE9 and AP9 models [8], respectively, and the solar protons from the JPL91 model [9].

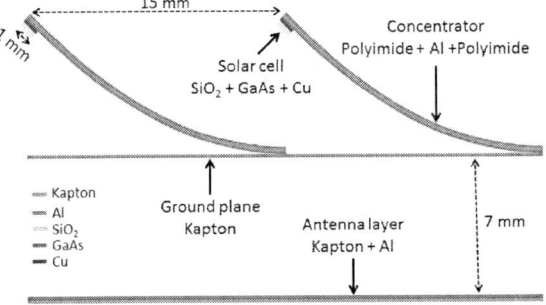

Figure 2. Concentrator system simulated with NOVICE.

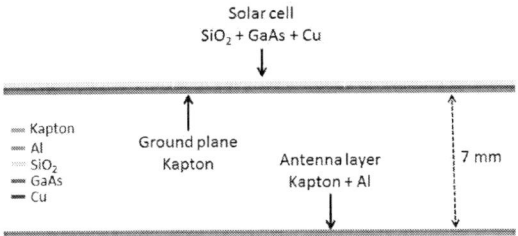

Figure 3. Flat plate system simulated with NOVICE.

The concentrator system that we are designing is a 1D reflective concentrator, depicted in Figure 1 and 2. Each reflector element consists of a half-parabola trough mirror, which focuses the light onto solar cells that are located on the top edge of the adjacent trough mirror. In our design, the concentrating mirrors are also the heat spreaders and thermal radiators for the system. Therefore, they need to incorporate a material with a high thermal conductivity (such as aluminum), and must have highly emissive surfaces on both sides, in order to maintain the solar cell at a reasonable temperature (~80 °C). As optical concentrators, they must also efficiently reflect sunlight towards the solar cells. In order to achieve a high specific power and fulfill the thermal and optical requirements, for a solar cell width of 1 mm, the optimum concentration is around 15x. In that scenario the amount of aluminum required to spread the heat along the mirror is in the order of 10 μm.

Therefore, the specifics of the concentrator system taken into account in the 3D simulations in NOVICE were (see Figure 2):

- Half-parabola trough mirrors with an aperture width of 15 mm (concentrating axis) and length of 100 mm (non-concentrating axis)

- The mirrors are made of 10 μm thick aluminum (for thermal conductivity) with 5 μm of polyimide on both sides (for emissivity)
- The ground plane simulated is made of 10 μm of kapton
- The antenna layer is made of 10 μm of kapton and 10 μm of aluminum and it is separated 7 mm from the ground layer

The row of solar cells is modeled as 10 separate 1 mm (concentrating axis) x 10 mm (non-concentrating axis) cells, placed along the top edge of the adjacent concentrator as shown in Figure 1 and 2. The solar cells are modeled after high-performance epitaxial lift-off (ELO) cells. Thus, instead of a semiconductor substrate, we assume the cells have 50 μm of copper as a handle layer. The front surface of the cells is protected by SiO_2 cover glass; we simulated five scenarios with different SiO_2 thicknesses: 12 μm, 25 μm, 36 μm, 50 μm and 75 μm as front radiation shield. Thus, we obtain the radiation environment that the solar cell experiences.

The flat plate simulated comprises the same solar cells, cover glass, ground plane and antenna layer than the CPV system as shown in Figure 3.

We simulated the integral fluences experienced by the solar cells, over 10 years of operation in GEO, for each of three particle types: trapped electrons using the AE9 model (Figure 4), trapped protons using the AP9 model (Figure 5), and solar protons using the JPL91 model (Figure 6). The figures show the results for both the flat plate scenarios (red dashed lines) and the concentrator system scenarios (grey solid lines), for three different values of front radiation shield SiO_2 thickness: 12 μm, 25 μm, and 75 μm.

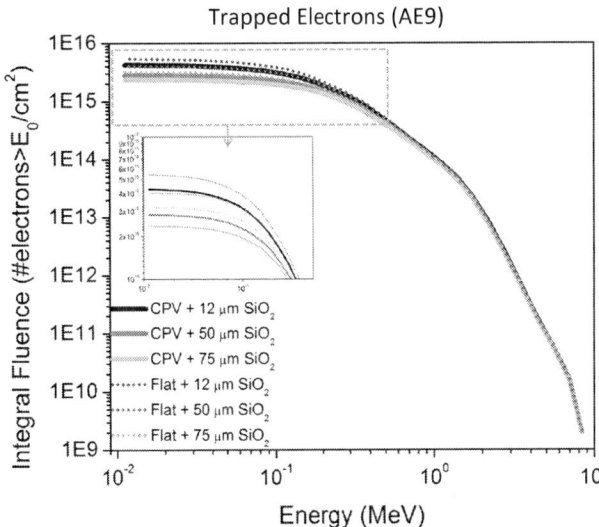

Figure 4. Integral fluences of trapped electrons experienced by the solar cells, after 10 years operation in GEO, in a flat plate module and in the concentrator photovoltaic system (CPV), for different thicknesses of cover glass on top of the solar cell.

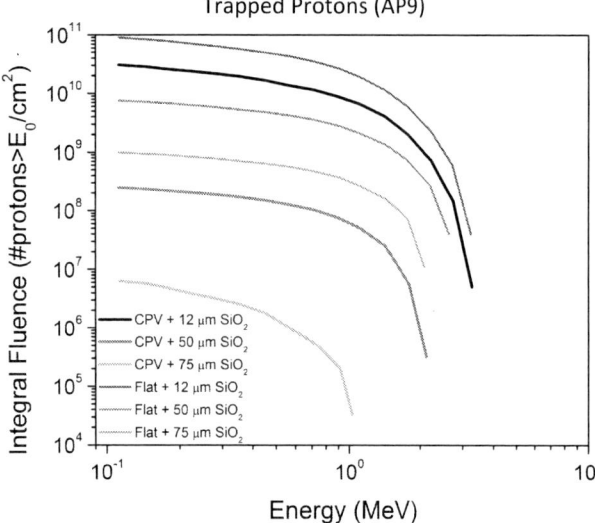

Figure 5. Integral fluences of trapped protons in a flat plate module and in the CPV system.

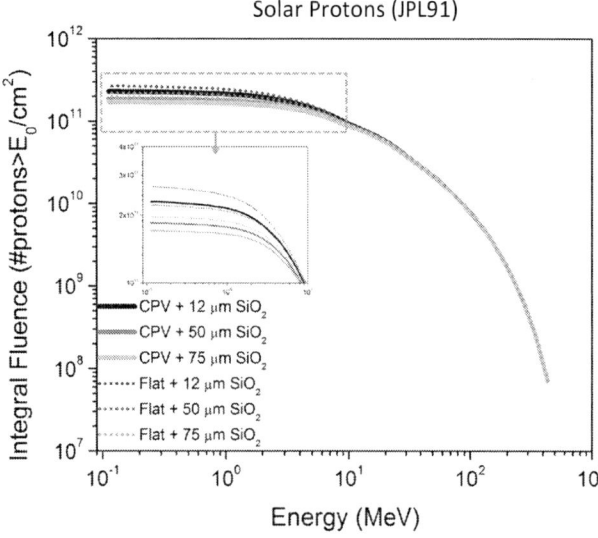

Figure 6. Integral fluences of solar protons experienced by the solar cells, after 10 years operation in GEO, in a flat plate module and in the concentrator photovoltaic system (CPV), for different thicknesses of cover glass on top of the solar cell.

Once the radiation environments have been simulated, the equivalent 1 MeV electron fluence can been calculated for each case using the Anspaugh relative damage coefficients of a dual junction GaAs/Ge [7]. The results are presented in Figure 7.

In order to estimate the impact that the calculated equivalent 1 MeV electron fluences would have on the performance of state of the art triple-junction (3J) solar cells, we have interpolated the 1 MeV electron degradation data presented in the datasheet for Spectrolab XTJ Prime triple-

junction solar cells [8]. This yields the remaining power factors presented in Figure 8, corresponding to predicted end-of-life (EOL) performance after 10 years in GEO. As expected, the concentrator design provides an additional barrier against protons and low energy electrons, which allows the amount of cover glass to be reduced for any given degradation threshold vs. the flat-plate configuration. For example, in order to maintain 85% of the beginning of life (BOL) power after operating 10 years in GEO, in the case of the flat plate system we need to protect the solar cells with ~60 μm of SiO_2, whereas we would need only ~25 μm of SiO_2 in the case of the 15 x concentrator system.

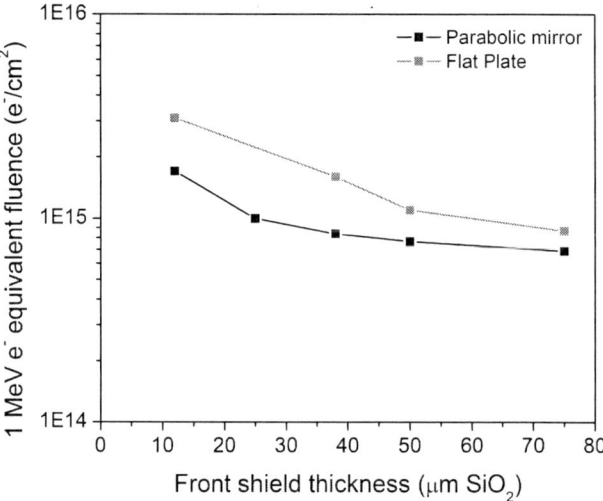

Figure 7. 1 MeV e⁻ equivalent fluence experienced by the solar cells for different thicknesses of cover glass, in the flat panel scenario and in the concentrator system, after operating 10 years in GEO.

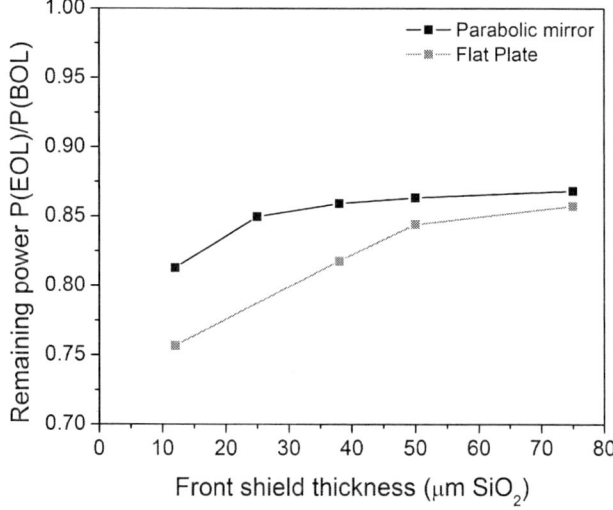

Figure 8. Remaining power after operating 10 years in GEO for solar cells with different SiO_2 shielding in a flat plate module and in the parabolic mirror concentrator system.

III. IMPACT OF HIGH ENERGY PARTICLE RADIATION ON THE PERFORMANCE OF SMALL SOLAR CELLS

Commercial multijunction solar cells for space applications typically have sizes ranging from ~25 cm^2 to 85 cm^2. However, the solar cells we are manufacturing for our 1D concentrators are approx. 1 mm x 10 mm in area. Furthermore, as can be observed in Figure 1, due to the geometry of the parabolic concentrator, the perimeter of the solar cells is particularly exposed to the radiation environment. If the performance degradation due to perimeter radiation damage is significant, designs that feature a high perimeter-to-area ratio, such as our concentrators, will experience an exaggerated decline in efficiency versus flat-plate configurations with larger cell areas.

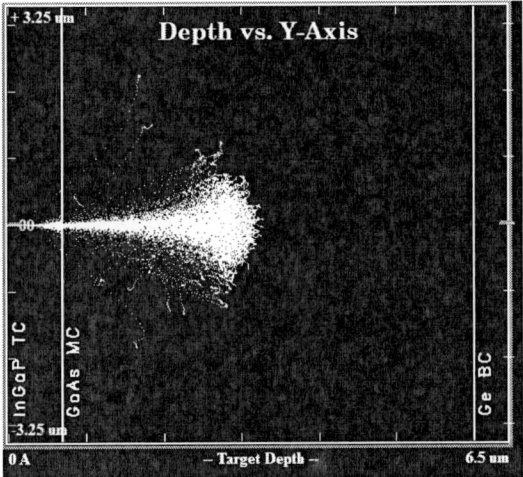

Figure 9. Damage track caused by 350 keV protons on their path through the InGaP/GaAs/Ge triple-junction solar cell. Irradiation is from the front side of the cell. Simulations carried out with SRIM (Stopping and Range of Ions in Matter) software.

In order to study the degradation caused by perimeter radiation in small solar cells, and thus to evaluate whether additional perimeter shielding might be needed for our concentrators, we have fabricated small solar cells and performed proton radiation testing on them. We manufactured 1 mm x 10 mm 3J solar cells from commercial space-grade InGaP/GaAs/Ge semiconductor wafers. The solar cells were irradiated at The Aerospace Corporation with 350 keV protons at a fluence of 10^{12} p$^+$cm^{-2}. Six solar cells were tested; two each in the following three configurations: (A), the protons were incident upon the back side of the solar cells, in which case the protons stop entirely within the >200 μm Ge substrate and negligible performance impact is expected. (B), the protons were incident upon the front side of the solar cells, in which case damage is expected to occur in the InGaP top cell and the GaAs middle cell, as shown in Figure 9. (C), the protons were incident upon one of the long edges of the solar cells, in a direction parallel to the front surface. In this case,

all three subcells are expected to suffer damage within a 3–4 micron depth from the irradiated perimeter edge, as shown in Figure 10.

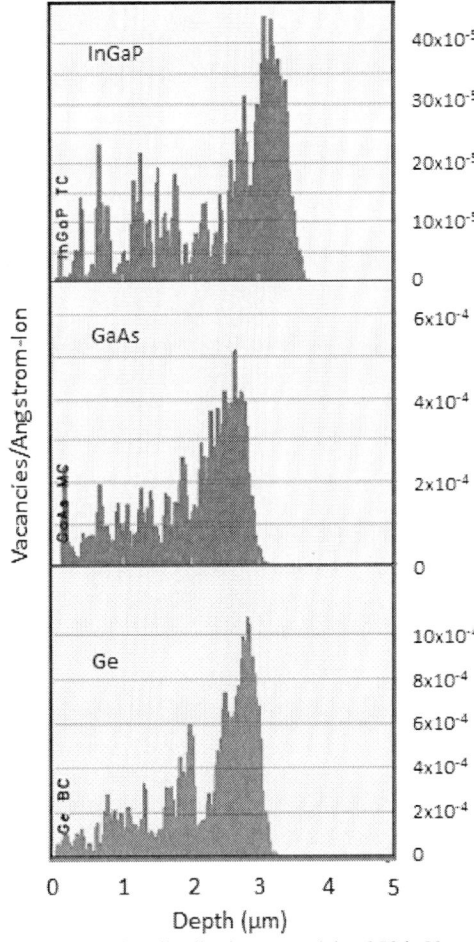

Figure 10. Vacancies distribution caused by 350 keV protons on their paths through the different constituent materials of the 3JSC: InGaP (top), GaAs (middle), and Ge (bottom). Irradiation is from the side of the cell, parallel to the junction interfaces. Simulations carried out with SRIM.

Figure 11 shows the I–V curves of the six solar cells after the irradiation, measured under approximate AM 1.5G illumination (~100 mW cm^{-2}). As expected, the solar cells subject to back-side irradiation do not exhibit any detectable performance deterioration versus non-irradiated cells (not shown). On the contrary, the cells subjected to front-side irradiation exhibit a significant performance degradation. The open circuit voltage decreases by ~670 mV, or by a factor of $V_{oc\ after}/V_{oc\ before} = 0.73$. The short circuit current degradation in these cells is also likely underestimated by the measurements in Figure 11, since the light source used was lacking in UV spectral content versus AM0 illumination, whereas the middle subcell is expected to suffer most of the degradation.

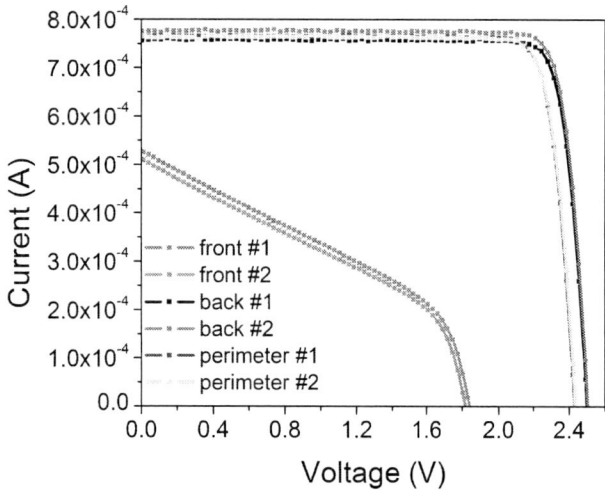

Figure 11. I–V curves of the solar cells after radiation exposure with 350 keV protons at a fluence of $1 \cdot 10^{12}$ p$^+$cm^{-2}. Note that the solar simulator used for the measurements had a low UV content vs. AM0 conditions.

Finally, for the solar cells where the perimeter had been irradiated, we observe only a slight degradation of the open circuit voltage, by around 70 mV, or $V_{oc\ after}/V_{oc\ before} = 0.97$. Therefore, we conclude that there is no significant degradation due to the perimeter radiation exposure, which is essential for concentrator solar systems that use solar cells with a size in the range of millimeter or hundreds of microns. In order to prove that small solar cells do not present higher degradation rates than standard cells, additional testing should be performed, such as with high energy electrons which will penetrate deeper in the semiconductor structure. Lower energy protons can be avoided with a reasonable shield. However, the high energy electrons are unavoidable in space.

IV. SUMMARY

In this paper we have presented radiation environment simulations for solar cells operating in GEO, comparing shielding performance for a flat plate solar module vs. a 1D reflective concentrator system. We have observed that the concentrator system provides a noteworthy degree of shielding for the cells, which allows for significantly reducing the amount of cover glass required.

A peculiarity of the concentrator system that we are designing is that it makes use of relatively small solar cells (1 mm wide), which have a relatively high perimeter-to-area ratio. In order to study the possible impact that the radiation of high energy particles could have on the performance of the cell perimeter, we irradiated some of our solar cells with 350-keV protons at a fluence of 10^{12} p$^+$cm^{-2}. We varied from which direction the cells were subject to irradiation: front, side (perimeter), and back. As expected, the degradation resulting from front-side irradiation was severe, and cover glass is necessary to stop the protons of the energy tested. On the contrary, no degradation was perceived when the protons were incident on the back side of the cells, since they stop in the inactive substrate region. Finally, only a slight degradation of the open circuit voltage was measured when the perimeter was irradiated, suggesting that the small solar cells do not have significantly higher degradation rates than the standard (larger) solar cells, and importantly, that no extra mass of shielding is necessary to protect the perimeter.

ACKNOWLEDGEMENT

We acknowledge financial support from Northrop Grumman. This effort made use of facilities provided by the Kavli Nanoscience Institute, the Material Molecular Research Center (MMRC), the Resnick Institute, and the Joint Center for Artificial Photosynthesis, at Caltech.

REFERENCES

[1] M. A. Brown, B. Whalen, *Retractable thin film solar concentrator for spacecraft*, US5885367 A, March, 23, 1999

[2] T. G. Stern "Interim results of the SLATS concentrator experiment on LIPS-II (space vehicle power plants)"; *Conference Record of the Twentieth IEEE Photovoltaic Specialists Conference*, Las Vegas, NV, USA, 1988, pp. 837-840 vol.2.

[3] O'Neill, M.J.; McDanal, A J.; Piszczor, M.F.; Edwards, D.L.; Eskenazi, M.I; Brandhorst, H.W., "Recent technology advances for the stretched lens array (SLA), a space solar array offering state of the art performance at low cost and ultra-light mass", *Photovoltaic Specialists Conference*, 2005. Conference Record of the Thirty-first IEEE , vol., no., pp.810,813, 3-7 Jan. 2005

[4] H. Curtis and D. Marvin, , "Final results from the PASP Plus flight experiment," *Photovoltaic Specialists Conference*, 1996., Conference Record of the Twenty Fifth IEEE , vol., no., pp.195,198, 13-17 May 1996 doi: 10.1109/PVSC.1996.563980

[5] Marc D. Rayman, Philip Varghese, David H. Lehman, Leslie L. Livesay, "Results from the Deep Space 1 technology validation mission", *Acta Astronautica*, Volume 47, Issues 2–9, July–November 2000, Pages 475-487, ISSN 0094-5765

[6] B. Hoang, S. White, B. Spence and S. Kiefer, "Commercialization of Deployable Space Systems' roll-out solar array (ROSA) technology for Space Systems Loral (SSL) solar arrays," 2016 *IEEE Aerospace Conference*, Big Sky, MT, 2016, pp.1-12.

[7] I. Jun, S. Kang, R. Evans, M. Cherng and R. Swimm "Radiation transport tools for space applications: a review" *5th Geant4 Space Users' Workshop*

[8] G. P. Ginet, T. P. O'Brien, S. L. Huston, W. R. Johnston, T. B. Guild, R. Friedel, C. D. Lindstrom, C. J. Roth, P. Whelan, R. A. Quinn, D. Madden, S. Morley and Yi-Jiun Su "AE9, AP9 and SPM: New Models for Specifying the Trapped Energetic Particle and Space Plasma Environment" *Space Science Reviews* November 2013, Volume 179, Issue 1–4, pp 579–615

[9] J. Feyman, G. Spitale and J. Wang "Interplanetary proton fluence model: JPL 1991" *Journal of geophysical research*, vol. 98, NO. A8, pages 13, 281-13, 294, August 1, 1993

[7] B. E. Anspaugh, *GaAs Solar Cell Radiation Handbook*, July 1996, JPL Publication 96-9

[8]http://www.spectrolab.com/DataSheets/cells/XTJ_Prime_Data_S heet_7-28-2016.pdf

Extracting the fixed charge density in HfO$_x$ films grown on highly-doped p-Si samples

Alexander To[1], Jie Cui[2], Bram Hoex[1*]

[1]University of New South Wales, Sydney, NSW, 2052 Australia, [2]Australian National University, Canberra, ACT, 2601, Australia

Abstract — A recently developed, novel method for the extraction of fixed interface charge, Q_f, and the surface recombination parameters, S_{n0} and S_{p0}, from the injection-level dependent effective minority carrier lifetime is applied to HfO$_x$ passivated highly doped c-Si samples. This contactless technique can, unlike conventional capacitance-voltage measurements, be applied to highly doped surfaces provided the surface becomes heavily depleted or inverted as the injection level increases. By fitting the measured injection level dependent effective lifetime curve before and after annealing, Q_f and the surface recombination velocity parameters, S_{n0} and S_{p0} are independently resolved and extracted. It was shown that the significantly higher effective lifetime for 15 nm and 30 nm thick HfO$_x$ films was a result of differing passivation mechanisms; the 15 nm film improves field effect passivation after annealing, whereas the surface passivation mechanism of the 30 nm film after anneal was predominantly a chemical passivation effect. This work presents interface parameter measurements of doped surfaces not accessible by the more common capacitance-voltage techniques.

I. Introduction

The recombination of charge carriers at the surfaces of silicon solar cells is a critical loss mechanism which needs to be minimised to achieve high-performance devices. This is typically referred to as surface or interface passivation and is commonly achieved via the deposition of thin-film dielectrics on the semiconductor surfaces. Dielectrics such as silicon nitride (SiN$_x$), aluminium oxide (AlO$_x$) and silicon dioxide (SiO$_x$), are widely used on both diffused and undiffused surfaces to reduce minority carrier recombination[1, 2].

With a wide bandgap of $E_q \approx 5.6$ [3], high refractive index (n=2.4), and high dielectric constant [4], hafnium oxide (HfO$_x$) films have been the subject of a broad range of research efforts. These efforts focused on applications in optical coatings [5], semiconductor devices such as DRAM [4], and dye-sensitized solar cells [6, 7]. Recently, excellent passivation results of undiffused silicon surfaces were reported for atomic-layer-deposition (ALD) deposited HfO$_x$ [8, 9]. This follows the work of previous researchers, who have demonstrated surface passivation results on low-moderately doped p-type and n-type silicon surfaces [3, 9, 10], and the application of HfO$_x$ films as an intervening layer between silicon and AlO$_x$ to tune the interface charge density [11, 12].

ALD grown HfO$_x$ films have relevance to c-Si photovoltaic devices as HfO$_x$ layers can be deposited at low temperatures, exhibit a low interface defect density, and a broad range of fixed charge densities [8, 10]. Furthermore, from corona oxide characterisation of semiconductor (COCOS) measurements, researchers have separately reported both positive and negative charge for ALD grown HfO$_x$ thin-films [8-10], suggesting that the fixed charge at the interface can be tailored depending on the underlying surface doping polarity.

The polarity and magnitude of fixed charge of thin films at the surface interface, in addition to the density of interface defects, D_{it}, is of interest to researchers and engineers as these parameters give insight into the underlying surface passivation mechanism. Surface passivation is achieved as a combination of both field-effect and chemical passivation mechanisms. Chemical passivation mechanisms are related to the reduction of D_{it}, as it affects the Shockley-Read-Hall (SRH) surface recombination velocity parameters, S_{n0} and S_{p0}, defined as

$$S_{p0} = D_{it}\sigma_p v_{th,} \qquad (1.1)$$
$$S_{n0} = D_{it}\sigma_n v_{th}. \qquad (1.2)$$

Where σ_n and σ_p are the electron and hole capture cross sections respectively, and v_{th} is the related carrier thermal velocity. Field-effect passivation mechanisms are techniques which reduce the concentration of one carrier species, while increasing the other; such is the case with doping or by the introduction of an electric field at the surface. Therefore, the values of surface parameters at the interface of the silicon substrate and passivating dielectric film give added insight over commonly reported effective parameters such as the surface saturation current (J_{0s}) and effective surface recombination velocity, which do not give insight into the passivation mechanism.

The polarity and magnitude of fixed charge can be measured with capacitance-voltage techniques on undiffused surfaces. These measurements can be applied on metal-oxide-semiconductor (MOS) stacks with a rear electrode, or contactless via the COCOS method [13, 14]. However these techniques are limited in the sense that 'leaky' dielectrics cannot hold sufficient deposited charge to induce flat-band conditions, and neither technique can be applied to samples with heavily diffused surfaces. An alternative method, such as Electric Field-Induced Second Harmonic Generation (EFISH) are rarely performed due to the high costs and complex analysis required to extract measurements [15].

To overcome the limitations of the previous methods, this work applies a recently developed method of extracting the interface parameter Q_f and surface recombination velocity parameters S_{n0} and S_{p0} on doped surfaces to study the properties of HfO$_x$ films deposited on heavily diffused p^+ surfaces [16].

978-1-5090-5606-4/17 $31.00 © 2017 IEEE

This technique relies on the fitting of the non-linear lifetime trend of surfaces which move into depletion and/or inversion as the sample moves into high injection to resolve individual components of Q_f and S_{n0}/S_{p0}. In this work, this method was applied to both annealed and as deposited HfO_x films to discern the resulting effect on the interface properties, and therefore the overall passivation mechanism on diffused p^+ surfaces.

II. EXPERIMENTAL

A. Sample Fabrication

Samples were fabricated on 280 μm thick, 100 Ω·cm, phosphorous doped, FZ <100>, polished c-Si wafers. After cleaning in a standard RCA procedure followed by a short HF dip, the samples were boron diffused in a tube diffusion process using a liquid BBr_3 source. Following boron-silicate glass (BSG) removal, thin HfO_x layers were deposited via atomic layer deposition (ALD, TFS200, BENEQ) on both sides, using a liquid tetrakis hafnium (TEMAH) precursor and deionized water at 250 °C. Layers approximately 15 and 30 nm thick were grown with 200 and 400 number of ALD cycles, respectively. The samples were then annealed in forming gas at 350 °C for 15 minutes. The electronic properties of these HfO_x film are discussed in detail in ref. [17].

B. Modelling and Characterisation Techniques

The interface extraction method relies on the simulation of the Auger-corrected lifetime curve as a function of injection level in Sentaurus TCAD, to match the measured photoconductance measurements of the fabricated test samples. To accurately simulate the fabricated samples, the HfO_x samples were extensively characterised. Thickness measurements were taken using a micrometer, and the active boron profile was determined using electrochemical capacitance-voltage (ECV, WEP, CVP21) measurements, which were calibrated using the sheet resistance determined by a 4-point probe measurement. Reflectance measurements were taken using a UV-VIS spectrophotometer and input into OPAL 2 [18], where the generation profile was simulated for a planar wafer of identical thickness.

The simulation method in Sentaurus TCAD is covered in previous work [16], and therefore covered briefly herein. A quasi-steady-state photoconductance measurement was simulated by solving the simulated structure at a range of generation levels in a steady-state condition. The excess conductance is calculated using according to the expression [19]

$$\sigma_L = q\Delta n(\mu_n + \mu_p)W, \qquad (2)$$

where μ_n and μ_p are the electron and hole mobility's was calculated using the model outlined by Klaassen [20, 21], Δn is the excess carrier concentration at each mesh point in the

sample and W is the sample width. From this value, the effective lifetime was calculated from

$$\tau_{eff} = \frac{\sigma_L}{J_{ph}(\mu_n+\mu_p)W}. \qquad (3)$$

Correction for Auger recombination in the bulk was performed using the Richter parameterisation [22]. The parameters S_{n0} and S_{p0} were calculated by varying D_{it} and using constant $v_{th} = 10^7$ cm/s and $\sigma_p = \sigma_n = 10^{-17}$ cm^2. The parameters D_{it} and Q_f were used as fitting parameters, however since the capture cross sections cannot be resolved in this method, S_{n0} and S_{p0} values are quoted.

III. RESULTS

The measured and simulated Auger-corrected inverse lifetime curves for the passivated p^+np^+ structures before and after annealing for the 15 and 30 nm thicknesses are shown in Fig. 1. Good agreement between measured and simulated values were achieved for all thicknesses and conditions. The surface parameters of the simulated structures are listed in Table I, and shows different trends in Q_f, S_{p0} and S_{n0} before and after annealing for the 15 and 30 nm films.

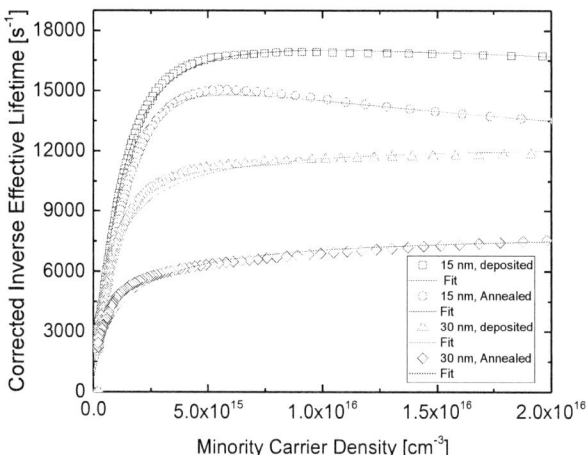

Fig. 1. Measured and simulated inverse corrected lifetime curves as a function of injection level for HfO_x passivated samples.

The lifetime measurement for the sample passivated by the 15 nm film was reproduced using both a higher S_{n0} and S_{p0} as well as Q_f value, suggesting the improvement in passivation was a result of improved field-effect passivation. In comparison, the improvement in lifetime observed for the 30 nm passivated sample after annealing was attributed primarily to the decrease in S_{n0} and S_{p0}, since the fitted Q_f value decreased slightly between treatments. The fact that the decrease in Q_f for the 30 nm film was larger than the increase in Q_f for the 15 nm film suggests that the sample was less reliant on field-effect passivation since the number of interface defect sites was low and, hence, the overall lifetime was still higher in comparison to the as-deposited state. These extracted values for

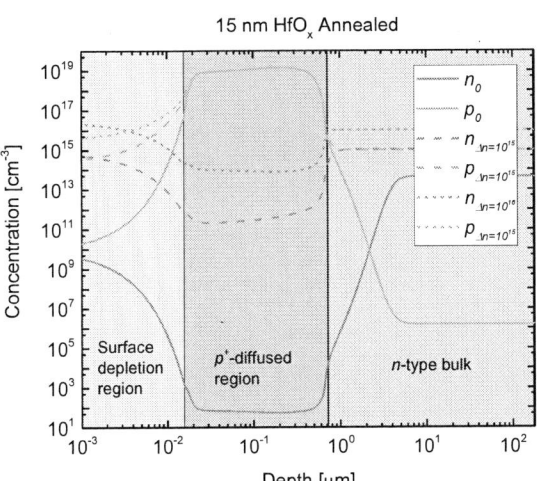 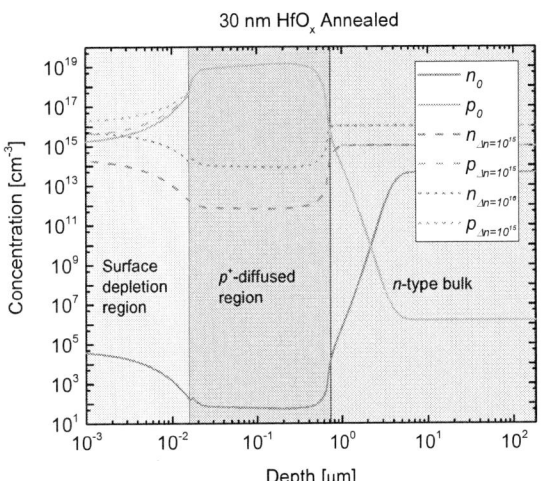

Fig. 2. Sentaurus simulated concentration of electrions and holes as a function of distance from the Si-HfO$_x$ surface interface for varying injections levels in the sample passivated with 15 nm HfO$_x$ (left) and 30 nm HfO$_x$ (right). The extracted interface parameters listed in Table I were used in the simulations.

Q_f and S_{n0} and S_{p0} measurements show good agreement with the identically processed layers on undiffused samples, measured with the COCOS technique, adding validity to this method and quoted in the work by Wang *et al* [10].

Analysis of the carrier concentrations at the sample surface as they vary with injection level reveals the underlying cause of the non-linear lifetime trend. Figure 2 shows the surface concentration of carriers as a function of depth into the sample for various injection levels for the 15 nm and 30 nm HfO$_x$ samples in the annealed state. In both samples, a significant depletion of the surface concentration of holes was evident due to the presence of fixed positive charge at the surface; this depletion was more pronounced for the 15 nm sample due to the higher Q_f in this sample. As the injection level increases, the surface electron concentration increases, and the difference between electrons and hole concentrations in the surface depletion region decreases, such that in the 15 nm case, for $\Delta n = 10^{16}$ cm^{-3}, the majority carrier had changed from holes to electrons. Thus the 15 nm sample had moved from depletion into inversion, whereas the 30 nm sample had moved into strong depletion. The resulting effect was a non-constant scaling of J_{0s} - which presents as a non-linear scaling of effective lifetime with injection level - as the recombination statistics become affected by the approach to parity of the carriers in the 30 nm case, and a transition of minority carrier species in the 15 nm case. Figure 5 plots the simulated J_{0s} as a function of injection level for both samples pictured in Figure 4, demonstrating the varying J_{0s} in the injection level range of interest.

TABLE I
EXTRACTED INTERFACE PARAMETERS Q_F AND S_{N0}=S_{P0}.

Thickness	Parameter	Deposited	After Anneal
15 nm	Q_f[cm^{-2}]	1.4×10^{12}	1.8×10^{12}
	S_{n0} [cm/s]	135	152
30 nm	Q_f[cm^{-2}]	3×10^{11}	2×10^{11}
	S_{n0} [cm/s]	65	40

The fitting results in Fig. 3 show that the method can be applied to HfO$_x$ passivated c-Si samples, due to the fixed positive charge inherent in the films. The extracted parameters in Table I show that even though the surface passivation was improved for both the 15 nm and 30 nm films, the underlying passivation mechanism was different. This highlights the importance of insights into the surface passivation mechanisms when developing surface passivation processes, as the magnitude and polarity of fixed charge is an important consideration when designing surface passivation technologies for silicon surfaces.

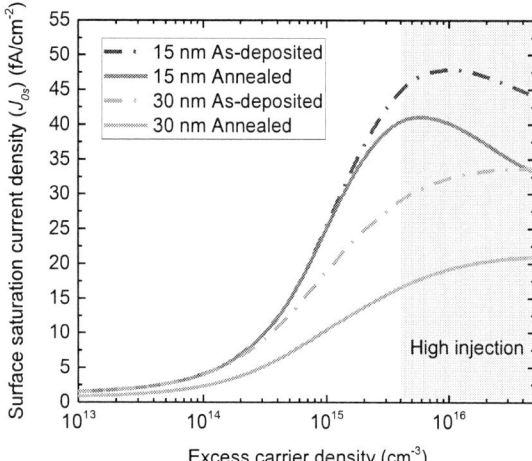

Fig. 3. Simulated J_{0s} vs. excess carrier density for c-Si samples passivated by15 nm and 30 nm thick HfO$_x$ layers in as-deposited and annealed state. The varying J_{0s} in the high injection region (shaded) is a result of the sample surfaces approaching inversion and strong depletion with increased illumination.

IV. CONCLUSIONS

The interface parameters, Q_f and S$_{n0}$ of boron diffused c-Si samples passivated with 15 and 30 nm thick ALD deposited HfO$_x$ layers were extracted. This was achieved by fitting the measured injection-level dependent, Auger-corrected inverse lifetime curves from photoconductance measurements with simulated test structures in Sentaurus TCAD. It was revealed that the interface characteristics of HfO$_x$ films varied significantly between the investigated thicknesses, as the 15 nm film exhibits a higher fixed charge and S$_{n0}$ and S$_{p0}$ value after annealing than the 30 nm film, which had lower Q_f and S$_{n0}$ and S$_{p0}$. This highlights the differing passivation mechanisms between the two samples of the same material. This work demonstrates how Q_f and S$_{n0}$ or S$_{p0}$ can be resolved independently from the inverse lifetime curve of diffused samples passivated with HfO$_x$.

REFERENCES

1. Hoex, B., et al., *Ultralow surface recombination of c-Si substrates passivated by plasma-assisted atomic layer deposited Al 2 O 3.* Applied Physics Letters, 2006. **89**(4): p. 042112.
2. Allongue, P., V. Kieling, and H. Gerischer, *Etching mechanism and atomic structure of H.Si(111) surfaces prepared in NH4F.* Electrochimica Acta, 1995. **40**(10): p. 1353-1360.
3. Dingemans, G. and W.M.M. Kessels, *(Invited) Aluminum Oxide and Other ALD Materials for Si Surface Passivation.* ECS Transactions, 2011. **41**(2): p. 293-301.
4. Wilk, G.D., R.M. Wallace, and J.M. Anthony, *High-κ gate dielectrics: Current status and materials properties considerations.* Journal of Applied Physics, 2001. **89**(10): p. 5243-5275.
5. Waldorf, A.J., et al., *Optical coatings deposited by reactive ion plating.* Applied Optics, 1993. **32**(28): p. 5583-5593.
6. Shanmugam, M., M.F. Baroughi, and D. Galipeau, *Effect of atomic layer deposited ultra thin HfO2 and Al2O3 interfacial layers on the performance of dye sensitized solar cells.* Thin Solid Films, 2010. **518**(10): p. 2678-2682.
7. Bills, B., M. Shanmugam, and M.F. Baroughi, *Effects of atomic layer deposited HfO2 compact layer on the performance of dye-sensitized solar cells.* Thin Solid Films, 2011. **519**(22): p. 7803-7808.
8. Cui, J., et al., *Highly effective electronic passivation of silicon surfaces by atomic layer deposited hafnium oxide.* Applied Physics Letters, 2017. **110**(2): p. 021602.
9. Lin, F., et al., *Low-Temperature Surface Passivation of Moderately Doped Crystalline Silicon by Atomic-Layer-Deposited Hafnium Oxide Films.* ECS Journal of Solid State Science and Technology, 2013. **2**(1): p. N11-N14.
10. Wang, J., S.S. Mottaghian, and M.F. Baroughi, *Passivation Properties of Atomic-Layer-Deposited Hafnium and Aluminum Oxides on Si Surfaces.* IEEE Transactions on Electron Devices, 2012. **59**(2): p. 342-348.
11. Simon, D.K., et al., *On the Control of the Fixed Charge Densities in Al2O3-Based Silicon Surface Passivation Schemes.* ACS Applied Materials and Interfaces, 2015. **7**(51): p. 28215-28222.
12. Dirnstorfer, I., et al., *Near surface inversion layer recombination in Al2O3 passivated n -type silicon.* Journal of Applied Physics, 2014. **116**(4).
13. Schroder, D.K., *Semiconductor material and device characterization.* 2006: John Wiley & Sons.
14. Wilson, M., et al., *COCOS (corona oxide characterization of semiconductor) non-contact metrology for gate dielectrics.* AIP Conference Proceedings, 2001. **550**(1): p. 220-225.
15. Gielis, J., et al., *Negative charge and charging dynamics in Al 2 O 3 films on Si characterized by second-harmonic generation.* J. Appl. Phys, 2008. **104**: p. 073701.
16. To, A., F.J. Ma, and B. Hoex, *Improved understanding of the recombination rate at inverted p+ silicon surfaces.* Japanese Journal of Applied Physics, 2017.
17. Yao, Y., et al., *Uniform Plating of Thin Nickel Layers for Silicon Solar Cells.* Energy Procedia, 2013. **38**: p. 807-815.
18. McIntosh, K.R. and S.C. Baker-Finch. *OPAL 2: Rapid optical simulation of silicon solar cells.* in *Photovoltaic Specialists Conference (PVSC), 2012 38th IEEE.* 2012.
19. Sinton, R.A. and A. Cuevas, *Contactless determination of current‐voltage characteristics and minority‐carrier lifetimes in semiconductors from quasi‐steady‐state photoconductance data.* Applied Physics Letters, 1996. **69**(17): p. 2510-2512.
20. Klaassen, D.B.M., *A unified mobility model for device simulation—I. Model equations and concentration dependence.* Solid-State Electronics, 1992. **35**(7): p. 953-959.
21. Klaassen, D.B.M., *A unified mobility model for device simulation—II. Temperature dependence of carrier mobility and lifetime.* Solid-State Electronics, 1992. **35**(7): p. 961-967.
22. Richter, A., et al., *Improved Parameterization of Auger Recombination in Silicon.* Energy Procedia, 2012. **27**: p. 88-94.

Near-unity ultra-wideband thermal infrared emission for space solar power radiative cooling

Ali Naqavi[1,*], Samuel P. Loke[1], Michael D. Kelzenberg[1], Emily C. Warmann[1], Pilar Espinet-González[1], Nina Vaidya[1], Jing-Shun Huang[1], Tatiana A. Roy[1], Alexander J. Messer[1,2], Tatiana G. Vinogradova[1,2], Ali Hajimiri[1], Sergio Pellegrino[1], and Harry A. Atwater[1]

[1] California Institute of Technology, Pasadena, CA 91125

[2] Northrop Grumman Aerospace Systems, Azusa, CA 91702

* Corresponding author e-mail address: naqavi@caltech.edu

Abstract — **We report the design, fabrication and characterization of ultrathin metasurfaces that exhibit wideband 300 K thermal emissivity. The emissive behavior of these structures is almost independent of the emission angle. Our ultralight subwavelength-thickness metasurfaces can be fabricated relatively easily and are excellent candidates for radiative cooling in space applications.**

Index Terms — **thermal radiation, radiative cooling, vibrons, polaritons, space solar power.**

I. INTRODUCTION

Space-based solar power can provide reliable clean energy during the 24 hours per day, in contrast to terrestrial solar power which has a daytime limitation on power generation. One of the main criteria for solar cells in space is rejecting the heat that is produced due to the optical loss and non-ideal optical-to-electrical energy conversion. Convective processes that help the thermal management of terrestrial solar cells are not applicable outside the atmosphere. In contrast, radiative cooling can be applied [1] since the electromagnetic waves can travel over space in the absence of a transfer material. This calls for the design and fabrication of thermal emissive surfaces that are very lightweight. Here we present the analysis, design, fabrication and characterization of metasurfaces based on the concept of Salisbury screen [2,3] for efficient emission of thermal radiation. Our metasurfaces have a thickness between 0.2 and 0.5 of λ, where λ is the free space wavelength of the peak of the blackbody spectrum at 300 K and weigh between 3 and 10 g/m². Besides, they have high thermal emissivity (0.75 to 0.9) over the wavelength range from 2 to 50 microns, that is unchanged from normal up to 60 degrees of emission angle.

Multiple phenomena contribute to the high emissivity in our structures. Similar to a conventional Salisbury screen, our simplest metasurface benefits from the existence of wave interference effects. Besides, due to vibronic resonances of the polyimide layer [4] and quantum confinement effects in its metallic inclusions [5], it exhibits enhanced emission over an extremely wide wavelength range. Similar polaritonic

emission occurs also for our other metasurfaces, which have more layers.

II. THERMAL EMISSIVE METASURFACE DESIGN

For a complete description of thermal emission from an object, a fluctuational electrodynamics approach can be applied [6], which is computationally expensive, and thus impractical in many cases. To calculate the emission in the far field, however, simplifying assumptions can be used. Based on Kirchhoff's law of thermal radiation, the absorption and the emission coefficients at a specific wavelength, angle and polarization are equal. Therefore, it is possible to replace the thermal emission problem by a conventional scattering (absorption) problem, which is much simpler to solve. Here we use this method to calculate the emissive properties of our metasurfaces.

Kirchhoff's law not only simplifies the far-field thermal emission problem significantly, but also provides guidelines for the design of emissive structures. Since emission and absorption of the electromagnetic waves are reciprocal, a structure that emits thermal radiation efficiently, also absorbs electromagnetic waves at the corresponding wavelengths very well. Therefore, methods of designing wave absorbers can be used to design efficient thermal emitters.

Inspired by this intuition, we designed metasurfaces based on the Salisbury screen, which is an electromagnetic wave absorber [2,3]. A generic Salisbury screen is shown schematically in Fig. 1. It consists of a thin metallic sheet that is separated from a metallic back reflector by a quarter-wavelength dielectric spacer. At resonance, the Salisbury screen satisfies impedance matching to free space and absorbs the incident wave completely. However, this condition is met for a single wavelength; thus the design is typically useful only for narrowband applications. For broadband emission, a more complex design is required, which can be realized by using more layers. Here we also consider structures with up to three consecutive dielectric-metal stacks, which will be addressed in section III.

978-1-5090-5606-4/17 $31.00 © 2017 IEEE

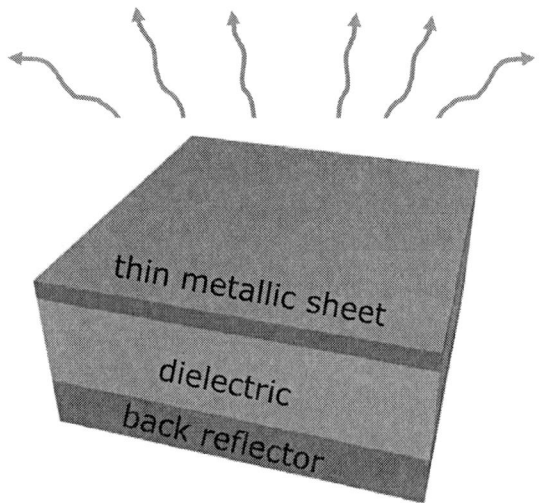

Fig. 1. A thermal emissive Salisbury screen. From top to bottom it consists of a thin metal sheet, a dielectric spacer and a back reflector.

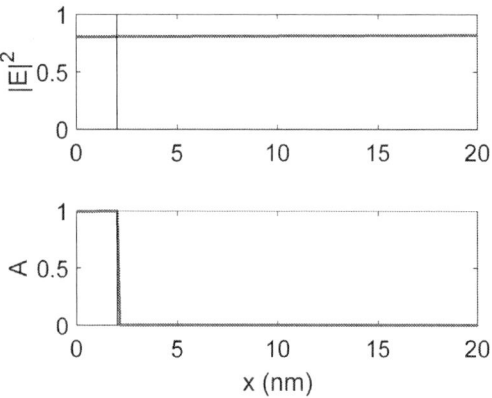

Fig. 2. Normalized intensity and absorption profiles as a function of depth for the Salisbury screen structure. The Cr top layer starts at x=0 mm and is 2 nm thick.

We considered CP1 polyimide (Nexolve) [7] as the dielectric spacer layer, and Cr as the metal for both the top conductive sheet and the back reflector. The optical data of CP1 were obtained through ellipsometry. Full wave electromagnetic simulations were performed to obtain optimal layer thicknesses for the three different structures.

Fig. 2 shows the normalized intensity and absorption profiles obtained by transfer matrix calculations for the Salisbury screen at resonance, based on which, several important points can be addressed. Most of the absorption happens in the top Cr sheet, which is only 2 nm thick. The absorption profile is plotted to the depth of 20 nm for better illustration, but we observed that the absorption/emission is very close to zero also for deeper parts of the structure (x>20nm). Actually, our transfer matrix calculations show that

at resonance, more than 85 % of the absorption occurs in the Cr layer which has a deeply-subwavelength thickness. Despite the continuity of the tangential electric field across the metal-dielectric interface, the absorption is dramatically higher in the metal, due to the high contrast between the permittivity of the CP1 and Cr. Besides, the peak of the electric field does not occur within the Cr layer, which is evidenced by the intensity values smaller than unity over the Cr layer. Actually, the electric field peak occurs within the CP1 layer, as this layer also plays a role in absorption, although marginally.

III. EXPERIMENTAL

We fabricated the Salisbury screen by ebeam evaporating 100 nm Cr onto a clean Si wafer, then spinning CP1 polyimide resin onto it to obtain a thickness of 2.1 microns, followed by ebeam evaporation of 2 nm Cr and then 10 nm SiO_2 to prevent oxidation of the Cr. We also fabricated multi-layer metasurfaces comprising additional alternating layers of polyimide, Cr and SiO_2. To fabricate the multilayer structures, the thickness of the SiO_2 layers topping the Cr was increased to 50 nm, to ensure that the solvent from subsequent CP1 layers did not penetrate the underlying layers during spin coating. We then measured wavelength-resolved reflection from the structures by using two different methods. First, we performed infrared ellipsometery to measure the specular reflection from the surface. We also measured reflection by using a Fourier transform infrared (FTIR) microscope that illuminates and collects from the surface of the sample with a Cassegrain lens within the annular range from about 15 to 35 degrees over the wavelengths from 2.5 to 15 microns. Since the reflection from our structure is almost independent of angle, excellent agreement in observed between the two measurements (see Fig. 3).

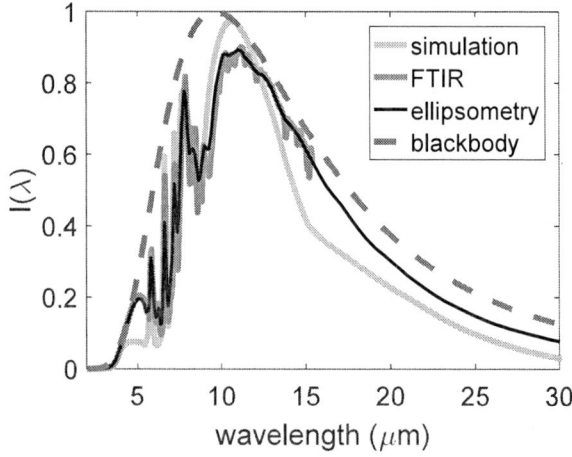

Fig. 3. Normalized spectral radiance of the Salisbury screen inferred from simulation, and reflection measurements by using FTIR and infrared ellipsometry at 30°, compared to the blackbody radiation spectrum at 300 K.

Comparison of the simulations and the measured reflection values (Fig. 3) shows a considerable amount of thermal radiation emission beyond the predictions of the transfer matrix calculations which use bulk optical data of materials. We attribute the large long-wavelength measured emissivity values to different effects including quantum confinement in the metallic particulates in the thin metal layer [5], vibronic resonances of the polyimide layer, and in particular coupling of these two effects. As electrons get hot in the very thin metal layers, their energy is elevated to discrete energy levels in the conduction band of the nanoparticle. At the metal-polyimide interface, this energy is transferred to the vibrons in the polyimide, which leads to the very broadband enhancement in emission.

Fig. 4 shows the angle-dependence of the emissivity of the three metasurfaces inferred from reflection measurements by applying Kirchhoff's law. Expectedly, adding the number of layers increases the emissivity, particularly at large angles. We observed that the measured emissivity values are larger than the values predicted by the simulations for the planar structure. Interestingly, the emissivity is almost independent of the incident angle up to around 60 degrees. The metasurface with more layers provides larger emissivity expectedly.

We also fabricated free-standing Salisbury screens by using CP1 films with a thickness of 1.8 microns, as shown in Fig. 5. The 300 K emissivity of these films is about 0.6 in the range from 5-30 microns, which can be improved by having a more appropriate film thickness.

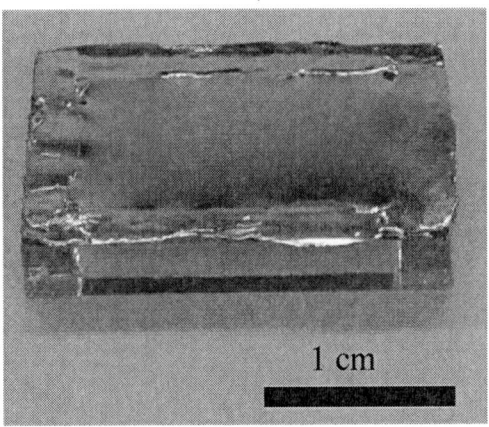

Fig. 5. The fabricated free-standing Salisbury screen. The flat internal part is the Salisbury screen and the surrounding area is the frame on which it is installed.

IV. CONCLUSION

We have designed, fabricated and characterized ultralight highly-emissive subwavelength metasurfaces with a measured emissivity of 0.75 to 0.9 for normal emission and an areal mass density of less than 10 g/m^2. The very high emissivity of these metasurfaces originates from different physical phenomena; Apart from classical optical interference effects, discreteness of conduction bands of metal nanoparticles in the very thin metal layers, vibronic resonances of the polyimide layers and transfer of energy of hot electrons in the metal to vibrons in the polyimide layer enhance emission over a broad spectrum. These phenomena provide large emissivity that is almost unchanged over a very wide angular range, from normal to 60 degrees. Our ultralight emissive metasurfaces are alternatives for thermal management in space technology.

ACKNOWLEDGEMENT

We acknowledge financial support from the Northrop Grumman Corporation. A.N. is supported from the Swiss Science National Foundation. The authors acknowledge Mark Kruer of Northrop Grumman Corporation for help in characterization of samples, Tom Tiwald of J. A. Woollam Co. for analyzing the ellipsometry measurements of the polyimide layers, and Lynn Rodman of Nexolve for providing materials and guidance in fabricating the thin polyimide layers.

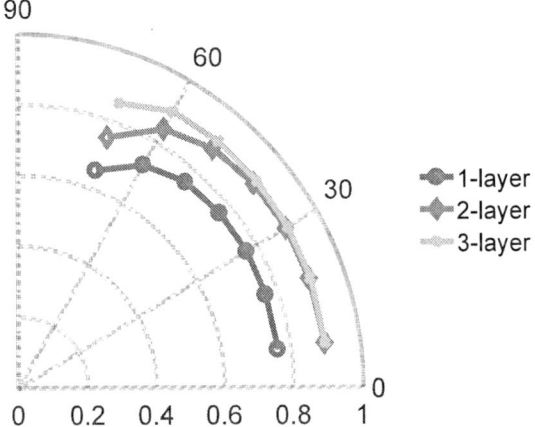

Fig. 4. 300 K emissivity over the wavelength from 2 to 50 microns versus the emission angle, obtained from the reflection measurements for the one-layer, two-layer and three-layer metasurfaces.

REFERENCES

[1] L. Zhu, A. Raman, and S. Fan, "Radiative cooling of solar absorbers using a visibly transparent photonic crystal thermal blackbody," *Proc. Natl. Acad. Sci. USA*, vol. 112, pp. 12282–12287, 2015.

[2] W. W. Salisbury, "Absorbent body for electromagnetic waves." U.S. Patent 2 599 944, June 10, 1952.

[3] R. L. Fante, and M. T. McCormack, "Reflection properties of the Salisbury screen," *IEEE Trans. Antenna. Propag.*, vol. 36, pp. 1443-1454, 1988.

[4] T. C. Preston and R. Signorell, "From plasmon spectra of metallic to vibron spectra of dielectric nanoparticles," *Accounts Chem. Res.*, vol. 45, pp. 1501-1509, 2012.

[5] J. A. Scholl, A. L. Koh and J. A. Dionne, "Quantum plasmon resonances of individual metallic nanoparticles," *Nature*, vol. 483, pp. 421-428, 2012.

[6] M. Krüger, G. Bimonte, T. Emig and M. Kardar, "Trace formulas for nonequilibrium Casimir interactions, heat radiation, and heat transfer for arbitrary objects," *Phys. Rev. B*, vol. 86, p. 115423, 2012.

[7] Nexolve Materials. (8/1/2017). [Online]. Available: http://www.nexolvematerials.com/.

Line-Focus and Point-Focus Space Photovoltaic Concentrators Using Robust Fresnel Lenses, 4-Junction Cells, & Graphene Radiators

Mark O'Neill[1], A.J. McDanal[1], Michael Piszczor[2], Matt Myers[2],

Paul Sharps[3], Claiborne McPheeters[3], Jeff Steinfeldt[3]

[1]Mark O'Neill, LLC, Keller, TX 76248, [2]NASA Glenn Research Center, Cleveland, OH 44135,

[3]SolAero Technologies, Albuquerque, NM 87123

Abstract — **At the past two PVSCs, our team presented recent advances in our line-focus space photovoltaic concentrator technology. In the past year, under Phase II and Phase II E SBIR programs and other programs funded by NASA, our team has extended the technology to point-focus concentrators as well, and has made much additional progress in the development of this new family of space photovoltaic concentrators. Both versions use flat, robust Fresnel lenses, high-efficiency 4-junction cells, and innovative graphene radiators. This paper presents the latest advances in this technology.**

Index Terms — **concentrator, Fresnel lens, multijunction cells, ultralight, graphene.**

I. INTRODUCTION AND SUMMARY

As discussed in our team's papers at the last two PVSCs [1]-[2], we have been working for the past several years on advanced space photovoltaic concentrator technology using three key elements:

- Ultralight, robust, color-mixing, flat Fresnel lens optical elements to collect and focus the sunlight onto multijunction solar cells. The latest lenses are strengthened with either:

  1. robust superstrates to support the silicone prisms forming the lens, or

  2. embedded metal mesh in the silicone lens itself.

- Advanced inverted metamorphic (IMM) solar cells with at least 4 junctions to enhance conversion efficiency.

- Waste heat radiators made from graphene, a new material with unprecedented in-plane thermal conductivity. The latest radiators also have silicone coatings on both surfaces to maximize emittance for waste heat radiation to deep space.

This paper presents technology advances in the past year for both the line-focus and the point-focus concentrators, which use different prismatic patterns in the lenses, different configurations for the cells and receivers, and different thicknesses for the radiators.

II. DESCRIPTION OF THE CONCENTRATOR MODULES

Fig. 1 shows the 4X line-focus concentrator module and Fig. 2 shows the 25X point-focus concentrator module. The robust lenses can use either a thin glass superstrate supporting the silicone prisms on the bottom side of the lens, or an embedded mesh in the silicone lens. Properly designed 4-junction IMM cells are used in the focal line or focal spot of each lens. Coated graphene radiators provide waste heat rejection in space.

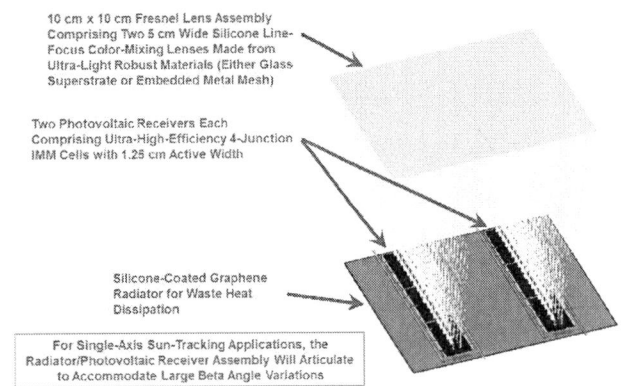

Fig. 1. 4X line-focus concentrator module.

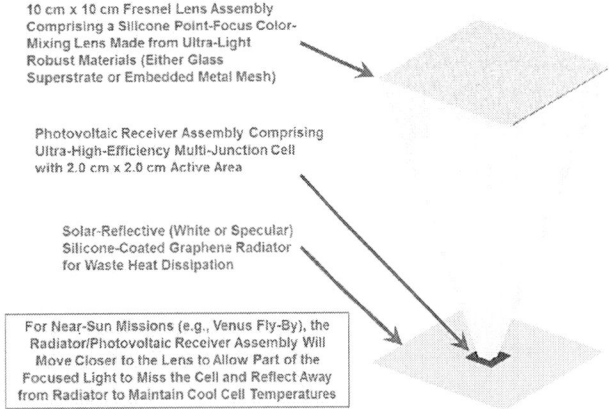

Fig. 2. 25X point-focus concentrator module.

978-1-5090-5606-4/17 $31.00 © 2017 IEEE

III. IMPROVED APPROACH FOR COLOR-MIXING LENSES

The prismatic patterns of the new lenses (both line-focus and point-focus) are optimized to provide exceptional color-mixing to avoid chromatic aberration power losses in the 4-junction IMM cells used in the new concentrator modules. In the original patented color-mixing lens approach [3], every other prism across the lens had its angle tweaked to overlap the short wavelengths of its refracted rays with the long wavelengths of refracted rays of its adjacent prisms. This two-part, every other prism, color mixing was adequate to minimize chromatic aberration power losses for 2-junction and 3-junction cells. But more sophisticated 4-junction and 6-junction cells require better color mixing, and the latest approach uses triplets of prisms across the lens to overlap short, medium, and long wavelengths of refracted solar rays to provide much better color mixing and much smaller chromatic aberration power losses. Fig. 3 shows the well matched current generation profiles for all four junctions in the latest point-focus lens.

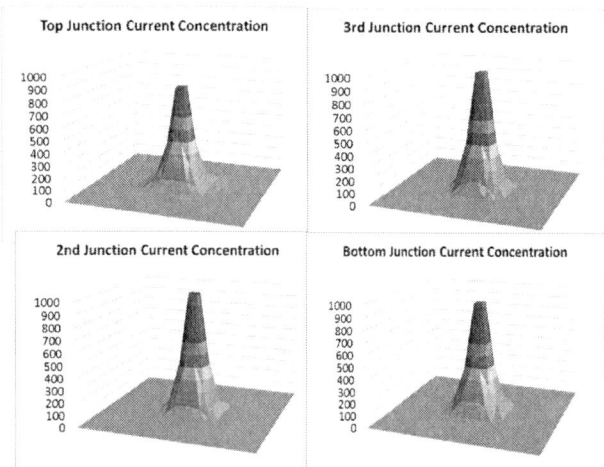

Fig. 3. Current concentration profiles for each of the 4 junctions in the IMM cell for the point-focus lens.

Tooling to produce the point-focus lens described above has recently been developed, and prototype lenses have been produced and evaluated. Fig. 4 shows a prototype lens focusing light from a distant lamp onto a simulated solar cell. The dark square represents the 2 cm x 2 cm solar cell, which also corresponds to the blue squares in Fig. 3. Note that the focal spot is intentionally much smaller than the cell to accommodate sun-pointing errors up to 2 degrees in any direction. The focal spot will drift toward the edge of the active area of the cell when such sun-pointing errors are present, but negligible amounts of light will fall off the cell below this 2 degree error tolerance. This same point-focus lens can be used with a smaller solar cell for missions requiring smaller sun-pointing tolerance. For example, a 1.4 cm x 1.4 cm cell can be used to provide 1 degree of sun-pointing tolerance at 50X geometric concentration ratio.

Fig. 4. Prototype point-focus lens with embedded stainless mesh in silicone lens.

Fig. 5. Prototype line-focus concentrator module with 2 lenses and 2 receivers using IMM cells.

As described in our last PVSC paper [2], we have already produced and tested line-focus lenses which perform well (over 90% net optical efficiency with 4-junction IMM cells in outdoor testing with direct terrestrial sunlight). Fig. 5 shows a recent prototype of the line-focus concentrator module of the basic building-block size shown previously in Fig. 1.

The line-focus concentrator of Fig. 1 and Fig. 5 at 4X geometric concentration ratio tolerates sun-pointing errors of ±2 degrees in the critical lateral direction and ±50 degrees in the longitudinal direction, with a slight seasonal receiver articulation to accommodate the latter with single-axis sun-tracking. A new U.S. Patent [4] describes the novel space concentrator technology in greater detail.

IV. GRAPHENE RADIATOR DEVELOPMENT

The latest graphene radiator for the line-focus concentrator uses a 25-micron-thick graphene sheet, strengthened on both sides with mesh, and coated on both sides with 25 microns of silicone to increase the emittance of the sheet from 33% bare to 70% coated, as discussed in our last PVSC paper [2]. The same approach is used for the point-focus concentrator, except the graphene sheet thickness is increased to 40 microns to better spread the heat over the radiator area from the central focal spot. Both 25-micron and 40-micron graphene sheets are commercially available at low cost. Fig. 6 shows the temperature profile over half of the line-focus radiator on GEO with the waste heat deposited in the central 1/10th of the radiator width. The temperature profile over the other half of the radiator is identical. Fig. 7 shows the temperature profile over one quadrant of the point-focus radiator on GEO with the waste heat deposited in the central 1/25th of the radiator area. The temperature profile over the other three quadrants of the radiator is identical.

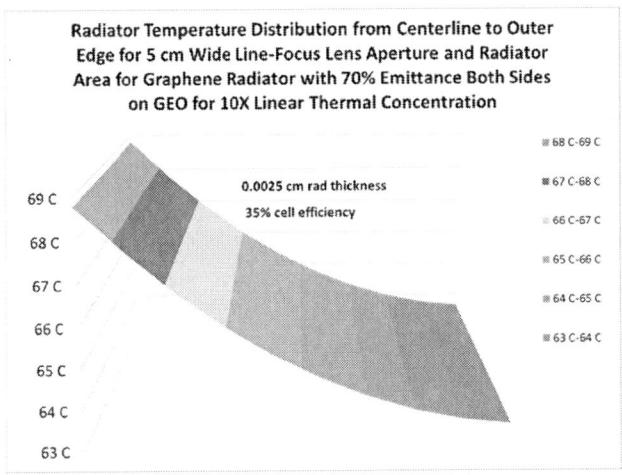

Fig. 6.   Radiator temperature profile for line-focus module.

In the mid-1990s, we developed a point-focus space concentrator using mini-dome Fresnel lenses to focus onto 2-junction mechanically stacked solar cells (GaAs over GaSb, made by Boeing), which flew on PASP+ [5]. At the time, the best radiator material identified by Boeing was aluminum. Its thermal conductivity limited the size of the lens aperture (and therefore the waste heat amount) to 3.7 cm x 3.7 cm, for an aluminum radiator thickness of 250 microns. With the new graphene material, which has a thermal conductivity almost a full order of magnitude higher than aluminum, we can now

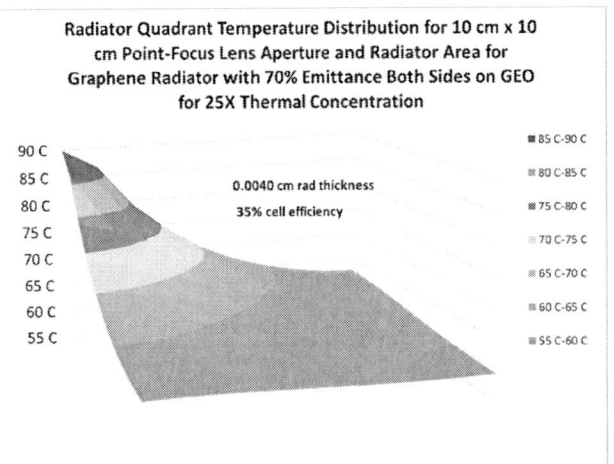

Fig. 7.   Radiator temperature profile for the point-focus module.

make the lenses and radiators 7.3 times (100 sq.cm./13.7 sq.cm.) larger in area while using a radiator thickness 84% (40μ/250μ) smaller. This improvement dramatically reduces both parts count and mass per unit power.

V. IMM CELL DEVELOPMENT

Development of the point-focus IMM cell has not yet started, but development of the line-focus IMM cell has been successfully completed. On a cell active area basis (not including the edge busbar and tabs), the prototype cells measured about 33% efficient at one-sun (AMO) before interconnect tabs were attached. Three-cell receiver circuits measured slightly lower about 31% after interconnecting the cells. We saw similar slightly lower cell performance in the lens-cell concentrator module discussed below. We are not yet certain of the cause for the reduction in cell performance after adding interconnects.

VI. SMALL CONCENTRATOR MODULE TESTING

Our team has done much testing over the past year, including outdoor testing of lens optical performance with 4-junction IMM cells under terrestrial direct sunlight. One of our prototype concentrator module test units is shown in Fig. 8. This small module is sized (with less than 11 cm diameter) to fit inside the cylindrical test tube on the NASA airplane flight test facilities, which fly above 80% of the atmosphere for near-space measurements. Results of these innovative tests include extrapolation from higher air mass (AM) values (between AM0 and AM1) at different altitudes to AM0 above the atmosphere in space. Prior to delivery to NASA, the prototype test unit shown in Fig. 8 was tested in large-area pulsed solar simulators (LAPSS) at SolAero, as shown in Fig. 9. This LAPSS testing proved to be very challenging due to the need for both a well collimated light source for the concentrator to focus well, and exceptional solar spectrum fidelity for each junction in the 4-junction IMM cell.

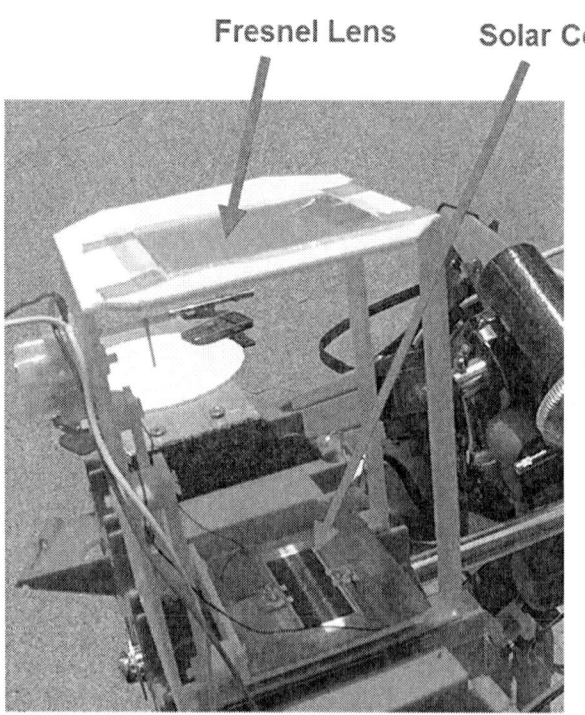

Fresnel Lens   Solar Cell

Fig. 8.   Small concentrator module testing outdoors.

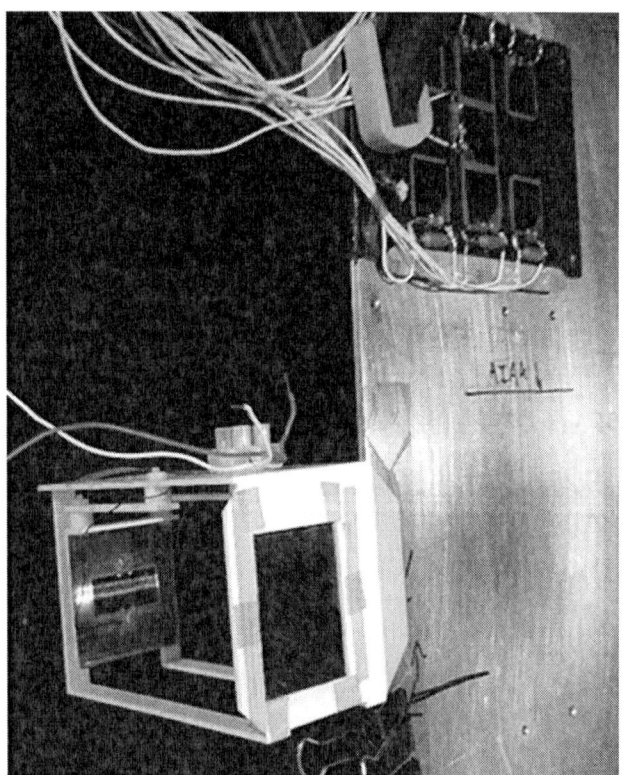

Fig. 9.   Small concentrator module LAPSS testing.

SolAero first performed testing in one LAPSS facility, but this first test provided unusable results, with clearly problematic current-voltage (IV) curves. Fortunately, the small concentrator module configuration allowed diagnosis of the problem. The cell was rotated outside of the lens and tested at 1 sun irradiance in the LAPSS facility, again with clearly problematic results in the IV curve. We at first thought the cell had been damaged. But we had earlier IV measurements in a cell solar simulator on the cell and we decided to remove the cell from the concentrator module and test it again in the cell simulator, where it performed well with a normal IV curve. These two tests showed that cell was good but that the LAPSS spectrum was inadequate for 4-junction cell testing. A second LAPSS test in a different simulator provided good results, summarized in Fig. 10. Separate IV curves were measured with the cell in the focus of the lens (red curves) and with the cell rotated out of the concentrator module and exposed to 1 sun irradiance (blue curves). These test results are close to expected results based on the calculated geometric concentration ratio of the cell using the total cell area less the combined area of busbar and interconnect tabs. The alignment, collimation, and spectral fidelity of the concentrator module and light source were not perfect, and we would not be surprised to see slightly better results in a future NASA aircraft test facility flight test.

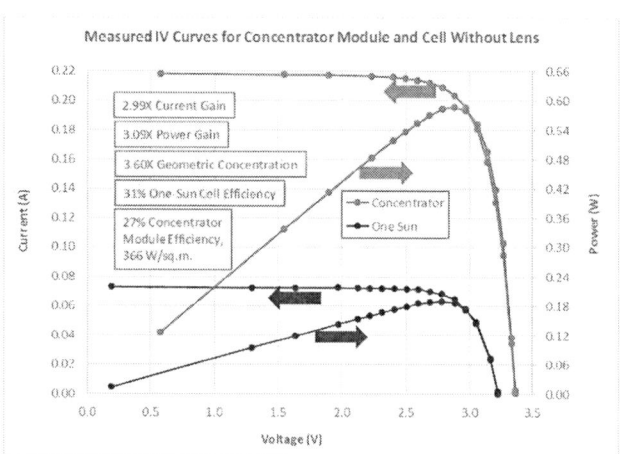

Fig. 10.   Measurements in SolAero's LAPSS test facility.

The prototype test unit shown in Fig. 8 and Fig. 9 was equipped with a one-sun reference cell prior to delivery to NASA for later testing on one of NASA's airplane test facilities, as shown in Fig. 11. This will allow the relative performance of both cells to be measured under the same near-space sunlight conditions. We used a similar approach several years ago to obtain good NASA Lear Jet measurements of the lens optical efficiency for the earlier mini-dome lenses. The module is also equipped with an offset 4-bar-linkage mechanism to allow for additional testing to demonstrate receiver articulation for large beta angle tolerance, or to rotate the lower cell beyond the lens and structure for direct one-sun irradiance in ground simulator testing. This feature has already proven useful in testing at SolAero, as discussed above.

Fig. 11. Small concentrator module delivered to NASA.

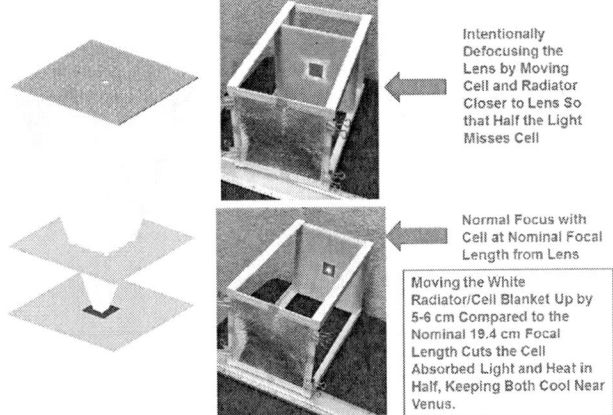

Intentionally Defocusing the Lens by Moving Cell and Radiator Closer to Lens So that Half the Light Misses Cell

Normal Focus with Cell at Nominal Focal Length from Lens

Moving the White Radiator/Cell Blanket Up by 5-6 cm Compared to the Nominal 19.4 cm Focal Length Cuts the Cell Absorbed Light and Heat in Half, Keeping Both Cool Near Venus.

Fig. 12. Solving the HIHT problem by moving the radiator and cell closer to the lens during near-sun portion of mission.

The 25X point-focus lens concentrator also offers outstanding advantages for the HIHT portion of the mission. Since the lightest type of platform for deployment and support of the point-focus concentrator array is a dual flexible blanket array with the top blanket comprising the lenses and the bottom blanket comprising the radiators (with solar cells mounted on the radiators), the distance between these blankets can be changed during the mission. By changing this distance, some of the focused sunlight can be directed to miss the solar cell and instead intercept solar-reflective radiator area, thereby reducing the absorbed solar irradiance and lowering the operating temperatures of the radiator and the cell. This simple yet effective way of overcoming the HIHT problem is shown in Fig. 12.

VII. LILT AND HIHT MITIGATION FOR DEEP SPACE MISSIONS

Over the past year, NASA has been supporting the development of the point-focus concentrator shown in Fig. 2 and Fig. 4 for deep space missions under the Game-Changing Technology Development Program for Extreme Environments Solar Power [6]. These future missions envision a spacecraft trajectory that involves of course a launch from Earth, followed by solar electric propulsion (SEP) flight toward Venus, and then a fly-by of Venus to provide a gravity-assist acceleration toward Jupiter or Saturn or one of the moons of these planets. The near-sun portion of the trajectory offers an extremely challenging high-intensity high-temperature (HIHT) environment since the solar irradiance at Venus is approximately twice the solar irradiance at Earth. When the spacecraft moves outward to Jupiter, the irradiance falls to less than 1/25th the solar irradiance at Earth, introducing an extremely challenging low-intensity low-temperature (LILT) environment.

The 25X point-focus lens concentrator offers outstanding advantages for the LILT portion of the mission, since the 25X concentration results in a solar irradiance on the small solar cell equivalent to 1 sun at 1 astronomical unit (AU) near Earth. Furthermore, the lens serves as a thermal radiation shield on the front of the concentrator, raising the operating temperature of the radiator and cell by about 20°C compared to one-sun cells near Jupiter, mitigating the low temperature problem.

VIII. SPACE ENVIRONMENTAL EFFECTS TESTING

At the last PVSC, we provided information about high-energy and low-energy charged particle exposure testing, which showed the new robust lens to be durable for space missions from 15 years on GEO to 1 year on the high-radiation TacSat 4 orbit. Eight different lens material approaches were tested with high-energy (2.7 MeV) protons through the backside (to deposit the protons near the frontside) at 5×10^{12} p$^+$/cm^2 with no loss in transmittance. Samples of the silicone lens with embedded metal mesh were also tested with low-energy protons (30 KeV) at various fluences (from 1×10^{13} p$^+$/cm^2 to 1×10^{16} p$^+$/cm^2) through the frontside. While some yellowing occurred only at the very highest fluence, these samples were uncoated and our team believes that the UV-rejection coating (multiple layers of ceramic materials), which we have successfully flown in space in the past [5], would absorb most of these low-energy particles to mitigate such yellowing of the silicone below.

In the past year, we have also conducted thermal cycling testing of the silicone lens with embedded metal mesh, again with positive results (no tearing of the silicone lens). The thermal cycling test comprised 516 cycles from -175°C to

+125°C plus 1484 cycles from -160°C to +125°C (to allow test completion on schedule). One of the samples had previously been exposed to 1×10^{15} p$^+$/cm^2 of 30 KeV protons, and none of the samples suffered tearing of the silicone. We believe that the radiation testing and thermal cycling testing show that the new robust lenses will be durable in the space environment.

IX. PERFORMANCE METRICS AND COST SAVINGS

As we gather experimental performance data and as we refine our performance and cost models, we are confirming that the new concentrator technology offers exceptional performance metrics and cost savings compared to conventional one-sun solar arrays. For example, the 25X point-focus concentrator saves about 95% of the sophisticated and relatively expensive IMM cell area and cost compared to a one-sun array. The mass of cell radiation shielding (front and back) is also reduced by about 95% for the 25X point-focus concentrator. For the 4X line-focus concentrator, the cell area and cost savings are about 75%, and the cell radiation shielding mass is also about 75% less for the same amount of shielding compared to a one-sun array. These cost and mass savings are particularly important for high-power missions (e.g., solar electric propulsion) and deep space missions (to Jupiter and beyond).

Fig. 13 and Fig 14 summarize the mass breakdowns for the key optical, thermal, and electrical elements of the 4X and 25X concentrator blankets, respectively, including the lenses (silicone with embedded stainless mesh), the silicone-coated graphene radiators, and the photovoltaic receiver assemblies. These mass breakdowns do not include the wiring harnesses, deployment and support structures, or the solar array drive assemblies. Note that the expected specific power for the combined mass of these three elements is excellent at > 1,200 W/kg for the 4X concentrator with 150 micron (6 mil) equivalent cover glass cell shielding (front and back), and > 1,400 W/kg for the 25X concentrator with 350 micron (14 mil) equivalent cover glass cell shielding (front and back).

Dual Blanket Areal Mass Density for 4X Line-Focus Silicone Lens with Stainless Mesh, Silicone-Coated Graphene Radiator Sheet, and Photovoltaic Receiver Elements for IMM Cell with 150 micron (6 mil) Equivalent Cover Glass Shielding Front and Back

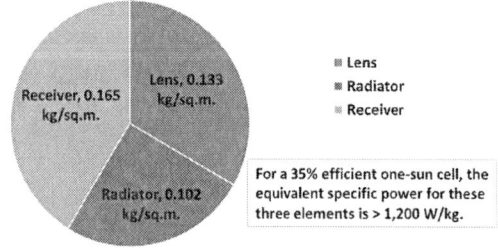

Fig. 13. Areal mass density of 4X concentrator blankets.

Dual Blanket Areal Mass Density for 25X Point-Focus Silicone Lens with Stainless Mesh, Silicone-Coated Graphene Radiator Sheet, and Photovoltaic Receiver Elements for IMM Cell with 350 micron (14 mil) Equivalent Cover Glass Shielding Front and Back

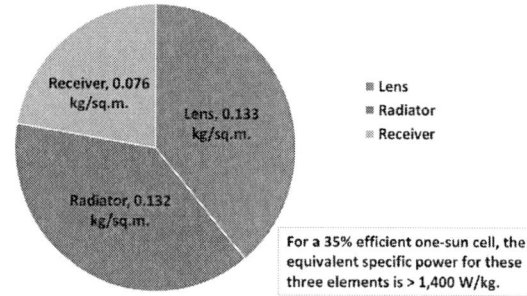

Fig. 14. Areal mass density of 25X concentrator blankets.

ACKNOWLEDGEMENT

The authors gratefully acknowledge NASA's continuing support of the advancement of the space photovoltaic concentrator technology discussed in this paper. This work has been supported under multiple NASA contracts, including Small Business Innovation Research (SBIR) contracts and Game-Changing Technology Development contracts.

REFERENCES

[1] Mark O'Neill, A.J. McDanal, Henry Brandhorst, Kevin Schmid, Peter LaCorte, Michael Piszczor, Matt Myers, "Recent space PV concentrator advances: more robust, lighter, and easier to track," *42nd IEEE Photovoltaic Specialists Conference, 2015, New Orleans.*

[2] Mark O'Neill, A.J. McDanal, Henry Brandhorst, Brian Spence, Shawn Iqbal, Paul Sharps, Clay McPheeters, Jeff Steinfeldt, Michael Piszczor, Matt Myers, "Space photovoltaic concentrator using robust fresnel lenses, 4-junction cells, graphene radiators, and articulating receivers," *43rd IEEE Photovoltaic Specialists Conference, 2016, Portland.*

[3] Mark O'Neill, "Color-mixing lens for solar concentrator system and methods of manufacture and operation thereof," *U.S. Patent No. 6,031,179, 2000.*

[4] Mark O'Neill, "Fresnel lens solar concentrator configured to focus sunlight at large longitudinal incidence angles onto an articulating energy receiver," *U.S. Patent No. 9,660,123, 2017.*

[5] H. Curtis and D. Marvin, "Final results from the PASP Plus flight experiment," *25th IEEE Photovoltaic Specialists Conference, 1996, Washington.*

[6] Frederick Elliott and Michael Piszczor, NASA Fact Sheet No. FS-2016-07-048-GRC, "Solar power generation in extreme space environments," 2016.
Accessed May 2017 at this NASA web site:
gameon.nasa.gov/gcd/files/2016/08/FS_EESP_160816.pdf

Simulation of Light Trapping Structures for Enhancing Radiation Hardness in Space Solar Cells

Nizami Z. Vagidov, Kyle H. Montgomery, Geoffrey K. Bradshaw, David M. Wilt

Air Force Research Laboratory, Space Vehicles Directorate, Kirtland AFB, NM, 87117, USA

Abstract — In an effort to reduce the radiation degradation of triple-junction solar cells, imbedded photon management is investigated as a means for reducing the middle subcell thickness while maintaining subcell's output. Three types of (In)GaAs subcells of the conventional InGaP/(In)GaAs/Ge triple-junction solar cells with different back-side gratings as light-reflecting devices were simulated. The best results were observed for the subcells with two-dimensional moth-eye back-side grating. More than 70 % of incident light is reflected into the non-zero diffraction orders which allows the thickness of the subcell to decrease from 3.5 to 1.25 µm. This shrinkage of the subcell substantially increases the radiation hardness of the multi-junction solar cell.

Index Terms — radiation hardness, triple-junction solar cell, moth-eye grating, blaze grating, lamellar grating, back reflector.

I. INTRODUCTION

Solar cells used for space applications suffer from power losses due to the radiation environment they are exposed to on-orbit. The energetic protons and electrons continuously bombard the cell, resulting in increased point defect formation and coinciding reduction of minority carrier diffusion length (MCDL). Cell designers must take this fact into consideration and optimize their cell design to maximize its performance at end of life (EOL) (in other words, after a given radiation dose). This consideration limits the ultimate performance that can be achieved, as there are trade-offs that must be made in the design, such as thinning the top subcell to allow for excess current generation in the radiation-soft middle subcell of a triple-junction solar cell (TJSC).

Given that the reduced MCDL is the cause of power loss at EOL, one option is to reduce the thickness of a given subcell sufficiently to minimize the impact of a reduced MCDL. In the case of a typical (In)GaAs middle subcell in a TJSC, this would mean reducing the thickness below 1 µm, to minimize the impact for the reduced MCDL under a typical 10^{15} cm^{-2} fluence of 1 MeV electrons [1]. However, given the absorption coefficient of (In)GaAs this would result in insufficient optical absorption.

In this work, we examine various light reflecting features that may be integrated into the architecture of the multi-junction solar cell to allow for increased optical absorption in subcell needing to be thinned to accommodate the reduction in MCDL. In the generic case, we will consider the (In)GaAs middle subcell of a standard TJSC, given that this subcell is the weakest in terms of its radiation hardness. Here we will focus on an integrated diffraction grating to increase the

optical path length, in particular for photons near the band edge of (In)GaAs. The goal herein is to determine the necessary parameters for given grating design to be integrated within the monolithic structure of the III-V-based solar cell.

II. OPTICAL SIMULATION

The design of TJSC includes a number of important components. One of them is antireflection coating (ARC). The design of an ARC for a TJSC is more challenging than for a single-junction solar cell since the spectral range which covers TJSC is much wider. The lattice-matched $In_{0.49}Ga_{0.51}P/In_{0.01}Ga_{0.99}As/Ge$ solar cell absorbs light in the range from 0.3 µm to 1.8 µm. The problem of choosing an appropriate ARC is simplified because the short-circuit current in the bottom Ge subcell is significantly higher than in InGaP and (In)GaAs subcells. Thus, the critical spectral region for an ARC design becomes much smaller: from 0.3 µm to 0.9 µm. An appropriate ARC for this spectral region is a double-layer anti-reflecting coating (DLAR) such as MgF_2/ZnS or SiO_2/Ta_2O_5 [2]. Simulations done using Lumerical software [3] show that the reflectance from MgF_2/ZnS DLAR is very low (see Fig. 1).

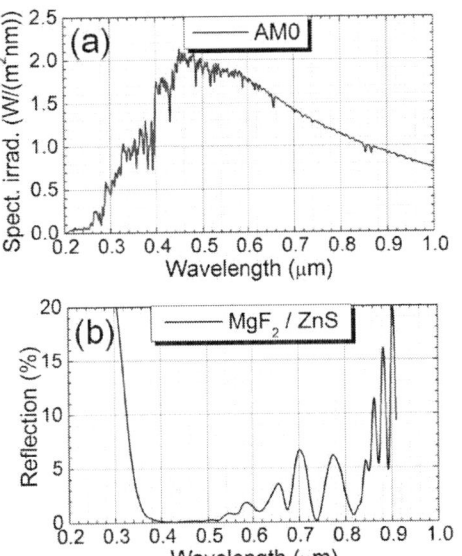

Fig. 1. (a) AM0 solar spectrum and (b) power reflectance from InGaP/(In)GaAs tandem solar cell with MgF_2/ZnS DLAR. The thickness of MgF_2 layer is 82 nm and of ZnS - 44 nm.

978-1-5090-5606-4/17 $31.00 © 2017 IEEE

One of the characteristics that defines reflectance from an ARC is the solar-weighted reflection (SWR) [4] which is defined as

$$SWR = \frac{\int_{\lambda_1}^{\lambda_2} \lambda\, R\,(\lambda)\, P_{AM0}\,(\lambda)}{\int_{\lambda_1}^{\lambda_2} \lambda\, P_{AM0}\,(\lambda)}. \qquad (1)$$

Here $R(\lambda)$ is wavelength-dependent surface reflectance and $P_{AM0}(\lambda)$ is AM0 solar power spectral density. In the spectral range from $\lambda_1 = 300$ nm to $\lambda_2 = 650$ nm the SWR from the surface covered by MgF_2/ZnS DLAR is about 1.5% which is close to the values from [2] and in the spectral range from 300 to 900 nm it is about 3% which corresponds to the results from Ref. [4].

The second important component of TJSC design is the thickness of the base of InGaP subcell. Depending on its thickness the top subcell can be made current rich or poor in comparison with the middle (In)GaAs subcell. As it was mentioned earlier, the middle subcell is made current rich because of its weak radiation hardness. Thus, the current of TJSC is limited by the current of the top subcell. In such configuration the thickness of the top subcell's base is small and is about 600 nm (see Fig. 2).

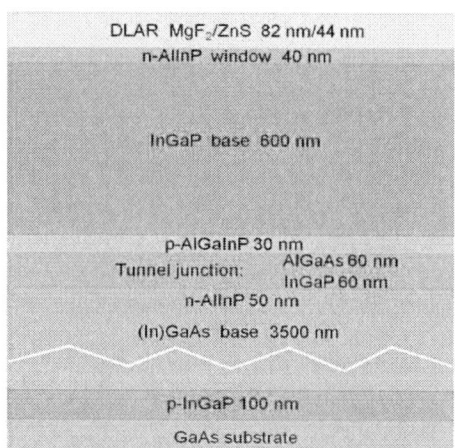

Fig. 2. The schematic structure of the simulated InGaP/(In)GaAs tandem solar cell with MgF_2/ZnS DLAR.

The parameters for the simulated $In_{0.49}Ga_{0.51}P/In_{0.01}Ga_{0.99}As$ tandem solar cell were taken from [4]. The base thickness of the middle subcell is large enough to provide higher current than in top subcell. The numerical integration of the external quantum efficiencies (EQE) of the above-mentioned tandem solar cell gives us the values of the short-circuit current of the top and the middle subcells equal approximately to 17 mA/cm^2 and 18 mA/cm^2, respectively. As it is seen from Fig. 3, the photocurrent in the top subcell is generated by absorption of solar light in the spectral region from 300 nm to 700 nm and in the middle cell from 650 nm to 900 nm. This fact makes the simulation of TJSC much easier. Instead of simulating the

whole structure, the optical and electrical models of single-junction InGaP and (In)GaAs solar cells can be analyzed.

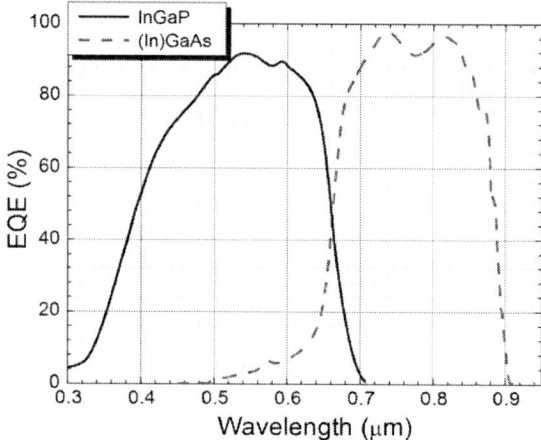

Fig. 3. The EQE of the simulated InGaP/(In)GaAs tandem solar cell with MgF_2/ZnS DLAR.

In order to find I-V characteristics of single-junction cells optical simulations were carried out using Lumerical FDTD SOLUTIONS tool which implements finite-difference time domain (FDTD) method. The photo-generation data obtained by these optical simulations was used as an input for electrical simulations by Lumerical DEVICE CT tool [3].

The optical model of the simulated device includes a plane wave source which is situated over the DLAR of the solar cell. Periodic boundary conditions in the x-, y-directions and perfectly matched layer (PML) condition in the z-direction substantially simplifies the simulation of the device. These assumptions allowed three-dimensional simulation of the structures that cannot be reduced to two-dimensional models.

As a result: (a) an InGaP single-junction solar cell with base thickness of 600 nm was analyzed, (b) two (In)GaAs single-junction solar cells with base thickness of 3.5 and 1.75 µm were simulated, and (c) (In)GaAs single-junction solar cells with four different light-reflecting structures (LRS) such as lamellar grating (Fig. 4(a)), blaze grating (Fig. 4(b)), moth-eye grating (Fig. 4(c) and 4(d)), and distributed Bragg reflector (DBR) (Fig. 4(e)) were analyzed. Further, we will consider (In)GaAs single-junction solar cell with 3.5 µm base thickness as a reference cell. All solar cell structures were illuminated by the normally incident light from FDTD source with a Gaussian-like pulse.

The first (lamellar) structure contains one-dimensional (1D) rectangular dielectric grating as a back-side LRS (Fig. 4(a)). Four parameters define this grating: 1) the grating period $\Lambda = 700$ nm, 2) the duty cycle $w = 50$ %, 3) the groove depth $h_1 = 250$ nm, and the thickness of dielectric layer $h_2 = 250$ nm. The second structure (Fig. 4(b)) contains a 1D asymmetric blaze grating with the following parameters: the period $\Lambda = 700$ nm, the blaze angle $\theta = 30°$, the height of the grating $h_1 = 303$ nm, and the thickness of dielectric layer $h_2 = 300$ nm.

978-1-5090-5606-4/17 $31.00 © 2017 IEEE

The third structure contains the so-called moth-eye grating [5] which can be described as asymmetric two-dimensional (2D) grating with periods $\Lambda_x = 500$ nm and $\Lambda_y = 800$ nm (Figs. 4(c) and 4(d)). This grating is an array of closely-packed half-ellipsoids with the height $h_1 = 360$ nm and the length of semi-axes $a = 500$ nm and $b = 800$ nm. The height of dielectric layer, where these half-ellipsoids are placed, is $h_2 = 250$ nm. The refraction index of (In)GaAs in all types of solar cells is equal to $n_1 = 3.6$ and the refraction index of dielectric n_2 is equal to 1.4. The parameters for gratings were chosen after analysis of literature where solar cells with back-side gratings were considered as light-reflecting structures [6-17].

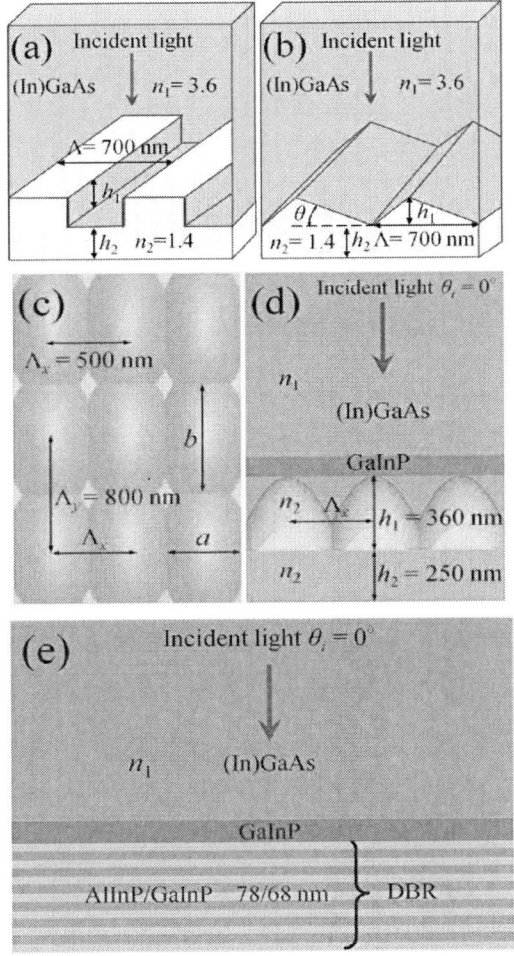

Fig. 4. (In)GaAs single-junction solar cell with three different dielectric gratings and one DBR as back-side light reflecting structures (LRS): (a) 1D rectangular grating with period $\Lambda = 700$ nm, duty cycle $w = 50$ %, groove depth $h_1 = 250$ nm, and dielectric layer with thickness $h_2 = 250$ nm; (b) 1D asymmetric blaze grating with period $\Lambda = 700$ nm, blaze angle $\theta = 30°$, height $h_1 = 303$ nm, and $h_2 = 300$ nm; (c), (d) 2D asymmetric moth-eye grating with periods $\Lambda_x = 500$ nm and $\Lambda_y = 800$ nm, $h_1 = 360$ nm, $h_2 = 250$ nm, $a = 500$ nm, and $b = 800$ nm. (e) 1D AlInP/GaInP DBR reflector with AlInP layer thickness 78 nm and GaInP layer thickness 68 nm. For all structures with gratings the refraction indices are the same and equal to $n_1 = 3.6$ and $n_2 = 1.4$.

In order to have all diffraction orders to be reflected, the period of grating should satisfy the following inequality:

$$\Lambda < \frac{\lambda}{m} \qquad (2)$$

where λ is the wavelength of incident light and m is the number of diffracted orders [6]. Duty cycle was taken equal to $w = 50$ % and aspect ratio h/Λ (groove depth/period) was about 0.25 [7-14].

The highest power reflection and the highest optical path length enhancement are achieved with the moth-eye grating. Fig. 5 shows that more than 70% of incident light is reflected into non-zero diffraction orders.

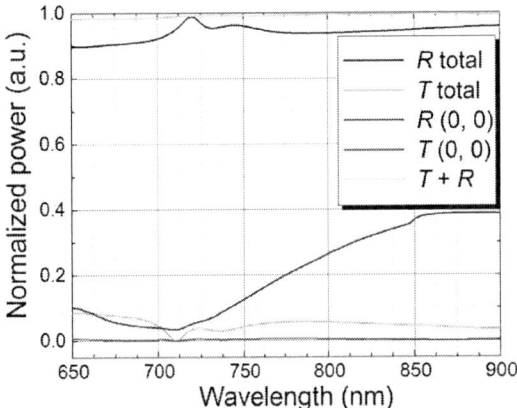

Fig. 5. Normalized power of reflected, R, and transmitted, T, waves for solar cell with moth-eye grating. R (0,0) and T (0,0) show the power of the $(0,0)^{th}$ order of reflected and transmitted waves, respectively.

The angles of reflection of non-zero orders are greater than the angle of total internal reflection ($\sim 16°$) as it can be seen from Fig. 6(a). The number of diffraction orders linearly increases with the decrease of wavelength (see Fig. 6(b)) as it can be expected from the diffraction grating equations for 2D grating [15]:

$$n_1 \sin\theta_{mn} \cos\theta_{mn} = m\frac{\lambda}{\Lambda_x} \qquad (3)$$

$$n_1 \sin\theta_{mn} \sin\theta_{mn} = m\frac{\lambda}{\Lambda_y}. \qquad (4)$$

The results for asymmetric 1D blaze grating were the second best. As it can be seen from Fig. 7(a) almost 55% of incident power is reflected into non-zero diffraction orders. The angles of reflection of non-zero diffraction orders are greater than the angle of total internal reflection (see Fig. 7(b)). As it was expected [16, 17], the reflection by a symmetric rectangular

978-1-5090-5606-4/17 $31.00 © 2017 IEEE

grating into non-zero diffraction order was small: only 30 % of incident power was reflected into these diffraction orders.

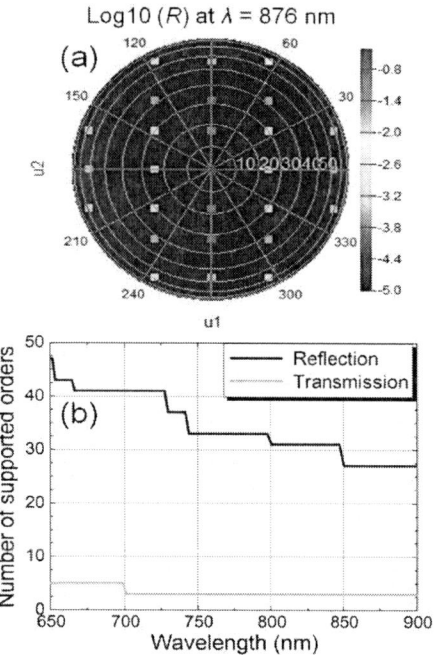

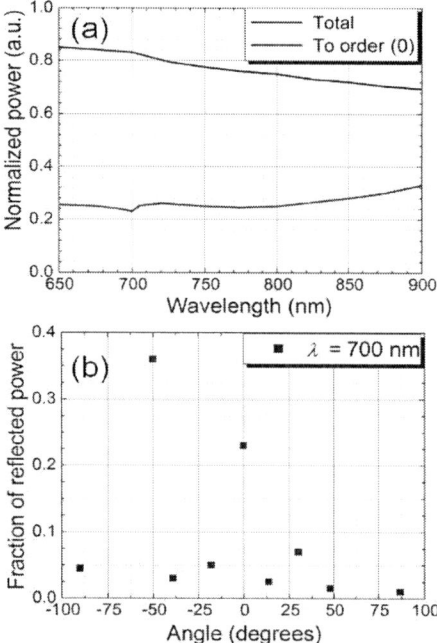

Fig. 6. (a) The angles of reflection and power of different diffraction orders and (b) the number of diffraction orders for the structure with 2D moth-eye grating as LRS.

Fig. 7. (a) The normalized power of reflected waves for solar cell with blaze grating LRS. (b) Fraction of reflected power in different diffraction orders of incident wave with wavelength $\lambda = 700$ nm in the case of blaze grating as back reflector.

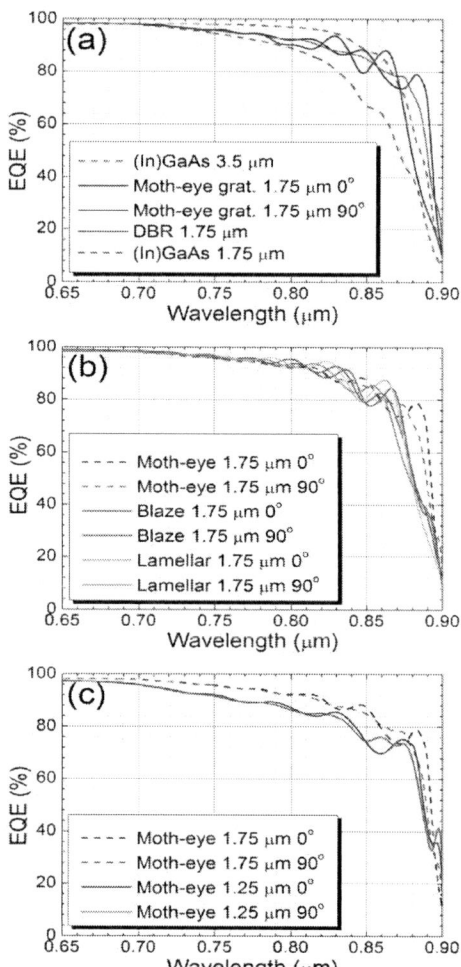

Fig. 8. EQE of (In)GaAs single-junction solar cells with different LRS: (a) Dashed lines represent EQE of solar cells with base thickness of 3.5 and 1.75 μm without any LRS. Short-circuit current densities limits for these two cells are 18.2 and 16.4 mA/cm², respectively. EQE of solar cells with moth-eye grating as back reflector are shown for two angles of polarization: 0° and 90°. EQE of the solar cell with AlInP/GaInP DBR as back reflector and with 1.75 μm base thickness is also shown. (b) EQE of solar cells with base thickness of 1.75 μm for three LRS and two polarizations are shown. (c) EQE of two solar cells with base thickness of 1.75 μm and 1.25 μm and with moth-eye grating as back reflectors for two polarizations are shown.

The EQE dependences found by optical simulation of the single-junction solar cells reflect the efficiency of the above-mentioned LRS. As one can see from Fig. 8(a) the EQE (and thus the short-circuit current) of the middle (In)GaAs subcell substantially drops when the thickness of its base decreases twice: from 3.5 to 1.75 μm. This significant drop can be compensated by introducing LRS as back reflectors. In the case of moth-eye grating the compensation of EQE takes place in the whole spectral range from 650 to 900 nm, whereas AlInP/GaInP distributed Bragg reflector (DBR) increases EQE only in a narrow spectral range from 810 to 900 nm (the

AlInP/GaInP DBR reflection bandwidth is about 75.8 nm [18]). As it is clearly seen from Fig. 8(b) the reflecting efficiency of moth-eye grating (especially in (In)GaAs band edge region) is higher than the reflectance of blaze and lamellar gratings. Finally, in Fig. 8(c) the EQEs of (In)GaAs single-junction cells with moth-eye grating as back reflector for two base thicknesses (1.75 and 1.25 µm) are shown. Assuming that internal quantum efficiency (IQE) is equal to 100% we can estimate the short-circuit density limit for the (In)GaAs subcell with 1.25 µm base thickness and InGaP subcell with 600 nm base thickness (compare EQE in Fig. 3 and Fig. 8(c)). The calculations show that the short-circuit currents of these two subcells are close to each other.

III. ELECTRICAL SIMULATION

The photo-generation data obtained by optical simulations was used as an input to find I-V characteristics of the single-junction solar cells considered above under different levels of radiation fluence. The radiation exposure by high-energy electrons results in decrease of MCDL which in its turn decreases short-circuit current and open-circuit voltage of the solar cell. Using the analysis done by M. Yamaguchi [1] the dependence of the MDCL in p-type InGaP and (In)GaAs single-junction solar cells on 1 MeV electron fluence can be found. The material parameters used for calculations of MCDL (L_n) in InGaP were taken from [19] and in (In)GaAs - from [20]. Both materials in the EOL (under 1 MeV electron fluence 10^{15} cm^{-2}) have L_n about 1µm (see Fig. 9).

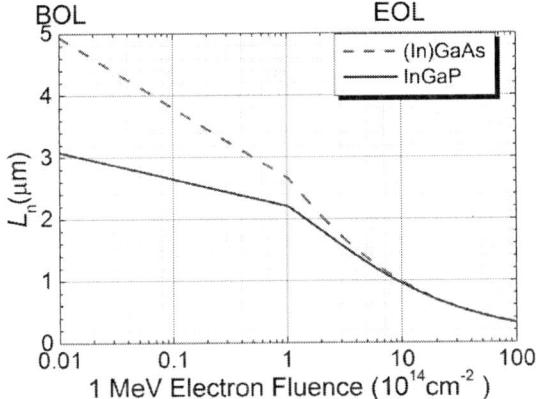

Fig. 9. The dependences of electron MCDL, L_n, in p-type (In)GaAs and InGaP bases on 1 MeV electron fluence are shown by dashed and solid lines, respectively. For p-type (In)GaAs and InGaP bases under zero electron fluence $L_n = 5$ µm and $L_n = 3.1$ µm, respectively.

The radiation hardness of several single-junction solar cells were analyzed: InGaP solar cell with 600 nm base thickness, reference (In)GaAs solar cell with 3.5 µm base thickness, and (In)GaAs solar cells with four different LRS with base thickness of 1.75 µm or less. The radiation hardness of all solar cells with an integrated LRS has been substantially improved by thinning the solar cell base. The highest radiation hardness cells have moth-eye gratings. As it is seen from

Fig. 10, such a cell with base thickness half of the (In)GaAs reference cell base has J_{SC} 9% higher than the reference cell in the EOL. This is why the solar cell with moth-eye grating was chosen as a middle (In)GaAs subcell.

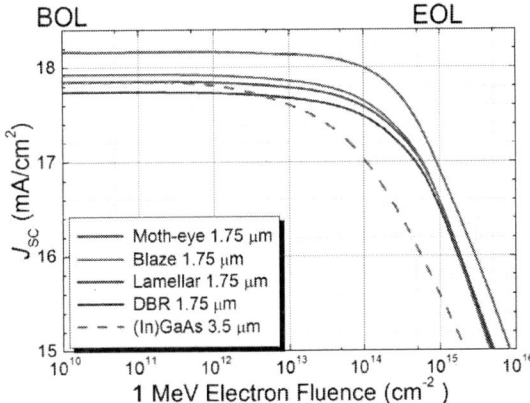

Fig. 10. The dependence of short-circuit current density, J_{SC}, of solar cells with four different back reflectors on 1 MeV electron fluence are shown by solid lines. All four solar cells have (In)GaAs base thickness of 1.75 µm. The dependence for the reference solar cell with (In)GaAs base thickness of 3.5 µm is shown by dashed line.

Normally, the manufacturers design the TJSC for operation in outer space by making the most radiation resistant subcell (which is InGaP top subcell) as a current limiter of the whole cell. The experimental values of J_{SC} of the highly-efficient InGaP/(In)GaAs/Ge cells are about 17.3 mA/cm^2 [21, 22]. Thus, the thickness of (In)GaAs base of the subcell with moth-eye grating can be made smaller than 1.75 µm (up to 1.25 µm) to have its J_{SC} slightly higher than the J_{SC} of the top subcell (see Fig. 11).

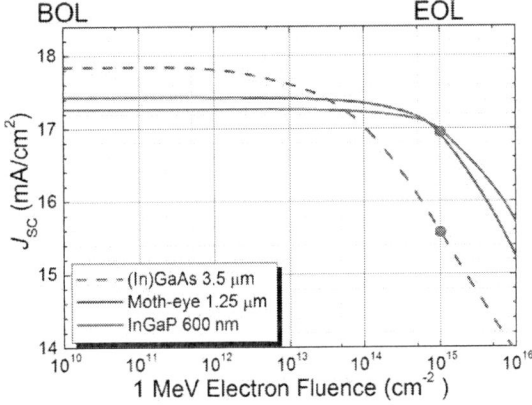

Fig. 11. The calculated dependences of short-circuit current density, J_{SC}, of solar cells with thin base thickness are shown by solid lines: the dependence with higher short-circuit current density at the beginning of life (BOL) corresponds to (In)GaAs solar cell with moth-eye grating as LRS and base thickness of 1.25 µm; the dependence with lower J_{SC} corresponds to InGaP subcell with base thickness of 600 nm. The dependence for reference (In)GaAs cell with base thickness of 3.5 µm is shown by dashed line.

978-1-5090-5606-4/17 $31.00 © 2017 IEEE

The design parameters for the simulation of top InGaP subcell were taken from the existing literature taking into account the experimental results of the highly-efficient InGaP/(In)GaAs/Ge cells [21, 22]. Note that the dependence of J_{SC} on 1 MeV electron fluence of (In)GaAs reference cell is very close to simulation results from Ref. [20]. The J_{SC} remaining factor of (In)GaAs subcell with moth-eye grating at the EOL is about 0.98 while the J_{SC} remaining factor of (In)GaAs reference subcell is much smaller ~ 0.87 (the remaining factor of J_{SC} is defined as a ratio of J_{SC} at EOL to J_{SC} at BOL of the solar cell). The J_{SC} of the top InGaP subcell and middle (In)GaAs subcell with moth-eye grating are equal to each other at EOL.

IV. CONCLUSION

We have examined three diffraction grating features for integration into a multi-junction solar cell architecture. It is evident that the moth-eye structure provides the greatest benefit in terms of increased optical path length. The output power degradation of a TJSC with a thin (In)GaAs subcell and an integrated light-reflecting structure at EOL after 15-20 years on Geo-stationary orbit is expected to be about 4 - 5 % instead of 11 - 15 % [21] for standard InGaP/(In)GaAs/Ge TJSC.

ACKNOWLEDGEMENT

This research was performed while N. Z. V. held an NRC Research Associateship award at Air Force Research Laboratory, Kirtland AFB.

REFERENCES

[1] M. Yamaguchi, "Radiation resistance of compound semiconductor solar cells," *Journal of Applied Physics*, vol. 78, pp. 1476-1480, 1995.

[2] R. Homier, A. Jaouad, A. Turala, C. E. Valdivia, D. Masson, S. G. Wallace, S. Fafard, R. Ares, and V. Aimez, "Antireflection coating design for triple-junction III–V/Ge high-efficiency solar cells using low absorption PECVD silicon nitride", *IEEE Journal of Photovoltaics*, vol. 2, pp. 393-397, 2012.

[3] https://www.lumerical.com/tcad-products.

[4] L. Yang, S. Pillai, and M. A. Green, "Can plasmonic Al nanoparticles improve absorption in triple junction solar cells?", *Scientific Reports*, vol. 5, pp. 11852-1-12, 2012.

[5] C.-H. Sun, P. Jiang, and B. Jiang, "Broadband moth-eye antireflection coatings on silicon," *Journal of Applied Physics*, vol. 92, pp. 061112 - 1 - 3, 2008.

[6] C. Eisele, C. E. Nebel, and M. Stutzmann, "Periodic light coupler gratings in amorphous thin film solar cells," *Journal of Applied Physics*, vol. 89, pp. 7722-7726, 2001.

[7] Ping Sheng, A. N. Bloch, and R. S. Stepleman, "Wavelength-selective absorption enhancement in thin-film solar cells," *Applied Physics Letters*, vol. 43, pp. 579-581, 1983.

[8] M. Peters, M. Rüdiger, H. Hauser, M. Hermle, and B. Bläsi, "Diffractive gratings for crystalline silicon solar cells—optimum parameters and loss mechanisms," *Progress in Photovoltaics: Research and Applications*, vol. 20, pp. 862–873, 2012.

[9] N.-N. Feng, J. Michel, L. Zeng, J. Liu, C.-Y. Hong, L. C. Kimerling, and X. Duan, "Design of highly efficient light-trapping structures for thin-film crystalline silicon solar cells," *IEEE Transactions on Electron Devices*, vol. 54, pp. 1926-1933, 2007.

[10] A. Mellor, I. Tobı́as, A. Martı́, and A. Luque, "A numerical study of Bi-periodic binary diffraction gratings for solar cell applications," *Solar Energy Materials and Solar Cells*, vol. 95, pp. 3527–3535, 2011.

[11] P. Bermel, C. Luo, L. Zeng, L. C. Kimerling, and J. D. Joannopoulos, "Improving thin-film crystalline silicon solar cell efficiencies with photonic crystals," *Optics Express*, vol. 15, pp. 16986- 17000, 2007.

[12] L. Zeng, Y. Yi, C. Hong, J. Liu, N. Feng, X. Duan, L. C. Kimerling, and B. A. Alamariu, "Efficiency enhancement in Si solar cells by textured photonic crystal back reflector," *Applied Physics Letters*, vol. 89, pp. 111111-1-3, 2006.

[13] M. Kroll, S. Fahr, C. Helgert, C. Rockstuhl, F. Lederer, and T. Pertsch, "Employing dielectric diffractive structures in solar cells – a numerical study," *Physica Status Solidi (a)*, vol. 205, pp. 2777–2795, 2008.

[14] L. Zhao, Y. H. Zuo, C. L. Zhou, H. L. Li, H. W. Diao, and W. J. Wang, "A highly efficient light-trapping structure for thin-film silicon solar cells," *Solar Energy*, vol. 84, pp. 110–115, 2010.

[15] H. Sai, Y. Kanamori, K. Arafune, Y. Ohshita, and M. Yamaguchi, "Light trapping effect of submicron surface textures in crystalline Si solar cells," *Progress in Photovoltaics: Research and Applications*, vol. 15, pp. 415–423, 2007.

[16] C. Heine and R. H. Morf, "Submicrometer gratings for solar energy applications," *Applied Optics*, vol. 34, pp. 2476-2482, 1995.

[17] S. Mokkapati and K. R. Catchpole, "Nanophotonic light trapping in solar cells," *Journal of Applied Physics*, vol. 112, pp. 101101 - 1 - 19, 2012.

[18] I. García, J. Geisz, M. Steiner, J. Olson, D. Friedman, and S. Kurtz, "Design of semiconductor-based back reflectors for high V_{oc} monolithic multijunction solar cells," in *39th IEEE Photovoltaic Specialists Conference*, 2012, pp. 002042-002047.

[19] N. M. Haegel, T. Christian, C. Scandrett, A. G. Norman, A. Mascarenhas, P. Misra, T. Liu, A. Sukiasyan, E. Pickett, and H. Yuen, "Doping dependence and anisotropy of minority electron mobility in molecular beam epitaxy-grown p-type GaInP," *Journal of Applied Physics*, vol. 105, pp. 202116-1-5, 2014.

[20] A. Mellor, N. P. Hylton, S. A. Maier, N. Ekins-Daukes, "Interstitial light-trapping design for multi-junction solar cells," *Solar Energy Materials and Solar Cells*, vol. 159, pp. 212–218, 2017.

[21] M. A. Stan, D. Aiken, P. R. Sharps, J. Hills, B. Clevenger, and N. S. Fatemi, "The development of >28% efficient triple-junction space solar cells at EMCORE PHOTOVOLTAICS," in *3rd World Conference on Photovoltaic Energy Conversion*, 2003, pp. 662-665.

[22] M. Yamaguchi, T. Takamoto, A. Khan, M. Imaizumi, S. Matsuda, and N. J. Ekins-Daukes, "Super-high-efficiency multi-junction solar cells," *Progress in Photovoltaics: Research and Applications*, vol. 13, pp. 125–132, 2005.

[23] M. Imaizumi, T. Nakamura, T. Takamoto, T. Ohshima, and M. Tajima, "Radiation degradation characteristics of component subcells in inverted metamorphic triple-junction solar cells irradiated with electrons and protons," *Progress in Photovoltaics: Research and Applications*, vol. 25, pp. 161–174, 2017.

978-1-5090-5606-4/17 $31.00 © 2017 IEEE

An Alternative Method for Solar Cell Integration

Jessica Buckner[1], Tracy Davis[2], Eric Muskovin[3], and Bernard Carpenter[4]

[1]Air Force Research Laboratory, Kirtland AFB, NM, 87117, USA, [2]Applied Technology Associates, Albuquerque, NM, 87123, USA, [3]Missouri University of Science and Technology, Rolla, MO, 65409, USA, [4]The Aerospace Corporation, Albuquerque, NM, 87110, USA

Keywords: Space technology, solar cell, coverglass, deep eutectic solvent

Abstract — A bonding alternative to silicone adhesives for solar cell integration has been explored. Efficacy and performance of traditional silicone adhesives is limited by manufacturability and UV degradation. A new method, utilizing deep eutectic solvents (DESs) been shown to mitigate obstacles inherent to silicone adhesives. The eutectic and ionic nature of DESs allows for freezing point depression and dissolution of metal oxides, creating a low temperature, graded, spinel-like structure when isothermally heated. This technique offers improved fracture toughness, strength, and UV exposure tolerance while preserving the optical and electrical properties of the cell. Understanding the structure and properties of the DES material is necessary for chemical manipulation and bond stability. In this study, the DESs and bond layers were examined using inductively coupled plasma-mass spectroscopy, optical microscopy, x-ray diffraction, and differential scanning calorimetry. Solubility of metal oxides in the DES are reported, the bond structures imply miscibility, and 0optimized bond behavior was achieved with the addition of hydrochloric acid to the DES.

I. INTRODUCTION

Silicone adhesives have historically been used to join coverglass to solar cells. While silicone adhesives exhibit high transmission and low joining temperatures, their efficacy is challenged by UV degradation and manufacturability. Additionally, the advent of flexible arrays has promoted the exploration of alternative bonding processes to improve joining reliability.

Deep eutectic solvents (DESs) provide a new, adhesiveless technique for solar cell integration. DESs are non-aqueous ionic liquid analogues formed by mixing a quaternary ammonium salt and a hydrogen bond donor [1]. DESs have many uses, including electrodeposition, metal extraction, and biological applications [2]. Relevant properties include high surface tension, conductivity, and viscosity [3], high solubility for metal oxides [1-4], and freezing point depression up to 200°C [5].

The eutectic and ionic nature of DESs can be exploited to form an inorganic bond that forms isothermally at low temperatures. This is possible by dissolving metal oxides into the solvent and utilizing the eutectic mixture to decrease normal processing temperature for formation of a spinel-like bond. Spinel is a hard oxide of the form AB_2O_4, where A is a divalent metal ion, and B represents a trivalent metal ion. The most common mineral form of spinel occurs when A is magnesium and B is aluminum. This joining layer offers improved fracture toughness, strength, and UV exposure tolerance compared to its silicone counterparts while maintaining optical and electric properties. This new joining method also decreases parasitic mass losses by implementation of a thinner joining layer than silicone adhesives and can be applied to both new and traditional solar cells.

Previous work [6] illustrated the potential success of adhesiveless bonding using choline chloride-malonic acid based DESs (Fig. 1). By manipulation of the composition factor (CF), equal to the molar ratio of choline chloride to molar ratio of oxalic acid, interfaces as thin as 8 microns were established. The re-melt temperature was higher than the initial room temperature melting of the solvent, and increased with reaction time. This suggests a stable bond in the temperature ranges expected from a space environment. A change in slope occurred at the transition from liquid to solid state diffusion, which are both necessary for stable bond formation. Long isothermal heating times (t<20 hours) allow for the initial liquid solute to diffuse into a glass sink, forming a graded spinel-like structure. Optical characteristics are also retained with this method, dependent on composition factor and bonding atmosphere. It is believed that high oxygen concentrations result in complete spinel formation while low concentrations do not allow for solidification to complete. Process feasibility was demonstrated and electrical properties measured with a calibrated X25 source at AM0. Comparable fill factor, short circuit current, and open circuit voltage were observed between the silicone bonded and DES bonded cell.

Previous work was replicated with choline chloride-oxalic acid based DESs, given the comparable physical and chemical properties. The aim of this investigation was to characterize the DESs and their bond properties to be able to better understand the chemistry of the bond layer and the role of the DES in bond formation. This knowledge can be used to manipulate and optimize bond properties.

Previous Work

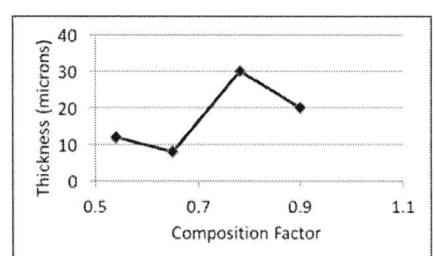

Bond Thickness

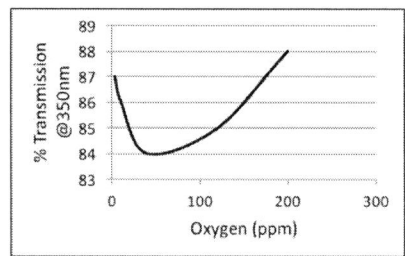

Atmosphere Dependence

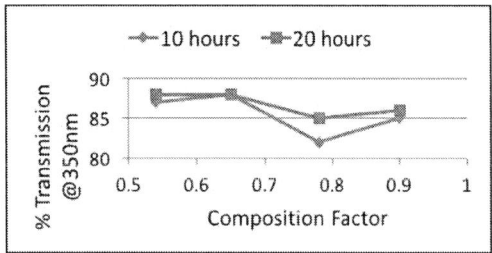

Optical Behavior

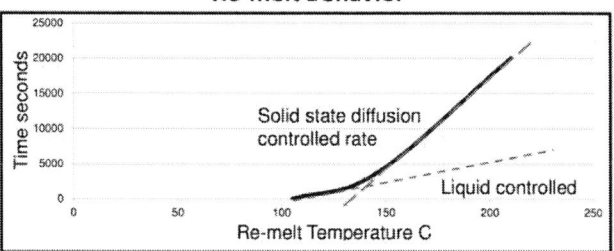

Re-melt Behavior

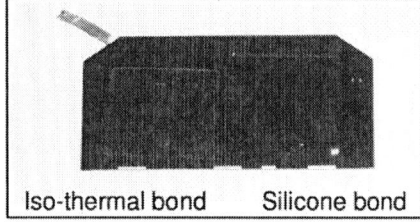

Demonstration of Process Feasibility

Iso-thermal bond   Silicone bond

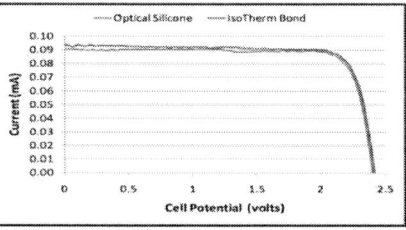

Electrical Performance

Fig. 1.   Summary of previous work on choline chloride-malonic acid DESs. Adapted from [6].

II. METHODS

Choline chloride-oxalic acid based DESs were synthesized at 60°C at different composition factors (CF=0.65,0.75,0.85). Excess metal oxides (Al_2O_3, SiO_2, CuO, Cu_2O) were mixed into different volumes and concentrations of HCl and added to the DES. The samples were then isothermally heated for 20 hours at variable temperatures and atmospheres. The DESs and bond layer were examined with inductively coupled plasma-mass spectrometry (ICP-MS), optical microscopy, differential scanning calorimetry (DSC), and x-ray diffraction (XRD).

III. RESULTS AND DISCUSSION

Initial work focused on bonding borosilicate glass slides together using the DES with dissolved metal oxides. Limited literature necessitated the need for solubility measurements of the mixtures. Solubility of Al_2O_3, SiO_2, CuO, and Cu_2O in varying CF of DES were measured (Table 1), indicating negligible dependence of solubility on composition factor and limited solubility of Al_2O_3 and SiO_2 in the DES. Early experiments were performed using copper oxides, but transparent bonds were difficult to achieve. Focus shifted to alumina and silica given the compatibility with glass. To increase the solubility of the metal oxides, hydrochloric acid (HCl) was added to the DESs. Future work will determine solubility of metal oxides in DES + HCl mixtures.

In the absence of HCl, DESs with dissolved metal oxides did not produce a solid bond, but remained liquid after heating. Upon addition of HCl, relatively strong, solid bonds were produced between 150-200°C at atmospheres of 100-300 ppm O_2. Bonding did not occur above 200°C; the constituents of the DES appeared to have volatilized at higher temperatures, leaving bare glass slides. Fig. 2 contains a photograph and optical microscopy of a bond produced after 20 hours at 150ºC in a 200 ppm O_2 atmosphere with a mixture of DES + HCl + 70%Al_2O_3/30% SiO_2. The porous appearance can be attributed to entrapped gases, likely due to vaporization of water or DES during reaction. Despite gas entrapment, the mixture appears to be miscible, forming structures similar to those published for ionic liquid-metal oxide microstructures [7]. Heat treated bonds were initially stable, but the hygroscopic nature of choline chloride remained, causing bond failure by absorption of water. Failure time varied as a function of heating parameters.

TABLE I. SOLUBILITY OF METAL OXIDES IN DES SOLUTION.

Sample	Composition Factor			
	0.65	0.75	0.85	1.0
SiO_2	4	4	4	2
Al_2O_3	9	8	9	7
CuO	196	2384	5702	8361
Cu_2O	4175	4538	9785	7135

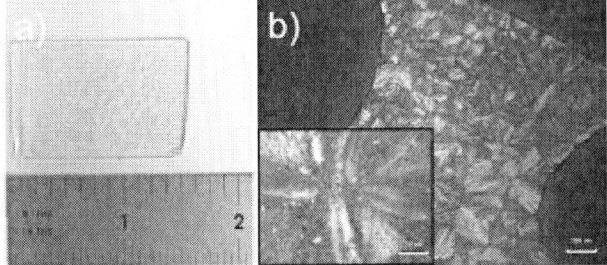

Fig. 2. Macrophoto (a) and optical images (b) of bond layer produced at 150ºC, 200 ppm O_2, 20 hours.

To confirm remaining constituents after boiling off of water, powder XRD was performed (Fig. 3) and confirmed that only choline chloride remained, with some unidentified peaks. None of these peaks corresponded to alumina or silica structures. Future work will focus on complete volatilization or transformation of choline chloride for a nonabsorbent bond. The concentration and volume of HCl added also affects the behavior of the DES and bond formation. DSC of a 0.65 CF DES with different volumes/concentrations of HCl is shown in Fig. 4, reflecting the freezing and melting behavior of the DES. Mixes 1 and 3 had ¼ volume of HCl added per the volume of DES, and mix 2 had equal volume HCl added per the volume of DES. DESs with equal volumes of 0.1 M HCl produced the most transparent macrostructures, strongest bonds, and had the least diffuse phase transformations. Higher concentrations and smaller volumes of HCl did not produce favorable bond behavior.

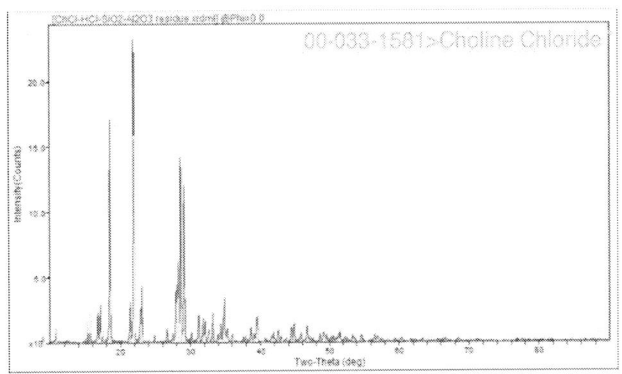

Fig. 3. Powder XRD of remaining DES material after boil off.

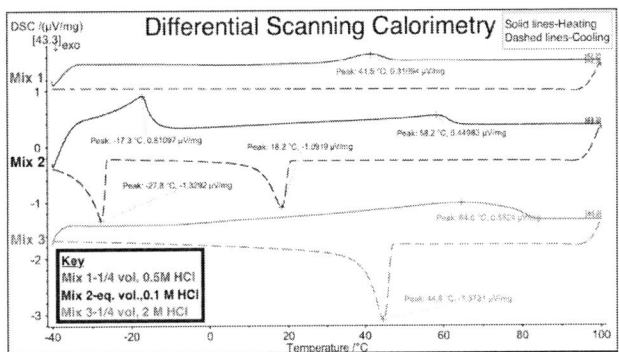

Fig 4. DSC of DES + HCl mixes 1,2 and 3.

IV. SUMMARY AND FUTURE WORK

In summary, the nature of the DESs were characterized to better understand the physical and chemical properties of the bond layer. Solubility of metal oxides that form optically desirable bonds in the DESs is limited, but solubility limits and bond performance can be increased with the addition of HCl to the solvent. Optical microscopy of the heat treated samples suggested miscibility, but detrimental pores formed during reaction due to entrapped gases. Bond performance was limited by the hygroscopic nature of the choline chloride remaining after heat treatment; failure times are dependent on atmosphere of heat treatment. HCl concentration and volume strongly influence the freezing and melting behavior of the DESs during heat treatment. Optimized bond behavior and structure were achieved with equal volumes of 0.1 M HCl.

Future work includes solubility measurements of metal oxides dissolved in the DESs with added HCl, additional heat treatments of the DESs with different process parameters, and thin film XRD of the bonds to detect crystallinity. Transmission as a function of atmosphere and temperature will be acquired. Thermogravimetric analysis of the DESs will be performed to understand the decomposition behavior, and alternate methods will be explored to fully volatilize the choline chloride from solution to maintain a stable bond.

Understanding the chemistry behind bond formation will aid in replication of the thermally matched, high strength, and versatile spinel-like bond, enabling its use as an alternative to silicone solar cell integration.

ACKNOWLEDGEMENTS

The authors would like to acknowledge the assistance of the Advanced Materials Laboratory and their staff with laboratory services. The authors would also like to acknowledge William Fahrenholtz and David Wilt for their support.

REFERENCES

[1] A. Abbott, D. Boothby, G. Capper, D. Davies, and R. Rasheed, Deep eutectic solvents formed between choline chloride and carboxylic acids: Versatile alternatives to ionic liquids, J. Am. Chem. Soc. 126, 9142-9147 (2004).

[2] E. Smith, A. Abbott, and K. Ryder, Deep Eutectic solvents (DESs) and their application, Chem. Rev. 114, 11060-11082 (2014).

[3] A. Abbott, G. Capper, D. Davies, R. Rasheed, and P. Shikotra, Selective extraction of metals from mixed oxide matrixes using choline-based ionic liquids, Inorg. Chem. 44, 6497-6499 (2005).

[4] A. Abbott, G. Capper, D. Davies, R. Rasheed, and V. Tambyrajah, Novel solvent properties of choline chloride/urea mixtures, Chem. Commun, 70-71 (2003).

[5] A. Abbott, G. Capper, D. Davies, K. McKenzie, and S> Obi, Solubility of metal oxides in deep eutectic solvents based on choline chloride, J. Chem. Eng. Data 51, 1280-1282 (2006).

[6] B. Carpenter, Adhesiveless Optical Bonding for Space Solar Cells, Presented at Space Power Workshop 2012, Manhattan Beach, April 16-19, 2012.

[7] J. Luczak, M. Paszkiewicz, A. Krukowska, A. Malankowska, and A. Zaleska-Medynska, Ionic liquids for nano- and microstructure preparation. Part 2: Application in Synthesis, Adv. Coll. Inter. Sci. 227, 1-52 (2016).

NIEL DOSE Analysis on Triple Junction Cells 30% Efficient and Related Single Junctions

Roberta Campesato [1], Erminio Greco[1], , Mariacristina Casale[1], Massimo Gervasi [2,3], P.G. Rancoita[2], Davide Rozza[2], Mauro Tacconi[2,3], Enos Gombia [4], Aldo Kingma[4], Carsten Baur[5]

[1]CESI, 20134 Milan, via Rubattino 54 , Italy, [2] INFN Milano-Bicocca, Italy, [3]and Univ. Milano Bicocca, Italy, [4] CNR, Bologna, Italy,[5] ESA/ ESTEC, Keplerlaan 1, 2201 AZ Noordwijk, The Netherlands

Abstract — **Space solar cells radiation hardness is of fundamental importance in view of the future missions towards harsh radiation environment (like the Jupiter missions) and for the new spacecraft using Electrical Propulsion. In this paper we report the radiation data for triple junction (TJ) solar cells and related component cells. Triple junction solar cell, InGaP top cell and GaAs middle cell degrade after electron radiation as expected. With proton irradiation, it was observed a high spread in the remaining factors, especially for the TJ and bottom cells. Very surprising was the germanium bottom junction that showed very high degradation after protons whereas it is quite stable against electrons. Radiation results have been analyzed by means of the Displacement Damage Dose method.**

Index Terms— **Photovoltaic cells, III-V semiconductor materials**

I. Introduction

In the last 10 years. Spacecraft are mainly powered by Triple junction solar cells based on III-V compound semiconductors. These solar cells are characterized by 30% conversion efficiency and a very high radiation hardness that allow to extend mission lifetime and to use electric propulsion.

The radiation analysis of solar cells is very important to predict the End Of Life (EOL) performances of the solar arrays.

The test of the solar cell radiation hardness is conducted on Earth by irradiating the solar cells using protons and electrons at different energies.

Of course, it is not possible to cover the full spectrum of charge particles energies in space, therefore the experiments on Earth are generally limited to a few energies for electrons and protons.

The evaluation of the radiation hardness of the solar cells is performed by means of two methods:
- The Equivalent Fluence method from JPL[1]
- The Displacement Damage Dose (DDD) from NRL [2].

The first method is the first implemented and used by the space actors; this method uses empirically determined relative damage coefficients (RDCs) and has the advantage of being immediately understandable but, to generate a good estimation, several particles energies shall be used (at least 8 for protons and 3 for electrons).

The DDD method uses calculated values of nonionizing energy loss (NIEL); the main advantage of this method is that it requires only 1 energy for proton and 2 energies for electrons in order to predict the EOL behavior of solar cells.

In this paper, we will present the results of electron and proton irradiation on triple junction solar cells and related component cells manufactured by CESI.

The analysis of the radiation hardness is conducted by the DDD method using an innovative approach for the NIEL calculation.

II. Triple Junction Solar Cells and Component Cells

InGaP/InGaAs/Ge triple junction solar cells and related component cells with a size of 4 cm2 and AM0 efficiency class 30% (CTJ30), have been manufactured and qualified following the ESA ECSS E ST20-08C standard.

These solar cells have been developed on a large MOCVD epitaxial reactor, VEECO E450G, suitable to simultaneously process up to thirteen 4-inch wafers per run.

The improvement in conversion efficiency was obtained by introducing Quantum Structures, mainly Bragg Reflectors, inside the solar cell stack and by a fine tuning of the electrical field inside the solar cell active regions [3].

These solar cells have standard thickness of 140 um and have been used on several satellites since 2013.

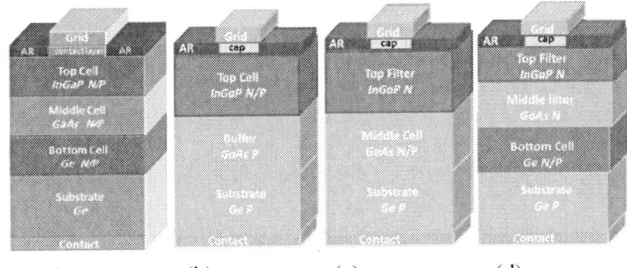

(a) (b) (c) (d)

Fig. 1. Scheme of a TJ cell (a), top cell (b), middle cell (c) and bottom cell (d).

The basic structure of the solar cells is reported in figure 1. The triple junction solar cell is composed by a germanium bottom junction obtained by diffusion into the germanium P-type substrate, a Middle junction of (In)GaAs, whose energy gap is around 1.38 eV and a top junction of InGaP with energy

978-1-5090-5606-4/17 $31.00 © 2017 IEEE

gap 1.85 eV. Component cells are single-junction cells which shall be an electrical and optical representation of the subcells inside the triple-junction cell. Therefore, to manufacture them, special attention was put to reproduce the optical thicknesses of all the upper layers present in the TJ structure. For example, for the Middle component cell, an n doped InGaP layer with the same thickness as the top cell was added to absorb the blue portion of the spectrum.

III. EXPERIMENTAL IRRADIATION RESULTS

TJ solar cells and component cells have been irradiated with protons and electrons at different fluences. Figure 2 shows the remaining power factors for TJ solar cells obtained with self annealing (solar cells kept in dry box for 1 month before measurement EOL) and after annealing (8 hours in AM0, 60°C).

The most important observation is related to the high spread of results in remaining factors obtained when solar cells are irradiated with low energy protons (0.7 MeV).

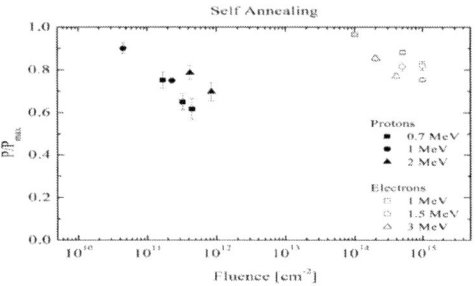

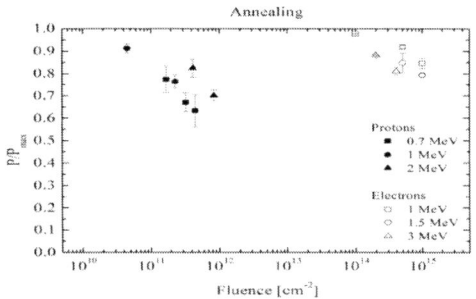

Fig. 2. Remaining power factors of TJ solar cells.

When component cells EOL behavior was analyzed, it turned out that the weakest and more unstable junction is the bottom one.

Germanium is high resistant against electrons [4] but it shows a bad radiation resistance against proton irradiation, especially if low energy protons are concerned.

Table 1 shows the main results for TJ and component cells after proton irradiation at 0.7 MeV, fluence 4.5e11 p^+/cm^2.

The irradiation was performed by CSNSM on 3 samples for each solar cell type whose area was 4 cm^2.

The bottom junction is highly degraded after proton irradiation whereas it is highly radiation resistant when irradiated with electrons (table 2). At the dose of 4.5e11 p/cm^2 with 0.7 MeV protons, the bottom junction has a shunted I-V curve, for this reason the short circuit current of TJ is limited by the middle junction but probably the bad behavior of the germanium junction could explain the low fill factor observed in TJ solar cells.

After annealing, the Voc of Top and Mid cells increases thus improving the Voc of the TJ as expected. The bottom junction seems to recover a portion of the short circuit current (+10% after annealing) but the shunted I-V curve is still present.

TABLE I

ELECTRICAL PERFORMANCES OF TJ, TOP, MID, BOTTOM (4 CM2 AREA) BEFORE AND AFTER IRRADIATION WITH 0.7 MEV PROTONS, 4.5E11 P/CM^2

	Isc [A]	Voc [V]	Pmax [W]	F.F.	Eff. [%]
BOL MEASUREMENT					
TJ	0.070	2.608	0.152	0.84	28.3
Top	0.071	1.365	0.082	0.85	15.03
Mid	0.077	1.012	0.065	0.84	11.91
Bot	0.105	0.220	0.010	0.46	2.01
Self annealing					
TJ	0.056	2.207	0.094	0.75	17.12
Top	0.069	1.197	0.065	0.78	11.91
Mid	0.057	0.806	0.034	0.75	6.24
Bot	0.039	0.158	0.002	0.35	0.39
After annealing					
TJ	0.057	2.232	0.097	0.77	17.70
Top	0.069	1.219	0.067	0.80	12.31
Mid	0.057	0.815	0.035	0.75	6.42
Bot	0.045	0.166	0.003	0.35	0.48

TABLE II

REMAINING FACTORS OF BOTTOM CELLS (4 CM^2 AREA) IRRADIATED WITH 1 MEV ELECTRONS

1 MeV e- fluence	RF Isc	RF Voc	RF Pm
1.00E+14	0.94	0.99	0.92
5.00E+14	0.90	0.96	0.84
1.00E+15	0.88	0.96	0.82

Smaller diodes of 1 mm^2 with the same epitaxial structure of Middle and top component cells, have been irradiated in order to measure the DLTS spectra on them.

To irradiate the samples inside the proton and electron accelerator chambers, the 1 mm² diodes were mounted on holder that allow to measure I-V curves and C-V curves versus frequency before and after irradiation (figure 3).

Fig. 3. Holder for 1 mm² diodes for DLTS and irradiation analysis

To avoid edge current due to cutting defect, the diodes are separated through mesa etching technique.
In this way it was possible to obtain good I-V curves from middle and top junction as reported in figure 4.

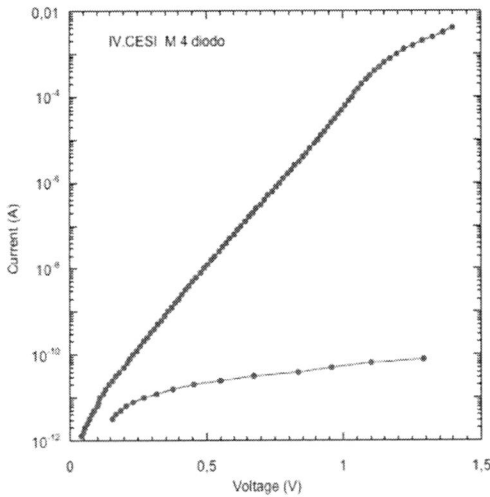

Figure 4. I-V curve of a 1 mm² diode having the middle junction structure

IV. NIEL ANALYSIS

In this section the photovoltaic parameters of the 3J solar cell, and single junction cells are investigated as a function of displacement damage dose (DDD) which is the product of the particle fluence Φ and the so-called NIEL (non-ionizing energy loss) $dE_{de}/d\chi$:

$$D^{NIEL}(E_d) = \Phi \frac{dE_{de}}{d\chi}, \quad (1)$$

With

$$\frac{dE_{de}}{d\chi} = \frac{N}{A} \int_{E_d}^{E_R^{max}} E_R L(E_R) \frac{d\sigma(E,E_R)}{dE_R} dE_R, \quad (2)$$

where N is the Avogadro constant; A is the atomic weight of the medium; E is the kinetic energy of the incoming particle; ER and E_{Rmax} are the recoil kinetic energy and the maximum energy transferred to the recoil nucleus respectively; Ed the displacement threshold energy; $L(E_R)$ is the Lindhard partition function; $d\sigma(E,E_R)/dE_R$ is the differential cross section for elastic Coulomb scattering for electrons or protons on nuclei. By inspection of equation (2) one can remark that DDD(Ed) depends on the displacement threshold energy Ed.

Furthermore, one can note that, for electrons, there is no relevant kinetic energy variation along the path inside the absorber (i.e. the 3J solar cell). On the contrary, such a change occurs for protons. In fact, their actual energy, in each junction, could be estimated by means of SRIM [5]. Thus, the doses were computed for the corresponding proton kinetic energies. For example a proton of 0.7 MeV loses about 30% of its initial before reaching the center of middle junction, and 40% before the center of bottom junction.

In the current study, the displacement damage doses for the 3J solar cells (see figure 5a) are those evaluated for the middle GaAs cell.

In addition, the relative degradation of Pmax, Isc, and Voc obtained after irradiation for the bottom cell exhibit an expected sudden drop. This was already observed in [2]. Therefore, the three sets (Pmax/Pmax(0), Isc/Isc(0), and Voc/Voc(0)) of experimental data were interpolated using the expression:

$$(1 - C_1) - C \cdot log_{10} \left[1 + \frac{D^{NIEL}(E_d)}{D_x} \right], \quad (3)$$

where C_1, C and D_x are obtained by a fit to the data in which the NIEL threshold energy, E_d, was moved to minimize the differences among electrons and protons data with respect the corresponding curve obtained by means of eq. (3).

It has also to be noted that C_1 is only relevant for the bottom cell, while it is negligible in the cases of 3J and GaInP, GaAs single cells. The optimal fit for the 3J cell was obtained using Ed $\approx$ 24 eV for Ga and As, while we obtained Ed $\approx$ 40 eV for Ge in the bottom junction (see figure 5b).

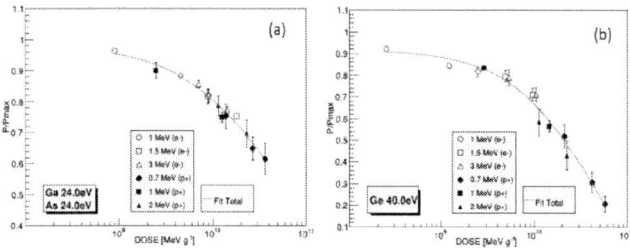

Fig. 5. (a) Optimal fit of Pmax (EOL)/Pmax(BOL) as function of the dose for the 3J solar cell; (b) Optimal fit of P/Pmax as function of the dose for single junction bottom cell.

IV. CONCLUSIONS AND FUTURE WORK

Triple junction InGaP/GaAs/Ge solar cells and related component cells have been irradiated with protons and electrons at different energies.

The data have been analyzed using the DDD methods by the evaluation of the NIEL. Very peculiar results were obtained for the bottom junction that showed high degradation after proton irradiation.

The analysis of the bottom junction by DDD requires a different computational expression.

Top cell and middle cell are going to be analyzed by DLTS to determine the different defects introduction after electrons and protons radiation.

REFERENCES

[1] Anspaugh "GaAs Solar Cell Radiation Handbook" NASA 1996
[2] S.R. Messengers, et al.: "Modeling solar cell degradation in space: A comparison of the NRL displacement damage dose and the JPL equivalent fluence approaches" PIP 2001, 10.1002/pip.357View
[3] G. Gori, R. Campesato : Photovoltaic Cell Having a high Conversion Efficiency PCT I09111-WO (2009)
[4] C. Baur et al.,"Investigation of Ge component cells", Photovoltaic Specialists Conference, Conference Record of the Thirty-first IEEE, 2005
[5] M.J. Boschini, P.G. Rancoita and M. Tacconi (2014), SR-NIEL Calculator: Screened Relativistic (SR) Treatment for Calculating the Displacement Damage and Nuclear Stopping Powers for Electrons, Protons, Light- and Heavy- Ions in Materials (version 3.5.4); [Online] available at INFN sez. Milano-Bicocca, Italy [2017, January]: http://www.sr-niel.org/.

Thin and Flexible Triple Junction Cells 30% Efficient: Qualification Results and Future Space Applications

Roberta Campesato [1], Mariacristina Casale[1], Giuseppe Gabetta[1], Emilio Fernandez Lisbona[2], Laurent D'Abrigeon[3]

[1]CESI, 20134 Milan, via Rubattino 54 , Italy, [2]ESA/ ESTEC, 2200 AG Noordwijk, Keplerlaan 1, The Netherlands, [3] Thales Alenia Space, 5 Allée des Gabians, 06156 Cannes la Bocca, France

Abstract — **InGaP/InGaAs/Ge triple junction solar cells with AM0 efficiency class 30% are the baseline for all the space missions since 2010. Next spacecraft generations demand low cost power ($/W), more specific power (W/g) and bendable/flexible solar arrays. For these reasons, new solar cells, based on multi junction III-V architecture characterized by a reduced thickness to make them bendable and flexible, have been developed and qualified. The main advantage of the developed technology lies in the possibility to strongly decrease the thickness of the solar cell without altering the high efficiency, their space reliability and the manufacturing yield and cost. In this paper we will present the qualification results of triple junction solar cells named CTJ30-80, characterized by a reduced thickness of 80 um.**

Index Terms— **Photovoltaic cells, III-V semiconductor materials**

I. INTRODUCTION

InGaP/InGaAs/Ge triple junction solar cells with a size of 26.5 cm2 and AM0 efficiency class 30% (CTJ30), have been manufactured and qualified at bare solar cell level in accordance with ESA ECSS E ST20-08C Rev. 1 standard and in the frame of ESA GSTP 6.2 contract 4000114125/15/NL/CBi/fk.

The improvement in conversion efficiency was obtained by introducing a Bragg Reflectors inside the solar cell stack and by a fine tuning of the electrical field inside the solar cell active regions [1]. These solar cells have standard thickness of 140 um and have been used on several satellites since 2013.

Next spacecraft generations demand more specific power (W/g) and bendable/flexible solar arrays. For these reasons, new solar cell approaches based on multi junction III-V solar cells and thin substrates (80 micron) have been developed. The main advantage of these technologies lies in the possibility to strongly decrease the mass with beneficial effects also on the cost of the III-V solar cells for space applications.

In this paper we describe the results of the qualification of CESI CTJ30 triple-junction space solar cells with a thickness of 80 um (CTJ30-80). Efficiency of 29% and bending radius of 3 cm have been proved on these devices that reach a specific power of about 1W/gr.

Their applicability to future flexible solar arrays will also be discussed.

II. THIN TRIPLE JUNCTION SOLAR CELLS STRUCTURE AND MANUFACTURING

Triple-junction space solar cells with a thickness of 80 um (CTJ30-80). were manufactured by CESI with an InGaP/InGaAs/Ge epitaxial structure entirely proprietary. The epitaxial structure was grown using the Veeco E450G MOCVD industrial reactor equipped with carriers that can simultaneously hold up to 13 Ge substrates per run. The solar cell structures were grown on 100-mm diameter and 140 μm-thick Ge substrates. After the epitaxial growth the Ge substrate was thinned down to 80 μm using a suitable wet etch process. The selected conditions allowed to obtain thin substrates with good thickness uniformity, without degradation of the electrical performances with respect to the standard 140μm thick solar cells. Two solar cells, with 26.5 cm^2 size and typical 1.3g mass were manufactured on each wafer. The front and rear contacts were made by evaporated silver 5 μm-thick, capped with 200nm Gold. The broad band AR coating was an optimized dual-layer Ta_2O_5/SiO_2.

About 500 CTJ30_80 solar cells (Fig. 1) were manufactured with an average efficiency of about 29%. The electrical performances were measured with respect to Secondary Working Standard sets that were calibrated by INTA-Spasolab. A batch of 125 cells was selected at random for the qualification campaign at bare solar cell level.

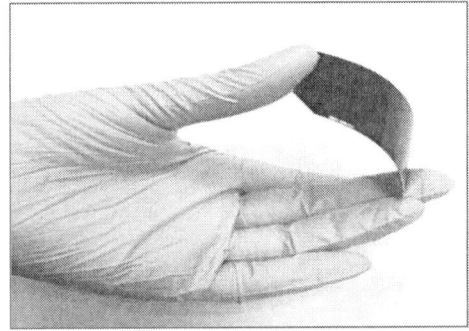

Figure 1. Picture of a CTJ30-80 solar cell in bent configuration

III. QUALIFICATION TEST PLAN

The thin GaInP/GaInAs/Ge triple junction bare solar cells (CTJ30-80) have been submitted to a qualification test campaign, according to the ESA generic specification ECSS-E-ST-20-08C Rev.1. The test campaign was designed to cover the application in space missions both in the Low Earth Orbit (LEO) and in the Geostationary Orbit (GEO).

The qualification tests were organised in the five test sequences of table 1:

TABLE I

ECSS TEST SEQUENCES APPLIED TO THE CTJ30-80 BARE SOLAR CELLS

Test	Subgroup
Contact Adherence	A
BOL performances	B
Electron Irradiation	C1
Extended storage simulation	O
Proton Irradiation	P

In addition to the qualification at bare cell level, some preliminary tests were carried out at solar cell assembly (SCA) level and in bent configuration to provide a preliminary performance assessment under operating conditions in terms of weldability and reliability to the thermal cycles during operation.

In this framework has emerged the demand of the space industry for a more detailed characterization of the flexible aspects of the bare solar cells in view of their application on flexible arrays. The discussions on this point focused on the following set of non-conventional tests on which an engineering phase is under evaluation:

Ultimate bending: a destructive test in which the solar cells get wrapped around cylinders with decreasing diameter, to identify the bending condition in which the cell breakage occurs.

Cells shrinking/ Cell-on-blanket test: whether a solar cell is coverglassed and/or glued to a flexible blanket the whole system adapts differently to the temperature changes. The lower inertia of the flexible/blanket solar arrays may lead to operating temperatures lower than -200°C. The implementation of a suitable test will be discussed .

Long term bending: If solar cells are held in a bent configuration for long time they may modify their electrical and mechanical properties (memory effects). The preliminary results on a few solar cells kept in bent configuration will be discussed.

The outcome of the testing activity on such nonconventional tests may constitute the background to a proposal for the addition of a specific test sequence in a future revision of the ECSS generic specification.

IV. QUALIFICATION RESULTS

The solar cells were successfully submitted to the test campaign of table 1 and the main results are reported herein (Table2).

TABLE II

ELECTRICAL PERFORMANCES OF CTJ30-80 SOLAR CELLS

CTJ30-80 bare	Area (cm2)	Isc (mA)	Voc (V)	FF	Eff (%)
Best BOL	26.5	473	2.61	0.85	29.0
Typical Avg	26.5	463	2.64	0.85	28.6
EOL (1MeV, 5e14 e/cm2)	26.5	445	2.51	0.84	26.0

Even with the reduced thickness of 80 um, CTJ30-80 solar cells were successfully welded by Parallel gap resistance welding without degradation in electrical performances. Pull tests were performed on both the front and rear contact. The pull force was applied at 0° angle: no metal delamination was observed and pull forces, higher than 1500g were consistently obtained at interconnect level on both front and rear sides.

Spectral responses have been performed BOL on samples of subgroup C1 and P. The results are reported in figure 2. With respect to standard CTJ30, EQE (External Quantum Efficiency) are similar demonstrating that the thinning process does not affect solar cell performances.

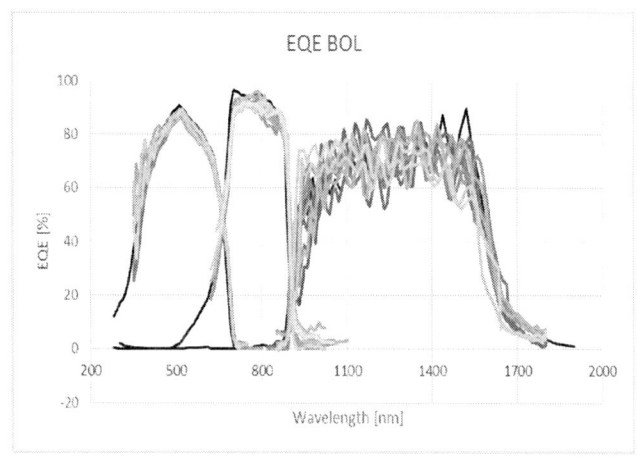

Figure 2. External Quantum efficiency of CTJ30-80 solar cells BOL

The CTJ30-80 solar cells have been irradiated with protons and electrons at different fluencies and energies. Tables 3 and 4 show the remaining factors for CTJ30-80 solar cells obtained after photon and thermal annealing.

Radiation resistance of thin solar cells is in line with the results obtained on CTJ30 solar cells [2] with standard thickness with just 1% less in Remaining Power Factors that is inside experimental accuracy.

TABLE III
RADIATION PERFORMANCE AT 1 MeV ELECTRON IRRADIATION EOL/BOL RATIO.

Dose e-/cm2	RF Isc	RF Voc	RF Pm
$5*10^{13}$	0.99	0.98	0.97
$1*10^{14}$	0.99	0.97	0.96
$5*10^{14}$	0.98	0.95	0.91
$1*10^{15}$	0.93	0.93	0.84
$3*10^{15}$	0.85	0.90	0.72

TABLE IV
RADIATION PERFORMANCE AT 1 MeV AND 2 MeV PROTON IRRADIATION EOL/BOL RATIO.

Dose	RF Isc	RF Voc	RF Pm
1 MeV proton			
1 E10 p+/cm^2	1.00	0.98	0.97
2 E10 p+/cm^2	0.99	0.96	0.95
5 E10 p+/cm^2	0.99	0.95	0.90
1 E11 p+/cm^2	0.97	0.92	0.85
3 E11 p+/cm^2	0.89	0.88	0.73
Dose	**R Isc**	**R Voc**	**RPm**
2 MeV proton			
1.5E11 p+/cm^2	0.98	0.94	0.88

Thin solar cells, solar cell assemblies (using standard coverglass and flexible coverglass from QOptic) and the first coupon (figure 3) were submitted to thermal cycling to test the mechanical robustness of the CTJ30-80 and their behaviour against thermal stress when mounted in bent configuration.

Figure 3. Picture of coupon with CTJ30-80 in bent configuration submitted to thermal cycling test

In the Thermal Cycling (CY) test the following samples have been tested:
- 20+3 CTJ30-80 bare solar cells (for qualification at bare level);
- 1 CTJ30-80-SCA (solar cell assembly) using standard CMG100 coverglass ;

- 4 CTJ30-80-SCA (solar cell assembly) using flexible QOptic coverglass;
- 5 CTJ30-80 in bent configuration on a 10 cm diameter cylinder (figure 3).

Thermal cycling consisted of 6000 cycles in the temperature range -130°C/+150°C in inert atmosphere.

Solar cells were measured at AM0, 25°C before and after thermal cycling.

The variation in electrical performances before and after thermal cycling is inside measurement errors as reported in table 5 for bare and SCA samples.

TABLE V
ELECTRICAL PERFORMANCES VARIATION BEFORE AND AFTER THERMAL CYCLING ON THIN SOLAR CELLS

	DIsc (%)	DVoc (%)	DPm (%)
Average	-0.4%	-0.1%	-0.4%
Max	0.3%	0.8%	1.4%
Min	-0.9%	-0.3%	-1.0%

After thermal cycling, the ultimate bending radius was measured. Generally an increase from 3 cm up to 5 cm was observed on CTJ30-80 solar cell bending radius after thermal cycling.

Solar cells mounted in bent configuration survived the thermal cycling, no visual defects have been detected after this test a part of one samples that resulted cracked after shipment. On these samples it was not possible to measure the complete I-V curve, only Isc and Voc were measured, results are reported in table 6.

TABLE VI
ISC AND VOC BEFORE AND AFTER THERMAL CYCLING IN BENT CONFIGURATION

Before thermal cycling		After thermal cycling	
Isc(mA)	Voc(V)	Isc (mA)	Voc(V)
438	2.61	440	2.57

A degradation of -1.5% was detected on Voc, whereas the Isc is stable.

IV. INNOVATIVE FLEXIBLE SOLAR ARRAYS

The next path identified toward very high power range of 35KW and above with limited mass increase and still improved competitiveness is the use of flexible arrays. Those flexible arrays are using PVA laid down on flexible dielectric sheet and rolled for launch inside the minimum volume and so with the minimum bending radius. Thin cells use will help to save mass and volume. On the other hand, their thickness shall be optimized versus next coming harsh radiative environment with electrical orbit raising strategies. The minimum remaining cell wafer and associated coverglassing should lead to optimum W/kg ratio at solar array level with about 160µm SCA, with potential criticality versus bending capability.

The bending capability of the bare cell and then of the SCA becomes mandatory characterization to secure rolling and then vibrations. Introduction of specific test sequence for the bending ability is considered as mandatory inside the TAT campaign.

Historically, this kind of tests are already performed by Thales Alenia Space since the introduction of thin silicon cell of 75µm to check their capability to withstand local bending under global panel distortion (to secure cell strength before acoustic test at panel level), but also to check their robustness with regard to very local imposed distortion on the panel sandwich.

For flexible solar array the test sequence will be based on the lessons learnt of those initial tests and will be specified to get an as much as possible extensive validation of the different feared events with:

- Static bending tests with digressive radius. Those tests will have to be performed on the used direction since difference between both directions are already observed. Objective is to go down to ultimate radius for electrical health and then down to mechanical breakage.
- Surrounding cell inspection (before and after each environmental or load case) including electrical check and ELM.
- Fatigue tests under bent conditions to check that no long term creeping are observed on cell, adhesive or coverglass.
- Combination of worst case stress with thermal cycling under bent conditions to combine bending stresses with thermo-elastic stresses.

All those tests will be defined with possible sequences in the same philosophy as ECSS-E-ST-20-08.

On CTJ30-80 solar cells the following tests were performed to verify their specifically robustness if applied to flexible arrays:

- Ultimate bending: the minimum radius before solar cells break is 3 cm for BOL solar cells and 5 cm for EOL samples (figure 4);
- Thermal shock test: bare CTJ30-80 and SCA using CMG100 coverglass were cycled between -193°C/ +150°C (2 minutes for cycle, total 100 cycles). No degradation in visual inspection and electrical performances were noticed (table 7);
- Long duration bending: solar cells are wrapped around a cylinder 10 mm radius since January 2017. Every month solar cells are unwrapped and the electrical performances are measured. Up to now no degradation is noticeable. Test will be considered finished End of September 2017).

TABLE VII

VARIATION OF SOLAR CELL ELECTRICAL PERFORMANCES AFTER THERMAL SHOCK TEST

	D Isc [%]	D Voc [%]	Delta Pm [%]
	after Thermal Shock		
CTJ30-80	0.3%	-0.1%	0.1%
SCA	0.2%	-0.1%	-0.4%

Figure 4. Bending radius for BOL CTJ30-80 bare solar cells

V. CONCLUSION

Triple junction solar cells with a thin substrates have been fully qualified following ECSS standards at bare level.

Front and back weldability has been proved and good adhesion of metals was obtained.

Irradiation results show good agreement with standard CTJ30 solar cells, only 1% less in remaining power factor was observed.

During qualification, thermal cycling at SCA level and on bent configuration have been done to verify the thermal stress resistance of CTJ30-80.

CTJ30-80 solar cells have been developed for all the solar arrays where lightweight, bendable and flexibility are requested.

To test their applicability to flexible solar arrays, more tests with respect to ECSS are necessary, some of these tests with experimental results confirm their suitability for flexible arrays.

REFERENCES

[1] G. Gori, R. Campesato : Photovoltaic Cell Having a high Conversion Efficiency PCT I09111-WO (2009)

[2] Roberta Campesato, Mariacristina Casale, Gabriele Gori, Giuseppe Gabetta: " Electron and Proton Irradiation on High Efficiency III-V Solar Cells Based on Three and Four Junctions", 28th EUPVSEC Paris 2013

Printed Assemblies of Microscale Triple-Junction (3J) Inverted Metamorphic (IMM) GaInP/GaAs/InGaAs Solar Cells

Boju Gai [1], John Geisz [3], Daniel Friedman [3], Jongseung Yoon [1,2,*]

[1]Department of Chemical Engineering and Materials Science, [2]Department of Electrical Engineering, University of Southern California, Los Angeles, California 90089, USA,

[3]National Renewable Energy Laboratory, Golden, Colorado 80401, USA.

Abstract —Transfer printing of microscale devices has been shown to be a manufacturable pathway to low-profile, low cost, high concentration photovoltaic modules, where extremely high efficiency solar cells are desirable to achieve competitive costs. Inverted metamorphic multijunction (IMM) solar cells represent a promising platform for ultrahigh efficiency concentrator photovoltaic systems with a clear pathway to 50% efficiency. While transfer printing of microscale solar cell devices has been demonstrated with traditional upright solar cell structures, IMM structures present additional processing challenges. We demonstrate, for the first time, the transfer printing of fully-functional microscale triple-junction (3J) IMM solar cells.

Index Terms — III-V; Photovoltaics, Inverted metamorphic multijunction (IMM); Transfer printing

I. INTRODUCTION

Inverted metamorphic multijunction (IMM) solar cells [1-5] are an enabling technology for ultrahigh efficiency concentrator photovoltaic systems due to their ability to monolithically integrate materials with different lattice constants into high performance two-terminal devices by incorporating compositionally-graded buffer layers. Substrate removal and optical enhancement [5] of solar cells with a metal back-surface reflector (BSR) are important consequence of processed *inverted* solar cells. Four-junction IMM concentrator cells have demonstrated 45.7% efficiency [2] and six-junction IMM devices are being developed to surpass 50%. In the current IMM PV technology, however, the device fabrication requires cm-scale dies and involves complete removal of the growth wafer by wet chemical etching, thereby substantially deteriorating the cost-effectiveness but also imposing severe restrictions in achievable cell size, spatial layout, and type of module substrate [1, 2]. In this regard, transfer printing has been demonstrated to be a manufacturable fabrication route that can circumvent these limitations for assembling more traditional *upright* solar cells with dimensions smaller than a millimeter (hereafter, microscale) into concentrator photovoltaic (CPV) systems [6-11]. Here we present materials design and fabrication strategies for transfer printing of *inverted* metamorphic multijunction cell fabrication by exploiting specialized etching schemes that results in print-ready microscale devices with a BSR and leave the GaAs substrate intact for possible reuse. Printable forms of GaInP/GaAs/InGaAs triple-junction IMM solar cells were

successfully fabricated and integrated onto a glass substrate with one-sun photovoltaic performance comparable to that in conventional processing methods as described in detail below.

II. RESULTS

Epitaxial materials for IMM 3J solar cells composed of GaInP/GaAs/GaInAs solar cells were grown by metal organic chemical vapor deposition (MOCVD) on a GaAs substrate

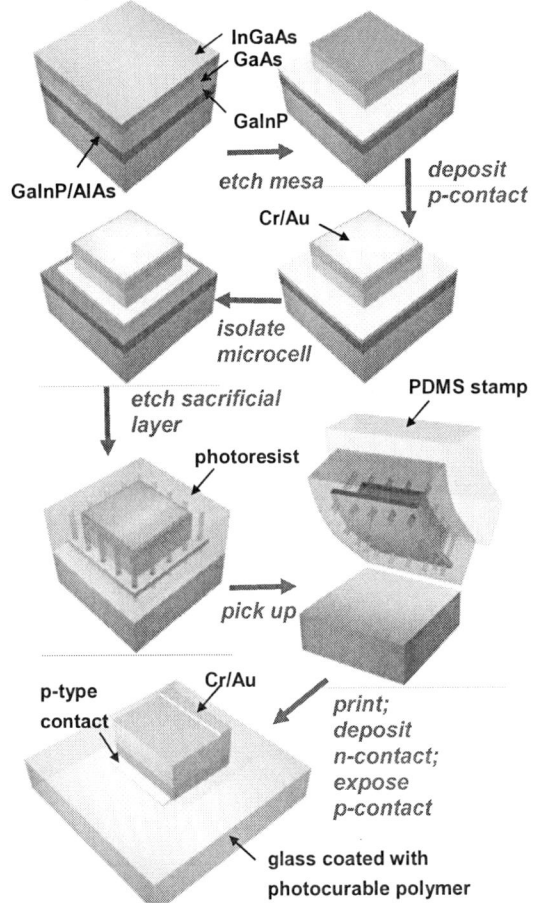

Figure 1. Schematic illustration of processing steps to fabricate printed assemblies of GaInP/GaAs/InGaAs 3J IMM microcells on a glass substrate.

following previously reported methods [3, 4]. A printable form of microscale (~500 x 500 μm²) solar cells (i.e. microcells) were fabricated using procedures illustrated in Figure 1. Corresponding optical images are depicted in Figure 2. The cell processing begins with the formation of mesa (~500 x 500 μm²) structure by photolithography and successive steps of wet chemical etching (Figure 2(a)). A mixture (1:1 by volume) hydrochloric acid (HCl) and deionized (DI) water was used for etching layers of $Ga_xIn_{1-x}P$, while a mixture (1:12:13 by volume) of phosphoric acid (H_3PO_4), DI water, and hydrogen peroxide (H_2O_2) was employed to etch layers of GaAs and $Ga_xIn_{1-x}As$. A second mesa (~580 x 580 μm²) was formed by photolithography and wet chemical etching using the mixture of phosphoric acid to produce isolated arrays of IMM microcells (Figure 2(b)). Subsequently, p-type metal contact (Cr/Au) was deposited by electron beam evaporation (Figure 2(c)). To enable the defect-free release of IMM microcells from the growth wafer, several critically important features were introduced. Etch holes necessary to deliver wet chemical etchants to the sacrificial layer were made on the second mesa

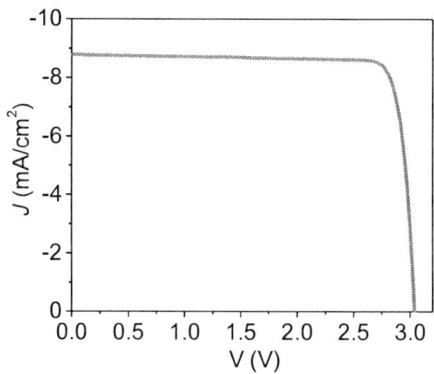

Figure 3. Representative current density (J)-voltage (V) characteristics of transfer-printed 3J IMM solar cells, measured under simulated AM1.5G illumination (1000 W/m²)

front surface. After the printing, the GaAs contact layer was etched except the region where n-type metal contact is deposited, followed by the deposition of metal contact (Cr/Au) and exposure of p-type contact by wet chemical etching.

The photovoltaic performance of printed 3J IMM solar cells were measured under AM1.5G standard solar illumination (1000 W/m²) as depicted in Figure 3. The average open-circuit voltage (V_{oc}), short-circuit current density (J_{sc}), fill factor (FF), and solar-to-electric energy conversion efficiency (η) were 3.039V, 8.92 mA/cm², 0.863, and 23.4%, respectively. Further studies for the measurement under concentrated illumination, assessment of perimeter recombination effects, and refinement of processing schemes to fabricate print-ready, fully-functional cells are currently underway.

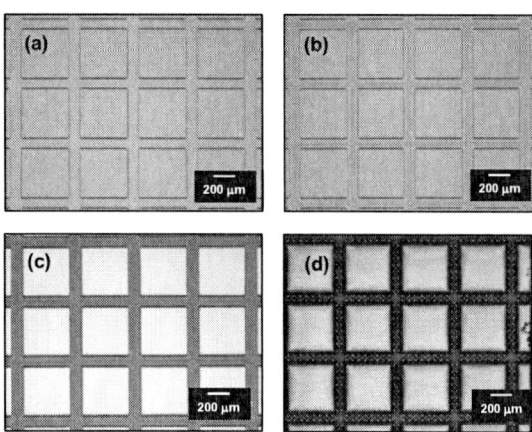

Figure 2. Optical micrographs of arrays of IMM microcells after fabrication steps of (a) mesa etching, (b) isolation, (c) p-contact metal deposition, and (d) undercut etching

outside the cell region to avoid unintentional etching of active junction area and associated surface recombination of photogenerated carriers [11]. As a sacrificial material for epitaxial liftoff, we employed a bilayer of $Ga_{0.51}In_{0.49}P$ and AlAs to protect the overlying active region from etching and contamination during the undercut etching. The GaInP and AlAs were partially etched by concentrated HCl through the etch holes, followed by the etching in the mixture of HCl and DI water (1:1 by volume) (Figure 2(d)). After the completion of undercut etching, arrays of IMM microcells were picked up by an elastomeric stamp made of poly(dimethyl siloxane) (PDMS) and printed onto a glass substrate using a thin layer of photocurable polymer as an adhesive [11, 12]. To flip over the cell and expose the GaInP cell to the illumination side, we employed a temporary carrier substrate made of PDMS such that the metal-deposited GaInAs side of IMM solar cells is faced down, while the surface of GaInP cell is exposed to the

IV. CONCLUSION

We have developed materials design and fabrication strategies of microscale IMM solar cells to circumvent existing limitations of conventional fabrication methods involving the complete etching of a growth wafer. We have demonstrated, for the first time, the transfer printing of fully-functional microscale three-junction IMM solar cells whose performance is comparable to the devices prepared by conventional fabrication processes.

REFERENCES

[1]   R. M. France, J. F. Geisz, I. Garcia, M. A. Steiner, W. E. McMahon, D. J. Friedman, *et al.*, "Quadruple-Junction Inverted Metamorphic Concentrator Devices," *IEEE Journal of Photovoltaics,* vol. 5, pp. 432-437, Jan 2015.

[2] R. M. France, J. F. Geisz, I. Garcia, M. A. Steiner, W. E. McMahon, D. J. Friedman, *et al.*, "Design Flexibility of Ultra-High Efficiency Four-Junction Inverted Metamorphic Solar Cells," *IEEE Journal of Photovoltaics,* vol. 6, pp. 578-583, 2016.

[3] J. F. Geisz, D. J. Friedman, J. S. Ward, A. Duda, W. J. Olavarria, T. E. Moriarty, *et al.*, "40.8% efficient inverted triple-junction solar cell with two independently metamorphic junctions," *Applied Physics Letters,* vol. 93, Sep 2008.

[4] J. F. Geisz, S. R. Kurtz, M. W. Wanlass, J. S. Ward, A. Duda, D. J. Friedman, *et al.*, "High-efficiency GaInP/GaAs/InGaAs triple-junction solar cells grown inverted with a metamorphic bottom junction," *Appl. Phys. Lett.,* vol. 91, p. 023502, 2007.

[5] M. A. Steiner, J. F. Geisz, I. Garcia, D. J. Friedman, A. Duda, and S. R. Kurtz, "Optical enhancement of the open-circuit voltage in high quality GaAs solar cells," *Journal of Applied Physics,* vol. 113, p. 123109, Mar 28 2013.

[6] X. Sheng, C. A. Bower, S. Bonafede, J. W. Wilson, B. Fisher, M. Meitl, *et al.*, "Printing-based assembly of quadruple-junction four-terminal microscale solar cells and their use in high-efficiency modules," *Nature Materials,* vol. 13, pp. 593-598, 2014.

[7] J. Yoon, A. J. Baca, S. I. Park, P. Elvikis, J. B. Geddes, L. F. Li, *et al.*, "Ultrathin silicon solar microcells for semitransparent, mechanically flexible and microconcentrator module designs," *Nature Materials,* vol. 7, pp. 907-915, 2008.

[8] J. Yoon, S. Jo, I. S. Chun, I. Jung, H. S. Kim, M. Meitl, *et al.*, "GaAs photovoltaics and optoelectronics using releasable multilayer epitaxial assemblies," *Nature,* vol. 465, pp. 329-U80, May 20 2010.

[9] B. Furman, E. Menard, A. Gray, M. Meitl, S. Bonafede, D. Kneeburg, *et al.*, "A HIGH CONCENTRATION PHOTOVOLTAIC MODULE UTILIZING MICRO-TRANSFER PRINTING AND SURFACE MOUNT TECHNOLOGY," *35th IEEE Photovoltaic Specialists Conference,* pp. 475-480, 2010.

[10] J. Justice, C. Bower, M. Meitl, M. B. Mooney, M. A. Gubbins, and B. Corbett, "Wafer-scale integration of group III-V lasers on silicon using transfer printing of epitaxial layers," *Nature Photonics,* vol. 6, pp. 610-614, Sep 2012.

[11] B. Gai, Y. Sun, H. Lim, H. Chen, J. Faucher, M. L. Lee, *et al.*, "Multilayer-Grown Ultrathin Nanostructured GaAs Solar Cells as a Cost-Competitive Materials Platform for III–V Photovoltaics," *ACS Nano,* vol. 11, pp. 992-999, 2017/01/24 2017.

[12] D. S. Kang, S. Arab, S. B. Cronin, X. L. Li, J. A. Rogers, and J. Yoon, "Carbon-doped GaAs single junction solar microcells grown in multilayer epitaxial assemblies," *Applied Physics Letters,* vol. 102, Jun 24 2013.

Comparative Study on Nonradiative Recombination Centers in Proton Irradiated InAs/GaAs Quantum Dot Structure by Two Wavelength Excited Photoluminescence

M. D. Haque[1], N. Kamata[1], S-I. Sato[2], and S. M. Hubbard[3]

[1]Department of Functional Materials Science, Saitama University, Saitama 338-8570, Japan

[2]Quantum Beam Science Research Directorate, National Institutes for Quantum and Radiological Science and Technology (QST), Takasaki 370-1292, Japan

[3]NanoPower Research Labs., Rochester Institute of Technology, Rochester, New York 14623, USA

Abstract — **Comparative study on nonradiative recombination (NRR) centers in InAs/GaAs quantum dot (QD) structure generated by 3 MeV proton irradiation is performed by two wavelength excited photoluminescence (TWEPL). The above-gap excitation (AGE) source of 2.33 eV or 1.26 eV is used to excite GaAs host material or InAs QDs, respectively. The QD PL intensity decreased after irradiation of below-gap excitation (BGE) of 0.75 eV over AGE, indicating a pair of NRR centers activated. The proton irradiation at fluence of $7 \times 10^{11} \text{cm}^{-2}$ reduces the NRR density, while that of $4 \times 10^{12} \text{cm}^{-2}$ increases it. Defect formation, carrier injection and their fluence dependence explain experimental results.**

Index Terms — **Intermediate band, NRR, proton irradiations, Quantum dots, TWEPL.**

I. INTRODUCTION

InAs/GaAs quantum dot (QD) structure has drawn considerable interest due to its band alignment and usage for high-efficiency intermediate band solar cell (IBSC). The bandgap engineering technique of III-V solar cell utilizes the benefits of broadened absorption spectrum of below band gap hetero-structures (InAs QDs) embedded into the host material of solar cell [1]. Highly lattice-mismatched narrow band gap material is deposited on the surface of host material before being capped in Stranski-Krastanov (S-K) growth mode. This allows us to make QDs in which carriers are confined in three dimensions. InAs QDs form an isolated intermediate band (IB) within the bandgap of host material and the photon absorption would take place from the valence band to the IB and from the IB to the conduction band of host material [2]. By using detailed balance approach, it is possible to obtain conversion efficiency over 60% for this kind of IBSC at 100 sun concentration under AM1.5 illumination [3]. The incorporated QDs in the active region of p-i-n solar cells are needed to be homogeneous, regular in pattern, small in size and densely packed so that individual wave functions begin to overlap, thus forming isolated IB [4]. For realizing high-efficiency IBSC, multilayer QDs are necessary to increase QD density for enhancing photon absorption. The advantage of multilayer QD structure is to attain greater active volume than that of single layer QD system [5]. However, the multiple QD layers make strain induced dislocations and result in the device degradation [6]. To resolve this, GaP layers have been introduced in the QD layers for strain balancing among the superlattices [2, 3, 7]. InAs QD based IBSCs are also attractive for space applications because of the superior radiation tolerance in addition to the high conversion efficiency. Since exposure of high energy particles such as electrons and protons introduces significant performance degradation of solar cell for spacecraft [2, 6, 8], the conversion efficiency of spacecraft solar cells at the end of mission (End of Life: EOL) is more important than the efficiency at the beginning of life (BOL). The performance degradation is caused by the increase in radiation induced defects in the active layer of solar cells and the effect of radiation induced defects as well as as-grown defects on the electronic/ optical properties should be studied. The most commonly used techniques for defect characterization are Deep Level Transient Spectroscopy (DLTS) and photoluminescence (PL). It been clarified from the results of DLTS measurements that different defect levels are introduced depending on radiation species (protons or electrons) and the defect density raises with increasing the irradiation fluences [2, 6]. Recently, R. Sreekumar et al., C.Y. Cheng et al. and S. Upadhyay et al. have found that PL efficiency of single and multiple InAs QD layer(s) in GaAs, which have no GaP strain balancing layer, enhances from low to intermediate fluence and quenches at higher fluence [9]-[12]. However, the effects of radiation defects on the performance of QD devices remain to be clarified. In particular, it has not been known how the defect levels affect the IBs.

In this paper, we report on the comprehensive analysis of the radiation defects as well as as-grown defects in InAs/GaAs QD devices by the spectroscopic method of two-wavelength excited photoluminescence (TWEPL) and propose a model for explaining carrier dynamics in detail. It allows us to determine spatial and energy distribution of defect levels which act as nonradiative recombination (NRR) centers through an efficient radiative channel of QD PL. Therefore, information of NRR centers in the QDs, QDs/GaAs interfaces and/or GaAs spacer layers can be obtained in spite of a relatively small volume of QDs in the device. The obtained results are complementary to the previous studies by DLTS.

II. SAMPLE STRUCTURE AND EXPERIMENT

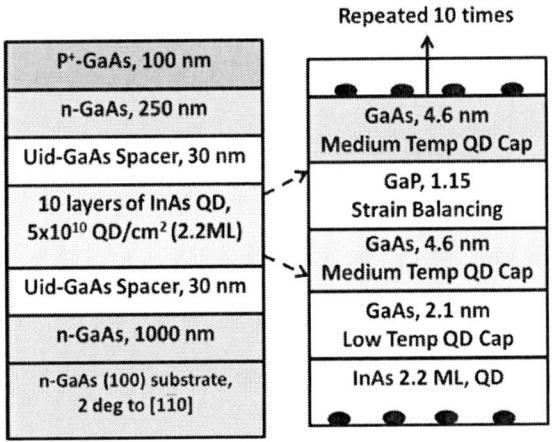

Fig. 1. Sample structure of InAs/GaAs QDs.

Figure 1 shows the schematic diagram of samples. The samples used in this study were GaAs p^+n diodes with embedded 10 periods of InAs QDs. They were grown by Metal Organic Vapor Phase Epitaxy (MOVPE) in Stranki-Krastanov mode on 350 μm thick Si doped GaAs substrate with the misorientation of $2°$ toward the [110] direction. The QD region was capped by unintentionally doped (uid) GaAs with the thickness of 30 nm. A single QD period consists of 2.2 monolayer of InAs followed by 2.1 nm GaAs low temperature cap, 4.6 nm GaAs spacer layer, 1.15 nm GaP strain balancing layer and finally a second 4.6 nm GaAs spacer layer. The GaP strain compensating layer allows multiple layers of high quality QDs to be grown. Each QD period was repeated 10 times. The QD height and diameter were 2 nm and 15 nm, respectively. The samples were irradiated with 3 MeV protons at the fluences of 7×10^{11} and 4×10^{12} protons/cm^2 at the Takasaki Advanced Radiation Research Institute, QST. Unirradiated, 7×10^{11}, and 4×10^{12} protons/cm^2 irradiated samples are identified by the letters of A, B and C, respectively. Details of device growth parameters and QD characteristics are presented elsewhere [6].

The detailed experimental layout is shown in Fig. 2. In the TWEPL, both the AGE ($h\nu_A = 2.33$ eV or 1.26 eV $> E_g$) and the BGE light ($h\nu_B = 0.75$ eV $< E_g$) are combined to excite the sample placed in a temperature controlled cryostat. Here E_g denotes the band-gap of GaAs or InAs QDs. The penetration depth of the 2.33 eV (532 nm wavelength) AGE light is roughly 300 nm, restricting only GaAs layers to be excited. The 1.26 eV (980 nm) AGE light is absorbed only in InAs

QDs, allowing us to define observation volume inside and/ or at the interface of the QDs. The photoluminescence of InAs/GaAs QDs were measured at a temperature of 9.5 K. The output power of laser was constant and neutral density filters were used for changing the excitation density.

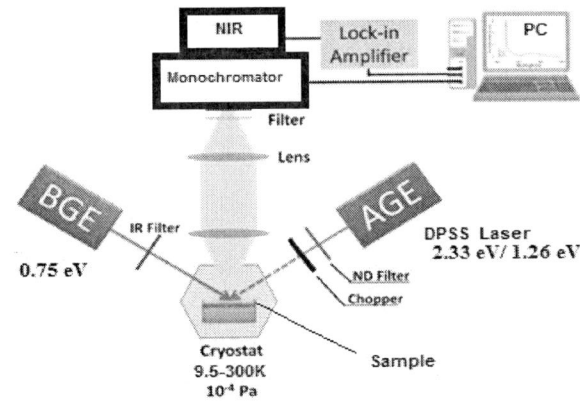

Fig. 2. Experimental set up of the two wavelength excited photoluminescence.

We investigated BGE power dependency by using 0.75 eV (1650 nm) as the BGE. The PL from the sample was fed to a monochromator through objective lenses, converted to photocurrent by NIR photo-multiplier and recorded by a computer after lock-in amplification. Here, an optical chopper was used with the digital lock-in-amplifier for improving S/N ratio of the PL signal. By measuring the QD PL intensity with and without the BGE, $I_{AGE+BGE}$, and I_{AGE}, respectively, the normalized PL intensity $I_N = I_{AGE+BGE} / I_{AGE}$ was determined. The deviation of normalized PL intensity from unity implies the presence of NRR centers [13, 14].

III. RESULTS AND DISCUSSION

Figure 3 shows the conventional QD PL spectra measured at 9.5 K under the AGE ($h\nu_{A1} = 2.33$ eV) for GaAs and the AGE ($h\nu_{A2} = 1.26$ eV) for QDs, respectively. Comparing the PL intensity, it was observed that the QD AGE resulted in higher PL intensity by a factor of 4 to that of GaAs AGE at the same excitation density of 19.85 mW/mm^2. This can be attributed to the fact that under GaAs AGE, the electrons are excited from valence band to the GaAs conduction band from which the electrons tunnel into QDs by two mechanisms: the direct capture of electrons from conduction band of GaAs barrier to defect levels and electrons in the conduction band of GaAs barrier relax into the QDs by phonon assisted tunneling [15]. The PL intensity at fluence of 7×10^{11} protons/cm^2 (sample B) was maximum compared to the other samples. It has been reported that the PL efficiency enhances from low to

978-1-5090-5606-4/17 $31.00 © 2017 IEEE

intermediate irradiation fluence due to the reduction of phonon bottleneck effect by defect assisted phonon emission [10, 16]. In other words, the point defect coupled with the QD whenever a point defect was located near the QD at distance of diameter of QD. This allows rapid energy relaxation through multi-phonon emission and transports electron from upper energy level to lower energy level. Thus the phonon bottleneck effect is reduced and the PL efficiency increases. Furthermore, the shallow electron trap center named PR4" (E_a = 0.30 eV, with the density, $N_T = 1 \times 10^{13}$ cm^{-3}) appears at the fluence of 7×10^{11} cm^{-2} in the GaAs layer or at the interface of InAs/GaAs by arsenic vacancy. PR4" trap centers have lower potential compared to other deep trap centers such as EL2, misfit dislocation and PR1. The photo-generated carriers are trapped to PR4" trap centers with higher probability than the other traps, and participate in direct tunneling from PR4" trap centers into QDs by quantum mechanical process [11]. Therefore, the PR4" trap centers can help carrier capture into QDs, in addition to the reduction of phonon bottleneck effect due to point defect [6, 9, 11, 12]. The PL intensity was the lowest for sample C with the fluence of 4×10^{12} protons/cm^2. The density of PR4" trap centers increased to 4×10^{14} cm^{-3} but the other deep trap center, PR1 (E_a = 0.81 eV, $N_T = 2 \times 10^{14}$ cm^{-3}) was also introduced [6]. In the higher fluence irradiation, complex defects specific to multiple displacement events in addition to point defects in GaAs contribute to the decrease in PL intensity [6, 11, 12, 16]. There are reports on defects in GaAs such as EL2 defects or arsenic antisite (As$_{Ga}$), gallium vacancy (V$_{Ga}$) arsenic interstitial (As$_i$) and associated pair defects that influences conduction mechanism [6, 18, 19].

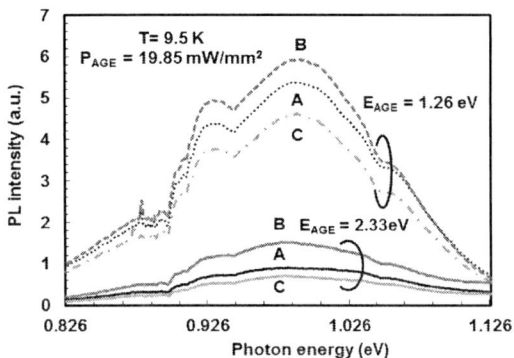

Fig. 3. PL spectra of InAs/GaAs QDs for 1.26eV and 2.33eV AGE. A (black), B (red), and C (green) are the data of unirradiated, 7×10^{11} and 4×10^{12} cm^{-2} irradiated samples, respectively. The same is true in Figs. 4, 6, and 7.

For the QD AGE, the tunneling of optically generated carriers into higher defect density GaAs barrier layer is not possible with selective excitation of the QD region. Since the number of NRR centers in the QDs and wetting layers is smaller than that in GaAs, the carrier trapping might be dominant when carriers moved out from QD levels. Hence, the carrier trapping does not take place from the QD region. P.C. Sercel reported that electrons could move out from the QD into a trap and they could also move back into the QD [16]. Thus, equilibrium is established in the density of electrons inside the QDs. These may cause the enhancement of PL efficiency for QD AGE compared to GaAs AGE excitation. The PL efficiency of unirradiated sample (A), 7×10^{11} protons/cm^2 irradiated sample (B) and 4×10^{12} protons/cm^2 irradiated sample (C) for QD AGE excitation represent same tendency as explained for GaAs AGE excitation.

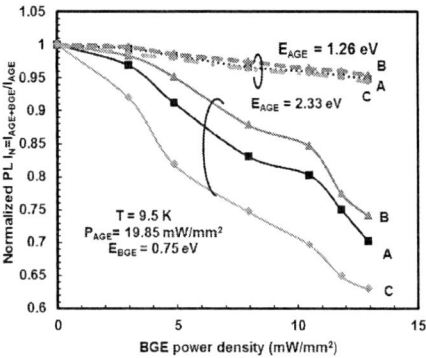

Fig. 4. TWEPL BGE Power dependency of normalized PL intensity for 1.26 eV and 2.33 eV AGE.

We investigated TWEPL BGE power dependency at the same AGE density to determine energy distribution of NRR centers in GaAs and InAs QDs, respectively. The results are shown in Fig. 4. The normalized PL intensity as a function of the BGE density was measured by combination of the AGE (either hν_{A1} = 2.33 eV or hν_{A2} = 1.26 eV) and the BGE (hν_B = 0.75 eV). For the GaAs AGE at the excitation density of 19.85 mW/mm^2, the normalized PL intensity quenches down to 0.70, 0.83 and 0.63 from unity up to BGE power density of 12.92 mW/mm^2 for the samples A, B, and C, respectively. The corresponding values for the QD AGE and BGE power density reduces to 0.949, 0.954 and 0.944 for the samples A, B, and C, respectively. The quenching of the normalized PL intensity from unity implies the existence of a pair of NRR centers that are activated by 0.75 eV BGE. There are reports on the same type of PL quenching in different QDs and quantum wells which is explained by two levels model [13, 14].

The quenching of normalized PL intensity due to superposition of BGE over AGE can be interpreted by two levels model as shown in Fig. 5. Under the GaAs AGE, electrons are excited from the valence band to the conduction band of GaAs and they are captured by the QDs through the wetting layer. The electrons at the GaAs conduction band are also captured by defect levels which act as NRR centers in the GaAs layers. Under the QD AGE, on the other hand, the electrons are excited from the valence band to the conduction band of QDs only. For both GaAs AGE and QD AGE, a part of photo-generated carriers recombine non-radiatively through the NRR levels which are located in the band gap of InAs QD and at the interface of InAs/GaAs. When the BGE energy corresponds to the energy difference between two sets of coexisting below gap NRR levels (L1, L2) in GaAs and (L1', L2') in InAs QD. The electrons in level L1 and L1' are excited to the level L2 and L2' from which they recombine non-radiatively with holes in the valence band of GaAs and InAs QDs, respectively. Hence, the hole density in the valence band (p) decreases by an amount Δp. Correspondingly, the electron vacancies in level L1 and L1' allow an increase of NRR from conduction band of GaAs and InAs QDs, respectively. Thus, the electron density in the conduction band (n) decreases by an amount Δn. The combination of both effects reduces the number of electron-hole pairs available for radiative recombination and the normalized PL intensity quenches. At fluence 7×10^{11} protons/cm^2 (sample B), the electrons in the GaAs conduction band are also captured by PR4" trap centers and then tunnel into the QDs layer. Therefore, the QD PL intensity enhances and the decrease in normalized PL intensity is lower than the other samples (A and C). While, at the fluence of 4×10^{12} protons/cm^2, (sample C), the introduction of 0.81 eV deep trap center increases NRR process which results in dominant quenching of normalized PL intensity compared to the sample A and B. According to the proposed model shown in Fig. 5 two pairs of NRR levels (L1, L2) and (L1', L2') for GaAs AGE while only one pair of NRR levels (L1', L2') for QD AGE are activated by the BGE. The level L1 represents the defects at 0.75 eV and 0.78 eV related to EL2 [6, 27]. The PR1 (0.81 eV) deep trap center is also equivalent to L1 [6]. The defects due to arsenic antisite interacting with As$_{Ga}$-V$_{As}$, threading dislocations, misfit dislocations, relaxation induced defects and point defects at 0.67 eV, 0.53 eV and 0.37 eV can form L2 [6, 23-26]. The level L1' may be combination of 0.32 eV and 0.40 eV for the cluster of point defects [16, 21]. The level L2' comprises of defect levels at 0.52 eV and 0.84 eV at the interface of GaAs cap layer/QD [6, 26]. Furthermore, the dominant defects levels are distributed in the GaAs layer and the quenching of normalized PL for GaAs AGE excitation is very high compared to QD AGE excitation. The degree of quenching of normalized PL at low temperature is also consistent with the relation of PL intensity only by AGE excitation for three different samples (A, B and C) for both AGE sources. Since no distinct tendency of saturation is observed with increasing BGE density, the density of NRR centers are high and no trap filling effect takes place in the present BGE density.

The existence of well-known EL2 family defects, gallium vacancy, arsenic interstitials, intrinsic point defects-oxygen inside GaAs barrier layers has already been reported [18, 19, 36]. Recently, Sato *et al.* also confirmed the presence of GaAs native defect, arsenic antisite (As$_{Ga}$) interacting with other defects forming a complex, defects due to incorporation of

InAs into GaAs, proton irradiated trap centers of 0.30eV and 0.81 eV at the fluences of 7×10^{11} protons/cm^2 and 4×10^{12} protons/cm^2, respectively by DLTS measurement. However, no distinct defect level was found inside QDs [6]. The threading and misfit dislocations at the InAs QDs/GaAs interface are originated due to lattice mismatch between InAs QDs and GaAs and the point defects and its complex inside the InAs QDs also exist [20-27]. It is reported that proton irradiation on InAs/GaAs QDs from low to intermediate fluence reduces as-grown defects. On the other hand at higher fluence, it introduces additional defects or NRR centers such as point defects, As and Ga vacancies, interstitials, antisites and combination of them [6, 9, 11, 12, 16]. Since most of the native defects as well as proton irradiated defects are originated in GaAs and InAs/GaAs interface, there is a possibility that a small amount of point defects might be generated due to proton irradiation inside QDs.

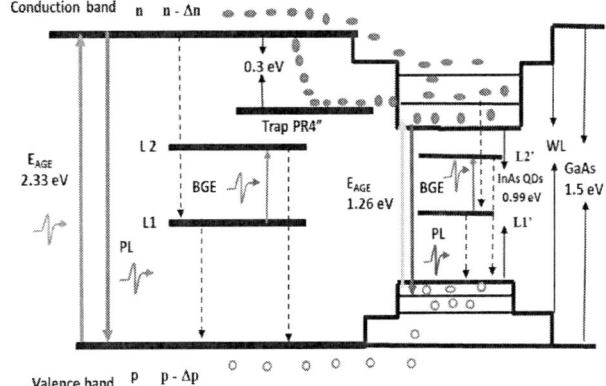

Fig. 5. Proposed model of carrier recombination for InAs/GaAs QD structures.

Due to the above phenomena, the decreased amount of the normalized PL intensity is larger in the GaAs AGE (hv_{A1}=2.33 eV) when compared with that in the QDs AGE (hv_{A2}=1.26 eV). To understand the origins of NRR centers in detail, i. e., to identify the native defects and also additional defects due to proton irradiation, it is important to determine the energy distribution of NRR centers by measuring the BGE energy dependence of the normalized PL intensity. A decreased contribution of NRR centers at 7×10^{11} protons/cm^2 fluence (sample B) together with the formation of the defect center PR4'' (0.3eV) which assists carrier flow to the QDs enhances the PL intensity at low temperature.

We also studied the AGE power dependence of the normalized PL intensity for the QD AGE and the GaAs AGE at fixed BGE power density 12.92 mW/mm^2 as shown in Figs. 6 and 7, respectively. For each case the normalized PL intensity quenches with decreasing the AGE density. The competition between radiative recombination and NRR of carriers generated by the AGE determines the QD PL intensity.

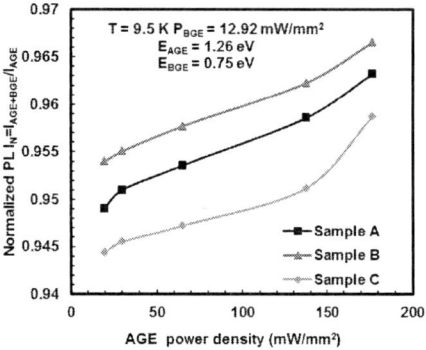

Fig. 6. TWEPL AGE Power dependency of normalized PL intensity for 1.26 eV AGE.

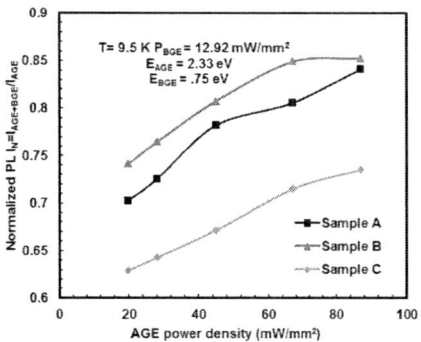

Fig. 7. TWEPL AGE Power dependency of normalized PL intensity for 2.33 eV AGE.

The strength of the BGE effect becomes large with decreasing the AGE density. The normalized PL intensity depends on the interaction between the below-gap states excited by the BGE and the AGE density. Therefore, at lower AGE density, the excitation via the below-gap level increases (BGE effect increases) relative to band to band excitation. The observed AGE density dependence is consistent with the rate equation analysis and previous results in QDs and quantum wells [13-14], [28]-[31]. For GaAs AGE (2.33 eV), the normalized PL intensity enhances from 0.70 to 0.84, 0.74 to 0.85, and 0.62 to 0.73 with increasing AGE power density from 19.85 mW/mm^2 to 87 mW/mm^2 for the samples A, B and C, respectively. The normalized PL intensity increased from 0.949 to 0.963, 0.953 to 0.966 and 0.944 to 0.958 with increasing 1.26 eV AGE power density from 19.85 mW/mm^2 to 177.07 mW/mm^2 for the samples A, B and C, respectively. Selecting excitation volume results in the different contribution of NRR centers, reflecting these higher distributions in the GaAs layers in contrast to that inside QDs and at their interfaces. It should be noted that such lower contribution of NRR centers in the smaller total volume of QDs can also be detected by the present TWEPL scheme under the QD AGE since the method obtains defect information through the QD PL.

IV. CONCLUSION

We studied NRR centers in InAs/GaAs QD structures and their influence against 3 MeV proton irradiation by TWEPL. The TWEPL method enabled us to selectively observe carrier dynamics in the GaAs layers and the QDs by choosing the appropriate AGE energy. The QD PL intensity under the QD AGE was higher than that under the GaAs AGE by a factor of four. Comparison of the normalized PL intensity by adding the BGE ($h\nu_B = 0.75$ eV) was also performed and the results showed dominant NRR process under GaAs AGE compared to that QD AGE. The density of NRR centers decreases due to irradiation of the 7×10^{11} protons/cm^2 irradiation whereas it increases due to the irradiation of 4×10^{12} protons/cm^2. Combination of defect formation and defect assisted carrier injection into QDs explains the result. We also detected NRR centers inside the QDs and/or at the QD interface in the former case. This shows a fundamental advantage of the TWEPL method by monitoring the QD PL.

ACKNOWLEDGEMENT

The authors would like to thanks David V. Forbes for fabricating samples at Nanopower Research Laboratory of Rochester Institute of Technology in United State of America. This research work was supported by Japan Society of promotional Science KAKENHI grant No. 25600087 and JP15638900.

REFERENCES

[1] Z. S. Bittner, S. Hellstroem, S. J. Polly, R. B. Laughumavarapu, B. Liang, D. L. Huffaker and S. M. Hubbard , "Investigation of optical transitions in InAs/GaAs(Sb)/AlAsSb quantum dots using modulation spectroscopy," *Applied Physics Letter*, vol. 105, pp. 253903-1-2539-4, 2014.

[2] S. M. Hubbard, S-I. Sato, K. Schmieder, W. Strong, D. Forbes C.G.Bailey, R.Hoheisel and R. J. Walters, "Impact of Nano-struc-tures and radiation environment on defect levels in III-V solar cell," in *40th IEEE, photovoltaic specialist conference*, 2014, p. 1045.

[3] S. M. Hubbard, A. Podell, C. Mackos, S.Polly, C. G. Bailey, D. V. Forbes, "Effect of vicinal substrates on the growth and device performance of quantum dot solar cells," *Solar Energy Materials and Solar Cells*, vol. 108, pp. 256-262, 2013.

[4] Y.Okada, N.J. Ekins-Daukes, T. kita, R. Tamaki, M.Yoshida, A. Pusch, O.Hess C.C Philips, D.J Farrel, K. Yoshida, N. Ahsan, Y. Shoji, T. Sogabe, and J-F. Guillemoles, " band solar cells: recent progress and future directions," *Applied Physics Reviews*, vol. 2, pp. 021302-1-02130247, 2015.

[5] A. Mandal, U.Verma, N. Halder, S. Chakrabarti, "Impact of monlayer coverage barrier thickness and growth rate on the thermal stability of photoluminescence of coupled InAs/ GaAs quantum dot hetero-structure with quaternary capping of InAlGaAs," *Materials Research Bulletin*, vol. 47, pp. 551-556, 2012.

[6] S-I. Sato, Kenneth J. Schmieder, S.M. Hubbard, D. V.Forbes, J.H. Warner, T. Ohshima and R. J.Walters," Defect characterization of proton irradiated GaAs pn-junction diodes with layers of InAs quantum dots," *Journal of Applied Physics*, vol. 119, pp. 185702- 1-185702-8, 2016.

[7] Kristina Driscoll, Mitchell F. Benntt, Stephen J. Polly, David V. Forbes and Seth M Hubbard, "Effect of quantum dot position and background doping on the performance of quantum dot enhanced GaAs solar cell," *Applied Physics Letter*, vol. 104, pp. 023119-1- 023119-5, 2014.

[8] T. Ohshima, S. Sato, M. Imaizumi, T. Nakamura, T. Sugaya, K. Matsubara, S. Niki, " Change in the electrical performance of GaAs solar cells with InGaAs quantum dots layers by electron irradiation," *Solar Energy Material and Solar cells*, vol. 108, pp. 023119-1- 023119-5, 2014.

[9] R. Sreekumar, A Mandal , S.K Gupta, S. Chakrabarti, "Effect of high energy proton irradiation on InAs/GaAs quantum dots: Enhance of photoluminescence efficiency (up to ~7 times) with minimum spectral shift," *Material Research Bulletin*, vol. 46, pp. 1786 -1793, 2011.

[10] C.Y Cheng , H Niu , C.H Chen ,T.N Yang, H.Y Wang, C.P Lee, "Effect of proton-irradiation on photoluminescence emission from self-assembled InAs/GaAs quantum dots," *Nuclear Instrument and Methods in Physics Research B*, vol. 261, pp.1171- 1175, 2007.

[11] S. Upadhyay, A. Mandal, N.B.V Subrahmanyam, P.Singh, P.Shete, B. Tongbram, S.Chakrabarti, "effect of high-energy proton implantation on the luminescence properties of InAs submonolayer quantum dots," *Journal of Luminescence*, vol. 171, pp. 27-32, 2016.

[12] S. Upadhyay , A. Mandal, H. Ghadi, D. Pal, A. basu, A. Agarwal, N.B.V. Subrahmanyam, P. Singh, S. Chakrabarti, "Effect of high energy proton implantation on the optical and electrical properties of In(Ga)As/GaAs QD heterostructures with variation in the capping layer," *Journal of Luminescence*, vol. 161, pp. 129 -134, 2015.

[13] N. Kamata, S. Saravanan ,J.M Zanardi Ocampo, P.O Vaccaro, Y. Arakawa, "Nonradiative centers in InAs quantum dots revealed by two wavelength excited photoluminescence," *Physica B*, vol. 376 (1), pp. 849-852, 2006.

[14] N. Kamata, J.M. Zanardi Ocampo, K. Hoshino, K Yamada, M. Nishioka, T. Someya and Y. Arakawa, "Below-gap spectroscopy of semiconductor quantum wells by two-wavelength excited photoluminescence(TWEPL)," *Recent Research Development on Quantum Electronics*, vol. 1, pp. 123-135, 1999.

[15] D. Sreenivasan, J.E.M. Havekort, T.J. Eijkemans, and R. Notzel, "Photoluminescence from low temperature grown InAs/GaAs quantum dots," *Applied Physics Letters*, vol. 90, pp.112109-1-112109-3, 2007.

[16] P.C. Sercel, "Multi-phonon assisted tunneling through deep level: A rapid energy relaxation mechanism in non-ideal quantum dots," *Physical Review B*, vol. 51, pp.14532-14541, 1995.

[17] Y. Ji, G. Chen, N. Tang, Q. Wang, X.G. Wang, J. Shao, X.S. Chen, and W. Lu, "Proton implantation-induced photoluminescence enhancement in self-assembled InAs/GaAs quantum dots *Applied. Physics Letter*, vol. 82, pp. 2802-2804, (2003).

[18] P.W.Yu, G.D. Robinson,J.R. Sizelove, and C. E. Stutz, "0.8eV photoluminescence of GaAs grown by molecular- beam epitaxy at low temperatures," *Physical Review B*, vol. 49, pp. 4689-4694, 1994.

[19] P. N. Brunkov, V.S. Kalinovsky, V.G. Nikitin and M.M. Sobolev,"Generation of the EL$_2$ defect in n-GaAs irradiated by high energy protons," *Semiconductor Science and Technology*, vol. 7, pp. 1237-1240, 1992.

[20] S.W. Lin, C. Balocco, M.Missous, A.R Peaker and A.M Song, "Coexistence of deep level with optically active InAs quantum dots," *Physical Review B*, vol.72, pp. 165302-7, 2005.

[21] C. Walther, J. Bollmann, H. Kissel, H. Kirmse, W. Neumann, and W. T. Masselink, "Characterization of electron trap states due to InAs quantum dots in GaAs," *Applied physics Letter*, vol. 76, pp. 2916-2918, 2000.

[22] Ken-ichi Shiramine, Yasunobo Horisaki, Dai Suzuki, Satoru Itoh, Yoshiki Ebiko, Shunichi Muto, Yoshiaki Nakata and Naoki Yokoyama, "Threading dislocations in multilayer structure of InAs self-assemble quantum dots," *Japanese Journal Applied Physics*, vol. 3, pp. 5493-5496, 1998.

[23] J. F. Chen and J. S. Wang, " Electron emission properties of relaxation-induced traps in InAs/GaAs quantum dots and the effect of electronic band structure," *Journal of Applied Physic*, vol. 102, pp. 043705-1-043705-6, 2007.

[24] S. Huang, S. J. Kim, X. Q.Pan and R.S. Goldman, "Origins of interlayer formation and misfit dislocation displacement in the vicinity of InAs/GaAs quantum dots," *Applied Physics Letter*, vol. 105, pp. 032107-1-032107-4, 2014.

[25] J.S wang , J.F Chen, J. L Huang and P.Y Wang, "Carrier distribution and relaxation-induced defects of InAs/GaAs quantum dots," *Applied Physics Letter*, vol. 77, pp. 3027-3029, 2000.

[26] L. Nasi, C. Bocchi, F. Germini, M. Prezioso, E. Gombia, R. Mosca, P. Frigeri, G. Trevisi, L . Seravalli and S. Franchi, " Defects in nanostructures with ripened InAs/GaAs quantum dots,"*Journal of Mater Science : Mater Electron*, vol. 19, pp. 96-100, 2008.

[27] M. Kaniewska, O. Engstrom, A. Barcz, M. Pacholak-Cybulska, "Deep level induced by InAs/GaAs quantum dots," *Material Science and Engineering C*, vol. 26, pp. 871-875, 2006.

[28] K. Hoshino, H. Kimura, T. Uchida, N. Kamata, K. Yamada M. Nishioka, Y. Arakawa, "Distribution of below- gap states in undoped GaAs/AlGaAs quantum wells revealed by two wavelength excited photoluminescence," *Journal of Luminescence* , vol. 79, pp.39-46, 1998.

[30] A.Z.M Touhidul Islam, K. hatta, N Murakoshi, T Fukuda, T. Takada, T Itatani and N. Kamata, "Detection of NRR centers in InGaAs/AlGaAs HEMTS: Two-wavelength excited photoluminescence studies," *Global Science and Technology Journal*, vol. 1, pp 1-11, 2013.

[31] N. Kamata, E. Kanoh, K. Hoshino, K. Yamada, M. Nishioka, and Y. Arakawa, "Saturation of luminescence quenching due to nonradiative centers in a GaAs/AlGaAs quantum well," *Material Science Forum*, vol. 196-20, 1 pp. 431-436, 1995.

Design and Prototyping Efforts for the Space Solar Power Initiative

Michael D. Kelzenberg[1*], Pilar Espinet-Gonzalez[1], Nina Vaidya[1], Tatiana A. Roy[1], Emily C. Warmann[1], Ali Naqvi[1], Samuel P. Loke[1], Jing-Shun Huang[1], Tatiana G. Vinogradova[2,1], Alexander J. Messer[2,1], Christophe Leclerc[1], Eleftherios E. Gdoutos[1], Fabien Royer[1], Ali Hajimiri[1], Sergio Pellegrino[1], and Harry A. Atwater[1]

[1] California Institute of Technology, Pasadena, CA 91125

[2] Northrop Grumman Aerospace Systems, Azusa, CA 91702

* Corresponding author e-mail address: mdk@caltech.edu

Abstract — The Space Solar Power Initiative (SSPI) seeks to enable reliable, cost-effective baseload power generation from large-scale solar power stations in space. We propose an ultralight, modular power station, having specific power in the range of 1–10 kW/kg for the photovoltaic (PV) collection subsystem. The building block of the power station is the 'tile,' a self-contained element that performs PV energy collection, conversion to radio frequency (RF), and transmission to earth. To minimize PV mass, we select a 1D, 10–20X parabolic trough concentrator geometry, which provides cooling and radiation shielding for the cells, and which folds flat for deployment. Here, we discuss the design, fabrication, and testing of the initial PV tile prototypes.

Index Terms — space solar power, concentrator photovoltaics, thermal management.

I. BACKGROUND

The placement of solar cells in space affords several advantages in terms of energy production as compared to terrestrial deployment. In particular, at a position in geostationary orbit, incident sunlight is nearly continuously available, and is not subject to attenuation by earth's atmosphere or weather patterns. A solar cell placed in space will receive approximately 5–10x more energy from the sun, on average, than at typical terrestrial locations (depending on location), and thus when coupled with wireless power transmission to earth, can reliably produce 24/7 baseload power without the need for large-scale energy storage.

Utilizing space-based solar power to meet terrestrial energy needs, however, requires solutions to several unique technical obstacles. First, the solar cells must be capable of operating in space, reliably and efficiently, for 10+ years lifetime, without maintenance or repair. They must not overheat despite the lack of air for convective cooling, and must survive exposure to ionizing radiation. Additionally, a mechanism is required to deliver the energy to earth. Finally, and most significantly, the system must be lightweight, compact, and efficient enough to justify the cost of launching it into orbit.

The concept of space-based solar power dates to the early years of the PV industry. It is interesting to note that when modern silicon PV cells were first developed in the 1950s, their extremely high cost precluded their use for nearly any terrestrial application; thus, the powering of early space satellites emerged as their first commercially successful application. [1] The rectenna was invented in the following decade, enabling efficient far-field wireless power transmission using beamed microwave energy. [2] Combining these technologies, space-based solar power was formally proposed in 1968, [3] leading to extensive feasibility studies in the 1970s. [4, 5]

These early studies proposed large, monolithic power stations, to be assembled on-orbit, comprising large-area PV arrays and high-power vacuum-tube-based RF transmitters, connected by high-voltage DC power conductors and rigid structural elements. High specific power (1–10 kW/kg) PV panels were identified as a key technical requirement, whereas optimistic estimates of then-achievable performance was ~0.25 kW/kg. [4] This, combined with the high mass, cost, and complexity of the monolithic structures, led to the conclusion that space-based solar power would not be economically viable until at least the 2005-2015 time frame. [5]

Although space-based solar power has not yet been realized, numerous technological advancements have occurred since the time of the initial studies. This includes the development of high-efficiency multijunction PV cells, ultralight structural materials, solid-state RF integrated circuits (ICs), and phased-array antennas. Proposals for more advanced space-based solar power systems have emerged, including modular systems with distributed solid-state RF transmission and simplified on-orbit assembly. [6-9] Lightweight PV concentrator panels have also been developed, including the stretched lens array (SLA) refractive concentrator, which achieved over 300 W/kg specific power and is expected to exceed 1 kW/kg, [10, 11] and "venetian blinds"-style reflective concentrators such as the SLATS [12] and FAST [13] projects. Space launch costs have also decreased in recent years with the introduction of new commercial rocket platforms.

Terrestrial PV has now grown to a multi-billion-dollar global industry, and the levelized cost of energy from PV installations is becoming less than that from fossil fuel sources. [14] Continued growth of terrestrial solar to meet baseload energy needs will increasingly require costly energy storage and/or grid improvements, which could be circumvented using space solar power. Considering these factors, there has never been a more favorable time to pursue space-based solar power.

978-1-5090-5606-4/17 $31.00 © 2017 IEEE

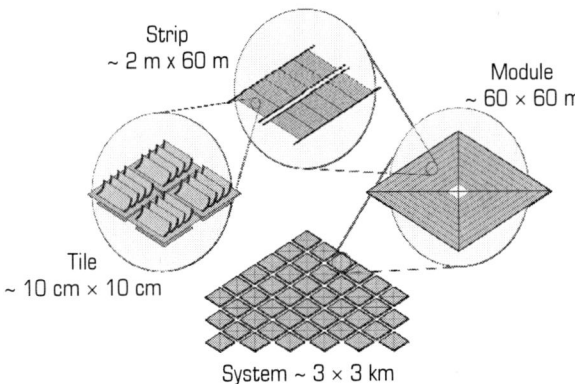

Fig. 1.   Conceptual illustration of the proposed space solar power station. The complete system has an area of ~3 x 3 km, and delivers 1–2 GW (peak) to the grid. Each tile is on the order of 10 x 10 cm, and is responsible for delivering 1–2 W (peak) to the grid.

II. SYSTEM DESIGN

We propose a modular space solar power station for geostationary orbit as depicted in Figure 1. The station has a spatial extent on the order of 3 x 3 km, and comprises a multitude of free-flying spacecraft, or "modules." Each module consists of a central hub containing the primary propulsion, control, communication, and deployment equipment. When launched, the solar array is folded and wrapped within the hub. High volumetric packing efficiency is possible because the PV concentrators and the patch antenna spacers can be flattened for storage. Upon deployment, four coilable booms extend radially from the hub, unfurling and tensioning the solar array. Position and attitude sensors are distributed throughout the array.

The building block of the power station is the 'tile,' a self-contained element that performs PV energy collection, conversion to RF via solid-state ICs, and transmission via a phased array of antennas, as illustrated in Fig. 2. To minimize mass associated with the PV cells and their radiation shielding, we use a 1D, 10–20x parabolic trough concentrator geometry, in which each reflector also serves as the mechanical mount, electrical interconnect, heat spreader, and radiative cooler.

Like other recent "sandwich module" designs for space solar power, [9] a key feature of the tile concept is that the DC electrical power produced by the solar cells is locally converted to RF energy, then radiated towards earth by nearby antennas. This avoids the need to collect or convey DC electrical power over large areas or long distances, which is a major source of mass and complexity in designs that separate the PV and RF apertures. Furthermore, the electrical power can be conveyed to the ICs at low voltage, with all cells operating in parallel, which simplifies the electrostatic design and obviates the need for bypass diodes. Finally, distributing the RF power conversion losses throughout the array area greatly reduces the mass required for thermal management.

This design decision does, however, cause there to be times during the daily geostationary orbit at which energy cannot be effectively collected and/or transmitted towards earth. We plan to address this issue in future work.

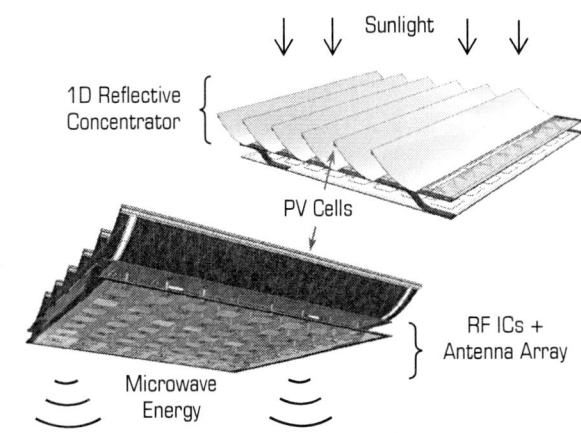

Fig. 2.   Illustration of tile concept; view from above and below.

III. PV CONCENTRATOR DESIGN

Commercial triple junction (3J) space PV cells are capable of 30% AM0 efficiency, and emerging technologies such as inverted metamorphic (IMM) and/or 4+ junction cells are projected to exceed 35%. [15] The active layer thickness of these cells is relatively thin (up to 10s of μm), suggesting that we could achieve our specific power target by thinning or removing the growth substrate. However, III-V solar cells require shielding to protect against damage from ionizing radiation in space; typically, 50+ μm of material on both sides. [16] Typically, front shielding is provided by cover glass, and rear shielding by the growth substrate or module backing. Thus to achieve specific power exceeding 1 kW/kg we must employ optical concentration.

Using a 1D concentrator affords a wide acceptance angle along the long axis of the concentrator, enabling a wider range of orbital maneuvering. We have selected a "venetian blinds"-style parabolic trough concentrator geometry, which is known to have favorable properties for space applications. [12, 13]

The key geometrical factors affecting the concentrator design are the spacing (pitch) of the reflector elements, and size (width) of the solar cells. The ratio of the former to the latter gives the nominal concentration ratio. We have chosen to work with 1 mm wide cells, which to our knowledge, is substantially smaller than used for prior space concentrators of this type. Using smaller concentrator elements reduces dramatically the amount of material required for heat spreading, at the cost of increased sensitivity to edge effects and alignment errors.

978-1-5090-5606-4/17 $31.00 © 2017 IEEE

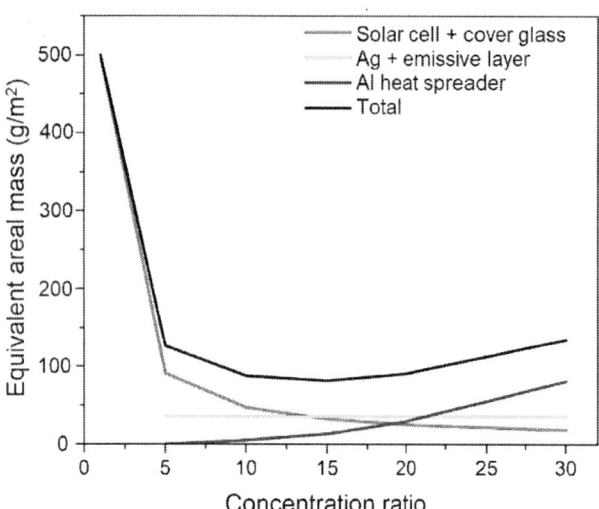

Fig. 3. Equivalent areal mass vs. concentration ratio for a simplified concentrator design utilizing 1 mm-wide solar cells. For each concentration value, the thickness of the Al heat-spreader layer is selected so as to yield a peak cell temperature of 100 °C.

As an example, Fig. 3 shows the calculated areal mass vs. concentration ratio for a simple concentrator design. Note that, because we have fixed the cell width at 1 mm, the concentration ratio is equivalent to the concentrator pitch (in millimeters). We consider only the mass of the cell (10 μm GaAs with 75 μm cover glass on both sides) and the reflector membrane in this calculation; structural support for the membrane is not included. The reflector membrane comprises a thin Ag mirror layer on top of an Al heat-spreading layer. The bottom side of the membrane is coated with a 25 μm thick carbon-loaded polyimide layer for improved emissivity. [17] We assume the reflector membrane has a top-side (sun facing) emissivity of 0.1 and a bottom-side emissivity of 0.8. We simulated the steady-state cell temperature using COMSOL, based on thermal loading for 3J IMM solar cells at a geostationary environment. For each simulation, the thickness of the Al heat-spreading layer was adjusted to produce a peak cell temperature of 100°C.

As shown in Fig. 3, higher concentration ratios decrease the areal mass associated with the solar cells and cover glass, but also increase the amount of aluminum required for heat spreading. The optimal concentration ratio is 10–20x. The corresponding acceptance angle is approximately 4° (10x) to 2° (20x).

IV. PROTOTYPING RESULTS

We have fabricated initial prototypes of ultralight PV concentrator tiles. Solar cells were prepared from commercial 3J Ge wafers (Spectrolab XTJ) and diced to approximately 10 mm x 1 mm size. Reflector membranes were fabricated from 25 μm thick aluminized Kapton sheeting, onto which we deposited a thin SiO_2-protected Ag reflective layer. Metal

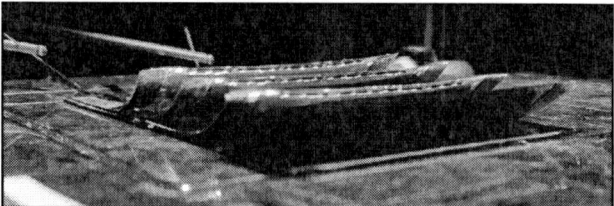

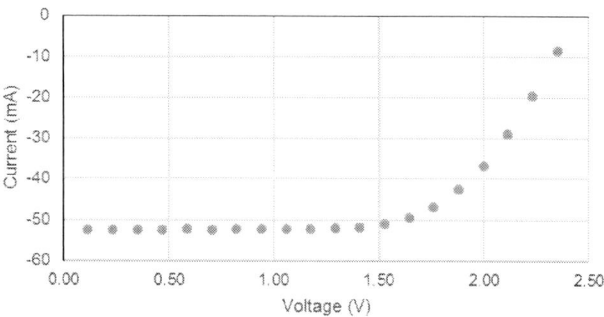

Fig 4. Photographs (top, center) and I-V data (bottom) of an ultralight PV concentrator prototype.

pads and traces were deposited on the back of each reflector membrane, then cells were attached using a combination of conductive and insulating adhesives. Cerium-doped, anti-reflective-coated cover glass (75 μm thick) was diced to size, then attached to the cells using PDMS (Dow Corning 93-500). Parabolic reflector supports, or "springs," were prepared by casting sheets of carbon-fiber referenced polymer (8-ply, T-800) against a parabolic mold, then cutting to 5 mm width.

Two springs were attached to each reflector membrane, one at each edge, using structural epoxy and an assembly jig. To improve the tendency of the reflector membranes to conform to the parabolic profile imparted by the spring supports, we applied a thin layer of polyimide (Nexolve CP1) to the front side of the mirrors prior to assembly, using a wire-wrapped drawdown rod. This caused the reflectors to curl naturally in the direction of the springs, and further improved the front-side emissivity.

The completed reflector subassemblies were then aligned by eye and glued together at the bases of the spring tabs. Thin strips of metalized Kapton were attached to electrically connect the cell banks in parallel. I-V data were measured under a solar simulator (Fig. 4, below) which used an AM0 filter set, but which only operated at 1 kW/m² intensity.

The prototype pictured in Figure 4 (top) consists of four reflectors, each approximately 1.5 x 6 cm in projected area. There are three active rows of solar cells, each containing four 1 cm long cells. The assembly weighs 1.00 g. Based on the active aperture area of ~18 cm², the efficiency is ~4.5%. Based on the total projected reflector area of ~36 cm², the areal mass is ~280 g/m². The mass breakdown was approximately 25% cells and cover glass, 25% adhesives, 15% reflector membranes, and 35% springs. The optical efficiency was extremely low (~20%) due to shape errors and alignment errors. Additional fill factor losses were caused by excessive series resistance in the metal traces and ribbons, and the failure of the finger electrodes during photolithography for our first batch of cells. This batch of cells also suffered from lower V_{OC} because the bottom junctions were inadvertently damaged during the mesa etch.

The completion of our first prototypes has provided an important proof-of-concept demonstration of the ultralight PV concentrator concept. Our design addressed first-order optical, electrical, and thermal performance, and also included radiation shielding. It was made using only space-grade materials, and achieved a low areal mass of <300 g/m². Efforts are now underway to improve the efficiency of the concentrators and solar cells. More recent prototypes have achieved dramatically higher specific power.

ACKNOWLEDGEMENTS

We acknowledge financial support from Northrop Grumman. This effort made use of facilities provided by the Kavli Nanoscience Institute, the Molecular Materials Research Center, the Resnick Institute, and the Joint Center for Artificial Photosynthesis at Caltech. We acknowledge the helpful contributions of Mark Kruer, Mike Levesque, and Erik Kurman at Northrop Grumman; Lynn Rodman at Nexolve; and Allen Smith at ABET.

REFERENCES

[1] J. Perlin, *From Space to Earth: the Story of Solar Electricity*. Ann Arbor, MI: Aatec Publications, 1999.

[2] W. C. Brown, "The History of Power Transmission by Radio Waves," *IEEE Transactions on Microwave Theory and Techniques,* vol. 32, pp. 1230-1242, 1984.

[3] P. E. Glaser, "Power from the Sun: Its Future," *Science,* vol. 162, pp. 857-861, 1968.

[4] P. E. Glaser, O. E. Maynard, J. J. R. Mackovciak, and E. I. Ralph, "Feasibility Study of a Satellite Solar Power Station," NASA Final Report, ADL-C-74830, 1974.

[5] "Solar Power Satellites," Office of Technology Assessment, NTIS order #PB82-108846, August 1981.

[6] J. C. Mankins, "A fresh look at space solar power: New architectures, concepts and technologies," *Acta Astronautica,* vol. 41, pp. 347-359, 1997.

[7] J. C. Mankins, "SPS-ALPHA: The First Practical Solar Power Satellite via Arbitrarily Large Phased Array," Final Report, 2011-2012 NASA NIAC Phase 1 Project 2012.

[8] S. Sasaki, "It's always sunny in space," *IEEE Spectrum,* vol. 51, pp. 46-51, 2014.

[9] P. Jaffe and J. McSpadden, "Energy Conversion and Transmission Modules for Space Solar Power," *Proceedings of the IEEE,* vol. 101, pp. 1424-1437, 2013.

[10] M. O. Neill, A. J. McDanal, H. Brandhorst, K. Schmid, P. LaCorte, M. Piszczor, *et al.*, "Recent space PV concentrator advances: More robust, lighter, and easier to track," in *2015 IEEE 42nd Photovoltaic Specialist Conference (PVSC),* 2015, pp. 1-6.

[11] M. O'Neill, A. McDanal, M. Piszczor, P. George, M. Eskenazi, M. Botke, *et al.*, "Recent progress on the stretched lens array (SLA)," presented at the 18th Space Photovoltaic Research and Technology Conference, 2005.

[12] T. G. Stern, "Interim results of the SLATS concentrator experiment on LIPS-II (space vehicle power plants)," in *Conference Record of the Twentieth IEEE Photovoltaic Specialists Conference,* 1988, pp. 837-840 vol.2.

[13] "Boeing Team to Develop Revolutionary Spacecraft Power System for DARPA," Press Release. Boeing, July 1 2009.

[14] "Renewable Infrastructure Investment Handbook: A Guide for Institutional Investors," World Economic Forum, Dec 2016.

[15] D. C. Law, X. Q. Liu, J. C. Boisvert, E. M. Redher, C. M. Fetzer, S. Mesropian, *et al.*, "Recent progress of Spectrolab high-efficiency space solar cells," in *2012 38th IEEE Photovoltaic Specialists Conference,* 2012, pp. 003146-003149.

[16] S. Bailey and R. Raffaelle, *Space Solar Cells and Arrays,* 2011.

[17] *Conductive Black CP1 Polyimide.* Available: http://www.nexolvematerials.com/low-cure-polyimides/conductive-black-cp1-polyimide

Defect Characterization of III-V Quantum Structure Solar Cells Using Photo-Induced Current Transient Spectroscopy

Shin-ichiro Sato[1], Takeyoshi Sugaya[2], Tetsuya Nakamura[3], and Takeshi Ohshima[1]

[1] National Institutes for Quantum and Radiological Science and Technology (QST), Takasaki, Gunma 370-1292, Japan

[2] National Institute of Advanced Industrial Science and Technology (AIST), 1-1-1 Umezono, Tsukuba, Ibaraki 305-8568, Japan

[3] Japan Aerospace Exploration Agency (JAXA), Tsukuba, Ibaraki, 305-8505, Japan

Abstract — Radiation degradation characterization method for solar cells embedded with quantum dot (QD) layers using Photo-Induced Current Transient Spectroscopy (PICTS) is proposed in this study. Contrary to Deep Level Transient Spectroscopy (DLTS), PICTS is capable of comparing directly photo-current degradation to radiation induced defects for p-i-n structure solar cells, which is a basic structure of QD solar cells. GaAs p-i-n solar cells with 50 $In_{0.4}Ga_{0.6}As$ QD layers are fabricated by Molecular Beam Epitaxy (MBE) and the radiation degradation is investigated by PICTS measurement. The current-voltage characteristics under AM0, 1 sun condition is also investigated and the degradation of solar cell performance is discussed by comparison to defect levels obtained from PICTS spectra. GaAs p-i-n solar cells without QDs are also investigated for comparison.

I. INTRODUCTION

Self-assembled semiconductor quantum dots (QDs), grown by the Stranski–Krastanov (S–K) growth mode, have attracted much attention because of their wide variety of applications including solar cells. With intermediate band solar cells, the intermediate-band formed by the QD superlattice is located in the bandgap of a matrix semiconductor, which is used to absorb the sub-bandgap photons by employing two-step photo-absorption via the intermediate-band state. The intermediate-band solar cells (or QD solar cells) are quite attractive not only for terrestrial use, but also for space use, since they have the potential to improve the radiation tolerance in addition to the conversion efficiency [1]. However, radiation effects on QD solar cells are less well understood and electrically active radiation induced defects should be investigated in order to develop space QD solar cells.

In General, the identification of electrically active defects (deep level traps) and their locations in devices is important in order to gain insight toward strategies for reducing defect density and improving device quality. One of the most effective methods to characterize defect levels is Deep Level Transient Spectroscopy (DLTS) [2]. For example, radiation degradation of solar cells based on the defect characterization using DLTS have been studied in order to understand the

degradation mechanism [3, 4]. However, QD layers are embedded into intrinsic layer of p-i-n structure solar cells. DLTS technique which measures transients of junction capacitance after pulse bias cannot characterize defect levels in the intrinsic layer of p-i-n devices. Therefore, special design is required to characterize defect levels in the QD layers using DLTS technique [1], and this fact makes the direct comparison of radiation induced defects to performance degradation of QD solar cells difficult.

In this study, we propose a radiation induced defect characterization method using Photo-Induced Current Transient Spectroscopy (PICTS) [5, 6]. PICTS is a method to characterize defect levels from transient currents in devices generated after pulse light illumination and its temperature dependence. This is the similar method to DLTS, and the rate window analysis, which activation energy and capture cross section of defect levels are derived by, is applicable for PICTS signals. PICTS technique permits the direct comparison of solar cell degradation to defect creation due to radiation exposure, since behavior of carriers in solar cells generated by light illumination is directly investigated.

II. EXPERIMENTAL

Samples used in this study were GaAs p-i-n solar cells where $In_{0.4}Ga_{0.6}As$ quantum dot layers are embedded into the i-layer. GaAs p-i-n solar cells were also fabricated for comparison. These samples were represented as 50xQD and baseline solar cells, respectively. A schematic diagram of the sample structure is shown in Fig. 1. They were grown by Molecular Beam Epitaxy (MBE) at the National Institute of Advanced Industrial Science and Technology (AIST). The QD region was capped above by 20 nm of unintentionally doped (uid) GaAs and was repeated 50 times. The diameter and height of QDs were 30 nm and 5 nm, respectively, and the density was 2.3×10^{10} cm^{-2} according to cross-sectional scanning transmission electron microscope (STEM). Details regarding the device growth parameters and QD characteristics can be found elsewhere [7-9].

978-1-5090-5606-4/17 $31.00 © 2017 IEEE

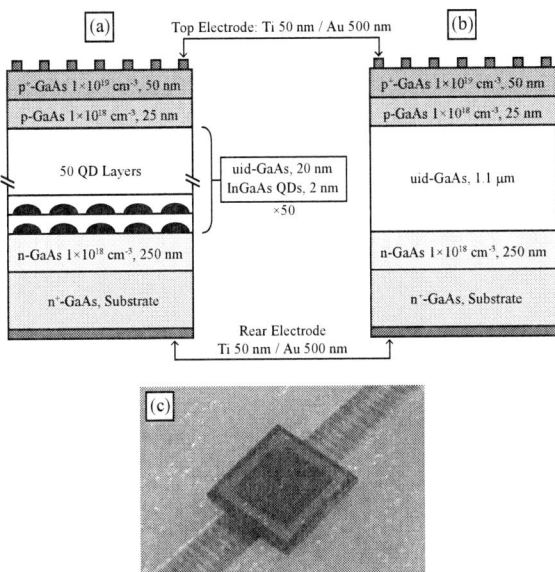

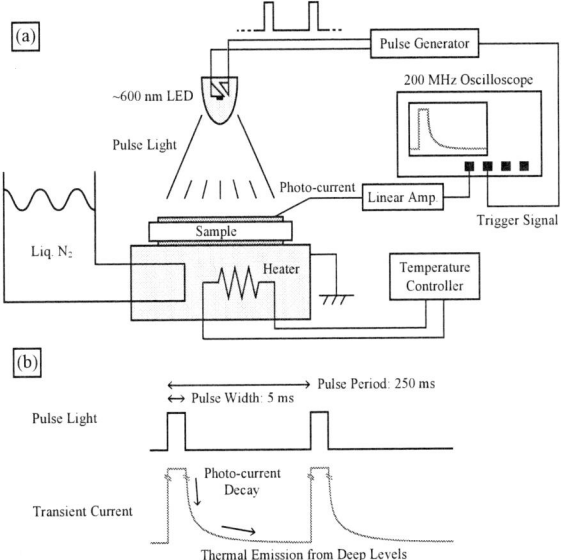

Fig. 1. A schematic diagram of (a) the 50xQD and (b) the baseline solar cells. (c) A photograph of the samples. "uid" in the figure represents "unintentionally doped".

Figure 2(a) shows a schematic diagram of PICTS measurement setup. LEDs with the peak wavelength of 600 nm were used for PICTS measurement. The pulse width and period were 5 ms and 250 ms as shown in Fig. 2(b), respectively. The light intensity of LED was sufficiently high enough to fill deep level traps in the active layer of solar cells. The rear electrode was grounded and the photo-current by pulsed lights from the LEDs was recorded from the front electrode using the 200 MHz Oscilloscope. No bias voltage was applied to the sample during PICTS measurement, corresponding to short-circuit condition. The sample temperature was scanned from 140 K to 400 K using the heater and liquid nitrogen. PICTS signals are defined as the difference between transient currents at times t_1 and t_2 after the LED light illumination turned off, and several signals with a constant t_1/t_2 ratio are recorded to perform the Arrhenius analysis (the rate window method). Activation energy and apparent capture cross section of deep trap levels are obtained from the Arrhenius analysis.

In order to intentionally induce defects in the samples, 1 MeV electrons were irradiated at room temperature and the variation with increasing electron fluence was investigated. The light current-voltage (LIV) characteristics under AM0, 1 sun condition was also investigated and the effect of radiation defects on the degradation of solar cell performance was discussed.

Fig. 2. (a) A schematic diagram of PICTS measurement setup. (b) Current transient caused by thermal emission from deep levels subsequent to photo-current decay is recorded.

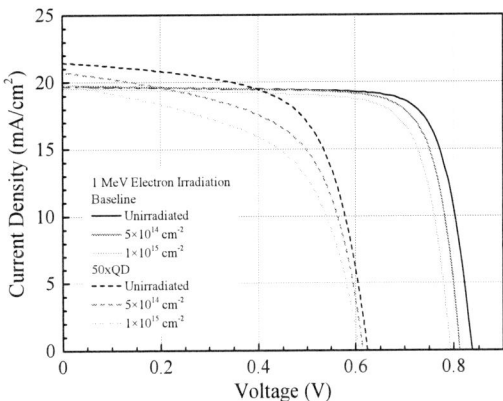

Fig. 3. LIV characteristics of the baseline (solid lines) and the 50xQD samples (dashed lines) at room temperature. The irradiation fluences of 1 MeV electrons are shown in the figure.

III. RESULTS AND DISCUSSION

The LIV characteristics of baseline and 50xQD samples and their variations with increasing irradiation fluence of 1 MeV electrons are shown in Fig. 3 and the degradation curves of short-circuit current (Isc), open-circuit voltage (Voc), output maximum (Pmax), and fill factor (FF) are shown in Fig. 4. Significant degradation of Voc appeared in the case of baseline sample, whereas Isc and FF were significantly reduced in the case of 50xQD sample, indicating the increase

978-1-5090-5606-4/17 $31.00 © 2017 IEEE 563

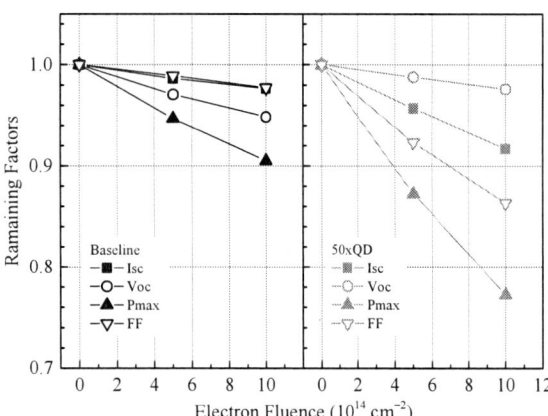

Fig. 4. Degradation curves of the baseline (left figure, black) and 50xQD samples (right figure, red). All the values are normalized by the values before irradiation.

in recombination current in the QD layer [10]. This trend corresponded to the previous study [11].

Figures 5 and 6 show the PICTS spectra of the baseline and 50xQD samples with different electron fluences, respectively. These PICTS signals were obtained by using $t_1 = 12$ ms, $t_2 = 180$ ms, and the emission rate = 83.3 s^{-1}. Although no distinct peak was found in the unirradiated baseline sample, at least 3 peaks appeared in the irradiated sample. The peak temperatures were 150 K, 200 K, and 350 K, which were represented as Traps 1, 2, and 3 in Fig. 5, respectively. All peak intensities increased with increasing electron fluence. This fact indicates that these peaks were originated from defect levels induced by 1 MeV electron irradiation. On the other hand, at least 3 peaks appeared in the unirradiated 50xQD samples and these peak intensities significantly increased due to 1 MeV electron irradiation, as shown in Fig. 6. The peak temperatures were 175 K, 225 K, and 300 K, represented as Traps 2, 3, and 4, respectively. A significant peak could appear at around 145 K after 1 MeV electron irradiation (Trap 1), although the peak temperature was not clearly identified.

Seven different PICTS spectra with a constant t_2/t_1 ratio ($t_2/t_1 = 3$) were analyzed to perform the Arrhenius analysis based on the rate window method. The results are shown in Fig. 7. These values were obtained from the results after 1 MeV electron irradiation at 1×10^{15} cm^{-2}. Traps 1 to 3 in the baseline sample in addition to Traps 3 and 4 in the 50xQD sample (which are shown as QD Trap 3 and QD Trap 4 in Fig. 7) were successfully characterized. The summary of obtained traps is given in Table I displaying the activation energy and the capture cross section as well as the tentative defect identification. Since signals from both electron and hole traps are detected in PICTS measurement, comparison to previous literatures is necessary to identify the observed trap levels.

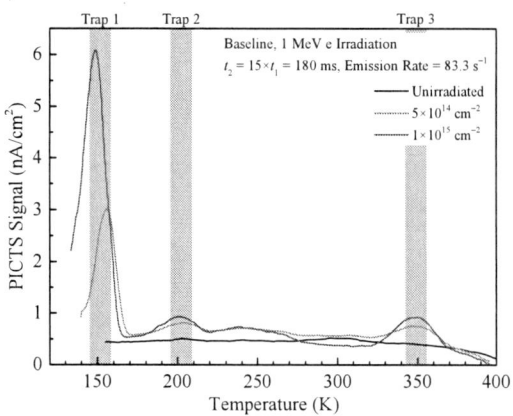

Fig. 5. PICTS spectra of the baseline sample with different electron fluences. The peak positions (Traps 1 to 3) are identified by shaded zones. The emission rate was 83.3 s^{-1}. The same is true in Fig. 6.

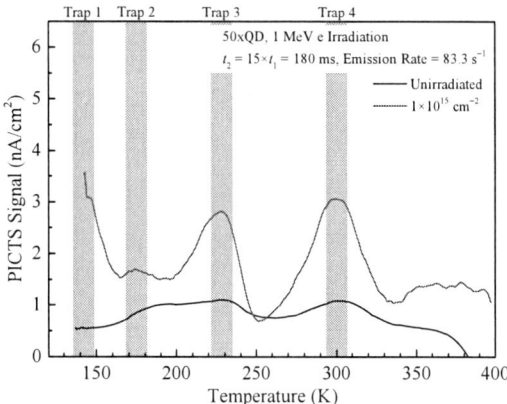

Fig. 6. PICTS spectra of the 50xQD sample with different electron fluences.

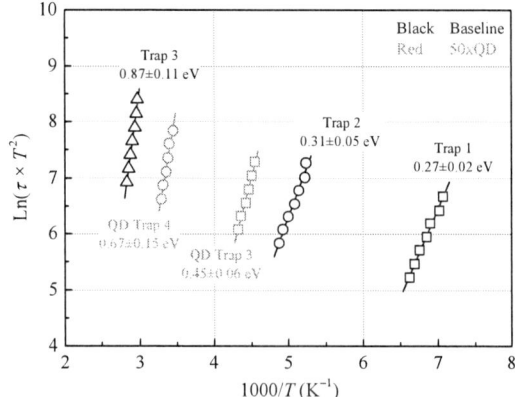

Fig. 7. Arrhenius plot of the baseline (black) and 50xQD samples (red) irradiated with 1 MeV electrons at 1×10^{15} cm^{-2}. τ in the ordinate is the inverse of emission rate.

TABLE I
SUMMARY OF OBSERVED TRAP LEVELS

Sample	Trap No.	Activation Energy (eV)	Capture Cross Section (cm^2)	Potential ID
Baseline	Trap 1	0.27	4×10^{-15}	E3 [13]
	Trap 2	0.31	7×10^{-17}	H2 [14]
	Trap 3	0.87	1×10^{-12}	E5 [13]
50xQD	QD Trap 1	–	–	E3
	QD Trap 2	–	–	unknown
	QD Trap 3	0.45	8×10^{-15}	unknown
	QD Trap 4	0.67	1×10^{-13}	E4 [13]

Traps 1 and 3 in the baseline sample are thought to be electron traps labeled E3 and E5, respectively, and Trap 2 is thought to be a hole trap labeled H2 [12, 13]. E3 and E5 have been identified As vacancy - As interstitial (V_{As}-As_i) pairs with different distances separating the vacancy and interstitial. Also, H2 has been identified a complex defect involving As_i [14]. The peak intensity of E3 was much higher than that of H2 and E5, suggesting that E3 was the dominant trap centers in the baseline sample. In the case of 50xQD sample, QD Trap 1 is thought to be the same origin as Trap 1 in the baseline sample (E3) because both peaks showed similar trends. Also, QD Trap 4 is thought to be E4 which has also been identified V_{As}-As_i pair [12, 13]. Although the reason that the intensity of E4 was more pronounced than that of E5 in the 50xQD sample is not clear at the present stage, the distance between V_{As} and As_i to stabilize in the QD layers might be different from the baseline sample. Since the peak intensity of QD Traps 3 and 4 was significantly high, it is likely that these defect levels act as recombination centers and make the Isc degradation of 50xQD sample more serious than that of baseline sample. Further investigation is necessary for deeper understanding.

IV. SUMMARY

We proposed the radiation degradation characterization method of QD solar cells, which has p-i-n structure, using PICTS technique and demonstrated the defect characterization. By analyzing PICTS spectra based on the rate window method, 3 defect levels in the baseline sample and 2 defect levels in the 50xQD sample were successfully characterized. We also discussed the origin of observed defect levels by reference to previous literatures about defects in GaAs. All trap levels in the baseline sample was successfully identified, whereas QD Traps 2 and 3 in the 50xQD sample was not identified. Relatively high peak intensities appeared in QD Traps 3 and 4, indicating the possibility that these deep defect levels contributed to the serious Isc degradation of 50xQD sample.

ACKNOWLEDGEMENT

We would like to thank Mr. Jiro Harada and Mr. Mitsunobu Sugai of Advanced Engineering Services Co., Ltd. (AES) for technical supports.

REFERENCES

[1] S-I. Sato, K. J. Schmieder, S. M. Hubbard, D. V. Forbes, J. H. Warner, T. Ohshima, and R. J. Walters, "Defect characterization of proton irradiated GaAs pn-junction diodes with layers of InAs quantum dots," *Journal of Applied Physics*, vol. 119, pp. 185702-1-8, 2016.

[2] D. V. Lang, "Deep level transient spectroscopy: A new method to characterize traps in semiconductors," *Journal of Applied Physics*, vol. 45, pp. 3023-3032, 1974.

[3] J. H. Warner, R. J. Walters, S. R. Messenger, G. P. Summers, S. M. Khanna, D. Estan, L. S. Erhardt, and A. Houdayer, "High-Energy Proton Irradiation Effects in GaAs Devices," *IEEE Transactions on Nuclear Science*, vol. 51, pp. 2887-2895, 2004.

[4] J. H. Warner, S. R. Messenger, R. J. Walters, G. P. Summers, M. J. Romero, and E. A Burke, "Displacement Damage Evolution in GaAs Following Electron, Proton and Silicon Ion Irradiation," *IEEE Transactions on Nuclear Science*, vol. 54, pp. 1961-1968, 2007.

[5] Ch. Hurtes, M. Boulou, A. Mitonneau, and D. Bois, "Deep-level spectroscopy in high-resistivity materials," Applied Physics Letters, vol. 32, pp. 821-823, 1978.

[6] P. Blood and J. W. Orton, *The Electrical Characterization of Semiconductors: Majority Carriers and Electron States*, London, London: Academic Press, 1992.

[7] T. Sugaya, T. Amano, M. Mori, S. Niki, and M. Kondo, "Highly Stacked and High-Quality Quantum Dots Fabricated by Intermittent Deposition of InGaAs," *Japanese Journal of Applied Physics*, vol. 49, pp. 030211-1-3, 2010.

[8] T. Sugaya, Y. Kamikawa, S. Furue, T. Amano, M. Mori, and S. Niki, "Multistacked quantum dot solar cells fabricated by intermittent deposition of InGaAs," *Solar Energy Materials & Solar Cells*, vol. 95, pp. 163-166, 2011.

[9] T. Sugaya, T. Amano, M. Mori, and S. Niki, "Highly stacked InGaAs quantum dot structures grown with two species of As,"

Journal of Vacuum Science & Technology B, vol. 28, pp. C3-C4-C8, 2010.

[10] T. Ohshima, S. Sato, M. Imaizumi, T. Nakamura, T. Sugaya, K. Matsubara, S. Niki, "Change in the electrical performance of GaAs solar cells with InGaAs quantum dot layers by electron irradiation," *Solar Energy Materials & Solar Cells*, vol. 108 pp. 263–268, 2013.

[11] T. Nakamura, T. Sumita, M. Imaizumi, T. Sugaya, K. Matsubara, S. Niki, A. Takeda, Y. Okano, S-I. Sato, and T. Ohshima, "Radiation Response of the Fill-Factor for GaAs Solar Cells with InGaAs Quantum Dot Layers," in 40th IEEE Photovoltaic Specialist Conference, 2014, pp. 2886-2891.

[12] F. H. Eisen, K. Bachem, E. Klausman, K. Koehler, and R. Haddad, "Ion irradiation damage in ntype GaAs in comparison with its electron irradiation damage," Journal of Applied Physics, vol. 72, pp 5593-5601, 1992.

[13] D. Pons and J. C. Bourgoin, "Irradiation-induced defects in GaAs," *Journal of Physics C: Solid State Physics*, vol. 18, pp. 3839-3871, 1985.

[14] D. Stievenard, X. Boddaert, J.C. Bourgoin, and H. J. von Bardeleben, "Behavior of electron-irradiation-induced defects in GaAs," *Physical Review B*, vol. 41, pp. 5271-5279, 1990.

Effect of luminescence coupling between InGaP and GaAs subcells to external quantum efficiency in triple-junction solar cells

Mitsunobu Sugai [1], Mitsuru Imaizumi [2], Tetsuya Nakamura [2], and Takeshi Ohshima [3]

[1] Advanced Engineering Services Co., Ltd. (AES), 1-6-1, Tsukuba, Ibaraki, 305-0032, Japan.

[2] Japan Aerospace Exploration Agency (JAXA), Tsukuba, Ibaraki, 305-8505, Japan.

[3] National Institute for Quantum Science and Technology (QST), Takasaki, Gunma, 370-1292, Japan.

Abstract — Luminescence coupling, which is reabsorption of recombination radiation in multi-junction solar cell, induces artifact response on external quantum efficiency (EQE). We have been investigated the relation between luminescence coupling and EQE in triple-junction solar cells (3J) for space applications. In this study, we focused the luminescence coupling induced by the emission from InGaP-top cells. LED light sources were introduced as bias light. The strength of the luminescence coupling between InGaP-top and GaAs-middle cells in newer-design 3J cells is higher than that of older-design 3J cells. The strength of "series coupling", which is induced in Ge-bottom cells by the coupling between InGaP-top and GaAs-middle cells, is about an order of magnitude lower than the strength of coupling between InGaP-top and GaAs-middle cells. In 3J cells that have the high intensity of emission from InGaP-top cell, the EQE of Ge-bottom cells only with the activation of InGaP-top cells does not need artifact correction.

Index term – Multi-junction solar cells, External quantum efficiency, Luminescence coupling, series coupling.

I. INTRODUCTION

In multi-junction solar cells, lower subcells absorb recombination radiation from upper subcells and produce photocurrent. It is called "luminescence coupling" [1], [2], [3]. The luminescence coupling causes the artifact response on external quantum efficiency (EQE) spectra [2], [4], [5]. The GaAs subcell layers emit strong luminescence; thus, the EQE of lower subcells under GaAs subcell shows high artifact in the GaAs absorption wavelength region. For example, Ge-bottom cells in lattice-matched triple-junction solar cells (LM-3J) exhibit such artifact in EQE. The artifact EQE makes the EQE analyses of subcells in multi-junction solar cells difficult. Therefore, various correction methods have been proposed [2], [3], [4].

In principle, the artifact EQE on GaAs-middle cells can be induced by luminescence from InGaP-top cell. However, the luminescence intensity of InGaP-top cells is lower than that of GaAs-middle cells in present LM-3J cells.

We studied the luminescence coupling between InGaP-top and GaAs-middle cells (LC$_{\text{InGaP-GaAs}}$) on relatively new design, high efficiency LM-3J. In addition, we also investigated the effect of LC$_{\text{InGaP-GaAs}}$ to the artifact EQE on the Ge-bottom cell.

II. EXPERIMENTS

We selected the LM-3J cells for space applications that have the conversion efficiency of about 30% for this study. The samples were classified into three groups according to the design generation. The older design is as Group I, the newer as Group III, and the intermediate as Group II.

We utilized the spectral response measuring equipment at Tsukuba space center to measure EQE. We set the bias voltage at 1.5V to measure the EQE of GaAs-middle cell. We controlled the bias voltage for EQE of Ge-bottom cell to avoid the effect of leak current on Ge-bottom cell [6].

LED light sources were introduced as bias lights. We utilized the blue LED (center wavelength of emission: 470nm, rated current: 200mA) for the InGaP-top bias light and infrared LED (center wavelength of emission: 1200nm, rated current: 600mA) for the Ge-bottom bias light. Fig. 1 shows the radiation spectra of LEDs at the rated currents.

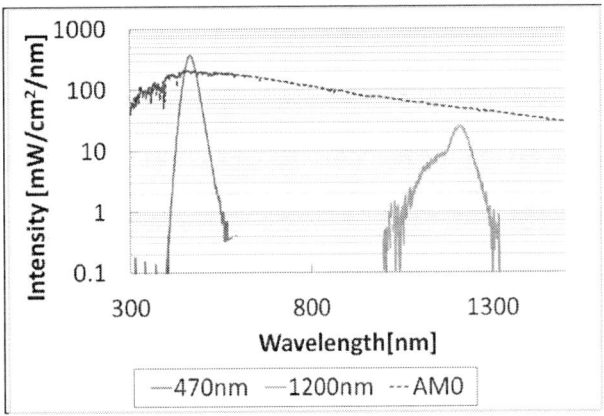

Fig. 1. Radiation spectra of LEDs for bias lights at the rated current. (rated current:470nm- 200mA, 1200nm- 600mA)

We introduced "response coefficient" for expressing the strength of luminescence coupling [6], [7]. When GaAs-middle cell is kept current-limiting as the targeted cell, the response coefficient is defined as the increasing rate of the output current against the power of the blue LED bias light to InGaP-top cell. It is expressed as

$$k = \frac{hc}{q\lambda}\frac{dJ_{OUT}}{dU_{IN}} \qquad (1)$$

Where J_{OUT} is the output current density of the 3J cell, U_{IN} is the input power of the blue LED bias light, h is plank constant, c is light speed, q is the elementary charge, and λ is the center wavelength of the blue LED bias light. The k from the equation (1) is the dimensionless value as the quantum efficiency.

III. RESULTS

1) InGaP-GaAs coupling & GaAs-EQE

We estimated the strength of $LC_{InGaP\text{-}GaAs}$. Fig. 2 shows the comparison of the response coefficient from $LC_{InGaP\text{-}GaAs}$ (k_{470}^{MID}). Group I indicated the lowest k_{470}^{MID} among the three groups. The highest k_{470}^{MID} was observed on Group III. The average of k_{470}^{MID} of Group II and III was about 7 and 14 times higher than those of Group I. Fig. 3 shows the electroluminescence spectra of the representative samples from each cell group when the injection current is 100mA. The peak intensity of the InGaP-top cells of Group II and Group III was about 7 and 19 times higher than that of Group I. The magnitude correlation of k_{470}^{MID} is same to that of EL intensity of the InGaP-top cell. Namely, the difference of the response coefficient from $LC_{InGaP\text{-}GaAs}$ reflected the difference of the intensity of the recombination radiation in InGaP-top cell. It means that the effect of $LC_{InGaP\text{-}GaAs}$ is enhanced with improvement of quality of InGaP-top cell.

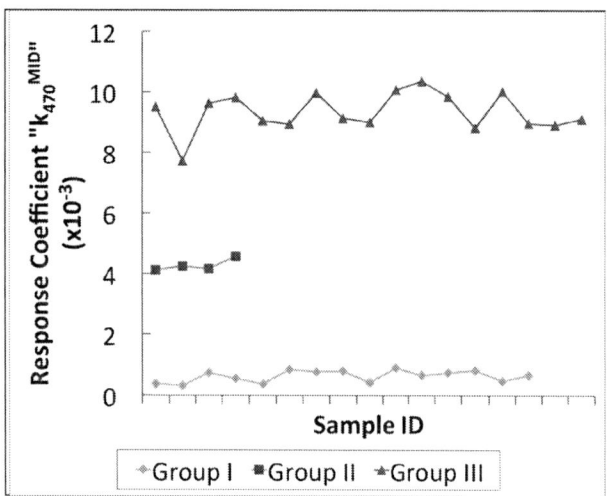

Fig. 2. Comparison of the response coefficient of GaAs-middle cell on 30%-3J of LM-3J by 470nm blue LED light.

We examined the change of the artifact EQE of GaAs-middle cell (GaAs-artifact) against the intensity of the blue LED bias light. Fig. 4 shows the comparison of GaAs-EQE of Group I and III by the blue LED bias light. Group III cell indicated higher GaAs-artifact and lower GaAs-EQE correspond to those of Group I. Fig. 5 shows the change of

GaAs-artifact with the blue LED bias light. The vertical axis shows the averaged GaAs-artifact in the wavelength region of 400-500nm. The horizontal axis shows the power input of the blue LED bias light. Group I did not indicate any relation of the GaAs-artifact with the bias light power, while the GaAs-artifact of Group II and III increased with the bias light power. Group III that has larger k_{470}^{MID} than that of Group II indicated the greater increase of the GaAs-artifact.

From the result, it was verified that the effect of $LC_{InGaP\text{-}GaAs}$ in older-design generation LM-3J to the EQE measurement is negligible. It is expected that the strength of $LC_{InGaP\text{-}GaAs}$ is getting higher with the quality improvement of LM-3J and the GaAs-artifact appears. In other words, the newer designs LM-3J need same correction on GaAs-EQE as Ge-EQE [4], [8].

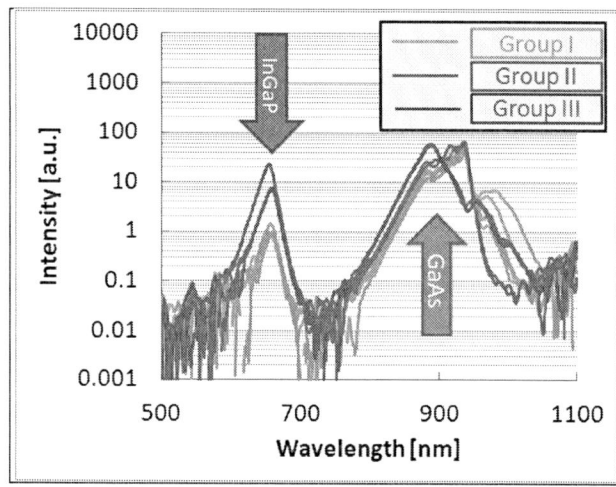

Fig. 3. Comparison of electroluminescence spectra at current input of 100mA.

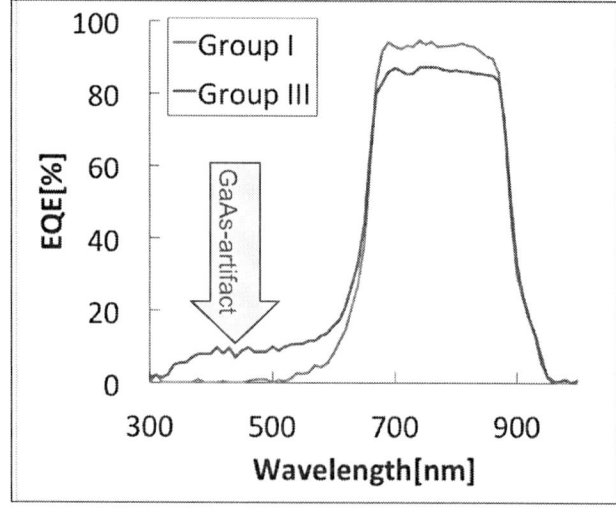

Fig. 4. Comparison of GaAs-EQE between Group I and III by the blue LED bias light.

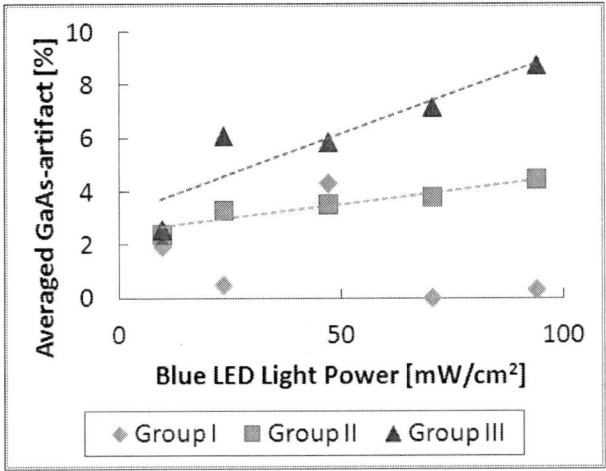

Fig. 5. Change of GaAs-artifact on intensity of 470nm LED bias light.

2) Series Coupling & Ge-EQE

In LM-3J, "series coupling" that the coupling photocarriers in GaAs-middle cell induced $LC_{GaAs-Ge}$ was expected. In particular, the strong $LC_{InGaP-GaAs}$ in Group III cells induces the greater number of photocarriers than that in Group I and II. We evaluated the series coupling effect induced by the blue LED bias light on Ge-bottom cell. Fig. 6 shows the comparison between k_{470}^{MID} and the response coefficient of the Ge-bottom cell induced by the blue LED bias light to the InGaP-top cell (k_{470}^{BOT}). The obtained k_{470}^{BOT} are an order of magnitude smaller than k_{470}^{MID}. Although k_{470}^{MID} showed the clear differences in each cell group, k_{470}^{BOT} did not exhibit significant difference among the three groups. In other words, the strength of the series coupling in the Ge-bottom cell did not depend on the intensity of the recombination radiation in InGaP-top cell. We think the effect of the series coupling can be negligible on EQE correction.

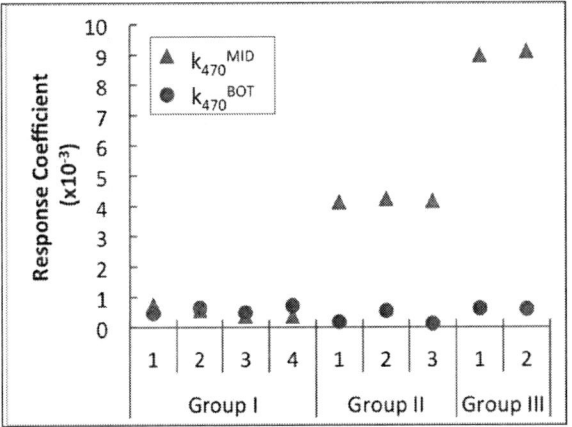

Fig. 6. Comparison of response coefficient on Ge-bottom cell induced by blue LED bias light.

From above, we expected that Ge-EQE without the artifact effect could be measured by only InGaP-top cell activation using single blue LED bias light (single-bias method). Figs. 7 and 8 show the comparison between the present method, which is our usual method using Xenon light and color filters, and the single-bias method. The red dot line shows the uncorrected result with the single-bias method. The correction is based on the luminescence coupling model [8]. The injection current of blue LED light was set at rated current, 200mA. It corresponds to the maximum light power shown in Fig. 5. The bias voltage was optimized for each sample cell.

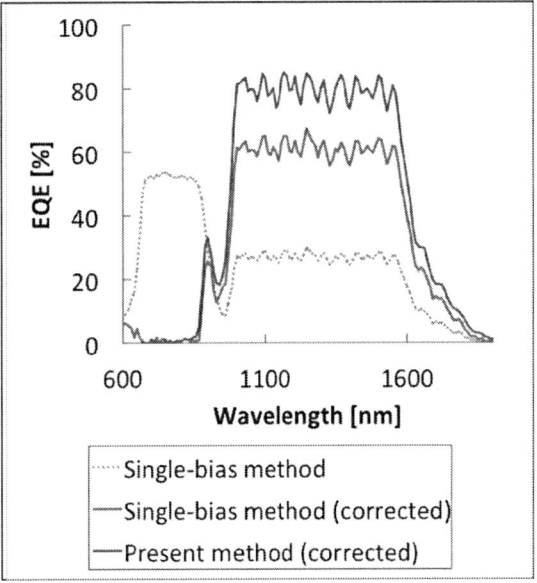

Fig. 7. Comparison of EQE measurement method on Group I.

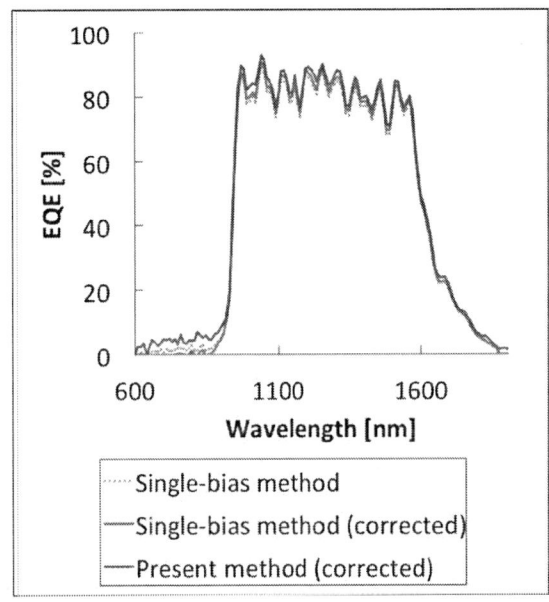

Fig. 8. Comparison of EQE measurement method on Group III.

The Ge-artifact of Group I with the single-bias method shown in Fig 7 was larger than that with the present method. After the correction, the Ge-EQE with the single-bias method is lower than the Ge-EQE with the present method. The Group III shown in Fig. 8 indicates lower Ge-artifact and higher Ge-EQE with the single-bias method compared to those with present method. In addition, the measured Ge-EQE with single-bias method is equivalent to the corrected Ge-EQE with the-present method. Because the intensity of color bias light for the present method was sufficient to make the Ge-bottom cells current limiting, the Ge-artifact with the present method is caused by luminescence coupling. Namely, the corrected Ge-EQE with the present method is considered to be the proper value. The result shown in Fig. 8 indicates that the single-bias method could measures the proper Ge-EQE on Group III cell. On the other hand, the single-bias method cannot measure the proper Ge-EQE on Group I shown in Fig. 7. It shows that the Ge-bottom cell of the Group I cell was not current limiting by the single-bias method.

We discuss the condition for the single-bias method. The single-bias method can be applied in the case the coupling current of GaAs-middle cells is larger than the leak current of Ge-bottom cells. If the intensity of the bias light for InGaP-top cell is constant, k_{470}^{MID} and the shunt resistance of Ge-bottom cell, R_{SH}^{BOT}, determine the condition for current limiting of Ge-bottom cell. Fig. 9 shows the relation between the Ge-EQE with the single-bias method and the product of k_{470}^{MID} and R_{SH}^{BOT} on the samples shown in Fig. 6. This relation shows that the single-bias method can be applied to the cells that satisfy the threshold, $k_{470}^{MID} \times R_{SH}^{BOT} > 10$, in the case under 100mW/cm^2 intensity of 470nm blue LED light. The threshold depends on the intensity of the bias light. Higher intensity of the blue bias light reduces the threshold and expands the application range.

We clarified that single-bias method is effective to measure more accurate Ge-EQE if the luminescence coupling between InGaP-top and GaAs-middle cell is strong.

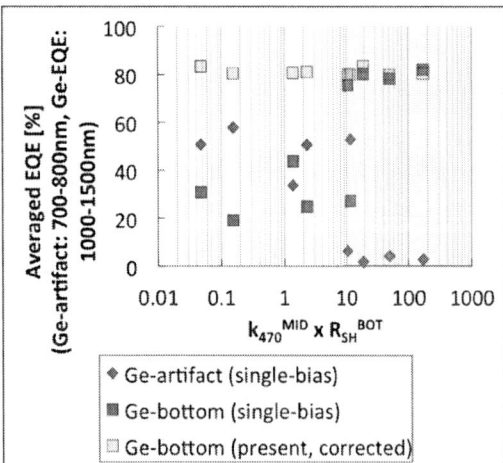

Fig. 9. Corelation between the Ge-EQE measured by single-bias measurement and the product of k_{470}^{MID} and R_{SH}^{BOT}.

IV. SUMMARY

We investigated the relation between artifact EQE of GaAs-middle cells and the luminescence coupling induced by the emission from InGaP-top cells in space 3J cells.

The InGaP-top cells in newer-design LM-3J indicated stronger luminescence than those in older-design LM-3J cells. The behavior of the artifact EQE of the GaAs-middle cells depended on the strength of the luminescence coupling between the InGaP-top and GaAs-middle cells.

The effect of series coupling, which is induced by the luminescence coupling between InGaP-top and GaAs-middle cell, to Ge-bottom cell was small compared with the coupling between InGaP and GaAs. In case of 3J cells that have high intensity of the emission of InGaP-top cell, the Ge-EQE can be measured only with the activation of InGaP-top cells, called single-bias method. The Ge-EQE with single-bias method does not need artifact correction. The threshold of the single-bias method is given by the product of the strength of luminescence coupling and the shunt resistance of Ge-bottom cell.

REFERENCES

[1] Vasiliki Paraskeva, Matthew Norton, Maria Hdjipanayi, Mauro Pravettoni, George E. Georghiou, "Luminescent emission of multi-junction InGaP/InGaAs/Ge PV cells under high intensity irradiation", Solar Energy Materials & Solar Cells 134 (2015) 175-184.

[2] Swee Hoe Lim, Jing-Jing Li, Elizabeth H. Steenbergen, Yong-Hang Zhang, "Luminescence coupling effects on multijunction solar cell external quantum efficiency measurement", Progress in Photovoltaics, Volume 21, Issue 3, May 2013.

[3] Tomah Sogabe, Akio Ogura, Mitsuyoshi Ohba, and Yoshitaka Okada, "Self-consistent modeling and novel approach for luminescence coupling measurement in III-V multi-junction solar cells", 23rd PVSEC, Oct 28-Nov 1, 2013, Taipei, Taiwan.

[4] Gerald Siefer, Carsten Baur, and Andreas W. Bett, "External Quantum Efficiency Measurements of Germanium Bottom Subcells: Measurement Artifacts and Correction Procedures", 35th PVSC, June 20 - 25, 2010, Honolulu, Hawaii, USA.

[5] Vasiliki Paraskeva, Maria Hdjipanayi, Matthew Norton, Mauro Pravettoni, George E. Georghiou, "Voltage and light bias dependent quantum efficiency measuremnts of GaInP/GaInAs/Ge triple junction devices", Solar Energy Materials & Solar Cells 116 (2015) 55-60.

[6] Mitsunobu Sugai, Jiro Harada, Mitsuru Imaizumi, and Takeshi Ohshima, "The study of dependency of external quantum efficiency of triple-junction solar cells on measurement conditions", 40th PVSC, June 9-13, 2014, Denver, Colorado, USA.

[7] Mitsunobu Sugai, Mitsuru Imaizumi, Tetsuya Nakamura, and Takeshi Ohshima, "The effect of luminescence coupling in external quantum efficiency measurement of multi-junction solar cells", 42nd PVSC, June 14-19, 2015, New Orleans, Louisiana, USA.

[8] Myles A. Steiner, John F. Geisz, Tom E. Moriarty, Ryan M. France, William E. McMahon, Jerry M. Olson, Sarah R. Kurtz, and Daniel J. Friedman, "Measuring IV curves and Subcell Photocurrents in the Presence of Luminescent Coupling", IEEE, Journal of Photovoltaics, Vol. 3, No. 2, Apr 2013.

[9] Tetsuya Nakamura, Mitsuru Imaizumi, Shin-ichiro Sato, Takesihi Ohshima, "Estimation of Subcell Photocurrent in IMM3J using LED Bias Light", 39th PVSC, June 16-21, 2013, Tampa, Florida, USA.

Lightweight Carbon Fiber Mirrors for Solar Concentrator Applications

Nina Vaidya[1], Michael D. Kelzenberg[1], Pilar Espinet-Gonzalez[1], Tatiana G. Vinogradova[1, 2], Jing-Shun Huang[1], Christophe Leclerc[1], Ali Naqavi[1], Emily C. Warmann[1], Sergio Pellegrino[1], Harry A. Atwater[1]

[1]California Institute of Technology, Pasadena, CA 91125, United States
[2]Northrop Grumman Aerospace Systems, Azusa, CA 91702, United States

Abstract — Lightweight parabolic mirrors for solar concentrators have been fabricated using carbon fiber reinforced polymer (CFRP) and a nanometer scale optical surface smoothing technique. The smoothing technique improved the surface roughness of the CFRP surface from ~3 μm root mean square (RMS) for as-cast to ~5 nm RMS after smoothing. The surfaces were then coated with metal, which retained the sub-wavelength surface roughness, to produce a high-quality specular reflector. The mirrors were tested in an 11x geometrical concentrator configuration and achieved an optical efficiency of 78% under an AM0 solar simulator. With further development, lightweight CFRP mirrors will enable dramatic improvements in the specific power, power per unit mass, achievable for concentrated photovoltaics in space.

Index Terms — mirrors, solar energy, optical device fabrication, polymers, ray tracing, space solar, optical design.

I. INTRODUCTION

We are seeking to develop technologies that enable cost-effective space-based solar power (SSP). SSP has long been proposed to meet earth's baseload electrical power needs with solar energy, by operating large-scale solar power stations in space and beaming the energy wirelessly to earth [1] or utilizing the DC power in space. The building block of our proposed power station is the 'tile,' depicted in Figure 1: a ~10 x 10 cm modular element which performs solar photovoltaic energy collection, conversion to radio frequency energy, and transmission of the energy towards earth-based receivers.

Due to high space launch costs, the key challenge is to increase the specific power, or power per unit mass, of SSP technology. Lightweight concentrating optics can increase the

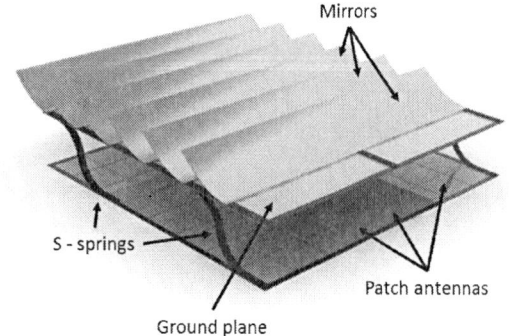

Fig. 1.   Conceptual rendering of space solar power tile.

specific power of space photovoltaic (PV) energy converters many fold, because reflective or refractive optics can generally be realized at dramatically lower area density (mass per unit area) than can solar cells and their radiation shielding [2]. Parabolic mirrors have been used extensively for concentrated photovoltaics, including in space applications [3, 4]. Here, we report a lightweight parabolic mirror fabrication process based on using cast carbon fiber reinforced polymer (CFRP) parabolas with a surface smoothing technique [5] to produce specular and highly reflective mirror surfaces.

II. DESIGN & FABRICATION

The PV concentrator shown in Figure 1 comprises parallel parabolic mirror troughs (parabolic in one plane and linear along the length) each having a focal line at the top back edge of the neighboring mirror. A row of multi-junction solar cells is attached at this point to collect the focused sunlight. This geometry is particularly well suited for space applications because the mirror troughs are foldable for efficient packing prior to launch, and also provide heatsinking, radiative cooling, and radiation shielding for the cells [4, 6].

This paper focuses on the fabrication and testing of the parabolic reflectors for this tile concentrator concept. We have investigated two fabrication approaches for the parabolic reflectors. Initially, we used metalized Kapton membranes as the reflectors, with parabolic supports placed at either end of the trough to impart the correct shape. Kapton is a well-known space-grade polymer and is commercially available in thin sheets with relatively smooth and specular surfaces. However, it was difficult to fabricate the relatively thick metal layer (2–10 μm Al), which is required for thermal conductivity, without degrading the specular reflectance and shape accuracy of the Kapton membrane reflectors. Furthermore, once the cells were mounted at the back edge of the kapton parabolic troughs, the shape deformed from the attachment and thermal stressed of the cells.

To improve the shape accuracy and optical efficiency, we used thin CFRP to fabricate the reflectors. Carbon fiber composites have excellent strength to mass ratio and find much use in aerospace. For our application, thin CFRP sheets are particularly promising because (a) they can be cast to the desired shape, (b) they are flexible but spring back to shape, (c) bare CFRP typically has high thermal emissivity, and by correct choice of fiber type and orientation, they can offer high in-plane thermal conductivity [7]. However, thin CFRP castings are difficult to form into precision shapes for optical applications,

978-1-5090-5606-4/17 $31.00 © 2017 IEEE

and typically have rough and non-specular surfaces which are unsuitable for direct use as a mirror substrate [8].

The composite parabolic reflectors were manufactured from unidirectional tape (T800 prepreg, 17 g/m²), arranged in 8 plies with stacking sequence [0/90/+45/−45]ₛ. A steel mold was machined, providing a convex surface of the desired parabolic profile extended by tangential flats on both sides. After lay-up, the composite was vacuum-bagged and cured in an autoclave furnace. Due to the elevated temperatures involved, thin composite materials tended to deform after curing because of imbalanced thermal stresses. Thus an iterative process was used to create a mold which yielded the desired parabolic profile in the castings. The shape of each casting was measured with a FARO ScanArm 3D scanner tool, and if necessary, another mold was machined to correct for any systematic shape errors observed.

Although this process produced CFRP sheets of the desired shape, their surfaces were rough, and they exhibited a diffuse optical appearance. The next challenge was to create a high-quality specular mirror on the surface of the CFRP castings, without deforming the parabolic shape. The surface roughness issue was solved by a novel smoothing technique [5], in which a resin mixture is applied to the surface and allowed to settle. Surface tension produces a smooth and conformal surface. The ultraviolet (UV) cure process minimizes shrinkage of the polymer, which maximizes smoothness and shape accuracy. A completed mirror is shown in Figure 2(a).

The procedure for creating mirrors from CFRP castings was:

1. Clean part, then oven dry.
2. Apply a thin layer of polymer on the surface.
3. Degas in vacuum chamber.
4. Brush off excess polymer and degas again, if necessary.
5. Cure with UV exposure.
6. Deposit reflector layers onto the UV-cured surface

The reflector layers applied to the smoothing polymer comprised a 10 nm Cr adhesion layer, a 120 nm Ag reflector layer, and a 10 nm SiO₂ protective layer. All were deposited by electron beam evaporation. Prior to smoothing or metallization, the CFRP average thickness was about 180 μm.

III. SHAPE CHARACTERIZATION

To determine the shape accuracy and performance potential of the CFRP reflectors, we scanned their shape using a FARO ScanArm 3D scanner. This produced point cloud data describing the mirror surfaces with ~25 μm accuracy, which were recorded with a point density of ~2500/cm². A typical point cloud data set for the fabricated mirror is plotted in Figure 2(b).

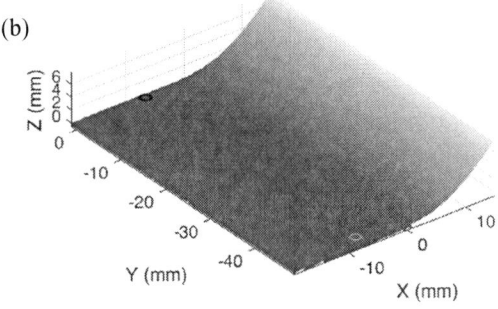

Fig. 2. (a) Photograph of the fabricated CRFP mirror after smoothing and Ag deposition. (b) Point cloud data for the same specimen acquired using a 3D scanner. The individual points are not distinguishable at this resolution. The coordinate system is aligned such that the nominal vertex of the parabolic profile occurs at $x = 0$ and $z = 0$. The raw data was cropped at $x = 15$ mm. The colormap is indexed to the z-value of each point. The black and red circles indicate the position of the left and right reference features, respectively, which were used to define the measurement coordinate system i.e., same as the raytracing coordinate system.

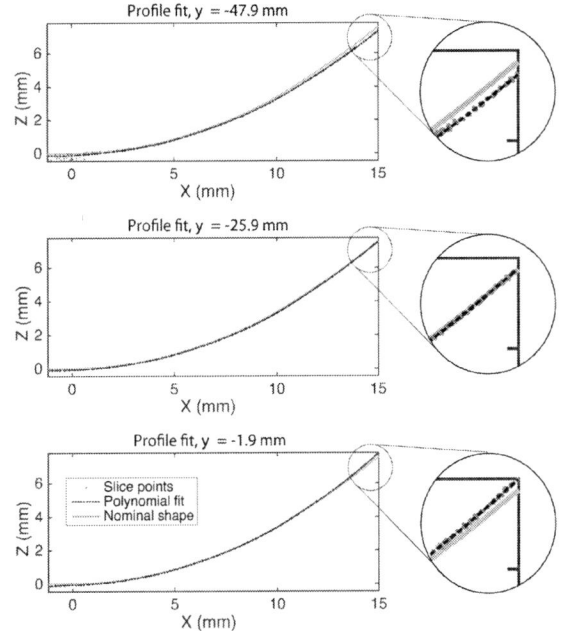

Fig. 3. Polynomial fit to point cloud data, and comparison to nominal parabolic profile, for three selected slices within the point cloud data set: the rightmost slice (top), the center slice (middle), and the leftmost slice (bottom). Plots use coordinate system of Figure 2(b).

978-1-5090-5606-4/17 $31.00 © 2017 IEEE

(a)

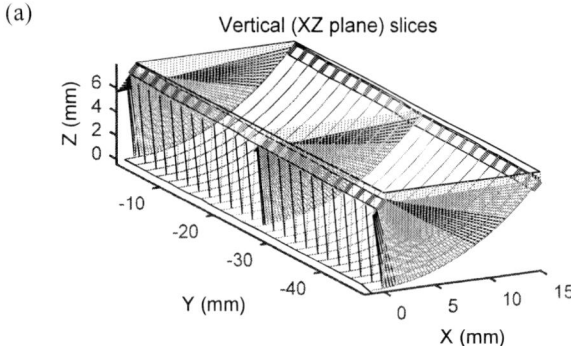

Vertical (XZ plane) slices

(b)

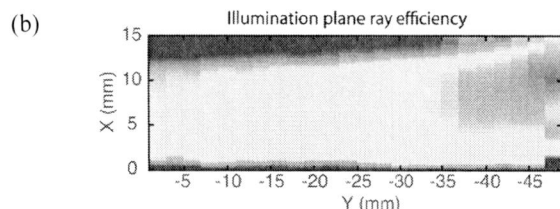

Illumination plane ray efficiency

(c)

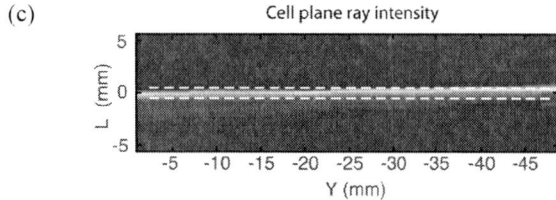

Cell plane ray intensity

Fig. 4. Ray tracing analysis for 2° incidence angle. All plots use the coordinate system of Figures 2(b) and 3.

(a) 3D plot showing the shape ('slices') used for ray tracing. The receiver cell positions (1 mm width) are also shown. A y-indexed color gradient is applied to better illustrate depth. Ray tracing paths are illustrated for array of rays the leftmost, center, and rightmost analysis planes, and for just the first ray of each slice (that is, nearest x=0). Incident rays are gray, reflected rays which strike the receiver cell are red, and reflected rays which miss the receiver cell are **black**.

(b) Illumination plane ray efficiency plot, indicating which areas of the reflector successfully reflect incident light to the receiver cell (yellow), and which areas of the reflector cause the light to miss the receiver cell (blue). Intermediate values occur due to the finite angular width of the source considered (1.5° disk*).

(c) Cell plane ray intensity plot, showing the intensity of light reaching the cell plane. The extent of the cell (1 mm) is indicated by the white dashed lines at $L = \pm 0.5$ mm.

To evaluate the accuracy of the mirror shape, we compared cross-sections of the surface data to the desired (nominal) parabolic profile. To assess the impact of the shape on the mirror's utility as a PV concentrator, we performed 2.5D raytracing to calculate the potential optical efficiency.

Following coordinate system alignment, the point cloud data were split and flattened into 2D x-z cross sections ('slices')

---

*We use a source width i.e., angular radius of 1.5° because it best approximates the illumination from our solar simulator. The sun's angular radius is ~0.25°.

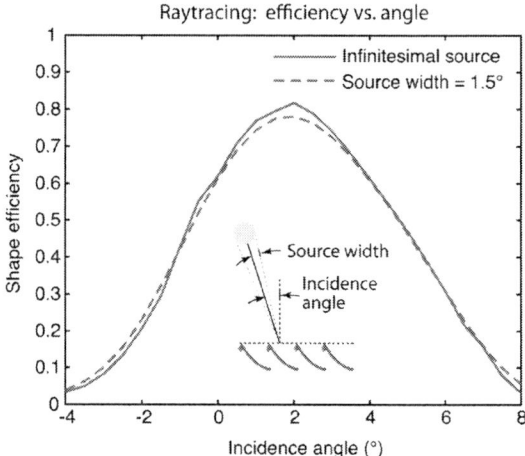

Fig. 5. Angular acceptance versus shape efficiency calculated by ray tracing.

along the length of the reflector. In the illustrated case, the slice width was 2 mm. Then, high-order polynomials were fit to the 2D point data for each slice (8th order fits were used). Figure 3 shows the slice point data, the polynomial fit, and the nominal parabolic profile, for three selected slices within the data set: the rightmost slice (top), the center slice (middle), and the leftmost slice (bottom).

Figure 4(a) shows the polynomial fits and receiver cell position for all slices in the data set. The data is displayed in 3D coordinates by plotting each 2D profile at the y-value corresponding to that slice's center plane. The receiver cell positions were determined by calculating where cells would be mounted on the back side of the measured reflector, then the cell positions were translated by a fixed x offset corresponding to the reflector pitch in the concentrator design (here, 15 mm) which positioned the cells at the focal line of the reflector.

Shown in Figure 4(a) is a down-sampled ray diagram for each of three slices featured in Figure 3. Red colored rays reach the receiver cell, while black colored rays miss. Figure 4(b) shows the ray efficiency map for all slices as an intensity plot. The correspondence of the left, center, and right-most columns of the image in Figure 4(b), to the ray diagrams above in Figure 4(a), is apparent.

Figure 4(c) shows the intensity distribution of rays reaching the cell plane, relative to the centerline of the cells in said plane (which defines $L=0$ in this image). It is observed that the position of peak intensity shifts slightly, relative to the cell position, over the length of the concentrator. The cause of this shift is evident from examining the three shape cross-sections plotted in Figure 3. A slight twist in the reflector shape has caused the focal line to become misaligned with the nominal focal line i.e., cell centerline position.

For this CFRP mirror, at an incidence angle of 2°, raytracing predicts that ~80% of incident light upon specular reflection from the mirror should reach the receiver cell (see Figure 5). We call this value the *shape efficiency* to distinguish it from true optical efficiency. Note that shape efficiency does not include scattering and absorption losses in the reflector, skewed rays, nor any shading or reflectance losses at the cell; furthermore, several simplifying assumptions have been made in its calculation. Nevertheless, shape measurements and shape efficiency calculations have proven to be useful analysis techniques in our pursuit of improved concentrator design and performance.

IV. SURFACE ROUGHNESS CHARACTERIZATION

An as-cast CFRP surface was characterized using laser scanning confocal microscopy due to the scale of the roughness of the CF surface. The RMS surface roughness was found to be 3 μm (Figure 6). A similarly made CFRP parabolic sample was smoothed using UV curable polymers and reflective layers deposited as described above to create the mirror that is demonstrated in this paper. An edge piece from the same finished mirror was characterized with an atomic force microscope (AFM) (Figure 7). The measured RMS surface roughness of the mirror was 4.5 nm, giving us 3 orders of magnitude improvement in surface roughness. Similar values of RMS roughness were measured before and after metallization.

For surfaces with subwavelength roughness, assuming Gaussian distribution of surface height, fraction of light scattered upon reflection is given by [8, 9] as

$$1 - \exp\left[-\left(\frac{4\pi\sigma\cos\theta}{\lambda}\right)^2\right]$$

where σ is the RMS roughness, θ is the incidence angle, and λ is the wavelength. If we desire to limit scattering loss to 2% at normal incidence, a mirror with 4.5 nm surface roughness is suitable for wavelengths above 400 nm.

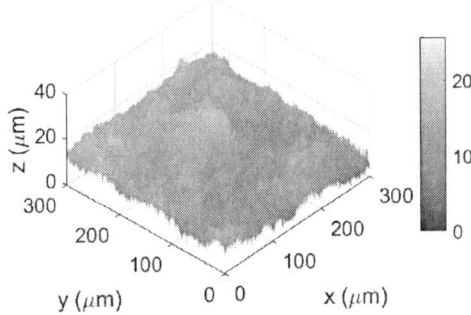

Fig. 6. Surface topography rendering of as-cast carbon fiber surface, measured by laser scanning confocal microscopy. The RMS surface roughness calculated from data is 3 μm.

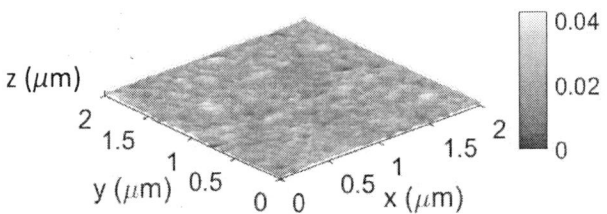

Fig. 7. Surface topography rendering of a finished mirror, measured by AFM. The RMS surface roughness is 4.5 nm.

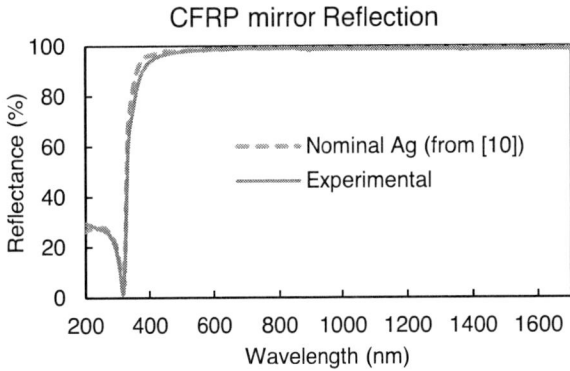

Fig. 8. Experimental reflectance data for a CFRP mirror, measured using a spectrophotometer. Also plotted is the nominal reflectance of polished Ag.

Figure 8 shows the experimental spectral reflectivity data for a CFRP reflector, taken using a Cary 5000 spectrophotometer, which agrees well with expectation for a polished Ag surface [10]. The roughness measurements and reflectance data confirm that our concentrator mirrors are adequately smooth for use over the solar spectrum down to wavelengths of about 400 nm, and that the primary limitation at shorter wavelengths is due to the Ag itself rather than surface scattering. Extending the usable range to ultraviolet wavelengths will require the use of a different reflective layer such as Al or dielectric-enhanced Ag, and may benefit from further reductions in surface roughness.

V. PERFORMANCE VALIDATION

The optimal alignment was determined by mounting the concentrator on a translation stage under an AM0 solar simulator, and adjusting the distance to the receiver to maximize short circuit current (I_{SC}). We determined the optical efficiency, as defined by:

$$\eta = \left(\frac{I_{SC\_mirror}}{I_{SC\_cell}} \frac{1}{C}\right)$$

where I_{SC_mirror} is the short circuit current recorded from the cell with the parabolic mirror under illumination, and I_{SC_cell} is short circuit current recorded from the cell alone under illumination without the concentrator. C is the geometric concentration ratio, defined as the ratio of the concentrator aperture area to the cell area, for a 2.5D concentrator this translates to a ratio of aperture width to cell width. The mirror has a nominal 15 mm optical width and the receiver cells used for this demonstration were 1.4 mm wide giving us a geometric concentration of 11x. Figure 9 shows a photograph of a concentrator pair during the alignment process. The reflected image of the cells (width of 1.4 mm) is enlarged and distributed across the entire focusing mirror (width of 15 mm), which indicates good optical performance. Figure 10 shows the measured optical efficiency of this concentrator pair at different aperture values. Here the peak optical efficiency of 77.5% corresponds to the point of zero position offset, which is the designed 15 mm aperture width. The peak optical efficiency was achieved at an incidence angle of 2 degrees, as previously predicted by ray tracing analysis of the same shape (Figure 5). The difference between the optical efficiency of 80% predicted by the ray trace analysis and the measured value can be accounted for by alignment errors and reflectance losses in the mirror coating.

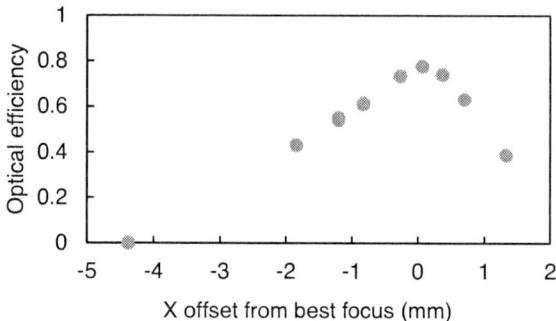

Fig. 10. CFRP mirror experimental optical efficiency curve. Peak optical efficiency is 77.5%

VII. CONCLUSIONS

A proof-of-concept 11x concentrator made with lightweight carbon fiber parabolic mirrors achieved up to 77.5% optical efficiency. The primary loss of efficiency was due to shape deviation from the nominal parabola, as the smoothed surface of the mirrors provided excellent specular reflectance over the visible and near infrared wavelengths. Further optimization of this system will include improving the shape, reducing the thickness of the polymer smoothing layer, and making thinner CFRP composite for overall mass reduction. In addition, we will investigate the stability of the system in vacuum and under elevated temperatures and thermal cycling consistent with operation in space. Overall, this UV curable nano-meter scale smoothing process for making CFRP mirrors offers promising performance for an ultra-light concentrated photovoltaic system intended for space applications.

ACKNOWLEDGEMENTS

We acknowledge funding from Northrop Grumman Corporation. This effort made use of facilities provided by the Kavli Nanoscience Institute, the Molecular Materials Research Center, the Resnick Institute, and the Joint Center for Artificial Photosynthesis at Caltech. We acknowledge the helpful contributions of Mark Kruer, Mike Levesque, and Erik Kurman at Northrop Grumman.

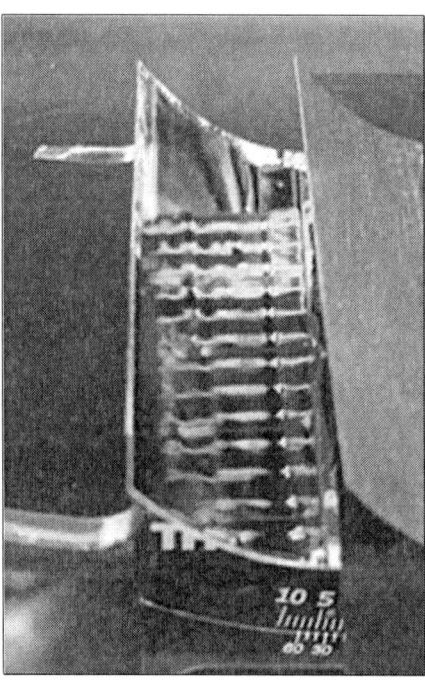

Fig. 9. Photograph of the image of the cells stretched clearly across the mirror indicating that the cells are positioned at the focus of this high quality mirror. Experimental configuration for testing the concentrator. The front CFRP parabolic shape supports a 1.4 mm cell at the top back edge.

REFERENCES

[1] P. Jaffe and J. McSpadden, "Energy Conversion and Transmission Modules for Space Solar Power," *Proceedings of the IEEE,* vol. 101, pp. 1424-1437, 2013.

[2] M. O'Neill, A. McDanal, M. Piszczor, P. George, M. Eskenazi, M. Botke, *et al.*, "Recent progress on the stretched lens array (SLA)," presented at the 18th Space Photovoltaic Research and Technology Conference, 2005.

[3] A. Rabl, "Optical and thermal properties of compound parabolic concentrators," *Solar Energy,* vol. 18, pp. 497-511, 1976.

[4] T. G. Stern, "Interim results of the SLATS concentrator experiment on LIPS-II (space vehicle power plants)," in *Conference Record of the Twentieth IEEE Photovoltaic Specialists Conference*, pp. 837-840 vol.2, 1988.

[5] N. Vaidya, T. E. Carver, and O. Solgaard, "Device fabrication using 3D printing," U.S. Provisional Patent Application No. 62/267,175, 2015.

[6] P. Espinet-Gonzalez, T. Vinogradova, M. D. Kelzenberg, A. Messer, E. Warmann, C. Peterson, *et al.*, "Impact of Space Radiation Environment on Concentrator Photovoltaic Systems," presented at the 44th IEEE PVSC, Washington DC, 2017.

[7] C. A. Silva, E. Marotta, M. Schuller, L. Peel, and M. O'Neill, "In-Plane Thermal Conductivity in Thin Carbon Fiber Composites," *Journal of Thermophysics and Heat Transfer,* vol. 21, pp. 460-467, 2007.

[8] J. B. Steeves, "Multilayer Active Shell Mirrors," Ph.D. Thesis, California Institute of Technology, Pasadena, CA, 2015.

[9] H. Davies, "The reflection of electromagnetic waves from a rough surface," *Proceedings of the IEE - Part IV: Institution Monographs,* vol. 101, pp. 209-214, 1954.

[10] P. B. Johnson and R. W. Christy, "Optical Constants of Noble Metals," *Physical Review B,* vol. 6, pp. 4370-4379, 1972.

GaAs Solar Cells on V-Grooved Silicon via Selective Area Growth

Michelle Vaisman, Nikhil Jain, Qiang Li, Kei May Lau, Adele C. Tamboli, and Emily L. Warren

National Renewable Energy Laboratory, Golden, CO 80401 USA; Yale University, New Haven, CT 06520;
Hong Kong University of Science and Technology, Clear Water Bay, Kowloon, Hong Kong

Abstract — Interest in integrating III-Vs onto Si has recently resurged as a promising pathway towards high-efficiency, low-cost tandem photovoltaics. Here, we present single-junction GaAs solar cells grown monolithically on on-axis Si (001) substrates using V-grooves, selective area growth, and aspect ratio trapping to mitigate defect formation without the use of expensive, thick graded buffers, along with homoepitaxially grown GaAs solar cell test structures. The GaAs buffer grown directly on Si is free of antiphase domains and exhibits a relatively low TDD of 4×10^7 cm^{-2}, despite the lack of a graded buffer. These demonstration solar cells show promise for further improvements to III-V/Si tandems to enable cost-competitive photovoltaics.

Index Terms – III-V on Si, GaAs, selective area growth, photovoltaic cell, solar energy, III-V semiconductor materials

I. INTRODUCTION

In order to achieve high efficiency photovoltaics at a low cost, research on III-V integration onto Si has resurged in recent years [1]. Currently, bonding approaches lack the benefit of low-cost, large-area Si substrates, while current epitaxial approaches are disadvantaged by the need for expensive, thick graded buffers to achieve good material quality, thus undermining their commercial viability. Current metamorphic approaches utilize GaAs$_y$P$_{1-y}$ or Si$_{1-x}$Ge$_x$ graded buffers as thick as 3.6-10 μm [2, 3]. Cost modeling has predicted that even thinner buffers, 3.0 μm in thickness, would constitute nearly 25% of total monolithic III-V/Si dual-junction manufacturing cost, as seen in Fig. 1. Recently, heteroepitaxy of III-Vs on nano-patterned Si via selective area growth (SAG) was proposed as a pathway to achieve low-cost monolithic tandem devices [4].

III-V growth on Si without the use of graded buffers faces challenges with material polarity, as well as large lattice mismatch, both of which can cause defects detrimental to device performance. As graded buffers help control strain relief to minimize threading dislocation density (TDD) [5], large TDD is expected to pose a challenge which predominantly hinders the device open-circuit voltage (V_{OC}) and can also reduce short-circuit current density (J_{SC}) and fill factor (FF) [6-8]. Another material challenge is crack formation induced by the large difference in thermal expansion coefficients between GaAs and Si, which can cause issues such as electrical shunting and additional in-plane electrical resistance [9]. Recently, a GaAs on Si solar cell exhibited controlled crack formation through the use of millimeter-scale notches, achieving a V_{OC} of 0.87 V and efficiency (η) of 18% [10].

While such millimeter-scale approaches have been investigated, nano-scale approaches still remain unexplored for III-V/Si solar cell defect mitigation. Nano-patterns with aspect ratios >1 have been shown to mitigate challenges with lattice mismatch by drastically reducing TDD by confining threads to annihilate on pattern sidewalls [11]. Furthermore, antiphase domains (APDs) can be suppressed by growth on V-grooved Si (111) surfaces [12]. In this work, we grow a monolithic single-junction GaAs solar cell on a GaAs/Si template formed via SAG on nano-patterned V-grooved Si, achieving 5.0% efficiency. The GaAs/Si templates used in this work exhibit a TDD of ~4×10^7 cm^{-2} and have been employed as a promising platform for high-performance laser diodes [13]. Our work demonstrates progress towards a commercially viable III-V/Si tandem for low-cost, high efficiency photovoltaics (PV).

II. EXPERIMENTAL

The high quality GaAs/Si template was prepared as previously described [14]. An on-axis n-Si (001) substrate was prepared by patterning SiO$_2$ stripes along the [110] direction, followed by RCA and dilute 1% HF wet chemical cleanings, and a 45% KOH etch at 70°C for the formation of V-grooves that exposed (111) facets. Initial GaAs nucleation and growth of a 1.5 μm GaAs buffer was conducted with a two-step growth process [14] in a low-pressure 0.1 atm metal-organic

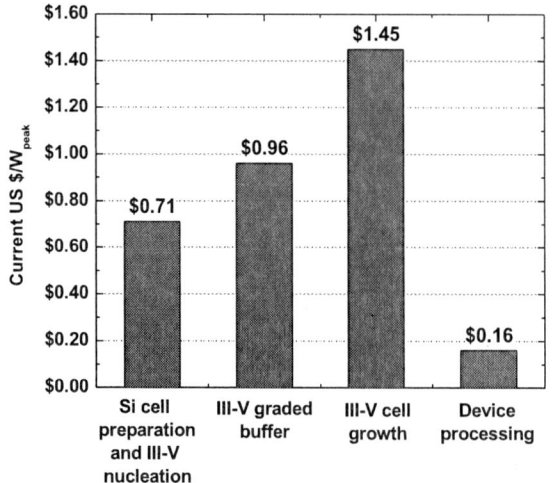

Figure 1. Techno-economic cost analysis conducted in 2013 predicts that the III-V graded buffer layer accounts for ~25% of the cost of a metamorphic III-V/Si tandem cell [15].

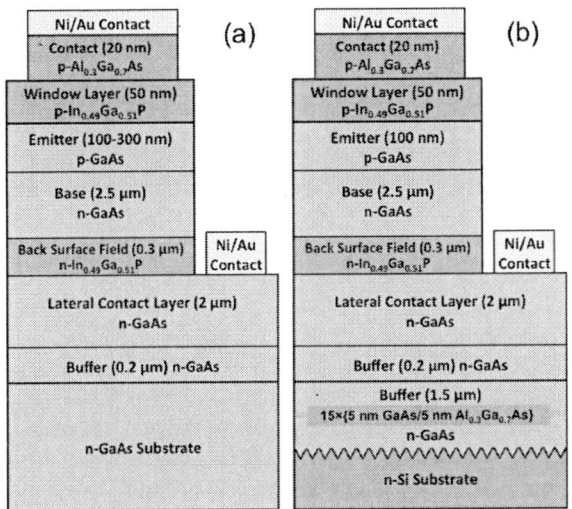
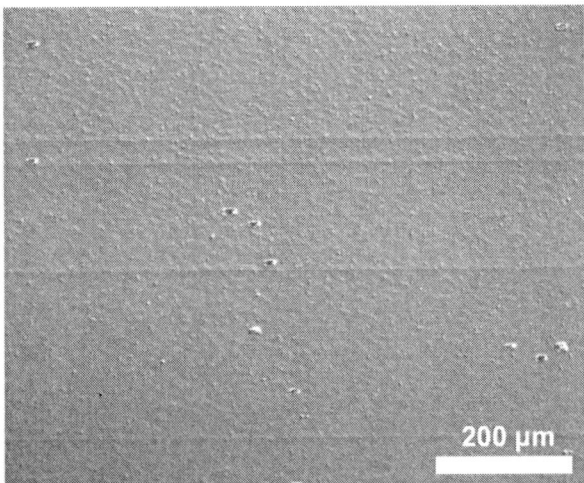

Figure 2. Schematic of (a) the homoepitaxially grown GaAs solar cells (MP423, MP425) and (b) the GaAs solar cell grown on a Si substrate (MP567).

Figure 3. Nomarski microscopy image of our GaAs solar cell grown on a GaAs/Si template showing noticeable cracking.

chemical vapor deposition (MOCVD) system. An AlGaAs/GaAs superlattice was utilized to help control TDD [16]. The template was then shipped from Hong Kong to the National Renewable Energy Laboratory for growth in an atmospheric MOCVD chamber.

Device schematics for the individually grown GaAs solar cells can be seen in Fig. 2. The two initial GaAs solar cell test structures in Fig. 2(a) were compared to determine an appropriate cell design to grow on the GaAs/Si template. The first cell, MP423, was grown with a 300 nm emitter and lower Zn doping level compared to MP425's 100 nm emitter doping of p ~ 1×10^{19} cm^{-3}. A p-on-n structure was chosen in this work to account for the anticipated high defect density in the GaAs grown on Si, enabling slightly higher defect tolerance in the device due to the inherently lower hole mobilities, and thus diffusion lengths, of n-type GaAs [17]. Furthermore, a two top contact design was used to isolate the top cell performance from potential influence of the GaAs-Si interface, such as from high n-doping caused by Si diffusion into the GaAs. The GaAs cell grown on the GaAs/Si template (MP567) used the same structure as MP425 [Fig. 2(b)].

Solar cells were fabricated using conventional photolithography, metal electroplating, and solution-based mesa etching. No anti-reflection (AR) coatings were applied to our devices. Nomarski microscopy and atomic force microscopy were used to characterize surface morphology of our samples, while X-ray diffraction (XRD) and Ayer's model [18] was utilized for the determination of GaAs/Si template TDD. Lighted current-voltage (LIV) measurements were taken under 1-sun, AM1.5G illumination conditions with a custom-built Xe lamp solar simulator. External quantum efficiency (EQE) was characterized using a custom-built tool under chopped, monochromatic light.

III. RESULTS AND DISCUSSION

The GaAs/Si template exhibited a smooth surface morphology with RMS roughness of 0.4-0.9 nm. Cracks were not found on the GaAs/Si templates, but crack formation was observed on the as-grown epitaxial material and GaAs solar cells on Si after processing (Fig. 3).

Our first cells grown homoepitaxially on GaAs substrates demonstrated high V_{OC} values of 1.03-1.04 V, indicative of good material quality (Table 1 and Fig. 4). The reduced emitter thickness in MP425 proved more optimal, enabling a larger J_{SC}. Despite the reduced emitter thickness, the increased emitter doping in MP425 was able to improved FF to 82.5% compared to MP423's FF of 79.7%. Given MP425's overall improved cell performance, the device on a GaAs/Si template (MP567) was grown with the same design.

The solar cell on a GaAs/Si template (MP567) exhibited a J_{SC} of 12.9 mA/cm^2 (Table 1), which is significantly lower than that of the corresponding homoepitaxially grown cell (MP425), which is also reflected in the EQE (Fig. 4). The V_{OC} and FF of the cell on Si were also lower than that on the GaAs substrate, contributing to a lower uncoated efficiency of 5.03% on Si compared to 17.9% on GaAs (Table 1 and Fig. 4). High TDD likely contributed to the reduced device V_{OC} and long wavelength EQE performance compared to the control device on GaAs. However, modeled efficiencies of GaAs cells on Si at this TDD can be as high as ~17% [8], suggesting other factors may be limiting cell performance, such as cracking. The low FF of the device on Si is suspected to be caused by a combination of shunting due to cracking, as well as processing issues.

While low compared with our homoepitaxially grown devices, the GaAs on Si cell exhibited an encouraging 5.03% efficiency, demonstrating promise as a working device with

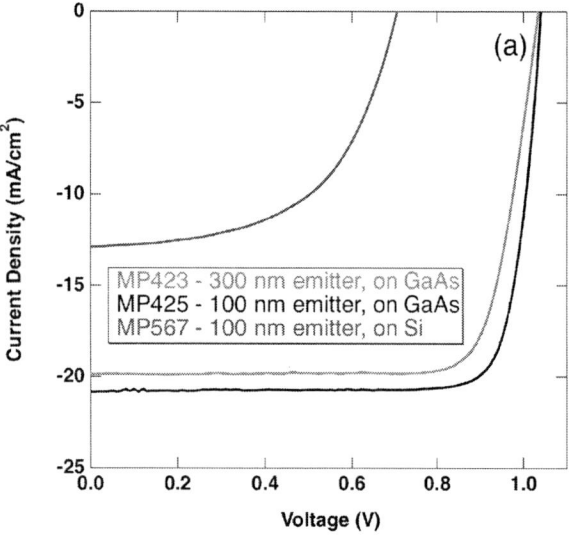

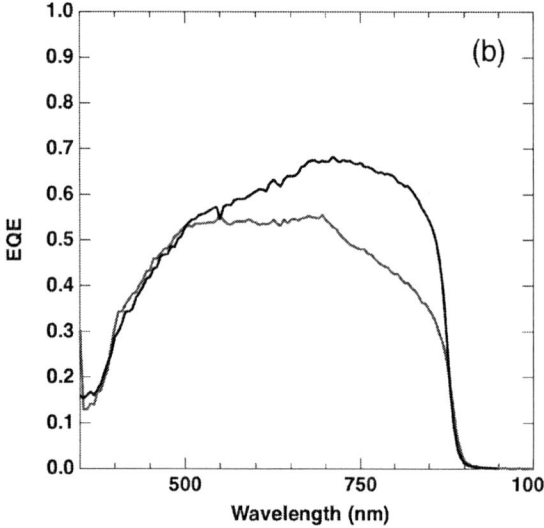

Figure 4. (a) LIV and (b) EQE of the uncoated GaAs solar cells grown homoepitaxially on GaAs with a thicker emitter (MP423, red) and a thinner emitter (MP425, black), and grown on Si with a thinner emitter (MP567, blue).

TABLE I
GaAs SOLAR CELL PARAMETERS

Sample	Substrate	V_{OC} (V)	J_{SC} (mA/cm^2)	FF (%)	η (%)
MP423	GaAs	1.03	19.9	79.7	16.4
MP425	GaAs	1.04	20.8	82.5	17.9
MP567	Si	0.705	12.9	55.3	5.03

the GaAs and Si subcells, and in a single-junction structure, this layer could likely be significantly thinned to reduce cost. We anticipate that thinner buffer layers may be used in the future to further cut costs, and potentially aid in mitigating crack formation.

Device and material improvements may help increase cell efficiency in future work. Using a high-performance AR coating will significantly improve cell J_{SC} with a concomitant boost to V_{OC}. Utilizing a p-i-n device design could help boost J_{SC} through field-assisted current collection. In addition, adjusting device polarity to a more standard n-on-p design could be an interesting option once TDD and cracking issues have been addressed; this design may cater to improved base diffusion lengths and thus improved long wavelength EQE and V_{OC} since electron mobilities are greater than that of holes in GaAs [17]. We further aim to improve upon cracking and TDD in future devices through exploration of the effect of growth parameters and fabrication techniques on material quality.

IV. CONCLUSIONS

We have demonstrated a GaAs solar cell grown epitaxially on nano-patterned Si, achieving 5.0% efficiency despite the large lattice-mismatch between these materials and lack of a graded buffer. The use of a two-step selective area growth process controlled TDD to 4×10^7 cm^{-2} on the GaAs/Si templates enabling a V_{OC} of 0.71 V. The results presented here show promise towards achieving the goal of commercially-viable III-V/Si multijunction devices for low-cost, high-efficiency PV.

ACKNOWLEDGEMENTS

We thank Myles Steiner, Waldo Olavarria, and Michelle Young for help with cell design, growth, and fabrication. We also thank Kelsey Horowitz for discussions on cost modeling. This work was supported by DOE EERE under contract DE-EE-00028394 as well as by grants from the Research Grants Council (No 16212115) and Innovation Technology Fund (ITS/320/14) of Hong Kong. M.V. was supported by a National Aeronautics and Space Administration (NASA) Space Technology Research Fellowship.

numerous pathways available for future improvement. While the current best GaAs on Si single-junction solar cell achieved 20% efficiency with the use of thermal cycle annealing and an InGaAs/GaAs superlattice buffer [19], our approach in this work offers cost benefits over these previous approaches by minimizing reactor growth time (Fig. 1).

The initial GaAs cell on Si investigated here utilizes a 1.5 μm thick GaAs buffer to mitigate dislocations, less than half of the thickness of the GaAs$_y$P$_{1-y}$ metamorphic graded buffers used for GaAs$_{0.75}$P$_{0.25}$ cells [2] and significantly thinner than the ~10 μm Si$_{1-x}$Ge$_x$ buffers used for GaAs cells on Ge-SiGe-Si [3]. The additional 2 μm GaAs lateral conduction layer seen in our device design would not be present in an actual tandem structure, as instead a thin tunnel junction would interconnect

REFERENCES

[1] N. Jain and M. K. Hudait, "III-V multijunction solar cell integration with silicon: present status, challenges and future outlook," *Energy Harvest. Syst.*, vol. 1, pp. 121-145, 2014.

[2] K. Nay Yaung, *et al.*, "GaAsP solar cells on GaP/Si with low threading dislocation density," *Appl. Phys. Lett.*, vol. 109, p. 032107, 2016.

[3] C. L. Andre, *et al.*, "Investigations of high-performance GaAs solar cells grown on Ge-Si$_{1-x}$Ge$_x$-Si substrates," *IEEE Trans. Electron Dev.*, vol. 52, pp. 1055-1060, 2005.

[4] E. L. Warren, *et al.*, "Selective area growth of GaAs on Si patterned using nanoimprint lithography," in *Proceedings of the 43rd IEEE Photovoltaic Specialists Conference*, 2016, pp. 1938-1941.

[5] E. A. Fitzgerald, *et al.*, "Dislocation dynamics in relaxed graded composition semiconductors," *Mat. Sci. Eng. B*, vol. 67, pp. 53-61, 1999.

[6] C. L. Andre, *et al.*, "Impact of dislocation densities on n$^+$ / p and p$^+$ / n junction GaAs diodes and solar cells on SiGe virtual substrates," *J. Appl. Phys.*, vol. 98, p. 014502, 2005.

[7] J. R. Lang, *et al.*, "Comparison of GaAsP solar cells on GaP and GaP/Si," *Appl. Phys. Lett.*, vol. 103, p. 092102, 2013.

[8] N. Jain and M. K. Hudait, "Impact of threading dislocations on the design of GaAs and InGaP/GaAs solar cells on Si using finite element analysis," *J. Photovolt.*, vol. 3, pp. 528-534, 2013.

[9] V. K. Yang, *et al.*, "Crack formation in GaAs heteroepitaxial films on Si and SiGe virtual substrates," *J. Appl. Phys.*, vol. 93, pp. 3859-3865, 2003.

[10] S. Oh, *et al.*, "Control of crack formation for the fabrication of crack-free and self-isolated high-efficiency gallium arsenide photovoltaic cells on silicon substrate," *J. Photovolt.*, vol. 6, pp. 1031-1035, 2016.

[11] J. Z. Li, *et al.*, "Defect reduction of GaAs epitaxy on Si (001) using selective aspect ratio trapping," *Appl. Phys. Lett.*, vol. 91, p. 021114, 2007.

[12] W. Guo, *et al.*, "Selective metal-organic chemical vapor deposition growth of high quality GaAs on Si(001)," *Appl. Phys. Lett.*, vol. 105, p. 062101, 2014.

[13] Y. Wan, *et al.*, "Optically pumped 1.3 μm room-temperature InAs quantum-dot micro-disk lasers directly grown on (001) silicon," *Opt. Lett.*, vol. 41, pp. 1664-1667, 2016.

[14] Q. Li, K. W. Ng, and K. M. Lau, "Growing antiphase-domain-free GaAs thin films out of highly ordered planar nanowire arrays on exact (001) silicon," *Appl. Phys. Lett.*, vol. 106, p. 072105, 2015.

[15] M. Woodhouse and A. Goodrich, National Renewable Energy Laboratory Report No. PR-6A20-601262013.

[16] N. Hayafuji, *et al.*, "Effectiveness of AlGaAs/GaAs superlattices in reducing dislocation density," *J. Cryst. Growth*, vol. 93, pp. 494-498, 1988.

[17] S. M. Sze and J. C. Irvin, "Resistivity, mobility, and impurity levels in GaAs, Ge, and Si at 300°K," *Solid-State Electron.*, vol. 11, pp. 599-602, 1968.

[18] J. E. Ayers, "The measurement of threading dislocation densities in semiconductor crystals by X-ray diffraction," *J. Cryst. Growth*, vol. 135, pp. 71-77, 1994.

[19] Y. Ohmachi, *et al.*, "High quality GaAs on Si and its application to a solar cell," in *MRS Proceedings*, 1988, p. 297.

High Temperature Annealing of $In_{1-x}Ga_xN$ MQW Solar Cells

Joshua J. Williams,[1] Heather McFavilen,[2] Steven Young,[2] Christiana B. Honsberg, and Stephen M. Goodnick[1]

[1]Arizona State University, Tempe, AZ, 85281, USA

[2]Photonitride Devices, Inc., Tempe, AZ, 85282, USA

Abstract — We demonstrate the reliability of $In_{1-x}Ga_xN$ MQW solar cells through the use of incrementally increased temperature anneals. Through use of room temperature IV measurements between 1 hour long anneals at increasing temperatures (400 °C to 800 °C), we begin to probe different failure mechanisms in $In_{1-x}Ga_xN$ solar cells. Slow degradation up to 700 °C is seen for high quality cells, while steady degradation with increasing temperatures is demonstrated for lower quality cells. Finally, at temperatures of 800 °C light-biased J-V curves almost completely collapse.

I. INTRODUCTION

Indium-gallium-nitride alloys have shown great promise for constructing high-efficiency solar cells ever since the true band gap on InN was discovered circa 2002-2003 [1]–[3]. The first $In_{1-x}Ga_xN$ solar cells were constructed in 2007 [4]–[6], and at that time they had efficiencies of only ~1%. Ten years have passed and in that time many research groups have attempted to improve the performance of $In_{1-x}Ga_xN$ solar cells [7], however efficiencies are still below 5%. That is largely due to material challenges associated with growing solar relevant compositions of $In_{1-x}Ga_xN$ ($0.2 < x < 0.8$).

In compositions where $x < 0.15$, high quality $In_{1-x}Ga_xN$ has been achieved and has even become a commodity as evidenced by the ubiquity of the "white" LED revolution. Building upon this technology many recent solar cells have demonstrated band gap-voltage offsets (W_{OC}) below 0.5 eV and fill factors above 75%, albeit at band gaps over 2.5 eV [8]–[10]. These band gaps limit the maximum achievable current density with 1-sun illumination to <10 mA/cm². However, these wide band gaps in conjunction with the strong nature of nitride chemical bonds make $In_{1-x}Ga_xN$ uniquely suited for niche applications such as a refractory solar cell [9].

Research efforts to build and operate an $In_{1-x}Ga_xN$ solar cell at extreme temperatures (>300 °C) have yielded promising results. Zhao *et. al* were the first to demonstrate the refractory nature of an $In_{1-x}Ga_xN$ solar cell with operational performance up to 400 °C [11]. More recently, Williams *et. al* have demonstrated performance with a similar cell up to 600 °C [9]. As research continues to improve the cell performance at all temperatures and maximum operational temperature, it is prudent to investigate failure limits and mechanisms of such solar cells.

In this paper we have taken the first steps in the long process of investigating failure mechanisms of $In_{1-x}Ga_xN$ solar cells. By thermally stressing the devices in a furnace and measuring the I-V characteristics, it is possible to determine when different temperature dependent phenomena take effect within the cell.

II. EXPERIMENTAL

The $In_{1-x}Ga_xN$ nitride solar cells used herein are a p-i-n structure with the p- and n-regions being composed of GaN and the i-region being a multiple quantum well structure of $In_{0.12}Ga_{0.88}N$/GaN with 40x periods. This composition of $In_{1-x}Ga_xN$ has a measured band gap of ~2.8 eV. Since the layers were grown on an insulating sapphire substrate, devices were fabricated in the form of mesa structures. Mesas (2 mm x 2 mm) are comprised of p-GaN on i-MQW, while the n-GaN is continuous across the entire wafer. Indium-tin-oxide (ITO) is used as a spreading layer on the p-GaN due to p-GaN's extremely high sheet resistance (>10,000 $\Omega/\square$). Contact layer stacks of Ti/Al/Ni/Au and Ti/Pt/Au are evaporated onto the n-GaN and p-GaN/ITO, respectively. A cross-sectional schematic is shown in Fig. 1. A detailed explanation of growth, fabrication, and initial device performance has been previously published [9].

Seven devices with a mixture of high and low quality performance were selected to be measured and thermally stressed in this work. Device I-V curves were measured with a 2-point probe setup under an Oriel solar simulator with a Xenon arc lamp light source calibrated to power output equivalent to 1-sun.

The wafer containing all the cell die was then loaded into a

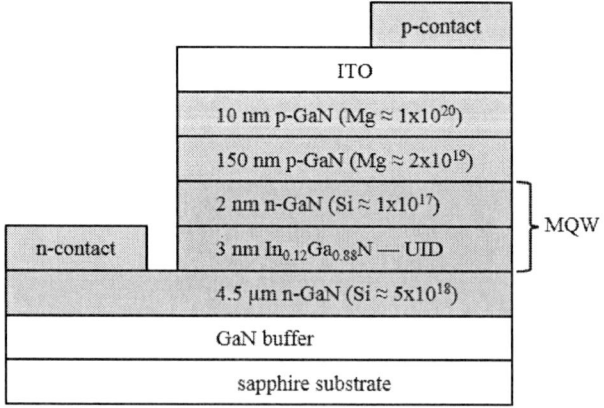

Fig. 1. Cross-sectional schematic of the In1-xGaxN solar cell. The p- region is represented by blue and the n-region is represented by red.

muffle tube furnace. The tube was flushed with ultra-high purity N_2 gas for 5 minutes. N_2 flow rate was then reduced to approximately 10 SCCM, and the sample was heated to the designated temperature. Sample heating was completed within 15 min, after which samples were held at temperature for 60 min and promptly removed. Upon cooling, samples' I-V curves were measured at room temperature and the process was repeated. The heat treatment was in increments of 100 °C from 400 °C to 800 °C, all for 1 hr.

III. RESULTS AND DISCUSSION

The effect of one hour long heat treatments at different temperatures produced affected the device J-V curves mostly in predictable ways, however there were a few counter-intuitive trends. The example J-V curves in Fig. 2 show the performance of one of the lesser cells and the performance of one of the champion cells. The champion cell shows negligible difference in performance as a result of temperature all the way up to 600 °C heat treatments. At 700 °C there is a perceivable increase in J_{SC}, this is likely due to an improvement in the ITO layer since anneals in nitrogen can increase the crystal quality. Additionally, at 700 °C, small but perceivable decreases to R_{Shunt} and increases to R_{Series} lead to an overall decrease in fill factor. In the low quality sample, the trends from the champion cell show up in addition to a strong decrease in V_{OC} as a function of temperature. This is probably due to further activation of the original defect (an unknown defect which reduced the V_{OC} relative to the neighboring champion cell). These trends are representative of the small sample set, i.e. high performance cells behavior similar to the graph on the right. Defective cells behave similar to the graph on the left.

After annealing at 800 °C, nearly all samples showed similar light IV curves. At first glance the cells seem nearly dead, however one can see that there is a very slight negative current in the 4th quadrant of the J-V graph. This indicates that there is a small photovoltaic response. It is interesting to note

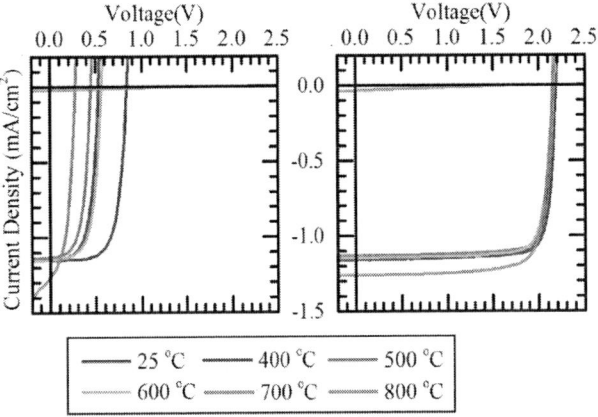

Fig. 2. Example J-V curves for two of the devices in the sample set. The two curves share a y-axis label on the left and y-axis numerals in the center. The curve on the left shows one the lesser cells. The curve on the right shows one of the champion cells. Measurements are taken at room temperature. The different curves represent J-V measurements after one hour anneals at the listed temperature.

that the V_{OC} in nearly all samples for the J-V curves at 800 °C is on par with the original V_{OC} of the champion cell (~2.2 V). Thus the 800 °C heat treatment has an effect of destroying a component of carrier collection, but it has not destroyed the junction nor significantly decreased minority carrier lifetimes near the junction. Therefore, we hypothesize that the collapse of the J-V curve is almost entirely due to the destruction of the contacting layers. This hypothesis makes sense as the ITO spreading layer is the least thermally robust of the compounds, followed by the metals in the contacting layers, and finally the nitrides. Finally, the increase in V_{OC} in the defective cells indicates that the unknown defect in some but not all of the devices can be thermally cured.

In order to better represent the performance of the entire sample over different measurements the key J-V parameters are plotted in the whisker-box plots in Fig. 3. These plots show the *normalized* parameters of the J-V curve. Each parameter for each cell is divided by its initial value, hence all

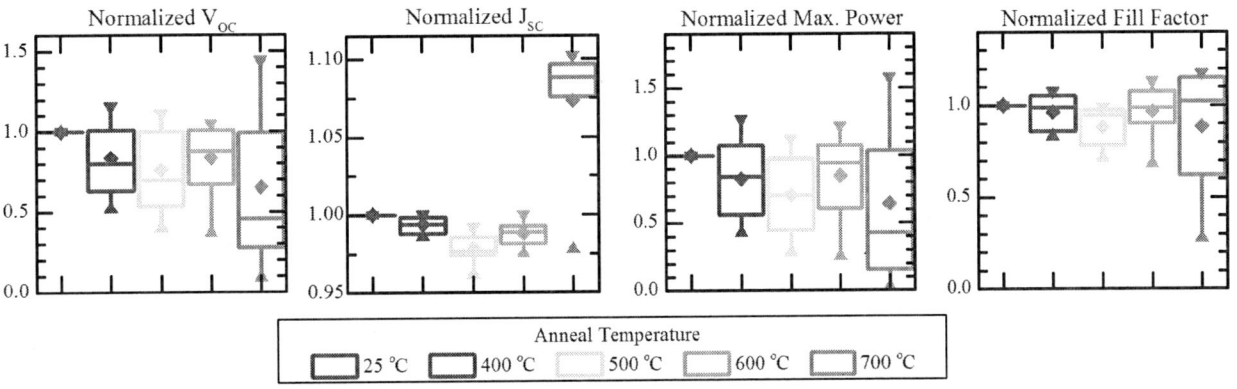

Fig. 3. Box-whisker plots for normalized J-V curve parameters. Data points for each cell are normalized against the 25 °C (Initial) metric for that cell. This allows comparison of trends between cells with drastically different performances. Diamonds show the mean value for a data set. Box levels are 25, 50, and 75 percentiles from bottom to top. End points are data outliers. Using the mean or 50 percentile line one can visualize the trends in each parameter.

values for the 25 °C box are 1. This method shows interesting trends in different performance metrics at different temperatures regardless of initial device performance.

The V_{OC} shows a downward trend with temperature, except for at 600 °C which shows a definitive rise. This is attributed to an increase in activation of Mg dopants in p-GaN (Mg has a notoriously low activation percentage, ~2%). Thermal anneals have been shown to activate more Mg through liberation of hydrogen [12]. At 700 °C the V_{OC} drops back down, probably due to inter-diffusion between layers creating a higher J_0. J_{SC} shows a trend similar to what was explained previously, i.e. the sudden jump in J_{SC} at 700 °C attributed to increased ITO transparency. The fill factor has a slight downward trend which is surprising given the tendency for R_{Shunt} and R_{Series} to get lower and higher, respectively, with temperature. Lastly, the maximum power is a convolution of all of these effects.

IV. CONCLUSION

This work is a first step towards understanding failure in these cells. Future work should be carried out to measure effects of time vs. elevated temperature, 600 °C (where the samples showed improvement) and 700 °C (where samples showed noticeable degradation but not complete collapse). Another unique measurement, if possible, would be to build an IV tester integrated with a furnace to measure the device performances while at elevated temperatures over a long period of time. It would be interesting to study the effect of dopant diffusion on various parameters at high temperatures. Similarly, an "electron wind" has been shown to increase the time to failure in semiconductor devices versus simple heating [13].

While all of these effects will be important to study, this work has further proven the extraordinary resiliency of GaN and $In_{1-x}Ga_xN$ to applications in high temperatures. Solar cells capable of withstanding one hour long anneals at temperatures of 700 °C with minimal change were demonstrated. Resilience at such high temperatures is the sign of a device with longer reliability when operating at lower temperature between 400 °C and 600 °C.

REFERENCES

[1] J. Wu, W. Walukiewicz, K. M. Yu, W. Shan, J. W. Ager, E. E. Haller, H. Lu, W. J. Schaff, W. K. Metzger, and S. Kurtz, "Superior radiation resistance of In 1-xGa xN alloys: Full-solar-spectrum photovoltaic material system," *J. Appl. Phys.*, vol. 94, no. 10, pp. 6477–6482, 2003.

[2] J. Wu, W. Walukiewicz, K. Yu, and J. A. Iii, "Indium

nitride: A narrow gap semiconductor," *Info:*, pp. 1–8, 2002.

[3] V. Y. Davydov, a. a. Klochikhin, R. P. Seisyan, V. V. Emtsev, S. V. Ivanov, F. Bechstedt, J. Furthmüller, H. Harima, a. V. Mudryi, J. Aderhold, O. Semchinova, and J. Graul, "Absorption and emission of hexagonal InN. Evidence of narrow fundamental band gap," *phys. stat. sol.*, vol. 229, no. 3, pp. 1972–1974, 2002.

[4] O. Jani, I. Ferguson, C. Honsberg, and S. Kurtz, "Design and characterization of GaN/InGaN solar cells," *Appl. Phys. Lett.*, vol. 91, no. 13, p. 132117, 2007.

[5] O. Jani, H. Yu, E. Trybus, B. Jampana, I. Ferguson, A. Doolittle, and C. Honsberg, "Effect of Phase Seperation on Performance on III-V Nitride Solar Cells," *22nd Eur. Photovolt. Sol. Energy Conf.*, no. September, pp. 64–67, 2007.

[6] X. Chen, K. D. Matthews, D. Hao, W. J. Schaff, and L. F. Eastman, "Growth, fabrication, and characterization of InGaN solar cells," *Phys. status solidi*, vol. 205, no. 5, pp. 1103–1105, May 2008.

[7] A. G. Bhuiyan, K. Sugita, A. Hashimoto, and A. Yamamoto, "InGaN Solar Cells: Present State of the Art and Important Challenges," *IEEE J. Photovoltaics*, vol. 2, no. 3, pp. 276–293, Jul. 2012.

[8] K. Y. Lai, G. J. Lin, Y.-R. Wu, M.-L. Tsai, and J.-H. He, "Efficiency dip observed with InGaN-based multiple quantum well solar cells.," *Opt. Express*, vol. 22 Suppl 7, no. December, pp. A1753-60, 2014.

[9] J. J. Williams, H. McFavilen, A. M. Fischer, D. Ding, S. R. Young, E. Vadiee, F. A. Ponce, C. Arena, C. B. Honsberg, and S. M. Goodnick, "Development of a high-band gap high temperature III-nitride solar cell for integration with concentrated solar power technology," in *2016 IEEE 43rd Photovoltaic Specialists Conference (PVSC)*, 2016, pp. 0193–0195.

[10] M.-J. Jeng, Y.-L. Lee, and L.-B. Chang, "Temperature dependences of In x Ga 1− x N multiple quantum well solar cells," *J. Phys. D. Appl. Phys.*, vol. 42, no. 10, p. 105101, May 2009.

[11] L. Zhao, T. Detchprohm, and C. Wetzel, "High 400 °C operation temperature blue spectrum concentration solar junction in GaInN/GaN," *Appl. Phys. Lett.*, vol. 105, pp. 243903-1-243903–4, 2014.

[12] J. Huang, T. Kuech, H. Lu, and I. Bhat, "Electrical characterization of Mg-doped GaN grown by metalorganic vapor phase epitaxy," *Appl. Phys. Lett.*, vol. 68, no. 17, pp. 2392–2394, 1996.

[13] J. R. Black, "Electromigration-A brief survey and some recent results," *IEEE Trans. Electron Devices*, vol. 16, no. 4, pp. 338–347, Apr. 1969.

Solar Probe Plus Array Reliability

Anton Yanchilin, Edward Gaddy

Johns Hopkins University Applied Physics Laboratory, Laurel, Maryland, 20723, USA

Abstract—The Solar Probe Plus will follow a trajectory passing the sun in the closest orbit at approximately 9.86 solar radii. During this mission the probe will encounter a high irradiance high temperature (HIHT) environment, making it necessary for a more intricate solar array configuration. This requires the standard Failure Modes Effects and array reliability calculations to be modified. Failure modes include solder deterioration, VDA Kapton evaporation, solar cell material deterioration, coverglass to cell adhesive degradation, and also typical interconnect and wire failures. Using statistics suited for two different types of strings, the reliability of the solar array is calculated when the probe is at aphelion and perihelion. The array contains not only two different types of strings, but also cell sizes and panel sections, making the true reliability more complex.

Through statistical analysis of individual components reliabilities, the reliability of the complete array with no, one and two string failures is calculated to be 89.55%, 5.22%, 0.13% respectively during perihelion mode and 80.17%, 13.64%, and 1.09% during aphelion mode. In result, the probe has a reliability of 94.89% with at most two string failures during perihelion mode and 94.91% during aphelion mode. After calculating the number of exposure hours per particular string exposed it was visible that the most inboard string, No. 22, experiences 39,994.875 exposure hours with strings No. 9 through No. 1 experiencing the full exposure time of the planned mission, 61,370.766 hours. With this, the reliability of 42 double-interconnected strings is 99.9% but with the inclusion of the two single-interconnected strings, this goes down to 94.9%. Due to the Solar Space Probe having a much different solar array configuration than that of most space systems it necessitates different statistical computation, in this case producing a unique reliability over the course of its 7-year mission.

I. INTRODUCTION

Undergoing much higher irradiances and temperatures than other spacecraft, the Solar Probe Plus is subject to various potential failure modes. These include solder deterioration, VDA Kapton evaporation, solar cell material deterioration as well as the typical interconnect and wire failures. The failure rates of each respective item mentioned will provide a basis for overall reliabilities of solar array configurations.

II. RELIABILITY COMPUTATION

As the Solar Probe Plus transitions from aphelion to perihelion, the flap angle of the array changes accordingly. The once fully open angle changes to a more restrictive one, allowing only about 8 of the outermost strings to be in the penumbra formed by the heat shield. As the mission progresses, the time spent for the aphelion and perihelion in each consecutive orbit changes. The reliability of the panel is mainly determined by two key items: time and strings present. The full 44 strings are illuminated during aphelion whereas there are only 16 during perihelion. The following calculations determine the chances of a string failure during the 7 year flight (approx. 61370.766 hours). At aphelion, since both primary and secondary wings are present, there will be 42 strings with double interconnects and 2 strings with one interconnect. Table I lists all of the relevant variables for the consequent calculations.

TABLE I

Information Regarding Array Reliability

Array Property	Value
Cell Failure Rate	$1*10^{-9}$ fail. per op. hour [2]
Mission Time	61370.66 hours
Solder Joint Fail Rate	$5*10^{-9}$ fail. per op. hour
Interconnect Fail Rate	$1*10^{-9}$ fail. per op. hour
No. of Cells in Series	34 cells
Failure Rate of Wire	$3.50*10^{-11}$ fail. per op. hour
Failure Rate of Crimp	$1*10^{-8}$ fail. per op. hour
No. Strings Pri. Panel	15 strings / side
No. Strings Sec. Panel	7 strings / side

A. Component Reliabilities

The given failure rate for a covered cell is $a_c = 10^{-9}$ failures per hour. Reliability over the mission time is found to be:

$$R_{cell} = e^{-a_c * t} = 0.99993863 \qquad (1)$$

Using the above equation with a given solder joint failure rate of $5*10^{-9}$ failures per hour gives a reliability of:

$$R_{sj} = 0.99969319 \qquad (2)$$

Due to standard welds being unable to sustain the high currents present in an HIHT environment, solder joints were used. Given an interconnect failure rate of 10^{-9} failures per hour allows for a reliability of:

$$R_{ic} = 0.99993863 \qquad (3)$$

A single connection will consist of two solder joints and one interconnect, allowing for the reliability between these components to be:

$$R_{sc} = R_{ic} * (R_{sj})^2 = 0.99932515 \qquad (4)$$

Using this value, the reliability of single and double interconnected cells can now be determined. The value for a cell with a single interconnector will have the same reliability as Equation 4. A cell with two interconnectors will have a reliability of:

$$R_{tc} = 1 - (1 - R_{sc})^2 = 0.99999954 \qquad (5)$$

978-1-5090-5606-4/17 $31.00 © 2017 IEEE

Using these two values, the following procedure is used to find the respective string values for strings with single and double interconnected cells. With 34 cells per string, the reliability of an unconnected string of each type would produce:

$$R_{sc*str} = R_{sc}^{35} * R_{cell}^{34} = 0.97461349 \qquad (6)$$

$$R_{tc*str} = R_{tc}^{35} * R_{cell}^{34} = 0.99789967 \qquad (7)$$

Now that the individual string reliabilities are known, connected string reliabilities are necessary. Given a wire failure rate and crimp failure rate of $3.5*10^{-11}$ and 10^{-8} respectively, the reliabilities of the wire and crimp over 7 years are:

$$R_w = 0.99999785 \qquad (8)$$

$$R_c = 0.99938648 \qquad (9)$$

Composed of the wire, crimp and solder joint, the harness was found to have a reliability of:

$$R_h = (R_{sj} * R_w * R_c)^2 = 0.99815629 \qquad (10)$$

Using this value, the respective reliabilities for connected strings with interconnects can be found by using the harness reliability and the string reliability.

$$R_{sc*str*h} = R_{sc*str} * R_h = 0.97281658 \qquad (11)$$

$$R_{tc*str*h} = R_{tc*str} * R_h = 0.99605982 \qquad (12)$$

B. Aphelion Reliability Calculations

Figure 1. SPP Array configuration in Aphelion mode [1]
The arrays reliability for 0, 1, or 2 failures is now able to be calculated using the above values. There are 2 strings with single interconnects (most outboard string) and 42 strings with two interconnects. Therefore the reliability of the array with 0 failures would be:

$$R_{array} = R_{sc*str*h}^2 * R_{tc*str*h}^{42} = 0.80176985 \qquad (13)$$

[1]Courtesy of Johns Hopkins University Applied Physics Laboratory

The probability of 1 failure is able to be calculated on both types of strings by incorporating combinations and using a standard probability rule when it comes to non-mutually exclusive events. The failure rates with respect to the number of interconnects are $F_{1*sc*str*h}$, $F_{1*tc*str*h}$, respectively:

$$(\frac{2}{44}) * \frac{2!}{1!1!} * (1 - R_{sc*str*h})^1 * (R_{sc*str*h})^1 = 0.00240404 \qquad (14)$$

$$(\frac{42}{44}) * \frac{42!}{41!1!} * (1 - R_{tc*str*h})^1 * (R_{tc*str*h})^{41} = 0.13435808 \qquad (15)$$

$$F_{1*sc*str*h} * F_{1*tc*str*h} = 0.00032300 \qquad (16)$$

The chances of this occurrence are not mutually exclusive and the probability of 1 failure overall is the sum of these two separate probabilities with the product of the two being subtracted:

$$F_{1*array} = 0.13643912 \qquad (17)$$

These values show the reliability of an array with at most one failed string to be 0.93820965. The same concept is applied to calculating the probability of 2 failed strings. The two different values required ($F_{2*sc*str*h}$, $F_{2*tc*str*h}$) are:

$$(\frac{2}{44}) * \frac{2!}{0!2!} * (1 - R_{sc*str*h})^2 * (R_{sc*str*h})^0 = 0.00003359 \qquad (18)$$

$$(\frac{42}{44}) * \frac{42!}{40!2!} * (1 - R_{tc*str*h})^2 * (R_{tc*str*h})^{42} = 0.01089551 \qquad (19)$$

$$F_{2*sc*str*h} * F_{2*tc*str*h} = 0.00000037 \qquad (20)$$

Therefore the total probability of having 2 failures within the Solar Probe Plus array in aphelion mode is:

$$F_{2*array} = 0.01092874 \qquad (21)$$

This makes the probability of having at most two failures within the array to be 0.94913817.

C. Perihelion Reliability Calculations

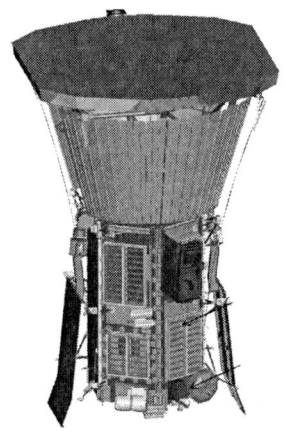

Figure 2. SPP Array configuration in Perihelion mode (88° angle) [2]

As the Solar Probe Plus enters Perihelion, the number of double interconnected strings will drop to 14. This will change the calculations starting with Equation 13. The new value for 0 failures at Perihelion is:

$$R_{array} = R_{sc*str*h}^2 * R_{tc*str*h}^{14} = 0.89548406 \qquad (22)$$

Due to having fewer strings involved in the calculation, it is inevitable that the reliability of this mode will be higher. Performing the same calculations above for this new configuration produces the reliabilities of an array with 1 failed string as well as an array with 2 failed strings of:

$$F_{1*array} = 0.05216039 \qquad (23)$$

$$F_{2*array} = 0.00127124 \qquad (24)$$

This makes the probability of having at most 2 string failures to be 0.948916 for the 7 year mission in perihelion. Even though the reliabilities for the two modes are very similar, the probability of not failing at all is significantly higher in perihelion mode than aphelion.

III. RESULTS

A summary of the results for the two main configurations found above are depicted in Table II. The calculations above and the results in Table II assume no cross coupling between aphelion and perihelion.

TABLE II

Overall Reliability Estimations for SPP Array

[2]Courtesy of Johns Hopkins University Applied Physics Laboratory

	Aphelion	Perihelion
0 failures	0.80176985	0.89548406
1 failure	0.13643912	0.05216039
2 failures	0.01092874	0.00127124
0,1, or 2	0.94913771	0.94891569

Rauschenbach establishes that the failure rate is failures per part operating hour [1]. This indicates the possibility that a different and better array reliability can be computed using this definition. This was done by creating a calculator using array geometry and umbra line predictions throughout the journey. After determining the specific time of operation per string and assuming a zero failure rate for cells during the intervals they do not operate, the reliability computation is as follows. Table III depicts the approximate amount of time that each string sees the sun, with strings 1-8 (most outboard) always in sunlight with string 9 experiencing partial shading. For example, the inboard string only experiences two thirds of the exposure time used for previous calculation implying that the true reliability of that inner section might especially be high.

TABLE III

Array Exposure Approximations

String no.	Exp. Time (years)	Exp. Time (Hours)
22	4.5625	39 995
21	4.8585	42 590
20	5.1095	44 790
19	5.324	46 670
18	5.512	48 318
17	5.6795	49 786
16	5.8315	51 119
15	5.9705	52 337
14	6.1005	53 477
13	6.23	54 612
12	6.372	55 857
11	6.5165	57 124
10	6.737	59 057
9	7.001	61 371

The probability of having at most 2 failures increases due to less time spent under the sun. This shows that the reliability of having no more than two failures for the entire array for 7 years will be in a range starting at 0.9489 to be the minimum reliability. The only thing that changes in these calculations is aphelion probabilities for 0-2 failures. The reason for this is because strings 1-8 are always under the sun therefore the operating time does not change.

The 2 strings that have single interconnects have a very prominent effect on the overall reliability of the solar array. The primary panels, 30 strings of double interconnected cells would have the following minimum reliability values. By minimum it means that it is possible with less interaction with radiation that some of the most inboard strings will demonstrate higher reliabilities. The total for 42 strings is also shown- that the inclusion of the two outermost strings in the solar array configuration brings about a sizable decrease in reliability values. Table VI displays the said reliabilities for entirety of the mission.

TABLE VI

Various String Arrangements and Corresponding Reliabilities

	0 fail	1 fails	2 fail
42 d.i.	0.8472	0.987 96	0.999 37
All	0.801 77	0.938 21	0.949 14
Prim.	0.888 31	0.993 72	0.999 77
Second.	0.902 58	0.948 65	0.9496

It also stands to mention that the formal reliability measurement for the Solar Probe Plus differs from the values in this paper in part due to the different conditions between aphelion/perihelion and the single interconnects between series cells in the most outboard string.

IV. CONCLUSION

Due to the secondary section having less overall strings than the primary section, it has a higher reliability for 0 failures over the lifetime. This, however, is accompanied by the fact that there are a higher proportion of strings with low reliability (single interconnect); the overall reliability is strongly affected, showing the two reliabilities of having 0, 1, or 2 failed strings to be close to one another. Being a different configuration than most standard spacecraft solar arrays, this one has a somewhat lower reliability over the course of a 7 year mission. This is to be expected due to the integration of cells with single interconnects. There is a probability of at least 94.9138% that the Solar Space Probe at aphelion will at most fail twice in aphelion mode and 94.8916% in perihelion mode.

REFERENCES

[1] Rauschenbach, H. S. *Solar Cell Array Design Handbook: The Principles and Technology of Photovoltaic Energy Conversion.* New York: Van Nostrand Reinhold, 1980. Print.

Photovoltaic Temperature Estimation Model for Rapid Irradiance Change Conditions in Tropical Regions Using Heuristic Algorithms

R.Srivatsan*, Lian L. Jiang*, and Douglas L.Maskell[†]

*Energy Research Institute @ NTU, Singapore, 637553

[†]School of Computer Science and Engineering, Nanyang Technological University, Singapore, 637553

Abstract—The knowledge of module temperature is necessary to implement any energy management technique that requires the prediction of solar power output. Current PV temperature estimation models are generally divided into three categories: Empirical models, Physical Steady State models and Physical Dynamic models. Each of these methods have their own disadvantages. In this paper, particle swarm optimization (PSO) is used to improve the accuracy of a simple physical dynamic model, the two parameter Resistance and Capacitance (RC) circuit model. The effectiveness of the proposed PSO-based parameter estimation for the RC circuit model is verified using an experimental dataset measured from a CIGS PV module at a 1-sec sampling interval. The performance of our proposed RC circuit model is then compared with two empirical models and a physical steady state model, the NOCT-standard model, the Veldhuis model and the Mattei model, respectively. The proposed model provides a significantly better temperature estimation than the NOCT and Mattei models, which do not effectively account for the thermal inertia of a PV module and hence are oversensitive to rapid irradiance variations. The proposed model provides a similar, but slightly better, temperature estimation to that of the Veldhuis model but is significantly easier to implement.

I. INTRODUCTION

The power output of a Photovoltaic (PV) System is significantly influenced by the irradiance and temperature [1]. This is shown in Fig. 1, where a change of average temeprature from 50 to 60°C leads to a 13% reduction in the power of the PV system. An accurate estimation of the temperature of a PV module plays an important role in solar yield forecasting models and in the operation of the latest MPPT algorithms [2] [3] [4]. Large variations in irradiance could result in inaccurate temperature estimation. This is particularly problematic for a city-state like Singapore, where the tropical climate causes large and rapid variations in solar insolation [5].

Many existing models can estimate module temperature [6] – [8]. They can be divided into three categories: empirical models, physical steady state models, and physical dynamic models, and are reviewed in detail in Section II. The main challenge faced by existing temperature estimation methods when applied under rapid irradiance change conditions, is their inability to simulate thermal inertia adequately. The thermal inertia of a PV module is represented by its delayed temperature response to heat flow [8]. The simpler empirical methods, underestimate this effect, which leads to over sensitivity to irradiance changes. The more complex methods based on the heat transfer physics of the PV module require knowledge of the internal material parameters and their dependence on temperature for accurate prediction.

Measurements taken in Singapore given in Fig. 2 show that the peak CIGS PV module temperate commonly exceeds 50C, which will impact the efficiency of the PV system. However, it is impractical and costly to install and maintain temperature sensors and the associated logging hardware for the different modules in a PV system. Therefore, it is necessary to build a temperature estimation model, which should be a compromise between representing the underlying physics of the PV module, including the thermal inertia, while at the same time not use too many parameters that would make it hard to implement.

A simple thermal resistance and capacitance (RC) circuit model was proposed to estimate PV module temperature in [9]. The parameters R and C, relating to the characteristics of the PV module, were identified using gradient methods, which could converge to a local minimum depending on the initial conditions. In this paper, we propose to use the thermal RC circuit model to predict the temperature, along with a heuristic global optimization algorithm the particle swarm optimization (PSO) method to identify the R and C values relating to the parameters of the PV module. Compared to the existing methods, such as the NOCT [10], Mattei [11], and Veldhuis models [12], the proposed RC circuit model gives a smaller normalized root mean square error (nRMSE) and a more accurate prediction over different weather conditions, particularly in conditions with rapid irradiance change. It is also very easy to implement.

II. EXISTING TEMPERATURE ESTIMATION MODELS

Photovoltaic temperature estimation models are grouped into three broad categories: empirical models, physical steady state models, and physical dynamic models. In this section, we review these three model categories.

A. Empirical Model

Empirical models usually propose simple relationships between the temperature and the ambient conditions and use fitting methods to identify the parameters that define these relationships. The proposed relationships are usually based on observation and do not simulate the underlying physical process that controls the temeprature of a PV module.

978-1-5090-5606-4/17 $31.00 © 2017 IEEE

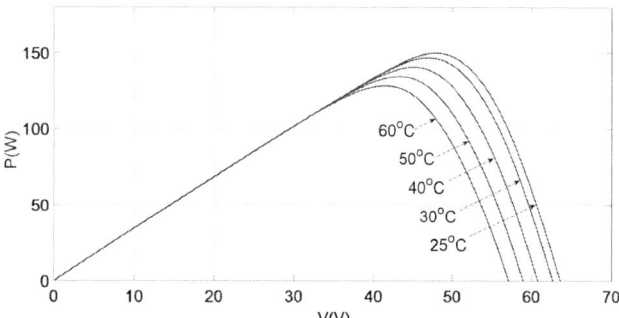

Fig. 1. Power Change with Temeprature for CIGS PV Module

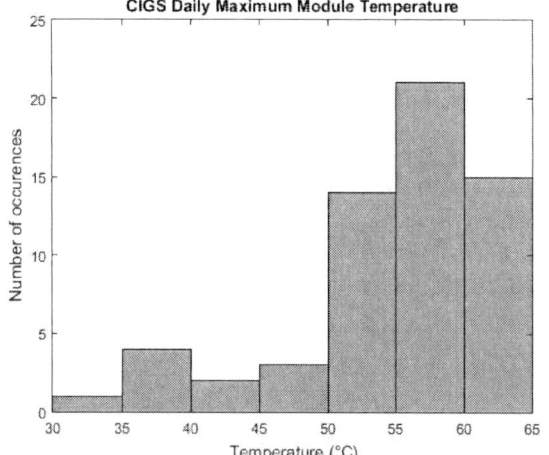

Fig. 2. Distribution of Maximum Module Temperature

A simple expample of an empirical model is the NOCT-Standard formula [12], given in Eq. 1.

$$T_m = T_a + \frac{G}{G_{NOCT}} \cdot (T_{NOCT} - T_{a,NOCT}) \quad (1)$$

T_a is the ambient temperature and G is the in-plane irradiance. T_{NOCT} is the technology dependent nominal operating cell temperature, which is the cell temperature at irradiance G_{NOCT} is 800 W/m^2, ambient temperature T_a, T_{NOCT} is 20°C and wind speed = 1m/s. T_{NOCT} depends on the PV technology and for the CIGS PV module used in this paper, it is around 45°C.

The main issue with the empirical models is that there is no consistency in the factors used to estimate temperature. Additionally, these factors can also vary based on the location and mounting methods, affecting accuracy. Similarly, the static nature of the equations causes the models to be extremely sensitive to rapid irradiance change, which is prevalent in Singapore. In order to tackle these problems, an empirical model taking into account all the factors that might affect the module temperature, such as radiation, convection, irradiance, wind speed, and humidity was proposed by Veldhuis et al [12] for Singapore. An exponential moving average was proposed to simulate the effect of thermal inertia on the estimation

process thus reducing the sensitivity of the model to rapid irradiance change. Eqs. 2-4 describe the calculations involved in this model. Eq. 2 represents the radiation contribution that results in the increase in the PV module temperature. Eq. 3 is the correction to the Eq. 2 that adds the effect of wind speed on the module temperature and Eq. 4 is the exponential moving average of the module temperature calculated by Eq.3. In these equations, T_r represents the module temperature contribution due to radiative heat transfer. T_m is the PV module back-surface temperature, k is the Ross coefficient, is the humidity coefficient, RH is the relative humidity, G is the irradiance, r is difference in temperature due to radiative cooling, h is the convective heat transfer coefficient, V_m is the average wind speed and α is the empirical exponential factor that affects the impact of wind speed on the convection heat transfer.

$$T_r = T_a + (k + (1 - R_H))G - r \quad (2)$$

$$T_r = T_r + (T_r - T_a)hv_m^a \quad (3)$$

$$T_m = T_{m,t}(\frac{2}{t+1}) + T_{m,EMA,t-1}(1 - \frac{2}{t+1}) \quad (4)$$

However, while the accuracy of the empirical model is improved, this method introduced a number of new disadvantages to the estimation process, which we will discuss in detail in Section IV.

B. Physical Steady State Models

A module in the field is susceptible to heat transfer through conduction, convection, and radiation. The main source of temperature rise of a PV panel is due to its exposure to solar radiation and the ambient (air) temperature. The module racks also act as a source of conductive heat loss for a PV module. Wind flow around the panels cause heat loss due to convection and there is a small amount of radiative heat loss from the panels to the surroundings and from the electricity generated from the panels. Fig.1 illustrates these mechanisms and Eqs. 5-6 puts them into a mathematical form.

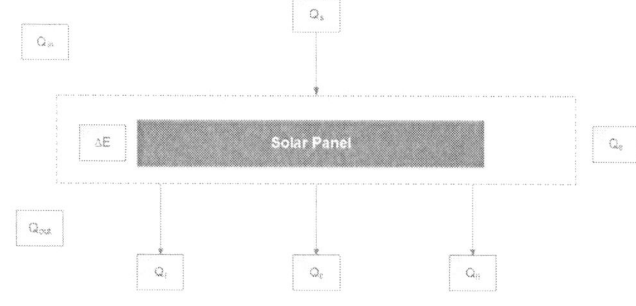

Fig. 3. PV Module Heat Exchange

$$Q_{in} = Q_s; Q_{out} = Q_c + Q_r + Q_n + Q_e; \quad (5)$$

$$\delta E = Q_{in} - Q_{out} \qquad (6)$$

In Fig.3, Q_s represents the input heat source to the solar panels from the surroundings. Q_c, Q_r, Q_n, and Q_e are the heat loss due to convection, conduction, radiation, and the heat dissipated by electricity generation, respectively.

Steady state models generally assume that input and output heat fluxes are the same and solve for the module temperature. As an example, the Mattei model [11] uses the mathematical equation shown in Eq. 7 to estimate the temperature of the PV module, as:

$$T_m = \frac{U_{PV}Ta + I(\tau\alpha - \eta_{STC}(1 - \beta_{STC}T_{STC}))}{U_{PV} + \beta_{STC}\eta_{STC}I} \qquad (7)$$

In this equation, T_m is the module temperature, T_a is the ambient temperature, G is the irradiance, is the efficiency of the solar panel at STC, which for the CIGS module used in this study is 13.8%, taken from the datasheet. τ is the transmittance of the cover system and α is the absorption coefficient. β_{STC} is the temperature coefficient at STC and U_{PV} is the heat exchange coefficient defined by Eq. 8, which depends on the wind speed.

$$U_{PV}(v) = 24.1 + 2.9v_w \qquad (8)$$

These models are generally more accurate than simple empirical models as they consider the internal physics of the modules; however, they are unable to simulate the thermal inertia of the PV module. These models also require information about several internal parameters of the system, which might not be available from the modules datasheet.

C. Physical Dynamic Models

The other popular way to deal with thermal inertia is to use the thermal equation of the PV module and not assume steady state [13] [14]. This results in a differential equation, as represented by Eq. 8 that can be solved using the Backward Euler method.

$$\frac{dT_{module}}{dt} = \frac{\alpha GA - P_{out} - (Q_{conv}^{front} + Q_{conv}^{rear})}{C_{module}} - \frac{(Q_{rad}^{front} + Q_{rad}^{rear})}{C_{module}} \qquad (9)$$

In the above equation, the L.H.S represents the rate of change of the temperature of the module. α is the absorptivity of the PV module, A is the area. Q_{conv}^{front} and Q_{conv}^{rear} are heat losses from the front and rear due to convection. Q_{rad}^{front} and Q_{rad}^{rear} are heat losses from the front and rear due to radiation. Finally, C_{module} is the heat capacity of the PV module. The major problem with this model is that the values for the parameters are usually unknown and need to be estimated. For example, the PSO algorithm [16] was used in [14] to identify the unknown parameters of the equation, based on four days of test data with a sampling period of 10 minutes. This

model was then used to estimate the temperature on different days. The results presented in [13] show that the temperature prediction error was reduced. However, several issues with this method make it unsuitable for easy application to other PV sites, particularly for sites with rapid irradiance change. Firstly, the data used to determine the model parameters used a sampling interval of 10 minutes. In tropical regions, the irradiance can change significantly in just a few seconds [17] and sampling at a 10 minute interval results in significant aliasing (the loss of high frequency components during the downsampling process) of the irradiance. Additionally, Eq. 9 depends on five parameters, which need to be known beforehand or determined through the application of PSO or some other technique.

A simpler *RC* circuit model [6] [9] can be used to model the thermal characteristics of a PV module, where *R* and *C* are the thermal resistance and capacitance of the module. In [6] the authors proposed an *RC* circuit for the PV module with individual *R* and *C* values for each of the layers. However, this method uses gradient based methods to calculate the *R* and *C* values, which have the tendency to converge to local minimas.

III. PROPOSED METHOD

In this work, we use the RC thermal model proposed in [9] as the basis for our PV module temperature estimation under rapid irradiance change. This model has the advantage of having the ability to simulate thermal inertia while using just two parameters. The most dominant heat exchange mechanism for a PV module is at the surface, where the radiative and convective heat transfer takes place, and these mechanisms are represented through a thermal circuit model shown in Fig. 4.

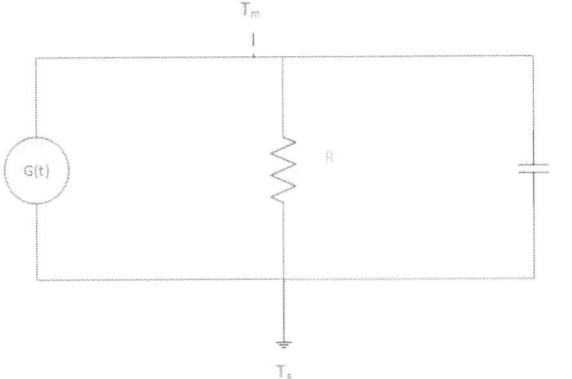

Fig. 4. RC Thermal Model

The iterative representation of the mathematical solution to the PV RC circuit model is shown in Eq. 10 . In Fig.4 (and Eq. 10), $G(t)$ represents the irradiance as a variable current source. *R* and *C* represent thermal resistance and thermal capacitance, while T_c, and T_a represent the module temperature and ambient temperature, respectively.

$$T_m(i) = T_a(i) + (1 - \frac{1}{RC})(T_m(i-1) - T_a(i-1)) + \frac{G(i-1)}{C} \quad (10)$$

Before the RC model can be used, the two parameters (R and C) need to be determined. To identify the unknown R and C parameters of Eq. 9, we use the PSO algorithm [16], because of its easy implementation and global optimization characteristic. In the optimization process we use the measured module temperature T_m, along with the irradiance $G(t)$ and the ambient temperature T_a, as inputs, and the normalized root mean square error (nRMSE) between the estimated and measured temperature values on the PV module as the fitness function to be optimized. The nRMSE is defined as:

$$nRMSE = \sqrt{\frac{\sum(\frac{y_{measured} - y_{estimated}}{y_{measured}})\Box}{n}} \quad (11)$$

IV. RESULTS AND DISCUSSIONS

To verify the effectiveness of the proposed temperature estimation method, we use 60 days of GHI irradiance data (measured using a CMP 11 Pyranometer) and air temperature data (measured using a Hobolink weather station) at Nanyang Technological University, Singapore (103.6829°E, 1.34653°N) over the period 06 May 2016 to 30th July 2016. The GHI and temperature are recorded at a one second sampling resolution using an OMRON ZR-RX 40 data logger. The GHI was converted to a plane of array (POA) irradiance to match the PV array module tilt of 10°E. For parameter estimation, and to determine the accuracy of the proposed temperature estimation method, we also use the actual PV module temperature measured at the back of one of the CIGS solar panels using a Pt100 temperature sensor from Ingenieurbüro Mencke and Tegtmeyer (using the same data logger as described above). That is, the ambient temperature and GHI are the inputs to the model in Eq. 10, and the PV temperature from a CIGS PV module is the desired output.

The sixty days chosen for the study cover a wide variety of weather conditions, varying from clear to cloudy and is representative of the rapidly changing weather conditions that are prevalent in Singapore. A scatter plot showing the variability index (VI) versus average clear sky index (CI), calculated according to [15], for the 60 days is shown in Fig. 5. The days with high CI but low VI are clear days, the days with low CI and low VI are overcast days, while the days with values in the middle are days with high irradiance variability. We then cluster the data based on CI and VI as shown in Fig. 5.

To illustrate the variability seen in tropical climates, the irradiance and ambient temperature versus time for four different days are shown in Fig. 6. The characteristics of these 4 days are summarised in table 1 and represent typical days with different VI and CI taken from clusters B, C, E and F in Fig. 5. On a clear day in Singapore (which is very rare), the

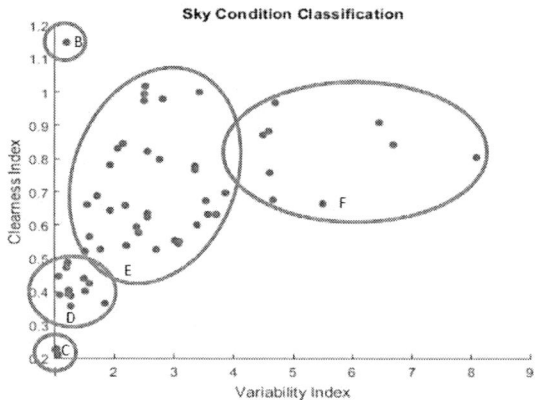

Fig. 5. Scatter Plot of Variability and Clearness Index

GHI shows a typical cosine characteristic with an irradiance greater than 800 W/m^2 in the middle of the day, and on a highly variable day, the irradiance can change rapidly from a value near the clear sky reading down to around 100-200 W/m2 within a matter of a few seconds.

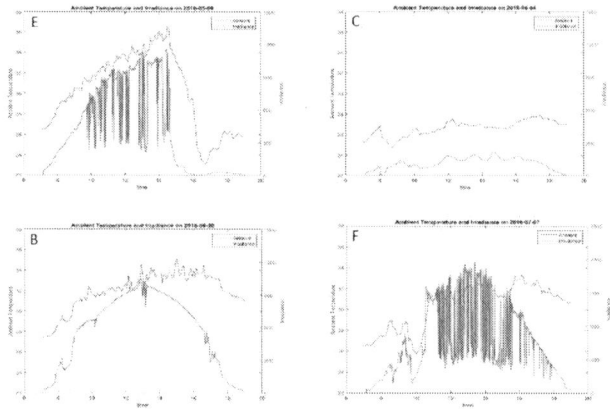

Fig. 6. Scatter Plot of Ambient Temperature and Solar Irradiance on different days

We then determine six data sets for the parameter estimation, as different weather conditions will influence the R and C values chosen, and as a result could affect the temperature prediction. These clusters are labelled from B – F as shown in Fig. 5, and cluster A is formed by considering all the sixty days.

The PSO algorithm is then used to determine the R and C parameter values of the model described by Eq. 10 for each day in a cluster. The parameters w, p1 and p2 are set as 1.1, 1.49 and 1.49, respectively. The number of particles and the number of iterations for the PSO algorithm are set as 400 and 20, respectively. During the parameter estimation, different days result in different R and C values. For example, the distribution of R and C values when the PSO algorithm is applied to set A (the full 60 days of data) is shown in Fig. 7. The individual R and C values for each day, determined by PSO, are then averaged to give the representative R and C value for a CIGS PV module. The final R and C values for each training cluster are given in Table 1.

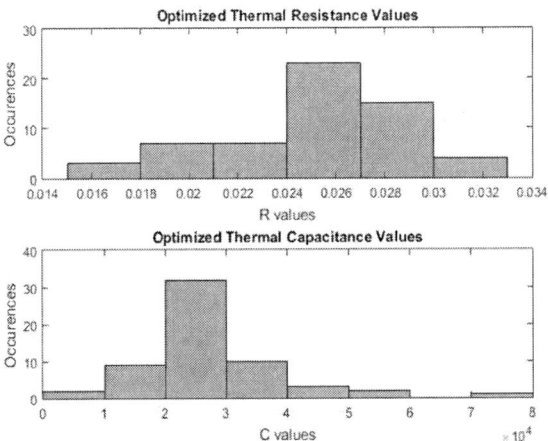

Fig. 7. Estimated RC Values

TABLE I
AVERAGE NRMSE AND VARIANCE FOR DIFFERENT METHODS

Model Name	Parameter Estimation		Predicted Temperature	
	R	C	nRMSE	Variance
NOCT Model	NA	NA	0.133681	0.001036
Mattei Model	NA	NA	0.086527	0.000361
Veldhuis Model	NA	NA	0.052186	0.000472
Set A(60 days)	0.024724	29658.65	0.053701	0.000236
Set B	0.025958	31.970334	0.09177	0.000445
Set C	0.015527	154238.08	0.103094	0.00087
Set D	0.023976	38125.16	0.054485088	0.000180695
Set E	0.025104	27946.85	0.05395	0.000269
Set F	0.027183	21728.31	0.059372	0.000494

Once the R and C values are determined, we use Eq. 10 to predict the module temperature from the GHI and the ambient temperature using a 1 second time step. The initial temperature is the ambient temperature in the morning when the GHI is equal to zero. The R and C values for each of the clusters is applied to the full 60 days of data and the average nRMSE and the variance between the actual temperature and the predicted temperature is determined. The results for the NOCT [10], Mattei [11] and Veldhuis [12] models are also shown in Table 1. The major problem with the NOCT and Mattei models is that they do not effectively account for the thermal inertia of a PV module. This can be seen in Fig. 8, which shows the temperature estimations for the NOCT, Mattei, Veldhuis and the proposed model (Set A) for a single day (07/07/2016). The NOCT and Mattei models are over sensitive to irradiance variations and the predicted

temperature is analogous to the irradiance (shown in Fig. 6) and hence will not be considered further.

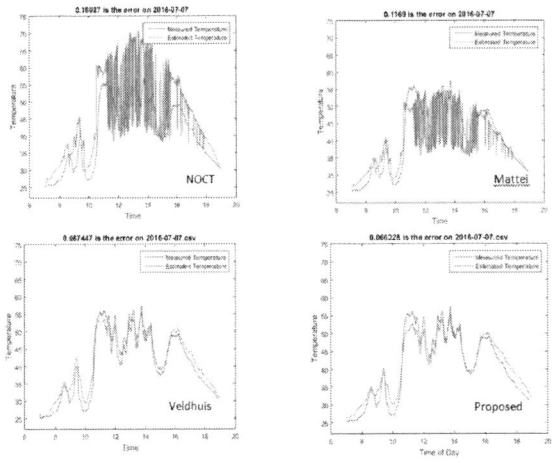

Fig. 8. Performance of different methods on a day with high irradiance variability

The temperature predictions using the Veldhuis model and the proposed model for four typical days, a clear day, a highly variable day, a less variable day and an overcast day, are given in Fig. 9. Fig. 9 shows that the proposed model and the Veldhuis model exhibit similar prediction trends, with the proposed model performing slightly better on the highly variable days (a nRMSE of 0.066228 compared to the 0.067447 of the Veldhuis model). A more detailed analysis showing the nRMSE for each of the 60 days for the 4 best training data sets (Sets A, D, E and F) and the Veldhuis model are shown in Fig. 10. Here we see that there is no real trend, and the daily variability for each model is considerable.

In terms of implementation, the Veldhuis model requires the identification of six parameters using a sensitivity analysis in order to estimate the module temperature. In order to identify these parameters, more information like the relative humidity and wind speed near the modules are required. Then, the parameter identification process also depends on the initial values chosen for the five parameters. Compared to that the Veldhuis model, our method works based on R and C values and only requires irradiance and ambient temperature as the inputs. Our model also runs three times faster than Veldhuis model because of a moving-average-free estimation process. The performance of the proposed model could also be further improved by including additional environmental inputs (such as humidity and windspeed) into the RC model, however this is left as future work.

V. CONCLUSIONS

In this work, we proposed a heuristic optimization algorithm based PV thermal model. The existing models were reviewed and classified into three categories, namely empirical models, physical steady state models and physical

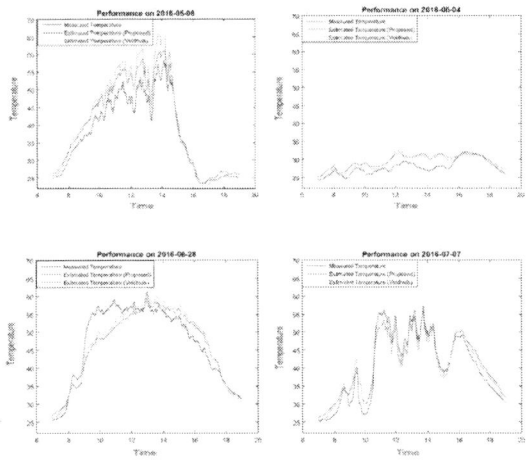

Fig. 9. Comparative performance of Veldhuis and Our Proposed Method on four days with different weather conditions

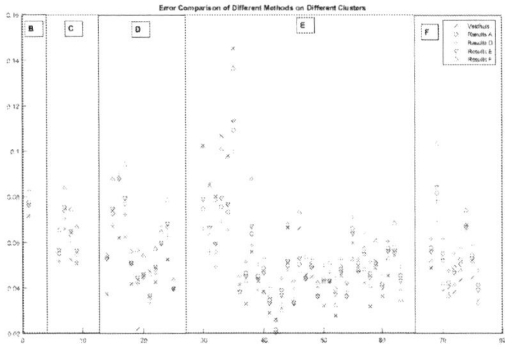

Fig. 10. Sixty day nRMSE values for each clusters

dynamic models. The disadvantages of existing methods includes vulnerability to rapid changing irradiance conditions, the parameter dependency, and difficult implementation, etc. After an initial analysis, we proposed an improvement to the RC circuit model, by which we estimated the R and C parameters using PSO algorithm. We chose three methods from the existing set of temperature models, named Mattei model, Veldhuis model and RC circuit model for analysing their applicability to a CIGS PV module placed on the field in Singapore under rapid varying conditions for a period of sixty days. The proposed RC circuit model provides a significantly better temperature estimation than the NOCT and Mattei models, which do not effectively account for the thermal inertia of a PV module and hence are oversensitive to rapid irradiance variations. The proposed model provides a similar, but slightly better, temperature estimation to that of the Veldhuis model but is significantly easier to implement. These results show that our model is highly suitable for

tropical areas where rapid changes in irradiance conditions are a common occurrence. Our future studies would include several topics such as improvement to the model with dynamic parameters, and online temperature prediction etc.

REFERENCES

[1] McCrone, Angus, et al. "Global trends in renewable energy investment 2016." Frankfurt School UNEP Collaborating Centre for Climate and Sustainable Energy Finance (2016).

[2] "Jiang, Lian; Maskell, Douglas L; Patra, Jagdish C; ",A novel ant colony optimization-based maximum power point tracking for photovoltaic systems under partially shaded conditions,Energy and Buildings,58,,227-236,2013,Elsevier

[3] Pelland, Sophie, et al. "Photovoltaic and solar forecasting: state of the art."IEA PVPS, Task 14 (2013): 1-36.

[4] Letendre, S., M. Makhyoun, and M. Taylor. "Predicting solar power production: irradiance forecasting models, applications and future prospects." Solar Electric Power Association, Tech. Rep (2014).

[5] Jayaraman, R. and Maskell, D.L., Temporal and spatial variations of the solar radiation observed in Singapore, Energy Procedia, vol. 25(2012), pp 108117, 2012.

[6] S. Armstrong, W.G. Hurley, A thermal model for photovoltaic panels under varying atmospheric conditions, Applied Thermal Engineering, Volume 30, Issues 1112, August 2010, Pages 1488-1495, ISSN 1359-4311, http://dx.doi.org/10.1016/j.applthermaleng.2010.03.012.

[7] Skoplaki, E., and J. A. Palyvos. "Operating temperature of photovoltaic modules: A survey of pertinent correlations." Renewable Energy 34.1 (2009): 23-29.

[8] Koehl, Michael, et al. "Modeling of the nominal operating cell temperature based on outdoor weathering." Solar Energy Materials and Solar Cells 95.7 (2011): 1638-1646.

[9] Mara, W., and M. Piotrowicz. "Extraction of thermal model parameters for field-installed photovoltaic module." Microelectronics Proceedings (MIEL), 2010 27th International Conference on. IEEE, 2010.

[10] Markvart T (editor). Solar electricity. 2nd edition. Chichester: Wiley; 2000.

[11] Mattei M, Notton G, Cristofari G, Muselli M, Poggi P. Calculation of the polycrystalline PV module temperature using a simple method of energy balance. Renew Energ 2006; 31: p. 553-567

[12] Veldhuis, Anton J., et al. "An empirical model for rack-mounted PV module temperatures for Southeast Asian Locations evaluated for Minute Time Scales." IEEE journal of photovoltaics 5.3 (2015): 774-782.

[13] Luketa-Hanlin, Amanda, and Joshua Stein. "Improvement and validation of a transient model to predict photovoltaic module temperature." Sandia National Laboratories SAND2012-4307 (2012).

[14] Chopde, Abhay, et al. "Parameter extraction for dynamic PV thermal model using particle swarm optimisation." Applied Thermal Engineering 100 (2016): 508-517.

[15] Stein, Joshua S., Clifford W. Hansen, and Matthew J. Reno. "The variability index: a new and novel metric for quantifying irradiance and PV output variability." World Renewable Energy Forum. 2012.

[16] J. Kennedy, R.C. Eberhart, Particle swarm optimization. Proc. IEEE Int. Conf. Neural Networks, New Jersey, USA, pp. 19421948, 1995.

[17] D. L. Maskell, S. Ramasubramanian and Xu Qing, "Module-based storage for regulating PV power intermittency at the point of generation," Photovoltaic Specialist Conference (PVSC), 2015 IEEE 42nd, New Orleans, LA, 2015, pp. 1-5. doi: 10.1109/PVSC.2015.7356279

978-1-5090-5606-4/17 $31.00 © 2017 IEEE

Accuracy of CdTe PV Energy Predictions Using Spectral Corrections

Mitchell Lee, Kendra Passow, and Paul Wolffersdorff

First Solar, San Francisco, CA, 94105, United States

Abstract — Outdoor Cadmium Telluride (CdTe) photovoltaic (PV) array performance is analyzed to evaluate the accuracy of a previously proposed spectral correction method and the impact of its inclusion on PV performance modeling. The spectral correction is of simple functional form, requiring air mass and precipitable water as inputs. Data is analyzed from four arrays, each containing twenty First Solar Series 4-2 CdTe PV modules. Two of the arrays are located in the desert climate of Mesa, Arizona. The other two arrays are located in the tropical climate of Kulim, Malaysia. Results demonstrate that the spectral correction method can improve the accuracy of PV performance modeling. It reduced seasonal biases when modeling the energy output of the arrays in Arizona, and it reduced large persistent biases when modeling the energy output of the arrays in Malaysia.

I. INTRODUCTION

The composition of the atmosphere varies on daily, seasonal, and annual time scales. Since different atmospheric constituents absorb and/or reflect irradiance at different wavelengths, these variations result in changes in outdoor spectrum. The performance of photovoltaic (PV) modules is often evaluated under a static spectral irradiance defined in the ASTM G173 standard [1], often referred to as "AM1.5". As the outdoor spectrum varies from AM1.5, so too does the performance of a PV module vary from its nameplate by a magnitude governed by its spectral response (SR) [1-2].

A metric often used to quantify the effects of spectrum on PV performance is called spectral shift (M), also known as spectral mismatch. M is defined as the ratio of irradiance-specific current generated under a particular spectrum to the irradiance-specific current generated under a reference spectrum [3], such as the G173 standard. M can also be defined using (1), where λ is wavelength, E is the spectral irradiance under consideration, and SR is the spectral responsivity of the PV device.

$$M = \frac{\int E(\lambda) \cdot SR(\lambda) d\lambda}{\int E(\lambda) d\lambda} \cdot \frac{\int E_{G173}(\lambda) d\lambda}{\int E_{G173}(\lambda) \cdot SR(\lambda) d\lambda} \quad (1)$$

Several correction models have been proposed [4-8], and many energy simulation tools have integrated or plan to integrate these methods [12,16]. However, spectral corrections are not universally applied within the PV industry. Although the theory is well-established, further evidence is necessary to illustrate that the spectral corrections will improve PV energy prediction accuracy. This paper will evaluate whether a spectral model proposed by Lee and Panchula [8] improves the accuracy of PV performance simulations. While the model is applicable to both cadmium telluride (CdTe) and crystalline silicon (c-Si)

technologies, this paper only investigates its impact on CdTe performance modeling.

The proposed model corrects for changes in spectrum due to precipitable water content (P_{wat}) and air mass (AM_a). P_{wat} and AM_a have been shown to significantly impact CdTe and c-Si PV technologies, with CdTe being more sensitive to P_{wat} and c-Si being more sensitive to AM_a [8-10]. The model is of the function form shown in (2), where coefficients b_0 through b_5 are dependent on the module SR.

$$M = b_0 + b_1 AM_a + b_2 P_{wat} + b_3 \sqrt{AM_a} + b_4 \sqrt{P_{wat}} + b_5 \frac{AM_a}{\sqrt{P_{wat}}} \quad (2)$$

In this study, the impact of the spectral model on the accuracy of PV performance modeling is evaluated. Data is analyzed from four arrays, each containing twenty First Solar Series 4-2A CdTe PV modules. Two of the arrays are located in the desert climate of Mesa, Arizona, and the other two arrays in the tropical climate of Kulim, Malaysia. For the Mesa arrays, two years of data was available. One year of data was available for the Kulim arrays.

For each test array, performance is estimated using power performance index (*PPI*) [11] and compared to M. In addition, energy output of the arrays is compared to PV simulations conducted with and without the inclusion of spectral effects using PlantPredict version 4.6.0 [12]. The PlantPredict simulations use measured on-site meteorological data.

Unlike the initial field-testing [8], this study includes a larger module sample size. It also compares the spectral model to measured power and energy production instead of I_{sc}, which are more important to the bankability of solar projects.

The accuracy of the spectral model was also evaluated by Schweiger and Hermann [13], who compared it to an approach where M was estimated using measured spectral irradiance and a module's spectral response curve. They found that the spectral model would have improved energy predictions for Series 4-2 in five diverse climates. In general, the research presented agrees with their findings, but with the added advantage of using measured PV array performance.

II. OUTDOOR PV PERFORMANCE DATA

The modules in Mesa, Arizona were installed in July of 2014. In order to remove the effects of initial module stabilization and data quality issues, this study analyzed data from October 1,

978-1-5090-5606-4/17 $31.00 © 2017 IEEE

2014 through September 30, 2016. The modules in Malaysia were installed in August of 2014; however, this study only analyzes data from March 1, 2016 to February 28, 2017 due to meteorological data availability.

Maximum power (P_{max}) at standard test conditions (STC) was measured using a solar simulator near the start of the analysis period for both locations. The average module P_{max} for Mesa and Kulim was 112.1W and 109.3W respectively. These P_{max} values were used instead of standard module bins, in an effort to improve the accuracy of this analysis. The two arrays in Mesa were constructed identically, with a fixed 25° tilt, south facing azimuth, and 50.4% ground coverage ratio. Similarly, the two arrays in Kulim were constructed identically with a fixed 10° tilt, south facing azimuth, and 70.3% ground coverage ratio. A DC:AC ratio below one was selected for all arrays to avoid clipping.

Co-located with the test site is a meteorological station with broadband global horizontal irradiance (*GHI*), plane of array (*POA*) irradiance, ambient temperature, and relative humidity. Representative module temperature measurements are also recorded. All measurements are recorded at five minute intervals. Because direct measurements of P_{wat} were not available, temperature and relative humidity were used to estimate P_{wat} according to the Gueymard method [15-16]. *M* was calculated for all timestamps using (2) with the relevant module coefficients as published in [17]: $b_0 = 0.86273$, $b_1 = -0.038948$, $b_2 = -0.012506$, $b_3 = 0.098871$, $b_4 = 0.084658$, and $b_5 = -0.0042948$.

III. Spectral Shift and PPI

M was compared with array performance as measured using *PPI*. *PPI* translates measured power to standard test conditions at 25 °C and 1000 W/m^2. As a result, this metric quantifies array performance while removing the effects of irradiance and temperature [11].

PPI was calculated on a weekly basis using five-minute resolution input data, and full weeks with poor *PPI* regressions, R^2 below 0.6, were excluded. For the Mesa arrays, input data was limited to when *POA* was between 500 W/m^2 and 1200 W/m^2. The transient irradiance conditions characteristic of the Kulim, Malaysia location have a destabilizing effect on the *PPI* metric. In an effort to mitigate these effects, a narrow range of *POA*, from 500 W/m^2 to 800 W/m^2, was used. Moreover, points

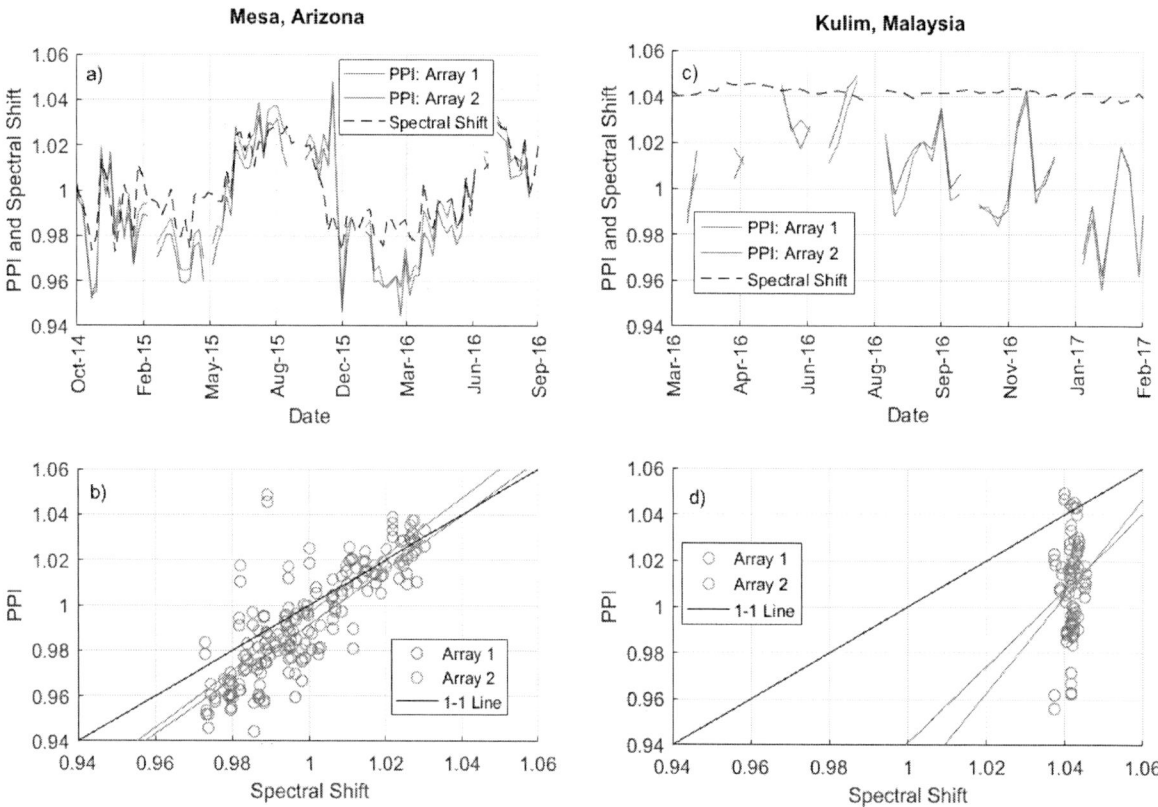

Fig. 1.   Comparison of weekly *PPI* and *M*. Subplot a) shows the data from Mesa, Arizona as a time series. Also included for comparison is spectral shift. Subplot b) shows *PPI* plotted against *M* for the Mesa site. Also included are a 1-1 line, and linear trend lines for both arrays. Subplots c) and d) are analogous plots, but for the Kulim, Malaysia location.

more than one standard deviation off of the power vs. irradiance regression line were excluded. For both locations, the adjusted module P_{max} from the previous section was used in the *PPI* calculations. A linear degradation rate of 0.5% per year was applied starting at the end of year one in accordance with First Solar energy prediction guidance [18]. For all time steps, array power was also adjusted by 0.2% to account for wiring and mismatch losses. Given the small scale of the PV system, these losses were estimated to be significantly less than those of utility scale solar farms. Module soiling data was not available and assumed negligible for both sites. In order to facilitate comparison, all filtering to *PPI* was also applied to *M*. *M* was then aggregated to weekly *POA*-weighted averages.

The results of the *PPI* and *M* comparison are illustrated in Fig. 1. Subplots 1a and 1b contain data from Mesa, Arizona, and subplots 1c and 1d contain data from Kulim, Malaysia. Fig. 1a shows that for the site in Mesa, Arizona there is a strong temporal relationship between *M* and *PPI*. It also illustrates that spectral effects can have a significant impact on seasonal performance. Corroborating the findings of [8], Fig. 1a shows spectrum positively impacting performance in summer months (when P_{wat} is high and AM_a is low), and it shows spectrum negatively impacting performance in winter months (when P_{wat} is low and AM_a is high). Fig. 1b illustrates that for the Mesa, Arizona system, *PPI* and *M* have a strong linear relationship, with R^2 values of 0.67 and 0.70 for Arrays 1 and 2, respectively. The high R^2 values suggest that most of the variation in STC-corrected array performance is due to spectral effects. Fig. 1b also illustrates that there is more variation in *PPI* than in *M*. For Array 1, the trend line of *PPI* versus *M* has a slope of 1.2041 and an intercept of -0.2126. Similarly, for Array 2, the trend line had a slope of 1.2664 and an intercept of -0.2697. The results suggest that the model may be underestimating the effects of spectrum on performance. Another possibility is that an unaccounted for variable, such as soiling, is influencing the *PPI* metric.

As seen in Fig. 1c and Fig. 1d, there was little agreement between *PPI* and *M* for the Kulim arrays. The R^2 between *PPI* and *M* was less than 0.05 for both arrays, suggesting that the relationship was statistically insignificant. This lack of relationship may be explained by two factors. First, all of the weekly *M* values occur over a relatively small range, 1.037 to 1.046, making a linear fit difficult. Second, and more importantly, the stability of the *PPI* metric can be heavily affected by cloudy conditions like those in Kulim, Malaysia. Ordinary least squares was used in this analysis for consistency despite not being recommended for cloudy data sets, in favor of more robust regression techniques [11].

IV. Spectral Shift and Performance Modeling

Energy production of the test arrays with and without spectral corrections was simulated in PlantPredict. The energy output of the simulations was then compared to actual PV array performance to determine whether or not the spectral model improved energy prediction accuracy. Because they are virtually identical, both arrays in Mesa were compared against the same PlantPredict simulation. Similarly, both arrays in Kulim were compared against the same PlantPredict simulation. All simulations used ground measured *GHI*, ambient temperature, and relative humidity. The globally available FS-4110A CdTe Sept2014 module file was used for all arrays and module quality adjustments were applied to make the effective nameplates match the solar simulator results. These values were used in lieu of sticker P_{max} in order to tease out any potential biases in the energy prediction and isolate the effects of spectrum. Mismatch and wiring losses were each estimated to be 0.1%, and the arrays were assumed to have no soiling losses. For the Mesa arrays, analysis started soon after module installation. As a result, a linear degradation rate of 0.5% was applied starting at the beginning of the second year. The Kulim arrays were installed more than a year in advance of the analysis period. As a result, a linear degradation rate was applied starting at the beginning of the simulation. All other parameters in the simulations were set to PlantPredict defaults [12,18].

A. Mesa, Arizona Test Arrays

Table I compares the measured array performance to the PlantPredict models. The table illustrates that the inclusion of spectral effects improved the energy prediction accuracy for both Mesa arrays. On an annual time scale the improvement was relatively small because the spectral gain during the summer is a similar magnitude compared to the spectral loss during the winter. For Array 1, the energy prediction error was reduced from 0.99% to 0.69%, and for Array 2 the error was reduced from 0.50% to 0.20%. However, for the Mesa arrays, the improvements to *seasonal* energy predication accuracy were more significant. Table I also contains the Mean Absolute Percentage Error (*MAPE*) of the energy predictions on a monthly basis. *MAPE* results illustrate that the average monthly energy prediction error was reduced from roughly two percent to one percent for both arrays. This suggests that the simulation including spectral effects was significantly better at modeling seasonal variation in PV performance at the Mesa location.

Fig. 2. shows the monthly energy prediction error with and without spectral shift. Also contained in each subplot are the *POA*-weighted monthly spectral losses. Note that the spectral loss convention was used for this plot in order to facilitate

TABLE I: Measured vs Predicted Energy

Array	Monthly Energy MAPE (%)		Total Energy Prediction Error (%)	
	Spectral On	Spectral Off	Spectral On	Spectral Off
Mesa, Array 1	1.04%	2.07%	0.69%	0.99%
Mesa, Array 2	1.04%	1.99%	0.20%	0.50%
Kulim, Array 1	1.45%	4.05%	-0.34%	-3.97%
Kulim, Array 2	1.25%	3.82%	-0.09%	-3.72%

visual comparison with the energy prediction error. In Fig. 2., a negative spectral loss is the equivalent of a spectral gain. Fig 2. illustrates that for most months in Mesa, the absolute error in predicted energy was smaller for the simulation that included the spectral model. When compared to both arrays, the addition of the spectral model improved energy prediction accuracy. However, the simulation with spectral effects included tends to overpredict when there is a spectral loss, and underpredict when there is a spectral gain. This suggests that (2) slightly underestimated the effects of spectrum on these test arrays.

B. Kulim, Malaysia Test Array

The inclusion of spectral effects improved the energy prediction accuracy for both arrays in Kulim, Malaysia. Annual energy prediction error was reduced by over 3.5% for both arrays.

However, as illustrated in Fig. 2c and Fig. 2d, the spectral model predicts minimal seasonal effects. The *POA* weighted spectral gain was between 3.7% and 4.4% for all months. As a result, the blue lines (spectral model on) maintain the same overall shape as the orange lines (spectral model off). Seasonal

variation in performance relative to the PlantPredict simulation remain unexplained by the spectral model. With the spectral model enabled, the PlantPredict simulation overpredicted the energy generation of both arrays in March by roughly 2.7%, and underpredicted energy generation in October by roughly 2%. This variation could be the result of unaccounted for spectral effects, uncertainty in other modeling steps, such as module temperature, or an unaccounted for loss factor such as soiling. Intermittent cloud coverage, like the conditions in Kulim Malaysia, will introduce more uncertainty into M estimates, as the model proposed by [8] is based on clear-sky conditions. Moreover, under tropical conditions, there will be more uncertainty in the other loss factors that comprise an energy prediction. A majority of the models within an energy prediction were developed for, and tested against, temperate conditions in Europe and North America.

V. DISCUSSION AND CONCLUSIONS

The findings suggest that the spectral model proposed by Lee and Panchula can improve the energy prediction accuracy of

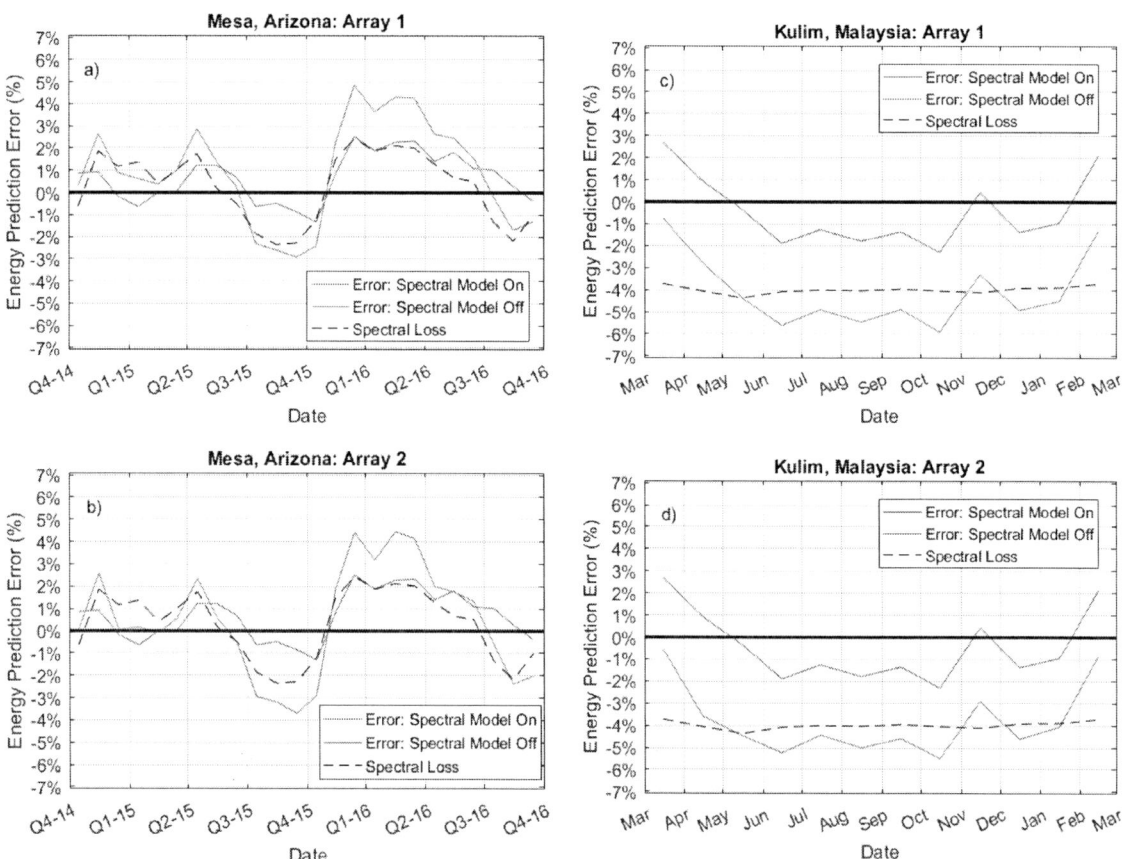

Fig. 2.   Monthly energy prediction error of PlantPredict simulation with spectral shift on and spectral shift off for a) Mesa, Array 1 performance, b) Mesa, Array 2, c) Kulim, Array 1, and d) Kulim, Array 2. Monthly spectral losses are included as dashed lines. The line indicating zero energy prediction error and zero spectral loss is in bold.

PV simulations. For both climates, the spectral correction model improved energy prediction accuracy when compared to the base case of no spectral correction.

For the desert climate in Mesa, Arizona, the inclusion of the model resulted in a small improvement in aggregate energy prediction on a multiyear time scale. Additionally, it considerably reduced seasonal biases in predicted energy output for this climate. Based on the model, these results match expectations, given that annual spectral effects in Mesa Arizona are small and seasonal effects are more substantial.

For the tropical climate of Kulim, Malaysia, the addition of the spectral model greatly improved the accuracy of the energy prediction on an annual time scale. For both arrays, the annual energy prediction error was reduced by over 3.5%. Nevertheless, some seasonal variation in performance remained unexplained by the spectral model.

Results from both climates corroborate the findings of [13], who found that the spectral model matched seasonal variation in First Solar Series 4-2 module performance for a location in Arizona, USA. They also found that the spectral model would improve energy prediction accuracy for a tropical climate in southeast, India, but that it did not match seasonal trends in performance.

VI. FUTURE WORK

This analysis, combined with the work of [13], suggests that the spectral model proposed by [8] will generally improve energy prediction accuracy over the base case of no spectral correction. However, there is room for additional research with respect to the impact of spectrum on PV performance. Future work could include comparison of the model to measured plant performance in a wider range of climates. Improvements may be made by the inclusion of cloud coverage or other atmospheric parameters in the spectral model. Note that spectral model development must balance accuracy with the need for straightforward implementation and meteorological inputs that are generally available on a global scale.

REFERENCES

[1] Nat. Renewable Energy Lab., Golden, CO. Reference solar spectral irradiance: ASTM G-173 (2012). [Online]. Available: http://rredc.nrel.gov/solar/spectra/am1.5/ASTMG173/ASTMG173.html.

[2] IEC 60904-3: Photovoltaic devices - Part 3: Measurement principles for terrestrial photovoltaic (PV) solar devices with reference spectral irradiance data (2009).

[3] C. R. Osterwald, "Translation of Device Performance Measurements to Reference Conditions," *Solar Cells*, vol. 18, no. 3-4, pp. 269-279, 1986.

[4] D. King, J. Kratochvill, and W. Boyson, "Measuring solar spectral and angle-of-incidence effects on photovoltaic modules and solar irradiance sensors," in 26th IEEE Photovoltaic Specialists Conference, 1997, pp. 1113 – 1116.

[5] T. R. Betts, R. Gottschalg, and D. G. Infield. "Spectral Irradiance Correction for PV System Yield Calculations." Proceedings of the 19 World Conference on PV Solar Energy Conversion, Vienna, 1998, pp. 1947-1952.

[6] B. C. Duck, and C. J. Fell, "Improving the Spectral Correction Function." in 43rd IEEE Photovoltaic Specialists Conference, 2016.

[7] L. Nelson, M. Frichtl, and A. Panchula, "Changes in cadmium telluride photovoltaic performance due to spectrum," *IEEE Journal of Photovoltaics*, vol. 3, No. 1, pp. 488-493, 2013.

[8] M. Lee, and A. Panchula, "Spectral Correction for Photovoltaic Module Performance Based on Air Mass and Precipitable Water", in 43rd IEEE Photovoltaic Specialists Conference, 2016.

[9] M. Lee, L. Ngan, W. Hayes, and A.F. Panchula, "Comparison of the Effects of Spectrum on Cadmium Telluride and Monocrystalline Silicon Photovoltaic Module Performance", in 42nd IEEE Photovoltaic Specialists Conference, 2015.

[10] E. F. Fernández, A. Soria-Moya, F. Almonacid, and J. Aguilera, "Comparative Assessment of the Spectral Impact on the Energy Yield of High Concentrator and Conventional Photovoltaic Technology," *Solar Energy Materials & Solar Cells*, vol. 147, pp. 185-197, 2016.

[11] L. Nelson, and C. Hansen, "Evaluation of Photovoltaic System Power Rating Methods for a Cadmium Telluride Array," in 37th IEEE Photovoltaics Specialists Conference, 2011.

[12] PlantPredict 4.6.0, "Resource Center," (2017). [Online]. Available: https://app.plantpredict.com/#/resource-center/user-manual.

[13] M. Schweiger, and W. Hermann, "Influence of Spectral Effects on Energy Yield of Different PV Modules: Comparison of P_{wat} and MMF Approach," TÜV Rheinland Energy GmbH Solar Energy., Cologne, Germany, Report No. 21237296.003, 2017.

[14] First Solar Application Notes PD-5-800 Rev. 6.0, "First Solar PV Module Product Bins and Distribution" 2016.

[15] C. Gueymard, "Analysis of Monthly Average Atmospheric Precipitable Water and Turbidity in Canada and Northern United States," Solar Energy, vol. 53, No.1, pp. 57-71, 1994.

[16] C. Gueymard, "Assessment of the Accuracy and Computing speed of Simplified Saturation Vapor Equations Using a New Reference Dataset," Journal of Applied Meteorology, vol. 32, pp 1294-1300, 1993.

[17] Sandia National Laboratories, PV Performance Modeling Collaborative, PV_LIB Toolbox, (2017). [Online]. Available: https://github.com/sandialabs/MATLAB_PV_LIB

[18] First Solar Application Notes PD-5-390 Rev. 1.0, "Module-related parameters for PVSYST Simulations of PV Systems Constructed with FS Series 4 PV Modules," 2015.

978-1-5090-5606-4/17 $31.00 © 2017 IEEE

PlantPredict: Solar Performance Modeling Made Simple

Kendra Passow, Lauren Ngan, Geoffrey Rich, Mitch Lee, and Stephen Kaplan

First Solar, San Francisco, CA, USA

Abstract — **First Solar has developed an energy prediction software called PlantPredict to model the generation of utility-scale photovoltaic power plants. PlantPredict focuses on efficiently and accurately modeling complex utility-scale photovoltaic power plants in a single energy prediction. Algorithm options are extensive and transparently documented. This includes important models that are not currently available in industry-standard comparable software such as spectral shift, non-linear temperature coefficient, and inverter thermal clipping. Through partnerships with weather file vendors, PlantPredict offers one-click download of weather data to further improve the efficiency of utility-scale energy predictions. Future releases will include important modeling capabilities such as module file generation, sloped terrain modeling, and plane of array irradiance input. With over 900 users across more than 300 companies, PlantPredict has seen steady growth since its initial release in late 2016.**

I. INTRODUCTION

Existing solar modeling software packages estimate energy generation of photovoltaic (PV) power plants with varying degrees of accuracy; however, the industry previously lacked modeling software that can aggregate large and often complex plant designs into a single energy prediction. First Solar has developed a free cloud-based web application to streamline the prediction of common PV technologies, allowing the ability to refine the energy prediction as a project develops from site prospecting through the verification of operating plant energy.

II. ORGANIZATION AND LAYOUT

Each organization using the software is assigned a unique account, where individual clients control their organization's permitted users. The Project Dashboard shows all available simulations from users within the same organization (see Fig. 1), allowing users within a company to easily share and/or compare simulation results. The program is hosted on Microsoft Window's Azure Cloud, with data security provided by Microsoft under Trust Center standards.

The general layout of the software allows users to quickly access component libraries. The inverter and module libraries are preloaded with common components from a range of manufacturers. Alternatively, users can easily create new components that can then be shared within their organization. The weather library is prepopulated with publicly available data sources such as NREL TMY3 in the United States [1], and weather data from Meteonorm [2] and Clean Power Research [3] are also available on a trial basis through their respective Application Programming Interfaces (APIs). The NREL National Solar Radiation Database resources are also available for query, including the Physical Solar Model [4] (United States and Mexico/Central America/northern South America) and SUNY [5] (South Asia and India). Weather files of any format may also be uploaded using an intuitive file parser and data quality check, where common formats are automatically recognized. Useful graphical and tabular visuals are available to examine or compare weather files of interest.

Figure 1.   Visual example of project map dashboard in PlantPredict.

The power plant can be modeled in detail as a nested structure of sub-models. The block is comprised of a subset of different array types, where each unique configuration can have a defined number of instances. The array also defines the medium voltage transformer and electrical losses including AC line loss, data acquisition system loss, and power conversion shelter cooling loss. Each array configuration similarly has one or more inverters associated with it, where the inverter power factor and set-point are defined. The inverter model is customizable, where the efficiency curves at various operating voltages and the temperature-elevation derate curves can be defined with an unlimited number of points. Each inverter then has one or more DC fields assigned, defining the module file, field nameplate power, field geometry, and DC losses.

Once the power plant configuration is defined, the power plant system dialogue allows the user to set a plant output limit and add up to six transmission lines and/or high voltage transformers in series between the power plant and the energy meter.

If the user has any difficulty with running a simulation, help dialogues are available at every step of the process, including video demonstrations. There is also a detailed user manual available in the resource center, including a transparent description of all available model algorithms. Suggestions and questions are also welcome in the "Contact Us" dialogue or by e-mailing support@plantpredict.com directly.

II. ALGORITHMS AND FEATURES

First Solar has invested heavily in understanding the field performance of utility-scale PV systems through a series of models describing spectral response, module temperature, degradation, etc. The culmination of this work is the implementation of models in a toolset that can be used to assess the value of power plants in all aspects of the business as well as in its customer base.

The core PV module is represented by a single-diode equivalent-circuit model, with a recombination term in order to better match module I–V curve characteristics. The model used is nearly identical to that implemented in PVsyst [6], but with additional numerical precision extracted for sensitive coefficients during the curve-matching process; in consequence, the module characterizations are interchangeable.

The ability to model spectral shift accurately is a major focus; the latest spectral shift model proposed by Lee et al. [7] is included, which allows accurate corrections for both CdTe and c-Si modules by considering the effects of both precipitable water and air mass. Other previously proposed models, including the precipitable water-only spectral response of CdTe [7][9] and the Sandia Air Mass Model [11], are also included for reference. These models drive improved accuracy of the energy estimation.

Generally, effort was made to include multiple validated model options in the optional advanced settings, such as for the diffuse decomposition and transposition models. Collectively, these algorithms represent the best known mathematical description of a PV power plant. Algorithms are well aligned with PV_LIB [12]. Addition and validation of new advanced algorithms will be an ongoing effort.

There are many other advantages of PlantPredict compared to currently available software. The program has the ability to model sub-hourly time scale, removing the uncertainty associated with hourly averaging. The user can also run multi-year performance estimates, including AC or DC degradation. Other advanced features include custom incidence angle modifier (IAM) profiles, non-linear temperature coefficient, and built-in record management to track project status. Finally, setup of an API for external users is available to allow for direct feed-in of operating plant data and automated prediction runs.

In June 2017, PlantPredict will incorporate a simple layout functionality to aid users who do not have a layout but want to understand how much energy is likely to be generated with a typical system on a given plot of land. Users are able to draw a land area on a map and choose from PlantPredict component libraries and specify mounting types after which they see estimates of tables and arrays of components. They are then led into the main powerplant builder to finish out their prediction.

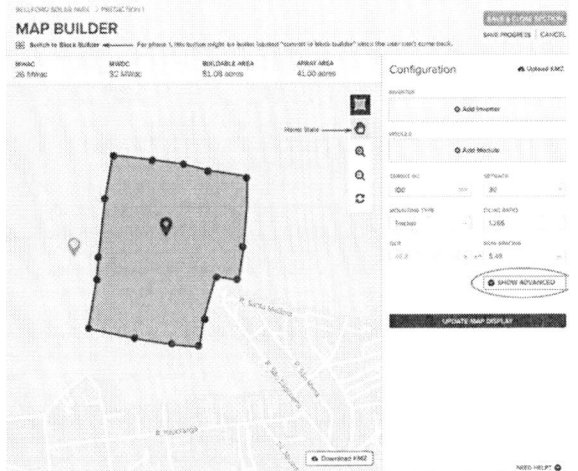

Figure 2. Mock-up of upcoming layout feature.

III. EXAMPLE OF API UTILITY

The API feature enables quick, easily altered predictions without using the user interface. This is particularly useful for operations and maintenance, optimization, or automated performance analysis. For more information and documentation on the API, please contact support@plantpredict.com.

One example of the capabilities of the PlantPredict API is presented here, showing a world map comparing the energy output of CdTe and c-Si PV module technologies (see Fig. 3).

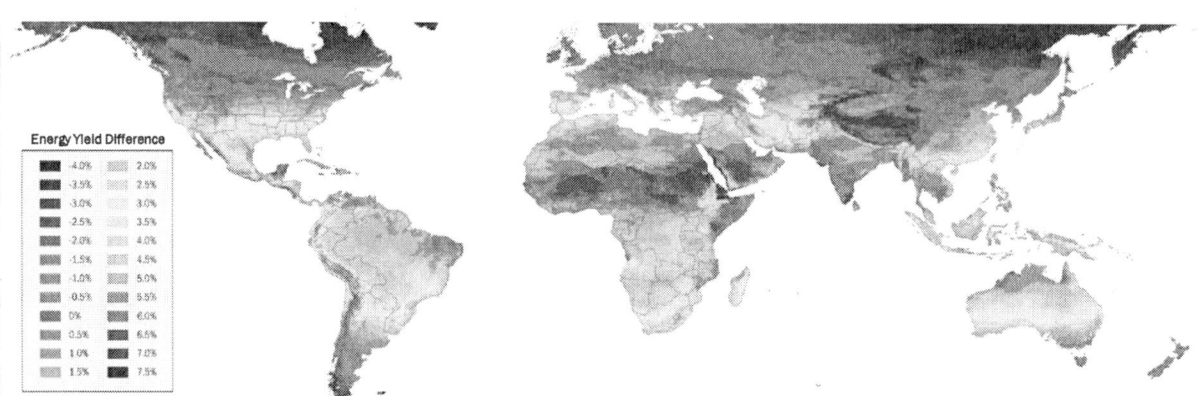

Figure 3.   Visual example of an energy prediction application enabled by use of the API, demonstrating the difference in energy production between CdTe and c-Si PV technologies.

One hundred well-distributed energy predictions were run representing a wide variety of climate types, which finished simulating in under an hour using the API. Efforts were made to ensure an apples to apples comparison between arrays, including using a constant ground coverage ratio (45%), DC:AC ratio (1.2), block DC capacity, and appropriate loss factors. The results of these predictions were used to create two functions that can estimate energy difference at any global location using latitude and average weather data for fixed tilt and tracker arrays. A worldwide grid of weather data was then used as input to calculate relative energy difference for each 0.25^0 x 0.25^0 grid cell, creating color maps for fixed tilt and tracker arrays.

III. SOFTWARE VALIDATION

Previous work has been presented on the accuracy of the software, both in comparison to PVsyst simulations and to measured data for nearly 1 GW of operating utility-scale PV assets [10]. This latter analysis was updated through the end of

2015, including the addition of two power plants to bring the total number of sites analyzed to 22. The energy meter error is illustrated in Fig. 4, showing an average underprediction of 0.12% with a standard deviation of ±1.92%. In addition to internal validation, external review has been completed by four major independent engineering firms. Reports and findings from these reviews are in the PlantPredict Resource Center. These reports validated the work comparing PlantPredict to PVsyst, indicating a close match for spectrally neutral locations; however, in locations with moderate or high spectral shift, PlantPredict is expected to lower uncertainty.

IV. PLANTPREDICT GROWTH AND INDUSTRY ACCEPTANCE

PlantPredict was released in September 2016. Since that time, the software has seen steady growth and industry adoption (see Fig. 5). Currently, more than 900 users representing nearly 400 companies are using the software. Through the addition of new algorithms, best-in-class prediction speed and accuracy, technical transparency and

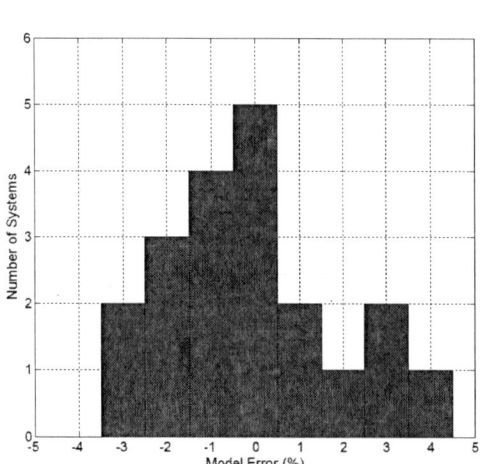

Figure 4.   Comparison of measured vs. predicted energy meter production for 22 sites using PlantPredict.

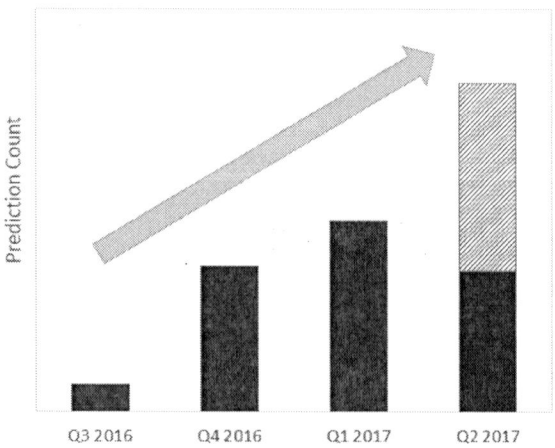

Figure 5. PlantPredict prediction count growth since release in 2016 as of May 2017 (hashed section represents projected Q2 2017 total).

high-quality customer service, we aim to become the utility-scale PV industry's choice for both bankable and convenient energy predictions.

IV. FUTURE IMPROVEMENTS

PlantPredict is run on a two-week agile sprint process which enables the continuous release of new features. Currently, many additions to the model library are in the works to keep pace with the state of the industry and user requests. Future work includes creation of a module file generator, with the ability to derive single diode model parameters using either datasheet values as in [13] or measured IV curves as in [14]. One particular focus of this work is the automated optimization of input parameters for a single diode model extended with a recombination current term in order to match efficiency-irradiance response curves provided by thin film manufacturer data.

The ability to input plane of array irradiance to avoid the errors associated with decomposition and transposition when measured data is available using the GTI-DIRINT model is also on the roadmap [15]. Other algorithmic improvements include tilted and dual-axis tracking, electrical shading, slope terrain modeling, non-linear degradation and electrical shading effects.

REFERENCES

[1] S. Wilcox, W. Marion, "Users Manual for TMY3 Data Sets", NREL/TP-581-43156, May 2008.

[2] J. Remund, S. Müller, Meteonorm Global Meteorological Database, Meteotest 2014.

[3] SolarAnywhere®, Clean Power Research

[4] NREL NSRDB PSM, https://nsrdb.nrel.gov/current-version#psm.

[5] NREL NSRDB SUNY, https://nsrdb.nrel.gov/international-datasets.

[6] PVsyst www.pvsyst.com

[7] M. Lee and A. Panchula, "Spectral Correction for PV Performance Based on Air Mass and Precipitable Water," *43rd IEEE Photovoltaic Specialists Conference* (2016).

[8] L. Nelson, et al, "Changes in Cadmium Telluride Photovoltaic System Performance due to Spectrum", *IEEE Journal of Photovoltaics*, Volume 3 Issue 1, Jan 2013. DOI 10.1109/JPHOTOV.2012.2226868.

[9] M. Lee, L. Ngan, W. Hayes, J. Sorensen, A. Panchula. "Understanding Next Generation Cadmium Telluride Photovoltaic Performance due to Spectrum," *42nd IEEE Photovoltaic Specialists Conference* (2015).

[10] K. Passow, L. Ngan, B. Littmann, M. Lee, and A. Panchula, "Accuracy of Energy Assessments in Utility Scale PV Power Plant using PlantPredict," *42nd IEEE Photovoltaic Specialists Conference* (2015).

[11] D. King, J. Kratochvil, and W. Boyson, "Measured Solar Spectral and Angle-of-Incidence Effects on Photovoltaic Modules and Solar Irradiance Sensors," *26th IEEE Photovoltaic Specialists Conference* (1997).

[12] Sandia National Laboratories, PV Performance Modeling Collaborative, PV_LIB Toolbox, (2015). [Online]. Available: https://pvpmc.sandia.gov/applications/pv_lib-toolbox/

[13] W. De Soto, S.A. Klein, and W.A. Beckman, "Improvement and validation of a model for photovoltaic array performance," *Solar Energy*, Volume 80 Issue 1, Jan 2007.

[14] C. Hansen, "Parameter Estimation for Single Diode Models of Photovoltaic Modules," Sandia Report 2015-2065, March 2015.

[15] B. Marion, "A model for deriving the direct normal and diffuse horizontal irradiance from the global tilted irradiance," *Solar Energy*, v. 122, pp. 1037-1046, 2015.

Integrability Comparison between BIPV and BAPV in Tropical Conditions: A Bangalore Case-study

Gayathri Aaditya, Roshan R Rao and Monto Mani

Indian Institute of Science, Bangalore – 560012, India

Abstract — Building Integrated/Applied Photovoltaic (BIPV/BAPV) technology is a unique building configuration integrating energy generation into a building's functional performance. BIPV comprises building envelope elements (wall, façade, fenestration) of PV while BAPV comprises PV applied on/in building elements. The uniqueness in this configuration lies in PV system design requiring maximum exposure to solar insolation, while building elements are designed to control and regulate solar exposure. PV performance is primarily governed by its output and efficiency, while building performance is primarily governed by its passive climatic-response, indoor thermal comfort and energy consumption. Integrability is a methodology evolved to aid in assessing a BIPV/BAPV configuration for its integrated building-PV performance. Tropical regions are gifted with ample sunny days, but due to the heat from the sun traditionally buildings have relied on high thermal-mass, shading devices and natural ventilation. However, with low-thermal mass in PV, its integration in buildings as BIPV results in high heat gain and loss, while integration as BAPV results in shading and altered building-heat transfer mechanisms. The current paper investigates the applicability of BIPV and BAPV, adopting the evolved integrability methodology, for tropical conditions, with Bangalore as case-study. Insights based on the study involving real-time measurements and simulation models have been presented in this paper.

Keywords – BIPV, BAPV, integrability index, PV performance, Tropical location, Design Builder, System Advisor Model.

I. Introduction

In India, BIPV installations are still nascent as compared to roof-top (BAPV) PV. Most installation of roof-top systems are driven more by energy security and sustainability rather than economic viability. BIPV systems are more an element of architectural style and identity, but given the urbanizing population and their humungous energy and building-resource demand [1], integration of PV in buildings is well envisaged. Currently, BIPV designs are influenced by the PV performance alone as done in the case of ground mounted systems, but needs to include impact on indoor thermal comfort, dust and wind regimes, maintainability, structural reliability, durability and economic viability. A performance assessment for BIPV must integrate these building performance functionalities of a typical dwelling.

In the case of BIPV, PV forms part of the building envelope material and requisite building-codes are yet to evolve. The available BIPV performance indices (like performance ratio, payback period) are typically one dimensional (PV), and do not reveal the whole building BIPV systems from a long-term sustainability point of view.

BIPV products have to comply with both the standards of buildings and PV.

Many researchers have already pointed out that the cell temperatures of a roof integrated BIPV system surfaces up to 50%~ 95°C [2, 3], and increased heating and cooling loads [4]. However, consequent impact on climatic performance and indoor thermal comfort have not integrated. Amongst few others [5] has tried to develop a comprehensive building index including aesthetics and ecology, but requires real-time testing and validation. In India, BIPV installations (not BAPV) have been really low (Frost and Sullivan n.d.). However, financial incentives like feed-in tariff, net metering and accelerated depreciation may assist faster dissemination. Currently, BIPV system designs are influenced by the PV performance alone as done in the case of ground mounted systems usually carried out using indicators like performance ratio and system efficiency. This does not consider important factors (like the provision of thermal comfort through passive means or being climate responsive). These indicators are typically one dimensional, as in only a certain aspect of the entire system is evaluated, and are not very effective in analysing the performance of the BIPV systems from a long-term sustainability point of view.

BIPV products have to comply with both the standards of buildings and PV. Benchmarking performance is presently quite difficult and individual building or PV performance indicators may not provide a correct perspective about the system. BIPV, as a technology, can be considered as an energy generator, nonetheless, without the essence of climate responsiveness embedded into its design, it can turn out to be an energy drainer in the Indian climatic conditions. In the case of BIPV, PV forms part of the building envelope it is only logical to expect an evaluation that spans the building aspects also. PV as a building material is yet to be ascertained and building codes are yet to be developed inclusive of PV. The framework of Integrability is a preliminary attempt in the direction of evaluation of BIPV systems. Integrability as a concept has been discussed in detail in Aaditya & Mani (2017) [6]. The present paper explains this framework in the context of comparing BIPV and BAPV system in the composite climate of Bangalore.

II. Integrability Methodology and Index

Integrability is a decision-making tool which helps us to evaluate, rate and compare BIPV systems. The methodology has been developed integrating three fundamental spheres of a BIPV system (building, PV and people). Corresponding performance parameters have been identified and devised to deliver a better performance evaluation scheme. The primary functionality of a dwelling is safety (elements of nature), comfort and productivity (thermal, light, acoustics) and aesthetics. The functionality of a BIPV system is to provide thermally comfortable indoors, generation of maximum electricity from the system by catering to the energy demands in a climate responsive building environment and ensuring an economically viable system [7].

The general formulation of Integrability Index is a geometric mean integrating the respective performances of the various functionalities of a building can be given in the form of an equation as showed below.

$$\text{INTEGRABILITY INDEX} = \sqrt[n]{(f1 * f2 * f3 \ldots \ldots * fn)}$$

The letter 'f' and 'n' stand for the normalized functionality values and the number of functionalities respectively. The Integrability Index is a number (as expected) between 0 and 1 with the maximum value of 1. The higher the value of the index, the better is the overall performance of the BIPV system. The intended outcome of the Integrability methodology is to maximize the functionalities involved in order to receive a high-performance index. Each functionality is also scaled to a corresponding 0~1 value. An approach very similar to that of the computation of a complex parameter like HDI (Human Development Index) has been taken in Integrability [8].

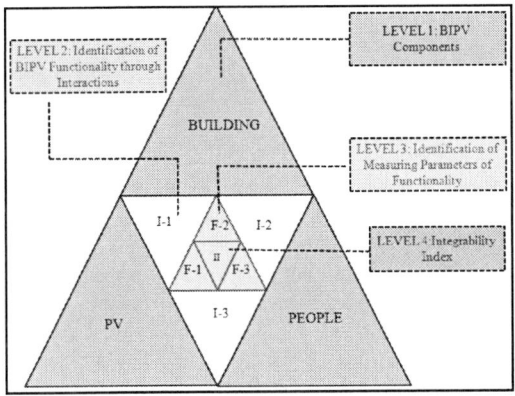

Fig. 1 . BIPV/BAPV Integrability Methodology [7]

The functionalities of a BIPV/BAPV system has been classified as electricity generation and provision of thermal comfort through climate-responsiveness and energy management for efficient consumption and economic feasibility which correspond to the five quantifiable parameters [6]. The electricity generation from PV is denoted as PV energy, PVE while the energy required for thermal comfort denoted as thermal comfort energy, TCE. The building actual energy demand has been represented as TEC (total energy consumed) and the total costs and benefits as C and B respectively. Apart from B and C, all

other values can be either calculated theoretically or measured real-time. Four parameters have been categorized from these five indicators. These four functionalities are normalized in the sense that for each of them the best and the worst-case scenarios have been considered. The first functionality indicates the compromise of the BIPV system from its ideal tilt and orientation. It is not restricted to single façade systems but can be extended to multi-façade BIPV systems. The second functionality is an equivalent indicator of the excess energy that needs to be expended to achieve thermal comfort for the occupants. The third functionality indicates the sufficiency of the BIPV system in meeting the electricity requirements of the building while the fourth functionality provides an indication of the economic viability of the system. As of now equal weightages are given for all the four parameters and thus it is expected that a BIPV system should perform well in all the four areas in order to be rated best. The computation of the four functionalities utilizing these five parameters has been discussed in detail in [6]. The formulas have been mentioned below as follows.

$$PVE(normalized) = \frac{Actual\ PVE - Minimum\ PVE}{Maximum\ PVE - Minimum\ PVE} \qquad (1)$$

$$TCE(normalized) = \frac{Maximum\ TCE - Actual\ TCE}{Maximum\ TCE - Minimum\ TCE} \qquad (2)$$

$$PV\ loading\ ratio(normalized) = \frac{\left\{\frac{PVE - TECe}{PVE + TECe} + 1\right\}}{2} \qquad (3)$$

III. COMPARATIVE STUDY: BIPV VS BAPV

BIPV and BAPV are often loosely interchanged and it would be interesting to understand which of these installations suits Bangalore climate. In order to facilitate this comparison and understand the utility of the Integrability Index building simulations have been employed. The first and foremost step in understanding performance of a BIPV system is to define the context of performance. Conventional performances only look at the electrical aspects of the system. In the present study, of the four functionalities of a BIPV system, the first two functionalities viz. tilt and orientation deviation and the amount of thermal discomfort alone have been considered. The intent of the paper is to convey the utility of the concept of Integrability. A single room (cuboid shaped) test-case dwelling has been devised to provide a comparative performance evaluation between a BIPV and BAPV since 32 % of the urban residential households (~ 330 million households surveyed) have been found to be single-room dwellings [9]. As a representative of the Indian urban conditions thus, a one room naturally ventilated building (geometrically a cube) has been modelled in Design-Builder (a CFD based dynamic building simulation software).

BAPV installation has a U-value (effective heat transfer co-efficient) of 2.66 W/m²K for roof BAPV and 2.46 W/m²K for façade BAPV systems as against BIPV (6.89 W/m²K for the roof BIPV and 5.86 W/m²K for façade BIPV). Given the fact that single-room dwellings are not built in isolation, but part of a larger building and building cluster, a simple cuboid dwelling is likely to have various configurations of solar exposure. 31 PV integration configurations have thus been tested to account for variations possible. The Integrability Index has been computed for all the possible 31 configurations using the computation methods discussed above. The performance of the systems has been evaluated only on the energy output and the thermal comfort aspects in this paper. System Advisory Model (SAM) simulations have also been carried out and the PV output energy have been computed for a BIPV and BAPV installation. A comparison has been inferred with the Integrability Methodology for the 31 cases in terms of PV output and thermal comfort aspects. The SAM modelling was used to understand the operating temperature impact on the PV output for BIPV and BIPV also.

IV. RESULTS AND DISCUSSION

The measured monthly average daytime PV cell temperatures in the BIPV setup have been plotted against that simulated using Design Builder (see Fig. 2). The low R^2 0.4 can be attributed to the soiling, water based cleaning intervals (lower cell temperature) and shadowing of adjacent vegetation not accounted for in the simulation model. Building on the measured BIPV DC output, and relying on the cell temperature variations derived from the simulation model, Fig. 3 illustrates the DC output of BIPV and BAPV.

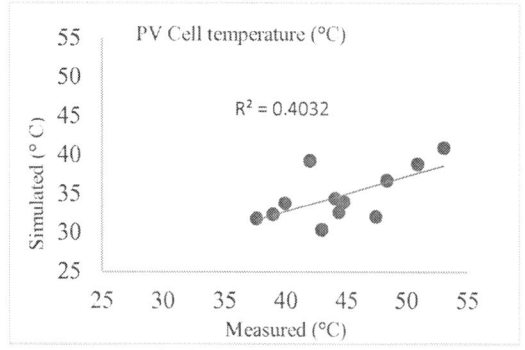

Fig. 2 Scatter plot between measured and simulated monthly average daytime cell temperatures

Further, the System Advisor Model (SAM) has been adopted to compute the PV performance of the same BIPV and BAPV system. (see Fig. 4). In both the cases, BAPV output has been found to be higher than that of a BIPV for a roof integrated system. In tropical conditions the roof accounts for the largest share of thermal transmittance, and in the case of the BAPV the shadowing of the roof and improved ventilation below the PV lowers the thermal transmittance through the roof and PV output respectively.

When comparisons are made only based on the PV output alone then results might favour BAPV systems predominantly. However, to truly ascertain that BAPV systems are the most suited ones for Bangalore climate, they should also be evaluated on the thermal comfort aspects. For naturally ventilated systems, thermal comfort energy has

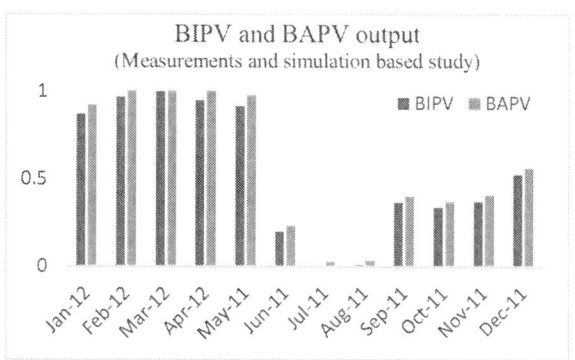

Fig. 3 Normalised DC output from BIPV and BAPV based on the measurements and simulation

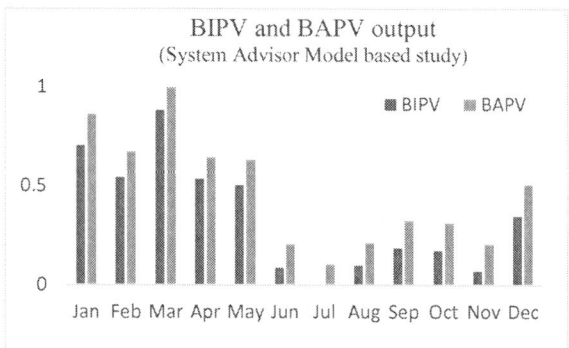

Fig. 4 Normalised DC output from BIPV and BAPV computed using System Advisor Model (SAM), NREL

been evaluated. Thermal comfort energy is considered as compared to Degree Discomfort Hours where the intensity of the temperature difference is not captured.

Fig. 5 and 6 illustrate the variation in TCE on a monthly normalized basis for single façade BIPV and BAPV systems. It can be seen that roof BIPV systems peaks twice (May and September) while the other façade systems peak once only in the month of August. On the other hand, BAPV systems follow same trend for both roof and façade systems and peak only during August. The peaking of TCE values in the month of August may be attributed to the need for de-humidification (monsoon season) along-with respective temperature controls. Fig. 7 illustrates the Integrability Index for the 31 cases for BAPV and BIPV systems and it can be clearly seen that BAPV systems for the case considered is always better than corresponding BIPV systems for all the configurations. The index has been a geometric mean of the chosen two functionalities. Thus, it is to be understood that BAPV is better than BIPV systems for Bangalore climate not only from the electrical perspective but also from the thermal perspective. The better thermal insulation provided by BAPV systems may be the rationale behind this. Roof BAPV system is the best with an index value of 0.63. The highlight of the study is that there is more evidence apart from just the electrical parameters to showcase that BAPV systems are better than similar BIPV systems for moderate climate. The methodology of

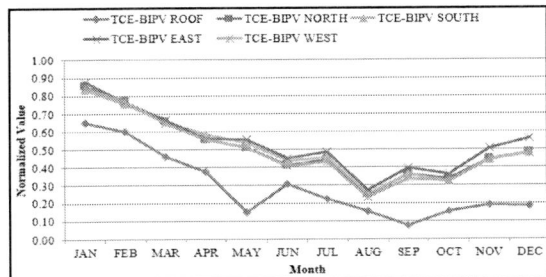

Fig. 5 Monthly simulated TCE normalized: BAPV single façade systems

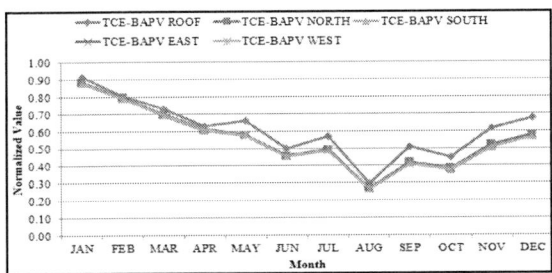

Fig. 6 Monthly simulated TCE normalized: BIPV single façade systems

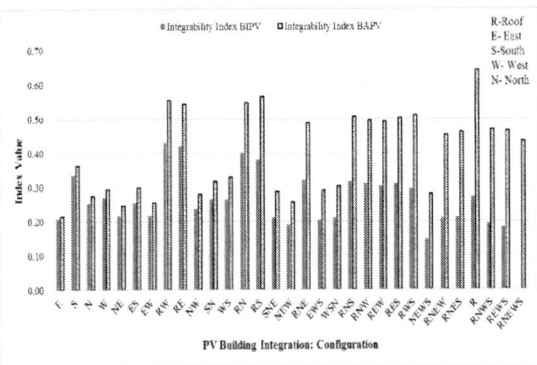

Fig. 7 Integrability Index BAPV vs BIPV systems

Integrability also underlines the fact that as the number of functionalities change, the results are also bound to change. For instance, brining in the cost benefit analysis for the 31 cases may change the scenario of comparison between the two. BIPV systems may displace traditional roofing materials and therefore might be cheaper compared to BAPV systems. If the economic feasibility of the system would also have been chosen as a functionality then a change in the performance indices would have been evident. The methodology thus gives ample flexibility and rates the systems contextually. This reinforces that there is lacuna with the traditional indicators which are hugely electrically oriented and fail to provide a holistic picture of the system. Many a times, the electrical data might seem to provide a rosy scenario about the system and only by evaluating the other functionalities of the system will it be appropriate to comment on the system performance.

V. CONCLUSIONS

The paper discusses Integrability for BIPV and BAPV as an integrated assessment methodology comprising various building functionalities. The study provides a comparative assessment of a single-room BIPV vs BAPV dwelling in Bangalore, India. The study finds that a roof integrated BIPV system may not be the defacto best configuration in tropical climates given overall building performance involving thermal comfort studies and PV performance.

ACKNOWLEDGEMENT

This research is partly based upon work supported by the Solar Energy Research Institute for India and the U.S. (SERIIUS) funded jointly by the U.S. Department of Energy subcontract DE AC36-08G028308 (Office of Science, Office of Basic Energy Sciences, and Energy Efficiency and Renewable Energy, Solar Energy Technology Program, with support from the Office of International Affairs) and the Government of India subcontract IUSSTF/JCERDC-SERIIUS/2012 dated 22nd Nov. 2012.

REFERENCES

[1]   M. Mani and B. V. Venkatarama Reddy, "Sustainability in Human Settlements: Imminent Material and Energy Challenges for Buildings in India". Journal of the Indian Institute of Science, Vol. 92(1), p145−162, 2012.

[2]   E. M. Saber, S. E. Lee, S. Manthapuri, W. Yi and C. Deb, "PV (photovoltaics) performance evaluation and simulation-based energy yield prediction for tropical buildings". Energy, vol. 71, p. 588−595.

[3]   B. Schams and G. TamizhMani, "BAPV modules with different air gaps: Effect of temperature on relative energy yield and lifetime". In 37th IEEE Photovoltaic Specialists Conference, p. 003213−003217, Washington, USA, 2011.

[4]   Y. Wang, W. Tian, J. Ren, J., Zhu and Q. Wang, "Influence of a building's integrated-photovoltaics on heating and cooling loads". Applied Energy, vol. 83(9), p. 989−1003, 2006.

[5]   T. Schuetze, W. Willkomm, and M. Roos, "Development of a Holistic Evaluation System for BIPV Façades". Energies, vol. 8(6), p. 6135−6152, 2015.

[6]   Gayathri Aaditya and Monto Mani, "Integration of Photovoltaics in Buildings", In Reference Module in Earth Systems and Environmental Sciences", Elsevier, 2013, doi.org/10.1016/B978-0-12-409548-9.10201-5.

[7]   A. Gayathri and M. Mani, "Climate-responsive Integrability of Building-Integrated Photovoltaics". International Journal of Low Carbon Technologies, 8(4), pp. 271-281, 2012.

[8]   UNDP (n.d.) "Why is geometric mean used for the HDI rather than the arithmetic mean?" http://hdr.undp.org/en/content/why-geometric-mean-used-hdi-rather-arithmetic-mean, 14 April 2016.

[9]   Government of India, "Census of India" http://censusindia.gov.in 2011, 14 April 2016.

978-1-5090-5606-4/17 $31.00 © 2017 IEEE

A New Photovoltaic System Topology Through Load Management

Joseph A. Azzolini and Meng Tao

School of Electrical, Computer and Energy Engineering, Arizona State University, Tempe, Arizona, 85287-5706, USA

Abstract — Conventional photovoltaic (PV) systems are designed to compensate for the intermittent nature of solar energy, and therefore require multiple power-managing components like storage batteries, maximum power point trackers, and charge controllers. Each of these components adds an additional cost and has an associated power loss. In this paper, we present a new topology for PV systems. Instead of managing the power from the PV array, our system manages the loads and matches them to the instantaneous maximum power available from the PV array. This results in a far simpler system than conventional systems with a higher efficiency and lower cost for various applications.

I. INTRODUCTION

Conventional photovoltaic (PV) systems can be designed in a variety of arrangements based on what is required from the system. Since the current-voltage (I-V) characteristics of a PV array are nonlinear, there exists only one optimal operating point under a given set of conditions. This point is referred to as the maximum power point. To ensure the system is operating at this optimal point, a maximum power point tracker (MPPT) is employed using various methods [1]. Energy storage devices, such as batteries, can be added to a PV system to compensate for the intermittent nature of solar energy. In order to maintain a healthy operation of the batteries, charge controllers must also be added to the system for safety reasons, and to extend the lifetime of the batteries by preventing over-charging/over-discharging. Batteries and power-managing devices all have energy conversion losses that must be taken into account as well [2]. The expected loads of a PV system also dictate which components are necessary. If all loads require DC power, then they can either be directly coupled to the PV array (if they can tolerate a range of voltages and currents), or connected through a DC-DC converter that adjusts and maintains voltage and current levels. If the loads require AC power, then a DC-AC inverter must be used. PV systems can also be connected to the electric grid for the added reliability of continuous access to power, and to sell back electricity in cases of excess PV power generation. In grid-connected systems, inverters are required to comply with all standards of the electrical utility, and be able to handle bi-directional power flows, which increases their complexity [3].

The components described above can be used to increase the flexibility of PV systems and allow for more applications of solar energy. However, each time the PV power is conditioned, managed, or stored, there is an associated cost and power loss incurred. For commercial-scale PV systems, the power management devices account for 10–35% of the total system cost depending on the complexity of the system [4], and lead to a typical system efficiency of 77% [5]. In this paper, we demonstrate a new topology for PV systems. Instead of managing the power from the PV array, our system manages the loads and matches them to the instantaneous maximum power available from the PV array. This results in a far simpler system with an efficiency approaching 100%. The simultaneous lower cost and higher efficiency promises a 15–40% lower levelized cost of electricity (LCOE) than conventional PV systems. The load-managing system can be configured to handle a large number of loads, without a significant cost increase. The system efficiency also increases with the number of loads being managed. Although the load-managing PV system is not a general-purpose concept, it enables solar electricity in several impactful applications.

II. SYSTEM OVERVIEW

The schematic for the basic load-managing system is shown in Fig.1. The major components of the basic system include a PV array, a photodetector, a programmable logic controller (PLC), a set of relays, and a number of DC loads to manage.

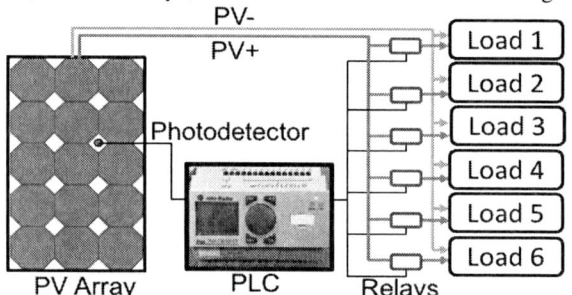

Fig. 1. Stand-alone load-managing PV system architecture with DC loads.

During a typical day, the solar irradiance will increase throughout the morning. When the solar irradiance increases to the point where enough power is available, the first load is connected directly to the PV array. As the available power increases with the solar irradiance, the PLC determines the number of parallel loads it can connect to the PV array based on the information provided by the photodetector. When the output power from the PV array reaches its maximum around midday, the PLC will connect the maximum number of loads to the PV array. In the afternoon when the solar irradiance decreases, the PLC begins to disconnect loads from the PV array until the sun sets, at which point there are no loads

978-1-5090-5606-4/17 $31.00 © 2017 IEEE

connected to the PV array. The system will also respond to any weather conditions throughout the day and adjust the number of loads connected accordingly. Overall, this system utilizes the maximum available power from the PV array while mitigating the amount of power losses compared to conventional systems.

The photodetector is made up of a small silicon solar cell and a transimpedance amplifier. The transimpedance amplifier converts the short-circuit current of the solar cell into a voltage signal from 0–10 V for the PLC. Since what is measured is the short-circuit current and what is needed is the maximum available power from the PV array, simulation was performed for the relationship between short-circuit current and maximum power of an ideal silicon solar cell, as shown in Fig. 2. The relationship is linear and thus short-circuit current can represent maximum power.

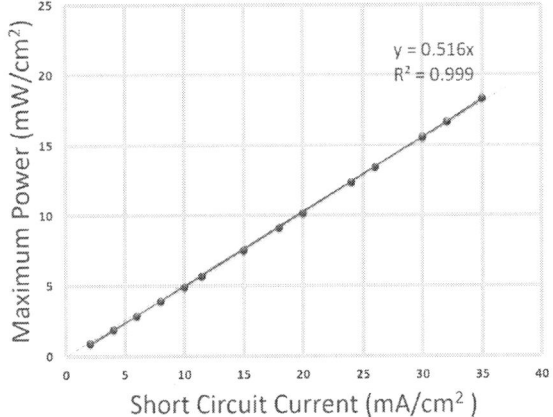

Fig. 2.   Relationship between short-circuit current and maximum power for ideal silicon cell.

Fig. 3 is the load-line analysis to further explain how the system operates. Several I-V curves of a commercial 72-cell silicon panel at different levels of irradiance are plotted. The I-V curve of a PV array is nonlinear, especially around the maximum power point (this region is often referred to as the "knee" of the curve). The intersection of the I-V curve and a load line is the operating point of the system, and only one optimal operating point exists for a given temperature and irradiance level. As the irradiance increases throughout the morning, the I-V curve gradually shifts up. To simplify the illustration, we assume that there are only three loads with a fixed, equal resistance. The fixed-resistance loads appear as straight lines in Fig. 3. The slope of each line is equal to the reciprocal of the combined resistance of the loads. The system can only operate along these lines. At point 1, the first load is connected to the PV array (in our case the array is just one panel). As the irradiance increases, the operating point will move along the load 1 line, until it reaches point 2. The system will switch from point 2 to point 3 by connecting load 2 in parallel with load 1. This critical point is referred to as a switch point (SP). The optimal switch point occurs when the power at point 3 is equal to the power at point 2. The system will then

operate along the load 1&2 line until it reaches the next switch point, and so on. In the afternoon, this process will happen in reverse as the irradiance decreases.

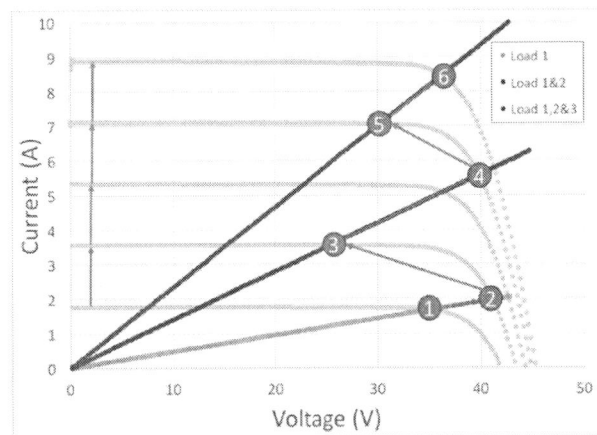

Fig. 3. Visualization of system operation for 3 equal loads.

It is noted that the system in Fig. 1 is the core of a load-managing system and it can be modified for various applications. For example, multiple photodetectors can be used for a large PV array. A microprocessor can replace the PLC for more complex systems. Variable-resistance loads such as batteries can be managed by the system with a second feedback. AC loads can be managed by adding a DC-AC inverter for each load. More importantly, the load-managing system can be backed up by the electric grid, eliminating its intermittency.

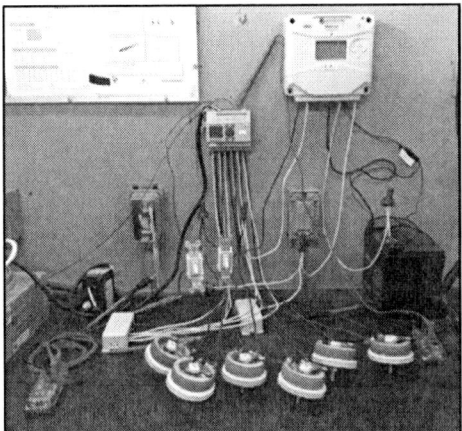

Fig. 4.   Load-managing PV system constructed at ASU

III. PHYSICAL IMPLEMENTATION

The core of the load-managing system has been implemented to explore the practical challenges of this system topology, pictured in Fig. 4. This system includes a PV panel, a photodetector, a PLC with six built-in relays, and six variable resistors to represent the loads. The PV panel and photodetector are installed on the roof, and thus are not pictured below. Data-logging digital multimeters take measurements of the total power delivered to the loads throughout the day, and the voltage signal from the photodetector. The maximum power point tracker (MPPT) and battery in the picture are only being used to take initial measurements of the PV panel, and thus are not part of the load-managing system.

IV. EXPERIMENTAL RESULTS AND DISCUSSION

Fig. 5 shows the experimental performance of the system in terms of power delivered to the loads as a function of time. There are four loads in this experiment, each set to 12 Ω, and the switch points can be found in Table I. In Fig. 5, the switch points are visualized four times (SP1 through SP4) in the morning. As the solar irradiance increases throughout the morning, the PLC subsequently connects loads to the PV panel based on the voltage signal received from the photodetector. By noon, all four loads are powered by the PV panel. As the solar irradiance decreases throughout the afternoon, the loads are disconnected from the PV panel accordingly.

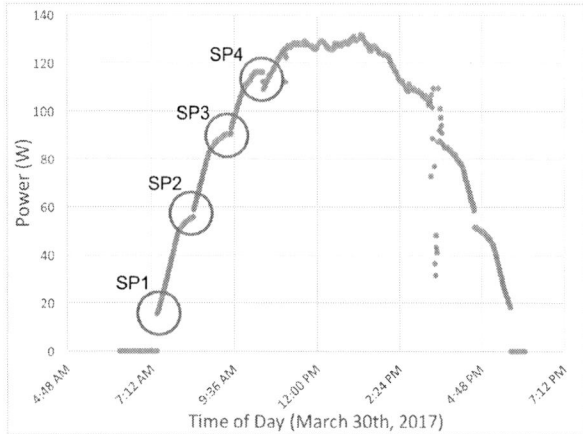

Fig. 5.   Total power delivered to 4 loads of 12 Ω each.

TABLE I
SWITCH POINTS FOR FIG. 5

Load Number	Switch Point (unit: V)
Load 1	1.25
Load 2	4.00
Load 3	6.75
Load 4	8.20

A dip in power is observed near SP4 in Fig. 5. This dip is attributed to a sub-optimal switch point setting. Optimal switch point settings result in a continuous power curve, like SP2 and SP3. The approximately flat top in Fig. 5 is due to a slight mismatch between the combined resistance of the four loads and the characteristic resistance of the PV panel at noon. The combined load resistance needs to match the characteristic resistance of the PV array around noon for maximum power to be extracted. For loads of equal resistance, the ohmic value can be calculated using the following equation

$$R_{PV} = \frac{V_{MP}}{I_{MP}} = \frac{R_{Load}}{n} \qquad (1)$$

where n is the number of loads.

The load-managing system also responds to changes in weather conditions in real time. At ~3:30 p.m. on March 30, 2017 (Fig. 5), the system briefly experienced cloud coverage. The photodetector responded to the drop in irradiance, and the rest of the system reacted accordingly. This is visible in the voltage signal from the photodetector on this day, as shown in Fig. 6.

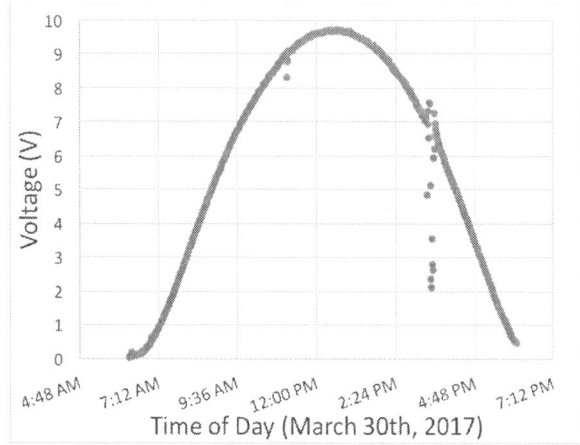

Fig. 6.   Voltage signal from photodetector on same day as Fig. 5.

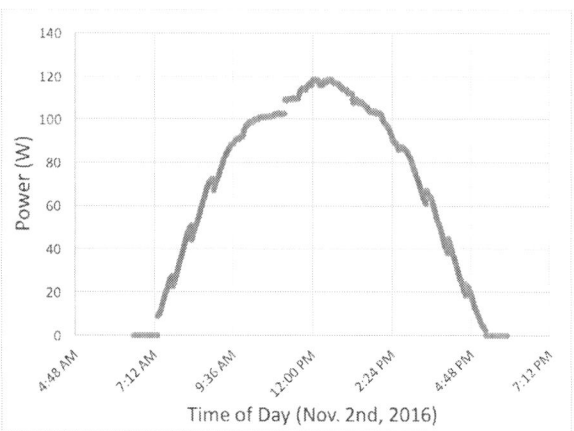

Fig. 7.   Total power delivered to 6 loads of 21 Ω each.

The experiment in Fig. 5 was repeated, but with the system managing six equal loads instead of four. The results can be seen in Fig. 7. With the two additional loads, the system has more available operating points to match the PV panel. Because of this, the power delivered to the loads fits more closely under a sine function and therefore represents a more efficient system. This positive correlation between number of loads and system efficiency provides a cost incentive for larger systems.

The proposed PV system can also work for managing loads of unequal resistances. In Fig. 8, loads 1, 2, and 4 have equal resistances of 17.75 Ω while load 3 has a resistance 8.87 Ω. The data can be fitted closely under a sine function. Thus, the load-managing system can maintain a high efficiency when managing unequal loads.

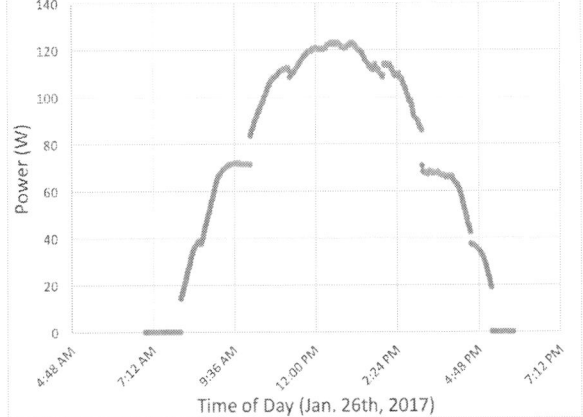

Fig. 8.   Total power delivered to 4 loads of unequal values.

V. SIMULATION RESULTS AND DISCUSSION

The daily efficiency of the system can be obtained by first encompassing the measured data with a sine function. The sine curve represents the maximum available power from the PV panel during the day. The system efficiency is then the ratio of the area under the data to the area under the sine curve:

$$\eta = \frac{Energy\ Delivered\ to\ Loads}{Total\ Energy\ Available} \times 100 \qquad (2)$$

To simulate the theoretical efficiency of the load-managing PV system, an ideal silicon panel, with zero series resistance and infinite shunt resistance, of 100 Wp was used to model the power being delivered to ten loads. It is assumed that all ten loads are of equal, fixed resistance. At peak power, the characteristic resistance of the panel is 14.92 Ω-cm^2. Following Eq. (1), the resistance of the ten loads is 149.2 Ω-cm^2 each. A new load is connected to the panel when the power delivered to n loads is equal to the power delivered to n+1 loads (Fig. 3).

Fig. 9 shows the simulated daily system efficiency with ten equal loads. The blue line is the maximum available PV power and the orange line is the total power delivered to the ten loads. The losses in the system are represented by the triangular regions between the two lines. These regions decrease in size as the number of loads being managed is increased. With ten loads, the daily system efficiency is ~99%. With more loads being managed, there are more available optimal operating points of the system that increase the total power delivered to the loads. The system efficiency thus approaches 100%.

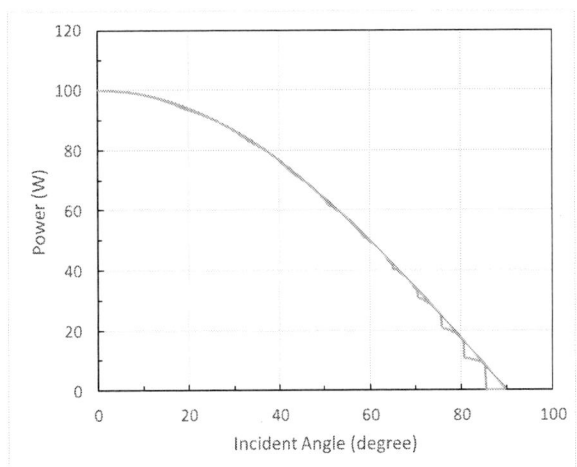

Fig. 9.   Simulated system efficiency when managing 10 equal loads.

Significant voltage fluctuations are noticed in the simulation, especially at low irradiance levels with fewer loads connected. The SPs are critical in limiting the voltage fluctuation. The theoretical system efficiency can reach 99.3% if the voltage is allowed to fluctuate freely for an ideal PV panel. If the voltage fluctuation is limited to 12%, the system efficiency drops to 98.9% for an ideal panel. For real panels, the voltage fluctuation will be larger. If a tighter voltage range is needed, DC-DC converters can be added to the first few loads in the system to maintain their voltage levels.

VI. APPLICATIONS

The load-managing PV system described above can be used to power a variety of applications. The ideal application for a load-managing PV system would have a total electrical load that can be divided into multiple sub-loads to be managed. Two key applications that meet this requirement are electric vehicle charging and industrial electrolysis.

Since cars and trucks contribute ~22% to total carbon emissions in the U.S. and ~17% globally [6], significant carbon reduction can be achieved if electric vehicles are charged with solar electricity. Several types of PV charging systems for electric vehicles have been explored [7]. A load-managing PV system that is backed by the grid, like in Fig. 10, would allow for a cost effective means of charging electric vehicles with solar electricity. Electrolytic processes such as hydrogen production from water can be powered by solar electricity using this system. Electrolysis is also used in the production of caustic soda, zinc and copper which can be powered by this system. Many solar-hydrogen systems have been proposed [8-10]. The total number of electrolyzers can be divided into sub-loads to be matched to the instantaneous maximum power of the PV array.

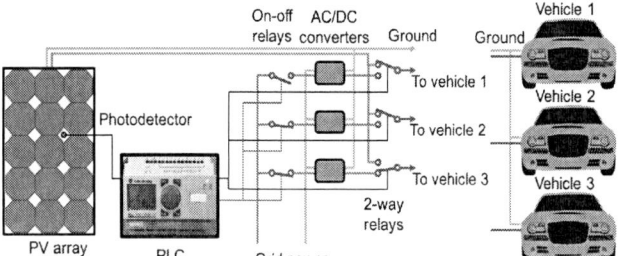

Fig. 10. A grid-backed, load-managing system for charging electric vehicles.

VII. System Costs

The cost of the load-managing system is really low. A 30-A 36-V relay is $11.90 on eBay. A PLC with eight outputs is $169 on Google. A silicon photodetector is $2. For less than $270 we can build the control electronics for a load-managing system to handle a 7.4-kWp PV array, i.e., the load-managing control electronics costs less than 4¢/Wp for small residential systems. For large-scale systems of ~1 MWp, the cost of the control electronics is estimated to be ~1¢/Wp. Power-managing devices such as charge controllers and power converters account for ~20% of the cost for conventional PV systems [4]. The load-managing system almost eliminates the cost for power-managing devices in conventional systems. The lower system cost is coupled with a higher system efficiency approaching 100%. The combined lower cost and higher efficiency results in a 15–40% LCOE as compared to conventional PV systems.

VIII. Conclusions

The load-managing PV system presented here has several key advantages over conventional PV systems. By managing the loads instead of the power, the system cost is significantly reduced while the system efficiency improves. The simultaneous benefit of reduced cost and improved efficiency leads to a 15-40% lower levelized cost of electricity as compared to conventional PV systems. It also has an excellent scalability and minimizes the negative impact of PV on the electric grid. The core of the system contains a PV array, a photodetector, a programmable logic controller, and a set of relays, without any power-managing components. The system performs maximum power point tracking by varying the number of loads in parallel to match the combined resistance of the loads with the characteristic resistance of the PV array. With a system efficiency approaching 100%, and the option of grid-backing, this system offers a cost effective way to incorporate solar electricity into several impactful applications.

IX. Acknowledgment

This work was partially supported by National Science Foundation under grant number CBET-1336297.

References

[1] J. Prasanth Ram, T. Sudhakar Babu, and N. Rajasekar, "A comprehensive review on solar PV maximum power point tracking techniques," 2016.

[2] S. Wenham, "Stand-Alone Photovoltaic System Components" in Applied Photovoltaics, Ch. 6, pp. 98-106, 2007.

[3] S. B. Kjaer, J. K. Pedersen, and F. Blaabjerg, "A review of single-phase grid-connected inverters for photovoltaic modules," *IEEE Trans. Ind. Appl.*, vol. 41, no. 5, pp. 1292–1306, 2005.

[4] D. Chung, C. Davidson, R. Fu, K. Ardani, and R. Margolis, U.S. Photovoltaic Prices and Cost Breakdowns: Q1 2015 Benchmarks for Residential, Commercial, and Utility-Scale Systems, http://www.nrel.gov/docs/fy15osti/64746.pdf.

[5] K. Ardani and R. Margolis, "2010 Solar Technologies Market Report," *NREL*, no. November, pp. 1–136, 2011.

[6] E. Roston, Electric Cars Can Help Clean Up the Grid - Electricity Needs to Come from Renewable Sources in Order to Arrest Climate Change, http://www.bloomberg.com/news/articles/2016-04-29/electric-cars-can- help-clean-up-the-grid.

[7] A. R. Bhatti, Z. Salam, M. J. B. A. Aziz, and K. P. Yee, "A critical review of electric vehicle charging using solar photovoltaic," *Int. J. Energy Res.*, pp. 439–461, 2016.

[8] L. G. Arriaga, W. Martínez, U. Cano, and H. Blud, "Direct coupling of a solar-hydrogen system in Mexico," *Int. J. Hydrogen Energy*, vol. 32, no. 13, pp. 2247–2252, 2007.

[9] R. E. Clarke, S. Giddey, F. T. Ciacchi, S. P. S. Badwal, B. Paul, and J. Andrews, "Direct coupling of an electrolyser to a solar PV system for generating hydrogen," *Int. J. Hydrogen Energy*, vol. 34, no. 6, pp. 2531–2542, 2009.

[10] T. Maeda, H. Ito, Y. Hasegawa, Z. Zhou, and M. Ishida, "Study on control method of the stand-alone direct-coupling photovoltaic - Water electrolyzer," *Int. J. Hydrogen Energy*, vol. 37, no. 6, pp. 4819–4828, 2012.

First Step for Power Generation Amount Estimation of Solar Matching System

Kazuya Hosokawa*, Toshiaki Yachi*, Yoichi Hirata** and Yasuyuki Watanabe**

*Tokyo University of Science, 6-3-1 Niijuku, Katushika-ku, Tokyo 125-8585, JAPAN
**Suwa Tokyo University of Science

Abstract — The solar matching photovoltaic (PV) system has already been proposed for the spectrum of sunlight except the spectrum for growing agricultural crops. The solar matching PV system needs an estimation of the amount of power generation to effectively use solar energy. In this study, the first step of estimating the power generation is shown.

Index Terms — photovoltaic system, solar matching, estimation of the amount of power generation.

I. INTRODUCTION

Flexible photovoltaic (PV) modules can be installed directly on the free curved surface of a rooftop or wall without an installation stand. They are expected to be used for various applications. Solar matching using flexible see-through organic thin-film PV (OPV) modules is one of the applications attracting attention [1]. It is important to estimate the power generation amount generated by the organic thin-film PV module to effectively utilize the electric power generation in solar matching and to determine the capacity of the power conditioner and the battery in the system design [2-4].

In this study, the actual characteristics of the see-through organic thin-film PV module have been measured as the first step. It has been compared compare the estimated power generation and the measured power generation of a see-through organic thin-film PV module installed on a flat surface.

II. SOLAR MATCHING SYSTEM AND POWER GENERATION ESTIMATION

A. Solar matching system

Figure 1 shows a schematic diagram of the solar matching system for power generation (PG). In solar matching, a PV module that transmits light, such as a see-through (OPV) module, is installed on a green house. The module transmits light in the wavelength range (red, blue) necessary for agricultural crops and generates electricity from the other light (green) that does not interfere with agricultural crop cultivation. This system is an attempt to separate the light for agricultural crops and PV modules to wavelength regions that do not affect each other.

In solar matching, it has been proved that the plant harvest amount and vegetable taste similar to that under cultivation in ordinary houses and sunlight-utilizing plant factories can be obtained [1]. Further study is now underway to improve the conversion efficiency of organic thin-film solar cells and plant production efficiency.

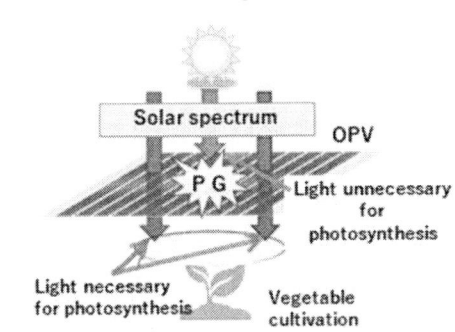

Fig. 1. Schematic diagram of solar matching [1].

B. Estimation of power generation of flexible PV module

When a flexible PV module is installed on a curved surface such as a cylinder or a dome, the module must have a plurality of azimuth angles and inclination angles. Therefore, equations to determine the power generation estimation have been proposed for a conventional PV module and for a flexible PV module. The equation for the flexible PV module must account for approximations of the curved surface of multiple polyhedrons [5].

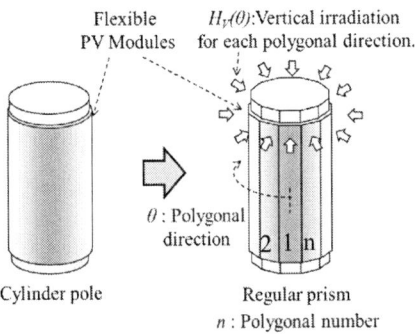

Fig. 2. Approximate polygonal column of a cylindrical PV module.

The amount of power generation is estimated individually from the irradiation on each surface constituting the polyhedron, and the total amount of power generation is the sum of the irradiation of each surface. Figure 2 shows an approximate polygonal column of a cylindrical PV module that can be applied to a solar matching house. The cylindrical shape model is a power generation amount estimation model having an inclination angle constant at 90° so that only the azimuth angle changes. When the model is applied to a solar matching house, the azimuth angle is constant and the inclination angle changes.

The estimation of the cylindrical flexible PV module power generation amount based on the power generation estimation equation of the conventional flat plate type solar module is shown below.

$$W_{cyl} = H_{Vave} \cdot A \cdot \eta$$
$$+ H_{Vrms} \cdot A \cdot \eta \cdot \alpha_w \cdot \left(T_{Mcyl}(H_{Vrms}) - T_{STC} \right) \quad (1)$$

$$H_{Vave} = \frac{1}{n} \cdot \sum_{i=1}^{n} H_V(\theta_i) \quad (2)$$

$$H_{Vrms} = \sqrt{\frac{1}{n} \cdot \sum_{i=1}^{n} \left(H_V(\theta_i) \right)^2} \quad (3)$$

where H_{Vave} [Wh/m^2] is the average and H_{rms} [Wh/m^2] is the root mean square of the irradiation over the entire area of the PV module. The product $T_{Mcyl}(H_{Vrms})$ indicates that H$_{Vrms}$ is used to calculate the temperature of the PV module T$_{Mcyl}$.

To estimate the power generation of the solar matching system, the conversion efficiency η [%] and the power generation temperature coefficient α_w [%/°C] of the organic thin-film PV module are required newly.

III. THE SEE-THROUGH OPV AND EXPERIMENTAL CONDITIONS

In this study, a see-through OPV module manufactured for solar matching was used. Figure 3 shows a see-through OPV module. In the module, the unit cell is 1.27 cm × 22.3 cm, and 20 cells are connected in series. The module also has a region where light is directly transmitted from the front surface to the back surface. This region was installed southward and parallel to the ground for the measurement of the amount of power generation.

Fig. 3.   The see through OPV module.

The voltage across the shunt resistor and the variable resistor was measured with a data logger (midi LOGGER GL800, Graphtec) and the output current and output voltage were measured. The shunt resistance was 5 mΩ and the variable resistance was used in the range of 0.5 to 500 Ω. The maximum power generation was calculated from the measured current and voltage. In addition, the module temperature was measured by attaching a thermocouple to the surface of the PV module, and the total irradiation was measured with a Pyranometer (MS-402, Eikoseiki) installed near the PV module.

The measurement period for autumn was September 25 to October 12, and the measurement period for winter was December 15 to January 6. The weather on the measurement days was cloudy or sunny. The data on October 15 was used for the verification of autumn, and the data on January 13 was used for winter verification.

IV. MEASUREMENT RESULTS AND ESTIMATION EQUATION

A. Solar radiation dependence and temperature dependence of conversion efficiency

The conversion efficiency of the PV module depends on the irradiation intensity and the module temperature. The relationship between solar radiation and module temperature was confirmed. Figure 4 shows the relationship between solar radiation intensity and module temperature in autumn and winter.

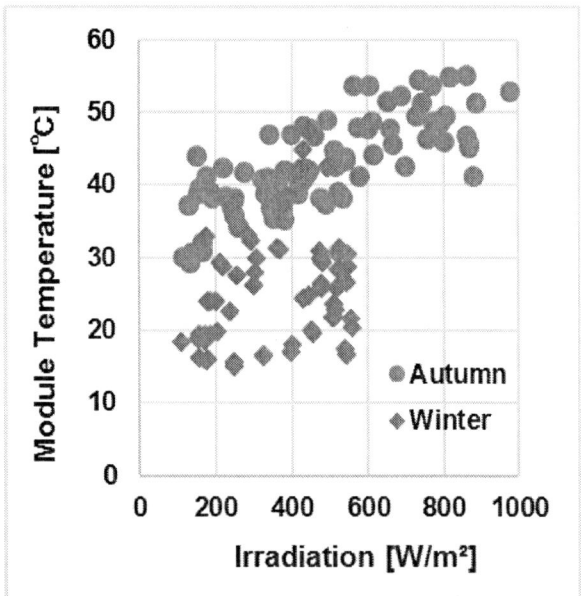

Fig. 4.   Relationship between solar radiation and module temperature.

There was a strong correlation between solar radiation intensity and module temperature in autumn (correlation coefficient equal to 0.76). When obtaining the conversion efficiency $\eta_{(H)}$ and the power generation temperature coefficient α_w, it is necessary to find out the module temperature and the solar radiation intensity to a narrow range to eliminate the correlation. On the other hand, there was a weak correlation between solar radiation intensity and module temperature in winter.

B. Conversion efficiency characteristic $\eta_{(H)}$ and power generation temperature coefficient α_w

First, It is obtained the conversion efficiency characteristics of the see-through organic thin-film PV module and the power generation temperature coefficient from the measurement data of autumn and winter. Figure 5 shows the relationship between the module temperature of the autumn and winter data and measured conversion efficiency. Figure 6 shows the relationship between solar radiation and measured conversion efficiency.

The power generation temperature coefficient is acquired in the range of high solar radiation intensity (400-600W/m²) and low solar radiation intensity (100-300W/m²). The conversion efficiency characteristic is acquired in the range of high module temperature (35-42°C) and low module temperature (26-33°C).

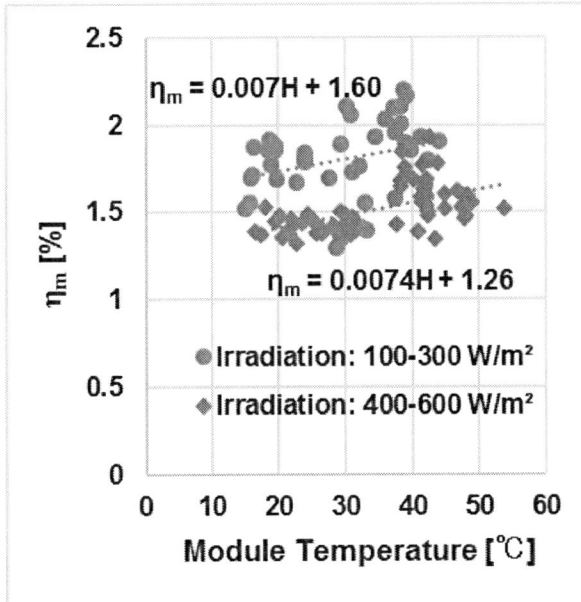

Fig. 5.   Relationship between module temperature and measured conversion efficiency in autumn and winter.

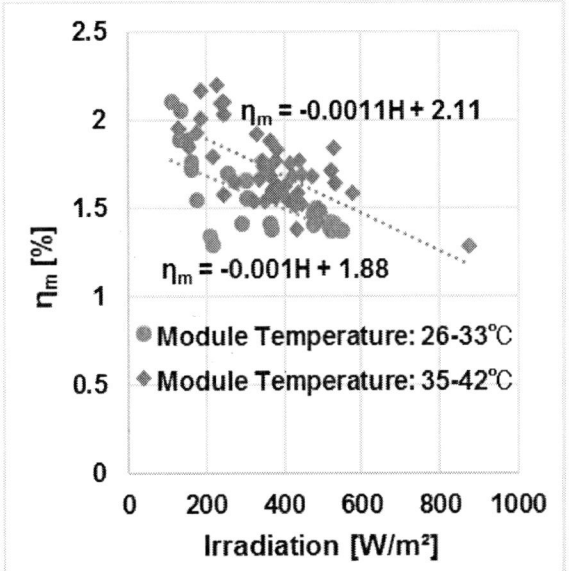

Fig. 6.   Relationship between solar radiation intensity and measured conversion efficiency in autumn and winter.

Combine the conversion efficiencies $\eta_{(H)}$ and the power generation temperature coefficients α_w to create an estimation equation of four patterns. Table 1 shows combinations of selected ranges.

Table 1. Selection pattern.

	High module temperature	Low module temperature
High irradiance	Pattern1	Pattern2
Low irradiance	Pattern3	Pattern4

By acquiring the power generation temperature coefficient by dividing the solar radiation intensity range, it is possible to make an estimation equation with small error. By obtaining the conversion efficiency characteristic by dividing the module temperature range, it is possible to make an estimation equation that is divided into temperature, wind speed, etc.

Table 2. Power generation estimation equations coefficients using autumn and winter data.

	$\eta_{(H)}$	α_w
Pattern1	-1.1H+2.11	0.51
Pattern2	-1.0H+1.88	0.51
Pattern3	-1.1H+2.11	0.40
Pattern4	-1.0H+1.88	0.40

Similarly, Table 3 shows estimation equations obtained from only autumn data. Table 4 shows data obtained only from winter data. Compare the power generation estimation equation obtained from the autumn and winter data with the coefficients of the obtained estimation equation divided into seasons.

Table 3. Power generation estimation equations coefficients using autumn data.

	$\eta_{(H)}$	α_w
Pattern1	$-0.6H+1.83$	-0.60
Pattern2	$-1.1H+2.20$	-0.60
Pattern3	$-0.6H+1.83$	-0.14
Pattern4	$-1.1H+2.20$	-0.14

The autumn estimation equations coefficients have following features.

- The slope of the conversion efficiency characteristic $\eta_{(H)}$ is different within the range of the module temperature.
- Both power generation temperature coefficient becomes negative.

Table 4. Power generation estimation equations coefficients using winter data.

	$\eta_{(H)}$	α_w
Pattern1	$-0.08H+1.46$	0.36
Pattern2	$-1.1H+1.97$	0.36
Pattern3	$-0.08H+1.46$	-0.99
Pattern4	$-1.1H+1.97$	-0.99

The winter estimation equations coefficients have following features.

- The slope of the conversion efficiency characteristic is very small in the high module temperature range.
- The power generation temperature coefficient becomes a positive value in the high irradiation range, but becomes negative in a low irradiation range.

V. VERIFICATION OF ESTIMATION EQUATION

A. Verification of estimation equation using autumn and winter data

Estimated autumn and winter validation days were estimated by estimation equations using autumn and winter data. Figure 7 and 8 show the results of the verification on the autumn verification day and on the winter validation day using the autumn and winter estimation equations.

The autumn verification day was sunny and the module temperature was relatively high. For this reason, Pattern 1 was expected to have the smallest error. However, the absolute error was the greatest, about 20% at each hour. The overall error rate was about 10%-20%.

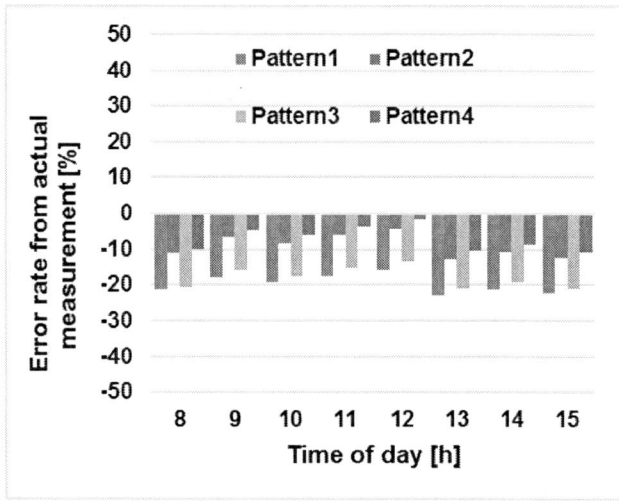

Fig. 7.　Error rate per hour on autumn verification day using autumn and winter estimation equations.

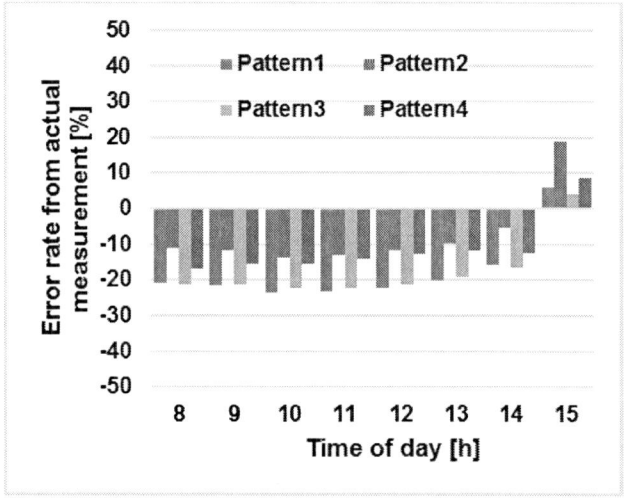

Fig. 8.　Error rate per hour on winter validation day using autumn and winter estimation equations.

The winter validation day was sunny until 13:00 and the module temperature was relatively high. For this reason, Pattern 1 was expected to have the smallest error. However, the absolute error was the greatest, about 20% at each hour. The absolute error among all patterns ranged from 10% to 20%.

Based on these verification results, It is guessed that the power generation temperature coefficient and conversion efficiency characteristics may depend on each season in autumn and winter. For that reason, the estimation equations were divided for each season.

B. Validation of estimation equation using seasonal data

The autumn verification day was estimated by the estimation equation using autumn data. Figures 9 and 10 show the verification results for the autumn verification day when using the autumn estimation equations.

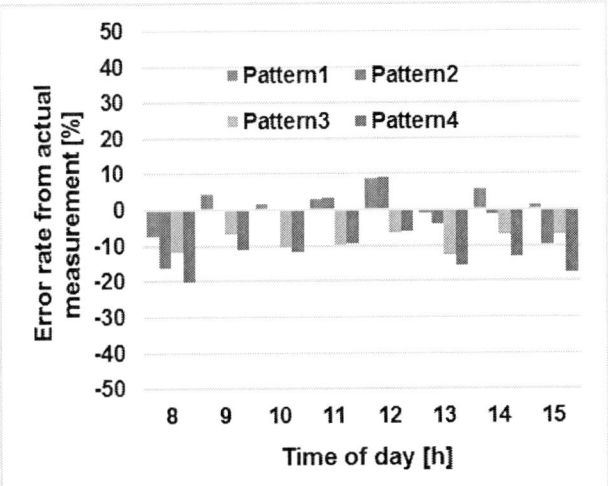

Fig. 9.   Error rate per hour on autumn verification day using equations by autumn data.

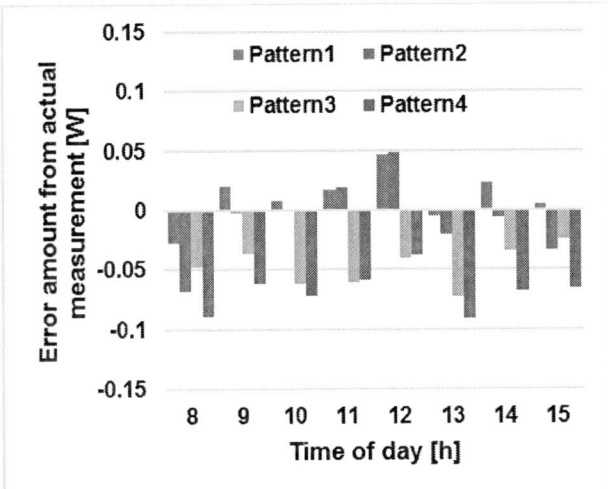

Fig. 10.   Error amount per hour on autumn verification day using equations by autumn data.

From the verification result, Pattern 1 had the smallest absolute error rate at each hour, the largest being 8.7% at 12:00. with the largest error. Among all patterns, the absolute error ranged from about 5% to 10%.

The winter verification date was estimated by the estimation formulas using the winter data. Figures 11 and 12 show the verification results for the winter validation day when using the winter estimation equations.

From the verification result, Pattern 3 has the smallest absolute error at each hour, and in the case of sunny, it was 12.9% at 10 o'clock, which has the largest error. In the case of sunny, among all patterns, the absolute error was about 5%-15%.

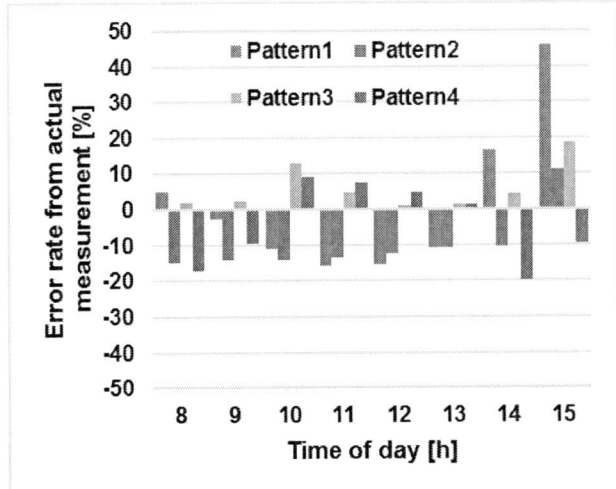

Fig. 11.   Error rate per hour on winter validation day using winter estimation equations.

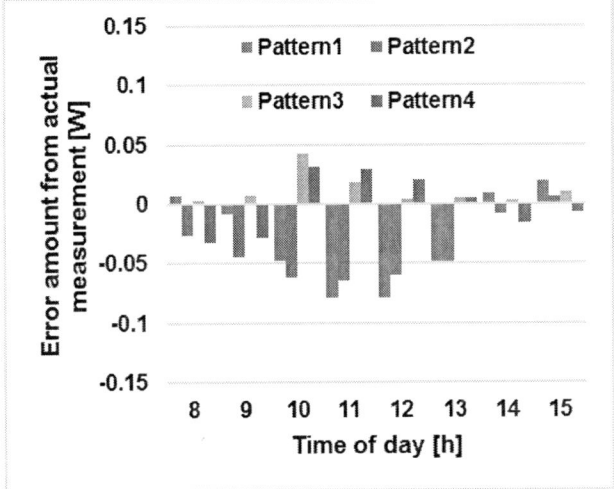

Fig. 12.   Error amount per hour on winter validation day using winter estimation equations.

VI. CONCLUSIONS

To estimate the power generation amount of the solar matching PV system, the power generation amount of the see-through OPV module installed on the flat surface has been measured. The following results has been obtained.

(1) By narrowing the solar radiation intensity and the module temperature range, estimation equations of four patterns were created for each season. As a result, it is possible to suppress the error by estimating by the estimation equation of the condition close to the weather on the verification day.

(2) The conversion efficiency characteristics and the power generation temperature coefficient of the see-through OPV are seasonally dependent. It is necessary to calculate an estimation equation by measurement for each season.

REFERENCES

[1] Y. Watanabe, MATERIAL STAGE, Vol.13, No.8, pp.1-6 ,2013(Japanese).

[2] A. M. Noorian, I. Moradi, and G. A. Kamali, "Evaluation of 12 models to estimate hourly diffuse irradiation on inclined surfaces", Renewable Energy, vol.33, no.6, pp.1406–1412, Jun 2008.

[3] M. Gulin, M. Vasak, and M. Baotic, "Estimation of the global solar irradiance on tilted surfaces," 17th International Conference on Electrical Drives and Power Electronics, Oct 2013.

[4] U.A. Yusufoglu, T.M. Pletzer, L.J. Koduvelikulathu, C. Comparotto, R. Kopecek, and H. Kurz, "Analysis of the Annual Performance of Bifacial Modules and Optimization Methods", IEEE Journal of Photovoltaics, vol.5, issue.1, pp. 320-328, Nov 2014.

[5] M. Shibasaki, T. Yachi, The Proc. of IEEE 40th PVSC, Denver, pp. 2759-2764, 2014.

Irradiance and temperature distributions at high latitudes: Design implications for photovoltaic systems

Anne Gerd Imenes[1,2] and Josefine Selj[3,4]

[1]Teknova AS, 4612 Kristiansand, Norway; [2]University of Agder, Department of Engineering Sciences, 4879 Grimstad, Norway;
[3]Institute for Energy Technology, 2007 Kjeller, Norway; [4]University of Oslo, Department of Technology Systems, 2007 Kjeller, Norway

Abstract — Irradiance and temperature distributions collected over several years at high-latitude locations in Norway (58-70°N) are presented. Measurements are coupled with standard efficiency data for selected commercial photovoltaic (PV) modules to evaluate the effect on relative efficiency. The modules display some performance variation caused by the different irradiance distributions, but the main effect is linked with temperature. The results may have implications for the choice of components and accuracy of yield estimates during the PV system design phase. The findings will provide input for the development of adapted building integrated (BIPV) modules with high performance and durability at high latitudes.

Index Terms — solar energy, temperature, modeling, energy efficiency, performance evaluation, photovoltaic systems.

I. INTRODUCTION

High latitude locations are subject to long winters and lower annual solar irradiation compared with regions closer to the equator, and the number of grid-connected photovoltaic (PV) installations in these regions are still modest. However, as the PV market grows worldwide, so also in the Nordic countries. Entrepreneurs and investors are increasingly focusing on field performance to maximize energy output over the system lifetime and minimize investment costs. Mono- and poly-crystalline Si technology dominates the world market, and is by far the most installed technology in Norway. A few examples exist using alternative technologies (CIS and thin film Si), but most are new installations with unsatisfactory data series to perform technology comparisons. However, based on irradiation measurements, module performance data, and available data from existing systems, we will evaluate the realistic performance of different technologies under Nordic irradiance and climatic conditions.

The work is part of a research project that aims to develop robust BIPV-solutions and develop new materials, components and solutions that are tailor-made for Nordic conditions [1,2]. Performance data are collected from existing PV installations and analysed to build knowledge about different technical solutions and typical yields observed in the field [1,3]. Both irradiance conditions and environmental stress factors that affect degradation and lifetime of PV modules are evaluated, such as ice/snow formation and wind-driven rain. Low operating temperatures are generally beneficial for PV efficiency and reduced degradation rates [19], but other stress factors caused by a cold climate may counteract this. High durability is especially important in building integrated photovoltaic (BIPV) applications, where the PV modules are considered as structural elements and must comply with building technical lifetime requirements [4].

Irradiance and module temperature are key parameters that influence the behaviour of PV systems, as they modify system efficiency and output energy. PV manufacturers currently rate their modules at standard test conditions (STC) and report the nominal operating cell temperature (NOCT). The objective of the recent international standard IEC 61853 [5] is to describe test requirements for evaluating PV module performance that better represent real operating conditions. These include local irradiance and temperatures, wind, solar spectrum, and climatic long term degradation [6]. IEC 61853 Part 1 prescribes a matrix for PV power assessment according to irradiance values (100-1100) W/m^2 and module temperatures (15-75) °C. This may be used to define a relative efficiency as a function of irradiance and temperature [7]. Using actual data from different geographical locations, this relative efficiency can improve yield estimation if implemented in design simulation software.

In [7], a generic power-rating model for crystalline silicon PV modules was developed based on 18 modules (mono- and poly-Si). The model was used to calculate the annual average relative efficiency for five locations in Europe, ranging from 38°N to 62°N. The results showed that the variation in simulated output between modules increased as the geographical location was moved further north. This was attributed to more cloudy climate conditions in the north, introducing larger uncertainties in the modelling compared to the sunnier south. This uncertainty should be considered when assessing PV performance simulations.

The documentation of actual irradiance and temperature conditions at high latitudes is therefore needed to improve system modelling accuracy and help PV system designers to evaluate how well the manufacturer's data represent the real operation conditions at different locations. In addition to improved accuracy of yield estimates, this may ultimately also affect the choice of PV technology and BOS components based on their 'climate-weighted' efficiencies.

II. BACKGROUND

PV installers usually predict the system yield of planned installations by running software simulations with synthetic weather data, as the availability of measured data is scarce. Synthetic data is generated from satellite maps and combined with measured data from discrete ground stations, then interpolated to nearby locations [8]. A major challenge is the diminishing coverage and accuracy of irradiation data at high latitudes [9-10]. Regions with long coastlines, mountain ranges, frequent clouds and snow coverage introduce uncertainties in the interpretation of satellite data. Ground measurements are therefore valuable to evaluate the actual irradiance received at a given location. However, as it is not recommended to extract data from ground measurements if the distance is more than 50 km away [11], a large number of sensors and dataloggers would be required to adequately serve the need. Cost and maintenance requirements prevent the widespread use of ground based irradiance sensors.

For high-precision PV simulations, both global and diffuse irradiance is required, ideally at high temporal resolution (<1 hr) and continuous time series spanning several years. In the Nordic regions, very few stations exist that provide such data, especially diffuse horizontal irradiance (DHI) data are rare. Hourly data of global horizontal irradiance (GHI) are, or have been, recorded at more than 50 meteorological stations across Norway [12-13]. However, the data quality and length of time series is greatly varying. Missing data and poor data quality generally represent a problem with most existing datasets in Norway today.

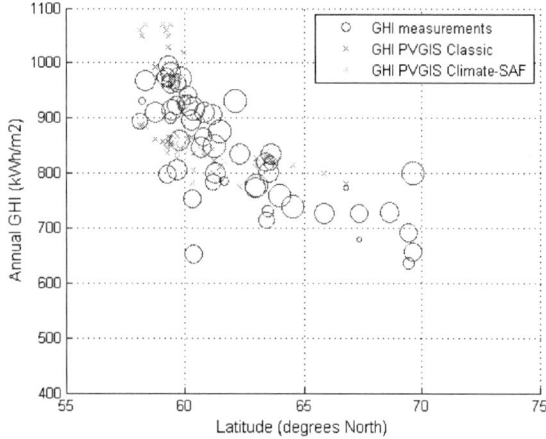

Fig. 1.   Average annual GHI resource in Norway in the range (58-70)°N and (0-40)°E based on long-time measurements of GHI (circle size proportional to length of time series available). PVGIS Classic and Climate-SAF data included for reference (crosses, up to 67°N).

With the increase in grid-installed PV systems, the number of in-plane reference cells have increased for performance monitoring purposes. These are generally not suited for solar radiation resource assessment, but combined with PV module temperatures provide valuable information about the real operating conditions for PV systems. Performance data can be combined with ambient temperature data widely available from a large number of meteorological stations. PV module efficiency versus solar irradiance for various temperature regimes and PV technologies has been extensively discussed in the literature, e.g. [14-17].

III. METHOD

The relative efficiency is defined as [7]:

$$\eta_{rel} = \frac{P(G,T')}{P_{STC} \cdot (G/G_{STC})} \ , \ where \ T' = T_{mod} - T_{STC}. \quad (1)$$

The relative efficiency is the ratio of the module efficiency under given conditions of irradiance G and module temperature T_{mod} to the efficiency at STC. The value of P_{STC} is here assumed to be the rated maximum power from manufacturer's specifications, but should ideally be measured on-site as there could be significant deviation caused by fabrication tolerances and initial degradation. The module temperature may be estimated from the rated NOCT temperature, which assumes a well-ventilated module and light air, together with irradiance and ambient temperature T_{amb}:

$$T_{mod} = T_{amb} \left(\frac{NOCT - 20°C}{800 \ W/m^2} \right) * G \quad (2)$$

Alternatively, a constant k may be derived from on-site measurements of module temperature versus irradiance in accordance with the linear expression $T_{mod} = T_{amb} + k*G$, see e.g. [18].

We here present irradiance and temperature distributions from five locations at high latitudes (58-70 °N) based on data recorded over several years. These distributions are analysed with respect to PV performances using manufacturer's data for commercially available PV modules (Table I), representing different PV technologies from the PVsyst database [19]. From this we calculate a relative efficiency weighted by the mean annual solar energy intensity distributions, evaluated for a set of realistic module operating temperatures.

TABLE I
EVALUATED PV MODULE TECHNOLOGIES

	Module type	P_{mp} [W_p]	η_{STC} [%]	NOCT [°C]	T.coeff. [%/°C]
µCSi-aSi:H	Sharp NA-F135 (GK)	135	9.7	44	-0.24
CdTe	FirstSolar FS-395	95	13.2	45	-0.25
poly-Si	Suntech STP230-20/Wd	230	14.1	45	-0.43
mono-Si	SunPower 230E-BLK-D	230	18.5	46	-0.38
HIT	Panasonic VBHN230SE51	230	18.3	44	-0.29
CIS	SolarFrontier SF170-S	170	13.9	47	-0.31
III-a-Si:H	Uni-Solar PVL-144	144	6.8	46	-0.21

Fig. 2 displays differences in normalized efficiencies relative to STC irradiance of 1000 W/m², for the selected PV technologies and at module temperatures of 10, 25 and 40 °C,

978-1-5090-5606-4/17 $31.00 © 2017 IEEE

based on module specifications in PVsyst. The T_{mod} = 25 °C curves correspond to STC temperature conditions, which have been found representative of real operating conditions in Southern Norway. In [20], the authors found temperatures centered around T_{mod} ~19 °C for 50 % of the time in a year, and around T_{mod} ~25 °C for 50 % of the annual energy production. This analysis was limited to mono-Si modules in an open-rack configuration, based on one year of data at the given location. Locations further north characterized by long cold winters may experience lower typical temperatures, whereas poorly ventilated systems or building integrated photovoltaics expect higher operating temperatures. Efficiency curves are therefore also evaluated at T_{mod} = 10 °C (wintertime conditions) and 40° C (BIPV applications).

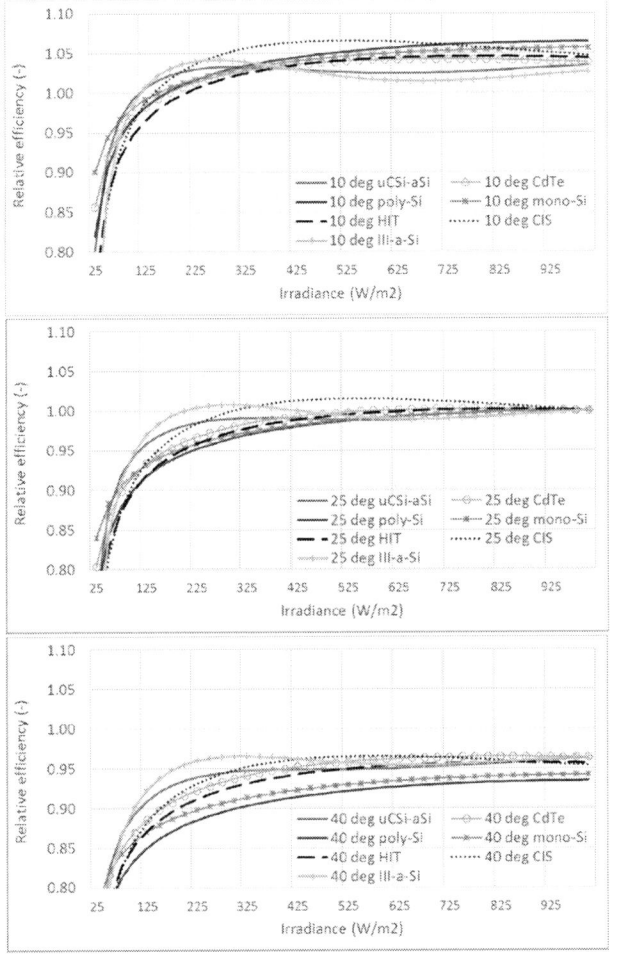

Fig. 2. Normalized efficiencies for various examples of PV module technologies, as a function of irradiance, based on manufacturer's specifications from the PVsyst database [19].

IV. RESULTS

Table II displays the locations evaluated and their corresponding annual GHI and DHI values based on the number of years of data available in this analysis. (Longer time series of global hourly irradiation exist for some locations.) Selected results from the irradiance and temperature distributions, derived from ground measurements at each location, are included below. By applying the GHI distributions to the PV module efficiency curves, the resulting relative efficiency can be seen for two examples in Fig. 8-9.

TABLE II
SUMMARY OF EVALUATED LOCATIONS

Location	Degr. North	Years meas.	GHI (kWh/m2)	DHI (kWh/m2)	DHI/GHI (%)	T_{amb} (°C)
Kristiansand	58	4	1054	*525*	*50*	7
Bergen	60	10	802	409	51	9
Aas	60	5	938	356	38	2
Trondheim	63	5	674	(300)	(44)	3
Tromsoe	70	5	680	*N/A*	*N/A*	1

Kristiansand (58 °N, 8 °E)

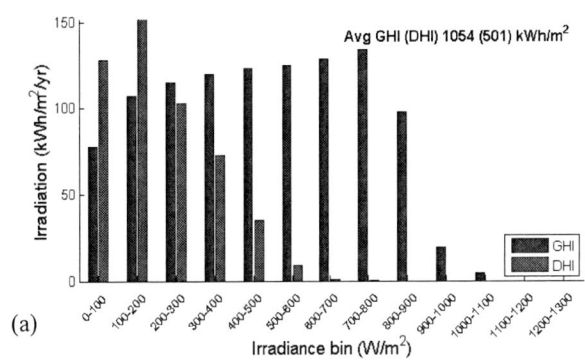

(a)

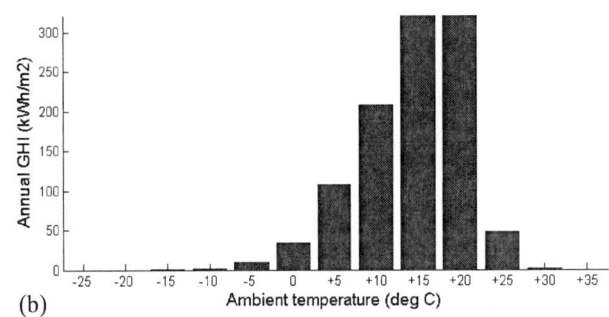

(b)

Fig. 3. (a) Annual distribution of GHI and DHI irradiances based on the average of N=4 years of 1-min measurements (2013-2016) in Kristiansand. (b) GHI distribution versus ambient temperature.

Bergen (60 °N, 5 °E)

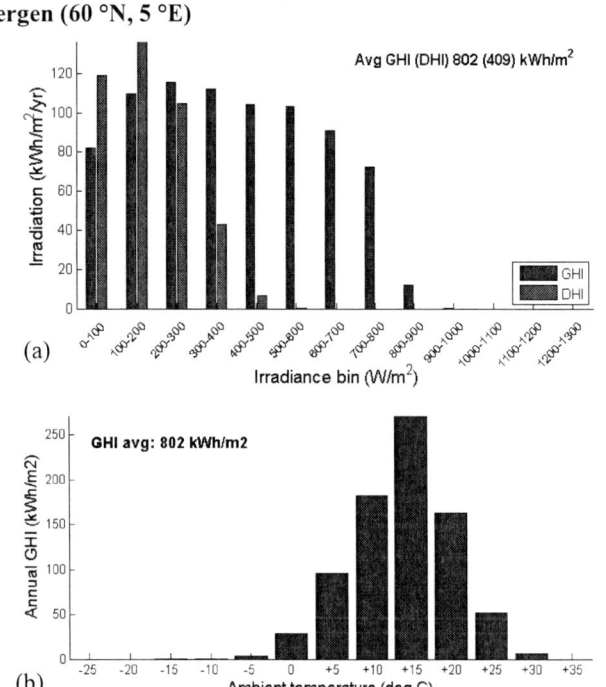

(a)

(b)

Fig. 4.   (a) Annual distribution of GHI and DHI irradiances based on the average of N=10 years of 1-hr measurements (2003-2013) in Bergen. (b) GHI distribution versus ambient temperature.

Trondheim (63 °N, 10 °E)

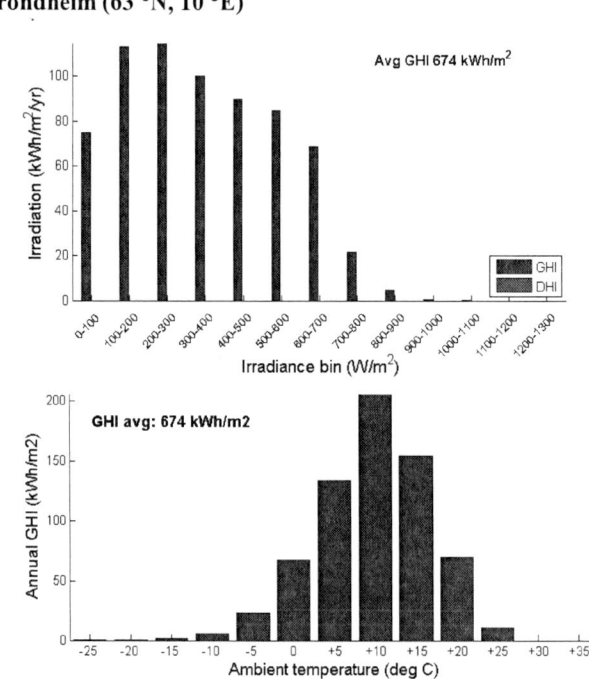

Fig. 6.   (a) Annual distribution of GHI and DHI irradiances based on the average of N=5 years of 1-hr measurements (2010-2014) in Trondheim. (b) GHI distribution versus ambient temperature.

Aas (60 °N, 11 °E)

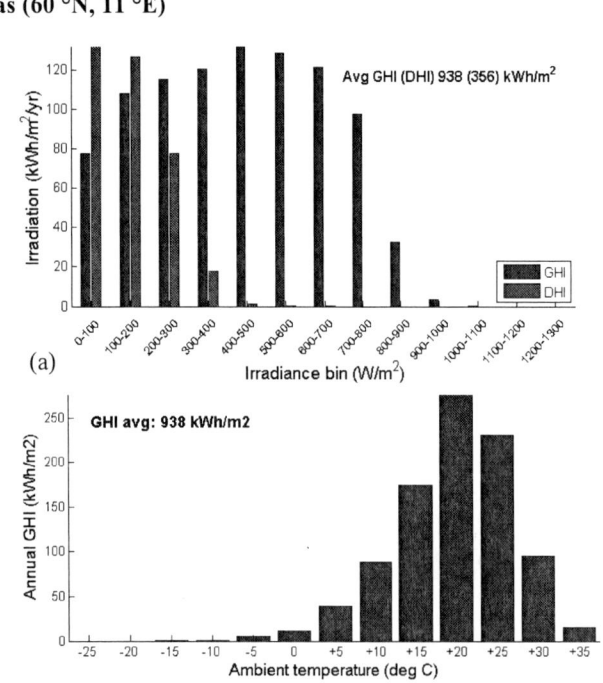

(a)

(b)

Fig. 5.   (a) Annual distribution of GHI and DHI irradiances based on the average of N=5 years of 10-min measurements (2011-2015) in Aas. (b) GHI distribution versus ambient temperature.

Tromsoe (70 °N, 19 °E)

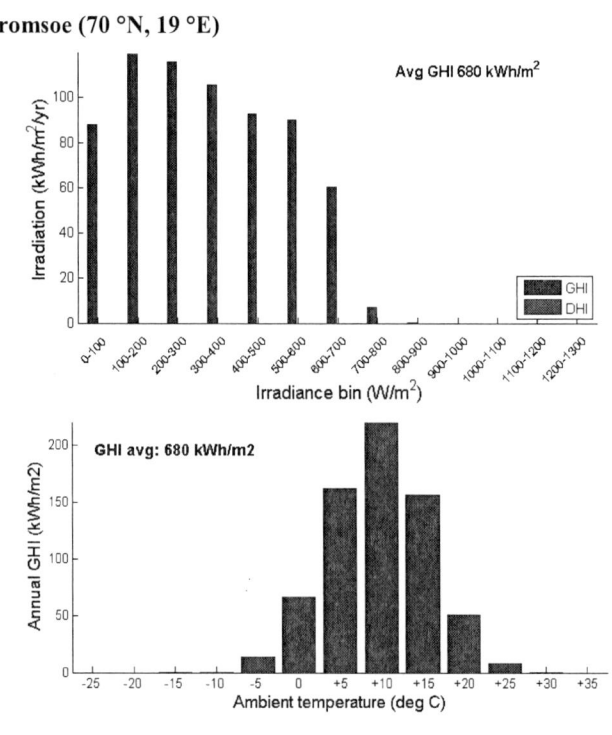

Fig. 7.   (a) Annual distribution of GHI irradiances based on the average of N=5 years of 1-hr measurements (2012-2016) in Tromsoe. (b) GHI distribution versus ambient temperature.

978-1-5090-5606-4/17 $31.00 © 2017 IEEE

Fig. 8-9 show that with the real GHI distribution, the resulting relative annual module efficiencies are lowered compared to the STC value, except at colder temperatures. Comparing the relative efficiencies at $T_{mod} = 25$ °C, the thin film modules appear to have slightly less performance drop than the crystalline silicon modules, but the overall main effect is due to differing temperature coefficients, as seen by comparing the three temperature ranges. Series and shunt resistance of the cells is of high importance with respect to low light efficiency, and there may be significant variation between different producers of the same technology. Hence, general conclusions on technology choice cannot be made based on the limited number of module types evaluated. Comparing Fig. 8 with Fig. 9, the trend is similar for both locations with no difference in ranking between the various modules. However, the relative efficiency levels are somewhat lower for the highest latitude (Tromsoe).

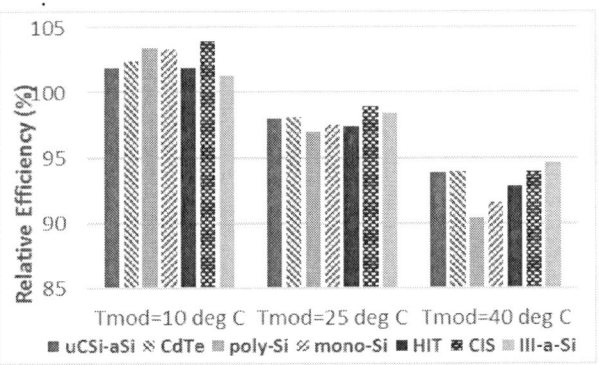

Fig. 8. Distribution of GHI-distribution weighted efficiency (average of N=4 years, 2013-2016, Kristiansand 58 °N), relative to STC efficiency, for three modelled PV module temperatures.

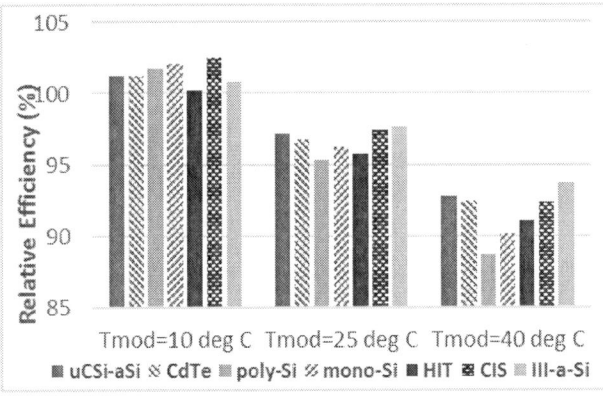

Fig. 9. Distribution of GHI-distribution weighted efficiency (average of N=5 years, 2012-2016, Tromsoe 70 °N), relative to STC efficiency, for three modelled PV module temperatures.

V. ANALYSIS OF RESULTS, MAIN FINDINGS

The results from the southern regions Kristiansand, Aas and Bergen show relatively even distributions of GHI across irradiance bins in the range 200-800 W/m². Most of the diffuse light is within the range 100-400 W/m², with a peak fraction at low intensities (100-200 W/m²). Additional data analysis from Kristiansand, evaluating the distribution of in-plane global irradiation at 20° tilt, show that the irradiance-distribution rises steadily from zero to a peak in the 900-1000 W/m² bin. Hence, for the tilted PV system (facing the equator), the STC conditions at irradiance 1000 W/m² is near the peak in terms of annual irradiation received, whereas for a horizontal system the distribution across the whole intensity range is important. The latter would favor PV technologies with stable high efficiency across all irradiance levels, including low light.

The GHI-distributions received in Bergen and Kristiansand are dominated by a high diffuse share. The annual GHI is centered around T_{amb} ~15° C. Higher mean ambient temperature is found for Aas (T_{amb} ~20°C), and lower mean temperatures in Trondheim and Tromsoe (T_{amb} ~10° C). The latter northern locations display GHI-distributions skewed toward lower intensity levels, in which case the STC conditions would be less suitable for design purposes.

There is a need for a larger set of reference sites and measurements of solar radiation properties, also including spectral effects and angular dependence of light over the hemisphere [21]. The angular distribution of diffuse light has not been measured, but using the isotropic sky model by Liu and Jordan [22] the distribution of off-normal incidence angles for a horizontal plane is calculated and shown in Table III. The resulting distribution is broad and cannot be represented by STC conditions. Generally, the diffuse fraction is high in the Nordic countries. Low-light performance and off-normal incidence angles are therefore design parameters that offer opportunities in terms of optical design improvements of PV cells and modules adapted to real operating conditions.

TABLE III
ANGULAR DISTRIBUTION OF DIFFUSE LIGHT (LIU AND JORDAN)

Angular interval	Mean angle of incidence	Fraction of isotropoic diffuse radiation
0°-10°	5°	0.0302
10°-20°	15°	0.0868
20°-30°	25°	0.1330
30°-40°	35°	0.1632
40°-50°	45°	0.1736
50°-60°	55°	0.1632
60°-70°	65°	0.1330
70°-80°	75°	0.0868
80°-90°	85°	0.0302

VI. Summary of the Work

Irradiance and temperature distributions of mean annual global and diffuse solar irradiation collected for five high-latitude locations (58-70 °N) over several years are presented. Some of these locations receive up to around 50 % diffuse light, with global intensity levels relatively evenly distributed across the irradiance range. The GHI-distributions are centered around ambient temperatures in the range 10-20°C. Results indicate that in terms of annual production, STC temperature may be representative of typical Norwegian conditions whereas irradiance levels are significantly lower than STC. The measured distributions are coupled with efficiency data for selected commercial PV modules, which display some performance differences. However, it is well known that particularly the shunt resistance is of high importance with respect to low light efficiency and there may be significant variation between different producers of the same technology. The work illustrates how actual irradiance and temperature conditions may be used to evaluate the choice of components and improve accuracy of yield estimates during PV system design. The model results will be followed up with experimental data from real systems in operation. This provides valuable input for the development of PV modules adapted to high latitude conditions. For a full analysis, it is also necessary to include spectral and angular distributions. Building knowledge about the available solar resource enables research into improved light utilization and thermal management, as well as new surface treatments for enhanced light absorption and the prevention of ice/snow formation on module surfaces [1].

Acknowledgements

Irradiance data received from Jan Asle Olseth at Geophysical Institute, University of Bergen, is gratefully acknowledged. The authors also acknowledge financial and project support from the Research Council of Norway and the partners in the BIPVNO project, *"Building Integrated Photovoltaics for Norway"*: NTNU, SINTEF, IFE, Teknova, Statsbygg, Omsorgsbygg, Undervisningsbygg, Glass og fasadeforeningen, Asplan Viak, Backegruppen, Isola, Rambøll, NorDan, Getek, and FUSen.

References

[1] *Building Integrated Photovoltaics for Norway*, EnergiX project no. 244031, The Research Council of Norway, www.bipvno.no .

[2] B. P. Jelle, "Building integrated photovoltaics: A concise description of the current state of the art and possible research pathways", *Energies*, vol. 9, issue 1, 2016.

[3] A. G. Imenes, "Performance of BIPV and BAPV Installations in Norway," in *43rd IEEE PVSC*, 2016, p. 3147.

[4] EN 50583:2016 *Photovoltaics in buildings, BIPV modules and systems.* Standard published 18.Jan.16, BSI.

[5] IEC 61853 Photovoltaic (PV) module performance testing and energy rating (Part 1:2011, Part 2:2016) International Electrotechnical Commission, available from the IEC Webstore (https://webstore.iec.ch).

[6] T. Huld, A. G. Amillo, E. Dunlop, "IEC 61853-3 Standard for calculating the energy rating of PV modules". Presentation at the 7th *Energy Rating and Module Performance Modeling Workshop*, 30-31 March 2017, Lugano. EU Commission, Joint Research Centre, Ispra.

[7] T. Huld, G. Friesen, A. Skoczek,, R. P. Kenny, T. Sample, M. Field, E. D. Dunlop, "A power-rating model for crystalline silicon PV modules", *Solar Energy Materials and Solar Cells,* vol. 95 (2011), pp. 3359-3369.

[8] Meteonorm Software, Meteotest, Bern, Switzerland (2016) (http://www.meteonorm.com/).

[9] T. Huld, R. Müller, A. Gambardella, "A new solar radiation database for estimating PV performance in Europe and Africa", *Solar Energy,* vol. 86, pp. 1803-1815.

[10] Ø. Kleven, H. Persson, "Solar Power Plants in the North", Technical Report no. 2013/12, Norut Narvik. Norway.

[11] R. Perez, R. Seals, A. Zelenka, "Comparing Satellite Remote Sensing and Ground Network Measurements for the Production of Site/Time Specific Irradiance Data", *Solar Energy,* vol. 60 (2), pp. 89-96.

[12] T. Haumann, *A Brief Look at the Performance of PV in Norway*, Master thesis, Dept. Physics and Technology, The Arctic University of Norway, December 2016.

[13] Norwegian Meteorological Institute (www.met.no), "*eKlima*" online database.

[14] W. Durisch, et al., "Efficiency model for photovoltaic modules and demonstration of its application to energy yield estimation", *Solar Energy Materials and Solar Cells,* vol. 91, pp. 79-84, 2007.

[15] G. Notton, V. Lazarov and L. Stoyanov, "Optimal sizing of a grid-connected PV system for various PV module technologies and inclinations, inverter efficiency characteristics and locations"*, Renewable Energy,* vol. 35, pp. 541-554, 2010.

[16] A. Louwen, et al., Comprehensive characterization and analysis of PV module performance under real operating conditions, *Progress in Photovoltaics: Research and Applications,* vol. 25 (2017), pp. 218-232.

[17] G. H. Yordanov, M. Tayyib, O.-M. Midtgård, J.-O. Odden, T. O. Sætre, "Test of the European Joint Research Centre Performance Model for c-Si PV Modules", in: Proc. 37th *IEEE Photovoltaic Specialists Conference,* (2011), pp. 2382-2387.

[18] L. Maturi, G. Belluardo, D. Moser, M. Buono, "BiPV system performance and efficiency drops: overview on PV module temperature conditions of different module types", *Energy Procedia,* vol. 48 (2014), pp. 1311-1319.

[19] PVsyst version 6 (Premium), Photovoltaic Software, PVsyst SA, Switzerland (2016) (http://www.pvsyst.com/).

[20] O. Dupre, R. Vaillon, M. A. Green, *Thermal Behaviour of Photovoltaic Devices*, Springer (2017), p.2-7.

[21] M. D. Lysko, *Measurement and Models of Solar Irradiance*, Doctoral thesis, Norwegian University of Science and Technology, Trondheim, August 2006. ISBN 82-471-8069-3 (electronic).

[22] B. Y. H. Liu and R. C. Jordan, "The long-term average performance of flat-plate solar energy collectors», *Solar Energy,* vol. 7, p. 53, 1963.

Step-by-step evaluation of photovoltaic module performance related to outdoor parameters: evaluation of the uncertainty

Anne Migan Dubois[a], Jordi Badosa[b], Fausto Calderón-Obaldía[a,b], Olivier Atlan[b], Vincent Bourdin[c], Marko Pavlov[b], Dae Young Kim[b], and Yvan Bonnassieux[d]

[a] GeePs; CNRS – CentraleSupelec – U-PSud – UPMC; 11 rue Joliot-Curie – F-91192 Gif-sur-Yvette
[b] LMD; École Polytechnique ; Route de Saclay – F-91128 Palaiseau
[c] LIMSI; CNRS ; Rue John von Neumann – F-91405 Orsay cedex
[d] LPICM; CNRS – École Polytechnique ; Route de Saclay – F-91128 Palaiseau

Abstract — Knowing the uncertainty in the PV production forecast is crucial in the optimization of smart-grid operations and in its stability. In this paper, we present uncertainty calculation in the PV energy production forecast process, calculated step by step. From horizontal irradiance and ambient temperature, all the steps needed to evaluate the PV production are studied with a special focus on the comparison between calculation and measurements. The uncertainty is evaluated by computing the relative mean bias error and the relative mean absolute error.

Index Terms — PV forecast, modelling, outdoor characterization, smart-grids.

I. INTRODUCTION

PV production mainly depends on the solar radiation incident on the PV modules. Solar resource variability and the uncertainty associated with the forecast of PV energy production are one of the most important factors that influence the grid stability, regardless of the size of the power grid.

The ability to precisely forecast the energy produced by PV systems is of great importance and has been identified as one of the key challenges for massive PV integration [1], [2].

Our approach is similar to indirect forecasts: firstly, we predict solar irradiance and ambient temperature, and then, using a PV performance model of the module, we calculate the PV power produced. The different stages of the PV forecast are summarized in Fig. 1, together with the methods used in this study.

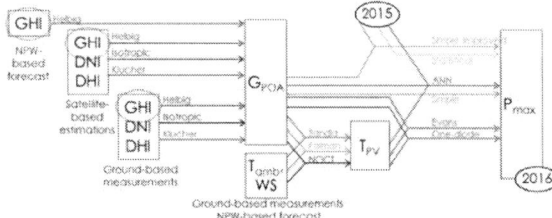

Fig. 1. Schematic of the protocol of PV production forecast and studied models.

This study focuses on the evaluation of the uncertainty on PV production estimation, step by step, starting from horizontal irradiances and ambient temperature.

The meteorological forecast step is not considered here, even though we know that it may carry the largest part of uncertainty.

The estimation of the uncertainty is done by comparison of calculated and measured values, with different methods of calculations for each step shown in Fig 1.

For this purpose, we firstly present the experimental measurements: atmospheric-related and from a PV-module platform.

In section III we consider diverse models to estimate G_{POA}, T_{PV} and P_{MPP}. All models are evaluated against local measurements at 1-hour time step.

All the results are grouped in section IV.

II. EXPERIMENTAL PLATFORMS DESCRIPTION

The experimental platforms are installed at the SIRTA observatory [3] located in Palaiseau (France, 48.7N, 2.2E), on the campus of École Polytechnique (Université Paris-Saclay), 15 km South-East of Paris.

A. Instrumental atmospheric measurements

This study uses two types of atmospheric measurements as input: ground-based measurements and satellite images.

The ground-based measurements are realized at the SIRTA observatory. It is a reference meteorological and climate observatory with more than 150 remote sensing and in-situ instruments. In terms of radiometric measurements, the site is part of the Baseline Surface Radiation Network (BSRN, http://bsrn.awi.de/) since 2003. GHI, DHI and DNI, as well as ground albedo measurements are realized following BSRN standards (with Kipp & Zonen CMP22 and CHP1 radiometers).

This study considers also estimations of GHI, DHI and DNI from Meteosat geostationary satellite observations computed by CAMS [4], [5].

978-1-5090-5606-4/17 $31.00 © 2017 IEEE

B. Outdoor photovoltaic characterization platform

A test bench PV platform was installed at SIRTA in 2014 and is composed of six commercial PV modules issued from different technologies (Fig. 2). In this paper, we only consider the c-Si PV module (see Table 1).

Fig. 2.   Outdoor characterization PV platform located at SIRTA.

The current-voltage characteristics are measured with Agilent DC electronic loads (6060B), each minute from sunrise to sunset. The P_{MPP} is derived from these characteristics. T_{PV} is measured with 4-wired class A platinum sensors (Pt100) and G_{POA} is measured with a solar radiometer (Hukseflux NR01) and a reference PV cell (SOLEMS RG100) installed in the same plane as the PV module.

TABLE I
FRANCEWATTS FL60-250MBP TECHNICAL PARAMETERS

Maximum power	P_{MPP}	250 W
Open circuit voltage	V_{OC}	37.67 V
Short circuit current	I_{SC}	8.64 A
Power temperature coefficient	TC_P	-0.48%/°C
Current temperature coefficient	TC_I	0.02%/°C
Module area	A	1.6285 m²

III. THEORETICAL MODELLING

A. POA irradiance calculation

In this part, the input data used are GHI, DHI and DNI solar irradiances given either by SIRTA ground measurements or by CAMS, as well as the tilt angle and the ground albedo.

The tilt irradiance is calculated using equation (1).

$$G_{POA} = B_{POA} + D_{POA} + A_{POA} \qquad (1)$$

The beam irradiance is calculated using the DNI as follow:

$$B_{POA} = DNI \times \cos(AOI) \qquad (2)$$

The angle of incidence between the sun's rays and the PV array can be determined as:

$$AOI = \cos^{-1}\left[\begin{array}{l} \cos(\theta_z)\cos(\theta_{tilt}) + \\ \sin(\theta_z)\sin(\theta_{tilt})\cos(\theta_A - \theta_{A,array}) \end{array} \right] \qquad (3)$$

where θ_A and θ_z are the solar azimuth and zenith angles, respectively. θ_{tilt} and $\theta_{A,array}$ are the tilt and azimuth angles of the array, respectively.

The albedo arriving in the plane of array is calculated thanks to the fill factor in front of the array:

$$A_{POA} = GHI \times Albedo \times \frac{1 - \cos(\theta_{tilt})}{2} \qquad (4)$$

Albedo corresponds to the coefficient of reflection of the ground and θ_{tilt} is the tilt angle of the PV module array.

Three models have been considered representing different ways to estimate the diffuse irradiance arriving on PV module:

1) Helbig Model [6]:

The fraction of diffuse irradiance is calculated only from GHI using an empirical relationship, giving an estimated value of DHI and DNI. The diffuse irradiance is then calculated in the same way as described in the isotropic hypothesis.

2) Isotropic hypothesis [7]:

The isotropic sky diffuse model assumes that the diffuse radiation from the sky vault is uniform across the sky. The diffuse irradiance in the plane of array is calculated by equation (5):

$$D_{POA} = DHI \times \frac{1 + \cos(\theta_{tilt})}{2} \qquad (5)$$

3) Non isotropic irradiance [8]:

Klucher found that the isotopic model gave good results for overcast skies but underestimates irradiance under clear and partly overcast conditions, when there is increased intensity near the horizon and in the circumsolar region of the sky.

$$D_{POA} = DHI \times \frac{1 + \cos(\theta_{tilt})}{2} \times$$
$$\left[1 + \left(1 - \frac{DHI}{GHI}\right)\sin^3\left(\frac{\theta_{tilt}}{2}\right) \right] \times \qquad (6)$$
$$\left[1 + \left(1 - \frac{DHI}{GHI}\right)\cos^2(AOI)\sin^3(\theta_z) \right]$$

B. PV module temperature calculation

In this part, we evaluate two models to calculate T_{PV} from T_{amb}, G_{POA} and WS local measurements.

1) Sandia Model [9]:

$$T_{PV} = G_{POA} \times \exp(a + b.WS) + T_{amb} \qquad (7)$$

a and b are empirical coefficients establishing the upper limit for module temperature at low wind speeds and high solar irradiance and the rate at which T_{PV} drops as WS increases, respectively.

2) Faiman Model [10]:

$$T_{PV} = T_{amb} + \frac{G_{POA}}{U_0 + U_1.WS} \qquad (8)$$

To evaluate the models, we first consider parameter values proposed by the authors (a = -3.47, b = -0.0594 s.m⁻¹ and U_0 = 25.0 W.m⁻².K⁻¹, U_1 = 6.84 W.m⁻³.s.K⁻¹). In a second step, we fit these coefficients to one year of measurements (2015) using the Levenberg-Marquardt method. The obtained coefficients were a = -3.1398, b = -0.305 s.m⁻¹ for Sandia model and U_0 = 21.777 W.m⁻².K⁻¹, U_1 = 9.855 W.m⁻³.s.K⁻¹ for Faiman model.

C. PV power modeling

In this study, we consider six different models to calculate P_{MPP} from the ground-based measurements of G_{POA} and T_{PV}:

1) Simple model:

It considers a constant CE measured by the manufacturer during a flash test and equal to 0.15086 for our studied PV module.

$$P_{MPP} = CE_{STC} \times G_{POA} \times A \qquad (9)$$

2) Simple improved model:

For this case, the conversion efficiency is taken equal to the average of the measured one during 2015: CE = 0.14339.

3) Evans model [11]:

It takes into account the linear variation of the conversion efficiency with temperature and the low light effect.

$$P_{MPP} = CE_{STC} \times G_{POA} \times A \times \left[1 - TC_1 \times \left(T_{PV} - T_{STC} \right) + \gamma \times \log_{10} \left(\frac{G_{POA}}{G_{STC}} \right) \right] \quad (10)$$

With TC1 = 0.0048 from manufacturer data and γ = 0.1 deduced from measurement during 2015.

4) Statistical model:

This model does not need internal information from the system to describe its performance. It is a data –driven approach which is able to extract relations on past data to predict the future behavior of the PV module. Here, the past is the year of 2015. Thus, quality of historical data is essential for accurate forecast.

5) One-diode electrical model [12]:

This model is based on the Shockley diode equation, with a current source to model the photo-current (I_{ph}), a single-diode junction (n is the ideality factor and I_0 the saturation current) and a series resistance (R_s), as shown in Fig. 3.

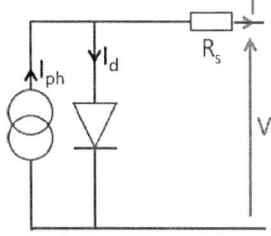

Fig. 3. One-diode electrical model.

The equation that drives this model is equation (11):

$$I = I_{ph} - I_0 \left[\exp\left(\frac{q(V + IR_s)}{nkT} \right) - 1 \right] \qquad (11)$$

$$P_{MPP} = \max(I \times V) \qquad (12)$$

I_{ph} depends on T_{PV} and G_{POA}, I_0 and R_s are temperature dependent and n is constant. All of the constants used in the above equation are determined by fitting the manufacturer flash test and ratings listed in Table 1, as shown in Fig 4.

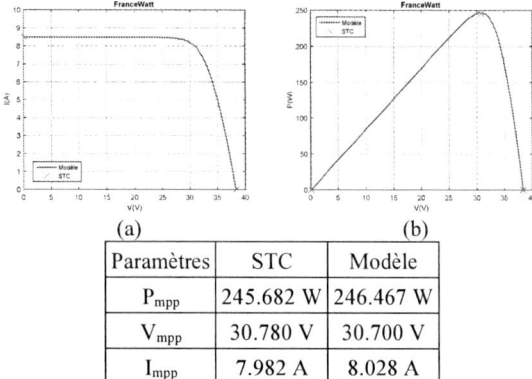

(a)   (b)

Paramètres	STC	Modèle
P_{mpp}	245.682 W	246.467 W
V_{mpp}	30.780 V	30.700 V
I_{mpp}	7.982 A	8.028 A

Fig. 4. Determination of the 4 parameters of the one-diode electrical model by fitting manufacturer data of Table I, I(V) in (a) and P(V) in (b).

6) Artificial neurons network [16]:

The ANN used was built using the feed forward neural network structure with a weighted linear combination and sigmoid function. The architecture chosen is one output (P_{MPP}), three inputs (G_{POA}, T_{amb} and WS) and one hidden layer of five neurons. The training period was the year 2015.

D. Evaluation indicators

In order to compare measurements and calculated values, we compute rMBE and rMAE, as defined in equations (13) and (14).

$$rMBE = \frac{\sum_{i=1}^{N} \left[P_{mpp,calc}(i) - P_{mpp,meas}(i) \right]}{\sum_{i=1}^{N} P_{mpp,meas}(i)} \qquad (13)$$

$$rMAE = \frac{\sum_{i=1}^{N} \left| P_{mpp,calc}(i) - P_{mpp,meas}(i) \right|}{\sum_{i=1}^{N} P_{mpp,meas}(i)} \qquad (14)$$

IV. RESULTS

All presented models in Section II are here evaluated for one independent year of data (2016). Fig. 4 to 6 compare the calculated values to the measured ones. Tables II to IV summarize the rMBE and rMAE obtained results.

A. POA irradiance calculation

In this part, we compare the G_{POA} calculation methods with in plane measurements.

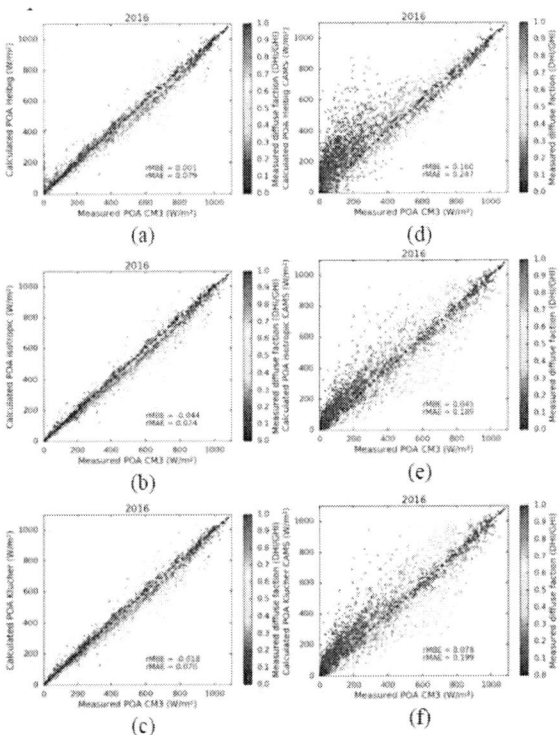

Fig. 4. Comparison of the calculated and the measured global irradiance in the plane of array using ground measurements (a), (b) and (c) and satellite estimation (e), (f) and (g). The models are Helbig (a) and (d), isotropic (b) and (e) and Klucher (c) and (f).

The error estimations are summarized in Table II.

TABLE II
UNCERTAINTY ESTIMATION IN THE CALCULATION OF G_{POA}

Input data	Model	rMBE	rMAE
CAMS	Helbig	0.160	0.247
	Isotropic	0.045	0.189
	Klucher	0.078	0.199
SIRTA	Helbig	0.001	0.079
	Isotropic	-0.044	0.074
	Klucher	-0.018	0.070

Table II shows that all methods using satellite irradiances have a positive bias and absolute error more than double than

the results with ground measurements. Isotropic model is the best if data comes from satellite.

B. PV module temperature calculation

The empirical coefficients of the PV module temperature models are first taken as literature values. Then, they were fitted using with the Levenberg-Marquardt method with data from 2015.

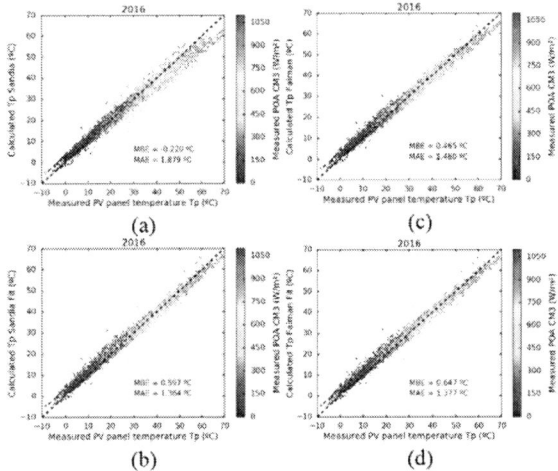

Fig. 5. Comparison of the calculated and the measured PV module temperature using Sandia model (a) and (b) and Faiman model (c) and (d). The empirical coefficient are those found in the literature (a) and (c) and fitted with data of 2015 (b) and (d).

The error estimation is summarized in Table III.

TABLE III
UNCERTAINTY ESTIMATION IN THE CALCULATION OF T_{PV}

Model	Coef. from	MBE (°C)	MAE (°C)
Sandia	Literature	-0.193	1.883
	2015 meas.	0.597	1.364
Faiman	Literature	0.485	1.471
	2015 meas.	0.647	1.377

Table III shows slightly better results when the model parameters are fitted to the measurements from 2015.

C. PV power modeling

The next figure compare the uncertainty of the model used to simulate the photoelectric effect.

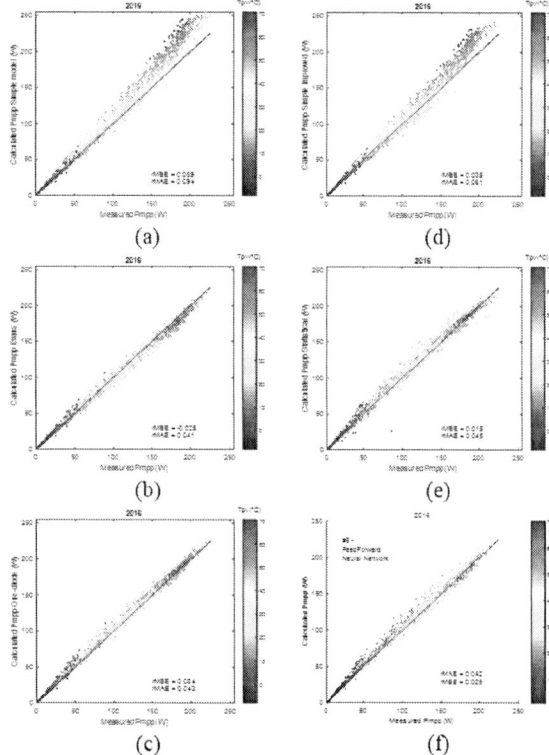

Fig. 6. Comparison of the calculated and the measured PV output maximum power using simple model (a), simple model with CE equal to the average CE of 2015 (d), Evans model (b), satstic model (e), one-diode electrical model (c) and ANN (f).

TABLE IV

UNCERTAINTY ESTIMATION IN THE CALCULATION OF P_{MPP}

Model	rMBE	rMAE
Simple	0.089	0.094
Improved simple	0.035	0.061
Evans	-0.025	0.041
Statistical	0.015	0.045
One-diode	0.034	0.043
ANN	0.0258	0.0425

Table IV shows the improvement of the accuracy of the models, from the simplest to the most complex. A good compromise seems to be the model of Evans.

Interestingly, the lowest rMAE value in the calculated G_{POA} (0.070) is 70% larger than for the best P_{MPP} calculation (4.1%).

V. CONCLUSION

In this paper, we have studied, step by step, the process of the estimation of the PV energy production from GHI, DHI, DNI and T_{amb}, with a special focus on the calculation of the uncertainty. Different models have been studied at each step of the calculation.

The next steps will be to show the link between all the models and the error propagation, to consider the step of meteorological and P_{MPP} forecast at different time horizons and to go through PV plants instead of a unique PV module.

ACKNOWLEDGMENTS

The authors thank the funding from IDEX of Université Paris Saclay, the research program TREND-X from École Polytechnique and the University of Costa-Rica.

GLOSSARY

ANN: Artificial neurons network
AOI: Angle Of Incidence
A_{POA}: Albedo in the plan of array (W.m^{-2})
B_{POA}: Beam irradiance in the plan of array (W.m^{-2})
CAMS: Copernicus Atmosphere Monitoring Service
CE: Conversion efficiency
DHI: Diffuse horizontal irradiance (W.m^{-2})
DNI: Direct normal irradiance (W.m^{-2})
D_{POA}: Diffuse irradiance in the plan of array (W.m^{-2})
GHI: Global horizontal irradiance (W.m^{-2})
G_{POA}: Plan of array irradiance (W.m^{-2})
P_{MPP}: Maximum power point (W)
rMAE: Relative mean absolute error
rMBE: Relative mean bias error
T_{amb}: Ambient temperature (°C)
T_{PV}: PV module temperature (°C)
WS: Wind speed (m.s^{-1})

REFERENCES

[1] EPIA, "*Connecting the Sun. Solar photovoltaic on the road to largescale grid integration*", 2012.
[2] PV GRID, "*Final Project Report*", 2014.
[3] Haeffelin, M., et al (2005). SIRTA, "*A ground-based atmospheric observatory for cloud and aerosol research*". Annales Geophysicae (Vol. 23, No. 2, pp. 253-275).
[4] Z. Qu, A. Oumbe, P. Blanc, B. Espinar, G. Gesell, B. Gschwind, L. Klüser, M. Lefèvre, L. Saboret, M. Schroedter-Homscheidt, and L. Wald L, "*Fast radiative transfer parameterisation for assessing the surface solar irradiance: The Heliosat-4 method*", Meteorologische Zeitschrift, 2016.
[5] M. Schroedter-Homscheidt A. Arola, N. Killius, M. Lefèvre, L. Saboret, W. Wandji, L. Wald, and E. We *Evaluating* y, "*The Copernicus Atmosphere Monitoring Service (CAMS) Radiation Service in a nutshell*", SolarPACES, Abu Dhabi, UAE, 2016.
[6] N. Helbig, "*Apllication of the radiosity approach to the radiation balance in complex terrain*" Thesis at University of Zurich, 2009.

[7] H. C. Hottel B. B. Woertz, *"Evaluation of flat-plate solar heat collector"* Trans. ASME 64, p. 91, 1942.

[8] T. M. Klucher, *"Evaluation of models to predict insolation on tilted surfaces"*, Solar Ener gy ,23 (2), pp. 111-114, 1979.

[9] D.L. King, W.E. Boyson, J.A. Kratochvil, "Photovoltaic Array Performance Model", Sandia National Laboratories, SAND2004-3535, 2004.

[10] D. Faiman, *"Assessing the outdoor operating temperature of photovoltaic modules"*, Prog. Photovolt.: Res. Appl. 16, pp. 307–315, 2008.

[11] D. L. Evans, *"Solar energy simplified method for predicting photovoltaic array output"*, Vol. 27, No. 6, pp. 555-560, 1981

[12] Antonanzas J., Osorio N., Escobar R., Urraca R., Martinez-de-Pison F.J., Antonanzas-Torres F., *"Review of photovoltaic power forecasting"*, Solar Energy 136, p. 78 (2016)

[13] Bishop J. W., *"Computer simulation of the effects of electrical mismatches in photovoltaic cell interconnection circuits"*, Solar Cells, 25 (1), p. 73 (1988)

[14] Karamirad, M., Omid, M., Alimardani, R., Mousazadeh, H., Heidari, S.N., *"ANN based simulation and experimental verification of analytical 4 and 5 parameters models of PV modules"*. Simul. Model. Pract. Theory 34, p. 86 (2013)

[15] W. R. Geoffrey, " *MPPT converter topologies using matlab PV model*", AUPEC: Innovation for Secure Power, Queensland University of Technology, Brisbane, Australia, pp. 138-143, 2000.

[16] A. Mellit., S. A. Kalogirou, *"Artificial intelligence techniques for photovoltaic applications: a review"*. Prog. Energy Combust. Sci. 34, pp. 574–632, 2008.33

Performance comparisons of a PV system by monitoring Solar irradiance with different pyranometers

Yasuhiro Matsumoto[1], J. Antonio Urbano[1], Ramón Peña[1], María de la Luz Olvera[1], Nun Pitalúa[2], Miguel A. Luna[1] and René Asomoza[1]

[1]Electrical Engineering Department, Centro de Investigación y de Estudios Avanzados del IPN
Av. IPN 2508, Colonia Zacatenco, C.P. 07360, México D.F
[2]Sustainable Development Group, Universidad de Sonora, Hermosillo, Sonora, México

Abstract — **Performance ratio (PR) of the 60 kWp photovoltaic-system (PVS) was evaluated in north Mexico City. The PVS energy PR is reported using two different instruments for detecting solar irradiation. Both, crystalline silicon (c-Si) based sensor (so-called, photovoltaic-pyranometer) and thermopile-pyranometer were mounted in the same plane of array (POA) of the photovoltaic modules. The average daily PV generated energy in the measured period was 256.67 kWh/day. The average PR was 85.7% by using silicon-pyranometer and 78.8% for thermopile-pyranometer. Also has measured during two months of March and April, 2017. The measured PR was 75.57% and 77.81% for the corresponding months, respectively.**

Index Terms — **PV system performance ratio, pyranometers, solar irradiation, temperature and wind dependences.**

I. INTRODUCTION

Solar photovoltaic systems (PVS) are becoming one of the important clean and alternative-energy technology in several countries. The PV systems are safe, reliable with a low-maintenance cost without any on-site pollutant emissions. Nowadays, the utility grid-connected PVS are increasing rapidly in the world and estimated global PV market grew to over 70 GW annually and about 300 GW cumulatively installed capacity at the end of 2016 [1]. In Mexico, the off-grid installed PV capacities was about 30 MW, which represented more than 85% of the total PV systems in 2010. But since the net-metering mechanism which was implemented in 2007, the total PV capacity has achieved about 370 to 400 MW in 2016 [2].

The Mexican energy reform implementation is expected to have a strong impact in the development of the PV market [3]. The International Energy Agency (IEA) cited the country's energy reform known as "Reforma Energética" as a major factor in the significant projected increase in renewables capacity. The reform sought to end the monopolies held on oil and gas by PEMEX the mexican National oil company, and on the electricity sector represented by the CFE (Comisión Federal de Electricidad). In doing so, it would open up

Mexico's energy sector to new players investment and new technology. The first two auctions of 2016 for new power supply did demonstrate that there is an appetite for investment in new solar PV and wind generation. Mexico is set to hit between 3 to 4 GW for 2018 [4] and 30-40GW of solar PV installations by 2040 under various scenarios projected by the IEA [5].

The present study evaluates the generated electricity from a 60kw PV system in terms of performance ratio (PR), taking in consideration Mexico City´s weather conditions. The PR were evaluated based upon two different solar irradiation sensors at the plane of array (POA), firstly during one month, from December 23, 2016 to January 22, 2017. A second set of PR-measurements were done from March 1 to April 30, 2017, during two complete months.

II. SYSTEM DESCRIPTION

The 60 kWp PV system (PVS) which consists of 240 single-crystalline silicon PV modules with a 250 Watt-peak (Wp) each. The detailed description of PVS can be found elsewhere [6]. The PV module-arrays were installed on the Institution´s building on the fifth-floor roof and fixed on aluminum framed structures oriented 30° East-faced from the geographical South. The system is located at 19° 30' 38" N, 99° 07' 50" W, and the module arrays were installed at about the latitude angle of 20°. This PV module arrays are subdivided (the electric-connection) into five sections. Each section is composed of a string of 48 PV modules that consist of 12-series and 4-parall connections.

For each of the five array arrangements is connected to the corresponding inverter; Fronius model IG-Plus-V 11.4-3 DELTA with the capacity of about 11.4 kW/each. The solar irradiance was measured using; a) reference crystalline-silicon (c-Si) solar cell, so-called photovoltaic-pyranometer [7] and b) second-class thermopile-pyranometer EKO MS-602, installed in the same PV-module plane of array (POA) with the angle of approximately 20 degrees from the horizontal. Also in a site, a

Fig. 1. The photovoltaic (upper left) arranged at POA and thermopile-pyranometers (bottoms) arrenged at POA and horizontal, are shown.

thermopile-pyranometer Yankee Environmental Systems; Model TSP-1, was installed to monitor the global horizontal solar irradiance, see Fig. 1.

III. MONITORING AND PERFORMACE CONCEPTS

A. System monitoring

Data monitoring is one of the important requirements for diverse PVS. Without an accurate data monitoring, the PVS performances cannot reliably be compared to the generated energy. An effective data monitoring not only helps to identify system performance but it also helps to resolve possible troubles [8]. The Fronius inverter system, integrates all of the monitored data every 5 minutes and logged, including solar irradiance using photovoltaic-pyranometer. In addition, the EKO MS-602 and the TSP-1 pyranometers were monitored by Campbell CR300 data logger every one minute.

B. General performance

Three of the IEC standard 61724 performance parameters have to be used to define the overall system performance with respect to the energy production, the solar resource and overall effect of system losses [9].

The performance ratio (PR) or so-called "quality factor", is the ratio between actual yield (i.e. annual production of electricity delivered at AC) and the ideal yield:

$$PR = \frac{\mathrm{Re}\,al \cdot Yield\,(AC)}{Ideal \cdot Yield\,(DC)}$$

In the present job, firstly, the measurements were done form December 23, 2016 to January 22, 2017, during 31 consecutive days, and then from March 1, to April 30, 2017.

C. Energy losses

Under normal PV system operating conditions, the measured data contains deviations caused by malfunctions such as string defects, shadings, module or inverter malfunctions that influence the measured performance of a PV system. Sometimes it is intuitive to think in terms of energy losses that occur at every step of the way, rather than component efficiencies. Both concepts are related as:

$$Losses = 1 - Efficiency$$

IV. SYSTEM GENERAL PERFORMANCE

In the first stage, the PV system was monitored for 31 consecutive days. Figure 2 shows the detected daily based solar irradiance in the POA through c-Si photovoltaic- and MS-602 thermopile-pyranometers. The average daily solar irradiance was 4.996 kWh/m^2 for c-Si and 5.452 kWh/m^2 for MS-602. The average irradiance detected for both sensors are shown in Fig. 2. The given units are in peak-hour or kWh/m^2. At the end of the figure, it is indicated by black-dot the detected average energy of the days.

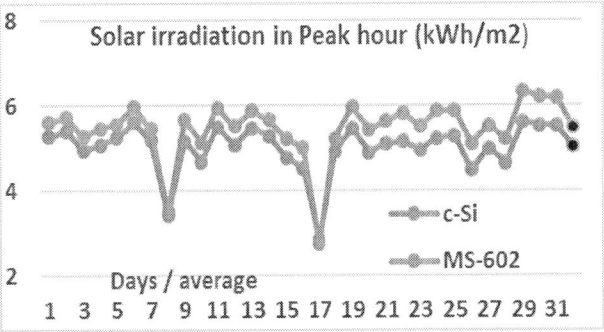

Fig. 2. Measured solar irradiance at the POA by using a) c-Si solar cell; photovoltaic-pyronometer and b) EKO MS-602 thermopile-pyranometer.

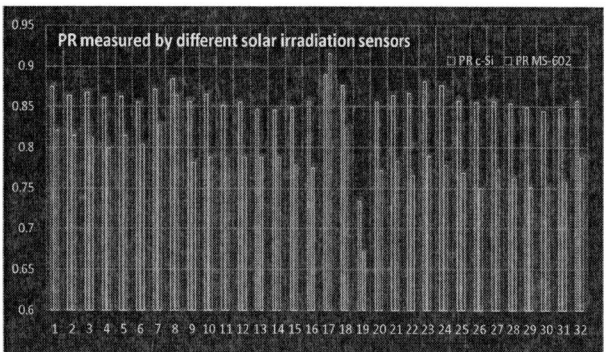

Fig. 3. Performance ratio (1=100%) for the PV system monitored using c-Si and MS-602 pyranometers. The last pair of measurements corresponds to the average PR of each pyranometers during the days.

Fig. 3 shows the performance ratio (PR) of the PVS in daily basis, starting day-1 on December 22th, 2016. The day 32th corresponds to the average daily PR. The obtained average PR was 85.7% for the data calculated by using c-Si pyranometer, and 78.8% for MS thermopile-pyranometer. There are near 7 % differences for the detected solar irradiances between the sensors, and this difference could be due to the c-Si detector dirt accumulation by a natural environmental dusts.

Fig. 4 shows the generated energy from the PVS and the solar energy detected by pyranometers in kWh. At the end, the black-points shows the daily-average energy during the days. From the same figure, it is possible to observe the detected solar energy differences and the tendencies throughout the days amongst the pyranometers. The corresponding thermopile-pyranometer detected average solar irradiation was 327.11 kWh, while the c-Si pyranometer detected 299.82 kWh. It means, c-Si pyranometer detected only 91.65% of the energy which MS-602 detected. Finally, the average daily PV generated energy in the measured period was 256.67 kWh/day.

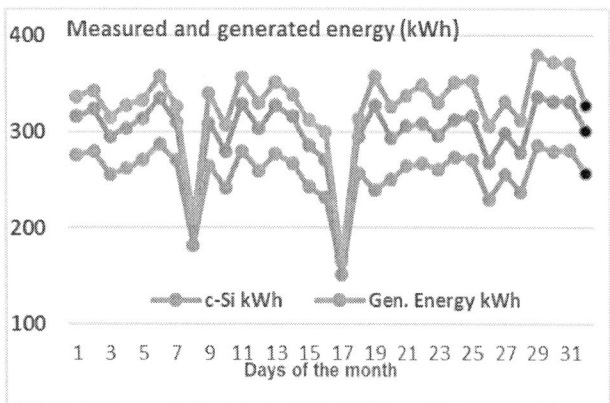

Fig. 4 shows the daily-based average generated energy from PV system and the measured solar irradiation by using two different pyranometers.

V. PERFORMANCE DIAGNOSIS-I

The purpose of monitoring PVS performance ratio (PR) is to determine whether the system is working as expected as to the incident solar irradiation. To do this, it requires measurement of the system output in real time and its operating conditions. Solar irradiance in a POA is by far, the most important data, and it is the base to calculate PR. Certainly, the differences in a spectral and directional responses between the pyranometers, the c-Si sensor lead to intraday and seasonal fluctuations [10]. For the precise generated electric power and so the PVS yield, it seems, the c-Si-based pyranometer measurements are not sufficiently precise. The c-Si sensors provide the required stability, with the spectral range from 400 to 1150 nm with a relatively quick response-time to the irradiance changes, but it has a temperature effects in the generated current to measure irradiance. Also, the angular distribution, the shift of transfer function over time, etc. The c-Si reference sensors are calibrated under indoor and outdoor conditions which should comply with IEC 60904-2 and -4, respectively. On the other hand, thermopile-pyranometer is based on a thermocouple device with a wider wavelength sensibility in a range of 300 to 3,000 nm [11]. The parameters that influence the uncertainty of pyranometers are irradiance level and Solar spectral distribution. But furthermore, also are the irradiance change rates during the measurement; cosine effect; the tilt angle; ambient and pyranometer´s dome temperature [12]. The known overall uncertainty of the instantaneous irradiance measurement based on secondary standard pyranometers is approximately 3% [13, 14].

Now, it is difficult to asseverate and to confirm whether the analyzed data can explain the obtained PR´s differences, and by using these averaged parameters, in any case, the c-Si based photovoltaic-pyranmeter´s temperature coefficient, also could influence in some extent. In our case, the photovoltaic module arrays are fixed but in the near future we are expecting to start with the tracking system to harvest greater amount of solar irradiation [15].

VI. PERFORMANCE DIAGNOSIS-II

Fig. 5 shows the PVS generated energy during two months; March and April, 2017. The produced average daily energy in March and April are 278.78 and 290.43 kWh, respectively. The average detected solar irradiations are 356.76 and 355.32 kWh for c-Si sensor, while for MS-602 sensor, 368.89 and 373.28 kWh for March and April, respectively. Table I summarizes different measured parameters together with the performance ratio (PR) related to each of the irradiance detectors. From Fig. 5. it can appreciate different performance ratios for the c-Si and MS-602 detected energies. The calculated PR increases from March to April for both type of pyranometers. In March, c-Si based irradiance was 356.76 kWh but it was only 368.89 kWh for MS-602. While in April,

978-1-5090-5606-4/17 $31.00 © 2017 IEEE

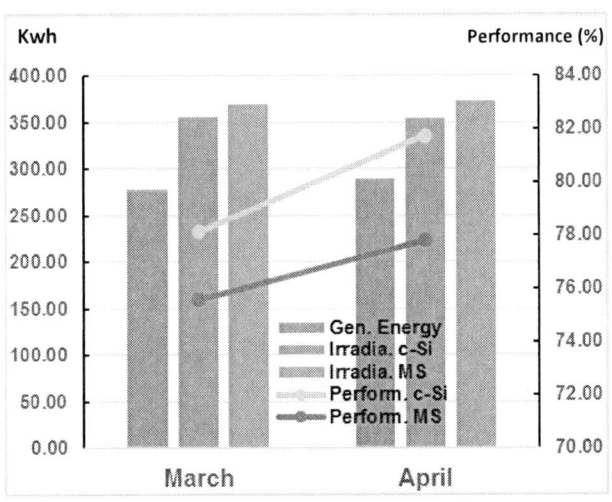

Fig. 5. Daily-based average generated energy from PV system. Measurements done from March to April 2017, also for the solar irradiation by using c-Si and MS-602 pyranometers.

TABLE I

The generated energy, the detected solar irradiance for both pyranometers and the corresponding performance ratio.

Month	Energy (kWh)	Irradia. c-Si (kWh)	Irradia. MS (kWh)	Perform. c-Si (%)	Perform. MS (%)
March	278.78	356.76	368.89	78.14	75.57
April	290.43	355.32	373.28	81.74	77.81

the irradiace was 355.32 kWh and 373.28 kWh, for c-Si and MS-602, repectively. The PR increased from 78.14 to 81.74% for c-Si based calculation, but the MS-602 based gives 75.57 and 77.81%, for March and April, respectively. We have to mention that the PV array was very dirty up to March 3rd due to the natural atmospheric dust accumulations during a couple of months, however, at the night time March 3rd, it was rained and cleaned up all of the PV array. This is one of the possible reason of PR increment from March to April. We have also compared ambient and PV module temperatures throughout the months and also wind velocity during the sun-shining times, even though, no other apparent variants might have

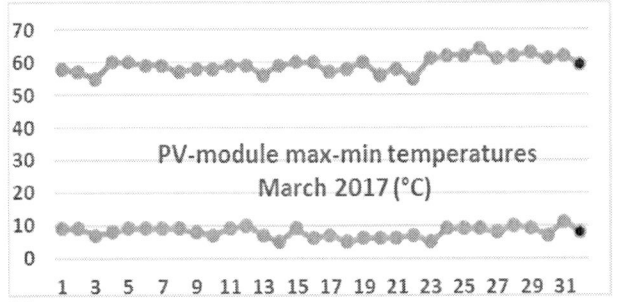

Fig. 6. Photovoltaic module maximum and minimum temperatures during March. The last points are the average of max. and minimum.

Fig. 7. One section of the 60 kWp PV module array installed in CINVESTAV, Mexico City. The annemometer is also shown.

influenced for the increased PR in April. Fig. 6 shows the maximum and minimum PV-module temperatures for March, which is almost equivalent for the hole month of April.

Fig. 7, shows 2/5 part of the installed 60 kWp PV system. As can see, it consists of two strings (upper and lower sides of the array) of 12-series with 4-parallel connected PV modules.

VII. DISCUSSIONS

In regards to the performance of a grid-connected 60 kWp photovoltaic system which was monitored from Dec. 23, 2016 to Jan. 22, 2017, the PR differences of 85.7% and 78.8% for c-Si and MS-602 which is about a 7% could be in part due to some of the dust accumulation on c-Si based sensor.

Even though, for March and April, and after cleaning the c-Si based photovoltaic pyranometer, the PR measurement differences from 3 to 5% could be due to the spectral limitation, and the lack of cosine sensibility of c-Si based sensor, which could be traduced as solar incidence angle limitation for the photovoltaic-sensor, but in addition, it may also influenced the c-Si sensor´s calibration.

The slight differences on the PR between the months of March and April is difficult to delucidate, because the ambient temperatures were almost constant during both months and also the wind velocities were almost with the same tendencies and seems does not affect substantially for the generated power.

VI. CONCLUSIONS

The performance of a grid-tied 60 kWp photovoltaic system at the north of Mexico City was evaluated and monitored from Dec. 23, 2016 to Jan. 22, and March-April, 2017. The daily produced energy was analyzed and interpreted together the incident solar irradiation. We have found big differences in the calculated performance ratio between both type of pyranometers. The c-Si phoptovoltaic-pyranometer detects about 3 to 5% less energy than that of MS-602 thermopile-pyranometer. This difference seems to be due to the spectral and the incidence angle limitation of c-Si-based sensor.

ACKNOWLEDGEMENT

We would acknowledge Mr. Martín Jiménez and Mr. Miguel Galván for their kind support to overcome logistics and to facilitate measurements.

REFERENCES

[1] PV activities in Japan, and global PV highlights Volume 23, No. 1, January 2017, RTS Japan.

[2] Solar power in Mexico https://en.wikipedia.org/wiki/Solar _power_in_Mexico

[3] CREARA; Energy Experts, PV GRID PARITY MONITOR Residential Sector 3rd issue *February 2015*

[4]. PV activities in Japan, and global PV highlights Volume 27, No. 5, May 2017, RTS Japan.

[5] http://www.pv-tech.org/news/mexico-to-hit-30-40gw-of-solar-pv-by-2040-iea. By Tom Kenning, Oct 27, 2016

[6] 2013 ISES Solar World Congress, "One-year 60 kWp photovoltaic system energy performance at CINVESTAV, Mexico City", to be published in Elsevier Energy Procedia.

[7] Photovoltaic pyranometer: https://en.wikipedia.org/wiki/Pyranometer

[8] http://www.southern-energy.com/files/43/64155.pdf

[9] B. Marion, J. Adelstein, K. Boyle, H. Hayden, et.al. February 2005, NREL/CP-520-37358 "Performance Parameters for Grid-Connected PV Systems", *31-IEEE PVSC, Lake Buena Vista, Florida, January 3-7, 2005.*

[10] Anton Driesse, Daniela Dirnberger, Christian Reise, Nils Reich, "Spectrally Selective Sensors for PV System Performance Monitoring", Fraunhofer ISE, Freiburg, Germany 38th IEEE PVSC Austin Texas (2012) 3294-3299

[11] Yasuhiro Matsumoto, Mauro Valdés, J. Antonio Urbano, Tomonao Kobayashi, Gabriela López and Ramón Peña, "Global solar irradiation in north Mexico city and some comparisons with the south", Energy Procedia (2014) 57, 1179 – 1188

[12] A. Guerin de Montgareuil, "A new accurate method for outdoor calibration of field pyranometers", 19th EUPVSEC, Paris France (2004).

[13] T.Betts, M. Bliss, R.Gottshlg and D. Infield, "Consideration of error sources for outdoor performance testing of photovoltaic modules", 20th EUPVSEC, Barcelona, Spain (2005) 2127-2130.

[14] A. Spena, C Cornaro, G. Inteccialagli and D. Chianese, "Data validation and uncertainty evaluation of the ester outdoor facility for tesitng of photovoltaic modules", 24th EUPVSEC, Hamburg, Germany, (2009)

[15] Hadis Moradi, Amir Abtahi and Roger Messenger, "Annual Performance Comparison Between Tracking and Fixed Photovoltaic Arrays" IEEE 43rd Photovoltaic Specialists Conference (PVSC), (2016), Portland, OR, USA. 978-1-5090-2724-8/16/$31.00 ©2016 IEEE

Financial Analysis of a Grid-connected Photovoltaic System in South Florida

Hadis Moradi, Amir Abtahi, and Ali Zilouchian

Florida Atlantic University, Boca Raton, FL, 33431, USA

Abstract —In this paper the performance and financial analysis of a grid-connected photovoltaic system installed at Florida Atlantic University (FAU) is evaluated. The power plant has the capacity of 14.8 kW and has been under operation since August 2014. This solar PV system is composed of two 7.4 kW sub-arrays, one fixed and one with single axis tracking. First, an overview of the system followed by local weather characteristics in Boca Raton, Florida is presented.

In addition, monthly averaged daily solar radiation in Boca Raton as well as system AC are calculated utilizing the PVwatts simulation calculator. Inputs such as module and inverter specifications are applied to the System Advisor Model (SAM) to design and optimize the system. Finally, the estimated local load demand as well as simulation results are extracted and analyzed.

Index Terms— Financial Analysis, Photovoltaic, System Advisor Model, Grid-tied Solar Unit.

I. INTRODUCTION

While concerns over fossil fuel depletion were the primary drivers for renewable energy development in the latter part of the 20th Century, climate change has been the main impetus behind wind and solar use and propagation in the last 2 decades. The steep decline in the cost of large wind generators and the more dramatic drop in the cost of PV and the associated components such as inverters and power conditioners, have made these power sources competitive in numerous countries. Advances in power electronics industry have also contributed to more efficient and reliable integration of these renewable resources into electric power grids [1-3]. The electric energy industry restructuring and the introduction of the concept of a smart grid has also led to new technologies such as distributed energy resources (DER) and distributed generation (DG) become more widely utilized. Generally, DG units could be defined as a local generation units that can be directly connected to the distribution utility grid [4-8].

PV electric power generation is economically feasible and environmentally sustainable, while requiring a relatively low maintenance for its operation. PV systems are often marketed and described in terms of the DC power rating of their modules, expressed in $'s/Watt. However, the value of a grid-tied solar system is a function of the energy generated, expressed in the scale of kWh [9,10]. The PV systems can be operated in both stand-alone and grid-connected modes.

A stand–alone PV system is an autonomous system that can operate without any connection to the grid and satisfy the design load. These systems are ideal when the grid is either not available or the cost of electricity is too high. Grid-tie and hybrid systems make up the majority of PV systems currently installed and operating. Hybrid systems allow the addition of batteries for either back-up in case of utility failure, or for demand control such as peak-shaving or utilities offer Time of Use (TOU) rates.

The performance ratio (PR), often called Quality Factor (QF), is independent from the irradiation, and mostly used to compare PV system performances. The intermittent nature of solar irradiation, especially in semi-tropical regions such as Florida, does not allow any PV operator to offer a power generation guarantee. However, it's possible to predict the energy generated with a high level of certainty based on reliable solar irradiation data [11].

In Florida, a net-metering mechanism was approved in 2007 for renewable energy systems under 500 kW capacities. It allows the users to feed a portion of the electricity generated into the grid and to receive credit per kWh supplied at the same rate as the utility kWh charges. Since 2012, net metering has also been available for multi-family housing. Based on different utility companies such as FPL [20] instructions, various types of renewable energy systems such as solar energy, wind energy, biomass, ocean energy, waste heat, hydroelectric power and geothermal energy are potentially eligible for net metering.

Each tenant will pay the difference between its individual consumption and the specific PV-generated-electricity; this difference is allocated to the electric utility company (CFE) to that tenant's utility account, according to a pre-arranged share. The PV Levelized cost of energy (LCOE) has experienced a significant decrease from 2009 to 2014, which is estimated at -18.4% compound annual growth rate, even though, for the average electricity consumer PV investment is still not competitive with grid electricity prices [12].

For the PV distributed generation (PVDG), there have been numerous studies to achieve the optimum allocation of the system. As mentioned, the optimal site and size of DG reflects the maximum loss reduction and improvement in voltage profile of distribution system. Different methodologies have been developed to determine the optimum location and optimum size of the DG. These methodologies are either based on analytical tools or on optimization programming methods [13].

Photovoltaic serves as a fundamental source to harness solar energy. Accompanied with receding prices, solar leasing and other innovative financing methods PV market is spreading widely. As per statistics specified in Renewables-2013 Global Status Report by Renewable Energy Policy Network for 21st

Century, the PV industry hit a 100 GigaWatt power production in 2012 [14]. Also International Renewable Energy Agency (IRENA) estimates the global weighted average LCOE of solar PV could fall by 59 percent by 2025 from 2015 [21], which makes it more competitive compared with conventional energy solutions. For the engineering project of a large-scale PV system, the economic analysis should be performed to evaluate the profitability of the PV system to ensure the investment cost can be recovered over the life cycle. It is concluded that the main factors affecting the PV system deployment are the initial capital cost of the system, the feed-in tariff and the PV system capital cost subsidization rate [15].

The System Advisor Model (SAM) is software used in renewable energy project analysis that integrates a detailed system performance model with a financial model and cost analysis [16]. The aim of this work is to study the financial performance of a 14.8 kW solar PV system installed at FAU campus in SAM and analyze the cost parameters and obtained simulation results. The present research, evaluates the generated electricity from the 14.8kw PV system, taking in consideration the local weather conditions. The PV module temperature, wind velocity and the solar irradiations are the main parameters for PV system performance evaluation.

II. SYSTEM OVERVIEW

The performance of a solar PV unit located at FAU in Boca Raton, Florida is studied and evaluated. This photovoltaic system consists of one fixed solar array and one tracker array. Each solar array has 12 modules in series and 2 strings in parallel. The tracking array has been designed to be installed on a North-South axis and tracking from East to West throughout the day with a broad turn range of 90°. The south facing fixed array has been designed on an East-West axis with the azimuth of 180° and tilt angle of 23° for "near optimum" annual generation [17]. The configuration of each installed array is shown in Fig .1.

Fig. 1. The configuration of one array installed at FAU

III. LOCAL METROLOGICAL DATA

Solar irradiation is one of the major parameters in designing solar PV systems. Also, in order to obtain the maximum power output from the PV arrays, the tilt angle must be set up correctly. Those parameters help to design the system based on the lowest solar irradiation of the year, which is going to be in the winter for south Florida location. The PVWatts software tools [22] is employed to calculate the solar irradiation based on the Typical Metrological Year (TMY) as the weather database for the location. Table. 1 shows the estimated solar radiation in (KWh/m²/day), and AC Energy in (KWh) for Boca Raton, Florida, USA.

TABLE I

MONTHLY AVERAGED DAILY SOLAR RADIATION AND SYSTEM'S AC ENERGY

Month	Solar Radiation (KWh/ /day)	AC Energy (KWh)
January	4.11	1,463
February	4.90	1,589
March	5.78	2,010
April	4.55	1,556
May	5.24	1,820
June	4.29	1,440
July	5.98	2,033
August	4.99	1,736
September	4.61	1,535
October	3.69	1,304
November	5.01	1,700
December	4.48	1,590
Annual	*4.80*	*19,776*

By selecting the location of the project, the required data such as global irradiance, diffuse irradiance, wind speed, relative humidity, snow depth and dry bulb temperature are obtained in the model. Some of the station specifications are provided in Table 2. Also solar global irradiance (W/m²) in 2014 when the system was installed at Boca Raton, FL is shown in Fig. 2.

TABLE II

LOCAL WEATHER DATA

Weather identification	Value
Latitude	26.37 °N
Longitude	-80.1 °E
Elevation	7.3 m
Global horizontal	5.26 kWh/m²/day
Direct normal (beam)	5.64 kWh/m²/day
Diffuse horizontal	1.68 kWh/m²/day
Average temperature	25.7 °C
Maximum snow depth	0 cm
Average wind speed	3 m/s

978-1-5090-5606-4/17 $31.00 © 2017 IEEE

Regarding the solar radiation curve in Fig .2, the solar irradiation in summertime is higher compared to the other seasons and the peak of solar potential occurs in May and minimum global irradiance happens in January.

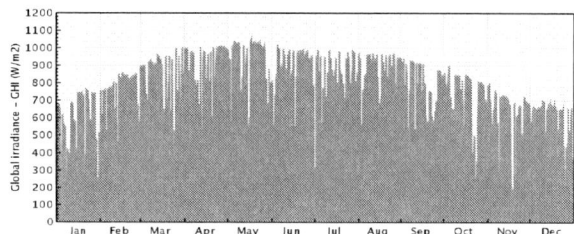

Fig. 2. Solar radiation curve in 2014, Boca Raton, FL

IV. MODULE SPECIFICATION

The solar panel used in the system is Trina TSM 285 PA14. It composed of multicrystalline 156mm by 156mm solar cells. Each panel has 72 (6 by 12) cells and module dimensions are 1956 × 992 × 46mm. The I-V curve of the module at Standard Test Condition (STC) with total irradiance of 1000 W/m² and cell temperature of 25 °C is shown in Fig .3 (right). According to the I-V curve the maximum power point is 285.31 Wdc. Also, the electrical specifications of the module are presented in Table 3.

TABLE III
MODULE ELECTRICAL DATA

Electrical specs	Value
Peak power watts-P_{max}	285 W
Power output tolerance	0/+3 %
Maximum power voltage	35.6 V
Maximum power current	8.02 A
Open circuit voltage-V_{oc}	44.7 V
Short circuit current	8.5 A
Module efficiency-η_m	14.7 %

V. INVERTER SPECIFICATION

In this section the employed inverter and its specifications are introduced. A Sunny Boy 7000-US inverters have been used for each solar array. The efficiency curve of the inverter is shown in Fig .3(left) and the technical data are presented in Table 4.

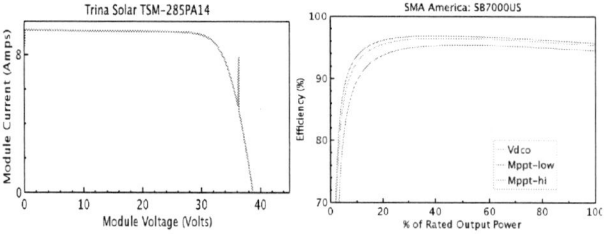

Fig. 3. (Right) Trina solar panel I-V curve at STC, (Left) Sunny Boy 7000-US efficiency curve

TABLE IV
INVERTER TECHNICAL DATA

Technical specs	Value
Max usable PV power at STC	8750 W
Max. DC power (@cos φ=1)	7400 W
Max. DC voltage	600 V
DC nominated voltage	310 V
MPP voltage rate	250-480 V
AC nominal power	7000 W
Nominal AC voltage	208/240/277 V
Max output current	34/29/25 A
Efficiency	97%

VI. SHADING AND LOSSES

In this section the shading, snow and possible losses, which decrease the system performance efficiency, are described. Various losses that system might experience due to the environmental conditions have significant effects on the system proper operation. Shading losses come from trees or buildings nearby or even due to self-shading from its own solar system structure and back-to-back PV rows. This system has no shading effects. Also, due to the weather condition in Boca Raton, the possibility of snow is zero, thus snow losses won't be applied to the system modeling. Other losses such as monthly soiling losses, module mismatch, diodes and connections and DC wiring are assumed 5%, 2%, 0.5% and 2% respectively. Thus the total DC power loss of 4.44% is considered as the inputs of the modeling. Also, AC wiring losses which comes from the electrical output of the inverter is assumed 1%. System performance degradation is assumed 0.5% per year, which means that total annual output of the PV system in financial modeling calculation will be degraded by half a present.

VII. SYSTEM COSTS AND FINANCIAL PARAMETERS

In this section the direct and indirect capital costs as well as operation and maintenance cost of the system is calculated and presented in Tables 5 and 6.

TABLE V
DIRECT CAPITAL COSTS

	Unit	kWdc/kWac per unit	$/Wdc, $/Wac	Cost
Module	48	0.3	1.0	$ 13,695.26
Inverter	2	7	0.21	$ 2,876.01

	$/Wdc	Cost
Balance of system equipment	0.36	$ 4,930.29
Installer margin and overhead	1.25	$ 17,119.08
Installation labor	0.30	$ 4,108.58

TABLE VI
INDIRECT CAPITAL COSTS

	$/Wdc	Cost
Permitting and environmental studies	0.1	$ 1,369.53

	%	Cost
Sale tax rate (present of direct cost)	6.0	$ 1,110.96

Based on the calculations the total direct costs, indirect costs and installed costs are $ 42,729.22, $ 2,480.49 and $ 45,209.71 respectively. Also, total installed cost per capacity is $ 3.30/Wdc. Also it can be assumed some fixed costs as operation and maintenance costs. In this study 25$/kW-yr is assumed as fixed cost by capacity and $1500 is assumed as a fixed annual cost for inverter replacements by considering manufacture warranty after 10 years.

Also in terms of financial parameters it is assumed that the installation has been done using standard loan with the inflation rate of 2% per year and Florida state tax rate of 6%. In addition, the analysis period is assumed 25 years. As incentives parameters, tax credits and direct cash incentives are considered. It is assumed that federal Investment Tax Credit (ITC) is 30% [18].

VIII. ELECTRICITY RATES

Various electricity rate and metering structures can be employed in SAM such as TOU rates, tiered rates, demand charges and TOU tiered rates. The rates for energy charges for are available in [19]. Florida Power and Light TOU tiered energy rates for weekdays are shown in Fig 4 and Table 7. The weekends are also placed in period one. It is assumed that monthly total excess rolled over to next month bill in kWh.

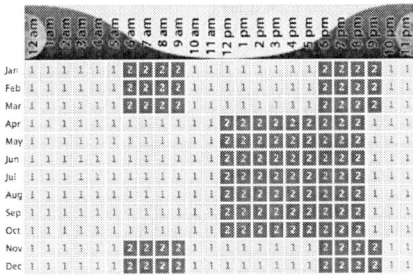

Fig. 4. FPL different periods of energy rates

TABLE VII
RATES FOR ENERGY CHARGES

Period	Tier	Max. Usage	Max. Usage units	Buy ($/kWh)
1	1	1000	kWh	0.18491
1	2	1e+38	kWh	0.06059
2	1	1000	kWh	0.17525
2	2	1e+38	kWh	0.06635

IX. LOAD DATA

The system performance is studied under a typical load demand using load data estimator. A simple residential load model is employed in this study. We assume a building with the floor area of 4,000 sq. ft has been built in 1985. Also the cooling and heating set points is considered 64 and 72 °F. The studied hourly electric load is shown in Fig. 5. It can be seen from the load profile that the demand peak occurs in summer time when the cooling demand is high in Florida.

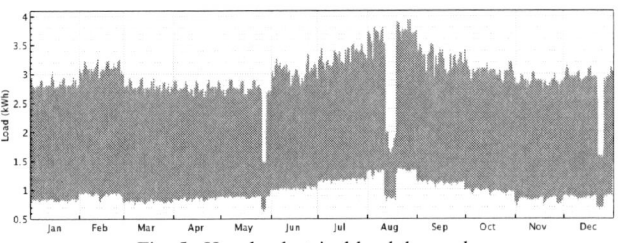

Fig. 5. Hourly electrical load demand

X. FINANCIAL MODEL RESULTS

The model's graphical and numerical results are shown in Fig. 6 and Table 8. It is concluded from the Fig .6(a) that the system has its maximum production in springtime when the solar irradiation is high. Fig .6(c) and 6(d) show how PV production is higher in May compared to November. Also, Fig .6(b) shows the system energy generation per year, which is decreasing over the time due to the system degradation.

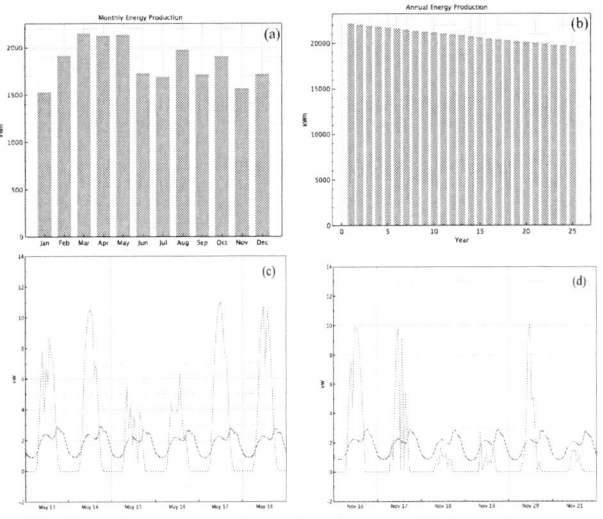

Fig. 6. (a) System monthly produced energy (b) Annual system energy generation (c) load demand vs. PV production sample in May (d) load demand vs. PV production sample in November

TABLE VIII
NUMERICAL RESULTS

Metric	Value
Annual energy (year1)	22,155 kWh
Capacity factor (year1)	18.5%
Energy yield (year1)	1,618 kWh/kW
Performance ratio (year1)	0.77
Eclectic bill without system (year1)	$2,637
Eclectic bill with system (year1)	$112
Net saving with system (year1)	$2,525
Payback period	13.9 years
Net Capital cost	$45,210

XX. CONCLUSION

In this paper, design, implementation and financial analysis of a grid-tied PV unit were carried out to fulfill the local load demand based on corresponding meteorological parameters. The 14.8 kW system comprised of 48 modules including fixed and tracker arrays connected to two inverters to support the load at FAU campus. The simulation results showed that although installation cost is relatively high, the system would operate within disbursement period after payback time. Cost savings on electric bill would be 95.7% in the first year of operation. Additionally, the data analysis demonstrated that the unit power output is maximized in May that is around 30% more than minimum generated power in January in south Florida. To improve the system performance in terms of technical and financial applications by smoothing or shifting the profile of energy output, a set of battery storage can be integrated to existing PV system as a potential future work.

REFERENCES

[1] A. Park, P. Lappas, "Evaluating demand charge reduction for commercial-scale solar PV coupled with battery storage," *Renewable Energy*, vol. 108, pp. 523-532, August 2017.

[2] M. H. Athari and Z. Wang, "Modeling the uncertainties in renewable generation and smart grid loads for the study of the grid vulnerability," *2016 IEEE Power & Energy Society Innovative Smart Grid Technologies Conference (ISGT)*, pp. 1–5, 2016.

[3] T. Arai and S. Wakao, "Computational analysis of battery operation in photovoltaic systems with varying charging and discharging rates," *2016 IEEE 43rd Photovoltaic Specialists Conference (PVSC)*, pp. 1773-1779, Portland, OR, 2016.

[4] A. S. Mobarakeh, A. Rajabi-Ghahnavieh and A. Zahedian, "A game theoretic framework for DG optimal contract pricing," *IEEE PES ISGT Europe 2013*, pp. 1-5, Lyngby, 2013.

[5] H. Khazaei, B. Vahidi, H. S. Hosseinian, and H. Rastegar, "Two-level decision-making model for a distribution company in day-ahead market," *IET Gen. Transm. Distrib.*, vol. 9, no. 12, pp. 1308–1315, Sep. 2015.

[6] A. Shahsavari, A. Fereidunian, and S. M. Mazhari, "A joint automatic and manual switch placement within distribution systems considering operational probabilities of control

sequences," *International Transactions on Electrical Energy Systems*, 2014.

[7] S. Najafi, M. Shafie-khah, N. Hajibandeh, G. J. Osório, J. PS. Catalão, "A New DG Planning Approach to Maximize Renewable-Based DG Penetration Level and Minimize Annual Loss," *Doctoral Conference on Computing, Electrical and Industrial Systems*, vol. 499, pp. 269-276, March 2017.

[8] M. Farajollahi, M. Fotuhi-Firuzabad and A. Safdarian, "Impact of erroneous measurements on power system real-time security analysis," *2015 23rd Iranian Conference on Electrical Engineering*, pp. 1630-1635, Tehran, 2015.

[9] T. Bano and K. Rao, "Performance analysis of 1MW grid connected photovoltaic power plant in Jaipur, India, " *2016 International Conference on Energy Efficient Technologies for Sustainability (ICEETS)*, pp. 165-170, Nagercoil, 2016.

[10] C. P. Cameron, W. E. Boyson and D. M. Riley, "Comparison of PV system performance-model predictions with measured PV system performance," *2008 33rd IEEE Photovoltaic Specialists Conference*, San Diego, pp. 1-6. CA, USA, 2008.

[11] Y. Matsumoto, J. A. Urbano, O. I. Gómez, R. Asomoza, J. García and R. Peña, "Seasonal quality factor; 60 kWp PV system at north Mexico City," *2014 IEEE 40th Photovoltaic Specialist Conference (PVSC)*, pp. 1948-1952, Denver, CO, 2014.

[12] Y. Matsumoto, C. Norberto, J. A. Urbano, M. Ortega and R. Asomoza, "Three-year PV system performance in Mexico City," *2016 IEEE 43rd Photovoltaic Specialists Conference (PVSC)*, pp. 3168-3172, Portland, OR, 2016.

[13] H. Sadeghian, M. H. Athari and Z. Wang, "Optimized Solar Photovoltaic Generation in a Real Local Distribution Network, " *2017 IEEE Innovative Smart Grid Technologies Conference (ISGT)*, 2017.

[14] Renewables-2013 Global Status Report by Renewable Energy Policy Network for 21st Century.

[15] C. H. Lin, W. L. Hsieh, C. S. Chen, C. T. Hsu, T. T. Ku and C. T. Tsai, "Financial analysis of a large-scale photovoltaic system and its impact on distribution feeders," in *IEEE Transactions on Industry Applications*, vol. 47, no. 4, pp. 1884-1891, July-Aug. 2011.

[16] N. J. Blair, P. Gilman and A. P. Dobos "Comparsion of photovoltaic models in the system advisor model," *National Renewable Energy Laboratory*.

[17] H. Moradi, A. Abtahi and R. Messenger, "Annual performance comparison between tracking and fixed photovoltaic arrays, " *2016 IEEE 43rd Photovoltaic Specialists Conference (PVSC)*, pp. 3179-3183, Portland, OR, 2016.

[18] http://www.dsireusa.org/

[19] http://en.openei.org/wiki/Utility_Rate_Database

[20] https://www.fpl.com/clean-energy/net-metering.html

[21] Solar and Wind Cost Reduction Potential to 2025 by IREA available at http://www.irena.org/DocumentDownloads/Publications/IRENA _Power_to_Change_2016.pdf

[22] PVWatts calculator by National Renewable Energy Laboratory available at http://pvwatts.nrel.gov/

Study of Photovoltaic Systems Monitoring Methods

E. Ortega [1], G. Aranguren [1], M.J. Sáenz[2], R. Gutiérrez[2] and J.C. Jimeno [2]

[1] Electronic Design Group. University of the Basque Country. Alameda Urquijo S/N, 48013-Bilbao, Bizkaia, Spain.

[2] Technological Institute of Microelectronics. University of the Basque Country. Alameda Urquijo S/N, 48013-Bilbao, Bizkaia, Spain.

Abstract — To maximize photovoltaic systems performance the monitoring of them is essential. There are several faults that produce energy losses in the system. When a fault happens the monitoring system has to detect it and give an indication about which is the component with the fault. For this, in most system a set of parameters are measured, usually four: current, voltage, temperature and irradiance. In some monitoring systems these measures are obtained for the whole plant whereas in others, the modules are measured one by one. In this work, a study of these methods is presented.

Index Terms — photovoltaic systems, performance ratio, condition monitoring

I. INTRODUCTION

Due to a worldwide increasing energy demand and the depletion of non renewable energy sources, renewable energy sources have been growing sharply during last decades. The renewable energy source which has grown more is solar energy. The installed capacity of photovoltaic (PV) power is increasing approximately 50GW per year [1] with a market size in 2014 of around 100 billions US dolars. For the year 2030 photovoltaic energy is expected to cover the 13% of the global energy demand.

The growth of the PV market has been accompanied with a reduction of the cost of the modules of more than the 70% on the last 8 years [2] and the increase in the PV modules efficiency.

PV systems performance has been growing in the same way, from a 50% on 1980 [3] until the current systems, usually with different types of monitoring systems, with a performance around 80% [4]. This means that, nowadays, PV systems have energy losses of about the 20% of the maximun output, around 20.000 milion US dolar per year.

However, full performance is still not possible due to several causes such as encapsulation failures, cells cracking, dust, snow or soiling accumulation or growth of the vegetation surrounding the PV modules. Faults that if detected, are easily fixable. Most of these failures could be avoided with appropriate monitoring systems that detect these faults in a fast and precise way.

Within all the PV systems performance measure indicators, the most used are final yield (Y_F), reference yield (Y_R) and performance ratio (PR) [3].

For a defined period of time, Y_F measures the energy delivered to the load divided by the rated output power.

$$Y_F = E / P_O \qquad (1)$$

Y_R represents the value of the hypothetically available energy on the field measured on a defined period of time and kW. This indicator is the ratio between the total solar isolation and the reference irradiance.

$$Y_R = H_t / G \qquad (2)$$

Finally, PR is the ratio between Y_F and Y_R, i.e. represents from the theoretically available energy the energy used. This ratio normalizes PV system performance to the received solar radiation, giving information about the total effect of losses in the PV plant.

$$PR = Y_F / Y_R \qquad (3)$$

There are significant differences in the PR of PV systems in function of the solar cells efficiency, the module temperature dependence and energy losses due to malfunctions, shading or dust among others. Huang et al. [4] in 2011 measured 202 PV systems in Taiwan to collect data of operational performance of PV systems, obtaining a PR of 74%. Matsumoto et al. [5] evaluated in 2016 the electricity generated from a 60kWp PV system in Mexico City obtaining a PR of 86.8%. Both measures were during a three years period. With this results PR could be used as an indicator of PV systems monitoring quality.

There are several failure causes that can be sorted out if they are detected. Because of that, monitoring is the most effective way to maximize the PR, minimizing the time until the fault is detected and it is fixed.

In this work, some of the most common PV faults are introduced in section II, most of them correctable if they are detected, the most used PV systems monitoring techniques main advantages and disadvantages are studied in section III and finally the need of developing a new single module monitoring system is exposed in section IV.

978-1-5090-5606-4/17 $31.00 © 2017 IEEE

II. PV MODULE FAILURE CAUSES

Failures in PV modules, which imply a loss of productivity, may be caused by several reasons, such as:

- Corrosion failures [6]. Corrosion has a direct influence in electrical performance of the PV module. Conductive parts corrosion generates an increment of the ohmic resistance, which implies a decrease of the module voltage and a decrease of the shunt resistance, which generate bigger current losses.
- Cell cracking failures [7]. Due to different causes during module lifetime cracking of cells happen with a rate of 1% per year. A cell crack causes open circuit on between the 1% and 10% of the cells.
- Hot-spots [8]. Caused by the increase of temperature on a cell or group of cells because their current capacity is lower than the current of the string. If this situation is prolonged during large periods of time, it contributes to the degradation of the PV module.
- Encapsulation failures [9]. The main cause for encapsulation failures is discoloration which makes that part of the incoming solar radiation is blocked.
- Electrical or mechanical interconnections [10]. Due to mechanical stress or repeated thermal expansion and contraction the interconnections between the cells or between the modules can be broken.
- Dust [11]. The accumulation of dust on PV modules can reduce the output power of the system on more than a 20%. Different works, estimated the effect of dust in PR in a decrease of up to the 50% in some environments.
- Partial shadows due to growth of vegetation [12]. If they are not detected growth of vegetation can generate permanent shadows on PV modules reducing output power.
- Potential Induced Degradation [13]. This effect appears on the modules with higher voltage if they are on negative potential against ground. This happens in strings with transformerless inverters. This effect is caused by modules surface humidity, which becomes the module surface electrically conductive.

The failure causes listed above imply a decrease in the PR of the PV systems and in addition, some of them speed up the PV module degradation, shortening the PV modules lifespan.

Furthermore, some of these failures are fixable if they are detected quickly but if the failure persists during a large period of time the effects of the failures on the module can become permanent and the module.

Other failures, such as accumulation of dust or partial shadows suppose considerable energy losses but are very easily correctable if detected. For these reasons, having an efficient monitoring system is vital in PV systems.

III. MONITORING SYSTEMS

There are several parameters which could be useful to measure the PV system performance and to monitor the systems but usually 4 main parameters are used. These parameters, which are common for all systems, 2 electrical, current and voltage, and another 2 environmental, temperature and solar irradiance. These parameters can be measured either at PV plant level, string level or PV module level (Fig. 1).

With these measures, three main goals must be achieved for optimal PV system monitoring. On the one hand, to estimate the PV modules ageing, secondly, to measure the PV system performance and in addition, to detect in a fast and reliable way the faults that the system may be suffering.

Many methods had been proposed to measure these parameters at plant, string or array level, combining a PV system (or string or array) model with temperature and irradiance measures. Some of the more remarkable ones are the method proposed by Drews et al. [14], which obtained these measures from satellite data. The aim of these approaches is to detect as soon as possible PV system faults using an automated failure detection routine, which searched for the cause of the PV system failure. Although these systems are able of detecting some energy losses due to faults, was not able to detect relatively small energy losses. The proposed system was not always able to detect a string error of 17%, only on summer, which implies that the time between the fault happens and it is detected can be very long, with the subsequent decrease on PR.

Other method, proposed by Chouder et al. [15] used a similar method but adding temperature and irradiance sensors instead of using satellite data. Using a PV array model they predicted the array output for the temperature and irradiance data and developed a system which measured PV system losses and was able to determine the fault type.

Other plant level methods [16] are based in statistical analysis of the PV system performance. Taking into account the natural PV modules degradation, through statistical analysis the expected output power of the plant is predicted and measuring the real output, failures could be detected.

At string level, different methods are used such as the method proposed by Firth et al. [17]. In their study 27 domestic single string PV systems performance was evaluated measuring modules irradiance and temperature and inverter input and output energy. The overall energy losses caused by faults varied from 3.6% to 18.9% in different PV systems. With this system measuring at string level, it was possible to detect energy losses due to the shading from a single tree, which were of 6.9%, being able to detect energy losses much more smaller than system level monitoring systems. Even the precision was much bigger than the system level monitoring methods the PR was still around 80%, with energy losses around 20%.

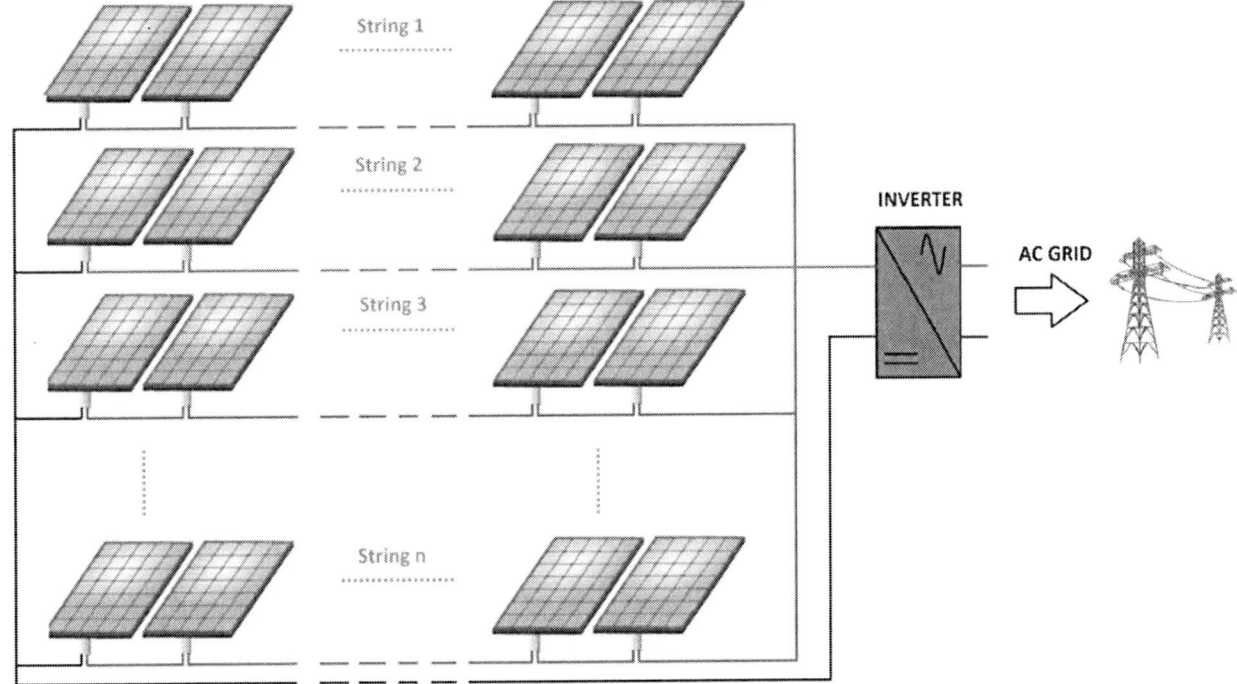

Fig. 1. Basic PV system block diagram.

These system, string or array level monitoring methods have the advantage that are cheaper methods than the module level monitoring methods but in addition to the lack of precision, they have other important disadvantage: Even if the fault is detected, there is not a clear indication of which is the module or the group of modules with the failure. This implies a loss of time between the failure detection and it solution, which decreases the PR.

The module level monitoring methods, on the other hand, are more precise methods which measure individual PV modules such as Hirata el al.[18] who proposed a method for PV systems failure detection based on single I-V curve measuring, With these methods, each PV module is fully characterized. Any small fault can be detected, but it is necessary to disconnect the module from the rest of the system before measuring it. Because of that, while the PV module is being measured it is not producing energy, which implies a decrease on PR.

Sera et al. [19] proposed also a module level method based on the measure of the series resistance, the temperature and the output power.

These methods are able to measure individual modules but present also some disadvantages. These monitoring devices are connected in the junction box in series to the module (Fig.2) so all the current of the string goes through the monitoring devices and they work at the module voltage.

With this structure, the monitoring device works at high power which makes more expensive the electronic components needed for the device and also increases the risk of failure of the device. The major problem is that, with this monitoring method, a failure on the monitoring device can put all the string in open circuit.

Many authors had proposed the use of sensor networks (wired and wireless) for PV systems monitoring deploying several sensors across the field for temperature, irradiance, voltage and current measuring [20] [21]. In these works several sensors for temperature and irradiance measures are deployed on a network. With this information and the current and voltage measures, usually measuring string DC powers or inverters output it is possible to estimate the energy losses due to faults.

In other works [22] the use of thermal imaging and unmanned aerial vehicles has been proposed. It has proven to be an useful method for visual faults detection but are not reliable enough as the only monitoring method.

IV. CONCLUSIONS

There is a consensus in that PV systems monitoring is the best way to maximize PV systems performance, with a direct effect in PR and availability of PV plants. However, each monitoring system affects in a different way to PV system performance. Based on previous works results, although

978-1-5090-5606-4/17 $31.00 © 2017 IEEE

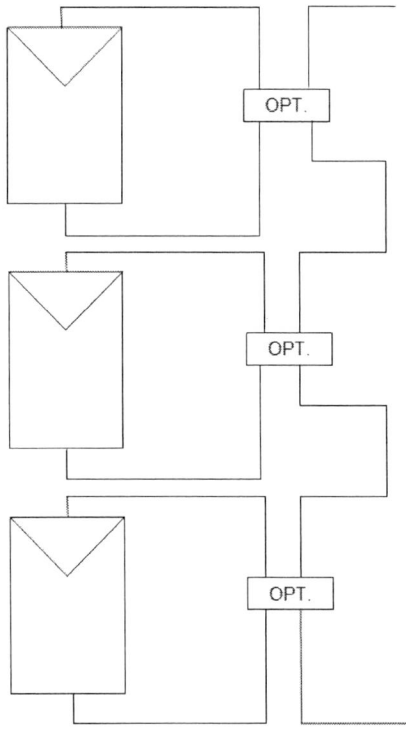

Fig. 2. PV Module level monitoring system.

system level monitoring systems are cheaper, are unable of detecting small energy losses and even if a bigger fault is detected, are not capable of giving a clear indication of which is the module failing. In conclusion, the lower the monitoring level is, higher the monitoring system resolution is and easier to detect small energy losses due to faults and to identify the failing components.

Ideally, to maximize PR, all the PV systems should be monitored at module level and this way, even the smaller energy losses will be detected in a fast and precise way, with a clear indication of which is the module or group of modules failing and the cause of it. For this, various disadvantages need to be solved.

The PV modules need to be measured without disconnecting them from the rest of the system, so the monitoring process can be done in an automatic way and without affecting to PV modules availability.

A failure on the monitoring device can not generate a failure on the PV module or string.

The cost of the monitoring system for each PV module must be smaller than the cost of the energy losses.

Related with the cost, the monitoring system needs to be low-power so the cost of the components and the power consumption is minimum.

In conclusion, with the current economically viable monitoring devices there is around a 20% of energy losses on the PV systems. To reduce energy losses and make PV energy even a more profitable and efficient energy source it is necessary a new monitoring system able to detect individual failures in PV modules in an automatic, cheap and efficient way.

ACKNOWLEDGEMENTS

This work received financial support from the Basque Government through the grant PRE_2016_1_0016.

REFERENCES

[1] S. Kurtz, N. Haegel, R. Sinton y R. Margolis, «A new era for solar,» *Nature photonics,* vol. 11, pp. 3-5, 2017.

[2] J. Weiner, «U:S: Distributed Solar Prices Fell 10 to 20 Percent in 0214, with trends continuing into 2015» *Berkeley Lab.*

[3] V. Sharma and S. Chandel, "Performance and degradation analysis for long term reliability of solar photovoltaic systems: A review," *Renewable and Sustainable Energy Reviews,* vol. 27, pp. 753-767, 2013.

[4] H. S. Huang, J. C. Jao, K. L. Yen and C. T. Tsai, "Performance and Availability Analyses of PV Generation Systems in Taiwan," *Interantional Journal of Electrical, Computer, Energetic, Electronic and Communication Engineering,* vol. 5, no. 6, pp. 731-735, 2011.

[5] Y. Matsumoto, C. Norberto, J. A. Urbano, M. Ortega and R. Asomoza, "Three-year PV system performance in Mexico City," in *Photovoltaic Specialists Conference (PVSC). 2016 IEEE 43rd,* 2016.

[6] C. Loredana, Faifer, M., M. Lazzaroni, M. M. Abdel Fattah Khalil, M. Catelani and L. Ciani, "Diagnostic architecture: A procedure based on the analysis of the failure applied to photovoltaic plants," *Elsevier Measuremet,* vol. 67, pp. 99-107, 2015.

[7] S. Kajari-Schröder, I. Kunze y M. Kšntges, «Criticality of Cracks in PV Modules,» *Proceedings of the 2nd International Conference on Crystalline Silicon Photovoltaics SiliconPV658-663,* vol. 27, pp. 658-663, 2012.

[8] K. A. Kim and P. T. Krein, "Hot spotting and second breakdown effects on reverse I-V characteristics for mono-crystalline Si photovoltaics," in *Energy Conversion Congress,* 2013.

[9] N. C. Park, J. S. Jeong, B. L. Kang and D. H. Kim, "The effect of encapsulant discoloration and delamination on the electrical characteristics of photovoltaic module," in *European Symposium on Reliablility of Electron Devices,* 2013.

[10] G. B. Alers, Solar Photovoltaic Module Failure Analysis, Microelectronics Failure Analysis: Desk reference, 2011.

[11] M. Catelani, L. Ciani, L. Cristaldi, M. Faifer, M. Lazzaroni y M. Rossi, «Characterization of photovoltaic panels: The effects of dust,» de *2nd IEEE Energycon. Conference & Exhibition. Advances in Energy Conversion Symp*, 2012.

[12] M. Seyedmahmoudian, S. Mekhilef, R. Rahmani, R. Yusof and E. T. Renani, "Analytical Modeling of Partially Shaded Photovoltaic Systems," *MDPI Energies,* vol. 6, no. 1, pp. 128-144, 2013.

[13] M. Oprea, S. Spataru, D. Sera, P. B. Poulsen, S. Thorsteinsson, R. Basu y A. R. Andersen, «Detection of potential induced degradation in c-Si PV panels using electrical impedance spectroscopy,» de *43rd Photovoltaic Specialists Conference (PVSC)*, 2016.

[14] A. Drews, A. C. Keizer, H. G. Beyer, E. Lorenz, J. Betcke, W. G. van Sark, W. Heydenreich, E. Wiemken, S. Stettler, P. Toggweiler, S. Bofinger, M. Schneider, G. heilscher y D. Heinemann, «Monitoring and remote failure detection of grid-connected PV systems based on satellite observations,» *Solar Energy,* vol. 81, pp. 548-564, 2007.

[15] A. Chouder y S. Silvestre, «Automatic supervision and fault detection of PV systems based on power losses analysis,» *Energy Conversion,* vol. 51, pp. 1929-1937, 2010.

[16] S. Vergura, G. Acciani, V. Amoruso, G. Patrono y F. Vacca, «Descriptive and Inferential Statistics for Supervising and Monitoring the Operation of PV Plants,» *IEEE Transactions on industrial electronics,* vol. 56, n° 11, pp. 4456-4464, 2009.

[17] S. K. Firth, K. J. Lomas y S. J. Rees, «A simple model of

PV system performance and its use in fault detection,» *Solar Energy,* vol. 84, pp. 624-635, 2010.

[18] Y. Hirata, S. Noro, T. Aoki and S. Shiwa, "Simple method to measureI-V curve of photovoltaic failures compared with other conventional methods," in *28th European Photovoltaic Solar Energy Conference and Exhibition*, Paris, 2013.

[19] D. Sera, R. Teodorescu y P. Rodriguez, «Photovoltaic module diagnostics by series resistance monitoring and temperature and rated power estimation,» de *Industrial Electronics, 2008. IECON 2008. 34th Annual Conference of IEEE*, 2008.

[20] G. Bayrak y M. Cebecí, «Monitoring A Grid Connected PV Power Generation System with Labview,» de *Int Conf on Renewable Energy Research and Applications*, Madrid, 2013.

[21] P. Papageorgas, D. Piromalis, K. Antonakouglou, G. Vokas, D. Tseles y K. G. Arvanitis, «Smart Solar Panels: In-situ monitoring of photovoltaic panels based on wired and wireless sensor networks,» de *TerraGreen 13 Int Conference*, 2013.

[22] F. A. M. M. e. a. Grimaccia, «Planning for PV plant performance monitoring by means of unmanned aerial systems (UAS),» *Int J Energy Environ Eng,* vol. 6, n° 1, pp. 47-54, 2015.

Global Design Aspects of Persistent and Autonomous PV Powered Systems

I. M. Peters[1], S. Watson[1], N. Sahraei[2], T. Buonassisi[1],

[1] Massachusetts Institute of Technology, Cambridge, MA 02141, United States of America

[2] Singapore MIT Alliance for Research and Technology, 1 CREATE Way, #10-01 CREATE Tower
Singapore 138602

Abstract — One of the major challenges for solar cell technology is that PV panels generate intermittent power, yet electrical appliances are designed for a fix power input. Addressing this challenge requires transforming the time-dependent output power of a PV panel into the required time-dependent input power of an appliance. This transformation is essential for the deployment of large amounts of solar power and can be accomplished, for example, by combining solar power with storage, complementary power sources, and/or by long-range grid connections. In this paper we look at the local variations in the design for a stand-alone solar/battery system with a constant power output. We find that the cost-optimum design of such a system depends mostly on the latitude, due to seasonal variations in the available solar resource. We also find that, with the used parameters, the ideal number of solar panels in the cost optimum design is between 1.5 and eight times larger than the number of solar panels necessary to deliver the integrated annual power needed.

Index Terms — PV systems, load management, global system design.

I. INTRODUCTION

Solar panels, like wind turbines, provide power depending on weather conditions and time of day. Due to variations in the weather, most prominently due to clouds, the output of a PV panel can oscillate rapidly over a day. Moreover, daily and seasonal variations cause oscillations with longer periods. Managing these intermittencies in electricity generation is one of the main challenges when integrating solar panels into an existing power infrastructure, be it a power grid or an electrical appliance, and has major implications, not least on the value of solar electricity [1]. Load matching and assessing and matching loads for PV power is, hence, a long-standing, yet very active field of research [2, 3].

In this study we take a general approach to assess what is needed to generate a stable and reliable electricity output using solar cells. The method used in this paper is sketched in figure 1. The oscillating output of a solar panel is used in conjunction with a storage medium, in this case a battery, to generate a constant power output that a consumer can use. This simple approach stands exemplarily for any method that allows achieving a constant output, be it by using a different storage medium, e.g. a fuel cell or pumped hydro power, a supplementing power source in the same or in another

location, like wind [4] or conventional power plants, or by integration into a grid that averages out generation and demand of different applications.

For any application, though, a PV system needs to perform this transformation in order to be useful. Designing a system that considers this transformation will, in general lead to different designs than a system that only considers average power generation. As more solar panels are installed, this principle design issues will become ever more pressing, and it marks a significant challenge, when considering the amount of solar panels that need to be installed to fulfill, for example, formulated climate goals [5,6].

The PV system considered in this study is a simple, localized PV system. It consists of solar panels with variable area and batteries of variable capacity. The design goal was that the system should produce a constant power output of 500W over the course of one year, and should do so for the smallest system price possible. The results of the presented calculation include the PV panel area, the battery capacity, and the total system capital costs needed to fulfill the two specified conditions for any given location in the world.

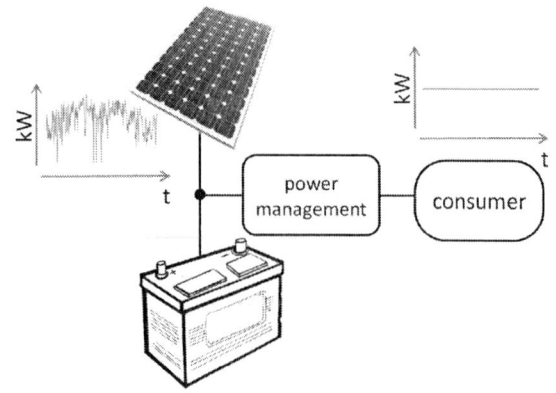

Fig. 1. Sketch of the problem addressed in this work. A solar panel delivers intermittent power with strong oscillations over time. Combining the solar panel with a battery, the power generation profile can be transformed into the time-dependent power requirement of a consumer. In the case discussed here, requirement of a constant power output is assumed.

978-1-5090-5606-4/17 $31.00 © 2017 IEEE

II. MODEL AND INPUT PARAMETERS

The main model input is a time series of daily average solar insolation throughout the year 2015 on a grid of 1 x 1 deg. This data was obtained from the NASA Clouds and Earth's Radiant Energy System (CERES) instrument [7]. A summary of this data is shown in figure 2. Insolation data and solar cell area are used to calculate power generation; generation and consumption are balanced, surplus is used to charge the battery, missing supply power is discharged from the battery. An optimization algorithm is used to find the lowest-priced system that provides a constant power output with 100% reliability. We assumed a price of 162 $/m² for solar panels with a 16% conversion efficiency, and 350 $/kWh for the batteries, crresponding to Li ion technology.

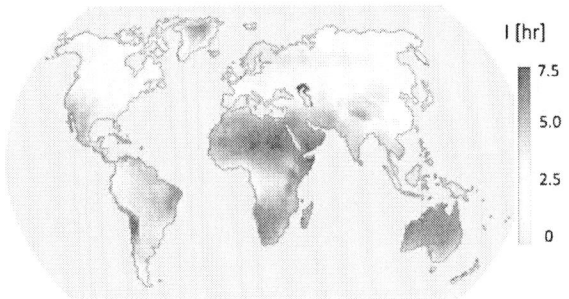

Fig. 2. Available solar resource. The figure shows the average insolation in units of sun hours per day for the year 2015. As the figure shows average values, seasonal variations are not visible. Seasonal variations provide an important contribution, and are increasingly strong the closer a location is to either of the poles. Note that data for Greenland is erroneous in this plot.

III   RESULTS & DISCUSSION

The primary results of the presented calculations are the battery capacity and the solar panel area required to fulfill the specified conditions. The obtained numbers are shown in figure 3. Battery capacity is given in units of days of autonomy, i.e. the number of days that the batteries would have to be able to supply the system all by themselves. The advantage of this unit is that it is independent of the actual load size. A strong dependence on latitude can be observed.

The smallest battery capacities are required in the tropics, but outside the equator. In these regions, insolation is high and consistent, without strong seasonal influence. The required battery capacity increases in more southern or, more noticeable in the plots, northern regions. Towards the poles, battery days of autonomy approach or even exceed values of 200 days – a feature caused by the arctic winter. A similar trend can be observed for the required solar panel area, although on a smaller scale. Whereas battery capacity varies over two orders of magnitude, solar panel area varies only over one.

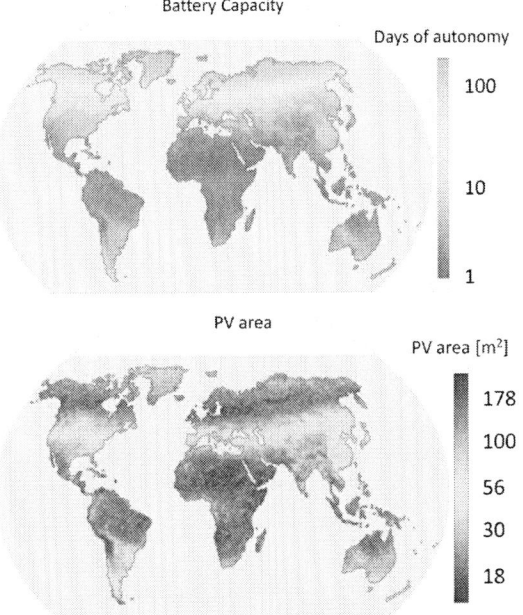

Fig. 3. Battery capacity and PV area for a cost optimum design and a constant power output of 500W, calculated for the year 2015. Battery capacity is given in units of days of autonomy, PV area in m².

The variation correlates strongly and inversely with the available solar resource; however, another component also plays a role: given the price ratio between solar cells and batteries, it can be advantageous to overdesign the system with respect to solar panel area, in order to save battery capacity. To illustrate this point, we also calculated a metric we called "PV area – multiplier", which is shown in figure 4. The values shown in this figure describe the ratio between the solar panel area in the cost-optimized system A_{opt} and the solar panel area required to fulfill the average, integrated power requirement over the course of the year A_{min}. The latter corresponds to the smallest possible solar panel area that would allow supplying the system with sufficient power continuously, if sufficient storage was available. It is worth noting that this value also corresponds to the solar panel area that is typically quoted when the necessary area to fulfill a certain power requirement is sought.

The main observation from this figure is that the minimum value of the defined multiplayer is about 1.4, meaning that, for the system costs specified here, the required solar panel area for continuous operation is at least 40% greater than the minimum area. Furthermore, values in the Tropics are below two, but the multiplier can be as high as eight in regions in the far North or South. The multiplier varies especially strong in Europe, North America and large parts of Russia. It is worth noting that the multiplier depends on the ratio between solar panel and battery price.

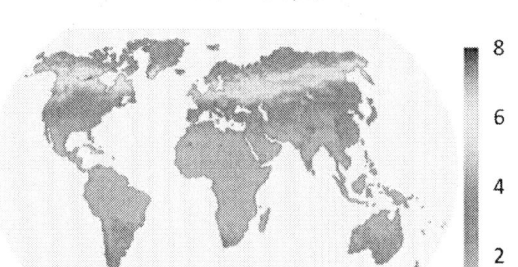

PV area – multiplier

Fig. 4. Required PV area given in units of the minimum area A_{min} needed to generate the required electricity on average over the course of a year. This factor gives an indication of how much PV systems are under-designed in size, if only average consumption is considered rather than a cost-optimized design that includes storage.

Finally, the total capital cost for the specified solar/battery system is shown in figure 5. With the given values, the total cost is mostly determined by the contribution from the batteries; solar panels add a much smaller contribution. Correspondingly, the total system cost varies over almost two orders of magnitude.

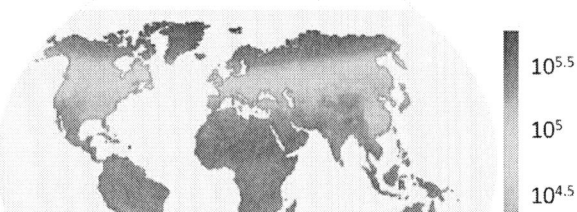

Total system cost

Fig. 5. Calculated total system cost for the system that fulfils the requirements. The cost is predominantly driven by the battery capacity.

Obviously, a power system with demand for constant power output at very high reliability and solar panels as its only power source is not practical in regions with long periods of little or no sunlight. But also in regions with moderate seasonal changes, such as North America, Europe and Northern Asia, a significant buffer is required to provide a constant power output with high reliability. In these regions, wind power provides a valuable option to complement solar power and reduce the required capacity for storage. Investigations how much an addition of wind can reduce the requirement for storage and the total system costs are ongoing, and have been the topic of numerous studies [8].

For systems in which electricity is not the final product, another possibility exists. Systems that use solar power to generate water - for example via desalination - or heat or cold - for example via air-conditioning - have the possibility to store the respective end product as an additional buffer. The two mentioned examples have the additional advantage that the demand is often correlated with the availability of sun light. With such options, the need for storage can be significantly reduced.

IV DISCUSSION

The objective of the presented study is to put into perspective how PV system design changes if integration into a wider context is taken into account. Cost reduction is the figure of merit used. The study presented here uses a rather simplistic model in which only isolated solar panels integrated with battery storage are considered. Synergy effects with wind or from extended grids are neglected, creating a scenario for island systems, which can be considered a worst case for an extended integration. While addition of wind, other supplementing power sources or grid integration will reduce the amount of solar cells required, we believe that some principle features shown in this study will remain valid. For once, designing a PV power system just on the average power it generates will underestimate the amount of PV needed to reliably and cost-effectively power an appliance.

Furthermore, designing any system that uses solar power will have a strong design component depending on the latitude of its application. Stationary systems should be specifically designed for the location in which they are used, mobile systems will have to be designed for the most Northern or Southern point where they should function reliably. Systems that use only solar energy are clearly most beneficial in tropical and sub-tropical regions. This offers an opportunity for invigorating solar businesses by designing solar powered applications for these regions. As the tropics and sub-tropics include many regions that are vulnerable, this analysis offers a further motivation to develop solar-powered solutions in the water-food-energy nexus.

Finally, reduction in battery prices offer a huge opportunity for dealing with intermittencies. However, purely relying on batteries to solve the intermittency challenge is unlikely to be prudent. The presented study was also motivated by the wish to find better ways in addressing this challenge.

V SUMMARY

Solar power is intermittent by nature. In order to provide reliable power, electricity generated by solar panels needs to be buffered or stored. Rather than only considering the solar panel area that is needed to generate a certain average power, in this paper we investigate designs of solar/battery systems that additionally generate a constant power output over the course of one year. To find a unique design solution, we perform a cost-optimization, to additionally find the system that fulfils the formulated requirement with the lowest possible capital cost. As a result of this investigation we obtain the battery capacity and the solar panel area as a function of location. Battery capacity is analysed in units of days of autonomy, and a strong dependence with latitude is found. The necessary capacity varies between two days in tropical areas outside the equator and 200 days close to the poles.

The obtained solar panel area is compared to the minimum area that would, for the same year, supply just the needed integrated average power. We find that in our design, the optimum solar panel area is at least 40% larger than this minimum value, for tropical regions, but can be up to eight times larger in the far north or south. This result is partially caused by the comparably high price for battery storage. The cost for storage dominates the system cost, consequently replacing battery capacity by adding more solar panels is, up to a point, advantageous. Generalizing this finding, it can be concluded that the amount of overcapacity in solar panels needed, will depend on the price with which energy can be stored.

In accordance with expectations, the results show that solar panels alone can be used in tropical regions, but need to be complemented by other energy sources in areas with a significant seasonal variation. Considering wind as a complementary source will be a next step.

ACKNOWLEDGEMENT

This work was financially supported by the U.S. Department of Energy SunShot NextGenIII program, Award Number DE-EE0006707, by funding from Singapore's National Research Foundation through the Singapore MIT Alliance for Research and Technology's "Low energy electronic systems (LEES) IRG", and by the National University of Singapore (NUS) and Singapore's National Research Foundation (NRF) through the Singapore Economic Development Board (EDB).

REFERENCES

[1] Burke et al. "The reliability of distributed solar in critical peak demand: A capital value assessment", Renewable Energy 68 (2014).

[2] Perez, Richard, Robert Seals, and Ronald Stewart. "Assessing the load matching capability of photovoltaics for US utilities based upon satellite-derived insolation data." Photovoltaic Specialists Conference, 1993., Conference Record of the Twenty Third IEEE. IEEE, 1993.

[3] Al-Hasan, A. Y., A. A. Ghoneim, and A. H. Abdullah. "Optimizing electrical load pattern in Kuwait using grid connected photovoltaic systems." Energy conversion and management 45.4 (2004): 483-494.

[4] 24 Waite & Modi "Potential for increased wind-generated electricity utilization using heat pumps in urban areas" Applied Energy 135 (2014).

[5] D. B. Needleman, J. R. Poindexter, R. C. Kurchin, I. M. Peters, G. Wilson, T. Buonassisi, "Economically sustainable scaling of photovoltaics to meet climate targets", EES 6 (2016), 2122–2129.

[6] N. Haegel et al. "Terawatt-scale photovoltaics: Trajectories and challenges" Science 356 (2017).

[7] https://ceres.larc.nasa.gov/

[8] For example: M. G. Rasmussen, G. B. Andresen, M. Greiner, "Storage and balancing synergies in a fully or highly renewable pan-European power system", Energy Policy, 51 (2012), 642-651.

How to Choose the best Empirical Model for Optimum Energy Yield Predictions

Steve Ransome[1] and Juergen Sutterlueti[2]

[1] Steve Ransome Consulting Ltd, KT2 6AF #99, Kingston upon Thames, UK
[2] Gantner Instruments Environment Solutions GmbH, 08297 Zwoenitz/ Germany

Abstract — **Many different empirical models have been used by the PV industry to characterize both indoor and outdoor module performance as a function of irradiance and module temperature to then predict energy yields vs. climate data. This paper has fitted 11 different empirical models to outdoor PV monitored data from Gantner Instruments' OTF in Arizona (GI) and also IEC 61853 matrix measurements from various 3rd parties (either by using in data from publications or from private communications). Some of the empirical models were found not to be able to fit PV performance very well at lowest and highest irradiance or with respect to temperature due to their choice of unphysical coefficient dependencies. The best features from all these models were combined to form a new optimised mechanistic performance model "MPM" which is being validated with GI data as well as IEC 61853 matrix data and can fit many matrix measurements with an rms <0.5%.**

Index Terms — **Energy, Modeling, Photovoltaic systems, Power, Simulation, Degradation**

I. INTRODUCTION

PV energy yield EY (also called YA for DC or YF for AC) is roughly proportional to the total of "plane of array insolation H_I (kWh/m²/y)" times "normalised module efficiency PR_{DC}" at all occurring irradiances (G_I kW/m²) and module temperatures (T_{MOD} C) as in equation (1). Many losses e.g. soiling, mismatch, wiring resistance and inverter efficiency have been assumed constant in this simplified equation.

$$EY \propto \sum_{G_I, T_{MOD}} H_I(G_I, T_{MOD}) \times PR_{DC}(G_I, T_{MOD}) \quad (1)$$

Figure 1 illustrates a plane of array insolation distribution H_I against plane of array irradiance G_I and module temperature T_{MOD} for a typical sunny site (Cairo $H_I \sim 2200$kWh/m²/y). The peak amount of insolation here occurs under bright and hot conditions (~800-900W/m² and 55-65C), with lower amounts of insolation occurring in cooler and duller weather.

Data for graphs such as figure 1 are calculated as follows:

1) Horizontal plane irradiance G_H and T_{AMB} (e.g. TMY)
2) Plane of array irradiance translation to GI
3) Module thermal performance modelling to T_{MOD}
4) Calculate $H_I(G_I, T_{MOD})$

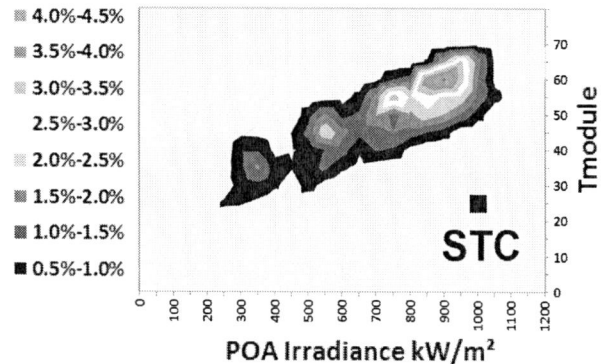

Fig. 1. Typical plot of plane of array insolation distribution (H_I) vs. POA irradiance G_I and module temperature T_{MOD} for a sunny site (Cairo).

Figure 2 shows a typical PR_{DC} against irradiance (x-axis) and module temperature (lines) for a standard c-Si module. The shape of the curves will vary slightly with different module manufacturers and PV technologies.

The overall shapes and values of the curves are characterised by the values of the 5 parameters defined in table I [1].

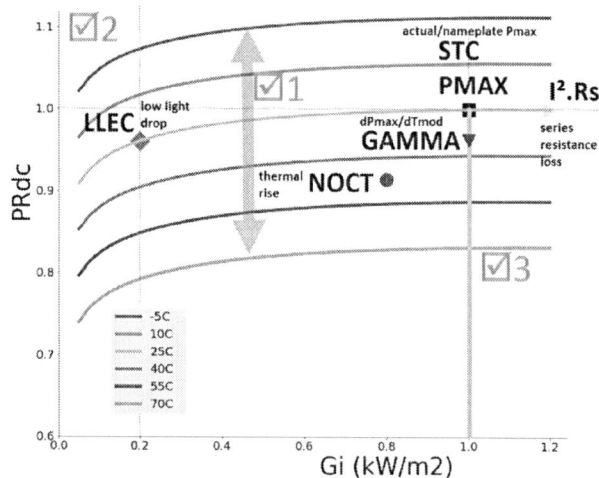

Fig. 2. Typical module PR_{DC} (%) vs. irradiance and module temperature.

978-1-5090-5606-4/17 $31.00 © 2017 IEEE

TABLE I

FIVE PARAMETERS THAT DETERMINE PR_{DC} VS. IRRADIANCE AND TEMPERATURE

(1) ■ PMAX POWER RATING = Manufacturer tolerance Value of "$P_{MAX.ACTUAL}/P_{MAX.NAMEPLATE}$ @ STC"
(2) ▼ GAMMA = $1/P_{MAX.STC} * dP_{MAX}/dT_{MOD}$ Vertical separation of curves at different temperatures (Usual range ~ -0.25 best to -0.50%/K)
(3) ♦ LLEC = Low light efficiency coefficient Value of "Eff@ 0.2kW/m² / Eff@ 1kW/m²" (Usual range ~ 85% worst to 105%)
(4) ● NOCT = Nominal operating cell temperature T_{MOD} @ 0.8kW/m² AM1.5 direct, 20C T_{AMB}, WS 1ms^{-1} (Usual range ~ 47±5C for ventilated modules)
(5) ▶ $I^2.R_S$ = R_{SERIES} loss = $I_{MP.STC}^2 * R_{SERIES}/P_{MAX.STC}$ Determines slope of dPR_{DC}/dG_I at high G_I (Usual range ~85% worst to 100% at 1kW/m2)

II. EMPIRICAL MODELLING

Empirical PV models usually calculate expected performance by summing several empirical coefficients (usually 3-7) each multiplied by different input dependencies i.e. functions of meteorological inputs such as G_I, T_{MOD} and Windspeed WS.

A simple example PVUSA is given in equation (2)

$$P_{CALC} = G_I * (C_1 + C_2 \times G_I + C_3 \times T_{AMB} + C_4 \times WS) \quad (2)$$

The PVUSA equation uses four empirical coefficients C_1 to C_4. Their input dependencies are as shown C_1= constant, C_2=G_I, C_3=T_{AMB} and C_4=WS.

The best fit is found by varying the empirical coefficients C_1 to C_4 optimising for the minimum RMS error between $P_{MAX.MEASURED}$ and $P_{MAX.CALCULATED}$.

Eleven empirical models that have been studied in this work (including PVUSA) are listed in alphabetical order in Table II.

TABLE II

ELEVEN EMPIRICAL MODELS

\<A\> Model Name [Reference]	\<B\> Number of coefficients	\<C\> Normalised coefficients?	\<D\> Is T_{MOD} modelled?	\<E\> Comments
CREST [2]	5	☒No	☑Yes	
HEYDENRICH [3]	3	☒No	☒No	
IEC 60891[4]	5?	☒No	☑Yes	Complicated calculation procedure
LFM 2013[5]	6	☑Yes	☒No	Simplified LFM (only 2 loss factors nI_{DC} and nV_{DC})
Mother PV[6]	5	☒No	☒No	Self-referenced Isc
POLYNOMIAL	4	☒No	☒No	Simple = $G_I * (C_1 + C_2*G_I + C_3*G_I^2 + C_4*G_I^3) * T_{CORR}$
PVCOMPARE[7]	7	☒No	☑Yes	Includes T_{AMB} and SolAlt terms
PVGIS[8][9]	6	☒No	☑Yes	
PVUSA[10]	4	☒No	☑Y	For use at high G_I only (>0.7kW/m² recommended)
PVUSA+[11]	5	☑Yes	☑Yes	5th coefficient to PVUSA gives improved fit at low light
SRCL2014 [1]	5	☑Yes	☑Yes	Physically significant coefficients e.g. LLEC, R_S
…				
MPM[12] Model"N"	6	☑Yes	☑Yes	New model in this paper

NOTE: These models have been randomly assigned ID letters A - N which are **not** in the same order as models appear above.

Column \<D\> indicates if the model has its own coefficients to model temperature dependence. If "No" then a separate multiplication is done to temperature correct as in equation (3).

$$PR_{DC}(T_{CORRECT}) = PR_{DC} \times (1 - \gamma * (T_{MOD} - 25)) \quad (3)$$

The empirical models coded A to N use a selection of input dependencies for their coefficients including the following terms:-

Irradiance: G_I, G_I^2, G_I^{-1} ·ln(G_I), ln(G_I)² …
Temperature: T_{MOD}, T_{MOD}^2, T_{AMB}, T_{AMB}^2 …
Combinations: T_{MOD}*ln(G_I), $T_{MOD} \cdot T_{AMB}$ …
Other: SolarAltitude, AirMass …

III. IEC 61853 VS. OUTDOOR MEASUREMENTS

IEC 61853 "the matrix method" defines the measurement of PV performance at 23 different (G_I, T_{MOD}) pairs of POA irradiance (0.1 to 1.0kW/m^2) and module temperature (15, 25, 50 and 75C) [13].

In this study, outdoor data was taken from GI's OTF in Arizona. Weather inputs included: date+time, POA irradiance (pyranometer and reference cell); ambient and module temperature.

Spectral distribution, beam fraction (=beam/global irradiance), horizontal plane clearness index kTh and relative humidity are measured by GI but have not been analysed here.

Solar altitude, azimuth and angle of incidence are calculated and are used for valid data selection (e.g. "solar altitude > 5 degrees and angle of incidence < 75 degrees").

The irradiance and module temperatures in outdoor data measurements are limited to combinations that naturally occur e.g. there are unlikely to be measurements with G_I=1.1kW/m^2 and T_{MOD}=10C. Figure 3 shows the proportion of POA insolation H_I (circle area) in Tempe, AZ vs. PR_{DC} (y axis) against "temperature corrected to the nearest 61853 module temperatures" (colours) and irradiances (x axis) as used in the IEC 61853 matrix method. The sizes of the circles show that a large proportion of the energy yield in AZ is generated between about 25C and 75C and between 0.2 and 1.0 kW/m^2. This indicates where the model fit needs to be good for an accurate energy yield calculation.

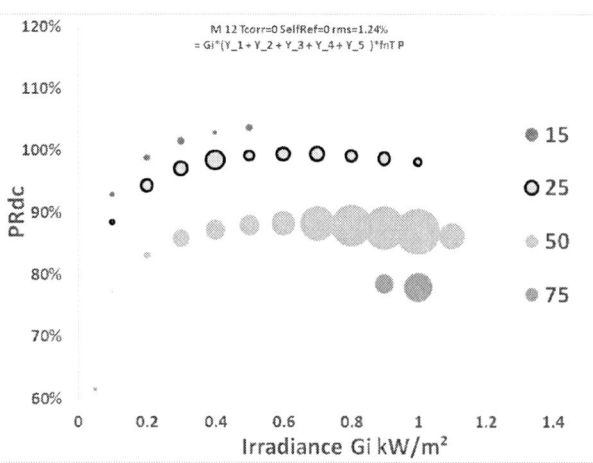

Fig. 3. Proportion of POA insolation ($\propto$ circle area) in each G_I,T_{MOD} bin from IEC 61853 for a c-Si module at GI's Tempe OTF.

IV. COMPARING THE MODELS

It was found that most of the 11 empirical models could fit the most commonly occurring weather conditions (e.g. 20C<T_{MOD}<60C and 0.2kW/m^2<G_I<1kW/m^2 reasonably well. However, some models with "unphysical coefficient dependencies" would diverge and have increasing errors at more extreme weather outside these conditions.

The fit by model D to a matrix of mc-Si module data #27 (dots) is shown in Figure 4. The fit (lines) has unphysical

temperature coefficients shown as "⊠1" that depend on module temperature (i.e. gamma ~ the change in PR_{DC} from 15-25C is much lower than from 55-65C), not what is usually measured as in Figure 2. Energy yield predictions using Model D would therefore be wrong at both predominantly cold and hot conditions due to the curves' non-linear separation.

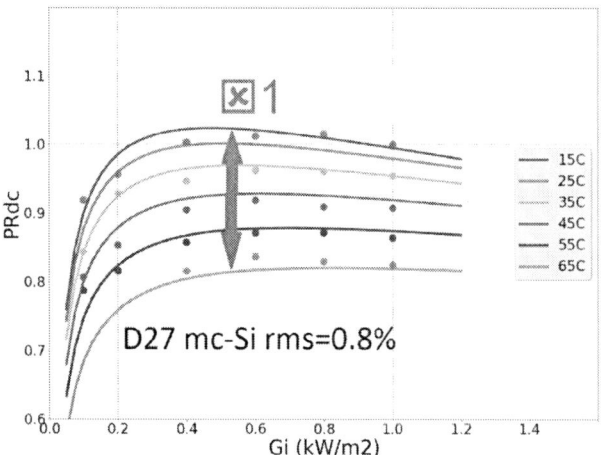

Fig. 4. Best fit for Model D with unphysical temperature behaviour ⊠1 for a mc-Si module #27 (contrast with figure 2)

Model A is shown in figure 5 for a matrix for a different CIS module (having poor low light performance #28 and so a different y-axis scale). Here we can clearly see a reasonably correct shape of curves but only from G_I = 0.2 to 1.0kW/m^2 with unphysically high modelled performance <0.1 "⊠2" and curving upwards rather than downwards >1.0kW/m^2 "⊠3". Energy yield predictions using Model A would be wrong at both low light and high light conditions.

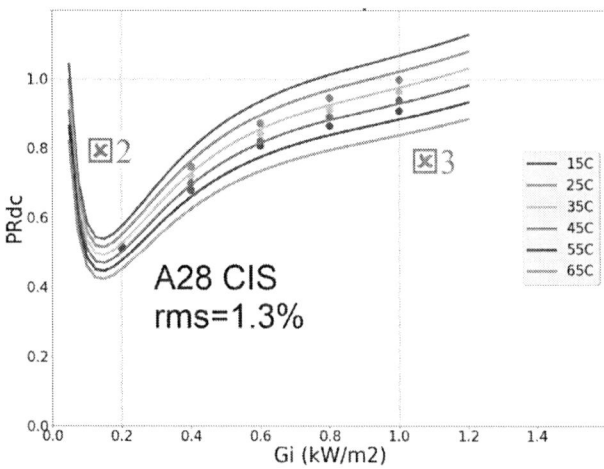

Fig. 5. Best fit for model A with unphysical low light behaviour ⊠2 and curving upwards high light performance ⊠3 CIS #28 (contrast with figure 2).

V. EMPIRICAL VS. MECHANISTIC MODELS

A simple mathematical study of the functions used in some of the 3rd party empirical models show why these errors happen. Not all dependencies listed can be used to give sensible model performance – any model containing an input dependency such as "$G_I * T_{MOD}$" will not be able to have a constant temperature coefficient with irradiance.

Therefore, a better type of model is needed. Table IV details the differences between empirical and mechanistic models.

TABLE IV
EMPIRICAL VS. MECHANISTIC MODELS

Empirical Model	Mechanistic Model
Not normalised – Coefficients values scale with array size	Normalised – Values independent of array size
Nonphysical coefficients included? e.g. $T_{AMB}*T_{MOD}$	Physically significant dependencies are used wherever possible
Not easy to use to compare and contrast arrays of different sizes	Easy to validate and compare different sized arrays

A new, optimised, **Mechanistic Performance Model** "MPM" [011] has been developed using the best features of existing models. It has 6 coefficients C_1 to C_6 defined in equation (4).

$$PR_{DC} = C_1 + C_2 \times dT_{MOD} + C_3 \times Log_{10}(G_I) + C_4 \times G_I + C_5 \times WS + C_6/G_I \qquad (4)$$

$dT_{MOD} = (T_{MOD} - 25)$; G_I (kW/m^2) and WS wind speed (ms^{-1}).

The MPM can be used to fit both IEC 61853 or outdoor data.

The 11 empirical models were also tested against the MPM by fitting to up to 3 years of measured data at GI's Tempe outdoor test facility (OTF) in Arizona for several PV technologies. Figure 6 shows the fit from the MPM for 500 random records from a c-Si mod at the AZ OTF Jun 2012 to Jun 2015. The scatter is mainly due to soiling effects but the rms fit error is only 1.2%. The MPM predicts curve shapes as smooth, uniform and parallel as expected from figure 2.

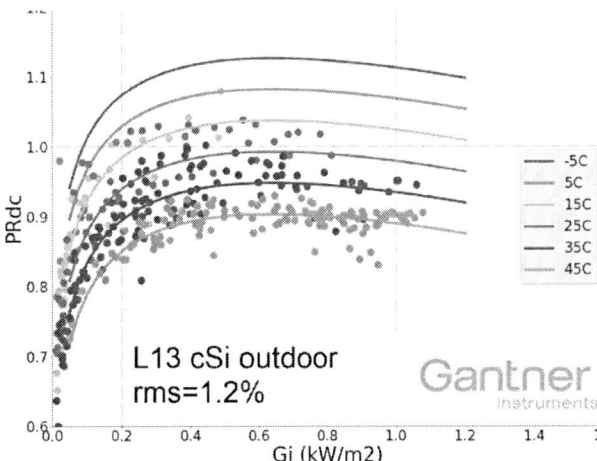

Fig. 6. Best fit for MPM model for typical outdoor c-Si #13, GI data Tempe with an rms error of 1.2%.

The MPM was also used to fit matrix data either provided or downloaded from publications from several third parties including ASU, ESTI, CFV and TUV Rheinland.

Note that not all IEC 61853 specified irradiances (e.g. 0.1 and 1.1kW/m^2 may be missing) or module temperatures (there may be 3-5 measured) are used by different third parties but the MPM interpolates and fits all the data well. Figures 7-10 show MPM fits to a variety of Matrix measurements from different sources where the rms error is <0.5% for well measured or modelled PV. Figure 11 fits the MPM to c-Si #23 data predicted by the SAPM [14] and figure 12 fits the modified 1-diode model used by PVSYST [15] for a CdTe device #24. Note rms error fits are always better than 0.5% and that one seems limited by measurement scatter (maybe mesh transmission for example all the 0.8kW/m^2 are lower than expected).

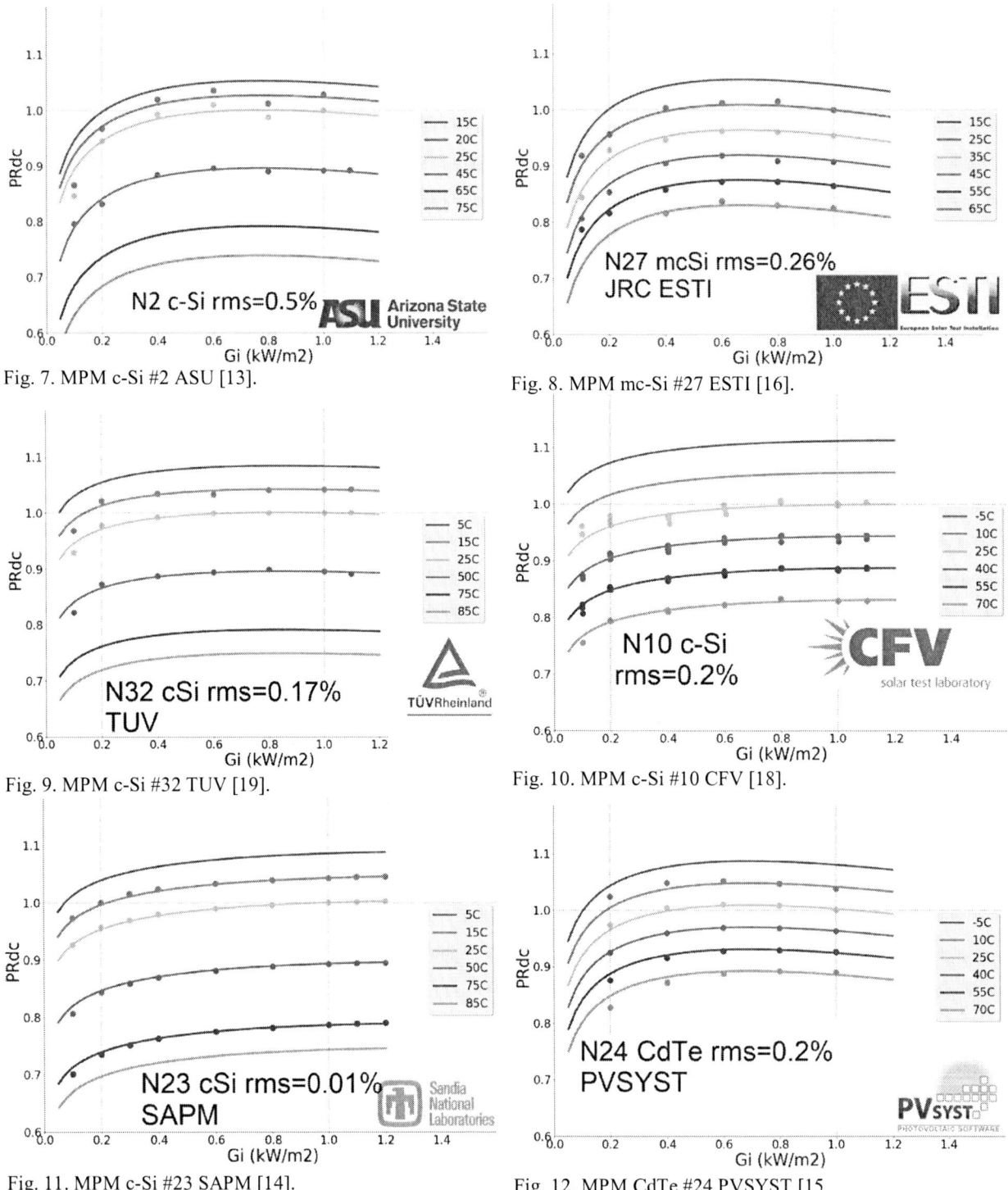

Fig. 7. MPM c-Si #2 ASU [13].

Fig. 8. MPM mc-Si #27 ESTI [16].

Fig. 9. MPM c-Si #32 TUV [19].

Fig. 10. MPM c-Si #10 CFV [18].

Fig. 11. MPM c-Si #23 SAPM [14].

Fig. 12. MPM CdTe #24 PVSYST [15

MPM smooths out some of the inevitable measurement scatter in matrix data. Table V gives a summary of normalised coefficients used in figs 7 to 12. Note all were indoor data so

C_5 was 0, also all data could be fitted well with $C_6=0$ but this sensitivity will be studied further [20]. These are all well behaved and meaningful values i.e. C_2 = Gamma.

TABLE V
EXAMPLE MPM COEFFICIENTS.

	Quality	Gamma	LogGi	Gi	WS	1/Gi	rms
	C_1	C_2	C_3	C_4	C_5	C_6	err
2) ASU_cSi	111.7%	-0.52%	21.2%	-11.9%	0.0%	0.0%	0.5%
27) ESTI_mcSi	115.5%	-0.45%	23.9%	-15.4%	0.0%	0.0%	0.3%
32) TUV cSi	105.2%	-0.42%	10.1%	-5.2%	0.0%	0.0%	0.2%
10) CFV_cSi	103.3%	-0.37%	9.4%	-3.4%	0.0%	0.0%	0.2%
23) SAPM_cSi	98.5%	-0.41%	9.4%	-2.6%	0.0%	0.0%	0.1%
24) PVSYST_CdTe	112.3%	-0.26%	19.2%	-12.0%	0.0%	0.0%	0.2%

VI. FURTHER WORK

For outdoor measurements, this study has so far just taken data from higher beam fraction and lower angle of incidence effects for its measured data against a reference cell rather than a pyranometer to do the simplest model fits but this will be expanded to cover mostly diffuse vs. mostly direct irradiance measurements and include angle of incidence and spectral corrections against a pyranometer [20].

VII. CONCLUSIONS

- Eleven in house and 3rd party empirical performance models have been compared and contrasted with outdoor data from GI and IEC 61853 data from 3rd parties.
- Several 3rd party empirical models fitted measured matrix data badly away from normal conditions due to their use of non-optimum dependencies.
- A new optimized mechanistic performance model MPM has been introduced and validated against GI outdoor data, 3rd party indoor matrix data and models such as the SAPM and PVSYST's implementation of the modified 1 diode model.
- The MPM can fit Matrix data well and could be considered as the standard method to interpolate and check matrix data for IEC 61853. It reduces 23 measurements to between 4 and 6 normalised coefficients with good accuracy.
- Gantner Instruments have implemented the MPM in their analysis software for both individual module measurements and large power plants [17].

ACKNOWLEDGEMENT

Thanks to many staff at ASU, CFV, CREST, GI, JRC ESTI, PVGIS, SANDIA and TUV for their help, data and discussions

REFERENCES

[1] Ransome "How Simulation Program kWh/kWp Predictions Depend on PV Model Discrepancies" 29th PVSEC 2014 Amsterdam

[2] Alhusna "Influence of Spectral Variations on Photovoltaic Module Energy Rating" *PVSAT 13* Bangor 2017

[3] Heydenrich "Describing the world with 3 parameters" *23rd PVSEC* Valencia 2008

[4] IEC 60891:2009 "Photovoltaic devices - Procedures for temperature and irradiance corrections to measured I-V characteristics"

[5] Stein et al "OUTDOOR PV PERFORMANCE EVALUATION OF THREE DIFFERENT MODELS: SINGLE-DIODE, SAPM AND LOSS FACTOR MODEL" *28th PVSEC 2013* Paris, France

[6] Montgareuil et al "MotherPV method From watt-peak to watt-hours" , *1st Sophia Workshop on PV Performance Modelling*

[7] Jardine et al "PV-COMPARE: Direct Comparison of Eleven PV Technologies at Two Locations in Northern and Southern Europe" *17th PVSEC Munich 2001*,

[8] PVGIS Yordanov thesis 2012

[9] Thomas Huld, PVGIS, private communication

[10] Whitaker et al "Application and validation of a new PV performance characterization method" *26th PVSC* 1997

[11] Ransome, "Comparing PV simulation models and methods with outdoor measurements" *Hawaii PVSC 35, 2010*

[12] Ransome "Choosing the best Empirical Model for predicting energy yield" *7th PVPMC Workshop*, Canobbio, Switzerland 2017

[13] Mani et al "Photovoltaic Module Power Rating per IEC 61853-1 Standard: A Study Under Natural Sunlight" *www.solarabcs.org*

[14] Mani et al "Performance Matrices per IEC 61853 Standards: Their Importance for the Energy Estimation Models" *energy.sandia.gov*

[15] PVSYST http://www.pvsyst.com/en/

[16] Kenny et al "Power rating of photovoltaic modules including validation of procedures to implement IEC 61853-1 on solar simulators and under natural sunlight" *27th PVSEC*, Frankfurt 2012

[17] Sutterlueti "Optimized PV Performance using State of the Art Monitoring for Increased Asset Value" 8th PVPMC 2017

[18] Cliff Hansen, SANDIA. Measurements by CFV , private communication http://cfvsolar.com/

[19] Markus Schweiger, TUV, private communication

[20] Ransome "A Systematic Comparison of >7 Empirical Models Used for Energy Yield Predictions vs PV Technology 5CO.7.6" to be presented at 33rd PVSEC Amsterdam Sep 2017

Modeling and Analysis of Photovoltaic Electrochemical System using Module-Level Power Electronics

Gowri M. Sriramagiri[1,2], Nuha Ahmed[1,2], Kevin D. Dobson[1], Steven S. Hegedus[1,2]

[1]Institute of Energy Conversion, University of Delaware, Newark, DE, 19716, USA

[2] Dept. of Electrical and Computer Engineering, University of Delaware, Newark, DE, 19716, USA

Abstract — The design of a model PV-electrolysis system using a practical water electrolyzer and a commercial solar module is presented and its day-long operation simulated. The fact that photovoltaic electrochemical (PV-EC) devices do not always operate under 1-Sun illumination is often overlooked, which implies that the PV source may not always provide sufficient voltage for electrolyzer operation. The energy transferred between the source and load of such a system is studied quantitatively under two coupling configurations. Modeling shows that voltage-matching with module-level power-electronics (MLPE) increases the net energy transferred between the source and the load by 14% compared to direct coupling. The benefit of MLPE will increase as difference between the operating voltage of the system and the maximum power point of the source increases.

Index Terms — solar electrolyzers, hydrogen, module-level power electronics, practical systems, balance-of-systems

I. INTRODUCTION

Artificial photosynthetic generation of solar fuels such as H_2 and methanol from water and CO_2 electrolysis, respectively, has gained significant attention lately. A considerable amount of research performed is directed towards improving the various components of a photovoltaic-driven (PV) electrolysis system, such as the electrodes, catalysts, electrolytes, ion-transfer membranes etc. All of these elements directly or indirectly affect its performance, measured by a figure-of-merit, solar-to-fuel efficiency (SFE). A PV electrolysis device can be achieved through several architectures ranging from a photoelectrochemical cell (PEC), where the catalyst material is coated onto the photo-electrode that is immersed in the electrolyte, to a photovoltaic-electrochemical (PV-EC) system, where the solar source is completely separated from the electrolyzer. [1] Both ends of this design pattern carry their own advantages.

The PEC avails a simpler design as all of its components fit into a single compact setup. However, it is limited by restrictions such as its requirement of semiconductor electrodes that have to withstand the corrosive effects of the electrolyte, high optical losses, and limited materials selection. On the other hand, while the PV-EC architecture entails with wiring losses and a bulky setup, it facilitates independent optimization of the solar and the electrolyzer systems [2] Each of the components, solar PV and electrochemical cells, have already been separately optimized and are commercially available, justifying the need for research to provide efficient coupling between them.

When a PV source is used to drive an electrolyzer load, it can be done so using different coupling mechanisms as shown in Fig. 1. The source can be directly connected to the load, in which case, the operating currents, current densities and the voltages of the source and the load are equal: $I_{PV} = I_{EC}$, $J_{PV} = J_{EC}$ and $V_{PV} = V_{EC}$, which is the simplest of all the configurations. However, this doesn't necessarily ensure maximum power transfer to the load at all PV source illumination levels, which is why indirect connection using coupling devices should be considered. Such devices can be designed similar to PV module level power electronic (MLPE) components, such as charge controllers, maximum power point trackers (MPPT) etc., or a combination of several such devices.

In this work we analyze a practical PV-EC system designed using commercially available solar module and water electrolysis devices. We simulate day-long performance of the resulting PV-EC system using real solar illumination data for sunny and cloudy days in Newark, DE, USA.

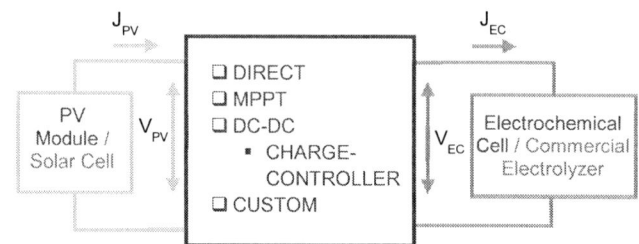

Fig.1: Schematic of various coupling methods for a PV-EC system

II. SYSTEM DESIGN

A. Choice of Source and Load

In order to demonstrate the simulated performance of a practical PV-EC system, we chose commercially available sources and loads, whose current-voltage behaviors are readily accessible and relatively closely matched under standard 1 sun conditions. A 50W, 7-cell proton-exchange membrane (PEM) hydrogen electrolyzer from H-TEC Education [3] made for demonstrational purposes, was chosen as the load. Keeping voltage and current characteristics in mind, a suitable 70W c-Si solar module, Siemens SP70 [4] was chosen from available modules. The

978-1-5090-5606-4/17 $31.00 © 2017 IEEE

operational parameters of both the electrolyzer and PV module are listed in Table 1.

The I-V characteristics of both the source and the load are given in Fig. 2. From this plot, it can be seen that the electrolyzer requires at least 10 V for any current to flow into the load. The operating point of the system under direct coupling can be calculated by taking the point of intersection of these two curves. From these parameters, the figure-of-merit of a solar electrolysis system, solar-to-fuel efficiency (SFE) is calculated using (1), [5]

$$SFE = \frac{J_{OP} \times V_{TH} \times FE}{P_{Solar}} \qquad (1)$$

Where J_{OP} is the operating current density, V_{TH} is the thermodynamic voltage of reacting species, FE is the faradaic efficiency of the reaction, P_{Solar} is the solar illumination intensity, which is 100 mW/cm^2 at for AM1.5. In this case, since the FE of the electrolyzer is not directly available, we do not report the SFE of such a system. We emphasize that this work only intends to focus on modeling quantitatively the differences between various coupling strategies rather than device efficiency figures.

Table 1: Specifications of *H-TEC 230* electrolyzer and *Siemens SP70* solar module

H-TEC 230	
Electrode Area	7 cells of 16 cm^2 each
Power	50 W @ 14 V DC
Permissible Voltage	10.5-14.0 V DC
Permissible Current	0-4.0 A DC
H$_2$ Production	230 cm^3/min
Siemens SP70	
P$_{MAX}$ [W$_P$]	70
η [%]	11.1
V$_{OC}$ [V]	21.4
J$_{SC}$ [A]	4.7
I$_{MPP}$ [A]	4.25
V$_{MPP}$ [V]	16.5

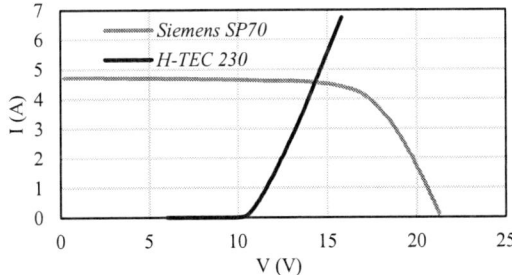

Fig. 2: I-V curves of the selected solar source and the electrolyzer load.

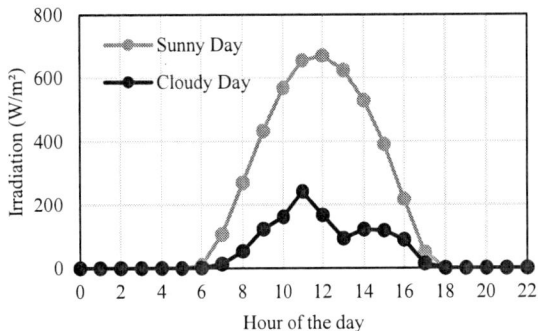

Fig. 3: Solar Illumination of selected sunny and cloudy days in 2016 at Newark, DE, USA. Data obtained from Newark metrological institute for February 28th, 2016 and Feb 25th, 2016.

B. Coupling Mechanisms and Need for MLPE

Most literature on solar fuel generation reports the performance of systems under 1-Sun illumination. Electrolyzers have a turn-on voltage similar to that of a diode, which is the thermodynamic voltage, V_{TH}, of the reacting species, required for its reduction into respective products. This is 1.23V for water reduction to hydrogen and 1.34 V for CO$_2$ reduction to CO. In addition to this, there is an overpotential, ΔV, required to overcome voltage losses within the device and electrolyte and ensure the kinetic transport of the species within the electrolyte, analogous to parasitic resistances in a diode. [5]

This means that under low illumination conditions, the voltage output from the PV system can drop below the minimum voltages required for the electrolyzer operation. This behavior is very similar to that of a battery charging process with a PV source. This often-overlooked notion calls for tapping the potential of using custom-designed electronics for specific systems. For instance, a device such as a MPPTs can be used in conjunction with a charge-controller-type device used for battery chargers. An MPPT tracks the maximum power point (MPP) of a solar array's I-V curve for every changing solar intensity and sets it output voltage and current to V$_{MP}$ and I$_{MP}$ respectively. Charge controllers designed typically for battery charging, maintain the output voltage suitable to the charging voltage of a battery.

The power electronic device desirable for an electrolyzer load with the solar array, would ideally have an MPPT to trace the array's MPP with changing solar irradiation throughout the day. Following this, that power is delivered to the load at a voltage that is always of sufficient magnitude to drive the electrolysis, through the voltage-regulator/charge-controller-type device, made specifically for the electrochemical cell under consideration. This ensures that the load is almost always operational, regardless of terrestrial illumination conditions.

978-1-5090-5606-4/17 $31.00 © 2017 IEEE

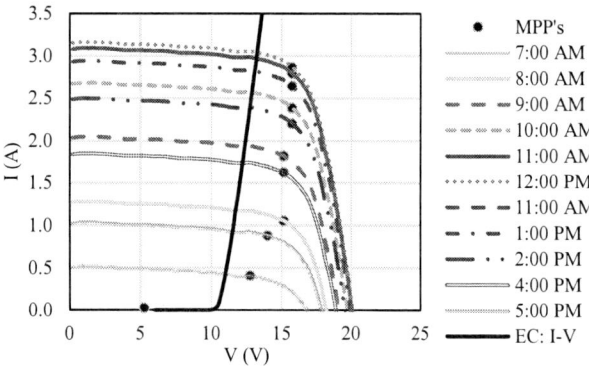

Fig. 4: Modeled I-V curves for *Siemens S70* module under different solar illumination intensities for each hour of the sunny day conditions.

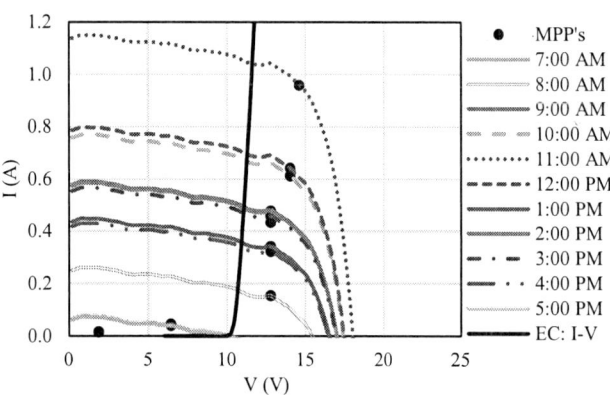

Fig. 5: Modeled I-V curves for Siemens S70 module under different solar illumination intensities for each hour of the cloudy day conditions.

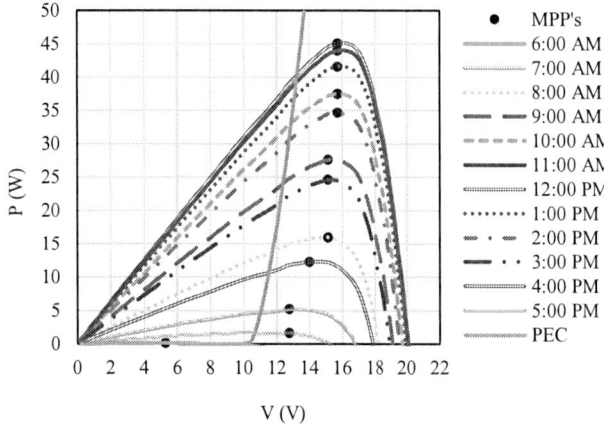

Fig. 6: Modeled P(V) curves for Siemens S70 module under different solar illumination intensities for each hour of the sunny day condition.

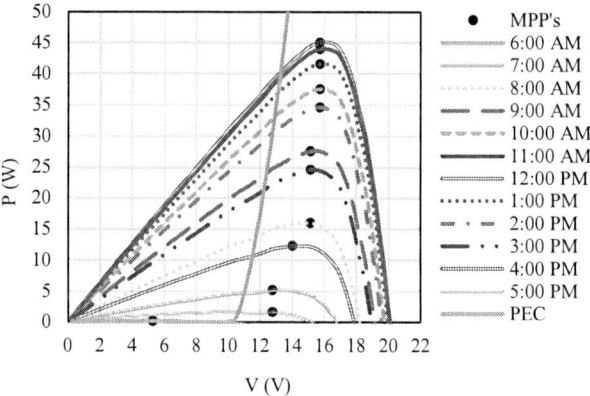

Fig. 7: Modeled P(V) curves for Siemens S70 module under different solar illumination intensities for each hour of the cloudy day condition.

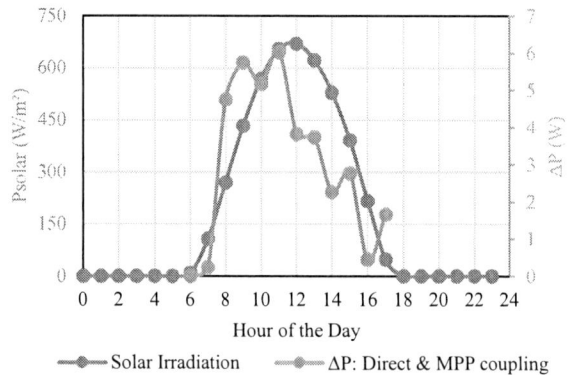

Fig. 8: Difference in the power transferred to the load in direct-coupled and MPPT-coupled configurations for the sunny day condition.

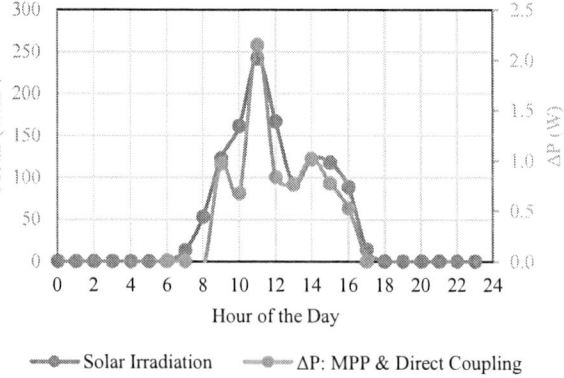

Fig. 9: Difference in the power transferred to the load in direct-coupled and MPPT-coupled configurations for the cloudy day condition.

978-1-5090-5606-4/17 $31.00 © 2017 IEEE

III. MODELING

The theoretical power output of the *Siemens SP70* module was calculated using a lumped circuit model, for two days in February 2016 with different solar irradiations as shown in Figure 3: sunny and cloudy. This was done using real hourly illumination data for Newark, DE, assuming a 0° tilt. The I-V curves of the solar module is used and the current as a function of voltage is scaled for varying solar intensity. The voltages are then corrected for operating cell temperatures with respect to changing ambient temperatures. While the V_{MP} of the module is reported for 25°C, in real applications the module will operate hotter, which will drive the V_{MP} lower. Fig. 4 and 5 show the I-V curves of the module for every hour of solar illumination, from 9 AM to 5 PM, on a sunny day and cloudy day, respectively. We can see how the maximum power points (indicated by the black circle) on the solar cell power curves shift away from that of the electrolyzer power curve with increasing illumination intensity every hour throughout the day under sunny and cloudy day conditions in the P(V) plots of Fig. 6 and 7. Two of the simple system coupling configurations described above, direct and MPPT, have been analyzed using this modeling approach.

IV. RESULTS AND DISCUSSION

The tabulated data of energy transferred for both days in each of these configurations is given in Table 2. The difference between the power transferred for each of the configurations hourly is presented in Fig. 8 and 9 for sunny and cloudy days, respectively. From these results, it can be seen that, throughout the day, the *power* transferred to the electrolyzer in the MPPT-coupling is higher than that of direct coupling, accounting for a 14% difference in *energy* transferred between the two different coupling configurations. Fig. 8 and 9 also show that the amount of extra power transferred through MPPT-coupling increases with higher solar irradiation. This follows the trend observed in Fig. 6 and 7, where, as the solar intensity goes up, a greater ΔP between the operating (intersecting) power point and the maximum power point is observed, increasing the difference between power transferred in the direct and MPPT coupling configurations. However, the overall gain in daily energy transfer through MPPT coupling over direct coupling is over 14% for both days of operation that were simulated in this work. This means that, with MPPT coupling, not only does the electrolyzer stay on under low illumination conditions, but overall, it receives more energy, enabling it to produce a higher gas output.

It could be argued that selecting a different PV module, having less rated power at 1 sun, would have provided a better match and thus minimized the gains with the MLPE, it would run the risk of having insufficient voltage to drive the electrolyzer at low sunlight conditions. This would significantly reduce the H_2 production.

Table 2: Energy transferred to the electrolyzer on 3 days of different levels of irradiation

Parameter	Sunny Day	Cloudy Day
Total Irradiation Per Day (W-hr)	411	109
MPPT Coupling Energy Transfer (W-hr/day)	290	60
Direct Coupling Energy Transfer (W-hr/day)	253	52
Δ Energy Transfer (%)	14.6	14.5

V. CONCLUSION

We present simulated PV-EC system performance during operation over an entire day and modeled two different coupling mechanisms: *direct* and *MPPT*. We highlight the need for development and usage of MLPE devices in the field of solar fuels generation by showing that the energy transferred between the source and the load of such a model configuration is higher when the voltage and current supplied to the load are electronically optimized for maximum power transfer. It was shown that on 2 days of very different solar irradiation: *cloudy* and *sunny*, the net energy transferred between the source and the load can be improved by 14% with MLPE devices used for coupling. Thus, using already existing power electronic device technologies such as MPP trackers or charge-controller-type devices to mediate the power transfer between PV and electrolyzer will ensure extracting the maximum performance of the solar modules in these systems and getting the maximum economic payback.

ACKNOWLEDGEMENT

This work was partially funded by a grant from the University of Delaware Energy Institute.

REFERENCES

[1] Jacobsson, T. Jesper, Viktor Fjällström, Marika Edoff, and Tomas Edvinsson. "Sustainable solar hydrogen production: from photoelectrochemical cells to PV-electrolyzers and back again." *Energy & Environmental Science* vol. 7, pp. 2056-2070, 2014

[2] White, J.L., Herb, J.T., Kaczur, J.J., Majsztrik, P.W. and Bocarsly, A.B., "Photons to formate: Efficient electrochemical solar energy conversion via reduction of carbon dioxide". *Journal of CO_2 Utilization*, vol. 7, pp.1-5, 2014.

[3] http://fuelcellstore.com/h-tec-education/electrolyzer-230-e107

[4] http://www.solardirect.com/PDF/SolarElectric/sp70.pdf

[5] Winkler, Mark T., Casandra R. Cox, Daniel G. Nocera, and Tonio Buonassisi. "Modeling integrated photovoltaic–electrochemical devices using steady-state equivalent circuits." *Proceedings of the National Academy of Sciences,* vol.110, no. 12: pp. E1076-E1082, 2013.

Betavoltaic Generation Function in Silicon

A.V. Sachenko[1], I.O. Sokolovskyi[1,2], and M. Evstigneev[2]

[1]V. Lashkaryov Institute of Semiconductor Physics, NAS of Ukraine, 03028 Kiev, Ukraine
[2] Department of Physics and Physical Oceanography, Memorial University of Newfoundland, St. John's, NL, A1B 3X7 Canada

Abstract — **Betavoltaics refers to direct conversion of radioactive beta-decay energy into electricity. This effect is rather close to photovoltaics both in underlying physics and mathematical modeling. In the present contribution, the electron-hole pair correlation function in silicon by beta-electrons of energy < 30 keV is obtained theoretically. A good agreement of our results with the experimental findings is demonstrated. The scattering of beta-particles of different energies and their electron-hole pair generation function depending on the angle of entrance into the sample are analysed. An accurate analytical approximation for the electron-hole pair generation function in silicon coupled to a tritium beta-source is shown to be given by a stretched exponential expression instead of the simple Beer-Lambert law.**

Index Terms — **betavoltaics, electron-hole pair generation function, silicon, tritium**

I. INTRODUCTION

Beta batteries are low-power long-lifetime energy sources that use beta-particles, i.e. high-energy electrons produced by a radioactive material in a beta-decay reaction. When a beta-particle with the typical kinetic energy of several keV enters a semiconductor diode, it creates thousands of electron-hole pairs by ionization. Those electron-hole pairs are separated by the built-in field of the p-n junction, and thus the kinetic energy of the incident beta-flux is converted into electricity [1].

Betavoltaic batteries have several attractive features that conventional photovoltaic elements and electrochemical cells lack. Beta-batteries are completely autonomous, i.e. they do not depend on the presence of solar radiation and do not rely on human intervention to replace or recharge them. Their service time is determined by the decay half-time of the radioactive source and amounts to years or even decades. Due to their longevity and robustness, beta-elements can be used in inaccessible or hostile environments, such as outer space, bottom of the ocean, or inside the human body, and can endure extreme temperature conditions from about -50 to 150 °C.

When modeling a beta-battery, of fundamental importance is the electron-hole pair generation function, i.e. the mean number of electron-hole pairs generated by a single beta particle per unit distance from the semiconductor surface. In spite of the fact that betavoltaics has been known for a long time (see [1] and references therein), we are not aware of an accurate expression for the generation function for particular semiconductor/source combinations published in the literature.

Here, we attempt to fill in this gap, focusing on Si, which is the most widely used semiconductor material, and tritium, which is the cheapest and the least destructive beta-source.

II. THE MODEL

The most efficient method to investigate the influence of beta-radiation on semiconductor materials is based on Monte Carlo (MC) calculations [2], [3].

Electron-hole pair generation in semiconductors is well described by the models, where the betas are assumed to be scattered elastically and, in addition, are acted upon by a constant stopping power [2]. Joy and Luo [4] specify the stopping power as

$$\frac{dE}{ds} = -785 \frac{\rho Z}{AE} \ln\left(\frac{1.166E}{J'}\right). \tag{1}$$

Here E is the instantaneous energy of a beta-electron (in eV), s is the path length along its curved trajectory (in Å); Z, A and ρ are atomic number, atomic weight and density (in g/cm^3) of the target material, and $J' = J/(1+kJ/E)$ is the modified mean ionization potential of material. For silicon, $J = 172$ eV and $k = 0.822$.

The differential Mott cross-sections of elastic scattering were calculated numerically by Czyżewski et al. [5]. The mean free path, i.e. the average distance between two elastic scattering events can be written as [6]

$$\lambda = \frac{10^{21}}{N_A} \frac{A}{\rho \sigma}, \tag{2}$$

where N_A is Avogadro number and σ is total collision cross-section (in nm^2).

To determine the generation function, we used the fact that the energy to produce a single electron-hole pair is related to the bandgap as [7]

$$E_{ehp} = 2.8E_g + 0.5\,\text{eV}, \tag{3}$$

i.e. the differential number of electron-hole pairs generated by an electron is related to its differential energy loss by $dn = dE/E_{ehp}$. The total generation function can be obtained by averaging this result with respect to the trajectories, energies, and incident angles of the incoming betas.

978-1-5090-5606-4/17 $31.00 © 2017 IEEE

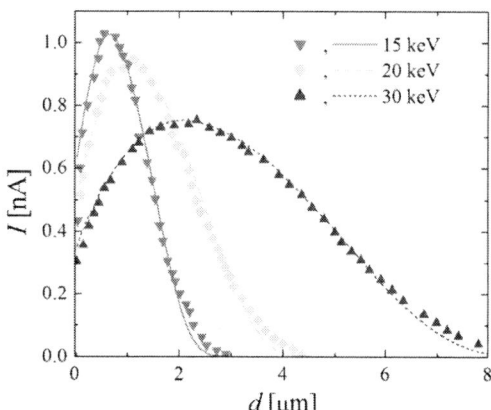

Fig. 1. Theoretical (solid lines) and experimental [8] (symbols) current generated by a monoenergetic electron beam at normal incidence in a p-n junction vs. junction depth. The three curves are obtained for three different incident electron energies.

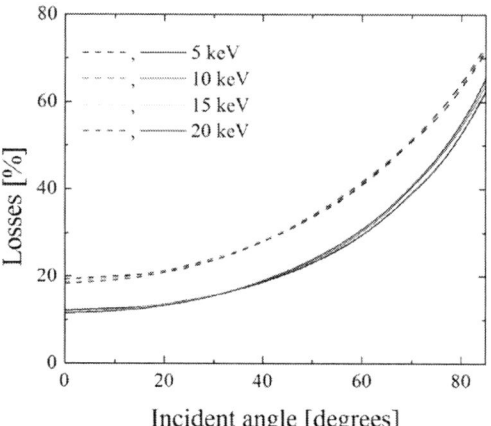

Fig. 2. Relative exit losses vs. input angle. Solid lines: exit probability; dashed lines: energy lost due to exit of betas from the material.

III. MODEL VALIDATION

Our program for MC calculations of the incident beta trajectories is very similar to the Casino code published in the literature. We used this code as a standard against which the validity of our own code was confirmed.

In addition, to verify our simulation results for the generation function, a comparison of generation current with the experiment [8] was made. In this experiment, a beam of monoenergetic electrons was incident normally to the surface of a p-n junction diode, where the flat boundary between the p- and n-regions formed a small angle with the frontal surface. Therefore, depending on the position of entrance of the thin incident electron beam, the p-n junction generated a different current value. The generated current is proportional to the generation function $g(z)$ at the distance z between the p-n boundary and the front surface along the surface normal. More precisely, in the experiment, the generation was effectively integrated by p-n-junction depth (about 400 nm) [8]. We have performed this averaging when calculating the theoretical curves as well. Fig. 1 shows a good agreement between our simulation results and the experiment.

IV. RESULTS AND DISCUSSION

A. Losses Due to the Exit of Betas from a Semi-Infinite Semiconductor

At a first glance, to calculate the generation function by betas from a particular source, characterized by a given energy and angular distribution, it may seems sufficient to proceed as follows. First, one determines the generation function for normal incidence of a monoenergetic electron beam for different energies present in the spectrum. Then, one averages this generation function with respect to the energies. Finally,

the result, weighted with an appropriate trigonometric function of the incidence angle, is averaged with respect to all incidence angles.

It turns out that this simple procedure does not work. The reason lies in the angular dependence of the fraction of the incident betas that leave the sample due to the elastic scattering, as well as the energy that those back-scattered betas carry away. These angular dependences for different energies of the incident electrons are shown in Fig. 2. We see that the fraction of the "lost" betas is non-negligible, constituting about 20% for normal incidence and raising up to over 70% for the 80 degrees incidence relative to the normal. The energy lost in this way varies from about 11% for normal incidence to over 60% at 80 degrees. Interestingly, both types of losses do not depend on the energy of the incident electrons. Therefore, in our Monte Carlo simulations, both energy and angular distribution of betas was present from the beginning. The energy distribution of the incident betas produced by a tritium source is taken from [9]; the angular distribution is assumed to be isotropic.

B. One-Sided Generation

First, we consider the situation when the beta source is brought in contact with a semi-infinite piece of silicon with a flat surface in the xy-plane. The simulated generation function in silicon by tritium beta-particles is shown in Fig. 3. The average number of electron-hole pairs generated by a single incident electron turned out to be $N_p = 1093$.

To approximately describe the resulting generation function by a simple analytical formula, we proposed a stretched exponential expression:

$$g(z) = g_0 e^{-(|z|/z_p)^\beta} . \qquad (4)$$

978-1-5090-5606-4/17 $31.00 © 2017 IEEE

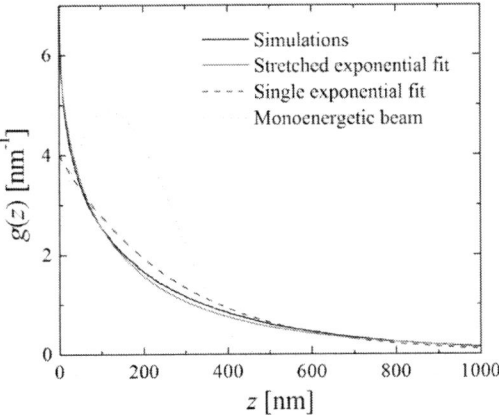

Fig. 3. Generation function vs. depth, as obtained from the simulations (solid black line), stretched exponential approximation (4) (solid line), and simple exponential approximation [i.e. (4) with β = 1] (blue dashed line). For comparison, generation function by a monoenergetic electron beam with the energy of 5.74 keV (the mean energy of a beta-particle coming from a tritium source) is shown as a green dotted line.

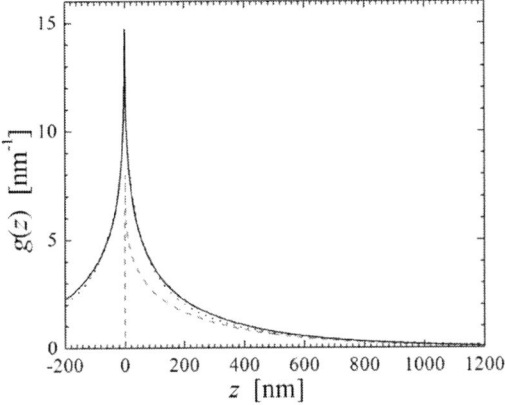

Fig. 4. Solid line: generation function vs. depth for a two-sided configuration. For comparison, the "one-sided" generation function from Fig. 3 is shown as a red dashed line. Blue dotted line is the fit with the stretched exponential expression (4).

The three fit parameters in this expression – g_0, z_p, and β – are found by the least square fitting the simulation curve under the constraint that the area of the fitting curve be equal to N_p:

$$\int_0^\infty dz\, g(z) = g_0 z_p \beta^{-1} \Gamma(\beta^{-1}) = N_p \,, \qquad (5)$$

where $\Gamma(\ldots)$ is the gamma function. In other words, only two out of the three parameters are independent. As a result of this procedure, the following parameter values are obtained:

$$g_0 = 6.51\,\text{nm}^{-1};\ z_p = 111.5\,\text{nm};\ \beta = 0.6\,. \qquad (6)$$

We note that a single-exponential approximation (with β set to 1), i.e. the familiar Beer-Lambert expression for the generation function corresponds to the photovoltaic counterpart, but is significantly less accurate then the stretched exponential approximation with the parameters above.

It is interesting to note that the generation function due to a monoenergetic electron beam of the energy 5.74 keV, the average beta energy of a tritium source, at normal incidence is very-different from the generation function obtained if one takes into account energy and angular distribution of the incident betas.

C. Two-Sided Generation

The one-sided configuration considered above is not a very good one, because half of the beta-particles produced by the source simply move away from the sample. A better use of betas would obviously be to sandwich a thin layer of the beta-source, e.g. tritiated amorphous silicon [10], between two semi-infinite silicon slabs. Then, all betas are used, and, besides, those betas that leave one slab (see discussion in section IVA) will enter the opposite slab to continue generating electron-hole pairs there. The resulting generation function is shown in Fig. 4 as solid black curve. It can be fitted quite accurately with the stretched exponential expression (4) with the fit parameters

$$g_0 = 13.69\,\text{nm}^{-1};\ z_p = 55\,\text{nm};\ \beta = 0.5\,. \qquad (6)$$

In the case of two-sided generation, each beta produces on average 1506 electron-hole pairs. This is about 50% more than in the case of one-sided generation.

D. Limit Efficiency

Assuming that all beta-generated electron-hole pairs contribute to the output power, an upper efficiency limit of a beta-battery can be obtained as the ratio of the energy extracted from an electron-hole pair to the energy necessary to produce a pair [1]

$$\eta_{\max} = \frac{E_g}{E_{ehp}} = \frac{E_g}{2.8 E_g + 0.5\,\text{eV}}\,, \qquad (7)$$

which equals 35.7% for wide-bandgap semiconductors and 30.8% for silicon. As our calculations show, this estimate is too optimistic for a silicon/tritium one-sided combination. If all electron-hole pairs generated contributed to the output power, then the upper efficiency limit is

$$\eta_{\max} = \frac{N_p E_g}{E_\beta}\,, \qquad (8)$$

where E_β is the mean energy of a single beta-particle and N_p is the mean number of pairs produced by one beta. For a silicon/tritium combination, the upper efficiency limit is 21.5% in one-sided configuration. For the two-sided generation, the limit efficiency is 29.4%, a bit lower than the result obtained from Olsen's expression (7).

V. CONCLUSIONS

Our simulations have shown that the electron-hole pair generation in silicon strongly depends on the energy and angular distribution of the incident beta-particles. The overall electron-hole pair generation function in silicon by betas coming from a tritium source can be approximated rather accurately by a stretched exponential function (4) with the parameters given by (5) for one-sided and (6) for two-sided configuration. Betaconversion efficiency turns out to be 21.5% for one-sided and 29.4% for two-sided configuration.

ACKNOWLEDGEMENT

M.E. is grateful to the Natural Sciences and Engineering Research Council of Canada (NSERC) and to the Research and Development Corporation of Newfoundland and Labrador (RDC) for financial support.

REFERENCES

[1] L.C. Olsen, P. Cabauy, and B.J. Elkind, "Betavoltaic power sources," *Physics Today*, vol. 65, pp. 35-38, 2012.

[2] R. Shimizu and Ding Ze-Jun, "Monte Carlo modelling of electron-solid interactions," *Reports on Progress in Physics*, vol. 55, pp. 487-531, 1992.

[3] H. Demers, N. Poirier-Demers, A. Réal Couture, D. Joly, M. Guilmain, N. de Jonge, D. Drouin, "Three-dimensional electron microscopy simulation with the CASINO Monte Carlo software," *Scanning*, vol. 33, pp. 135-146, 2011.

[4] D.C. Joy and S. Luo, "An empirical stopping power relationship for low-energy electrons," *Scanning*, vol. 11, pp. 176-180, 1989.

[5] Z. Czyżewski, D. O'Neill MacCallum, A. Romig, and D.C. Joy, "Calculations of Mott scattering cross section," *Journal of Applied Physics*, vol. 68, pp. 3066-3072, 1990.

[6] P. Hovington, D. Drouin, R. Gauvin, "CASINO: A new Monte Carlo code in C language for electron beam interaction —Part I: description of the program," *Scanning*, vol. 19, pp. 1-14, 1997.

[7] C.A. Klein, "Bandgap dependence and related features of radiation ionization energies in semiconductors", *Journal of Applied Physics*, vol. 39, pp. 2029-2038, 1968.

[8] U. Werner, F. Koch and G. Oelgart, "Kilovolt electron energy loss distribution in Si," *Journal of Physics D: Applied Physics,* vol. 21, pp. 116-124, 1988.

[9] *KATRIN design report – 2004.* Forschungszentrum Karlsruhe. https://publikationen.bibliothek.kit.edu/270060419.

[10] T. Kosteski, N.P. Kherani, P. Stradins, F. Gaspari, W.T. Shmayda, L.S. Sidhu and S. Zukotynski, "Tritiated amorphous silicon betavoltaic devices", *IEE Proceedings - Circuits Devices and Systems*, vol. 150, pp. 274-281, 2003.

Multi-objective optimization for color-tunability and transparency in colloidal quantum dot solar cells

Ebuka S. Arinze,[1] Botong Qiu,[1] Nathan Palmquist,[2] Yan Cheng,[1] Yida Lin,[1] Gabrielle Nyirjesy,[2] Gary Qian,[1] and Susanna M. Thon[1,*]

[1]Department of Electrical and Computer Engineering, Johns Hopkins University, 3400 N. Charles Street, Baltimore, MD, 21218, USA
[2] Department of Material Science and Engineering, Johns Hopkins University, 3400 N. Charles Street, Baltimore, MD, 21218, USA

ABSTRACT — Owing to their solution-processing flexibility, band gap tunability and infrared responsivity, colloidal quantum dots (CQDs) are favorable materials for building solar cells with tunable spectral profiles. Here, we design, optimize and fabricate multicolored and semitransparent CQD devices using thin-film interference and multiobjective optimization algorithms. We obtain a target color or transparency level by maximizing reflection or transmission over a relevant wavelength range while simultaneously maximizing device photocurrent. We fabricate color-tuned devices with photocurrents of 10-15 mA/cm² and semitransparent devices with ~ 30% visible transparency. Our optimization technique creates a foundation for the custom-design of spectrally selective optoelectronic devices.

Index Terms —photovoltaic cells, interference, thin films, optical devices, color and vision.

I. INTRODUCTION

Color-tuned and semi-transparent solar cells, photovoltaic devices with controlled and tunable reflection and transmission spectra, are of significant interest due to their potential applications in building-integrated photovoltaics, vehicular heat and power management, and multijunction photovoltaics. Solution-processed solar cells are ideal for these large-area applications due to their ease and flexibility of fabrication, thin film and lightweight nature, associated low costs, and high efficiency potential.

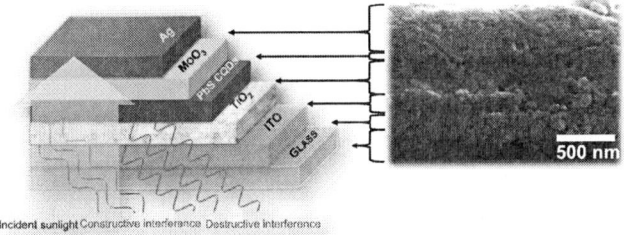

Fig. 1. CQD-based solar cell diagram illustrating the spectrally-dependent optical interference patterns that can result from thickness manipulation of the stuctural layers. As incident broadband sunlight passes through the device, constructive or destructive interference occurs, resulting in wavelength-dependent reflection and transmission, giving the cell its apparent color or transparency. Right: Cross-sectional scanning electron microscope (SEM) image of a device structure.

Strategic and selective reflection or transmission in the visible spectral region leads to tunable apparent device color or visual transparency. Color-tuning and semitransparency are typically achieved at the expense of useful absorption and photocurrent in solar cells based on solution-processed materials such as perovskites and organic polymers that mostly absorb visible light [1,2]. Two benefits of colloidal quantum dots (CQDs) for color-tuned photovoltaics are their tunable band gap, a result of quantum confinement effects, and their large spectral absorption range that can extend into and beyond the near infrared (NIR) portion of the spectrum [3]. If suitable spectral engineering techniques could be employed in CQD solar cell development, the optical loss in the visible portion of the spectrum could be compensated for by strong absorption in the NIR.

II. OPTIMIZATION TECHNIQUE

CQD solar cells are typically composed of several different layers with optical thicknesses on the order of the relevant solar wavelengths. If the films are smooth enough, this results in complex multilayer interference, illustrated in Fig.1 for our model PbS CQD heterojunction solar cell. The structure consists of an optically thick glass substrate and thin layers of indium tin oxide (ITO, bottom contact), TiO_2 (n-type semiconductor), PbS CQD film (p-type semiconductor), MoO_3 (buffer layer), and Ag (top electrode).

We use the Transfer Matrix Method (TMM) [4], which inputs layer thicknesses and refractive indices and outputs normalized electric field profiles within the multilayer stack, to calculate the optical properties of our devices. Device "transparency" is computed by averaging transmittance data over the visible wavelength range (420 nm – 680 nm) output by the TMM calculations.

The first step in our method is to use the TMM to calculate the expected reflectance spectrum of our multilayered device stack by inputting the corresponding thicknesses and associated optical constants. This calculated reflectance spectrum is a unique property of the device, and by combining it with an illuminating spectrum, we obtain the associated reflectance spectrum. In our application, we use the

978-1-5090-5606-4/17 $31.00 © 2017 IEEE

AM1.5G spectrum as our illuminating spectrum for computation.

To obtain the equivalent color, the objective is to find the XYZ tristimulus values via application of color matching functions. Here, we use the 1931 CIE color matching functions [5] in our computations. By normalizing these XYZ tristimulus values, we are able to obtain the xyz chromaticity coordinates. These xyz coordinates are then, in turn, converted to rgb color space values via a transformation matrix. The calculated xy coordinates are plotted on a 2-dimensional chromaticity plot, as shown in Fig. 3b, and are represented by their associated rgb values.

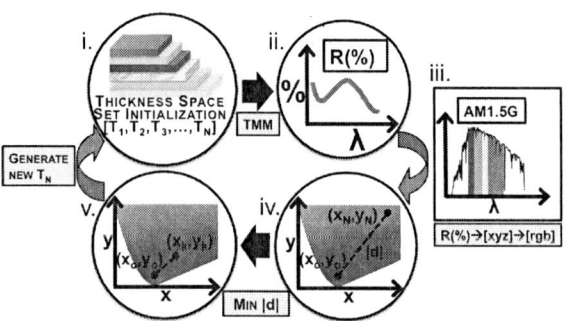

Fig. 2. Graphical depiction of the optimization technique used to generate cells with specific reflection and absorption spectra. A space set of thickness combinations is (i) initialized, and each combination is transformed to (ii) a reflection spectrum via the Transfer Matrix Method. These spectra in combination with incident (iii) AM1.5G spectrum and color-matching functions are translated to rgb colors on (iv) chromaticity plots where the distance to the intended color is (v) minimized. This optimization cycle repeats until a global minimum is found.

In order to optimize the apparent color as well as the performance of our devices, we employ particle swarm optimization (PSO), a population-based algorithm [6], tailored for our specific application, as illustrated in Fig. 2. A combination of solution thickness sets, based on the initial specified "swarm size", is initialized and fed into the TMM to generate associated reflection spectra, which are then converted to apparent color. These [rgb] co-ordinates are then optimized for a specific reflected color (wavelength) response by minimizing the distance between the target point and solution point on the chromaticity plot, thus obtaining a global solution via multiple iteration cycles. This cycle is illustrated in Fig 2. Additionally, we specify practical experimental lower and upper bounds and photocurrent constraints in our

multi-objective optimization algorithm [7] sequence to navigate for global solutions.

III. FABRICATION RESULTS AND DISCUSSION

Based on designs generated using our optimization technique, we fabricated several proof-of-principle CQD solar cell devices using PbS CQDs with exciton peak wavelengths near 950 nm. We employed commercial ITO-coated glass substrates with ITO thicknesses of 28 nm for our "red" and "green" cell designs. For the "blue" device, we deposited ITO on a glass substrate using e-beam evaporation, followed by an annealing process, to achieve our simulated target optical thickness. The TiO$_2$ layer was also deposited using e-beam evaporation for accurate thickness control. The PbS CQD layer was deposited using a layer-by-layer solid state ligand exchange process. The top contact was composed of a thin MoO$_3$ buffer layer and Ag, both of which were deposited via e-beam evaporation.

Additionally, we fabricated semi-transparent devices based on our optimization technique. The top contact of these devices was a composite transparent electrode consisting of spin-coated Ag nanowires and ITO nanoparticles. Our test devices had measured average visible transparencies (AVTs) ranging from 27.3% to 32.2%. The measured transmittance spectrum of the highest efficiency device is plotted in Fig. 3(a).

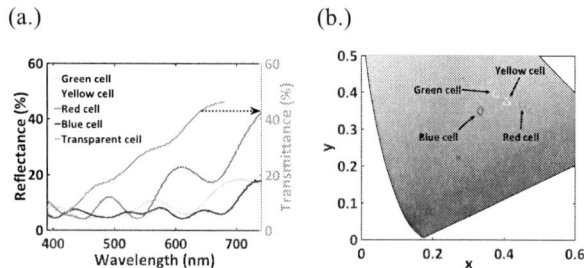

Fig. 3. (a) Experimental reflectance and transmittance spectra for colored and semi-transparent solar cells, respectively. (b) Chromaticity plot showing the computed coordinates for different colored devices. Crosses indicate design points while the corresponding colored shapes indicate experimental points.

The reflection spectra and calculated apparent colors of our fabricated devices are shown in Fig. 3(a)-(b). Photographs of five different types of CQD solar cells are shown in Fig. 4, and detailed electrical performances and design parameters are shown in Table 1.

Cell Type	V_{OC} (V)	J_{SC} (mA/cm^2)	FF	PCE (%)	Design Parameters ITO/TiO$_2$/PbS/MoO$_3$/Ag (nm)
Blue	0.56±0.01	14.6±0.6	0.44±0.01	3.6.±0.1	240/113/400/100/475
Green	0.55±0.02	12.1±0.7	0.42±0.02	2.8±0.1	28/150/297/12/60
Yellow	0.53±0.01	12.6±0.5	0.41±0.01	2.7±0.2	28/150/297/12/60
Red	0.50±0.03	10.3±0.8	0.41±0.03	2.1±0.3	28/166/222/22/274
Transparent	0.46±0.05	5.2±0.7	0.31±0.01	0.8±0.1	28/349/170/17/85(ITO)

Table 1. Average performance characteristics of colored and transparent CQD solar cell devices showing open-circuit voltage (V_{OC}), short-circuit current (J_{SC}), fill factor (FF) and power conversion efficiency (PCE). All measurements are for at least 5 devices.

IV. OUTLOOK

Using thin film interference methods and optimization algorithms, we developed a technique for generating arbitrary spectral profiles in multilayered solar cell structures [8]. To obtain a target apparent color or transparency level, our model maximizes reflection or transmission over a relevant wavelength range while maximizing photocarrier generation. Our analysis indicated that designs with minimum transparency do not necessarily correspond to the highest attainable device photocurrent, providing a route for achieving high efficiency in color-tuned devices. Experimentally, we fabricated proof-of-principle blue, green, yellow, red and semi-transparent devices. The measured reflectance and transmittance spectra matched well with the predicted color and transparency levels.

Future studies will emphasize expanding the application of our model to hybrid materials systems (single junction and tandem design structures based on other materials) and explicitly including additional (electronic) loss mechanisms. Lastly, combined with the progressive development of more efficient room-temperature-processed transparent electrode materials, this study should lead to diverse applications of flexible optoelectronic devices.

REFERENCES

[1] W. Zhang, M. Anaya, G. Lozano, M. E. Calvo, M. B. Johnston, H. Míguez, and H. J. Snaith, "Highly Efficient Perovskite Solar Cells with Tunable Structural Color," *Nano Lett*, vol. **15**, pp. 1698–1702, 2015.

[2] H. J. Park, T. Xu, J. Y. Lee, A. Ledbetter, and L. J. Guo, "Photonic Color Filters Integrated with Organic Solar Cells for Energy Harvesting," *ACS Nano*, vol. **5**, pp. 7055–7060, 2011.

[3] G. H. Carey, A. L. Abdelhady, Z. Ning, S. M. Thon, O. M. Bakr, and E. H. Sargent, "Colloidal Quantum Dot Solar Cells," *Chem. Rev*, vol. **115**, pp. 12732–12763, 2015.

[4] G. F. Burkhard, E. T. Hoke, and M. D. McGehee, "Accounting for Interference, Scattering, and Electrode Absorption to Make Accurate Internal Quantum Efficiency Measurements in Organic and Other Thin Solar Cells," *Adv. Mater*, vol. **22**, pp. 3293–3297, 2010.

[5] T. Smith and J. Guild, "The C.I.E. colorimetric standards and their use," *Trans. Opt. Soc*, vol. **33**, pp. 73, 1931.

[6] J. Kennedy, "Particle Swarm Optimization," in *Encyclopedia of Machine Learning*, C. Sammut and G. I. Webb, eds. (Springer US, 2011), pp. 760–766.

[7] K. Deb, *Multi-Objective Optimization Using Evolutionary Algorithms* (John Wiley & Sons, 2001).

[8] E. S. Arinze, B. Qiu, N. Palmquist, Y. Cheng, Y. Lin, G. Nyirjesy, G. Qian, and S. M. Thon, "Color-tuned and transparent colloidal quantum dot solar cells via optimized multilayer interference," *Opt. Express*, vol. **25**, pp. A101-A112, 2017.

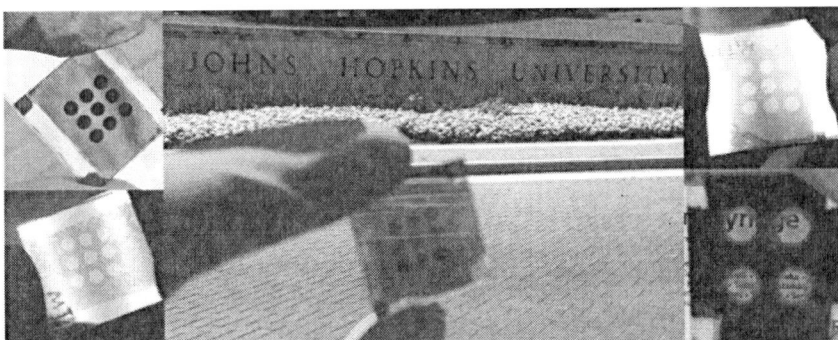

Fig. 4. Photographs of blue (top left), green (bottom left), red (center), yellow (top right), and semi-transparent (bottom right) CQD solar cells with layer structures based on the designs from our computational optimization method.

978-1-5090-5606-4/17 $31.00 © 2017 IEEE

Cubic phase In$_x$Ga$_{1-x}$N/GaN quantum wells for their application to tandem Solar Cells

C. A. Hernández-Gutiérrez[1], Y. L. Casallas-Moreno[2], Dagoberto. Cardona[2], Yu. Kudriavtsev[3], A. Morales-Acevedo[3], G. Santana-Rodríguez[4], M. López-López[2].

1. Cinvestav-IPN, Nanotechnology program, DF, 07360, México,

2. Cinvestav-IPN, Physics department, DF, 07360, México,

3. Cinvestav-IPN, Electrical engineering department SEES, DF, 07360, México,

4. National Autonomous University of Mexico UNAM, Coyoacán, DF, 04510, México,

*Corresponding author: chernandez@fis.cinvestav.mx

Abstract — **In$_x$Ga$_{1-x}$N/GaN QWs were grown by MBE for tandem solar cell applications. High Indium concentration in the In$_x$Ga$_{1-x}$N ternary alloy was measured by SIMS, reaching an atomic concentration ~27% corresponding to PL emission at 2.29 eV from the well. The error between the modeled emission of the well and the experimental PL emission was only 30 meV for QW1 and 90 meV for a second well (QW2). Therefore, because of high Indium incorporation, the structure is an attractive candidate to be incorporated as a top material in a tandem solar cell.**

Index Terms — **cubic-phase, GaN, In$_x$Ga$_{1-x}$N, quantum-well, SIMS, PL.**

I. INTRODUCTION

In$_x$Ga$_{1-x}$N based solar cells present technological challenges due to the difficulty of Indium incorporation. Typically, when Indium concentration exceeds 20%, the obtained ternary alloy tends to show phase separation and this problem gets worse as the thickness of the In$_x$Ga$_{1-x}$N layer increases [1-3]. Fischer et al. [4] have estimated the critical thickness of stressed In$_x$Ga$_{1-x}$N before phase separation, obtaining 12 nm for Indium concentrations of 25% and as the concentration increases the critical thickness is reduced. Hence, although the In$_x$Ga$_{1-x}$N has a high absorption coefficient, thick films are difficult to achieve without the phase separation occurring.

In addition, it is difficult to achieve p-type doping for In$_x$Ga$_{1-x}$N layers so that p-n junctions are difficult to obtain too. In order to overcome these technological issues, several approaches have been attempted such as the realization of (PIN) double heterostructures p-GaN/In$_x$Ga$_{1-x}$N/n-GaN and solar cells based on In$_x$Ga$_{1-x}$N/GaN multi-quantum wells (MQWs), but poor efficiencies have been reported, less than 3.5% [5-7]. Due to low Indium concentration, high bandgaps are achieved and a poor photocurrent is generated. Furthermore, when the Indium incorporation increases in the active layer, the crystalline and surface quality reduce and the short circuit current density is also reduced by a recombination process. Moreover, low Indium incorporation also affects the external quantum efficiency (EQE) reducing the range of absorbed photons with a cutoff wavelength of less than 480 nm.

To improve the poor photocurrent due to the low Indium incorporation avoiding phase separation, we propose the growth of multi-quantum wells (MQWs) in the thermodynamic metastable cubic phase of III-Nitrides. Cubic phase Nitrides have some advantages in comparison with the more familiar hexagonal phase, such as requiring less Indium in the InxGa1-xN layers to achieve lower bandgap, the absence of spontaneous and piezoelectric polarization fields, the ability to cleave cubic (001) on the perpendicular {110} cleavage planes for device fabrication, and the enhanced mobility of holes [8, 9]. Furthermore, one of the most important advantages of cubic phase nitrides for solar cell application is the possibility to be integrated with GaAs(001) or Si(001) for a tandem solar cell.

Therefore, in this work, In$_x$Ga$_{1-x}$N/GaN quantum wells were grown by plasma-assisted MBE tuned to absorb radiation around 2 eV, as the top material for a tandem solar cell. The growth temperatures of the quantum wells were chosen to achieve an appropriate distribution of Indium with a maximum concentration higher than 20%.

II. EXPERIMENT

The samples were carried out by plasma-assisted MBE on a (001) GaAs substrate. After the substrate thermal desorption, a GaAs film was grown as a buffer layer. Then, GaN was grown at low temperature (LT< 700°C). After this GaN layer was grown, the temperature was raised up to 720°C to deposit a 30 nm of the GaN barrier layer. Then, a 10 nm thick In$_x$Ga$_{1-x}$N quantum well was grown at 600°C, with an Indium beam equivalent pressure (BEP) of 1.8x10^{-7} Torr. The barrier and the quantum well deposition processes were repeated varying only the well growth temperature, keeping constant the GaN barrier deposition temperature. The wells growth temperatures for sample S1 were 600°C and 650°C, for the sample S2 were 600°C, 620°C and 640°C. Figure 1 illustrates the schematic diagram for sample S1 (S2 is not shown). The elemental depth distribution was analyzed using a TOF-SIMS-5 secondary ion mass spectrometer from ION-TOF GmbH using Cs+ as primary ions with an energy of 500 eV to avoid ion implantation. Photoluminescence was measured employing a 325 nm HeCd LASER at a low temperature of ~11 K.

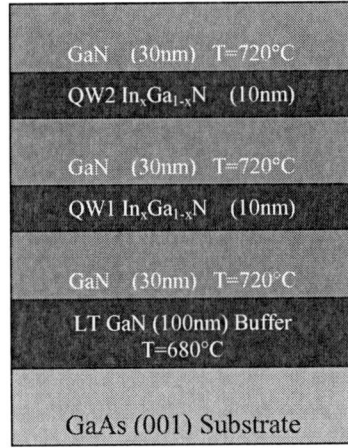

Fig. 1. Schematic diagram of In$_x$Ga$_{1-x}$N/GaN quantum wells for sample S1. The schema of sample S2 is not shown here.

III. RESULTS AND DISCUSSION

To quantify the Indium concentration incorporation, the SIMS analysis was performed as is shown in fig 2. From SIMS analysis Indium incorporation as a function of temperature was observed, due to the temperature-dependent sticking coefficient. For sample S1 the Indium atomic concentrations were ~27% and ~20% for each well, respectively (the SIMS analysis of S2 is not shown but is very similar to that of sample S1).

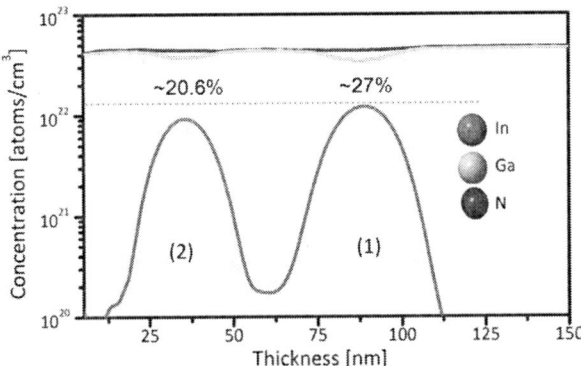

Fig 2. SIMS depth profile for sample S2.

After estimation of Indium concentration by SIMS, the quantum energy levels of each well can be calculated by solving the finite potential square well Schrödinger equation using the effective masses approximation. The conduction and the valence band discontinuity were obtained by Anderson's rule, moreover, and the In$_x$Ga$_{1-x}$N bandgap Eg of the wells was calculated by an empirical relation, employing the bowing parameter [10]. As a result, the estimated basal energy for each well of sample S1 was 2.59 eV corresponding to the well grown at 650 ° C and 2.32 eV for the well grown at 600 ° C. The band diagram for sample S1 is illustrated in fig 3.

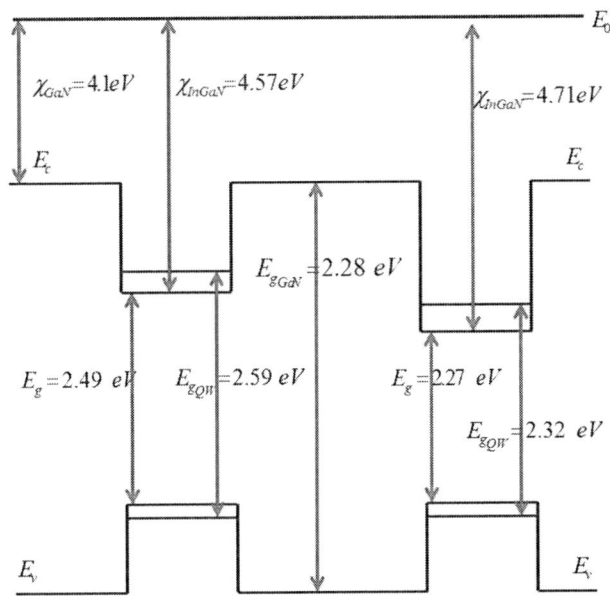

Fig 3. QWs band diagram of sample S1.

To demonstrated that the above methodology is adequate to study our QWs, low-temperature photoluminescence (PL) measurements were done. PL shows a broad spectra centered at green wavelengths due to the high indium incorporation. The spectra widening is because the variation of Indium composition from high to low as is shown in fig 1. After deconvolution analysis of S1 each emission contribution was discriminated. The basal energy transition of the well grown at 600°C is approximately at 2.29 eV and of the well grown at 650°C is 2.68 eV with an error from our calculations of 30 meV for QW1 and 90 meV for QW2. Then, our energy calculations are very accurate and the error could come from the empirical estimation of Eg and the effective masses of electrons and holes. In addition, the PL of sample S2 is similar to S1, but due to three QWs, the GaN emission was suppressed attributed to the high absorption coefficient of the QWs. So the PL spectra demonstrates the high Indium incorporation into the cubic In$_x$Ga$_{1-x}$N as measured by SIMS and thus it can be used as a high energy photon-absorbing layer in a tandem solar cell.

Very recently (2017), Valdueza-Felip at al [5] reported high Indium incorporation of 40% for the more familiar hexagonal phase obtaining an Eg of 2.25 eV. Hence, its worth to note here that we are reporting 13% less Indium than for their case (40%), but with a badgap about 40 meV below. Therefore, the metastable cubic phase should be considered as a promising material, since less Indium incorporation is required and also it will be easier to integrate with GaAs(001) or Si(001) in a tandem solar cell.

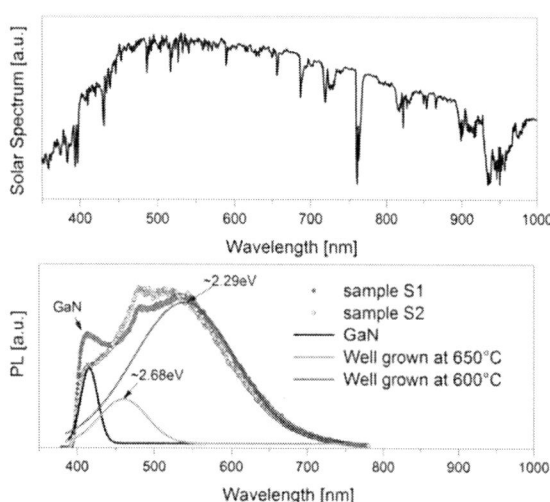

Fig 4. PL emsion of samples S1 and S2. The sample S1 was deconvoluted to discriminate the origin of each emission. The solar spectrum was added to show qualitatively what part of this spectrum could be absorbed by our quantum wells.

VI. CONCLUSION

$In_xGa_{1-x}N$/GaN QWs were grown by MBE, modeled and characterized by SIMS and PL. SIMS depth profile show a high Indium concentration of ~27%, while photoluminescence measurements analysis corroborated the high In incorporation with a basal energy emission in the well around 2.29 eV. Therefore, the cubic phase $InxGa1_{-x}N$/GaN QWs showed an attractive reduction of band gap from 3.28 eV to 2.29 eV without phase separation, making this structure suitable to absorb photons from the high-energy region of the solar spectrum for a tandem-type solar cell.

ACKNOWLEDGEMENT

The authors would like to thank CONACYT Mexico for financial support.

REFERENCES

[1] J. Wu, "When group-III nitrides go infrared: New properties and perspectives", *Journal of applied physics 106, 011101, 2009.*

[2] A. Yamamoto, M. R. Islam, T.-T. Kang, and A. Hashimoto, "Recent advances in InN-based solar cells: Status and challenges in InGaN and InAlN solar cells", physica status solidi C, vol. 7, no. 5, pp. 1309–1316, 2010.

[3] F. A. Ponce et al., "Microstructure and electronic properties of InGaN alloys", physica status solidi B, vol. 240, no. 2, pp. 273–284, 2003.

[4] A. Fischer, H. Kühne, H. Richter, Physical Review Letters, "New Approach in Equilibrium Theory for Strained Layer Relaxation", 73 (20) 2712, (1994).

[5] S. Valdueza-Felip, A. Ajay, L. Redaelli, M.P. Chauvat, P. Ruterana, T. Cremel, M. Jiménez-Rodríguez, K. Kheng, E. Monroy, "P-i-n InGaN homojunctions (10–40% In) synthesized by plasma-assisted molecular beam epitaxy with extended photoresponse to 600 nm", Solar Energy Materials & Solar Cells 160 355–360 (2017).

[6] Chloe A. M. Fabien, Member, IEEE, Aymeric Maros, Student Member, IEEE, Christiana B. Honsberg, Senior Member, IEEE, and William Alan Doolittle, IEEE, "III-Nitride Double-Heterojunction Solar Cells With High In-Content InGaN Absorbing Layers: Comparison of Large-Area and Small-Area Devices", IEEE journal of photovoltaics Vol. 6, 2, 460-464, (2016).

[7] Sirona Valdueza-Felip, Anna Mukhtarova, Louis Grenet, Catherine Bougerol, Christophe Durand, Joel Eymery, and Eva Monroy, "Improved conversion efficiency of as-grown InGaN/GaN quantum-well solar cells for hybrid integration", Applied Physics Express 7 032301(1)-032301(4) (2014).

[8] Shunfeng L, Schomann J, As D J and Lischka K, "Room temperature green light emission from nonpolar cubic InGaN/GaNInGaN/GaN multi-quantum-wells", Applied physics letters, 90 071903 071903(1)- 071903(3), (2007).

[9] As D J, Schikora D, Greiner A, Lubbers M, Mimkes J and Lischka K, "p- and n-type cubic GaN epilayers on GaAs", Physical Review B 54 R11118 (1996).

[10] J. Wu, W. Walukiewicz, K. M. Yu, and J. W. Ager III, E. E. Haller, Hai Lu and William J. Schaff, "Small band gap bowing in In1-xGaxN alloys", Applied physics letters. 80, 25 (2002).

Modeling of p-i-n GaAsPN/GaP MQWs solar cell: towards lattice matched III-V/Si tandem

Khim Kharel and Alexandre Freundlich

Center for Advanced Materials, University of Houston

Houston, Texas, 77096, USA

Abstract — Dual junction thin film III-V/Si solar cell's efficiencies are limited to 20% because of high dislocation densities (> $10^6/cm^2$) due to lattice mismatch. GaAsPN is a promising dilute nitride semiconductor lattice matched to silicon and suitable for the optimal bandgap for III-V/Si tandem solar cells. We designed defect tolerant, resonantly coupled asymmetric GaAsPN/GaP quantum wells in the intrinsic region of p-i-n MQWs solar cell for faster collection of photo-carriers (few picoseconds). Optimum number of resonantly coupled quantum wells of GaAsPN/GaP projects short circuit current of 16mA/cm^2 under AM1.5G spectrum using realistic drift-diffusion model. We have also simulated spectral response, JV characteristics and maximum efficiency of GaAsPN/GaP p-i-n MQWs solar cell where efficiency surpass the Shockley–Queisser limit for a single-junction device.

Index Terms — III-V-N semiconductor materials, Resonant thermo-tunneling design, Quantum well, Drift-diffusion model.

I. INTRODUCTION

III-V/Si tandem solar cells create huge attention for the reduction of levelized cost of solar electricity by increasing cell's efficiency more than 30%. Radiative efficiency of silicon based tandem maximum at 41.9% (1.74eV/1.12eV) under AM 1.5G spectrum [1]. Similarly, optimized top cell thickness of series connected III-V/Si (1.7eV/1.12eV) tandem efficiency maximum at 37% under AM 1.5G spectrum [2]. In monolithic integration of III-V/Si, realistic efficiencies of dual junction solar cells are limited to 20% because of high dislocation densities (> $10^6 cm^{-2}$) due to lattice mismatch which degrades minority carrier's properties. Dilute nitride based lattice matched III-V/Si solar cells are more promising but less explored. GaAsyP1-x-yNx is a direct band gap semiconductor (N>0.6%) and having a lattice constant matched with silicon at y=4.7*x-0.1. Geisz et al. in 2005 [3] has reported for the first time, a dual junction GaAsPN /Si tandems solar cell with an efficiency of 5.2% under AM1.5G using GaP as a nucleation layer. Similarly, Geisz et al. in 2002 [4] and M.da Silva et al. in 2015 [5] have also reported 3.8% and 2.3% efficiencies under AM 1.5G spectrum using p-i-n GaAsPN based solar cell on GaP substrate taking different thickness of GaAsPN absorber. Deficient performance of solar cell is because of degradation of minority carrier properties [6, 7] and difficulties in doping [8] in the bulk dilute nitride semiconductors. Based on our previous worked on a dilute nitride MQWs solar cell, we have applied defect tolerant design of quantum wells in the intrinsic region of p-i-n solar cells whose thickness is much smaller than that of minority carrier diffusion length. To facilitate carrier extraction in a deep quantum well of the conduction band, we have used thermoresonant tunneling design purposed by A.Freundlich and A. Alemu (US pattern 20130186458A1). In a coupled quantum well, carriers are thermally excited to higher bound states and then resonantly coupled with a weaker bound state of adjacent coupled well till complete extraction.

II. MODELING APPROACH

Bandgap evolution of purposed design is calculated using 8 band k.p Hamiltonian with Band Anti-crossing model to the conduction band to account for the lesser amounts of nitrogen impurities. Confinement energies of electrons and holes are from transfer matrix approach and envelop function approximation [9]. Taking account of all possible transitions from electrons to holes' energy levels, optical absorption in a coupled quantum wells system are evaluated using Fermi golden rule including the effect of excitons. We have also evaluated thermionic escape time for carriers excited to higher confined states and tunneling escape time for resonantly coupled states using confinement energies, potential barrier height and electric field associated across depletion region of p-i-n device [10]. Internal quantum efficiency is calculated from total thermionic escape, tunneling and recombination time i.e.

$$IQE \equiv \frac{\tau_{tot}}{\tau_{esp}} \quad (1) \quad \text{Where,}$$

escape and total times are $\dfrac{1}{\tau_{esp}} \equiv \dfrac{1}{\tau_{thermal}} + \dfrac{1}{\tau_{tun}}$ and

$$\frac{1}{\tau_{tot}} \equiv \frac{1}{\tau_{esp}} + \frac{1}{\tau_{recombinaton}} \quad (2)$$

Where tunneling escape time for potential of V (z) and thermal escape time with electric field are given by [10]

$$\frac{1}{\tau_{tun}} \equiv \frac{n\hbar}{2m_w L^2} \exp\left(-\frac{2}{h}\int_0^b \sqrt{2m_b\big(qV(z)-E_n-qFz\big)}dz\right)$$

(3)

$$\frac{1}{\tau_{thermal}} \equiv \frac{1}{L}\sqrt{\frac{kT}{2\pi m_w}}\exp\left(-\frac{E_{barrier}(F)}{kT}\right)$$

(4)

Where, L is length of quantum well, b is barrier width m_w and m_b are the carrier effective masses in well and barrier region having effective barrier height for nth sub band with energy E_n and CB or VB offset $\Delta E_{C,V}$ given by

$$E_{barrier}(F) \equiv \Delta E_{C,V} - E_n - \frac{qFL}{2}$$

(5)

From experience, recombination time in a MQWs solar cell is taken as 100ps. For N number of coupled quantum wells, we have used IQE=$(IQE)^N$ where (IQE) is the internal quantum efficiency for intrinsic region of single coupled quantum well. Finally, we have evaluated spectral response, JV characteristics and photo-conversion efficiencies using absorption as well as IQE calculation in i region adopting drift- diffusion approach.

III. RESULT AND DISCUSSION:

Example of the calculation of confinement energies of GaAsPN/GaP for a given thickness of quantum well system as-

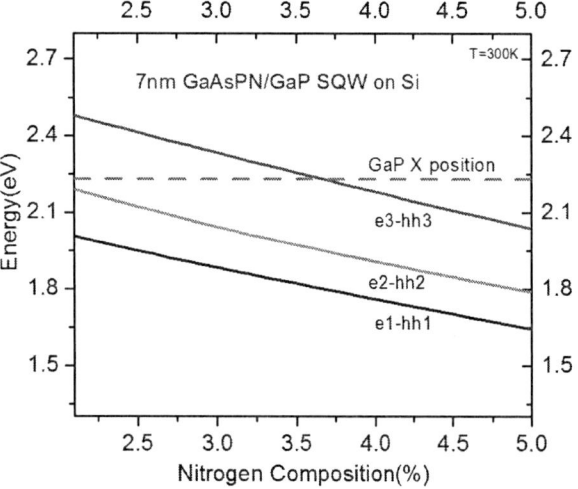

Fig 1: Confinement energies of 7 nm GaAsPN/GaP as a function of nitrogen compositions.

a function of nitrogen compositions is as shown in the figure [1]. Effective bandgap i.e. 1.72 eV can be obtained by choosing 12 nm thickness of $GaAs_{0.09}P_{0.87}N_{0.04}$/GaP quantum well. Based on the band structure calculation, we have purposed the thermo-resonant design which consists of three asymmetric coupled quantum wells of $GaAs_{0.09}P_{0.87}N_{0.04}$ of thickness 12 nm, 5.9 nm and 2.5 nm separated by 4 nm Gallium phosphide barrier as shown in the figure [2]. Similarly, for a given range of nitrogen composition, we have evaluated optical absorption coefficient at room temperature of such coupled quantum well system using band structure calculation and Fermi Golden rule taking account of excitonic effect. The enhancement of the absorption can be observed in the performance of internal quantum efficiency in the intrinsic region as shown in figure [3]. The major causes for that enhancement: increase in the effective mass of dilute nitride alloys, there by density of states and we have also taken excitonic absorption along with thermal broadening effect.

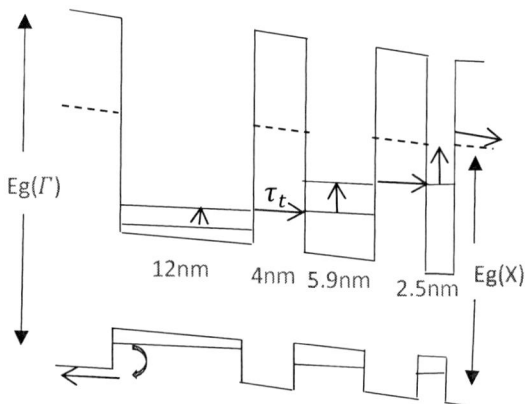

Fig 2: Thermoresonant tunneling design of GaAsPN/GaP quantum well

Based on our design, using calculated values of tunneling escape times, thermionic escape times in each route for escape mechanism and recombination time, IQE for a single coupled quantum well is about 0.9983 and 0.95 for thirty number of such coupled quantum wells. However, for substantial number of coupled quantum wells, IQE decreases in a large ratio because of the reduction of electric field associated with the depletion region of p-i-n cell due to increase in the thickness of the intrinsic region. Using absorption coefficient and IQE of the intrinsic region with respect to the number of coupled quantum wells, internal quantum efficiency of the p-i-n MQWs solar cell is obtained using the realistic drift-diffusion framework. Device parameters for Gallium Phosphide based solar cell are extracted

978-1-5090-5606-4/17 $31.00 © 2017 IEEE

from reference [11] as well as other past experiments in the literature. Fig [3] shows the calculation of internal quantum efficiencies in different regions such as emitter, base, collector

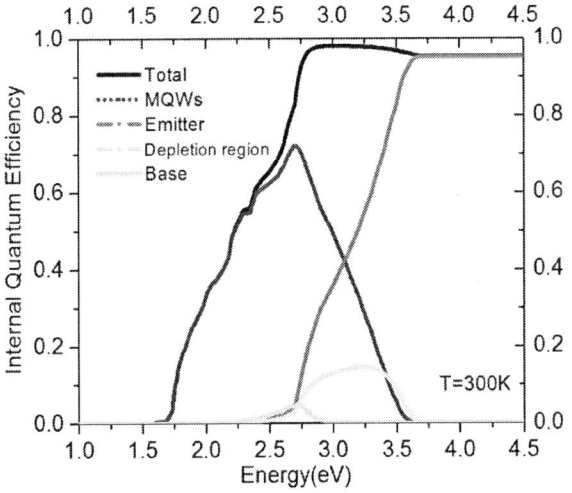

Fig 3: Internal quantum efficiencies of ten periods resonantly coupled quantum well in p-i-n GaAsPN/GaP MQWs solar cell having contribution in different regions.

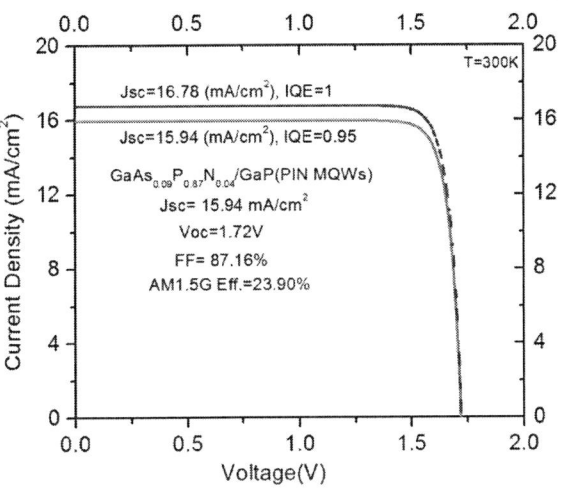

Fig 4: Calculated current versus voltage characteristics of purposed p-i-n GaAsPN/GaP MQWs based solar cell with 100% and 95% extraction of photo-carriers from coupled quantum wells.

and intrinsic region for ten periods of resonantly coupled quantum wells.

The calculated current versus voltage (JV) Characteristics of the 1.72 eV purposed p-i-n $GaAs_{0.09}P_{0.87}N_{0.04}$/GaP coupled

quantum well solar cell is as shown in the figure [4]. Dark currents were derived using the parameters extracted from past experiments of Gallium Phosphide (GaP) based solar cells. Similarly, minority charge carrier's diffusion lengths are derived as a function of doping concentrations. We have neglected any reflection or shadowing losses in the calculation. Details of the calculation is based on the dilute nitride multi quantum well solar cells calculation reported in reference [12]. Faster carrier collection prevents from recombination losses, hence improve in the performance of solar cell properties. However, in a convectional p-i-n MQWs, large recombination loses degrade both Jsc and Voc as well as efficiency. Similarly, resonant thermotunneling design uses the excessive amounts heat energy to excite carriers (otherwise wasted) to convert into electrical energy. Based on the calculation of the IQE of the intrinsic region for thirty periods of coupled quantum wells, maximum short circuit current under 1 Sun, AM 1.5G spectrum is about 16mA/cm^2 maintaining open circuit voltage in the photo-current line of the Gallium Phosphide based cells. The expected efficiency surpasses the Shockley-Queisser efficiency limit for a single junction cell. Obtaining promising preliminary projected values of the 1.72 eV p-i-n $GaAs_{0.09}P_{0.87}N_{0.04}$/GaP MQWs solar cells support monolithic integration of lattice matching III-V on silicon solar cells.

IV. CONCLUSION

Modeling approach of the design device and promising projected solar cell properties in this paper demonstrates the feasibility of the inclusion of resonantly coupled multi-quantum wells in the intrinsic region of the Gallium Phosphide based solar cell. This is a very important motivation for monolithic integration of lattice matching III-V on silicon solar cells and getting practical value of cell's efficiency above 30%. Designing of tunnel junction between III-V-N and silicon tandem solar cells is explained in reference [13].

REFERENCES

[1] James P. Connolly, Denis Mencaraglia, Charles Renard and Daniel Bouchier, Prog. Photovolt: Res. Appl. 2014; 22:810–820.

[2] S.R. Kurtz, P. Faine, J.M. Olson, J.Appl.Phys.68, 4(1990)

[3] J. F. Geisz, J. M. Olson, D. J. Friedman, K. M. Jones, R. C. Reedy, and M. J. Romero, Conference Record of the Thirty first IEEE Photovoltaic Specialists Conference, 2005, pp. 695–698.

[4] J. F. Geisz, D. J. Friedman, and S. Kurtz, Conference Record of the Twenty-Ninth IEEE Photovoltaic Specialists Conference, 2002 pp. 864–867.

[5] Mickael Da Silva, Samy Almosni, C. Cornet, Antoine L´etoublon, Christophe Levallois, SPIE, Proc. 2015, Physics, Simulation, and Photonic Engineering of Photovoltaic Devices IV.

[6] S. R. Kurtz, E. D. Jones, Appl. Phys Lett 74, 19 (1999).

[7] S.Y. Xie, S.F. Yoon, S. Wang, J. Appl. Phys 97, 7(2005).

[8] W. Li, M. Pessa, J. Toivonen, H. Lipsanen, Phys. Rev B 64, (2001) 113308.

[9] L. Bhusal and A. Freundlich, Physical Review B 75, 075321(2007).

[10] A. Freundlich, A. Alemu, IEEE Journal of Photovoltaics 2, 3 (2012).

[11] Xuesong Lu, Susan Huang, Martin B. Diaz, Nicole Kotulak, Ruiying Hao, Robert Opila, and Allen Barnett, IEEE Journal of Photovoltaic, 2, 2 (2012).

[12] A. Freundlich and A. Alemu, phys. stat. sol. (c) 2, 8, 2978–2981 (2005).

[13] Alain Rolland, Laurent Pedesseau, Jacky Even,Samy Almosni ,Cedric Robert , Charles Cornet, Jean Marc Jancu, Jamal Benhlal , Olivier Durand,Alain Le Corre,Pierre Rale,Laurent Lombez ,Jean-Francois Guillemoles , Eric Tea ,Sana Laribi, Opt Quant Electron (2014) 46:1397–1403.

InP Quantum Dot Intermediate Band Solar Cell Grown via MOCVD

Hyun Kum, Yushuai Dai, Michael Slocum, Zachary Bittner, Seth Hubbard

Rochester Institute of Technology, Rochester NY 14623

Abstract—Growth of type-II InP quantum dots in wide band-gap InGaP (E_G = 1.85 eV) for intermediate band solar cell applications is studied, grown via Aixtron close-coupled showerhead metal-organic chemical vapor deposition system. Theoretical calculation and optical characterizations verify type-II band alignment, showing approximately 300 meV electron confinement in the conduction band and around 5 meV potential barrier for holes in the valence band. AFM measurements predominantly show type-II dots (dot height greater than 15 nm), which is further verified by a single QD peak with narrow full width at half maximum obtained by photoluminescence measurements. Dot density of 0.7 x 10^{10} cm^{-2} is measured with diameter and height of 56 ± 10 nm and 18 ± 2.8 nm, respectively. This quantum dot system promises a more ideal band alignment for intermediate band assisted photon absorption and higher two-step photon absorption (TSPA) temperatures. Here, we observe TSPA response up to 200 K for a solar cell with 5x layers of InP QDs in the intrinsic region.

I. INTRODUCTION

Intermediate band solar cells (IBSC) have been proposed as a means of exceeding the detailed balance homo-junction limiting efficiency, in which photons at a lower energy than the bandgap can be absorbed via an intermediate band (IB) formed in the forbidden gap. This can be implemented by utilizing quantum dots to form the IB within a single junction solar cell. A majority of the quantum dots (QD) for IBSC work completed to date has been with the InAs QD in a GaAs bulk material system. Although this has been demonstrated to show two-step photo-absorption (TSPA) [1] and voltage preservation [2], it is only achievable at cryogenic temperatures due to the shallow conduction band offset of 0.2 eV in the QD. A material system that is a more radical departure from the InAs QDs in GaAs are InP QD in lattice matched $In_{0.49}Ga_{0.51}P$ (hereafter just InGaP) with type-II band alignment. Type-II InP QDs in InGaP have a band structure with a conduction band confinement of 300 meV with relatively small valence band energy off-set of around 5 meV [3], which acts as a hole barrier. This is a desirable band off-set for multi-step photon absorption and may allow two-step photon absorption at higher temperatures than InAs QDs. Due to this particular band energy profile, the electron and hole wavefunctions are separated, leading to longer carrier lifetimes in the IB. This is advantageous as it provides enough time for carriers populating the IB to absorb a second photon and produce current before it has a chance to recombine radiatively (or non-radiatively). Unfortunately, growth of type-II InP QDs via MOCVD has not been investigated in detail for photovoltaic applications. It has been reported by several groups that InP

QDs grow in two varieties, type A and type B. Type A is smaller and shorter with type-I band alignment, while type B is larger and taller (height typically greater than 15 nm) and exhibits the type-II band alignment that is desired for IBSC [4]. Although the relative densities of each type can be controlled, no one has reported being able to grow fully type A or type B InP QDs to date. Simulated band alignment of a single type B QD with a diameter and height of 50 nm and 20 nm, respectively, at 300 K using the nextnano simulation software (6-band $k \cdot p$) is shown in Figure 1. As can be seen, besides being able to absorb photons with energy greater than the band gap ($E > E_G$), photons with energy less than the band gap (E_L and E_H) can also be absorbed for two-step (VB to IB, then IB to CB) e-h pair generation. Theoretical max efficiency ($\sim$60%) is obtained for systems with EL of 0.7 eV and E_G of 1.9 eV. However, the InP/InGaP QD system is also expected to have a much higher efficiency limit ($\sim$50%) than conventional homo-junction solar cells ($\sim$30%, estimated from [5]).

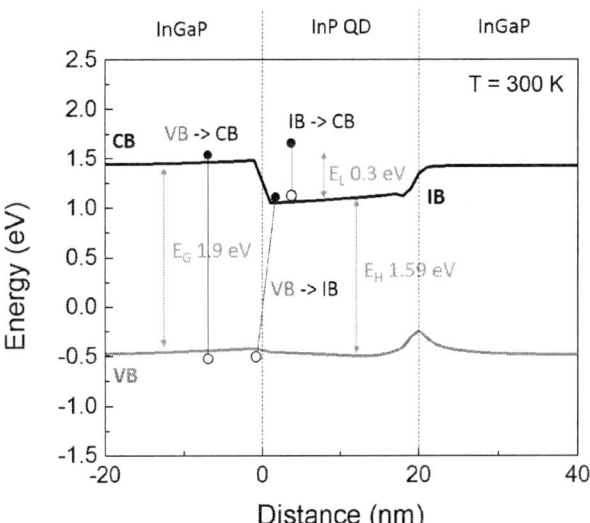

Figure 1. Band energy simulation of type-II InP QDs using nextnano (6-band $k \cdot p$ for T = 300 K. E_L is the lower energy sub-bandgap and E_H is the higher energy sub-bandgap.

II. EXPERIMENTAL

Growth was carried out in a commercial Aixtron close-coupled showerhead metal-organic chemical vapor deposition system (MOCVD) using trimethylgallium (TMGa), trimethylindium (TMIn), arsine (AsH3), and phosphine (PH3) precursors on Si-doped GaAs substrates with a 2° offcut towards the (110). All samples had an initial 50 nm unintentionally doped GaAs buffer layer grown at 580 °C at a growth rate of 2 μm/h, followed by a 100 nm InGaP buffer layer grown at the same temperature at a rate of 1.05 μm/h. Temperature was then lowered to the InP QD growth temperature of 530 °C. Photoluminescence (PL) measurements were made using a 532 nm (2.33 eV) diode laser with samples mounted on a liquid helium cryostat for cooling. For PL measurements, the dots (single layer) were capped with 5 nm low-temperature InGaP layer grown at 530 °C, followed by a high temperature 100 nm $In_{0.49}Ga_{0.51}P$ grown at 580 °C. Growth rates, temperature, and thicknesses were measured in-situ via a LayTec EpiTT monitor. Three solar cells (a baseline n-i-p InGaP, 3x InP QD, and 5x InP QD solar cells) were fabricated using standard lithography, deposition, and wet etch techniques for III-V semiconductors. Backside p-contact metal was thermally evaporated, consisting of Au/Zn/Au (20/20/300 nm), and annealed at 405 °C for 6 minutes. Ge/Au/Ni/Au (25/50/35/400 nm) metal scheme was used for the n-contact. The fabricated wafers were cleaved into 1×1 cm^2 cells for standard solar cell measurements.

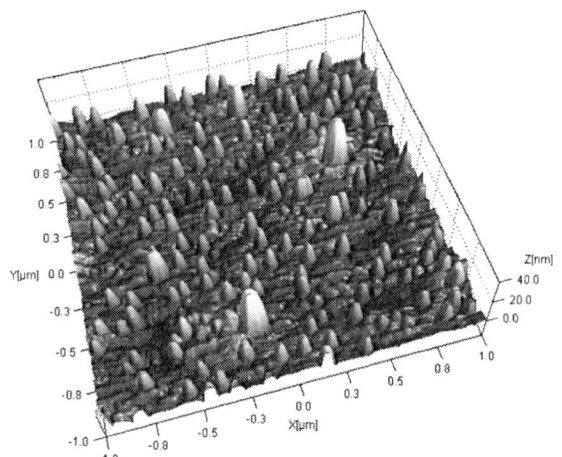

Figure 2. AFM image of InP surface QDs grown on InGaP lattice matched to GaAs.

Light IV measurements at AM0 illumination were carried out in a TS Space System close-match solar simulator with samples mounted on a cooling chuck set at 25 °C. External quantum efficiency (EQE) measurements were taken using a Newport IQE 200 system. Two-step photon absorption measurements were done using two light sources consisting of a white light source through a monochromator (primary light source for excitation from the CB to IB or VB) and a tunable IR laser with a wavelength of 1.3 μm (secondary light source for IB to CB excitation). Only the secondary light source was chopped to detect the resulting photocurrent arising from the IB to CB excitation using a preamp (SR570) and lock-in amplifier (SR830).

III. RESULTS AND DISCUSSION

Initial growth conditions were based on literature reports [3], [6]. Typical parameters that determine the size and density of QDs are growth temperature, rate, and time. Other techniques, such as growth interrupts, may also play a role. Optimal growth temperature was found to be 530 °C at a growth rate of approximately 0.10 ML/s (total nominal thickness of 5.0 ML) with a growth interrupt of 60 s. During growth interrupt, all precursors are turned off, including phosphine. We were able to achieve a maximum dot density of 0.7×10^{10} cm^{-2}. Atomic force microscopy (AFM) images are shown in Figure 2 for a scan area of 1×1 μm^2. Scanning probe image processor (SPIP) analysis of a 5×5 μm^2 AFM image (not shown) show the mean dot diameter and height of 56 ± 10 nm and 18 ± 2.8 nm, respectively.

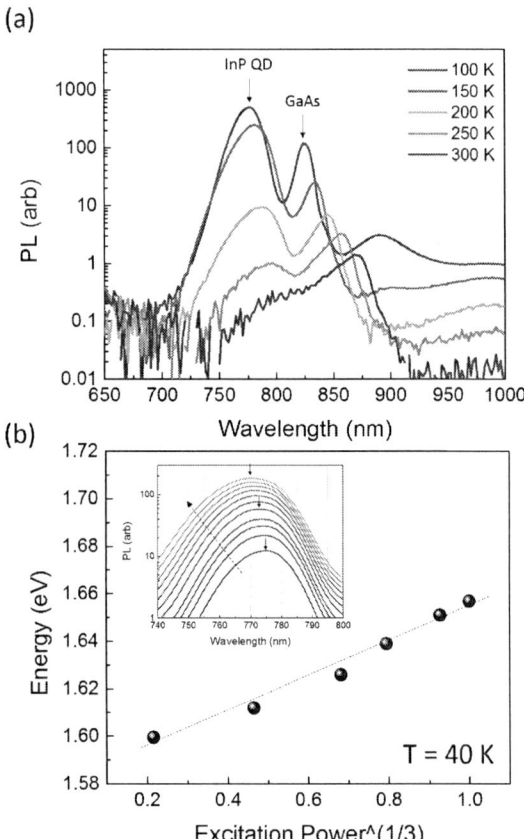

Figure 3. (a) Temperature dependent PL measurement of InP QDs in InGaP. (b) PL spectrum as a function of laser excitation power, showing blue shift of the peak wavelength, characteristic of type-II confinement structures.

It has been theoretically shown by Pryor et al. [4] that InP dots with a height of approximately 15 nm or larger in an InGaP host would have a type-II band alignment with emission near 1.6 eV. This was confirmed by simulation in Figure 1 and is experimentally verified by temperature dependent PL measurements taken from 100 K to 300 K (Figure 3 (a)) for a single layer of buried dots shown in Figure 2. Strong InP QD luminescence can be seen at 775 nm (1.6 eV) at 100 K, consistent with the energy of type-II InP QDs studied by Tagayaki et al. [3] and simulation results by nextnano (Figure 1). We were not able to observe luminescence from type-I QDs, which have been reported to appear at a higher energy of ∼1.7 eV. Again, this is an indication of growth of predominantly type B QDs. The full width at half maximum of the QD peak is around 40 meV at 40 K, indicating uniform size distribution of the InP QDs leading to reduced inhomogeneous broadening in the PL. The PL emission at ∼1.55 eV corresponds with a conduction band confinement of approximately 300 meV at RT. To verify type-II band alignment, power dependent PL was measured. Typically, there will be a linear dependence of the emission energy to the cube root of the excitation power of the laser due to the spatial separation of electrons and holes caused by type-II band alignment, leading to band bending at the QD/host interface and increasing the ground state energy. Large blue-shifts in PL have been observed for type-II GaSb QD systems due to the large conduction band barrier off-set. For InP QDs, we were able to observe a blue-shift of about ∼12 meV at 40 K, shown Figure 3 (b). This value is again consistent with values obtained by Tagayaki et al. and is an unambiguous indication of type-II confinement.

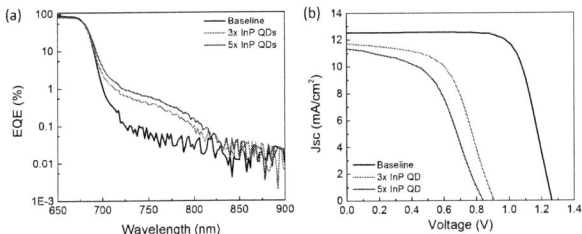

Figure 4. (a) External quantum efficieny and (b) light I-V (at AM0) of InGaP baseline and InP QD solar cells at 300 K.

Three n-i-p solar cells were grown consisting of (1) baseline InGaP cell, (2) 3x InP QDs, and (3) 5x InP QDs embedded in the intrinsic region. Figure 4(a) shows the sub-bandgap external quantum efficiency of the three cells. As expected, as the QD number of quantum dot layers are increased, so does the sub-band EQE response in the wavelength range of ∼ 680 nm to 800 nm, which matches the PL peaks of the band edge of the InGaP and InP QDs measured at 300 K. The efficiency is quite small, in the order of 1%, but for IBSC applications, this is desirable since it means the carriers are confined in the quantum dots to absorb low energy photons instead of being collected via thermal escape. Figure 4(b) shows the Light I-V characteristics of the three cells with no ARC under an AM0 illumination. Reduction in V_{OC} and I_{SC} is observed, which is due to a combination of reduced quality of growth after the InP QD layers and well as carrier trapping in the QDs that are generated in the emitter region. This was verified by applying a negative voltage bias to the cell, which significantly improved the EQE by allowing efficient carrier tunneling out of the QD states.

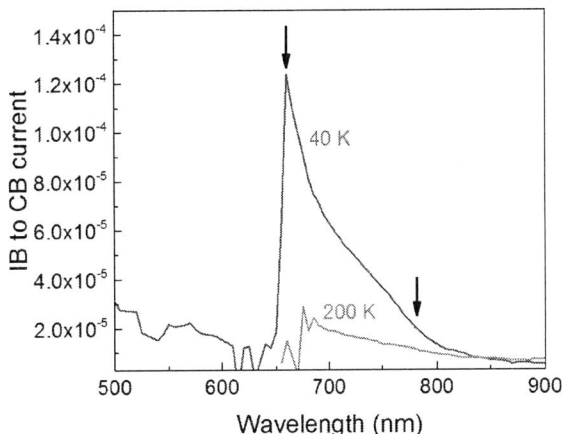

Figure 5. Photocurrent induced by IB to CB transition with 1300 nm laser excitation at 40 and 200 K. The arrows indicate the band energy of InGaP (650 nm) and E_H (770 nm) at 40 K.

Finally, two-step photon absorption dynamics was measured using the setup described in the experimental section. We were able to clearly measure photocurrent produced by the 1300 nm laser up to 200 K, as shown in Figure 5. The wavelength of the generated photocurrent matches the experimental and simulated band edges of the InGaP and InP QDs. As the monochromater photon energy is decreased below the bandgap of InGaP (1.91 eV at 40 K), carriers start to accumulate in the IB which is readily excited and collected by the 1300 nm laser. When the photon energy of the monochromater goes below the E_H energy (with the VB energy set as the reference of 0 eV), there is no longer any carriers in the IB for the laser to excite, quenching the photocurrent. The wavelength range that we see photocurrent induced by the 1300 nm laser matches the E_L energy level of ∼300 meV.

IV. CONCLUSIONS

In conclusion, we demonstrate the growth of predominantly type-II InP QDs on an InGaP matrix using MOCVD. Type-II confinement is confirmed by excitation power dependent PL, with 6-band $k \cdot p$ simulation showing approximately 300 meV conduction band confinement and 5 meV valence band barrier. TSPA measurements confirm absorption of sub-band energy photons, which clearly show the E_G, E_H, and E_L energy levels up to 200 K. Further optimization and development will allow InP QDs in InGaP to be an attractive system for highly efficient and radiation hard IBSC for extraterrestrial applications.

ACKNOWLEDGEMENTS

This work was supported by the Air Force Research Laboratory through STTR FA9453-15-C-0404.

REFERENCES

[1] A. Mart, E. Antoln, C. R. Stanley, C. D. Farmer, N. Lpez, P. Daz, E. Cnovas, P. G. Linares, and A. Luque, "Production of Photocurrent due to Intermediate-to-Conduction-Band Transitions: A Demonstration of a Key Operating Principle of the Intermediate-Band Solar Cell,"

Physical Review Letters, vol. 97, no. 24, Dec. 2006. [Online]. Available: https://link.aps.org/doi/10.1103/PhysRevLett.97.247701

[2] P. G. Linares, A. Mart, E. Antoln, C. D. Farmer, i. Ramiro, C. R. Stanley, and A. Luque, "Voltage recovery in intermediate band solar cells," *Solar Energy Materials and Solar Cells*, vol. 98, pp. 240–244, Mar. 2012. [Online]. Available: http://linkinghub.elsevier.com/retrieve/pii/S0927024811006295

[3] T. Tayagaki and T. Sugaya, "Type-II InP quantum dots in wide-bandgap InGaP host for intermediate-band solar cells," *Applied Physics Letters*, vol. 108, no. 15, p. 153901, Apr. 2016. [Online]. Available: http://scitation.aip.org/content/aip/journal/apl/108/15/10.1063/1.4946761

[4] C. Pryor, M. E. Pistol, and L. Samuelson, "Electronic structure of strained I n P/G a 0.51 In 0.49 P quantum dots," *Physical Review B*, vol. 56, no. 16, p. 10404, 1997. [Online]. Available: http://journals.aps.org/prb/abstract/10.1103/PhysRevB.56.10404

[5] A. Luque, A. Mart, and C. Stanley, "Understanding intermediate-band solar cells," *Nature Photonics*, vol. 6, no. 3, pp. 146–152, Feb. 2012. [Online]. Available: http://www.nature.com/doifinder/10.1038/nphoton.2012.1

[6] R. Rdel, A. Bauer, S. Kremling, S. Reitzenstein, S. Hfling, M. Kamp, L. Worschech, and A. Forchel, "Density and size control of InP/GaInP quantum dots on GaAs substrate grown by gas source molecular beam epitaxy," *Nanotechnology*, vol. 23, no. 1, p. 015605, Jan. 2012.

Modified Limiting Efficiency for

Multiple Exciton Generation Solar Cells

Jongwon Lee and Christiana B. Honsbnerg

School of Electrical, Computer and Energy Engineering, Arizona State University, Tempe, AZ, 85287, USA

Abstract — The thermodynamic limit of multiple exciton generation (MEG) solar cells have been have reviewed due to considering the mathematical expression of recombination process. The chemical potential (CP) of MEG should be lower than its bandgap energy so that we have analyzed that mathematical discrepancy of CP while with and without including Auger recombination (AR). Thus, we conclude the meaning of quantum yield in the recombination current in the detailed balance of MEG. Furthermore, after analyzing open circuit voltage (Voc) limit for MEG, theoretical results between with and without AR are nearly identical due to similar CP which only one radiative recombination process can be considerable under restrictions of Voc.

I. INTRODUCTION

Multiple exciton generation solar cells (MEGSC) are the promising third generation photovoltaic devices to surpass the single junction(SJ) limit due to multiple electron and hole pairs (EHP) per an incoming photon [1]. After developing experimental possibility and detailed balance, it has been good guide for future generation solar cells [2,3]. In the detailed balance of MEGSC, the initial calculation results without considering multiple carrier recombination (MR) or Auger recombination (AR) showed 98% which is higher than Carnot efficiency (=95%) under the full light concentrate (=46200 suns) [4]. Thus, it can state that MEGSC is the most efficient thermal engine due to negligible loss of solar cells. But, without any stringent limit, its results are too ideal to demonstrate even considering the loss of solar cells. After including Auger recombination (AR) effect in the detailed balance of MEGSC, it has used as the references to calculate the limiting efficiencies of MEG [4-7].

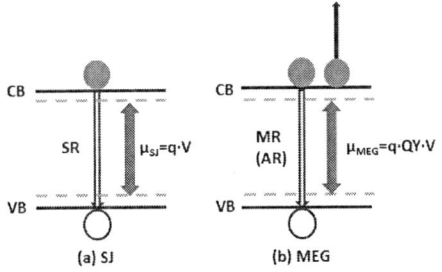

Fig.1. The chemical potential of SJ and MEG (a) one EHP recombination (b) two EHPs recombination by Auger recombination

Fig.1 describes the chemical potential (CP) between SJ solar cell and MEGSC with explaining recombination processes. SJ shows the one-radiative recombination and MEGSC demonstrates the MR or AR process. Both CPs from SJ and MEG are defined within the bandgap energy (=E_g) that each CP are defined as μ=q·V and μ_{MEG}=q·QY·V for SJ and MEG respectively where q is the element of charge, QY is the quantum yield and V is the operating voltage. μ_{MEG} is from ref [4-7] including AR impact. But, μ_{MEG} can lead the mathematical confusion that its CP can be greater than E_g due to including QY($\geq$1). A MEGSC is a part of single junction solar cell so that its CP has to be defined within E_g. Thus, the meaning of QY can be differently interpreted for the recombination process.

Multiple EHPs recombination in MEGSC is important to emit single photon AR. Therefore, including AR or MR effect into the detailed balance is critical to give the impact of results [4-7]. To avoid the mathematical confusion due to q·QY·V, alternative mathematical expression of CP for MEGSC is also necessary to develop and adjust CP of MEG. Thus, we will focus on the recombination current of MEG between multiple EHPs recombination (MR) and single EHP recombination (SR, one-radiative recombination) case with/without restrictions.

In this paper, we will discuss two points for the detailed balance for MEGSC about (1) the meaning of QY in the recombination process of MEG and (2) simplified expression of recombination current with constringent condition such as the limit of open circuit voltage (=V_{OC}) to adjust CP of MEG. Therefore, we will revisit the detailed balance equations for conventional MEG theory and research about the recombination current.

II. THEORY

2.1 Detailed balance of MEG

The first MEG observation and theory of bulk-Si [2,3] was a key to develop the limiting efficiency of MEGSC while considering one radiative recombination. Then, the current models are included AR to explain the multiple EHPs recombination in the recombination current terms [4-7]. The QY is expressed as the number of generated EHPs per one incoming photon energy in the semiconductor material. It is shown in Eq. (1) and its related equations are from Eq. (2) to

978-1-5090-5606-4/17 $31.00 © 2017 IEEE

Eq. (5). Eq. (1) is the ideal quantum yield (IQY) and non-ideal QY (NQY) respectively. IQY is increased step-like instead of linearly increasing NQY. NQY is from the experiment extractions based on pump-probe measurements and it is theoretically modeled [9-11]. QY is also shown in Fig.2.

$$\text{Ideal } QY(E) = \begin{cases} 0 & 0 < E < E_g \\ m & m \cdot E_g < E < (m+1) \cdot E_g \quad m = 1, 2, 3, . \\ M & E \geq M \cdot E_g \end{cases}$$

$$\text{Non Ideal } QY(E) = \begin{cases} 0 & 0 < E < E_g \\ 1 & E_g < E < E_{th} \cdot E_g \,\cdot\cdot \\ 1 + A \cdot \left(\frac{E - E_{th}}{E_g}\right) & E \geq E_{th} \cdot E_g \end{cases} \quad (1)$$

where m is the number of multiple EHPs generated, M is the maximum number of EHPs which are generated, E_g is the bandgap, and E is the photon energy. A(=1) is slope of the linearized QY; and E_{th} is the threshold energy for a MEG event, $E_{th} = r (\geq 2)$ where r is a positive real number.

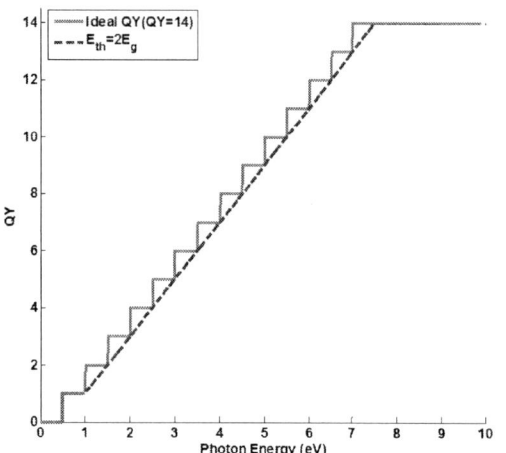

Fig. 2. IQY (maximum QY=14) and seven the NQY with seven different E_{th}s where E_g is 0.5 eV.

$$\phi(E_1, E_2, T, \mu) = \frac{2\pi}{h^3 c^2} \int_{E_1}^{E_2} \frac{E^2}{\exp[(E - \mu)/kT] - 1} dE \quad (2)$$

$$\phi_{MEG}(E_1, E_2, T, \mu) = \frac{2\pi}{h^3 c^2} \int_{E_1}^{E_2} \frac{QY(E) \cdot E^2}{\exp[(E - \mu_{MEG})/kT] - 1} dE \quad (3)$$

$$J_{BB} = q \cdot Conc \cdot f_S \cdot \phi_{MEG}(E_g, \infty, T_S, 0)$$
$$+ q \cdot Conc \cdot (1 - f_S) \cdot \phi_{MEG}(E_g, \infty, T_C, 0) \quad (4)$$
$$- q \cdot \phi_{MEG}(E_g, \infty, T_C, \mu_{MEG})$$

$$\mu_{MEG} = q \cdot QY(E) \cdot V \quad (5)$$

where ϕ is the particle flux given by Planck's equation for a temperature T with a CP μ in the photon energy range

between E_1 and E_2 (from Eq 3); h is Planck's constant; c is the speed of light; μ is the CP of single junction solar cell (=q·V) where V is the operating voltage, μ_{MEG} is the CP of MEG (=q·QY(E)·V); k is the Boltzmann constant; J is the current density of the solar cell; q is the element of charge; Conc is the optical concentration; f_S is the geometry factor (=1/46200), TS is the temperature of sun (=6000K); T_C is the solar cell's temperature (=300K).

2.2 The mathematical approaches of chemical potential

The CP of MEGSC is mathematically represented as q·QY·V and this value can be higher than E_g due to including QY≥1. For instance, from $2E_g$ to $3E_g$, if the maximum QY is 2, the CP is 2·q·V. Thus, it can intuitively state that the optimum CP should be two times of V. But, this value can be violated because optimum CP of single junction solar cell should be lower than E_g.

$$\int_{E_g}^{\infty} \frac{QY \cdot E^2}{\exp[(E - q \cdot QY \cdot V)/kT_C] - 1} dE =$$
$$\int_{E_g}^{2E_g} \frac{E^2}{\exp[(E - q \cdot V)/kT_C] - 1} dE + \int_{2E_g}^{3E_g} \frac{2 \cdot E^2}{\exp[(E - q \cdot 2 \cdot V)/kT_C] - 1} dE + \cdots + \int_{M \cdot E_g}^{\infty} \frac{M \cdot E^2}{\exp[(E - q \cdot M \cdot V)/kT_C] - 1} dE$$

Fig.3 The IQY case and its mathematical approaches to calculate the recombination current and obtain the effective chemical potential of MEGSC

With regarding of recombination current in IQY case, we can divide the photon energy range from $m \cdot E_g$ to $(m+1) \cdot E_g$ where m is 1,2,3... In the integral calculations, the beginning point of each integrand is $m \cdot E_g$ and its corresponding CP is $m \cdot q \cdot V$. Thus, its exponential term can be expressed during calculations as $\exp[m \cdot (E_g - q \cdot V)/(k \cdot T_C)]$ that its CP(=q·V) can be defined within E_g. It is shown in Fig.3 and Eq. 6.

It is a specific case to explain effective CP of MEGSC due to IQY case. In other words, the most terms of MEGSC are explained by the number of generated EHPs per an incoming photon that IQY case should be the standard explanations to explain MEG process. NQY case is experimentally obtained by the pump-probe measurement. Its non-linearity is from the imperfections of nanostructures owing not to showing the zero-dimensional structures (delta-function) of density of state. If the nanostructure has the near-perfect materials, its QY can display the near IQY case. Further, MEGSC is a part of SJ solar cell that its CP should be q·V instead of q·QY·V. Thus, we use the effective CP from IQY case. Therefore, electrons having energy QY·E_g will contribute to do AR for single photon emission and QY of CP in the recombination current of MEGSC is a term for explanations of AR.

$$\mu_{MEG, effective} = q \cdot V \quad (6)$$

While comparing (1) CP of one radiative recombination and (2) effective CP of MEGSC, these two concepts for CP are necessary to adjust mathematical approaches for MEG. In the one-radiative recombination process, we previously mentioned that without restricting V_{OC} under the full concentration, it demonstrated the highest efficiency over Carnot efficiency (=95%). Therefore, to avoid this conflict, it is important to provide the other assumption such as limit of open circuit voltage (V_{OC}). One of assumptions of solar cells is 'the operating voltage cannot pass the bandgap of materials" and its mathematical expression is '$E-\mu$ ($=q\cdot V$) $\gg$ $k\cdot T_C$'. Thus, we would use this assumption that the limit of $q\cdot V_{OC}$ is less than $E_g-k\cdot T_C$. It is shown in Eq. (7) and Eq. (8).

$$
\begin{aligned}
J_{BB} = &\ q\cdot C\cdot f_S\cdot \phi_{MEG}(E_g,\infty,T_S,0) \\
&+ q\cdot C\cdot(1-f_S)\cdot\phi_{MEG}(E_g,\infty,T_C,0) \\
&- q\cdot\phi(E_g,\infty,T_C,\mu\ (=q\cdot V))
\end{aligned} \tag{7}
$$

$$
\mu\ (=q\cdot V)\ \leq\ E_g-k\cdot T_C \tag{8}
$$
$$
\text{where } T_C=300K.
$$

Furthermore, we will use this assumption for MR case (see Eq. (5)) to compare between MR and SR case. Thus, the summary of these two cases are shown in Eq. 9.

$$
\begin{aligned}
MR: &\ \phi_{MEG}(E_g,\infty,T_C,\mu_{MEG}) \quad \mu_{MEG},\text{effective} \leq E_g-k\cdot T_C \\
SR: &\ \phi(E_g,\infty,T_C,\mu) \quad \mu \leq E_g-k\cdot T_C
\end{aligned} \tag{9}
$$

III. RESULTS

3.1 The conventional approaches of MEGSC

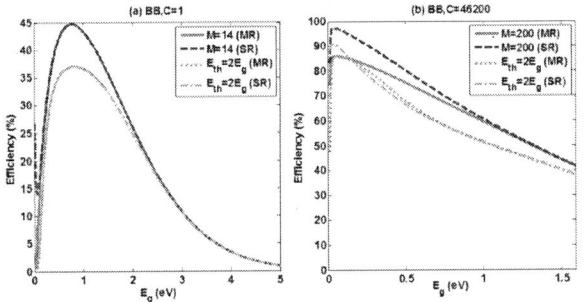

Fig.3.The comparison between SR and MR without restrictions of V_{OC}. The maximum QY is set as 14 for one sun illumination (Conc=1) and 200 for full light concentration (Conc=46200)

Table 1. Efficiency vs optimum E_g (eV) for IQY and NQY between MR and SR without limit of V_{OC} for Blackbody Radiation where η (%) is the maximum conversion efficiency, M is the maximum IQY and E_{th} is the threshold energy of NQY.

One sun (Conc=1)					
M	14 MR	14 SR	E_{th}	$2E_g$ MR	$2E_g$ SR
E_g(eV)	0.77	0.77	E_g	0.82	0.82
η (%)	44.7	44.7	η (%)	37.1	37.1
Full concentration (Conc=46200)					
M	200 MR	200 SR	E_{th}	$2E_g$ MR	$2E_g$ SR
E_g(eV)	0.05	0.04	E_g	0.07	0.04
η (%)	85.9	97.3	η (%)	85.8	90.8

For this simulation, we choose 14 and 200 of maximum QY(=QY_{max}) for one sun illumination (Conc=1) and full concentration (Conc=46200) respectively to compare MR and SR cases. The calculations without restrictions (SR case) shows 97.3% of theoretical efficiency under QY_{max}=200 with BB spectrum at full light concentration (Conc=46200) (see Table. 1) [4]. Thus, MEG can be the most ideal thermodynamic engine and we can neglect the most of thermalization loss through MEGSC. But, its value is too ideal to present due to higher value than Carnot efficiency (1-T_C/T_{SUN}=0.95, 95%). Thus, later publication including AR impact (MR case) demonstrated 85.9% under the same conditions [4] and it has been the reference for the detailed balance of MEGSC [4-7]. It shows about 11.4% efficiency difference between two cases. But, the results between MR/SR cases under one sun illumination are nearly identical because of similar optimum power point under the moderate maximum QY (QY_{max}=14). Typically, SR case including non-ideal QY(=NQY, E_{th}=2E_g) shows the similar tendencies to IQY case that its theoretical maximum efficiency (=90.8%) under maximum light concentration is higher than 85.8% of MR case. We summarized these results in Fig.3 and Table.1. The large efficiency discrepancies are shown at low bandgap energy region (below 0.5eV) under the full light concentration (Conc=46200).

3.2 Results with restrictions of V_{OC}

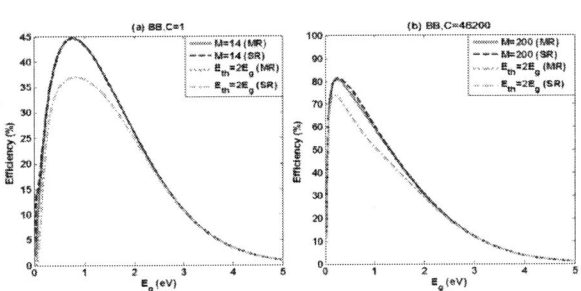

Fig.4.The comparison between SR and MR with restrictions of V_{OC}.

Table 2. Efficiency vs optimum E_g (eV) for IQY and NQY between MR and SR without limit of V_{OC} for Blackbody Radiation.

One sun (Conc=1)					
M	14 MR	14 SR	E_{th}	$2E_g$ MR	$2E_g$ SR
E_g(eV)	0.77	0.77	E_g(eV)	0.82	0.82
η (%)	44.7	44.7	η (%)	37.1	37.1
Full concentration (Conc=46200)					
M	200 MR	200 SR	E_{th}	$2E_g$ MR	$2E_g$ SR
E_g(eV)	0.25	0.27	E_g(eV)	0.2	0.2
η (%)	81.0	81.5	η (%)	74.1	74.1

In the previous chapter, we have discussed the large discrepancies between MR and SR under full light concentration without constringent restriction of V_{OC}. In this chapter, we will discuss the impact of limiting V_{OC} between

MR and SR. With limit of V_{OC} ($\leq E_g$-k·T_C.), the entire results between MR and SR are similar under both one sun and full concentration due to similar maximum power points. In contrast to no limit of V_{OC} case, for instance, the efficiencies between MR and SR for ideal QY(QY=200) are 81.0% and 81.5% are respectively and its optimum bandgaps are also similar under the full light concentration. Even, NQY cases of MR/SR are shown the same results that its maximum power points are also the same. Thus, the limit of V_{OC} provide the crucial role that the impact of both one-radiative recombination process (SR case) and MR case can be similar under the certain restriction of V_{OC}. In other words, the impact between MR and SR case are quite similar under the restriction of V_{OC} that using SR (one-radiative recombination) case can be available into the detailed balance of MEGSC even if MR is still considerable process. Typically, this impact significantly shows at low E_g region (below 0.5 eV).

IV. RESULTS

We demonstrate and review the limiting efficiencies of MEGSC (1) for verifying the meaning of QY in the recombination currents and (2) for researching the limit of V_{OC} in comparison to between MR/SR effects. Without any restrictions, the one radiative recombination provides too ideal results that its theoretical efficiency is higher than Carnot engine in IQY case under the full light concentration. After inclusion of AR effects in the recombination process, the modified calculations show the conventional thermodynamic limit of MEG. While analyzing the detailed balance equations, we have found that QY in the recombination current is for explaining reverse MEG process that effective CP is q·V instead of q·QY·V.

We have further researched about the restrictions of V_{OC} to compare the effect MR/SR. While comparing between MR and SR cases, both theoretical efficiencies show similar results due to near-identical CPs. Therefore, the impact of SR can be acceptable under the certain restriction even if MR case can contribute to provide significant impact for MEGSC.

REFERENCE

[1] W. Shockley and H. J. Queisser, "Detailed Balance Limit of Efficiency of p-n Junction Solar Cells," Journal of Applied Physics, vol. 32, pp. 510-519, 1961.

[2] S. Kolodinski, J. H. Werner, T. Wittchen, and H. J. Queisser, "Quantum efficiencies exceeding unity due to impact ionization in silicon solar cells," Applied Physics Letters., vol. 63, pp 2405-2407, 1993.

[3] J.H. Werner, S. Kolodinski, and H.J. Queisser, "Novel optimization principles and efficiency limits for semiconductor solar cells," Physical Review Letters, vol. 72, pp. 3851-3854, 1994.

[4] J.H. Werner, R. Brendel, H.J. Queisser, "New upper efficiency limits for semiconductor solar cells", in Proc. of the IEEE 1st

World Conference on Photovoltaic Energy Conversion, Waikoloa, HI, USA, 1994, pp. 1742–1745.

[5] J. H. Werner, R. Brendel, and H. J. Queisser, "Radiative efficiency limit of terrestrial solar cells with internal carrier multiplication," Applied Physics Letters, vol. 67, no. 7, pp. 1028-1030, 1995.

[6] R. Brendel, J. H. Werner and H. J. Queisser, "Thermodynamic efficiency limits for semiconductor solar cells with carrier multiplication", Solar Energy Materials and Solar Cells., vol. 41/42, pp. 419-425, 1996.

[7] A. De Vos and B. Desoete, "On the ideal performance of solar cells with larger-than-unity quantum efficiency", Solar Energy Materials and Solar Cells., vol. 51, pp 413-424, 1998.

[8] A. Luque, A. Marti and L. Cuadra., "Thermodynamics of solar energy conversion in novel structures," Physica E: Low-dimensional Systems and Nanostructures, vol. 14, pp. 107-114, 2002.

[9] R. D. Schaller, M. Sykora, J. M. Pietryga, and V. I. Klimov, "Seven Excitons at a Cost of One: Redefining the limits for conversion efficiency of photons into charge carriers," Nano Letters, vol. 6, no. 3, pp. 424-429, 2006.

[10] M. C. Beard, and R. J. Ellingson, "Multiple exciton generation in semiconductor nanocrystals: Toward efficient solar energy conversion," Laser & Photonics Reviews, vol. 2, no. 5, pp. 377-399, 2008.

[11] M. C. Hanna, and A. J. Nozik, "Solar conversion efficiency of photovoltaic and photoelectrolysis cells with carrier multiplication absorbers," Journal of Applied Physics, vol. 100, no. 7, pp. 074510, 2006.

A Simple Monte Carlo Model of a Hot Carrier Cell

Tor Oskar Saetre

University of Agder, Department of Engineering Sciences, 4898 Grimstad, Norway

Abstract — In the present work, a simplified model of a hot carrier cell is examined at different energy levels of carrier collection. Incident photons, Monte Carlo generated by employing the ASTM G173-03 data set, are accounted for individually as they interact with the cell. The hot carrier cells are examined in a transient state caused by a flash of photons with AM1.5 spectrum. Excess photon energy is recycled giving a form of thermalization of hot carriers. The sensitivity of contact variables and the device band gap on cell efficiency, has been studied.

Index Terms — carrier collection, energy recycling, hot carrier cells, Monte Carlo simulation.

I. INTRODUCTION

Conventional solar cells collect most of the carriers after they thermalize with the lattice. Hot carrier cells are in principle able to extract more energy than conventional solar cells by collecting the carriers before the thermalization at the cell temperature. The field of hot carrier solar cells has attracted a lot of attention and is one of the promising future approaches to improved solar cell efficiency, cf. e.g. [1]-[19].

In the present paper, a simplified model of a hot carrier cell is studied. Incident photons are generated by Monte Carlo methods [20]-[24] and individually accounted for in the model. The energy levels at which the carriers are collected, are varied to examine the sensitivity and performance of the cells at different contact energy levels and thus different open circuit cell voltages.

The distribution of excited electrons and holes will vary over time in a transient state approach. After a monochromatic pulse, the stages can generally be listed as [1]:

1. Coherent distributions
2. Carrier scattering
3. Thermalization of hot carriers
4. Carrier cooling
5. Lattice thermalized carriers
6. Recombination
7. Return to thermal equilibrium

The stages are distributions in energy-momentum space, but particle momenta are not implemented in the present model. Stage 4, carrier cooling, is prevented in hot carrier cells to extract more of the incident energy and thus avoid some losses.

In the present model, the carriers will initially be generated to Stage 1 caused by a polychromatic pulse. A simple Monte Carlo procedure enables the simulation of thermalization of hot carriers from Stage 1 to Stage 3. Carrier scattering is not considered. The device is idealized by assuming that there is no losses due to Stages 4-7, which are not considered in the present model.

II. MODEL DESCRIPTION

A. Photon generation

The data set provided by ASTM G173-03 for the direct normal AM1.5 spectrum have been employed in the Monte Carlo computer program to generate the distribution of photons. This is achieved by using an intrinsic function random number generator which generates a uniform distribution between 0 and 1. These pseudo-random numbers can be converted into any distribution with well-known Monte Carlo techniques. A typical generated distribution of photons giving an AM1.5 distribution of radiation is seen in Fig. 1. The photons are then individually accounted for in the device model. At every simulation run, 10^6 photons were generated and accounted for.

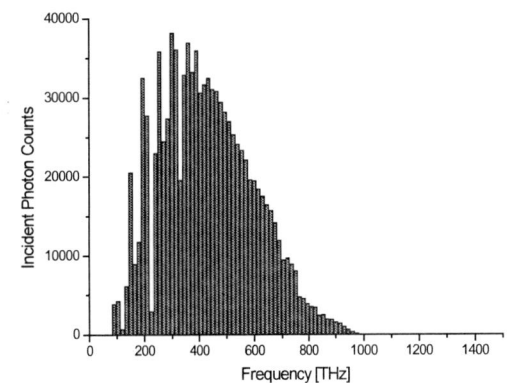

Fig. 1. Typical photon distribution generated by employing the ASTM G173-03 data set for direct normal AM1.5 spectrum, downloaded from NREL. This sample contains 10^6 photons binned into 100 interval-bins for this plot.

B. Device model

The device model studied in the present paper is a simple and idealistic model, cf. Fig. 2.

Firstly, an ensemble of photons is generated. Any spectrum can be generated.

Secondly, each successful photon is exciting an electron from the valence band to the conduction band, thus creating a hole in the valence band.

Thirdly, the electrons and holes with energies at or better than the contact levels are withdrawn. Excess energy of the carriers collected, which normally would be lost, are accounted for and

978-1-5090-5606-4/17 $31.00 © 2017 IEEE

distributed randomly amongst the electrons and holes which cannot reach the contacts.

Fourthly, if there are some remaining electrons or holes which can reach the contacts after redistributing the excess energy, they are withdrawn over the contacts. This iterative process is continued until no more carriers can reach the contacts.

The simulation procedure is thus not for a steady-state operation, but emulates a controlled pulse-of-photons event and the effects thereof.

The energy level of the hole in the valence band is determined as a distance from the band edge of the valence band by a number drawn from a uniform random number distribution. The energy level of the corresponding electron in the conduction band is then given by the energy of the photon absorbed.

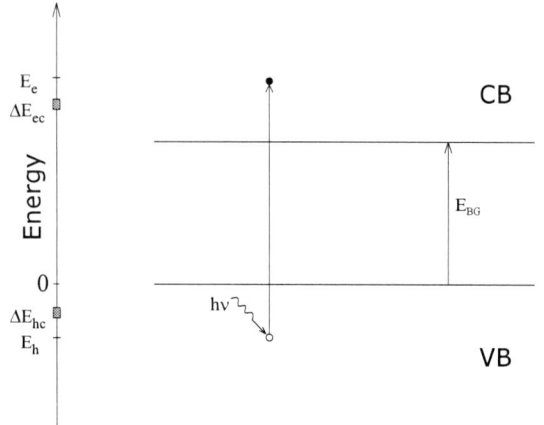

Fig. 2. Between the valence band (VB) and the conduction band (CB) there is a band gap of energy E_{BG}. When a photon of energy $h\nu$ excites an electron from energy level E_h, a hole is created at that level of energy. The electron obtains the energy E_e after the excitation process. Contacts normally withdraw carriers across a small range of energies, ΔE_{hc} and ΔE_{ec}. In most of the present simulations, only the highest energy limit in ΔE_{hc} and lowest energy limit in ΔE_{ec} for carrier withdrawal are used, and not the intervals. A zero-energy reference point is defined at the upper band edge of the valence band.

The cell characteristics are ideal with a fill factor of 1 and the open circuit voltage, V_{oc}, given by:

$$V_{oc} = \frac{(E_{ec} - E_{hc})}{q} \qquad (1)$$

where q is the charge of an electron, E_{ec} is the lower energy limit of the range ΔE_{ec}, and E_{hc} is the higher energy limit of the range ΔE_{hc}. All energies in this model are relative to the valence band edge which is defined as the zero electron volt level, cf. Fig. 2. The design of contacts is of importance in hot carrier cells since they can cool the hot carriers. The contact energy intervals are thus important in applications and should be small to avoid loss of energy. In the present model, the carriers are withdrawn at levels E_{ec} and E_{hc} for electrons and holes respectively. Excess energy is the energy differences ($E_e - E_{ec}$ - ΔE_{ec}) and ($E_{hc} - E_h - \Delta E_{hc}$). The excess energy differences are recycled to the ensemble of excited electrons and holes that could not reach the contact levels. In the present model, there are three loss mechanisms:

1. Photons with energies $h\nu < (E_{BG} - E_h)$ cannot generate an electron-hole pair due to lack of energy.
2. Holes at $E_h > E_{hc}$ and excited electrons at $E_e < E_{ec}$ that remains after the iterative energy thermalization, cannot contribute to the device's DC-current.
3. The effect of the direct contact losses due to the intervals ΔE_{ec} and ΔE_{hc} are examined in one of the simulated cases.

III. RESULTS AND DISCUSSION

Some results from model simulations can be seen in Figs. 3-7. The curves are plotted as linear point-to-point plots. The simulation method gives some statistical variations which to some degree can cause deviations from smoothness of the curves. In the simulations plotted in Figs. 3-6, the third loss mechanism was not active by assigning ΔE_{ec}=0 and ΔE_{hc}=0.

A. The effect of introducing Stage 3: Thermalization.

Fig. 3 shows the effect of varying the absorption limit. The absorption limit is the energy difference between the edge of the valence band (zero reference level) and the minimum level in the valence band where photons are absorbed by generating electron-hole pairs. In the present work, the probability of absorption within this range is uniform. At the lowest absorption limit values, all the holes are generated at energies greater than the contact level. Hence, no current can pass.

When the minimum energy level for absorption is less than the contact energy level, power is generated. Increasing the absorption limit results in a quick rise in power until the peak value. Further increasing the absorption limit results in reduced power levels. This is due to the loss mechanisms described above. Two curves are presented in Fig. 3. The black curve was published previously [24]. This is the result with only Stage 1 implemented. The added benefit of recycling the energy is shown by the upper red curve's higher efficiency values.

B. The effect of the contact energy for electrons.

With reference to the loss mechanisms stated above, it would be expected that a variation of the contact energies would change the power output of the device. In Fig. 4, simulation results with different values of the energy level of the contact in the conduction band are seen. With variations from 0.6eV to 3.0eV, the optimum device efficiency is given by a value of 1.8eV for E_{BG}=0.5eV and E_{hc}=-1.0eV.

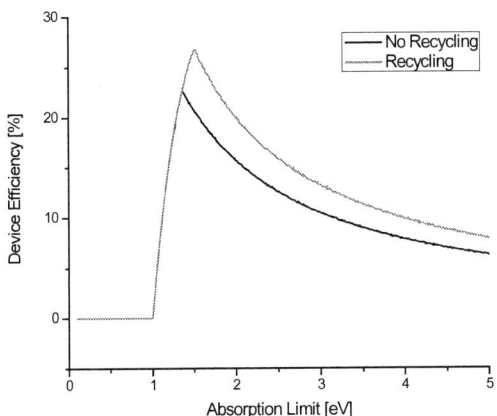

Fig. 3. The efficiency of the device resulting from variations in the photon absorption limit in the valence band. The absorption limit is the energy difference between the edge of the valence band (zero reference level) and the minimum level in the valence band where photons are absorbed by generating electron-hole pairs. When the limit is at higher energies than E_{hc}, no current is generated. Black curve: excess carrier energies are lost. Red curve: excess carrier energies are recycled.
Device parameters: E_{BG}=0.5eV, E_{ec}=2.0eV, E_{hc}=-1.0eV, ΔE_{ec}=0, ΔE_{hc}=0.

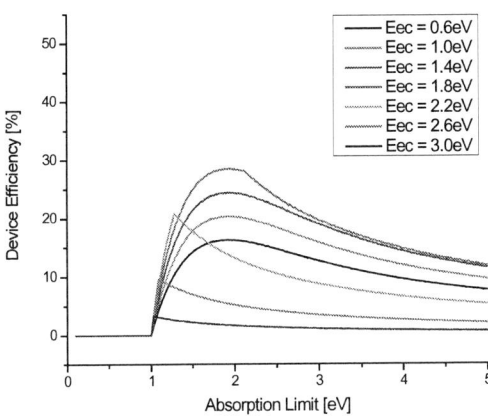

Fig. 4. The efficiency resulting from variations in the photon absorption limit in the valence band. The curves show results from simulations with different energy levels at the contact in the conduction band.
Device parameters: E_{BG}=0.5eV, E_{ec}: several values, E_{hc}=-1.0eV, ΔE_{ec}=0, ΔE_{hc}=0.

C. The effect of the contact energy for holes.

In Fig. 5, simulation results with different values of the energy level of the contact in the valence band are seen. With variations from -0.2eV to -3.0eV, the optimum device efficiency is given by a value of -0.2eV for E_{BG}=0.5eV and E_{ec}=2.0eV. This figure also demonstrates the device efficiency's dependence on the combination of the values of the absorption limit and the contact level for holes.

D. The effect of the band gap energy.

In Fig. 6, simulation results with different values of the band gap energy are seen. With variations from 0.2eV to 1.8eV, the optimum device efficiency is given by a value of 0.2eV for E_{BG}=0.5eV, E_{ec}=2.0eV and E_{hc}=-1.0eV. This result confirms the well-known fact that the band gap in these devices should be small.

E. Contact losses.

The energy span of the contacts given by ΔE_{hc} and ΔE_{ec}, give rise to loss mechanism 3. The wider the intervals, the more loss occurs due to contact cooling of charge carriers. Fig. 7 shows some results where the widening of the energy span of the contacts reduces the device efficiency, as would be expected.

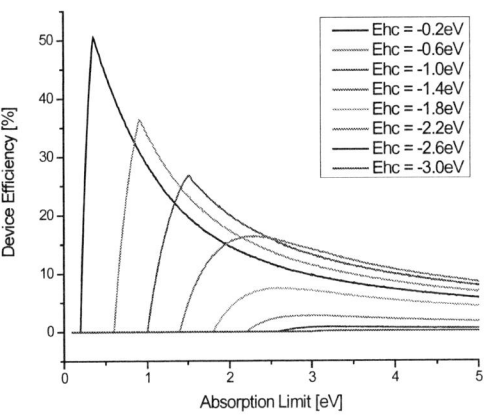

Fig. 5. The efficiency resulting from variations in the photon absorption limit in the valence band. The curves show results from simulations with different energy levels at the contact in the valence band.
Device parameters: E_{BG}=0.5eV, E_{ec}=2.0eV, E_{hc}: several values, ΔE_{ec}=0, ΔE_{hc}=0.

F. General considerations.

The generally high valence band effective mass will probably result in holes being generated close to the band edge. Hence, more of the surplus photon energy should be absorbed in the conduction band than in the valence band in real devices. A simplified device could thus be constructed with an ordinary contact for the holes and a hot carrier contact in the conduction band only. Special attention could thus be raised to materials

978-1-5090-5606-4/17 $31.00 © 2017 IEEE

which may enable most photon absorptions very close to the band edge of the valence band. The benefit of this proposal will be studied in future research.

The implementation of a full energy-momentum space in future may increase the usefulness of these simulations by enabling direct comparison with selected materials.

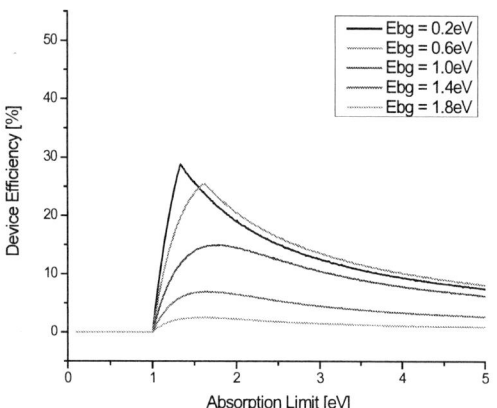

Fig. 6. The efficiency resulting from variations in the photon absorption limit in the valence band. The curves show results from simulations with different energy levels of the band gap.
Device parameters: E_{BG}: several values, E_{ec}=2.0eV, E_{hc}=-1.0eV, ΔE_{ec}=0, ΔE_{hc}=0.

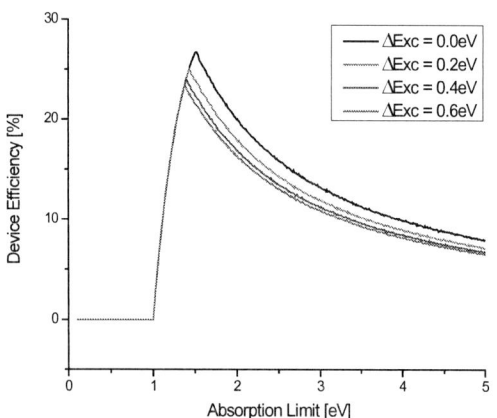

Fig. 7. Efficiency losses at the contacts at the conduction band and valence band. ΔE_{xc} are the different energy spans of the contacts. The black upper curve is identical to the red upper curve of Fig. 3.
Device parameters: E_{BG}=0.5eV, E_{ec}=2.0eV, E_{hc}=-1.0eV. ΔE_{xc}=ΔE_{hc}=ΔE_{ec}, cf. Fig. 2.

IV. CONCLUSION

A Monte Carlo modelling approach to hot carrier solar cells has been presented. The effects of a pulse of photons with an AM1.5 standard spectrum have been studied under different conditions. Excess carrier energy has been recycled randomly to the ensemble of remaining carriers. The energy level of the contact in the valence band is the most sensitive variable studied in the present work.

REFERENCES

[1] M.A. Green, Third Generation Photovoltaics, Springer 2003, pp. 69-79.

[2] G.J. Conibeer, C.-W. Jiang, D. König, S. Shrestha, T. Walsh, M.A. Green, Thin Solid Films, 516 (2008), 6968.

[3] G. Conibeer, N. Ekins-Daukes, J.-F. Guillemoles, D. König, E.-C. Cho, C.-W. Jiang, S. Shrestha, M. Green, Solar Energy Materials & Solar Cells, 93 (2009), 713.

[4] G. Conibeer, R. Patterson, L. Huang, J.-F. Guillemoles, D. König, S. Shrestha, M. Green, Solar Energy Materials & Solar Cells, 94 (2010), 1516.

[5] D. König, K. Casalenuovo, Y. Takeda, G. Conibeer, J.F. Guillemoles, R. Patterson, L.M. Huang, M.A. Green, Physica E, 42 (2010), 2862.

[6] A. Le Bris, L. Lombez, J.F. Guillemoles, R. Esteban, M. Laroche, J.J. Greffet, G. Boissier, P. Christol, S. Collin, J.L. Pelouard, P. Ascheoug, F. Pellé, Proc. 25th European Solar Energy Conference and Exhibition, 6-10 Sept. 2010, Valencia, Spain, 683.

[7] A. Luque, A. Marti, Solar Energy Materials & Solar Cells, 94 (2010), 287.

[8] A. Mart, A. Luque, IEEE Journal of Photovoltaics, 3 (2013), 1298.

[9] L.C. Hirst, M.K. Yakes, C.G. Bailey, J.G. Tischler, M.P. Lumb, M. Gonzalez, M.F. Fuhrer, N.J. Ekins-Daukes, R.J. Walters, IEEE Journal of Photovoltaics, 4 (2014), 1526, DOI: 10.1109/JPHOTOV.2014.2355412.

[10] D. Konig, Y. Yao, R. Patterson, Japanese Journal of Applied Physics, 53 (2014), 05FV04, DOI: 10.7567/JJAP.53.05FV04.

[11] D. Watanabe, N. Kasamatsu, Y. Harada, T. Kita, Applied Physics Letters, 105 (2014), 171904, DOI: 10.1063/1.4900947.

[12] Y. Takeda, A. Ichiki, Y. Kusano, N. Sugimoto, T. Motohiro, Journal of Applied Physics, 118 (2015), 124510, DOI: 10.1063/1.4931888.

[13] Su, S.H., T.J. Liao, X.H. Chen, G.Z. Su, J.C. Chen, IEEE Journal of Quantum Electronics, 51 (2015), 4800208, DOI: 10.1109/JQE.2015.2469152.

[14] D. Konig, Y. Yao, Japanese Journal of Applied Physics, 54 (2015), 08KA03, DOI: 10.7567/JJAP.54.08KA03.

[15] G. Conibeer, S. Shresta, S.J. Huang, R. Patterson, H.Z. Xia, Y. Feng, P.F. Zhang, N. Gupta, M. Tayebjee, S. Smyth, Y.X. Liao, S. Lin, P. Wang, X. Dai, S.M. Chung, Solar Energy Materials & Solar Cells, 135 (2015), 124.

[16] H. Esmaielpour, V.R. Whiteside, J. Tang, S. Vijeyaragunathan, T.D. Mishima, S. Cairns, M.B. Santos, B. Wang, I.R. Sellers, Prog. Photovolt: Res. Appl., 24 (2016), 591.

[17] Y. Zhang, M.J.Y. Tayebjee, S. Smyth, M. Dvorak, Z.M. Wen, H.Z. Xia, M. Heilmann, Y.X. Liao, Z.W. Zhang, T. Williamson, J. Williams, S. Bremner, S. Shrestha, S.J. Huang, T.W. Schmidt, G.J. Conibeer, Applied Physics Letter, 108 (2016), 131904, DOI: 10.1063/1.4945594.

[18] S.M. Chung, S. Shresta, X.M. Wen, Y. Feng, N. Gupta, H.Z. Xia, P. Yu, J. Tang, G. Conibeer, Solar Energy Materials & Solar Cells, 144 (2016), 781.

[19] M. Li, S. Bhaumik, T.W. Goh, M.S. Kumar, N. Yantara, M. Grätzel, S. Mhaisalkar, N. Mathews, T.C. Sum, Nature Communications, 2017, DOI: 10.1038/ncomms14350.

[20] T.O. Saetre, Proc. 29th European Photovoltaic Solar Energy Conf. and Exib., WIP Renewable Energies, 2014, 73.

[21] T.O. Saetre, Photovoltaic Specialist Conference (PVSC), 2015 IEEE 42nd, DOI: 10.1109/PVSC.2015.7356110.

[22] T.O. Saetre, Photovoltaic Specialist Conference (PVSC), 2015 IEEE 42nd, DOI: 10.1109/PVSC.2015.7355803.

[23] T.O. Saetre, Photovoltaic Specialist Conference (PVSC), 2016 IEEE 43rd, DOI: 10.1109/PVSC.2016.7749609.

[24] T.O. Saetre, Photovoltaic Specialist Conference (PVSC), 2016 IEEE 43rd, DOI: 10.1109/PVSC.2016.7749610.

Optimization of Semiconductor Quantum Dots for Luminescent Solar Concentrators: Minimizing Reabsorption Losses

Anatoli I. Shkrebtii[1*], Anatoliy V. Sachenko[2], Igor O. Sokolovskyi[2,3], Vitaliy P. Kostylyov[2], Mykola R. Kulish[2], Denis V. Khomenko[2] and Mykhaylo A. Evstigneev[3]

[1] University of Ontario Institute of Technology, 2000 Simcoe St. N., Oshawa, Ontario, L1H 7K4, Canada

[2] V. Lashkaryov Institute of Semiconductor Physics, NAS of Ukraine, 45 Pr. Nauky, Kyiv 03028, Ukraine

[3] Memorial University of Newfoundland, St. John's, Newfoundland and Labrador, A1B 3X7 Canada

Abstract — We investigate theoretically dependence of the reabsorption in semiconductor quantum dots (QDs) on their geometric and electron parameters, namely the radius $\bar{r}$, its dispersion $\Delta\bar{r}$ and bulk semiconductor band gap E_{g0}. QDs are promising as luminophores in luminescence solar concentrators (LSCs). The photo-luminescence (PL) process in QDs is influenced by the reabsorption. To understand how to minimize the detrimental reabsorption losses, we considered six semiconductors, typically used to fabricate QDs, with a wide range of their bulk bandgaps: CdS (E_{g0} = 2.42 eV), CdSe (E_{g0} = 1.67 eV), CdTe (E_{g0} = 1.5 eV), InP (E_{g0} = 1.27 eV), InAs (E_{g0} = 0.36 eV), and PbSe (E_{g0} = 0.27 eV). We prove that by adjusting the QD radius $\bar{r}$ and dispersion $\Delta\bar{r}$, it is possible to optimize nanocrystal dimensions in order to minimize the reabsorption. It was shown that for the semiconductor bulk band gap in the range between 1.27 eV to 2.42 eV the optimum QD size and dispersion can always be chosen, at which the reabsorption is below the total experimental error of the measured normalized both absorption coefficient and luminescence intensity. Further reduction of E_{g0}, however, increases the reabsorption at any values of $\bar{r}$ and $\Delta\bar{r}$: for instance, for PbSe based QD of 1 nm radius and its 1% dispersion, the reabsorption reaches 54%. We estimate the fragment of the solar spectrum, from which the photons, involved in the luminescence processes, originate. This is important for stacked LSCs application.

Index Terms — luminescent solar concentrators, solar cells, quantum dots, reabsorption, radius dispersion

I. INTRODUCTION

Currently, an intensive search for alternative approaches to diversity the methods of the energy harvesting is underway. Luminescent solar concentrators (LSCs) are attracting significant attention from both fundamental research and their application as a part of the integrated photovoltaics [1] (see, also, Fig. 20 there), and [2]-[4]. LSC based photoconvertors consists of a transparent plate (made of organic or inorganic materials) and doped with luminescent materials. The luminescent photons are propagating inside the plate toward solar cells, attached at the ends of the plates (details of the LSC design can be found, *e.g.*, in the review [1]). Among several types of phosphors, the most attractive are semiconductor quantum

dots (QDs) with their radius ranging from 1 to 20 nm [4]-[10]. Through the band-to-band excitations, the luminescence in semiconductor QDs converts a broad solar spectrum into photons of particular wavelengths, that form a narrow band. Changing the optical bandgap of the QDs by choosing a proper semiconductor bulk gap, varying the core size, and adjusting their radial distribution, the luminescence band can be formed to fit the spectral position of the luminescence to the maximum sensitivity of the solar cell. The different types of LSCs can be assembled in a stack to improve the energy harvesting of the wide solar spectrum range.

There are several mechanisms of loses in LSCs, and the reabsorption process is among them. In this research, we combine the effects of QD size, inhomogeneity and thermal energy level smearing on the formation of the luminescent spectra and the resulting reabsorption. That is, the corresponding contributing factors considered are the thermal effects, the nanoparticle size and their dispersion.

II. FORMULATION OF THE PROBLEM AND THE GOALS

To analyze and quantify the reabsorption, we will start with the established formalisms, used in the literature. We first define how to quantify the reabsorption, which is depicted schematically in Fig. 1. There we normalize both the amplitude of the first maximum of the QD absorbance (blue solid line) and the QD luminescent spectrum (red dashed line) and point to their crossing point. The luminescent photons, emitted within the energy range Δ, can be reabsorbed. The heights of the overlap between the two normalized spectra, at the point where the absorption and luminescence lines are crossing, can be used as a measure of the reabsorption.

To quantify the reabsorption, we have to calculate first the overlap of the normalized narrow luminescent QD band with its wide absorption band. Electrons and holes in such QDs can be considered as quasiparticles in three-dimensional (3D) quantum well. Measuring the energy position of the quantized levels for hole and electron from the bottom of the conduction band and the top of the valence band, respectively, and following [11] (p. 155, Problem 62, see also [4]) the energy of the quantized levels in QD can be calculated as:

$$E_{n,l}^{e,h} = E_{g0} + \frac{\hbar^2 \varphi_{l,n}^2}{2r^2}\left[\frac{1}{m_e} + \frac{1}{m_h}\right] - \frac{1.786e^2}{4\pi\varepsilon\varepsilon_0 r} - 0.248E_{Ry}^*, \quad (1)$$

where E_{Ry}^* is the Rydberg energy, ε and ε_0 are the relative dielectric constants for semiconductor and its absolute vacuum value respectively, $\hbar$ is the reduced Planck's constant, m_e and m_h are respectively the effective masses of the electron and hole, e is the electron charge, r is the nanocrystal core radius, $n = 1, 2, 3, \ldots$ are the principal quantum numbers, $l = 0, 1, 2, \ldots$ are the orbital quantum numbers, while $\varphi_{l,n}$ is a universal set of numbers given in the table ([11], p. 155, Problem 62, and in the appendix).

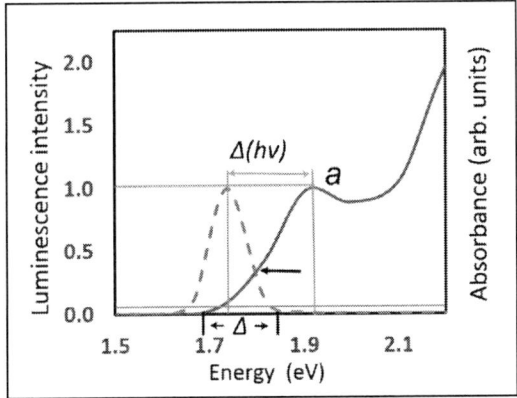

Fig. 1. Color online. Spectra of QD absorption (solid line) and luminescence (dashes). Luminescent peak maximum and the first maximum of the absorbance (with the energy *a*) are normalized to unity. The red horizontal line indicates the total accuracy of absorption and luminescence measurements. The energy separation between the first peak of absorption spectrum and the peak of QD luminescence is $\Delta(h\nu)$. The horizontal arrow indicates the point where the absorption and luminescence spectra are crossing. The spectral range of the reabsorption is Δ.

The first term in Eq. 1 is the semiconductor bulk bandgap, the second term describes the energy of quantized levels due to the carriers' confinement in nanocrystal; the third term describes the energy reduction due to interaction of electrons and holes. The last fourth term is the Rydberg energy, which does not depend on the size of the nanoparticles and can be usually neglected, except the cases of semiconductor with the dielectric constant below 10, such as, *e.g.*, in InSb.

The QD absorption spectrum is formed by transitions between the size-quantized electron and hole levels with the same quantum numbers n and l (see Fig. 2a). In the real nanocrystals two main mechanisms are responsible for the broad absorption band: (i) at nonzero absolute temperature

there always is a thermal broadening of the quantized energy levels due to atomic vibrations and (ii) the nanocrystal size dispersion leads to the additional inhomogeneous broadening, which is temperature independent (Fig. 2b). The optical band gap E_{abs} is also shown in Fig. 2b.

When calculating the absorption spectra of QD, as a good approximation the thermal energy level broadening is usually neglected ([11]-[12]), that is, only the contribution from nanocrystal size dispersion is considered. In this case, the absorption spectrum is composed of a series of Gaussian peaks with energy E_i and width ΔE_i, and the spectrum can be written as following [13]:

$$\alpha(E_{ph}) = 10^4 \frac{2\pi\hbar e^2}{m_0^2 c n_r \varepsilon_0} \frac{1}{V_{QD}} \sum_i \frac{|P_i|^2}{3E_{ph}\sqrt{2\pi}\Delta E_i} \exp\left(-\frac{(E_{ph}-E_i)^2}{2\Delta E_i^2}\right), \quad (2)$$

where E_{ph} is the photon energy, m_0 is a mass of a free electron, c - the speed of light in a vacuum, n_r is the real part of the refractive index of the bulk semiconductor, V_{QD} is the volume of the quantum dot, the line width of the optical transition ΔE_i is the standard deviation. P_i and E_i are the matrix element of momentum and energy with respect to each of the i-th exciton transition.

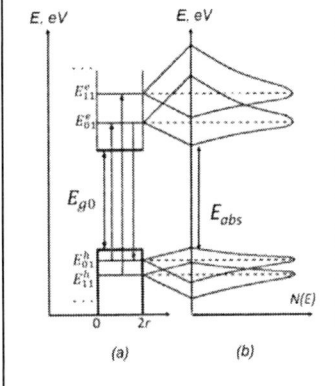

Fig. 2. Quantized energy levels of electron and hole in QDs in the case of: (a) 3D confined nanoparticle with its $2r$ core fixed size and discrete energy levels $E_{n,l}^{e,h}$ (see text for the explanation). In the case (b) the QDs size dispersion and the thermal broadening of the energy levels are present. E_{g0} is the bulk semiconductor gap and $N(E)$ is the carriers' density of states.

Equation (2) has been used in [13] to compare with the experimental absorption spectra of CdSe QDs. A good agreement of the calculated and experimental absorption spectra has been demonstrated for the QDs with their radius, ranging from 1.5 to 8 nm.

When illuminated by sunlight, electrons are excited from the quantized levels of the valence band to the corresponding quantized energy levels of the conduction band. Both electrons and holes relax quickly (within a picosecond time interval) to the E_{01}^e and E_{01}^h energy levels respectively, then recombining through the transition from E_{01}^h to E_{01}^e, emitting the luminescent photon.

Majority of the research on nanocrystals properties consider ensembles of QDs with 10% size dispersion (see, *e.g.*, [6]) with the Gaussian distribution [5] of their radius r around its mean value:

$$P(r,\bar{r}) = \frac{1}{\sigma_r \sqrt{2\pi}} e^{-(r-\bar{r})^2 / 2\sigma_r^2}. \tag{3}$$

In (3) $\bar{r}$ is the mean QD radius, the standard deviation σ_r can be determined using electron microscopy data.

Each QD emits luminescence photons with its characteristic energy $h\nu_i$, the combined emission of the set of QDs forms a luminescent band. The intensity of the luminescence band of QDs is a sum of luminescence intensities $I_{PL}(h\nu_i, r_i)$, where r_i is the variable radius of i-th QD, emitting photons of energy $h\nu_i$. Following [5], the intensity of the luminescence band of the QD set can be described as:

$$I_{PL}^{En} = \sum_d \alpha(h\nu_{ex}, r) I_{PL}(h\nu_i, r_i) P(r_i, \bar{r}). \tag{4}$$

The energy separation between the first maximum of the absorption band and the luminescence band maximum of QDs can be extracted from (1) as:

$$\Delta(h\nu) = \frac{1.786e^2}{4\pi\varepsilon_0\varepsilon\,\bar{r}} - 0.248 E_{Ry}^*, \tag{5}$$

where $\bar{r}$ is the average radius of the ensemble of nanoparticles. As it has been found [5], the intensity of luminescence band, calculated by (4), correctly describes the corresponding experimental dependence for the ensemble of InP quantum dots.

Usually an ensemble of quantum dots contains particles of different, statistically distributed sizes and together with the thermal smearing of QDs energy levels, which result in both absorption and luminescence spectra, dependent on characteristics of the QD ensemble and the temperature. This modifies the luminescence and absorption bands overlap, determining the reabsorption extent. The larger the dispersion of the quantum dot size, the stronger is the reabsorption.

While the above expressions in general correctly describe the evolution of luminescence and absorption spectra of the size dispersed QDs, they do not allow estimating the evolution of the reabsorption with their average radius. The purpose of this work is to derive expressions that properly calculate the reabsorption by taking into account the distribution of nanoparticle size and the thermal spreading of the quantized energy levels. Next, analyzing the derived expressions, we quantify the impact of the QDs size and their dispersion on the absorption, luminescence and reabsorption.

III. ABSORPTION AND LUMINESCENT SPECTRA OF QD CORE

Usually, QDs with high luminescence quantum yield contain a core surrounded by one or more shells of semiconductor materials and a shell of organic material (see Fig. 3). Typically, the core is made of narrow-gap semiconductor that transforms the broad solar spectrum into a narrow luminescence band. The first (inorganic) shell ensures the dangling bonds passivation at the surface of the QD core. The bandgap of this inorganic shell is bigger than the core gap. If several inorganic shells are present, each successive shell differs from the previous by increased bandgap. The main purpose of the multi-shell design is to align semiconductor lattice constants, which reduces formation of dislocations. The outer organic shell consists of organic molecules on the top of inorganic shell, which prevents aggregation of the QDs.

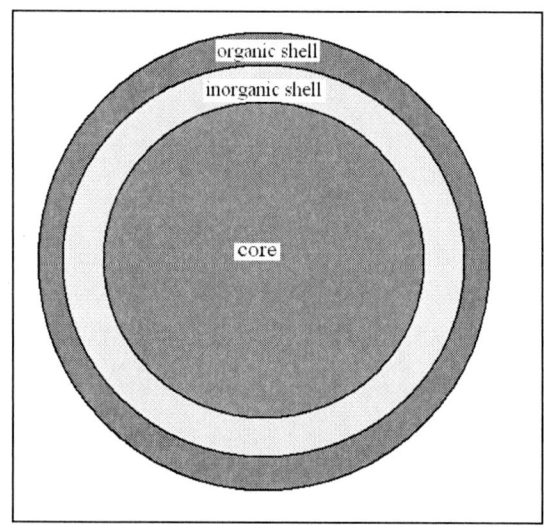

Fig. 3. Schematic representation of the structure of a multi-shell spherical quantum dot. QD core is composed of semiconducting material (its bulk energy gap is E_{g0}), covered by inorganic shell(s) to passivate the core surface. The outer organic shell ensures that QDs do not aggregate.

Geometry of the QDs and their size dispersion are usually characterized by the electron microscopy, the average core size $\bar{r}$ can be extracted from the absorption and luminescence spectra. To investigate the evolution of the reabsorption when QD size is changing, it is sufficient to determine the average QD core radius. This can be done by comparing a known experimental position of the absorption spectrum maximum, formed by electron transitions from the energy levels E_{01}^h on E_{01}^e with the location of the theoretical maximum. To estimate the mean radius, we introduce in (2) the term that takes into account the thermal broadening of the quantized energy levels and after a simple transformations, we obtain formula for the $\alpha_1(E_{ph})$ in the form:

$$\alpha_1(E_{ph}) = A \int_0^\infty \frac{1}{r^3} \exp\left(-\frac{\left(E_{ph} - E_1(r)\right)^2}{2\sigma_E^2}\right) \exp\left(-\frac{(r-\bar{r})^2}{2\sigma_r^2}\right) dr. \tag{6}$$

Here A is the QD radius independent constant, determined from normalization of the absorption to unity. The first exponent (Gaussian distribution on the energy) describes the contribution of the thermal smearing σ_E of the quantized energy levels. When calculating the absorption spectrum from (6), it is sufficiently accurate using $\sigma_E = 2kT$ (k is the Boltzmann constant and T is the absolute temperature). The second exponential (Gaussian QD size distribution) relates QDs radius dispersion σ_r and the optical spectra broadening. Here the energy $E_l(r_i)$ corresponds to the absorption maximum of the QD with its core radius r_i. Next, we calculate shape of the absorption band that formed by the electron transition from level E_{01}^e to level E_{01}^h.

The energy position of the maximum the luminescence peak is estimated from (5). We derived the expression for a shape of the luminescence spectrum as following:

$$I_{PL}\left(E_{ph}\right) = B\int_0^\infty \frac{1}{r_i^3}\exp\left(-\frac{\left(E_{ph}-E_{PL}(r_i)\right)^2}{2\sigma_E^2}\right)\exp\left(-\frac{(r_i-\bar{r})^2}{2\sigma_r^2}\right)dr, \quad (7)$$

where B is the QD radius independent constant, determined from the normalization condition of the luminescence intensity $I_{PL}(E_{ph})$ to unity, where $E_{PL}(r_i)$ is the energy of the luminescence intensity maximum of the QD of radius r_i.

According to [14], which treats interaction of semiconductor band states with photonic field, the low-energy part of PL spectra is determined by the density of states, while the high-energy part is defined by the distribution function. The density of states, which contributes to formation of the absorption spectra, is much higher than the density of states, contributing to formation of luminescence spectra. Therefore, the standard deviation σ_E for the absorption spectra is larger than for the luminescence spectra.

IV. EFFECT OF THE QUANTUM DOT SIZE AND THERMAL BROADENING OF THE QUANTUM LEVEL ON THE REABSORPTION

We evaluate the combined effects of QD size (namely its core radius), QD size dispersion and the thermal broadening on the reabsorption using as an example InP QDs. Their absorption and luminescence spectra, as discussed in [5], are shown in Fig. 4. To calculate the absorption, originated due to electron transitions from E_{01}^h to E_{01}^e, and the luminescence intensity we use (6) and (7) and take into account the dependence of the optical bandgap on the QD core radius [15].

To estimate the reabsorption magnitude, we follow several well-defined steps: Firstly, using the experimental values of the first absorption maximum energy and the peak of the luminescence (shown in Fig. 1), we find the mean radius of the quantum dot $\bar{r}$ from Eq. (5). Secondly, using Eqs. (6) and (7), we calculate the absorption and luminescence spectra,

normalized to unity. The height of the point, where these two curves intersect (see Fig. 1), is a measure of reabsorption. It is known that transmittance measurement error does not exceed 0.4% (see, for example, [16] and the error of the luminescence intensity measurement does not exceed 2.6% (see, *e.g.*, [17]-[18]). Therefore, when choosing QDs for solar fluorescent concentrators with minimal reabsorption, we can consider only those semiconductors, in which the value of reabsorption is less than 3%.

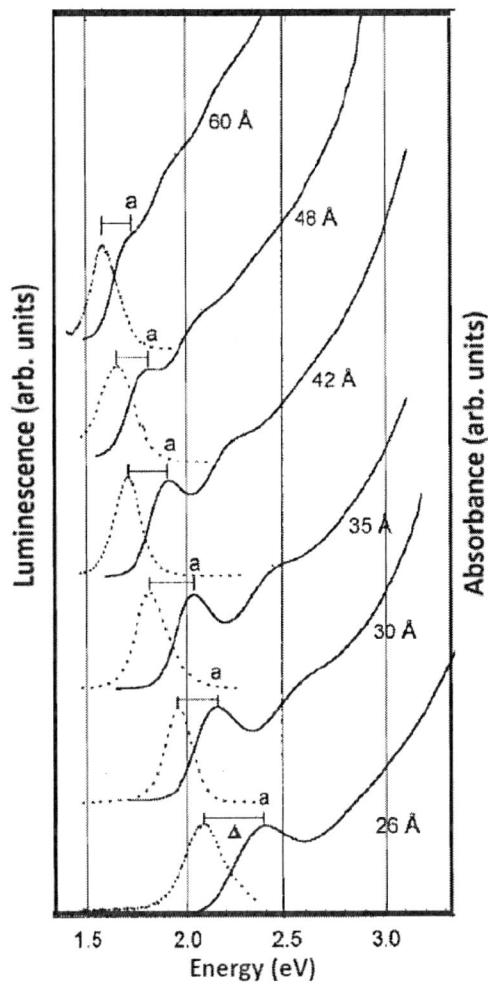

Fig. 4. Absorption (solid line) and PL (dotted line) spectra at 298 K for colloidal ensembles of InP QDs with different mean diameters. (Adapted from the original Fig. 1 in [5]). The energy separation Δ between the first peak of absorption spectrum and the peak of QD luminescence is $\Delta(h\nu)$, the notations used in Fig. 1.

In the majority of research on the efficiency of photovoltaic solar luminescent concentrators, the PMMA plate with QDs is used. Since the plastic plate is transparent in the wavelength range of $0.4 - 1$ μm, (or $3.1 - 1.2$ eV), the analysis of the effect of the QD radius and its dispersion on the reabsorption has been performed for the semiconductor QD core with the optical bandgap in the range of 1.2 – 3.1 eV.

Using Eqs. (6) and (7) we find a dependence of the nanoparticles reabsorption on their radius and dispersion (10%, 5%, 2% and 1% QDs dispersion values are considered) and this is shown in Fig. 5. As a distinct feature of the theoretical curves in Fig. 5, the presence of the minimum indicates that the optimum QD core sizes r_{min}, which minimizes the reabsorption, can always be chosen. Indeed, QD size increase above r_{min} reduces the energy separation between the peaks of the absorption and luminescence, thus increasing the reabsorption. When QD size decreases below r_{min}, the effect of thermal broadening of the quantized energy levels results in the reabsorption growth. According to Fig. 5, with a decrease of the QD size dispersion, the optimum r_{min} becomes smaller. It is clear from Fig. 5 that for the CdS based quantum dots r_{min} is always above 1.6 nm at any value of the QD size dispersion. On the other hand, in PbSe quantum dots even for their core radius (or diameter?) of 1 nm and very low QD size dispersion of 1%, the reabsorption cannot be below 54%.

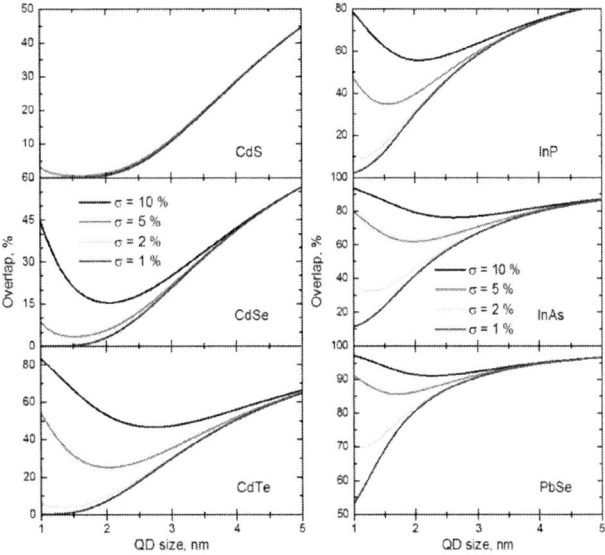

Fig. 5. Dependence of the reabsorption on the QD core radius for four values of the size dispersion. In constructing graphs used experimental dependence of absorption spectra and luminescence. To prepare the graphs, experimental dependences of the absorption and luminescence has been considered. The following energy gap values have been considered: CdS (E_{g0} = 2.49 eV), CdSe (E_{g0}= 1.74 eV), CdTe (E_{g0} = 1.43 eV) InP (E_{g0}= 1.27 eV), InAs (E_{g0}= 0.355 eV), PbSe (E_{g0}= 0.27 eV).

Using the results from Fig. 5 and known values of the semiconductor bulk gap, we plot in Fig. 6 the gap dependence of the reabsorption for several values of QD core size

dispersion, namely 10%, 5%, 2% and 1%. According to Fig. 6, reduction of the QD core size dispersion results in the reabsorption decrease. It is also shifting down with increasing the optical band gap of the QD. Consequently, the value of the band gap, at which the reabsorption becomes negligible, shifts to the lower energy side. Current QD technology offers highly homogeneous nanoparticles with 5% size dispersion [9]-[10]). At 5% size dispersion, the reabsorption can be negligible (that is below experimental error) if the QD core semiconductor bulk band gap exceeds 1.8 eV.

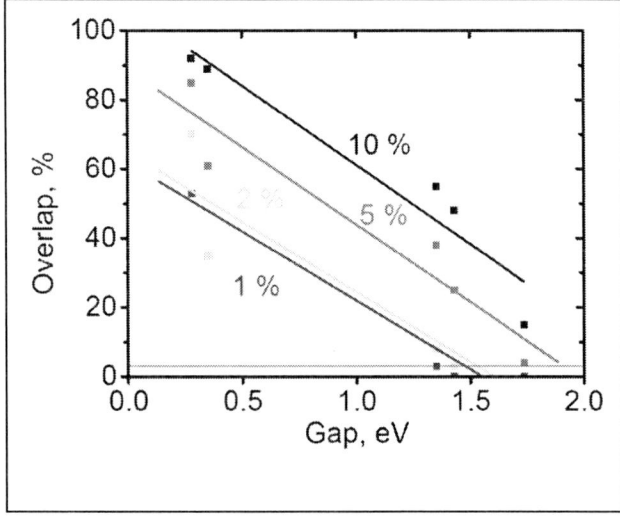

Fig 6. Dependence of the reabsorption losses on the QD optical band gap E_{abs} for different size dispersion values. Black horizontal line, close to the x-axis, aims to compare the normalized experimental accuracy of the measured QDs spectra to the reabsorption losses.

To determine the width of the solar spectrum, from which the incoming photons are transformed into luminescent photons, consider the dependence of the QDs reabsorption on the optical bandgap. For this, we substitute r_{min} values from Fig. 5 into Eq. (1) and calculate the optical band gap $E_{abs} = E_{01}^h - E_{01}^h$ of the QD core. Alternatively, we can use the gaps, due to the quantum confinement in the nanoparticles, given in [15]. The resulting reabsorption dependence on the optical bandgap of QD is shown in Fig. 7, which demonstrates that the slope of the reabsorption dependence decreases with reduction of the QD size dispersion. This means that with the CD size dispersion decrease, the sunlight absorption bandwidth also decreases. In particular, for the QD cores with a 5% size variation, the range of absorption of solar light photons extends the high side starting from 2.7 eV.

978-1-5090-5606-4/17 $31.00 © 2017 IEEE

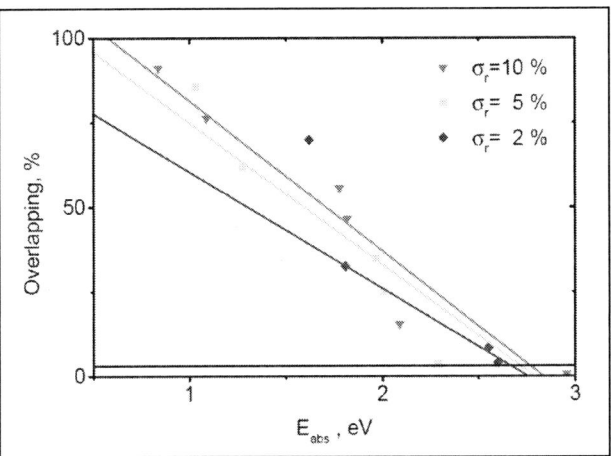

Fig. 7. Dependence of the QD reabsorption on the optical bandgap of the nanoparticle. $T = 300°$ C

V. CONCLUSION

We investigated the reabsorption processes in quantum dots (QDs), including the effect of the QD core size, its dispersion and the semiconductor core bulk band gap. It is shown that decreasing the core size, the optimum QD radius, which minimizes the reabsorption, can be found. Decreasing or increasing the QD core size below or above the optimal value always increase the reabsorption. We demonstrated that decreasing the semiconductor core bulk bandgap E_{g0} (while keeping it sufficiently large, i.e., $E_{g0} \geq 1.8$ eV) and low QD size dispersion (5% or less), the reabsorption can be negligibly small (that is, being below the experimental detection limit). In contrast, when $E_{g0} < 1.8$ eV, further band gap E_g decrease only leads to the reabsorption growth. For instance, even for highly homogeneous nanoparticles with only 1% size dispersion made from PbSe QDs with $E_g = 0.27$ eV the lowest reabsorption extends to 54%.

REFERENCES

[1] X. Huang, S. Han, W. Huang, and X. Liu, "Enhancing solar cell efficiency: the search for luminescent materials as spectral converters," *Chem. Soc. Rev.* vol. 42, pp. 173-201, 2013.

[2] N. Aste, L. C. Tagliabue, C. D. Pero, D. Testa, and R. Fusco, "Performance analysis of a large-area luminescent solar concentrator module", *Renew. Ener.* vol. 76, pp. 330-337, 2015.

[3] C. Li, W. Chen, D. Wu, D. Quan, Z. Zhou, J. Hao, J. Qin, Y. Li, Z. He, and K. Wang, "Large Stokes Shift and High Efficiency Luminescent Solar Concentrator Incorporated with CuInS2/ZnS Quantum Dots Scientific Reports 5, 17777 (2015).

[4] M. R. Kulish, V. P. Kostylyov, A. V. Sachenko, I. O. Sokolovskyi, D. V. Khomenko, and A. I. Shkrebtii, "Luminescent converter of solar light into electrical energy. Review", *Semicond. Phys, Quantum Electronics & Optoelectronics*, vol. 19, 229-247, 2016.

[5] O. I. Mićić, H. M. Cheong, H. Fu, A. Zunger, J. R. Sprague, A. Mascarenhas, A. J. Nozik, "Size-Dependent Spectroscopy InP Quantum Dots", *J. Ph. Chem. B*, vol. 101, pp. 4904-4912, 1997.

[6] A. Irman, "Modification of Spontaneous Emission of Quantum Dots by Photonic Crystals". PhD Thesis, Complex Photonic Systems Group and Institute Faculty of Science and Technology University of Twente Enschede, The Netherlands (2003). http://cops.nano-cops.com/sites/default/files/irman.pdf

[7] A. Kitai, *Luminescent materials and applications,* Chichester, England, Hoboken, NJ: John Wiley, 2008.

[8] O. Chen, J. Zhao, V. P. Chauhan, J. Cui, C. Wong, D. K. Harris, H. Wei, H. Han, *et al.*, "Compact high-quality CdSe–CdS core–shell nanocrystals with narrow emission linewidths and suppressed blinking," *Nat. Mater,* vol. 12, pp. 445-451, 2013.

[9] M. Aliofkhazraei, *Handbook of nanoparticles*, London: Springer International Publishing, 2015.

[10] D. Segets, "Analysis of Particle Size Distributions of Quantum Dots: From Theory to Application," *KONA Powder and Particle Journal*, vol. 33, pp. 48-62, 2016.

[11] S. Flügge, *Practical quantum mechanics,* Berlin: Springer, 1999.

[12] D. Bera, L. Qian, T. Tseng, and P. H. Holloway, "Quantum Dots and Their Multimodal Applications: A Review," *Materials,* vol. 3, pp. 2260-2345, 2010.

[13] J. Jasieniak, L. Smith, J. van Embden, P. Mulvaney, and M. Califano, "Re-examination of the Size-Dependent Absorption Properties of CdSe Quantum Dots," *J. Phys. Chem. C*, vol. 113, pp. 19468-19474, 2009.

[14] W. van Roosbroeck and W. Shockley, "Photon - Radiative Recombination of Electrons and Holes in Germanium" *Phys.Rev.* vol. 94, pp. 1558-1560, 1954.

[15] S. Baskoutas and A. F. Terzis, "Size-dependent band gap of colloidal quantum dots," *J. Appl. Phys.* vol. 99, pp. 13708-1, 2006.

[16] A. Ruiz, J. Zwinikels, I. Bougleux, S. Bruce, E. Early and P. Y. Barnes. *Inter-American metrology system (Sim 2.2). Intercomparison of wavelength scale and photometric scale of spectrophotometry laboratories* CENAM-NRC-INMETRO-NIST (2000), http://www.cenam.mx/comparaciones /ComparacionSIM2_2CENAM-NRC-INMETRO-NIS T.pdf

[17] K. Firosz, Scope of *Accreditation to ISO/IEC 17025:2005. Molecular Devices,* Inc. 1311 Orleans Drive Sunnyvale, CA 94089, (2005). http://www.ophiropt.com/user_files/laser/Ophir-SpiriconScope.pdf.

[18] B. J. Palmer, K. Winterton, M. Jense, *Scope of Accreditation to ISO/IEC 17025:2005.* Ophir-Spiricon LLC 3050 N 300 W North Logan, UT 84431. ANSI-ASQ National Accreditation Board/ACLASS, http://www.ophiropt.com/user_files/laser/ Ophir-SpiriconScope.pdf

APPENDIX

Table 1.Values of the parameter $\varphi_{l,n}$ used in Eq. 1 (from [11]).

l/n	1	2	3	4
0	3.142	6.263	9.425	12.566
1	4.493	7.725	10.904	14.066
2	5.764	9.095	12.323	
3	6.988	10.417	13.698	
4	8.183	11.705		
5	9.356	12.967		
6	10.513	14.207		
7	11.657			
8	12.791			
9	13.916			

Development of absorber and energy selective contacts for hot carrier solar cells

Santosh Shrestha, Simon Chung, Yuanxun Liao, Wenkai Cao, Neeti Gupta, Yi Zhang, Xiaoming Wen, Gavin Conibeer

School of Photovoltaic and Renewable Energy Engineering, UNSW Sydney, NSW 2052, Australia

Abstract —The hot carrier solar cell is one of the most promising advanced PV concepts with theoretical efficiency of over 65% at one sun. Two crucial components of the HC solar cell are: (i) Absorber which can sufficiently reduce the rate of hot carrier cooling so that they can be collected at higher energies, and (ii) Energy selective contacts (ESCs) which allow extraction of hot carriers only through a narrow energy range. In this paper, potential of HfN and ZrN thin films as hot carrier absorber has been investigated. Al_2O_3/Ge QW/ Al_2O_3, Al_2O_3/Si QW/Al_2O_3, and Al_2O_3/PbS QDs/Al_2O_3 double barrier structures have been investigated as energy selective contacts.

Index Terms — carrier cooling, double barrier, energy selective contacts, hafnium nitride, hot carrier, photovoltaic cells, solar energy, zirconium nitride.

I. INTRODUCTION

In conventional single junction solar cells about half of the incident solar energy is lost due to the non-absorption of the below band gap photons and thermalisation of above bandgap photons. The Hot carrier solar cell (HCSC) aims to minimise these loses by minimising these losses. Efficiency for this technology is predicted to be over 65% at one sun and 85% for maximal concentration [1-3]. The concept of

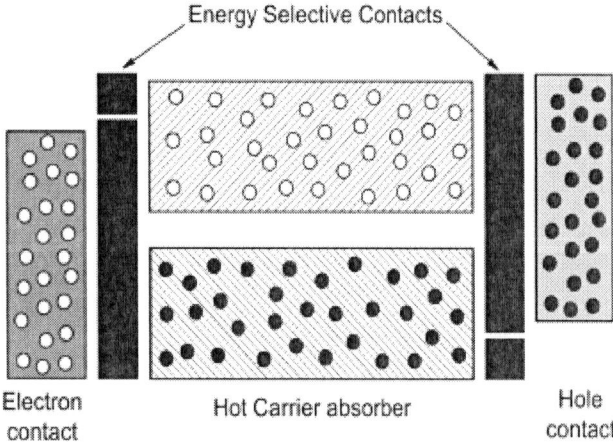

Fig. 1. Schematic diagram of an ideal H_C solar cell. The absorber has a hot carrier distribution at temperature T_H. The hot carriers are collected through the ESCs at temperature T_A.

the HC solar cells is illustrated in Fig. 1. It consists of an absorber and two energy selective contacts (ESCs).

For the absorber, a low band gap material is desirable so that a large fraction of the solar spectrum can be absorbed. The critical requirement of the absorber is, however, to have slow rate of carrier cooling- of the order of nanoseconds. This allows the collection of photo-generated carriers at higher energies (hot carriers) and thus enables extraction of the carriers at higher voltages. The hot carriers also need to be collected within a narrow energy range through ESCs. Carriers with other energies are reflected back to the absorber where their energies are re-normalised through elastic carrier-carrier scattering thus re-filling the depleted carriers at the extraction energy. Only a small fraction of excess energy above the band edge is lost when the hot carriers come in contact with cold carriers in the external metal contacts through ESCs [4].

In this paper, hafnium nitride and zirconium nitride films are presented as potential absorber materials based on their large phononic band gap, processibility and long carrier cooling time. MQW-based absorber is presented elsewhere [5]. For energy selective contacts, double barrier resonant (DBR) tunneling structures employing Ge quantum well (QW) and PbS quantum dots (QDs) will be discussed.

II. POTENTIAL OF HAFNIUM NITRIDE AND ZIRCONIUM NITRIDE AS ABSORBER FOR HOT CARRIER SOLAR CELLS

Requisites of HC absorber are described by Conibeer et al. [6]. Slow carrier cooling of hot carriers in the absorber is a critical requirement. It is necessary for the collection of the carriers before they thermailise on the band edges, i.e. whilst they are at higher energies. The efficiency of HCSC critically depends on the hot carrier thermilisaiton time in the absorber. In most semiconductors, hot carriers cooling time is of the order of a few picoseconds. For an efficient

HCSC, the carrier cooing rate needs to be significantly slower- of the order of nanoseconds. Calculation by Takeda et al. shows that the conversion efficiency (at 1000 concentration) is about 40% for the thermalisation time of 100 ps but it increases to 55% for the thermalisation time of 1 ns [7].

The hot carriers primarily lose their energies by scattering of hot electrons with zone centre optical phonons and optical phonons lose their energies, predominantly, by decaying into two acoustic phonons of half the energy and opposite momenta of the original phonon, i.e. via Klemens decay mechanism [8,9]. Another decay route, which is much less efficient, is via Ridley mechanism where an optical phonon decays into optical phonon with a lower energy and an acoustic phonon [10]. If the lifetime of optical phonons is long enough, they can transfer their energies to the carriers and return them to 'hot' state.

Klemens mechanism is restricted in materials where the minimum energy gap between the optical and acoustic phonons, i.e. phononic band gap, is higher than the maximum acoustic phonon energy [11]. Ridley mechanism is also suppressed in such materials with large phonon band gap and/or narrow dispersion of optical modes. For binary compounds optical and acoustic phonon energies can be approximated from a 1-D force constant model treating the atoms as simple harmonic oscillators [12]. It can be calculated that a minimum of $M/m > 4$ is necessary to produce a large phononic band gap to prevent Klemens decay. Potential absorber materials should also have a small dispersion of optical modes to suppress Ridley decay and small electronic bandgap so that a large fraction of the incident solar radiation can be absorbed. Calculation shows that HfN and ZrN have very large phonon gaps and small dispersion of optical modes. Modelling work has also shown that these materials have very large phononic band gap and small dispersion of optical modes [13]. The abundance of these materials is also relatively large. Hence these materials can be good candidates for the HC solar cell absorber. While HfN and ZrN are metallic, Hf_3N_4 and Zr_3N_4 are semiconductors with band gap of ~0.9 eV and 1 eV, respectively [14].

X-ray diffraction (XRD) spectra of selected HfN films grown on a silicon and MgO substrates by RF sputtering are shown in Fig. 2(a) and 2(b), respectively. During the film growth the substrates were heated to 450 °C. The films were about 100 nm. XRD were performed using a PANalytical Empyrean system using CuK_α (λ = 1.5418 Å). Peak

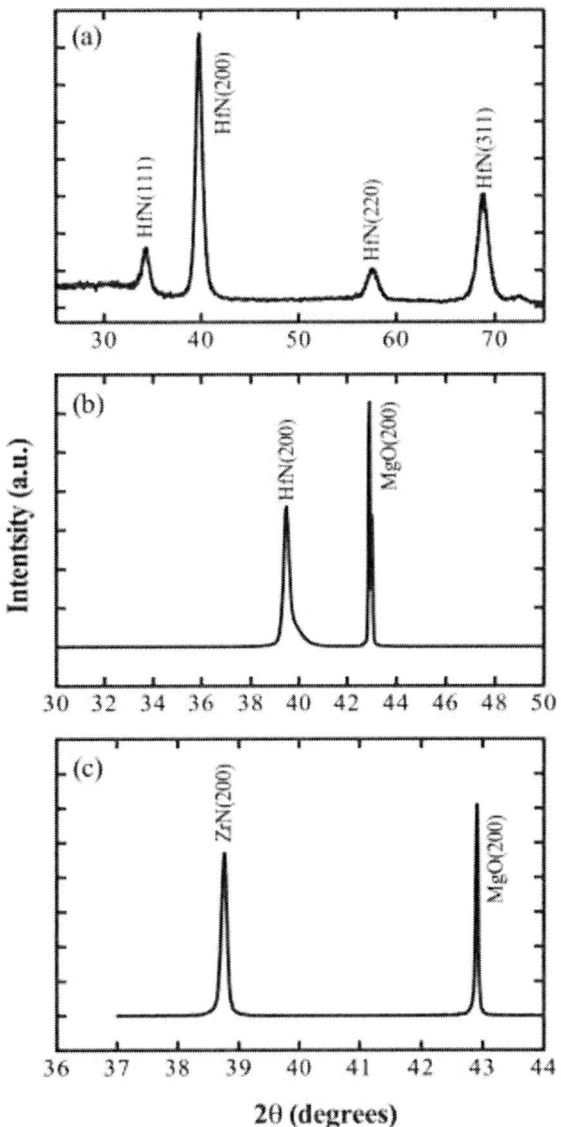

Fig. 2. XRD spectra of (a) HfN film deposited on Si (100) substrate, (b) HfN film deposited on MgO(100) substrate, and (c) ZrN film deposited on MgO(100) substrate.

corresponding to the HfN and substrates are labelled. It can be deduced that the HfN film preferentially grow in [100] direction on Si (100) and MgO(100) substrate. Growth of relatively better crystal quality on MgO substrate can be attributed to smaller lattice mismatch between HfN and MgO. Lattice mismatch between HfN and MgO is about 7.5% whereas it is 17% between HfN and Si. XRD spectrum of a typical ZrN film deposited at 500 °C is shown in Fig 2(c) which also show preferential growth in (100) direction on MgO(100) substrate. The compositional analysis of these films with X-ray photoelectron spectroscopy and Rutherford backscattering spectroscopy [15] have shown that most of the films are metal-rich.

Carrier cooling properties of selected HfN and ZrN films were investigated by ultrafast transient absorption (TA) spectroscopy. Time resolved photoluminescence [16], which is typically used for the charaterisation of hot carrier dynamics, could not be used as these materials were non-luminescent. Details of TA spectroscopy are described in other references [17]. In short, the sample of interest is excited by a pump pulse and a weak probe with different delays with respect to the pump is used to monitor the change in optical density (ΔOD) before and after the pump.

The ΔOD is directly related to the carrier dynamics and thus can be studied as carriers relax back to steady state conditions. This method has been used to study carrier dynamics in many materials including in semiconductor nanoparticles [18-21]. HfN and ZrN films grown on quartz substrates were used as transmitted signal is detected in this method. The TA measurements were performed using 400 nm excitation pump source with 100 fs duration and 1 kHz repetition rate. White light continuum was used as the probe beam and detected by a polychromator CCD.

TA results of typical HfN and ZrN films are shown in Fig. 3. On the left TA spectra at various time delays are shown for HfN and ZrN films. The changes in optical density, ΔOD, before and after the pump are plotted as a function of the

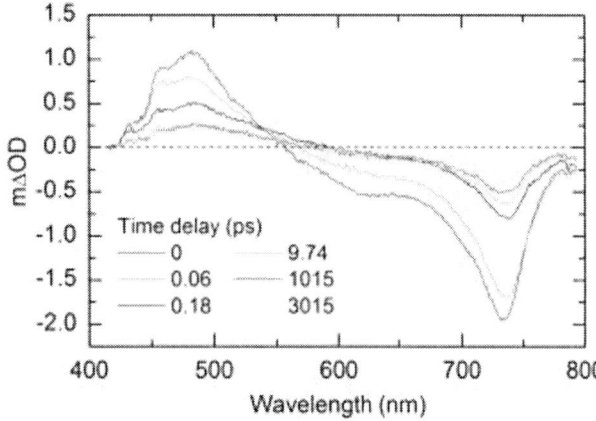

Fig. 3(a) TA spectra of a HfN sample deposited at 450 °C in visible range.

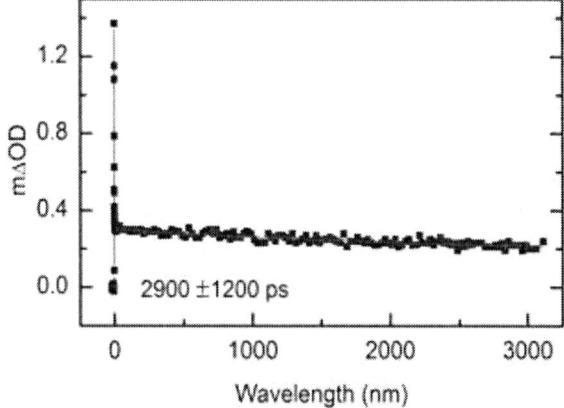

Fig. 3 (b) ΔOD as a function of delay time measured at 730 nm for the sample in (a). The red curve is the exponential fit.

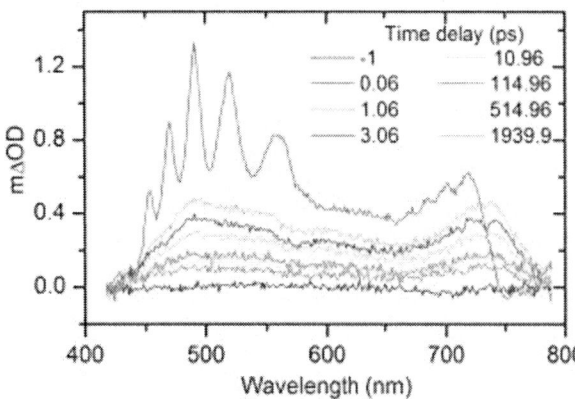

Fig. 3 (c) TA spectra of a ZrN sample deposited at 500 °C in visible range.

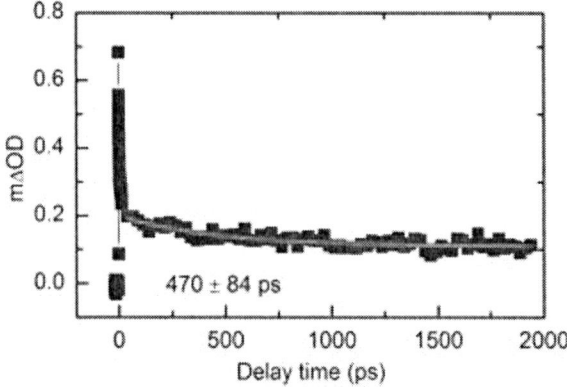

Fig. 3 (d) ΔOD as a function of delay time measured at 730 nm for the sample in (c). The red curve is the exponential fit.

probe wavelength. TA spectra for HfN show an excited state absorption peak around 440 nm and a bleaching peak around 730 nm whereas bleaching peak is not observed in the case of ZrN. On the right column, time evolution of ΔOD at a particular wavelength is plotted as a function of time delays. Immediately after the pump a rapid change in ΔOD is observed in both HfN and ZrN spectra, which may be due to electron-electron scattering. Then it slowly decreases with longer delay time. The data has been fitted with a single exponential fit which is shown by the red curves. From these fits decay time constant of about 2900 ± 1200 ps, and 470 ± 84 ps can be extracted for HfN and ZrN, respectively.

II. POTENTIAL OF AL_2O_3/GE QW/AL_2O_3 AND AL_2O_3/PBS QD/ AL_2O_3 AS ENERGY SELECTIVE CONTACTS

For energy selective contacts, double barrier resonant tunneling structures (DBRs) consisting of quantum well (QW) and quantum dot (QD) have been investigated. Such structures are expected to give conduction of carriers strongly peaked at the discrete energy levels, and lower at other energies. Evidence of resonance tunneling can be demonstrated by negative differential resistance (NDR) in I–V measurements [22]. Energy at which this occurs can be tuned by the choice of barrier and well thickness as well as material combinations. This allows optimisation of ESCs to operate HC devices at optimal energy extraction point for a particular absorber and ESC combination.

Ge QW-based structures (Al_2O_3/Ge QW/Al_2O_3) were fabricated on highly doped n-type silicon substrate (0.001-0.006 Ω.cm) by sputtering. The thickness of Al_2O_3 barriers was typically 0.5 nm, and Ge QW was typically 4 nm thick. From TEM an interface roughness of about 0.2 nm has been observed between the Ge and Al_2O_3 layers which is good for minimizing degradation caused by nonuniformity. Au was used at the back contact and for the front contact pattered Al was used. For the PbS QD-based structures, Al_2O_3 layers were fabricated by atomic layer deposition (ALD). PbS QDs were chemically synthesized as discussed in literature [23]. A mono-layer of PbS QDs was transferred on to the Al_2O_3 film using Langmuir-Blodgett method [24]. The samples were prepared on highly doped n-type silicon wafer (0.001-0.006 Ω.cm). Top Pt contacts were fabricated using a focused ion beam (FIB) with in-situ SEM to precisely locate the contacts over single-layer of QDs. Au back contact on the Si substrate was fabricated by sputtering gold on the rear surface.

Figure 4 shows results of I-V measurements on a Al_2O_3/Ge QW/Al_2O_3 structure at 300K and 90K. The Al_2O_3 barriers were 0.5 nm and Ge QW was 4 nm thick. NDR features were also be observed in on Al_2O_3/ PbS QDs/ Al_2O_3 structure at 300K and 190K. Full width at half maximum (FWHM), peak to valley current ratio (PVCR) and quality factor (QF = PVCR/FWHM) of the NDR peaks are given in the respective figures. PVCR and QF increase at lower temperature, which is likely to be primarily due to the reduction of background current.

II. DISCUSSION AND CONCLUSIONS

TA measurements have shown that there is carrier activity for several nanoseconds in HfN and several hundred picoseconds in ZrN. The observed decay time constants in these materials are more than two orders of magnitude larger than that for Si and GaAs [25,26]. Slow carrier cooling in these materials can be attributed to a large gap between

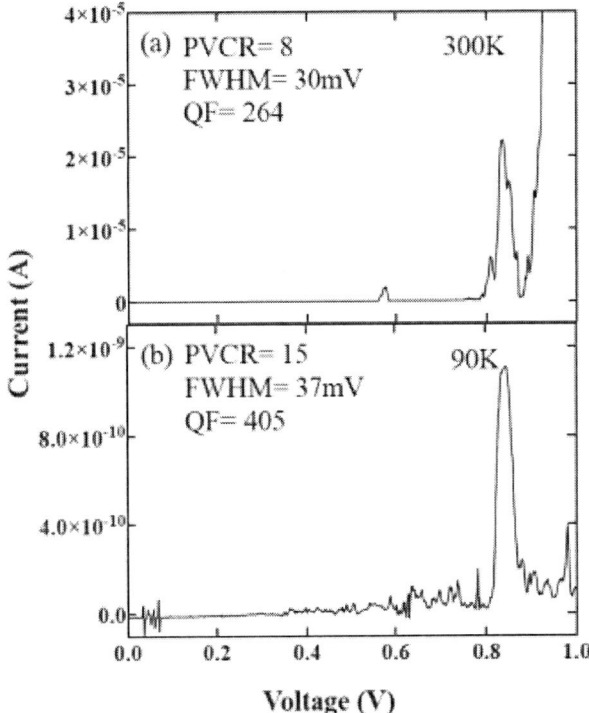

Fig. 4. I-V curves measurement on Al_2O_3/ Ge QW/ Al_2O_3 at (a) 300K and (b) 90K.

optical and acoustic modes. This is strongly suggestive of hot carrier behavior at long lifetimes in these materials and thus are potential hot carrier absorber materials. As mentioned earlier, for TA measurements HfN and ZrN films grown on quartz substrate were used. These films were non-stoichiometric and the crystal quality was inferior to that grown on silicon and MgO substrates. Non-stoichiometry and poor crystal quality introduce additional phonon modes in the phonon gap which can allow decay process to occur and thus decrease the hot carrier lifetime. Hence stoichiometric, better quality materials are expected to further increase the carrier lifetime.

IV measurements on A_2O_3/Ge QW/A_2O_3 and A_2O_3/PbS QDs/A_2O_3 DBRT structures show NDR behavior both at low temperature and room temperature. FWHM of the NDR peaks is about 30 mV. Ideally hot carriers need to be extracted at a discrete energy through mono-energetic contacts for optimum efficiency [2], although in this case the power output from the device would be minimum. Therefore, in practical HC solar cell devices ∂E should be kept as small as possible. A rough estimate of $\partial E \sim 25$ mV has been recommended as a reasonable value [27]. The observed FWHM is only slightly larger than the recommended value which indicates that these structures can be used as ESCs for hot carrier solar cells.

Further work is planned to improve the quality of HfN and ZrN, for example, by atomic layer deposition. Study of DBRT devices to control the position of the resonance peak to match with the optimum energy extraction level for a given absorber will also be conducted. Integration of these ESC structures with HfN and ZrN, absorbers to fabricate complete hot carrier solar cell devices is also planned.

ACKNOWLEDGEMENT

This program has been supported by the Australian Government through the Australian Renewable Energy Agency (ARENA). Responsibility for the views, information, or advice expressed herein is not accepted by the Australian Government.

REFERENCES

[1] R. Ross, A.J. Nozik, J. Appl Phys 53, 3318 (1982).

[2] P. Würfel, Sol. Energy Mats. and Sol. Cells. 46, 43 (1997).

[3] M.A. Green, Third Generation Photovoltaics (Springer-Verlag, 2003).

[4] G.J. Conibeer, C-W. Jiang, D. König, S. Shrestha, T. Walsh, M.A. Green, Thin Solid Films, 516, 6968 (2008).

[5] G. Conibeer, Y. Zhang, S. Chung, Y. Liao, S. Brember, S. Shrestha, "Multiple quantum wells as slowed hot carrier cooling absorbers in hot carrier cells", Proc. 44th IEEE Photovoltaic Specialist Conference, June 2017, Washington DC.

[6] G. Conibeer, S. Shrestha, S. Huang, R. Patterson, H. Xia, Y. Feng, P. Zhang, N. Gupta, M. Tayebjee, S. Smyth, Y. Liao, S. Lin, P. Wang, X. Dai, S. Chung, Sol. Energy Mater. Sol. Cells 135, 124 (2015).

[7] Y. Takeda, T. Ito, T. Motohiro, D. König, S. Shrestha, G. Conibeer, J. Appl. Phys. 105, 074905 (2009).

[8] A. Othonos, J Appl Phys. Reviews, 83, 1789 (1998).

[9] P.G. Klemens, Phys. Rev. 148, 845 (1966).

[10] J.W. Pomeroy, M. Kuball, H. Lu, W.J. Schaff, X. Wang, A. Yoshikawa, Appl. Phys. Lett. 86, 223501 (2005).

[11] G. Conibeer, R. Patterson, L. Huang, J.F. Guillemoles, D. König, S. Shrestha, M.A. Green, Solar Energy Mater. Solar Cells 94, 1516 (2010).

[12] P. Misra, Physics of Condensed Matter (Academic Press, 2010), p. 44.

[13] B. Saha, J. Acharya, T.D. Sands, U.V. Waghmare, J. Applied Physics 107, 033715 (2010).

[14] D.I. Bazhanov, A.A. Knizhnik, A.A. Safonov, A.A. Bagaturyants, M.W. Stoker, A.A. Korkin, Journal of Applied Physics 97, 044108 (2005).

[15] L.C. Feldman, J.W. Mayer, Fundamentals of Surface and Thin Film Analysis (North-Holland, NewYork,1986).

[16] Y. Zhang, M.J.Y. Tayebjee, S. Smyth, M. Dvořák, X. Wen, H. Xia, M. Heilmann, Y. Liao, Z. Zhang, T. Williamson, J. Williams, S. Bremner, S. Shrestha, S. Huang, T.W. Schmidt, G.J. Conibeer, Applied Physics Letters 108, 131904 (2016).

[17] G.V. Hartland, Chemical Science 1, 303 (2010).

[18] A. Pandey, P. Guyot-Sionnest, Science 322, 929 (2008).

[19] M. L. Mueller, X. Yan, B. Dragnea, L. Li, Nano Letters 11, 56 (2011).

[20] X. Wen, P. Yu, Y.-R. Toh, Y.-C. Lee, K.-Y. Huang, S. Huang, S. Shrestha, G. Conibeer, J. Tang, Journal of Materials Chemistry C 2, 3826 (2014).

[21] V.I. Klimov, Journal of Physical Chemistry B 104, 6112 (2000).

[22] S.K. Shrestha, P. Aliberti, G.J. Conibeer, Sol. Energy Mater. Sol. Cells. 94, 1546 (2010).

[23] W. Cao, Z. Zhang, R. Patterson, Y. Lin, X. Wen, B.P. Veetil, P. Zhang, Q. Zhang, S. Shrestha, G. Conibeer, S. Huang, RSC Advances 6, 90846 (2016).

[24] J. Orbulescu, R.M. Leblanc, American Chemical Society 996, 172 (2008).

[25] A. J. Sabbah, D. M. Riffe, Physical Review B 66, 165217 (2002).

[26] P. Langot, N. Del Fatti, D. Christofilos, R. Tommasi, F. Vallée, Phys. Rev. B 54, 14487 (1996).

[27] G.J. Conibeer et al, Selective energy contacts for potential application too hot carrier solar PV cells, Proc. 3rd World Conference on Photovoltaic Energy Conversion, 2003, p. 2730.

GaAsBi Devices for Thermal Energy Conversion

Margaret Stevens, Abigail Licht, Nicole Pfiester, Emily Carlson, Kevin Grossklaus, and Thomas E. Vandervelde

Renewable Energy and Applied Photonics Laboratory, Electrical and Computer Engineering Department, Tufts University Medford, Massachusetts, 02155

Abstract — Small band gap thermophotovoltaics can be engineered to match the thermal spectrum of many heat sources. III-V-Bismides offer new band gap combinations at the InP lattice constant. Through simulation, we explore homojunction and heterojunction GaAs$_{0.76}$Bi$_{0.24}$ devices, with a band gap of 0.25eV, tuned for a 1350K source. Homojunction devices, simulated with the Silvaco-Atlas software package, demonstrate high short-circuit current but also high dark current density. Heterojunction structures using In$_{0.53}$Ga$_{0.47}$As do not demonstrate as significant gains in short-circuit current, but greatly reduce dark current. This leads to increases in open-circuit voltage, fill factor, and efficiency.

Index Terms — thermophotovoltaics, III-V semiconductors, bismuth compounds, epitaxial layers, device simulation.

I. INTRODUCTION

Waste heat power generation is a relatively untapped source of electricity. In 2015 as much as 59% of all energy generated for consumption and transportation in the United States was dissipated as heat [1]. This massive energy waste stems from two main sources: heat losses from converting thermal energy to mechanical energy in power plants and exhaust from internal-combustion engines in vehicles. Thermophotovoltaic diodes (TPVs), which convert infrared radiation into electricity by the photovoltaic effect, can be used to recover some of this wasted energy. By engineering TPV diodes to have a spectral response optimized for the thermal spectrum of the heat, this formerly wasted energy can be recaptured and recycled to improve the efficiency of our preexisting power structures.

TPVs are suited for a variety of different power applications ranging from the aforementioned waste heat recovery, to mobile power sources [2], to deep space power generation [3]. To date, TPV systems have implemented diodes with band gaps ranging from 0.5-0.8eV, corresponding to source temperatures from 1000-2000K or IR radiation from 1-3μm [4]. Although the 0.5-0.8eV range is ideal for integrating TPVs with multijunction photovoltaics optimized for the solar spectrum, the high source temperatures can contribute to thermal stresses on the equipment and require additional insulation, ultimately limiting the potential applications.

By utilizing small band gap materials, a longer-wavelength TPV cell can be used with an existing thermal configuration to harvest a larger portion of the infrared spectrum, ultimately generating more power. In this work, we explore TPV devices tuned for a 1350K blackbody thermal source, to match those employed in radioisotope power systems. Detailed balance limit calculations show the maximum theoretical TPV diode efficiency with this source temperature is 43.9%, predicted to occur for a single junction diode band gap of 0.25eV [5]. This is a significant improvement over the state-of-the-art strained InGaAs on InP device (0.6eV), which has a maximum theoretical diode efficiency of 17.8%. Long-wavelength, small band gap TPVs would allow for better harvesting of the 1350K thermal spectrum. However, there are many challenges associated with growing small band gap III-V materials such as miscibility gaps and lattice matching constraints.

III-V-Bismide alloys are presently under investigation for many mid-infrared optoelectronics devices [6][7][8] and multijunction photovoltaics [9][10]. Incorporation of bismuth, the largest and least studied nonradioactive group-V element, has a dramatic effect on the electrical and optical properties of a III-V alloy. Due to the valence band anticrossing effect, a small incorporation of bismuth can significantly narrow the band gap [11]. This allows for fine-tuning of the band gap of a semiconductor without causing significant changes to the lattice constant. Density function theory (DFT) calculations, shown in Figure 1, have predicted GaAs$_{1-x}$Bi$_x$ could achieve a band gap of 0.25eV on an InP substrate with a 0.95% lattice

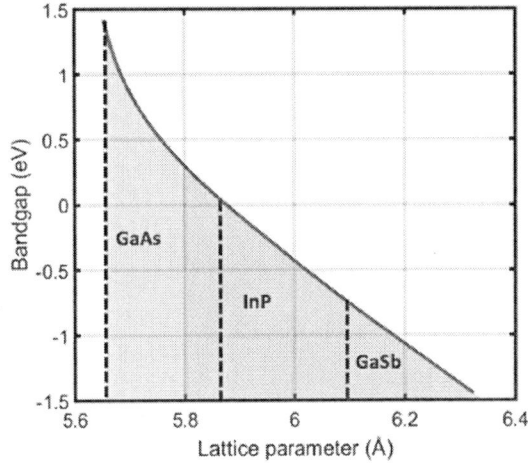

Fig1: Solid blue curve depicts the calculated band gap of GaAs$_{1-x}$Bi$_x$ from GaAs (5.65 Å) to GaBi (6.33 Å) [10]. Dashed lines indicate the lattice constants of common III-V substrates. Shading marks the incorporations of bismuth that have been achieved experimentally (blue) and have yet to be achieved experimentally (green). Our target bismuth incorporation, 24%, will require advanced epitaxial techniques to achieve high incorporations with good crystalline quality.

978-1-5090-5606-4/17 $31.00 © 2017 IEEE

mismatch [10]. This mismatch is comparable to $In_{0.66}Ga_{0.34}As$ strained on InP at 0.6eV.

Previous simulations have shown promising results for bismide-based devices for multijunction photovoltaic purposes [9][10]. In this work, we applied similar simulation techniques to explore III-V-Bismide alloys for long-wavelength TPVs. We employed the Silvaco-Atlas software package to design homojunction and heterojunction TPV diodes targeting a 1350K heat source. Conventional designs utilize InGaAs strained on InP as the TPV diode material, which has achieved preliminary system efficiencies of 19% utilizing monolithically integrated modules [12]. To increase the power conversion efficiency of TPVs employed with a 1350K heat source, we simulated a 0.25eV $GaAs_{0.76}Bi_{0.24}$ TPV diode to explore device designs that increase V_{OC} and reduce dark current.

II. SILVACO SIMULATIONS

A. Homojunction Devices

Using the device simulation program Atlas by Silvaco, we optimized the structure of our TPV diodes to maximize efficiency when paired with a 1350K blackbody heat source. To simulate bismide-based devices, we fed in material parameters including band gap [10], carrier mobilities [13][14], recombination lifetimes [15], and absorption coefficients [16]. These parameters were obtained from density functional theory (DFT) simulations of GaAsBi [10][17] and from experimental data in the literature [13-16]. The InGaAs TPV diodes were n-on-p devices modeled after single junction 0.6eV InGaAs devices in the literature [18]. The GaAsBi devices were p-on-n, as we found that structure yielded higher V_{OC}. This could be attributed to lower Auger recombination in the main absorption regions of the device, leading to overall lower dark current density. Layer thicknesses and doping profiles for all devices were optimized by maximizing the EQE response. The doping values for the GaAsBi device are higher than preferred for a practical device (p: 1e19 cm^{-3} and n: 5e18 cm^{-3}); however, such high doping profiles are necessary to generate a response at room temperature (300K) for the simple single junction device with no spectral control [19].

Figure 2(a) shows the EQE response of both TPV devices compared to the normalized power density of the blackbody source in question. The 0.25eV GaAsBi TPV diode has lower overall EQE than the InGaAs device, but is far better matched to the 1350K thermal source as predicted by the detailed-balance calculation. This indicates the 0.25eV GaAsBi device will produce higher short circuit current than the 0.6eV InGaAs device. These gains are confirmed in Figure 2(b), which demonstrates the simulated light IV response of the InGaAs and GaAsBi-based TPV diodes when paired with a 1350K blackbody source. The room temperature performance metrics of each device are compared in Table 1.

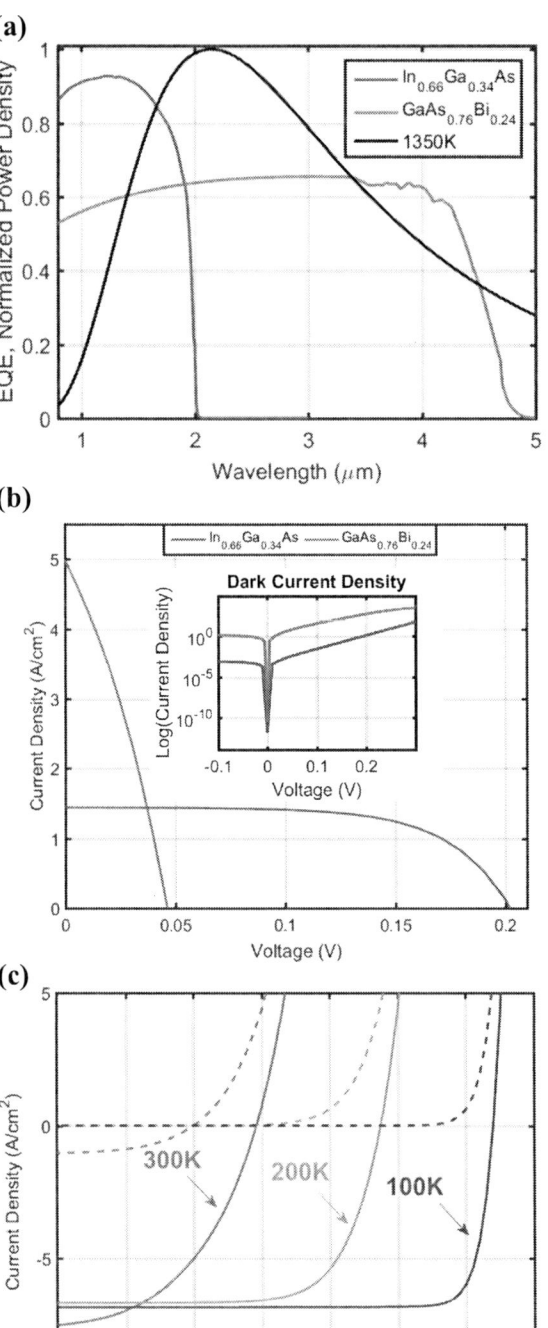

Fig2: (a) Simulated EQE of TPV devices compared to the normalized power density of a 1350K blackbody source. (b) Simulated light IV curves of TPV diodes operating at room temperature. Inset shows logarithmic plot of the dark current for both devices. (c) Dark and light IV simulations of the $GaAs_{0.76}Bi_{0.24}$ diode with decreasing operating temperature.

978-1-5090-5606-4/17 $31.00 © 2017 IEEE   702

TABLE I
SIMULATED PERFORMANCE METRICS OF SINGLE JUNCTION TPV DIODES (1350K) OPERATING AT ROOM TEMPERATURE (300K)

Material	J_{SC} (A/cm^2)	V_{OC} (V)	Efficiency (%)
$In_{0.66}Ga_{0.34}As$	1.442	0.202	3.146
$GaAs_{0.76}Bi_{0.24}$	4.986	0.046	1.228
Theoretical 0.25eV diode [21]	6.837	0.09	4.776
$In_{0.53}Ga_{0.47}As/$ $GaAs_{0.76}Bi_{0.24}$	1.450	0.353	3.835

B. High Dark Current Density

Although the 0.25eV TPV device far outperforms the InGaAs TPV in terms of J_{SC}, the overall efficiency suffers due to low V_{OC} and low fill factor. The inset in Figure 2(b) shows the GaAsBi device suffers from high dark current density (J_0), a common problem for small band gap TPV diodes operating under room temperature conditions [20]. V_{OC} is related to J_0 by the following equation,

$$V_{OC} = \frac{nk_B T_C}{q} \ln\left(1 + \frac{J_{SC}}{J_0}\right),\qquad(1)$$

where n is the diode ideality factor, k_B is Boltzmann's constant, and T_C is the operating temperature of the cell. To translate the improvements in J_{SC} due to the well-matched diode into overall efficiency improvements, we must design device architectures that reduce J_0. To demonstrate potential improvements in V_{OC}, we reduce the operating temperature of the diode from 300K to 100K as shown in Figure 2(c). As the operating temperature decreases, the magnitude of the dark current decreases, resulting in an increase in V_{OC} and fill factor. At 100K, the V_{OC} and power conversion efficiency are simulated to be 0.202eV and 20.96%, respectively. However, cooling the devices in such a manner is not practical for most power generation applications. Therefore, it is imperative to employ structures that reduce dark current to access the benefits of small band gap TPVs.

Analyzing the upper limits of J_0 as a function of semiconductor band gap allows us to look at the prospects for V_{OC} and FF for our small band gap homojunction TPVs. Combined with the predicted J_{SC} from low operating temperature simulations, we can predict more realistic upper bounds on power conversion efficiency for small band gap homojunction devices. An analysis laid out by MG Mauk (2006) for band gaps >0.5eV, extrapolated to 0.25eV, suggests an upper limit of 4.8% for power conversion efficiency for our single junction devices [21]. Adding in photon recycling effects can increase minority carrier lifetimes, and raise the upper limit on power conversion efficiency to 8.3%. These calculations assume a diode ideality factor of n=2/3. Although promising, Mauk notes that it remains to be seen if the trends outlined in this analysis will continue experimentally for band gaps

<0.5eV, as there are few fabricated small band gap TPV homojunctions with band gaps <0.5eV. However, even a small increase in V_{OC} due to backside reflector, or the addition of spectral control [19] would have a large impact on the power conversion efficiency of the homojunction devices.

C. Heterojunction Devices

If J_0 cannot be substantially decreased due to severe recombination effects, an alternative method of increasing the V_{OC} of a device is to employ a heterojunction structure. Previous work on heterojunction TPVs have shown an improvement in V_{OC} and temperature stability of the device due to the presence of a wider band gap emitter [22]. Ideally, photogeneration would occur in the small band gap n-type base, allowing holes to diffuse efficiently along the valence band, while dark current charge carriers would be hindered by the large built-in potential across the conduction band.

We initially explored $In_{0.53}Ga_{0.47}As$ lattice matched to InP (0.74eV) as our wide band gap p-type emitter. This would be desirable from an epitaxial standpoint, as we intend to grow our small band gap bismide materials on InP substrates to achieve high bismuth incorporation without introducing global strain effects. We optimized our device design to produce the highest J_{SC} (p-InGaAs: 7e17 cm^{-3} and n-GaAsBi: 5e18 cm^{-3}). The band structure of our heterojunction is shown in Figure 3.

Figure 4(a) shows the room temperature IV response of our $In_{0.53}Ga_{0.47}As/GaAs_{0.76}Bi_{0.24}$ device, compared to the homojunction InGaAs and GaAsBi devices previously discussed. Table 1 summarizes the performance metrics of all three TPV devices along with the theoretical upper limits discussed in Section B. The heterojunction architecture significantly outperforms both homojunctions in terms of V_{OC}, resulting in the highest power conversion efficiency. The inset shows that the heterojunction device has the lowest J_0 of all three devices, which could indicate improved temperature stability above 300K.

However, the heterojunction device does not demonstrate the increased J_{SC} characteristic of the GaAsBi homojunction,

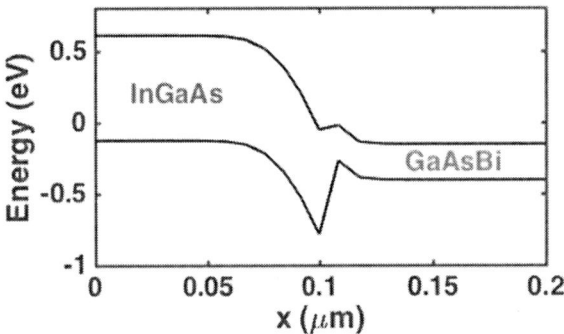

Fig3: Band structure of our heterojunction device using $In_{0.53}Ga_{0.47}As/GaAs_{0.76}Bi_{0.24}$ as the emitter/base. Schematic is zoomed in to highlight the interface between the two materials. Full device thickness is 3μm.

978-1-5090-5606-4/17 $31.00 © 2017 IEEE

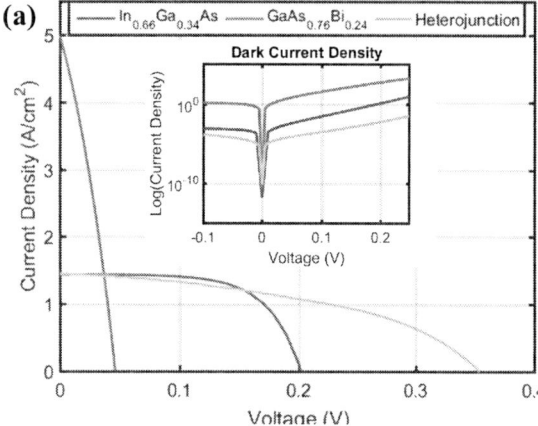

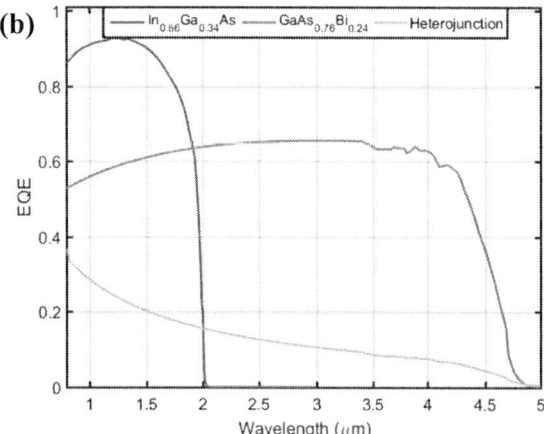

Fig4: (a) IV simulations of InGaAs/GaAsBi heterojunction device, demonstrating increased V_{OC} when compared to either the InGaAs or GaAsBi device. Inset shows reduced dark current when compared to either device as well. (b) EQE response of all devices discussed in this study.

even though the $In_{0.53}Ga_{0.47}As$ emitter is thin (100nm). Figure 4(b) shows the EQE response of our heterojunction device between 1-5μm is significantly reduced compared to the homojunction device and cannot be improved by changing device design parameters. The large recombination rate at the heterojunction between the two materials is most likely responsible for this decrease in charge carrier generation due to valence band mismatch between the InGaAs and the GaAsBi materials as well as tunneling effects from the short InGaAs emitter. Future work will involve simulating other wide band gap quaternary GaAsBi-based alloys, containing aluminum or indium, to create a better heterojunction while maintaining a lattice constant around the InP substrate.

III. CONCLUSIONS

In this work, we presented device simulations for TPV diodes tuned for a 1350K heat source as an alternative to

InGaAs strained on InP. We demonstrated increased J_{SC} associated from using a material with a better matched band gap; however, we note significant reductions in V_{OC} and FF that lead to lower overall power conversion efficiency. We explored the fundamental limitations on the dark current density of a small band gap device, and note higher power conversion efficiencies could be achievable by adding photon recycling mechanisms such as spectral control or backside reflectors. Additionally, we presented an $In_{0.53}Ga_{0.47}As/GaAs_{0.76}Bi_{0.24}$ heterojunction device as an alternative method for increasing V_{OC} and reducing dark current density. Future simulation work involves exploring better matched emitters to reduce recombination at the heterojunction and increase charge carrier generation from the small band gap bismide.

IV. ACKNOWLEDGEMENTS

This work was supported by a NASA Space Technology Research Fellowship (NNX15AQ79H), the NSF STEM Leader Program (NSF EEC-1444926), and ONR grant N00014-15-1-2946.

REFERENCES

[1] L. L. N. Laboratory, "Estimated U.S. Energy Consumption in 2015: 97.5 Quads," 2016. [Online]. Available: https://flowcharts.llnl.gov/.

[2] L. Fraas, J. Avery, and L. Minkin, "Design of a Portable Fuel Fired Cylindrical TPV Battery Replacement," *Am. Inst. Aeronaut. Astronaut.*, pp. 1–15, 2007.

[3] A. Datas and A. Martí, "Thermophotovoltaic energy in space applications: Review and future potential," *Sol. Energy Mater. Sol. Cells*, vol. 161, no. December 2016, pp. 285–296, 2017.

[4] B. Wernsman, R. R. Siergiej, S. D. Link, R. G. Mahorter, M. N. Palmisiano, R. J. Wehrer, R. W. Schultz, G. P. Schmuck, R. L. Messham, S. Murray, C. S. Murray, F. Newman, D. Taylor, D. M. DePoy, and T. Rahmlow, "Greater than 20% radiant heat conversion efficiency of a thermophotovoltaic radiator/module system using reflective spectral control," *IEEE Trans. Electron Devices*, vol. 51, no. 3, pp. 512–515, 2004.

[5] D. L. Chubb, *Fundamentals of Thermophotovoltaic Energy Conversion*, First. Oxford: Elsevier, 2007.

[6] O. Delorme, L. Cerutti, E. Tournié, and J.-B. Rodriguez, "Molecular beam epitaxy and characterization of high Bi content GaSbBi alloys," *J. Cryst. Growth*, 2017.

[7] S. J. Sweeney, Z. Batool, K. Hild, S. R. Jin, and T. J. C. Hosea, "The potential role of Bismide alloys in future photonic devices," *Int. Conf. Transparent Opt. Networks*, no. Figure 3, pp. 4–7, 2011.

[8] R. D. Richards, F. Harun, J. S. Cheong, A. Mellor, N. P. Hylton, T. Wilson, T. Thomas, J. P. R. David, and S. Yorkshire, "GaAsBi : An Alternative to InGaAs Based Multiple Quantum Well Photovoltaics," *43rd IEEE Photovoltaics Spec. Conf. Proc.*, pp. 1135–1137, 2016.

[9] T. Thomas, A. Mellor, N. P. Hylton, M. Fuhrer, D. Alonso-Alvarez, A. Braun, N. J. Ekins-Daukes, J. P. R. David, and S. J.

Sweeney, "Requirements for a GaAsBi 1 eV sub-cell in a GaAs-based multi-junction solar cell," *Semicond. Sci. Technol.*, vol. 30, no. 9, p. 94010, 2015.

[10] A. Zayan, M. Stevens, and T. E. Vandervelde, "GaAsBi Alloys for Photovoltaic and Thermophotovoltaic Applications," in *43rd IEEE Photovoltaics Specialists Conference Proceedings*, 2016, pp. 2839–2843.

[11] K. Alberi, J. Wu, W. Walukiewicz, K. M. Yu, O. . Dubon, W. S.P, C. X. Wang, X. Liu, Y. . Cho, and J. Furdyna, "Valence-band anticrossing in mismatched III-V semiconductor alloys," *Phys. Rev. B*, vol. 75, no. 45203, 2007.

[12] C. J. Crowley, N. A. Elkouh, S. Murray, and D. L. Chubb, "Thermophotovoltaic Converter Performance for Radioisotope Power Systems Thermophotovoltaic Converter Performance for Radioisotope Power Systems," *AIP Conf. Proc.*, vol. 601, no. 746, 2005.

[13] R. N. Kini, L. Bhusal, A. J. Ptak, R. France, and A. Mascarenhas, "Electron hall mobility in GaAsBi," *J. Appl. Phys.*, vol. 106, no. 4, 2009.

[14] R. N. Kini, A. J. Ptak, B. Fluegel, R. France, R. C. Reedy, and A. Mascarenhas, "Effect of Bi alloying on the hole transport in the dilute bismide alloy GaAs1-xBix," *Phys. Rev. B - Condens. Matter Mater. Phys.*, vol. 83, no. 7, pp. 1–6, 2011.

[15] S. Nargelas, K. Jarašiunas, K. Bertulis, and V. Pačebutas, "Hole diffusivity in GaAsBi alloys measured by a picosecond transient grating technique," *Appl. Phys. Lett.*, vol. 98, no. 8, pp. 98–101, 2011.

[16] M. Masnadi-Shirazi, R. B. Lewis, V. Bahrami-Yekta, T. Tiedje, M. Chicoine, and P. Servati, "Band gap and optical absorption edge of GaAs1-xBix alloys with 0<x<17.8%," *J. Appl. Phys.*, vol. 116, no. 223506, 2014.

[17] A. H. Reshak, H. Kamarudin, S. Auluck, and I. V Kityk, "Bismuth in gallium arsenide Structural and electronic properties of GaAs1−xBix alloys," *J. Solid State Chem.*, vol. 186, no. C, pp. 47–53, 2012.

[18] M. K. Hudait, Y. Lin, M. N. Palmisiano, and S. A. Ringel, "0.6-eV bandgap In0.69Ga0.31As thermophotovoltaic devices grown on InAsyP1-y step-graded buffers by molecular beam epitaxy," *IEEE Electron Device Lett.*, vol. 24, no. 9, pp. 538–540, 2003.

[19] N. A. Pfiester and T. E. Vandervelde, "Selective emitters for thermophotovoltaic applications," *Phys. Status Solidi Appl. Mater. Sci.*, vol. 1600410, no. 1, pp. 1–24, 2016.

[20] A. S. Licht, D. F. Demeo, J. B. Rodriguez, and T. E. Vandervelde, "Decreasing Dark Current in Long Wavelength InAs / GaSb Thermophotovoltaics via Bandgap Engineering," in *40th IEEE Photovoltaics Specialist Conference (PVSC)*, 2014, pp. 482–486.

[21] M. G. Mauk, "Survey of thermophotovoltaic (TPV) devices," *Springer Ser. Opt. Sci.*, vol. 118, pp. 673–738, 2006.

[22] R. K. Huang, R. J. Ram, M. J. Manfra, M. K. Connors, L. J. Missaggia, and G. W. Turner, "Heterojunction thermophotovoltaic devices with high voltage factor," *J. Appl. Phys.*, vol. 101, no. 4, pp. 1–4, 2007.

AUTHOR INDEX

Aaditya, Gayathri 604
Abbas, A. 1691, 2457, 3430
Abbas, Ahmed E. 1888
Abbas, Ali 186, 752, 1674
Abbott, Malcolm D.......... 1322, 2576, 2600
Abdalla, L.B 1245
Abdallah, Amir A. 3435
Abdallah, Shaimaa A. 219
Abdellaoui, Imane.....................900
Abdullah, Ahmad 2128
Aberle, Armin.....................2318
Aberle, Armin G. 284, 496, 499, 1922
Ablekim, Tursun.....................3422
Aboubakr, Benazzouz.....................487
Abouelkhair, Hussain M.....................2324
Abtahi, Amir 638
Abudayyeh, Omar K.....................88
Acebo, Laura 155
Addamane, S. J.....................281
Adewoyin, Adeyinka.....................2381
Adhikari, Dipendra.....................2582
Affouda, Chaffra A.....................259
Agarwal, Mohit 2330
Agarwal, Sumit.....................1777
Agarwal, Vivek.................. 2952, 2981, 2986, 3050
Agbo, Solomon N.....................2114
Ager, Joel.....................3410
Agrawal, Rakesh 1449
Aguiar, Jeff.....................2702
Aguiar, Jeffrey 2467
Aguirre, Rodolfo 2419
Ahamioje, Joseph A. 2931
Ahanzhamejhad, Ramez H.....................170
Ahlswede, E. 3260
Ahlswede, Erik 791
Ahmad, Jawad.....................3096
Ahmed, Benlarabi 487
Ahmed, Nuha.....................658, 2667
Aho, Arto 297, 2520
Aho, T. 1189
Aho, Timo 297
Ahrenkiel, P.....................869
Ahrenkiel, Phil 206, 831
Ahrenkiel, Richard K. 3448
Ahrenkiel, S. Phillip.....................2514
Ahsan, Nazmul 2334
Aierken, Abuduwayiti.....................226
Aindow, Mark 1522
Aïssa, Brahim 3435
Akaki, Yoji.....................2338
Akari, Shunsuke.....................2385
Akarm, Muhammad Nadeem 2776
Ake-Sultan, Bernt 2864
Akiki, Tilda.....................1968
Akimoto, Katsuhiro 33, 160, 900

Akimoto, Naoki.....................712
Akiyama, Hidefumi 721, 2781, 3528
Akwari, Chinedum 735, 2446
Al Mahmud, Abdullah 1067
Alahmed, Ahmed.....................1110
Alam, Giri Wahyu 1498
Alam, Muhammad A. 1055, 1259
Alam, Muhammad Ashraful.....................1904
Alberi, Kirstin.....................2506
Albin, David.....................1196, 3305, 3319
Alcubilla, R. 1781
Alcubilla, Ramón 944
Aleman, Monica.....................2227, 3435
Alexander, Jessica A. 966
Alfadhili, Fadhil K.....................730, 815
Al-Fadhili, Fadhil K. 2462
Al-Ghzaiwat, Mutaz 2593
Algora, C. 1210
Algora, Carlos.....................1204
Alharbi, Fahhad H. 963
Alharthi, Yahya Z. 1018, 1110
Ali, Asad.....................1228
Ali, Jaffar Moideen Yacob.....................2318
Ali, Waqar.....................1228
Alivisatos, A. Paul.....................1737
Al-Jassim, M. 1196
Al-Jassim, M.M. 1312, 2280, 2785
Al-Jassim, Mowafak 62, 1371, 1381, 1400, 2789, 2887, 3305, 3319
Al-Jassim, Mowafak M. 3147
Aljaziri, Marwa 2011
Alkhayat, Rabee B.....................815
Allebé, C. 50, 2073
Allebé, Christophe.....................3254, 3256
Allen, Thomas 2076
Almheiri, Anwar 1946
Almonacid, Florencia 2858
Alrashidi, Hameed 2858
Altermatt, Pietro P. 1922, 2220, 3304
Alvarez, Diego Alonso 1339
Alvarez, Genesis 2941
Alvarez, José 2453, 2528
Aly, Shahzada P.....................963
Alzahmi, Wadhah.....................1946
Amdemeskel, Mekbib W.....................2672
Anctil, Annick.....................2124
Anderberg, A. 467
Andler, Joseph 1449
Ando, Daisuke.....................931
Ando, Yasutaka 970
Ando, Yuta 192
Andreani, Lucio.....................290
Angeles-Ordóñez, G. 142
Annigoni, Eleonora.....................1395, 2794
Anoma, Marc Abou 1549

AUTHOR INDEX

Anselmo, Andrew 74, 2839, 2897
Antony, Aldrin 1755
Anttu, Nicklas...2502
Anyadiegwu, Ifeanacho 970
Anyanwu, Uchechi 319
Araki, Hideaki...2338
Araki, Kenji.........................359, 412, 1479, 1711,
    1714, 1743, 2498, 2548, 2566
Aranguren, G. 643
Archer, Alexander 771
Arehart, A. R. 30, 2414
Arehart, Aaron R. 215, 2446, 3139
Arinze, Ebuka S. 667
Armour, Eric 827
Armour, Eric A. 210, 2506
Armoush, Maher 1058
Arnold, Daniel B. 3002
Arnou, Panagiota 146, 186
Arora, B. M. 396, 1995, 2716
Arora, Brij M. 3478
Arp, Juergen 1411
Arredondo, C. A...2031
Artegiani, Elisa................752, 1669, 2372
Aryal, Krishna...182
Asadirad, Mojtaba..866
Asahi, Shigeo 23
Asgharzadeh, Amir 1537, 1543, 3333
Ashrafee, Tasnuva 735
Aslam, Aasma...2355
Asomoza, René 632
Astakhov, Oleksandr 2114
Aswani, U 1898
Athresh, Eashwer...................................2395, 2399
Atia, Adam A...3230
Atkins, R. 229
Atlan, Olivier...626
Atwater, Harry A. 512, 521, 558, 572,
    1248, 1589, 1737, 2236
Augarten, Yael 1651
Augusto, André 1589, 2596
Avasthi, Sushobhan........ 251, 837, 841, 986, 2395, 2399
Avenet, Julien 1933
Avery, J. E. 1863
Awadallah, Osama 3473
Awasthi, Vishnu 2345
Ayala, Orlando 735
Azkona, N. 2740
Azkona, Nekane 2677
Azzolini, Joseph A..608
Baba, Masaaki..1724
Babbe, Finn 151, 2054
Babcock, Sean J...2298
Bachman, Benjamin F. 3381
Badel, N..50
Badosa, Jordi 626

Badr, Ikken 487
Bae, Soohyun 935
Baggu, Murali 2991
Baik, Sungsun 2242
Bailey, C. 845
Bailey, Christopher G. 2298
Bailey, J. 2414
Bailey, Jeff...1686, 3327
Baines, Tom 742, 1445
Baka, Maro 3343
Baker, Rupesh 3172
Bakhshi, Sara 322
Bakker, Klaas 2875
Balaji, Pradeep 2596
Balakrishnan, G...281
Balasubramaniam, Kavaipatti R. 1704
Baldus-Jeursen, Christopher...............................1908
Ball, Greg 2263
Ballif, C. 50, 2073
Ballif, Christophe 55, 1220, 1395,
    2104, 2794, 3254, 3256, 3435
Baloch, Ahmer A.B...............................963, 1058
Banda, Pedro 1946
Banerje, Rangan 1151
Banerjee, Sanjay K. 363
Barahman, Gil 2285
Barakel, Damien 2255
Barnes, T. M. 138
Barnes, Teresa M..3422
Barnett, Allen 315
Barraud, L. 50
Barraud, Loris 3254, 3256
Barrigon, Enrique 2502
Barth, Kurt 424
Bartolo, Robert E. 195
Bartsch, J. 884
Basore, Paul A. 2163
Bastide, Stéphane 3402
Bastola, Ebin 738, 781
Basu, Prabir K. 396
Battaglia, A. 1747
Baudrit, Mathieu 2492, 2562
Bauer, Andreas...791, 2058
Bauer, Jan 1376
Bauhuis, G. 1189
Baumann, Thomas 1077
Baumgartner, Franz..1077
Baur, Carsten...541, 2087
Baxter, Jason B. 3143
Bearda, Twan 1233
Beauchemin, Ryan D. 102
Becerril-Romero, Ignacio 155
Becker, Jacob J. 3366, 3410
Bedair, Salah M. 2195
Belanger, Ted..1427

AUTHOR INDEX

Belletête, Marc 1579
Belluardo, Giorgio 3360, 3482
Bemrrr, Andreas..3500
Benamara, Mourad..3370
Benatto, Gisele A. Dos Reis 2672, 2682
Benick, J...2064
Benick, Jan 2511
Bennett, Dirk 2042
Bennett, Mitchell.......................................247
Bennett, Mitchell F........................ 210, 259, 873, 2091
Berardone, Irene 402
Berg, Alexander 1773
Berg, Morgann 3417
Bermel, Peter.....................................1904, 2467
Bernard, Annie...................................2870, 2891
Berry, Joseph J...2176
Bert, J...1733
Bertoni, Mariana......................944, 2610, 2854
Bertoni, Mariana I. 2179, 3309
Besanger, Yvon..3102
Bett, Alexander J.......................................1253
Bett, Andreas W...2511
Bettenwort, Gerd 1965
Beutel, Paul...2511
Beutner, Volker...1855
Bhaduri, Sonali....................................2799, 3478
Bhan, Mohan Krishan 496
Bhandari, Khagendra P..................738, 748, 781, 815
Bhatia, A..1656
Bhatia, Swasti..1755
Bhattacharya, Indranil................................3083
Bhattacharya, Sitangshu 2376
Bheemreddy, Venkata 2688
Bialek, Tom 2991
Bidiville, Adrien 1333
Biedenham, Richard E...................................3245
Biegelsen, D. K. 1733
Biiss, M. 2457
Binetti, Simona..1669
Birch, Max T. 2423
Birkmirc, Robert..726
Birkmire, Robert W.....................................2637
Bishop, Doug 3275
Bishop, Douglas..726
Bishop, Douglas M. 1441
Bissels, G...1189
Biswas, R..1350
Bittau, F..3430
Bittau, Francesco 752
Bittner, Zachary.......................................677
Bittner, Zachary S.....................18, 202, 2084
Bivour, M..2064
Bivour, Martin 1253
Blakely, Logan...1573
Blanche, Pierre-Alexandre 1147

Bläsi, Benedikt...352
Blasi, David 1531
Blum, Adrienne 2692
Blum, Adrienne L. 2765
Bob, Brion 2258
Bobela, David C..2506
Bobyl, A.V. 1025, 1811
Boca, Andreea 2099
Boccard, Mathieu 55, 1220, 1317, 1790, 3366
Boeck, Torta...3396
Bohra, Rakesh 1912
Boizot, Bruno 83, 2087
Bolaji, Adewumi 2381
Boley, Allison 2573
Bolke, J. G. 1656
Bonnassieux, Yvan 626
Bonomo, Pierluigi 2118
Book, Felix 1824
Bora, Birinchi 3478
Borgers, Tom 3343
Borgström, Magnus......................................2502
Borgström, Magnus T 1286
Borland, John 2947
Borne, Axel 2864
Borowik, Lukasz..1516
Bosco, Nick 3190, 3200
Bosco, Nick S..2864
Bosson, Christopher J..................................2423
Bostock, Peter 2267
Bothe, Karsten 2692
Bourcois, Jérôme 2087
Bourdin, Vincent 626
Bourgoin, Jacques C. 2087
Bourne, Ben C. 1549
Bousselham, Abdelkader 1058
Bouttcmy, Muriel 2711
Bowden, Stuart 240, 925, 1797, 2719
Bowden, Stuart G. 1589, 2596
Bowen, Leon..1445
Bowers, J.W. 2457, 3430
Bowers, Jake W. 146, 186, 752, 2349
Boyce, Ken 2000
Boyce, Kenneth 1933
Boyd, Matthew..1933
Boyer, Jacob...215
Boyer, Jacob T. 2079, 2554
Boyer-Richard, Soline 2192
Brabec, Christoph J. 1346, 3500
Bradshaw, Geoffrey K. 88, 301, 531
Brady, Brendan 3388
Braga, Daniel Sena 2307
Braid, Jennifer L. 1927, 2697, 3456
Brammertz, G...3260
Brand, A.A. 884
Brates, Nanu 1728

AUTHOR INDEX

Bräuninger, Matthias 3256
Breitenstein, Otwin 1376
Breitwieser, M. 3135
Bremner, S. P. 953
Bremner, Stephen 858, 1215, 1845, 2186, 2569
Bremner, Stephen P. 948
Brendel, Rolf 1366, 3371
Breus, V. 1752
Bright, Jamie M. 1405
Brinnig, Samuel..2622
Brito, Pedro P. 2307
Britt, Jeffrey 1455
Brittman, Sarah 2245
Broderick, Robert 3008
Broderick, Robert J. 1435, 1555, 1567, 1573, 3025, 3031
Brolo, Alexander G. 3388
Bruckman, Laura.......................................1933
Bruckman, Laura S. 2000
Brückner, Sebastian..................................2538
Brughera, Céline 2492
Brule, Carlton...1728
Brulo, Gregory S. 1469
Bryan, Jonathan 1317
Buchanan, Wayne 1196
Büchler, A. 884, 3135
Buckner, Jessica..537
Buerhop, Claudia 3500
Bukowsky, Colton R. 1737
Bulkin, P. 1781
Bulkin, Pavel 1237
Bullock, James...................................59, 2076
Buonassisi, T. 648, 1140, 3295
Buonassisi, Tonio 284, 1264, 1491, 2242, 2532, 2744, 3236, 3290, 3300
Burgers, A.R. 3150
Burgers, Antonius R. 917
Burkhardt, S...1752
Burnham, Laurie 1435
Burroughs, Scott 272, 1469
Busquet, Severine 1061
Butt, Isaac...182
Cabarrocas, Pere Roca I..............464, 1237, 2528, 2593
Cachet-Vivier, Christine.............................3402
Caffy, Florent 1516
Calderón-Obaldía, Fausto 626
Calle, Eric..944
Calvo-Barrio, Lorenzo...............................3285
Campa, Andrej 1346
Campanelli, Mark......................................437
Campbell, Calli M. 3366, 3410
Campesato, Roberta 76, 541, 545
Campos, Cláudio Dias 2307
Camus, Christian 3500
Cañadillas, David................................429, 1116

Canino, A. 1747
Caño, P. 1210
Cao, Huihui 1619
Cao, Wenkai...696
Cao, Xin..2427
Cao, Yunxue....................392, 1430, 1873, 2918
Cappelluti, F. 1189
Cardona, Dagoberto..................................670
Cardwell, D. 3511
Cariou, Romain 2511, 2528
Carlin, John A. 215
Carlson, David E. 3442
Carlson, Emily 701
Carneiro, Lucas M. 417
Carolus, Jome 2875
Carpenter, Bernard 537
Carr, Anna J. 1081
Carriere, Jarrett 2833
Carruthers, Steve 3514
Carter, Catrice M. 3393
Carter, Cedric 2135
Carter, Sam 3514
Casale, Mariacristina 76, 541, 545
Casallas-Moreno, Y. L. 670
Casper, Chadwick 1476
Cassini, Denio A. 1917
Castañeda, Carlos A. Rodríguez 1858
Catthoor, Francky 3343
Cattin, Jean...3435
Cattoni, Andrea 1289
Cavani, Olivier...2087
Cédola, A. P. 1189
Cendagorta-Galarza, Manuel.....................429
Cepeda, Kyle 876
Cesar, I. 3150
Cesar, Ilkay...917
Chai, Gaoda 976
Chai, Jing 1922
Chakraborty, Sagnik.................................3300
Chamarthi, Phani Kumar...........................2952
Chamberlin, Charles.................................1271
Champliaud, J. 50
Champliaud, Jonathan 3435
Champness, C. H. 2388
Chan, Calvin 3417
Chan, Catherine E............................2576, 2600
Chan, Mandy 2808
Chan, Maria K.Y. 6, 1256, 2759
Chan, R. 3511
Chandralal, Sreeram 1674
Chandran, Deepak 2986
Chang, Jipeng 1873, 2823
Chang, Sheng-Hao 1051
Chang, Via-Chung....................................1051
Chantana, Jakapan 757, 2385

AUTHOR INDEX

Chaporr, Patrick 2711
Chapuis, Valentin 2104
Chattopadhyay, Kamanio 2811
Chattopadhyay, Shashwata....... 1850, 1858, 2849, 3478
Chaudhry, Ghulam M. 1018, 1110
Chaujar, Rishu 377
Chaurasia, Saloni 837, 841
Chausseau, Matthieu 2711
Chavali, Raghu Vamsi Krishna 1904
Chavez, Jose J. 2419
Chemisana, Daniel 1339
Chen, Benjamin 2358
Chen, Chien-Hsun 911
Chen, Chun-Chi 1635
Chen, Daniel 2576
Chen, Eric Y. 1598, 3384
Chen, Haiyan 2220
Chen, Hung-Ling 1289
Chen, Junyan...1835, 2732
Chen, Kaifeng 2185
Chen, Kunji 2656
Chen, Lung-Chien 367
Chen, Meixi 326, 999, 2035
Chen, Peng-Wei 2660
Chen, Ran 2576
Chen, Renfang 1241
Chen, Shi-Wei 1627
Chen, Sung-Yu 911
Chen, Tsung-Cheng 329
Chen, Tzu-Yu 1627
Chen, Wanghua 2593
Chen, Weijian 2392
Chen, Y. 2785
Chen, Yang 2502
Chen, Yao- Hui 893, 2664
Chen, Yifeng..1922, 2220
Chen, Yunfei...761, 2427
Chen, Yusi 1835
Chen, Zhi David 1044
Chen, Zihan 2392
Chendo, Michael 2381
Cheng, Y. 14
Cheng, Yan 667
Cheng, Yuh-Jen 1610
Cheng, Zhe..3473
Cheng, Zhongkai 3393
Chenna, Shiva Tarun 1674
Chiang, Cho-Chun 893, 2664
Chiang, Fu-Kuo 198
Chikhalkar, Abhinav......................................823, 827
Chin, Ken K...761, 2427
Chinnusamy, Saravanan 980
Chiu, Chun-Yu 1169
Chiu, P. 2094
Chiu, Philip...2099

Chmielewski, Daniel J..........................215, 2079, 2554
Cho, Eunhwan 333, 1824, 1838
Cho, Junsik 810
Cho, Yasuo 3323
Choi, Gyu-Seok 2723
Choi, J. -K 2019
Choi, Rae-Won 2723
Choi, Seungkeun 1037
Choi, Sungjin 1758
Chong, Cheemun 2600
Choubisa, Hitarth 1022
Choudhury, K. R. 2312
Chouhan, Arun Singh 986
Choulat, Patrick 2227, 3435
Chow, E.M. 1733
Chowdhury, Ahrar Ahmed 888
Christians, Jeffrey A. 2176
Christmann, G. 50
Chu, Chi-Wei 1051
Chu, Haifeng 1222
Chu, Sheng 1299
Chua, Soo Jin 284
Chuang, Ta-Wei 343, 367, 893, 2664
Chung, Daniel 2707, 3304
Chung, Haejun 1904
Chung, Simon 696, 2186
Ciesla, Alison M. 2576, 2600
Cifuentes, L. 1210
Cifuentes, Luis 1204
Ciocia, Alessandro 3096
Cirino, Daniel A. Merced 3044
Clayton-Warwick, D. 138
Cleveland, Erin 247, 2091
Clinton, Evan A. 305
Cobo-Yepes, Nicolás 2963
Codd, Daniel S. 3245
Cohen, Bat-El 2170
Cole, Wesley J. 2163
Colegrove, E. 1312
Colegrove, Eric 3147, 3319
Coll, Pablo Guimera 2610
Collin, Stéphane 1289, 3147
Collins, Robert W..............807, 2462, 2582, 2646, 3426
Collins, Shamara...........................802, 1638, 2449, 3413
Comagliotti, Emanuele 3435
Condorelli, G. 1747
Conibeer, Gavin 696, 2186, 2392
Conlon, Benjamin P. 219
Conrad, Brianna 315
Cordeiro, Patricia 2135
Cordova, Adam 1965
Cornagliotti, Emanuele 1804, 2227
Cornaro, Cristina 3482
Cornell, Robert..1275
Correa, J.M. 433

AUTHOR INDEX

Correa-Baena, Juan-Pablo 3300
Cossio, Gabriel 1181
Costa, Sara...1979
Costa, Suellen C. 2307
Côté, Alexandre..1908
Cousar, Larry C. 921
Cravens, R. 2094
Crawford, L. 1733
Crupi, F. 2073
Cruz, José Ortega 1959, 1990
Cruz, Leila R. De Oliveira..............................2307
Cruz-Campa, Jose Luis 337
Cuevas, Andres...2076
Cui, Jie 2076
Cui, Min 1765
Cunningham, Daniel W. 1463
Cunningham, Joseph 3161
Cur, Jie 517
Curran, Alan J. 1927, 2697, 3488
Curvat, L. 50
Cushing, Scott K. 417
Da Fonseca, Jérémy 2492, 2562
Dabney, M. S. 138
D'Abrigeon, Laurent..545
Daenen, Michael 2875
Dagenais, Mario 195, 1048
Dagyte, Vilgaile 2502
Dahal, Saroj..................................309, 3123
Dahal, Som...240
Dai, Yushuai 18, 222, 677, 1184
Dalal, V L 1350
Dalal, Vikram 2247
Dalpian, G..1245
Dam-Hansen, Carsten.........................2672, 2682
Danel, A. 1747
Dang, Hongmei...2432
Dangate, Milind S. 980
Daniil, Andreana...944
Danzl, F.J.K. 3150
Darbali-Zamora, Rachid 2957, 2963
Das, Ujjwal 408, 1473, 1761, 1828, 2667
Das, Ujjwal K. 2637
Datas, Alejandro 2562
Dauskardt, Reinhold.........................3190, 3200
Davidsen, Rasmus S. 2672
Davies, J. I. 1210
Davis, Kristopher O. 74, 322, 1804, 3448
Davis, Tracy..537
De Coux, Patricia..464
De Melo, O. 2342
De Nicolas, S. Martin...50
De Oliveira, Michele C. C.................................1917
De Villers, Bertrand J. Tremolet 1354
De Wolf, Stefaan 55, 3256, 3435
De, F. C. Lins Vanessa.....................................1917

Debnath, M. C. 14
Debnath, Tanmoy 1067
Deboever, Jeremiah.........................1555, 1567
Debrot, F. 50
Debucquoy, Maarten......................................1233
Debusschere, Vincent 3102
Deceglie, Michael 2771, 2789
Deceglie, Michael G....................2488, 2804, 3452
Deckerl, D. 1752
Decobert, Jean 2528
Deer, Tanya...1908
Deitz, Julia I...3139
Delahoy, Alan E. 761, 2427
Delhotal, J. 3224
Deligiannls, D. 3150
Deline, Chris 116, 1537, 1922, 3184, 3333
Demadrille, Renaud 1516
Demirkan, Korhan 820
Deng, Changhong 1158
Deng, Weiwei 2220
Denk, Patrick 1360
Descoeudres, A. 50
Descoeudres, Antoine 3254
Despeisse, M. 50, 2073
Despeisse, Matthieu.........................3254, 3256, 3435
Desrues, Thibaut 2492, 2562
Deutsch, Todd G. 47
Devos, Arnaud 464
Dewitt, Daniel 1835
Dey, Anamika 1034
Dhakal, Tara P. 989
Dhere, N. 389, 1701
Di Leo, Paolo 3096
Di Mare, Simone..2372
Di Napoli, Simone 2205
Diaz, Liliana Ruiz 1147
Diercks, David R. 46
Dimitrievska, Mirjana......................................3285
Dimopoulos, Theodoros 178
Dimroth, Frank 2511
Ding, Jie 1937, 2823
Dinger, Justin..............................2692, 2765
Diniz, Antonia Sônia A. C..................1917, 2307
Dirriwachter, Antonius B. 3448
Dise, John 132, 1104
Dise, Skip 1427
Dobrich, Anja 2538
Dobroliubov, Aleksandr....................................2776
Dobson, Kevin 315
Dobson, Kevin D. 658
Dobson, Weston 2692, 2765
Dogan, Yusuf 229
Doi, T. 441
Dominguez, A. 2342
Dong, Jianfei 2605

AUTHOR INDEX

Doolittle, William A. 305
Dooraghi, Michael.................1169
Döscher, Henning....................47
Doty, Matthew F. 1598, 3384
Dougher, Chris.....................3245
Dougherty, Brian 1933
Drahi, Etienne 464
Drayton, Jennifer A. 164
Drees, M. 3511
Dréon, Julie 55
D'Rozario, Julia 18
Drummy, Lawrence F. 966
Du, Chen-Hsun......................911
Du, Xingzhi........................2558
Du, Zhongming.............198, 767, 1707
Duan, Baosong................392, 2823
Duan, Wenqi........................346
Dubey, R. 1995
Dubey, Rajiv.............. 1704, 2849, 3478
Dubois, Anne Migan.................626
Duenow, J.N. 1312
Duenow, Joel.................1196, 3147
Duerinckx, Filip...............2227, 3435
Dugan, Roger C. 3055
Dugdill, Brian......................2014
Dumbrell, Robert................420, 3315
Dunham, Scott T. 3119
Dupré, Cécilia....................2492, 2562
Durand, Olivier....................2192
Durose, Ken......................742, 1445
Durstock, Michael F 966
Durygin, Andriy....................3473
Dusane, Rajiv O 2330
Dussarrat, Christian 326
Dutt, A............................370
Dutt, Ateet........................2342
Dutta, P...........................869
Dutta, Pavel.....................866, 2368
Duttagupta, S.P....................1898
Eafanti, Joshua....................3190
Ebe, Falko.........................2996
Ebert, Matthieu....................1531
Ebong, Abasifreke..................888
Ediger, E..........................2364
Edinger, Stefan.....................178
Edoff, Marika.......................796
Edwards, Daniel J..................3514
Eeles, Alexander................146, 186
Efthymiou, Venizelos...............3107
Egbe, Daniel Ayuk Mbi 1360
Eggink, Wouter.....................2109
Ekins-Daukes, Ned..................1339
El Assimi, Taha....................3402
Elangovan, Hemaprabha...............2811
Elanzeery, Hossam...............151, 2054

Eldho, T.I. 1898
El-Henawey, Mohamed...............2247
Elkhatib, Mohamed............2141, 2969
Elleuch, Omar 359
Ellibee, Donald....................1543
Ellingson, Randall.................2926
Ellingson, Randall J. 1030
Ellingson, Randy J..........738, 748, 781, 815
Ellis, Chase T. 873
Elnosh, Ammar.....................1946
Elsehrawy, Farid...................1189
Emery, K.A. 490
Engerer, Nicholas A. 1405
Eriksen, Ryan.....................2870
Eriksen, Ryan S....................2891
Ermer, J. 2094
Ermer, James......................2099
Ermer, Jim H. 37
Escarra, Matthew D. 37, 3245
Escobar, D. Martínez 1959, 1990
Esfandiari, Parichehr 178
Espinct-Gonzalez, Pilar.............558
Espíndola-Rodríguez, Moisés.........155, 512, 572, 3265
Espinet-González, Pilar............521, 1248
Essa, Gharibah 2011
Essig, Stephanie 55, 3254, 3371
Etcheberry, Arnaud.................2711
Etgar, Lioz........................2170
Eugen, Rene.......................2864
Evani, Vamsi................802, 1638, 2449
Evans, Garrett Z. 921
Evstigneev, M..............663, 1025, 1811
Evstigneev, Mykhaylo A. 690
Eylers, Katharina 3396
Fada, Justin S. 2697, 3456, 3488
Faes, A. 50
Fairbrother, Andrew..........1933, 2000, 3204
Faleev, Nikolai.................1215, 2573
Fan, S. 3376
Fan, Shanhui....................2185, 2732
Fang, Liang 226
Fang, Y. 1603
Fang, Yi 305
Fano, V. 2740
Fano, Vanesa......................2677
Faraj, Abudul 2014
Farnung, Boris....................2267
Farré, Laia Arqués 3285
Farshchi, Rouin 1459, 1686
Faur, Maria 896
Faur, Orry.........................896
Favre, W. 1747
Fedina, Maria 2070
Fejfar, A. 2073
Felder, T..........................2312

AUTHOR INDEX

Feldmann, F. 2064
Feng, Sheng-Kai 343, 367, 893, 2664
Feng, Shien-Ping 1012
Feng, Zhiqiang 1922, 2220
Fenning, David P. 1494, 2245
Ferekides, Chris 802, 1638, 2449, 2467, 3413
Ferekides, Chris S. 1511
Ferekides, Christos 175
Ferguson, Andrew J. 1354
Ferguson, L. 1863
Fernández, Eduardo F. 2858
Fernandez, R. Mis 1691, 2457
Fetzer, C. 2094
Fiducia, Thomas 424
Fields, Brian J. 2618
Filipic, Miha 1233
Filonovich, Sergej 464, 1237
Firth, Peter 1317
Fischer, A. 1603
Fischer, Alec 823
Fischer, Alec M. 305
Fisher, Brent 210, 272, 1469
Fisher, Dallas 989
Fitzgerald, Eugene A. 213
Fleming, Robert A. 1869
Flicker, J. D. 3224
Flicker, Jack 1280
Florides, Michalis 1941
Foldyna, Martin 2528, 2593
Forberich, Karen 1346
Forbes, David V. 3468
Forchhammer, Soren 2682
Forsh, P.A. 1811
Foster, Robert 2014
Fouchier, Marin 3402
Fournel, Frank 2492, 2562
Fraas, L. M. 1863
Fraas, Lewis 2042
France, Ryan M. 47, 232
Fraser, Ray 337
Frederiksen, Kenn H. B. 2682
Freeman, Janine M. 3494
Freiburger, Brennen M. 1869
French, Roger 1933
French, Roger H. 1927, 2000, 2697, 3456, 3488
Freundlich, Alexandre 236, 673, 1452
Fridman, Lucas 2000
Friedman, Daniel 549, 2543
Friedman, Daniel J. 42, 268, 1201
Friend, Mari Paz 429
Fritzsche, M. 1752
Frontini, Francesco 2118
Fthenakis, V. 2019
Fthenakis, V. M. 3230
Fthenakis, Vasilis 3077

Fu, Ran 1259, 1463
Fuhrich, Alexander 3396
Fuhrmann, Bianca 83
Fujiwara, Koji 1973
Fukuda, Tetsuya 931
Funabiki, Shigeyuki 2906
Fung, Tsun H. 2576
Fuyuki, Takashi 2593
Gabetta, Giuseppe 76, 545
Gabor, Andrew M. 74, 2839, 2897
Gaddy, Edward 585
Gahr, Stefan 178
Gai, Boju 549, 2291
Gaiaschi, Sofia 2711
Gallon, Joshua B. 3448
Galtieri, Jason 2975, 3214
Gambogi, W. 2312
Gao, Hui 226
Gao, Peng 1648
Gao, Wei 226
Gao, Y. 869
Gao, Yijun 2392
Gao, Ying 2368
Gao, Yuan 2048, 2605
Gao, Yujie 2870, 2891
García, I. 1210
Garcia, Iván 1204
Garcia, Juan Lopez 402
Garcia-Linares, Pablo 2562
Garg, Vivek 2345
Garner, Sean 2870
Garner, Sean M. 2891
Garnett, Erik C. 2245
Garreau-Iles, L. 2312
Garrillo, Pablo A. Fernández 1516
Garuz, Richard 2255
Gaury, Benoit 1303, 2438
Gdoutos, Eleftherios E. 558
Geelan-Small, Peter 3304
Gehre, Simon 3500
Geissbiihler, J. 50
Geissbuehler, Jonas 3256
Geisz, John 549, 3371
Geisz, John F. 232, 268, 1737, 2195, 3254
Georghiou, George E. 276, 1163, 1941, 1954, 3107
Geraghty, Paul 1342
Gerardi, C. 1747
Gerber, Andreas 1400, 1651
Gerdimenes, Anne 619
Gervasi, Massimo 541
Ghaisas, S.V 389, 1701
Ghimire, Kiran 993
Ghosh, Kunal 716
Gibbs, Jacob M. 730
Gibelli, François 2192

AUTHOR INDEX

Giebink, Noel C. 1469
Giguère, Jean-Benoit 1360
Gilchrist, James B. 966
Gillispie, Kellen 2762
Giordano, Francesco 3096
Giraldo, Sergio 3265, 3285
Giussani, A. 845
Giussani, Alessandro 206, 831, 2514
Givot, Bradley L. 2864
Gladden, Christopher 1476
Glasgow, Nate 1427
Glatthaar, M. 884, 3135
Gloeckler, Markus 1193
Glunz, S. 3135
Glunz, S.W. 2064
Glunz, Stefan W. 1253, 2511
Gokkaya, Huseyin Cem 958
Goldschmidt, Jan Christoph 1253
Golembeski, Andrew A. 3143
Goma, Elias Garcia 3462
Gombia, Enos 541
Gona, Michael N. 2349
Gong, Chen 1585
Gong, Jue 2251
Gonzálcz-Díaz, Benjamín 3240
Gonzalez, Maria 259
Gonzalez, S. 3224
Gonzalez, Sigifredo 2147, 3002, 3020
Gonzalez-Díaz, Benjamín 429, 1116
Goodarzi, Mohsen 2707
Gooding, Renee 1280, 1543
Goodnick, S. M. 1603
Goodnick, Stephen 1790
Goodnick, Stephen M. 305, 582, 1797
Gordillo, G. 433, 503
Gordon, Ivan 1233
Gori, Gabriele 76
Gorman, Brian 62, 1371, 1381
Górnez-González, L. A. 2614
Gostein, Michael 2808, 2923
Goswamy, Naveen 1908
Gotoh, Kazuhiro 1765, 1794
Gottschalg, Ralph 1411, 2827, 3208
Govaerts, Jonathan 3343
Goverde, Hans 3343
Gowda, Ramesh Rame 1912
Graf, Martin 2511
Graham, Kenneth 1044
Grandidier, Jonathan 2099
Grassman, Tyler J 182, 215, 2079, 2554, 3139
Greco, Erminio 76, 541
Grede, Alex J. 1469
Green, Martin 2213, 2403
Green, Martin A. 858
Green, Michael 2926

Greenhalgh, R.C. 3430
Gregory, Geoffrey 74
Grévin, Benjamin 1516
Grice, Corey R. 771, 1643, 2473, 3426
Grieco, William J. 2618
Griffin, Alecia 2870
Griffin, Alecia C. 2891
Grijalva, Santiago 1555, 1567
Grini, S. 3269
Große, T. 1752
Großer, Stephan 2232
Grossklaus, Kevin 701
Großschädl, Bettina 1329
Grovenor, Chris 424
Grover, Sachit 1193, 2473
Grübel, B. 884
Gu, Fei 1346
Gu, Tian 1473
Gu, Tingyi 1828
Gu, Xiaohong 1933, 2000, 2844, 3195, 3204
Guarracino, Ilaria 1339
Guay, Nathan 1543
Gudla, Sushanth 1389
Guerrero-Lemus, Ricardo 429
Guillemoles, Jean-François 1289, 2192
Guillevin, N. 3150
Guillevin, Nicolas 917
Guina, M. 1189
Guina, Mircea 297, 2520
Guischard, Felix 2836
Gunawan, Oki 1441, 3275
Gunnarsson, William B. 2443
Guo, D. 1603, 2816
Guo, Hong 1299
Guo, Q. 3
Guo, Qi 226
Guo, Shuwen 1430, 1873, 2918
Guo, Yongjie 1719
Gupta, Amit Kumar 2952, 2981, 2986, 3050
Gupta, Mool C. 937
Gupta, Neeti 696
Gupta, Ritesh Kant 1034
Gupta, Shivam 377
Gupta, V. 1733
Gustafsson, Mattias 2025
Guthrey, Harvey 1400, 2887
Gutiérrez, J. R. 2740
Gutiérrez, R. 643
Gutscher, S. 884
Guwaeder, Abdulmunim 1122
Gwak, Jihye 810
Ha, Dongheon 1585
Habermann, D. 1752
Hack, James 999
Hack, James H. 326

AUTHOR INDEX

Hacke, Peter 1371, 1381, 1421, 1922, 2819, 2854, 3305
Hackl, Wolfeanz 178
Haddad, M. 2094
Haddadian, Rojiar 1927
Hadi, Sabina Abdul 213, 1741
Hadjipanayi, Maria 276
Hadke, Shreyash 986
Hadley, Wendy 2014
Haegel, Nancy M. 62
Hagendorf, Christian 1376, 2232
Hägglund, Carl 796
Hahn, Carina E. 175
Hai, Hoang Tri 931
Haight, Richard 1441
Hajimiri, Ali 521, 558
Hajizadeh, Amin 3092
Halbwax, Mathieu 3402
Hall, Allen 1511
Hallam, Brett J. 2576
Halliday, Douglas P. 2423
Hamadani, Behrang H. 263, 437, 508
Hameiri, Ziv 66, 420, 3290, 3315
Hamon, Gwénaëlle 2528
Hamui, L. 2614
Hamzaoui, Saad 900
Hamzavy, Babak T. 2618
Han, Sang M. 88
Han, Xinyue 1719
Han, Youngsik 2242
Hanada, Toru 940
Handwerker, Carol A. 1449
Haney, Paul 1303
Haney, Paul M. 2438
Hanley, J. 2094
Hanna, Amir 1055
Hannappel, Thomas 2524, 2538, 2538
Hanriot, Sergio De Morais 2307
Hansen, Clifford 1127, 1537, 3184, 3333, 3348
Hansen, Clifford W. 110, 1543, 1549
Hansen, Ole 2672
Hansen, Richard 2042
Hansen, Shirley 2042
Hao, Xia 160
Hao, Xiaojing 858, 2213, 2403
Haohui, L. 1140
Haohui, Liu 2744
Haque, K A S M Ehteshamul 346
Haque, M. D. 552
Hara, Shigeomi 1950, 3339
Hara, Tomoya 2548
Harari, Joseph 3402
Hardikar, Kedar 2688
Häring, Adrian 2263
Hariskos, Dimitrios 2058

Harmand, Jean-Christophe 1289
Harris, Christian 319
Harris, James 1835, 2732
Harris, Tom 2991
Harvey, Steven 2887
Harvey, Steven P. 1371, 1381, 2702, 3305, 3319
Haschke, Jan 3435
Haslinger, Michael 1804
Hassan, Ibrahim A. I. 2858
Hatch, S. 14
Hatton, Peter D. 2423
Hauch, Jens 3500
Haug, F.-J. 2073
Hausgen, Paul E. 102
Hausmann, J. 1752
Havu, Ville 2070
Haysom, Joan E. 1094
He, Junwen 1469, 1737
He, Qiuxiang 3304
He, Wenshuang 392
Hea, Wenshuang 2823
Heben, Michael 2926
Heben, Michael J. 170, 730, 748, 815, 1030, 2462
Hegedus, Steven 408, 1473, 1761, 1828, 2667
Hegedus, Steven S. 658
Heidmann, Berit 3396
Heilbrunner, Herwig 1360
Heilscher, Gerd 2996
Heinz, F. D. 3135
Heinze, Matthias 2263
Heller, Dominic 1077
Henes, Dan 1094
Hentz, Sandrine 966
Hermle, M. 2064
Hermle, Martin 1253, 2511
Hernandez, J. A. 2031
Hernández, Johan 1143
Hernandez, Joseph 3473
Hernandez-Alvidrez, Javier 2153
Hernández-Gutiérrez, C. A. 670
Hernández-Rodríguez, Cecilio 429
Herrera, Daniel J. 219
Herrmann, W. 107
Herz, Magnus 3360
Heta, Y. 2312
Hetterich, Michael 1682, 2216
Hettick, Mark 59, 823, 2076
Heurlin, Magnus 1286
Hickey, Benjamin 1459
Hidaka, Kazuyuki 1973
Higa, M. 441
Hilfiker, M. 2364
Hill, Alex 1893
Himwas, Chalermchai 1289
Hindi, Basel 1058

AUTHOR INDEX

Hinken, David 2692
Hinojosa, M. 1210
Hinzer, Karin 1094
Hirai, Masakazu.....................................1769
Hirata, Yoichi.....................................613
Hirose, Kotaro.....................................3323
Hirstl, Louise C. 2091
Hishikawa, Y. 441, 1003
Hishikawa, Yoshihiro 480, 2781
Ho, Jian Wei.....................................496
Ho, Wen-Jeng.....................................343, 367, 893, 2664
Hoang, Bao.....................................96
Ho-Baillie, Anita.....................................858, 1845, 2569
Hobbs, William B. 2618
Hoerteis, Matthias.....................................914
Hoex, Bram.....................................517
Hoff, Thomas.....................................132, 1104
Hofmann, Johannes 2407
Hoheisel, Raymond 247, 3514
Höhn, Oliver.....................................352
Holman, Z. C. 3376
Holman, Zachary 1790, 3366
Holman, Zachary C. 1220, 1228, 1317, 1322, 1820, 3250
Holmgren, William F. 110, 1127
Holzmann, Daniel.....................................914
Hong, Chung-Yu.....................................294
Hong, Keunkee.....................................399
Honsberg, Christiana.....................................827, 3088
Honsberg, Christiana B. 240, 305, 582, 681, 1215, 1841, 2573
Hopf, Markus.....................................1965
Horenstein, Mark.....................................2870
Horenstein, N Mark.....................................2891
Horner, Greg S. 3448
Horowitz, Kelsey A.W. 1259, 1463
Horzel, J.....................................50, 2073
Hoshii, Takuya.....................................2334
Hosokawa, Kazuya.....................................613
Hossain, Istiaque 2247
Hossain, Mohammad A. 3456
Hossain, Mohammad I.....................................963
Howard, John M. 2443
Hsi, Edward.....................................1275
Hsu, Chia-Jhe.....................................1623
Hsu, Chih An 1638, 2449, 3413
Hsu, Lung-Hsing.....................................1610
Hsu, Shun-Chieh.....................................1606, 1623
Hsu, Shu-Tsung.....................................445, 448, 476
Hsu, Wei-Lun.....................................1048
Hsu, Yu -Chen.....................................888
Hu, Chehao.....................................229
Hu, Cheng-Shun.....................................329
Hu, Hailin.....................................1858
Hu, Juejun.....................................1473

Hu, Lilei.....................................3129
Hu, Long 2392
Hu, Yang 1927
Hu, Yicong 2392
Huang, Jialiang.....................................2213
Huang, Jingsheng 1937, 2823
Huang, Jing-Shun.....................................512, 521, 558, 572, 1248
Huang, Shujuan 2392
Huang, Vi-Wen 1631
Huang, Weijing 1873, 2918
Huang, Wei-Ming 1627, 1631
Huang, Wen-Hsi 385
Huang, Ying-Yuan 1807
Huang, Yi-Wen 1627
Huang, Yu-Ming 1606
Huang, Yu-Ting 1012
Huang, Z. 3260
Huayamave, Victor.....................................2839
Hubbard, S. M. 552, 845, 2755
Hubbard, Seth 677
Hubbard, Seth M.....................................18, 202, 206, 222, 831, 1184, 2084, 2298, 2514
Huber, Christian 2216
Hudson, A.I. 2755
Huey, Bryan D. 1522
Huffaker, D.L. 2755
Huffaker, Diana.....................................202
Huhn, Vito.....................................1651
Huld, Thomas 2167
Hunault, Philippe 2711
Hung, Yung-Jr.....................................1606
Huo, Yijie 1835
Husein, Sebastian 944
Huss, Alexandra M.....................................164
Hussain, Babar 451, 2355
Hussain, Muhammad M. 1055
Hutchings, Douglas.....................................1869
Hutchings, Douglas A 921
Hutter, Oliver S. 1445
Hwang, James.....................................333
Hyvl, M. 2073
Iandolo, Beniamino 2672
Ianno, N.J. 2364
Ichikawa, Yukimi 1769
Idlbi, Basem 2996
Ikki, Osamu 2159
Ilic, Ognjen 1737
Imai, Jun 2906
Imaizumi, Mitsuru 567, 3506
Imtiaz, Syed N.....................................1067
Ingenhoven, Philip.....................................3482
Ingenito, A.....................................2073
Inns, Daniel 3113
Isabella, Olindo.....................................2605
Isbilir, Kenan 2827

AUTHOR INDEX

Isherwood, Patrick J. M. 2349
Ishii, Tomoaki 455
Ishino, Yuya 757
Ishizuka, Shogo 33
Islam, Kazi 37
Islam, Muhammad Monirul......................33, 900
Islam, Raisul 1835
Isoaho, Riku 2520
Iwasaki, Kazuya 2338
Iwata, Naotaka 2642
Iwuoha, Emmanuel 1360
Iyer, Abhishek....................326, 999, 2035
Iyer, Parameswar K. 1034
Izquierdo-Roca, Victor 3265, 3285
Jackson, Christine.................................215
Jackson, Philip 2205, 2453
Jacob, David 1549
Jacobson, Arne 1271
Jadkar, S.R. 1701
Jaeckel, Bengt 1411
Jae-Yun, Fa-Jun Ma, ...1845
Jagdish, A K 2811
Jäger-Waldau, Arnulf............................2167
Jahn, Ulrike 3360
Jain, Aditi 333
Jain, Nikhil 42, 46, 232, 578, 2195, 3371
Janoch, Rob 74, 2839, 2897
Jansen, Mark J. 1081
Jany, Christophe 2492, 2562
Janz, Stefan 83, 2407
Jaramillo, Adolfo 1143
Jared, Bradley 1473
Jarmar, T. 30
Jasti, Naga Prathibha 986
Jaswal, Rohit 3172
Javed, Mehwish Azher 1317
Javey, Ali 59, 823, 2076
Jeangros, Q. 2073
Jenkins, P. P. 845
Jenkins, Phillip P................. 247, 373, 1838, 2091, 3514
Jensen, Brian 2014
Jensen, M. A. 3295
Jensen, Mallory A. 1491, 3290, 3300
Jensen, Soren..............................1196, 3147
Jeong, Woo-Lim 777, 1665
Jhaveri, Janam 1773
Ji, Liang 1933, 2000
Ji, Yaping 37
Ji, Yaping Vera 3245
Jia, Jieyang 1835, 2732
Jian, Ding-Rung 1627, 1631
Jiang, C. S. 1312, 2789
Jiang, C.-S. 2280, 2785
Jiang, Chun-Sheng 62, 1371
Jiang, Lian L. 589

Jiang, Lian Lian 120
Jiang, Xuefang 1937
Jiang, Yu 3220
Jimeno, J. C. 643, 2740
Jimeno, Juan Carlos.............................2677
Jin, C. 1781
Jin, Yu 3119
John, Jim J. 1946
John, Joachim 1804, 2227
John, Suru Vivian 1360
Johnson, A. D. 1210
Johnson, E.V. 1781
Johnson, Erik V. 2593
Johnson, J. L. 1656, 1661
Johnson, Jay 2135, 2141, 2153, 2969, 3002, 3008
Johnston, S. 2785
Johnston, Steve.....................62, 202, 459, 1371, 1381, 1400, 2213, 2819, 2887, 3305, 3452
Jones, C. Birk............2618, 3008, 3155, 3488
Jones, David 1342
Joonwichien, Supawan 904
Joshi, Madhuwanti S.............2952, 2981, 2986, 3050
Joshi, Pranav 2247
Jošt, Marko 1346
Jovanovic, Raka 963
Juang, B.C. 2755
Juang, Bor-Chau 202
Juárez, A. Sánchez 1990
Juárez, Aarón Sánchez...........................1959
Juhl, Mattias K. 420, 3315
Juhl, Mattias Klaus 66
Julien, Scott.................................1933, 2000
Junci, Wang 496
Junda, Maxwell 771
Junda, Maxwell M....................2462, 2582, 3426, 3468
Jung, Jae Hak 487
Jung, Jiirgen 2864
Jung, Sang Hoon 2723
Jung, Sang Hyun 244
Juso, Hiroyuki 3506
Kabalan, Amal.....................................2358
Kaczynski, Ryan 1455
Kaizu, Toshiyuki 23
Kaizuka, Izumi 2159
Kakosimos, Konstantinos E......................1888
Kalainatharr, Sivaperuman 2334
Kalb, J. 1733
Kale, Abhijit.......................................1801
Kale, Abhijit S. 1777
Kallickal, Johnson 1543, 3348
Kalt, Heinz 1682, 2216
Kamata, N. 552
Kamevama, Satoshi 2642
Kamino, Brett 3256
Kamins, Ted 1835

AUTHOR INDEX

Kaminski, P.M. 3430
Kaminski, Piotr 1674
Kaminski-Cachopo, Anne 2562
Kamioka, Takefumi 2498, 2548, 2566, 2642
Kanemitsu, Yoshihiko..............................721, 2781
Kanevce, A. 1312
Kanevce, Ana...3147
Kang, Ho Kwan 244
Kang, Min Gu 356, 1758, 2723
Kang, Yoonmook 935
Kankiewicz, Adam 132, 1104, 1132, 1427
Kannan, C. V. 2716
Kao, Ming-Hsuan 1627
Kaplan, Stephen 600, 1071
Kaplar, R. 3224
Karas, Joseph 925
Karki, Shankar... 182, 735, 807, 2298, 2446, 2646, 3139
Karmarkar, M. 1661
Karpowich, Lindsey 914
Karthik, Shravan 3172
Kashkoush, Ismail...322
Kaslin, Remo 1077
Kasry, Amal 2858
Kasu, Makoto 1950, 3339
Kato, Takekazu 1175
Kato, Takuya 160
Katsube, Ryoji 2361
Kaule, Felix 2622
Kausika, Bala Bhavya..............................3014, 3167
Kavaipatti, Balasubramaniam 2799
Kawatsu, Tomoyuki 381, 2588
Kazmerski, Lawrence L. 2799
Kazmerski, L.L. 1245
Kazmerski, Lawrence L. 1917, 2307
Kazumi, Kenji 2361
Keeler, Gordon 1473
Keller, Nico 1077
Kelly, George..1275, 2263
Kelly, Matthew 3514
Kelzenberg, Michael D. 512, 521, 558, 572, 1248
Kempe, M.D. 138
Kempe, Michael 1933, 2000
Kempe, Michael D..3208
Kephart, Jason 785
Kephart, Jason M. 3417
Kern, Gregory 2147
Kern, Gregory A. 3020
Kesavan, Arul Varman..1614
Kessels, Wilhelmus M.M. 1817
Kessler, Emily..206
Khadimallah, A. 869
Khalili, A. 1189
Khan, Imran 802, 1638, 3413
Khan, Imran S. 2449
Khan, Mohammad R. 1055

Khan, Taj M. 451
Khanna, Raghav...2926
Kharait, Rounak A..2833
Kharel, Khim 236, 673, 1452
Khatavkar, Sanchit 2716
Khatiwada, D. 869
Khatiwada, Devendra 866
Khatri, Ishwor 192
Khatri, Trijul 377
Khomenko, Denis V..690
Khoo, Yong Sheng 1922
Khor, Alan 3172
Khoram, Parisa 2245
Khorenko, Victor 83, 2087
Khoury, R. 1781
Kiefer, Fabian 1366
Killam, Alex 2719
Killinger, Sven 126, 1405
Kim, Boram 2201, 2524, 2538
Kim, Chang Zoo 244
Kim, D. 1189
Kim, Dae Young 626
Kim, Dong Seop 399
Kim, Dong-Ho 2631, 2634
Kim, Donghwan 935, 1758
Kim, Hae-Sun 777, 1665
Kim, Hyo Jin 849
Kim, In-Young 777
Kim, Jae Hyun..............................363, 2844, 3195, 3204
Kim, Jin-Hyeok 777
Kim, Jisun 399
Kim, Ka-Hyun 1758
Kim, Kangho 244
Kim, Kihwan 810
Kim, Kyoung- Tae 1037
Kim, Min-Soo..487
Kim, Moon 2759
Kim, Sangpyeong...240
Kim, Soo Min 2723
Kim, Woo Kyoung 487
Kim, Yeongho 827
Kim, Yong Bae...2723
Kim, Yong Whan 849
Kim, Youngjo...244
Kimbal, Gregory M. 110
Kimura, Daiki 854
Kindole, Dickson 970
Kindvall, Anna..785
King, Bruce H. 3155, 3488
King, Richard 827
King, Richard R. 301, 823, 1215, 1841
Kingma, Aldo 541
Kini, Roshan 2926
Kinoshita, Kosuke 1504, 2588
Kirk, A..3511

AUTHOR INDEX

Kita, Takashi 23
Kleider, Jean-Paul 2528
Klein, Talysa R. 2482, 3371, 3439
Kleinschmidt, Peter....................2538
Klemm, Hagen W. 3396
Klenk, Markus 1077
Klie, Robert F. 2759
Klimm, Elisabeth 2836
Klise, Katherine A. 3161
Klisel, Geoffrey T. 3494
Kluska, S. 884, 3135
Knight, Bruce 2014
Knopf, Hannes 1965
Ko, Changhee 326
Kobayashi, Jonathan 1061
Koehl, Michael 3488
Koepgel, Ringo 2622
Kogler, Willi 791
Kohlstädt, Markus 1253
Koike, Junichi 931
Koirala, Prakash 2462, 3426
Kojima, Nobuaki....................359, 2498, 2566
Kojima, Takuto 1504, 2588
Komsa, Hannu-Pekka 2070
Konagai, Makoto 1769, 2627
König, M. 1752
Konstantinou, Georgios 1941
Kontges, Marc 1366
Kopecek, Radovan 1222
Koschny, T. 1350
Kostylyov, V.P. 1025, 1811
Kostylyov, Vitaliy P. 690
Kotipalli, Ratan 2209
Kottantharayil, A. 1995
Kottantharayil, Anil 396, 716, 1850, 2799, 3478
Kottokkaran, Ranjith 2247
Kotulak, Nicole 999, 1838
Kotulak, Nicole A. 247
Koyama, Koichi 1765, 1787
Kozodoy, Peter 1476
Krabb, Peter 178
Krantz, Patrick W. 730
Krasikov, D. 2816
Krc, Janez 1346
Krein, Philip T. 3214
Krich, Jacob J. 1294
Krishnan, Mani R. 1912
Krishnan, Sheeja 76
Krishnaswami, Hariharan....................2931, 2936
Krogen, J. 2094
Krügener, Jan....................1494
Krut, Dimitri D. 37
Ku, Chen-Hao....................329
Kubiniec, Alex 132, 1132, 1427
Kuciauskas, Darius 1679

Kudriavtsev, Yu....................670
Kuitche, Joseph 1877, 1883
Kulish, Mykola R....................690
Kum, Hyun 18, 222, 677, 2084
Kumar, Rajesh....................3478
Kumar, Shailendra....................2345
Kumar, Sukanya Santhosh....................980
Kumar, Vijay....................2716
Kumari, Khushboo 251
Kuo, Hao-Chung 1610, 1627
Kuo, Po-Tsun 1006
Kuo, Ting-Wei 329
Kurdgelashvili, Lado....................2035
Kurihara, Risa 2159
Kurimoto, Yuji 931
Kurokawa, Yasuyoshi 1765
Kurstjens, Rufi 83
Kurtz, Sarah 1275, 1922, 2263, 3190
Kusaki, Kazuki 23
Kuthanazhi, Vivek....................3478
Kwon, Jung-Dae 2631, 2634
Kwon, Sang Jik 195
Kyureghian, H. 2364
La Centra, Ricci 2870, 2891
Lachaurne, Raphaël....................2528, 3402
Lachowicz, A. 50
Lackner, David 2511
Lacroix, Jean-Sébastien 1579
Lafleur-Lambert, Antoine 1360
Lafont, Ombline 2453
Lagumavarapu, Ramesh B. 202
Lai, B. 3295
Lai, Barry 1494, 2170, 2179, 2245, 3300, 3309
Lai, Yi....................1009
Laine, Hannu S. 1491, 1494, 3236
Lakshmanan, Ramakrishnan 2870, 2891
Landgraf, D. 1752
Lang, Mario 2216
Lapierre, Ray R. 1294
Larrey, Vincent 2492
Larsen, Ross E. 1354
Larson, Bryon W. 1354
Lasalvia, Vincenzo....................881, 1491, 1801, 2242, 3439
Laschinski, Joachim 1965
Lassise, Maxwell 3410
Latham, Joseph 1086
Lau, Derwin 3220
Lau, Kei May 578
Lave, Matthew 1435, 3008, 3025, 3031, 3184, 3348
Lavrova, Olga 1280, 2618, 3488, 3494
Law, D. 2094
Lazarou, Constantinos 276
Le Corre, Alain 2192
Le Donne, Alessia 1669
Le Gall, Sylvain 3402

AUTHOR INDEX

Le Guen, Vincent 70
Le Rouzo, Judikaël 2255
Lebreton, Fabien 464, 1237
Leclerc, Christophe.............................558, 572
Lecouvey, Christophe 2492, 2562
Ledinek, Dorothea 796
Ledinsky, M...2073
Lee, Angela 417
Lee, Benjamin G. 881, 1737, 1801, 1832, 2482, 3439
Lee, Calvin..1342
Lee, Dong-Seon..................................777, 1665
Lee, Eunjoo..399
Lee, Eunsang..2124
Lee, Hae-Seok..935
Lee, Hyeonseok..1012
Lee, Jaejin..244
Lee, Jeong In 356, 1758
Lee, Ji-Hoon...2631, 2634
Lee, Jihwan...1181
Lee, Jinwoo...1455
Lee, Jongwon...681, 1215
Lee, Kan-Hua 359, 412, 1479, 1711,
    1714, 1743, 2498, 2566
Lee, Kyumin...1526
Lee, Kyu-Tae..1469
Lee, M. L..3376
Lee, Minjoo..2291
Lee, Minjoo L..42
Lee, Mitch..600
Lee, Mitchell..595
Lee, Seunghun..1253
Lee, Soonil..363
Lee, Yeonbae..1204
Lee, Yun Seog..1441
Lefebvre, Amy..1933
Lefebvre, Amy L...2000
Lehman, Peter..1271
Lehr, J..3224
Leilaeioun, M..3376
Leilaeioun, Mehdi 1322, 1790
Leite, Marina S...................1508, 1585, 2443
Lekx, David..1094
Lemaître, Aristide...1289
Lemus, Ricardo Guerrero 1116, 3240
Lennon, Alison..3220
Lennon, Kyle..3384
Lennon, Kyle R..1598
Leone, Stephen R...417
Leonhardt, M..1752
Leow, Shin Woei..3275
Lepkowski, Daniel...215
Lepkowski, Daniel L................................2079, 2554
Lester, Luke F...219
Leto, Riccardo..1728
Leu, S..1752

Levcenco, Sergiu..3396
Levi, D.H. 467, 490
Levi, Dean 483
Levrat, J. 50
Levrat, Jacques 3435
Levy, David H. 3442
Li, Chu Tu 1094
Li, Duanhui 1473
Li, Guan-Yi 343, 367, 893, 2664
Li, Jian 771, 1643
Li, Jian V. 2473, 2728, 2749
Li, Joel B. 3300
Li, Kexue 424
Li, L. 3295
Li, Lan 1473
Li, Li 1175
Li, Lu 1619
Li, Mengjie 3315
Li, Qiang 578
Li, Rui 1094
Li, Siming 3143
Li, Wenjie 3275
Li, Xiaoping 1193
Li, Xinyi 255
Li, Xueying 2170
Li, Y. 869
Li, Yongkuan 2368
Li, Yunjun 907
Li, Yunpeng 2220
Li, Zhanhang 226
Li, Zhuohui 1598, 3384
Liang, B.L. 2755
Liang, Jianbo 2548, 2551
Liao, Anqi 948
Liao, Yuanxun 696
Liao, Yuaxun 2186
Libby, Cara S. 2618
Licht, Abigail 701
Lichty, Marlene L.. 2298
Lie, Stener 3275
Lim, Bianca 2318
Lin, Albert... 294, 1631
Lin, Albert S. 1627
Lin, Cheng-Shian 1006
Lin, Chien-Chung...............1606, 1610, 1623, 1627
Lin, Ching-Fuh 1006, 1009
Lin, Fen 284
Lin, Ming-Yi 1051
Lin, Shang-Pang 1006
Lin, Yan 1100
Lin, Yandan 2048
Lin, Yan-Zhang 1623
Lin, Yida 667
Lin, Yu-Hsuan 911
Lin, Yung-Sheng... 329

AUTHOR INDEX

Lin, Zong-Xian 367
Lincoln, Jason 2897
Lincoln, Jason L. 2839
Lincot, Daniel 2453
Linton, John 337
Lipovšek, Benjamin 1346
Lipski, Michael V. 1469
Lisbona, Emilio Fernandez 545
Lisco, F. 1691, 2457
Litjens, Geert..3014
Liu, A. Y. 1485
Liu, Chenxi 3172
Liu, Fang Fang 1648
Liu, Fangyang 2213
Liu, H. 1189
Liu, H.Y. 14
Liu, Haitao...472
Liu, Han-Wen...2660
Liu, Haohui..284, 2532
Liu, Hsiang-Yu...2637
Liu, Huiyun..3370
Liu, Jheng-Jie...............................343, 893, 2664
Liu, Kanglin...1100
Liu, Mengxia...3129
Liu, Qihang..1245
Liu, Ruimin 2220
Liu, Simon H. 93
Liu, X.Q. 2094
Liu, Xiangxin.......................198, 767, 1707
Liu, Xinbing...1728
Liu, Xing-Quan..2099
Liu, Zhe...284
Liu, Zhen..2532
Liu, Zhengjun..1494
Liu, Zhengxin..1241
Livera, Andreas......................................276, 1954
Liyanage, Geethika K. 170, 730, 815, 2462
Llin, Lourdes Ferre.......................................1339
Lloyd, Alexis..2870
Lloyd, Michael A...............................726, 3143
Lnr, Yiming..2558
Loach, Andrew J..2697
Lodha, Saurabh...716
Lokanath, Sumanth 1275
Loke, Samuel P.......................................512, 521, 558
Loke, W.K. 1210
Lombardero, I..1210
Lombez, Laurent.....................70, 1289, 2192
Lonergan, Mark...802
Long, Yean-San.......................................448, 476
Looney, Erin E. 1491, 3236, 3290, 3300
Löper, P. 2073
Löper, Philipp...55
Lopez, Cristina S. Polo 2118
López, G. 1781

Lopez, Roberto 2728, 2749
López-González, J.M.....................................1781
López-López, M. 670
Lopez-Marino, Simón 155
Lorentzen, Justin 3514
Lorenzo, Antonio T. 1127
Loser, Ulrich 2272
Lossen, Jan 1222
Lotshaw, W.T. 2755
Lou, Chaogang...1619
Loubar, Anais 2711
Loyer, Camille 2000
Lu, Ching-Ying 1835
Lu, Hongbo 255
Lu, J.P. 1733
Lu, Jiawen 2656
Lu, Kyle B. 3448
Lu, Zhou...1728
Lubenow, Tomas 3333
Lujan, R. 1733
Luka, Tabea 2232
Lumb, Matthew P.............210, 247, 259, 272, 873, 2506
Luna, Miguel A. 632
Lunacek, Monte 3008
Lunt, Richard R. 2124
Luo, Shiqiang..976
Luo, Wei 1922
Luo, Yanqi 2170, 2245
Luria, Justin L. 1522
Luther, Joseph M. 2176
Lynn, Kelvin G. 3422
Lyons, Alan 2285
Lyu, Yadong 1933, 2844, 3195, 3204
Lyu, Zheng 1835, 2732
Ma, D. 229
Ma, Fa-Jun 2569
Ma, Xiaokun 1469
Macalpine, Sara...1537
Macco, Bart..1817
Macdonald, D..1485, 3295
Macdonald, Daniel 2707, 3300
Mack, C. 490
Mack, I. 2073
Mack, Shawn 259, 873
Mackie, Neil 820
Maclaren, Scott...1511
Macmaster, Steven W. 2864
Madani, Keeya 940, 1824
Madsen, C. K. 229
Maeda, P.Y. 1733
Magaña, Ernesto 1494
Magdaleno, R. Santos....................................1990
Magdaleno, Rocío De La Luz Santos 1959
Magnin, Vincent 3402
Magnone, Lydie 1415

AUTHOR INDEX

Mahadik, N. A. 845
Mahapatra, Chiranjibi 2849
Maia, Cristiana Brasil 2307
Maidaniuk, Yurii 3370
Mailoal, Jonathan 1264
Major, Jonathan D. 742, 1445
Makita, Kikuo 861, 1724
Makoutz, Emily A. 3381
Makrides, George 1163, 1941, 1954, 3107
Malhotra, Raghav 3172
Malik, Roger 1193
Maliya, Heini 226
Malkov, Andrei V. 146, 186
Mallick, Tapas K. 2858
Manda, Surya 761
Mandelis, Andreas 3129
Manganiello, Patrizio 3343
Mangelinck-Noël, Nathalie 1498
Mani, Monto 604
Maniscalco, B. 2457
Maniscalco, Biancamaria 2827
Mann, Colin 1248
Mann, Colin J. 93, 512
Mansfield, Lorelle 1400, 2473
Mansoori, A. 281
Mantel, Claire 2682
Manzoor, Salman 1228, 1322
Marie, Benoit 1498
Marion, Bill 1134, 1537, 1543, 3333, 3348
Markevich, V. P. 1485
Markides, Christos N. 1339
Maros, Aymeric 1215
Marsh, Brett M. 417
Marsillac, Sylvain 182, 735, 807, 2298, 2446, 2646, 3139
Marsillac, Sylvain X. 2582
Marti, Shilpa 2936
Martín, I. 1781
Martin, Mickaël 2492
Martinez, Aaron D. 2536, 3406
Martinez-Morales, Alfredo A. 2881
Martínez-Pérez, Alejandro 3285
Martín-Martín, D. 3376
Martins, Ana C. 2104
Martinson, Alex B.F. 6, 1256
Masada, Isao 1504
Mascarenhas, Angelo 2506
Maser, Jörg 3309
Maser, Jörg M. 2179
Maskell, Douglas L. 120, 589
Mastroianni, Simone 1253
Masuda, Atsushi 1268
Masuda, Shota 1794
Masutomi, Yasuki 3339
Matei, I. 1733

Mather, Barry 1561
Mathew, Leo 363
Mathew, X. 142
Mathews, N. R. 142
Matsubara, Koji 381, 1333
Matsui, Takuya 381, 1333
Matsumoto, Yasuhiro 632
Matsumura, Hideki 1765, 1787
Matsuo, K. 3
Matthew, Leo 2506
Maximenko, S. I. 845
Maximenko, Sergey 2091
Maximenko, Sergey I. 873
May, Matthias M. 2538
Mayberry, Ryan 914
Mazumder, Malay 2870
Mazumder, Malay K. 2891
Mazur, Yuriy I. 3370
Mccandless, Brian 1196, 3319
Mccandless, Brian E. 726
Mcclary, Scott A. 1449
Mcclung, Larry 2833
Mcclure, E. L. 845
Mcclure, Elisabeth L. 2298
Mcclure, Harumi 2947
Mccndless, Brian E. 3143
Mccomb, David W. 966, 3139
Mcdanal, A.J. 525
Mcdanold, Byron K. 2864
Mcfavilen, Heather 305, 582
Mcintosh, Keith R. 1322
Mcintyre, Maxwell 1040
Mcintyre, Michael 1086
Mckenna, Russell 126
Mcmahon, William E. 268, 3381, 3406
Mcmeans, Philip A. 921
Mcpheeters, Claiborne 42, 525, 2099
Meakin, David 1927
Medic, V. 2364
Medici, Vasco 2118
Meeker, Michael A. 873
Mehlich, H. 1752
Mehta, Hitesh K. 3038
Meier, Florian 2996
Meissner, Dieter 178
Meitl, Matt 272
Meitl, Matthew 873
Melamed, Celeste L. 3406
Melchiorre, Michele 151, 2054
Meleco, A. J. 14
Mellor, Alexander 1339
Melnikov, Alexander 3129
Melvin, Andrew 1
Méndez, Juan A. 3240
Meng, Fanying 1241

AUTHOR INDEX

Meng, Hsin-Fei 1635
Meng, Xiaodong 2854
Menossi, Daniele 752, 1669, 2372
Menozzi, Roberto 2205
Men-Pérez, E. 370
Meot, Jacques 2593
Merdzhanova, Tsvetelina 2114
Merghcim, Julia 3500
Merkle, Agnes 3371
Merz, Christopher 1965
Merzlic, Sebastien 1933, 2000
Messer, Alexander 512
Messer, Alexander J. 521, 558
Messerschmidt, Michael 2682
Messmer, C. 2064
Metzger, W.K. 1312
Metzger, Wyatt K. 1196, 3147, 3305, 3319
Meuris, M. 3260
Mewe, Agnes A. 917
Meyers, Bennet 3354
Mi, Z. 2388
Mi, Zetian 1299
Mia, Md Dalim 2749
Michaelson, Lynne 925
Micheli, Leonardo 2301, 2789, 2804, 2858, 2881
Michl, Bernhard 1329
Mihailetchi, Valentin D. 1222
Mihaylov, Blagovest 1411
Mikofski, Mark 3354
Mikofski, Mark M. 110
Milakovich, Timothy 213
Miller, Bill 1473
Miller, David C. 2789, 2864, 3195, 3208
Miller, Elisa M. 2536
Milleville, Christopher C. 1598, 3384
Mil'shtein, S. 2411
Min, Jung-Hong 777
Minemoto, Takashi 455, 757, 2385
Minkin, L. 1863
Mints, Paula 2039
Miryala, Tejaswini 2646
Mishima, T. D. 14
Mishra, Himani 2376
Misra, Sudhajit 175, 802, 2467
Mitchell, Bernhard 2707, 3304
Mittag, Max 1531
Miyajima, Sakutaro 480
Miyashita, Naoya 854, 2334
Mizuno, Hidenori 1724
Moffett, C.E. 2736
Mohammed, Khaja H. 921
Mohapatra, Soumya Ranjan 3050
Mohr, Christian 83
Monnard, Raphäel 3256
Montenegro, Davis 3055

Montes, Carlos 429
Montgomery, Kyle H. 531
Montiel-Chicharro, Daniel 3208
Moon, Soo-Jin 3256
Moore, A. 2816
Moore, Andrew 1522
Moore, James 1838, 2091
Moore, James E. 259, 272, 373, 2506
Moore, Jay 2947
Moosa, Hassa 2011
Moosa, Maitha 2011
Moradi, Hadis 638, 2941
Moraitis, Panagiotis 3167
Morales, Christophe 2492, 2562
Morales, Cristian 2870, 2891
Morales-Acevedo, A. 670
Morel, Don 802, 1638, 3413
Morgado-Dias, F. 3178
Moriarty, T. 490
Moriarty, Tom 483
Moriki, Akinori 2906
Morin, Jean-Francois 1360
Morishige, Ashley E. 1494, 3236, 3290, 3300
Morita, Hiroshi 1973
Morral, Anna Fontcuberta I. 944
Morris, Jeromie 2996
Morrison, Matthew 229
Mortazavi, Soheyl 2875
Moseley, J. 1312
Moseley, John 62, 1196, 1381, 2887, 3123, 3147
Moser, David 3360, 3482
Moustafa, A. 1747
Moutinho, H.R. 1312, 2280, 2785
Moutinho, Helio 62, 2789, 3305
M'sirdi, Nacer K. 1968
Muaddi, Saad 1110
Mueller, Thomas 496, 2318
Mukherjee, Shaibal 2345
Mulder, P. 1189
Muller, Bjorn 126, 2267
Muller, M. 2280
Muller, Matthew 2294, 2301, 2789, 2804, 2858, 2881
Müller, R. 2064
Munasinghe, Anjali 2124
Munday, Jeremy N. 1585
Mundt, Laura 1253
Mundus, Markus 1253
Munkhammar, Joakim 3067
Muñoz, D. 1747
Munoz, Krystal 925
Munshi, Amit 1674
Munshi, Amita 980
Mur, Pierre 2562
Muralidharan, Pradyumna 1790, 1797
Muramatsu, Kazuo 2642

AUTHOR INDEX

Murphy, J. D. 1485
Murugesan, Arumugam 2172
Muskovin, Eric 537
Mutitu, James 315
Mwove, Johnson Kyalo 2014
Myers, Matt 525
Nærland, Tine Uberg 2610
Nagaoka, Akira 1679
Nagarajan, Adarsh 2991
Nage, M. 3150
Nagel, H. 3135
Nägelein, Andreas 2538
Nair, P. R. 2716
Nair, Pradeep R. 1015, 1022, 1755
Nakada, Tokio 192
Nakamur, Tetsuya 567
Nakamura, Kyotaro 1504, 1794, 2498, 2566, 2588, 2642
Nakamura, Shigeyuki 2338
Nakamura, Tetsuya 562, 3506
Nakano, Yoshiaki 854, 2201, 2524, 2538, 3528
Nakata, Tatsuya 854
Nakatsuka, Shigeru 2385
Nam, Wooseok 2242
Nanda, A. 229
Nandal, Vikas 1015
Nanduri, Sai Naga Raghuram 1018
Naqavi, Ali 512, 521, 558, 572, 1248
Narasimhan, K.L. 396, 1850, 1995, 3478
Nardone, Marco 309, 3123
Naseem, Hameed A. 921
Natsheh, Ammar 2011
Naumann, Volker 1376
Nawara, Witek 3462
Nawaz, Syed F. 1067
Nayfeh, Ammar 213, 1741
Naylor, Mark 914
Nayshevsky, Illya 2285
Ndione, Paul 1253
Needell, David R. 1737
Neely, J. 3224
Neely, Jason 2141
Neergat, Manoj 1704
Nehme, Bechara 1968
Nelson, George T. 202, 206, 222, 1184, 2084
Nemeth, William 1777, 1801, 1817, 1832, 2242, 2702, 3439
Nespoli, Lorenzo 2118
Nett, Zach 1737
Neuschitzer, Markus 155
Neuwirth, Markus 1682
Newlands, Allan 2042
Ng, Annie 958
Ngan, Lauren 600
Nguven, Dac-Trung 2192

Nguyen, H. T. 3295
Nguyen, Tinh 3204
Nickel, Benedikt 3388
Nicolay, S. 50
Nicolay, Sylvain 3256
Niemi, T. 1189
Niesen, Bjoern 3256
Nietzold, Tara 944, 2179, 3309
Nii, Kohdai 85
Niki, Shigeru 33
Nilsson, Ulf H. 2864
Nishikawa, Naoyuki 1385
Nishio, M. 3
Nishioka, Kensuke 480, 1479
Noack, Max 2247
Nobre, André M. 3172
Nobuhara, Shohei 1175
Nocerino, John 93
Noda, Naoto 326
Noda, Yoshimasa 970
Nofuentes, Gustavo 2858
Nogay, G. 2073
Noh, Shinyoung 858
Nonnenmacher, H. J. 1752
Norman, Andrew 1381, 2887
Norman, Andrew G. 2536, 3406
Norwood, Robert A. 1147
Nose, Yoshitaro 1679, 2361, 2385
Nowakowski, Marilyn L 3524
Nsofor, Ugochukwu 1828
Nukala, Tejeswar 3061
Nunomura, Shota 381
Nurdin, Muhammad 3102
Nussbaumer, Hartmut 1077
Nuzzo, Ralph G. 1469, 1737
Nyirjesy, Gabrielle 667
Oberbeck, Lars 3370
O'Brien, Greg 1933
O'Brien, Gregory 2000
Ocaña, Luis 429
O'Carroll, Deirdre M. 3393
Ochoa, M. 1210
Odden, Jan Ove 2651, 2776
Oehler, Fabrice 1289
Ogawa, Tomoki 2548
Ogura, Atsushi 1504, 2588, 2642
Ogutman, Kortan 1804, 3448
Oh, Jaewon 1858, 1877, 1883, 2912
Oh, Seung Kyu 866
Oh, Soo-Young 487
Ohdaira, Keisuke 1385, 1787
Ohigashi, Takashi 2159
Ohshima, H. 441
Ohshima, Takeshi 562, 567
Ohshita, Yoshio 1504, 1794, 2498, 2566, 2588, 2642

AUTHOR INDEX

Ohta, Taisuke 3417
Ok, Young-Woo 333, 1807, 1838
Oka, Naotaka 1973
Okada, Yoshitaka.....................10, 85, 854, 2334
Okafor, Jonathan O. 219
Okano, Y. 3
Okel, Lars A.G. 1081
Oliva, Florian 3265, 3285
Olopade, Muteeu......................................2381
Olvera, María De La Luz 632
O'Neill, Mark 525
Oney, Michael F. T 2176
Onno, Arthur...3370
Onunkwo, Ifeoma......................................2135
Oo, W.M. Hlaing.......................................1661
Opila, Robert...................................999, 2035
Opila, Robert L.................................315, 326
Oreski, Gemot 178
Orlovskaya, Nina A..................................2324
Ortega, E..643
Ortega, Pablo 944
Ortiz, Brenden R......................................3406
Ortiz-Rivera, Eduardo I. 2957, 2963
Ory, Daniel 70
Oshima, Ryuji...861
Ososanya, Esther......................................2432
Osowski, M. 3511
Osterwald, C.R.................................467, 490
Ota, Yasuyuki...1479
Otaegi, A. 2740
Otaegi, Aloña 2677
Otnes, Gaute 1286, 2502
Ottoson, L...467, 490
Ouyang, Zi 2403, 3220
Oviedo, Felipe 2744
Ozanne, A. -S...1747
Paap, Scott 1473
Packard, Corinne E. 46
Page, Matthew......................................1777, 2242
Paggi, Marco 402
Palekis, Vasilios 175, 802, 1511, 1638, 2467, 3413
Palekis, Vasilis 2449
Palitzsch, Wolfram....................................2272
Palmer, Evan 496
Palmiotti, Elizabeth..................................1400
Palmquist, Nathan 667
Pan, Hui 1100
Pan, N..3511
Pan, Zhen..226
Panchal, A. K...3061
Panchal, Ashish K.....................................3038
Pandey, Rahul 377
Paolone, Mario 1415
Parashar, Parag 1627, 1631
Paraskeva, Vasiliki....................................276

Parenti, Robert C. 3520
Parikh, Anuja V. 3123
Parikh, Harsh 2682
Park, Chinho..487
Park, Ji-Sang 6, 1256
Park, Joo Hyung 810
Park, Kyung Ho 244
Park, S. 2388
Park, Seonyong 2087
Park, Somin 1044
Park, Sungeun 935
Park, Won-Kyu 244
Partain, Larry..2042
Passow, Kendra 595, 600, 1071
Paszuk, Agnieszka 2524, 2538
Patra, Payal...761
Patterson, Robert J 2392
Paudel, Naba R. 2443
Paul, Douglas 1339
Paul, Nicolas..70
Paul, P. K. 30
Paul, Pran K. 2446, 3139
Paul, Sanjoy 2473, 2749
Paulauskas, Tadas 2759
Paull, P. K. 2414
Paull, Sanjoy 2728
Paulsen, Andrew 3514
Pavgi, Ashwini 1877, 1883
Paviet-Salomon, B. 50
Paviet-Salomon, Bertrand 3256
Pavilonis, Michael.....................................1476
Pavlov, Marko 626
Pavlovsky, Igor 907
Pawar, Vaibhav 2986
Payne, David 315
Payne, David N.R. 2576
Peaker, A. R. 1485
Peale, Robert E. 2324
Peharz, Gerhard................................178, 1329
Peibst, Robby 1366, 3371
Pellegrino, Sergio..................512, 521, 558, 572
Peña, J.L. 1691, 2457
Pena, Juan Luis 1669, 2372
Peña, Ramón 632
Peng, Jun 2076
Peng, Shou 761, 2427
Penning, David P 2170
Peppanen, Jouni 3025
Pera, David 1979
Peraca, Nicolás Márquez 263
Perez, Richard 132, 1104
Pérez-Rodríguez, Alejandro 3265, 3285
Perez-Wurfl, Ivan......................................315
Perkins, C. 2280
Perkins, Craig 2702, 2789

AUTHOR INDEX

Perkins, Craig L. 2294
Perl, E. 3376
Perl, Emmett E. 42, 1201
Perna, Allison 2467
Pesala, Bala 2858
Peschel, Gina 3396
Peshek, Timothy J. 1927, 2697, 3456
Peters, I. M. 648, 1140
Peters, Ian Marius 284, 1264, 2532, 2744
Peters, Marius 3236
Petersen, Michael 2682
Peterson, Chris 512
Peterson, Josh 1169
Petoukhoff, Christopher E. 3393
Pfiester, Nicole 701
Phillips, Adam 748
Phillips, Adam B. 170, 730, 815, 1030, 2462
Phillips, Laurie J. 1445
Phillips, Nancy H. 2864
Phinikarides, Alexander 1954
Picard, Sandrine 2087
Piccinelli, Fabio 1669, 2372
Pickel, Tobias 3500
Pierro, Marco 3482
Pieters, Bart E. 1651
Pihan, Etienne 1498
Pillai, Supriya 2403
Pistor, Paul 155, 3285
Piszczor, Michael 525
Pitalúa, Nun 632
Platzer-Björkman, C. 3269
Plessing, Lukas 178
Pleus, Albert 1835
Plochowietz, A. 1733
Podraza, Nikolas 2646
Podraza, Nikolas J. 2462, 2582, 2771, 3426, 3468
Poindexter, Jeremy 3300
Poissant, Yves 1908
Pokharel, Nikhil 831, 2514
Polojärvi, Ville 297
Poncho, Corpuz 2947
Poortmans, J. 3260
Poortmans, Jef 1233
Pop, Sergiu C. 921, 1869
Poplavskyy, Dmitry 1459, 1686
Porter, Ilana J. 417
Potamialis, C. 2457
Pötz, Sandra 178
Pouladi, S. 869
Pouladi, Sara 866
Poulsen, Peter B. 2672, 2682
Powalla, Michael 791
Previtali, Jonathan 1275
Price, Jared S. 1469
Prietl, Christine 1329

Printraza, Nikolas J. 993
Procel, P. 2073
Ptak, Aaron J. 46, 62, 2275
Puska, Martti J. 2070
Puthanveettil, Suresh E. 76
Qazi, Farah 1317
Qian, Gary 667
Qian, Shen 958
Qin, Xuefei 1594
Qiu, Botong 667
Qudsia, Syeda 1317
Quinto, Carlos 429
Quiroz, Jimmy E. 1280
Rada, Jacob 1271
Radhakrishnan, Hariharsudan
Sivaramakrishnan 1233
Raghavan, Srinivasan 837, 841, 986, 2395, 2399
Ragunathan, Gautham 1181
Rahman, Mosaddequr 1067
Rahn, Christopher D. 1469
Raiker, Gautam A. 3073
Raj, Samuel 284, 496, 499
Rajan, Grace 182, 735, 807, 2298, 2446, 2646
Rajbhandari, Pravakar P. 989
Rajput, Amit Singh 499
Raju, T. Bhim 1034
Raker, David 2926
Rale, Pierre 1289, 3147
Ramakumar, Rama 1122
Ramamurthy, Praveen C. 1614, 2811
Rambabu, Sugguna 3478
Ramic, Zekija 2776
Ramírez, A. 503
Ramirez, A.A. 433
Ramírez, E. A. 433, 503
Ramos, Helena Geirinhas 3178
Ramos, Javier 2255
Ramprasad, Sumukh 496
Ramu, Govind 1275, 2263
Rancoita, P.G. 541
Rand, James 925
Ranjan, Rajeev 2395, 2399
Ranjan, Upasna 2811
Ranjbar, S. 3260
Ransome, Steve 652
Rao, Arun D 1614
Rao, B.V. 1898
Rao, Rajesh 363, 2506
Rao, Roshan R 604
Raorane, Neha 1755
Raote, Yojak 1022
Rashkin, L. 3224
Rastogi, A.C. 3279
Rathi, M. 869
Rathi, Monika 866, 2368

AUTHOR INDEX

Rathore, Sudharm 2902
Rau, Uwe 1651, 2114
Raupp, Christopher 1984
Ravindra, M. 76
Ravindra, Pramod 2395, 2399
Raychaudhuri, S...1733
Razooqi, Mohammed A. 2462
Recart, Federico 2677
Reddy, Anurag 3528
Reddy, K.S. 2858
Reddy, Rekha 3524
Reed, S. 869
Reedy, Robert C. 881
Reese, M. O. 138
Regalado-Pérez, E. 142
Reichel, C. 2064
Reichert, Andreas 2407
Reinders, Angèle 2109
Reindl, T. 1140
Reise, Christian 2267
Rejon, V. 1691, 2457
Ren, Zekun 284, 2532, 2744
Ren, Zhiwei..958
Reno, Matthew J. 1555, 1567, 1573,
    1579, 2975, 3025, 3031, 3055
Renteria, E. J. 281
Repins, Ingrid....................................2728, 3452
Reusser, Jean 2255
Reyes-Banda, M.G. 142
Rey-Stolle, I. 1210
Rey-Stolle, Ignacio 1204
Rhodes, Christopher 1476
Riaz, Hiba 1741
Ribeyron, P. -J...1747
Ricardo, Julian Do Nascimento 3077
Rich, Geoffrey 600
Richards, J..3224
Richardson, Walter 1116
Richter, A. 1752, 2064
Richter, Mauricio...3360
Riedel, Nicholas 2672, 2682
Rienacker, Michael 3371
Riesen, Yannick..3435
Rigdon, Terry B. 3448
Riggs, Brian 37
Riggs, Brian C. 3245
Riley, Daniel 1537, 3155, 3184, 3348
Riley, Daniel M..................................1543, 1549
Rimmaudo, I. 1691, 2457
Rimmaudo, Ivan.....................................1669, 2372
Rincon-Charris, Amilcar A. 2963
Ringel, Steven A. 215, 2079, 2446, 2554
Ringleb, Franziska 3396
Rivera, Eduardo I. Ortiz 3044
Riverola, Alberto..1339

Robert, Sofie 1804
Roberts, Jesse...3083
Robertson, John 37
Robertson, Kyle W. 1294
Robinson, Charles D. 3155
Rochat, Raphael 326
Rocheleau, Richard E...1061
Rockett, A. 30
Rockett, Angus.......................182, 1400, 2446
Rockett, Angus A. 1511
Rodrigues, Sandy..3178
Rodriguez, D. J. 2031
Rodríguez, Diego J. 1143
Rodríguez, Pedro 2677
Rodríguez-Gallegos, Carlos D. 2318
Roest, Stefan 3462
Roeth, A. J. 14
Rogers, John A. 1469
Rohatgi, Ajeet 333, 940, 1807, 1824, 1838
Roland, Paul J. 1030
Roller, John 508
Romanin, Vince 37, 3245
Romeo, Alessandro.....................752, 1669, 2372
Ronoh, Geoffrey Kibiegon 970
Rooijakkers, Tom T.H. 1081
Ropp, Michael..2147, 3020
Rosales-Ascnsio, Enrique 3240
Rose, Volker 2179, 3300
Ross, N. 3269
Rotoli, P...1747
Rounsaville, Brian 940, 1807, 1824
Routhier, Alexander F. 3088
Rowell, David..3524
Roy, Sam 2358
Roy, Tatiana A... 521, 558
Royer, Fabien..558
Rozza, Davide 541
Rubbard, Seth M. 3468
Ruffini, Leia 2453
Ruiz, Carmen M. 2255
Ruiz, E. O. Ángel 1990
Rummel, S. 467
Rupp, B. 1733
Ruppalt, Laura B..873
Russell, Annie 1094
Russell, Richard..2227, 3435
Russell, Thomas C.R. 2236
Ruth, Daniel 2301
Ryou, J. 869
Ryou, Jae-Hyun 866, 2368
Saavedra, Michael 1473
Sablon, Kimberly..1181
Sabnis, Sanjeev 2849
Sacchetto, Davide 3256
Sachenko, A.V......................................663, 1025, 1811

AUTHOR INDEX

Sachenko, Anatoliy V.................................690
Sáenz, M.J...643
Saetre, Tor Oskar 685
Sahayaraj, S...3260
Sahli, Florent...3256
Sahraei, N. 648
Sai, Hitoshi.....................................381, 1333
Saifullah, Muhammad 810
Sainsbury, Cassidy......................2692, 2765
Saito, K. 3
Saito, Tomohiro.......................................931
Saive, Rebecca.............................1589, 2236
Sakamoto, Katsuyoshi..............................85
Sakamoto, Norihiko...............................1268
Sakurai, Takeaki...................33, 160, 900
Salamo, Gregory J. 3370
Salavei, Andrei.......................................2372
Salazar, J. 370
Salazar-Duque, John E. 2963
Salo, Kristian..1494
Salome, Pedro...796
Salpakari, Jyri...3236
Salvetat, Thierry 2492
Samoilenko, Yegor...................................1697
Sampath, W.S. 980, 2736
Sampath, Walajabad 424, 785, 1674
Sampath, Walajabad S. 3417
Sample, Tony...1275
Sampson, Matthew D...........................6, 1256
Samuelson, Lars.......................................2502
Samundsett, C. 3295
Samundsett, Chris...................................2076
Sánchez, Yudania......................................155
Sánchez-Pérez, P. A. 1959, 1990
Sanchiz, Joaquín.......................................429
Sandeep, K. 396
Sang, Baosheng.......................................1455
Sang, Shiyu 472, 1430, 2918
Sangjeong, Myeong...................................356
Sankaran, M. 76
Sankin, I. 2816
Santana, G. 370, 2342, 2614
Santana-Rodríguez, G. 670
Santbergen, Rudi....................................2605
Santhanam, Parthiban.............................2185
Santos, M. B. 14
Santoyo-Salazar, J....................................370
Saraf, Akash 761
Saraswat, Krishna...................................1835
Sargent, Edward H. 3129
Sarmah, Nabin 2858
Sarvari, Hojjatollah 1044, 2432
Sarwar, Jawad...1888
Sasaki, A. 1003
Sastry, O. S. 3478

Sato, Daisuke 1743
Sato, Shin-Ichiro 562
Sato, S-I. 552
Satzinger, Valentin 178
Saucedo, Edgardo...............155, 3265, 3285
Savin, Hele 944, 1494, 3236
Sawallich, S. 3150
Sayed, Islam E.H. 2195
Sayyah, Arash 2891
Scaccabarozzi, Andrea 1289
Scarpulla, M.A. 1656, 1661
Scarpulla, Michael A............175, 802, 1679, 2467
Schäfer, Nicolas 2216
Schaller, Richard D. 6
Scheiman, David 1838, 3514
Schelhasl, Laura T...................................2176
Scheltens, Frank J. 966
Schenller, E.J. 1701
Schermer, J. 1189
Schindler, F. 2064
Schitthelm, F. 1752
Schlemmer, James 1104
Schmid, Martina.....................................3396
Schmidt, Jan 3371
Schmidt, Thomas....................................3396
Schmieder, Kenneth J. ...210, 259, 272, 873, 2091, 2506
Schnabe, Thomas....................................2216
Schnabel, Erdmut 3488
Schnabel, Manuel ...1817, 2482, 2543, 3254, 3371, 3439
Schnabel, T. 3260
Schnabel, Thomas 791
Schneider, Kevin 1476
Schneller, Ej 389
Schneller, Eric J....................2839, 2897, 3448
Schoenfeld, Winston 2839
Schoenfeld, Winston V. 322, 1804
Schoenfelder, Stephan............................2622
Schoenwald, David 2969
Scholl, Jonathan A..................................1549
Schoop, Urs 1455
Schorch, M. 1752
Schriemer, Henry P................................1094
Schubert, M. C. 3135
Schubert, Martin C. 1329
Schulte, Kevin 62
Schulte, Kevin L. 46, 232, 2275
Schulte-Huxel, Henning 1366, 2543, 3371
Schulz, Gerd 914
Schulze, Patricia S.C. 1253
Schwabe, Hartmut..................................2622
Schweiger, M. 107
Sclj, Josefine 619
Scofield, A.C. 2755
Scolari, Enrica 1415
Sculati-Meillaud, Fanny.........................2794

AUTHOR INDEX

Seif, Johannes P. 3435
Seigneur, Hubert 2839, 2897
Sellami, Nazmi 2858
Sellers, Andrew 2926
Sellers, Diane G. 3384
Sellers, I. R. 14
Selvamanickam, V. 869
Selvamanickam, Venkat 866, 2368
Semichaevsky, Andrey 319
Sen, Fatih G. 2759
Senaud, L.-L. 50
Sengar, Brajendra S. 2345
Sengupta, Manajit 116, 1169
Senthilarasu, S. 2858
Sepeher, Mohsen M. 1094
Sera, Dezso 1421, 2682
Serra, João M. 1979
Sethia, Saurabh 2902
Seydel, Elisabeth 1682
Shafarman, William N. 26
Shah, S. 1350
Shahirinia, Amir 3092
Shanmugam, Vinodh 2318
Sharma, Ashok K. 396
Sharma, Romika 3300
Sharma, S. 2094
Sharps, Paul 42, 525, 2099
She, Hui 1863
Shen, Chang-Hong 1627
Shen, Zeqing 3393
Shephard, Les E. 1116
Shervin, Kaveh 1452
Shervin, Shahab 866
Shetty, Nishit 876
Shi, Jianwei 1820
Shi, Jiatiwei 1322
Shi, Xuanyi 3220
Shi, Zhan 1037
Shibata, Hajime 33, 1268
Shieh, Jia-Min 1627
Shigekawa, Naoteru 2548, 2551
Shih, Cheng-Hao 2035
Shih, I. 2388
Shih, Ishiang 1299
Shima, D. M. 281
Shimura, H. 1003
Shin, Hyun-Beom 244
Shin, Myunghun 2631
Shin, Seunghyun 935
Shin, Woo Jung 385
Shinde, O.S. 389, 1701
Shirasawa, Katsuhiko 904, 931
Shkrebtii, Anatoli I. 690
Shoji, Yasushi 10
Shore, Andrew 437
Shrestha, Niraj 1030
Shrestha, Santosh 696, 2186
Shu, Chia-Jhe 1606
Shu, Jinn-Kong 1606
Shubhrant, Abhishek 2902
Si, Fai Tong 2605
Siddiki, Mahbube K. 1018, 1110
Sidhu, Navjot Kaur 3279
Siebentritt, Susanne 151, 2054, 2205, 2478
Siepchen, Bastian 761
Sikchang, Hyo 356
Silva, Francois 464, 1237
Silva, José A. 1979
Silvaggio, Amber C. 2554
Silverman, Timothy 1259, 1893
Silverman, Timothy J. 1400, 2771, 3452
Simon, John 42, 46, 62, 1201, 2275
Simon, Kirby 876
Simpson, L. 2280
Simpson, Lin 2294
Simpson, Lin J. 1893, 2789
Sinapis, Kostas 1081, 1090
Singh, Aparna 2902
Singh, Ashish 1034
Singh, Ashish K. 1704
Singh, Hemant K. 1995, 3478
Singh, Rajeev 2762
Singh, Rhythm 1151
Singh, Rubina 1855
Singh, Sukvhinder 2227
Singlr, Vijay P. 2432
Sinha, Archana 3478
Sinha, Parikhit 2005
Sinisuka, Ngapuli I 3102
Sink, Joseph 3333
Sinton, Ronald 2707
Sinton, Ronald A. 2692, 2765
Sio, H. C. 3295
Sio, Hang Cheong 3300
Sites, James R. 164, 1308
Slocum, Michael 677
Slocum, Michael A. 18, 202, 206, 222,
      831, 1184, 2084, 2514, 3468
Slooff, Lenneke H. 1081
Smaglik, Nathan 831, 2514
Smestad, Greg P. 2858
Smith, Benjamin 1134
Smith, Brittany L. 18, 1184
Smith, David J. 2573
Smith, Mathew 2941
So, Won-Shup 487
Soares, Gabriela De Amorim 2875
Sodabanlu, Hassanet 854
Söderström, T. 1752
Sofia, Sarah E. 1264

AUTHOR INDEX

Sogabe, Tomah 85, 712
Sokolovskyi, I.O. 663, 1025, 1811
Sokolovskyi, Igor O. 690
Solanki, Chetan S. 3478
Solanki, Chetan Singh 1850
Soltanmohammad, Sina..26
Soman, Anishkumar 1828
Søndergaard, Sissel Tind..2651
Song, Dengyuan..1430
Song, Hee-Eun 356, 1758
Song, Myungkwan 2631, 2634
Song, Tao 1308
Song, Zhaoning.................... 170, 730, 748, 815, 1030
Sonp, Hee-Eun 2723
Sood, Neeru 2858
Sossan, Fabrizio 1415
Soudachanh, A. L. 281
Sozzi, Giovanna..2205
Spandana, B..396
Spataru, Sergiu..................................... 1421, 2682, 2819
Spaulding, David 820, 1686
Spertino, Filippo..3096
Spiering, Stefanie 791
Spinelli, P..3150
Spooner, Ted 1275, 2263
Sreekumar, Nimisha...1755
Sridharan, Akirt..999
Sriramagiri, Gowri 658, 1196
Srivatsan, R. 120, 589
Stark, Cameron 1855
Starkl, Hannes 178
Steeman, Rob...337
Steenhoff, Volker 3388
Stefancich, Marco 1946
Steijvers, Henk 2875
Stein, Joshua....................................... 1537, 3333, 3348
Stein, Joshua S................. 1543, 3155, 3161, 3184, 3488
Steiner, Myles A. 42, 47, 232, 1201, 2195, 3254
Steinfedt, Jeff..525
Stender, C. 3511
Stender, Christopher L...3524
Stephan, Jack 2124
Stevens, Margaret 701
Steward, Malia 1037
Stewart, J. 3224
Stika, K. 2312
Stiles, Phil 2833
Stoddard, Nathan 2610
Stokes, Adam 1381, 2887
Stolt, L. 30
Stone, Kevin H. 2176
Stradins, Paul881, 1491, 1777, 1801, 1817, 1832, 2242, 24
Stradins, Pauls...2482, 2702
Strandberg, Rune 706, 2651
Stride, John A. 2392

Stuart, Thomas 2926
Stuckelberger, J..2073
Stuckelberger, Michael 2610, 2854, 3309
Stuckelberger, Michael E......................................2179
Stueve, Bill 2808, 2923
Sturm, James C. 1773
Stutz, Elias Z. 944
Su, Bojie...392, 1430, 1873, 2918
Su, Chengfeng 392, 1873
Subbiah, Jegadesan 1342
Subedi, Indra 2771, 3468
Subedi, Kamala Khanal 781
Sudbury, Benjamin A. 1322
Suga, Mitsunobu 567
Sugaya, Takeyoshi.............................. 562, 861, 1724
Sugimoto, Hiroki 160
Sugiyama, Masakazu 854, 2201, 2524, 2538, 3528
Sugiyama, Mutsumi 192
Sugiyama, Ryo 712
Suhana, Hadi 3102
Sumita, Taishi 3506
Sun, C. 1485
Sun, Ce 2759
Sun, Chang 3300
Sun, Chenguang..1241
Sun, Kaiwen 2213
Sun, Qiang 1648
Sun, Qiming 3129
Sun, S. 869
Sun, Sicong...2368
Sun, Wen-Cheng 2227
Sun, Xiaolin. 2656
Sun, Xingshu..1055, 1259, 1904
Sun, Yaojie 2048
Sun, Yubo...2467
Sun, Yukun 2291
Sun, Zeming 937
Supplie, Oliver 2524, 2538
Surya, Charles 958
Sutou, Yuji 931
Sutterlueti, Juergen 652
Suzuki, Ryota...1504, 2588
Swain, Santosh K. 3422
Swartz, Craig H. 2473, 2749
Sweatt, William 1473
Syu, Hong-Jhang 1009
Szabo, Sandor 2167
Szlufcik, Jozef..................................1233, 2227, 3435
Tabet, Nouar......................................963, 1058, 3435
Tacconi, Mauro 541
Tachibana, Shoji...1504
Tadese, Alemu 1104
Tadesse, Alemu...............................132, 1132, 1427
Tae, Christian 1835
Taekjeong, Kyung 356

AUTHOR INDEX

Takahashi, Akiko 2906
Takahashi, Isao 1765, 1794
Takahashi, Takuji 455
Takahashi, Yasuhito 1973
Takamoto, Tatsuya 3506
Takato, Hidetaka 381, 904, 1724, 3323
Takenouchi, T..441
Tamaki, Ryo 10
Tamboli, Adele 3254, 3371
Tamboli, Adele C. 578, 2482, 2488,
 2536, 2543, 3381, 3406
Tamizhmani, Govindasamy 1389, 1850,
 1858, 1877, 1883, 1959, 1984, 2789, 2912
Tan, Jin 1158
Tan, Joel M. R.......................................3275
Tan, K.H. 1210
Tan, Xuehai 761
Tanahashi, Katsuto 3323
Tanahashi, Tadanori...............................1268
Tanaka, Aki 2642
Tanaka, T..3
Tanaka, Takahiro 2947
Tang, Chiu C. 2423
Tang, Houjun 1100
Tang, Mingchu 3370
Tang, Tao 2558
Tanke-Pedretti, Anna 1473
Tao, Meng..385, 608
Tao, Yuguo 1824
Tappan, Ian A. 2864
Tassone, Christopher J............................2176
Tatapudi, Sai 1850, 1858, 1877, 1959, 2912
Tatapudi, Sai Ravi Vasista 2789
Tatavarti, Rao 1184, 2084
Tatavarti, Sudersena Rao.........................1181
Tate, John Keith.......................................333
Tayagaki, T..3
Tayyib, Muhammad 2776
Tchemycheva, Maria 1289
Tedeschi, Giampiero 2372
Teena, Percis 3113
Tennyson, Elizabeth M............................1508, 2443
Terheiden, Barbara.................................1824
Terukov, E.I. 1025, 1811
Teubner, Thomas....................................3396
Teymouri, Arastoo 2403
Thanh, Nguyen Cong 1765
Thankalekshmi, Ratheesh R. 3279
Theelen, Mirjam.....................................2875
Theigi, San 881
Theingi, San...1832
Theocharides, Spyros.................1163, 3107
Therrien, Francis....................................1579
Thibeault, Brian.......................................315
Thimsen, Elijah......................................876

Thompson, Christopher............................1196
Thompson, Corey S..................................1869
Thon, Susanna M.....................................667
Thorseth, Anders............................2672, 2682
Thorsteinsson, Sune.........................2672, 2682
Thway, Maung 284, 2744
Tidwell, Steven......................................1086
Timò, Gianluca 290
Tirumalai, Tejas 2923
Tischler, Joseph G...................................873
Titus, Jochen..820
To, Alexander 517
To, B. 2280
Toberer, Eric S. 2536, 3406
Todorov, Teodor 1441
Togay, M. 2457
Togay, Mustafa 146, 186
Tomasi, A. 50
Tomasi, Andrea 3435
Tomasulo, Stephanie 2091
Tonic, Marko 1346
Toor, Fatima 346, 1537, 1543, 3184, 3333, 3348
Toprasertpong, Kasidit...................2201, 2524
Torralba, Encarnacion.............................3402
Tous, Loïc 2227, 3435
Tracy, Jared....................3190, 3200
Traverse, Christopher 2124
Trout, T. John..2312
Trupke, Thorsten 66, 420, 2707, 3304, 3315
Tsafarakis, Odysseas..............................1090
Tsai, Cheng- Ying 3366
Tsai, Jia-Lin 1606
Tsai, Jia-Ling 294
Tseng, Zong-Liang 367
Tsutsumi, S. 3
Tu, Wei-Chen 1051
Tucher, Nico.................352, 1253, 2511
Tukiainen, Antti 297, 2520
Tuminello, F. 3511
Tummala, Abhishiktha 2912
Turek, Marko 2232
Turner, John A. 47
Tuteja, Mohit 1511
Tyagi, Astha 716
Tyler, Kevin 301
Tyson, Tom 925
Tzolov, Marian 1040
Ubukata, Akinori.....................................861
Ueda, Kohsuke.......................................3506
Ueda, T. 1003
Uematsu, Takumi 1950
Ulbricht, Christoph 1360
Ulicná, Sona 146, 186
Uma, B. R. 76
Umishio, Hiroshi......................................381

AUTHOR INDEX

Unold, Thomas 3396
Unsur, Veysel...888
Upadhyaya, Ajay D.............................940, 1807
Upadhyaya, Vijay D..333
Upadhyaya, Vijaykumar 940, 1807, 1824
Uprety, Prakash 3468
Urbano, J. Antonio...632
Uruena, Angel..2227, 3435
Usami, Noritaka 1765, 1794
Utsunomiya, Satoshi 904
Vadiee, E. 1603
Vadiee, Ehsan........................305, 827, 1841
Vagidov, Nizami Z. 531
Vähänissi, Ville...1494
Vaida, Mihai E. 417
Vaidya, Nina..................512, 521, 558, 572, 1248
Vaisman, M. 3376
Vaisman, Michelle...............................578, 3381
Vaissiére, Nicolas 2528
Valderrama, Nicolas 1893
Valdivia, Christopher E.................................1094
Van Aken, Bas B. 3462
Van Alsburg, Jane..1455
Van De Loo, Bas W.H.....................................1817
Van Der Heide, Arvid 3343
Van Hest, Maikel F.A.M.............2482, 3371, 3439
Van Sark, Wilfried...3014
Van Sark, Wilfried G.J.H.M. 1090, 3167
Vandamme, Nicolas 2453
Vandervelde, Thomas E. 701
Vanka, S. 2388
Vanka, Srinivas 1299
Vansant, Kaitlyn 1922, 3452
Vargas, Carlos..3290
Vasi, J. 1995
Vasi, Juzer 1850, 3478
Vasileska, D. 1603, 2816
Vasileska, Dragica 1790, 1797
Vasilyev, Leonid A. 3448
Vasudevan, Saravanan 2172
Vauche, Laura 2492, 2562
Vedde, Jan 2682
Veettil, Binesh Puthen 2392
Vehse, Martin 3388
Veinberg-Vidal, Elias............................2492, 2562
Veith-Wolf, Boris 1366
Velappan, Krishnakumar.................................761
Venizelou, Venizelos 276, 1163, 1941, 3107
Verbitskiy, V.N..1811
Verlinden, Pierre J................................1922, 2220
Vermang, B. 3260
Vermang, Bart 2209
Verschac, Rodrigo...1175
Vetter, E. 1752
Viana, Marcelo Machado 1917, 2307

Vignola, Frank...1169
Vijh, Aarohi 3520
Vilcot, Jean-Pierre 3402
Vincent, Nina 1893
Vines, L. 3269
Vinogradova, Tatiana 512
Vinogradova, Tatiana G. 521, 558, 572
Virtuani, Alessandro 1395, 2104, 2794
Vlasiuk, V.M. 1025
Vlasyuk, V.M. 1811
Vleugels, J. 3260
Voarino, Philippe 2492, 2562
Vogt, Malte Ruben 1366
Von Gastrow, Guillaume 944
Voroshazi, Eszter 3343
Voss, Henrik 2682
Waddle, John M. 309, 3123
Wade, Andreas 2005
Wagner, Sigurd 1773
Waiis, J.M. 2457
Waldhauser, Wolfgang 1329
Walker, Don 93, 512, 1248
Walls, J.M. 1691, 3430
Walls, John 1674
Walls, John M..................146, 186, 752, 2349
Walls, John Michael 2827
Walls, Michael 424
Walter, Arnaud 3256
Walters, Joseph 2839, 2897
Walters, R. J. 845
Walters, Robert 3514
Walters, Robert J. 210, 247, 259, 272,
        373, 873, 1838, 2091, 2506
Waltmger, A. 1752
Walukiewicz, W. 3
Walukiewicz, Wladek....................................1204
Wan, Kai-Tak 1933, 2000
Wan, Ronghua 226
Wan, Yimao 59, 2076
Wang, Ao......................................1937, 2823
Wang, Baomin 1469
Wang, Changlei 993
Wang, Da-Wei 3220
Wang, Deng...2048
Wang, Feng 1044
Wang, Fumei...........................392, 1937, 2823
Wang, Haotian 1342
Wang, He. 226, 392, 1430, 1648, 1873, 1937, 2823, 2918
Wang, Hongfeng 1215
Wang, Laidong 385
Wang, Mu 3370
Wang, Q. 1733
Wang, Rui 1100
Wang, Shenghao 160
Wang, Shizhen 976

AUTHOR INDEX

Wang, Sisi 2600
Wang, Teng-Yu 2660
Wang, Xiaohui...2432
Wang, Y. 1733
Wang, Y. D. 1733
Wang, Yan 1922
Wang, Yichen 1299
Wang, Yiwang 1100
Wang, Yongqian 2220
Wang, Yu 1933, 2000
Wang, Yu-Cian 2498, 2566
Wang, Zigang 1922
Ward, J. Scott...3254
Warmann, Emily...1248
Warmann, Emily C.................512, 521, 558, 572
Warner, Jeffery. H..2091
Warren, Emily...3371
Warren, Emily L. 578, 2482, 2488, 2543, 3381
Washington, Lori 3520
Washio, Hidetoshi...3506
Watanabe, Kentaroh.............................854, 3528
Watanabe, Yasuyuki.......................................613
Waters, Martin...2923
Watson, S. 648
Watthage, Suneth C...............170, 730, 748, 815, 1030
Watts, John L.R. 2762
Weeber, Arthur...2875
Weick, Clément 2492, 2562
Weigand, William 1790
Weiss, Charlotte....................................83, 2407
Weiss, Dirk 1264
Weiss, Karl-Anders.......................................2836
Wen, Ching-Chang 329
Wen, Xiaoming..696
Wenham, Stuart R. 2576, 2600
Werner, Florian 2205, 2478
Werner, Jérémie.....................................55, 3256
West, Bradley M. 2179, 3309
Western, N. J...953
Western, Ned J...948
Wheeler, Tobias 1476
Whipple, Steven 88
Whiteside, V. R. 14
Wibowo, A. 3511
Wibowo, Andre 1181, 1184, 2084
Wicaksono, S. 1210
Widén, Joakim 3067
Wieghold, Sarah 3300
Wienands, Karl 1253
Wiese, Martin 1531
Wille-Haussmann, Bernhard 126
Williams, J. 1603
Williams, Joshua J. 305, 582
Williams, R. 490
Wilson, Gregory 3236

Wilson, Marshall 322
Wilt, David M. 88, 102, 301, 531
Wilt, Sam..301
Wilterdink, Harrison 2692, 2765
Winkler, Kristina 1253
Winkler, Thilo 3500
Wirsching, Sven 3500
Wirth, Harry...1531
Wischmann, Wiltraud 2058
Wissen, A...1752
Witte, Wolfram 2205
Witteck, Robert 1366
Wohlgemuth, John 1275
Wojtowicz, Anna 164
Wolden, C. A. 138
Wolden, Colin A. 1697
Wolf, Martin 2692
Wolffersdorff, Paul.......................................595
Wong, Johnson 499, 3113
Wong, Johnson Kai Chi 496
Wong, Lydia H. 3275
Woodhouse, Michael 1259
Woods, Jason 1893
Woods-Robinson, Rachel 3410
Worrell, Ernst 3014
Wright, Lewis D. 146, 186
Wu, Gordon 96
Wu, J. 1189
Wu, Jiang 3370
Wu, Kuen-Yi 911
Wu, Po-Ching 1623
Wu, Ruei-Ying 1635
Wu, Shang-Hsuan 1051
Wu, Teng-Chun 448, 476
Wu, Yonggang 1594
Wu, Yuh-Renn 294
Wu, Zhuopeng 1241
Würfel, Uli 1253
Wyrsch, Nicolas 3435
Wyss, P. 2073
Xia, Hongze 2392
Xia, Zihuan 1594
Xiao, C. 1312, 2785
Xiao, Chuanxiao 62, 1371
Xiao, T. Patrick 2185
Xiao, Zhi Bin 1648
Xie, Yu 116
Xiong, Gang 1193, 2473
Xiong, Zhen 2220, 3304
Xu, Jun 2656
Xu, Ling...2656
Xu, Lu 1737
Xu, Menglei 1233
Xu, Qi 37, 3245
Xu, Qianfeng 2285

AUTHOR INDEX

Xu, Tao 2251
Xu, Xiaojie 3410
Xu, Zhaoran 59
Xue, Muyu 1835, 2732
Yablonovitch, Eli 2185
Yachi, Toshiaki 613
Yadav, Karan Shishir 2902
Yadav, Tarun S. 396
Yakes, Michael K. 873, 2091
Yamada, Noboru 1724, 1743
Yamada, Nobuyuki 2906
Yamagami, Takeru 192
Yamagoe, K. 441
Yamaguchi, Hiroshi 3506
Yamaguchi, Koichi 712
Yamaguchi, Masafumi 359, 412, 1479,
    1711, 1714, 1743, 2498, 2548, 2566
Yamaguchi, Seira 1385
Yamamichi, Masaaki 1275, 2263
Yamaya, Haruki 2159
Yan, Chang 2213
Yan, Di 2076
Yan, Yanfa 771, 993, 1643, 2443, 2473, 3426
Yancey, Billy 2128
Yanchilin, Anton 585
Yang, Fan 2656
Yang, Guangtao 2605
Yang, Hao-Yu 343, 367, 893, 2664
Yang, Hong 392, 1430, 1873, 1937, 2823, 2918
Yang, Jianfeng 2392
Yang, Mohshi 907
Yang, Peter 1100
Yang, Shuying 2697, 3456
Yang, X. 2785
Yang, Yang 2220
Yang, Yi Tong 1648
Yang, Yun-Chie 893, 2664
Yang, Zhihao 74
Yao, Li You 1648
Yao, Y. 869, 1752
Yao, Yangyi 1048
Yao, Yao 866, 2368
Yarnaquchi, Koichi 85
Yates, Peter 1445
Yaung, K. Nay 3376
Yaung, Kevin Nay 284, 2744
Ye, Feng 2220
Ye, J. 2785
Ye, Qilin 948
Yeh, Chun-Ming 911
Yellowhair, Julius 2870
Yellowhair, Julius E. 2891
Yi, Chuqi 2569
Yilmaz, S. 3430
Yoo, Chang Youn 399
Yoon, Howard W. 437
Yoon, Jongseung 549, 2291
Yoon, S. F. 1210
Yoon, Woojun 373, 1838
Yoshiba, Shuhei 1769
Yoshino, Kenji 1679
Yoshita, M. 1003
Yoshita, Masahiro 2781
You, Bang-Jin 367
You, Liang-Chian 1635
Young, David 46
Young, David L. 1817, 1832, 2275, 3254
Young, James L. 47
Young, Steven 582
Youssef, Amanda 1491, 2242, 3300
Youtsey, Christopher 3524
Yu, Edward T. 363, 1181
Yu, Jia 2453
Yu, K. M. 3
Yu, Kin Man 1204
Yu, Li-Chieh 3204
Yu, Linwei 2656
Yu, Ming 1193
Yu, Peichen 294, 1610, 1635
Yu, Pei-Chen 1606
Yu, Sun 1522
Yu, Zhengshan J. 1228, 1317, 1322, 2039, 3250
Yuan, Bo 315
Yuan, Lin 2392
Yue, Yao 93
Yun, Jae Ho 810
Zachariah, S. 1995
Zachariah, Sachin 2799, 2849, 3478
Zahler, James 1463
Zahler, James M. 3245
Zakaria, Naimi 487
Zamora, Rachid Darbali 3044
Zang, Kai 1835
Zapalac, G. 2414
Zapalac, Geordie 820, 3327
Zauner, Andy 1237
Zech, Tobias 1531
Zelenina, Anastasiya 2054, 2478
Zeman, Miro 2605
Zeng, Guoping 907
Zeng, Xulu 1286
Zeyu, L. 1781
Zhai, Yonghui 472
Zhan, Tien-Chien 294
Zhang, Bao 226
Zhang, C. 1603
Zhang, Chaomin 240, 827, 1215, 1841, 2573
Zhang, Guoqi 2605
Zhang, Hua 3304
Zhang, Huan 2558

AUTHOR INDEX

Zhang, Jili..1100
Zhang, Jing..3384
Zhang, Junjun.............................1937, 2823
Zhang, Lei................408, 1761, 1828, 2667
Zhang, Liang..2247
Zhang, Liping.......................................1241
Zhang, Nian...2432
Zhang, Qiming..226
Zhang, Wei...................................255, 1193
Zhang, Weijie...820
Zhang, X...2094
Zhang, Xiaochen 1567
Zhang, Xue 392, 1430, 1873, 2918
Zhang, Yang...195
Zhang, Yi......................................696, 2186
Zhang, Yong-Hang.........................3366, 3410
Zhang, Yufeng..................198, 767, 1707
Zhang, Z...2019
Zhang, Zhilong......................................2392
Zhang, Zongyi.......................................1594
Zhangl, Xiaochen 1555
Zhao, Dewei...993
Zhao, Hui....................392, 1430, 1873, 2918
Zhao, J...1752
Zhao, Jing..1845
Zhao, Pan....................1430, 1873, 2918
Zhao, Xin-Hao..3366
Zhao, Yuan...3366
Zhao, Yuetao..1044
Zhe, Liu...2744
Zheng, N..869
Zhigunov, D.M. 1811
Zhongbiao, Ye.......................................1044
Zhou, Guomin...472
Zhou, Hang..................................976, 2558
Zhou, Jian..1594
Zhou, Xiao W. 2419
Zhu, Jiang..3208
Zhu, Lin.......................................721, 3528
Zhu, Yan..66, 3290
Zhu, Ziyao...198
Zide, Joshua M. O.................................3384
Zielnik, Allen..3208
Zilles, Roberto......................................1917
Zilouchian, Ali...............................638, 2941
Zimmerman, Jeramy D. 3381
Zin, Ngwe..322
Zinaddinov, M.......................................2411
Zoppi, Guillaume. 742
Zubia, David...2419
Zunger, Alex...1245

IEEE
445 Hoes Lane
Piscataway, NJ 08854-4141

ISBN 978-1-5090-5606-4